BREAK THROUGH
To improving results

Available in MyMathLab® for Your Course

Achieve Your Potential

Success in math can make a difference in your life. MyMathLab is a learning experience with resources to help you achieve your potential in this course and beyond. MyMathLab will help you learn the new skills required, and also help you learn the concepts and make connections for future courses and careers.

Visualization and Conceptual Understanding

These MyMathLab resources will help you think visually and connect the concepts.

NEW! Guided Visualizations

These engaging interactive figures bring mathematical concepts to life, helping students visualize the concepts through directed explorations and purposeful manipulation. Guided Visualizations are assignable in MyMathLab and encourage active learning, critical thinking, and conceptual learning.

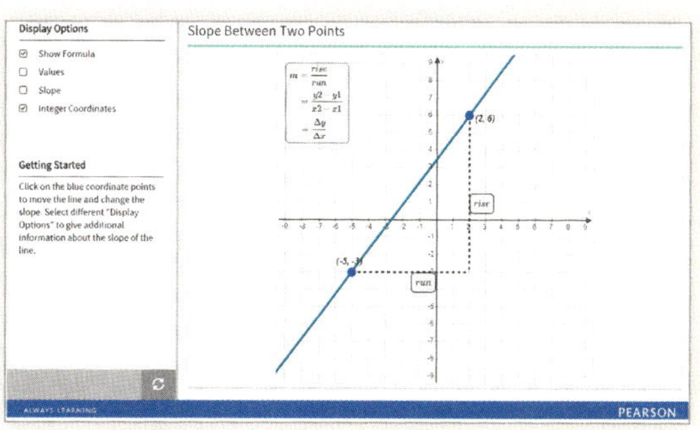

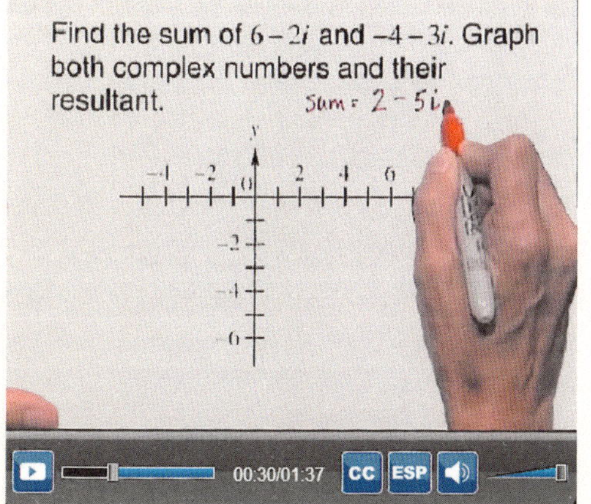

Video Assessment Exercises

Assignable in MyMathLab, Example Solution Videos present the detailed solution process for every example in the text. Additional Quick Reviews cover definitions and procedures for each section. Assessment exercises check conceptual understanding of the mathematics.

www.mymathlab.com

Preparedness and Study Skills

MyMathLab® gives access to many learning resources which refresh knowledge of topics previously learned. Integrated Review MyMathLab Courses, Getting Ready material and Skills for Success Modules are some of the tools available.

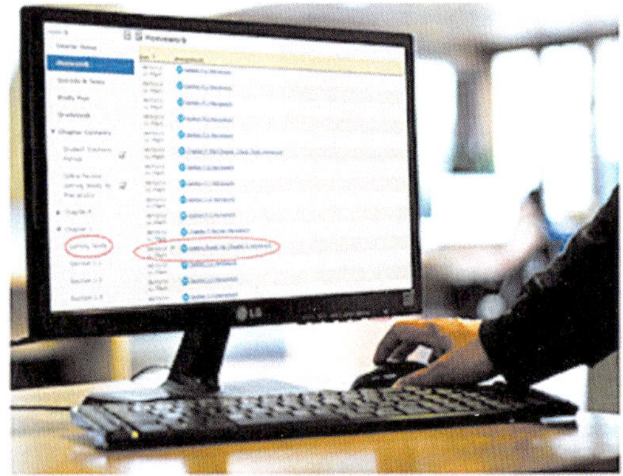

Getting Ready

Students refresh prerequisite topics through skill review quizzes and personalized homework integrated in MyMathLab. With Getting Ready content in MyMathLab students get just the help they need to be prepared to learn the new material.

Updated Videos

Updated videos are available to support students outside of the classroom and cover all topics in the text. Quick Review videos cover key definitions and procedures. Example Solution videos offer a detailed solution process for every example in the textbook.

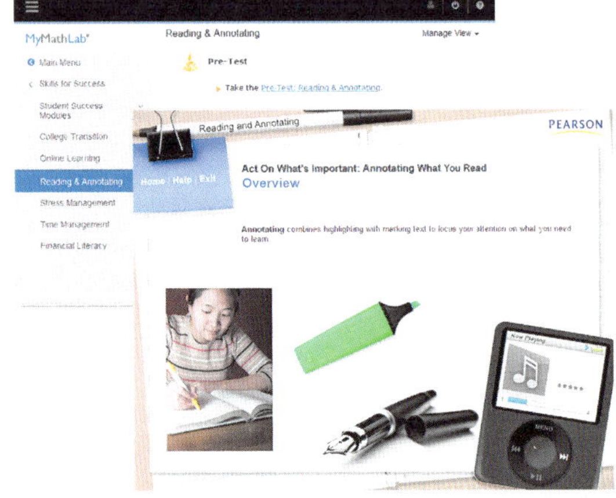

Skills for Success Modules

Skills for Success Modules help foster success in collegiate courses and prepare students for future professions. Topics such as "Time Management," "Stress Management" and "Financial Literacy" are available within the MyMathLab course.

Precalculus

SIXTH EDITION

Margaret L. Lial
American River College

John Hornsby
University of New Orleans

David I. Schneider
University of Maryland

Callie J. Daniels
St. Charles Community College

PEARSON

Boston Columbus Indianapolis New York San Francisco
Amsterdam Cape Town Dubai London Madrid Milan Munich Paris Montréal Toronto
Delhi Mexico City São Paulo Sydney Hong Kong Seoul Singapore Taipei Tokyo

Editorial Director: Chris Hoag
Editor in Chief: Anne Kelly
Editorial Assistant: Ashley Gordon
Program Manager: Danielle Simbajon
Project Manager: Christine O'Brien
Program Management Team Lead: Karen Wernholm
Project Management Team Lead: Peter Silvia
Media Producer: Jonathan Wooding
TestGen Content Manager: John Flanagan
MathXL Content Manager: Kristina Evans
Marketing Manager: Claire Kozar
Marketing Assistant: Fiona Murray
Senior Author Support/Technology Specialist: Joe Vetere
Rights and Permissions Project Manager: Gina Cheselka
Procurement Specialist: Carol Melville
Associate Director of Design: Andrea Nix
Program Design Lead: Beth Paquin
Text Design: Cenveo® Publisher Services
Composition: Cenveo® Publisher Services
Illustrations: Cenveo® Publisher Services
Cover Design: Cenveo® Publisher Services
Cover Image: MakiEni's photo/Getty Images

Library of Congress Cataloging-in-Publication Data

Names: Lial, Margaret L., author.

Title: Precalculus / Margaret L. Lial, American River College [and three others]..

Description: 6th edition. | Boston : Pearson, [2017]

Identifiers: LCCN 2015045217 | ISBN 9780134217420 (hardcover) | ISBN 013421742X (hardcover) | ISBN 9780134217635 (hardcover) | ISBN 0134217632 (hardcover)

Subjects: LCSH: Precalculus. | Algebra. | Trigonometry.

Classification: LCC QA154.3 .L56 2017 | DDC 512--dc23

LC record available at http://lccn.loc.gov/2015045217

www.pearsonhighered.com

ISBN 13: 978-0-13-421742-0
ISBN 10: 0-13-421742-X

To Rhonda, Sandy, and Betty
Johnny

To my MS & T professors, Gus Garver, Troy Hicks, and Jagdish Patel
C.J.D.

Contents

Preface

WELCOME TO THE 6TH EDITION

In the sixth edition of *Precalculus,* we continue our ongoing commitment to providing the best possible text to help instructors teach and students succeed. In this edition, we have remained true to the pedagogical style of the past while staying focused on the needs of today's students. Support for all classroom types (traditional, hybrid, and online) may be found in this classic text and its supplements backed by the power of Pearson's MyMathLab.

In this edition, we have drawn upon the extensive teaching experience of the Lial team, with special consideration given to reviewer suggestions. General updates include enhanced readability with improved layout of examples, better use of color in displays, and language written with students in mind. All calculator screenshots have been updated and now provide color displays to enhance students' conceptual understanding. Each homework section now begins with a group of *Concept Preview* exercises, assignable in MyMathLab, which may be used to ensure students' understanding of vocabulary and basic concepts prior to beginning the regular homework exercises.

Further enhancements include numerous current data examples and exercises that have been updated to reflect current information. Additional real-life exercises have been included to pique student interest; answers to writing exercises have been provided; better consistency has been achieved between the directions that introduce examples and those that introduce the corresponding exercises; and better guidance for rounding of answers has been provided in the exercise sets.

The Lial team believes this to be our best *Precalculus* edition yet, and we sincerely hope that you enjoy using it as much as we have enjoyed writing it. Additional textbooks in this series are

College Algebra, Twelfth Edition
Trigonometry, Eleventh Edition
College Algebra & Trigonometry, Sixth Edition.

HIGHLIGHTS OF NEW CONTENT

- In **Chapter R,** more detail has been added to set-builder notation, illustrations of the rules for exponents have been provided, and many exercises have been updated to better match section examples.

- Several new and updated application exercises have been inserted into the **Chapter 1** exercise sets. New objectives have been added to **Section 1.4** outlining the four methods for solving a quadratic equation, along with guidance suggesting when each method may be used efficiently.

- **Chapters 2 and 3** contain numerous new and updated application exercises, along with many updated calculator screenshots that are now provided in color. In response to reviewer suggestions, the discussion on increasing, decreasing, and constant functions in **Section 2.3** has been written to apply to open intervals of the domain. Also as a response to reviewers, intercepts of graphs are now defined in terms of coordinates rather than a single number. This notation continues throughout the text.

- In **Chapter 4,** greater emphasis is given to the concept of exponential and logarithmic functions as inverses, there is a new table providing descriptions of the additional properties of exponents, and additional exercises requiring graphing logarithmic functions with translations have been included. There are also many new and updated real-life applications of exponential and logarithmic functions.

- In **Chapter 5,** we now include historical material for students to see how trigonometry developed as a means to solve applied problems involving right triangles. In this chapter and the others that cover trigonometry, we have reorganized exercise sets to correspond to the flow of the examples when necessary.
- **Chapter 6** continues to focus on the periodic nature of the circular functions. To illustrate, we have added exercises that involve data of average monthly temperatures of regions that lie below the equator, as well as data that describe the fractional part of the moon illuminated for each day of a particular month. A new example (and corresponding exercises) for analyzing damped oscillatory motion has been included in **Section 6.7.**
- **Chapter 7** now includes a derivation of the product-to-sum identity for the product $\sin A \cos B$, as well as new figures illustrating periodic functions associated with music tones and frequencies.
- In **Chapter 8,** we have reorganized the two sections dealing with vectors. The material covered has not changed, but we have rewritten the sections so that **Section 8.3** first covers geometrically defined vectors and applications, while **Section 8.4** then introduces algebraically defined vectors and the dot product.
- In **Chapter 9,** special attention has been given to finding partial fraction decompositions in **Section 9.4** and to linear programing in **Section 9.6.** Examples have been rewritten to promote student understanding of these very difficult topics.
- In **Chapter 10,** greater emphasis is given to analyzing the specific aspects of conic sections, such as finding the equation of the axis of symmetry of a parabola, finding the coordinates of the foci of ellipses and hyperbolas, and finding the equations of the asymptotes of hyperbolas.
- Throughout **Chapter 11,** examples have been carefully updated to ensure that students are able to understand each step of the solutions. Special consideration was given to mathematical induction in **Section 11.5** by providing numerous additional side comments for the steps in the solution of examples in this difficult section.
- For visual learners, numbered **Figure** and **Example** references within the text are set using the same typeface as the figure number itself and bold print for the example. This makes it easier for the students to identify and connect them. We also have increased our use of a "drop down" style, when appropriate, to distinguish between simplifying expressions and solving equations, and we have added many more explanatory side comments. Guided Visualizations, with accompanying exercises and explorations, are now available and assignable in MyMathLab.
- *Precalculus* is widely recognized for the quality of its exercises. In the sixth edition, nearly 1500 are new or modified, and hundreds present updated real-life data. Furthermore, the MyMathLab course has expanded coverage of all exercise types appearing in the exercise sets, as well as the mid-chapter Quizzes and Summary Exercises.

FEATURES OF THIS TEXT

SUPPORT FOR LEARNING CONCEPTS

We provide a variety of features to support students' learning of the essential topics of precalculus. Explanations that are written in understandable terms, figures and graphs that illustrate examples and concepts, graphing technology that supports and enhances algebraic manipulations, and real-life applications that enrich the topics with meaning all provide opportunities for students to deepen their understanding of mathematics. These features help students make mathematical connections and expand their own knowledge base.

- **Examples** Numbered examples that illustrate the techniques for working exercises are found in every section. We use traditional explanations, side comments, and pointers to describe the steps taken—and to warn students about common pitfalls. Some examples provide additional graphing calculator solutions, although these can be omitted if desired.
- **Now Try Exercises** Following each numbered example, the student is directed to try a corresponding odd-numbered exercise (or exercises). This feature allows for quick feedback to determine whether the student has understood the principles illustrated in the example.
- **Real-Life Applications** We have included hundreds of real-life applications, many with data updated from the previous edition. They come from fields such as business, entertainment, sports, biology, astronomy, geology, music, highway design, and environmental studies.
- **Function Boxes** Beginning in Chapter 2, functions provide a unifying theme throughout the text. Special function boxes offer a comprehensive, visual introduction to each type of function and also serve as an excellent resource for reference and review. Each function box includes a table of values, traditional and calculator-generated graphs, the domain, the range, and other special information about the function. These boxes are assignable in MyMathLab.
- **Figures and Photos** Today's students are more visually oriented than ever before, and we have updated the figures and photos in this edition to promote visual appeal. Guided Visualizations with accompanying exercises and explorations are now available and assignable in MyMathLab.
- **Use of Graphing Technology** We have integrated the use of graphing calculators where appropriate, although ***this technology is completely optional and can be omitted without loss of continuity.*** We continue to stress that graphing calculators support understanding but that students must first master the underlying mathematical concepts. Exercises that require the use of a graphing calculator are marked with the icon.
- **Cautions and Notes** Text that is marked CAUTION warns students of common errors, and NOTE comments point out explanations that should receive particular attention.
- **Looking Ahead to Calculus** These margin notes offer glimpses of how the topics currently being studied are used in calculus.

SUPPORT FOR PRACTICING CONCEPTS

This text offers a wide variety of exercises to help students master precalculus. The extensive exercise sets provide ample opportunity for practice, and the exercise problems increase in difficulty so that students at every level of understanding are challenged. The variety of exercise types promotes understanding of the concepts and reduces the need for rote memorization.

- NEW **Concept Preview** Each exercise set now begins with a group of **CONCEPT PREVIEW** exercises designed to promote understanding of vocabulary and basic concepts of each section. These new exercises are assignable in MyMathLab and will provide support especially for hybrid, online, and flipped courses.

- **Exercise Sets** In addition to traditional drill exercises, this text includes writing exercises, optional graphing calculator problems, and multiple-choice, matching, true/false, and completion exercises. Those marked ***Concept Check*** focus on conceptual thinking. ***Connecting Graphs with Equations*** exercises challenge students to write equations that correspond to given graphs.

- **Relating Concepts Exercises** Appearing at the end of selected exercise sets, these groups of exercises are designed so that students who work them in numerical order will follow a line of reasoning that leads to an understanding of how various topics and concepts are related. All answers to these exercises appear in the student answer section, and these exercises are assignable in MyMathLab.

- **Complete Solutions to Selected Exercises** Exercise numbers marked indicate that a full worked-out solution appears in the eText. These are often exercises that extend the skills and concepts presented in the numbered examples.

SUPPORT FOR REVIEW AND TEST PREP

Ample opportunities for review are found within the chapters and at the ends of chapters. Quizzes that are interspersed within chapters provide a quick assessment of students' understanding of the material presented up to that point in the chapter. Chapter "Test Preps" provide comprehensive study aids to help students prepare for tests.

- **Quizzes** Students can periodically check their progress with in-chapter quizzes that appear in all chapters, beginning with Chapter 1. All answers, with corresponding section references, appear in the student answer section. These quizzes are assignable in MyMathLab.

- **Summary Exercises** These sets of in-chapter exercises give students the all-important opportunity to work *mixed* review exercises, requiring them to synthesize concepts and select appropriate solution methods. The summary exercises are assignable in MyMathLab.

- **End-of-Chapter Test Prep** Following the final numbered section in each chapter, the Test Prep provides a list of **Key Terms,** a list of **New Symbols** (if applicable), and a two-column **Quick Review** that includes a section-by-section summary of concepts and examples. This feature concludes with a comprehensive set of **Review Exercises** and a **Chapter Test.** The Test Prep, Review Exercises, and Chapter Test are assignable in MyMathLab.

Get the most out of MyMathLab®

MyMathLab is the world's leading online resource for teaching and learning mathematics. MyMathLab helps students and instructors improve results, and it provides engaging experiences and personalized learning for each student so learning can happen in any environment. Plus, it offers flexible and time-saving course management features to allow instructors to easily manage their classes while remaining in complete control, regardless of course format.

Personalized Support for Students

- MyMathLab comes with many learning resources–eText, animations, videos, and more–all designed to support your students as they progress through their course.
- The Adaptive Study Plan acts as a personal tutor, updating in real time based on student performance to provide personalized recommendations on what to work on next. With the new Companion Study Plan assignments, instructors can now assign the Study Plan as a prerequisite to a test or quiz, helping to guide students through concepts they need to master.
- Personalized Homework enables instructors to create homework assignments tailored to each student's specific needs and focused on the topics they have not yet mastered.

Used by nearly 4 million students each year, the MyMathLab and MyStatLab family of products delivers consistent, measurable gains in student learning outcomes, retention, and subsequent course success.

BREAKTHROUGH
To improving results

Resources for Success

MyMathLab® Online Course for Lial, Hornsby, Schneider, Daniels *Precalculus*

MyMathLab delivers proven results in helping individual students succeed. The authors Lial, Hornsby, Schneider, and Daniels have developed specific content in MyMathLab to give students the practice they need to develop a conceptual understanding of precalculus and the analytical skills necessary for success in mathematics. The MyMathLab features described here support precalculus students in a variety of classroom formats (traditional, hybrid, and online).

Concept Preview Exercises

Each Homework section now begins with a group of Concept Preview Exercises, assignable in MyMathLab and also available in Learning Catalytics. These may be used to ensure that students understand the related vocabulary and basic concepts before beginning the regular homework problems. Learning Catalytics is a "bring your own device" system of prebuilt questions designed to enhance student engagement and facilitate assessment.

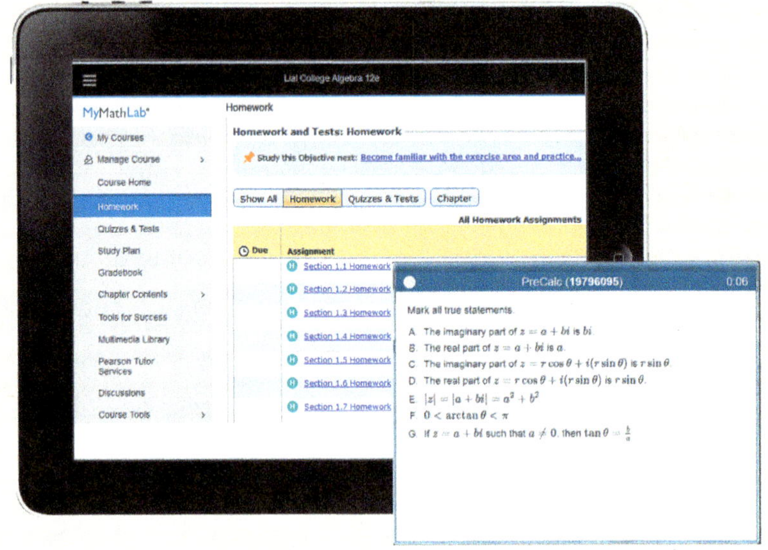

MyNotes and MyClassroomExamples

MyNotes provide a note-taking structure for students to use while they read the text or watch the MyMathLab videos. MyClassroom Examples offer structure for notes taken during lecture and are for use with the Classroom Examples found in the Annotated Instructor Edition.

Both sets of notes are available in MyMathLab and can be customized by the instructor.

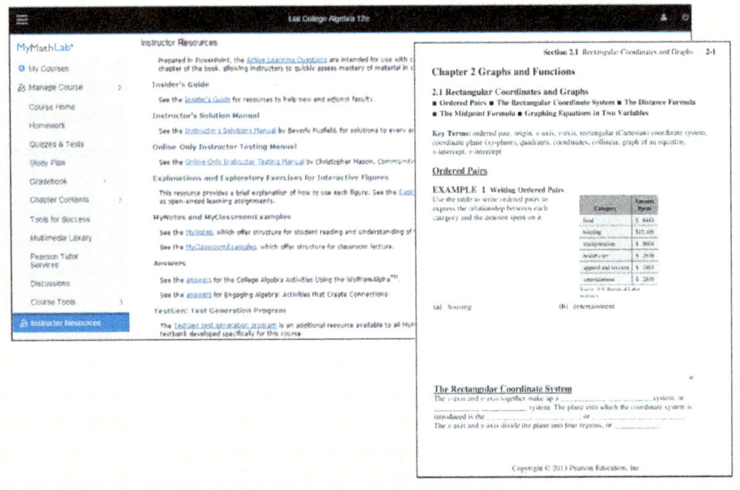

www.mymathlab.com

BREAKTHROUGH
To improving results

Resources for Success

Student Supplements

Student's Solutions Manual

By Beverly Fusfield

- Provides detailed solutions to all odd-numbered text exercises

ISBN: 0-13-431434-4 & 978-0-13-431434-1

Video Lectures with Optional Captioning

- Feature Quick Reviews and Example Solutions:
  Quick Reviews cover key definitions and procedures from each section.
  Example Solutions walk students through the detailed solution process for every example in the textbook.
- Ideal for distance learning or supplemental instruction at home or on campus
- Include optional text captioning
- Available in MyMathLab®

MyNotes

- Available in MyMathLab and offer structure for students as they watch videos or read the text
- Include textbook examples along with ample space for students to write solutions and notes
- Include key concepts along with prompts for students to read, write, and reflect on what they have just learned
- **Customizable** so that instructors can add their own examples or remove examples that are not covered in their courses

MyClassroomExamples

- Available in MyMathLab and offer structure for classroom lecture
- Include Classroom Examples along with ample space for students to write solutions and notes
- Include key concepts along with fill in the blank opportunities to keep students engaged
- **Customizable** so that instructors can add their own examples or remove Classroom Examples that are not covered in their courses

Instructor Supplements

Annotated Instructor's Edition

- Provides answers in the margins to almost all text exercises, as well as helpful Teaching Tips and Classroom Examples
- Includes sample homework assignments indicated by problem numbers underlined in blue within each end-of-section exercise set
- Sample homework problems assignable in MyMathLab

ISBN: 0-13-419239-7 & 978-0-13-419239-0

Online Instructor's Solutions Manual

By Beverly Fusfield

- Provides complete solutions to all text exercises
- Available in MyMathLab or downloadable from Pearson Education's online catalog

Online Instructor's Testing Manual

By David Atwood

- Includes diagnostic pretests, chapter tests, final exams, and additional test items, grouped by section, with answers provided
- Available in MyMathLab or downloadable from Pearson Education's online catalog

TestGen®

- Enables instructors to build, edit, print, and administer tests
- Features a computerized bank of questions developed to cover all text objectives
- Available in MyMathLab or downloadable from Pearson Education's online catalog

Online PowerPoint Presentation and Classroom Example PowerPoints

- Written and designed specifically for this text
- Include figures and examples from the text
- Provide Classroom Example PowerPoints that include full worked-out solutions to all Classroom Examples
- Available in MyMathLab or downloadable from Pearson Education's online catalog

ACKNOWLEDGMENTS

We wish to thank the following individuals who provided valuable input into this edition of the text.

Barbara Aramenta – Pima Community College
Robert Bates – Honolulu Community College
Troy Brachey – Tennessee Tech University
Hugh Cornell – University of North Florida
John E. Daniels – Central Michigan University
Dan Fahringer – HACC Harrisburg
Doug Grenier – Rogers State University
Mary Hill – College of DuPage
Keith Hubbard – Stephen Austin State University
Christine Janowiak – Arapahoe Community College
Tarcia Jones – Austin Community College, Rio Grande
Rene Lumampao – Austin Community College
Rosana Maldonado – South Texas Community College, Pecan
Nilay S. Manzagol – Georgia State University
Marianna McClymonds – Phoenix College
Randy Nichols – Delta College
Preeti Parikh – SUNY Maritime
Deanna Robinson-Briedel – McLennan Community College
Sutandra Sarkar – Georgia State University
Patty Schovanec – Texas Tech University
Jimmy Vincente – El Centro College
Deanna M. Welsch – Illinois Central College
Amanda Wheeler – Amarillo College
Li Zhou – Polk State College

Our sincere thanks to those individuals at Pearson Education who have supported us throughout this revision: Anne Kelly, Christine O'Brien, Joe Vetere, and Danielle Simbajon. Terry McGinnis continues to provide behind-the-scenes guidance for both content and production. We have come to rely on her expertise during all phases of the revision process. Marilyn Dwyer of Cenveo® Publishing Services, with the assistance of Carol Merrigan, provided excellent production work. Special thanks go out to Paul Lorczak, John Morin, and Perian Herring for their excellent accuracy-checking. We thank Lucie Haskins, who provided an accurate index, and Jack Hornsby, who provided assistance in creating calculator screens, researching data updates, and proofreading. We appreciate the valuable suggestions for Chapter 9 that Mary Hill of *College of Dupage* made during our meeting with her in March 2010.

As an author team, we are committed to providing the best possible precalculus course to help instructors teach and students succeed. As we continue to work toward this goal, we welcome any comments or suggestions you might send, via e-mail, to math@pearson.com.

Margaret L. Lial
John Hornsby
David I. Schneider
Callie J. Daniels

R Review of Basic Concepts

Positive and negative numbers, used to represent gains and losses on a board such as this one, are examples of *real numbers* encountered in applications of mathematics.

R.1 Sets

- Basic Definitions
- Operations on Sets

Basic Definitions

A **set** is a collection of objects. The objects that belong to a set are its **elements,** or **members.** In algebra, the elements of a set are usually numbers. Sets are commonly written using **set braces,** $\{\ \}$.

$$\{1, 2, 3, 4\} \quad \text{The set containing the elements 1, 2, 3, and 4}$$

The order in which the elements are listed is not important. As a result, this same set can also be written as $\{4, 3, 2, 1\}$ or with any other arrangement of the four numbers.

To show that 4 is an element of the set $\{1, 2, 3, 4\}$, we use the symbol $\in$.

$$4 \in \{1, 2, 3, 4\}$$

Since 5 is *not* an element of this set, we place a slash through the symbol $\in$.

$$5 \notin \{1, 2, 3, 4\}$$

It is customary to name sets with capital letters.

$$S = \{1, 2, 3, 4\} \quad S \text{ is used to name the set.}$$

Set S was written above by listing its elements. Set S might also be described as

"the set containing the first four counting numbers."

The set F, consisting of all fractions between 0 and 1, is an example of an **infinite set**—one that has an unending list of distinct elements. A **finite set** is one that has a limited number of elements. The process of counting its elements comes to an end.

Some infinite sets can be described by listing. For example, the set of numbers N used for counting, which are the **natural numbers** or the **counting numbers,** can be written as follows.

$$N = \{1, 2, 3, 4, \ldots\} \quad \text{Natural (counting) numbers}$$

The three dots (*ellipsis points*) show that the list of elements of the set continues according to the established pattern.

Sets are often written in **set-builder notation,** which uses a variable, such as x, to describe the elements of the set. The following set-builder notation represents the set $\{3, 4, 5, 6\}$ and is read "the set of all elements x such that x is a natural number between 2 and 7." The numbers 2 and 7 are *not* between 2 and 7.

$$\{x \mid x \text{ is a natural number between 2 and 7}\} = \{3, 4, 5, 6\} \quad \text{Set-builder notation}$$

The set of all elements x — such that — x is a natural number between 2 and 7

EXAMPLE 1 Using Set Notation and Terminology

Identify each set as *finite* or *infinite*. Then determine whether 10 is an element of the set.

(a) $\{7, 8, 9, \ldots, 14\}$

(b) $\left\{1, \frac{1}{4}, \frac{1}{16}, \frac{1}{64}, \ldots\right\}$

(c) $\{x \mid x \text{ is a fraction between 1 and 2}\}$

(d) $\{x \mid x \text{ is a natural number between 9 and 11}\}$

SOLUTION

(a) The set is finite, because the process of counting its elements 7, 8, 9, 10, 11, 12, 13, and 14 comes to an end. The number 10 belongs to the set.

$$10 \in \{7, 8, 9, \ldots, 14\}$$

(b) The set is infinite, because the ellipsis points indicate that the pattern continues indefinitely. In this case,

$$10 \notin \left\{1, \tfrac{1}{4}, \tfrac{1}{16}, \tfrac{1}{64}, \ldots\right\}.$$

(c) Between any two distinct natural numbers there are infinitely many fractions, so this set is infinite. The number 10 is not an element.

(d) There is only one natural number between 9 and 11, namely 10. So the set is finite, and 10 is an element.

✔ **Now Try Exercises 11, 13, 15, and 17.**

EXAMPLE 2 Listing the Elements of a Set

Use set notation, and list all the elements of each set.

(a) $\{x \mid x \text{ is a natural number less than } 5\}$

(b) $\{x \mid x \text{ is a natural number greater than 7 and less than } 14\}$

SOLUTION

(a) The natural numbers less than 5 form the set $\{1, 2, 3, 4\}$.

(b) This is the set $\{8, 9, 10, 11, 12, 13\}$. ✔ **Now Try Exercise 25.**

When we are discussing a particular situation or problem, the **universal set** (whether expressed or implied) contains all the elements included in the discussion. The letter U is used to represent the universal set. The **null set, or empty set,** is the set containing no elements. We write the null set by either using the special symbol $\emptyset$, or else writing set braces enclosing no elements, $\{ \; \}$.

CAUTION Do not combine these symbols. **$\{\emptyset\}$ *is not the null set.*** It is the set containing the symbol $\emptyset$.

Every element of the set $S = \{1, 2, 3, 4\}$ is a natural number. S is an example of a *subset* of the set N of natural numbers. This relationship is written using the symbol $\subseteq$.

$$S \subseteq N$$

By definition, set A is a **subset** of set B if every element of set A is also an element of set B. For example, if $A = \{2, 5, 9\}$ and $B = \{2, 3, 5, 6, 9, 10\}$, then $A \subseteq B$. However, there are some elements of B that are not in A, so B is not a subset of A. This relationship is written using the symbol $\nsubseteq$.

$$B \nsubseteq A$$

Every set is a subset of itself. Also, $\emptyset$ is a subset of every set.

If A is any set, then $A \subseteq A$ and $\emptyset \subseteq A$.

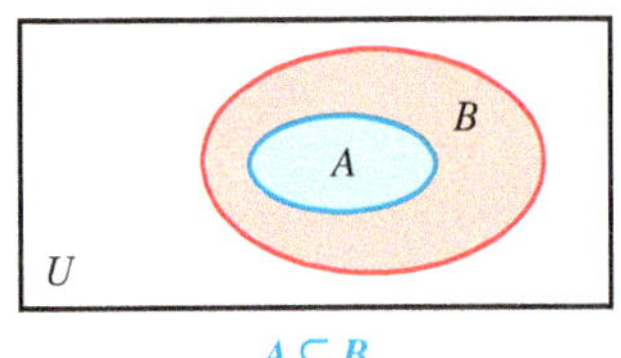

$A \subseteq B$

Figure 1

Figure 1 shows a set A that is a subset of set B. The rectangle in the drawing represents the universal set U. Such a diagram is a **Venn diagram.**

Two sets A and B are equal whenever $A \subseteq B$ and $B \subseteq A$. Equivalently, $A = B$ if the two sets contain exactly the same elements. For example,

$$\{1, 2, 3\} = \{3, 1, 2\}$$

is true because both sets contain exactly the same elements. However,

$$\{1, 2, 3\} \neq \{0, 1, 2, 3\}$$

because the set $\{0, 1, 2, 3\}$ contains the element 0, which is not an element of $\{1, 2, 3\}$.

EXAMPLE 3 Examining Subset Relationships

Let $U = \{1, 3, 5, 7, 9, 11, 13\}$, $A = \{1, 3, 5, 7, 9, 11\}$, $B = \{1, 3, 7, 9\}$, $C = \{3, 9, 11\}$, and $D = \{1, 9\}$. Determine whether each statement is *true* or *false*.

(a) $D \subseteq B$ **(b)** $B \subseteq D$ **(c)** $C \nsubseteq A$ **(d)** $U = A$

SOLUTION

(a) All elements of D, namely 1 and 9, are also elements of B, so D is a subset of B, and $D \subseteq B$ is true.

(b) There is at least one element of B (for example, 3) that is not an element of D, so B is *not* a subset of D. Thus, $B \subseteq D$ is false.

(c) C is a subset of A, because every element of C is also an element of A. Thus, $C \subseteq A$ is true, and as a result, $C \nsubseteq A$ is false.

(d) U contains the element 13, but A does not. Therefore, $U = A$ is false.

✔ **Now Try Exercises 53, 55, 63, and 65.**

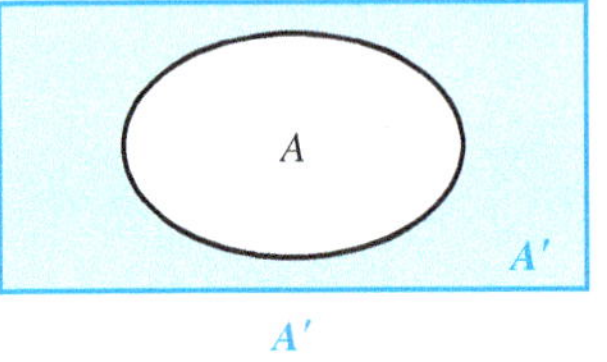

A'

Figure 2

Operations on Sets

Given a set A and a universal set U, the set of all elements of U that do ***not*** belong to set A is the **complement** of set A. For example, if set A is the set of all students in a class 30 years old or older, and set U is the set of all students in the class, then the complement of A would be the set of all students in the class younger than age 30.

The complement of set A is written $\mathbf{A'}$ (read **"A-prime"**). The Venn diagram in **Figure 2** shows a set A. Its complement, A', is in color. Using set-builder notation, the complement of set A is described as follows.

$$A' = \{x \mid x \in U,\ \ x \notin A\}$$

EXAMPLE 4 Finding Complements of Sets

Let $U = \{1, 2, 3, 4, 5, 6, 7\}$, $A = \{1, 3, 5, 7\}$, and $B = \{3, 4, 6\}$. Find each set.

(a) A' **(b)** B' **(c)** $\varnothing'$ **(d)** U'

SOLUTION

(a) Set A' contains the elements of U that are not in A. Thus, $A' = \{2, 4, 6\}$.

(b) $B' = \{1, 2, 5, 7\}$ **(c)** $\varnothing' = U$ **(d)** $U' = \varnothing$

✔ **Now Try Exercise 89.**

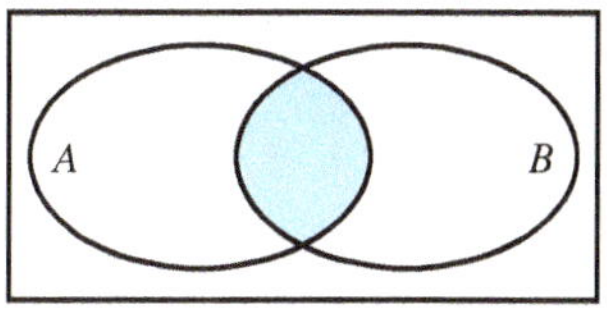

$A \cap B$

Figure 3

Given two sets A and B, the set of all elements belonging both to set A ***and*** to set B is the **intersection** of the two sets, written $A \cap B$. For example, if $A = \{1, 2, 4, 5, 7\}$ and $B = \{2, 4, 5, 7, 9, 11\}$, then we have the following.

$$A \cap B = \{1, 2, 4, 5, 7\} \cap \{2, 4, 5, 7, 9, 11\} = \{2, 4, 5, 7\}$$

The Venn diagram in **Figure 3** shows two sets A and B. Their intersection, $A \cap B$, is in color. Using set-builder notation, the intersection of sets A and B is described as follows.

$$\mathbf{A \cap B = \{x \mid x \in A \text{ and } x \in B\}}$$

Two sets that have no elements in common are **disjoint sets.** If A and B are any two disjoint sets, then $A \cap B = \emptyset$. For example, there are no elements common to both $\{50, 51, 54\}$ and $\{52, 53, 55, 56\}$, so these two sets are disjoint.

$$\{50, 51, 54\} \cap \{52, 53, 55, 56\} = \emptyset$$

EXAMPLE 5 Finding Intersections of Two Sets

Find each of the following. Identify any disjoint sets.

(a) $\{9, 15, 25, 36\} \cap \{15, 20, 25, 30, 35\}$

(b) $\{2, 3, 4, 5, 6\} \cap \{1, 2, 3, 4\}$

(c) $\{1, 3, 5\} \cap \{2, 4, 6\}$

SOLUTION

(a) $\{9, 15, 25, 36\} \cap \{15, 20, 25, 30, 35\} = \{15, 25\}$

The elements 15 and 25 are the only ones belonging to both sets.

(b) $\{2, 3, 4, 5, 6\} \cap \{1, 2, 3, 4\} = \{2, 3, 4\}$

(c) $\{1, 3, 5\} \cap \{2, 4, 6\} = \emptyset$ Disjoint sets

✔ **Now Try Exercises 69, 75, and 85.**

The set of all elements belonging to set A ***or*** to set B (or to both) is the **union** of the two sets, written $A \cup B$. For example, if $A = \{1, 3, 5\}$ and $B = \{3, 5, 7, 9\}$, then we have the following.

$$A \cup B = \{1, 3, 5\} \cup \{3, 5, 7, 9\} = \{1, 3, 5, 7, 9\}$$

The Venn diagram in **Figure 4** shows two sets A and B. Their union, $A \cup B$, is in color.

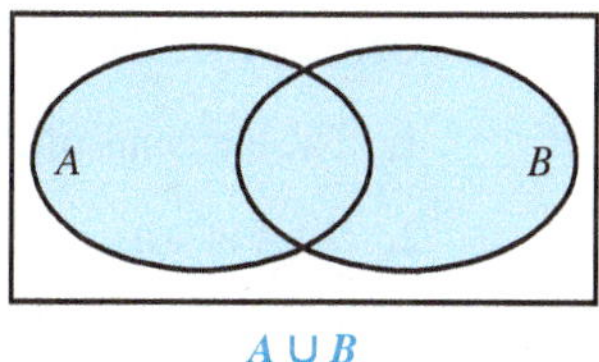

$A \cup B$

Figure 4

Using set-builder notation, the union of sets A and B is described as follows.

$$\mathbf{A \cup B = \{x \mid x \in A \text{ or } x \in B\}}$$

EXAMPLE 6 Finding Unions of Two Sets

Find each of the following.

(a) $\{1, 2, 5, 9, 14\} \cup \{1, 3, 4, 8\}$

(b) $\{1, 3, 5, 7\} \cup \{2, 4, 6\}$

(c) $\{1, 3, 5, 7, \ldots\} \cup \{2, 4, 6, \ldots\}$

SOLUTION

(a) Begin by listing the elements of the first set, $\{1, 2, 5, 9, 14\}$. Then include any elements from the second set that are not already listed.

$$\{1, 2, 5, 9, 14\} \cup \{1, 3, 4, 8\} = \{1, 2, 3, 4, 5, 8, 9, 14\}$$

(b) $\{1, 3, 5, 7\} \cup \{2, 4, 6\} = \{1, 2, 3, 4, 5, 6, 7\}$

(c) $\{1, 3, 5, 7, \ldots\} \cup \{2, 4, 6, \ldots\} = N$ Natural numbers

✔ Now Try Exercises 71 and 83.

The **set operations** are summarized below.

Set Operations

Let A and B define sets, with universal set U.

The **complement** of set A is the set A' of all elements in the universal set that do *not* belong to set A.

$$A' = \{x \mid x \in U, \quad x \notin A\}$$

The **intersection** of sets A and B, written $A \cap B$, is made up of all the elements belonging to both set A *and* set B.

$$A \cap B = \{x \mid x \in A \text{ and } x \in B\}$$

The **union** of sets A and B, written $A \cup B$, is made up of all the elements belonging to set A *or* set B.

$$A \cup B = \{x \mid x \in A \text{ or } x \in B\}$$

R.1 Exercises

CONCEPT PREVIEW *Fill in the blank to correctly complete each sentence.*

1. The elements of the set of natural numbers are ________.
2. Set A is a(n) ________ of set B if every element of set A is also an element of set B.
3. The set of all elements of the universal set U that do not belong to set A is the ________ of set A.
4. The ________ of sets A and B is made up of all the elements belonging to both set A *and* set B.
5. The ________ of sets A and B is made up of all the elements belonging to set A *or* set B (or both).

CONCEPT PREVIEW *Work each problem.*

6. Identify the set $\left\{1, \frac{1}{3}, \frac{1}{9}, \frac{1}{27}, \ldots\right\}$ as finite or infinite.

7. Use set notation and write the elements belonging to the set $\{x \mid x$ is a natural number less than $6\}$.

8. Let $U = \{1, 2, 3, 4, 5\}$ and $A = \{1, 2, 3\}$. Find A'.

9. Find $\{16, 18, 21, 50\} \cap \{15, 16, 17, 18\}$.

10. Find $\{16, 18, 21, 50\} \cup \{15, 16, 17, 18\}$.

Identify each set as finite *or* infinite. *Then determine whether* 10 *is an element of the set.* ***See Example 1.***

11. $\{4, 5, 6, \ldots, 15\}$

12. $\{1, 2, 3, 4, 5, \ldots, 75\}$

13. $\left\{1, \frac{1}{2}, \frac{1}{4}, \frac{1}{8}, \ldots\right\}$

14. $\{4, 5, 6, \ldots\}$

15. $\{x \mid x$ is a natural number greater than $11\}$

16. $\{x \mid x$ is a natural number greater than or equal to $10\}$

17. $\{x \mid x$ is a fraction between 1 and $2\}$

18. $\{x \mid x$ is an even natural number$\}$

Use set notation, and list all the elements of each set. ***See Example 2.***

19. $\{12, 13, 14, \ldots, 20\}$

20. $\{8, 9, 10, \ldots, 17\}$

21. $\left\{1, \frac{1}{2}, \frac{1}{4}, \ldots, \frac{1}{32}\right\}$

22. $\{3, 9, 27, \ldots, 729\}$

23. $\{17, 22, 27, \ldots, 47\}$

24. $\{74, 68, 62, \ldots, 38\}$

25. $\{x \mid x$ is a natural number greater than 8 and less than $15\}$

26. $\{x \mid x$ is a natural number not greater than $4\}$

Insert $\in$ *or* $\notin$ *in each blank to make the resulting statement true.* ***See Examples 1 and 2.***

27. 6 ____ $\{3, 4, 5, 6\}$

28. 9 ____ $\{2, 3, 5, 9, 8\}$

29. 5 ____ $\{4, 6, 8, 10\}$

30. 13 ____ $\{3, 5, 12, 14\}$

31. 0 ____ $\{0, 2, 3, 4\}$

32. 0 ____ $\{0, 5, 6, 7, 8, 10\}$

33. $\{3\}$ ____ $\{2, 3, 4, 5\}$

34. $\{5\}$ ____ $\{3, 4, 5, 6, 7\}$

35. $\{0\}$ ____ $\{0, 1, 2, 5\}$

36. $\{2\}$ ____ $\{2, 4, 6, 8\}$

37. 0 ____ $\varnothing$

38. $\varnothing$ ____ $\varnothing$

Determine whether each statement is true *or* false. ***See Examples 1–3.***

39. $3 \in \{2, 5, 6, 8\}$

40. $6 \in \{2, 5, 8, 9\}$

41. $1 \in \{11, 5, 4, 3, 1\}$

42. $12 \in \{18, 17, 15, 13, 12\}$

43. $9 \notin \{8, 5, 2, 1\}$

44. $3 \notin \{7, 6, 5, 4\}$

45. $\{2, 5, 8, 9\} = \{2, 5, 9, 8\}$

46. $\{3, 0, 9, 6, 2\} = \{2, 9, 0, 3, 6\}$

47. $\{5, 8, 9\} = \{5, 8, 9, 0\}$

48. $\{3, 7, 12, 14\} = \{3, 7, 12, 14, 0\}$

49. $\{x \mid x$ is a natural number less than $3\} = \{1, 2\}$

50. $\{x \mid x$ is a natural number greater than $10\} = \{11, 12, 13, \ldots\}$

Let $A = \{2, 4, 6, 8, 10, 12\}$, $B = \{2, 4, 8, 10\}$, $C = \{4, 10, 12\}$, $D = \{2, 10\}$, *and* $U = \{2, 4, 6, 8, 10, 12, 14\}$.

Determine whether each statement is true *or* false. **See Example 3.**

51. $A \subseteq U$
52. $C \subseteq U$
53. $D \subseteq B$
54. $D \subseteq A$
55. $A \subseteq B$
56. $B \subseteq C$
57. $\varnothing \subseteq A$
58. $\varnothing \subseteq \varnothing$
59. $\{4, 8, 10\} \subseteq B$
60. $\{0, 2\} \subseteq D$
61. $B \subseteq D$
62. $A \subseteq C$

Insert $\subseteq$ *or* $\nsubseteq$ *in each blank to make the resulting statement true.* **See Example 3.**

63. $\{2, 4, 6\}$ _____ $\{2, 3, 4, 5, 6\}$
64. $\{1, 5\}$ _____ $\{0, 1, 2, 3, 5\}$
65. $\{0, 1, 2\}$ _____ $\{1, 2, 3, 4, 5\}$
66. $\{5, 6, 7, 8\}$ _____ $\{1, 2, 3, 4, 5, 6, 7\}$
67. $\varnothing$ _____ $\{1, 4, 6, 8\}$
68. $\varnothing$ _____ $\varnothing$

Determine whether each statement is true *or* false. **See Examples 4–6.**

69. $\{5, 7, 9, 19\} \cap \{7, 9, 11, 15\} = \{7, 9\}$
70. $\{8, 11, 15\} \cap \{8, 11, 19, 20\} = \{8, 11\}$
71. $\{1, 2, 7\} \cup \{1, 5, 9\} = \{1\}$
72. $\{6, 12, 14, 16\} \cup \{6, 14, 19\} = \{6, 14\}$
73. $\{2, 3, 5, 9\} \cap \{2, 7, 8, 10\} = \{2\}$
74. $\{6, 8, 9\} \cup \{9, 8, 6\} = \{8, 9\}$
75. $\{3, 5, 9, 10\} \cap \varnothing = \{3, 5, 9, 10\}$
76. $\{3, 5, 9, 10\} \cup \varnothing = \{3, 5, 9, 10\}$
77. $\{1, 2, 4\} \cup \{1, 2, 4\} = \{1, 2, 4\}$
78. $\{1, 2, 4\} \cap \{1, 2, 4\} = \varnothing$
79. $\varnothing \cup \varnothing = \varnothing$
80. $\varnothing \cap \varnothing = \varnothing$

Let $U = \{0, 1, 2, 3, 4, 5, 6, 7, 8, 9, 10, 11, 12, 13\}$, $M = \{0, 2, 4, 6, 8\}$, $N = \{1, 3, 5, 7, 9, 11, 13\}$, $Q = \{0, 2, 4, 6, 8, 10, 12\}$, *and* $R = \{0, 1, 2, 3, 4\}$.

Use these sets to find each of the following. Identify any disjoint sets. **See Examples 4–6.**

81. $M \cap R$
82. $M \cap U$
83. $M \cup N$
84. $M \cup R$
85. $M \cap N$
86. $U \cap N$
87. $N \cup R$
88. $M \cup Q$
89. N'
90. Q'
91. $M' \cap Q$
92. $Q \cap R'$
93. $\varnothing \cap R$
94. $\varnothing \cap Q$
95. $N \cup \varnothing$
96. $R \cup \varnothing$
97. $(M \cap N) \cup R$
98. $(N \cup R) \cap M$
99. $(Q \cap M) \cup R$
100. $(R \cup N) \cap M'$
101. $(M' \cup Q) \cap R$
102. $Q \cap (M \cup N)$
103. $Q' \cap (N' \cap U)$
104. $(U \cap \varnothing') \cup R$
105. $\{x \mid x \in U,\ \ x \notin M\}$
106. $\{x \mid x \in U,\ \ x \notin R\}$
107. $\{x \mid x \in M \text{ and } x \in Q\}$
108. $\{x \mid x \in Q \text{ and } x \in R\}$
109. $\{x \mid x \in M \text{ or } x \in Q\}$
110. $\{x \mid x \in Q \text{ or } x \in R\}$

R.2 Real Numbers and Their Properties

- Sets of Numbers and the Number Line
- Exponents
- Order of Operations
- Properties of Real Numbers
- Order on the Number Line
- Absolute Value

Sets of Numbers and the Number Line

As mentioned previously, the set of **natural numbers** is written in set notation as follows.

$$\{1, 2, 3, 4, \ldots\} \quad \text{Natural numbers}$$

Including 0 with the set of natural numbers gives the set of **whole numbers.**

$$\{0, 1, 2, 3, 4, \ldots\} \quad \text{Whole numbers}$$

Including the negatives of the natural numbers with the set of whole numbers gives the set of **integers.**

$$\{\ldots, -3, -2, -1, 0, 1, 2, 3, \ldots\} \quad \text{Integers}$$

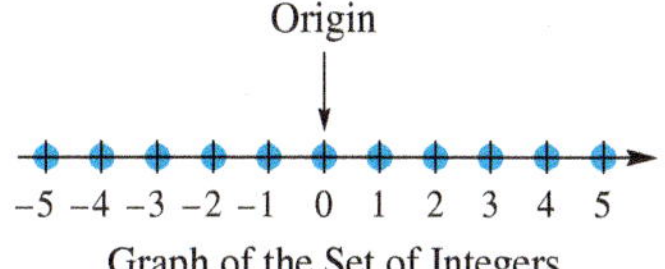

Graph of the Set of Integers

Figure 5

Integers can be graphed on a **number line.** See **Figure 5.** Every number corresponds to one and only one point on the number line, and each point corresponds to one and only one number. The number associated with a given point is the **coordinate** of the point. This correspondence forms a **coordinate system.**

The result of dividing two integers (with a nonzero divisor) is a *rational number*, or *fraction.* A **rational number** is an element of the set defined as follows.

$$\left\{\frac{p}{q} \,\middle|\, p \text{ and } q \text{ are integers and } q \neq 0\right\} \quad \text{Rational numbers}$$

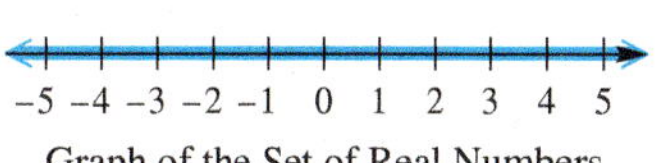

Graph of the Set of Real Numbers

Figure 6

The set of rational numbers includes the natural numbers, the whole numbers, and the integers. For example, the integer -3 is a rational number because it can be written as $\frac{-3}{1}$. Numbers that can be written as repeating or terminating decimals are also rational numbers. For example, $0.\overline{6} = 0.66666\ldots$ represents a rational number that can be expressed as the fraction $\frac{2}{3}$.

The set of all numbers that correspond to points on a number line is the **real numbers,** shown in **Figure 6.** Real numbers can be represented by decimals. Because every fraction has a decimal form—for example, $\frac{1}{4} = 0.25$—real numbers include rational numbers.

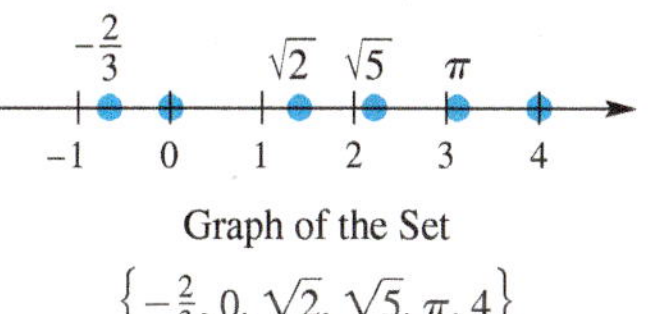

Graph of the Set $\left\{-\frac{2}{3}, 0, \sqrt{2}, \sqrt{5}, \pi, 4\right\}$

$\sqrt{2}$, $\sqrt{5}$, and π are irrational. Because $\sqrt{2}$ is approximately equal to 1.41, it is located between 1 and 2, slightly closer to 1.

Figure 7

Some real numbers cannot be represented by quotients of integers. These numbers are **irrational numbers.** The set of irrational numbers includes $\sqrt{2}$ and $\sqrt{5}$. Another irrational number is π, which is *approximately* equal to 3.14159. Some rational and irrational numbers are graphed in **Figure 7.**

The sets of numbers discussed so far are summarized as follows.

Sets of Numbers

Set	Description
Natural numbers	$\{1, 2, 3, 4, \ldots\}$
Whole numbers	$\{0, 1, 2, 3, 4, \ldots\}$
Integers	$\{\ldots, -3, -2, -1, 0, 1, 2, 3, \ldots\}$
Rational numbers	$\left\{\frac{p}{q} \,\middle\|\, p \text{ and } q \text{ are integers and } q \neq 0\right\}$
Irrational numbers	$\{x \mid x \text{ is real but not rational}\}$
Real numbers	$\{x \mid x \text{ corresponds to a point on a number line}\}$

EXAMPLE 1 Identifying Sets of Numbers

Let $A = \left\{-8, -6, -\frac{12}{4}, -\frac{3}{4}, 0, \frac{3}{8}, \frac{1}{2}, 1, \sqrt{2}, \sqrt{5}, 6\right\}$. List all the elements of A that belong to each set.

(a) Natural numbers **(b)** Whole numbers **(c)** Integers

(d) Rational numbers **(e)** Irrational numbers **(f)** Real numbers

SOLUTION

(a) Natural numbers: 1 and 6 **(b)** Whole numbers: 0, 1, and 6

(c) Integers: $-8, -6, -\frac{12}{4}$ (or -3), 0, 1, and 6

(d) Rational numbers: $-8, -6, -\frac{12}{4}$ (or -3), $-\frac{3}{4}, 0, \frac{3}{8}, \frac{1}{2}$, 1, and 6

(e) Irrational numbers: $\sqrt{2}$ and $\sqrt{5}$

(f) All elements of A are real numbers.

✔ Now Try Exercises 11, 13, and 15.

Figure 8 shows the relationships among the subsets of the real numbers. As shown, the natural numbers are a subset of the whole numbers, which are a subset of the integers, which are a subset of the rational numbers. The union of the rational numbers and irrational numbers is the set of real numbers.

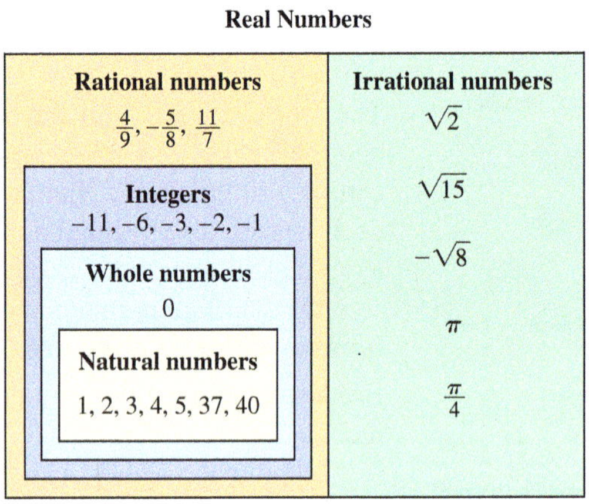

Figure 8

Exponents

Any collection of numbers or variables joined by the basic operations of addition, subtraction, multiplication, or division (except by 0), or the operations of raising to powers or taking roots, formed according to the rules of algebra, is an **algebraic expression.**

$$-2x^2 + 3x, \quad \frac{15y}{2y - 3}, \quad \sqrt{m^3 - 64}, \quad (3a + b)^4 \qquad \text{Algebraic expressions}$$

The expression 2^3 is an **exponential expression,** or **exponential,** where the 3 indicates that three factors of 2 appear in the corresponding product. The number 2 is the **base,** and the number 3 is the **exponent.**

$$2^3 = \underbrace{2 \cdot 2 \cdot 2}_{\text{Three factors of 2}} = 8$$

Exponent: 3

Base: 2

Exponential Notation

If n is any positive integer and a is any real number, then the nth power of a is written using exponential notation as follows.

$$a^n = \underbrace{a \cdot a \cdot a \cdot \ldots \cdot a}_{n \text{ factors of } a}$$

Read a^n as **"a to the nth power"** or simply **"a to the nth."**

EXAMPLE 2 Evaluating Exponential Expressions

Evaluate each exponential expression, and identify the base and the exponent.

(a) 4^3 **(b)** $(-6)^2$ **(c)** -6^2 **(d)** $4 \cdot 3^2$ **(e)** $(4 \cdot 3)^2$

SOLUTION

(a) $4^3 = \underbrace{4 \cdot 4 \cdot 4}_{3 \text{ factors of } 4} = 64$ The base is 4 and the exponent is 3.

(b) $(-6)^2 = (-6)(-6) = 36$ The base is -6 and the exponent is 2.

(c) $-6^2 = -(6 \cdot 6) = -36$ Notice that parts (b) and (c) are different.
The base is 6 and the exponent is 2.

(d) $4 \cdot 3^2 = 4 \cdot 3 \cdot 3 = 36$ The base is 3 and the exponent is 2.
$3^2 = 3 \cdot 3$, NOT $3 \cdot 2$

(e) $(4 \cdot 3)^2 = 12^2 = 144$ $(4 \cdot 3)^2 \neq 4 \cdot 3^2$
The base is $4 \cdot 3$, or 12, and the exponent is 2.

✔ **Now Try Exercises 17, 19, 21, and 23.**

Order of Operations When an expression involves more than one operation symbol, such as $5 \cdot 2 + 3$, we use the following order of operations.

Order of Operations

If grouping symbols such as parentheses, square brackets, absolute value bars, or fraction bars are present, begin as follows.

Step 1 Work separately above and below each **fraction bar.**

Step 2 Use the rules below within each set of **parentheses** or **square brackets.** Start with the innermost set and work outward.

If no grouping symbols are present, follow these steps.

Step 1 Simplify all **powers** and **roots.** ***Work from left to right.***

Step 2 Do any **multiplications** or **divisions** in order. ***Work from left to right.***

Step 3 Do any **negations, additions,** or **subtractions** in order. ***Work from left to right.***

EXAMPLE 3 Using Order of Operations

Evaluate each expression.

(a) $6 \div 3 + 2^3 \cdot 5$

(b) $(8 + 6) \div 7 \cdot 3 - 6$

(c) $\dfrac{4 + 3^2}{6 - 5 \cdot 3}$

(d) $\dfrac{-(-3)^3 + (-5)}{2(-8) - 5(3)}$

SOLUTION

(a)

$$
\begin{aligned}
& 6 \div 3 + 2^3 \cdot 5 \\
&= 6 \div 3 + 8 \cdot 5 && \text{Evaluate the exponential.} \\
&= 2 + 8 \cdot 5 && \text{Divide.} \\
&= 2 + 40 && \text{Multiply.} \\
&= 42 && \text{Add.}
\end{aligned}
$$

Multiply or divide *in order from left to right.*

(b)

$$
\begin{aligned}
& (8 + 6) \div 7 \cdot 3 - 6 \\
&= 14 \div 7 \cdot 3 - 6 && \text{Work inside the parentheses.} \\
&= 2 \cdot 3 - 6 && \text{Divide.} \\
&= 6 - 6 && \text{Multiply.} \\
&= 0 && \text{Subtract.}
\end{aligned}
$$

Be careful to divide *before* multiplying here.

(c) Work separately above and below the fraction bar, and then divide as a last step.

$$
\begin{aligned}
& \frac{4 + 3^2}{6 - 5 \cdot 3} \\
&= \frac{4 + 9}{6 - 15} && \text{Evaluate the exponential and multiply.} \\
&= \frac{13}{-9} && \text{Add and subtract.} \\
&= -\frac{13}{9} && \tfrac{a}{-b} = -\tfrac{a}{b}
\end{aligned}
$$

(d)

$$
\begin{aligned}
& \frac{-(-3)^3 + (-5)}{2(-8) - 5(3)} \\
&= \frac{-(-27) + (-5)}{2(-8) - 5(3)} && \text{Evaluate the exponential.} \\
&= \frac{27 + (-5)}{-16 - 15} && \text{Multiply.} \\
&= \frac{22}{-31} && \text{Add and subtract.} \\
&= -\frac{22}{31} && \tfrac{a}{-b} = -\tfrac{a}{b}
\end{aligned}
$$

✔ **Now Try Exercises 25, 27, and 33.**

EXAMPLE 4 Using Order of Operations

Evaluate each expression for $x = -2$, $y = 5$, and $z = -3$.

(a) $-4x^2 - 7y + 4z$ **(b)** $\dfrac{2(x-5)^2 + 4y}{z+4}$ **(c)** $\dfrac{\frac{x}{2} - \frac{y}{5}}{\frac{3z}{9} + \frac{8y}{5}}$

SOLUTION

(a)

$$-4x^2 - 7y + 4z$$

$$= -4(-2)^2 - 7(5) + 4(-3) \quad \text{Substitute: } x = -2,\ y = 5, \text{ and } z = -3.$$

Use parentheses around substituted values to avoid errors.

$$= -4(4) - 7(5) + 4(-3) \quad \text{Evaluate the exponential.}$$

$$= -16 - 35 - 12 \quad \text{Multiply.}$$

$$= -63 \quad \text{Subtract.}$$

(b)

$$\frac{2(x-5)^2 + 4y}{z+4}$$

$$= \frac{2(-2-5)^2 + 4(5)}{-3+4} \quad \text{Substitute: } x = -2,\ y = 5, \text{ and } z = -3.$$

$$= \frac{2(-7)^2 + 4(5)}{-3+4} \quad \text{Work inside the parentheses.}$$

$$= \frac{2(49) + 4(5)}{-3+4} \quad \text{Evaluate the exponential.}$$

$$= \frac{98 + 20}{1} \quad \text{Multiply in the numerator. Add in the denominator.}$$

$$= 118 \quad \text{Add; } \tfrac{a}{1} = a.$$

(c) This is a *complex fraction*. Work separately above and below the main fraction bar, and then divide as a last step.

$$\frac{\frac{x}{2} - \frac{y}{5}}{\frac{3z}{9} + \frac{8y}{5}}$$

$$= \frac{\frac{-2}{2} - \frac{5}{5}}{\frac{3(-3)}{9} + \frac{8(5)}{5}} \quad \text{Substitute: } x = -2,\ y = 5, \text{ and } z = -3.$$

$$= \frac{-1 - 1}{-1 + 8} \quad \text{Simplify the fractions.}$$

$$= -\frac{2}{7} \quad \text{Subtract and add; } \tfrac{-a}{b} = -\tfrac{a}{b}.$$

✔ **Now Try Exercises 35, 43, and 45.**

Properties of Real Numbers Recall the following basic properties.

Properties of Real Numbers

Let a, b, and c represent real numbers.

Property	Description
Closure Properties $a + b$ is a real number. ab is a real number.	The sum or product of two real numbers is a real number.
Commutative Properties $a + b = b + a$ $ab = ba$	The sum or product of two real numbers is the same regardless of their order.
Associative Properties $(a + b) + c = a + (b + c)$ $(ab)c = a(bc)$	The sum or product of three real numbers is the same no matter which two are added or multiplied first.
Identity Properties There exists a unique real number 0 such that $a + 0 = a$ and $0 + a = a$. There exists a unique real number 1 such that $a \cdot 1 = a$ and $1 \cdot a = a$.	The sum of a real number and 0 is that real number, and the product of a real number and 1 is that real number.
Inverse Properties There exists a unique real number $-a$ such that $a + (-a) = 0$ and $-a + a = 0$. If $a \neq 0$, there exists a unique real number $\frac{1}{a}$ such that $a \cdot \frac{1}{a} = 1$ and $\frac{1}{a} \cdot a = 1$.	The sum of any real number and its negative is 0, and the product of any nonzero real number and its reciprocal is 1.
Distributive Properties $a(b + c) = ab + ac$ $a(b - c) = ab - ac$	The product of a real number and the sum (or difference) of two real numbers equals the sum (or difference) of the products of the first number and each of the other numbers.
Multiplication Property of Zero $0 \cdot a = a \cdot 0 = 0$	The product of a real number and 0 is 0.

CAUTION With the commutative properties, the ***order*** changes, but with the associative properties, the ***grouping*** changes.

Commutative Properties	Associative Properties
$(x + 4) + 9 = (4 + x) + 9$	$(x + 4) + 9 = x + (4 + 9)$
$7 \cdot (5 \cdot 2) = (5 \cdot 2) \cdot 7$	$7 \cdot (5 \cdot 2) = (7 \cdot 5) \cdot 2$

EXAMPLE 5 Simplifying Expressions

Use the commutative and associative properties to simplify each expression.

(a) $6 + (9 + x)$ **(b)** $\frac{5}{8}(16y)$ **(c)** $-10p\left(\frac{6}{5}\right)$

SOLUTION

(a) $6 + (9 + x)$

$= (6 + 9) + x$ Associative property

$= 15 + x$ Add.

(b) $\frac{5}{8}(16y)$

$= \left(\frac{5}{8} \cdot 16\right)y$ Associative property

$= 10y$ Multiply.

(c) $-10p\left(\frac{6}{5}\right)$

$= \frac{6}{5}(-10p)$ Commutative property

$= \left[\frac{6}{5}(-10)\right]p$ Associative property

$= -12p$ Multiply.

✔ **Now Try Exercises 63 and 65.**

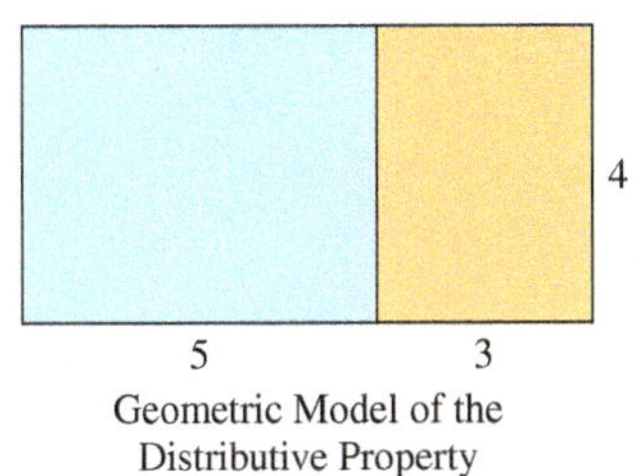

Geometric Model of the Distributive Property

Figure 9

Figure 9 helps to explain the distributive property. The area of the entire region shown can be found in two ways, as follows.

$$4(5 + 3) = 4(8) = 32$$

or

$$4(5) + 4(3) = 20 + 12 = 32$$

The result is the same. This means that

$$4(5 + 3) = 4(5) + 4(3).$$

EXAMPLE 6 Using the Distributive Property

Rewrite each expression using the distributive property and simplify, if possible.

(a) $3(x + y)$ **(b)** $-(m - 4n)$ **(c)** $7p + 21$ **(d)** $\frac{1}{3}\left(\frac{4}{5}m - \frac{3}{2}n - 27\right)$

SOLUTION

(a) $3(x + y)$

$= 3x + 3y$

Distributive property

(b) $-(m - 4n)$

$= -1(m - 4n)$

$= -1(m) + (-1)(-4n)$ Be careful with the negative signs.

$= -m + 4n$

(c) $7p + 21$

$= 7p + 7 \cdot 3$

$= 7 \cdot p + 7 \cdot 3$

$= 7(p + 3)$

Distributive property in reverse

(d) $\frac{1}{3}\left(\frac{4}{5}m - \frac{3}{2}n - 27\right)$

$= \frac{1}{3}\left(\frac{4}{5}m\right) + \frac{1}{3}\left(-\frac{3}{2}n\right) + \frac{1}{3}(-27)$

$= \frac{4}{15}m - \frac{1}{2}n - 9$

✔ **Now Try Exercises 67, 69, and 71.**

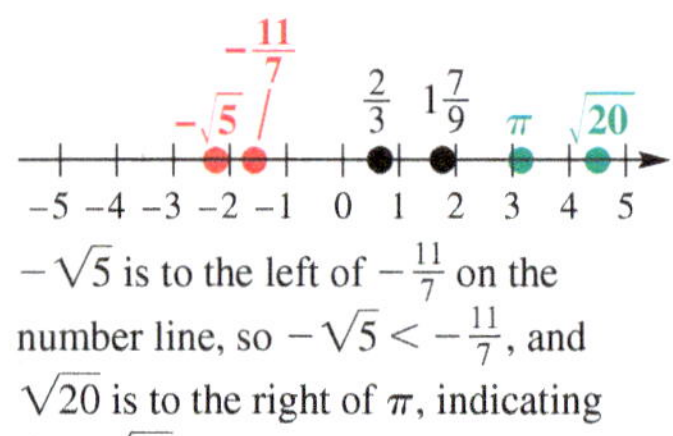

$-\sqrt{5}$ is to the left of $-\frac{11}{7}$ on the number line, so $-\sqrt{5} < -\frac{11}{7}$, and $\sqrt{20}$ is to the right of π, indicating that $\sqrt{20} > \pi$.

Figure 10

Order on the Number Line If the real number a is to the left of the real number b on a number line, then

***a* is less than *b*,** written $a < b$.

If a is to the right of b, then

***a* is greater than *b*,** written $a > b$.

The inequality symbol must point toward the lesser number.

See **Figure 10.** Statements involving these symbols, as well as the symbols less than or equal to, $\leq$, and greater than or equal to, $\geq$, are **inequalities.** The inequality $a < b < c$ says that b is ***between*** a and c because $a < b$ and $b < c$.

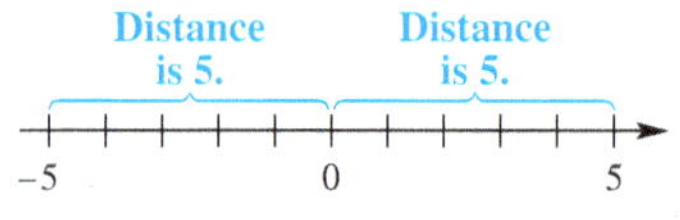

Figure 11

Absolute Value The undirected distance on a number line from a number to 0 is the **absolute value** of that number. The absolute value of the number a is written $|a|$. For example, the distance on a number line from 5 to 0 is 5, as is the distance from -5 to 0. See **Figure 11.** Therefore, both of the following are true.

$$|5| = 5 \quad \text{and} \quad |-5| = 5$$

NOTE ***Because distance cannot be negative, the absolute value of a number is always positive or* 0.**

The algebraic definition of absolute value follows.

Absolute Value

Let a represent a real number.

$$|a| = \begin{cases} a & \text{if } a \geq 0 \\ -a & \text{if } a < 0 \end{cases}$$

That is, the absolute value of a positive number or* 0 *equals that number, while the absolute value of a negative number equals its negative (or opposite).

EXAMPLE 7 Evaluating Absolute Values

Evaluate each expression.

(a) $\left|-\frac{5}{8}\right|$ **(b)** $-|8|$ **(c)** $-|-2|$ **(d)** $|2x|$, for $x = \pi$

SOLUTION

(a) $\left|-\frac{5}{8}\right| = \frac{5}{8}$

(b) $-|8| = -(8) = -8$

(c) $-|-2| = -(2) = -2$

(d) $|2\pi| = 2\pi$

✔ **Now Try Exercises 83 and 87.**

Absolute value is useful in applications where only the *size* (or magnitude), not the *sign*, of the difference between two numbers is important.

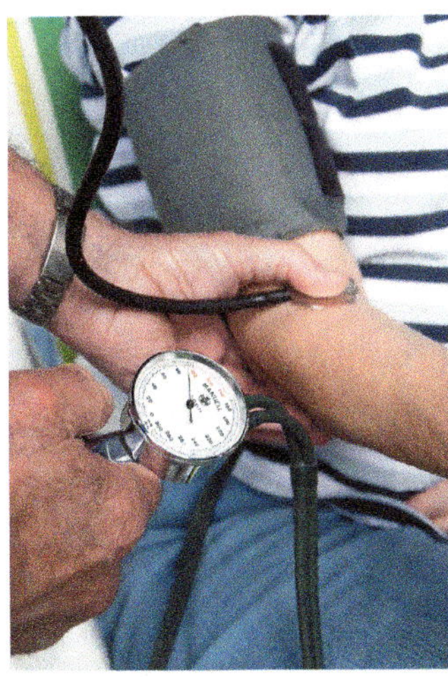

EXAMPLE 8 **Measuring Blood Pressure Difference**

Systolic blood pressure is the maximum pressure produced by each heartbeat. Both low blood pressure and high blood pressure may be cause for medical concern. Therefore, health care professionals are interested in a patient's "pressure difference from normal," or P_d.

If 120 is considered a normal systolic pressure, then

$$P_d = |P - 120|, \quad \text{where } P \text{ is the patient's recorded systolic pressure.}$$

Find P_d for a patient with a systolic pressure, P, of 113.

SOLUTION

$$\begin{aligned} P_d &= |P - 120| & \\ &= |113 - 120| & \text{Let } P = 113. \\ &= |-7| & \text{Subtract.} \\ &= 7 & \text{Definition of absolute value} \end{aligned}$$

✔ **Now Try Exercise 89.**

Properties of Absolute Value

Let a and b represent real numbers.

Property	Description
1. $\|a\| \geq 0$	The absolute value of a real number is positive or 0.
2. $\|-a\| = \|a\|$	The absolute values of a real number and its opposite are equal.
3. $\|a\| \cdot \|b\| = \|ab\|$	The product of the absolute values of two real numbers equals the absolute value of their product.
4. $\frac{\|a\|}{\|b\|} = \left\|\frac{a}{b}\right\|$ $(b \neq 0)$	The quotient of the absolute values of two real numbers equals the absolute value of their quotient.
5. $\|a + b\| \leq \|a\| + \|b\|$ (the triangle inequality)	The absolute value of the sum of two real numbers is less than or equal to the sum of their absolute values.

LOOKING AHEAD TO CALCULUS

One of the most important definitions in calculus, that of the **limit,** uses absolute value. The symbols ϵ (epsilon) and δ (delta) are often used to represent small quantities in mathematics.

Suppose that a function f is defined at every number in an open interval I containing a, except perhaps at a itself. Then the limit of $f(x)$ as x approaches a is L, written

$$\lim_{x \to a} f(x) = L,$$

if for every $\epsilon > 0$ there exists a $\delta > 0$ such that $|f(x) - L| < \epsilon$ whenever $0 < |x - a| < \delta$.

Examples of Properties 1–4:

$|-15| = 15$ and $15 \geq 0$. — Property 1

$|-10| = 10$ and $|10| = 10$, so $|-10| = |10|$. — Property 2

$|5| \cdot |-4| = 5 \cdot 4 = 20$ and $|5(-4)| = |-20| = 20$, so $|5| \cdot |-4| = |5(-4)|$. — Property 3

$\frac{|2|}{|3|} = \frac{2}{3}$ and $\left|\frac{2}{3}\right| = \frac{2}{3}$, so $\frac{|2|}{|3|} = \left|\frac{2}{3}\right|$. — Property 4

Example of the triangle inequality:

$$|a + b| = |3 + (-7)| = |-4| = 4$$
$$|a| + |b| = |3| + |-7| = 3 + 7 = 10$$

Let $a = 3$ and $b = -7$.

Thus, $|a + b| \leq |a| + |b|$. — Property 5

NOTE As seen in **Example 9(b),** absolute value bars can also act as grouping symbols. Remember this when applying the rules for order of operations.

EXAMPLE 9 Evaluating Absolute Value Expressions

Let $x = -6$ and $y = 10$. Evaluate each expression.

(a) $|2x - 3y|$ **(b)** $\frac{2|x| - |3y|}{|xy|}$

SOLUTION

(a) $|2x - 3y|$

$= |2(-6) - 3(10)|$ Substitute: $x = -6$, $y = 10$.

$= |-12 - 30|$ Work inside the absolute value bars. Multiply.

$= |-42|$ Subtract.

$= 42$ Definition of absolute value

(b) $\frac{2|x| - |3y|}{|xy|}$

$= \frac{2|-6| - |3(10)|}{|-6(10)|}$ Substitute: $x = -6$, $y = 10$.

$= \frac{2 \cdot 6 - |30|}{|-60|}$ $|-6| = 6$; Multiply.

$= \frac{12 - 30}{60}$ Multiply; $|30| = 30$, $|-60| = 60$.

$= \frac{-18}{60}$ Subtract.

$= -\frac{3}{10}$ Write in lowest terms; $\frac{-a}{b} = -\frac{a}{b}$.

✔ **Now Try Exercises 93 and 95.**

Distance between Points on a Number Line

If P and Q are points on a number line with coordinates a and b, respectively, then the distance $d(P, Q)$ between them is given by the following.

$$d(P, Q) = |b - a| \quad \text{or} \quad d(P, Q) = |a - b|$$

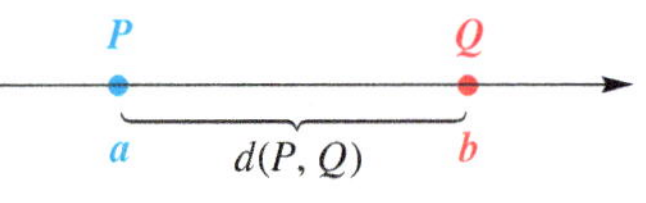

Figure 12

That is, the distance between two points on a number line is the absolute value of the difference between their coordinates in either order. See **Figure 12.**

EXAMPLE 10 **Finding the Distance between Two Points**

Find the distance between -5 and 8.

SOLUTION Use the first formula in the preceding box, with $a = -5$ and $b = 8$.

$$d(P, Q) = |b - a| = |8 - (-5)| = |8 + 5| = |13| = 13$$

Using the second formula in the box, we obtain the same result.

$$d(P, Q) = |a - b| = |(-5) - 8| = |-13| = 13$$

✔ **Now Try Exercise 105.**

R.2 Exercises

CONCEPT PREVIEW *Fill in the blank(s) to correctly complete each sentence.*

1. $\{0, 1, 2, 3, \ldots\}$ describes the set of ________.

2. $\{\ldots, -3, -2, -1, 0, 1, 2, 3, \ldots\}$ describes the set of ________.

3. In the expression 6^3, 6 is the ________, and 3 is the ________.

4. If the real number a is to the left of the real number b on a number line, then a ________ b.

5. The distance on a number line from a number to 0 is the ________ of that number.

6. **CONCEPT PREVIEW** Match each number from Column I with the letter or letters of the sets of numbers from Column II to which the number belongs. There may be more than one choice, so give all choices.

I		II	
(a) 0	**(b)** 34	**A.** Natural numbers	**B.** Whole numbers
(c) $-\frac{9}{4}$	**(d)** $\sqrt{36}$	**C.** Integers	**D.** Rational numbers
(e) $\sqrt{13}$	**(f)** 2.16	**E.** Irrational numbers	**F.** Real numbers

CONCEPT PREVIEW *Evaluate each expression.*

7. 10^3

8. $2 \cdot 5 - 10 \div 2$

9. $|-4|$

10. **CONCEPT PREVIEW** Simplify the expression $-7(x - 4y)$.

Let $A = \left\{-6, -\frac{12}{4}, -\frac{5}{8}, -\sqrt{3}, 0, \frac{1}{4}, 1, 2\pi, 3, \sqrt{12}\right\}$. *List all the elements of A that belong to each set.* ***See Example 1.***

11. Natural numbers

12. Whole numbers

13. Integers

14. Rational numbers

15. Irrational numbers

16. Real numbers

Evaluate each expression. ***See Example 2.***

17. -2^4

18. -3^5

19. $(-2)^4$

20. $(-2)^6$

21. $(-3)^5$

22. $(-2)^5$

23. $-2 \cdot 3^4$

24. $-4 \cdot 5^3$

Evaluate each expression. **See Example 3.**

25. $-2 \cdot 5 + 12 \div 3$

26. $9 \cdot 3 - 16 \div 4$

27. $-4(9 - 8) + (-7)(2)^3$

28. $6(-5) - (-3)(2)^4$

29. $(4 - 2^3)(-2 + \sqrt{25})$

30. $(5 - 3^2)(\sqrt{16} - 2^3)$

31. $\left(-\frac{2}{9} - \frac{1}{4}\right) - \left[-\frac{5}{18} - \left(-\frac{1}{2}\right)\right]$

32. $\left[-\frac{5}{8} - \left(-\frac{2}{5}\right)\right] - \left(\frac{3}{2} - \frac{11}{10}\right)$

33. $\dfrac{-8 + (-4)(-6) \div 12}{4 - (-3)}$

34. $\dfrac{15 \div 5 \cdot 4 \div 6 - 8}{-6 - (-5) - 8 \div 2}$

Evaluate each expression for $p = -4$, $q = 8$, *and* $r = -10$. **See Example 4.**

35. $-p^2 - 7q + r^2$

36. $-p^2 - 2q + r$

37. $\dfrac{q + r}{q + p}$

38. $\dfrac{p + r}{p + q}$

39. $\dfrac{3q}{r} - \dfrac{5}{p}$

40. $\dfrac{3r}{q} - \dfrac{2}{r}$

41. $\dfrac{5r}{2p - 3r}$

42. $\dfrac{3q}{3p - 2r}$

43. $\dfrac{\frac{q}{2} - \frac{r}{3}}{\frac{3p}{4} + \frac{q}{8}}$

44. $\dfrac{\frac{q}{4} - \frac{r}{5}}{\frac{p}{2} + \frac{q}{2}}$

45. $\dfrac{-(p + 2)^2 - 3r}{2 - q}$

46. $\dfrac{-(q - 6)^2 - 2p}{4 - p}$

47. $\dfrac{3p + 3(4 + p)^3}{r + 8}$

48. $\dfrac{5q + 2(1 + p)^3}{r + 3}$

Identify the property illustrated in each statement. Assume all variables represent real numbers. **See Examples 5 and 6.**

49. $6 \cdot 12 + 6 \cdot 15 = 6(12 + 15)$

50. $8(m + 4) = 8m + 32$

51. $(t - 6) \cdot \left(\frac{1}{t - 6}\right) = 1$, if $t - 6 \neq 0$

52. $\frac{2 + m}{2 - m} \cdot \frac{2 - m}{2 + m} = 1$, if $m \neq 2$ or -2

53. $(7.5 - y) + 0 = 7.5 - y$

54. $1 \cdot (3x - 7) = 3x - 7$

55. $5(t + 3) = (t + 3) \cdot 5$

56. $-7 + (x + 3) = (x + 3) + (-7)$

57. $(5x)\left(\frac{1}{x}\right) = 5\left(x \cdot \frac{1}{x}\right)$

58. $(38 + 99) + 1 = 38 + (99 + 1)$

59. $5 + \sqrt{3}$ is a real number.

60. 5π is a real number.

Write a short answer to each question.

61. Is there a commutative property for subtraction? That is, in general, is $a - b$ equal to $b - a$? Support your answer with an example.

62. Is there an associative property for subtraction? That is, does $(a - b) - c$ equal $a - (b - c)$ in general? Support your answer with an example.

Simplify each expression. **See Examples 5 and 6.**

63. $\frac{10}{11}(22z)$

64. $\left(\frac{3}{4}r\right)(-12)$

65. $(m + 5) + 6$

66. $8 + (a + 7)$

67. $\frac{3}{8}\left(\frac{16}{9}y + \frac{32}{27}z - \frac{40}{9}\right)$

68. $-\frac{1}{4}(20m + 8y - 32z)$

Use the distributive property to rewrite sums as products and products as sums. ***See Example 6.***

69. $8p - 14p$ **70.** $15x - 10x$ **71.** $-4(z - y)$ **72.** $-3(m + n)$

Concept Check *Use the distributive property to calculate each value mentally.*

73. $72 \cdot 17 + 28 \cdot 17$

74. $32 \cdot 80 + 32 \cdot 20$

75. $123\frac{5}{8} \cdot 1\frac{1}{2} - 23\frac{5}{8} \cdot 1\frac{1}{2}$

76. $17\frac{2}{5} \cdot 14\frac{3}{4} - 17\frac{2}{5} \cdot 4\frac{3}{4}$

Concept Check *Decide whether each statement is* true *or* false. *If false, correct the statement so it is true.*

77. $|6 - 8| = |6| - |8|$

78. $|(-3)^3| = -|3^3|$

79. $|-5| \cdot |6| = |-5 \cdot 6|$

80. $\frac{|-14|}{|2|} = \left|\frac{-14}{2}\right|$

81. $|a - b| = |a| - |b|$, if $b > a > 0$

82. If a is negative, then $|a| = -a$.

Evaluate each expression. ***See Example 7.***

83. $|-10|$ **84.** $|-15|$ **85.** $-\left|\frac{4}{7}\right|$

86. $-\left|\frac{7}{2}\right|$ **87.** $-|-8|$ **88.** $-|-12|$

(Modeling) Blood Pressure Difference *Use the formula for determining blood pressure difference from normal,* $P_d = |P - 120|$, *to solve each problem.* ***See Example 8.***

89. Calculate the P_d value for a woman whose actual systolic blood pressure is 116.

90. Determine two possible values for a person's systolic blood pressure if his P_d value is 17.

Let $x = -4$ *and* $y = 2$. *Evaluate each expression.* ***See Example 9.***

91. $|3x - 2y|$ **92.** $|2x - 5y|$ **93.** $|-3x + 4y|$ **94.** $|-5y + x|$

95. $\frac{2|y| - 3|x|}{|xy|}$ **96.** $\frac{4|x| + 4|y|}{|x|}$ **97.** $\frac{|-8y + x|}{-|x|}$ **98.** $\frac{|x| + 2|y|}{-|x|}$

Determine whether each statement is true *or* false.

99. $|25| = |-25|$

100. $|-8| \geq 0$

101. $|5 + (-13)| = |5| + |-13|$

102. $|8 - 12| = |8| - |12|$

103. $|11| \cdot |-6| = |-66|$

104. $\left|\frac{10}{-2}\right| = \frac{|10|}{|-2|}$

Find the given distances between points P, Q, R, and S on a number line, with coordinates −4, −1, 8, *and* 12, *respectively.* ***See Example 10.***

105. $d(P, Q)$ **106.** $d(P, R)$ **107.** $d(Q, R)$ **108.** $d(Q, S)$

Concept Check *Determine what signs on values of x and y would make each statement true. Assume that x and y are not 0. (You should be able to work these mentally.)*

109. $xy > 0$

110. $x^2y > 0$

111. $\frac{x}{y} < 0$

112. $\frac{y^2}{x} < 0$

113. $\frac{x^3}{y} > 0$

114. $-\frac{x}{y} > 0$

Solve each problem.

115. ***Golf Scores*** Jordan Spieth won the 2015 Masters Golf Tournament with a total score that was 18 under par, and Zach Johnson won the 2007 tournament with a total score that was 1 above par. Using -18 to represent 18 below par and $+1$ to represent 1 over par, find the difference between these scores (in either order) and take the absolute value of this difference. What does this final number represent? (*Source:* www.masters.org)

116. ***Total Football Yardage*** As of 2015, Emmitt Smith of the Dallas Cowboys was the NFL career leader for rushing. During his 15 years in the NFL, he gained 18,355 yd rushing, 3224 yd receiving, and -15 yd returning fumbles. Find his total yardage (called *all-purpose yards*). Is this the same as the sum of the absolute values of the three categories? Explain. (*Source:* www.pro-football-reference.com)

(Modeling) Blood Alcohol Concentration *The blood alcohol concentration (BAC) of a person who has been drinking is approximated by the following formula.*

$$\text{BAC} = \text{number of oz} \times \%\text{ alcohol} \times 0.075 \div \text{body wt in lb} - \text{hr of drinking} \times 0.015$$

(Source: Lawlor, J., *Auto Math Handbook: Mathematical Calculations, Theory, and Formulas for Automotive Enthusiasts,* HP Books.*)*

117. Suppose a policeman stops a 190-lb man who, in 2 hr, has ingested four 12-oz beers (48 oz), each having a 3.2% alcohol content. Calculate the man's BAC to the nearest thousandth. Follow the order of operations.

118. Calculate the BAC to the nearest thousandth for a 135-lb woman who, in 3 hr, has consumed three 12-oz beers (36 oz), each having a 4.0% alcohol content.

119. Calculate the BAC to the nearest thousandth for a 200-lb man who, in 4 hr, has consumed three 20-oz beers, each having a 3.8% alcohol content. If the man's weight were greater and all other variables remained the same, how would that affect his BAC?

120. Calculate the BAC to the nearest thousandth for a 150-lb woman who, in 2 hr, has consumed two 6-oz glasses of wine, each having a 14% alcohol content. If the woman drank the same two glasses of wine over a longer period of time, how would that affect her BAC?

Archie Manning, father of NFL quarterbacks Peyton and Eli, signed this photo for author Hornsby's son, Jack.

(Modeling) Passer Rating for NFL Quarterbacks *The current system of rating passers in the National Football League is based on four performance components: completions, touchdowns, yards gained, and interceptions, as percentages of the number of passes attempted. It uses the following formula.*

$$\text{Rating} = \frac{\left(250 \cdot \frac{C}{A}\right) + \left(1000 \cdot \frac{T}{A}\right) + \left(12.5 \cdot \frac{Y}{A}\right) + 6.25 - \left(1250 \cdot \frac{I}{A}\right)}{3},$$

where A = attempted passes, C = completed passes, T = touchdown passes, Y = yards gained passing, and I = interceptions.

In addition to the weighting factors appearing in the formula, the four category ratios are limited to nonnegative values with the following maximums.

$$0.775 \text{ for } \frac{C}{A}, \quad 0.11875 \text{ for } \frac{T}{A}, \quad 12.5 \text{ for } \frac{Y}{A}, \quad 0.095 \text{ for } \frac{I}{A}$$

Exercises 121–132 give the 2014 regular-season statistics for the top twelve quarterbacks. Use the formula to determine the rating to the nearest tenth for each.

Quarterback, Team	A Att	C Comp	T TD	Y Yards	I Int
121. Tony Romo, DAL	435	304	34	3705	9
122. Aaron Rodgers, GB	520	341	38	4381	5
123. Ben Roethlisberger, PIT	608	408	32	4952	9
124. Peyton Manning, DEN	597	395	39	4727	15
125. Tom Brady, NE	582	373	33	4109	9
126. Drew Brees, NO	659	456	33	4952	17
127. Andrew Luck, IND	616	380	40	4761	16
128. Carson Palmer, ARI	224	141	11	1626	3
129. Ryan Fitzpatrick, HOU	312	197	17	2483	8
130. Russell Wilson, SEA	452	285	20	3475	7
131. Matt Ryan, ATL	628	415	28	4694	14
132. Philip Rivers, SD	570	379	31	4286	18

Source: www.nfl.com

Solve each problem using the passer rating formula above.

133. Peyton Manning, when he played for the Indianapolis Colts, set a full-season rating record of 121.1 in 2004 and held that record until Aaron Rodgers, of the Green Bay Packers, surpassed it in 2011. (As of 2014, Rodgers's all-time record held.) If Rodgers had 343 completions, 45 touchdowns, 6 interceptions, and 4643 yards, for 502 attempts, what was his rating in 2011?

134. Steve Young, of the San Francisco 49ers, set a full season rating record of 112.8 in 1994 and held that record until Peyton Manning surpassed it in 2004. If Manning had 336 completions, 49 touchdowns, 10 interceptions, and 4557 yards, for 497 attempts, what was his rating in 2004?

135. If Tom Brady, of the New England Patriots, during the 2010 regular season, had 324 completions, 36 touchdowns, 4 interceptions, and 3900 yards, what was his rating in 2010 for 492 attempts?

136. Refer to the passer rating formula and determine the highest rating possible (considered a "perfect" passer rating).

R.3 Polynomials

- Rules for Exponents
- Polynomials
- Addition and Subtraction
- Multiplication
- Division

Rules for Exponents

Recall that the notation a^m (where m is a positive integer and a is a real number) means that a appears as a factor m times. In the same way, a^n (where n is a positive integer) means that a appears as a factor n times.

Rules for Exponents

For all positive integers m and n and all real numbers a and b, the following rules hold.

Rule	Example	Description
Product Rule $a^m \cdot a^n = a^{m+n}$	$2^2 \cdot 2^3 = (2 \cdot 2)(2 \cdot 2 \cdot 2) = 2^{2+3} = 2^5$	When multiplying powers of like bases, keep the base and add the exponents.
Power Rule 1 $(a^m)^n = a^{mn}$	$(4^5)^3 = 4^5 \cdot 4^5 \cdot 4^5 = 4^{5+5+5} = 4^{5 \cdot 3} = 4^{15}$	To raise a power to a power, multiply the exponents.
Power Rule 2 $(ab)^m = a^m b^m$	$(7x)^3 = (7x)(7x)(7x) = (7 \cdot 7 \cdot 7)(x \cdot x \cdot x) = 7^3x^3$	To raise a product to a power, raise each factor to that power.
Power Rule 3 $\left(\frac{a}{b}\right)^m = \frac{a^m}{b^m}$ $(b \neq 0)$	$\left(\frac{3}{5}\right)^4 = \left(\frac{3}{5}\right)\left(\frac{3}{5}\right)\left(\frac{3}{5}\right)\left(\frac{3}{5}\right) = \frac{3 \cdot 3 \cdot 3 \cdot 3}{5 \cdot 5 \cdot 5 \cdot 5} = \frac{3^4}{5^4}$	To raise a quotient to a power, raise the numerator and the denominator to that power.

EXAMPLE 1 Using the Product Rule

Simplify each expression.

(a) $y^4 \cdot y^7$ **(b)** $(6z^5)(9z^3)(2z^2)$

SOLUTION

(a) $y^4 \cdot y^7 = y^{4+7} = y^{11}$ Product rule: Keep the base and add the exponents.

(b) $(6z^5)(9z^3)(2z^2)$

$= (6 \cdot 9 \cdot 2) \cdot (z^5z^3z^2)$ Commutative and associative properties

$= 108z^{5+3+2}$ Multiply. Apply the product rule.

$= 108z^{10}$ Add.

✔ Now Try Exercises 13 and 17.

EXAMPLE 2 Using the Power Rules

Simplify. Assume all variables represent nonzero real numbers.

(a) $(5^3)^2$ **(b)** $(3^4x^2)^3$ **(c)** $\left(\frac{2^5}{b^4}\right)^3$ **(d)** $\left(\frac{-2m^6}{t^2z}\right)^5$

SOLUTION

(a) $(5^3)^2 = 5^{3(2)} = 5^6$ Power rule 1

(b) $(3^4x^2)^3$

$= (3^4)^3(x^2)^3$ Power rule 2

$= 3^{4(3)}x^{2(3)}$ Power rule 1

$= 3^{12}x^6$

(c) $\left(\frac{2^5}{b^4}\right)^3$

$= \frac{(2^5)^3}{(b^4)^3}$ Power rule 3

$= \frac{2^{15}}{b^{12}}$ Power rule 1

(d) $\left(\frac{-2m^6}{t^2z}\right)^5$

$= \frac{(-2m^6)^5}{(t^2z)^5}$ Power rule 3

$= \frac{(-2)^5(m^6)^5}{(t^2)^5z^5}$ Power rule 2

$= \frac{-32m^{30}}{t^{10}z^5}$ Evaluate $(-2)^5$. Then use Power rule 1.

$= -\frac{32m^{30}}{t^{10}z^5}$ $\frac{-a}{b} = -\frac{a}{b}$

✔ **Now Try Exercises 23, 25, 29, and 31.**

CAUTION The expressions mn^2 and $(mn)^2$ are ***not*** equivalent. The second power rule can be used only with the second expression:

$$(mn)^2 = m^2n^2.$$

A zero exponent is defined as follows.

Zero Exponent

For any nonzero real number a, $\mathbf{a^0 = 1.}$

***That is, any nonzero number with a zero exponent equals* 1.**

To illustrate why a^0 is defined to equal 1, consider the product

$$a^n \cdot a^0, \quad \text{for} \quad a \neq 0.$$

We want the definition of a^0 to be consistent so that the product rule applies. Now apply this rule.

$$a^n \cdot a^0 = a^{n+0} = a^n$$

The product of a^n and a^0 must be a^n, and thus a^0 is acting like the identity element 1. So, for consistency, we *define* a^0 to equal 1. **(0^0 is undefined.)**

EXAMPLE 3 Using the Definition of a^0

Evaluate each power.

(a) 4^0 **(b)** $(-4)^0$ **(c)** -4^0

(d) $-(-4)^0$ **(e)** $(7r)^0$

SOLUTION

(a) $4^0 = 1$ Base is 4. **(b)** $(-4)^0 = 1$ Base is -4.

(c) $-4^0 = -(4^0) = -1$ Base is 4.

(d) $-(-4)^0 = -(1) = -1$ Base is -4.

(e) $(7r)^0 = 1,\ r \neq 0$ Base is $7r$.

✔ **Now Try Exercise 35.**

Polynomials The product of a number and one or more variables raised to powers is a **term.** The number is the **numerical coefficient,** or just the **coefficient,** of the variables. The coefficient of the variable in $-3m^4$ is -3, and the coefficient in $-p^2$ is -1. **Like terms** are terms with the same variables each raised to the same powers.

$$-13x^3,\quad 4x^3,\quad -x^3 \quad \text{Like terms} \qquad 6y,\quad 6y^2,\quad 4y^3 \quad \text{Unlike terms}$$

A **polynomial** is a term or a finite sum of terms, with only positive or zero integer exponents permitted on the variables. If the terms of a polynomial contain only the variable x, then the polynomial is a **polynomial in x.**

$$5x^3 - 8x^2 + 7x - 4, \qquad 9p^5 - 3, \qquad 8r^2, \qquad 6 \quad \text{Polynomials}$$

The terms of a polynomial cannot have variables in a denominator.

$$9x^2 - 4x + \frac{6}{x} \quad \text{Not a polynomial}$$

The **degree of a term** with one variable is the exponent on the variable. For example, the degree of $2x^3$ is 3, and the degree of $17x$ (that is, $17x^1$) is 1. The greatest degree of any term in a polynomial is the **degree of the polynomial.** For example, the polynomial

$$4x^3 - 2x^2 - 3x + 7 \quad \text{has degree } 3$$

because the greatest degree of any term is 3. A nonzero constant such as -6, equivalent to $-6x^0$, has degree 0. (The polynomial 0 has no degree.)

A polynomial can have more than one variable. A term containing more than one variable has degree equal to the sum of all the exponents appearing on the variables in the term. For example,

$$-3x^4y^3z^5 \quad \text{has degree} \quad 4 + 3 + 5 = 12.$$

$$5xy^2z^7 \quad \text{has degree} \quad 1 + 2 + 7 = 10.$$

The degree of a polynomial in more than one variable is equal to the greatest degree of any term appearing in the polynomial. By this definition, the polynomial

$$2x^4y^3 - 3x^5y + x^6y^2 \quad \text{has degree } 8$$

because the x^6y^2 term has the greatest degree, 8.

A polynomial containing exactly three terms is a **trinomial.** A two-term polynomial is a **binomial.** A single-term polynomial is a **monomial.**

EXAMPLE 4 Classifying Expressions as Polynomials

Identify each as a *polynomial* or *not a polynomial*. For each polynomial, give the degree and identify it as a *monomial, binomial, trinomial*, or *none of these*.

(a) $9x^7 - 4x^3 + 8x^2$ **(b)** $2t^4 - \frac{1}{t^2}$ **(c)** $-\frac{4}{5}x^3y^2$

SOLUTION

(a) $9x^7 - 4x^3 + 8x^2$ is a polynomial. The first term, $9x^7$, has greatest degree, so this a polynomial of degree 7. Because it has three terms, it is a trinomial.

(b) $2t^4 - \frac{1}{t^2}$ is not a polynomial because it has a variable in the denominator.

(c) $-\frac{4}{5}x^3y^2$ is a polynomial. Add the exponents $3 + 2 = 5$ to determine that it is of degree 5. Because there is one term, it is a monomial.

✔ **Now Try Exercises 37, 39, and 45.**

Addition and Subtraction Polynomials are added by adding coefficients of like terms. They are subtracted by subtracting coefficients of like terms.

EXAMPLE 5 Adding and Subtracting Polynomials

Add or subtract, as indicated.

(a) $(2y^4 - 3y^2 + y) + (4y^4 + 7y^2 + 6y)$

(b) $(-3m^3 - 8m^2 + 4) - (m^3 + 7m^2 - 3)$

(c) $(8m^4p^5 - 9m^3p^5) + (11m^4p^5 + 15m^3p^5)$

(d) $4(x^2 - 3x + 7) - 5(2x^2 - 8x - 4)$

SOLUTION

(a) $(2y^4 - 3y^2 + y) + (4y^4 + 7y^2 + 6y)$ ($y = 1y$)

$= (2 + 4)y^4 + (-3 + 7)y^2 + (1 + 6)y$ Add coefficients of like terms.

$= 6y^4 + 4y^2 + 7y$ Work inside the parentheses.

(b) $(-3m^3 - 8m^2 + 4) - (m^3 + 7m^2 - 3)$

$= (-3 - 1)m^3 + (-8 - 7)m^2 + [4 - (-3)]$ Subtract coefficients of like terms.

$= -4m^3 - 15m^2 + 7$ Simplify.

(c) $(8m^4p^5 - 9m^3p^5) + (11m^4p^5 + 15m^3p^5)$

$= 19m^4p^5 + 6m^3p^5$

(d) $4(x^2 - 3x + 7) - 5(2x^2 - 8x - 4)$

$= 4x^2 - 4(3x) + 4(7) - 5(2x^2) - 5(-8x) - 5(-4)$ Distributive property

$= 4x^2 - 12x + 28 - 10x^2 + 40x + 20$ Multiply.

$= -6x^2 + 28x + 48$ Add like terms.

✔ **Now Try Exercises 49 and 51.**

As shown in **Examples 5(a), (b), and (d),** polynomials in one variable are often written with their terms in **descending order** (or descending degree). The term of greatest degree is first, the one of next greatest degree is next, and so on.

Multiplication One way to find the product of two polynomials, such as $3x - 4$ and $2x^2 - 3x + 5$, is to distribute each term of $3x - 4$, multiplying by each term of $2x^2 - 3x + 5$.

$$(3x - 4)(2x^2 - 3x + 5)$$

$$= 3x(2x^2 - 3x + 5) - 4(2x^2 - 3x + 5) \quad \text{Distributive property}$$

$$= 3x(2x^2) + 3x(-3x) + 3x(5) - 4(2x^2) - 4(-3x) - 4(5) \quad \text{Distributive property again}$$

$$= 6x^3 - 9x^2 + 15x - 8x^2 + 12x - 20 \quad \text{Multiply.}$$

$$= 6x^3 - 17x^2 + 27x - 20 \quad \text{Combine like terms.}$$

Another method is to write such a product vertically, similar to the method used in arithmetic for multiplying whole numbers.

$$\begin{array}{rrrrl} & 2x^2 & - 3x & + 5 & \\ & & 3x & - 4 & \\ \hline & -8x^2 & + 12x & - 20 & \leftarrow -4(2x^2 - 3x + 5) \\ 6x^3 & - 9x^2 & + 15x & & \leftarrow 3x(2x^2 - 3x + 5) \\ \hline 6x^3 & - 17x^2 & + 27x & - 20 & \text{Add in columns.} \end{array}$$

Place like terms in the same column.

EXAMPLE 6 Multiplying Polynomials

Multiply $(3p^2 - 4p + 1)(p^3 + 2p - 8)$.

SOLUTION

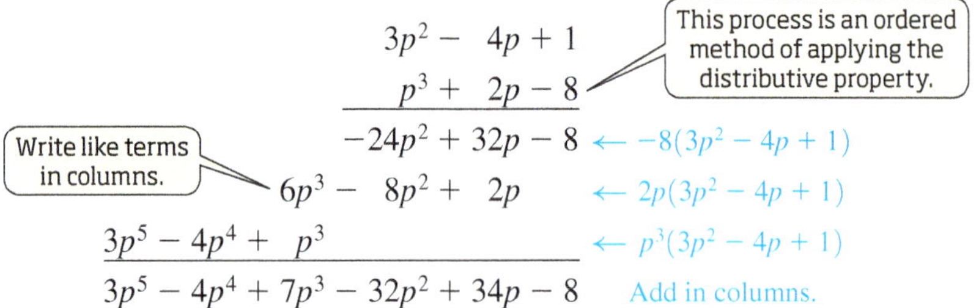

$$\begin{array}{rrrrrrl} & & & 3p^2 & - 4p & + 1 & \\ & & p^3 & & + 2p & - 8 & \\ \hline & & & -24p^2 & + 32p & - 8 & \leftarrow -8(3p^2 - 4p + 1) \\ & & 6p^3 & - 8p^2 & + 2p & & \leftarrow 2p(3p^2 - 4p + 1) \\ 3p^5 & - 4p^4 & + p^3 & & & & \leftarrow p^3(3p^2 - 4p + 1) \\ \hline 3p^5 & - 4p^4 & + 7p^3 & - 32p^2 & + 34p & - 8 & \text{Add in columns.} \end{array}$$

✔ **Now Try Exercise 63.**

The **FOIL method** is a convenient way to find the product of two binomials. The memory aid **FOIL** (for **F**irst, **O**uter, **I**nner, **L**ast) gives the pairs of terms to be multiplied when distributing each term of the first binomial, multiplying by each term of the second binomial.

EXAMPLE 7 Using the FOIL Method to Multiply Two Binomials

Find each product.

(a) $(6m + 1)(4m - 3)$ **(b)** $(2x + 7)(2x - 7)$ **(c)** $r^2(3r + 2)(3r - 2)$

SOLUTION

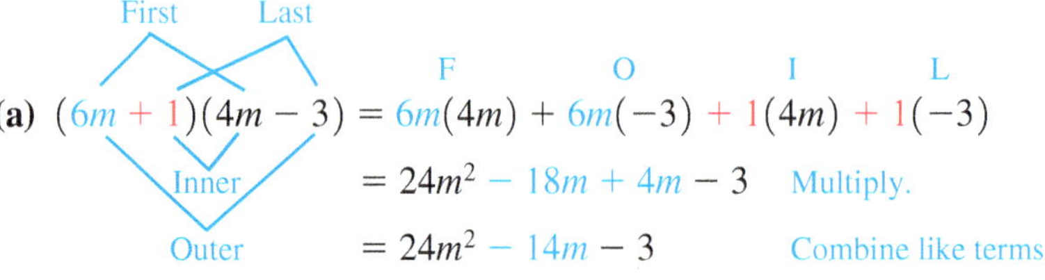

(a) $(6m + 1)(4m - 3) = 6m(4m) + 6m(-3) + 1(4m) + 1(-3)$

$$= 24m^2 - 18m + 4m - 3 \quad \text{Multiply.}$$

$$= 24m^2 - 14m - 3 \quad \text{Combine like terms.}$$

(b) $(2x + 7)(2x - 7)$

$= 4x^2 - 14x + 14x - 49$ FOIL method

$= 4x^2 - 49$ Combine like terms.

(c) $r^2(3r + 2)(3r - 2)$

$= r^2(9r^2 - 6r + 6r - 4)$ FOIL method

$= r^2(9r^2 - 4)$ Combine like terms.

$= 9r^4 - 4r^2$ Distributive property

✔ **Now Try Exercises 55 and 57.**

In **Example 7(a),** the product of two binomials is a trinomial, while in **Examples 7(b) and (c)**, the product of two binomials is a binomial. The product of two binomials of the forms $x + y$ and $x - y$ is a special product form called a **difference of squares.** The squares of binomials, $(x + y)^2$ and $(x - y)^2$, are also special product forms called **perfect square trinomials.**

Special Products

Product of the Sum and Difference of Two Terms $(x + y)(x - y) = x^2 - y^2$ (Difference of squares)

Square of a Binomial

$(x + y)^2 = x^2 + 2xy + y^2$

$(x - y)^2 = x^2 - 2xy + y^2$ (Perfect square trinomials)

EXAMPLE 8 Using the Special Products

Find each product.

(a) $(3p + 11)(3p - 11)$

(b) $(5m^3 - 3)(5m^3 + 3)$

(c) $(9k - 11r^3)(9k + 11r^3)$

(d) $(2m + 5)^2$

(e) $(3x - 7y^4)^2$

SOLUTION

(a) $(3p + 11)(3p - 11)$

$= (3p)^2 - 11^2$ $(x + y)(x - y) = x^2 - y^2$

$= 9p^2 - 121$ Power rule 2

$(3p)^2 = 3^2p^2$, not $3p^2$

(b) $(5m^3 - 3)(5m^3 + 3)$

$= (5m^3)^2 - 3^2$ $(x - y)(x + y) = x^2 - y^2$

$= 25m^6 - 9$ Power rules 2 and 1

(c) $(9k - 11r^3)(9k + 11r^3)$

$= (9k)^2 - (11r^3)^2$

$= 81k^2 - 121r^6$

Be careful applying the power rules.

(d) $(2m + 5)^2$

$= (2m)^2 + 2(2m)(5) + 5^2$ $\quad (x + y)^2 = x^2 + 2xy + y^2$

$= 4m^2 + 20m + 25$ $\quad$ Power rule 2; Multiply.

(e) $(3x - 7y^4)^2$

$= (3x)^2 - 2(3x)(7y^4) + (7y^4)^2$ $\quad (x - y)^2 = x^2 - 2xy + y^2$

$= 9x^2 - 42xy^4 + 49y^8$ $\quad$ Power rule 2; Multiply.

✔ **Now Try Exercises 69, 71, 73, and 75.**

CAUTION See **Examples 8(d) and (e).** ***The square of a binomial has three terms.*** Do **not** give $x^2 + y^2$ as the result of expanding $(x + y)^2$, or $x^2 - y^2$ as the result of expanding $(x - y)^2$.

$$(x + y)^2 = x^2 + 2xy + y^2$$
$$(x - y)^2 = x^2 - 2xy + y^2$$

Remember to include the middle term.

EXAMPLE 9 Multiplying More Complicated Binomials

Find each product.

(a) $[(3p - 2) + 5q][(3p - 2) - 5q]$ **(b)** $(x + y)^3$ **(c)** $(2a + b)^4$

SOLUTION

(a) $[(3p - 2) + 5q][(3p - 2) - 5q]$

$= (3p - 2)^2 - (5q)^2$ $\quad$ Product of the sum and difference of two terms

$= 9p^2 - 12p + 4 - 25q^2$ $\quad$ Square both quantities.

(b) $(x + y)^3$ — This does *not* equal $x^3 + y^3$.

$= (x + y)^2(x + y)$ $\quad a^3 = a^2 \cdot a$

$= (x^2 + 2xy + y^2)(x + y)$ $\quad$ Square $x + y$.

$= x^3 + x^2y + 2x^2y + 2xy^2 + xy^2 + y^3$ $\quad$ Multiply.

$= x^3 + 3x^2y + 3xy^2 + y^3$ $\quad$ Combine like terms.

(c) $(2a + b)^4$

$= (2a + b)^2(2a + b)^2$ $\quad a^4 = a^2 \cdot a^2$

$= (4a^2 + 4ab + b^2)(4a^2 + 4ab + b^2)$ $\quad$ Square each $2a + b$.

$= 16a^4 + 16a^3b + 4a^2b^2 + 16a^3b + 16a^2b^2 + 4ab^3 + 4a^2b^2 + 4ab^3 + b^4$ $\quad$ Distributive property

$= 16a^4 + 32a^3b + 24a^2b^2 + 8ab^3 + b^4$ $\quad$ Combine like terms.

✔ **Now Try Exercises 79, 83, and 85.**

Division The quotient of two polynomials can be found with an algorithm (that is, a step-by-step procedure) for long division similar to that used for dividing whole numbers. ***Both polynomials must be written in descending order to use this algorithm.***

EXAMPLE 10 Dividing Polynomials

Divide $4m^3 - 8m^2 + 5m + 6$ by $2m - 1$.

SOLUTION

$4m^3$ divided by $2m$ is $2m^2$.
$-6m^2$ divided by $2m$ is $-3m$.
$2m$ divided by $2m$ is 1.

$$
\begin{array}{r}
2m^2 - 3m + 1 \\
2m - 1 \overline{)4m^3 - 8m^2 + 5m + 6} \\
\underline{4m^3 - 2m^2} \qquad\qquad \\
-6m^2 + 5m \qquad \\
\underline{-6m^2 + 3m} \qquad \\
2m + 6 \\
\underline{2m - 1} \\
7
\end{array}
$$

- $4m^3 - 2m^2 \leftarrow 2m^2(2m - 1) = 4m^3 - 2m^2$
- $-6m^2 + 5m \leftarrow$ Subtract. Bring down the next term.
- $-6m^2 + 3m \leftarrow -3m(2m - 1) = -6m^2 + 3m$
- $2m + 6 \leftarrow$ Subtract. Bring down the next term.
- $2m - 1 \leftarrow 1(2m - 1) = 2m - 1$
- $7 \leftarrow$ Subtract. The remainder is 7.

To subtract, add the opposite.

Thus, $\dfrac{4m^3 - 8m^2 + 5m + 6}{2m - 1} = 2m^2 - 3m + 1 + \dfrac{7}{2m - 1}$.

Remember to add $\dfrac{\text{remainder}}{\text{divisor}}$.

✔ Now Try Exercise 101.

When a polynomial has a missing term, we allow for that term by inserting a term with a 0 coefficient for it. For example,

$3x^2 - 7$ is equivalent to $3x^2 + 0x - 7$,

and $2x^3 + x + 10$ is equivalent to $2x^3 + 0x^2 + x + 10$.

EXAMPLE 11 Dividing Polynomials with Missing Terms

Divide $3x^3 - 2x^2 - 150$ by $x^2 - 4$.

SOLUTION Both polynomials have missing first-degree terms. Insert each missing term with a 0 coefficient.

$$
\begin{array}{r}
3x - 2 \\
x^2 + 0x - 4 \overline{)3x^3 - 2x^2 + 0x - 150} \\
\underline{3x^3 + 0x^2 - 12x} \qquad\quad \\
-2x^2 + 12x - 150 \\
\underline{-2x^2 + 0x + 8} \\
12x - 158
\end{array}
$$

- Missing term ($0x$ in the divisor)
- Missing term ($0x$ in the dividend)
- $12x - 158 \leftarrow$ Remainder

The division process ends when the remainder is 0 or the degree of the remainder is less than that of the divisor. Because $12x - 158$ has lesser degree than the divisor $x^2 - 4$, it is the remainder. Thus, the entire quotient is written as follows.

$$\frac{3x^3 - 2x^2 - 150}{x^2 - 4} = 3x - 2 + \frac{12x - 158}{x^2 - 4}$$

✔ Now Try Exercise 103.

R.3 Exercises

CONCEPT PREVIEW *Fill in the blank to correctly complete each sentence.*

1. The polynomial $2x^5 - x + 4$ is a trinomial of degree ________.
2. A polynomial containing exactly one term is a(n) ________.
3. A polynomial containing exactly two terms is a(n) ________.
4. In the term $-6x^2y$, -6 is the ________.
5. A convenient way to find the product of two binomials is to use the ________ method.

CONCEPT PREVIEW *Decide whether each is* true *or* false. *If false, correct the right side of the equation.*

6. $5^0 = 1$
7. $y^2 \cdot y^5 = y^7$
8. $(a^2)^3 = a^5$
9. $(x + y)^2 = x^2 + y^2$
10. $x^2 + x^2 = x^4$

Simplify each expression. **See Example 1.**

11. $(-4x^5)(4x^2)$
12. $(3y^4)(-6y^3)$
13. $n^6 \cdot n^4 \cdot n$
14. $a^8 \cdot a^5 \cdot a$
15. $9^3 \cdot 9^5$
16. $4^2 \cdot 4^8$
17. $(-3m^4)(6m^2)(-4m^5)$
18. $(-8t^3)(2t^6)(-5t^4)$
19. $(5x^2y)(-3x^3y^4)$
20. $(-4xy^3)(7x^2y)$
21. $\left(\frac{1}{2}mn\right)(8m^2n^2)$
22. $(35m^4n)\left(-\frac{2}{7}mn^2\right)$

Simplify each expression. Assume all variables represent nonzero real numbers. **See Examples 1–3.**

23. $(2^2)^5$
24. $(6^4)^3$
25. $(-6x^2)^3$
26. $(-2x^5)^5$
27. $-(4m^3n^0)^2$
28. $-(2x^0y^4)^3$
29. $\left(\frac{r^8}{s^2}\right)^3$
30. $\left(\frac{p^4}{q}\right)^2$
31. $\left(\frac{-4m^2}{tp^2}\right)^4$
32. $\left(\frac{-5n^4}{r^2}\right)^3$
33. $-\left(\frac{x^3y^5}{z}\right)^0$
34. $-\left(\frac{p^2q^3}{r^3}\right)^0$

Match each expression in Column I with its equivalent in Column II. **See Example 3.**

	I	II
35.	(a) 6^0	A. 0
	(b) -6^0	B. 1
	(c) $(-6)^0$	C. -1
	(d) $-(-6)^0$	D. 6
		E. -6

	I	II
36.	(a) $3p^0$	A. 0
	(b) $-3p^0$	B. 1
	(c) $(3p)^0$	C. -1
	(d) $(-3p)^0$	D. 3
		E. -3

Identify each expression as a polynomial *or* not a polynomial. *For each polynomial, give the degree and identify it as a* monomial, binomial, trinomial, *or* none of these. ***See Example 4.***

37. $-5x^{11}$

38. $-4y^5$

39. $6x + 3x^4$

40. $-9y + 5y^3$

41. $-7z^5 - 2z^3 + 1$

42. $-9t^4 + 8t^3 - 7$

43. $15a^2b^3 + 12a^3b^8 - 13b^5 + 12b^6$

44. $-16x^5y^7 + 12x^3y^8 - 4xy^9 + 18x^{10}$

45. $\frac{3}{8}x^5 - \frac{1}{x^2} + 9$

46. $\frac{2}{3}t^6 + \frac{3}{t^5} + 1$

47. 5

48. 9

Add or subtract, as indicated. ***See Example 5.***

49. $(5x^2 - 4x + 7) + (-4x^2 + 3x - 5)$

50. $(3m^3 - 3m^2 + 4) + (-2m^3 - m^2 + 6)$

51. $2(12y^2 - 8y + 6) - 4(3y^2 - 4y + 2)$

52. $3(8p^2 - 5p) - 5(3p^2 - 2p + 4)$

53. $(6m^4 - 3m^2 + m) - (2m^3 + 5m^2 + 4m) + (m^2 - m)$

54. $-(8x^3 + x - 3) + (2x^3 + x^2) - (4x^2 + 3x - 1)$

Find each product. ***See Examples 6–8.***

55. $(4r - 1)(7r + 2)$

56. $(5m - 6)(3m + 4)$

57. $x^2\left(3x - \frac{2}{3}\right)\left(5x + \frac{1}{3}\right)$

58. $m^3\left(2m - \frac{1}{4}\right)\left(3m + \frac{1}{2}\right)$

59. $4x^2(3x^3 + 2x^2 - 5x + 1)$

60. $2b^3(b^2 - 4b + 3)$

61. $(2z - 1)(-z^2 + 3z - 4)$

62. $(3w + 2)(-w^2 + 4w - 3)$

63. $(m - n + k)(m + 2n - 3k)$

64. $(r - 3s + t)(2r - s + t)$

65. $(2x + 3)(2x - 3)(4x^2 - 9)$

66. $(3y - 5)(3y + 5)(9y^2 - 25)$

67. $(x + 1)(x + 1)(x - 1)(x - 1)$

68. $(t + 4)(t + 4)(t - 4)(t - 4)$

Find each product. ***See Examples 8 and 9.***

69. $(2m + 3)(2m - 3)$

70. $(8s - 3t)(8s + 3t)$

71. $(4x^2 - 5y)(4x^2 + 5y)$

72. $(2m^3 + n)(2m^3 - n)$

73. $(4m + 2n)^2$

74. $(a - 6b)^2$

75. $(5r - 3t^2)^2$

76. $(2z^4 - 3y)^2$

77. $[(2p - 3) + q]^2$

78. $[(4y - 1) + z]^2$

79. $[(3q + 5) - p][(3q + 5) + p]$

80. $[(9r - s) + 2][(9r - s) - 2]$

81. $[(3a + b) - 1]^2$

82. $[(2m + 7) - n]^2$

83. $(y + 2)^3$

84. $(z - 3)^3$

85. $(q - 2)^4$

86. $(r + 3)^4$

Perform the indicated operations. ***See Examples 5–9.***

87. $(p^3 - 4p^2 + p) - (3p^2 + 2p + 7)$

88. $(x^4 - 3x^2 + 2) - (-2x^4 + x^2 - 3)$

89. $(7m + 2n)(7m - 2n)$

90. $(3p + 5)^2$

91. $-3(4q^2 - 3q + 2) + 2(-q^2 + q - 4)$

92. $2(3r^2 + 4r + 2) - 3(-r^2 + 4r - 5)$

93. $p(4p - 6) + 2(3p - 8)$

94. $m(5m - 2) + 9(5 - m)$

95. $-y(y^2 - 4) + 6y^2(2y - 3)$

96. $-z^3(9 - z) + 4z(2 + 3z)$

Perform each division. ***See Examples 10 and 11.***

97. $\dfrac{-4x^7 - 14x^6 + 10x^4 - 14x^2}{-2x^2}$

98. $\dfrac{-8r^3s - 12r^2s^2 + 20rs^3}{-4rs}$

99. $\dfrac{4x^3 - 3x^2 + 1}{x - 2}$

100. $\dfrac{3x^3 - 2x + 5}{x - 3}$

101. $\dfrac{6m^3 + 7m^2 - 4m + 2}{3m + 2}$

102. $\dfrac{10x^3 + 11x^2 - 2x + 3}{5x + 3}$

103. $\dfrac{x^4 + 5x^2 + 5x + 27}{x^2 + 3}$

104. $\dfrac{k^4 - 4k^2 + 2k + 5}{k^2 + 1}$

(Modeling) Solve each problem.

105. ***Geometric Modeling*** Consider the figure, which is a square divided into two squares and two rectangles.

(a) The length of each side of the largest square is $x + y$. Use the formula for the area of a square to write the area of the largest square as a power.

(b) Use the formulas for the area of a square and the area of a rectangle to write the area of the largest square as a trinomial that represents the sum of the areas of the four figures that make it up.

(c) Explain why the expressions in parts (a) and (b) must be equivalent.

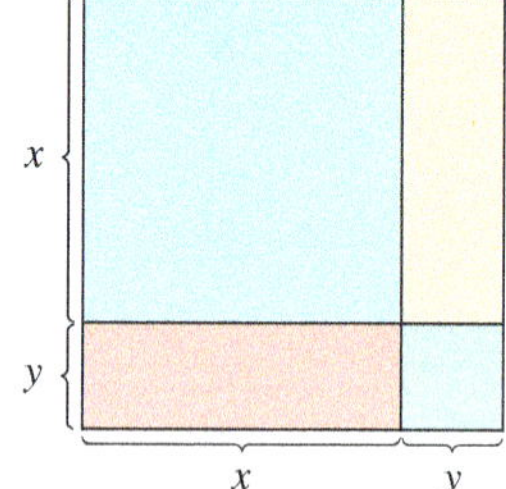

(d) What special product formula from this section does this exercise reinforce geometrically?

106. ***Geometric Modeling*** Use the figure to geometrically support the distributive property. Write a short paragraph explaining this process.

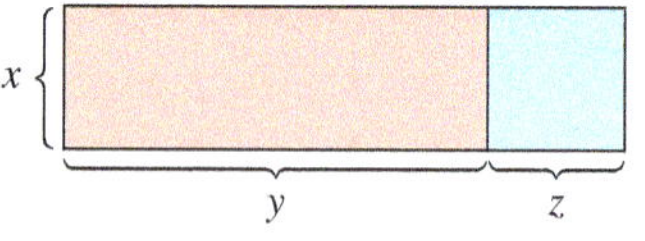

107. ***Volume of the Great Pyramid*** An amazing formula from ancient mathematics was used by the Egyptians to find the volume of the frustum of a square pyramid, as shown in the figure. Its volume is given by

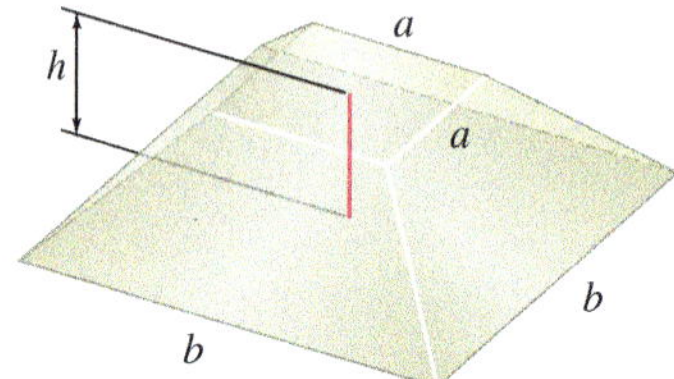

$$V = \frac{1}{3}h(a^2 + ab + b^2),$$

where b is the length of the base, a is the length of the top, and h is the height. (*Source:* Freebury, H. A., *A History of Mathematics*, Macmillan Company, New York.)

(a) When the Great Pyramid in Egypt was partially completed to a height h of 200 ft, b was 756 ft, and a was 314 ft. Calculate its volume at this stage of construction to the nearest thousand feet.

(b) Try to visualize the figure if $a = b$. What is the resulting shape? Find its volume.

(c) Let $a = b$ in the Egyptian formula and simplify. Are the results the same?

108. *Volume of the Great Pyramid* Refer to the formula and the discussion in **Exercise 107.**

(a) Use $V = \frac{1}{3}h(a^2 + ab + b^2)$ to determine a formula for the volume of a pyramid with square base of length b and height h by letting $a = 0$.

(b) The Great Pyramid in Egypt had a square base of length 756 ft and a height of 481 ft. Find the volume of the Great Pyramid to the nearest tenth million cubic feet. Compare it with the 273-ft-tall Superdome in New Orleans, which has an approximate volume of 100 million ft^3. (*Source: Guinness Book of World Records.*)

(c) The Superdome covers an area of 13 acres. How many acres, to the nearest tenth, does the Great Pyramid cover? (*Hint:* 1 acre = 43,560 ft^2)

(Modeling) Number of Farms in the United States *The graph shows the number of farms in the United States for selected years since 1950. We can use the formula*

$$\textit{Number of farms} = 0.001259x^2 - 5.039x + 5044$$

to get a good approximation of the number of farms for these years by substituting the year for x and evaluating the polynomial. For example, if $x = 1960$, *the value of the polynomial is approximately* 4.1, *which differs from the data in the bar graph by only* 0.1.

Evaluate the polynomial for each year, and then give the difference from the value in the graph.

109. 1950

110. 1970

111. 1990

112. 2012

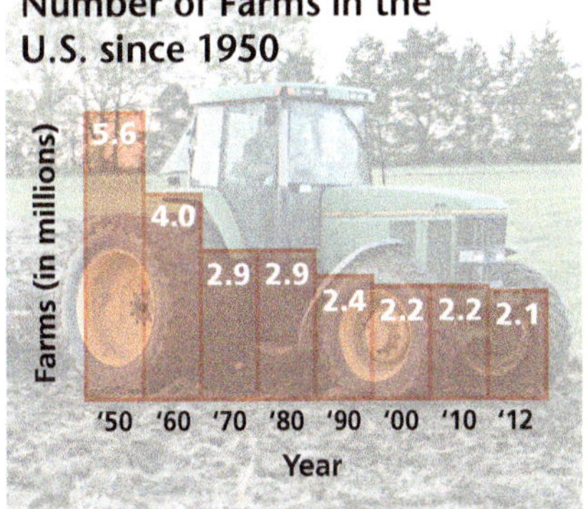

Source: U.S. Department of Agriculture.

Concept Check *Perform each operation mentally.*

113. $(0.25^3)(400^3)$ **114.** $(24^2)(0.5^2)$ **115.** $\dfrac{4.2^5}{2.1^5}$ **116.** $\dfrac{15^4}{5^4}$

Relating Concepts

For individual or collaborative investigation *(Exercises 117–120)*

The special products can be used to perform selected multiplications. On the left, we use $(x + y)(x - y) = x^2 - y^2$. *On the right,* $(x - y)^2 = x^2 - 2xy + y^2$.

$51 \times 49 = (50 + 1)(50 - 1)$	$47^2 = (50 - 3)^2$
$= 50^2 - 1^2$	$= 50^2 - 2(50)(3) + 3^2$
$= 2500 - 1$	$= 2500 - 300 + 9$
$= 2499$	$= 2209$

Use special products to evaluate each expression.

117. 99×101 **118.** 63×57 **119.** 102^2 **120.** 71^2

R.4 Factoring Polynomials

- Factoring Out the Greatest Common Factor
- Factoring by Grouping
- Factoring Trinomials
- Factoring Binomials
- Factoring by Substitution

The process of finding polynomials whose product equals a given polynomial is called **factoring.** Unless otherwise specified, we consider only integer coefficients when factoring polynomials. For example, because

$$4x + 12 = 4(x + 3),$$

both 4 and $x + 3$ are **factors** of $4x + 12$, and $4(x + 3)$ is a **factored form** of $4x + 12$.

A polynomial with variable terms that cannot be written as a product of two polynomials of lesser degree is a **prime polynomial.** A polynomial is **factored completely** when it is written as a product of prime polynomials.

Factoring Out the Greatest Common Factor

To factor a polynomial such as $6x^2y^3 + 9xy^4 + 18y^5$, we look for a monomial that is the **greatest common factor (GCF)** of the three terms.

$$6x^2y^3 + 9xy^4 + 18y^5$$
$$= 3y^3(2x^2) + 3y^3(3xy) + 3y^3(6y^2) \quad \text{GCF} = 3y^3$$
$$= 3y^3(2x^2 + 3xy + 6y^2) \quad \text{Distributive property}$$

EXAMPLE 1 Factoring Out the Greatest Common Factor

Factor out the greatest common factor from each polynomial.

(a) $9y^5 + y^2$

(b) $6x^2t + 8xt - 12t$

(c) $14(m + 1)^3 - 28(m + 1)^2 - 7(m + 1)$

SOLUTION

(a) $9y^5 + y^2$

$$= y^2(9y^3) + y^2(1) \quad \text{GCF} = y^2$$
$$= y^2(9y^3 + 1)$$

Remember to include the 1.

CHECK Multiply out the factored form: $y^2(9y^3 + 1) = \underbrace{9y^5 + y^2}_{\text{Original polynomial}}$. ✓

(b) $6x^2t + 8xt - 12t$

$$= 2t(3x^2 + 4x - 6) \quad \text{GCF} = 2t$$

CHECK $2t(3x^2 + 4x - 6) = 6x^2t + 8xt - 12t$ ✓

(c) $14(m + 1)^3 - 28(m + 1)^2 - 7(m + 1)$

$$= 7(m + 1)[2(m + 1)^2 - 4(m + 1) - 1] \quad \text{GCF} = 7(m + 1)$$
$$= 7(m + 1)[2(m^2 + 2m + 1) - 4m - 4 - 1] \quad \text{Square } m + 1\text{; distributive property}$$

Remember the middle term.

$$= 7(m + 1)(2m^2 + 4m + 2 - 4m - 4 - 1) \quad \text{Distributive property}$$
$$= 7(m + 1)(2m^2 - 3) \quad \text{Combine like terms.}$$

✓ Now Try Exercises 13, 19, and 25.

CAUTION In **Example 1(a),** the 1 is essential in the answer because

$$y^2(9y^3) \neq 9y^5 + y^2.$$

Factoring can always be checked by multiplying.

Factoring by Grouping

When a polynomial has more than three terms, it can sometimes be factored using **factoring by grouping.** Consider this example.

$$ax + ay + 6x + 6y$$

Terms with common factor a — Terms with common factor 6

$$= (ax + ay) + (6x + 6y)$$ Group the terms so that each group has a common factor.

$$= a(x + y) + 6(x + y)$$ Factor each group.

$$= (x + y)(a + 6)$$ Factor out $x + y$.

It is not always obvious which terms should be grouped. In cases like the one above, group in pairs. Experience and repeated trials are the most reliable tools.

EXAMPLE 2 Factoring by Grouping

Factor each polynomial by grouping.

(a) $mp^2 + 7m + 3p^2 + 21$ **(b)** $2y^2 + az - 2z - ay^2$

(c) $4x^3 + 2x^2 - 2x - 1$

SOLUTION

(a) $mp^2 + 7m + 3p^2 + 21$

$= (mp^2 + 7m) + (3p^2 + 21)$ Group the terms.

$= m(p^2 + 7) + 3(p^2 + 7)$ Factor each group.

$= (p^2 + 7)(m + 3)$ $p^2 + 7$ is a common factor.

CHECK $(p^2 + 7)(m + 3)$

$= mp^2 + 3p^2 + 7m + 21$ FOIL method

$= mp^2 + 7m + 3p^2 + 21$ ✓ Commutative property

(b) $2y^2 + az - 2z - ay^2$

$= 2y^2 - 2z - ay^2 + az$ Rearrange the terms.

$= (2y^2 - 2z) + (-ay^2 + az)$ Group the terms.

$= 2(y^2 - z) - a(y^2 - z)$ Factor out 2 and $-a$ so that $y^2 - z$ is a common factor.

Be careful with signs here.

$= (y^2 - z)(2 - a)$ Factor out $y^2 - z$.

(c) $4x^3 + 2x^2 - 2x - 1$

$= (4x^3 + 2x^2) + (-2x - 1)$ Group the terms.

$= 2x^2(2x + 1) - 1(2x + 1)$ Factor each group.

$= (2x + 1)(2x^2 - 1)$ Factor out $2x + 1$.

✔ **Now Try Exercises 29 and 31.**

Factoring Trinomials As shown in the diagram below, factoring is the opposite of multiplication.

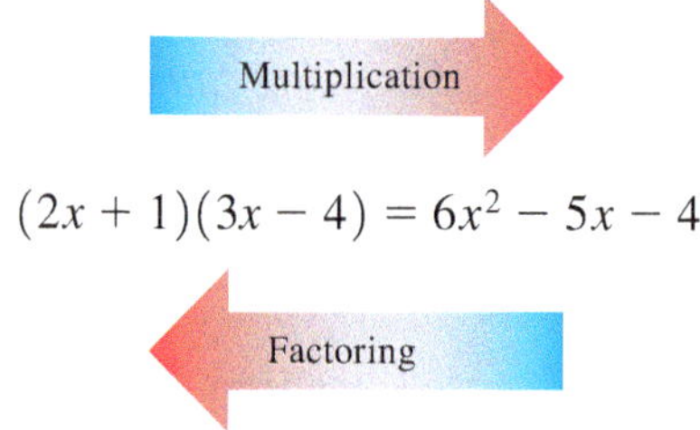

One strategy when factoring trinomials uses the FOIL method in reverse. This strategy requires trial-and-error to find the correct arrangement of coefficients of the binomial factors.

EXAMPLE 3 Factoring Trinomials

Factor each trinomial, if possible.

(a) $4y^2 - 11y + 6$ **(b)** $6p^2 - 7p - 5$

(c) $2x^2 + 13x - 18$ **(d)** $16y^3 + 24y^2 - 16y$

SOLUTION

(a) To factor this polynomial, we must find values for integers a, b, c, and d in such a way that

$$4y^2 - 11y + 6$$
$$= (ay + b)(cy + d). \quad \text{FOIL method}$$

Using the FOIL method, we see that $ac = 4$ and $bd = 6$. The positive factors of 4 are 4 and 1 or 2 and 2. Because the middle term has a negative coefficient, we consider only negative factors of 6. The possibilities are -2 and -3 or -1 and -6.

Now we try various arrangements of these factors until we find one that gives the correct coefficient of y.

$(2y - 1)(2y - 6)$	$(2y - 2)(2y - 3)$	$(y - 2)(4y - 3)$
$= 4y^2 - 14y + 6$	$= 4y^2 - 10y + 6$	$= 4y^2 - 11y + 6$
Incorrect	Incorrect	Correct

Therefore, $4y^2 - 11y + 6$ factors as $(y - 2)(4y - 3)$.

CHECK $(y - 2)(4y - 3)$

$$= 4y^2 - 3y - 8y + 6 \quad \text{FOIL method}$$
$$= 4y^2 - 11y + 6 \checkmark \quad \text{Original polynomial}$$

(b) Again, we try various possibilities to factor $6p^2 - 7p - 5$. The positive factors of 6 could be 2 and 3 or 1 and 6. As factors of -5 we have only -1 and 5 or -5 and 1.

$(2p - 5)(3p + 1)$	$(3p - 5)(2p + 1)$
$= 6p^2 - 13p - 5$ Incorrect	$= 6p^2 - 7p - 5$ Correct

Thus, $6p^2 - 7p - 5$ factors as $(3p - 5)(2p + 1)$.

(c) If we try to factor $2x^2 + 13x - 18$, we find that none of the pairs of factors gives the correct coefficient of x. Additional trials are also unsuccessful.

$(2x+9)(x-2)$	$(2x-3)(x+6)$	$(2x-1)(x+18)$
$= 2x^2 + 5x - 18$	$= 2x^2 + 9x - 18$	$= 2x^2 + 35x - 18$
Incorrect	Incorrect	Incorrect

This trinomial cannot be factored with integer coefficients and is prime.

(d)

$$16y^3 + 24y^2 - 16y$$
$$= 8y(2y^2 + 3y - 2) \quad \text{Factor out the GCF, } 8y.$$
$$= 8y(2y-1)(y+2) \quad \text{Factor the trinomial.}$$

Remember to include the common factor in the final form.

✔ **Now Try Exercises 35, 37, 39, and 41.**

NOTE In **Example 3,** we chose positive factors of the positive first term (instead of two negative factors). This makes the work easier.

Each of the special patterns for multiplication can be used in reverse to obtain a pattern for factoring. Perfect square trinomials can be factored as follows.

Factoring Perfect Square Trinomials

$$x^2 + 2xy + y^2 = (x + y)^2$$
$$x^2 - 2xy + y^2 = (x - y)^2$$

EXAMPLE 4 Factoring Perfect Square Trinomials

Factor each trinomial.

(a) $16p^2 - 40pq + 25q^2$ **(b)** $36x^2y^2 + 84xy + 49$

SOLUTION

(a) Because $16p^2 = (4p)^2$ and $25q^2 = (5q)^2$, we use the second pattern shown in the box, with $4p$ replacing x and $5q$ replacing y.

$$16p^2 - 40pq + 25q^2$$
$$= (4p)^2 - 2(4p)(5q) + (5q)^2$$
$$= (4p - 5q)^2$$

Make sure that the middle term of the trinomial being factored, $-40pq$ here, is twice the product of the two terms in the binomial $4p - 5q$.

$$-40pq = 2(4p)(-5q)$$

Thus, $16p^2 - 40pq + 25q^2$ factors as $(4p - 5q)^2$.

CHECK $(4p - 5q)^2 = 16p^2 - 40pq + 25q^2$ ✓ Multiply.

(b) $36x^2y^2 + 84xy + 49$ factors as $(6xy + 7)^2$. $2(6xy)(7) = 84xy$

CHECK Square $6xy + 7$: $(6xy + 7)^2 = 36x^2y^2 + 84xy + 49$. ✓

✔ **Now Try Exercises 51 and 55.**

Factoring Binomials Check first to see whether the terms of a binomial have a common factor. If so, factor it out. The binomial may also fit one of the following patterns.

Factoring Binomials

Difference of Squares	$x^2 - y^2 = (x + y)(x - y)$
Difference of Cubes	$x^3 - y^3 = (x - y)(x^2 + xy + y^2)$
Sum of Cubes	$x^3 + y^3 = (x + y)(x^2 - xy + y^2)$

CAUTION ***There is no factoring pattern for a sum of squares in the real number system.*** In particular, for real numbers x and y,

$$x^2 + y^2 \textbf{ does not factor as } (x + y)^2.$$

EXAMPLE 5 Factoring Differences of Squares

Factor each polynomial.

(a) $4m^2 - 9$

(b) $256k^4 - 625m^4$

(c) $(a + 2b)^2 - 4c^2$

(d) $x^2 - 6x + 9 - y^4$

(e) $y^2 - x^2 + 6x - 9$

SOLUTION

(a) $4m^2 - 9$

$= (2m)^2 - 3^2$ Write as a difference of squares.

$= (2m + 3)(2m - 3)$ Factor.

Check by multiplying.

(b) $256k^4 - 625m^4$

$= (16k^2)^2 - (25m^2)^2$ Write as a difference of squares.

Don't stop here. $= (16k^2 + 25m^2)(16k^2 - 25m^2)$ Factor.

$= (16k^2 + 25m^2)(4k + 5m)(4k - 5m)$ Factor $16k^2 - 25m^2$.

CHECK $(16k^2 + 25m^2)(4k + 5m)(4k - 5m)$

$= (16k^2 + 25m^2)(16k^2 - 25m^2)$ Multiply the last two factors.

$= 256k^4 - 625m^4$ ✓ Original polynomial

(c) $(a + 2b)^2 - 4c^2$

$= (a + 2b)^2 - (2c)^2$ Write as a difference of squares.

$= [(a + 2b) + 2c][(a + 2b) - 2c]$ Factor.

$= (a + 2b + 2c)(a + 2b - 2c)$ Check by multiplying.

(d) $x^2 - 6x + 9 - y^4$

$= (x^2 - 6x + 9) - y^4$ Group terms.

$= (x - 3)^2 - y^4$ Factor the trinomial.

$= (x - 3)^2 - (y^2)^2$ Write as a difference of squares.

$= [(x - 3) + y^2][(x - 3) - y^2]$ Factor.

$= (x - 3 + y^2)(x - 3 - y^2)$

(e) $y^2 - x^2 + 6x - 9$

Be careful with signs. This is a perfect square trinomial.

$= y^2 - (x^2 - 6x + 9)$ Factor out the negative sign, and group the last three terms.

$= y^2 - (x - 3)^2$ Write as a difference of squares.

$= [y - (x - 3)][y + (x - 3)]$ Factor.

$= (y - x + 3)(y + x - 3)$ Distributive property

✔ **Now Try Exercises 59, 61, 65, and 69.**

CAUTION When factoring as in **Example 5(e),** be careful with signs. Inserting an open parenthesis following the minus sign requires changing the signs of all of the following terms.

EXAMPLE 6 Factoring Sums or Differences of Cubes

Factor each polynomial.

(a) $x^3 + 27$ **(b)** $m^3 - 64n^3$ **(c)** $8q^6 + 125p^9$

SOLUTION

(a) $x^3 + 27$

$= x^3 + 3^3$ Write as a sum of cubes.

$= (x + 3)(x^2 - 3x + 3^2)$ Factor.

$= (x + 3)(x^2 - 3x + 9)$ Apply the exponent.

(b) $m^3 - 64n^3$

$= m^3 - (4n)^3$ Write as a difference of cubes.

$= (m - 4n)[m^2 + m(4n) + (4n)^2]$ Factor.

$= (m - 4n)(m^2 + 4mn + 16n^2)$ Multiply; $(4n)^2 = 4^2n^2$.

(c) $8q^6 + 125p^9$

$= (2q^2)^3 + (5p^3)^3$ Write as a sum of cubes.

$= (2q^2 + 5p^3)[(2q^2)^2 - 2q^2(5p^3) + (5p^3)^2]$ Factor.

$= (2q^2 + 5p^3)(4q^4 - 10q^2p^3 + 25p^6)$ Simplify.

✔ **Now Try Exercises 73, 75, and 77.**

Factoring by Substitution More complicated polynomials may be factored using substitution.

EXAMPLE 7 Factoring by Substitution

Factor each polynomial.

(a) $10(2a-1)^2 - 19(2a-1) - 15$ **(b)** $(2a-1)^3 + 8$

(c) $6z^4 - 13z^2 - 5$

SOLUTION

(a) $10(2a-1)^2 - 19(2a-1) - 15$

$= 10u^2 - 19u - 15$ Replace $2a-1$ with u so that $(2a-1)^2$ becomes u^2.

$= (5u+3)(2u-5)$ Factor.

Don't stop here. Replace u with $2a-1$.

$= [5(2a-1)+3][2(2a-1)-5]$ Replace u with $2a-1$.

$= (10a-5+3)(4a-2-5)$ Distributive property

$= (10a-2)(4a-7)$ Simplify.

$= 2(5a-1)(4a-7)$ Factor out the common factor.

(b) $(2a-1)^3 + 8$

$= u^3 + 2^3$ Replace $2a-1$ with u. Write as a sum of cubes.

$= (u+2)(u^2-2u+4)$ Factor.

$= [(2a-1)+2][(2a-1)^2 - 2(2a-1)+4]$ Replace u with $2a-1$.

$= (2a+1)(4a^2-4a+1-4a+2+4)$ Add, and then multiply.

$= (2a+1)(4a^2-8a+7)$ Combine like terms.

(c) $6z^4 - 13z^2 - 5$

$= 6u^2 - 13u - 5$ Replace z^2 with u.

$= (2u-5)(3u+1)$ Factor the trinomial.

Remember to make the final substitution.

$= (2z^2-5)(3z^2+1)$ Replace u with z^2.

✔ **Now Try Exercises 83, 87, and 91.**

R.4 Exercises

CONCEPT PREVIEW *Fill in the blank(s) to correctly complete each sentence.*

1. The process of finding polynomials whose product equals a given polynomial is called ________.

2. A polynomial is factored completely when it is written as a product of ________.

3. Factoring is the opposite of ________.

4. When a polynomial has more than three terms, it can sometimes be factored using ________.

5. There is no factoring pattern for a ________ in the real number system. In particular, $x^2 + y^2$ does not factor as $(x+y)^2$, for real numbers x and y.

CONCEPT PREVIEW *Work each problem.*

6. Match each polynomial in Column I with its factored form in Column II.

I	II
(a) $x^2 + 10xy + 25y^2$	**A.** $(x + 5y)(x - 5y)$
(b) $x^2 - 10xy + 25y^2$	**B.** $(x + 5y)^2$
(c) $x^2 - 25y^2$	**C.** $(x - 5y)^2$
(d) $25y^2 - x^2$	**D.** $(5y + x)(5y - x)$

7. Match each polynomial in Column I with its factored form in Column II.

I	II
(a) $8x^3 - 27$	**A.** $(3 - 2x)(9 + 6x + 4x^2)$
(b) $8x^3 + 27$	**B.** $(2x - 3)(4x^2 + 6x + 9)$
(c) $27 - 8x^3$	**C.** $(2x + 3)(4x^2 - 6x + 9)$

8. Which of the following is the correct factorization of $6x^2 + x - 12$?

A. $(3x + 4)(2x + 3)$ **B.** $(3x - 4)(2x - 3)$
C. $(3x + 4)(2x - 3)$ **D.** $(3x - 4)(2x + 3)$

9. Which of the following is the correct complete factorization of $x^4 - 1$?

A. $(x^2 - 1)(x^2 + 1)$ **B.** $(x^2 + 1)(x + 1)(x - 1)$
C. $(x^2 - 1)^2$ **D.** $(x - 1)^2(x + 1)^2$

10. Which of the following is the correct factorization of $x^3 + 8$?

A. $(x + 2)^3$ **B.** $(x + 2)(x^2 + 2x + 4)$
C. $(x + 2)(x^2 - 2x + 4)$ **D.** $(x + 2)(x^2 - 4x + 4)$

Factor out the greatest common factor from each polynomial. ***See Examples 1 and 2.***

11. $12m + 60$ **12.** $15r - 27$ **13.** $8k^3 + 24k$

14. $9z^4 + 81z$ **15.** $xy - 5xy^2$ **16.** $5h^2j + hj$

17. $-4p^3q^4 - 2p^2q^5$ **18.** $-3z^5w^2 - 18z^3w^4$

19. $4k^2m^3 + 8k^4m^3 - 12k^2m^4$ **20.** $28r^4s^2 + 7r^3s - 35r^4s^3$

21. $2(a + b) + 4m(a + b)$ **22.** $6x(a + b) - 4y(a + b)$

23. $(5r - 6)(r + 3) - (2r - 1)(r + 3)$

24. $(4z - 5)(3z - 2) - (3z - 9)(3z - 2)$

25. $2(m - 1) - 3(m - 1)^2 + 2(m - 1)^3$ **26.** $5(a + 3)^3 - 2(a + 3) + (a + 3)^2$

27. *Concept Check* When directed to completely factor the polynomial $4x^2y^5 - 8xy^3$, a student wrote $2xy^3(2xy^2 - 4)$. When the teacher did not give him full credit, he complained because when his answer is multiplied out, the result is the original polynomial. Give the correct answer.

28. *Concept Check* Kurt factored $16a^2 - 40a - 6a + 15$ by grouping and obtained $(8a - 3)(2a - 5)$. Callie factored the same polynomial and gave an answer of $(3 - 8a)(5 - 2a)$. Which answer is correct?

Factor each polynomial by grouping. ***See Example 2.***

29. $6st + 9t - 10s - 15$ **30.** $10ab - 6b + 35a - 21$

31. $2m^4 + 6 - am^4 - 3a$ **32.** $4x^6 + 36 - x^6y - 9y$

33. $p^2q^2 - 10 - 2q^2 + 5p^2$ **34.** $20z^2 - 8x + 5pz^2 - 2px$

Factor each trinomial, if possible. **See Examples 3 and 4.**

35. $6a^2 - 11a + 4$
36. $8h^2 - 2h - 21$
37. $3m^2 + 14m + 8$
38. $9y^2 - 18y + 8$
39. $15p^2 + 24p + 8$
40. $9x^2 + 4x - 2$
41. $12a^3 + 10a^2 - 42a$
42. $36x^3 + 18x^2 - 4x$
43. $6k^2 + 5kp - 6p^2$
44. $14m^2 + 11mr - 15r^2$
45. $5a^2 - 7ab - 6b^2$
46. $12s^2 + 11st - 5t^2$
47. $12x^2 - xy - y^2$
48. $30a^2 + am - m^2$
49. $24a^4 + 10a^3b - 4a^2b^2$
50. $18x^5 + 15x^4z - 75x^3z^2$
51. $9m^2 - 12m + 4$
52. $16p^2 - 40p + 25$
53. $32a^2 + 48ab + 18b^2$
54. $20p^2 - 100pq + 125q^2$
55. $4x^2y^2 + 28xy + 49$
56. $9m^2n^2 + 12mn + 4$
57. $(a - 3b)^2 - 6(a - 3b) + 9$
58. $(2p + q)^2 - 10(2p + q) + 25$

Factor each polynomial. **See Examples 5 and 6.**

59. $9a^2 - 16$
60. $16q^2 - 25$
61. $x^4 - 16$
62. $y^4 - 81$
63. $25s^4 - 9t^2$
64. $36z^2 - 81y^4$
65. $(a + b)^2 - 16$
66. $(p - 2q)^2 - 100$
67. $p^4 - 625$
68. $m^4 - 1296$
69. $x^2 - 8x + 16 - y^2$
70. $m^2 + 10m + 25 - n^4$
71. $y^2 - x^2 + 12x - 36$
72. $9m^2 - n^2 - 2n - 1$
73. $8 - a^3$
74. $27 - r^3$
75. $125x^3 - 27$
76. $8m^3 - 27n^3$
77. $27y^9 + 125z^6$
78. $27z^9 + 64y^{12}$
79. $(r + 6)^3 - 216$
80. $(b + 3)^3 - 27$
81. $27 - (m + 2n)^3$
82. $125 - (4a - b)^3$

Factor each polynomial. **See Example 7.**

83. $7(3k - 1)^2 + 26(3k - 1) - 8$
84. $6(4z - 3)^2 + 7(4z - 3) - 3$
85. $9(a - 4)^2 + 30(a - 4) + 25$
86. $4(5x + 7)^2 + 12(5x + 7) + 9$
87. $(a + 1)^3 + 27$
88. $(x - 4)^3 + 64$
89. $(3x + 4)^3 - 1$
90. $(5x - 2)^3 - 8$
91. $m^4 - 3m^2 - 10$
92. $a^4 - 2a^2 - 48$
93. $12t^4 - t^2 - 35$
94. $10m^4 + 43m^2 - 9$

Factor by any method. **See Examples 1–7.**

95. $4b^2 + 4bc + c^2 - 16$
96. $(2y - 1)^2 - 4(2y - 1) + 4$
97. $x^2 + xy - 5x - 5y$
98. $8r^2 - 3rs + 10s^2$
99. $p^4(m - 2n) + q(m - 2n)$
100. $36a^2 + 60a + 25$
101. $4z^2 + 28z + 49$
102. $6p^4 + 7p^2 - 3$
103. $1000x^3 + 343y^3$
104. $b^2 + 8b + 16 - a^2$
105. $125m^6 - 216$
106. $q^2 + 6q + 9 - p^2$
107. $64 + (3x + 2)^3$
108. $216p^3 + 125q^3$
109. $(x + y)^3 - (x - y)^3$
110. $100r^2 - 169s^2$
111. $144z^2 + 121$
112. $(3a + 5)^2 - 18(3a + 5) + 81$
113. $(x + y)^2 - (x - y)^2$
114. $4z^4 - 7z^2 - 15$

115. *Concept Check* Are there any conditions under which a sum of squares can be factored? If so, give an example.

116. *Geometric Modeling* Explain how the figures give geometric interpretation to the formula $x^2 + 2xy + y^2 = (x + y)^2$.

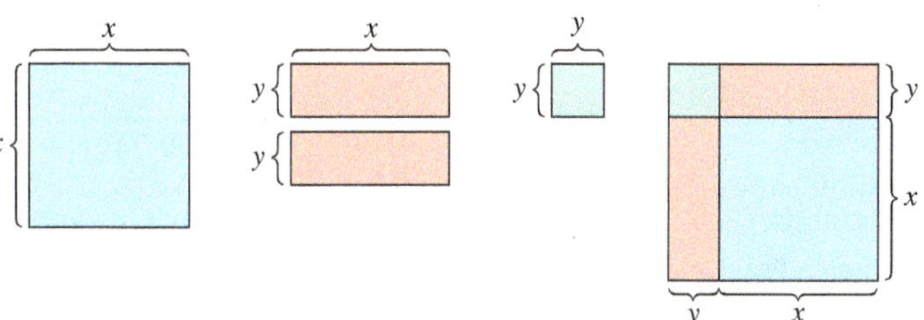

Factor each polynomial over the set of rational number coefficients.

117. $49x^2 - \frac{1}{25}$ **118.** $81y^2 - \frac{1}{49}$ **119.** $\frac{25}{9}x^4 - 9y^2$ **120.** $\frac{121}{25}y^4 - 49x^2$

Concept Check Find all values of b or c that will make the polynomial a perfect square trinomial.

121. $4z^2 + bz + 81$

122. $9p^2 + bp + 25$

123. $100r^2 - 60r + c$

124. $49x^2 + 70x + c$

Relating Concepts

For individual or collaborative investigation *(Exercises 125–130)*

The polynomial $x^6 - 1$ can be considered either a difference of squares or a difference of cubes. Work ***Exercises 125–130 in order,*** *to connect the results obtained when two different methods of factoring are used.*

125. Factor $x^6 - 1$ by first factoring as a difference of squares, and then factor further by using the patterns for a sum of cubes and a difference of cubes.

126. Factor $x^6 - 1$ by first factoring as a difference of cubes, and then factor further by using the pattern for a difference of squares.

127. Compare the answers in **Exercises 125 and 126.** Based on these results, what is the factorization of $x^4 + x^2 + 1$?

128. The polynomial $x^4 + x^2 + 1$ cannot be factored using the methods described in this section. However, there is a technique that enables us to factor it, as shown here. Supply the reason why each step is valid.

$$
\begin{aligned}
&x^4 + x^2 + 1 \\
&\quad = x^4 + 2x^2 + 1 - x^2 && \underline{\qquad\qquad\qquad} \\
&\quad = (x^4 + 2x^2 + 1) - x^2 && \underline{\qquad\qquad\qquad} \\
&\quad = (x^2 + 1)^2 - x^2 && \underline{\qquad\qquad\qquad} \\
&\quad = (x^2 + 1 - x)(x^2 + 1 + x) && \underline{\qquad\qquad\qquad} \\
&\quad = (x^2 - x + 1)(x^2 + x + 1) && \underline{\qquad\qquad\qquad}
\end{aligned}
$$

129. How does the answer in **Exercise 127** compare with the final line in **Exercise 128?**

130. Factor $x^8 + x^4 + 1$ using the technique outlined in **Exercise 128.**

R.5 Rational Expressions

- Rational Expressions
- Lowest Terms of a Rational Expression
- Multiplication and Division
- Addition and Subtraction
- Complex Fractions

Rational Expressions

The quotient of two polynomials P and Q, with $Q \neq 0$, is a **rational expression.**

$$\frac{x+6}{x+2}, \quad \frac{(x+6)(x+4)}{(x+2)(x+4)}, \quad \frac{2p^2+7p-4}{5p^2+20p} \qquad \text{Rational expressions}$$

The **domain** of a rational expression is the set of real numbers for which the expression is defined. Because the denominator of a fraction cannot be 0, the domain consists of all real numbers except those that make the denominator 0. We find these numbers by setting the denominator equal to 0 and solving the resulting equation. For example, in the rational expression

$$\frac{x+6}{x+2},$$

the solution to the equation $x+2=0$ is excluded from the domain. The solution is -2, so the domain is the set of all real numbers x not equal to -2.

$$\{x \mid x \neq -2\} \qquad \text{Set-builder notation}$$

If the denominator of a rational expression contains a product, we determine the domain with the **zero-factor property,** which states that $ab = 0$ if and only if $a = 0$ or $b = 0$.

EXAMPLE 1 Finding the Domain

Find the domain of the rational expression.

$$\frac{(x+6)(x+4)}{(x+2)(x+4)}$$

SOLUTION

$$(x+2)(x+4) = 0 \qquad \text{Set the denominator equal to zero.}$$

$$x+2=0 \quad \text{or} \quad x+4=0 \qquad \text{Zero-factor property}$$

$$x=-2 \quad \text{or} \quad x=-4 \qquad \text{Solve each equation.}$$

The domain is the set of real numbers *not equal to* -2 or -4, written

$$\{x \mid x \neq -2, -4\}.$$

✔ **Now Try Exercises 11 and 13.**

Lowest Terms of a Rational Expression

A rational expression is written in **lowest terms** when the greatest common factor of its numerator and its denominator is 1. We use the following **fundamental principle of fractions** to write a rational expression in lowest terms by dividing out common factors.

Fundamental Principle of Fractions

$$\frac{ac}{bc} = \frac{a}{b} \quad (b \neq 0, c \neq 0)$$

EXAMPLE 2 Writing Rational Expressions in Lowest Terms

Write each rational expression in lowest terms.

(a) $\dfrac{2x^2 + 7x - 4}{5x^2 + 20x}$ (b) $\dfrac{6 - 3x}{x^2 - 4}$

SOLUTION

(a) $\dfrac{2x^2 + 7x - 4}{5x^2 + 20x}$

$$= \frac{(2x - 1)(x + 4)}{5x(x + 4)} \quad \text{Factor.}$$

$$= \frac{2x - 1}{5x} \quad \text{Divide out the common factor.}$$

To determine the domain, we find values of x that make the *original* denominator $5x^2 + 20x$ equal to 0, and exclude them.

$$5x^2 + 20x = 0 \quad \text{Set the denominator equal to 0.}$$

$$5x(x + 4) = 0 \quad \text{Factor.}$$

$$5x = 0 \quad \text{or} \quad x + 4 = 0 \quad \text{Zero-factor property}$$

$$x = 0 \quad \text{or} \quad x = -4 \quad \text{Solve each equation.}$$

The domain is $\{x \mid x \neq 0, -4\}$. ***From now on, we will assume such restrictions when writing rational expressions in lowest terms.***

(b) $\dfrac{6 - 3x}{x^2 - 4}$

$$= \frac{3(2 - x)}{(x + 2)(x - 2)} \quad \text{Factor.}$$

$$= \frac{3(2 - x)(-1)}{(x + 2)(x - 2)(-1)} \quad \begin{array}{l}2 - x \text{ and } x - 2 \text{ are opposites.}\\ \text{Multiply numerator and denominator by } -1.\end{array}$$

$$= \frac{3(2 - x)(-1)}{(x + 2)(2 - x)} \quad (x - 2)(-1) = -x + 2 = 2 - x$$

Be careful with signs.

$$= \frac{-3}{x + 2} \quad \text{Divide out the common factor.}$$

Working in an alternative way would lead to the equivalent result $\frac{3}{-x - 2}$.

✔ Now Try Exercises 23 and 27.

LOOKING AHEAD TO CALCULUS

A standard problem in calculus is investigating what value an expression such as $\frac{x^2 - 1}{x - 1}$ approaches as x approaches 1. We cannot do this by simply substituting 1 for x in the expression since the result is the indeterminate form $\frac{0}{0}$. When we factor the numerator and write the expression in lowest terms, it becomes $x + 1$. Then, by substituting 1 for x, we obtain $1 + 1 = 2$, which is the **limit** of $\frac{x^2 - 1}{x - 1}$ as x approaches 1.

CAUTION The fundamental principle requires a pair of common *factors*, one in the numerator and one in the denominator. ***Only after a rational expression has been factored can any common factors be divided out.*** For example,

$$\frac{2x + 4}{6} = \frac{2(x + 2)}{2 \cdot 3} = \frac{x + 2}{3}. \quad \text{Factor first, and then divide.}$$

Multiplication and Division We now multiply and divide fractions.

Multiplication and Division

For fractions $\frac{a}{b}$ and $\frac{c}{d}$ $(b \neq 0, d \neq 0)$, the following hold.

$$\frac{a}{b} \cdot \frac{c}{d} = \frac{ac}{bd} \quad \text{and} \quad \frac{a}{b} \div \frac{c}{d} = \frac{a}{b} \cdot \frac{d}{c} \quad (c \neq 0)$$

That is, to find the product of two fractions, multiply their numerators to find the numerator of the product. Then multiply their denominators to find the denominator of the product.

To divide two fractions, multiply the* dividend *(the first fraction) by the reciprocal of the* divisor *(the second fraction).

EXAMPLE 3 Multiplying or Dividing Rational Expressions

Multiply or divide, as indicated.

(a) $\dfrac{2y^2}{9} \cdot \dfrac{27}{8y^5}$ **(b)** $\dfrac{3m^2 - 2m - 8}{3m^2 + 14m + 8} \cdot \dfrac{3m + 2}{3m + 4}$

(c) $\dfrac{3p^2 + 11p - 4}{24p^3 - 8p^2} \div \dfrac{9p + 36}{24p^4 - 36p^3}$ **(d)** $\dfrac{x^3 - y^3}{x^2 - y^2} \cdot \dfrac{2x + 2y + xz + yz}{2x^2 + 2y^2 + zx^2 + zy^2}$

SOLUTION

(a) $\dfrac{2y^2}{9} \cdot \dfrac{27}{8y^5}$

$$= \frac{2y^2 \cdot 27}{9 \cdot 8y^5} \quad \text{Multiply fractions.}$$

$$= \frac{2 \cdot 9 \cdot 3 \cdot y^2}{9 \cdot 2 \cdot 4 \cdot y^2 \cdot y^3} \quad \text{Factor.}$$

$$= \frac{3}{4y^3} \quad \text{Lowest terms}$$

Although we usually factor first and then multiply the fractions (see parts (b)–(d)), we did the opposite here. Either order is acceptable.

(b) $\dfrac{3m^2 - 2m - 8}{3m^2 + 14m + 8} \cdot \dfrac{3m + 2}{3m + 4}$

$$= \frac{(m - 2)(3m + 4)}{(m + 4)(3m + 2)} \cdot \frac{3m + 2}{3m + 4} \quad \text{Factor.}$$

$$= \frac{(m - 2)(3m + 4)(3m + 2)}{(m + 4)(3m + 2)(3m + 4)} \quad \text{Multiply fractions.}$$

$$= \frac{m - 2}{m + 4} \quad \text{Lowest terms}$$

(c) $\dfrac{3p^2 + 11p - 4}{24p^3 - 8p^2} \div \dfrac{9p + 36}{24p^4 - 36p^3}$

$$= \frac{(p + 4)(3p - 1)}{8p^2(3p - 1)} \div \frac{9(p + 4)}{12p^3(2p - 3)} \quad \text{Factor.}$$

$$= \frac{(p + 4)(3p - 1)}{8p^2(3p - 1)} \cdot \frac{12p^3(2p - 3)}{9(p + 4)} \quad \text{Multiply by the reciprocal of the divisor.}$$

$$= \frac{12p^3(2p - 3)}{9 \cdot 8p^2} \quad \text{Divide out common factors. Multiply fractions.}$$

$$= \frac{3 \cdot 4 \cdot p^2 \cdot p(2p - 3)}{3 \cdot 3 \cdot 4 \cdot 2 \cdot p^2} \quad \text{Factor.}$$

$$= \frac{p(2p - 3)}{6} \quad \text{Lowest terms}$$

(d) $\dfrac{x^3 - y^3}{x^2 - y^2} \cdot \dfrac{2x + 2y + xz + yz}{2x^2 + 2y^2 + zx^2 + zy^2}$

$$= \frac{(x - y)(x^2 + xy + y^2)}{(x + y)(x - y)} \cdot \frac{2(x + y) + z(x + y)}{2(x^2 + y^2) + z(x^2 + y^2)} \quad \text{Factor. Group terms and factor.}$$

$$= \frac{(x - y)(x^2 + xy + y^2)}{(x + y)(x - y)} \cdot \frac{(x + y)(2 + z)}{(x^2 + y^2)(2 + z)} \quad \text{Factor by grouping.}$$

$$= \frac{x^2 + xy + y^2}{x^2 + y^2} \quad \text{Divide out common factors. Multiply fractions.}$$

✔ **Now Try Exercises 33, 43, and 47.**

Addition and Subtraction We add and subtract rational expressions in the same way that we add and subtract fractions.

Addition and Subtraction

For fractions $\frac{a}{b}$ and $\frac{c}{d}$ $(b \neq 0, d \neq 0)$, the following hold.

$$\frac{a}{b} + \frac{c}{d} = \frac{ad + bc}{bd} \quad \text{and} \quad \frac{a}{b} - \frac{c}{d} = \frac{ad - bc}{bd}$$

That is, to add (or subtract) two fractions in practice, find their least common denominator (LCD) and change each fraction to one with the LCD as denominator. The sum (or difference) of their numerators is the numerator of their sum (or difference), and the LCD is the denominator of their sum (or difference).

Finding the Least Common Denominator (LCD)

Step 1 Write each denominator as a product of prime factors.

Step 2 Form a product of all the different prime factors. Each factor should have as exponent the ***greatest*** exponent that appears on that factor.

EXAMPLE 4 Adding or Subtracting Rational Expressions

Add or subtract, as indicated.

(a) $\frac{5}{9x^2} + \frac{1}{6x}$

(b) $\frac{y}{y-2} + \frac{8}{2-y}$

(c) $\frac{3}{(x-1)(x+2)} - \frac{1}{(x+3)(x-4)}$

SOLUTION

(a) $\frac{5}{9x^2} + \frac{1}{6x}$

Step 1 Write each denominator as a product of prime factors.

$$9x^2 = 3^2 \cdot x^2$$

$$6x = 2^1 \cdot 3^1 \cdot x^1$$

Step 2 For the LCD, form the product of all the prime factors, with each factor having the greatest exponent that appears on it.

Greatest exponent on 3 is 2. Greatest exponent on x is 2.

$$\text{LCD} = 2^1 \cdot 3^2 \cdot x^2$$

$$= 18x^2$$

Write the given expressions with this denominator, and then add.

$$\frac{5}{9x^2} + \frac{1}{6x}$$

$$= \frac{5 \cdot 2}{9x^2 \cdot 2} + \frac{1 \cdot 3x}{6x \cdot 3x} \qquad \text{LCD} = 18x^2$$

$$= \frac{10}{18x^2} + \frac{3x}{18x^2} \qquad \text{Multiply.}$$

$$= \frac{10 + 3x}{18x^2} \qquad \text{Add the numerators.}$$

Always check to see that the answer is in lowest terms.

(b) $\frac{y}{y-2} + \frac{8}{2-y}$ — We arbitrarily choose $y - 2$ as the LCD.

$$= \frac{y}{y-2} + \frac{8(-1)}{(2-y)(-1)} \qquad \text{Multiply the second expression by } -1 \text{ in both the numerator and the denominator.}$$

$$= \frac{y}{y-2} + \frac{-8}{y-2} \qquad \text{Simplify.}$$

$$= \frac{y-8}{y-2} \qquad \text{Add the numerators.}$$

We could use $2 - y$ as the common denominator instead of $y - 2$.

$$\frac{y(-1)}{(y-2)(-1)} + \frac{8}{2-y}$$ Multiply the first expression by -1 in both the numerator and the denominator.

$$= \frac{-y}{2-y} + \frac{8}{2-y}$$ Simplify.

$$= \frac{8-y}{2-y}$$ This equivalent expression results.

(c) $$\frac{3}{(x-1)(x+2)} - \frac{1}{(x+3)(x-4)}$$ The LCD is $(x-1)(x+2)(x+3)(x-4)$.

$$= \frac{3(x+3)(x-4)}{(x-1)(x+2)(x+3)(x-4)} - \frac{1(x-1)(x+2)}{(x+3)(x-4)(x-1)(x+2)}$$

$$= \frac{3(x^2-x-12)-(x^2+x-2)}{(x-1)(x+2)(x+3)(x-4)}$$ Multiply in the numerators, and then subtract them.

Be careful with signs.

$$= \frac{3x^2-3x-36-x^2-x+2}{(x-1)(x+2)(x+3)(x-4)}$$ Distributive property

$$= \frac{2x^2-4x-34}{(x-1)(x+2)(x+3)(x-4)}$$ Combine like terms in the numerator.

✔ **Now Try Exercises 57, 63, and 69.**

CAUTION ***When subtracting fractions where the second fraction has more than one term in the numerator, as in Example 4(c), be sure to distribute the negative sign to each term.*** Use parentheses as in the second step to avoid an error.

Complex Fractions

The quotient of two rational expressions is a **complex fraction**. There are two methods for simplifying a complex fraction.

EXAMPLE 5 Simplifying Complex Fractions

Simplify each complex fraction. In part (b), use two methods.

(a) $$\frac{6-\frac{5}{k}}{1+\frac{5}{k}}$$ **(b)** $$\frac{\frac{a}{a+1}+\frac{1}{a}}{\frac{1}{a}+\frac{1}{a+1}}$$

SOLUTION

(a) Method 1 for simplifying uses the identity property for multiplication. We multiply both numerator and denominator by the LCD of all the fractions, k.

$$\frac{6-\frac{5}{k}}{1+\frac{5}{k}} = \frac{k\left(6-\frac{5}{k}\right)}{k\left(1+\frac{5}{k}\right)} = \frac{6k-k\left(\frac{5}{k}\right)}{k+k\left(\frac{5}{k}\right)} = \frac{6k-5}{k+5}$$

Distribute k to *all* terms within the parentheses.

(b) $$\frac{\frac{a}{a+1}+\frac{1}{a}}{\frac{1}{a}+\frac{1}{a+1}}=\frac{\left(\frac{a}{a+1}+\frac{1}{a}\right)a(a+1)}{\left(\frac{1}{a}+\frac{1}{a+1}\right)a(a+1)}$$

For Method 1, multiply both numerator and denominator by the LCD of all the fractions, $a(a+1)$.

(Method 1)

$$=\frac{\frac{a}{a+1}(a)(a+1)+\frac{1}{a}(a)(a+1)}{\frac{1}{a}(a)(a+1)+\frac{1}{a+1}(a)(a+1)}$$

Distributive property

$$=\frac{a^2+(a+1)}{(a+1)+a}$$

Multiply.

$$=\frac{a^2+a+1}{2a+1}$$

Combine like terms.

$$\frac{\frac{a}{a+1}+\frac{1}{a}}{\frac{1}{a}+\frac{1}{a+1}}=\frac{\frac{a^2+1(a+1)}{a(a+1)}}{\frac{1(a+1)+1(a)}{a(a+1)}}$$

For Method 2, find the LCD, and add terms in the numerator and denominator of the complex fraction.

(Method 2)

$$=\frac{\frac{a^2+a+1}{a(a+1)}}{\frac{2a+1}{a(a+1)}}$$

Combine terms in the numerator and denominator.

$$=\frac{a^2+a+1}{a(a+1)}\cdot\frac{a(a+1)}{2a+1}$$

Multiply by the reciprocal of the divisor.

$$=\frac{a^2+a+1}{2a+1}$$

The result is the same as in Method 1.

Multiply fractions, and write in lowest terms.

✔ **Now Try Exercises 71 and 83.**

R.5 Exercises

CONCEPT PREVIEW *Fill in the blank(s) to correctly complete each sentence.*

1. The quotient of two polynomials in which the denominator is not equal to zero is a ________.

2. The domain of a rational expression consists of all real numbers except those that make the ________ equal to 0.

3. In the rational expression $\frac{x+1}{x-5}$, the domain cannot include the number ________.

4. A rational expression is in lowest terms when the greatest common factor of its numerator and its denominator is ________.

CONCEPT PREVIEW *Perform the indicated operation, and write each answer in lowest terms.*

5. $\frac{2x}{5}\cdot\frac{10}{x^2}$

6. $\frac{y^3}{8}\div\frac{y}{4}$

7. $\frac{3}{x}+\frac{7}{x}$

8. $\frac{4}{x-y} - \frac{9}{x-y}$ **9.** $\frac{2x}{5} + \frac{x}{4}$ **10.** $\frac{7}{x^2} - \frac{8}{y}$

Find the domain of each rational expression. ***See Example 1.***

11. $\frac{x+3}{x-6}$ **12.** $\frac{2x-4}{x+7}$ **13.** $\frac{3x+7}{(4x+2)(x-1)}$

14. $\frac{9x+12}{(2x+3)(x-5)}$ **15.** $\frac{12}{x^2+5x+6}$ **16.** $\frac{3}{x^2-5x-6}$

17. $\frac{x^2-1}{x+1}$ **18.** $\frac{x^2-25}{x-5}$ **19.** $\frac{x^3-1}{x-1}$

20. *Concept Check* Use specific values for x and y to show that in general, $\frac{1}{x} + \frac{1}{y}$ is not equivalent to $\frac{1}{x+y}$.

Write each rational expression in lowest terms. ***See Example 2.***

21. $\frac{8x^2+16x}{4x^2}$ **22.** $\frac{36y^2+72y}{9y^2}$ **23.** $\frac{3(3-t)}{(t+5)(t-3)}$

24. $\frac{-8(4-y)}{(y+2)(y-4)}$ **25.** $\frac{8k+16}{9k+18}$ **26.** $\frac{20r+10}{30r+15}$

27. $\frac{m^2-4m+4}{m^2+m-6}$ **28.** $\frac{r^2-r-6}{r^2+r-12}$ **29.** $\frac{8m^2+6m-9}{16m^2-9}$

30. $\frac{6y^2+11y+4}{3y^2+7y+4}$ **31.** $\frac{x^3+64}{x+4}$ **32.** $\frac{y^3-27}{y-3}$

Multiply or divide, as indicated. ***See Example 3.***

33. $\frac{15p^3}{9p^2} \div \frac{6p}{10p^2}$ **34.** $\frac{8r^3}{6r} \div \frac{5r^2}{9r^3}$ **35.** $\frac{2k+8}{6} \div \frac{3k+12}{2}$

36. $\frac{5m+25}{10} \div \frac{6m+30}{12}$ **37.** $\frac{x^2+x}{5} \cdot \frac{25}{xy+y}$ **38.** $\frac{y^3+y^2}{7} \cdot \frac{49}{y^4+y^3}$

39. $\frac{4a+12}{2a-10} \div \frac{a^2-9}{a^2-a-20}$ **40.** $\frac{6r-18}{9r^2+6r-24} \div \frac{4r-12}{12r-16}$

41. $\frac{p^2-p-12}{p^2-2p-15} \cdot \frac{p^2-9p+20}{p^2-8p+16}$ **42.** $\frac{x^2+2x-15}{x^2+11x+30} \cdot \frac{x^2+2x-24}{x^2-8x+15}$

43. $\frac{m^2+3m+2}{m^2+5m+4} \div \frac{m^2+5m+6}{m^2+10m+24}$ **44.** $\frac{y^2+y-2}{y^2+3y-4} \div \frac{y^2+3y+2}{y^2+4y+3}$

45. $\frac{x^3+y^3}{x^3-y^3} \cdot \frac{x^2-y^2}{x^2+2xy+y^2}$ **46.** $\frac{x^2-y^2}{(x-y)^2} \cdot \frac{x^2-xy+y^2}{x^2-2xy+y^2} \div \frac{x^3+y^3}{(x-y)^4}$

47. $\frac{xz-xw+2yz-2yw}{z^2-w^2} \cdot \frac{4z+4w+xz+wx}{16-x^2}$

48. $\frac{ac+ad+bc+bd}{a^2-b^2} \cdot \frac{a^3-b^3}{2a^2+2ab+2b^2}$

49. *Concept Check* Which of the following rational expressions is equivalent to -1? In choices A, B, and D, $x \neq -4$, and in choice C, $x \neq 4$. (*Hint:* There may be more than one answer.)

A. $\frac{x-4}{x+4}$ **B.** $\frac{-x-4}{x+4}$ **C.** $\frac{x-4}{4-x}$ **D.** $\frac{x-4}{-x-4}$

50. Explain how to find the least common denominator of several fractions.

Add or subtract, as indicated. **See Example 4.**

51. $\frac{3}{2k} + \frac{5}{3k}$

52. $\frac{8}{5p} + \frac{3}{4p}$

53. $\frac{1}{6m} + \frac{2}{5m} + \frac{4}{m}$

54. $\frac{8}{3p} + \frac{5}{4p} + \frac{9}{2p}$

55. $\frac{1}{a} - \frac{b}{a^2}$

56. $\frac{3}{z} + \frac{x}{z^2}$

57. $\frac{5}{12x^2y} - \frac{11}{6xy}$

58. $\frac{7}{18a^3b^2} - \frac{2}{9ab}$

59. $\frac{17y+3}{9y+7} - \frac{-10y-18}{9y+7}$

60. $\frac{7x+8}{3x+2} - \frac{x+4}{3x+2}$

61. $\frac{1}{x+z} + \frac{1}{x-z}$

62. $\frac{m+1}{m-1} + \frac{m-1}{m+1}$

63. $\frac{3}{a-2} - \frac{1}{2-a}$

64. $\frac{4}{p-q} - \frac{2}{q-p}$

65. $\frac{x+y}{2x-y} - \frac{2x}{y-2x}$

66. $\frac{m-4}{3m-4} - \frac{5m}{4-3m}$

67. $\frac{4}{x+1} + \frac{1}{x^2-x+1} - \frac{12}{x^3+1}$

68. $\frac{5}{x+2} + \frac{2}{x^2-2x+4} - \frac{60}{x^3+8}$

69. $\frac{3x}{x^2+x-12} - \frac{x}{x^2-16}$

70. $\frac{p}{2p^2-9p-5} - \frac{2p}{6p^2-p-2}$

Simplify each complex fraction. **See Example 5.**

71. $\dfrac{1+\frac{1}{x}}{1-\frac{1}{x}}$

72. $\dfrac{2-\frac{2}{y}}{2+\frac{2}{y}}$

73. $\dfrac{\frac{1}{x+1}-\frac{1}{x}}{\frac{1}{x}}$

74. $\dfrac{\frac{1}{y+3}-\frac{1}{y}}{\frac{1}{y}}$

75. $\dfrac{1+\frac{1}{1-b}}{1-\frac{1}{1+b}}$

76. $\dfrac{2+\frac{2}{1+x}}{2-\frac{2}{1-x}}$

77. $\dfrac{\frac{1}{a^3+b^3}}{\frac{1}{a^2+2ab+b^2}}$

78. $\dfrac{\frac{1}{x^3-y^3}}{\frac{1}{x^2-y^2}}$

79. $\dfrac{m-\frac{1}{m^2-4}}{\frac{1}{m+2}}$

80. $\dfrac{y+\frac{1}{y^2-9}}{\frac{1}{y+3}}$

81. $\dfrac{\frac{3}{p^2-16}+p}{\frac{1}{p-4}}$

82. $\dfrac{\frac{6}{x^2-25}+x}{\frac{1}{x-5}}$

83. $\dfrac{\frac{y+3}{y}-\frac{4}{y-1}}{\frac{y}{y-1}+\frac{1}{y}}$

84. $\dfrac{\frac{x+4}{x}-\frac{3}{x-2}}{\frac{x}{x-2}+\frac{1}{x}}$

85. $\dfrac{\frac{1}{x+h}-\frac{1}{x}}{h}$

86. $\dfrac{\frac{-2}{x+h}-\frac{-2}{x}}{h}$

87. $\dfrac{\frac{1}{(x+h)^2+9}-\frac{1}{x^2+9}}{h}$

88. $\dfrac{\frac{2}{(x+h)^2+16}-\frac{2}{x^2+16}}{h}$

(Modeling) Distance from the Origin of the Nile River *The Nile River in Africa is about* 4000 mi *long. The Nile begins as an outlet of Lake Victoria at an altitude of* 7000 ft *above sea level and empties into the Mediterranean Sea at sea level* (0 ft). *The distance from its origin in thousands of miles is related to its height above sea level in thousands of feet (x) by the following formula.*

$$\text{Distance} = \frac{7 - x}{0.639x + 1.75}$$

For example, when the river is at an altitude of 600 ft, $x = 0.6$ *(thousand), and the distance from the origin is*

$$\text{Distance} = \frac{7 - 0.6}{0.639(0.6) + 1.75} \approx 3, \quad \textit{which represents } 3000 \text{ mi.}$$

(Source: World Almanac and Book of Facts.)

89. What is the distance from the origin of the Nile when the river has an altitude of 7000 ft?

90. What is the distance, to the nearest mile, from the origin of the Nile when the river has an altitude of 1200 ft?

(Modeling) Cost-Benefit Model for a Pollutant *In situations involving environmental pollution, a* ***cost-benefit model*** *expresses cost in terms of the percentage of pollutant removed from the environment. Suppose a cost-benefit model is expressed as*

$$y = \frac{6.7x}{100 - x},$$

where y is the cost in thousands of dollars of removing x percent of a certain pollutant. Find the value of y for each given value of x.

91. $x = 75\ (75\%)$

92. $x = 95\ (95\%)$

R.6 Rational Exponents

- Negative Exponents and the Quotient Rule
- Rational Exponents
- Complex Fractions Revisited

Negative Exponents and the Quotient Rule

Suppose that n is a positive integer, and we wish to define a^{-n} to be consistent with the application of the product rule. Consider the product $a^n \cdot a^{-n}$, and apply the rule.

$$
\begin{aligned}
& a^n \cdot a^{-n} \\
&= a^{n+(-n)} && \text{Product rule: Add exponents.} \\
&= a^0 && n \text{ and } -n \text{ are additive inverses.} \\
&= 1 && \text{Definition of } a^0
\end{aligned}
$$

The expression a^{-n} acts as the *reciprocal* of a^n, which is written $\frac{1}{a^n}$. Thus, these two expressions must be equivalent.

Negative Exponent

Let a be a nonzero real number and n be any integer.

$$a^{-n} = \frac{1}{a^n}$$

EXAMPLE 1 **Using the Definition of a Negative Exponent**

Write each expression without negative exponents, and evaluate if possible. Assume all variables represent nonzero real numbers.

(a) 4^{-2} **(b)** -4^{-2} **(c)** $\left(\frac{2}{5}\right)^{-3}$ **(d)** $(xy)^{-3}$ **(e)** xy^{-3}

SOLUTION

(a) $4^{-2} = \frac{1}{4^2} = \frac{1}{16}$

(b) $-4^{-2} = -\frac{1}{4^2} = -\frac{1}{16}$

(c) $\left(\frac{2}{5}\right)^{-3} = \frac{1}{\left(\frac{2}{5}\right)^3} = \frac{1}{\frac{8}{125}} = 1 \div \frac{8}{125} = 1 \cdot \frac{125}{8} = \frac{125}{8}$

Multiply by the reciprocal of the divisor.

(d) $(xy)^{-3} = \frac{1}{(xy)^3} = \frac{1}{x^3y^3}$

Base is xy.

(e) $xy^{-3} = x \cdot \frac{1}{y^3} = \frac{x}{y^3}$

Base is y.

✔ **Now Try Exercises 11, 13, 15, 17, and 19.**

CAUTION ***A negative exponent indicates a reciprocal, not a sign change of the expression.***

Example 1(c) showed the following.

$$\left(\frac{2}{5}\right)^{-3} = \frac{125}{8} = \left(\frac{5}{2}\right)^3$$

We can generalize this result. If $a \neq 0$ and $b \neq 0$, then for any integer n, the following is true.

$$\left(\frac{a}{b}\right)^{-n} = \left(\frac{b}{a}\right)^n$$

The following example suggests the **quotient rule** for exponents.

$$\frac{5^6}{5^2} = \frac{5 \cdot 5 \cdot 5 \cdot 5 \cdot 5 \cdot 5}{5 \cdot 5} = 5^4 \longleftarrow$$

This exponent is the result of dividing common factors, or, essentially, subtracting the original exponents.

Quotient Rule

Let m and n be integers and a be a nonzero real number.

$$\frac{a^m}{a^n} = a^{m-n}$$

That is, when dividing powers of like bases, keep the same base and subtract the exponent of the denominator from the exponent of the numerator.

CAUTION When applying the quotient rule, be sure to subtract the exponents in the correct order. Be careful especially when the exponent in the denominator is negative, and avoid sign errors.

EXAMPLE 2 Using the Quotient Rule

Simplify each expression. Assume all variables represent nonzero real numbers.

(a) $\dfrac{12^5}{12^2}$ **(b)** $\dfrac{a^5}{a^{-8}}$

(c) $\dfrac{16m^{-9}}{12m^{11}}$ **(d)** $\dfrac{25r^7z^5}{10r^9z}$

SOLUTION

(a) $\dfrac{12^5}{12^2} = 12^{5-2} = 12^3$

(b) $\dfrac{a^5}{a^{-8}} = a^{5-(-8)} = a^{13}$ Use parentheses to avoid errors.

(c) $\dfrac{16m^{-9}}{12m^{11}}$

$$= \frac{16}{12} \cdot m^{-9-11}$$

$$= \frac{4}{3}m^{-20}$$

$$= \frac{4}{3} \cdot \frac{1}{m^{20}}$$

$$= \frac{4}{3m^{20}}$$

(d) $\dfrac{25r^7z^5}{10r^9z}$

$$= \frac{25}{10} \cdot \frac{r^7}{r^9} \cdot \frac{z^5}{z^1}$$

$$= \frac{5}{2}r^{7-9}z^{5-1}$$

$$= \frac{5}{2}r^{-2}z^4$$

$$= \frac{5z^4}{2r^2}$$

✔ **Now Try Exercises 23, 29, 31, and 33.**

The previous rules for exponents were stated for positive integer exponents and for zero as an exponent. Those rules continue to apply in expressions involving negative exponents, as seen in the next example.

EXAMPLE 3 Using the Rules for Exponents

Simplify each expression. Write answers without negative exponents. Assume all variables represent nonzero real numbers.

(a) $3x^{-2}(4^{-1}x^{-5})^2$ **(b)** $\dfrac{12p^3q^{-1}}{8p^{-2}q}$ **(c)** $\dfrac{(3x^2)^{-1}(3x^5)^{-2}}{(3^{-1}x^{-2})^2}$

SOLUTION

(a) $3x^{-2}(4^{-1}x^{-5})^2$

$= 3x^{-2}(4^{-2}x^{-10})$ Power rules

$= 3 \cdot 4^{-2} \cdot x^{-2+(-10)}$ Rearrange factors; product rule

$= 3 \cdot 4^{-2} \cdot x^{-12}$ Simplify the exponent on x.

$= \dfrac{3}{16x^{12}}$ Write with positive exponents, and multiply.

(b) $\dfrac{12p^3q^{-1}}{8p^{-2}q}$

$= \dfrac{12}{8} \cdot \dfrac{p^3}{p^{-2}} \cdot \dfrac{q^{-1}}{q^1}$ Write as separate factors.

$= \dfrac{3}{2} \cdot p^{3-(-2)}q^{-1-1}$ Write $\frac{12}{8}$ in lowest terms; quotient rule

$= \dfrac{3}{2}p^5q^{-2}$ Simplify the exponents.

$= \dfrac{3p^5}{2q^2}$ Write with positive exponents, and multiply.

(c) $\dfrac{(3x^2)^{-1}(3x^5)^{-2}}{(3^{-1}x^{-2})^2}$

$= \dfrac{3^{-1}x^{-2}3^{-2}x^{-10}}{3^{-2}x^{-4}}$ Power rules

$= \dfrac{3^{-1+(-2)}x^{-2+(-10)}}{3^{-2}x^{-4}}$ Product rule

$= \dfrac{3^{-3}x^{-12}}{3^{-2}x^{-4}}$ Simplify the exponents.

$= 3^{-3-(-2)}x^{-12-(-4)}$ Quotient rule

Be careful with signs.

$= 3^{-1}x^{-8}$ Simplify the exponents.

$= \dfrac{1}{3x^8}$ Write with positive exponents, and multiply.

✔ **Now Try Exercises 37, 43, and 45.**

CAUTION Notice the use of the power rule $(ab)^n = a^nb^n$ in **Example 3(c).** ***Remember to apply the exponent to the numerical coefficient* 3.**

$$(3x^2)^{-1} = 3^{-1}(x^2)^{-1} = 3^{-1}x^{-2}$$

Rational Exponents The definition of a^n can be extended to rational values of n by defining $a^{1/n}$ to be the nth root of a. By one of the power rules of exponents (extended to a rational exponent), we have the following.

$$(a^{1/n})^n = a^{(1/n)n} = a^1 = a$$

This suggests that $a^{1/n}$ is a number whose nth power is a.

The Expression $a^{1/n}$

$a^{1/n}$, n Even If n is an *even* positive integer, and if $a > 0$, then $a^{1/n}$ is the positive real number whose nth power is a. That is, $(a^{1/n})^n = a$. (In this case, $a^{1/n}$ is the principal nth root of a.)

$a^{1/n}$, n Odd If n is an *odd* positive integer, and a is *any nonzero real number*, then $a^{1/n}$ is the positive or negative real number whose nth power is a. That is, $(a^{1/n})^n = a$.

For all positive integers n, $0^{1/n} = 0$.

EXAMPLE 4 **Using the Definition of $a^{1/n}$**

Evaluate each expression.

(a) $36^{1/2}$ (b) $-100^{1/2}$ (c) $-(225)^{1/2}$ (d) $625^{1/4}$

(e) $(-1296)^{1/4}$ (f) $-1296^{1/4}$ (g) $(-27)^{1/3}$ (h) $-32^{1/5}$

SOLUTION

(a) $36^{1/2} = 6$ because $6^2 = 36$.

(b) $-100^{1/2} = -10$

(c) $-(225)^{1/2} = -15$

(d) $625^{1/4} = 5$

(e) $(-1296)^{1/4}$ is not a real number.

(f) $-1296^{1/4} = -6$

(g) $(-27)^{1/3} = -3$

(h) $-32^{1/5} = -2$

✔ **Now Try Exercises 47, 49, and 53.**

The notation $a^{m/n}$ must be defined in such a way that all the previous rules for exponents still hold. For the power rule to hold, $(a^{1/n})^m$ must equal $a^{m/n}$. Therefore, $a^{m/n}$ is defined as follows.

The Expression $a^{m/n}$

Let m be any integer, n be any positive integer, and a be any real number for which $a^{1/n}$ is a real number.

$$a^{m/n} = (a^{1/n})^m$$

EXAMPLE 5 **Using the Definition of $a^{m/n}$**

Evaluate each expression.

(a) $125^{2/3}$ (b) $32^{7/5}$ (c) $-81^{3/2}$ (d) $(-27)^{2/3}$ (e) $16^{-3/4}$ (f) $(-4)^{5/2}$

SOLUTION

(a) $$\begin{aligned} 125^{2/3} &= (125^{1/3})^2 \\ &= 5^2 \\ &= 25 \end{aligned}$$

(b) $$\begin{aligned} 32^{7/5} &= (32^{1/5})^7 \\ &= 2^7 \\ &= 128 \end{aligned}$$

(c) $$\begin{aligned} -81^{3/2} &= -(81^{1/2})^3 \\ &= -9^3 \\ &= -729 \end{aligned}$$

(d) $$\begin{aligned} (-27)^{2/3} &= [(-27)^{1/3}]^2 \\ &= (-3)^2 \\ &= 9 \end{aligned}$$

(e) $$\begin{aligned} 16^{-3/4} &= \frac{1}{16^{3/4}} \\ &= \frac{1}{(16^{1/4})^3} \\ &= \frac{1}{2^3} \\ &= \frac{1}{8} \end{aligned}$$

(f) $(-4)^{5/2}$ is not a real number. This is because $(-4)^{1/2}$ is not a real number.

✔ **Now Try Exercises 55, 59, and 61.**

NOTE For all real numbers a, integers m, and positive integers n for which $a^{1/n}$ is a real number, $a^{m/n}$ can be interpreted as follows.

$$a^{m/n} = (a^{1/n})^m \quad \text{or} \quad a^{m/n} = (a^m)^{1/n}$$

So $a^{m/n}$ can be evaluated either as $(a^{1/n})^m$ or as $(a^m)^{1/n}$.

$$27^{4/3} = (27^{1/3})^4 = 3^4 = 81$$

or

$$27^{4/3} = (27^4)^{1/3} = 531{,}441^{1/3} = 81$$

The result is the same.

The earlier results for integer exponents also apply to rational exponents.

Definitions and Rules for Exponents

Let r and s be rational numbers. The following results are valid for all positive numbers a and b.

Product rule $\quad a^r \cdot a^s = a^{r+s}$

Quotient rule $\quad \dfrac{a^r}{a^s} = a^{r-s}$

Negative exponent $\quad a^{-r} = \dfrac{1}{a^r}$

Power rules $\quad (a^r)^s = a^{rs} \qquad (ab)^r = a^r b^r \qquad \left(\dfrac{a}{b}\right)^r = \dfrac{a^r}{b^r}$

EXAMPLE 6 Using the Rules for Exponents

Simplify each expression. Assume all variables represent positive real numbers.

(a) $\dfrac{27^{1/3} \cdot 27^{5/3}}{27^3}$ **(b)** $81^{5/4} \cdot 4^{-3/2}$ **(c)** $6y^{2/3} \cdot 2y^{1/2}$

(d) $\left(\dfrac{3m^{5/6}}{y^{3/4}}\right)^2 \left(\dfrac{8y^3}{m^6}\right)^{2/3}$ **(e)** $m^{2/3}(m^{7/3} + 2m^{1/3})$

SOLUTION

(a) $\dfrac{27^{1/3} \cdot 27^{5/3}}{27^3}$

$= \dfrac{27^{1/3+5/3}}{27^3}$ Product rule

$= \dfrac{27^2}{27^3}$ Simplify.

$= 27^{2-3}$ Quotient rule

$= 27^{-1}$ Simplify the exponent.

$= \dfrac{1}{27}$ Negative exponent

(b) $81^{5/4} \cdot 4^{-3/2}$

$= (81^{1/4})^5(4^{1/2})^{-3}$

$= 3^5 \cdot 2^{-3}$

$= \dfrac{3^5}{2^3}$

$= \dfrac{243}{8}$

(c) $6y^{2/3} \cdot 2y^{1/2}$

$= 6 \cdot 2y^{2/3+1/2}$ Product rule

$= 12y^{7/6}$ Multiply. Simplify the exponent.

(d) $\left(\frac{3m^{5/6}}{y^{3/4}}\right)^2\left(\frac{8y^3}{m^6}\right)^{2/3}$

$= \frac{9m^{5/3}}{y^{3/2}} \cdot \frac{4y^2}{m^4}$ Power rules

$= 36m^{5/3-4}y^{2-3/2}$ Quotient rule

$= 36m^{-7/3}y^{1/2}$ Simplify the exponents.

$= \frac{36y^{1/2}}{m^{7/3}}$ Simplify.

(e) $m^{2/3}(m^{7/3} + 2m^{1/3})$

Do *not* multiply the exponents.

$= m^{2/3} \cdot m^{7/3} + m^{2/3} \cdot 2m^{1/3}$ Distributive property

$= m^{2/3+7/3} + 2m^{2/3+1/3}$ Product rule

$= m^3 + 2m$ Simplify the exponents.

✔ **Now Try Exercises 65, 67, 75, and 81.**

EXAMPLE 7 Factoring Expressions with Negative or Rational Exponents

Factor out the least power of the variable or variable expression. Assume all variables represent positive real numbers.

(a) $12x^{-2} - 8x^{-3}$

(b) $4m^{1/2} + 3m^{3/2}$

(c) $(y-2)^{-1/3} + (y-2)^{2/3}$

SOLUTION

(a) The least exponent of the variable x in $12x^{-2} - 8x^{-3}$ is -3. Because 4 is a common numerical factor, factor out $4x^{-3}$.

$12x^{-2} - 8x^{-3}$

$= 4x^{-3}(3x^{-2-(-3)} - 2x^{-3-(-3)})$ Factor.

$= 4x^{-3}(3x - 2)$ Simplify the exponents.

CHECK $4x^{-3}(3x-2) = 12x^{-2} - 8x^{-3}$ ✓ Multiply.

(b) $4m^{1/2} + 3m^{3/2}$

$= m^{1/2}(4 + 3m)$ The least exponent is 1/2. Factor out $m^{1/2}$.

(c) $(y-2)^{-1/3} + (y-2)^{2/3}$

$= (y-2)^{-1/3}[1 + (y-2)]$ Factor out $(y-2)^{-1/3}$.

$= (y-2)^{-1/3}(y-1)$ Simplify.

✔ **Now Try Exercises 89, 95, and 99.**

LOOKING AHEAD TO CALCULUS

The technique of **Example 7(c)** is used often in calculus.

Complex Fractions Revisited Negative exponents are sometimes used to write complex fractions. Recall that complex fractions are simplified either by first multiplying the numerator and denominator by the LCD of all the denominators, or by performing any indicated operations in the numerator and the denominator and then using the definition of division for fractions.

EXAMPLE 8 Simplifying a Fraction with Negative Exponents

Simplify $\frac{(x+y)^{-1}}{x^{-1}+y^{-1}}$. Write the result with only positive exponents.

SOLUTION

$$\frac{(x+y)^{-1}}{x^{-1}+y^{-1}}$$

$$= \frac{\frac{1}{x+y}}{\frac{1}{x}+\frac{1}{y}}$$ Definition of negative exponent

$$= \frac{\frac{1}{x+y}}{\frac{y+x}{xy}}$$ Add fractions in the denominator.

$$= \frac{1}{x+y} \cdot \frac{xy}{x+y}$$ Multiply by the reciprocal of the denominator of the complex fraction.

$$= \frac{xy}{(x+y)^2}$$ Multiply fractions.

✔ Now Try Exercise 105.

CAUTION Remember that if $r \neq 1$, then $(x+y)^r \neq x^r + y^r$. In particular, this means that $(x+y)^{-1} \neq x^{-1} + y^{-1}$.

R.6 Exercises

CONCEPT PREVIEW *Decide whether each statement is* true *or* false. *If false, correct the right side of the equation.*

1. $5^{-2} = \frac{1}{5^2}$

2. $\left(\frac{2}{3}\right)^{-2} = \left(\frac{3}{2}\right)^2$

3. $\frac{a^6}{a^4} = a^{-2}$

4. $(3x^2)^{-1} = 3x^{-2}$

5. $(x+y)^{-1} = x^{-1} + y^{-1}$

6. $m^{2/3} \cdot m^{1/3} = m^{2/9}$

CONCEPT PREVIEW *Match each expression in Column I with its equivalent expression in Column II. Choices may be used once, more than once, or not at all.*

	I	II
7.	**(a)** 4^{-2}	**A.** 16
	(b) -4^{-2}	**B.** $\frac{1}{16}$
	(c) $(-4)^{-2}$	**C.** -16
	(d) $-(-4)^{-2}$	**D.** $-\frac{1}{16}$

	I	II
8.	**(a)** 5^{-3}	**A.** 125
	(b) -5^{-3}	**B.** -125
	(c) $(-5)^{-3}$	**C.** $\frac{1}{125}$
	(d) $-(-5)^{-3}$	**D.** $-\frac{1}{125}$

I		II	
9. (a) $\left(\frac{4}{9}\right)^{3/2}$	**10. (a)** $\left(\frac{8}{27}\right)^{2/3}$	**A.** $\frac{9}{4}$	**B.** $-\frac{9}{4}$
(b) $\left(\frac{4}{9}\right)^{-3/2}$	**(b)** $\left(\frac{8}{27}\right)^{-2/3}$	**C.** $-\frac{4}{9}$	**D.** $\frac{4}{9}$
(c) $-\left(\frac{9}{4}\right)^{3/2}$	**(c)** $-\left(\frac{27}{8}\right)^{2/3}$	**E.** $\frac{8}{27}$	**F.** $-\frac{27}{8}$
(d) $-\left(\frac{4}{9}\right)^{-3/2}$	**(d)** $-\left(\frac{27}{8}\right)^{-2/3}$	**G.** $\frac{27}{8}$	**H.** $-\frac{8}{27}$

Write each expression without negative exponents, and evaluate if possible. Assume all variables represent nonzero real numbers. ***See Example 1.***

11. $(-4)^{-3}$ **12.** $(-5)^{-2}$ **13.** -5^{-4}

14. -7^{-2} **15.** $\left(\frac{1}{3}\right)^{-2}$ **16.** $\left(\frac{4}{3}\right)^{-3}$

17. $(4x)^{-2}$ **18.** $(5t)^{-3}$ **19.** $4x^{-2}$

20. $5t^{-3}$ **21.** $-a^{-3}$ **22.** $-b^{-4}$

Simplify each expression. Write answers without negative exponents. Assume all variables represent nonzero real numbers. ***See Examples 2 and 3.***

23. $\frac{4^8}{4^6}$ **24.** $\frac{5^9}{5^7}$ **25.** $\frac{x^{12}}{x^8}$ **26.** $\frac{y^{14}}{y^{10}}$

27. $\frac{r^7}{r^{10}}$ **28.** $\frac{y^8}{y^{12}}$ **29.** $\frac{6^4}{6^{-2}}$ **30.** $\frac{7^5}{7^{-3}}$

31. $\frac{4r^{-3}}{6r^{-6}}$ **32.** $\frac{15s^{-4}}{5s^{-8}}$ **33.** $\frac{16m^{-5}n^4}{12m^2n^{-3}}$ **34.** $\frac{15a^{-5}b^{-1}}{25a^{-2}b^4}$

35. $-4r^{-2}(r^4)^2$ **36.** $-2m^{-1}(m^3)^2$ **37.** $(5a^{-1})^4(a^2)^{-3}$ **38.** $(3p^{-4})^2(p^3)^{-1}$

39. $\frac{(p^{-2})^0}{5p^{-4}}$ **40.** $\frac{(m^4)^0}{9m^{-3}}$ **41.** $\frac{(3pq)q^2}{6p^2q^4}$ **42.** $\frac{(-8xy)y^3}{4x^5y^4}$

43. $\frac{4a^5(a^{-1})^3}{(a^{-2})^{-2}}$ **44.** $\frac{12k^{-2}(k^{-3})^{-4}}{6k^5}$ **45.** $\frac{(5x)^{-2}(5x^3)^{-3}}{(5^{-2}x^{-3})^3}$ **46.** $\frac{(8y^2)^{-4}(8y^5)^{-2}}{(8^{-3}y^{-4})^2}$

Evaluate each expression. ***See Example 4.***

47. $169^{1/2}$ **48.** $121^{1/2}$ **49.** $16^{1/4}$ **50.** $625^{1/4}$

51. $\left(-\frac{64}{27}\right)^{1/3}$ **52.** $\left(-\frac{8}{27}\right)^{1/3}$ **53.** $(-4)^{1/2}$ **54.** $(-64)^{1/4}$

Simplify each expression. Write answers without negative exponents. Assume all variables represent positive real numbers. ***See Examples 5 and 6.***

55. $8^{2/3}$ **56.** $27^{4/3}$ **57.** $100^{3/2}$

58. $64^{3/2}$ **59.** $-81^{3/4}$ **60.** $(-32)^{-4/5}$

61. $\left(\frac{27}{64}\right)^{-4/3}$ **62.** $\left(\frac{121}{100}\right)^{-3/2}$ **63.** $3^{1/2} \cdot 3^{3/2}$

64. $6^{4/3} \cdot 6^{2/3}$ **65.** $\frac{64^{5/3}}{64^{4/3}}$ **66.** $\frac{125^{7/3}}{125^{5/3}}$

67. $y^{7/3} \cdot y^{-4/3}$

68. $r^{-8/9} \cdot r^{17/9}$

69. $\dfrac{k^{1/3}}{k^{2/3} \cdot k^{-1}}$

70. $\dfrac{z^{3/4}}{z^{5/4} \cdot z^{-2}}$

71. $\dfrac{(x^{1/4}y^{2/5})^{20}}{x^2}$

72. $\dfrac{(r^{1/5}s^{2/3})^{15}}{r^2}$

73. $\dfrac{(x^{2/3})^2}{(x^2)^{7/3}}$

74. $\dfrac{(p^3)^{1/4}}{(p^{5/4})^2}$

75. $\left(\dfrac{16m^3}{n}\right)^{1/4}\left(\dfrac{9n^{-1}}{m^2}\right)^{1/2}$

76. $\left(\dfrac{25^4a^3}{b^2}\right)^{1/8}\left(\dfrac{4^2b^{-5}}{a^2}\right)^{1/4}$

77. $\dfrac{p^{1/5}p^{7/10}p^{1/2}}{(p^3)^{-1/5}}$

78. $\dfrac{z^{1/3}z^{-2/3}z^{1/6}}{(z^{-1/6})^3}$

(Modeling) Solve each problem.

79. ***Holding Time of Athletes*** A group of ten athletes were tested for isometric endurance by measuring the length of time they could resist a load pulling on their legs while seated. The approximate amount of time (called the **holding time**) that they could resist the load was given by the formula

$$t = 31{,}293w^{-1.5},$$

where w is the weight of the load in pounds and the holding time t is measured in seconds. (*Source:* Townend, M. Stewart, *Mathematics in Sport*, Chichester, Ellis Horwood Limited.)

(a) Determine the holding time, to the nearest second, for a load of 25 lb.

(b) When the weight of the load is doubled, by what factor is the holding time changed?

80. ***Duration of a Storm*** Suppose that meteorologists approximate the duration of a particular storm by using the formula

$$T = 0.07D^{3/2},$$

where T is the time (in hours) that a storm of diameter D (in miles) lasts.

(a) The National Weather Service reports that a storm 4 mi in diameter is headed toward New Haven. How many minutes is the storm expected to last?

(b) A thunderstorm is predicted for a farming community. The crops need at least 1.5 hr of rain. Local radar shows that the storm is 7 mi in diameter. Will it rain long enough to meet the farmers' need?

Find each product. Assume all variables represent positive real numbers. ***See Example 6(e).***

81. $y^{5/8}(y^{3/8} - 10y^{11/8})$

82. $p^{11/5}(3p^{4/5} + 9p^{19/5})$

83. $-4k(k^{7/3} - 6k^{1/3})$

84. $-5y(3y^{9/10} + 4y^{3/10})$

85. $(x + x^{1/2})(x - x^{1/2})$

86. $(2z^{1/2} + z)(z^{1/2} - z)$

87. $(r^{1/2} - r^{-1/2})^2$

88. $(p^{1/2} - p^{-1/2})(p^{1/2} + p^{-1/2})$

Factor out the least power of the variable or variable expression. Assume all variables represent positive real numbers. ***See Example 7.***

89. $4k^{-1} + k^{-2}$

90. $y^{-5} - 3y^{-3}$

91. $4t^{-2} + 8t^{-4}$

92. $5r^{-6} - 10r^{-8}$

93. $9z^{-1/2} + 2z^{1/2}$

94. $3m^{2/3} - 4m^{-1/3}$

95. $p^{-3/4} - 2p^{-7/4}$

96. $6r^{-2/3} - 5r^{-5/3}$

97. $-4a^{-2/5} + 16a^{-7/5}$

98. $-3p^{-3/4} - 30p^{-7/4}$

99. $(p+4)^{-3/2} + (p+4)^{-1/2} + (p+4)^{1/2}$

100. $(3r+1)^{-2/3} + (3r+1)^{1/3} + (3r+1)^{4/3}$

101. $2(3x+1)^{-3/2} + 4(3x+1)^{-1/2} + 6(3x+1)^{1/2}$

102. $7(5t+3)^{-5/3} + 14(5t+3)^{-2/3} - 21(5t+3)^{1/3}$

103. $4x(2x+3)^{-5/9} + 6x^2(2x+3)^{4/9} - 8x^3(2x+3)^{13/9}$

104. $6y^3(4y-1)^{-3/7} - 8y^2(4y-1)^{4/7} + 16y(4y-1)^{11/7}$

Perform all indicated operations, and write each answer with positive integer exponents. ***See Example 8.***

105. $\dfrac{a^{-1}+b^{-1}}{(ab)^{-1}}$

106. $\dfrac{p^{-1}-q^{-1}}{(pq)^{-1}}$

107. $\dfrac{r^{-1}+q^{-1}}{r^{-1}-q^{-1}} \cdot \dfrac{r-q}{r+q}$

108. $\dfrac{x^{-2}+y^{-2}}{x^{-2}-y^{-2}} \cdot \dfrac{x+y}{x-y}$

109. $\dfrac{x-9y^{-1}}{(x-3y^{-1})(x+3y^{-1})}$

110. $\dfrac{a-16b^{-1}}{(a+4b^{-1})(a-4b^{-1})}$

Simplify each rational expression. Assume all variable expressions represent positive real numbers. (Hint: Use factoring and divide out any common factors as a first step.)

111. $\dfrac{(x^2+1)^4(2x) - x^2(4)(x^2+1)^3(2x)}{(x^2+1)^8}$

112. $\dfrac{(y^2+2)^5(3y) - y^3(6)(y^2+2)^4(3y)}{(y^2+2)^7}$

113. $\dfrac{4(x^2-1)^3 + 8x(x^2-1)^4}{16(x^2-1)^3}$

114. $\dfrac{10(4x^2-9)^2 - 25x(4x^2-9)^3}{15(4x^2-9)^6}$

115. $\dfrac{2(2x-3)^{1/3} - (x-1)(2x-3)^{-2/3}}{(2x-3)^{2/3}}$

116. $\dfrac{7(3t+1)^{1/4} - (t-1)(3t+1)^{-3/4}}{(3t+1)^{3/4}}$

Concept Check *Answer each question.*

117. If the lengths of the sides of a cube are tripled, by what factor will the volume change?

118. If the radius of a circle is doubled, by what factor will the area change?

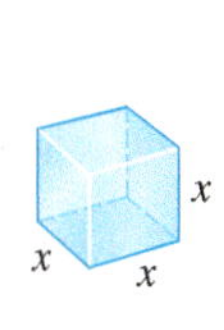

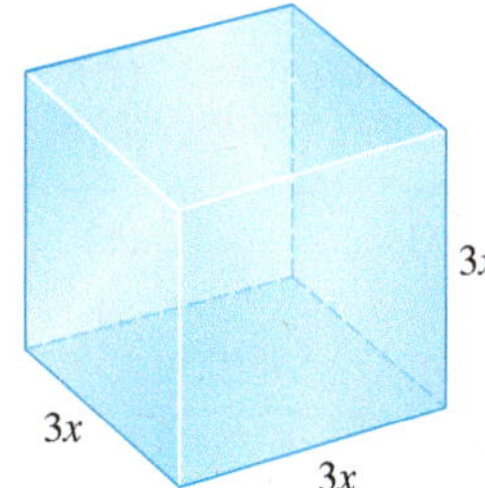

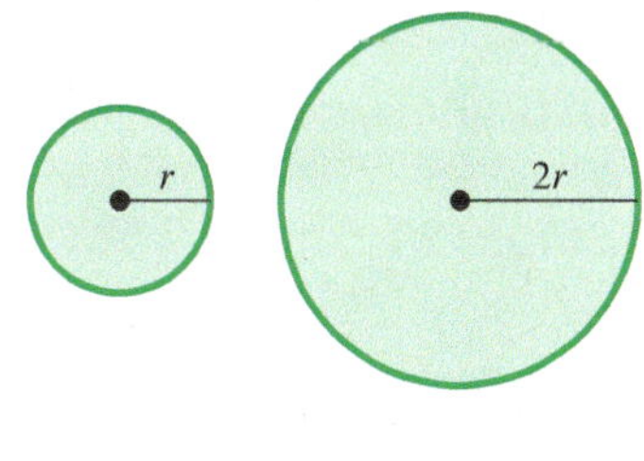

Concept Check *Calculate each value mentally.*

119. $0.2^{2/3} \cdot 40^{2/3}$

120. $0.1^{3/2} \cdot 90^{3/2}$

121. $\dfrac{2^{2/3}}{2000^{2/3}}$

122. $\dfrac{20^{3/2}}{5^{3/2}}$

R.7 Radical Expressions

- Radical Notation
- Simplified Radicals
- Operations with Radicals
- Rationalizing Denominators

Radical Notation

Previously, we used rational exponents to express roots. An alternative notation for roots is **radical notation.**

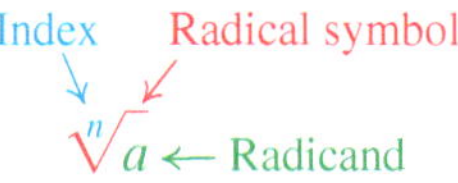

Radical Notation for $a^{1/n}$

Let a be a real number, n be a positive integer, and $a^{1/n}$ be a real number.

$$\sqrt[n]{a} = a^{1/n}$$

Radical Notation for $a^{m/n}$

Let a be a real number, m be an integer, n be a positive integer, and $\sqrt[n]{a}$ be a real number.

$$a^{m/n} = \left(\sqrt[n]{a}\right)^m = \sqrt[n]{a^m}$$

We use the familiar notation $\sqrt{a}$ instead of $\sqrt[2]{a}$ for the square root. For even values of n (square roots, fourth roots, and so on), when a is positive, there are two nth roots, one positive and one negative.

$\sqrt[n]{a}$ represents the positive root, the **principal nth root.**

$-\sqrt[n]{a}$ represents the **negative root.**

EXAMPLE 1 Evaluating Roots

Write each root using exponents and evaluate.

(a) $\sqrt[4]{16}$ **(b)** $-\sqrt[4]{16}$ **(c)** $\sqrt[5]{-32}$

(d) $\sqrt[3]{1000}$ **(e)** $\sqrt[6]{\dfrac{64}{729}}$ **(f)** $\sqrt[4]{-16}$

SOLUTION

(a) $\sqrt[4]{16} = 16^{1/4} = 2$

(b) $-\sqrt[4]{16} = -16^{1/4} = -2$

(c) $\sqrt[5]{-32} = (-32)^{1/5} = -2$

(d) $\sqrt[3]{1000} = 1000^{1/3} = 10$

(e) $\sqrt[6]{\dfrac{64}{729}} = \left(\dfrac{64}{729}\right)^{1/6} = \dfrac{2}{3}$

(f) $\sqrt[4]{-16}$ is not a real number.

✔ **Now Try Exercises 11, 13, 17, and 21.**

EXAMPLE 2 Converting from Rational Exponents to Radicals

Write in radical form and simplify. Assume all variable expressions represent positive real numbers.

(a) $8^{2/3}$ **(b)** $(-32)^{4/5}$ **(c)** $-16^{3/4}$ **(d)** $x^{5/6}$

(e) $3x^{2/3}$ **(f)** $2p^{1/2}$ **(g)** $(3a+b)^{1/4}$

SOLUTION

(a) $8^{2/3} = \left(\sqrt[3]{8}\right)^2 = 2^2 = 4$ **(b)** $(-32)^{4/5} = \left(\sqrt[5]{-32}\right)^4 = (-2)^4 = 16$

(c) $-16^{3/4} = -\left(\sqrt[4]{16}\right)^3 = -(2)^3 = -8$

(d) $x^{5/6} = \sqrt[6]{x^5}$ **(e)** $3x^{2/3} = 3\sqrt[3]{x^2}$

(f) $2p^{1/2} = 2\sqrt{p}$ **(g)** $(3a+b)^{1/4} = \sqrt[4]{3a+b}$

✔ **Now Try Exercises 23 and 25.**

LOOKING AHEAD TO CALCULUS

In calculus, the "power rule" for derivatives requires converting radicals to rational exponents.

CAUTION It is not possible to "distribute" exponents over a sum, so in **Example 2(g),** $(3a+b)^{1/4}$ *cannot be written as* $(3a)^{1/4} + b^{1/4}$.

$$\sqrt[n]{x^n + y^n} \quad \textit{is not equivalent to} \quad x + y.$$

(For example, let $n = 2$, $x = 3$, and $y = 4$ to see this.)

EXAMPLE 3 Converting from Radicals to Rational Exponents

Write in exponential form. Assume all variable expressions represent positive real numbers.

(a) $\sqrt[4]{x^5}$ **(b)** $\sqrt{3y}$ **(c)** $10\left(\sqrt[5]{z}\right)^2$

(d) $5\sqrt[3]{(2x^4)^7}$ **(e)** $\sqrt{p^2+q}$

SOLUTION

(a) $\sqrt[4]{x^5} = x^{5/4}$ **(b)** $\sqrt{3y} = (3y)^{1/2}$ **(c)** $10\left(\sqrt[5]{z}\right)^2 = 10z^{2/5}$

(d) $5\sqrt[3]{(2x^4)^7} = 5(2x^4)^{7/3} = 5 \cdot 2^{7/3}x^{28/3}$

(e) $\sqrt{p^2+q} = (p^2+q)^{1/2}$

✔ **Now Try Exercises 27 and 29.**

We *cannot* simply write $\sqrt{x^2} = x$ for all real numbers x. For example, what if x represents a negative number such as -5?

$$\sqrt{x^2} = \sqrt{(-5)^2} = \sqrt{25} = 5 \neq x$$

To take care of the fact that a negative value of x can produce a positive result, we use absolute value. For any real number a, the following holds.

$$\sqrt{a^2} = |a|$$

Examples: $\sqrt{(-9)^2} = |-9| = 9$ and $\sqrt{13^2} = |13| = 13$

We can generalize this result to any *even* nth root.

Evaluating $\sqrt[n]{a^n}$

If n is an *even* positive integer, then $\sqrt[n]{a^n} = |a|$.

If n is an *odd* positive integer, then $\sqrt[n]{a^n} = a$.

EXAMPLE 4 Using Absolute Value to Simplify Roots

Simplify.

(a) $\sqrt{p^4}$ (b) $\sqrt[4]{p^4}$ (c) $\sqrt{16m^8r^6}$

(d) $\sqrt[6]{(-2)^6}$ (e) $\sqrt[5]{m^5}$ (f) $\sqrt{(2k+3)^2}$

(g) $\sqrt{x^2-4x+4}$

SOLUTION

(a) $\sqrt{p^4} = \sqrt{(p^2)^2} = |p^2| = p^2$

(b) $\sqrt[4]{p^4} = |p|$

(c) $\sqrt{16m^8r^6} = |4m^4r^3| = 4m^4|r^3|$

(d) $\sqrt[6]{(-2)^6} = |-2| = 2$

(e) $\sqrt[5]{m^5} = m$

(f) $\sqrt{(2k+3)^2} = |2k+3|$

(g) $\sqrt{x^2-4x+4} = \sqrt{(x-2)^2} = |x-2|$

✔ **Now Try Exercises 35, 37, and 39.**

NOTE When working with variable radicands, we will *usually* assume that all variables in radicands represent only nonnegative real numbers.

The following rules for working with radicals are simply the power rules for exponents written in radical notation.

Rules for Radicals

Suppose that a and b represent real numbers, and m and n represent positive integers for which the indicated roots are real numbers.

Rule	Description
Product rule $\sqrt[n]{a} \cdot \sqrt[n]{b} = \sqrt[n]{ab}$	The product of two roots is the root of the product.
Quotient rule $\sqrt[n]{\frac{a}{b}} = \frac{\sqrt[n]{a}}{\sqrt[n]{b}} \quad (b \neq 0)$	The root of a quotient is the quotient of the roots.
Power rule $\sqrt[m]{\sqrt[n]{a}} = \sqrt[mn]{a}$	The index of the root of a root is the product of their indexes.

EXAMPLE 5 Simplifying Radical Expressions

Simplify. Assume all variable expressions represent positive real numbers.

(a) $\sqrt{6} \cdot \sqrt{54}$ **(b)** $\sqrt[3]{m} \cdot \sqrt[3]{m^2}$

(c) $\sqrt{\frac{7}{64}}$ **(d)** $\sqrt[4]{\frac{a}{b^4}}$

(e) $\sqrt[7]{\sqrt[3]{2}}$ **(f)** $\sqrt[4]{\sqrt{3}}$

SOLUTION

(a) $\sqrt{6} \cdot \sqrt{54}$

$= \sqrt{6 \cdot 54}$ Product rule

$= \sqrt{324}$ Multiply.

$= 18$

(b) $\sqrt[3]{m} \cdot \sqrt[3]{m^2}$

$= \sqrt[3]{m^3}$

$= m$

(c) $\sqrt{\frac{7}{64}} = \frac{\sqrt{7}}{\sqrt{64}} = \frac{\sqrt{7}}{8}$ Quotient rule

(d) $\sqrt[4]{\frac{a}{b^4}} = \frac{\sqrt[4]{a}}{\sqrt[4]{b^4}} = \frac{\sqrt[4]{a}}{b}$

(e) $\sqrt[7]{\sqrt[3]{2}} = \sqrt[21]{2}$ Power rule

(f) $\sqrt[4]{\sqrt{3}} = \sqrt[4 \cdot 2]{3} = \sqrt[8]{3}$

✔ **Now Try Exercises 45, 49, 53, and 73.**

NOTE Converting to rational exponents shows why these rules work.

$$\sqrt[7]{\sqrt[3]{2}} = (2^{1/3})^{1/7} = 2^{1/3(1/7)} = 2^{1/21} = \sqrt[21]{2} \quad \text{Example 5(e)}$$

$$\sqrt[4]{\sqrt{3}} = (3^{1/2})^{1/4} = 3^{1/2(1/4)} = 3^{1/8} = \sqrt[8]{3} \quad \text{Example 5(f)}$$

Simplified Radicals In working with numbers, we generally prefer to write a number in its simplest form. For example, $\frac{10}{2}$ is written as 5, and $-\frac{9}{6}$ is written as $-\frac{3}{2}$. Similarly, expressions with radicals can be written in their simplest forms.

Simplified Radicals

An expression with radicals is simplified when all of the following conditions are satisfied.

1. The radicand has no factor raised to a power greater than or equal to the index.
2. The radicand has no fractions.
3. No denominator contains a radical.
4. Exponents in the radicand and the index of the radical have greatest common factor 1.
5. All indicated operations have been performed (if possible).

EXAMPLE 6 Simplifying Radicals

Simplify each radical.

(a) $\sqrt{175}$ **(b)** $-3\sqrt[5]{32}$ **(c)** $\sqrt[3]{81x^5y^7z^6}$

SOLUTION

(a) $\sqrt{175}$

$= \sqrt{25 \cdot 7}$ Factor.

$= \sqrt{25} \cdot \sqrt{7}$ Product rule

$= 5\sqrt{7}$ Square root

(b) $-3\sqrt[5]{32}$

$= -3\sqrt[5]{2^5}$ Exponential form

$= -3 \cdot 2$ $\sqrt[n]{a^n} = a$ if n is odd.

$= -6$ Multiply.

(c) $\sqrt[3]{81x^5y^7z^6}$

$= \sqrt[3]{27 \cdot 3 \cdot x^3 \cdot x^2 \cdot y^6 \cdot y \cdot z^6}$ Factor.

$= \sqrt[3]{27x^3y^6z^6(3x^2y)}$ Group all perfect cubes.

$= 3xy^2z^2\sqrt[3]{3x^2y}$ Remove all perfect cubes from the radical.

✔ **Now Try Exercises 41 and 59.**

Operations with Radicals Radicals with the same radicand and the same index, such as $3\sqrt[4]{11pq}$ and $-7\sqrt[4]{11pq}$, are **like radicals.** On the other hand, examples of **unlike radicals** are as follows.

$2\sqrt{5}$ and $2\sqrt{3}$ Radicands are different.

$2\sqrt{3}$ and $2\sqrt[3]{3}$ Indexes are different.

We add or subtract like radicals using the distributive property. ***Only like radicals can be combined.***

EXAMPLE 7 Adding and Subtracting Radicals

Add or subtract, as indicated. Assume all variables represent positive real numbers.

(a) $3\sqrt[4]{11pq} - 7\sqrt[4]{11pq}$ **(b)** $\sqrt{98x^3y} + 3x\sqrt{32xy}$

(c) $\sqrt[3]{64m^4n^5} - \sqrt[3]{-27m^{10}n^{14}}$

SOLUTION

(a) $3\sqrt[4]{11pq} - 7\sqrt[4]{11pq}$

$= (3 - 7)\sqrt[4]{11pq}$ Distributive property

$= -4\sqrt[4]{11pq}$ Subtract.

(b) $\sqrt{98x^3y} + 3x\sqrt{32xy}$

$= \sqrt{49 \cdot 2 \cdot x^2 \cdot x \cdot y} + 3x\sqrt{16 \cdot 2 \cdot x \cdot y}$ Factor.

$= 7x\sqrt{2xy} + 3x(4)\sqrt{2xy}$ Remove all perfect squares from the radicals.

$= 7x\sqrt{2xy} + 12x\sqrt{2xy}$ Multiply.

$= (7x + 12x)\sqrt{2xy}$ Distributive property

$= 19x\sqrt{2xy}$ Add.

(c) $\sqrt[3]{64m^4n^5} - \sqrt[3]{-27m^{10}n^{14}}$

$= \sqrt[3]{64m^3n^3(mn^2)} - \sqrt[3]{-27m^9n^{12}(mn^2)}$ Factor.

$= 4mn\sqrt[3]{mn^2} - (-3m^3n^4)\sqrt[3]{mn^2}$ Remove all perfect cubes from the radicals.

$= 4mn\sqrt[3]{mn^2} + 3m^3n^4\sqrt[3]{mn^2}$ $-(-a) = a$

$= (4mn + 3m^3n^4)\sqrt[3]{mn^2}$ ← This *cannot* be simplified further.

✔ **Now Try Exercises 77 and 81.**

CAUTION The terms $4mn\sqrt[3]{mn^2}$ and $3m^3n^4\sqrt[3]{mn^2}$ in **Example 7(c)** are rewritten in the final line using the distributive property.

If the index of the radical and an exponent in the radicand have a common factor, we can simplify the radical by first writing it in exponential form. We simplify the rational exponent, and then write the result as a radical again, as shown in **Example 8**.

EXAMPLE 8 Simplifying Radicals

Simplify each radical. Assume all variables represent positive real numbers.

(a) $\sqrt[6]{3^2}$ **(b)** $\sqrt[6]{x^{12}y^3}$ **(c)** $\sqrt[9]{\sqrt{6^3}}$

SOLUTION

(a) $\sqrt[6]{3^2}$

$= 3^{2/6}$

$= 3^{1/3}$

$= \sqrt[3]{3}$

(b) $\sqrt[6]{x^{12}y^3}$

$= (x^{12}y^3)^{1/6}$

$= x^{12/6}y^{3/6}$

$= x^2y^{1/2}$

$= x^2\sqrt{y}$

(c) $\sqrt[9]{\sqrt{6^3}}$

$= \sqrt[9]{6^{3/2}}$

$= (6^{3/2})^{1/9}$

$= 6^{1/6}$

$= \sqrt[6]{6}$

✔ **Now Try Exercises 71 and 75.**

In **Example 8(a),** we simplified $\sqrt[6]{3^2}$ as $\sqrt[3]{3}$. However, to simplify $\left(\sqrt[6]{x}\right)^2$, the variable x must represent a nonnegative number. For example, consider the statement

$$(-8)^{2/6} = [(-8)^{1/6}]^2.$$

This result is not a real number because $(-8)^{1/6}$ is not a real number. On the other hand,

$$(-8)^{1/3} = -2.$$

Here, even though $\frac{2}{6} = \frac{1}{3}$,

$$\left(\sqrt[6]{x}\right)^2 \neq \sqrt[3]{x}.$$

If a is nonnegative, then it is always true that $a^{m/n} = a^{(mp)/(np)}$. Simplifying rational exponents on negative bases should be considered case by case.

EXAMPLE 9 **Multiplying Radical Expressions**

Find each product.

(a) $\left(\sqrt{7}-\sqrt{10}\right)\left(\sqrt{7}+\sqrt{10}\right)$ **(b)** $\left(\sqrt{2}+3\right)\left(\sqrt{8}-5\right)$

SOLUTION

(a) $\left(\sqrt{7}-\sqrt{10}\right)\left(\sqrt{7}+\sqrt{10}\right)$

$= \left(\sqrt{7}\right)^2 - \left(\sqrt{10}\right)^2$ Product of the sum and difference of two terms

$= 7 - 10$ $\left(\sqrt{a}\right)^2 = a$

$= -3$ Subtract.

(b) $\left(\sqrt{2}+3\right)\left(\sqrt{8}-5\right)$

$= \sqrt{2}\left(\sqrt{8}\right) - \sqrt{2}(5) + 3\sqrt{8} - 3(5)$ FOIL method

$= \sqrt{16} - 5\sqrt{2} + 3\left(2\sqrt{2}\right) - 15$ Multiply; $\sqrt{8} = 2\sqrt{2}$.

$= 4 - 5\sqrt{2} + 6\sqrt{2} - 15$ Simplify.

$= -11 + \sqrt{2}$ Combine like terms.

✔ **Now Try Exercises 85 and 91.**

Rationalizing Denominators Condition 3 for a simplified radical requires that no denominator contain a radical. We achieve this by **rationalizing the denominator**—that is, multiplying by a form of 1.

EXAMPLE 10 **Rationalizing Denominators**

Rationalize each denominator.

(a) $\dfrac{4}{\sqrt{3}}$ **(b)** $\sqrt[4]{\dfrac{3}{5}}$

SOLUTION

(a) $\dfrac{4}{\sqrt{3}} = \dfrac{4}{\sqrt{3}} \cdot \dfrac{\sqrt{3}}{\sqrt{3}} = \dfrac{4\sqrt{3}}{3}$ Multiply by $\frac{\sqrt{3}}{\sqrt{3}}$ (which equals 1). In the denominator, $\sqrt{a} \cdot \sqrt{a} = a$.

(b) $\sqrt[4]{\dfrac{3}{5}} = \dfrac{\sqrt[4]{3}}{\sqrt[4]{5}}$ Quotient rule

The denominator will be a rational number if it equals $\sqrt[4]{5^4}$. That is, four factors of 5 are needed under the radical. We multiply by $\frac{\sqrt[4]{5^3}}{\sqrt[4]{5^3}}$.

$\dfrac{\sqrt[4]{3}}{\sqrt[4]{5}}$ Because $\sqrt[4]{5}$ has just one factor of 5, three additional factors are needed.

$= \dfrac{\sqrt[4]{3} \cdot \sqrt[4]{5^3}}{\sqrt[4]{5} \cdot \sqrt[4]{5^3}}$ Multiply by $\frac{\sqrt[4]{5^3}}{\sqrt[4]{5^3}}$.

$= \dfrac{\sqrt[4]{3 \cdot 5^3}}{\sqrt[4]{5^4}}$ Product rule

$= \dfrac{\sqrt[4]{375}}{5}$ Simplify.

✔ **Now Try Exercises 63 and 67.**

EXAMPLE 11 Simplifying Radical Expressions with Fractions

Simplify each expression. Assume all variables represent positive real numbers.

(a) $\dfrac{\sqrt[4]{xy^3}}{\sqrt[4]{x^3y^2}}$ **(b)** $\sqrt[3]{\dfrac{5}{x^6}} - \sqrt[3]{\dfrac{4}{x^9}}$

SOLUTION

(a) $\dfrac{\sqrt[4]{xy^3}}{\sqrt[4]{x^3y^2}}$

$= \sqrt[4]{\dfrac{xy^3}{x^3y^2}}$ Quotient rule

$= \sqrt[4]{\dfrac{y}{x^2}}$ Simplify the radicand.

$= \dfrac{\sqrt[4]{y}}{\sqrt[4]{x^2}}$ Quotient rule

$= \dfrac{\sqrt[4]{y}}{\sqrt[4]{x^2}} \cdot \dfrac{\sqrt[4]{x^2}}{\sqrt[4]{x^2}}$ Rationalize the denominator.

$= \dfrac{\sqrt[4]{x^2y}}{x}$ $\sqrt[4]{x^2} \cdot \sqrt[4]{x^2} = \sqrt[4]{x^4} = x$

(b) $\sqrt[3]{\dfrac{5}{x^6}} - \sqrt[3]{\dfrac{4}{x^9}}$

$= \dfrac{\sqrt[3]{5}}{\sqrt[3]{x^6}} - \dfrac{\sqrt[3]{4}}{\sqrt[3]{x^9}}$ Quotient rule

$= \dfrac{\sqrt[3]{5}}{x^2} - \dfrac{\sqrt[3]{4}}{x^3}$ Simplify the denominators.

$= \dfrac{x\sqrt[3]{5}}{x^3} - \dfrac{\sqrt[3]{4}}{x^3}$ Write with a common denominator.

$= \dfrac{x\sqrt[3]{5} - \sqrt[3]{4}}{x^3}$ Subtract the numerators.

✔ **Now Try Exercises 93 and 95.**

LOOKING AHEAD TO CALCULUS

Another standard problem in calculus is investigating the value that an expression such as

$$\frac{\sqrt{x^2 + 9} - 3}{x^2}$$

approaches as x approaches 0. This cannot be done by simply substituting 0 for x because the result is $\frac{0}{0}$. However, by **rationalizing the numerator,** we can show that for $x \neq 0$ the expression is equivalent to

$$\frac{1}{\sqrt{x^2 + 9} + 3}.$$

Then, by substituting 0 for x, we find that the original expression approaches $\frac{1}{6}$ as x approaches 0.

In **Example 9(a),** we saw that the product

$$(\sqrt{7} - \sqrt{10})(\sqrt{7} + \sqrt{10}) \quad \text{equals} \quad -3, \quad \text{a rational number.}$$

This suggests a way to rationalize a denominator that is a binomial in which one or both terms is a square root radical. The expressions $a - b$ and $a + b$ are **conjugates.**

EXAMPLE 12 Rationalizing a Binomial Denominator

Rationalize the denominator of $\dfrac{1}{1 - \sqrt{2}}$.

SOLUTION $\dfrac{1}{1 - \sqrt{2}}$

$\dfrac{1 + \sqrt{2}}{1 + \sqrt{2}} = 1$

$= \dfrac{1(1 + \sqrt{2})}{(1 - \sqrt{2})(1 + \sqrt{2})}$ Multiply numerator and denominator by the conjugate of the denominator, $1 + \sqrt{2}$.

$= \dfrac{1 + \sqrt{2}}{1 - 2}$ $(x - y)(x + y) = x^2 - y^2$

$= \dfrac{1 + \sqrt{2}}{-1}$ Subtract.

$= -1 - \sqrt{2}$ Divide by -1.

✔ **Now Try Exercise 101.**

R.7 Exercises

CONCEPT PREVIEW *Work each problem.*

1. Write $\sqrt[3]{64}$ using exponents and evaluate.
2. Write $27^{2/3}$ in radical form and simplify.

CONCEPT PREVIEW *Match the rational exponent expression in Column I with the equivalent radical expression in Column II. Assume that x is not 0.*

I		II	
3. (a) $(-3x)^{1/3}$	**4. (a)** $-3x^{1/3}$	**A.** $\frac{3}{\sqrt[3]{x}}$	**B.** $-3\sqrt[3]{x}$
(b) $(-3x)^{-1/3}$	**(b)** $-3x^{-1/3}$	**C.** $\frac{1}{\sqrt[3]{3x}}$	**D.** $\frac{-3}{\sqrt[3]{x}}$
(c) $(3x)^{1/3}$	**(c)** $3x^{-1/3}$	**E.** $3\sqrt[3]{x}$	**F.** $\sqrt[3]{-3x}$
(d) $(3x)^{-1/3}$	**(d)** $3x^{1/3}$	**G.** $\sqrt[3]{3x}$	**H.** $\frac{1}{\sqrt[3]{-3x}}$

CONCEPT PREVIEW *Perform the operation and/or simplify each of the following. Assume all variables represent positive real numbers.*

5. $\sqrt[5]{t^5}$
6. $\sqrt{6} \cdot \sqrt{24}$
7. $\sqrt{50}$
8. $\sqrt{\frac{7}{36}}$
9. $3\sqrt{xy} - 8\sqrt{xy}$
10. $(2 + \sqrt{3})(2 - \sqrt{3})$

Write each root using exponents and evaluate. ***See Example 1.***

11. $\sqrt[3]{125}$
12. $\sqrt[3]{216}$
13. $\sqrt[4]{81}$
14. $\sqrt[4]{256}$
15. $\sqrt[3]{-125}$
16. $\sqrt[3]{-343}$
17. $\sqrt[4]{-81}$
18. $\sqrt[4]{-256}$
19. $\sqrt[5]{32}$
20. $\sqrt[7]{128}$
21. $-\sqrt[5]{-32}$
22. $-\sqrt[3]{-343}$

If the expression is in exponential form, write it in radical form. If it is in radical form, write it in exponential form. Assume all variables represent positive real numbers. ***See Examples 2 and 3.***

23. $m^{2/3}$
24. $p^{5/4}$
25. $(2m + p)^{2/3}$
26. $(5r + 3t)^{4/7}$
27. $\sqrt[5]{k^2}$
28. $\sqrt[4]{z^5}$
29. $-3\sqrt{5p^3}$
30. $-m\sqrt{2y^5}$

Concept Check Answer each question.

31. For which of the following cases is $\sqrt{ab} = \sqrt{a} \cdot \sqrt{b}$ a true statement?

    A. a and b both positive **B.** a and b both negative

32. For which positive integers n greater than or equal to 2 is $\sqrt[n]{a^n} = a$ always a true statement?

33. For what values of x is $\sqrt{9ax^2} = 3x\sqrt{a}$ a true statement? Assume $a \geq 0$.

34. Which of the following expressions is *not* simplified? Give the simplified form.

    A. $\sqrt[3]{2y}$ **B.** $\frac{\sqrt{5}}{2}$ **C.** $\sqrt[4]{m^3}$ **D.** $\sqrt{\frac{3}{4}}$

Simplify. ***See Example 4.***

35. $\sqrt[4]{x^4}$ 36. $\sqrt[6]{x^6}$ 37. $\sqrt{25k^4m^2}$

38. $\sqrt[4]{81p^{12}q^4}$ 39. $\sqrt{(4x-y)^2}$ 40. $\sqrt[4]{(5+2m)^4}$

Simplify each expression. Assume all variables represent positive real numbers. ***See Examples 1, 4–6, and 8–11.***

41. $\sqrt[3]{81}$ 42. $\sqrt[3]{250}$ 43. $-\sqrt[4]{32}$ 44. $-\sqrt[4]{243}$

45. $\sqrt{14}\cdot\sqrt{3pqr}$ 46. $\sqrt{7}\cdot\sqrt{5xt}$ 47. $\sqrt[3]{7x}\cdot\sqrt[3]{2y}$ 48. $\sqrt[3]{9x}\cdot\sqrt[3]{4y}$

49. $-\sqrt{\frac{9}{25}}$ 50. $-\sqrt{\frac{16}{49}}$ 51. $-\sqrt[3]{\frac{5}{8}}$ 52. $-\sqrt[4]{\frac{3}{16}}$

53. $\sqrt[4]{\frac{m}{n^4}}$ 54. $\sqrt[6]{\frac{r}{s^6}}$ 55. $3\sqrt[5]{-3125}$ 56. $5\sqrt[3]{-343}$

57. $\sqrt[3]{16(-2)^4(2)^8}$ 58. $\sqrt[3]{25(-3)^4(5)^3}$ 59. $\sqrt{8x^5z^8}$ 60. $\sqrt{24m^6n^5}$

61. $\sqrt[4]{x^4+y^4}$ 62. $\sqrt[3]{27+a^3}$ 63. $\sqrt{\frac{2}{3x}}$ 64. $\sqrt{\frac{5}{3p}}$

65. $\sqrt{\frac{x^5y^3}{z^2}}$ 66. $\sqrt{\frac{g^3h^5}{r^3}}$ 67. $\sqrt[3]{\frac{8}{x^4}}$ 68. $\sqrt[3]{\frac{9}{16p^4}}$

69. $\sqrt[4]{\frac{g^3h^5}{9r^6}}$ 70. $\sqrt[4]{\frac{32x^5}{y^5}}$ 71. $\sqrt[8]{3^4}$ 72. $\sqrt[9]{5^3}$

73. $\sqrt[3]{\sqrt{4}}$ 74. $\sqrt[4]{\sqrt{25}}$ 75. $\sqrt[4]{\sqrt[3]{2}}$ 76. $\sqrt[5]{\sqrt[3]{9}}$

Perform the indicated operations. Assume all variables represent positive real numbers. ***See Examples 7, 9, and 11.***

77. $8\sqrt{2x}-\sqrt{8x}+\sqrt{72x}$ 78. $4\sqrt{18k}-\sqrt{72k}+\sqrt{50k}$

79. $2\sqrt[3]{3}+4\sqrt[3]{24}-\sqrt[3]{81}$ 80. $\sqrt[3]{32}-5\sqrt[3]{4}+2\sqrt[3]{108}$

81. $\sqrt[4]{81x^6y^3}-\sqrt[4]{16x^{10}y^3}$ 82. $\sqrt[4]{256x^5y^6}+\sqrt[4]{625x^9y^2}$

83. $5\sqrt{6}+2\sqrt{10}$ 84. $3\sqrt{11}-5\sqrt{13}$

85. $(\sqrt{2}+3)(\sqrt{2}-3)$ 86. $(\sqrt{5}+\sqrt{2})(\sqrt{5}-\sqrt{2})$

87. $(\sqrt[3]{11}-1)(\sqrt[3]{11^2}+\sqrt[3]{11}+1)$ 88. $(\sqrt[3]{7}+3)(\sqrt[3]{7^2}-3\sqrt[3]{7}+9)$

89. $(\sqrt{3}+\sqrt{8})^2$ 90. $(\sqrt{5}+\sqrt{10})^2$

91. $(3\sqrt{2}+\sqrt{3})(2\sqrt{3}-\sqrt{2})$ 92. $(4\sqrt{5}+\sqrt{2})(3\sqrt{2}-\sqrt{5})$

93. $\frac{\sqrt[3]{mn}\cdot\sqrt[3]{m^2}}{\sqrt[3]{n^2}}$ 94. $\frac{\sqrt[4]{8m^2n^3}\cdot\sqrt[4]{2m^2}}{\sqrt[4]{32m^4n^3}}$

95. $\sqrt[3]{\frac{2}{x^6}}-\sqrt[3]{\frac{5}{x^9}}$ 96. $\sqrt[4]{\frac{7}{t^{12}}}+\sqrt[4]{\frac{9}{t^4}}$

97. $\frac{1}{\sqrt{2}}+\frac{3}{\sqrt{8}}+\frac{1}{\sqrt{32}}$ 98. $\frac{2}{\sqrt{12}}-\frac{1}{\sqrt{27}}-\frac{5}{\sqrt{48}}$

99. $\frac{-4}{\sqrt[3]{3}}+\frac{1}{\sqrt[3]{24}}-\frac{2}{\sqrt[3]{81}}$ 100. $\frac{5}{\sqrt[3]{2}}-\frac{2}{\sqrt[3]{16}}+\frac{1}{\sqrt[3]{54}}$

Rationalize each denominator. Assume all variables represent nonnegative numbers and that no denominators are 0. ***See Example 12.***

101. $\frac{\sqrt{3}}{\sqrt{5}+\sqrt{3}}$ **102.** $\frac{\sqrt{7}}{\sqrt{3}-\sqrt{7}}$ **103.** $\frac{\sqrt{7}-1}{2\sqrt{7}+4\sqrt{2}}$

104. $\frac{1+\sqrt{3}}{3\sqrt{5}+2\sqrt{3}}$ **105.** $\frac{p-4}{\sqrt{p}+2}$ **106.** $\frac{9-r}{3-\sqrt{r}}$

107. $\frac{3m}{2+\sqrt{m+n}}$ **108.** $\frac{a}{\sqrt{a+b}-1}$ **109.** $\frac{5\sqrt{x}}{2\sqrt{x}+\sqrt{y}}$

110. *Concept Check* By what number should the numerator and denominator of

$$\frac{1}{\sqrt[3]{3}-\sqrt[3]{5}}$$

be multiplied in order to rationalize the denominator? Write this fraction with a rationalized denominator.

(Modeling) Solve each problem.

111. ***Rowing Speed*** Olympic rowing events have one-, two-, four-, or eight-person crews, with each person pulling a single oar. Increasing the size of the crew increases the speed of the boat. An analysis of Olympic rowing events concluded that the approximate speed, s, of the boat (in feet per second) was given by the formula

$$s = 15.18\sqrt[9]{n},$$

where n is the number of oarsmen. Estimate the speed of a boat with a four-person crew. (*Source:* Townend, M. Stewart, *Mathematics in Sport,* Chichester, Ellis Horwood Limited.)

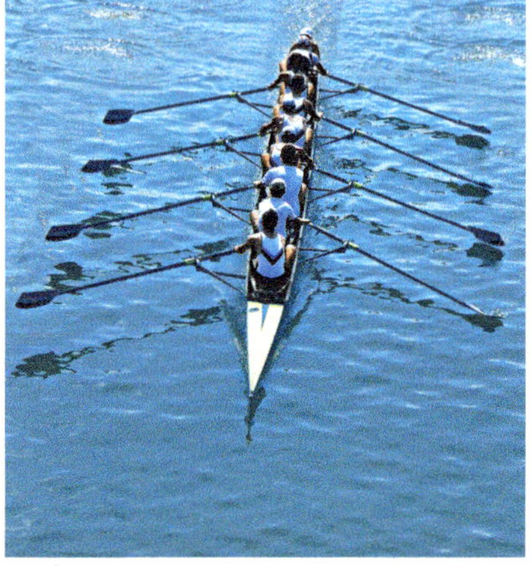

112. ***Rowing Speed*** **See Exercise 111.** Estimate the speed of a boat with an eight-person crew.

*(Modeling) **Windchill** The National Weather Service has used the formula*

$$\text{Windchill temperature} = 35.74 + 0.6215T - 35.75V^{0.16} + 0.4275TV^{0.16},$$

where T is the temperature in °F and V is the wind speed in miles per hour, to calculate windchill. (Source: National Oceanic and Atmospheric Administration, National Weather Service.) Use the formula to calculate the windchill to the nearest degree given the following conditions.

113. 10°F, 30 mph wind **114.** 30°F, 15 mph wind

Concept Check Simplify each expression mentally.

115. $\sqrt[4]{8}\cdot\sqrt[4]{2}$ **116.** $\frac{\sqrt[3]{54}}{\sqrt[3]{2}}$ **117.** $\frac{\sqrt[5]{320}}{\sqrt[5]{10}}$

118. $\sqrt{0.1}\cdot\sqrt{40}$ **119.** $\frac{\sqrt[3]{15}}{\sqrt[3]{5}}\cdot\sqrt[3]{9}$ **120.** $\sqrt[6]{2}\cdot\sqrt[6]{4}\cdot\sqrt[6]{8}$

The screen in **Figure A** *seems to indicate that π and $\sqrt[4]{\frac{2143}{22}}$ are exactly equal, since the eight decimal values given by the calculator agree. However, as shown in* **Figure B** *using one more decimal place in the display, they differ in the ninth decimal place.* ***The radical expression is a very good approximation for π, but it is still only an approximation.***

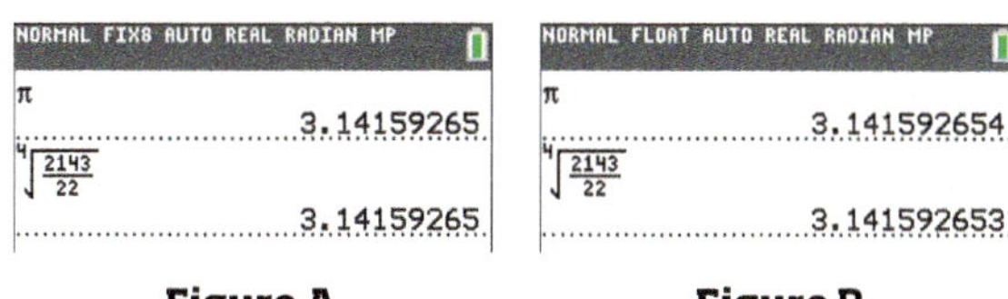

Figure A **Figure B**

Use your calculator to answer each question. Refer to the display for π in **Figure B.**

121. The Chinese of the fifth century used $\frac{355}{113}$ as an approximation for π. How many decimal places of accuracy does this fraction give?

122. A value for π that the Greeks used circa 150 CE is equivalent to $\frac{377}{120}$. In which decimal place does this value first differ from π?

123. The Hindu mathematician Bhaskara used $\frac{3927}{1250}$ as an approximation for π circa 1150 CE. In which decimal place does this value first differ from π?

Chapter R Test Prep

Key Terms

R.1 set, elements (members), infinite set, finite set, Venn diagram, disjoint sets

R.2 number line, coordinate, coordinate system, algebraic expression, exponential expression (exponential), base, exponent, absolute value

R.3 term, numerical coefficient (coefficient), like terms, polynomial, polynomial in x, degree of a term, degree of a polynomial, trinomial, binomial, monomial, descending order, FOIL method

R.4 factoring, factored form, prime polynomial, factored completely, greatest common factor (GCF)

R.5 rational expression, domain of a rational expression, lowest terms, complex fraction

R.7 radicand, principal nth root, like radicals, unlike radicals, conjugates

New Symbols

Symbol	Meaning
$\{\ \}$	set braces
$\in$	is an element of
$\notin$	is not an element of
$\{x \mid x$ **has property** $p\}$	set-builder notation
U	universal set
$\varnothing$, or $\{\ \}$	null (empty) set
$\subseteq$	is a subset of
$\nsubseteq$	is not a subset of
A'	complement of a set A
$\cap$	set intersection
$\cup$	set union
a^n	n factors of a
$<$	is less than
$>$	is greater than
$\leq$	is less than or equal to
$\geq$	is greater than or equal to
$\lvert a \rvert$	absolute value of a
$\sqrt[n]{\ }$	radical symbol with index n

Quick Review

Concepts	Examples
R.1 Sets	
Set Operations	
For all sets A and B, with universal set U: The **complement** of set A is the set A' of all elements in U that do *not* belong to set A.	Let $U = \{1, 2, 3, 4, 5, 6\}$, $A = \{1, 2, 3, 4\}$, and $B = \{3, 4, 6\}$.
$A' = \{x \mid x \in U,\ \ x \notin A\}$	$A' = \{5, 6\}$
The **intersection** of sets A and B, written $A \cap B$, is made up of all the elements belonging to both set A *and* set B.	
$A \cap B = \{x \mid x \in A \text{ and } x \in B\}$	$A \cap B = \{3, 4\}$
The **union** of sets A and B, written $A \cup B$, is made up of all the elements belonging to set A *or* set B.	
$A \cup B = \{x \mid x \in A \text{ or } x \in B\}$	$A \cup B = \{1, 2, 3, 4, 6\}$
R.2 Real Numbers and Their Properties	
Sets of Numbers	
Natural numbers	
$\{1, 2, 3, 4, \ldots\}$	$5, 17, 142$
Whole numbers	
$\{0, 1, 2, 3, 4, \ldots\}$	$0, 27, 96$
Integers	
$\{\ldots, -3, -2, -1, 0, 1, 2, 3, \ldots\}$	$-24, 0, 19$
Rational numbers	
$\left\{\frac{p}{q} \middle\vert p \text{ and } q \text{ are integers and } q \neq 0\right\}$	$-\frac{3}{4}, -0.28, 0, 7, \frac{9}{16}, 0.66\overline{6}$
Irrational numbers	
$\{x \mid x \text{ is real but not rational}\}$	$-\sqrt{15}, 0.101101110\ldots, \sqrt{2}, \pi$
Real numbers	
$\{x \mid x \text{ corresponds to a point on a number line}\}$	$-46, 0.7, \pi, \sqrt{19}, \frac{8}{5}$
Properties of Real Numbers	
For all real numbers a, b, and c, the following hold.	
Closure Properties	
$a + b$ **is a real number.**	$1 + \sqrt{2}$ is a real number.
ab **is a real number.**	$3\sqrt{7}$ is a real number.
Commutative Properties	
$a + b = b + a$	$5 + 18 = 18 + 5$
$ab = ba$	$-4 \cdot 8 = 8 \cdot (-4)$
Associative Properties	
$(a + b) + c = a + (b + c)$	$[6 + (-3)] + 5 = 6 + (-3 + 5)$
$(ab)c = a(bc)$	$(7 \cdot 6)20 = 7(6 \cdot 20)$

Concepts	Examples
Identity Properties There exists a unique real number 0 such that $\mathbf{a + 0 = a}$ and $\mathbf{0 + a = a.}$	$145 + 0 = 145$ and $0 + 145 = 145$
There exists a unique real number 1 such that $\mathbf{a \cdot 1 = a}$ and $\mathbf{1 \cdot a = a.}$	$-60 \cdot 1 = -60$ and $1 \cdot (-60) = -60$
Inverse Properties There exists a unique real number $-a$ such that $\mathbf{a + (-a) = 0}$ and $\mathbf{-a + a = 0.}$	$17 + (-17) = 0$ and $-17 + 17 = 0$
If $a \neq 0$, there exists a unique real number $\frac{1}{a}$ such that $\mathbf{a \cdot \frac{1}{a} = 1}$ and $\mathbf{\frac{1}{a} \cdot a = 1.}$	$22 \cdot \frac{1}{22} = 1$ and $\frac{1}{22} \cdot 22 = 1$
Distributive Properties $\mathbf{a(b + c) = ab + ac}$ $\mathbf{a(b - c) = ab - ac}$	$3(5 + 8) = 3 \cdot 5 + 3 \cdot 8$ $6(4 - 2) = 6 \cdot 4 - 6 \cdot 2$
Multiplication Property of Zero $\mathbf{0 \cdot a = a \cdot 0 = 0}$	$0 \cdot 4 = 4 \cdot 0 = 0$
Order $a > b$ if a is to the right of b on a number line. $a < b$ if a is to the left of b on a number line.	$7 > -5$ $0 < 15$
Absolute Value $\|a\| = \begin{cases} a & \text{if } a \geq 0 \\ -a & \text{if } a < 0 \end{cases}$	$\|3\| = 3$ and $\|-3\| = 3$

R.3 Polynomials

Concepts	Examples
Operations To add or subtract polynomials, add or subtract the coefficients of like terms.	$(2x^2 + 3x + 1) - (x^2 - x + 2)$ $= (2 - 1)x^2 + (3 + 1)x + (1 - 2)$ $= x^2 + 4x - 1$
To multiply polynomials, distribute each term of the first polynomial, multiplying by each term of the second polynomial.	$(x - 5)(x^2 + 5x + 25)$ $= x^3 + 5x^2 + 25x - 5x^2 - 25x - 125$ $= x^3 - 125$
To divide polynomials when the divisor has two or more terms, use a process of long division similar to that for dividing whole numbers.	$\begin{array}{r} 4x^2 - 10x + 21 \\ x + 2 \overline{)4x^3 - 2x^2 + x - 1} \\ \underline{4x^3 + 8x^2} \\ -10x^2 + x \\ \underline{-10x^2 - 20x} \\ 21x - 1 \\ \underline{21x + 42} \\ -43 \end{array}$ ← Remainder $\frac{4x^3 - 2x^2 + x - 1}{x + 2} = 4x^2 - 10x + 21 + \frac{-43}{x + 2}$

Concepts	Examples
Special Products	
Product of the Sum and Difference of Two Terms	
$(x+y)(x-y)=x^2-y^2$	$(7-x)(7+x)=7^2-x^2$ $=49-x^2$
Square of a Binomial	
$(x+y)^2=x^2+2xy+y^2$	$(3a+b)^2=(3a)^2+2(3a)(b)+b^2$ $=9a^2+6ab+b^2$
$(x-y)^2=x^2-2xy+y^2$	$(2m-5)^2=(2m)^2-2(2m)(5)+5^2$ $=4m^2-20m+25$

R.4 Factoring Polynomials

Concepts	Examples
Factoring Patterns	
Perfect Square Trinomial	
$x^2+2xy+y^2=(x+y)^2$ $x^2-2xy+y^2=(x-y)^2$	$p^2+4pq+4q^2=(p+2q)^2$ $9m^2-12mn+4n^2=(3m-2n)^2$
Difference of Squares	
$x^2-y^2=(x+y)(x-y)$	$4t^2-9=(2t+3)(2t-3)$
Difference of Cubes	
$x^3-y^3=(x-y)(x^2+xy+y^2)$	$r^3-8=(r-2)(r^2+2r+4)$
Sum of Cubes	
$x^3+y^3=(x+y)(x^2-xy+y^2)$	$27x^3+64=(3x+4)(9x^2-12x+16)$

R.5 Rational Expressions

Concepts	Examples
Operations Let $\frac{a}{b}$ and $\frac{c}{d}$ $(b \neq 0, d \neq 0)$ represent fractions.	
$\frac{a}{b} \pm \frac{c}{d} = \frac{ad \pm bc}{bd}$	$\frac{2}{x}+\frac{5}{y}=\frac{2y+5x}{xy}$ $\quad \frac{x}{6}-\frac{2y}{5}=\frac{5x-12y}{30}$
$\frac{a}{b} \cdot \frac{c}{d} = \frac{ac}{bd}$ and $\frac{a}{b} \div \frac{c}{d} = \frac{ad}{bc}$ $(c \neq 0)$	$\frac{3}{q} \cdot \frac{3}{2p} = \frac{9}{2pq}$ $\quad \frac{z}{4} \div \frac{z}{2t} = \frac{z}{4} \cdot \frac{2t}{z} = \frac{2zt}{4z} = \frac{t}{2}$

R.6 Rational Exponents

Concepts	Examples
Rules for Exponents Let r and s be rational numbers. The following results are valid for all positive numbers a and b.	
Product rule $\quad a^r \cdot a^s = a^{r+s}$	$6^2 \cdot 6^3 = 6^5$
Quotient rule $\quad \frac{a^r}{a^s} = a^{r-s}$	$\frac{p^5}{p^2} = p^3$
Negative exponent $\quad a^{-r} = \frac{1}{a^r}$	$4^{-3} = \frac{1}{4^3}$
Power rules $\quad (a^r)^s = a^{rs}$ $\quad \left(\frac{a}{b}\right)^r = \frac{a^r}{b^r}$ $(ab)^r = a^r b^r$	$(m^2)^3 = m^6$ $\quad \left(\frac{x}{3}\right)^2 = \frac{x^2}{3^2}$ $(3x)^4 = 3^4 x^4$

Concepts	Examples
R.7 Radical Expressions	
Radical Notation Let a be a real number, n be a positive integer, and $a^{1/n}$ be a real number. $$\sqrt[n]{a} = a^{1/n}$$	$\sqrt[4]{16} = 16^{1/4} = 2$
Let a be a real number, m be an integer, n be a positive integer, and $\sqrt[n]{a}$ be a real number. $$a^{m/n} = \left(\sqrt[n]{a}\right)^m = \sqrt[n]{a^m}$$	$8^{2/3} = \left(\sqrt[3]{8}\right)^2 = \sqrt[3]{8^2} = 4$
Operations Operations with radical expressions are performed like operations with polynomials.	$\sqrt{8x} + \sqrt{32x}$ $= 2\sqrt{2x} + 4\sqrt{2x}$ $= 6\sqrt{2x}$ $\left(\sqrt{5} - \sqrt{3}\right)\left(\sqrt{5} + \sqrt{3}\right)$ $= 5 - 3$ $= 2$ $\left(\sqrt{2} + \sqrt{7}\right)\left(\sqrt{3} - \sqrt{6}\right)$ $= \sqrt{6} - 2\sqrt{3} + \sqrt{21} - \sqrt{42}$ FOIL method; $\sqrt{12} = 2\sqrt{3}$
Rationalizing the Denominator Rationalize the denominator by multiplying numerator and denominator by a form of 1.	$\dfrac{\sqrt{7y}}{\sqrt{5}} = \dfrac{\sqrt{7y}}{\sqrt{5}} \cdot \dfrac{\sqrt{5}}{\sqrt{5}} = \dfrac{\sqrt{35y}}{5}$

Chapter R Review Exercises

1. Use set notation to list all the elements of the set $\{6, 8, 10, \ldots, 20\}$.
2. Is the set $\{x \mid x \text{ is a decimal between 0 and 1}\}$ finite or infinite?
3. *Concept Check* *True* or *false:* The set of negative integers and the set of whole numbers are disjoint sets.
4. *Concept Check* *True* or *false:* 9 is an element of the set $\{999\}$.

Determine whether each statement is true *or* false.

5. $1 \in \{6, 2, 5, 1\}$
6. $7 \notin \{1, 3, 5, 7\}$
7. $\{8, 11, 4\} = \{8, 11, 4, 0\}$
8. $\{0\} = \varnothing$

Let $A = \{1, 3, 4, 5, 7, 8\}$, $B = \{2, 4, 6, 8\}$, $C = \{1, 3, 5, 7\}$, $D = \{1, 2, 3\}$, $E = \{3, 7\}$, *and* $U = \{1, 2, 3, 4, 5, 6, 7, 8, 9, 10\}$.

Determine whether each statement is true *or* false.

9. $\varnothing \subseteq A$
10. $E \subseteq C$
11. $D \nsubseteq B$
12. $E \nsubseteq A$

Refer to the sets given for Exercises 9–12. Specify each set.

13. A' **14.** $B \cap A$ **15.** $B \cap E$

16. $C \cup E$ **17.** $D \cap \emptyset$ **18.** $B \cup \emptyset$

19. $(C \cap D) \cup B$ **20.** $(D' \cap U) \cup E$ **21.** $\emptyset'$

22. ***Concept Check*** *True* or *false:* For all sets A and B, $(A \cap B) \subseteq (A \cup B)$.

Let $K = \left\{-12, -6, -0.9, -\sqrt{7}, -\sqrt{4}, 0, \frac{1}{8}, \frac{\pi}{4}, 6, \sqrt{11}\right\}$. *List all the elements of K that belong to each set.*

23. Integers **24.** Rational numbers

Choose all words from the following list that apply to each number.

natural number whole number integer
rational number irrational number real number

25. $\frac{4\pi}{5}$ **26.** $\frac{\pi}{0}$ **27.** 0 **28.** $-\sqrt{36}$

Write each algebraic identity (true statement) as a complete English sentence without using the names of the variables. For instance, $z(x + y) = zx + zy$ can be stated as "The multiple of a sum is the sum of the multiples."

29. $\frac{1}{xy} = \frac{1}{x} \cdot \frac{1}{y}$ **30.** $a(b - c) = ab - ac$

31. $(ab)^n = a^n b^n$ **32.** $a^2 - b^2 = (a + b)(a - b)$

33. $\left(\frac{a}{b}\right)^n = \frac{a^n}{b^n}$ **34.** $|st| = |s| \cdot |t|$

Identify the property illustrated in each statement.

35. $8(5 + 9) = (5 + 9)8$ **36.** $4 \cdot 6 + 4 \cdot 12 = 4(6 + 12)$

37. $3 \cdot (4 \cdot 2) = (3 \cdot 4) \cdot 2$ **38.** $-8 + 8 = 0$

39. $(9 + p) + 0 = 9 + p$ **40.** $\frac{1}{\sqrt{2}} \cdot \frac{\sqrt{2}}{\sqrt{2}} = \frac{\sqrt{2}}{2}$

41. ***(Modeling) Online College Courses*** The number of students (in millions) taking at least one online college course between the years 2002 and 2012 can be approximated by the formula

$$\text{Number of students} = 0.0112x^2 + 0.4663x + 1.513,$$

where $x = 0$ corresponds to 2002, $x = 1$ corresponds to 2003, and so on. According to this model, how many students took at least one online college course in 2012? (*Source:* Babson Survey Research Group.)

42. ***Counting Marshmallows*** Recently, there were media reports about students providing a correction to the following question posed on boxes of Swiss Miss Chocolate: *On average, how many mini-marshmallows are in one serving?*

$$3 + 2 \times 4 \div 2 - 3 \times 7 - 4 + 47 = \underline{\qquad}$$

The company provided 92 as the answer. What is the *correct* calculation provided by the students? (*Source:* Swiss Miss Chocolate box.)

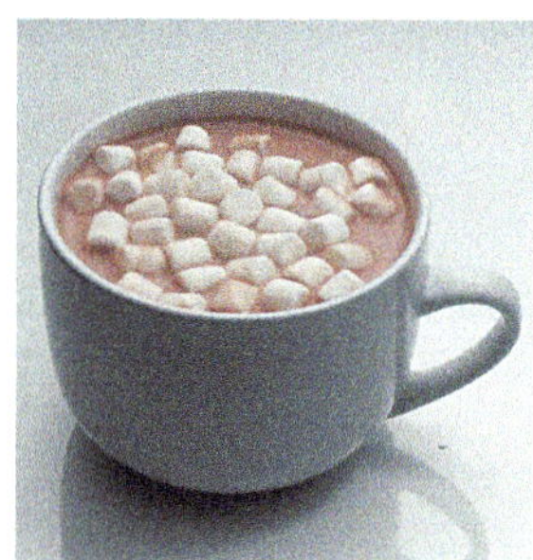

Simplify each expression.

43. $(-4-1)(-3-5)-2^3$

44. $(6-9)(-2-7)\div(-4)$

45. $\left(-\frac{5}{9}-\frac{2}{3}\right)-\frac{5}{6}$

46. $\left(-\frac{2^3}{5}-\frac{3}{4}\right)-\left(-\frac{1}{2}\right)$

47. $\frac{6(-4)-3^2(-2)^3}{-5[-2-(-6)]}$

48. $\frac{(-7)(-3)-(-2^3)(-5)}{(-2^2-2)(-1-6)}$

Evaluate each expression for $a=-1$, $b=-2$, and $c=4$.

49. $-c(2a-5b)$

50. $(a-2)\div 5\cdot b+c$

51. $\frac{9a+2b}{a+b+c}$

52. $\frac{3|b|-4|c|}{|ac|}$

Perform the indicated operations.

53. $(3q^3-9q^2+6)+(4q^3-8q+3)$

54. $2(3y^6-9y^2+2y)-(5y^6-4y)$

55. $(8y-7)(2y^2+7y-3)$

56. $(2r+11s)(4r-9s)$

57. $(3k-5m)^2$

58. $(4a-3b)^2$

Perform each division.

59. $\frac{30m^3-9m^2+22m+5}{5m+1}$

60. $\frac{72r^2+59r+12}{8r+3}$

61. $\frac{3b^3-8b^2+12b-30}{b^2+4}$

62. $\frac{5m^3-7m^2+14}{m^2-2}$

Factor as completely as possible.

63. $3(z-4)^2+9(z-4)^3$

64. $7z^2-9z^3+z$

65. $z^2-6zk-16k^2$

66. $r^2+rp-42p^2$

67. $48a^8-12a^7b-90a^6b^2$

68. $6m^2-13m-5$

69. $49m^8-9n^2$

70. $169y^4-1$

71. $6(3r-1)^2+(3r-1)-35$

72. $8y^3-1000z^6$

73. $xy+2x-y-2$

74. $15mp+9mq-10np-6nq$

Factor each expression. (These expressions arise in calculus from a technique called the product rule that is used to determine the shape of a curve.)

75. $(3x-4)^2+(x-5)(2)(3x-4)(3)$

76. $(5-2x)(3)(7x-8)^2(7)+(7x-8)^3(-2)$

Perform the indicated operations.

77. $\frac{k^2+k}{8k^3}\cdot\frac{4}{k^2-1}$

78. $\frac{3r^3-9r^2}{r^2-9}\div\frac{8r^3}{r+3}$

79. $\frac{x^2+x-2}{x^2+5x+6}\div\frac{x^2+3x-4}{x^2+4x+3}$

80. $\frac{27m^3-n^3}{3m-n}\div\frac{9m^2+3mn+n^2}{9m^2-n^2}$

81. $\frac{p^2-36q^2}{p^2-12pq+36q^2}\cdot\frac{p^2-5pq-6q^2}{p^2+2pq+q^2}$

82. $\frac{1}{4y}+\frac{8}{5y}$

83. $\frac{m}{4-m}+\frac{3m}{m-4}$

84. $\frac{3}{x^2-4x+3}-\frac{2}{x^2-1}$

85. $\frac{p^{-1}+q^{-1}}{1-(pq)^{-1}}$

86. $\frac{3+\frac{2m}{m^2-4}}{\frac{5}{m-2}}$

Simplify each expression. Write answers without negative exponents. Assume all variables represent positive real numbers.

87. $\left(-\frac{5}{4}\right)^{-2}$

88. $3^{-1}-4^{-1}$

89. $(5z^3)(-2z^5)$

90. $(8p^2q^3)(-2p^5q^{-4})$

91. $(-6p^5w^4m^{12})^0$

92. $(-6x^2y^{-3}z^2)^{-2}$

93. $\frac{-8y^7p^{-2}}{y^{-4}p^{-3}}$

94. $\frac{a^{-6}(a^{-8})}{a^{-2}(a^{11})}$

95. $\frac{(p+q)^4(p+q)^{-3}}{(p+q)^6}$

96. $\frac{[p^2(m+n)^3]^{-2}}{p^{-2}(m+n)^{-5}}$

97. $(7r^{1/2})(2r^{3/4})(-r^{1/6})$

98. $(a^{3/4}b^{2/3})(a^{5/8}b^{-5/6})$

99. $\frac{y^{5/3}\cdot y^{-2}}{y^{-5/6}}$

100. $\left(\frac{25m^3n^5}{m^{-2}n^6}\right)^{-1/2}$

101. $\frac{(p^{15}q^{12})^{-4/3}}{(p^{24}q^{16})^{-3/4}}$

102. Simplify the product $-m^{3/4}(8m^{1/2}+4m^{-3/2})$. Assume the variable represents a positive real number.

Simplify each expression. Assume all variables represent positive real numbers.

103. $\sqrt{200}$

104. $\sqrt[3]{16}$

105. $\sqrt[4]{1250}$

106. $-\sqrt{\frac{16}{3}}$

107. $-\sqrt[3]{\frac{2}{5p^2}}$

108. $\sqrt{\frac{2^7y^8}{m^3}}$

109. $\sqrt[4]{\sqrt[3]{m}}$

110. $\frac{\sqrt[4]{8p^2q^5}\cdot\sqrt[4]{2p^3q}}{\sqrt[4]{p^5q^2}}$

111. $\left(\sqrt[3]{2}+4\right)\left(\sqrt[3]{2^2}-4\sqrt[3]{2}+16\right)$

112. $\frac{3}{\sqrt{5}}-\frac{2}{\sqrt{45}}+\frac{6}{\sqrt{80}}$

113. $\sqrt{18m^3}-3m\sqrt{32m}+5\sqrt{m^3}$

114. $\frac{2}{7-\sqrt{3}}$

115. $\frac{6}{3-\sqrt{2}}$

116. $\frac{k}{\sqrt{k}-3}$

Concept Check *Correct each* ***INCORRECT*** *statement by changing the right side of the equation.*

117. $x(x^2+5)=x^3+5$

118. $-3^2=9$

119. $(m^2)^3=m^5$

120. $(3x)(3y)=3xy$

121. $\frac{\left(\frac{a}{b}\right)}{2}=\frac{2a}{b}$

122. $\frac{m}{r}\cdot\frac{n}{r}=\frac{mn}{r}$

123. $\frac{1}{(-2)^3}=2^{-3}$

124. $(-5)^2=-5^2$

125. $\left(\frac{8}{7}+\frac{a}{b}\right)^{-1}=\frac{7}{8}+\frac{b}{a}$

Chapter R Test

Let $U = \{1, 2, 3, 4, 5, 6, 7, 8\}$, $A = \{1, 2, 3, 4, 5, 6\}$, $B = \{1, 3, 5\}$, $C = \{1, 6\}$, *and* $D = \{4\}$. *Decide whether each statement is* true *or* false.

1. $B' = \{2, 4, 6, 8\}$

2. $C \subseteq A$

3. $(B \cap C) \cup D = \{1, 3, 4, 5, 6\}$

4. $(A' \cup C) \cap B' = \{6, 7, 8\}$

5. Let $A = \left\{-13, -\frac{12}{4}, 0, \frac{3}{5}, \frac{\pi}{4}, 5.9, \sqrt{49}\right\}$. List all the elements of A that belong to each set.

(a) Integers **(b)** Rational numbers **(c)** Real numbers

6. Evaluate the expression $\left|\frac{x^2 + 2yz}{3(x + z)}\right|$ for $x = -2$, $y = -4$, and $z = 5$.

7. Identify each property illustrated. Let a, b, and c represent any real numbers.

(a) $a + (b + c) = (a + b) + c$

(b) $a + (c + b) = a + (b + c)$

(c) $a(b + c) = ab + ac$

(d) $a + [b + (-b)] = a + 0$

8. ***(Modeling) Passer Rating for NFL Quarterbacks*** Approximate the quarterback rating (to the nearest tenth) of Drew Brees of the New Orleans Saints during the 2013 regular season. He attempted 650 passes, completed 446, had 5162 total yards, threw for 39 touchdowns, and had 12 interceptions. (*Source:* www.nfl.com)

$$\text{Rating} = \frac{\left(250 \cdot \frac{C}{A}\right) + \left(1000 \cdot \frac{T}{A}\right) + \left(12.5 \cdot \frac{Y}{A}\right) + 6.25 - \left(1250 \cdot \frac{I}{A}\right)}{3},$$

where A = attempted passes, C = completed passes, T = touchdown passes, Y = yards gained passing, and I = interceptions.

In addition to the weighting factors that appear in the formula, the four category ratios are limited to nonnegative values with the following maximums.

$$0.775 \text{ for } \frac{C}{A}, \quad 0.11875 \text{ for } \frac{T}{A}, \quad 12.5 \text{ for } \frac{Y}{A}, \quad 0.095 \text{ for } \frac{I}{A}$$

Perform the indicated operations.

9. $(x^2 - 3x + 2) - (x - 4x^2) + 3x(2x + 1)$

10. $(6r - 5)^2$

11. $(t + 2)(3t^2 - t + 4)$

12. $\frac{2x^3 - 11x^2 + 28}{x - 5}$

(Modeling) Adjusted Poverty Threshold *The adjusted poverty threshold for a single person between the years 1999 and 2013 can be approximated by the formula*

$$y = 2.719x^2 + 196.1x + 8718,$$

where $x = 0$ *corresponds to 1999,* $x = 1$ *corresponds to 2000, and so on, and the adjusted poverty threshold amount,* y*, is in dollars. According to this model, what was the adjusted poverty threshold, to the nearest dollar, in each given year? (Source: U.S. Census Bureau.)*

13. 2005

14. 2012

Factor completely.

15. $6x^2 - 17x + 7$

16. $x^4 - 16$

17. $24m^3 - 14m^2 - 24m$

18. $x^3y^2 - 9x^3 - 8y^2 + 72$

19. $(a - b)^2 + 2(a - b)$

20. $1 - 27x^6$

Perform the indicated operations.

21. $\dfrac{5x^2 - 9x - 2}{30x^3 + 6x^2} \div \dfrac{x^4 - 3x^2 - 4}{2x^8 + 6x^7 + 4x^6}$

22. $\dfrac{x}{x^2 + 3x + 2} + \dfrac{2x}{2x^2 - x - 3}$

23. $\dfrac{a + b}{2a - 3} - \dfrac{a - b}{3 - 2a}$

24. $\dfrac{y - 2}{y - \dfrac{4}{y}}$

Simplify or evaluate as appropriate. Assume all variables represent positive real numbers.

25. $\sqrt{18x^5y^8}$

26. $\sqrt{32x} + \sqrt{2x} - \sqrt{18x}$

27. $\left(\sqrt{x} - \sqrt{y}\right)\left(\sqrt{x} + \sqrt{y}\right)$

28. $\dfrac{14}{\sqrt{11} - \sqrt{7}}$

29. $\left(\dfrac{x^{-2}y^{-1/3}}{x^{-5/3}y^{-2/3}}\right)^3$

30. $\left(-\dfrac{64}{27}\right)^{-2/3}$

31. ***Concept Check*** *True* or *false*: For all real numbers x, $\sqrt{x^2} = x$.

32. ***(Modeling) Period of a Pendulum*** The period t, in seconds, of the swing of a pendulum is given by the formula

$$t = 2\pi\sqrt{\frac{L}{32}},$$

where L is the length of the pendulum in feet. Find the period of a pendulum 3.5 ft long. Use a calculator, and round the answer to the nearest tenth.

1 Equations and Inequalities

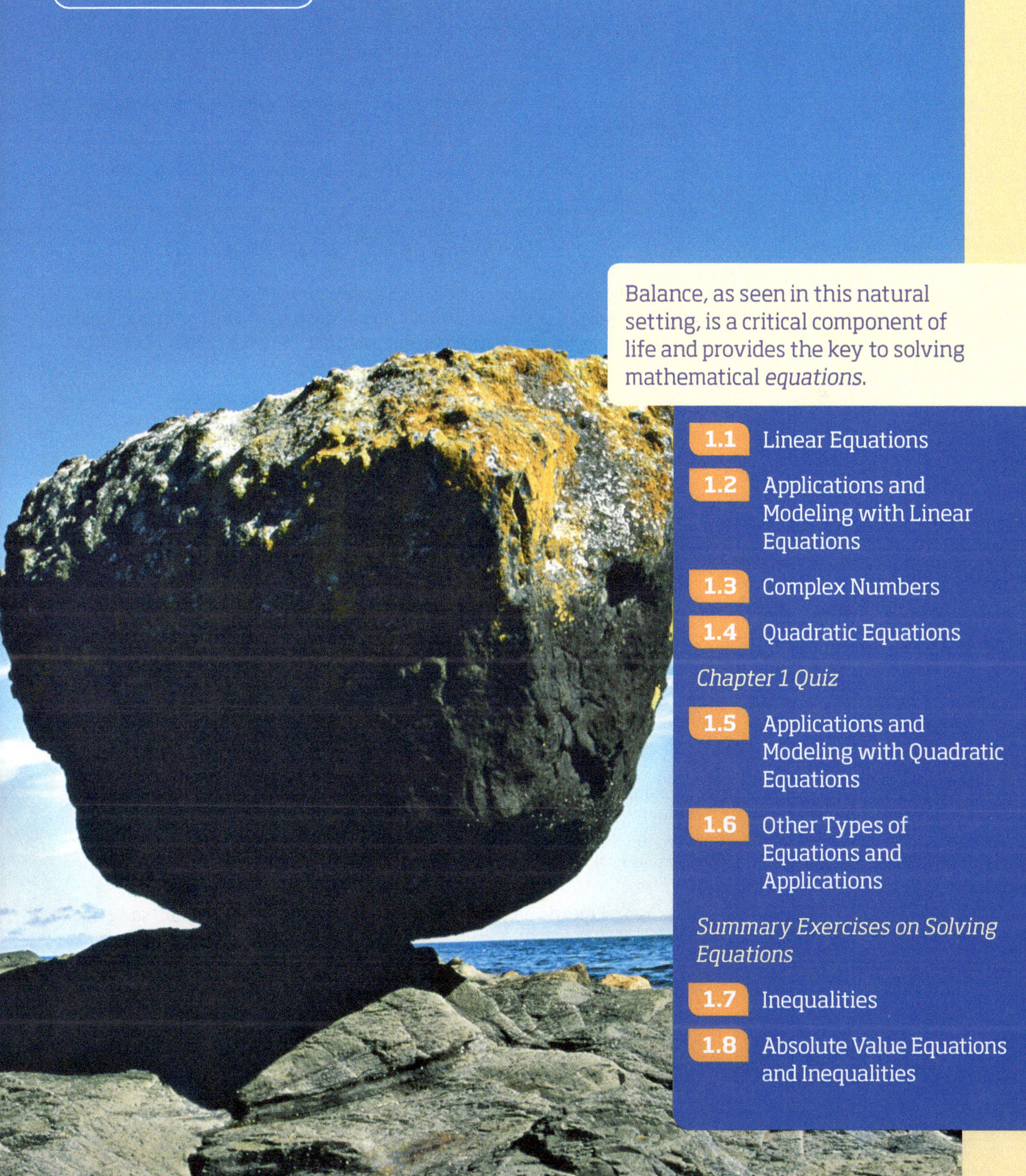

Balance, as seen in this natural setting, is a critical component of life and provides the key to solving mathematical *equations*.

1.1 Linear Equations

- Basic Terminology of Equations
- Linear Equations
- Identities, Conditional Equations, and Contradictions
- Solving for a Specified Variable (Literal Equations)

Basic Terminology of Equations

An **equation** is a statement that two expressions are equal.

$$x + 2 = 9, \quad 11x = 5x + 6x, \quad x^2 - 2x - 1 = 0 \quad \text{Equations}$$

To *solve* an equation means to find all numbers that make the equation a true statement. These numbers are the **solutions,** or **roots,** of the equation. A number that is a solution of an equation is said to *satisfy* the equation, and the solutions of an equation make up its **solution set.** Equations with the same solution set are **equivalent equations.** For example,

$$x = 4, \quad x + 1 = 5, \quad \text{and} \quad 6x + 3 = 27 \quad \text{are equivalent equations}$$

because they have the same solution set, $\{4\}$. However, the equations

$$x^2 = 9 \quad \text{and} \quad x = 3 \quad \text{are } \textit{not} \text{ equivalent}$$

because the first has solution set $\{-3, 3\}$ while the solution set of the second is $\{3\}$.

One way to solve an equation is to rewrite it as a series of simpler equivalent equations using the **addition and multiplication properties of equality.**

Addition and Multiplication Properties of Equality

Let a, b, and c represent real numbers.

$$\textbf{If } a = b, \textbf{ then } a + c = b + c.$$

That is, the same number may be added to each side of an equation without changing the solution set.

$$\textbf{If } a = b \textbf{ and } c \neq 0, \textbf{ then } ac = bc.$$

That is, each side of an equation may be multiplied by the same nonzero number without changing the solution set. (Multiplying each side by zero leads to $0 = 0$.)

These properties can be extended: The same number may be subtracted from each side of an equation, and each side may be divided by the same nonzero number, without changing the solution set.

Linear Equations

We use the properties of equality to solve *linear equations.*

Linear Equation in One Variable

A **linear equation in one variable** is an equation that can be written in the form

$$ax + b = 0,$$

where a and b are real numbers and $a \neq 0$.

A linear equation is a **first-degree equation** because the greatest degree of the variable is 1.

$$3x + \sqrt{2} = 0, \quad \frac{3}{4}x = 12, \quad 0.5(x+3) = 2x - 6 \qquad \text{Linear equations}$$

$$\sqrt{x} + 2 = 5, \quad \frac{1}{x} = -8, \quad x^2 + 3x + 0.2 = 0 \qquad \text{Nonlinear equations}$$

EXAMPLE 1 Solving a Linear Equation

Solve $3(2x - 4) = 7 - (x + 5)$.

SOLUTION

$3(2x - 4) = 7 - (x + 5)$

Be careful with signs.

$6x - 12 = 7 - x - 5$ Distributive property

$6x - 12 = 2 - x$ Combine like terms.

$6x - 12 + x = 2 - x + x$ Add x to each side.

$7x - 12 = 2$ Combine like terms.

$7x - 12 + 12 = 2 + 12$ Add 12 to each side.

$7x = 14$ Combine like terms.

$\frac{7x}{7} = \frac{14}{7}$ Divide each side by 7.

$x = 2$

CHECK

A check of the solution is recommended.

$3(2x - 4) = 7 - (x + 5)$ Original equation

$3(2 \cdot 2 - 4) \stackrel{?}{=} 7 - (2 + 5)$ Let $x = 2$.

$3(4 - 4) \stackrel{?}{=} 7 - (7)$ Work inside the parentheses.

$0 = 0$ ✓ True

Replacing x with 2 results in a true statement, so 2 is a solution of the given equation. The solution set is $\{2\}$.

✓ **Now Try Exercise 13.**

EXAMPLE 2 Solving a Linear Equation with Fractions

Solve $\frac{2x+4}{3} + \frac{1}{2}x = \frac{1}{4}x - \frac{7}{3}$.

SOLUTION

$\frac{2x+4}{3} + \frac{1}{2}x = \frac{1}{4}x - \frac{7}{3}$

Distribute to *all* terms within the parentheses.

$12\left(\frac{2x+4}{3} + \frac{1}{2}x\right) = 12\left(\frac{1}{4}x - \frac{7}{3}\right)$ Multiply by 12, the LCD of the fractions.

$12\left(\frac{2x+4}{3}\right) + 12\left(\frac{1}{2}x\right) = 12\left(\frac{1}{4}x\right) - 12\left(\frac{7}{3}\right)$ Distributive property

$4(2x + 4) + 6x = 3x - 28$ Multiply.

$8x + 16 + 6x = 3x - 28$ Distributive property

$14x + 16 = 3x - 28$ Combine like terms.

$11x = -44$ Subtract $3x$. Subtract 16.

$x = -4$ Divide each side by 11.

CHECK

$$\frac{2x+4}{3}+\frac{1}{2}x=\frac{1}{4}x-\frac{7}{3}$$ Original equation

$$\frac{2(-4)+4}{3}+\frac{1}{2}(-4)\stackrel{?}{=}\frac{1}{4}(-4)-\frac{7}{3}$$ Let $x=-4$.

$$\frac{-4}{3}+(-2)\stackrel{?}{=}-1-\frac{7}{3}$$ Simplify on each side.

$$-\frac{10}{3}=-\frac{10}{3}\ ✓$$ True

The solution set is $\{-4\}$.

✔ **Now Try Exercise 21.**

Identities, Conditional Equations, and Contradictions An equation satisfied by every number that is a meaningful replacement for the variable is an **identity.**

$$3(x+1)=3x+3$$ Identity

An equation that is satisfied by some numbers but not others is a **conditional equation.**

$$2x=4$$ Conditional equation

The equations in **Examples 1 and 2** are conditional equations. An equation that has no solution is a **contradiction.**

$$x=x+1$$ Contradiction

EXAMPLE 3 Identifying Types of Equations

Determine whether each equation is an *identity*, a *conditional equation*, or a *contradiction.* Give the solution set.

(a) $-2(x+4)+3x=x-8$ **(b)** $5x-4=11$ **(c)** $3(3x-1)=9x+7$

SOLUTION

(a) $-2(x+4)+3x=x-8$

$$-2x-8+3x=x-8$$ Distributive property

$$x-8=x-8$$ Combine like terms.

$$0=0$$ Subtract x. Add 8.

When a *true* statement such as $0=0$ results, the equation is an identity, and the solution set is **{all real numbers}.**

(b) $5x-4=11$

$$5x=15$$ Add 4 to each side.

$$x=3$$ Divide each side by 5.

This is a conditional equation, and its solution set is $\{3\}$.

(c) $3(3x-1)=9x+7$

$$9x-3=9x+7$$ Distributive property

$$-3=7$$ Subtract $9x$.

When a *false* statement such as $-3=7$ results, the equation is a contradiction, and the solution set is the **empty set,** or **null set,** symbolized $\varnothing$.

✔ **Now Try Exercises 31, 33, and 35.**

Identifying Types of Linear Equations

1. If solving a linear equation leads to a true statement such as $0 = 0$, the equation is an **identity.** Its solution set is **{all real numbers}. (See Example 3(a).)**

2. If solving a linear equation leads to a single solution such as $x = 3$, the equation is **conditional.** Its solution set consists of a single element. **(See Example 3(b).)**

3. If solving a linear equation leads to a false statement such as $-3 = 7$, the equation is a **contradiction.** Its solution set is $\emptyset$. **(See Example 3(c).)**

Solving for a Specified Variable (Literal Equations) A formula is an example of a **literal equation** (an equation involving letters).

EXAMPLE 4 Solving for a Specified Variable

Solve each formula or equation for the specified variable.

(a) $I = Prt$, for t **(b)** $A - P = Prt$, for P

(c) $3(2x - 5a) + 4b = 4x - 2$, for x

SOLUTION

(a) This is the formula for **simple interest** I on a principal amount of P dollars at an annual interest rate r for t years. To solve for t, we treat t as if it were the only variable, and the other variables as if they were constants.

$$I = Prt \quad \text{Goal: Isolate } t \text{ on one side.}$$

$$\frac{I}{Pr} = \frac{Prt}{Pr} \quad \text{Divide each side by } Pr.$$

$$\frac{I}{Pr} = t, \quad \text{or} \quad t = \frac{I}{Pr}$$

(b) The formula $A = P(1 + rt)$, which can also be written $A - P = Prt$, gives the **future value,** or **maturity value,** A of P dollars invested for t years at annual simple interest rate r.

$$A - P = Prt \quad \text{Goal: Isolate } P\text{, the specified variable.}$$

$$A = P + Prt \quad \text{Transform so that all terms involving } P \text{ are on one side.}$$

Pay close attention to this step.

$$A = P(1 + rt) \quad \text{Factor out } P.$$

$$\frac{A}{1 + rt} = P, \quad \text{or} \quad P = \frac{A}{1 + rt} \quad \text{Divide by } 1 + rt.$$

(c)

$$3(2x - 5a) + 4b = 4x - 2 \quad \text{Solve for } x.$$

$$6x - 15a + 4b = 4x - 2 \quad \text{Distributive property}$$

$$6x - 4x = 15a - 4b - 2 \quad \text{Isolate the } x\text{-terms on one side.}$$

$$2x = 15a - 4b - 2 \quad \text{Combine like terms.}$$

$$x = \frac{15a - 4b - 2}{2} \quad \text{Divide each side by 2.}$$

✔ **Now Try Exercises 39, 47, and 49.**

EXAMPLE 5 Applying the Simple Interest Formula

A woman borrowed \$5240 for new furniture. She will pay it off in 11 months at an annual simple interest rate of 4.5%. How much interest will she pay?

SOLUTION Use the simple interest formula $I = Prt$.

$$I = 5240(0.045)\left(\frac{11}{12}\right) = \$216.15 \qquad P = 5240,\ r = 0.045,\ \text{and } t = \tfrac{11}{12} \text{ (year)}$$

She will pay \$216.15 interest on her purchase.

✔ **Now Try Exercise 59.**

1.1 Exercises

CONCEPT PREVIEW *Fill in the blank to correctly complete each sentence.*

1. A(n) ________ is a statement that two expressions are equal.
2. To ________ an equation means to find all numbers that make the equation a true statement.
3. A linear equation is a(n) ________ because the greatest degree of the variable is 1.
4. A(n) ________ is an equation satisfied by every number that is a meaningful replacement for the variable.
5. A(n) ________ is an equation that has no solution.

CONCEPT PREVIEW *Decide whether each statement is* true *or* false.

6. The solution set of $2x + 5 = x - 3$ is $\{-8\}$.
7. The equation $5(x - 8) = 5x - 40$ is an example of an identity.
8. The equation $5x = 4x$ is an example of a contradiction.
9. Solving the literal equation $A = \frac{1}{2}bh$ for the variable h gives $h = \frac{A}{2b}$.

10. **CONCEPT PREVIEW** Which one is *not* a linear equation?

A. $5x + 7(x - 1) = -3x$ B. $9x^2 - 4x + 3 = 0$

C. $7x + 8x = 13x$ D. $0.04x - 0.08x = 0.40$

Solve each equation. ***See Examples 1 and 2.***

11. $5x + 4 = 3x - 4$
12. $9x + 11 = 7x + 1$
13. $6(3x - 1) = 8 - (10x - 14)$
14. $4(-2x + 1) = 6 - (2x - 4)$
15. $\frac{5}{6}x - 2x + \frac{4}{3} = \frac{5}{3}$
16. $\frac{7}{4} + \frac{1}{5}x - \frac{3}{2} = \frac{4}{5}x$
17. $3x + 5 - 5(x + 1) = 6x + 7$
18. $5(x + 3) + 4x - 3 = -(2x - 4) + 2$
19. $2[x - (4 + 2x) + 3] = 2x + 2$
20. $4[2x - (3 - x) + 5] = -6x - 28$
21. $\frac{1}{14}(3x - 2) = \frac{x + 10}{10}$
22. $\frac{1}{15}(2x + 5) = \frac{x + 2}{9}$
23. $0.2x - 0.5 = 0.1x + 7$
24. $0.01x + 3.1 = 2.03x - 2.96$
25. $-4(2x - 6) + 8x = 5x + 24 + x$
26. $-8(3x + 4) + 6x = 4(x - 8) + 4x$

27. $0.5x + \frac{4}{3}x = x + 10$

28. $\frac{2}{3}x + 0.25x = x + 2$

29. $0.08x + 0.06(x + 12) = 7.72$

30. $0.04(x - 12) + 0.06x = 1.52$

Determine whether each equation is an identity, *a* conditional equation, *or a* contradiction. *Give the solution set.* ***See Example 3.***

31. $4(2x + 7) = 2x + 22 + 3(2x + 2)$

32. $\frac{1}{2}(6x + 20) = x + 4 + 2(x + 3)$

33. $2(x - 8) = 3x - 16$

34. $-8(x + 5) = -8x - 5(x + 8)$

35. $4(x + 7) = 2(x + 12) + 2(x + 1)$

36. $-6(2x + 1) - 3(x - 4) = -15x + 1$

37. $0.3(x + 2) - 0.5(x + 2) = -0.2x - 0.4$

38. $-0.6(x - 5) + 0.8(x - 6) = 0.2x - 1.8$

Solve each formula for the specified variable. Assume that the denominator is not 0 if variables appear in the denominator. ***See Examples 4(a) and (b).***

39. $V = lwh$, for l (volume of a rectangular box)

40. $I = Prt$, for P (simple interest)

41. $P = a + b + c$, for c (perimeter of a triangle)

42. $P = 2l + 2w$, for w (perimeter of a rectangle)

43. $\mathcal{A} = \frac{1}{2}h(B + b)$, for B (area of a trapezoid)

44. $\mathcal{A} = \frac{1}{2}h(B + b)$, for h (area of a trapezoid)

45. $S = 2\pi rh + 2\pi r^2$, for h (surface area of a right circular cylinder)

46. $s = \frac{1}{2}gt^2$, for g (distance traveled by a falling object)

47. $S = 2lw + 2wh + 2hl$, for h (surface area of a rectangular box)

48. $z = \frac{x - \mu}{\sigma}$, for x (standardized value)

Solve each equation for x. ***See Example 4(c).***

49. $2(x - a) + b = 3x + a$

50. $5x - (2a + c) = 4(x + c)$

51. $ax + b = 3(x - a)$

52. $4a - ax = 3b + bx$

53. $\frac{x}{a - 1} = ax + 3$

54. $\frac{x - 1}{2a} = 2x - a$

55. $a^2x + 3x = 2a^2$

56. $ax + b^2 = bx - a^2$

57. $3x = (2x - 1)(m + 4)$

58. $-x = (5x + 3)(3k + 1)$

Simple Interest *Work each problem.* ***See Example 5.***

59. Elmer borrowed \$3150 from his brother Julio to pay for books and tuition. He agreed to repay Julio in 6 months with simple annual interest at 4%.

(a) How much will the interest amount to?

(b) What amount must Elmer pay Julio at the end of the 6 months?

60. Levada borrows \$30,900 from her bank to open a florist shop. She agrees to repay the money in 18 months with simple annual interest of 5.5%.

(a) How much must she pay the bank in 18 months?

(b) How much of the amount in part (a) is interest?

Celsius and Fahrenheit Temperatures *In the metric system of weights and measures, temperature is measured in degrees Celsius (°C) instead of degrees Fahrenheit (°F). To convert between the two systems, we use the equations*

$$C = \frac{5}{9}(F - 32) \quad \text{and} \quad F = \frac{9}{5}C + 32.$$

In each exercise, convert to the other system. Round answers to the nearest tenth of a degree if necessary.

61. 20°C **62.** 200°C **63.** 50°F

64. 77°F **65.** 100°F **66.** 350°F

Work each problem. Round to the nearest tenth of a degree, if necessary.

67. ***Temperature of Venus*** Venus is the hottest planet, with a surface temperature of 867°F. What is this temperature in Celsius? (*Source: World Almanac and Book of Facts.*)

68. ***Temperature at Soviet Antarctica Station*** A record low temperature of −89.4°C was recorded at the Soviet Antarctica Station of Vostok on July 21, 1983. Find the corresponding Fahrenheit temperature. (*Source: World Almanac and Book of Facts.*)

69. ***Temperature in South Carolina*** A record high temperature of 113°F was recorded for the state of South Carolina on June 29, 2012. What is the corresponding Celsius temperature? (*Source:* U.S. National Oceanic and Atmospheric Administration.)

70. ***Temperature in Haiti*** The average annual temperature in Port-au-Prince, Haiti, is approximately 28.1°C. What is the corresponding Fahrenheit temperature? (*Source:* www.haiti.climatemps.com)

1.2 Applications and Modeling with Linear Equations

- Solving Applied Problems
- Geometry Problems
- Motion Problems
- Mixture Problems
- Modeling with Linear Equations

Solving Applied Problems

One of the main reasons for learning mathematics is to be able use it to solve application problems. While there is no one method that enables us to solve all types of applied problems, the following six steps provide a useful guide.

Solving an Applied Problem

Step 1 **Read** the problem carefully until you understand what is given and what is to be found.

Step 2 **Assign a variable** to represent the unknown value, using diagrams or tables as needed. Write down what the variable represents. If necessary, express any other unknown values in terms of the variable.

Step 3 **Write an equation** using the variable expression(s).

Step 4 **Solve** the equation.

Step 5 **State the answer** to the problem. Does it seem reasonable?

Step 6 **Check** the answer in the words of the original problem.

Geometry Problems

EXAMPLE 1 Finding the Dimensions of a Square

If the length of each side of a square is increased by 3 cm, the perimeter of the new square is 40 cm more than twice the length of each side of the original square. Find the dimensions of the original square.

SOLUTION

Step 1 **Read** the problem. We must find the length of each side of the original square.

Step 2 **Assign a variable.** Since the length of a side of the original square is to be found, let the variable represent this length.

Let x = the length of a side of the original square in centimeters.

The length of a side of the new square is 3 cm more than the length of a side of the old square.

Then $x + 3$ = the length of a side of the new square.

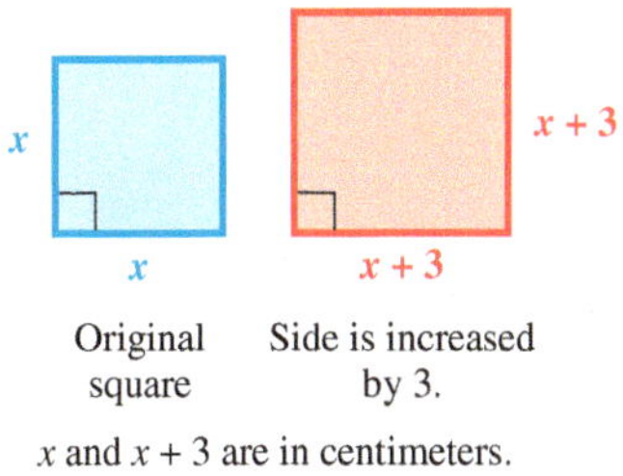

See **Figure 1.** Now write a variable expression for the perimeter of the new square. The perimeter of a square is 4 times the length of a side.

Thus, $4(x + 3)$ = the perimeter of the new square.

Original square | Side is increased by 3.

x and $x + 3$ are in centimeters.

Figure 1

Step 3 **Write an equation.** Translate the English sentence that follows into its equivalent algebraic equation.

The new perimeter	is	40	more than	twice the length of each side of the original square.
$4(x+3)$	$=$	40	$+$	$2x$

Step 4 **Solve** the equation.

$4x + 12 = 40 + 2x$ Distributive property

$2x = 28$ Subtract $2x$ and 12.

$x = 14$ Divide by 2.

Step 5 **State the answer.** Each side of the original square measures 14 cm.

Step 6 **Check.** Go back to the words of the original problem to see that all necessary conditions are satisfied. The length of a side of the new square would be $14 + 3 = 17$ cm. The perimeter of the new square would be $4(17) = 68$ cm. Twice the length of a side of the original square would be $2(14) = 28$ cm. Because $40 + 28 = 68$, the answer checks.

✔ Now Try Exercise 15.

Motion Problems

LOOKING AHEAD TO CALCULUS

In calculus the concept of the **definite integral** is used to find the distance traveled by an object traveling at a *non-constant* velocity.

PROBLEM SOLVING HINT In a motion problem, the three components *distance, rate,* and *time* are denoted by the letters d, r, and t, respectively. (The *rate* is also called the *speed* or *velocity.* Here, rate is understood to be constant.) These variables are related by the following equations.

$$d = rt, \quad \text{and its related forms} \quad r = \frac{d}{t} \quad \text{and} \quad t = \frac{d}{r}$$

EXAMPLE 2 Solving a Motion Problem

Maria and Eduardo are traveling to a business conference. The trip takes 2 hr for Maria and 2.5 hr for Eduardo because he lives 40 mi farther away. Eduardo travels 5 mph faster than Maria. Find their average rates.

SOLUTION

Step 1 **Read** the problem. We must find Maria's and Eduardo's average rates.

Step 2 **Assign a variable.** Because average rates are to be found, we let the variable represent one of these rates.

Let x = Maria's rate.

Because Eduardo travels 5 mph faster than Maria, we can express his average rate using the same variable.

Then $x + 5$ = Eduardo's rate.

Make a table. The expressions in the last column were found by multiplying the corresponding rates and times.

	r	t	d
Maria	x	2	$2x$
Eduardo	$x+5$	2.5	$2.5(x+5)$

Summarize the given information in a table.

Use $d = rt$.

Step 3 **Write an equation.** Eduardo's distance traveled exceeds Maria's distance by 40 mi. Translate this into an equation.

$$\underbrace{2.5(x+5)}_{\text{Eduardo's distance}} \underbrace{=}_{\text{is}} \underbrace{2x + 40}_{\text{40 more than Maria's.}}$$

Step 4 **Solve.**

$$2.5x + 12.5 = 2x + 40 \quad \text{Distributive property}$$

$$0.5x = 27.5 \quad \text{Subtract } 2x \text{ and } 12.5.$$

$$x = 55 \quad \text{Divide by } 0.5.$$

Step 5 **State the answer.** Maria's rate of travel is 55 mph, and Eduardo's rate is

$$55 + 5 = 60 \text{ mph.}$$

Step 6 **Check.** The conditions of the problem are satisfied, as shown below.

Distance traveled by Maria: $2(55) = 110$ mi

Distance traveled by Eduardo: $2.5(60) = 150$ mi

$150 - 110 = 40$ as required.

✔ **Now Try Exercise 19.**

Mixture Problems Problems involving mixtures of two types of the same substance, salt solution, candy, and so on, often involve percentages.

PROBLEM-SOLVING HINT In mixture problems involving solutions,

$$\begin{matrix}\text{rate (percent)} \\ \text{of concentration}\end{matrix} \cdot \text{quantity} = \begin{matrix}\text{amount of pure} \\ \text{substance present.}\end{matrix}$$

The concentration of the final mixture must be between the concentrations of the two solutions making up the mixture.

George Polya (1887–1985)

Polya, a native of Budapest, Hungary, wrote more than 250 papers and a number of books. He proposed a general outline for solving applied problems in his classic book *How to Solve It.*

EXAMPLE 3 Solving a Mixture Problem

A chemist needs a 20% solution of alcohol. She has a 15% solution on hand, as well as a 30% solution. How many liters of the 15% solution should she add to 3 L of the 30% solution to obtain the 20% solution?

SOLUTION

Step 1 **Read** the problem. We must find the required number of liters of 15% alcohol solution.

Step 2 **Assign a variable.**

Let x = the number of liters of 15% solution to be added.

Figure 2 and the table show what is happening in the problem. The numbers in the last column were found by multiplying the strengths and the numbers of liters.

Figure 2

Strength	Liters of Solution	Liters of Pure Alcohol
15%	x	$0.15x$
30%	3	$0.30(3)$
20%	$x + 3$	$0.20(x + 3)$

Sum must equal

Step 3 **Write an equation.** The number of liters of pure alcohol in the 15% solution plus the number of liters in the 30% solution must equal the number of liters in the final 20% solution.

Liters of pure alcohol in 15% solution + Liters of pure alcohol in 30% solution = Liters of pure alcohol in 20% solution

$$0.15x + 0.30(3) = 0.20(x + 3)$$

Step 4 **Solve.**

$$0.15x + 0.90 = 0.20x + 0.60 \quad \text{Distributive property}$$

$$0.30 = 0.05x \quad \text{Subtract 0.60 and } 0.15x.$$

$$6 = x \quad \text{Divide by 0.05.}$$

Step 5 **State the answer.** Thus, 6 L of 15% solution should be mixed with 3 L of 30% solution, giving $6 + 3 = 9$ L of 20% solution.

Step 6 **Check.** The answer checks because the amount of alcohol in the two solutions is equal to the amount of alcohol in the mixture.

$$0.15(6) + 0.9 = 0.9 + 0.9 = 1.8 \quad \text{Solutions}$$

$$0.20(6 + 3) = 0.20(9) = 1.8 \quad \text{Mixture}$$

✔ **Now Try Exercise 29.**

PROBLEM-SOLVING HINT In mixed investment problems, multiply the principal amount P by the interest rate r, expressed as a decimal, and the time t, in years, to find the amount of interest earned I.

$$I = Prt \quad \text{Simple interest formula}$$

EXAMPLE 4 Solving an Investment Problem

An artist has sold a painting for \$410,000. He invests a portion of the money for 6 months at 2.65% and the rest for a year at 2.91%. His broker tells him the two investments will earn a total of \$8761. How much should be invested at each rate to obtain that amount of interest?

SOLUTION

Step 1 **Read** the problem. We must find the amount to be invested at each rate.

Step 2 **Assign a variable.**

Let $x =$ the dollar amount to be invested for 6 months at 2.65%.

$410{,}000 - x =$ the dollar amount to be invested for 1 yr at 2.91%.

P Invested Amount	r Interest Rate (%)	t Time (in years)	I Interest Earned
x	2.65	0.5	$x(0.0265)(0.5)$
$410{,}000 - x$	2.91	1	$(410{,}000 - x)(0.0291)(1)$

Summarize the information in a table using the formula $I = Prt$.

Step 3 **Write an equation.** The sum of the two interest amounts must equal the total interest earned.

$$\underbrace{0.5x(0.0265)}_{\text{Interest from 2.65\% investment}} + \underbrace{0.0291(410{,}000 - x)}_{\text{Interest from 2.91\% investment}} = \underbrace{8761}_{\text{Total interest}}$$

Step 4 **Solve.**

$$0.01325x + 11{,}931 - 0.0291x = 8761 \quad \text{Distributive property}$$

$$11{,}931 - 0.01585x = 8761 \quad \text{Combine like terms.}$$

$$-0.01585x = -3170 \quad \text{Subtract 11,931.}$$

$$x = 200{,}000 \quad \text{Divide by } -0.01585.$$

Step 5 **State the answer.** The artist should invest \$200,000 at 2.65% for 6 months and

$$\$410{,}000 - \$200{,}000 = \$210{,}000$$

at 2.91% for 1 yr to earn \$8761 in interest.

Step 6 **Check.** The 6-month investment earns

$$\$200{,}000(0.0265)(0.5) = \$2650,$$

and the 1-yr investment earns

$$\$210{,}000(0.0291)(1) = \$6111.$$

The total amount of interest earned is

$$\$2650 + \$6111 = \$8761, \quad \text{as required.}$$

✔ **Now Try Exercise 35.**

Modeling with Linear Equations A **mathematical model** is an equation (or inequality) that describes the relationship between two quantities. A **linear model** is a linear equation. The next example shows how a linear model is applied.

EXAMPLE 5 Modeling Prevention of Indoor Pollutants

If a vented range hood removes contaminants such as carbon monoxide and nitrogen dioxide from the air at a rate of F liters of air per second, then the percent P of contaminants that are also removed from the surrounding air can be modeled by the linear equation

$$P = 1.06F + 7.18, \quad \text{where} \quad 10 \le F \le 75.$$

What flow F (to the nearest hundredth) must a range hood have to remove 50% of the contaminants from the air? (*Source: Proceedings of the Third International Conference on Indoor Air Quality and Climate.*)

SOLUTION Replace P with 50 in the linear model, and solve for F.

$$\begin{aligned} P &= 1.06F + 7.18 && \text{Given model} \\ 50 &= 1.06F + 7.18 && \text{Let } P = 50. \\ 42.82 &= 1.06F && \text{Subtract 7.18.} \\ F &\approx 40.40 && \text{Divide by 1.06.} \end{aligned}$$

Therefore, to remove 50% of the contaminants, the flow rate must be 40.40 L of air per second.

Now Try Exercise 41.

EXAMPLE 6 Modeling Health Care Costs

The projected per capita health care expenditures in the United States, where y is in dollars, and x is years after 2000, are given by the following linear equation.

$$y = 331x + 5091 \quad \text{Linear model}$$

(*Source:* Centers for Medicare and Medicaid Services.)

(a) What were the per capita health care expenditures in the year 2010?

(b) If this model continues to describe health care expenditures, when will the per capita expenditures reach \$11,000?

SOLUTION In part (a) we are given information to determine a value for x and asked to find the corresponding value of y, whereas in part (b) we are given a value for y and asked to find the corresponding value of x.

(a) The year 2010 is 10 yr after the year 2000. Let $x = 10$ and find the value of y.

$$\begin{aligned} y &= 331x + 5091 && \text{Given model} \\ y &= 331(10) + 5091 && \text{Let } x = 10. \\ y &= 8401 && \text{Multiply and then add.} \end{aligned}$$

In 2010, the estimated per capita health care expenditures were \$8401.

(b) Let $y = 11{,}000$ in the given model, and find the value of x.

$$\begin{aligned} 11{,}000 &= 331x + 5091 && \text{Let } y = 11{,}000. \\ 5909 &= 331x && \text{Subtract 5091.} \\ x &\approx 17.9 && \text{Divide by 331.} \end{aligned}$$

17 corresponds to 2000 + 17 = 2017.

The x-value of 17.9 indicates that per capita health care expenditures are projected to reach \$11,000 during the 17th year after 2000—that is, 2017.

Now Try Exercise 45.

1.2 Exercises

CONCEPT PREVIEW *Solve each problem.*

1. ***Time Traveled*** How long will it take a car to travel 400 mi at an average rate of 50 mph?

2. ***Distance Traveled*** If a train travels at 80 mph for 15 min, what is the distance traveled?

3. ***Investing*** If a person invests \$500 at 2% simple interest for 4 yr, how much interest is earned?

4. ***Value of Coins*** If a jar of coins contains 40 half-dollars and 200 quarters, what is the monetary value of the coins?

5. ***Acid Mixture*** If 120 L of an acid solution is 75% acid, how much pure acid is there in the mixture?

6. ***Sale Price*** Suppose that a computer that originally sold for x dollars has been discounted 60%. Which one of the following expressions does not represent its sale price?

   A. $x - 0.60x$ **B.** $0.40x$ **C.** $\frac{4}{10}x$ **D.** $x - 0.60$

7. ***Acid Mixture*** Suppose two acid solutions are mixed. One is 26% acid and the other is 34% acid. Which one of the following concentrations cannot possibly be the concentration of the mixture?

   A. 24% **B.** 30% **C.** 31% **D.** 33%

8. ***Unknown Numbers*** Consider the following problem.

   The difference between seven times a number and 9 is equal to five times the sum of the number and 2. Find the number.

   If x represents the number, which equation is correct for solving this problem?

   A. $7x - 9 = 5(x + 2)$ **B.** $9 - 7x = 5(x + 2)$

   C. $7x - 9 = 5x + 2$ **D.** $9 - 7x = 5x + 2$

9. ***Unknown Numbers*** Consider the following problem.

   One number is 3 less than 6 times a second number. Their sum is 46. Find the numbers.

   If x represents the second number, which equation is correct for solving this problem?

   A. $46 - (x + 3) = 6x$ **B.** $(3 - 6x) + x = 46$

   C. $46 - (3 - 6x) = x$ **D.** $(6x - 3) + x = 46$

10. ***Dimensions of a Rectangle*** Which one or more of the following cannot be a correct equation to solve a geometry problem, if x represents the length of a rectangle? (*Hint:* Solve each equation and consider the solution.)

    A. $2x + 2(x - 1) = 14$ **B.** $-2x + 7(5 - x) = 52$

    C. $5(x + 2) + 5x = 10$ **D.** $2x + 2(x - 3) = 22$

Note: **Geometry formulas can be found on the back inside cover of this book.**

Solve each problem. ***See Example 1.***

11. ***Perimeter of a Rectangle*** The perimeter of a rectangle is 294 cm. The width is 57 cm. Find the length.

12. ***Perimeter of a Storage Shed*** Michael must build a rectangular storage shed. He wants the length to be 6 ft greater than the width, and the perimeter will be 44 ft. Find the length and the width of the shed.

13. *Dimensions of a Puzzle Piece* A puzzle piece in the shape of a triangle has perimeter 30 cm. Two sides of the triangle are each twice as long as the shortest side. Find the length of the shortest side. (Side lengths in the figure are in centimeters.)

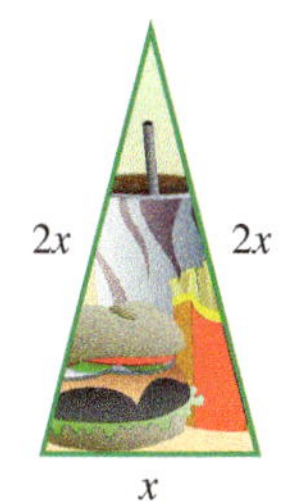

14. *Dimensions of a Label* The length of a rectangular label is 2.5 cm less than twice the width. The perimeter is 40.6 cm. Find the width. (Side lengths in the figure are in centimeters.)

15. *Perimeter of a Plot of Land* The perimeter of a triangular plot of land is 2400 ft. The longest side is 200 ft less than twice the shortest. The middle side is 200 ft less than the longest side. Find the lengths of the three sides of the triangular plot.

16. *World Largest Ice Cream Cake* The world's largest ice cream cake, a rectangular cake made by a Dairy Queen in Toronto, Ontario, Canada, on May 10, 2011, had length 0.39 m greater than its width. Its perimeter was 17.02 m. What were the length and width of this 10-ton cake? (*Source:* www.guinnessworldrecords.com)

17. *Storage Bin Dimensions* A storage bin is in the shape of a rectangular box. Find the height of the box if its length is 18 ft, its width is 8 ft, and its surface area is 496 ft^2. (In the figure, h = height. Assume that the given surface area includes that of the top lid of the box.)

18. *Cylinder Dimensions* A right circular cylinder has radius 6 in. and volume 144π in.^{3}. What is its height? (In the figure, h = height.)

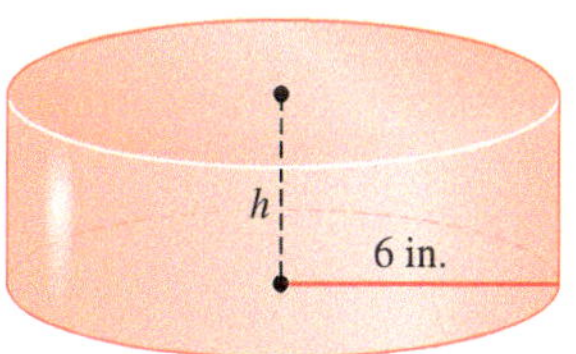

h is in inches.

Solve each problem. **See Example 2.**

19. *Distance to an Appointment* Margaret drove to a business appointment at 50 mph. Her average speed on the return trip was 40 mph. The return trip took $\frac{1}{4}$ hr longer because of heavy traffic. How far did she travel to the appointment?

	r	t	d
Morning	50	x	
Afternoon	40	$x + \frac{1}{4}$	

20. *Distance between Cities* Elwyn averaged 50 mph traveling from Denver to Minneapolis. Returning by a different route that covered the same number of miles, he averaged 55 mph. What is the distance between the two cities to the nearest ten miles if his total traveling time was 32 hr?

	r	t	d
Going	50	x	
Returning	55	$32 - x$	

21. *Distance to Work* David gets to work in 20 min when he drives his car. Riding his bike (by the same route) takes him 45 min. His average driving speed is 4.5 mph greater than his average speed on his bike. How far does he travel to work?

22. *Speed of a Plane* Two planes leave Los Angeles at the same time. One heads south to San Diego, while the other heads north to San Francisco. The San Diego plane flies 50 mph slower than the San Francisco plane. In $\frac{1}{2}$ hr, the planes are 275 mi apart. What are their speeds?

23. *Running Times* Mary and Janet are running in the Apple Hill Fun Run. Mary runs at 7 mph, Janet at 5 mph. If they start at the same time, how long will it be before they are 1.5 mi apart?

24. *Running Times* If the run in **Exercise 23** has a staggered start, and Janet starts first, with Mary starting 10 min later, how long will it be before Mary catches up with Janet?

25. *Track Event Speeds* At the 2008 Summer Olympics in Beijing, China, Usain Bolt (Jamaica) set a new Olympic and world record in the 100-m dash with a time of 9.69 sec. If this pace could be maintained for an entire 26-mi marathon, what would his time be? How would this time compare to the fastest time for a marathon, which is 2 hr, 3 min, 23 sec, set in 2013? (*Hint:* 1 m $\approx$ 3.281 ft.) (*Source: Sports Illustrated Almanac.*)

26. *Track Event Speeds* On August 16, 2009, at the World Track and Field Championship in Berlin, Usain Bolt set a new world record in the 100-m dash with a time of 9.58 sec. Refer to **Exercise 25** and answer the questions using Bolt's 2009 time. (*Source: Sports Illustrated Almanac.*)

27. *Boat Speed* Callie took 20 min to drive her boat upstream to water-ski at her favorite spot. Coming back later in the day, at the same boat speed, took her 15 min. If the current in that part of the river is 5 km per hr, what was her boat speed?

28. *Wind Speed* Joe traveled against the wind in a small plane for 3 hr. The return trip with the wind took 2.8 hr. Find the speed of the wind to the nearest tenth if the speed of the plane in still air is 180 mph.

Solve each problem. ***See Example 3.***

29. *Acid Mixture* How many gallons of a 5% acid solution must be mixed with 5 gal of a 10% solution to obtain a 7% solution?

Strength	Gallons of Solution	Gallons of Pure Acid
5%	x	
10%	5	
7%	$x + 5$	

30. *Acid Mixture* A student needs 10% hydrochloric acid for a chemistry experiment. How much 5% acid should she mix with 60 mL of 20% acid to get a 10% solution?

Strength	mL of Solution	mL of Pure Acid
5%	x	
20%	60	
10%	$x + 60$	

31. *Alcohol Mixture* Beau wishes to strengthen a mixture from 10% alcohol to 30% alcohol. How much pure alcohol should be added to 7 L of the 10% mixture?

32. *Alcohol Mixture* How many gallons of pure alcohol should be mixed with 20 gal of a 15% alcohol solution to obtain a mixture that is 25% alcohol?

33. *Saline Solution* How much water should be added to 8 mL of 6% saline solution to reduce the concentration to 4%?

34. *Acid Mixture* How much pure acid should be added to 18 L of 30% acid to increase the concentration to 50% acid?

Solve each problem. ***See Example 4.***

35. *Real Estate Financing* Cody wishes to sell a piece of property for \$240,000. He wants the money to be paid off in two ways: a short-term note at 2% interest and a long-term note at 2.5%. Find the amount of each note if the total annual interest paid is \$5500.

P Note Amount	r Interest Rate (%)	t Time (in years)	I Interest Paid
x	2	1	$x(0.02)(1)$
$240{,}000 - x$	2.5	1	$(240{,}000 - x)(0.025)(1)$

36. *Buying and Selling Land* Roger bought two plots of land for a total of \$120,000. When he sold the first plot, he made a profit of 15%. When he sold the second, he lost 10%. His total profit was \$5500. How much did he pay for each piece of land?

Land Price	Rate of Return (%)	Profit (or Loss)
x	15%	
$120{,}000 - x$	10%	

37. *Retirement Planning* In planning her retirement, Janet deposits some money at 2.5% interest, with twice as much deposited at 3%. Find the amount deposited at each rate if the total annual interest income is \$850.

38. *Investing a Building Fund* A church building fund has invested some money in two ways: part of the money at 3% interest and four times as much at 2.75%. Find the amount invested at each rate if the total annual income from interest is \$2800.

39. *Lottery Winnings* Linda won \$200,000 in a state lottery. She first paid income tax of 30% on the winnings. She invested some of the rest at 1.5% and some at 4%, earning \$4350 interest per year. How much did she invest at each rate?

40. *Cookbook Royalties* Becky earned \$48,000 from royalties on her cookbook. She paid a 28% income tax on these royalties. The balance was invested in two ways, some of it at 3.25% interest and some at 1.75%. The investments produced \$904.80 interest per year. Find the amount invested at each rate.

(Modeling) *Solve each problem.* ***See Examples 5 and 6.***

41. *Warehouse Club Membership* If the annual fee for a warehouse club membership is \$100 and the reward rate is 2% on club purchases for the year, then the linear equation

$$y = 100 - 0.02x$$

models the actual annual cost of the membership y, in dollars. Here x represents the annual amount of club purchases, also in dollars.

(a) Determine the actual annual cost of the membership if club purchases for the year are \$2400.

(b) What amount of club purchases would reduce the actual annual cost of the membership to \$50?

(c) How much would a member have to spend in annual club purchases to reduce the annual membership cost to \$0?

42. *Warehouse Club Membership* Suppose that the annual fee for a warehouse club membership is \$50 and that the reward rate on club purchases for the year is 1.6%. Then the actual annual cost of a membership y, in dollars, for an amount of annual club purchases x, in dollars, can be modeled by the following linear equation.

$$y = 50 - 0.016x$$

(a) Determine the actual annual cost of the membership if club purchases for the year are \$1500.

(b) What amount of club purchases would reduce the actual annual cost of the membership to \$0?

(c) If club purchases for the year exceed \$3125, how is the actual annual membership cost affected?

43. *Indoor Air Pollution* Formaldehyde is an indoor air pollutant formerly found in plywood, foam insulation, and carpeting. When concentrations in the air reach 33 micrograms per cubic foot ($\mu g/ft^3$), eye irritation can occur. One square foot of new plywood could emit 140 μg per hr. (*Source:* A. Hines, *Indoor Air Quality & Control.*)

(a) A room has 100 ft^2 of new plywood flooring. Find a linear equation F that computes the amount of formaldehyde, in micrograms, emitted in x hours.

(b) The room contains 800 ft^3 of air and has no ventilation. Determine how long it would take for concentrations to reach 33 $\mu g/ft^3$. (Round to the nearest tenth.)

44. *Classroom Ventilation* According to the American Society of Heating, Refrigerating and Air-Conditioning Engineers, Inc. (ASHRAE), a nonsmoking classroom should have a ventilation rate of 15 ft^3 per min for each person in the room.

(a) Write an equation that models the total ventilation V (in cubic feet per hour) necessary for a classroom with x students.

(b) A common unit of ventilation is air change per hour (ach). One ach is equivalent to exchanging all the air in a room every hour. If x students are in a classroom having volume 15,000 ft^3, determine how many air exchanges per hour (A) are necessary to keep the room properly ventilated.

(c) Find the necessary number of ach (A) if the classroom has 40 students in it.

(d) In areas like bars and lounges that allow smoking, the ventilation rate should be increased to 50 ft^3 per min per person. Compared to classrooms, ventilation should be increased by what factor in heavy smoking areas?

45. *College Enrollments* The graph shows the projections in total enrollment at degree-granting institutions from fall 2014 to fall 2021.

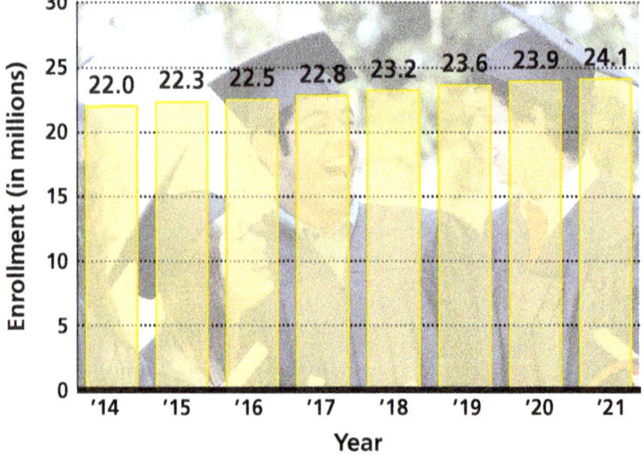

Source: U.S. Department of Education, National Center for Education Statistics.

The following linear model provides the approximate enrollment, in millions, between the years 2014 and 2021, where $x = 0$ corresponds to 2014, $x = 1$ to 2015, and so on, and y is in millions of students.

$$y = 0.3143x + 21.95$$

(a) Use the model to determine projected enrollment for fall 2018.

(b) Use the model to determine the year in which enrollment is projected to reach 24 million.

(c) How do your answers to parts (a) and (b) compare to the corresponding values shown in the graph?

(d) The actual enrollment in fall 2000 was 15.3 million. The model here is based on data from 2014 to 2021. If we were to use the model for 2000, what would the projected enrollment be?

(e) Compare the actual value and the value based on the model in part (d). Discuss the pitfalls of using the model to predict enrollment for years preceding 2014.

46. ***Baby Boom*** U.S. population during the years between 1946 and 1964, commonly known as the Baby Boom, can be modeled by the following linear equation.

$$y = 2.8370x + 140.83$$

Here y represents the population in millions as of July 1 of a given year, and x represents number of years after 1946. Thus, $x = 0$ corresponds to 1946, $x = 1$ corresponds to 1947, and so on. (*Source:* U.S. Census Bureau.)

(a) According to the model, what was the U.S. population on July 1, 1952?

(b) In what year did the U.S. population reach 150 million?

1.3 Complex Numbers

- Basic Concepts of Complex Numbers
- Operations on Complex Numbers

Basic Concepts of Complex Numbers

The set of real numbers does not include all the numbers needed in algebra. For example, there is no real number solution of the equation

$$x^2 = -1$$

because no real number, when squared, gives -1. To extend the real number system to include solutions of equations of this type, the number i is defined.

Imaginary Unit *i*

$$i = \sqrt{-1}, \quad \text{and therefore,} \quad i^2 = -1.$$

(Note that $-i$ is also a square root of -1.)

Complex numbers are formed by adding real numbers and multiples of i.

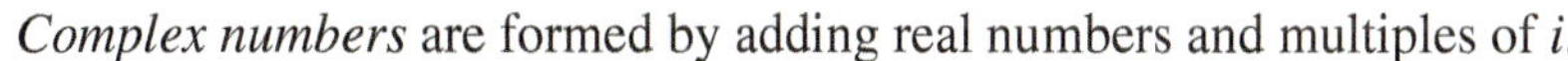

Complex Number

If a and b are real numbers, then any number of the form

$$a + bi$$

is a **complex number.** In the complex number $a + bi$, a is the **real part** and b is the **imaginary part.***

Square roots of negative numbers were not incorporated into an integrated number system until the 16th century. They were then used as solutions of equations and later (in the 18th century) in surveying. Today, such numbers are used extensively in science and engineering.

Two complex numbers $a + bi$ and $c + di$ are equal provided that their real parts are equal and their imaginary parts are equal—that is, they are equal if and only if $a = c$ and $b = d$.

*In some texts, the term bi is defined to be the imaginary part.

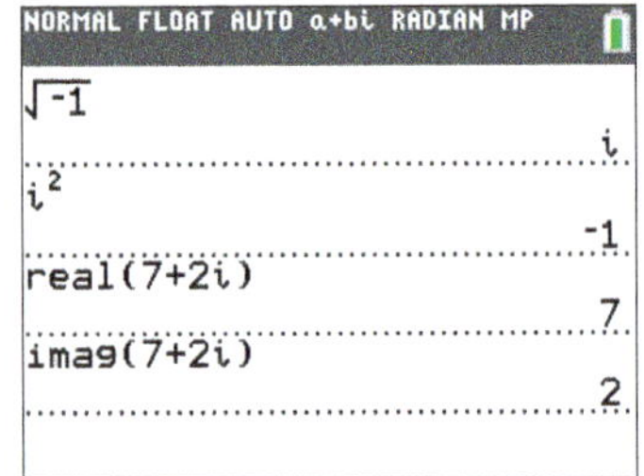

The calculator is in complex number mode. This screen supports the definition of i. It also shows how the calculator returns the real and imaginary parts of the complex number $7 + 2i$.

Figure 3

Some graphing calculators, such as the TI-84 Plus, are capable of working with complex numbers, as seen in **Figure 3.** ■

The following important concepts apply to a complex number $a + bi$.

1. If $b = 0$, then $a + bi = a$, which is a real number. (This means that the set of real numbers is a subset of the set of complex numbers. See **Figure 4.**)
2. If $b \neq 0$, then $a + bi$ is a **nonreal complex number.**
   Examples: $7 + 2i, \quad -1 - i$
3. If $a = 0$ and $b \neq 0$, then the nonreal complex number is a **pure imaginary number.**
   Examples: $3i, \quad -16i$

The form $a + bi$ (or $a + ib$) is **standard form.** (The form $a + ib$ is used to write expressions such as $i\sqrt{5}$, because $\sqrt{5}i$ could be mistaken for $\sqrt{5i}$.)

The relationships among the subsets of the complex numbers are shown in **Figure 4.**

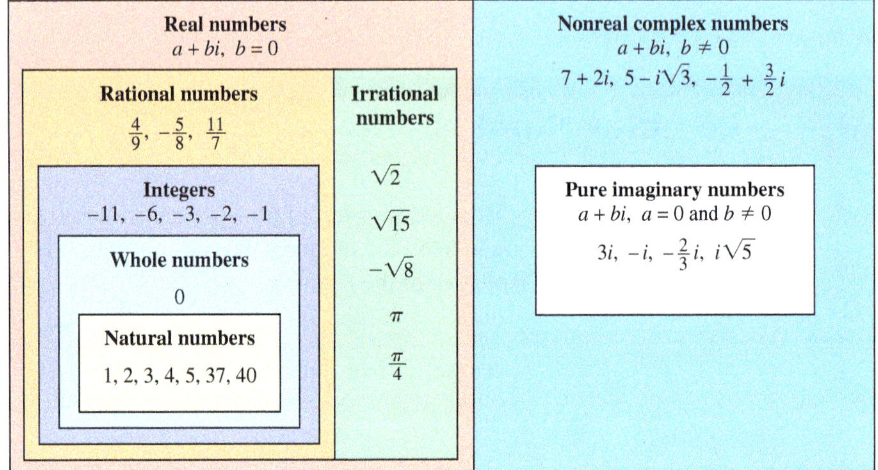

Figure 4

For a positive real number a, the expression $\sqrt{-a}$ is defined as follows.

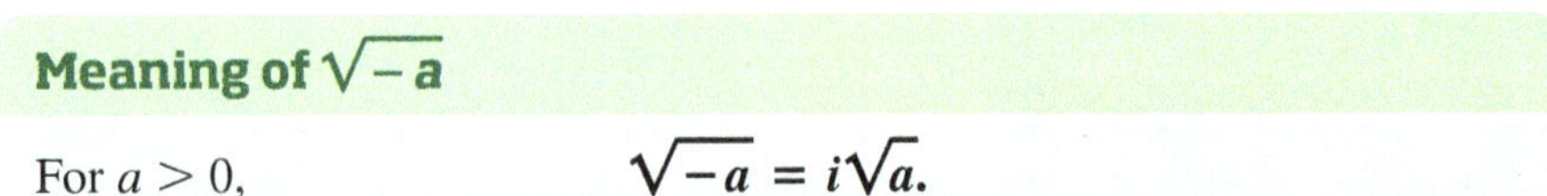

Meaning of $\sqrt{-a}$

For $a > 0$, $\quad \sqrt{-a} = i\sqrt{a}.$

EXAMPLE 1 Writing $\sqrt{-a}$ as $i\sqrt{a}$

Write each number as the product of a real number and i.

(a) $\sqrt{-16}$ **(b)** $\sqrt{-70}$ **(c)** $\sqrt{-48}$

SOLUTION

(a) $\sqrt{-16} = i\sqrt{16} = 4i$

(b) $\sqrt{-70} = i\sqrt{70}$

(c) $\sqrt{-48} = i\sqrt{48} = i\sqrt{16 \cdot 3} = 4i\sqrt{3}$ Product rule for radicals

✔ **Now Try Exercises 21, 23, and 25.**

Operations on Complex Numbers Products or quotients with negative radicands are simplified by first rewriting $\sqrt{-a}$ as $i\sqrt{a}$ for a positive number a.

CAUTION ***When working with negative radicands, use the definition $\sqrt{-a} = i\sqrt{a}$ before using any of the other rules for radicals.*** In particular, the rule $\sqrt{c} \cdot \sqrt{d} = \sqrt{cd}$ is valid only when c and d are *not* both negative. For example, consider the following.

$$\sqrt{-4} \cdot \sqrt{-9} = 2i \cdot 3i = 6i^2 = -6 \quad \text{Correct}$$

$$\sqrt{-4} \cdot \sqrt{-9} = \sqrt{(-4)(-9)} = \sqrt{36} = 6 \quad \text{Incorrect}$$

EXAMPLE 2 Finding Products and Quotients Involving $\sqrt{-a}$

Find each product or quotient. Simplify the answers.

(a) $\sqrt{-7} \cdot \sqrt{-7}$ **(b)** $\sqrt{-6} \cdot \sqrt{-10}$ **(c)** $\dfrac{\sqrt{-20}}{\sqrt{-2}}$ **(d)** $\dfrac{\sqrt{-48}}{\sqrt{24}}$

SOLUTION

(a) $\sqrt{-7} \cdot \sqrt{-7}$

$= i\sqrt{7} \cdot i\sqrt{7}$ First write all square roots in terms of i.

$= i^2 \cdot \left(\sqrt{7}\right)^2$

$= -1 \cdot 7$ $\quad i^2 = -1;\ \left(\sqrt{a}\right)^2 = a$

$= -7$ $\quad$ Multiply.

(b) $\sqrt{-6} \cdot \sqrt{-10}$

$= i\sqrt{6} \cdot i\sqrt{10}$

$= i^2 \cdot \sqrt{60}$

$= -1\sqrt{4 \cdot 15}$

$= -1 \cdot 2\sqrt{15}$

$= -2\sqrt{15}$

(c) $\dfrac{\sqrt{-20}}{\sqrt{-2}} = \dfrac{i\sqrt{20}}{i\sqrt{2}} = \sqrt{\dfrac{20}{2}} = \sqrt{10}$ $\quad$ Quotient rule for radicals

(d) $\dfrac{\sqrt{-48}}{\sqrt{24}} = \dfrac{i\sqrt{48}}{\sqrt{24}} = i\sqrt{\dfrac{48}{24}} = i\sqrt{2}$ $\quad$ Quotient rule for radicals

✔ **Now Try Exercises 29, 31, 33, and 35.**

EXAMPLE 3 Simplifying a Quotient Involving $\sqrt{-a}$

Write $\dfrac{-8 + \sqrt{-128}}{4}$ in standard form $a + bi$.

SOLUTION

$$\frac{-8 + \sqrt{-128}}{4}$$

$= \dfrac{-8 + \sqrt{-64 \cdot 2}}{4}$ $\quad$ Product rule for radicals

$= \dfrac{-8 + 8i\sqrt{2}}{4}$ $\quad$ $\sqrt{-64} = 8i$

Be sure to factor before simplifying.

$= \dfrac{4\left(-2 + 2i\sqrt{2}\right)}{4}$ $\quad$ Factor.

$= -2 + 2i\sqrt{2}$ $\quad$ Lowest terms; standard form

✔ **Now Try Exercise 41.**

With the definitions $i^2 = -1$ and $\sqrt{-a} = i\sqrt{a}$ for $a > 0$, all properties of real numbers are extended to complex numbers. As a result, complex numbers are added, subtracted, multiplied, and divided using real number properties and the following definitions.

Addition and Subtraction of Complex Numbers

For complex numbers $a + bi$ and $c + di$,

$$(a + bi) + (c + di) = (a + c) + (b + d)i$$

and

$$(a + bi) - (c + di) = (a - c) + (b - d)i.$$

That is, to add or subtract complex numbers, add or subtract the real parts and add or subtract the imaginary parts.

EXAMPLE 4 Adding and Subtracting Complex Numbers

Find each sum or difference. Write answers in standard form.

(a) $(3 - 4i) + (-2 + 6i)$ **(b)** $(-4 + 3i) - (6 - 7i)$

SOLUTION

(a) $(3 - 4i) + (-2 + 6i)$

Add real parts. Add imaginary parts.

$= \underbrace{[3 + (-2)]} + \underbrace{[-4 + 6]}i$ Commutative, associative, distributive properties

$= 1 + 2i$ Standard form

(b) $(-4 + 3i) - (6 - 7i)$

$= (-4 - 6) + [3 - (-7)]i$ Subtract real parts. Subtract imaginary parts.

$= -10 + 10i$ Standard form

✔ **Now Try Exercises 47 and 49.**

The product of two complex numbers is found by multiplying as though the numbers were binomials and using the fact that $i^2 = -1$, as follows.

$(a + bi)(c + di)$

$= ac + adi + bic + bidi$ FOIL method

$= ac + adi + bci + bdi^2$ Commutative property; Multiply.

$= ac + (ad + bc)i + bd(-1)$ Distributive property; $i^2 = -1$

$= (ac - bd) + (ad + bc)i$ Group like terms.

Multiplication of Complex Numbers

For complex numbers $a + bi$ and $c + di$,

$$(a + bi)(c + di) = (ac - bd) + (ad + bc)i.$$

To find a given product in routine calculations, it is often easier just to multiply as with binomials and use the fact that $i^2 = -1$.

EXAMPLE 5 Multiplying Complex Numbers

Find each product. Write answers in standard form.

(a) $(2 - 3i)(3 + 4i)$ **(b)** $(4 + 3i)^2$ **(c)** $(6 + 5i)(6 - 5i)$

SOLUTION

(a) $(2 - 3i)(3 + 4i)$

$= 2(3) + 2(4i) - 3i(3) - 3i(4i)$ FOIL method

$= 6 + 8i - 9i - 12i^2$ Multiply.

$= 6 - i - 12(-1)$ Combine like terms; $i^2 = -1$

$= 18 - i$ Standard form

(b) $(4 + 3i)^2$

$= 4^2 + 2(4)(3i) + (3i)^2$ Square of a binomial

Remember to add twice the product of the two terms.

$= 16 + 24i + 9i^2$ Multiply.

$= 16 + 24i + 9(-1)$ $i^2 = -1$

$= 7 + 24i$ Standard form

(c) $(6 + 5i)(6 - 5i)$

$= 6^2 - (5i)^2$ Product of the sum and difference of two terms

$= 36 - 25(-1)$ Square 6; $(5i)^2 = 5^2i^2 = 25(-1)$.

$= 36 + 25$ Multiply.

$= 61$, or $61 + 0i$ Standard form

```
NORMAL FLOAT AUTO REAL RADIAN MP
(2-3i)(3+4i)
                    18-i
(4+3i)²
                   7+24i
(6+5i)(6-5i)
                      61
```

This screen shows how the TI-84 Plus displays the results found in **Example 5.**

✔ **Now Try Exercises 55, 59, and 63.**

Example 5(c) showed that $(6 + 5i)(6 - 5i) = 61$. The numbers $6 + 5i$ and $6 - 5i$ differ only in the sign of their imaginary parts and are **complex conjugates.** ***The product of a complex number and its conjugate is always a real number.*** This product is the sum of the squares of the real and imaginary parts.

Property of Complex Conjugates

For real numbers a and b,

$$(a + bi)(a - bi) = a^2 + b^2.$$

To find the quotient of two complex numbers in standard form, we multiply both the numerator and the denominator by the complex conjugate of the denominator.

EXAMPLE 6 Dividing Complex Numbers

Find each quotient. Write answers in standard form.

(a) $\dfrac{3+2i}{5-i}$ **(b)** $\dfrac{3}{i}$

SOLUTION

(a) $\dfrac{3+2i}{5-i}$

$= \dfrac{(3+2i)(5+i)}{(5-i)(5+i)}$ Multiply by the complex conjugate of the denominator in both the numerator and the denominator.

$= \dfrac{15+3i+10i+2i^2}{25-i^2}$ Multiply.

$= \dfrac{13+13i}{26}$ Combine like terms; $i^2 = -1$

$= \dfrac{13}{26} + \dfrac{13i}{26}$ $\dfrac{a+bi}{c} = \dfrac{a}{c} + \dfrac{bi}{c}$

$= \dfrac{1}{2} + \dfrac{1}{2}i$ Write in lowest terms and standard form.

CHECK $\left(\dfrac{1}{2} + \dfrac{1}{2}i\right)(5-i) = 3+2i$ ✓ Quotient × Divisor = Dividend

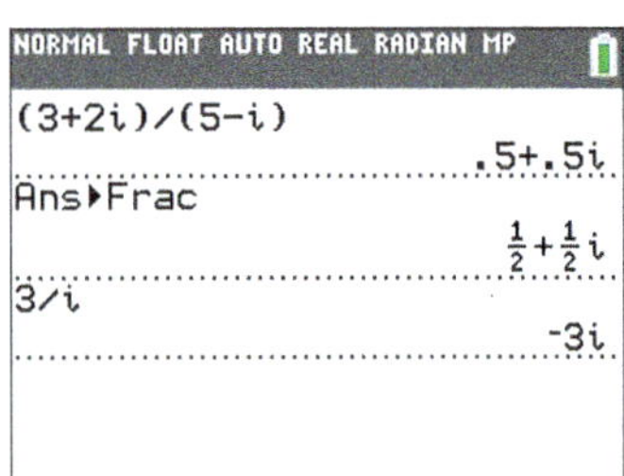

This screen supports the results in **Example 6.**

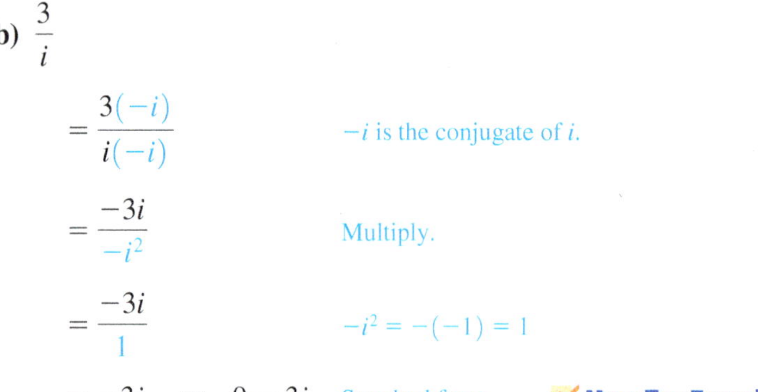

(b) $\dfrac{3}{i}$

$= \dfrac{3(-i)}{i(-i)}$ $-i$ is the conjugate of i.

$= \dfrac{-3i}{-i^2}$ Multiply.

$= \dfrac{-3i}{1}$ $-i^2 = -(-1) = 1$

$= -3i$, or $0 - 3i$ Standard form

✔ **Now Try Exercises 73 and 79.**

Powers of i can be simplified using the facts

$$i^2 = -1 \quad \text{and} \quad i^4 = (i^2)^2 = (-1)^2 = 1.$$

Consider the following powers of i.

$i^1 = i$	$i^5 = i^4 \cdot i = 1 \cdot i = i$
$i^2 = -1$	$i^6 = i^4 \cdot i^2 = 1(-1) = -1$
$i^3 = i^2 \cdot i = (-1) \cdot i = -i$	$i^7 = i^4 \cdot i^3 = 1 \cdot (-i) = -i$
$i^4 = i^2 \cdot i^2 = (-1)(-1) = 1$	$i^8 = i^4 \cdot i^4 = 1 \cdot 1 = 1$ and so on.

NORMAL FLOAT AUTO REAL RADIAN MP
i²
-1
i³
-i
i⁴
1

Powers of i can be found on the TI-84 Plus calculator.

Powers of i cycle through the same four outcomes $(i, -1, -i,$ and $1)$ because i^4 has the same multiplicative property as 1. Also, any power of i with an exponent that is a multiple of 4 has value 1. As with real numbers, $i^0 = 1$.

EXAMPLE 7 Simplifying Powers of *i*

Simplify each power of i.

(a) i^{15} **(b)** i^{-3}

SOLUTION

(a) Because $i^4 = 1$, write the given power as a product involving i^4.

$$i^{15} = i^{12} \cdot i^3 = (i^4)^3 \cdot i^3 = 1^3(-i) = -i$$

(b) Multiply i^{-3} by 1 in the form of i^4 to create the least positive exponent for i.

$$i^{-3} = i^{-3} \cdot 1 = i^{-3} \cdot i^4 = i \qquad i^4 = 1$$

✔ **Now Try Exercises 89 and 97.**

1.3 Exercises

CONCEPT PREVIEW *Fill in the blank to correctly complete each sentence.*

1. By definition, $i =$ ________, and therefore, $i^2 =$ ________.
2. If a and b are real numbers, then any number of the form $a + bi$ is a(n) ________.
3. The numbers $6 + 5i$ and $6 - 5i$, which differ only in the sign of their imaginary parts, are ________.
4. The product of a complex number and its conjugate is always a(n) ________.
5. To find the quotient of two complex numbers in standard form, multiply both the numerator and the denominator by the complex conjugate of the ________.

CONCEPT PREVIEW *Decide whether each statement is* true *or* false. *If false, correct the right side of the equation.*

6. $\sqrt{-25} = 5i$
7. $\sqrt{-4} \cdot \sqrt{-9} = -6$
8. $i^{12} = 1$
9. $(-2 + 7i) - (10 - 6i) = -12 + i$
10. $(5 + 3i)^2 = 16$

Concept Check *Identify each number as* real, complex, pure imaginary, *or* nonreal complex. *(More than one of these descriptions will apply.)*

11. -4
12. 0
13. $13i$
14. $-7i$
15. $5 + i$
16. $-6 - 2i$
17. π
18. $\sqrt{24}$
19. $\sqrt{-25}$
20. $\sqrt{-36}$

Write each number as the product of a real number and i. ***See Example 1.***

21. $\sqrt{-25}$
22. $\sqrt{-36}$
23. $\sqrt{-10}$
24. $\sqrt{-15}$
25. $\sqrt{-288}$
26. $\sqrt{-500}$
27. $-\sqrt{-18}$
28. $-\sqrt{-80}$

Find each product or quotient. Simplify the answers. ***See Example 2.***

29. $\sqrt{-13} \cdot \sqrt{-13}$
30. $\sqrt{-17} \cdot \sqrt{-17}$
31. $\sqrt{-3} \cdot \sqrt{-8}$
32. $\sqrt{-5} \cdot \sqrt{-15}$
33. $\dfrac{\sqrt{-30}}{\sqrt{-10}}$
34. $\dfrac{\sqrt{-70}}{\sqrt{-7}}$

35. $\dfrac{\sqrt{-24}}{\sqrt{8}}$ **36.** $\dfrac{\sqrt{-54}}{\sqrt{27}}$ **37.** $\dfrac{\sqrt{-10}}{\sqrt{-40}}$

38. $\dfrac{\sqrt{-8}}{\sqrt{-72}}$ **39.** $\dfrac{\sqrt{-6}\cdot\sqrt{-2}}{\sqrt{3}}$ **40.** $\dfrac{\sqrt{-12}\cdot\sqrt{-6}}{\sqrt{8}}$

Write each number in standard form $a + bi$. ***See Example 3.***

41. $\dfrac{-6-\sqrt{-24}}{2}$ **42.** $\dfrac{-9-\sqrt{-18}}{3}$ **43.** $\dfrac{10+\sqrt{-200}}{5}$

44. $\dfrac{20+\sqrt{-8}}{2}$ **45.** $\dfrac{-3+\sqrt{-18}}{24}$ **46.** $\dfrac{-5+\sqrt{-50}}{10}$

Find each sum or difference. Write answers in standard form. ***See Example 4.***

47. $(3+2i)+(9-3i)$ **48.** $(4-i)+(8+5i)$

49. $(-2+4i)-(-4+4i)$ **50.** $(-3+2i)-(-4+2i)$

51. $(2-5i)-(3+4i)-(-2+i)$ **52.** $(-4-i)-(2+3i)+(-4+5i)$

53. $-i\sqrt{2}-2-\left(6-4i\sqrt{2}\right)-\left(5-i\sqrt{2}\right)$

54. $3\sqrt{7}-\left(4\sqrt{7}-i\right)-4i+\left(-2\sqrt{7}+5i\right)$

Find each product. Write answers in standard form. ***See Example 5.***

55. $(2+i)(3-2i)$ **56.** $(-2+3i)(4-2i)$ **57.** $(2+4i)(-1+3i)$

58. $(1+3i)(2-5i)$ **59.** $(3-2i)^2$ **60.** $(2+i)^2$

61. $(3+i)(3-i)$ **62.** $(5+i)(5-i)$ **63.** $(-2-3i)(-2+3i)$

64. $(6-4i)(6+4i)$ **65.** $\left(\sqrt{6}+i\right)\left(\sqrt{6}-i\right)$ **66.** $\left(\sqrt{2}-4i\right)\left(\sqrt{2}+4i\right)$

67. $i(3-4i)(3+4i)$ **68.** $i(2+7i)(2-7i)$ **69.** $3i(2-i)^2$

70. $-5i(4-3i)^2$ **71.** $(2+i)(2-i)(4+3i)$ **72.** $(3-i)(3+i)(2-6i)$

Find each quotient. Write answers in standard form. ***See Example 6.***

73. $\dfrac{6+2i}{1+2i}$ **74.** $\dfrac{14+5i}{3+2i}$ **75.** $\dfrac{2-i}{2+i}$ **76.** $\dfrac{4-3i}{4+3i}$

77. $\dfrac{1-3i}{1+i}$ **78.** $\dfrac{-3+4i}{2-i}$ **79.** $\dfrac{-5}{i}$ **80.** $\dfrac{-6}{i}$

81. $\dfrac{8}{-i}$ **82.** $\dfrac{12}{-i}$ **83.** $\dfrac{2}{3i}$ **84.** $\dfrac{5}{9i}$

(Modeling) Alternating Current *Complex numbers are used to describe current I, voltage E, and impedance Z (the opposition to current). These three quantities are related by the equation*

$E = IZ$, *which is known as* ***Ohm's Law.***

Thus, if any two of these quantities are known, the third can be found. In each exercise, solve the equation $E = IZ$ for the remaining value.

85. $I=5+7i,\ Z=6+4i$ **86.** $I=20+12i,\ Z=10-5i$

87. $I=10+4i,\ E=88+128i$ **88.** $E=57+67i,\ Z=9+5i$

Simplify each power of i. ***See Example 7.***

89. i^{25} **90.** i^{29} **91.** i^{22} **92.** i^{26}

93. i^{23} **94.** i^{27} **95.** i^{32} **96.** i^{40}

97. i^{-13} **98.** i^{-14} **99.** $\frac{1}{i^{-11}}$ **100.** $\frac{1}{i^{-12}}$

Work each problem.

101. Show that $\frac{\sqrt{2}}{2} + \frac{\sqrt{2}}{2}i$ is a square root of i.

102. Show that $-\frac{\sqrt{2}}{2} - \frac{\sqrt{2}}{2}i$ is a square root of i.

103. Show that $\frac{\sqrt{3}}{2} + \frac{1}{2}i$ is a cube root of i.

104. Show that $-\frac{\sqrt{3}}{2} + \frac{1}{2}i$ is a cube root of i.

105. Show that $-2 + i$ is a solution of the equation $x^2 + 4x + 5 = 0$.

106. Show that $-2 - i$ is a solution of the equation $x^2 + 4x + 5 = 0$.

107. Show that $-3 + 4i$ is a solution of the equation $x^2 + 6x + 25 = 0$.

108. Show that $-3 - 4i$ is a solution of the equation $x^2 + 6x + 25 = 0$.

1.4 Quadratic Equations

- The Zero-Factor Property
- The Square Root Property
- Completing the Square
- The Quadratic Formula
- Solving for a Specified Variable
- The Discriminant

A *quadratic equation* is defined as follows.

Quadratic Equation in One Variable

An equation that can be written in the form

$$ax^2 + bx + c = 0,$$

where a, b, and c are real numbers with $a \neq 0$, is a **quadratic equation.** The given form is called **standard form.**

A quadratic equation is a **second-degree equation**—that is, an equation with a squared variable term and no terms of greater degree.

$$x^2 = 25, \quad 4x^2 + 4x - 5 = 0, \quad 3x^2 = 4x - 8 \quad \text{Quadratic equations}$$

The Zero-Factor Property

When the expression $ax^2 + bx + c$ in a quadratic equation is easily factorable over the real numbers, it is efficient to factor and then apply the following **zero-factor property.**

Zero-Factor Property

If a and b are complex numbers with $ab = 0$, then $a = 0$ or $b = 0$ or both equal zero.

EXAMPLE 1 Using the Zero-Factor Property

Solve $6x^2 + 7x = 3$.

SOLUTION

$$
\begin{aligned}
6x^2 + 7x &= 3 && \text{Don't factor out } x \text{ here.}\\
6x^2 + 7x - 3 &= 0 && \text{Standard form}\\
(3x - 1)(2x + 3) &= 0 && \text{Factor.}\\
3x - 1 = 0 \quad \text{or} \quad 2x + 3 &= 0 && \text{Zero-factor property}\\
3x = 1 \quad \text{or} \quad 2x &= -3 && \text{Solve each equation.}\\
x = \frac{1}{3} \quad \text{or} \quad x &= -\frac{3}{2}
\end{aligned}
$$

CHECK $6x^2 + 7x = 3$ Original equation

$$6\left(\frac{1}{3}\right)^2 + 7\left(\frac{1}{3}\right) \stackrel{?}{=} 3 \quad \text{Let } x = \tfrac{1}{3}.$$

$$\frac{6}{9} + \frac{7}{3} \stackrel{?}{=} 3$$

$$3 = 3 \checkmark \quad \text{True}$$

$$6\left(-\frac{3}{2}\right)^2 + 7\left(-\frac{3}{2}\right) \stackrel{?}{=} 3 \quad \text{Let } x = -\tfrac{3}{2}.$$

$$\frac{54}{4} - \frac{21}{2} \stackrel{?}{=} 3$$

$$3 = 3 \checkmark \quad \text{True}$$

Both values check because true statements result. The solution set is $\left\{\frac{1}{3}, -\frac{3}{2}\right\}$.

✔ **Now Try Exercise 15.**

The Square Root Property When a quadratic equation can be written in the form $x^2 = k$, where k is a constant, the equation can be solved as follows.

$$
\begin{aligned}
x^2 &= k\\
x^2 - k &= 0 && \text{Subtract } k.\\
\left(x - \sqrt{k}\right)\left(x + \sqrt{k}\right) &= 0 && \text{Factor.}\\
x - \sqrt{k} = 0 \quad \text{or} \quad x + \sqrt{k} &= 0 && \text{Zero-factor property}\\
x = \sqrt{k} \quad \text{or} \quad x &= -\sqrt{k} && \text{Solve each equation.}
\end{aligned}
$$

This proves the **square root property.**

Square Root Property

If $x^2 = k$, then $\boldsymbol{x = \sqrt{k}}$ **or** $\boldsymbol{x = -\sqrt{k}}$.

That is, the solution set of $x^2 = k$ is

$$\left\{\sqrt{k}, -\sqrt{k}\right\}, \quad \textit{which may be abbreviated} \quad \left\{\pm\sqrt{k}\right\}.$$

Both solutions $\sqrt{k}$ and $-\sqrt{k}$ of $x^2 = k$ are real if $k > 0$. Both are pure imaginary if $k < 0$. If $k < 0$, then we write the solution set as

$$\left\{\pm i\sqrt{|k|}\right\}.$$

If $k = 0$, then there is only one distinct solution, 0, sometimes called a **double solution.**

EXAMPLE 2 **Using the Square Root Property**

Solve each quadratic equation.

(a) $x^2 = 17$ **(b)** $x^2 = -25$ **(c)** $(x-4)^2 = 12$

SOLUTION

(a) $x^2 = 17$

$x = \pm\sqrt{17}$ Square root property

The solution set is $\{\pm\sqrt{17}\}$.

(b) $x^2 = -25$

$x = \pm\sqrt{-25}$ Square root property

$x = \pm 5i$ $\sqrt{-1} = i$

The solution set is $\{\pm 5i\}$.

(c)

$$(x-4)^2 = 12$$

$x - 4 = \pm\sqrt{12}$ Generalized square root property

$x = 4 \pm \sqrt{12}$ Add 4.

$x = 4 \pm 2\sqrt{3}$ $\sqrt{12} = \sqrt{4 \cdot 3} = 2\sqrt{3}$

CHECK $(x-4)^2 = 12$ Original equation

$\left(4 + 2\sqrt{3} - 4\right)^2 \stackrel{?}{=} 12$ Let $x = 4 + 2\sqrt{3}$.

$\left(2\sqrt{3}\right)^2 \stackrel{?}{=} 12$

$2^2 \cdot \left(\sqrt{3}\right)^2 \stackrel{?}{=} 12$

$12 = 12$ ✓ True

$\left(4 - 2\sqrt{3} - 4\right)^2 \stackrel{?}{=} 12$ Let $x = 4 - 2\sqrt{3}$.

$\left(-2\sqrt{3}\right)^2 \stackrel{?}{=} 12$

$(-2)^2 \cdot \left(\sqrt{3}\right)^2 \stackrel{?}{=} 12$

$12 = 12$ ✓ True

The solution set is $\{4 \pm 2\sqrt{3}\}$.

✓ **Now Try Exercises 27, 29, and 31.**

Completing the Square Any quadratic equation can be solved by the method of **completing the square,** summarized in the box below. While this method may seem tedious, it has several useful applications, including analyzing the graph of a parabola and developing a general formula for solving quadratic equations.

Solving a Quadratic Equation Using Completing the Square

To solve $ax^2 + bx + c = 0$, where $a \neq 0$, using completing the square, follow these steps.

Step 1 If $a \neq 1$, divide each side of the equation by a.

Step 2 Rewrite the equation so that the constant term is alone on one side of the equality symbol.

Step 3 Square half the coefficient of x, and add this square to each side of the equation.

Step 4 Factor the resulting trinomial as a perfect square and combine like terms on the other side.

Step 5 Use the square root property to complete the solution.

EXAMPLE 3 **Using Completing the Square ($a = 1$)**

Solve $x^2 - 4x - 14 = 0$.

SOLUTION $x^2 - 4x - 14 = 0$

Step 1 This step is not necessary because $a = 1$.

Step 2 $x^2 - 4x = 14$ Add 14 to each side.

Step 3 $x^2 - 4x + 4 = 14 + 4$ $\left[\frac{1}{2}(-4)\right]^2 = 4$; Add 4 to each side.

Step 4 $(x - 2)^2 = 18$ Factor. Combine like terms.

Step 5 $x - 2 = \pm\sqrt{18}$ Square root property

Take *both* roots.

$x = 2 \pm \sqrt{18}$ Add 2 to each side.

$x = 2 \pm 3\sqrt{2}$ Simplify the radical.

The solution set is $\left\{2 \pm 3\sqrt{2}\right\}$.

✔ **Now Try Exercise 41.**

EXAMPLE 4 **Using Completing the Square ($a \neq 1$)**

Solve $9x^2 - 12x + 9 = 0$.

SOLUTION $9x^2 - 12x + 9 = 0$

Step 1 $x^2 - \frac{4}{3}x + 1 = 0$ Divide by 9 so that $a = 1$.

Step 2 $x^2 - \frac{4}{3}x = -1$ Subtract 1 from each side.

Step 3 $x^2 - \frac{4}{3}x + \frac{4}{9} = -1 + \frac{4}{9}$ $\left[\frac{1}{2}\left(-\frac{4}{3}\right)\right]^2 = \frac{4}{9}$; Add $\frac{4}{9}$ to each side.

Step 4 $\left(x - \frac{2}{3}\right)^2 = -\frac{5}{9}$ Factor. Combine like terms.

Step 5 $x - \frac{2}{3} = \pm\sqrt{-\frac{5}{9}}$ Square root property

$x - \frac{2}{3} = \pm\frac{\sqrt{5}}{3}i$ $\sqrt{-\frac{5}{9}} = \frac{\sqrt{-5}}{\sqrt{9}} = \frac{i\sqrt{5}}{3}$, or $\frac{\sqrt{5}}{3}i$

$x = \frac{2}{3} \pm \frac{\sqrt{5}}{3}i$ Add $\frac{2}{3}$ to each side.

The solution set is $\left\{\frac{2}{3} \pm \frac{\sqrt{5}}{3}i\right\}$.

✔ **Now Try Exercise 47.**

The Quadratic Formula If we start with the equation $ax^2 + bx + c = 0$, for $a > 0$, and complete the square to solve for x in terms of the constants a, b, and c, the result is a general formula for solving any quadratic equation.

$$ax^2 + bx + c = 0$$

$$x^2 + \frac{b}{a}x + \frac{c}{a} = 0 \quad \text{Divide each side by } a. \text{ (Step 1)}$$

$$x^2 + \frac{b}{a}x = -\frac{c}{a} \quad \text{Subtract } \tfrac{c}{a} \text{ from each side. (Step 2)}$$

Square half the coefficient of x: $\left[\frac{1}{2}\left(\frac{b}{a}\right)\right]^2 = \left(\frac{b}{2a}\right)^2 = \frac{b^2}{4a^2}$.

$$x^2 + \frac{b}{a}x + \frac{b^2}{4a^2} = -\frac{c}{a} + \frac{b^2}{4a^2}$$ Add $\frac{b^2}{4a^2}$ to each side. (Step 3)

$$\left(x + \frac{b}{2a}\right)^2 = \frac{b^2}{4a^2} + \frac{-c}{a}$$ Factor. Use the commutative property. (Step 4)

$$\left(x + \frac{b}{2a}\right)^2 = \frac{b^2}{4a^2} + \frac{-4ac}{4a^2}$$ Write fractions with a common denominator.

$$\left(x + \frac{b}{2a}\right)^2 = \frac{b^2 - 4ac}{4a^2}$$ Add fractions.

$$x + \frac{b}{2a} = \pm\sqrt{\frac{b^2 - 4ac}{4a^2}}$$ Square root property (Step 5)

$$x + \frac{b}{2a} = \frac{\pm\sqrt{b^2 - 4ac}}{2a}$$ Since $a > 0$, $\sqrt{4a^2} = 2a$.

$$x = \frac{-b}{2a} \pm \frac{\sqrt{b^2 - 4ac}}{2a}$$ Subtract $\frac{b}{2a}$ from each side.

Quadratic Formula This result is also true for $a < 0$. ⟶

$$x = \frac{-b \pm \sqrt{b^2 - 4ac}}{2a}$$ Combine terms on the right.

Quadratic Formula

The solutions of the quadratic equation $ax^2 + bx + c = 0$, where $a \neq 0$, are given by the quadratic formula.

$$x = \frac{-b \pm \sqrt{b^2 - 4ac}}{2a}$$

EXAMPLE 5 Using the Quadratic Formula (Real Solutions)

Solve $x^2 - 4x = -2$.

SOLUTION

$$x^2 - 4x + 2 = 0$$ Write in standard form. Here $a = 1$, $b = -4$, and $c = 2$.

$$x = \frac{-b \pm \sqrt{b^2 - 4ac}}{2a}$$ Quadratic formula

$$x = \frac{-(-4) \pm \sqrt{(-4)^2 - 4(1)(2)}}{2(1)}$$ Substitute $a = 1$, $b = -4$, and $c = 2$.

The fraction bar extends *under* $-b$.

$$x = \frac{4 \pm \sqrt{16 - 8}}{2}$$ Simplify.

$$x = \frac{4 \pm 2\sqrt{2}}{2}$$ $\sqrt{16 - 8} = \sqrt{8} = \sqrt{4 \cdot 2} = 2\sqrt{2}$

$$x = \frac{2(2 \pm \sqrt{2})}{2}$$ Factor out 2 in the numerator.

Factor first, then divide.

$$x = 2 \pm \sqrt{2}$$ Lowest terms

The solution set is $\{2 \pm \sqrt{2}\}$.

✔ **Now Try Exercise 53.**

CAUTION ***Remember to extend the fraction bar in the quadratic formula under the $-b$ term in the numerator.***

Throughout this text, unless otherwise specified, we use the set of complex numbers as the domain when solving equations of degree 2 or greater.

EXAMPLE 6 Using the Quadratic Formula (Nonreal Complex Solutions)

Solve $2x^2 = x - 4$.

SOLUTION

$$2x^2 - x + 4 = 0 \quad \text{Write in standard form.}$$

$$x = \frac{-(-1) \pm \sqrt{(-1)^2 - 4(2)(4)}}{2(2)} \quad \text{Quadratic formula with } a = 2,\ b = -1,\ c = 4$$

Use parentheses and substitute carefully to avoid errors.

$$x = \frac{1 \pm \sqrt{1 - 32}}{4}$$

$$x = \frac{1 \pm \sqrt{-31}}{4} \quad \text{Simplify.}$$

$$x = \frac{1 \pm i\sqrt{31}}{4} \quad \sqrt{-1} = i$$

The solution set is $\left\{\frac{1}{4} \pm \frac{\sqrt{31}}{4}i\right\}$.

✔ **Now Try Exercise 57.**

The equation $x^3 + 8 = 0$ is a **cubic equation** because the greatest degree of the terms is 3. While a quadratic equation (degree 2) can have as many as two solutions, a cubic equation (degree 3) can have as many as three solutions. The maximum possible number of solutions corresponds to the degree of the equation.

EXAMPLE 7 Solving a Cubic Equation

Solve $x^3 + 8 = 0$ using factoring and the quadratic formula.

SOLUTION

$$x^3 + 8 = 0$$

$$(x + 2)(x^2 - 2x + 4) = 0 \quad \text{Factor as a sum of cubes.}$$

$$x + 2 = 0 \quad \text{or} \quad x^2 - 2x + 4 = 0 \quad \text{Zero-factor property}$$

$$x = -2 \quad \text{or} \quad x = \frac{-(-2) \pm \sqrt{(-2)^2 - 4(1)(4)}}{2(1)} \quad \text{Quadratic formula with } a = 1,\ b = -2,\ c = 4$$

$$x = \frac{2 \pm \sqrt{-12}}{2} \quad \text{Simplify.}$$

$$x = \frac{2 \pm 2i\sqrt{3}}{2} \quad \text{Simplify the radical.}$$

$$x = \frac{2(1 \pm i\sqrt{3})}{2} \quad \text{Factor out 2 in the numerator.}$$

$$x = 1 \pm i\sqrt{3} \quad \text{Divide out the common factor.}$$

The solution set is $\{-2, 1 \pm i\sqrt{3}\}$.

✔ **Now Try Exercise 67.**

Solving for a Specified Variable To solve a quadratic equation for a specified variable, we usually apply the square root property or the quadratic formula.

EXAMPLE 8 Solving for a Quadratic Variable in a Formula

Solve each equation for the specified variable. Use $\pm$ when taking square roots.

(a) $\mathscr{A} = \frac{\pi d^2}{4}$, for d **(b)** $rt^2 - st = k \ (r \neq 0)$, for t

SOLUTION

(a)

$$\mathscr{A} = \frac{\pi d^2}{4}$$ Goal: Isolate d, the specified variable.

$$4\mathscr{A} = \pi d^2$$ Multiply each side by 4.

$$\frac{4\mathscr{A}}{\pi} = d^2$$ Divide each side by π.

$$d = \pm\sqrt{\frac{4\mathscr{A}}{\pi}}$$ Interchange sides; square root property

See the Note following this example.

$$d = \frac{\pm\sqrt{4\mathscr{A}}}{\sqrt{\pi}} \cdot \frac{\sqrt{\pi}}{\sqrt{\pi}}$$ Multiply by $\frac{\sqrt{\pi}}{\sqrt{\pi}}$.

$$d = \frac{\pm\sqrt{4\mathscr{A}\pi}}{\pi}$$ Multiply numerators. Multiply denominators.

$$d = \frac{\pm 2\sqrt{\mathscr{A}\pi}}{\pi}$$ Simplify the radical.

(b) Because $rt^2 - st = k$ has terms with t^2 and t, use the quadratic formula.

$$rt^2 - st - k = 0$$ Write in standard form.

$$t = \frac{-b \pm \sqrt{b^2 - 4ac}}{2a}$$ Quadratic formula

$$t = \frac{-(-s) \pm \sqrt{(-s)^2 - 4(r)(-k)}}{2(r)}$$ Here, $a = r$, $b = -s$, and $c = -k$.

$$t = \frac{s \pm \sqrt{s^2 + 4rk}}{2r}$$ Simplify.

✔ **Now Try Exercises 71 and 77.**

NOTE In **Example 8,** we took both positive and negative square roots. However, if the variable represents time or length in an application, we consider only the *positive* square root.

The Discriminant The quantity under the radical in the quadratic formula, $b^2 - 4ac$, is the **discriminant.**

$$x = \frac{-b \pm \sqrt{b^2 - 4ac}}{2a}$$ ← Discriminant

When the numbers a, b, and c are *integers* (but not necessarily otherwise), the value of the discriminant $b^2 - 4ac$ can be used to determine whether the solutions of a quadratic equation are rational, irrational, or nonreal complex numbers. The number and type of solutions based on the value of the discriminant are shown in the following table.

Solutions of Quadratic Equations

Discriminant	Number of Solutions	Type of Solutions	
Positive, perfect square	Two	Rational	
Positive, but not a perfect square	Two	Irrational	← As seen in Example 5
Zero	One (a double solution)	Rational	
Negative	Two	Nonreal complex	← As seen in Example 6

CAUTION ***The restriction on a, b, and c is important.*** For example,

$$x^2 - \sqrt{5}x - 1 = 0 \quad \text{has discriminant} \quad b^2 - 4ac = 5 + 4 = 9,$$

which would indicate two rational solutions *if the coefficients were integers.* By the quadratic formula, the two solutions $\frac{\sqrt{5} \pm 3}{2}$ are *irrational* numbers.

EXAMPLE 9 Using the Discriminant

Evaluate the discriminant for each equation. Then use it to determine the number of distinct solutions, and tell whether they are *rational, irrational*, or *nonreal complex* numbers.

(a) $5x^2 + 2x - 4 = 0$ **(b)** $x^2 - 10x = -25$ **(c)** $2x^2 - x + 1 = 0$

SOLUTION

(a) For $5x^2 + 2x - 4 = 0$, use $a = 5$, $b = 2$, and $c = -4$.

$$b^2 - 4ac = 2^2 - 4(5)(-4) = 84 \leftarrow \text{Discriminant}$$

The discriminant 84 is positive and not a perfect square, so there are two distinct irrational solutions.

(b) First, write the equation in standard form as

$$x^2 - 10x + 25 = 0.$$

Thus, $a = 1$, $b = -10$, and $c = 25$.

$$b^2 - 4ac = (-10)^2 - 4(1)(25) = 0 \leftarrow \text{Discriminant}$$

There is one distinct rational solution, a double solution.

(c) For $2x^2 - x + 1 = 0$, use $a = 2$, $b = -1$, and $c = 1$.

$$b^2 - 4ac = (-1)^2 - 4(2)(1) = -7 \leftarrow \text{Discriminant}$$

There are two distinct nonreal complex solutions. (They are complex conjugates.)

✔ **Now Try Exercises 83, 85, and 89.**

1.4 Exercises

CONCEPT PREVIEW *Match the equation in Column I with its solution(s) in Column II.*

I		II	
1. $x^2 = 25$	**2.** $x^2 = -25$	**A.** $\pm 5i$	**B.** $\pm 2\sqrt{5}$
3. $x^2 + 5 = 0$	**4.** $x^2 - 5 = 0$	**C.** $\pm i\sqrt{5}$	**D.** 5
5. $x^2 = -20$	**6.** $x^2 = 20$	**E.** $\pm\sqrt{5}$	**F.** -5
7. $x - 5 = 0$	**8.** $x + 5 = 0$	**G.** ± 5	**H.** $\pm 2i\sqrt{5}$

CONCEPT PREVIEW *Use Choices A–D to answer each question.*

A. $3x^2 - 17x - 6 = 0$ **B.** $(2x + 5)^2 = 7$

C. $x^2 + x = 12$ **D.** $(3x - 1)(x - 7) = 0$

9. Which equation is set up for direct use of the zero-factor property? Solve it.

10. Which equation is set up for direct use of the square root property? Solve it.

11. Only one of the equations does not require Step 1 of the method for completing the square described in this section. Which one is it? Solve it.

12. Only one of the equations is set up so that the values of a, b, and c can be determined immediately. Which one is it? Solve it.

Solve each equation using the zero-factor property. ***See Example 1.***

13. $x^2 - 5x + 6 = 0$ **14.** $x^2 + 2x - 8 = 0$ **15.** $5x^2 - 3x - 2 = 0$

16. $2x^2 - x - 15 = 0$ **17.** $-4x^2 + x = -3$ **18.** $-6x^2 + 7x = -10$

19. $x^2 - 100 = 0$ **20.** $x^2 - 64 = 0$ **21.** $4x^2 - 4x + 1 = 0$

22. $9x^2 - 12x + 4 = 0$ **23.** $25x^2 + 30x + 9 = 0$ **24.** $36x^2 + 60x + 25 = 0$

Solve each equation using the square root property. ***See Example 2.***

25. $x^2 = 16$ **26.** $x^2 = 121$ **27.** $27 - x^2 = 0$

28. $48 - x^2 = 0$ **29.** $x^2 = -81$ **30.** $x^2 = -400$

31. $(3x - 1)^2 = 12$ **32.** $(4x + 1)^2 = 20$ **33.** $(x + 5)^2 = -3$

34. $(x - 4)^2 = -5$ **35.** $(5x - 3)^2 = -3$ **36.** $(-2x + 5)^2 = -8$

Solve each equation using completing the square. ***See Examples 3 and 4.***

37. $x^2 - 4x + 3 = 0$ **38.** $x^2 - 7x + 12 = 0$ **39.** $2x^2 - x - 28 = 0$

40. $4x^2 - 3x - 10 = 0$ **41.** $x^2 - 2x - 2 = 0$ **42.** $x^2 - 10x + 18 = 0$

43. $2x^2 + x = 10$ **44.** $3x^2 + 2x = 5$ **45.** $-2x^2 + 4x + 3 = 0$

46. $-3x^2 + 6x + 5 = 0$ **47.** $-4x^2 + 8x = 7$ **48.** $-3x^2 + 9x = 7$

Concept Check Answer each question.

49. Francisco claimed that the equation

$$x^2 - 8x = 0$$

cannot be solved by the quadratic formula since there is no value for c. Is he correct?

50. Francesca, Francisco's twin sister, claimed that the equation

$$x^2 - 19 = 0$$

cannot be solved by the quadratic formula since there is no value for b. Is she correct?

Solve each equation using the quadratic formula. **See Examples 5 and 6.**

51. $x^2 - x - 1 = 0$ **52.** $x^2 - 3x - 2 = 0$ **53.** $x^2 - 6x = -7$

54. $x^2 - 4x = -1$ **55.** $x^2 = 2x - 5$ **56.** $x^2 = 2x - 10$

57. $-4x^2 = -12x + 11$ **58.** $-6x^2 = 3x + 2$ **59.** $\frac{1}{2}x^2 + \frac{1}{4}x - 3 = 0$

60. $\frac{2}{3}x^2 + \frac{1}{4}x = 3$ **61.** $0.2x^2 + 0.4x - 0.3 = 0$

62. $0.1x^2 - 0.1x = 0.3$ **63.** $(4x - 1)(x + 2) = 4x$

64. $(3x + 2)(x - 1) = 3x$ **65.** $(x - 9)(x - 1) = -16$

66. ***Concept Check*** Why do the following two equations have the same solution set? (Do not solve.)

$$-2x^2 + 3x - 6 = 0 \quad \text{and} \quad 2x^2 - 3x + 6 = 0$$

Solve each cubic equation using factoring and the quadratic formula. **See Example 7.**

67. $x^3 - 8 = 0$ **68.** $x^3 - 27 = 0$

69. $x^3 + 27 = 0$ **70.** $x^3 + 64 = 0$

Solve each equation for the specified variable. (Assume no denominators are 0.) **See Example 8.**

71. $s = \frac{1}{2}gt^2$, for t **72.** $\mathscr{A} = \pi r^2$, for r

73. $F = \frac{kMv^2}{r}$, for v **74.** $E = \frac{e^2k}{2r}$, for e

75. $r = r_0 + \frac{1}{2}at^2$, for t **76.** $s = s_0 + gt^2 + k$, for t

77. $h = -16t^2 + v_0 t + s_0$, for t **78.** $S = 2\pi rh + 2\pi r^2$, for r

For each equation, **(a)** *solve for x in terms of y, and* **(b)** *solve for y in terms of x.* **See Example 8.**

79. $4x^2 - 2xy + 3y^2 = 2$ **80.** $3y^2 + 4xy - 9x^2 = -1$

81. $2x^2 + 4xy - 3y^2 = 2$ **82.** $5x^2 - 6xy + 2y^2 = 1$

Evaluate the discriminant for each equation. Then use it to determine the number of distinct solutions, and tell whether they are rational, irrational, *or* nonreal complex numbers. *(Do not solve the equation.)* **See Example 9.**

83. $x^2 - 8x + 16 = 0$ **84.** $x^2 + 4x + 4 = 0$ **85.** $3x^2 + 5x + 2 = 0$

86. $8x^2 = -14x - 3$ **87.** $4x^2 = -6x + 3$ **88.** $2x^2 + 4x + 1 = 0$

89. $9x^2 + 11x + 4 = 0$ **90.** $3x^2 = 4x - 5$ **91.** $8x^2 - 72 = 0$

Concept Check *Answer each question.*

92. Show that the discriminant for the equation

$$\sqrt{2}x^2 + 5x - 3\sqrt{2} = 0$$

is 49. If this equation is completely solved, it can be shown that the solution set is $\left\{-3\sqrt{2}, \frac{\sqrt{2}}{2}\right\}$. We have a discriminant that is positive and a perfect square, yet the two solutions are irrational. Does this contradict the discussion in this section?

93. Is it possible for the solution set of a quadratic equation with integer coefficients to consist of a single irrational number?

94. Is it possible for the solution set of a quadratic equation with real coefficients to consist of one real number and one nonreal complex number?

Find the values of a, b, and c for which the quadratic equation

$$ax^2 + bx + c = 0$$

has the given numbers as solutions. (Hint: Use the zero-factor property in reverse.)

95. 4, 5 **96.** $-3, 2$ **97.** $1 + \sqrt{2}, 1 - \sqrt{2}$ **98.** $i, -i$

Chapter 1 Quiz (Sections 1.1–1.4)

1. Solve the linear equation $3(x - 5) + 2 = 1 - (4 + 2x)$.

2. Determine whether each equation is an *identity*, a *conditional equation*, or a *contradiction*. Give the solution set.

(a) $4x - 5 = -2(3 - 2x) + 3$

(b) $5x - 9 = 5(-2 + x) + 1$

(c) $5x - 4 = 3(6 - x)$

3. Solve the equation $ay + 2x = y + 5x$ for y. (Assume $a \neq 1$.)

4. ***Earning Interest*** Johnny deposits some money at 2.5% annual interest and twice as much at 3.0%. Find the amount deposited at each rate if his total annual interest income is \$850.

5. ***(Modeling) Minimum Hourly Wage*** One model for the minimum hourly wage in the United States for the period 1979–2014 is

$$y = 0.128x - 250.43,$$

where x represents the year and y represents the wage, in dollars. (*Source:* Bureau of Labor Statistics.) The actual 2008 minimum wage was \$6.55. What does this model predict as the wage? What is the difference between the actual wage and the predicted wage?

6. Write $\frac{-4 + \sqrt{-24}}{8}$ in standard form $a + bi$.

7. Write the quotient $\frac{7 - 2i}{2 + 4i}$ in standard form $a + bi$.

Solve each equation.

8. $3x^2 - x = -1$ **9.** $x^2 - 29 = 0$ **10.** $\mathcal{A} = \frac{1}{2}r^2\theta$, for r

1.5 Applications and Modeling with Quadratic Equations

- Geometry Problems
- The Pythagorean Theorem
- Height of a Projected Object
- Modeling with Quadratic Equations

Geometry Problems

To solve these applications, we continue to use a six-step problem-solving strategy.

EXAMPLE 1 Solving a Problem Involving Volume

A piece of machinery produces rectangular sheets of metal such that the length is three times the width. Equal-sized squares measuring 5 in. on a side can be cut from the corners so that the resulting piece of metal can be shaped into an open box by folding up the flaps. If specifications call for the volume of the box to be 1435 in.^{3}, find the dimensions of the original piece of metal.

SOLUTION

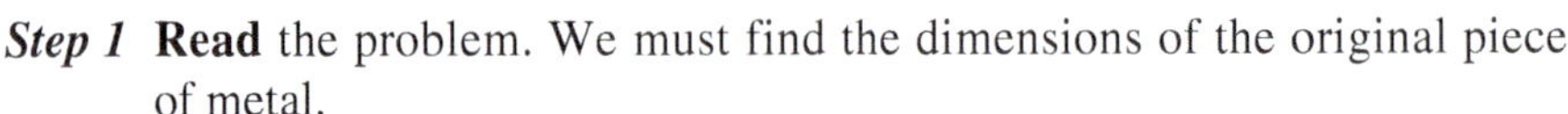

Step 1 **Read** the problem. We must find the dimensions of the original piece of metal.

Step 2 **Assign a variable.** We know that the length is three times the width.

Let x = the width (in inches) and thus, $3x$ = the length.

The box is formed by cutting $5 + 5 = 10$ in. from both the length and the width. See **Figure 5.** The width of the bottom of the box is $x - 10$, the length of the bottom of the box is $3x - 10$, and the height is 5 in. (the length of the side of each cut-out square). See **Figure 6.**

Figure 5

Step 3 **Write an equation.** The formula for volume of a box is $V = lwh$.

$$\underset{1435}{\text{Volume}} = \underset{(3x-10)}{\text{length}} \times \underset{(x-10)}{\text{width}} \times \underset{(5)}{\text{height}}$$

$$1435 = (3x - 10)(x - 10)(5)$$

Figure 6

(Note that the dimensions of the box must be positive numbers, so $3x - 10$ and $x - 10$ must be greater than 0, which implies $x > \frac{10}{3}$ and $x > 10$. These are both satisfied when $x > 10$.)

Step 4 **Solve** the equation from Step 3.

$$1435 = 15x^2 - 200x + 500 \quad \text{Multiply.}$$
$$0 = 15x^2 - 200x - 935 \quad \text{Subtract 1435 from each side.}$$
$$0 = 3x^2 - 40x - 187 \quad \text{Divide each side by 5.}$$
$$0 = (3x + 11)(x - 17) \quad \text{Factor.}$$
$$3x + 11 = 0 \quad \text{or} \quad x - 17 = 0 \quad \text{Zero-factor property}$$
$$x = -\frac{11}{3} \quad \text{or} \quad x = 17 \quad \text{Solve each equation.}$$

The width cannot be negative.

Step 5 **State the answer.** Only 17 satisfies the restriction $x > 10$. Thus, the dimensions of the original piece should be 17 in. by $3(17) = 51$ in.

Step 6 **Check.** The length and width of the bottom of the box are

$$51 - 2(5) = 41 \text{ in.} \quad \text{Length}$$

and

$$17 - 2(5) = 7 \text{ in.} \quad \text{Width}$$

The height is 5 in. (the amount cut on each corner), so the volume is

$$V = lwh = 41 \times 7 \times 5 = 1435 \text{ in.}^3, \quad \text{as required.}$$

✔ Now Try Exercise 27.

PROBLEM-SOLVING HINT As seen in **Example 1,** discard any solution that does not satisfy the physical constraints of a problem.

The Pythagorean Theorem **Example 2** requires the use of the **Pythagorean theorem** for right triangles. Recall that the **legs** of a right triangle form the right angle, and the **hypotenuse** is the side opposite the right angle.

Pythagorean Theorem

In a right triangle, the sum of the squares of the lengths of the legs is equal to the square of the length of the hypotenuse.

$$a^2 + b^2 = c^2$$

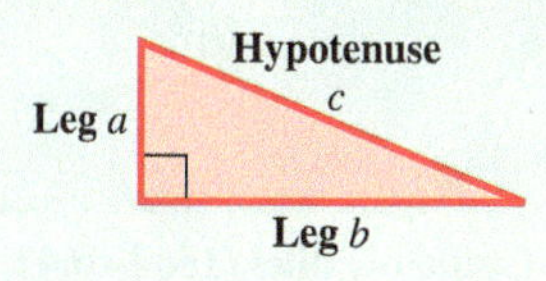

EXAMPLE 2 Applying the Pythagorean Theorem

A piece of property has the shape of a right triangle. The longer leg is 20 m longer than twice the length of the shorter leg. The hypotenuse is 10 m longer than the length of the longer leg. Find the lengths of the sides of the triangular lot.

SOLUTION

Step 1 **Read** the problem. We must find the lengths of the three sides.

Step 2 **Assign a variable.**

Let x = the length of the shorter leg (in meters).

Then $2x + 20$ = the length of the longer leg, and

$(2x + 20) + 10$, or $2x + 30$ = the length of the hypotenuse.

See **Figure 7.**

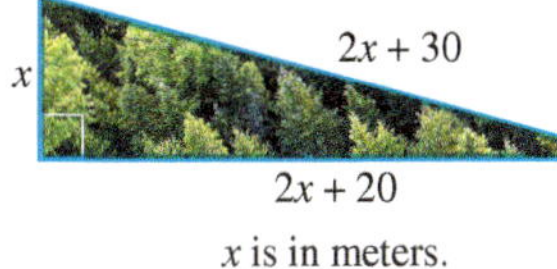

x is in meters.

Figure 7

Step 3 **Write an equation.**

$$\begin{array}{ccccc} a^2 & + & b^2 & = & c^2 \\ \downarrow & & \downarrow & & \downarrow \\ x^2 & + & (2x+20)^2 & = & (2x+30)^2 \end{array}$$

The hypotenuse is c. Substitute into the Pythagorean theorem.

Step 4 **Solve** the equation.

$$x^2 + (4x^2 + 80x + 400) = 4x^2 + 120x + 900$$ Square the binomials. Remember the middle terms.

$$x^2 - 40x - 500 = 0$$ Standard form

$$(x - 50)(x + 10) = 0$$ Factor.

$$x - 50 = 0 \quad \text{or} \quad x + 10 = 0$$ Zero-factor property

$$x = 50 \quad \text{or} \quad x = -10$$ Solve each equation.

Step 5 **State the answer.** Because x represents a length, -10 is not reasonable. The lengths of the sides of the triangular lot are

$$50 \text{ m}, \quad 2(50) + 20 = 120 \text{ m}, \quad \text{and} \quad 2(50) + 30 = 130 \text{ m}.$$

Step 6 **Check.** The lengths 50, 120, and 130 satisfy the words of the problem and also satisfy the Pythagorean theorem.

✔ **Now Try Exercise 35.**

Galileo Galilei (1564–1642)

According to legend, Galileo dropped objects of different weights from the Leaning Tower of Pisa to disprove the Aristotelian view that heavier objects fall faster than lighter objects. He developed the formula $d = 16t^2$ for freely falling objects, where d is the distance in feet that an object falls (neglecting air resistance) in t seconds, regardless of weight.

Height of a Projected Object If air resistance is neglected, the height s (in feet) of an object projected directly upward from an initial height of s_0 feet, with initial velocity v_0 feet per second, is given by the following equation.

$$s = -16t^2 + v_0t + s_0$$

Here t represents the number of seconds after the object is projected. The coefficient of t^2, -16, is a constant based on the gravitational force of Earth. This constant varies on other surfaces, such as the moon and other planets.

EXAMPLE 3 Solving a Problem Involving Projectile Height

If a projectile is launched vertically upward from the ground with an initial velocity of 100 ft per sec, neglecting air resistance, its height s (in feet) above the ground t seconds after projection is given by

$$s = -16t^2 + 100t.$$

(a) After how many seconds will it be 50 ft above the ground?

(b) How long will it take for the projectile to return to the ground?

SOLUTION

(a) We must find value(s) of t so that height s is 50 ft.

$$s = -16t^2 + 100t$$

$$50 = -16t^2 + 100t \qquad \text{Let } s = 50.$$

$$0 = -16t^2 + 100t - 50 \qquad \text{Standard form}$$

$$0 = 8t^2 - 50t + 25 \qquad \text{Divide by } -2.$$

$$t = \frac{-b \pm \sqrt{b^2 - 4ac}}{2a} \qquad \text{Quadratic formula}$$

Substitute carefully.

$$t = \frac{-(-50) \pm \sqrt{(-50)^2 - 4(8)(25)}}{2(8)} \qquad \text{Substitute } a = 8,\ b = -50, \text{ and } c = 25.$$

$$t = \frac{50 \pm \sqrt{1700}}{16} \qquad \text{Simplify.}$$

$$t \approx 0.55 \quad \text{or} \quad t \approx 5.70 \qquad \text{Use a calculator.}$$

Both solutions are acceptable. The projectile reaches 50 ft twice—once on its way up (after 0.55 sec) and once on its way down (after 5.70 sec).

(b) When the projectile returns to the ground, the height s will be 0 ft.

$$s = -16t^2 + 100t$$

$$0 = -16t^2 + 100t \qquad \text{Let } s = 0.$$

$$0 = -4t(4t - 25) \qquad \text{Factor.}$$

$$-4t = 0 \quad \text{or} \quad 4t - 25 = 0 \qquad \text{Zero-factor property}$$

$$t = 0 \quad \text{or} \quad t = 6.25 \qquad \text{Solve each equation.}$$

The first solution, 0, represents the time at which the projectile was on the ground prior to being launched, so it does not answer the question. The projectile will return to the ground 6.25 sec after it is launched.

✔ **Now Try Exercise 47.**

LOOKING AHEAD TO CALCULUS

In calculus, you will need to be able to write an algebraic expression from the description in a problem like those in this section. Using calculus techniques, you will be asked to find the value of the variable that produces an optimum (a maximum or minimum) value of the expression.

Modeling with Quadratic Equations

EXAMPLE 4 Analyzing Trolley Ridership

The I-Ride Trolley service carries passengers along the International Drive resort area of Orlando, Florida. The bar graph in **Figure 8** shows I-Ride Trolley ridership data in millions. The quadratic equation

$$y = -0.00525x^2 + 0.0913x + 1.64$$

models ridership from 2000 to 2013, where y represents ridership in millions, and $x = 0$ represents 2000, $x = 1$ represents 2001, and so on.

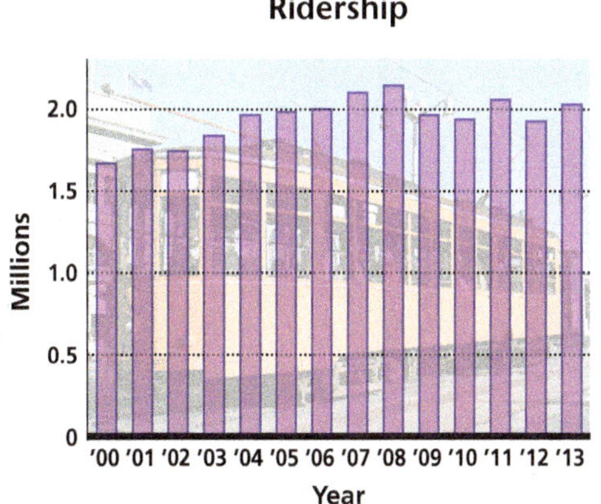

Source: I-Ride Trolley, International Drive Master Transit, www.itrolley.com

Figure 8

(a) Use the model to determine ridership in 2011. Compare the result to the actual ridership figure of 2.1 million.

(b) According to the model, in what year did ridership reach 1.8 million?

SOLUTION

(a) Because $x = 0$ represents the year 2000, use $x = 11$ to represent 2011.

$$y = -0.00525x^2 + 0.0913x + 1.64 \quad \text{Given model}$$

$$y = -0.00525(11)^2 + 0.0913(11) + 1.64 \quad \text{Let } x = 11.$$

$$y \approx 2.0 \text{ million} \quad \text{Use a calculator.}$$

The prediction is about 0.1 million (that is, 100,000) less than the actual figure of 2.1 million.

(b)

$$y = -0.00525x^2 + 0.0913x + 1.64 \quad \text{Given model}$$

$$1.8 = -0.00525x^2 + 0.0913x + 1.64 \quad \text{Let } y = 1.8.$$

Solve this equation for x.

$$0 = -0.00525x^2 + 0.0913x - 0.16 \quad \text{Standard form}$$

$$x = \frac{-0.0913 \pm \sqrt{(0.0913)^2 - 4(-0.00525)(-0.16)}}{2(-0.00525)} \quad \text{Quadratic formula}$$

$$x \approx 2.0 \quad \text{or} \quad x \approx 15.4 \quad \text{Use a calculator.}$$

The year 2002 corresponds to $x = 2.0$. Thus, according to the model, ridership reached 1.8 million in the year 2002. This outcome closely matches the bar graph and seems reasonable.

The year 2015 corresponds to $x = 15.4$. Round down to the year 2015 because 15.4 yr from 2000 occurs during 2015. There is no value on the bar graph to compare this to, because the last data value is for the year 2013. Always view results that are *beyond* the data in a model with skepticism, and realistically consider whether the model will continue as given. The model *predicts* that ridership will be 1.8 million again in the year 2015.

✔ **Now Try Exercise 49.**

1.5 Exercises

CONCEPT PREVIEW *Answer each question.*

1. *Area of a Parking Lot* For the rectangular parking area of the shopping center shown, with x in yards, which one of the following equations says that the area is 40,000 yd^2?

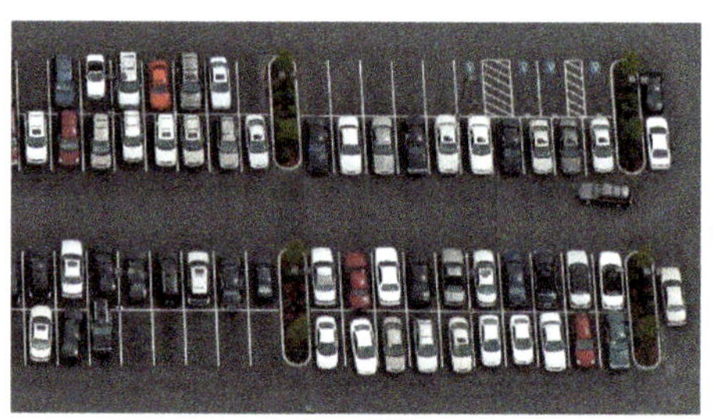

A. $x(2x + 200) = 40{,}000$ **B.** $2x + 2(2x + 200) = 40{,}000$

C. $x + (2x + 200) = 40{,}000$ **D.** $x^2 + (2x + 200)^2 = 40{,}000^2$

2. *Diagonal of a Rectangle* If a rectangle is r feet long and s feet wide, which expression represents the length of its diagonal in terms of r and s?

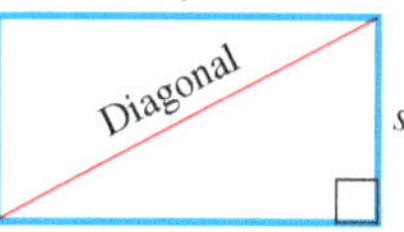

A. $\sqrt{rs}$  **B.** $r + s$ **C.** $\sqrt{r^2 + s^2}$ **D.** $r^2 + s^2$

3. *Sides of a Right Triangle* To solve for the lengths of the right triangle sides, which equation is correct?

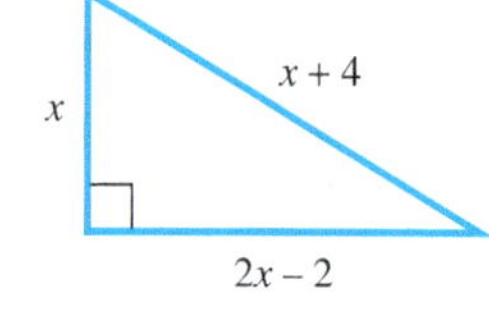

A. $x^2 = (2x - 2)^2 + (x + 4)^2$

B. $x^2 + (x + 4)^2 = (2x - 2)^2$

C. $x^2 = (2x - 2)^2 - (x + 4)^2$

D. $x^2 + (2x - 2)^2 = (x + 4)^2$

4. *Area of a Picture* The mat and frame around the picture shown measure x inches across. Which equation says that the area of the picture itself is 600 in.^{2}?

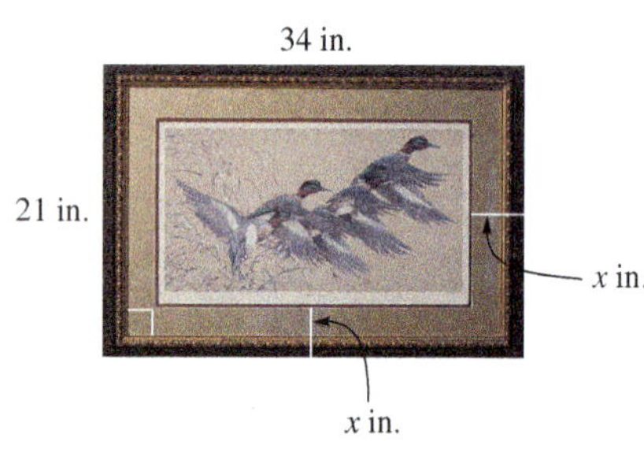

A. $2(34 - 2x) + 2(21 - 2x) = 600$

B. $(34 - 2x)(21 - 2x) = 600$

C. $(34 - x)(21 - x) = 600$

D. $x(34)(21) = 600$

5. *Volume of a Box* A rectangular piece of metal is 5 in. longer than it is wide. Squares with sides 2 in. long are cut from the four corners, and the flaps are folded upward to form an open box. Which equation indicates that the volume of the box is 64 in.^{3}?

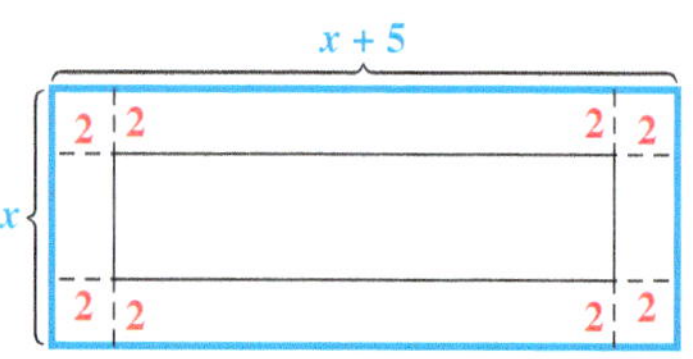

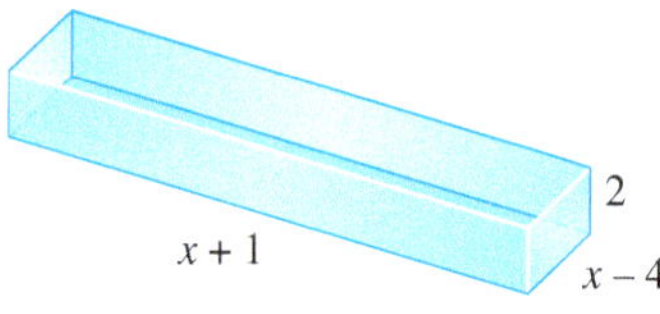

A. $(x + 1)(x - 4)(2) = 64$ **B.** $x(x + 5)(2) = 64$

C. $(x + 1)(x - 4) = 64$ **D.** $x(x + 5) = 64$

6. *Height of a Projectile* If a projectile is launched vertically upward from the ground with an initial velocity of 60 ft per sec, neglecting air resistance, its height s (in feet) above the ground t seconds after projection is given by

$$s = -16t^2 + 60t.$$

Which equation should be used to determine the time at which the height of the projectile reaches 40 ft?

A. $s = -16(40)^2 + 60$ **B.** $s = -16(40)^2 + 60(40)$

C. $40 = -16t^2 + 60t$ **D.** $40 = -16t^2$

7. *Height of a Projectile* If a projectile is launched vertically upward from the ground with an initial velocity of 45 ft per sec, neglecting air resistance, its height s (in feet) above the ground t seconds after projection is given by

$$s = -16t^2 + 45t.$$

Which equation should be used to determine the height of the projectile after 2 sec?

A. $s = 2(-16t^2 + 45t)$ **B.** $s = -16(2)^2 + 45(2)$

C. $2 = -16t^2 + 45t$ **D.** $2 = -16t^2$

8. *New Car Sales* Suppose that the quadratic equation

$$S = 0.0538x^2 - 0.807x + 8.84$$

models sales of new cars, where S represents sales in millions, and $x = 0$ represents 2000, $x = 1$ represents 2001, and so on. Which equation should be used to determine sales in 2010?

A. $10 = 0.0538x^2 - 0.807x + 8.84$

B. $2010 = 0.0538x^2 - 0.807x + 8.84$

C. $S = 0.0538(10)^2 - 0.807(10) + 8.84$

D. $S = 0.0538(2010)^2 - 0.807(2010) + 8.84$

To prepare for the applications that come later, work the following basic problems that lead to quadratic equations.

Unknown Numbers *In Exercises 9–18, use the following facts.*

If x represents an integer, then $x + 1$ represents the next consecutive integer.

If x represents an even integer, then $x + 2$ represents the next consecutive even integer.

If x represents an odd integer, then $x + 2$ represents the next consecutive odd integer.

9. Find two consecutive integers whose product is 56.

10. Find two consecutive integers whose product is 110.

11. Find two consecutive even integers whose product is 168.

12. Find two consecutive even integers whose product is 224.

13. Find two consecutive odd integers whose product is 63.

14. Find two consecutive odd integers whose product is 143.

15. The sum of the squares of two consecutive odd integers is 202. Find the integers.

16. The sum of the squares of two consecutive even integers is 52. Find the integers.

17. The difference of the squares of two positive consecutive even integers is 84. Find the integers.

18. The difference of the squares of two positive consecutive odd integers is 32. Find the integers.

Solve each problem. **See Examples 1 and 2.**

19. *Dimensions of a Right Triangle* The lengths of the sides of a right triangle are consecutive even integers. Find these lengths. (*Hint:* Use the Pythagorean theorem.)

20. *Dimensions of a Right Triangle* The lengths of the sides of a right triangle are consecutive positive integers. Find these lengths. (*Hint:* Use the Pythagorean theorem.)

21. *Dimensions of a Square* The length of each side of a square is 3 in. more than the length of each side of a smaller square. The sum of the areas of the squares is 149 in.^{2}. Find the lengths of the sides of the two squares.

22. *Dimensions of a Square* The length of each side of a square is 5 in. more than the length of each side of a smaller square. The difference of the areas of the squares is 95 in.^{2}. Find the lengths of the sides of the two squares.

Solve each problem. **See Example 1.**

23. *Dimensions of a Parking Lot* A parking lot has a rectangular area of 40,000 yd^2. The length is 200 yd more than twice the width. Find the dimensions of the lot.

24. *Dimensions of a Garden* An ecology center wants to set up an experimental garden using 300 m of fencing to enclose a rectangular area of 5000 m^2. Find the dimensions of the garden.

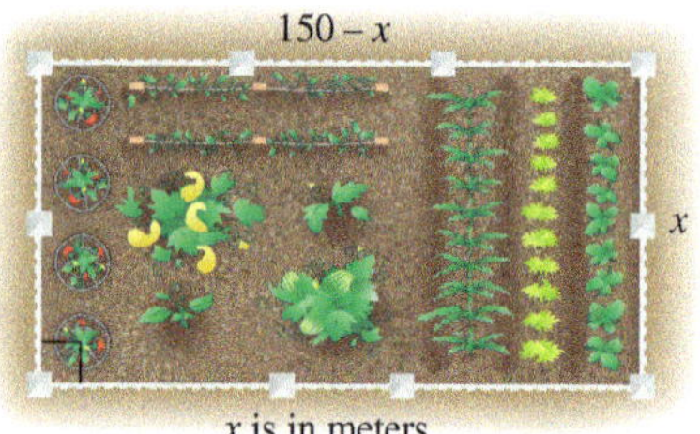

25. *Dimensions of a Rug* Zachary wants to buy a rug for a room that is 12 ft wide and 15 ft long. He wants to leave a uniform strip of floor around the rug. He can afford to buy 108 ft^2 of carpeting. What dimensions should the rug have?

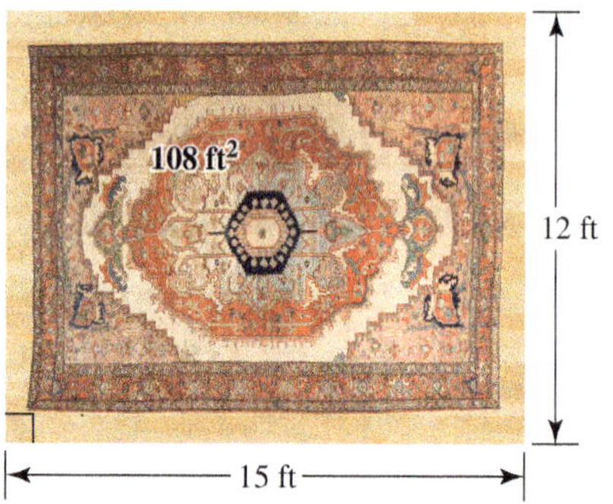

26. *Width of a Flower Border* A landscape architect has included a rectangular flower bed measuring 9 ft by 5 ft in her plans for a new building. She wants to use two colors of flowers in the bed: one in the center and the other for a border of the same width on all four sides. If she has enough plants to cover 24 ft^2 for the border, how wide can the border be?

27. *Volume of a Box* A rectangular piece of metal is 10 in. longer than it is wide. Squares with sides 2 in. long are cut from the four corners, and the flaps are folded upward to form an open box. If the volume of the box is 832 in.^{3}, what were the original dimensions of the piece of metal?

28. *Volume of a Box* In **Exercise 27,** suppose that the piece of metal has length twice the width, and 4-in. squares are cut from the corners. If the volume of the box is 1536 in.^{3}, what were the original dimensions of the piece of metal?

29. *Manufacturing to Specifications* A manufacturing firm wants to package its product in a cylindrical container 3 ft high with surface area 8π ft^2. What should the radius of the circular top and bottom of the container be? (*Hint:* The surface area consists of the circular top and bottom and a rectangle that represents the side cut open vertically and unrolled.)

30. *Manufacturing to Specifications* In **Exercise 29,** what radius would produce a container with a volume of π times the radius? (*Hint:* The volume is the area of the circular base times the height.)

31. *Dimensions of a Square* What is the length of the side of a square if its area and perimeter are numerically equal?

32. *Dimensions of a Rectangle* A rectangle has an area that is numerically twice its perimeter. If the length is twice the width, what are its dimensions?

33. *Radius of a Can* A can of Blue Runner Red Kidney Beans has surface area 371 cm^2. Its height is 12 cm. What is the radius of the circular top? Round to the nearest hundredth.

34. *Dimensions of a Cereal Box* The volume of a 15-oz cereal box is 180.4 in.^{3}. The length of the box is 3.2 in. less than the height, and its width is 2.3 in. Find the height and length of the box to the nearest tenth.

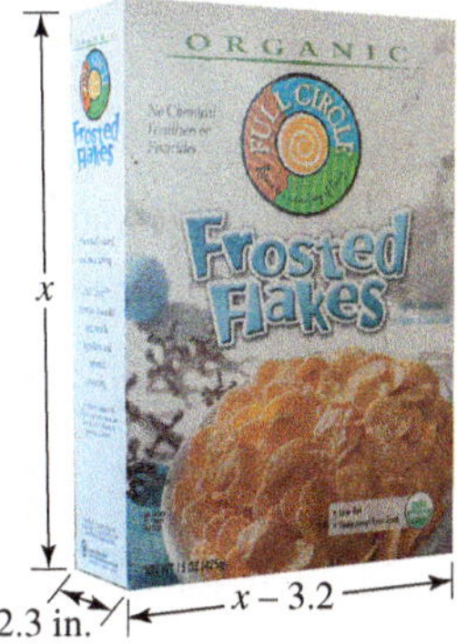

Solve each problem. ***See Example 2.***

35. *Height of a Dock* A boat is being pulled into a dock with a rope attached to the boat at water level. When the boat is 12 ft from the dock, the length of the rope from the boat to the dock is 3 ft longer than twice the height of the dock above the water. Find the height of the dock.

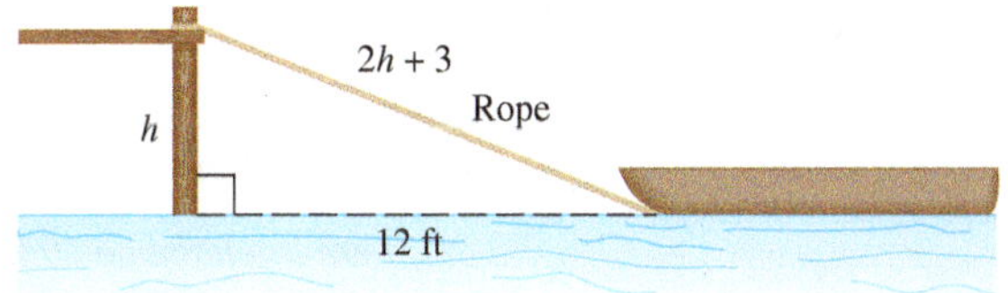

36. *Height of a Kite* Grady is flying a kite on 50 ft of string. Its vertical distance from his hand is 10 ft more than its horizontal distance from his hand. Assuming that the string is being held 5 ft above ground level, find its horizontal distance from Grady and its vertical distance from the ground.

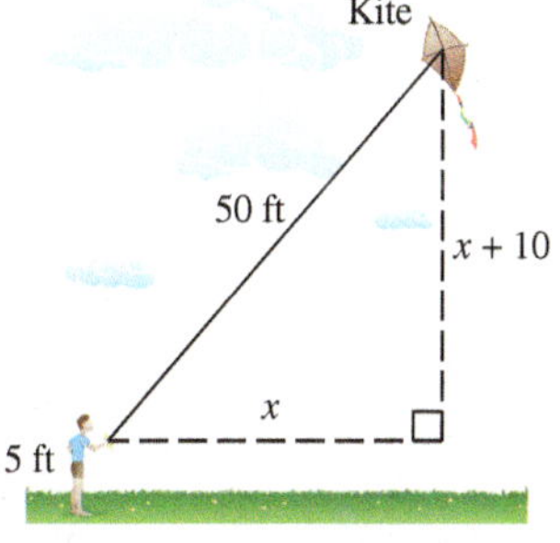

37. *Radius Covered by a Circular Lawn Sprinkler* A square lawn has area 800 ft^2. A sprinkler placed at the center of the lawn sprays water in a circular pattern as shown in the figure. What is the radius of the circle?

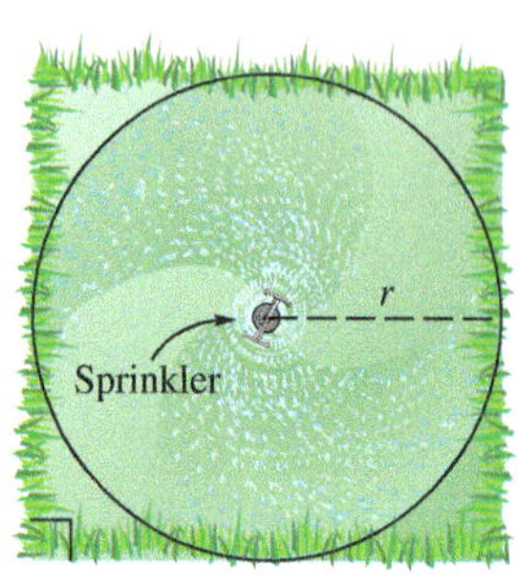

38. *Dimensions of a Solar Panel Frame* Molly has a solar panel with a width of 26 in. To get the proper inclination for her climate, she needs a right triangular support frame that has one leg twice as long as the other. To the nearest tenth of an inch, what dimensions should the frame have?

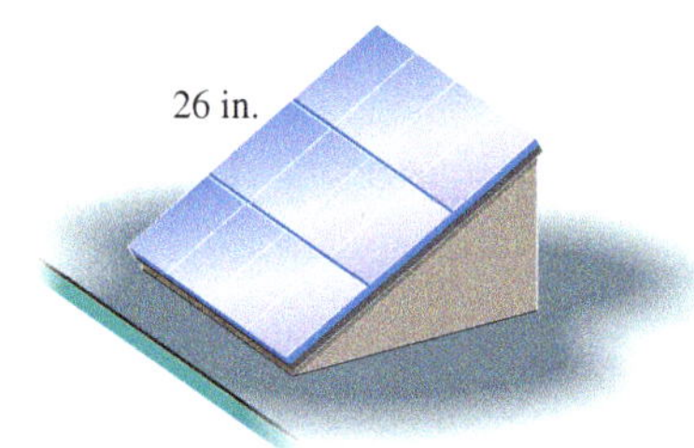

39. *Length of a Ladder* A building is 2 ft from a 9-ft fence that surrounds the property. A worker wants to wash a window in the building 13 ft from the ground. He plans to place a ladder over the fence so it rests against the building. (See the figure.) He decides he should place the ladder 8 ft from the fence for stability. To the nearest tenth of a foot, how long a ladder will he need?

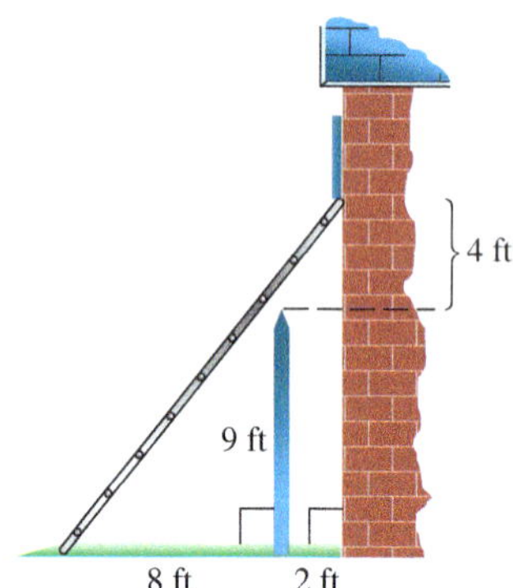

40. *Range of Receivers* Tanner and Sheldon have received communications receivers for Christmas. If they leave from the same point at the same time, Tanner walking north at 2.5 mph and Sheldon walking east at 3 mph, how long will they be able to talk to each other if the range of the communications receivers is 4 mi? Round the answer to the nearest minute.

41. *Length of a Walkway* A nature conservancy group decides to construct a raised wooden walkway through a wetland area. To enclose the most interesting part of the wetlands, the walkway will have the shape of a right triangle with one leg 700 yd longer than the other and the hypotenuse 100 yd longer than the longer leg. Find the total length of the walkway.

42. *Broken Bamboo* Problems involving the Pythagorean theorem have appeared in mathematics for thousands of years. This one is taken from the ancient Chinese work *Arithmetic in Nine Sections:*

> *There is a bamboo 10 ft high, the upper end of which, being broken, reaches the ground 3 ft from the stem. Find the height of the break.*

(Modeling) Solve each problem. **See Example 3.**

Height of a Projectile A projectile is launched from ground level with an initial velocity of v_0 feet per second. Neglecting air resistance, its height in feet t seconds after launch is given by

$$s = -16t^2 + v_0 t.$$

*In Exercises 43–46, find the time(s) that the projectile will **(a)** reach a height of* 80 ft *and **(b)** return to the ground for the given value of v_0. Round answers to the nearest hundredth if necessary.*

43. $v_0 = 96$ **44.** $v_0 = 128$ **45.** $v_0 = 32$ **46.** $v_0 = 16$

47. *Height of a Projected Ball* An astronaut on the moon throws a baseball upward. The astronaut is 6 ft, 6 in. tall, and the initial velocity of the ball is 30 ft per sec. The height s of the ball in feet is given by the equation

$$s = -2.7t^2 + 30t + 6.5,$$

where t is the number of seconds after the ball was thrown.

(a) After how many seconds is the ball 12 ft above the moon's surface? Round to the nearest hundredth.

(b) How many seconds will it take for the ball to hit the moon's surface? Round to the nearest hundredth.

48. *Concept Check* The ball in **Exercise 47** will never reach a height of 100 ft. How can this be determined algebraically?

(Modeling) Solve each problem. ***See Example 4.***

49. *NFL Salary Cap* In 1994, the National Football League introduced a salary cap that limits the amount of money spent on players' salaries. The quadratic model

$$y = 0.2313x^2 + 2.600x + 35.17$$

approximates this cap in millions of dollars for the years 1994–2009, where $x = 0$ represents 1994, $x = 1$ represents 1995, and so on. (*Source:* www.businessinsider.com)

(a) Approximate the NFL salary cap in 2007 to the nearest tenth of a million dollars.

(b) According to the model, in what year did the salary cap reach 90 million dollars?

50. *NFL Rookie Wage Scale* Salaries, in millions of dollars, for rookies selected in the first round of the NFL 2014 draft can be approximated by the quadratic model

$$y = 0.0258x^2 - 1.30x + 23.3,$$

where x represents draft pick order. Players selected earlier in the round have higher salaries than those selected later in the round. (*Source:* www.forbes.com)

(a) Use the model to estimate the salary of the player selected first overall to the nearest tenth of a million dollars.

(b) What is the estimated salary of the player selected 10th overall? Round to the nearest tenth of a million dollars.

51. *Carbon Monoxide Exposure* Carbon monoxide (CO) combines with the hemoglobin of the blood to form carboxyhemoglobin (COHb), which reduces transport of oxygen to tissues. Smokers routinely have a 4% to 6% COHb level in their blood. The quadratic model

$$T = 0.00787x^2 - 1.528x + 75.89$$

approximates the exposure time in hours necessary to reach this 4% to 6% level, where $50 \le x \le 100$ is the amount of carbon monoxide present in the air in parts per million (ppm). (*Source: Indoor Air Quality Environmental Information Handbook: Combustion Sources.*)

(a) A kerosene heater or a room full of smokers is capable of producing 50 ppm of carbon monoxide. How long would it take for a nonsmoking person to start feeling the above symptoms? Round to the nearest tenth.

(b) Find the carbon monoxide concentration necessary for a person to reach the 4% to 6% COHb level in 3 hr. Round to the nearest tenth.

52. *Carbon Monoxide Exposure* Refer to **Exercise 51.** High concentrations of carbon monoxide (CO) can cause coma and death. The time required for a person to reach a COHb level capable of causing a coma can be approximated by the quadratic model

$$T = 0.0002x^2 - 0.316x + 127.9,$$

where T is the exposure time in hours necessary to reach this level and $500 \leq x \leq 800$ is the amount of carbon monoxide present in the air in parts per million (ppm). (*Source: Indoor Air Quality Environmental Information Handbook: Combustion Sources.*)

(a) What is the exposure time when $x = 600$ ppm?

(b) Find the concentration of CO necessary to produce a coma in 4 hr. Round to the nearest tenth part per million.

53. *Methane Gas Emissions* The table gives methane gas emissions from all sources in the United States, in millions of metric tons. The quadratic model

$$y = 0.0429x^2 - 9.73x + 606$$

approximates the emissions for these years. In the model, x represents the number of years since 2008, so $x = 0$ represents 2008, $x = 1$ represents 2009, and so on.

Year	Millions of Metric Tons of Methane
2008	606.0
2009	596.6
2010	585.5
2011	578.4
2012	567.3

Source: U.S. Environmental Protection Agency.

(a) According to the model, what would emissions be in 2014? Round to the nearest tenth of a million metric tons.

(b) Find the nearest year beyond 2008 for which this model predicts that emissions will reach 500 million metric tons.

54. *Cost of Public Colleges* The average cost, in dollars, for tuition and fees for in-state students at four-year public colleges over the period 2000–2014 can be modeled by the equation

$$y = 4.065x^2 + 370.1x + 3450$$

where $x = 0$ corresponds to 2000, $x = 1$ corresponds to 2001, and so on. Based on this model, for what year after 2000 was the average cost \$8605? (*Source:* The College Board, *Annual Survey of Colleges.*)

55. *Internet Publishing* Estimated revenue from Internet publishing and web search portals in the United States during the years 2007 through 2012 can be modeled by the equation

$$y = 710.55x^2 + 1333.7x + 32{,}399$$

where $x = 0$ corresponds to the year 2007, $x = 1$ corresponds to 2008, and so on, and y is in millions of dollars. Approximate the revenue from these services in 2010 to the nearest million. (*Source:* U.S. Census Bureau.)

56. *Cable's Top Internet Speeds* The top cable Internet speeds during the years 2007 through 2013 can be modeled by the equation

$$y = 23.09x^2 - 62.12x + 32.78,$$

where $x = 0$ corresponds to 2007, $x = 1$ corresponds to 2008, and so on, and y is in megabits per second (MBS). Based on this model, what was cable TV's top Internet speed in 2012? (*Source:* National Cable & Telecommunications Association.)

Relating Concepts

For individual or collaborative investigation *(Exercises 57–60)*

If p units of an item are sold for x dollars per unit, the revenue is $R = px$. Use this idea to analyze the following problem, ***working Exercises 57–60 in order.***

Number of Apartments Rented *The manager of an 80-unit apartment complex knows from experience that at a rent of* $300*, all the units will be full. On the average, one additional unit will remain vacant for each* $20 *increase in rent over* $300*. Furthermore, the manager must keep at least* 30 *units rented due to other financial considerations. Currently, the revenue from the complex is* $35,000*. How many apartments are rented?*

57. Suppose that x represents the number of $20 increases over $300. Represent the number of apartment units that will be rented in terms of x.

58. Represent the rent per unit in terms of x.

59. Use the answers in **Exercises 57 and 58** to write an equation that defines the revenue generated when there are x increases of $20 over $300.

60. According to the problem, the revenue currently generated is $35,000. Substitute this value for revenue into the equation from **Exercise 59.** Solve for x to answer the question in the problem.

Solve each problem. ***(See Exercises 57–60.)***

61. *Number of Airline Passengers* The cost of a charter flight to Miami is $225 each for 75 passengers, with a refund of $5 per passenger for each passenger in excess of 75. How many passengers must take the flight to produce a revenue of $16,000?

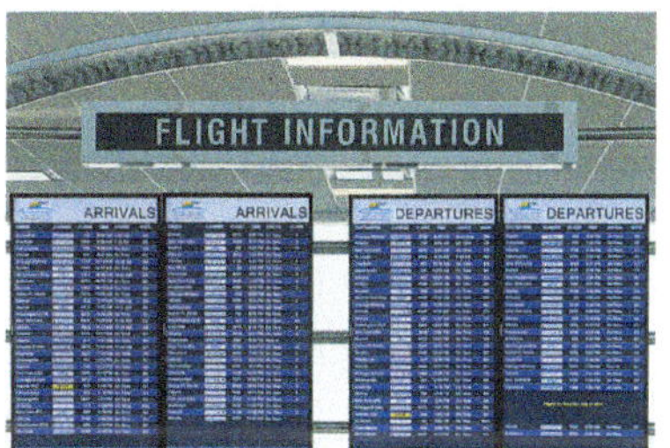

62. *Number of Bus Passengers* A charter bus company charges a fare of $40 per person, plus $2 per person for each unsold seat on the bus. If the bus holds 100 passengers and x represents the number of unsold seats, how many passengers must ride the bus to produce revenue of $5950? (*Note:* Because of the company's commitment to efficient fuel use, the charter will not run unless filled to at least half-capacity.)

63. *Harvesting a Cherry Orchard* The manager of a cherry orchard wants to schedule the annual harvest. If the cherries are picked now, the average yield per tree will be 100 lb, and the cherries can be sold for 40 cents per pound. Past experience shows that the yield per tree will increase about 5 lb per week, while the price will decrease about 2 cents per pound per week. How many weeks should the manager wait to get an average revenue of $38.40 per tree?

64. *Recycling Aluminum Cans* A local group of scouts has been collecting old aluminum cans for recycling. The group has already collected 12,000 lb of cans, for which they could currently receive $4 per hundred pounds. The group can continue to collect cans at the rate of 400 lb per day. However, a glut in the old-can market has caused the recycling company to announce that it will lower its price, starting immediately, by $0.10 per hundred pounds per day. The scouts can make only one trip to the recycling center. How many days should they wait in order to receive $490 for their cans?

1.6 Other Types of Equations and Applications

- Rational Equations
- Work Rate Problems
- Equations with Radicals
- Equations with Rational Exponents
- Equations Quadratic in Form

Rational Equations

A **rational equation** is an equation that has a rational expression for one or more terms. To solve a rational equation, multiply each side by the least common denominator (LCD) of the terms of the equation to eliminate fractions, and then solve the resulting equation.

A value of the variable that appears to be a solution after each side of a rational equation is multiplied by a variable expression (the LCD) is called a **proposed solution.** Because a rational expression is not defined when its denominator is 0, **proposed solutions for which any denominator equals 0 are excluded from the solution set.**

Be sure to check all proposed solutions in the original equation.

EXAMPLE 1 Solving Rational Equations That Lead to Linear Equations

Solve each equation.

(a) $\frac{3x-1}{3} - \frac{2x}{x-1} = x$ **(b)** $\frac{x}{x-2} = \frac{2}{x-2} + 2$

SOLUTION

(a) The least common denominator is $3(x-1)$, which is equal to 0 if $x = 1$. Therefore, 1 cannot possibly be a solution of this equation.

$$\frac{3x-1}{3} - \frac{2x}{x-1} = x$$

$$3(x-1)\left(\frac{3x-1}{3}\right) - 3(x-1)\left(\frac{2x}{x-1}\right) = 3(x-1)x$$ Multiply by the LCD, $3(x-1)$, where $x \neq 1$.

$$(x-1)(3x-1) - 3(2x) = 3x(x-1)$$ Divide out common factors.

$$3x^2 - 4x + 1 - 6x = 3x^2 - 3x$$ Multiply.

$$1 - 10x = -3x$$ Subtract $3x^2$. Combine like terms.

$$1 = 7x$$ Solve the linear equation.

$$x = \frac{1}{7}$$ Proposed solution

The proposed solution $\frac{1}{7}$ meets the requirement that $x \neq 1$ and does not cause any denominator to equal 0. Substitute to check for correct algebra.

CHECK

$$\frac{3x-1}{3} - \frac{2x}{x-1} = x$$ Original equation

$$\frac{3\left(\frac{1}{7}\right)-1}{3} - \frac{2\left(\frac{1}{7}\right)}{\frac{1}{7}-1} \stackrel{?}{=} \frac{1}{7}$$ Let $x = \frac{1}{7}$.

$$-\frac{4}{21} - \left(-\frac{1}{3}\right) \stackrel{?}{=} \frac{1}{7}$$ Simplify the complex fractions.

$$\frac{1}{7} = \frac{1}{7} \checkmark$$ True

The solution set is $\left\{\frac{1}{7}\right\}$.

(b)

$$\frac{x}{x-2} = \frac{2}{x-2} + 2$$

$$(x-2)\left(\frac{x}{x-2}\right) = (x-2)\left(\frac{2}{x-2}\right) + (x-2)2$$ Multiply by the LCD, $x - 2$, where $x \neq 2$.

$$x = 2 + 2(x-2)$$ Divide out common factors.

$$x = 2 + 2x - 4$$ Distributive property

$$-x = -2$$ Solve the linear equation.

$$x = 2$$ Proposed solution

The proposed solution is 2. However, the variable is restricted to real numbers except 2. If $x = 2$, then not only does it cause a zero denominator, but also multiplying by $x - 2$ in the first step is multiplying both sides by 0, which is not valid. Thus, the solution set is $\emptyset$.

✔ **Now Try Exercises 17 and 19.**

EXAMPLE 2 Solving Rational Equations That Lead to Quadratic Equations

Solve each equation.

(a) $\frac{3x+2}{x-2} + \frac{1}{x} = \frac{-2}{x^2-2x}$ **(b)** $\frac{-4x}{x-1} + \frac{4}{x+1} = \frac{-8}{x^2-1}$

SOLUTION

(a)

$$\frac{3x+2}{x-2} + \frac{1}{x} = \frac{-2}{x^2-2x}$$

$$\frac{3x+2}{x-2} + \frac{1}{x} = \frac{-2}{x(x-2)}$$ Factor the last denominator.

$$x(x-2)\left(\frac{3x+2}{x-2}\right) + x(x-2)\left(\frac{1}{x}\right) = x(x-2)\left(\frac{-2}{x(x-2)}\right)$$ Multiply by $x(x-2)$, $x \neq 0, 2$.

$$x(3x+2) + (x-2) = -2$$ Divide out common factors.

$$3x^2 + 2x + x - 2 = -2$$ Distributive property

$$3x^2 + 3x = 0$$ Standard form

$$3x(x+1) = 0$$ Factor.

Set *each* factor equal to 0. → $3x = 0$ or $x + 1 = 0$ Zero-factor property

$x = 0$ or $x = -1$ Proposed solutions

Because of the restriction $x \neq 0$, the only valid proposed solution is -1. Check -1 in the original equation. The solution set is $\{-1\}$.

(b)

$$\frac{-4x}{x-1} + \frac{4}{x+1} = \frac{-8}{x^2-1}$$

$$\frac{-4x}{x-1} + \frac{4}{x+1} = \frac{-8}{(x+1)(x-1)}$$ Factor.

The restrictions on x are $x \neq \pm 1$. Multiply by the LCD, $(x+1)(x-1)$.

$$(x+1)(x-1)\left(\frac{-4x}{x-1}\right)+(x+1)(x-1)\left(\frac{4}{x+1}\right)=(x+1)(x-1)\left(\frac{-8}{(x+1)(x-1)}\right)$$

$$-4x(x+1)+4(x-1)=-8 \quad \text{Divide out common factors.}$$

$$-4x^2-4x+4x-4=-8 \quad \text{Distributive property}$$

$$-4x^2+4=0 \quad \text{Standard form}$$

$$x^2-1=0 \quad \text{Divide by } -4.$$

$$(x+1)(x-1)=0 \quad \text{Factor.}$$

$$x+1=0 \quad \text{or} \quad x-1=0 \quad \text{Zero-factor property}$$

$$x=-1 \quad \text{or} \quad x=1 \quad \text{Proposed solutions}$$

Neither proposed solution is valid, so the solution set is $\emptyset$.

✔ **Now Try Exercises 25 and 27.**

Work Rate Problems If a job can be completed in 3 hr, then the rate of work is $\frac{1}{3}$ of the job per hr. After 1 hr the job would be $\frac{1}{3}$ complete, and after 2 hr the job would be $\frac{2}{3}$ complete. In 3 hr the job would be $\frac{3}{3}$ complete, meaning that 1 complete job had been accomplished.

PROBLEM-SOLVING HINT If a job can be completed in t units of time, then the rate of work, r, is $\frac{1}{t}$ of the job per unit time.

$$r=\frac{1}{t}$$

The amount of work completed, A, is found by multiplying the rate of work, r, and the amount of time worked, t. This formula is similar to the distance formula $d = rt$.

Amount of work completed = rate of work × amount of time worked

or $$A = rt$$

EXAMPLE 3 Solving a Work Rate Problem

One printer can do a job twice as fast as another. Working together, both printers can do the job in 2 hr. How long would it take each printer, working alone, to do the job?

SOLUTION

Step 1 **Read** the problem. We must find the time it would take each printer, working alone, to do the job.

Step 2 **Assign a variable.** Let x represent the number of hours it would take the faster printer, working alone, to do the job. The time for the slower printer to do the job alone is then $2x$ hours.

Therefore, $\frac{1}{x}$ = the rate of the faster printer (job per hour)

and $\frac{1}{2x}$ = the rate of the slower printer (job per hour).

The time for the printers to do the job together is 2 hr. Multiplying each rate by the time will give the fractional part of the job completed by each.

	Rate	Time	Part of the Job Completed
Faster Printer	$\frac{1}{x}$	2	$2\left(\frac{1}{x}\right) = \frac{2}{x}$
Slower Printer	$\frac{1}{2x}$	2	$2\left(\frac{1}{2x}\right) = \frac{1}{x}$

$A = rt$

Step 3 **Write an equation.** The sum of the two parts of the job completed is 1 because one whole job is done.

$$\underbrace{\text{Part of the job done by the faster printer}}_{\frac{2}{x}} + \underbrace{\text{Part of the job done by the slower printer}}_{\frac{1}{x}} = \underbrace{\text{One whole job}}_{1}$$

Step 4 **Solve.**

$$x\left(\frac{2}{x} + \frac{1}{x}\right) = x(1) \quad \text{Multiply each side by } x, \text{ where } x \neq 0.$$

$$x\left(\frac{2}{x}\right) + x\left(\frac{1}{x}\right) = x(1) \quad \text{Distributive property}$$

$$2 + 1 = x \quad \text{Multiply.}$$

$$3 = x \quad \text{Add.}$$

Step 5 **State the answer.** The faster printer would take 3 hr to do the job alone. The slower printer would take $2(3) = 6$ hr. Give *both* answers here.

Step 6 **Check.** The answer is reasonable because the time working together (2 hr, as stated in the problem) is less than the time it would take the faster printer working alone (3 hr, as found in Step 4).

✔ **Now Try Exercise 39.**

NOTE **Example 3** can also be solved by using the fact that the sum of the rates of the individual printers is equal to their rate working together. Because the printers can complete the job together in 2 hr, their combined rate is $\frac{1}{2}$ of the job per hr.

$$\frac{1}{x} + \frac{1}{2x} = \frac{1}{2}$$

$$2x\left(\frac{1}{x} + \frac{1}{2x}\right) = 2x\left(\frac{1}{2}\right) \quad \text{Multiply each side by } 2x.$$

$$2 + 1 = x \quad \text{Distributive property}$$

$$3 = x \quad \text{Same solution found earlier}$$

Equations with Radicals To solve an equation such as

$$x - \sqrt{15 - 2x} = 0,$$

in which the variable appears in a radicand, we use the following **power property** to eliminate the radical.

Power Property

If P and Q are algebraic expressions, then every solution of the equation $P = Q$ is also a solution of the equation $P^n = Q^n$, for any positive integer n.

When the power property is used to solve equations, the new equation may have *more* solutions than the original equation. For example, the equation

$$x = -2 \quad \text{has solution set} \quad \{-2\}.$$

If we square each side of the equation $x = -2$, we obtain the new equation

$$x^2 = 4, \quad \text{which has solution set} \quad \{-2, 2\}.$$

Because the solution sets are not equal, the equations are not equivalent. ***When we use the power property to solve an equation, it is essential to check all proposed solutions in the original equation.***

CAUTION ***Be very careful when using the power property.*** It does *not* say that the equations $P = Q$ and $P^n = Q^n$ are equivalent. It says only that each solution of the original equation $P = Q$ is also a solution of the new equation $P^n = Q^n$.

Solving an Equation Involving Radicals

Step 1 Isolate the radical on one side of the equation.

Step 2 Raise each side of the equation to a power that is the same as the index of the radical so that the radical is eliminated.

If the equation still contains a radical, repeat Steps 1 and 2.

Step 3 Solve the resulting equation.

Step 4 Check each proposed solution in the *original* equation.

EXAMPLE 4 Solving an Equation Containing a Radical (Square Root)

Solve $x - \sqrt{15 - 2x} = 0$.

SOLUTION

$$x - \sqrt{15 - 2x} = 0$$

Step 1 $x = \sqrt{15 - 2x}$ Isolate the radical.

Step 2 $x^2 = \left(\sqrt{15 - 2x}\right)^2$ Square each side.

$x^2 = 15 - 2x$ $\left(\sqrt{a}\right)^2 = a$, for $a \geq 0$.

Step 3 $x^2 + 2x - 15 = 0$ Write in standard form.

$(x + 5)(x - 3) = 0$ Factor.

$x + 5 = 0$ or $x - 3 = 0$ Zero-factor property

$x = -5$ or $x = 3$ Proposed solutions

Step 4

CHECK $\quad x - \sqrt{15 - 2x} = 0 \quad$ Original equation

$-5 - \sqrt{15 - 2(-5)} \stackrel{?}{=} 0$ Let $x = -5$.	$3 - \sqrt{15 - 2(3)} \stackrel{?}{=} 0$ Let $x = 3$.
$-5 - \sqrt{25} \stackrel{?}{=} 0$	$3 - \sqrt{9} \stackrel{?}{=} 0$
$-5 - 5 \stackrel{?}{=} 0$	$3 - 3 \stackrel{?}{=} 0$
$-10 = 0$ False	$0 = 0$ ✓ True

As the check shows, only 3 is a solution, so the solution set is $\{3\}$.

✓ **Now Try Exercise 45.**

EXAMPLE 5 Solving an Equation Containing Two Radicals

Solve $\sqrt{2x + 3} - \sqrt{x + 1} = 1$.

SOLUTION

$$\sqrt{2x + 3} - \sqrt{x + 1} = 1$$ Isolate one of the radicals on one side of the equation.

Step 1 $\quad \sqrt{2x + 3} = 1 + \sqrt{x + 1} \quad$ Isolate $\sqrt{2x + 3}$.

Step 2 $\quad \left(\sqrt{2x + 3}\right)^2 = \left(1 + \sqrt{x + 1}\right)^2 \quad$ Square each side.

$$2x + 3 = 1 + 2\sqrt{x + 1} + (x + 1)$$ **Be careful:** $(a + b)^2 = a^2 + 2ab + b^2$

Don't forget this term when squaring.

Step 1 $\quad x + 1 = 2\sqrt{x + 1} \quad$ Isolate the remaining radical.

Step 2 $\quad (x + 1)^2 = \left(2\sqrt{x + 1}\right)^2 \quad$ Square again.

$x^2 + 2x + 1 = 4(x + 1)$ $\quad (ab)^2 = a^2b^2 \quad$ Apply the exponents.

$x^2 + 2x + 1 = 4x + 4 \quad$ Distributive property

Step 3 $\quad x^2 - 2x - 3 = 0 \quad$ Write in standard form.

$(x - 3)(x + 1) = 0 \quad$ Factor.

$x - 3 = 0$ or $x + 1 = 0 \quad$ Zero-factor property

$x = 3$ or $x = -1 \quad$ Proposed solutions

Step 4

CHECK $\quad \sqrt{2x + 3} - \sqrt{x + 1} = 1 \quad$ Original equation

$\sqrt{2(3) + 3} - \sqrt{3 + 1} \stackrel{?}{=} 1$ Let $x = 3$.	$\sqrt{2(-1) + 3} - \sqrt{-1 + 1} \stackrel{?}{=} 1$ Let $x = -1$.
$\sqrt{9} - \sqrt{4} \stackrel{?}{=} 1$	$\sqrt{1} - \sqrt{0} \stackrel{?}{=} 1$
$3 - 2 \stackrel{?}{=} 1$	$1 - 0 \stackrel{?}{=} 1$
$1 = 1$ ✓ True	$1 = 1$ ✓ True

Both 3 and -1 are solutions of the original equation, so $\{-1, 3\}$ is the solution set.

✓ **Now Try Exercise 57.**

CAUTION Remember to isolate a radical in Step 1. It would be incorrect to square each term individually as the first step in **Example 5.**

EXAMPLE 6 Solving an Equation Containing a Radical (Cube Root)

Solve $\sqrt[3]{4x^2 - 4x + 1} - \sqrt[3]{x} = 0$.

SOLUTION

$$\sqrt[3]{4x^2 - 4x + 1} - \sqrt[3]{x} = 0$$

Step 1 $\sqrt[3]{4x^2 - 4x + 1} = \sqrt[3]{x}$ Isolate a radical.

Step 2 $\left(\sqrt[3]{4x^2 - 4x + 1}\right)^3 = \left(\sqrt[3]{x}\right)^3$ Cube each side.

$4x^2 - 4x + 1 = x$ Apply the exponents.

Step 3 $4x^2 - 5x + 1 = 0$ Write in standard form.

$(4x - 1)(x - 1) = 0$ Factor.

$4x - 1 = 0$ or $x - 1 = 0$ Zero-factor property

$x = \frac{1}{4}$ or $x = 1$ Proposed solutions

Step 4

CHECK $\sqrt[3]{4x^2 - 4x + 1} - \sqrt[3]{x} = 0$ Original equation

$\sqrt[3]{4\left(\frac{1}{4}\right)^2 - 4\left(\frac{1}{4}\right) + 1} - \sqrt[3]{\frac{1}{4}} \stackrel{?}{=} 0$ Let $x = \frac{1}{4}$.

$\sqrt[3]{\frac{1}{4}} - \sqrt[3]{\frac{1}{4}} \stackrel{?}{=} 0$

$0 = 0$ ✓ True

$\sqrt[3]{4(1)^2 - 4(1) + 1} - \sqrt[3]{1} \stackrel{?}{=} 0$ Let $x = 1$.

$\sqrt[3]{1} - \sqrt[3]{1} \stackrel{?}{=} 0$

$0 = 0$ ✓ True

Both are valid solutions, and the solution set is $\left\{\frac{1}{4}, 1\right\}$.

✓ **Now Try Exercise 69.**

Equations with Rational Exponents An equation with a rational exponent contains a variable, or variable expression, raised to an exponent that is a rational number. For example, the radical equation

$$\left(\sqrt[5]{x}\right)^3 = 27 \quad \text{can be written with a rational exponent as} \quad x^{3/5} = 27$$

and solved by raising each side to the reciprocal of the exponent, with care taken regarding signs as seen in **Example 7(b).**

EXAMPLE 7 Solving Equations with Rational Exponents

Solve each equation.

(a) $x^{3/5} = 27$ **(b)** $(x - 4)^{2/3} = 16$

SOLUTION

(a) $x^{3/5} = 27$

$\left(x^{3/5}\right)^{5/3} = 27^{5/3}$ Raise each side to the power $\frac{5}{3}$, the reciprocal of the exponent of x.

$x = 243$ $27^{5/3} = \left(\sqrt[3]{27}\right)^5 = 3^5 = 243$

CHECK Let $x = 243$ in the original equation.

$$x^{3/5} = 243^{3/5} = \left(\sqrt[5]{243}\right)^3 = 3^3 = 27 \quad ✓ \text{ True}$$

The solution set is $\{243\}$.

(b)

$$(x-4)^{2/3} = 16$$

$$\left[(x-4)^{2/3}\right]^{3/2} = \pm 16^{3/2} \qquad \text{Raise each side to the power } \tfrac{3}{2}\text{. Insert } \pm \text{ because this involves an even root, as indicated by the 2 in the denominator.}$$

$$x - 4 = \pm 64 \qquad \pm 16^{3/2} = \pm\left(\sqrt{16}\right)^3 = \pm 4^3 = \pm 64$$

$$x = 4 \pm 64 \qquad \text{Add 4 to each side.}$$

$$x = -60 \quad \text{or} \quad x = 68 \qquad \text{Proposed solutions}$$

CHECK $(x-4)^{2/3} = 16$ Original equation

$(-60-4)^{2/3} \stackrel{?}{=} 16$ Let $x = -60$.	$(68-4)^{2/3} \stackrel{?}{=} 16$ Let $x = 68$.
$(-64)^{2/3} \stackrel{?}{=} 16$	$64^{2/3} \stackrel{?}{=} 16$
$\left(\sqrt[3]{-64}\right)^2 \stackrel{?}{=} 16$	$\left(\sqrt[3]{64}\right)^2 \stackrel{?}{=} 16$
$16 = 16$ ✓ True	$16 = 16$ ✓ True

Both proposed solutions check, so the solution set is $\{-60, 68\}$.

✔ **Now Try Exercises 75 and 79.**

Equations Quadratic in Form Many equations that are not quadratic equations can be solved using similar methods. The equation

$$(x+1)^{2/3} - (x+1)^{1/3} - 2 = 0$$

is not a quadratic equation in x. However, with the substitutions

$$u = (x+1)^{1/3} \quad \text{and} \quad u^2 = \left[(x+1)^{1/3}\right]^2 = (x+1)^{2/3},$$

the equation becomes

$$u^2 - u - 2 = 0,$$

which is a quadratic equation in u. This quadratic equation can be solved to find u, and then $u = (x+1)^{1/3}$ can be used to find the values of x, the solutions to the original equation.

Equation Quadratic in Form

An equation is **quadratic in form** if it can be written as

$$au^2 + bu + c = 0,$$

where $a \neq 0$ and u is some algebraic expression.

EXAMPLE 8 Solving Equations Quadratic in Form

Solve each equation.

(a) $(x+1)^{2/3} - (x+1)^{1/3} - 2 = 0$ **(b)** $6x^{-2} + x^{-1} = 2$

SOLUTION

(a)

$$(x+1)^{2/3} - (x+1)^{1/3} - 2 = 0 \quad (x+1)^{2/3} = [(x+1)^{1/3}]^2, \text{ so}$$

$$u^2 - u - 2 = 0 \quad \text{let } u = (x+1)^{1/3}.$$

$$(u-2)(u+1) = 0 \quad \text{Factor.}$$

$$u - 2 = 0 \quad \text{or} \quad u + 1 = 0 \quad \text{Zero-factor property}$$

$$u = 2 \quad \text{or} \quad u = -1 \quad \text{Solve each equation.}$$

Don't forget this step.

$$(x+1)^{1/3} = 2 \quad \text{or} \quad (x+1)^{1/3} = -1 \quad \text{Replace } u \text{ with } (x+1)^{1/3}.$$

$$[(x+1)^{1/3}]^3 = 2^3 \quad \text{or} \quad [(x+1)^{1/3}]^3 = (-1)^3 \quad \text{Cube each side.}$$

$$x + 1 = 8 \quad \text{or} \quad x + 1 = -1 \quad \text{Apply the exponents.}$$

$$x = 7 \quad \text{or} \quad x = -2 \quad \text{Proposed solutions}$$

CHECK $(x+1)^{2/3} - (x+1)^{1/3} - 2 = 0$ Original equation

$(7+1)^{2/3} - (7+1)^{1/3} - 2 \stackrel{?}{=} 0$ Let $x = 7$.	$(-2+1)^{2/3} - (-2+1)^{1/3} - 2 \stackrel{?}{=} 0$ Let $x = -2$.
$8^{2/3} - 8^{1/3} - 2 \stackrel{?}{=} 0$	$(-1)^{2/3} - (-1)^{1/3} - 2 \stackrel{?}{=} 0$
$4 - 2 - 2 \stackrel{?}{=} 0$	$1 + 1 - 2 \stackrel{?}{=} 0$
$0 = 0$ ✓ True	$0 = 0$ ✓ True

Both proposed solutions check, so the solution set is $\{-2, 7\}$.

(b)

$$6x^{-2} + x^{-1} = 2$$

$$6x^{-2} + x^{-1} - 2 = 0 \quad \text{Subtract 2 from each side.}$$

$$6u^2 + u - 2 = 0 \quad \text{Let } u = x^{-1}. \text{ Then } u^2 = x^{-2}.$$

$$(3u+2)(2u-1) = 0 \quad \text{Factor.}$$

$$3u + 2 = 0 \quad \text{or} \quad 2u - 1 = 0 \quad \text{Zero-factor property}$$

Don't stop here. Substitute for u.

$$u = -\frac{2}{3} \quad \text{or} \quad u = \frac{1}{2} \quad \text{Solve each equation.}$$

$$x^{-1} = -\frac{2}{3} \quad \text{or} \quad x^{-1} = \frac{1}{2} \quad \text{Replace } u \text{ with } x^{-1}.$$

$$x = -\frac{3}{2} \quad \text{or} \quad x = 2 \quad x^{-1} \text{ is the reciprocal of } x.$$

Both proposed solutions check, so the solution set is $\left\{-\frac{3}{2}, 2\right\}$.

✔ **Now Try Exercises 93 and 99.**

CAUTION ***When using a substitution variable in solving an equation that is quadratic in form, do not forget the step that gives the solution in terms of the original variable.***

EXAMPLE 9 Solving an Equation Quadratic in Form

Solve $12x^4 - 11x^2 + 2 = 0$.

SOLUTION

$$12x^4 - 11x^2 + 2 = 0$$

$$12(x^2)^2 - 11x^2 + 2 = 0 \quad x^4 = (x^2)^2$$

$$12u^2 - 11u + 2 = 0 \quad \text{Let } u = x^2. \text{ Then } u^2 = x^4.$$

$$(3u - 2)(4u - 1) = 0 \quad \text{Solve the quadratic equation.}$$

$$3u - 2 = 0 \quad \text{or} \quad 4u - 1 = 0 \quad \text{Zero-factor property}$$

$$u = \frac{2}{3} \quad \text{or} \quad u = \frac{1}{4} \quad \text{Solve each equation.}$$

$$x^2 = \frac{2}{3} \quad \text{or} \quad x^2 = \frac{1}{4} \quad \text{Replace } u \text{ with } x^2.$$

$$x = \pm\sqrt{\frac{2}{3}} \quad \text{or} \quad x = \pm\sqrt{\frac{1}{4}} \quad \text{Square root property}$$

$$x = \frac{\pm\sqrt{2}}{\sqrt{3}} \cdot \frac{\sqrt{3}}{\sqrt{3}} \quad \text{or} \quad x = \pm\frac{1}{2} \quad \text{Simplify radicals.}$$

$$x = \pm\frac{\sqrt{6}}{3}$$

Check that the solution set is $\left\{\pm\frac{\sqrt{6}}{3}, \pm\frac{1}{2}\right\}$.

✔ **Now Try Exercise 87.**

NOTE To solve the equation from **Example 9,**

$$12x^4 - 11x^2 + 2 = 0,$$

we could factor $12x^4 - 11x^2 + 2$ directly as $(3x^2 - 2)(4x^2 - 1)$, set each factor equal to zero, and then solve the resulting two quadratic equations. ***Which method to use is a matter of personal preference.***

1.6 Exercises

CONCEPT PREVIEW *Fill in the blank to correctly complete each sentence.*

1. A(n) ________ is an equation that has a rational expression for one or more terms.

2. Proposed solutions for which any denominator equals ________ are excluded from the solution set of a rational equation.

3. If a job can be completed in 4 hr, then the rate of work is ________ of the job per hour.

4. When the power property is used to solve an equation, it is essential to check all proposed solutions in the ________.

5. An equation such as $x^{3/2} = 8$ is an equation with a(n) ________, because it contains a variable raised to an exponent that is a rational number.

CONCEPT PREVIEW *Match each equation in Column I with the correct first step for solving it in Column II.*

I	II
6. $\frac{2x+3}{x} + \frac{5}{x+5} = 7$	**A.** Cube each side of the equation.
7. $\sqrt{x+5} = 7$	**B.** Multiply each side of the equation by $x(x+5)$.
8. $(x+5)^{5/2} = 32$	**C.** Raise each side of the equation to the power $\frac{2}{5}$.
9. $(x+5)^{2/3} - (x+5)^{1/3} - 6 = 0$	**D.** Square each side of the equation.
10. $\sqrt[3]{x(x+5)} = \sqrt[3]{-6}$	**E.** Let $u = (x+5)^{1/3}$ and $u^2 = (x+5)^{2/3}$.

Decide what values of the variable cannot possibly be solutions for each equation. Do not solve. ***See Examples 1 and 2.***

11. $\frac{5}{2x+3} - \frac{1}{x-6} = 0$

12. $\frac{2}{x+1} + \frac{3}{5x-2} = 0$

13. $\frac{3}{x-2} + \frac{1}{x+1} = \frac{3}{x^2-x-2}$

14. $\frac{2}{x+3} - \frac{5}{x-1} = \frac{-5}{x^2+2x-3}$

15. $\frac{1}{4x} - \frac{2}{x} = 3$

16. $\frac{5}{2x} + \frac{2}{x} = 6$

Solve each equation. ***See Example 1.***

17. $\frac{2x+5}{2} - \frac{3x}{x-2} = x$

18. $\frac{4x+3}{4} - \frac{2x}{x+1} = x$

19. $\frac{x}{x-3} = \frac{3}{x-3} + 3$

20. $\frac{x}{x-4} = \frac{4}{x-4} + 4$

21. $\frac{-2}{x-3} + \frac{3}{x+3} = \frac{-12}{x^2-9}$

22. $\frac{3}{x-2} + \frac{1}{x+2} = \frac{12}{x^2-4}$

23. $\frac{4}{x^2+x-6} - \frac{1}{x^2-4} = \frac{2}{x^2+5x+6}$

24. $\frac{3}{x^2+x-2} - \frac{1}{x^2-1} = \frac{7}{2x^2+6x+4}$

Solve each equation. ***See Example 2.***

25. $\frac{2x+1}{x-2} + \frac{3}{x} = \frac{-6}{x^2-2x}$

26. $\frac{4x+3}{x+1} + \frac{2}{x} = \frac{1}{x^2+x}$

27. $\frac{x}{x-1} - \frac{1}{x+1} = \frac{2}{x^2-1}$

28. $\frac{-x}{x+1} - \frac{1}{x-1} = \frac{-2}{x^2-1}$

29. $\frac{5}{x^2} - \frac{43}{x} = 18$

30. $\frac{7}{x^2} + \frac{19}{x} = 6$

31. $2 = \frac{3}{2x-1} + \frac{-1}{(2x-1)^2}$

32. $6 = \frac{7}{2x-3} + \frac{3}{(2x-3)^2}$

33. $\frac{2x-5}{x} = \frac{x-2}{3}$

34. $\frac{x+4}{2x} = \frac{x-1}{3}$

35. $\frac{2x}{x-2} = 5 + \frac{4x^2}{x-2}$

36. $\frac{3x^2}{x-1} + 2 = \frac{x}{x-1}$

Solve each problem. ***See Example 3.***

37. ***Painting a House*** (This problem appears in the 1994 movie *Little Big League*.) If Joe can paint a house in 3 hr, and Sam can paint the same house in 5 hr, how long does it take them to do it together?

38. ***Painting a House*** Repeat **Exercise 37,** but assume that Joe takes 6 hr working alone, and Sam takes 8 hr working alone.

39. ***Pollution in a River*** Two chemical plants are polluting a river. If plant A produces a predetermined maximum amount of pollutant twice as fast as plant B, and together they produce the maximum pollutant in 26 hr, how long will it take plant B alone?

	Rate	Time	Part of Job Completed
Pollution from A	$\frac{1}{x}$	26	$26\left(\frac{1}{x}\right)$
Pollution from B		26	

x represents the number of hours it takes plant A, working alone, to produce the maximum pollutant.

40. ***Filling a Settling Pond*** A sewage treatment plant has two inlet pipes to its settling pond. One pipe can fill the pond 3 times as fast as the other pipe, and together they can fill the pond in 12 hr. How long will it take the faster pipe to fill the pond alone?

	Rate	Time	Part of Job Completed
Faster Inlet Pipe	$\frac{1}{x}$	12	
Slower Inlet Pipe	$\frac{1}{3x}$	12	

x represents the number of hours it takes the faster inlet pipe, working alone, to fill the settling pond.

41. ***Filling a Pool*** An inlet pipe can fill Blake's pool in 5 hr, and an outlet pipe can empty it in 8 hr. In his haste to surf the Internet, Blake left both pipes open. How long did it take to fill the pool?

42. ***Filling a Pool*** Suppose Blake discovered his error (see **Exercise 41**) after an hour-long surf. If he then closed the outlet pipe, how much more time would be needed to fill the pool?

43. ***Filling a Sink*** With both taps open, Robert can fill his kitchen sink in 5 min. When full, the sink drains in 10 min. How long will it take to fill the sink if Robert forgets to put in the stopper?

44. ***Filling a Sink*** If Robert (see **Exercise 43**) remembers to put in the stopper after 1 min, how much longer will it take to fill the sink?

Solve each equation. ***See Examples 4–6.***

45. $x - \sqrt{2x + 3} = 0$

46. $x - \sqrt{3x + 18} = 0$

47. $\sqrt{3x + 7} = 3x + 5$

48. $\sqrt{4x + 13} = 2x - 1$

49. $\sqrt{4x + 5} - 6 = 2x - 11$

50. $\sqrt{6x + 7} - 9 = x - 7$

51. $\sqrt{4x} - x + 3 = 0$

52. $\sqrt{2x} - x + 4 = 0$

53. $\sqrt{x} - \sqrt{x-5} = 1$

54. $\sqrt{x} - \sqrt{x-12} = 2$

55. $\sqrt{x+7} + 3 = \sqrt{x-4}$

56. $\sqrt{x+5} + 2 = \sqrt{x-1}$

57. $\sqrt{2x+5} - \sqrt{x+2} = 1$

58. $\sqrt{4x+1} - \sqrt{x-1} = 2$

59. $\sqrt{3x} = \sqrt{5x+1} - 1$

60. $\sqrt{2x} = \sqrt{3x+12} - 2$

61. $\sqrt{x+2} = 1 - \sqrt{3x+7}$

62. $\sqrt{2x-5} = 2 + \sqrt{x-2}$

63. $\sqrt{2\sqrt{7x+2}} = \sqrt{3x+2}$

64. $\sqrt{3\sqrt{2x+3}} = \sqrt{5x-6}$

65. $3 - \sqrt{x} = \sqrt{2\sqrt{x} - 3}$

66. $\sqrt{x} + 2 = \sqrt{4 + 7\sqrt{x}}$

67. $\sqrt[3]{4x+3} = \sqrt[3]{2x-1}$

68. $\sqrt[3]{2x} = \sqrt[3]{5x+2}$

69. $\sqrt[3]{5x^2 - 6x + 2} - \sqrt[3]{x} = 0$

70. $\sqrt[3]{3x^2 - 9x + 8} = \sqrt[3]{x}$

71. $\sqrt[4]{x-15} = 2$

72. $\sqrt[4]{3x+1} = 1$

73. $\sqrt[4]{x^2 + 2x} = \sqrt[4]{3}$

74. $\sqrt[4]{x^2 + 6x} = 2$

Solve each equation. **See Example 7.**

75. $x^{3/2} = 125$

76. $x^{5/4} = 32$

77. $(x^2 + 24x)^{1/4} = 3$

78. $(3x^2 + 52x)^{1/4} = 4$

79. $(x-3)^{2/5} = 4$

80. $(x+200)^{2/3} = 36$

81. $(2x+5)^{1/3} - (6x-1)^{1/3} = 0$

82. $(3x+7)^{1/3} - (4x+2)^{1/3} = 0$

83. $(2x-1)^{2/3} = x^{1/3}$

84. $(x-3)^{2/5} = (4x)^{1/5}$

85. $x^{2/3} = 2x^{1/3}$

86. $3x^{3/4} = x^{1/2}$

Solve each equation. **See Examples 8 and 9.**

87. $2x^4 - 7x^2 + 5 = 0$

88. $4x^4 - 8x^2 + 3 = 0$

89. $x^4 + 2x^2 - 15 = 0$

90. $3x^4 + 10x^2 - 25 = 0$

91. $(x-1)^{2/3} + (x-1)^{1/3} - 12 = 0$

92. $(2x-1)^{2/3} + 2(2x-1)^{1/3} - 3 = 0$

93. $(x+1)^{2/5} - 3(x+1)^{1/5} + 2 = 0$

94. $(x+5)^{2/3} + (x+5)^{1/3} - 20 = 0$

95. $4(x+1)^4 - 13(x+1)^2 = -9$

96. $25(x-5)^4 - 116(x-5)^2 = -64$

97. $6(x+2)^4 - 11(x+2)^2 = -4$

98. $8(x-4)^4 - 10(x-4)^2 = -3$

99. $10x^{-2} + 33x^{-1} - 7 = 0$

100. $7x^{-2} - 10x^{-1} - 8 = 0$

101. $x^{-2/3} + x^{-1/3} - 6 = 0$

102. $2x^{-2/5} - x^{-1/5} - 1 = 0$

103. $16x^{-4} - 65x^{-2} + 4 = 0$

104. $625x^{-4} - 125x^{-2} + 4 = 0$

Solve each equation for the specified variable. (Assume all denominators are nonzero.)

105. $d = k\sqrt{h}$, for h

106. $x^{2/3} + y^{2/3} = a^{2/3}$, for y

107. $m^{3/4} + n^{3/4} = 1$, for m

108. $\frac{1}{R} = \frac{1}{r_1} + \frac{1}{r_2}$, for R

109. $\frac{E}{e} = \frac{R + r}{r}$, for e

110. $a^2 + b^2 = c^2$, for b

Relating Concepts

For individual or collaborative investigation *(Exercises 111–114)*

In this section we introduced methods of solving equations quadratic in form by substitution and solving equations involving radicals by raising each side of the equation to a power. Suppose we wish to solve

$$x - \sqrt{x} - 12 = 0.$$

We can solve this equation using either of the two methods. ***Work Exercises 111–114 in order,*** *to see how both methods apply.*

111. Let $u = \sqrt{x}$ and solve the equation by substitution.

112. Solve the equation by isolating $\sqrt{x}$ on one side and then squaring.

113. Which one of the methods used in **Exercises 111 and 112** do you prefer? Why?

114. Solve $3x - 2\sqrt{x} - 8 = 0$ using one of the two methods described.

Summary Exercises on Solving Equations

This section of miscellaneous equations provides practice in solving all the types introduced in this chapter so far. Solve each equation.

1. $4x - 3 = 2x + 3$

2. $5 - (6x + 3) = 2(2 - 2x)$

3. $x(x + 6) = 9$

4. $x^2 = 8x - 12$

5. $\sqrt{x + 2} + 5 = \sqrt{x + 15}$

6. $\frac{5}{x + 3} - \frac{6}{x - 2} = \frac{3}{x^2 + x - 6}$

7. $\frac{3x + 4}{3} - \frac{2x}{x - 3} = x$

8. $\frac{x}{2} + \frac{4}{3}x = x + 5$

9. $5 - \frac{2}{x} + \frac{1}{x^2} = 0$

10. $(2x + 1)^2 = 9$

11. $x^{-2/5} - 2x^{-1/5} - 15 = 0$

12. $\sqrt{x + 2} + 1 = \sqrt{2x + 6}$

13. $x^4 - 3x^2 - 4 = 0$

14. $1.2x + 0.3 = 0.7x - 0.9$

15. $\sqrt[3]{2x + 1} = \sqrt[3]{9}$

16. $3x^2 - 2x = -1$

17. $3[2x - (6 - 2x) + 1] = 5x$

18. $\sqrt{x} + 1 = \sqrt{11 - \sqrt{x}}$

19. $(14 - 2x)^{2/3} = 4$

20. $-x^{-2} + 2x^{-1} = 1$

21. $\frac{3}{x - 3} = \frac{3}{x - 3}$

22. $a^2 + b^2 = c^2$, for a

1.7 Inequalities

- Linear Inequalities
- Three-Part Inequalities
- Quadratic Inequalities
- Rational Inequalities

An **inequality** says that one expression is greater than, greater than or equal to, less than, or less than or equal to another. As with equations, a value of the variable for which the inequality is true is a solution of the inequality, and the set of all solutions is the solution set of the inequality. Two inequalities with the same solution set are equivalent.

Inequalities are solved with the properties of inequality, which are similar to the properties of equality.

Properties of Inequality

Let a, b, and c represent real numbers.

1. If $a < b$, then $a + c < b + c$.

2. If $a < b$ and if $c > 0$, then $ac < bc$.

3. If $a < b$ and if $c < 0$, then $ac > bc$.

Replacing $<$ with $>$, $\leq$, or $\geq$ results in similar properties. (Restrictions on c remain the same.)

NOTE Multiplication may be replaced by division in Properties 2 and 3. ***Always remember to reverse the direction of the inequality symbol when multiplying or dividing by a negative number.***

Linear Inequalities

The definition of a *linear inequality* is similar to the definition of a linear equation.

Linear Inequality in One Variable

A **linear inequality in one variable** is an inequality that can be written in the form

$$ax + b > 0,^*$$

where a and b are real numbers and $a \neq 0$.

*The symbol $>$ can be replaced with $<$, $\leq$, or $\geq$.

EXAMPLE 1 Solving a Linear Inequality

Solve $-3x + 5 > -7$.

SOLUTION

$$-3x + 5 > -7$$

$$-3x + 5 - 5 > -7 - 5 \quad \text{Subtract 5.}$$

$$-3x > -12 \quad \text{Combine like terms.}$$

$$\frac{-3x}{-3} < \frac{-12}{-3} \quad \text{Divide by } -3. \text{ Reverse the direction of the inequality symbol when multiplying or dividing by a negative number.}$$

Don't forget to reverse the inequality symbol here.

$$x < 4$$

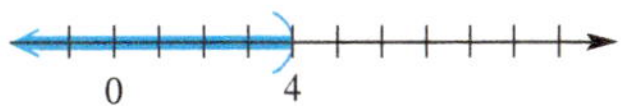

Figure 9

Thus, the original inequality $-3x + 5 > -7$ is satisfied by any real number less than 4. The solution set can be written $\{x \mid x < 4\}$.

A graph of the solution set is shown in **Figure 9,** where the parenthesis is used to show that 4 itself does not belong to the solution set. As shown below, testing values from the solution set in the original inequality will produces true statements. Testing values outside the solution set produces false statements.

CHECK $\quad -3x + 5 > -7 \quad$ Original inequality

$-3(0) + 5 \overset{?}{>} -7 \quad$ Let $x = 0$.	$-3(5) + 5 \overset{?}{>} -7 \quad$ Let $x = 5$.
$5 > -7$ ✓ True	$-10 > -7 \quad$ False

The solution set of the inequality,

$$\{x \mid x < 4\}, \quad \text{Set-builder notation}$$

is an example of an **interval.** We use **interval notation** to write intervals. With this notation, we write the above interval as

$$(-\infty, 4). \quad \text{Interval notation}$$

The symbol $-\infty$ does not represent an actual number. Rather, it is used to show that the interval includes all real numbers less than 4. The interval $(-\infty, 4)$ is an example of an **open interval** because the endpoint, 4, is not part of the interval. An interval that includes both its endpoints is a **closed interval.** A square bracket indicates that a number *is* part of an interval, and a parenthesis indicates that a number *is not* part of an interval.

✔ **Now Try Exercise 13.**

In the table that follows, we assume that $a < b$.

Summary of Types of Intervals

Type of Interval	Set	Interval Notation	Graph
Open interval	$\{x \mid x > a\}$	(a, ∞)	a
	$\{x \mid a < x < b\}$	(a, b)	a b
	$\{x \mid x < b\}$	$(-\infty, b)$	b
Other intervals	$\{x \mid x \geq a\}$	$[a, \infty)$	a
	$\{x \mid a < x \leq b\}$	$(a, b]$	a b
	$\{x \mid a \leq x < b\}$	$[a, b)$	a b
	$\{x \mid x \leq b\}$	$(-\infty, b]$	b
Closed interval	$\{x \mid a \leq x \leq b\}$	$[a, b]$	a b
Disjoint interval	$\{x \mid x < a \text{ or } x > b\}$	$(-\infty, a) \cup (b, \infty)$	a b
All real numbers	$\{x \mid x \text{ is a real number}\}$	$(-\infty, \infty)$	

EXAMPLE 2 **Solving a Linear Inequality**

Solve $4 - 3x \leq 7 + 2x$. Give the solution set in interval notation.

SOLUTION

$$4 - 3x \leq 7 + 2x$$

$$4 - 3x - 4 \leq 7 + 2x - 4 \quad \text{Subtract 4.}$$

$$-3x \leq 3 + 2x \quad \text{Combine like terms.}$$

$$-3x - 2x \leq 3 + 2x - 2x \quad \text{Subtract } 2x.$$

$$-5x \leq 3 \quad \text{Combine like terms.}$$

$$\frac{-5x}{-5} \geq \frac{3}{-5} \quad \text{Divide by } -5. \text{ Reverse the direction of the inequality symbol.}$$

$$x \geq -\frac{3}{5} \quad \frac{a}{-b} = -\frac{a}{b}$$

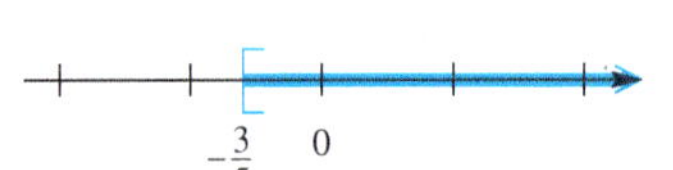

Figure 10

In interval notation, the solution set is $\left[-\frac{3}{5}, \infty\right)$. See **Figure 10** for the graph.

✔ **Now Try Exercise 15.**

A product will break even, or begin to produce a profit, only if the revenue from selling the product at least equals the cost of producing it. If R represents revenue and C is cost, then the **break-even point** is the point where $R = C$.

EXAMPLE 3 **Finding the Break-Even Point**

If the revenue and cost of a certain product are given by

$$R = 4x \quad \text{and} \quad C = 2x + 1000,$$

where x is the number of units produced and sold, at what production level does *R at least equal C*?

SOLUTION Set $R \geq C$ and solve for x.

At least equal to translates as $\geq$.

$$R \geq C$$

$$4x \geq 2x + 1000 \quad \text{Substitute.}$$

$$2x \geq 1000 \quad \text{Subtract } 2x.$$

$$x \geq 500 \quad \text{Divide by 2.}$$

The break-even point is at $x = 500$. This product will at least break even if the number of units produced and sold is in the interval $[500, \infty)$.

✔ **Now Try Exercise 25.**

Three-Part Inequalities

The inequality $-2 < 5 + 3x < 20$ says that

$5 + 3x$ is *between* -2 and 20.

This inequality is solved using an extension of the properties of inequality given earlier, working with all three expressions at the same time.

EXAMPLE 4 **Solving a Three-Part Inequality**

Solve $-2 < 5 + 3x < 20$. Give the solution set in interval notation.

SOLUTION

$$-2 < 5 + 3x < 20$$

$$-2 - 5 < 5 + 3x - 5 < 20 - 5 \quad \text{Subtract 5 from each part.}$$

$$-7 < 3x < 15 \quad \text{Combine like terms in each part.}$$

$$\frac{-7}{3} < \frac{3x}{3} < \frac{15}{3} \quad \text{Divide each part by 3.}$$

$$-\frac{7}{3} < x < 5$$

Figure 11

The solution set, graphed in **Figure 11,** is the interval $\left(-\frac{7}{3}, 5\right)$.

✔ **Now Try Exercise 29.**

Quadratic Inequalities We can distinguish a *quadratic inequality* from a linear inequality by noticing that it is of degree 2.

Quadratic Inequality

A **quadratic inequality** is an inequality that can be written in the form

$$ax^2 + bx + c < 0,^*$$

where a, b, and c are real numbers and $a \neq 0$.

*The symbol $<$ can be replaced with $>$, $\leq$, or $\geq$.

One method of solving a quadratic inequality involves finding the solutions of the corresponding quadratic equation and then testing values in the intervals on a number line determined by those solutions.

Solving a Quadratic Inequality

Step 1 Solve the corresponding quadratic equation.

Step 2 Identify the intervals determined by the solutions of the equation.

Step 3 Use a test value from each interval to determine which intervals form the solution set.

EXAMPLE 5 **Solving a Quadratic Inequality**

Solve $x^2 - x - 12 < 0$.

SOLUTION

Step 1 Find the values of x that satisfy $x^2 - x - 12 = 0$.

$$x^2 - x - 12 = 0 \quad \text{Corresponding quadratic equation}$$

$$(x + 3)(x - 4) = 0 \quad \text{Factor.}$$

$$x + 3 = 0 \quad \text{or} \quad x - 4 = 0 \quad \text{Zero-factor property}$$

$$x = -3 \quad \text{or} \quad x = 4 \quad \text{Solve each equation.}$$

Step 2 The two numbers -3 and 4 cause the expression $x^2 - x - 12$ to ***equal*** zero and can be used to divide the number line into three intervals, as shown in **Figure 12.** The expression $x^2 - x - 12$ will take on a value that is either ***less than*** zero or ***greater than*** zero on each of these intervals. We are looking for x-values that make the expression ***less than*** zero, so we use open circles at -3 and 4 to indicate that they are not included in the solution set.

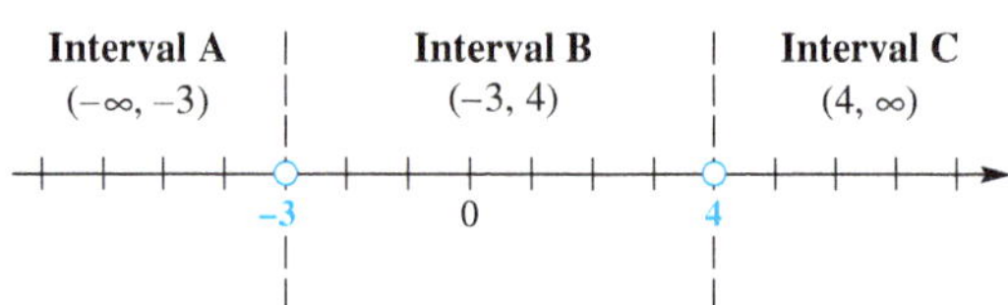

Use open circles because the inequality symbol does not include *equality*. -3 and 4 do not satisfy the *inequality*.

Figure 12

Step 3 Choose a test value in each interval to see whether it satisfies the original inequality, $x^2 - x - 12 < 0$. If the test value makes the statement true, then the entire interval belongs to the solution set.

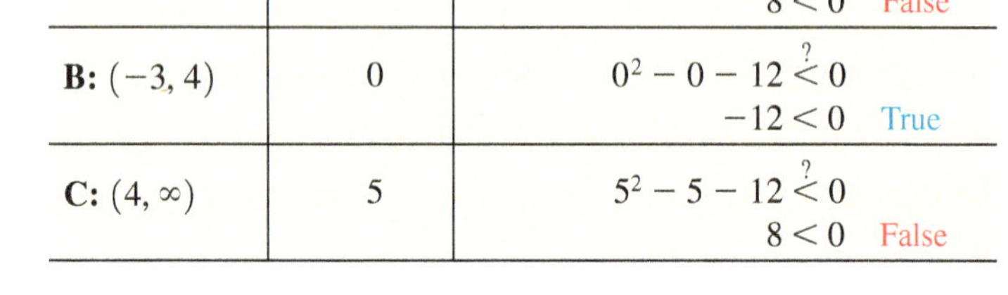

Interval	Test Value	Is $x^2 - x - 12 < 0$ True or False?
A: $(-\infty, -3)$	-4	$(-4)^2 - (-4) - 12 \overset{?}{<} 0$ $8 < 0$ False
B: $(-3, 4)$	0	$0^2 - 0 - 12 \overset{?}{<} 0$ $-12 < 0$ True
C: $(4, \infty)$	5	$5^2 - 5 - 12 \overset{?}{<} 0$ $8 < 0$ False

Because the values in Interval B make the inequality true, the solution set is the interval $(-3, 4)$. See **Figure 13.** ✔ **Now Try Exercise 41.**

-3 0 4

Figure 13

EXAMPLE 6 Solving a Quadratic Inequality

Solve $2x^2 + 5x - 12 \geq 0$.

SOLUTION

Step 1 Find the values of x that satisfy $2x^2 + 5x - 12 = 0$.

$$2x^2 + 5x - 12 = 0 \quad \text{Corresponding quadratic equation}$$

$$(2x - 3)(x + 4) = 0 \quad \text{Factor.}$$

$$2x - 3 = 0 \quad \text{or} \quad x + 4 = 0 \quad \text{Zero-factor property}$$

$$x = \frac{3}{2} \quad \text{or} \quad x = -4 \quad \text{Solve each equation.}$$

Step 2 The values $\frac{3}{2}$ and -4 cause the expression $2x^2 + 5x - 12$ to equal 0 and can be used to form the intervals $(-\infty, -4)$, $\left(-4, \frac{3}{2}\right)$, and $\left(\frac{3}{2}, \infty\right)$ on the number line, as seen in **Figure 14.**

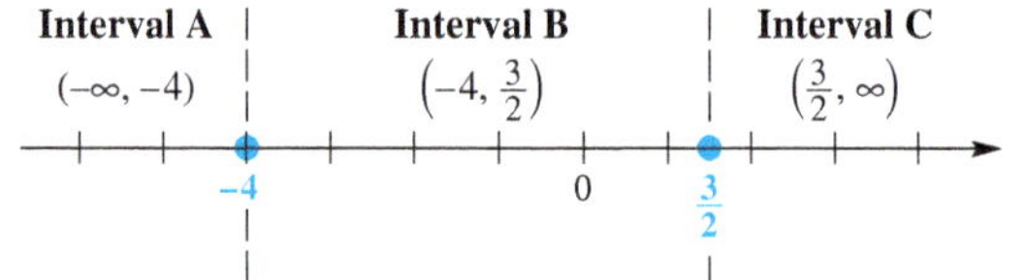

Use closed circles because the inequality symbol includes *equality*. -4 and $\frac{3}{2}$ satisfy the *inequality*.

Figure 14

Step 3 Choose a test value in each interval.

Interval	Test Value	Is $2x^2 + 5x - 12 \ge 0$ True or False?
A: $(-\infty, -4)$	-5	$2(-5)^2 + 5(-5) - 12 \overset{?}{\ge} 0$ $13 \ge 0$ True
B: $\left(-4, \frac{3}{2}\right)$	0	$2(0)^2 + 5(0) - 12 \overset{?}{\ge} 0$ $-12 \ge 0$ False
C: $\left(\frac{3}{2}, \infty\right)$	2	$2(2)^2 + 5(2) - 12 \overset{?}{\ge} 0$ $6 \ge 0$ True

The values in Intervals A and C make the inequality true, so the solution set is a disjoint interval: the *union* of the two intervals, written

$$(-\infty, -4] \cup \left[\frac{3}{2}, \infty\right).$$

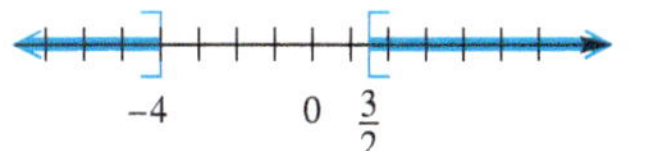

Figure 15

The graph of the solution set is shown in **Figure 15.**

Now Try Exercise 39.

NOTE Inequalities that use the symbols $<$ and $>$ are **strict inequalities**, while $\le$ and $\ge$ are used in **nonstrict inequalities.** The solutions of the equation in **Example 5** were not included in the solution set because the inequality was a *strict* inequality. In **Example 6,** the solutions of the equation *were* included in the solution set because of the nonstrict inequality.

EXAMPLE 7 Finding Projectile Height

If a projectile is launched from ground level with an initial velocity of 96 ft per sec, its height s in feet t seconds after launching is given by the following equation.

$$s = -16t^2 + 96t$$

When will the projectile be greater than 80 ft above ground level?

SOLUTION

$$-16t^2 + 96t > 80 \quad \text{Set } s \text{ greater than 80.}$$

$$-16t^2 + 96t - 80 > 0 \quad \text{Subtract 80.}$$

$$t^2 - 6t + 5 < 0 \quad \text{Divide by } -16.$$

Reverse the direction of the inequality symbol.

Now solve the corresponding *equation.*

$$t^2 - 6t + 5 = 0$$

$$(t-1)(t-5) = 0 \quad \text{Factor.}$$

$$t - 1 = 0 \quad \text{or} \quad t - 5 = 0 \quad \text{Zero-factor property}$$

$$t = 1 \quad \text{or} \quad t = 5 \quad \text{Solve each equation.}$$

Use these values to determine the intervals

$$(-\infty, 1), \quad (1, 5), \quad \text{and} \quad (5, \infty).$$

We are solving a strict inequality, so solutions of the equation $t^2 - 6t + 5 = 0$ are ***not*** included. Choose a test value from each interval to see whether it satisfies the inequality $t^2 - 6t + 5 < 0$. See **Figure 16** on the next page.

Interval A $(-\infty, 1)$	Interval B $(1, 5)$	Interval C $(5, \infty)$
Test Value 0	Test Value 3	Test Value 7
$0^2 - 6(0) + 5 \overset{?}{<} 0$	$3^2 - 6(3) + 5 \overset{?}{<} 0$	$7^2 - 6(7) + 5 \overset{?}{<} 0$
$5 < 0$	$-4 < 0$	$12 < 0$
False	True	False

Figure 16

The values in Interval B, $(1, 5)$, make the inequality true. The projectile is greater than 80 ft above ground level between 1 and 5 sec after it is launched.

✔ **Now Try Exercise 81.**

Rational Inequalities Inequalities involving one or more rational expressions are **rational inequalities.**

$$\frac{5}{x+4} \geq 1 \quad \text{and} \quad \frac{2x-1}{3x+4} < 5 \qquad \text{Rational inequalities}$$

Solving a Rational Inequality

Step 1 Rewrite the inequality, if necessary, so that 0 is on one side and there is a single fraction on the other side.

Step 2 Determine the values that will cause either the numerator or the denominator of the rational expression to equal 0. These values determine the intervals on the number line to consider.

Step 3 Use a test value from each interval to determine which intervals form the solution set.

A value causing a denominator to equal zero will never be included in the solution set. If the inequality is strict, any value causing the numerator to equal zero will be excluded. If the inequality is nonstrict, any such value will be included.

CAUTION Solving a rational inequality such as $\frac{5}{x+4} \geq 1$ by multiplying each side by $x + 4$ requires considering *two cases*, because the sign of $x + 4$ depends on the value of x. If $x + 4$ is negative, then the inequality symbol must be reversed. The procedure described in the preceding box eliminates the need for considering separate cases.

EXAMPLE 8 Solving a Rational Inequality

Solve $\frac{5}{x+4} \geq 1$.

SOLUTION

Step 1

$$\frac{5}{x+4} - 1 \geq 0 \qquad \text{Subtract 1 so that 0 is on one side.}$$

$$\frac{5}{x+4} - \frac{x+4}{x+4} \geq 0 \qquad \text{Use } x+4 \text{ as the common denominator.}$$

Note the careful use of parentheses.

$$\frac{5-(x+4)}{x+4} \geq 0 \qquad \text{Write as a single fraction.}$$

$$\frac{1-x}{x+4} \geq 0 \qquad \text{Combine like terms in the numerator, being careful with signs.}$$

Step 2 The quotient possibly changes sign only where x-values make the numerator or denominator 0. This occurs at

$$1 - x = 0 \quad \text{or} \quad x + 4 = 0$$

$$x = 1 \quad \text{or} \quad x = -4.$$

These values form the intervals $(-\infty, -4)$, $(-4, 1)$, and $(1, \infty)$ on the number line, as seen in **Figure 17.**

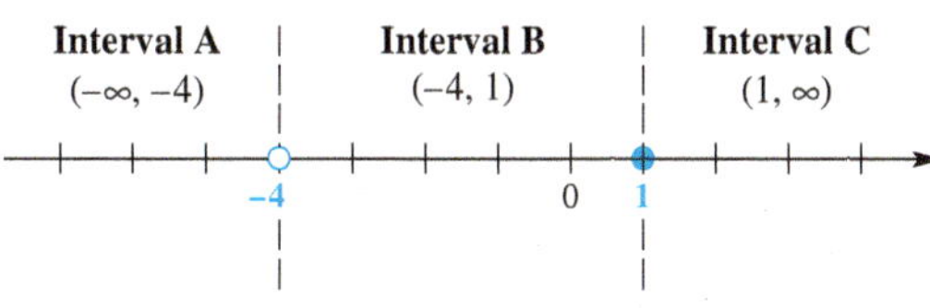

Figure 17

Use a solid circle on 1 because the symbol is $\geq$. The value -4 cannot be in the solution set because it causes the denominator to equal 0. Use an open circle on -4.

Step 3 Choose test values.

Interval	Test Value	Is $\frac{5}{x+4} \geq 1$ True or False?
A: $(-\infty, -4)$	-5	$\frac{5}{-5+4} \stackrel{?}{\geq} 1$ $-5 \geq 1$ False
B: $(-4, 1)$	0	$\frac{5}{0+4} \stackrel{?}{\geq} 1$ $\frac{5}{4} \geq 1$ True
C: $(1, \infty)$	2	$\frac{5}{2+4} \stackrel{?}{\geq} 1$ $\frac{5}{6} \geq 1$ False

The values in Interval B, $(-4, 1)$, satisfy the original inequality. The value 1 makes the nonstrict inequality true, so it must be included in the solution set. Because -4 makes the denominator 0, it must be excluded. The solution set is the interval $(-4, 1]$.

✔ **Now Try Exercise 59.**

CAUTION ***Be careful with the endpoints of the intervals when solving rational inequalities.***

EXAMPLE 9 Solving a Rational Inequality

Solve $\frac{2x-1}{3x+4} < 5$.

SOLUTION

$$\frac{2x-1}{3x+4} - 5 < 0 \quad \text{Subtract 5.}$$

$$\frac{2x-1}{3x+4} - \frac{5(3x+4)}{3x+4} < 0 \quad \text{The common denominator is } 3x+4.$$

$$\frac{2x-1-5(3x+4)}{3x+4} < 0 \quad \text{Write as a single fraction.}$$

Be careful with signs.

$$\frac{2x-1-15x-20}{3x+4} < 0 \quad \text{Distributive property}$$

$$\frac{-13x-21}{3x+4} < 0 \quad \text{Combine like terms in the numerator.}$$

Set the numerator and denominator of $\frac{-13x-21}{3x+4}$ equal to 0 and solve the resulting equations to find the values of x where sign changes may occur.

$$-13x - 21 = 0 \quad \text{or} \quad 3x + 4 = 0$$

$$x = -\frac{21}{13} \quad \text{or} \quad x = -\frac{4}{3}$$

Use these values to form intervals on the number line. Use an open circle at $-\frac{21}{13}$ because of the strict inequality, and use an open circle at $-\frac{4}{3}$ because it causes the denominator to equal 0. See **Figure 18.**

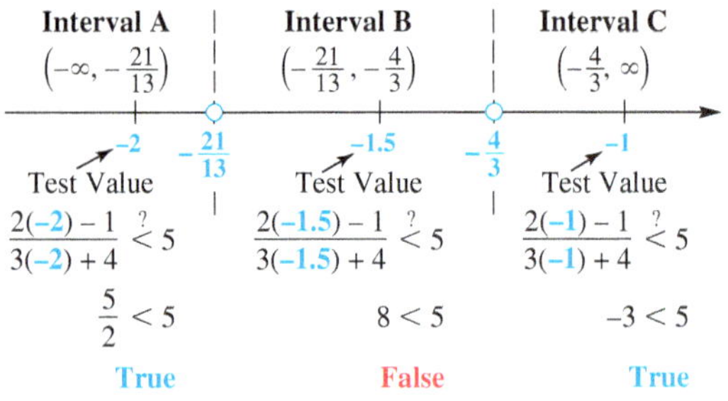

Figure 18

Choosing a test value from each interval shows that the values in Intervals A and C satisfy the original inequality, $\frac{2x-1}{3x+4} < 5$. So the solution set is the union of these intervals.

$$\left(-\infty, -\frac{21}{13}\right) \cup \left(-\frac{4}{3}, \infty\right)$$

✔ **Now Try Exercise 71.**

1.7 Exercises

CONCEPT PREVIEW *Match the inequality in each exercise in Column I with its equivalent interval notation in Column II.*

I	II
1. $x < -6$	**A.** $(-2, 6]$
2. $x \leq 6$	**B.** $[-2, 6)$
3. $-2 < x \leq 6$	**C.** $(-\infty, -6]$
4. $x^2 \geq 0$	**D.** $[6, \infty)$
5. $x \geq -6$	**E.** $(-\infty, -3) \cup (3, \infty)$
6. $6 \leq x$	**F.** $(-\infty, -6)$
7. 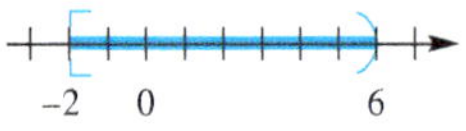	**G.** $(0, 8)$
8. 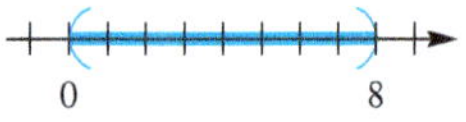	**H.** $(-\infty, \infty)$
9. 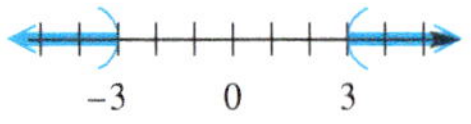	**I.** $[-6, \infty)$
10. 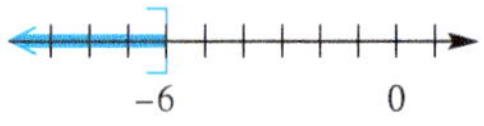	**J.** $(-\infty, 6]$

11. Explain how to determine whether to use a parenthesis or a square bracket when writing the solution set of a linear inequality in interval notation.

12. *Concept Check* The three-part inequality $a < x < b$ means "a is less than x and x is less than b." Which inequality is *not* satisfied by some real number x?

A. $-3 < x < 10$ B. $0 < x < 6$

C. $-3 < x < -1$ D. $-8 < x < -10$

Solve each inequality. Give the solution set in interval notation. **See Examples 1 and 2.**

13. $-2x + 8 \le 16$

14. $-3x - 8 \le 7$

15. $-2x - 2 \le 1 + x$

16. $-4x + 3 \ge -2 + x$

17. $3(x + 5) + 1 \ge 5 + 3x$

18. $6x - (2x + 3) \ge 4x - 5$

19. $8x - 3x + 2 < 2(x + 7)$

20. $2 - 4x + 5(x - 1) < -6(x - 2)$

21. $\dfrac{4x + 7}{-3} \le 2x + 5$

22. $\dfrac{2x - 5}{-8} \le 1 - x$

23. $\dfrac{1}{3}x + \dfrac{2}{5}x - \dfrac{1}{2}(x + 3) \le \dfrac{1}{10}$

24. $-\dfrac{2}{3}x - \dfrac{1}{6}x + \dfrac{2}{3}(x + 1) \le \dfrac{4}{3}$

Break-Even Interval Find all intervals where each product will at least break even. **See Example 3.**

25. The cost to produce x units of picture frames is $C = 50x + 5000$, while the revenue is $R = 60x$.

26. The cost to produce x units of baseball caps is $C = 100x + 6000$, while the revenue is $R = 500x$.

27. The cost to produce x units of coffee cups is $C = 105x + 900$, while the revenue is $R = 85x$.

28. The cost to produce x units of briefcases is $C = 70x + 500$, while the revenue is $R = 60x$.

Solve each inequality. Give the solution set in interval notation. **See Example 4.**

29. $-5 < 5 + 2x < 11$

30. $-7 < 2 + 3x < 5$

31. $10 \le 2x + 4 \le 16$

32. $-6 \le 6x + 3 \le 21$

33. $-11 > -3x + 1 > -17$

34. $2 > -6x + 3 > -3$

35. $-4 \le \dfrac{x + 1}{2} \le 5$

36. $-5 \le \dfrac{x - 3}{3} \le 1$

37. $-3 \le \dfrac{3x - 4}{-5} < 4$

38. $1 \le \dfrac{4x - 5}{-2} < 9$

Solve each quadratic inequality. Give the solution set in interval notation. **See Examples 5 and 6.**

39. $x^2 - x - 6 > 0$

40. $x^2 - 7x + 10 > 0$

41. $2x^2 - 9x \le 18$

42. $3x^2 + x \le 4$

43. $-x^2 - 4x - 6 \le -3$

44. $-x^2 - 6x - 16 > -8$

45. $x(x - 1) \le 6$

46. $x(x + 1) < 12$

47. $x^2 \le 9$

48. $x^2 > 16$

49. $x^2 + 5x + 7 < 0$

50. $4x^2 + 3x + 1 \leq 0$

51. $x^2 - 2x \leq 1$

52. $x^2 + 4x > -1$

53. *Concept Check* Which inequality has solution set $(-\infty, \infty)$?

A. $(x-3)^2 \geq 0$ **B.** $(5x-6)^2 \leq 0$

C. $(6x+4)^2 > 0$ **D.** $(8x+7)^2 < 0$

54. *Concept Check* Which inequality in **Exercise 53** has solution set $\varnothing$?

Solve each rational inequality. Give the solution set in interval notation. ***See Examples 8 and 9.***

55. $\frac{x-3}{x+5} \leq 0$

56. $\frac{x+1}{x-4} > 0$

57. $\frac{1-x}{x+2} < -1$

58. $\frac{6-x}{x+2} > 1$

59. $\frac{3}{x-6} \leq 2$

60. $\frac{3}{x-2} < 1$

61. $\frac{-4}{1-x} < 5$

62. $\frac{-6}{3x-5} \leq 2$

63. $\frac{10}{3+2x} \leq 5$

64. $\frac{1}{x+2} \geq 3$

65. $\frac{7}{x+2} \geq \frac{1}{x+2}$

66. $\frac{5}{x+1} > \frac{12}{x+1}$

67. $\frac{3}{2x-1} > \frac{-4}{x}$

68. $\frac{-5}{3x+2} \geq \frac{5}{x}$

69. $\frac{4}{2-x} \geq \frac{3}{1-x}$

70. $\frac{4}{x+1} < \frac{2}{x+3}$

71. $\frac{x+3}{x-5} \leq 1$

72. $\frac{x+2}{3+2x} \leq 5$

Solve each rational inequality. Give the solution set in interval notation.

73. $\frac{2x-3}{x^2+1} \geq 0$

74. $\frac{9x-8}{4x^2+25} < 0$

75. $\frac{(5-3x)^2}{(2x-5)^3} > 0$

76. $\frac{(5x-3)^3}{(25-8x)^2} \leq 0$

77. $\frac{(2x-3)(3x+8)}{(x-6)^3} \geq 0$

78. $\frac{(9x-11)(2x+7)}{(3x-8)^3} > 0$

(Modeling) Solve each problem.

79. *Box Office Receipts* U.S. movie box office receipts, in billions of dollars, are shown in 5-year increments from 1993 to 2013. (*Source:* www.boxofficemojo.com)

Year	Receipts
1993	5.154
1998	6.949
2003	9.240
2008	9.631
2013	10.924

These receipts R are reasonably approximated by the linear model

$$R = 0.2844x + 5.535,$$

where $x = 0$ corresponds to 1993, $x = 5$ corresponds to 1998, and so on. Using the model, calculate the year in which the receipts first exceed each amount.

(a) \$7.6 billion **(b)** \$10 billion

80. *Recovery of Solid Waste* The percent W of municipal solid waste recovered is shown in the bar graph. The linear model

$$W = 0.33x + 33.1,$$

where $x = 1$ represents 2008, $x = 2$ represents 2009, and so on, fits the data reasonably well.

(a) Based on this model, when did the percent of waste recovered first exceed 34%?

(b) In what years was it between 33.9% and 34.5%?

Municipal Solid Waste Recovered

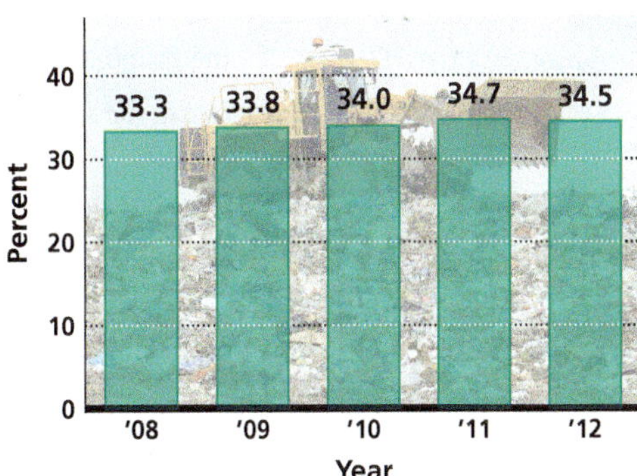

Source: U.S. Environmental Protection Agency.

Solve each problem. ***See Example 7.***

81. *Height of a Projectile* A projectile is fired straight up from ground level. After t seconds, its height above the ground is s feet, where

$$s = -16t^2 + 220t.$$

For what time period is the projectile at least 624 ft above the ground?

82. *Height of a Projectile* See **Exercise 81.** For what time period is the projectile at least 744 ft above the ground?

83. *Height of a Baseball* A baseball is hit so that its height, s, in feet after t seconds is

$$s = -16t^2 + 44t + 4.$$

For what time period is the ball at least 32 ft above the ground?

84. *Height of a Baseball* See **Exercise 83.** For what time period is the ball greater than 28 ft above the ground?

85. *Velocity of an Object* Suppose the velocity, v, of an object is given by

$$v = 2t^2 - 5t - 12,$$

where t is time in seconds. (Here t can be positive or negative.) Find the intervals where the velocity is negative.

86. *Velocity of an Object* The velocity of an object, v, after t seconds is given by

$$v = 3t^2 - 18t + 24.$$

Find the interval where the velocity is negative.

Relating Concepts

For individual or collaborative investigation *(Exercises 87–90)*

Inequalities that involve more than two factors, such as

$$(3x - 4)(x + 2)(x + 6) \le 0,$$

can be solved using an extension of the method shown in ***Examples 5 and 6. Work Exercises 87–90 in order,*** *to see how the method is extended.*

87. Use the zero-factor property to solve $(3x - 4)(x + 2)(x + 6) = 0$.

88. Plot the three solutions in **Exercise 87** on a number line, using closed circles because of the nonstrict inequality, $\le$.

89. The number line from **Exercise 88** should show four intervals formed by the three points. For each interval, choose a test value from the interval and decide whether it satisfies the original inequality.

90. On a single number line, do the following.

(a) Graph the intervals that satisfy the inequality, including endpoints. This is the graph of the solution set of the inequality.

(b) Write the solution set in interval notation.

Use the technique described in ***Exercises 87–90*** *to solve each inequality. Write each solution set in interval notation.*

91. $(2x - 3)(x + 2)(x - 3) \ge 0$

92. $(x + 5)(3x - 4)(x + 2) \ge 0$

93. $4x - x^3 \ge 0$

94. $16x - x^3 \ge 0$

95. $(x + 1)^2(x - 3) < 0$

96. $(x - 5)^2(x + 1) < 0$

97. $x^3 + 4x^2 - 9x \ge 36$

98. $x^3 + 3x^2 - 16x \le 48$

99. $x^2(x + 4)^2 \ge 0$

100. $-x^2(2x - 3)^2 \le 0$

1.8 Absolute Value Equations and Inequalities

- Basic Concepts
- Absolute Value Equations
- Absolute Value Inequalities
- Special Cases
- Absolute Value Models for Distance and Tolerance

Basic Concepts

Recall that the **absolute value** of a number a, written $|a|$, gives the undirected distance from a to 0 on a number line. By this definition, the equation $|x| = 3$ can be solved by finding all real numbers at a distance of 3 units from 0. As shown in **Figure 19,** two numbers satisfy this equation, -3 and 3, so the solution set is $\{-3, 3\}$.

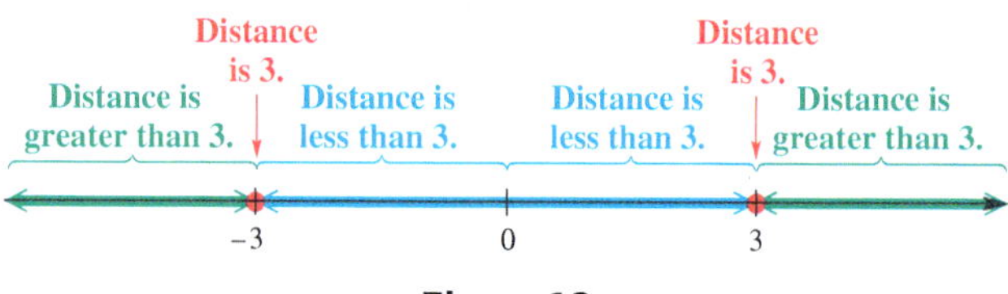

Figure 19

Similarly, $|x| < 3$ is satisfied by all real numbers whose undirected distances from 0 are less than 3. As shown in **Figure 19,** this is the interval

$$-3 < x < 3, \quad \text{or} \quad (-3, 3).$$

Finally, $|x| > 3$ is satisfied by all real numbers whose undirected distances from 0 are greater than 3. These numbers are less than -3 or greater than 3, so the solution set is

$$(-\infty, -3) \cup (3, \infty).$$

Notice in **Figure 19** that the union of the solution sets of $|x| = 3$, $|x| < 3$, and $|x| > 3$ is the set of real numbers.

These observations support the cases for solving absolute value equations and inequalities summarized in the table that follows. If the equation or inequality fits the form of Case 1, 2, or 3, change it to its equivalent form and solve. The solution set and its graph will look similar to those shown.

Solving Absolute Value Equations and Inequalities

Absolute Value Equation or Inequality*	Equivalent Form	Graph of the Solution Set	Solution Set
Case 1: $\lvert x\rvert = k$	$x = k$ or $x = -k$		$\{-k, k\}$
Case 2: $\lvert x\rvert < k$	$-k < x < k$		$(-k, k)$
Case 3: $\lvert x\rvert > k$	$x < -k$ or $x > k$		$(-\infty, -k) \cup (k, \infty)$

*For each equation or inequality in Cases 1–3, assume that $k > 0$.

In Cases 2 and 3, the strict inequality may be replaced by its nonstrict form. Additionally, if an absolute value equation takes the form $|a| = |b|$, then a and b must be equal in value or opposite in value.

Thus, the equivalent form of $|a| = |b|$ is $a = b$ or $a = -b$.

Absolute Value Equations Because absolute value represents undirected distance from 0 on a number line, solving an absolute value equation requires solving two possibilities, as shown in the examples that follow.

EXAMPLE 1 Solving Absolute Value Equations (Case 1 and the Special Case $|a| = |b|$)

Solve each equation.

(a) $|5 - 3x| = 12$ **(b)** $|4x - 3| = |x + 6|$

SOLUTION

(a) For the given expression $5 - 3x$ to have absolute value 12, it must represent either 12 or -12. This equation fits the form of Case 1.

$$|5 - 3x| = 12$$

Don't forget this second possibility.

$$5 - 3x = 12 \quad \text{or} \quad 5 - 3x = -12 \qquad \text{Case 1}$$

$$-3x = 7 \quad \text{or} \quad -3x = -17 \qquad \text{Subtract 5.}$$

$$x = -\frac{7}{3} \quad \text{or} \quad x = \frac{17}{3} \qquad \text{Divide by } -3.$$

Check the solutions $-\frac{7}{3}$ and $\frac{17}{3}$ by substituting them in the original absolute value equation. The solution set is $\left\{-\frac{7}{3}, \frac{17}{3}\right\}$.

LOOKING AHEAD TO CALCULUS

The precise definition of a **limit** in calculus requires writing absolute value inequalities.

A standard problem in calculus is to find the "interval of convergence" of a **power series** by solving the following inequality.

$$|x - a| < r$$

This inequality says that x can be any number within r units of a on the number line, so its solution set is indeed an interval—namely the interval $(a - r, a + r)$.

(b) If the absolute values of two expressions are equal, then those expressions are either equal in value or opposite in value.

$$|4x - 3| = |x + 6|$$

$4x - 3 = x + 6$ or $4x - 3 = -(x + 6)$ Consider both possibilities.

$3x = 9$ or $4x - 3 = -x - 6$ Solve each linear equation.

$x = 3$ or $5x = -3$

$x = -\frac{3}{5}$

CHECK $|4x - 3| = |x + 6|$ Original equation

$\left|4\left(-\frac{3}{5}\right) - 3\right| \stackrel{?}{=} \left|-\frac{3}{5} + 6\right|$ Let $x = -\frac{3}{5}$.

$\left|-\frac{12}{5} - 3\right| \stackrel{?}{=} \left|-\frac{3}{5} + 6\right|$

$\left|-\frac{27}{5}\right| = \left|\frac{27}{5}\right|$ ✓ True

$|4(3) - 3| \stackrel{?}{=} |3 + 6|$ Let $x = 3$.

$|12 - 3| \stackrel{?}{=} |3 + 6|$

$|9| = |9|$ ✓ True

Both solutions check. The solution set is $\left\{-\frac{3}{5}, 3\right\}$.

✔ **Now Try Exercises 9 and 19.**

Absolute Value Inequalities

EXAMPLE 2 Solving Absolute Value Inequalities (Cases 2 and 3)

Solve each inequality.

(a) $|2x + 1| < 7$ **(b)** $|2x + 1| > 7$

SOLUTION

(a) This inequality fits Case 2. If the absolute value of an expression is less than 7, then the value of the expression is *between* -7 and 7.

$$|2x + 1| < 7$$

$-7 < 2x + 1 < 7$ Case 2

$-8 < 2x < 6$ Subtract 1 from each part.

$-4 < x < 3$ Divide each part by 2.

The final inequality gives the solution set $(-4, 3)$ in interval notation.

(b) This inequality fits Case 3. If the absolute value of an expression is greater than 7, then the value of the expression is either less than -7 *or* greater than 7.

$$|2x + 1| > 7$$

$2x + 1 < -7$ or $2x + 1 > 7$ Case 3

$2x < -8$ or $2x > 6$ Subtract 1 from each side.

$x < -4$ or $x > 3$ Divide each side by 2.

The solution set written in interval notation is $(-\infty, -4) \cup (3, \infty)$.

✔ **Now Try Exercises 27 and 29.**

Cases 1, 2, and 3 require that the absolute value expression be isolated on one side of the equation or inequality.

EXAMPLE 3 Solving an Absolute Value Inequality (Case 3)

Solve $|2 - 7x| - 1 > 4$.

SOLUTION

$$|2 - 7x| - 1 > 4$$

$$|2 - 7x| > 5 \qquad \text{Add 1 to each side.}$$

$$2 - 7x < -5 \quad \text{or} \quad 2 - 7x > 5 \qquad \text{Case 3}$$

$$-7x < -7 \quad \text{or} \quad -7x > 3 \qquad \text{Subtract 2 from each side.}$$

$$x > 1 \quad \text{or} \quad x < -\frac{3}{7} \qquad \text{Divide by } -7. \text{ Reverse the direction of each inequality.}$$

The solution set written in interval notation is $\left(-\infty, -\frac{3}{7}\right) \cup (1, \infty)$.

✔ **Now Try Exercise 51.**

Special Cases The three cases given in this section require the constant k to be positive. ***When $k \leq 0$, use the fact that the absolute value of any expression must be nonnegative, and consider the conditions necessary for the statement to be true.***

EXAMPLE 4 Solving Special Cases

Solve each equation or inequality.

(a) $|2 - 5x| \geq -4$ **(b)** $|4x - 7| < -3$ **(c)** $|5x + 15| = 0$

SOLUTION

(a) Since the absolute value of a number is always nonnegative, the inequality

$$|2 - 5x| \geq -4 \text{ is always true.}$$

The solution set includes all real numbers, written $(-\infty, \infty)$.

(b) There is no number whose absolute value is less than -3 (or less than *any* negative number).

$$\text{The solution set of } |4x - 7| < -3 \text{ is } \varnothing.$$

(c) The absolute value of a number will be 0 only if that number is 0. Therefore, $|5x + 15| = 0$ is equivalent to

$$5x + 15 = 0, \quad \text{which has solution set} \quad \{-3\}.$$

CHECK Substitute -3 into the original equation.

$$|5x + 15| = 0 \qquad \text{Original equation}$$

$$|5(-3) + 15| \stackrel{?}{=} 0 \qquad \text{Let } x = -3.$$

$$0 = 0 \checkmark \qquad \text{True}$$

✔ **Now Try Exercises 55, 57, and 59.**

Absolute Value Models for Distance and Tolerance If a and b represent two real numbers, then the absolute value of their difference,

$$\text{either} \quad |a - b| \quad \text{or} \quad |b - a|,$$

represents the undirected distance between them.

EXAMPLE 5 Using Absolute Value Inequalities with Distances

Write each statement using an absolute value inequality.

(a) k is no less than 5 units from 8. **(b)** n is within 0.001 unit of 6.

SOLUTION

(a) Since the distance from k to 8, written $|k - 8|$ or $|8 - k|$, is no less than 5, the distance is greater than or equal to 5. This can be written as

$$|k - 8| \geq 5, \quad \text{or, equivalently,} \quad |8 - k| \geq 5. \quad \text{Either form is acceptable.}$$

(b) This statement indicates that the distance between n and 6 is less than 0.001.

$$|n - 6| < 0.001, \quad \text{or, equivalently,} \quad |6 - n| < 0.001$$

✔ **Now Try Exercises 69 and 71.**

EXAMPLE 6 Using Absolute Value to Model Tolerance

In quality control situations, such as filling bottles on an assembly line, we often wish to keep the difference between two quantities within some predetermined amount, called the **tolerance.**

Suppose $y = 2x + 1$ and we want y to be within 0.01 unit of 4. For what values of x will this be true?

SOLUTION

$$|y - 4| < 0.01 \quad \text{Write an absolute value inequality.}$$

$$|2x + 1 - 4| < 0.01 \quad \text{Substitute } 2x + 1 \text{ for } y.$$

$$|2x - 3| < 0.01 \quad \text{Combine like terms.}$$

$$-0.01 < 2x - 3 < 0.01 \quad \text{Case 2}$$

$$2.99 < 2x < 3.01 \quad \text{Add 3 to each part.}$$

$$1.495 < x < 1.505 \quad \text{Divide each part by 2.}$$

Reversing these steps shows that keeping x in the interval $(1.495, 1.505)$ ensures that the difference between y and 4 is within 0.01 unit.

✔ **Now Try Exercise 75.**

1.8 Exercises

CONCEPT PREVIEW *Match each equation or inequality in Column I with the graph of its solution set in Column II.*

I	II
1. $\lvert x \rvert = 7$	A. −7 0 7
2. $\lvert x \rvert = -7$	B. −7 0 7 ∅
3. $\lvert x \rvert > -7$	C. −7 0 7
4. $\lvert x \rvert > 7$	D. −7 0 7
5. $\lvert x \rvert < 7$	E. −7 0 7
6. $\lvert x \rvert \geq 7$	F. −7 0 7
7. $\lvert x \rvert \leq 7$	G. −7 0 7
8. $\lvert x \rvert \neq 7$	H. −7 0 7

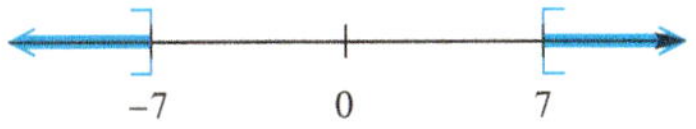

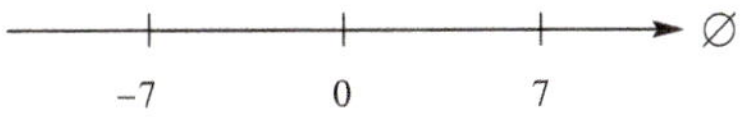

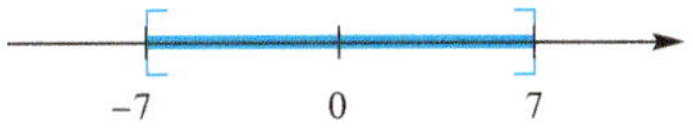

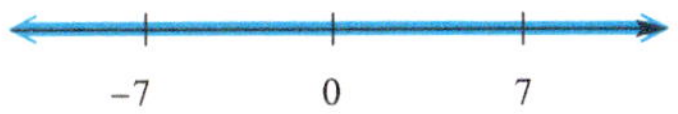

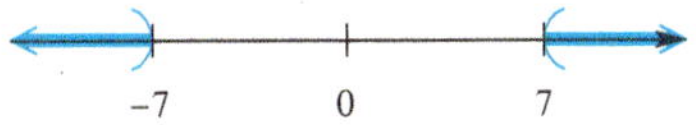

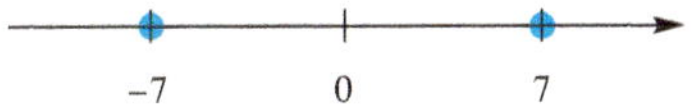

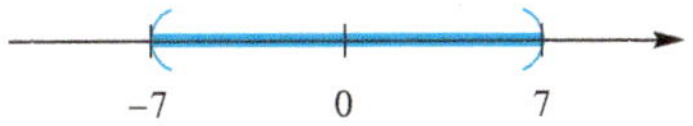

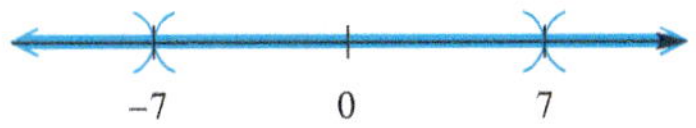

Solve each equation. ***See Example 1.***

9. $\lvert 3x - 1 \rvert = 2$

10. $\lvert 4x + 2 \rvert = 5$

11. $\lvert 5 - 3x \rvert = 3$

12. $\lvert 7 - 3x \rvert = 3$

13. $\left\lvert \frac{x-4}{2} \right\rvert = 5$

14. $\left\lvert \frac{x+2}{2} \right\rvert = 7$

15. $\left\lvert \frac{5}{x-3} \right\rvert = 10$

16. $\left\lvert \frac{3}{2x-1} \right\rvert = 4$

17. $\left\lvert \frac{6x+1}{x-1} \right\rvert = 3$

18. $\left\lvert \frac{2x+3}{3x-4} \right\rvert = 1$

19. $\lvert 2x - 3 \rvert = \lvert 5x + 4 \rvert$

20. $\lvert x + 1 \rvert = \lvert 1 - 3x \rvert$

21. $\lvert 4 - 3x \rvert = \lvert 2 - 3x \rvert$

22. $\lvert 3 - 2x \rvert = \lvert 5 - 2x \rvert$

23. $\lvert 5x - 2 \rvert = \lvert 2 - 5x \rvert$

24. The equation $\lvert 5x - 6 \rvert = 3x$ cannot have a negative solution. Why?

25. The equation $\lvert 7x + 3 \rvert = -5x$ cannot have a positive solution. Why?

26. *Concept Check* Determine the solution set of each equation by inspection.

(a) $-\lvert x \rvert = \lvert x \rvert$ **(b)** $\lvert -x \rvert = \lvert x \rvert$ **(c)** $\lvert x^2 \rvert = \lvert x \rvert$ **(d)** $-\lvert x \rvert = 9$

Solve each inequality. Give the solution set in interval notation. ***See Example 2.***

27. $\lvert 2x + 5 \rvert < 3$

28. $\lvert 3x - 4 \rvert < 2$

29. $\lvert 2x + 5 \rvert \geq 3$

30. $\lvert 3x - 4 \rvert \geq 2$

31. $\left\lvert \frac{1}{2} - x \right\rvert < 2$

32. $\left\lvert \frac{3}{5} + x \right\rvert < 1$

33. $4\lvert x - 3 \rvert > 12$

34. $5\lvert x + 1 \rvert > 10$

35. $\lvert 5 - 3x \rvert > 7$

36. $\lvert 7 - 3x \rvert > 4$

37. $\lvert 5 - 3x \rvert \leq 7$

38. $\lvert 7 - 3x \rvert \leq 4$

39. $\left\lvert \frac{2}{3}x + \frac{1}{2} \right\rvert \leq \frac{1}{6}$

40. $\left\lvert \frac{5}{3} - \frac{1}{2}x \right\rvert > \frac{2}{9}$

41. $\lvert 0.01x + 1 \rvert < 0.01$

42. Explain why the equation $\lvert x \rvert = \sqrt{x^2}$ has infinitely many solutions.

Solve each equation or inequality. ***See Examples 3 and 4.***

43. $|4x + 3| - 2 = -1$

44. $|8 - 3x| - 3 = -2$

45. $|6 - 2x| + 1 = 3$

46. $|4 - 4x| + 2 = 4$

47. $|3x + 1| - 1 < 2$

48. $|5x + 2| - 2 < 3$

49. $\left|5x + \frac{1}{2}\right| - 2 < 5$

50. $\left|2x + \frac{1}{3}\right| + 1 < 4$

51. $|10 - 4x| + 1 \geq 5$

52. $|12 - 6x| + 3 \geq 9$

53. $|3x - 7| + 1 < -2$

54. $|-5x + 7| - 4 < -6$

Solve each equation or inequality. ***See Example 4.***

55. $|10 - 4x| \geq -4$

56. $|12 - 9x| \geq -12$

57. $|6 - 3x| < -11$

58. $|18 - 3x| < -13$

59. $|8x + 5| = 0$

60. $|7 + 2x| = 0$

61. $|4.3x + 9.8| < 0$

62. $|1.5x - 14| < 0$

63. $|2x + 1| \leq 0$

64. $|3x + 2| \leq 0$

65. $|3x + 2| > 0$

66. $|4x + 3| > 0$

67. *Concept Check* Write an equation involving absolute value that says the distance between p and q is 2 units.

68. *Concept Check* Write an equation involving absolute value that says the distance between r and s is 6 units.

Write each statement using an absolute value equation or inequality. ***See Example 5.***

69. m is no more than 2 units from 7.

70. z is no less than 5 units from 4.

71. p is within 0.0001 unit of 9.

72. k is within 0.0002 unit of 10.

73. r is no less than 1 unit from 29.

74. q is no more than 8 units from 22.

(Modeling) Solve each problem. ***See Example 6.***

75. *Tolerance* Suppose that $y = 5x + 1$ and we want y to be within 0.002 unit of 6. For what values of x will this be true?

76. *Tolerance* Repeat **Exercise 75**, but let $y = 10x + 2$.

77. *Weights of Babies* Dr. Tydings has found that, over the years, 95% of the babies he has delivered weighed x pounds, where

$$|x - 8.2| \leq 1.5.$$

What range of weights corresponds to this inequality?

78. *Temperatures on Mars* The temperatures on the surface of Mars in degrees Celsius approximately satisfy the inequality $|C + 84| \leq 56$. What range of temperatures corresponds to this inequality?

79. *Conversion of Methanol to Gasoline* The industrial process that is used to convert methanol to gasoline is carried out at a temperature range of 680°F to 780°F. Using F as the variable, write an absolute value inequality that corresponds to this range.

80. *Wind Power Extraction Tests* When a model kite was flown in crosswinds in tests to determine its limits of power extraction, it attained speeds of 98 to 148 ft per sec in winds of 16 to 26 ft per sec. Using x as the variable in each case, write absolute value inequalities that correspond to these ranges.

(Modeling) Carbon Dioxide Emissions When humans breathe, carbon dioxide is emitted. In one study, the emission rates of carbon dioxide by college students were measured during both lectures and exams. The average individual rate R_L (in grams per hour) during a lecture class satisfied the inequality

$$|R_L - 26.75| \leq 1.42,$$

whereas during an exam the rate R_E satisfied the inequality

$$|R_E - 38.75| \leq 2.17.$$

(*Source:* Wang, T. C., *ASHRAE Trans.*, 81 (Part 1), 32.)

Use this information to solve each problem.

81. Find the range of values for R_L and R_E.

82. The class had 225 students. If T_L and T_E represent the total amounts of carbon dioxide in grams emitted during a 1-hour lecture and a 1-hour exam, respectively, write inequalities that model the ranges for T_L and T_E.

Relating Concepts

For individual or collaborative investigation *(Exercises 83–86)*

To see how to solve an equation that involves the absolute value of a quadratic polynomial, such as $|x^2 - x| = 6$, ***work Exercises 83–86 in order.***

83. For $x^2 - x$ to have an absolute value equal to 6, what are the two possible values that it may be? (*Hint:* One is positive and the other is negative.)

84. Write an equation stating that $x^2 - x$ is equal to the positive value found in **Exercise 83,** and solve it using the zero-factor property.

85. Write an equation stating that $x^2 - x$ is equal to the negative value found in **Exercise 83,** and solve it using the quadratic formula. (*Hint:* The solutions are not real numbers.)

86. Give the complete solution set of $|x^2 - x| = 6$, using the results from **Exercises 84 and 85.**

Use the method described in ***Exercises 83–86,*** *if applicable, and properties of absolute value to solve each equation or inequality. (Hint: Exercises 93 and 94 can be solved by inspection.)*

87. $|3x^2 + x| = 14$

88. $|2x^2 - 3x| = 5$

89. $|4x^2 - 23x - 6| = 0$

90. $|6x^3 + 23x^2 + 7x| = 0$

91. $|x^2 + 1| - |2x| = 0$

92. $\left|\dfrac{x^2 + 2}{x}\right| - \dfrac{11}{3} = 0$

93. $|x^4 + 2x^2 + 1| < 0$

94. $|x^2 + 10| < 0$

95. $\left|\dfrac{x - 4}{3x + 1}\right| \geq 0$

96. $\left|\dfrac{9 - x}{7 + 8x}\right| \geq 0$

Chapter 1 Test Prep

Key Terms

1.1 equation, solution (root), solution set, equivalent equations, linear equation in one variable, first-degree equation, identity, conditional equation, contradiction, literal equation, simple interest, future value (maturity value)

1.2 mathematical model, linear model

1.3 imaginary unit, complex number, real part, imaginary part, pure imaginary number, nonreal complex number, standard form, complex conjugate

1.4 quadratic equation, standard form, second-degree equation, double solution, cubic equation, discriminant

1.5 leg, hypotenuse

1.6 rational equation, proposed solution, equation quadratic in form

1.7 inequality, linear inequality in one variable, interval, interval notation, open interval, closed interval, break-even point, quadratic inequality, strict inequality, nonstrict inequality, rational inequality

1.8 tolerance

New Symbols

Symbol	Meaning
$\varnothing$	empty or null set
i	imaginary unit
∞	infinity
$-\infty$	negative infinity
(a, b), $(-\infty, a]$, $[a, b)$	interval notation
$\lvert a \rvert$	absolute value of a

Quick Review

Concepts	Examples
1.1 Linear Equations	
Addition and Multiplication Properties of Equality Let a, b, and c represent real numbers. **If $a = b$, then $a + c = b + c$.** **If $a = b$ and $c \neq 0$, then $ac = bc$.**	Solve. $5(x+3) = 3x+7$ $5x + 15 = 3x + 7$ Distributive property $2x = -8$ Subtract $3x$. Subtract 15. $x = -4$ Divide by 2. The solution set is $\{-4\}$.
1.2 Applications and Modeling with Linear Equations	
Solving an Applied Problem *Step 1* Read the problem.	How many liters of 30% alcohol solution and 80% alcohol solution must be mixed to obtain 50 L of 50% alcohol solution?

Concepts	Examples

Step 2 Assign a variable.

Let $x =$ the number of liters of 30% solution.
$50 - x =$ the number of liters of 80% solution.
Summarize the information of the problem in a table.

Strength	Liters of Solution	Liters of Pure Alcohol
30%	x	$0.30x$
80%	$50 - x$	$0.80(50 - x)$
50%	50	$0.50(50)$

Step 3 Write an equation.

The equation is $0.30x + 0.80(50 - x) = 0.50(50)$.

Step 4 Solve the equation.

Solve the equation to obtain $x = 30$.

Step 5 State the answer.

Therefore, 30 L of the 30% solution and $50 - 30 = 20$ L of the 80% solution must be mixed.

Step 6 Check.

CHECK

$$0.30(30) + 0.80(50 - 30) \stackrel{?}{=} 0.50(50)$$
$$25 = 25 \quad ✓ \text{ True}$$

1.3 Complex Numbers

Definition of *i*

$$i = \sqrt{-1}, \quad \text{and therefore,} \quad i^2 = -1$$

Definition of Complex Number (*a* and *b* real)

$$\underset{\substack{\uparrow \\ \text{Real part}}}{a} + \underset{\substack{\uparrow \\ \text{Imaginary part}}}{bi}$$

In the complex number $-6 + 2i$, the real part is -6 and the imaginary part is 2.

Definition of $\sqrt{-a}$

For $a > 0$, $\sqrt{-a} = i\sqrt{a}$.

Simplify.

$$\sqrt{-4} = 2i$$
$$\sqrt{-12} = i\sqrt{12} = i\sqrt{4 \cdot 3} = 2i\sqrt{3}$$

Adding and Subtracting Complex Numbers
Add or subtract the real parts, and add or subtract the imaginary parts.

$$\begin{aligned}&(2 + 3i) + (3 + i) - (2 - i)\\&= (2 + 3 - 2) + (3 + 1 + 1)i\\&= 3 + 5i\end{aligned}$$

Multiplying and Dividing Complex Numbers
Multiply complex numbers as with binomials, and use the fact that $i^2 = -1$.

$$\begin{aligned}&(6 + i)(3 - 2i)\\&= 18 - 12i + 3i - 2i^2 && \text{FOIL method}\\&= (18 + 2) + (-12 + 3)i && i^2 = -1\\&= 20 - 9i\end{aligned}$$

Concepts	Examples
Divide complex numbers by multiplying the numerator and denominator by the complex conjugate of the denominator.	$\frac{3+i}{1+i}$ $= \frac{(3+i)(1-i)}{(1+i)(1-i)}$ Multiply by $\frac{1-i}{1-i}$. $= \frac{3-3i+i-i^2}{1-i^2}$ Multiply. $= \frac{4-2i}{2}$ Combine like terms; $i^2 = -1$ $= \frac{2(2-i)}{2}$ Factor in the numerator. $= 2 - i$ Divide out the common factor.
1.4 Quadratic Equations **Zero-Factor Property** If a and b are complex numbers with $ab = 0$, then $a = 0$ or $b = 0$ or both equal zero.	Solve. $6x^2 + x - 1 = 0$ $(3x-1)(2x+1) = 0$ Factor. $3x - 1 = 0$ or $2x + 1 = 0$ Zero-factor property $x = \frac{1}{3}$ or $x = -\frac{1}{2}$ The solution set is $\left\{-\frac{1}{2}, \frac{1}{3}\right\}$.
Square Root Property The solution set of $x^2 = k$ is $\left\{\sqrt{k}, -\sqrt{k}\right\}$, abbreviated $\left\{\pm\sqrt{k}\right\}$.	$x^2 = 12$ $x = \pm\sqrt{12} = \pm 2\sqrt{3}$ The solution set is $\left\{\pm 2\sqrt{3}\right\}$.
Quadratic Formula The solutions of the quadratic equation $ax^2 + bx + c = 0$, where $a \neq 0$, are given by the quadratic formula. $x = \frac{-b \pm \sqrt{b^2 - 4ac}}{2a}$	$x^2 + 2x + 3 = 0$ $x = \frac{-2 \pm \sqrt{2^2 - 4(1)(3)}}{2(1)}$ $a = 1, b = 2, c = 3$ $x = \frac{-2 \pm \sqrt{-8}}{2}$ Simplify. $x = \frac{-2 \pm 2i\sqrt{2}}{2}$ Simplify the radical. $x = \frac{2\left(-1 \pm i\sqrt{2}\right)}{2}$ Factor out 2 in the numerator. $x = -1 \pm i\sqrt{2}$ Divide out the common factor. The solution set is $\left\{-1 \pm i\sqrt{2}\right\}$.

Concepts | **Examples**

1.5 Applications and Modeling with Quadratic Equations

Pythagorean Theorem
In a right triangle, the sum of the squares of the lengths of legs a and b is equal to the square of the length of hypotenuse c.

$$a^2 + b^2 = c^2$$

In a right triangle, the shorter leg is 7 in. less than the longer leg, and the hypotenuse is 2 in. greater than the longer leg. What are the lengths of the sides?

Let $x =$ the length of the longer leg.

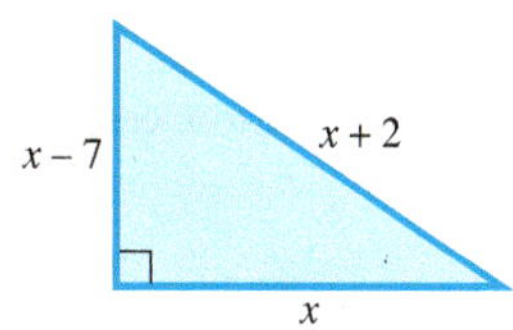

$(x-7)^2 + x^2 = (x+2)^2$ — Substitute into the Pythagorean theorem.

$x^2 - 14x + 49 + x^2 = x^2 + 4x + 4$ — Square the binomials.

$x^2 - 18x + 45 = 0$ — Standard form

$(x-15)(x-3) = 0$ — Factor.

$x - 15 = 0$ or $x - 3 = 0$ — Zero-factor property

$x = 15$ or $x = 3$ — Solve each equation.

The value 3 must be rejected because the height would be negative. The lengths of the sides are 15 in., 8 in., and 17 in. Check to see that the conditions of the problem are satisfied.

Height of a Projected Object
The height s (in feet) of an object projected directly upward from an initial height of s_0 feet, with initial velocity v_0 feet per second, is

$$s = -16t^2 + v_0t + s_0,$$

where t is the number of seconds after the object is projected.

The height of an object projected upward from ground level with an initial velocity of 64 ft per sec is given by

$$s = -16t^2 + 64t.$$

Find the time(s) that the projectile will reach a height of 56 ft.

$56 = -16t^2 + 64t$ — Let $s = 56$.

$0 = -16t^2 + 64t - 56$ — Subtract 56.

$0 = 2t^2 - 8t + 7$ — Divide by -8.

$t = \dfrac{-(-8) \pm \sqrt{(-8)^2 - 4(2)(7)}}{2(2)}$ — Quadratic formula

$t = \dfrac{8 \pm \sqrt{8}}{4}$ — Simplify.

$t \approx 1.29$ or $t \approx 2.71$ — Use a calculator.

The object reaches a height of 56 ft twice—once on its way up (after 1.29 sec) and once on its way down (after 2.71 sec).

Concepts	Examples

1.6 Other Types of Equations and Applications

Power Property

If P and Q are algebraic expressions, then every solution of the equation $P = Q$ is also a solution of the equation $P^n = Q^n$, for any positive integer n.

Quadratic in Form

An equation in the form $au^2 + bu + c = 0$, where $a \neq 0$ and u is an algebraic expression, can be solved by using a substitution variable.

If the power property is applied, or if both sides of an equation are multiplied by a variable expression, check all proposed solutions.

Solve.

$$(x+1)^{2/3} + (x+1)^{1/3} - 6 = 0$$

$$u^2 + u - 6 = 0 \quad \text{Let } u = (x+1)^{1/3}.$$

$$(u+3)(u-2) = 0$$

$$u + 3 = 0 \quad \text{or} \quad u - 2 = 0$$

$$u = -3 \quad \text{or} \quad u = 2$$

$$(x+1)^{1/3} = -3 \quad \text{or} \quad (x+1)^{1/3} = 2$$

$$x + 1 = -27 \quad \text{or} \quad x + 1 = 8 \quad \text{Cube.}$$

$$x = -28 \quad \text{or} \quad x = 7 \quad \text{Subtract 1.}$$

Both solutions check. The solution set is $\{-28, 7\}$.

1.7 Inequalities

Properties of Inequality

Let a, b, and c represent real numbers.

1. **If $a < b$, then $a + c < b + c$.**
2. **If $a < b$ and if $c > 0$, then $ac < bc$.**
3. **If $a < b$ and if $c < 0$, then $ac > bc$.**

Solve.

$$-3(x+4) + 2x < 6$$

$$-3x - 12 + 2x < 6$$

$$-x < 18$$

$$x > -18 \quad \text{Multiply by } -1. \text{ Change } < \text{ to } >.$$

The solution set is $(-18, \infty)$.

Solving a Quadratic Inequality

Step 1 Solve the corresponding quadratic equation.

$$x^2 + 6x \leq 7$$

$$x^2 + 6x - 7 = 0 \quad \text{Corresponding equation}$$

$$(x+7)(x-1) = 0 \quad \text{Factor.}$$

$$x + 7 = 0 \quad \text{or} \quad x - 1 = 0 \quad \text{Zero-factor property}$$

$$x = -7 \quad \text{or} \quad x = 1 \quad \text{Solve each equation.}$$

Step 2 Identify the intervals determined by the solutions of the equation.

Step 3 Use a test value from each interval to determine which intervals form the solution set.

The intervals formed are $(-\infty, -7)$, $(-7, 1)$, and $(1, \infty)$. Test values show that values in the intervals $(-\infty, -7)$ and $(1, \infty)$ do not satisfy the original inequality, while those in $(-7, 1)$ do. Because the symbol $\leq$ includes equality, the endpoints are included.

The solution set is $[-7, 1]$.

Solving a Rational Inequality

Step 1 Rewrite the inequality so that 0 is on one side and a single fraction is on the other.

$$\frac{x}{x+3} \geq \frac{5}{x+3}$$

$$\frac{x}{x+3} - \frac{5}{x+3} \geq 0$$

$$\frac{x-5}{x+3} \geq 0$$

Step 2 Find the values that make either the numerator or the denominator 0.

The values -3 and 5 make either the numerator or the denominator 0. The intervals formed are

$$(-\infty, -3),\ (-3, 5),\ \text{and } (5, \infty).$$

Step 3 Use a test value from each interval to determine which intervals form the solution set.

The value -3 must be excluded and 5 must be included. Test values show that values in the intervals $(-\infty, -3)$ and $(5, \infty)$ yield true statements.

The solution set is $(-\infty, -3) \cup [5, \infty)$.

Concepts	Examples

1.8 Absolute Value Equations and Inequalities

Solving Absolute Value Equations and Inequalities

For each equation or inequality in Cases 1–3, assume that $k > 0$.

Solve.

Case 1: To solve $|x| = k$, use the equivalent form

$$x = k \quad \text{or} \quad x = -k.$$

$$|5x - 2| = 3$$

$$5x - 2 = 3 \quad \text{or} \quad 5x - 2 = -3$$

$$5x = 5 \quad \text{or} \quad 5x = -1$$

$$x = 1 \quad \text{or} \quad x = -\frac{1}{5}$$

The solution set is $\left\{-\frac{1}{5}, 1\right\}$.

Case 2: To solve $|x| < k$, use the equivalent form

$$-k < x < k.$$

$$|5x - 2| < 3$$

$$-3 < 5x - 2 < 3$$

$$-1 < 5x < 5$$

$$-\frac{1}{5} < x < 1$$

The solution set is $\left(-\frac{1}{5}, 1\right)$.

Case 3: To solve $|x| > k$, use the equivalent form

$$x < -k \quad \text{or} \quad x > k.$$

$$|5x - 2| \geq 3$$

$$5x - 2 \leq -3 \quad \text{or} \quad 5x - 2 \geq 3$$

$$5x \leq -1 \quad \text{or} \quad 5x \geq 5$$

$$x \leq -\frac{1}{5} \quad \text{or} \quad x \geq 1$$

The solution set is $\left(-\infty, -\frac{1}{5}\right] \cup [1, \infty)$.

Chapter 1 Review Exercises

Solve each equation.

1. $2x + 8 = 3x + 2$

2. $\frac{1}{6}x - \frac{1}{12}(x - 1) = \frac{1}{2}$

3. $5x - 2(x + 4) = 3(2x + 1)$

4. $9x - 11(k + p) = x(a - 1)$, for x

5. $A = \frac{24f}{B(p + 1)}$, for f (approximate annual interest rate)

Solve each problem.

6. *Concept Check* Which of the following cannot be a correct equation to solve a geometry problem, if x represents the measure of a side of a rectangle? (*Hint:* Solve the equations and consider the solutions.)

A. $2x + 2(x + 2) = 20$

B. $2x + 2(5 + x) = -2$

C. $8(x + 2) + 4x = 16$

D. $2x + 2(x - 3) = 10$

7. *Concept Check* If x represents the number of pennies in a jar in an applied problem, which of the following equations cannot be a correct equation for finding x? (*Hint:* Solve the equations and consider the solutions.)

A. $5x + 3 = 11$
B. $12x + 6 = -4$
C. $100x = 50(x + 3)$
D. $6(x + 4) = x + 24$

8. *Airline Carry-On Baggage Size* Carry-on rules for domestic economy-class travel differ from one airline to another, as shown in the table.

Airline	Size (linear inches)
Alaska	51
American	45
Delta	45
Southwest	50
United	45
USAirways	45

Source: Individual airline websites.

To determine the number of linear inches for a carry-on, add the length, width, and height of the bag.

(a) One Samsonite rolling bag measures 9 in. by 12 in. by 21 in. Are there any airlines that would not allow it as a carry-on?

(b) A Lark wheeled bag measures 10 in. by 14 in. by 22 in. On which airlines does it qualify as a carry-on?

Solve each problem.

9. *Dimensions of a Square* If the length of each side of a square is decreased by 4 in., the perimeter of the new square is 10 in. more than half the perimeter of the original square. What are the dimensions of the original square?

10. *Distance from a Library* Becky can ride her bike to the university library in 20 min. The trip home, which is all uphill, takes her 30 min. If her rate is 8 mph faster on her trip there than her trip home, how far does she live from the library?

11. *Alcohol Mixture* Alan wishes to strengthen a mixture that is 10% alcohol to one that is 30% alcohol. How much pure alcohol should he add to 12 L of the 10% mixture?

12. *Loan Interest Rates* A realtor borrowed \$90,000 to develop some property. He was able to borrow part of the money at 5.5% interest and the rest at 6%. The annual interest on the two loans amounts to \$5125. How much was borrowed at each rate?

13. *Speed of a Plane* Mary Lynn left by plane to visit her mother in Louisiana, 420 km away. Fifteen minutes later, her mother left to meet her at the airport. She drove the 20 km to the airport at 40 km per hr, arriving just as the plane taxied in. What was the speed of the plane?

14. *Toxic Waste* Two chemical plants are releasing toxic waste into a holding tank. Plant I releases waste twice as fast as Plant II. Together they fill the tank in 3 hr. How long would it take the slower plant to fill the tank working alone?

15. *(Modeling) Lead Intake* As directed by the "Safe Drinking Water Act" of December 1974, the EPA proposed a maximum lead level in public drinking water of 0.05 mg per liter. This standard assumed an individual consumption of two liters of water per day.

(a) If EPA guidelines are followed, write an equation that models the maximum amount of lead A ingested in x years. Assume that there are 365.25 days in a year.

(b) If the average life expectancy is 72 yr, find the EPA maximum lead intake from water over a lifetime.

16. *(Modeling) Online Retail Sales* Projected e-commerce sales (in billions of dollars) for the years 2010–2018 can be modeled by the equation

$$y = 40.892x + 150.53,$$

where $x = 0$ corresponds to 2010, $x = 1$ corresponds to 2011, and so on. Based on this model, what would expected retail e-commerce sales be in 2018? (*Source:* Statistics Portal.)

17. *(Modeling) Minimum Wage* U.S. minimum hourly wage, in dollars, for selected years from 1956–2009 is shown in the table. The linear model

$$y = 0.1132x + 0.4609$$

approximates the minimum wage during this time period, where x is the number of years after 1956 and y is the minimum wage in dollars.

Year	Minimum Wage	Year	Minimum Wage
1956	1.00	1996	4.75
1963	1.25	1997	5.15
1975	2.10	2007	5.85
1981	3.35	2008	6.55
1990	3.80	2009	7.25

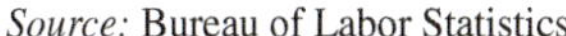

Source: Bureau of Labor Statistics.

(a) Use the model to approximate the minimum wage in 1990. How does it compare to the data in the table?

(b) Use the model to approximate the year in which the minimum wage was \$5.85. How does the answer compare to the data in the table?

18. *(Modeling) New York State Population* The U.S. population, in millions, for selected years is given in the table. The bar graph shows the percentages of the U.S. population that lived in New York State during those years.

Year	U.S. Population (in millions)
1980	226.5
1990	248.7
2000	281.4
2010	308.7
2014	318.9

Source: U.S. Census Bureau.

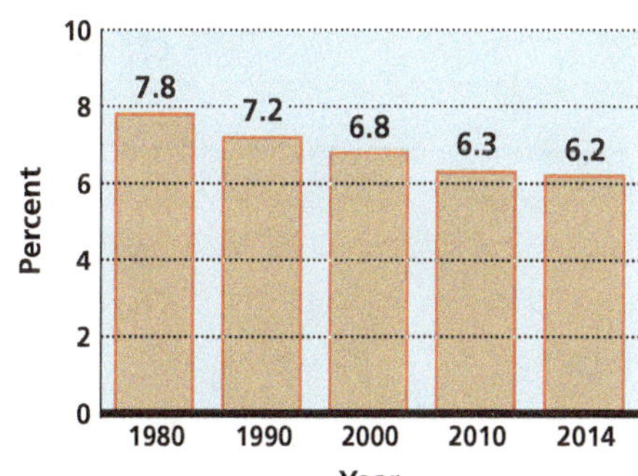

Source: U.S. Census Bureau.

(a) Find the number of Americans, to the nearest tenth of a million, living in New York State for each year given in the table.

(b) The percentages given in the bar graph decrease each year, while the populations given in the table increase each year. From the answers to part (a), is the number of Americans living in New York State increasing or decreasing?

Perform each operation. Write answers in standard form.

19. $(6 - i) + (7 - 2i)$

20. $(-11 + 2i) - (8 - 7i)$

21. $15i - (3 + 2i) - 11$

22. $-6 + 4i - (8i - 2)$

23. $(5 - i)(3 + 4i)$

24. $(-8 + 2i)(-1 + i)$

25. $(5 - 11i)(5 + 11i)$

26. $(4 - 3i)^2$

27. $-5i(3 - i)^2$

28. $4i(2 + 5i)(2 - i)$

29. $\dfrac{-12 - i}{-2 - 5i}$

30. $\dfrac{-7 + i}{-1 - i}$

Simplify each power of i.

31. i^{11}

32. i^{60}

33. i^{1001}

34. i^{110}

35. i^{-27}

36. $\dfrac{1}{i^{17}}$

Solve each equation.

37. $(x + 7)^2 = 5$

38. $(2 - 3x)^2 = 8$

39. $2x^2 + x - 15 = 0$

40. $12x^2 = 8x - 1$

41. $-2x^2 + 11x = -21$

42. $-x(3x + 2) = 5$

43. $(2x + 1)(x - 4) = x$

44. $\sqrt{2}x^2 - 4x + \sqrt{2} = 0$

45. $x^2 - \sqrt{5}\,x - 1 = 0$

46. $(x + 4)(x + 2) = 2x$

47. *Concept Check* Which equation has two real, distinct solutions? Do not actually solve.

A. $(3x - 4)^2 = -9$

B. $(4 - 7x)^2 = 0$

C. $(5x - 9)(5x - 9) = 0$

D. $(7x + 4)^2 = 11$

48. *Concept Check* See **Exercise 47.**

(a) Which equations have only one distinct real solution?

(b) Which equation has two nonreal complex solutions?

Evaluate the discriminant for each equation. Then use it to determine the number and type of solutions.

49. $-6x^2 + 2x = -3$

50. $8x^2 = -2x - 6$

51. $-8x^2 + 10x = 7$

52. $16x^2 + 3 = -26x$

53. $x(9x + 6) = -1$

54. $25x^2 + 110x + 121 = 0$

Solve each problem.

55. *(Modeling) Height of a Projectile* A projectile is fired straight up from ground level. After t seconds its height s, in feet above the ground, is given by

$$s = -16t^2 + 220t.$$

At what times is the projectile exactly 750 ft above the ground?

56. *Dimensions of a Picture Frame* Zach went into a frame-it-yourself shop. He wanted a frame 3 in. longer than it was wide. The frame he chose extended 1.5 in. beyond the picture on each side. Find the outside dimensions of the frame if the area of the unframed picture is 70 in.^{2}.

57. *Kitchen Flooring* Paula plans to replace the vinyl floor covering in her 10-ft by 12-ft kitchen. She wants to have a border of even width of a special material. She can afford only 21 ft^2 of this material. How wide a border can she have?

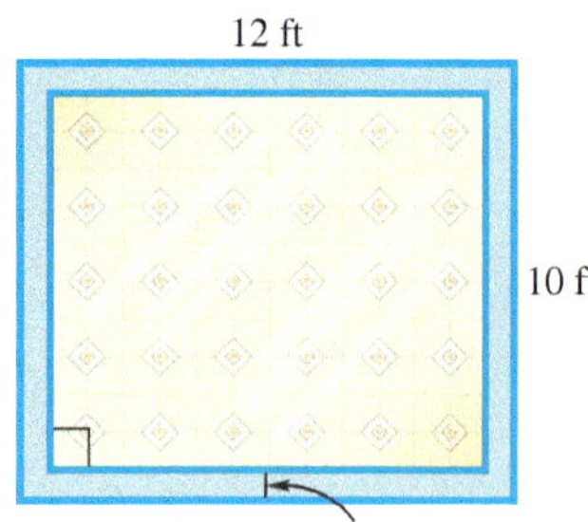

58. *(Modeling) Airplane Landing Speed* To determine the appropriate landing speed of a small airplane, the formula

$$D = 0.1s^2 - 3s + 22$$

is used, where s is the initial landing speed in feet per second and D is the length of the runway in feet. If the landing speed is too fast, the pilot may run out of runway. If the speed is too slow, the plane may stall. If the runway is 800 ft long, what is the appropriate landing speed? Round to the nearest tenth.

59. *(Modeling) U.S. Government Spending on Medical Care* The amount spent in billions of dollars by the U.S. government on medical care during the period 1990–2013 can be approximated by the equation

$$y = 1.016x^2 + 12.49x + 197.8$$

where $x = 0$ corresponds to 1990, $x = 1$ corresponds to 1991, and so on. According to this model, about how much was spent by the U.S. government on medical care in 2009? Round to the nearest tenth of a billion. (*Source*: *U.S. Office of Management and Budget*.)

60. *Dimensions of a Right Triangle* The lengths of the sides of a right triangle are such that the shortest side is 7 in. shorter than the middle side, while the longest side (the hypotenuse) is 1 in. longer than the middle side. Find the lengths of the sides.

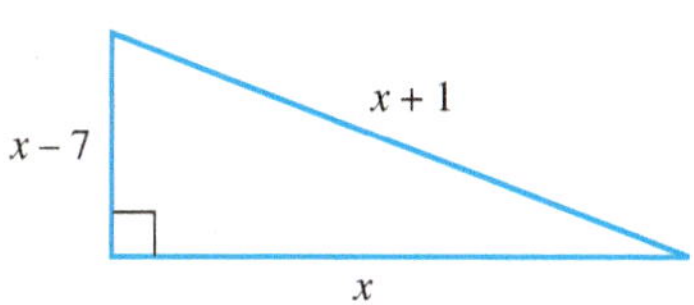

Solve each equation.

61. $4x^4 + 3x^2 - 1 = 0$

62. $x^2 - 2x^4 = 0$

63. $\frac{2}{x} - \frac{4}{3x} = 8 + \frac{3}{x}$

64. $2 - \frac{5}{x} = \frac{3}{x^2}$

65. $\frac{10}{4x - 4} = \frac{1}{1 - x}$

66. $\frac{13}{x^2 + 10} = \frac{2}{x}$

67. $(x - 4)^{2/5} = 9$

68. $(x^2 - 6x)^{1/4} = 2$

69. $(x - 2)^{2/3} = x^{1/3}$

70. $\sqrt{2x + 3} = x + 2$

71. $\sqrt{x + 2} - x = 2$

72. $\sqrt{x} - \sqrt{x + 3} = -1$

73. $\sqrt{4x - 2} = \sqrt{3x + 1}$

74. $\sqrt{5x - 15} - \sqrt{x + 1} = 2$

75. $\sqrt{x + 3} - \sqrt{3x + 10} = 1$

76. $\sqrt[5]{2x} = \sqrt[5]{3x + 2}$

77. $\sqrt[3]{6x + 2} - \sqrt[3]{4x} = 0$

78. $\sqrt{x^2 + 3x} - 2 = 0$

79. $\frac{x}{x + 2} + \frac{1}{x} + 3 = \frac{2}{x^2 + 2x}$

80. $\frac{2}{x + 2} + \frac{1}{x + 4} = \frac{4}{x^2 + 6x + 8}$

81. $(2x + 3)^{2/3} + (2x + 3)^{1/3} - 6 = 0$

82. $(x + 3)^{-2/3} - 2(x + 3)^{-1/3} = 3$

Solve each inequality. Give the solution set using interval notation.

83. $-9x + 3 < 4x + 10$

84. $11x \geq 2(x - 4)$

85. $-5x - 4 \geq 3(2x - 5)$

86. $7x - 2(x - 3) \leq 5(2 - x)$

87. $5 \leq 2x - 3 \leq 7$

88. $-8 > 3x - 5 > -12$

89. $x^2 + 3x - 4 \leq 0$

90. $x^2 + 4x - 21 > 0$

91. $6x^2 - 11x < 10$

92. $x^2 - 3x \geq 5$

93. $x^3 - 16x \leq 0$

94. $2x^3 - 3x^2 - 5x < 0$

95. $\frac{3x + 6}{x - 5} > 0$

96. $\frac{x + 7}{2x + 1} \leq 1$

97. $\frac{3x - 2}{x} - 4 > 0$

98. $\frac{5x + 2}{x} + 1 < 0$

99. $\frac{3}{x - 1} \leq \frac{5}{x + 3}$

100. $\frac{3}{x + 2} > \frac{2}{x - 4}$

(Modeling) Solve each problem.

101. ***Ozone Concentration*** Guideline levels for indoor ozone are less than 50 parts per billion (ppb). In a scientific study, a Purafil air filter was used to reduce an initial ozone concentration of 140 ppb. The filter removed 43% of the ozone. (*Source:* Parmar and Grosjean, *Removal of Air Pollutants from Museum Display Cases*, Getty Conservation Institute, Marina del Rey, CA.)

(a) What is the ozone concentration after the Purafil air filter is used?

(b) What is the maximum initial concentration of ozone that this filter will reduce to an acceptable level? Round the answer to the nearest tenth part per billion.

102. ***Break-Even Interval*** A company produces earbuds. The revenue from the sale of x units of these earbuds is

$$R = 8x.$$

The cost to produce x units of earbuds is

$$C = 3x + 1500.$$

In what interval will the company at least break even?

103. ***Height of a Projectile*** A projectile is launched upward from the ground. Its height s in feet above the ground after t seconds is given by the following equation.

$$s = -16t^2 + 320t$$

(a) After how many seconds in the air will it hit the ground?

(b) During what time interval is the projectile more than 576 ft above the ground?

104. ***Social Security*** The total amount paid by the U.S. government to individuals for Social Security retirement and disability insurance benefits during the period 2004–2013 can be approximated by the linear model

$$y = 35.7x + 486,$$

where $x = 0$ corresponds to 2004, $x = 1$ corresponds to 2005, and so on. The variable y is in billions of dollars. Based on this model, during what year did the amount paid by the government first exceed \$800 billion? Round the answer to the nearest year. Compare the answer to the bar graph.

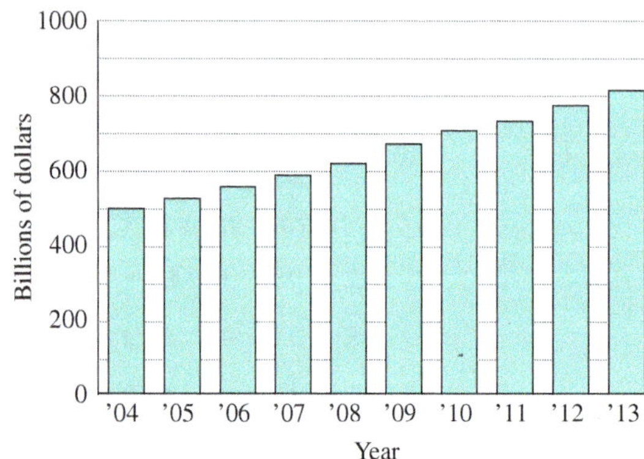

Source: U.S. Office of Management and Budget.

105. Without actually solving the inequality, explain why 3 cannot be in the solution set of $\frac{14x + 9}{x - 3} < 0$.

106. Without actually solving the inequality, explain why -4 must be in the solution set of $\frac{x + 4}{2x + 1} \geq 0$.

Solve each equation or inequality.

107. $|x + 4| = 7$

108. $|2 - x| = 3$

109. $\left|\dfrac{7}{2 - 3x}\right| - 9 = 0$

110. $\left|\dfrac{8x - 1}{3x + 2}\right| - 7 = 0$

111. $|5x - 1| = |2x + 3|$

112. $|x + 10| = |x - 11|$

113. $|2x + 9| \le 3$

114. $|8 - 5x| \ge 2$

115. $|7x - 3| > 4$

116. $\left|\dfrac{1}{2}x + \dfrac{2}{3}\right| < 3$

117. $|3x + 7| - 5 < 5$

118. $|7x + 8| - 6 > -3$

119. $|4x - 12| \ge -3$

120. $|7 - 2x| \le -9$

121. $|x^2 + 4x| \le 0$

122. $|x^2 + 4x| > 0$

Write each statement using an absolute value equation or inequality.

123. k is 12 units from 6.

124. p is at least 3 units from 1.

125. t is no less than 0.01 unit from 5.

Chapter 1 Test

Solve each equation.

1. $3(x - 4) - 5(x + 2) = 2 - (x + 24)$

2. $\dfrac{2}{3}x + \dfrac{1}{2}(x - 4) = x - 4$

3. $6x^2 - 11x - 7 = 0$

4. $(3x + 1)^2 = 8$

5. $3x^2 + 2x = -2$

6. $\dfrac{12}{x^2 - 9} = \dfrac{2}{x - 3} - \dfrac{3}{x + 3}$

7. $\dfrac{4x}{x - 2} + \dfrac{3}{x} = \dfrac{-6}{x^2 - 2x}$

8. $\sqrt{3x + 4} + 5 = 2x + 1$

9. $\sqrt{-2x + 3} + \sqrt{x + 3} = 3$

10. $\sqrt[3]{3x - 8} = \sqrt[3]{9x + 4}$

11. $x^4 - 17x^2 + 16 = 0$

12. $(x + 3)^{2/3} + (x + 3)^{1/3} - 6 = 0$

13. $|4x + 3| = 7$

14. $|2x + 1| = |5 - x|$

15. ***Surface Area of a Rectangular Solid*** The formula for the surface area of a rectangular solid is

$$S = 2HW + 2LW + 2LH,$$

where S, H, W, and L represent surface area, height, width, and length, respectively. Solve this formula for W.

16. Perform each operation. Write answers in standard form.

(a) $(9 - 3i) - (4 + 5i)$

(b) $(4 + 3i)(-5 + 3i)$

(c) $(8 + 3i)^2$

(d) $\dfrac{3 + 19i}{1 + 3i}$

17. Simplify each power of i.

(a) i^{42}

(b) i^{-31}

(c) $\dfrac{1}{i^{19}}$

Solve each problem.

18. ***(Modeling) Water Consumption for Snowmaking*** Ski resorts require large amounts of water in order to make snow. Snowmass Ski Area in Colorado plans to pump between 1120 and 1900 gal of water per minute at least 12 hr per day from Snowmass Creek between mid-October and late December. (*Source:* York Snow Incorporated.)

(a) Determine an equation that will calculate the *minimum* amount of water A (in gallons) pumped after x days during mid-October to late December.

(b) Find the minimum amount of water pumped in 30 days.

(c) Suppose the water being pumped from Snowmass Creek was used to fill swimming pools. The average backyard swimming pool holds 20,000 gal of water. Determine an equation that will give the minimum number of pools P that could be filled after x days. How many pools could be filled each day (to the nearest whole number)?

(d) To the nearest day, in how many days could a minimum of 1000 pools be filled?

19. ***Dimensions of a Rectangle*** The perimeter of a rectangle is 620 m. The length is 20 m less than twice the width. What are the length and width?

20. ***Nut Mixture*** To make a special mix, the owner of a fruit and nut stand wants to combine cashews that sell for \$7.00 per lb with walnuts that sell for \$5.50 per lb to obtain 35 lb of a mixture that sells for \$6.50 per lb. How many pounds of each type of nut should be used in the mixture?

21. ***Speed of an Excursion Boat*** An excursion boat travels upriver to a landing and then returns to its starting point. The trip upriver takes 1.2 hr, and the trip back takes 0.9 hr. If the average speed on the return trip is 5 mph faster than on the trip upriver, what is the boat's speed upriver?

22. ***(Modeling) Cigarette Use*** The percentage of college freshmen who smoke declined substantially from the year 2004 to the year 2014 and can be modeled by the linear equation

$$y = -0.461x + 6.32,$$

where x represents the number of years since 2004. Thus, $x = 0$ represents 2004, $x = 1$ represents 2005, and so on, (*Source:* Higher Education Research Institute, UCLA.)

(a) Use the model to determine the percentage of college freshmen who smoked in the year 2014. Round the answer to the nearest tenth of a percent.

(b) According to the model, in what year did 4.9% of college freshmen smoke?

23. ***(Modeling) Height of a Projectile*** A projectile is launched straight up from ground level with an initial velocity of 96 ft per sec. Its height in feet, s, after t seconds is given by the equation

$$s = -16t^2 + 96t.$$

(a) At what time(s) will it reach a height of 80 ft?

(b) After how many seconds will it return to the ground?

Solve each inequality. Give the answer using interval notation.

24. $-2(x - 1) - 12 < 2(x + 1)$

25. $-3 \le \frac{1}{2}x + 2 \le 3$

26. $2x^2 - x \ge 3$

27. $\frac{x + 1}{x - 3} < 5$

28. $|2x - 5| < 9$

29. $|2x + 1| - 11 \ge 0$

30. $|3x + 7| \le 0$

2 Graphs and Functions

The fact that the left and right sides of this butterfly mirror each other is an example of *symmetry*, a phenomenon found throughout nature and interpreted mathematically in this chapter.

2.1 Rectangular Coordinates and Graphs

- Ordered Pairs
- The Rectangular Coordinate System
- The Distance Formula
- The Midpoint Formula
- Equations in Two Variables

Ordered Pairs

The idea of pairing one quantity with another is often encountered in everyday life.

- A numerical score in a mathematics course is paired with a corresponding letter grade.
- The number of gallons of gasoline pumped into a tank is paired with the amount of money needed to purchase it.
- Expense categories are paired with dollars spent by the average American household in 2013. (See the table in the margin.)

Category	Amount Spent
food	\$ 8506
housing	\$21,374
transportation	\$12,153
health care	\$ 4917
apparel and services	\$ 2076
entertainment	\$ 3240

Source: U.S. Bureau of Labor Statistics.

Pairs of related quantities, such as a 96 determining a grade of A, 3 gallons of gasoline costing \$10.50, and 2013 spending on food of \$8506, can be expressed as *ordered pairs:* (96, A), (3, \$10.50), (food, \$8506). An **ordered pair** consists of two components, written inside parentheses.

EXAMPLE 1 Writing Ordered Pairs

Use the table to write ordered pairs to express the relationship between each category and the amount spent on it.

(a) housing **(b)** entertainment

SOLUTION

(a) Use the data in the second row: (housing, \$21,374).

(b) Use the data in the last row: (entertainment, \$3240).

✔ Now Try Exercise 13.

In mathematics, we are most often interested in ordered pairs whose components are numbers. The ordered pairs (a, b) and (c, d) are equal provided that $a = c$ *and* $b = d$.

NOTE Notation such as (2, 4) is used to show an interval on a number line, and the same notation is used to indicate an ordered pair of numbers. The intended use is usually clear from the context of the discussion.

The Rectangular Coordinate System

Each real number corresponds to a point on a number line. This idea is extended to ordered pairs of real numbers by using two perpendicular number lines, one horizontal and one vertical, that intersect at their zero-points. The point of intersection is the **origin.** The horizontal line is the ***x*-axis,** and the vertical line is the ***y*-axis.**

The x-axis and y-axis together make up a **rectangular coordinate system,** or **Cartesian coordinate system** (named for one of its coinventors, René Descartes. The other coinventor was Pierre de Fermat). The plane into which the coordinate system is introduced is the **coordinate plane,** or ***xy*-plane.** See **Figure 1.** The x-axis and y-axis divide the plane into four regions, or **quadrants,** labeled as shown. The points on the x-axis or the y-axis belong to no quadrant.

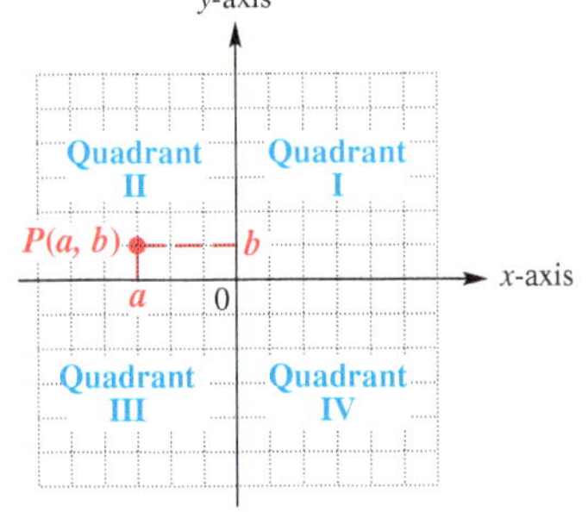

Rectangular (Cartesian) Coordinate System

Figure 1

Each point P in the xy-plane corresponds to a unique ordered pair (a, b) of real numbers. The point P corresponding to the ordered pair (a, b) often is written $P(a, b)$ as in **Figure 1** and referred to as "the point (a, b)." The numbers a and b are the **coordinates** of point P.

Figure 2

To locate on the xy-plane the point corresponding to the ordered pair $(3, 4)$, for example, start at the origin, move 3 units in the positive x-direction, and then move 4 units in the positive y-direction. See **Figure 2.** Point A corresponds to the ordered pair $(3, 4)$.

The Distance Formula Recall that the distance on a number line between points P and Q with coordinates x_1 and x_2 is

$$d(P, Q) = |x_1 - x_2| = |x_2 - x_1|. \quad \text{Definition of distance}$$

By using the coordinates of their ordered pairs, we can extend this idea to find the distance between any two points in a plane.

Figure 3 shows the points $P(-4, 3)$ and $R(8, -2)$. If we complete a right triangle that has its 90° angle at $Q(8, 3)$ as in the figure, the legs have lengths

$$d(P, Q) = |8 - (-4)| = 12$$

and

$$d(Q, R) = |3 - (-2)| = 5.$$

By the Pythagorean theorem, the hypotenuse has length

$$\sqrt{12^2 + 5^2} = \sqrt{144 + 25} = \sqrt{169} = 13.$$

Thus, the distance between $(-4, 3)$ and $(8, -2)$ is 13.

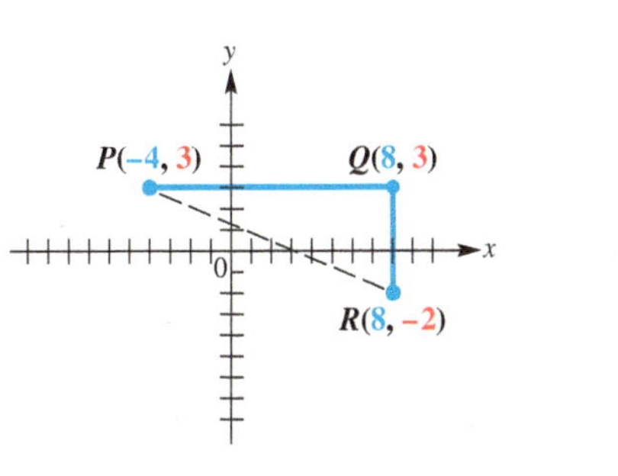

Figure 3

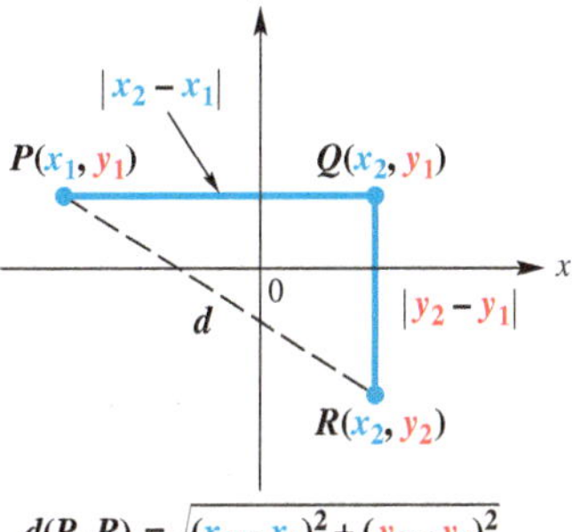

Figure 4

To obtain a general formula, let $P(x_1, y_1)$ and $R(x_2, y_2)$ be any two distinct points in a plane, as shown in **Figure 4.** Complete a triangle by locating point Q with coordinates (x_2, y_1). The Pythagorean theorem gives the distance between P and R.

$$d(P, R) = \sqrt{(x_2 - x_1)^2 + (y_2 - y_1)^2}$$

Absolute value bars are not necessary in this formula because, for all real numbers a and b,

$$|a - b|^2 = (a - b)^2.$$

The **distance formula** can be summarized as follows.

Distance Formula

Suppose that $P(x_1, y_1)$ and $R(x_2, y_2)$ are two points in a coordinate plane. The distance between P and R, written $d(P, R)$, is given by the following formula.

$$d(P, R) = \sqrt{(x_2 - x_1)^2 + (y_2 - y_1)^2}$$

René Descartes (1596–1650)

The initial flash of *analytic geometry* may have come to Descartes as he was watching a fly crawling about on the ceiling near a corner of his room. It struck him that the path of the fly on the ceiling could be described if only one knew the relation connecting the fly's distances from two adjacent walls.

Source: An Introduction to the History of Mathematics by Howard Eves.

LOOKING AHEAD TO CALCULUS

In analytic geometry and calculus, the distance formula is extended to two points in space. Points in space can be represented by **ordered triples.** The distance between the two points

$$(x_1, y_1, z_1) \quad \text{and} \quad (x_2, y_2, z_2)$$

is given by the following expression.

$$\sqrt{(x_2 - x_1)^2 + (y_2 - y_1)^2 + (z_2 - z_1)^2}$$

The distance formula can be stated in words.

The distance between two points in a coordinate plane is the square root of the sum of the square of the difference between their x-coordinates and the square of the difference between their y-coordinates.

Although our derivation of the distance formula assumed that P and R are not on a horizontal or vertical line, the result is true for any two points.

EXAMPLE 2 Using the Distance Formula

Find the distance between $P(-8, 4)$ and $Q(3, -2)$.

SOLUTION Use the distance formula.

$$d(P, Q) = \sqrt{(x_2 - x_1)^2 + (y_2 - y_1)^2} \quad \text{Distance formula}$$

$$= \sqrt{[3 - (-8)]^2 + (-2 - 4)^2} \quad x_1 = -8,\ y_1 = 4,\ x_2 = 3,\ y_2 = -2$$

Be careful when subtracting a negative number.

$$= \sqrt{11^2 + (-6)^2}$$

$$= \sqrt{121 + 36}$$

$$= \sqrt{157}$$

✔ **Now Try Exercise 15(a).**

A statement of the form "If p, then q" is a **conditional statement.** The related statement "If q, then p" is its **converse.** The *converse* of the Pythagorean theorem is also a true statement.

If the sides a, b, and c of a triangle satisfy $a^2 + b^2 = c^2$, then the triangle is a right triangle with legs having lengths a and b and hypotenuse having length c.

EXAMPLE 3 Applying the Distance Formula

Determine whether the points $M(-2, 5)$, $N(12, 3)$, and $Q(10, -11)$ are the vertices of a right triangle.

SOLUTION A triangle with the three given points as vertices, shown in **Figure 5,** is a right triangle if the square of the length of the longest side equals the sum of the squares of the lengths of the other two sides. Use the distance formula to find the length of each side of the triangle.

$$d(M, N) = \sqrt{[12 - (-2)]^2 + (3 - 5)^2} = \sqrt{196 + 4} = \sqrt{200}$$

$$d(M, Q) = \sqrt{[10 - (-2)]^2 + (-11 - 5)^2} = \sqrt{144 + 256} = \sqrt{400} = 20$$

$$d(N, Q) = \sqrt{(10 - 12)^2 + (-11 - 3)^2} = \sqrt{4 + 196} = \sqrt{200}$$

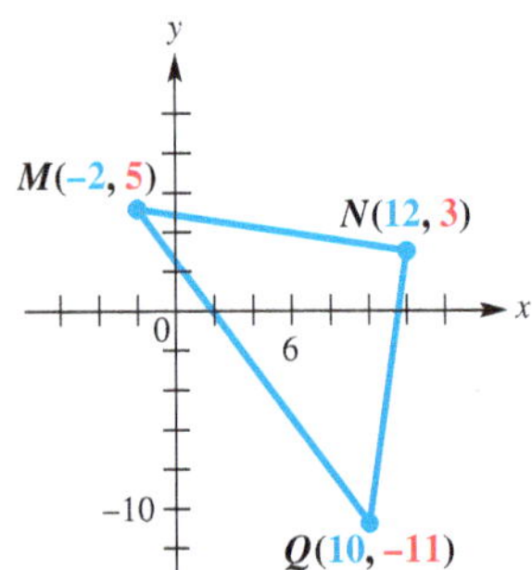

Figure 5

The longest side, of length 20 units, is chosen as the hypotenuse. Because

$$\left(\sqrt{200}\right)^2 + \left(\sqrt{200}\right)^2 = 400 = 20^2$$

is true, the triangle is a right triangle with hypotenuse joining M and Q.

✔ **Now Try Exercise 23.**

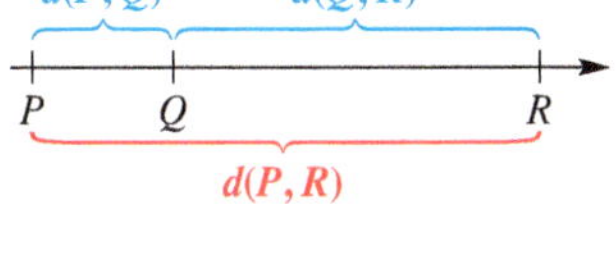

Figure 6

Using a similar procedure, we can tell whether three points are **collinear** (that is, lying on a straight line). See **Figure 6.**

Three points are collinear if the sum of the distances between two pairs of the points is equal to the distance between the remaining pair of points.

EXAMPLE 4 Applying the Distance Formula

Determine whether the points $P(-1, 5)$, $Q(2, -4)$, and $R(4, -10)$ are collinear.

SOLUTION Use the distance formula.

$$d(P, Q) = \sqrt{(-1-2)^2 + [5-(-4)]^2} = \sqrt{9+81} = \sqrt{90} = 3\sqrt{10}$$

$$\sqrt{90} = \sqrt{9 \cdot 10} = 3\sqrt{10}$$

$$d(Q, R) = \sqrt{(2-4)^2 + [-4-(-10)]^2} = \sqrt{4+36} = \sqrt{40} = 2\sqrt{10}$$

$$d(P, R) = \sqrt{(-1-4)^2 + [5-(-10)]^2} = \sqrt{25+225} = \sqrt{250} = 5\sqrt{10}$$

Because $3\sqrt{10} + 2\sqrt{10} = 5\sqrt{10}$ is true, the three points are collinear.

✔ **Now Try Exercise 29.**

The Midpoint Formula The midpoint of a line segment is equidistant from the endpoints of the segment. The **midpoint formula** is used to find the coordinates of the midpoint of a line segment. To develop the midpoint formula, let $P(x_1, y_1)$ and $Q(x_2, y_2)$ be any two distinct points in a plane. (Although **Figure 7** shows $x_1 < x_2$, no particular order is required.) Let $M(x, y)$ be the midpoint of the segment joining P and Q. Draw vertical lines from each of the three points to the x-axis, as shown in **Figure 7.**

Figure 7

The ordered pair $M(x, y)$ is the midpoint of the line segment joining P and Q, so the distance between x and x_1 equals the distance between x and x_2.

$$x_2 - x = x - x_1$$

$$x_2 + x_1 = 2x \quad \text{Add } x \text{ and } x_1 \text{ to each side.}$$

$$x = \frac{x_1 + x_2}{2} \quad \text{Divide by 2 and rewrite.}$$

Similarly, the y-coordinate is $\frac{y_1 + y_2}{2}$, yielding the following formula.

Midpoint Formula

The coordinates of the midpoint M of the line segment with endpoints $P(x_1, y_1)$ and $Q(x_2, y_2)$ are given by the following.

$$M = \left(\frac{x_1 + x_2}{2}, \frac{y_1 + y_2}{2}\right)$$

That is, the x-coordinate of the midpoint of a line segment is the* average *of the x-coordinates of the segment's endpoints, and the y-coordinate is the* average *of the y-coordinates of the segment's endpoints.

EXAMPLE 5 **Using the Midpoint Formula**

Use the midpoint formula to do each of the following.

(a) Find the coordinates of the midpoint M of the line segment with endpoints $(8, -4)$ and $(-6, 1)$.

(b) Find the coordinates of the other endpoint Q of a line segment with one endpoint $P(-6, 12)$ and midpoint $M(8, -2)$.

SOLUTION

(a) The coordinates of M are found using the midpoint formula.

$$M = \left(\frac{8 + (-6)}{2}, \frac{-4 + 1}{2}\right) = \left(1, -\frac{3}{2}\right) \quad \text{Substitute in } M = \left(\frac{x_1 + x_2}{2}, \frac{y_1 + y_2}{2}\right).$$

The coordinates of midpoint M are $\left(1, -\frac{3}{2}\right)$.

(b) Let (x, y) represent the coordinates of Q. Use both parts of the midpoint formula.

Substitute carefully.

x-value of P ↓, x-value of M ↓	y-value of P ↓, y-value of M ↓
$\frac{x + (-6)}{2} = 8$	$\frac{y + 12}{2} = -2$
$x - 6 = 16$	$y + 12 = -4$
$x = 22$	$y = -16$

The coordinates of endpoint Q are $(22, -16)$.

✔ **Now Try Exercises 15(b) and 35.**

EXAMPLE 6 **Applying the Midpoint Formula**

Figure 8 depicts how a graph might indicate the increase in the revenue generated by fast-food restaurants in the United States from \$69.8 billion in 1990 to \$195.1 billion in 2014. Use the midpoint formula and the two given points to estimate the revenue from fast-food restaurants in 2002, and compare it to the actual figure of \$138.3 billion.

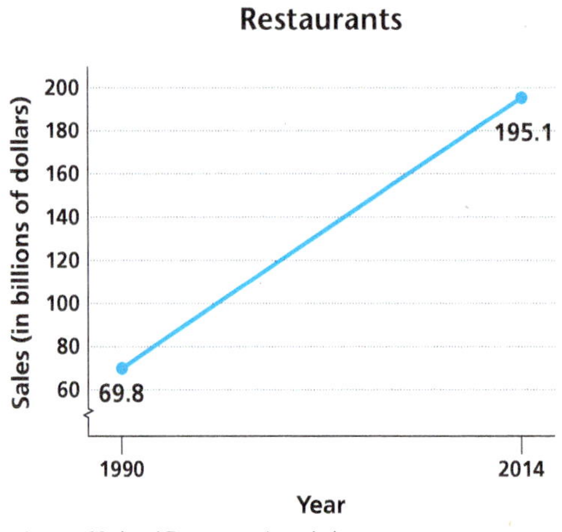

Source: National Restaurant Association.

Figure 8

SOLUTION The year 2002 lies halfway between 1990 and 2014, so we must find the coordinates of the midpoint of the line segment that has endpoints

$$(1990, 69.8) \quad \text{and} \quad (2014, 195.1).$$

(Here, the second component is in billions of dollars.)

$$M = \left(\frac{1990 + 2014}{2}, \frac{69.8 + 195.1}{2}\right) = (2002, 132.5) \quad \text{Use the midpoint formula.}$$

Our estimate is \$132.5 billion, which is less than the actual figure of \$138.3 billion. Models are used to predict outcomes. They rarely give exact values.

✔ **Now Try Exercise 41.**

Equations in Two Variables Ordered pairs are used to express the solutions of equations in two variables. When an ordered pair represents the solution of an equation with the variables x and y, the x-value is written first. For example, we say that

$$(1, 2) \quad \text{is a solution of} \quad 2x - y = 0.$$

Substituting 1 for x and 2 for y in the equation gives a true statement.

$$2x - y = 0$$
$$2(1) - 2 \stackrel{?}{=} 0 \quad \text{Let } x = 1 \text{ and } y = 2.$$
$$0 = 0 \checkmark \quad \text{True}$$

EXAMPLE 7 Finding Ordered-Pair Solutions of Equations

For each equation, find at least three ordered pairs that are solutions.

(a) $y = 4x - 1$ **(b)** $x = \sqrt{y - 1}$ **(c)** $y = x^2 - 4$

SOLUTION

(a) Choose any real number for x or y, and substitute in the equation to obtain the corresponding value of the other variable. For example, let $x = -2$ and then let $y = 3$.

$$y = 4x - 1$$
$$y = 4(-2) - 1 \quad \text{Let } x = -2.$$
$$y = -8 - 1 \quad \text{Multiply.}$$
$$y = -9 \quad \text{Subtract.}$$

$$y = 4x - 1$$
$$3 = 4x - 1 \quad \text{Let } y = 3.$$
$$4 = 4x \quad \text{Add 1.}$$
$$1 = x \quad \text{Divide by 4.}$$

This gives the ordered pairs $(-2, -9)$ and $(1, 3)$. Verify that the ordered pair $(0, -1)$ is also a solution.

(b)

$$x = \sqrt{y - 1} \quad \text{Given equation}$$
$$1 = \sqrt{y - 1} \quad \text{Let } x = 1.$$
$$1 = y - 1 \quad \text{Square each side.}$$
$$2 = y \quad \text{Add 1.}$$

One ordered pair is $(1, 2)$. Verify that the ordered pairs $(0, 1)$ and $(2, 5)$ are also solutions of the equation.

(c) A table provides an organized method for determining ordered pairs. Here, we let x equal -2, -1, 0, 1, and 2 in $y = x^2 - 4$ and determine the corresponding y-values.

x	y	
-2	0	$(-2)^2 - 4 = 4 - 4 = 0$
-1	-3	$(-1)^2 - 4 = 1 - 4 = -3$
0	-4	$0^2 - 4 = -4$
1	-3	$1^2 - 4 = -3$
2	0	$2^2 - 4 = 0$

Five ordered pairs are $(-2, 0)$, $(-1, -3)$, $(0, -4)$, $(1, -3)$, and $(2, 0)$.

✔ **Now Try Exercises 47(a), 51(a), and 53(a).**

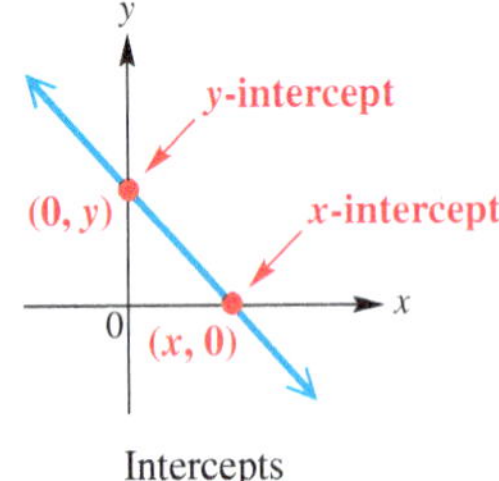

Intercepts

The **graph** of an equation is found by plotting ordered pairs that are solutions of the equation. The **intercepts** of the graph are good points to plot first. An ***x*-intercept** is a point where the graph intersects the x-axis. A ***y*-intercept** is a point where the graph intersects the y-axis. In other words, the x-intercept is represented by an ordered pair with y-coordinate 0, and the y-intercept is an ordered pair with x-coordinate 0.

A general algebraic approach for graphing an equation using intercepts and point-plotting follows.

Graphing an Equation by Point Plotting

Step 1 Find the intercepts.

Step 2 Find as many additional ordered pairs as needed.

Step 3 Plot the ordered pairs from Steps 1 and 2.

Step 4 Join the points from Step 3 with a smooth line or curve.

EXAMPLE 8 Graphing Equations

Graph each of the equations here, from **Example 7.**

(a) $y = 4x - 1$ **(b)** $x = \sqrt{y - 1}$ **(c)** $y = x^2 - 4$

SOLUTION

(a) ***Step 1*** Let $y = 0$ to find the x-intercept, and let $x = 0$ to find the y-intercept.

$y = 4x - 1$	$y = 4x - 1$
$0 = 4x - 1$ Let $y = 0$.	$y = 4(0) - 1$ Let $x = 0$.
$1 = 4x$	$y = 0 - 1$
$\frac{1}{4} = x$	$y = -1$

The intercepts are $\left(\frac{1}{4}, 0\right)$ and $(0, -1)$.* Note that the y-intercept is one of the ordered pairs we found in **Example 7(a).**

*Intercepts are sometimes defined as numbers, such as x-intercept $\frac{1}{4}$ and y-intercept -1. In this text, we define them as ordered pairs, such as $\left(\frac{1}{4}, 0\right)$ and $(0, -1)$.

Step 2 We use the other ordered pairs found in **Example 7(a):**

$$(-2, -9) \quad \text{and} \quad (1, 3).$$

Step 3 Plot the four ordered pairs from Steps 1 and 2 as shown in **Figure 9.**

Step 4 Join the points plotted in Step 3 with a straight line. This line, also shown in **Figure 9,** is the graph of the equation $y = 4x - 1$.

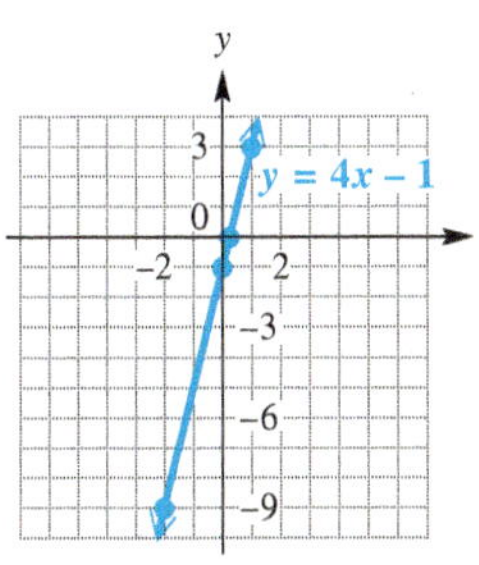

Figure 9

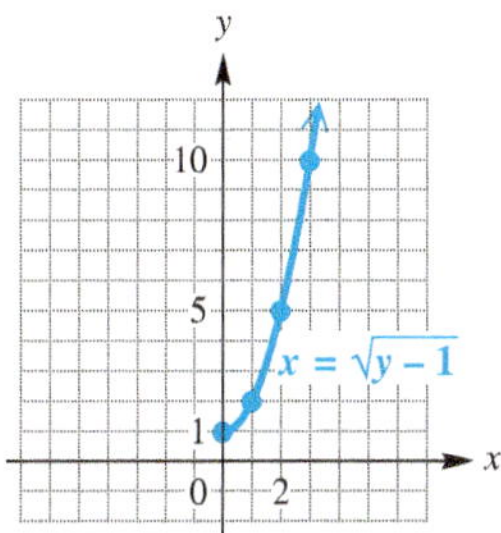

Figure 10

(b) For $x = \sqrt{y-1}$, the y-intercept $(0, 1)$ was found in **Example 7(b).** Solve

$$x = \sqrt{0 - 1} \quad \text{Let } y = 0.$$

to find the x-intercept. When $y = 0$, the quantity under the radical symbol is negative, so there is no x-intercept. In fact, $y - 1$ must be greater than or equal to 0, so y must be greater than or equal to 1.

We start by plotting the ordered pairs from **Example 7(b)** and then join the points with a smooth curve as in **Figure 10.** To confirm the direction the curve will take as x increases, we find another solution, $(3, 10)$. (Point plotting for graphs other than lines is often inefficient. We will examine other graphing methods later.)

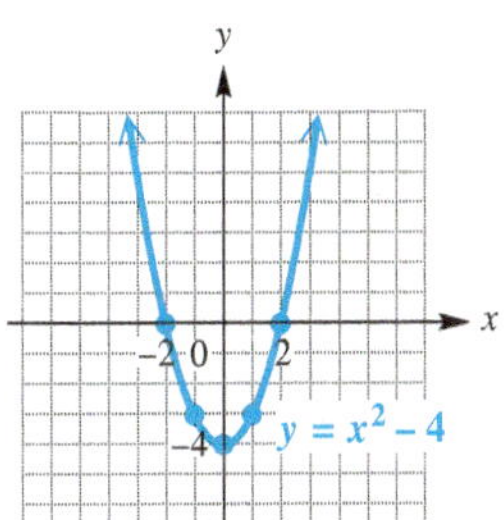

Figure 11

(c) In **Example 7(c),** we made a table of five ordered pairs that satisfy the equation $y = x^2 - 4$.

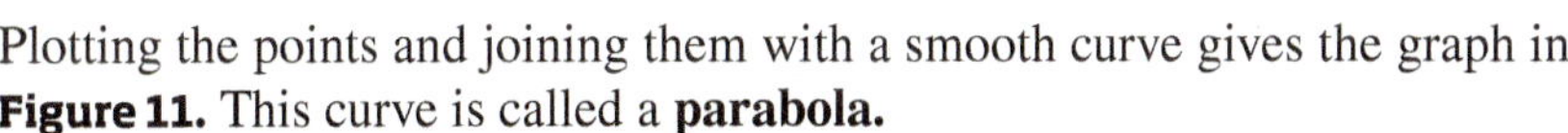

$(-2, 0)$, $(-1, -3)$, $(0, -4)$, $(1, -3)$, $(2, 0)$

↑ x-intercept ↑ y-intercept ↑ x-intercept

Plotting the points and joining them with a smooth curve gives the graph in **Figure 11.** This curve is called a **parabola.**

Now Try Exercises 47(b), 51(b), and 53(b).

To graph an equation on a calculator, such as

$$y = 4x - 1, \quad \text{Equation from **Example 8(a)**}$$

we must first solve it for y (if necessary). Here the equation is already in the correct form, $y = 4x - 1$, so we enter $4x - 1$ for y_1.*

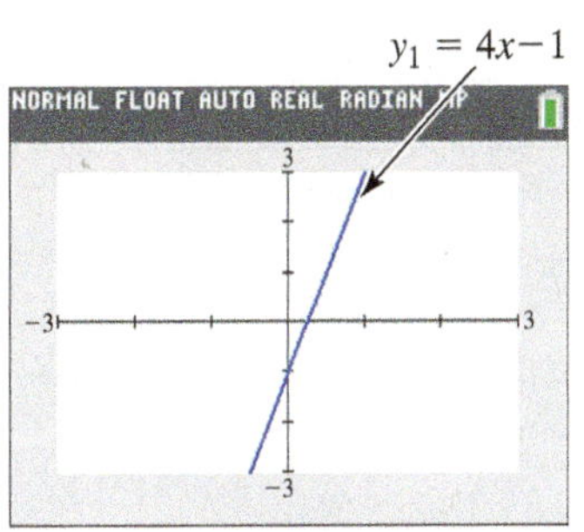

Figure 12

The intercepts can help determine an appropriate window, since we want them to appear in the graph. A good choice is often the **standard viewing window** for the TI-84 Plus, which has x minimum $= -10$, x maximum $= 10$, y minimum $= -10$, y maximum $= 10$, with x scale $= 1$ and y scale $= 1$. (The x and y scales determine the spacing of the tick marks.) Because the intercepts here are very close to the origin, we have chosen the x and y minimum and maximum to be -3 and 3 instead. See **Figure 12.** ■

*In this text, we use lowercase letters for variables when referencing graphing calculators. (Some models use uppercase letters.)

2.1 Exercises

CONCEPT PREVIEW *Fill in the blank to correctly complete each sentence.*

1. The point $(-1, 3)$ lies in quadrant ______ in the rectangular coordinate system.
2. The point $(4, \_\_)$ lies on the graph of the equation $y = 3x - 6$.
3. Any point that lies on the x-axis has y-coordinate equal to ______.
4. The y-intercept of the graph of $y = -2x + 6$ is ______.
5. The x-intercept of the graph of $2x + 5y = 10$ is ______.
6. The distance from the origin to the point $(-3, 4)$ is ______.

CONCEPT PREVIEW *Determine whether each statement is* true *or* false. *If false, explain why.*

7. The graph of $y = x^2 + 2$ has no x-intercepts.
8. The graph of $y = x^2 - 2$ has two x-intercepts.
9. The midpoint of the segment joining $(0, 0)$ and $(4, 4)$ is 2.
10. The distance between the points $(0, 0)$ and $(4, 4)$ is 4.

Give three ordered pairs from each table. **See Example 1.**

11.

x	y
2	−5
−1	7
3	−9
5	−17
6	−21

12.

x	y
3	3
−5	−21
8	18
4	6
0	−6

13. *Percent of High School Students Who Smoke*

Year	Percent
1999	35
2001	29
2003	22
2005	23
2007	20
2009	20

Source: Centers for Disease Control and Prevention.

14. *Number of U.S. Viewers of the Super Bowl*

Year	Viewers (millions)
2002	86.8
2004	89.8
2006	90.7
2008	97.4
2010	106.5
2012	111.4
2014	111.5

Source: www.tvbythenumbers.com

For the points P and Q, find **(a)** *the distance $d(P, Q)$ and* **(b)** *the coordinates of the midpoint M of line segment PQ.* **See Examples 2 and 5(a).**

15. $P(-5, -6), Q(7, -1)$

16. $P(-4, 3), Q(2, -5)$

17. $P(8, 2), Q(3, 5)$

18. $P(-8, 4), Q(3, -5)$

19. $P(-6, -5), Q(6, 10)$

20. $P(6, -2), Q(4, 6)$

21. $P(3\sqrt{2}, 4\sqrt{5}), Q(\sqrt{2}, -\sqrt{5})$

22. $P(-\sqrt{7}, 8\sqrt{3}), Q(5\sqrt{7}, -\sqrt{3})$

Determine whether the three points are the vertices of a right triangle. ***See Example 3.***

23. $(-6, -4), (0, -2), (-10, 8)$

24. $(-2, -8), (0, -4), (-4, -7)$

25. $(-4, 1), (1, 4), (-6, -1)$

26. $(-2, -5), (1, 7), (3, 15)$

27. $(-4, 3), (2, 5), (-1, -6)$

28. $(-7, 4), (6, -2), (0, -15)$

Determine whether the three points are collinear. ***See Example 4.***

29. $(0, -7), (-3, 5), (2, -15)$

30. $(-1, 4), (-2, -1), (1, 14)$

31. $(0, 9), (-3, -7), (2, 19)$

32. $(-1, -3), (-5, 12), (1, -11)$

33. $(-7, 4), (6, -2), (-1, 1)$

34. $(-4, 3), (2, 5), (-1, 4)$

Find the coordinates of the other endpoint of each line segment, given its midpoint and one endpoint. ***See Example 5(b).***

35. midpoint $(5, 8)$, endpoint $(13, 10)$

36. midpoint $(-7, 6)$, endpoint $(-9, 9)$

37. midpoint $(12, 6)$, endpoint $(19, 16)$

38. midpoint $(-9, 8)$, endpoint $(-16, 9)$

39. midpoint (a, b), endpoint (p, q)

40. midpoint $(6a, 6b)$, endpoint $(3a, 5b)$

Solve each problem. ***See Example 6.***

41. *Bachelor's Degree Attainment* The graph shows a straight line that approximates the percentage of Americans 25 years and older who had earned bachelor's degrees or higher for the years 1990–2012. Use the midpoint formula and the two given points to estimate the percent in 2001. Compare the answer with the actual percent of 26.2.

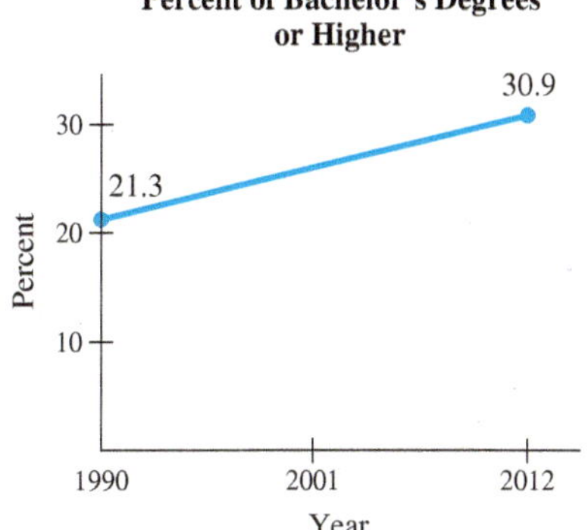

Source: U.S. Census Bureau.

42. *Newspaper Advertising Revenue* The graph shows a straight line that approximates national advertising revenue, in millions of dollars, for newspapers in the United States for the years 2006–2012. Use the midpoint formula and the two given points to estimate revenue in 2009. Compare the answer with the actual figure of 4424 million dollars.

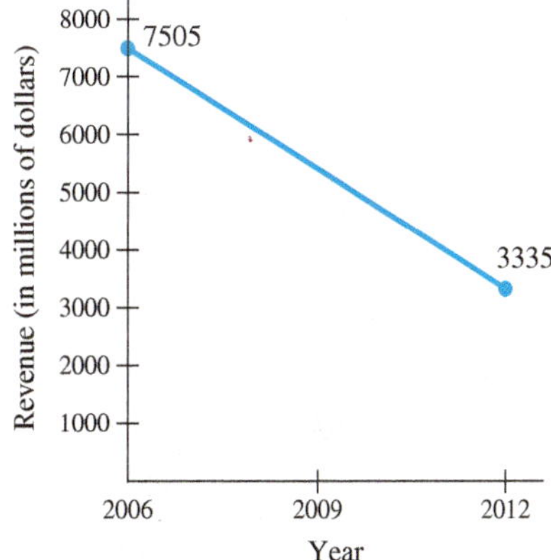

Source: Newspaper Association of America.

43. *Poverty Level Income Cutoffs* The table lists how poverty level income cutoffs (in dollars) for a family of four have changed over time. Use the midpoint formula to approximate the poverty level cutoff in 2012 to the nearest dollar.

Year	Income (in dollars)
1990	13,359
2000	17,604
2010	22,315
2011	23,021
2013	23,834

Source: U.S. Census Bureau.

44. *Public College Enrollment* Enrollments in public colleges for recent years are shown in the table. Assuming a linear relationship, estimate the enrollments for **(a)** 2003 and **(b)** 2009. Give answers to the nearest tenth of thousands if applicable.

Year	Enrollment (in thousands)
2000	11,753
2006	13,180
2012	14,880

Source: National Center for Education Statistics.

45. Show that if M is the midpoint of the line segment with endpoints $P(x_1, y_1)$ and $Q(x_2, y_2)$, then

$$d(P, M) + d(M, Q) = d(P, Q) \quad \text{and} \quad d(P, M) = d(M, Q).$$

46. Write the distance formula $d = \sqrt{(x_2 - x_1)^2 + (y_2 - y_1)^2}$ using a rational exponent.

*For each equation, **(a)** give a table with at least three ordered pairs that are solutions, and **(b)** graph the equation.* ***See Examples 7 and 8.***

47. $y = \frac{1}{2}x - 2$

48. $y = -\frac{1}{2}x + 2$

49. $2x + 3y = 5$

50. $3x - 2y = 6$

51. $y = x^2$

52. $y = x^2 + 2$

53. $y = \sqrt{x - 3}$

54. $y = \sqrt{x} - 3$

55. $y = |x - 2|$

56. $y = -|x + 4|$

57. $y = x^3$

58. $y = -x^3$

Concept Check Answer the following.

59. If a vertical line is drawn through the point $(4, 3)$, at what point will it intersect the x-axis?

60. If a horizontal line is drawn through the point $(4, 3)$, at what point will it intersect the y-axis?

61. If the point (a, b) is in the second quadrant, then in what quadrant is $(a, -b)$? $(-a, b)$? $(-a, -b)$? (b, a)?

62. Show that the points $(-2, 2)$, $(13, 10)$, $(21, -5)$, and $(6, -13)$ are the vertices of a rhombus (all sides equal in length).

63. Are the points $A(1, 1)$, $B(5, 2)$, $C(3, 4)$, and $D(-1, 3)$ the vertices of a parallelogram (opposite sides equal in length)? of a rhombus (all sides equal in length)?

64. Find the coordinates of the points that divide the line segment joining $(4, 5)$ and $(10, 14)$ into three equal parts.

2.2 Circles

- Center-Radius Form
- General Form
- An Application

Center-Radius Form

By definition, a **circle** is the set of all points in a plane that lie a given distance from a given point. The given distance is the **radius** of the circle, and the given point is the **center.**

We can find the equation of a circle from its definition using the distance formula. Suppose that the point (h, k) is the center and the circle has radius r, where $r > 0$. Let (x, y) represent any point on the circle. See **Figure 13.**

LOOKING AHEAD TO CALCULUS

The circle $x^2 + y^2 = 1$ is called the **unit circle.** It is important in interpreting the *trigonometric* or *circular* functions that appear in the study of calculus.

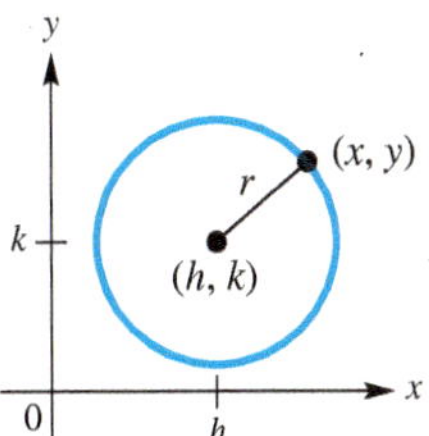

Figure 13

$$\sqrt{(x_2 - x_1)^2 + (y_2 - y_1)^2} = d \quad \text{Distance formula}$$

$$\sqrt{(x - h)^2 + (y - k)^2} = r \quad (x_1, y_1) = (h, k),\ (x_2, y_2) = (x, y), \text{ and } d = r$$

$$(x - h)^2 + (y - k)^2 = r^2 \quad \text{Square each side.}$$

Center-Radius Form of the Equation of a Circle

A circle with center (h, k) and radius r has equation

$$(x - h)^2 + (y - k)^2 = r^2,$$

which is the **center-radius form** of the equation of the circle. As a special case, a circle with center $(0, 0)$ and radius r has the following equation.

$$x^2 + y^2 = r^2$$

EXAMPLE 1 Finding the Center-Radius Form

Find the center-radius form of the equation of each circle described.

(a) center $(-3, 4)$, radius 6 **(b)** center $(0, 0)$, radius 3

SOLUTION

(a)

$$(x - h)^2 + (y - k)^2 = r^2 \quad \text{Center-radius form}$$

$$[x - (-3)]^2 + (y - 4)^2 = 6^2 \quad \text{Substitute. Let } (h, k) = (-3, 4) \text{ and } r = 6.$$

$$(x + 3)^2 + (y - 4)^2 = 36 \quad \text{Simplify.}$$

Be careful with signs here.

(b) The center is the origin and $r = 3$.

$$x^2 + y^2 = r^2 \quad \text{Special case of the center-radius form}$$

$$x^2 + y^2 = 3^2 \quad \text{Let } r = 3.$$

$$x^2 + y^2 = 9 \quad \text{Apply the exponent.}$$

✔ **Now Try Exercises 11(a) and 17(a).**

EXAMPLE 2 Graphing Circles

Graph each circle discussed in **Example 1.**

(a) $(x+3)^2+(y-4)^2=36$ **(b)** $x^2+y^2=9$

SOLUTION

(a) Writing the given equation in center-radius form

$$[x-(-3)]^2+(y-4)^2=6^2$$

gives $(-3, 4)$ as the center and 6 as the radius. See **Figure 14.**

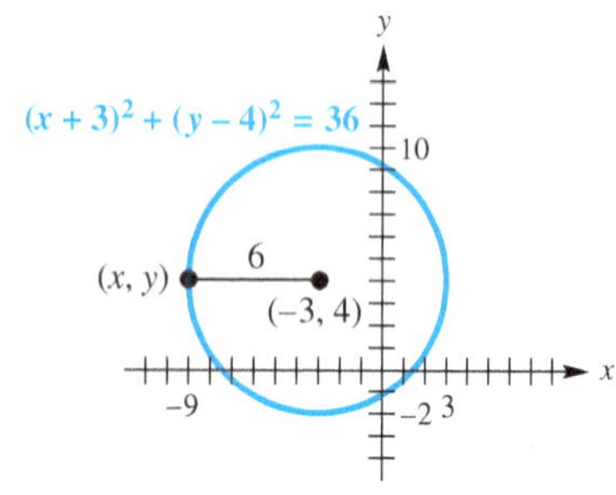

Figure 14

Figure 15

(b) The graph with center $(0, 0)$ and radius 3 is shown in **Figure 15.**

✔ **Now Try Exercises 11(b) and 17(b).**

The circles graphed in **Figures 14 and 15** of **Example 2** can be generated on a graphing calculator by first solving for y and then entering two expressions for y_1 and y_2. See **Figures 16 and 17.** In both cases, the plot of y_1 yields the top half of the circle, and that of y_2 yields the bottom half. It is necessary to use a **square viewing window** to avoid distortion when graphing circles.

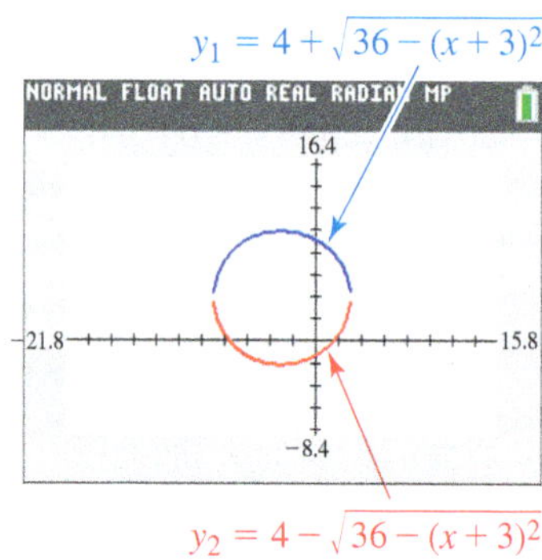

The graph of this circle has equation $(x+3)^2+(y-4)^2=36$.

Figure 16

The graph of this circle has equation $x^2+y^2=9$.

Figure 17

General Form Consider the center-radius form of the equation of a circle, and rewrite it so that the binomials are expanded and the right side equals 0.

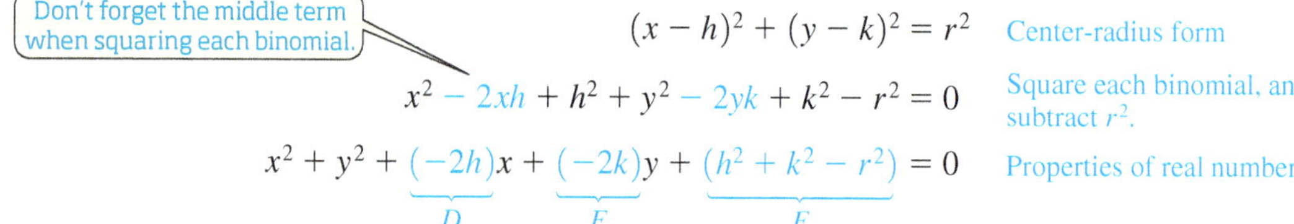

$$(x-h)^2+(y-k)^2=r^2 \quad \text{Center-radius form}$$

$$x^2-2xh+h^2+y^2-2yk+k^2-r^2=0 \quad \text{Square each binomial, and subtract } r^2.$$

$$x^2+y^2+\underbrace{(-2h)}_{D}x+\underbrace{(-2k)}_{E}y+\underbrace{(h^2+k^2-r^2)}_{F}=0 \quad \text{Properties of real numbers}$$

If $r>0$, then the graph of this equation is a circle with center (h, k) and radius r.

This form is the **general form of the equation of a circle.**

General Form of the Equation of a Circle

For some real numbers D, E, and F, the equation

$$x^2 + y^2 + Dx + Ey + F = 0$$

can have a graph that is a circle or a point, or is nonexistent.

Starting with an equation in this general form, we can complete the square to get an equation of the form

$$(x - h)^2 + (y - k)^2 = c, \quad \text{for some number } c.$$

There are three possibilities for the graph based on the value of c.

1. If $c > 0$, then $r^2 = c$, and the graph of the equation is a circle with radius $\sqrt{c}$.
2. If $c = 0$, then the graph of the equation is the single point (h, k).
3. If $c < 0$, then no points satisfy the equation, and the graph is nonexistent.

The next example illustrates the procedure for finding the center and radius.

EXAMPLE 3 Finding the Center and Radius by Completing the Square

Show that $x^2 - 6x + y^2 + 10y + 18 = 0$ has a circle as its graph. Find the center and radius.

SOLUTION We complete the square twice, once for x and once for y. Begin by subtracting 18 from each side.

$$x^2 - 6x + y^2 + 10y + 18 = 0$$

$$(x^2 - 6x \quad\quad) + (y^2 + 10y \quad\quad) = -18$$

Think: $\left[\frac{1}{2}(-6)\right]^2 = (-3)^2 = 9$ and $\left[\frac{1}{2}(10)\right]^2 = 5^2 = 25$

Add 9 and 25 on the left to complete the two squares, and to compensate, add 9 and 25 on the right.

$$(x^2 - 6x + 9) + (y^2 + 10y + 25) = -18 + 9 + 25 \quad \text{Complete the square.}$$

Add 9 and 25 on *both* sides.

$$(x - 3)^2 + (y + 5)^2 = 16 \quad \text{Factor. Add on the right.}$$

$$(x - 3)^2 + [y - (-5)]^2 = 4^2 \quad \text{Center-radius form}$$

Because $4^2 = 16$ and $16 > 0$, the equation represents a circle with center $(3, -5)$ and radius 4.

✔ **Now Try Exercise 27.**

EXAMPLE 4 Finding the Center and Radius by Completing the Square

Show that $2x^2 + 2y^2 - 6x + 10y = 1$ has a circle as its graph. Find the center and radius.

SOLUTION To complete the square, the coefficient of the x^2-term and that of the y^2-term must be 1. In this case they are both 2, so begin by dividing each side by 2.

$$2x^2 + 2y^2 - 6x + 10y = 1$$

$$x^2 + y^2 - 3x + 5y = \frac{1}{2} \quad \text{Divide by 2.}$$

$$(x^2 - 3x \qquad) + (y^2 + 5y \qquad) = \frac{1}{2} \quad \text{Rearrange and regroup terms in anticipation of completing the square.}$$

$$\left(x^2 - 3x + \frac{9}{4}\right) + \left(y^2 + 5y + \frac{25}{4}\right) = \frac{1}{2} + \frac{9}{4} + \frac{25}{4} \quad \text{Complete the square for } \textit{both } x \text{ and } y\text{; } \left[\tfrac{1}{2}(-3)\right]^2 = \tfrac{9}{4} \text{ and } \left[\tfrac{1}{2}(5)\right]^2 = \tfrac{25}{4}.$$

$$\left(x - \frac{3}{2}\right)^2 + \left(y + \frac{5}{2}\right)^2 = 9 \quad \text{Factor and add.}$$

$$\left(x - \frac{3}{2}\right)^2 + \left[y - \left(-\frac{5}{2}\right)\right]^2 = 3^2 \quad \text{Center-radius form}$$

The equation has a circle with center $\left(\frac{3}{2}, -\frac{5}{2}\right)$ and radius 3 as its graph.

✔ **Now Try Exercise 31.**

EXAMPLE 5 Determining Whether a Graph Is a Point or Nonexistent

The graph of the equation $x^2 + 10x + y^2 - 4y + 33 = 0$ either is a point or is nonexistent. Which is it?

SOLUTION

$$x^2 + 10x + y^2 - 4y + 33 = 0$$

$$x^2 + 10x + y^2 - 4y = -33 \quad \text{Subtract 33.}$$

Think: $\left[\frac{1}{2}(10)\right]^2 = 25$ and $\left[\frac{1}{2}(-4)\right]^2 = 4$ Prepare to complete the square for both x and y.

$$(x^2 + 10x + 25) + (y^2 - 4y + 4) = -33 + 25 + 4 \quad \text{Complete the square.}$$

$$(x + 5)^2 + (y - 2)^2 = -4 \quad \text{Factor and add.}$$

Because $-4 < 0$, there are *no* ordered pairs (x, y), with x and y both real numbers, satisfying the equation. The graph of the given equation is nonexistent—it contains no points. (If the constant on the right side were 0, the graph would consist of the single point $(-5, 2)$.)

✔ **Now Try Exercise 33.**

An Application Seismologists can locate the epicenter of an earthquake by determining the intersection of three circles. The radii of these circles represent the distances from the epicenter to each of three receiving stations. The centers of the circles represent the receiving stations.

EXAMPLE 6 **Locating the Epicenter of an Earthquake**

Suppose receiving stations A, B, and C are located on a coordinate plane at the points

$$(1, 4), \quad (-3, -1), \quad \text{and} \quad (5, 2).$$

Let the distances from the earthquake epicenter to these stations be 2 units, 5 units, and 4 units, respectively. Where on the coordinate plane is the epicenter located?

SOLUTION Graph the three circles as shown in **Figure 18.** From the graph it appears that the epicenter is located at $(1, 2)$. To check this algebraically, determine the equation for each circle and substitute $x = 1$ and $y = 2$.

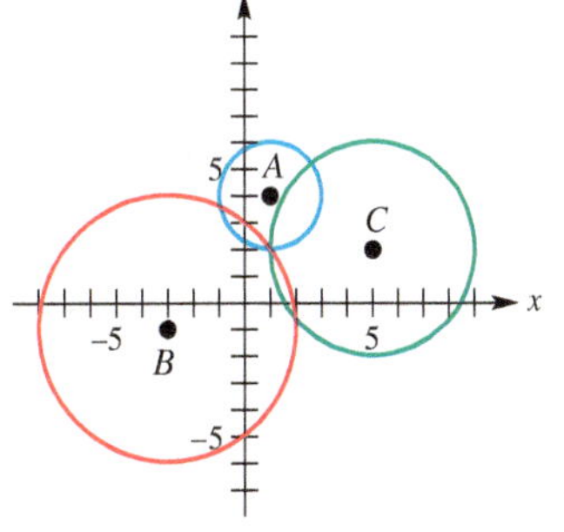

Figure 18

Station A:

$(x-1)^2+(y-4)^2=2^2$ Equation of a circle with center $(1, 4)$ and radius 2

$(1-1)^2+(2-4)^2 \stackrel{?}{=} 4$ Let $x = 1$ and $y = 2$.

$0+4 \stackrel{?}{=} 4$

$4 = 4$ ✓

Station B:

$(x+3)^2+(y+1)^2=5^2$ Equation of a circle with center $(-3, -1)$ and radius 5

$(1+3)^2+(2+1)^2 \stackrel{?}{=} 25$ Let $x = 1$ and $y = 2$.

$16+9 \stackrel{?}{=} 25$

$25 = 25$ ✓

Station C:

$(x-5)^2+(y-2)^2=4^2$ Equation of a circle with center $(5, 2)$ and radius 4

$(1-5)^2+(2-2)^2 \stackrel{?}{=} 16$ Let $x = 1$ and $y = 2$.

$16+0 \stackrel{?}{=} 16$

$16 = 16$ ✓

The point $(1, 2)$ lies on all three graphs. Thus, we can conclude that the epicenter of the earthquake is at $(1, 2)$.

✓ **Now Try Exercise 39.**

2.2 Exercises

CONCEPT PREVIEW *Fill in the blank(s) to correctly complete each sentence.*

1. The circle with equation $x^2 + y^2 = 49$ has center with coordinates ________ and radius equal to ________.
2. The circle with center $(3, 6)$ and radius 4 has equation ________.
3. The graph of $(x-4)^2 + (y+7)^2 = 9$ has center with coordinates ________.
4. The graph of $x^2 + (y-5)^2 = 9$ has center with coordinates ________.

CONCEPT PREVIEW *Match each equation in Column I with its graph in Column II.*

I	II
5. $(x-3)^2+(y-2)^2=25$	**A.**
6. $(x-3)^2+(y+2)^2=25$	**B.**
7. $(x+3)^2+(y-2)^2=25$	**C.**
8. $(x+3)^2+(y+2)^2=25$	**D.**

A.

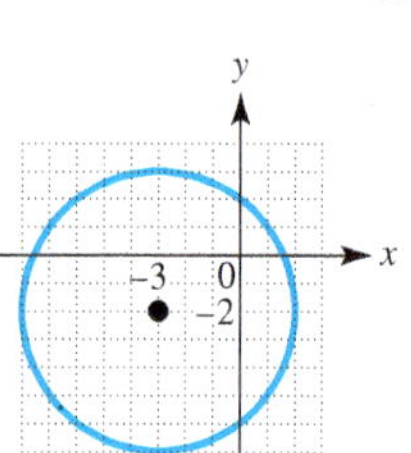

B.

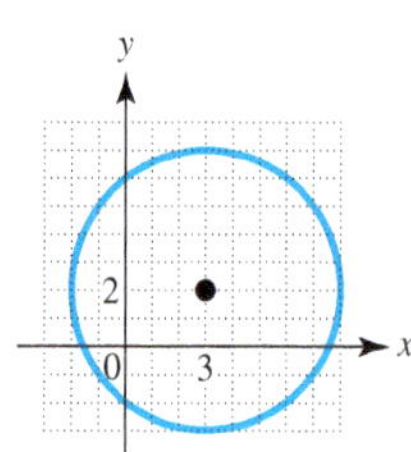

C.

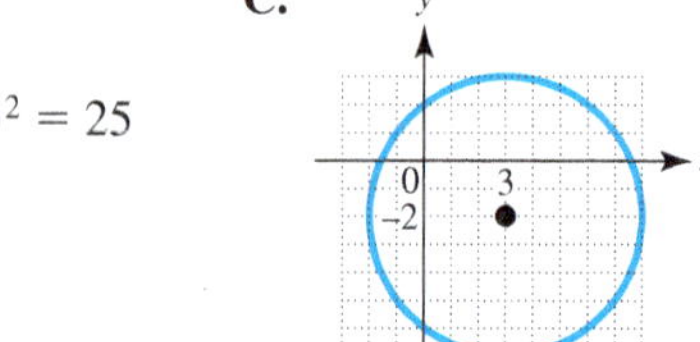

D.

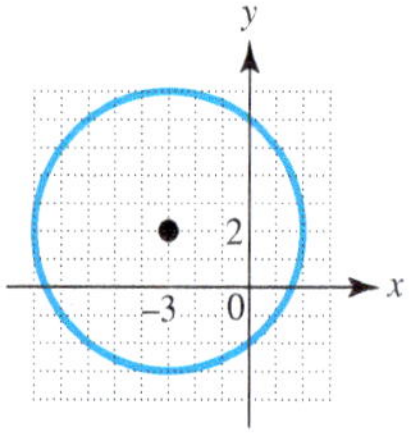

CONCEPT PREVIEW *Answer each question.*

9. How many points lie on the graph of $x^2+y^2=0$?

10. How many points lie on the graph of $x^2+y^2=-100$?

*In the following exercises, **(a)** find the center-radius form of the equation of each circle described, and **(b)** graph it. See **Examples 1 and 2.***

11. center $(0, 0)$, radius 6

12. center $(0, 0)$, radius 9

13. center $(2, 0)$, radius 6

14. center $(3, 0)$, radius 3

15. center $(0, 4)$, radius 4

16. center $(0, -3)$, radius 7

17. center $(-2, 5)$, radius 4

18. center $(4, 3)$, radius 5

19. center $(5, -4)$, radius 7

20. center $(-3, -2)$, radius 6

21. center $\left(\sqrt{2}, \sqrt{2}\right)$, radius $\sqrt{2}$

22. center $\left(-\sqrt{3}, -\sqrt{3}\right)$, radius $\sqrt{3}$

Connecting Graphs with Equations *Use each graph to determine an equation of the circle in **(a)** center-radius form and **(b)** general form.*

23.

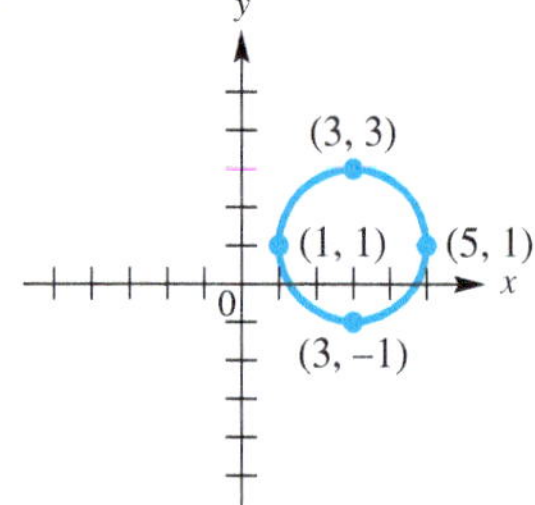

24.

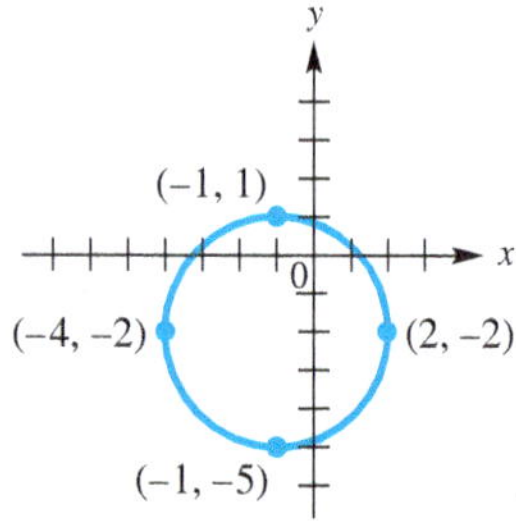

25.

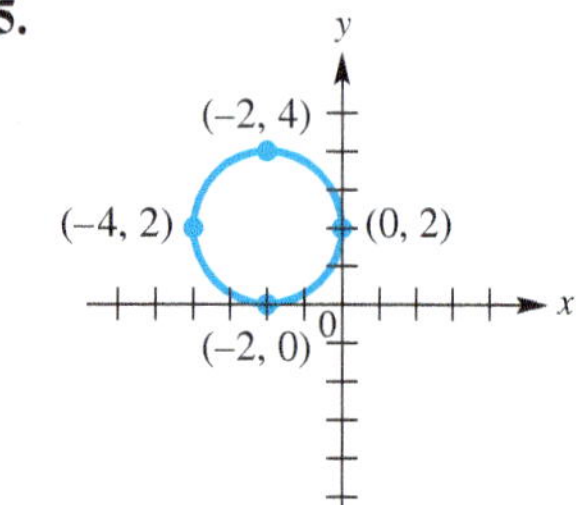

26.

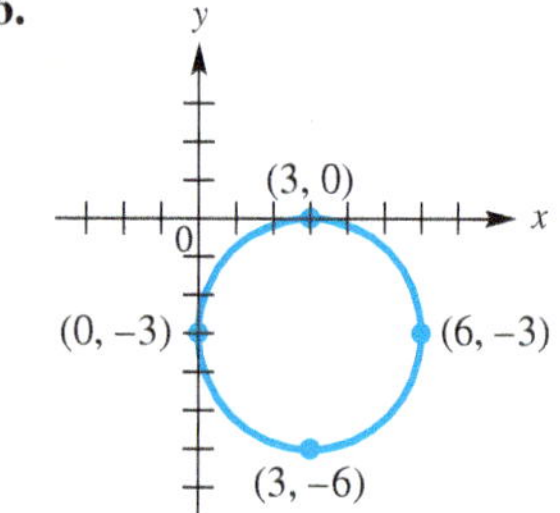

Decide whether or not each equation has a circle as its graph. If it does, give the center and radius. If it does not, describe the graph. **See Examples 3–5.**

27. $x^2 + y^2 + 6x + 8y + 9 = 0$

28. $x^2 + y^2 + 8x - 6y + 16 = 0$

29. $x^2 + y^2 - 4x + 12y = -4$

30. $x^2 + y^2 - 12x + 10y = -25$

31. $4x^2 + 4y^2 + 4x - 16y - 19 = 0$

32. $9x^2 + 9y^2 + 12x - 18y - 23 = 0$

33. $x^2 + y^2 + 2x - 6y + 14 = 0$

34. $x^2 + y^2 + 4x - 8y + 32 = 0$

35. $x^2 + y^2 - 6x - 6y + 18 = 0$

36. $x^2 + y^2 + 4x + 4y + 8 = 0$

37. $9x^2 + 9y^2 - 6x + 6y - 23 = 0$

38. $4x^2 + 4y^2 + 4x - 4y - 7 = 0$

Epicenter of an Earthquake *Solve each problem. To visualize the situation, use graph paper and a compass to carefully graph each circle.* **See Example 6.**

39. Suppose that receiving stations *X*, *Y*, and *Z* are located on a coordinate plane at the points

$$(7, 4), \quad (-9, -4), \quad \text{and} \quad (-3, 9),$$

respectively. The epicenter of an earthquake is determined to be 5 units from *X*, 13 units from *Y*, and 10 units from *Z*. Where on the coordinate plane is the epicenter located?

40. Suppose that receiving stations *P*, *Q*, and *R* are located on a coordinate plane at the points

$$(3, 1), \quad (5, -4), \quad \text{and} \quad (-1, 4),$$

respectively. The epicenter of an earthquake is determined to be $\sqrt{5}$ units from *P*, 6 units from *Q*, and $2\sqrt{10}$ units from *R*. Where on the coordinate plane is the epicenter located?

41. The locations of three receiving stations and the distances to the epicenter of an earthquake are contained in the following three equations:

$$(x - 2)^2 + (y - 1)^2 = 25, \quad (x + 2)^2 + (y - 2)^2 = 16,$$
$$\text{and} \quad (x - 1)^2 + (y + 2)^2 = 9.$$

Determine the location of the epicenter.

42. The locations of three receiving stations and the distances to the epicenter of an earthquake are contained in the following three equations:

$$(x - 2)^2 + (y - 4)^2 = 25, \quad (x - 1)^2 + (y + 3)^2 = 25,$$
$$\text{and} \quad (x + 3)^2 + (y + 6)^2 = 100.$$

Determine the location of the epicenter.

Concept Check *Work each of the following.*

43. Find the center-radius form of the equation of a circle with center $(3, 2)$ and tangent to the x-axis. (*Hint:* A line **tangent** to a circle touches it at exactly one point.)

44. Find the equation of a circle with center at $(-4, 3)$, passing through the point $(5, 8)$. Write it in center-radius form.

45. Find all points (x, y) with $x = y$ that are 4 units from $(1, 3)$.

46. Find all points satisfying $x + y = 0$ that are 8 units from $(-2, 3)$.

47. Find the coordinates of all points whose distance from $(1, 0)$ is $\sqrt{10}$ and whose distance from $(5, 4)$ is $\sqrt{10}$.

48. Find the equation of the circle of least radius that contains the points $(1, 4)$ and $(-3, 2)$ within or on its boundary.

49. Find all values of y such that the distance between $(3, y)$ and $(-2, 9)$ is 12.

50. Suppose that a circle is tangent to both axes, is in the third quadrant, and has radius $\sqrt{2}$. Find the center-radius form of its equation.

51. Find the shortest distance from the origin to the graph of the circle with equation

$$x^2 - 16x + y^2 - 14y + 88 = 0.$$

52. Phlash Phelps, the morning radio personality on SiriusXM Satellite Radio's *Sixties on Six* Decades channel, is an expert on U.S. geography. He loves traveling around the country to strange, out-of-the-way locations. The photo shows Phlash seated in front of a sign in a small Arizona settlement called *Nothing*. (Nothing is so small that it's not named on current maps.) The sign indicates that Nothing is 50 mi from Wickenburg, AZ, 75 mi from Kingman, AZ, 105 mi from Phoenix, AZ, and 180 mi from Las Vegas, NV. Explain how the concepts of **Example 6** can be used to locate Nothing, AZ, on a map of Arizona and southern Nevada.

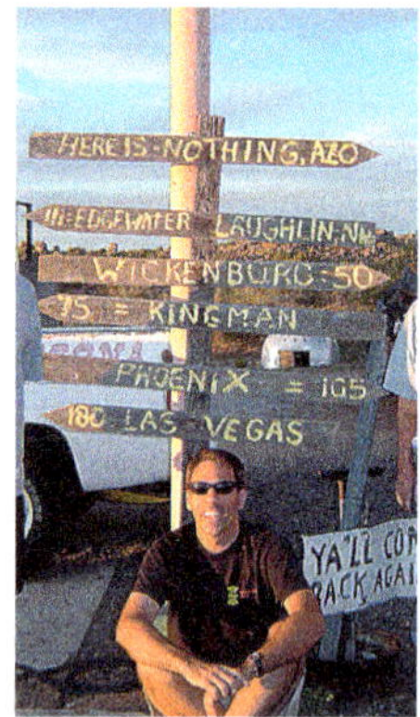

Relating Concepts

For individual or collaborative investigation *(Exercises 53–58)*

The distance formula, midpoint formula, and center-radius form of the equation of a circle are closely related in the following problem.

A circle has a diameter with endpoints $(-1, 3)$ and $(5, -9)$. Find the center-radius form of the equation of this circle.

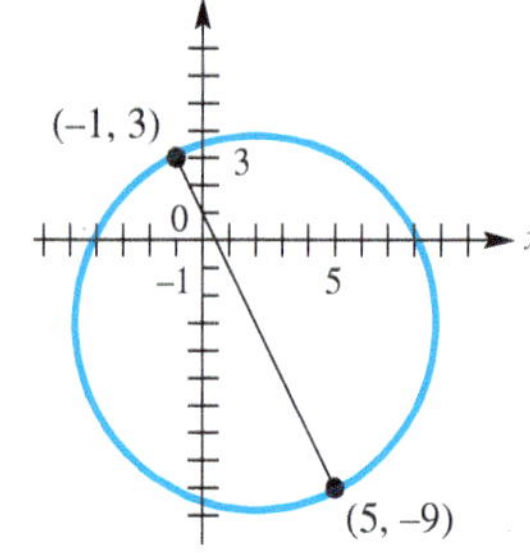

Work Exercises 53–58 in order, *to see the relationships among these concepts.*

53. To find the center-radius form, we must find both the radius and the coordinates of the center. Find the coordinates of the center using the midpoint formula. (The center of the circle must be the midpoint of the diameter.)

54. There are several ways to find the radius of the circle. One way is to find the distance between the center and the point $(-1, 3)$. Use the result from **Exercise 53** and the distance formula to find the radius.

55. Another way to find the radius is to repeat **Exercise 54,** but use the point $(5, -9)$ rather than $(-1, 3)$. Do this to obtain the same answer found in **Exercise 54.**

56. There is yet another way to find the radius. Because the radius is half the diameter, it can be found by finding half the length of the diameter. Using the endpoints of the diameter given in the problem, find the radius in this manner. The same answer found in **Exercise 54** should be obtained.

57. Using the center found in **Exercise 53** and the radius found in **Exercises 54–56,** give the center-radius form of the equation of the circle.

58. Use the method described in **Exercises 53–57** to find the center-radius form of the equation of the circle with diameter having endpoints $(3, -5)$ and $(-7, 3)$.

Find the center-radius form of the circle described or graphed. **(See Exercises 53–58.)**

59. a circle having a diameter with endpoints $(-1, 2)$ and $(11, 7)$

60. a circle having a diameter with endpoints $(5, 4)$ and $(-3, -2)$

61.

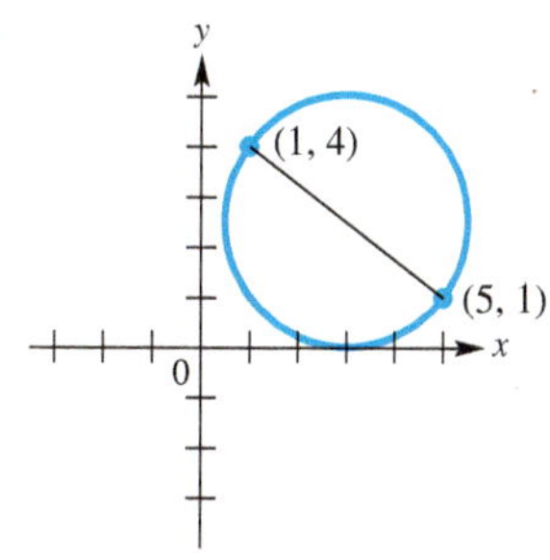

62.

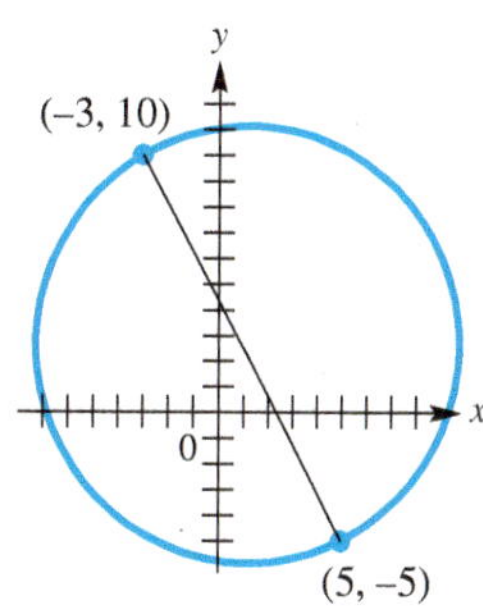

2.3 Functions

- Relations and Functions
- Domain and Range
- Determining Whether Relations Are Functions
- Function Notation
- Increasing, Decreasing, and Constant Functions

Relations and Functions

As we saw previously, one quantity can sometimes be described in terms of another.

- The letter grade a student receives in a mathematics course depends on a numerical score.
- The amount paid (in dollars) for gas at a gas station depends on the number of gallons pumped.
- The dollars spent by the average American household depends on the expense category.

We used ordered pairs to represent these corresponding quantities. For example, (3, \$10.50) indicates that we pay \$10.50 for 3 gallons of gas. Since the amount we pay *depends* on the number of gallons pumped, the amount (in dollars) is called the *dependent variable*, and the number of gallons pumped is called the *independent variable.*

Generalizing, if the value of the second component y depends on the value of the first component x, then y is the **dependent variable** and x is the **independent variable.**

Independent variable → x; Dependent variable → y

$$(x, y)$$

A set of ordered pairs such as $\{(3, 10.50), (8, 28.00), (10, 35.00)\}$ is a *relation.* A *function* is a special kind of relation.

Relation and Function

A **relation** is a set of ordered pairs. A **function** is a relation in which, for each distinct value of the first component of the ordered pairs, there is *exactly one* value of the second component.

NOTE The relation from the beginning of this section representing the number of gallons of gasoline and the corresponding cost is a function because each x-value is paired with exactly one y-value.

EXAMPLE 1 Deciding Whether Relations Define Functions

Decide whether each relation defines a function.

$$F = \{(1, 2), (-2, 4), (3, 4)\}$$

$$G = \{(1, 1), (1, 2), (1, 3), (2, 3)\}$$

$$H = \{(-4, 1), (-2, 1), (-2, 0)\}$$

SOLUTION Relation F is a function because for each different x-value there is exactly one y-value. We can show this correspondence as follows.

$$\{1, -2, 3\} \quad x\text{-values of } F$$
$$\downarrow \quad \downarrow \quad \downarrow$$
$$\{2, 4, 4\} \quad y\text{-values of } F$$

As the correspondence below shows, relation G is not a function because one first component corresponds to *more than one* second component.

$$\{1, 2\} \quad x\text{-values of } G$$
$$\{1, 2, 3\} \quad y\text{-values of } G$$

In relation H the last two ordered pairs have the same x-value paired with two different y-values (-2 is paired with both 1 and 0), so H is a relation but not a function. ***In a function, no two ordered pairs can have the same first component and different second components.***

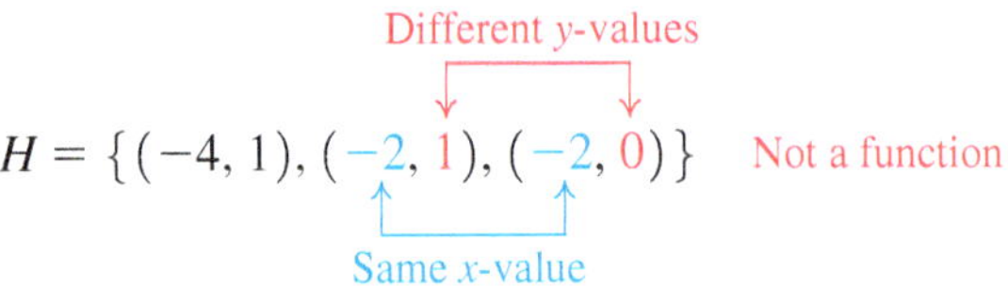

✔ Now Try Exercises 11 and 13.

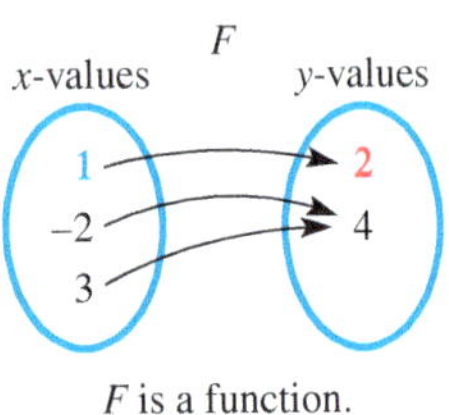

F is a function.

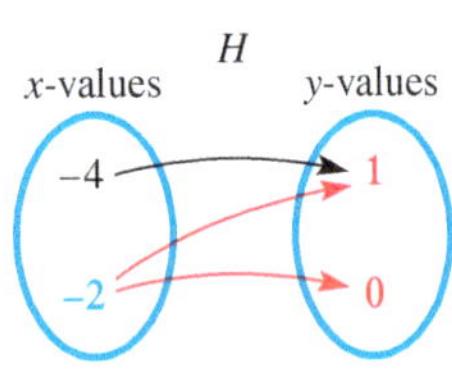

H is not a function.

Figure 19

Relations and functions can also be expressed as a correspondence or *mapping* from one set to another, as shown in **Figure 19** for function F and relation H from **Example 1.** The arrow from 1 to 2 indicates that the ordered pair $(1, 2)$ belongs to F—each first component is paired with exactly one second component. In the mapping for relation H, which is not a function, the first component -2 is paired with two different second components, 1 and 0.

Because relations and functions are sets of ordered pairs, we can represent them using tables and graphs. A table and graph for function F are shown in **Figure 20.**

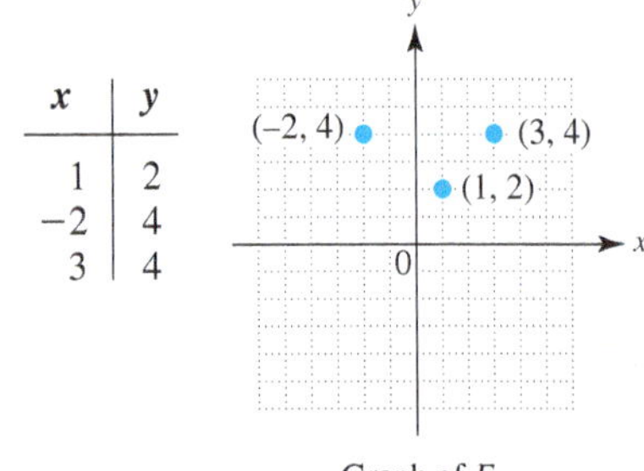

x	y
1	2
−2	4
3	4

Graph of F

Figure 20

Finally, we can describe a relation or function using a rule that tells how to determine the dependent variable for a specific value of the independent variable. The rule may be given in words: for instance, "the dependent variable is twice the independent variable." Usually the rule is an equation, such as the one below.

$$\text{Dependent variable} \longrightarrow y = 2x \longleftarrow \text{Independent variable}$$

On this particular day, an *input* of pumping 7.870 gallons of gasoline led to an *output* of \$29.58 from the purchaser's wallet. This is an example of a function whose domain consists of numbers of gallons pumped, and whose range consists of amounts from the purchaser's wallet. Dividing the dollar amount by the number of gallons pumped gives the exact price of gasoline that day. Use a calculator to check this. Was this pump fair? (Later we will see that this price is an example of the slope m of a linear function of the form $y = mx$.)

In a function, there is exactly one value of the dependent variable, the second component, for each value of the independent variable, the first component.

NOTE Another way to think of a function relationship is to think of the independent variable as an **input** and of the dependent variable as an **output.** This is illustrated by the **input-output (function) machine** for the function $y = 2x$.

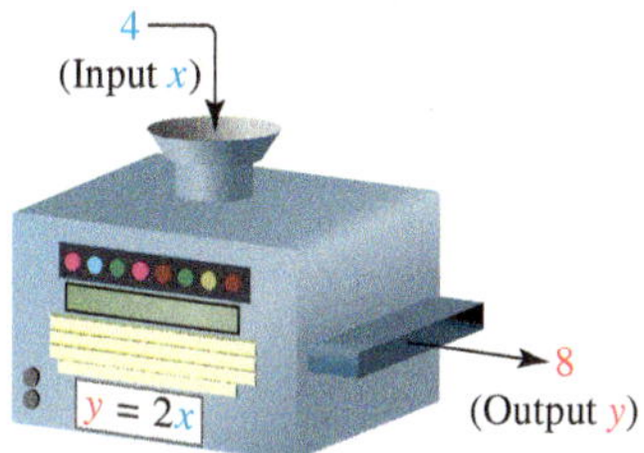

Function machine

Domain and Range We now consider two important concepts concerning relations.

Domain and Range

For every relation consisting of a set of ordered pairs (x, y), there are two important sets of elements.

- The set of all values of the independent variable (x) is the **domain.**
- The set of all values of the dependent variable (y) is the **range.**

EXAMPLE 2 Finding Domains and Ranges of Relations

Give the domain and range of each relation. Tell whether the relation defines a function.

(a) $\{(3, -1), (4, 2), (4, 5), (6, 8)\}$

(b)

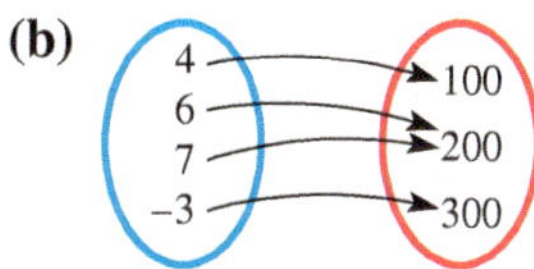

(c)

x	y
−5	2
0	2
5	2

SOLUTION

(a) The domain is the set of x-values, $\{3, 4, 6\}$. The range is the set of y-values, $\{-1, 2, 5, 8\}$. This relation is not a function because the same x-value, 4, is paired with two different y-values, 2 and 5.

(b) The domain is $\{4, 6, 7, -3\}$ and the range is $\{100, 200, 300\}$. This mapping defines a function. Each x-value corresponds to exactly one y-value.

(c) This relation is a set of ordered pairs, so the domain is the set of x-values $\{-5, 0, 5\}$ and the range is the set of y-values $\{2\}$. The table defines a function because each different x-value corresponds to exactly one y-value (even though it is the same y-value).

✔ **Now Try Exercises 19, 21, and 23.**

EXAMPLE 3 **Finding Domains and Ranges from Graphs**

Give the domain and range of each relation.

(a)

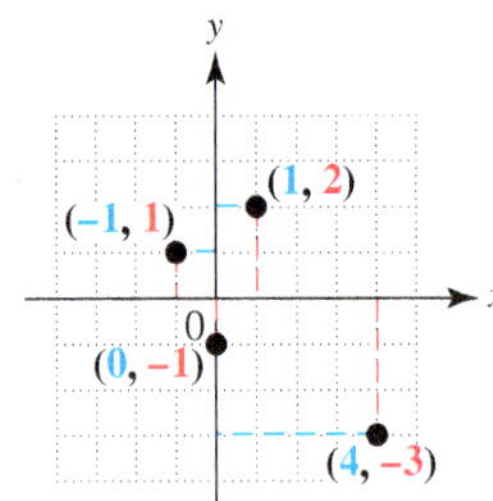

(b)

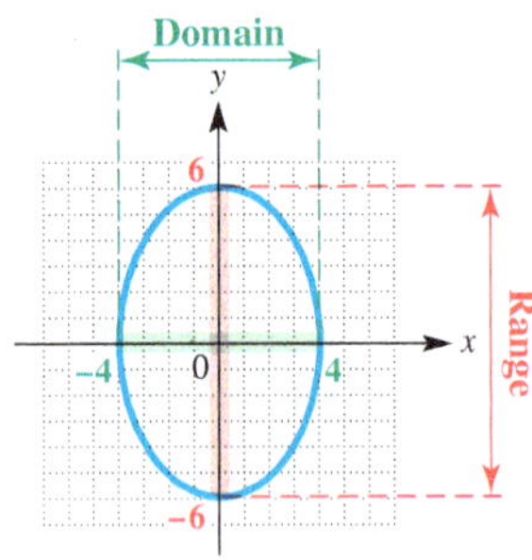

(c)

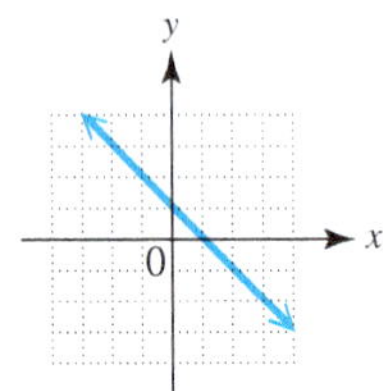

(d)

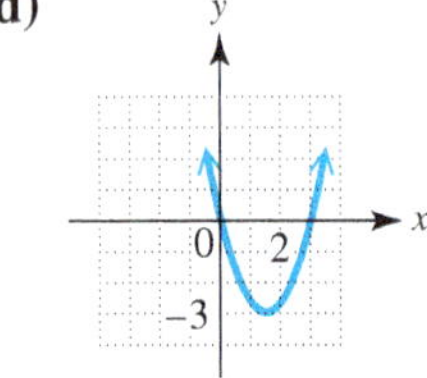

SOLUTION

(a) The domain is the set of x-values, $\{-1, 0, 1, 4\}$. The range is the set of y-values, $\{-3, -1, 1, 2\}$.

(b) The x-values of the points on the graph include all numbers between -4 and 4, inclusive. The y-values include all numbers between -6 and 6, inclusive.

The domain is $[-4, 4]$.
The range is $[-6, 6]$.　　Use interval notation.

(c) The arrowheads indicate that the line extends indefinitely left and right, as well as up and down. Therefore, both the domain and the range include all real numbers, which is written $(-\infty, \infty)$.

(d) The arrowheads indicate that the graph extends indefinitely left and right, as well as upward. The domain is $(-\infty, \infty)$. Because there is a least y-value, -3, the range includes all numbers greater than or equal to -3, written $[-3, \infty)$.

✔ **Now Try Exercises 27 and 29.**

Relations are often defined by equations, such as $y = 2x + 3$ and $y^2 = x$, so we must sometimes determine the domain of a relation from its equation. In this book, we assume the following agreement on the domain of a relation.

Agreement on Domain

Unless specified otherwise, the domain of a relation is assumed to be all real numbers that produce real numbers when substituted for the independent variable.

To illustrate this agreement, because any real number can be used as a replacement for x in $y = 2x + 3$, the domain of this function is the set of all real numbers.

As another example, the function $y = \frac{1}{x}$ has the set of all real numbers *except* 0 as domain because y is undefined if $x = 0$.

In general, the domain of a function defined by an algebraic expression is the set of all real numbers, except those numbers that lead to division by 0 or to an even root of a negative number.

(There are also exceptions for logarithmic and trigonometric functions. They are covered in further treatment of precalculus mathematics.)

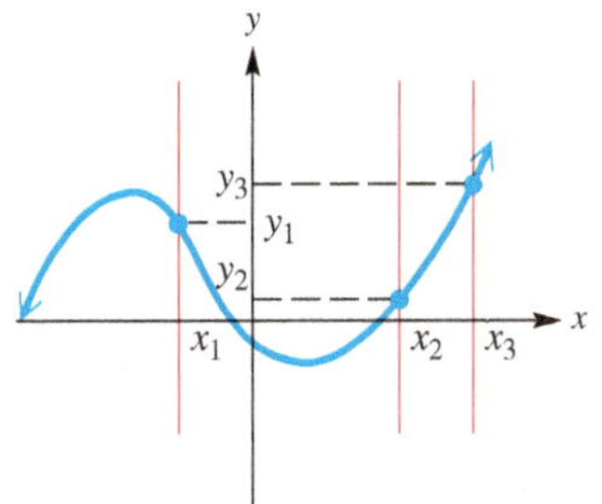

This is the graph of a function. Each x-value corresponds to only one y-value.

(a)

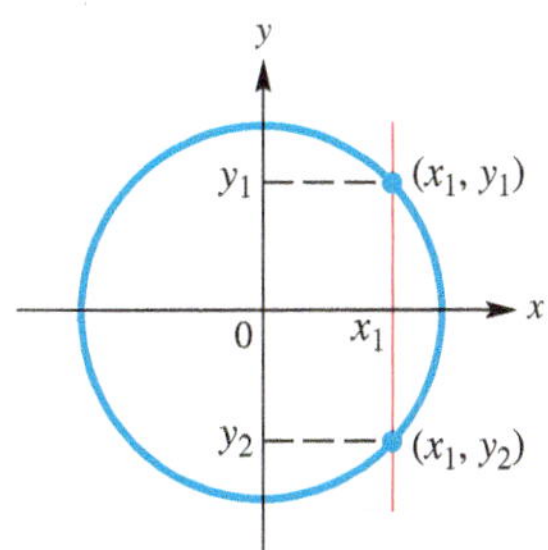

This is not the graph of a function. The same x-value corresponds to two different y-values.

(b)

Figure 21

Determining Whether Relations Are Functions Because each value of x leads to only one value of y in a function, any vertical line must intersect the graph in at most one point. This is the **vertical line test** for a function.

Vertical Line Test

If every vertical line intersects the graph of a relation in no more than one point, then the relation is a function.

The graph in **Figure 21(a)** represents a function because each vertical line intersects the graph in no more than one point. The graph in **Figure 21(b)** is not the graph of a function because there exists a vertical line that intersects the graph in more than one point.

EXAMPLE 4 Using the Vertical Line Test

Use the vertical line test to determine whether each relation graphed in **Example 3** is a function.

SOLUTION We repeat each graph from **Example 3,** this time with vertical lines drawn through the graphs.

(a)

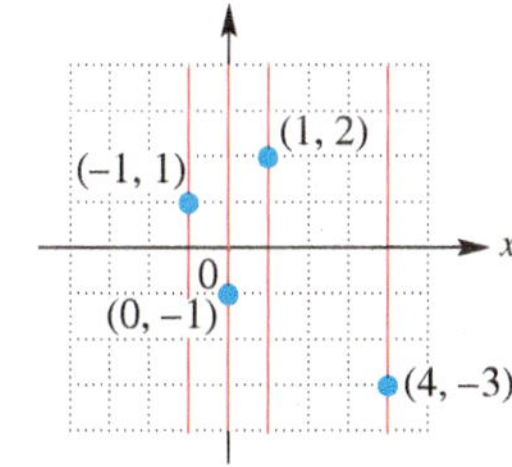

(b)

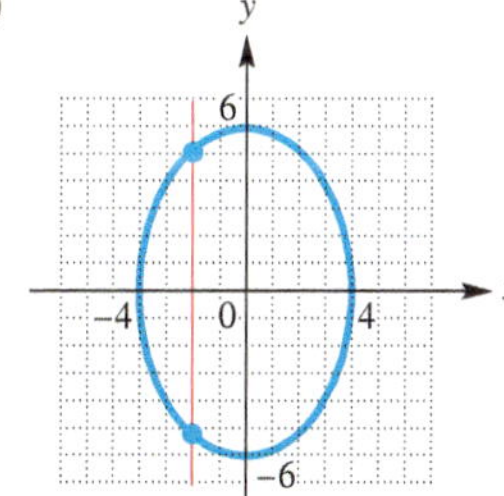

(c)

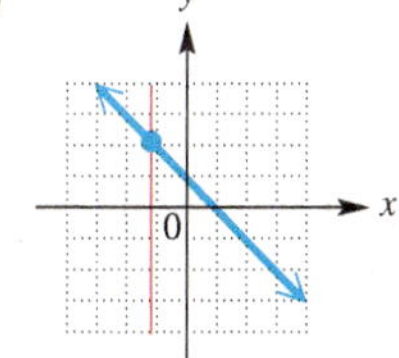

(d)

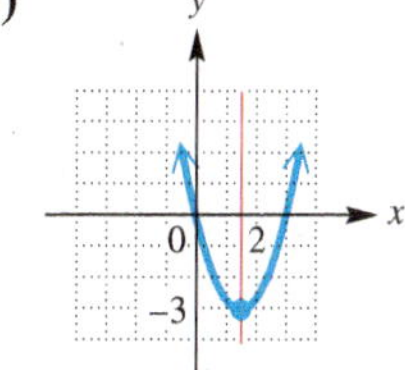

- The graphs of the relations in parts (a), (c), and (d) pass the vertical line test because every vertical line intersects each graph no more than once. Thus, these graphs represent functions.
- The graph of the relation in part (b) fails the vertical line test because the same x-value corresponds to two different y-values. Therefore, it is not the graph of a function.

✔ **Now Try Exercises 27 and 29.**

The vertical line test is a simple method for identifying a function defined by a graph. Deciding whether a relation defined by an equation or an inequality is a function, as well as determining the domain and range, is more difficult. The next example gives some hints that may help.

EXAMPLE 5 Identifying Functions, Domains, and Ranges

Decide whether each relation defines y as a function of x, and give the domain and range.

(a) $y = x + 4$ **(b)** $y = \sqrt{2x - 1}$ **(c)** $y^2 = x$

(d) $y \le x - 1$ **(e)** $y = \dfrac{5}{x - 1}$

SOLUTION

(a) In the defining equation (or rule), $y = x + 4$, y is always found by adding 4 to x. Thus, each value of x corresponds to just one value of y, and the relation defines a function. The variable x can represent any real number, so the domain is

$$\{x \mid x \text{ is a real number}\}, \quad \text{or} \quad (-\infty, \infty).$$

Because y is always 4 more than x, y also may be any real number, and so the range is $(-\infty, \infty)$.

(b) For any choice of x in the domain of $y = \sqrt{2x - 1}$, there is exactly one corresponding value for y (the radical is a nonnegative number), so this equation defines a function. The equation involves a square root, so the quantity under the radical sign cannot be negative.

$$2x - 1 \ge 0 \quad \text{Solve the inequality.}$$

$$2x \ge 1 \quad \text{Add 1.}$$

$$x \ge \frac{1}{2} \quad \text{Divide by 2.}$$

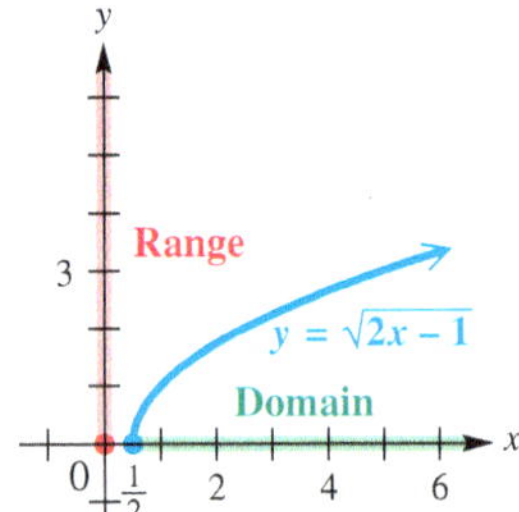

Figure 22

The domain of the function is $\left[\frac{1}{2}, \infty\right)$. Because the radical must represent a nonnegative number, as x takes values greater than or equal to $\frac{1}{2}$, the range is $\{y \mid y \ge 0\}$, or $[0, \infty)$. See **Figure 22.**

(c) The ordered pairs $(16, 4)$ and $(16, -4)$ both satisfy the equation $y^2 = x$. There exists at least one value of x—for example, 16—that corresponds to two values of y, 4 and -4, so this equation does not define a function.

Because x is equal to the square of y, the values of x must always be nonnegative. The domain of the relation is $[0, \infty)$. Any real number can be squared, so the range of the relation is $(-\infty, \infty)$. See **Figure 23.**

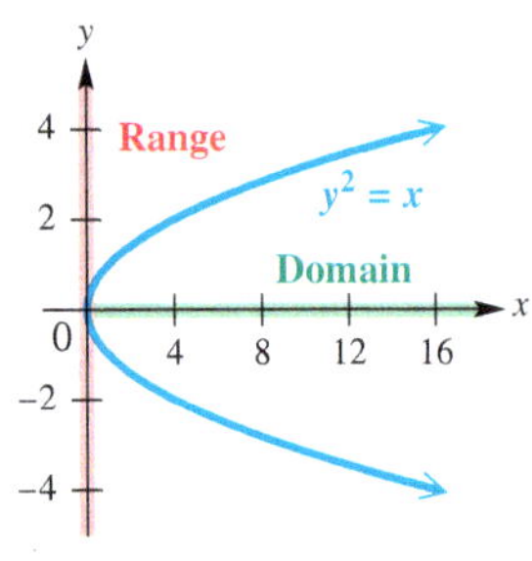

Figure 23

(d) By definition, y is a function of x if every value of x leads to exactly one value of y. Substituting a particular value of x, say 1, into $y \le x - 1$ corresponds to many values of y. The ordered pairs

$$(1, 0), \quad (1, -1), \quad (1, -2), \quad (1, -3), \quad \text{and} \quad \text{so on}$$

all satisfy the inequality, so y is not a function of x here. Any number can be used for x or for y, so the domain and the range of this relation are both the set of real numbers, $(-\infty, \infty)$.

(e) Given any value of x in the domain of

$$y = \frac{5}{x - 1},$$

we find y by subtracting 1 from x, and then dividing the result into 5. This process produces exactly one value of y for each value in the domain, so this equation defines a function.

The domain of $y = \frac{5}{x-1}$ includes all real numbers except those that make the denominator 0. We find these numbers by setting the denominator equal to 0 and solving for x.

$$x - 1 = 0$$

$$x = 1 \quad \text{Add 1.}$$

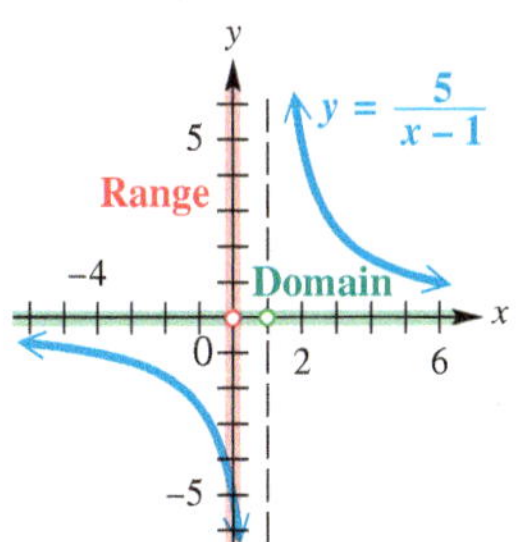

Figure 24

Thus, the domain includes all real numbers except 1, written as the interval $(-\infty, 1) \cup (1, \infty)$. Values of y can be positive or negative, but never 0, because a fraction cannot equal 0 unless its numerator is 0. Therefore, the range is the interval $(-\infty, 0) \cup (0, \infty)$, as shown in **Figure 24**.

✔ **Now Try Exercises 37, 39, and 45.**

Variations of the Definition of Function

1. A **function** is a relation in which, for each distinct value of the first component of the ordered pairs, there is exactly one value of the second component.
2. A **function** is a set of ordered pairs in which no first component is repeated.
3. A **function** is a rule or correspondence that assigns exactly one range value to each distinct domain value.

LOOKING AHEAD TO CALCULUS

One of the most important concepts in calculus, that of the **limit of a function,** is defined using function notation:

$$\lim_{x \to a} f(x) = L$$

(read "the limit of $f(x)$ as x approaches a is equal to L") means that the values of $f(x)$ become as close as we wish to L when we choose values of x sufficiently close to a.

Function Notation When a function f is defined with a rule or an equation using x and y for the independent and dependent variables, we say, "*y is a function of x*" to emphasize that y *depends on* x. We use the notation

$$y = f(x),$$

called **function notation,** to express this and read $f(x)$ as **"f of x,"** or **"f at x."** The letter f is the name given to this function.

For example, if $y = 3x - 5$, we can name the function f and write

$$f(x) = 3x - 5.$$

Note that $f(x)$ is just another name for the dependent variable y. For example, if $y = f(x) = 3x - 5$ and $x = 2$, then we find y, or $f(2)$, by replacing x with 2.

$$f(2) = 3 \cdot 2 - 5 \quad \text{Let } x = 2.$$

$$f(2) = 1 \quad \text{Multiply, and then subtract.}$$

The statement "In the function f, if $x = 2$, then $y = 1$" represents the ordered pair $(2, 1)$ and is abbreviated with function notation as follows.

$$f(2) = 1$$

The symbol $f(2)$ is read "f of 2" or "f at 2."

Function notation can be illustrated as follows.

Name of the function

Defining expression

$$y = f(x) = 3x - 5$$

Value of the function   Name of the independent variable

CAUTION ***The symbol $f(x)$ does not indicate "f times x,"*** but represents the y-value associated with the indicated x-value. As just shown, $f(2)$ is the y-value that corresponds to the x-value 2.

EXAMPLE 6 Using Function Notation

Let $f(x) = -x^2 + 5x - 3$ and $g(x) = 2x + 3$. Find each of the following.

(a) $f(2)$ **(b)** $f(q)$ **(c)** $g(a + 1)$

SOLUTION

(a)
$$f(x) = -x^2 + 5x - 3$$
$$f(2) = -2^2 + 5 \cdot 2 - 3 \quad \text{Replace } x \text{ with 2.}$$
$$f(2) = -4 + 10 - 3 \quad \text{Apply the exponent and multiply.}$$
$$f(2) = 3 \quad \text{Add and subtract.}$$

Thus, $f(2) = 3$, and the ordered pair $(2, 3)$ belongs to f.

(b)
$$f(x) = -x^2 + 5x - 3$$
$$f(q) = -q^2 + 5q - 3 \quad \text{Replace } x \text{ with } q.$$

(c)
$$g(x) = 2x + 3$$
$$g(a + 1) = 2(a + 1) + 3 \quad \text{Replace } x \text{ with } a + 1.$$
$$g(a + 1) = 2a + 2 + 3 \quad \text{Distributive property}$$
$$g(a + 1) = 2a + 5 \quad \text{Add.}$$

The replacement of one variable with another variable or expression, as in parts (b) and (c), is important in later courses.

✔ Now Try Exercises 51, 59, and 65.

Functions can be evaluated in a variety of ways, as shown in **Example 7.**

EXAMPLE 7 Using Function Notation

For each function, find $f(3)$.

(a) $f(x) = 3x - 7$ **(b)** $f = \{(-3, 5), (0, 3), (3, 1), (6, -1)\}$

(c)

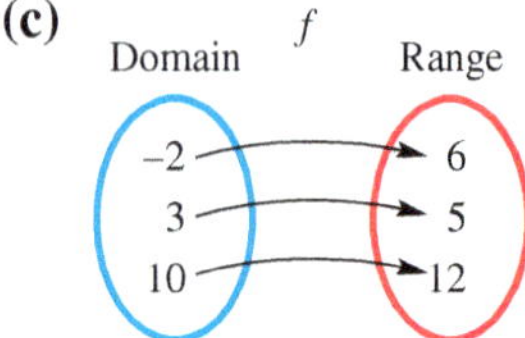

(d)

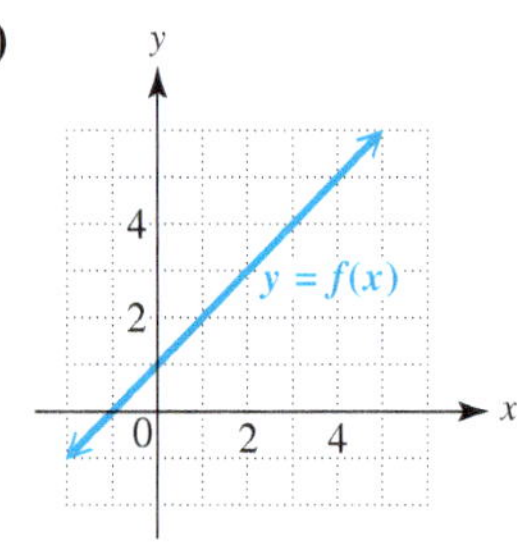

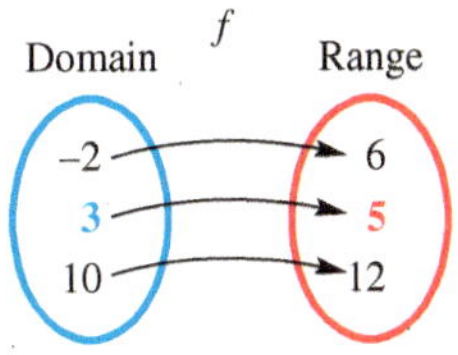

Figure 25

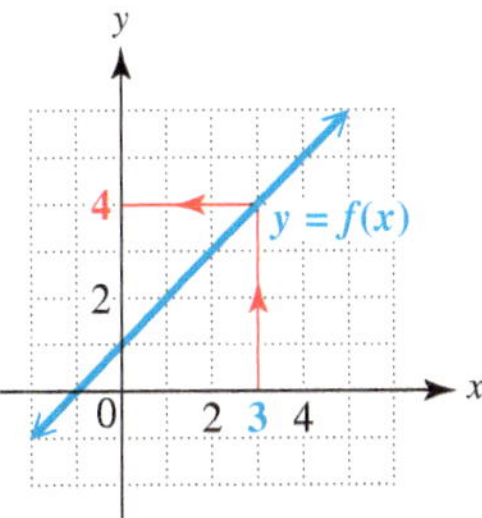

Figure 26

SOLUTION

(a) $f(x) = 3x - 7$

$f(3) = 3(3) - 7$ Replace x with 3.

$f(3) = 2$ Simplify.

$f(3) = 2$ indicates that the ordered pair $(3, 2)$ belongs to f.

(b) For $f = \{(-3, 5), (0, 3), (3, 1), (6, -1)\}$, we want $f(3)$, the y-value of the ordered pair where $x = 3$. As indicated by the ordered pair $(3, 1)$, when $x = 3$, $y = 1$, so $f(3) = 1$.

(c) In the mapping, repeated in **Figure 25,** the domain element 3 is paired with 5 in the range, so $f(3) = 5$.

(d) To evaluate $f(3)$ using the graph, find 3 on the x-axis. See **Figure 26.** Then move up until the graph of f is reached. Moving horizontally to the y-axis gives 4 for the corresponding y-value. Thus, $f(3) = 4$.

✔ **Now Try Exercises 67, 69, and 71.**

If a function f is defined by an equation with x and y (and not with function notation), use the following steps to find $f(x)$.

Finding an Expression for $f(x)$

Consider an equation involving x and y. Assume that y can be expressed as a function f of x. To find an expression for $f(x)$, use the following steps.

Step 1 Solve the equation for y.

Step 2 Replace y with $f(x)$.

EXAMPLE 8 Writing Equations Using Function Notation

Assume that y is a function f of x. Rewrite each equation using function notation. Then find $f(-2)$ and $f(p)$.

(a) $y = x^2 + 1$ **(b)** $x - 4y = 5$

SOLUTION

(a) ***Step 1*** $y = x^2 + 1$ This equation is already solved for y.

Step 2 $f(x) = x^2 + 1$ Let $y = f(x)$.

Now find $f(-2)$ and $f(p)$.

$f(-2) = (-2)^2 + 1$ Let $x = -2$.

$f(-2) = 4 + 1$

$f(-2) = 5$

$f(p) = p^2 + 1$ Let $x = p$.

(b) ***Step 1*** $x - 4y = 5$ Given equation.

$-4y = -x + 5$ Add $-x$.

$y = \frac{x - 5}{4}$ Multiply by -1. Divide by 4.

Step 2 $f(x) = \frac{1}{4}x - \frac{5}{4}$ Let $y = f(x)$; $\frac{a - b}{c} = \frac{a}{c} - \frac{b}{c}$.

Now find $f(-2)$ and $f(p)$.

$$f(x) = \frac{1}{4}x - \frac{5}{4}$$

$$f(-2) = \frac{1}{4}(-2) - \frac{5}{4} \quad \text{Let } x = -2. \qquad\qquad f(p) = \frac{1}{4}p - \frac{5}{4} \quad \text{Let } x = p.$$

$$f(-2) = -\frac{7}{4}$$

✔ **Now Try Exercises 77 and 81.**

Increasing, Decreasing, and Constant Functions Informally speaking, a function *increases* over an open interval of its domain if its graph rises from left to right on the interval. It *decreases* over an open interval of its domain if its graph falls from left to right on the interval. It is *constant* over an open interval of its domain if its graph is horizontal on the interval.

For example, consider **Figure 27.**

y is increasing. y is constant. y is decreasing.

Figure 27

- The function increases over the open interval $(-2, 1)$ because the y-values continue to get larger for x-values in that interval.
- The function is constant over the open interval $(1, 4)$ because the y-values are always 5 for all x-values there.
- The function decreases over the open interval $(4, 6)$ because in that interval the y-values continuously get smaller.

The intervals refer to the x-values where the y-values either increase, decrease, or are constant.

The formal definitions of these concepts follow.

Increasing, Decreasing, and Constant Functions

Suppose that a function f is defined over an *open* interval I and x_1 and x_2 are in I.

(a) f **increases** over I if, whenever $x_1 < x_2$, $f(x_1) < f(x_2)$.

(b) f **decreases** over I if, whenever $x_1 < x_2$, $f(x_1) > f(x_2)$.

(c) f is **constant** over I if, for every x_1 and x_2, $f(x_1) = f(x_2)$.

Figure 28 illustrates these ideas.

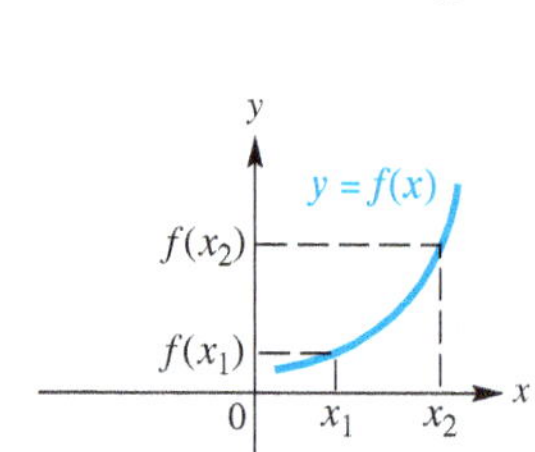

Whenever $x_1 < x_2$, and $f(x_1) < f(x_2)$, f is increasing.

(a)

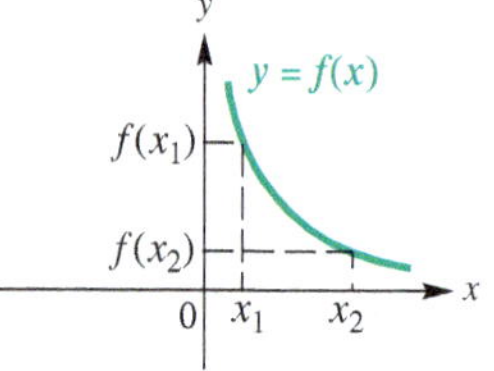

Whenever $x_1 < x_2$, and $f(x_1) > f(x_2)$, f is decreasing.

(b)

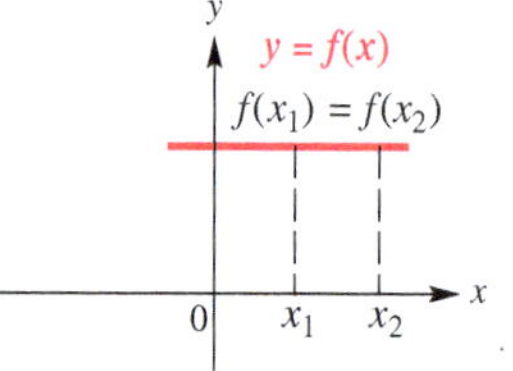

For every x_1 and x_2, if $f(x_1) = f(x_2)$, then f is constant.

(c)

Figure 28

NOTE To decide whether a function is increasing, decreasing, or constant over an interval, ask yourself, ***"What does y do as x goes from left to right?"*** Our definition of *increasing, decreasing,* and *constant* function behavior applies to open intervals of the domain, not to individual points.

EXAMPLE 9 Determining Increasing, Decreasing, and Constant Intervals

Figure 29 shows the graph of a function. Determine the largest open intervals of the domain over which the function is increasing, decreasing, or constant.

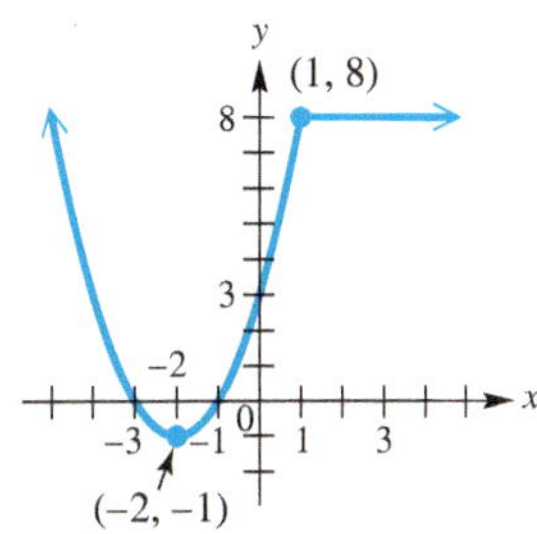

Figure 29

SOLUTION We observe the domain and ask, "*What is happening to the y-values as the x-values are getting larger?*" Moving from left to right on the graph, we see the following:

- On the open interval $(-\infty, -2)$, the y-values are *decreasing*.
- On the open interval $(-2, 1)$, the y-values are *increasing*.
- On the open interval $(1, \infty)$, the y-values are *constant* (and equal to 8).

Therefore, the function is decreasing on $(-\infty, -2)$, increasing on $(-2, 1)$, and constant on $(1, \infty)$.

✔ **Now Try Exercise 91.**

EXAMPLE 10 Interpreting a Graph

Figure 30 shows the relationship between the number of gallons, $g(t)$, of water in a small swimming pool and time in hours, t. By looking at this graph of the function, we can answer questions about the water level in the pool at various times. For example, at time 0 the pool is empty. The water level then increases, stays constant for a while, decreases, and then becomes constant again.

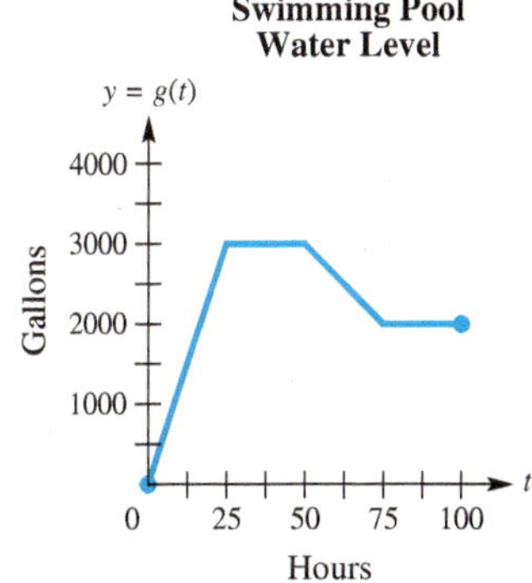

Figure 30

Use the graph to respond to the following.

(a) What is the maximum number of gallons of water in the pool? When is the maximum water level first reached?

(b) For how long is the water level increasing? decreasing? constant?

(c) How many gallons of water are in the pool after 90 hr?

(d) Describe a series of events that could account for the water level changes shown in the graph.

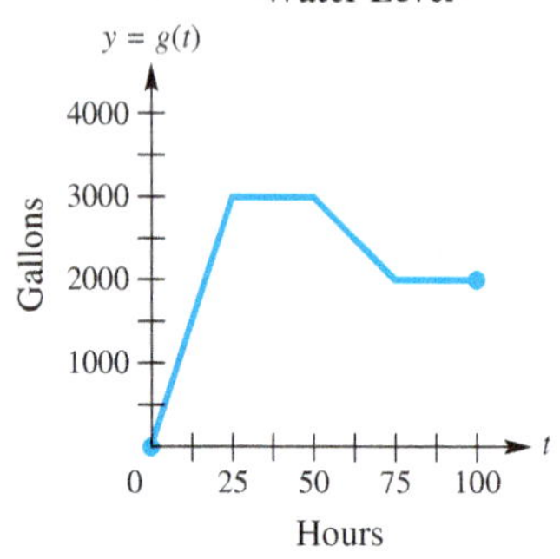

Figure 30 (repeated)

SOLUTION

(a) The maximum range value is 3000, as indicated by the horizontal line segment for the hours 25 to 50. This maximum number of gallons, 3000, is first reached at $t = 25$ hr.

(b) The water level is increasing for $25 - 0 = 25$ hr. The water level is decreasing for $75 - 50 = 25$ hr. It is constant for

$$\begin{aligned}&(50 - 25) + (100 - 75)\\&= 25 + 25\\&= 50 \text{ hr.}\end{aligned}$$

(c) When $t = 90$, $y = g(90) = 2000$. There are 2000 gal after 90 hr.

(d) Looking at the graph in **Figure 30,** we might write the following description.

The pool is empty at the beginning and then is filled to a level of 3000 gal during the first 25 hr. For the next 25 hr, the water level remains the same. At 50 hr, the pool starts to be drained, and this draining lasts for 25 hr, until only 2000 gal remain. For the next 25 hr, the water level is unchanged.

✔ **Now Try Exercise 93.**

2.3 Exercises

CONCEPT PREVIEW *Fill in the blank(s) to correctly complete each sentence.*

1. The domain of the relation $\{(3, 5), (4, 9), (10, 13)\}$ is ________.

2. The range of the relation in **Exercise 1** is ________.

3. The equation $y = 4x - 6$ defines a function with independent variable ________ and dependent variable ________.

4. The function in **Exercise 3** includes the ordered pair $(6, \_\_)$.

5. For the function $f(x) = -4x + 2$, $f(-2) =$ ________.

6. For the function $g(x) = \sqrt{x}$, $g(9) =$ ________.

7. The function in **Exercise 6** has domain ________.

8. The function in **Exercise 6** has range ________.

9. The largest open interval over which the function graphed here increases is ________.

10. The largest open interval over which the function graphed here decreases is ________.

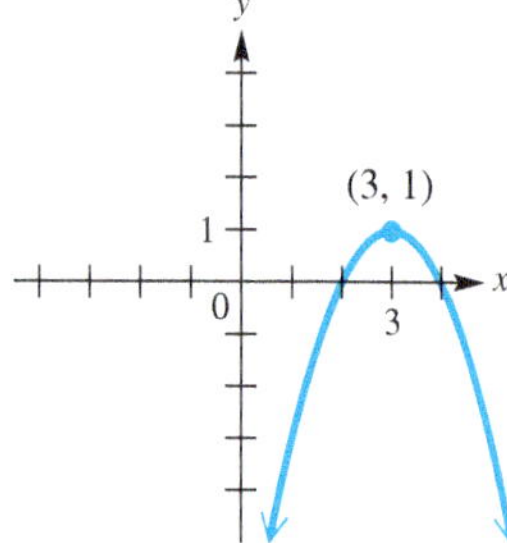

Decide whether each relation defines a function. ***See Example 1.***

11. $\{(5, 1), (3, 2), (4, 9), (7, 8)\}$

12. $\{(8, 0), (5, 7), (9, 3), (3, 8)\}$

13. $\{(2, 4), (0, 2), (2, 6)\}$

14. $\{(9, -2), (-3, 5), (9, 1)\}$

15. $\{(-3, 1), (4, 1), (-2, 7)\}$

16. $\{(-12, 5), (-10, 3), (8, 3)\}$

17.

x	y
3	-4
7	-4
10	-4

18.

x	y
-4	$\sqrt{2}$
0	$\sqrt{2}$
4	$\sqrt{2}$

Decide whether each relation defines a function, and give the domain and range. ***See Examples 1–4.***

19. $\{(1, 1), (1, -1), (0, 0), (2, 4), (2, -4)\}$

20. $\{(2, 5), (3, 7), (3, 9), (5, 11)\}$

21.

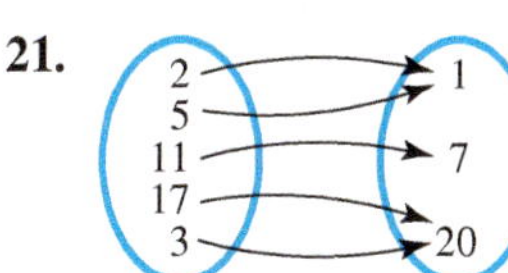

22.

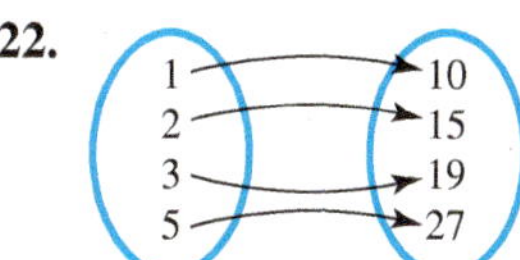

23.

x	y
0	0
-1	1
-2	2

24.

x	y
0	0
1	-1
2	-2

25. ***Number of Visits to U.S. National Parks***

Year (x)	Number of Visits (y) (millions)
2010	64.9
2011	63.0
2012	65.1
2013	63.5

Source: National Park Service.

26. ***Attendance at NCAA Women's College Basketball Games***

Season* (x)	Attendance (y)
2011	11,159,999
2012	11,210,832
2013	11,339,285
2014	11,181,735

Source: NCAA.

*Each season overlaps the given year with the previous year.

27.

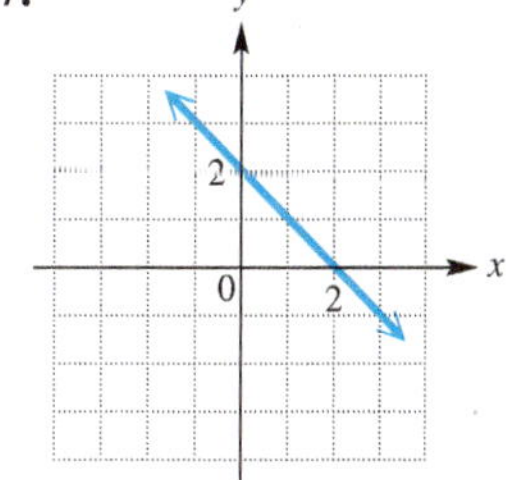

28.

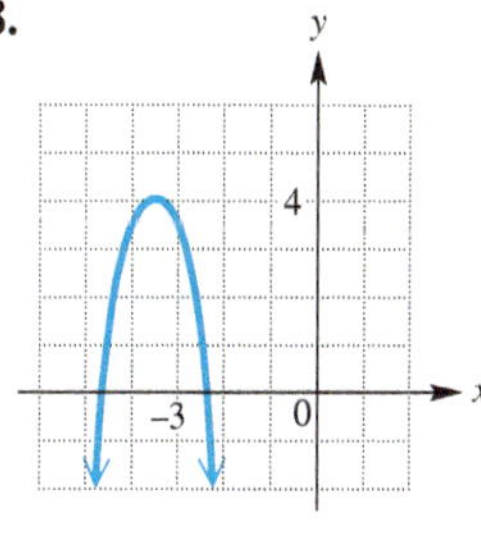

29.

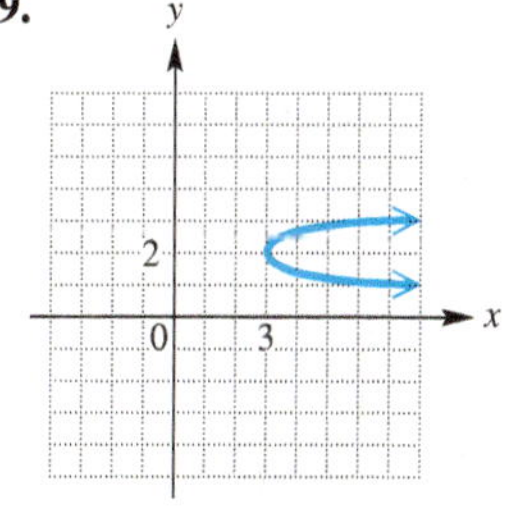

30.

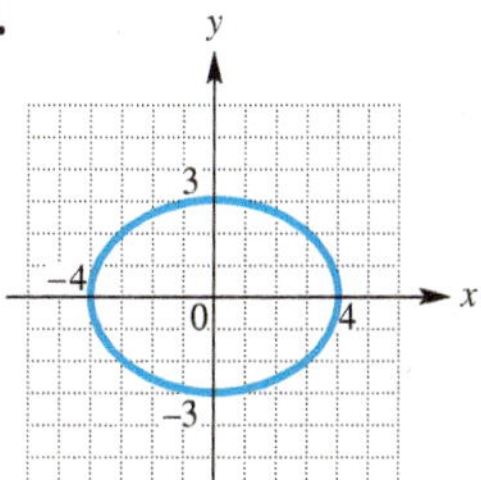

31.

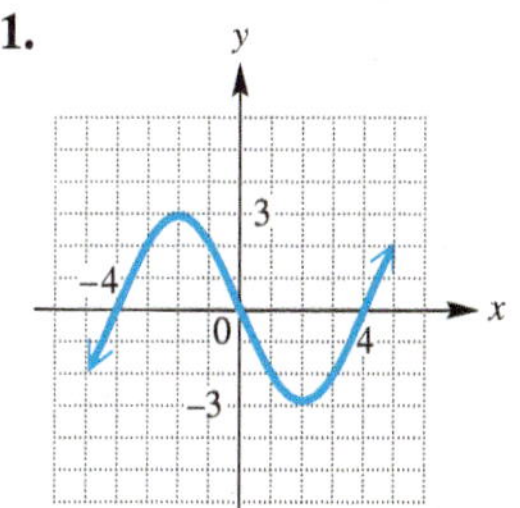

32.

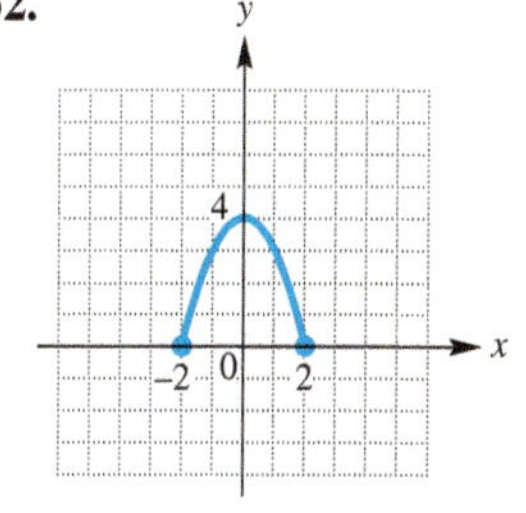

Decide whether each relation defines y as a function of x. Give the domain and range. ***See Example 5.***

33. $y = x^2$ **34.** $y = x^3$ **35.** $x = y^6$

36. $x = y^4$ **37.** $y = 2x - 5$ **38.** $y = -6x + 4$

39. $x + y < 3$ **40.** $x - y < 4$ **41.** $y = \sqrt{x}$

42. $y = -\sqrt{x}$ **43.** $xy = 2$ **44.** $xy = -6$

45. $y = \sqrt{4x + 1}$ **46.** $y = \sqrt{7 - 2x}$

47. $y = \dfrac{2}{x - 3}$ **48.** $y = \dfrac{-7}{x - 5}$

49. ***Concept Check*** Choose the correct answer: For function f, the notation $f(3)$ means

A. the variable f times 3, or $3f$.

B. the value of the dependent variable when the independent variable is 3.

C. the value of the independent variable when the dependent variable is 3.

D. f equals 3.

50. ***Concept Check*** Give an example of a function from everyday life. (*Hint:* Fill in the blanks: ________ depends on ________, so ________ is a function of ________.)

Let $f(x) = -3x + 4$ *and* $g(x) = -x^2 + 4x + 1$. *Find each of the following. Simplify if necessary.* ***See Example 6.***

51. $f(0)$ **52.** $f(-3)$ **53.** $g(-2)$

54. $g(10)$ **55.** $f\left(\frac{1}{3}\right)$ **56.** $f\left(-\frac{7}{3}\right)$

57. $g\left(\frac{1}{2}\right)$ **58.** $g\left(-\frac{1}{4}\right)$ **59.** $f(p)$

60. $g(k)$ **61.** $f(-x)$ **62.** $g(-x)$

63. $f(x + 2)$ **64.** $f(a + 4)$ **65.** $f(2m - 3)$ **66.** $f(3t - 2)$

For each function, find ***(a)*** $f(2)$ *and* ***(b)*** $f(-1)$. ***See Example 7.***

67. $f = \{(-1, 3), (4, 7), (0, 6), (2, 2)\}$ **68.** $f = \{(2, 5), (3, 9), (-1, 11), (5, 3)\}$

69.

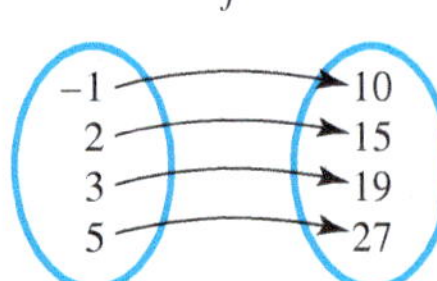

70.

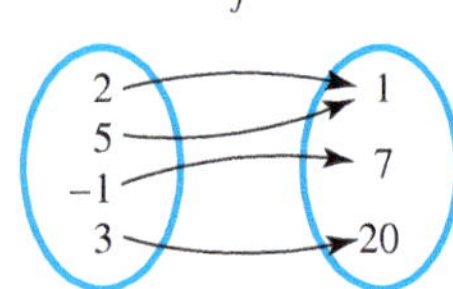

71.

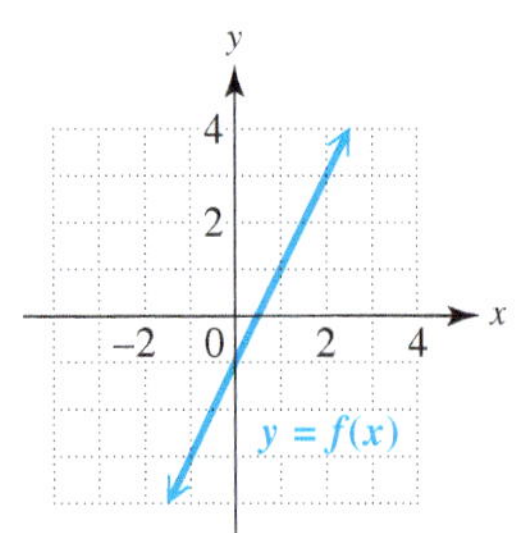

72.

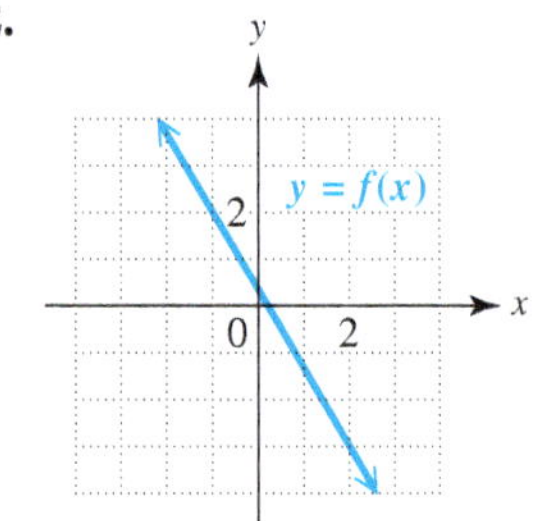

Use the graph of $y = f(x)$ to find each function value: ***(a)*** $f(-2)$, ***(b)*** $f(0)$, ***(c)*** $f(1)$, *and* ***(d)*** $f(4)$. ***See Example 7(d).***

73.

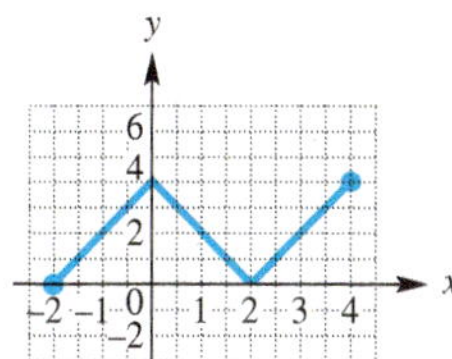

74.

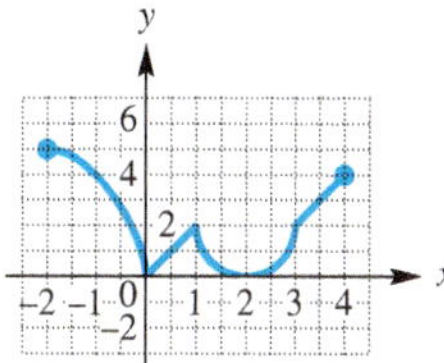

75.

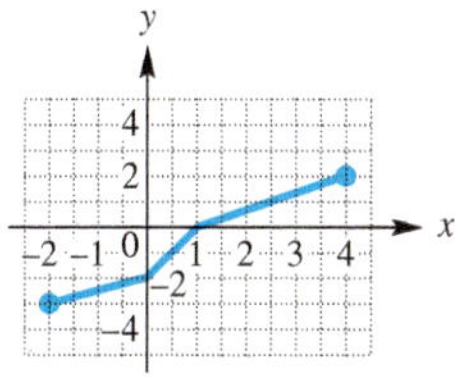

76.

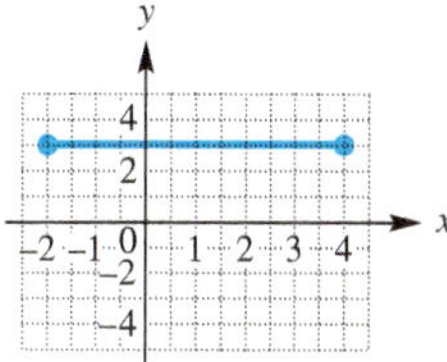

An equation that defines y as a function of x is given. ***(a)*** *Rewrite each equation using function notation $f(x)$.* ***(b)*** *Find $f(3)$.* ***See Example 8.***

77. $x + 3y = 12$

78. $x - 4y = 8$

79. $y + 2x^2 = 3 - x$

80. $y - 3x^2 = 2 + x$

81. $4x - 3y = 8$

82. $-2x + 5y = 9$

Concept Check *Answer each question.*

83. If $(3, 4)$ is on the graph of $y = f(x)$, which one of the following must be true: $f(3) = 4$ or $f(4) = 3$?

84. The figure shows a portion of the graph of

$$f(x) = x^2 + 3x + 1$$

and a rectangle with its base on the x-axis and a vertex on the graph. What is the area of the rectangle? (*Hint:* $f(0.2)$ is the height.)

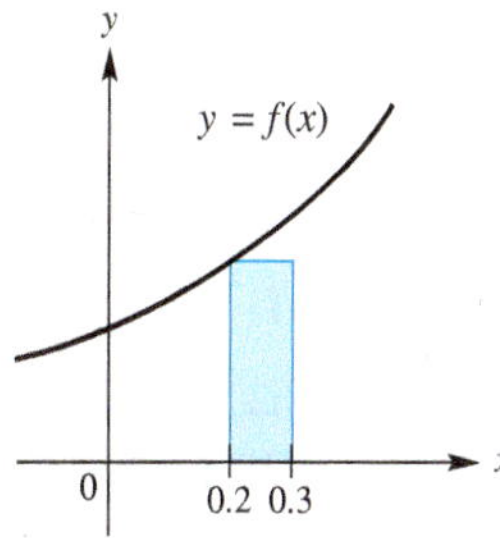

85. The graph of $y_1 = f(x)$ is shown with a display at the bottom. What is $f(3)$?

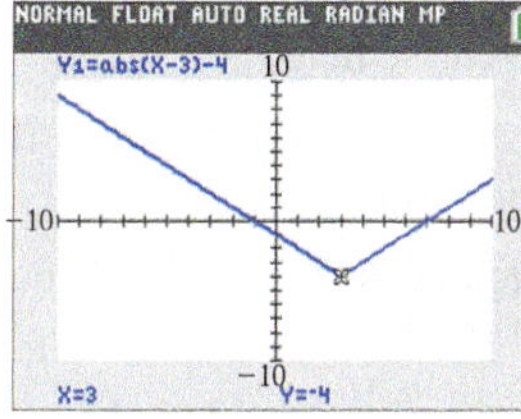

86. The graph of $y_1 = f(x)$ is shown with a display at the bottom. What is $f(-2)$?

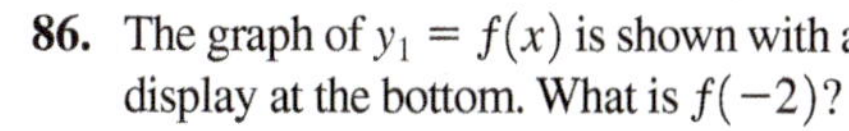

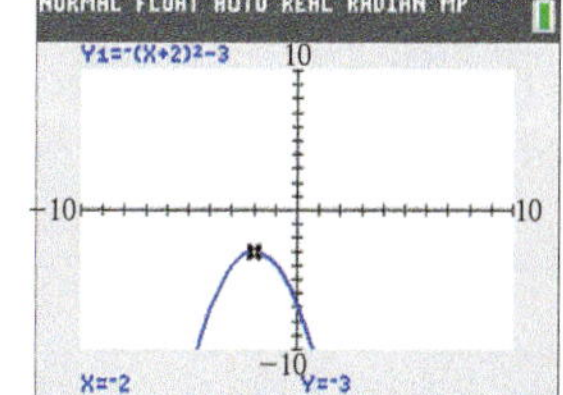

Determine the largest open intervals of the domain over which each function is ***(a)*** *increasing,* ***(b)*** *decreasing, and* ***(c)*** *constant.* ***See Example 9.***

87.

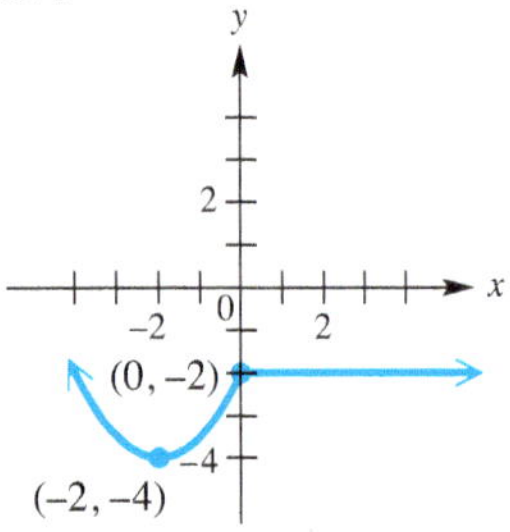

88.

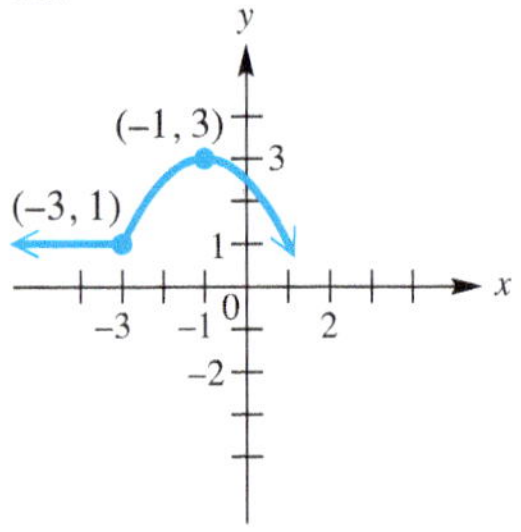

89.

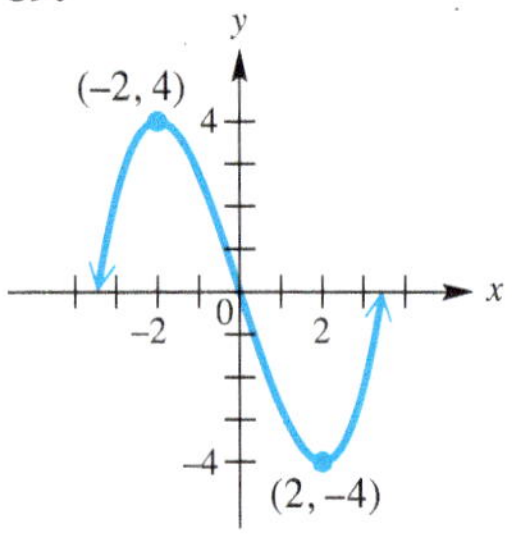

90.

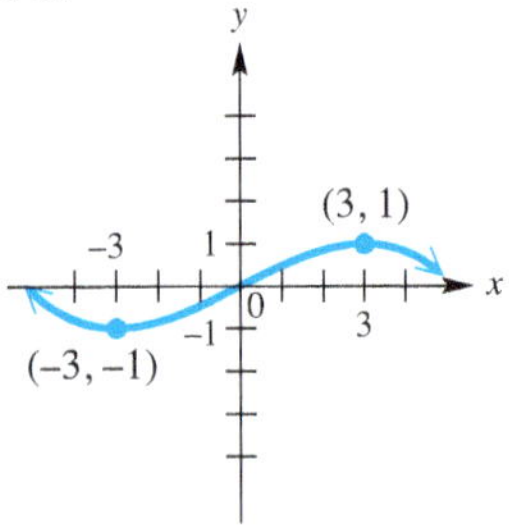

91.

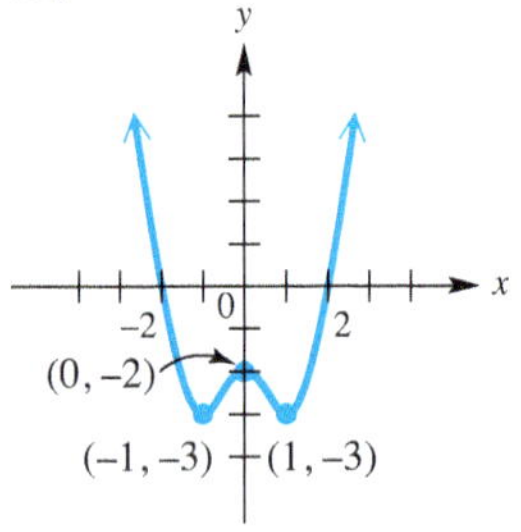

92.

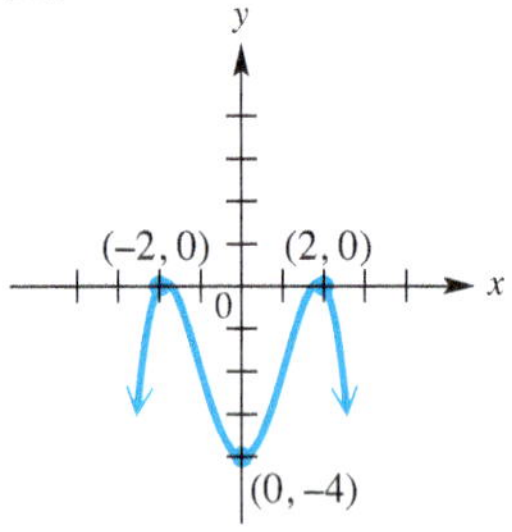

Solve each problem. ***See Example 10.***

93. ***Electricity Usage*** The graph shows the daily megawatts of electricity used on a record-breaking summer day in Sacramento, California.

(a) Is this the graph of a function?

(b) What is the domain?

(c) Estimate the number of megawatts used at 8 A.M.

(d) At what time was the most electricity used? the least electricity?

(e) Call this function f. What is $f(12)$? Interpret your answer.

(f) During what time intervals is usage increasing? decreasing?

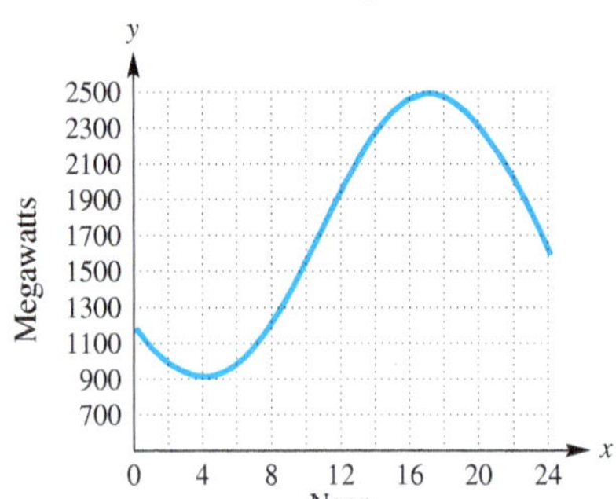

Source: Sacramento Municipal Utility District.

94. ***Height of a Ball*** A ball is thrown straight up into the air. The function $y = h(t)$ in the graph gives the height of the ball (in feet) at t seconds. (*Note:* The graph does *not* show the path of the ball. The ball is rising straight up and then falling straight down.)

(a) What is the height of the ball at 2 sec?

(b) When will the height be 192 ft?

(c) During what time intervals is the ball going up? down?

(d) How high does the ball go? When does the ball reach its maximum height?

(e) After how many seconds does the ball hit the ground?

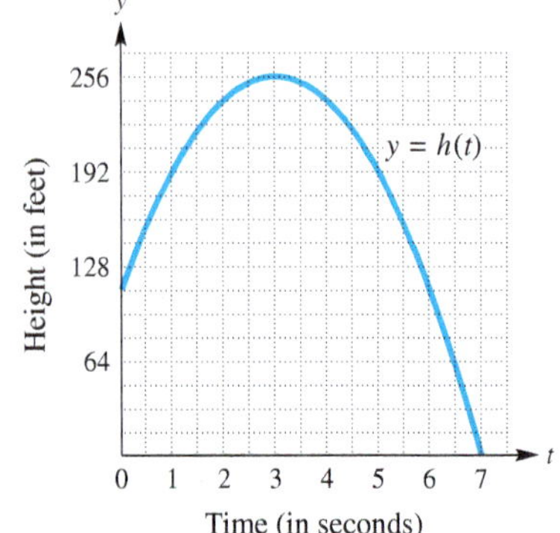

95. ***Temperature*** The graph shows temperatures on a given day in Bratenahl, Ohio.

(a) At what times during the day was the temperature over 55°?

(b) When was the temperature at or below 40°?

(c) Greenville, South Carolina, is 500 mi south of Bratenahl, Ohio, and its temperature is 7° higher all day long. At what time was the temperature in Greenville the same as the temperature at noon in Bratenahl?

(d) Use the graph to give a word description of the 24-hr period in Bratenahl.

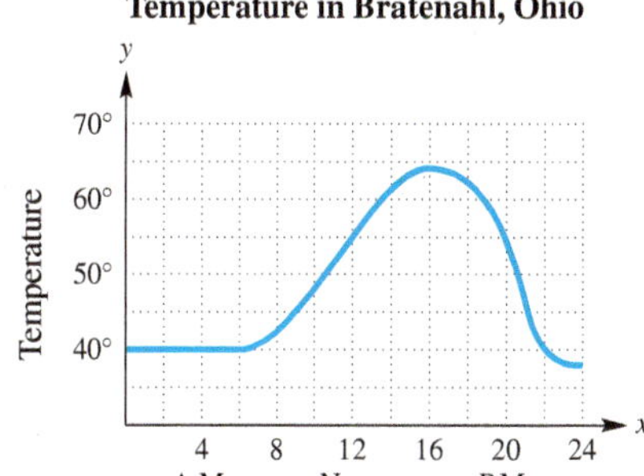

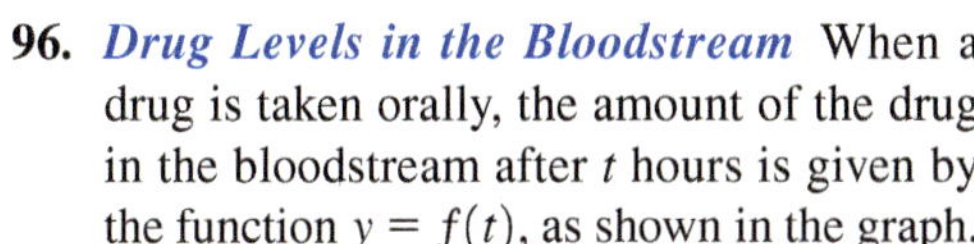

96. ***Drug Levels in the Bloodstream*** When a drug is taken orally, the amount of the drug in the bloodstream after t hours is given by the function $y = f(t)$, as shown in the graph.

(a) How many units of the drug are in the bloodstream at 8 hr?

(b) During what time interval is the level of the drug in the bloodstream increasing? decreasing?

(c) When does the level of the drug in the bloodstream reach its maximum value, and how many units are in the bloodstream at that time?

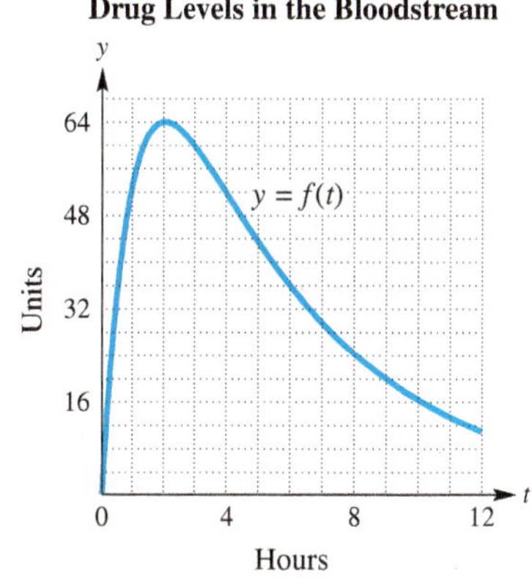

(d) When the drug reaches its maximum level in the bloodstream, how many additional hours are required for the level to drop to 16 units?

(e) Use the graph to give a word description of the 12-hr period.

2.4 Linear Functions

- Basic Concepts of Linear Functions
- Standard Form $Ax + By = C$
- Slope
- Average Rate of Change
- Linear Models

Basic Concepts of Linear Functions

We begin our study of specific functions by looking at *linear* functions.

Linear Function

A function f is a **linear function** if, for real numbers a and b,

$$f(x) = ax + b.$$

If $a \neq 0$, then the domain and the range of f are both $(-\infty, \infty)$.

Lines can be graphed by finding ordered pairs and plotting them. Although only two points are necessary to graph a linear function, we usually plot a third point as a check. The intercepts are often good points to choose for graphing lines.

EXAMPLE 1 Graphing a Linear Function Using Intercepts

Graph $f(x) = -2x + 6$. Give the domain and range.

SOLUTION The x-intercept is found by letting $f(x) = 0$ and solving for x.

$$f(x) = -2x + 6$$

$$0 = -2x + 6 \quad \text{Let } f(x) = 0.$$

$$x = 3 \quad \text{Add } 2x \text{ and divide by 2.}$$

We plot the x-intercept $(3, 0)$. The y-intercept is found by evaluating $f(0)$.

$$f(0) = -2(0) + 6 \quad \text{Let } x = 0.$$

$$f(0) = 6 \quad \text{Simplify.}$$

Therefore, another point on the graph is the y-intercept, $(0, 6)$. We plot this point and join the two points with a straight-line graph. We use the point $(2, 2)$ as a check. See **Figure 31.** The domain and the range are both $(-\infty, \infty)$.

The corresponding calculator graph with $f(x) = y_1$ is shown in **Figure 32.**

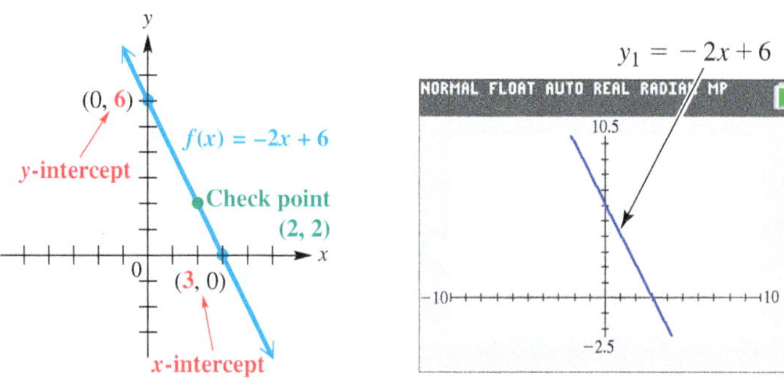

Figure 31 **Figure 32**

✔ **Now Try Exercise 13.**

If $a = 0$ in the definition of linear function, then the equation becomes $f(x) = b$. In this case, the domain is $(-\infty, \infty)$ and the range is $\{b\}$. A function of the form $\boldsymbol{f(x) = b}$ is a **constant function,** and its graph is a horizontal line.

EXAMPLE 2 Graphing a Horizontal Line

Graph $f(x) = -3$. Give the domain and range.

SOLUTION Because $f(x)$, or y, always equals -3, the value of y can never be 0 and the graph has no x-intercept. If a straight line has no x-intercept then it must be parallel to the x-axis, as shown in **Figure 33.** The domain of this linear function is $(-\infty, \infty)$ and the range is $\{-3\}$. **Figure 34** shows the calculator graph.

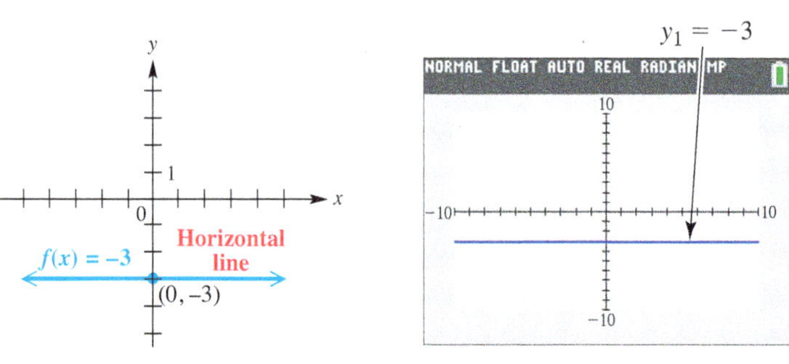

Figure 33 **Figure 34**

✔ **Now Try Exercise 17.**

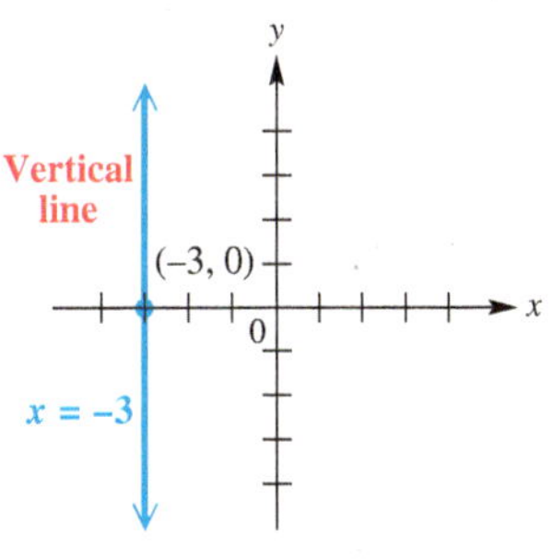

Figure 35

EXAMPLE 3 Graphing a Vertical Line

Graph $x = -3$. Give the domain and range of this relation.

SOLUTION Because x always equals -3, the value of x can never be 0, and the graph has no y-intercept. Using reasoning similar to that of **Example 2,** we find that this graph is parallel to the y-axis, as shown in **Figure 35.** The domain of this relation, which is *not* a function, is $\{-3\}$, while the range is $(-\infty, \infty)$.

✔ **Now Try Exercise 25.**

Standard Form $Ax + By = C$ Equations of lines are often written in the form $Ax + By = C$, known as **standard form.**

NOTE The definition of "standard form" is, ironically, not standard from one text to another. Any linear equation can be written in infinitely many different, but equivalent, forms. For example, the equation $2x + 3y = 8$ can be written equivalently as

$$2x + 3y - 8 = 0, \quad 3y = 8 - 2x, \quad x + \frac{3}{2}y = 4, \quad 4x + 6y = 16,$$

and so on. In this text we will agree that if the coefficients and constant in a linear equation are rational numbers, then we will consider the standard form to be $Ax + By = C$, where $A \geq 0$, A, B, and C are integers, and the greatest common factor of A, B, and C is 1. If $A = 0$, then we choose $B > 0$. (If two or more integers have a greatest common factor of 1, they are said to be **relatively prime.**)

EXAMPLE 4 Graphing $Ax + By = C$ ($C = 0$)

Graph $4x - 5y = 0$. Give the domain and range.

SOLUTION Find the intercepts.

$$4x - 5y = 0$$
$$4(0) - 5y = 0 \quad \text{Let } x = 0.$$
$$y = 0 \quad \text{The } y\text{-intercept is } (0, 0).$$

$$4x - 5y = 0$$
$$4x - 5(0) = 0 \quad \text{Let } y = 0.$$
$$x = 0 \quad \text{The } x\text{-intercept is } (0, 0).$$

The graph of this function has just one intercept—the origin $(0, 0)$. We need to find an additional point to graph the function by choosing a different value for x (or y).

$$4(5) - 5y = 0 \quad \text{We choose } x = 5.$$
$$20 - 5y = 0 \quad \text{Multiply.}$$
$$4 = y \quad \text{Add } 5y. \text{ Divide by 5.}$$

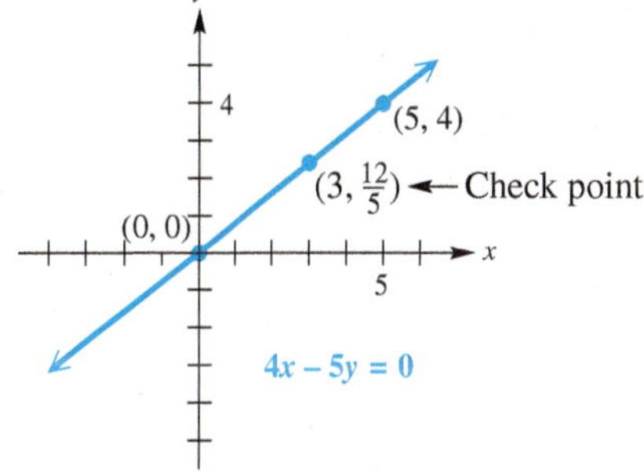

Figure 36

This leads to the ordered pair $(5, 4)$. Complete the graph using the two points $(0, 0)$ and $(5, 4)$, with a third point as a check. The domain and range are both $(-\infty, \infty)$. See **Figure 36.**

✔ **Now Try Exercise 23.**

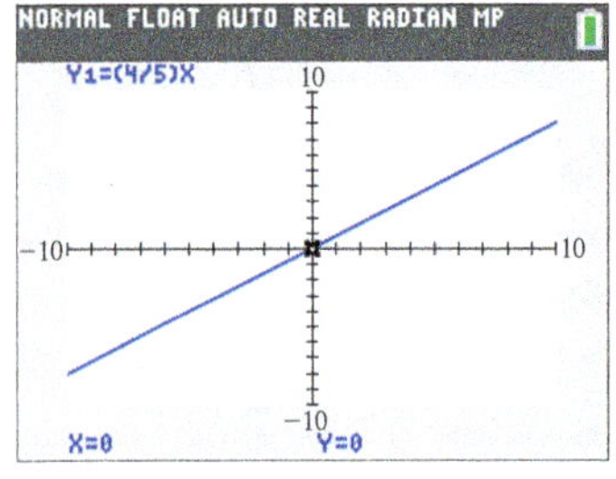

Figure 37

To use a graphing calculator to graph a linear function, as in **Figure 37,** we must first solve the defining equation for y.

$$4x - 5y = 0 \qquad \text{Equation from Example 4}$$

$$y = \frac{4}{5}x \qquad \text{Subtract } 4x\text{. Divide by } -5.$$

Slope

Slope is a numerical measure of the steepness and orientation of a straight line. (Geometrically, this may be interpreted as the ratio of **rise** to **run.**) The slope of a highway (sometimes called the *grade*) is often given as a percent. For example, a 10% $\left(\text{or } \frac{10}{100} = \frac{1}{10}\right)$ slope means the highway rises 1 unit for every 10 horizontal units.

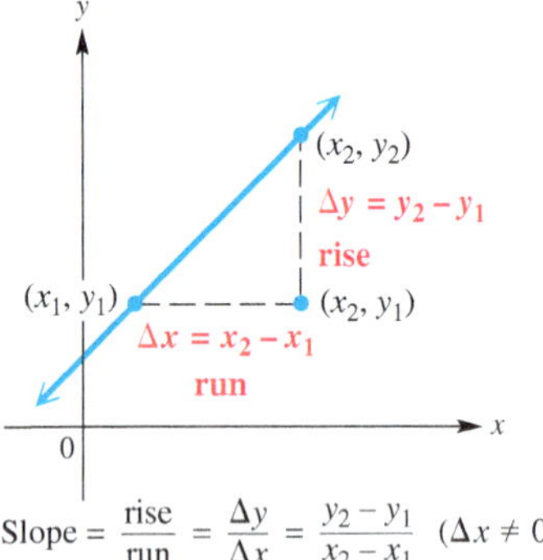

$$\text{Slope} = \frac{\text{rise}}{\text{run}} = \frac{\Delta y}{\Delta x} = \frac{y_2 - y_1}{x_2 - x_1} \quad (\Delta x \neq 0)$$

Figure 38

To find the slope of a line, start with two distinct points (x_1, y_1) and (x_2, y_2) on the line, as shown in **Figure 38,** where $x_1 \neq x_2$. As we move along the line from (x_1, y_1) to (x_2, y_2), the horizontal difference

$$\Delta x = x_2 - x_1$$

is the **change in *x*,** denoted by Δx (read **"delta *x*"**), where Δ is the Greek capital letter **delta.** The vertical difference, the **change in *y*,** can be written

$$\Delta y = y_2 - y_1.$$

The *slope* of a nonvertical line is defined as the quotient (ratio) of the change in y and the change in x, as follows.

Slope

The **slope** m of the line through the points (x_1, y_1) and (x_2, y_2) is given by the following.

$$m = \frac{\textbf{rise}}{\textbf{run}} = \frac{\Delta y}{\Delta x} = \frac{y_2 - y_1}{x_2 - x_1}, \quad \text{where } \Delta x \neq 0$$

That is, the slope of a line is the change in y divided by the corresponding change in x, where the change in x is not 0.

LOOKING AHEAD TO CALCULUS

The concept of slope of a line is extended in calculus to general curves. The **slope of a curve at a point** is understood to mean the slope of the line tangent to the curve at that point.

The line in the figure is tangent to the curve at point P.

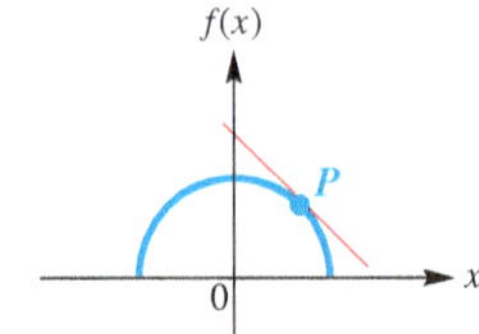

CAUTION ***When using the slope formula, it makes no difference which point is (x_1, y_1) or (x_2, y_2). However, be consistent.*** Start with the x- and y-values of *one* point (either one), and subtract the corresponding values of the *other* point.

$$\textit{Use} \quad \frac{y_2 - y_1}{x_2 - x_1} \quad \text{or} \quad \frac{y_1 - y_2}{x_1 - x_2}, \quad \textit{not} \quad \frac{y_2 - y_1}{x_1 - x_2} \quad \text{or} \quad \frac{y_1 - y_2}{x_2 - x_1}.$$

Be sure to write the difference of the y-values in the numerator and the difference of the x-values in the denominator.

The slope of a line can be found only if the line is nonvertical. This guarantees that $x_2 \neq x_1$ so that the denominator $x_2 - x_1 \neq 0$.

Undefined Slope

The slope of a vertical line is undefined.

EXAMPLE 5 Finding Slopes with the Slope Formula

Find the slope of the line through the given points.

(a) $(-4, 8), (2, -3)$ **(b)** $(2, 7), (2, -4)$ **(c)** $(5, -3), (-2, -3)$

SOLUTION

(a) Let $x_1 = -4$, $y_1 = 8$, and $x_2 = 2$, $y_2 = -3$.

$$m = \frac{\text{rise}}{\text{run}} = \frac{\Delta y}{\Delta x}$$ Definition of slope

$$= \frac{-3 - 8}{2 - (-4)}$$ Substitute carefully.

$$= \frac{-11}{6}, \quad \text{or} \quad -\frac{11}{6}$$ Subtract; $\frac{-a}{b} = -\frac{a}{b}$.

We can also subtract in the opposite order, letting $x_1 = 2$, $y_1 = -3$ and $x_2 = -4$, $y_2 = 8$. The same slope results.

$$m = \frac{8 - (-3)}{-4 - 2} = \frac{11}{-6}, \quad \text{or} \quad -\frac{11}{6}$$

(b) If we attempt to use the slope formula with the points $(2, 7)$ and $(2, -4)$, we obtain a zero denominator.

$$m = \frac{-4 - 7}{2 - 2} = \frac{-11}{0}$$ Undefined

The formula is not valid here because $\Delta x = x_2 - x_1 = 2 - 2 = 0$. A sketch would show that the line through $(2, 7)$ and $(2, -4)$ is vertical. As mentioned above, the slope of a vertical line is undefined.

(c) For $(5, -3)$ and $(-2, -3)$, the slope equals 0.

$$m = \frac{-3 - (-3)}{-2 - 5} = \frac{0}{-7} = 0$$

A sketch would show that the line through $(5, -3)$ and $(-2, -3)$ is horizontal.

✔ **Now Try Exercises 41, 47, and 49.**

The results in **Example 5(c)** suggest the following generalization.

LOOKING AHEAD TO CALCULUS

The **derivative** of a function provides a formula for determining the slope of a line tangent to a curve. If the slope is positive on a given interval, then the function is increasing there. If it is negative, then the function is decreasing. If it is 0, then the function is constant.

Slope Equal to Zero

The slope of a horizontal line is 0.

Theorems for similar triangles can be used to show that the slope of a line is independent of the choice of points on the line. ***That is, slope is the same no matter which pair of distinct points on the line are used to find it.***

If the equation of a line is in the form

$$y = ax + b,$$

we can show that the slope of the line is a. To do this, we use function notation and the definition of slope.

$$m = \frac{f(x_2) - f(x_1)}{x_2 - x_1} \quad \text{Slope formula}$$

$$m = \frac{[a(x+1) + b] - (ax + b)}{(x+1) - x} \quad \text{Let } f(x) = ax + b,\ x_1 = x, \text{ and } x_2 = x + 1.$$

$$m = \frac{ax + a + b - ax - b}{x + 1 - x} \quad \text{Distributive property}$$

$$m = \frac{a}{1} \quad \text{Combine like terms.}$$

$$m = a \quad \text{The slope is } a.$$

This discussion enables us to find the slope of the graph of any linear equation by solving for y and identifying the coefficient of x, which is the slope.

EXAMPLE 6 Finding Slope from an Equation

Find the slope of the line $4x + 3y = 12$.

SOLUTION Solve the equation for y.

$$4x + 3y = 12$$

$$3y = -4x + 12 \quad \text{Subtract } 4x.$$

$$y = -\frac{4}{3}x + 4 \quad \text{Divide by 3.}$$

Be careful to divide *each* term by 3.

The slope is $-\frac{4}{3}$, which is the coefficient of x when the equation is solved for y.

✔ **Now Try Exercise 55(a).**

Because the slope of a line is the ratio of vertical change (rise) to horizontal change (run), if we know the slope of a line and the coordinates of a point on the line, we can draw the graph of the line.

EXAMPLE 7 Graphing a Line Using a Point and the Slope

Graph the line passing through the point $(-1, 5)$ and having slope $-\frac{5}{3}$.

SOLUTION First locate the point $(-1, 5)$ as shown in **Figure 39.** The slope of this line is $\frac{-5}{3}$, so a change of -5 units vertically (that is, 5 units *down*) corresponds to a change of 3 units horizontally (that is, 3 units to the *right*). This gives a second point, $(2, 0)$, which can then be used to complete the graph.

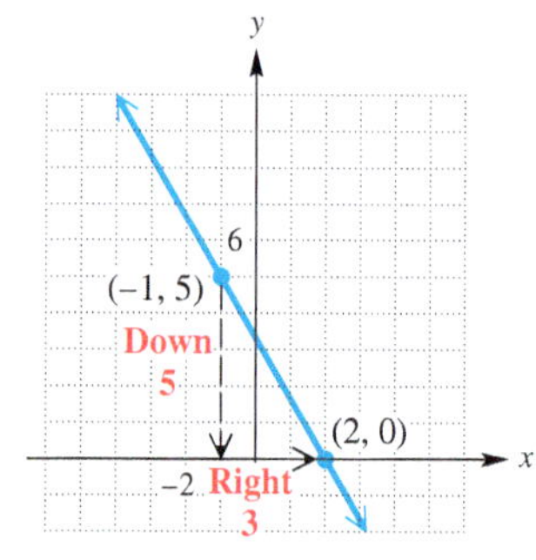

Figure 39

Because $\frac{-5}{3} = \frac{5}{-3}$, another point could be obtained by starting at $(-1, 5)$ and moving 5 units *up* and 3 units to the *left*. We would reach a different second point, $(-4, 10)$, but the graph would be the same line. Confirm this in **Figure 39.**

✔ **Now Try Exercise 59.**

Figure 40 shows lines with various slopes.

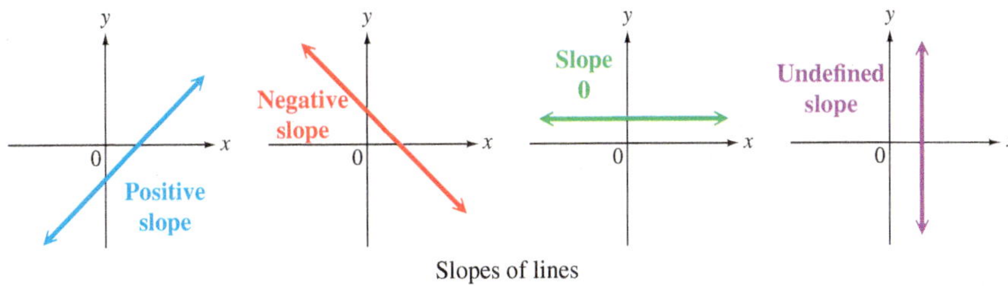

Slopes of lines

Figure 40

Notice the following important concepts.

- A line with a positive slope rises from left to right. The corresponding linear function is increasing on its entire domain.
- A line with a negative slope falls from left to right. The corresponding linear function is decreasing on its entire domain.
- A line with slope 0 neither rises nor falls. The corresponding linear function is constant on its entire domain.
- The slope of a vertical line is undefined.

Average Rate of Change We know that the slope of a line is the ratio of the vertical change in y to the horizontal change in x. ***Thus, slope gives the average rate of change in y per unit of change in x, where the value of y depends on the value of x.*** If f is a linear function defined on the interval $[a, b]$, then we have the following.

$$\textbf{Average rate of change on } [a, b] = \frac{f(b) - f(a)}{b - a}$$

This is simply another way to write the slope formula, using function notation.

EXAMPLE 8 Interpreting Slope as Average Rate of Change

In 2009, Google spent $2800 million on research and development. In 2013, Google spent $8000 million on research and development. Assume a linear relationship, and find the average rate of change in the amount of money spent on R&D per year. Graph as a line segment, and interpret the result. (*Source:* MIT Technology Review.)

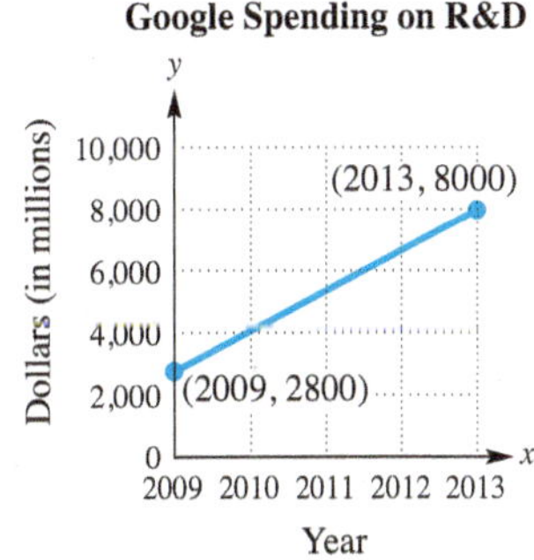

Figure 41

SOLUTION To use the slope formula, we need two ordered pairs. Here, If $x = 2009$, then $y = 2800$, and if $x = 2013$, then $y = 8000$. This gives the two ordered pairs $(2009, 2800)$ and $(2013, 8000)$. (Here y is in millions of dollars.)

$$\text{Average rate of change} = \frac{8000 - 2800}{2013 - 2009} = \frac{5200}{4} = 1300$$

The graph in **Figure 41** confirms that the line through the ordered pairs rises from left to right and therefore has positive slope. Thus, the annual amount of money spent by Google on R&D *increased* by an average of about of $1300 million each year from 2009 to 2013.

Now Try Exercise 75.

Linear Models In **Example 8,** we used the graph of a line to approximate real data, a process known as **mathematical modeling.** Points on the straight-line graph model, or approximate, the actual points that correspond to the data.

A **linear cost function** has the form

$$C(x) = mx + b, \quad \text{Linear cost function}$$

where x represents the number of items produced, m represents the **cost** per item, and b represents the **fixed cost.** The fixed cost is constant for a particular product and does not change as more items are made. The value of mx, which increases as more items are produced, covers labor, materials, packaging, shipping, and so on.

The **revenue function** for selling a product depends on the price per item p and the number of items sold x. It is given by the following function.

$$R(x) = px \quad \text{Revenue function}$$

Profit is found by subtracting cost from revenue and is described by the **profit function.**

$$P(x) = R(x) - C(x) \quad \text{Profit function}$$

In applications we are often interested in values of x that will assure that profit is a positive number. In such cases we solve $R(x) - C(x) > 0$.

EXAMPLE 9 Writing Linear Cost, Revenue, and Profit Functions

Assume that the cost to produce an item is a linear function and all items produced are sold. The fixed cost is \$1500, the variable cost per item is \$100, and the item sells for \$125. Write linear functions to model each of the following.

(a) cost **(b)** revenue **(c)** profit

(d) How many items must be sold for the company to make a profit?

SOLUTION

(a) The cost function is linear, so it will have the following form.

$$C(x) = mx + b \quad \text{Cost function}$$

$$C(x) = 100x + 1500 \quad \text{Let } m = 100 \text{ and } b = 1500.$$

(b) The revenue function is defined by the product of 125 and x.

$$R(x) = px \quad \text{Revenue function}$$

$$R(x) = 125x \quad \text{Let } p = 125.$$

(c) The profit function is found by subtracting the cost function from the revenue function.

$$P(x) = R(x) - C(x)$$

$$P(x) = 125x - (100x + 1500) \quad \text{Use parentheses here.}$$

$$P(x) = 125x - 100x - 1500 \quad \text{Distributive property}$$

$$P(x) = 25x - 1500 \quad \text{Combine like terms.}$$

ALGEBRAIC SOLUTION

(d) To make a profit, $P(x)$ must be positive.

$P(x) = 25x - 1500$ Profit function from part (c)

Set $P(x) > 0$ and solve.

$P(x) > 0$

$25x - 1500 > 0$ $P(x) = 25x - 1500$

$25x > 1500$ Add 1500 to each side.

$x > 60$ Divide by 25.

The number of items must be a whole number, so at least 61 items must be sold for the company to make a profit.

GRAPHING CALCULATOR SOLUTION

(d) Define y_1 as $25x - 1500$ and graph the line. Use the capability of a calculator to locate the x-intercept. See **Figure 42.** As the graph shows, y-values for x less than 60 are negative, and y-values for x greater than 60 are positive, so at least 61 items must be sold for the company to make a profit.

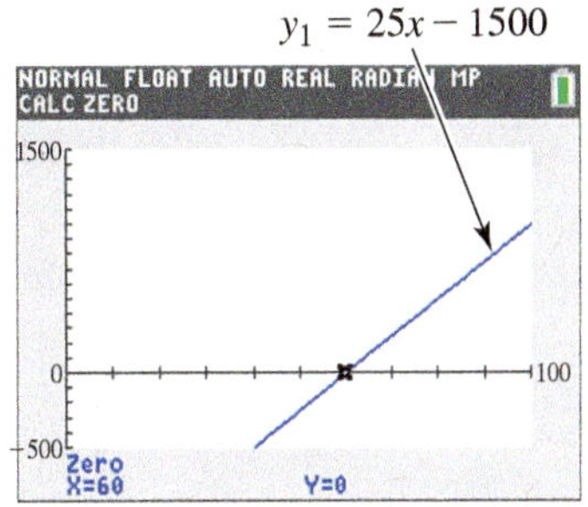

Figure 42

✔ **Now Try Exercise 77.**

CAUTION In problems involving $R(x) - C(x)$ like **Example 9(c),** pay attention to the use of parentheses around the expression for $C(x)$.

2.4 Exercises

CONCEPT PREVIEW *Match the description in Column I with the correct response in Column II. Some choices may not be used.*

I	II
1. a linear function whose graph has y-intercept $(0, 6)$	**A.** $f(x) = 5x$
2. a vertical line	**B.** $f(x) = 3x + 6$
3. a constant function	**C.** $f(x) = -8$
4. a linear function whose graph has x-intercept $(-2, 0)$ and y-intercept $(0, 4)$	**D.** $f(x) = x^2$
	E. $x + y = -6$
5. a linear function whose graph passes through the origin	**F.** $f(x) = 3x + 4$
	G. $2x - y = -4$
6. a function that is not linear	**H.** $x = 9$

CONCEPT PREVIEW *For each given slope, identify the line in A–D that could have this slope.*

7. -3 **8.** 0 **9.** 3 **10.** undefined

A.

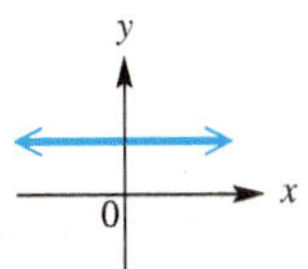

B.

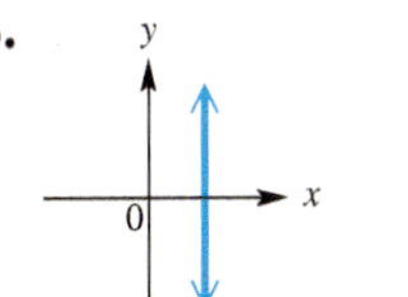

C.

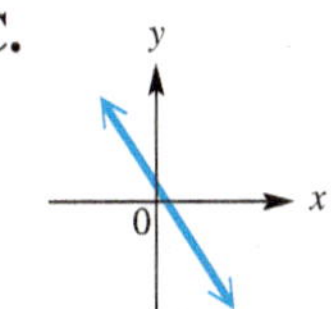

D.

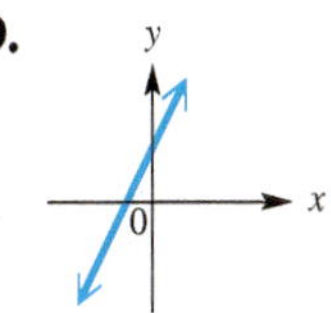

Graph each linear function. Give the domain and range. Identify any constant functions. ***See Examples 1 and 2.***

11. $f(x) = x - 4$ **12.** $f(x) = -x + 4$ **13.** $f(x) = \frac{1}{2}x - 6$

14. $f(x) = \frac{2}{3}x + 2$ **15.** $f(x) = 3x$ **16.** $f(x) = -2x$

17. $f(x) = -4$ **18.** $f(x) = 3$ **19.** $f(x) = 0$

20. *Concept Check* Write the equation of the linear function f with graph having slope 9 and passing through the origin. Give the domain and range.

Graph each line. Give the domain and range. ***See Examples 3 and 4.***

21. $-4x + 3y = 12$ **22.** $2x + 5y = 10$ **23.** $3y - 4x = 0$

24. $3x + 2y = 0$ **25.** $x = 3$ **26.** $x = -4$

27. $2x + 4 = 0$ **28.** $-3x + 6 = 0$

29. $-x + 5 = 0$ **30.** $3 + x = 0$

Match each equation with the sketch that most closely resembles its graph. ***See Examples 2 and 3.***

31. $y = 5$ **32.** $y = -5$ **33.** $x = 5$ **34.** $x = -5$

A. **B.** **C.** **D.**

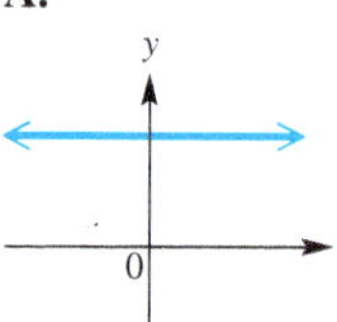

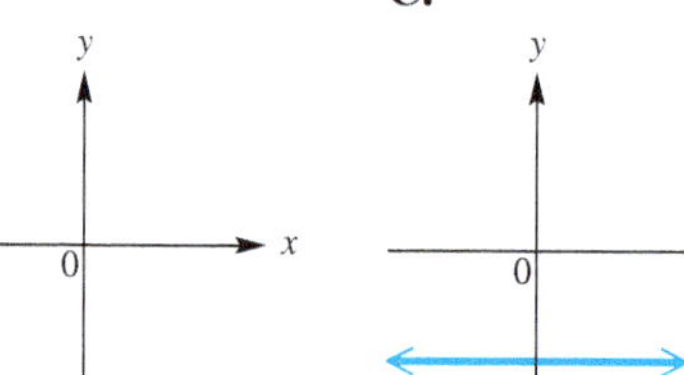

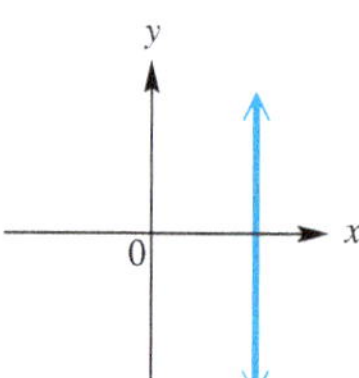

Use a graphing calculator to graph each equation in the standard viewing window. ***See Examples 1, 2, and 4.***

35. $y = 3x + 4$ **36.** $y = -2x + 3$

37. $3x + 4y = 6$ **38.** $-2x + 5y = 10$

39. *Concept Check* If a walkway rises 2.5 ft for every 10 ft on the horizontal, which of the following express its slope (or grade)? (There are several correct choices.)

A. 0.25 **B.** 4 **C.** $\frac{2.5}{10}$ **D.** 25%

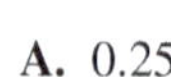

E. $\frac{1}{4}$ **F.** $\frac{10}{2.5}$ **G.** 400% **H.** 2.5%

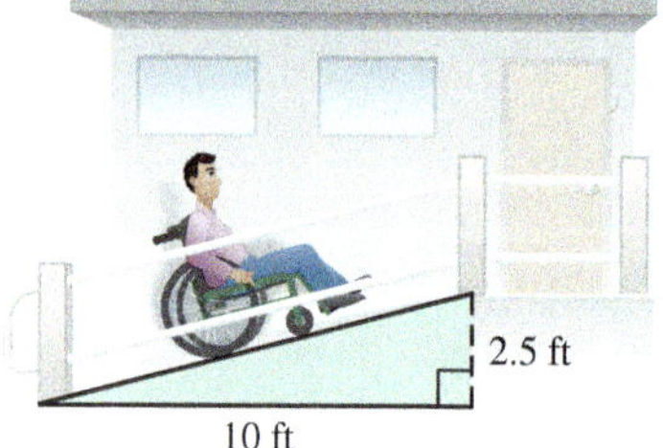

40. *Concept Check* If the pitch of a roof is $\frac{1}{4}$, how many feet in the horizontal direction correspond to a rise of 4 ft?

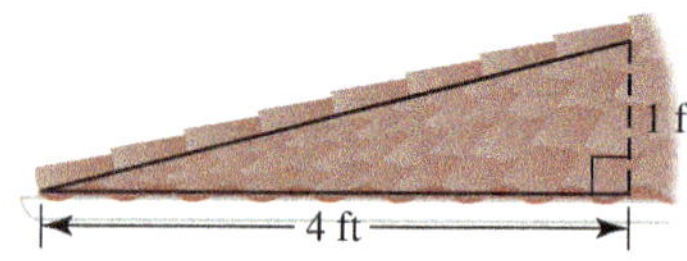

Find the slope of the line satisfying the given conditions. ***See Example 5.***

41. through $(2, -1)$ and $(-3, -3)$

42. through $(-3, 4)$ and $(2, -8)$

43. through $(5, 8)$ and $(3, 12)$

44. through $(5, -3)$ and $(1, -7)$

45. through $(5, 9)$ and $(-2, 9)$

46. through $(-2, 4)$ and $(6, 4)$

47. horizontal, through $(5, 1)$

48. horizontal, through $(3, 5)$

49. vertical, through $(4, -7)$

50. vertical, through $(-8, 5)$

For each line, ***(a)*** *find the slope and* ***(b)*** *sketch the graph.* ***See Examples 6 and 7.***

51. $y = 3x + 5$

52. $y = 2x - 4$

53. $2y = -3x$

54. $-4y = 5x$

55. $5x - 2y = 10$

56. $4x + 3y = 12$

Graph the line passing through the given point and having the indicated slope. Plot two points on the line. ***See Example 7.***

57. through $(-1, 3)$, $m = \frac{3}{2}$

58. through $(-2, 8)$, $m = \frac{2}{5}$

59. through $(3, -4)$, $m = -\frac{1}{3}$

60. through $(-2, -3)$, $m = -\frac{3}{4}$

61. through $\left(-\frac{1}{2}, 4\right)$, $m = 0$

62. through $\left(\frac{3}{2}, 2\right)$, $m = 0$

63. through $\left(-\frac{5}{2}, 3\right)$, undefined slope

64. through $\left(\frac{9}{4}, 2\right)$, undefined slope

Concept Check *Find and interpret the average rate of change illustrated in each graph.*

65.

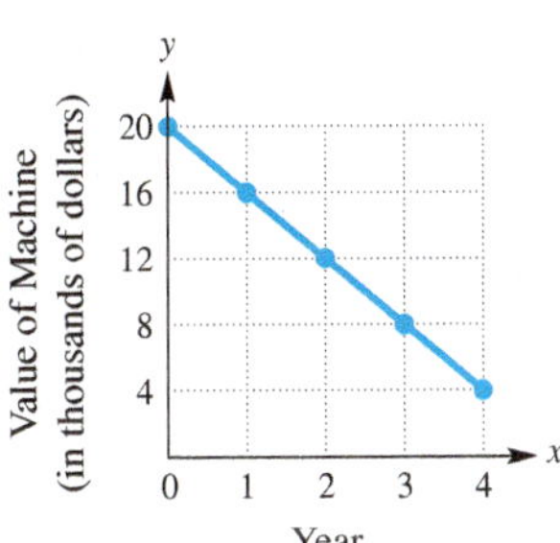

66.

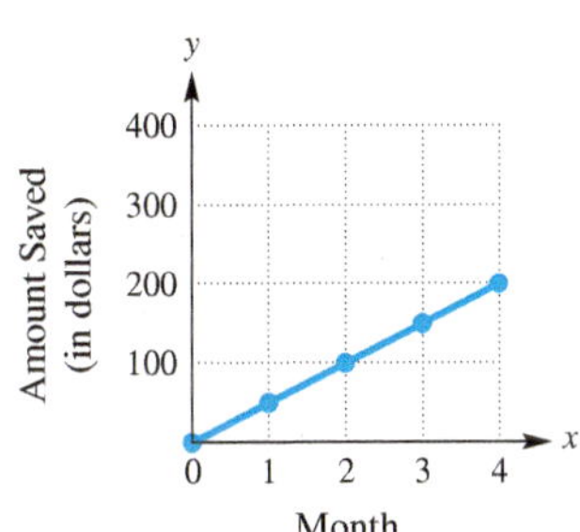

67.

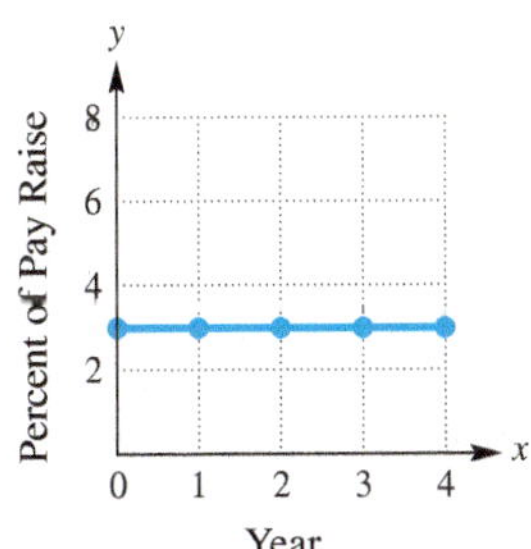

68.

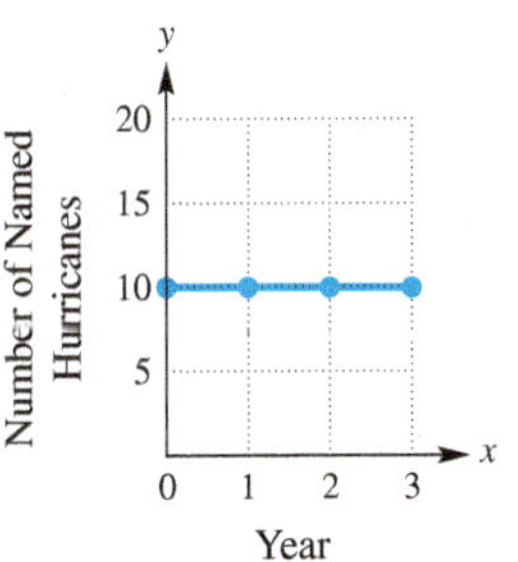

Solve each problem. ***See Example 8.***

69. ***Dropouts*** In 1980, the number of high school dropouts in the United States was 5085 thousand. By 2012, this number had decreased to 2562 thousand. Find and interpret the average rate of change per year in the number of high school dropouts. Round the answer to the nearest tenth. (*Source: 2013 Digest of Education Statistics.*)

70. ***Plasma Flat-Panel TV Sales*** The total amount spent on plasma flat-panel TVs in the United States changed from \$5302 million in 2006 to \$1709 million in 2013. Find and interpret the average rate of change in sales, in millions of dollars per year. Round the answer to the nearest hundredth. (*Source:* Consumer Electronics Association.)

71. *(Modeling) Olympic Times for 5000-Meter Run* The graph shows the winning times (in minutes) at the Olympic Games for the men's 5000-m run, together with a linear approximation of the data.

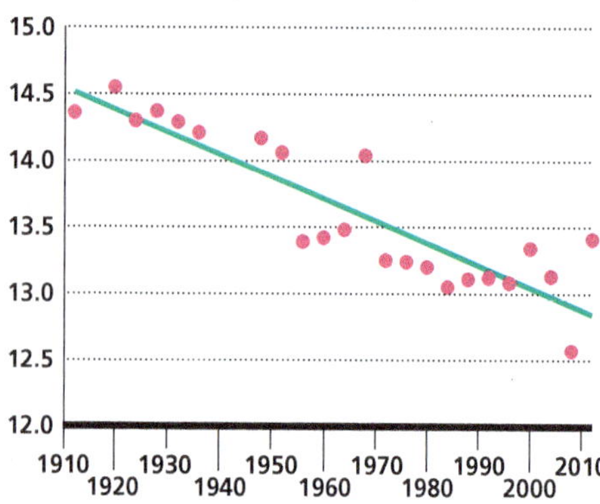

Source: World Almanac and Book of Facts.

(a) An equation for a linear model, based on data from 1912–2012 (where x represents the year), is

$$y = -0.0167x + 46.45.$$

Determine the slope. (**See Example 6.**) What does the slope of this line represent? Why is the slope negative?

(b) What reason might explain why there are no data points for the years 1916, 1940, and 1944?

(c) The winning time for the 2000 Olympic Games was 13.35 min. What does the model predict to the nearest hundredth? How far is the prediction from the actual value?

72. *(Modeling) U.S. Radio Stations* The graph shows the number of U.S. radio stations on the air, along with the graph of a linear function that models the data.

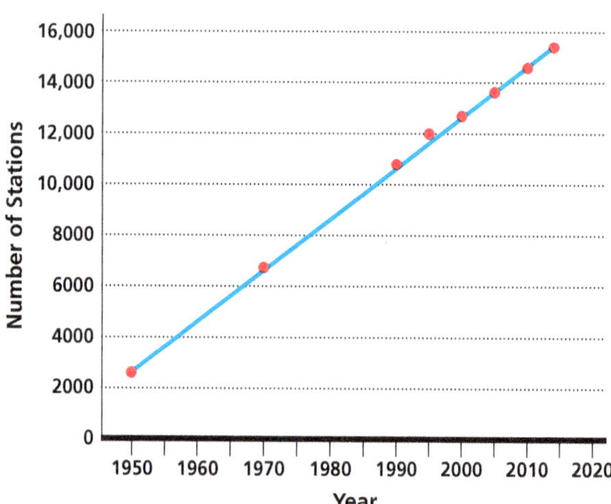

Source: Federal Communications Commission.

(a) An equation for a linear model, based on data from 1950–2014 (where $x = 0$ represents 1950, $x = 10$ represents 1960, and so on) is

$$y = 200.02x + 2727.7.$$

Determine the slope. (**See Example 6.**) What does the slope of this line represent? Why is the slope positive?

(b) Use the model in part (a) to predict the number of stations in 2018.

73. *Cellular Telephone Subscribers* The table gives the number of cellular telephone subscribers in the U.S. (in thousands) from 2008 through 2013. Find the average annual rate of change during this time period. Round to the nearest unit.

Year	Subscribers (in thousands)
2008	270,334
2009	285,646
2010	296,286
2011	315,964
2012	326,475
2013	335,652

Source: CTIA-The Wireless Association.

74. *Earned Run Average* In 2006, in an effort to end the so-called "steroid era," Major League Baseball introduced a strict drug-testing policy in order to discourage players from using performance-enhancing drugs. The table shows how overall earned run average, or ERA, changed from 2006 through 2014. Find the average annual rate of change, to the nearest thousandth, during this period.

Year	ERA
2006	4.53
2007	4.47
2008	4.32
2009	4.32
2010	4.08
2011	3.94
2012	4.01
2013	3.87
2014	3.74

Source: www.baseball-reference.com

75. *Mobile Homes* The graph provides a good approximation of the number of mobile homes (in thousands) placed in use in the United States from 2003 through 2013.

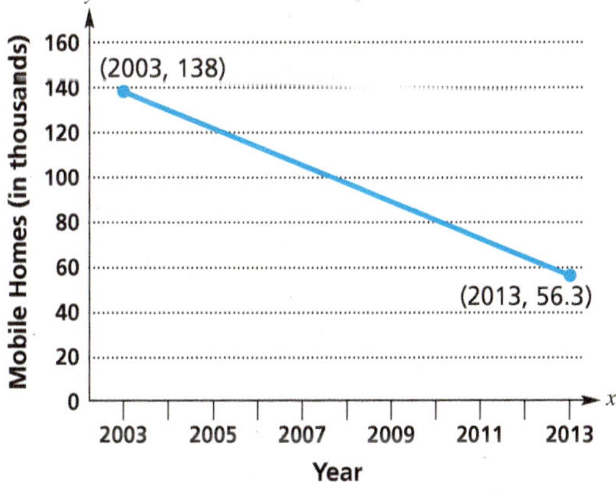

Source: U.S. Census Bureau.

(a) Use the given ordered pairs to find the average rate of change in the number of mobile homes per year during this period.

(b) Interpret what a negative slope means in this situation.

76. ***Teen Birth Rates*** In 1991, there were 61.8 births per thousand for adolescent females aged 15–19. By 2013, this number had decreased to 26.6 births per thousand. Find and interpret the average annual rate of change in teen births per year for this period. Round the answer to the nearest tenth. (*Source:* U.S. Department of Health and Human Services.)

(Modeling) Cost, Revenue, and Profit Analysis *A firm will break even (no profit and no loss) as long as revenue just equals cost. The value of x (the number of items produced and sold) where $C(x) = R(x)$ is the* ***break-even point.*** *Assume that each of the following can be expressed as a linear function. Find*

(a) *the cost function,* ***(b)*** *the revenue function, and* ***(c)*** *the profit function.*

(d) *Find the break-even point and decide whether the product should be produced, given the restrictions on sales.*

See Example 9.

	Fixed Cost	Variable Cost	Price of Item	
77.	\$ 500	\$ 10	\$ 35	No more than 18 units can be sold.
78.	\$2700	\$150	\$280	No more than 25 units can be sold.
79.	\$1650	\$400	\$305	All units produced can be sold.
80.	\$ 180	\$ 11	\$ 20	No more than 30 units can be sold.

(Modeling) Break-Even Point *The manager of a small company that produces roof tile has determined that the total cost in dollars, $C(x)$, of producing x units of tile is given by*

$$C(x) = 200x + 1000,$$

while the revenue in dollars, $R(x)$, from the sale of x units of tile is given by

$$R(x) = 240x.$$

81. Find the break-even point and the cost and revenue at the break-even point.

82. Suppose the variable cost is actually \$220 per unit, instead of \$200. How does this affect the break-even point? Is the manager better off or not?

Relating Concepts

For individual or collaborative investigation *(Exercises 83–92)*

The table shows several points on the graph of a linear function. ***Work Exercises 83–92 in order,*** *to see connections between the slope formula, distance formula, midpoint formula, and linear functions.*

x	y
0	−6
1	−3
2	0
3	3
4	6
5	9
6	12

83. Use the first two points in the table to find the slope of the line.

84. Use the second and third points in the table to find the slope of the line.

85. Make a conjecture by filling in the blank: If we use any two points on a line to find its slope, we find that the slope is ______ in all cases.

86. Find the distance between the first two points in the table. (*Hint:* Use the distance formula.)

87. Find the distance between the second and fourth points in the table.

88. Find the distance between the first and fourth points in the table.

89. Add the results in **Exercises 86 and 87,** and compare the sum to the answer found in **Exercise 88.** What do you notice?

90. Fill in each blank, basing the answers on observations in **Exercises 86–89:**

If points A, B, and C lie on a line in that order, then the distance between A and B added to the distance between ______ and ______ is equal to the distance between ______ and ______.

91. Find the midpoint of the segment joining $(0, -6)$ and $(6, 12)$. Compare the answer to the middle entry in the table. What do you notice?

92. If the table were set up to show an x-value of 4.5, what would be the corresponding y-value?

Chapter 2 Quiz (Sections 2.1-2.4)

1. For $A(-4, 2)$ and $B(-8, -3)$, find $d(A, B)$, the distance between A and B.

2. *Two-Year College Enrollment* Enrollments in two-year colleges for selected years are shown in the table. Use the midpoint formula to estimate the enrollments for 2006 and 2010.

Year	Enrollment (in millions)
2004	6.55
2008	6.97
2012	7.50

Source: National Center for Education Statistics.

3. Graph $y = -x^2 + 4$ by plotting points.

4. Graph $x^2 + y^2 = 16$.

5. Determine the radius and the coordinates of the center of the circle with equation

$$x^2 + y^2 - 4x + 8y + 3 = 0.$$

For Exercises 6–8, refer to the graph of $f(x) = |x + 3|$.

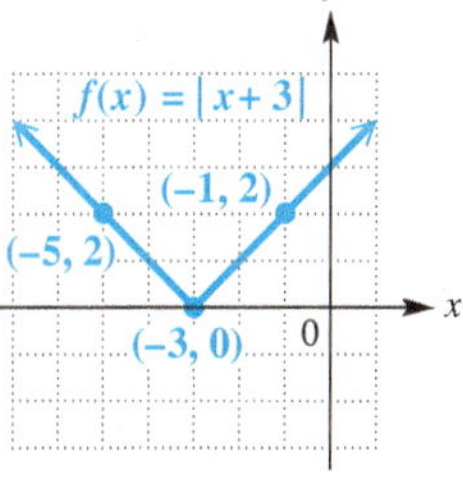

6. Find $f(-1)$.

7. Give the domain and the range of f.

8. Give the largest open interval over which the function f is

(a) decreasing, **(b)** increasing, **(c)** constant.

9. Find the slope of the line through the given points.

(a) $(1, 5)$ and $(5, 11)$ **(b)** $(-7, 4)$ and $(-1, 4)$ **(c)** $(6, 12)$ and $(6, -4)$

10. *Motor Vehicle Sales* The graph shows a straight line segment that approximates new motor vehicle sales in the United States from 2009 to 2013. Determine the average rate of change from 2009 to 2013, and interpret the results.

New Vehicle Sales

Number of Vehicles (in thousands): 20,000; 15,000; 10,000; 5000; 0
10,602
15,884
Year: 2009 2010 2011 2012 2013

Source: U.S. Bureau of Economic Analysis.

2.5 Equations of Lines and Linear Models

- Point-Slope Form
- Slope-Intercept Form
- Vertical and Horizontal Lines
- Parallel and Perpendicular Lines
- Modeling Data
- Graphical Solution of Linear Equations in One Variable

Point-Slope Form

The graph of a linear function is a straight line. We now develop various forms for the equation of a line.

Figure 43 shows the line passing through the fixed point (x_1, y_1) having slope m. (Assuming that the line has a slope guarantees that it is not vertical.) Let (x, y) be any other point on the line. Because the line is not vertical, $x - x_1 \neq 0$. Now use the definition of slope.

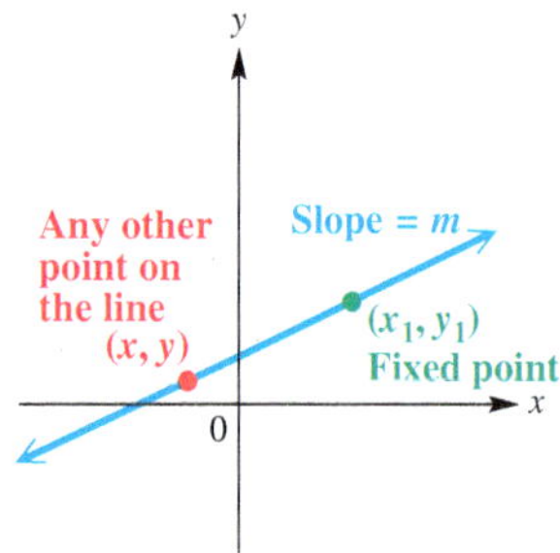

Figure 43

$$m = \frac{y - y_1}{x - x_1} \quad \text{Slope formula}$$

$$m(x - x_1) = y - y_1 \quad \text{Multiply each side by } x - x_1.$$

$$y - y_1 = m(x - x_1) \quad \text{Interchange sides.}$$

This result is the *point-slope form* of the equation of a line.

LOOKING AHEAD TO CALCULUS

A standard problem in calculus is to find the equation of the line tangent to a curve at a given point. The derivative (see *Looking Ahead to Calculus* earlier in this chapter) is used to find the slope of the desired line, and then the slope and the given point are used in the point-slope form to solve the problem.

Point-Slope Form

The **point-slope form** of the equation of the line with slope m passing through the point (x_1, y_1) is given as follows.

$$y - y_1 = m(x - x_1)$$

EXAMPLE 1 Using the Point-Slope Form (Given a Point and the Slope)

Write an equation of the line through the point $(-4, 1)$ having slope -3.

SOLUTION Here $x_1 = -4$, $y_1 = 1$, and $m = -3$.

$$y - y_1 = m(x - x_1) \quad \text{Point-slope form}$$

$$y - 1 = -3[x - (-4)] \quad x_1 = -4,\ y_1 = 1,\ m = -3$$

Be careful with signs.

$$y - 1 = -3(x + 4)$$

$$y - 1 = -3x - 12 \quad \text{Distributive property}$$

$$y = -3x - 11 \quad \text{Add 1.}$$

✔ Now Try Exercise 29.

EXAMPLE 2 Using the Point-Slope Form (Given Two Points)

Write an equation of the line through the points $(-3, 2)$ and $(2, -4)$. Write the result in standard form $Ax + By = C$.

SOLUTION Find the slope first.

$$m = \frac{-4 - 2}{2 - (-3)} = -\frac{6}{5} \quad \text{Definition of slope}$$

The slope m is $-\frac{6}{5}$. Either the point $(-3, 2)$ or the point $(2, -4)$ can be used for (x_1, y_1). We choose $(-3, 2)$.

$$y - y_1 = m(x - x_1) \quad \text{Point-slope form}$$

$$y - 2 = -\frac{6}{5}[x - (-3)] \quad x_1 = -3,\ y_1 = 2,\ m = -\frac{6}{5}$$

$$5(y - 2) = -6(x + 3) \quad \text{Multiply by 5.}$$

$$5y - 10 = -6x - 18 \quad \text{Distributive property}$$

$$6x + 5y = -8 \quad \text{Standard form}$$

Verify that we obtain the same equation if we use $(2, -4)$ instead of $(-3, 2)$ in the point-slope form.

✔ **Now Try Exercise 19.**

NOTE The lines in **Examples 1 and 2** both have negative slopes. Keep in mind that a slope of the form $-\frac{A}{B}$ may be interpreted as either $\frac{-A}{B}$ or $\frac{A}{-B}$.

Slope-Intercept Form As a special case of the point-slope form of the equation of a line, suppose that a line has y-intercept $(0, b)$. If the line has slope m, then using the point-slope form with $x_1 = 0$ and $y_1 = b$ gives the following.

$$y - y_1 = m(x - x_1) \quad \text{Point-slope form}$$

$$y - b = m(x - 0) \quad x_1 = 0,\ y_1 = b$$

$$y - b = mx \quad \text{Distributive property}$$

$$y = mx + b \quad \text{Solve for } y.$$

Slope ⟶ m; b ⟵ The y-intercept is $(0, b)$.

Because this result shows the slope of the line and indicates the y-intercept, it is known as the *slope-intercept form* of the equation of the line.

Slope-Intercept Form

The **slope-intercept form** of the equation of the line with slope m and y-intercept $(0, b)$ is given as follows.

$$y = mx + b$$

EXAMPLE 3 Finding Slope and y-Intercept from an Equation of a Line

Find the slope and y-intercept of the line with equation $4x + 5y = -10$.

SOLUTION Write the equation in slope-intercept form.

$$4x + 5y = -10$$

$$5y = -4x - 10 \quad \text{Subtract } 4x.$$

$$y = -\frac{4}{5}x - 2 \quad \text{Divide by 5.}$$

($-\frac{4}{5}$ is m; -2 is b.)

The slope is $-\frac{4}{5}$, and the y-intercept is $(0, -2)$.

✔ **Now Try Exercise 37.**

NOTE Generalizing from **Example 3,** we see that the slope m of the graph of the equation

$$Ax + By = C$$

is $-\frac{A}{B}$, and the y-intercept is $\left(0, \frac{C}{B}\right)$.

EXAMPLE 4 Using the Slope-Intercept Form (Given Two Points)

Write an equation of the line through the points $(1, 1)$ and $(2, 4)$. Then graph the line using the slope-intercept form.

SOLUTION In **Example 2,** we used the *point-slope form* in a similar problem. Here we show an alternative method using the *slope-intercept form.* First, find the slope.

$$m = \frac{4-1}{2-1} = \frac{3}{1} = 3 \quad \text{Definition of slope}$$

Now substitute 3 for m in $y = mx + b$ and choose one of the given points, say $(1, 1)$, to find the value of b.

$$y = mx + b \quad \text{Slope-intercept form}$$

$$1 = 3(1) + b \quad m = 3, x = 1, y = 1$$

$$\text{The } y\text{-intercept is } (0, b). \rightarrow b = -2 \quad \text{Solve for } b.$$

The slope-intercept form is

$$y = 3x - 2.$$

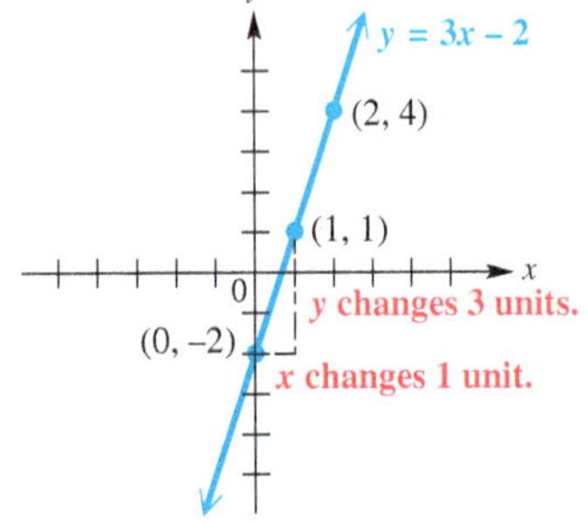

Figure 44

The graph is shown in **Figure 44.** We can plot $(0, -2)$ and then use the definition of slope to arrive at $(1, 1)$. Verify that $(2, 4)$ also lies on the line.

✔ **Now Try Exercise 19.**

EXAMPLE 5 Finding an Equation from a Graph

Use the graph of the linear function f shown in **Figure 45** to complete the following.

(a) Identify the slope, y-intercept, and x-intercept.

(b) Write an equation that defines f.

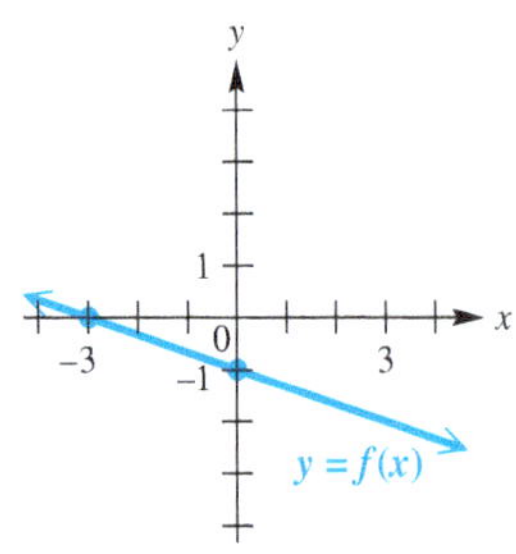

Figure 45

SOLUTION

(a) The line falls 1 unit each time the x-value increases 3 units. Therefore, the slope is $\frac{-1}{3} = -\frac{1}{3}$. The graph intersects the y-axis at the y-intercept $(0, -1)$ and the x-axis at the x-intercept $(-3, 0)$.

(b) The slope is $m = -\frac{1}{3}$, and the y-intercept is $(0, -1)$.

$$y = f(x) = mx + b \quad \text{Slope-intercept form}$$

$$f(x) = -\frac{1}{3}x - 1 \quad m = -\frac{1}{3}, b = -1$$

✔ **Now Try Exercise 45.**

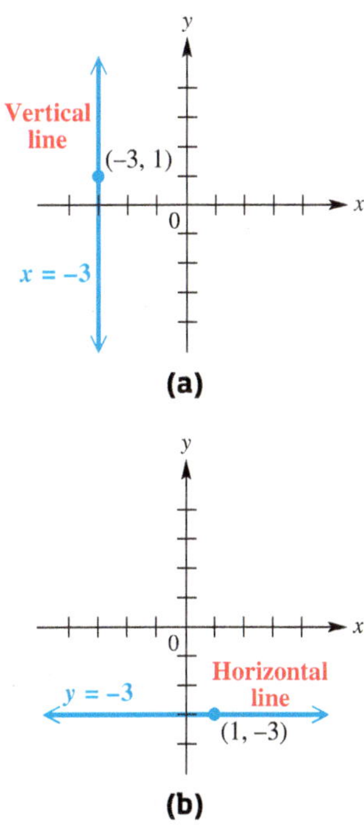

Figure 46

Vertical and Horizontal Lines The vertical line through the point (a, b) passes through all points of the form (a, y), for any value of y. Consequently, the equation of a vertical line through (a, b) is $x = a$. For example, the vertical line through $(-3, 1)$ has equation $x = -3$. See **Figure 46(a).** Because each point on the y-axis has x-coordinate 0, ***the equation of the y-axis is*** $x = 0$.

The horizontal line through the point (a, b) passes through all points of the form (x, b), for any value of x. Therefore, the equation of a horizontal line through (a, b) is $y = b$. For example, the horizontal line through $(1, -3)$ has equation $y = -3$. See **Figure 46(b).** Because each point on the x-axis has y-coordinate 0, ***the equation of the x-axis is*** $y = 0$.

Equations of Vertical and Horizontal Lines

An equation of the **vertical line** through the point (a, b) is $x = a$.

An equation of the **horizontal line** through the point (a, b) is $y = b$.

Parallel and Perpendicular Lines Two parallel lines are equally "steep," so they should have the same slope. Also, two distinct lines with the same "steepness" are parallel. The following result summarizes this discussion. (The statement "p if and only if q" means "if p then q *and* if q then p.")

Parallel Lines

Two distinct nonvertical lines are parallel if and only if they have the same slope.

When two lines have slopes with a product of -1, the lines are perpendicular.

Perpendicular Lines

Two lines, neither of which is vertical, are perpendicular if and only if their slopes have a product of -1. Thus, the slopes of perpendicular lines, neither of which is vertical, are *negative reciprocals.*

Example: If the slope of a line is $-\frac{3}{4}$, then the slope of any line perpendicular to it is $\frac{4}{3}$ because

$$-\frac{3}{4}\left(\frac{4}{3}\right) = -1.$$

(Numbers like $-\frac{3}{4}$ and $\frac{4}{3}$ are **negative reciprocals** of each other.) A proof of this result is outlined in **Exercises 79–85.**

NOTE Because a vertical line has *undefined* slope, it does not follow the *mathematical* rules for parallel and perpendicular lines. We intuitively know that all vertical lines are parallel and that a vertical line and a horizontal line are perpendicular.

EXAMPLE 6 Finding Equations of Parallel and Perpendicular Lines

Write an equation in both slope-intercept and standard form of the line that passes through the point $(3, 5)$ and satisfies the given condition.

(a) parallel to the line $2x + 5y = 4$

(b) perpendicular to the line $2x + 5y = 4$

SOLUTION

(a) We know that the point $(3, 5)$ is on the line, so we need only find the slope to use the point-slope form. We find the slope by writing the equation of the given line in slope-intercept form. (That is, we solve for y.)

$$2x + 5y = 4$$

$$5y = -2x + 4 \quad \text{Subtract } 2x.$$

$$y = -\frac{2}{5}x + \frac{4}{5} \quad \text{Divide by 5.}$$

The slope is $-\frac{2}{5}$. Because the lines are parallel, $-\frac{2}{5}$ is also the slope of the line whose equation is to be found. Now substitute this slope and the given point $(3, 5)$ in the point-slope form.

$$y - y_1 = m(x - x_1) \quad \text{Point-slope form}$$

$$y - 5 = -\frac{2}{5}(x - 3) \quad m = -\tfrac{2}{5},\ x_1 = 3,\ y_1 = 5$$

$$y - 5 = -\frac{2}{5}x + \frac{6}{5} \quad \text{Distributive property}$$

$$\text{Slope-intercept form} \longrightarrow y = -\frac{2}{5}x + \frac{31}{5} \quad \text{Add } 5 = \tfrac{25}{5}.$$

$$5y = -2x + 31 \quad \text{Multiply by 5.}$$

$$\text{Standard form} \longrightarrow 2x + 5y = 31 \quad \text{Add } 2x.$$

(b) There is no need to find the slope again—in part (a) we found that the slope of the line $2x + 5y = 4$ is $-\frac{2}{5}$. The slope of any line perpendicular to it is $\frac{5}{2}$.

$$y - y_1 = m(x - x_1) \quad \text{Point-slope form}$$

$$y - 5 = \frac{5}{2}(x - 3) \quad m = \tfrac{5}{2},\ x_1 = 3,\ y_1 = 5$$

$$y - 5 = \frac{5}{2}x - \frac{15}{2} \quad \text{Distributive property}$$

$$\text{Slope-intercept form} \longrightarrow y = \frac{5}{2}x - \frac{5}{2} \quad \text{Add } 5 = \tfrac{10}{2}.$$

$$2y = 5x - 5 \quad \text{Multiply by 2.}$$

$$-5x + 2y = -5 \quad \text{Subtract } 5x.$$

$$\text{Standard form} \longrightarrow 5x - 2y = 5 \quad \text{Multiply by } -1 \text{ so that } A > 0.$$

✔ **Now Try Exercises 51 and 53.**

We can use a graphing calculator to support the results of **Example 6.** In **Figure 47(a),** we graph the equations of the parallel lines

$$y_1 = -\frac{2}{5}x + \frac{4}{5} \quad \text{and} \quad y_2 = -\frac{2}{5}x + \frac{31}{5}. \qquad \text{See Example 6(a).}$$

The lines appear to be parallel, giving visual support for our result. We must use caution, however, when viewing such graphs, as the limited resolution of a graphing calculator screen may cause two lines to *appear* to be parallel even when they are not. For example, **Figure 47(b)** shows the graphs of the equations

$$y_1 = 2x + 6 \quad \text{and} \quad y_2 = 2.01x - 3$$

in the standard viewing window, and they appear to be parallel. This is not the case, however, because their slopes, 2 and 2.01, are different.

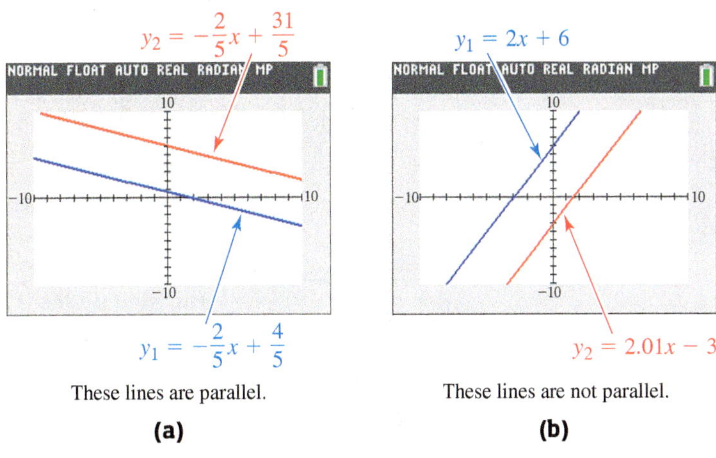

These lines are parallel.

(a)

These lines are not parallel.

(b)

Figure 47

Now we graph the equations of the perpendicular lines

$$y_1 = -\frac{2}{5}x + \frac{4}{5} \quad \text{and} \quad y_2 = \frac{5}{2}x - \frac{5}{2}. \qquad \text{See Example 6(b).}$$

If we use the standard viewing window, the lines do not appear to be perpendicular. See **Figure 48(a).** To obtain the correct perspective, we must use a square viewing window, as in **Figure 48(b).**

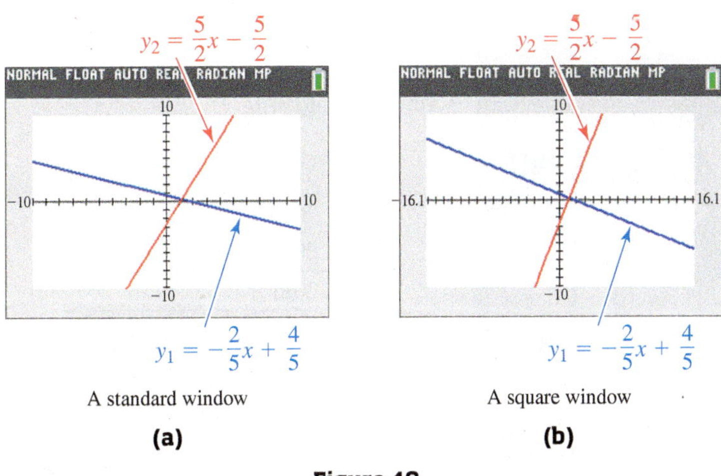

A standard window

(a)

A square window

(b)

Figure 48

A summary of the various forms of linear equations follows.

Summary of Forms of Linear Equations

Equation	Description	When to Use
$y = mx + b$	**Slope-Intercept Form** Slope is m. y-intercept is $(0, b)$.	The slope and y-intercept can be easily identified and used to quickly graph the equation. This form can also be used to find the equation of a line given a point and the slope.
$y - y_1 = m(x - x_1)$	**Point-Slope Form** Slope is m. Line passes through (x_1, y_1).	This form is ideal for finding the equation of a line if the slope and a point on the line or two points on the line are known.
$Ax + By = C$	**Standard Form** (If the coefficients and constant are rational, then A, B, and C are expressed as relatively prime integers, with $A \geq 0$.) Slope is $-\frac{A}{B}$ $(B \neq 0)$. x-intercept is $\left(\frac{C}{A}, 0\right)$ $(A \neq 0)$. y-intercept is $\left(0, \frac{C}{B}\right)$ $(B \neq 0)$.	The x- and y-intercepts can be found quickly and used to graph the equation. The slope must be calculated.
$y = b$	**Horizontal Line** Slope is 0. y-intercept is $(0, b)$.	If the graph intersects only the y-axis, then y is the only variable in the equation.
$x = a$	**Vertical Line** Slope is undefined. x-intercept is $(a, 0)$.	If the graph intersects only the x-axis, then x is the only variable in the equation.

Modeling Data We can write equations of lines that mathematically describe, or model, real data if the data change at a fairly constant rate. In this case, the data fit a linear pattern, and the rate of change is the slope of the line.

EXAMPLE 7 Finding an Equation of a Line That Models Data

Average annual tuition and fees for in-state students at public four-year colleges are shown in the table for selected years and graphed as ordered pairs of points in **Figure 49,** where $x = 0$ represents 2009, $x = 1$ represents 2010, and so on, and y represents the cost in dollars. This graph of ordered pairs of data is a **scatter diagram.**

Year	Cost (in dollars)
2009	6312
2010	6695
2011	7136
2012	7703
2013	8070

Source: National Center for Education Statistics.

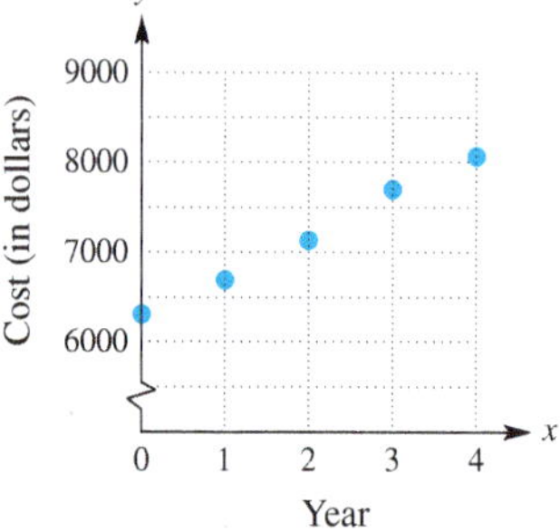

Figure 49

(a) Find an equation that models the data.

(b) Use the equation from part (a) to estimate the cost of tuition and fees at public four-year colleges in 2015.

SOLUTION

(a) The points in **Figure 49** lie approximately on a straight line, so we can write a linear equation that models the relationship between year x and cost y. We choose two data points, $(0, 6312)$ and $(4, 8070)$, to find the slope of the line.

$$m = \frac{8070 - 6312}{4 - 0} = \frac{1758}{4} = 439.5$$

The slope 439.5 indicates that the cost of tuition and fees increased by about \$440 per year from 2009 to 2013. We use this slope, the y-intercept $(0, 6312)$, and the slope-intercept form to write an equation of the line.

$$y = mx + b \qquad \text{Slope-intercept form}$$

$$y = 439.5x + 6312 \qquad \text{Substitute for } m \text{ and } b.$$

(b) The value $x = 6$ corresponds to the year 2015, so we substitute 6 for x.

$$y = 439.5x + 6312 \qquad \text{Model from part (a)}$$

$$y = 439.5(6) + 6312 \qquad \text{Let } x = 6.$$

$$y = 8949 \qquad \text{Multiply, and then add.}$$

The model estimates that average tuition and fees for in-state students at public four-year colleges in 2015 were about \$8949.

✔ **Now Try Exercise 63(a) and (b).**

NOTE In **Example 7,** if we had chosen different data points, we would have obtained a slightly different equation.

Guidelines for Modeling

Step 1 Make a scatter diagram of the data.

Step 2 Find an equation that models the data. For a line, this involves selecting two data points and finding the equation of the line through them.

Linear regression is a technique from statistics that provides the line of "best fit." **Figure 50** shows how a TI-84 Plus calculator accepts the data points, calculates the equation of this line of best fit (in this case, $y_1 = 452.4x + 6278.4$), and plots the data points and line on the same screen.

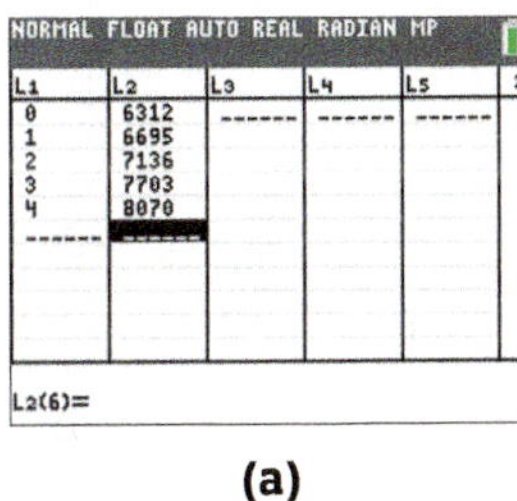

(a)

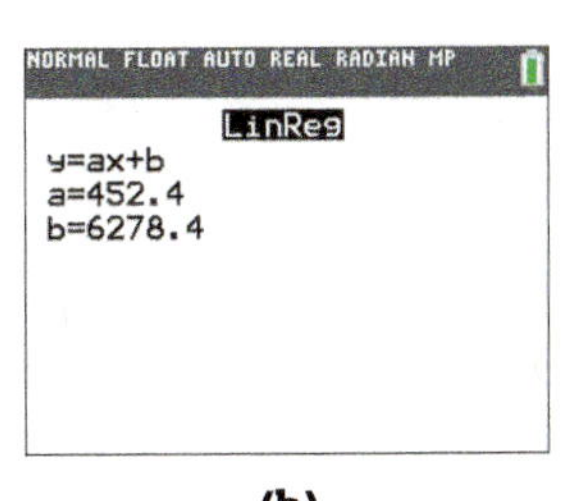

(b)

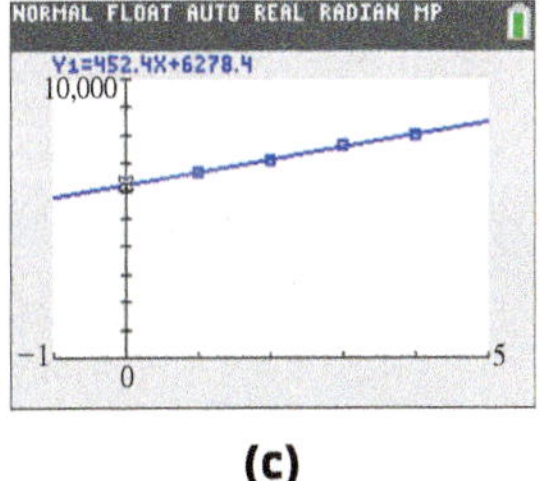

(c)

Figure 50

Graphical Solution of Linear Equations in One Variable Suppose that y_1 and y_2 are linear expressions in x. We can solve the equation $y_1 = y_2$ graphically as follows (assuming it has a unique solution).

1. Rewrite the equation as $y_1 - y_2 = 0$.
2. Graph the linear function $y_3 = y_1 - y_2$.
3. Find the x-intercept of the graph of the function y_3. This x-value is the solution of $y_1 = y_2$.

Some calculators use the term *zero* to identify the x-value of an x-intercept. In general, if $f(a) = 0$, then a is a **zero** of f.

$y_1 = -2x - 4(2 - x) - 3x - 4$

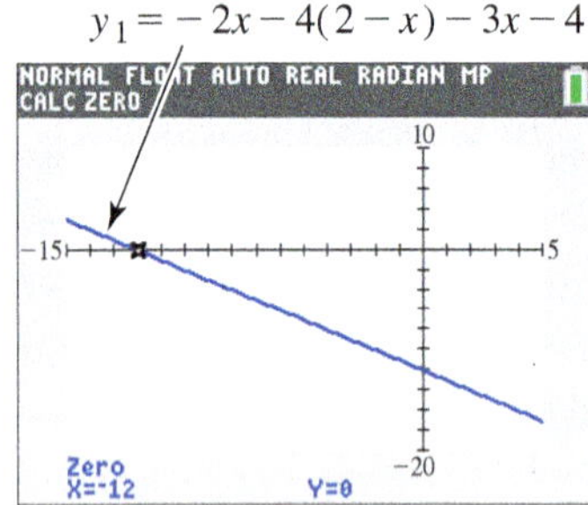

Figure 51

EXAMPLE 8 Solving an Equation with a Graphing Calculator

Use a graphing calculator to solve $-2x - 4(2 - x) = 3x + 4$.

SOLUTION We write an equivalent equation with 0 on one side.

$$-2x - 4(2 - x) - 3x - 4 = 0 \quad \text{Subtract } 3x \text{ and } 4.$$

Then we graph $y = -2x - 4(2 - x) - 3x - 4$ to find the x-intercept. The standard viewing window cannot be used because the x-intercept does not lie in the interval $[-10, 10]$. As seen in **Figure 51,** the solution of the equation is -12, and the solution set is $\{-12\}$.

✔ **Now Try Exercise 69.**

2.5 Exercises

CONCEPT PREVIEW *Fill in the blank(s) to correctly complete each sentence.*

1. The graph of the line $y - 3 = 4(x - 8)$ has slope _____ and passes through the point $(8, \_\_)$.
2. The graph of the line $y = -2x + 7$ has slope _____ and y-intercept _____.
3. The vertical line through the point $(-4, 8)$ has equation _____ $= -4$.
4. The horizontal line through the point $(-4, 8)$ has equation _____ $= 8$.
5. Any line parallel to the graph of $6x + 7y = 9$ must have slope _____.
6. Any line perpendicular to the graph of $6x + 7y = 9$ must have slope _____.

CONCEPT PREVIEW *Match each equation with its graph in A–D.*

7. $y = \frac{1}{4}x + 2$

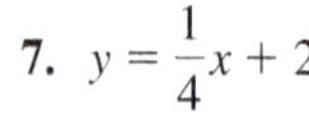

8. $4x + 3y = 12$
9. $y - (-1) = \frac{3}{2}(x - 1)$
10. $y = 4$

A.

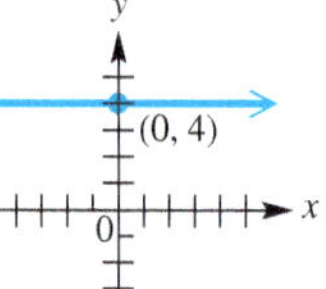

B.

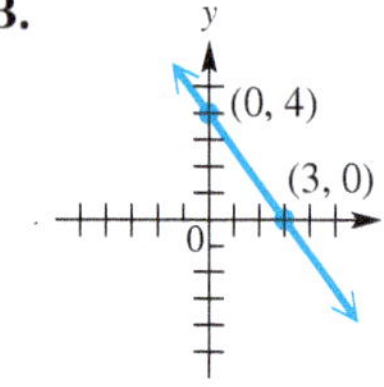

C.

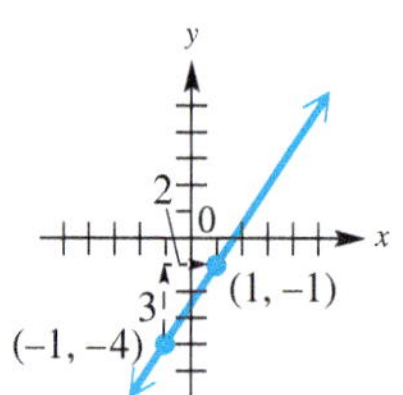

D.

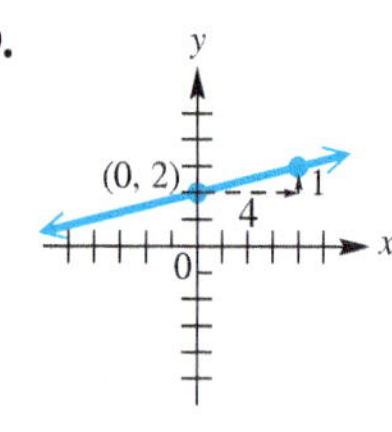

Write an equation for each line described. Give answers in standard form for Exercises 11–20 and in slope-intercept form (if possible) for Exercises 21–32. ***See Examples 1–4.***

11. through $(1, 3)$, $m = -2$

12. through $(2, 4)$, $m = -1$

13. through $(-5, 4)$, $m = -\frac{3}{2}$

14. through $(-4, 3)$, $m = \frac{3}{4}$

15. through $(-8, 4)$, undefined slope

16. through $(5, 1)$, undefined slope

17. through $(5, -8)$, $m = 0$

18. through $(-3, 12)$, $m = 0$

19. through $(-1, 3)$ and $(3, 4)$

20. through $(2, 3)$ and $(-1, 2)$

21. x-intercept $(3, 0)$, y-intercept $(0, -2)$

22. x-intercept $(-4, 0)$, y-intercept $(0, 3)$

23. vertical, through $(-6, 4)$

24. vertical, through $(2, 7)$

25. horizontal, through $(-7, 4)$

26. horizontal, through $(-8, -2)$

27. $m = 5$, $b = 15$

28. $m = -2$, $b = 12$

29. through $(-2, 5)$ having slope -4

30. through $(4, -7)$ having slope -2

31. slope 0, y-intercept $\left(0, \frac{3}{2}\right)$

32. slope 0, y-intercept $\left(0, -\frac{5}{4}\right)$

33. *Concept Check* Fill in each blank with the appropriate response:

The line $x + 2 = 0$ has x-intercept ________. It ________ (does/does not) have a y-intercept. The slope of this line is ________ (0/undefined).

The line $4y = 2$ has y-intercept ________. It ________ (does/does not) have an x-intercept. The slope of this line is ________ (0/undefined).

34. *Concept Check* Match each equation with the line that would most closely resemble its graph. (*Hint:* Consider the signs of m and b in the slope-intercept form.)

(a) $y = 3x + 2$ **(b)** $y = -3x + 2$ **(c)** $y = 3x - 2$ **(d)** $y = -3x - 2$

A.

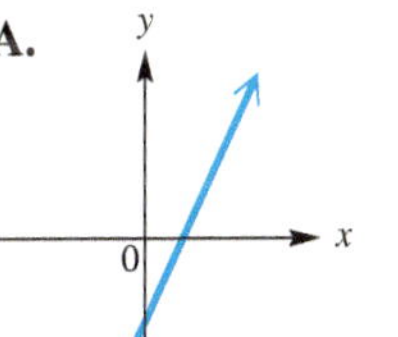

B.

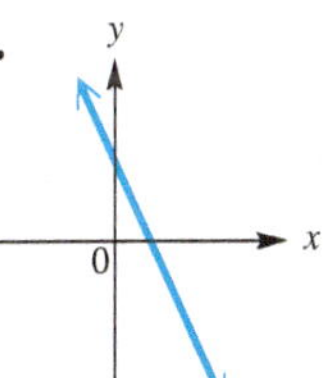

C.

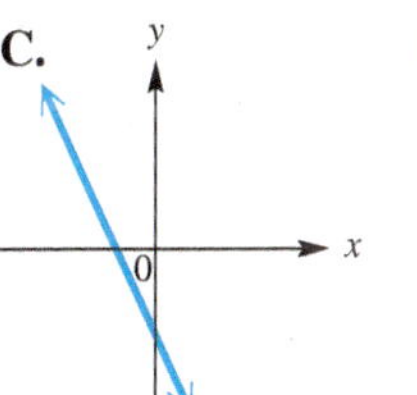

D.

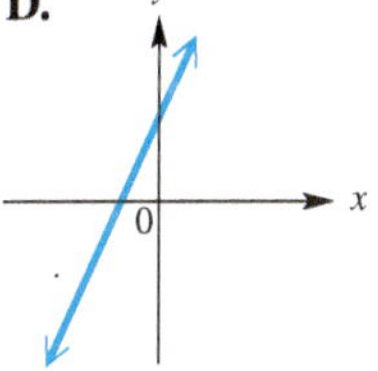

Find the slope and y-intercept of each line, and graph it. ***See Example 3.***

35. $y = 3x - 1$

36. $y = -2x + 7$

37. $4x - y = 7$

38. $2x + 3y = 16$

39. $4y = -3x$

40. $2y = x$

41. $x + 2y = -4$

42. $x + 3y = -9$

43. $y - \frac{3}{2}x - 1 = 0$

44. *Concept Check* The table represents a linear function f.

(a) Find the slope of the graph of $y = f(x)$.

(b) Find the y-intercept of the line.

(c) Write an equation for this line in slope-intercept form.

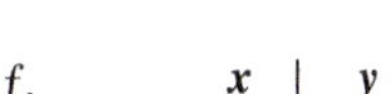

x	y
-2	-11
-1	-8
0	-5
1	-2
2	1
3	4

Connecting Graphs with Equations *The graph of a linear function f is shown.* ***(a)*** *Identify the slope, y-intercept, and x-intercept.* ***(b)*** *Write an equation that defines f.* ***See Example 5.***

45.

46.

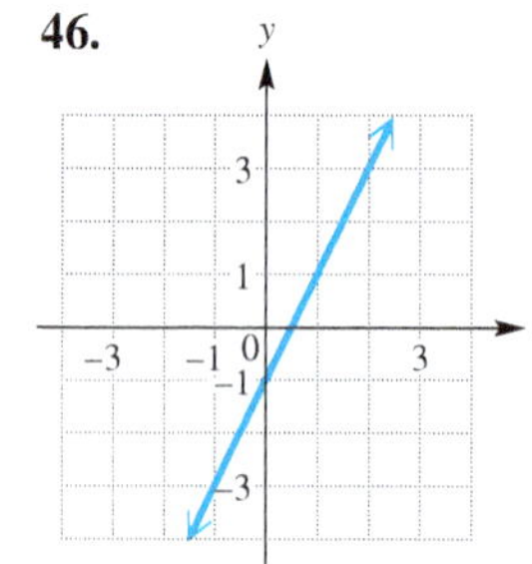

47.

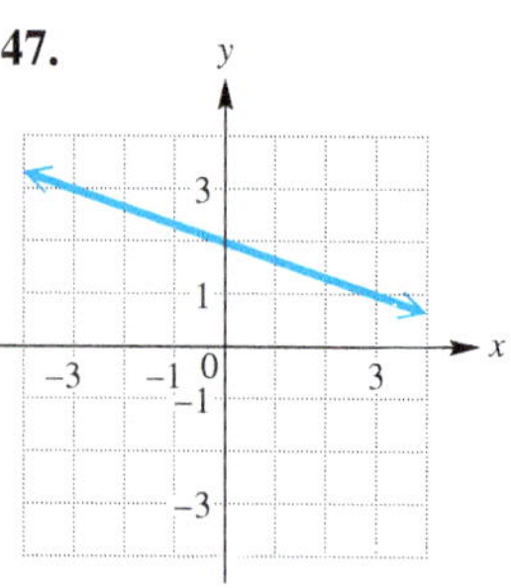

48.

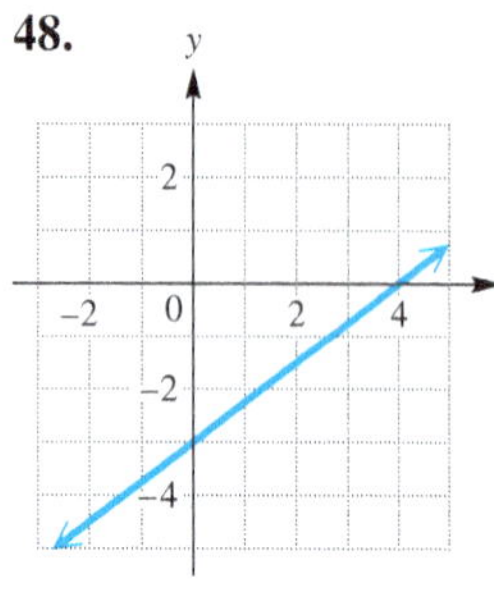

49.

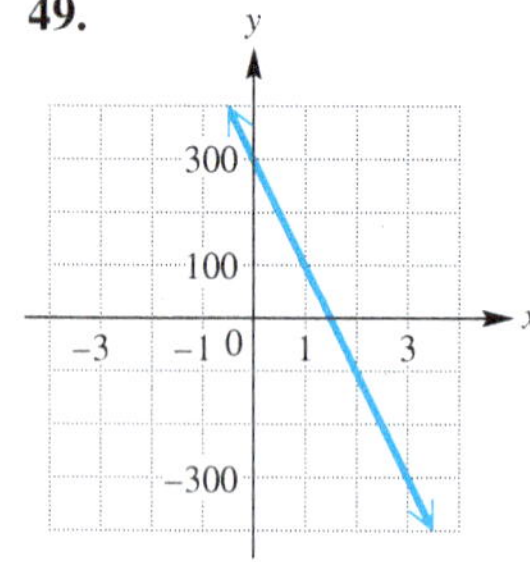

50.

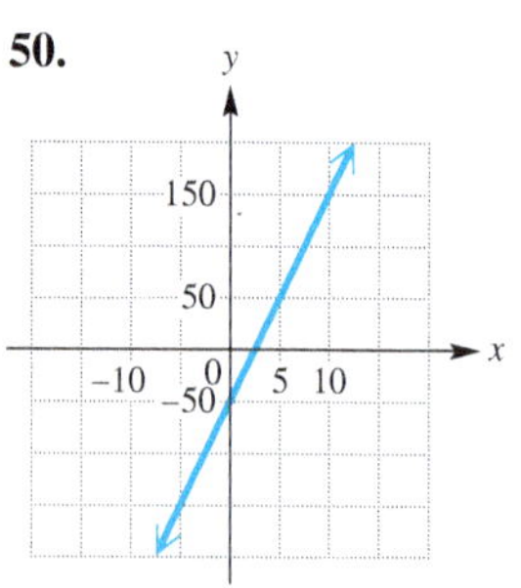

Write an equation ***(a)*** *in standard form and* ***(b)*** *in slope-intercept form for each line described.* ***See Example 6.***

51. through $(-1, 4)$, parallel to $x + 3y = 5$

52. through $(3, -2)$, parallel to $2x - y = 5$

53. through $(1, 6)$, perpendicular to $3x + 5y = 1$

54. through $(-2, 0)$, perpendicular to $8x - 3y = 7$

55. through $(4, 1)$, parallel to $y = -5$

56. through $(-2, -2)$, parallel to $y = 3$

57. through $(-5, 6)$, perpendicular to $x = -2$

58. through $(4, -4)$, perpendicular to $x = 4$

Work each problem.

59. Find k so that the line through $(4, -1)$ and $(k, 2)$ is

(a) parallel to $3y + 2x = 6$ **(b)** perpendicular to $2y - 5x = 1$.

60. Find r so that the line through $(2, 6)$ and $(-4, r)$ is

(a) parallel to $2x - 3y = 4$ **(b)** perpendicular to $x + 2y = 1$.

(Modeling) *Solve each problem.* ***See Example 7.***

61. *Annual Tuition and Fees* Refer to the table that accompanies **Figure 49** in **Example 7.**

(a) Use the data points $(0, 6312)$ and $(3, 7703)$ to find a linear equation that models the data.

(b) Use the equation from part (a) to estimate average tuition and fees for in-state students at public four-year colleges in 2013. How does the result compare to the actual figure given in the table, $8070?

62. *Annual Tuition and Fees* Refer to the table that accompanies **Figure 49** in **Example 7.**

(a) Use the data points for the years 2009 and 2011 to find a linear equation that models the data.

(b) Use the equation from part (a) to estimate average tuition and fees for in-state students at public four-year colleges in 2013. How does the result compare to the actual figure given in the table, \$8070?

63. *Cost of Private College Education* The table lists average annual cost (in dollars) of tuition and fees at private four-year colleges for selected years.

Year	Cost (in dollars)
2009	22,036
2010	21,908
2011	22,771
2012	23,460
2013	24,525

Source: National Center for Education Statistics.

(a) Determine a linear function $f(x) = ax + b$ that models the data, where $x = 0$ represents 2009, $x = 1$ represents 2010, and so on. Use the points $(0, 22{,}036)$ and $(4, 24{,}525)$ to graph f and a scatter diagram of the data on the same coordinate axes. (Use a graphing calculator if desired.) What does the slope of the graph indicate?

(b) Use the function from part (a) to approximate average tuition and fees in 2012. Compare the approximation to the actual figure given in the table, \$23,460.

(c) Use the linear regression feature of a graphing calculator to find the equation of the line of best fit.

64. *Distances and Velocities of Galaxies* The table lists the distances (in megaparsecs; 1 megaparsec $= 3.085 \times 10^{24}$ cm, and 1 megaparsec $= 3.26$ million light-years) and velocities (in kilometers per second) of four galaxies moving rapidly away from Earth.

Galaxy	Distance	Velocity
Virgo	15	1600
Ursa Minor	200	15,000
Corona Borealis	290	24,000
Bootes	520	40,000

Source: Acker, A., and C. Jaschek, *Astronomical Methods and Calculations,* John Wiley and Sons. Karttunen, H. (editor), *Fundamental Astronomy,* Springer-Verlag.

(a) Plot the data using distances for the x-values and velocities for the y-values. What type of relationship seems to hold between the data?

(b) Find a linear equation in the form $y = mx$ that models these data using the points $(520, 40{,}000)$ and $(0, 0)$. Graph the equation with the data on the same coordinate axes.

(c) The galaxy Hydra has a velocity of 60,000 km per sec. How far away, to the nearest megaparsec, is it according to the model in part (b)?

(d) The value of m is the **Hubble constant.** The Hubble constant can be used to estimate the age of the universe A (in years) using the formula

$$A = \frac{9.5 \times 10^{11}}{m}.$$

Approximate A using the value of m. Round to the nearest hundredth of a billion years.

(e) Astronomers currently place the value of the Hubble constant between 50 and 100. What is the range for the age of the universe A?

65. *Celsius and Fahrenheit Temperatures* When the Celsius temperature is 0°, the corresponding Fahrenheit temperature is 32°. When the Celsius temperature is 100°, the corresponding Fahrenheit temperature is 212°. Let C represent the Celsius temperature and F the Fahrenheit temperature.

(a) Express F as an exact linear function of C.

(b) Solve the equation in part (a) for C, thus expressing C as a function of F.

(c) For what temperature is $F = C$ a true statement?

66. *Water Pressure on a Diver* The pressure P of water on a diver's body is a linear function of the diver's depth, x. At the water's surface, the pressure is 1 atmosphere. At a depth of 100 ft, the pressure is about 3.92 atmospheres.

(a) Find a linear function that relates P to x.

(b) Compute the pressure, to the nearest hundredth, at a depth of 10 fathoms (60 ft).

67. *Consumption Expenditures* In Keynesian macroeconomic theory, total consumption expenditure on goods and services, C, is assumed to be a linear function of national personal income, I. The table gives the values of C and I for 2009 and 2013 in the United States (in billions of dollars).

Year	2009	2013
Total consumption (C)	\$10,089	\$11,484
National income (I)	\$12,026	\$14,167

Source: U.S. Bureau of Economic Analysis.

(a) Find a formula for C as a function of I.

(b) The slope of the linear function found in part (a) is the **marginal propensity to consume.** What is the marginal propensity to consume for the United States from 2009–2013?

68. *Concept Check* A graph of $y = f(x)$ is shown in the standard viewing window. Which is the only value of x that could possibly be the solution of the equation $f(x) = 0$?

A. -15 **B.** 0 **C.** 5 **D.** 15

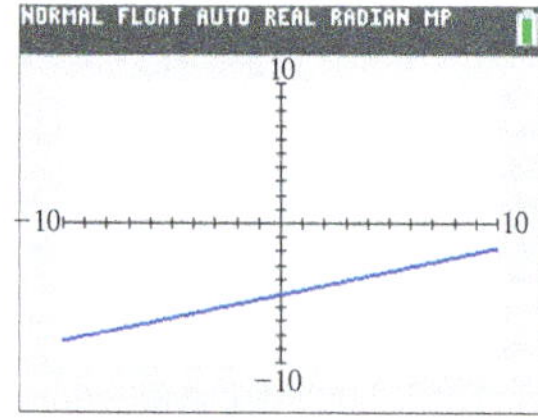

Use a graphing calculator to solve each linear equation. ***See Example 8.***

69. $2x + 7 - x = 4x - 2$

70. $7x - 2x + 4 - 5 = 3x + 1$

71. $3(2x + 1) - 2(x - 2) = 5$

72. $4x - 3(4 - 2x) = 2(x - 3) + 6x + 2$

73. **(a)** Solve $-2(x - 5) = -x - 2$ using traditional paper-and-pencil methods.

(b) Explain why the standard viewing window of a graphing calculator cannot graphically support the solution found in part (a). What minimum and maximum x-values would make it possible for the solution to be seen?

74. Use a graphing calculator to try to solve

$$-3(2x + 6) = -4x + 8 - 2x.$$

Explain what happens. What is the solution set?

If three distinct points A, B, and C in a plane are such that the slopes of nonvertical line segments AB, AC, and BC are equal, then A, B, and C are collinear. Otherwise, they are not. Use this fact to determine whether the three points given are collinear.

75. $(-1, 4), (-2, -1), (1, 14)$

76. $(0, -7), (-3, 5), (2, -15)$

77. $(-1, -3), (-5, 12), (1, -11)$

78. $(0, 9), (-3, -7), (2, 19)$

Relating Concepts

For individual or collaborative investigation *(Exercises 79-85)*

In this section we state that two lines, neither of which is vertical, are perpendicular if and only if their slopes have a product of −1. In Exercises 79–85, we outline a partial proof of this for the case where the two lines intersect at the origin. ***Work these exercises in order****, and refer to the figure as needed.*

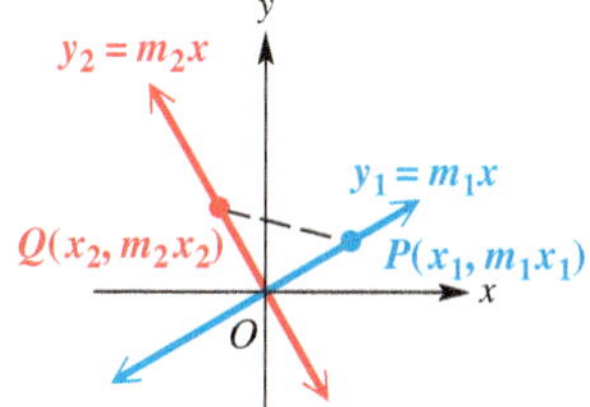

By the converse of the Pythagorean theorem, if

$$[d(O, P)]^2 + [d(O, Q)]^2 = [d(P, Q)]^2,$$

then triangle POQ is a right triangle with right angle at O.

79. Find an expression for the distance $d(O, P)$.

80. Find an expression for the distance $d(O, Q)$.

81. Find an expression for the distance $d(P, Q)$.

82. Use the results from **Exercises 79–81,** and substitute into the equation from the Pythagorean theorem. Simplify to show that this leads to the equation

$$-2m_1m_2x_1x_2 - 2x_1x_2 = 0.$$

83. Factor $-2x_1x_2$ from the final form of the equation in **Exercise 82.**

84. Use the property that if $ab = 0$ then $a = 0$ or $b = 0$ to solve the equation in **Exercise 83,** showing that $m_1m_2 = -1$.

85. State a conclusion based on the results of **Exercises 79–84.**

Summary Exercises on Graphs, Circles, Functions, and Equations

These summary exercises provide practice with some of the concepts covered so far in this chapter.

For the points P and Q, find ***(a)*** *the distance d(P, Q),* ***(b)*** *the coordinates of the midpoint of the segment PQ, and* ***(c)*** *an equation for the line through the two points. Write the equation in slope-intercept form if possible.*

1. $P(3, 5), Q(2, -3)$

2. $P(-1, 0), Q(4, -2)$

3. $P(-2, 2), Q(3, 2)$

4. $P(2\sqrt{2}, \sqrt{2}), Q(\sqrt{2}, 3\sqrt{2})$

5. $P(5, -1), Q(5, 1)$

6. $P(1, 1), Q(-3, -3)$

7. $P(2\sqrt{3}, 3\sqrt{5}), Q(6\sqrt{3}, 3\sqrt{5})$

8. $P(0, -4), Q(3, 1)$

Write an equation for each of the following, and sketch the graph.

9. the line through $(-2, 1)$ and $(4, -1)$

10. the horizontal line through $(2, 3)$

11. the circle with center $(2, -1)$ and radius 3

12. the circle with center $(0, 2)$ and tangent to the x-axis

13. the line through $(3, -5)$ with slope $-\frac{5}{6}$

14. the vertical line through $(-4, 3)$

15. the line through $(-3, 2)$ and parallel to the line $2x + 3y = 6$

16. the line through the origin and perpendicular to the line $3x - 4y = 2$

Decide whether or not each equation has a circle as its graph. If it does, give the center and the radius.

17. $x^2 + y^2 - 4x + 2y = 4$

18. $x^2 + y^2 + 6x + 10y + 36 = 0$

19. $x^2 + y^2 - 12x + 20 = 0$

20. $x^2 + y^2 + 2x + 16y = -61$

21. $x^2 + y^2 - 2x + 10 = 0$

22. $x^2 + y^2 - 8y - 9 = 0$

Solve each problem.

23. Find the coordinates of the points of intersection of the line $y = 2$ and the circle with center at $(4, 5)$ and radius 4.

24. Find the shortest distance from the origin to the graph of the circle with equation

$$x^2 + y^2 - 10x - 24y + 144 = 0.$$

*For each relation, **(a)** find the domain and range, and **(b)** if the relation defines y as a function f of x, rewrite the relation using function notation and find $f(-2)$.*

25. $x - 4y = -6$

26. $y^2 - x = 5$

27. $(x + 2)^2 + y^2 = 25$

28. $x^2 - 2y = 3$

2.6 Graphs of Basic Functions

- Continuity
- The Identity, Squaring, and Cubing Functions
- The Square Root and Cube Root Functions
- The Absolute Value Function
- Piecewise-Defined Functions
- The Relation $x = y^2$

Continuity The graph of a linear function—a straight line—may be drawn by hand over any interval of its domain without picking the pencil up from the paper. In mathematics we say that a function with this property is *continuous* over any interval. The formal definition of continuity requires concepts from calculus, but we can give an informal definition at the college algebra level.

Continuity (Informal Definition)

A function is **continuous** over an interval of its domain if its hand-drawn graph over that interval can be sketched without lifting the pencil from the paper.

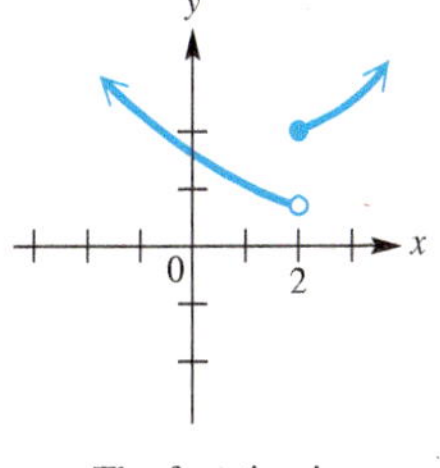

The function is discontinuous at $x = 2$.

Figure 52

If a function is not continuous at a *point,* then it has a *discontinuity* there. **Figure 52** shows the graph of a function with a discontinuity at the point where $x = 2$.

EXAMPLE 1 Determining Intervals of Continuity

Describe the intervals of continuity for each function in **Figure 53.**

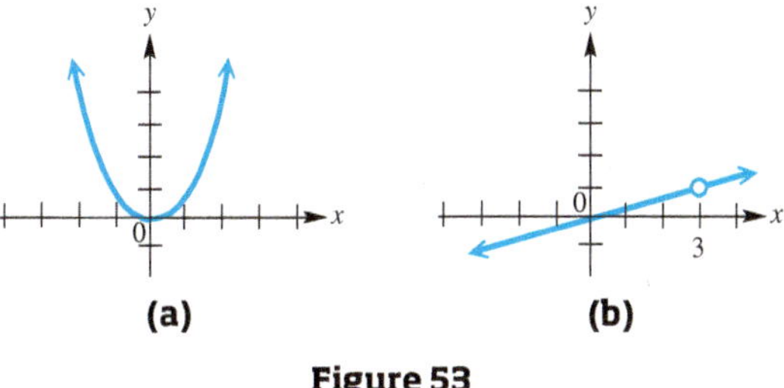

Figure 53

SOLUTION The function in **Figure 53(a)** is continuous over its entire domain, $(-\infty, \infty)$. The function in **Figure 53(b)** has a point of discontinuity at $x = 3$. Thus, it is continuous over the intervals

$$(-\infty, 3) \quad \text{and} \quad (3, \infty).$$

✔ **Now Try Exercises 11 and 15.**

Graphs of the basic functions studied in college algebra can be sketched by careful point plotting or generated by a graphing calculator. As you become more familiar with these graphs, you should be able to provide quick rough sketches of them.

The Identity, Squaring, and Cubing Functions The **identity function** $f(x) = x$ pairs every real number with itself. See **Figure 54.**

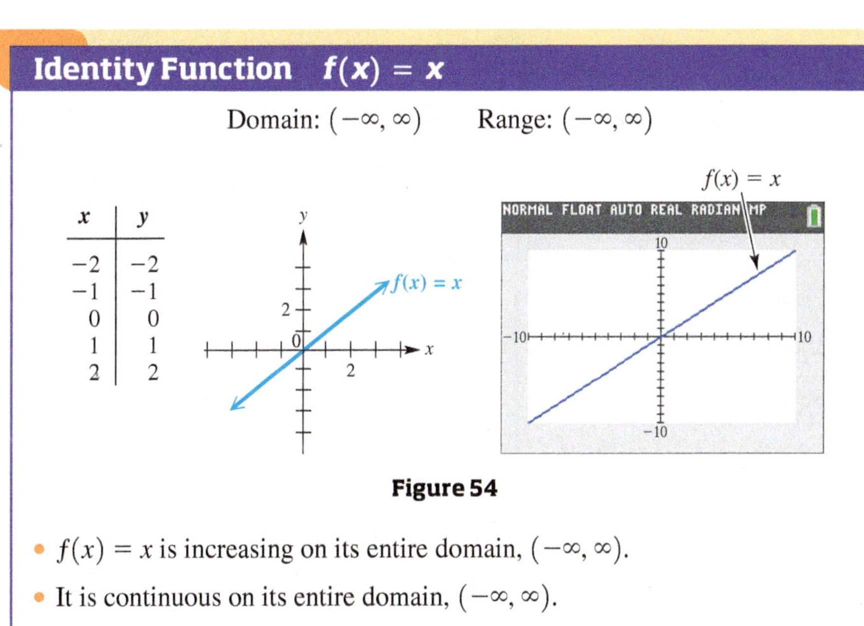

x	y
−2	−2
−1	−1
0	0
1	1
2	2

Figure 54

- $f(x) = x$ is increasing on its entire domain, $(-\infty, \infty)$.
- It is continuous on its entire domain, $(-\infty, \infty)$.

LOOKING AHEAD TO CALCULUS
Many calculus theorems apply only to continuous functions.

The **squaring function** $f(x) = x^2$ pairs each real number with its square. Its graph is a **parabola.** The point $(0, 0)$ at which the graph changes from decreasing to increasing is the **vertex** of the parabola. See **Figure 55.** (For a parabola that opens downward, the vertex is the point at which the graph changes from increasing to decreasing.)

Squaring Function $f(x) = x^2$

Domain: $(-\infty, \infty)$ Range: $[0, \infty)$

x	y
-2	4
-1	1
0	0
1	1
2	4

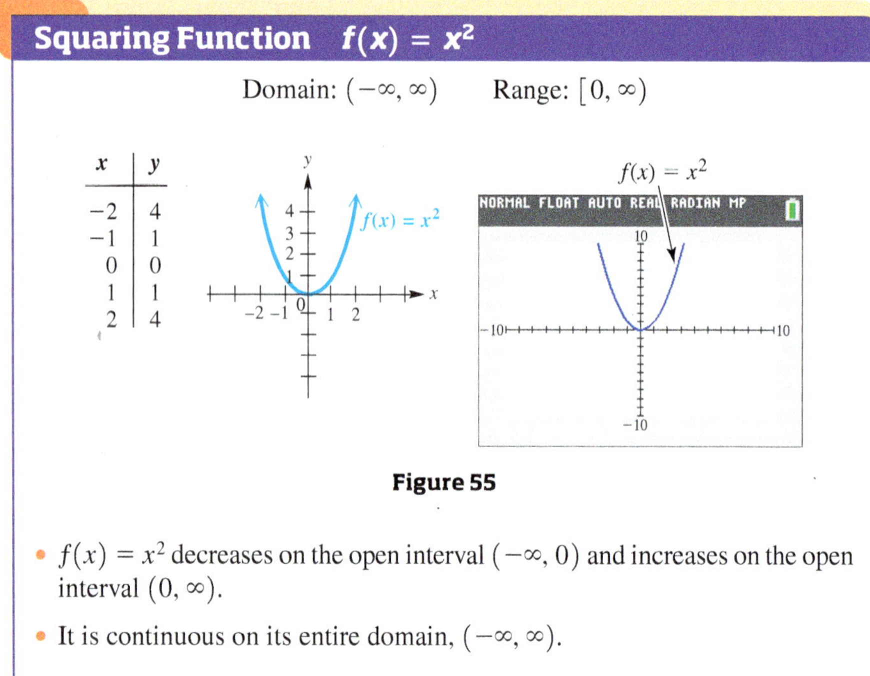

Figure 55

- $f(x) = x^2$ decreases on the open interval $(-\infty, 0)$ and increases on the open interval $(0, \infty)$.
- It is continuous on its entire domain, $(-\infty, \infty)$.

The function $f(x) = x^3$ is the **cubing function.** It pairs each real number with the cube of the number. See **Figure 56.** The point $(0, 0)$ at which the graph changes from "opening downward" to "opening upward" is an **inflection point.**

Cubing Function $f(x) = x^3$

Domain: $(-\infty, \infty)$ Range: $(-\infty, \infty)$

x	y
-2	-8
-1	-1
0	0
1	1
2	8

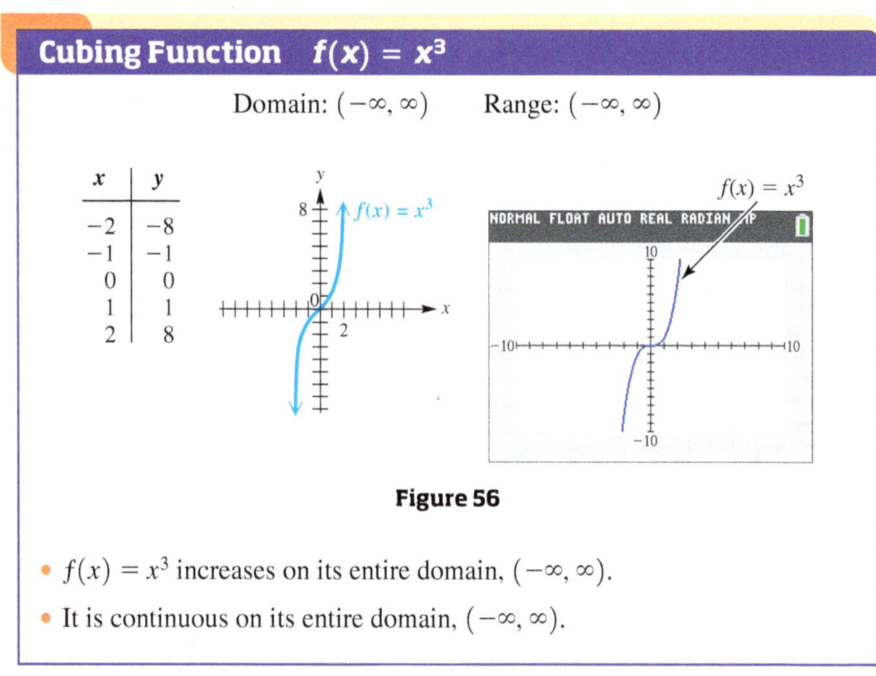

Figure 56

- $f(x) = x^3$ increases on its entire domain, $(-\infty, \infty)$.
- It is continuous on its entire domain, $(-\infty, \infty)$.

The Square Root and Cube Root Functions The function $f(x) = \sqrt{x}$ is the **square root function.** It pairs each real number with its principal square root. See **Figure 57.** For the function value to be a real number, the domain must be restricted to $[0, \infty)$.

Square Root Function $f(x) = \sqrt{x}$

Domain: $[0, \infty)$ Range: $[0, \infty)$

x	y
0	0
1	1
4	2
9	3
16	4

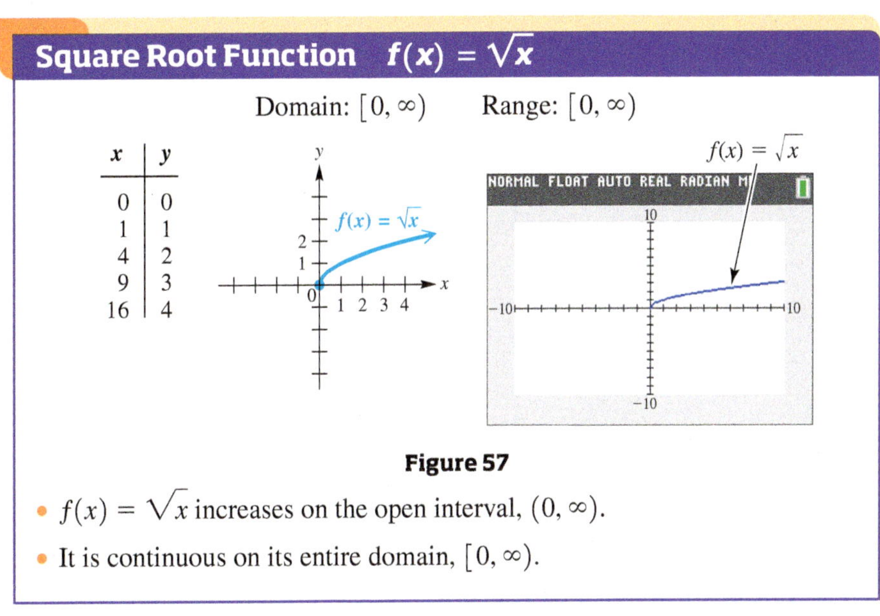

Figure 57

- $f(x) = \sqrt{x}$ increases on the open interval, $(0, \infty)$.
- It is continuous on its entire domain, $[0, \infty)$.

The **cube root function** $f(x) = \sqrt[3]{x}$ pairs each real number with its cube root. See **Figure 58.** The cube root function differs from the square root function in that *any* real number has a real number cube root. Thus, the domain is $(-\infty, \infty)$.

Cube Root Function $f(x) = \sqrt[3]{x}$

Domain: $(-\infty, \infty)$ Range: $(-\infty, \infty)$

x	y
−8	−2
−1	−1
0	0
1	1
8	2

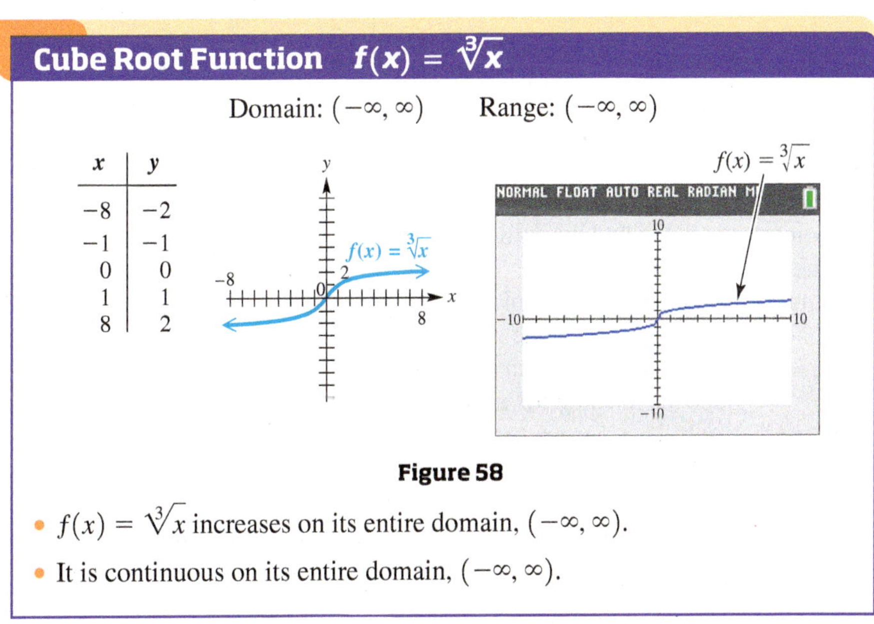

Figure 58

- $f(x) = \sqrt[3]{x}$ increases on its entire domain, $(-\infty, \infty)$.
- It is continuous on its entire domain, $(-\infty, \infty)$.

The Absolute Value Function The **absolute value function,** $f(x) = |x|$, which pairs every real number with its absolute value, is graphed in **Figure 59** on the next page and is defined as follows.

$$f(x) = |x| = \begin{cases} x & \text{if } x \geq 0 \\ -x & \text{if } x < 0 \end{cases} \quad \text{Absolute value function}$$

That is, we use $|x| = x$ if x is positive or 0, and we use $|x| = -x$ if x is negative.

Absolute Value Function $f(x) = |x|$

Domain: $(-\infty, \infty)$ Range: $[0, \infty)$

x	y
-2	2
-1	1
0	0
1	1
2	2

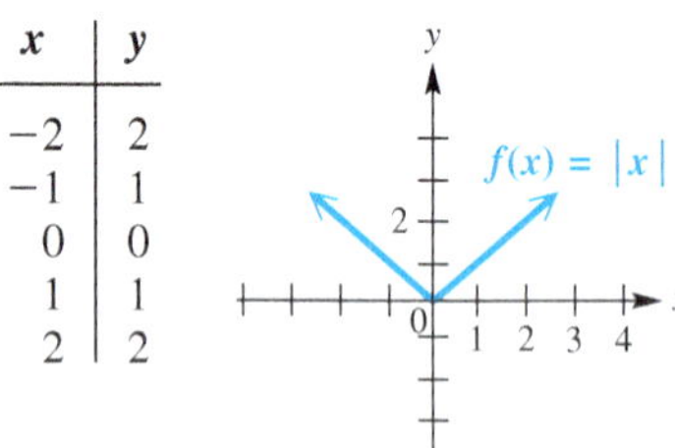

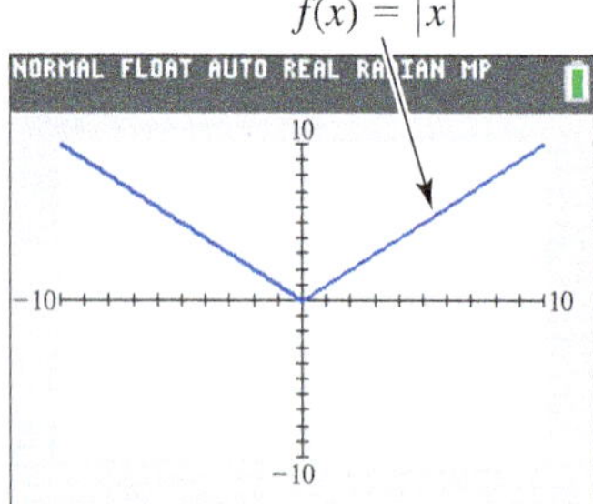

Figure 59

- $f(x) = |x|$ decreases on the open interval $(-\infty, 0)$ and increases on the open interval $(0, \infty)$.
- It is continuous on its entire domain, $(-\infty, \infty)$.

Piecewise-Defined Functions

The absolute value function is a **piecewise-defined function.** It is defined by different rules over different intervals of its domain.

EXAMPLE 2 Graphing Piecewise-Defined Functions

Graph each function.

(a) $f(x) = \begin{cases} -2x + 5 & \text{if } x \leq 2 \\ x + 1 & \text{if } x > 2 \end{cases}$ **(b)** $f(x) = \begin{cases} 2x + 3 & \text{if } x \leq 0 \\ -x^2 + 3 & \text{if } x > 0 \end{cases}$

ALGEBRAIC SOLUTION

(a) We graph each interval of the domain separately. If $x \leq 2$, the graph of $f(x) = -2x + 5$ has an endpoint at $x = 2$. We find the corresponding y-value by substituting 2 for x in $-2x + 5$ to obtain $y = 1$. To find another point on this part of the graph, we choose $x = 0$, so $y = 5$. We draw the graph through $(2, 1)$ and $(0, 5)$ as a partial line with endpoint $(2, 1)$.

We graph the function for $x > 2$ similarly, using $f(x) = x + 1$. This partial line has an open endpoint at $(2, 3)$. We use $y = x + 1$ to find another point with x-value greater than 2 to complete the graph. See **Figure 60.**

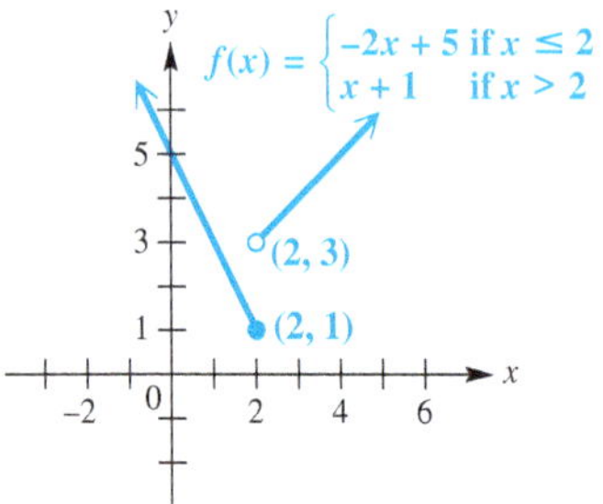

Figure 60

GRAPHING CALCULATOR SOLUTION

(a) We use the TEST feature of the TI-84 Plus to graph the piecewise-defined function. (Press 2ND MATH to display a list containing inequality symbols.) The result of a true statement is 1, and the result of a false statement is 0. We choose x with the appropriate inequality based on how the function is defined. Next we multiply each defining expression by the test condition result. We then add these products to obtain the complete function.

The expression for the function in part (a) is shown at the top of the screen in **Figure 61.**

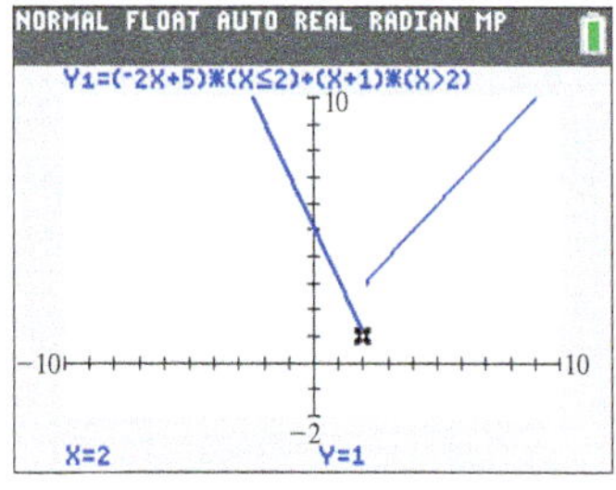

Figure 61

(b) First graph $f(x) = 2x + 3$ for $x \le 0$. Then for $x > 0$, graph $f(x) = -x^2 + 3$. The two graphs meet at the point $(0, 3)$. See **Figure 62.**

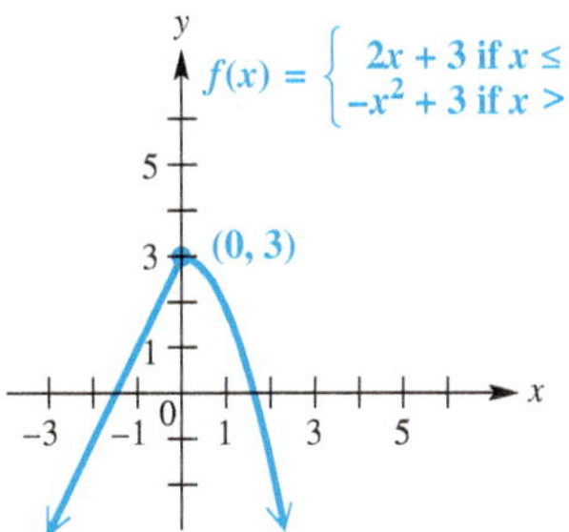

Figure 62

(b) Use the procedure described in part (a). The expression for the function is shown at the top of the screen in **Figure 63.**

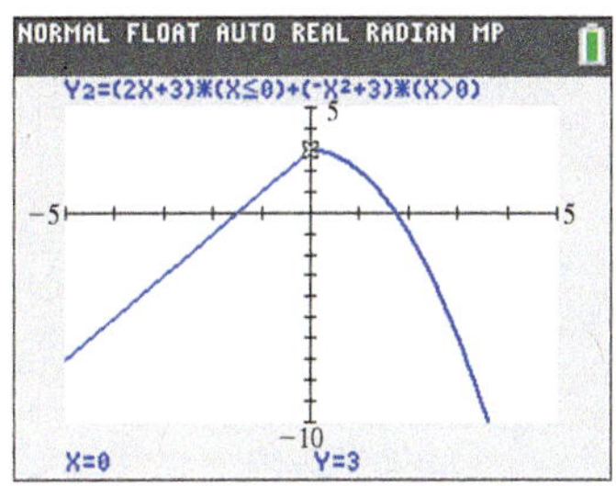

Figure 63

✔ **Now Try Exercises 23 and 29.**

Another piecewise-defined function is the *greatest integer function.*

LOOKING AHEAD TO CALCULUS

The **greatest integer function** is used in calculus as a classic example of how the limit of a function may not exist at a particular value in its domain. For a limit to exist, the left- and right-hand limits must be equal. We can see from the graph of the greatest integer function that for an integer value such as 3, as x approaches 3 from the left, function values are all 2, while as x approaches 3 from the right, function values are all 3. Because the left- and right-hand limits are different, the limit as x approaches 3 does not exist.

$f(x) = [\![x]\!]$

The **greatest integer function** $f(x) = [\![x]\!]$ pairs every real number x with the greatest integer less than or equal to x.

Examples: $[\![8.4]\!] = 8$, $[\![-5]\!] = -5$, $[\![\pi]\!] = 3$, $[\![-6.9]\!] = -7$.

The graph is shown in **Figure 64.** In general, if $f(x) = [\![x]\!]$, then

for $-2 \le x < -1$, $f(x) = -2$,

for $-1 \le x < 0$, $f(x) = -1$,

for $0 \le x < 1$, $f(x) = 0$,

for $1 \le x < 2$, $f(x) = 1$,

for $2 \le x < 3$, $f(x) = 2$, and so on.

Greatest Integer Function $f(x) = [\![x]\!]$

Domain: $(-\infty, \infty)$

Range: $\{y \mid y \text{ is an integer}\} = \{\ldots, -3, -2, -1, 0, 1, 2, 3, \ldots\}$

x	y
−2	−2
−1.5	−2
−0.99	−1
0	0
0.001	0
3	3
3.99	3

$f(x) = [\![x]\!]$

The dots indicate that the graph continues indefinitely in the same pattern.

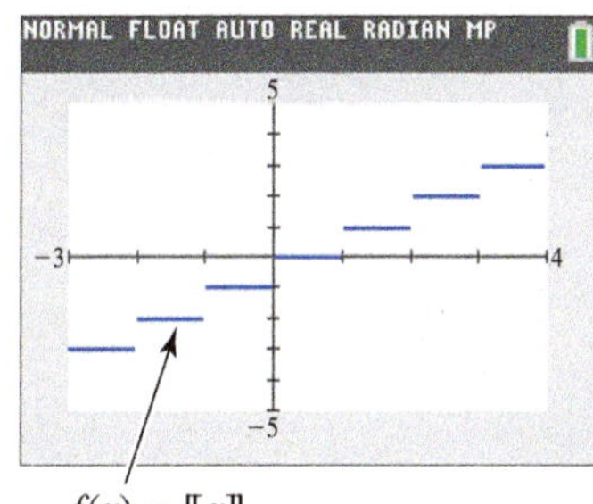

Figure 64

- $f(x) = [\![x]\!]$ is constant on the open intervals $\ldots, (-2, -1), (-1, 0), (0, 1), (1, 2), (2, 3), \ldots$.
- It is discontinuous at all integer values in its domain, $(-\infty, \infty)$.

EXAMPLE 3 Graphing a Greatest Integer Function

Graph $f(x) = [\![\frac{1}{2}x + 1]\!]$.

SOLUTION If x is in the interval $[0, 2)$, then $y = 1$. For x in $[2, 4)$, $y = 2$, and so on. Some sample ordered pairs are given in the table.

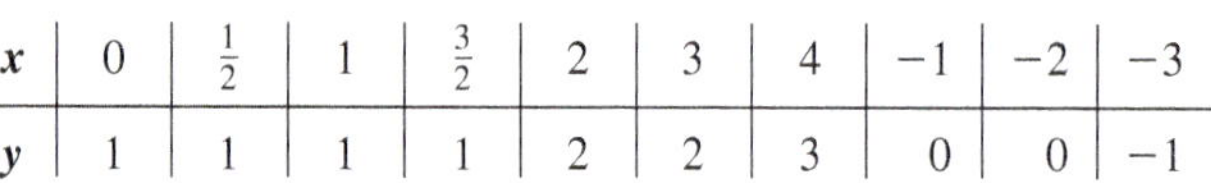

x	0	$\frac{1}{2}$	1	$\frac{3}{2}$	2	3	4	-1	-2	-3
y	1	1	1	1	2	2	3	0	0	-1

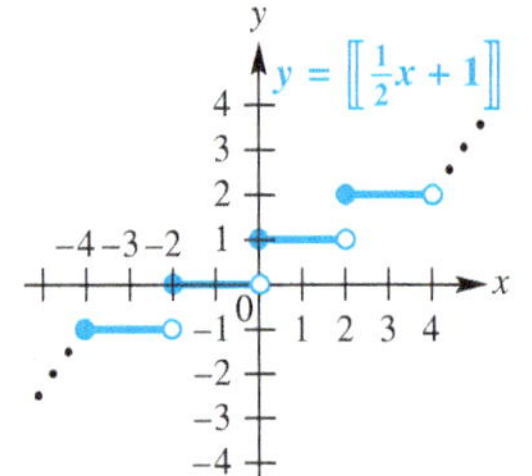

Figure 65

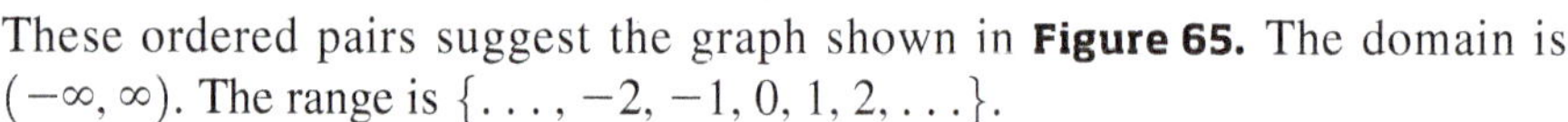

These ordered pairs suggest the graph shown in **Figure 65.** The domain is $(-\infty, \infty)$. The range is $\{\ldots, -2, -1, 0, 1, 2, \ldots\}$.

✔ Now Try Exercise 45.

The greatest integer function is an example of a **step function,** a function with a graph that looks like a series of steps.

EXAMPLE 4 Applying a Greatest Integer Function

An express mail company charges \$25 for a package weighing up to 2 lb. For each additional pound or fraction of a pound, there is an additional charge of \$3. Let $y = D(x)$ represent the cost to send a package weighing x pounds. Graph $y = D(x)$ for x in the interval $(0, 6]$.

SOLUTION For x in the interval $(0, 2]$, we obtain $y = 25$. For x in $(2, 3]$, $y = 25 + 3 = 28$. For x in $(3, 4]$, $y = 28 + 3 = 31$, and so on. The graph, which is that of a step function, is shown in **Figure 66.** In this case, the first step has a different length.

Figure 66

✔ Now Try Exercise 47.

The Relation $x = y^2$ ***Recall that a function is a relation where every domain value is paired with one and only one range value.*** Consider the relation defined by the equation $x = y^2$, which is not a function. Notice from the table of selected ordered pairs on the next page that this relation has two different y-values for each positive value of x.

If we plot the points from the table and join them with a smooth curve, we find that the graph of $x = y^2$ is a parabola opening to the right with vertex $(0, 0)$. See **Figure 67(a).** The domain is $[0, \infty)$ and the range is $(-\infty, \infty)$.

To use a calculator in function mode to graph the relation $x = y^2$, we graph the two functions $y_1 = \sqrt{x}$ (to generate the top half of the parabola) and $y_2 = -\sqrt{x}$ (to generate the bottom half). See **Figure 67(b).** ■

Selected Ordered Pairs for $x = y^2$

x	y
0	0
1	± 1
4	± 2
9	± 3

There are two different y-values for the same x-value.

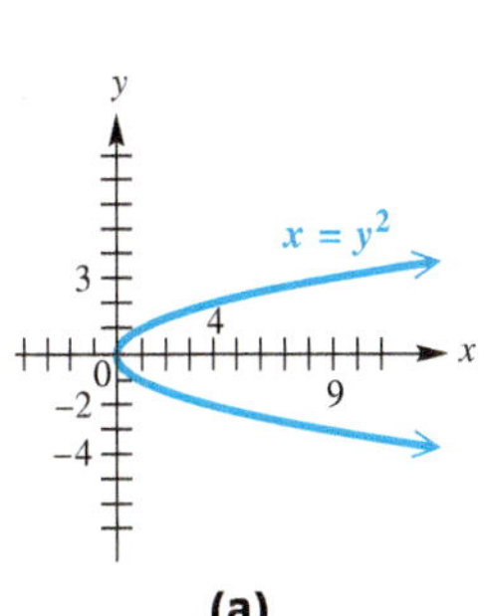

(a)

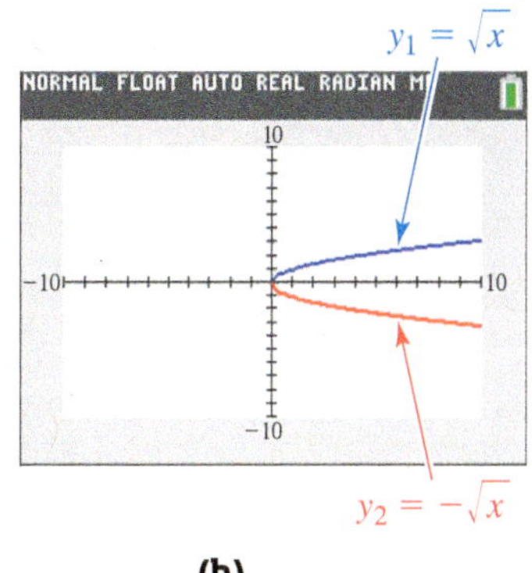

(b)

Figure 67

2.6 Exercises

CONCEPT PREVIEW *To answer each question, refer to the following basic graphs.*

A.

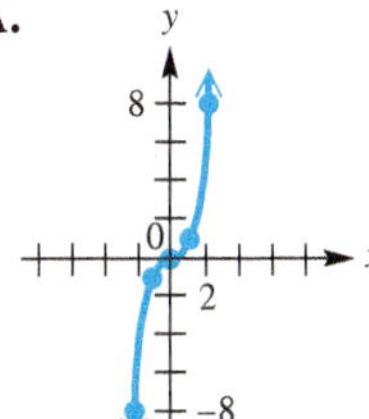

B.

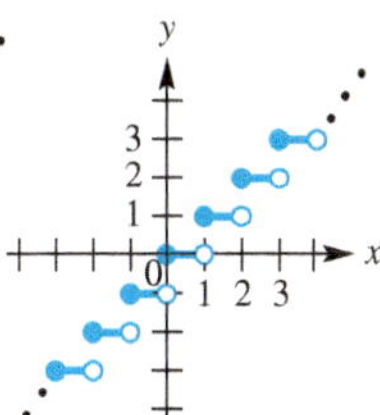

C.

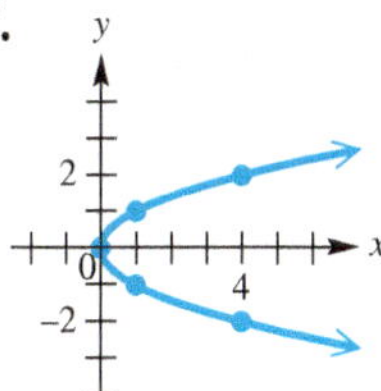

D.

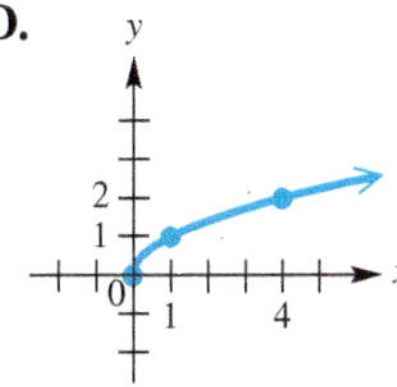

E.

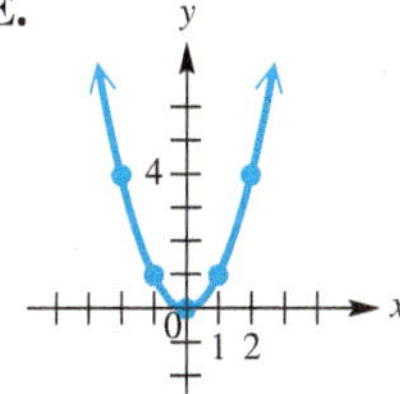

F.

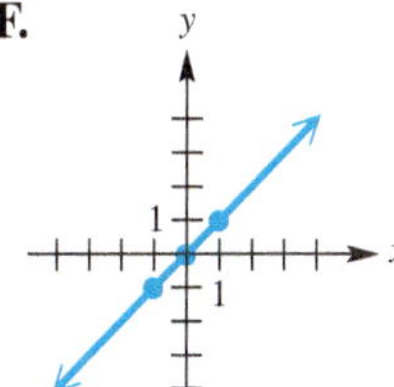

G.

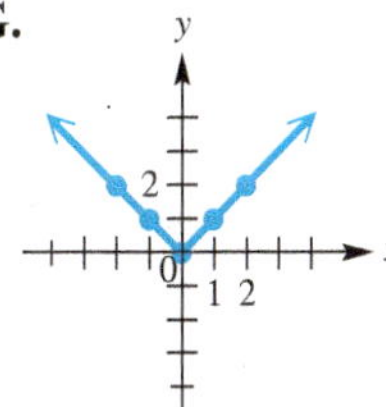

H.

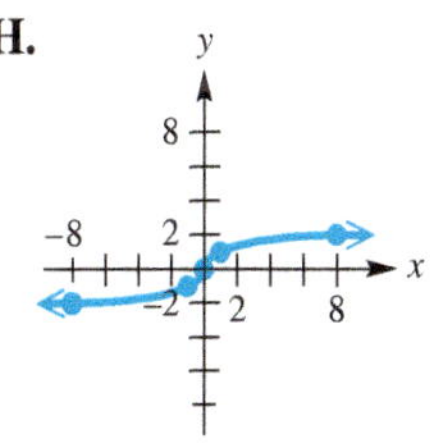

I.

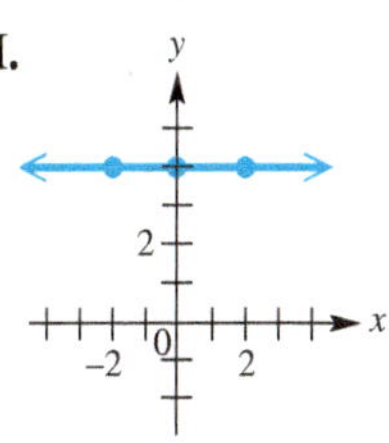

1. Which one is the graph of $f(x) = x^2$? What is its domain?
2. Which one is the graph of $f(x) = |x|$? On what open interval is it increasing?
3. Which one is the graph of $f(x) = x^3$? What is its range?
4. Which one is not the graph of a function? What is its equation?
5. Which one is the identity function? What is its equation?
6. Which one is the graph of $f(x) = [\![x]\!]$? What is the function value when $x = 1.5$?
7. Which one is the graph of $f(x) = \sqrt[3]{x}$? Is there any open interval over which the function is decreasing?
8. Which one is the graph of $f(x) = \sqrt{x}$? What is its domain?
9. Which one is discontinuous at many points? What is its range?
10. Which graphs of functions decrease over part of the domain and increase over the rest of the domain? On what open intervals do they increase? decrease?

Determine the intervals of the domain over which each function is continuous. ***See Example 1.***

11.

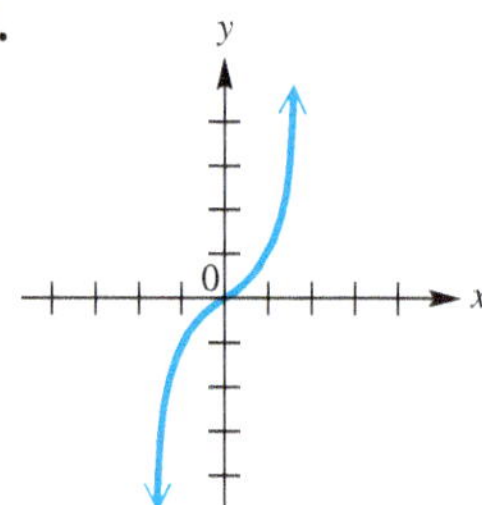

12.

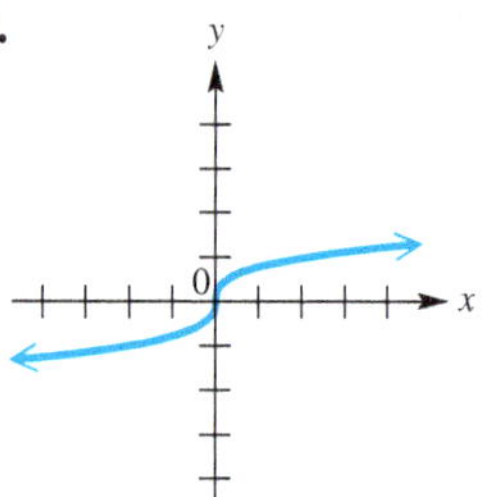

13.

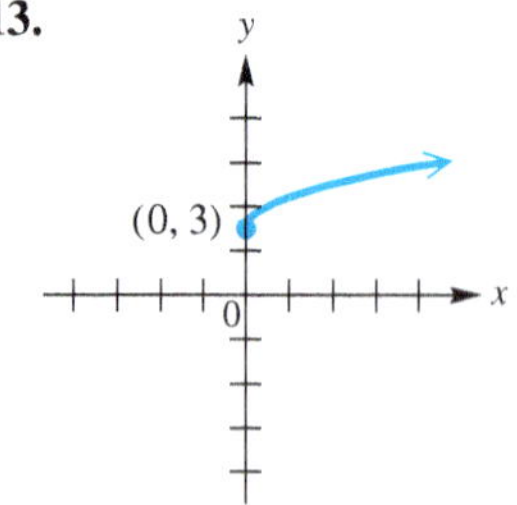

14.

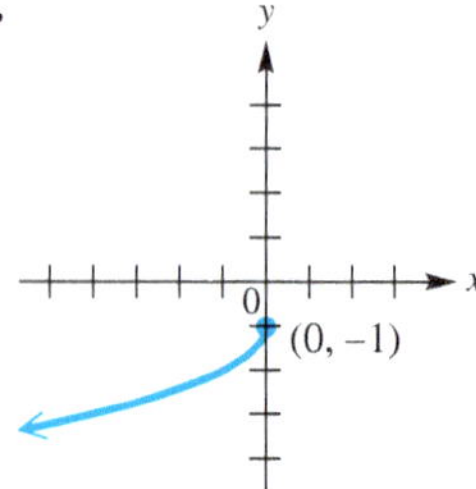

15.

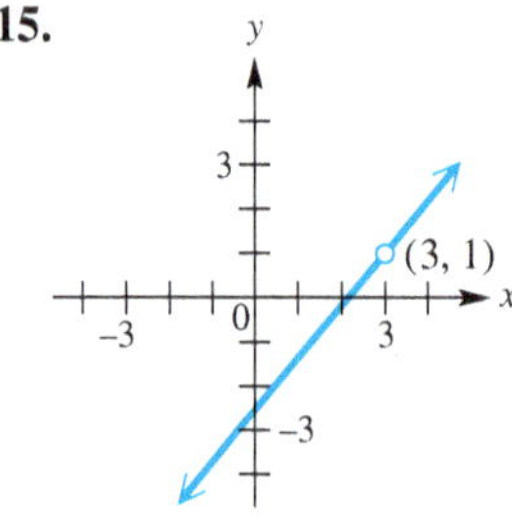

16.

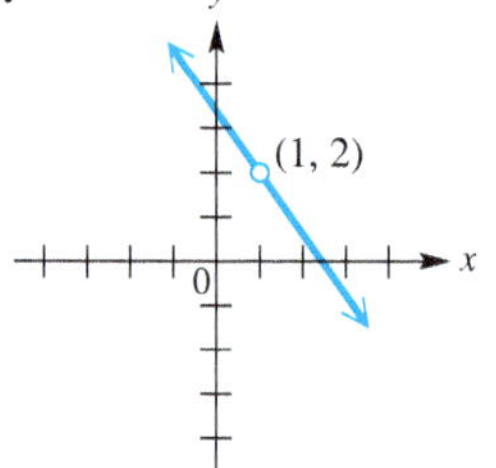

For each piecewise-defined function, find ***(a)*** $f(-5)$, ***(b)*** $f(-1)$, ***(c)*** $f(0)$, *and* ***(d)*** $f(3)$. ***See Example 2.***

17. $f(x) = \begin{cases} 2x & \text{if } x \le -1 \\ x - 1 & \text{if } x > -1 \end{cases}$

18. $f(x) = \begin{cases} x - 2 & \text{if } x < 3 \\ 5 - x & \text{if } x \ge 3 \end{cases}$

19. $f(x) = \begin{cases} 2 + x & \text{if } x < -4 \\ -x & \text{if } -4 \le x \le 2 \\ 3x & \text{if } x > 2 \end{cases}$

20. $f(x) = \begin{cases} -2x & \text{if } x < -3 \\ 3x - 1 & \text{if } -3 \le x \le 2 \\ -4x & \text{if } x > 2 \end{cases}$

Graph each piecewise-defined function. ***See Example 2.***

21. $f(x) = \begin{cases} x - 1 & \text{if } x \le 3 \\ 2 & \text{if } x > 3 \end{cases}$

22. $f(x) = \begin{cases} 6 - x & \text{if } x \le 3 \\ 3 & \text{if } x > 3 \end{cases}$

23. $f(x) = \begin{cases} 4 - x & \text{if } x < 2 \\ 1 + 2x & \text{if } x \ge 2 \end{cases}$

24. $f(x) = \begin{cases} 2x + 1 & \text{if } x \ge 0 \\ x & \text{if } x < 0 \end{cases}$

25. $f(x) = \begin{cases} -3 & \text{if } x \le 1 \\ -1 & \text{if } x > 1 \end{cases}$

26. $f(x) = \begin{cases} -2 & \text{if } x \le 1 \\ 2 & \text{if } x > 1 \end{cases}$

27. $f(x) = \begin{cases} 2 + x & \text{if } x < -4 \\ -x & \text{if } -4 \le x \le 5 \\ 3x & \text{if } x > 5 \end{cases}$

28. $f(x) = \begin{cases} -2x & \text{if } x < -3 \\ 3x - 1 & \text{if } -3 \le x \le 2 \\ -4x & \text{if } x > 2 \end{cases}$

29. $f(x) = \begin{cases} -\frac{1}{2}x^2 + 2 & \text{if } x \leq 2 \\ \frac{1}{2}x & \text{if } x > 2 \end{cases}$

30. $f(x) = \begin{cases} x^3 + 5 & \text{if } x \leq 0 \\ -x^2 & \text{if } x > 0 \end{cases}$

31. $f(x) = \begin{cases} 2x & \text{if } -5 \leq x < -1 \\ -2 & \text{if } -1 \leq x < 0 \\ x^2 - 2 & \text{if } \ 0 \leq x \leq 2 \end{cases}$

32. $f(x) = \begin{cases} 0.5x^2 & \text{if } -4 \leq x \leq -2 \\ x & \text{if } -2 < x < 2 \\ x^2 - 4 & \text{if } \ 2 \leq x \leq 4 \end{cases}$

33. $f(x) = \begin{cases} x^3 + 3 & \text{if } -2 \leq x \leq 0 \\ x + 3 & \text{if } \ 0 < x < 1 \\ 4 + x - x^2 & \text{if } \ 1 \leq x \leq 3 \end{cases}$

34. $f(x) = \begin{cases} -2x & \text{if } -3 \leq x < -1 \\ x^2 + 1 & \text{if } -1 \leq x \leq 2 \\ \frac{1}{2}x^3 + 1 & \text{if } \ 2 < x \leq 3 \end{cases}$

Connecting Graphs with Equations *Give a rule for each piecewise-defined function. Also give the domain and range.*

35.

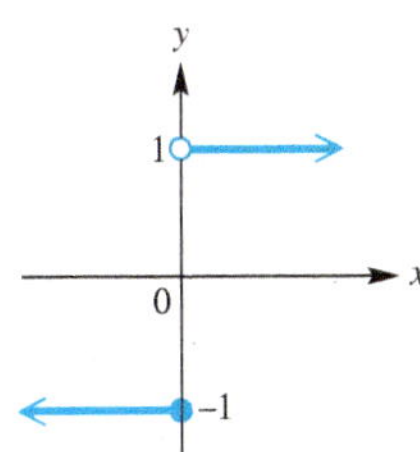

36.

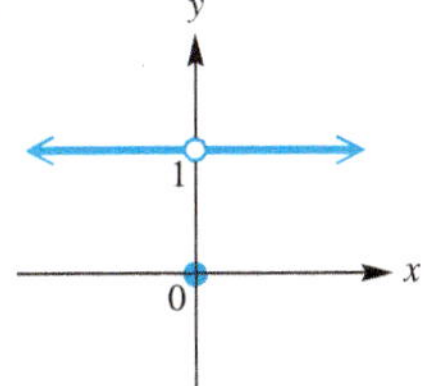

37.

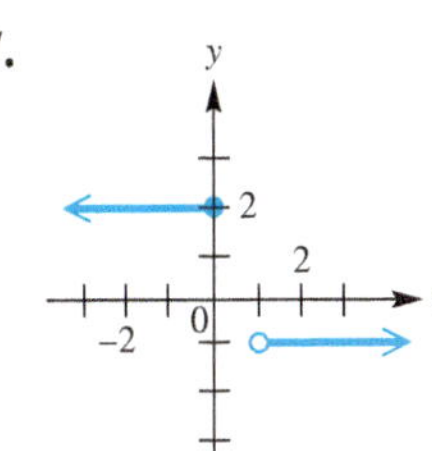

38.

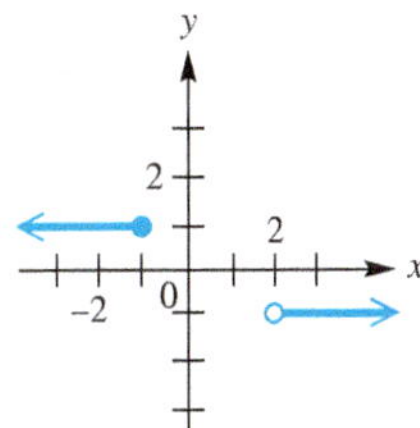

39.

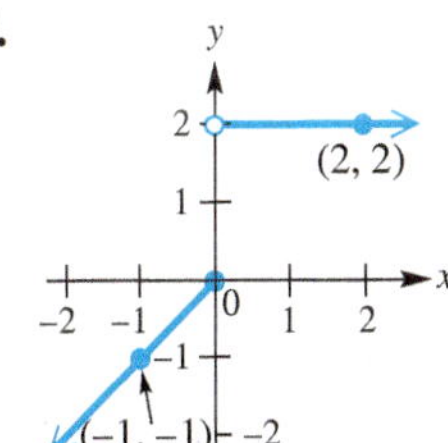

40.

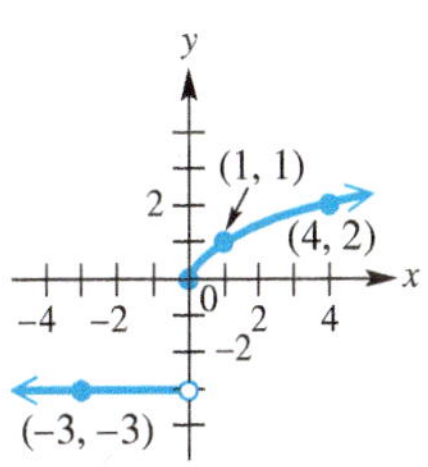

41.

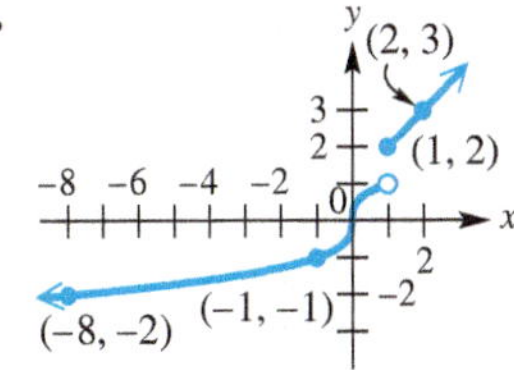

42.

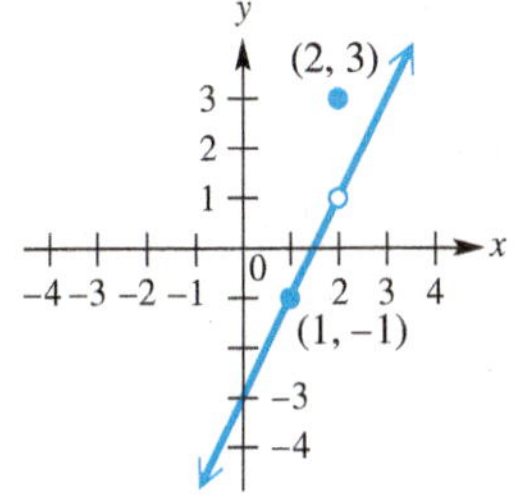

Graph each function. Give the domain and range. **See Example 3.**

43. $f(x) = [\![-x]\!]$

44. $f(x) = -[\![x]\!]$

45. $f(x) = [\![2x]\!]$

46. $g(x) = [\![2x - 1]\!]$

(Modeling) Solve each problem. ***See Example 4.***

47. *Postage Charges* Assume that postage rates are \$0.49 for the first ounce, plus \$0.21 for each additional ounce, and that each letter carries one \$0.49 stamp and as many \$0.21 stamps as necessary. Graph the function f that models the number of stamps on a letter weighing x ounces over the interval $(0, 5]$.

48. *Parking Charges* The cost of parking a car at an hourly parking lot is \$3 for the first half-hour and \$2 for each additional half-hour or fraction of a half-hour. Graph the function f that models the cost of parking a car for x hours over the interval $(0, 2]$.

49. *Water in a Tank* Sketch a graph that depicts the amount of water in a 100-gal tank. The tank is initially empty and then filled at a rate of 5 gal per minute. Immediately after it is full, a pump is used to empty the tank at 2 gal per minute.

50. *Distance from Home* Sketch a graph showing the distance a person is from home after x hours if he or she drives on a straight road at 40 mph to a park 20 mi away, remains at the park for 2 hr, and then returns home at a speed of 20 mph.

51. *New Truck Market Share* The new vehicle market share (in percent) in the United States for trucks is shown in the graph. Let $x = 0$ represent 2000, $x = 8$ represent 2008, and so on.

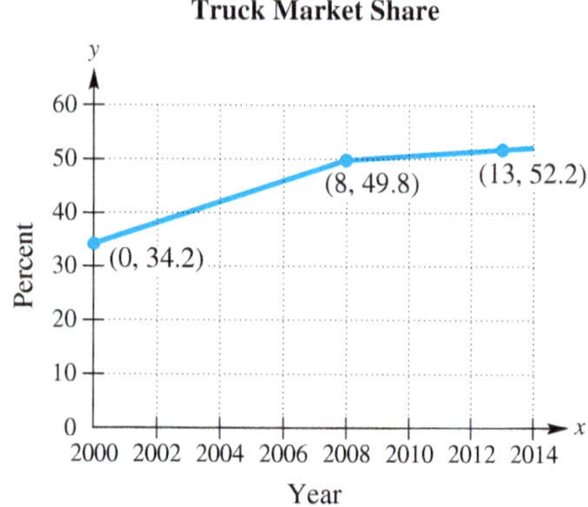

Source: U.S. Bureau of Economic Analysis.

(a) Use the points on the graph to write equations for the graphs in the intervals $[0, 8]$ and $(8, 13]$.

(b) Define this graph as a piecewise-defined function f.

52. *Flow Rates* A water tank has an inlet pipe with a flow rate of 5 gal per minute and an outlet pipe with a flow rate of 3 gal per minute. A pipe can be either closed or completely open. The graph shows the number of gallons of water in the tank after x minutes. Use the concept of slope to interpret each piece of this graph.

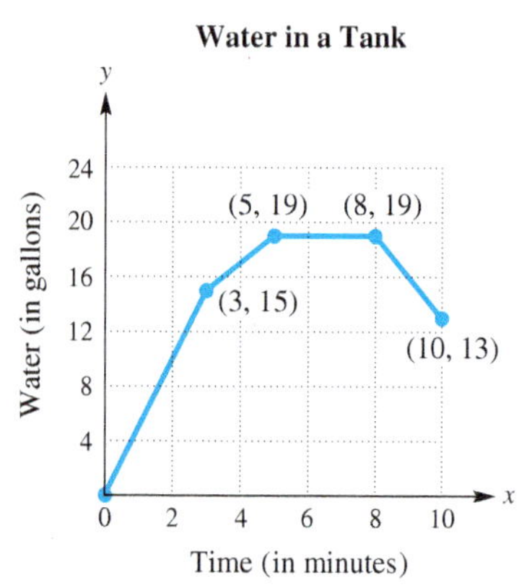

53. ***Swimming Pool Levels*** The graph of $y = f(x)$ represents the amount of water in thousands of gallons remaining in a swimming pool after x days.

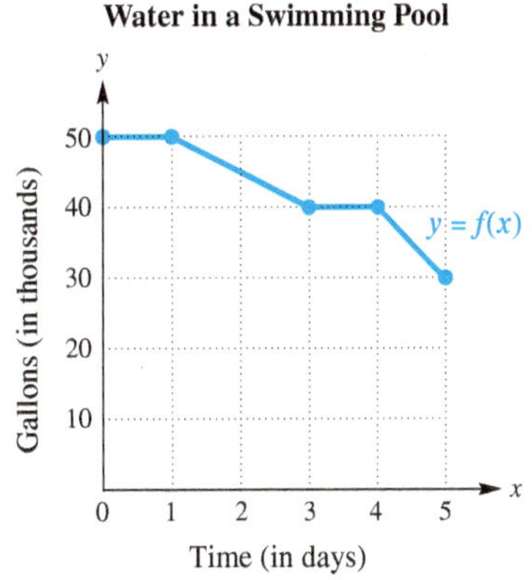

(a) Estimate the initial and final amounts of water contained in the pool.

(b) When did the amount of water in the pool remain constant?

(c) Approximate $f(2)$ and $f(4)$.

(d) At what rate was water being drained from the pool when $1 \le x \le 3$?

54. ***Gasoline Usage*** The graph shows the gallons of gasoline y in the gas tank of a car after x hours.

(a) Estimate how much gasoline was in the gas tank when $x = 3$.

(b) When did the car burn gasoline at the greatest rate?

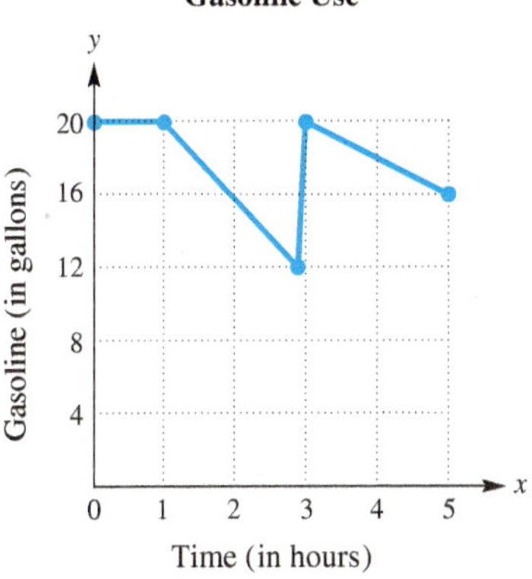

55. ***Lumber Costs*** Lumber that is used to frame walls of houses is frequently sold in multiples of 2 ft. If the length of a board is not exactly a multiple of 2 ft, there is often no charge for the additional length. For example, if a board measures at least 8 ft, but less than 10 ft, then the consumer is charged for only 8 ft.

(a) Suppose that the cost of lumber is \$0.80 every 2 ft. Find a formula for a function f that computes the cost of a board x feet long for $6 \le x \le 18$.

(b) Determine the costs of boards with lengths of 8.5 ft and 15.2 ft.

56. ***Snow Depth*** The snow depth in a particular location varies throughout the winter. In a typical winter, the snow depth in inches might be approximated by the following function.

$$f(x) = \begin{cases} 6.5x & \text{if } 0 \le x \le 4 \\ -5.5x + 48 & \text{if } 4 < x \le 6 \\ -30x + 195 & \text{if } 6 < x \le 6.5 \end{cases}$$

Here, x represents the time in months with $x = 0$ representing the beginning of October, $x = 1$ representing the beginning of November, and so on.

(a) Graph $y = f(x)$.

(b) In what month is the snow deepest? What is the deepest snow depth?

(c) In what months does the snow begin and end?

2.7 Graphing Techniques

- Stretching and Shrinking
- Reflecting
- Symmetry
- Even and Odd Functions
- Translations

Graphing techniques presented in this section show how to graph functions that are defined by altering the equation of a basic function.

Stretching and Shrinking

We begin by considering how the graphs of

$$y = af(x) \quad \text{and} \quad y = f(ax)$$

compare to the graph of $y = f(x)$, where $a > 0$.

EXAMPLE 1 Stretching or Shrinking Graphs

Graph each function.

(a) $g(x) = 2|x|$ **(b)** $h(x) = \frac{1}{2}|x|$ **(c)** $k(x) = |2x|$

SOLUTION

(a) Comparing the tables of values for $f(x) = |x|$ and $g(x) = 2|x|$ in **Figure 68,** we see that for corresponding x-values, the y-values of g are each twice those of f. The graph of $f(x) = |x|$ is *vertically stretched.* The graph of $g(x)$, shown in blue in **Figure 68,** is narrower than that of $f(x)$, shown in red for comparison.

x	$f(x) = \|x\|$	$g(x) = 2\|x\|$
−2	2	4
−1	1	2
0	0	0
1	1	2
2	2	4

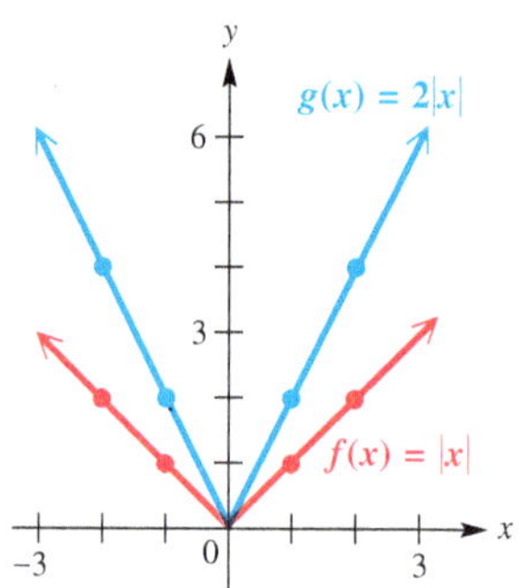

Figure 68

(b) The graph of $h(x) = \frac{1}{2}|x|$ is also the same general shape as that of $f(x)$, but here the coefficient $\frac{1}{2}$ is between 0 and 1 and causes a *vertical shrink.* The graph of $h(x)$ is wider than the graph of $f(x)$, as we see by comparing the tables of values. See **Figure 69.**

x	$f(x) = \|x\|$	$h(x) = \frac{1}{2}\|x\|$
−2	2	1
−1	1	$\frac{1}{2}$
0	0	0
1	1	$\frac{1}{2}$
2	2	1

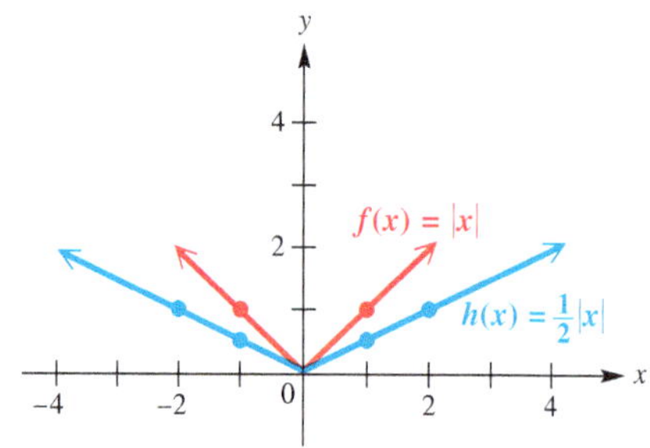

Figure 69

(c) Use the property of absolute value that states $|ab| = |a| \cdot |b|$ to rewrite $|2x|$.

$$k(x) = |2x| = |2| \cdot |x| = 2|x| \quad \text{Property 3}$$

Therefore, the graph of $k(x) = |2x|$ is the same as the graph of $g(x) = 2|x|$ in part (a). This is a *horizontal shrink* of the graph of $f(x) = |x|$. See **Figure 68** on the previous page.

✔ **Now Try Exercises 17 and 19.**

Vertical Stretching or Shrinking of the Graph of a Function

Suppose that $a > 0$. If a point (x, y) lies on the graph of $y = f(x)$, then the point (x, ay) lies on the graph of $\boldsymbol{y = af(x)}$.

(a) If $a > 1$, then the graph of $y = af(x)$ is a **vertical stretching** of the graph of $y = f(x)$.

(b) If $0 < a < 1$, then the graph of $y = af(x)$ is a **vertical shrinking** of the graph of $y = f(x)$.

Figure 70 shows graphical interpretations of vertical stretching and shrinking. ***In both cases, the x-intercepts of the graph remain the same but the y-intercepts are affected.***

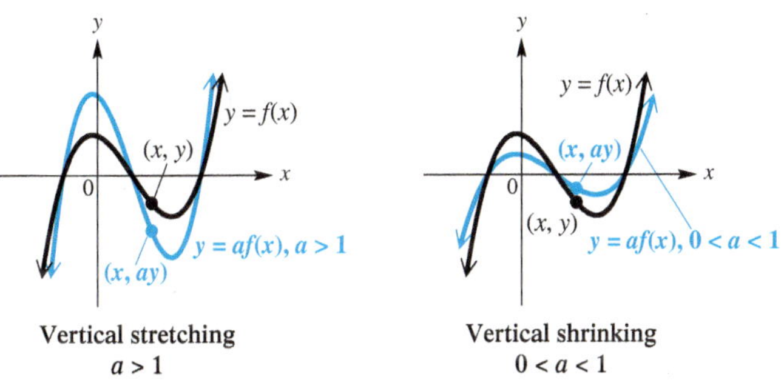

Figure 70

Graphs of functions can also be stretched and shrunk horizontally.

y = f(x)

(x/a, y)

(x, y)

y = f(ax), 0 < a < 1

Horizontal stretching
0 < a < 1

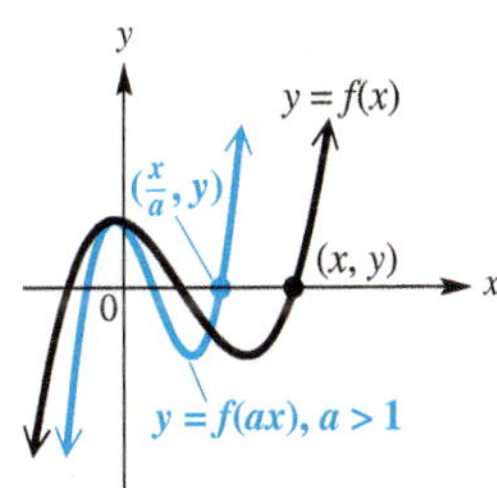

Horizontal shrinking
a > 1

Figure 71

Horizontal Stretching or Shrinking of the Graph of a Function

Suppose that $a > 0$. If a point (x, y) lies on the graph of $y = f(x)$, then the point $\left(\frac{x}{a}, y\right)$ lies on the graph of $\boldsymbol{y = f(ax)}$.

(a) If $0 < a < 1$, then the graph of $y = f(ax)$ is a **horizontal stretching** of the graph of $y = f(x)$.

(b) If $a > 1$, then the graph of $y = f(ax)$ is a **horizontal shrinking** of the graph of $y = f(x)$.

See **Figure 71** for graphical interpretations of horizontal stretching and shrinking. ***In both cases, the y-intercept remains the same but the x-intercepts are affected.***

Reflecting Forming the mirror image of a graph across a line is called ***reflecting the graph across the line.***

EXAMPLE 2 Reflecting Graphs across Axes

Graph each function.

(a) $g(x) = -\sqrt{x}$ **(b)** $h(x) = \sqrt{-x}$

SOLUTION

(a) The tables of values for $g(x) = -\sqrt{x}$ and $f(x) = \sqrt{x}$ are shown with their graphs in **Figure 72.** As the tables suggest, every y-value of the graph of $g(x) = -\sqrt{x}$ is the negative of the corresponding y-value of $f(x) = \sqrt{x}$. This has the effect of reflecting the graph across the x-axis.

x	$f(x) = \sqrt{x}$	$g(x) = -\sqrt{x}$
0	0	0
1	1	-1
4	2	-2

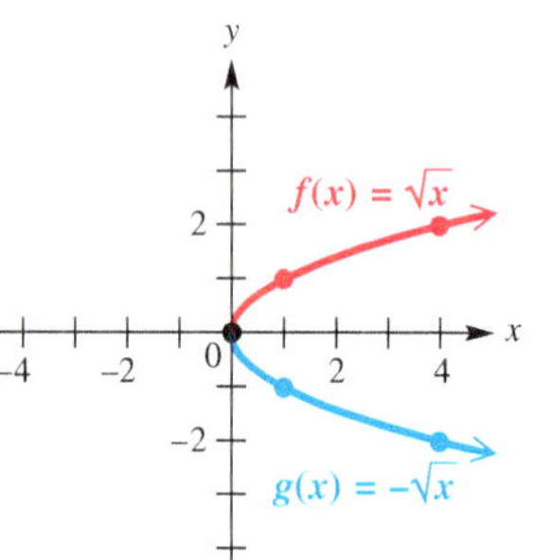

Figure 72

(b) The domain of $h(x) = \sqrt{-x}$ is $(-\infty, 0]$, while the domain of $f(x) = \sqrt{x}$ is $[0, \infty)$. Choosing x-values for $h(x)$ that are negatives of those used for $f(x)$, we see that corresponding y-values are the same. The graph of h is a reflection of the graph of f across the y-axis. See **Figure 73.**

x	$f(x) = \sqrt{x}$	$h(x) = \sqrt{-x}$
-4	undefined	2
-1	undefined	1
0	0	0
1	1	undefined
4	2	undefined

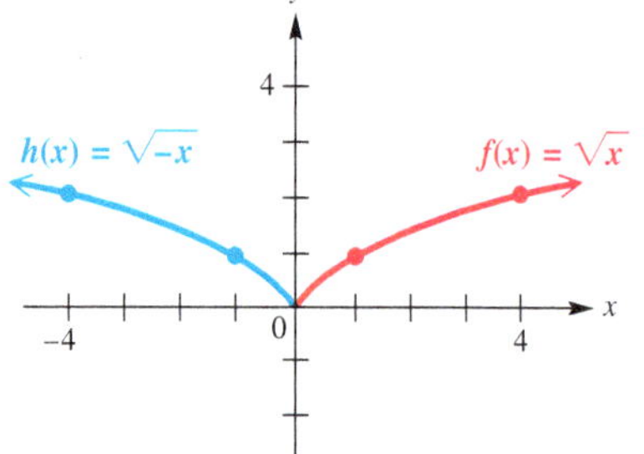

Figure 73

✔ **Now Try Exercises 27 and 33.**

The graphs in **Example 2** suggest the following generalizations.

Reflecting across an Axis

The graph of $y = -f(x)$ is the same as the graph of $y = f(x)$ reflected across the x-axis. (If a point (x, y) lies on the graph of $y = f(x)$, then $(x, -y)$ lies on this reflection.)

The graph of $y = f(-x)$ is the same as the graph of $y = f(x)$ reflected across the y-axis. (If a point (x, y) lies on the graph of $y = f(x)$, then $(-x, y)$ lies on this reflection.)

Symmetry The graph of f shown in **Figure 74(a)** is cut in half by the y-axis, with each half the mirror image of the other half. Such a graph is *symmetric with respect to the y-axis.* ***In general, for a graph to be symmetric with respect to the y-axis, the point $(-x, y)$ must be on the graph whenever the point (x, y) is on the graph.***

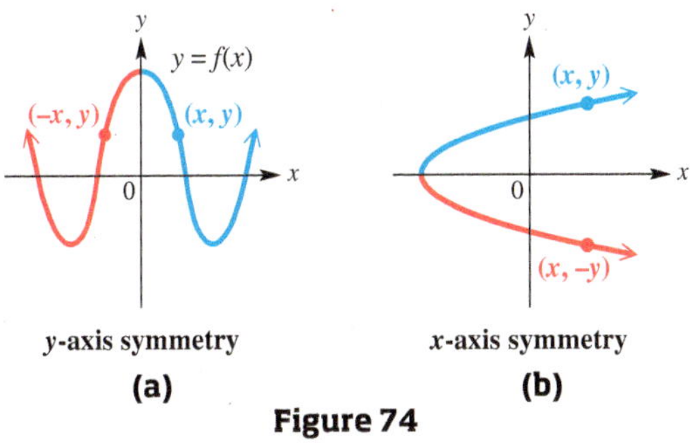

y-axis symmetry
(a)

x-axis symmetry
(b)

Figure 74

Similarly, if the graph in **Figure 74(b)** were folded in half along the x-axis, the portion at the top would exactly match the portion at the bottom. Such a graph is *symmetric with respect to the x-axis.* ***In general, for a graph to be symmetric with respect to the x-axis, the point $(x, -y)$ must be on the graph whenever the point (x, y) is on the graph.***

Symmetry with Respect to an Axis

The graph of an equation is **symmetric with respect to the y-axis** if the replacement of x with $-x$ results in an equivalent equation.

The graph of an equation is **symmetric with respect to the x-axis** if the replacement of y with $-y$ results in an equivalent equation.

Examples: Of the basic functions in the previous section, graphs of the squaring and absolute value functions are symmetric with respect to the y-axis.

EXAMPLE 3 Testing for Symmetry with Respect to an Axis

Test for symmetry with respect to the x-axis and the y-axis.

(a) $y = x^2 + 4$ **(b)** $x = y^2 - 3$ **(c)** $x^2 + y^2 = 16$ **(d)** $2x + y = 4$

SOLUTION

(a) In $y = x^2 + 4$, replace x with $-x$.

Use parentheses around $-x$.

$$y = x^2 + 4$$
$$y = (-x)^2 + 4$$
$$y = x^2 + 4$$

The result is equivalent to the original equation.

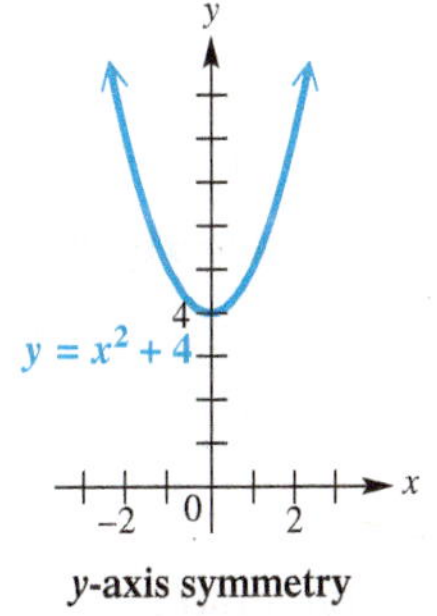

y-axis symmetry

Figure 75

Thus the graph, shown in **Figure 75,** is symmetric with respect to the y-axis. The y-axis cuts the graph in half, with the halves being mirror images.

Now replace y with $-y$ to test for symmetry with respect to the x-axis.

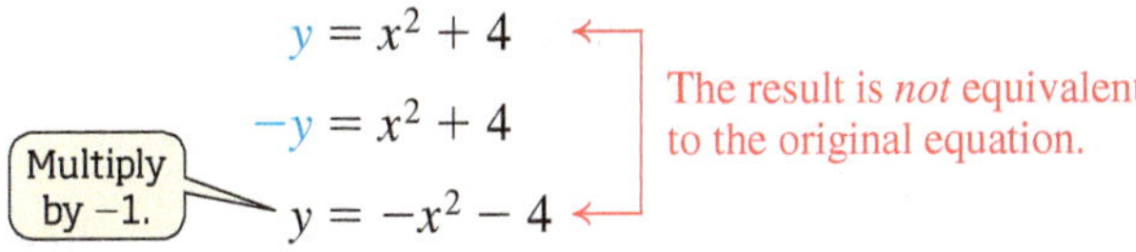

The graph is *not* symmetric with respect to the x-axis. See **Figure 75.**

LOOKING AHEAD TO CALCULUS

The tools of calculus enable us to find areas of regions in the plane. To find the area of the region below the graph of $y = x^2$, above the x-axis, bounded on the left by the line $x = -2$ and on the right by $x = 2$, draw a sketch of this region. Because of the symmetry of the graph of $y = x^2$, the desired area is twice the area to the right of the y-axis. Thus, symmetry can be used to reduce the original problem to an easier one by simply finding the area to the right of the y-axis and then doubling the answer.

(b) In $x = y^2 - 3$, replace y with $-y$.

$$x = (-y)^2 - 3 = y^2 - 3 \quad \text{Same as the original equation}$$

The graph is symmetric with respect to the x-axis, as shown in **Figure 76.** It is *not* symmetric with respect to the y-axis.

(c) Substitute $-x$ for x and then $-y$ for y in $x^2 + y^2 = 16$.

$$(-x)^2 + y^2 = 16 \quad \text{and} \quad x^2 + (-y)^2 = 16$$

Both simplify to the original equation,

$$x^2 + y^2 = 16.$$

The graph, a circle of radius 4 centered at the origin, is symmetric with respect to *both* axes. See **Figure 77.**

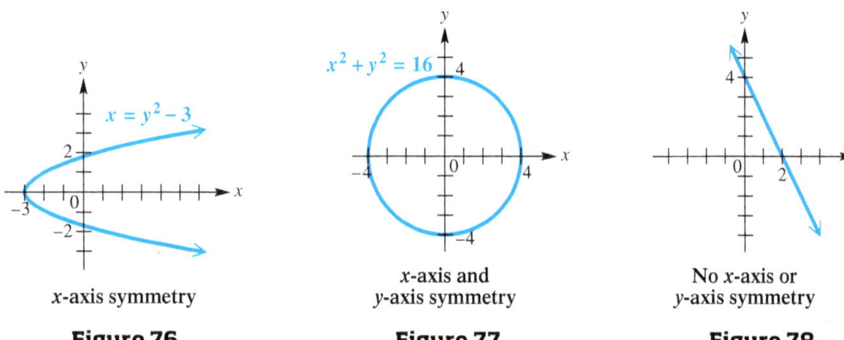

x-axis symmetry — **Figure 76**

x-axis and *y*-axis symmetry — **Figure 77**

No *x*-axis or *y*-axis symmetry — **Figure 78**

(d) $2x + y = 4$

Replace x with $-x$ and then replace y with $-y$.

$2x + y = 4$			$2x + y = 4$	
$2(-x) + y = 4$	Not equivalent		$2x + (-y) = 4$	Not equivalent
$-2x + y = 4$			$2x - y = 4$	

See **Figure 78.**

Now Try Exercise 45.

Another kind of symmetry occurs when a graph can be rotated 180° about the origin, with the result coinciding exactly with the original graph. Symmetry of this type is *symmetry with respect to the origin. **In general, for a graph to be symmetric with respect to the origin, the point $(-x, -y)$ is on the graph whenever the point (x, y) is on the graph.***

Figure 79 shows two such graphs.

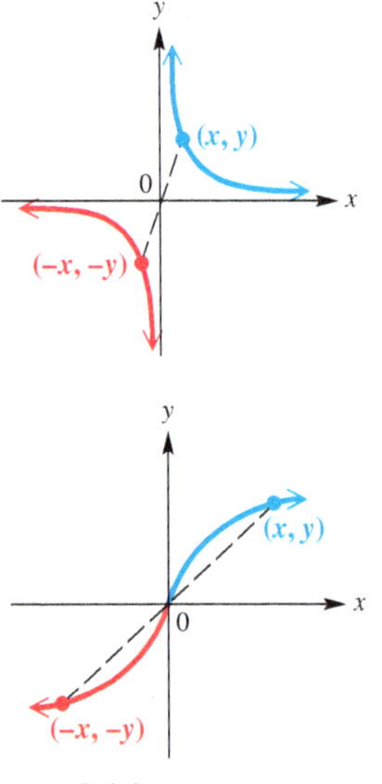

Origin symmetry

Figure 79

Symmetry with Respect to the Origin

The graph of an equation is **symmetric with respect to the origin** if the replacement of both x with $-x$ and y with $-y$ at the same time results in an equivalent equation.

Examples: Of the basic functions in the previous section, graphs of the cubing and cube root functions are symmetric with respect to the origin.

EXAMPLE 4 Testing for Symmetry with Respect to the Origin

Determine whether the graph of each equation is symmetric with respect to the origin.

(a) $x^2 + y^2 = 16$ **(b)** $y = x^3$

SOLUTION

(a) Replace x with $-x$ and y with $-y$.

$$x^2 + y^2 = 16$$
$$(-x)^2 + (-y)^2 = 16 \quad \text{Equivalent}$$
$$x^2 + y^2 = 16$$

Use parentheses around $-x$ and $-y$.

The graph, which is the circle shown in **Figure 77** in **Example 3(c),** is symmetric with respect to the origin.

(b) Replace x with $-x$ and y with $-y$.

$$y = x^3$$
$$-y = (-x)^3 \quad \text{Equivalent}$$
$$-y = -x^3$$
$$y = x^3$$

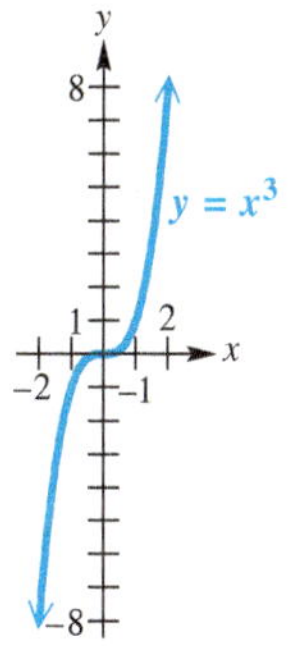

Origin symmetry

Figure 80

The graph, which is that of the cubing function, is symmetric with respect to the origin and is shown in **Figure 80.**

Now Try Exercise 49.

Notice the following important concepts regarding symmetry:

- A graph symmetric with respect to both the x- and y-axes is automatically symmetric with respect to the origin. (See **Figure 77.**)
- A graph symmetric with respect to the origin need *not* be symmetric with respect to either axis. (See **Figure 80.**)
- Of the three types of symmetry—with respect to the x-axis, with respect to the y-axis, and with respect to the origin—a graph possessing any two types must also exhibit the third type of symmetry.
- A graph symmetric with respect to the x-axis does not represent a function. (See **Figures 76 and 77.**)

Summary of Tests for Symmetry

	Symmetry with Respect to:		
	x-axis	**y-axis**	**Origin**
Equation is unchanged if:	y is replaced with $-y$	x is replaced with $-x$	x is replaced with $-x$ and y is replaced with $-y$
Example:	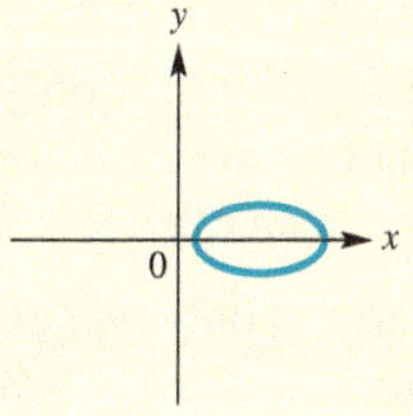	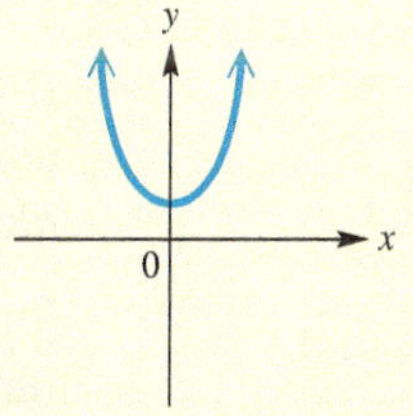	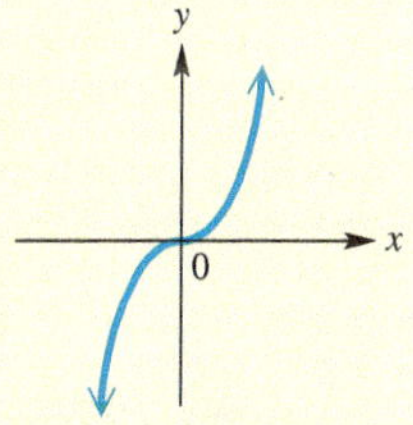

Even and Odd Functions The concepts of symmetry with respect to the y-axis and symmetry with respect to the origin are closely associated with the concepts of *even* and *odd functions.*

Even and Odd Functions

A function f is an **even function** if $f(-x) = f(x)$ for all x in the domain of f. (Its graph is symmetric with respect to the y-axis.)

A function f is an **odd function** if $f(-x) = -f(x)$ for all x in the domain of f. (Its graph is symmetric with respect to the origin.)

EXAMPLE 5 Determining Whether Functions Are Even, Odd, or Neither

Determine whether each function defined is *even, odd,* or *neither.*

(a) $f(x) = 8x^4 - 3x^2 + 1$ **(b)** $f(x) = 6x^3 - 9x$ **(c)** $f(x) = 3x^2 + 5x$

SOLUTION

(a) Replacing x with $-x$ gives the following.

$$f(x) = 8x^4 - 3x^2 + 1$$

$$f(-x) = 8(-x)^4 - 3(-x)^2 + 1 \quad \text{Replace } x \text{ with } -x.$$

$$f(-x) = 8x^4 - 3x^2 + 1 \quad \text{Apply the exponents.}$$

$$f(-x) = f(x) \quad 8x^4 - 3x^2 + 1 = f(x)$$

Because $f(-x) = f(x)$ for each x in the domain of the function, f is even.

(b) $f(x) = 6x^3 - 9x$

$$f(-x) = 6(-x)^3 - 9(-x) \quad \text{Replace } x \text{ with } -x.$$

$$f(-x) = -6x^3 + 9x \quad \text{Be careful with signs.}$$

$$f(-x) = -f(x) \quad -6x^3 + 9x = -(6x^3 - 9x) = -f(x)$$

The function f is odd because $f(-x) = -f(x)$.

(c) $f(x) = 3x^2 + 5x$

$$f(-x) = 3(-x)^2 + 5(-x) \quad \text{Replace } x \text{ with } -x.$$

$$f(-x) = 3x^2 - 5x \quad \text{Simplify.}$$

Because $f(-x) \neq f(x)$ and $f(-x) \neq -f(x)$, the function f is neither even nor odd.

✔ **Now Try Exercises 53, 55, and 57.**

NOTE Consider a function defined by a polynomial in x.

- If the function has only *even* exponents on x (including the case of a constant where x^0 is understood to have the even exponent 0), it will *always* be an even function.
- Similarly, if only *odd* exponents appear on x, the function will be an odd function.

Translations The next examples show the results of horizontal and vertical shifts, or **translations,** of the graph of $f(x) = |x|$.

EXAMPLE 6 Translating a Graph Vertically

Graph $g(x) = |x| - 4$.

SOLUTION Comparing the table shown with **Figure 81,** we see that for corresponding x-values, the y-values of g are each 4 *less* than those for f. The graph of $g(x) = |x| - 4$ is the same as that of $f(x) = |x|$, but translated 4 units down. The lowest point is at $(0, -4)$. The graph is symmetric with respect to the y-axis and is therefore the graph of an even function.

x	$f(x) = \|x\|$	$g(x) = \|x\| - 4$
−4	4	0
−1	1	−3
0	0	−4
1	1	−3
4	4	0

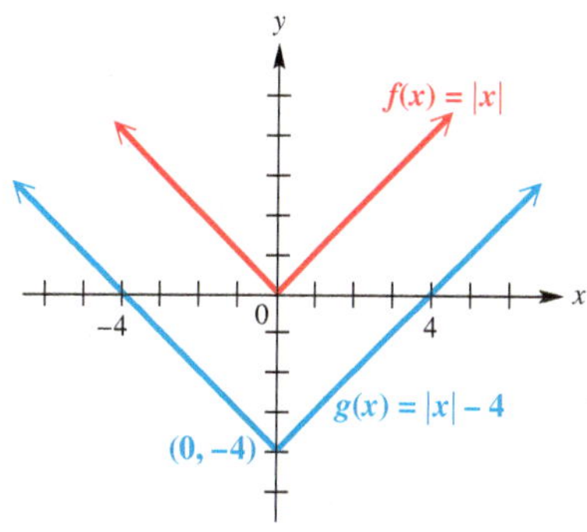

Figure 81

Now Try Exercise 67.

The graphs in **Example 6** suggest the following generalization.

Vertical Translations

Given a function g defined by $\mathbf{g(x) = f(x) + c}$, where c is a real number:

- For every point (x, y) on the graph of f, there will be a corresponding point $(x, y + c)$ on the graph of g.
- The graph of g will be the same as the graph of f, but translated c units up if c is positive or $|c|$ units down if c is negative.

The graph of g is a **vertical translation** of the graph of f.

Figure 82 shows a graph of a function f and two vertical translations of f. **Figure 83** shows two vertical translations of $y_1 = x^2$ on a TI-84 Plus calculator screen.

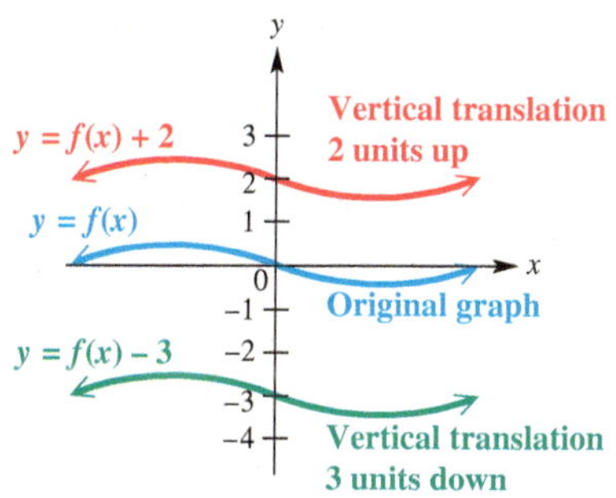

Figure 82

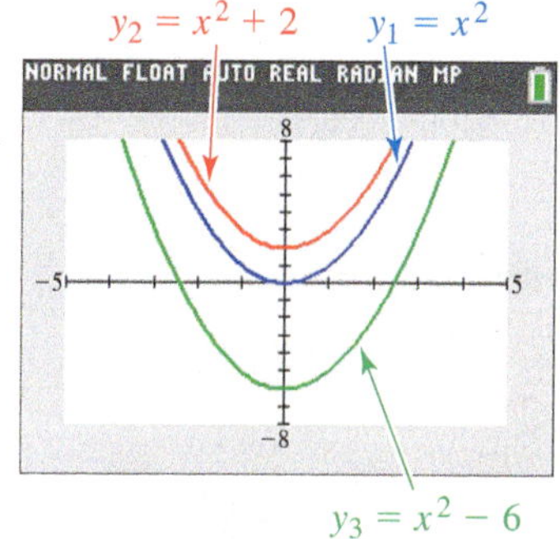

y_2 is the graph of $y_1 = x^2$ translated 2 units *up*. y_3 is that of y_1 translated 6 units *down*.

Figure 83

EXAMPLE 7 Translating a Graph Horizontally

Graph $g(x) = |x - 4|$.

SOLUTION Comparing the tables of values given with **Figure 84** shows that for corresponding y-values, the x-values of g are each 4 *more* than those for f. The graph of $g(x) = |x - 4|$ is the same as that of $f(x) = |x|$, but translated 4 units to the right. The lowest point is at $(4, 0)$. As suggested by the graphs in **Figure 84,** this graph is symmetric with respect to the line $x = 4$.

x	$f(x) = \|x\|$	$g(x) = \|x - 4\|$
−2	2	6
0	0	4
2	2	2
4	4	0
6	6	2

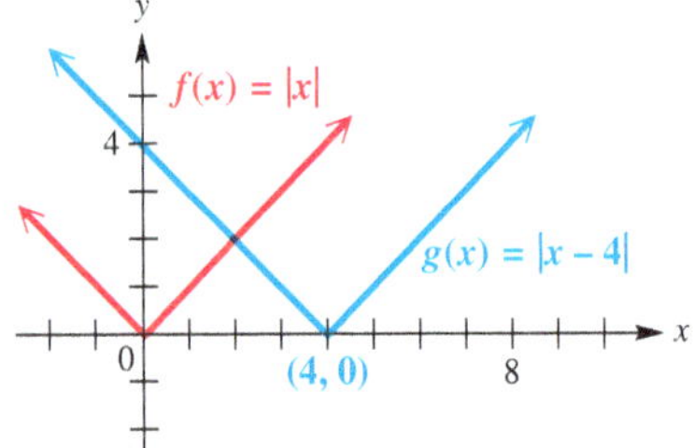

Figure 84

✔ **Now Try Exercise 65.**

The graphs in **Example 7** suggest the following generalization.

Horizontal Translations

Given a function g defined by $\mathbf{g(x) = f(x - c)}$, where c is a real number:

- For every point (x, y) on the graph of f, there will be a corresponding point $(x + c, y)$ on the graph of g.
- The graph of g will be the same as the graph of f, but translated c units to the right if c is positive or $|c|$ units to the left if c is negative.

The graph of g is a **horizontal translation** of the graph of f.

Figure 85 shows a graph of a function f and two horizontal translations of f. **Figure 86** shows two horizontal translations of $y_1 = x^2$ on a TI-84 Plus calculator screen.

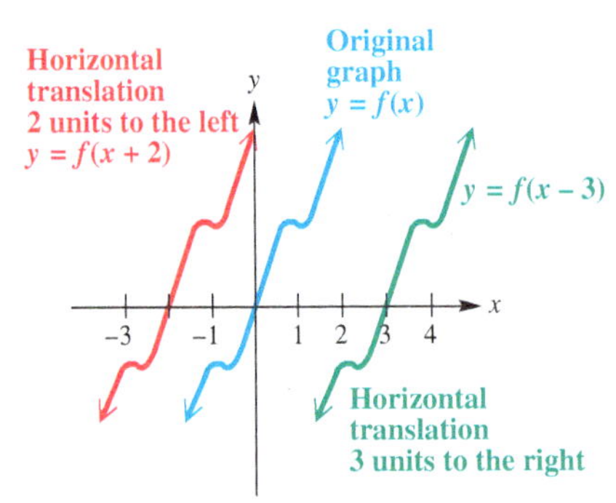

Figure 85

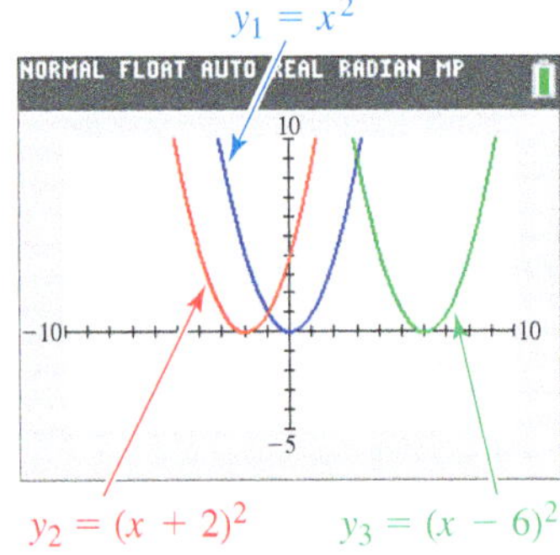

y_2 is the graph of $y_1 = x^2$ translated 2 units to the *left*. y_3 is that of y_1 translated 6 units to the *right*.

Figure 86

Summary of Translations

$(c > 0)$ To Graph:	Shift the Graph of $y = f(x)$ by c Units:
$y = f(x) + c$	up
$y = f(x) - c$	down
$y = f(x + c)$	left
$y = f(x - c)$	right

Vertical and horizontal translations are summarized in the table, where f is a function and c is a positive number.

CAUTION ***Errors frequently occur when horizontal shifts are involved.*** To determine the direction and magnitude of a horizontal shift, find the value that causes the expression $x - h$ to equal 0, as shown below.

$F(x) = (x - 5)^2$	$F(x) = (x + 5)^2$
Because $+5$ causes $x - 5$ to equal 0, the graph of $F(x)$ illustrates a shift of	Because -5 causes $x + 5$ to equal 0, the graph of $F(x)$ illustrates a shift of
5 units to the right.	**5 units to the left.**

EXAMPLE 8 Using More Than One Transformation

Graph each function.

(a) $f(x) = -|x + 3| + 1$ **(b)** $h(x) = |2x - 4|$ **(c)** $g(x) = -\frac{1}{2}x^2 + 4$

SOLUTION

(a) To graph $f(x) = -|x + 3| + 1$, the *lowest* point on the graph of $y = |x|$ is translated 3 units to the left and 1 unit up. The graph opens down because of the negative sign in front of the absolute value expression, making the lowest point now the highest point on the graph, as shown in **Figure 87.** The graph is symmetric with respect to the line $x = -3$.

Figure 87

(b) To determine the horizontal translation, factor out 2.

$$h(x) = |2x - 4|$$

$$h(x) = |2(x - 2)| \quad \text{Factor out 2.}$$

$$h(x) = |2| \cdot |x - 2| \quad |ab| = |a| \cdot |b|$$

$$h(x) = 2|x - 2| \quad |2| = 2$$

The graph of h is the graph of $y = |x|$ translated 2 units to the right, and vertically stretched by a factor of 2. Horizontal shrinking gives the same appearance as vertical stretching for this function. See **Figure 88.**

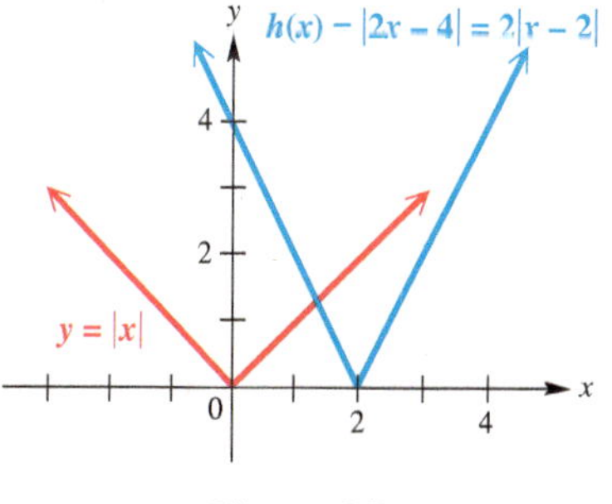

Figure 88

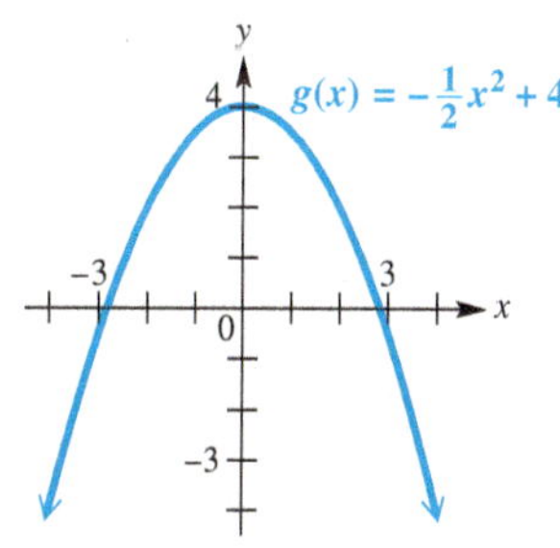

Figure 89

(c) The graph of $g(x) = -\frac{1}{2}x^2 + 4$ has the same shape as that of $y = x^2$, but it is wider (that is, shrunken vertically), reflected across the x-axis because the coefficient $-\frac{1}{2}$ is negative, and then translated 4 units up. See **Figure 89.**

✔ **Now Try Exercises 71, 73, and 81.**

EXAMPLE 9 **Graphing Translations and Reflections of a Given Graph**

A graph of a function $y = f(x)$ is shown in **Figure 90.** Use this graph to sketch each of the following graphs.

(a) $g(x) = f(x) + 3$

(b) $h(x) = f(x + 3)$

(c) $k(x) = f(x - 2) + 3$

(d) $F(x) = -f(x)$

SOLUTION In each part, pay close attention to how the plotted points in **Figure 90** are translated or reflected.

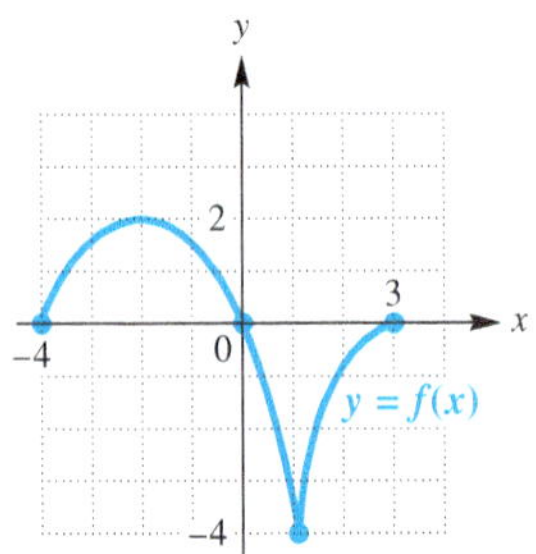

Figure 90

(a) The graph of $g(x) = f(x) + 3$ is the same as the graph in **Figure 90,** translated 3 units up. See **Figure 91(a).**

(b) To obtain the graph of $h(x) = f(x + 3)$, the graph of $y = f(x)$ must be translated 3 units to the left because $x + 3 = 0$ when $x = -3$. See **Figure 91(b).**

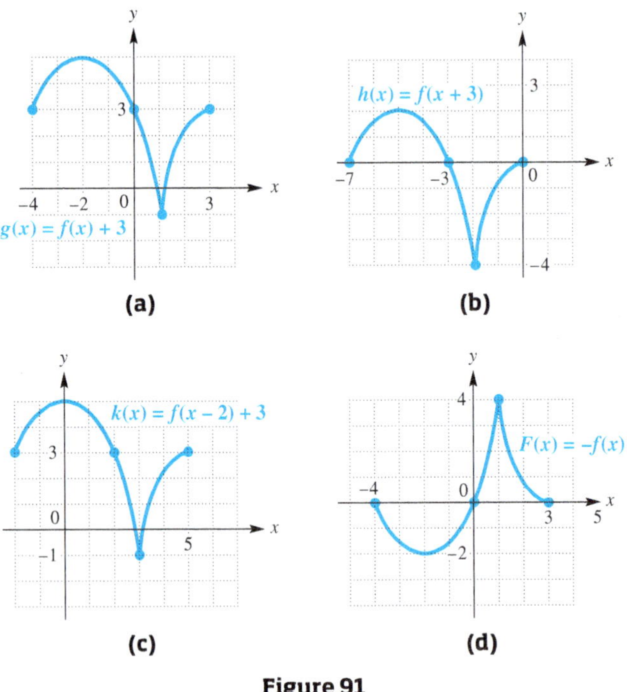

Figure 91

(c) The graph of $k(x) = f(x - 2) + 3$ will look like the graph of $f(x)$ translated 2 units to the right and 3 units up, as shown in **Figure 91(c).**

(d) The graph of $F(x) = -f(x)$ is that of $y = f(x)$ reflected across the x-axis. See **Figure 91(d).**

✔ **Now Try Exercise 87.**

Summary of Graphing Techniques

In the descriptions that follow, assume that $a > 0$, $h > 0$, and $k > 0$. In comparison with the graph of $y = f(x)$:

1. The graph of $\boldsymbol{y = f(x) + k}$ is translated k units up.
2. The graph of $\boldsymbol{y = f(x) - k}$ is translated k units down.
3. The graph of $\boldsymbol{y = f(x + h)}$ is translated h units to the left.
4. The graph of $\boldsymbol{y = f(x - h)}$ is translated h units to the right.
5. The graph of $\boldsymbol{y = af(x)}$ is a vertical stretching of the graph of $y = f(x)$ if $a > 1$. It is a vertical shrinking if $0 < a < 1$.
6. The graph of $\boldsymbol{y = f(ax)}$ is a horizontal stretching of the graph of $y = f(x)$ if $0 < a < 1$. It is a horizontal shrinking if $a > 1$.
7. The graph of $\boldsymbol{y = -f(x)}$ is reflected across the x-axis.
8. The graph of $\boldsymbol{y = f(-x)}$ is reflected across the y-axis.

2.7 Exercises

CONCEPT PREVIEW *Fill in the blank(s) to correctly complete each sentence.*

1. To graph the function $f(x) = x^2 - 3$, shift the graph of $y = x^2$ down ______ units.
2. To graph the function $f(x) = x^2 + 5$, shift the graph of $y = x^2$ up ______ units.
3. The graph of $f(x) = (x + 4)^2$ is obtained by shifting the graph of $y = x^2$ to the ______ 4 units.
4. The graph of $f(x) = (x - 7)^2$ is obtained by shifting the graph of $y = x^2$ to the ______ 7 units.
5. The graph of $f(x) = -\sqrt{x}$ is a reflection of the graph of $y = \sqrt{x}$ across the ______-axis.
6. The graph of $f(x) = \sqrt{-x}$ is a reflection of the graph of $y = \sqrt{x}$ across the ______-axis.
7. To obtain the graph of $f(x) = (x + 2)^3 - 3$, shift the graph of $y = x^3$ to the left ______ units and down ______ units.
8. To obtain the graph of $f(x) = (x - 3)^3 + 6$, shift the graph of $y = x^3$ to the right ______ units and up ______ units.
9. The graph of $f(x) = |-x|$ is the same as the graph of $y = |x|$ because reflecting it across the ______-axis yields the same ordered pairs.
10. The graph of $x = y^2$ is the same as the graph of $x = (-y)^2$ because reflecting it across the ______-axis yields the same ordered pairs.
11. *Concept Check* Match each equation in Column I with a description of its graph from Column II as it relates to the graph of $y = x^2$.

I	II
(a) $y = (x - 7)^2$	**A.** a translation 7 units to the left
(b) $y = x^2 - 7$	**B.** a translation 7 units to the right
(c) $y = 7x^2$	**C.** a translation 7 units up
(d) $y = (x + 7)^2$	**D.** a translation 7 units down
(e) $y = x^2 + 7$	**E.** a vertical stretching by a factor of 7

12. *Concept Check* Match each equation in Column I with a description of its graph from Column II as it relates to the graph of $y = \sqrt[3]{x}$.

I	II
(a) $y = 4\sqrt[3]{x}$	**A.** a translation 4 units to the right
(b) $y = -\sqrt[3]{x}$	**B.** a translation 4 units down
(c) $y = \sqrt[3]{-x}$	**C.** a reflection across the x-axis
(d) $y = \sqrt[3]{x - 4}$	**D.** a reflection across the y-axis
(e) $y = \sqrt[3]{x} - 4$	**E.** a vertical stretching by a factor of 4

13. *Concept Check* Match each equation with the sketch of its graph in A–I.

(a) $y = x^2 + 2$ **(b)** $y = x^2 - 2$ **(c)** $y = (x + 2)^2$

(d) $y = (x - 2)^2$ **(e)** $y = 2x^2$ **(f)** $y = -x^2$

(g) $y = (x - 2)^2 + 1$ **(h)** $y = (x + 2)^2 + 1$ **(i)** $y = (x + 2)^2 - 1$

A.

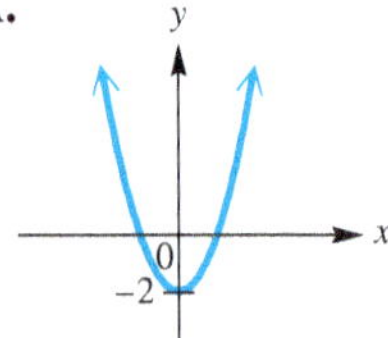

B.

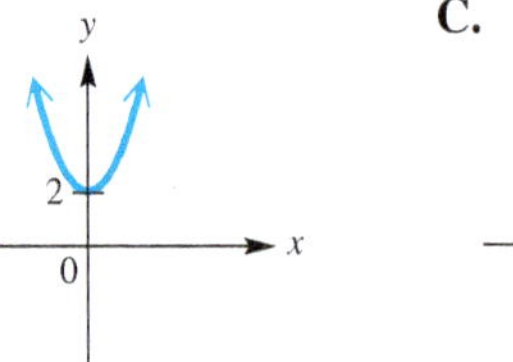

C.

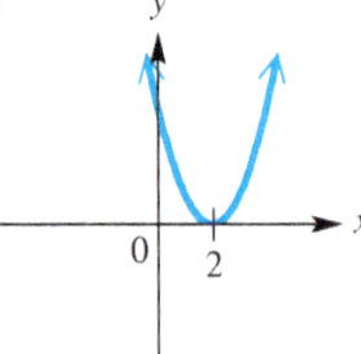

D.

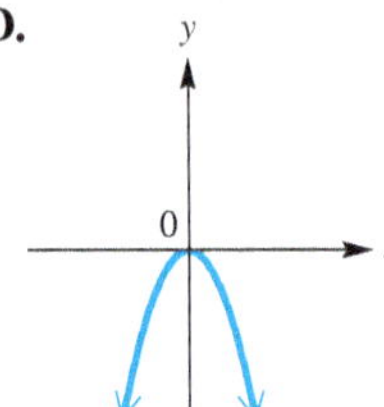

E.

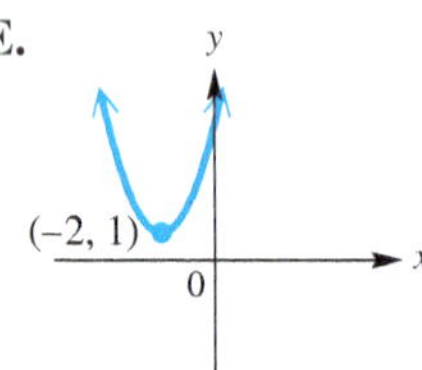

F.

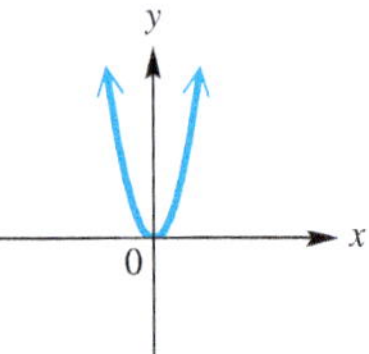

G.

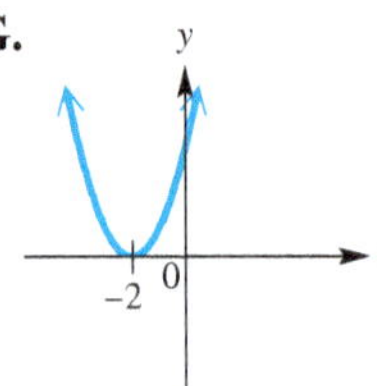

H.

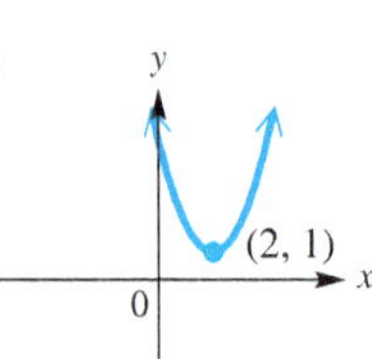

I.

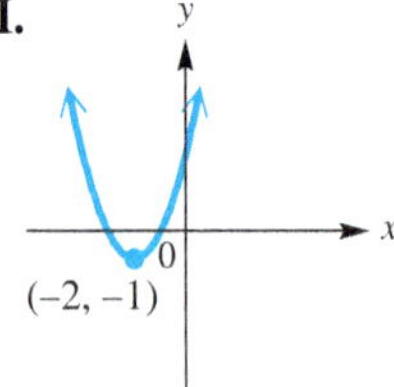

14. *Concept Check* Match each equation with the sketch of its graph in A–I.

(a) $y = \sqrt{x + 3}$ **(b)** $y = \sqrt{x} - 3$ **(c)** $y = \sqrt{x} + 3$

(d) $y = 3\sqrt{x}$ **(e)** $y = -\sqrt{x}$ **(f)** $y = \sqrt{x - 3}$

(g) $y = \sqrt{x - 3} + 2$ **(h)** $y = \sqrt{x + 3} + 2$ **(i)** $y = \sqrt{x - 3} - 2$

A.

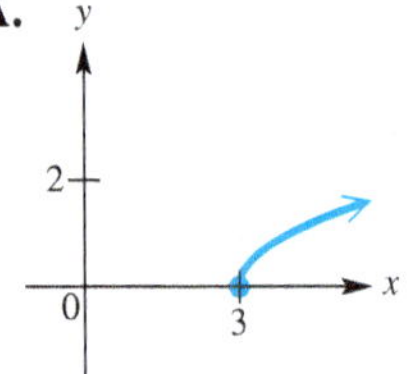

B.

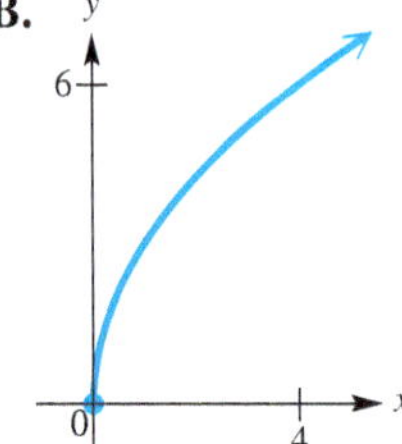

C.

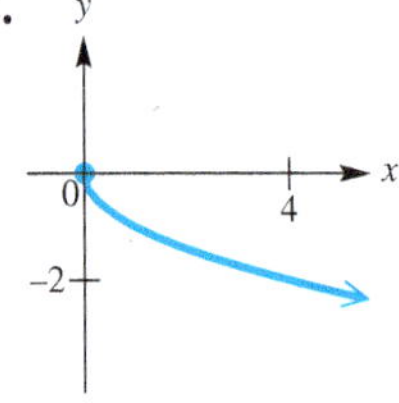

D.

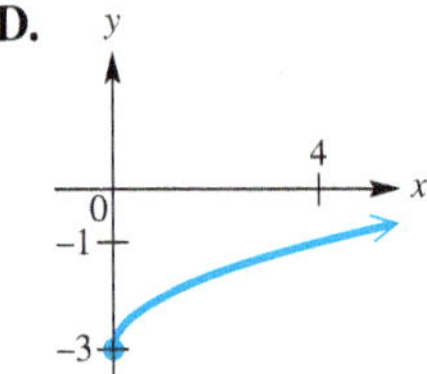

E.

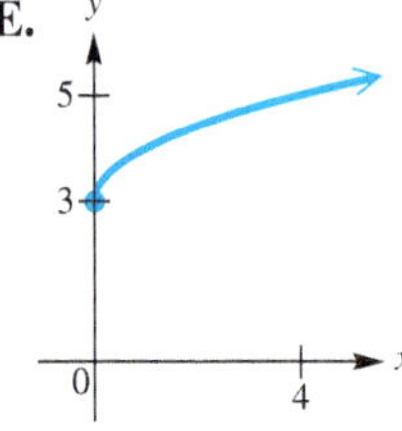

F.

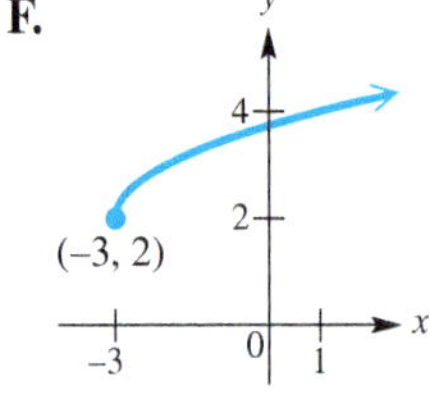

G.

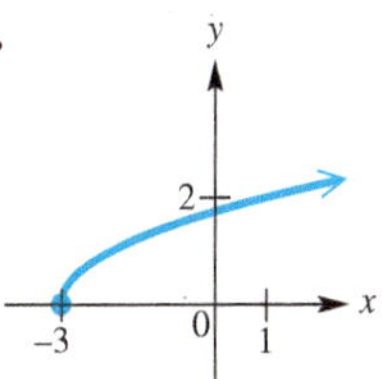

H.

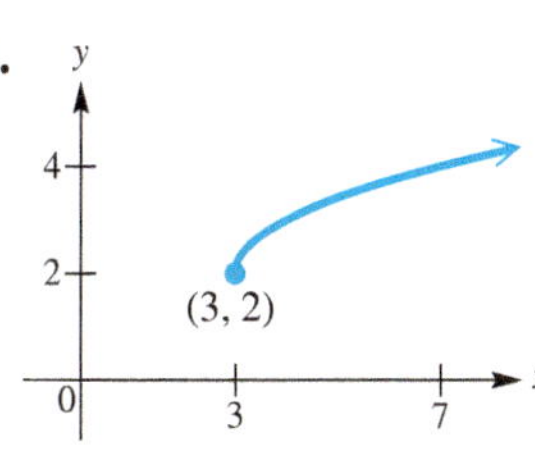

I.

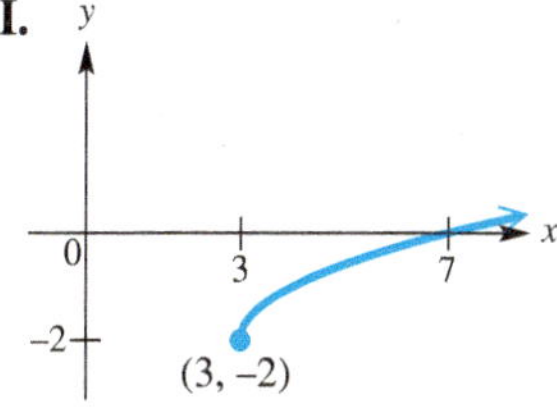

15. *Concept Check* Match each equation with the sketch of its graph in A–I.

(a) $y = |x - 2|$ **(b)** $y = |x| - 2$ **(c)** $y = |x| + 2$

(d) $y = 2|x|$ **(e)** $y = -|x|$ **(f)** $y = |-x|$

(g) $y = -2|x|$ **(h)** $y = |x - 2| + 2$ **(i)** $y = |x + 2| - 2$

A.

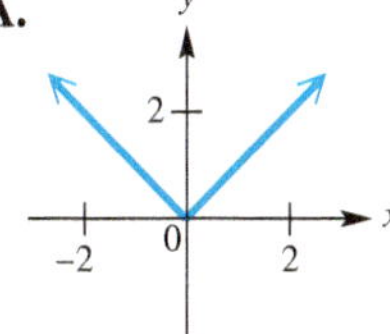

B.

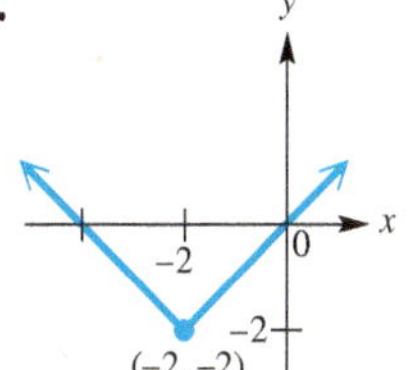

C.

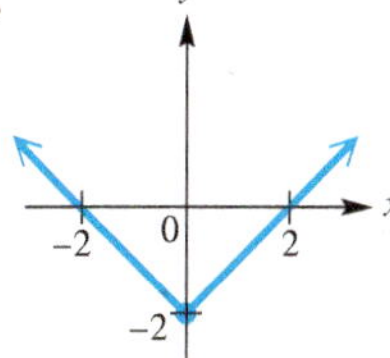

D.

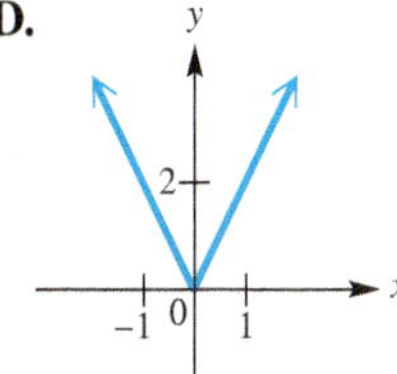

E.

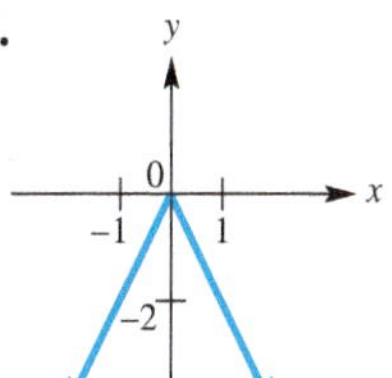

F.

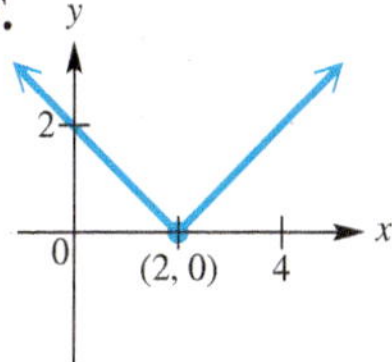

G.

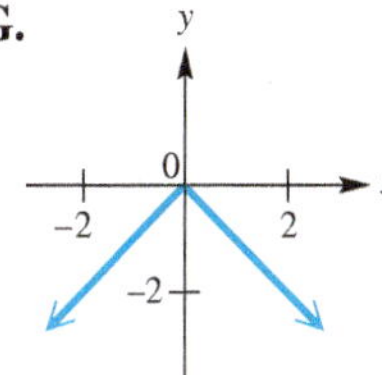

H.

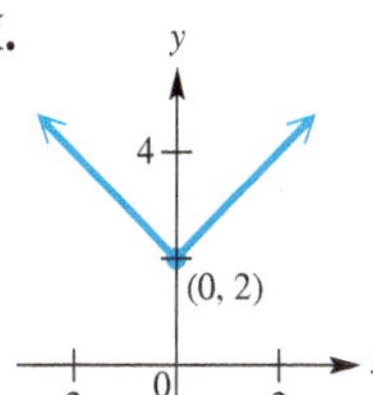

I.

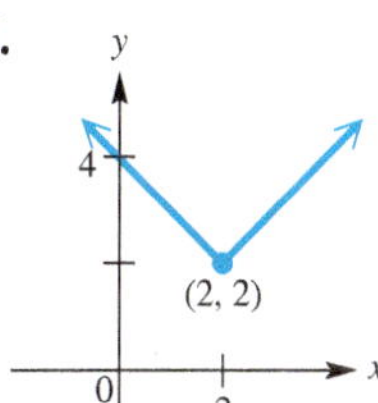

16. *Concept Check* Describe how the graph of $f(x) = 2(x + 1)^3 - 6$ compares to the graph of $y = x^3$.

Graph each function. ***See Examples 1 and 2.***

17. $f(x) = 3|x|$ **18.** $f(x) = 4|x|$ **19.** $f(x) = \frac{2}{3}|x|$

20. $f(x) = \frac{3}{4}|x|$ **21.** $g(x) = 2x^2$ **22.** $g(x) = 3x^2$

23. $g(x) = \frac{1}{2}x^2$ **24.** $g(x) = \frac{1}{3}x^2$ **25.** $f(x) = -\frac{1}{2}x^2$

26. $f(x) = -\frac{1}{3}x^2$ **27.** $f(x) = -3|x|$ **28.** $f(x) = -2|x|$

29. $h(x) = \left|-\frac{1}{2}x\right|$ **30.** $h(x) = \left|-\frac{1}{3}x\right|$ **31.** $h(x) = \sqrt{4x}$

32. $h(x) = \sqrt{9x}$ **33.** $f(x) = -\sqrt{-x}$ **34.** $f(x) = -|-x|$

Concept Check Suppose the point $(8, 12)$ *is on the graph of* $y = f(x)$.

35. Find a point on the graph of

(a) $y = f(x + 4)$ **(b)** $y = f(x) + 4$.

36. Find a point on the graph of

(a) $y = \frac{1}{4}f(x)$ **(b)** $y = 4f(x)$.

37. Find a point on the graph of

(a) $y = f(4x)$ **(b)** $y = f\left(\frac{1}{4}x\right)$.

38. Find a point on the graph of the reflection of $y = f(x)$

(a) across the x-axis **(b)** across the y-axis.

*Concept Check Plot each point, and then plot the points that are symmetric to the given point with respect to the **(a)** x-axis, **(b)** y-axis, and **(c)** origin.*

39. $(5, -3)$ **40.** $(-6, 1)$ **41.** $(-4, -2)$ **42.** $(-8, 0)$

43. *Concept Check* The graph of $y = |x - 2|$ is symmetric with respect to a vertical line. What is the equation of that line?

44. *Concept Check* Repeat **Exercise 43** for the graph of $y = -|x + 1|$.

Without graphing, determine whether each equation has a graph that is symmetric with respect to the x-axis, *the* y-axis, *the* origin, *or* none of these. ***See Examples 3 and 4.***

45. $y = x^2 + 5$ **46.** $y = 2x^4 - 3$

47. $x^2 + y^2 = 12$ **48.** $y^2 - x^2 = -6$

49. $y = -4x^3 + x$ **50.** $y = x^3 - x$

51. $y = x^2 - x + 8$ **52.** $y = x + 15$

Determine whether each function is even, odd, *or* neither. ***See Example 5.***

53. $f(x) = -x^3 + 2x$ **54.** $f(x) = x^5 - 2x^3$

55. $f(x) = 0.5x^4 - 2x^2 + 6$ **56.** $f(x) = 0.75x^2 + |x| + 4$

57. $f(x) = x^3 - x + 9$ **58.** $f(x) = x^4 - 5x + 8$

Graph each function. ***See Examples 6–8 and the Summary of Graphing Techniques box following Example 9.***

59. $f(x) = x^2 - 1$ **60.** $f(x) = x^2 - 2$ **61.** $f(x) = x^2 + 2$

62. $f(x) = x^2 + 3$ **63.** $g(x) = (x - 4)^2$ **64.** $g(x) = (x - 2)^2$

65. $g(x) = (x + 2)^2$ **66.** $g(x) = (x + 3)^2$ **67.** $g(x) = |x| - 1$

68. $g(x) = |x + 3| + 2$ **69.** $h(x) = -(x + 1)^3$ **70.** $h(x) = -(x - 1)^3$

71. $h(x) = 2x^2 - 1$ **72.** $h(x) = 3x^2 - 2$ **73.** $f(x) = 2(x - 2)^2 - 4$

74. $f(x) = -3(x - 2)^2 + 1$ **75.** $f(x) = \sqrt{x + 2}$ **76.** $f(x) = \sqrt{x - 3}$

77. $f(x) = -\sqrt{x}$ **78.** $f(x) = \sqrt{x} - 2$ **79.** $f(x) = 2\sqrt{x} + 1$

80. $f(x) = 3\sqrt{x} - 2$ **81.** $g(x) = \frac{1}{2}x^3 - 4$ **82.** $g(x) = \frac{1}{2}x^3 + 2$

83. $g(x) = (x + 3)^3$ **84.** $f(x) = (x - 2)^3$ **85.** $f(x) = \frac{2}{3}(x - 2)^2$

86. *Concept Check* What is the relationship between the graphs of $f(x) = |x|$ and $g(x) = |-x|$?

Work each problem. ***See Example 9.***

87. Given the graph of $y = g(x)$ in the figure, sketch the graph of each function, and describe how it is obtained from the graph of $y = g(x)$.

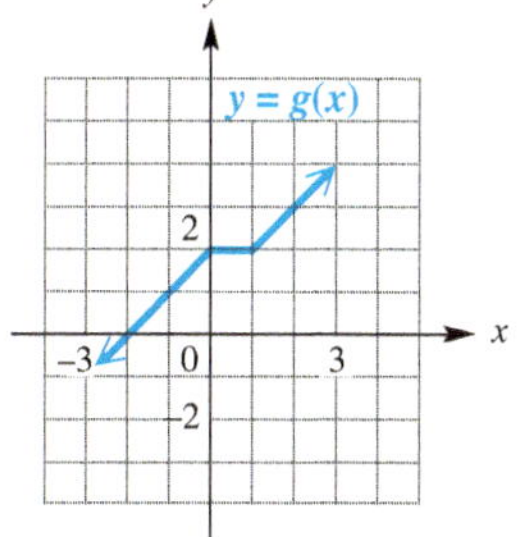

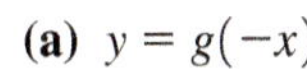

(a) $y = g(-x)$

(b) $y = g(x - 2)$

(c) $y = -g(x)$

(d) $y = -g(x) + 2$

88. Given the graph of $y = f(x)$ in the figure, sketch the graph of each function, and describe how it is obtained from the graph of $y = f(x)$.

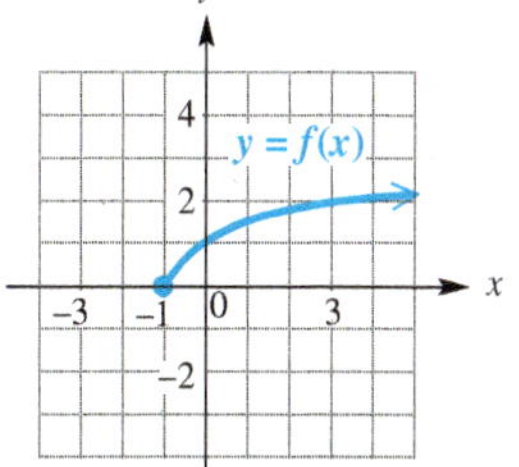

(a) $y = -f(x)$

(b) $y = 2f(x)$

(c) $y = f(-x)$

(d) $y = \frac{1}{2}f(x)$

Connecting Graphs with Equations *Each of the following graphs is obtained from the graph of $f(x) = |x|$ or $g(x) = \sqrt{x}$ by applying several of the transformations discussed in this section. Describe the transformations and give an equation for the graph.*

89.

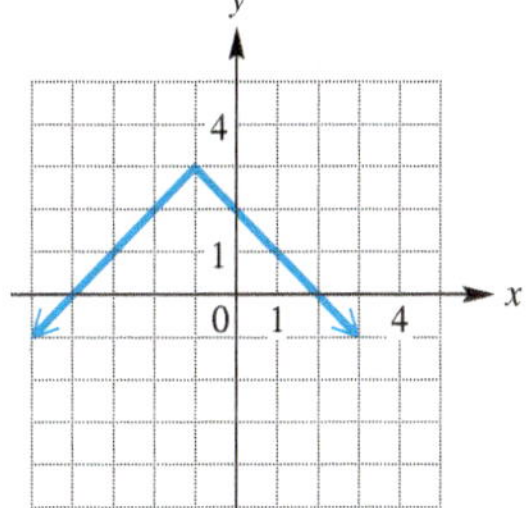

90.

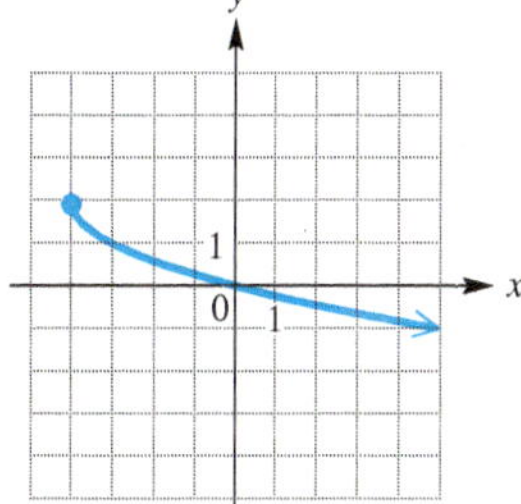

91.

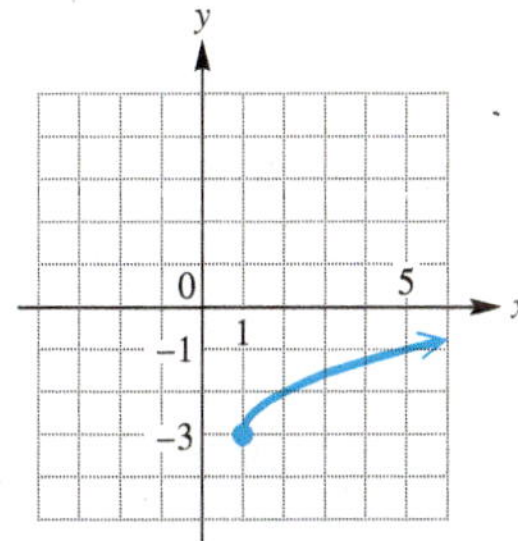

92.

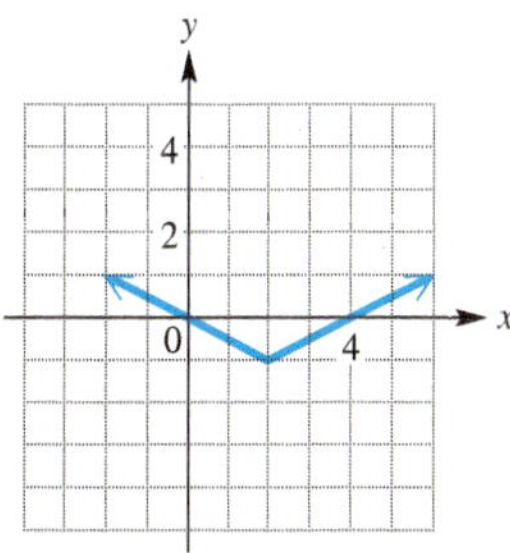

93.

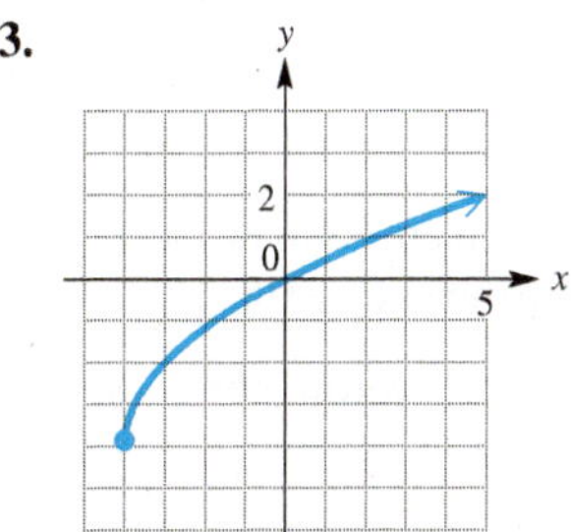

94.

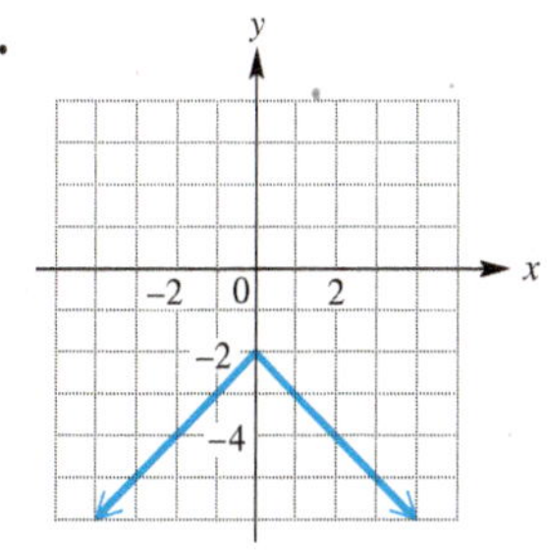

Concept Check Suppose that for a function f,

$$f(3) = 6.$$

For the given assumptions, find another function value.

95. The graph of $y = f(x)$ is symmetric with respect to the origin.

96. The graph of $y = f(x)$ is symmetric with respect to the y-axis.

97. The graph of $y = f(x)$ is symmetric with respect to the line $x = 6$.

98. For all x, $f(-x) = f(x)$.

99. For all x, $f(-x) = -f(x)$.

100. f is an odd function.

Work each problem.

101. Find a function $g(x) = ax + b$ whose graph can be obtained by translating the graph of $f(x) = 2x + 5$ up 2 units and 3 units to the left.

102. Find a function $g(x) = ax + b$ whose graph can be obtained by translating the graph of $f(x) = 3 - x$ down 2 units and 3 units to the right.

103. *Concept Check* Complete the left half of the graph of $y = f(x)$ in the figure for each condition.

(a) $f(-x) = f(x)$ **(b)** $f(-x) = -f(x)$

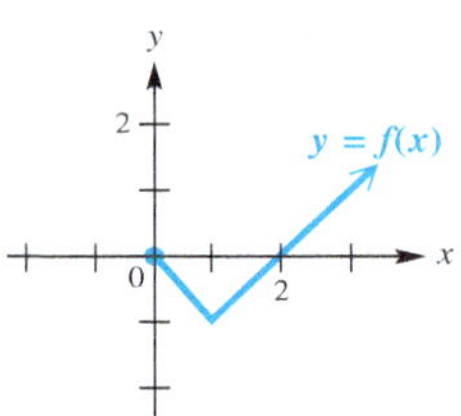

104. *Concept Check* Complete the right half of the graph of $y = f(x)$ in the figure for each condition.

(a) f is odd. **(b)** f is even.

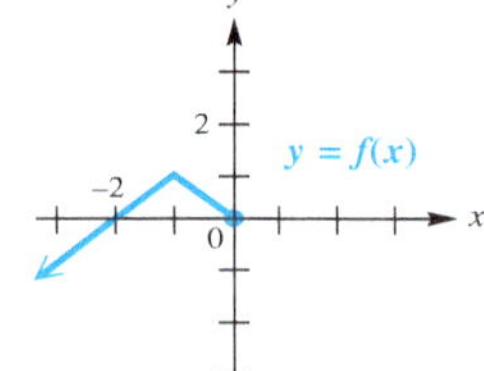

Chapter 2 Quiz (Sections 2.5-2.7)

1. For the line passing through the points $(-3, 5)$ and $(-1, 9)$, find the following.

(a) the slope-intercept form of its equation **(b)** its x-intercept

2. Find the slope-intercept form of the equation of the line passing through the point $(-6, 4)$ and perpendicular to the graph of $3x - 2y = 6$.

3. Suppose that P has coordinates $(-8, 5)$. Find the equation of the line through P that is

(a) vertical **(b)** horizontal.

4. For each basic function graphed, give the name of the function, the domain, the range, and open intervals over which it is decreasing, increasing, or constant.

(a) (b) (c)

5. *(Modeling) Long-Distance Call Charges* A certain long-distance carrier provides service between Podunk and Nowheresville. If x represents the number of minutes for the call, where $x > 0$, then the function

$$f(x) = 0.40[\![x]\!] + 0.75$$

gives the total cost of the call in dollars. Find the cost of a 5.5-min call.

Graph each function.

6. $f(x) = \begin{cases} \sqrt{x} & \text{if } x \geq 0 \\ 2x + 3 & \text{if } x < 0 \end{cases}$ 7. $f(x) = -x^3 + 1$ 8. $f(x) = 2|x - 1| + 3$

9. *Connecting Graphs with Equations* The function $g(x)$ graphed here is obtained by stretching, shrinking, reflecting, and/or translating the graph of $f(x) = \sqrt{x}$. Give the equation that defines this function.

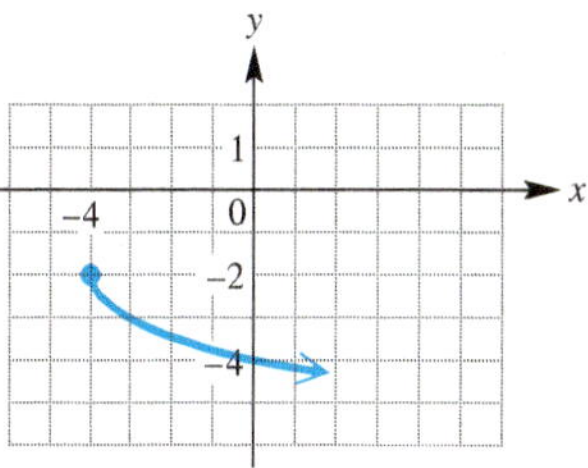

10. Determine whether each function is *even*, *odd*, or *neither*.

(a) $f(x) = x^2 - 7$ (b) $f(x) = x^3 - x - 1$ (c) $f(x) = x^{101} - x^{99}$

2.8 Function Operations and Composition

- Arithmetic Operations on Functions
- The Difference Quotient
- Composition of Functions and Domain

Arithmetic Operations on Functions **Figure 92** shows the situation for a company that manufactures DVDs. The two lines are the graphs of the linear functions for

$$\text{revenue } R(x) = 168x \quad \text{and} \quad \text{cost } C(x) = 118x + 800,$$

where x is the number of DVDs produced and sold, and x, $R(x)$, and $C(x)$ are given in thousands. When 30,000 (that is, 30 thousand) DVDs are produced and sold, profit is found as follows.

$$P(x) = R(x) - C(x) \quad \text{Profit function}$$

$$P(30) = R(30) - C(30) \quad \text{Let } x = 30.$$

$$P(30) = 5040 - 4340 \quad R(30) = 168(30);\ C(30) = 118(30) + 800$$

$$P(30) = 700 \quad \text{Subtract.}$$

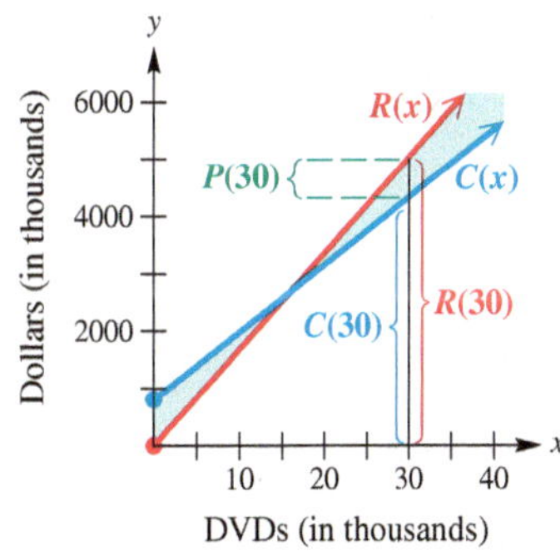

Figure 92

Thus, the profit from the sale of 30,000 DVDs is $700,000.

The profit function is found by *subtracting* the cost function from the revenue function. New functions can be formed by using other operations as well.

Operations on Functions and Domains

Given two functions f and g, then for all values of x for which both $f(x)$ and $g(x)$ are defined, the functions $f + g$, $f - g$, fg, and $\frac{f}{g}$ are defined as follows.

$$(f + g)(x) = f(x) + g(x) \quad \text{Sum function}$$

$$(f - g)(x) = f(x) - g(x) \quad \text{Difference function}$$

$$(fg)(x) = f(x) \cdot g(x) \quad \text{Product function}$$

$$\left(\frac{f}{g}\right)(x) = \frac{f(x)}{g(x)}, \quad g(x) \neq 0 \quad \text{Quotient function}$$

The **domains of $f + g$, $f - g$,** and **fg** include all real numbers in the intersection of the domains of f and g, while the **domain of $\frac{f}{g}$** includes those real numbers in the intersection of the domains of f and g for which $g(x) \neq 0$.

NOTE The condition $g(x) \neq 0$ in the definition of the quotient means that the domain of $\left(\frac{f}{g}\right)(x)$ is restricted to all values of x for which $g(x)$ is not 0. The condition does *not* mean that $g(x)$ is a function that is never 0.

EXAMPLE 1 Using Operations on Functions

Let $f(x) = x^2 + 1$ and $g(x) = 3x + 5$. Find each of the following.

(a) $(f + g)(1)$ **(b)** $(f - g)(-3)$ **(c)** $(fg)(5)$ **(d)** $\left(\frac{f}{g}\right)(0)$

SOLUTION

(a) First determine $f(1) = 2$ and $g(1) = 8$. Then use the definition.

$$\begin{aligned}
&(f + g)(1) \\
&= f(1) + g(1) && (f + g)(x) = f(x) + g(x) \\
&= 2 + 8 && f(1) = 1^2 + 1;\ g(1) = 3(1) + 5 \\
&= 10 && \text{Add.}
\end{aligned}$$

(b) $(f - g)(-3)$

$$\begin{aligned}
&= f(-3) - g(-3) && (f - g)(x) = f(x) - g(x) \\
&= 10 - (-4) && f(-3) = (-3)^2 + 1;\ g(-3) = 3(-3) + 5 \\
&= 14 && \text{Subtract.}
\end{aligned}$$

(c) $(fg)(5)$

$$\begin{aligned}
&= f(5) \cdot g(5) && (fg)(x) = f(x) \cdot g(x) \\
&= (5^2 + 1)(3 \cdot 5 + 5) && f(x) = x^2 + 1;\ g(x) = 3x + 5 \\
&= 26 \cdot 20 && f(5) = 26;\ g(5) = 20 \\
&= 520 && \text{Multiply.}
\end{aligned}$$

(d) $\left(\frac{f}{g}\right)(0)$

$= \frac{f(0)}{g(0)}$ $\quad \left(\frac{f}{g}\right)(x) = \frac{f(x)}{g(x)}$

$= \frac{0^2 + 1}{3(0) + 5}$ $\quad f(x) = x^2 + 1$, $g(x) = 3x + 5$

$= \frac{1}{5}$ $\quad$ Simplify.

✔ **Now Try Exercises 11, 13, 15, and 17.**

EXAMPLE 2 Using Operations on Functions and Determining Domains

Let $f(x) = 8x - 9$ and $g(x) = \sqrt{2x - 1}$. Find each function in (a)–(d).

(a) $(f + g)(x)$ **(b)** $(f - g)(x)$ **(c)** $(fg)(x)$ **(d)** $\left(\frac{f}{g}\right)(x)$

(e) Give the domains of the functions in parts (a)–(d).

SOLUTION

(a) $(f + g)(x)$

$= f(x) + g(x)$

$= 8x - 9 + \sqrt{2x - 1}$

(b) $(f - g)(x)$

$= f(x) - g(x)$

$= 8x - 9 - \sqrt{2x - 1}$

(c) $(fg)(x)$

$= f(x) \cdot g(x)$

$= (8x - 9)\sqrt{2x - 1}$

(d) $\left(\frac{f}{g}\right)(x)$

$= \frac{f(x)}{g(x)}$

$= \frac{8x - 9}{\sqrt{2x - 1}}$

(e) To find the domains of the functions in parts (a)–(d), we first find the domains of f and g.

The domain of f is the set of all real numbers $(-\infty, \infty)$.

Because g is defined by a square root radical, the radicand must be nonnegative (that is, greater than or equal to 0).

$g(x) = \sqrt{2x - 1}$ $\quad$ Rule for $g(x)$

$2x - 1 \geq 0$ $\quad$ $2x - 1$ must be nonnegative.

$x \geq \frac{1}{2}$ $\quad$ Add 1 and divide by 2.

Thus, the domain of g is $\left[\frac{1}{2}, \infty\right)$.

The domains of $f + g$, $f - g$, and fg are the intersection of the domains of f and g, which is

$$(-\infty, \infty) \cap \left[\frac{1}{2}, \infty\right) = \left[\frac{1}{2}, \infty\right).$$

The intersection of two sets is the set of all elements common to both sets.

The domain of $\frac{f}{g}$ includes those real numbers in the intersection above for which $g(x) = \sqrt{2x - 1} \neq 0$—that is, the domain of $\frac{f}{g}$ is $\left(\frac{1}{2}, \infty\right)$.

✔ **Now Try Exercises 19 and 23.**

EXAMPLE 3 Evaluating Combinations of Functions

If possible, use the given representations of functions f and g to evaluate

$$(f+g)(4),\quad (f-g)(-2),\quad (fg)(1),\quad \text{and}\quad \left(\frac{f}{g}\right)(0).$$

(a)

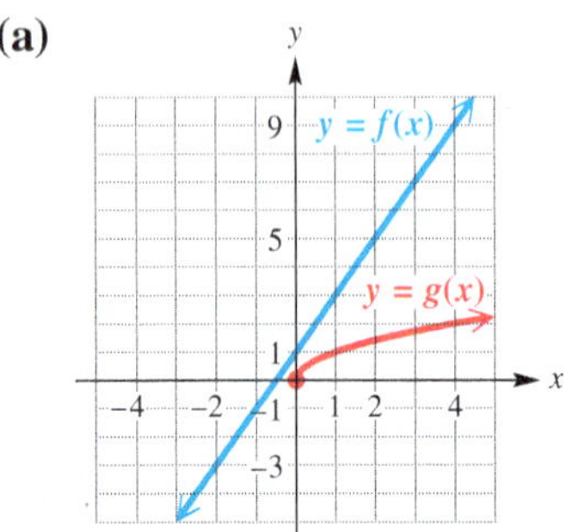

(b)

x	$f(x)$	$g(x)$
-2	-3	undefined
0	1	0
1	3	1
4	9	2

(c) $f(x) = 2x + 1,\quad g(x) = \sqrt{x}$

SOLUTION

(a) From the figure, $f(4) = 9$ and $g(4) = 2$.

$$\begin{aligned}&(f+g)(4)\\&= f(4) + g(4) && (f+g)(x) = f(x) + g(x)\\&= 9 + 2 && \text{Substitute.}\\&= 11 && \text{Add.}\end{aligned}$$

For $(f-g)(-2)$, although $f(-2) = -3$, $g(-2)$ is undefined because -2 is not in the domain of g. Thus $(f-g)(-2)$ is undefined.

The domains of f and g both include 1.

$$\begin{aligned}&(fg)(1)\\&= f(1) \cdot g(1) && (fg)(x) = f(x) \cdot g(x)\\&= 3 \cdot 1 && \text{Substitute.}\\&= 3 && \text{Multiply.}\end{aligned}$$

The graph of g includes the origin, so $g(0) = 0$. Thus $\left(\frac{f}{g}\right)(0)$ is undefined.

(b) From the table, $f(4) = 9$ and $g(4) = 2$.

$$\begin{aligned}&(f+g)(4)\\&= f(4) + g(4) && (f+g)(x) = f(x) + g(x)\\&= 9 + 2 && \text{Substitute.}\\&= 11 && \text{Add.}\end{aligned}$$

In the table, $g(-2)$ is undefined, and thus $(f-g)(-2)$ is also undefined.

$$\begin{aligned}&(fg)(1)\\&= f(1) \cdot g(1) && (fg)(x) = f(x) \cdot g(x)\\&= 3 \cdot 1 && f(1) = 3 \text{ and } g(1) = 1\\&= 3 && \text{Multiply.}\end{aligned}$$

The quotient function value $\left(\frac{f}{g}\right)(0)$ is undefined because the denominator, $g(0)$, equals 0.

(c) Using $f(x) = 2x + 1$ and $g(x) = \sqrt{x}$, we can find $(f + g)(4)$ and $(fg)(1)$. Because -2 is not in the domain of g, $(f - g)(-2)$ is not defined.

$(f+g)(4)$	$(fg)(1)$
$= f(4) + g(4)$	$= f(1) \cdot g(1)$
$= (2 \cdot 4 + 1) + \sqrt{4}$	$= (2 \cdot 1 + 1) \cdot \sqrt{1}$
$= 9 + 2$	$= 3(1)$
$= 11$	$= 3$

$\left(\frac{f}{g}\right)(0)$ is undefined since $g(0) = 0$.

✔ **Now Try Exercises 33 and 37.**

The Difference Quotient Suppose a point P lies on the graph of $y = f(x)$ as in **Figure 93,** and suppose h is a positive number. If we let $(x, f(x))$ denote the coordinates of P and $(x + h, f(x + h))$ denote the coordinates of Q, then the line joining P and Q has slope as follows.

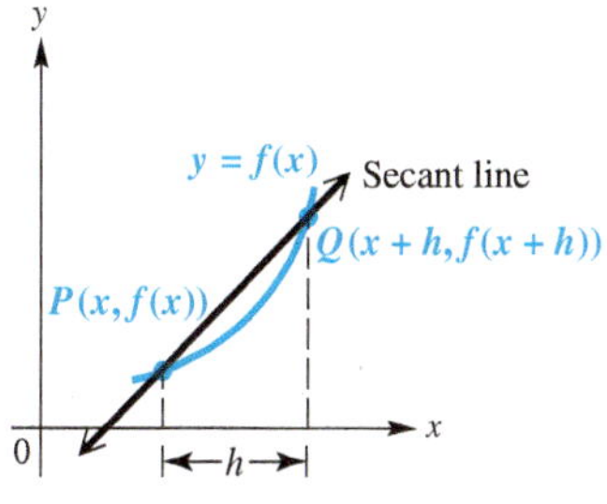

Figure 93

$$m = \frac{f(x+h) - f(x)}{(x+h) - x} \quad \text{Slope formula}$$

$$m = \frac{\mathbf{f(x+h) - f(x)}}{\mathbf{h}}, \quad \mathbf{h \neq 0} \quad \text{Difference quotient}$$

This boldface expression is the **difference quotient.**

Figure 93 shows the graph of the line PQ (called a **secant line**). As h approaches 0, the slope of this secant line approaches the slope of the line tangent to the curve at P. Important applications of this idea are developed in calculus.

EXAMPLE 4 Finding the Difference Quotient

Let $f(x) = 2x^2 - 3x$. Find and simplify the expression for the difference quotient,

$$\frac{f(x+h) - f(x)}{h}.$$

SOLUTION We use a three-step process.

Step 1 Find the first term in the numerator, $f(x + h)$. Replace x in $f(x)$ with $x + h$.

$$f(x+h)$$
$$= 2(x+h)^2 - 3(x+h) \quad f(x) = 2x^2 - 3x$$

Step 2 Find the entire numerator, $f(x + h) - f(x)$.

$$f(x+h) - f(x)$$

From Step 1

$$= [2(x+h)^2 - 3(x+h)] - (2x^2 - 3x) \quad \text{Substitute.}$$

$$= 2(x^2 + 2xh + h^2) - 3(x+h) - (2x^2 - 3x) \quad \text{Square } x + h.$$

Remember this term when squaring $x + h$

$$= 2x^2 + 4xh + 2h^2 - 3x - 3h - 2x^2 + 3x \quad \text{Distributive property}$$

$$= 4xh + 2h^2 - 3h \quad \text{Combine like terms.}$$

Step 3 Find the difference quotient by dividing by h.

$$\frac{f(x+h)-f(x)}{h}$$

$$= \frac{4xh + 2h^2 - 3h}{h} \quad \text{Substitute } 4xh + 2h^2 - 3h \text{ for } f(x+h) - f(x) \text{, from Step 2.}$$

$$= \frac{h(4x + 2h - 3)}{h} \quad \text{Factor out } h.$$

$$= 4x + 2h - 3 \quad \text{Divide out the common factor.}$$

✔ **Now Try Exercises 45 and 55.**

LOOKING AHEAD TO CALCULUS

The difference quotient is essential in the definition of the **derivative of a function** in calculus. The derivative provides a formula, in function form, for finding the slope of the tangent line to the graph of the function at a given point.

To illustrate, it is shown in calculus that the derivative of

$$f(x) = x^2 + 3$$

is given by the function

$$f'(x) = 2x.$$

Now, $f'(0) = 2(0) = 0$, meaning that the slope of the tangent line to $f(x) = x^2 + 3$ at $x = 0$ is 0, which implies that the tangent line is horizontal. If you draw this tangent line, you will see that it is the line $y = 3$, which is indeed a horizontal line.

CAUTION In **Example 4,** notice that the expression $f(x + h)$ is not equivalent to $f(x) + f(h)$. These expressions differ by $4xh$.

$$f(x+h) = 2(x+h)^2 - 3(x+h) = 2x^2 + 4xh + 2h^2 - 3x - 3h$$

$$f(x) + f(h) = (2x^2 - 3x) + (2h^2 - 3h) = 2x^2 - 3x + 2h^2 - 3h$$

In general, for a function f, $f(x + h)$ is *not* equivalent to $f(x) + f(h)$.

Composition of Functions and Domain The diagram in **Figure 94** shows a function g that assigns to each x in its domain a value $g(x)$. Then another function f assigns to each $g(x)$ in its domain a value $f(g(x))$. This two-step process takes an element x and produces a corresponding element $f(g(x))$.

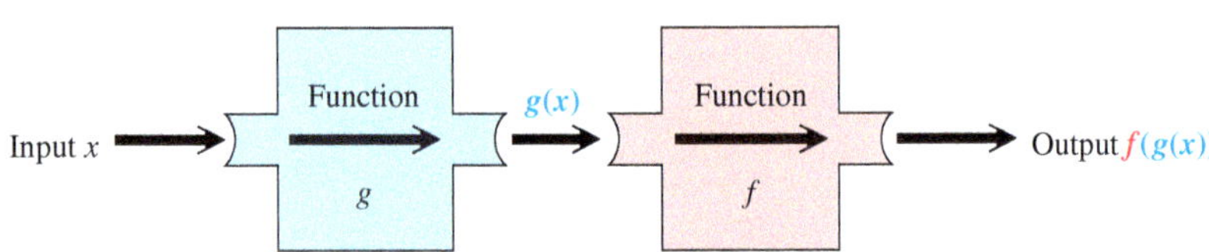

Figure 94

The function with y-values $f(g(x))$ is the *composition* of functions f and g, which is written $\boldsymbol{f \circ g}$ and read **"f of g"** or **"f compose g".**

Composition of Functions and Domain

If f and g are functions, then the **composite function,** or **composition,** of f and g is defined by

$$(f \circ g)(x) = f(g(x)).$$

The **domain of $f \circ g$** is the set of all numbers x in the domain of g such that $g(x)$ is in the domain of f.

As a real-life example of how composite functions occur, consider the following retail situation.

A \$40 pair of blue jeans is on sale for 25% off. If we purchase the jeans before noon, they are an additional 10% off. What is the final sale price of the jeans?

We might be tempted to say that the jeans are 35% off and calculate \$40(0.35) = \$14, giving a final sale price of

$$\$40 - \$14 = \$26. \quad \text{Incorrect}$$

\$26 *is not correct.* To find the final sale price, we must first find the price after taking 25% off and then take an additional 10% off *that* price. See **Figure 95.**

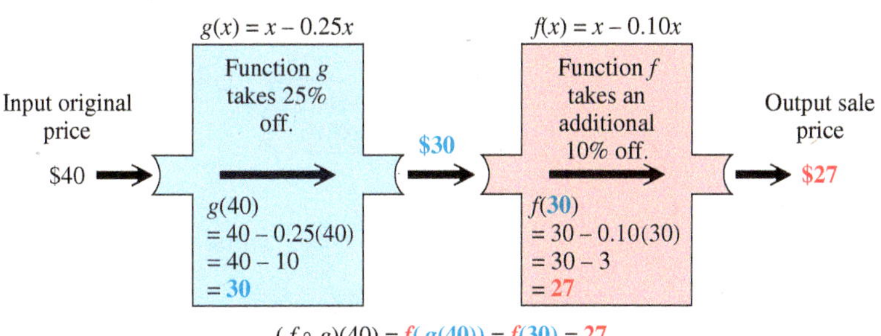

Figure 95

EXAMPLE 5 Evaluating Composite Functions

Let $f(x) = 2x - 1$ and $g(x) = \frac{4}{x-1}$.

(a) Find $(f \circ g)(2)$. **(b)** Find $(g \circ f)(-3)$.

SOLUTION

(a) First find $g(2)$: $g(2) = \frac{4}{2-1} = \frac{4}{1} = 4.$

Now find $(f \circ g)(2)$.

$$\begin{aligned}
&(f \circ g)(2) \\
&= f(g(2)) && \text{Definition of composition} \\
&= f(4) && g(2) = 4 \\
&= 2(4) - 1 && \text{Definition of } f \\
&= 7 && \text{Simplify.}
\end{aligned}$$

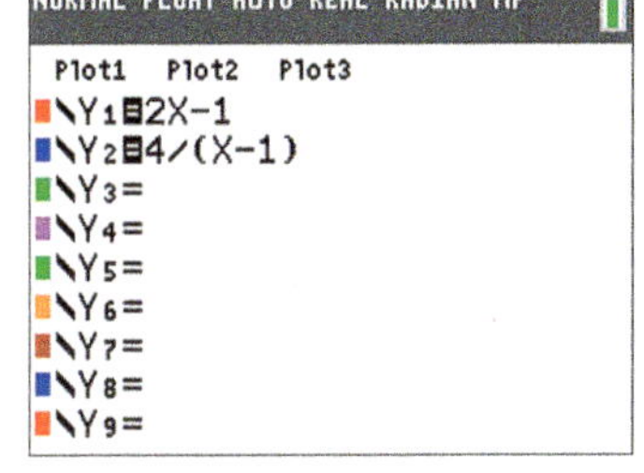

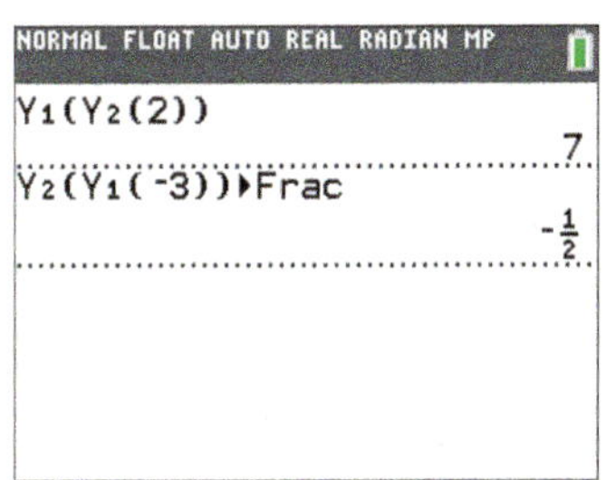

The screens show how a graphing calculator evaluates the expressions in **Example 5.**

(b) $(g \circ f)(-3)$

$$\begin{aligned}
&= g(f(-3)) && \text{Definition of composition} \\
&= g[2(-3) - 1] && f(-3) = 2(-3) - 1 \\
&= g(-7) && \text{Multiply, and then subtract.} \\
&= \frac{4}{-7-1} && g(x) = \frac{4}{x+1} \\
&= -\frac{1}{2} && \text{Subtract in the denominator. Write in lowest terms.}
\end{aligned}$$

✔ **Now Try Exercise 57.**

EXAMPLE 6 Determining Composite Functions and Their Domains

Given that $f(x) = \sqrt{x}$ and $g(x) = 4x + 2$, find each of the following.

(a) $(f \circ g)(x)$ and its domain **(b)** $(g \circ f)(x)$ and its domain

SOLUTION

(a) $(f \circ g)(x)$

$$\begin{aligned}
&= f(g(x)) && \text{Definition of composition} \\
&= f(4x + 2) && g(x) = 4x + 2 \\
&= \sqrt{4x + 2} && f(x) = \sqrt{x}
\end{aligned}$$

The domain and range of g are both the set of all real numbers, $(-\infty, \infty)$. The domain of f is the set of all nonnegative real numbers, $[0, \infty)$. Thus, $g(x)$, which is defined as $4x + 2$, must be greater than or equal to zero.

The radicand must be nonnegative.

$$4x + 2 \geq 0 \quad \text{Solve the inequality.}$$

$$x \geq -\frac{1}{2} \quad \text{Subtract 2. Divide by 4.}$$

Therefore, the domain of $f \circ g$ is $\left[-\frac{1}{2}, \infty\right)$.

(b) $(g \circ f)(x)$

$$= g(f(x)) \quad \text{Definition of composition}$$

$$= g\left(\sqrt{x}\right) \quad f(x) = \sqrt{x}$$

$$= 4\sqrt{x} + 2 \quad g(x) = 4x + 2$$

The domain and range of f are both the set of all nonnegative real numbers, $[0, \infty)$. The domain of g is the set of all real numbers, $(-\infty, \infty)$. Therefore, the domain of $g \circ f$ is $[0, \infty)$.

✔ **Now Try Exercise 75.**

EXAMPLE 7 Determining Composite Functions and Their Domains

Given that $f(x) = \frac{6}{x-3}$ and $g(x) = \frac{1}{x}$, find each of the following.

(a) $(f \circ g)(x)$ and its domain

(b) $(g \circ f)(x)$ and its domain

SOLUTION

(a) $(f \circ g)(x)$

$$= f(g(x)) \quad \text{By definition}$$

$$= f\left(\frac{1}{x}\right) \quad g(x) = \frac{1}{x}$$

$$= \frac{6}{\frac{1}{x} - 3} \quad f(x) = \frac{6}{x-3}$$

$$= \frac{6x}{1 - 3x} \quad \text{Multiply the numerator and denominator by } x.$$

The domain of g is the set of all real numbers *except* 0, which makes $g(x)$ undefined. The domain of f is the set of all real numbers *except* 3. The expression for $g(x)$, therefore, cannot equal 3. We determine the value that makes $g(x) = 3$ and *exclude* it from the domain of $f \circ g$.

$$\frac{1}{x} = 3 \quad \text{The solution must be excluded.}$$

$$1 = 3x \quad \text{Multiply by } x.$$

$$x = \frac{1}{3} \quad \text{Divide by 3.}$$

Therefore, the domain of $f \circ g$ is the set of all real numbers *except* 0 and $\frac{1}{3}$, written in interval notation as

$$(-\infty, 0) \cup \left(0, \frac{1}{3}\right) \cup \left(\frac{1}{3}, \infty\right).$$

LOOKING AHEAD TO CALCULUS

Finding the derivative of a function in calculus is called **differentiation.** To differentiate a composite function such as

$$h(x) = (3x + 2)^4,$$

we interpret $h(x)$ as $(f \circ g)(x)$, where

$$g(x) = 3x + 2 \quad \text{and} \quad f(x) = x^4.$$

The **chain rule** allows us to differentiate composite functions. Notice the use of the composition symbol and function notation in the following, which comes from the chain rule.

If $h(x) = (f \circ g)(x)$, then
$$h'(x) = f'(g(x)) \cdot g'(x).$$

(b) $(g \circ f)(x)$

$= g(f(x))$ By definition

$= g\left(\dfrac{6}{x-3}\right)$ $f(x) = \frac{6}{x-3}$

$= \dfrac{1}{\frac{6}{x-3}}$ Note that this is meaningless if $x = 3$; $g(x) = \frac{1}{x}$.

$= \dfrac{x-3}{6}$ $\dfrac{1}{\frac{a}{b}} = 1 \div \frac{a}{b} = 1 \cdot \frac{b}{a} = \frac{b}{a}$

The domain of f is the set of all real numbers *except* 3. The domain of g is the set of all real numbers *except* 0. The expression for $f(x)$, which is $\frac{6}{x-3}$, is never zero because the numerator is the nonzero number 6. The domain of $g \circ f$ is the set of all real numbers *except* 3, written $(-\infty, 3) \cup (3, \infty)$.

✔ **Now Try Exercise 87.**

NOTE It often helps to consider the *unsimplified* form of a composition expression when determining the domain in a situation like **Example 7(b).**

EXAMPLE 8 Showing That $(g \circ f)(x)$ Is Not Equivalent to $(f \circ g)(x)$

Let $f(x) = 4x + 1$ and $g(x) = 2x^2 + 5x$. Show that $(g \circ f)(x) \neq (f \circ g)(x)$. (This is sufficient to prove that this inequality is true in general.)

SOLUTION First, find $(g \circ f)(x)$. Then find $(f \circ g)(x)$.

$(g \circ f)(x)$

$= g(f(x))$ By definition

$= g(4x + 1)$ $f(x) = 4x + 1$

$= 2(4x + 1)^2 + 5(4x + 1)$ $g(x) = 2x^2 + 5x$

$= 2(16x^2 + 8x + 1) + 20x + 5$ Square $4x + 1$ and apply the distributive property.

$= 32x^2 + 16x + 2 + 20x + 5$ Distributive property

$= 32x^2 + 36x + 7$ Combine like terms.

$(f \circ g)(x)$

$= f(g(x))$ By definition

$= f(2x^2 + 5x)$ $g(x) = 2x^2 + 5x$

$= 4(2x^2 + 5x) + 1$ $f(x) = 4x + 1$

$= 8x^2 + 20x + 1$ Distributive property

Thus, $(g \circ f)(x) \neq (f \circ g)(x)$.

✔ **Now Try Exercise 91.**

As **Example 8** shows, ***it is not always true that $f \circ g = g \circ f$.*** One important circumstance in which equality holds occurs when f and g are *inverses* of each other, a concept discussed later in the text.

In calculus it is sometimes necessary to treat a function as a composition of two functions. The next example shows how this can be done.

EXAMPLE 9 Finding Functions That Form a Given Composite

Find functions f and g such that

$$(f \circ g)(x) = (x^2 - 5)^3 - 4(x^2 - 5) + 3.$$

SOLUTION Note the repeated quantity $x^2 - 5$. If we choose $g(x) = x^2 - 5$ and $f(x) = x^3 - 4x + 3$, then we have the following.

$$\begin{aligned}
&(f \circ g)(x) \\
&\quad = f(g(x)) && \text{By definition} \\
&\quad = f(x^2 - 5) && g(x) = x^2 - 5 \\
&\quad = (x^2 - 5)^3 - 4(x^2 - 5) + 3 && \text{Use the rule for } f.
\end{aligned}$$

There are other pairs of functions f and g that also satisfy these conditions. For instance,

$$f(x) = (x - 5)^3 - 4(x - 5) + 3 \quad \text{and} \quad g(x) = x^2.$$

✔ Now Try Exercise 99.

2.8 Exercises

CONCEPT PREVIEW *Without using paper and pencil, evaluate each expression given the following functions.*

$$f(x) = x + 1 \quad \text{and} \quad g(x) = x^2$$

1. $(f + g)(2)$ **2.** $(f - g)(2)$ **3.** $(fg)(2)$

4. $\left(\frac{f}{g}\right)(2)$ **5.** $(f \circ g)(2)$ **6.** $(g \circ f)(2)$

CONCEPT PREVIEW *Refer to functions f and g as described in* ***Exercises 1–6,*** *and find the following.*

7. domain of f **8.** domain of g **9.** domain of $f + g$ **10.** domain of $\frac{f}{g}$

Let $f(x) = x^2 + 3$ and $g(x) = -2x + 6$. Find each of the following. ***See Example 1.***

11. $(f + g)(3)$ **12.** $(f + g)(-5)$ **13.** $(f - g)(-1)$ **14.** $(f - g)(4)$

15. $(fg)(4)$ **16.** $(fg)(-3)$ **17.** $\left(\frac{f}{g}\right)(-1)$ **18.** $\left(\frac{f}{g}\right)(5)$

For the pair of functions defined, find $(f + g)(x)$, $(f - g)(x)$, $(fg)(x)$, and $\left(\frac{f}{g}\right)(x)$. Give the domain of each. ***See Example 2.***

19. $f(x) = 3x + 4$, $g(x) = 2x - 5$

20. $f(x) = 6 - 3x$, $g(x) = -4x + 1$

21. $f(x) = 2x^2 - 3x$, $g(x) = x^2 - x + 3$

22. $f(x) = 4x^2 + 2x$, $g(x) = x^2 - 3x + 2$

23. $f(x) = \sqrt{4x - 1}$, $g(x) = \frac{1}{x}$

24. $f(x) = \sqrt{5x - 4}$, $g(x) = -\frac{1}{x}$

Associate's Degrees Earned *The graph shows the number of associate's degrees earned (in thousands) in the United States from 2004 through 2012.*

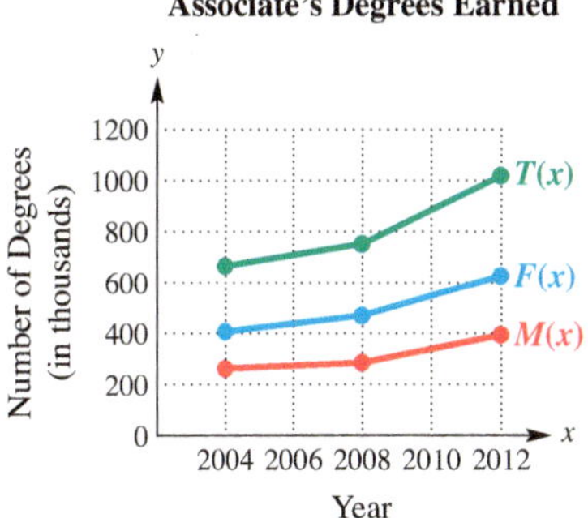

Source: National Center for Education Statistics.

$M(x)$ gives the number of degrees earned by males.

$F(x)$ gives the number earned by females.

$T(x)$ gives the total number for both groups.

25. Estimate $M(2008)$ and $F(2008)$, and use the results to estimate $T(2008)$.

26. Estimate $M(2012)$ and $F(2012)$, and use the results to estimate $T(2012)$.

27. Use the slopes of the line segments to decide in which period (2004–2008 or 2008–2012) the total number of associate's degrees earned increased more rapidly.

28. ***Concept Check*** Refer to the graph of Associate's Degrees Earned.

If $2004 \leq k \leq 2012$, $T(k) = r$, and $F(k) = s$, then $M(k) =$ ______.

Science and Space/Technology Spending *The graph shows dollars (in billions) spent for general science and for space/other technologies in selected years.*

$G(x)$ represents the dollars spent for general science.

$S(x)$ represents the dollars spent for space and other technologies.

$T(x)$ represents the total expenditures for these two categories.

29. Estimate $(T - S)(2000)$. What does this function represent?

30. Estimate $(T - G)(2010)$. What does this function represent?

31. In which category and which period(s) does spending decrease?

32. In which period does spending for $T(x)$ increase most?

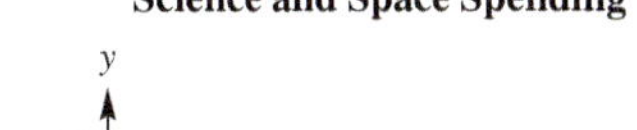
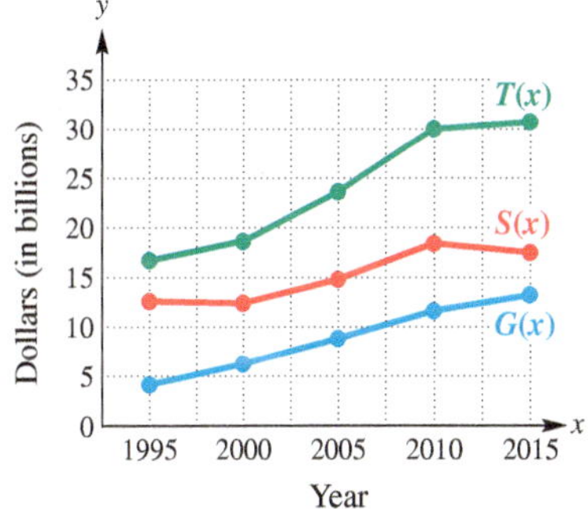

Source: U.S. Office of Management and Budget.

Use the graph to evaluate each expression. ***See Example 3(a).***

33. (a) $(f + g)(2)$ **(b)** $(f - g)(1)$ **(c)** $(fg)(0)$ **(d)** $\left(\frac{f}{g}\right)(1)$

34. (a) $(f + g)(0)$ **(b)** $(f - g)(-1)$ **(c)** $(fg)(1)$ **(d)** $\left(\frac{f}{g}\right)(2)$

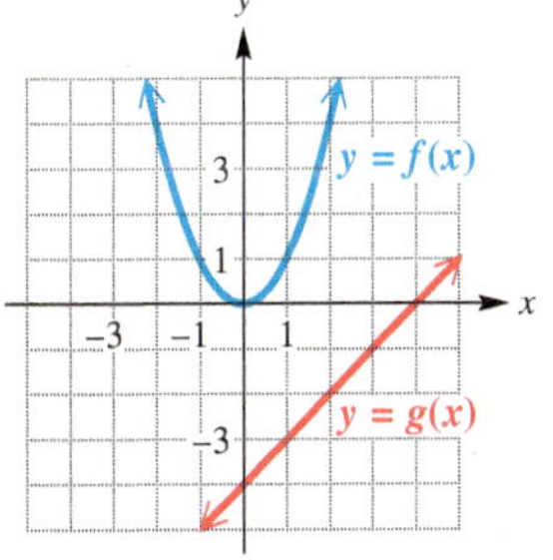

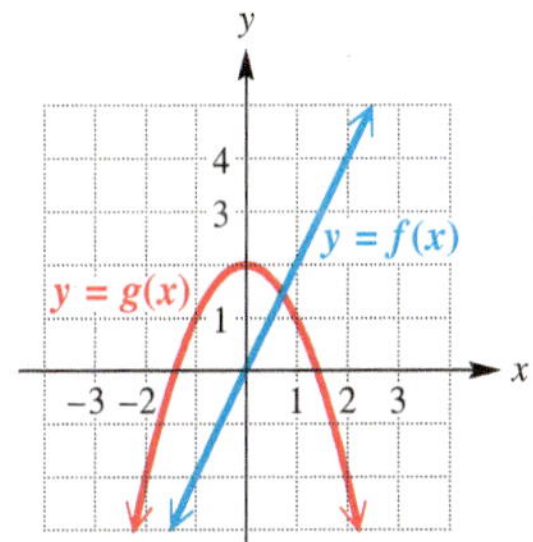

35. **(a)** $(f+g)(-1)$ **(b)** $(f-g)(-2)$ **(c)** $(fg)(0)$ **(d)** $\left(\frac{f}{g}\right)(2)$

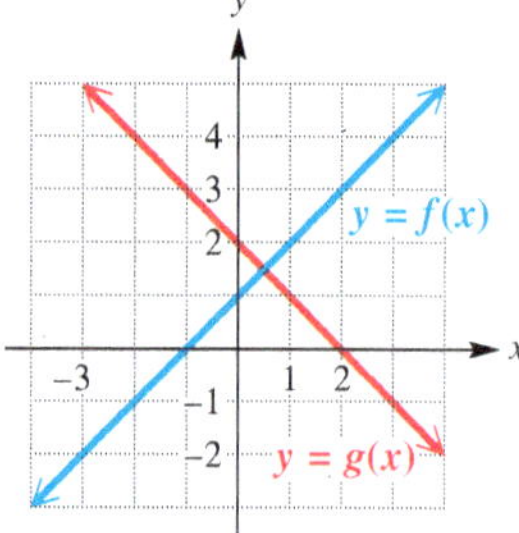

36. **(a)** $(f+g)(1)$ **(b)** $(f-g)(0)$ **(c)** $(fg)(-1)$ **(d)** $\left(\frac{f}{g}\right)(1)$

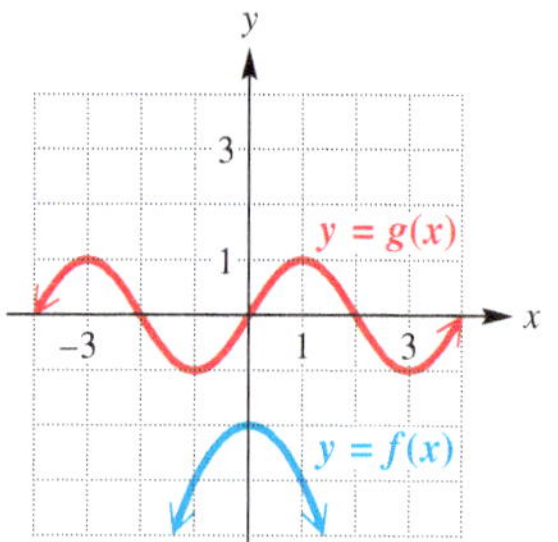

Use the table to evaluate each expression in parts (a)–(d), if possible. ***See Example 3(b).***

(a) $(f+g)(2)$ **(b)** $(f-g)(4)$ **(c)** $(fg)(-2)$ **(d)** $\left(\frac{f}{g}\right)(0)$

37.

x	$f(x)$	$g(x)$
-2	0	6
0	5	0
2	7	-2
4	10	5

38.

x	$f(x)$	$g(x)$
-2	-4	2
0	8	-1
2	5	4
4	0	0

39. Use the table in **Exercise 37** to complete the following table.

x	$(f+g)(x)$	$(f-g)(x)$	$(fg)(x)$	$\left(\frac{f}{g}\right)(x)$
-2				
0				
2				
4				

40. Use the table in **Exercise 38** to complete the following table.

x	$(f+g)(x)$	$(f-g)(x)$	$(fg)(x)$	$\left(\frac{f}{g}\right)(x)$
-2				
0				
2				
4				

41. *Concept Check* How is the difference quotient related to slope?

42. *Concept Check* Refer to **Figure 93.** How is the secant line PQ related to the tangent line to a curve at point P?

For each function, find ***(a)*** $f(x+h)$, ***(b)*** $f(x+h)-f(x)$, *and* ***(c)*** $\frac{f(x+h)-f(x)}{h}$.

See Example 4.

43. $f(x)=2-x$ **44.** $f(x)=1-x$ **45.** $f(x)=6x+2$

46. $f(x)=4x+11$ **47.** $f(x)=-2x+5$ **48.** $f(x)=-4x+2$

49. $f(x) = \frac{1}{x}$ **50.** $f(x) = \frac{1}{x^2}$ **51.** $f(x) = x^2$

52. $f(x) = -x^2$ **53.** $f(x) = 1 - x^2$ **54.** $f(x) = 1 + 2x^2$

55. $f(x) = x^2 + 3x + 1$ **56.** $f(x) = x^2 - 4x + 2$

Let $f(x) = 2x - 3$ and $g(x) = -x + 3$. Find each function value. ***See Example 5.***

57. $(f \circ g)(4)$ **58.** $(f \circ g)(2)$ **59.** $(f \circ g)(-2)$ **60.** $(g \circ f)(3)$

61. $(g \circ f)(0)$ **62.** $(g \circ f)(-2)$ **63.** $(f \circ f)(2)$ **64.** $(g \circ g)(-2)$

Concept Check The tables give some selected ordered pairs for functions f and g.

x	3	4	6
$f(x)$	1	3	9

x	2	7	1	9
$g(x)$	3	6	9	12

Find each of the following.

65. $(f \circ g)(2)$ **66.** $(f \circ g)(7)$ **67.** $(g \circ f)(3)$

68. $(g \circ f)(6)$ **69.** $(f \circ f)(4)$ **70.** $(g \circ g)(1)$

71. *Concept Check* Why can we not determine $(f \circ g)(1)$ given the information in the tables for **Exercises 65–70?**

72. *Concept Check* Extend the concept of composition of functions to evaluate $(g \circ (f \circ g))(7)$ using the tables for **Exercises 65–70.**

*Given functions f and g, find **(a)** $(f \circ g)(x)$ and its domain, and **(b)** $(g \circ f)(x)$ and its domain.* ***See Examples 6 and 7.***

73. $f(x) = -6x + 9, \quad g(x) = 5x + 7$ **74.** $f(x) = 8x + 12, \quad g(x) = 3x - 1$

75. $f(x) = \sqrt{x}, \quad g(x) = x + 3$ **76.** $f(x) = \sqrt{x}, \quad g(x) = x - 1$

77. $f(x) = x^3, \quad g(x) = x^2 + 3x - 1$ **78.** $f(x) = x + 2, \quad g(x) = x^4 + x^2 - 4$

79. $f(x) = \sqrt{x - 1}, \quad g(x) = 3x$ **80.** $f(x) = \sqrt{x - 2}, \quad g(x) = 2x$

81. $f(x) = \frac{2}{x}, \quad g(x) = x + 1$ **82.** $f(x) = \frac{4}{x}, \quad g(x) = x + 4$

83. $f(x) = \sqrt{x + 2}, \quad g(x) = -\frac{1}{x}$ **84.** $f(x) = \sqrt{x + 4}, \quad g(x) = -\frac{2}{x}$

85. $f(x) = \sqrt{x}, \quad g(x) = \frac{1}{x + 5}$ **86.** $f(x) = \sqrt{x}, \quad g(x) = \frac{3}{x + 6}$

87. $f(x) = \frac{1}{x - 2}, \quad g(x) = \frac{1}{x}$ **88.** $f(x) = \frac{1}{x + 4}, \quad g(x) = -\frac{1}{x}$

89. *Concept Check* Fill in the missing entries in the table.

x	$f(x)$	$g(x)$	$g(f(x))$
1	3	2	7
2	1	5	
3	2		

90. *Concept Check* Suppose $f(x)$ is an odd function and $g(x)$ is an even function. Fill in the missing entries in the table.

x	-2	-1	0	1	2
$f(x)$			0	-2	
$g(x)$	0	2	1		
$(f \circ g)(x)$		1	-2		

91. Show that $(f \circ g)(x)$ is not equivalent to $(g \circ f)(x)$ for

$$f(x) = 3x - 2 \quad \text{and} \quad g(x) = 2x - 3.$$

92. Show that for the functions

$$f(x) = x^3 + 7 \quad \text{and} \quad g(x) = \sqrt[3]{x - 7},$$

both $(f \circ g)(x)$ and $(g \circ f)(x)$ equal x.

For certain pairs of functions f and g, $(f \circ g)(x) = x$ and $(g \circ f)(x) = x$. Show that this is true for each pair in ***Exercises 93–96.***

93. $f(x) = 4x + 2, \quad g(x) = \frac{1}{4}(x - 2)$

94. $f(x) = -3x, \quad g(x) = -\frac{1}{3}x$

95. $f(x) = \sqrt[3]{5x + 4}, \quad g(x) = \frac{1}{5}x^3 - \frac{4}{5}$

96. $f(x) = \sqrt[3]{x + 1}, \quad g(x) = x^3 - 1$

Find functions f and g such that $(f \circ g)(x) = h(x)$. (There are many possible ways to do this.) ***See Example 9.***

97. $h(x) = (6x - 2)^2$

98. $h(x) = (11x^2 + 12x)^2$

99. $h(x) = \sqrt{x^2 - 1}$

100. $h(x) = (2x - 3)^3$

101. $h(x) = \sqrt{6x} + 12$

102. $h(x) = \sqrt[3]{2x + 3} - 4$

Solve each problem.

103. *Relationship of Measurement Units* The function $f(x) = 12x$ computes the number of inches in x feet, and the function $g(x) = 5280x$ computes the number of feet in x miles. What is $(f \circ g)(x)$, and what does it compute?

104. The function $f(x) = 3x$ computes the number of feet in x yards, and the function $g(x) = 1760x$ computes the number of yards in x miles. What is $(f \circ g)(x)$, and what does it compute?

105. *Area of an Equilateral Triangle* The area of an equilateral triangle with sides of length x is given by the function $\mathcal{A}(x) = \frac{\sqrt{3}}{4}x^2$.

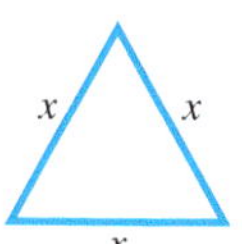

(a) Find $\mathcal{A}(2x)$, the function representing the area of an equilateral triangle with sides of length twice the original length.

(b) Find the area of an equilateral triangle with side length 16. Use the formula $\mathcal{A}(2x)$ found in part (a).

106. *Perimeter of a Square* The perimeter x of a square with side of length s is given by the formula $x = 4s$.

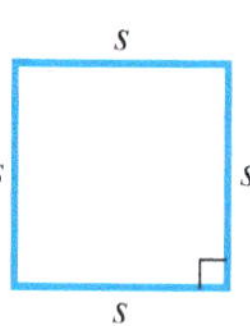

(a) Solve for s in terms of x.

(b) If y represents the area of this square, write y as a function of the perimeter x.

(c) Use the composite function of part (b) to find the area of a square with perimeter 6.

107. ***Oil Leak*** An oil well off the Gulf Coast is leaking, with the leak spreading oil over the water's surface as a circle. At any time t, in minutes, after the beginning of the leak, the radius of the circular oil slick on the surface is $r(t) = 4t$ feet. Let $\mathcal{A}(r) = \pi r^2$ represent the area of a circle of radius r.

(a) Find $(\mathcal{A} \circ r)(t)$.

(b) Interpret $(\mathcal{A} \circ r)(t)$.

(c) What is the area of the oil slick after 3 min?

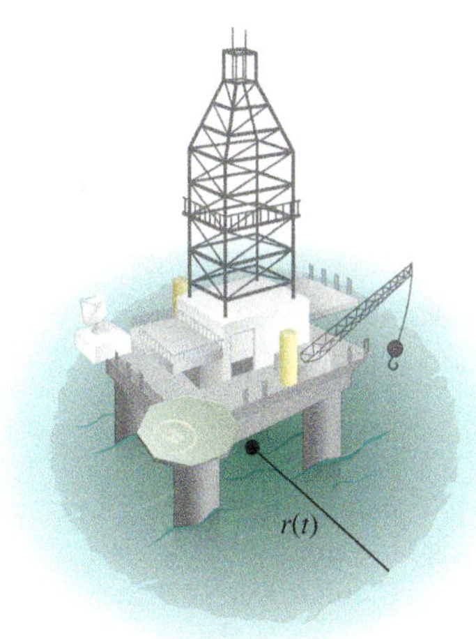

108. ***Emission of Pollutants*** When a thermal inversion layer is over a city (as happens in Los Angeles), pollutants cannot rise vertically but are trapped below the layer and must disperse horizontally. Assume that a factory smokestack begins emitting a pollutant at 8 A.M. Assume that the pollutant disperses horizontally over a circular area. If t represents the time, in hours, since the factory began emitting pollutants ($t = 0$ represents 8 A.M.), assume that the radius of the circle of pollutants at time t is $r(t) = 2t$ miles. Let $\mathcal{A}(r) = \pi r^2$ represent the area of a circle of radius r.

(a) Find $(\mathcal{A} \circ r)(t)$. **(b)** Interpret $(\mathcal{A} \circ r)(t)$.

(c) What is the area of the circular region covered by the layer at noon?

109. ***(Modeling) Product of Two Factors*** Suppose that x represents a real number between 0 and 100. Consider two factors of the following forms: $N(x)$ represents the number that is x less than 100 and $G(x)$ represents the number that is 20 greater than the product of 5 and x.

(a) Express the factor $N(x)$ in terms of x.

(b) Express the factor $G(x)$ in terms of x.

(c) Express their product $C(x)$.

(d) Suppose that in an application $C(x)$ represents a cost in dollars and the value of x is 20. Evaluate $C(20)$.

110. ***Software Author Royalties*** A software author invests his royalties in two accounts for 1 yr.

(a) The first account pays 2% simple interest. If he invests x dollars in this account, write an expression for y_1 in terms of x, where y_1 represents the amount of interest earned.

(b) He invests in a second account \$500 more than he invested in the first account. This second account pays 1.5% simple interest. Write an expression for y_2, where y_2 represents the amount of interest earned.

(c) What does $y_1 + y_2$ represent?

(d) How much interest will he receive if \$250 is invested in the first account?

111. ***Sale Pricing*** In the sale room at a clothing store, every item is on sale for half the original price, plus 1 dollar.

(a) Write a function g that finds half of x.

(b) Write a function f that adds 1 to x.

(c) Write and simplify the function $(f \circ g)(x)$.

(d) Use the function from part (c) to find the sale price of a shirt at this store that has original price \$60.

112. ***Area of a Square*** The area of a square is x^2 square inches. Suppose that 3 in. is added to one dimension and 1 in. is subtracted from the other dimension. Express the area $\mathcal{A}(x)$ of the resulting rectangle as a product of two functions.

Chapter 2 Test Prep

Key Terms

2.1 ordered pair, origin, x-axis, y-axis, rectangular (Cartesian) coordinate system, coordinate plane (xy-plane), quadrants, coordinates, conditional statement, collinear, graph of an equation, x-intercept, y-intercept

2.2 circle, radius, center of a circle

2.3 dependent variable, independent variable, relation, function, input, output, input-output (function) machine, domain, range, increasing function, decreasing function, constant function

2.4 linear function, constant function, standard form, relatively prime, slope, average rate of change, mathematical modeling, linear cost function, cost, fixed cost, revenue function, profit function

2.5 point-slope form, slope-intercept form, negative reciprocals, scatter diagram, linear regression, zero (of a function)

2.6 continuous function, parabola, vertex, piecewise-defined function, step function

2.7 symmetry, even function, odd function, vertical translation, horizontal translation

2.8 difference quotient, secant line, composite function (composition)

New Symbols

(a, b)	ordered pair
$f(x)$	function f evaluated at x (read "f of x" or "f at x")
Δx	change in x
Δy	change in y
m	slope
$[\![x]\!]$	the greatest integer less than or equal to x
$f \circ g$	composite function

Quick Review

Concepts	Examples

2.1 Rectangular Coordinates and Graphs

Distance Formula

Suppose that $P(x_1, y_1)$ and $R(x_2, y_2)$ are two points in a coordinate plane. The distance between P and R, written $d(P, R)$, is given by the following formula.

$$d(P, R) = \sqrt{(x_2 - x_1)^2 + (y_2 - y_1)^2}$$

Example: Find the distance between the points $P(-1, 4)$ and $R(6, -3)$.

$$\begin{aligned} d(P, R) &= \sqrt{[6 - (-1)]^2 + (-3 - 4)^2} \\ &= \sqrt{49 + 49} \\ &= \sqrt{98} \\ &= 7\sqrt{2} \qquad \sqrt{98} = \sqrt{49 \cdot 2} = 7\sqrt{2} \end{aligned}$$

Midpoint Formula

The coordinates of the midpoint M of the line segment with endpoints $P(x_1, y_1)$ and $Q(x_2, y_2)$ are given by the following.

$$M = \left(\frac{x_1 + x_2}{2}, \frac{y_1 + y_2}{2}\right)$$

Example: Find the coordinates of the midpoint M of the line segment with endpoints $(-1, 4)$ and $(6, -3)$.

$$M = \left(\frac{-1 + 6}{2}, \frac{4 + (-3)}{2}\right) = \left(\frac{5}{2}, \frac{1}{2}\right)$$

Concepts	Examples
2.2 Circles	
Center-Radius Form of the Equation of a Circle The equation of a circle with center (h, k) and radius r is given by the following. $$(x - h)^2 + (y - k)^2 = r^2$$	Find the center-radius form of the equation of the circle with center $(-2, 3)$ and radius 4. $$[x - (-2)]^2 + (y - 3)^2 = 4^2$$ $$(x + 2)^2 + (y - 3)^2 = 16$$
General Form of the Equation of a Circle $$x^2 + y^2 + Dx + Ey + F = 0$$	The general form of the equation of the preceding circle is $$x^2 + y^2 + 4x - 6y - 3 = 0.$$
2.3 Functions	
A **relation** is a set of ordered pairs. A **function** is a relation in which, for each value of the first component of the ordered pairs, there is *exactly one* value of the second component. The set of first components is the **domain.** The set of second components is the **range.**	The relation $y = x^2$ defines a function because each choice of a number for x corresponds to one and only one number for y. The domain is $(-\infty, \infty)$, and the range is $[0, \infty)$. The relation $x = y^2$ does *not* define a function because a number x may correspond to two numbers for y. The domain is $[0, \infty)$, and the range is $(-\infty, \infty)$.
Vertical Line Test If every vertical line intersects the graph of a relation in no more than one point, then the relation is a function.	Determine whether each graph is that of a function. **A.** 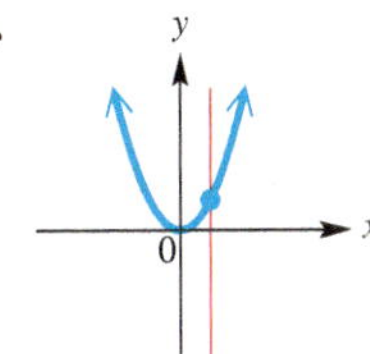 **B.** 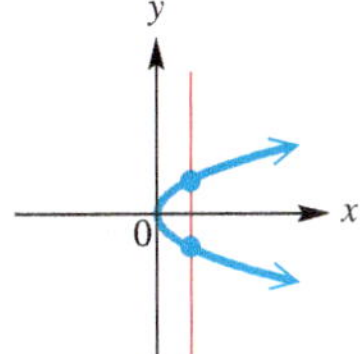 By the vertical line test, graph A is the graph of a function, but graph B is not.
Increasing, Decreasing, and Constant Functions 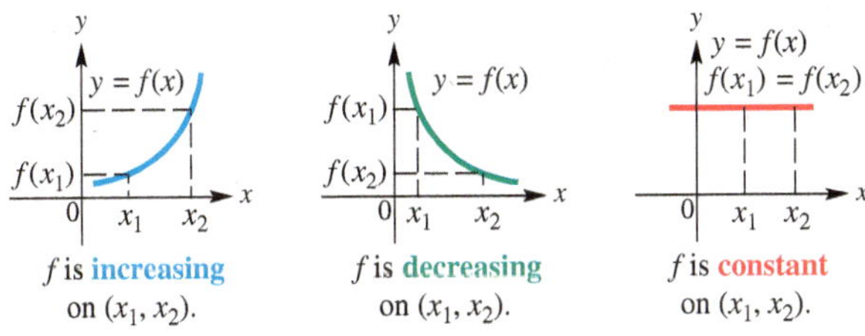	Discuss the function in graph A in terms of whether it is increasing, decreasing, or constant. The function in graph A is decreasing on the open interval $(-\infty, 0)$ and increasing on the open interval $(0, \infty)$.
2.4 Linear Functions	
A function f is a **linear function** if, for real numbers a and b, $$f(x) = ax + b.$$ The graph of a linear function is a line.	The equation $$y = f(x) = \frac{1}{2}x - 4$$ defines y as a linear function f of x.
Definition of Slope The slope m of the line through the points (x_1, y_1) and (x_2, y_2) is given by the following. $$m = \frac{\text{rise}}{\text{run}} = \frac{\Delta y}{\Delta x} = \frac{y_2 - y_1}{x_2 - x_1}, \quad \text{where } \Delta x \neq 0$$	Find the slope of the line through the points $(2, 4)$ and $(-1, 7)$. $$m = \frac{7 - 4}{-1 - 2} = \frac{3}{-3} = -1$$

Concepts	Examples

2.5 Equations of Lines and Linear Models

Summary of Forms of Linear Equations

Equation	Description
$y = mx + b$	**Slope-Intercept Form** Slope is m. y-intercept is $(0, b)$.
$y - y_1 = m(x - x_1)$	**Point-Slope Form** Slope is m. Line passes through (x_1, y_1).
$Ax + By = C$	**Standard Form** (A, B, and C integers, $A \geq 0$.) Slope is $-\frac{A}{B}$ $(B \neq 0)$. x-intercept is $\left(\frac{C}{A}, 0\right)$ $(A \neq 0)$. y-intercept is $\left(0, \frac{C}{B}\right)$ $(B \neq 0)$.
$y = b$	**Horizontal Line** Slope is 0. y-intercept is $(0, b)$.
$x = a$	**Vertical Line** Slope is undefined. x-intercept is $(a, 0)$.

Consider the following equations.

$$y = 3x + \frac{2}{3} \quad \text{Slope-intercept form}$$

The slope m is 3, and the y-intercept is $\left(0, \frac{2}{3}\right)$.

$$y - 3 = -2(x - 4) \quad \text{Point-slope form}$$

The slope m is -2. The line passes through the point $(4, 3)$.

$$4x + 5y = 7 \quad \text{Standard form with } A = 4, B = 5, C = 7$$

The slope is $m = -\frac{A}{B} = -\frac{4}{5}$.

The x-intercept has $\frac{C}{A} = \frac{7}{4}$ and is $\left(\frac{7}{4}, 0\right)$.

The y-intercept has $\frac{C}{B} = \frac{7}{5}$ and is $\left(0, \frac{7}{5}\right)$.

$$y = -6 \quad \text{Horizontal line}$$

The slope is 0. The y-intercept is $(0, -6)$.

$$x = 3 \quad \text{Vertical line}$$

The slope is undefined. The x-intercept is $(3, 0)$.

2.6 Graphs of Basic Functions

Basic Functions

Identity Function $f(x) = x$

Squaring Function $f(x) = x^2$

Cubing Function $f(x) = x^3$

Square Root Function $f(x) = \sqrt{x}$

Cube Root Function $f(x) = \sqrt[3]{x}$

Absolute Value Function $f(x) = |x|$

Greatest Integer Function $f(x) = [\![x]\!]$

Refer to the function boxes in **Section 2.6.** Graphs of the basic functions are also shown on the back inside cover of the print text.

2.7 Graphing Techniques

Stretching and Shrinking

If $a > 1$, then the graph of $y = af(x)$ is a **vertical stretching** of the graph of $y = f(x)$.

If $0 < a < 1$, then the graph of $y = af(x)$ is a **vertical shrinking** of the graph of $y = f(x)$.

If $0 < a < 1$, then the graph of $y = f(ax)$ is a **horizontal stretching** of the graph of $y = f(x)$.

If $a > 1$, then the graph of $y = f(ax)$ is a **horizontal shrinking** of the graph of $y = f(x)$.

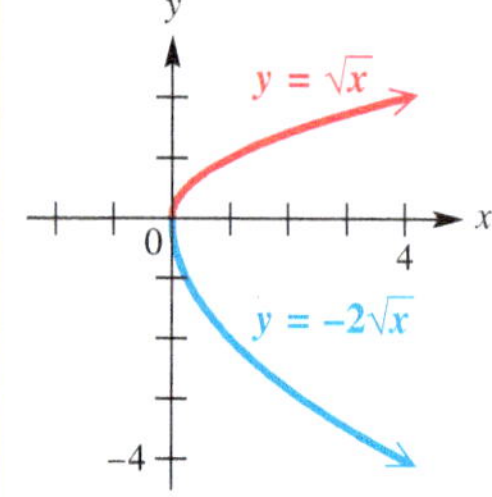

The graph of

$$y = -2\sqrt{x}$$

is the graph of $y = \sqrt{x}$ stretched vertically by a factor of 2 and reflected across the x-axis.

Concepts	Examples

Reflection across an Axis

The graph of $y = -f(x)$ is the same as the graph of $y = f(x)$ reflected across the x-axis.

The graph of $y = f(-x)$ is the same as the graph of $y = f(x)$ reflected across the y-axis.

Symmetry

The graph of an equation is **symmetric with respect to the y-axis** if the replacement of x with $-x$ results in an equivalent equation.

The graph of an equation is **symmetric with respect to the x-axis** if the replacement of y with $-y$ results in an equivalent equation.

The graph of an equation is **symmetric with respect to the origin** if the replacement of both x with $-x$ and y with $-y$ at the same time results in an equivalent equation.

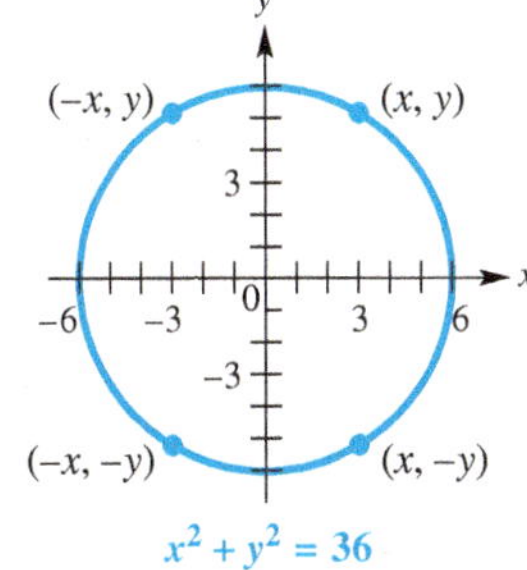

The graph of

$$x^2 + y^2 = 36$$

is symmetric with respect to the y-axis, the x-axis, and the origin.

Translations

Let f be a function and c be a positive number.

To Graph:	Shift the Graph of $y = f(x)$ by c Units:
$y = f(x) + c$	up
$y = f(x) - c$	down
$y = f(x + c)$	left
$y = f(x - c)$	right

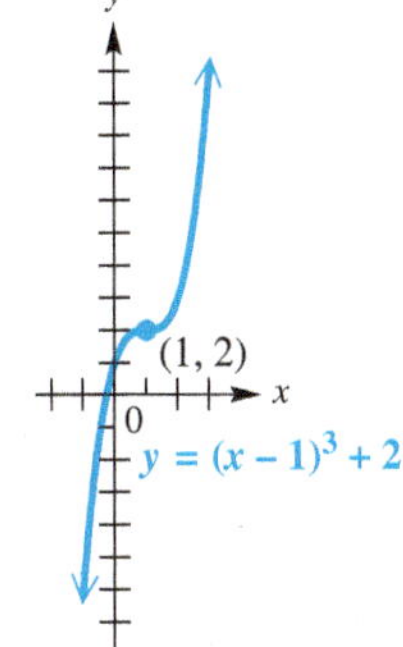

The graph of

$$y = (x - 1)^3 + 2$$

is the graph of $y = x^3$ translated 1 unit to the right and 2 units up.

2.8 Function Operations and Composition

Operations on Functions

Given two functions f and g, then for all values of x for which both $f(x)$ and $g(x)$ are defined, the following operations are defined.

$$(f + g)(x) = f(x) + g(x) \quad \text{Sum function}$$

$$(f - g)(x) = f(x) - g(x) \quad \text{Difference function}$$

$$(fg)(x) = f(x) \cdot g(x) \quad \text{Product function}$$

$$\left(\frac{f}{g}\right)(x) = \frac{f(x)}{g(x)}, \quad g(x) \neq 0 \quad \text{Quotient function}$$

The **domains of $f + g$, $f - g$,** and **fg** include all real numbers in the intersection of the domains of f and g, while the **domain of $\frac{f}{g}$** includes those real numbers in the intersection of the domains of f and g for which $g(x) \neq 0$.

Let $f(x) = 2x - 4$ and $g(x) = \sqrt{x}$.

$$(f + g)(x) = 2x - 4 + \sqrt{x}$$

$$(f - g)(x) = 2x - 4 - \sqrt{x}$$

$$(fg)(x) = (2x - 4)\sqrt{x}$$

The domain is $[0, \infty)$.

$$\left(\frac{f}{g}\right)(x) = \frac{2x - 4}{\sqrt{x}}$$

The domain is $(0, \infty)$.

Difference Quotient

The line joining $P(x, f(x))$ and $Q(x + h, f(x + h))$ has slope

$$m = \frac{\mathbf{f(x + h) - f(x)}}{\mathbf{h}}, \quad h \neq 0.$$

The boldface expression is the difference quotient.

Refer to **Example 4** in **Section 2.8.**

Concepts	Examples
Composition of Functions If f and g are functions, then the composite function, or composition, of f and g is defined by $(f \circ g)(x) = f(g(x))$. The domain of $f \circ g$ is the set of all x in the domain of g such that $g(x)$ is in the domain of f.	Let $f(x) = 2x - 4$ and $g(x) = \sqrt{x}$. $(f \circ g)(x) = 2\sqrt{x} - 4$ The domain is all x such that $x \geq 0$, represented by the interval $[0, \infty)$.

Chapter 2 Review Exercises

Find the distance between each pair of points, and give the coordinates of the midpoint of the line segment joining them.

1. $P(3, -1)$, $Q(-4, 5)$ **2.** $M(-8, 2)$, $N(3, -7)$ **3.** $A(-6, 3)$, $B(-6, 8)$

4. Are the points $(5, 7)$, $(3, 9)$, and $(6, 8)$ the vertices of a right triangle? If so, at what point is the right angle?

5. Determine the coordinates of B for line segment AB, given that A has coordinates $(-6, 10)$ and the coordinates of its midpoint M are $(8, 2)$.

6. Use the distance formula to determine whether the points $(-2, -5)$, $(1, 7)$, and $(3, 15)$ are collinear.

Find the center-radius form of the equation for each circle.

7. center $(-2, 3)$, radius 15 **8.** center $\left(\sqrt{5}, -\sqrt{7}\right)$, radius $\sqrt{3}$

9. center $(-8, 1)$, passing through $(0, 16)$ **10.** center $(3, -6)$, tangent to the x-axis

Connecting Graphs with Equations *Use each graph to determine an equation of the circle. Express in center-radius form.*

11.

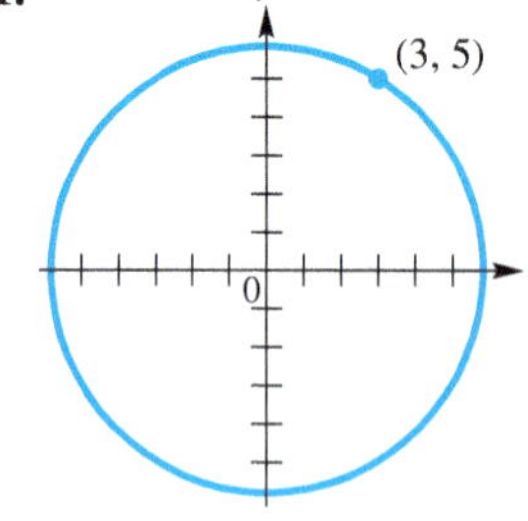

12.

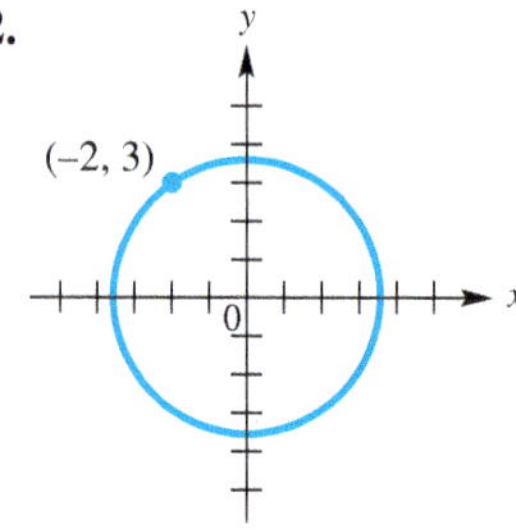

13.

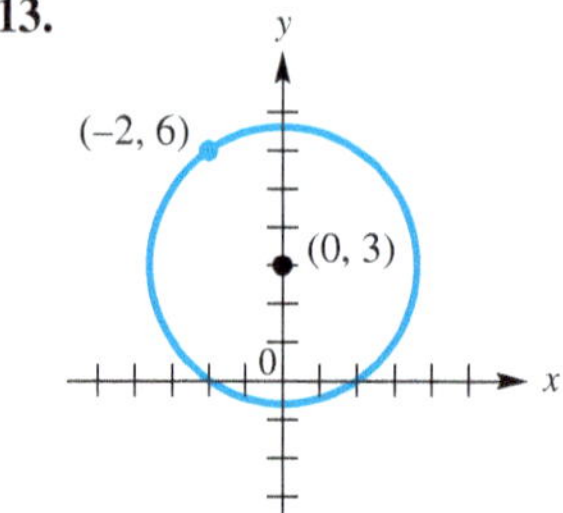

14.

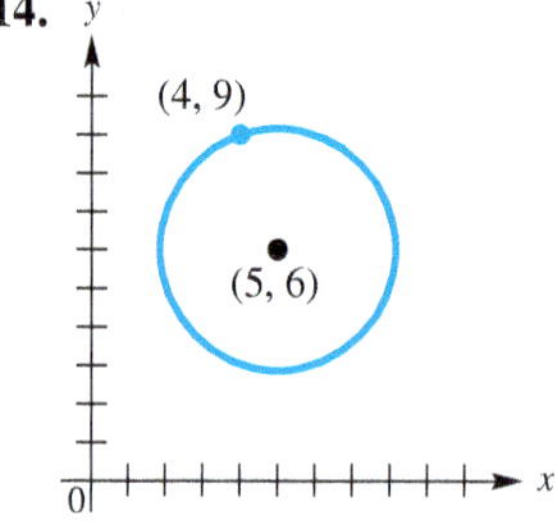

Find the center and radius of each circle.

15. $x^2 + y^2 - 4x + 6y + 12 = 0$

16. $x^2 + y^2 - 6x - 10y + 30 = 0$

17. $2x^2 + 2y^2 + 14x + 6y + 2 = 0$

18. $3x^2 + 3y^2 + 33x - 15y = 0$

For each graph, decide whether y is a function of x. Give the domain and range of each relation.

19.

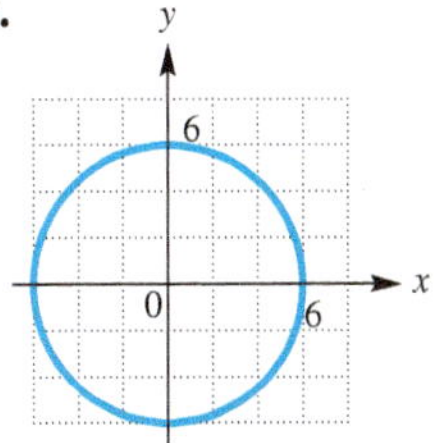

20.

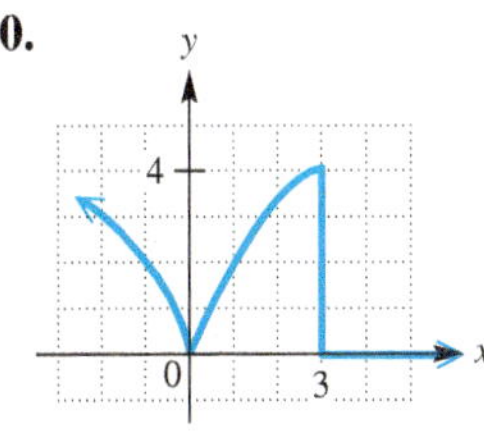

21.

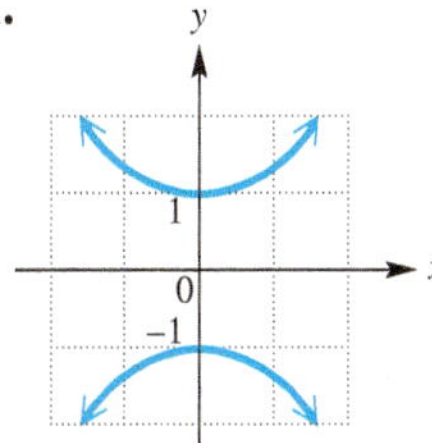

22.

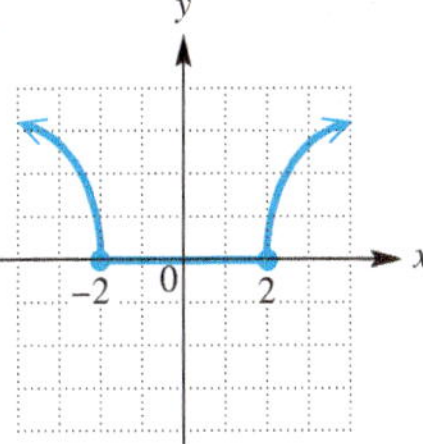

23.

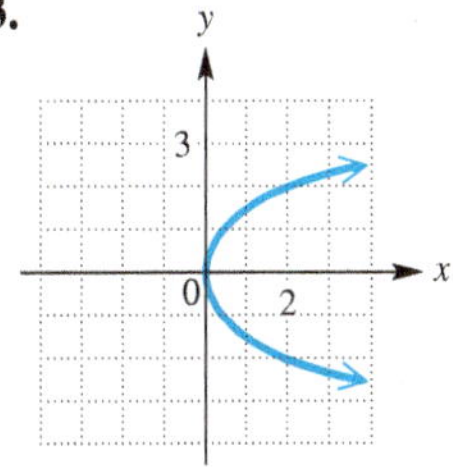

24.

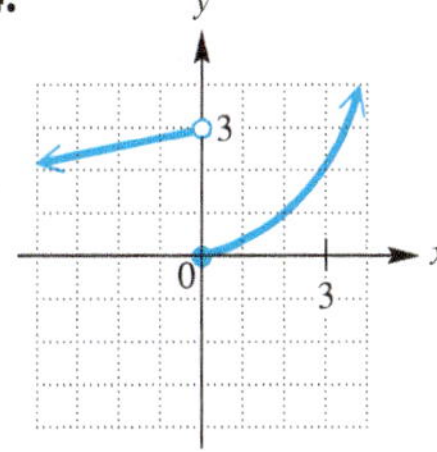

Decide whether each equation defines y as a function of x.

25. $y = 6 - x^2$

26. $x = \frac{1}{3}y^2$

27. $y = \pm\sqrt{x - 2}$

28. $y = -\frac{4}{x}$

Give the domain of each function.

29. $f(x) = -4 + |x|$

30. $f(x) = \frac{8 + x}{8 - x}$

31. $f(x) = \sqrt{6 - 3x}$

32. For the function graphed in **Exercise 22**, determine the largest open intervals over which it is **(a)** increasing, **(b)** decreasing, and **(c)** constant.

Let $f(x) = -2x^2 + 3x - 6$. *Find each of the following.*

33. $f(3)$

34. $f(-0.5)$

35. $f(0)$

36. $f(k)$

Graph each equation.

37. $2x - 5y = 5$

38. $3x + 7y = 14$

39. $2x + 5y = 20$

40. $3y = x$

41. $f(x) = x$

42. $x - 4y = 8$

43. $x = -5$

44. $f(x) = 3$

45. $y + 2 = 0$

46. ***Concept Check*** The equation of the line that lies along the x-axis is ________.

Graph the line satisfying the given conditions.

47. through $(0, 5)$, $m = -\frac{2}{3}$

48. through $(2, -4)$, $m = \frac{3}{4}$

Find the slope of each line, provided that it has a slope.

49. through $(2, -2)$ and $(3, -4)$

50. through $(8, 7)$ and $\left(\frac{1}{2}, -2\right)$

51. through $(0, -7)$ and $(3, -7)$

52. through $(5, 6)$ and $(5, -2)$

53. $11x + 2y = 3$

54. $9x - 4y = 2$

55. $x - 2 = 0$

56. $x - 5y = 0$

Work each problem.

57. *(Modeling) Distance from Home* The graph depicts the distance y that a person driving a car on a straight road is from home after x hours. Interpret the graph. At what speeds did the car travel?

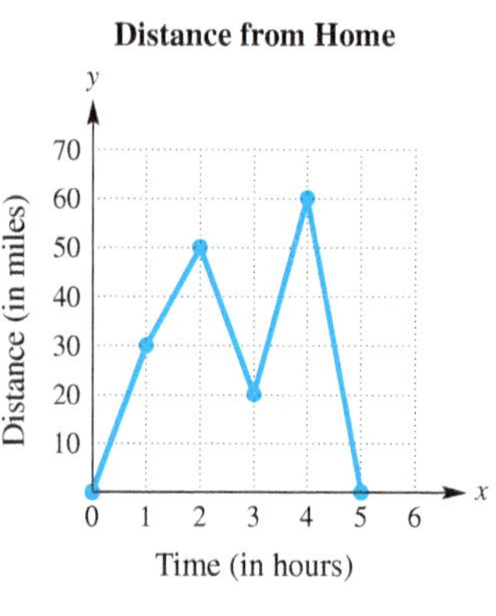

58. *(Modeling) Job Market* The figure shows the number of jobs gained or lost in a recent period from September to May.

(a) Is this the graph of a function?

(b) In what month were the most jobs lost? the most gained?

(c) What was the largest number of jobs lost? of jobs gained?

(d) Do these data show an upward or a downward trend? If so, which is it?

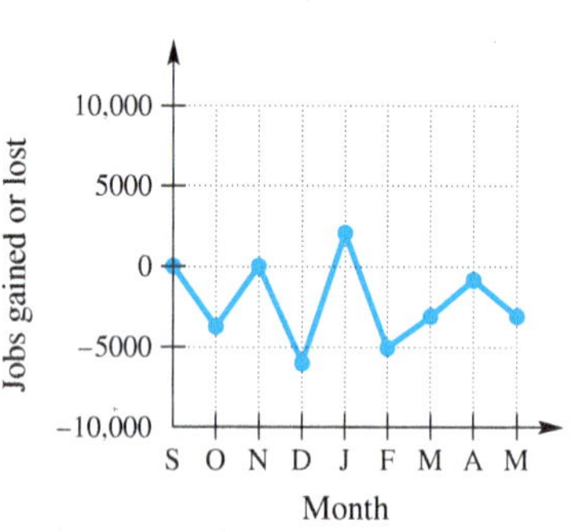

59. *(Modeling) E-Filing Tax Returns* The percent of tax returns filed electronically in 2001 was 30.7%. In 2013, the figure was 82.9%. (*Source:* Internal Revenue Service.)

(a) Use the information given for the years 2001 and 2013, letting $x = 0$ represent 2001, $x = 12$ represent 2013, and y represent the percent of returns filed electronically, to find a linear equation that models the data. Write the equation in slope-intercept form. Interpret the slope of the graph of this equation.

(b) Use the equation from part (a) to approximate the percent of tax returns that were filed electronically in 2009.

60. *Family Income* In 1980 the median family income in the United States was about \$21,000 per year. In 2013 it was about \$63,800 per year. Find the average annual rate of change of median family income to the nearest dollar over that period. (*Source:* U.S. Census Bureau.)

*For each line described, write an equation in **(a)** slope-intercept form, if possible, and **(b)** standard form.*

61. through $(3, -5)$ with slope -2

62. through $(-2, 4)$ and $(1, 3)$

63. through $(2, -1)$, parallel to $3x - y = 1$

64. x-intercept $(-3, 0)$, y-intercept $(0, 5)$

65. through $(2, -10)$, perpendicular to a line with undefined slope

66. through $(0, 5)$, perpendicular to $8x + 5y = 3$

67. through $(-7, 4)$, perpendicular to $y = 8$

68. through $(3, -5)$, parallel to $y = 4$

Graph each function.

69. $f(x) = |x| - 3$

70. $f(x) = -|x|$

71. $f(x) = -(x + 1)^2 + 3$

72. $f(x) = -\sqrt{x} - 2$

73. $f(x) = [\![x - 3]\!]$

74. $f(x) = 2\sqrt[3]{x + 1} - 2$

75. $f(x) = \begin{cases} -4x + 2 & \text{if } x \leq 1 \\ 3x - 5 & \text{if } x > 1 \end{cases}$

76. $f(x) = \begin{cases} x^2 + 3 & \text{if } x < 2 \\ -x + 4 & \text{if } x \geq 2 \end{cases}$

77. $f(x) = \begin{cases} |x| & \text{if } x < 3 \\ 6 - x & \text{if } x \geq 3 \end{cases}$

78. *Concept Check* If x represents an integer, then what is the simplest form of the expression $[\![x]\!] + x$?

Concept Check Decide whether each statement is true *or* false. *If false, tell why.*

79. The graph of an even function is symmetric with respect to the y-axis.

80. The graph of a nonzero function cannot be symmetric with respect to the x-axis.

81. If (a, b) is on the graph of an even function, then so is $(a, -b)$.

82. The graph of an odd function is symmetric with respect to the origin.

83. The constant function $f(x) = 0$ is both even and odd.

84. If (a, b) is on the graph of an odd function, then so is $(-a, b)$.

Decide whether each equation has a graph that is symmetric with respect to the x-axis, *the* y-axis, *the* origin, *or* none of these.

85. $x + y^2 = 10$

86. $5y^2 + 5x^2 = 30$

87. $x^2 = y^3$

88. $y^3 = x + 4$

89. $6x + y = 4$

90. $|y| = -x$

91. $y = 1$

92. $|x| = |y|$

93. $x^2 - y^2 = 0$

94. $x^2 + (y - 2)^2 = 4$

Describe how the graph of each function can be obtained from the graph of $f(x) = |x|$.

95. $g(x) = -|x|$

96. $h(x) = |x| - 2$

97. $k(x) = 2|x - 4|$

Let $f(x) = 3x - 4$. *Find an equation for each reflection of the graph of* $f(x)$.

98. across the x-axis

99. across the y-axis

100. across the origin

101. *Concept Check* The graph of a function f is shown in the figure. Sketch the graph of each function defined as follows.

(a) $y = f(x) + 3$

(b) $y = f(x - 2)$

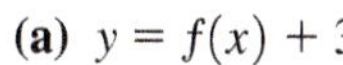

(c) $y = f(x + 3) - 2$

(d) $y = |f(x)|$

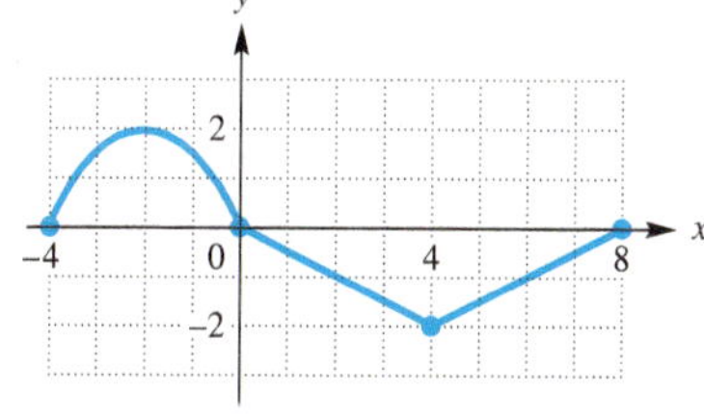

102. *Concept Check* Must the domain of g be a subset of the domain of $f \circ g$?

Let $f(x) = 3x^2 - 4$ *and* $g(x) = x^2 - 3x - 4$. *Find each of the following.*

103. $(fg)(x)$

104. $(f - g)(4)$

105. $(f + g)(-4)$

106. $(f + g)(2k)$

107. $\left(\frac{f}{g}\right)(3)$

108. $\left(\frac{f}{g}\right)(-1)$

109. the domain of $(fg)(x)$

110. the domain of $\left(\frac{f}{g}\right)(x)$

For each function, find and simplify $\frac{f(x + h) - f(x)}{h}$, $h \neq 0$.

111. $f(x) = 2x + 9$

112. $f(x) = x^2 - 5x + 3$

Let $f(x) = \sqrt{x-2}$ and $g(x) = x^2$. Find each of the following, if possible.

113. $(g \circ f)(x)$ **114.** $(f \circ g)(x)$ **115.** $(g \circ f)(3)$

116. $(f \circ g)(-6)$ **117.** $(g \circ f)(-1)$ **118.** the domain of $f \circ g$

Use the table to evaluate each expression, if possible.

119. $(f + g)(1)$

120. $(f - g)(3)$

121. $(fg)(-1)$

122. $\left(\frac{f}{g}\right)(0)$

x	$f(x)$	$g(x)$
-1	3	-2
0	5	0
1	7	1
3	9	9

Use the tables for f and g to evaluate each expression.

123. $(g \circ f)(-2)$

124. $(f \circ g)(3)$

x	$f(x)$
-2	1
0	4
2	3
4	2

x	$g(x)$
1	2
2	4
3	-2
4	0

Concept Check *The graphs of two functions f and g are shown in the figures.*

125. Find $(f \circ g)(2)$.

126. Find $(g \circ f)(3)$.

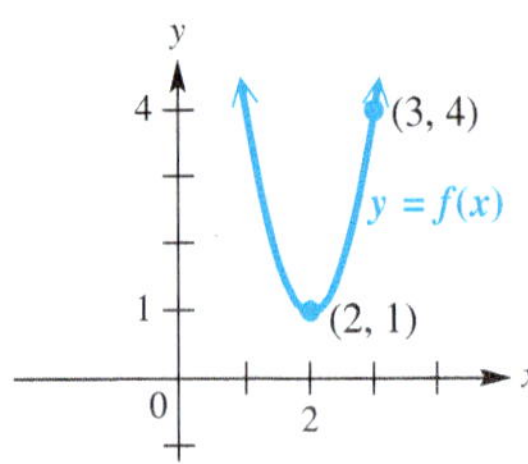

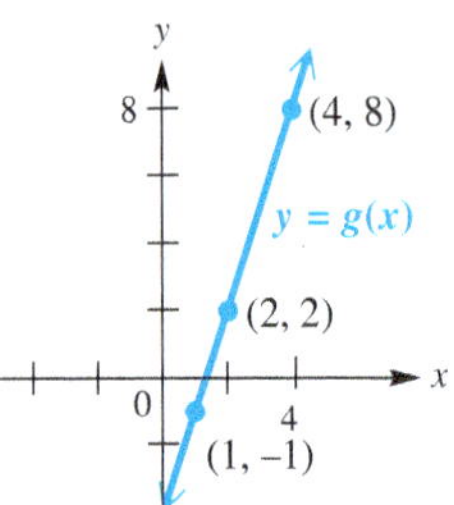

Solve each problem.

127. ***Relationship of Measurement Units*** There are 36 in. in 1 yd, and there are 1760 yd in 1 mi. Express the number of inches in x miles by forming two functions and then considering their composition.

128. ***(Modeling) Perimeter of a Rectangle*** Suppose the length of a rectangle is twice its width. Let x represent the width of the rectangle. Write a formula for the perimeter P of the rectangle in terms of x alone. Then use $P(x)$ notation to describe it as a function. What type of function is this?

129. ***(Modeling) Volume of a Sphere*** The formula for the volume of a sphere is $V(r) = \frac{4}{3}\pi r^3$, where r represents the radius of the sphere. Construct a model function V representing the amount of volume gained when the radius r (in inches) of a sphere is increased by 3 in.

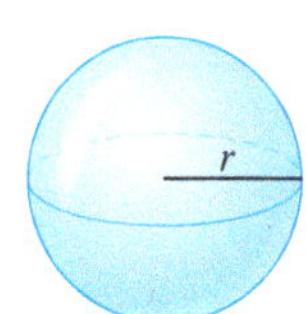

130. ***(Modeling) Dimensions of a Cylinder*** A cylindrical can makes the most efficient use of materials when its height is the same as the diameter of its top.

(a) Express the volume V of such a can as a function of the diameter d of its top.

(b) Express the surface area S of such a can as a function of the diameter d of its top. (*Hint:* The curved side is made from a rectangle whose length is the circumference of the top of the can.)

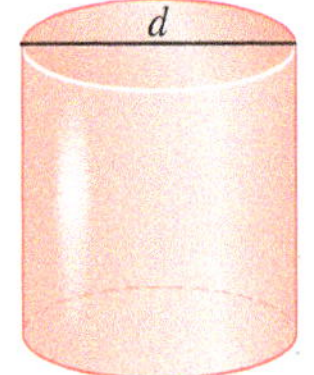

Chapter 2 Test

1. Match the set described in Column I with the correct interval notation from Column II. Choices in Column II may be used once, more than once, or not at all.

I	II
(a) Domain of $f(x) = \sqrt{x} + 3$	A. $[-3, \infty)$
(b) Range of $f(x) = \sqrt{x - 3}$	B. $[3, \infty)$
(c) Domain of $f(x) = x^2 - 3$	C. $(-\infty, \infty)$
(d) Range of $f(x) = x^2 + 3$	D. $[0, \infty)$
(e) Domain of $f(x) = \sqrt[3]{x - 3}$	E. $(-\infty, 3)$
(f) Range of $f(x) = \sqrt[3]{x} + 3$	F. $(-\infty, 3]$
(g) Domain of $f(x) = \lvert x \rvert - 3$	G. $(3, \infty)$
(h) Range of $f(x) = \lvert x + 3 \rvert$	H. $(-\infty, 0]$
(i) Domain of $x = y^2$	
(j) Range of $x = y^2$	

The graph shows the line that passes through the points $(-2, 1)$ and $(3, 4)$. Refer to it to answer Exercises 2–6.

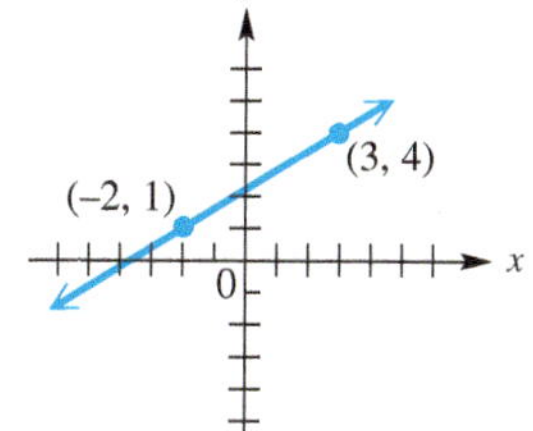

2. What is the slope of the line?

3. What is the distance between the two points shown?

4. What are the coordinates of the midpoint of the line segment joining the two points?

5. Find the standard form of the equation of the line.

6. Write the linear function $f(x) = ax + b$ that has this line as its graph.

7. ***Connecting Graphs with Equations*** Use each graph to determine an equation of the circle. Express it in center-radius form.

(a)

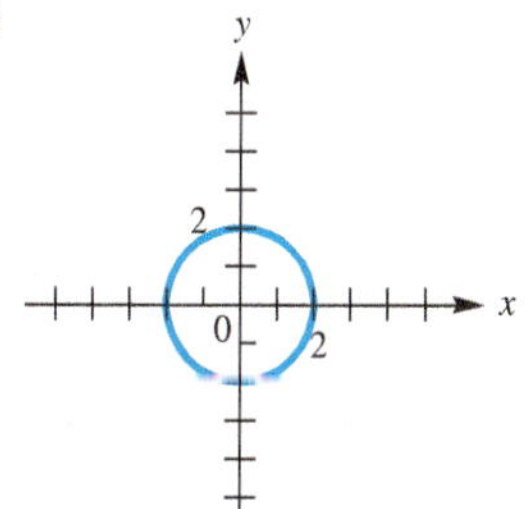

(b)

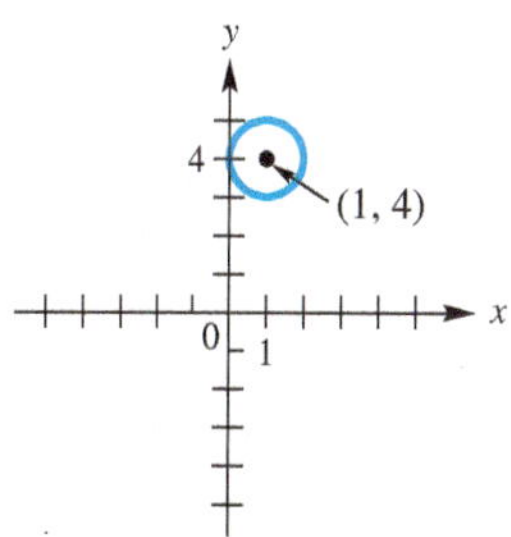

8. Graph the circle with equation $x^2 + y^2 + 4x - 10y + 13 = 0$.

9. In each case, determine whether y is a function of x. Give the domain and range. If it is a function, give the largest open intervals over which it is increasing, decreasing, or constant.

(a)

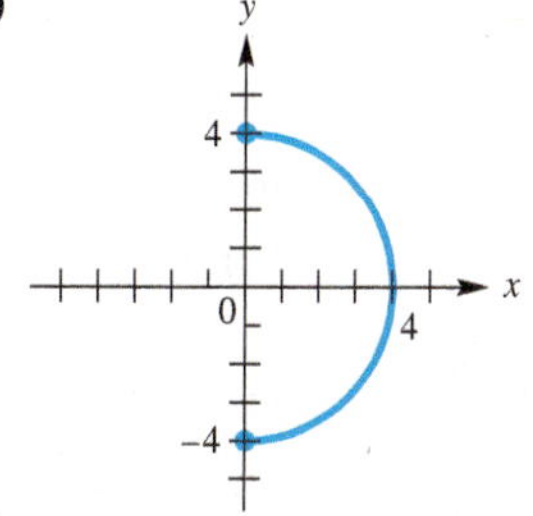

(b)

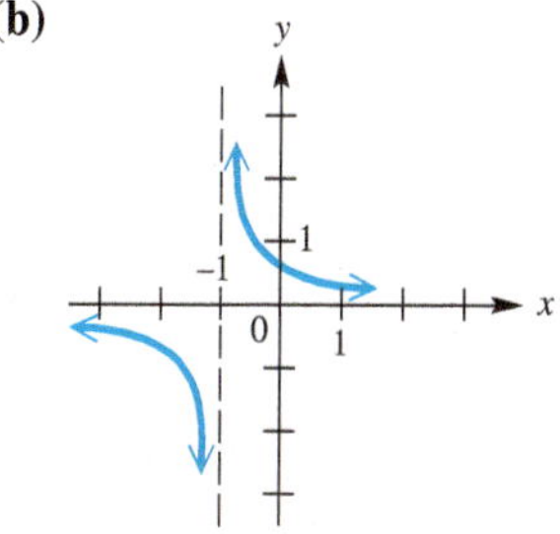

10. Suppose point A has coordinates $(5, -3)$. What is the equation of the

(a) vertical line through A? **(b)** horizontal line through A?

11. Find the slope-intercept form of the equation of the line passing through $(2, 3)$ and

(a) parallel to the graph of $y = -3x + 2$

(b) perpendicular to the graph of $y = -3x + 2$.

12. Consider the graph of the function shown here. Give the open interval(s) over which the function is

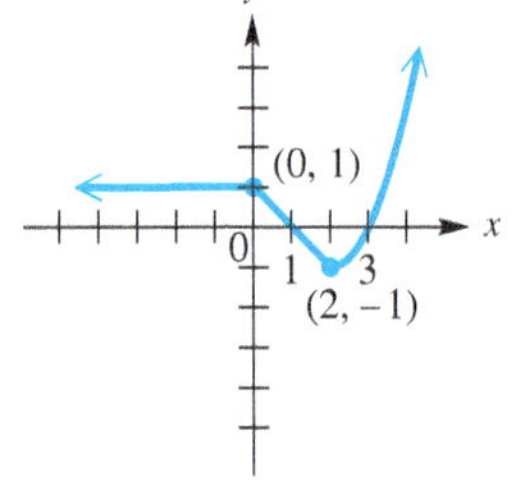

(a) increasing **(b)** decreasing

(c) constant **(d)** continuous.

(e) What is the domain of this function?

(f) What is the range of this function?

Graph each function.

13. $f(x) = |x - 2| - 1$ **14.** $f(x) = [\![x + 1]\!]$ **15.** $f(x) = \begin{cases} 3 & \text{if } x < -2 \\ 2 - \frac{1}{2}x & \text{if } x \geq -2 \end{cases}$

16. The graph of $y = f(x)$ is shown here. Sketch the graph of each of the following. Use ordered pairs to indicate three points on the graph.

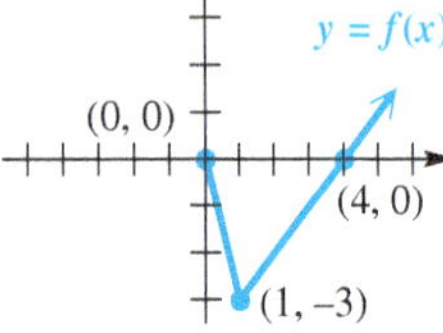

(a) $y = f(x) + 2$ **(b)** $y = f(x + 2)$

(c) $y = -f(x)$ **(d)** $y = f(-x)$

(e) $y = 2f(x)$

17. Describe how the graph of $f(x) = -2\sqrt{x + 2} - 3$ can be obtained from the graph of $y = \sqrt{x}$.

18. Determine whether the graph of $3x^2 - 2y^2 = 3$ is symmetric with respect to the

(a) x-axis **(b)** y-axis **(c)** origin.

19. Let $f(x) = 2x^2 - 3x + 2$ and $g(x) = -2x + 1$. Find each of the following. Simplify the expressions when possible.

(a) $(f - g)(x)$ **(b)** $\left(\frac{f}{g}\right)(x)$

(c) domain of $\frac{f}{g}$ **(d)** $\frac{f(x + h) - f(x)}{h}$ $(h \neq 0)$

(e) $(f + g)(1)$ **(f)** $(fg)(2)$ **(g)** $(f \circ g)(0)$

Let $f(x) = \sqrt{x + 1}$ *and* $g(x) = 2x - 7$. *Find each of the following.*

20. $(f \circ g)(x)$ and its domain **21.** $(g \circ f)(x)$ and its domain

22. ***(Modeling) Cost, Revenue, and Profit Analysis*** Dotty starts up a small business manufacturing bobble-head figures of famous soccer players. Her initial cost is \$3300. Each figure costs \$4.50 to manufacture.

(a) Write a cost function C, where x represents the number of figures manufactured.

(b) Find the revenue function R if each figure in part (a) sells for \$10.50.

(c) Give the profit function P.

(d) How many figures must be produced and sold for Dotty to earn a profit?

3 Polynomial and Rational Functions

Polynomial functions are used as models in many practical applications including the height of a thrown ball, the volume of a box, and, as seen in the photo here, the trajectories of water spouts.

3.1 Quadratic Functions and Models

- Polynomial Functions
- Quadratic Functions
- Graphing Techniques
- Completing the Square
- The Vertex Formula
- Quadratic Models

Polynomial Functions

Polynomial Function

A **polynomial function** f of degree n, where n is a nonnegative integer, is given by

$$f(x) = a_n x^n + a_{n-1}x^{n-1} + \cdots + a_1 x + a_0,$$

where $a_n, a_{n-1}, \ldots, a_1$, and a_0 are complex numbers, with $a_n \neq 0$.

In this chapter we primarily consider polynomial functions having real coefficients. When analyzing a polynomial function, the degree n and the **leading coefficient** a_n are important. These are both given in the **dominating term** $a_n x^n$.

LOOKING AHEAD TO CALCULUS
In calculus, polynomial functions are used to approximate more complicated functions. For example, the trigonometric function $\sin x$ is approximated by the polynomial

$$x - \frac{x^3}{6} + \frac{x^5}{120} - \frac{x^7}{5040}.$$

Polynomial Function	Function Type	Degree n	Leading Coefficient a_n
$f(x) = 2$	Constant	0	2
$f(x) = 5x - 1$	Linear	1	5
$f(x) = 4x^2 - x + 1$	Quadratic	2	4
$f(x) = 2x^3 - \frac{1}{2}x + 5$	Cubic	3	2
$f(x) = x^4 + \sqrt{2}x^3 - 3x^2$	Quartic	4	1

The function $f(x) = 0$ is the **zero polynomial** and has no degree.

Quadratic Functions

Polynomial functions of degree 2 are *quadratic functions*. Again, we are most often concerned with real coefficients.

Quadratic Function

A function f is a **quadratic function** if

$$f(x) = ax^2 + bx + c,$$

where a, b, and c are complex numbers, with $a \neq 0$.

The simplest quadratic function is

$$f(x) = x^2. \quad \text{Squaring function}$$

See **Figure 1.** This graph is a **parabola.** Every quadratic function with real coefficients defined over the real numbers has a graph that is a parabola. The domain of $f(x) = x^2$ is $(-\infty, \infty)$, and the range is $[0, \infty)$. The lowest point on the graph occurs at the origin $(0, 0)$. Thus, the function decreases on the open interval $(-\infty, 0)$ and increases on the open interval $(0, \infty)$. (Remember that these intervals indicate x-values.)

x	$f(x)$
-2	4
-1	1
0	0
1	1
2	4

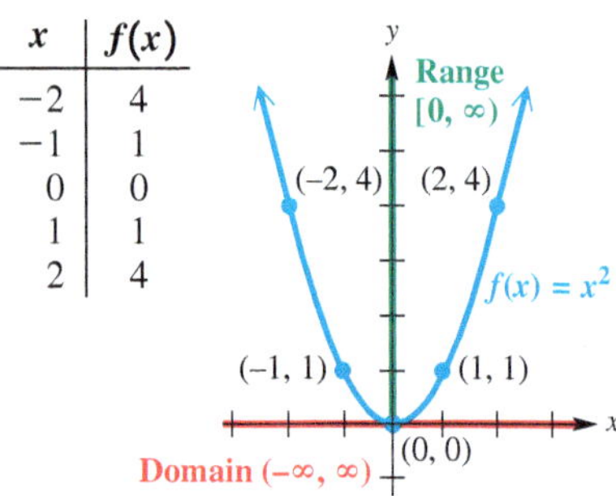

Figure 1

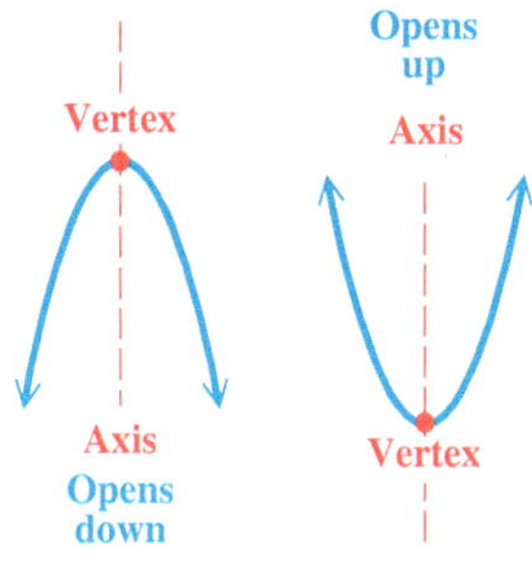

Figure 2

Parabolas are symmetric with respect to a line (the y-axis in **Figure 1**). This line is the **axis of symmetry,** or **axis,** of the parabola. The point where the axis intersects the parabola is the **vertex** of the parabola. As **Figure 2** shows, the vertex of a parabola that opens down is the highest point of the graph, and the vertex of a parabola that opens up is the lowest point of the graph.

Graphing Techniques Graphing techniques may be applied to the graph of $f(x) = x^2$ to give the graph of a different quadratic function. Compared to the basic graph of $f(x) = x^2$, the graph of $F(x) = a(x - h)^2 + k$, with $a \neq 0$, has the following characteristics.

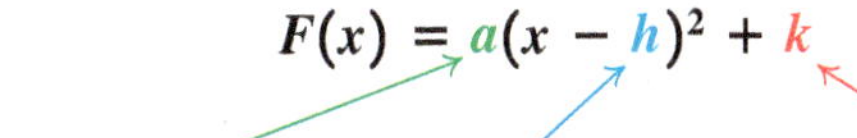

- Opens up if $a > 0$
- Opens down if $a < 0$
- Vertically stretched (narrower) if $|a| > 1$
- Vertically shrunk (wider) if $0 < |a| < 1$

Horizontal shift:

- h units right if $h > 0$
- $|h|$ units left if $h < 0$

Vertical shift:

- k units up if $k > 0$
- $|k|$ units down if $k < 0$

EXAMPLE 1 Graphing Quadratic Functions

Graph each function. Give the domain and range.

(a) $f(x) = x^2 - 4x - 2$ (by plotting points)

(b) $g(x) = -\frac{1}{2}x^2$ (and compare to $y = x^2$ and $y = \frac{1}{2}x^2$)

(c) $F(x) = -\frac{1}{2}(x - 4)^2 + 3$ (and compare to the graph in part (b))

SOLUTION

(a) See the table with **Figure 3.** The domain of $f(x) = x^2 - 4x - 2$ is $(-\infty, \infty)$, the range is $[-6, \infty)$, the vertex is the point $(2, -6)$, and the axis has equation $x = 2$. **Figure 4** shows how a graphing calculator displays this graph.

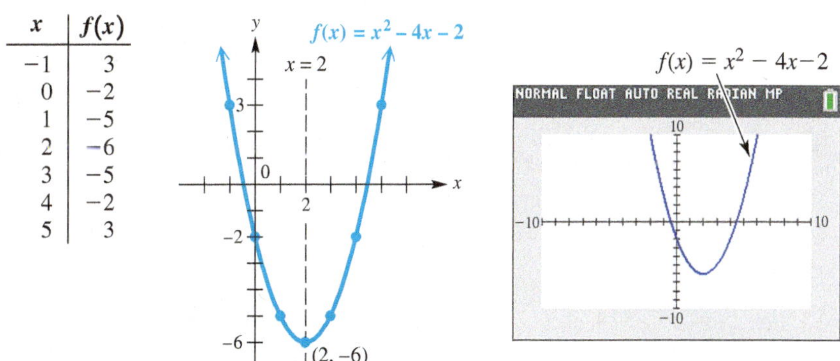

x	$f(x)$
-1	3
0	-2
1	-5
2	-6
3	-5
4	-2
5	3

Figure 3 **Figure 4**

(b) Think of $g(x) = -\frac{1}{2}x^2$ as $g(x) = -\left(\frac{1}{2}x^2\right)$. The graph of $y = \frac{1}{2}x^2$ is a wider version of the graph of $y = x^2$, and the graph of $g(x) = -\left(\frac{1}{2}x^2\right)$ is a reflection of the graph of $y = \frac{1}{2}x^2$ across the x-axis. See **Figure 5** on the next page. The vertex is the point $(0, 0)$, and the axis of the parabola is the line $x = 0$ (the y-axis). The domain is $(-\infty, \infty)$, and the range is $(-\infty, 0]$.

Calculator graphs are shown in **Figure 6.**

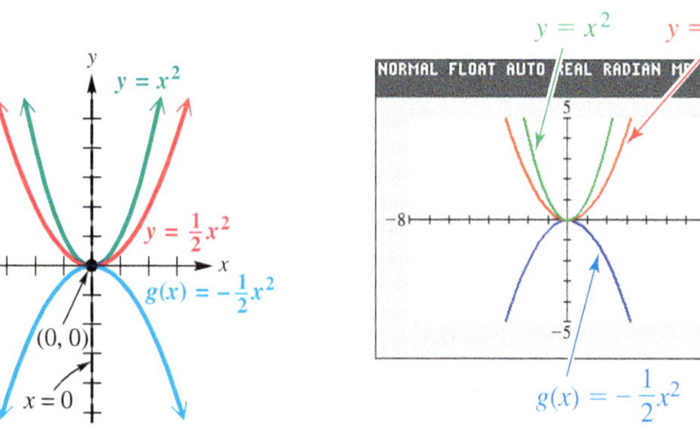

Figure 5 **Figure 6**

(c) Notice that $F(x) = -\frac{1}{2}(x-4)^2 + 3$ is related to $g(x) = -\frac{1}{2}x^2$ from part (b). The graph of $F(x)$ is the graph of $g(x)$ translated 4 units to the right and 3 units up. See **Figure 7.** The vertex is the point $(4, 3)$, which is also shown in the calculator graph in **Figure 8,** and the axis of the parabola is the line $x = 4$. The domain is $(-\infty, \infty)$, and the range is $(-\infty, 3]$.

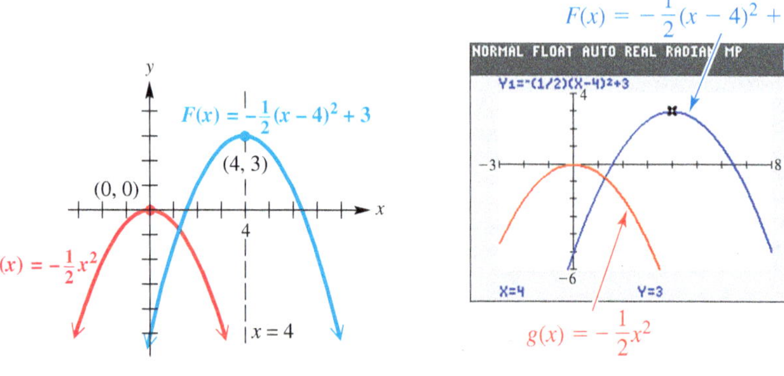

Figure 7 **Figure 8**

✔ **Now Try Exercises 19 and 21.**

Completing the Square

In general, the graph of the quadratic function

$$f(x) = a(x-h)^2 + k \quad (a \neq 0)$$

is a parabola with ***vertex*** (h, k) and ***axis of symmetry*** $x = h$. The parabola opens up if a is positive and down if a is negative. With these facts in mind, we *complete the square* to graph the general quadratic function

$$f(x) = ax^2 + bx + c.$$

EXAMPLE 2 Graphing a Parabola ($a = 1$)

Graph $f(x) = x^2 - 6x + 7$. Find the largest open intervals over which the function is increasing or decreasing.

SOLUTION We express $x^2 - 6x + 7$ in the form $(x-h)^2 + k$ by completing the square. In preparation for this, we first write

$$f(x) = (x^2 - 6x \qquad\quad) + 7. \quad \text{Prepare to complete the square.}$$

We must add a number inside the parentheses to obtain a perfect square trinomial. Find this number by taking half the coefficient of x and squaring the result.

$f(x) = x^2 - 6x + 7$

$f(x) = (x - 3)^2 - 2$

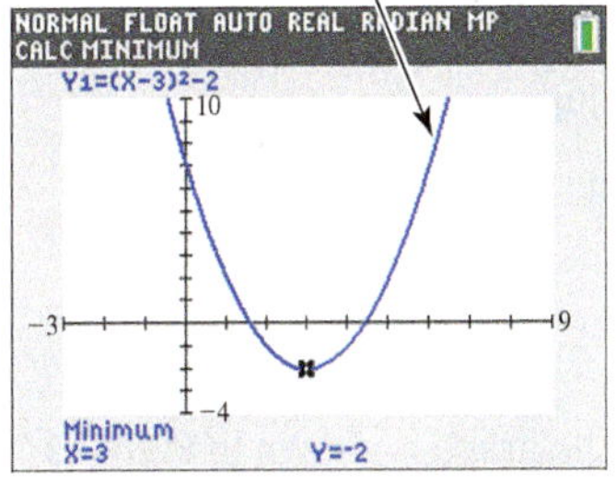

This screen shows that the vertex of the graph in **Figure 9** is the point $(3, -2)$. Because it is the *lowest* point on the graph, we direct the calculator to find the *minimum*.

$$\left[\frac{1}{2}(-6)\right]^2 = (-3)^2 = 9 \quad \text{Take half the coefficient of } x. \text{ Square the result.}$$

$$f(x) = (x^2 - 6x + 9 - 9) + 7 \quad \text{Add and subtract 9.}$$

This is the same as adding 0.

$$f(x) = (x^2 - 6x + 9) - 9 + 7 \quad \text{Regroup terms.}$$

$$f(x) = (x - 3)^2 - 2 \quad \text{Factor and simplify.}$$

The vertex of the parabola is the point $(3, -2)$, and the axis is the line $x = 3$. We find additional ordered pairs that satisfy the equation, as shown in the table, and plot and join these points to obtain the graph in **Figure 9.**

x	y	
0	7	← y-intercept
1	2	
3	−2	← Vertex
5	2	← Find using symmetry
6	7	← about the axis.

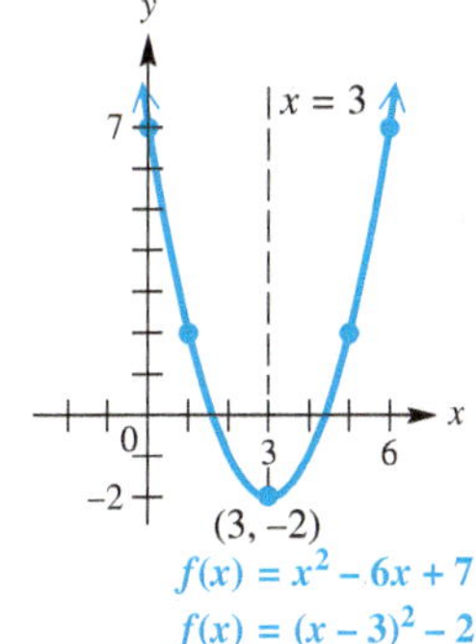

Figure 9

The domain of this function is $(-\infty, \infty)$, and the range is $[-2, \infty)$. Because the lowest point on the graph is the vertex $(3, -2)$, the function is decreasing on $(-\infty, 3)$ and increasing on $(3, \infty)$.

✔ **Now Try Exercise 31.**

NOTE In **Example 2** we added and subtracted 9 *on the same side* of the equation to complete the square. This differs from adding the same number to *each side of the equation*, as is sometimes done in the procedure. We want $f(x)$—that is, y—alone on one side of the equation, so we adjusted that step in the process of completing the square here.

EXAMPLE 3 Graphing a Parabola ($a \neq 1$)

Graph $f(x) = -3x^2 - 2x + 1$. Identify the intercepts of the graph.

SOLUTION To complete the square, the coefficient of x^2 must be 1.

$$f(x) = -3\left(x^2 + \frac{2}{3}x \qquad\right) + 1 \quad \text{Factor } -3 \text{ from the first two terms.}$$

$$f(x) = -3\left(x^2 + \frac{2}{3}x + \frac{1}{9} - \frac{1}{9}\right) + 1 \quad \left[\frac{1}{2}\left(\frac{2}{3}\right)\right]^2 = \left(\frac{1}{3}\right)^2 = \frac{1}{9}, \text{ so add and subtract } \frac{1}{9}.$$

$$f(x) = -3\left(x^2 + \frac{2}{3}x + \frac{1}{9}\right) - 3\left(-\frac{1}{9}\right) + 1 \quad \text{Distributive property}$$

Be careful here.

$$f(x) = -3\left(x + \frac{1}{3}\right)^2 + \frac{4}{3} \quad \text{Factor and simplify.}$$

The vertex is the point $\left(-\frac{1}{3}, \frac{4}{3}\right)$. The intercepts are good additional points to find. The y-intercept is found by evaluating $f(0)$.

$$f(0) = -3(0)^2 - 2(0) + 1 \quad \text{Let } x = 0 \text{ in } f(x) = -3x^2 - 2x + 1.$$

$$f(0) = 1 \longleftarrow \text{The } y\text{-intercept is } (0, 1).$$

The x-intercepts are found by setting $f(x)$ equal to 0 and solving for x.

$$0 = -3x^2 - 2x + 1 \quad \text{Set } f(x) = 0.$$

$$0 = 3x^2 + 2x - 1 \quad \text{Multiply by } -1.$$

$$0 = (3x - 1)(x + 1) \quad \text{Factor.}$$

$$x = \frac{1}{3} \quad \text{or} \quad x = -1 \quad \text{Zero-factor property}$$

Therefore, the x-intercepts are $\left(\frac{1}{3}, 0\right)$ and $(-1, 0)$. The graph is shown in **Figure 10.**

$f(x) = -3x^2 - 2x + 1$

$f(x) = -3\left(x + \frac{1}{3}\right)^2 + \frac{4}{3}$

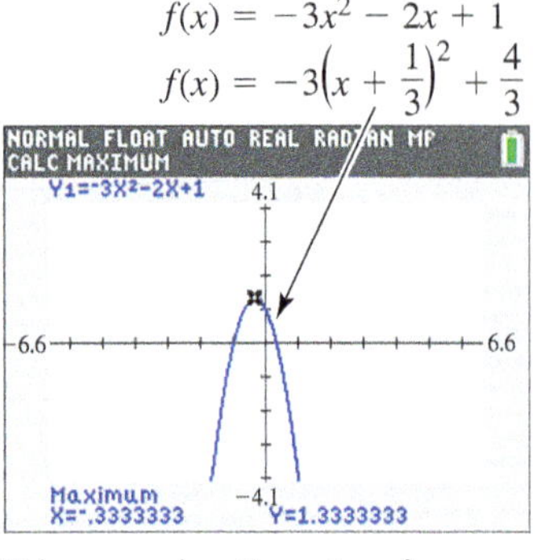

This screen gives the vertex of the graph in **Figure 10** as the point $\left(-\frac{1}{3}, \frac{4}{3}\right)$. (The display shows decimal approximations.) We want the highest point on the graph, so we direct the calculator to find the *maximum*.

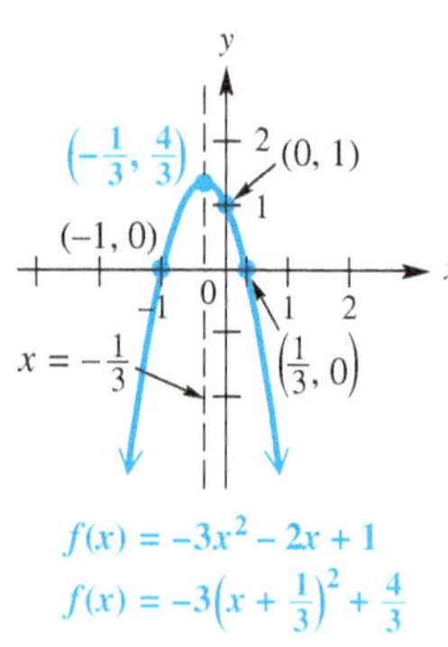

Figure 10

✔ **Now Try Exercise 33.**

NOTE It is possible to reverse the process of **Example 3** and write the quadratic function from its graph in **Figure 10** if the vertex and any other point on the graph are known. Because quadratic functions take the form

$$f(x) = a(x - h)^2 + k,$$

we can substitute the x- and y-values of the vertex, $\left(-\frac{1}{3}, \frac{4}{3}\right)$, for h and k.

$$f(x) = a\left[x - \left(-\frac{1}{3}\right)\right]^2 + \frac{4}{3} \quad \text{Let } h = -\tfrac{1}{3} \text{ and } k = \tfrac{4}{3}.$$

$$f(x) = a\left(x + \frac{1}{3}\right)^2 + \frac{4}{3} \quad \text{Simplify.}$$

We find the value of a by substituting the x- and y-coordinates of any other point on the graph, say $(0, 1)$, into this function and solving for a.

$$1 = a\left(0 + \frac{1}{3}\right)^2 + \frac{4}{3} \quad \text{Let } x = 0 \text{ and } y = 1.$$

$$1 = a\left(\frac{1}{9}\right) + \frac{4}{3} \quad \text{Square.}$$

$$-\frac{1}{3} = \frac{1}{9}a \quad \text{Subtract } \tfrac{4}{3}.$$

$$a = -3 \quad \text{Multiply by 9. Interchange sides.}$$

Verify in **Example 3** that the vertex form of the quadratic function is

$$f(x) = -3\left(x + \frac{1}{3}\right)^2 + \frac{4}{3}.$$

Exercises of this type are labeled ***Connecting Graphs with Equations***.

LOOKING AHEAD TO CALCULUS

An important concept in calculus is the **definite integral.** If the graph of f lies above the x-axis, the symbol

$$\int_a^b f(x)\,dx$$

represents the area of the region above the x-axis and below the graph of f from $x = a$ to $x = b$. For example, in **Figure 10** with

$$f(x) = -3x^2 - 2x + 1,$$

$a = -1$, and $b = \frac{1}{3}$, calculus provides the tools for determining that the area enclosed by the parabola and the x-axis is $\frac{32}{27}$ (square units).

The Vertex Formula We can generalize the earlier work to obtain a formula for the vertex of a parabola.

$$f(x) = ax^2 + bx + c \quad \text{General quadratic form}$$

$$= a\left(x^2 + \frac{b}{a}x \qquad \right) + c \quad \text{Factor } a \text{ from the first two terms.}$$

$$= a\left(x^2 + \frac{b}{a}x + \frac{b^2}{4a^2}\right) + c - a\left(\frac{b^2}{4a^2}\right) \quad \text{Add } \left[\tfrac{1}{2}\left(\tfrac{b}{a}\right)\right]^2 = \tfrac{b^2}{4a^2} \text{ inside the parentheses. Subtract } a\left(\tfrac{b^2}{4a^2}\right) \text{ outside the parentheses.}$$

$$= a\left(x + \frac{b}{2a}\right)^2 + c - \frac{b^2}{4a} \quad \text{Factor and simplify.}$$

$$f(x) = a\Bigg[x - \underbrace{\left(-\frac{b}{2a}\right)}_{h}\Bigg]^2 + \underbrace{\frac{4ac - b^2}{4a}}_{k} \quad \text{Vertex form of } f(x) = a(x-h)^2 + k$$

Thus, the vertex (h, k) can be expressed in terms of a, b, and c. ***It is not necessary to memorize the expression for k because it is equal to $f(h) = f\left(-\frac{b}{2a}\right)$.***

Graph of a Quadratic Function

The quadratic function $f(x) = ax^2 + bx + c$ can be written as

$$y = f(x) = a(x - h)^2 + k, \quad \text{with } a \neq 0,$$

where $h = -\dfrac{b}{2a}$ and $k = f(h)$. Vertex formula

The graph of f has the following characteristics.

1. It is a parabola with vertex (h, k) and the vertical line $x = h$ as axis.
2. It opens up if $a > 0$ and down if $a < 0$.
3. It is wider than the graph of $y = x^2$ if $|a| < 1$ and narrower if $|a| > 1$.
4. The y-intercept is $(0, f(0)) = (0, c)$.
5. The x-intercepts are found by solving the equation $ax^2 + bx + c = 0$.
   - If $b^2 - 4ac > 0$, then the x-intercepts are $\left(\frac{-b \pm \sqrt{b^2 - 4ac}}{2a}, 0\right)$.
   - If $b^2 - 4ac = 0$, then the x-intercept is $\left(-\frac{b}{2a}, 0\right)$.
   - If $b^2 - 4ac < 0$, then there are no x-intercepts.

EXAMPLE 4 Using the Vertex Formula

Find the axis and vertex of the parabola having equation $f(x) = 2x^2 + 4x + 5$.

SOLUTION The axis of the parabola is the vertical line

$$x = h = -\frac{b}{2a} = -\frac{4}{2(2)} = -1. \quad \text{Use the vertex formula. Here } a = 2 \text{ and } b = 4.$$

The vertex is $(-1, f(-1))$. Evaluate $f(-1)$.

$$f(-1) = 2(-1)^2 + 4(-1) + 5 = 3$$

The vertex is $(-1, 3)$.

✔ **Now Try Exercise 31(a).**

Quadratic Models Because the vertex of a vertical parabola is the highest or lowest point on the graph, equations of the form

$$y = ax^2 + bx + c$$

are important in certain problems where we must find the maximum or minimum value of some quantity.

- When $a < 0$, the y-coordinate of the vertex gives the maximum value of y.
- When $a > 0$, the y-coordinate of the vertex gives the minimum value of y.

The x-coordinate of the vertex tells *where* the maximum or minimum value occurs.

If air resistance is neglected, the height s (in feet) of an object projected directly upward from an initial height s_0 feet with initial velocity v_0 feet per second is

$$\mathbf{s(t) = -16t^2 + v_0 t + s_0,}$$

where t is the number of seconds after the object is projected. The coefficient of t^2 (that is, -16) is a constant based on the gravitational force of Earth. This constant is different on other surfaces, such as the moon and the other planets.

EXAMPLE 5 Solving a Problem Involving Projectile Motion

A ball is projected directly upward from an initial height of 100 ft with an initial velocity of 80 ft per sec.

(a) Give the function that describes the height of the ball in terms of time t.

(b) After how many seconds does the ball reach its maximum height? What is this maximum height?

(c) For what interval of time is the height of the ball greater than 160 ft?

(d) After how many seconds will the ball hit the ground?

ALGEBRAIC SOLUTION

(a) Use the projectile height function.

$$s(t) = -16t^2 + v_0 t + s_0 \qquad \text{Let } v_0 = 80 \text{ and}$$

$$s(t) = -16t^2 + 80t + 100 \qquad s_0 = 100.$$

(b) The coefficient of t^2 is -16, so the graph of the projectile function is a parabola that opens down. Find the coordinates of the vertex to determine the maximum height and when it occurs. Let $a = -16$ and $b = 80$ in the vertex formula.

$$t = -\frac{b}{2a} = -\frac{80}{2(-16)} = 2.5$$

$$s(t) = -16t^2 + 80t + 100$$

$$s(2.5) = -16(2.5)^2 + 80(2.5) + 100$$

$$s(2.5) = 200$$

Therefore, after 2.5 sec the ball reaches its maximum height of 200 ft.

GRAPHING CALCULATOR SOLUTION

(a) Use the projectile height function as in the algebraic solution, with $v_0 = 80$ and $s_0 = 100$.

$$s(t) = -16t^2 + 80t + 100$$

(b) Using the capabilities of a calculator, we see in **Figure 11** that the vertex coordinates are indeed $(2.5, 200)$.

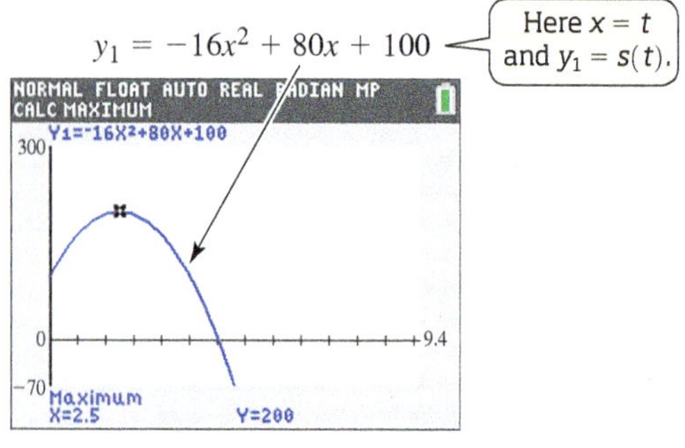

Figure 11

Be careful not to misinterpret the graph in* Figure 11*. It does not show the path followed by the ball. It defines height as a function of time.

(c) We must solve the related quadratic *inequality*.

$$-16t^2 + 80t + 100 > 160$$

$$-16t^2 + 80t - 60 > 0 \quad \text{Subtract 160.}$$

$$4t^2 - 20t + 15 < 0 \quad \text{Divide by } -4. \text{ Reverse the inequality symbol.}$$

Use the quadratic formula to find the solutions of $4t^2 - 20t + 15 = 0$.

$$t = \frac{-(-20) \pm \sqrt{(-20)^2 - 4(4)(15)}}{2(4)} \quad \text{Here } a = 4,\ b = -20, \text{ and } c = 15.$$

$$t = \frac{5 - \sqrt{10}}{2} \approx 0.92 \quad \text{or} \quad t = \frac{5 + \sqrt{10}}{2} \approx 4.08$$

These numbers divide the number line into three intervals:

$$(-\infty, 0.92), \quad (0.92, 4.08), \quad \text{and} \quad (4.08, \infty).$$

Using a test value from each interval shows that $(0.92, 4.08)$ satisfies the *inequality*. The ball is greater than 160 ft above the ground between 0.92 sec and 4.08 sec.

(d) The height is 0 when the ball hits the ground. We use the quadratic formula to find the *positive* solution of the equation

$$-16t^2 + 80t + 100 = 0.$$

Here, $a = -16$, $b = 80$, and $c = 100$.

$$t = \frac{-80 \pm \sqrt{80^2 - 4(-16)(100)}}{2(-16)}$$

$$t \approx -1.04 \text{ (Reject)} \quad \text{or} \quad t \approx 6.04$$

The ball hits the ground after about 6.04 sec.

(c) If we graph

$$y_1 = -16x^2 + 80x + 100 \quad \text{and} \quad y_2 = 160,$$

as shown in **Figures 12 and 13,** and locate the two points of intersection, we find that the x-coordinates for these points are approximately

$$0.92 \quad \text{and} \quad 4.08.$$

Therefore, between 0.92 sec and 4.08 sec, y_1 is greater than y_2, and the ball is greater than 160 ft above the ground.

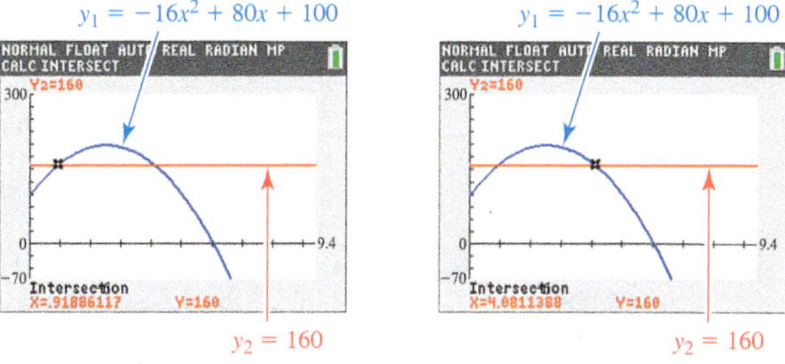

Figure 12 **Figure 13**

(d) **Figure 14** shows that the x-intercept of the graph of $y = -16x^2 + 80x + 100$ in the given window is approximately $(6.04, 0)$, which means that the ball hits the ground after about 6.04 sec.

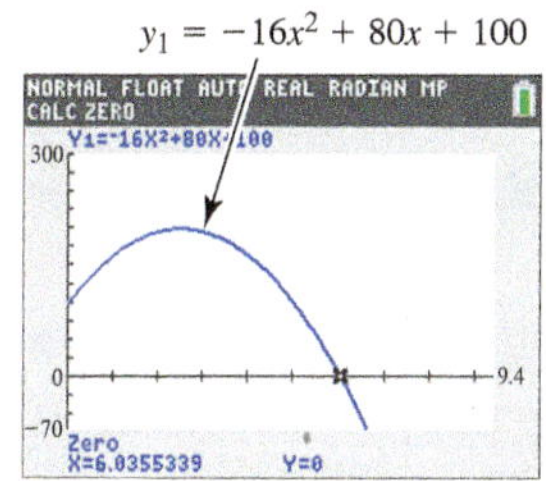

Figure 14

✔ **Now Try Exercise 57.**

EXAMPLE 6 Modeling the Number of Hospital Outpatient Visits

The number of hospital outpatient visits (in millions) for selected years is shown in the table.

Year	Visits	Year	Visits
99	573.5	106	690.4
100	592.7	107	693.5
101	612.0	108	710.0
102	640.5	109	742.0
103	648.6	110	750.4
104	662.1	111	754.5
105	673.7	112	778.0

Source: American Hospital Association.

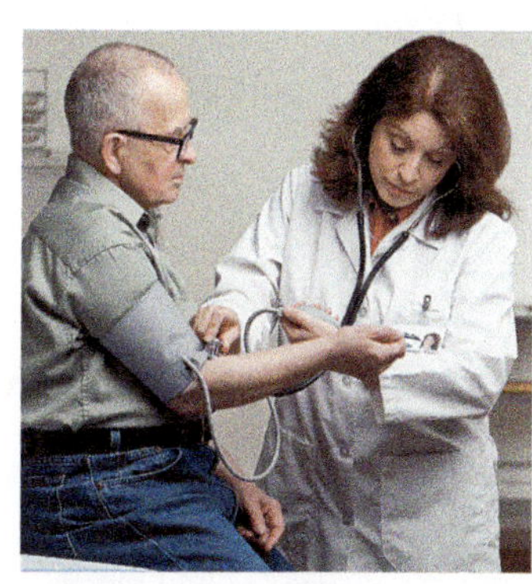

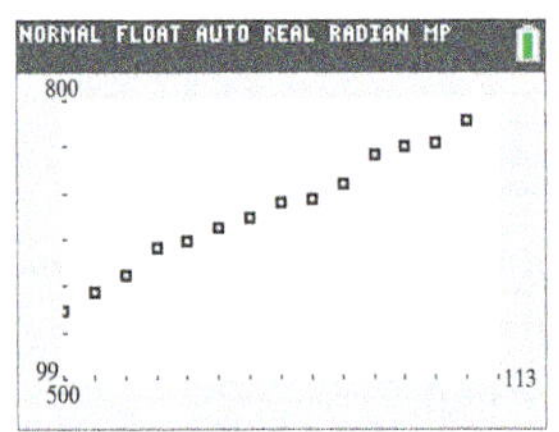

(a)

$f(x) = -0.16161x^2 + 49.071x - 2695.5$

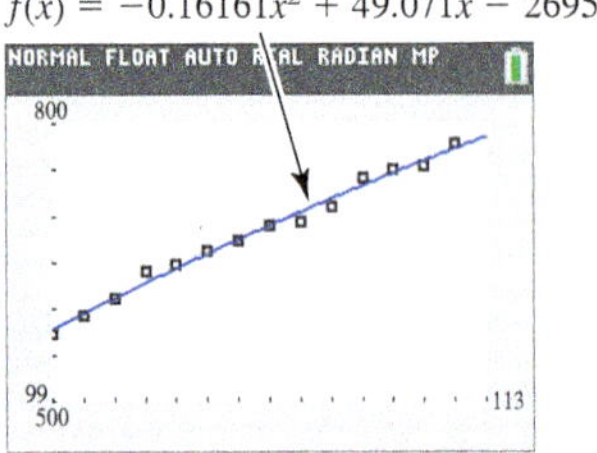

(b)

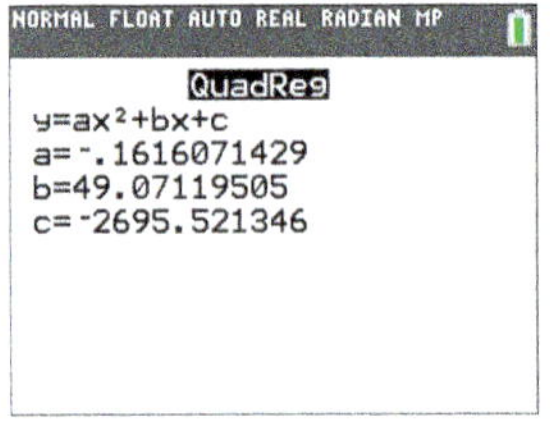

(c)

Figure 15

In the table on the preceding page, 99 represents 1999, 100 represents 2000, and so on, and the number of outpatient visits is given in millions.

(a) Prepare a scatter diagram, and determine a quadratic model for these data.

(b) Use the model from part (a) to predict the number of visits in 2016.

SOLUTION

(a) Linear regression is used to determine linear equations that model data. With a graphing calculator, we can use **quadratic regression** to find quadratic equations that model data.

The scatter diagram in **Figure 15(a)** suggests that a quadratic function with a negative value of a (so the graph opens down) would be a reasonable model for the data. Using quadratic regression, the quadratic function

$$f(x) = -0.16161x^2 + 49.071x - 2695.5$$

approximates the data well. See **Figure 15(b).** The quadratic regression values of a, b, and c are displayed in **Figure 15(c).**

(b) The year 2016 corresponds to $x = 116$. The model predicts that there will be 822 million visits in 2016.

$$f(x) = -0.16161x^2 + 49.071x - 2695.5$$

$$f(116) = -0.16161(116)^2 + 49.071(116) - 2695.5$$

$$f(116) \approx 822 \text{ million}$$

✔ **Now Try Exercise 73.**

3.1 Exercises

CONCEPT PREVIEW *Fill in the blank(s) to correctly complete each sentence.*

1. A polynomial function with leading term $3x^5$ has degree ______.

2. The lowest point on the graph of a parabola that opens up is the ________ of the parabola.

3. The highest point on the graph of a parabola that opens down is the ________ of the parabola.

4. The axis of symmetry of the graph of $f(x) = 2(x + 4)^2 - 6$ has equation $x =$ ______.

5. The vertex of the graph of $f(x) = x^2 + 2x + 4$ has x-coordinate ______.

6. The graph of $f(x) = -2x^2 - 6x + 5$ opens down with y-intercept $(0, \text{\_\_\_\_\_\_})$, so it has ____________ x-intercept(s).
(no/one/two)

CONCEPT PREVIEW *Match each equation in Column I with the description of the parabola that is its graph in Column II.*

I	II
7. $y = (x + 4)^2 + 2$	**A.** vertex $(-2, 4)$, opens up
8. $y = (x + 2)^2 + 4$	**B.** vertex $(-2, 4)$, opens down
9. $y = -(x + 4)^2 + 2$	**C.** vertex $(-4, 2)$, opens up
10. $y = -(x + 2)^2 + 4$	**D.** vertex $(-4, 2)$, opens down

Consider the graph of each quadratic function. Do the following. **See Examples 1–4.**

(a) *Give the domain and range.* ***(b)*** *Give the coordinates of the vertex.*

(c) *Give the equation of the axis.* ***(d)*** *Find the y-intercept.*

(e) *Find the x-intercepts.*

11.

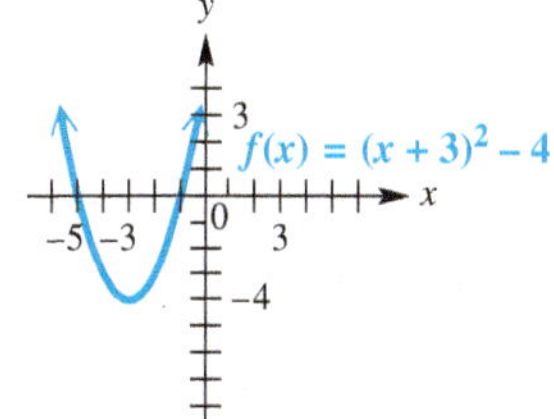

12.

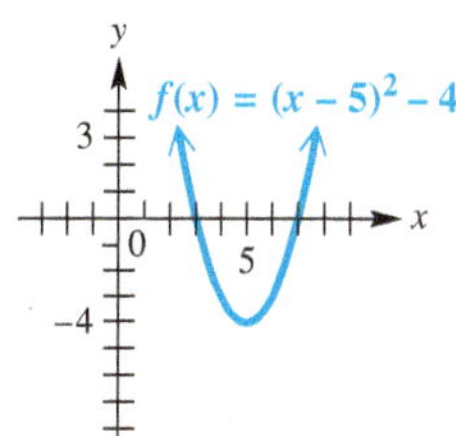

13.

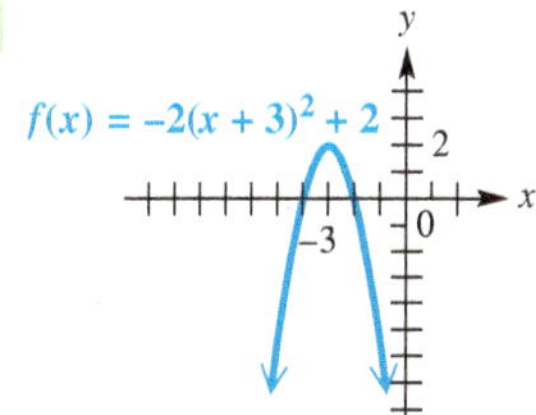

14.

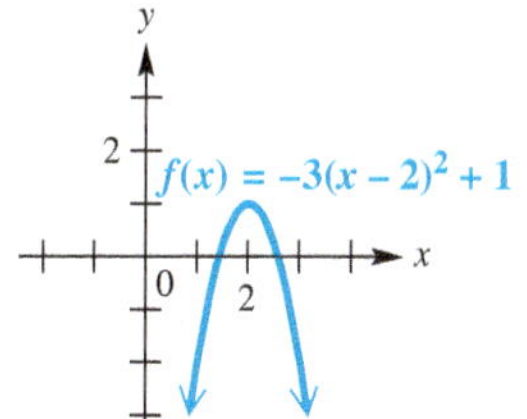

Concept Check *Match each function with its graph without actually entering it into a calculator. Then, after completing the exercises, check the answers with a calculator. Use the standard viewing window.*

15. $f(x) = (x - 4)^2 - 3$

16. $f(x) = -(x - 4)^2 + 3$

17. $f(x) = (x + 4)^2 - 3$

18. $f(x) = -(x + 4)^2 + 3$

A.

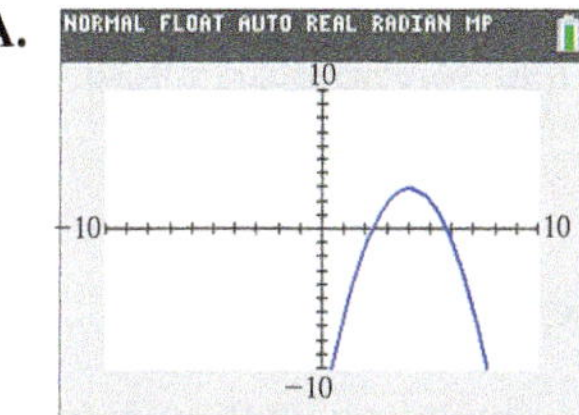

B.

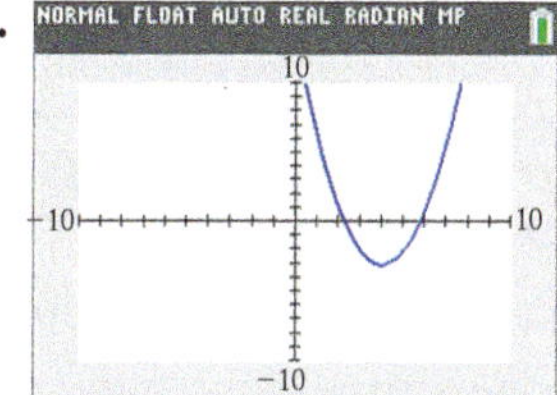

C.

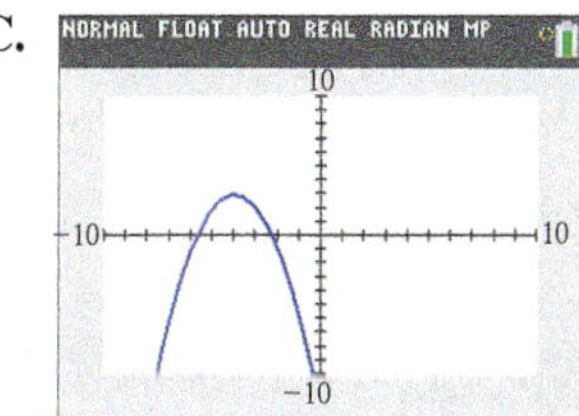

D.

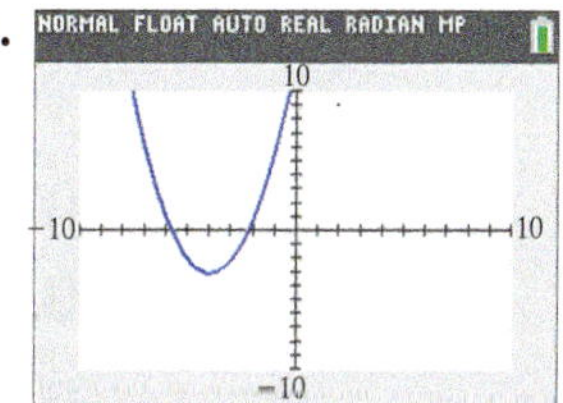

19. Graph the following on the same coordinate system.

(a) $y = x^2$ **(b)** $y = 3x^2$ **(c)** $y = \frac{1}{3}x^2$

(d) How does the coefficient of x^2 affect the shape of the graph?

20. Graph the following on the same coordinate system.

(a) $y = x^2$ **(b)** $y = x^2 - 2$ **(c)** $y = x^2 + 2$

(d) How do the graphs in parts (b) and (c) differ from the graph of $y = x^2$?

21. Graph the following on the same coordinate system.

(a) $y = (x - 2)^2$ **(b)** $y = (x + 1)^2$ **(c)** $y = (x + 3)^2$

(d) How do these graphs differ from the graph of $y = x^2$?

22. *Concept Check* A quadratic function $f(x)$ has vertex $(0, 0)$, and all of its intercepts are the same point. What is the general form of its equation?

*Graph each quadratic function. Give the **(a)** vertex, **(b)** axis, **(c)** domain, and **(d)** range. Then determine **(e)** the largest open interval of the domain over which the function is increasing and **(f)** the largest open interval over which the function is decreasing.* ***See Examples 1–4.***

23. $f(x) = (x - 2)^2$

24. $f(x) = (x + 4)^2$

25. $f(x) = (x + 3)^2 - 4$

26. $f(x) = (x - 5)^2 - 4$

27. $f(x) = -\frac{1}{2}(x + 1)^2 - 3$

28. $f(x) = -3(x - 2)^2 + 1$

29. $f(x) = x^2 - 2x + 3$

30. $f(x) = x^2 + 6x + 5$

31. $f(x) = x^2 - 10x + 21$

32. $f(x) = 2x^2 - 4x + 5$

33. $f(x) = -2x^2 - 12x - 16$

34. $f(x) = -3x^2 + 24x - 46$

35. $f(x) = -\frac{1}{2}x^2 - 3x - \frac{1}{2}$

36. $f(x) = \frac{2}{3}x^2 - \frac{8}{3}x + \frac{5}{3}$

Concept Check The figure shows the graph of a quadratic function $y = f(x)$. Use it to answer each question.

37. What is the minimum value of $f(x)$?

38. For what value of x is $f(x)$ as small as possible?

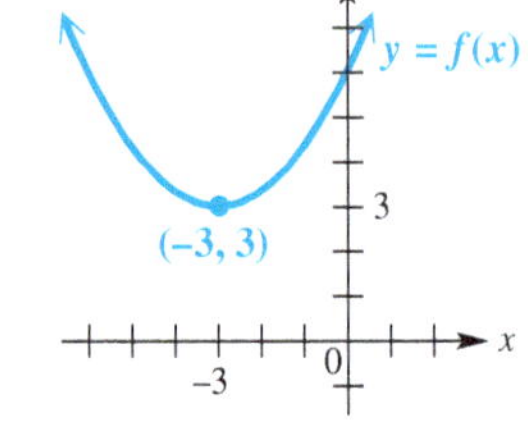

39. How many real solutions are there to the equation $f(x) = 1$?

40. How many real solutions are there to the equation $f(x) = 4$?

Concept Check Several graphs of the quadratic function

$$f(x) = ax^2 + bx + c$$

are shown below. For the given restrictions on a, b, and c, select the corresponding graph from choices A–F. (Hint: Use the discriminant.)

41. $a < 0;\ b^2 - 4ac = 0$

42. $a > 0;\ b^2 - 4ac < 0$

43. $a < 0;\ b^2 - 4ac < 0$

44. $a < 0;\ b^2 - 4ac > 0$

45. $a > 0;\ b^2 - 4ac > 0$

46. $a > 0;\ b^2 - 4ac = 0$

A.

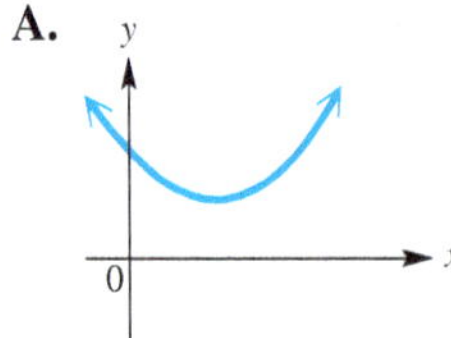

B.

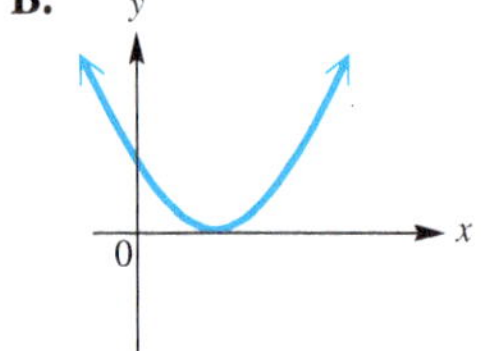

C.

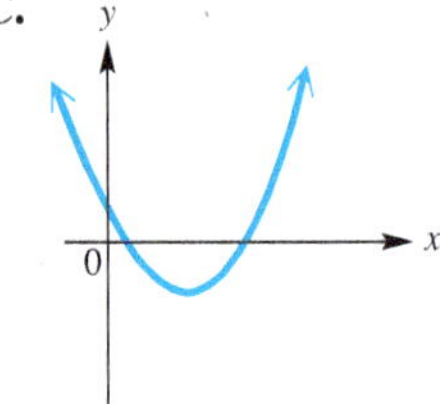

D.

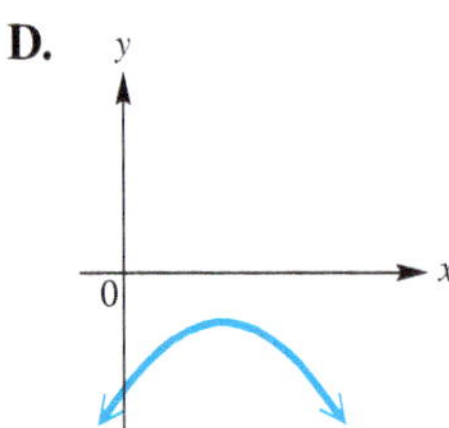

E.

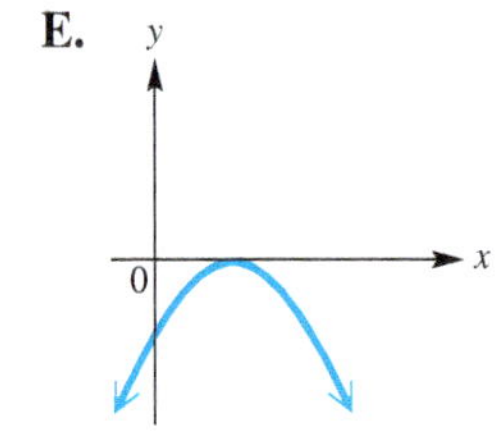

F.

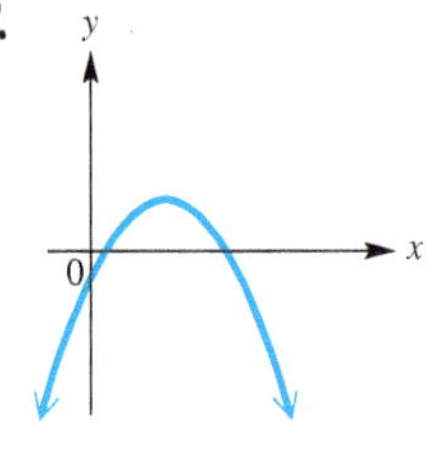

Connecting Graphs with Equations *Find a quadratic function f having the graph shown. (Hint: See the Note following* ***Example 3.****)*

47.

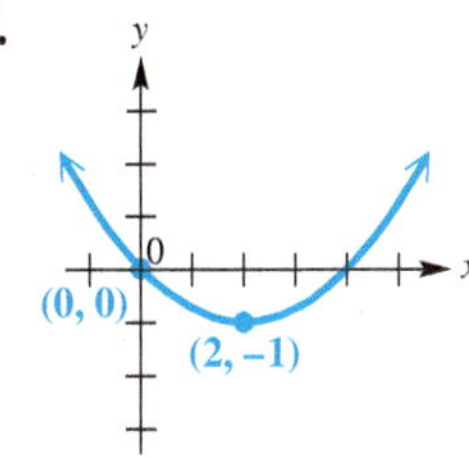

48.

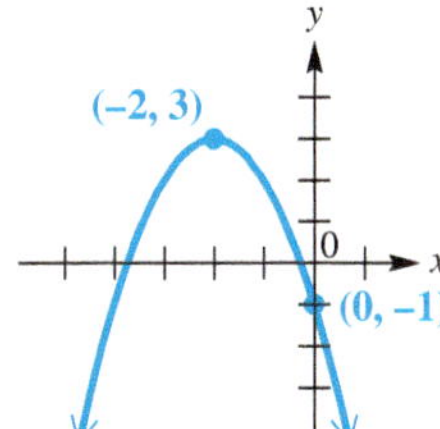

49.

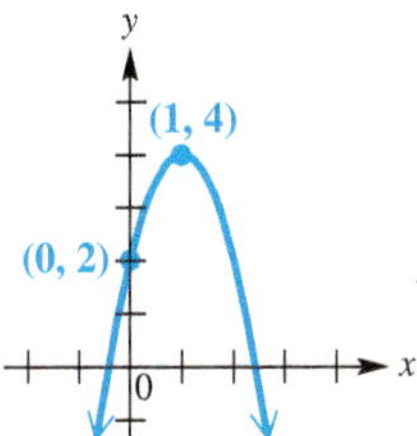

50.

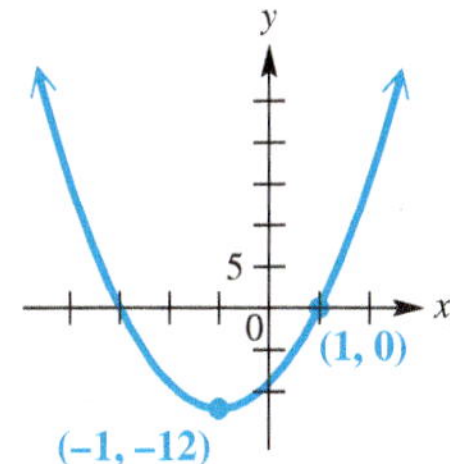

(Modeling) *In each scatter diagram, tell whether a linear or a quadratic model is appropriate for the data. If linear, tell whether the slope should be positive or negative. If quadratic, tell whether the leading coefficient of x^2 should be positive or negative.*

51. number of shopping centers as a function of time

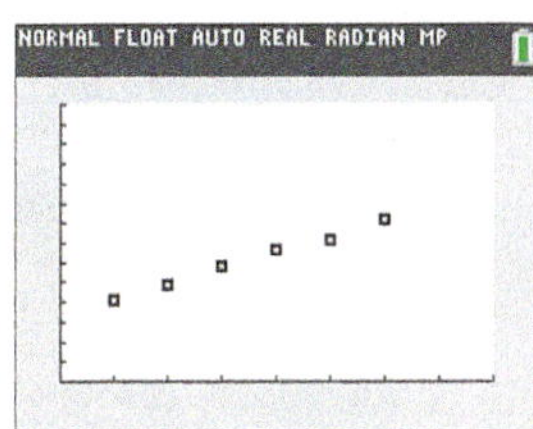

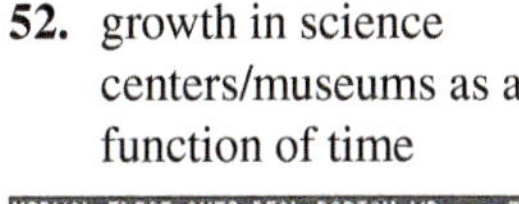

52. growth in science centers/museums as a function of time

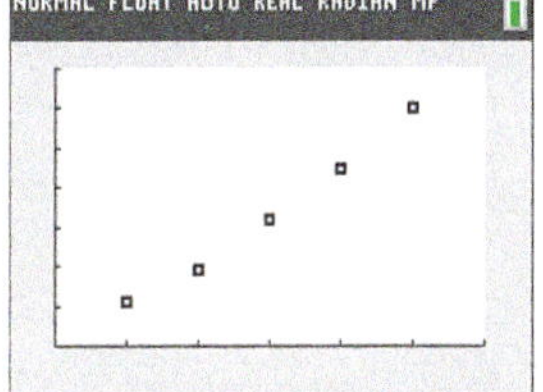

53. value of U.S. salmon catch as a function of time

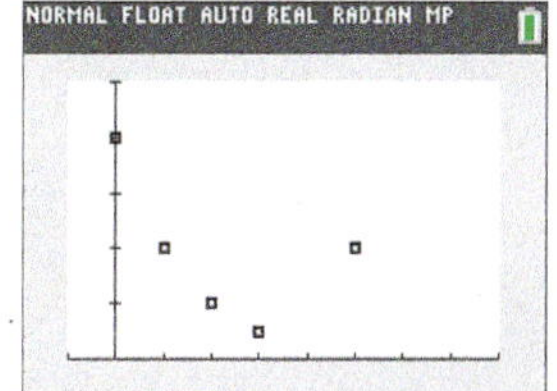

54. height of an object projected upward as a function of time

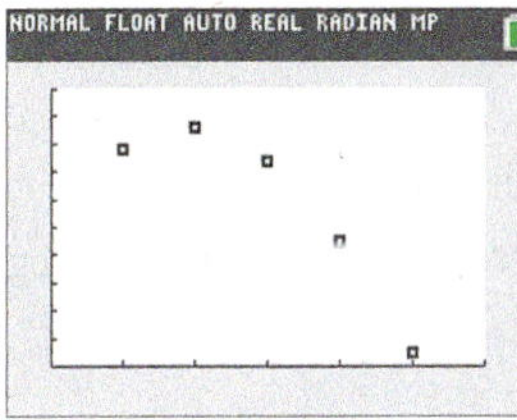

55. Social Security assets as a function of time

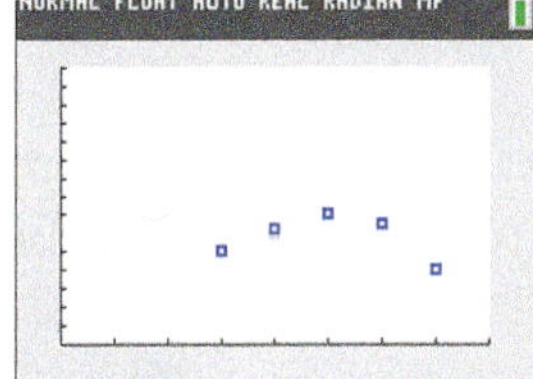

56. newborns with AIDS as a function of time

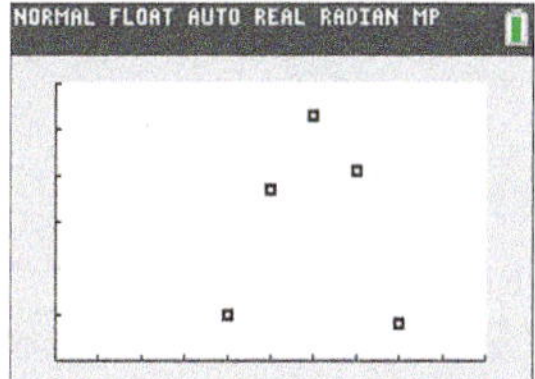

(Modeling) *Solve each problem. Give approximations to the nearest hundredth.* ***See Example 5.***

57. ***Height of a Toy Rocket*** A toy rocket (not internally powered) is launched straight up from the top of a building 50 ft tall at an initial velocity of 200 ft per sec.

(a) Give the function that describes the height of the rocket in terms of time t.

(b) Determine the time at which the rocket reaches its maximum height and the maximum height in feet.

(c) For what time interval will the rocket be more than 300 ft above ground level?

(d) After how many seconds will it hit the ground?

58. ***Height of a Projected Rock*** A rock is projected directly upward from ground level with an initial velocity of 90 ft per sec.

(a) Give the function that describes the height of the rock in terms of time t.

(b) Determine the time at which the rock reaches its maximum height and the maximum height in feet.

(c) For what time interval will the rock be more than 120 ft above ground level?

(d) After how many seconds will it hit the ground?

59. ***Area of a Parking Lot*** One campus of Houston Community College has plans to construct a rectangular parking lot on land bordered on one side by a highway. There are 640 ft of fencing available to fence the other three sides. Let x represent the length of each of the two parallel sides of fencing.

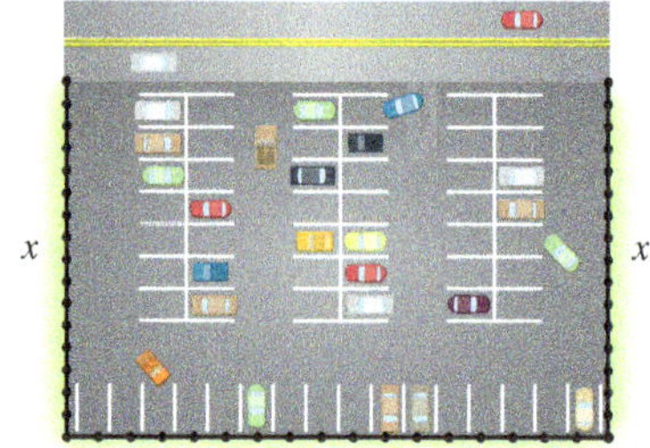

(a) Express the length of the remaining side to be fenced in terms of x.

(b) What are the restrictions on x?

(c) Determine a function $\mathscr{A}$ that represents the area of the parking lot in terms of x.

(d) Determine the values of x that will give an area between 30,000 and 40,000 ft^2.

(e) What dimensions will give a maximum area, and what will this area be?

60. ***Area of a Rectangular Region*** A farmer wishes to enclose a rectangular region bordering a river with fencing, as shown in the diagram. Suppose that x represents the length of each of the three parallel pieces of fencing. She has 600 ft of fencing available.

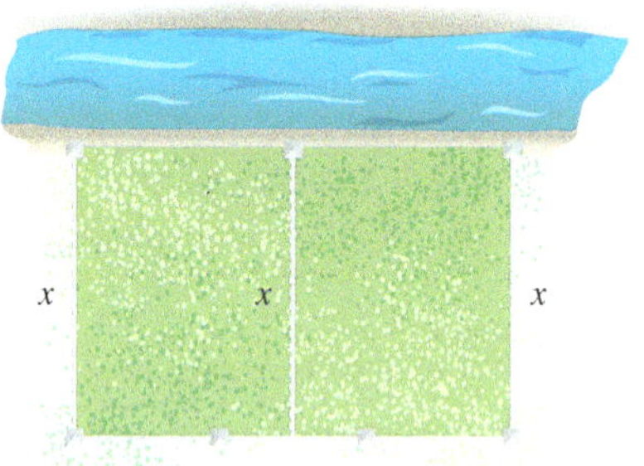

(a) What is the length of the remaining piece of fencing in terms of x?

(b) Determine a function $\mathscr{A}$ that represents the total area of the enclosed region. Give any restrictions on x.

(c) What dimensions for the total enclosed region would give an area of 22,500 ft^2?

(d) What is the maximum area that can be enclosed?

61. ***Volume of a Box*** A piece of cardboard is twice as long as it is wide. It is to be made into a box with an open top by cutting 2-in. squares from each corner and folding up the sides. Let x represent the width (in inches) of the original piece of cardboard.

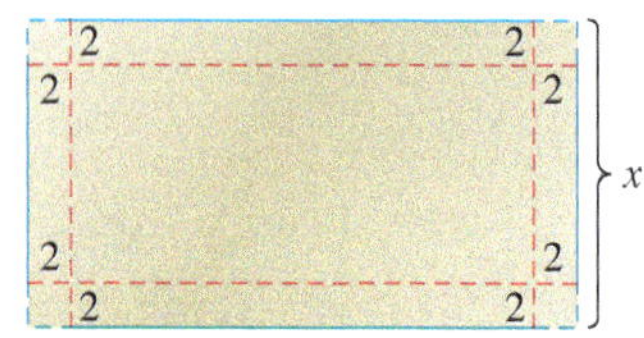

(a) Represent the length of the original piece of cardboard in terms of x.

(b) What will be the dimensions of the bottom rectangular base of the box? Give the restrictions on x.

(c) Determine a function V that represents the volume of the box in terms of x.

(d) For what dimensions of the bottom of the box will the volume be 320 in.^3?

(e) Find the values of x if such a box is to have a volume between 400 and 500 in.^3.

62. *Volume of a Box* A piece of sheet metal is 2.5 times as long as it is wide. It is to be made into a box with an open top by cutting 3-in. squares from each corner and folding up the sides. Let x represent the width (in inches) of the original piece.

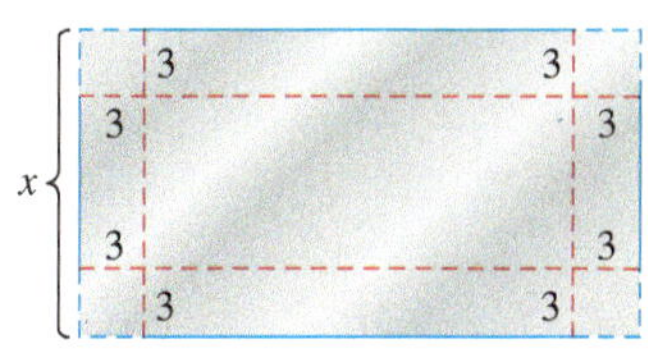

(a) Represent the length of the original piece of sheet metal in terms of x.

(b) What are the restrictions on x?

(c) Determine a function V that represents the volume of the box in terms of x.

(d) For what values of x (that is, original widths) will the volume of the box be between 600 and 800 in.^{3}?

63. *Shooting a Free Throw* If a person shoots a free throw from a position 8 ft above the floor, then the path of the ball may be modeled by the parabola

$$y = \frac{-16x^2}{0.434v^2} + 1.15x + 8,$$

where v is the initial velocity of the ball in feet per second, as illustrated in the figure. (*Source:* Rist, C., "The Physics of Foul Shots," *Discover.*)

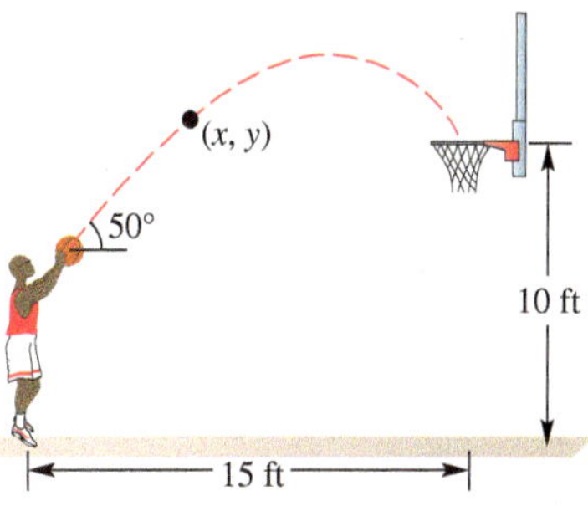

(a) If the basketball hoop is 10 ft high and located 15 ft away, what initial velocity v should the basketball have?

(b) What is the maximum height of the basketball?

64. *Shooting a Free Throw* See **Exercise 63.** If a person shoots a free throw from an underhand position 3 ft above the floor, then the path of the ball may be modeled by

$$y = \frac{-16x^2}{0.117v^2} + 2.75x + 3.$$

Repeat parts (a) and (b) from **Exercise 63.** Then compare the paths for the overhand shot and the underhand shot.

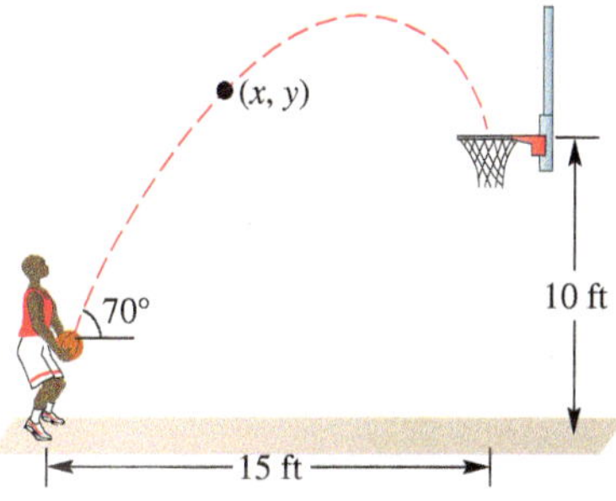

65. *Sum and Product of Two Numbers* Find two numbers whose sum is 20 and whose product is the maximum possible value. (*Hint: Let x be one number. Then $20 - x$ is the other number. Form a quadratic function by multiplying them, and then find the maximum value of the function.*)

66. *Sum and Product of Two Numbers* Find two numbers whose sum is 32 and whose product is the maximum possible value.

67. *Height of an Object* If an object is projected upward from ground level with an initial velocity of 32 ft per sec, then its height in feet after t seconds is given by

$$s(t) = -16t^2 + 32t.$$

Find the number of seconds it will take the object to reach its maximum height. What is this maximum height?

68. *Height of an Object* If an object is projected upward from an initial height of 100 ft with an initial velocity of 64 ft per sec, then its height in feet after t seconds is given by

$$s(t) = -16t^2 + 64t + 100.$$

Find the number of seconds it will take the object to reach its maximum height. What is this maximum height?

(Modeling) Solve each problem. ***See Example 6.***

69. ***Price of Chocolate Chip Cookies*** The average price in dollars of a pound of chocolate chip cookies from 2002 to 2013 is shown in the table.

Year	Price per Pound	Year	Price per Pound
2002	2.59	2008	2.88
2003	2.81	2009	3.17
2004	2.65	2010	3.25
2005	2.67	2011	3.35
2006	2.88	2012	3.62
2007	2.70	2013	3.64

Source: Consumer Price Index.

The data are modeled by the quadratic function

$$f(x) = 0.0095x^2 - 0.0076x + 2.660,$$

where $x = 0$ corresponds to 2002 and $f(x)$ is the price in dollars. If this model continues to apply, what will it predict for the price of a pound of chocolate chip cookies in 2018?

70. ***Concentration of Atmospheric CO_2*** The quadratic function

$$f(x) = 0.0118x^2 + 0.8633x + 317$$

models the worldwide atmospheric concentration of carbon dioxide in parts per million (ppm) over the period 1960–2013, where $x = 0$ represents the year 1960. If this model continues to apply, what will be the atmospheric CO_2 concentration in 2020? Round to the nearest unit. (*Source*: U.S. Department of Energy.)

71. ***Spending on Shoes and Clothing*** The total amount spent by Americans on shoes and clothing from 2000 to 2013 can be modeled by

$$f(x) = 0.7714x^2 - 3.693x + 297.9,$$

where $x = 0$ represents 2000 and $f(x)$ is in billions of dollars. Based on this model, in what year did spending on shoes and clothing reach a minimum? (*Source*: Bureau of Economic Analysis.)

72. ***Accident Rate*** According to data from the National Highway Traffic Safety Administration, the accident rate as a function of the age of the driver in years x can be approximated by the function

$$f(x) = 0.0232x^2 - 2.28x + 60.0, \quad \text{for } 16 \le x \le 85.$$

Find both the age at which the accident rate is a minimum and the minimum rate to the nearest hundredth.

73. ***College Enrollment*** The table lists total fall enrollments in degree-granting postsecondary colleges in the United States for selected years.

(a) Plot the data. Let $x = 0$ correspond to the year 2008.

(b) Find a quadratic function $f(x) = ax^2 + bx + c$ that models the data.

(c) Plot the data together with f in the same window. How well does f model enrollment?

(d) Use f to estimate total enrollment in 2013 to the nearest tenth of a million.

Year	Enrollment (in millions)
2008	19.1
2009	20.4
2010	21.0
2011	21.0
2012	20.6

Source: National Center for Education Statistics.

74. *Two-Year College Enrollment* The table lists total fall enrollments in degree-granting two-year colleges in the United States for selected years.

(a) Plot the data. Let $x = 0$ correspond to the year 2008.

(b) Find a quadratic function $g(x) = ax^2 + bx + c$ that models the data.

(c) Plot the data together with g in the same window. How well does g model enrollment?

(d) Use g to estimate total enrollment in 2013 to the nearest tenth of a million.

Year	Enrollment (in millions)
2008	6.9
2009	7.5
2010	7.7
2011	7.5
2012	7.2

Source: National Center for Education Statistics.

75. *Foreign-Born Americans* The table lists the percent of the U.S. population that was foreign-born for selected years.

(a) Plot the data. Let $x = 0$ correspond to the year 1930, $x = 10$ correspond to 1940, and so on.

(b) Find a quadratic function $f(x) = a(x - h)^2 + k$ that models the data. Use $(40, 4.7)$ as the vertex and $(20, 6.9)$ as the other point to determine a.

(c) Plot the data together with f in the same window. How well does f model the percent of the U.S. population that is foreign-born?

(d) Use the quadratic regression feature of a graphing calculator to determine the quadratic function g that provides the best fit for the data.

(e) Use functions f and g to predict the percent, to the nearest tenth, of the U.S. population in 2019 that will be foreign-born.

Year	Percent
1930	11.6
1940	8.8
1950	6.9
1960	5.4
1970	4.7
1980	6.2
1990	7.9
2000	11.1
2010	12.4
2012	12.9

Source: U.S. Census Bureau.

76. *Automobile Stopping Distance* Selected values of the stopping distance y, in feet, of a car traveling x miles per hour are given in the table.

(a) Plot the data.

(b) The quadratic function

$$f(x) = 0.056057x^2 + 1.06657x$$

is one model that has been used to approximate stopping distances. Find $f(45)$ to the nearest foot, and interpret this result.

(c) How well does f model the car's stopping distance?

Speed (in mph)	Stopping Distance (in feet)
20	46
30	87
40	140
50	240
60	282
70	371

Source: National Safety Institute Student Workbook.

Concept Check *Work each problem.*

77. Find a value of c so that $y = x^2 - 10x + c$ has exactly one x-intercept.

78. For what values of a does $y = ax^2 - 8x + 4$ have no x-intercepts?

79. Define the quadratic function f having x-intercepts $(2, 0)$ and $(5, 0)$ and y-intercept $(0, 5)$.

80. Define the quadratic function f having x-intercepts $(1, 0)$ and $(-2, 0)$ and y-intercept $(0, 4)$.

81. The distance between the two points $P(x_1, y_1)$ and $R(x_2, y_2)$ is

$$d(P, R) = \sqrt{(x_1 - x_2)^2 + (y_1 - y_2)^2}. \quad \text{Distance formula}$$

Find the closest point on the line $y = 2x$ to the point $(1, 7)$. (*Hint:* Every point on $y = 2x$ has the form $(x, 2x)$, and the closest point has the minimum distance.)

82. A quadratic equation $f(x) = 0$ has a solution $x = 2$. Its graph has vertex $(5, 3)$. What is the other solution of the equation?

Relating Concepts

For individual or collaborative investigation *(Exercises 83-86)*

A quadratic inequality such as

$$x^2 + 2x - 8 < 0$$

can be solved by first solving the related quadratic equation

$$x^2 + 2x - 8 = 0,$$

identifying intervals determined by the solutions of this equation, and then using a test value from each interval to determine which intervals form the solution set. ***Work Exercises 83–86 in order*** *to learn a graphical method of solving inequalities.*

83. Graph $f(x) = x^2 + 2x - 8$.

84. The real solutions of $x^2 + 2x - 8 = 0$ are the x-values of the x-intercepts of the graph in **Exercise 83.** These are values of x for which $f(x) = 0$. What are these values? What is the solution set of this equation?

85. The real solutions of $x^2 + 2x - 8 < 0$ are the x-values for which the graph in **Exercise 83** lies *below* the x-axis. These are values of x for which $f(x) < 0$ is true. What interval of x-values represents the solution set of this inequality?

86. The real solutions of $x^2 + 2x - 8 > 0$ are the x-values for which the graph in **Exercise 83** lies *above* the x-axis. These are values of x for which $f(x) > 0$ is true. What intervals of x-values represent the solution set of this inequality?

Use the technique described in ***Exercises 83–86*** *to solve each inequality. Write the solution set in interval notation.*

87. $x^2 - x - 6 < 0$

88. $x^2 - 9x + 20 < 0$

89. $2x^2 - 9x \geq 18$

90. $3x^2 + x \geq 4$

91. $-x^2 + 4x + 1 \geq 0$

92. $-x^2 + 2x + 6 > 0$

3.2 Synthetic Division

- Synthetic Division
- Remainder Theorem
- Potential Zeros of Polynomial Functions

The outcome of a division problem can be written using multiplication, even when the division involves polynomials. The **division algorithm** illustrates this.

Division Algorithm

Let $f(x)$ and $g(x)$ be polynomials with $g(x)$ of lesser degree than $f(x)$ and $g(x)$ of degree 1 or more. There exist unique polynomials $q(x)$ and $r(x)$ such that

$$f(x) = g(x) \cdot q(x) + r(x),$$

where either $r(x) = 0$ or the degree of $r(x)$ is less than the degree of $g(x)$.

This is the long-division process for the result shown to the right.

$$\begin{array}{r} 3x - 2 \\ x^2 + 0x - 4\overline{)3x^3 - 2x^2 + 0x - 150} \\ \underline{3x^3 + 0x^2 - 12x} \phantom{{}-150} \\ -2x^2 + 12x - 150 \\ \underline{-2x^2 - 0x + 8} \\ 12x - 158 \end{array}$$

Consider the result shown here.

$$\frac{3x^3 - 2x^2 - 150}{x^2 - 4} = 3x - 2 + \frac{12x - 158}{x^2 - 4}.$$

We can express it using the preceding division algorithm.

$$\underbrace{3x^3 - 2x^2 - 150}_{f(x)} = \underbrace{(x^2 - 4)}_{g(x)}\underbrace{(3x - 2)}_{q(x)} + \underbrace{12x - 158}_{r(x)}$$

Dividend (original polynomial) = Divisor · Quotient + Remainder

Synthetic Division When a given polynomial in x is divided by a first-degree binomial of the form $x - k$, a shortcut method called **synthetic division** may be used. The example on the left below is simplified by omitting all variables and writing only coefficients, with 0 used to represent the coefficient of any missing terms. The coefficient of x in the divisor is always 1 in these divisions, so it too can be omitted. These omissions simplify the problem, as shown on the right.

$$\begin{array}{r} 3x^2 + 10x + 40 \\ x - 4\overline{)3x^3 - 2x^2 + 0x - 150} \\ \underline{3x^3 - 12x^2} \phantom{{}+0x-150} \\ 10x^2 + 0x \phantom{{}-150} \\ \underline{10x^2 - 40x} \phantom{{}-150} \\ 40x - 150 \\ \underline{40x - 160} \\ 10 \end{array}
\qquad
\begin{array}{rrrr} & 3 & 10 & 40 \\ -4\overline{)3} & -2 & 0 & -150 \\ \underline{3} & \underline{-12} & & \\ & 10 & 0 & \\ & \underline{10} & \underline{-40} & \\ & & 40 & -150 \\ & & \underline{40} & \underline{-160} \\ & & & 10 \end{array}$$

The numbers in color that are repetitions of the numbers directly above them can also be omitted, as shown on the left below.

$$\begin{array}{rrrr} & 3 & 10 & 40 \\ -4\overline{)3} & -2 & 0 & -150 \\ & \underline{-12} & & \\ & 10 & 0 & \\ & & \underline{-40} & \\ & & 40 & -150 \\ & & & \underline{-160} \\ & & & 10 \end{array}
\qquad
\begin{array}{rrrr} & 3 & 10 & 40 \\ -4\overline{)3} & -2 & 0 & -150 \\ & \underline{-12} & & \\ & 10 & & \\ & & \underline{-40} & \\ & & 40 & \\ & & & \underline{-160} \\ & & & 10 \end{array}$$

The numbers in color are again repetitions of those directly above them. They may be omitted, as shown on the right above.

The entire process can now be condensed vertically.

$$\begin{array}{rrrr} -4\overline{)3} & -2 & 0 & -150 \\ & \underline{-12} & \underline{-40} & \underline{-160} \\ 3 & 10 & 40 & 10 \end{array}$$

The top row of numbers can be omitted since it duplicates the bottom row if the 3 is brought down.

The rest of the bottom row is obtained by subtracting -12, -40, and -160 from the corresponding terms above them.

To simplify the arithmetic, we replace subtraction in the second row by addition and compensate by changing the -4 at the upper left to its additive inverse, 4.

$$\begin{array}{rrrr} 4\overline{)3} & -2 & 0 & -150 \\ & \underline{12} & \underline{40} & \underline{160} \\ 3 & 10 & 40 & 10 \\ \downarrow & \downarrow & \downarrow & \downarrow \end{array}$$

Additive inverse → 4; Signs changed ← 12, 40, 160

Quotient → $3x^2 + 10x + 40 + \dfrac{10}{x - 4}$ ← Remainder

Synthetic division provides an efficient process for dividing a polynomial in x by a binomial of the form $x - k$. Begin by writing the coefficients of the polynomial in decreasing powers of the variable, using 0 as the coefficient of any missing powers. The number k is written to the left in the same row. The answer is found in the bottom row with the remainder farthest to the right and the coefficients of the quotient on the left when written in order of decreasing degree.

CAUTION ***To avoid errors, use 0 as the coefficient for any missing terms, including a missing constant, when setting up the division.***

EXAMPLE 1 Using Synthetic Division

Use synthetic division to perform the division.

$$\frac{5x^3 - 6x^2 - 28x - 2}{x + 2}$$

SOLUTION Express $x + 2$ in the form $x - k$ by writing it as $x - (-2)$.

x + 2 leads to −2.

$$-2\overline{)5 \quad -6 \quad -28 \quad -2} \longleftarrow \text{Coefficients of the polynomial}$$

Bring down the 5, and multiply: $-2(5) = -10$.

$$\begin{array}{r|rrrr} -2 & 5 & -6 & -28 & -2 \\ & \downarrow & -10 & & \\ \hline & 5 & & & \end{array}$$

Add -6 and -10 to obtain -16. Multiply: $-2(-16) = 32$.

$$\begin{array}{r|rrrr} -2 & 5 & -6 & -28 & -2 \\ & & -10 & 32 & \\ \hline & 5 & -16 & & \end{array}$$

Add -28 and 32, obtaining 4. Multiply: $-2(4) = -8$.

$$\begin{array}{r|rrrr} -2 & 5 & -6 & -28 & -2 \\ & & -10 & 32 & -8 \\ \hline & 5 & -16 & 4 & \end{array}$$

Add columns. Be careful with signs.

Add -2 and -8 to obtain -10.

$$\begin{array}{r|rrrr} -2 & 5 & -6 & -28 & -2 \\ & & -10 & 32 & -8 \\ \hline & 5 & -16 & 4 & -10 \end{array} \longleftarrow \text{Remainder}$$

Quotient (5, −16, 4)

Because the divisor $x - k$ has degree 1, the degree of the quotient will always be one less than the degree of the polynomial to be divided.

$$\frac{5x^3 - 6x^2 - 28x - 2}{x + 2} = 5x^2 - 16x + 4 + \frac{-10}{x + 2}$$

Remember to add $\frac{\text{remainder}}{\text{divisor}}$.

✔ **Now Try Exercise 15.**

The result of the division in **Example 1** can be written as

$$5x^3 - 6x^2 - 28x - 2 = (x + 2)(5x^2 - 16x + 4) + (-10). \quad \text{Multiply by } x + 2.$$

The theorem that follows is a generalization of this product form.

Special Case of the Division Algorithm

For any polynomial $f(x)$ and any complex number k, there exists a unique polynomial $q(x)$ and number r such that the following holds.

$$f(x) = (x - k)q(x) + r$$

We can illustrate this connection using the mathematical statement

$$\underbrace{5x^3 - 6x^2 - 28x - 2}_{f(x)} \;\underset{=}{=}\; \underbrace{(x + 2)}_{(x-k)\,\cdot}\underbrace{(5x^2 - 16x + 4)}_{q(x)} \underset{+}{+} \underbrace{(-10)}_{r}.$$

This form of the division algorithm is used to develop the *remainder theorem.*

Remainder Theorem Suppose $f(x)$ is written as $f(x) = (x - k)q(x) + r$. This equality is true for all complex values of x, so it is true for $x = k$.

$$f(k) = (k - k)q(k) + r, \quad \text{or} \quad f(k) = r \qquad \text{Replace } x \text{ with } k.$$

This proves the following **remainder theorem,** which gives a new method of evaluating polynomial functions.

Remainder Theorem

If the polynomial $f(x)$ is divided by $x - k$, then the remainder is equal to $f(k)$.

In **Example 1,** when $f(x) = 5x^3 - 6x^2 - 28x - 2$ was divided by $x + 2$, or $x - (-2)$, the remainder was -10. Substitute -2 for x in $f(x)$.

$$f(-2) = 5(-2)^3 - 6(-2)^2 - 28(-2) - 2$$
$$f(-2) = -40 - 24 + 56 - 2$$
$$f(-2) = -10$$

Use parentheses around substituted values to avoid errors.

An alternative way to find the value of a polynomial is to use synthetic division. By the remainder theorem, instead of replacing x by -2 to find $f(-2)$, divide $f(x)$ by $x + 2$ as in **Example 1.** Then $f(-2)$ is the remainder, -10.

$$\begin{array}{r|rrrr} -2 & 5 & -6 & -28 & -2 \\ & & -10 & 32 & -8 \\ \hline & 5 & -16 & 4 & -10 \end{array} \leftarrow f(-2)$$

EXAMPLE 2 **Applying the Remainder Theorem**

Let $f(x) = -x^4 + 3x^2 - 4x - 5$. Use the remainder theorem to find $f(-3)$.

SOLUTION Use synthetic division with $k = -3$.

$$\begin{array}{r|rrrrr} -3 & -1 & 0 & 3 & -4 & -5 \\ & & 3 & -9 & 18 & -42 \\ \hline & -1 & 3 & -6 & 14 & -47 \end{array} \leftarrow \text{Remainder}$$

$f(-3)$ is equal to the remainder when dividing by $x + 3$.

By this result, $f(-3) = -47$.

✔ **Now Try Exercise 37.**

Potential Zeros of Polynomial Functions A **zero** of a polynomial function $f(x)$ is a number k such that $f(k) = 0$. ***Real number zeros are the x-values of the x-intercepts of the graph of the function.***

The remainder theorem gives a quick way to decide whether a number k is a zero of a polynomial function $f(x)$, as follows.

1. Use synthetic division to find $f(k)$.
2. If the remainder is 0, then $f(k) = 0$ and k is a zero of $f(x)$. If the remainder is not 0, then k is not a zero of $f(x)$.

A zero of $f(x)$ is a **root,** or **solution,** of the equation $f(x) = 0$.

EXAMPLE 3 Deciding Whether a Number Is a Zero

Decide whether the given number k is a zero of $f(x)$.

(a) $f(x) = x^3 - 4x^2 + 9x - 6; \quad k = 1$

(b) $f(x) = x^4 + x^2 - 3x + 1; \quad k = -1$

(c) $f(x) = x^4 - 2x^3 + 4x^2 + 2x - 5; \quad k = 1 + 2i$

SOLUTION

(a) To decide whether 1 is a zero of $f(x) = x^3 - 4x^2 + 9x - 6$, use synthetic division.

$$\text{Proposed zero} \longrightarrow 1\overline{)\begin{array}{rrrr} 1 & -4 & 9 & -6 \\ & 1 & -3 & 6 \\ \hline 1 & -3 & 6 & 0 \end{array}} \quad \begin{array}{l} \longleftarrow f(x) = x^3 - 4x^2 + 9x - 6 \\ \\ \longleftarrow \text{Remainder} \end{array}$$

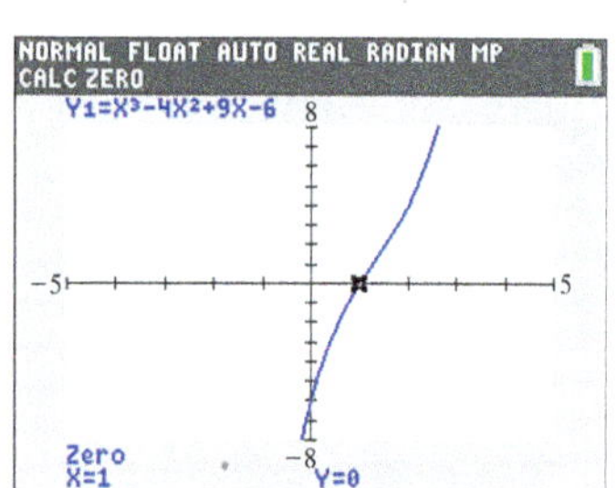

Figure 16

Because the remainder is 0, $f(1) = 0$, and 1 is a zero of the given polynomial function. An x-intercept of the graph of $f(x) = x^3 - 4x^2 + 9x - 6$ is the point $(1, 0)$. The graph in **Figure 16** supports this.

(b) For $f(x) = x^4 + x^2 - 3x + 1$, remember to use 0 as coefficient for the missing x^3-term in the synthetic division.

$$\text{Proposed zero} \longrightarrow -1\overline{)\begin{array}{rrrrr} 1 & 0 & 1 & -3 & 1 \\ & -1 & 1 & -2 & 5 \\ \hline 1 & -1 & 2 & -5 & 6 \end{array}} \longleftarrow \text{Remainder}$$

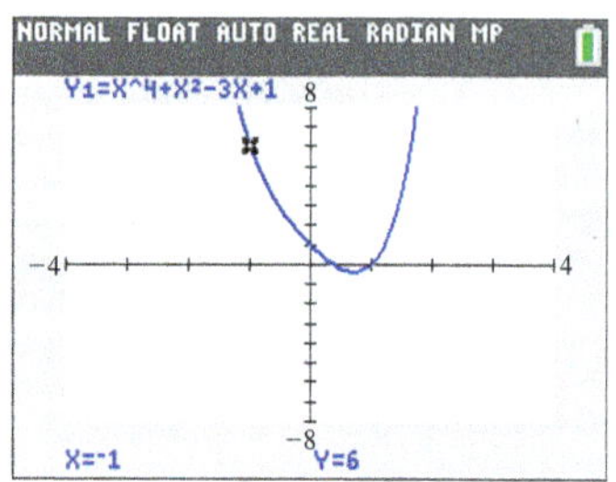

Figure 17

The remainder is not 0, so -1 is not a zero of $f(x) = x^4 + x^2 - 3x + 1$. In fact, $f(-1) = 6$, indicating that $(-1, 6)$ is on the graph of $f(x)$. The graph in **Figure 17** supports this.

(c) Use synthetic division and operations with complex numbers to determine whether $1 + 2i$ is a zero of $f(x) = x^4 - 2x^3 + 4x^2 + 2x - 5$.

$$1 + 2i\overline{)\begin{array}{rrrrr} 1 & -2 & 4 & 2 & -5 \\ & 1 + 2i & -5 & -1 - 2i & 5 \\ \hline 1 & -1 + 2i & -1 & 1 - 2i & 0 \end{array}} \quad \begin{array}{l} \\ i^2 = -1 \\ \longleftarrow \text{Remainder} \end{array}$$

$$\begin{aligned}(1 + 2i)(-1 + 2i) &= -1 + 4i^2 \\ &= -5\end{aligned}$$

The remainder is 0, so $1 + 2i$ is a zero of the given polynomial function. Notice that $1 + 2i$ is *not* a real number zero. Therefore, it is not associated with an x-intercept on the graph of $f(x)$.

✔ **Now Try Exercises 49 and 59.**

3.2 Exercises

CONCEPT PREVIEW *Fill in the blank(s) to correctly complete each sentence.*

1. In arithmetic, the result of the division

$$\begin{array}{r} 3 \\ 5\overline{)19} \\ \underline{15} \\ 4 \end{array}$$

can be written

$19 = 5 \cdot$ _____ + _____.

2. In algebra, the result of the division

$$\begin{array}{r} x + 3 \\ x - 1\overline{)x^2 + 2x + 3} \\ \underline{x^2 - x} \\ 3x + 3 \\ \underline{3x - 3} \\ 6 \end{array}$$

can be written $x^2 + 2x + 3 =$
$(x - 1)($_____$) +$ _____.

3. To perform the division in **Exercise 2** using synthetic division, we begin by writing the following.

$$\_\_\_\_)\overline{\_\_\_\_ \quad 2 \quad 3}$$

4. To perform the division

$$x + 2\overline{)x^3 + 4x + 2}$$

using synthetic division, we begin by writing the following.

$$\_\_\_\_)\overline{1 \quad \_\_\_\_ \quad 4 \quad 2}$$

5. To perform the division

$$x - 3\overline{)x^3 + 6x^2 + 2x}$$

using synthetic division, we begin by writing the following.

$$\_\_\_\_)\overline{1 \quad \_\_\_\_ \quad 2 \quad \_\_\_\_}$$

6. Consider the following function.

$$f(x) = 2x^4 + 6x^3 - 5x^2 + 3x + 8$$
$$f(x) = (x - 2)(2x^3 + 10x^2 + 15x + 33) + 74$$

By inspection, we can state that $f(2) =$ _____.

Use synthetic division to perform each division. **See Example 1.**

7. $\dfrac{x^3 + 3x^2 + 11x + 9}{x + 1}$

8. $\dfrac{x^3 + 7x^2 + 13x + 6}{x + 2}$

9. $\dfrac{5x^4 + 5x^3 + 2x^2 - x - 3}{x + 1}$

10. $\dfrac{2x^4 - x^3 - 7x^2 + 7x - 10}{x - 2}$

11. $\dfrac{x^4 + 4x^3 + 2x^2 + 9x + 4}{x + 4}$

12. $\dfrac{x^4 + 5x^3 + 4x^2 - 3x + 9}{x + 3}$

13. $\dfrac{x^5 + 3x^4 + 2x^3 + 2x^2 + 3x + 1}{x + 2}$

14. $\dfrac{x^6 - 3x^4 + 2x^3 - 6x^2 - 5x + 3}{x + 2}$

15. $\dfrac{-9x^3 + 8x^2 - 7x + 2}{x - 2}$

16. $\dfrac{-11x^4 - 2x^3 - 8x^2 - 4}{x + 1}$

17. $\dfrac{\frac{1}{3}x^3 - \frac{2}{9}x^2 + \frac{2}{27}x - \frac{1}{81}}{x - \frac{1}{3}}$

18. $\dfrac{x^3 + x^2 + \frac{1}{2}x + \frac{1}{8}}{x + \frac{1}{2}}$

19. $\dfrac{x^4 - 3x^3 - 4x^2 + 12x}{x - 2}$

20. $\dfrac{x^4 - x^3 - 5x^2 - 3x}{x + 1}$

21. $\dfrac{x^3 - 1}{x - 1}$

22. $\dfrac{x^4 - 1}{x - 1}$

23. $\dfrac{x^5 + 1}{x + 1}$

24. $\dfrac{x^7 + 1}{x + 1}$

Use synthetic division to divide $f(x)$ by $x - k$ for the given value of k. Then express $f(x)$ in the form $f(x) = (x - k)q(x) + r$.

25. $f(x) = 2x^3 + x^2 + x - 8; \quad k = -1$

26. $f(x) = 2x^3 + 3x^2 - 16x + 10; \quad k = -4$

27. $f(x) = x^3 + 4x^2 + 5x + 2; \quad k = -2$

28. $f(x) = -x^3 + x^2 + 3x - 2; \quad k = 2$

29. $f(x) = 4x^4 - 3x^3 - 20x^2 - x; \quad k = 3$

30. $f(x) = 2x^4 + x^3 - 15x^2 + 3x; \quad k = -3$

31. $f(x) = 3x^4 + 4x^3 - 10x^2 + 15; \quad k = -1$

32. $f(x) = -5x^4 + x^3 + 2x^2 + 3x + 1; \quad k = 1$

For each polynomial function, use the remainder theorem to find $f(k)$. **See Example 2.**

33. $f(x) = x^2 + 5x + 6; \quad k = -2$

34. $f(x) = x^2 - 4x - 5; \quad k = 5$

35. $f(x) = 2x^2 - 3x - 3; \quad k = 2$

36. $f(x) = -x^3 + 8x^2 + 63; \quad k = 4$

37. $f(x) = x^3 - 4x^2 + 2x + 1; \quad k = -1$

38. $f(x) = 2x^3 - 3x^2 - 5x + 4; \quad k = 2$

39. $f(x) = x^2 - 5x + 1; \quad k = 2 + i$

40. $f(x) = x^2 - x + 3; \quad k = 3 - 2i$

41. $f(x) = x^2 + 4; \quad k = 2i$

42. $f(x) = 2x^2 + 10; \quad k = i\sqrt{5}$

43. $f(x) = 2x^5 - 10x^3 - 19x^2 - 50; \quad k = 3$

44. $f(x) = x^4 + 6x^3 + 9x^2 + 3x - 3; \quad k = 4$

45. $f(x) = 6x^4 + x^3 - 8x^2 + 5x + 6; \quad k = \frac{1}{2}$

46. $f(x) = 6x^3 - 31x^2 - 15x; \quad k = -\frac{1}{2}$

Use synthetic division to decide whether the given number k is a zero of the polynomial function. If it is not, give the value of $f(k)$. **See Examples 2 and 3.**

47. $f(x) = x^2 + 2x - 8; \quad k = 2$

48. $f(x) = x^2 + 4x - 5; \quad k = -5$

49. $f(x) = x^3 - 3x^2 + 4x - 4; \quad k = 2$

50. $f(x) = x^3 + 2x^2 - x + 6; \quad k = -3$

51. $f(x) = 2x^3 - 6x^2 - 9x + 4; \quad k = 1$

52. $f(x) = 2x^3 + 9x^2 - 16x + 12; \quad k = 1$

53. $f(x) = x^3 + 7x^2 + 10x; \quad k = 0$

54. $f(x) = 2x^3 - 3x^2 - 5x; \quad k = 0$

55. $f(x) = 5x^4 + 2x^3 - x + 3; \quad k = \frac{2}{5}$

56. $f(x) = 16x^4 + 3x^2 - 2; \quad k = \frac{1}{2}$

57. $f(x) = x^2 - 2x + 2; \quad k = 1 - i$

58. $f(x) = x^2 - 4x + 5; \quad k = 2 - i$

59. $f(x) = x^2 + 3x + 4; \quad k = 2 + i$

60. $f(x) = x^2 - 3x + 5; \quad k = 1 - 2i$

61. $f(x) = 4x^4 + x^2 + 17x + 3; \quad k = -\frac{3}{2}$

62. $f(x) = 3x^4 + 13x^3 - 10x + 8; \quad k = -\frac{4}{3}$

63. $f(x) = x^3 + 3x^2 - x + 1; \quad k = 1 + i$

64. $f(x) = 2x^3 - x^2 + 3x - 5; \quad k = 2 - i$

Relating Concepts

For individual or collaborative investigation *(Exercises 65–74)*

The remainder theorem indicates that when a polynomial $f(x)$ is divided by $x - k$, the remainder is equal to $f(k)$. For

$$f(x) = x^3 - 2x^2 - x + 2,$$

use the remainder theorem to find each of the following. Then determine the coordinates of the corresponding point on the graph of $f(x)$.

65. $f(-2)$ **66.** $f(-1)$ **67.** $f\left(-\frac{1}{2}\right)$ **68.** $f(0)$

69. $f(1)$ **70.** $f\left(\frac{3}{2}\right)$ **71.** $f(2)$ **72.** $f(3)$

73. Use the results from **Exercises 65–72** to plot eight points on the graph of $f(x)$. Join these points with a smooth curve.

74. Apply the method above to graph $f(x) = -x^3 - x^2 + 2x$. Use x-values -3, -1, $\frac{1}{2}$, and 2 and the fact that $f(0) = 0$.

3.3 Zeros of Polynomial Functions

- Factor Theorem
- Rational Zeros Theorem
- Number of Zeros
- Conjugate Zeros Theorem
- Zeros of a Polynomial Function
- Descartes' Rule of Signs

Factor Theorem

Consider the polynomial function

$$f(x) = x^2 + x - 2,$$

which is written in factored form as

$$f(x) = (x - 1)(x + 2).$$

For this function, $f(1) = 0$ and $f(-2) = 0$, and thus 1 and -2 are zeros of $f(x)$. Notice the special relationship between each linear factor and its corresponding zero. The **factor theorem** summarizes this relationship.

Factor Theorem

For any polynomial function $f(x)$, $x - k$ is a factor of the polynomial if and only if $f(k) = 0$.

EXAMPLE 1 Deciding Whether $x - k$ Is a Factor

Determine whether $x - 1$ is a factor of each polynomial.

(a) $f(x) = 2x^4 + 3x^2 - 5x + 7$

(b) $f(x) = 3x^5 - 2x^4 + x^3 - 8x^2 + 5x + 1$

SOLUTION

(a) By the factor theorem, $x - 1$ will be a factor if $f(1) = 0$. Use synthetic division and the remainder theorem to decide.

$$\begin{array}{r|rrrrr} 1 & 2 & 0 & 3 & -5 & 7 \\ & & 2 & 2 & 5 & 0 \\ \hline & 2 & 2 & 5 & 0 & 7 \end{array} \leftarrow f(1) = 7$$

Use a zero coefficient for the missing term.

The remainder is 7, not 0, so $x - 1$ is not a factor of $2x^4 + 3x^2 - 5x + 7$.

(b)

$$\begin{array}{r|rrrrrr} 1 & 3 & -2 & 1 & -8 & 5 & 1 \\ & & 3 & 1 & 2 & -6 & -1 \\ \hline & 3 & 1 & 2 & -6 & -1 & 0 \end{array}$$

$\leftarrow f(x) = 3x^5 - 2x^4 + x^3 - 8x^2 + 5x + 1$

$\leftarrow f(1) = 0$

Because the remainder is 0, $x - 1$ is a factor. Additionally, we can determine from the coefficients in the bottom row that the other factor is

$$3x^4 + 1x^3 + 2x^2 - 6x - 1.$$

Thus, we can express the polynomial in factored form.

$$f(x) = (x - 1)(3x^4 + x^3 + 2x^2 - 6x - 1)$$

✔ **Now Try Exercises 9 and 11.**

We can use the factor theorem to factor a polynomial of greater degree into linear factors of the form $ax - b$.

EXAMPLE 2 Factoring a Polynomial Given a Zero

Factor $f(x) = 6x^3 + 19x^2 + 2x - 3$ into linear factors given that -3 is a zero.

SOLUTION Because -3 is a zero of f, $x - (-3) = x + 3$ is a factor.

$$\begin{array}{r|rrrr} -3 & 6 & 19 & 2 & -3 \\ & & -18 & -3 & 3 \\ \hline & 6 & 1 & -1 & 0 \end{array}$$

Use synthetic division to divide $f(x)$ by $x + 3$.

The quotient is $6x^2 + x - 1$, which is the factor that accompanies $x + 3$.

$$f(x) = (x + 3)(6x^2 + x - 1)$$

$$f(x) = \underbrace{(x + 3)(2x + 1)(3x - 1)}_{\text{These factors are all linear.}} \quad \text{Factor } 6x^2 + x - 1.$$

✔ **Now Try Exercise 21.**

LOOKING AHEAD TO CALCULUS

Finding the derivative of a polynomial function is one of the basic skills required in a first calculus course. For the functions

$$f(x) = x^4 - x^2 + 5x - 4,$$
$$g(x) = -x^6 + x^2 - 3x + 4,$$
and $$h(x) = 3x^3 - x^2 + 2x - 4,$$

the derivatives are

$$f'(x) = 4x^3 - 2x + 5,$$
$$g'(x) = -6x^5 + 2x - 3,$$
and $$h'(x) = 9x^2 - 2x + 2.$$

Notice the use of the "prime" notation. For example, the derivative of $f(x)$ is denoted $f'(x)$.

Look for the pattern among the exponents and the coefficients. Using this pattern, what is the derivative of

$$F(x) = 4x^4 - 3x^3 + 6x - 4?$$

The answer is at the top of the next page.

Rational Zeros Theorem

The **rational zeros theorem** gives a method to determine all possible candidates for rational zeros of a polynomial function with integer coefficients.

Rational Zeros Theorem

If $\frac{p}{q}$ is a rational number written in lowest terms, and if $\frac{p}{q}$ is a zero of f, a polynomial function with integer coefficients, then p is a factor of the constant term and q is a factor of the leading coefficient.

Proof $f\left(\frac{p}{q}\right) = 0$ because $\frac{p}{q}$ is a zero of $f(x)$.

$$a_n\left(\frac{p}{q}\right)^n + a_{n-1}\left(\frac{p}{q}\right)^{n-1} + \cdots + a_1\left(\frac{p}{q}\right) + a_0 = 0 \quad \text{Definition of zero of } f$$

$$a_n\left(\frac{p^n}{q^n}\right) + a_{n-1}\left(\frac{p^{n-1}}{q^{n-1}}\right) + \cdots + a_1\left(\frac{p}{q}\right) + a_0 = 0 \quad \text{Power rule for exponents}$$

$$a_np^n + a_{n-1}p^{n-1}q + \cdots + a_1pq^{n-1} = -a_0q^n \quad \text{Multiply by } q^n. \text{ Subtract } a_0q^n.$$

$$p(a_np^{n-1} + a_{n-1}p^{n-2}q + \cdots + a_1q^{n-1}) = -a_0q^n \quad \text{Factor out } p.$$

Answer to Looking Ahead to Calculus:
$F'(x) = 16x^3 - 9x^2 + 6$

This result shows that $-a_0q^n$ equals the product of the two factors p and $(a_np^{n-1} + \cdots + a_1q^{n-1})$. For this reason, p must be a factor of $-a_0q^n$. Because it was assumed that $\frac{p}{q}$ is written in lowest terms, p and q have no common factor other than 1, so p is not a factor of q^n. Thus, p must be a factor of a_0. In a similar way, it can be shown that q is a factor of a_n.

EXAMPLE 3 Using the Rational Zeros Theorem

Consider the polynomial function.

$$f(x) = 6x^4 + 7x^3 - 12x^2 - 3x + 2$$

(a) List all possible rational zeros.

(b) Find all rational zeros and factor $f(x)$ into linear factors.

SOLUTION

(a) For a rational number $\frac{p}{q}$ to be a zero, p must be a factor of $a_0 = 2$, and q must be a factor of $a_4 = 6$. Thus, p can be ± 1 or ± 2, and q can be ± 1, ± 2, ± 3, or ± 6. The possible rational zeros $\frac{p}{q}$ are ± 1, ± 2, $\pm \frac{1}{2}$, $\pm \frac{1}{3}$, $\pm \frac{1}{6}$, and $\pm \frac{2}{3}$.

(b) Use the remainder theorem to show that 1 is a zero.

Use "trial and error" to find zeros.

$$\begin{array}{r|rrrrr} 1 & 6 & 7 & -12 & -3 & 2 \\ & & 6 & 13 & 1 & -2 \\ \hline & 6 & 13 & 1 & -2 & 0 \end{array} \leftarrow f(1) = 0$$

The 0 remainder shows that 1 is a zero. The quotient is $6x^3 + 13x^2 + x - 2$.

$$f(x) = (x - 1)(6x^3 + 13x^2 + x - 2) \quad \text{Begin factoring } f(x).$$

Now, use the quotient polynomial and synthetic division to find that -2 is a zero.

$$\begin{array}{r|rrrr} -2 & 6 & 13 & 1 & -2 \\ & & -12 & -2 & 2 \\ \hline & 6 & 1 & -1 & 0 \end{array} \leftarrow f(-2) = 0$$

The new quotient polynomial is $6x^2 + x - 1$. Therefore, $f(x)$ can now be completely factored as follows.

$$f(x) = (x - 1)(x + 2)(6x^2 + x - 1)$$

$$f(x) = (x - 1)(x + 2)(3x - 1)(2x + 1)$$

Setting $3x - 1 = 0$ and $2x + 1 = 0$ yields the zeros $\frac{1}{3}$ and $-\frac{1}{2}$. In summary, the rational zeros are 1, -2, $\frac{1}{3}$, and $-\frac{1}{2}$. These zeros correspond to the x-intercepts of the graph of $f(x)$ in **Figure 18.** The linear factorization of $f(x)$ is as follows.

$$f(x) = 6x^4 + 7x^3 - 12x^2 - 3x + 2$$

$$f(x) = (x - 1)(x + 2)(3x - 1)(2x + 1)$$

Check by multiplying these factors.

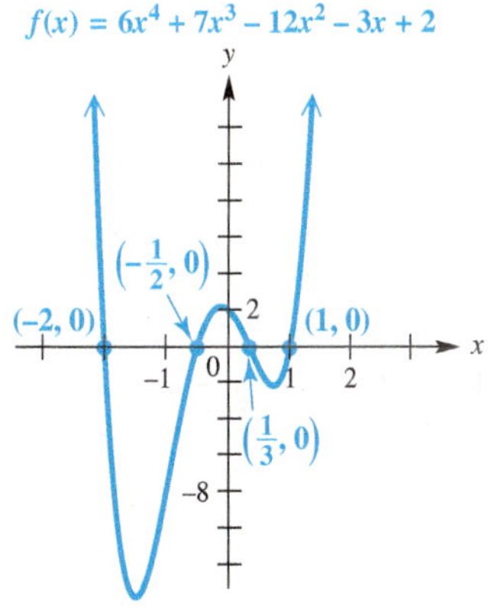

Figure 18

✔ **Now Try Exercise 39.**

NOTE Once we obtained the quadratic factor

$$6x^2 + x - 1$$

in **Example 3,** we were able to complete the work by factoring it directly. Had it not been easily factorable, we could have used the quadratic formula to find the other two zeros (and factors).

CAUTION ***The rational zeros theorem gives only possible rational zeros. It does not tell us whether these rational numbers are actual zeros.*** We must rely on other methods to determine whether or not they are indeed zeros. Furthermore, the polynomial must have integer coefficients.

To apply the rational zeros theorem to a polynomial with fractional coefficients, multiply through by the least common denominator of all the fractions. For example, any rational zeros of $p(x)$ defined below will also be rational zeros of $q(x)$.

$$p(x) = x^4 - \frac{1}{6}x^3 + \frac{2}{3}x^2 - \frac{1}{6}x - \frac{1}{3}$$

$$q(x) = 6x^4 - x^3 + 4x^2 - x - 2 \quad \text{Multiply the terms of } p(x) \text{ by 6.}$$

Carl Friedrich Gauss (1777–1855)

The **fundamental theorem of algebra** was first proved by Carl Friedrich Gauss in his doctoral thesis in 1799, when he was 22 years old.

Number of Zeros The **fundamental theorem of algebra** says that every function defined by a polynomial of degree 1 or more has a zero, which means that every such polynomial can be factored.

Fundamental Theorem of Algebra

Every function defined by a polynomial of degree 1 or more has at least one complex zero.

From the fundamental theorem, if $f(x)$ is of degree 1 or more, then there is some number k_1 such that $f(k_1) = 0$. By the factor theorem,

$$f(x) = (x - k_1)q_1(x), \quad \text{for some polynomial } q_1(x).$$

If $q_1(x)$ is of degree 1 or more, the fundamental theorem and the factor theorem can be used to factor $q_1(x)$ in the same way. There is some number k_2 such that $q_1(k_2) = 0$, so

$$q_1(x) = (x - k_2)q_2(x)$$

and

$$f(x) = (x - k_1)(x - k_2)q_2(x).$$

Assuming that $f(x)$ has degree n and repeating this process n times gives

$$f(x) = a(x - k_1)(x - k_2) \cdots (x - k_n). \quad a \text{ is the leading coefficient.}$$

Each of these factors leads to a zero of $f(x)$, so $f(x)$ has the n zeros $k_1, k_2, k_3, \ldots, k_n$. This result suggests the **number of zeros theorem.**

Number of Zeros Theorem

A function defined by a polynomial of degree n has *at most* n distinct zeros.

For example, a polynomial function of degree 3 has *at most* three distinct zeros but can have as few as one zero. Consider the following polynomial.

$$f(x) = x^3 + 3x^2 + 3x + 1$$
$$f(x) = (x + 1)^3 \qquad \text{Factored form of } f(x)$$

The function f is of degree 3 but has one distinct zero, -1. Actually, the zero -1 occurs *three* times because there are three factors of $x + 1$. The number of times a zero occurs is referred to as the **multiplicity of the zero.**

EXAMPLE 4 Finding a Polynomial Function That Satisfies Given Conditions (Real Zeros)

Find a polynomial function $f(x)$ of degree 3 with real coefficients that satisfies the given conditions.

(a) Zeros of -1, 2, and 4; $f(1) = 3$

(b) -2 is a zero of multiplicity 3; $f(-1) = 4$

SOLUTION

(a) These three zeros give $x - (-1) = x + 1$, $x - 2$, and $x - 4$ as factors of $f(x)$. Because $f(x)$ is to be of degree 3, these are the only possible factors by the number of zeros theorem. Therefore, $f(x)$ has the form

$$f(x) = a(x + 1)(x - 2)(x - 4), \quad \text{for some real number } a.$$

To find a, use the fact that $f(1) = 3$.

$$f(1) = a(1 + 1)(1 - 2)(1 - 4) \qquad \text{Let } x = 1.$$
$$3 = a(2)(-1)(-3) \qquad f(1) = 3$$
$$3 = 6a \qquad \text{Multiply.}$$
$$a = \frac{1}{2} \qquad \text{Divide by 6.}$$

Thus, $$f(x) = \frac{1}{2}(x + 1)(x - 2)(x - 4), \qquad \text{Let } a = \tfrac{1}{2}.$$

or, $$f(x) = \frac{1}{2}x^3 - \frac{5}{2}x^2 + x + 4. \qquad \text{Multiply.}$$

(b) The polynomial function $f(x)$ has the following form.

$$f(x) = a(x + 2)(x + 2)(x + 2) \qquad \text{Factor theorem}$$
$$f(x) = a(x + 2)^3 \qquad (x + 2) \text{ is a factor three times.}$$

To find a, use the fact that $f(-1) = 4$.

$$f(-1) = a(-1 + 2)^3 \qquad \text{Let } x = -1.$$
$$4 = a(1)^3 \qquad f(-1) = 4$$
$$a = 4 \qquad \text{Solve for } a.$$

Thus, $f(x) = 4(x + 2)^3$,

Remember: $(x + 2)^3 \neq x^3 + 2^3$

or, $$f(x) = 4x^3 + 24x^2 + 48x + 32. \qquad \text{Multiply.}$$

✔ **Now Try Exercises 53 and 57.**

NOTE In **Example 4(a),** we cannot clear the denominators in $f(x)$ by multiplying each side by 2 because the result would equal $2 \cdot f(x)$, not $f(x)$.

Conjugate Zeros Theorem The following properties of complex conjugates are needed to prove the **conjugate zeros theorem.** We use a simplified notation for conjugates here. If $z = a + bi$, then the conjugate of z is written $\bar{z}$, where $\bar{z} = a - bi$. For example, if $z = -5 + 2i$, then $\bar{z} = -5 - 2i$.

Properties of Conjugates

For any complex numbers c and d, the following properties hold.

$$\overline{c + d} = \bar{c} + \bar{d}, \quad \overline{c \cdot d} = \bar{c} \cdot \bar{d}, \quad \text{and} \quad \overline{c^n} = (\bar{c})^n$$

In general, if z is a zero of a polynomial function with *real* coefficients, then so is $\bar{z}$. For example, the remainder theorem can be used to show that both $2 + i$ and $2 - i$ are zeros of $f(x) = x^3 - x^2 - 7x + 15$.

Conjugate Zeros Theorem

If $f(x)$ defines a polynomial function ***having only real coefficients*** and if $z = a + bi$ is a zero of $f(x)$, where a and b are real numbers, then

$\bar{z} = a - bi$ is also a zero of $f(x)$.

Proof Start with the polynomial function

$$f(x) = a_n x^n + a_{n-1}x^{n-1} + \cdots + a_1 x + a_0,$$

where all coefficients are real numbers. If the complex number z is a zero of $f(x)$, then we have the following.

$$f(z) = a_n z^n + a_{n-1}z^{n-1} + \cdots + a_1 z + a_0 = 0$$

Take the conjugate of both sides of this equation.

$$\overline{a_n z^n + a_{n-1}z^{n-1} + \cdots + a_1 z + a_0} = \bar{0}$$

$$\overline{a_n z^n} + \overline{a_{n-1}z^{n-1}} + \cdots + \overline{a_1 z} + \overline{a_0} = \bar{0}$$ Use generalizations of the properties $\overline{c + d} = \bar{c} + \bar{d}$ and $\overline{c \cdot d} = \bar{c} \cdot \bar{d}$.

$$\overline{a_n}\,\overline{z^n} + \overline{a_{n-1}}\,\overline{z^{n-1}} + \cdots + \overline{a_1}\,\bar{z} + \overline{a_0} = \bar{0}$$

$$a_n(\bar{z})^n + a_{n-1}(\bar{z})^{n-1} + \cdots + a_1(\bar{z}) + a_0 = 0$$ Use the property $\overline{c^n} = (\bar{c})^n$ and the fact that for any real number a, $\bar{a} = a$.

$$f(\bar{z}) = 0$$

Hence $\bar{z}$ is also a zero of $f(x)$, which completes the proof.

CAUTION ***When the conjugate zeros theorem is applied, it is essential that the polynomial have only real coefficients. For example,***

$$f(x) = x - (1 + i)$$

has $1 + i$ as a zero, but the conjugate $1 - i$ is not a zero.

EXAMPLE 5 Finding a Polynomial Function That Satisfies Given Conditions (Complex Zeros)

Find a polynomial function $f(x)$ of least degree having only real coefficients and zeros 3 and $2 + i$.

SOLUTION The complex number $2 - i$ must also be a zero, so the polynomial has at least three zeros: 3, $2 + i$, and $2 - i$. For the polynomial to be of least degree, these must be the only zeros. By the factor theorem there must be three factors: $x - 3$, $x - (2 + i)$, and $x - (2 - i)$.

$$f(x) = (x-3)[x-(2+i)][x-(2-i)] \quad \text{Factor theorem}$$

$$f(x) = (x-3)(x-2-i)(x-2+i) \quad \text{Distribute negative signs.}$$

The multiplication steps are not shown here.

$$f(x) = (x-3)(x^2-4x+5) \quad \text{Multiply and combine like terms; } i^2 = -1.$$

$$f(x) = x^3 - 7x^2 + 17x - 15 \quad \text{Multiply again.}$$

Any nonzero multiple of $x^3 - 7x^2 + 17x - 15$ also satisfies the given conditions on zeros. The information on zeros given in the problem is not sufficient to give a specific value for the leading coefficient.

✔ **Now Try Exercise 69.**

Zeros of a Polynomial Function The theorem on conjugate zeros helps predict the number of real zeros of polynomial functions with real coefficients.

- A polynomial function with real coefficients of *odd* degree n, where $n \geq 1$, must have at least one real zero (because zeros of the form $a + bi$, where $b \neq 0$, occur in conjugate pairs).
- A polynomial function with real coefficients of *even* degree n may have no real zeros.

EXAMPLE 6 Finding All Zeros Given One Zero

Find all zeros of $f(x) = x^4 - 7x^3 + 18x^2 - 22x + 12$, given that $1 - i$ is a zero.

SOLUTION Because the polynomial function has only real coefficients and $1 - i$ is a zero, by the conjugate zeros theorem $1 + i$ is also a zero. To find the remaining zeros, first use synthetic division to divide the original polynomial by $x - (1 - i)$.

$(1-i)(-6-i) = -6 - i + 6i + i^2 = -7 + 5i$

$$\begin{array}{r|rrrrr} 1-i & 1 & -7 & 18 & -22 & 12 \\ & & 1-i & -7+5i & 16-6i & -12 \\ \hline & 1 & -6-i & 11+5i & -6-6i & 0 \end{array}$$

By the factor theorem, because $x = 1 - i$ is a zero of $f(x)$, $x - (1 - i)$ is a factor, and $f(x)$ can be written as follows.

$$f(x) = [x-(1-i)][x^3 + (-6-i)x^2 + (11+5i)x + (-6-6i)]$$

We know that $x = 1 + i$ is also a zero of $f(x)$. Continue to use synthetic division and divide the quotient polynomial above by $x - (1 + i)$.

$$\begin{array}{r|rrrr} 1+i & 1 & -6-i & 11+5i & -6-6i \\ & & 1+i & -5-5i & 6+6i \\ \hline & 1 & -5 & 6 & 0 \end{array}$$

Using the result of the synthetic division, $f(x)$ can be written in the following factored form.

$$f(x) = [x - (1 - i)][x - (1 + i)](x^2 - 5x + 6)$$
$$f(x) = [x - (1 - i)][x - (1 + i)](x - 2)(x - 3)$$

The remaining zeros are 2 and 3. The four zeros are $1 - i$, $1 + i$, 2, and 3.

✔ **Now Try Exercise 35.**

NOTE If we had been unable to factor $x^2 - 5x + 6$ into linear factors, in **Example 6,** we would have used the quadratic formula to solve the equation $x^2 - 5x + 6 = 0$ to find the remaining two zeros of the function.

Descartes' Rule of Signs The following rule helps to determine the number of positive and negative real zeros of a polynomial function. A **variation in sign** is a change from positive to negative or from negative to positive in successive terms of the polynomial when they are written in order of descending powers of the variable. ***Missing terms (those with 0 coefficients) are counted as no change in sign and can be ignored.***

Descartes' Rule of Signs

Let $f(x)$ define a polynomial function with real coefficients and a nonzero constant term, with terms in descending powers of x.

(a) The number of positive real zeros of f either equals the number of variations in sign occurring in the coefficients of $f(x)$, or is less than the number of variations by a positive even integer.

(b) The number of negative real zeros of f either equals the number of variations in sign occurring in the coefficients of $f(-x)$, or is less than the number of variations by a positive even integer.

EXAMPLE 7 Applying Descartes' Rule of Signs

Determine the different possibilities for the numbers of positive, negative, and nonreal complex zeros of

$$f(x) = x^4 - 6x^3 + 8x^2 + 2x - 1.$$

SOLUTION We first consider the possible number of positive zeros by observing that $f(x)$ has three variations in signs.

$$f(x) = \underbrace{+x^4 - 6x^3}_{1} \underbrace{+ 8x^2}_{2} \underbrace{+ 2x - 1}_{3}$$

Thus, $f(x)$ has either three or one (because $3 - 2 = 1$) positive real zeros.

For negative zeros, consider the variations in signs for $f(-x)$.

$$f(-x) = (-x)^4 - 6(-x)^3 + 8(-x)^2 + 2(-x) - 1$$
$$f(-x) = x^4 + 6x^3 \underbrace{+ 8x^2 - 2x}_{1} - 1$$

There is only one variation in sign, so $f(x)$ has exactly one negative real zero.

Because $f(x)$ is a fourth-degree polynomial function, it must have four complex zeros, some of which may be repeated. Descartes' rule of signs has indicated that exactly one of these zeros is a negative real number.

- One possible combination of the zeros is one negative real zero, three positive real zeros, and no nonreal complex zeros.
- Another possible combination of the zeros is one negative real zero, one positive real zero, and two nonreal complex zeros.

By the conjugate zeros theorem, any possible nonreal complex zeros must occur in conjugate pairs because $f(x)$ has real coefficients. The table below summarizes these possibilities.

Possible Number of Zeros		
Positive	**Negative**	**Nonreal Complex**
3	1	0
1	1	2

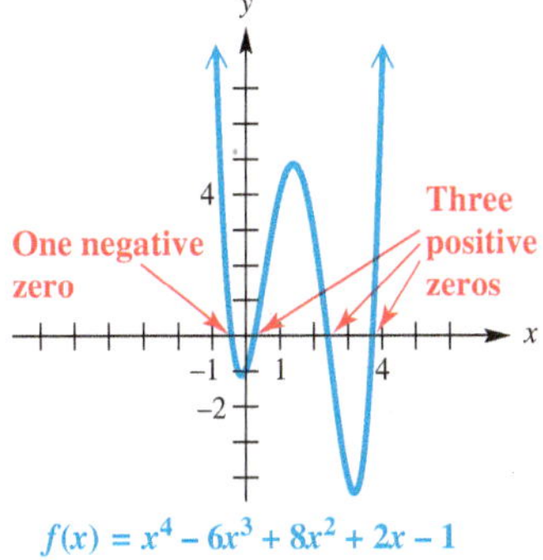

Figure 19

The graph of $f(x)$ in **Figure 19** verifies the correct combination of three positive real zeros with one negative real zero, as seen in the first row of the table.*

✔ **Now Try Exercise 79.**

NOTE ***Descartes' rule of signs does not identify the multiplicity of the zeros of a function.*** For example, if it indicates that a function $f(x)$ has exactly two positive real zeros, then $f(x)$ may have two distinct positive real zeros or one positive real zero of multiplicity 2.

3.3 Exercises

CONCEPT PREVIEW *Determine whether each statement is* true *or* false. *If false, explain why.*

1. Because $x - 1$ is a factor of $f(x) = x^6 - x^4 + 2x^2 - 2$, we can also conclude that $f(1) = 0$.
2. Because $f(1) = 0$ for $f(x) = x^6 - x^4 + 2x^2 - 2$, we can conclude that $x - 1$ is a factor of $f(x)$.
3. For $f(x) = (x + 2)^4(x - 3)$, the number 2 is a zero of multiplicity 4.
4. Because $2 + 3i$ is a zero of $f(x) = x^2 - 4x + 13$, we can conclude that $2 - 3i$ is also a zero.

* The authors would like to thank Mary Hill of College of DuPage for her input into **Example 7.**

5. A polynomial function having degree 6 and only real coefficients may have no real zeros.

6. The polynomial function $f(x) = 2x^5 + 3x^4 - 8x^3 - 5x + 6$ has three variations in sign.

7. If $z = 7 - 6i$, then $\bar{z} = -7 + 6i$.

8. The product of a complex number and its conjugate is always a real number.

Use the factor theorem and synthetic division to decide whether the second polynomial is a factor of the first. **See Example 1.**

9. $x^3 - 5x^2 + 3x + 1;\quad x - 1$

10. $x^3 + 6x^2 - 2x - 7;\quad x + 1$

11. $2x^4 + 5x^3 - 8x^2 + 3x + 13;\quad x + 1$

12. $-3x^4 + x^3 - 5x^2 + 2x + 4;\quad x - 1$

13. $-x^3 + 3x - 2;\quad x + 2$

14. $-2x^3 + x^2 - 63;\quad x + 3$

15. $4x^2 + 2x + 54;\quad x - 4$

16. $5x^2 - 14x + 10;\quad x + 2$

17. $x^3 + 2x^2 + 3;\quad x - 1$

18. $2x^3 + x + 2;\quad x + 1$

19. $2x^4 + 5x^3 - 2x^2 + 5x + 6;\quad x + 3$

20. $5x^4 + 16x^3 - 15x^2 + 8x + 16;\quad x + 4$

Factor $f(x)$ into linear factors given that k is a zero. **See Example 2.**

21. $f(x) = 2x^3 - 3x^2 - 17x + 30;\quad k = 2$

22. $f(x) = 2x^3 - 3x^2 - 5x + 6;\quad k = 1$

23. $f(x) = 6x^3 + 13x^2 - 14x + 3;\quad k = -3$

24. $f(x) = 6x^3 + 17x^2 - 63x + 10;\quad k = -5$

25. $f(x) = 6x^3 + 25x^2 + 3x - 4;\quad k = -4$

26. $f(x) = 8x^3 + 50x^2 + 47x - 15;\quad k = -5$

27. $f(x) = x^3 + (7 - 3i)x^2 + (12 - 21i)x - 36i;\quad k = 3i$

28. $f(x) = 2x^3 + (3 + 2i)x^2 + (1 + 3i)x + i;\quad k = -i$

29. $f(x) = 2x^3 + (3 - 2i)x^2 + (-8 - 5i)x + (3 + 3i);\quad k = 1 + i$

30. $f(x) = 6x^3 + (19 - 6i)x^2 + (16 - 7i)x + (4 - 2i);\quad k = -2 + i$

31. $f(x) = x^4 + 2x^3 - 7x^2 - 20x - 12;\quad k = -2$ (multiplicity 2)

32. $f(x) = 2x^4 + x^3 - 9x^2 - 13x - 5;\quad k = -1$ (multiplicity 3)

For each polynomial function, one zero is given. Find all other zeros. **See Examples 2 and 6.**

33. $f(x) = x^3 - x^2 - 4x - 6;\quad 3$

34. $f(x) = x^3 + 4x^2 - 5;\quad 1$

35. $f(x) = x^3 - 7x^2 + 17x - 15;\quad 2 - i$

36. $f(x) = 4x^3 + 6x^2 - 2x - 1;\quad \frac{1}{2}$

37. $f(x) = x^4 + 5x^2 + 4;\quad -i$

38. $f(x) = x^4 + 26x^2 + 25;\quad i$

For each polynomial function, ***(a)*** *list all possible rational zeros,* ***(b)*** *find all rational zeros, and* ***(c)*** *factor $f(x)$ into linear factors.* **See Example 3.**

39. $f(x) = x^3 - 2x^2 - 13x - 10$

40. $f(x) = x^3 + 5x^2 + 2x - 8$

41. $f(x) = x^3 + 6x^2 - x - 30$

42. $f(x) = x^3 - x^2 - 10x - 8$

43. $f(x) = 6x^3 + 17x^2 - 31x - 12$

44. $f(x) = 15x^3 + 61x^2 + 2x - 8$

45. $f(x) = 24x^3 + 40x^2 - 2x - 12$

46. $f(x) = 24x^3 + 80x^2 + 82x + 24$

For each polynomial function, find all zeros and their multiplicities.

47. $f(x) = (x - 2)^3(x^2 - 7)$

48. $f(x) = (x + 1)^2(x - 1)^3(x^2 - 10)$

49. $f(x) = 3x(x - 2)(x + 3)(x^2 - 1)$

50. $f(x) = 5x^2(x^2 - 16)(x + 5)$

51. $f(x) = (x^2 + x - 2)^5\left(x - 1 + \sqrt{3}\right)^2$

52. $f(x) = (2x^2 - 7x + 3)^3\left(x - 2 - \sqrt{5}\right)$

Find a polynomial function $f(x)$ of degree 3 with real coefficients that satisfies the given conditions. **See Example 4.**

53. Zeros of -3, 1, and 4; $f(2) = 30$

54. Zeros of 1, -1, and 0; $f(2) = 3$

55. Zeros of -2, 1, and 0; $f(-1) = -1$

56. Zeros of 2, -3, and 5; $f(3) = 6$

57. Zero of -3 having multiplicity 3; $f(3) = 36$

58. Zero of 2 and zero of 4 having multiplicity 2; $f(1) = -18$

59. Zero of 0 and zero of 1 having multiplicity 2; $f(2) = 10$

60. Zero of -4 and zero of 0 having multiplicity 2; $f(-1) = -6$

Find a polynomial function $f(x)$ of least degree having only real coefficients and zeros as given. Assume multiplicity 1 unless otherwise stated. **See Examples 4–6.**

61. $5 + i$ and $5 - i$

62. $7 - 2i$ and $7 + 2i$

63. 0, i, and $1 + i$

64. 0, $-i$, and $2 + i$

65. $1 + \sqrt{2}$, $1 - \sqrt{2}$, and 1

66. $1 - \sqrt{3}$, $1 + \sqrt{3}$, and 1

67. $2 - i$, 3, and -1

68. $3 + 2i$, -1, and 2

69. 2 and $3 + i$

70. -1 and $4 - 2i$

71. $1 - \sqrt{2}$, $1 + \sqrt{2}$, and $1 - i$

72. $2 + \sqrt{3}$, $2 - \sqrt{3}$, and $2 + 3i$

73. $2 - i$ and $6 - 3i$

74. $5 + i$ and $4 - i$

75. 4, $1 - 2i$, and $3 + 4i$

76. -1, $5 - i$, and $1 + 4i$

77. $1 + 2i$ and 2 (multiplicity 2)

78. $2 + i$ and -3 (multiplicity 2)

Determine the different possibilities for the numbers of positive, negative, and nonreal complex zeros of each function. **See Example 7.**

79. $f(x) = 2x^3 - 4x^2 + 2x + 7$

80. $f(x) = x^3 + 2x^2 + x - 10$

81. $f(x) = 4x^3 - x^2 + 2x - 7$

82. $f(x) = 3x^3 + 6x^2 + x + 7$

83. $f(x) = 5x^4 + 3x^2 + 2x - 9$

84. $f(x) = 3x^4 + 2x^3 - 8x^2 - 10x - 1$

85. $f(x) = -8x^4 + 3x^3 - 6x^2 + 5x - 7$

86. $f(x) = 6x^4 + 2x^3 + 9x^2 + x + 5$

87. $f(x) = x^5 + 3x^4 - x^3 + 2x + 3$

88. $f(x) = 2x^5 - x^4 + x^3 - x^2 + x + 5$

89. $f(x) = 2x^5 - 7x^3 + 6x + 8$

90. $f(x) = 11x^5 - x^3 + 7x - 5$

91. $f(x) = 5x^6 - 6x^5 + 7x^3 - 4x^2 + x + 2$

92. $f(x) = 9x^6 - 7x^4 + 8x^2 + x + 6$

93. $f(x) = 7x^5 + 6x^4 + 2x^3 + 9x^2 + x + 5$

94. $f(x) = -2x^5 + 10x^4 - 6x^3 + 8x^2 - x + 1$

*Find all complex zeros of each polynomial function. Give exact values. List multiple zeros as necessary.**

95. $f(x) = x^4 + 2x^3 - 3x^2 + 24x - 180$

96. $f(x) = x^3 - x^2 - 8x + 12$

97. $f(x) = x^4 + x^3 - 9x^2 + 11x - 4$

98. $f(x) = x^3 - 14x + 8$

99. $f(x) = 2x^5 + 11x^4 + 16x^3 + 15x^2 + 36x$

100. $f(x) = 3x^3 - 9x^2 - 31x + 5$

101. $f(x) = x^5 - 6x^4 + 14x^3 - 20x^2 + 24x - 16$

102. $f(x) = 9x^4 + 30x^3 + 241x^2 + 720x + 600$

103. $f(x) = 2x^4 - x^3 + 7x^2 - 4x - 4$

104. $f(x) = 32x^4 - 188x^3 + 261x^2 + 54x - 27$

105. $f(x) = 5x^3 - 9x^2 + 28x + 6$

106. $f(x) = 4x^3 + 3x^2 + 8x + 6$

107. $f(x) = x^4 + 29x^2 + 100$

108. $f(x) = x^4 + 4x^3 + 6x^2 + 4x + 1$

109. $f(x) = x^4 + 2x^2 + 1$

110. $f(x) = x^4 - 8x^3 + 24x^2 - 32x + 16$

111. $f(x) = x^4 - 6x^3 + 7x^2$

112. $f(x) = 4x^4 - 65x^2 + 16$

113. $f(x) = x^4 - 8x^3 + 29x^2 - 66x + 72$

114. $f(x) = 12x^4 - 43x^3 + 50x^2 + 38x - 12$

115. $f(x) = x^6 - 9x^4 - 16x^2 + 144$

116. $f(x) = x^6 - x^5 - 26x^4 + 44x^3 + 91x^2 - 139x + 30$

If c and d are complex numbers, prove each statement. (Hint: Let $c = a + bi$ and $d = m + ni$ and form all the conjugates, the sums, and the products.)

117. $\overline{c + d} = \overline{c} + \overline{d}$

118. $\overline{c \cdot d} = \overline{c} \cdot \overline{d}$

119. $\overline{a} = a$ for any real number a

120. $\overline{c^2} = (\overline{c})^2$

In 1545, a method of solving a cubic equation of the form

$$x^3 + mx = n,$$

developed by Niccolo Tartaglia, was published in the Ars Magna, *a work by Girolamo Cardano. The formula for finding the one real solution of the equation is*

$$x = \sqrt[3]{\frac{n}{2} + \sqrt{\left(\frac{n}{2}\right)^2 + \left(\frac{m}{3}\right)^3}} - \sqrt[3]{\frac{-n}{2} + \sqrt{\left(\frac{n}{2}\right)^2 + \left(\frac{m}{3}\right)^3}}.$$

(Source: Gullberg, J., *Mathematics from the Birth of Numbers,* W.W. Norton & Company.) *Use the formula to solve each equation for the one real solution.*

121. $x^3 + 9x = 26$

122. $x^3 + 15x = 124$

---

* The authors would like to thank Aileen Solomon of Trident Technical College for preparing and suggesting the inclusion of **Exercises 95–108.**

3.4 Polynomial Functions: Graphs, Applications, and Models

- Graphs of $f(x) = ax^n$
- Graphs of General Polynomial Functions
- Behavior at Zeros
- Turning Points and End Behavior
- Graphing Techniques
- Intermediate Value and Boundedness Theorems
- Approximations of Real Zeros
- Polynomial Models

Graphs of $f(x) = ax^n$

We can now graph polynomial functions of degree 3 or greater with real number coefficients and domains (because the graphs are in the real number plane). We begin by inspecting the graphs of several functions of the form

$$f(x) = ax^n, \quad \text{with } a = 1.$$

The identity function $f(x) = x$, the squaring function $f(x) = x^2$, and the cubing function $f(x) = x^3$ were graphed earlier using a general point-plotting method.

Each function in **Figure 20** has odd degree and is an odd function exhibiting symmetry about the origin. Each has domain $(-\infty, \infty)$ and range $(-\infty, \infty)$ and is continuous on its entire domain $(-\infty, \infty)$. Additionally, these odd functions are increasing on their entire domain $(-\infty, \infty)$, appearing as though they fall to the left and rise to the right.

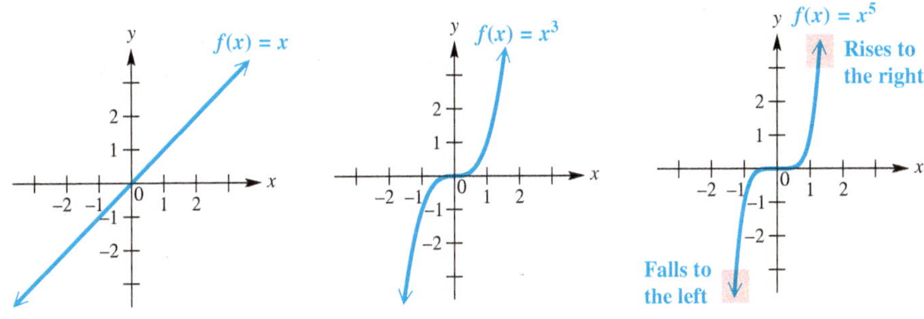

Figure 20

Each function in **Figure 21** has even degree and is an even function exhibiting symmetry about the y-axis. Each has domain $(-\infty, \infty)$ but restricted range $[0, \infty)$. These even functions are also continuous on their entire domain $(-\infty, \infty)$. However, they are decreasing on $(-\infty, 0)$ and increasing on $(0, \infty)$, appearing as though they rise both to the left and to the right.

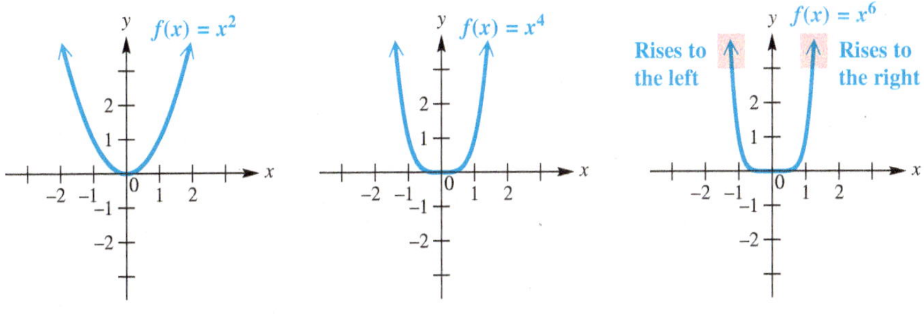

Figure 21

The behaviors in the graphs of these basic polynomial functions as x increases (decreases) without bound also apply to more complicated polynomial functions.

Graphs of General Polynomial Functions

As with quadratic functions, the absolute value of a in $f(x) = ax^n$ determines the width of the graph.

- When $|a| > 1$, the graph is stretched vertically, making it narrower.
- When $0 < |a| < 1$, the graph is shrunk or compressed vertically, making it wider.

Compared to the graph of $f(x) = ax^n$, the following also hold true.

- The graph of $f(x) = -ax^n$ is reflected across the x-axis.
- The graph of $f(x) = ax^n + k$ is translated (shifted) k units up if $k > 0$ and $|k|$ units down if $k < 0$.
- The graph of $f(x) = a(x - h)^n$ is translated h units to the right if $h > 0$ and $|h|$ units to the left if $h < 0$.
- The graph of $f(x) = a(x - h)^n + k$ shows a combination of these translations.

EXAMPLE 1 Examining Vertical and Horizontal Translations

Graph each polynomial function. Determine the largest open intervals of the domain over which each function is increasing or decreasing.

(a) $f(x) = x^5 - 2$ **(b)** $f(x) = (x + 1)^6$ **(c)** $f(x) = -2(x - 1)^3 + 3$

SOLUTION

(a) The graph of $f(x) = x^5 - 2$ will be the same as that of $f(x) = x^5$, but translated 2 units down. See **Figure 22.** This function is increasing on its entire domain $(-\infty, \infty)$.

(b) In $f(x) = (x + 1)^6$, function f has a graph like that of $f(x) = x^6$, but because

$$x + 1 = x - (-1),$$

it is translated 1 unit to the left. See **Figure 23.** This function is decreasing on $(-\infty, -1)$ and increasing on $(-1, \infty)$.

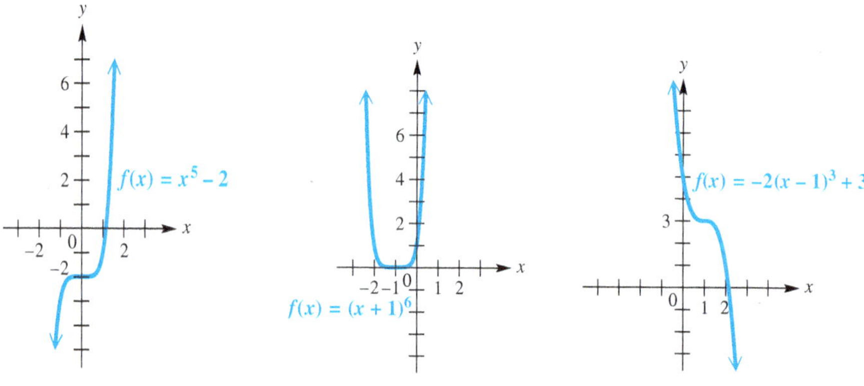

Figure 22 **Figure 23** **Figure 24**

(c) The negative sign in -2 causes the graph of

$$f(x) = -2(x - 1)^3 + 3$$

to be reflected across the x-axis when compared with the graph of $f(x) = x^3$. Because $|-2| > 1$, the graph is stretched vertically when compared to the graph of $f(x) = x^3$. As shown in **Figure 24,** the graph is also translated 1 unit to the right and 3 units up. This function is decreasing on its entire domain $(-\infty, \infty)$.

✔ **Now Try Exercises 13, 15, and 19.**

Unless otherwise restricted, the domain of a polynomial function is the set of all real numbers. Polynomial functions are smooth, continuous curves on the interval $(-\infty, \infty)$. The range of a polynomial function of odd degree is also the set of all real numbers.

The graphs in **Figure 25** suggest that for every polynomial function f of odd degree there is at least one real value of x that satisfies $f(x) = 0$. The real zeros correspond to the x-intercepts of the graph and can be determined by inspecting the factored form of each polynomial.

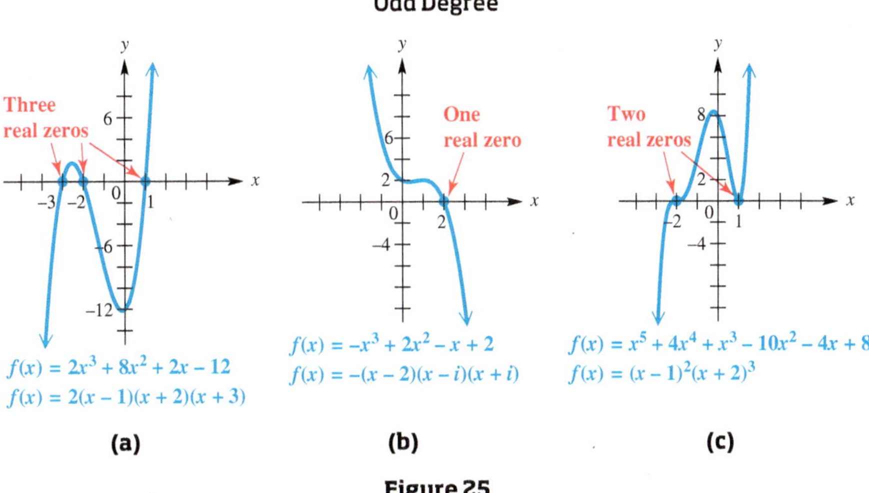

Figure 25

A polynomial function of even degree has a range of the form $(-\infty, k]$ or $[k, \infty)$, for some real number k. **Figure 26** shows two typical graphs.

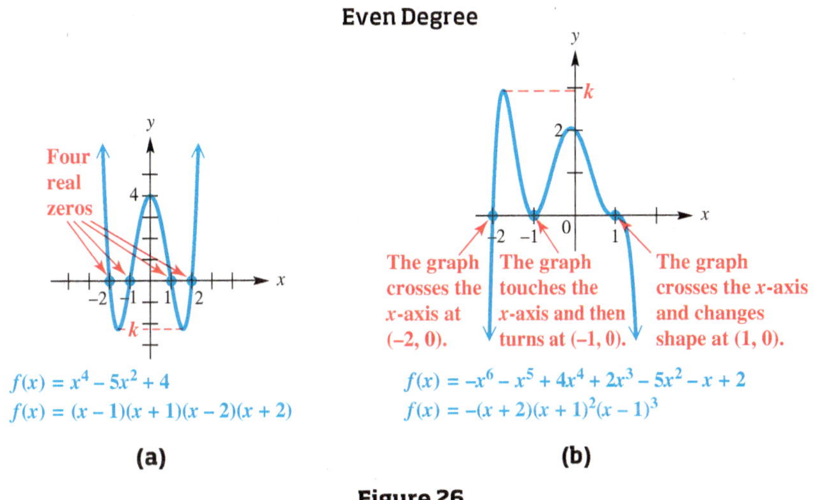

Figure 26

Behavior at Zeros **Figure 26(b)** shows a sixth-degree polynomial function with three distinct zeros, yet the behavior of the graph at each zero is different. This behavior depends on the multiplicity of the zero as determined by the exponent on the corresponding factor. The factored form of the polynomial function $f(x)$ is

$$-(x + 2)^1(x + 1)^2(x - 1)^3.$$

- $(x + 2)$ is a factor of multiplicity 1. Therefore, the graph crosses the x-axis at $(-2, 0)$.
- $(x + 1)$ is a factor of multiplicity 2. Therefore, the graph is tangent to the x-axis at $(-1, 0)$. This means that it touches the x-axis, then turns and changes behavior from decreasing to increasing similar to that of the squaring function $f(x) = x^2$ at its zero.
- $(x - 1)$ is a factor of multiplicity 3. Therefore, the graph crosses the x-axis ***and*** is tangent to the x-axis at $(1, 0)$. This causes a change in concavity (that is, how the graph opens upward or downward) at this x-intercept with behavior similar to that of the cubing function $f(x) = x^3$ at its zero.

Figure 27 generalizes the behavior of such graphs at their zeros.

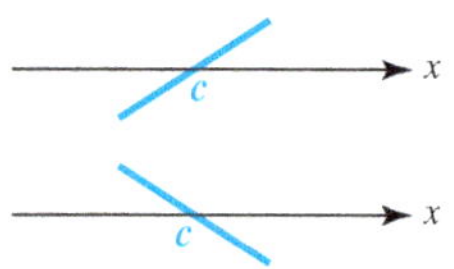

The graph crosses the x-axis at $(c, 0)$ if c is a zero of multiplicity 1.

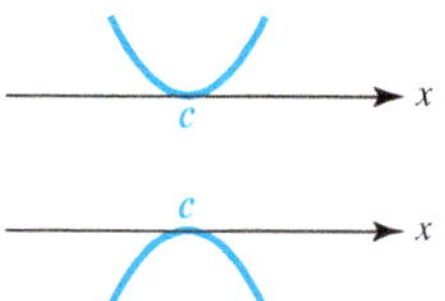

The graph is tangent to the x-axis at $(c, 0)$ if c is a zero of even multiplicity. The graph bounces, or turns, at c.

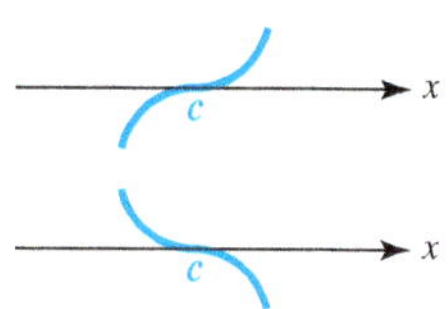

The graph crosses ***and*** is tangent to the x-axis at $(c, 0)$ if c is a zero of odd multiplicity greater than 1. The graph wiggles at c.

Figure 27

Turning Points and End Behavior The graphs in **Figures 25 and 26** show that polynomial functions often have **turning points** where the function changes from increasing to decreasing or from decreasing to increasing.

Turning Points

A polynomial function of degree n has at most $n - 1$ turning points, with at least one turning point between each pair of successive zeros.

The **end behavior** of a polynomial graph is determined by the *dominating term*—that is, the term of greatest degree. A polynomial of the form

$$f(x) = a_n x^n + a_{n-1}x^{n-1} + \cdots + a_0$$

has the same end behavior as $f(x) = a_n x^n$. For example,

$$f(x) = 2x^3 + 8x^2 + 2x - 12$$

has the same end behavior as $f(x) = 2x^3$. It is large and positive for large positive values of x, while it is large and negative for negative values of x with large absolute value. That is, it rises to the right and falls to the left.

Figure 25(a) shows that as x increases without bound, y does also. For the same graph, as x decreases without bound, y does also.

$$\text{As} \quad x \to \infty, \quad y \to \infty \quad \text{and} \quad \text{as} \quad x \to -\infty, \quad y \to -\infty.$$

LOOKING AHEAD TO CALCULUS

To find the x-coordinates of the two turning points of the graph of

$$f(x) = 2x^3 + 8x^2 + 2x - 12,$$

we can use the "maximum" and "minimum" capabilities of a graphing calculator and determine that, to the nearest thousandth, they are -0.131 and -2.535. In calculus, their exact values can be found by determining the zeros of the derivative function of $f(x)$,

$$f'(x) = 6x^2 + 16x + 2,$$

because the turning points occur precisely where the tangent line has slope 0. Using the quadratic formula would show that the zeros are

$$\frac{-4 \pm \sqrt{13}}{3},$$

which agree with the calculator approximations.

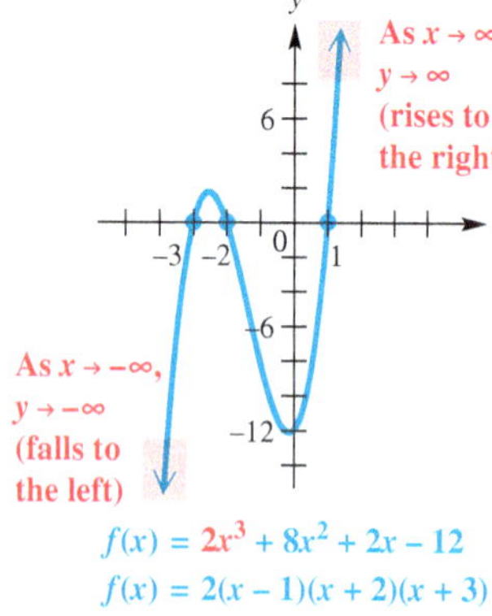

Figure 25(a) (repeated)

As $x \to -\infty$, $y \to \infty$ (rises to the left)

As $x \to \infty$, $y \to -\infty$ (falls to the right)

$f(x) = -x^3 + 2x^2 - x + 2$

$f(x) = -(x - 2)(x - i)(x + i)$

Figure 25(b) (repeated)

The graph in **Figure 25(b)** has the same end behavior as $f(x) = -x^3$.

$$\text{As} \quad x \to \infty, \quad y \to -\infty \quad \text{and} \quad \text{as} \quad x \to -\infty, \quad y \to \infty.$$

The graph of a polynomial function with a dominating term of even degree will show end behavior in the same direction. See **Figure 26.**

End Behavior of Graphs of Polynomial Functions

Suppose that ax^n is the dominating term of a polynomial function f of ***odd degree.***

1. If $a > 0$, then as $x \to \infty$, $f(x) \to \infty$, and as $x \to -\infty$, $f(x) \to -\infty$. Therefore, the end behavior of the graph is of the type shown in **Figure 28(a).** We symbolize it as ↙↗.
2. If $a < 0$, then as $x \to \infty$, $f(x) \to -\infty$, and as $x \to -\infty$, $f(x) \to \infty$. Therefore, the end behavior of the graph is of the type shown in **Figure 28(b).** We symbolize it as ↖↘.

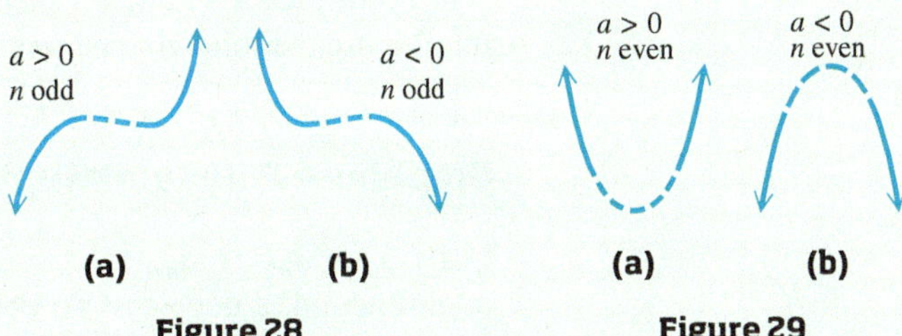

(a) **(b)**

Figure 28

(a) **(b)**

Figure 29

Suppose that ax^n is the dominating term of a polynomial function f of ***even degree.***

1. If $a > 0$, then as $|x| \to \infty$, $f(x) \to \infty$. Therefore, the end behavior of the graph is of the type shown in **Figure 29(a).** We symbolize it as ↖↗.
2. If $a < 0$, then as $|x| \to \infty$, $f(x) \to -\infty$. Therefore, the end behavior of the graph is of the type shown in **Figure 29(b).** We symbolize it as ↙↘.

EXAMPLE 2 Determining End Behavior

The graphs of the polynomial functions defined as follows are shown in A–D.

$$f(x) = x^4 - x^2 + 5x - 4, \qquad g(x) = -x^6 + x^2 - 3x - 4,$$

$$h(x) = 3x^3 - x^2 + 2x - 4, \quad \text{and} \quad k(x) = -x^7 + x - 4$$

Based on the discussion of end behavior, match each function with its graph.

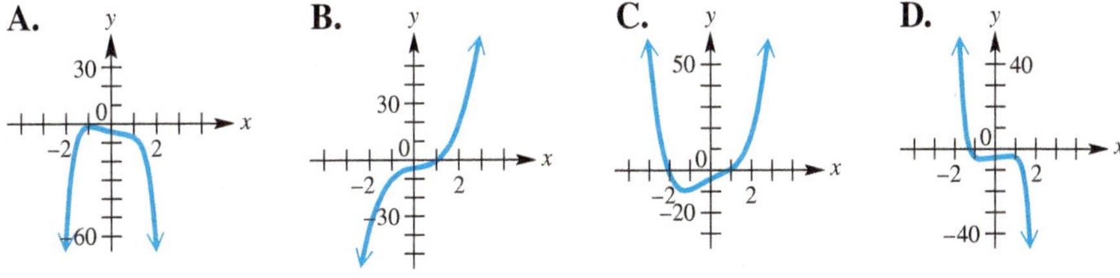

SOLUTION

- Function f has even degree and a dominating term with positive leading coefficient, as in C.
- Function g has even degree and a dominating term with negative leading coefficient, as in A.
- Function h has odd degree and a dominating term with positive coefficient, as in B.
- Function k has odd degree and a dominating term with negative coefficient, as in D.

✔ **Now Try Exercises 21, 23, 25, and 27.**

Graphing Techniques We have discussed several characteristics of the graphs of polynomial functions that are useful for graphing the function by hand. A **comprehensive graph** of a polynomial function $f(x)$ will show the following characteristics.

- all x-intercepts (indicating the real zeros) and the behavior of the graph at these zeros
- the y-intercept
- the sign of $f(x)$ within the intervals formed by the x-intercepts
- enough of the domain to show the end behavior

In **Examples 3 and 4,** we sketch the graphs of two polynomial functions by hand. We use the following general guidelines.

Graphing a Polynomial Function

Let $f(x) = a_nx^n + a_{n-1}x^{n-1} + \cdots + a_1x + a_0$, with $a_n \neq 0$, be a polynomial function of degree n. To sketch its graph, follow these steps.

Step 1 Find the real zeros of f. Plot the corresponding x-intercepts.

Step 2 Find $f(0) = a_0$. Plot the corresponding y-intercept.

Step 3 Use end behavior, whether the graph crosses, bounces on, or wiggles through the x-axis at the x-intercepts, and selected points as necessary to complete the graph.

EXAMPLE 3 Graphing a Polynomial Function

Graph $f(x) = 2x^3 + 5x^2 - x - 6$.

SOLUTION

Step 1 The possible rational zeros are ±1, ±2, ±3, ±6, $\pm\frac{1}{2}$, and $\pm\frac{3}{2}$. Use synthetic division to show that 1 is a zero.

$$\begin{array}{r|rrrr} 1 & 2 & 5 & -1 & -6 \\ & & 2 & 7 & 6 \\ \hline & 2 & 7 & 6 & 0 \end{array} \leftarrow f(1) = 0$$

We use the results of the synthetic division to factor as follows.

$$f(x) = (x-1)(2x^2 + 7x + 6)$$

$$f(x) = (x-1)(2x+3)(x+2) \quad \text{Factor again.}$$

Set each linear factor equal to 0, and then solve for x to find zeros. The three zeros of f are 1, $-\frac{3}{2}$, and -2. Plot the corresponding x-intercepts. See **Figure 30.**

Step 2 $f(0) = -6$, so plot $(0, -6)$. See **Figure 30.**

Step 3 The dominating term of $f(x)$ is $2x^3$, so the graph will have end behavior similar to that of $f(x) = x^3$. It will rise to the right and fall to the left as ↙↗. See **Figure 30.** Each zero of $f(x)$ occurs with multiplicity 1, meaning that the graph of $f(x)$ will cross the x-axis at each of its zeros. Because the graph of a polynomial function has no breaks, gaps, or sudden jumps, we now have sufficient information to sketch the graph of $f(x)$.

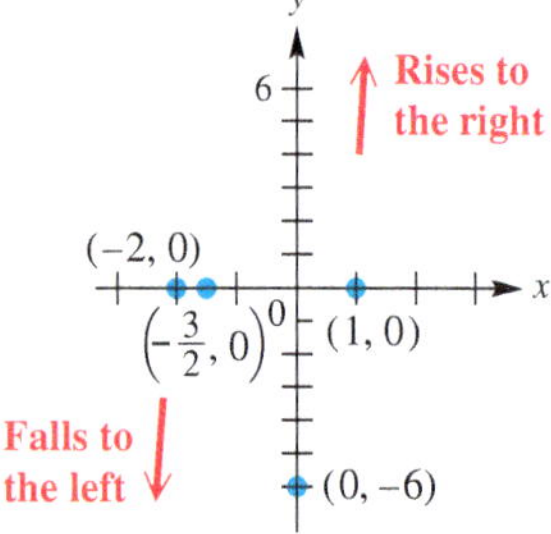

Figure 30

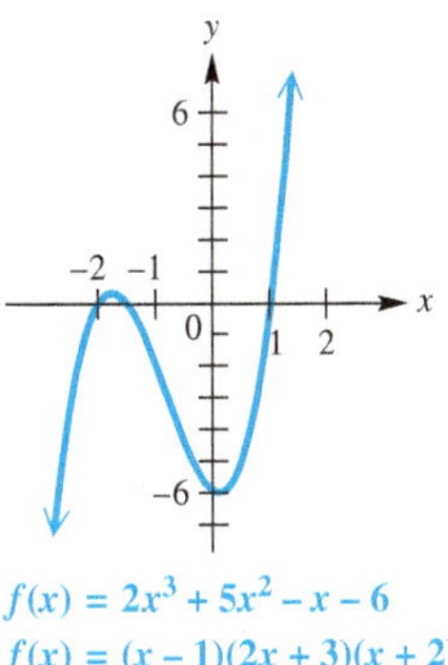

$f(x) = 2x^3 + 5x^2 - x - 6$
$f(x) = (x - 1)(2x + 3)(x + 2)$

Figure 31

Begin sketching at either end of the graph with the appropriate end behavior, and draw a smooth curve that crosses the x-axis at each zero, has a turning point between successive zeros, and passes through the y-intercept as shown in **Figure 31.**

Additional points may be used to verify whether the graph is above or below the x-axis between the zeros and to add detail to the sketch of the graph. The zeros divide the x-axis into four intervals:

$$(-\infty, -2), \quad \left(-2, -\frac{3}{2}\right), \quad \left(-\frac{3}{2}, 1\right), \quad \text{and} \quad (1, \infty).$$

Select an x-value as a test point in each interval, and substitute it into the equation for $f(x)$ to determine additional points on the graph. A typical selection of test points and the results of the tests are shown in the table.

Interval	Test Point x	Value of $f(x)$	Sign of $f(x)$	Graph Above or Below x-Axis
$(-\infty, -2)$	-3	-12	Negative	Below
$\left(-2, -\frac{3}{2}\right)$	$-\frac{7}{4}$	$\frac{11}{32}$	Positive	Above
$\left(-\frac{3}{2}, 1\right)$	0	-6	Negative	Below
$(1, \infty)$	2	28	Positive	Above

✔ **Now Try Exercise 29.**

EXAMPLE 4 Graphing a Polynomial Function

Graph $f(x) = -(x - 1)(x - 3)(x + 2)^2$.

SOLUTION

Step 1 Because the polynomial is given in factored form, the zeros can be determined by inspection. They are 1, 3, and -2. Plot the corresponding x-intercepts of the graph of $f(x)$. See **Figure 32.**

Step 2

$$f(0) = -(0 - 1)(0 - 3)(0 + 2)^2 \quad \text{Find } f(0).$$
$$f(0) = -(-1)(-3)(2)^2 \quad \text{Simplify in parentheses.}$$
$$f(0) = -12 \longleftarrow \text{The } y\text{-intercept is } (0, -12).$$

Plot the y-intercept $(0, -12)$. See **Figure 32.**

Step 3 The dominating term of $f(x)$ can be found by multiplying the factors and identifying the term of greatest degree. Here it is $-(x)(x)(x)^2 = -x^4$, indicating that the end behavior of the graph is ↙ ↘. Because 1 and 3 are zeros of multiplicity 1, the graph will cross the x-axis at these zeros. The graph of $f(x)$ will touch the x-axis at -2 and then turn and change direction because it is a zero of even multiplicity.

Begin at either end of the graph with the appropriate end behavior and draw a smooth curve that crosses the x-axis at 1 and 3 and that touches the x-axis at -2, then turns and changes direction. The graph will also pass through the y-intercept $(0, -12)$. See **Figure 32.**

Using test points within intervals formed by the x-intercepts is a good way to add detail to the graph and verify the accuracy of the sketch. A typical selection of test points is $(-3, -24)$, $(-1, -8)$, $(2, 16)$, and $(4, -108)$.

✔ **Now Try Exercise 33.**

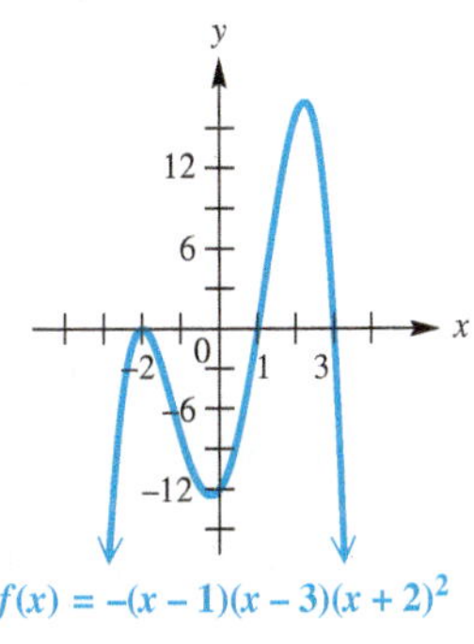

$f(x) = -(x - 1)(x - 3)(x + 2)^2$

Figure 32

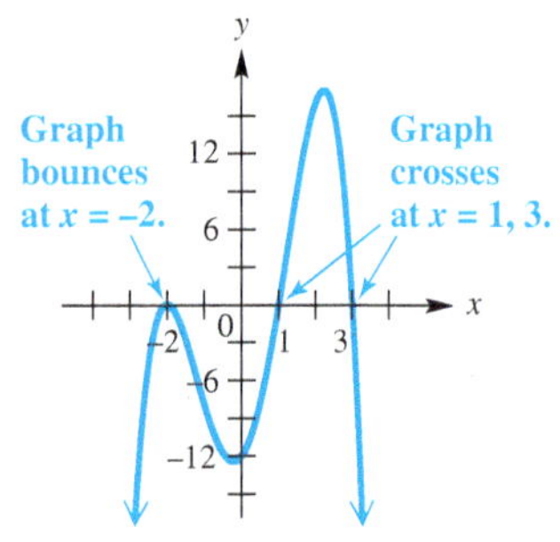

Figure 32 (repeated)

NOTE It is possible to reverse the process of **Example 4** and write the polynomial function from its graph if the zeros and any other point on the graph are known. Suppose that we are asked to find a polynomial function of least degree having the graph shown in **Figure 32** (repeated in the margin). Because the graph crosses the x-axis at 1 and 3 and bounces at -2, we know that the factored form of the function is as follows.

Multiplicity one — Multiplicity two

$$f(x) = a(x-1)^1(x-3)^1(x+2)^2$$

Now find the value of a by substituting the x- and y-values of any other point on the graph, say $(0, -12)$, into this function and solving for a.

$$f(x) = a(x-1)(x-3)(x+2)^2$$

$$-12 = a(0-1)(0-3)(0+2)^2 \quad \text{Let } x = 0 \text{ and } y = -12.$$

$$-12 = a(12) \quad \text{Simplify.}$$

$$a = -1 \quad \text{Divide by 12. Interchange sides.}$$

Verify in **Example 4** that the polynomial function is

$$f(x) = -(x-1)(x-3)(x+2)^2.$$

Exercises of this type are labeled ***Connecting Graphs with Equations***.

We emphasize the important relationships among the following concepts.

- the x-intercepts of the graph of $y = f(x)$
- the zeros of the function f
- the solutions of the equation $f(x) = 0$
- the factors of $f(x)$

For example, the graph of the function

$$f(x) = 2x^3 + 5x^2 - x - 6 \quad \text{Example 3}$$

$$f(x) = (x-1)(2x+3)(x+2) \quad \text{Factored form}$$

has x-intercepts $(1, 0)$, $\left(-\frac{3}{2}, 0\right)$, and $(-2, 0)$ as shown in **Figure 31** on the previous page. Because $1, -\frac{3}{2}$, and -2 are the x-values where the function is 0, they are the zeros of f. Also, $1, -\frac{3}{2}$, and -2 are the solutions of the polynomial equation

$$2x^3 + 5x^2 - x - 6 = 0.$$

This discussion is summarized as follows.

Relationships among x-Intercepts, Zeros, Solutions, and Factors

If f is a polynomial function and $(c, 0)$ is an x-intercept of the graph of $y = f(x)$, then

c is a zero of f, c is a solution of $f(x) = 0$,

and **$x - c$ is a factor of $f(x)$.**

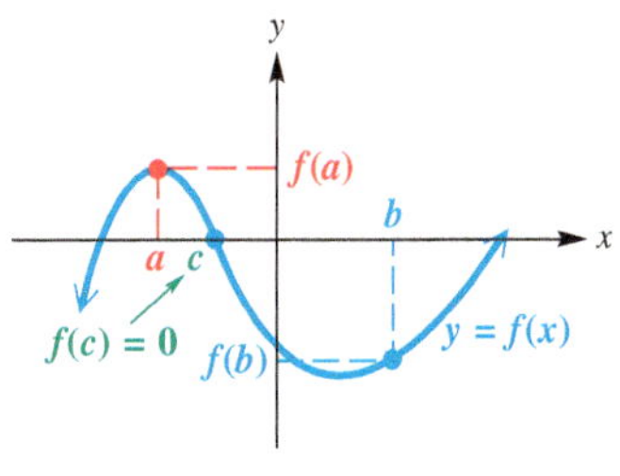

Figure 33

Intermediate Value and Boundedness Theorems As **Examples 3 and 4** show, one key to graphing a polynomial function is locating its zeros. In the special case where the potential zeros are rational numbers, the zeros are found by the rational zeros theorem.

Occasionally, irrational zeros can be found by inspection. For instance, $f(x) = x^3 - 2$ has the irrational zero $\sqrt[3]{2}$.

The next two theorems apply to the zeros of every polynomial function with real coefficients. The first theorem uses the fact that graphs of polynomial functions are continuous curves. The proof requires advanced methods, so it is not given here. **Figure 33** illustrates the theorem.

Intermediate Value Theorem

If $f(x)$ is a polynomial function with only *real coefficients,* and if for real numbers a and b the values $f(a)$ and $f(b)$ are opposite in sign, then there exists at least one real zero between a and b.

This theorem helps identify intervals where zeros of polynomial functions are located. If $f(a)$ and $f(b)$ are opposite in sign, then 0 is between $f(a)$ and $f(b)$, and so there must be a number c between a and b where $f(c) = 0$.

EXAMPLE 5 Locating a Zero

Use synthetic division and a graph to show that $f(x) = x^3 - 2x^2 - x + 1$ has a real zero between 2 and 3.

ALGEBRAIC SOLUTION

Use synthetic division to find $f(2)$ and $f(3)$.

$$\begin{array}{r|rrrr} 2 & 1 & -2 & -1 & 1 \\ & & 2 & 0 & -2 \\ \hline & 1 & 0 & -1 & -1 = f(2) \end{array}$$

$$\begin{array}{r|rrrr} 3 & 1 & -2 & -1 & 1 \\ & & 3 & 3 & 6 \\ \hline & 1 & 1 & 2 & 7 = f(3) \end{array}$$

Because $f(2)$ is negative and $f(3)$ is positive, by the intermediate value theorem there must be a real zero between 2 and 3.

GRAPHING CALCULATOR SOLUTION

The graphing calculator screen in **Figure 34** indicates that this zero is approximately 2.2469796. (Notice that there are two other zeros as well.)

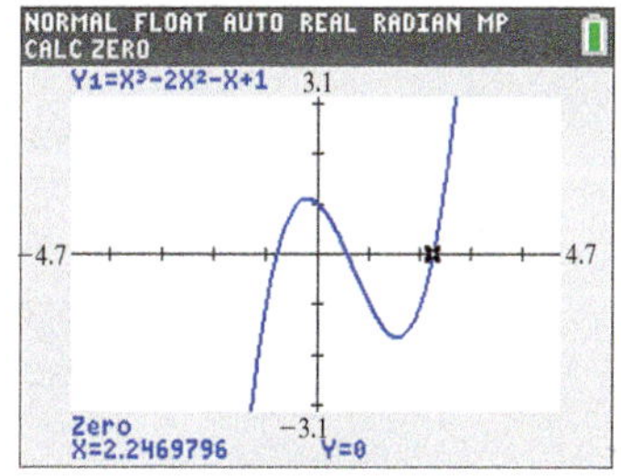

Figure 34

✔ **Now Try Exercise 49.**

CAUTION ***Be careful when interpreting the intermediate value theorem.*** If $f(a)$ and $f(b)$ are ***not*** opposite in sign, it does not necessarily mean that there is no zero between a and b. In **Figure 35,** $f(a)$ and $f(b)$ are both negative, but -3 and -1, which are between a and b, are zeros of $f(x)$.

Figure 35

The intermediate value theorem for polynomials helps limit the search for real zeros to smaller and smaller intervals. In **Example 5,** we used the theorem to verify that there is a real zero between 2 and 3. To locate the zero more accurately, we can use the theorem repeatedly. (Prior to modern-day methods involving calculators and computers, this was done by hand.)

The **boundedness theorem** shows how the bottom row of a synthetic division is used to place upper and lower bounds on possible real zeros of a polynomial function.

Boundedness Theorem

Let $f(x)$ be a polynomial function of degree $n \geq 1$ with *real coefficients* and with a *positive* leading coefficient. Suppose $f(x)$ is divided synthetically by $x - c$.

(a) If $c > 0$ and all numbers in the bottom row of the synthetic division are nonnegative, then $f(x)$ has no zero greater than c.

(b) If $c < 0$ and the numbers in the bottom row of the synthetic division alternate in sign (with 0 considered positive or negative, as needed), then $f(x)$ has no zero less than c.

Proof We outline the proof of part (a). The proof for part (b) is similar.

By the division algorithm, if $f(x)$ is divided by $x - c$, then for some $q(x)$ and r,

$$f(x) = (x - c)q(x) + r,$$

where all coefficients of $q(x)$ are nonnegative, $r \geq 0$, and $c > 0$. If $x > c$, then $x - c > 0$. Because $q(x) > 0$ and $r \geq 0$,

$$f(x) = (x - c)q(x) + r > 0.$$

This means that $f(x)$ will never be 0 for $x > c$.

EXAMPLE 6 Using the Boundedness Theorem

Show that the real zeros of $f(x) = 2x^4 - 5x^3 + 3x + 1$ satisfy these conditions.

(a) No real zero is greater than 3. **(b)** No real zero is less than -1.

SOLUTION

(a) Because $f(x)$ has real coefficients and the leading coefficient, 2, is positive, use the boundedness theorem. Divide $f(x)$ synthetically by $x - 3$.

$$\begin{array}{r|rrrrr} 3 & 2 & -5 & 0 & 3 & 1 \\ & & 6 & 3 & 9 & 36 \\ \hline & 2 & 1 & 3 & 12 & 37 \end{array} \leftarrow \text{All are nonnegative.}$$

Here $3 > 0$ and all numbers in the last row of the synthetic division are nonnegative, so $f(x)$ has no real zero greater than 3.

(b) We use the boundedness theorem again and divide $f(x)$ synthetically by $x - (-1)$, or $x + 1$.

$$\begin{array}{r|rrrrr} -1 & 2 & -5 & 0 & 3 & 1 \\ & & -2 & 7 & -7 & 4 \\ \hline & 2 & -7 & 7 & -4 & 5 \end{array} \leftarrow \text{These numbers alternate in sign.}$$

Here $-1 < 0$ and the numbers in the last row alternate in sign, so $f(x)$ has no real zero less than -1.

✔ **Now Try Exercises 57 and 59.**

Approximations of Real Zeros

We can approximate the irrational real zeros of a polynomial function using a graphing calculator.

EXAMPLE 7 Approximating Real Zeros of a Polynomial Function

Approximate the real zeros of $f(x) = x^4 - 6x^3 + 8x^2 + 2x - 1$.

SOLUTION The dominating term is x^4, so the graph will have end behavior similar to the graph of $f(x) = x^4$, which is positive for all values of x with large absolute values. That is, the end behavior is up at the left and the right, ↖↗. There are at most four real zeros because the polynomial is fourth-degree.

Since $f(0) = -1$, the y-intercept is $(0, -1)$. Because the end behavior is positive on the left and the right, by the intermediate value theorem f has at least one real zero on either side of $x = 0$. To approximate the zeros, we use a graphing calculator. The graph in **Figure 36** shows that there are four real zeros, and the table indicates that they are between

$$-1 \text{ and } 0, \quad 0 \text{ and } 1, \quad 2 \text{ and } 3, \quad \text{and} \quad 3 \text{ and } 4$$

because there is a sign change in $f(x) = y_1$ in each case.

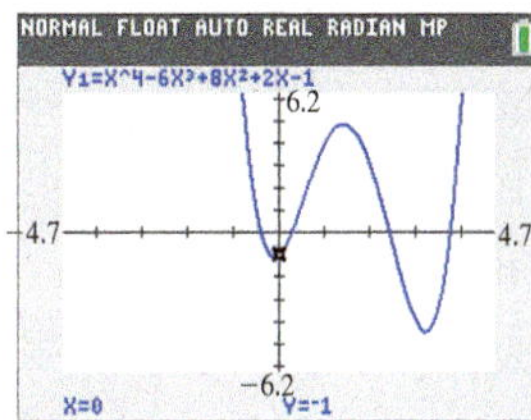

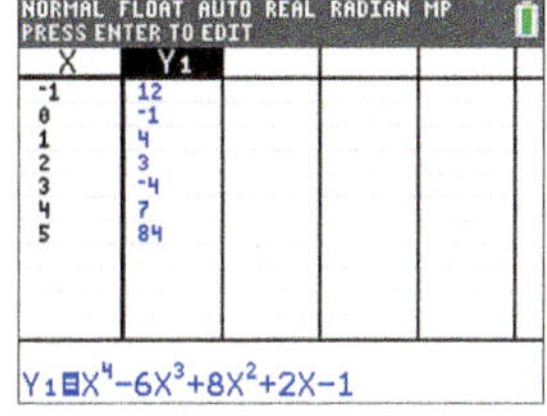

Figure 36

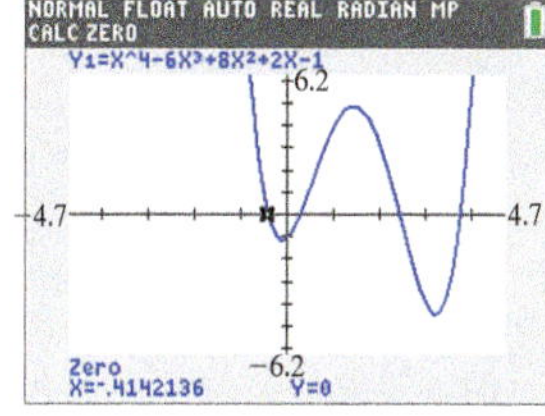

Figure 37

Using a calculator, we can find zeros to a great degree of accuracy. **Figure 37** shows that the negative zero is approximately -0.4142136. Similarly, we find that the other three zeros are approximately

$$0.26794919, \quad 2.4142136, \quad \text{and} \quad 3.7320508.$$

✔ **Now Try Exercise 77.**

Polynomial Models

EXAMPLE 8 Examining a Polynomial Model

The table shows the number of transactions, in millions, by users of bank debit cards for selected years.

Year	Transactions (in millions)
1995	829
1998	3765
2000	5290
2004	14,106
2008	28,464
2012	44,351

Source: Statistical Abstract of the United States.

(a) Using $x = 0$ to represent 1995, $x = 3$ to represent 1998, and so on, use the regression feature of a calculator to determine the quadratic function that best fits the data. Plot the data and the graph.

(b) Repeat part (a) for a cubic function (degree 3).

(c) Repeat part (a) for a quartic function (degree 4).

(d) The **correlation coefficient,** R, is a measure of the strength of the relationship between two variables. The values of R and R^2 are used to determine how well a regression model fits a set of data. The closer the value of R^2 is to 1, the better the fit. Compare R^2 for the three functions found in parts (a)–(c) to decide which function best fits the data.

SOLUTION

(a) The best-fitting quadratic function for the data is

$$y = 131.4x^2 + 342.9x + 901.4.$$

The regression coordinates screen and the graph are shown in **Figure 38.**

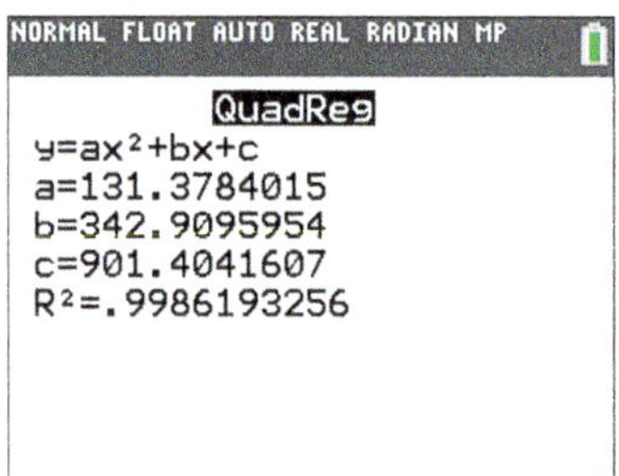

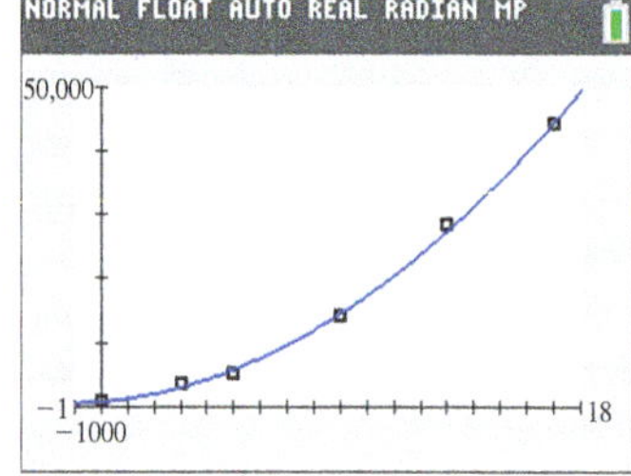

Figure 38

(b) The best-fitting cubic function is shown in **Figure 39** and is

$$y = -1.606x^3 + 172.1x^2 + 92.33x + 1119.$$

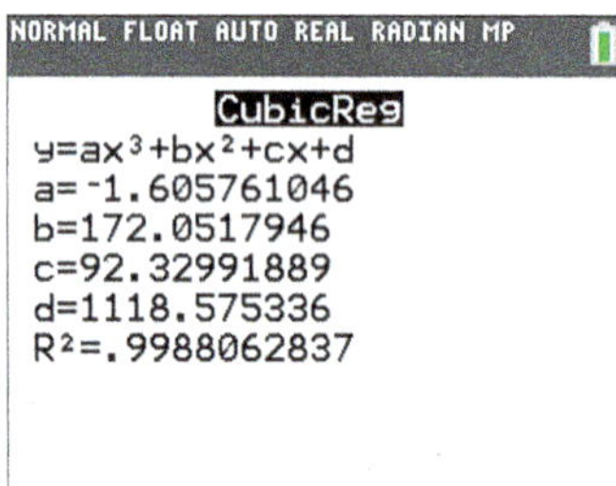

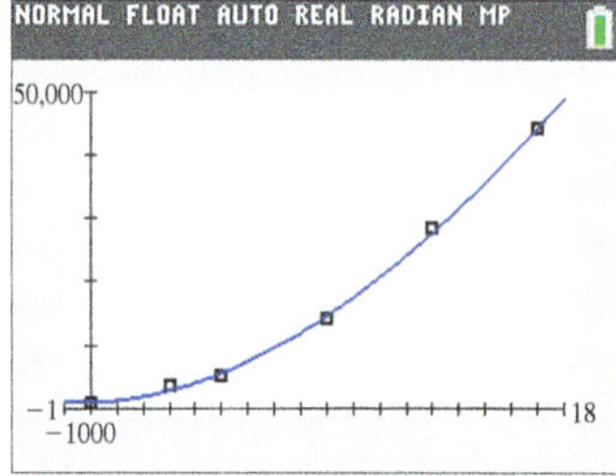

Figure 39

(c) The best-fitting quartic function is shown in **Figure 40** and is

$$y = -1.088x^4 + 34.71x^3 - 195.1x^2 + 1190x + 868.6.$$

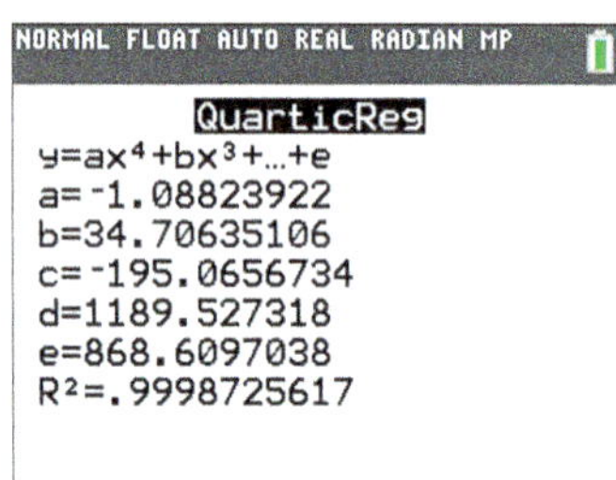

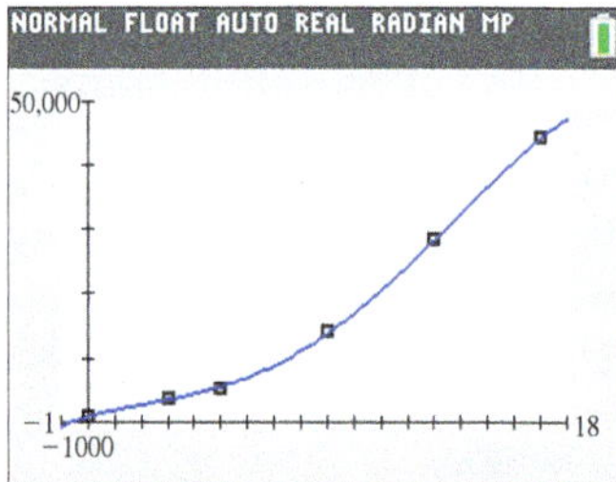

Figure 40

(d) With the statistical diagnostics turned on, the value of R^2 is displayed with the regression results on the TI-84 Plus each time that a regression model is executed. By inspecting the R^2 value for each model above, we see that the quartic function provides the best fit because it has the largest R^2 value of 0.9998725617.

✔ **Now Try Exercise 99.**

NOTE In **Example 8(d),** we selected the quartic function as the best model based on the comparison of R^2 values of the models. In practice, however, the best choice of a model should also depend on the set of data being analyzed as well as analysis of its trends and attributes.

3.4 Exercises

CONCEPT PREVIEW *Comprehensive graphs of four polynomial functions are shown in A–D. They represent the graphs of functions defined by these four equations, but not necessarily in the order listed.*

$y = x^3 - 3x^2 - 6x + 8$ $\qquad$ $y = x^4 + 7x^3 - 5x^2 - 75x$

$y = -x^3 + 9x^2 - 27x + 17$ $\qquad$ $y = -x^5 + 36x^3 - 22x^2 - 147x - 90$

Apply the concepts of this section to work each problem.

A.

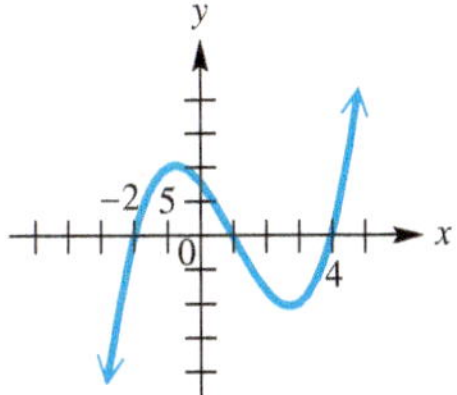

B.

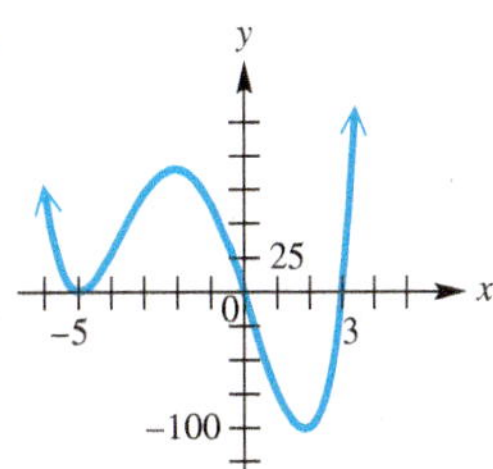

C.

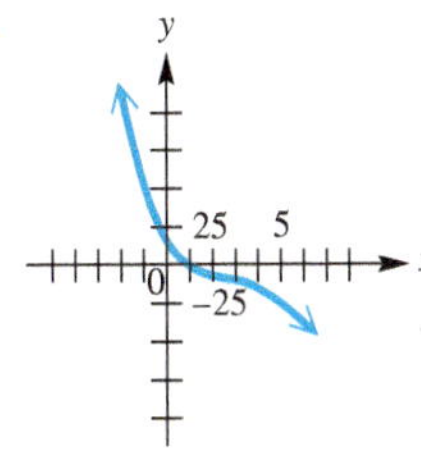

D.

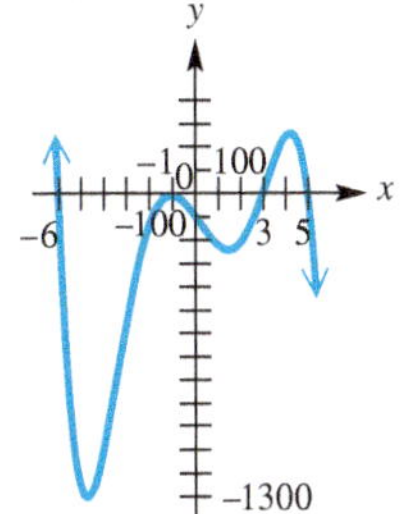

1. Which one of the graphs is that of $y = x^3 - 3x^2 - 6x + 8$?
2. Which one of the graphs is that of $y = x^4 + 7x^3 - 5x^2 - 75x$?
3. How many real zeros does the function graphed in C have?
4. Which one of C and D is the graph of $y = -x^3 + 9x^2 - 27x + 17$?
5. Which of the graphs cannot be that of a cubic polynomial function?
6. Which one of the graphs is that of a function whose range is *not* $(-\infty, \infty)$?
7. The function $f(x) = x^4 + 7x^3 - 5x^2 - 75x$ has the graph shown in B. Use the graph to factor the polynomial.
8. The function $f(x) = -x^5 + 36x^3 - 22x^2 - 147x - 90$ has the graph shown in D. Use the graph to factor the polynomial.

Graph each function. Determine the largest open intervals of the domain over which each function is **(a)** *increasing or* **(b)** *decreasing.* ***See Example 1.***

9. $f(x) = 2x^4$
10. $f(x) = \frac{1}{4}x^6$
11. $f(x) = -\frac{2}{3}x^5$
12. $f(x) = -\frac{5}{4}x^5$
13. $f(x) = \frac{1}{2}x^3 + 1$
14. $f(x) = -x^4 + 2$
15. $f(x) = -(x + 1)^3 + 1$
16. $f(x) = (x + 2)^3 - 1$

17. $f(x) = (x - 1)^4 + 2$

18. $f(x) = \frac{1}{3}(x + 3)^4 - 3$

19. $f(x) = \frac{1}{2}(x - 2)^2 + 4$

20. $f(x) = \frac{1}{3}(x + 1)^3 - 3$

Use an end behavior diagram, ↖ ↗, ↙ ↘, ↖ ↘, *or* ↙ ↗, *to describe the end behavior of the graph of each polynomial function.* ***See Example 2.***

21. $f(x) = 5x^5 + 2x^3 - 3x + 4$

22. $f(x) = -x^3 - 4x^2 + 2x - 1$

23. $f(x) = -4x^3 + 3x^2 - 1$

24. $f(x) = 4x^7 - x^5 + x^3 - 1$

25. $f(x) = 9x^6 - 3x^4 + x^2 - 2$

26. $f(x) = 10x^6 - x^5 + 2x - 2$

27. $f(x) = 3 + 2x - 4x^2 - 5x^{10}$

28. $f(x) = 7 + 2x - 5x^2 - 10x^4$

Graph each polynomial function. Factor first if the polynomial is not in factored form. ***See Examples 3 and 4.***

29. $f(x) = x^3 + 5x^2 + 2x - 8$

30. $f(x) = x^3 + 3x^2 - 13x - 15$

31. $f(x) = 2x(x - 3)(x + 2)$

32. $f(x) = x(x + 1)(x - 1)$

33. $f(x) = x^2(x - 2)(x + 3)^2$

34. $f(x) = x^2(x - 5)(x + 3)(x - 1)$

35. $f(x) = (3x - 1)(x + 2)^2$

36. $f(x) = (4x + 3)(x + 2)^2$

37. $f(x) = x^3 + 5x^2 - x - 5$

38. $f(x) = x^3 + x^2 - 36x - 36$

39. $f(x) = x^3 - x^2 - 2x$

40. $f(x) = 3x^4 + 5x^3 - 2x^2$

41. $f(x) = 2x^3(x^2 - 4)(x - 1)$

42. $f(x) = x^2(x - 3)^3(x + 1)$

43. $f(x) = 2x^3 - 5x^2 - x + 6$

44. $f(x) = 2x^4 + x^3 - 6x^2 - 7x - 2$

45. $f(x) = 3x^4 - 7x^3 - 6x^2 + 12x + 8$

46. $f(x) = x^4 + 3x^3 - 3x^2 - 11x - 6$

Use the intermediate value theorem to show that each polynomial function has a real zero between the numbers given. ***See Example 5.***

47. $f(x) = 2x^2 - 7x + 4$; 2 and 3

48. $f(x) = 3x^2 - x - 4$; 1 and 2

49. $f(x) = 2x^3 - 5x^2 - 5x + 7$; 0 and 1

50. $f(x) = 2x^3 - 9x^2 + x + 20$; 2 and 2.5

51. $f(x) = 2x^4 - 4x^2 + 4x - 8$; 1 and 2

52. $f(x) = x^4 - 4x^3 - x + 3$; 0.5 and 1

53. $f(x) = x^4 + x^3 - 6x^2 - 20x - 16$; 3.2 and 3.3

54. $f(x) = x^4 - 2x^3 - 2x^2 - 18x + 5$; 3.7 and 3.8

55. $f(x) = x^4 - 4x^3 - 20x^2 + 32x + 12$; -1 and 0

56. $f(x) = x^5 + 2x^4 + x^3 + 3$; -1.8 and -1.7

Show that the real zeros of each polynomial function satisfy the given conditions. ***See Example 6.***

57. $f(x) = x^4 - x^3 + 3x^2 - 8x + 8$; no real zero greater than 2

58. $f(x) = 2x^5 - x^4 + 2x^3 - 2x^2 + 4x - 4$; no real zero greater than 1

59. $f(x) = x^4 + x^3 - x^2 + 3$; no real zero less than -2

60. $f(x) = x^5 + 2x^3 - 2x^2 + 5x + 5$; no real zero less than -1

61. $f(x) = 3x^4 + 2x^3 - 4x^2 + x - 1$; no real zero greater than 1

62. $f(x) = 3x^4 + 2x^3 - 4x^2 + x - 1$; no real zero less than -2

63. $f(x) = x^5 - 3x^3 + x + 2$; no real zero greater than 2

64. $f(x) = x^5 - 3x^3 + x + 2$; no real zero less than -3

Connecting Graphs with Equations *Find a polynomial function f of least degree having the graph shown. (Hint: See the Note following* ***Example 4.****)*

65.

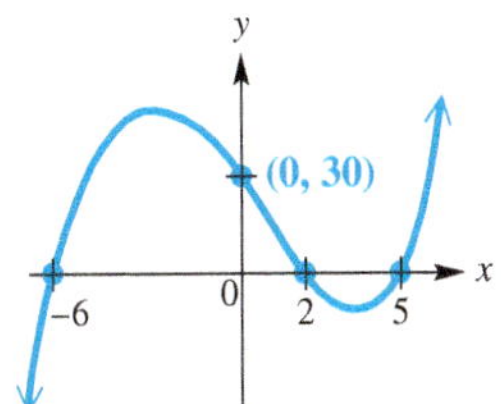

66.

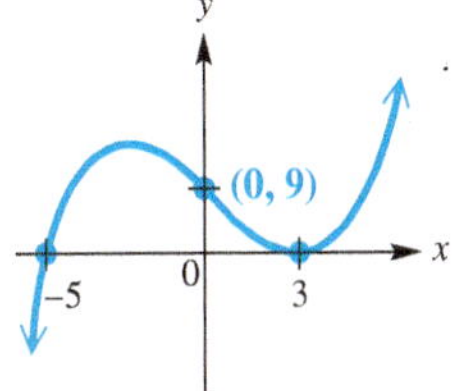

67.

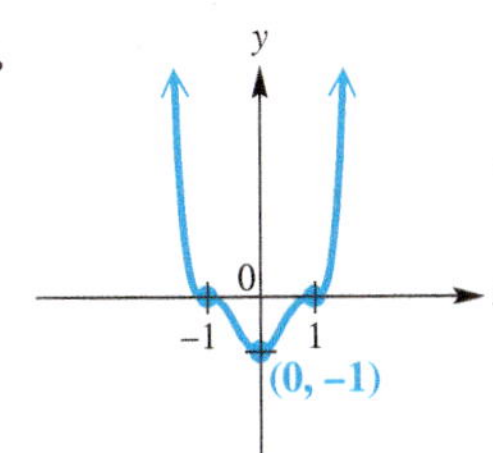

68.

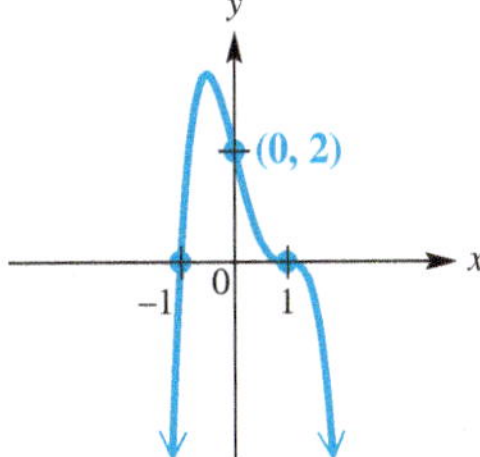

69.

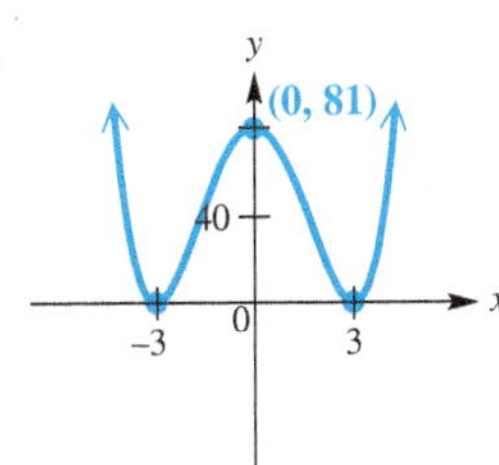

70.

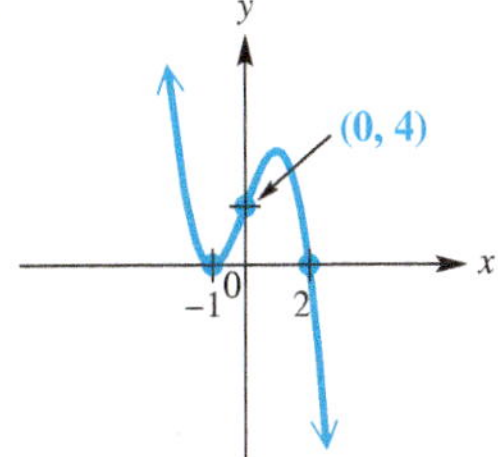

Graph each function in the viewing window specified. Compare the graph to the one shown in the answer section of this text. Then use the graph to find $f(1.25)$.

71. $f(x) = 2x(x - 3)(x + 2)$; window: $[-3, 4]$ by $[-20, 12]$
Compare to **Exercise 31.**

72. $f(x) = x^2(x - 2)(x + 3)^2$; window: $[-4, 3]$ by $[-24, 4]$
Compare to **Exercise 33.**

73. $f(x) = (3x - 1)(x + 2)^2$; window: $[-4, 2]$ by $[-15, 15]$
Compare to **Exercise 35.**

74. $f(x) = x^3 + 5x^2 - x - 5$; window: $[-6, 2]$ by $[-30, 30]$
Compare to **Exercise 37.**

Approximate the real zero discussed in each specified exercise. ***See Example 7.***

75. Exercise 47

76. Exercise 49

77. Exercise 51

78. Exercise 50

For the given polynomial function, approximate each zero as a decimal to the nearest tenth. ***See Example 7.***

79. $f(x) = x^3 + 3x^2 - 2x - 6$

80. $f(x) = x^3 - 3x + 3$

81. $f(x) = -2x^4 - x^2 + x + 5$

82. $f(x) = -x^4 + 2x^3 + 3x^2 + 6$

Use a graphing calculator to find the coordinates of the turning points of the graph of each polynomial function in the given domain interval. Give answers to the nearest hundredth.

83. $f(x) = 2x^3 - 5x^2 - x + 1; \quad [-1, 0]$

84. $f(x) = x^3 + 4x^2 - 8x - 8; \quad [0.3, 1]$

85. $f(x) = 2x^3 - 5x^2 - x + 1; \quad [1.4, 2]$

86. $f(x) = x^3 - x + 3; \quad [-1, 0]$

87. $f(x) = x^3 + 4x^2 - 8x - 8; \quad [-3.8, -3]$

88. $f(x) = x^4 - 7x^3 + 13x^2 + 6x - 28; \quad [-1, 0]$

Solve each problem.

89. *(Modeling) Social Security Numbers* Your Social Security number (SSN) is unique, and with it you can construct your own personal Social Security polynomial. Let the polynomial function be defined as follows, where a_i represents the ith digit in your SSN:

$$\text{SSN}(x) = (x - a_1)(x + a_2)(x - a_3)(x + a_4)(x - a_5) \cdot (x + a_6)(x - a_7)(x + a_8)(x - a_9).$$

For example, if the SSN is 539-58-0954, the polynomial function is

$$\text{SSN}(x) = (x - 5)(x + 3)(x - 9)(x + 5)(x - 8)(x + 0)(x - 9)(x + 5)(x - 4).$$

A comprehensive graph of this function is shown in **Figure A.** In **Figure B,** we show a screen obtained by zooming in on the positive zeros, as the comprehensive graph does not show the local behavior well in this region. Use a graphing calculator to graph your own "personal polynomial."

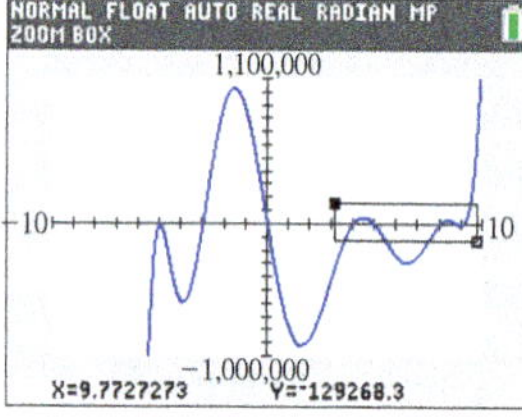

Figure A

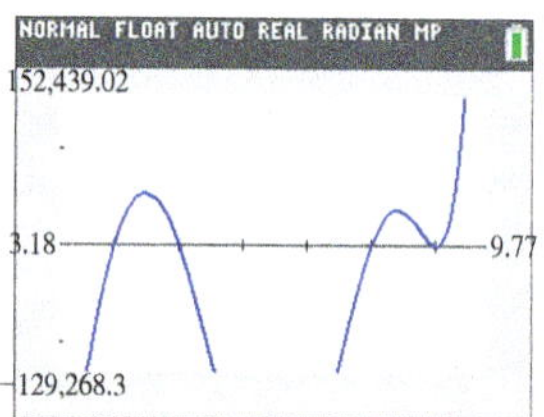

Figure B

90. A comprehensive graph of

$$f(x) = x^4 - 7x^3 + 18x^2 - 22x + 12$$

is shown in the two screens, along with displays of the two real zeros. Find the two remaining nonreal complex zeros.

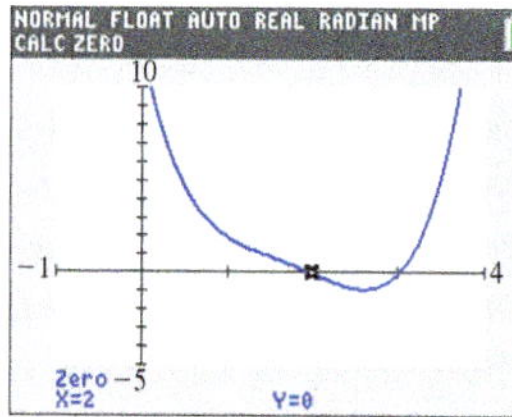

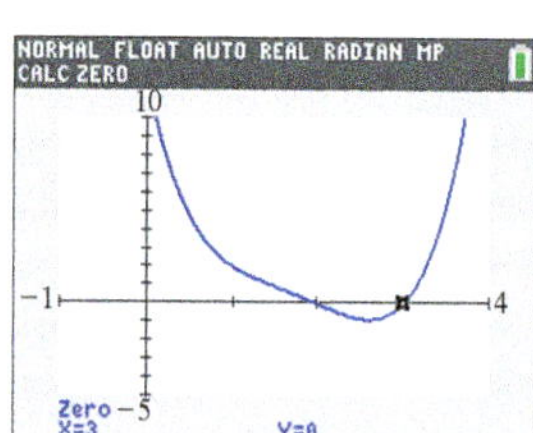

(Modeling) The following exercises are geometric in nature and lead to polynomial models. Solve each problem.

91. ***Volume of a Box*** A rectangular piece of cardboard measuring 12 in. by 18 in. is to be made into a box with an open top by cutting equal-size squares from each corner and folding up the sides. Let x represent the length of a side of each such square in inches. Give approximations to the nearest hundredth.

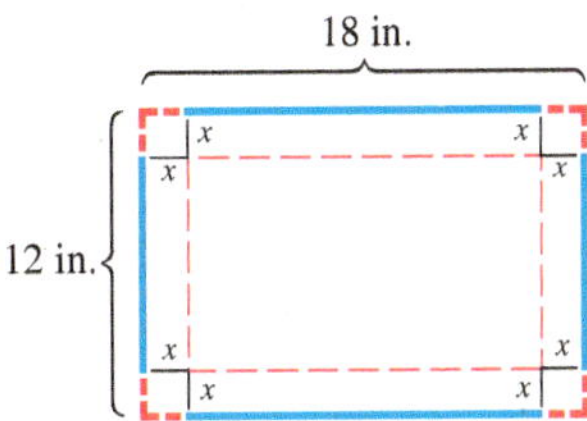

(a) Give the restrictions on x.

(b) Determine a function V that gives the volume of the box as a function of x.

(c) For what value of x will the volume be a maximum? What is this maximum volume? (*Hint:* Use the function of a graphing calculator that allows us to determine a maximum point within a given interval.)

(d) For what values of x will the volume be greater than 80 in.^{3}?

92. ***Construction of a Rain Gutter*** A piece of rectangular sheet metal is 20 in. wide. It is to be made into a rain gutter by turning up the edges to form parallel sides. Let x represent the length of each of the parallel sides. Give approximations to the nearest hundredth.

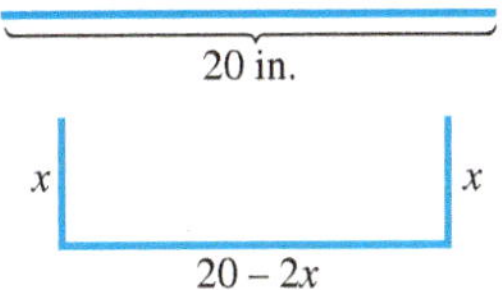

(a) Give the restrictions on x.

(b) Determine a function $\mathcal{A}$ that gives the area of a cross section of the gutter.

(c) For what value of x will $\mathcal{A}$ be a maximum (and thus maximize the amount of water that the gutter will hold)? What is this maximum area?

(d) For what values of x will the area of a cross section be less than 40 in.^{2}?

93. ***Sides of a Right Triangle*** A certain right triangle has area 84 in.^{2}. One leg of the triangle measures 1 in. less than the hypotenuse. Let x represent the length of the hypotenuse.

(a) Express the length of the leg mentioned above in terms of x. Give the domain of x.

(b) Express the length of the other leg in terms of x.

(c) Write an equation based on the information determined thus far. Square both sides and then write the equation with one side as a polynomial with integer coefficients, in descending powers, and the other side equal to 0.

(d) Solve the equation in part (c) graphically. Find the lengths of the three sides of the triangle.

94. ***Area of a Rectangle*** Find the value of x in the figure that will maximize the area of rectangle $ABCD$. Round to the nearest thousandth.

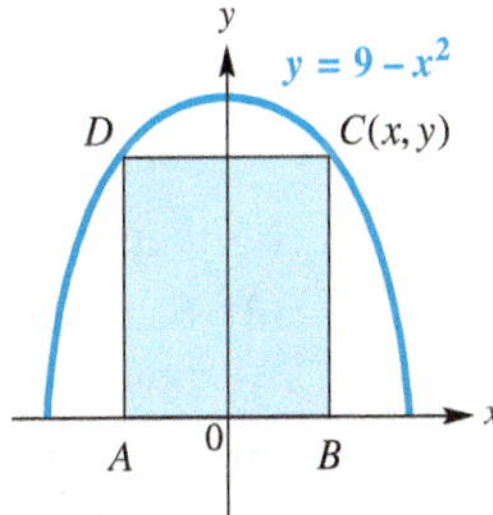

95. ***Butane Gas Storage*** A storage tank for butane gas is to be built in the shape of a right circular cylinder of altitude 12 ft, with a half sphere attached to each end. If x represents the radius of each half sphere, what radius should be used to cause the volume of the tank to be 144π ft^3?

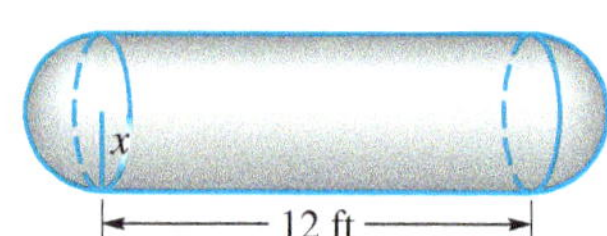

96. ***Volume of a Box*** A standard piece of notebook paper measuring 8.5 in. by 11 in. is to be made into a box with an open top by cutting equal-size squares from each corner and folding up the sides. Let x represent the length of a side of each such square in inches. Use the table feature of a graphing calculator to do the following. Round to the nearest hundredth.

(a) Find the maximum volume of the box.

(b) Determine when the volume of the box will be greater than 40 in.^{3}.

97. ***Floating Ball*** The polynomial function

$$f(x) = \frac{\pi}{3}x^3 - 5\pi x^2 + \frac{500\pi d}{3}$$

can be used to find the depth that a ball 10 cm in diameter sinks in water. The constant d is the density of the ball, where the density of water is 1. The smallest *positive* zero of $f(x)$ equals the depth that the ball sinks. Approximate this depth for each material and interpret the results.

(a) A wooden ball with $d = 0.8$ (to the nearest hundredth)

(b) A solid aluminum ball with $d = 2.7$

(c) A spherical water balloon with $d = 1$

98. ***Floating Ball*** Refer to **Exercise 97.** If a ball has a 20-cm diameter, then the function becomes

$$f(x) = \frac{\pi}{3}x^3 - 10\pi x^2 + \frac{4000\pi d}{3}.$$

This function can be used to determine the depth that the ball sinks in water. Find the depth that this size ball sinks when $d = 0.6$. Round to the nearest hundredth.

(Modeling) *Solve each problem.* ***See Example 8.***

99. ***Highway Design*** To allow enough distance for cars to pass on two-lane highways, engineers calculate minimum sight distances between curves and hills. The table shows the minimum sight distance y in feet for a car traveling at x miles per hour.

x (in mph)	20	30	40	50	60	65	70
y (in feet)	810	1090	1480	1840	2140	2310	2490

Source: Haefner, L., *Introduction to Transportation Systems*, Holt, Rinehart and Winston.

(a) Make a scatter diagram of the data.

(b) Use the regression feature of a calculator to find the best-fitting linear function for the data. Graph the function with the data.

(c) Repeat part (b) for a cubic function.

(d) Estimate the minimum sight distance for a car traveling 43 mph using the functions from parts (b) and (c).

(e) By comparing graphs of the functions in parts (b) and (c) with the data, decide which function best fits the given data.

100. *Water Pollution* Copper in high doses can be lethal to aquatic life. The table lists copper concentrations in freshwater mussels after 45 days at various distances downstream from an electroplating plant. The concentration C is measured in micrograms of copper per gram of mussel x kilometers downstream.

x	5	21	37	53	59
C	20	13	9	6	5

Source: Foster, R., and J. Bates, "Use of mussels to monitor point source industrial discharges," *Environ. Sci. Technol.*; Mason, C., *Biology of Freshwater Pollution*, John Wiley & Sons.

(a) Make a scatter diagram of the data.

(b) Use the regression feature of a calculator to find the best-fitting quadratic function for the data. Graph the function with the data.

(c) Repeat part (b) for a cubic function.

(d) By comparing graphs of the functions in parts (b) and (c) with the data, decide which function best fits the given data.

(e) Concentrations above 10 are lethal to mussels. Find the values of x (using the cubic function) for which this is the case.

101. *Government Spending on Health Research and Training* The table lists the annual amount (in billions of dollars) spent by the federal government on health research and training programs over a 10-yr period.

Year	Amount (billions of $)	Year	Amount (billions of $)
2004	27.1	2009	30.6
2005	28.1	2010	34.2
2006	28.8	2011	36.2
2007	29.3	2012	34.5
2008	29.9	2013	32.9

Source: U.S. Office of Management and Budget.

Which one of the following provides the best model for these data, where x represents the year?

A. $f(x) = 0.2(x - 2004)^2 + 27.1$ **B.** $g(x) = (x - 2004) + 27.1$

C. $h(x) = 2.5\sqrt{x - 2004} + 27.1$ **D.** $k(x) = 0.1(x - 2004)^3 + 27.1$

102. *Swing of a Pendulum* Grandfather clocks use pendulums to keep accurate time. The relationship between the length of a pendulum L and the time T for one complete oscillation can be expressed by the equation

$$L = kT^n,$$

where k is a constant and n is a positive integer to be determined. The data in the table were taken for different lengths of pendulums.

L (ft)	T (sec)	L (ft)	T (sec)
1.0	1.11	3.0	1.92
1.5	1.36	3.5	2.08
2.0	1.57	4.0	2.22
2.5	1.76		

(a) As the length of the pendulum increases, what happens to T?

(b) Use the data to approximate k and determine the best value for n.

(c) Using the values of k and n from part (b), predict T for a pendulum having length 5 ft. Round to the nearest hundredth.

(d) If the length L of a pendulum doubles, what happens to the period T?

Relating Concepts

For individual or collaborative investigation *(Exercises 103-108)*

For any function $y = f(x)$, the following hold true.

(a) The real solutions of $f(x) = 0$ correspond to the x-intercepts of the graph.

(b) The real solutions of $f(x) < 0$ are the x-values for which the graph lies *below* the x-axis.

(c) The real solutions of $f(x) > 0$ are the x-values for which the graph lies *above* the x-axis.

In each exercise, a polynomial function $f(x)$ is given in both expanded and factored forms. Graph each function, and solve the equations and inequalities. Give multiplicities of solutions when applicable.

103. $f(x) = x^3 - 3x^2 - 6x + 8$
$f(x) = (x - 4)(x - 1)(x + 2)$
(a) $f(x) = 0$ **(b)** $f(x) < 0$
(c) $f(x) > 0$

104. $f(x) = x^3 + 4x^2 - 11x - 30$
$f(x) = (x - 3)(x + 2)(x + 5)$
(a) $f(x) = 0$ **(b)** $f(x) < 0$
(c) $f(x) > 0$

105. $f(x) = 2x^4 - 9x^3 - 5x^2 + 57x - 45$
$f(x) = (x - 3)^2(2x + 5)(x - 1)$
(a) $f(x) = 0$ **(b)** $f(x) < 0$
(c) $f(x) > 0$

106. $f(x) = 4x^4 + 27x^3 - 42x^2 - 445x - 300$
$f(x) = (x + 5)^2(4x + 3)(x - 4)$
(a) $f(x) = 0$ **(b)** $f(x) < 0$
(c) $f(x) > 0$

107. $f(x) = -x^4 - 4x^3 + 3x^2 + 18x$
$f(x) = x(2 - x)(x + 3)^2$
(a) $f(x) = 0$ **(b)** $f(x) \geq 0$
(c) $f(x) \leq 0$

108. $f(x) = -x^4 + 2x^3 + 8x^2$
$f(x) = x^2(4 - x)(x + 2)$
(a) $f(x) = 0$ **(b)** $f(x) \geq 0$
(c) $f(x) \leq 0$

Summary Exercises on Polynomial Functions, Zeros, and Graphs

We use all of the theorems for finding complex zeros of polynomial functions in the next example.

EXAMPLE Finding All Zeros of a Polynomial Function

Find all zeros of $f(x) = x^4 - 3x^3 + 6x^2 - 12x + 8$.

SOLUTION We consider the number of positive zeros by observing the variations in signs for $f(x)$.

$$f(x) = +x^4 - 3x^3 + 6x^2 - 12x + 8$$

1 2 3 4

Because $f(x)$ has four sign changes, we can use Descartes' rule of signs to determine that there are four, two, or zero positive real zeros. For negative zeros, we consider the variations in signs for $f(-x)$.

$$f(-x) = (-x)^4 - 3(-x)^3 + 6(-x)^2 - 12(-x) + 8$$
$$f(-x) = x^4 + 3x^3 + 6x^2 + 12x + 8$$

Because $f(-x)$ has no sign changes, there are no negative real zeros. The function has degree 4, so it has a maximum of four zeros with possibilities summarized in the table on the next page.

Positive	Negative	Nonreal Complex
4	0	0
2	0	2
0	0	4

We can now use the rational zeros theorem to determine that the possible rational zeros are ± 1, ± 2, ± 4, and ± 8. Based on Descartes' rule of signs, we discard the negative rational zeros from this list and try to find a positive rational zero. We start by using synthetic division to check 4.

$$\begin{array}{r|rrrrr} \text{Proposed zero} \rightarrow 4) & 1 & -3 & 6 & -12 & 8 \\ & & 4 & 4 & 40 & 112 \\ \hline & 1 & 1 & 10 & 28 & 120 \leftarrow f(4) = 120 \end{array}$$

We find that 4 is not a zero. However, $4 > 0$, and the numbers in the bottom row of the synthetic division are nonnegative. Thus, the boundedness theorem indicates that there are no zeros greater than 4. We can discard 8 as a possible rational zero and use synthetic division to show that 1 and 2 are zeros.

$$\begin{array}{r|rrrrr} 1) & 1 & -3 & 6 & -12 & 8 \\ & & 1 & -2 & 4 & -8 \\ \hline 2) & 1 & -2 & 4 & -8 & 0 \leftarrow f(1) = 0 \\ & & 2 & 0 & 8 & \\ \hline & 1 & 0 & 4 & 0 & \leftarrow f(2) = 0 \end{array}$$

The polynomial now factors as

$$f(x) = (x-1)(x-2)(x^2+4).$$

We find the remaining two zeros using algebra to solve for x in the quadratic factor of the following equation.

$$(x-1)(x-2)(x^2+4) = 0$$

$$x - 1 = 0 \quad \text{or} \quad x - 2 = 0 \quad \text{or} \quad x^2 + 4 = 0 \qquad \text{Zero-factor property}$$

$$x = 1 \quad \text{or} \quad x = 2 \quad \text{or} \quad x^2 = -4$$

$$x = \pm 2i \qquad \text{Square root property}$$

The linear factored form of the polynomial is

$$f(x) = (x-1)(x-2)(x-2i)(x+2i),$$

and the corresponding zeros are 1, 2, $2i$, and $-2i$. ✓ **Now Try Exercise 3.**

EXERCISES

For each polynomial function, complete the following in order.

(a) Use Descartes' rule of signs to determine the different possibilities for the numbers of positive, negative, and nonreal complex zeros.

(b) Use the rational zeros theorem to determine the possible rational zeros.

(c) Use synthetic division with the boundedness theorem where appropriate and/or factoring to find the rational zeros, if any.

(d) Find all other complex zeros (both real and nonreal), if any.

1. $f(x) = 6x^3 - 41x^2 + 26x + 24$

2. $f(x) = 2x^3 - 5x^2 - 4x + 3$

3. $f(x) = 3x^4 - 5x^3 + 14x^2 - 20x + 8$

4. $f(x) = 2x^4 - 3x^3 + 16x^2 - 27x - 18$

5. $f(x) = 6x^4 - 5x^3 - 11x^2 + 10x - 2$

6. $f(x) = 5x^4 + 8x^3 - 19x^2 - 24x + 12$

7. $f(x) = x^5 - 6x^4 + 16x^3 - 24x^2 + 16x$ (*Hint:* Factor out x first.)

8. $f(x) = 2x^4 + 8x^3 - 7x^2 - 42x - 9$

9. $f(x) = 8x^4 + 8x^3 - x - 1$ (*Hint:* Factor the polynomial.)

10. $f(x) = 2x^5 + 5x^4 - 9x^3 - 11x^2 + 19x - 6$

For each polynomial function, complete the following in order.

(a) *Use Descartes' rule of signs to determine the different possibilities for the numbers of positive, negative, and nonreal complex zeros.*

(b) *Use the rational zeros theorem to determine the possible rational zeros.*

(c) *Find the rational zeros, if any.*

(d) *Find all other real zeros, if any.*

(e) *Find any other complex zeros (that is, zeros that are not real), if any.*

(f) *Find the x-intercepts of the graph, if any.*

(g) *Find the y-intercept of the graph.*

(h) *Use synthetic division to find $f(4)$, and give the coordinates of the corresponding point on the graph.*

(i) *Determine the end behavior of the graph.*

(j) *Sketch the graph.*

11. $f(x) = x^4 + 3x^3 - 3x^2 - 11x - 6$

12. $f(x) = -2x^5 + 5x^4 + 34x^3 - 30x^2 - 84x + 45$

13. $f(x) = 2x^5 - 10x^4 + x^3 - 5x^2 - x + 5$

14. $f(x) = 3x^4 - 4x^3 - 22x^2 + 15x + 18$

15. $f(x) = -2x^4 - x^3 + x + 2$

16. $f(x) = 4x^5 + 8x^4 + 9x^3 + 27x^2 + 27x$ (*Hint:* Factor out x first.)

17. $f(x) = 3x^4 - 14x^2 - 5$ (*Hint:* Factor the polynomial.)

18. $f(x) = -x^5 - x^4 + 10x^3 + 10x^2 - 9x - 9$

19. $f(x) = -3x^4 + 22x^3 - 55x^2 + 52x - 12$

20. For the polynomial functions in **Exercises 11–19** that have irrational zeros, find approximations to the nearest thousandth.

3.5 Rational Functions: Graphs, Applications, and Models

- **The Reciprocal Function $f(x) = \frac{1}{x}$**
- **The Function $f(x) = \frac{1}{x^2}$**
- **Asymptotes**
- **Graphing Techniques**
- **Rational Models**

A rational expression is a fraction that is the quotient of two polynomials. A *rational function* is defined by a quotient of two polynomial functions.

Rational Function

A function f of the form

$$f(x) = \frac{p(x)}{q(x)},$$

where $p(x)$ and $q(x)$ are polynomial functions, with $q(x) \neq 0$, is a **rational function.**

$$f(x) = \frac{1}{x}, \quad f(x) = \frac{x+1}{2x^2+5x-3}, \quad f(x) = \frac{3x^2-3x-6}{x^2+8x+16} \qquad \text{Rational functions}$$

Any values of x such that $q(x) = 0$ are excluded from the domain of a rational function, so this type of function often has a **discontinuous graph**—that is, a graph that has one or more breaks in it.

The Reciprocal Function $f(x) = \frac{1}{x}$ The simplest rational function with a variable denominator is the **reciprocal function.**

$$f(x) = \frac{1}{x} \qquad \text{Reciprocal function}$$

The domain of this function is the set of all nonzero real numbers. The number 0 cannot be used as a value of x, but it is helpful to find values of $f(x)$ for some values of x very close to 0. We use the table feature of a graphing calculator to do this. The tables in **Figure 41** suggest that $|f(x)|$ increases without bound as x gets closer and closer to 0, which is written in symbols as

$$|f(x)| \to \infty \quad \text{as} \quad x \to 0.$$

(The symbol $x \to 0$ means that x approaches 0, without necessarily ever being equal to 0.) Because x cannot equal 0, the graph of $f(x) = \frac{1}{x}$ will never intersect the vertical line $x = 0$. This line is a **vertical asymptote.**

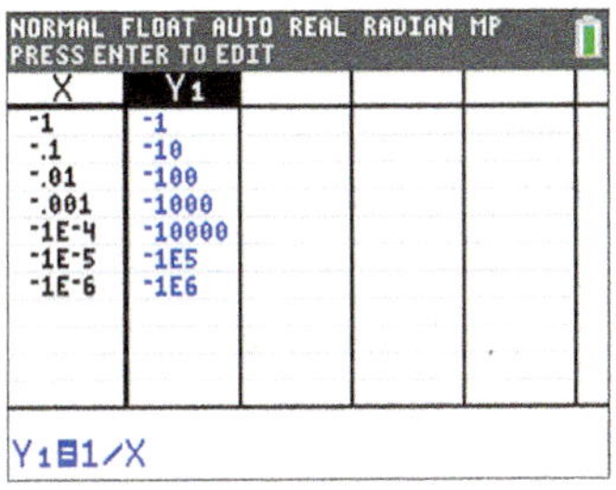

NORMAL FLOAT AUTO REAL RADIAN MP
PRESS ENTER TO EDIT

X	Y1
-1	-1
-.1	-10
-.01	-100
-.001	-1000
-1E-4	-10000
-1E-5	-1E5
-1E-6	-1E6

Y1=1/X

As x approaches 0 from the left, $y_1 = \frac{1}{x}$ approaches $-\infty$. (–1E–6 means -1×10^{-6}.)

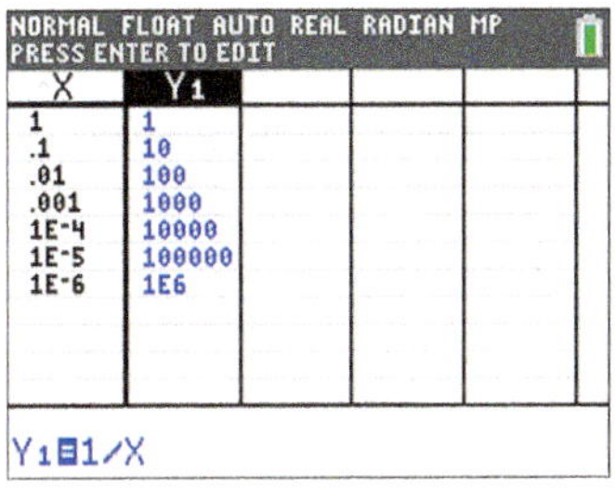

NORMAL FLOAT AUTO REAL RADIAN MP
PRESS ENTER TO EDIT

X	Y1
1	1
.1	10
.01	100
.001	1000
1E-4	10000
1E-5	100000
1E-6	1E6

Y1=1/X

As x approaches 0 from the right, $y_1 = \frac{1}{x}$ approaches ∞.

Figure 41

As $|x|$ increases without bound (written $|x| \to \infty$), the values of $f(x) = \frac{1}{x}$ get closer and closer to 0, as shown in the tables in **Figure 42.** Letting $|x|$ increase without bound causes the graph of $f(x) = \frac{1}{x}$ to move closer and closer to the horizontal line $y = 0$. This line is a **horizontal asymptote.**

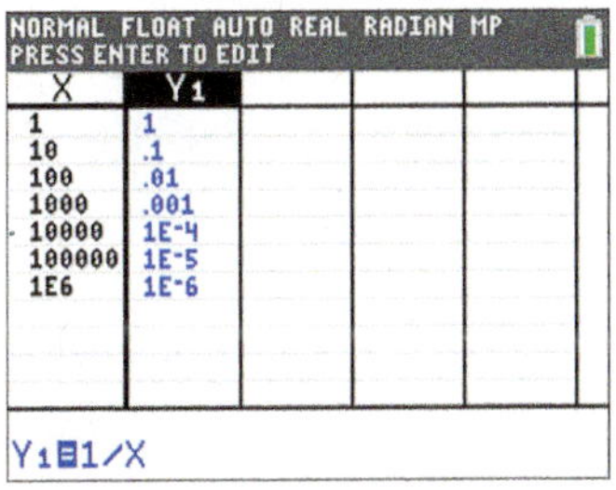

NORMAL FLOAT AUTO REAL RADIAN MP
PRESS ENTER TO EDIT

X	Y1
1	1
10	.1
100	.01
1000	.001
10000	1E-4
100000	1E-5
1E6	1E-6

Y1=1/X

As x approaches ∞, $y_1 = \frac{1}{x}$ approaches 0 through positive values.

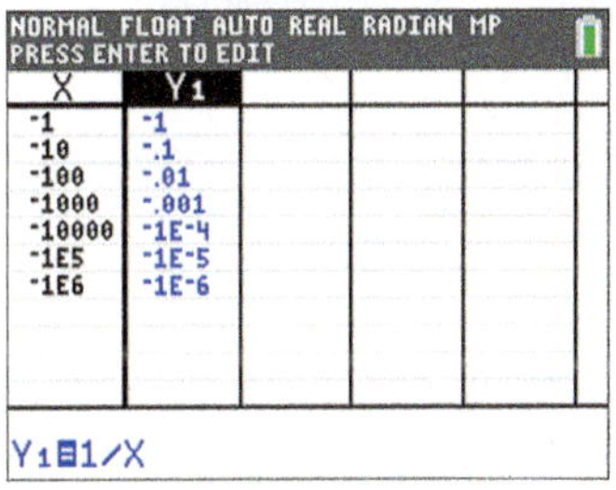

NORMAL FLOAT AUTO REAL RADIAN MP
PRESS ENTER TO EDIT

X	Y1
-1	-1
-10	-.1
-100	-.01
-1000	-.001
-10000	-1E-4
-1E5	-1E-5
-1E6	-1E-6

Y1=1/X

As x approaches $-\infty$, $y_1 = \frac{1}{x}$ approaches 0 through negative values.

Figure 42

The graph of $f(x) = \frac{1}{x}$ is shown in **Figure 43.**

Reciprocal Function $f(x) = \frac{1}{x}$

Domain: $(-\infty, 0) \cup (0, \infty)$ Range: $(-\infty, 0) \cup (0, \infty)$

x	y
-2	$-\frac{1}{2}$
-1	-1
$-\frac{1}{2}$	-2
0	undefined
$\frac{1}{2}$	2
1	1
2	$\frac{1}{2}$

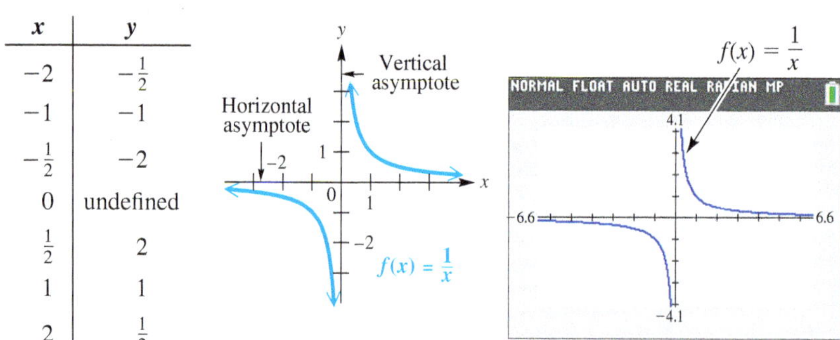

Figure 43

- $f(x) = \frac{1}{x}$ decreases on the open intervals $(-\infty, 0)$ and $(0, \infty)$.
- It is discontinuous at $x = 0$.
- The y-axis is a vertical asymptote, and the x-axis is a horizontal asymptote.
- It is an odd function, and its graph is symmetric with respect to the origin.

The graph of $y = \frac{1}{x}$ can be translated and/or reflected.

EXAMPLE 1 Graphing a Rational Function

Graph $y = -\frac{2}{x}$. Give the domain and range and the largest open intervals of the domain over which the function is increasing or decreasing.

SOLUTION The expression $-\frac{2}{x}$ can be written as $-2\left(\frac{1}{x}\right)$ or $2\left(\frac{1}{-x}\right)$, indicating that the graph may be obtained by stretching the graph of $y = \frac{1}{x}$ vertically by a factor of 2 and reflecting it across either the x-axis or the y-axis. The x- and y-axes remain the horizontal and vertical asymptotes. The domain and range are both still $(-\infty, 0) \cup (0, \infty)$. See **Figure 44.**

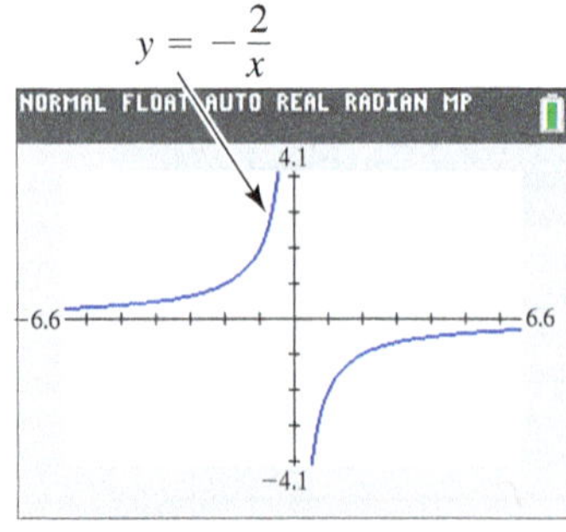

The graph in **Figure 44** is shown here using a **decimal window.** Using a nondecimal window *may* produce an extraneous vertical line that is not part of the graph.

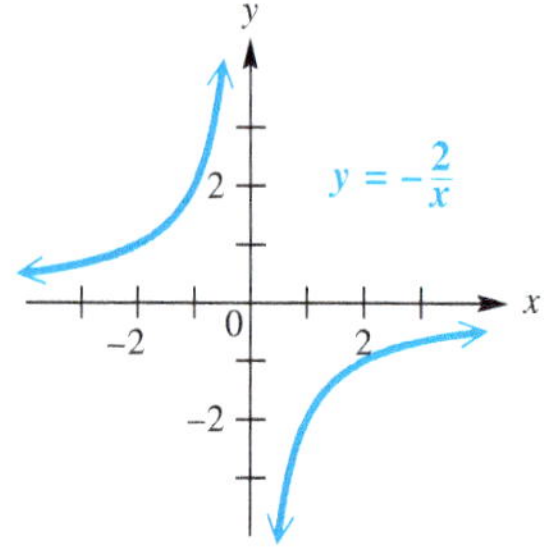

Figure 44

The graph shows that $f(x)$ is increasing on both sides of its vertical asymptote. Thus, it is increasing on $(-\infty, 0)$ and $(0, \infty)$.

✔ **Now Try Exercise 17.**

EXAMPLE 2 Graphing a Rational Function

Graph $f(x) = \frac{2}{x+1}$. Give the domain and range and the largest open intervals of the domain over which the function is increasing or decreasing.

ALGEBRAIC SOLUTION

The expression $\frac{2}{x+1}$ can be written as $2\left(\frac{1}{x+1}\right)$, indicating that the graph may be obtained by shifting the graph of $y = \frac{1}{x}$ to the left 1 unit and stretching it vertically by a factor of 2. See **Figure 45.**

The horizontal shift affects the domain, which is now $(-\infty, -1) \cup (-1, \infty)$. The line $x = -1$ is the vertical asymptote, and the line $y = 0$ (the x-axis) remains the horizontal asymptote. The range is still $(-\infty, 0) \cup (0, \infty)$. The graph shows that $f(x)$ is decreasing on both sides of its vertical asymptote. Thus, it is decreasing on $(-\infty, -1)$ and $(-1, \infty)$.

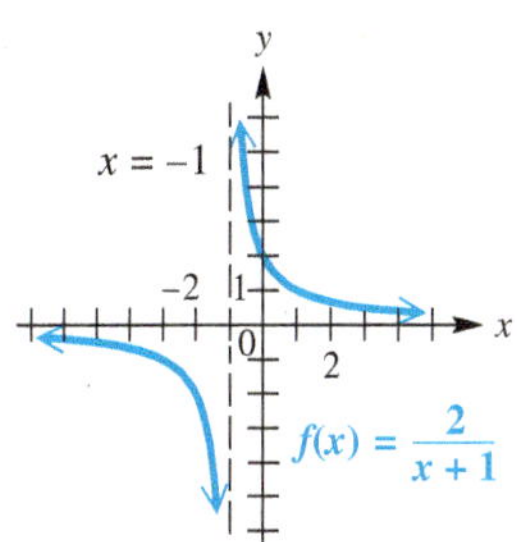

Figure 45

GRAPHING CALCULATOR SOLUTION

When entering this rational function into the function editor of a calculator, make sure that the numerator is 2 and the denominator is the entire expression $(x + 1)$.

The graph of this function has a vertical asymptote at $x = -1$ and a horizontal asymptote at $y = 0$, so it is reasonable to choose a viewing window that contains the locations of both asymptotes as well as enough of the graph to determine its basic characteristics. See **Figure 46.**

$f(x) = \frac{2}{x+1}$

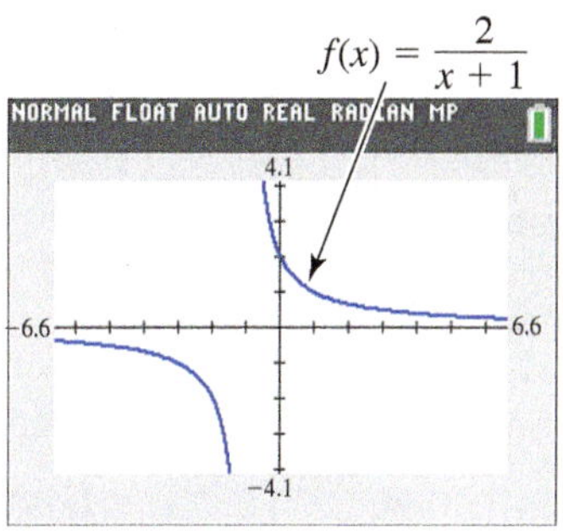

Figure 46

✔ **Now Try Exercise 19.**

The Function $f(x) = \frac{1}{x^2}$

The rational function

$$f(x) = \frac{1}{x^2} \quad \text{Rational function}$$

also has domain $(-\infty, 0) \cup (0, \infty)$. We can use the table feature of a graphing calculator to examine values of $f(x)$ for some x-values close to 0. See **Figure 47.**

NORMAL FLOAT AUTO REAL RADIAN MP
PRESS ENTER TO EDIT

X	Y1
-1	1
-.1	100
-.01	10000
-.001	1E6
-1E-4	1E8
-1E-5	1E10
-1E-6	1E12

Y1=1/X²

As x approaches 0 from the left, $y_1 = \frac{1}{x^2}$ approaches ∞.

NORMAL FLOAT AUTO REAL RADIAN MP
PRESS ENTER TO EDIT

X	Y1
1	1
.1	100
.01	10000
.001	1E6
1E-4	1E8
1E-5	1E10
1E-6	1E12

Y1=1/X²

As x approaches 0 from the right, $y_1 = \frac{1}{x^2}$ approaches ∞.

Figure 47

The tables suggest that $f(x)$ increases without bound as x gets closer and closer to 0. Notice that as x approaches 0 from *either* side, function values are all positive and there is symmetry with respect to the y-axis. Thus, $f(x) \to \infty$ as $x \to 0$. The y-axis $(x = 0)$ is the vertical asymptote.

NORMAL FLOAT AUTO REAL RADIAN MP
PRESS ENTER TO EDIT

X	Y₁
1	1
10	.01
100	1E-4
1000	1E-6
10000	1E-8
100000	1E-10
1E6	1E-12

Y₁=1/X²

As x approaches ∞, $y_1 = \frac{1}{x^2}$ approaches 0 through positive values.

NORMAL FLOAT AUTO REAL RADIAN MP
PRESS ENTER TO EDIT

X	Y₁
-1	1
-10	.01
-100	1E-4
-1000	1E-6
-10000	1E-8
-1E5	1E-10
-1E6	1E-12

Y₁=1/X²

As x approaches $-\infty$, $y_1 = \frac{1}{x^2}$ approaches 0 through positive values.

Figure 48

As $|x|$ increases without bound, $f(x)$ approaches 0, as suggested by the tables in **Figure 48.** Again, function values are all positive. The x-axis is the horizontal asymptote of the graph.

The graph of $f(x) = \frac{1}{x^2}$ is shown in **Figure 49.**

Rational Function $f(x) = \frac{1}{x^2}$

Domain: $(-\infty, 0) \cup (0, \infty)$ Range: $(0, \infty)$

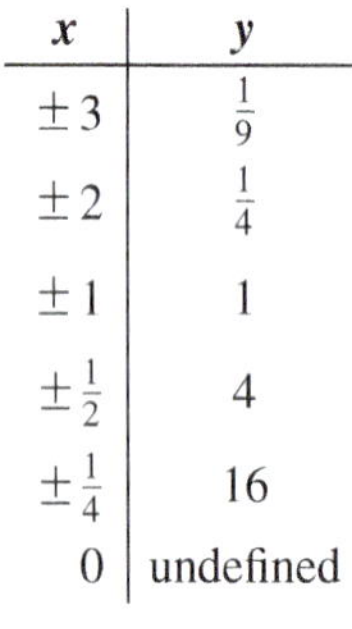

x	y
± 3	$\frac{1}{9}$
± 2	$\frac{1}{4}$
± 1	1
$\pm \frac{1}{2}$	4
$\pm \frac{1}{4}$	16
0	undefined

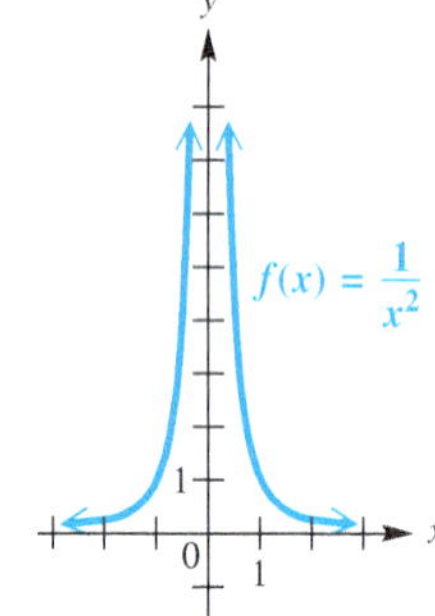

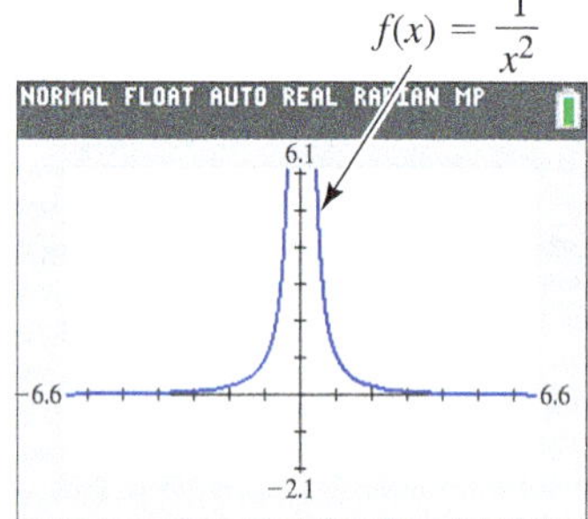

Figure 49

- $f(x) = \frac{1}{x^2}$ increases on the open interval $(-\infty, 0)$ and decreases on the open interval $(0, \infty)$.
- It is discontinuous at $x = 0$.
- The y-axis is a vertical asymptote, and the x-axis is a horizontal asymptote.
- It is an even function, and its graph is symmetric with respect to the y-axis.

EXAMPLE 3 Graphing a Rational Function

Graph $g(x) = \frac{1}{(x+2)^2} - 1$. Give the domain and range and the largest open intervals of the domain over which the function is increasing or decreasing.

SOLUTION The function $g(x) = \frac{1}{(x+2)^2} - 1$ is equivalent to

$$g(x) = f(x+2) - 1, \quad \text{where} \quad f(x) = \frac{1}{x^2}.$$

This indicates that the graph will be shifted 2 units to the left and 1 unit down. The horizontal shift affects the domain, now $(-\infty, -2) \cup (-2, \infty)$. The vertical shift affects the range, now $(-1, \infty)$.

The vertical asymptote has equation $x = -2$, and the horizontal asymptote has equation $y = -1$. A traditional graph is shown in **Figure 50,** with a calculator graph in **Figure 51.** Both graphs show that this function is increasing on $(-\infty, -2)$ and decreasing on $(-2, \infty)$.

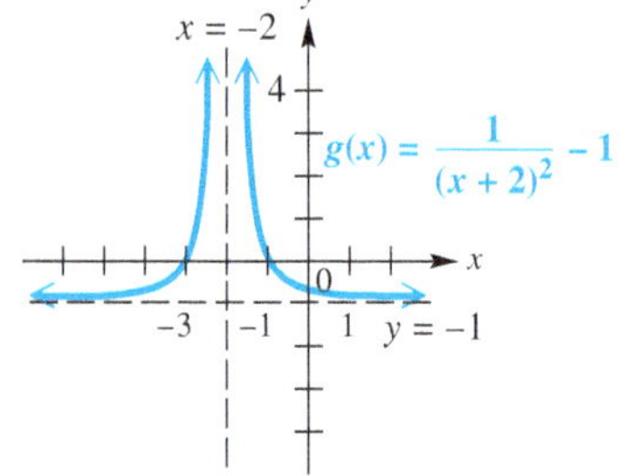

This is the graph of $y = \frac{1}{x^2}$ shifted 2 units to the left and 1 unit down.

Figure 50

Now Try Exercise 27.

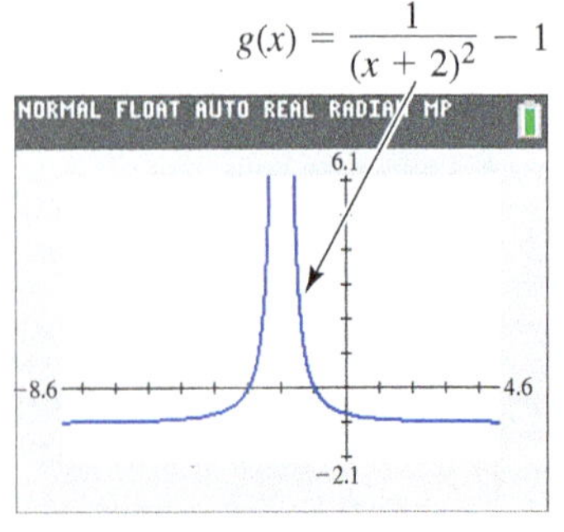

Figure 51

LOOKING AHEAD TO CALCULUS

The rational function

$$f(x) = \frac{2}{x+1}$$

in **Example 2** has a vertical asymptote at $x = -1$. In calculus, the behavior of the graph of this function for values close to -1 is described using **one-sided limits.** As x approaches -1 from the *left*, the function values decrease without bound. This is written

$$\lim_{x \to -1^-} f(x) = -\infty.$$

As x approaches -1 from the *right*, the function values increase without bound. This is written

$$\lim_{x \to -1^+} f(x) = \infty.$$

Asymptotes The preceding examples suggest the following definitions of vertical and horizontal asymptotes.

Asymptotes

Let $p(x)$ and $q(x)$ define polynomial functions. Consider the rational function $f(x) = \frac{p(x)}{q(x)}$, written in lowest terms, and real numbers a and b.

1. If $|f(x)| \to \infty$ as $x \to a$, then the line $x = a$ is a **vertical asymptote.**
2. If $f(x) \to b$ as $|x| \to \infty$, then the line $y = b$ is a **horizontal asymptote.**

Locating asymptotes is important when graphing rational functions.

- We find *vertical asymptotes* by determining the values of x that make the denominator equal to 0.
- We find *horizontal asymptotes* (and, in some cases, *oblique asymptotes*), by considering what happens to $f(x)$ as $|x| \to \infty$. These asymptotes determine the end behavior of the graph.

Determining Asymptotes

To find the asymptotes of a rational function defined by a rational expression in *lowest terms,* use the following procedures.

1. Vertical Asymptotes
Find any vertical asymptotes by setting the denominator equal to 0 and solving for x. If a is a zero of the denominator, then **the line $x = a$ is a vertical asymptote.**

2. Other Asymptotes
Determine any other asymptotes by considering three possibilities:

(a) If the numerator has lesser degree than the denominator, then there is a **horizontal asymptote $y = 0$** (the x-axis).

(b) If the numerator and denominator have the same degree, and the function is of the form

$$f(x) = \frac{a_n x^n + \cdots + a_0}{b_n x^n + \cdots + b_0}, \quad \text{where} \quad a_n, b_n \neq 0,$$

then the **horizontal asymptote has equation $y = \frac{a_n}{b_n}$.**

(c) If the numerator is of degree exactly one more than the denominator, then there will be an **oblique (slanted) asymptote.** To find it, divide the numerator by the denominator and disregard the remainder. Set the rest of the quotient equal to y to obtain the equation of the asymptote.

NOTE The graph of a rational function may have more than one vertical asymptote, or it may have none at all.

> ***The graph cannot intersect any vertical asymptote. There can be at most one other (nonvertical) asymptote, and the graph may intersect that asymptote.*** **(See Example 7.)**

EXAMPLE 4 Finding Asymptotes of Rational Functions

Give the equations of any vertical, horizontal, or oblique asymptotes for the graph of each rational function.

(a) $f(x) = \dfrac{x+1}{(2x-1)(x+3)}$ **(b)** $f(x) = \dfrac{2x+1}{x-3}$ **(c)** $f(x) = \dfrac{x^2+1}{x-2}$

SOLUTION

(a) To find the vertical asymptotes, set the denominator equal to 0 and solve.

$$(2x-1)(x+3) = 0$$

$$2x - 1 = 0 \quad \text{or} \quad x + 3 = 0 \qquad \text{Zero-factor property}$$

$$x = \frac{1}{2} \quad \text{or} \quad x = -3 \qquad \text{Solve each equation.}$$

The equations of the vertical asymptotes are $x = \frac{1}{2}$ and $x = -3$.

To find the equation of the horizontal asymptote, begin by multiplying the factors in the denominator.

$$f(x) = \frac{x+1}{(2x-1)(x+3)} = \frac{x+1}{2x^2+5x-3}$$

Now divide each term in the numerator and denominator by x^2. We choose the exponent 2 because it is the greatest power of x in the entire expression.

$$f(x) = \frac{\frac{x}{x^2} + \frac{1}{x^2}}{\frac{2x^2}{x^2} + \frac{5x}{x^2} - \frac{3}{x^2}} = \frac{\frac{1}{x} + \frac{1}{x^2}}{2 + \frac{5}{x} - \frac{3}{x^2}}$$

Stop here. Leave the expression in complex form.

As $|x|$ increases without bound, the quotients $\frac{1}{x}$, $\frac{1}{x^2}$, $\frac{5}{x}$, and $\frac{3}{x^2}$ all approach 0, and the value of $f(x)$ approaches

$$\frac{0+0}{2+0-0} = 0. \qquad \frac{0}{2} = 0$$

The line $y = 0$ (that is, the x-axis) is therefore the horizontal asymptote. This supports procedure 2(a) of determining asymptotes on the previous page.

(b) Set the denominator $x - 3$ equal to 0 to find that the vertical asymptote has equation $x = 3$. To find the horizontal asymptote, divide each term in the rational expression by x since the greatest power of x in the expression is 1.

$$f(x) = \frac{2x+1}{x-3} = \frac{\frac{2x}{x} + \frac{1}{x}}{\frac{x}{x} - \frac{3}{x}} = \frac{2 + \frac{1}{x}}{1 - \frac{3}{x}}$$

As $|x|$ increases without bound, the quotients $\frac{1}{x}$ and $\frac{3}{x}$ both approach 0, and the value of $f(x)$ approaches

$$\frac{2+0}{1-0} = 2.$$

The line $y = 2$ is the horizontal asymptote. This supports procedure 2(b) of determining asymptotes on the previous page.

(c) Setting the denominator $x - 2$ equal to 0 shows that the vertical asymptote has equation $x = 2$. If we divide by the greatest power of x as before (x^2 in this case), we see that there is no horizontal asymptote because

$$f(x) = \frac{x^2 + 1}{x - 2} = \frac{\frac{x^2}{x^2} + \frac{1}{x^2}}{\frac{x}{x^2} - \frac{2}{x^2}} = \frac{1 + \frac{1}{x^2}}{\frac{1}{x} - \frac{2}{x^2}}$$

does not approach any real number as $|x| \to \infty$, due to the fact that $\frac{1 + 0}{0 - 0} = \frac{1}{0}$ is undefined. This happens whenever the degree of the numerator is greater than the degree of the denominator.

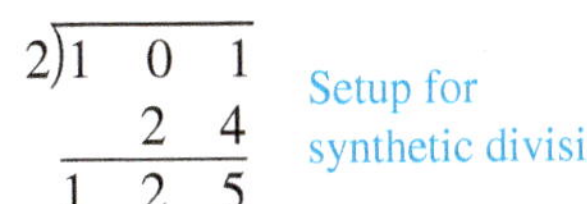
Setup for synthetic division

In such cases, divide the denominator into the numerator to write the expression in another form. We use synthetic division, as shown in the margin. The result enables us to write the function as follows.

$$f(x) = x + 2 + \frac{5}{x - 2}$$

For very large values of $|x|$, $\frac{5}{x-2}$ is close to 0, and the graph approaches the line $y = x + 2$. This line is an **oblique asymptote** (slanted, neither vertical nor horizontal) for the graph of the function. This supports procedure 2(c) of determining asymptotes.

✔ **Now Try Exercises 37, 39, and 41.**

Graphing Techniques A comprehensive graph of a rational function will show the following characteristics.

- all x- and y-intercepts
- all asymptotes: vertical, horizontal, and/or oblique
- the point at which the graph intersects its nonvertical asymptote (if there is any such point)
- the behavior of the function on each domain interval determined by the vertical asymptotes and x-intercepts

Graphing a Rational Function

Let $f(x) = \frac{p(x)}{q(x)}$ define a function where $p(x)$ and $q(x)$ are polynomial functions and the rational expression is written in lowest terms. To sketch its graph, follow these steps.

Step 1 Find any vertical asymptotes.

Step 2 Find any horizontal or oblique asymptotes.

Step 3 If $q(0) \neq 0$, plot the y-intercept by evaluating $f(0)$.

Step 4 Plot the x-intercepts, if any, by solving $f(x) = 0$. (These will correspond to the zeros of the numerator, $p(x)$.)

Step 5 Determine whether the graph will intersect its nonvertical asymptote $y = b$ or $y = mx + b$ by solving $f(x) = b$ or $f(x) = mx + b$.

Step 6 Plot selected points, as necessary. Choose an x-value in each domain interval determined by the vertical asymptotes and x-intercepts.

Step 7 Complete the sketch.

EXAMPLE 5 Graphing a Rational Function (x-Axis as Horizontal Asymptote)

Graph $f(x) = \frac{x+1}{2x^2+5x-3}$.

SOLUTION

Steps 1 and 2 In **Example 4(a),** we found that $2x^2 + 5x - 3 = (2x - 1)(x + 3)$, so the vertical asymptotes have equations $x = \frac{1}{2}$ and $x = -3$, and the horizontal asymptote is the x-axis.

Step 3 The y-intercept is $\left(0, -\frac{1}{3}\right)$, as justified below.

$$f(0) = \frac{0+1}{2(0)^2 + 5(0) - 3} = -\frac{1}{3} \quad \text{The } y\text{-intercept corresponds to the ratio of the constant terms.}$$

Step 4 The x-intercept is found by solving $f(x) = 0$.

$$\frac{x+1}{2x^2+5x-3} = 0 \quad \text{Set } f(x) = 0.$$

$$x + 1 = 0 \quad \text{If a rational expression is equal to 0, then its numerator must equal 0.}$$

$$x = -1 \quad \text{The } x\text{-intercept is } (-1, 0).$$

Step 5 To determine whether the graph intersects its horizontal asymptote, solve this equation.

$$f(x) = 0 \longleftarrow y\text{-value of horizontal asymptote}$$

The horizontal asymptote is the x-axis, so the solution of $f(x) = 0$ was found in Step 4. The graph intersects its horizontal asymptote at $(-1, 0)$.

Step 6 Plot a point in each of the intervals determined by the x-intercepts and vertical asymptotes, $(-\infty, -3)$, $(-3, -1)$, $\left(-1, \frac{1}{2}\right)$ and $\left(\frac{1}{2}, \infty\right)$, to get an idea of how the graph behaves in each interval.

Interval	Test Point x	Value of $f(x)$	Sign of $f(x)$	Graph Above or Below x-Axis
$(-\infty, -3)$	-4	$-\frac{1}{3}$	Negative	Below
$(-3, -1)$	-2	$\frac{1}{5}$	Positive	Above
$\left(-1, \frac{1}{2}\right)$	0	$-\frac{1}{3}$	Negative	Below
$\left(\frac{1}{2}, \infty\right)$	2	$\frac{1}{5}$	Positive	Above

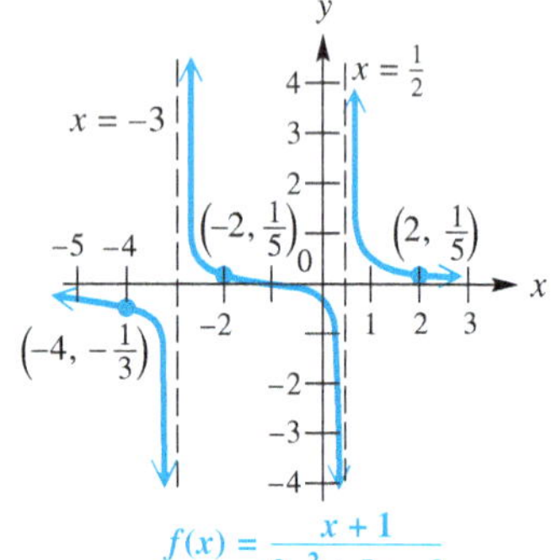

Figure 52

Step 7 Complete the sketch. See **Figure 52.** This function is decreasing on each interval of its domain—that is, on $(-\infty, -3)$, $\left(-3, \frac{1}{2}\right)$ and $\left(\frac{1}{2}, \infty\right)$.

✔ **Now Try Exercise 67.**

EXAMPLE 6 Graphing a Rational Function (Does Not Intersect Its Horizontal Asymptote)

Graph $f(x) = \frac{2x+1}{x-3}$.

SOLUTION

Steps 1 and 2 As determined in **Example 4(b),** the equation of the vertical asymptote is $x = 3$. The horizontal asymptote has equation $y = 2$.

Step 3 $f(0) = -\frac{1}{3}$, so the *y*-intercept is $\left(0, -\frac{1}{3}\right)$.

Step 4 Solve $f(x) = 0$ to find any *x*-intercepts.

$$\frac{2x+1}{x-3} = 0 \qquad \text{Set } f(x) = 0.$$

$$2x + 1 = 0 \qquad \text{If a rational expression is equal to 0, then its numerator must equal 0.}$$

$$x = -\frac{1}{2} \qquad \text{The } x\text{-intercept is } \left(-\tfrac{1}{2}, 0\right).$$

Step 5 The graph does not intersect its horizontal asymptote because $f(x) = 2$ has no solution.

$$\frac{2x+1}{x-3} = 2 \qquad \text{Set } f(x) = 2.$$

$$2x + 1 = 2x - 6 \qquad \text{Multiply each side by } x - 3.$$

A false statement results.

$$1 = -6 \qquad \text{Subtract } 2x.$$

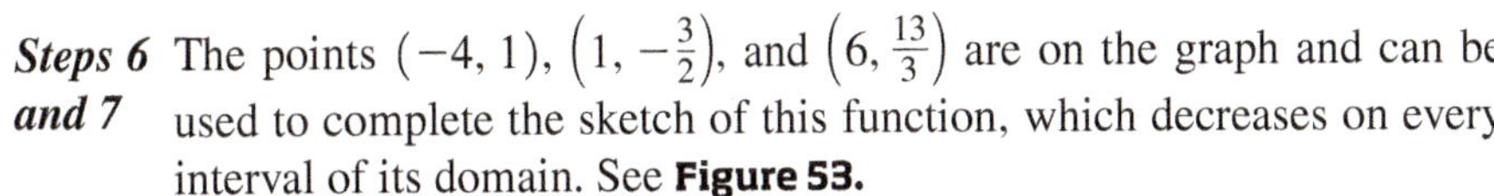

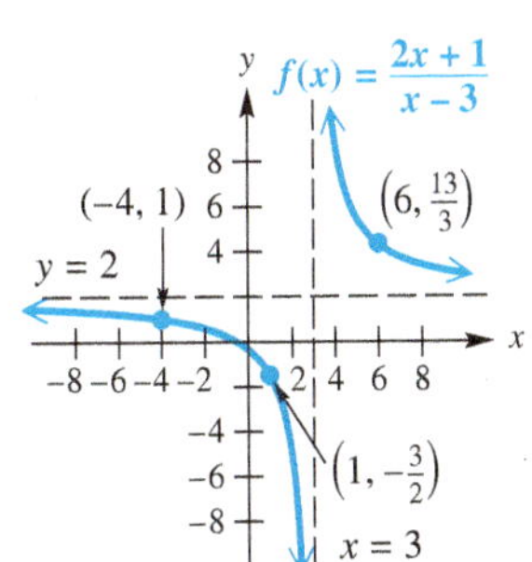

Figure 53

Steps 6 and 7 The points $(-4, 1)$, $\left(1, -\frac{3}{2}\right)$, and $\left(6, \frac{13}{3}\right)$ are on the graph and can be used to complete the sketch of this function, which decreases on every interval of its domain. See **Figure 53.**

✔ **Now Try Exercise 63.**

EXAMPLE 7 Graphing a Rational Function (Intersects Its Horizontal Asymptote)

Graph $f(x) = \dfrac{3x^2 - 3x - 6}{x^2 + 8x + 16}$.

SOLUTION

Step 1 To find the vertical asymptote(s), solve $x^2 + 8x + 16 = 0$.

$$x^2 + 8x + 16 = 0 \qquad \text{Set the denominator equal to 0.}$$

$$(x + 4)^2 = 0 \qquad \text{Factor.}$$

$$x = -4 \qquad \text{Zero-factor property}$$

The numerator is not 0 when $x = -4$.

The vertical asymptote has equation $x = -4$.

Step 2 We divide all terms by x^2 and consider the behavior of each term as $|x|$ increases without bound to get the equation of the horizontal asymptote,

$$y = \frac{3}{1}, \quad \begin{matrix}\longleftarrow \text{Leading coefficient of numerator} \\ \longleftarrow \text{Leading coefficient of denominator}\end{matrix} \quad \text{or} \quad y = 3.$$

Step 3 $f(0) = -\frac{3}{8}$, so the *y*-intercept is $\left(0, -\frac{3}{8}\right)$.

Step 4 Solve $f(x) = 0$ to find any *x*-intercepts.

$$\frac{3x^2 - 3x - 6}{x^2 + 8x + 16} = 0 \qquad \text{Set } f(x) = 0.$$

$$3x^2 - 3x - 6 = 0 \qquad \text{Set the numerator equal to 0.}$$

$$x^2 - x - 2 = 0 \qquad \text{Divide by 3.}$$

$$(x - 2)(x + 1) = 0 \qquad \text{Factor.}$$

$$x = 2 \quad \text{or} \quad x = -1 \qquad \text{Zero-factor property}$$

The *x*-intercepts are $(-1, 0)$ and $(2, 0)$.

Step 5 We set $f(x) = 3$ and solve to locate the point where the graph intersects the horizontal asymptote.

$$\frac{3x^2 - 3x - 6}{x^2 + 8x + 16} = 3 \qquad \text{Set } f(x) = 3.$$

$$3x^2 - 3x - 6 = 3x^2 + 24x + 48 \qquad \text{Multiply each side by } x^2 + 8x + 16.$$

$$-27x = 54 \qquad \text{Subtract } 3x^2 \text{ and } 24x. \text{ Add } 6.$$

$$x = -2 \qquad \text{Divide by } -27.$$

The graph intersects its horizontal asymptote at $(-2, 3)$.

Steps 6 and 7 Some other points that lie on the graph are $(-10, 9)$, $\left(-8, 13\frac{1}{8}\right)$, and $\left(5, \frac{2}{3}\right)$. These are used to complete the graph, as shown in **Figure 54.**

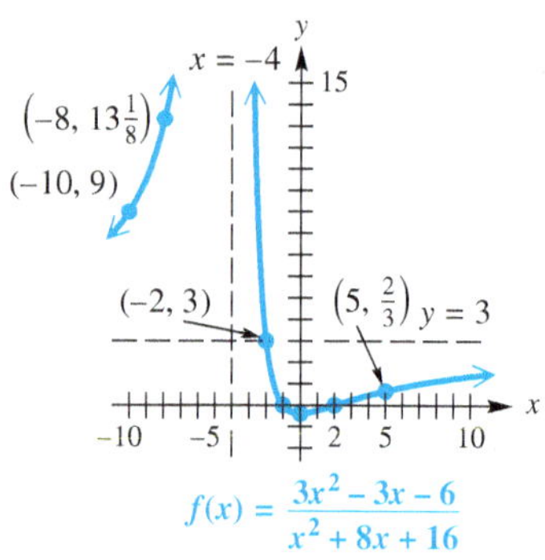

Figure 54

✔ **Now Try Exercise 83.**

Notice the behavior of the graph of the function in **Figure 54** near the line $x = -4$. As $x \to -4$ from either side, $f(x) \to \infty$.

If we examine the behavior of the graph of the function in **Figure 53** (on the previous page) near the line $x = 3$, we find that $f(x) \to -\infty$ as x approaches 3 from the left, while $f(x) \to \infty$ as x approaches 3 from the right. The behavior of the graph of a rational function near a vertical asymptote $x = a$ partially depends on the exponent on $x - a$ in the denominator.

LOOKING AHEAD TO CALCULUS

The rational function

$$f(x) = \frac{2x + 1}{x - 3},$$

seen in **Example 6,** has horizontal asymptote $y = 2$. In calculus, the behavior of the graph of this function as x approaches $-\infty$ and as x approaches ∞ is described using **limits at infinity.** As x approaches $-\infty$, $f(x)$ approaches 2. This is written

$$\lim_{x \to -\infty} f(x) = 2.$$

As x approaches ∞, $f(x)$ approaches 2. This is written

$$\lim_{x \to \infty} f(x) = 2.$$

Behavior of Graphs of Rational Functions near Vertical Asymptotes

Suppose that $f(x)$ is a rational expression in lowest terms. If n is the largest positive integer such that $(x - a)^n$ is a factor of the denominator of $f(x)$, then the graph will behave in the manner illustrated.

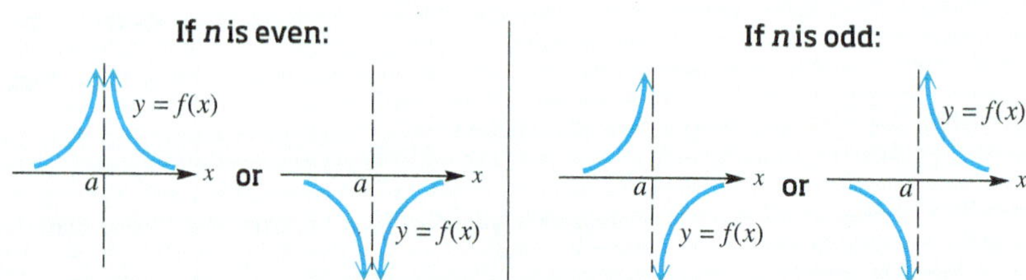

We have observed that the behavior of the graph of a polynomial function near its zeros is dependent on the multiplicity of the zero. The same statement can be made for rational functions.

Suppose that $f(x)$ is defined by a rational expression in lowest terms. If n is the greatest positive integer such that $(x - c)^n$ is a factor of the numerator of $f(x)$, then the graph will behave in the manner illustrated.

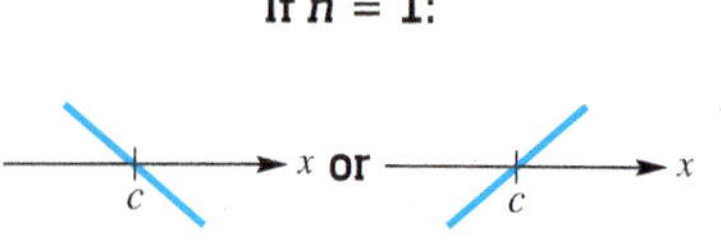

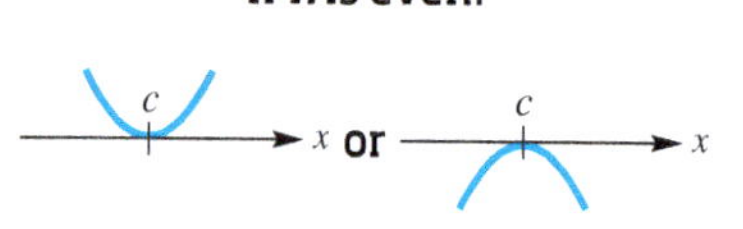

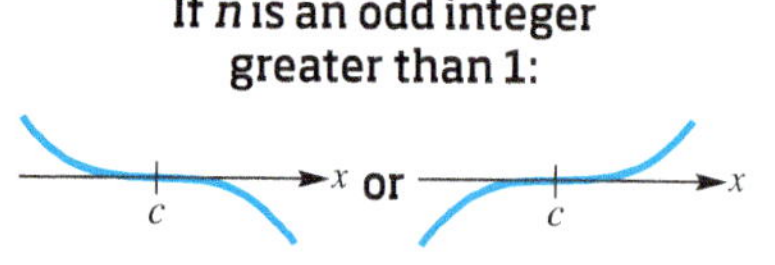

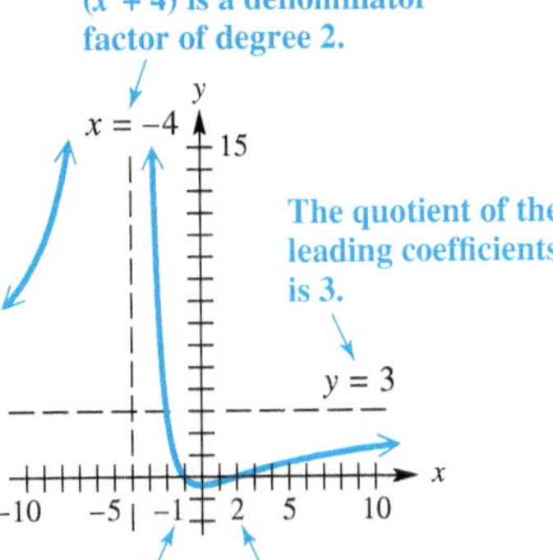

Figure 54 (repeated)

NOTE Suppose that we are asked to reverse the process of **Example 7** and find the equation of a rational function having the graph shown in **Figure 54** (repeated in the margin). Because the graph crosses the x-axis at its x-intercepts $(-1, 0)$ and $(2, 0)$, the numerator must have factors

$$(x + 1) \quad \text{and} \quad (x - 2),$$

each of degree 1.

The behavior of the graph at its vertical asymptote $x = -4$ suggests that there is a factor of $(x + 4)$ of even degree in the denominator. The horizontal asymptote at $y = 3$ indicates that the numerator and denominator have the same degree (both 2) and that the ratio of leading coefficients is 3.

Verify in **Example 7** that the rational function is

$$f(x) = \frac{3(x + 1)(x - 2)}{(x + 4)^2}$$

$$f(x) = \frac{3x^2 - 3x - 6}{x^2 + 8x + 16}. \quad \text{Multiply the factors in the numerator. Square in the denominator.}$$

Exercises of this type are labeled ***Connecting Graphs with Equations.***

EXAMPLE 8 Graphing a Rational Function with an Oblique Asymptote

Graph $f(x) = \dfrac{x^2 + 1}{x - 2}$.

SOLUTION As shown in **Example 4(c),** the vertical asymptote has equation $x = 2$, and the graph has an oblique asymptote with equation $y = x + 2$. The y-intercept is $\left(0, -\frac{1}{2}\right)$, and the graph has no x-intercepts because the numerator, $x^2 + 1$, has no real zeros. The graph does not intersect its oblique asymptote because the following has no solution.

$$\frac{x^2 + 1}{x - 2} = x + 2 \quad \text{Set the expressions defining the function and the oblique asymptote equal.}$$

$$x^2 + 1 = x^2 - 4 \quad \text{Multiply each side by } x - 2.$$

$$1 = -4 \quad \text{False}$$

Using the y-intercept, asymptotes, the points $\left(4, \frac{17}{2}\right)$ and $\left(-1, -\frac{2}{3}\right)$, and the general behavior of the graph near its asymptotes leads to the graph in **Figure 55.**

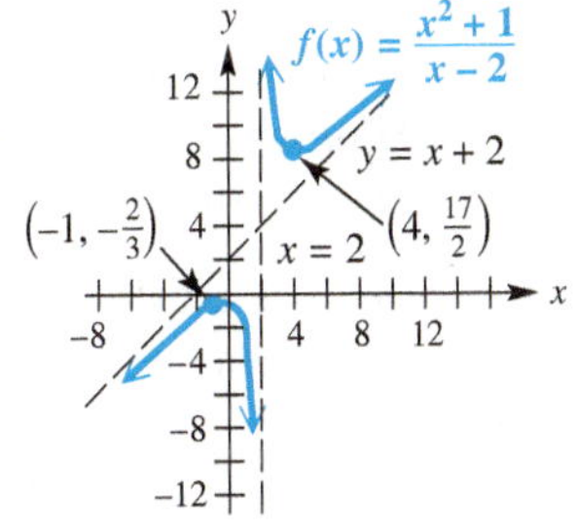

Figure 55

✔ **Now Try Exercise 87.**

A rational function that is not in lowest terms often has a **point of discontinuity** in its graph. Such a point is sometimes called a *hole*.

LOOKING AHEAD TO CALCULUS

Different types of discontinuity are discussed in calculus. The function in **Example 9,**

$$f(x) = \frac{x^2 - 4}{x - 2},$$

is said to have a **removable discontinuity** at $x = 2$, because the discontinuity can be removed by redefining f at 2. The function in **Example 8,**

$$f(x) = \frac{x^2 + 1}{x - 2},$$

has **infinite discontinuity** at $x = 2$, as indicated by the vertical asymptote there. The greatest integer function has **jump discontinuities** because the function values "jump" from one value to another for integer domain values.

EXAMPLE 9 Graphing a Rational Function Defined by an Expression That Is Not in Lowest Terms

Graph $f(x) = \dfrac{x^2 - 4}{x - 2}$.

ALGEBRAIC SOLUTION

The domain of this function cannot include 2. The expression $\frac{x^2 - 4}{x - 2}$ should be written in lowest terms.

$$f(x) = \frac{x^2 - 4}{x - 2} \quad \text{Factor and then divide.}$$

$$f(x) = \frac{(x + 2)(x - 2)}{x - 2} \quad \text{Factor.}$$

$$f(x) = x + 2, \quad x \neq 2$$

Therefore, the graph of this function will be the same as the graph of $y = x + 2$ (a straight line), with the exception of the point with x-value 2. A hole appears in the graph at $(2, 4)$. See **Figure 56.**

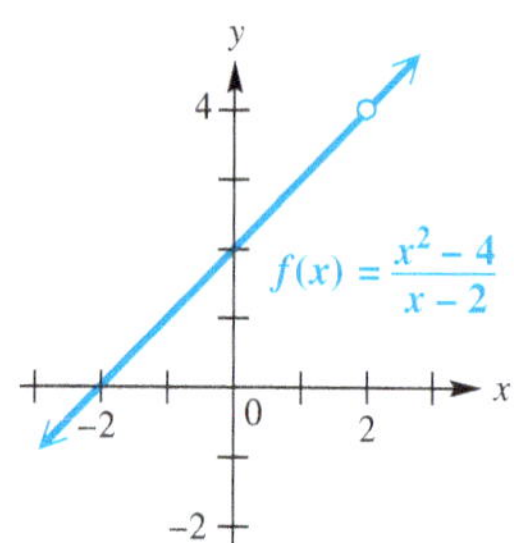

Figure 56

GRAPHING CALCULATOR SOLUTION

If we set the window of a graphing calculator so that an x-value of 2 is displayed, then we can see that the calculator cannot determine a value for y. We define

$$y_1 = \frac{x^2 - 4}{x - 2}$$

and graph it in such a window, as in **Figure 57.** The error message in the table further supports the existence of a discontinuity at $x = 2$. (For the table, $y_2 = x + 2$.)

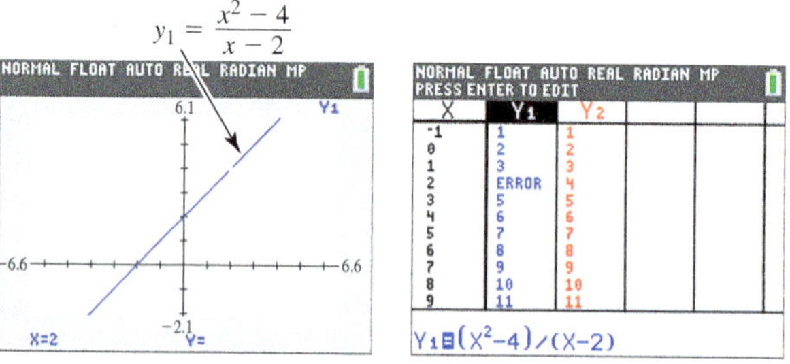

Figure 57

Notice the visible discontinuity at $x = 2$ in the graph for the chosen window. (If the standard viewing window is chosen, the discontinuity is not visible.)

✔ Now Try Exercise 91.

Rational Models

Rational functions have a variety of applications.

EXAMPLE 10 Modeling Traffic Intensity with a Rational Function

Vehicles arrive randomly at a parking ramp at an average rate of 2.6 vehicles per minute. The parking attendant can admit 3.2 vehicles per minute. However, since arrivals are random, lines form at various times. (*Source:* Mannering, F. and W. Kilareski, *Principles of Highway Engineering and Traffic Analysis,* 2nd ed., John Wiley & Sons.)

(a) The **traffic intensity** x is defined as the ratio of the average arrival rate to the average admittance rate. Determine x for this parking ramp.

(b) The average number of vehicles waiting in line to enter the ramp is given by

$$f(x) = \frac{x^2}{2(1 - x)},$$

where $0 \leq x < 1$ is the traffic intensity. Graph $f(x)$ and compute $f(0.8125)$ for this parking ramp.

(c) What happens to the number of vehicles waiting as the traffic intensity approaches 1?

SOLUTION

(a) The average arrival rate is 2.6 vehicles per minute and the average admittance rate is 3.2 vehicles per minute, so

$$x = \frac{2.6}{3.2} = 0.8125.$$

(b) A calculator graph of f is shown in **Figure 58.**

$$f(0.8125) = \frac{0.8125^2}{2(1 - 0.8125)} \approx 1.76 \text{ vehicles}$$

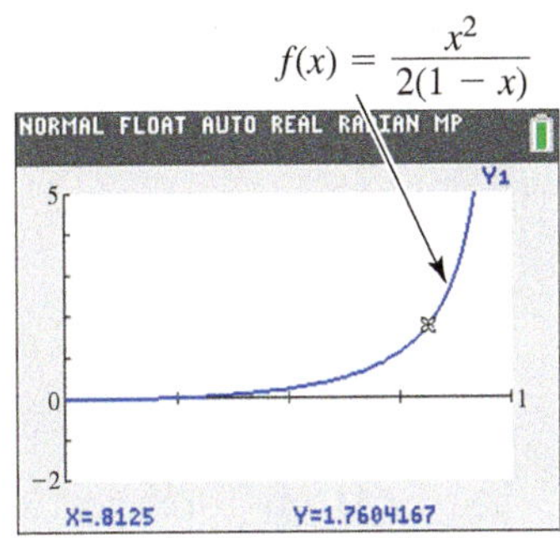

Figure 58

(c) From the graph we see that as x approaches 1, $y = f(x)$ gets very large. Thus, the average number of waiting vehicles gets very large. This is what we would expect.

✔ **Now Try Exercise 113.**

3.5 Exercises

CONCEPT PREVIEW *Provide a short answer to each question.*

1. What is the domain of the function $f(x) = \frac{1}{x}$? What is its range?

2. What is the domain of the function $f(x) = \frac{1}{x^2}$? What is its range?

3. What is the largest open interval of the domain over which the function $f(x) = \frac{1}{x}$ increases? decreases? is constant?

4. What is the largest open interval of the domain over which the function $f(x) = \frac{1}{x^2}$ increases? decreases? is constant?

5. What is the equation of the vertical asymptote of the graph of $y = \frac{1}{x-3} + 2$? Of the horizontal asymptote?

6. What is the equation of the vertical asymptote of the graph of $y = \frac{1}{(x+2)^2} - 4$? Of the horizontal asymptote?

7. Is $f(x) = \frac{1}{x^2}$ an even or an odd function? What symmetry does its graph exhibit?

8. Is $f(x) = \frac{1}{x}$ an even or an odd function? What symmetry does its graph exhibit?

Concept Check Use the graphs of the rational functions in choices A–D to answer each question. There may be more than one correct choice.

A.

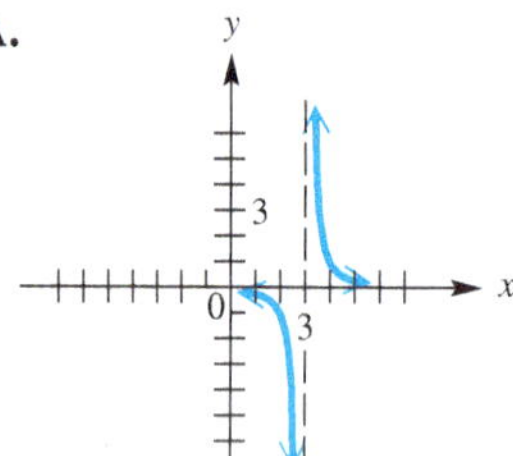

B.

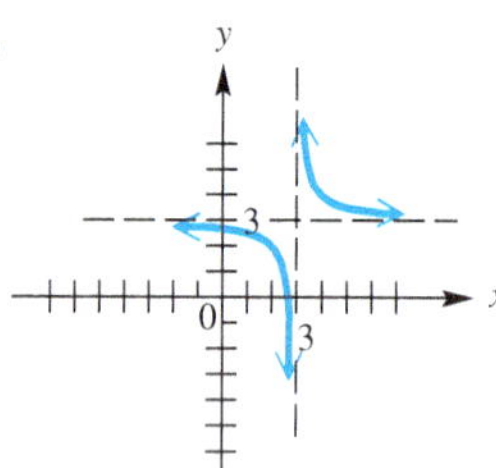

C.

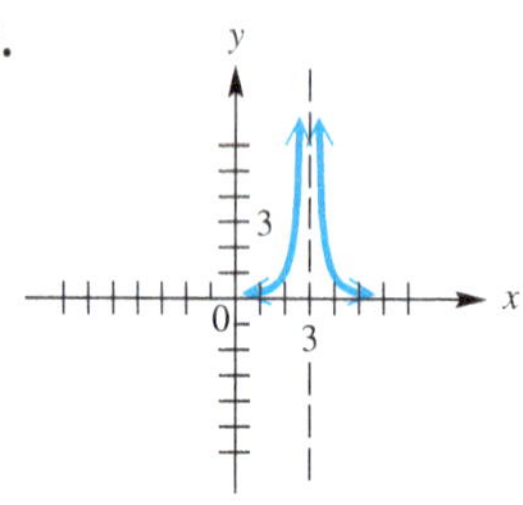

D.

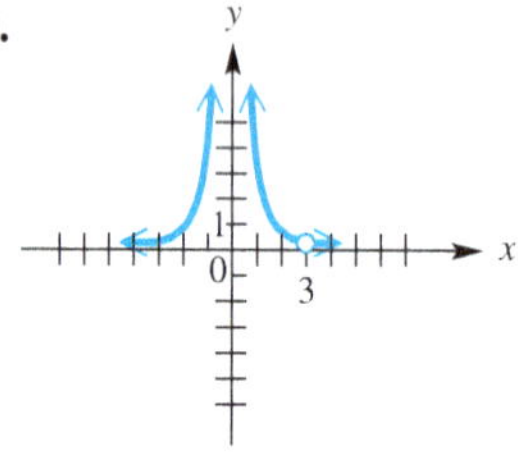

9. Which choices have domain $(-\infty, 3) \cup (3, \infty)$?

10. Which choices have range $(-\infty, 3) \cup (3, \infty)$?

11. Which choices have range $(-\infty, 0) \cup (0, \infty)$?

12. Which choices have range $(0, \infty)$?

13. If f represents the function, only one choice has a single solution to the equation $f(x) = 3$. Which one is it?

14. Which choices have domain $(-\infty, 0) \cup (0, 3) \cup (3, \infty)$?

15. Which choices have the x-axis as a horizontal asymptote?

16. Which choices are symmetric with respect to a vertical line?

*Explain how the graph of each function can be obtained from the graph of $y = \frac{1}{x}$ or $y = \frac{1}{x^2}$. Then graph f and give the **(a)** domain and **(b)** range. Determine the largest open intervals of the domain over which the function is **(c)** increasing or **(d)** decreasing.* ***See Examples 1–3.***

17. $f(x) = \dfrac{2}{x}$

18. $f(x) = -\dfrac{3}{x}$

19. $f(x) = \dfrac{1}{x+2}$

20. $f(x) = \dfrac{1}{x-3}$

21. $f(x) = \dfrac{1}{x} + 1$

22. $f(x) = \dfrac{1}{x} - 2$

23. $f(x) = -\dfrac{2}{x^2}$

24. $f(x) = \dfrac{1}{x^2} + 3$

25. $f(x) = \dfrac{1}{(x-3)^2}$

26. $f(x) = \dfrac{-2}{(x-3)^2}$

27. $f(x) = \dfrac{-1}{(x+2)^2} - 3$

28. $f(x) = \dfrac{-1}{(x-4)^2} + 2$

Concept Check *Match the rational function in Column I with the appropriate description in Column II. Choices in Column II can be used only once.*

I	II
29. $f(x) = \dfrac{x+7}{x+1}$	**A.** The x-intercept is $(-3, 0)$.
30. $f(x) = \dfrac{x+10}{x+2}$	**B.** The y-intercept is $(0, 5)$.
31. $f(x) = \dfrac{1}{x+4}$	**C.** The horizontal asymptote is $y = 4$.
32. $f(x) = \dfrac{-3}{x^2}$	**D.** The vertical asymptote is $x = -1$.
33. $f(x) = \dfrac{x^2-16}{x+4}$	**E.** There is a hole in its graph at $(-4, -8)$.
34. $f(x) = \dfrac{4x+3}{x-7}$	**F.** The graph has an oblique asymptote.
35. $f(x) = \dfrac{x^2+3x+4}{x-5}$	**G.** The x-axis is its horizontal asymptote, and the y-axis is not its vertical asymptote.
36. $f(x) = \dfrac{x+3}{x-6}$	**H.** The x-axis is its horizontal asymptote, and the y-axis is its vertical asymptote.

Give the equations of any vertical, horizontal, or oblique asymptotes for the graph of each rational function. ***See Example 4.***

37. $f(x) = \dfrac{3}{x-5}$ **38.** $f(x) = \dfrac{-6}{x+9}$ **39.** $f(x) = \dfrac{4-3x}{2x+1}$

40. $f(x) = \dfrac{2x+6}{x-4}$ **41.** $f(x) = \dfrac{x^2-1}{x+3}$ **42.** $f(x) = \dfrac{x^2+4}{x-1}$

43. $f(x) = \dfrac{x^2-2x-3}{2x^2-x-10}$ **44.** $f(x) = \dfrac{3x^2-6x-24}{5x^2-26x+5}$

45. $f(x) = \dfrac{x^2+1}{x^2+9}$ **46.** $f(x) = \dfrac{4x^2+25}{x^2+9}$

Concept Check *Work each problem.*

47. Let f be the function whose graph is obtained by translating the graph of $y = \frac{1}{x}$ to the right 3 units and up 2 units.

(a) Write an equation for $f(x)$ as a quotient of two polynomials.

(b) Determine the zero(s) of f.

(c) Identify the asymptotes of the graph of $f(x)$.

48. Repeat **Exercise 47** if f is the function whose graph is obtained by translating the graph of $y = -\frac{1}{x^2}$ to the left 3 units and up 1 unit.

49. After the numerator is divided by the denominator,

$$f(x) = \frac{x^5+x^4+x^2+1}{x^4+1} \quad \text{becomes} \quad f(x) = x + 1 + \frac{x^2-x}{x^4+1}.$$

(a) What is the oblique asymptote of the graph of the function?

(b) Where does the graph of the function intersect its asymptote?

(c) As $x \to \infty$, does the graph of the function approach its asymptote from above or below?

50. Choices A–D below show the four ways in which the graph of a rational function can approach the vertical line $x = 2$ as an asymptote. Identify the graph of each rational function defined in parts (a)–(d).

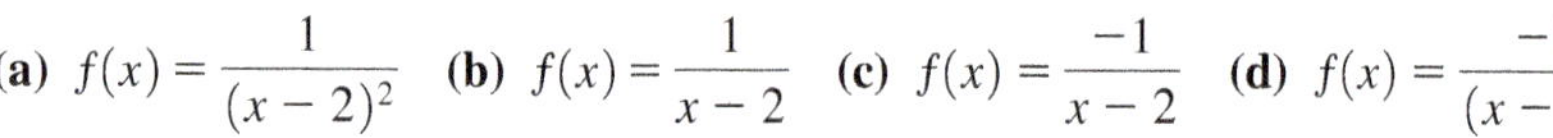

(a) $f(x) = \frac{1}{(x-2)^2}$ **(b)** $f(x) = \frac{1}{x-2}$ **(c)** $f(x) = \frac{-1}{x-2}$ **(d)** $f(x) = \frac{-1}{(x-2)^2}$

A.

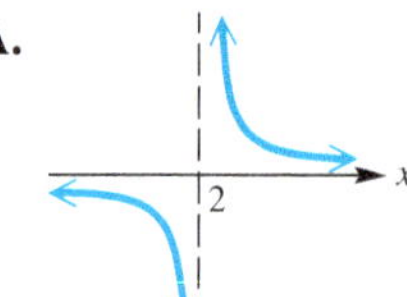

B.

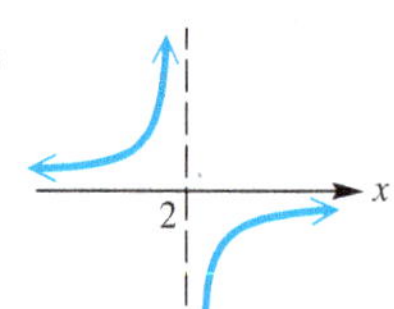

C.

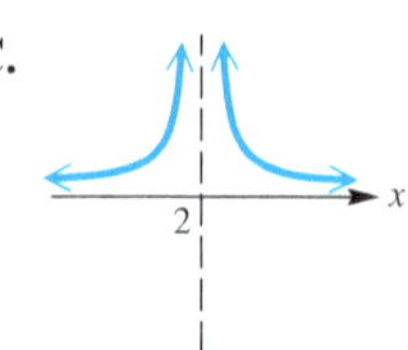

D.

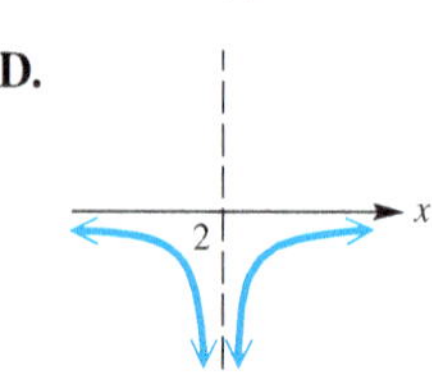

51. Which function has a graph that does not have a vertical asymptote?

A. $f(x) = \frac{1}{x^2 + 2}$ **B.** $f(x) = \frac{1}{x^2 - 2}$ **C.** $f(x) = \frac{3}{x^2}$ **D.** $f(x) = \frac{2x + 1}{x - 8}$

52. Which function has a graph that does not have a horizontal asymptote?

A. $f(x) = \frac{2x - 7}{x + 3}$ **B.** $f(x) = \frac{3x}{x^2 - 9}$

C. $f(x) = \frac{x^2 - 9}{x + 3}$ **D.** $f(x) = \frac{x + 5}{(x + 2)(x - 3)}$

Identify any vertical, horizontal, or oblique asymptotes in the graph of $y = f(x)$. State the domain of f.

53.

y, x, 10, 6, −10, −6, −2, 0, 6, 10, −4, −8

54.

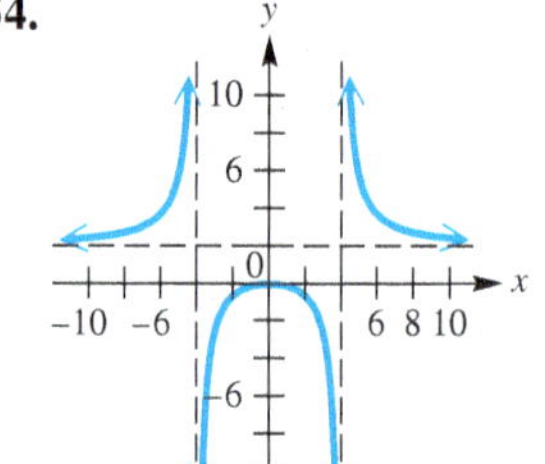

55.

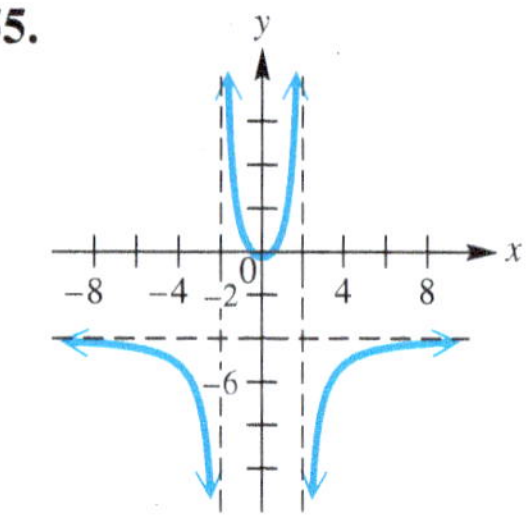

56.

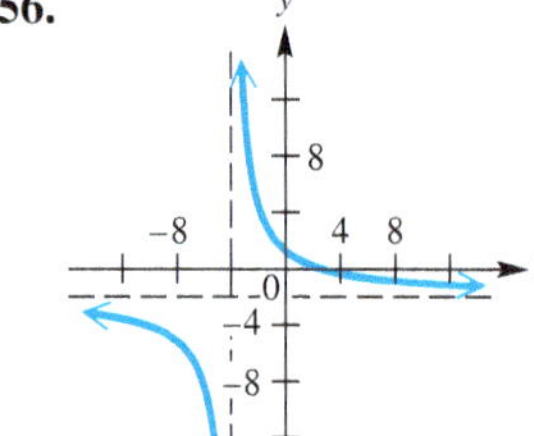

57.

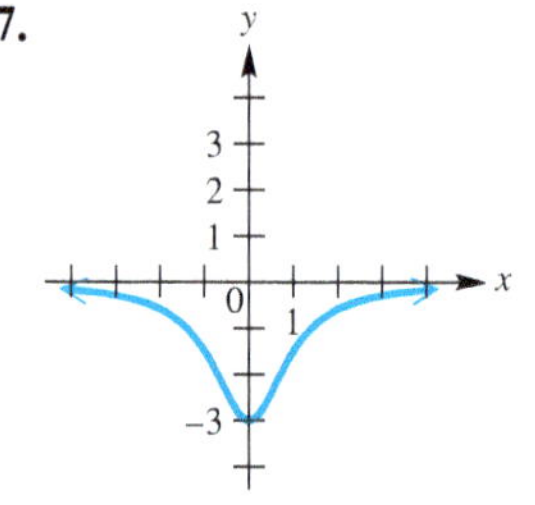

58.

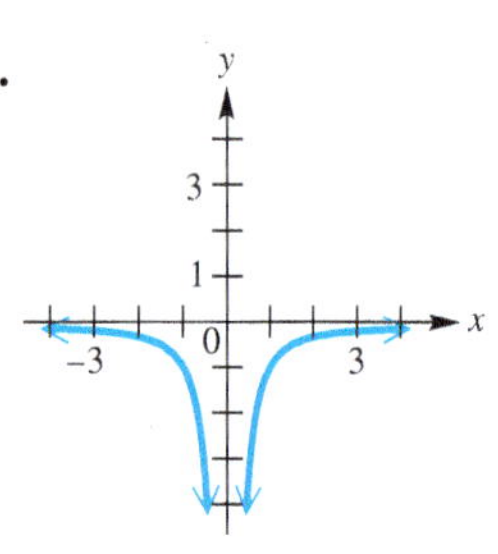

59.

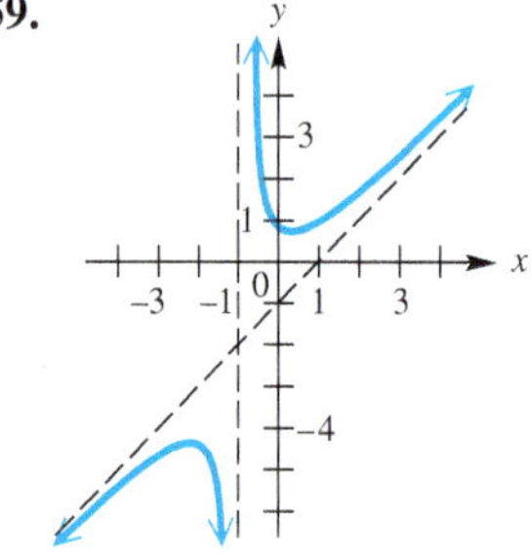

60.

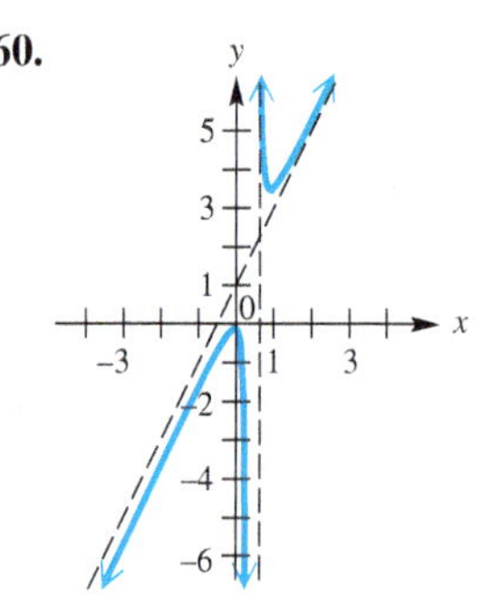

Graph each rational function. ***See Examples 5–9.***

61. $f(x) = \dfrac{x+1}{x-4}$

62. $f(x) = \dfrac{x-5}{x+3}$

63. $f(x) = \dfrac{x+2}{x-3}$

64. $f(x) = \dfrac{x-3}{x+4}$

65. $f(x) = \dfrac{4-2x}{8-x}$

66. $f(x) = \dfrac{6-3x}{4-x}$

67. $f(x) = \dfrac{3x}{x^2-x-2}$

68. $f(x) = \dfrac{2x+1}{x^2+6x+8}$

69. $f(x) = \dfrac{5x}{x^2-1}$

70. $f(x) = \dfrac{x}{4-x^2}$

71. $f(x) = \dfrac{(x+6)(x-2)}{(x+3)(x-4)}$

72. $f(x) = \dfrac{(x+3)(x-5)}{(x+1)(x-4)}$

73. $f(x) = \dfrac{3x^2+3x-6}{x^2-x-12}$

74. $f(x) = \dfrac{4x^2+4x-24}{x^2-3x-10}$

75. $f(x) = \dfrac{9x^2-1}{x^2-4}$

76. $f(x) = \dfrac{16x^2-9}{x^2-9}$

77. $f(x) = \dfrac{(x-3)(x+1)}{(x-1)^2}$

78. $f(x) = \dfrac{x(x-2)}{(x+3)^2}$

79. $f(x) = \dfrac{x}{x^2-9}$

80. $f(x) = \dfrac{-5}{2x+4}$

81. $f(x) = \dfrac{1}{x^2+1}$

82. $f(x) = \dfrac{(x-5)(x-2)}{x^2+9}$

83. $f(x) = \dfrac{(x+4)^2}{(x-1)(x+5)}$

84. $f(x) = \dfrac{(x+1)^2}{(x+2)(x-3)}$

85. $f(x) = \dfrac{20+6x-2x^2}{8+6x-2x^2}$

86. $f(x) = \dfrac{18+6x-4x^2}{4+6x+2x^2}$

87. $f(x) = \dfrac{x^2+1}{x+3}$

88. $f(x) = \dfrac{2x^2+3}{x-4}$

89. $f(x) = \dfrac{x^2+2x}{2x-1}$

90. $f(x) = \dfrac{x^2-x}{x+2}$

91. $f(x) = \dfrac{x^2-9}{x+3}$

92. $f(x) = \dfrac{x^2-16}{x+4}$

93. $f(x) = \dfrac{2x^2-5x-2}{x-2}$

94. $f(x) = \dfrac{x^2-5}{x-3}$

95. $f(x) = \dfrac{x^2-1}{x^2-4x+3}$

96. $f(x) = \dfrac{x^2-4}{x^2+3x+2}$

97. $f(x) = \dfrac{(x^2-9)(2+x)}{(x^2-4)(3+x)}$

98. $f(x) = \dfrac{(x^2-16)(3+x)}{(x^2-9)(4+x)}$

99. $f(x) = \dfrac{x^4-20x^2+64}{x^4-10x^2+9}$

100. $f(x) = \dfrac{x^4-5x^2+4}{x^4-24x^2+108}$

Connecting Graphs with Equations *Find a rational function f having the graph shown. (Hint: See the note preceding **Example 8.**)*

101. **102.** 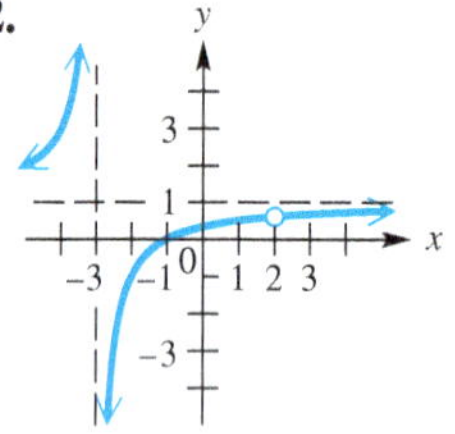 **103.**

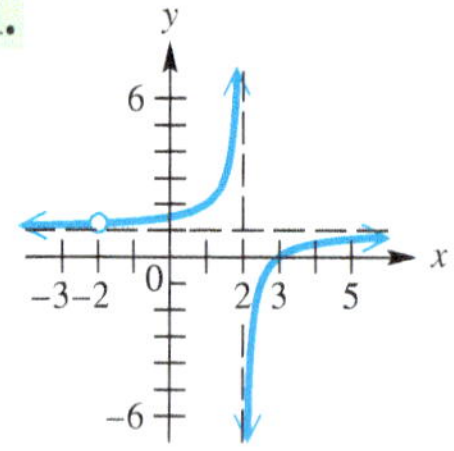

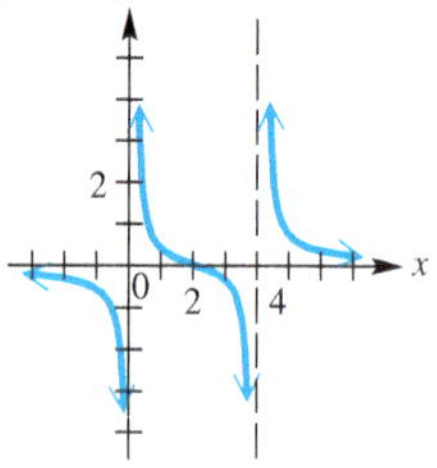

104. 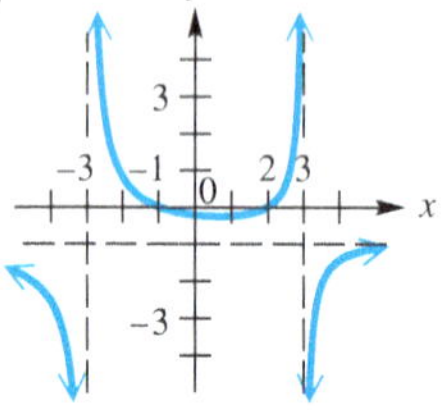 **105.** 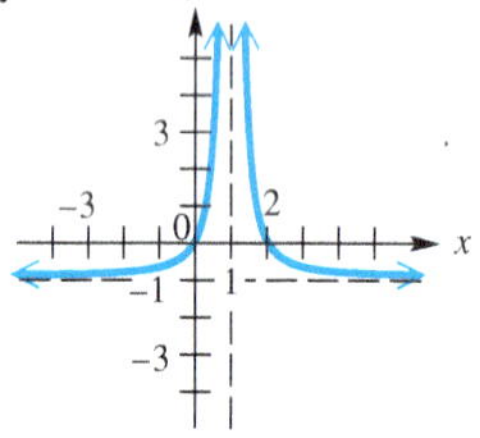 **106.**

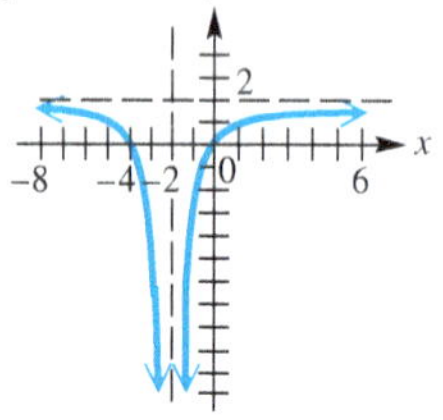

Concept Check *Find a rational function f having a graph with the given features.*

107. x-intercepts: $(-1, 0)$ and $(3, 0)$
y-intercept: $(0, -3)$
vertical asymptote: $x = 1$
horizontal asymptote: $y = 1$

108. x-intercepts: $(1, 0)$ and $(3, 0)$
y-intercept: none
vertical asymptotes: $x = 0$ and $x = 2$
horizontal asymptote: $y = 1$

Use a graphing calculator to graph the rational function in each specified exercise. Then use the graph to find $f(1.25)$.

109. Exercise 61 **110. Exercise 67** **111. Exercise 89** **112. Exercise 91**

(Modeling) Solve each problem. ***See Example 10.***

113. *Traffic Intensity* Let the average number of vehicles arriving at the gate of an amusement park per minute be equal to k, and let the average number of vehicles admitted by the park attendants be equal to r. Then the average waiting time T (in minutes) for each vehicle arriving at the park is given by the rational function

$$T(r) = \frac{2r - k}{2r^2 - 2kr},$$

where $r > k$. (*Source:* Mannering, F., and W. Kilareski, *Principles of Highway Engineering and Traffic Analysis*, 2nd ed., John Wiley & Sons.)

(a) It is known from experience that on Saturday afternoon $k = 25$. Use graphing to estimate the admittance rate r that is necessary to keep the average waiting time T for each vehicle to 30 sec.

(b) If one park attendant can serve 5.3 vehicles per minute, how many park attendants will be needed to keep the average wait to 30 sec?

114. *Waiting in Line* **Queuing theory** (also known as **waiting-line theory**) investigates the problem of providing adequate service economically to customers waiting in line. Suppose customers arrive at a fast-food service window at the rate of 9 people per hour. With reasonable assumptions, the average time (in hours) that a customer will wait in line before being served is modeled by

$$f(x) = \frac{9}{x(x - 9)},$$

where x is the average number of people served per hour. A graph of $f(x)$ for $x > 9$ is shown in the figure on the next page.

(a) Why is the function meaningless if the average number of people served per hour is less than 9?

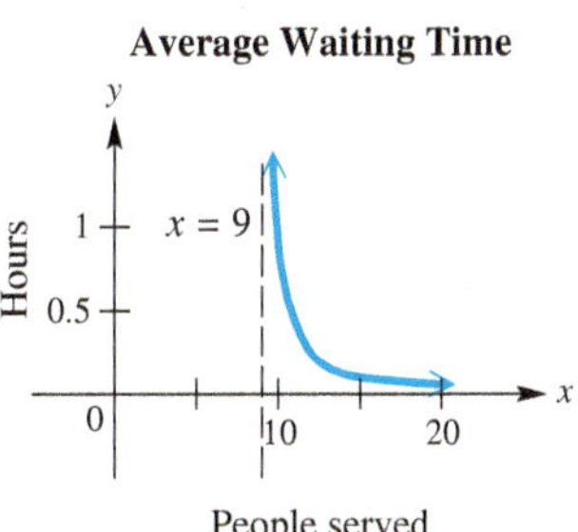

Suppose the average time to serve a customer is 5 min.

(b) How many customers can be served in an hour?

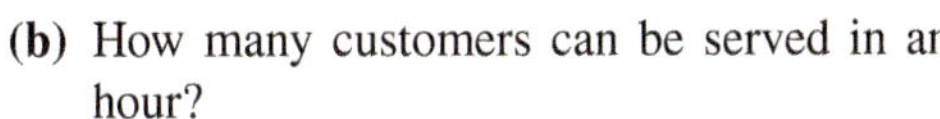

(c) How many minutes will a customer have to wait in line (on the average)?

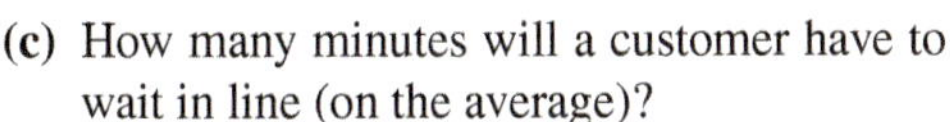

(d) Suppose we want to halve the average waiting time to 7.5 min $\left(\frac{1}{8}\text{ hr}\right)$. How fast must an employee work to serve a customer (on the average)? (*Hint:* Let $f(x) = \frac{1}{8}$ and solve the equation for x. Convert the answer to minutes and round to the nearest hundredth.) How might this reduction in serving time be accomplished?

115. ***Braking Distance*** Braking distance for automobiles traveling at x miles per hour, where $20 \le x \le 70$, can be modeled by the rational function

$$d(x) = \frac{8710x^2 - 69{,}400x + 470{,}000}{1.08x^2 - 324x + 82{,}200}.$$

(*Source:* Mannering, F., and W. Kilareski, *Principles of Highway Engineering and Traffic Analysis*, 2nd ed., John Wiley & Sons.)

(a) Use graphing to estimate x to the nearest unit when $d(x) = 300$.

(b) Complete the table for each value of x.

(c) If a car doubles its speed, does the braking distance double or more than double? Explain.

(d) Suppose that the automobile braking distance doubled whenever the speed doubled. What type of relationship would exist between the braking distance and the speed?

x	$d(x)$	x	$d(x)$
20		50	
25		55	
30		60	
35		65	
40		70	
45			

116. ***Braking Distance*** The **grade** x of a hill is a measure of its steepness. For example, if a road rises 10 ft for every 100 ft of horizontal distance, then it has an uphill grade of

$$x = \frac{10}{100}, \quad \text{or} \quad 10\%.$$

Grades are typically kept quite small—usually less than 10%. The braking distance D for a car traveling at 50 mph on a wet, uphill grade is given by

$$D(x) = \frac{2500}{30(0.3 + x)}.$$

(*Source:* Haefner, L., *Introduction to Transportation Systems*, Holt, Rinehart and Winston.)

(a) Evaluate $D(0.05)$ and interpret the result.

(b) Describe what happens to braking distance as the hill becomes steeper. Does this agree with your driving experience?

(c) Estimate the grade associated with a braking distance of 220 ft.

117. *Tax Revenue* Economist Arthur Laffer has been a center of controversy because of his **Laffer curve**, an idealized version of which is shown here.

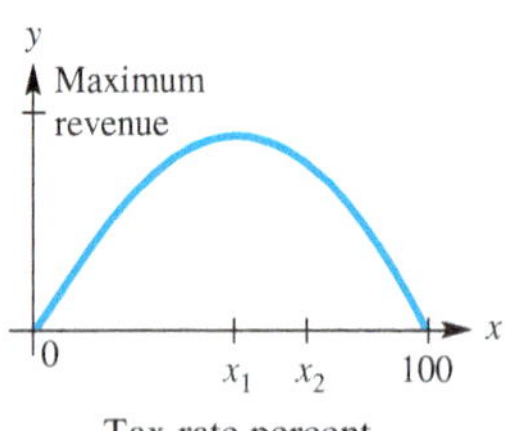

According to this curve, increasing a tax rate, say from x_1 percent to x_2 percent on the graph, can actually lead to a decrease in government revenue. All economists agree on the endpoints, 0 revenue at tax rates of both 0% and 100%, but there is much disagreement on the location of the rate x_1 that produces maximum revenue. Suppose an economist studying the Laffer curve produces the rational function

$$R(x) = \frac{80x - 8000}{x - 110},$$

where $R(x)$ is government revenue in tens of millions of dollars for a tax rate of x percent, with the function valid for $55 \le x \le 100$. Find the revenue for the following tax rates. Round to the nearest tenth if necessary.

(a) 55% **(b)** 60% **(c)** 70% **(d)** 90% **(e)** 100%

118. *Tax Revenue* **See Exercise 117.** Suppose an economist determines that

$$R(x) = \frac{60x - 6000}{x - 120},$$

where $y = R(x)$ is government revenue in tens of millions of dollars for a tax rate of x percent, with $y = R(x)$ valid for $50 \le x \le 100$. Find the revenue for each tax rate. Round to the nearest tenth if necessary.

(a) 50% **(b)** 60% **(c)** 80% **(d)** 100%

Relating Concepts

For individual or collaborative investigation *(Exercises 119–128)*

Consider the following "monster" rational function.

$$f(x) = \frac{x^4 - 3x^3 - 21x^2 + 43x + 60}{x^4 - 6x^3 + x^2 + 24x - 20}$$

Analyzing this function will synthesize many of the concepts of this and earlier sections. ***Work Exercises 119–128 in order.***

119. Find the equation of the horizontal asymptote.

120. Given that -4 and -1 are zeros of the numerator, factor the numerator completely.

121. **(a)** Given that 1 and 2 are zeros of the denominator, factor the denominator completely.

(b) Write the entire quotient for f so that the numerator and the denominator are in factored form.

122. **(a)** What is the common factor in the numerator and the denominator?

(b) For what value of x will there be a point of discontinuity (i.e., a hole)?

123. What are the x-intercepts of the graph of f?

124. What is the y-intercept of the graph of f?

125. Find the equations of the vertical asymptotes.

126. Determine the point or points of intersection of the graph of f with its horizontal asymptote.

127. Sketch the graph of f.

128. Use the graph of f to solve each inequality.

(a) $f(x) < 0$ (b) $f(x) > 0$

Chapter 3 Quiz (Sections 3.1-3.5)

1. Graph each quadratic function. Give the vertex, axis, domain, range, and largest open intervals of the domain over which the function is increasing or decreasing.

(a) $f(x) = -2(x+3)^2 - 1$ (b) $f(x) = 2x^2 - 8x + 3$

2. *(Modeling) Height of a Projected Object* A ball is projected directly upward from an initial height of 200 ft with an initial velocity of 64 ft per sec.

(a) Use the function $s(t) = -16t^2 + v_0 t + s_0$ to describe the height of the ball in terms of time t.

(b) For what interval of time is the height of the ball greater than 240 ft? Round to the nearest hundredth.

Use synthetic division to decide whether the given number k is a zero of the polynomial function. If it is not, give the value of f(k).

3. $f(x) = 2x^4 + x^3 - 3x + 4;\ k = 2$

4. $f(x) = x^2 - 4x + 5;\ k = 2 + i$

5. Find a polynomial function f of least degree having only real coefficients with zeros -2, 3, and $3 - i$.

Graph each polynomial function. Factor first if the polynomial is not in factored form.

6. $f(x) = x(x-2)^3(x+2)^2$

7. $f(x) = 2x^4 - 9x^3 - 5x^2 + 57x - 45$

8. $f(x) = -4x^5 + 16x^4 + 13x^3 - 76x^2 - 3x + 18$

Graph each rational function.

9. $f(x) = \dfrac{3x+1}{x^2 + 7x + 10}$

10. $f(x) = \dfrac{x^2 + 2x + 1}{x - 1}$

Summary Exercises on Solving Equations and Inequalities

A rational inequality can be solved by rewriting it so that 0 is on one side. Then we determine the values that cause either the numerator or the denominator to *equal* 0, and by using a test value from each interval determined by these values, we can find the solution set.

We now solve rational inequalities by inspecting the graph of a related function. The graphs can be obtained using technology or the steps for graphing a rational function.

EXAMPLE **Solving a Rational Inequality**

Solve the inequality.

$$\frac{1-x}{x+4} \geq 0$$

SOLUTION Graph the related rational function

$$y = \frac{1-x}{x+4}.$$

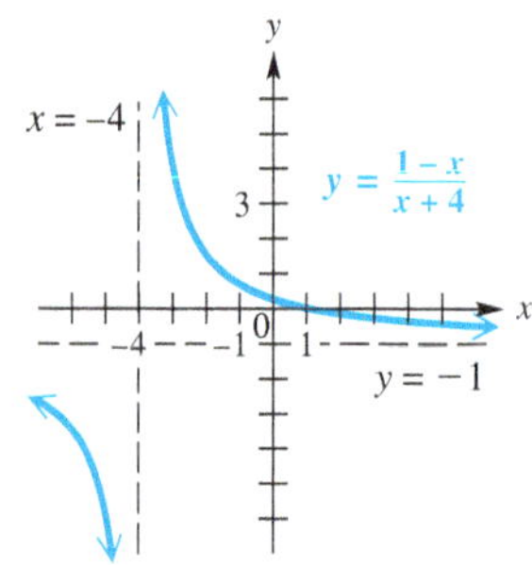

The real solutions of $\frac{1-x}{x+4} \geq 0$ are the x-values for which the graph lies above or on the x-axis. This is true for all x to the right of the vertical asymptote at $x = -4$, up to and including the x-intercept at $(1, 0)$. Therefore, the solution set of the inequality is $(-4, 1]$.

By inspecting the graph of the related function, we can also determine that the solution set of $\frac{1-x}{x+4} < 0$ is $(-\infty, -4) \cup (1, \infty)$ and that the solution set of the equation $\frac{1-x}{x+4} = 0$ is $\{1\}$, the x-value of the x-intercept. (This graphical method may be used to solve other equations and inequalities including those defined by polynomials.)

✔ **Now Try Exercise 19.**

EXERCISES

Concept Check *Use the graph of the function to solve each equation or inequality.*

1. **(a)** $f(x) > 0$ **(b)** $f(x) \leq 0$

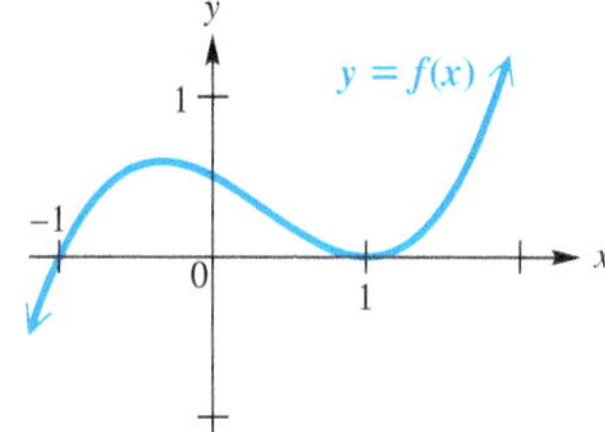

2. **(a)** $f(x) < 0$ **(b)** $f(x) > 0$

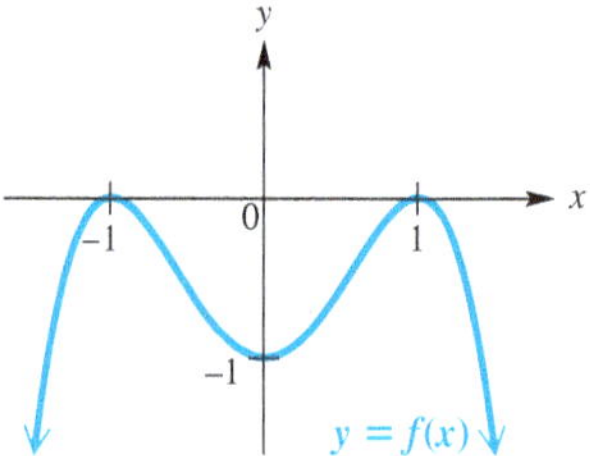

3. **(a)** $f(x) = 0$ **(b)** $f(x) > 0$

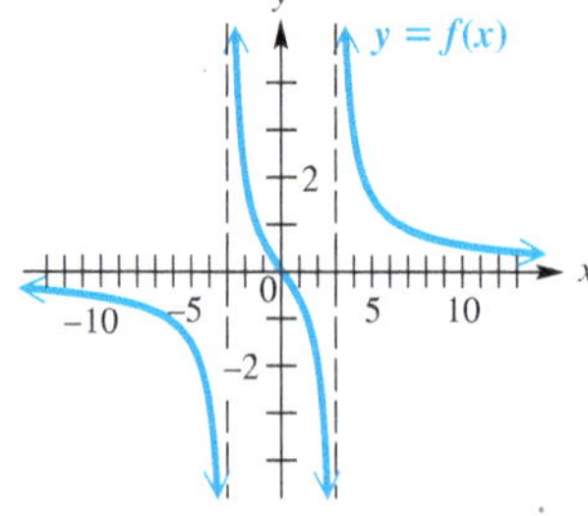

4. **(a)** $f(x) = 0$ **(b)** $f(x) \leq 0$

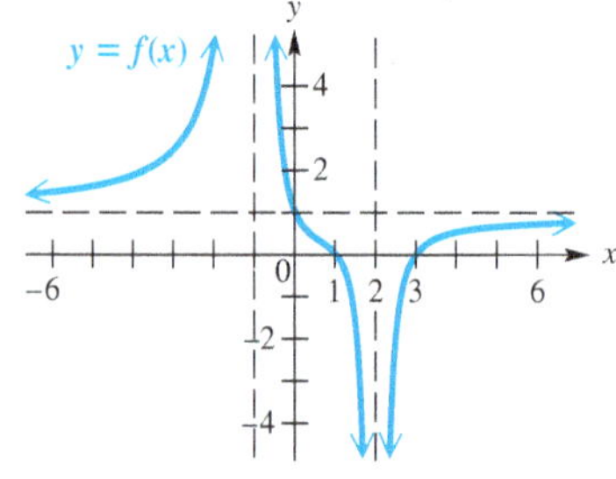

Solve each equation.

5. $\frac{5x+8}{-2} = 2x - 10$

6. $\frac{1}{5}x + 0.25x = \frac{1}{2}x - 1$

7. $(x-5)^{-4} - 13(x-5)^{-2} = -36$

8. $x = 13\sqrt{x} - 40$

9. $\sqrt{2x-5} - \sqrt{x-3} = 1$

10. $3 = \sqrt{x+2} + \sqrt{x-1}$

11. $x^{2/3} + \frac{1}{2} = \frac{3}{4}$

12. $27 - (x-4)^{3/2} = 0$

Sketch the graph of an appropriate function and then use the graph to solve each equation or inequality. (Note: First determine whether the expression is an equation or an inequality. When appropriate, use the steps for graphing polynomial functions or rational functions.)

13. $25x^2 - 20x + 4 > 0$

14. $3x + 4 \le x^2$

15. $x^4 - 2x^3 - 3x^2 + 4x + 4 = 0$

16. $x^3 + 5x^2 + 3x - 9 \ge 0$

17. $-x^4 - x^3 + 12x^2 = 0$

18. $-4x^4 + 13x^2 - 3 > 0$

19. $\dfrac{2x^2 - 13x + 15}{x^2 - 3x} \ge 0$

20. $\dfrac{x^2 + 3x - 1}{x + 1} > 3$

21. $\dfrac{x - 1}{(x - 3)^2} = 0$

22. $\dfrac{x}{x^2 - 4} > 0$

3.6 Variation

- Direct Variation
- Inverse Variation
- Combined and Joint Variation

Direct Variation To apply mathematics we often need to express relationships between quantities. For example,

- In chemistry, the ideal gas law describes how temperature, pressure, and volume are related.
- In physics, various formulas in optics describe the relationship between the focal length of a lens and the size of an image.

When one quantity is a constant multiple of another quantity, the two quantities are said to *vary directly.* For example, if you work for an hourly wage of \$10, then

$$[\text{pay}] = 10 \cdot [\text{hours worked}].$$

Doubling the hours doubles the pay. Tripling the hours triples the pay, and so on. This is stated more precisely as follows.

Direct Variation

y **varies directly** as x, or y is **directly proportional** to x, if there exists a nonzero real number k, called the **constant of variation,** such that for all x,

$$y = kx.$$

The direct variation equation $y = kx$ defines a linear function, where the constant of variation k is the slope of the line. For $k > 0$,

- As the value of x *increases*, the value of y *increases*.
- As the value of x *decreases*, the value of y *decreases*.

When used to describe a direct variation relationship, the phrase "directly proportional" is sometimes abbreviated to just "proportional."

The steps involved in solving a variation problem are summarized on the next page.

Solving a Variation Problem

Step 1 Write the general relationship among the variables as an equation. Use the constant k.

Step 2 Substitute given values of the variables and find the value of k.

Step 3 Substitute this value of k into the equation from Step 1, obtaining a specific formula.

Step 4 Substitute the remaining values and solve for the required unknown.

EXAMPLE 1 Solving a Direct Variation Problem

The area of a rectangle varies directly as its length. If the area is 50 m^2 when the length is 10 m, find the area when the length is 25 m. (See **Figure 59.**)

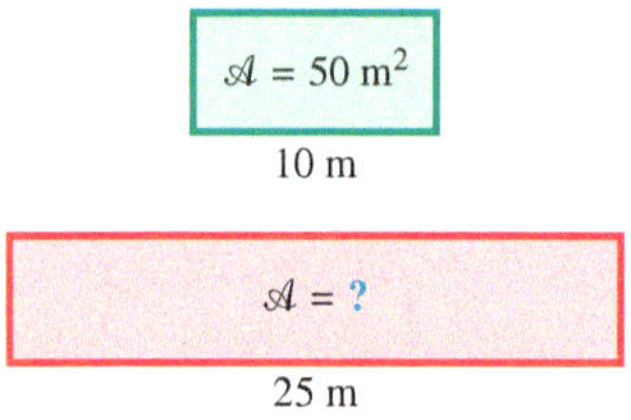

$\mathcal{A} = 50\text{ m}^2$

10 m

$\mathcal{A} = ?$

25 m

Figure 59

SOLUTION

Step 1 The area varies directly as the length, so

$$\mathcal{A} = kL,$$

where $\mathcal{A}$ represents the area of the rectangle, L is the length, and k is a nonzero constant.

Step 2 Because $\mathcal{A} = 50$ when $L = 10$, we can solve the equation $\mathcal{A} = kL$ for k.

$$50 = 10k \quad \text{Substitute for } \mathcal{A} \text{ and L.}$$

$$k = 5 \quad \text{Divide by 10. Interchange sides.}$$

Step 3 Using this value of k, we can express the relationship between the area and the length as follows.

$$\mathcal{A} = 5L \quad \text{Direct variation equation}$$

Step 4 To find the area when the length is 25, we replace L with 25.

$$\mathcal{A} = 5L$$

$$\mathcal{A} = 5(25) \quad \text{Substitute for } L.$$

$$\mathcal{A} = 125 \quad \text{Multiply.}$$

The area of the rectangle is 125 m^2 when the length is 25 m.

✔ **Now Try Exercise 27.**

Sometimes y varies as a power of x. If n is a positive integer greater than or equal to 2, then y is a greater-power polynomial function of x.

Direct Variation as nth Power

Let n be a positive real number. Then y **varies directly as the nth power** of x, or y is **directly proportional to the nth power** of x, if for all x there exists a nonzero real number k such that

$$y = kx^n.$$

For example, the area of a square of side x is given by the formula $\mathcal{A} = x^2$, so the area varies directly as the square of the length of a side. Here $k = 1$.

Inverse Variation Another type of variation is *inverse variation.* With inverse variation, where $k > 0$, as the value of one variable increases, the value of the other decreases. This relationship can be expressed as a rational function.

Inverse Variation as *n*th Power

Let n be a positive real number. Then y **varies inversely as the *n*th power** of x, or y is **inversely proportional to the *n*th power** of x, if for all x there exists a nonzero real number k such that

$$y = \frac{k}{x^n}.$$

If $n = 1$, then $y = \frac{k}{x}$, and y **varies inversely** as x.

EXAMPLE 2 Solving an Inverse Variation Problem

In a certain manufacturing process, the cost of producing a single item varies inversely as the square of the number of items produced. If 100 items are produced, each costs \$2. Find the cost per item if 400 items are produced.

SOLUTION

Step 1 Let x represent the number of items produced and y represent the cost per item. Then, for some nonzero constant k, the following holds.

$$y = \frac{k}{x^2} \qquad y \text{ varies inversely as the square of } x.$$

Step 2

$$2 = \frac{k}{100^2} \qquad \text{Substitute; } y = 2 \text{ when } x = 100.$$

$$k = 20{,}000 \qquad \text{Solve for } k.$$

Step 3 The relationship between x and y is $y = \frac{20{,}000}{x^2}$.

Step 4 When 400 items are produced, the cost per item is found as follows.

$$y = \frac{20{,}000}{x^2} = \frac{20{,}000}{400^2} = 0.125$$

The cost per item is \$0.125, or 12.5 cents.

✔ **Now Try Exercise 37.**

Combined and Joint Variation In **combined variation,** one variable depends on more than one other variable. Specifically, when a variable depends on the *product* of two or more other variables, it is referred to as *joint variation.*

Joint Variation

Let m and n be real numbers. Then y **varies jointly** as the nth power of x and the mth power of z if for all x and z, there exists a nonzero real number k such that

$$y = kx^nz^m.$$

CAUTION Note that *and* in the expression "*y* varies jointly as *x* and *z*" translates as the product $y = kxz$. The word "and" does not indicate addition here.

EXAMPLE 3 Solving a Joint Variation Problem

The area of a triangle varies jointly as the lengths of the base and the height. A triangle with base 10 ft and height 4 ft has area 20 ft^2. Find the area of a triangle with base 3 ft and height 8 ft. (See **Figure 60.**)

4 ft

10 ft

Area = 20 ft^2

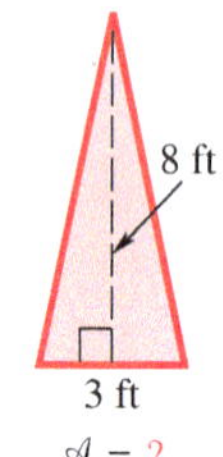

Figure 60

SOLUTION

Step 1 Let $\mathcal{A}$ represent the area, b the base, and h the height of the triangle. Then, for some number k,

$$\mathcal{A} = kbh. \quad \text{$\mathcal{A}$ varies jointly as b and h.}$$

Step 2 $\mathcal{A}$ is 20 when b is 10 and h is 4, so substitute and solve for k.

$$20 = k(10)(4) \quad \text{Substitute for $\mathcal{A}$, b, and h.}$$

$$\frac{1}{2} = k \quad \text{Solve for k.}$$

Step 3 The relationship among the variables is the familiar formula for the area of a triangle,

$$\mathcal{A} = \frac{1}{2}bh.$$

Step 4 To find $\mathcal{A}$ when $b = 3$ ft and $h = 8$ ft, substitute into the formula.

$$\mathcal{A} = \frac{1}{2}(3)(8) = 12 \text{ ft}^2$$

✔ **Now Try Exercise 39.**

EXAMPLE 4 Solving a Combined Variation Problem

The number of vibrations per second (the pitch) of a steel guitar string varies directly as the square root of the tension and inversely as the length of the string. If the number of vibrations per second is 50 when the tension is 225 newtons and the length is 0.60 m, find the number of vibrations per second when the tension is 196 newtons and the length is 0.65 m.

SOLUTION

Step 1 Let n represent the number of vibrations per second, T represent the tension, and L represent the length of the string. Then, from the information in the problem, write the variation equation.

$$n = \frac{k\sqrt{T}}{L} \quad \text{n varies directly as the square root of T and inversely as L.}$$

Step 2 Substitute the given values for n, T, and L and solve for k.

$$50 = \frac{k\sqrt{225}}{0.60} \quad \text{Let $n = 50$, $T = 225$, $L = 0.60$.}$$

$$30 = k\sqrt{225} \quad \text{Multiply by 0.60.}$$

$$30 = 15k \quad \sqrt{225} = 15$$

$$k = 2 \quad \text{Divide by 15. Interchange sides.}$$

Step 3 Substitute for k to find the relationship among the variables.

$$n = \frac{2\sqrt{T}}{L}$$

Step 4 Now use the second set of values for T and L to find n.

$$n = \frac{2\sqrt{196}}{0.65} \approx 43 \quad \text{Let } T = 196, L = 0.65.$$

The number of vibrations per second is approximately 43.

✔ Now Try Exercise 43.

3.6 Exercises

CONCEPT PREVIEW *Fill in the blank(s) to correctly complete each sentence, or answer the question as appropriate.*

1. For $k > 0$, if y varies directly as x, then when x increases, y ________, and when x decreases, y ________.
2. For $k > 0$, if y varies inversely as x, then when x increases, y ________, and when x decreases, y ________.
3. In the equation $y = 6x$, y varies directly as x. When $x = 5$, $y = 30$. What is the value of y when $x = 10$?
4. In the equation $y = \frac{12}{x}$, y varies inversely as x. When $x = 3$, $y = 4$. What is the value of y when $x = 6$?
5. Consider the two ordered pairs (x, y) from **Exercise 3.** Divide the y-value by the x-value. What is the result in each case?
6. Consider the two ordered pairs (x, y) from **Exercise 4.** Multiply the y-value by the x-value. What is the result in each case?

Solve each problem. ***See Examples 1–4.***

7. If y varies directly as x, and $y = 20$ when $x = 4$, find y when $x = -6$.
8. If y varies directly as x, and $y = 9$ when $x = 30$, find y when $x = 40$.
9. If m varies jointly as x and y, and $m = 10$ when $x = 2$ and $y = 14$, find m when $x = 21$ and $y = 8$.
10. If m varies jointly as z and p, and $m = 10$ when $z = 2$ and $p = 7.5$, find m when $z = 6$ and $p = 9$.
11. If y varies inversely as x, and $y = 10$ when $x = 3$, find y when $x = 20$.
12. If y varies inversely as x, and $y = 20$ when $x = \frac{1}{4}$, find y when $x = 15$.
13. Suppose r varies directly as the square of m, and inversely as s. If $r = 12$ when $m = 6$ and $s = 4$, find r when $m = 6$ and $s = 20$.
14. Suppose p varies directly as the square of z, and inversely as r. If $p = \frac{32}{5}$ when $z = 4$ and $r = 10$, find p when $z = 3$ and $r = 32$.
15. Let a be directly proportional to m and n^2, and inversely proportional to y^3. If $a = 9$ when $m = 4$, $n = 9$, and $y = 3$, find a when $m = 6$, $n = 2$, and $y = 5$.
16. Let y vary directly as x, and inversely as m^2 and r^2. If $y = \frac{5}{3}$ when $x = 1$, $m = 2$, and $r = 3$, find y when $x = 3$, $m = 1$, and $r = 8$.

Concept Check *Match each statement with its corresponding graph in choices A–D. In each case,* $k > 0$.

17. y varies directly as x. $(y = kx)$

18. y varies inversely as x. $\left(y = \frac{k}{x}\right)$

19. y varies directly as the second power of x. $(y = kx^2)$

20. x varies directly as the second power of y. $(x = ky^2)$

A. 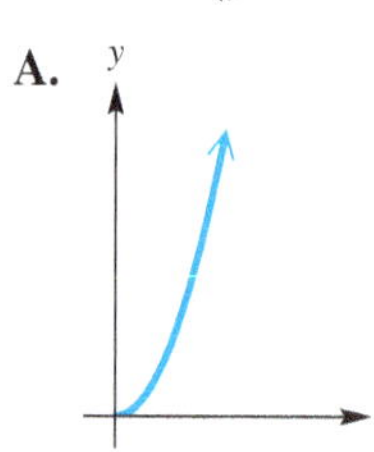

B. 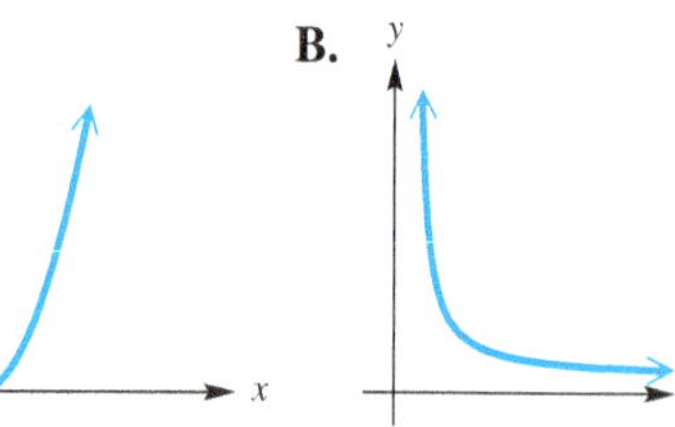

C. 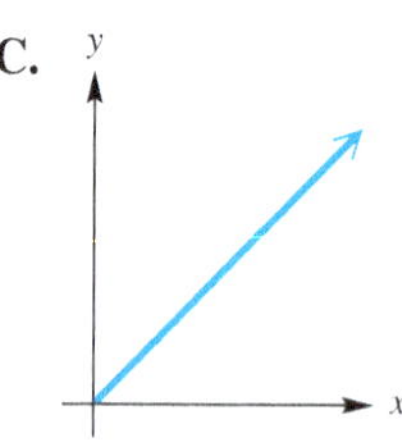

D. 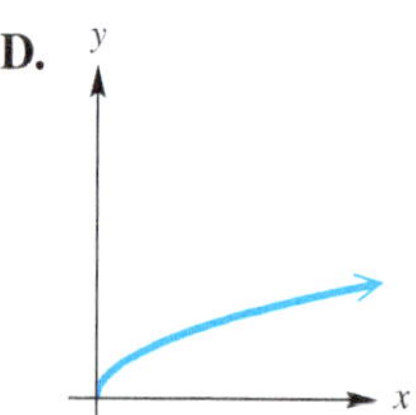

Concept Check *Write each formula as an English phrase using the word* varies *or* proportional.

21. $C = 2\pi r$, where C is the circumference of a circle of radius r

22. $d = \frac{1}{5}s$, where d is the approximate distance (in miles) from a storm, and s is the number of seconds between seeing lightning and hearing thunder

23. $r = \frac{d}{t}$, where r is the speed when traveling d miles in t hours

24. $d = \dfrac{1}{4\pi n r^2}$, where d is the distance a gas atom of radius r travels between collisions, and n is the number of atoms per unit volume

25. $s = kx^3$, where s is the strength of a muscle that has length x

26. $f = \dfrac{mv^2}{r}$, where f is the centripetal force of an object of mass m moving along a circle of radius r at velocity v

Solve each problem. ***See Examples 1–4.***

27. ***Circumference of a Circle*** The circumference of a circle varies directly as the radius. A circle with radius 7 in. has circumference 43.96 in. Find the circumference of the circle if the radius changes to 11 in.

28. ***Pressure Exerted by a Liquid*** The pressure exerted by a certain liquid at a given point varies directly as the depth of the point beneath the surface of the liquid. The pressure at 10 ft is 50 pounds per square inch (psi). What is the pressure at 15 ft?

29. ***Resistance of a Wire*** The resistance in ohms of a platinum wire temperature sensor varies directly as the temperature in kelvins (K). If the resistance is 646 ohms at a temperature of 190 K, find the resistance at a temperature of 250 K.

30. ***Weight on the Moon*** The weight of an object on Earth is directly proportional to the weight of that same object on the moon. A 200-lb astronaut would weigh 32 lb on the moon. How much would a 50-lb dog weigh on the moon?

31. ***Distance to the Horizon*** The distance that a person can see to the horizon on a clear day from a point above the surface of Earth varies directly as the square root of the height at that point. If a person 144 m above the surface of Earth can see 18 km to the horizon, how far can a person see to the horizon from a point 64 m above the surface?

32. ***Water Emptied by a Pipe*** The amount of water emptied by a pipe varies directly as the square of the diameter of the pipe. For a certain constant water flow, a pipe emptying into a canal will allow 200 gal of water to escape in an hour. The diameter of the pipe is 6 in. How much water would a 12-in. pipe empty into the canal in an hour, assuming the same water flow?

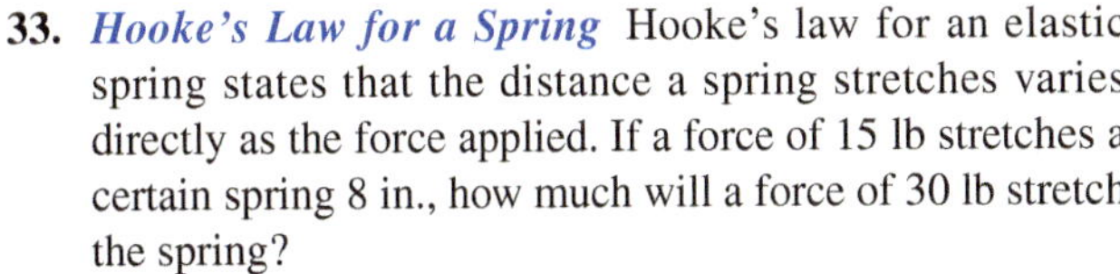

33. ***Hooke's Law for a Spring*** Hooke's law for an elastic spring states that the distance a spring stretches varies directly as the force applied. If a force of 15 lb stretches a certain spring 8 in., how much will a force of 30 lb stretch the spring?

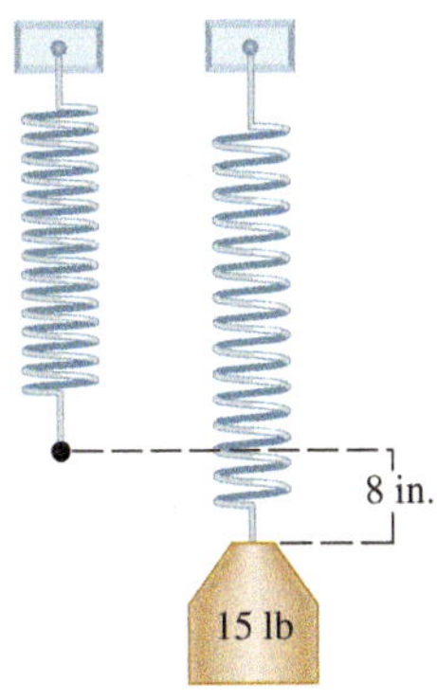

34. ***Current in a Circuit*** The current in a simple electrical circuit varies inversely as the resistance. If the current is 50 amps when the resistance is 10 ohms, find the current if the resistance is 5 ohms.

35. ***Speed of a Pulley*** The speed of a pulley varies inversely as its diameter. One kind of pulley, with diameter 3 in., turns at 150 revolutions per minute. Find the speed of a similar pulley with diameter 5 in.

36. ***Weight of an Object*** The weight of an object varies inversely as the square of its distance from the center of Earth. If an object 8000 mi from the center of Earth weighs 90 lb, find its weight when it is 12,000 mi from the center of Earth.

37. ***Current Flow*** In electric current flow, it is found that the resistance offered by a fixed length of wire of a given material varies inversely as the square of the diameter of the wire. If a wire 0.01 in. in diameter has a resistance of 0.4 ohm, what is the resistance of a wire of the same length and material with diameter 0.03 in., to the nearest ten-thousandth of an ohm?

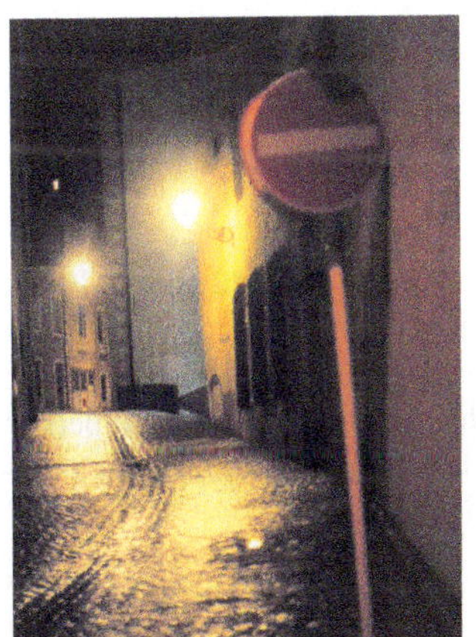

38. ***Illumination*** The illumination produced by a light source varies inversely as the square of the distance from the source. The illumination of a light source at 5 m is 70 candelas. What is the illumination 12 m from the source?

39. ***Simple Interest*** Simple interest varies jointly as principal and time. If $1000 invested for 2 yr earned $70, find the amount of interest earned by $5000 for 5 yr.

40. ***Volume of a Gas*** The volume of a gas varies inversely as the pressure and directly as the temperature in kelvins (K). If a certain gas occupies a volume of 1.3 L at 300 K and a pressure of 18 newtons, find the volume at 340 K and a pressure of 24 newtons.

41. ***Force of Wind*** The force of the wind blowing on a vertical surface varies jointly as the area of the surface and the square of the velocity. If a wind of 40 mph exerts a force of 50 lb on a surface of $\frac{1}{2}$ ft^2, how much force will a wind of 80 mph place on a surface of 2 ft^2?

42. ***Volume of a Cylinder*** The volume of a right circular cylinder is jointly proportional to the square of the radius of the circular base and to the height. If the volume is 300 cm^3 when the height is 10.62 cm and the radius is 3 cm, find the volume, to the nearest tenth, of a cylinder with radius 4 cm and height 15.92 cm.

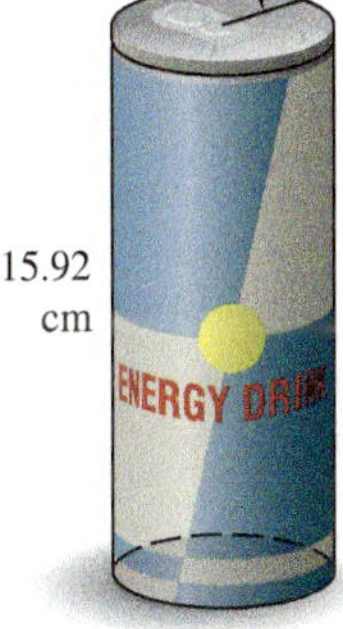

43. *Sports Arena Construction* The roof of a new sports arena rests on round concrete pillars. The maximum load a cylindrical column of circular cross section can hold varies directly as the fourth power of the diameter and inversely as the square of the height. The arena has 9-m-tall columns that are 1 m in diameter and will support a load of 8 metric tons. How many metric tons will be supported by a column 12 m high and $\frac{2}{3}$ m in diameter?

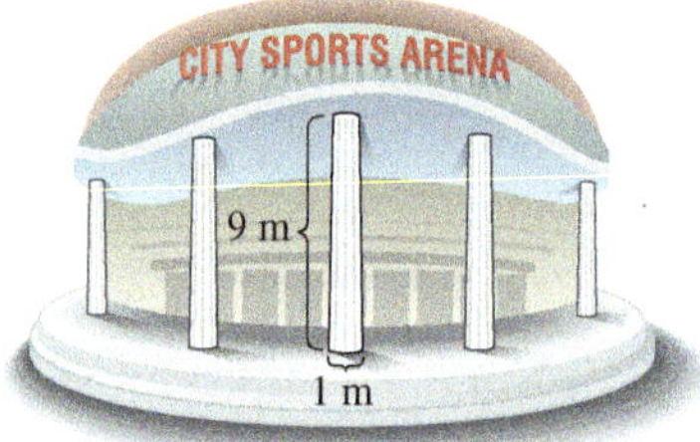

Load = 8 metric tons

44. *Sports Arena Construction* The sports arena in **Exercise 43** requires a horizontal beam 16 m long, 24 cm wide, and 8 cm high. The maximum load of such a horizontal beam that is supported at both ends varies directly as the width of the beam and the square of its height and inversely as the length between supports. If a beam of the same material 8 m long, 12 cm wide, and 15 cm high can support a maximum of 400 kg, what is the maximum load the beam in the arena will support?

45. *Period of a Pendulum* The period of a pendulum varies directly as the square root of the length of the pendulum and inversely as the square root of the acceleration due to gravity. Find the period when the length is 121 cm and the acceleration due to gravity is 980 cm per second squared, if the period is 6π seconds when the length is 289 cm and the acceleration due to gravity is 980 cm per second squared.

46. *Skidding Car* The force needed to keep a car from skidding on a curve varies inversely as the radius r of the curve and jointly as the weight of the car and the square of the speed. It takes 3000 lb of force to keep a 2000-lb car from skidding on a curve of radius 500 ft at 30 mph. What force will keep the same car from skidding on a curve of radius 800 ft at 60 mph?

47. *Body Mass Index* The federal government has developed the **body mass index** (BMI) to determine ideal weights. A person's BMI is directly proportional to his or her weight in pounds and inversely proportional to the square of his or her height in inches. (A BMI of 19 to 25 corresponds to a healthy weight.) A 6-foot-tall person weighing 177 lb has BMI 24. Find the BMI (to the nearest whole number) of a person whose weight is 130 lb and whose height is 66 in.

48. *Poiseuille's Law* According to Poiseuille's law, the resistance to flow of a blood vessel, R, is directly proportional to the length, l, and inversely proportional to the fourth power of the radius, r. If $R = 25$ when $l = 12$ and $r = 0.2$, find R, to the nearest hundredth, as r increases to 0.3, while l is unchanged.

49. *Stefan-Boltzmann Law* The Stefan-Boltzmann law says that the radiation of heat R from an object is directly proportional to the fourth power of the kelvin temperature of the object. For a certain object, $R = 213.73$ at room temperature (293 K). Find R, to the nearest hundredth, if the temperature increases to 335 K.

50. *Nuclear Bomb Detonation* Suppose the effects of detonating a nuclear bomb will be felt over a distance from the point of detonation that is directly proportional to the cube root of the yield of the bomb. Suppose a 100-kiloton bomb has certain effects to a radius of 3 km from the point of detonation. Find the distance to the nearest tenth that the effects would be felt for a 1500-kiloton bomb.

51. *Malnutrition Measure* A measure of malnutrition, called the **pelidisi**, varies directly as the cube root of a person's weight in grams and inversely as the person's sitting height in centimeters. A person with a pelidisi below 100 is considered undernourished, while a pelidisi greater than 100 indicates overfeeding. A person who weighs 48,820 g with a sitting height of 78.7 cm has a pelidisi of 100. Find the pelidisi (to the nearest whole number) of a person whose weight is 54,430 g and whose sitting height is 88.9 cm. Is this individual undernourished or overfed?

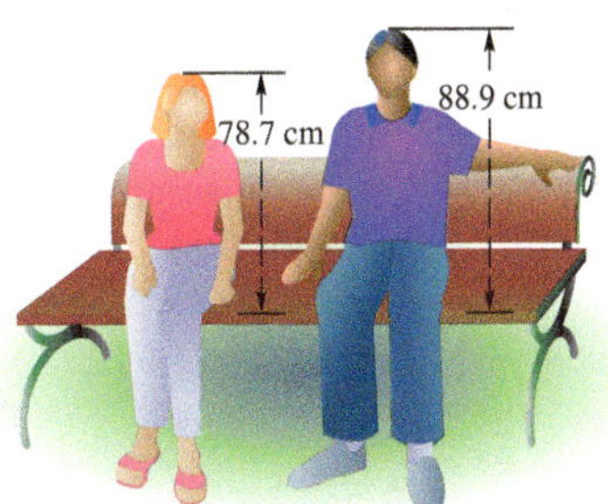

52. *Photography* Variation occurs in a formula from photography. In

$$L = \frac{25F^2}{st},$$

the luminance, L, varies directly as the square of the F-stop, F, and inversely as the product of the film ASA number, s, and the shutter speed, t.

(a) What would an appropriate F-stop be for 200 ASA film and a shutter speed of $\frac{1}{250}$ sec when 500 footcandles of light is available?

(b) If 125 footcandles of light is available and an F-stop of 2 is used with 200 ASA film, what shutter speed should be used?

Concept Check Work each problem.

53. What happens to y if y varies inversely as x, and x is doubled?

54. What happens to y if y varies directly as x, and x is halved?

55. Suppose y is directly proportional to x, and x is replaced by $\frac{1}{3}x$. What happens to y?

56. Suppose y is inversely proportional to x, and x is tripled. What happens to y?

57. Suppose p varies directly as r^3 and inversely as t^2. If r is halved and t is doubled, what happens to p?

58. Suppose m varies directly as p^2 and q^4. If p doubles and q triples, what happens to m?

Chapter 3 Test Prep

Key Terms

3.1 polynomial function
leading coefficient
dominating term
zero polynomial
quadratic function
parabola
axis of symmetry (axis)
vertex
quadratic regression

3.2 synthetic division
zero of a polynomial function
root (or solution) of an equation

3.3 multiplicity of a zero

3.4 turning points
end behavior

3.5 rational function
discontinuous graph
vertical asymptote
horizontal asymptote
oblique asymptote
point of discontinuity (hole)

3.6 varies directly (directly proportional to)
constant of variation
varies inversely (inversely proportional to)
combined variation
varies jointly

New Symbols

Symbol	Meaning	Symbol	Meaning
$\bar{z}$	conjugate of $z = a + bi$	$\lvert f(x)\rvert \to \infty$	absolute value of $f(x)$ increases without bound
	end behavior diagrams	$x \to a$	x approaches a

Quick Review

Concepts / Examples

3.1 Quadratic Functions and Models

1. The graph of
$$f(x) = a(x - h)^2 + k, \quad \text{with } a \neq 0,$$
is a parabola with vertex at (h, k) and the vertical line $x = h$ as axis.

2. The graph opens up if $a > 0$ and down if $a < 0$.

3. The graph is wider than the graph of $f(x) = x^2$ if $|a| < 1$ and narrower if $|a| > 1$.

Vertex Formula

The vertex of the graph of $f(x) = ax^2 + bx + c$, with $a \neq 0$, may be found by completing the square or using the vertex formula.

$$\left(-\frac{b}{2a}, f\left(-\frac{b}{2a}\right)\right) \quad \text{Vertex}$$

Graphing a Quadratic Function $f(x) = ax^2 + bx + c$

Step 1 Find the vertex either by using the vertex formula or by completing the square. Plot the vertex.

Step 2 Plot the y-intercept by evaluating $f(0)$.

Step 3 Plot any x-intercepts by solving $f(x) = 0$.

Step 4 Plot any additional points as needed, using symmetry about the axis.

The graph opens up if $a > 0$ and down if $a < 0$.

Examples

Graph $f(x) = -(x + 3)^2 + 1$.

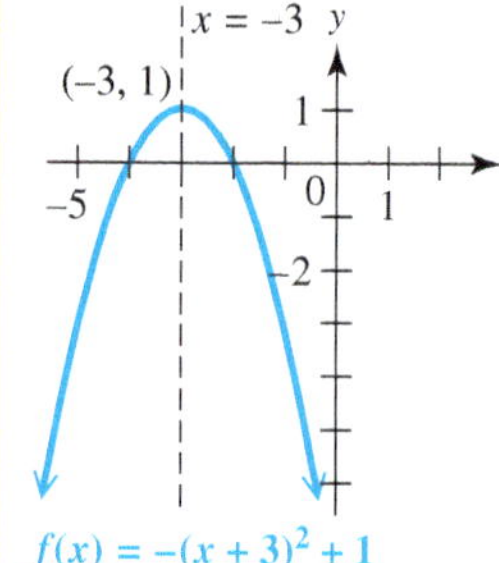

The graph opens down because $a < 0$. It is the graph of $y = -x^2$ shifted 3 units left and 1 unit up, so the vertex is $(-3, 1)$, with axis $x = -3$. The domain is $(-\infty, \infty)$, and the range is $(-\infty, 1]$. The function is increasing on $(-\infty, -3)$ and decreasing on $(-3, \infty)$.

Graph $f(x) = x^2 + 4x + 3$. The vertex of the graph is

$$\left(-\frac{b}{2a}, f\left(-\frac{b}{2a}\right)\right) = (-2, -1). \quad a = 1, b = 4, c = 3$$

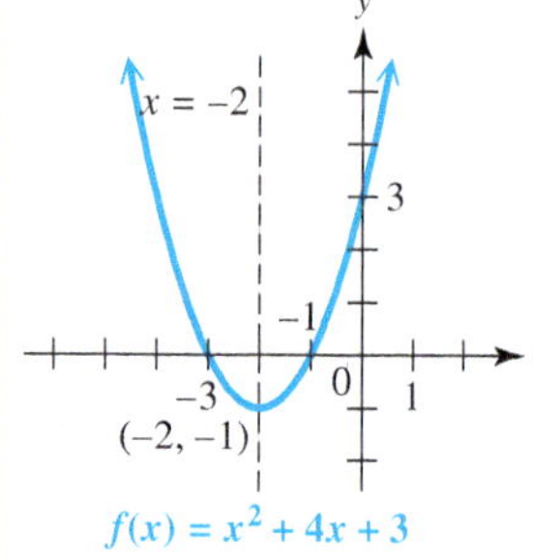

The graph opens up because $a > 0$. $f(0) = 3$, so the y-intercept is $(0, 3)$. The solutions of $x^2 + 4x + 3 = 0$ are -1 and -3, which correspond to the x-intercepts. The domain is $(-\infty, \infty)$, and the range is $[-1, \infty)$. The function is decreasing on $(-\infty, -2)$ and increasing on $(-2, \infty)$.

Concepts	Examples
3.2 Synthetic Division	
Division Algorithm Let $f(x)$ and $g(x)$ be polynomials with $g(x)$ of lesser degree than $f(x)$ and $g(x)$ of degree 1 or more. There exist unique polynomials $q(x)$ and $r(x)$ such that $f(x) = g(x) \cdot q(x) + r(x),$ where either $r(x) = 0$ or the degree of $r(x)$ is less than the degree of $g(x)$.	Use synthetic division to divide $f(x) = 2x^3 - 3x + 2 \quad \text{by} \quad x - 1,$ and write the result as $f(x) = g(x) \cdot q(x) + r(x)$.
Synthetic Division Synthetic division is a shortcut method for dividing a polynomial by a binomial of the form $x - k$.	$\begin{array}{r\|rrrr} 1 & 2 & 0 & -3 & 2 \\ & & 2 & 2 & -1 \\ \hline & 2 & 2 & -1 & 1 \end{array}$ Coefficients of the quotient (2, 2, −1); Remainder (1) $\underbrace{2x^3 - 3x + 2}_{f(x)} = \underbrace{(x-1)}_{g(x)} \cdot \underbrace{(2x^2 + 2x - 1)}_{q(x)} + \underbrace{1}_{r(x)}$
Remainder Theorem If the polynomial $f(x)$ is divided by $x - k$, the remainder is $f(k)$.	By the result above, for $f(x) = 2x^3 - 3x + 2$, $f(1) = 1.$
3.3 Zeros of Polynomial Functions	
Factor Theorem For any polynomial function $f(x)$, $x - k$ is a factor of the polynomial if and only if $f(k) = 0$.	For the polynomial functions $f(x) = x^3 + x + 2 \quad \text{and} \quad g(x) = x^3 - 1,$ $f(-1) = 0$. Therefore, $x - (-1)$, or $x + 1$, is a factor of $f(x)$. Because $x - 1$ is a factor of $g(x)$, $g(1) = 0$.
Rational Zeros Theorem If $\frac{p}{q}$ is a rational number written in lowest terms, and if $\frac{p}{q}$ is a zero of f, a polynomial function with integer coefficients, then p is a factor of the constant term and q is a factor of the leading coefficient.	The only rational numbers that can possibly be zeros of $f(x) = 2x^3 - 9x^2 - 4x - 5$ are ± 1, ± 5, $\pm \frac{1}{2}$, and $\pm \frac{5}{2}$. By synthetic division, it can be shown that the only rational zero of $f(x)$ is 5. $\begin{array}{r\|rrrr} 5 & 2 & -9 & -4 & -5 \\ & & 10 & 5 & 5 \\ \hline & 2 & 1 & 1 & 0 \end{array} \leftarrow f(5)$
Fundamental Theorem of Algebra Every function defined by a polynomial of degree 1 or more has at least one complex zero.	$f(x) = x^3 + x + 2$ has at least one and at most three distinct zeros.
Number of Zeros Theorem A function defined by a polynomial of degree n has at most n distinct zeros.	
Conjugate Zeros Theorem If $f(x)$ defines a polynomial function ***having only real coefficients*** and if $z = a + bi$ is a zero of $f(x)$, where a and b are real numbers, then the conjugate $\bar{z} = a - bi$ is also a zero of $f(x)$.	$1 + 2i$ is a zero of $f(x) = x^3 - 5x^2 + 11x - 15,$ and therefore its conjugate $1 - 2i$ is also a zero.

Concepts	Examples

Descartes' Rule of Signs

Let $f(x)$ define a polynomial function with real coefficients and a nonzero constant term, with terms in descending powers of x.

(a) The number of positive real zeros of f either equals the number of variations in sign occurring in the coefficients of $f(x)$ or is less than the number of variations by a positive even integer.

(b) The number of negative real zeros of f either equals the number of variations in sign occurring in the coefficients of $f(-x)$ or is less than the number of variations by a positive even integer.

There are three sign changes for

$$f(x) = +3x^3 - 2x^2 + x - 4,$$

1 2 3

so there will be three or one positive real zeros. Because

$$f(-x) = -3x^3 - 2x^2 - x - 4$$

has no sign changes, there will be no negative real zeros. The table shows the possibilities for the numbers of positive, negative, and nonreal complex zeros.

Positive	Negative	Nonreal Complex
3	0	0
1	0	2

3.4 Polynomial Functions: Graphs, Applications, and Models

Graphing Using Translations

The graph of the function

$$f(x) = a(x - h)^n + k$$

can be found by considering the effects of the constants a, h, and k on the graph of $f(x) = ax^n$.

- When $|a| > 1$, the graph is stretched vertically.
- When $0 < |a| < 1$, the graph is shrunk vertically.
- When $a < 0$, the graph is reflected across the x-axis.
- The graph is translated h units right if $h > 0$ and $|h|$ units left if $h < 0$.
- The graph is translated k units up if $k > 0$ and $|k|$ units down if $k < 0$.

Graph $f(x) = -(x + 2)^4 + 1$.

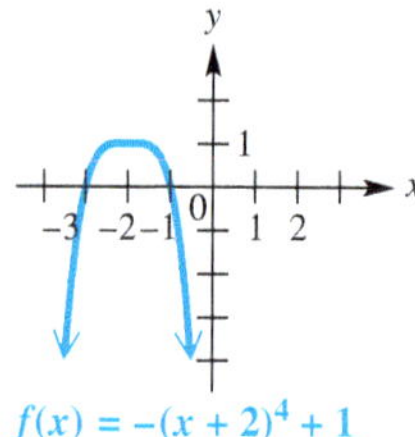

The negative sign causes the graph to be reflected across the x-axis compared to the graph of $f(x) = x^4$. The graph is translated 2 units to the left and 1 unit up. The function is increasing on $(-\infty, -2)$ and decreasing on $(-2, \infty)$.

Multiplicity of a Zero

The behavior of the graph of a polynomial function $f(x)$ near a zero depends on the multiplicity of the zero. If $(x - c)^n$ is a factor of $f(x)$, then the graph will behave in the following manner.

- For $n = 1$, the graph will cross the x-axis at $(c, 0)$.
- For n even, the graph will bounce, or turn, at $(c, 0)$.
- For n an odd integer greater than 1, the graph will wiggle through the x-axis at $(c, 0)$.

Determine the behavior of f near its zeros, and graph.

$$f(x) = (x - 1)(x - 3)^2(x + 1)^3$$

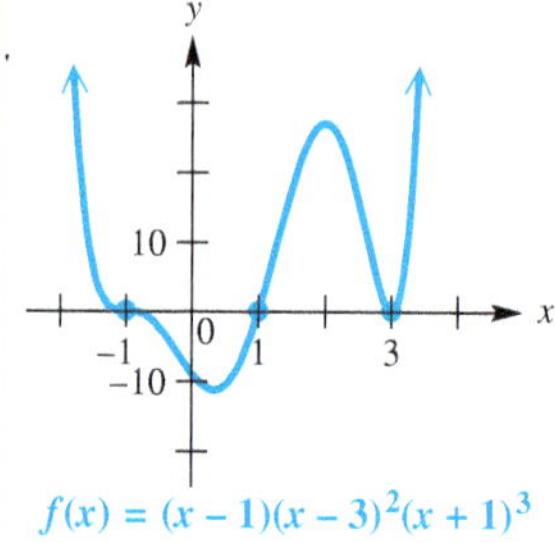

The graph will cross the x-axis at $x = 1$, bounce at $x = 3$, and wiggle through the x-axis at $x = -1$. Since the dominating term is x^6, the end behavior is ↖ ↗. The y-intercept is $(0, -9)$ because $f(0) = -9$.

Concepts	Examples

Turning Points

A polynomial function of degree n has at most $n - 1$ turning points, with at least one turning point between each pair of successive zeros.

The graph of

$$f(x) = 4x^5 - 2x^3 + 3x^2 + x - 10$$

has at most four turning points (because $5 - 1 = 4$).

End Behavior

The end behavior of the graph of a polynomial function $f(x)$ is determined by the dominating term, or term of greatest degree. If ax^n is the dominating term of $f(x)$, then the end behavior is as follows.

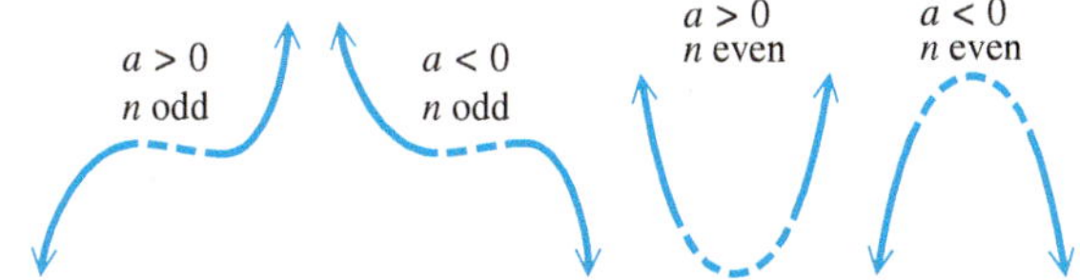

The end behavior of

$$f(x) = 3x^5 + 2x^2 + 7$$

is .

The end behavior of

$$f(x) = -x^4 - 3x^3 + 2x - 9$$

is .

Graphing Polynomial Functions

To graph a polynomial function f, first find the x-intercepts and y-intercept.

Then use end behavior, whether the graph crosses, bounces on, or wiggles through the x-axis at the x-intercepts, and selected points as necessary to complete the graph.

Graph $f(x) = (x + 2)(x - 1)(x + 3)$.

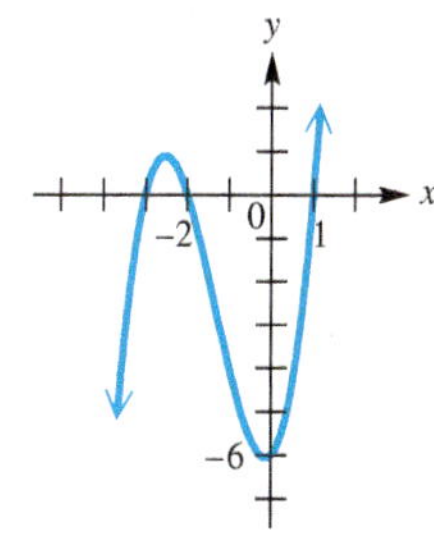

$f(x) = (x + 2)(x - 1)(x + 3)$

The x-intercepts correspond to the zeros of f, which are -2, 1, and -3. Because $f(0) = 2(-1)(3) = -6$, the y-intercept is $(0, -6)$. The dominating term is $x(x)(x)$, or x^3, so the end behavior is . Begin at either end of the graph with the correct end behavior, and draw a smooth curve that crosses the x-axis at each zero, has a turning point between successive zeros, and passes through the y-intercept.

Intermediate Value Theorem

If $f(x)$ is a polynomial function with only *real coefficients*, and if for real numbers a and b the values of $f(a)$ and $f(b)$ are opposite in sign, then there exists at least one real zero between a and b.

For the polynomial function

$$f(x) = -x^4 + 2x^3 + 3x^2 + 6,$$

$$f(3.1) = 2.0599 \quad \text{and} \quad f(3.2) = -2.6016.$$

Because $f(3.1) > 0$ and $f(3.2) < 0$, there exists at least one real zero between 3.1 and 3.2.

Boundedness Theorem

Let $f(x)$ be a polynomial function of degree $n \geq 1$ with *real coefficients* and with a *positive* leading coefficient. Suppose $f(x)$ is divided synthetically by $x - c$.

(a) If $c > 0$ and all numbers in the bottom row of the synthetic division are nonnegative, then $f(x)$ has no zero greater than c.

(b) If $c < 0$ and the numbers in the bottom row of the synthetic division alternate in sign (with 0 considered positive or negative, as needed), then $f(x)$ has no zero less than c.

Show that $f(x) = x^3 - x^2 - 8x + 12$ has no zero greater than 4 and no zero less than -4.

4)	1	−1	−8	12	
		4	12	16	
	1	3	4	28	← All signs positive

−4)	1	−1	−8	12	
		−4	20	−48	
	1	−5	12	−36	← Alternating signs

Concepts	Examples

3.5 Rational Functions: Graphs, Applications, and Models

Graphing Rational Functions

To graph a rational function in lowest terms, find the asymptotes and intercepts. Determine whether the graph intersects its nonvertical asymptote. Plot selected points, as necessary, to complete the sketch.

Graph $f(x) = \dfrac{x^2 - 1}{(x + 3)(x - 2)}$.

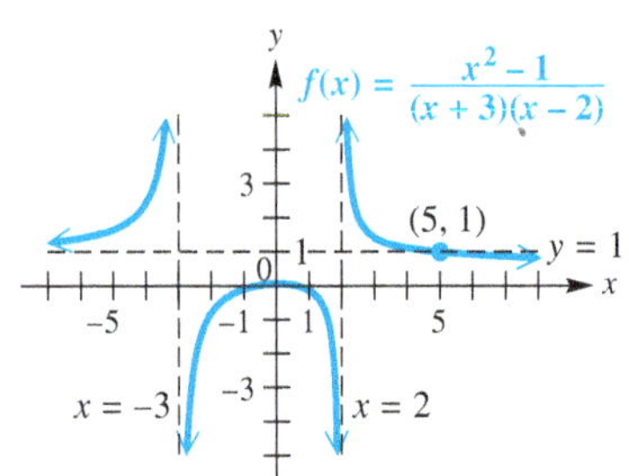

Point of Discontinuity

A rational function that is not in lowest terms often has a hole, or point of discontinuity, in its graph.

Graph $f(x) = \dfrac{x^2 - 1}{x + 1}$.

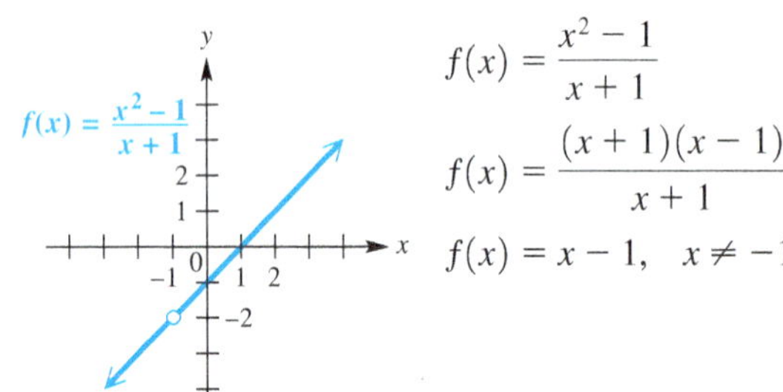

$$f(x) = \frac{x^2 - 1}{x + 1}$$

$$f(x) = \frac{(x + 1)(x - 1)}{x + 1}$$

$$f(x) = x - 1, \quad x \neq -1$$

The graph is that of $y = x - 1$, with a hole at $(-1, -2)$.

3.6 Variation

Direct Variation

y varies directly as the nth power of x if for all x there exists a nonzero real number k such that

$$y = kx^n.$$

The area of a circle varies directly as the square of the radius.

$$\mathcal{A} = kr^2 \quad (k = \pi)$$

Inverse Variation

y varies inversely as the nth power of x if for all x there exists a nonzero real number k such that

$$y = \frac{k}{x^n}.$$

Pressure of a gas varies inversely as volume.

$$P = \frac{k}{V}$$

Joint Variation

For real numbers m and n, y varies jointly as the nth power of x and the mth power of z if for all x and z, there exists a nonzero real number k such that

$$y = kx^nz^m.$$

The area of a triangle varies jointly as its base and its height.

$$\mathcal{A} = kbh \quad \left(k = \tfrac{1}{2}\right)$$

Chapter 3 Review Exercises

Graph each quadratic function. Give the vertex, axis, x-intercepts, y-intercept, domain, range, and largest open intervals of the domain over which each function is increasing or decreasing.

1. $f(x) = 3(x+4)^2 - 5$

2. $f(x) = -\frac{2}{3}(x-6)^2 + 7$

3. $f(x) = -3x^2 - 12x - 1$

4. $f(x) = 4x^2 - 4x + 3$

Concept Check *Consider the function*

$$f(x) = a(x-h)^2 + k, \quad \text{for } a > 0.$$

5. What are the coordinates of the lowest point of its graph?

6. What is the y-intercept of its graph?

7. Under what conditions will its graph have one or more x-intercepts? For these conditions, express the x-intercept(s) in terms of a, h, and k.

8. If a is positive, what is the least value of $ax^2 + bx + c$ in terms of a, b, and c?

(Modeling) Solve each problem.

9. ***Area of a Rectangle*** Use a quadratic function to find the dimensions of the rectangular region of maximum area that can be enclosed with 180 m of fencing, if no fencing is needed along one side of the region.

10. ***Height of a Projectile*** A projectile is fired vertically upward, and its height $s(t)$ in feet after t seconds is given by the function

$$s(t) = -16t^2 + 800t + 600.$$

(a) From what height was the projectile fired?

(b) After how many seconds will it reach its maximum height?

(c) What is the maximum height it will reach?

(d) Between what two times (in seconds, to the nearest tenth) will it be more than 5000 ft above the ground?

(e) After how many seconds, to the nearest tenth, will the projectile hit the ground?

11. ***Food Bank Volunteers*** During the course of a year, the number of volunteers available to run a food bank each month is modeled by $V(x)$, where

$$V(x) = 2x^2 - 32x + 150$$

between the months of January and August. Here x is time in months, with $x = 1$ representing January. From August to December, $V(x)$ is modeled by

$$V(x) = 31x - 226.$$

Find the number of volunteers in each of the following months.

(a) January **(b)** May **(c)** August **(d)** October **(e)** December

(f) Sketch a graph of $y = V(x)$ for January through December. In what month are the fewest volunteers available?

12. *Concentration of Atmospheric CO_2* In 1990, the International Panel on Climate Change (IPCC) stated that if current trends of burning fossil fuel and deforestation were to continue, then future amounts of atmospheric carbon dioxide in parts per million (ppm) would increase, as shown in the table.

Year	Carbon Dioxide
1990	353
2000	375
2075	590
2175	1090
2275	2000

Source: IPCC.

(a) Let $x = 0$ represent 1990, $x = 10$ represent 2000, and so on. Find a function of the form

$$f(x) = a(x - h)^2 + k$$

that models the data. Use $(0, 353)$ as the vertex and $(285, 2000)$ as another point to determine a.

(b) Use the function to predict the amount of carbon dioxide in 2300. Round to the nearest unit.

Consider the function $f(x) = -2.64x^2 + 5.47x + 3.54$.

13. Use the discriminant to explain how to determine the number of x-intercepts the graph of $f(x)$ will have before graphing it on a calculator.

14. Graph the function in the standard viewing window of a calculator, and use the calculator to solve the equation $f(x) = 0$. Express solutions as approximations to the nearest hundredth.

15. Use the answer to **Exercise 14** and the graph of f to solve the following. Give approximations to the nearest hundredth.

(a) $f(x) > 0$ (b) $f(x) < 0$

16. Use the capabilities of a calculator to find the coordinates of the vertex of the graph. Express coordinates to the nearest hundredth.

Use synthetic division to perform each division.

17. $\dfrac{x^3 + x^2 - 11x - 10}{x - 3}$

18. $\dfrac{3x^3 + 8x^2 + 5x + 10}{x + 2}$

19. $\dfrac{2x^3 - x + 6}{x + 4}$

20. $\dfrac{3x^3 + 6x^2 - 8x + 3}{x + 3}$

Use synthetic division to divide $f(x)$ by $x - k$ for the given value of k. Then express $f(x)$ in the form $f(x) = (x - k)q(x) + r$.

21. $f(x) = 5x^3 - 3x^2 + 2x - 6$; $k = 2$

22. $f(x) = -3x^3 + 5x - 6$; $k = -1$

Use synthetic division to find $f(2)$.

23. $f(x) = -x^3 + 5x^2 - 7x + 1$

24. $f(x) = 2x^3 - 3x^2 + 7x - 12$

25. $f(x) = 5x^4 - 12x^2 + 2x - 8$

26. $f(x) = x^5 + 4x^2 - 2x - 4$

Use synthetic division to determine whether k is a zero of the function.

27. $f(x) = x^3 + 2x^2 + 3x + 2$; $k = -1$

28. $f(x) = 2x^3 + 5x^2 + 30$; $k = -4$

29. *Concept Check* If $f(x)$ is a polynomial function with real coefficients, and if $7 + 2i$ is a zero of the function, then what other complex number must also be a zero?

30. *Concept Check* Suppose the polynomial function f has a zero at $x = -3$. Which of the following statements *must* be true?

A. $(3, 0)$ is an x-intercept of the graph of f.

B. $(0, 3)$ is a y-intercept of the graph of f.

C. $x - 3$ is a factor of $f(x)$.

D. $f(-3) = 0$

Find a polynomial function $f(x)$ of least degree with real coefficients having zeros as given.

31. $-1, 4, 7$

32. $8, 2, 3$

33. $\sqrt{3}, -\sqrt{3}, 2, 3$

34. $-2 + \sqrt{5}, -2 - \sqrt{5}, -2, 1$

35. $2, 4, -i$

36. $0, 5, 1 + 2i$

Find all rational zeros of each function.

37. $f(x) = 2x^3 - 9x^2 - 6x + 5$

38. $f(x) = 8x^4 - 14x^3 - 29x^2 - 4x + 3$

Show that each polynomial function has a real zero as described in parts (a) and (b). In Exercises 39 and 40, also work part (c).

39. $f(x) = 3x^3 - 8x^2 + x + 2$

(a) between -1 and 0 **(b)** between 2 and 3

(c) Find the zero in part (b) to three decimal places.

40. $f(x) = 4x^3 - 37x^2 + 50x + 60$

(a) between 2 and 3 **(b)** between 7 and 8

(c) Find the zero in part (b) to three decimal places.

41. $f(x) = 6x^4 + 13x^3 - 11x^2 - 3x + 5$

(a) no zero greater than 1 **(b)** no zero less than -3

Solve each problem.

42. Use Descartes' rule of signs to determine the different possibilities for the numbers of positive, negative, and nonreal complex zeros of

$$f(x) = x^3 + 3x^2 - 4x - 2.$$

43. Is $x + 1$ a factor of $f(x) = x^3 + 2x^2 + 3x + 2$?

44. Find a polynomial function f with real coefficients of degree 4 with 3, 1, and $-1 + 3i$ as zeros, and $f(2) = -36$.

45. Find a polynomial function f of degree 3 with -2, 1, and 4 as zeros, and $f(2) = 16$.

46. Find all zeros of $f(x) = x^4 - 3x^3 - 8x^2 + 22x - 24$, given that $1 + i$ is a zero.

47. Find all zeros of $f(x) = 2x^4 - x^3 + 7x^2 - 4x - 4$, given that 1 and $-2i$ are zeros.

48. Find a value of k such that $x - 4$ is a factor of $f(x) = x^3 - 2x^2 + kx + 4$.

49. Find a value of k such that when the polynomial $x^3 - 3x^2 + kx - 4$ is divided by $x - 2$, the remainder is 5.

50. Give the maximum number of turning points of the graph of each function.

(a) $f(x) = x^5 - 9x^2$ **(b)** $f(x) = 4x^3 - 6x^2 + 2$

51. ***Concept Check*** Give an example of a cubic polynomial function having exactly one real zero, and then sketch its graph.

52. ***Concept Check*** Give an example of a fourth-degree polynomial function having exactly two distinct real zeros, and then sketch its graph.

53. ***Concept Check*** If the dominating term of a polynomial function is $10x^7$, what can we conclude about each of the following features of the graph of the function?

(a) domain **(b)** range **(c)** end behavior **(d)** number of zeros

(e) number of turning points

54. ***Concept Check*** Repeat **Exercise 53** for a polynomial function with dominating term $-9x^6$.

Graph each polynomial function.

55. $f(x)=(x-2)^2(x+3)$

56. $f(x)=-2x^3+7x^2-2x-3$

57. $f(x)=2x^3+x^2-x$

58. $f(x)=x^4-3x^2+2$

59. $f(x)=x^4+x^3-3x^2-4x-4$

60. $f(x)=-2x^4+7x^3-4x^2-4x$

Concept Check *For each polynomial function, identify its graph from choices A–F.*

61. $f(x)=(x-2)^2(x-5)$

62. $f(x)=-(x-2)^2(x-5)$

63. $f(x)=(x-2)^2(x-5)^2$

64. $f(x)=(x-2)(x-5)$

65. $f(x)=-(x-2)(x-5)$

66. $f(x)=-(x-2)^2(x-5)^2$

A.

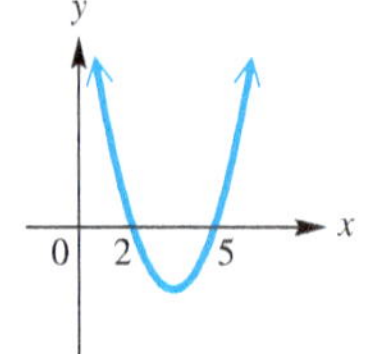

B.

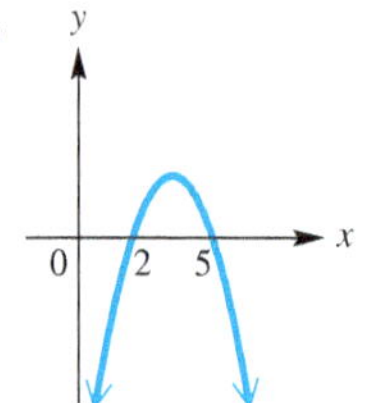

C.

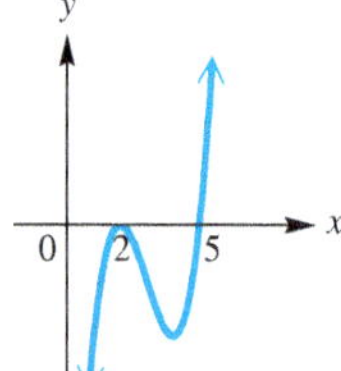

D.

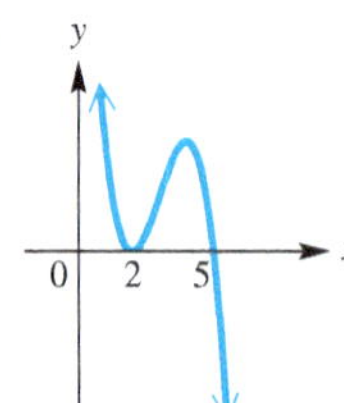

E.

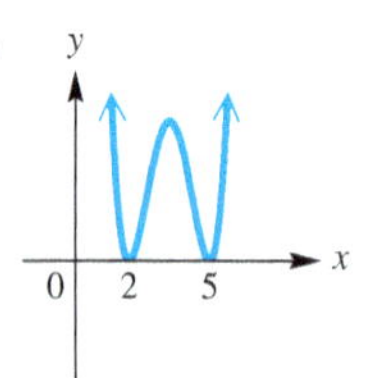

F.

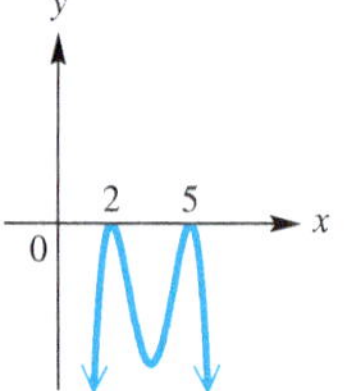

Graph each polynomial function in the viewing window specified. Then approximate the real zeros to as many decimal places as the calculator will provide.

67. $f(x)=x^3-8x^2+2x+5$; window: $[-10, 10]$ by $[-60, 60]$

68. $f(x)=x^4-4x^3-5x^2+14x-15$; window: $[-10, 10]$ by $[-60, 60]$

Solve each problem.

69. ***(Modeling) Medicare Beneficiary Spending*** Out-of-pocket spending projections for a typical Medicare beneficiary as a share of his or her income are given in the table. Let $x=0$ represent 1990, so $x=8$ represents 1998. Use a graphing calculator to do the following.

(a) Graph the data points.

(b) Find a quadratic function to model the data.

(c) Find a cubic function to model the data.

(d) Graph each function in the same viewing window as the data points.

(e) Compare the two functions. Which is a better fit for the data?

Year	Percent of Income
1998	18.6
2000	19.3
2005	21.7
2010	24.7
2015	27.5
2020	28.3
2025	28.6

Source: Urban Institute's Analysis of Medicare Trustees' Report.

70. ***Dimensions of a Cube*** After a 2-in. slice is cut off the top of a cube, the resulting solid has a volume of 32 in.^{3}. Find the dimensions of the original cube.

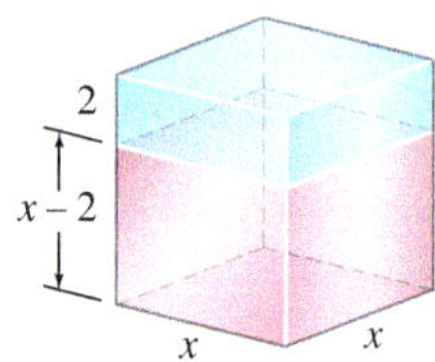

71. *Dimensions of a Box* The width of a rectangular box is three times its height, and its length is 11 in. more than its height. Find the dimensions of the box if its volume is 720 in.^{3}.

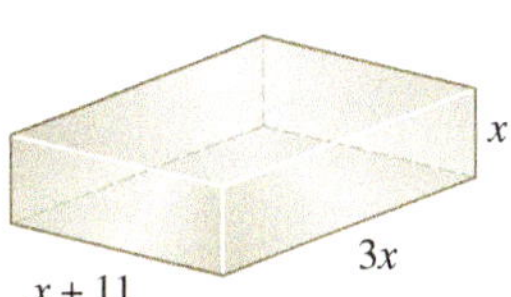

72. The function $f(x) = \frac{1}{x}$ is negative at $x = -1$ and positive at $x = 1$ but has no zero between -1 and 1. Explain why this does not contradict the intermediate value theorem.

Graph each rational function.

73. $f(x) = \dfrac{4}{x - 1}$

74. $f(x) = \dfrac{4x - 2}{3x + 1}$

75. $f(x) = \dfrac{6x}{x^2 + x - 2}$

76. $f(x) = \dfrac{2x}{x^2 - 1}$

77. $f(x) = \dfrac{x^2 + 4}{x + 2}$

78. $f(x) = \dfrac{x^2 - 1}{x}$

79. $f(x) = \dfrac{-2}{x^2 + 1}$

80. $f(x) = \dfrac{4x^2 - 9}{2x + 3}$

Solve each problem.

81. *Concept Check* Work each of the following.

(a) Sketch the graph of a function that does not intersect its horizontal asymptote $y = 1$, has the line $x = 3$ as a vertical asymptote, and has x-intercepts $(2, 0)$ and $(4, 0)$.

(b) Find an equation for a possible corresponding rational function.

82. *Concept Check* Work each of the following.

(a) Sketch the graph of a function that is never negative and has the lines $x = -1$ and $x = 1$ as vertical asymptotes, the x-axis as a horizontal asymptote, and the origin as an x-intercept.

(b) Find an equation for a possible corresponding rational function.

83. *Connecting Graphs with Equations* Find a rational function f having the graph shown.

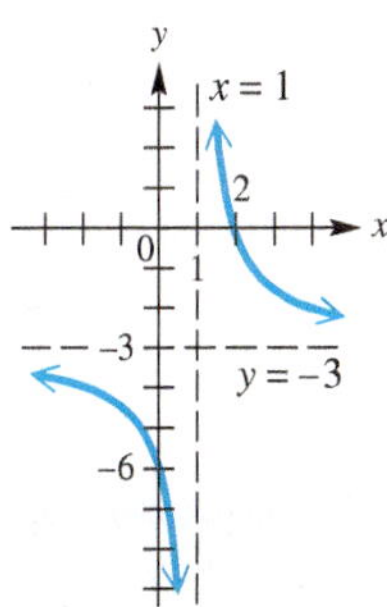

84. *Concept Check* The rational function

$$f(x) = \frac{x^3 + 7x^2 - 25x - 175}{x^3 + 3x^2 - 25x - 75}$$

has two holes and one vertical asymptote.

(a) What are the x-values of the holes?

(b) What is the equation of the vertical asymptote?

(Modeling) Solve each problem.

85. *Environmental Pollution* In situations involving environmental pollution, a cost-benefit model expresses cost as a function of the percentage of pollutant removed from the environment. Suppose a cost-benefit model is expressed as

$$C(x) = \frac{6.7x}{100 - x},$$

where $C(x)$ is cost in thousands of dollars of removing x percent of a pollutant.

(a) Graph the function in the window $[0, 100]$ by $[0, 100]$.

(b) How much would it cost to remove 95% of the pollutant? Round to the nearest tenth.

86. *Antique-Car Competition* Antique-car owners often enter their cars in a **concours d'elegance** in which a maximum of 100 points can be awarded to a particular car based on its attractiveness. The function

$$C(x) = \frac{10x}{49(101 - x)}$$

models the cost, in thousands of dollars, of restoring a car so that it will win x points.

(a) Graph the function in the window $[0, 101]$ by $[0, 10]$.

(b) How much would an owner expect to pay to restore a car in order to earn 95 points? Round to the nearest tenth.

Solve each problem.

87. If x varies directly as y, and $x = 20$ when $y = 14$, find y when $x = 50$.

88. If x varies directly as y, and $x = 12$ when $y = 4$, find x when $y = 12$.

89. If t varies inversely as s, and $t = 3$ when $s = 5$, find s when $t = 20$.

90. If z varies inversely as w, and $z = 10$ when $w = \frac{1}{2}$, find z when $w = 10$.

91. f varies jointly as g^2 and h, and $f = 50$ when $g = 5$ and $h = 4$. Find f when $g = 3$ and $h = 6$.

92. p varies jointly as q and r^2, and $p = 100$ when $q = 2$ and $r = 3$. Find p when $q = 5$ and $r = 2$.

93. *Power of a Windmill* The power a windmill obtains from the wind varies directly as the cube of the wind velocity. If a wind of 10 km per hr produces 10,000 units of power, how much power is produced by a wind of 15 km per hr?

94. *Pressure in a Liquid* The pressure on a point in a liquid is directly proportional to the distance from the surface to the point. In a certain liquid, the pressure at a depth of 4 m is 60 kg per m^2. Find the pressure at a depth of 10 m.

Chapter 3 Test

1. Graph the quadratic function $f(x) = -2x^2 + 6x - 3$. Give the intercepts, vertex, axis, domain, range, and the largest open intervals of the domain over which the function is increasing or decreasing.

2. *(Modeling) Height of a Projectile* A small rocket is fired directly upward, and its height s in feet after t seconds is given by the function

$$s(t) = -16t^2 + 88t + 48.$$

(a) Determine the time at which the rocket reaches its maximum height.

(b) Determine the maximum height.

(c) Between what two times (in seconds, to the nearest tenth) will the rocket be more than 100 ft above ground level?

(d) After how many seconds will the rocket hit the ground?

Use synthetic division to perform each division.

3. $\dfrac{3x^3 + 4x^2 - 9x + 6}{x + 2}$

4. $\dfrac{2x^3 - 11x^2 + 25}{x - 5}$

5. Use synthetic division to determine $f(5)$ for

$$f(x) = 2x^3 - 9x^2 + 4x + 8.$$

6. Use the factor theorem to determine whether the polynomial $x - 3$ is a factor of

$$6x^4 - 11x^3 - 35x^2 + 34x + 24.$$

If it is, what is the other factor? If it is not, explain why.

7. Given that -2 is a zero, find all zeros of

$$f(x) = x^3 + 8x^2 + 25x + 26.$$

8. Find a fourth degree polynomial function f having only real coefficients, -1, 2, and i as zeros, and $f(3) = 80$.

9. Why can't the polynomial function $f(x) = x^4 + 8x^2 + 12$ have any real zeros?

10. Consider the polynomial function

$$f(x) = x^3 - 5x^2 + 2x + 7.$$

(a) Use the intermediate value theorem to show that f has a zero between 1 and 2.

(b) Use Descartes' rule of signs to determine the different possibilities for the numbers of positive, negative, and nonreal complex zeros.

(c) Use a graphing calculator to find all real zeros to as many decimal places as the calculator will give.

11. Graph the polynomial functions

$$f(x) = x^4 \quad \text{and} \quad g(x) = -2(x + 5)^4 + 3$$

on the same axes. How can the graph of g be obtained by a transformation of the graph of f?

12. Use end behavior to determine which one of the following graphs is that of $f(x) = -x^7 + x - 4$.

A. 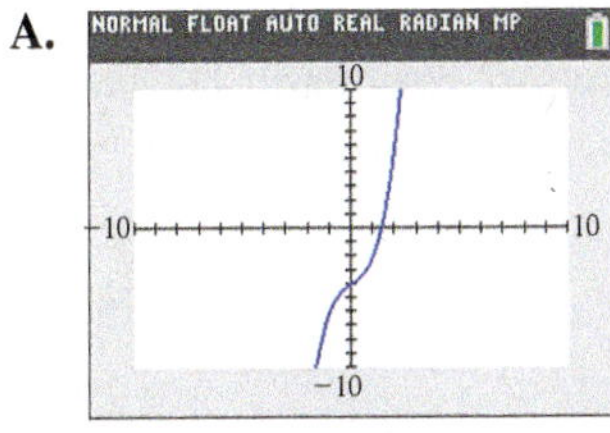

B. 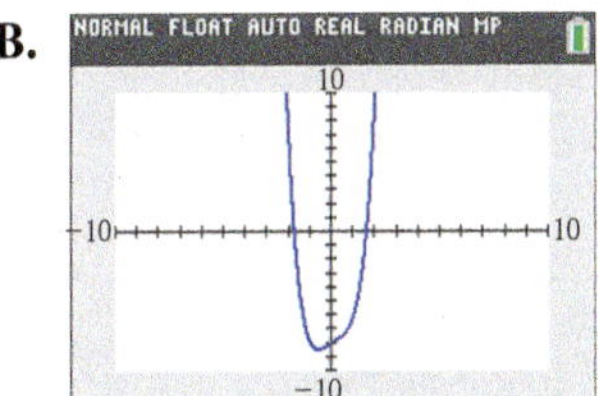

C. 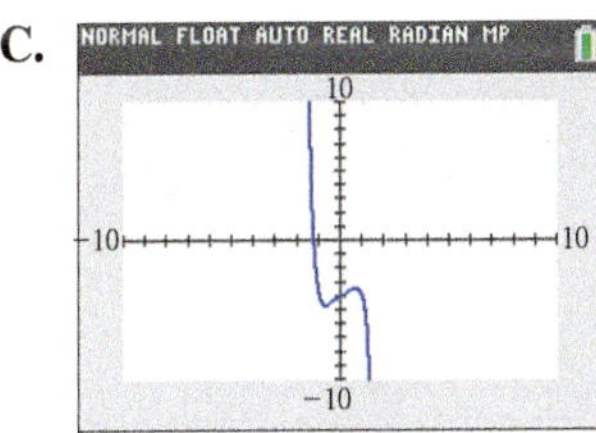

D. 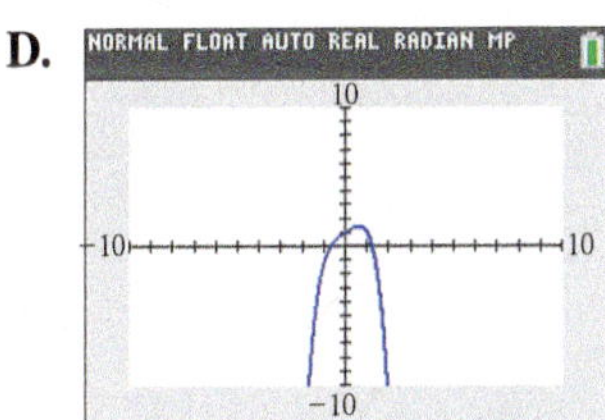

Graph each polynomial function.

13. $f(x) = x^3 - 5x^2 + 3x + 9$

14. $f(x) = 2x^2(x - 2)^2$

15. $f(x) = -x^3 - 4x^2 + 11x + 30$

16. ***Connecting Graphs with Equations*** Find a cubic polynomial function f having the graph shown.

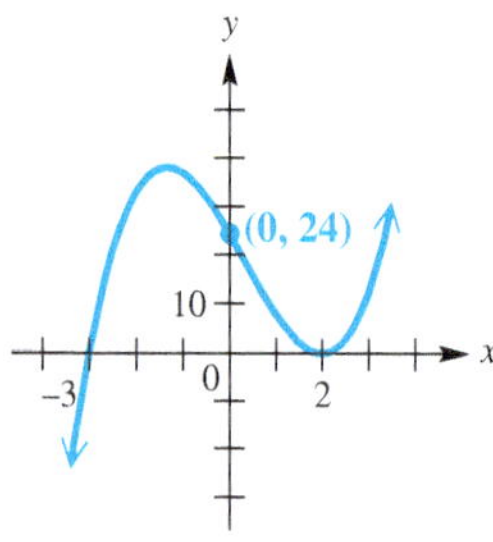

17. ***(Modeling) Oil Pressure*** The pressure of oil in a reservoir tends to drop with time. Engineers found that the change in pressure is modeled by

$$f(t) = 1.06t^3 - 24.6t^2 + 180t,$$

for t (in years) in the interval $[0, 15]$.

(a) What was the change after 2 yr?

(b) For what time periods, to the nearest tenth of a year, is the amount of change in pressure increasing? decreasing? Use a graph to decide.

Graph each rational function.

18. $f(x) = \dfrac{3x - 1}{x - 2}$

19. $f(x) = \dfrac{x^2 - 1}{x^2 - 9}$

20. Consider the rational function $f(x) = \dfrac{2x^2 + x - 6}{x - 1}$.

(a) Determine the equation of the oblique asymptote.

(b) Determine the x-intercepts.

(c) Determine the y-intercept.

(d) Determine the equation of the vertical asymptote.

(e) Sketch the graph.

21. If y varies directly as the square root of x, and $y = 12$ when $x = 4$, find y when $x = 100$.

22. ***Weight on and above Earth*** The weight w of an object varies inversely as the square of the distance d between the object and the center of Earth. If a man weighs 90 kg on the surface of Earth, how much would he weigh 800 km above the surface? (*Hint:* The radius of Earth is about 6400 km.)

4 Inverse, Exponential, and Logarithmic Functions

The magnitudes of earthquakes, the loudness of sounds, and the growth or decay of some populations are examples of quantities that are described by *exponential functions* and their inverses, *logarithmic functions*.

4.1 Inverse Functions

- One-to-One Functions
- Inverse Functions
- Equations of Inverses
- An Application of Inverse Functions to Cryptography

One-to-One Functions

Suppose we define the following function F.

$$F = \{(-2, 2), (-1, 1), (0, 0), (1, 3), (2, 5)\}$$

(We have defined F so that each *second* component is used only once.) We can form another set of ordered pairs from F by interchanging the x- and y-values of each pair in F. We call this set G.

$$G = \{(2, -2), (1, -1), (0, 0), (3, 1), (5, 2)\}$$

G is the *inverse* of F. Function F was defined with each *second* component used only once, so set G will also be a function. (Each *first* component must be used only once.) In order for a function to have an inverse that is also a function, it must exhibit this one-to-one relationship.

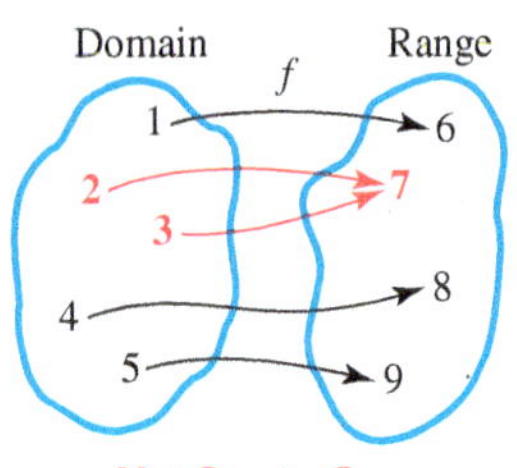

Not One-to-One

Figure 1

In a one-to-one function, each x-value corresponds to only one y-value, and each y-value corresponds to only one x-value.

The function f shown in **Figure 1** is not one-to-one because the y-value 7 corresponds to *two* x-values, 2 and 3. That is, the ordered pairs $(2, 7)$ and $(3, 7)$ both belong to the function. The function f in **Figure 2** is one-to-one.

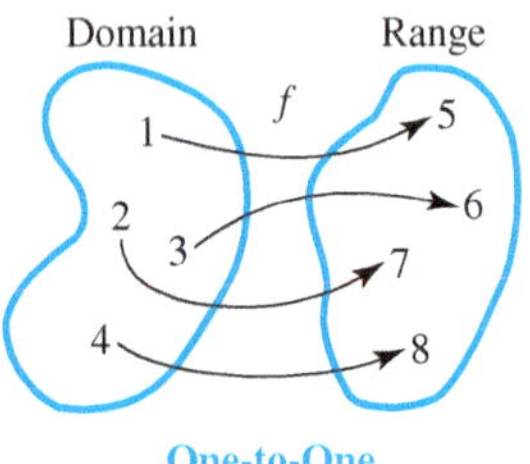

One-to-One

Figure 2

One-to-One Function

A function f is a **one-to-one function** if, for elements a and b in the domain of f,

$$\boldsymbol{a \neq b \quad \text{implies} \quad f(a) \neq f(b).}$$

That is, different values of the domain correspond to different values of the range.

Using the concept of the *contrapositive* from the study of logic, the boldface statement in the preceding box is equivalent to

$$\boldsymbol{f(a) = f(b) \quad \text{implies} \quad a = b.}$$

This means that if two range values are equal, then their corresponding domain values are equal. We use this statement to show that a function f is one-to-one in **Example 1(a)**.

EXAMPLE 1 Deciding Whether Functions Are One-to-One

Determine whether each function is one-to-one.

(a) $f(x) = -4x + 12$ **(b)** $f(x) = \sqrt{25 - x^2}$

SOLUTION

(a) We can determine that the function $f(x) = -4x + 12$ is one-to-one by showing that $f(a) = f(b)$ leads to the result $a = b$.

$$f(a) = f(b)$$
$$-4a + 12 = -4b + 12 \qquad f(x) = -4x + 12$$
$$-4a = -4b \qquad \text{Subtract 12.}$$
$$a = b \qquad \text{Divide by } -4.$$

By the definition, $f(x) = -4x + 12$ is one-to-one.

(b) We can determine that the function $f(x) = \sqrt{25 - x^2}$ is not one-to-one by showing that *different* values of the domain correspond to the *same* value of the range. If we choose $a = 3$ and $b = -3$, then $3 \neq -3$, but

$$f(3) = \sqrt{25 - 3^2} = \sqrt{25 - 9} = \sqrt{16} = 4$$

and $$f(-3) = \sqrt{25 - (-3)^2} = \sqrt{25 - 9} = 4.$$

Here, even though $3 \neq -3$, $f(3) = f(-3) = 4$. By the definition, f is *not* a one-to-one function.

✔ **Now Try Exercises 17 and 19.**

As illustrated in **Example 1(b),** a way to show that a function is *not* one-to-one is to produce a pair of different domain elements that lead to the same function value. There is a useful graphical test for this, the **horizontal line test.**

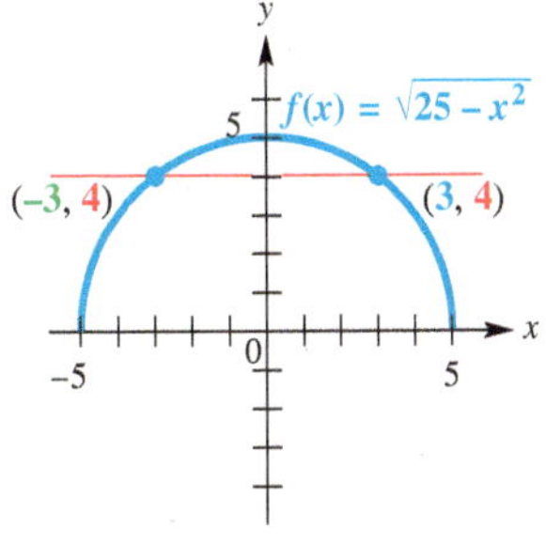

Figure 3

Horizontal Line Test

A function is one-to-one if every horizontal line intersects the graph of the function at most once.

NOTE In **Example 1(b),** the graph of the function is a semicircle, as shown in **Figure 3.** Because there is at least one horizontal line that intersects the graph in more than one point, this function is not one-to-one.

EXAMPLE 2 Using the Horizontal Line Test

Determine whether each graph is the graph of a one-to-one function.

(a)

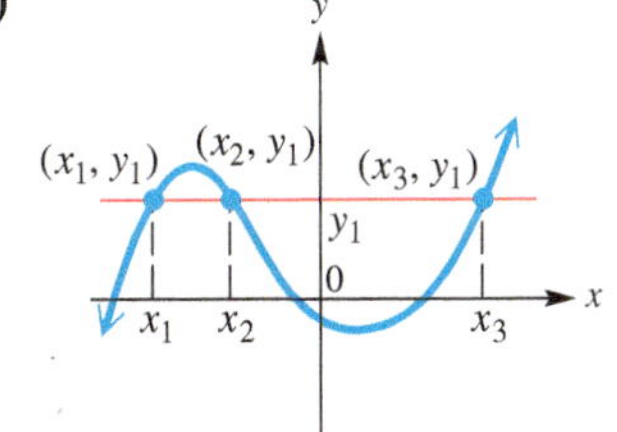

(b)

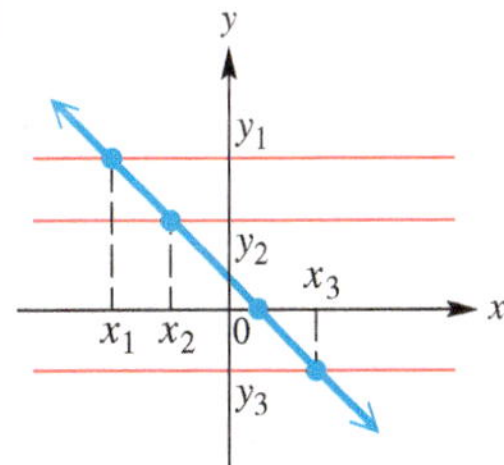

SOLUTION

(a) Each point where the horizontal line intersects the graph has the same value of y but a different value of x. Because more than one different value of x (here three) lead to the same value of y, the function is not one-to-one.

(b) Every horizontal line will intersect the graph at exactly one point, so this function is one-to-one.

✔ **Now Try Exercises 11 and 13.**

The function graphed in **Example 2(b)** decreases on its entire domain.

In general, a function that is either increasing or decreasing on its entire domain, such as $f(x) = -x$, $g(x) = x^3$, and $h(x) = \frac{1}{x}$, must be one-to-one.

Tests to Determine Whether a Function Is One-to-One

1. Show that $f(a) = f(b)$ implies $a = b$. This means that f is one-to-one. **(See Example 1(a).)**
2. In a one-to-one function, every y-value corresponds to no more than one x-value. To show that a function is not one-to-one, find at least two x-values that produce the same y-value. **(See Example 1(b).)**
3. Sketch the graph and use the horizontal line test. **(See Example 2.)**
4. If the function either increases or decreases on its entire domain, then it is one-to-one. A sketch is helpful here, too. **(See Example 2(b).)**

Inverse Functions Certain pairs of one-to-one functions "undo" each other. For example, consider the functions

$$g(x) = 8x + 5 \quad \text{and} \quad f(x) = \frac{1}{8}x - \frac{5}{8}.$$

We choose an arbitrary element from the domain of g, say 10. Evaluate $g(10)$.

$$g(x) = 8x + 5 \qquad \text{Given function}$$
$$g(10) = 8 \cdot 10 + 5 \qquad \text{Let } x = 10.$$
$$g(10) = 85 \qquad \text{Multiply and then add.}$$

Now, we evaluate $f(85)$.

$$f(x) = \frac{1}{8}x - \frac{5}{8} \qquad \text{Given function}$$
$$f(85) = \frac{1}{8}(85) - \frac{5}{8} \qquad \text{Let } x = 85.$$
$$f(85) = \frac{85}{8} - \frac{5}{8} \qquad \text{Multiply.}$$
$$f(85) = 10 \qquad \text{Subtract and then divide.}$$

Starting with 10, we "applied" function g and then "applied" function f to the result, which returned the number 10. See **Figure 4.**

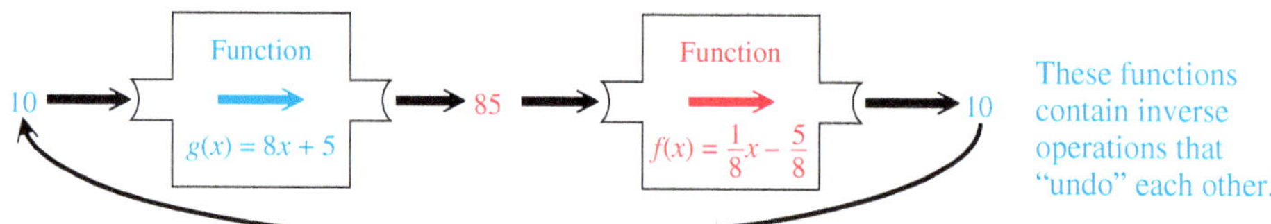

Figure 4

As further examples, confirm the following.

$$g(3) = 29 \quad \text{and} \quad f(29) = 3$$
$$g(-5) = -35 \quad \text{and} \quad f(-35) = -5$$
$$g(2) = 21 \quad \text{and} \quad f(21) = 2$$
$$f(2) = -\frac{3}{8} \quad \text{and} \quad g\left(-\frac{3}{8}\right) = 2$$

In particular, for the pair of functions $g(x) = 8x + 5$ and $f(x) = \frac{1}{8}x - \frac{5}{8}$,

$$f(g(2)) = 2 \quad \text{and} \quad g(f(2)) = 2.$$

In fact, for *any* value of x,

$$f(g(x)) = x \quad \text{and} \quad g(f(x)) = x.$$

Using the notation for composition of functions, these two equations can be written as follows.

$$(f \circ g)(x) = x \quad \text{and} \quad (g \circ f)(x) = x$$ The result is the identity function.

Because the compositions of f and g yield the *identity* function, they are *inverses* of each other.

Inverse Function

Let f be a one-to-one function. Then g is the **inverse function** of f if

$$(f \circ g)(x) = x \quad \text{for every } x \text{ in the domain of } g,$$

and

$$(g \circ f)(x) = x \quad \text{for every } x \text{ in the domain of } f.$$

The condition that f is one-to-one in the definition of inverse function is essential. Otherwise, g will not define a function.

EXAMPLE 3 Determining Whether Two Functions Are Inverses

Let functions f and g be defined respectively by

$$f(x) = x^3 - 1 \quad \text{and} \quad g(x) = \sqrt[3]{x + 1}.$$

Is g the inverse function of f?

SOLUTION As shown in **Figure 5,** the horizontal line test applied to the graph indicates that f is one-to-one, so the function has an inverse. Because it is one-to-one, we now find $(f \circ g)(x)$ and $(g \circ f)(x)$.

$(f \circ g)(x)$	$(g \circ f)(x)$
$= f(g(x))$	$= g(f(x))$
$= \left(\sqrt[3]{x+1}\right)^3 - 1$	$= \sqrt[3]{(x^3 - 1) + 1}$
$= x + 1 - 1$	$= \sqrt[3]{x^3}$
$= x$	$= x$

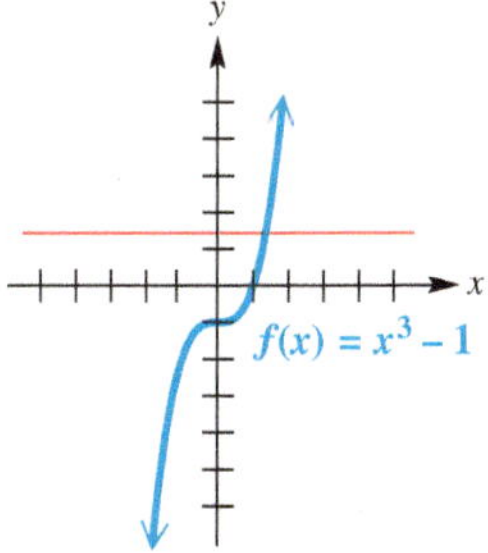

Figure 5

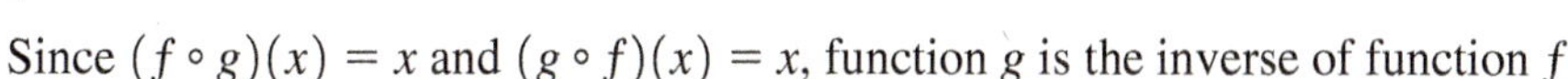
Since $(f \circ g)(x) = x$ and $(g \circ f)(x) = x$, function g is the inverse of function f.

✔ **Now Try Exercise 41.**

A special notation is used for inverse functions: If g is the inverse of a function f, then g is written as f^{-1} (read **"f-inverse"**).

$$f(x) = x^3 - 1 \quad \text{has inverse} \quad f^{-1}(x) = \sqrt[3]{x + 1}.$$ See **Example 3.**

CAUTION ***Do not confuse the -1 in f^{-1} with a negative exponent.*** The symbol $f^{-1}(x)$ represents the inverse function of f, *not* $\frac{1}{f(x)}$.

By the definition of inverse function, the domain of f is the range of f^{-1}, and the range of f is the domain of f^{-1}. See **Figure 6.**

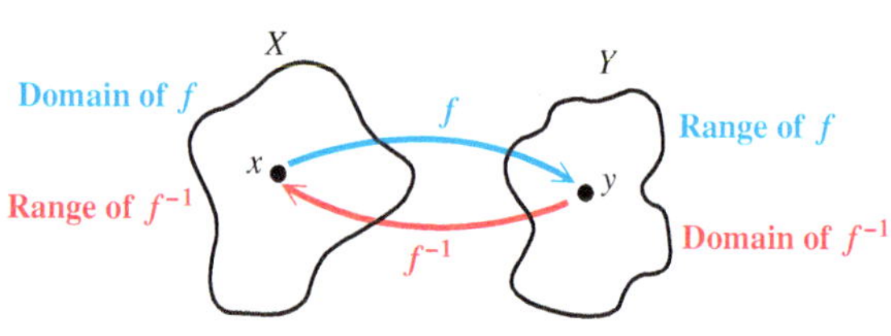

Figure 6

EXAMPLE 4 Finding Inverses of One-to-One Functions

Find the inverse of each function that is one-to-one.

(a) $F = \{(-2, 1), (-1, 0), (0, 1), (1, 2), (2, 2)\}$

(b) $G = \{(3, 1), (0, 2), (2, 3), (4, 0)\}$

(c) The table in the margin shows the number of hurricanes recorded in the North Atlantic during the years 2009–2013. Let f be the function defined in the table, with the years forming the domain and the numbers of hurricanes forming the range.

Year	Number of Hurricanes
2009	3
2010	12
2011	7
2012	10
2013	2

Source: www.wunderground.com

SOLUTION

(a) Each x-value in F corresponds to just one y-value. However, the y-value 2 corresponds to two x-values, 1 and 2. Also, the y-value 1 corresponds to both -2 and 0. Because at least one y-value corresponds to more than one x-value, F is not one-to-one and does not have an inverse.

(b) Every x-value in G corresponds to only one y-value, and every y-value corresponds to only one x-value, so G is a one-to-one function. The inverse function is found by interchanging the x- and y-values in each ordered pair.

$$G^{-1} = \{(1, 3), (2, 0), (3, 2), (0, 4)\}$$

Notice how the domain and range of G become the range and domain, respectively, of G^{-1}.

(c) Each x-value in f corresponds to only one y-value, and each y-value corresponds to only one x-value, so f is a one-to-one function. The inverse function is found by interchanging the x- and y-values in the table.

$$f^{-1}(x) = \{(3, 2009), (12, 2010), (7, 2011), (10, 2012), (2, 2013)\}$$

The domain and range of f become the range and domain of f^{-1}.

✔ **Now Try Exercises 37, 51, and 53.**

Equations of Inverses The inverse of a one-to-one function is found by interchanging the x- and y-values of each of its ordered pairs. The equation of the inverse of a function defined by $y = f(x)$ is found in the same way.

Finding the Equation of the Inverse of $y = f(x)$

For a one-to-one function f defined by an equation $y = f(x)$, find the defining equation of the inverse as follows. (If necessary, replace $f(x)$ with y first. Any restrictions on x and y should be considered.)

Step 1 Interchange x and y.

Step 2 Solve for y.

Step 3 Replace y with $f^{-1}(x)$.

EXAMPLE 5 Finding Equations of Inverses

Determine whether each equation defines a one-to-one function. If so, find the equation of the inverse.

(a) $f(x) = 2x + 5$ **(b)** $y = x^2 + 2$ **(c)** $f(x) = (x - 2)^3$

SOLUTION

(a) The graph of $y = 2x + 5$ is a nonhorizontal line, so by the horizontal line test, f is a one-to-one function. Find the equation of the inverse as follows.

$$f(x) = 2x + 5 \quad \text{Given function}$$

$$y = 2x + 5 \quad \text{Let } y = f(x).$$

Step 1 $$x = 2y + 5 \quad \text{Interchange } x \text{ and } y.$$

Step 2 $$x - 5 = 2y \quad \text{Subtract 5.}$$

$$y = \frac{x-5}{2} \quad \text{Divide by 2. Rewrite.}$$

(Subtract 5; Divide by 2; Rewrite: Solve for y.)

Step 3 $$f^{-1}(x) = \frac{1}{2}x - \frac{5}{2} \quad \text{Replace } y \text{ with } f^{-1}(x). \quad \frac{a-b}{c} = \left(\frac{1}{c}\right)a - \frac{b}{c}$$

Thus, the equation $f^{-1}(x) = \frac{x-5}{2} = \frac{1}{2}x - \frac{5}{2}$ represents a linear function. In the function $y = 2x + 5$, the value of y is found by starting with a value of x, multiplying by 2, and adding 5.

The equation $f^{-1}(x) = \frac{x-5}{2}$ for the inverse *subtracts* 5 and then *divides* by 2. An inverse is used to "undo" what a function does to the variable x.

(b) The equation $y = x^2 + 2$ has a parabola opening up as its graph, so some horizontal lines will intersect the graph at two points. For example, both $x = 3$ and $x = -3$ correspond to $y = 11$. Because of the presence of the x^2-term, there are many pairs of x-values that correspond to the same y-value. This means that the function defined by $y = x^2 + 2$ is not one-to-one and does not have an inverse.

Proceeding with the steps for finding the equation of an inverse leads to

$$y = x^2 + 2$$

$$x = y^2 + 2 \quad \text{Interchange } x \text{ and } y.$$

$$x - 2 = y^2 \quad \text{Solve for } y.$$

$$\pm\sqrt{x-2} = y. \quad \text{Square root property}$$

Remember both roots.

The last equation shows that there are two y-values for each choice of x greater than 2, indicating that this is not a function.

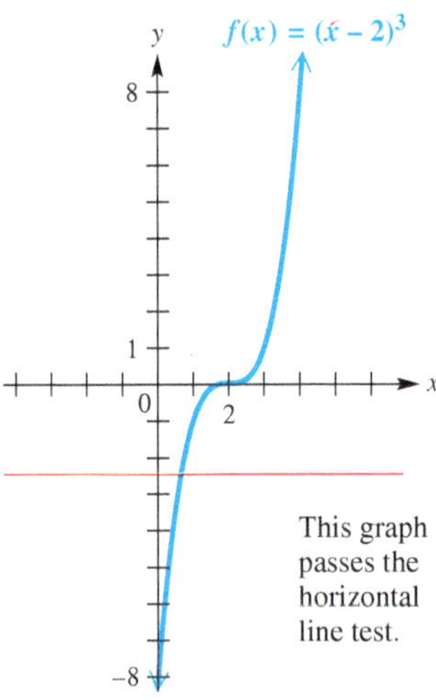

Figure 7

(c) **Figure 7** shows that the horizontal line test assures us that this horizontal translation of the graph of the cubing function is one-to-one.

	$f(x) = (x - 2)^3$	Given function
	$y = (x - 2)^3$	Replace $f(x)$ with y.
Step 1	$x = (y - 2)^3$	Interchange x and y.
Step 2	$\sqrt[3]{x} = \sqrt[3]{(y-2)^3}$	Take the cube root on each side. (Solve for y.)
	$\sqrt[3]{x} = y - 2$	$\sqrt[3]{a^3} = a$ (Solve for y.)
	$\sqrt[3]{x} + 2 = y$	Add 2. (Solve for y.)
Step 3	$f^{-1}(x) = \sqrt[3]{x} + 2$	Replace y with $f^{-1}(x)$. Rewrite.

✔ **Now Try Exercises 59(a), 63(a), and 65(a).**

EXAMPLE 6 Finding the Equation of the Inverse of a Rational Function

The following rational function is one-to-one. Find its inverse.

$$f(x) = \frac{2x+3}{x-4}, \quad x \neq 4$$

SOLUTION

	$f(x) = \frac{2x+3}{x-4}, \quad x \neq 4$	Given function
	$y = \frac{2x+3}{x-4}$	Replace $f(x)$ with y.
Step 1	$x = \frac{2y+3}{y-4}, \quad y \neq 4$	Interchange x and y.
Step 2	$x(y - 4) = 2y + 3$	Multiply by $y - 4$. (Solve for y.)
	$xy - 4x = 2y + 3$	Distributive property (Solve for y.)
Pay close attention here.	$xy - 2y = 4x + 3$	Add $4x$ and $-2y$. (Solve for y.)
	$y(x - 2) = 4x + 3$	Factor out y. (Solve for y.)
	$y = \frac{4x+3}{x-2}, \quad x \neq 2$	Divide by $x - 2$. (Solve for y.)

In the final line, we give the condition $x \neq 2$. (Note that 2 is not in the *range* of f, so it is not in the domain of f^{-1}.)

Step 3 $\quad f^{-1}(x) = \frac{4x+3}{x-2}, \quad x \neq 2 \quad$ Replace y with $f^{-1}(x)$.

✔ **Now Try Exercise 71(a).**

One way to graph the inverse of a function f whose equation is known follows.

Step 1 Find some ordered pairs that are on the graph of f.

Step 2 Interchange x and y to find ordered pairs that are on the graph of f^{-1}.

Step 3 Plot those points, and sketch the graph of f^{-1} through them.

Another way is to select points on the graph of f and use symmetry to find corresponding points on the graph of f^{-1}.

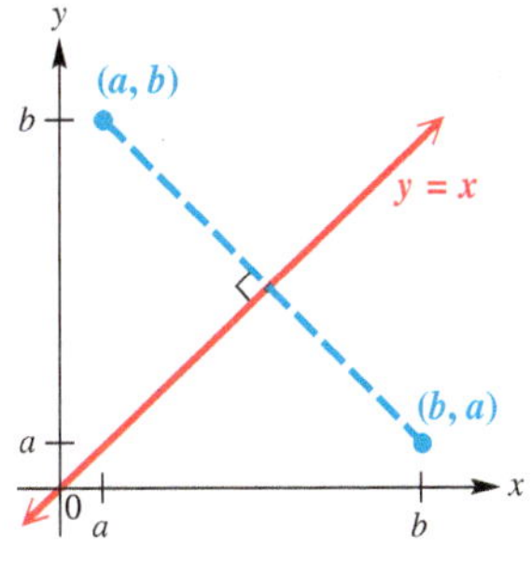

Figure 8

For example, suppose the point (a, b) shown in **Figure 8** is on the graph of a one-to-one function f. Then the point (b, a) is on the graph of f^{-1}. The line segment connecting (a, b) and (b, a) is perpendicular to, and cut in half by, the line $y = x$. The points (a, b) and (b, a) are "mirror images" of each other with respect to $y = x$.

Thus, we can find the graph of f^{-1} from the graph of f by locating the mirror image of each point in f with respect to the line $y = x$.

EXAMPLE 7 Graphing f^{-1} Given the Graph of f

In each set of axes in **Figure 9,** the graph of a one-to-one function f is shown in blue. Graph f^{-1} in red.

SOLUTION In **Figure 9,** the graphs of two functions f shown in blue are given with their inverses shown in red. In each case, the graph of f^{-1} is a reflection of the graph of f with respect to the line $y = x$.

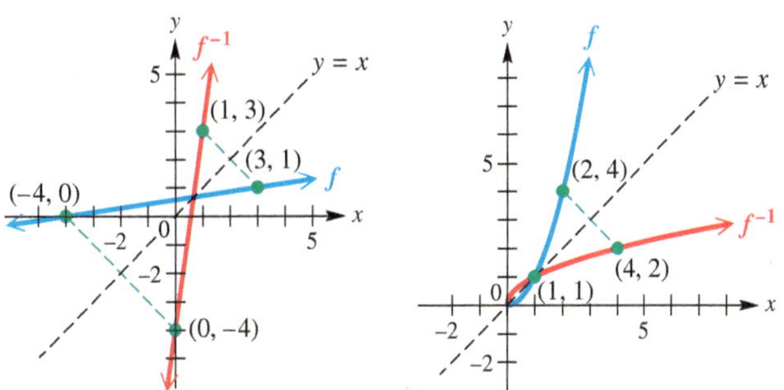

Figure 9

✔ **Now Try Exercises 77 and 81.**

EXAMPLE 8 Finding the Inverse of a Function (Restricted Domain)

Let $f(x) = \sqrt{x + 5}, \quad x \geq -5$. Find $f^{-1}(x)$.

SOLUTION The domain of f is restricted to the interval $[-5, \infty)$. Function f is one-to-one because it is an increasing function and thus has an inverse function. Now we find the equation of the inverse.

$$f(x) = \sqrt{x + 5}, \quad x \geq -5 \qquad \text{Given function}$$

$$y = \sqrt{x + 5}, \quad x \geq -5 \qquad \text{Replace } f(x) \text{ with } y.$$

Step 1 $$x = \sqrt{y + 5}, \quad y \geq -5 \qquad \text{Interchange } x \text{ and } y.$$

Step 2 $$x^2 = \left(\sqrt{y + 5}\right)^2 \qquad \text{Square each side.}$$

$$x^2 = y + 5 \qquad \left(\sqrt{a}\right)^2 = a \text{ for } a \geq 0$$

$$y = x^2 - 5 \qquad \text{Subtract 5. Rewrite.}$$

(Solve for y.)

However, we cannot define $f^{-1}(x)$ as $x^2 - 5$. The domain of f is $[-5, \infty)$, and its range is $[0, \infty)$. The range of f is the domain of f^{-1}, so f^{-1} must be defined as follows.

Step 3 $$f^{-1}(x) = x^2 - 5, \quad x \geq 0$$

As a check, the range of f^{-1}, $[-5, \infty)$, is the domain of f.

Graphs of f and f^{-1} are shown in **Figures 10 and 11.** The line $y = x$ is included on the graphs to show that the graphs of f and f^{-1} are mirror images with respect to this line.

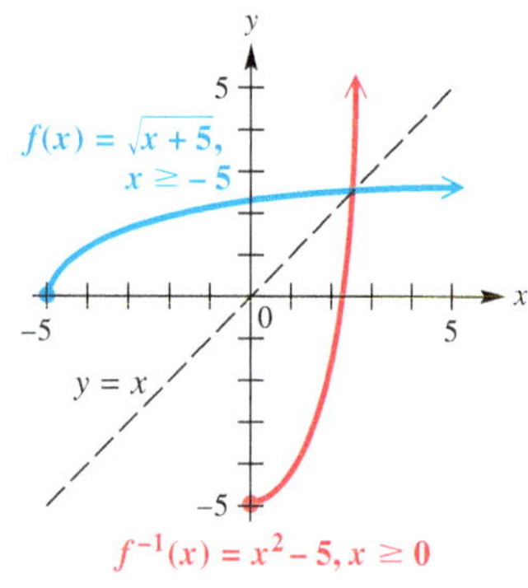

Figure 10

Figure 11

✔ **Now Try Exercise 75.**

Important Facts about Inverses

1. If f is one-to-one, then f^{-1} exists.
2. The domain of f is the range of f^{-1}, and the range of f is the domain of f^{-1}.
3. If the point (a, b) lies on the graph of f, then (b, a) lies on the graph of f^{-1}. The graphs of f and f^{-1} are reflections of each other across the line $y = x$.
4. To find the equation for f^{-1}, replace $f(x)$ with y, interchange x and y, and solve for y. This gives $f^{-1}(x)$.

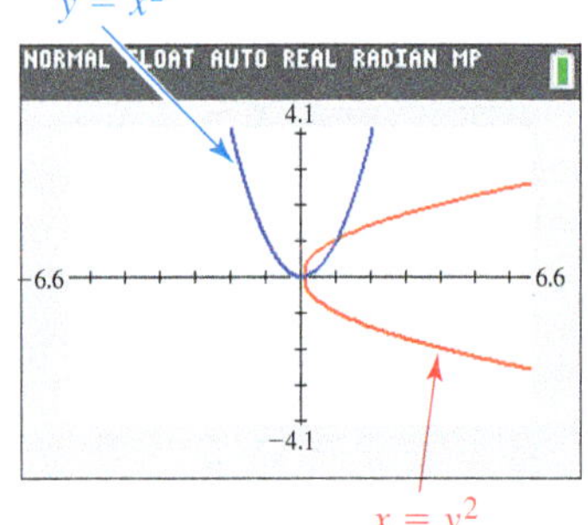

Despite the fact that $y = x^2$ is not one-to-one, the calculator will draw its "inverse," $x = y^2$.

Figure 12

Some graphing calculators have the capability of "drawing" the reflection of a graph across the line $y = x$. This feature does not require that the function be one-to-one, however, so the resulting figure may not be the graph of a function. See **Figure 12.** ***It is necessary to understand the mathematics to interpret results correctly.*** ■

An Application of Inverse Functions to Cryptography A one-to-one function and its inverse can be used to make information secure. The function is used to encode a message, and its inverse is used to decode the coded message. In practice, complicated functions are used.

EXAMPLE 9 Using Functions to Encode and Decode a Message

Use the one-to-one function $f(x) = 3x + 1$ and the following numerical values assigned to each letter of the alphabet to encode and decode the message BE MY FACEBOOK FRIEND.

A	1	**H**	8	**O**	15	**V**	22
B	2	**I**	9	**P**	16	**W**	23
C	3	**J**	10	**Q**	17	**X**	24
D	4	**K**	11	**R**	18	**Y**	25
E	5	**L**	12	**S**	19	**Z**	26
F	6	**M**	13	**T**	20		
G	7	**N**	14	**U**	21		

A	1	**N**	14
B	2	**O**	15
C	3	**P**	16
D	4	**Q**	17
E	5	**R**	18
F	6	**S**	19
G	7	**T**	20
H	8	**U**	21
I	9	**V**	22
J	10	**W**	23
K	11	**X**	24
L	12	**Y**	25
M	13	**Z**	26

SOLUTION The message BE MY FACEBOOK FRIEND would be encoded as

7 16 40 76 19 4 10 16 7
46 46 34 19 55 28 16 43 13

because

B corresponds to 2 and $f(2) = 3(2) + 1 = 7$,

E corresponds to 5 and $f(5) = 3(5) + 1 = 16$, and so on.

Using the inverse $f^{-1}(x) = \frac{1}{3}x - \frac{1}{3}$ to decode yields

$$f^{-1}(7) = \frac{1}{3}(7) - \frac{1}{3} = 2,$$ which corresponds to B,

$$f^{-1}(16) = \frac{1}{3}(16) - \frac{1}{3} = 5,$$ which corresponds to E, and so on.

✔ **Now Try Exercise 97.**

4.1 Exercises

CONCEPT PREVIEW *Determine whether the function represented in each table is one-to-one.*

1. The table shows the number of registered passenger cars in the United States for the years 2008–2012.

Year	Registered Passenger Cars (in thousands)
2008	137,080
2009	134,880
2010	139,892
2011	125,657
2012	111,290

Source: U.S. Federal Highway Administration.

2. The table gives the number of representatives currently in Congress from each of five New England states.

State	Number of Representatives
Connecticut	5
Maine	2
Massachusetts	9
New Hampshire	2
Vermont	1

Source: www.house.gov

CONCEPT PREVIEW *Fill in the blank(s) to correctly complete each sentence.*

3. For a function to have an inverse, it must be ________.

4. If two functions f and g are inverses, then $(f \circ g)(x) =$ ________ and ________ $= x$.

5. The domain of f is equal to the ________ of f^{-1}, and the range of f is equal to the ________ of f^{-1}.

6. If the point (a, b) lies on the graph of f, and f has an inverse, then the point ________ lies on the graph of f^{-1}.

7. If $f(x) = x^3$, then $f^{-1}(x) =$ ________.

8. If a function f has an inverse, then the graph of f^{-1} may be obtained by reflecting the graph of f across the line with equation ________.

9. If a function f has an inverse and $f(-3) = 6$, then $f^{-1}(6) =$ ________.

10. If $f(-4) = 16$ and $f(4) = 16$, then f ________ (does/does not) have an inverse because ________.

Determine whether each function graphed or defined is one-to-one. **See Examples 1 and 2.**

11.

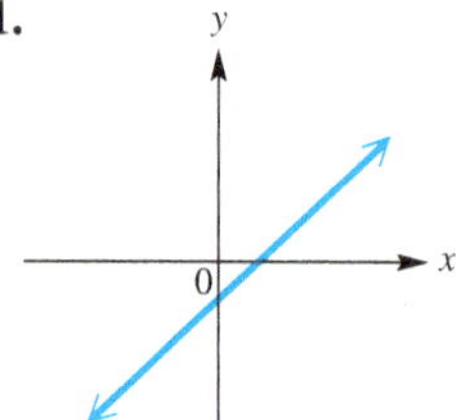

12.

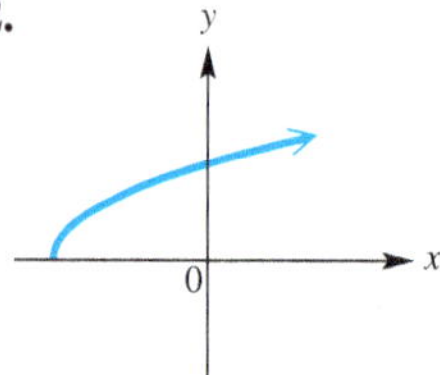

13.

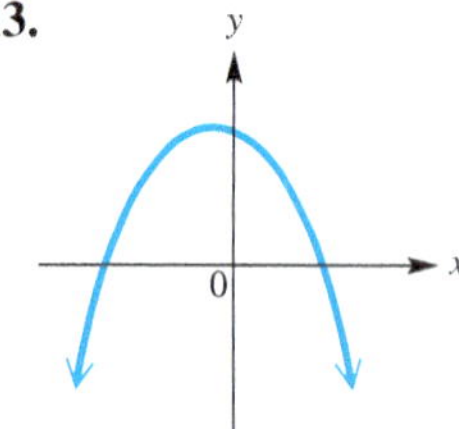

14.

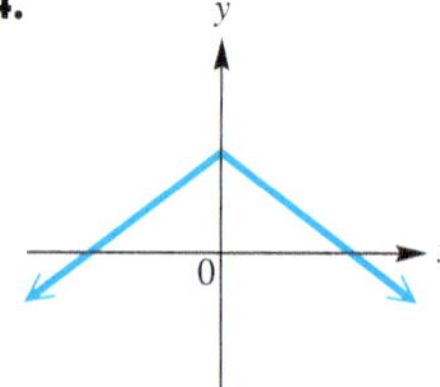

15.

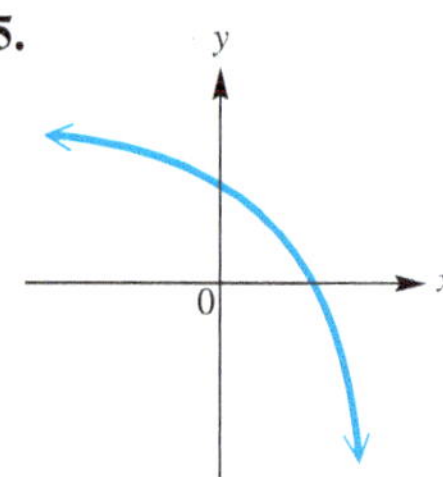

16.

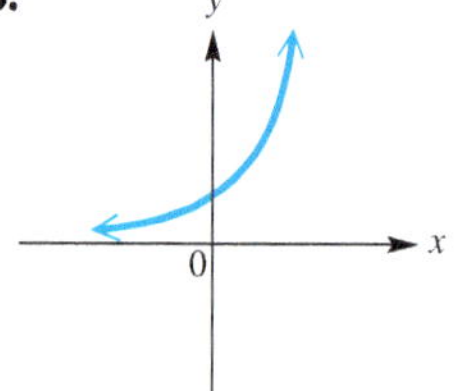

17. $y = 2x - 8$

18. $y = 4x + 20$

19. $y = \sqrt{36 - x^2}$

20. $y = -\sqrt{100 - x^2}$

21. $y = 2x^3 - 1$

22. $y = 3x^3 - 6$

23. $y = \dfrac{-1}{x + 2}$

24. $y = \dfrac{4}{x - 8}$

25. $y = 2(x + 1)^2 - 6$

26. $y = -3(x - 6)^2 + 8$

27. $y = \sqrt[3]{x + 1} - 3$

28. $y = -\sqrt[3]{x + 2} - 8$

Concept Check Answer each question.

29. Can a constant function, such as $f(x) = 3$, defined over the set of real numbers, be one-to-one?

30. Can a polynomial function of even degree defined over the set of real numbers have an inverse?

Concept Check An everyday activity is described. Keeping in mind that an inverse operation "undoes" what an operation does, describe each inverse activity.

31. tying your shoelaces

32. starting a car

33. entering a room

34. climbing the stairs

35. screwing in a light bulb

36. filling a cup

Determine whether the given functions are inverses. **See Example 4.**

37.

x	$f(x)$
3	-4
2	-6
5	8
1	9
4	3

x	$g(x)$
-4	3
-6	2
8	5
9	1
3	4

38.

x	$f(x)$
-2	-8
-1	-1
0	0
1	1
2	8

x	$g(x)$
8	-2
1	-1
0	0
-1	1
-8	2

39. $f = \{(2, 5), (3, 5), (4, 5)\}$; $g = \{(5, 2)\}$

40. $f = \{(1, 1), (3, 3), (5, 5)\}$; $g = \{(1, 1), (3, 3), (5, 5)\}$

Use the definition of inverses to determine whether f and g are inverses. **See Example 3.**

41. $f(x) = 2x + 4, \quad g(x) = \frac{1}{2}x - 2$

42. $f(x) = 3x + 9, \quad g(x) = \frac{1}{3}x - 3$

43. $f(x) = -3x + 12, \quad g(x) = -\frac{1}{3}x - 12$

44. $f(x) = -4x + 2, \quad g(x) = -\frac{1}{4}x - 2$

45. $f(x) = \frac{x + 1}{x - 2}, \quad g(x) = \frac{2x + 1}{x - 1}$

46. $f(x) = \frac{x - 3}{x + 4}, \quad g(x) = \frac{4x + 3}{1 - x}$

47. $f(x) = \frac{2}{x + 6}, \quad g(x) = \frac{6x + 2}{x}$

48. $f(x) = \frac{-1}{x + 1}, \quad g(x) = \frac{1 - x}{x}$

49. $f(x) = x^2 + 3, \quad x \geq 0; \quad g(x) = \sqrt{x - 3}, \quad x \geq 3$

50. $f(x) = \sqrt{x + 8}, \quad x \geq -8; \quad g(x) = x^2 - 8, \quad x \geq 0$

Find the inverse of each function that is one-to-one. **See Example 4.**

51. $\{(-3, 6), (2, 1), (5, 8)\}$

52. $\left\{(3, -1), (5, 0), (0, 5), \left(4, \frac{2}{3}\right)\right\}$

53. $\{(1, -3), (2, -7), (4, -3), (5, -5)\}$

54. $\{(6, -8), (3, -4), (0, -8), (5, -4)\}$

Determine whether each pair of functions graphed are inverses. **See Example 7.**

55.

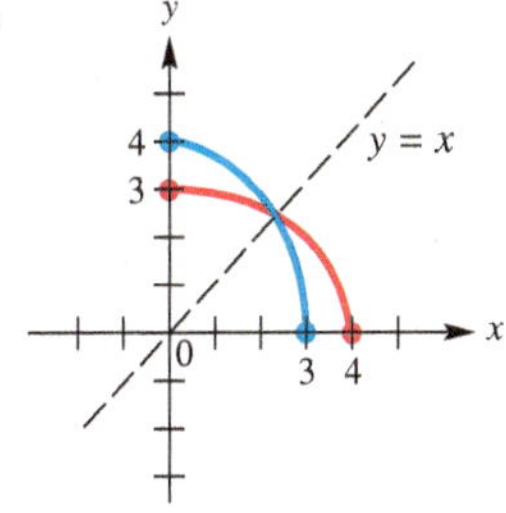

56.

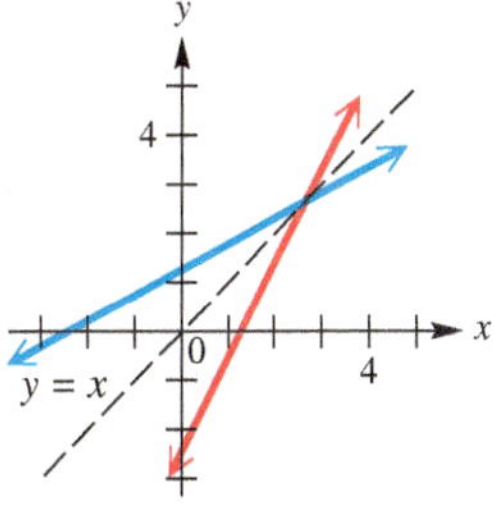

57.

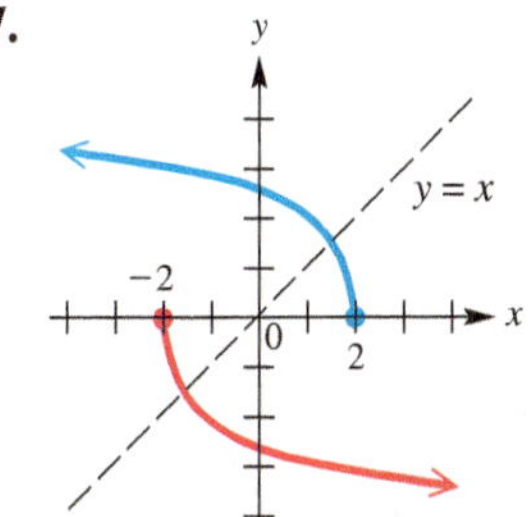

58.

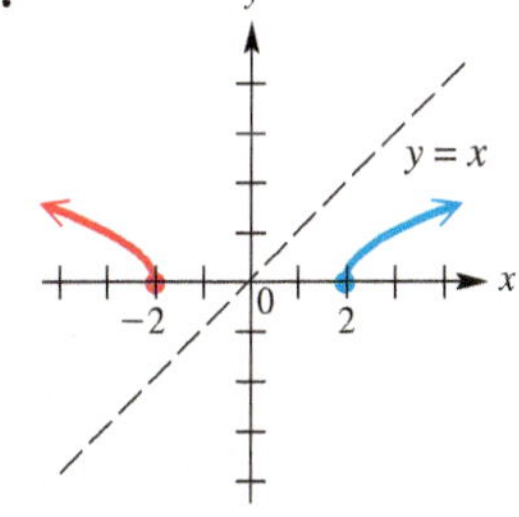

For each function that is one-to-one, **(a)** *write an equation for the inverse function,* **(b)** *graph f and f^{-1} on the same axes, and* **(c)** *give the domain and range of both f and f^{-1}. If the function is not one-to-one, say so.* ***See Examples 5–8.***

59. $f(x) = 3x - 4$

60. $f(x) = 4x - 5$

61. $f(x) = -4x + 3$

62. $f(x) = -6x - 8$

63. $f(x) = x^3 + 1$

64. $f(x) = -x^3 - 2$

65. $f(x) = x^2 + 8$

66. $f(x) = -x^2 + 2$

67. $f(x) = \frac{1}{x}, \quad x \neq 0$

68. $f(x) = \frac{4}{x}, \quad x \neq 0$

69. $f(x) = \frac{1}{x-3}, \quad x \neq 3$

70. $f(x) = \frac{1}{x+2}, \quad x \neq -2$

71. $f(x) = \frac{x+1}{x-3}, \quad x \neq 3$

72. $f(x) = \frac{x+2}{x-1}, \quad x \neq 1$

73. $f(x) = \frac{2x+6}{x-3}, \quad x \neq 3$

74. $f(x) = \frac{-3x+12}{x-6}, \quad x \neq 6$

75. $f(x) = \sqrt{x+6}, \quad x \geq -6$

76. $f(x) = -\sqrt{x^2-16}, \quad x \geq 4$

Graph the inverse of each one-to-one function. ***See Example 7.***

77.

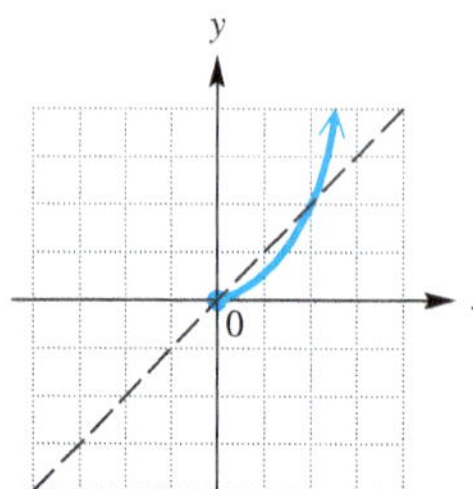

78.

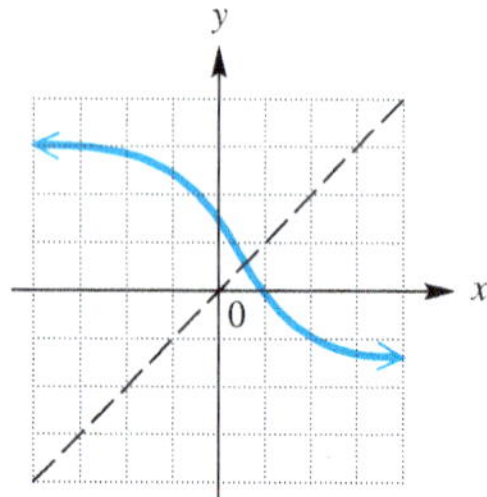

79.

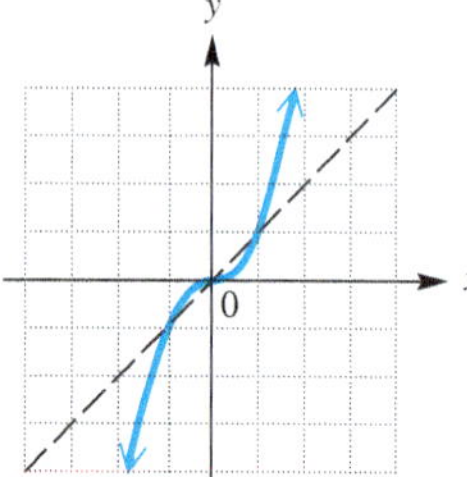

80.

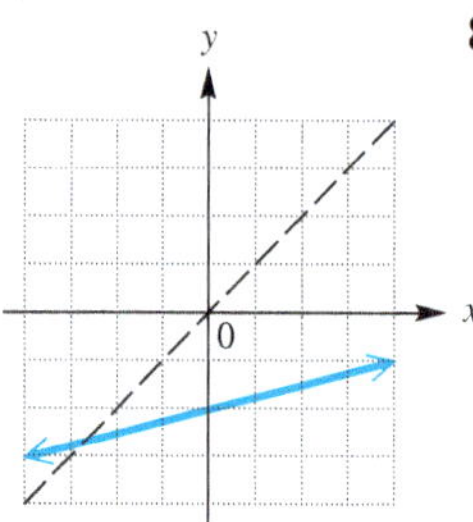

81.

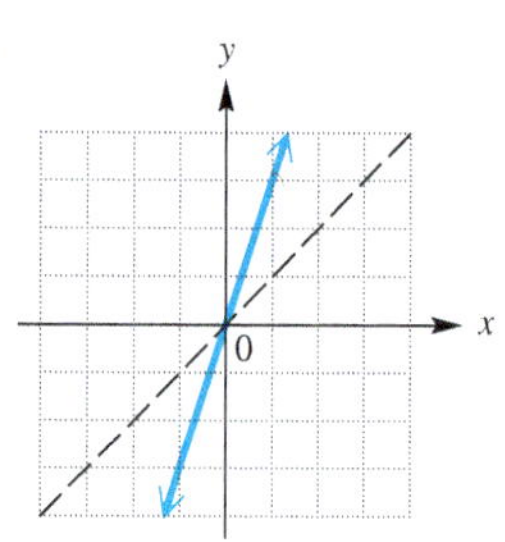

82.

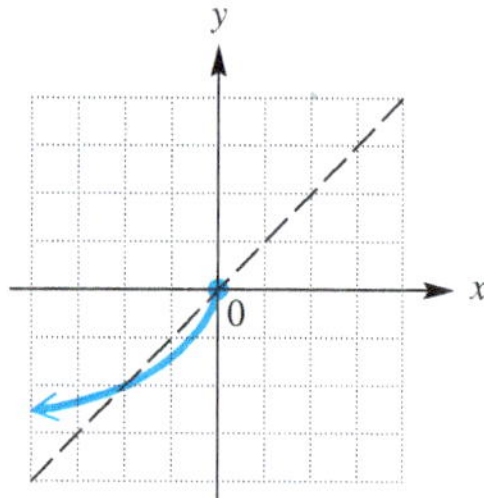

Concept Check *The graph of a function f is shown in the figure. Use the graph to find each value.*

83. $f^{-1}(4)$

84. $f^{-1}(2)$

85. $f^{-1}(0)$

86. $f^{-1}(-2)$

87. $f^{-1}(-3)$

88. $f^{-1}(-4)$

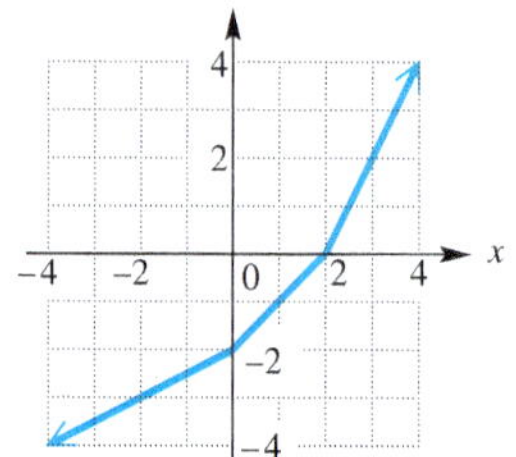

Concept Check *Answer each of the following.*

89. Suppose $f(x)$ is the number of cars that can be built for x dollars. What does $f^{-1}(1000)$ represent?

90. Suppose $f(r)$ is the volume (in cubic inches) of a sphere of radius r inches. What does $f^{-1}(5)$ represent?

91. If a line has slope a, what is the slope of its reflection across the line $y = x$?

92. For a one-to-one function f, find $(f^{-1} \circ f)(2)$, where $f(2) = 3$.

Use a graphing calculator to graph each function defined as follows, using the given viewing window. Use the graph to decide which functions are one-to-one. If a function is one-to-one, give the equation of its inverse.

93. $f(x) = 6x^3 + 11x^2 - 6;$
$[-3, 2]$ by $[-10, 10]$

94. $f(x) = x^4 - 5x^2;$
$[-3, 3]$ by $[-8, 8]$

95. $f(x) = \frac{x-5}{x+3}, \quad x \neq -3;$
$[-8, 8]$ by $[-6, 8]$

96. $f(x) = \frac{-x}{x-4}, \quad x \neq 4;$
$[-1, 8]$ by $[-6, 6]$

Use the following alphabet coding assignment to work each problem. ***See Example 9.***

A	1	**H**	8	**O**	15	**V**	22
B	2	**I**	9	**P**	16	**W**	23
C	3	**J**	10	**Q**	17	**X**	24
D	4	**K**	11	**R**	18	**Y**	25
E	5	**L**	12	**S**	19	**Z**	26
F	6	**M**	13	**T**	20		
G	7	**N**	14	**U**	21		

97. The function $f(x) = 3x - 2$ was used to encode a message as

37 25 19 61 13 34 22 1 55 1 52 52 25 64 13 10.

Find the inverse function and determine the message.

98. The function $f(x) = 2x - 9$ was used to encode a message as

−5 9 5 5 9 27 15 29 −1 21 19 31 −3 27 41.

Find the inverse function and determine the message.

99. Encode the message SEND HELP, using the one-to-one function

$$f(x) = x^3 - 1.$$

Give the inverse function that the decoder will need when the message is received.

100. Encode the message SAILOR BEWARE, using the one-to-one function

$$f(x) = (x + 1)^3.$$

Give the inverse function that the decoder will need when the message is received.

4.2 Exponential Functions

- Exponents and Properties
- Exponential Functions
- Exponential Equations
- Compound Interest
- The Number e and Continuous Compounding
- Exponential Models

Exponents and Properties

Recall the definition of $a^{m/n}$: If a is a real number, m is an integer, n is a positive integer, and $\sqrt[n]{a}$ is a real number, then

$$a^{m/n} = \left(\sqrt[n]{a}\right)^m.$$

For example,

$$16^{3/4} = \left(\sqrt[4]{16}\right)^3 = 2^3 = 8,$$

$$27^{-1/3} = \frac{1}{27^{1/3}} = \frac{1}{\sqrt[3]{27}} = \frac{1}{3}, \quad \text{and} \quad 64^{-1/2} = \frac{1}{64^{1/2}} = \frac{1}{\sqrt{64}} = \frac{1}{8}.$$

In this section, we extend the definition of a^r to include all *real* (not just rational) values of the exponent r. Consider the graphs of $y = 2^x$ for different domains in **Figure 13.**

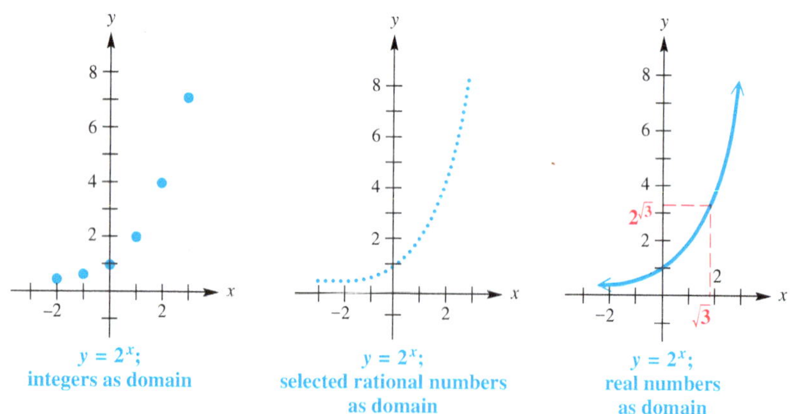

Figure 13

The equations that use just integers or selected rational numbers as domain in **Figure 13** leave holes in the graphs. In order for the graph to be continuous, we must extend the domain to include irrational numbers such as $\sqrt{3}$. We might evaluate $2^{\sqrt{3}}$ by approximating the exponent with the rational numbers 1.7, 1.73, 1.732, and so on. Because these values approach the value of $\sqrt{3}$ more and more closely, it is reasonable that $2^{\sqrt{3}}$ should be approximated more and more closely by the numbers $2^{1.7}$, $2^{1.73}$, $2^{1.732}$, and so on. These expressions can be evaluated using rational exponents as follows.

$$2^{1.7} = 2^{17/10} = \left(\sqrt[10]{2}\right)^{17} \approx 3.249009585$$

Because any irrational number may be approximated more and more closely using rational numbers, we can extend the definition of a^r to include all real number exponents and apply all previous theorems for exponents. In addition to the rules for exponents presented earlier, we use several new properties in this chapter.

Additional Properties of Exponents

For any real number $a > 0$, $a \neq 1$, the following statements hold.

Property	Description
(a) a^x **is a unique real number for all real numbers** x.	$y = a^x$ can be considered a function $f(x) = a^x$ with domain $(-\infty, \infty)$.
(b) $a^b = a^c$ **if and only if** $b = c$.	The function $f(x) = a^x$ is one-to-one.
(c) If $a > 1$ **and** $m < n$, **then** $a^m < a^n$.	*Example:* $2^3 < 2^4 \quad (a > 1)$ Increasing the exponent leads to a *greater* number. The function $f(x) = 2^x$ is an *increasing* function.
(d) If $0 < a < 1$ **and** $m < n$, **then** $a^m > a^n$.	*Example:* $\left(\frac{1}{2}\right)^2 > \left(\frac{1}{2}\right)^3 \quad (0 < a < 1)$ Increasing the exponent leads to a *lesser* number. The function $f(x) = \left(\frac{1}{2}\right)^x$ is a *decreasing* function.

Exponential Functions We now define a function $f(x) = a^x$ whose domain is the set of all real numbers. Notice how the independent variable x appears in the exponent in this function. In earlier chapters, this was not the case.

Exponential Function

If $a > 0$ and $a \neq 1$, then the **exponential function with base a** is

$$f(x) = a^x.$$

NOTE The restrictions on a in the definition of an exponential function are important. Consider the outcome of breaking each restriction.

If $a < 0$, say $a = -2$, and we let $x = \frac{1}{2}$, then $f\left(\frac{1}{2}\right) = (-2)^{1/2} = \sqrt{-2}$, which is *not* a real number.

If $a = 1$, then the function becomes the constant function $f(x) = 1^x = 1$, which is *not* an exponential function.

EXAMPLE 1 Evaluating an Exponential Function

For $f(x) = 2^x$, find each of the following.

(a) $f(-1)$ **(b)** $f(3)$ **(c)** $f\left(\frac{5}{2}\right)$ **(d)** $f(4.92)$

SOLUTION

(a) $f(-1) = 2^{-1} = \frac{1}{2}$ Replace x with -1. **(b)** $f(3) = 2^3 = 8$

(c) $f\left(\frac{5}{2}\right) = 2^{5/2} = (2^5)^{1/2} = 32^{1/2} = \sqrt{32} = \sqrt{16 \cdot 2} = 4\sqrt{2}$

(d) $f(4.92) = 2^{4.92} \approx 30.2738447$ Use a calculator.

✔ **Now Try Exercises 13, 19, and 23.**

We repeat the final graph of $y = 2^x$ (with real numbers as domain) from **Figure 13** and summarize important details of the function $f(x) = 2^x$ here.

- The y-intercept is $(0, 1)$.
- Because $2^x > 0$ for all x and $2^x \rightarrow 0$ as $x \rightarrow -\infty$, the x-axis is a horizontal asymptote.
- As the graph suggests, the domain of the function is $(-\infty, \infty)$ and the range is $(0, \infty)$.
- The function is increasing on its entire domain. Therefore, it is one-to-one.

These observations lead to the following generalizations about the graphs of exponential functions.

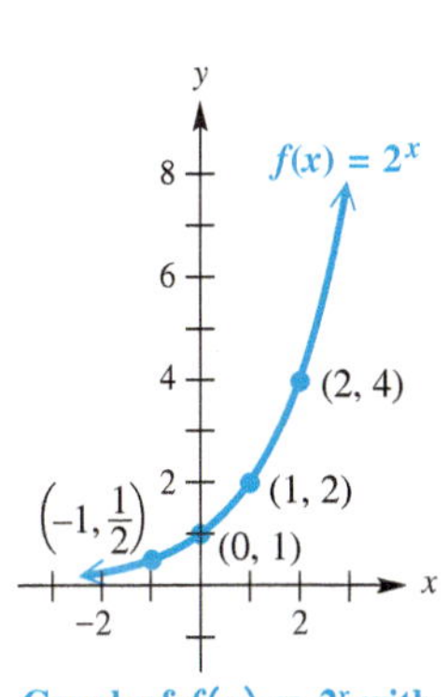

Graph of $f(x) = 2^x$ with domain $(-\infty, \infty)$

Figure 13 (repeated)

Exponential Function $f(x) = a^x$

Domain: $(-\infty, \infty)$ Range: $(0, \infty)$

For $f(x) = 2^x$:

x	$f(x)$
-2	$\frac{1}{4}$
-1	$\frac{1}{2}$
0	1
1	2
2	4
3	8

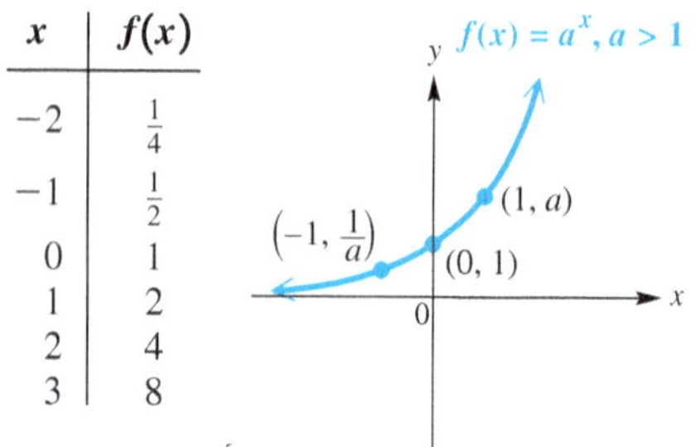

$f(x) = a^x, a > 1$

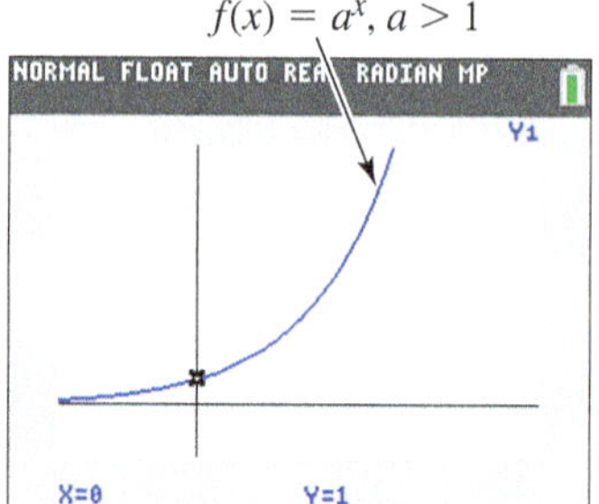

This is the general behavior seen on a calculator graph for **any base a, for $a > 1$.**

Figure 14

- **$f(x) = a^x$, for $a > 1$,** is increasing and continuous on its entire domain, $(-\infty, \infty)$.
- The x-axis is a horizontal asymptote as $x \to -\infty$.
- The graph passes through the points $\left(-1, \frac{1}{a}\right)$, $(0, 1)$, and $(1, a)$.

For $f(x) = \left(\frac{1}{2}\right)^x$:

x	$f(x)$
-3	8
-2	4
-1	2
0	1
1	$\frac{1}{2}$
2	$\frac{1}{4}$

y
$f(x) = a^x, 0 < a < 1$
$\left(-1, \frac{1}{a}\right)$
$(0, 1)$
$(1, a)$
0
x

$f(x) = a^x, 0 < a < 1$

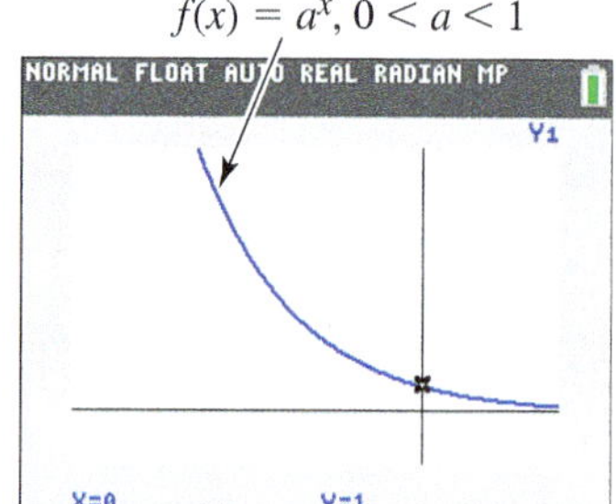

This is the general behavior seen on a calculator graph for **any base a, for $0 < a < 1$.**

Figure 15

- **$f(x) = a^x$, for $0 < a < 1$,** is decreasing and continuous on its entire domain, $(-\infty, \infty)$.
- The x-axis is a horizontal asymptote as $x \to \infty$.
- The graph passes through the points $\left(-1, \frac{1}{a}\right)$, $(0, 1)$, and $(1, a)$.

Recall that the graph of $y = f(-x)$ is the graph of $y = f(x)$ reflected across the y-axis. Thus, we have the following.

$$\text{If } f(x) = 2^x, \quad \text{then} \quad f(-x) = 2^{-x} = 2^{-1 \cdot x} = (2^{-1})^x = \left(\frac{1}{2}\right)^x.$$

This is supported by the graphs in **Figures 14 and 15.**

The graph of $f(x) = 2^x$ is typical of graphs of $f(x) = a^x$ where $a > 1$. For larger values of a, the graphs rise more steeply, but the general shape is similar to the graph in **Figure 14.** When $0 < a < 1$, the graph decreases in a manner similar to the graph of $f(x) = \left(\frac{1}{2}\right)^x$ in **Figure 15.**

In **Figure 16,** the graphs of several typical exponential functions illustrate these facts.

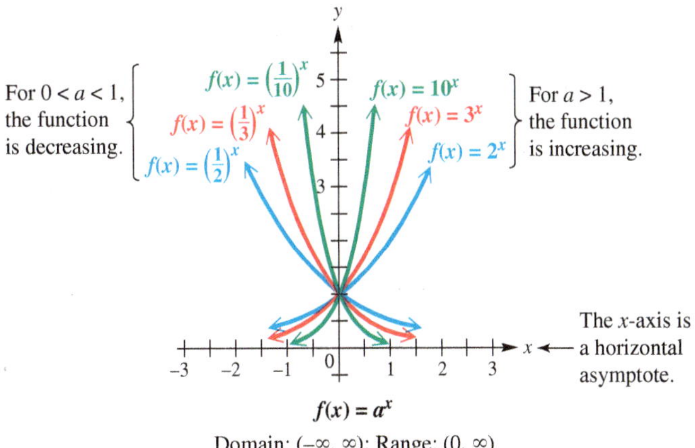

Figure 16

In summary, the graph of a function of the form $f(x) = a^x$ has the following features.

Characteristics of the Graph of $f(x) = a^x$

1. The points $\left(-1, \frac{1}{a}\right)$, $(0, 1)$, and $(1, a)$ are on the graph.
2. If $a > 1$, then f is an increasing function.

   If $0 < a < 1$, then f is a decreasing function.
3. The x-axis is a horizontal asymptote.
4. The domain is $(-\infty, \infty)$, and the range is $(0, \infty)$.

EXAMPLE 2 Graphing an Exponential Function

Graph $f(x) = \left(\frac{1}{5}\right)^x$. Give the domain and range.

SOLUTION The y-intercept is $(0, 1)$, and the x-axis is a horizontal asymptote. Plot a few ordered pairs, and draw a smooth curve through them as shown in **Figure 17.**

x	$f(x)$
-2	25
-1	5
0	1
1	$\frac{1}{5}$
2	$\frac{1}{25}$

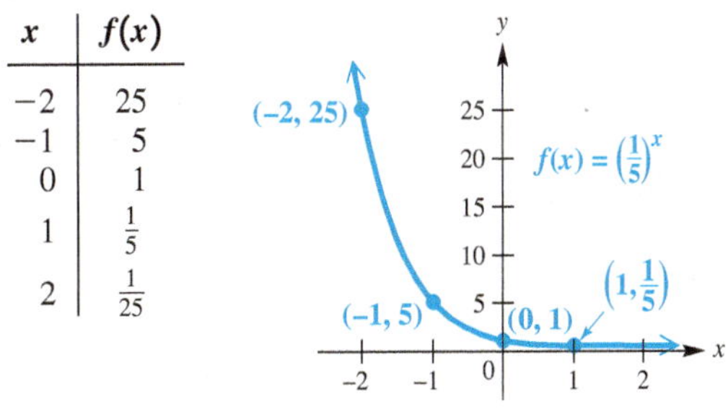

Figure 17

This function has domain $(-\infty, \infty)$, range $(0, \infty)$, and is one-to-one. It is decreasing on its entire domain.

✔ **Now Try Exercise 29.**

EXAMPLE 3 Graphing Reflections and Translations

Graph each function. Show the graph of $y = 2^x$ for comparison. Give the domain and range.

(a) $f(x) = -2^x$ **(b)** $f(x) = 2^{x+3}$ **(c)** $f(x) = 2^{x-2} - 1$

SOLUTION In each graph, we show in particular how the point $(0, 1)$ on the graph of $y = 2^x$ has been translated.

(a) The graph of $f(x) = -2^x$ is that of $f(x) = 2^x$ reflected across the x-axis. See **Figure 18.** The domain is $(-\infty, \infty)$, and the range is $(-\infty, 0)$.

(b) The graph of $f(x) = 2^{x+3}$ is the graph of $f(x) = 2^x$ translated 3 units to the left, as shown in **Figure 19.** The domain is $(-\infty, \infty)$, and the range is $(0, \infty)$.

(c) The graph of $f(x) = 2^{x-2} - 1$ is that of $f(x) = 2^x$ translated 2 units to the right and 1 unit down. See **Figure 20.** The domain is $(-\infty, \infty)$, and the range is $(-1, \infty)$.

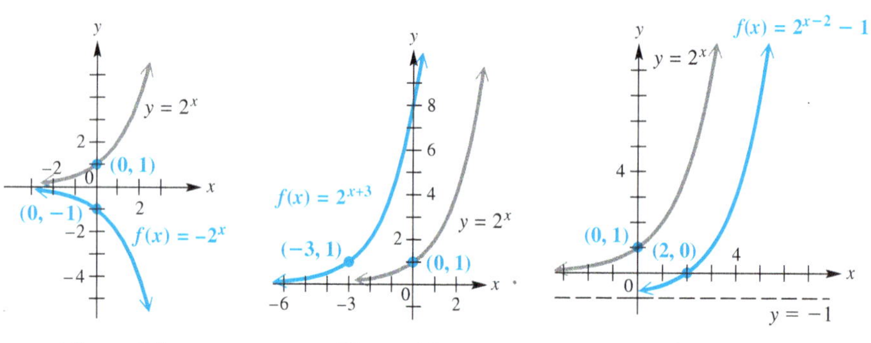

Figure 18 **Figure 19** **Figure 20**

✔ **Now Try Exercises 39, 41, and 47.**

Exponential Equations Because the graph of $f(x) = a^x$ is that of a one-to-one function, to solve $a^{x_1} = a^{x_2}$, we need only show that $x_1 = x_2$. This property is used to solve an **exponential equation,** which is an equation with a variable as exponent.

EXAMPLE 4 Solving an Exponential Equation

Solve $\left(\frac{1}{3}\right)^x = 81$.

SOLUTION ***Write each side of the equation using a common base.***

$$\left(\frac{1}{3}\right)^x = 81$$

$(3^{-1})^x = 81$ Definition of negative exponent

$3^{-x} = 81$ $(a^m)^n = a^{mn}$

$3^{-x} = 3^4$ Write 81 as a power of 3.

$-x = 4$ Set exponents equal (Property (b) given earlier).

$x = -4$ Multiply by -1.

Check by substituting -4 for x in the original equation. The solution set is $\{-4\}$.

✔ **Now Try Exercise 73.**

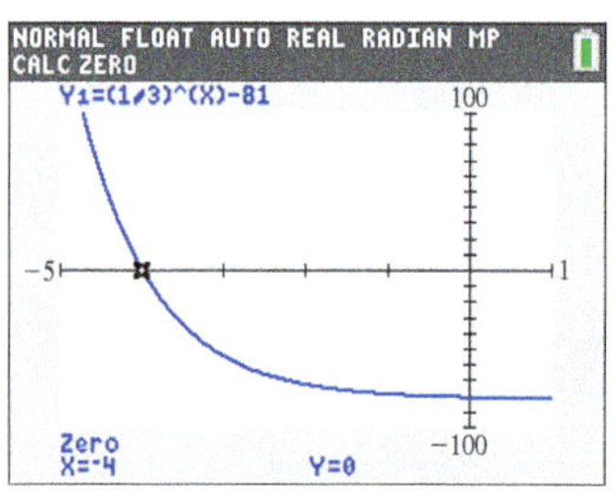

The x-intercept of the graph of $y = \left(\frac{1}{3}\right)^x - 81$ can be used to verify the solution in **Example 4.**

EXAMPLE 5 Solving an Exponential Equation

Solve $2^{x+4} = 8^{x-6}$.

SOLUTION Write each side of the equation using a common base.

$$2^{x+4} = 8^{x-6}$$

$$2^{x+4} = (2^3)^{x-6} \quad \text{Write 8 as a power of 2.}$$

$$2^{x+4} = 2^{3x-18} \quad (a^m)^n = a^{mn}$$

$$x + 4 = 3x - 18 \quad \text{Set exponents equal (Property (b)).}$$

$$-2x = -22 \quad \text{Subtract } 3x \text{ and 4.}$$

$$x = 11 \quad \text{Divide by } -2.$$

Check by substituting 11 for x in the original equation. The solution set is $\{11\}$.

✔ **Now Try Exercise 81.**

Later in this chapter, we describe a general method for solving exponential equations where the approach used in **Examples 4 and 5** is not possible. For instance, the above method could not be used to solve an equation like

$$7^x = 12$$

because it is not easy to express both sides as exponential expressions with the same base.

In **Example 6,** we solve an equation that has the variable as the base of an exponential expression.

EXAMPLE 6 Solving an Equation with a Fractional Exponent

Solve $x^{4/3} = 81$.

SOLUTION Notice that the variable is in the base rather than in the exponent.

$$x^{4/3} = 81$$

$$\left(\sqrt[3]{x}\right)^4 = 81 \quad \text{Radical notation for } a^{m/n}$$

$$\sqrt[3]{x} = \pm 3 \quad \text{Take fourth roots on each side. Remember to use } \pm.$$

$$x = \pm 27 \quad \text{Cube each side.}$$

Check *both* solutions in the original equation. Both check, so the solution set is $\{\pm 27\}$.

Alternative Method There may be more than one way to solve an exponential equation, as shown here.

$$x^{4/3} = 81$$

$$(x^{4/3})^3 = 81^3 \quad \text{Cube each side.}$$

$$x^4 = (3^4)^3 \quad \text{Write 81 as } 3^4.$$

$$x^4 = 3^{12} \quad (a^m)^n = a^{mn}$$

$$x = \pm\sqrt[4]{3^{12}} \quad \text{Take fourth roots on each side.}$$

$$x = \pm 3^3 \quad \text{Simplify the radical.}$$

$$x = \pm 27 \quad \text{Apply the exponent.}$$

The same solution set, $\{\pm 27\}$, results.

✔ **Now Try Exercise 83.**

Compound Interest Recall the formula for simple interest, $I = Prt$, where P is principal (amount deposited), r is annual rate of interest expressed as a decimal, and t is time in years that the principal earns interest. Suppose $t = 1$ yr. Then at the end of the year, the amount has grown to the following.

$$P + Pr = P(1 + r) \quad \text{Original principal plus interest}$$

If this balance earns interest at the same interest rate for another year, the balance at the end of *that* year will increase as follows.

$$[P(1 + r)] + [P(1 + r)]r = [P(1 + r)](1 + r) \quad \text{Factor.}$$
$$= P(1 + r)^2 \quad a \cdot a = a^2$$

After the third year, the balance will grow in a similar pattern.

$$[P(1 + r)^2] + [P(1 + r)^2]r = [P(1 + r)^2](1 + r) \quad \text{Factor.}$$
$$= P(1 + r)^3 \quad a^2 \cdot a = a^3$$

Continuing in this way produces a formula for interest compounded annually.

$$A = P(1 + r)^t$$

The general formula for compound interest can be derived in the same way.

Compound Interest

If P dollars are deposited in an account paying an annual rate of interest r compounded (paid) n times per year, then after t years the account will contain A dollars, according to the following formula.

$$A = P\left(1 + \frac{r}{n}\right)^{tn}$$

EXAMPLE 7 Using the Compound Interest Formula

Suppose \$1000 is deposited in an account paying 4% interest per year compounded quarterly (four times per year).

(a) Find the amount in the account after 10 yr with no withdrawals.

(b) How much interest is earned over the 10-yr period?

SOLUTION

(a)
$$A = P\left(1 + \frac{r}{n}\right)^{tn} \quad \text{Compound interest formula}$$
$$A = 1000\left(1 + \frac{0.04}{4}\right)^{10(4)} \quad \text{Let } P = 1000,\ r = 0.04,\ n = 4, \text{ and } t = 10.$$
$$A = 1000(1 + 0.01)^{40} \quad \text{Simplify.}$$
$$A = 1488.86 \quad \text{Round to the nearest cent.}$$

Thus, \$1488.86 is in the account after 10 yr.

(b) The interest earned for that period is

$$\$1488.86 - \$1000 = \$488.86.$$

✔ **Now Try Exercise 97(a).**

In the formula for compound interest

$$A = P\left(1 + \frac{r}{n}\right)^{tn},$$

A is sometimes called the **future value** and P the **present value.** A is also called the **compound amount** and is the balance *after* interest has been earned.

EXAMPLE 8 Finding Present Value

Becky must pay a lump sum of \$6000 in 5 yr.

(a) What amount deposited today (present value) at 3.1% compounded annually will grow to \$6000 in 5 yr?

(b) If only \$5000 is available to deposit now, what annual interest rate is necessary for the money to increase to \$6000 in 5 yr?

SOLUTION

(a)

$$A = P\left(1 + \frac{r}{n}\right)^{tn} \quad \text{Compound interest formula}$$

$$6000 = P\left(1 + \frac{0.031}{1}\right)^{5(1)} \quad \text{Let } A = 6000,\ r = 0.031,\ n = 1, \text{ and } t = 5.$$

$$6000 = P(1.031)^5 \quad \text{Simplify.}$$

$$P = \frac{6000}{(1.031)^5} \quad \text{Divide by } (1.031)^5 \text{ to solve for } P.$$

$$P \approx 5150.60 \quad \text{Use a calculator.}$$

If Becky leaves \$5150.60 for 5 yr in an account paying 3.1% compounded annually, she will have \$6000 when she needs it. Thus, \$5150.60 is the present value of \$6000 if interest of 3.1% is compounded annually for 5 yr.

(b)

$$A = P\left(1 + \frac{r}{n}\right)^{tn} \quad \text{Compound interest formula}$$

$$6000 = 5000(1 + r)^5 \quad \text{Let } A = 6000,\ P = 5000,\ n = 1, \text{ and } t = 5.$$

$$\frac{6}{5} = (1 + r)^5 \quad \text{Divide by 5000.}$$

$$\left(\frac{6}{5}\right)^{1/5} = 1 + r \quad \text{Take the fifth root on each side.}$$

$$\left(\frac{6}{5}\right)^{1/5} - 1 = r \quad \text{Subtract 1.}$$

$$r \approx 0.0371 \quad \text{Use a calculator.}$$

An interest rate of 3.71% will produce enough interest to increase the \$5000 to \$6000 by the end of 5 yr.

✔ **Now Try Exercises 99 and 103.**

CAUTION When performing the computations in problems like those in **Examples 7 and 8,** do not round off during intermediate steps. Keep all calculator digits and round at the end of the process.

n	$\left(1 + \frac{1}{n}\right)^n$ (rounded)
1	2
2	2.25
5	2.48832
10	2.59374
100	2.70481
1000	2.71692
10,000	2.71815
1,000,000	2.71828

The Number e and Continuous Compounding The more often interest is compounded within a given time period, the more interest will be earned. Surprisingly, however, there is a limit on the amount of interest, no matter how often it is compounded.

Suppose that \$1 is invested at 100% interest per year, compounded n times per year. Then the interest rate (in decimal form) is 1.00, and the interest rate per period is $\frac{1}{n}$. According to the formula (with $P = 1$), the compound amount at the end of 1 yr will be

$$A = \left(1 + \frac{1}{n}\right)^n.$$

A calculator gives the results in the margin for various values of n. The table suggests that as n increases, the value of $\left(1 + \frac{1}{n}\right)^n$ gets closer and closer to some fixed number. This is indeed the case. This fixed number is called e. ***(In mathematics, e is a real number and not a variable.)***

Value of e

$$e \approx 2.718281828459045$$

Figure 21 shows graphs of the functions

$$y = 2^x, \quad y = 3^x, \quad \text{and} \quad y = e^x.$$

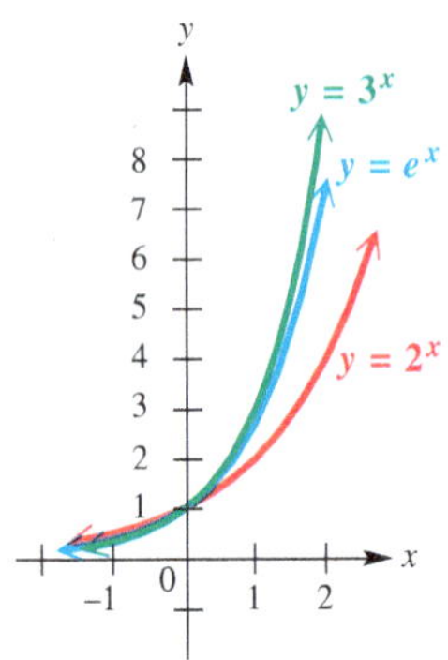

Figure 21

Because $2 < e < 3$, the graph of $y = e^x$ lies "between" the other two graphs.

As mentioned above, the amount of interest earned increases with the frequency of compounding, but the value of the expression $\left(1 + \frac{1}{n}\right)^n$ approaches e as n gets larger. Consequently, the formula for compound interest approaches a limit as well, called the compound amount from **continuous compounding.**

Continuous Compounding

If P dollars are deposited at a rate of interest r compounded continuously for t years, then the compound amount A in dollars on deposit is given by the following formula.

$$\boldsymbol{A = Pe^{rt}}$$

EXAMPLE 9 Solving a Continuous Compounding Problem

Suppose \$5000 is deposited in an account paying 3% interest compounded continuously for 5 yr. Find the total amount on deposit at the end of 5 yr.

SOLUTION

$A = Pe^{rt}$	Continuous compounding formula
$A = 5000e^{0.03(5)}$	Let $P = 5000$, $r = 0.03$, and $t = 5$.
$A = 5000e^{0.15}$	Multiply exponents.
$A \approx 5809.17$ or \$5809.17	Use a calculator.

Check that daily compounding would have produced a compound amount about \$0.03 less.

Now Try Exercise 97(b).

EXAMPLE 10 Comparing Interest Earned as Compounding Is More Frequent

In **Example 7,** we found that \$1000 invested at 4% compounded quarterly for 10 yr grew to \$1488.86. Compare this same investment compounded annually, semiannually, monthly, daily, and continuously.

SOLUTION Substitute 0.04 for r, 10 for t, and the appropriate number of compounding periods for n into the formulas

$$A = P\left(1 + \frac{r}{n}\right)^{tn} \quad \text{Compound interest formula}$$

and

$$A = Pe^{rt}. \quad \text{Continuous compounding formula}$$

The results for amounts of \$1 and \$1000 are given in the table.

Compounded	\$1	\$1000
Annually	$(1+0.04)^{10} \approx 1.48024$	\$1480.24
Semiannually	$\left(1+\frac{0.04}{2}\right)^{10(2)} \approx 1.48595$	\$1485.95
Quarterly	$\left(1+\frac{0.04}{4}\right)^{10(4)} \approx 1.48886$	\$1488.86
Monthly	$\left(1+\frac{0.04}{12}\right)^{10(12)} \approx 1.49083$	\$1490.83
Daily	$\left(1+\frac{0.04}{365}\right)^{10(365)} \approx 1.49179$	\$1491.79
Continuously	$e^{10(0.04)} \approx 1.49182$	\$1491.82

Comparing the results for a \$1000 investment, we notice the following.

- Compounding semiannually rather than annually increases the value of the account after 10 yr by \$5.71.
- Quarterly compounding grows to \$2.91 more than semiannual compounding after 10 yr.
- Daily compounding yields only \$0.96 more than monthly compounding.
- Continuous compounding yields only \$0.03 more than daily compounding.

Each increase in compounding frequency earns less additional interest.

✔ **Now Try Exercise 105.**

LOOKING AHEAD TO CALCULUS

In calculus, the derivative allows us to determine the slope of a tangent line to the graph of a function. For the function

$$f(x) = e^x,$$

the derivative is the function f itself:

$$f'(x) = e^x.$$

Therefore, in calculus the exponential function with base e is much easier to work with than exponential functions having other bases.

Exponential Models The number e is important as the base of an exponential function in many practical applications. In situations involving growth or decay of a quantity, the amount or number present at time t often can be closely modeled by a function of the form

$$y = y_0e^{kt},$$

where y_0 is the amount or number present at time $t = 0$ and k is a constant.

Exponential functions are used to model the growth of microorganisms in a culture, the growth of certain populations, and the decay of radioactive material.

EXAMPLE 11 Using Data to Model Exponential Growth

Data from recent years indicate that future amounts of carbon dioxide in the atmosphere may grow according to the table. Amounts are given in parts per million.

Year	Carbon Dioxide (ppm)
1990	353
2000	375
2075	590
2175	1090
2275	2000

Source: International Panel on Climate Change (IPCC).

(a) Make a scatter diagram of the data. Do the carbon dioxide levels appear to grow exponentially?

(b) One model for the data is the function

$$y = 0.001942e^{0.00609x},$$

where x is the year and $1990 \le x \le 2275$. Use a graph of this model to estimate when future levels of carbon dioxide will double and triple over the preindustrial level of 280 ppm.

SOLUTION

(a) We show a calculator graph for the data in **Figure 22(a).** The data appear to resemble the graph of an increasing exponential function.

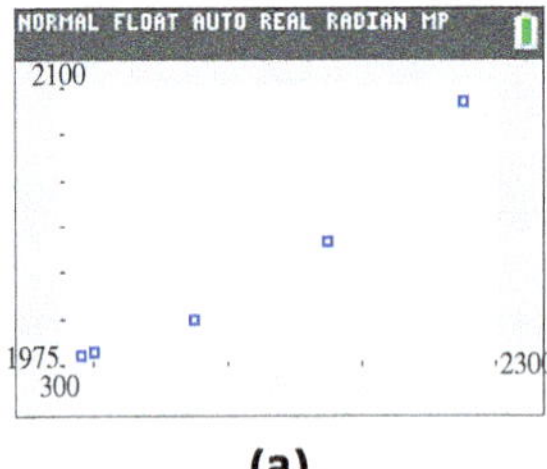

(a)

$y = 0.001942e^{0.00609x}$

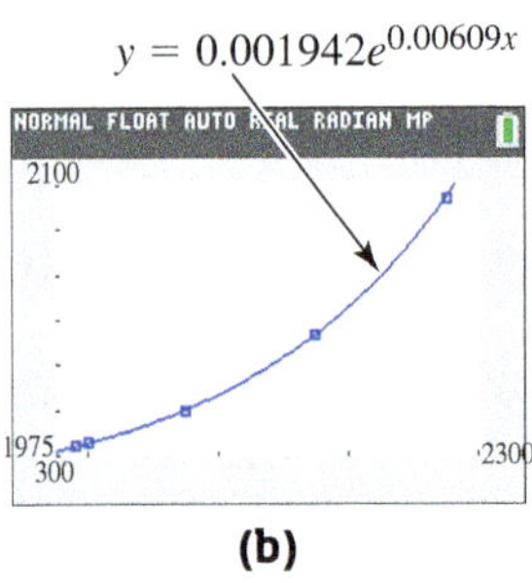

(b)

Figure 22

(b) A graph of $y = 0.001942e^{0.00609x}$ in **Figure 22(b)** shows that it is very close to the data points. We graph $y_2 = 2 \cdot 280 = 560$ in **Figure 23(a)** and $y_2 = 3 \cdot 280 = 840$ in **Figure 23(b)** on the same coordinate axes as the given function, and we use the calculator to find the intersection points.

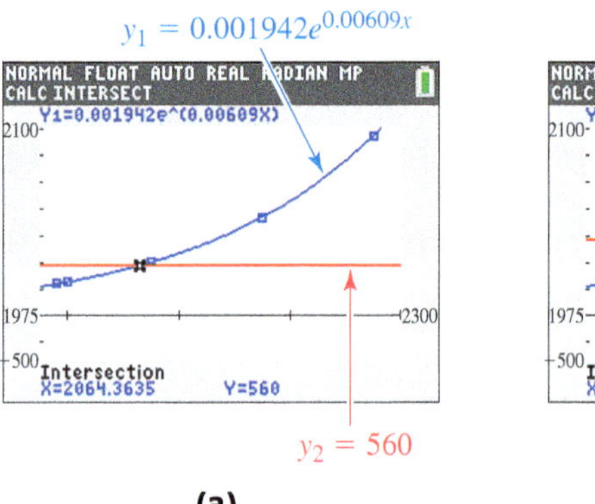

(a) (b)

Figure 23

The graph of the function intersects the horizontal lines at x-values of approximately 2064.4 and 2130.9. According to this model, carbon dioxide levels will have doubled during 2064 and tripled by 2131.

Now Try Exercise 107.

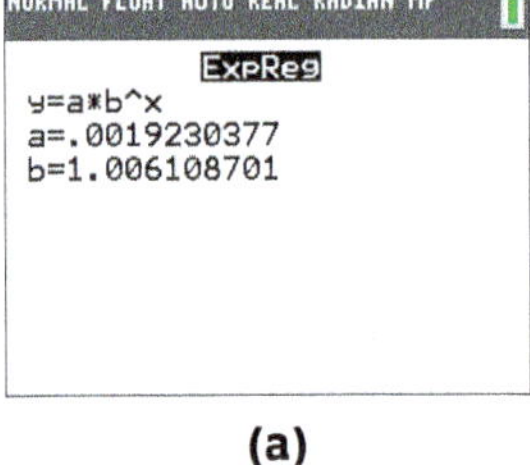

(a)

$y_1 = 0.001923(1.006109)^x$

(b)

Figure 24

Graphing calculators are capable of fitting exponential curves to scatter diagrams like the one found in **Example 11.** The TI-84 Plus displays another (different) equation in **Figure 24(a)** for the atmospheric carbon dioxide example, approximated as follows.

$$y = 0.001923(1.006109)^x$$

This calculator form differs from the model in **Example 11. Figure 24(b)** shows the data points and the graph of this exponential regression equation. ■

4.2 Exercises

CONCEPT PREVIEW *Fill in the blank(s) to correctly complete each sentence.*

1. If $f(x) = 4^x$, then $f(2) =$ ______ and $f(-2) =$ ______.
2. If $a > 1$, then the graph of $f(x) = a^x$ ______ (rises/falls) from left to right.
3. If $0 < a < 1$, then the graph of $f(x) = a^x$ ______ (rises/falls) from left to right.
4. The domain of $f(x) = 4^x$ is ______ and the range is ______.
5. The graph of $f(x) = 8^x$ passes through the points $(-1, \_\_)$, $(0, \_\_)$, and $(1, \_\_)$.
6. The graph of $f(x) = -\left(\frac{1}{3}\right)^{x+4} - 5$ is that of $f(x) = \left(\frac{1}{3}\right)^x$ reflected across the ______-axis, translated ______ units to the left and ______ units down.

CONCEPT PREVIEW *Solve each equation. Round answers to the nearest hundredth as needed.*

7. $\left(\frac{1}{4}\right)^x = 64$
8. $x^{2/3} = 36$
9. $A = 2000\left(1 + \frac{0.03}{4}\right)^{8(4)}$
10. $10{,}000 = 5000(1 + r)^{25}$

For $f(x) = 3^x$ and $g(x) = \left(\frac{1}{4}\right)^x$, find each of the following. Round answers to the nearest thousandth as needed. ***See Example 1.***

11. $f(2)$
12. $f(3)$
13. $f(-2)$
14. $f(-3)$
15. $g(2)$
16. $g(3)$
17. $g(-2)$
18. $g(-3)$
19. $f\left(\frac{3}{2}\right)$
20. $f\left(-\frac{5}{2}\right)$
21. $g\left(\frac{3}{2}\right)$
22. $g\left(-\frac{5}{2}\right)$
23. $f(2.34)$
24. $f(-1.68)$
25. $g(-1.68)$
26. $g(2.34)$

Graph each function. ***See Example 2.***

27. $f(x) = 3^x$
28. $f(x) = 4^x$
29. $f(x) = \left(\frac{1}{3}\right)^x$
30. $f(x) = \left(\frac{1}{4}\right)^x$
31. $f(x) = \left(\frac{3}{2}\right)^x$
32. $f(x) = \left(\frac{5}{3}\right)^x$
33. $f(x) = \left(\frac{1}{10}\right)^{-x}$
34. $f(x) = \left(\frac{1}{6}\right)^{-x}$
35. $f(x) = 4^{-x}$
36. $f(x) = 10^{-x}$
37. $f(x) = 2^{|x|}$
38. $f(x) = 2^{-|x|}$

Graph each function. Give the domain and range. ***See Example 3.***

39. $f(x) = 2^x + 1$
40. $f(x) = 2^x - 4$
41. $f(x) = 2^{x+1}$
42. $f(x) = 2^{x-4}$
43. $f(x) = -2^{x+2}$
44. $f(x) = -2^{x-3}$
45. $f(x) = 2^{-x}$
46. $f(x) = -2^{-x}$
47. $f(x) = 2^{x-1} + 2$
48. $f(x) = 2^{x+3} + 1$
49. $f(x) = 2^{x+2} - 4$
50. $f(x) = 2^{x-3} - 1$

Graph each function. Give the domain and range. **See Example 3.**

51. $f(x)=\left(\frac{1}{3}\right)^{x}-2$ **52.** $f(x)=\left(\frac{1}{3}\right)^{x}+4$ **53.** $f(x)=\left(\frac{1}{3}\right)^{x+2}$

54. $f(x)=\left(\frac{1}{3}\right)^{x-4}$ **55.** $f(x)=\left(\frac{1}{3}\right)^{-x+1}$ **56.** $f(x)=\left(\frac{1}{3}\right)^{-x-2}$

57. $f(x)=\left(\frac{1}{3}\right)^{-x}$ **58.** $f(x)=-\left(\frac{1}{3}\right)^{-x}$ **59.** $f(x)=\left(\frac{1}{3}\right)^{x-2}+2$

60. $f(x)=\left(\frac{1}{3}\right)^{x-1}+3$ **61.** $f(x)=\left(\frac{1}{3}\right)^{x+2}-1$ **62.** $f(x)=\left(\frac{1}{3}\right)^{x+3}-2$

Connecting Graphs with Equations *Write an equation for the graph given. Each represents an exponential function f with base 2 or 3, translated and/or reflected.*

63.

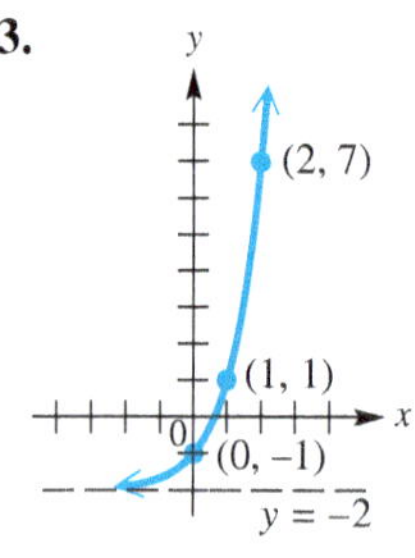

64.

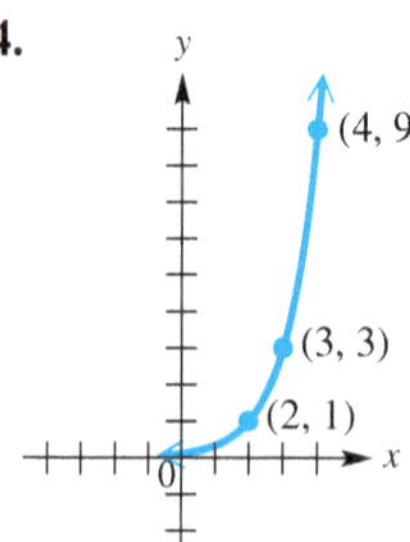

65.

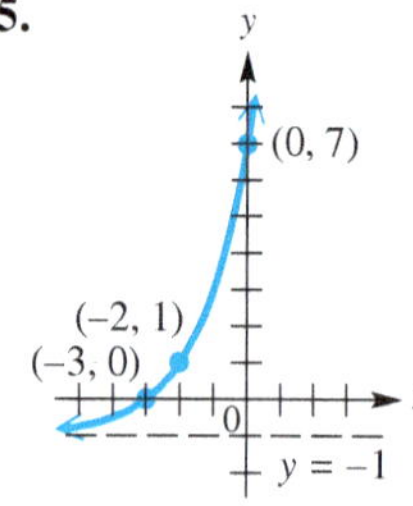

66.

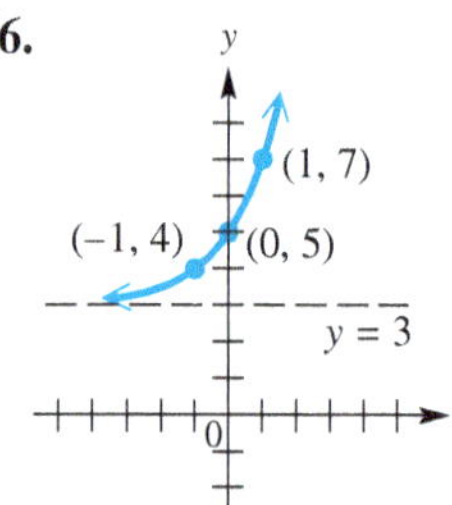

67.

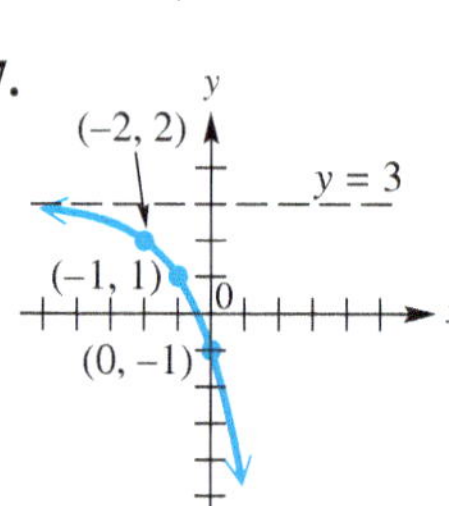

68.

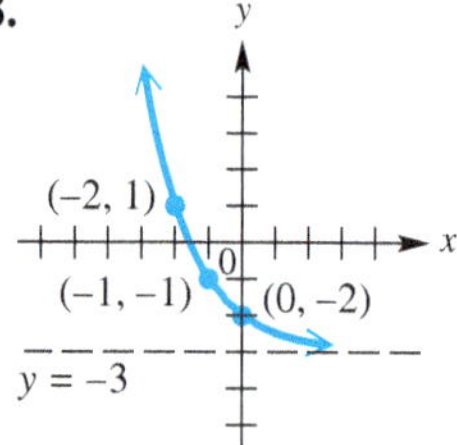

69.

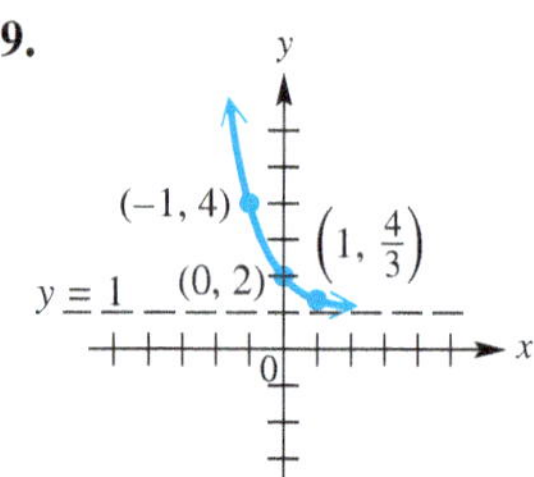

70.

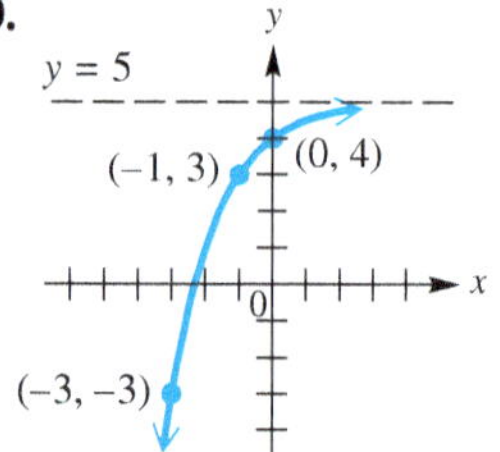

Solve each equation. **See Examples 4–6.**

71. $4^x=2$ **72.** $125^x=5$ **73.** $\left(\frac{5}{2}\right)^x=\frac{4}{25}$ **74.** $\left(\frac{2}{3}\right)^x=\frac{9}{4}$

75. $2^{3-2x}=8$ **76.** $5^{2+2x}=25$ **77.** $e^{4x-1}=(e^2)^x$ **78.** $e^{3-x}=(e^3)^{-x}$

79. $27^{4x}=9^{x+1}$ **80.** $32^{2x}=16^{x-1}$ **81.** $4^{x-2}=2^{3x+3}$ **82.** $2^{6-3x}=8^{x+1}$

83. $x^{2/3}=4$ **84.** $x^{2/5}=16$ **85.** $x^{5/2}=32$ **86.** $x^{3/2}=27$

87. $x^{-6}=\frac{1}{64}$ **88.** $x^{-4}=\frac{1}{256}$ **89.** $x^{5/3}=-243$ **90.** $x^{7/5}=-128$

91. $\left(\frac{1}{e}\right)^{-x}=\left(\frac{1}{e^2}\right)^{x+1}$ **92.** $e^{x-1}=\left(\frac{1}{e^4}\right)^{x+1}$ **93.** $\left(\sqrt{2}\right)^{x+4}=4^x$

94. $\left(\sqrt[3]{5}\right)^{-x}=\left(\frac{1}{5}\right)^{x+2}$ **95.** $\frac{1}{27}=x^{-3}$ **96.** $\frac{1}{32}=x^{-5}$

Solve each problem. **See Examples 7–9.**

97. *Future Value* Find the future value and interest earned if \$8906.54 is invested for 9 yr at 3% compounded

(a) semiannually **(b)** continuously.

98. *Future Value* Find the future value and interest earned if \$56,780 is invested at 2.8% compounded

(a) quarterly for 23 quarters **(b)** continuously for 15 yr.

99. *Present Value* Find the present value that will grow to \$25,000 if interest is 3.2% compounded quarterly for 11 quarters.

100. *Present Value* Find the present value that will grow to \$45,000 if interest is 3.6% compounded monthly for 1 yr.

101. *Present Value* Find the present value that will grow to \$5000 if interest is 3.5% compounded quarterly for 10 yr.

102. *Interest Rate* Find the required annual interest rate to the nearest tenth of a percent for \$65,000 to grow to \$65,783.91 if interest is compounded monthly for 6 months.

103. *Interest Rate* Find the required annual interest rate to the nearest tenth of a percent for \$1200 to grow to \$1500 if interest is compounded quarterly for 9 yr.

104. *Interest Rate* Find the required annual interest rate to the nearest tenth of a percent for \$5000 to grow to \$6200 if interest is compounded quarterly for 8 yr.

Solve each problem. **See Example 10.**

105. *Comparing Loans* Bank A is lending money at 6.4% interest compounded annually. The rate at Bank B is 6.3% compounded monthly, and the rate at Bank C is 6.35% compounded quarterly. At which bank will we pay the *least* interest?

106. *Future Value* Suppose \$10,000 is invested at an annual rate of 2.4% for 10 yr. Find the future value if interest is compounded as follows.

(a) annually **(b)** quarterly **(c)** monthly **(d)** daily (365 days)

(Modeling) Solve each problem. **See Example 11.**

107. *Atmospheric Pressure* The atmospheric pressure (in millibars) at a given altitude (in meters) is shown in the table.

Altitude	Pressure	Altitude	Pressure
0	1013	6000	472
1000	899	7000	411
2000	795	8000	357
3000	701	9000	308
4000	617	10,000	265
5000	541		

Source: Miller, A. and J. Thompson, *Elements of Meteorology*, Fourth Edition, Charles E. Merrill Publishing Company, Columbus, Ohio.

(a) Use a graphing calculator to make a scatter diagram of the data for atmospheric pressure P at altitude x.

(b) Would a linear or an exponential function fit the data better?

(c) The following function approximates the data.

$$P(x) = 1013e^{-0.0001341x}$$

Use a graphing calculator to graph P and the data on the same coordinate axes.

(d) Use P to predict the pressures at 1500 m and 11,000 m, and compare them to the actual values of 846 millibars and 227 millibars, respectively.

108. *World Population Growth* World population in millions closely fits the exponential function

$$f(x) = 6084e^{0.0120x},$$

where x is the number of years since 2000. (*Source:* U.S. Census Bureau.)

(a) The world population was about 6853 million in 2010. How closely does the function approximate this value?

(b) Use this model to predict world population in 2020 and 2030.

109. *Deer Population* The exponential growth of the deer population in Massachusetts can be approximated using the model

$$f(x) = 50{,}000(1 + 0.06)^x,$$

where 50,000 is the initial deer population and 0.06 is the rate of growth. $f(x)$ is the total population after x years have passed. Find each value to the nearest thousand.

(a) Predict the total population after 4 yr.

(b) If the initial population was 30,000 and the growth rate was 0.12, how many deer would be present after 3 yr?

(c) How many additional deer can we expect in 5 yr if the initial population is 45,000 and the current growth rate is 0.08?

110. *Employee Training* A person learning certain skills involving repetition tends to learn quickly at first. Then learning tapers off and skill acquisition approaches some upper limit. Suppose the number of symbols per minute that a person using a keyboard can type is given by

$$f(t) = 250 - 120(2.8)^{-0.5t},$$

where t is the number of months the operator has been in training. Find each value to the nearest whole number.

(a) $f(2)$ **(b)** $f(4)$ **(c)** $f(10)$

(d) What happens to the number of symbols per minute after several months of training?

Use a graphing calculator to find the solution set of each equation. Approximate the solution(s) to the nearest tenth.

111. $5e^{3x} = 75$ **112.** $6^{-x} = 1 - x$ **113.** $3x + 2 = 4^x$ **114.** $x = 2^x$

115. A function of the form $f(x) = x^r$, where r is a constant, is a **power function.** Discuss the difference between an exponential function and a power function.

116. *Concept Check* If $f(x) = a^x$ and $f(3) = 27$, determine each function value.

(a) $f(1)$ **(b)** $f(-1)$ **(c)** $f(2)$ **(d)** $f(0)$

Concept Check Give an equation of the form $f(x) = a^x$ to define the exponential function whose graph contains the given point.

117. $(3, 8)$ **118.** $(3, 125)$ **119.** $(-3, 64)$ **120.** $(-2, 36)$

Concept Check Use properties of exponents to write each function in the form $f(t) = ka^t$, where k is a constant. (Hint: Recall that $a^{x+y} = a^x \cdot a^y$.)

121. $f(t) = 3^{2t+3}$ **122.** $f(t) = 2^{3t+2}$ **123.** $f(t) = \left(\frac{1}{3}\right)^{1-2t}$ **124.** $f(t) = \left(\frac{1}{2}\right)^{1-2t}$

In calculus, the following can be shown.

$$e^x = 1 + x + \frac{x^2}{2 \cdot 1} + \frac{x^3}{3 \cdot 2 \cdot 1} + \frac{x^4}{4 \cdot 3 \cdot 2 \cdot 1} + \frac{x^5}{5 \cdot 4 \cdot 3 \cdot 2 \cdot 1} + \cdots$$

Using more terms, one can obtain a more accurate approximation for e^x.

125. Use the terms shown, and replace x with 1 to approximate $e^1 = e$ to three decimal places. Check the result with a calculator.

126. Use the terms shown, and replace x with -0.05 to approximate $e^{-0.05}$ to four decimal places. Check the result with a calculator.

Relating Concepts

For individual or collaborative investigation *(Exercises 127–132)*

Consider $f(x) = a^x$, where $a > 1$. ***Work these exercises in order.***

127. Is f a one-to-one function? If so, what kind of related function exists for f?

128. If f has an inverse function f^{-1}, sketch f and f^{-1} on the same set of axes.

129. If f^{-1} exists, find an equation for $y = f^{-1}(x)$. (You need not solve for y.)

130. If $a = 10$, what is the equation for $y = f^{-1}(x)$? (You need not solve for y.)

131. If $a = e$, what is the equation for $y = f^{-1}(x)$? (You need not solve for y.)

132. If the point (p, q) is on the graph of f, then the point ________ is on the graph of f^{-1}.

4.3 Logarithmic Functions

- **Logarithms**
- **Logarithmic Equations**
- **Logarithmic Functions**
- **Properties of Logarithms**

Logarithms

The previous section dealt with exponential functions of the form $y = a^x$ for all positive values of a, where $a \neq 1$. The horizontal line test shows that exponential functions are one-to-one and thus have inverse functions. The equation defining the inverse of a function is found by interchanging x and y in the equation that defines the function. Starting with $y = a^x$ and interchanging x and y yields

$$x = a^y.$$

Here y is the exponent to which a must be raised in order to obtain x. We call this exponent a **logarithm,** symbolized by the abbreviation **"log."** The expression $\log_a x$ represents the logarithm in this discussion. The number a is the **base** of the logarithm, and x is the **argument** of the expression. It is read **"logarithm with base a of x,"** or **"logarithm of x with base a,"** or **"base a logarithm of x."**

Logarithm

For all real numbers y and all positive numbers a and x, where $a \neq 1$,

$$y = \log_a x \quad \text{is equivalent to} \quad x = a^y.$$

The expression $\log_a x$ represents the exponent to which the base a must be raised in order to obtain x.

EXAMPLE 1 Writing Equivalent Logarithmic and Exponential Forms

The table shows several pairs of equivalent statements, written in both logarithmic and exponential forms.

SOLUTION

Logarithmic Form	Exponential Form
$\log_2 8 = 3$	$2^3 = 8$
$\log_{1/2} 16 = -4$	$\left(\frac{1}{2}\right)^{-4} = 16$
$\log_{10} 100{,}000 = 5$	$10^5 = 100{,}000$
$\log_3 \frac{1}{81} = -4$	$3^{-4} = \frac{1}{81}$
$\log_5 5 = 1$	$5^1 = 5$
$\log_{3/4} 1 = 0$	$\left(\frac{3}{4}\right)^0 = 1$

To remember the relationships among a, x, and y in the two equivalent forms $y = \log_a x$ and $x = a^y$, refer to these diagrams.

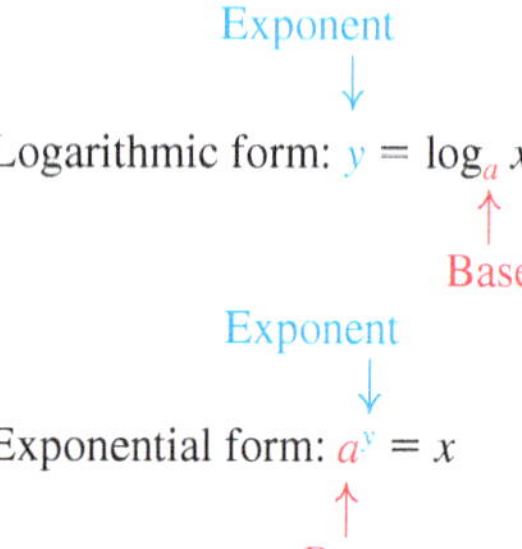

✔ **Now Try Exercises 11, 13, 15, and 17.**

Logarithmic Equations

The definition of logarithm can be used to solve a **logarithmic equation,** which is an equation with a logarithm in at least one term.

EXAMPLE 2 Solving Logarithmic Equations

Solve each equation.

(a) $\log_x \frac{8}{27} = 3$ **(b)** $\log_4 x = \frac{5}{2}$ **(c)** $\log_{49} \sqrt[3]{7} = x$

SOLUTION Many logarithmic equations can be solved by first writing the equation in exponential form.

(a)

$$\log_x \frac{8}{27} = 3$$

$$x^3 = \frac{8}{27} \quad \text{Write in exponential form.}$$

$$x^3 = \left(\frac{2}{3}\right)^3 \quad \frac{8}{27} = \left(\frac{2}{3}\right)^3$$

$$x = \frac{2}{3} \quad \text{Take cube roots.}$$

CHECK

$$\log_x \frac{8}{27} = 3 \quad \text{Original equation}$$

$$\log_{2/3} \frac{8}{27} \stackrel{?}{=} 3 \quad \text{Let } x = \tfrac{2}{3}.$$

$$\left(\frac{2}{3}\right)^3 \stackrel{?}{=} \frac{8}{27} \quad \text{Write in exponential form.}$$

$$\frac{8}{27} = \frac{8}{27} \ ✓ \quad \text{True}$$

The solution set is $\left\{\frac{2}{3}\right\}$.

(b)

$$\log_4 x = \frac{5}{2}$$

$4^{5/2} = x$ — Write in exponential form.

$(4^{1/2})^5 = x$ — $a^{mn} = (a^m)^n$

$2^5 = x$ — $4^{1/2} = (2^2)^{1/2} = 2$

$32 = x$ — Apply the exponent.

CHECK $\log_4 32 \stackrel{?}{=} \frac{5}{2}$ — Let $x = 32$.

$4^{5/2} \stackrel{?}{=} 32$

$2^5 \stackrel{?}{=} 32$ — $4^{5/2} = \left(\sqrt{4}\right)^5 = 2^5$

$32 = 32$ ✓ True

The solution set is $\{32\}$.

(c) $\log_{49} \sqrt[3]{7} = x$

$49^x = \sqrt[3]{7}$ — Write in exponential form.

$(7^2)^x = 7^{1/3}$ — Write with the same base.

$7^{2x} = 7^{1/3}$ — Power rule for exponents

$2x = \frac{1}{3}$ — Set exponents equal.

$x = \frac{1}{6}$ — Divide by 2.

A check shows that the solution set is $\left\{\frac{1}{6}\right\}$.

✔ **Now Try Exercises 19, 29, and 35.**

Logarithmic Functions

We define the logarithmic function with base a.

Logarithmic Function

If $a > 0$, $a \neq 1$, and $x > 0$, then the **logarithmic function with base *a*** is

$$f(x) = \log_a x.$$

Exponential and logarithmic functions are inverses of each other. To show this, we use the three steps for finding the inverse of a function.

	$f(x) = 2^x$	Exponential function with base 2
	$y = 2^x$	Let $y = f(x)$.
Step 1	$x = 2^y$	Interchange x and y.
Step 2	$y = \log_2 x$	Solve for y by writing in equivalent logarithmic form.
Step 3	$f^{-1}(x) = \log_2 x$	Replace y with $f^{-1}(x)$.

The graph of $f(x) = 2^x$ has the x-axis as horizontal asymptote and is shown in red in **Figure 25.** Its inverse, $f^{-1}(x) = \log_2 x$, has the y-axis as vertical asymptote and is shown in blue. The graphs are reflections of each other across the line $y = x$. As a result, their domains and ranges are interchanged.

x	$f(x) = 2^x$
-2	$\frac{1}{4}$
-1	$\frac{1}{2}$
0	1
1	2
2	4

x	$f^{-1}(x) = \log_2 x$
$\frac{1}{4}$	-2
$\frac{1}{2}$	-1
1	0
2	1
4	2

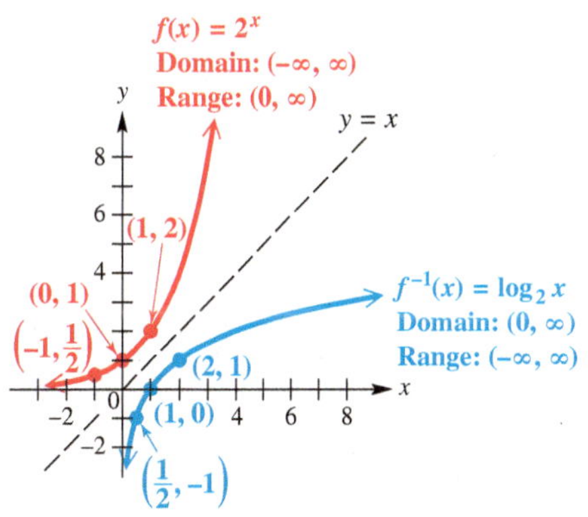

Figure 25

The domain of an exponential function is the set of all real numbers, so the range of a logarithmic function also will be the set of all real numbers. In the same way, both the range of an exponential function and the domain of a logarithmic function are the set of all positive real numbers.

Thus, logarithms can be found for positive numbers only.

Domain: $(0, \infty)$ Range: $(-\infty, \infty)$

For $f(x) = \log_2 x$:

x	$f(x)$
$\frac{1}{4}$	-2
$\frac{1}{2}$	-1
1	0
2	1
4	2
8	3

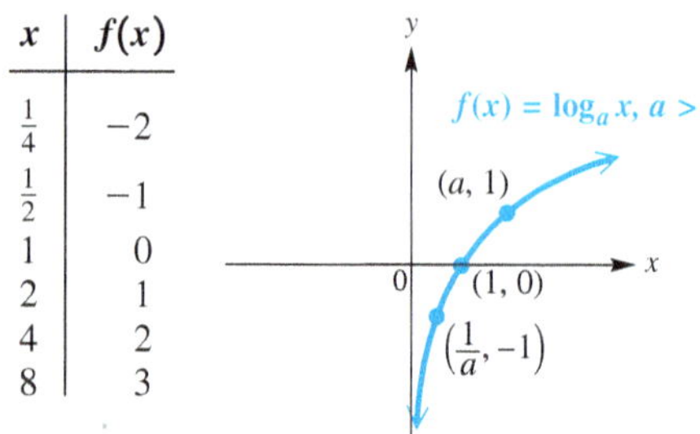

$f(x) = \log_a x$, $a > 1$

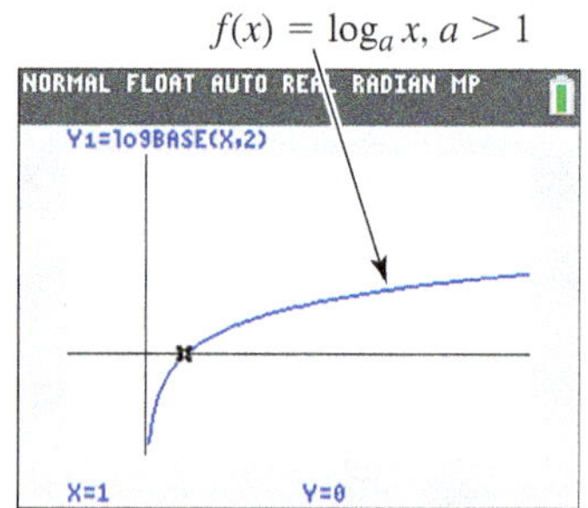

This is the general behavior seen on a calculator graph for **any base a, for $a > 1$.**

Figure 26

- **$f(x) = \log_a x$, for $a > 1$,** is increasing and continuous on its entire domain, $(0, \infty)$.
- The y-axis is a vertical asymptote as $x \to 0$ from the right.
- The graph passes through the points $\left(\frac{1}{a}, -1\right)$, $(1, 0)$, and $(a, 1)$.

For $f(x) = \log_{1/2} x$:

x	$f(x)$
$\frac{1}{4}$	2
$\frac{1}{2}$	1
1	0
2	-1
4	-2
8	-3

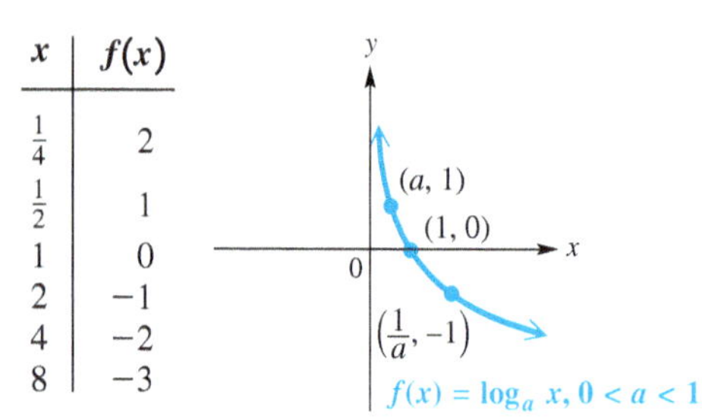

$f(x) = \log_a x$, $0 < a < 1$

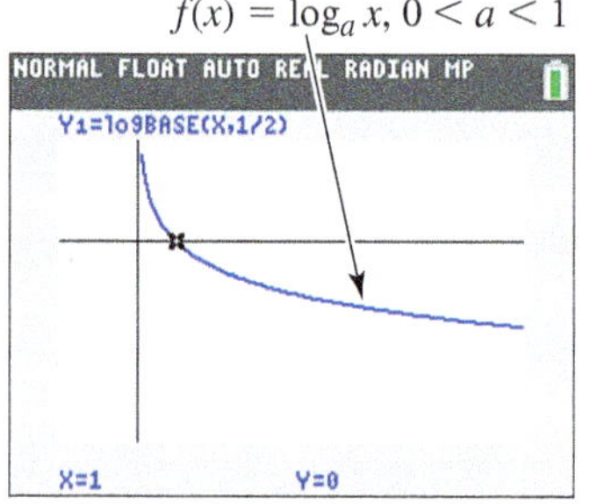

This is the general behavior seen on a calculator graph for **any base a, for $0 < a < 1$.**

Figure 27

- **$f(x) = \log_a x$, for $0 < a < 1$,** is decreasing and continuous on its entire domain, $(0, \infty)$.
- The y-axis is a vertical asymptote as $x \to 0$ from the right.
- The graph passes through the points $\left(\frac{1}{a}, -1\right)$, $(1, 0)$, and $(a, 1)$.

Calculator graphs of logarithmic functions sometimes do not give an accurate picture of the behavior of the graphs near the vertical asymptotes. While it may seem as if the graph has an endpoint, this is not the case. The resolution of the calculator screen is not precise enough to indicate that the graph approaches the vertical asymptote as the value of x gets closer to it. Do not draw incorrect conclusions just because the calculator does not show this behavior. ■

The graphs in **Figures 26 and 27** and the information with them suggest the following generalizations about the graphs of logarithmic functions of the form $f(x) = \log_a x$.

Characteristics of the Graph of $f(x) = \log_a x$

1. The points $\left(\frac{1}{a}, -1\right)$, $(1, 0)$, and $(a, 1)$ are on the graph.
2. If $a > 1$, then f is an increasing function.
   If $0 < a < 1$, then f is a decreasing function.
3. The y-axis is a vertical asymptote.
4. The domain is $(0, \infty)$, and the range is $(-\infty, \infty)$.

EXAMPLE 3 Graphing Logarithmic Functions

Graph each function.

(a) $f(x) = \log_{1/2} x$ **(b)** $f(x) = \log_3 x$

SOLUTION

(a) One approach is to first graph $y = \left(\frac{1}{2}\right)^x$, which defines the inverse function of f, by plotting points. Some ordered pairs are given in the table with the graph shown in red in **Figure 28.**

The graph of $f(x) = \log_{1/2} x$ is the reflection of the graph of $y = \left(\frac{1}{2}\right)^x$ across the line $y = x$. The ordered pairs for $y = \log_{1/2} x$ are found by interchanging the x- and y-values in the ordered pairs for $y = \left(\frac{1}{2}\right)^x$. See the graph in blue in **Figure 28.**

x	$y = \left(\frac{1}{2}\right)^x$
-2	4
-1	2
0	1
1	$\frac{1}{2}$
2	$\frac{1}{4}$
4	$\frac{1}{16}$

x	$f(x) = \log_{1/2} x$
4	-2
2	-1
1	0
$\frac{1}{2}$	1
$\frac{1}{4}$	2
$\frac{1}{16}$	4

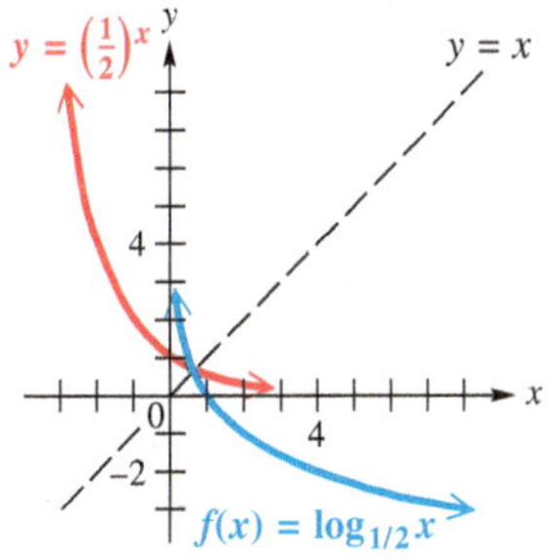

Figure 28

x	$f(x) = \log_3 x$
$\frac{1}{3}$	-1
1	0
3	1
9	2

Think: $x = 3^y$

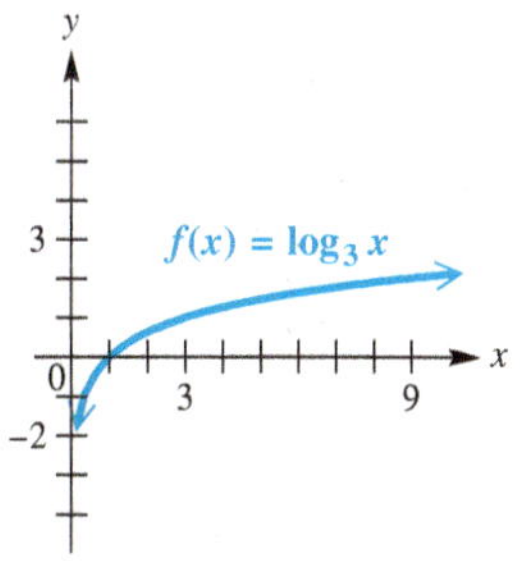

Figure 29

(b) Another way to graph a logarithmic function is to write $f(x) = y = \log_3 x$ in exponential form as $x = 3^y$, and then select y-values and calculate corresponding x-values. Several selected ordered pairs are shown in the table for the graph in **Figure 29.**

✔ **Now Try Exercise 55.**

CAUTION If we write a logarithmic function in exponential form in order to graph it, as in **Example 3(b),** we start *first* with y-values to calculate corresponding x-values. ***Be careful to write the values in the ordered pairs in the correct order.***

More general logarithmic functions can be obtained by forming the composition of $f(x) = \log_a x$ with a function $g(x)$. For example, if $f(x) = \log_2 x$ and $g(x) = x - 1$, then

$$(f \circ g)(x) = f(g(x)) = \log_2(x - 1).$$

The next example shows how to graph such functions.

EXAMPLE 4 Graphing Translated Logarithmic Functions

Graph each function. Give the domain and range.

(a) $f(x) = \log_2(x - 1)$ **(b)** $f(x) = (\log_3 x) - 1$

(c) $f(x) = \log_4(x + 2) + 1$

SOLUTION

(a) The graph of $f(x) = \log_2(x - 1)$ is the graph of $g(x) = \log_2 x$ translated 1 unit to the right. The vertical asymptote has equation $x = 1$. Because logarithms can be found only for positive numbers, we solve $x - 1 > 0$ to find the domain, $(1, \infty)$. To determine ordered pairs to plot, use the equivalent exponential form of the equation $y = \log_2(x - 1)$.

$$y = \log_2(x - 1)$$

$$x - 1 = 2^y \quad \text{Write in exponential form.}$$

$$x = 2^y + 1 \quad \text{Add 1.}$$

We first choose values for y and then calculate each of the corresponding x-values. The range is $(-\infty, \infty)$. See **Figure 30.**

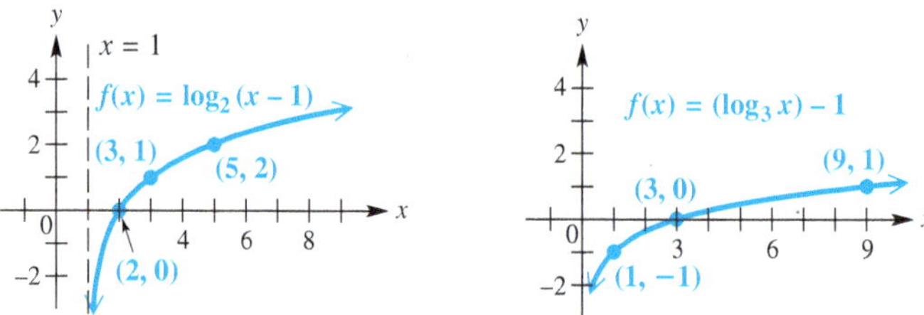

Figure 30 **Figure 31**

(b) The function $f(x) = (\log_3 x) - 1$ has the same graph as $g(x) = \log_3 x$ translated 1 unit down. We find ordered pairs to plot by writing the equation $y = (\log_3 x) - 1$ in exponential form.

$$y = (\log_3 x) - 1$$

$$y + 1 = \log_3 x \quad \text{Add 1.}$$

$$x = 3^{y+1} \quad \text{Write in exponential form.}$$

Again, choose y-values and calculate the corresponding x-values. The graph is shown in **Figure 31.** The domain is $(0, \infty)$, and the range is $(-\infty, \infty)$.

(c) The graph of $f(x) = \log_4(x + 2) + 1$ is obtained by shifting the graph of $y = \log_4 x$ to the left 2 units and up 1 unit. The domain is found by solving

$$x + 2 > 0,$$

which yields $(-2, \infty)$. The vertical asymptote has been shifted to the left 2 units as well, and it has equation $x = -2$. The range is unaffected by the vertical shift and remains $(-\infty, \infty)$. See **Figure 32.**

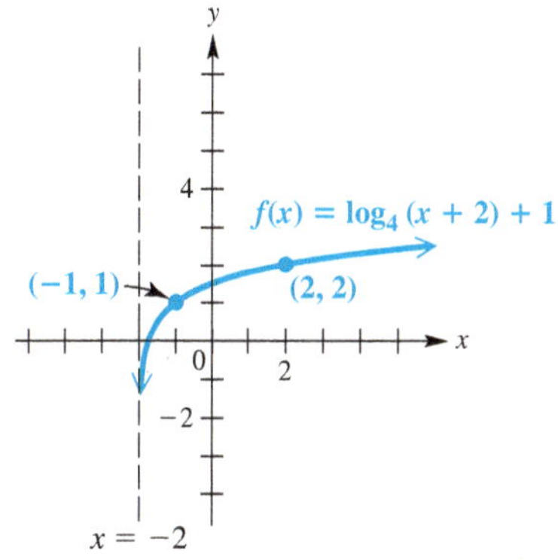

Figure 32

✔ **Now Try Exercises 43, 47, and 61.**

NOTE If we are given a graph such as the one in **Figure 31** and asked to find its equation, we could reason as follows: The point $(1, 0)$ on the basic logarithmic graph has been shifted *down* 1 unit, and the point $(3, 0)$ on the given graph is 1 unit lower than $(3, 1)$, which is on the graph of $y = \log_3 x$. Thus, the equation will be

$$y = (\log_3 x) - 1.$$

Properties of Logarithms

The properties of logarithms enable us to change the form of logarithmic statements so that products can be converted to sums, quotients can be converted to differences, and powers can be converted to products.

Properties of Logarithms

For $x > 0$, $y > 0$, $a > 0$, $a \neq 1$, and any real number r, the following properties hold.

Property	Description
Product Property $\log_a xy = \log_a x + \log_a y$	The logarithm of the product of two numbers is equal to the sum of the logarithms of the numbers.
Quotient Property $\log_a \frac{x}{y} = \log_a x - \log_a y$	The logarithm of the quotient of two numbers is equal to the difference between the logarithms of the numbers.
Power Property $\log_a x^r = r \log_a x$	The logarithm of a number raised to a power is equal to the exponent multiplied by the logarithm of the number.
Logarithm of 1 $\log_a 1 = 0$	The base a logarithm of 1 is 0.
Base *a* Logarithm of *a* $\log_a a = 1$	The base a logarithm of a is 1.

Proof To prove the product property, let $m = \log_a x$ and $n = \log_a y$.

$\log_a x = m$ means $a^m = x$

$\log_a y = n$ means $a^n = y$ Write in exponential form.

LOOKING AHEAD TO CALCULUS

A technique called **logarithmic differentiation,** which uses the properties of logarithms, can often be used to differentiate complicated functions.

Now consider the product xy.

$$xy = a^m \cdot a^n \qquad x = a^m \text{ and } y = a^n\text{; Substitute.}$$

$$xy = a^{m+n} \qquad \text{Product rule for exponents}$$

$$\log_a xy = m + n \qquad \text{Write in logarithmic form.}$$

$$\log_a xy = \log_a x + \log_a y \qquad \text{Substitute.}$$

The last statement is the result we wished to prove. The quotient and power properties are proved similarly and are left as exercises.

EXAMPLE 5 Using Properties of Logarithms

Use the properties of logarithms to rewrite each expression. Assume all variables represent positive real numbers, with $a \neq 1$ and $b \neq 1$.

(a) $\log_6 (7 \cdot 9)$ **(b)** $\log_9 \frac{15}{7}$ **(c)** $\log_5 \sqrt{8}$

(d) $\log_a \sqrt[3]{m^2}$ **(e)** $\log_a \frac{mnq}{p^2t^4}$ **(f)** $\log_b \sqrt[n]{\frac{x^3y^5}{z^m}}$

SOLUTION

(a) $\log_6 (7 \cdot 9)$

$$= \log_6 7 + \log_6 9 \qquad \text{Product property}$$

(b) $\log_9 \frac{15}{7}$

$$= \log_9 15 - \log_9 7 \qquad \text{Quotient property}$$

(c) $\log_5 \sqrt{8}$

$$= \log_5 (8^{1/2}) \qquad \sqrt{a} = a^{1/2}$$

$$= \frac{1}{2}\log_5 8 \qquad \text{Power property}$$

(d) $\log_a \sqrt[3]{m^2}$

$$= \log_a m^{2/3} \qquad \sqrt[n]{a^m} = a^{m/n}$$

$$= \frac{2}{3}\log_a m \qquad \text{Power property}$$

(e) $\log_a \frac{mnq}{p^2t^4}$

$$= \log_a m + \log_a n + \log_a q - (\log_a p^2 + \log_a t^4) \qquad \text{Product and quotient properties}$$

Use parentheses to avoid errors.

$$= \log_a m + \log_a n + \log_a q - (2\log_a p + 4\log_a t) \qquad \text{Power property}$$

$$= \log_a m + \log_a n + \log_a q - 2\log_a p - 4\log_a t \qquad \text{Distributive property}$$

Be careful with signs.

(f) $\log_b \sqrt[n]{\frac{x^3y^5}{z^m}}$

$$= \log_b \left(\frac{x^3y^5}{z^m}\right)^{1/n} \qquad \sqrt[n]{a} = a^{1/n}$$

$$= \frac{1}{n}\log_b \frac{x^3y^5}{z^m} \qquad \text{Power property}$$

$$= \frac{1}{n}(\log_b x^3 + \log_b y^5 - \log_b z^m) \qquad \text{Product and quotient properties}$$

$$= \frac{1}{n}(3\log_b x + 5\log_b y - m\log_b z) \qquad \text{Power property}$$

$$= \frac{3}{n}\log_b x + \frac{5}{n}\log_b y - \frac{m}{n}\log_b z \qquad \text{Distributive property}$$

✔ **Now Try Exercises 71, 73, and 77.**

EXAMPLE 6 Using Properties of Logarithms

Write each expression as a single logarithm with coefficient 1. Assume all variables represent positive real numbers, with $a \neq 1$ and $b \neq 1$.

(a) $\log_3(x+2) + \log_3 x - \log_3 2$ **(b)** $2\log_a m - 3\log_a n$

(c) $\frac{1}{2}\log_b m + \frac{3}{2}\log_b 2n - \log_b m^2 n$

SOLUTION

(a) $\log_3(x+2) + \log_3 x - \log_3 2$

$= \log_3 \frac{(x+2)x}{2}$ Product and quotient properties

(b) $2\log_a m - 3\log_a n$

$= \log_a m^2 - \log_a n^3$ Power property

$= \log_a \frac{m^2}{n^3}$ Quotient property

(c) $\frac{1}{2}\log_b m + \frac{3}{2}\log_b 2n - \log_b m^2 n$

$= \log_b m^{1/2} + \log_b (2n)^{3/2} - \log_b m^2 n$ Power property

$= \log_b \frac{m^{1/2}(2n)^{3/2}}{m^2 n}$ Use parentheses around $2n$. Product and quotient properties

$= \log_b \frac{2^{3/2} n^{1/2}}{m^{3/2}}$ Rules for exponents

$= \log_b \left(\frac{2^3 n}{m^3}\right)^{1/2}$ Rules for exponents

$= \log_b \sqrt{\frac{8n}{m^3}}$ Definition of $a^{1/n}$

✔ **Now Try Exercises 83, 87, and 89.**

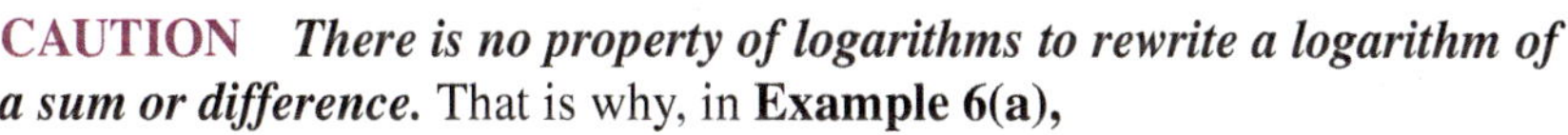

CAUTION ***There is no property of logarithms to rewrite a logarithm of a sum or difference.*** That is why, in **Example 6(a),**

$\log_3(x+2)$ cannot be written as $\log_3 x + \log_3 2$.

The distributive property does not apply here because $\log_3(x+y)$ is one term. The abbreviation "log" is a function name, *not* a factor.

Napier's Rods

The search for ways to make calculations easier has been a long, ongoing process. Machines built by Charles Babbage and Blaise Pascal, a system of "rods" used by John Napier, and slide rules were the forerunners of today's calculators and computers. The invention of logarithms by John Napier in the 16th century was a great breakthrough in the search for easier calculation methods.

Source: IBM Corporate Archives.

EXAMPLE 7 Using Properties of Logarithms with Numerical Values

Given that $\log_{10} 2 \approx 0.3010$, find each logarithm without using a calculator.

(a) $\log_{10} 4$ **(b)** $\log_{10} 5$

SOLUTION

(a) $\log_{10} 4$

$= \log_{10} 2^2$

$= 2\log_{10} 2$

$\approx 2(0.3010)$

≈ 0.6020

(b) $\log_{10} 5$

$= \log_{10} \frac{10}{2}$

$= \log_{10} 10 - \log_{10} 2$

$\approx 1 - 0.3010$

≈ 0.6990

✔ **Now Try Exercises 93 and 95.**

NOTE The values in **Example 7** are approximations of logarithms, so the final digit may differ from the actual 4-decimal-place approximation after properties of logarithms are applied.

Recall that for inverse functions f and g, $(f \circ g)(x) = (g \circ f)(x) = x$. We can use this property with exponential and logarithmic functions to state two more properties. If $f(x) = a^x$ and $g(x) = \log_a x$, then

$$(f \circ g)(x) = a^{\log_a x} = x \quad \text{and} \quad (g \circ f)(x) = \log_a (a^x) = x.$$

Theorem on Inverses

For $a > 0$, $a \neq 1$, the following properties hold.

$$a^{\log_a x} = x \ (\text{for } x > 0) \quad \text{and} \quad \log_a a^x = x$$

Examples: $7^{\log_7 10} = 10$, $\log_5 5^3 = 3$, and $\log_r r^{k+1} = k + 1$

The second statement in the theorem will be useful when we solve logarithmic and exponential equations.

4.3 Exercises

CONCEPT PREVIEW *Match the logarithm in Column I with its value in Column II. Remember that $\log_a x$ is the exponent to which a must be raised in order to obtain x.*

I	II
1. (a) $\log_2 16$	**A.** 0
(b) $\log_3 1$	**B.** $\frac{1}{2}$
(c) $\log_{10} 0.1$	**C.** 4
(d) $\log_2 \sqrt{2}$	**D.** -3
(e) $\log_e \frac{1}{e^2}$	**E.** -1
(f) $\log_{1/2} 8$	**F.** -2

I	II
2. (a) $\log_3 81$	**A.** -2
(b) $\log_3 \frac{1}{3}$	**B.** -1
(c) $\log_{10} 0.01$	**C.** 0
(d) $\log_6 \sqrt{6}$	**D.** $\frac{1}{2}$
(e) $\log_e 1$	**E.** $\frac{9}{2}$
(f) $\log_3 27^{3/2}$	**F.** 4

CONCEPT PREVIEW *Write each equivalent form.*

3. Write $\log_2 8 = 3$ in exponential form.

4. Write $10^3 = 1000$ in logarithmic form.

CONCEPT PREVIEW *Solve each logarithmic equation.*

5. $\log_x \frac{16}{81} = 2$

6. $\log_{36} \sqrt[3]{6} = x$

CONCEPT PREVIEW *Sketch the graph of each function. Give the domain and range.*

7. $f(x) = \log_5 x$

8. $g(x) = \log_{1/5} x$

CONCEPT PREVIEW *Use the properties of logarithms to rewrite each expression. Assume all variables represent positive real numbers.*

9. $\log_{10} \frac{2x}{7}$

10. $3 \log_4 x - 5 \log_4 y$

If the statement is in exponential form, write it in an equivalent logarithmic form. If the statement is in logarithmic form, write it in exponential form. ***See Example 1.***

11. $3^4 = 81$

12. $2^5 = 32$

13. $\left(\frac{2}{3}\right)^{-3} = \frac{27}{8}$

14. $10^{-4} = 0.0001$

15. $\log_6 36 = 2$

16. $\log_5 5 = 1$

17. $\log_{\sqrt{3}} 81 = 8$

18. $\log_4 \frac{1}{64} = -3$

Solve each equation. ***See Example 2.***

19. $x = \log_5 \frac{1}{625}$

20. $x = \log_3 \frac{1}{81}$

21. $\log_x \frac{1}{32} = 5$

22. $\log_x \frac{27}{64} = 3$

23. $x = \log_8 \sqrt[4]{8}$

24. $x = \log_7 \sqrt[5]{7}$

25. $x = 3^{\log_3 8}$

26. $x = 12^{\log_{12} 5}$

27. $x = 2^{\log_2 9}$

28. $x = 8^{\log_8 11}$

29. $\log_x 25 = -2$

30. $\log_x 16 = -2$

31. $\log_4 x = 3$

32. $\log_2 x = 3$

33. $x = \log_4 \sqrt[3]{16}$

34. $x = \log_5 \sqrt[4]{25}$

35. $\log_9 x = \frac{5}{2}$

36. $\log_4 x = \frac{7}{2}$

37. $\log_{1/2} (x + 3) = -4$

38. $\log_{1/3} (x + 6) = -2$

39. $\log_{(x+3)} 6 = 1$

40. $\log_{(x-4)} 19 = 1$

41. $3x - 15 = \log_x 1 \quad (x > 0, x \neq 1)$

42. $4x - 24 = \log_x 1 \quad (x > 0, x \neq 1)$

Graph each function. Give the domain and range. ***See Example 4.***

43. $f(x) = (\log_2 x) + 3$

44. $f(x) = \log_2 (x + 3)$

45. $f(x) = |\log_2 (x + 3)|$

Graph each function. Give the domain and range. ***See Example 4.***

46. $f(x) = (\log_{1/2} x) - 2$

47. $f(x) = \log_{1/2} (x - 2)$

48. $f(x) = |\log_{1/2} (x - 2)|$

Concept Check *In Exercises 49–54, match the function with its graph from choices A–F.*

49. $f(x) = \log_2 x$

50. $f(x) = \log_2 2x$

51. $f(x) = \log_2 \frac{1}{x}$

52. $f(x) = \log_2 \left(\frac{1}{2}x\right)$

53. $f(x) = \log_2 (x - 1)$

54. $f(x) = \log_2 (-x)$

A.

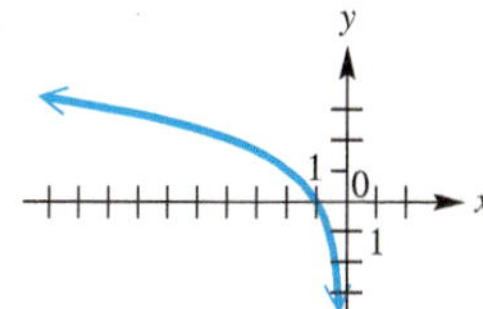

B.

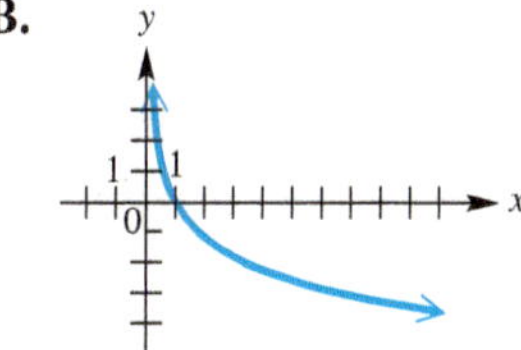

C.

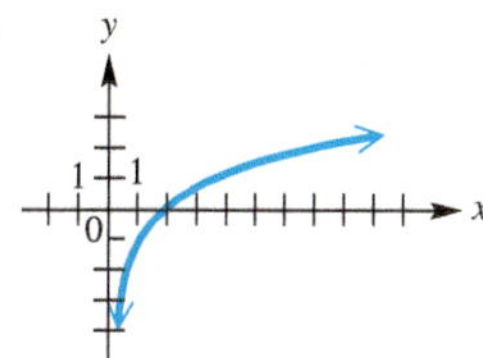

D.

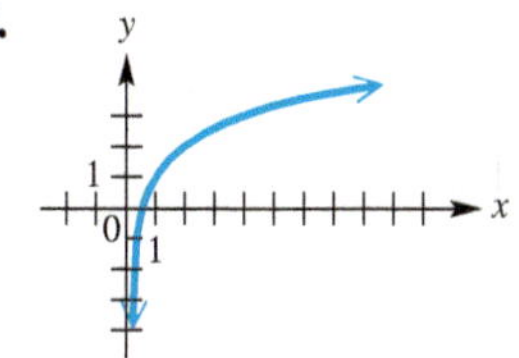

E.

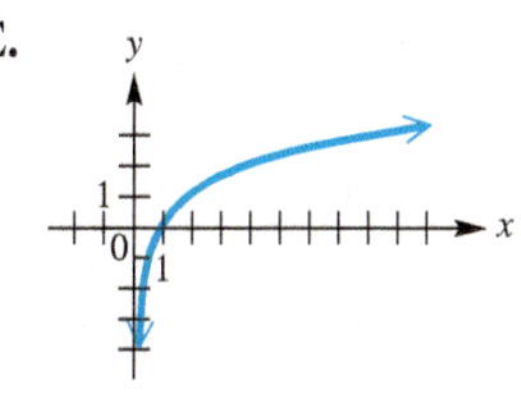

F.

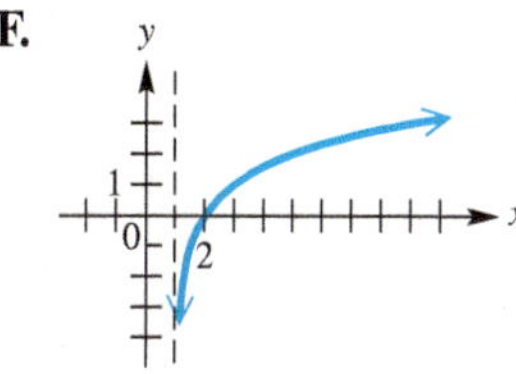

Graph each function. ***See Examples 3 and 4.***

55. $f(x) = \log_5 x$ **56.** $f(x) = \log_{10} x$ **57.** $f(x) = \log_5 (x + 1)$

58. $f(x) = \log_6 (x - 2)$ **59.** $f(x) = \log_{1/2} (1 - x)$ **60.** $f(x) = \log_{1/3} (3 - x)$

61. $f(x) = \log_3 (x - 1) + 2$ **62.** $f(x) = \log_2 (x + 2) - 3$ **63.** $f(x) = \log_{1/2} (x + 3) - 2$

64. *Concept Check* To graph the function $f(x) = -\log_5 (x - 7) - 4$, reflect the graph of $y = \log_5 x$ across the ______-axis, then shift the graph ______ units to the right and ______ units down.

Connecting Graphs with Equations *Write an equation for the graph given. Each represents a logarithmic function f with base 2 or 3, translated and/or reflected.* ***See the Note following Example 4.***

65.

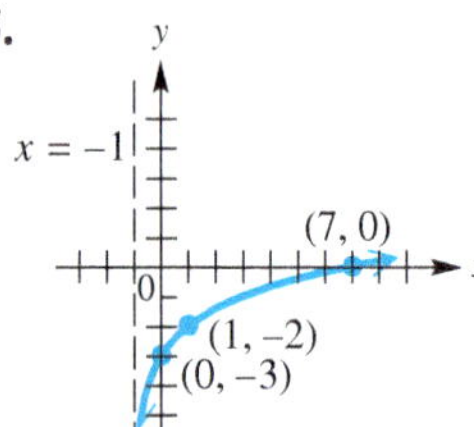

66.

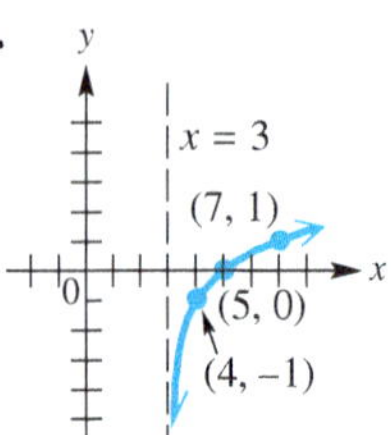

67.

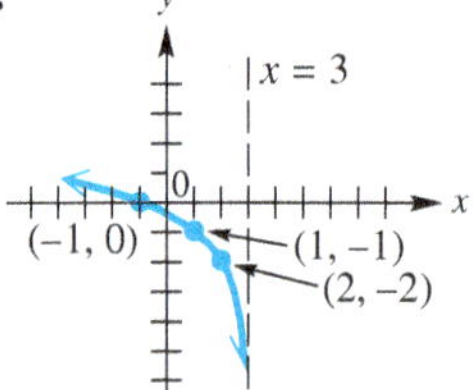

68.

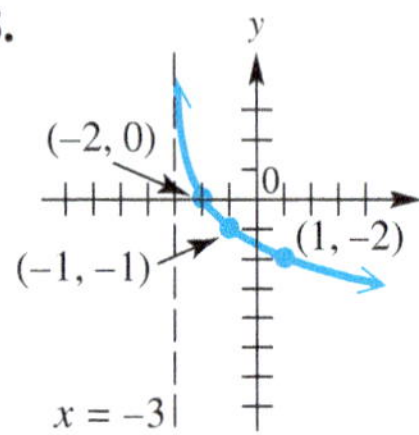

69.

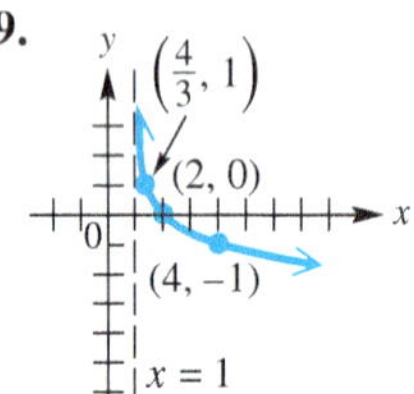

70.

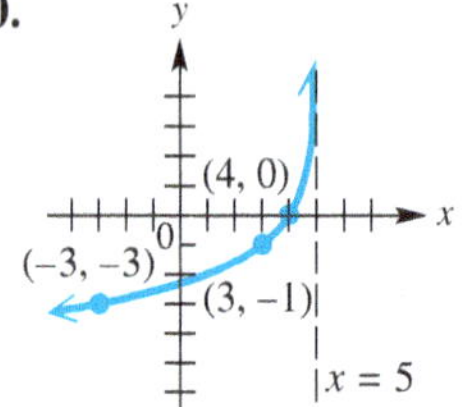

Use the properties of logarithms to rewrite each expression. Simplify the result if possible. Assume all variables represent positive real numbers. ***See Example 5.***

71. $\log_2 \frac{6x}{y}$ **72.** $\log_3 \frac{4p}{q}$ **73.** $\log_5 \frac{5\sqrt{7}}{3}$

74. $\log_2 \frac{2\sqrt{3}}{5}$ **75.** $\log_4 (2x + 5y)$ **76.** $\log_6 (7m + 3q)$

77. $\log_2 \sqrt{\frac{5r^3}{z^5}}$ **78.** $\log_3 \sqrt[3]{\frac{m^5 n^4}{t^2}}$ **79.** $\log_2 \frac{ab}{cd}$

80. $\log_2 \frac{xy}{tqr}$ **81.** $\log_3 \frac{\sqrt{x} \cdot \sqrt[3]{y}}{w^2 \sqrt{z}}$ **82.** $\log_4 \frac{\sqrt[3]{a} \cdot \sqrt[4]{b}}{\sqrt{c} \cdot \sqrt[3]{d^2}}$

Write each expression as a single logarithm with coefficient 1. Assume all variables represent positive real numbers, with a $\neq$ 1 *and b* $\neq$ 1. ***See Example 6.***

83. $\log_a x + \log_a y - \log_a m$ **84.** $\log_b k + \log_b m - \log_b a$

85. $\log_a m - \log_a n - \log_a t$ **86.** $\log_b p - \log_b q - \log_b r$

87. $\frac{1}{3} \log_b x^4 y^5 - \frac{3}{4} \log_b x^2 y$ **88.** $\frac{1}{2} \log_a p^3 q^4 - \frac{2}{3} \log_a p^4 q^3$

89. $2 \log_a (z + 1) + \log_a (3z + 2)$ **90.** $5 \log_a (z + 7) + \log_a (2z + 9)$

91. $-\frac{2}{3} \log_5 5m^2 + \frac{1}{2} \log_5 25m^2$ **92.** $-\frac{3}{4} \log_3 16p^4 - \frac{2}{3} \log_3 8p^3$

Given that $\log_{10} 2 \approx 0.3010$ *and* $\log_{10} 3 \approx 0.4771$, *find each logarithm without using a calculator.* ***See Example 7.***

93. $\log_{10} 6$ **94.** $\log_{10} 12$ **95.** $\log_{10} \frac{3}{2}$ **96.** $\log_{10} \frac{2}{9}$

97. $\log_{10} \frac{9}{4}$ **98.** $\log_{10} \frac{20}{27}$ **99.** $\log_{10} \sqrt{30}$ **100.** $\log_{10} 36^{1/3}$

Solve each problem.

101. ***(Modeling) Interest Rates of Treasury Securities*** The table gives interest rates for various U.S. Treasury Securities on January 2, 2015.

(a) Make a scatter diagram of the data.

(b) Which type of function will model this data best: linear, exponential, or logarithmic?

Time	Yield
3-month	0.02%
6-month	0.10%
2-year	0.66%
5-year	1.61%
10-year	2.11%
30-year	2.60%

Source: www.federalreserve.gov

102. ***Concept Check*** Use the graph to estimate each logarithm.

(a) $\log_3 0.3$ **(b)** $\log_3 0.8$

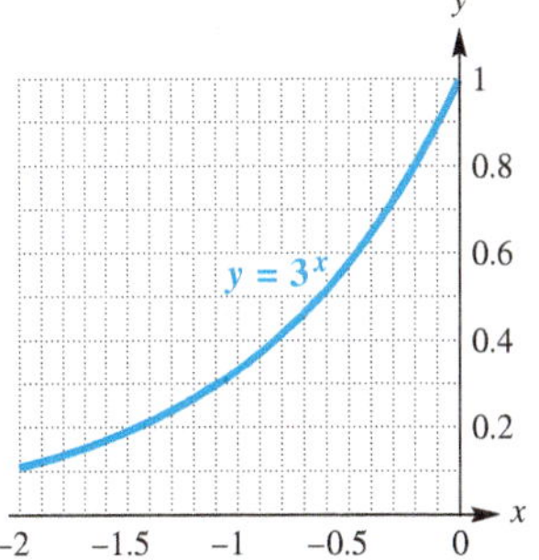

103. ***Concept Check*** Suppose $f(x) = \log_a x$ and $f(3) = 2$. Determine each function value.

(a) $f\left(\frac{1}{9}\right)$ **(b)** $f(27)$ **(c)** $f(9)$ **(d)** $f\left(\frac{\sqrt{3}}{3}\right)$

104. Use properties of logarithms to evaluate each expression.

(a) $100^{\log_{10} 3}$ **(b)** $\log_{10} (0.01)^3$ **(c)** $\log_{10} (0.0001)^5$ **(d)** $1000^{\log_{10} 5}$

105. Using the compound interest formula $A = P\left(1 + \frac{r}{n}\right)^{tn}$, show that the amount of time required for a deposit to double is

$$\frac{1}{\log_2 \left(1 + \frac{r}{n}\right)^n}.$$

106. ***Concept Check*** If $(5, 4)$ is on the graph of the logarithmic function with base a, which of the following statements is true:

$$5 = \log_a 4 \quad \text{or} \quad 4 = \log_a 5?$$

Use a graphing calculator to find the solution set of each equation. Give solutions to the nearest hundredth.

107. $\log_{10} x = x - 2$ **108.** $2^{-x} = \log_{10} x$

109. Prove the quotient property of logarithms: $\log_a \frac{x}{y} = \log_a x - \log_a y$.

110. Prove the power property of logarithms: $\log_a x^r = r \log_a x$.

Summary Exercises on Inverse, Exponential, and Logarithmic Functions

The following exercises are designed to help solidify your understanding of inverse, exponential, and logarithmic functions from **Sections 4.1–4.3.**

Determine whether the functions in each pair are inverses of each other.

1. $f(x) = 3x - 4, \quad g(x) = \frac{1}{3}x + \frac{4}{3}$

2. $f(x) = 8 - 5x, \quad g(x) = 8 + \frac{1}{5}x$

3. $f(x) = 1 + \log_2 x, \quad g(x) = 2^{x-1}$

4. $f(x) = 3^{x/5} - 2, \quad g(x) = 5 \log_3 (x + 2)$

Determine whether each function is one-to-one. If it is, then sketch the graph of its inverse function.

5.

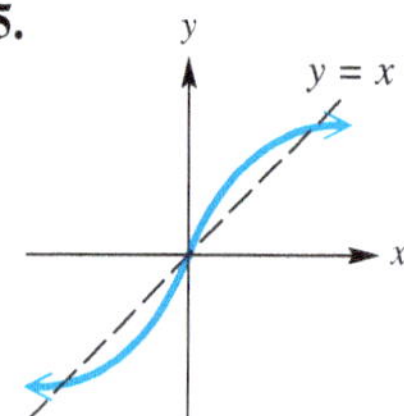

6.

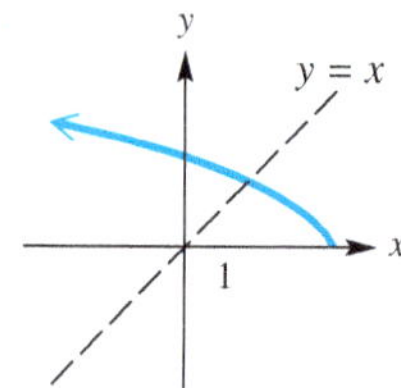

7.

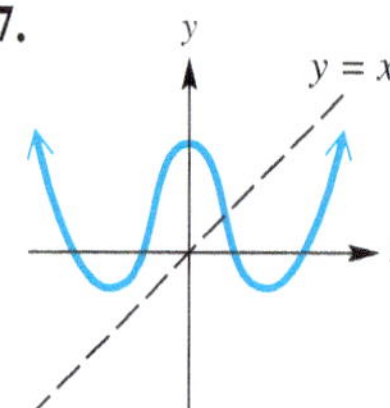

8.

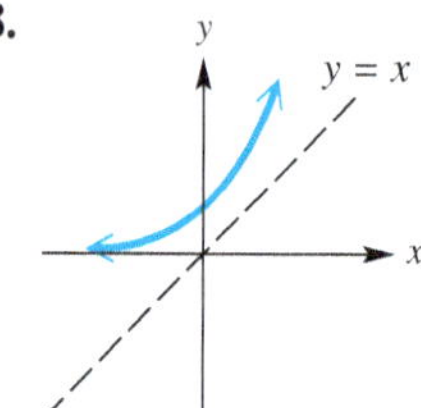

In Exercises 9–12, match each function with its graph from choices A–D.

9. $y = \log_3 (x + 2)$

10. $y = 5 - 2^x$

11. $y = \log_2 (5 - x)$

12. $y = 3^x - 2$

A.

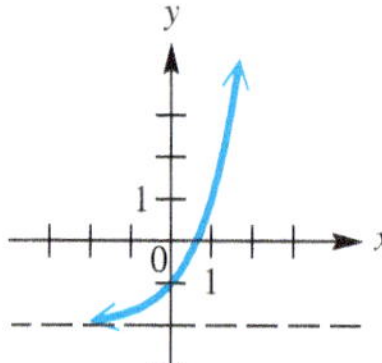

B.

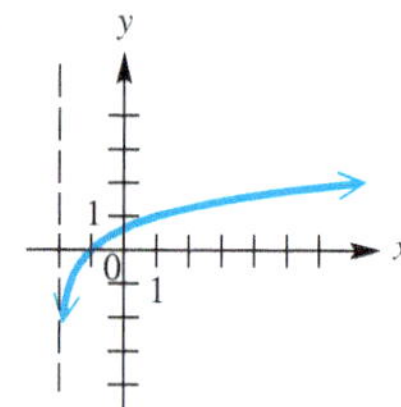

C.

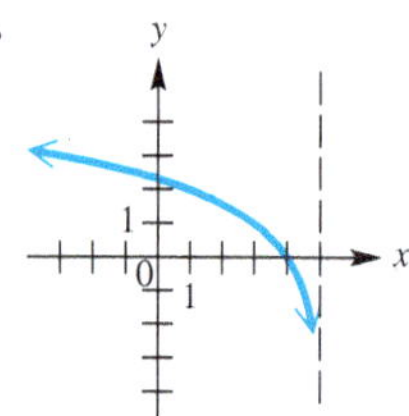

D.

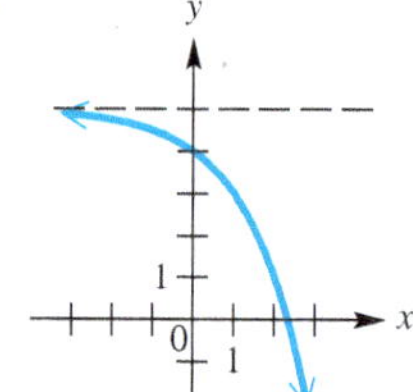

13. The functions in **Exercises 9–12** form two pairs of inverse functions. Determine which functions are inverses of each other.

14. Determine the inverse of the function $f(x) = \log_5 x$. (*Hint:* Replace $f(x)$ with y, and write in exponential form.)

For each function that is one-to-one, write an equation for the inverse function. Give the domain and range of both f and f^{-1}. If the function is not one-to-one, say so.

15. $f(x) = 3x - 6$

16. $f(x) = 2(x + 1)^3$

17. $f(x) = 3x^2$

18. $f(x) = \dfrac{2x - 1}{5 - 3x}$

19. $f(x) = \sqrt[3]{5 - x^4}$

20. $f(x) = \sqrt{x^2 - 9}, \quad x \geq 3$

Write an equivalent statement in logarithmic form.

21. $\left(\dfrac{1}{10}\right)^{-3} = 1000$

22. $a^b = c$

23. $\left(\sqrt{3}\right)^4 = 9$

24. $4^{-3/2} = \dfrac{1}{8}$

25. $2^x = 32$

26. $27^{4/3} = 81$

Solve each equation.

27. $3x = 7^{\log_7 6}$

28. $x = \log_{10} 0.001$

29. $x = \log_6 \dfrac{1}{216}$

30. $\log_x 5 = \dfrac{1}{2}$

31. $\log_{10} 0.01 = x$

32. $\log_x 3 = -1$

33. $\log_x 1 = 0$

34. $x = \log_2 \sqrt{8}$

35. $\log_x \sqrt[3]{5} = \dfrac{1}{3}$

36. $\log_{1/3} x = -5$

37. $\log_{10}(\log_2 2^{10}) = x$

38. $x = \log_{4/5} \dfrac{25}{16}$

39. $2x - 1 = \log_6 6^x$

40. $x = \sqrt{\log_{1/2} \dfrac{1}{16}}$

41. $2^x = \log_2 16$

42. $\log_3 x = -2$

43. $\left(\dfrac{1}{3}\right)^{x+1} = 9^x$

44. $5^{2x-6} = 25^{x-3}$

4.4 Evaluating Logarithms and the Change-of-Base Theorem

- Common Logarithms
- Applications and Models with Common Logarithms
- Natural Logarithms
- Applications and Models with Natural Logarithms
- Logarithms with Other Bases

Common Logarithms

Two of the most important bases for logarithms are 10 and e. Base 10 logarithms are **common logarithms.** The common logarithm of x is written **log x,** where the base is understood to be 10.

Common Logarithm

For all positive numbers x,

$$\log x = \log_{10} x.$$

A calculator with a log key can be used to find the base 10 logarithm of any positive number.

EXAMPLE 1 Evaluating Common Logarithms with a Calculator

Use a calculator to find the values of

$$\log 1000, \quad \log 142, \quad \text{and} \quad \log 0.005832.$$

SOLUTION **Figure 33** shows that the exact value of log 1000 is 3 (because $10^3 = 1000$), and that

$$\log 142 \approx 2.152288344$$

and $$\log 0.005832 \approx -2.234182485.$$

Most common logarithms that appear in calculations are approximations, as seen in the second and third displays.

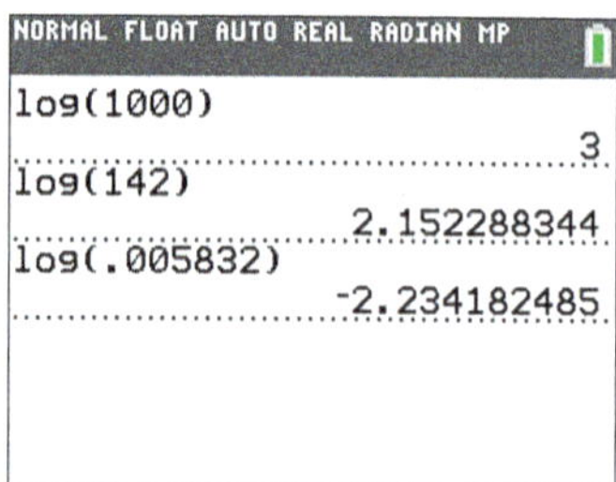

Figure 33

✔ **Now Try Exercises 11, 15, and 17.**

For $a > 1$, base a logarithms of numbers between 0 and 1 are always negative, and base a logarithms of numbers greater than 1 are always positive.

Applications and Models with Common Logarithms In chemistry, the **pH** of a solution is defined as

$$\mathbf{pH = -\log[H_3O^+]},$$

where $[H_3O^+]$ is the hydronium ion concentration in moles* per liter. The pH value is a measure of the acidity or alkalinity of a solution. Pure water has pH 7.0, substances with pH values greater than 7.0 are alkaline, and substances with pH values less than 7.0 are acidic. See **Figure 34.** It is customary to round pH values to the nearest tenth.

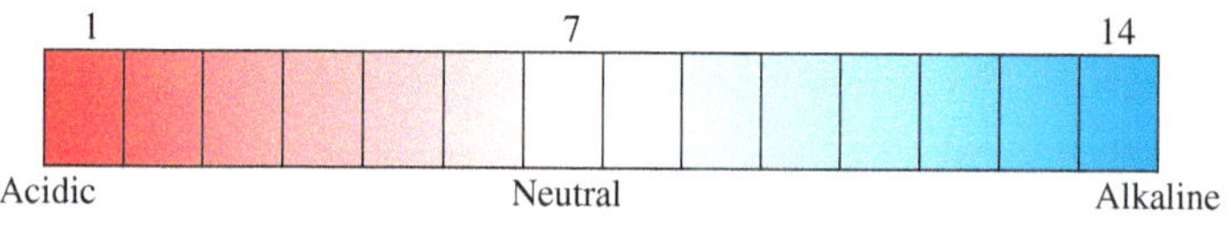

Figure 34

EXAMPLE 2 Finding pH

(a) Find the pH of a solution with $[H_3O^+] = 2.5 \times 10^{-4}$.

(b) Find the hydronium ion concentration of a solution with pH = 7.1.

SOLUTION

(a)

$pH = -\log[H_3O^+]$	
$pH = -\log(2.5 \times 10^{-4})$	Substitute $[H_3O^+] = 2.5 \times 10^{-4}$.
$pH = -(\log 2.5 + \log 10^{-4})$	Product property
$pH = -(0.3979 - 4)$	$\log 10^{-4} = -4$
$pH = -0.3979 + 4$	Distributive property
$pH \approx 3.6$	Add.

*A *mole* is the amount of a substance that contains the same number of molecules as the number of atoms in exactly 12 grams of carbon-12.

(b)

$$\begin{aligned} pH &= -\log[H_3O^+] & \\ 7.1 &= -\log[H_3O^+] & \text{Substitute pH} = 7.1. \\ -7.1 &= \log[H_3O^+] & \text{Multiply by } -1. \\ [H_3O^+] &= 10^{-7.1} & \text{Write in exponential form.} \\ [H_3O^+] &\approx 7.9 \times 10^{-8} & \text{Evaluate } 10^{-7.1} \text{ with a calculator.} \end{aligned}$$

✔ **Now Try Exercises 29 and 33.**

NOTE In the fourth line of the solution in **Example 2(a),** we use the equality symbol, =, rather than the approximate equality symbol, ≈, when replacing log 2.5 with 0.3979. This is often done for convenience, despite the fact that most logarithms used in applications are indeed approximations.

EXAMPLE 3 Using pH in an Application

Wetlands are classified as *bogs, fens, marshes,* and *swamps* based on pH values. A pH value between 6.0 and 7.5 indicates that the wetland is a "rich fen." When the pH is between 3.0 and 6.0, it is a "poor fen," and if the pH falls to 3.0 or less, the wetland is a "bog." (*Source:* R. Mohlenbrock, "Summerby Swamp, Michigan," *Natural History*.)

Suppose that the hydronium ion concentration of a sample of water from a wetland is 6.3×10^{-5}. How would this wetland be classified?

SOLUTION

$$\begin{aligned} pH &= -\log[H_3O^+] & \text{Definition of pH} \\ pH &= -\log(6.3 \times 10^{-5}) & \text{Substitute for } [H_3O^+]. \\ pH &= -(\log 6.3 + \log 10^{-5}) & \text{Product property} \\ pH &= -\log 6.3 - (-5) & \text{Distributive property; } \log 10^n = n \\ pH &= -\log 6.3 + 5 & \text{Definition of subtraction} \\ pH &\approx 4.2 & \text{Use a calculator.} \end{aligned}$$

The pH is between 3.0 and 6.0, so the wetland is a poor fen.

✔ **Now Try Exercise 37.**

EXAMPLE 4 Measuring the Loudness of Sound

The loudness of sounds is measured in **decibels.** We first assign an intensity of I_0 to a very faint **threshold sound.** If a particular sound has intensity I, then the decibel rating d of this louder sound is given by the following formula.

$$d = 10 \log \frac{I}{I_0}$$

Find the decibel rating d of a sound with intensity $10{,}000I_0$.

SOLUTION

$$\begin{aligned} d &= 10 \log \frac{10{,}000I_0}{I_0} & \text{Let } I = 10{,}000I_0. \\ d &= 10 \log 10{,}000 & \frac{I_0}{I_0} = 1 \\ d &= 10(4) & \log 10{,}000 = \log 10^4 = 4 \\ d &= 40 & \text{Multiply.} \end{aligned}$$

The sound has a decibel rating of 40.

✔ **Now Try Exercise 63.**

Natural Logarithms

In most practical applications of logarithms, the irrational number e is used as the base. Logarithms with base e are **natural logarithms** because they occur in the life sciences and economics in natural situations that involve growth and decay. The base e logarithm of x is written **ln** x (read **"el-en** x**"**). ***The expression* ln *x represents the exponent to which e must be raised in order to obtain x.***

Natural Logarithm

For all positive numbers x,

$$\ln x = \log_e x.$$

Figure 35

A graph of the natural logarithmic function $f(x) = \ln x$ is given in **Figure 35.**

EXAMPLE 5 Evaluating Natural Logarithms with a Calculator

Use a calculator to find the values of

$$\ln e^3, \quad \ln 142, \quad \text{and} \quad \ln 0.005832.$$

SOLUTION **Figure 36** shows that the exact value of $\ln e^3$ is 3, and that

$$\ln 142 \approx 4.955827058$$

and $$\ln 0.005832 \approx -5.144395284.$$

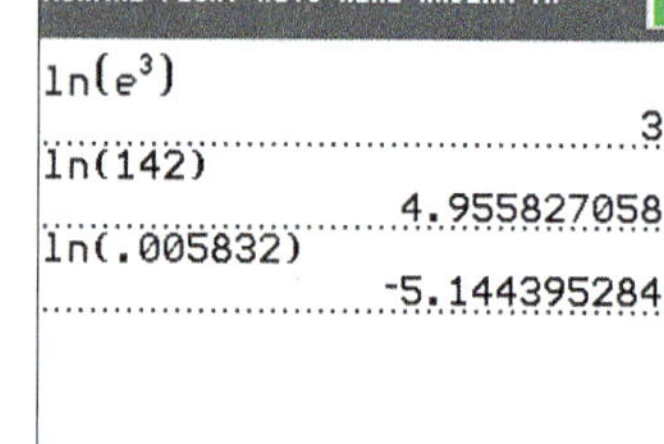

Figure 36

✔ **Now Try Exercises 45, 51, and 53.**

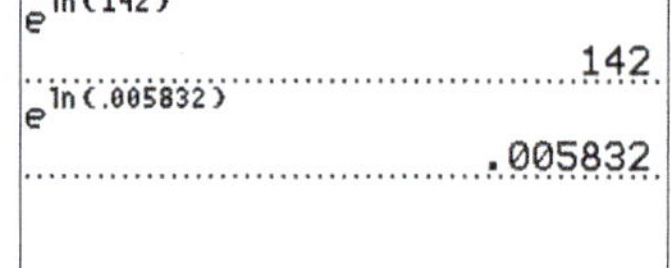

Figure 37

Figure 37 illustrates that **ln** x ***is the exponent to which e must be raised in order to obtain x.***

Applications and Models with Natural Logarithms

EXAMPLE 6 Measuring the Age of Rocks

Geologists sometimes measure the age of rocks by using "atomic clocks." By measuring the amounts of argon-40 and potassium-40 in a rock, it is possible to find the age t of the specimen in years with the formula

$$t = (1.26 \times 10^9)\frac{\ln\left(1 + 8.33\left(\frac{A}{K}\right)\right)}{\ln 2},$$

where A and K are the numbers of atoms of argon-40 and potassium-40, respectively, in the specimen.

(a) How old is a rock in which $A = 0$ and $K > 0$?

(b) The ratio $\frac{A}{K}$ for a sample of granite from New Hampshire is 0.212. How old is the sample?

LOOKING AHEAD TO CALCULUS

The natural logarithmic function $f(x) = \ln x$ and the reciprocal function $g(x) = \frac{1}{x}$ have an important relationship in calculus. The derivative of the natural logarithmic function is the reciprocal function. Using **Leibniz notation** (named after one of the co-inventors of calculus), we write this fact as $\frac{d}{dx}(\ln x) = \frac{1}{x}$.

SOLUTION

(a) If $A = 0$, then $\frac{A}{K} = 0$ and the equation is as follows.

$$t = (1.26 \times 10^9)\frac{\ln\left(1 + 8.33\left(\frac{A}{K}\right)\right)}{\ln 2} \quad \text{Given formula}$$

$$t = (1.26 \times 10^9)\frac{\ln 1}{\ln 2} \quad \frac{A}{K} = 0, \text{ so } \ln(1+0) = \ln 1$$

$$t = (1.26 \times 10^9)(0) \quad \ln 1 = 0$$

$$t = 0$$

The rock is new (0 yr old).

(b) Because $\frac{A}{K} = 0.212$, we have the following.

$$t = (1.26 \times 10^9)\frac{\ln(1 + 8.33(0.212))}{\ln 2} \quad \text{Substitute.}$$

$$t \approx 1.85 \times 10^9 \quad \text{Use a calculator.}$$

The granite is about 1.85 billion yr old.

✔ **Now Try Exercise 77.**

EXAMPLE 7 Modeling Global Temperature Increase

Carbon dioxide in the atmosphere traps heat from the sun. The additional solar radiation trapped by carbon dioxide is **radiative forcing.** It is measured in watts per square meter (w/m^2). In 1896 the Swedish scientist Svante Arrhenius modeled radiative forcing R caused by additional atmospheric carbon dioxide, using the logarithmic equation

$$R = k \ln \frac{C}{C_0},$$

where C_0 is the preindustrial amount of carbon dioxide, C is the current carbon dioxide level, and k is a constant. Arrhenius determined that $10 \le k \le 16$ when $C = 2C_0$. (*Source:* Clime, W., *The Economics of Global Warming*, Institute for International Economics, Washington, D.C.)

(a) Let $C = 2C_0$. Is the relationship between R and k linear or logarithmic?

(b) The average global temperature increase T (in °F) is given by $T(R) = 1.03R$. Write T as a function of k.

SOLUTION

(a) If $C = 2C_0$, then $\frac{C}{C_0} = 2$, so $R = k \ln 2$ is a linear relation, because $\ln 2$ is a constant.

(b)

$$T(R) = 1.03R$$

$$T(k) = 1.03k \ln \frac{C}{C_0} \quad \text{Use the given expression for } R.$$

✔ **Now Try Exercise 75.**

Logarithms with Other Bases We can use a calculator to find the values of either natural logarithms (base e) or common logarithms (base 10). However, sometimes we must use logarithms with other bases. The change-of-base theorem can be used to convert logarithms from one base to another.

LOOKING AHEAD TO CALCULUS

In calculus, natural logarithms are more convenient to work with than logarithms with other bases. The change-of-base theorem enables us to convert any logarithmic function to a *natural* logarithmic function.

Change-of-Base Theorem

For any positive real numbers x, a, and b, where $a \neq 1$ and $b \neq 1$, the following holds.

$$\log_a x = \frac{\log_b x}{\log_b a}$$

Proof Let $y = \log_a x$.

Then $a^y = x$ Write in exponential form.

$\log_b a^y = \log_b x$ Take the base b logarithm on each side.

$y \log_b a = \log_b x$ Power property

$y = \frac{\log_b x}{\log_b a}$ Divide each side by $\log_b a$.

$\log_a x = \frac{\log_b x}{\log_b a}$. Substitute $\log_a x$ for y.

Any positive number other than 1 can be used for base b in the change-of-base theorem, but usually the only practical bases are e and 10 since most calculators give logarithms for these two bases only.

Using the change-of-base theorem, we can graph an equation such as $y = \log_2 x$ by directing the calculator to graph $y = \frac{\log x}{\log 2}$, or, equivalently, $y = \frac{\ln x}{\ln 2}$. ■

EXAMPLE 8 Using the Change-of-Base Theorem

Use the change-of-base theorem to find an approximation to four decimal places for each logarithm.

(a) $\log_5 17$ **(b)** $\log_2 0.1$

SOLUTION

(a) We use natural logarithms to approximate this logarithm. Because $\log_5 5 = 1$ and $\log_5 25 = 2$, we can estimate $\log_5 17$ to be a number between 1 and 2.

$$\log_5 17 = \frac{\ln 17}{\ln 5} \approx 1.7604 \quad \text{Check: } 5^{1.7604} \approx 17$$

The first two entries in **Figure 38(a)** show that the results are the same whether natural or common logarithms are used.

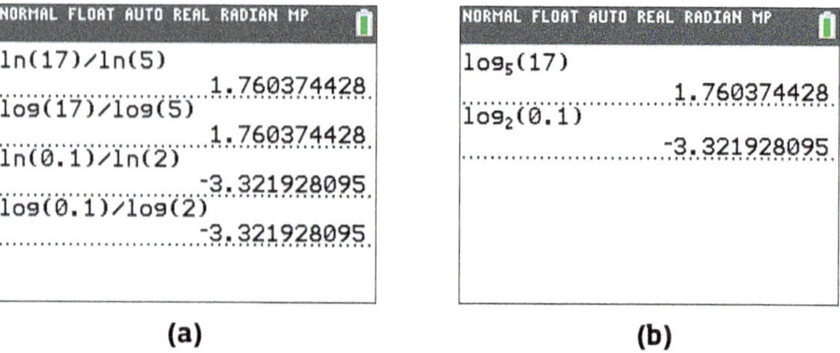

(a) (b)

Figure 38

(b) We use common logarithms for this approximation.

$$\log_2 0.1 = \frac{\log 0.1}{\log 2} \approx -3.3219 \quad \text{Check: } 2^{-3.3219} \approx 0.1$$

The last two entries in **Figure 38(a)** show that the results are the same whether natural or common logarithms are used.

Some calculators, such as the TI-84 Plus, evaluate these logarithms directly without using the change-of-base theorem. See **Figure 38(b).**

✔ **Now Try Exercises 79 and 81.**

EXAMPLE 9 Modeling Diversity of Species

One measure of the diversity of the species in an ecological community is modeled by the formula

$$H = -[P_1 \log_2 P_1 + P_2 \log_2 P_2 + \cdots + P_n \log_2 P_n],$$

where $P_1, P_2, \ldots, P_n$ are the proportions of a sample that belong to each of n species found in the sample. (*Source:* Ludwig, J., and J. Reynolds, *Statistical Ecology: A Primer on Methods and Computing*, © 1988, John Wiley & Sons, NY.)

Find the measure of diversity in a community with two species where there are 90 of one species and 10 of the other.

SOLUTION There are 100 members in the community, so $P_1 = \frac{90}{100} = 0.9$ and $P_2 = \frac{10}{100} = 0.1$.

$$H = -[0.9 \log_2 0.9 + 0.1 \log_2 0.1] \quad \text{Substitute for } P_1 \text{ and } P_2.$$

In **Example 8(b),** we found that $\log_2 0.1 \approx -3.32$. Now we find $\log_2 0.9$.

$$\log_2 0.9 = \frac{\log 0.9}{\log 2} \approx -0.152 \quad \text{Change-of-base theorem}$$

Now evaluate H.

$$H = -[0.9 \log_2 0.9 + 0.1 \log_2 0.1]$$
$$H \approx -[0.9(-0.152) + 0.1(-3.32)] \quad \text{Substitute approximate values.}$$
$$H \approx 0.469 \quad \text{Simplify.}$$

Verify that $H \approx 0.971$ if there are 60 of one species and 40 of the other. As the proportions of n species get closer to $\frac{1}{n}$ each, the measure of diversity increases to a maximum of $\log_2 n$.

✔ **Now Try Exercise 73.**

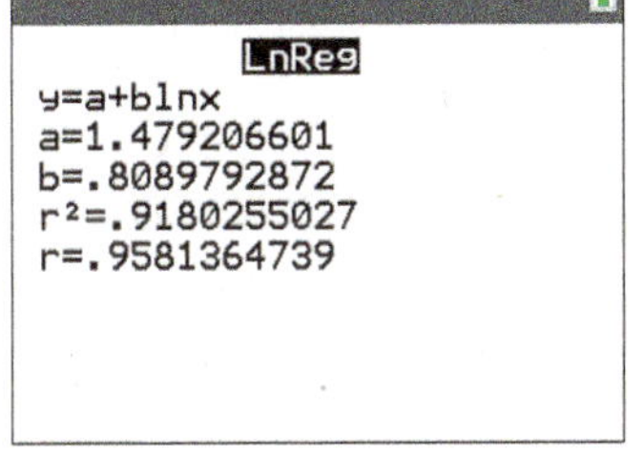

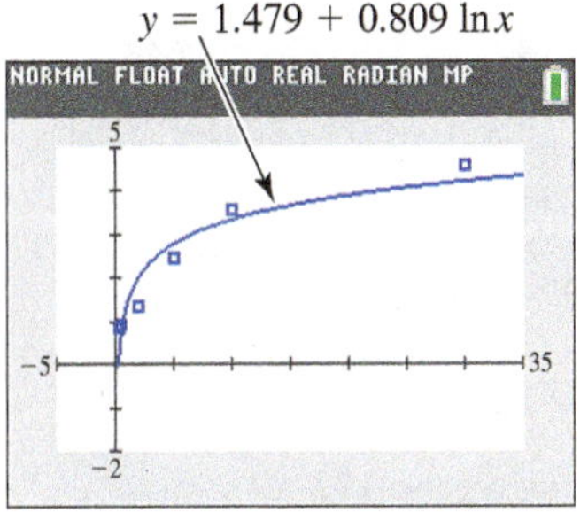

Figure 39

We saw previously that graphing calculators are capable of fitting exponential curves to data that suggest such behavior. The same is true for logarithmic curves. For example, during the early 2000s on one particular day, interest rates for various U.S. Treasury Securities were as shown in the table.

Time	3-mo	6-mo	2-yr	5-yr	10-yr	30-yr
Yield	0.83%	0.91%	1.35%	2.46%	3.54%	4.58%

Source: U.S. Treasury.

Figure 39 shows how a calculator gives the best-fitting natural logarithmic curve for the data, as well as the data points and the graph of this curve. ■

4.4 Exercises

CONCEPT PREVIEW *Answer each of the following.*

1. For the exponential function $f(x) = a^x$, where $a > 1$, is the function increasing or decreasing over its entire domain?

2. For the logarithmic function $g(x) = \log_a x$, where $a > 1$, is the function increasing or decreasing over its entire domain?

3. If $f(x) = 5^x$, what is the rule for $f^{-1}(x)$?

4. What is the name given to the exponent to which 4 must be raised to obtain 11?

5. A base e logarithm is called a(n) ________ logarithm, and a base 10 logarithm is called a(n) ________ logarithm.

6. How is $\log_3 12$ written in terms of natural logarithms using the change-of-base theorem?

7. Why is $\log_2 0$ undefined?

8. Between what two consecutive integers must $\log_2 12$ lie?

9. The graph of $y = \log x$ shows a point on the graph. Write the logarithmic equation associated with that point.

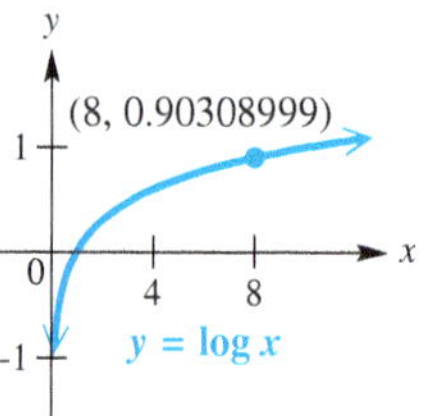

10. The graph of $y = \ln x$ shows a point on the graph. Write the logarithmic equation associated with that point.

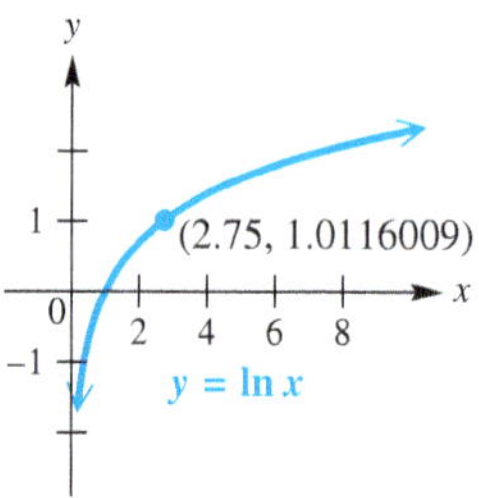

Find each value. If applicable, give an approximation to four decimal places. ***See Example 1.***

11. $\log 10^{12}$	**12.** $\log 10^7$	**13.** $\log 0.1$	**14.** $\log 0.01$
15. $\log 63$	**16.** $\log 94$	**17.** $\log 0.0022$	**18.** $\log 0.0055$
19. $\log(387 \times 23)$	**20.** $\log(296 \times 12)$	**21.** $\log \frac{518}{342}$	**22.** $\log \frac{643}{287}$

23. $\log 387 + \log 23$

24. $\log 296 + \log 12$

25. $\log 518 - \log 342$

26. $\log 643 - \log 287$

Answer each question.

27. Why is the result in **Exercise 23** the same as that in **Exercise 19?**

28. Why is the result in **Exercise 25** the same as that in **Exercise 21?**

For each substance, find the pH from the given hydronium ion concentration. ***See Example 2(a).***

29. grapefruit, 6.3×10^{-4}

30. limes, 1.6×10^{-2}

31. crackers, 3.9×10^{-9}

32. sodium hydroxide (lye), 3.2×10^{-14}

Find the $[H_3O^+]$ *for each substance with the given pH.* ***See Example 2(b).***

33. soda pop, 2.7

34. wine, 3.4

35. beer, 4.8

36. drinking water, 6.5

Suppose that water from a wetland area is sampled and found to have the given hydronium ion concentration. Determine whether the wetland is a rich fen, a poor fen, or a bog. ***See Example 3.***

37. 2.49×10^{-5}

38. 6.22×10^{-5}

39. 2.49×10^{-2}

40. 3.14×10^{-2}

41. 2.49×10^{-7}

42. 5.86×10^{-7}

Solve each problem.

43. Use a calculator to find an approximation for each logarithm.

(a) log 398.4 **(b)** log 39.84 **(c)** log 3.984

(d) From the answers to parts (a)–(c), make a conjecture concerning the decimal values in the approximations of common logarithms of numbers greater than 1 that have the same digits.

44. Given that $\log 25 \approx 1.3979$, $\log 250 \approx 2.3979$, and $\log 2500 \approx 3.3979$, make a conjecture for an approximation of log 25,000. Why does this pattern continue?

Find each value. If applicable, give an approximation to four decimal places. ***See Example 5.***

45. $\ln e^{1.6}$

46. $\ln e^{5.8}$

47. $\ln \frac{1}{e^2}$

48. $\ln \frac{1}{e^4}$

49. $\ln \sqrt{e}$

50. $\ln \sqrt[3]{e}$

51. ln 28

52. ln 39

53. ln 0.00013

54. ln 0.0077

55. $\ln(27 \times 943)$

56. $\ln(33 \times 568)$

57. $\ln \frac{98}{13}$

58. $\ln \frac{84}{17}$

59. $\ln 27 + \ln 943$

60. $\ln 33 + \ln 568$

61. $\ln 98 - \ln 13$

62. $\ln 84 - \ln 17$

Solve each problem. ***See Examples 4, 6, 7, and 9.***

63. ***Decibel Levels*** Find the decibel ratings of sounds having the following intensities.

(a) $100I_0$ **(b)** $1000I_0$ **(c)** $100{,}000I_0$ **(d)** $1{,}000{,}000I_0$

(e) If the intensity of a sound is doubled, by how much is the decibel rating increased? Round to the nearest whole number.

64. ***Decibel Levels*** Find the decibel ratings of the following sounds, having intensities as given. Round each answer to the nearest whole number.

(a) whisper, $115I_0$ **(b)** busy street, $9{,}500{,}000I_0$

(c) heavy truck, 20 m away, $1{,}200{,}000{,}000I_0$

(d) rock music, $895{,}000{,}000{,}000I_0$

(e) jetliner at takeoff, $109{,}000{,}000{,}000{,}000I_0$

65. ***Earthquake Intensity*** The magnitude of an earthquake, measured on the Richter scale, is $\log_{10} \frac{I}{I_0}$, where I is the amplitude registered on a seismograph 100 km from the epicenter of the earthquake, and I_0 is the amplitude of an earthquake of a certain (small) size. Find the Richter scale ratings for earthquakes having the following amplitudes.

(a) $1000I_0$ **(b)** $1{,}000{,}000I_0$ **(c)** $100{,}000{,}000I_0$

66. ***Earthquake Intensity*** On December 26, 2004, an earthquake struck in the Indian Ocean with a magnitude of 9.1 on the Richter scale. The resulting tsunami killed an estimated 229,900 people in several countries. Express this reading in terms of I_0 to the nearest hundred thousand.

67. ***Earthquake Intensity*** On February 27, 2010, a massive earthquake struck Chile with a magnitude of 8.8 on the Richter scale. Express this reading in terms of I_0 to the nearest hundred thousand.

68. ***Earthquake Intensity Comparison*** Compare the answers to **Exercises 66 and 67.** How many times greater was the force of the 2004 earthquake than that of the 2010 earthquake?

69. ***(Modeling) Bachelor's Degrees in Psychology*** The table gives the number of bachelor's degrees in psychology (in thousands) earned at U.S. colleges and universities for selected years from 1980 through 2012. Suppose x represents the number of years since 1950. Thus, 1980 is represented by 30, 1990 is represented by 40, and so on.

Year	Degrees Earned (in thousands)
1980	42.1
1990	54.0
2000	74.2
2010	97.2
2011	100.9
2012	109.0

Source: National Center for Education Statistics.

The following function is a logarithmic model for the data.

$$f(x) = -273 + 90.6 \ln x$$

Use this function to estimate the number of bachelor's degrees in psychology earned in the year 2016 to the nearest tenth thousand. What assumption must we make to estimate the number of degrees in years beyond 2012?

70. ***(Modeling) Domestic Leisure Travel*** The bar graph shows numbers of leisure trips within the United States (in millions of person-trips of 50 or more miles one-way) over the years 2009–2014. The function

$$f(t) = 1458 + 95.42 \ln t, \quad t \geq 1,$$

where t represents the number of years since 2008 and $f(t)$ is the number of person-trips, in millions, approximates the curve reasonably well.

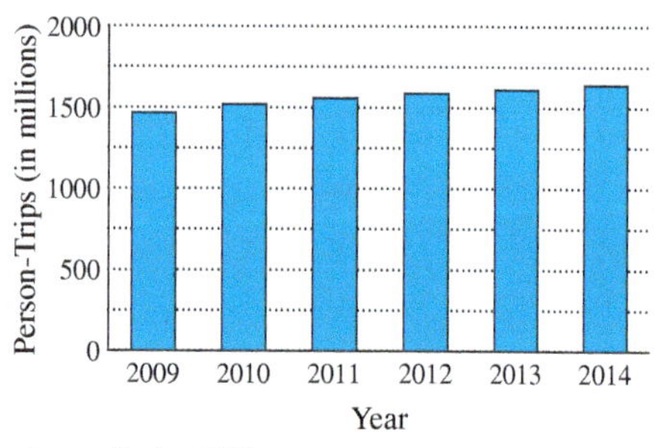

Source: Statista 2014.

Use the function to approximate the number of person-trips in 2012 to the nearest million. How does this approximation compare to the actual number of 1588 million?

71. *(Modeling) Diversity of Species* The number of species $S(n)$ in a sample is given by

$$S(n) = a \ln\left(1 + \frac{n}{a}\right),$$

where n is the number of individuals in the sample, and a is a constant that indicates the diversity of species in the community. If $a = 0.36$, find $S(n)$ for each value of n. (*Hint:* $S(n)$ must be a whole number.)

(a) 100 **(b)** 200 **(c)** 150 **(d)** 10

72. *(Modeling) Diversity of Species* In **Exercise 71,** find $S(n)$ if a changes to 0.88. Use the following values of n.

(a) 50 **(b)** 100 **(c)** 250

73. *(Modeling) Diversity of Species* Suppose a sample of a small community shows two species with 50 individuals each. Find the measure of diversity H.

74. *(Modeling) Diversity of Species* A virgin forest in northwestern Pennsylvania has 4 species of large trees with the following proportions of each:

hemlock, 0.521; beech, 0.324; birch, 0.081; maple, 0.074.

Find the measure of diversity H to the nearest thousandth.

75. *(Modeling) Global Temperature Increase* In **Example 7,** we expressed the average global temperature increase T (in °F) as

$$T(k) = 1.03k \ln \frac{C}{C_0},$$

where C_0 is the preindustrial amount of carbon dioxide, C is the current carbon dioxide level, and k is a constant. Arrhenius determined that $10 \le k \le 16$ when C was double the value C_0. Use $T(k)$ to find the range of the rise in global temperature T (rounded to the nearest degree) that Arrhenius predicted. (*Source:* Clime, W., *The Economics of Global Warming,* Institute for International Economics, Washington, D.C.)

76. *(Modeling) Global Temperature Increase* (Refer to **Exercise 75.**) According to one study by the IPCC, future increases in average global temperatures (in °F) can be modeled by

$$T(C) = 6.489 \ln \frac{C}{280},$$

where C is the concentration of atmospheric carbon dioxide (in ppm). C can be modeled by the function

$$C(x) = 353(1.006)^{x-1990},$$

where x is the year. (*Source:* International Panel on Climate Change (IPCC).)

(a) Write T as a function of x.

(b) Using a graphing calculator, graph $C(x)$ and $T(x)$ on the interval [1990, 2275] using different coordinate axes. Describe the graph of each function. How are C and T related?

(c) Approximate the slope of the graph of T. What does this slope represent?

(d) Use graphing to estimate x and $C(x)$ when $T(x) = 10$°F.

77. *Age of Rocks* Use the formula of **Example 6** to estimate the age of a rock sample having $\frac{A}{K} = 0.103$. Give the answer in billions of years, rounded to the nearest hundredth.

78. ***(Modeling) Planets' Distances from the Sun and Periods of Revolution*** The table contains the planets' average distances D from the sun and their periods P of revolution around the sun in years. The distances have been normalized so that Earth is one unit away from the sun. For example, since Jupiter's distance is 5.2, its distance from the sun is 5.2 times farther than Earth's.

Planet	D	P
Mercury	0.39	0.24
Venus	0.72	0.62
Earth	1	1
Mars	1.52	1.89
Jupiter	5.2	11.9
Saturn	9.54	29.5
Uranus	19.2	84.0
Neptune	30.1	164.8

Source: Ronan, C., *The Natural History of the Universe*, MacMillan Publishing Co., New York.

(a) Using a graphing calculator, make a scatter diagram by plotting the point $(\ln D, \ln P)$ for each planet on the xy-coordinate axes. Do the data points appear to be linear?

(b) Determine a linear equation that models the data points. Graph the line and the data on the same coordinate axes.

(c) Use this linear model to predict the period of Pluto if its distance is 39.5. Compare the answer to the actual value of 248.5 yr.

Use the change-of-base theorem to find an approximation to four decimal places for each logarithm. See ***Example 8.***

79. $\log_2 5$ **80.** $\log_2 9$ **81.** $\log_8 0.59$ **82.** $\log_8 0.71$

83. $\log_{1/2} 3$ **84.** $\log_{1/3} 2$ **85.** $\log_\pi e$ **86.** $\log_\pi \sqrt{2}$

87. $\log_{\sqrt{13}} 12$ **88.** $\log_{\sqrt{19}} 5$ **89.** $\log_{0.32} 5$ **90.** $\log_{0.91} 8$

Let $u = \ln a$ *and* $v = \ln b$. *Write each expression in terms of* u *and* v *without using the* ln *function.*

91. $\ln\left(b^4\sqrt{a}\right)$ **92.** $\ln \dfrac{a^3}{b^2}$ **93.** $\ln\sqrt{\dfrac{a^3}{b^5}}$ **94.** $\ln\left(\sqrt[3]{a} \cdot b^4\right)$

Concept Check *Use the various properties of exponential and logarithmic functions to evaluate the expressions in parts (a)–(c).*

95. Given $g(x) = e^x$, find **(a)** $g(\ln 4)$ **(b)** $g(\ln 5^2)$ **(c)** $g\left(\ln \frac{1}{e}\right)$.

96. Given $f(x) = 3^x$, find **(a)** $f(\log_3 2)$ **(b)** $f(\log_3(\ln 3))$ **(c)** $f(\log_3(2 \ln 3))$.

97. Given $f(x) = \ln x$, find **(a)** $f(e^6)$ **(b)** $f(e^{\ln 3})$ **(c)** $f(e^{2 \ln 3})$.

98. Given $f(x) = \log_2 x$, find **(a)** $f(2^7)$ **(b)** $f(2^{\log_2 2})$ **(c)** $f(2^{2 \log_2 2})$.

Work each problem.

99. ***Concept Check*** Which of the following is equivalent to $2 \ln(3x)$ for $x > 0$?

A. $\ln 9 + \ln x$ **B.** $\ln 6x$ **C.** $\ln 6 + \ln x$ **D.** $\ln 9x^2$

100. ***Concept Check*** Which of the following is equivalent to $\ln(4x) - \ln(2x)$ for $x > 0$?

A. $2 \ln x$ **B.** $\ln 2x$ **C.** $\dfrac{\ln 4x}{\ln 2x}$ **D.** $\ln 2$

101. The function $f(x) = \ln |x|$ plays a prominent role in calculus. Find its domain, its range, and the symmetries of its graph.

102. Consider the function $f(x) = \log_3 |x|$.

(a) What is the domain of this function?

(b) Use a graphing calculator to graph $f(x) = \log_3 |x|$ in the window $[-4, 4]$ by $[-4, 4]$.

(c) How might one easily misinterpret the domain of the function by merely observing the calculator graph?

Use properties of logarithms to rewrite each function, and describe how the graph of the given function compares to the graph of $g(x) = \ln x$.

103. $f(x) = \ln(e^2 x)$ **104.** $f(x) = \ln \frac{x}{e}$ **105.** $f(x) = \ln \frac{x}{e^2}$

Chapter 4 Quiz (Sections 4.1–4.4)

1. For the one-to-one function $f(x) = \sqrt[3]{3x-6}$, find $f^{-1}(x)$.

2. Solve $4^{2x+1} = 8^{3x-6}$.

3. Graph $f(x) = -3^x$. Give the domain and range.

4. Graph $f(x) = \log_4(x+2)$. Give the domain and range.

5. ***Future Value*** Suppose that \$15,000 is deposited in a bank certificate of deposit at an annual rate of 2.7% for 8 yr. Find the future value if interest is compounded as follows.

(a) annually **(b)** quarterly **(c)** monthly **(d)** daily (365 days)

6. Use a calculator to evaluate each logarithm to four decimal places.

(a) log 34.56 **(b)** ln 34.56

7. What is the meaning of the expression $\log_6 25$?

8. Solve each equation.

(a) $x = 3^{\log_3 4}$ **(b)** $\log_x 25 = 2$ **(c)** $\log_4 x = -2$

9. Assuming all variables represent positive real numbers, use properties of logarithms to rewrite

$$\log_3 \frac{\sqrt{x} \cdot y}{pq^4}.$$

10. Given $\log_b 9 = 3.1699$ and $\log_b 5 = 2.3219$, find the value of $\log_b 225$.

11. Find the value of $\log_3 40$ to four decimal places.

12. If $f(x) = 4^x$, what is the value of $f(\log_4 12)$?

4.5 Exponential and Logarithmic Equations

- Exponential Equations
- Logarithmic Equations
- Applications and Models

Exponential Equations We solved exponential equations in earlier sections. General methods for solving these equations depend on the property below, which follows from the fact that logarithmic functions are one-to-one.

Property of Logarithms

If $x > 0$, $y > 0$, $a > 0$, and $a \neq 1$, then the following holds.

$$x = y \quad \text{is equivalent to} \quad \log_a x = \log_a y.$$

EXAMPLE 1 Solving an Exponential Equation

Solve $7^x = 12$. Give the solution to the nearest thousandth.

SOLUTION The properties of exponents cannot be used to solve this equation, so we apply the preceding property of logarithms. While any appropriate base b can be used, the best practical base is base 10 or base e. We choose base e (natural) logarithms here.

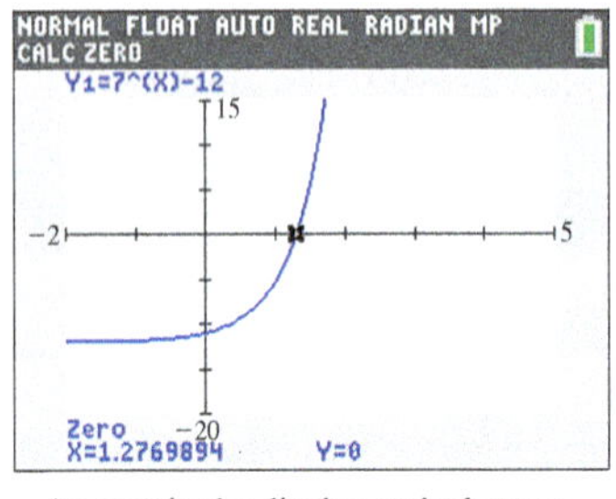

As seen in the display at the bottom of the screen, when rounded to three decimal places, the solution of $7^x - 12 = 0$ agrees with that found in **Example 1.**

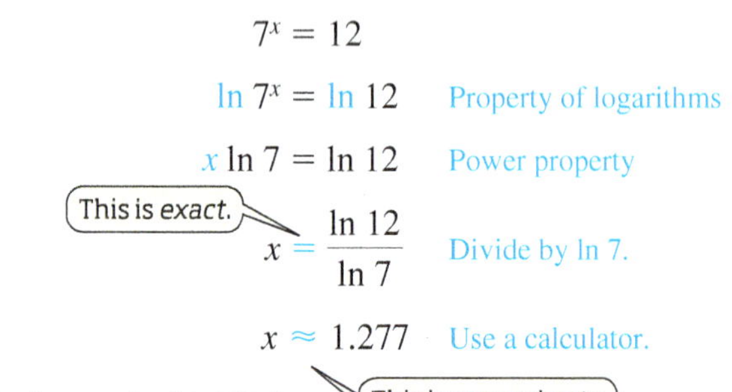

$$7^x = 12$$

$$\ln 7^x = \ln 12 \quad \text{Property of logarithms}$$

$$x \ln 7 = \ln 12 \quad \text{Power property}$$

$$x = \frac{\ln 12}{\ln 7} \quad \text{Divide by } \ln 7.$$

This is exact.

$$x \approx 1.277 \quad \text{Use a calculator.}$$

The solution set is $\{1.277\}$. This is approximate.

✔ **Now Try Exercise 11.**

CAUTION Do not confuse a quotient like $\frac{\ln 12}{\ln 7}$ in **Example 1** with $\ln \frac{12}{7}$, which can be written as $\ln 12 - \ln 7$. ***We cannot change the quotient of two logarithms to a difference of logarithms.***

$$\frac{\ln 12}{\ln 7} \neq \ln \frac{12}{7}$$

EXAMPLE 2 Solving an Exponential Equation

Solve $3^{2x-1} = 0.4^{x+2}$. Give the solution to the nearest thousandth.

SOLUTION

$$3^{2x-1} = 0.4^{x+2}$$

$$\ln 3^{2x-1} = \ln 0.4^{x+2} \quad \text{Take the natural logarithm on each side.}$$

$$(2x-1)\ln 3 = (x+2)\ln 0.4 \quad \text{Power property}$$

$$2x \ln 3 - \ln 3 = x \ln 0.4 + 2 \ln 0.4 \quad \text{Distributive property}$$

$$2x \ln 3 - x \ln 0.4 = 2 \ln 0.4 + \ln 3 \quad \text{Write so that the terms with } x \text{ are on one side.}$$

$$x(2 \ln 3 - \ln 0.4) = 2 \ln 0.4 + \ln 3 \quad \text{Factor out } x.$$

$$x = \frac{2 \ln 0.4 + \ln 3}{2 \ln 3 - \ln 0.4} \quad \text{Divide by } 2 \ln 3 - \ln 0.4.$$

$$x = \frac{\ln 0.4^2 + \ln 3}{\ln 3^2 - \ln 0.4} \quad \text{Power property}$$

$$x = \frac{\ln 0.16 + \ln 3}{\ln 9 - \ln 0.4} \quad \text{Apply the exponents.}$$

$$x = \frac{\ln 0.48}{\ln 22.5} \quad \text{Product and quotient properties}$$

This is exact.

$$x \approx -0.236 \quad \text{Use a calculator.}$$

The solution set is $\{-0.236\}$. This is approximate.

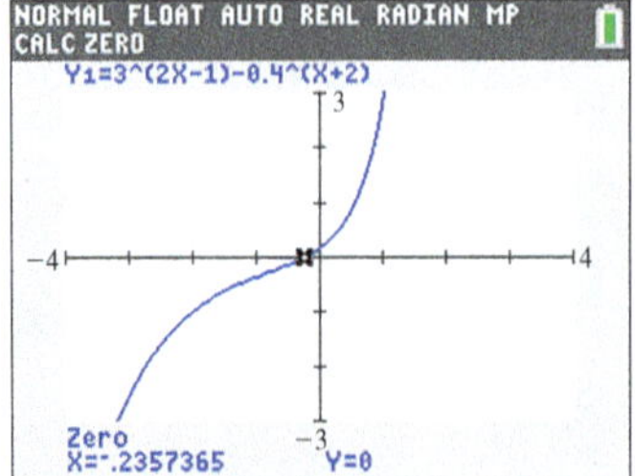

This screen supports the solution found in **Example 2.**

✔ **Now Try Exercise 19.**

EXAMPLE 3 Solving Base e Exponential Equations

Solve each equation. Give solutions to the nearest thousandth.

(a) $e^{x^2} = 200$ **(b)** $e^{2x+1} \cdot e^{-4x} = 3e$

SOLUTION

(a)

$$e^{x^2} = 200$$

$$\ln e^{x^2} = \ln 200 \quad \text{Take the natural logarithm on each side.}$$

$$x^2 = \ln 200 \quad \ln e^{x^2} = x^2$$

Remember both roots.

$$x = \pm\sqrt{\ln 200} \quad \text{Square root property}$$

$$x \approx \pm 2.302 \quad \text{Use a calculator.}$$

The solution set is $\{\pm 2.302\}$.

(b)

$$e^{2x+1} \cdot e^{-4x} = 3e$$

$$e^{-2x+1} = 3e \quad a^m \cdot a^n = a^{m+n}$$

$$e^{-2x} = 3 \quad \text{Divide by } e;\ \frac{e^{-2x+1}}{e^1} = e^{-2x+1-1} = e^{-2x}.$$

$$\ln e^{-2x} = \ln 3 \quad \text{Take the natural logarithm on each side.}$$

$$-2x \ln e = \ln 3 \quad \text{Power property}$$

$$-2x = \ln 3 \quad \ln e = 1$$

$$x = -\frac{1}{2}\ln 3 \quad \text{Multiply by } -\tfrac{1}{2}.$$

$$x \approx -0.549 \quad \text{Use a calculator.}$$

The solution set is $\{-0.549\}$.

✔ **Now Try Exercises 21 and 23.**

EXAMPLE 4 Solving an Exponential Equation (Quadratic in Form)

Solve $e^{2x} - 4e^x + 3 = 0$. Give exact value(s) for x.

SOLUTION If we substitute $u = e^x$, we notice that the equation is quadratic in form.

$$e^{2x} - 4e^x + 3 = 0$$

$$(e^x)^2 - 4e^x + 3 = 0 \quad a^{mn} = (a^n)^m$$

$$u^2 - 4u + 3 = 0 \quad \text{Let } u = e^x.$$

$$(u-1)(u-3) = 0 \quad \text{Factor.}$$

$$u - 1 = 0 \quad \text{or} \quad u - 3 = 0 \quad \text{Zero-factor property}$$

$$u = 1 \quad \text{or} \quad u = 3 \quad \text{Solve for } u.$$

$$e^x = 1 \quad \text{or} \quad e^x = 3 \quad \text{Substitute } e^x \text{ for } u.$$

$$\ln e^x = \ln 1 \quad \text{or} \quad \ln e^x = \ln 3 \quad \text{Take the natural logarithm on each side.}$$

$$x = 0 \quad \text{or} \quad x = \ln 3 \quad \ln e^x = x;\ \ln 1 = 0$$

Both values check, so the solution set is $\{0, \ln 3\}$.

✔ **Now Try Exercise 35.**

Logarithmic Equations The following equations involve logarithms of variable expressions.

EXAMPLE 5 Solving Logarithmic Equations

Solve each equation. Give exact values.

(a) $7 \ln x = 28$ **(b)** $\log_2(x^3 - 19) = 3$

SOLUTION

(a)

$$7 \ln x = 28$$

$$\log_e x = 4 \quad \ln x = \log_e x; \text{ Divide by 7.}$$

$$x = e^4 \quad \text{Write in exponential form.}$$

The solution set is $\{e^4\}$.

(b)

$$\log_2(x^3 - 19) = 3$$

$$x^3 - 19 = 2^3 \quad \text{Write in exponential form.}$$

$$x^3 - 19 = 8 \quad \text{Apply the exponent.}$$

$$x^3 = 27 \quad \text{Add 19.}$$

$$x = \sqrt[3]{27} \quad \text{Take cube roots.}$$

$$x = 3 \quad \sqrt[3]{27} = 3$$

The solution set is $\{3\}$.

✔ **Now Try Exercises 41 and 49.**

EXAMPLE 6 Solving a Logarithmic Equation

Solve $\log(x + 6) - \log(x + 2) = \log x$. Give exact value(s).

SOLUTION Recall that logarithms are defined only for nonnegative numbers.

$$\log(x + 6) - \log(x + 2) = \log x$$

$$\log \frac{x + 6}{x + 2} = \log x \quad \text{Quotient property}$$

$$\frac{x + 6}{x + 2} = x \quad \text{Property of logarithms}$$

$$x + 6 = x(x + 2) \quad \text{Multiply by } x + 2.$$

$$x + 6 = x^2 + 2x \quad \text{Distributive property}$$

$$x^2 + x - 6 = 0 \quad \text{Standard form}$$

$$(x + 3)(x - 2) = 0 \quad \text{Factor.}$$

$$x + 3 = 0 \quad \text{or} \quad x - 2 = 0 \quad \text{Zero-factor property}$$

$$x = -3 \quad \text{or} \quad x = 2 \quad \text{Solve for } x.$$

The proposed negative solution (-3) is not in the domain of $\log x$ in the original equation, so the only valid solution is the positive number 2. The solution set is $\{2\}$.

✔ **Now Try Exercise 69.**

CAUTION Recall that the domain of $y = \log_a x$ is $(0, \infty)$. ***For this reason, it is always necessary to check that proposed solutions of a logarithmic equation result in logarithms of positive numbers in the original equation.***

EXAMPLE 7 Solving a Logarithmic Equation

Solve $\log_2[(3x-7)(x-4)] = 3$. Give exact value(s).

SOLUTION

$$\log_2[(3x-7)(x-4)] = 3$$

$$(3x-7)(x-4) = 2^3 \quad \text{Write in exponential form.}$$

$$3x^2 - 19x + 28 = 8 \quad \text{Multiply. Apply the exponent.}$$

$$3x^2 - 19x + 20 = 0 \quad \text{Standard form}$$

$$(3x-4)(x-5) = 0 \quad \text{Factor.}$$

$$3x - 4 = 0 \quad \text{or} \quad x - 5 = 0 \quad \text{Zero-factor property}$$

$$x = \frac{4}{3} \quad \text{or} \quad x = 5 \quad \text{Solve for } x.$$

A check is necessary to be sure that the argument of the logarithm in the given equation is positive. In both cases, the product $(3x-7)(x-4)$ leads to 8, and $\log_2 8 = 3$ is true. The solution set is $\left\{\frac{4}{3}, 5\right\}$.

✔ **Now Try Exercise 53.**

EXAMPLE 8 Solving a Logarithmic Equation

Solve $\log(3x+2) + \log(x-1) = 1$. Give exact value(s).

SOLUTION

$$\log(3x+2) + \log(x-1) = 1$$

$$\log_{10}[(3x+2)(x-1)] = 1 \quad \log x = \log_{10} x;\ \text{product property}$$

$$(3x+2)(x-1) = 10^1 \quad \text{Write in exponential form.}$$

$$3x^2 - x - 2 = 10 \quad \text{Multiply; } 10^1 = 10.$$

$$3x^2 - x - 12 = 0 \quad \text{Subtract 10.}$$

$$x = \frac{-b \pm \sqrt{b^2 - 4ac}}{2a} \quad \text{Quadratic formula}$$

$$x = \frac{-(-1) \pm \sqrt{(-1)^2 - 4(3)(-12)}}{2(3)} \quad \text{Substitute } a = 3,\ b = -1,\ c = -12.$$

The two proposed solutions are

$$\frac{1 - \sqrt{145}}{6} \quad \text{and} \quad \frac{1 + \sqrt{145}}{6}.$$

The first proposed solution, $\frac{1 - \sqrt{145}}{6}$, is negative. Substituting for x in $\log(x-1)$ results in a negative argument, which is not allowed. Therefore, this solution must be rejected.

The second proposed solution, $\frac{1 + \sqrt{145}}{6}$, is positive. Substituting it for x in $\log(3x+2)$ results in a positive argument. Substituting it for x in $\log(x+1)$ also results in a positive argument. Both are necessary conditions. Therefore, the solution set is $\left\{\frac{1 + \sqrt{145}}{6}\right\}$.

✔ **Now Try Exercise 77.**

NOTE We could have replaced 1 with $\log_{10} 10$ in **Example 8** by first writing

$$\log(3x+2) + \log(x-1) = 1 \quad \text{Equation from Example 8}$$
$$\log_{10}[(3x+2)(x-1)] = \log_{10} 10 \quad \text{Substitute.}$$
$$(3x+2)(x-1) = 10, \quad \text{Property of logarithms}$$

and then continuing as shown on the preceding page.

EXAMPLE 9 Solving a Base *e* Logarithmic Equation

Solve $\ln e^{\ln x} - \ln(x-3) = \ln 2$. Give exact value(s).

SOLUTION This logarithmic equation differs from those in **Examples 7 and 8** because the expression on the right side involves a logarithm.

$$\ln e^{\ln x} - \ln(x-3) = \ln 2$$
$$\ln x - \ln(x-3) = \ln 2 \quad e^{\ln x} = x$$
$$\ln \frac{x}{x-3} = \ln 2 \quad \text{Quotient property}$$
$$\frac{x}{x-3} = 2 \quad \text{Property of logarithms}$$
$$x = 2(x-3) \quad \text{Multiply by } x-3.$$
$$x = 2x - 6 \quad \text{Distributive property}$$
$$x = 6 \quad \text{Solve for } x.$$

Check that the solution set is $\{6\}$.

✔ **Now Try Exercise 79.**

Solving an Exponential or Logarithmic Equation

To solve an exponential or logarithmic equation, change the given equation into one of the following forms, where a and b are real numbers, $a > 0$ and $a \neq 1$, and follow the guidelines.

1. $a^{f(x)} = b$

   Solve by taking logarithms on each side.

2. $\log_a f(x) = b$

   Solve by changing to exponential form $a^b = f(x)$.

3. $\log_a f(x) = \log_a g(x)$

   The given equation is equivalent to the equation $f(x) = g(x)$. Solve algebraically.

4. In a more complicated equation, such as

   $$e^{2x+1} \cdot e^{-4x} = 3e, \quad \text{See Example 3(b).}$$

   it may be necessary to first solve for $a^{f(x)}$ or $\log_a f(x)$ and then solve the resulting equation using one of the methods given above.

5. Check that each proposed solution is in the domain.

Applications and Models

EXAMPLE 10 Applying an Exponential Equation to the Strength of a Habit

The strength of a habit is a function of the number of times the habit is repeated. If N is the number of repetitions and H is the strength of the habit, then, according to psychologist C.L. Hull,

$$H = 1000(1 - e^{-kN}),$$

where k is a constant. Solve this equation for k.

SOLUTION

$$H = 1000(1 - e^{-kN})$$ First solve for e^{-kN}.

$$\frac{H}{1000} = 1 - e^{-kN}$$ Divide by 1000.

$$\frac{H}{1000} - 1 = -e^{-kN}$$ Subtract 1.

$$e^{-kN} = 1 - \frac{H}{1000}$$ Multiply by -1 and rewrite.

Now solve for k.

$$\ln e^{-kN} = \ln\left(1 - \frac{H}{1000}\right)$$ Take the natural logarithm on each side.

$$-kN = \ln\left(1 - \frac{H}{1000}\right)$$ $\ln e^x = x$

$$k = -\frac{1}{N}\ln\left(1 - \frac{H}{1000}\right)$$ Multiply by $-\frac{1}{N}$.

With the final equation, if one pair of values for H and N is known, k can be found, and the equation can then be used to find either H or N for given values of the other variable.

✔ **Now Try Exercise 91.**

EXAMPLE 11 Modeling PC Tablet Sales in the U.S.

Year	Sales (in millions)
2010	10.3
2011	24.1
2012	35.1
2013	39.8
2014	42.1

Source: Forrester Research.

The table gives U.S. tablet sales (in millions) for several years. The data can be modeled by the function

$$f(t) = 20.57 \ln t + 10.58, \quad t \geq 1,$$

where t is the number of years after 2009.

(a) Use the function to estimate the number of tablets sold in the United States in 2015.

(b) If this trend continues, approximately when will annual sales reach 60 million?

SOLUTION

(a) The year 2015 is represented by $t = 2015 - 2009 = 6$.

$$f(t) = 20.57 \ln t + 10.58$$ Given function

$$f(6) = 20.57 \ln 6 + 10.58$$ Let $t = 6$.

$$f(6) \approx 47.4$$ Use a calculator.

Based on this model, 47.4 million tablets were sold in 2015.

(b) Replace $f(t)$ with 60 and solve for t.

$$f(t) = 20.57 \ln t + 10.58 \quad \text{Given function}$$

$$60 = 20.57 \ln t + 10.58 \quad \text{Let } f(t) = 60.$$

$$49.42 = 20.57 \ln t \quad \text{Subtract 10.58.}$$

$$\ln t = \frac{49.42}{20.57} \quad \text{Divide by 20.57 and rewrite.}$$

$$t = e^{49.42/20.57} \quad \text{Write in exponential form.}$$

$$t \approx 11.05 \quad \text{Use a calculator.}$$

Adding 11 to 2009 gives the year 2020. Based on this model, annual sales will reach 60 million in 2020.

✔ **Now Try Exercise 111.**

4.5 Exercises

CONCEPT PREVIEW *Match each equation in Column I with the best first step for solving it in Column II.*

I

1. $10^x = 150$

2. $e^{2x-1} = 24$

3. $\log_4 (x^2 - 10) = 2$

4. $e^{2x} \cdot e^x = 2e$

5. $2e^{2x} - 5e^x - 3 = 0$

6. $\log (2x - 1) + \log (x + 4) = 1$

II

A. Use the product rule for exponents.

B. Take the common logarithm on each side.

C. Write the sum of logarithms as the logarithm of a product.

D. Let $u = e^x$ and write the equation in quadratic form.

E. Change to exponential form.

F. Take the natural logarithm on each side.

CONCEPT PREVIEW *An exponential equation such as*

$$5^x = 9$$

can be solved for its exact solution using the meaning of logarithm and the change-of-base theorem. Because x is the exponent to which 5 must be raised in order to obtain 9, the exact solution is

$$\log_5 9, \quad or \quad \frac{\log 9}{\log 5}, \quad or \quad \frac{\ln 9}{\ln 5}.$$

For each equation, give the exact solution in three forms similar to the forms above.

7. $7^x = 19$

8. $3^x = 10$

9. $\left(\frac{1}{2}\right)^x = 12$

10. $\left(\frac{1}{3}\right)^x = 4$

Solve each equation. In Exercises 11–34, give irrational solutions as decimals correct to the nearest thousandth. In Exercises 35–40, give solutions in exact form. ***See Examples 1–4.***

11. $3^x = 7$

12. $5^x = 13$

13. $\left(\frac{1}{2}\right)^x = 5$

14. $\left(\frac{1}{3}\right)^x = 6$

15. $0.8^x = 4$

16. $0.6^x = 3$

17. $4^{x-1} = 3^{2x}$

18. $2^{x+3} = 5^{2x}$

19. $6^{x+1} = 4^{2x-1}$

20. $3^{x-4} = 7^{2x+5}$ **21.** $e^{x^2} = 100$ **22.** $e^{x^4} = 1000$

23. $e^{3x-7} \cdot e^{-2x} = 4e$ **24.** $e^{1-3x} \cdot e^{5x} = 2e$ **25.** $\left(\frac{1}{3}\right)^x = -3$

26. $\left(\frac{1}{9}\right)^x = -9$ **27.** $0.05(1.15)^x = 5$ **28.** $1.2(0.9)^x = 0.6$

29. $3(2)^{x-2} + 1 = 100$ **30.** $5(1.2)^{3x-2} + 1 = 7$ **31.** $2(1.05)^x + 3 = 10$

32. $3(1.4)^x - 4 = 60$ **33.** $5(1.015)^{x-1980} = 8$ **34.** $6(1.024)^{x-1900} = 9$

35. $e^{2x} - 6e^x + 8 = 0$ **36.** $e^{2x} - 8e^x + 15 = 0$ **37.** $2e^{2x} + e^x = 6$

38. $3e^{2x} + 2e^x = 1$ **39.** $5^{2x} + 3(5^x) = 28$ **40.** $3^{2x} - 12(3^x) = -35$

Solve each equation. Give solutions in exact form. ***See Examples 5–9.***

41. $5 \ln x = 10$ **42.** $3 \ln x = 9$

43. $\ln 4x = 1.5$ **44.** $\ln 2x = 5$

45. $\log(2 - x) = 0.5$ **46.** $\log(3 - x) = 0.75$

47. $\log_6(2x + 4) = 2$ **48.** $\log_5(8 - 3x) = 3$

49. $\log_4(x^3 + 37) = 3$ **50.** $\log_7(x^3 + 65) = 0$

51. $\ln x + \ln x^2 = 3$ **52.** $\log x + \log x^2 = 3$

53. $\log_3[(x + 5)(x - 3)] = 2$ **54.** $\log_4[(3x + 8)(x - 6)] = 3$

55. $\log_2[(2x + 8)(x + 4)] = 5$ **56.** $\log_5[(3x + 5)(x + 1)] = 1$

57. $\log x + \log(x + 15) = 2$ **58.** $\log x + \log(2x + 1) = 1$

59. $\log(x + 25) = \log(x + 10) + \log 4$ **60.** $\log(3x + 5) - \log(2x + 4) = 0$

61. $\log(x - 10) - \log(x - 6) = \log 2$ **62.** $\log(x^2 - 9) - \log(x - 3) = \log 5$

63. $\ln(7 - x) + \ln(1 - x) = \ln(25 - x)$ **64.** $\ln(3 - x) + \ln(5 - x) = \ln(50 - 6x)$

65. $\log_8(x + 2) + \log_8(x + 4) = \log_8 8$ **66.** $\log_2(5x - 6) - \log_2(x + 1) = \log_2 3$

67. $\log_2(x^2 - 100) - \log_2(x + 10) = 1$ **68.** $\log_2(x - 2) + \log_2(x - 1) = 1$

69. $\log x + \log(x - 21) = \log 100$ **70.** $\log x + \log(3x - 13) = \log 10$

71. $\log(9x + 5) = 3 + \log(x + 2)$ **72.** $\log(11x + 9) = 3 + \log(x + 3)$

73. $\ln(4x - 2) - \ln 4 = -\ln(x - 2)$ **74.** $\ln(5 + 4x) - \ln(3 + x) = \ln 3$

75. $\log_5(x + 2) + \log_5(x - 2) = 1$ **76.** $\log_2(x - 7) + \log_2 x = 3$

77. $\log_2(2x - 3) + \log_2(x + 1) = 1$ **78.** $\log_5(3x + 2) + \log_5(x - 1) = 1$

79. $\ln e^x - 2 \ln e = \ln e^4$ **80.** $\ln e^x - \ln e^3 = \ln e^3$

81. $\log_2(\log_2 x) = 1$ **82.** $\log x = \sqrt{\log x}$

83. $\log x^2 = (\log x)^2$ **84.** $\log_2 \sqrt{2x^2} = \frac{3}{2}$

85. ***Concept Check*** Consider the following statement: "We must reject any negative proposed solution when we solve an equation involving logarithms." Is this correct? Why or why not?

86. ***Concept Check*** What values of x could not possibly be solutions of the following equation?

$$\log_a(4x - 7) + \log_a(x^2 + 4) = 0$$

Solve each equation for the indicated variable. Use logarithms with the appropriate bases. ***See Example 10.***

87. $p = a + \dfrac{k}{\ln x}$, for x

88. $r = p - k \ln t$, for t

89. $T = T_0 + (T_1 - T_0)10^{-kt}$, for t

90. $A = \dfrac{Pr}{1 - (1 + r)^{-n}}$, for n

91. $I = \dfrac{E}{R}(1 - e^{-Rt/2})$, for t

92. $y = \dfrac{K}{1 + ae^{-bx}}$, for b

93. $y = A + B(1 - e^{-Cx})$, for x

94. $m = 6 - 2.5 \log \dfrac{M}{M_0}$, for M

95. $\log A = \log B - C \log x$, for A

96. $d = 10 \log \dfrac{I}{I_0}$, for I

97. $A = P\left(1 + \dfrac{r}{n}\right)^{tn}$, for t

98. $D = 160 + 10 \log x$, for x

To solve each problem, refer to the formulas for compound interest.

$$A = P\left(1 + \frac{r}{n}\right)^{tn} \quad \text{and} \quad A = Pe^{rt}$$

99. ***Compound Amount*** If \$10,000 is invested in an account at 3% annual interest compounded quarterly, how much will be in the account in 5 yr if no money is withdrawn?

100. ***Compound Amount*** If \$5000 is invested in an account at 4% annual interest compounded continuously, how much will be in the account in 8 yr if no money is withdrawn?

101. ***Investment Time*** Kurt wants to buy a \$30,000 truck. He has saved \$27,000. Find the number of years (to the nearest tenth) it will take for his \$27,000 to grow to \$30,000 at 4% interest compounded quarterly.

102. ***Investment Time*** Find t to the nearest hundredth of a year if \$1786 becomes \$2063 at 2.6%, with interest compounded monthly.

103. ***Interest Rate*** Find the interest rate to the nearest hundredth of a percent that will produce \$2500, if \$2000 is left at interest compounded semiannually for 8.5 yr.

104. ***Interest Rate*** At what interest rate, to the nearest hundredth of a percent, will \$16,000 grow to \$20,000 if invested for 7.25 yr and interest is compounded quarterly?

(Modeling) Solve each application. ***See Example 11.***

105. In the central Sierra Nevada (a mountain range in California), the percent of moisture that falls as snow rather than rain is approximated reasonably well by

$$f(x) = 86.3 \ln x - 680,$$

where x is the altitude in feet and $f(x)$ is the percent of moisture that falls as snow. Find the percent of moisture, to the nearest tenth, that falls as snow at each altitude.

(a) 3000 ft **(b)** 4000 ft **(c)** 7000 ft

106. Northwest Creations finds that its total sales in dollars, $T(x)$, from the distribution of x thousand catalogues is approximated by

$$T(x) = 5000 \log(x + 1).$$

Find the total sales, to the nearest dollar, resulting from the distribution of each number of catalogues.

(a) 5000 **(b)** 24,000 **(c)** 49,000

107. *Average Annual Public University Costs* The table shows the cost of a year's tuition, room and board, and fees at 4-year public colleges for the years 2006–2014. Letting y represent the cost in dollars and x the number of years since 2006, the function

$$f(x) = 13{,}017(1.05)^x$$

models the data quite well. According to this function, in what year will the 2006 cost be doubled?

Year	Average Annual Cost
2006	\$12,837
2007	\$13,558
2008	\$14,372
2009	\$15,235
2010	\$16,178
2011	\$17,156
2012	\$17,817
2013	\$18,383
2014	\$18,943

Source: The College Board, *Annual Survey of Colleges*.

108. *Race Speed* At the World Championship races held at Rome's Olympic Stadium in 1987, American sprinter Carl Lewis ran the 100-m race in 9.86 sec. His speed in meters per second after t seconds is closely modeled by the function

$$f(t) = 11.65(1 - e^{-t/1.27}).$$

(*Source:* Banks, Robert B., *Towing Icebergs, Falling Dominoes, and Other Adventures in Applied Mathematics,* Princeton University Press.)

(a) How fast, to the nearest hundredth, was he running as he crossed the finish line?

(b) After how many seconds, to the nearest hundredth, was he running at the rate of 10 m per sec?

109. *Women in Labor Force* The percent of women in the U.S. civilian labor force can be modeled fairly well by the function

$$f(x) = \frac{67.21}{1 + 1.081e^{-x/24.71}},$$

where x represents the number of years since 1950. (*Source: Monthly Labor Review,* U.S. Bureau of Labor Statistics.)

(a) What percent, to the nearest whole number, of U.S. women were in the civilian labor force in 2014?

(b) In what year were 55% of U.S. women in the civilian labor force?

110. *Height of the Eiffel Tower* One side of the Eiffel Tower in Paris has a shape that can be approximated by the graph of the function

$$f(x) = -301 \ln \frac{x}{207}, \quad x > 0,$$

where x and $f(x)$ are both measured in feet. (*Source:* Banks, Robert B., *Towing Icebergs, Falling Dominoes, and Other Adventures in Applied Mathematics,* Princeton University Press.)

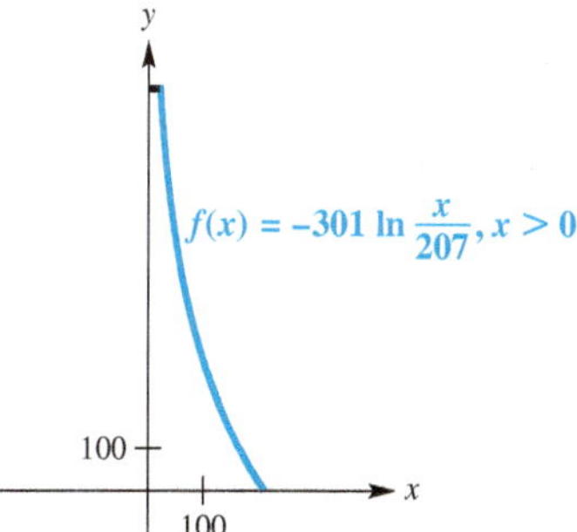

(a) Why does the shape of the left side of the graph of the Eiffel Tower have the formula given by $f(-x)$?

(b) The short horizontal segment at the top of the figure has length 7.8744 ft. How tall, to the nearest foot, is the Eiffel Tower?

(c) How far from the center of the tower is the point on the right side that is 500 ft above the ground? Round to the nearest foot.

111. *CO_2 Emissions Tax* One action that government could take to reduce carbon emissions into the atmosphere is to levy a tax on fossil fuel. This tax would be based on the amount of carbon dioxide emitted into the air when the fuel is burned. The **cost-benefit equation**

$$\ln(1 - P) = -0.0034 - 0.0053x$$

models the approximate relationship between a tax of x dollars per ton of carbon and the corresponding percent reduction P (in decimal form) of emissions of carbon dioxide. (*Source:* Nordhause, W., "To Slow or Not to Slow: The Economics of the Greenhouse Effect," Yale University, New Haven, Connecticut.)

(a) Write P as a function of x.

(b) Graph P for $0 \leq x \leq 1000$. Discuss the benefit of continuing to raise taxes on carbon.

(c) Determine P, to the nearest tenth, when $x = \$60$. Interpret this result.

(d) What value of x will give a 50% reduction in carbon emissions?

112. *Radiative Forcing* Radiative forcing, R, measures the influence of carbon dioxide in altering the additional solar radiation trapped in Earth's atmosphere. The International Panel on Climate Change (IPCC) in 1990 estimated k to be 6.3 in the radiative forcing equation

$$R = k \ln \frac{C}{C_0},$$

where C_0 is the preindustrial amount of carbon dioxide and C is the current level. (*Source:* Clime, W., *The Economics of Global Warming,* Institute for International Economics, Washington, D.C.)

(a) Use the equation $R = 6.3 \ln \frac{C}{C_0}$ to determine the radiative forcing R (in watts per square meter to the nearest tenth) expected by the IPCC if the carbon dioxide level in the atmosphere doubles from its preindustrial level.

(b) Determine the global temperature increase T, to the nearest tenth, that the IPCC predicted would occur if atmospheric carbon dioxide levels were to double, given $T(R) = 1.03R$.

Find $f^{-1}(x)$, and give the domain and range.

113. $f(x) = e^{x-5}$ **114.** $f(x) = e^x + 10$ **115.** $f(x) = e^{x+1} - 4$

116. $f(x) = \ln(x + 2)$ **117.** $f(x) = 2 \ln 3x$ **118.** $f(x) = \ln(x - 1) + 6$

Use a graphing calculator to solve each equation. Give irrational solutions correct to the nearest hundredth.

119. $e^x + \ln x = 5$ **120.** $e^x - \ln(x + 1) = 3$ **121.** $2e^x + 1 = 3e^{-x}$

122. $e^x + 6e^{-x} = 5$ **123.** $\log x = x^2 - 8x + 14$ **124.** $\ln x = -\sqrt[3]{x + 3}$

125. Find the **error** in the following **"proof"** that $2 < 1$.

$\frac{1}{9} < \frac{1}{3}$ True statement

$\left(\frac{1}{3}\right)^2 < \frac{1}{3}$ Rewrite the left side.

$\log\left(\frac{1}{3}\right)^2 < \log \frac{1}{3}$ Take the logarithm on each side.

$2 \log \frac{1}{3} < 1 \log \frac{1}{3}$ Property of logarithms; identity property

$2 < 1$ Divide each side by $\log \frac{1}{3}$.

4.6 Applications and Models of Exponential Growth and Decay

- The Exponential Growth or Decay Function
- Growth Function Models
- Decay Function Models

LOOKING AHEAD TO CALCULUS

The exponential growth and decay function formulas are studied in calculus in conjunction with the topic known as **differential equations.**

The Exponential Growth or Decay Function

In many situations in ecology, biology, economics, and the social sciences, a quantity changes at a rate proportional to the amount present. The amount present at time t is a special function of t called an **exponential growth or decay function.**

Exponential Growth or Decay Function

Let y_0 be the amount or number present at time $t = 0$. Then, under certain conditions, the amount y present at any time t is modeled by

$$y = y_0 e^{kt}, \quad \text{where } k \text{ is a constant.}$$

The constant k determines the type of function.

- When $k > 0$, the function describes growth. Examples of exponential growth include compound interest and atmospheric carbon dioxide.
- When $k < 0$, the function describes decay. One example of exponential decay is radioactive decay.

Growth Function Models

The amount of time it takes for a quantity that grows exponentially to become twice its initial amount is its **doubling time.**

EXAMPLE 1 Determining a Function to Model Exponential Growth

Earlier in this chapter, we discussed the growth of atmospheric carbon dioxide over time using a function based on the data from the table. Now we determine such a function from the data.

Year	Carbon Dioxide (ppm)
1990	353
2000	375
2075	590
2175	1090
2275	2000

Source: International Panel on Climate Change (IPCC).

(a) Find an exponential function that gives the amount of carbon dioxide y in year x.

(b) Estimate the year when future levels of carbon dioxide will be double the preindustrial level of 280 ppm.

SOLUTION

(a) The data points exhibit exponential growth, so the equation will take the form

$$y = y_0 e^{kx}.$$

We must find the values of y_0 and k. The data begin with the year 1990, so to simplify our work we let 1990 correspond to $x = 0$, 1991 correspond to $x = 1$, and so on. Here y_0 is the initial amount and $y_0 = 353$ in 1990 when $x = 0$. Thus the equation is

$$y = 353e^{kx}. \quad \text{Let } y_0 = 353.$$

From the last pair of values in the table, we know that in 2275 the carbon dioxide level is expected to be 2000 ppm. The year 2275 corresponds to $2275 - 1990 = 285$. Substitute 2000 for y and 285 for x, and solve for k.

$y = 353e^{kx}$	Solve for k.
$2000 = 353e^{k(285)}$	Substitute 2000 for y and 285 for x.
$\frac{2000}{353} = e^{285k}$	Divide by 353.
$\ln \frac{2000}{353} = \ln e^{285k}$	Take the natural logarithm on each side.
$\ln \frac{2000}{353} = 285k$	$\ln e^x = x$, for all x.
$k = \frac{1}{285} \cdot \ln \frac{2000}{353}$	Multiply by $\frac{1}{285}$ and rewrite.
$k \approx 0.00609$	Use a calculator.

A function that models the data is

$$y = 353e^{0.00609x}.$$

(b)

$y = 353e^{0.00609x}$	Solve the model from part (a) for the year x.
$560 = 353e^{0.00609x}$	To double the level 280, let $y = 2(280) = 560$.
$\frac{560}{353} = e^{0.00609x}$	Divide by 353.
$\ln \frac{560}{353} = \ln e^{0.00609x}$	Take the natural logarithm on each side.
$\ln \frac{560}{353} = 0.00609x$	$\ln e^x = x$, for all x.
$x = \frac{1}{0.00609} \cdot \ln \frac{560}{353}$	Multiply by $\frac{1}{0.00609}$ and rewrite.
$x \approx 75.8$	Use a calculator.

Since $x = 0$ corresponds to 1990, the preindustrial carbon dioxide level will double in the 75th year after 1990, or during 2065, according to this model.

✔ **Now Try Exercise 43.**

EXAMPLE 2 Finding Doubling Time for Money

How long will it take for money in an account that accrues interest at a rate of 3%, compounded continuously, to double?

SOLUTION

$A = Pe^{rt}$	Continuous compounding formula
$2P = Pe^{0.03t}$	Let $A = 2P$ and $r = 0.03$.
$2 = e^{0.03t}$	Divide by P.
$\ln 2 = \ln e^{0.03t}$	Take the natural logarithm on each side.
$\ln 2 = 0.03t$	$\ln e^x = x$
$\frac{\ln 2}{0.03} = t$	Divide by 0.03.
$23.10 \approx t$	Use a calculator.

It will take about 23 yr for the amount to double. ✔ **Now Try Exercise 31.**

EXAMPLE 3 Using an Exponential Function to Model Population Growth

According to the U.S. Census Bureau, the world population reached 6 billion people during 1999 and was growing exponentially. By the end of 2010, the population had grown to 6.947 billion. The projected world population (in billions of people) t years after 2010 is given by the function

$$f(t) = 6.947e^{0.00745t}.$$

(a) Based on this model, what will the world population be in 2025?

(b) If this trend continues, approximately when will the world population reach 9 billion?

SOLUTION

(a) Since $t = 0$ represents the year 2010, in 2025, t would be $2025 - 2010 = 15$ yr. We must find $f(t)$ when t is 15.

$$f(t) = 6.947e^{0.00745t} \quad \text{Given function}$$

$$f(15) = 6.947e^{0.00745(15)} \quad \text{Let } t = 15.$$

$$f(15) \approx 7.768 \quad \text{Use a calculator.}$$

The population will be 7.768 billion at the end of 2025.

(b)

$$f(t) = 6.947e^{0.00745t} \quad \text{Given function}$$

$$9 = 6.947e^{0.00745t} \quad \text{Let } f(t) = 9.$$

$$\frac{9}{6.947} = e^{0.00745t} \quad \text{Divide by 6.947.}$$

$$\ln \frac{9}{6.947} = \ln e^{0.00745t} \quad \text{Take the natural logarithm on each side.}$$

$$\ln \frac{9}{6.947} = 0.00745t \quad \ln e^x = x, \text{ for all } x.$$

$$t = \frac{\ln \frac{9}{6.947}}{0.00745} \quad \text{Divide by 0.00745 and rewrite.}$$

$$t \approx 34.8 \quad \text{Use a calculator.}$$

Thus, 34.8 yr after 2010, during the year 2044, world population will reach 9 billion.

✔ **Now Try Exercise 39.**

Decay Function Models **Half-life** is the amount of time it takes for a quantity that decays exponentially to become half its initial amount.

NOTE In **Example 4** on the next page, the initial amount of substance is given as 600 g. Because half-life is constant over the lifetime of a decaying quantity, starting with any initial amount, y_0, and substituting $\frac{1}{2}y_0$ for y in $y = y_0e^{kt}$ would allow the common factor y_0 to be divided out. The rest of the work would be the same.

EXAMPLE 4 Determining an Exponential Function to Model Radioactive Decay

Suppose 600 g of a radioactive substance are present initially and 3 yr later only 300 g remain.

(a) Determine an exponential function that models this decay.

(b) How much of the substance will be present after 6 yr?

SOLUTION

(a) We use the given values to find k in the exponential equation

$$y = y_0 e^{kt}.$$

Because the initial amount is 600 g, $y_0 = 600$, which gives $y = 600e^{kt}$. The initial amount (600 g) decays to half that amount (300 g) in 3 yr, so its half-life is 3 yr. Now we solve this exponential equation for k.

$$\begin{aligned} y &= 600e^{kt} && \text{Let } y_0 = 600. \\ 300 &= 600e^{3k} && \text{Let } y = 300 \text{ and } t = 3. \\ 0.5 &= e^{3k} && \text{Divide by 600.} \\ \ln 0.5 &= \ln e^{3k} && \text{Take the natural logarithm on each side.} \\ \ln 0.5 &= 3k && \ln e^x = x, \text{ for all } x. \\ \frac{\ln 0.5}{3} &= k && \text{Divide by 3.} \\ k &\approx -0.231 && \text{Use a calculator.} \end{aligned}$$

A function that models the situation is

$$y = 600e^{-0.231t}.$$

(b) To find the amount present after 6 yr, let $t = 6$.

$$\begin{aligned} y &= 600e^{-0.231t} && \text{Model from part (a)} \\ y &= 600e^{-0.231(6)} && \text{Let } t = 6. \\ y &= 600e^{-1.386} && \text{Multiply.} \\ y &\approx 150 && \text{Use a calculator.} \end{aligned}$$

After 6 yr, 150 g of the substance will remain. ✔ **Now Try Exercise 19.**

EXAMPLE 5 Solving a Carbon Dating Problem

Carbon-14, also known as radiocarbon, is a radioactive form of carbon that is found in all living plants and animals. After a plant or animal dies, the radiocarbon disintegrates. Scientists can determine the age of the remains by comparing the amount of radiocarbon with the amount present in living plants and animals. This technique is called **carbon dating.** The amount of radiocarbon present after t years is given by

$$y = y_0 e^{-0.0001216t},$$

where y_0 is the amount present in living plants and animals.

(a) Find the half-life of carbon-14.

(b) Charcoal from an ancient fire pit on Java contained $\frac{1}{4}$ the carbon-14 of a living sample of the same size. Estimate the age of the charcoal.

SOLUTION

(a) If y_0 is the amount of radiocarbon present in a living thing, then $\frac{1}{2}y_0$ is half this initial amount. We substitute and solve the given equation for t.

$$y = y_0 e^{-0.0001216t} \quad \text{Given equation}$$

$$\frac{1}{2}y_0 = y_0 e^{-0.0001216t} \quad \text{Let } y = \tfrac{1}{2}y_0.$$

$$\frac{1}{2} = e^{-0.0001216t} \quad \text{Divide by } y_0.$$

$$\ln \frac{1}{2} = \ln e^{-0.0001216t} \quad \text{Take the natural logarithm on each side.}$$

$$\ln \frac{1}{2} = -0.0001216t \quad \ln e^x = x, \text{ for all } x.$$

$$\frac{\ln \frac{1}{2}}{-0.0001216} = t \quad \text{Divide by } -0.0001216.$$

$$5700 \approx t \quad \text{Use a calculator.}$$

The half-life is 5700 yr.

(b) Solve again for t, this time letting the amount $y = \frac{1}{4}y_0$.

$$y = y_0 e^{-0.0001216t} \quad \text{Given equation}$$

$$\frac{1}{4}y_0 = y_0 e^{-0.0001216t} \quad \text{Let } y = \tfrac{1}{4}y_0.$$

$$\frac{1}{4} = e^{-0.0001216t} \quad \text{Divide by } y_0.$$

$$\ln \frac{1}{4} = \ln e^{-0.0001216t} \quad \text{Take the natural logarithm on each side.}$$

$$\frac{\ln \frac{1}{4}}{-0.0001216} = t \quad \ln e^x = x; \text{ Divide by } -0.0001216.$$

$$t \approx 11{,}400 \quad \text{Use a calculator.}$$

The charcoal is 11,400 yr old.

✔ **Now Try Exercise 23.**

EXAMPLE 6 Modeling Newton's Law of Cooling

Newton's law of cooling says that the rate at which a body cools is proportional to the difference in temperature between the body and the environment around it. The temperature $f(t)$ of the body at time t in appropriate units after being introduced into an environment having constant temperature T_0 is

$$f(t) = T_0 + Ce^{-kt}, \quad \text{where } C \text{ and } k \text{ are constants.}$$

A pot of coffee with a temperature of 100°C is set down in a room with a temperature of 20°C. The coffee cools to 60°C after 1 hr.

(a) Write an equation to model the data.

(b) Find the temperature after half an hour.

(c) How long will it take for the coffee to cool to 50°C?

SOLUTION

(a) We must find values for C and k in the given formula. As given, when $t = 0$, $T_0 = 20$, and the temperature of the coffee is $f(0) = 100$.

$$f(t) = T_0 + Ce^{-kt} \quad \text{Given function}$$

$$100 = 20 + Ce^{-0k} \quad \text{Let } t = 0,\ f(0) = 100, \text{ and } T_0 = 20.$$

$$100 = 20 + C \quad e^0 = 1$$

$$80 = C \quad \text{Subtract 20.}$$

The following function models the data.

$$f(t) = 20 + 80e^{-kt} \quad \text{Let } T_0 = 20 \text{ and } C = 80.$$

The coffee cools to 60°C after 1 hr, so when $t = 1$, $f(1) = 60$.

$$f(t) = 20 + 80e^{-kt} \quad \text{Above function with } T_0 = 20 \text{ and } C = 80$$

$$60 = 20 + 80e^{-1k} \quad \text{Let } t = 1 \text{ and } f(1) = 60.$$

$$40 = 80e^{-k} \quad \text{Subtract 20.}$$

$$\frac{1}{2} = e^{-k} \quad \text{Divide by 80.}$$

$$\ln\frac{1}{2} = \ln e^{-k} \quad \text{Take the natural logarithm on each side.}$$

$$\ln\frac{1}{2} = -k \quad \ln e^x = x, \text{ for all } x.$$

$$k \approx 0.693 \quad \text{Multiply by } -1, \text{ rewrite, and use a calculator.}$$

Thus, the model is $f(t) = 20 + 80e^{-0.693t}$.

(b) To find the temperature after $\frac{1}{2}$ hr, let $t = \frac{1}{2}$ in the model from part (a).

$$f(t) = 20 + 80e^{-0.693t} \quad \text{Model from part (a)}$$

$$f\left(\frac{1}{2}\right) = 20 + 80e^{(-0.693)(1/2)} \quad \text{Let } t = \tfrac{1}{2}.$$

$$f\left(\frac{1}{2}\right) \approx 76.6°\text{C} \quad \text{Use a calculator.}$$

(c) To find how long it will take for the coffee to cool to 50°C, let $f(t) = 50$.

$$f(t) = 20 + 80e^{-0.693t} \quad \text{Model from part (a)}$$

$$50 = 20 + 80e^{-0.693t} \quad \text{Let } f(t) = 50.$$

$$30 = 80e^{-0.693t} \quad \text{Subtract 20.}$$

$$\frac{3}{8} = e^{-0.693t} \quad \text{Divide by 80.}$$

$$\ln\frac{3}{8} = \ln e^{-0.693t} \quad \text{Take the natural logarithm on each side.}$$

$$\ln\frac{3}{8} = -0.693t \quad \ln e^x = x, \text{ for all } x.$$

$$t = \frac{\ln\frac{3}{8}}{-0.693} \quad \text{Divide by } -0.693 \text{ and rewrite.}$$

$t \approx 1.415$ hr, or about 1 hr, 25 min

✔ **Now Try Exercise 27.**

4.6 Exercises

CONCEPT PREVIEW *Population Growth* *A population is increasing according to the exponential function*

$$y = 2e^{0.02x},$$

where y is in millions and x is the number of years. Match each question in Column I with the correct procedure in Column II to answer the question.

I	II
1. How long will it take for the population to triple?	**A.** Evaluate $y = 2e^{0.02(1/3)}$.
2. When will the population reach 3 million?	**B.** Solve $2e^{0.02x} = 6$.
3. How large will the population be in 3 yr?	**C.** Evaluate $y = 2e^{0.02(3)}$.
4. How large will the population be in 4 months?	**D.** Solve $2e^{0.02x} = 3$.

CONCEPT PREVIEW *Radioactive Decay* *Strontium-90 decays according to the exponential function*

$$y = y_0 e^{-0.0241t},$$

where t is time in years. Match each question in Column I with the correct procedure in Column II to answer the question.

I	II
5. If the initial amount of Strontium-90 is 200 g, how much will remain after 10 yr?	**A.** Solve $0.75y_0 = y_0 e^{-0.0241t}$.
6. If the initial amount of Strontium-90 is 200 g, how much will remain after 20 yr?	**B.** Evaluate $y = 200e^{-0.0241(10)}$.
7. What is the half-life of Strontium-90?	**C.** Solve $\frac{1}{2}y_0 = y_0 e^{-0.0241t}$.
8. How long will it take for any amount of Strontium-90 to decay to 75% of its initial amount?	**D.** Evaluate $y = 200e^{-0.0241(20)}$.

(Modeling) The exercises in this set are grouped according to discipline. They involve exponential or logarithmic models. ***See Examples 1–6.***

Physical Sciences (Exercises 9–28)

An initial amount of a radioactive substance y_0 is given, along with information about the amount remaining after a given time t in appropriate units. For an equation of the form $y = y_0 e^{kt}$ that models the situation, give the exact value of k in terms of natural logarithms.

9. $y_0 = 60$ g; After 3 hr, 20 g remain.

10. $y_0 = 30$ g; After 6 hr, 10 g remain.

11. $y_0 = 10$ mg; The half-life is 100 days.

12. $y_0 = 20$ mg; The half-life is 200 days.

13. $y_0 = 2.4$ lb; After 2 yr, 0.6 lb remains.

14. $y_0 = 8.1$ kg; After 4 yr, 0.9 kg remains.

Solve each problem.

15. *Decay of Lead* A sample of 500 g of radioactive lead-210 decays to polonium-210 according to the function

$$A(t) = 500e^{-0.032t},$$

where t is time in years. Find the amount of radioactive lead remaining after

(a) 4 yr, **(b)** 8 yr, **(c)** 20 yr. **(d)** Find the half-life.

16. *Decay of Plutonium* Repeat **Exercise 15** for 500 g of plutonium-241, which decays according to the function $A(t) = A_0e^{-0.053t}$, where t is time in years.

17. *Decay of Radium* Find the half-life of radium-226, which decays according to the function $A(t) = A_0e^{-0.00043t}$, where t is time in years.

18. *Decay of Tritium* Find the half-life of tritium, a radioactive isotope of hydrogen, which decays according to the function $A(t) = A_0e^{-0.056t}$, where t is time in years.

19. *Radioactive Decay* If 12 g of a radioactive substance are present initially and 4 yr later only 6.0 g remain, how much of the substance will be present after 7 yr?

20. *Radioactive Decay* If 1 g of strontium-90 is present initially, and 2 yr later 0.95 g remains, how much strontium-90 will be present after 5 yr?

21. *Decay of Iodine* How long will it take any quantity of iodine-131 to decay to 25% of its initial amount, knowing that it decays according to the exponential function $A(t) = A_0e^{-0.087t}$, where t is time in days?

22. *Magnitude of a Star* The magnitude M of a star is modeled by

$$M = 6 - \frac{5}{2}\log\frac{I}{I_0},$$

where I_0 is the intensity of a just-visible star and I is the actual intensity of the star being measured. The dimmest stars are of magnitude 6, and the brightest are of magnitude 1. Determine the ratio of light intensities between a star of magnitude 1 and a star of magnitude 3.

23. *Carbon-14 Dating* Suppose an Egyptian mummy is discovered in which the amount of carbon-14 present is only about one-third the amount found in living human beings. How long ago did the Egyptian die?

24. *Carbon-14 Dating* A sample from a refuse deposit near the Strait of Magellan had 60% of the carbon-14 of a contemporary sample. How old was the sample?

25. *Carbon-14 Dating* Paint from the Lascaux caves of France contains 15% of the normal amount of carbon-14. Estimate the age of the paintings.

26. *Dissolving a Chemical* The amount of a chemical that will dissolve in a solution increases exponentially as the (Celsius) temperature t is increased according to the model

$$A(t) = 10e^{0.0095t}.$$

At what temperature will 15 g dissolve?

27. *Newton's Law of Cooling* Boiling water, at 100°C, is placed in a freezer at 0°C. The temperature of the water is 50°C after 24 min. Find the temperature of the water to the nearest hundredth after 96 min. (*Hint:* Change minutes to hours.)

28. *Newton's Law of Cooling* A piece of metal is heated to 300°C and then placed in a cooling liquid at 50°C. After 4 min, the metal has cooled to 175°C. Find its temperature to the nearest hundredth after 12 min. (*Hint:* Change minutes to hours.)

Finance (Exercises 29–34)

29. *Comparing Investments* Russ, who is self-employed, wants to invest \$60,000 in a pension plan. One investment offers 3% compounded quarterly. Another offers 2.75% compounded continuously.

    (a) Which investment will earn more interest in 5 yr?

    (b) How much more will the better plan earn?

30. *Growth of an Account* If Russ (see **Exercise 29**) chooses the plan with continuous compounding, how long will it take for his \$60,000 to grow to \$70,000?

31. *Doubling Time* Find the doubling time of an investment earning 2.5% interest if interest is compounded continuously.

32. *Doubling Time* If interest is compounded continuously and the interest rate is tripled, what effect will this have on the time required for an investment to double?

33. *Growth of an Account* How long will it take an investment to triple if interest is compounded continuously at 3%?

34. *Growth of an Account* Use the Table feature of a graphing calculator to find how long it will take \$1500 invested at 2.75% compounded daily to triple in value. Zoom in on the solution by systematically decreasing the increment for x. Find the answer to the nearest day. (Find the answer to the nearest day by eventually letting the increment of x equal $\frac{1}{365}$. The decimal part of the solution can be multiplied by 365 to determine the number of days greater than the nearest year. For example, if the solution is determined to be 16.2027 yr, then multiply 0.2027 by 365 to get 73.9855. The solution is then, to the nearest day, 16 yr, 74 days.) Confirm the answer algebraically.

Social Sciences (Exercises 35–44)

35. *Legislative Turnover* The turnover of legislators is a problem of interest to political scientists. It was found that one model of legislative turnover in a particular body was

$$M(t) = 434e^{-0.08t},$$

where $M(t)$ represents the number of continuously serving members at time t. Here, $t = 0$ represents 1965, $t = 1$ represents 1966, and so on. Use this model to approximate the number of continuously serving members in each year.

(a) 1969 **(b)** 1973 **(c)** 1979

36. *Legislative Turnover* Use the model in **Exercise 35** to determine the year in which the number of continuously serving members was 338.

37. *Population Growth* In 2000 India's population reached 1 billion, and it is projected to be 1.4 billion in 2025. (*Source:* U.S. Census Bureau.)

(a) Find values for P_0 and a so that $P(x) = P_0a^{x-2000}$ models the population of India in year x. Round a to five decimal places.

(b) Predict India's population in 2020 to the nearest tenth of a billion.

(c) In what year is India's population expected to reach 1.5 billion?

38. *Population Decline* A midwestern city finds its residents moving to the suburbs. Its population is declining according to the function

$$P(t) = P_0e^{-0.04t},$$

where t is time measured in years and P_0 is the population at time $t = 0$. Assume that $P_0 = 1{,}000{,}000$.

(a) Find the population at time $t = 1$ to the nearest thousand.

(b) How long, to the nearest tenth of a year, will it take for the population to decline to 750,000?

(c) How long, to the nearest tenth of a year, will it take for the population to decline to half the initial number?

39. ***Health Care Spending*** Out-of-pocket spending in the United States for health care increased between 2008 and 2012. The function

$$f(x) = 7446e^{0.0305x}$$

models average annual expenditures per household, in dollars. In this model, x represents the year, where $x = 0$ corresponds to 2008. (*Source*: U.S. Bureau of Labor Statistics.)

(a) Estimate out-of-pocket household spending on health care in 2012 to the nearest dollar.

(b) In what year did spending reach $7915 per household?

40. ***Recreational Expenditures*** Personal consumption expenditures for recreation in billions of dollars in the United States during the years 2000–2013 can be approximated by the function

$$A(t) = 632.37e^{0.0351t},$$

where $t = 0$ corresponds to the year 2000. Based on this model, how much were personal consumption expenditures in 2013 to the nearest billion? (*Source:* U.S. Bureau of Economic Analysis.)

41. ***Housing Costs*** Average annual per-household spending on housing over the years 2000–2012 is approximated by

$$H = 12{,}744e^{0.0264t},$$

where t is the number of years since 2000. Find H to the nearest dollar for each year. (*Source:* U.S. Bureau of Labor Statistics.)

(a) 2005 **(b)** 2009 **(c)** 2012

42. ***Evolution of Language*** The number of years, n, since two independently evolving languages split off from a common ancestral language is approximated by

$$n \approx -7600 \log r,$$

where r is the proportion of words from the ancestral language common to both languages. Find each of the following to the nearest year.

(a) Find n if $r = 0.9$. **(b)** Find n if $r = 0.3$.

(c) How many years have elapsed since the split if half of the words of the ancestral language are common to both languages?

43. ***School District Growth*** Student enrollment in the Wentzville School District, one of the fastest-growing school districts in the state of Missouri, has projected growth as shown in the graph.

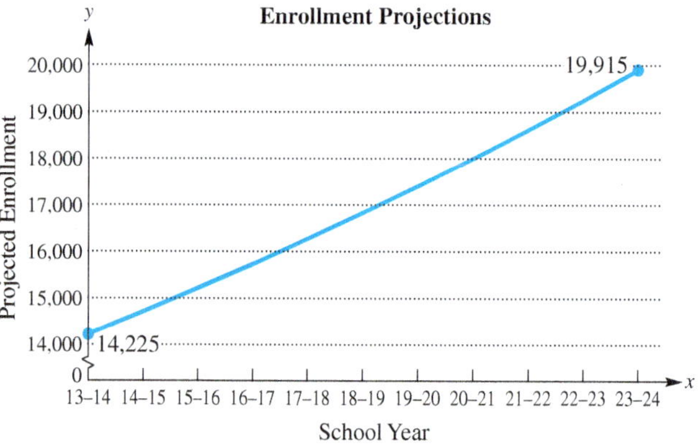

Source: Wentzville School District.

(a) Use the model $y = y_0e^{kx}$ to find an exponential function that gives the projected enrollment y in school year x. Let the school year 2013–14 correspond to $x = 0$, 2014–15 correspond to $x = 1$, and so on, and use the two points indicated on the graph.

(b) Estimate the school year for which projected enrollment will be 21,500 students.

44. *YouTube Views* The number of views of a YouTube video increases after the number of hours posted as shown in the table.

Hour	Number of Views
20	100
25	517
30	2015
35	10,248

(a) Use the model $y = y_0e^{kx}$ to find an exponential function that gives projected number of views y after number of hours x. Let hour 20 correspond to $x = 0$, hour 25 correspond to $x = 5$, and so on, and use the first and last data values given in the table.

(b) Estimate the number of views after 50 hr.

Life Sciences (Exercises 45–50)

45. *Spread of Disease* During an epidemic, the number of people who have never had the disease and who are not immune (they are *susceptible*) decreases exponentially according to the function

$$f(t) = 15{,}000e^{-0.05t},$$

where t is time in days. Find the number of susceptible people at each time.

(a) at the beginning of the epidemic **(b)** after 10 days **(c)** after 3 weeks

46. *Spread of Disease* Refer to **Exercise 45** and determine how long it will take, to the nearest day, for the initial number of people susceptible to decrease to half its amount.

47. *Growth of Bacteria* The growth of bacteria makes it necessary to time-date some food products so that they will be sold and consumed before the bacteria count is too high. Suppose for a certain product the number of bacteria present is given by

$$f(t) = 500e^{0.1t},$$

where t is time in days and the value of $f(t)$ is in millions. Find the number of bacteria, in millions, present at each time.

(a) 2 days **(b)** 4 days **(c)** 1 week

48. *Growth of Bacteria* How long will it take the bacteria population in **Exercise 47** to double? Round the answer to the nearest tenth.

49. *Medication Effectiveness* Drug effectiveness decreases over time. If, each hour, a drug is only 90% as effective as the previous hour, at some point the patient will not be receiving enough medication and must receive another dose. If the initial dose was 200 mg and the drug was administered 3 hr ago, the expression $200(0.90)^3$, which equals 145.8, represents the amount of effective medication still in the system. (The exponent is equal to the number of hours since the drug was administered.)

The amount of medication still available in the system is given by the function

$$f(t) = 200(0.90)^t.$$

In this model, t is in hours and $f(t)$ is in milligrams. How long will it take for this initial dose to reach the dangerously low level of 50 mg? Round the answer to the nearest tenth.

50. *Population Size* Many environmental situations place effective limits on the growth of the number of an organism in an area. Many such limited-growth situations are described by the **logistic function**

$$G(x) = \frac{MG_0}{G_0 + (M - G_0)e^{-kMx}},$$

where G_0 is the initial number present, M is the maximum possible size of the population, and k is a positive constant. The screens illustrate a typical logistic function calculation and graph.

NORMAL FLOAT AUTO REAL RADIAN MP

L1	L2	L3	L4	L5
10	14	------	------	------
20	25			
30	42			
40	50			
50	58			
60	60			
------	------			

L1={10,20,30,40,50,60}

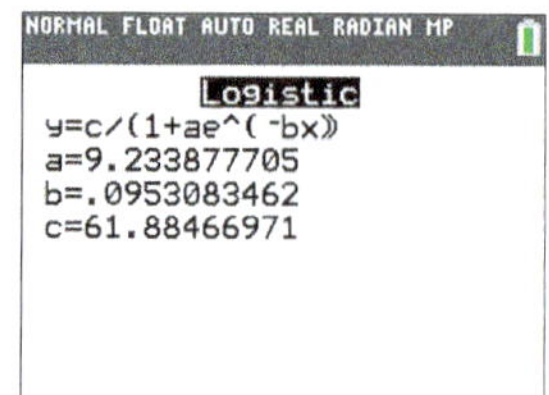

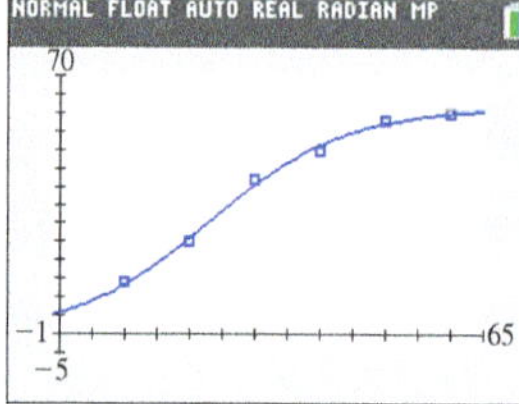

Assume that $G_0 = 100$, $M = 2500$, $k = 0.0004$, and $x =$ time in decades (10-yr periods).

(a) Use a calculator to graph the function, using $0 \le x \le 8$ and $0 \le y \le 2500$.

(b) Estimate the value of $G(2)$ from the graph. Then evaluate $G(2)$ algebraically to find the population after 20 yr.

(c) Find the x-coordinate of the intersection of the curve with the horizontal line $y = 1000$ to estimate the number of decades required for the population to reach 1000. Then solve $G(x) = 1000$ algebraically to obtain the exact value of x.

Economics (Exercises 51–56)

51. *Consumer Price Index* The U.S. Consumer Price Index for the years 1990–2013 is approximated by

$$A(t) = 100e^{0.0264t},$$

where t represents the number of years after 1990. (Since $A(16)$ is about 153, the amount of goods that could be purchased for \$100 in 1990 cost about \$153 in 2006.) Use the function to determine the year in which costs will be 125% higher than in 1990. (*Source:* U.S. Bureau of Labor Statistics.)

52. *Product Sales* Sales of a product, under relatively stable market conditions but in the absence of promotional activities such as advertising, tend to decline at a constant yearly rate. This rate of sales decline varies considerably from product to product, but it seems to remain the same for any particular product. The sales decline can be expressed by the function

$$S(t) = S_0 e^{-at},$$

where $S(t)$ is the rate of sales at time t measured in years, S_0 is the rate of sales at time $t = 0$, and a is the sales decay constant.

(a) Suppose the sales decay constant for a particular product is $a = 0.10$. Let $S_0 = 50{,}000$ and find $S(1)$ and $S(3)$ to the nearest thousand.

(b) Find $S(2)$ and $S(10)$ to the nearest thousand if $S_0 = 80{,}000$ and $a = 0.05$.

53. *Product Sales* Use the sales decline function given in **Exercise 52.** If $a = 0.10$, $S_0 = 50{,}000$, and t is time measured in years, find the number of years it will take for sales to fall to half the initial sales. Round the answer to the nearest tenth.

54. *Cost of Bread* Assume the cost of a loaf of bread is \$4. With continuous compounding, find the number of years, to the nearest tenth, it would take for the cost to triple at an annual inflation rate of 4%.

55. *Electricity Consumption* Suppose that in a certain area the consumption of electricity has increased at a continuous rate of 6% per year. If it continued to increase at this rate, find the number of years, to the nearest tenth, before twice as much electricity would be needed.

56. *Electricity Consumption* Suppose a conservation campaign, together with higher rates, caused demand for electricity to increase at only 2% per year. (See **Exercise 55.**) Find the number of years, to the nearest tenth, before twice as much electricity would be needed.

(Modeling) Solve each problem that uses a logistic function.

57. *Heart Disease* As age increases, so does the likelihood of coronary heart disease (CHD). The fraction of people x years old with some CHD is modeled by

$$f(x)=\frac{0.9}{1+271e^{-0.122x}}.$$

(*Source:* Hosmer, D., and S. Lemeshow, *Applied Logistic Regression,* John Wiley and Sons.)

(a) Evaluate $f(25)$ and $f(65)$ to the nearest hundredth. Interpret the results.

(b) At what age, to the nearest year, does this likelihood equal 50%?

58. *Tree Growth* The height of a certain tree in feet after x years is modeled by

$$f(x)=\frac{50}{1+47.5e^{-0.22x}}.$$

(a) Make a table for f starting at $x = 10$, and incrementing by 10. What appears to be the maximum height of the tree?

(b) Graph f and identify the horizontal asymptote. Explain its significance.

(c) After how many years was the tree 30 ft tall? Round to the nearest tenth.

Summary Exercises on Functions: Domains and Defining Equations

Finding the Domain of a Function: A Summary To find the domain of a function, given the equation that defines the function, remember that the value of x input into the equation must yield a real number for y when the function is evaluated. For the functions studied so far in this book, there are three cases to consider when determining domains.

Guidelines for Domain Restrictions

1. No input value can lead to 0 in a denominator, because division by 0 is undefined.
2. No input value can lead to an even root of a negative number, because this situation does not yield a real number.
3. No input value can lead to the logarithm of a negative number or 0, because this situation does not yield a real number.

Unless otherwise specified, we determine domains as follows.

- The domain of a **polynomial function** is the set of all real numbers.
- The domain of an **absolute value function** is the set of all real numbers for which the expression inside the absolute value bars (the argument) is defined.
- If a **function is defined by a rational expression,** the domain is the set of all real numbers for which the denominator is not zero.
- The domain of a **function defined by a radical with *even* root index** is the set of all real numbers that make the radicand greater than or equal to zero.

  If the root index is *odd*, the domain is the set of all real numbers for which the radicand is itself a real number.
- For an **exponential function** with constant base, the domain is the set of all real numbers for which the exponent is a real number.
- For a **logarithmic function,** the domain is the set of all real numbers that make the argument of the logarithm greater than zero.

Determining Whether an Equation Defines y as a Function of x

For y to be a function of x, it is necessary that every input value of x in the domain leads to one and only one value of y.

To determine whether an equation such as

$$x - y^3 = 0 \quad \text{or} \quad x - y^2 = 0$$

represents a function, solve the equation for y. In the first equation above, doing so leads to

$$y = \sqrt[3]{x}.$$

Notice that every value of x in the domain (that is, all real numbers) leads to one and only one value of y. So in the first equation, we can write y as a function of x. However, in the second equation above, solving for y leads to

$$y = \pm\sqrt{x}.$$

If we let $x = 4$, for example, we get two values of y: -2 and 2. Thus, in the second equation, we cannot write y as a function of x.

EXERCISES

Find the domain of each function. Write answers using interval notation.

1. $f(x) = 3x - 6$

2. $f(x) = \sqrt{2x - 7}$

3. $f(x) = |x + 4|$

4. $f(x) = \dfrac{x + 2}{x - 6}$

5. $f(x) = \dfrac{-2}{x^2 + 7}$

6. $f(x) = \sqrt{x^2 - 9}$

7. $f(x) = \dfrac{x^2 + 7}{x^2 - 9}$

8. $f(x) = \sqrt[3]{x^3 + 7x - 4}$

9. $f(x) = \log_5(16 - x^2)$

10. $f(x) = \log \dfrac{x + 7}{x - 3}$

11. $f(x) = \sqrt{x^2 - 7x - 8}$

12. $f(x) = 2^{1/x}$

13. $f(x) = \dfrac{1}{2x^2 - x + 7}$

14. $f(x) = \dfrac{x^2 - 25}{x + 5}$

15. $f(x) = \sqrt{x^3 - 1}$

16. $f(x) = \ln |x^2 - 5|$

17. $f(x) = e^{x^2+x+4}$

18. $f(x) = \dfrac{x^3 - 1}{x^2 - 1}$

19. $f(x) = \sqrt{\dfrac{-1}{x^3 - 1}}$

20. $f(x) = \sqrt[3]{\dfrac{1}{x^3 - 8}}$

21. $f(x) = \ln (x^2 + 1)$

22. $f(x) = \sqrt{(x - 3)(x + 2)(x - 4)}$

23. $f(x) = \log \left(\dfrac{x + 2}{x - 3} \right)^2$

24. $f(x) = \sqrt[12]{(4 - x)^2(x + 3)}$

25. $f(x) = e^{|1/x|}$

26. $f(x) = \dfrac{1}{|x^2 - 7|}$

27. $f(x) = x^{100} - x^{50} + x^2 + 5$

28. $f(x) = \sqrt{-x^2 - 9}$

29. $f(x) = \sqrt[4]{16 - x^4}$

30. $f(x) = \sqrt[3]{16 - x^4}$

31. $f(x) = \sqrt{\dfrac{x^2 - 2x - 63}{x^2 + x - 12}}$

32. $f(x) = \sqrt[5]{5 - x}$

33. $f(x) = |\sqrt{5 - x}|$

34. $f(x) = \sqrt{\dfrac{-1}{x - 3}}$

35. $f(x) = \log \left| \dfrac{1}{4 - x} \right|$

36. $f(x) = 6^{x^2-9}$

37. $f(x) = 6^{\sqrt{x^2-25}}$

38. $f(x) = 6^{\sqrt[3]{x^2-25}}$

39. $f(x) = \ln \left(\dfrac{-3}{(x + 2)(x - 6)} \right)$

40. $f(x) = \dfrac{-2}{\log x}$

Determine which one of the choices (A, B, C, or D) is an equation in which y can be written as a function of x.

41. **A.** $3x + 2y = 6$ **B.** $x = \sqrt{|y|}$ **C.** $x = |y + 3|$ **D.** $x^2 + y^2 = 9$

42. **A.** $3x^2 + 2y^2 = 36$ **B.** $x^2 + y - 2 = 0$ **C.** $x - |y| = 0$ **D.** $x = y^2 - 4$

43. **A.** $x = \sqrt{y^2}$ **B.** $x = \log y^2$ **C.** $x^3 + y^3 = 5$ **D.** $x = \dfrac{1}{y^2 + 3}$

44. **A.** $\dfrac{x^2}{4} + \dfrac{y^2}{4} = 1$ **B.** $x = 5y^2 - 3$ **C.** $\dfrac{x^2}{4} - \dfrac{y^2}{9} = 1$ **D.** $x = 10^y$

45. **A.** $x = \dfrac{2 - y}{y + 3}$ **B.** $x = \ln (y + 1)^2$ **C.** $\sqrt{x} = |y + 1|$ **D.** $\sqrt[4]{x} = y^2$

46. **A.** $e^{y^2} = x$ **B.** $e^{y+2} = x$ **C.** $e^{|y|} = x$ **D.** $10^{|y+2|} = x$

47. **A.** $x^2 = \dfrac{1}{y^2}$ **B.** $x + 2 = \dfrac{1}{y^2}$ **C.** $3x = \dfrac{1}{y^4}$ **D.** $2x = \dfrac{1}{y^3}$

48. **A.** $|x| = |y|$ **B.** $x = |y^2|$ **C.** $x = \dfrac{1}{y}$ **D.** $x^4 + y^4 = 81$

49. **A.** $\dfrac{x^2}{4} - \dfrac{y^2}{9} = 1$ **B.** $\dfrac{y^2}{4} - \dfrac{x^2}{9} = 1$ **C.** $\dfrac{x}{4} - \dfrac{y}{9} = 0$ **D.** $\dfrac{x^2}{4} - \dfrac{y^2}{9} = 0$

50. **A.** $y^2 - \sqrt{(x + 2)^2} = 0$ **B.** $y - \sqrt{(x + 2)^2} = 0$

C. $y^6 - \sqrt{(x + 1)^2} = 0$ **D.** $y^4 - \sqrt{x^2} = 0$

Chapter 4 Test Prep

Key Terms

4.1 one-to-one function
inverse function

4.2 exponential function
exponential equation
compound interest
future value
present value
compound amount
continuous compounding

4.3 logarithm
base
argument
logarithmic equation
logarithmic function

4.4 common logarithm
pH
natural logarithm

4.6 doubling time
half-life

New Symbols

$f^{-1}(x)$	inverse of $f(x)$
e	a constant, approximately 2.718281828459045
$\log_a x$	logarithm of x with the base a
$\log x$	common (base 10) logarithm of x
$\ln x$	natural (base e) logarithm of x

Quick Review

Concepts	Examples
4.1 Inverse Functions	
One-to-One Function In a one-to-one function, each x-value corresponds to only one y-value, and each y-value corresponds to only one x-value.	The function $y = f(x) = x^2$ is not one-to-one, because $y = 16$, for example, corresponds to both $x = 4$ and $x = -4$.
A function f is one-to-one if, for elements a and b in the domain of f, $a \neq b$ implies $f(a) \neq f(b)$.	
Horizontal Line Test A function is one-to-one if every horizontal line intersects the graph of the function at most once.	The graph of $f(x) = 2x - 1$ is a straight line with slope 2. f is a one-to-one function by the horizontal line test.
Inverse Functions Let f be a one-to-one function. Then g is the inverse function of f if $(f \circ g)(x) = x$ for every x in the domain of g and $(g \circ f)(x) = x$ for every x in the domain of f. To find $g(x)$, interchange x and y in $y = f(x)$, solve for y, and replace y with $g(x)$, which is $f^{-1}(x)$.	Find the inverse of f. $f(x) = 2x - 1$ Given function; $y = 2x - 1$ Let $y = f(x)$.; $x = 2y - 1$ Interchange x and y.; $y = \frac{x+1}{2}$ Solve for y.; $f^{-1}(x) = \frac{x+1}{2}$ Replace y with $f^{-1}(x)$.; $f^{-1}(x) = \frac{1}{2}x + \frac{1}{2}$ $\quad \frac{x+1}{2} = \frac{x}{2} + \frac{1}{2} = \frac{1}{2}x + \frac{1}{2}$

Concepts	Examples

4.2 Exponential Functions

Additional Properties of Exponents
For any real number $a > 0$, $a \neq 1$, the following hold true.

Concepts	Examples
(a) a^x is a unique real number for all real numbers x.	**(a)** 2^x is a unique real number for all real numbers x.
(b) $a^b = a^c$ if and only if $b = c$.	**(b)** $2^x = 2^3$ if and only if $x = 3$.
(c) If $a > 1$ and $m < n$, then $a^m < a^n$.	**(c)** $2^5 < 2^{10}$, because $2 > 1$ and $5 < 10$.
(d) If $0 < a < 1$ and $m < n$, then $a^m > a^n$.	**(d)** $\left(\frac{1}{2}\right)^5 > \left(\frac{1}{2}\right)^{10}$ because $0 < \frac{1}{2} < 1$ and $5 < 10$.

Exponential Function
If $a > 0$ and $a \neq 1$, then the exponential function with base a is $\boldsymbol{f(x) = a^x}$.

$f(x) = 3^x$ is the exponential function with base 3.

x	y
-1	$\frac{1}{3}$
0	1
1	3

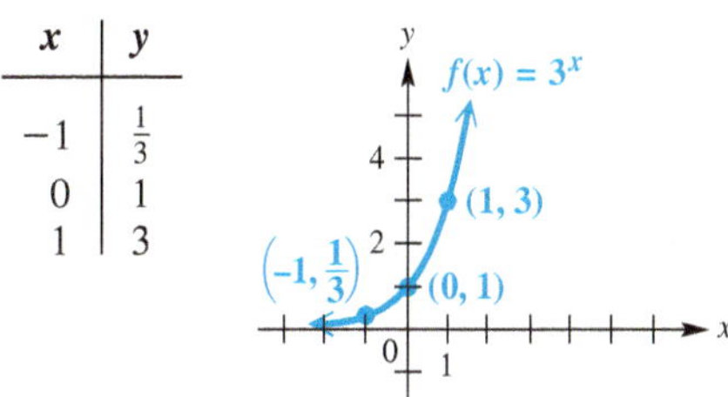

Graph of $f(x) = a^x$

1. The points $\left(-1, \frac{1}{a}\right)$, $(0, 1)$, and $(1, a)$ are on the graph.
2. If $a > 1$, then f is an increasing function.
   If $0 < a < 1$, then f is a decreasing function.
3. The x-axis is a horizontal asymptote.
4. The domain is $(-\infty, \infty)$, and the range is $(0, \infty)$.

4.3 Logarithmic Functions

Logarithm
For all real numbers y and all positive numbers a and x, where $a \neq 1$, $\boldsymbol{y = \log_a x}$ is equivalent to $\boldsymbol{x = a^y}$.

$\log_3 81 = 4$ is equivalent to $3^4 = 81$.

Logarithmic Function
If $a > 0$, $a \neq 1$, and $x > 0$, then the logarithmic function with base a is $\boldsymbol{f(x) = \log_a x}$.

$f(x) = \log_3 x$ is the logarithmic function with base 3.

x	y
$\frac{1}{3}$	-1
1	0
3	1

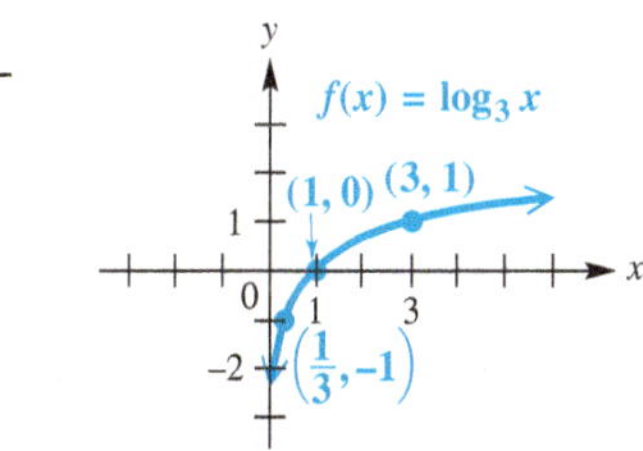

Graph of $f(x) = \log_a x$

1. The points $\left(\frac{1}{a}, -1\right)$, $(1, 0)$, and $(a, 1)$ are on the graph.
2. If $a > 1$, then f is an increasing function.
   If $0 < a < 1$, then f is a decreasing function.
3. The y-axis is a vertical asymptote.
4. The domain is $(0, \infty)$, and the range is $(-\infty, \infty)$.

Properties of Logarithms
For $x > 0$, $y > 0$, $a > 0$, $a \neq 1$, and any real number r, the following properties hold.

Property	Name	Example
$\log_a xy = \log_a x + \log_a y$	Product property	$\log_2 (3 \cdot 5) = \log_2 3 + \log_2 5$
$\log_a \frac{x}{y} = \log_a x - \log_a y$	Quotient property	$\log_2 \frac{3}{5} = \log_2 3 - \log_2 5$
$\log_a x^r = r \log_a x$	Power property	$\log_6 3^5 = 5 \log_6 3$
$\log_a 1 = 0$	Logarithm of 1	$\log_{10} 1 = 0$
$\log_a a = 1$	Base a logarithm of a	$\log_{10} 10 = 1$

Theorem on Inverses
For $a > 0$ and $a \neq 1$, the following properties hold.

$$a^{\log_a x} = x \quad (x > 0) \quad \text{and} \quad \log_a a^x = x$$

$2^{\log_2 5} = 5$ and $\log_2 2^5 = 5$

Concepts	Examples

4.4 Evaluating Logarithms and the Change-of-Base Theorem

Common and Natural Logarithms

For all positive numbers x, base 10 logarithms and base e logarithms are written as follows.

$$\log x = \log_{10} x \quad \text{Common logarithm}$$

$$\ln x = \log_e x \quad \text{Natural logarithm}$$

Approximate log 0.045 and ln 247.1.

$$\log 0.045 \approx -1.3468$$

$$\ln 247.1 \approx 5.5098 \quad \text{Use a calculator.}$$

Change-of-Base Theorem

For any positive real numbers x, a, and b, where $a \neq 1$ and $b \neq 1$, the following holds.

$$\log_a x = \frac{\log_b x}{\log_b a}$$

Approximate $\log_8 7$.

$$\log_8 7 = \frac{\log 7}{\log 8} = \frac{\ln 7}{\ln 8} \approx 0.9358 \quad \text{Use a calculator.}$$

4.5 Exponential and Logarithmic Equations

Property of Logarithms

If $x > 0$, $y > 0$, $a > 0$, and $a \neq 1$, then the following holds.

$$x = y \quad \text{is equivalent to} \quad \log_a x = \log_a y.$$

Solve.

$$e^{5x} = 10$$

$$\ln e^{5x} = \ln 10 \quad \text{Take natural logarithms.}$$

$$5x = \ln 10 \quad \ln e^x = x, \text{ for all } x.$$

$$x = \frac{\ln 10}{5} \quad \text{Divide by 5.}$$

$$x \approx 0.461 \quad \text{Use a calculator.}$$

The solution set can be written with the exact value, $\left\{\frac{\ln 10}{5}\right\}$, or with the approximate value, $\{0.461\}$.

$$\log_2 (x^2 - 3) = \log_2 6$$

$$x^2 - 3 = 6 \quad \text{Property of logarithms}$$

$$x^2 = 9 \quad \text{Add 3.}$$

$$x = \pm 3 \quad \text{Take square roots.}$$

Both values check, so the solution set is $\{\pm 3\}$.

4.6 Applications and Models of Exponential Growth and Decay

Exponential Growth or Decay Function

Let y_0 be the amount or number present at time $t = 0$. Then, under certain conditions, the amount present at any time t is modeled by

$$y = y_0 e^{kt}, \quad \text{where } k \text{ is a constant.}$$

The formula for continuous compounding,

$$A = Pe^{rt},$$

is an example of exponential growth. Here, A is the compound amount if P dollars are invested at an annual interest rate r for t years.

If $P = \$200$, $r = 3\%$, and $t = 5$ yr, find A.

$$A = Pe^{rt}$$

$$A = 200e^{0.03(5)} \quad \text{Substitute.}$$

$$A \approx \$232.37 \quad \text{Use a calculator.}$$

Chapter 4 Review Exercises

Determine whether each function as graphed or defined is one-to-one.

1.

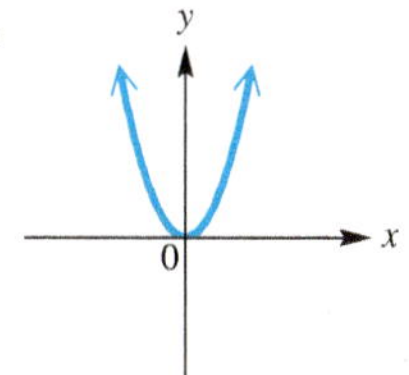

2.

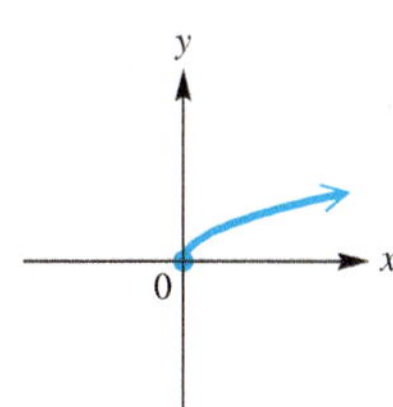

3.

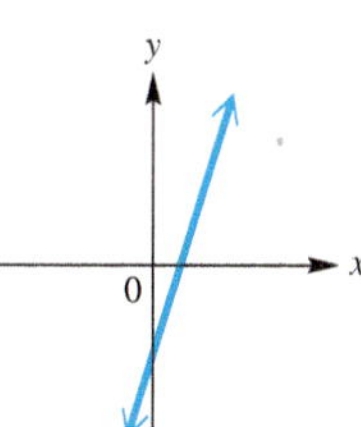

4. $y = x^3 + 1$ **5.** $y = (x + 3)^2$ **6.** $y = \sqrt{3x^2 + 2}$

Find the inverse of each function that is one-to-one.

7. $f(x) = x^3 - 3$ **8.** $f(x) = \sqrt{25 - x^2}$

Concept Check *Work each problem.*

9. Suppose $f(t)$ is the amount an investment will grow to t years after 2004. What does $f^{-1}(\$50{,}000)$ represent?

10. The graphs of two functions are shown. Based on their graphs, are these functions inverses?

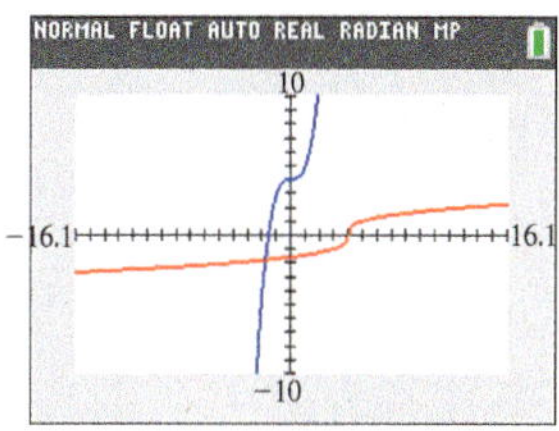

11. To have an inverse, a function must be a(n) ________ function.

12. *True* or *false*? The x-coordinate of the x-intercept of the graph of $y = f(x)$ is the y-coordinate of the y-intercept of the graph of $y = f^{-1}(x)$.

Match each equation with the figure that most closely resembles its graph.

13. $y = \log_{0.3} x$ **14.** $y = e^x$ **15.** $y = \ln x$ **16.** $y = 0.3^x$

A.

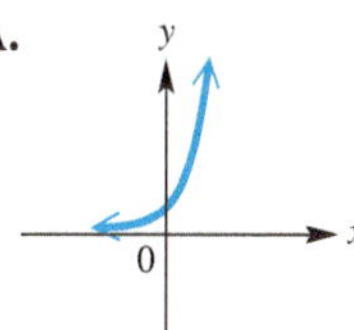

B.

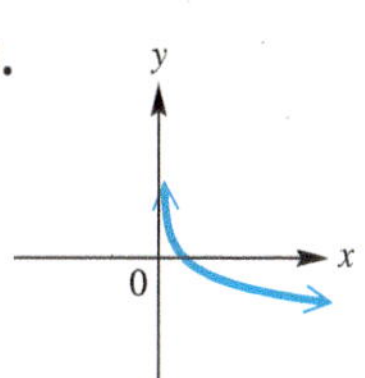

C. **D.**

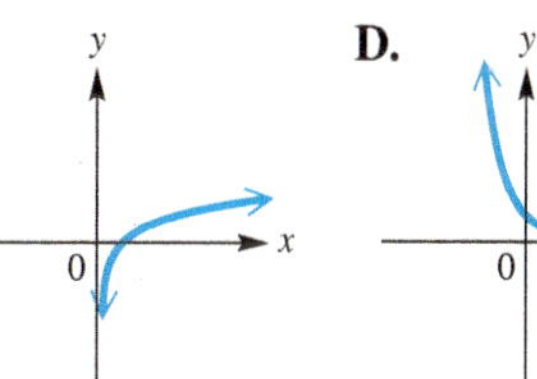

Write each equation in logarithmic form.

17. $2^5 = 32$ **18.** $100^{1/2} = 10$ **19.** $\left(\frac{3}{4}\right)^{-1} = \frac{4}{3}$

20. Graph $f(x) = \left(\frac{1}{5}\right)^{x+2} - 1$. Give the domain and range.

Write each equation in exponential form.

21. $\log 1000 = 3$ **22.** $\log_9 27 = \frac{3}{2}$ **23.** $\ln \sqrt{e} = \frac{1}{2}$

24. *Concept Check* What is the base of the logarithmic function whose graph contains the point $(81, 4)$?

25. *Concept Check* What is the base of the exponential function whose graph contains the point $\left(-4, \frac{1}{16}\right)$?

Use properties of logarithms to rewrite each expression. Simplify the result if possible. Assume all variables represent positive real numbers.

26. $\log_5\left(x^2y^4\sqrt[5]{m^3p}\right)$

27. $\log_3 \frac{mn}{5r}$

28. $\log_7(7k + 5r^2)$

Use a calculator to find an approximation to four decimal places for each logarithm.

29. $\log 0.0411$

30. $\log 45.6$

31. $\ln 144{,}000$

32. $\ln 470$

33. $\log_{2/3} \frac{5}{8}$

34. $\log_3 769$

Solve each equation. Unless otherwise specified, give irrational solutions as decimals correct to the nearest thousandth.

35. $16^{x+4} = 8^{3x-2}$

36. $4^x = 12$

37. $3^{2x-5} = 13$

38. $2^{x+3} = 5^x$

39. $6^{x+3} = 4^x$

40. $e^{x-1} = 4$

41. $e^{2-x} = 12$

42. $2e^{5x+2} = 8$

43. $10e^{3x-7} = 5$

44. $5^{x+2} = 2^{2x-1}$

45. $6^{x-3} = 3^{4x+1}$

46. $e^{8x} \cdot e^{2x} = e^{20}$

47. $e^{6x} \cdot e^x = e^{21}$

48. $100(1.02)^{x/4} = 200$

49. $2e^{2x} - 5e^x - 3 = 0$ (Give exact form.)

50. $\left(\frac{1}{2}\right)^x + 2 = 0$

51. $4(1.06)^x + 2 = 8$

52. *Concept Check* Which one or more of the following choices is the solution set of $5^x = 9$?

A. $\{\log_5 9\}$

B. $\{\log_9 5\}$

C. $\left\{\frac{\log 9}{\log 5}\right\}$

D. $\left\{\frac{\ln 9}{\ln 5}\right\}$

Solve each equation. Give solutions in exact form.

53. $3 \ln x = 13$

54. $\ln 5x = 16$

55. $\log(2x + 7) = 0.25$

56. $\ln x + \ln x^3 = 12$

57. $\log_2(x^3 + 5) = 5$

58. $\log_3(x^2 - 9) = 3$

59. $\log_4[(3x + 1)(x - 4)] = 2$

60. $\ln e^{\ln x} - \ln(x - 4) = \ln 3$

61. $\log x + \log(13 - 3x) = 1$

62. $\log_7(3x + 2) - \log_7(x - 2) = 1$

63. $\ln(6x) - \ln(x + 1) = \ln 4$

64. $\log_{16}\sqrt{x + 1} = \frac{1}{4}$

65. $\ln[\ln e^{-x}] = \ln 3$

66. $S = a \ln\left(1 + \frac{n}{a}\right)$, for n

67. $d = 10 \log \frac{I}{I_0}$, for I_0

68. $D = 200 + 100 \log x$, for x

69. Use a graphing calculator to solve the equation $e^x = 4 - \ln x$. Give solution(s) to the nearest thousandth.

Solve each problem.

70. *(Modeling) Decibel Levels* Decibel rating of the loudness of a sound is modeled by

$$d = 10 \log \frac{I}{I_0},$$

where I is the intensity of a particular sound, and I_0 is the intensity of a very faint threshold sound. A few years ago, there was a controversy about a proposed government limit on factory noise. One group wanted a maximum of 89 decibels, while another group wanted 86. Find the percent by which the 89-decibel intensity exceeds that for 86 decibels.

71. *Earthquake Intensity* The magnitude of an earthquake, measured on the Richter scale, is $\log \frac{I}{I_0}$, where I is the amplitude registered on a seismograph 100 km from the epicenter of the earthquake, and I_0 is the amplitude of an earthquake of a certain (small) size. On August 24, 2014, the Napa Valley in California was shaken by an earthquake that measured 6.0 on the Richter scale.

(a) Express this reading in terms of I_0.

(b) On April 1, 2014, a quake measuring 8.2 on the Richter scale struck off the coast of Chile. It was the largest earthquake in 2014. Express the magnitude of an 8.2 reading in terms of I_0 to the nearest hundred thousand.

(c) How much greater than the force of the 6.0 earthquake was the force of the earthquake that measured 8.2?

72. *Earthquake Intensity* The San Francisco earthquake of 1906 had a Richter scale rating of 8.3.

(a) Express the magnitude of this earthquake in terms of I_0 to the nearest hundred thousand.

(b) In 1989, the San Francisco region experienced an earthquake with a Richter scale rating of 7.1. Express the magnitude of this earthquake in terms of I_0 to the nearest hundred thousand.

(c) Compare the magnitudes of the two San Francisco earthquakes discussed in parts (a) and (b).

73. *Interest Rate* What annual interest rate, to the nearest tenth, will produce \$4700 if \$3500 is left at interest compounded annually for 10 yr?

74. *Growth of an Account* Find the number of years (to the nearest tenth) needed for \$48,000 to become \$53,647 at 2.8% interest compounded semiannually.

75. *Growth of an Account* Manuel deposits \$10,000 for 12 yr in an account paying 3% interest compounded annually. He then puts this total amount on deposit in another account paying 4% interest compounded semiannually for another 9 yr. Find the total amount on deposit after the entire 21-yr period.

76. *Growth of an Account* Anne deposits \$12,000 for 8 yr in an account paying 2.5% interest compounded annually. She then leaves the money alone with no further deposits at 3% interest compounded annually for an additional 6 yr. Find the total amount on deposit after the entire 14-yr period.

77. *Cost from Inflation* Suppose the inflation rate is 4%. Use the formula for continuous compounding to find the number of years, to the nearest tenth, for a \$1 item to cost \$2.

78. *(Modeling) Drug Level in the Bloodstream* After a medical drug is injected directly into the bloodstream, it is gradually eliminated from the body. Graph the following functions on the interval $[0, 10]$. Use $[0, 500]$ for the range of $A(t)$. Determine the function that best models the amount $A(t)$ (in milligrams) of a drug remaining in the body after t hours if 350 mg were initially injected.

(a) $A(t) = t^2 - t + 350$ **(b)** $A(t) = 350 \log(t + 1)$

(c) $A(t) = 350(0.75)^t$ **(d)** $A(t) = 100(0.95)^t$

79. *(Modeling) Chicago Cubs' Payroll* The table shows the total payroll (in millions of dollars) of the Chicago Cubs baseball team for the years 2010–2014.

Year	Total Payroll (millions of dollars)
2010	145.4
2011	134.3
2012	111.0
2013	107.4
2014	92.7

Source: www.baseballprospectus.com/compensation

Letting $f(x)$ represent the total payroll and x represent the number of years since 2010, we find that the function

$$f(x) = 146.02e^{-0.112x}$$

models the data quite well. According to this function, when will the total payroll halve its 2010 value?

80. *(Modeling) Transistors on Computer Chips* Computing power has increased dramatically as a result of the ability to place an increasing number of transistors on a single processor chip. The table lists the number of transistors on some popular computer chips made by Intel.

Year	Chip	Transistors
1989	486DX	1,200,000
1994	Pentium	3,300,000
2000	Pentium 4	42,000,000
2006	Core 2 Duo	291,000,000
2008	Core 2 Quad	820,000,000
2010	Core (2nd gen.)	1,160,000,000
2012	Core (3rd gen.)	1,400,000,000

Source: Intel.

(a) Make a scatter diagram of the data. Let the x-axis represent the year, where $x = 0$ corresponds to 1989, and let the y-axis represent the number of transistors.

(b) Decide whether a linear, a logarithmic, or an exponential function best describes the data.

(c) Determine a function f that approximates these data. Plot f and the data on the same coordinate axes.

(d) Assuming that this trend continues, use f to estimate the number of transistors on a chip, to the nearest million, in the year 2016.

81. *Financial Planning* The traditional IRA (individual retirement account) is a common tax-deferred saving plan in the United States. Earned income deposited into an IRA is not taxed in the current year, and no taxes are incurred on the interest paid in subsequent years. However, when the money is withdrawn from the account after age $59\frac{1}{2}$, taxes must be paid on the entire amount withdrawn.

Suppose we deposited \$5000 of earned income into an IRA, we can earn an annual interest rate of 4%, and we are in a 25% tax bracket. (*Note:* Interest rates and tax brackets are subject to change over time, but some assumptions must be made to evaluate the investment.) Also, suppose that we deposit the \$5000 at age 25 and withdraw it at age 60, and that interest is compounded continuously.

(a) How much money will remain after we pay the taxes at age 60?

(b) Suppose that instead of depositing the money into an IRA, we pay taxes on the money and the annual interest. How much money will we have at age 60? (*Note:* We effectively start with \$3750 (75% of \$5000), and the money earns 3% (75% of 4%) interest after taxes.)

(c) To the nearest dollar, how much additional money will we earn with the IRA?

(d) Suppose we pay taxes on the original \$5000 but are then able to earn 4% in a tax-free investment. Compare the balance at age 60 with the IRA balance.

82. Consider $f(x) = \log_4(2x^2 - x)$.

(a) Use the change-of-base theorem with base e to write $\log_4(2x^2 - x)$ in a suitable form to graph with a calculator.

(b) Graph the function using a graphing calculator. Use the window $[-2.5, 2.5]$ by $[-5, 2.5]$.

(c) What are the x-intercepts?

(d) Give the equations of the vertical asymptotes.

(e) Why is there no y-intercept?

Chapter 4 Test

1. Consider the function $f(x) = \sqrt[3]{2x - 7}$.

(a) What are the domain and range of f?

(b) Explain why f^{-1} exists.

(c) Write an equation for $f^{-1}(x)$.

(d) What are the domain and range of f^{-1}?

(e) Graph both f and f^{-1}. How are the two graphs related with respect to the line $y = x$?

2. Match each equation with its graph.

(a) $y = \log_{1/3} x$ (b) $y = e^x$ (c) $y = \ln x$ (d) $y = \left(\frac{1}{3}\right)^x$

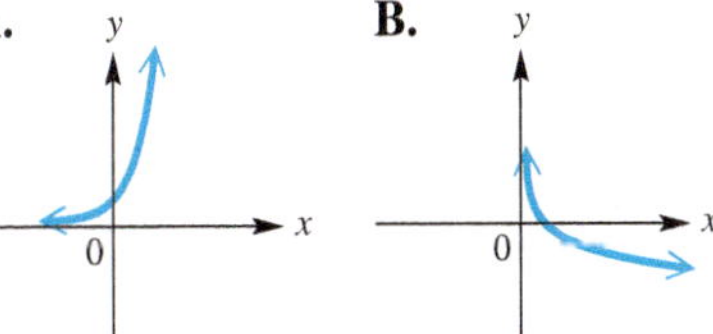

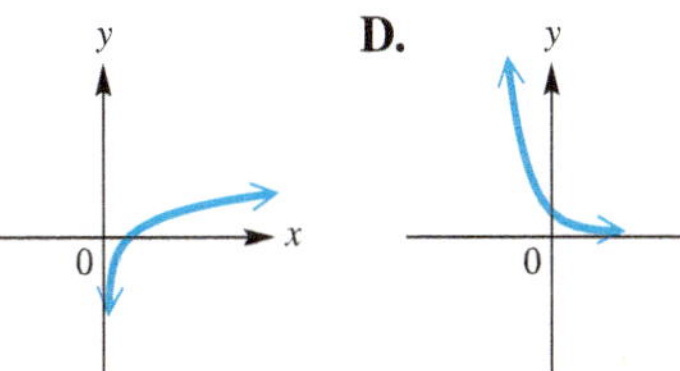

3. Solve $\left(\frac{1}{8}\right)^{2x-3} = 16^{x+1}$.

4. (a) Write $4^{3/2} = 8$ in logarithmic form.

(b) Write $\log_8 4 = \frac{2}{3}$ in exponential form.

5. Graph $f(x) = \left(\frac{1}{2}\right)^x$ and $g(x) = \log_{1/2} x$ on the same axes. What is their relationship?

6. Use properties of logarithms to rewrite the expression. Assume all variables represent positive real numbers.

$$\log_7 \frac{x^2\sqrt[4]{y}}{z^3}$$

Use a calculator to find an approximation to four decimal places for each logarithm.

7. $\log 2388$ **8.** $\ln 2388$ **9.** $\log_9 13$

10. Solve $x^{2/3} = 25$.

Solve each equation. Give irrational solutions as decimals correct to the nearest thousandth.

11. $12^x = 1$ **12.** $9^x = 4$ **13.** $16^{2x+1} = 8^{3x}$

14. $2^{x+1} = 3^{x-4}$ **15.** $e^{0.4x} = 4^{x-2}$

16. $2e^{2x} - 5e^x + 3 = 0$ (Give both exact and approximate values.)

Solve each equation. Give solutions in exact form.

17. $\log_x \frac{9}{16} = 2$ **18.** $\log_2 [(x-4)(x-2)] = 3$

19. $\log_2 x + \log_2 (x+2) = 3$ **20.** $\ln x - 4 \ln 3 = \ln \frac{1}{5} x$

21. $\log_3 (x+1) - \log_3 (x-3) = 2$

22. A friend is taking another mathematics course and says, "I have no idea what an expression like $\log_5 27$ really means." Write an explanation of what it means, and tell how we can find an approximation for it with a calculator.

Solve each problem.

23. ***(Modeling) Skydiver Fall Speed*** A skydiver in free fall travels at a speed modeled by

$$v(t) = 176(1 - e^{-0.18t})$$

feet per second after t seconds. How long, to the nearest second, will it take for the skydiver to attain a speed of 147 ft per sec (100 mph)?

24. ***Growth of an Account*** How many years, to the nearest tenth, will be needed for \$5000 to increase to \$18,000 at 3.0% annual interest compounded **(a)** monthly **(b)** continuously?

25. ***Tripling Time*** For *any* amount of money invested at 2.8% annual interest compounded continuously, how long, to the nearest tenth of a year, will it take to triple?

26. ***(Modeling) Radioactive Decay*** The amount of a certain radioactive material, in grams, present after t days is modeled by

$$A(t) = 600e^{-0.05t}.$$

(a) Find the amount present after 12 days, to the nearest tenth of a gram.

(b) Find the half-life of the material, to the nearest tenth of a day.

5 Trigonometric Functions

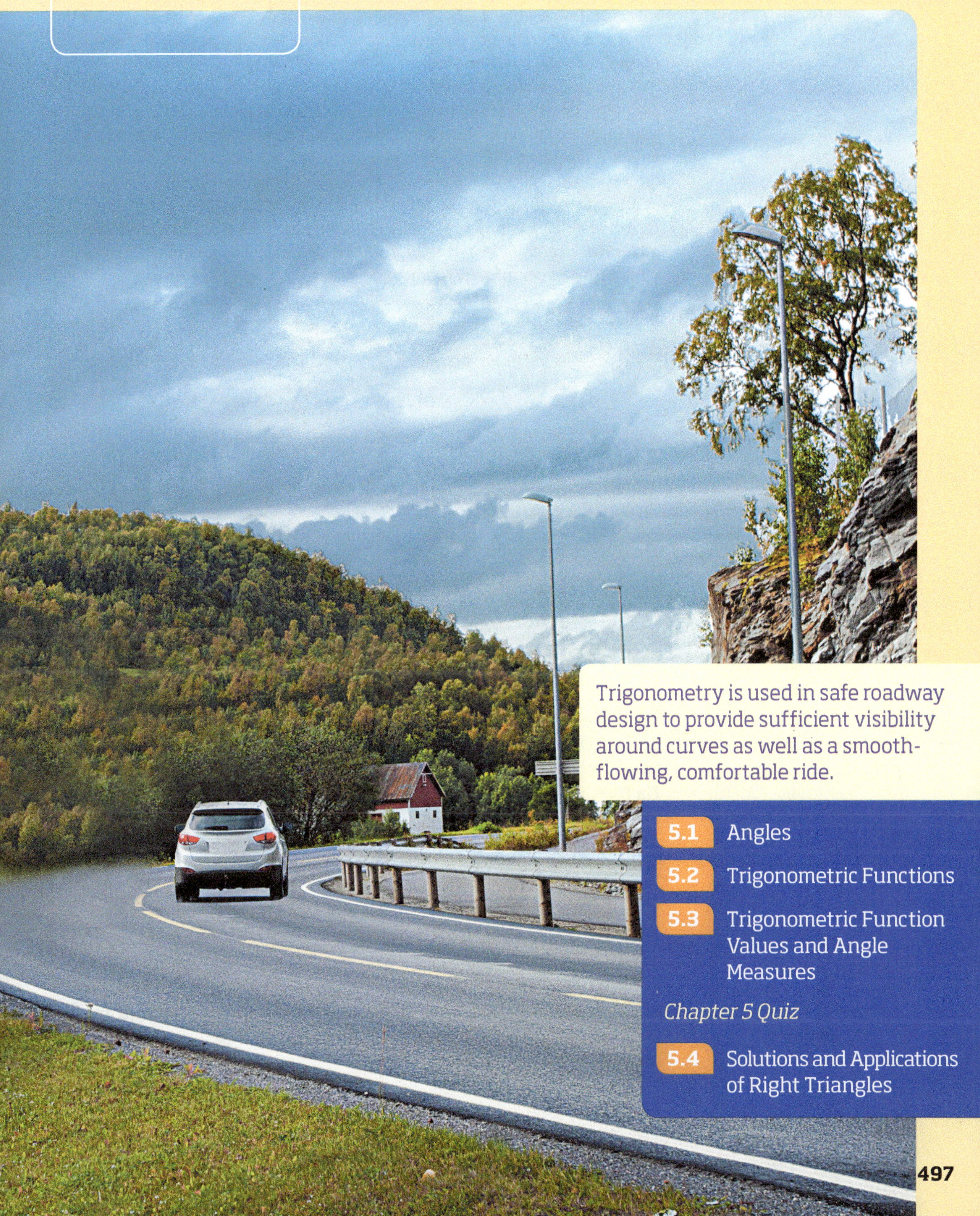

Trigonometry is used in safe roadway design to provide sufficient visibility around curves as well as a smooth-flowing, comfortable ride.

5.1 Angles

- Basic Terminology
- Degree Measure
- Standard Position
- Coterminal Angles

Basic Terminology

Two distinct points A and B determine a line called **line AB.** The portion of the line between A and B, including points A and B themselves, is **line segment AB,** or simply **segment AB.** The portion of line AB that starts at A and continues through B, and on past B, is the **ray AB.** Point A is the **endpoint of the ray.** See **Figure 1.**

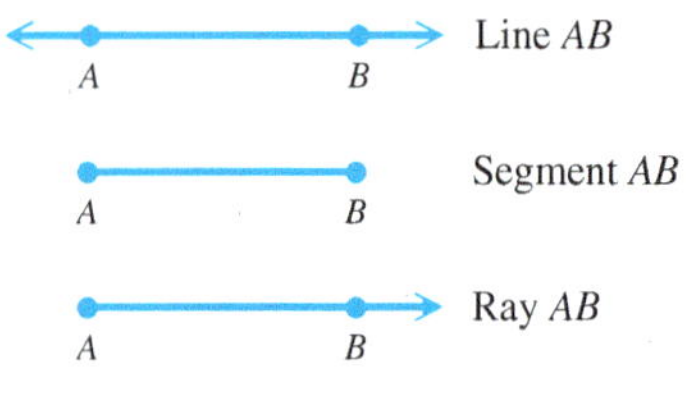

Figure 1

In trigonometry, an **angle** consists of two rays in a plane with a common endpoint, or two line segments with a common endpoint. These two rays (or segments) are the **sides** of the angle, and the common endpoint is the **vertex** of the angle. Associated with an angle is its measure, generated by a rotation about the vertex. See **Figure 2.** This measure is determined by rotating a ray starting at one side of the angle, the **initial side,** to the position of the other side, the **terminal side.** ***A counterclockwise rotation generates a positive measure, and a clockwise rotation generates a negative measure.*** The rotation can consist of more than one complete revolution.

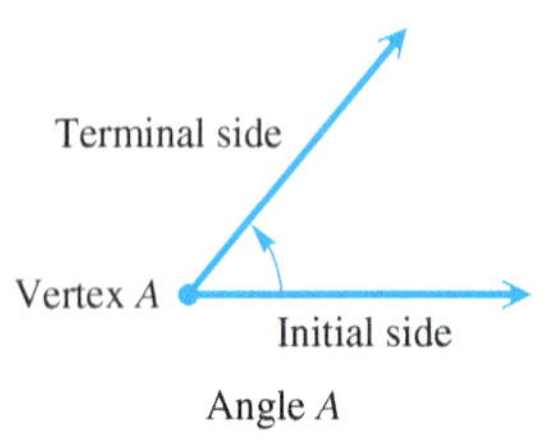

Figure 2

Figure 3 shows two angles, one **positive** and one **negative.**

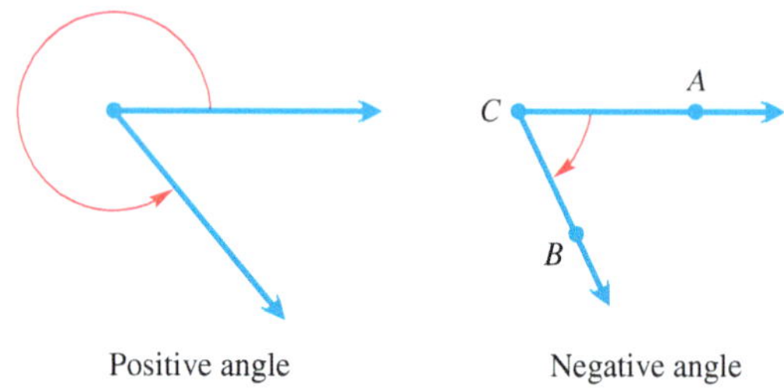

Figure 3

An angle can be named by using the name of its vertex. For example, the angle on the right in **Figure 3** can be named angle C. Alternatively, an angle can be named using three letters, with the vertex letter in the middle. Thus, the angle on the right also could be named angle ACB or angle BCA.

Degree Measure

The most common unit for measuring angles is the **degree.** Degree measure was developed by the Babylonians 4000 yr ago. To use degree measure, we assign 360 degrees to a complete rotation of a ray.* In **Figure 4,** notice that the terminal side of the angle corresponds to its initial side when it makes a complete rotation.

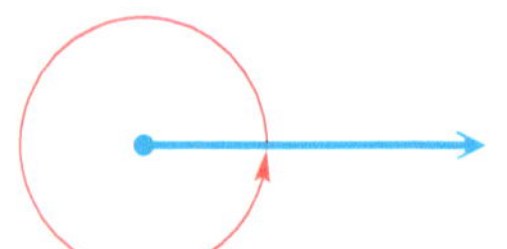

A complete rotation of a ray gives an angle whose measure is 360°. $\frac{1}{360}$ of a complete rotation gives an angle whose measure is 1°.

Figure 4

$$\text{One degree, written } 1^\circ\text{, represents } \frac{1}{360} \text{ of a complete rotation.}$$

Therefore, 90° represents $\frac{90}{360} = \frac{1}{4}$ of a complete rotation, and 180° represents $\frac{180}{360} = \frac{1}{2}$ of a complete rotation.

An angle measuring between 0° and 90° is an **acute angle.** An angle measuring exactly 90° is a **right angle.** The symbol ⌝ is often used at the vertex of a right angle to denote the 90° measure. An angle measuring more than 90° but less than 180° is an **obtuse angle,** and an angle of exactly 180° is a **straight angle.**

*The Babylonians were the first to subdivide the circumference of a circle into 360 parts. There are various theories about why the number 360 was chosen. One is that it is approximately the number of days in a year, and it has many divisors, which makes it convenient to work with in computations.

The Greek Letters		
Α	α	alpha
Β	β	beta
Γ	γ	gamma
Δ	δ	delta
Ε	ε	epsilon
Ζ	ζ	zeta
Η	η	eta
Θ	θ	theta
Ι	ι	iota
Κ	κ	kappa
Λ	λ	lambda
Μ	μ	mu
Ν	ν	nu
Ξ	ξ	xi
Ο	o	omicron
Π	π	pi
Ρ	ρ	rho
Σ	σ	sigma
Τ	τ	tau
Υ	υ	upsilon
Φ	ϕ	phi
Χ	χ	chi
Ψ	ψ	psi
Ω	ω	omega

In **Figure 5,** we use the **Greek letter θ (theta)*** to name each angle. The table in the margin lists the upper- and lowercase Greek letters, which are often used in trigonometry.

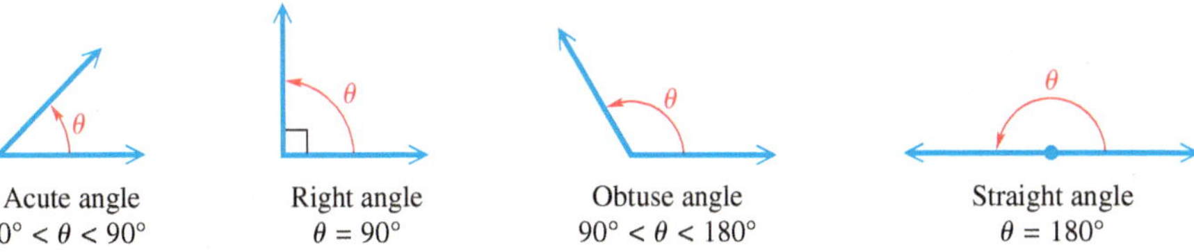

Figure 5

If the sum of the measures of two positive angles is 90°, the angles are **complementary** and the angles are **complements** of each other. Two positive angles with measures whose sum is 180° are **supplementary,** and the angles are **supplements.**

EXAMPLE 1 Finding the Complement and the Supplement of an Angle

Find the measure of **(a)** the complement and **(b)** the supplement of an angle measuring 40°.

SOLUTION

(a) To find the measure of its complement, subtract the measure of the angle from 90°.

$$90^\circ - 40^\circ = 50^\circ \quad \text{Complement of } 40^\circ$$

(b) To find the measure of its supplement, subtract the measure of the angle from 180°.

$$180^\circ - 40^\circ = 140^\circ \quad \text{Supplement of } 40^\circ$$

Now Try Exercise 11.

EXAMPLE 2 Finding Measures of Complementary and Supplementary Angles

Find the measure of each marked angle in **Figure 6.**

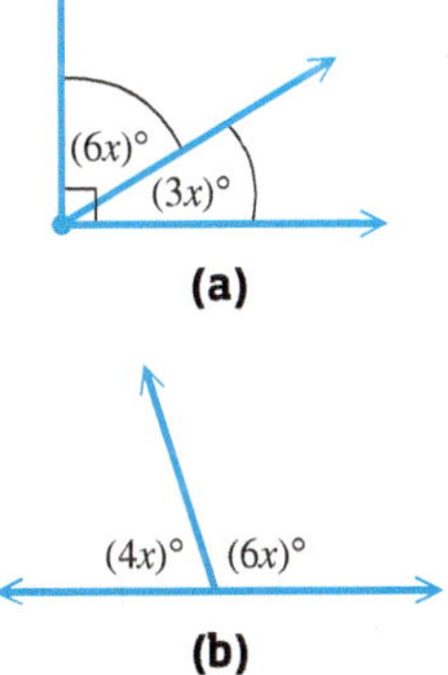

Figure 6

SOLUTION

(a) Because the two angles in **Figure 6(a)** form a right angle, they are complementary angles.

$$6x + 3x = 90 \quad \text{Complementary angles sum to } 90^\circ.$$

$$9x = 90 \quad \text{Combine like terms.}$$

Don't stop here.

$$x = 10 \quad \text{Divide by 9.}$$

Be sure to determine the measure of each angle by substituting 10 for x in $6x$ and $3x$. The two angles have measures of $6(10) = 60^\circ$ and $3(10) = 30^\circ$.

(b) The angles in **Figure 6(b)** are supplementary, so their sum must be 180°.

$$4x + 6x = 180 \quad \text{Supplementary angles sum to } 180^\circ.$$

$$10x = 180 \quad \text{Combine like terms.}$$

$$x = 18 \quad \text{Divide by 10.}$$

The angle measures are $4x = 4(18) = 72^\circ$ and $6x = 6(18) = 108^\circ$.

Now Try Exercises 23 and 25.

*In addition to θ (theta), other Greek letters such as α (alpha) and β (beta) are used to name angles.

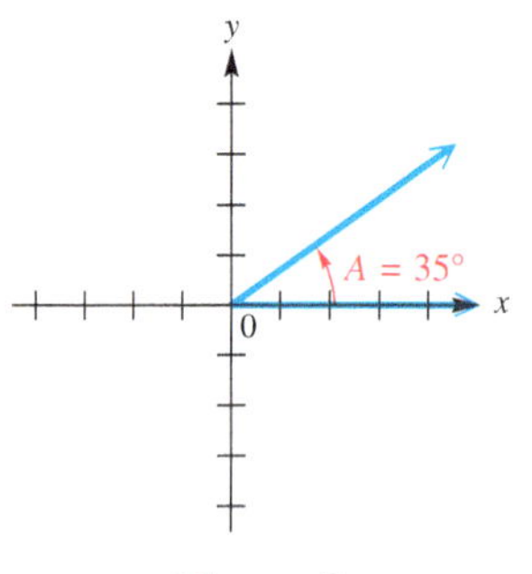

Figure 7

The measure of angle A in **Figure 7** is 35°. This measure is often expressed by saying that $\boldsymbol{m(\text{angle } A)}$ is 35°, where $m(\text{angle } A)$ is read **"the measure of angle A."** The symbolism $m(\text{angle } A) = 35°$ is abbreviated as $A = 35°$.

Traditionally, portions of a degree have been measured with minutes and seconds. One **minute,** written **1′**, is $\frac{1}{60}$ of a degree.

$$1' = \left(\frac{1}{60}\right)^\circ \quad \text{or} \quad 60' = 1^\circ$$

One **second, 1″**, is $\frac{1}{60}$ of a minute.

$$1'' = \left(\frac{1}{60}\right)' = \left(\frac{1}{3600}\right)^\circ \quad \text{or} \quad 60'' = 1' \text{ and } 3600'' = 1^\circ$$

The measure 12° 42′ 38″ represents 12 degrees, 42 minutes, 38 seconds.

EXAMPLE 3 Calculating with Degrees, Minutes, and Seconds

Perform each calculation.

(a) 51° 29′ + 32° 46′ **(b)** 90° − 73° 12′

SOLUTION

(a)
$$\begin{array}{r} 51^\circ\, 29' \\ +\, 32^\circ\, 46' \\ \hline 83^\circ\, 75' \end{array}$$
Add degrees and minutes separately.

The sum 83° 75′ can be rewritten as follows.

$$\begin{aligned} &83^\circ\, 75' \\ &= 83^\circ + 1^\circ\, 15' && 75' = 60' + 15' = 1^\circ\, 15' \\ &= 84^\circ\, 15' && \text{Add.} \end{aligned}$$

(b)
$$\begin{array}{r} 90^\circ \\ -\, 73^\circ\, 12' \\ \hline \end{array} \quad \text{can be written} \quad \begin{array}{r} 89^\circ\, 60' \\ -\, 73^\circ\, 12' \\ \hline 16^\circ\, 48' \end{array}$$
Write 90° as 89° 60′.

✔ **Now Try Exercises 41 and 45.**

An alternative way to measure angles involves decimal degrees. For example,

$$12.4238^\circ \quad \text{represents} \quad 12\frac{4238}{10{,}000}{}^\circ.$$

EXAMPLE 4 Converting between Angle Measures

(a) Convert 74° 08′ 14″ to decimal degrees to the nearest thousandth.

(b) Convert 34.817° to degrees, minutes, and seconds to the nearest second.

SOLUTION

(a) 74° 08′ 14″

$$\begin{aligned} &= 74^\circ + \left(\frac{8}{60}\right)^\circ + \left(\frac{14}{3600}\right)^\circ && 08' \cdot \frac{1^\circ}{60'} = \left(\frac{8}{60}\right)^\circ \text{ and } 14'' \cdot \frac{1^\circ}{3600''} = \left(\frac{14}{3600}\right)^\circ \\ &\approx 74^\circ + 0.1333^\circ + 0.0039^\circ && \text{Divide to express the fractions as decimals.} \\ &\approx 74.137^\circ && \text{Add and round to the nearest thousandth.} \end{aligned}$$

```
NORMAL FLOAT AUTO REAL DEGREE MP
74°8'14"
                    74.13722222
74°8'14"
                         74.137
34.817▶DMS
                      34°49'1.2"
```

This screen shows how the TI-84 Plus performs the conversions in **Example 4.** The ▶DMS option is found in the ANGLE Menu.

(b) $34.817°$

$= 34° + 0.817°$	Write as a sum.
$= 34° + 0.817(60')$	$0.817° \cdot \frac{60'}{1°} = 0.817(60')$
$= 34° + 49.02'$	Multiply.
$= 34° + 49' + 0.02'$	Write $49.02'$ as a sum.
$= 34° + 49' + 0.02(60'')$	$0.02' \cdot \frac{60''}{1'} = 0.02(60'')$
$= 34° + 49' + 1.2''$	Multiply.
$\approx 34°\, 49'\, 01''$	Approximate to the nearest second.

✔ **Now Try Exercises 61 and 71.**

Standard Position An angle is in **standard position** if its vertex is at the origin and its initial side lies on the positive x-axis. The angles in **Figures 8(a) and 8(b)** are in standard position. An angle in standard position is said to lie in the quadrant in which its terminal side lies. An acute angle is in quadrant I (**Figure 8(a)**) and an obtuse angle is in quadrant II (**Figure 8(b)**). **Figure 8(c)** shows ranges of angle measures for each quadrant when $0° < \theta < 360°$.

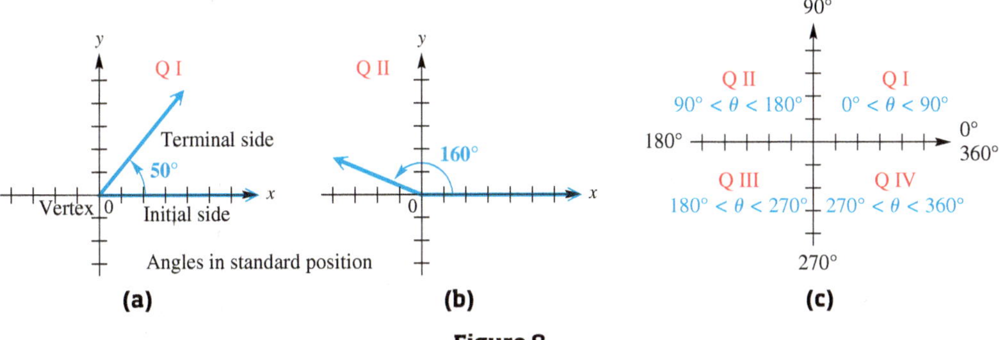

Figure 8

Quadrantal Angles

Angles in standard position whose terminal sides lie on the x-axis or y-axis, such as angles with measures 90°, 180°, 270°, and so on, are **quadrantal angles.**

Coterminal Angles A complete rotation of a ray results in an angle measuring 360°. By continuing the rotation, angles of measure larger than 360° can be produced. The angles in **Figure 9** with measures 60° and 420° have the same initial side and the same terminal side, but different amounts of rotation. Such angles are **coterminal angles.** ***Their measures differ by a multiple of 360°.*** As shown in **Figure 10,** angles with measures 110° and 830° are coterminal.

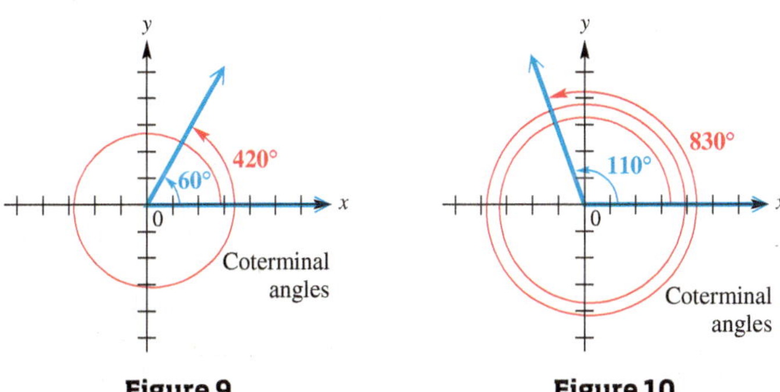

Figure 9 **Figure 10**

EXAMPLE 5 Finding Measures of Coterminal Angles

Find the angle of least positive measure that is coterminal with each angle.

(a) 908° **(b)** -75° **(c)** -800°

SOLUTION

(a) Subtract 360° as many times as needed to obtain an angle with measure greater than 0° but less than 360°.

$$908^\circ - 2 \cdot 360^\circ = 188^\circ \quad \text{Multiply } 2 \cdot 360^\circ\text{. Then subtract.}$$

An angle of 188° is coterminal with an angle of 908°. See **Figure 11.**

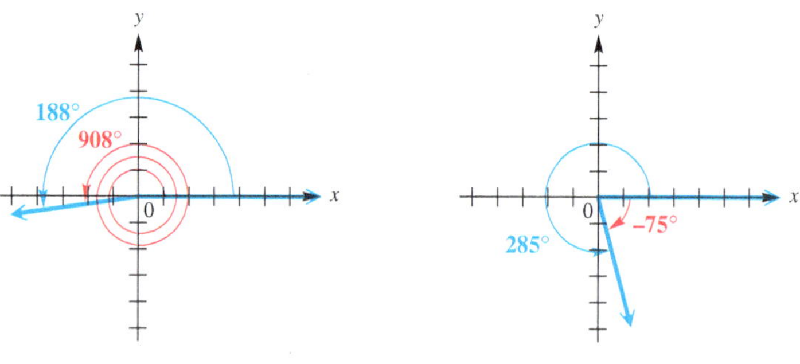

Figure 11 **Figure 12**

(b) Add 360° to the given negative angle measure to obtain the angle of least *positive* measure. See **Figure 12.**

$$-75^\circ + 360^\circ = 285^\circ$$

(c) The least integer multiple of 360° greater than 800° is

$$3 \cdot 360^\circ = 1080^\circ.$$

Add 1080° to -800° to obtain

$$-800^\circ + 1080^\circ = 280^\circ.$$

✔ **Now Try Exercises 81, 91, and 95.**

Sometimes it is necessary to find an expression that will generate all angles coterminal with a given angle. For example, we can obtain any angle coterminal with 60° by adding an integer multiple of 360° to 60°. Let n represent any integer. Then the following expression represents all such coterminal angles.

$$60^\circ + \boldsymbol{n \cdot 360^\circ} \quad \text{Angles coterminal with } 60^\circ$$

The table below shows a few possibilities.

Examples of Angles Coterminal with 60°

Value of n	Angle Coterminal with 60°
2	$60^\circ + 2 \cdot 360^\circ = 780^\circ$
1	$60^\circ + 1 \cdot 360^\circ = 420^\circ$
0	$60^\circ + 0 \cdot 360^\circ = 60^\circ$ (the angle itself)
-1	$60^\circ + (-1) \cdot 360^\circ = -300^\circ$
-2	$60^\circ + (-2) \cdot 360^\circ = -660^\circ$

This table shows some examples of coterminal quadrantal angles.

Examples of Coterminal Quadrantal Angles

Quadrantal Angle θ	Coterminal with θ
$0°$	$\pm 360°, \pm 720°$
$90°$	$-630°, -270°, 450°$
$180°$	$-180°, 540°, 900°$
$270°$	$-450°, -90°, 630°$

EXAMPLE 6 Analyzing Revolutions of a Disk Drive

A constant angular velocity disk drive spins a disk at a constant speed. Suppose a disk makes 480 revolutions per min. Through how many degrees will a point on the edge of the disk move in 2 sec?

SOLUTION The disk revolves 480 times in 1 min, or $\frac{480}{60}$ times $= 8$ times per sec (because 60 sec $= 1$ min). In 2 sec, the disk will revolve $2 \cdot 8 = 16$ times. Each revolution is $360°$, so in 2 sec a point on the edge of the disk will revolve

$$16 \cdot 360° = 5760°.$$

A unit analysis expression can also be used.

$$\frac{480 \text{ rev}}{1 \text{ min}} \times \frac{1 \text{ min}}{60 \text{ sec}} \times \frac{360°}{1 \text{ rev}} \times 2 \text{ sec} = 5760°$$ Divide out common units.

✔ Now Try Exercise 123.

5.1 Exercises

CONCEPT PREVIEW *Fill in the blank(s) to correctly complete each sentence.*

1. One degree, written $1°$, represents ______ of a complete rotation.

2. If the measure of an angle is $x°$, its complement can be expressed as ______ $- x°$.

3. If the measure of an angle is $x°$, its supplement can be expressed as ______ $- x°$.

4. The measure of an angle that is its own complement is ______.

5. The measure of an angle that is its own supplement is ______.

6. One minute, written $1'$, is ______ of a degree.

7. One second, written $1''$, is ______ of a minute.

8. $12° \, 30'$ written in decimal degrees is ______.

9. $55.25°$ written in degrees and minutes is ______.

10. If n represents any integer, then an expression representing all angles coterminal with $45°$ is $45° +$ ______.

Find the measure of (a) the complement and (b) the supplement of an angle with the given measure. ***See Examples 1 and 3.***

11. $30°$ **12.** $60°$ **13.** $45°$ **14.** $90°$

15. $54°$ **16.** $10°$ **17.** $1°$ **18.** $89°$

19. $14° \, 20'$ **20.** $39° \, 50'$ **21.** $20° \, 10' \, 30''$ **22.** $50° \, 40' \, 50''$

Find the measure of each marked angle. **See Example 2.**

23.

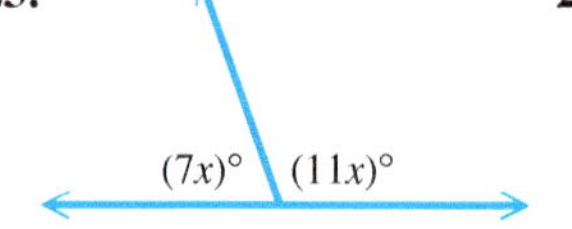

24.

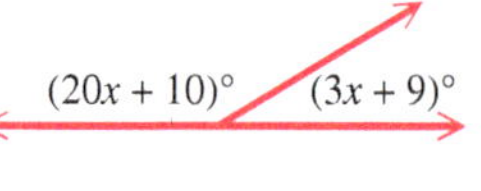

25.

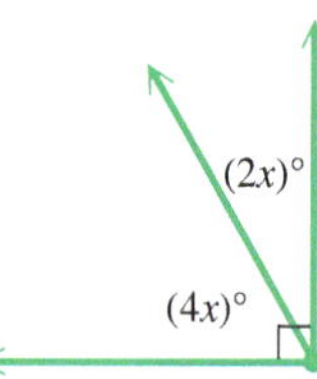

26.

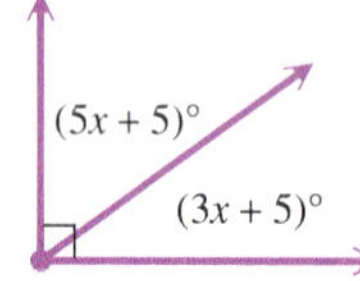

27.

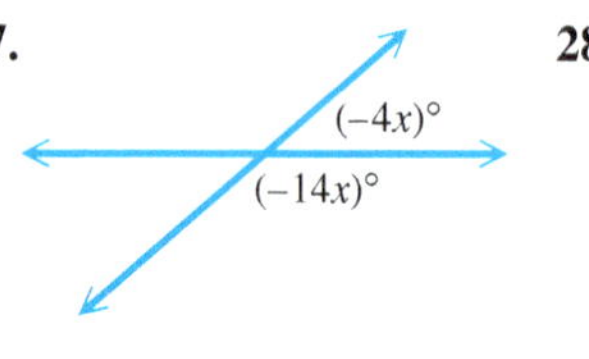

28.

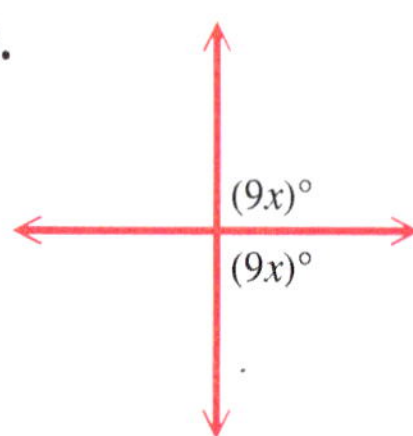

29. supplementary angles with measures $10x + 7$ and $7x + 3$ degrees

30. supplementary angles with measures $6x - 4$ and $8x - 12$ degrees

31. complementary angles with measures $9x + 6$ and $3x$ degrees

32. complementary angles with measures $3x - 5$ and $6x - 40$ degrees

Find the measure of the smaller angle formed by the hands of a clock at the following times.

33.

34.

35. 3:15 **36.** 9:45 **37.** 8:20 **38.** 6:10

Perform each calculation. **See Example 3.**

39. $62° 18' + 21° 41'$ **40.** $75° 15' + 83° 32'$ **41.** $97° 42' + 81° 37'$

42. $110° 25' + 32° 55'$ **43.** $47° 29' - 71° 18'$ **44.** $47° 23' - 73° 48'$

45. $90° - 51° 28'$ **46.** $90° - 17° 13'$ **47.** $180° - 119° 26'$

48. $180° - 124° 51'$ **49.** $90° - 72° 58' 11''$ **50.** $90° - 36° 18' 47''$

51. $26° 20' + 18° 17' - 14° 10'$ **52.** $55° 30' + 12° 44' - 8° 15'$

Convert each angle measure to decimal degrees. If applicable, round to the nearest thousandth of a degree. **See Example 4(a).**

53. $35° 30'$ **54.** $82° 30'$ **55.** $112° 15'$ **56.** $133° 45'$

57. $-60° 12'$ **58.** $-70° 48'$ **59.** $20° 54' 36''$ **60.** $38° 42' 18''$

61. $91° 35' 54''$ **62.** $34° 51' 35''$ **63.** $274° 18' 59''$ **64.** $165° 51' 09''$

Convert each angle measure to degrees, minutes, and seconds. If applicable, round to the nearest second. **See Example 4(b).**

65. 39.25° **66.** 46.75° **67.** 126.76° **68.** 174.255°

69. −18.515° **70.** −25.485° **71.** 31.4296° **72.** 59.0854°

73. 89.9004° **74.** 102.3771° **75.** 178.5994° **76.** 122.6853°

Find the angle of least positive measure (not equal to the given measure) that is coterminal with each angle. **See Example 5.**

77. 32° **78.** 86° **79.** 26° 30′ **80.** 58° 40′

81. −40° **82.** −98° **83.** −125° 30′ **84.** −203° 20′

85. 361° **86.** 541° **87.** −361° **88.** −541°

89. 539° **90.** 699° **91.** 850° **92.** 1000°

93. 5280° **94.** 8440° **95.** −5280° **96.** −8440°

Give two positive and two negative angles that are coterminal with the given quadrantal angle.

97. 90° **98.** 180° **99.** 0° **100.** 270°

Write an expression that generates all angles coterminal with each angle. Let n represent any integer.

101. 30° **102.** 45° **103.** 135° **104.** 225°

105. −90° **106.** −180° **107.** 0° **108.** 360°

109. Why do the answers to **Exercises 107 and 108** give the same set of angles?

110. *Concept Check* Which two of the following are not coterminal with $r°$?

A. $360° + r°$ **B.** $r° - 360°$ **C.** $360° - r°$ **D.** $r° + 180°$

Concept Check Sketch each angle in standard position. Draw an arrow representing the correct amount of rotation. Find the measure of two other angles, one positive and one negative, that are coterminal with the given angle. Give the quadrant of each angle, if applicable.

111. 75° **112.** 89° **113.** 174° **114.** 234°

115. 300° **116.** 512° **117.** −61° **118.** −159°

119. 90° **120.** 180° **121.** −90° **122.** −180°

Solve each problem. **See Example 6.**

123. *Revolutions of a Turntable* A turntable in a shop makes 45 revolutions per min. How many revolutions does it make per second?

124. *Revolutions of a Windmill* A windmill makes 90 revolutions per min. How many revolutions does it make per second?

125. *Rotating Tire* A tire is rotating 600 times per min. Through how many degrees does a point on the edge of the tire move in $\frac{1}{2}$ sec?

126. *Rotating Airplane Propeller* An airplane propeller rotates 1000 times per min. Find the number of degrees that a point on the edge of the propeller will rotate in 2 sec.

127. *Rotating Pulley* A pulley rotates through 75° in 1 min. How many rotations does the pulley make in 1 hr?

128. ***Surveying*** One student in a surveying class measures an angle as 74.25°, while another student measures the same angle as 74° 20′. Find the difference between these measurements, both to the nearest minute and to the nearest hundredth of a degree.

129. ***Viewing Field of a Telescope*** As a consequence of Earth's rotation, celestial objects such as the moon and the stars appear to move across the sky, rising in the east and setting in the west. As a result, if a telescope on Earth remains stationary while viewing a celestial object, the object will slowly move outside the viewing field of the telescope. For this reason, a motor is often attached to telescopes so that the telescope rotates at the same rate as Earth. Determine how long it should take the motor to turn the telescope through an angle of 1 min in a direction perpendicular to Earth's axis.

130. ***Angle Measure of a Star on the American Flag*** Determine the measure of the angle in each point of the five-pointed star appearing on the American flag. (*Hint:* Inscribe the star in a circle, and use the following theorem from geometry: *An angle whose vertex lies on the circumference of a circle is equal to half the central angle that cuts off the same arc.* See the figure.)

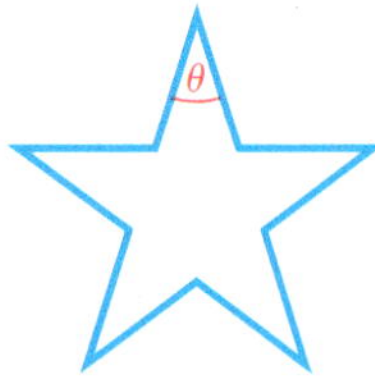

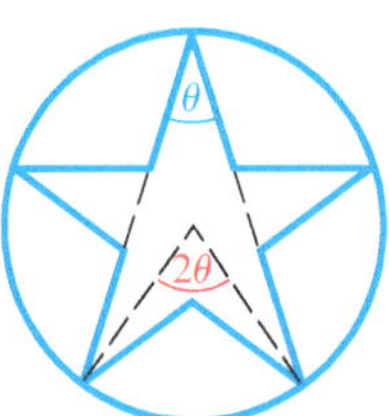

5.2 Trigonometric Functions

- Trigonometric Functions
- Quadrantal Angles
- Reciprocal Identities
- Signs and Ranges of Function Values
- Pythagorean Identities
- Quotient Identities

Trigonometric Functions

To define the six **trigonometric functions,** we start with an angle θ in standard position and choose any point P having coordinates (x, y) on the terminal side of angle θ. (The point P must not be the vertex of the angle.) See **Figure 13.** A perpendicular from P to the x-axis at point Q determines a right triangle, having vertices at O, P, and Q. We find the distance r from $P(x, y)$ to the origin, $(0, 0)$, using the distance formula.

$$d(O, P) = \sqrt{(x_2 - x_1)^2 + (y_2 - y_1)^2} \quad \text{Distance formula}$$

$$r = \sqrt{(x - 0)^2 + (y - 0)^2} \quad \text{Substitute } (x, y) \text{ for } (x_2, y_2) \text{ and } (0, 0) \text{ for } (x_1, y_1).$$

$$r = \sqrt{x^2 + y^2} \quad \text{Subtract.}$$

Notice that $r > 0$ because this is the undirected distance.

The six trigonometric functions of angle θ are

sine, cosine, tangent, cotangent, secant, and **cosecant,**

abbreviated **sin, cos, tan, cot, sec,** and **csc.**

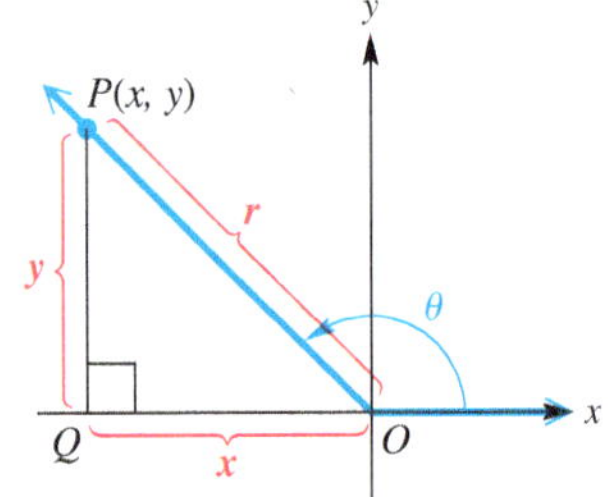

Figure 13

Trigonometric Functions

Let (x, y) be a point other than the origin on the terminal side of an angle θ in standard position. The distance from the point to the origin is $r = \sqrt{x^2 + y^2}$. The six trigonometric functions of θ are defined as follows.

$$\sin\theta = \frac{y}{r} \qquad \cos\theta = \frac{x}{r} \qquad \tan\theta = \frac{y}{x} \quad (x \neq 0)$$

$$\csc\theta = \frac{r}{y} \quad (y \neq 0) \qquad \sec\theta = \frac{r}{x} \quad (x \neq 0) \qquad \cot\theta = \frac{x}{y} \quad (y \neq 0)$$

EXAMPLE 1 Finding Function Values of an Angle

The terminal side of an angle θ in standard position passes through the point $(8, 15)$. Find the values of the six trigonometric functions of angle θ.

SOLUTION **Figure 14** shows angle θ and the triangle formed by dropping a perpendicular from the point $(8, 15)$ to the x-axis. The point $(8, 15)$ is 8 units to the right of the y-axis and 15 units above the x-axis, so $x = 8$ and $y = 15$. Now use $r = \sqrt{x^2 + y^2}$.

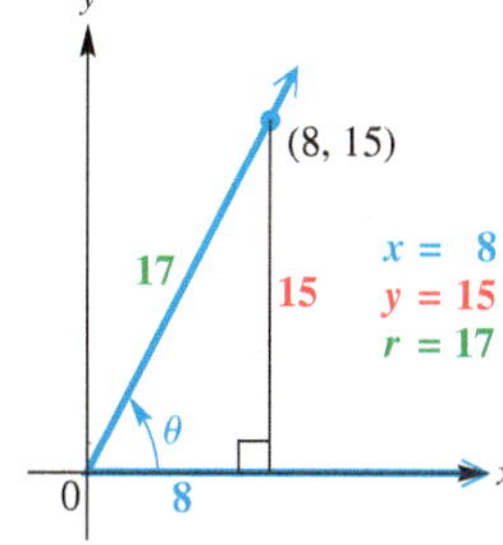

Figure 14

$$r = \sqrt{8^2 + 15^2} = \sqrt{64 + 225} = \sqrt{289} = 17$$

We can now use these values for x, y, and r to find the values of the six trigonometric functions of angle θ.

$$\sin\theta = \frac{y}{r} = \frac{15}{17} \qquad \cos\theta = \frac{x}{r} = \frac{8}{17} \qquad \tan\theta = \frac{y}{x} = \frac{15}{8}$$

$$\csc\theta = \frac{r}{y} = \frac{17}{15} \qquad \sec\theta = \frac{r}{x} = \frac{17}{8} \qquad \cot\theta = \frac{x}{y} = \frac{8}{15}$$

✔ **Now Try Exercise 13.**

EXAMPLE 2 Finding Function Values of an Angle

The terminal side of an angle θ in standard position passes through the point $(-3, -4)$. Find the values of the six trigonometric functions of angle θ.

SOLUTION As shown in **Figure 15,** $x = -3$ and $y = -4$.

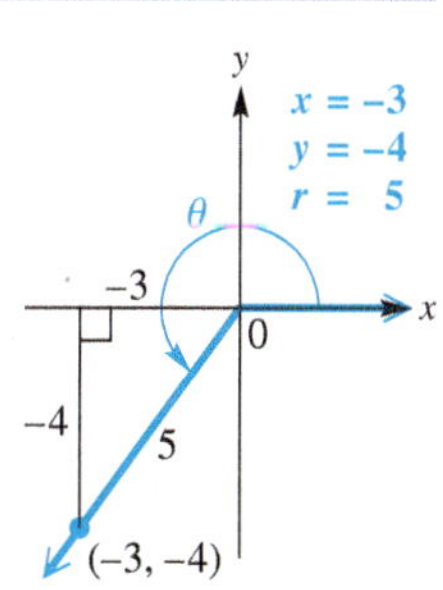

Figure 15

$$r = \sqrt{(-3)^2 + (-4)^2} \qquad r = \sqrt{x^2 + y^2}$$

$$r = \sqrt{25} \qquad \text{Simplify the radicand.}$$

$$r = 5 \qquad r > 0$$

Now we use the definitions of the trigonometric functions.

$$\sin\theta = \frac{-4}{5} = -\frac{4}{5} \qquad \cos\theta = \frac{-3}{5} = -\frac{3}{5} \qquad \tan\theta = \frac{-4}{-3} = \frac{4}{3}$$

$$\csc\theta = \frac{5}{-4} = -\frac{5}{4} \qquad \sec\theta = \frac{5}{-3} = -\frac{5}{3} \qquad \cot\theta = \frac{-3}{-4} = \frac{3}{4}$$

✔ **Now Try Exercise 17.**

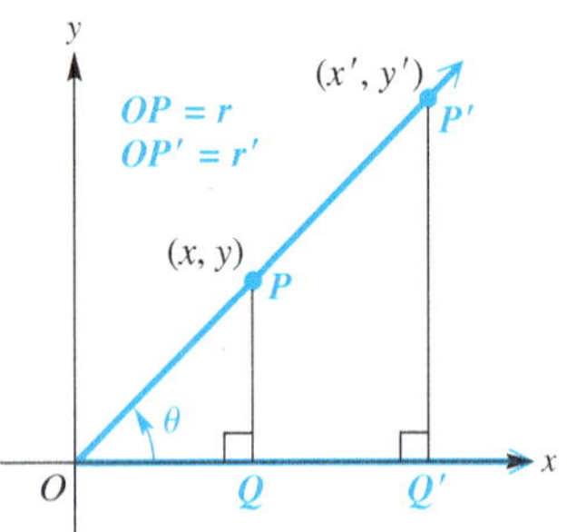

Figure 16

We can find the six trigonometric functions using *any* point other than the origin on the terminal side of an angle. To see why any point can be used, refer to **Figure 16,** which shows an angle θ and two distinct points on its terminal side. Point P has coordinates (x, y), and point P' (read "**P-prime**") has coordinates (x', y'). Let r be the length of the hypotenuse of triangle OPQ, and let r' be the length of the hypotenuse of triangle $OP'Q'$. Because corresponding sides of similar triangles are proportional, we have

$$\frac{y}{r} = \frac{y'}{r'}. \quad \text{Corresponding sides are proportional.}$$

Thus $\sin \theta = \frac{y}{r}$ is the same no matter which point is used to find it. A similar result holds for the other five trigonometric functions.

We can also find the trigonometric function values of an angle if we know the equation of the line coinciding with the terminal ray. Recall that the graph of the equation

$$Ax + By = 0 \quad \text{Linear equation in two variables}$$

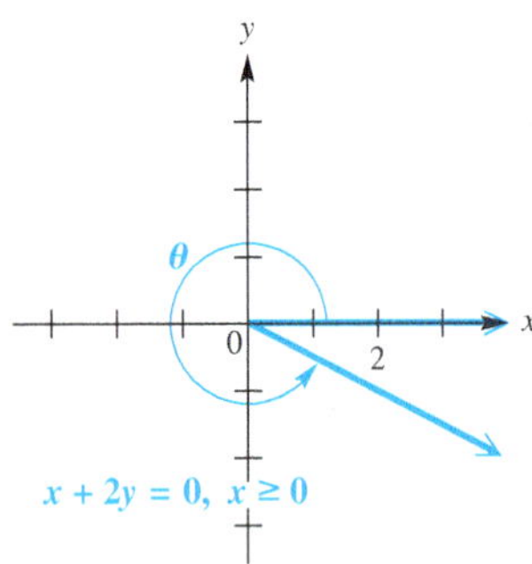

Figure 17

is a line that passes through the origin $(0, 0)$. If we restrict x to have only nonpositive or only nonnegative values, we obtain as the graph a ray with endpoint at the origin. For example, the graph of $x + 2y = 0$, $x \geq 0$, shown in **Figure 17,** is a ray that can serve as the terminal side of an angle θ in standard position. By choosing a point on the ray, we can find the trigonometric function values of the angle.

EXAMPLE 3 Finding Function Values of an Angle

Find the six trigonometric function values of an angle θ in standard position, if the terminal side of θ is defined by $x + 2y = 0$, $x \geq 0$.

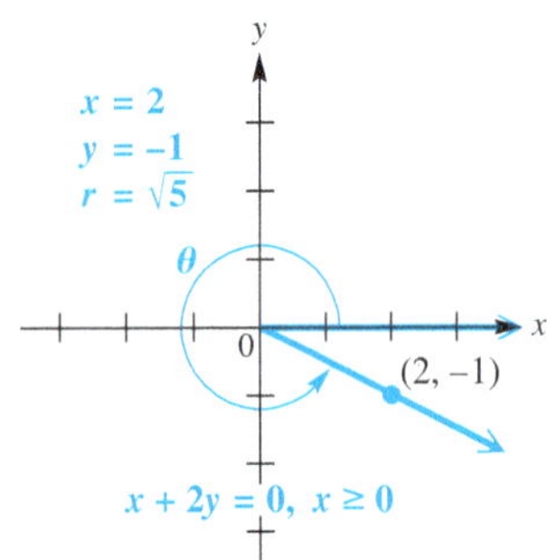

Figure 18

SOLUTION The angle is shown in **Figure 18.** We can use *any* point except $(0, 0)$ on the terminal side of θ to find the trigonometric function values. We choose $x = 2$ and find the corresponding y-value.

$$x + 2y = 0, \quad x \geq 0$$
$$2 + 2y = 0 \quad \text{Let } x = 2.$$
$$2y = -2 \quad \text{Subtract 2.}$$
$$y = -1 \quad \text{Divide by 2.}$$

The point $(2, -1)$ lies on the terminal side, and so

$$r = \sqrt{2^2 + (-1)^2} = \sqrt{5}.$$

Now we use the definitions of the trigonometric functions.

$$\sin \theta = \frac{y}{r} = \frac{-1}{\sqrt{5}} = \frac{-1}{\sqrt{5}} \cdot \frac{\sqrt{5}}{\sqrt{5}} = -\frac{\sqrt{5}}{5}$$

$$\cos \theta = \frac{x}{r} = \frac{2}{\sqrt{5}} = \frac{2}{\sqrt{5}} \cdot \frac{\sqrt{5}}{\sqrt{5}} = \frac{2\sqrt{5}}{5}$$

Multiply by $\frac{\sqrt{5}}{\sqrt{5}}$, a form of 1, to rationalize the denominators.

$$\tan \theta = \frac{y}{x} = \frac{-1}{2} = -\frac{1}{2}$$

$$\csc \theta = \frac{r}{y} = \frac{\sqrt{5}}{-1} = -\sqrt{5} \qquad \sec \theta = \frac{r}{x} = \frac{\sqrt{5}}{2} \qquad \cot \theta = \frac{x}{y} = \frac{2}{-1} = -2$$

✔ **Now Try Exercise 43.**

Recall that when the equation of a line is written in the form

$$y = mx + b, \quad \text{Slope-intercept form}$$

the coefficient m of x gives the slope of the line. In **Example 3,** the equation $x + 2y = 0$ can be written as $y = -\frac{1}{2}x$, so the slope of this line is $-\frac{1}{2}$. Notice that $\tan \theta = -\frac{1}{2}$.

In general, it is true that m = tan θ.

NOTE The trigonometric function values we found in **Examples 1–3** are *exact.* If we were to use a calculator to approximate these values, the decimal results would not be acceptable if exact values were required.

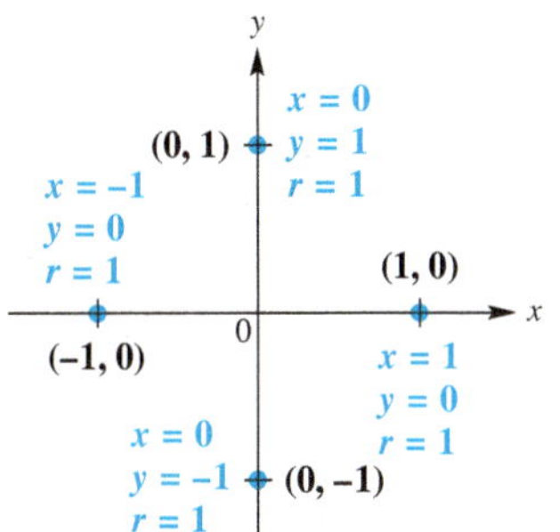

Figure 19

Quadrantal Angles

If the terminal side of an angle in standard position lies along the y-axis, any point on this terminal side has x-coordinate 0. Similarly, an angle with terminal side on the x-axis has y-coordinate 0 for any point on the terminal side. Because the values of x and y appear in the denominators of some trigonometric functions, and because a fraction is undefined if its denominator is 0, some trigonometric function values of quadrantal angles (i.e., those with terminal side on an axis) are undefined.

When determining trigonometric function values of quadrantal angles, **Figure 19** can help find the ratios. Because *any* point on the terminal side can be used, it is convenient to choose the point one unit from the origin, with $r = 1$. (Later we will extend this idea to the *unit circle.*)

To find the function values of a quadrantal angle, determine the position of the terminal side, choose the one of these four points that lies on this terminal side, and then use the definitions involving x, y, and r.

EXAMPLE 4 Finding Function Values of Quadrantal Angles

Find the values of the six trigonometric functions for each angle.

(a) an angle of 90°

(b) an angle θ in standard position with terminal side passing through $(-3, 0)$

SOLUTION

(a) **Figure 20** shows that the terminal side passes through $(0, 1)$. So $x = 0$, $y = 1$, and $r = 1$. Thus, we have the following.

$$\sin 90^\circ = \frac{1}{1} = 1 \qquad \cos 90^\circ = \frac{0}{1} = 0 \qquad \tan 90^\circ = \frac{1}{0} \quad \text{Undefined}$$

$$\csc 90^\circ = \frac{1}{1} = 1 \qquad \sec 90^\circ = \frac{1}{0} \quad \text{Undefined} \qquad \cot 90^\circ = \frac{0}{1} = 0$$

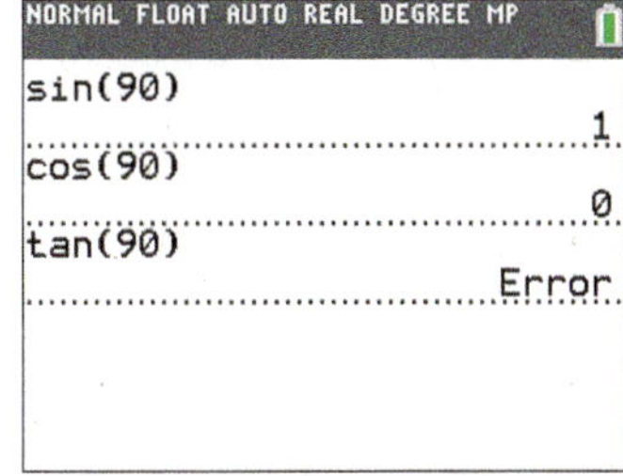

A calculator in degree mode returns the correct values for sin 90° and cos 90°. The screen shows an ERROR message for tan 90°, because 90° is not in the domain of the tangent function.

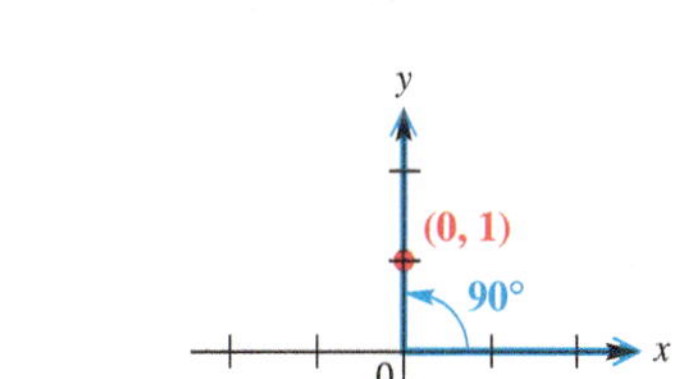

Figure 20

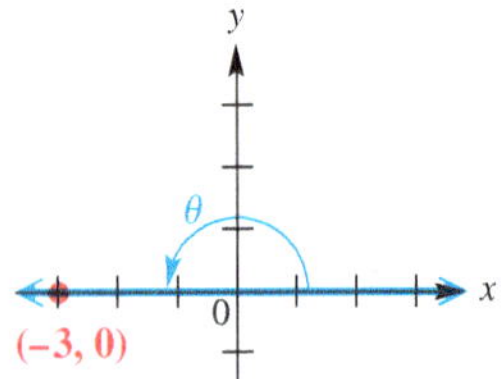

Figure 21

(b) **Figure 21** shows the angle. Here,

$$x = -3, \quad y = 0, \quad \text{and} \quad r = 3,$$

so the trigonometric functions have the following values.

$$\sin\theta = \frac{0}{3} = 0 \qquad \cos\theta = \frac{-3}{3} = -1 \qquad \tan\theta = \frac{0}{-3} = 0$$

$$\csc\theta = \frac{3}{0} \quad \text{Undefined} \qquad \sec\theta = \frac{3}{-3} = -1 \qquad \cot\theta = \frac{-3}{0} \quad \text{Undefined}$$

Verify that these values can also be found using the point $(-1, 0)$.

✔ **Now Try Exercises 21, 55, 57, and 59.**

The conditions under which the trigonometric function values of quadrantal angles are undefined are summarized here.

Conditions for Undefined Function Values

Identify the terminal side of a quadrantal angle.

- If the terminal side of the quadrantal angle lies along the y-axis, then the tangent and secant functions are undefined.
- If the terminal side of the quadrantal angle lies along the x-axis, then the cotangent and cosecant functions are undefined.

The function values of some commonly used quadrantal angles, 0°, 90°, 180°, 270°, and 360°, are summarized in the table. They can be determined when needed using **Figure 19** on the previous page and the method of **Example 4(a).**

For other quadrantal angles such as −90°, −270°, and 450°, first determine the coterminal angle that lies between 0° and 360°, and then refer to the table entries for that particular angle. For example, the function values of a −90° angle would correspond to those of a 270° angle.

Function Values of Quadrantal Angles

θ	$\sin\theta$	$\cos\theta$	$\tan\theta$	$\cot\theta$	$\sec\theta$	$\csc\theta$
0°	0	1	0	Undefined	1	Undefined
90°	1	0	Undefined	0	Undefined	1
180°	0	−1	0	Undefined	−1	Undefined
270°	−1	0	Undefined	0	Undefined	−1
360°	0	1	0	Undefined	1	Undefined

The values given in this table can be found with a calculator that has trigonometric function keys. ***Make sure the calculator is set to degree mode.***

NORMAL FLOAT AUTO REAL DEGREE MP
MATHPRINT CLASSIC
NORMAL SCI ENG
FLOAT 0 1 2 3 4 5 6 7 8 9
RADIAN DEGREE
FUNCTION PARAMETRIC POLAR SEQ
THICK DOT-THICK THIN DOT-THIN
SEQUENTIAL SIMUL
REAL a+bi re^(θi)
FULL HORIZONTAL GRAPH-TABLE

TI-84 Plus

Figure 22

CAUTION ***One of the most common errors involving calculators in trigonometry occurs when the calculator is set for radian measure, rather than degree measure.*** Be sure to set your calculator to degree mode. See **Figure 22.**

Reciprocal Identities An identity is an equation that is true for all values of the variables for which all expressions are defined.

$$(x+y)^2 = x^2 + 2xy + y^2 \qquad 2(x+3) = 2x + 6 \qquad \text{Identities}$$

Recall the definition of a reciprocal.

The **reciprocal** of a nonzero number x is $\frac{1}{x}$.

Examples: The reciprocal of 2 is $\frac{1}{2}$, and the reciprocal of $\frac{8}{11}$ is $\frac{11}{8}$. There is no reciprocal for 0 because $\frac{1}{0}$ is undefined.

The definitions of the trigonometric functions given earlier in this section were written so that functions in the same column were reciprocals of each other. Because $\sin\theta = \frac{y}{r}$ and $\csc\theta = \frac{r}{y}$,

$$\sin\theta = \frac{1}{\csc\theta} \quad \text{and} \quad \csc\theta = \frac{1}{\sin\theta}, \quad \text{provided } \sin\theta \neq 0.$$

Also, $\cos\theta$ and $\sec\theta$ are reciprocals, as are $\tan\theta$ and $\cot\theta$. The **reciprocal identities** hold for any angle θ that does not lead to a 0 denominator.

Reciprocal Identities

For all angles θ for which both functions are defined, the following identities hold.

$$\sin\theta = \frac{1}{\csc\theta} \qquad \cos\theta = \frac{1}{\sec\theta} \qquad \tan\theta = \frac{1}{\cot\theta}$$

$$\csc\theta = \frac{1}{\sin\theta} \qquad \sec\theta = \frac{1}{\cos\theta} \qquad \cot\theta = \frac{1}{\tan\theta}$$

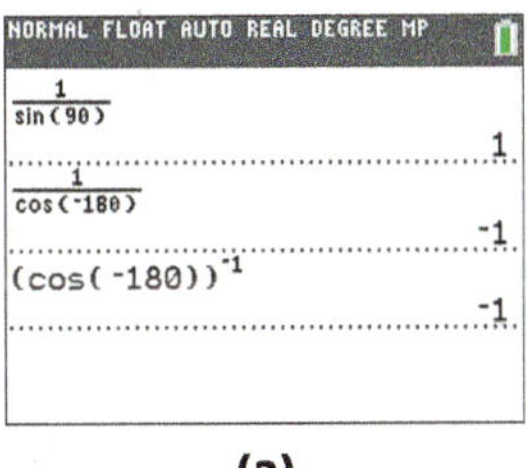

(a)

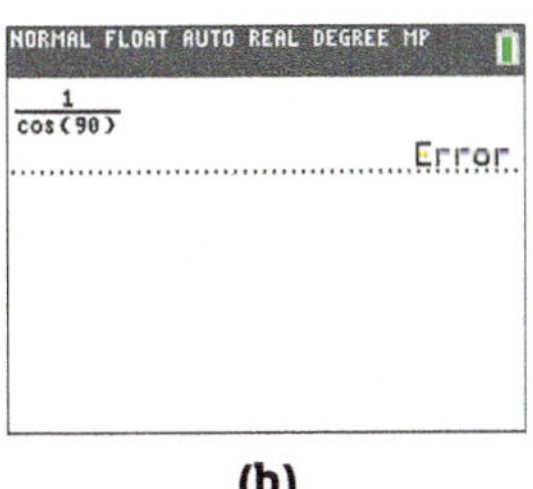

(b)

NORMAL FLOAT AUTO REAL DEGREE MP
ERROR: DIVIDE BY 0
1:Quit
2:Goto
Attempted calculation contains division by 0. Calculation fails.

(c)

Figure 23

The screen in **Figure 23(a)** shows that $\csc 90° = 1$ and $\sec(-180°) = -1$ using appropriate reciprocal identities. The third entry uses the reciprocal function key x^{-1} to evaluate $\sec(-180°)$. **Figure 23(b)** shows that attempting to find $\sec 90°$ by entering $\frac{1}{\cos 90°}$ produces an ERROR message, indicating that the reciprocal is undefined. See **Figure 23(c).** ■

CAUTION ***Be sure not to use the inverse trigonometric function keys to find reciprocal function values.*** For example, consider the following.

$$\cos^{-1}(-180°) \neq (\cos(-180°))^{-1}$$

This is the *inverse cosine function,* which will be discussed later in the text.

This is the *reciprocal function,* which correctly evaluates $\sec(-180°)$, as seen in **Figure 23(a).**

$$(\cos(-180°))^{-1} = \frac{1}{\cos(-180°)} = \sec(-180°)$$

The reciprocal identities can be written in different forms. For example,

$$\sin\theta = \frac{1}{\csc\theta} \quad \text{is equivalent to} \quad \csc\theta = \frac{1}{\sin\theta} \quad \text{and} \quad (\sin\theta)(\csc\theta) = 1.$$

EXAMPLE 5 Using the Reciprocal Identities

Find each function value.

(a) $\cos\theta$, given that $\sec\theta = \frac{5}{3}$

(b) $\sin\theta$, given that $\csc\theta = -\frac{\sqrt{12}}{2}$

SOLUTION

(a) We use the fact that $\cos\theta$ is the reciprocal of $\sec\theta$.

$$\cos\theta = \frac{1}{\sec\theta} = \frac{1}{\frac{5}{3}} = 1 \div \frac{5}{3} = 1 \cdot \frac{3}{5} = \frac{3}{5} \qquad \text{Simplify the complex fraction.}$$

(b)

$$\sin\theta = \frac{1}{\csc\theta} \qquad \sin\theta \text{ is the reciprocal of } \csc\theta.$$

$$= \frac{1}{-\frac{\sqrt{12}}{2}} \qquad \text{Substitute } \csc\theta = -\frac{\sqrt{12}}{2}.$$

$$= -\frac{2}{\sqrt{12}} \qquad \text{Simplify the complex fraction as in part (a).}$$

$$= -\frac{2}{2\sqrt{3}} \qquad \sqrt{12} = \sqrt{4\cdot 3} = 2\sqrt{3}$$

$$= -\frac{1}{\sqrt{3}} \qquad \text{Divide out the common factor 2.}$$

$$= -\frac{1}{\sqrt{3}} \cdot \frac{\sqrt{3}}{\sqrt{3}} \qquad \text{Rationalize the denominator.}$$

$$= -\frac{\sqrt{3}}{3} \qquad \text{Multiply.}$$

✔ **Now Try Exercises 87 and 93.**

Signs and Ranges of Function Values In the definitions of the trigonometric functions, r is the distance from the origin to the point (x, y). This distance is undirected, so $r > 0$. If we choose a point (x, y) in quadrant I, then both x and y will be positive, and the values of all six functions will be positive.

A point (x, y) in quadrant II satisfies $x < 0$ and $y > 0$. This makes the values of sine and cosecant positive for quadrant II angles, while the other four functions take on negative values. Similar results can be obtained for the other quadrants.

This important information is summarized here.

Signs of Trigonometric Function Values

θ in Quadrant	$\sin\theta$	$\cos\theta$	$\tan\theta$	$\cot\theta$	$\sec\theta$	$\csc\theta$
I	+	+	+	+	+	+
II	+	−	−	−	−	+
III	−	−	+	+	−	−
IV	−	+	−	−	+	−

NOTE Because numbers that are reciprocals always have the same sign, the sign of a function value automatically determines the sign of the reciprocal function value.

EXAMPLE 6 Determining Signs of Functions of Nonquadrantal Angles

Determine the signs of the trigonometric functions of an angle in standard position with the given measure.

(a) $87°$ **(b)** $300°$ **(c)** $-200°$

SOLUTION

(a) An angle of $87°$ is in the first quadrant, with x, y, and r all positive, so all of its trigonometric function values are positive.

(b) A $300°$ angle is in quadrant IV, so the cosine and secant are positive, while the sine, cosecant, tangent, and cotangent are negative.

(c) A $-200°$ angle is in quadrant II. The sine and cosecant are positive, and all other function values are negative.

✔ **Now Try Exercises 97, 99, and 103.**

EXAMPLE 7 Identifying the Quadrant of an Angle

Identify the quadrant (or possible quadrants) of an angle θ that satisfies the given conditions.

(a) $\sin \theta > 0, \tan \theta < 0$ **(b)** $\cos \theta < 0, \sec \theta < 0$

SOLUTION

(a) Because $\sin \theta > 0$ in quadrants I and II and $\tan \theta < 0$ in quadrants II and IV, both conditions are met only in quadrant II.

(b) The cosine and secant functions are both negative in quadrants II and III, so in this case θ could be in either of these two quadrants.

✔ **Now Try Exercises 113 and 119.**

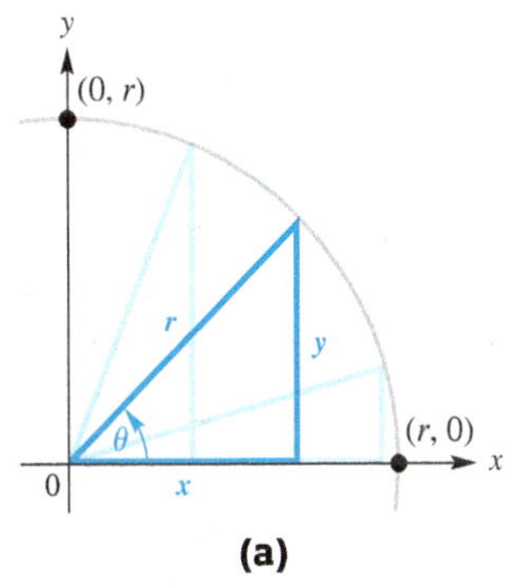

(a)

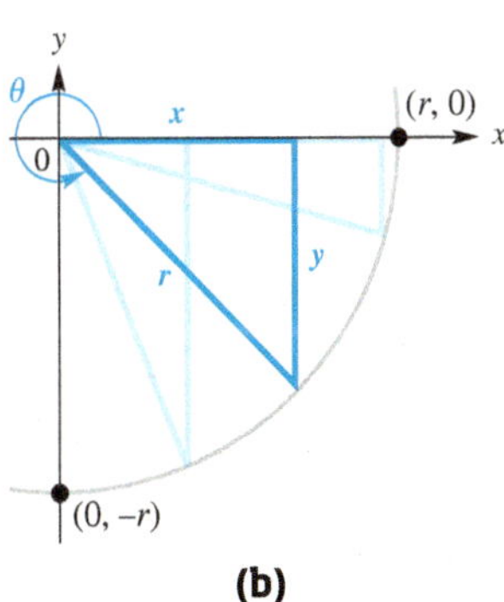

(b)

Figure 24

Figure 24(a) shows an angle θ as it increases in measure from near $0°$ toward $90°$. In each case, the value of r is the same. As the measure of the angle increases, y increases but never exceeds r, so $y \leq r$. Dividing both sides by the positive number r gives $\frac{y}{r} \leq 1$.

In a similar way, angles in quadrant IV as in **Figure 24(b)** suggest that

$$-1 \leq \frac{y}{r},$$

so

$$-1 \leq \frac{y}{r} \leq 1$$

and

$$\mathbf{-1 \leq \sin \theta \leq 1.} \quad \frac{y}{r} = \sin \theta \text{ for any angle } \theta.$$

Similarly,

$$\mathbf{-1 \leq \cos \theta \leq 1.} \quad \frac{x}{r} = \cos \theta \text{ for any angle } \theta.$$

The tangent of an angle is defined as $\frac{y}{x}$. It is possible that $x < y$, $x = y$, or $x > y$. Thus, $\frac{y}{x}$ can take any value, so

$\tan \theta$ can be any real number, as can $\cot \theta$.

The functions $\sec \theta$ and $\csc \theta$ are reciprocals of the functions $\cos \theta$ and $\sin \theta$, respectively, making

$$\sec \theta \leq -1 \text{ or } \sec \theta \geq 1 \text{ and } \csc \theta \leq -1 \text{ or } \csc \theta \geq 1.$$

In summary, the ranges of the trigonometric functions are as follows.

Ranges of Trigonometric Functions

Trigonometric Function of θ	Range (Set-Builder Notation)	Range (Interval Notation)
$\sin\theta$, $\cos\theta$	$\{y \mid \lvert y\rvert \le 1\}$	$[-1, 1]$
$\tan\theta$, $\cot\theta$	$\{y \mid y \text{ is a real number}\}$	$(-\infty, \infty)$
$\sec\theta$, $\csc\theta$	$\{y \mid \lvert y\rvert \ge 1\}$	$(-\infty, -1] \cup [1, \infty)$

EXAMPLE 8 Determining Whether a Value Is in the Range of a Trigonometric Function

Determine whether each statement is *possible* or *impossible*.

(a) $\sin\theta = 2.5$ **(b)** $\tan\theta = 110.47$ **(c)** $\sec\theta = 0.6$

SOLUTION

(a) For any value of θ, we know that

$$-1 \le \sin\theta \le 1.$$

Here $2.5 > 1$, so it is impossible to find a value of θ that satisfies $\sin\theta = 2.5$.

(b) The tangent function can take on any real number value. Thus, $\tan\theta = 110.47$ is possible.

(c) Because $\lvert \sec\theta \rvert \ge 1$ for all θ for which the secant is defined, the statement $\sec\theta = 0.6$ is impossible.

✔ **Now Try Exercises 121, 125, and 127.**

The six trigonometric functions are defined in terms of x, y, and r, where the Pythagorean theorem shows that

$$r^2 = x^2 + y^2 \quad \text{and} \quad r > 0.$$

With these relationships, knowing the value of only one function and the quadrant in which the angle lies makes it possible to find the values of the other trigonometric functions.

EXAMPLE 9 Finding All Function Values Given One Value and the Quadrant

Suppose that angle θ is in quadrant II and $\sin\theta = \frac{2}{3}$. Find the values of the five remaining trigonometric functions.

SOLUTION Choose any point on the terminal side of angle θ. For simplicity, since $\sin\theta = \frac{y}{r}$, choose the point with $r = 3$.

$$\sin\theta = \frac{2}{3} \quad \text{Given value}$$

$$\frac{y}{r} = \frac{2}{3} \quad \text{Substitute } \tfrac{y}{r} \text{ for } \sin\theta.$$

Because $\frac{y}{r} = \frac{2}{3}$ and $r = 3$, it follows that $y = 2$. We must find the value of x.

$$x^2 + y^2 = r^2 \quad \text{Pythagorean theorem}$$

$$x^2 + 2^2 = 3^2 \quad \text{Substitute.}$$

$$x^2 + 4 = 9 \quad \text{Apply exponents.}$$

$$x^2 = 5 \quad \text{Subtract 4.}$$

Remember *both* roots.

$$x = \sqrt{5} \quad \text{or} \quad x = -\sqrt{5} \quad \text{Square root property: If } x^2 = k, \text{ then } x = \sqrt{k} \text{ or } x = -\sqrt{k}.$$

Because θ is in quadrant II, x must be negative. Choose $x = -\sqrt{5}$ so that the point $(-\sqrt{5}, 2)$ is on the terminal side of θ. See **Figure 25.**

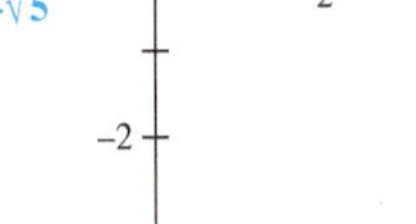

Figure 25

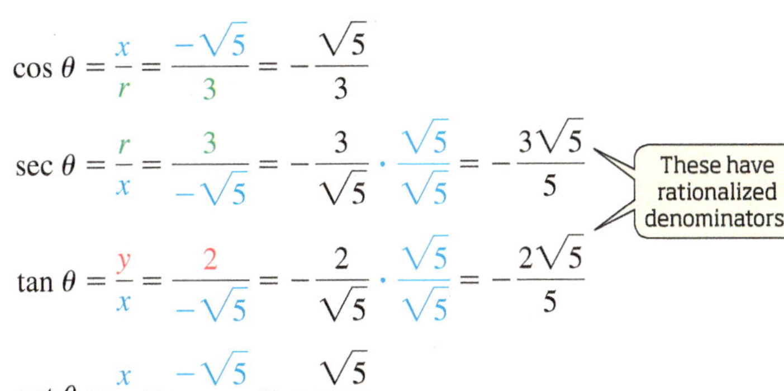

$$\cos\theta = \frac{x}{r} = \frac{-\sqrt{5}}{3} = -\frac{\sqrt{5}}{3}$$

$$\sec\theta = \frac{r}{x} = \frac{3}{-\sqrt{5}} = -\frac{3}{\sqrt{5}} \cdot \frac{\sqrt{5}}{\sqrt{5}} = -\frac{3\sqrt{5}}{5}$$

$$\tan\theta = \frac{y}{x} = \frac{2}{-\sqrt{5}} = -\frac{2}{\sqrt{5}} \cdot \frac{\sqrt{5}}{\sqrt{5}} = -\frac{2\sqrt{5}}{5}$$

These have rationalized denominators.

$$\cot\theta = \frac{x}{y} = \frac{-\sqrt{5}}{2} = -\frac{\sqrt{5}}{2}$$

$$\csc\theta = \frac{r}{y} = \frac{3}{2}$$

✔ **Now Try Exercise 141.**

Pythagorean Identities

We now derive three new identities.

$$x^2 + y^2 = r^2 \quad \text{Pythagorean theorem}$$

$$\frac{x^2}{r^2} + \frac{y^2}{r^2} = \frac{r^2}{r^2} \quad \text{Divide by } r^2.$$

$$\left(\frac{x}{r}\right)^2 + \left(\frac{y}{r}\right)^2 = 1 \quad \text{Power rule for exponents; } \frac{a^m}{b^m} = \left(\frac{a}{b}\right)^m$$

$(\cos\theta)^2$ and $\cos^2\theta$ are equivalent forms.

$$(\cos\theta)^2 + (\sin\theta)^2 = 1 \quad \cos\theta = \frac{x}{r},\ \sin\theta = \frac{y}{r}$$

$$\mathbf{\sin^2\theta + \cos^2\theta = 1} \quad \text{Apply exponents; commutative property}$$

Starting again with $x^2 + y^2 = r^2$ and dividing through by x^2 gives the following.

$$\frac{x^2}{x^2} + \frac{y^2}{x^2} = \frac{r^2}{x^2} \quad \text{Divide by } x^2.$$

$$1 + \left(\frac{y}{x}\right)^2 = \left(\frac{r}{x}\right)^2 \quad \text{Power rule for exponents}$$

$$1 + (\tan\theta)^2 = (\sec\theta)^2 \quad \tan\theta = \frac{y}{x},\ \sec\theta = \frac{r}{x}$$

$$\mathbf{\tan^2\theta + 1 = \sec^2\theta} \quad \text{Apply exponents; commutative property}$$

Similarly, dividing through by y^2 leads to another identity.

$$\mathbf{1 + \cot^2\theta = \csc^2\theta}$$

These three identities are the **Pythagorean identities** because the original equation that led to them, $x^2 + y^2 = r^2$, comes from the Pythagorean theorem.

Pythagorean Identities

For all angles θ for which the function values are defined, the following identities hold.

$$\sin^2\theta + \cos^2\theta = 1 \qquad \tan^2\theta + 1 = \sec^2\theta \qquad 1 + \cot^2\theta = \csc^2\theta$$

We give only one form of each identity. However, algebraic transformations produce equivalent forms. For example, by subtracting $\sin^2\theta$ from both sides of $\sin^2\theta + \cos^2\theta = 1$, we obtain an equivalent identity.

$$\cos^2\theta = 1 - \sin^2\theta \quad \text{Alternative form}$$

It is important to be able to transform these identities quickly and also to recognize their equivalent forms.

LOOKING AHEAD TO CALCULUS

The reciprocal, Pythagorean, and quotient identities are used in calculus to find derivatives and integrals of trigonometric functions. A standard technique of integration called **trigonometric substitution** relies on the Pythagorean identities.

Quotient Identities Consider the quotient of the functions $\sin\theta$ and $\cos\theta$, for $\cos\theta \neq 0$.

$$\frac{\sin\theta}{\cos\theta} = \frac{\frac{y}{r}}{\frac{x}{r}} = \frac{y}{r} \div \frac{x}{r} = \frac{y}{r} \cdot \frac{r}{x} = \frac{y}{x} = \tan\theta$$

Similarly, $\frac{\cos\theta}{\sin\theta} = \cot\theta$, for $\sin\theta \neq 0$. Thus, we have the **quotient identities.**

Quotient Identities

For all angles θ for which the denominators are not zero, the following identities hold.

$$\frac{\sin\theta}{\cos\theta} = \tan\theta \qquad \frac{\cos\theta}{\sin\theta} = \cot\theta$$

EXAMPLE 10 Using Identities to Find Function Values

Find $\sin\theta$ and $\tan\theta$, given that $\cos\theta = -\frac{\sqrt{3}}{4}$ and $\sin\theta > 0$.

SOLUTION Start with the Pythagorean identity that includes $\cos\theta$.

$$\sin^2\theta + \cos^2\theta = 1 \quad \text{Pythagorean identity}$$

$$\sin^2\theta + \left(-\frac{\sqrt{3}}{4}\right)^2 = 1 \quad \text{Replace } \cos\theta \text{ with } -\tfrac{\sqrt{3}}{4}.$$

$$\sin^2\theta + \frac{3}{16} = 1 \quad \text{Square } -\tfrac{\sqrt{3}}{4}.$$

$$\sin^2\theta = \frac{13}{16} \quad \text{Subtract } \tfrac{3}{16}.$$

$$\sin\theta = \pm\frac{\sqrt{13}}{4} \quad \text{Take square roots.}$$

Choose the correct sign here.

$$\sin\theta = \frac{\sqrt{13}}{4} \quad \text{Choose the positive square root because } \sin\theta \text{ is positive.}$$

To find tan θ, use the values of cos θ and sin θ and the quotient identity for tan θ.

$$\tan\theta = \frac{\sin\theta}{\cos\theta} = \frac{\frac{\sqrt{13}}{4}}{-\frac{\sqrt{3}}{4}} = \frac{\sqrt{13}}{4}\left(-\frac{4}{\sqrt{3}}\right) = -\frac{\sqrt{13}}{\sqrt{3}}$$

$$= -\frac{\sqrt{13}}{\sqrt{3}} \cdot \frac{\sqrt{3}}{\sqrt{3}} = -\frac{\sqrt{39}}{3} \quad \text{Rationalize the denominator.}$$

✔ **Now Try Exercise 145.**

CAUTION ***In exercises like Examples 9 and 10, be careful to choose the correct sign when square roots are taken.*** Refer as needed to the diagrams preceding **Example 6** that summarize the signs of the functions.

EXAMPLE 11 Using Identities to Find Function Values

Find sin θ and cos θ, given that tan $\theta = \frac{4}{3}$ and θ is in quadrant III.

SOLUTION Because θ is in quadrant III, sin θ and cos θ will both be negative. It is tempting to say that since $\tan\theta = \frac{\sin\theta}{\cos\theta}$ and $\tan\theta = \frac{4}{3}$, then $\sin\theta = -4$ and $\cos\theta = -3$. This is *incorrect,* however—both sin θ and cos θ must be in the interval $[-1, 1]$.

We use the Pythagorean identity $\tan^2\theta + 1 = \sec^2\theta$ to find sec θ, and then the reciprocal identity $\cos\theta = \frac{1}{\sec\theta}$ to find cos θ.

$$\tan^2\theta + 1 = \sec^2\theta \quad \text{Pythagorean identity}$$

$$\left(\frac{4}{3}\right)^2 + 1 = \sec^2\theta \quad \tan\theta = \frac{4}{3}$$

$$\frac{16}{9} + 1 = \sec^2\theta \quad \text{Square } \tfrac{4}{3}.$$

$$\frac{25}{9} = \sec^2\theta \quad \text{Add.}$$

Be careful to choose the correct sign here.

$$-\frac{5}{3} = \sec\theta \quad \text{Choose the negative square root because sec } \theta \text{ is negative when } \theta \text{ is in quadrant III.}$$

$$-\frac{3}{5} = \cos\theta \quad \text{Secant and cosine are reciprocals.}$$

Now we use this value of cos θ to find sin θ.

$$\sin^2\theta = 1 - \cos^2\theta \quad \text{Pythagorean identity (alternative form)}$$

$$\sin^2\theta = 1 - \left(-\frac{3}{5}\right)^2 \quad \cos\theta = -\tfrac{3}{5}$$

$$\sin^2\theta = 1 - \frac{9}{25} \quad \text{Square } -\tfrac{3}{5}.$$

$$\sin^2\theta = \frac{16}{25} \quad \text{Subtract.}$$

Again, be careful.

$$\sin\theta = -\frac{4}{5} \quad \text{Choose the negative square root.}$$

✔ **Now Try Exercise 143.**

NOTE **Example 11** can also be worked by sketching θ in standard position in quadrant III, finding r to be 5, and then using the definitions of $\sin \theta$ and $\cos \theta$ in terms of x, y, and r. See **Figure 26.**

When using this method, be sure to choose the correct signs for x and y as determined by the quadrant in which the terminal side of θ lies. This is analogous to choosing the correct signs after applying the Pythagorean identities.

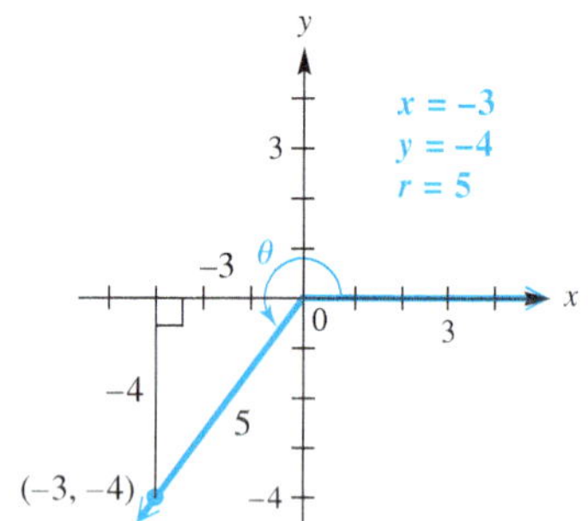

Figure 26

5.2 Exercises

CONCEPT PREVIEW *The terminal side of an angle θ in standard position passes through the point $(-3, -3)$. Use the figure to find the following values. Rationalize denominators when applicable.*

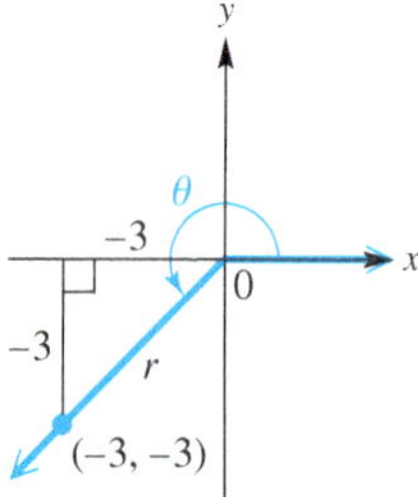

1.  r **2.** $\sin \theta$

3. $\cos \theta$ **4.** $\tan \theta$

CONCEPT PREVIEW *Determine whether each statement is* possible *or* impossible.

5. $\sin \theta = \frac{1}{2}, \ \csc \theta = 2$ **6.** $\tan \theta = 2, \ \cot \theta = -2$ **7.** $\sin \theta > 0, \ \csc \theta < 0$

8. $\cos \theta = 1.5$ **9.** $\cot \theta = -1.5$ **10.** $\sin^2 \theta + \cos^2 \theta = 2$

Sketch an angle θ in standard position such that θ has the least positive measure, and the given point is on the terminal side of θ. Then find the values of the six trigonometric functions for each angle. Rationalize denominators when applicable. ***See Examples 1, 2, and 4.***

11. $(5, -12)$ **12.** $(-12, -5)$ **13.** $(3, 4)$ **14.** $(-4, -3)$

15. $(-8, 15)$ **16.** $(15, -8)$ **17.** $(-7, -24)$ **18.** $(-24, -7)$

19. $(0, 2)$ **20.** $(0, 5)$ **21.** $(-4, 0)$ **22.** $(-5, 0)$

23. $(1, \sqrt{3})$ **24.** $(-1, \sqrt{3})$ **25.** $(-2\sqrt{3}, -2)$ **26.** $(-2\sqrt{3}, 2)$

Concept Check *Suppose that the point (x, y) is in the indicated quadrant. Determine whether the given ratio is positive or negative. Recall that $r = \sqrt{x^2 + y^2}$. (Hint: Drawing a sketch may help.)*

27. II, $\frac{x}{r}$ **28.** III, $\frac{y}{r}$ **29.** IV, $\frac{y}{x}$ **30.** IV, $\frac{x}{y}$

31. II, $\frac{y}{r}$ **32.** III, $\frac{x}{r}$ **33.** IV, $\frac{x}{r}$ **34.** IV, $\frac{y}{r}$

35. II, $\frac{x}{y}$ **36.** II, $\frac{y}{x}$ **37.** III, $\frac{y}{x}$ **38.** III, $\frac{x}{y}$

39. III, $\frac{r}{x}$ **40.** III, $\frac{r}{y}$ **41.** I, $\frac{x}{y}$ **42.** I, $\frac{y}{x}$

An equation of the terminal side of an angle θ in standard position is given with a restriction on x. Sketch the least positive such angle θ, and find the values of the six trigonometric functions of θ. **See Example 3.**

43. $2x + y = 0, x \geq 0$
44. $3x + 5y = 0, x \geq 0$
45. $-4x + 7y = 0, x \leq 0$
46. $6x - 5y = 0, x \geq 0$
47. $x + y = 0, x \geq 0$
48. $x - y = 0, x \geq 0$
49. $-\sqrt{3}x + y = 0, x \leq 0$
50. $\sqrt{3}x + y = 0, x \leq 0$

Find the indicated function value. If it is undefined, say so. **See Example 4.**

51. $\cos 90°$
52. $\sin 90°$
53. $\tan 180°$
54. $\cot 90°$
55. $\sec 180°$
56. $\csc 270°$
57. $\sin(-270°)$
58. $\cos(-90°)$
59. $\cot 540°$
60. $\tan 450°$
61. $\csc(-450°)$
62. $\sec(-540°)$
63. $\sin 1800°$
64. $\cos 1800°$
65. $\csc 1800°$
66. $\cot 1800°$
67. $\sec 1800°$
68. $\tan 1800°$
69. $\cos(-900°)$
70. $\sin(-900°)$

Use trigonometric function values of quadrantal angles to evaluate each expression. An expression such as $\cot^2 90°$ *means* $(\cot 90°)^2$, *which is equal to* $0^2 = 0$.

71. $\cos 90° + 3 \sin 270°$
72. $\tan 0° - 6 \sin 90°$
73. $3 \sec 180° - 5 \tan 360°$
74. $4 \csc 270° + 3 \cos 180°$
75. $\tan 360° + 4 \sin 180° + 5 \cos^2 180°$
76. $5 \sin^2 90° + 2 \cos^2 270° - \tan 360°$
77. $-2 \sin^4 0° + 3 \tan^2 0°$
78. $-3 \sin^4 90° + 4 \cos^3 180°$
79. $\sin^2(-90°) + \cos^2(-90°)$
80. $\cos^2(-180°) + \sin^2(-180°)$

If n is an integer, $n \cdot 180°$ *represents an integer multiple of* $180°$, $(2n + 1) \cdot 90°$ *represents an odd integer multiple of* $90°$, *and so on. Determine whether each expression is equal to* 0, 1, *or* -1, *or is* undefined.

81. $\cos[(2n + 1) \cdot 90°]$
82. $\sin[n \cdot 180°]$
83. $\tan[n \cdot 180°]$
84. $\sin[270° + n \cdot 360°]$
85. $\tan[(2n + 1) \cdot 90°]$
86. $\cot[n \cdot 180°]$

Use the appropriate reciprocal identity to find each function value. Rationalize denominators when applicable. **See Example 5.**

87. $\sec \theta$, given that $\cos \theta = \frac{2}{3}$
88. $\sec \theta$, given that $\cos \theta = \frac{5}{8}$
89. $\csc \theta$, given that $\sin \theta = -\frac{3}{7}$
90. $\csc \theta$, given that $\sin \theta = -\frac{8}{43}$
91. $\cot \theta$, given that $\tan \theta = 5$
92. $\cot \theta$, given that $\tan \theta = 18$
93. $\sin \theta$, given that $\csc \theta = \frac{\sqrt{8}}{2}$
94. $\sin \theta$, given that $\csc \theta = \frac{\sqrt{24}}{3}$
95. $\sin \theta$, given that $\csc \theta = 1.25$
96. $\cos \theta$, given that $\sec \theta = 8$

Determine the signs of the trigonometric functions of an angle in standard position with the given measure. **See Example 6.**

97. $74°$
98. $84°$
99. $218°$
100. $195°$
101. $178°$
102. $125°$
103. $-80°$
104. $-15°$
105. $855°$
106. $1005°$
107. $-345°$
108. $-640°$

Identify the quadrant (or possible quadrants) of an angle θ that satisfies the given conditions. ***See Example 7.***

109. $\sin \theta > 0, \ \csc \theta > 0$ **110.** $\cos \theta > 0, \ \sec \theta > 0$ **111.** $\cos \theta > 0, \ \sin \theta > 0$

112. $\sin \theta > 0, \ \tan \theta > 0$ **113.** $\tan \theta < 0, \ \cos \theta < 0$ **114.** $\cos \theta < 0, \ \sin \theta < 0$

115. $\sec \theta > 0, \ \csc \theta > 0$ **116.** $\csc \theta > 0, \ \cot \theta > 0$ **117.** $\sec \theta < 0, \ \csc \theta < 0$

118. $\cot \theta < 0, \ \sec \theta < 0$ **119.** $\sin \theta < 0, \ \csc \theta < 0$ **120.** $\tan \theta < 0, \ \cot \theta < 0$

Determine whether each statement is possible *or* impossible. ***See Example 8.***

121. $\sin \theta = 2$ **122.** $\sin \theta = 3$ **123.** $\cos \theta = -0.96$ **124.** $\cos \theta = -0.56$

125. $\tan \theta = 0.93$ **126.** $\cot \theta = 0.93$ **127.** $\sec \theta = -0.3$ **128.** $\sec \theta = -0.9$

129. $\csc \theta = 100$ **130.** $\csc \theta = -100$

Use identities to solve each of the following. Rationalize denominators when applicable. ***See Examples 9–11.***

131. Find $\cos \theta$, given that $\sin \theta = \frac{3}{5}$ and θ is in quadrant II.

132. Find $\sin \theta$, given that $\cos \theta = \frac{4}{5}$ and θ is in quadrant IV.

133. Find $\csc \theta$, given that $\cot \theta = -\frac{1}{2}$ and θ is in quadrant IV.

134. Find $\sec \theta$, given that $\tan \theta = \frac{\sqrt{7}}{3}$ and θ is in quadrant III.

135. Find $\tan \theta$, given that $\sin \theta = \frac{1}{2}$ and θ is in quadrant II.

136. Find $\cot \theta$, given that $\csc \theta = -2$ and θ is in quadrant III.

137. Find $\cot \theta$, given that $\csc \theta = -1.45$ and θ is in quadrant III.

138. Find $\tan \theta$, given that $\sin \theta = 0.6$ and θ is in quadrant II.

Give all six trigonometric function values for each angle θ. Rationalize denominators when applicable. ***See Examples 9–11.***

139. $\tan \theta = -\frac{15}{8}$, and θ is in quadrant II **140.** $\cos \theta = -\frac{3}{5}$, and θ is in quadrant III

141. $\sin \theta = \frac{\sqrt{5}}{7}$, and θ is in quadrant I **142.** $\tan \theta = \sqrt{3}$, and θ is in quadrant III

143. $\cot \theta = \frac{\sqrt{3}}{8}$, and θ is in quadrant I **144.** $\csc \theta = 2$, and θ is in quadrant II

145. $\sin \theta = \frac{\sqrt{2}}{6}$, and $\cos \theta < 0$ **146.** $\cos \theta = \frac{\sqrt{5}}{8}$, and $\tan \theta < 0$

147. $\sec \theta = -4$, and $\sin \theta > 0$ **148.** $\csc \theta = -3$, and $\cos \theta > 0$

149. $\sin \theta = 1$ **150.** $\cos \theta = 1$

Work each problem.

151. Derive the identity $1 + \cot^2 \theta = \csc^2 \theta$ by dividing $x^2 + y^2 = r^2$ by y^2.

152. Derive the quotient identity $\frac{\cos \theta}{\sin \theta} = \cot \theta$.

Concept Check *Suppose that $90° < \theta < 180°$. Find the sign of each function value.*

153. $\sin 2\theta$ **154.** $\csc 2\theta$ **155.** $\tan \frac{\theta}{2}$ **156.** $\cot \frac{\theta}{2}$

Concept Check *Suppose that $-90° < \theta < 90°$. Find the sign of each function value.*

157. $\cos \frac{\theta}{2}$ **158.** $\sec \frac{\theta}{2}$ **159.** $\sec(-\theta)$ **160.** $\cos(-\theta)$

5.3 Trigonometric Function Values and Angle Measures

- Right-Triangle-Based Definitions of the Trigonometric Functions
- Cofunctions
- Trigonometric Function Values of Special Angles
- Reference Angles
- Special Angles as Reference Angles
- Determination of Angle Measures with Special Reference Angles
- Calculator Approximations of Trigonometric Function Values
- Calculator Approximations of Angle Measures
- An Application

Right-Triangle-Based Definitions of the Trigonometric Functions

Angles in standard position can be used to define the trigonometric functions. There is also another way to approach them: as ratios of the lengths of the sides of right triangles.

Figure 27 shows an acute angle A in standard position. The definitions of the trigonometric function values of angle A require x, y, and r. As drawn in **Figure 27,** x and y are the lengths of the two legs of the right triangle ABC, and r is the length of the hypotenuse.

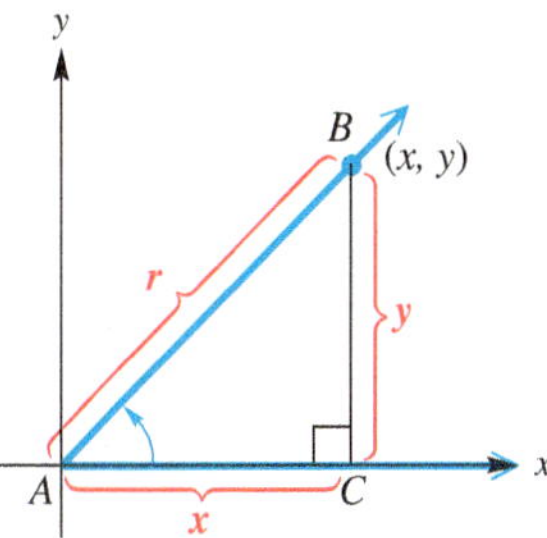

Figure 27

The side of length y is the **side opposite** angle A, and the side of length x is the **side adjacent** to angle A. We use the lengths of these sides to replace x and y in the definitions of the trigonometric functions, and the length of the hypotenuse to replace r, to obtain the following right-triangle-based definitions. In the definitions, we use the standard abbreviations for the sine, cosine, tangent, cosecant, secant, and cotangent functions.

Right-Triangle-Based Definitions of Trigonometric Functions

Let A represent any acute angle in standard position.

$$\sin A = \frac{y}{r} = \frac{\text{side opposite } A}{\text{hypotenuse}} \qquad \csc A = \frac{r}{y} = \frac{\text{hypotenuse}}{\text{side opposite } A}$$

$$\cos A = \frac{x}{r} = \frac{\text{side adjacent to } A}{\text{hypotenuse}} \qquad \sec A = \frac{r}{x} = \frac{\text{hypotenuse}}{\text{side adjacent to } A}$$

$$\tan A = \frac{y}{x} = \frac{\text{side opposite } A}{\text{side adjacent to } A} \qquad \cot A = \frac{x}{y} = \frac{\text{side adjacent to } A}{\text{side opposite } A}$$

NOTE We will sometimes shorten wording like "side opposite A" to just "side opposite" when the meaning is obvious.

EXAMPLE 1 Finding Trigonometric Function Values of an Acute Angle

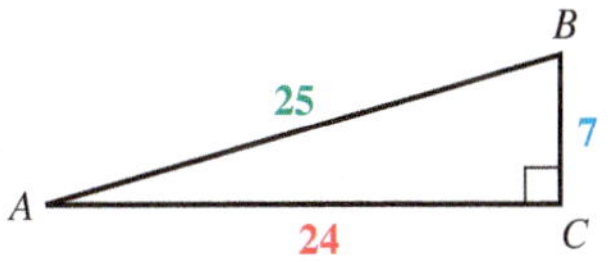

Find the sine, cosine, and tangent values for angles A and B in the right triangle in **Figure 28.**

Figure 28

SOLUTION The length of the side opposite angle A is 7, the length of the side adjacent to angle A is 24, and the length of the hypotenuse is 25.

$$\sin A = \frac{\text{side opposite}}{\text{hypotenuse}} = \frac{7}{25} \qquad \cos A = \frac{\text{side adjacent}}{\text{hypotenuse}} = \frac{24}{25} \qquad \tan A = \frac{\text{side opposite}}{\text{side adjacent}} = \frac{7}{24}$$

The length of the side opposite angle B is 24, and the length of the side adjacent to angle B is 7.

$$\sin B = \frac{24}{25} \qquad \cos B = \frac{7}{25} \qquad \tan B = \frac{24}{7}$$

Use the right-triangle-based definitions of the trigonometric functions.

✔ **Now Try Exercise 11.**

NOTE The cosecant, secant, and cotangent ratios are reciprocals of the sine, cosine, and tangent values, respectively, so in **Example 1** we have

$$\csc A = \frac{25}{7} \qquad \sec A = \frac{25}{24} \qquad \cot A = \frac{24}{7}$$

$$\csc B = \frac{25}{24} \qquad \sec B = \frac{25}{7} \qquad \text{and} \qquad \cot B = \frac{7}{24}.$$

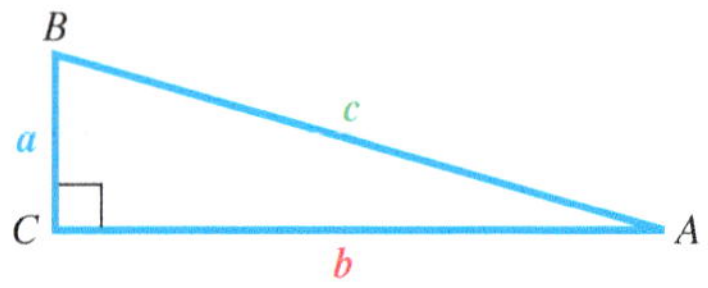

Whenever we use A, B, and C to name angles in a right triangle, C will be the right angle.

Figure 29

Cofunctions

Figure 29 shows a right triangle with acute angles A and B and a right angle at C. The length of the side opposite angle A is a, and the length of the side opposite angle B is b. The length of the hypotenuse is c. By the preceding definitions, $\sin A = \frac{a}{c}$. Also, $\cos B = \frac{a}{c}$. Thus, we have the following.

$$\sin A = \frac{a}{c} = \cos B$$

Similarly, $\tan A = \frac{a}{b} = \cot B$ and $\sec A = \frac{c}{b} = \csc B.$

In any right triangle, the sum of the two acute angles is 90°, so they are *complementary*. In **Figure 29,** A and B are thus complementary, and we have established that $\sin A = \cos B$. This can also be written as follows.

$$\sin A = \cos(90° - A) \qquad B = 90° - A$$

This is an example of a more general relationship between **cofunction** pairs.

sine, cosine
tangent, cotangent
secant, cosecant

} Cofunction pairs

Cofunction Identities

For any acute angle A, the following hold.

$$\sin A = \cos(90° - A) \quad \sec A = \csc(90° - A) \quad \tan A = \cot(90° - A)$$

$$\cos A = \sin(90° - A) \quad \csc A = \sec(90° - A) \quad \cot A = \tan(90° - A)$$

The cofunction identities state the following.

Cofunction values of complementary angles are equal.

EXAMPLE 2 Writing Functions in Terms of Cofunctions

Write each function in terms of its cofunction.

(a) cos 52° **(b)** tan 71° **(c)** sec 24°

SOLUTION

(a)

Cofunctions

$$\cos 52° = \sin(90° - 52°) = \sin 38° \qquad \cos A = \sin(90° - A)$$

Complementary angles

(b) $\tan 71° = \cot(90° - 71°) = \cot 19°$ **(c)** $\sec 24° = \csc 66°$

✔ **Now Try Exercises 27 and 29.**

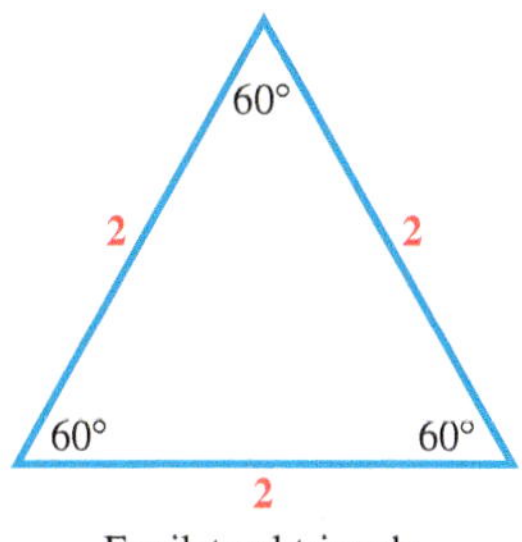

Equilateral triangle

(a)

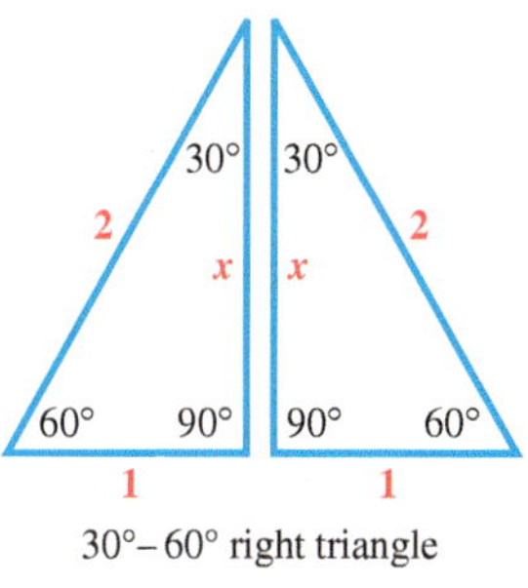

30°–60° right triangle

(b)

Figure 30

Trigonometric Function Values of Special Angles Certain special angles, such as 30°, 45°, and 60°, occur so often in trigonometry and in more advanced mathematics that they deserve special study. We start with an equilateral triangle, a triangle with all sides of equal length. Each angle of such a triangle measures 60°. Although the results we will obtain are independent of the length, for convenience we choose the length of each side to be 2 units. See **Figure 30(a).**

Bisecting one angle of this equilateral triangle leads to two right triangles, each of which has angles of 30°, 60°, and 90°, as shown in **Figure 30(b).** An angle bisector of an equilateral triangle also bisects the opposite side. Thus the shorter leg has length 1. Let x represent the length of the longer leg.

$$2^2 = 1^2 + x^2 \quad \text{Pythagorean theorem}$$
$$4 = 1 + x^2 \quad \text{Apply the exponents.}$$
$$3 = x^2 \quad \text{Subtract 1 from each side.}$$
$$\sqrt{3} = x \quad \text{Square root property; choose the positive root.}$$

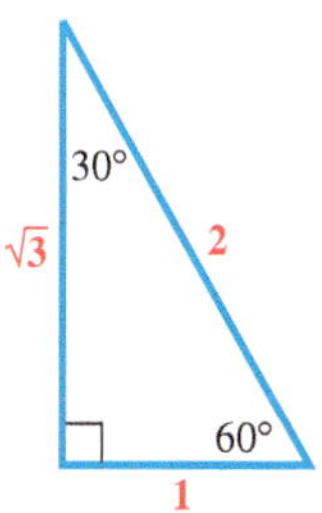

Figure 31

Figure 31 summarizes our results using a 30°–60° right triangle. As shown in the figure, the side opposite the 30° angle has length 1. For the 30° angle,

$$\text{hypotenuse} = 2, \quad \text{side opposite} = 1, \quad \text{side adjacent} = \sqrt{3}.$$

Now we use the definitions of the trigonometric functions.

$$\sin 30° = \frac{\text{side opposite}}{\text{hypotenuse}} = \frac{1}{2}$$

$$\cos 30° = \frac{\text{side adjacent}}{\text{hypotenuse}} = \frac{\sqrt{3}}{2}$$

$$\tan 30° = \frac{\text{side opposite}}{\text{side adjacent}} = \frac{1}{\sqrt{3}} = \frac{1}{\sqrt{3}} \cdot \frac{\sqrt{3}}{\sqrt{3}} = \frac{\sqrt{3}}{3}$$

$$\csc 30° = \frac{2}{1} = 2$$

Rationalize the denominators.

$$\sec 30° = \frac{2}{\sqrt{3}} = \frac{2}{\sqrt{3}} \cdot \frac{\sqrt{3}}{\sqrt{3}} = \frac{2\sqrt{3}}{3}$$

$$\cot 30° = \frac{\sqrt{3}}{1} = \sqrt{3}$$

EXAMPLE 3 Finding Trigonometric Function Values for 60°

Find the six trigonometric function values for a 60° angle.

SOLUTION Refer to **Figure 31** to find the following ratios.

$$\sin 60° = \frac{\sqrt{3}}{2} \qquad \cos 60° = \frac{1}{2} \qquad \tan 60° = \frac{\sqrt{3}}{1} = \sqrt{3}$$

$$\csc 60° = \frac{2}{\sqrt{3}} = \frac{2\sqrt{3}}{3} \qquad \sec 60° = \frac{2}{1} = 2 \qquad \cot 60° = \frac{1}{\sqrt{3}} = \frac{\sqrt{3}}{3}$$

✔ **Now Try Exercises 31, 33, and 35.**

NOTE The results in **Example 3** can also be found using the fact that cofunction values of the complementary angles 60° and 30° are equal.

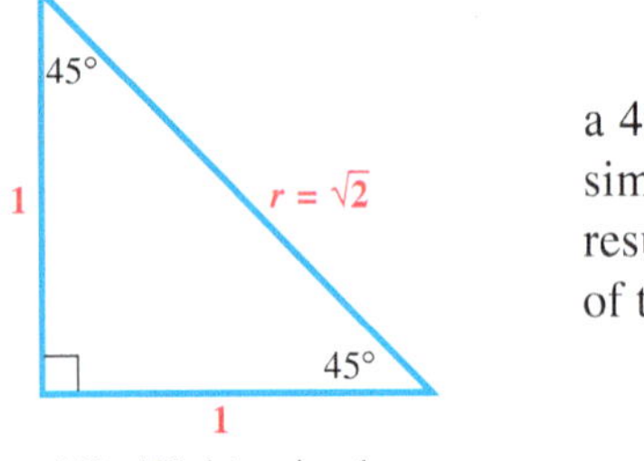

45°–45° right triangle

Figure 32

We find the values of the trigonometric functions for 45° by starting with a 45°–45° right triangle, as shown in **Figure 32.** This triangle is isosceles. For simplicity, we choose the lengths of the equal sides to be 1 unit. (As before, the results are independent of the length of the equal sides.) If r represents the length of the hypotenuse, then we can find its value using the Pythagorean theorem.

$$1^2 + 1^2 = r^2 \quad \text{Pythagorean theorem}$$

$$2 = r^2 \quad \text{Simplify.}$$

$$\sqrt{2} = r \quad \text{Choose the positive root.}$$

Now we use the measures indicated on the 45°–45° right triangle in **Figure 32.**

$$\sin 45^\circ = \frac{1}{\sqrt{2}} = \frac{\sqrt{2}}{2} \qquad \cos 45^\circ = \frac{1}{\sqrt{2}} = \frac{\sqrt{2}}{2} \qquad \tan 45^\circ = \frac{1}{1} = 1$$

$$\csc 45^\circ = \frac{\sqrt{2}}{1} = \sqrt{2} \qquad \sec 45^\circ = \frac{\sqrt{2}}{1} = \sqrt{2} \qquad \cot 45^\circ = \frac{1}{1} = 1$$

Function values for 30°, 45°, and 60° are summarized in the table that follows.

Function Values of Special Angles

θ	$\sin\theta$	$\cos\theta$	$\tan\theta$	$\cot\theta$	$\sec\theta$	$\csc\theta$
30°	$\frac{1}{2}$	$\frac{\sqrt{3}}{2}$	$\frac{\sqrt{3}}{3}$	$\sqrt{3}$	$\frac{2\sqrt{3}}{3}$	2
45°	$\frac{\sqrt{2}}{2}$	$\frac{\sqrt{2}}{2}$	1	1	$\sqrt{2}$	$\sqrt{2}$
60°	$\frac{\sqrt{3}}{2}$	$\frac{1}{2}$	$\sqrt{3}$	$\frac{\sqrt{3}}{3}$	2	$\frac{2\sqrt{3}}{3}$

NOTE You will be able to reproduce this table quickly if you learn the values of sin 30°, sin 45°, and sin 60°. Then you can complete the rest of the table using the reciprocal, cofunction, and quotient identities.

Reference Angles Associated with every nonquadrantal angle in standard position is an acute angle called its *reference angle*. A **reference angle** for an angle θ, written θ', is the acute angle made by the terminal side of angle θ and the x-axis.

NOTE Reference angles are always positive and are between 0° and 90°.

Figure 33 on the next page shows several angles θ (each less than one complete counterclockwise revolution) in quadrants II, III, and IV, respectively, with the reference angle θ' also shown. In quadrant I, angles θ and θ' are the same. If an angle θ is negative or has measure greater than 360°, its reference angle is found by first finding its coterminal angle that is between 0° and 360°, and then using the diagrams in **Figure 33.**

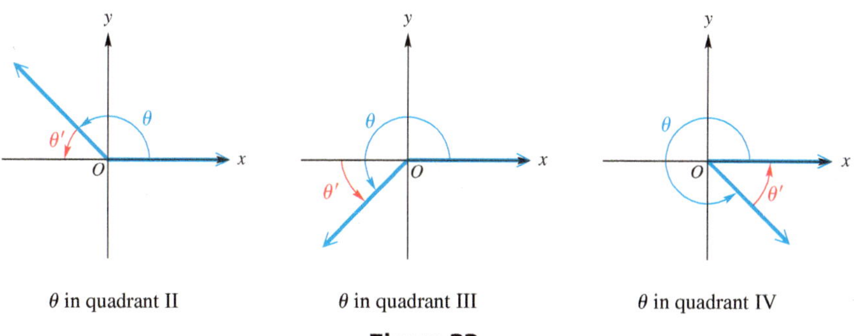

Figure 33

CAUTION A common error is to find the reference angle by using the terminal side of θ and the y-axis. ***The reference angle is always found with reference to the x-axis.***

EXAMPLE 4 Finding Reference Angles

Find the reference angle for each angle.

(a) 218° **(b)** 1387°

SOLUTION

(a) As shown in **Figure 34(a),** the positive acute angle made by the terminal side of this angle and the x-axis is

$$218^\circ - 180^\circ = 38^\circ.$$

For $\theta = 218^\circ$, the reference angle $\theta' = 38^\circ$.

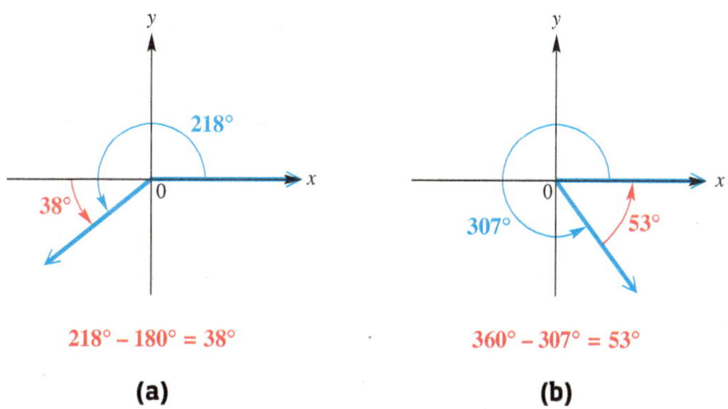

(a) **(b)**

Figure 34

(b) First find a coterminal angle between 0° and 360°. Divide 1387° by 360° to obtain a quotient of about 3.9. Begin by subtracting 360° three times (because of the whole number 3 in 3.9).

$$\begin{aligned} &1387^\circ - 3 \cdot 360^\circ \\ &= 1387^\circ - 1080^\circ \quad \text{Multiply.} \\ &= 307^\circ \quad \text{Subtract.} \end{aligned}$$

The reference angle for 307° (and thus for 1387°) is $360^\circ - 307^\circ = 53^\circ$. See **Figure 34(b).**

✔ **Now Try Exercises 57 and 61.**

The preceding example suggests the following table for finding the reference angle θ' for any angle θ between 0° and 360°.

Reference Angle θ' for θ, where $0° < \theta < 360°$*

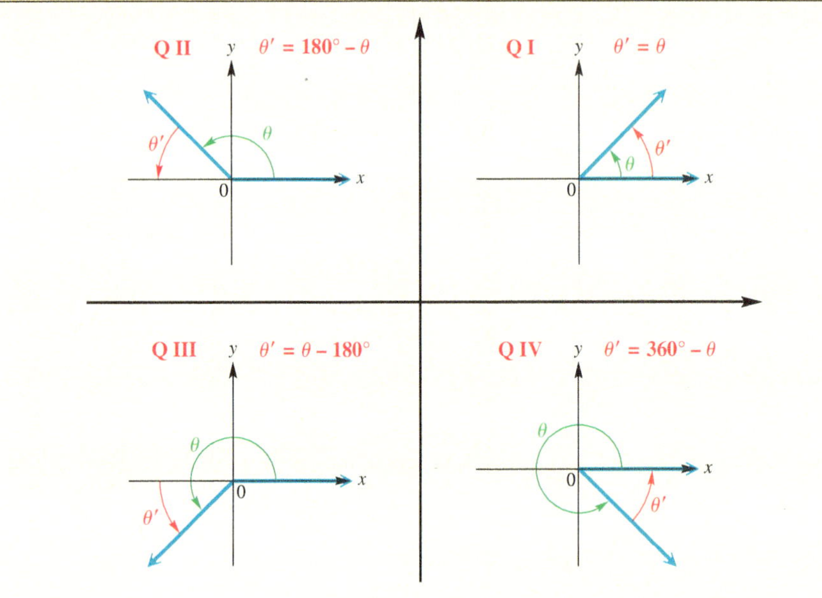

Special Angles as Reference Angles We can now find exact trigonometric function values of angles with reference angles of 30°, 45°, or 60°.

EXAMPLE 5 Finding Trigonometric Function Values of a Quadrant III Angle

Find exact values of the six trigonometric functions of 210°.

SOLUTION An angle of 210° is shown in **Figure 35.** The reference angle is

$$210° - 180° = 30°.$$

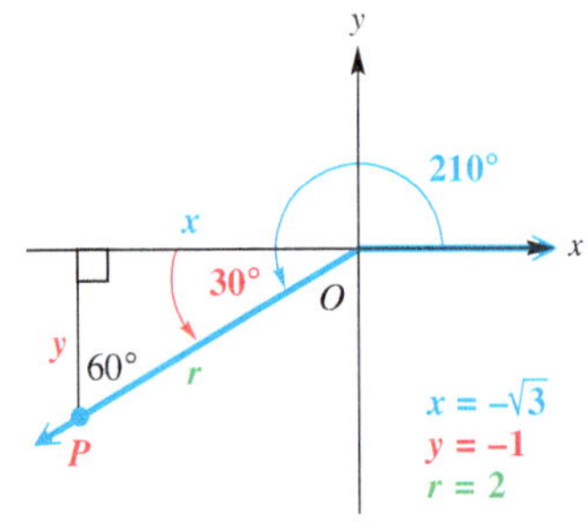

Figure 35

To find the trigonometric function values of 210°, choose point P on the terminal side of the angle so that the distance from the origin O to P is 2. (Any positive number would work, but 2 is most convenient.) By the results from 30°–60° right triangles, the coordinates of point P become $(-\sqrt{3}, -1)$, with $x = -\sqrt{3}$, $y = -1$, and $r = 2$. Then, by the definitions of the trigonometric functions, we obtain the following.

$$\sin 210° = \frac{-1}{2} = -\frac{1}{2} \qquad \csc 210° = \frac{2}{-1} = -2$$

$$\cos 210° = \frac{-\sqrt{3}}{2} = -\frac{\sqrt{3}}{2} \qquad \sec 210° = \frac{2}{-\sqrt{3}} = -\frac{2\sqrt{3}}{3}$$

$$\tan 210° = \frac{-1}{-\sqrt{3}} = \frac{\sqrt{3}}{3} \qquad \cot 210° = \frac{-\sqrt{3}}{-1} = \sqrt{3}$$

Rationalize denominators as needed.

✔ **Now Try Exercise 71.**

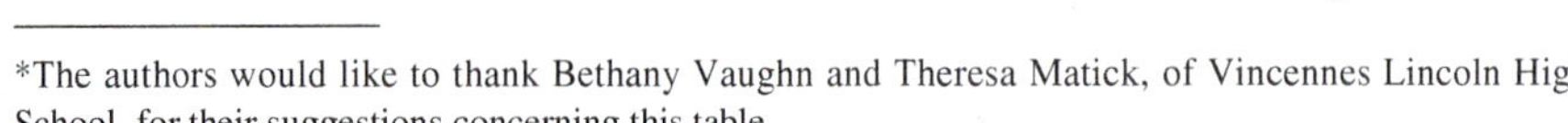

*The authors would like to thank Bethany Vaughn and Theresa Matick, of Vincennes Lincoln High School, for their suggestions concerning this table.

Notice in **Example 5** that the trigonometric function values of 210° correspond in absolute value to those of its reference angle 30°. The signs are different for the sine, cosine, secant, and cosecant functions because 210° is a quadrant III angle. These results suggest a shortcut for finding the trigonometric function values of a non-acute angle, using the reference angle.

In **Example 5,** the reference angle for 210° is 30°. Using the trigonometric function values of 30°, and choosing the correct signs for a quadrant III angle, we obtain the same results.

We determine the values of the trigonometric functions for any nonquadrantal angle θ as follows. Keep in mind that all function values are positive when the terminal side is in Quadrant I, the sine and cosecant are positive in Quadrant II, the tangent and cotangent are positive in Quadrant III, and the cosine and secant are positive in Quadrant IV. In other cases, the function values are negative.

Finding Trigonometric Function Values for Any Nonquadrantal Angle θ

Step 1 If $\theta > 360°$, or if $\theta < 0°$, then find a coterminal angle by adding or subtracting 360° as many times as needed to obtain an angle greater than 0° but less than 360°.

Step 2 Find the reference angle θ'.

Step 3 Find the trigonometric function values for reference angle θ'.

Step 4 Determine the correct signs for the values found in Step 3. (Use the table of signs given earlier in the text or the paragraph above, if necessary.) This gives the values of the trigonometric functions for angle θ.

NOTE To avoid sign errors when finding the trigonometric function values of an angle, sketch it in standard position. Include a reference triangle complete with appropriate values for x, y, and r as done in **Figure 35.**

EXAMPLE 6 Finding Trigonometric Function Values Using Reference Angles

Find the exact value of each expression.

(a) $\cos(-240°)$ **(b)** $\tan 675°$

SOLUTION

(a) Because an angle of $-240°$ is coterminal with an angle of

$$-240° + 360° = 120°,$$

the reference angle is $180° - 120° = 60°$, as shown in **Figure 36.** The cosine is negative in quadrant II.

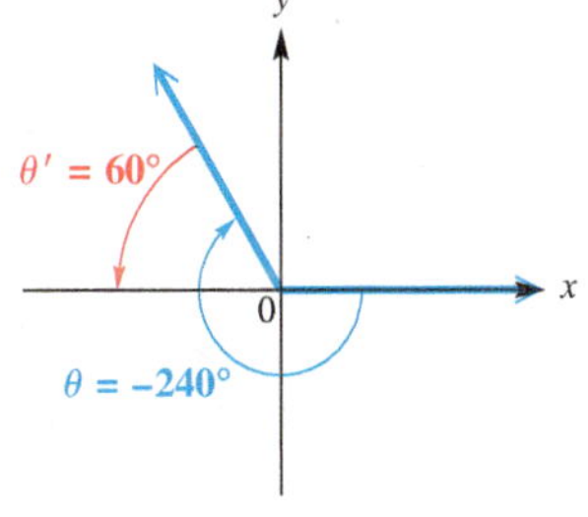

Figure 36

$$\cos(-240°)$$
$$= \cos 120° \quad \leftarrow \text{Coterminal angle}$$
$$= -\cos 60° \quad \leftarrow \text{Reference angle}$$
$$= -\frac{1}{2} \quad \text{Evaluate.}$$

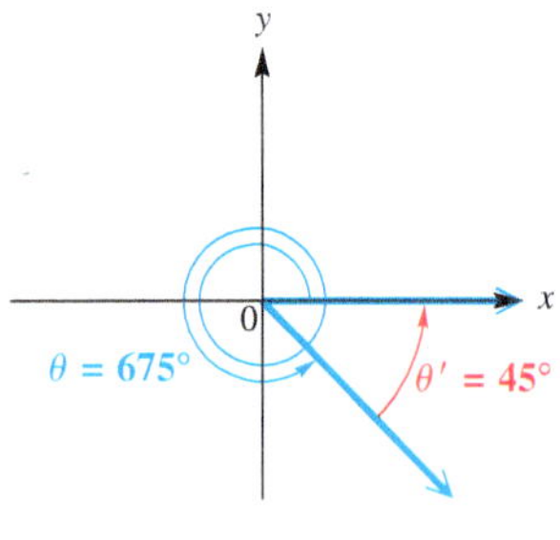

Figure 37

(b) Subtract 360° to find an angle between 0° and 360° coterminal with 675°.

$$675° - 360° = 315°$$

As shown in **Figure 37,** the reference angle is $360° - 315° = 45°$. An angle of 315° is in quadrant IV, so the tangent will be negative.

$$\begin{aligned} &\tan 675° \\ &= \tan 315° && \text{Coterminal angle} \\ &= -\tan 45° && \text{Reference angle; quadrant-based sign choice} \\ &= -1 && \text{Evaluate.} \end{aligned}$$

✔ **Now Try Exercises 89 and 91.**

Determination of Angle Measures with Special Reference Angles

The ideas discussed in this section can be used "in reverse" to find the measures of certain angles, given a trigonometric function value and an interval in which the angle must lie. We are most often interested in the interval $[0°, 360°)$.

EXAMPLE 7 Finding Angle Measures

Find all values of θ, if θ is in the interval $[0°, 360°)$ and $\cos \theta = -\frac{\sqrt{2}}{2}$.

SOLUTION The value of $\cos \theta$ is negative, so θ may lie in either quadrant II or III. Because the absolute value of $\cos \theta$ is $\frac{\sqrt{2}}{2}$, the reference angle θ' must be 45°. The two possible angles θ are sketched in **Figure 38.**

$$180° - 45° = 135° \quad \text{Quadrant II angle } \theta \quad \text{(from Figure 38(a))}$$

$$180° + 45° = 225° \quad \text{Quadrant III angle } \theta \quad \text{(from Figure 38(b))}$$

(a) **(b)**

Figure 38

✔ **Now Try Exercise 97.**

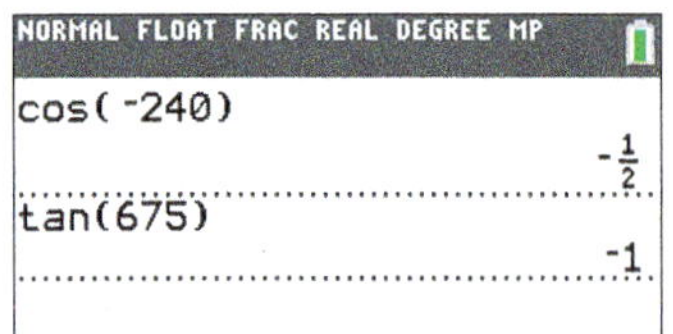

Degree mode

A calculator can be used to find exact values such as $\cos(-240°)$ and $\tan 675°$.

Calculator Approximations of Trigonometric Function Values

We have found exact function values for special angles and for angles having special reference angles. Calculators provide approximations for function values of angles that do not satisfy these conditions.

CAUTION ***When evaluating trigonometric functions of angles given in degrees, the calculator must be in degree mode.*** An easy way to check this is to enter sin 90. The displayed answer should be 1. Also, if the angle or the reference angle is not a special or quadrantal angle, then the value given by the calculator is an *approximation*. And even if the angle or reference angle *is* a special angle, the value given by the calculator will often be an approximation.

EXAMPLE 8 Finding Function Values with a Calculator

Approximate the value of each expression.

(a) $\sin 49° 12'$ **(b)** $\sec 97.977°$ **(c)** $\dfrac{1}{\cot 51.4283°}$ **(d)** $\sin(-246°)$

SOLUTION See **Figure 39.** We give values to eight decimal places below.

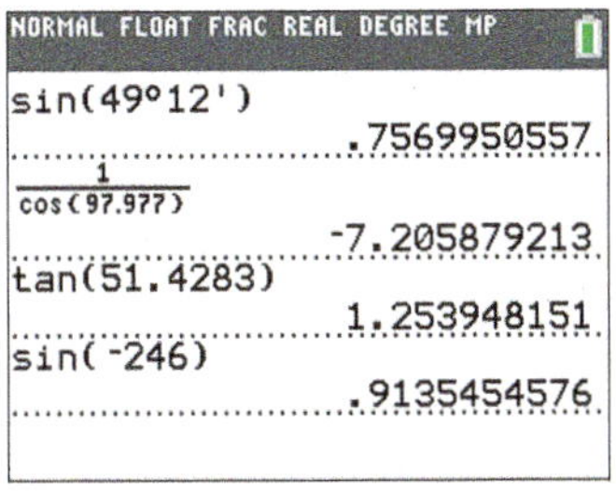

NORMAL FLOAT FRAC REAL DEGREE MP
sin(49°12')
.7569950557
1/cos(97.977)
-7.205879213
tan(51.4283)
1.253948151
sin(-246)
.9135454576

Degree mode

Figure 39

(a) We *may* begin by converting 49° 12′ to decimal degrees.

$$49°\,12' = 49\frac{12°}{60} = 49.2°$$

However, some calculators allow direct entry of degrees, minutes, and seconds. (The method of entry varies among models.) Entering either $\sin(49°\,12')$ or $\sin 49.2°$ gives the same approximation.

$$\sin 49°\,12' = \sin 49.2° \approx 0.75699506$$

(b) There are no dedicated calculator keys for the secant, cosecant, and cotangent functions. However, we can use reciprocal identities to evaluate them. Recall that $\sec\theta = \frac{1}{\cos\theta}$ for all angles θ, where $\cos\theta \neq 0$. Therefore, we use the reciprocal of the cosine function to evaluate the secant function.

$$\sec 97.977° = \frac{1}{\cos 97.977°} \approx -7.20587921$$

(c) Use the reciprocal identity $\frac{1}{\cot\theta} = \tan\theta$ to simplify the expression first.

$$\frac{1}{\cot 51.4283°} = \tan 51.4283° \approx 1.25394815$$

(d) $\sin(-246°) \approx 0.91354546$

✔ **Now Try Exercises 109, 111, 115, and 119.**

Calculator Approximations of Angle Measures To find the measure of an angle having a certain trigonometric function value, calculators have three *inverse functions* (denoted $\mathbf{sin^{-1}}$, $\mathbf{cos^{-1}}$, and $\mathbf{tan^{-1}}$).

If x is an appropriate number, then $\sin^{-1} x$, $\cos^{-1} x$, or $\tan^{-1} x$ gives the measure of an angle whose sine, cosine, or tangent, respectively, is x.

For applications in this chapter, these functions will return angles in quadrant I.

EXAMPLE 9 Using Inverse Trigonometric Functions to Find Angles

Find an angle θ in the interval $[0°, 90°)$ that satisfies each condition.

$\sin\theta = 0.96770915$ **(b)** $\sec\theta = 1.0545829$

SOLUTION

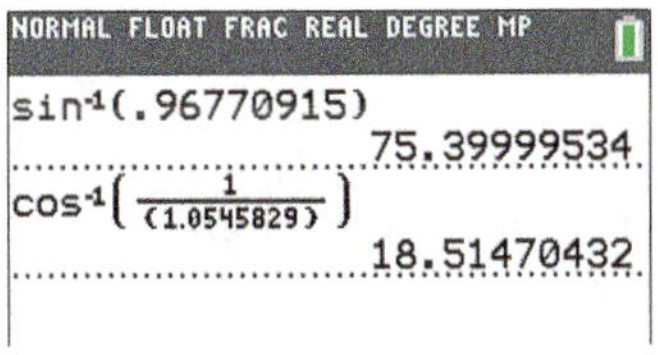

NORMAL FLOAT FRAC REAL DEGREE MP
sin⁻¹(.96770915)
75.39999534
cos⁻¹(1/(1.0545829))
18.51470432

Degree mode

Figure 40

(a) Using degree mode and the inverse sine function, we find that an angle θ having sine value 0.96770915 is 75.399995°. (There are infinitely many such angles, but the calculator gives only this one.)

$$\theta = \sin^{-1} 0.96770915 \approx 75.399995°$$

See **Figure 40.**

(b) Use the identity $\cos\theta = \frac{1}{\sec\theta}$. If $\sec\theta = 1.0545829$, then

$$\cos\theta = \frac{1}{1.0545829}.$$

Now, find θ using the inverse cosine function.

$$\theta = \cos^{-1}\left(\frac{1}{1.0545829}\right) \approx 18.514704°$$ See **Figure 40** on the previous page.

✔ **Now Try Exercises 125 and 129.**

CAUTION Compare **Examples 8(b) and 9(b).**

- To determine the secant of an angle, as in **Example 8(b),** we find the *reciprocal of the cosine* of the angle.
- To determine an angle with a given secant value, as in **Example 9(b),** we find the *inverse cosine of the reciprocal* of the value.

An Application

EXAMPLE 10 Finding Grade Resistance

When an automobile travels uphill or downhill on a highway, it experiences a force due to gravity. This force F in pounds is the **grade resistance** and is modeled by

$$F = W\sin\theta,$$

where θ is the grade and W is the weight of the automobile. If the automobile is moving uphill, then $\theta > 0°$; if downhill, then $\theta < 0°$. See **Figure 41.** (*Source:* Mannering, F. and W. Kilareski, *Principles of Highway Engineering and Traffic Analysis,* Second Edition, John Wiley and Sons.)

Figure 41

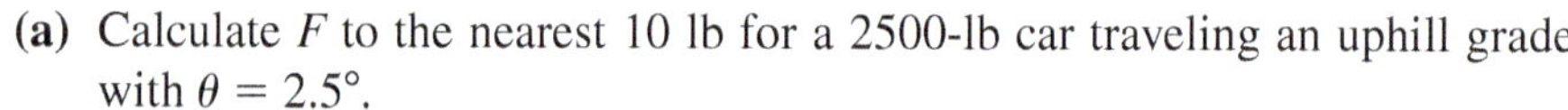

(a) Calculate F to the nearest 10 lb for a 2500-lb car traveling an uphill grade with $\theta = 2.5°$.

(b) Calculate F to the nearest 10 lb for a 5000-lb truck traveling a downhill grade with $\theta = -6.1°$.

(c) Calculate F for $\theta = 0°$ and $\theta = 90°$. Do these answers agree with intuition?

SOLUTION

(a)

$F = W\sin\theta$ — Given model for grade resistance

$F = 2500\sin 2.5°$ — Substitute given values.

$F \approx 110$ lb — Evaluate.

(b) $F = W\sin\theta = 5000\sin(-6.1°) \approx -530$ lb

F is negative because the truck is moving downhill.

(c) $F = W\sin\theta = W\sin 0° = W(0) = 0$ lb

$F = W\sin\theta = W\sin 90° = W(1) = W$ lb

This agrees with intuition because if $\theta = 0°$, then there is level ground and gravity does not cause the vehicle to roll. If θ were 90°, the road would be vertical and the full weight of the vehicle would be pulled downward by gravity, so $F = W$.

✔ **Now Try Exercises 135 and 137.**

5.3 Exercises

CONCEPT PREVIEW *Match each trigonometric function in Column I with its value in Column II. Choices may be used once, more than once, or not at all.*

I		II		
1. $\sin 30°$	**2.** $\cos 45°$	**A.** $\sqrt{3}$	**B.** 1	**C.** $\frac{1}{2}$
3. $\tan 45°$	**4.** $\sec 60°$	**D.** $\frac{\sqrt{3}}{2}$	**E.** $\frac{2\sqrt{3}}{3}$	**F.** $\frac{\sqrt{3}}{3}$
5. $\csc 60°$	**6.** $\cot 30°$	**G.** 2	**H.** $\frac{\sqrt{2}}{2}$	**I.** $\sqrt{2}$

CONCEPT PREVIEW *Fill in the blanks to correctly complete each sentence.*

7. The value of $\sin 240°$ is ______________ (positive/negative) because 240° is in quadrant _____. The reference angle is _____, and the *exact* value of $\sin 240°$ is _____.

8. The value of $\cos 390°$ is ______________ (positive/negative) because 390° is in quadrant _____. The reference angle is _____, and the *exact* value of $\cos 390°$ is _____.

9. The value of $\tan(-150°)$ is ______________ (positive/negative) because $-150°$ is in quadrant _____. The reference angle is _____, and the *exact* value of $\tan(-150°)$ is _____.

10. The value of $\sec 135°$ is ______________ (positive/negative) because 135° is in quadrant _____. The reference angle is _____, and the *exact* value of $\sec 135°$ is _____.

Find exact values or expressions for $\sin A$, $\cos A$, *and* $\tan A$. **See Example 1.**

11.

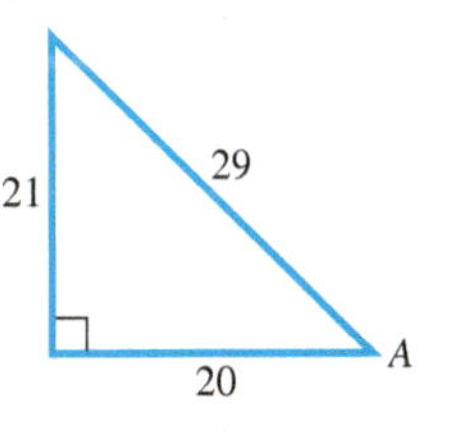

12.

53
45
28
A

13.

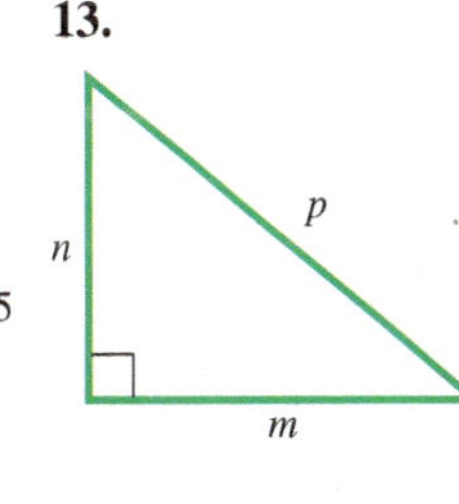

14.

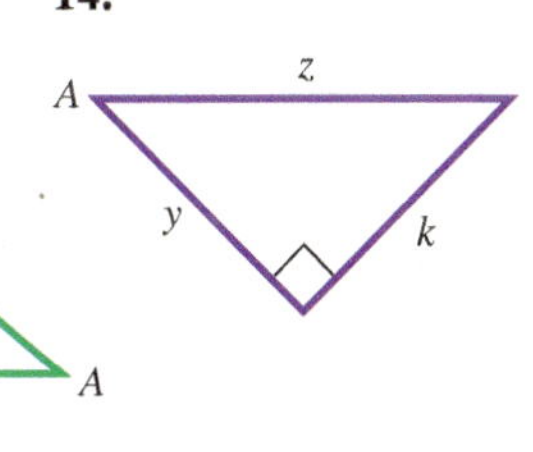

Suppose ABC is a right triangle with sides of lengths a, b, and c and right angle at C.

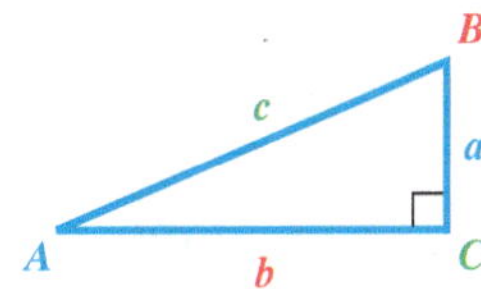

Use the Pythagorean theorem to find the unknown side length. Then find exact values of the six trigonometric functions for angle B. Rationalize denominators when applicable. **See Example 1.**

15. $a = 5, b = 12$ **16.** $a = 3, b = 4$ **17.** $a = 6, c = 7$ **18.** $b = 7, c = 12$

19. $a = 3, c = 10$ **20.** $b = 8, c = 11$ **21.** $a = 1, c = 2$ **22.** $a = \sqrt{2}, c = 2$

Write each function in terms of its cofunction. **See Example 2.**

23. $\cos 30°$ **24.** $\sin 45°$ **25.** $\csc 60°$ **26.** $\cot 73°$

27. $\sec 39°$ **28.** $\tan 25.4°$ **29.** $\sin 38.7°$ **30.** $\csc 49.9°$

Give the exact value of each expression. **See Example 3.**

31. $\tan 30°$ **32.** $\cot 30°$ **33.** $\sin 30°$ **34.** $\cos 30°$

35. $\sec 30°$ **36.** $\csc 30°$ **37.** $\csc 45°$ **38.** $\sec 45°$

39. $\cos 45°$ **40.** $\cot 45°$ **41.** $\tan 45°$ **42.** $\sin 45°$

43. $\sin 60°$ **44.** $\cos 60°$ **45.** $\tan 60°$ **46.** $\csc 60°$

Concept Check *Work each problem.*

47. Find the equation of the line that passes through the origin and makes a 30° angle with the x-axis.

48. Find the equation of the line that passes through the origin and makes a 60° angle with the x-axis.

49. What angle does the line $y = \sqrt{3}x$ make with the positive x-axis?

50. What angle does the line $y = \frac{\sqrt{3}}{3}x$ make with the positive x-axis?

Find the exact value of the variables in each figure.

51.

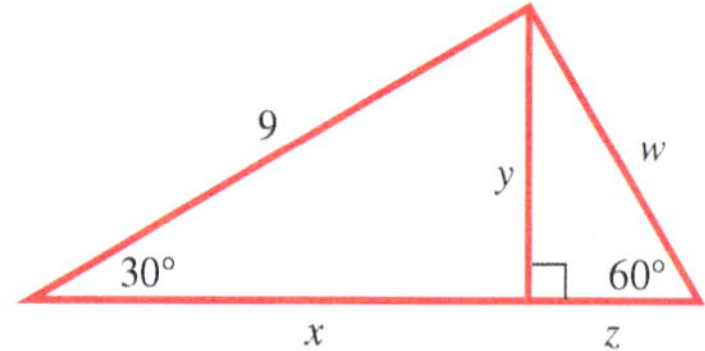

52.

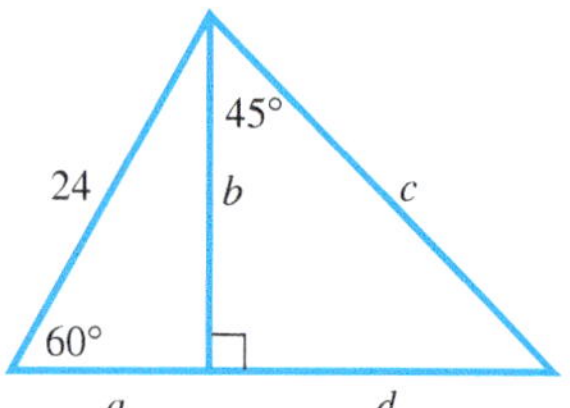

53.

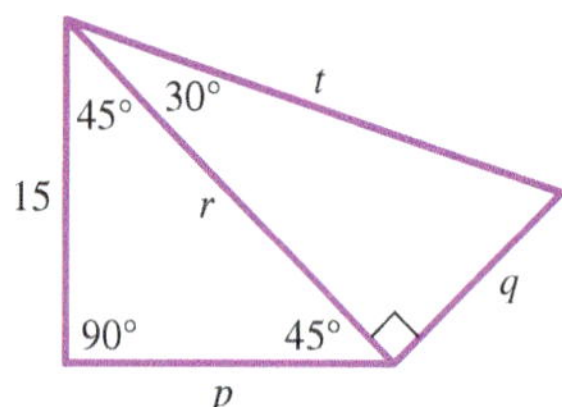

54.

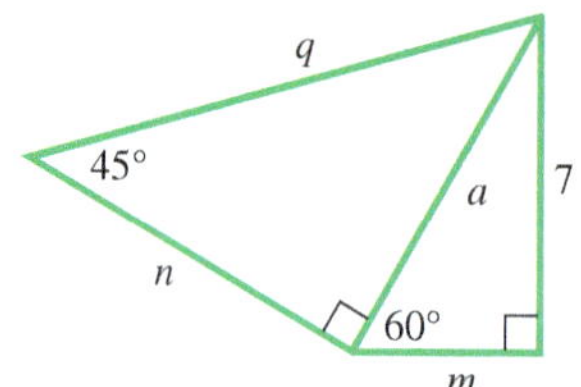

Find a formula for the area of each figure in terms of s.

55.

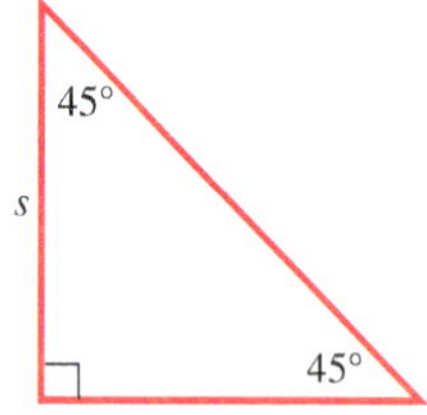

56.

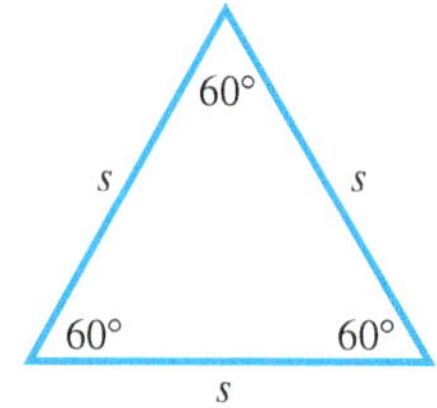

Concept Check *Match each angle in Column I with its reference angle in Column II. Choices may be used once, more than once, or not at all.* **See Example 4.**

I		II	
57. 98°	**58.** 212°	**A.** 45°	**B.** 60°
59. −135°	**60.** −60°	**C.** 82°	**D.** 30°
61. 750°	**62.** 480°	**E.** 38°	**F.** 32°

Complete the table with exact trigonometric function values. Do not use a calculator. ***See Examples 5 and 6.***

	θ	$\sin\theta$	$\cos\theta$	$\tan\theta$	$\cot\theta$	$\sec\theta$	$\csc\theta$
63.	30°	$\frac{1}{2}$	$\frac{\sqrt{3}}{2}$			$\frac{2\sqrt{3}}{3}$	2
64.	45°			1	1		
65.	60°		$\frac{1}{2}$	$\sqrt{3}$		2	
66.	120°	$\frac{\sqrt{3}}{2}$		$-\sqrt{3}$			$\frac{2\sqrt{3}}{3}$
67.	135°	$\frac{\sqrt{2}}{2}$	$-\frac{\sqrt{2}}{2}$			$-\sqrt{2}$	$\sqrt{2}$
68.	150°		$-\frac{\sqrt{3}}{2}$	$-\frac{\sqrt{3}}{3}$			2
69.	210°	$-\frac{1}{2}$		$\frac{\sqrt{3}}{3}$	$\sqrt{3}$		-2
70.	240°	$-\frac{\sqrt{3}}{2}$	$-\frac{1}{2}$			-2	$-\frac{2\sqrt{3}}{3}$

Find exact values of the six trigonometric functions of each angle. Rationalize denominators when applicable. ***See Examples 5 and 6.***

71. 300° **72.** 315° **73.** 405° **74.** 420° **75.** 480° **76.** 495°

77. 570° **78.** 750° **79.** 1305° **80.** 1500° **81.** −300° **82.** −390°

83. −510° **84.** −1020° **85.** −1290° **86.** −855° **87.** −1860° **88.** −2205°

Find the exact value of each expression. ***See Example 6.***

89. sin 1305° **90.** sin 1500° **91.** cos(−510°) **92.** tan(−1020°)

93. csc(−855°) **94.** sec(−495°) **95.** tan 3015° **96.** cot 2280°

Find all values of θ, if θ is in the interval $[0°, 360°)$ and has the given function value. ***See Example 7.***

97. $\sin\theta = \frac{1}{2}$ **98.** $\cos\theta = \frac{\sqrt{3}}{2}$ **99.** $\tan\theta = -\sqrt{3}$

100. $\sec\theta = -\sqrt{2}$ **101.** $\cos\theta = \frac{\sqrt{2}}{2}$ **102.** $\cot\theta = -\frac{\sqrt{3}}{3}$

103. $\csc\theta = -2$ **104.** $\sin\theta = -\frac{\sqrt{3}}{2}$ **105.** $\tan\theta = \frac{\sqrt{3}}{3}$

106. $\cos\theta = -\frac{1}{2}$ **107.** $\csc\theta = -\sqrt{2}$ **108.** $\cot\theta = -1$

Use a calculator to approximate the value of each expression. Give answers to six decimal places. In Exercises 119–122, simplify the expression before using the calculator. ***See Example 8.***

109. sin 38° 42′ **110.** cos 41° 24′ **111.** sec 13° 15′

112. $\csc 145° 45'$ **113.** $\cot 183° 48'$ **114.** $\tan 421° 30'$

115. $\sin(-312° 12')$ **116.** $\tan(-80° 06')$ **117.** $\csc(-317° 36')$

118. $\cot(-512° 20')$ **119.** $\dfrac{1}{\cot 23.4°}$ **120.** $\dfrac{1}{\sec 14.8°}$

121. $\dfrac{\cos 77°}{\sin 77°}$ **122.** $\dfrac{\sin 33°}{\cos 33°}$

Find a value of θ in the interval $[0°, 90°)$ that satisfies each statement. Write each answer in decimal degrees to six decimal places. ***See Example 9.***

123. $\tan \theta = 1.4739716$ **124.** $\tan \theta = 6.4358841$ **125.** $\sin \theta = 0.27843196$

126. $\sin \theta = 0.84802194$ **127.** $\cot \theta = 1.2575516$ **128.** $\csc \theta = 1.3861147$

129. $\sec \theta = 2.7496222$ **130.** $\sec \theta = 1.1606249$ **131.** $\cos \theta = 0.70058013$

132. $\cos \theta = 0.85536428$ **133.** $\csc \theta = 4.7216543$ **134.** $\cot \theta = 0.21563481$

(Modeling) Grade Resistance *Solve each problem.* ***See Example 10.***

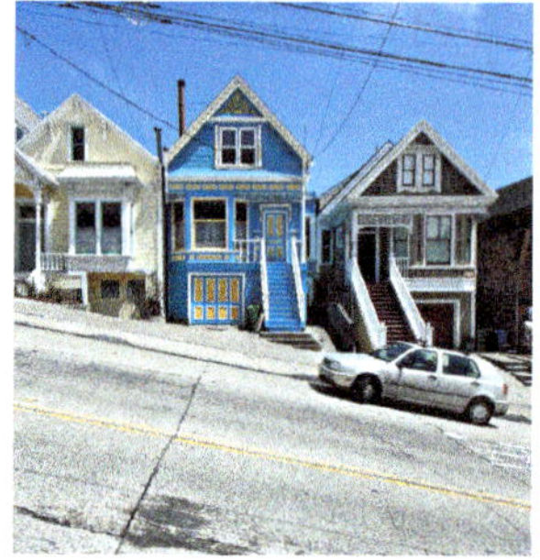

135. Find the grade resistance, to the nearest ten pounds, for a 2100-lb car traveling on a 1.8° uphill grade.

136. Find the grade resistance, to the nearest ten pounds, for a 2400-lb car traveling on a −2.4° downhill grade.

137. A 2600-lb car traveling downhill has a grade resistance of −130 lb. Find the angle of the grade to the nearest tenth of a degree.

138. A 3000-lb car traveling uphill has a grade resistance of 150 lb. Find the angle of the grade to the nearest tenth of a degree.

139. A car traveling on a 2.7° uphill grade has a grade resistance of 120 lb. Determine the weight of the car to the nearest hundred pounds.

140. A car traveling on a −3° downhill grade has a grade resistance of −145 lb. Determine the weight of the car to the nearest hundred pounds.

141. Which has the greater grade resistance: a 2200-lb car on a 2° uphill grade or a 2000-lb car on a 2.2° uphill grade?

142. Complete the table for values of $\sin \theta$, $\tan \theta$, and $\frac{\pi\theta}{180}$ to four decimal places.

θ	0°	0.5°	1°	1.5°	2°	2.5°	3°	3.5°	4°
$\sin \theta$									
$\tan \theta$									
$\dfrac{\pi\theta}{180}$									

(a) How do $\sin \theta$, $\tan \theta$, and $\frac{\pi\theta}{180}$ compare for small grades θ?

(b) Highway grades are usually small. Give two approximations of the grade resistance $F = W \sin \theta$ that do not use the sine function.

(c) A stretch of highway has a 4-ft vertical rise for every 100 ft of horizontal run. Use an approximation from part (b) to estimate the grade resistance, to the nearest pound, for a 2000-lb car on this stretch of highway.

(d) Without evaluating a trigonometric function, estimate the grade resistance, to the nearest pound, for an 1800-lb car on a stretch of highway that has a 3.75° grade.

(Modeling) Design of Highway Curves *When highway curves are designed, the outside of the curve is often slightly elevated or inclined above the inside of the curve. See the figure. This inclination is the* ***superelevation.***

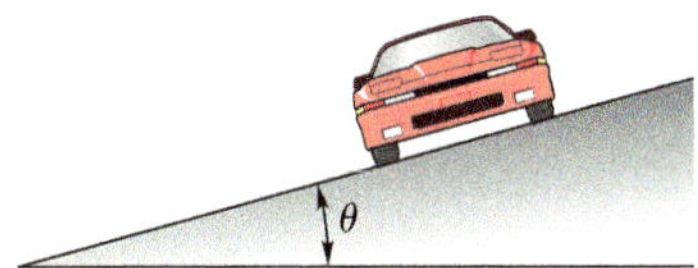

For safety reasons, it is important that both the curve's radius and superelevation be correct for a given speed limit. If an automobile is traveling at velocity V (in feet per second), the safe radius R, in feet, for a curve with superelevation θ is modeled by the formula

$$R = \frac{V^2}{g(f + \tan \theta)},$$

where f and g are constants. (*Source:* Mannering, F. and W. Kilareski, *Principles of Highway Engineering and Traffic Analysis,* Second Edition, John Wiley and Sons.)

143. A roadway is being designed for automobiles traveling at 45 mph. If $\theta = 3^\circ$, $g = 32.2$, and $f = 0.14$, calculate R to the nearest foot. (*Hint:* 45 mph = 66 ft per sec)

144. Determine the radius of the curve, to the nearest foot, if the speed in **Exercise 143** is increased to 70 mph.

145. How would increasing angle θ affect the results? Verify your answer by repeating **Exercises 143 and 144** with $\theta = 4^\circ$.

146. Refer to **Exercise 143** and use the same values for f and g. A highway curve has radius $R = 1150$ ft and a superelevation of $\theta = 2.1^\circ$. What should the speed limit (in miles per hour) be for this curve?

(Modeling) Measuring Speed by Radar *Any offset between a stationary radar gun and a moving target creates a "cosine effect" that reduces the radar reading by the cosine of the angle between the gun and the vehicle. That is, the radar speed reading is the product of the actual speed and the cosine of the angle.* (*Source:* Fischetti, M., "Working Knowledge," *Scientific American.*)

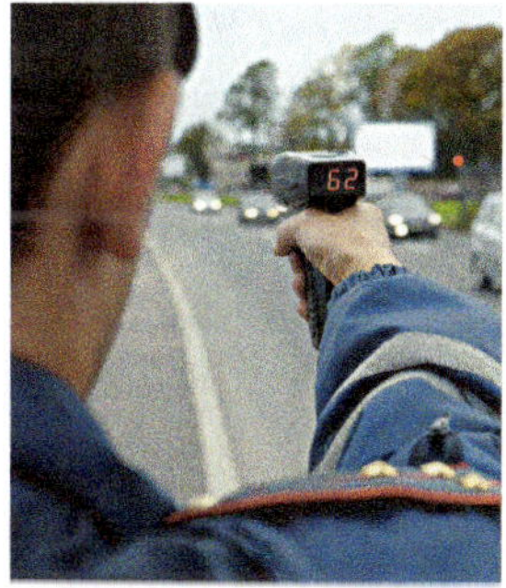

147. Find the radar readings, to the nearest unit, for Auto A and Auto B shown in the figure.

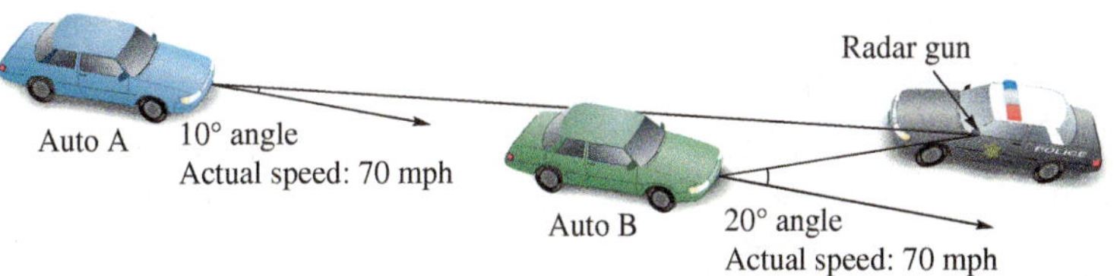

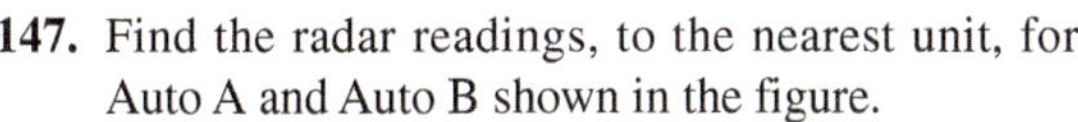

148. The speed reported by a radar gun is reduced by the cosine of angle θ, shown in the figure, where r represents reduced speed and a represents actual speed. Use the figure to show why this "cosine effect" occurs.

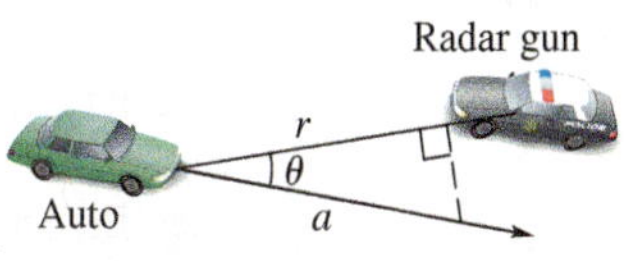

Chapter 5 Quiz (Sections 5.1–5.3)

1. Find the measure of **(a)** the complement and **(b)** the supplement of an angle measuring 19°.

Find the measure of each unknown angle.

2.

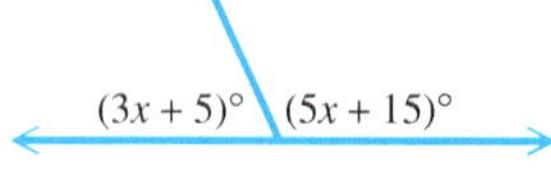

3.

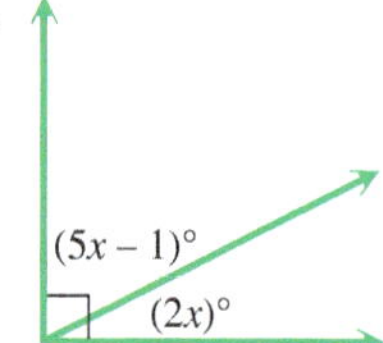

Solve each problem.

4. Perform each conversion.

(a) 77° 12′ 09″ to decimal degrees **(b)** 22.0250° to degrees, minutes, seconds

5. Find the angle of least positive measure (not equal to the given measure) that is coterminal with each angle.

(a) 410° **(b)** −60° **(c)** 890° **(d)** 57°

6. ***Rotating Flywheel*** A flywheel rotates 300 times per min. Through how many degrees does a point on the edge of the flywheel move in 1 sec?

Solve each problem.

7. The terminal side of an angle θ in standard position passes through the point $(-24, 7)$. Find the values of the six trigonometric functions of θ.

8. Find exact values of the six trigonometric functions for angle A in the right triangle.

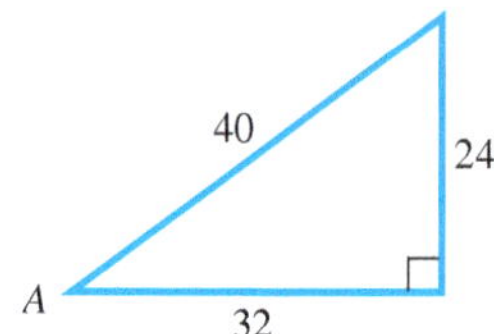

9. Complete the table with exact trigonometric function values.

θ	$\sin\theta$	$\cos\theta$	$\tan\theta$	$\cot\theta$	$\sec\theta$	$\csc\theta$
30°						
45°						
60°						

10. ***Area of a Solar Cell*** A solar cell converts the energy of sunlight directly into electrical energy. The amount of energy a cell produces depends on its area. Suppose a solar cell is hexagonal, as shown in the figure on the left.

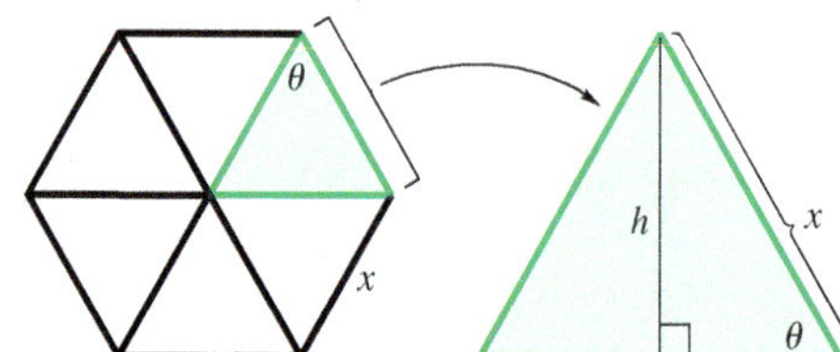

Express its area $\mathcal{A}$ in terms of $\sin\theta$ and any side x. (*Hint:* Consider one of the six equilateral triangles from the hexagon. See the figure on the right.) (*Source:* Kastner, B., *Space Mathematics,* NASA.)

11. Find the exact value of each variable in the figure.

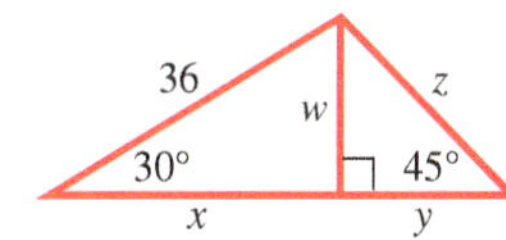

Find exact values of the six trigonometric functions for each angle. Rationalize denominators when applicable.

12. $135°$ 13. $-150°$ 14. $1020°$

Find all values of θ, if θ is in the interval [0°, 360°) and has the given function value.

15. $\sin \theta = \frac{\sqrt{3}}{2}$ 16. $\sec \theta = -\sqrt{2}$

Use a calculator to approximate the value of each expression. Give answers to six decimal places.

17. $\sin 42° 18'$ 18. $\sec(-212° 12')$

Find a value of θ in the interval [0°, 90°) that satisfies each statement. Write each answer in decimal degrees to six decimal places.

19. $\tan \theta = 2.6743210$ 20. $\csc \theta = 2.3861147$

5.4 Solutions and Applications of Right Triangles

- Historical Background
- Significant Digits
- Solving Triangles
- Angles of Elevation or Depression
- Bearing
- Further Applications

Historical Background The beginnings of trigonometry can be traced back to antiquity. **Figure 42** shows the Babylonian tablet **Plimpton 322,** which provides a table of secant values. The Greek mathematicians Hipparchus and Claudius Ptolemy developed a table of chords, which gives values of sines of angles between 0° and 90° in increments of 15 minutes. Until the advent of scientific calculators in the late 20th century, tables were used to find function values.

Applications of *spherical trigonometry* accompanied the study of astronomy for these ancient civilizations. Until the mid-20th century, spherical trigonometry was studied in undergraduate courses. See **Figure 43.**

An introduction to applications of the *plane trigonometry* studied in this text involves applying the ratios to sides of objects that take the shape of right triangles.

Plimpton 322

Figure 42

Figure 43

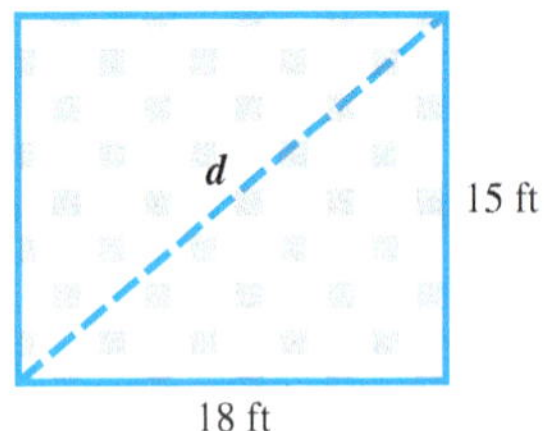

Figure 44

Significant Digits

A number that represents the result of counting, or that results from theoretical work and is not the result of measurement, is an **exact number.** There are 50 states in the U.S. In this statement, 50 is an exact number.

Most values obtained for trigonometric applications are measured values that are *not* exact. Suppose we quickly measure a room as 15 ft by 18 ft. See **Figure 44.** We calculate the length of a diagonal of the room as follows.

$d^2 = 15^2 + 18^2$ Pythagorean theorem

$d^2 = 549$ Apply the exponents and add.

$d = \sqrt{549}$ Square root property; Choose the positive root.

$d \approx 23.430749$

Should this answer be given as the length of the diagonal of the room? Of course not. The number 23.430749 contains six decimal places, while the original data of 15 ft and 18 ft are accurate only to the nearest foot. The results of a calculation can be no more accurate than the least accurate number in the calculation. Thus, the diagonal of the 15-by-18-ft room is approximately 23 ft.

If a wall measured to the nearest foot is 18 ft long, this actually means that the wall has length between 17.5 ft and 18.5 ft. If the wall is measured more accurately as 18.3 ft long, then its length is really between 18.25 ft and 18.35 ft. The results of physical measurement are only approximately accurate and depend on the precision of the measuring instrument as well as the aptness of the observer. The digits obtained by actual measurement are **significant digits.** The measurement 18 ft is said to have two significant digits; 18.3 ft has three significant digits.

In the following numbers, the significant digits are identified in color.

408   21.5   18.00   6.700   0.0025   0.09810   7300

Notice the following.

- 18.00 has four significant digits. The zeros in this number represent measured digits accurate to the nearest hundredth.
- The number 0.0025 has only two significant digits, 2 and 5, because the zeros here are used only to locate the decimal point.
- The number 7300 causes some confusion because it is impossible to determine whether the zeros are measured values. The number 7300 may have two, three, or four significant digits. When presented with this situation, we assume that the zeros are not significant, unless the context of the problem indicates otherwise.

To determine the number of significant digits for answers in applications of angle measure, use the following table.

Significant Digits for Angles

Angle Measure to Nearest	Examples	Write Answer to This Number of Significant Digits
Degree	$62°$, $36°$	two
Ten minutes, or nearest tenth of a degree	$52° 30'$, $60.4°$	three
Minute, or nearest hundredth of a degree	$81° 48'$, $71.25°$	four
Ten seconds, or nearest thousandth of a degree	$10° 52' 20''$, $21.264°$	five

To perform calculations with measured numbers, start by identifying the number with the least number of significant digits. Round the final answer to the same number of significant digits as this number. ***Remember that the answer is no more accurate than the least accurate number in the calculation.***

Solving Triangles ***To solve a triangle means to find the measures of all the angles and sides of the triangle.*** As shown in **Figure 45,** we use a to represent the length of the side opposite angle A, b for the length of the side opposite angle B, and so on. In a right triangle, the letter c is reserved for the hypotenuse.

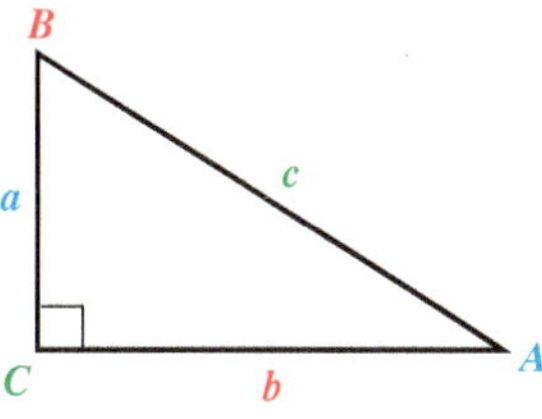

When we are solving triangles, a labeled sketch is an important aid.

Figure 45

EXAMPLE 1 Solving a Right Triangle Given an Angle and a Side

Solve right triangle ABC, if $A = 34° 30'$ and $c = 12.7$ in.

SOLUTION To solve the triangle, find the measures of the remaining sides and angles. See **Figure 46.** To find the value of a, use a trigonometric function involving the known values of angle A and side c. Because the sine of angle A is given by the quotient of the side opposite A and the hypotenuse, use $\sin A$.

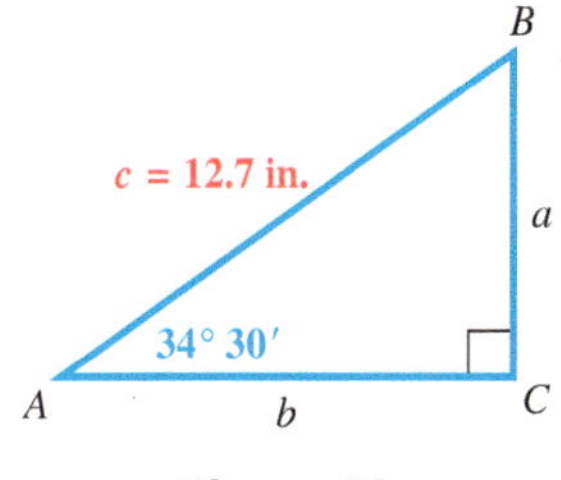

Figure 46

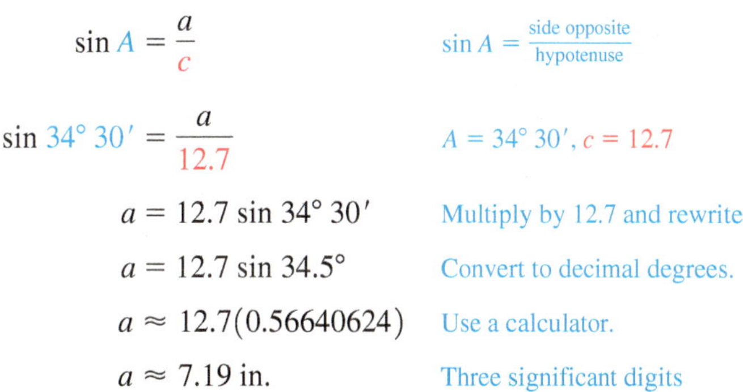

$$\sin A = \frac{a}{c} \qquad \sin A = \frac{\text{side opposite}}{\text{hypotenuse}}$$

$$\sin 34° 30' = \frac{a}{12.7} \qquad A = 34° 30',\ c = 12.7$$

$$a = 12.7 \sin 34° 30' \qquad \text{Multiply by 12.7 and rewrite.}$$

$$a = 12.7 \sin 34.5° \qquad \text{Convert to decimal degrees.}$$

$$a \approx 12.7(0.56640624) \qquad \text{Use a calculator.}$$

$$a \approx 7.19 \text{ in.} \qquad \text{Three significant digits}$$

To find the value of b, we could substitute the value of a just calculated and the given value of c in the Pythagorean theorem. It is better, however, to use the information given in the problem rather than a result just calculated. If an error is made in finding a, then b also would be incorrect. And, rounding more than once may cause the result to be less accurate. To find b, use $\cos A$.

$$\cos A = \frac{b}{c} \qquad \cos A = \frac{\text{side adjacent}}{\text{hypotenuse}}$$

$$\cos 34° 30' = \frac{b}{12.7} \qquad A = 34° 30',\ c = 12.7$$

$$b = 12.7 \cos 34° 30' \qquad \text{Multiply by 12.7 and rewrite.}$$

$$b \approx 10.5 \text{ in.} \qquad \text{Three significant digits}$$

Once b is found, the Pythagorean theorem can be used to verify the results.

LOOKING AHEAD TO CALCULUS

The derivatives of the **parametric equations** $x = f(t)$ and $y = g(t)$ often represent the rate of change of physical quantities, such as velocities. When x and y are related by an equation, the derivatives are **related rates** because a change in one causes a related change in the other. Determining these rates in calculus often requires solving a right triangle.

All that remains to solve triangle *ABC* is to find the measure of angle *B*.

$$A + B = 90^\circ \quad \text{A and B are complementary angles.}$$

$$34^\circ\, 30' + B = 90^\circ \quad A = 34^\circ 30'$$

$$B = 55^\circ\, 30' \quad 90^\circ = 89^\circ\, 60';\ \text{Subtract } 34^\circ\, 30'.$$

✔ **Now Try Exercise 23.**

NOTE In **Example 1,** we could have found the measure of angle *B* first and then used the trigonometric function values of *B* to find the lengths of the unknown sides. A right triangle can usually be solved in several ways.

To maintain accuracy, always use given information as much as possible, and avoid rounding in intermediate steps.

EXAMPLE 2 Solving a Right Triangle Given Two Sides

Solve right triangle *ABC*, if $a = 29.43$ cm and $c = 53.58$ cm.

SOLUTION We draw a sketch showing the given information, as in **Figure 47.** One way to begin is to find angle *A* using the sine function.

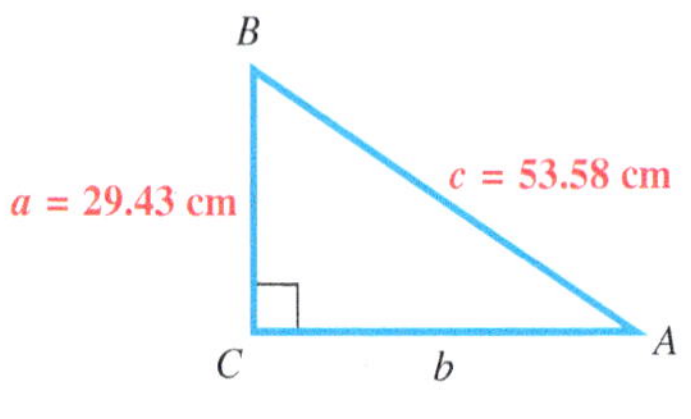

Figure 47

$$\sin A = \frac{a}{c} \quad \sin A = \frac{\text{side opposite}}{\text{hypotenuse}}$$

$$\sin A = \frac{29.43}{53.58} \quad a = 29.43,\ c = 53.58$$

$$\sin A \approx 0.5492721165 \quad \text{Use a calculator.}$$

$$A \approx \sin^{-1}(0.5492721165) \quad \text{Use the inverse sine function.}$$

$$A \approx 33.32^\circ \quad \text{Four significant digits}$$

$$A \approx 33^\circ\, 19' \quad 33.32^\circ = 33^\circ + 0.32(60')$$

The measure of *B* is approximately

$$90^\circ - 33^\circ\, 19' = 56^\circ\, 41'. \quad 90^\circ = 89^\circ\, 60'$$

We now find *b* using the Pythagorean theorem.

$$a^2 + b^2 = c^2 \quad \text{Pythagorean theorem}$$

$$29.43^2 + b^2 = 53.58^2 \quad a = 29.43,\ c = 53.58$$

$$b^2 = 53.58^2 - 29.43^2 \quad \text{Subtract } 29.43^2.$$

$$b = \sqrt{2004.6915} \quad \text{Simplify on the right; square root property}$$

$$b \approx 44.77 \text{ cm} \quad \text{Choose the positive square root.}$$

✔ **Now Try Exercise 33.**

Angles of Elevation or Depression In applications of right triangles, the **angle of elevation** from point X to point Y (above X) is the acute angle formed by ray XY and a horizontal ray with endpoint at X. See **Figure 48(a).** The **angle of depression** from point X to point Y (below X) is the acute angle formed by ray XY and a horizontal ray with endpoint X. See **Figure 48(b).**

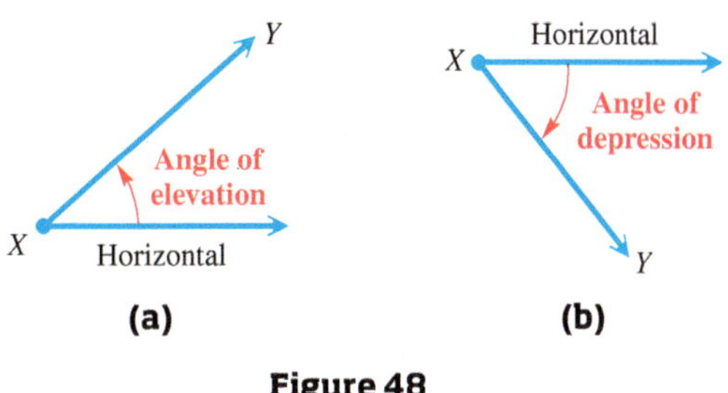

Figure 48

CAUTION Be careful when interpreting the angle of depression. ***Both the angle of elevation and the angle of depression are measured between the line of sight and a horizontal line.***

To solve applied trigonometry problems, follow the same procedure as solving a triangle. ***Drawing a sketch and labeling it correctly in Step 1 is crucial.***

Solving an Applied Trigonometry Problem

Step 1 Draw a sketch, and label it with the given information. Label the quantity to be found with a variable.

Step 2 Use the sketch to write an equation relating the given quantities to the variable.

Step 3 Solve the equation, and check that the answer makes sense.

EXAMPLE 3 Finding a Length Given the Angle of Elevation

At a point A, 123 ft from the base of a flagpole, the angle of elevation to the top of the flagpole is $26° 40'$. Find the height of the flagpole.

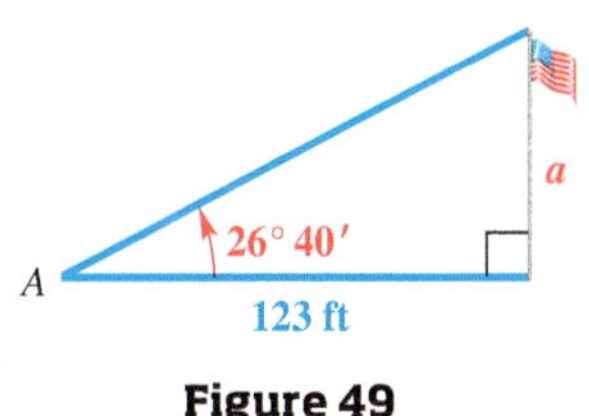

Figure 49

SOLUTION

Step 1 See **Figure 49.** The length of the side adjacent to A is known, and the length of the side opposite A must be found. We will call it a.

Step 2 The tangent ratio involves the given values. Write an equation.

$$\tan A = \frac{\text{side opposite}}{\text{side adjacent}} \qquad \text{Tangent ratio}$$

$$\tan 26° 40' = \frac{a}{123} \qquad A = 26° 40'; \text{ side adjacent} = 123$$

Step 3

$$a = 123 \tan 26° 40' \qquad \text{Multiply by 123 and rewrite.}$$

$$a \approx 123(0.50221888) \qquad \text{Use a calculator.}$$

$$a \approx 61.8 \text{ ft} \qquad \text{Three significant digits}$$

The height of the flagpole is 61.8 ft.

Now Try Exercise 47.

EXAMPLE 4 Finding an Angle of Depression

From the top of a 210-ft cliff, David observes a lighthouse that is 430 ft offshore. Find the angle of depression from the top of the cliff to the base of the lighthouse.

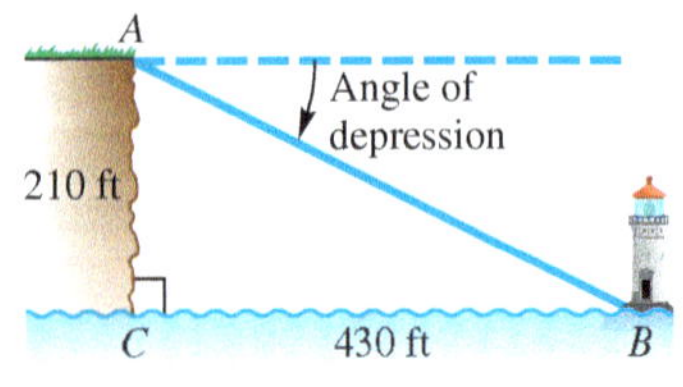

Figure 50

SOLUTION As shown in **Figure 50,** the angle of depression is measured from a horizontal line down to the base of the lighthouse. The angle of depression and angle B, in the right triangle shown, are alternate interior angles whose measures are equal. We use the tangent ratio to solve for angle B.

$$\tan B = \frac{\text{side opposite}}{\text{side adjacent}} \quad \text{Tangent ratio}$$

$$\tan B = \frac{210}{430} \quad \text{Side opposite} = 210\text{; side adjacent} = 430$$

$$B = \tan^{-1}\left(\frac{210}{430}\right) \quad \text{Use the inverse tangent function.}$$

$$B \approx 26° \quad \text{Two significant digits}$$

✔ **Now Try Exercise 49.**

Bearing We now investigate problems involving *bearing*, a term used in navigation. **Bearing** refers to the direction of motion of an object, such as a ship or airplane, or the direction of a second object at a distance relative to the ship or airplane.

We introduce two methods of measuring bearing.

Expressing Bearing (Method 1)

When a single angle is given, it is understood that bearing is measured in a clockwise direction from due north.

Several sample bearings using Method 1 are shown in **Figure 51.**

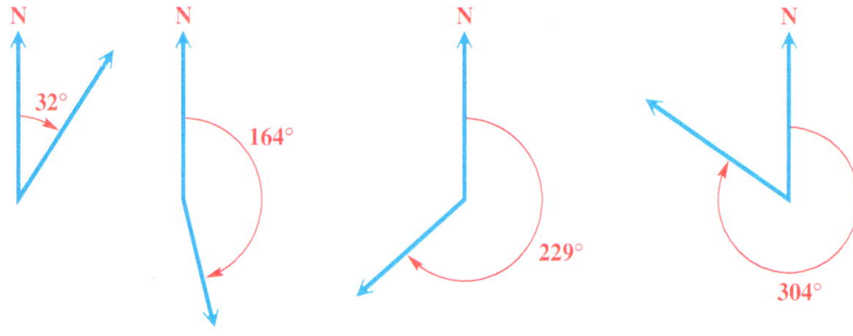

Bearings of 32°, 164°, 229°, and 304°

Figure 51

CAUTION ***A correctly labeled sketch is crucial*** when solving applications like those that follow. Some of the necessary information is often not directly stated in the problem and can be determined only from the sketch.

EXAMPLE 5 Solving a Problem Involving Bearing (Method 1)

Radar stations A and B are on an east-west line, 3.7 km apart. Station A detects a plane at C, on a bearing of 61°. Station B simultaneously detects the same plane, on a bearing of 331°. Find the distance from A to C.

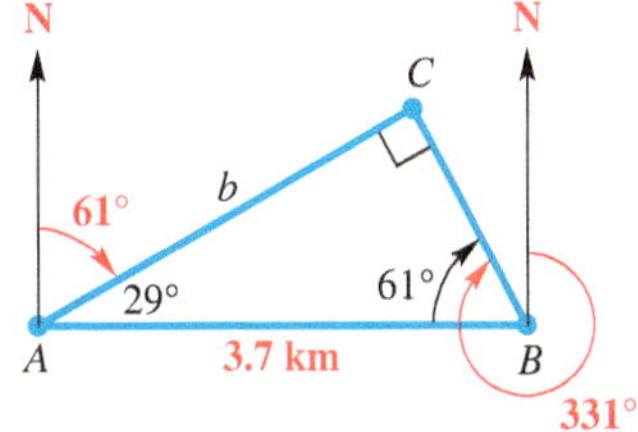

Figure 52

SOLUTION Begin with a sketch showing the given information. See **Figure 52.** A line drawn due north is perpendicular to an east-west line, so right angles are formed at A and B. Angles CBA and CAB can be found as follows.

$$\angle CBA = 331° - 270° = 61° \quad \text{and} \quad \angle CAB = 90° - 61° = 29°$$

A right triangle is formed. The distance from A to C, denoted b in the figure, can be found using the cosine function for angle CAB.

$$\cos 29° = \frac{b}{3.7} \qquad \text{Cosine ratio}$$

$$b = 3.7 \cos 29° \qquad \text{Multiply by 3.7 and rewrite.}$$

$$b \approx 3.2 \text{ km} \qquad \text{Two significant digits}$$

✔ **Now Try Exercise 69.**

Expressing Bearing (Method 2)

Start with a north-south line and use an acute angle to show the direction, either east or west, from this line.

Figure 53 shows several sample bearings using this method. Either N or S always comes first, followed by an acute angle, and then E or W.

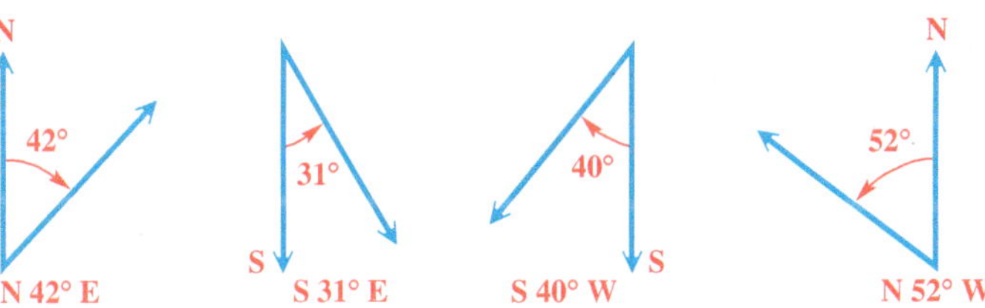

Figure 53

EXAMPLE 6 Solving a Problem Involving Bearing (Method 2)

A ship leaves port and sails on a bearing of N 47° E for 3.5 hr. It then turns and sails on a bearing of S 43° E for 4.0 hr. If the ship's rate is 22 knots (nautical miles per hour), find the distance that the ship is from port.

SOLUTION Draw and label a sketch as in **Figure 54.** Choose a point C on a bearing of N 47° E from port at point A. Then choose a point B on a bearing of S 43° E from point C. Because north-south lines are parallel, angle ACD measures 47° by alternate interior angles. The measure of angle ACB is

$$47° + 43° = 90°,$$

making triangle ABC a right triangle.

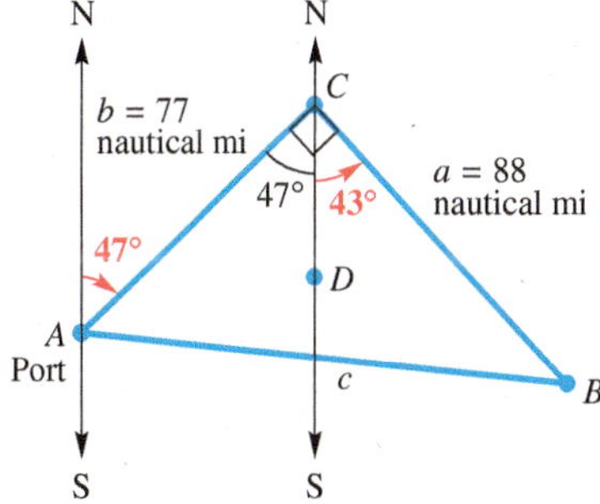

Figure 54

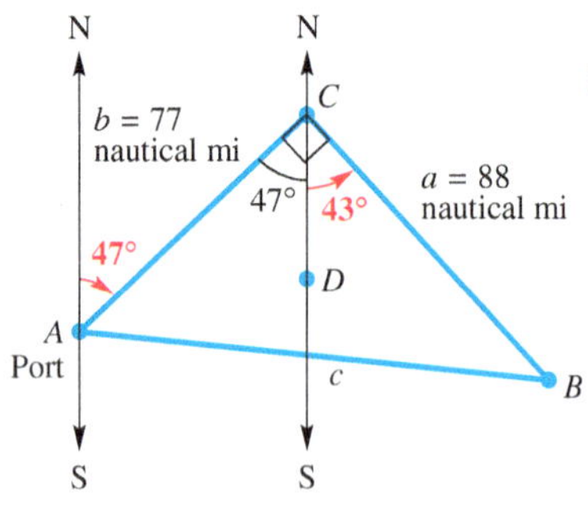

Figure 54 (repeated)

Use the formula relating distance, rate, and time to find the distances in **Figure 54** from A to C and from C to B.

$$b = 22 \times 3.5 = 77 \text{ nautical mi}$$
$$a = 22 \times 4.0 = 88 \text{ nautical mi}$$
Distance = rate × time

Now find c, the distance from port at point A to the ship at point B.

$$a^2 + b^2 = c^2 \quad \text{Pythagorean theorem}$$
$$88^2 + 77^2 = c^2 \quad a = 88,\ b = 77$$
$$c = \sqrt{88^2 + 77^2} \quad \text{If } a^2 + b^2 = c^2 \text{ and } c > 0, \text{ then } c = \sqrt{a^2 + b^2}.$$
$$c \approx 120 \text{ nautical mi} \quad \text{Two significant digits}$$

✔ **Now Try Exercise 75.**

Further Applications

EXAMPLE 7 Using Trigonometry to Measure a Distance

The **subtense bar method** is a method that surveyors use to determine a small distance d between two points P and Q. The subtense bar with length b is centered at Q and situated perpendicular to the line of sight between P and Q. See **Figure 55.** Angle θ is measured, and then the distance d can be determined.

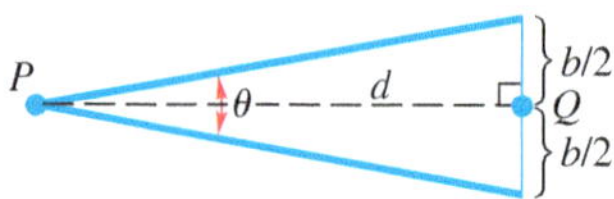

Figure 55

(a) Find d when $\theta = 1°\,23'\,12''$ and $b = 2.0000$ cm.

(b) How much change would there be in the value of d if θ measured $1''$ larger?

SOLUTION

(a) From **Figure 55,** we obtain the following.

$$\cot\frac{\theta}{2} = \frac{d}{\frac{b}{2}} \quad \text{Cotangent ratio}$$

$$d = \frac{b}{2}\cot\frac{\theta}{2} \quad \text{Multiply and rewrite.}$$

Let $b = 2$. To evaluate $\frac{\theta}{2}$, we change θ to decimal degrees.

$$1°\,23'\,12'' \approx 1.386666667°$$

Then $$d = \frac{2}{2}\cot\frac{1.386666667°}{2} \approx 82.634110 \text{ cm.}$$

Use $\cot\theta = \frac{1}{\tan\theta}$ to evaluate.

(b) If θ is $1''$ larger, then $\theta = 1°\,23'\,13'' \approx 1.386944444°$.

$$d = \frac{2}{2}\cot\frac{1.386944444°}{2} \approx 82.617558 \text{ cm}$$

The difference is $82.634110 - 82.617558 = 0.016552$ cm.

✔ **Now Try Exercise 87.**

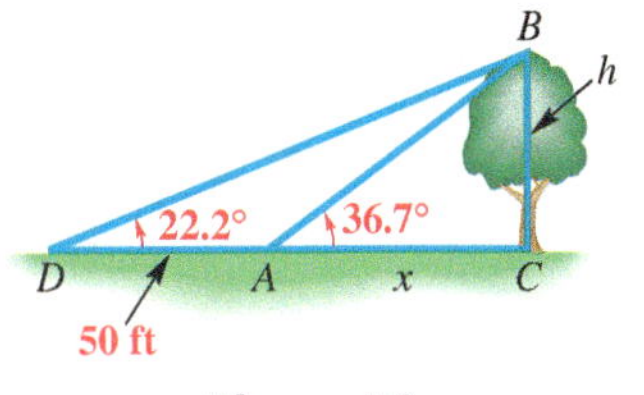

Figure 56

EXAMPLE 8 Solving a Problem Involving Angles of Elevation

Francisco needs to know the height of a tree. From a given point on the ground, he finds that the angle of elevation to the top of the tree is 36.7°. He then moves back 50 ft. From the second point, the angle of elevation to the top of the tree is 22.2°. See **Figure 56.** Find the height of the tree to the nearest foot.

ALGEBRAIC SOLUTION

Figure 56 shows two unknowns: x, the distance from the center of the trunk of the tree to the point where the first observation was made, and h, the height of the tree. See **Figure 57** in the Graphing Calculator Solution. Because nothing is given about the length of the hypotenuse of either triangle ABC or triangle BCD, we use a ratio that does not involve the hypotenuse—namely, the tangent.

In triangle ABC, $\tan 36.7° = \frac{h}{x}$ or $h = x \tan 36.7°$.

In triangle BCD, $\tan 22.2° = \frac{h}{50 + x}$ or $h = (50 + x) \tan 22.2°$.

Each expression equals h, so the expressions must be equal.

$$x \tan 36.7° = (50 + x) \tan 22.2°$$

Equate expressions for h.

$$x \tan 36.7° = 50 \tan 22.2° + x \tan 22.2°$$

Distributive property

$$x \tan 36.7° - x \tan 22.2° = 50 \tan 22.2°$$

Write the x-terms on one side.

$$x(\tan 36.7° - \tan 22.2°) = 50 \tan 22.2°$$

Factor out x.

$$x = \frac{50 \tan 22.2°}{\tan 36.7° - \tan 22.2°}$$

Divide by the coefficient of x.

We saw above that $h = x \tan 36.7°$. Substitute for x.

$$h = \left(\frac{50 \tan 22.2°}{\tan 36.7° - \tan 22.2°}\right) \tan 36.7°$$

Use a calculator.

$$\tan 36.7° = 0.74537703 \quad \text{and} \quad \tan 22.2° = 0.40809244$$

Thus,

$$\tan 36.7° - \tan 22.2° = 0.74537703 - 0.40809244 = 0.33728459$$

and $$h = \left(\frac{50(0.40809244)}{0.33728459}\right) 0.74537703 \approx 45.$$

To the nearest foot, the height of the tree is 45 ft.

GRAPHING CALCULATOR SOLUTION*

In **Figure 57,** we have superimposed **Figure 56** on coordinate axes with the origin at D. By definition, the tangent of the angle between the x-axis and the graph of a line with equation $y = mx + b$ is the slope of the line, m. For line DB, $m = \tan 22.2°$. Because b equals 0, the equation of line DB is

$$y_1 = (\tan 22.2°)x.$$

The equation of line AB is

$$y_2 = (\tan 36.7°)x + b.$$

Because $b \neq 0$ here, we use the point $A(50, 0)$ and the point-slope form to find the equation.

$$y_2 - y_0 = m(x - x_0) \quad \text{Point-slope form}$$

$$y_2 - 0 = m(x - 50) \quad x_0 = 50, y_0 = 0$$

$$y_2 = \tan 36.7°(x - 50)$$

Lines y_1 and y_2 are graphed in **Figure 58.** The y-coordinate of the point of intersection of the graphs gives the length of BC, or h. Thus, $h \approx 45$.

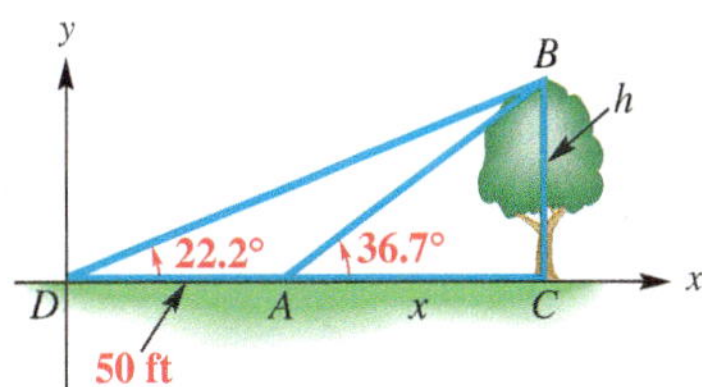

Figure 57

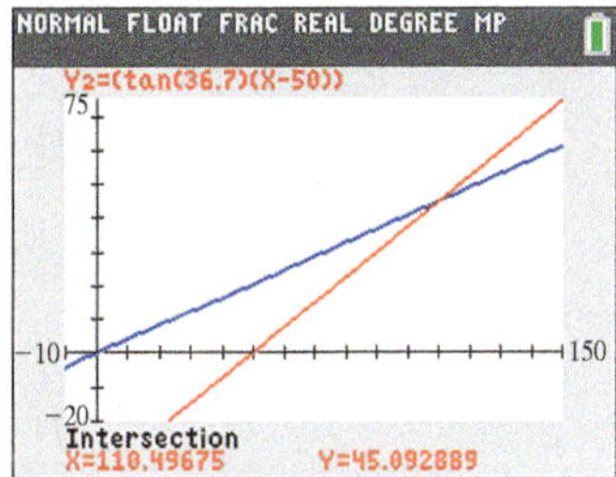

Figure 58

✔ Now Try Exercise 77.

*Source: Reprinted with permission from *The Mathematics Teacher*, copyright 1995 by the National Council of Teachers of Mathematics. All rights reserved.

5.4 Exercises

CONCEPT PREVIEW *Match each equation in Column I with the appropriate right triangle in Column II. In each case, the goal is to find the value of x.*

I

1. $x = 5 \cot 38°$
2. $x = 5 \cos 38°$
3. $x = 5 \tan 38°$
4. $x = 5 \csc 38°$
5. $x = 5 \sin 38°$
6. $x = 5 \sec 38°$

II

A.

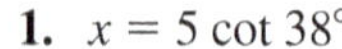

B.

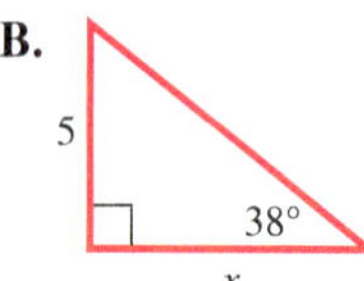

C.

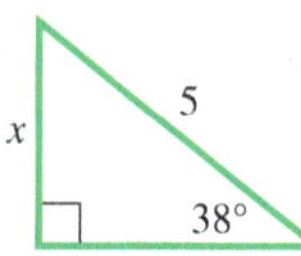

D.

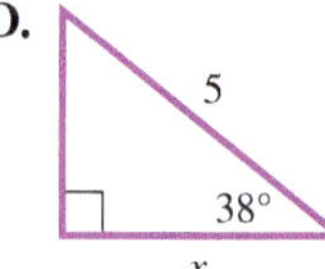

E.

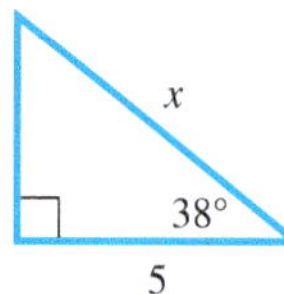

F.

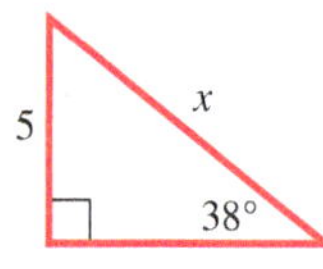

CONCEPT PREVIEW *Match the measure of bearing in Column I with the appropriate graph in Column II.*

I

7. 110°
8. S 20° W
9. N 70° W
10. 270°
11. 180°
12. N 70° E

II

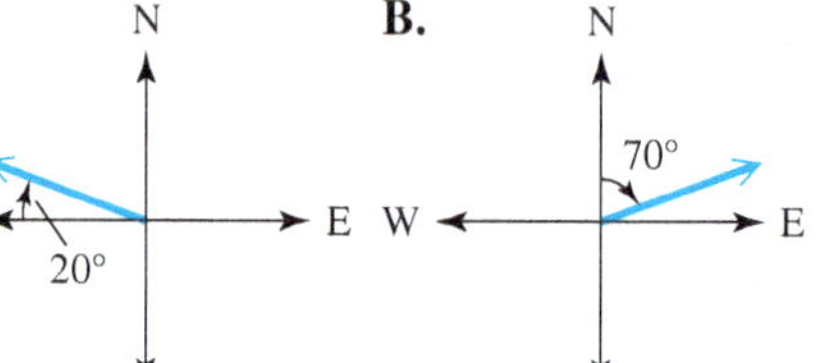

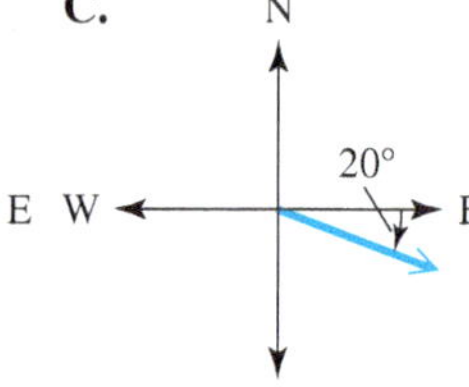

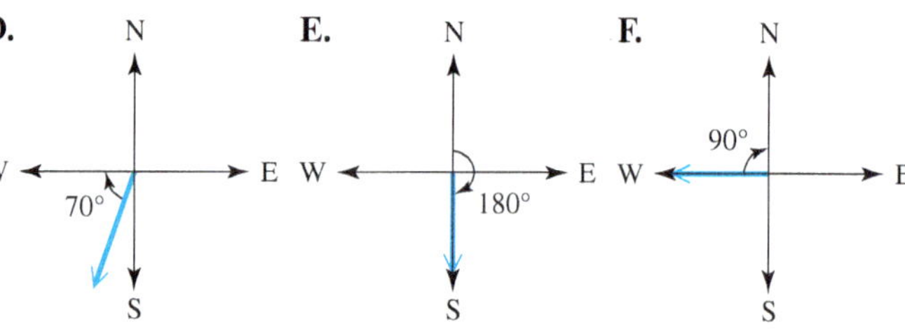

Concept Check Refer to the discussion of accuracy and significant digits in this section to answer the following.

13. *Lake Ponchartrain Causeway* The world's longest bridge over a body of water (continuous) is the causeway that joins the north and south shores of Lake Ponchartrain, a saltwater lake that lies north of New Orleans, Louisiana. It consists of two parallel spans. The longer of the spans measures 23.83 mi. State the range represented by this number. (*Source:* www.worldheritage.org)

14. *Mt. Everest* When Mt. Everest was first surveyed, the surveyors obtained a height of 29,000 ft to the nearest foot. State the range represented by this number. (The surveyors thought no one would believe a measurement of 29,000 ft, so they reported it as 29,002.) (*Source:* Dunham, W., *The Mathematical Universe,* John Wiley and Sons.)

Solve each right triangle. When two sides are given, give angles in degrees and minutes. ***See Examples 1 and 2.***

15.

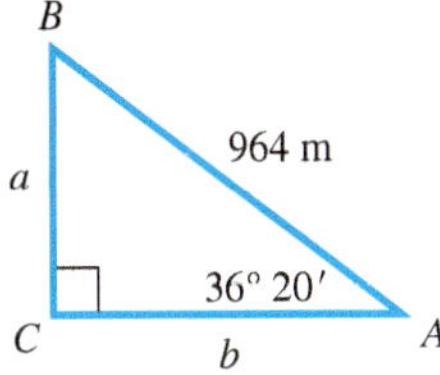

16.

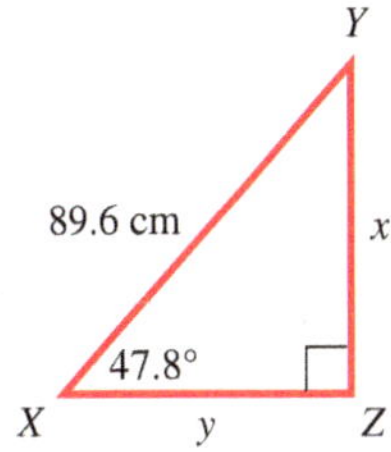

17.

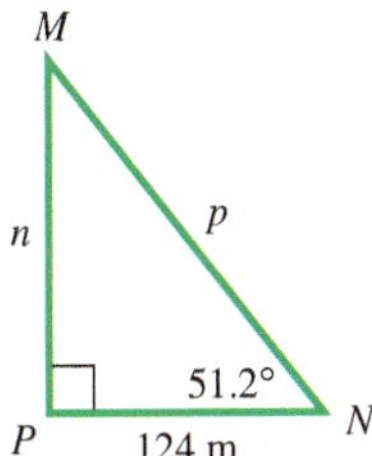

18.

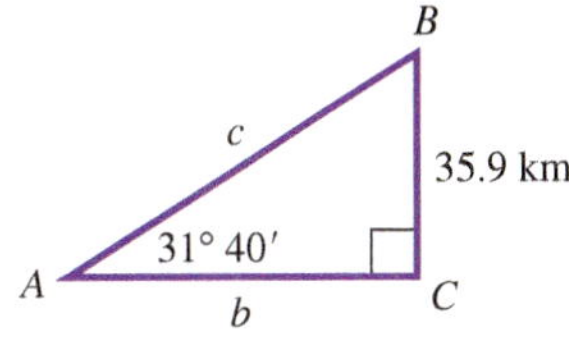

19.

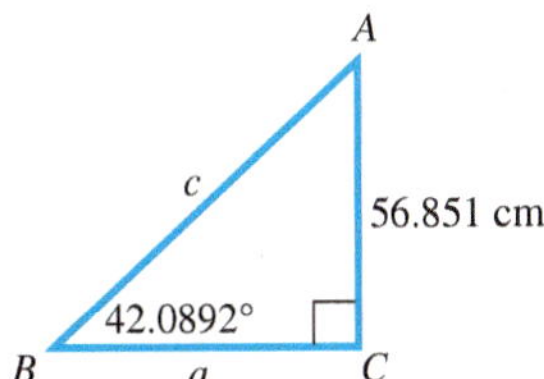

20.

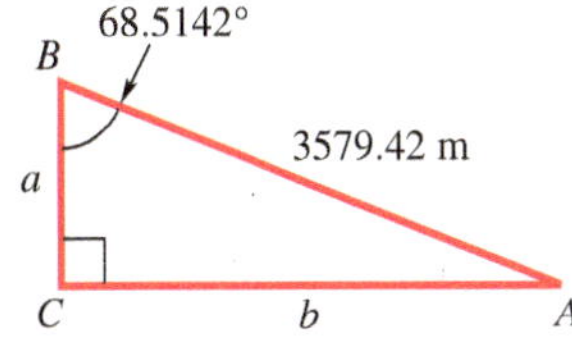

21.

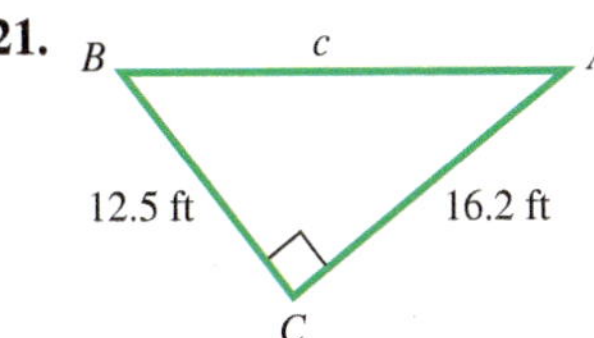

22.

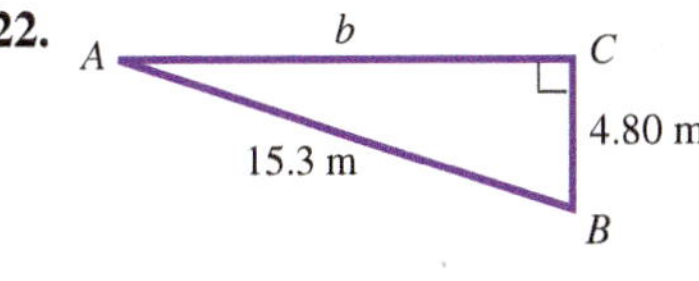

Solve each right triangle. In each case, $C = 90°$. If angle information is given in degrees and minutes, give answers in the same way. If angle information is given in decimal degrees, do likewise in answers. When two sides are given, give angles in degrees and minutes. ***See Examples 1 and 2.***

23. $A = 28.0°$, $c = 17.4$ ft

24. $B = 46.0°$, $c = 29.7$ m

25. $B = 73.0°$, $b = 128$ in.

26. $A = 62.5°$, $a = 12.7$ m

27. $A = 61.0°$, $b = 39.2$ cm

28. $B = 51.7°$, $a = 28.1$ ft

29. $a = 13$ m, $c = 22$ m

30. $b = 32$ ft, $c = 51$ ft

31. $a = 76.4$ yd, $b = 39.3$ yd

32. $a = 958$ m, $b = 489$ m

33. $a = 18.9$ cm, $c = 46.3$ cm

34. $b = 219$ m, $c = 647$ m

35. $A = 53°24'$, $c = 387.1$ ft

36. $A = 13°47'$, $c = 1285$ m

37. $B = 39°09'$, $c = 0.6231$ m

38. $B = 82°51'$, $c = 4.825$ cm

Solve each problem. ***See Examples 1–4.***

39. ***Height of a Ladder on a Wall*** A 13.5-m fire truck ladder is leaning against a wall. Find the distance d the ladder goes up the wall (above the top of the fire truck) if the ladder makes an angle of $43° 50'$ with the horizontal.

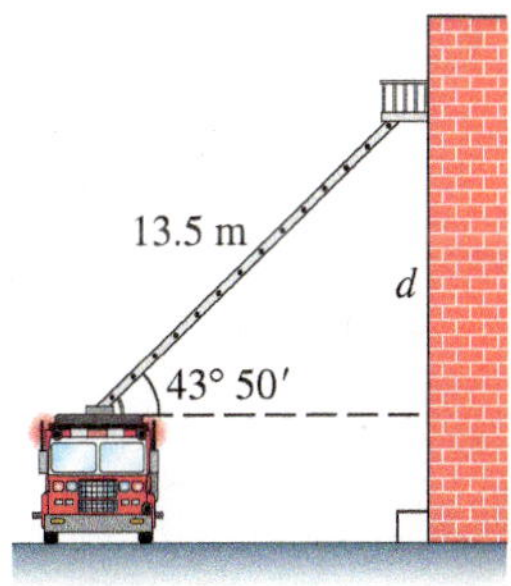

40. ***Distance across a Lake*** To find the distance RS across a lake, a surveyor lays off length $RT = 53.1$ m, so that angle $T = 32° 10'$ and angle $S = 57° 50'$. Find length RS.

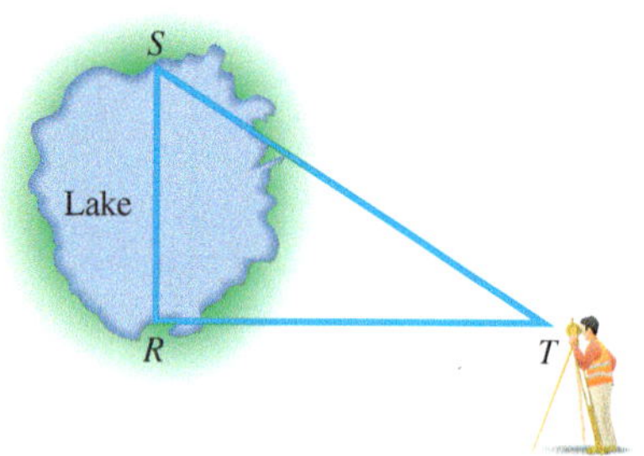

41. ***Height of a Building*** From a window 30.0 ft above the street, the angle of elevation to the top of the building across the street is 50.0° and the angle of depression to the base of this building is 20.0°. Find the height of the building across the street.

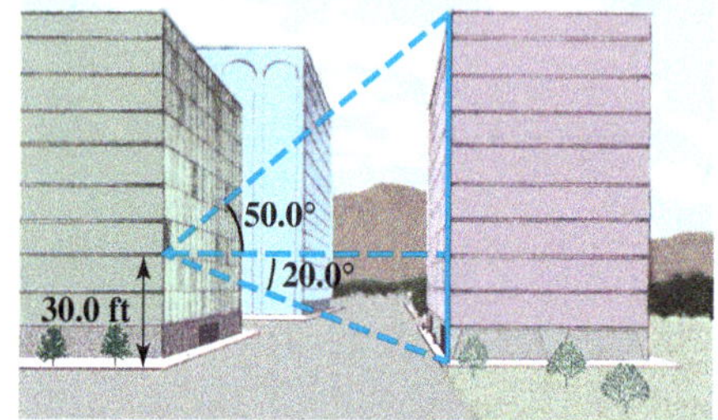

42. ***Diameter of the Sun*** To determine the diameter of the sun, an astronomer might sight with a **transit** (a device used by surveyors for measuring angles) first to one edge of the sun and then to the other, estimating that the included angle equals 32′. Assuming that the distance d from Earth to the sun is 92,919,800 mi, approximate the diameter of the sun.

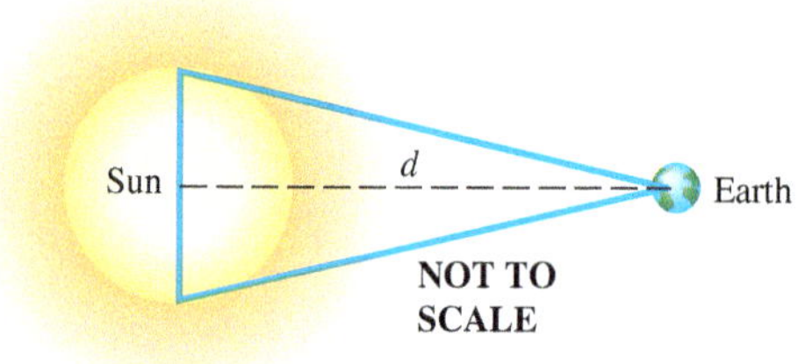

43. ***Side Lengths of a Triangle*** The length of the base of an isosceles triangle is 42.36 in. Each base angle is 38.12°. Find the length of each of the two equal sides of the triangle. (*Hint:* Divide the triangle into two right triangles.)

44. ***Altitude of a Triangle*** Find the altitude of an isosceles triangle having base 184.2 cm if the angle opposite the base is 68° 44′.

Solve each problem. ***See Examples 3 and 4.***

45. ***Height of a Tower*** The shadow of a vertical tower is 40.6 m long when the angle of elevation of the sun is 34.6°. Find the height of the tower.

46. ***Distance from the Ground to the Top of a Building*** The angle of depression from the top of a building to a point on the ground is 32° 30′. How far is the point on the ground from the top of the building if the building is 252 m high?

47. ***Length of a Shadow*** Suppose that the angle of elevation of the sun is 23.4°. Find the length of the shadow cast by a person who is 5.75 ft tall.

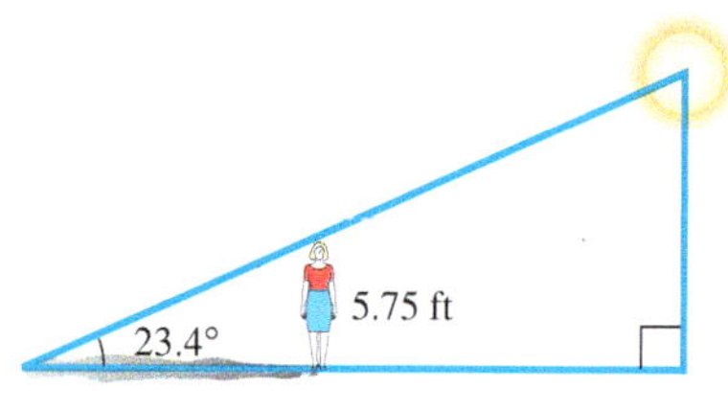

48. *Airplane Distance* An airplane is flying 10,500 ft above level ground. The angle of depression from the plane to the base of a tree is 13° 50′. How far horizontally must the plane fly to be directly over the tree?

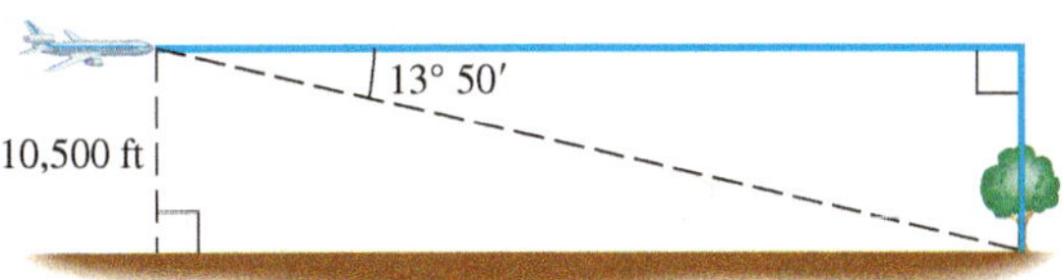

49. *Angle of Depression of a Light* A company safety committee has recommended that a floodlight be mounted in a parking lot so as to illuminate the employee exit, as shown in the figure. Find the angle of depression of the light to the nearest minute.

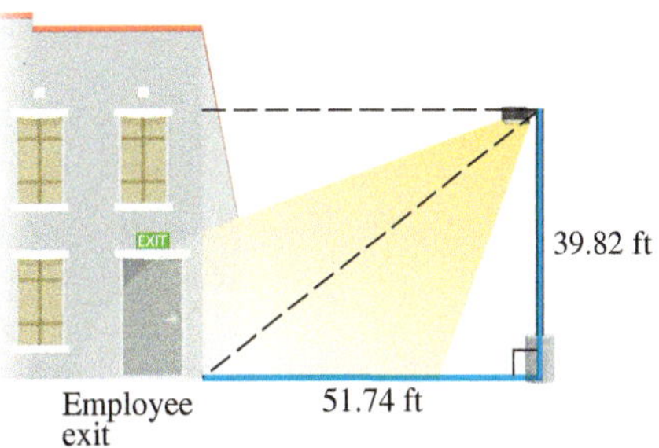

50. *Height of a Building* The angle of elevation from the top of a small building to the top of a nearby taller building is 46° 40′, and the angle of depression to the bottom is 14° 10′. If the shorter building is 28.0 m high, find the height of the taller building.

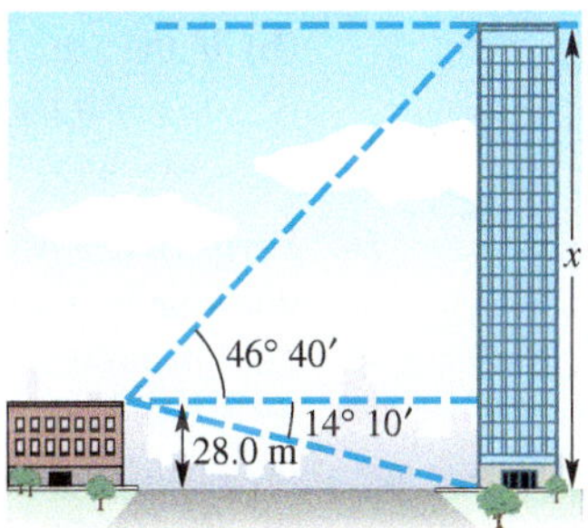

51. *Angle of Elevation of the Sun* The length of the shadow of a building 34.09 m tall is 37.62 m. Find the angle of elevation of the sun to the nearest hundredth of a degree.

52. *Angle of Elevation of the Sun* The length of the shadow of a flagpole 55.20 ft tall is 27.65 ft. Find the angle of elevation of the sun to the nearest hundredth of a degree.

53. *Angle of Elevation of the Pyramid of the Sun* The Pyramid of the Sun is in the ancient Mexican city of Teotihuacan. The base is a square with sides about 700 ft long. The height of the pyramid is about 200 ft. Find the angle of elevation θ of the edge indicated in the figure to two significant digits. (*Hint:* The base of the triangle in the figure has measure half the diagonal of the square base of the pyramid.) (*Source*: www.britannica.com)

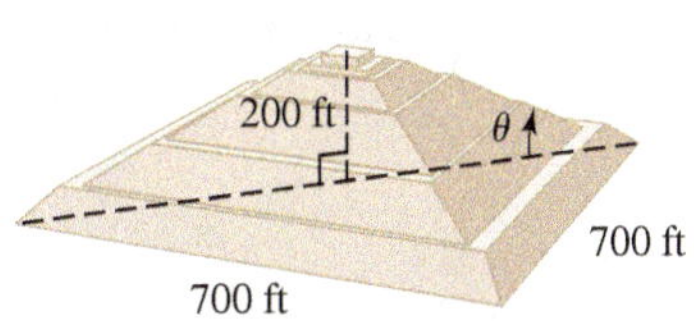

54. ***Cloud Ceiling*** The U.S. Weather Bureau defines a **cloud ceiling** as the altitude of the lowest clouds that cover more than half the sky. To determine a cloud ceiling, a powerful searchlight projects a circle of light vertically on the bottom of the cloud. An observer sights the circle of light in the crosshairs of a tube called a **clinometer.** A pendant hanging vertically from the tube and resting on a protractor gives the angle of elevation. Find the cloud ceiling if the searchlight is located 1000 ft from the observer and the angle of elevation is 30.0° as measured with a clinometer at eye-height 6 ft. (Assume three significant digits.)

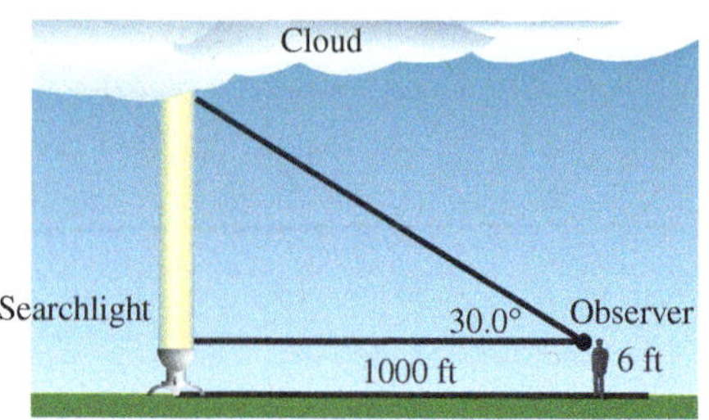

55. ***Height of Mt. Everest*** The highest mountain peak in the world is Mt. Everest, located in the Himalayas. The height of this enormous mountain was determined in 1856 by surveyors using trigonometry long before it was first climbed in 1953. This difficult measurement had to be done from a great distance. At an altitude of 14,545 ft on a different mountain, the straight-line distance to the peak of Mt. Everest is 27.0134 mi and its angle of elevation is $\theta = 5.82°$. (*Source:* Dunham, W., *The Mathematical Universe,* John Wiley and Sons.)

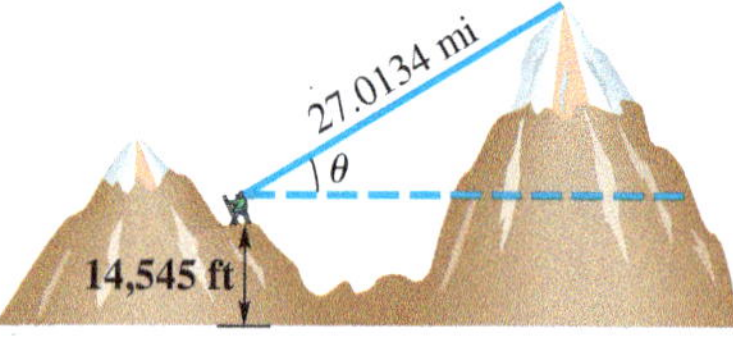

(a) Approximate the height (in feet) of Mt. Everest.

(b) In the actual measurement, Mt. Everest was over 100 mi away and the curvature of Earth had to be taken into account. Would the curvature of Earth make the peak appear taller or shorter than it actually is?

56. ***Error in Measurement*** A degree may seem like a very small unit, but an error of one degree in measuring an angle may be very significant. For example, suppose a laser beam directed toward the visible center of the moon misses its assigned target by 30.0″. How far is it (in miles) from its assigned target? Take the distance from the surface of Earth to that of the moon to be 234,000 mi. (*Source: A Sourcebook of Applications of School Mathematics* by Donald Bushaw et al.)

The two methods of expressing bearing can be interpreted using a rectangular coordinate system. Suppose that an observer for a radar station is located at the origin of a coordinate system. Find the bearing of an airplane located at each point. Express the bearing using both methods.

57. $(-4, 0)$ **58.** $(5, 0)$ **59.** $(0, 4)$ **60.** $(0, -2)$

61. $(-5, 5)$ **62.** $(-3, -3)$ **63.** $(2, -2)$ **64.** $(2, 2)$

Solve each problem. ***See Examples 5 and 6.***

65. ***Distance Flown by a Plane*** A plane flies 1.3 hr at 110 mph on a bearing of 38°. It then turns and flies 1.5 hr at the same speed on a bearing of 128°. How far is the plane from its starting point?

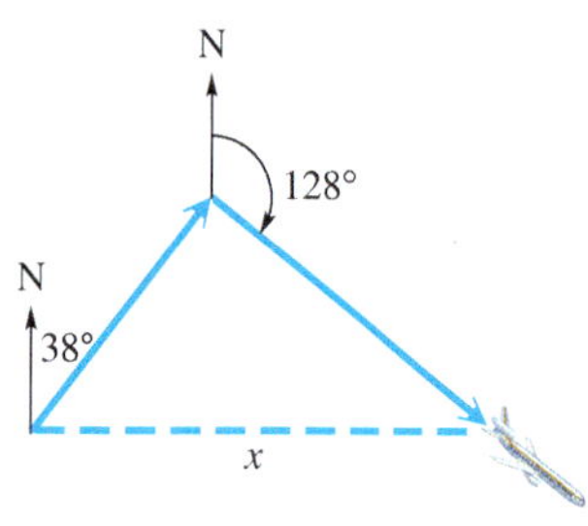

66. *Distance Traveled by a Ship* A ship travels 55 km on a bearing of 27° and then travels on a bearing of 117° for 140 km. Find the distance from the starting point to the ending point.

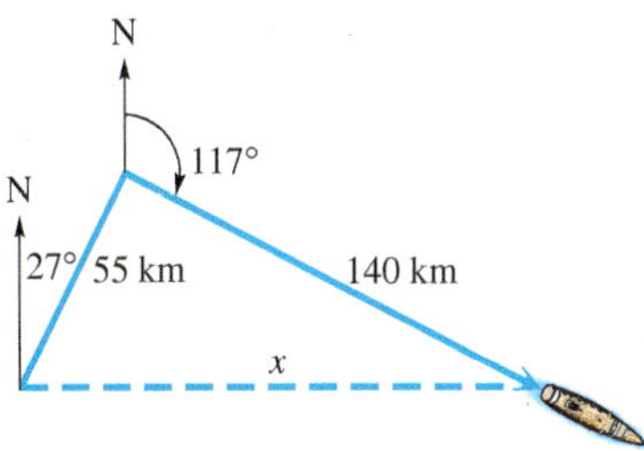

67. *Distance between Two Ships* Two ships leave a port at the same time. The first ship sails on a bearing of 40° at 18 knots (nautical miles per hour) and the second on a bearing of 130° at 26 knots. How far apart are they after 1.5 hr?

68. *Distance between Two Ships* Two ships leave a port at the same time. The first ship sails on a bearing of 52° at 17 knots and the second on a bearing of 322° at 22 knots. How far apart are they after 2.5 hr?

69. *Distance between Two Docks* Two docks are located on an east-west line 2587 ft apart. From dock A, the bearing of a coral reef is 58° 22′. From dock B, the bearing of the coral reef is 328° 22′. Find the distance from dock A to the coral reef.

70. *Distance between Two Lighthouses* Two lighthouses are located on a north-south line. From lighthouse A, the bearing of a ship 3742 m away is 129° 43′. From lighthouse B, the bearing of the ship is 39° 43′. Find the distance between the lighthouses.

71. *Distance between Two Ships* A ship leaves its home port and sails on a bearing of S 61°50′ E. Another ship leaves the same port at the same time and sails on a bearing of N 28°10′ E. If the first ship sails at 24.0 mph and the second ship sails at 28.0 mph, find the distance between the two ships after 4 hr.

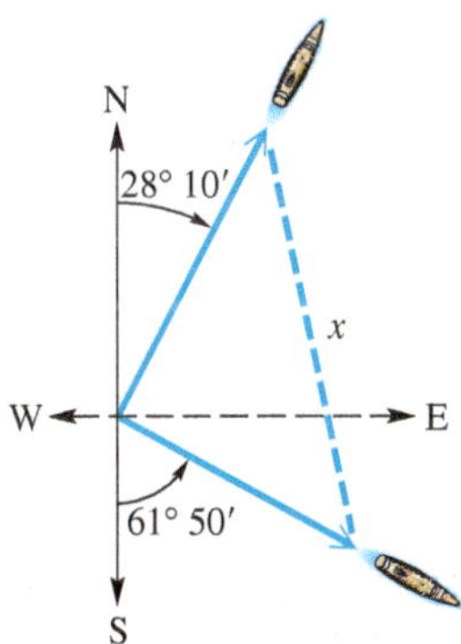

72. *Distance between Transmitters* Radio direction finders are set up at two points A and B, which are 2.50 mi apart on an east-west line. From A, it is found that the bearing of a signal from a radio transmitter is N 36°20′ E, and from B the bearing of the same signal is N 53°40′ W. Find the distance of the transmitter from B.

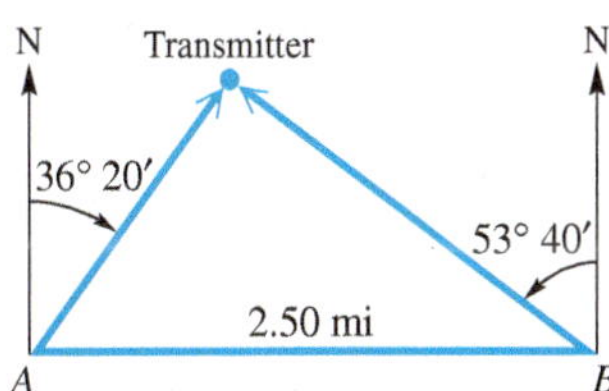

73. *Flying Distance* The bearing from A to C is S 52° E. The bearing from A to B is N 84° E. The bearing from B to C is S 38° W. A plane flying at 250 mph takes 2.4 hr to go from A to B. Find the distance from A to C.

74. *Flying Distance* The bearing from A to C is N 64° W. The bearing from A to B is S 82° W. The bearing from B to C is N 26° E. A plane flying at 350 mph takes 1.8 hr to go from A to B. Find the distance from B to C.

75. *Distance between Two Cities* The bearing from Winston-Salem, North Carolina, to Danville, Virginia, is N 42° E. The bearing from Danville to Goldsboro, North Carolina, is S 48° E. A car traveling at 65 mph takes 1.1 hr to go from Winston-Salem to Danville and 1.8 hr to go from Danville to Goldsboro. Find the distance from Winston-Salem to Goldsboro.

76. *Distance between Two Cities* The bearing from Atlanta to Macon is S 27° E, and the bearing from Macon to Augusta is N 63° E. An automobile traveling at 62 mph needs $1\frac{1}{4}$ hr to go from Atlanta to Macon and $1\frac{3}{4}$ hr to go from Macon to Augusta. Find the distance from Atlanta to Augusta.

Solve each problem. **See Examples 7 and 8.**

77. *Height of a Pyramid* The angle of elevation from a point on the ground to the top of a pyramid is 35° 30′. The angle of elevation from a point 135 ft farther back to the top of the pyramid is 21° 10′. Find the height of the pyramid.

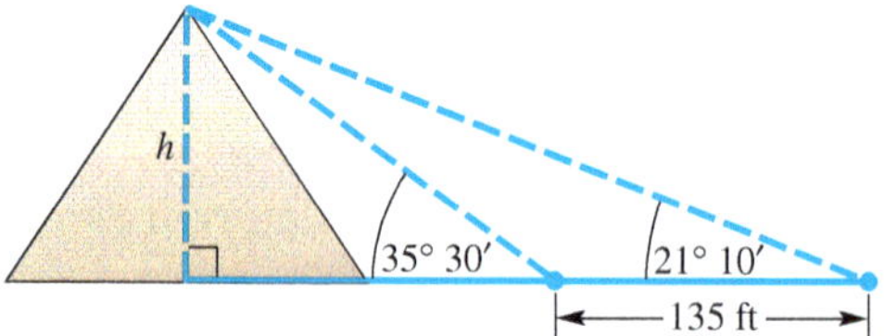

78. *Distance between a Whale and a Lighthouse* A whale researcher is watching a whale approach directly toward a lighthouse as she observes from the top of this lighthouse. When she first begins watching the whale, the angle of depression to the whale is 15° 50′. Just as the whale turns away from the lighthouse, the angle of depression is 35° 40′. If the height of the lighthouse is 68.7 m, find the distance traveled by the whale as it approached the lighthouse.

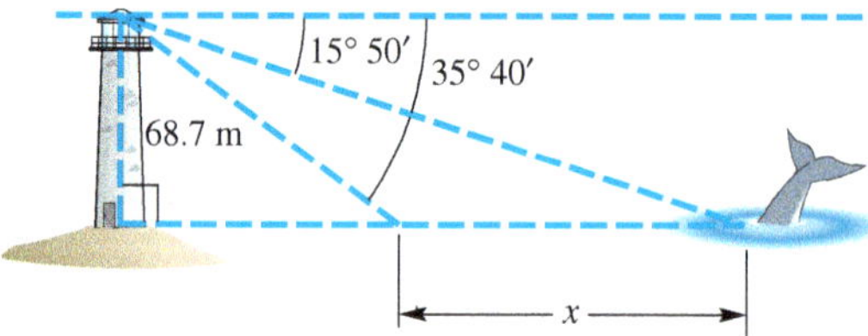

79. *Height of an Antenna* A scanner antenna is on top of the center of a house. The angle of elevation from a point 28.0 m from the center of the house to the top of the antenna is 27° 10′, and the angle of elevation to the bottom of the antenna is 18° 10′. Find the height of the antenna.

80. *Height of Mt. Whitney* The angle of elevation from Lone Pine to the top of Mt. Whitney is 10° 50′. A hiker, traveling 7.00 km from Lone Pine along a straight, level road toward Mt. Whitney, finds the angle of elevation to be 22° 40′. Find the height of the top of Mt. Whitney above the level of the road.

81. Find h as indicated in the figure.

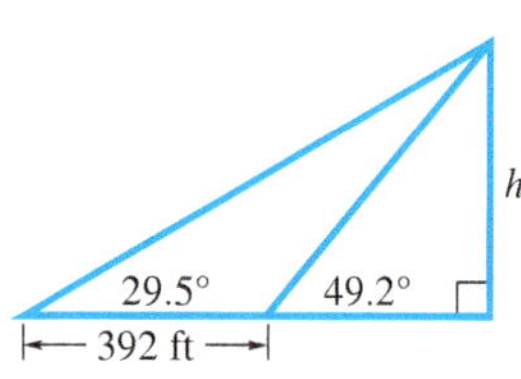

82. Find h as indicated in the figure.

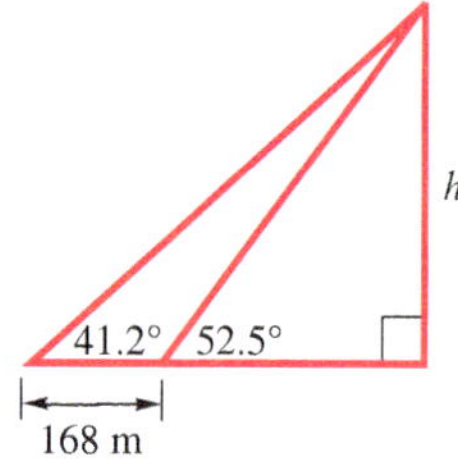

83. ***Distance of a Plant from a Fence*** In one area, the lowest angle of elevation of the sun in winter is $23^\circ 20'$. Find the minimum distance x that a plant needing full sun can be placed from a fence 4.65 ft high.

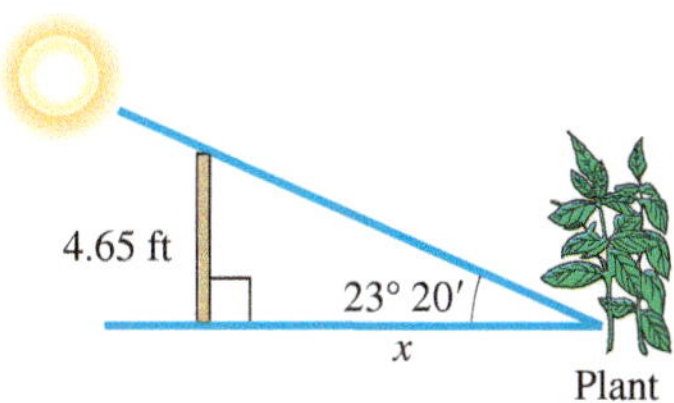

84. ***Distance through a Tunnel*** A tunnel is to be built from A to B. Both A and B are visible from C. If AC is 1.4923 mi and BC is 1.0837 mi, and if C is 90°, find the measures of angles A and B.

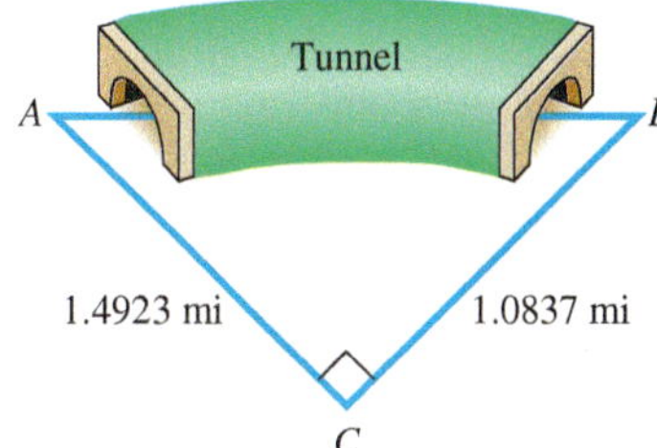

85. ***Height of a Plane above Earth*** Find the minimum height h above the surface of Earth so that a pilot at point A in the figure can see an object on the horizon at C, 125 mi away. Assume 4.00×10^3 mi as the radius of Earth.

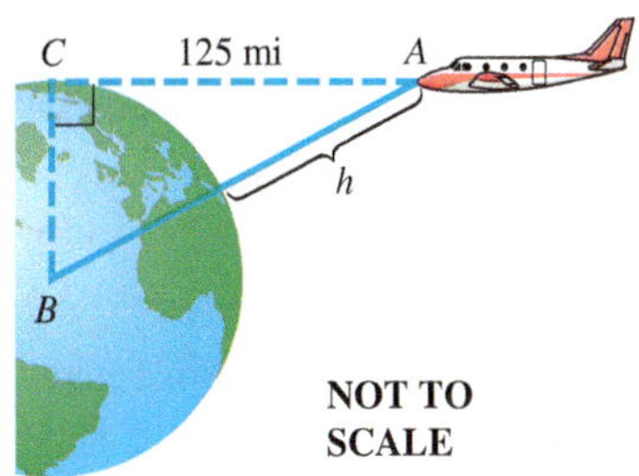

86. ***Length of a Side of a Piece of Land*** A piece of land has the shape shown in the figure. Find the length x.

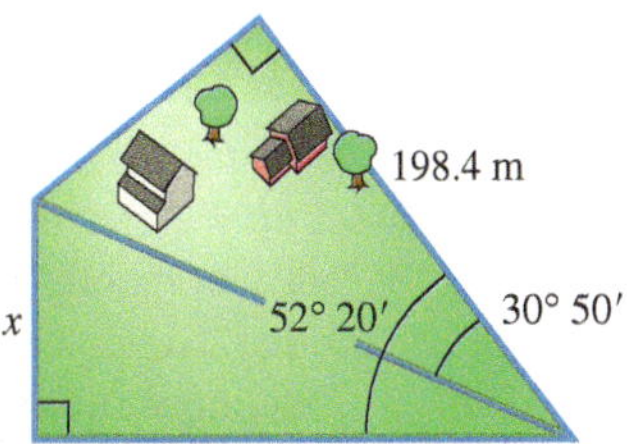

87. ***(Modeling) Distance between Two Points*** A variation of the subtense bar method that surveyors use to determine larger distances d between two points P and Q is shown in the figure. The subtense bar with length b is placed between points P and Q so that the bar is centered on and perpendicular to the line of sight between P and Q. Angles α and β are measured from points P and Q, respectively. (*Source:* Mueller, I. and K. Ramsayer, *Introduction to Surveying*, Frederick Ungar Publishing Co.)

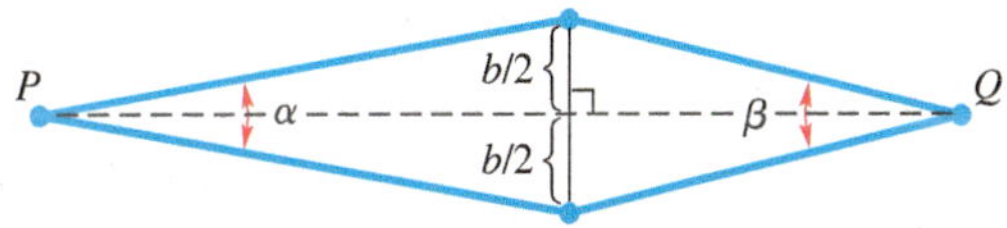

(a) Find a formula for d involving α, β, and b.

(b) Use the formula from part (a) to determine d if $\alpha = 37' \, 48''$, $\beta = 42' \, 03''$, and $b = 2.000$ cm.

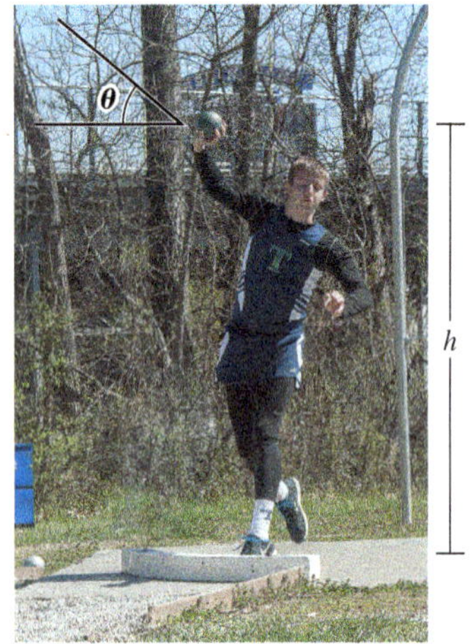

88. *(Modeling) Distance of a Shot Put* A shot-putter trying to improve performance may wonder whether there is an optimal angle to aim for, or whether the velocity (speed) at which the ball is thrown is more important. The figure shows the path of a steel ball thrown by a shot-putter. The distance D depends on initial velocity v, height h, and angle θ when the ball is released.

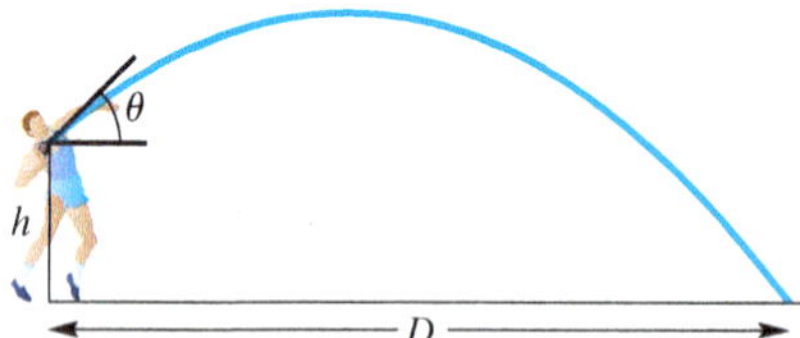

One model developed for this situation gives D as

$$D = \frac{v^2 \sin\theta \cos\theta + v\cos\theta\sqrt{(v\sin\theta)^2 + 64h}}{32}.$$

Typical ranges for the variables are v: 33–46 ft per sec; h: 6–8 ft; and θ: 40°–45°. (*Source:* Kreighbaum, E. and K. Barthels, *Biomechanics,* Allyn & Bacon.)

(a) To see how angle θ affects distance D, let $v = 44$ ft per sec and $h = 7$ ft. Calculate D, to the nearest hundredth, for $\theta = 40°$, 42°, and 45°. How does distance D change as θ increases?

(b) To see how velocity v affects distance D, let $h = 7$ and $\theta = 42°$. Calculate D, to the nearest hundredth, for $v = 43$, 44, and 45 ft per sec. How does distance D change as v increases?

(c) Which affects distance D more, v or θ? What should the shot-putter do to improve performance?

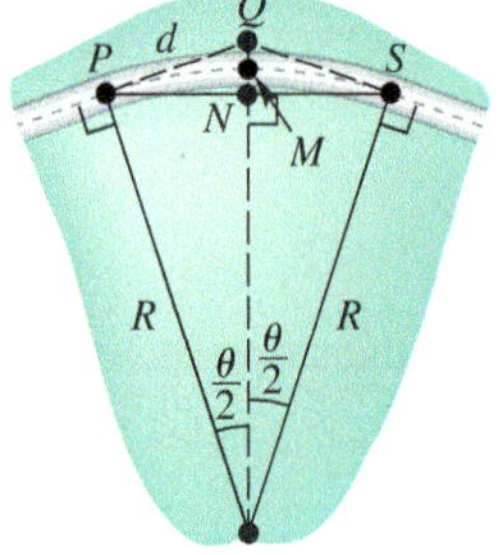

89. *(Modeling) Highway Curves* A basic highway curve connecting two straight sections of road may be circular. In the figure in the margin, the points P and S mark the beginning and end of the curve. Let Q be the point of intersection where the two straight sections of highway leading into the curve would meet if extended. The radius of the curve is R, and the central angle θ denotes how many degrees the curve turns. (*Source:* Mannering, F. and W. Kilareski, *Principles of Highway Engineering and Traffic Analysis*, Second Edition, John Wiley and Sons.)

(a) If $R = 965$ ft and $\theta = 37°$, find the distance d between P and Q.

(b) Find an expression in terms of R and θ for the distance between points M and N.

90. *(Modeling) Stopping Distance on a Curve* Refer to **Exercise 89.** When an automobile travels along a circular curve, objects like trees and buildings situated on the inside of the curve can obstruct the driver's vision. In the figure, the *minimum* distance d that should be cleared on the inside of the highway is modeled by the equation

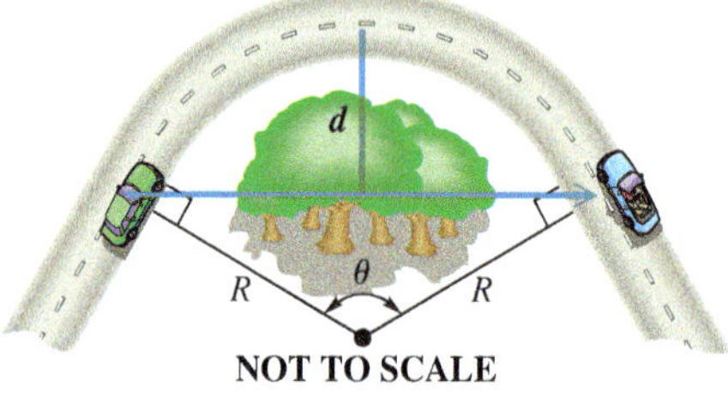

$$d = R\left(1 - \cos\frac{\theta}{2}\right).$$

(*Source:* Mannering, F. and W. Kilareski, *Principles of Highway Engineering and Traffic Analysis,* Second Edition, John Wiley and Sons.)

(a) It can be shown that if θ is measured in degrees, then $\theta \approx \frac{57.3S}{R}$, where S is the safe stopping distance for the given speed limit. Compute d to the nearest foot for a 55 mph speed limit if $S = 336$ ft and $R = 600$ ft.

(b) Compute d to the nearest foot for a 65 mph speed limit given $S = 485$ ft and $R = 600$ ft.

(c) How does the speed limit affect the amount of land that should be cleared on the inside of the curve?

Chapter 5 Test Prep

Key Terms

5.1 line, line segment (or segment), ray, endpoint of a ray, angle, side of an angle, vertex of an angle, initial side, terminal side, positive angle, negative angle, degree, acute angle, right angle, obtuse angle, straight angle, complementary angles (complements), supplementary angles (supplements), minute, second, angle in standard position, quadrantal angle, coterminal angles

5.2 sine (sin), cosine (cos), tangent (tan), cotangent (cot), secant (sec), cosecant (csc), degree mode, reciprocal

5.3 side opposite, side adjacent, cofunctions, reference angle

5.4 exact number, significant digits, angle of elevation, angle of depression, bearing

New Symbols

⅂	right angle symbol (for a right triangle)
θ	Greek letter theta
°	degree
′	minute
″	second

Quick Review

Concepts	Examples
5.1 Angles	
Types of Angles Two positive angles with a sum of 90° are **complementary angles.**	$70°$ and $90° - 70° = 20°$ are complementary.
Two positive angles with a sum of 180° are **supplementary angles.**	$70°$ and $180° - 70° = 110°$ are supplementary.
1 degree = 60 minutes **(1° = 60′)** 1 minute = 60 seconds **(1′ = 60″)**	$15°\ 30'\ 45''$ $= 15° + \frac{30°}{60} + \frac{45}{3600}°$ $30' \cdot \frac{1°}{60'} = \frac{30°}{60}$ and $45'' \cdot \frac{1°}{3600''} = \frac{45}{3600}°$. $= 15.5125°$ Decimal degrees
Coterminal angles have measures that differ by a multiple of 360°. Their terminal sides coincide when in standard position.	The acute angle θ in the figure is in standard position. If θ measures $46°$, find the measure of a positive and a negative coterminal angle. $46° + 360° = 406°$ $46° - 360° = -314°$ 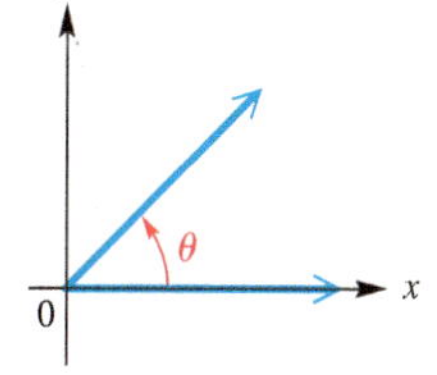

Concepts	Examples

5.2 Trigonometric Functions

Trigonometric Functions

Let (x, y) be a point other than the origin on the terminal side of an angle θ in standard position. The distance from the point to the origin is

$$r = \sqrt{x^2 + y^2}.$$

The six trigonometric functions of θ are defined as follows.

$$\sin\theta = \frac{y}{r} \qquad \cos\theta = \frac{x}{r} \qquad \tan\theta = \frac{y}{x}\ (x \neq 0)$$

$$\csc\theta = \frac{r}{y}\ (y \neq 0) \quad \sec\theta = \frac{r}{x}\ (x \neq 0) \quad \cot\theta = \frac{x}{y}\ (y \neq 0)$$

See the summary table of trigonometric function values for quadrantal angles in this section.

If the point $(-2, 3)$ is on the terminal side of an angle θ in standard position, find the values of the six trigonometric functions of θ.

Here $x = -2$ and $y = 3$, so

$$r = \sqrt{(-2)^2 + 3^2} = \sqrt{4 + 9} = \sqrt{13}.$$

$$\sin\theta = \frac{3\sqrt{13}}{13} \qquad \cos\theta = -\frac{2\sqrt{13}}{13} \qquad \tan\theta = -\frac{3}{2}$$

$$\csc\theta = \frac{\sqrt{13}}{3} \qquad \sec\theta = -\frac{\sqrt{13}}{2} \qquad \cot\theta = -\frac{2}{3}$$

Reciprocal Identities

$$\sin\theta = \frac{1}{\csc\theta} \qquad \cos\theta = \frac{1}{\sec\theta} \qquad \tan\theta = \frac{1}{\cot\theta}$$

$$\csc\theta = \frac{1}{\sin\theta} \qquad \sec\theta = \frac{1}{\cos\theta} \qquad \cot\theta = \frac{1}{\tan\theta}$$

If $\cot\theta = -\frac{2}{3}$, find $\tan\theta$.

$$\tan\theta = \frac{1}{\cot\theta} = \frac{1}{-\frac{2}{3}} = -\frac{3}{2}$$

Pythagorean Identities

$$\sin^2\theta + \cos^2\theta = 1 \qquad \tan^2\theta + 1 = \sec^2\theta$$

$$1 + \cot^2\theta = \csc^2\theta$$

Find $\sin\theta$ and $\tan\theta$, given that $\cos\theta = \frac{\sqrt{3}}{5}$ and $\sin\theta < 0$.

$$\sin^2\theta + \cos^2\theta = 1 \qquad \text{Pythagorean identity}$$

$$\sin^2\theta + \left(\frac{\sqrt{3}}{5}\right)^2 = 1 \qquad \text{Replace } \cos\theta \text{ with } \frac{\sqrt{3}}{5}.$$

$$\sin^2\theta + \frac{3}{25} = 1 \qquad \text{Square } \frac{\sqrt{3}}{5}.$$

$$\sin^2\theta = \frac{22}{25} \qquad \text{Subtract } \frac{3}{25}.$$

$$\sin\theta = -\frac{\sqrt{22}}{5} \qquad \text{Choose the negative root.}$$

Quotient Identities

$$\frac{\sin\theta}{\cos\theta} = \tan\theta \qquad \frac{\cos\theta}{\sin\theta} = \cot\theta$$

To find $\tan\theta$, use the values of $\sin\theta$ and $\cos\theta$ from above and the quotient identity $\tan\theta = \frac{\sin\theta}{\cos\theta}$.

$$\tan\theta = \frac{\sin\theta}{\cos\theta} = \frac{-\frac{\sqrt{22}}{5}}{\frac{\sqrt{3}}{5}} = -\frac{\sqrt{22}}{\sqrt{3}} \cdot \frac{\sqrt{3}}{\sqrt{3}} = -\frac{\sqrt{66}}{3}$$

Simplify the complex fraction, and rationalize the denominator.

Signs of the Trigonometric Functions

$x < 0, y > 0, r > 0$ II Sine and cosecant positive	$x > 0, y > 0, r > 0$ I All functions positive
$x < 0, y < 0, r > 0$ III Tangent and cotangent positive	$x > 0, y < 0, r > 0$ IV Cosine and secant positive

Identify the quadrant(s) of any angle θ that satisfies $\sin\theta < 0$, $\tan\theta > 0$.

Because $\sin\theta < 0$ in quadrants III and IV, and $\tan\theta > 0$ in quadrants I and III, both conditions are met only in quadrant III.

Concepts | **Examples**

5.3 Trigonometric Function Values and Angle Measures

Right-Triangle-Based Definitions of Trigonometric Functions

Let A represent any acute angle in standard position.

$$\sin A = \frac{y}{r} = \frac{\text{side opposite}}{\text{hypotenuse}} \qquad \csc A = \frac{r}{y} = \frac{\text{hypotenuse}}{\text{side opposite}}$$

$$\cos A = \frac{x}{r} = \frac{\text{side adjacent}}{\text{hypotenuse}} \qquad \sec A = \frac{r}{x} = \frac{\text{hypotenuse}}{\text{side adjacent}}$$

$$\tan A = \frac{y}{x} = \frac{\text{side opposite}}{\text{side adjacent}} \qquad \cot A = \frac{x}{y} = \frac{\text{side adjacent}}{\text{side opposite}}$$

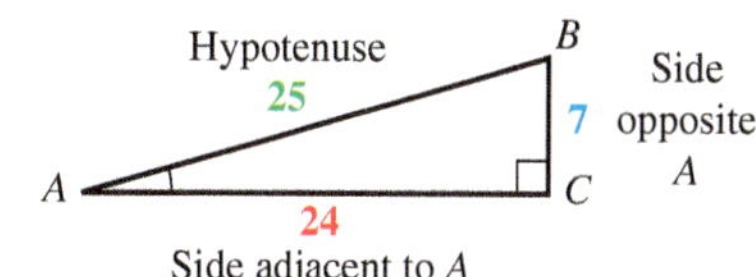

$$\sin A = \frac{7}{25} \qquad \cos A = \frac{24}{25} \qquad \tan A = \frac{7}{24}$$

$$\csc A = \frac{25}{7} \qquad \sec A = \frac{25}{24} \qquad \cot A = \frac{24}{7}$$

Cofunction Identities

For any acute angle A, cofunction values of complementary angles are equal.

$$\sin A = \cos(90^\circ - A) \qquad \cos A = \sin(90^\circ - A)$$

$$\sec A = \csc(90^\circ - A) \qquad \csc A = \sec(90^\circ - A)$$

$$\tan A = \cot(90^\circ - A) \qquad \cot A = \tan(90^\circ - A)$$

$$\sin 55^\circ = \cos(90^\circ - 55^\circ) = \cos 35^\circ$$

$$\sec 48^\circ = \csc(90^\circ - 48^\circ) = \csc 42^\circ$$

$$\tan 72^\circ = \cot(90^\circ - 72^\circ) = \cot 18^\circ$$

Function Values of Special Angles

θ	$\sin\theta$	$\cos\theta$	$\tan\theta$	$\cot\theta$	$\sec\theta$	$\csc\theta$
30°	$\frac{1}{2}$	$\frac{\sqrt{3}}{2}$	$\frac{\sqrt{3}}{3}$	$\sqrt{3}$	$\frac{2\sqrt{3}}{3}$	2
45°	$\frac{\sqrt{2}}{2}$	$\frac{\sqrt{2}}{2}$	1	1	$\sqrt{2}$	$\sqrt{2}$
60°	$\frac{\sqrt{3}}{2}$	$\frac{1}{2}$	$\sqrt{3}$	$\frac{\sqrt{3}}{3}$	2	$\frac{2\sqrt{3}}{3}$

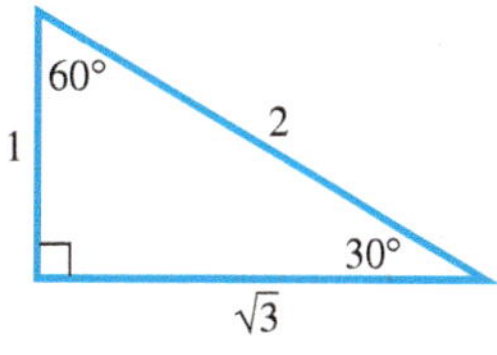

30°–60° right triangle

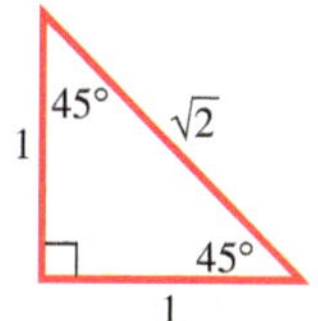

45°–45° right triangle

Reference Angle θ' for θ in (0°, 360°)

θ in Quadrant	I	II	III	IV
θ' is	θ	$180^\circ - \theta$	$\theta - 180^\circ$	$360^\circ - \theta$

Quadrant I: For $\theta = 25^\circ$, $\theta' = 25^\circ$

Quadrant II: For $\theta = 152^\circ$, $\theta' = 28^\circ$

Quadrant III: For $\theta = 200^\circ$, $\theta' = 20^\circ$

Quadrant IV: For $\theta = 320^\circ$, $\theta' = 40^\circ$

Finding Trigonometric Function Values for Any Nonquadrantal Angle θ

Step 1 Add or subtract 360° as many times as needed to obtain an angle greater than 0° but less than 360°.

Step 2 Find the reference angle θ'.

Step 3 Find the trigonometric function values for θ'.

Step 4 Determine the correct signs for the values found in Step 3.

Find sin 1050°.

$$1050^\circ - 2(360^\circ) = 330^\circ \quad \text{Coterminal angle in quadrant IV}$$

The reference angle for 330° is $\theta' = 30^\circ$.

$$\sin 1050^\circ$$
$$= -\sin 30^\circ \quad \text{Sine is negative in quadrant IV.}$$
$$= -\frac{1}{2} \quad \sin 30^\circ = \frac{1}{2}$$

To approximate a trigonometric function value of an angle in degrees, make sure the calculator is in degree mode.

Approximate each value.

$$\cos 50^\circ\, 15' = \cos 50.25^\circ \approx 0.63943900$$

$$\csc 32.5^\circ = \frac{1}{\sin 32.5^\circ} \approx 1.86115900 \quad \csc\theta = \frac{1}{\sin\theta}$$

Concepts	Examples
To find the corresponding angle measure given a trigonometric function value, use an appropriate inverse function.	Find an angle θ in the interval $[0°, 90°)$ that satisfies each condition in color. $\cos\theta \approx 0.73677482$ $\theta \approx \cos^{-1}(0.73677482)$ $\theta \approx 42.542600°$ $\csc\theta \approx 1.04766792$ $\sin\theta \approx \frac{1}{1.04766792}$ $\quad \sin\theta = \frac{1}{\csc\theta}$ $\theta \approx \sin^{-1}\left(\frac{1}{1.04766792}\right)$ $\theta \approx 72.65°$

5.4 Solutions and Applications of Right Triangles

Solving an Applied Trigonometry Problem

Step 1 Draw a sketch, and label it with the given information. Label the quantity to be found with a variable.

Step 2 Use the sketch to write an equation relating the given quantities to the variable.

Step 3 Solve the equation, and check that the answer makes sense.

Find the angle of elevation of the sun if a 48.6-ft flagpole casts a shadow 63.1 ft long.

Step 1 See the sketch. We must find θ.

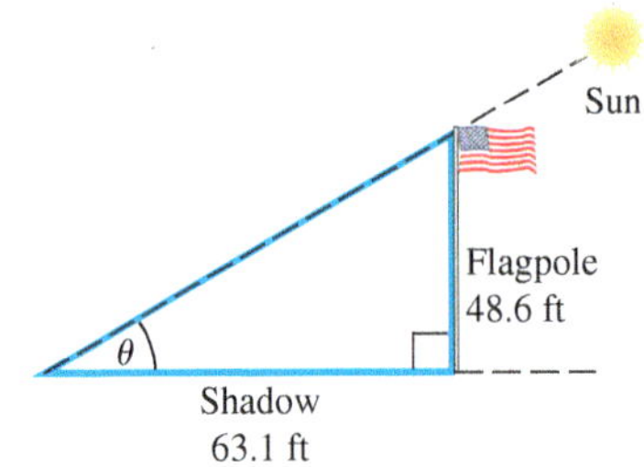

Step 2
$$\tan\theta = \frac{48.6}{63.1}$$
$$\tan\theta \approx 0.770206$$

Step 3
$$\theta = \tan^{-1} 0.770206$$
$$\theta \approx 37.6°$$

The angle of elevation rounded to three significant digits is 37.6°, or 37° 40′.

Expressing Bearing

Use one of the following methods.

Method 1 When a single angle is given, bearing is measured in a clockwise direction from due north.

Method 2 Start with a north-south line and use an acute angle to show direction, either east or west, from this line.

Example: 220° *Example:* S 40° W

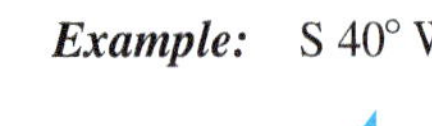

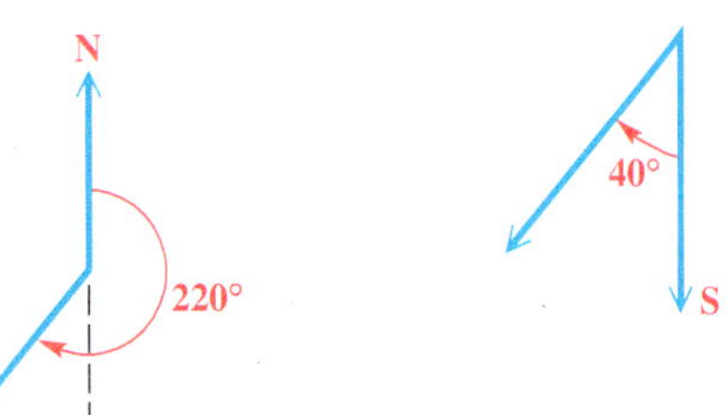

Chapter 5 Review Exercises

1. Give the measures of the complement and the supplement of an angle measuring 35°.

Find the angle of least positive measure that is coterminal with each angle.

2. $-51°$
3. $-174°$
4. $792°$

Work each problem.

5. **Rotating Propeller** The propeller of a speedboat rotates 650 times per min. Through how many degrees does a point on the edge of the propeller rotate in 2.4 sec?

6. **Rotating Pulley** A pulley is rotating 320 times per min. Through how many degrees does a point on the edge of the pulley move in $\frac{2}{3}$ sec?

Convert decimal degrees to degrees, minutes, seconds, and convert degrees, minutes, seconds to decimal degrees. If applicable, round to the nearest second or the nearest thousandth of a degree.

7. $119° \, 08' \, 03''$
8. $47° \, 25' \, 11''$
9. $275.1005°$
10. $-61.5034°$

Find the six trigonometric function values for each angle. Rationalize denominators when applicable.

11.

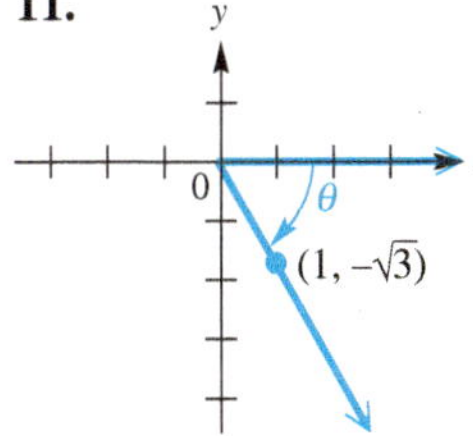

12.

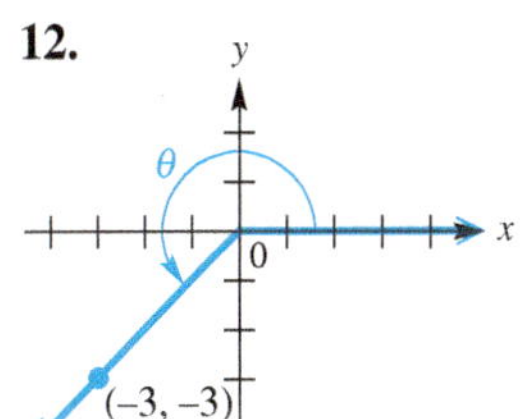

13.

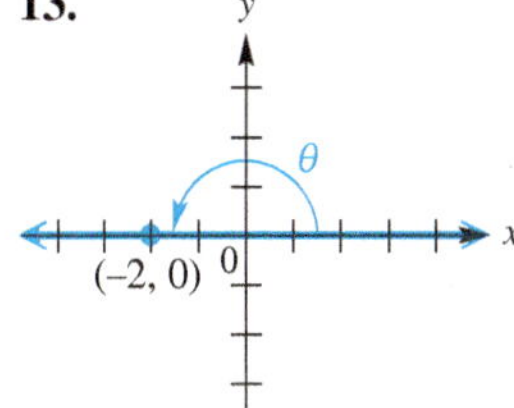

Find the values of the six trigonometric functions for an angle in standard position having each given point on its terminal side. Rationalize denominators when applicable.

14. $(9, -2)$
15. $(3, -4)$
16. $(1, -5)$
17. $(-8, 15)$
18. $\left(6\sqrt{3}, -6\right)$
19. $\left(-2\sqrt{2}, 2\sqrt{2}\right)$

An equation of the terminal side of an angle θ in standard position is given with a restriction on x. Sketch the least positive such angle θ, and find the values of the six trigonometric functions of θ.

20. $y = -5x, \ x \leq 0$
21. $5x - 3y = 0, \ x \geq 0$
22. $12x + 5y = 0, \ x \geq 0$

Complete the table with the appropriate function values of the given quadrantal angles. If the value is undefined, say so.

	θ	$\sin\theta$	$\cos\theta$	$\tan\theta$	$\cot\theta$	$\sec\theta$	$\csc\theta$
23.	180°						
24.	−90°						

Give all six trigonometric function values for each angle θ. Rationalize denominators when applicable.

25. $\cos\theta = -\frac{5}{8}$, and θ is in quadrant III

26. $\sin\theta = \frac{\sqrt{3}}{5}$, and $\cos\theta < 0$

27. $\sec\theta = -\sqrt{5}$, and θ is in quadrant II

28. $\tan\theta = 2$, and θ is in quadrant III

29. $\sec\theta = \frac{5}{4}$, and θ is in quadrant IV

30. $\sin\theta = -\frac{2}{5}$, and θ is in quadrant III

Find exact values of the six trigonometric functions for each angle A.

31.

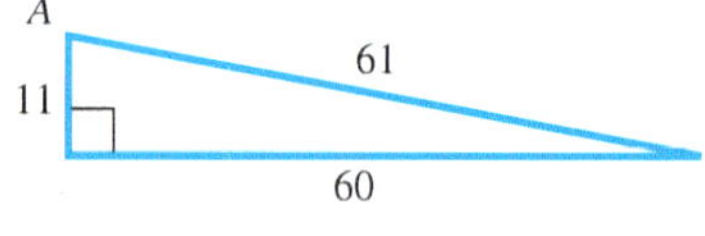

32.

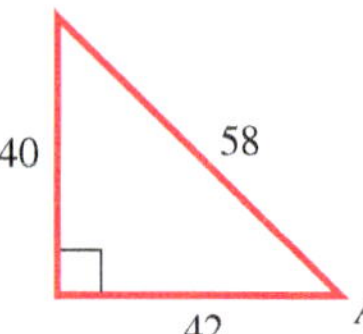

Find exact values of the six trigonometric functions for each angle. Do not use a calculator. Rationalize denominators when applicable.

33. $1020°$ **34.** $120°$ **35.** $-1470°$ **36.** $-225°$

Find all values of θ, if θ is in the interval $[0°, 360°)$ and θ has the given function value.

37. $\cos\theta = -\dfrac{1}{2}$

38. $\sin\theta = -\dfrac{1}{2}$

39. $\sec\theta = -\dfrac{2\sqrt{3}}{3}$

40. $\cot\theta = -1$

Find the sine, cosine, and tangent function values for each angle.

41.

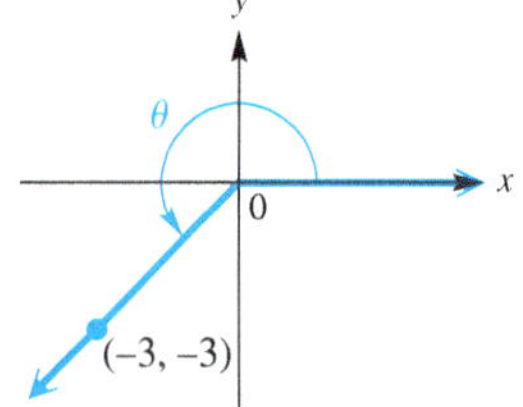

42.

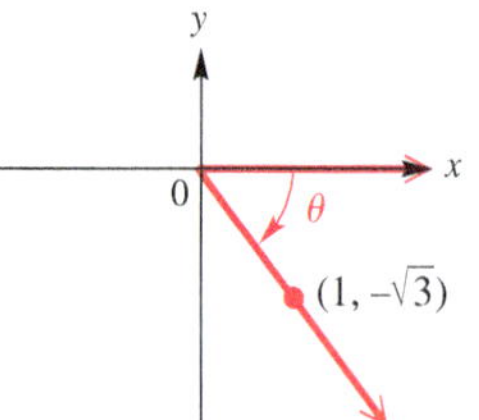

Use a calculator to approximate the value of each expression. Give answers to six decimal places.

43. $\sec 222° 30'$ **44.** $\sin 72° 30'$ **45.** $\csc 78° 21'$

46. $\cot 305.6°$ **47.** $\tan 11.7689°$ **48.** $\sec 58.9041°$

Use a calculator to find each value of θ, where θ is in the interval $[0°, 90°)$. Give answers in decimal degrees to six decimal places.

49. $\sin\theta = 0.82584121$ **50.** $\cot\theta = 1.1249386$ **51.** $\cos\theta = 0.97540415$

52. $\sec\theta = 1.2637891$ **53.** $\tan\theta = 1.9633124$ **54.** $\csc\theta = 9.5670466$

Find two angles in the interval $[0°, 360°)$ that satisfy each of the following. Round answers to the nearest degree.

55. $\sin\theta = 0.73135370$

56. $\tan\theta = 1.3763819$

Solve each problem.

57. A student wants to use a calculator to find the value of cot 25°. However, instead of entering $\frac{1}{\tan 25}$, he enters $\tan^{-1} 25$. Assuming the calculator is in degree mode, will this produce the correct answer? Explain.

58. Explain the process for using a calculator to find $\sec^{-1} 10$.

Solve each right triangle. In Exercise 60, give angles to the nearest minute. In Exercises 61 and 62, label the triangle ABC as in Exercises 59 and 60.

59.

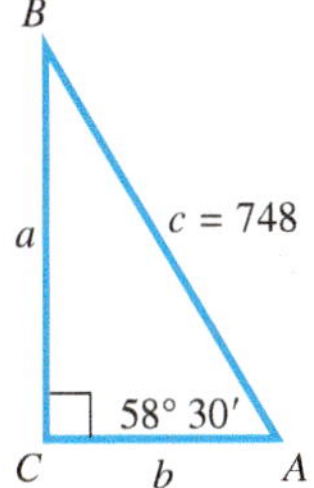

60.

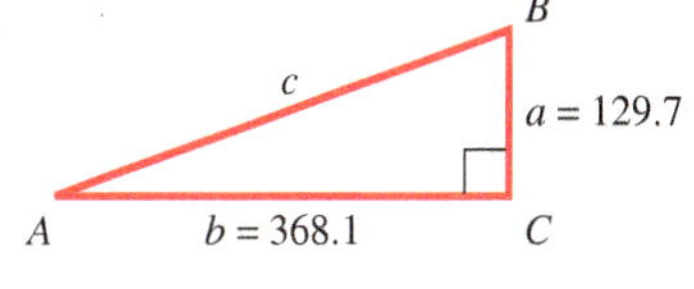

61. $A = 39.72°$, $b = 38.97$ m

62. $B = 47° 53'$, $b = 298.6$ m

Solve each problem.

63. ***Height of a Tower*** The angle of elevation from a point 93.2 ft from the base of a tower to the top of the tower is 38° 20′. Find the height of the tower.

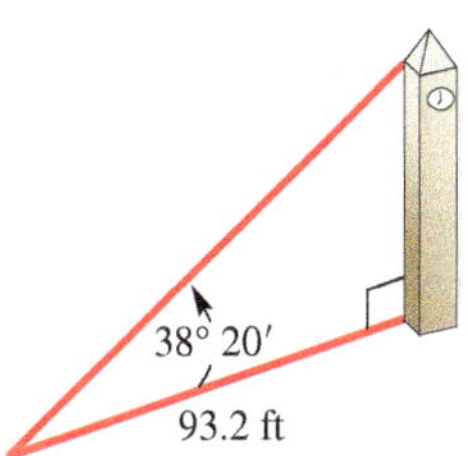

64. ***Height of a Tower*** The angle of depression from a television tower to a point on the ground 36.0 m from the bottom of the tower is 29.5°. Find the height of the tower.

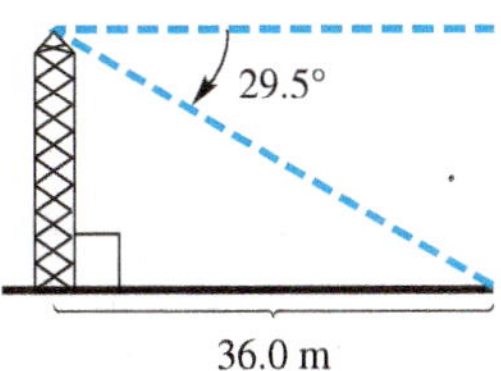

65. ***Length of a Diagonal*** One side of a rectangle measures 15.24 cm. The angle between the diagonal and that side is 35.65°. Find the length of the diagonal.

66. ***Length of Sides of an Isosceles Triangle*** An isosceles triangle has a base of length 49.28 m. The angle opposite the base is 58.746°. Find the length of each of the two equal sides.

67. ***Distance between Two Points*** The bearing of point B from point C is 254°. The bearing of point A from point C is 344°. The bearing of point A from point B is 32°. If the distance from A to C is 780 m, find the distance from A to B.

68. ***Distance a Ship Sails*** The bearing from point A to point B is S 55° E, and the bearing from point B to point C is N 35° E. If a ship sails from A to B, a distance of 81 km, and then from B to C, a distance of 74 km, how far is it from A to C?

69. ***Distance between Two Points*** Two cars leave an intersection at the same time. One heads due south at 55 mph. The other travels due west. After 2 hr, the bearing of the car headed west from the car headed south is 324°. How far apart are they at that time?

70. ***(Modeling) Height of a Satellite*** Artificial satellites that orbit Earth often use VHF signals to communicate with the ground. VHF signals travel in straight lines. The height h of the satellite above Earth and the time T that the satellite can communicate with a fixed location on the ground are related by the model

$$h = R\left(\frac{1}{\cos \frac{180T}{P}} - 1\right),$$

where $R = 3955$ mi is the radius of Earth and P is the period for the satellite to orbit Earth. (*Source:* Schlosser, W., T. Schmidt-Kaler, and E. Milone, *Challenges of Astronomy,* Springer-Verlag.)

(a) Find h to the nearest mile when $T = 25$ min and $P = 140$ min. (Evaluate the cosine function in degree mode.)

(b) What is the value of h to the nearest mile if T is increased to 30 min?

Chapter 5 Test

1. Give the measures of the complement and the supplement of an angle measuring 67°.

Find the measure of each marked angle.

2. $(7x + 19)°$ $(2x - 1)°$

3. $(-3x + 5)°$ $(-8x + 30)°$

Perform each conversion.

4. 74° 18′ 36″ to decimal degrees

5. 45.2025° to degrees, minutes, seconds

Solve each problem.

6. Find the angle of least positive measure that is coterminal with each angle.

(a) 390° **(b)** $-80°$ **(c)** 810°

7. ***Rotating Tire*** A tire rotates 450 times per min. Through how many degrees does a point on the edge of the tire move in 1 sec?

Sketch an angle θ in standard position such that θ has the least positive measure, and the given point is on the terminal side of θ. Then find the values of the six trigonometric functions for the angle. If any of these are undefined, say so.

8. $(2, -7)$

9. $(0, -2)$

Work each problem.

10. Draw a sketch of an angle in standard position having the line with the equation $3x - 4y = 0$, $x \leq 0$, as its terminal side. Indicate the angle of least positive measure θ, and find the values of the six trigonometric functions of θ.

11. Complete the table with the appropriate function values of the given quadrantal angles. If the value is undefined, say so.

θ	$\sin\theta$	$\cos\theta$	$\tan\theta$	$\cot\theta$	$\sec\theta$	$\csc\theta$
$90°$						
$-360°$						
$630°$						

12. If the terminal side of a quadrantal angle lies along the negative x-axis, which two of its trigonometric function values are undefined?

13. Identify the possible quadrant(s) in which θ must lie under the given conditions.

(a) $\cos\theta > 0,\ \tan\theta > 0$ **(b)** $\sin\theta < 0,\ \csc\theta < 0$ **(c)** $\cot\theta > 0,\ \cos\theta < 0$

14. Find the five remaining trigonometric function values of θ if $\sin\theta = \frac{3}{7}$ and θ is in quadrant II.

Solve each problem.

15. Find exact values of the six trigonometric functions for angle A in the right triangle.

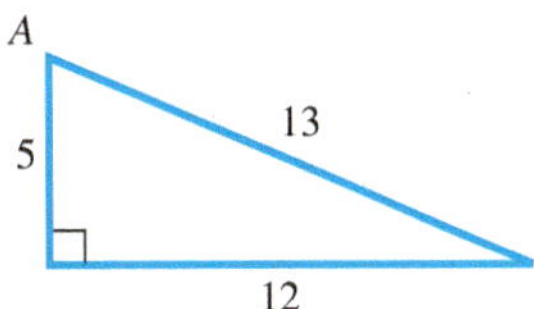

16. Find the exact value of each variable in the figure.

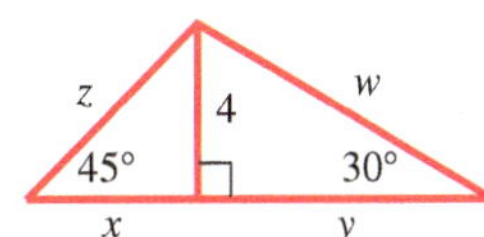

Find exact values of the six trigonometric functions for each angle. Rationalize denominators when applicable.

17. $240°$ **18.** $-135°$ **19.** $990°$

Find all values of θ, if θ is in the interval $[0°, 360°)$ and has the given function value.

20. $\cos\theta = -\frac{\sqrt{2}}{2}$ **21.** $\csc\theta = -\frac{2\sqrt{3}}{3}$ **22.** $\tan\theta = 1$

Solve each problem.

23. How would we find $\cot\theta$ using a calculator, if $\tan\theta = 1.6778490$? Evaluate $\cot\theta$.

24. Use a calculator to approximate the value of each expression. Give answers to six decimal places.

(a) $\sin 78° 21'$ **(b)** $\tan 117.689°$ **(c)** $\sec 58.9041°$

25. Find the value of θ in the interval $[0°, 90°]$ in decimal degrees, if

$$\sin\theta = 0.27843196.$$

Give the answer to six decimal places.

26. Solve the right triangle.

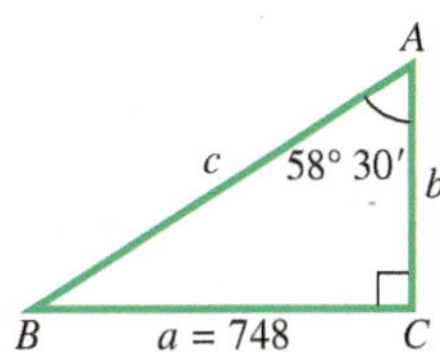

27. *Antenna Mast Guy Wire* A guy wire 77.4 m long is attached to the top of an antenna mast that is 71.3 m high. Find the angle that the wire makes with the ground.

28. *Height of a Flagpole* To measure the height of a flagpole, Jan Marie found that the angle of elevation from a point 24.7 ft from the base to the top is $32° 10'$. What is the height of the flagpole?

29. *Altitude of a Mountain* The highest point in Texas is Guadalupe Peak. The angle of depression from the top of this peak to a small miner's cabin at an approximate elevation of 2000 ft is 26°. The cabin is located 14,000 ft horizontally from a point directly under the top of the mountain. Find the altitude of the top of the mountain to the nearest hundred feet.

30. *Distance between Two Points* Two ships leave a port at the same time. The first ship sails on a bearing of 32° at 16 knots (nautical miles per hour) and the second on a bearing of 122° at 24 knots. How far apart are they after 2.5 hr?

31. *Distance of a Ship from a Pier* A ship leaves a pier on a bearing of S 62° E and travels for 75 km. It then turns and continues on a bearing of N 28° E for 53 km. How far is the ship from the pier?

32. Find h as indicated in the figure.

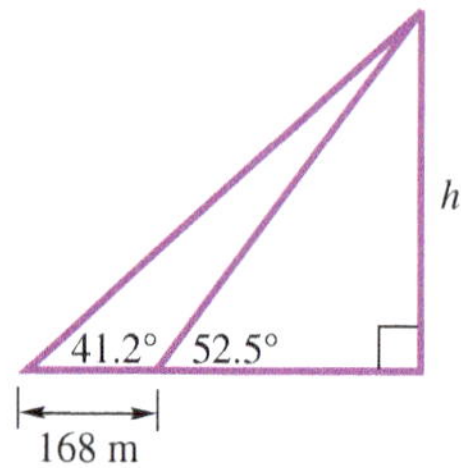

6 The Circular Functions and Their Graphs

Phenomena that repeat in a regular pattern, such as average monthly temperature, fractional part of the moon's illumination, and high and low tides, can be modeled by *periodic functions.*

6.1 Radian Measure

- Radian Measure
- Conversions between Degrees and Radians
- Arc Length on a Circle
- Area of a Sector of a Circle

Radian Measure

We have seen that angles can be measured in degrees. In more theoretical work in mathematics, *radian measure* of angles is preferred. Radian measure enables us to treat the trigonometric functions as functions with domains of *real numbers,* rather than angles.

Figure 1 shows an angle θ in standard position, along with a circle of radius r. The vertex of θ is at the center of the circle. Because angle θ intercepts an arc on the circle equal in length to the radius of the circle, we say that angle θ has a measure of *1 radian.*

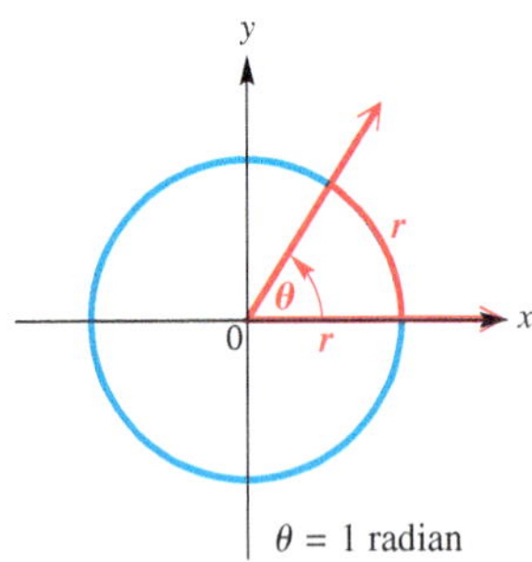

Figure 1

Radian

An angle with its vertex at the center of a circle that intercepts an arc on the circle equal in length to the radius of the circle has a measure of **1 radian.**

It follows that an angle of measure 2 radians intercepts an arc equal in length to twice the radius of the circle, an angle of measure $\frac{1}{2}$ radian intercepts an arc equal in length to half the radius of the circle, and so on. ***In general, if θ is a central angle of a circle of radius r, and θ intercepts an arc of length s, then the radian measure of θ is $\frac{s}{r}$. See* Figure 2.**

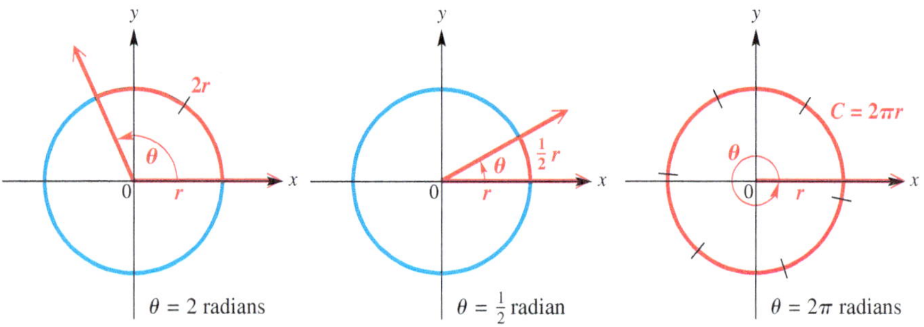

Figure 2

The ratio $\frac{s}{r}$ is a pure number, where s and r are expressed in the same units. ***Thus, "radians" is not a unit of measure like feet or centimeters.***

Conversions between Degrees and Radians

The **circumference** of a circle—the distance around the circle—is given by $C = 2\pi r$, where r is the radius of the circle. The formula $C = 2\pi r$ shows that the radius can be measured off 2π times around a circle. Therefore, an angle of 360°, which corresponds to a complete circle, intercepts an arc equal in length to 2π times the radius of the circle. Thus, an angle of 360° has a measure of 2π radians.

$$\mathbf{360^\circ = 2\pi \text{ radians}}$$

An angle of 180° is half the size of an angle of 360°, so an angle of 180° has half the radian measure of an angle of 360°.

$$\mathbf{180^\circ = \frac{1}{2}(2\pi) \text{ radians} = \pi \text{ radians}} \quad \text{Degree/radian relationship}$$

We can use the relationship $180° = \pi$ radians to develop a method for converting between degrees and radians as follows.

$$\mathbf{180° = \pi \text{ radians}} \quad \text{Degree/radian relationship}$$

$$\mathbf{1° = \frac{\pi}{180} \text{ radian}} \quad \text{Divide by 180.} \qquad \text{or} \qquad \mathbf{1 \text{ radian} = \frac{180°}{\pi}} \quad \text{Divide by } \pi.$$

NOTE Replacing π with its approximate integer value 3 in the fractions above and simplifying gives a couple of facts to help recall the relationship between degrees and radians. Remember that these are only approximations.

$$1° \approx \frac{1}{60} \text{ radian} \quad \text{and} \quad 1 \text{ radian} \approx 60°$$

Converting between Degrees and Radians

- Multiply a degree measure by $\frac{\pi}{180}$ radian and simplify to convert to radians.
- Multiply a radian measure by $\frac{180°}{\pi}$ and simplify to convert to degrees.

EXAMPLE 1 Converting Degrees to Radians

Convert each degree measure to radians.

(a) $45°$ **(b)** $-270°$ **(c)** $249.8°$

SOLUTION

(a) $45° = 45\left(\frac{\pi}{180} \text{ radian}\right) = \frac{\pi}{4} \text{ radian}$ Multiply by $\frac{\pi}{180}$ radian.

(b) $-270° = -270\left(\frac{\pi}{180} \text{ radian}\right) = -\frac{3\pi}{2} \text{ radians}$ Multiply by $\frac{\pi}{180}$ radian. Write in lowest terms.

(c) $249.8° = 249.8\left(\frac{\pi}{180} \text{ radian}\right) \approx 4.360 \text{ radians}$ Nearest thousandth

✔ **Now Try Exercises 11, 17, and 45.**

```
NORMAL FLOAT AUTO REAL RADIAN MP
45°
                    .7853981634
-270°
                    -4.71238898
249.8°
                    4.359832471
```

This radian mode screen shows TI-84 Plus conversions for **Example 1.** Verify that the first two results are *approximations* for the *exact* values of $\frac{\pi}{4}$ and $-\frac{3\pi}{2}$.

EXAMPLE 2 Converting Radians to Degrees

Convert each radian measure to degrees.

(a) $\frac{9\pi}{4}$ **(b)** $-\frac{5\pi}{6}$ **(c)** 4.25

SOLUTION

(a) $\frac{9\pi}{4} \text{ radians} = \frac{9\pi}{4}\left(\frac{180°}{\pi}\right) = 405°$ Multiply by $\frac{180°}{\pi}$.

(b) $-\frac{5\pi}{6} \text{ radians} = -\frac{5\pi}{6}\left(\frac{180°}{\pi}\right) = -150°$ Multiply by $\frac{180°}{\pi}$.

(c) $4.25 \text{ radians} = 4.25\left(\frac{180°}{\pi}\right) \approx 243.5°, \text{ or } 243° \, 30'$ $0.50706(60') \approx 30'$

✔ **Now Try Exercises 29, 33, and 57.**

```
NORMAL FLOAT AUTO REAL DEGREE MP
(9π/4)r
                    405
(-5π/6)r
                    -150
4.25r▸DMS
                    243°30'25.427"
```

This degree mode screen shows how a TI-84 Plus calculator converts the radian measures in **Example 2** to degree measures.

NOTE Another way to convert a radian measure that is a rational multiple of π, such as $\frac{9\pi}{4}$, to degrees is to substitute $180°$ for π. In **Example 2(a),** doing this would give the following.

$$\frac{9\pi}{4} \text{ radians} = \frac{9(180°)}{4} = 405°$$

One of the most important facts to remember when working with angles and their measures is summarized in the following statement.

Agreement on Angle Measurement Units

If no unit of angle measure is specified, then the angle is understood to be measured in radians.

For example, **Figure 3(a)** shows an angle of 30°, and **Figure 3(b)** shows an angle of 30 (which means 30 radians). An angle with measure 30 radians is coterminal with an angle of approximately 279°.

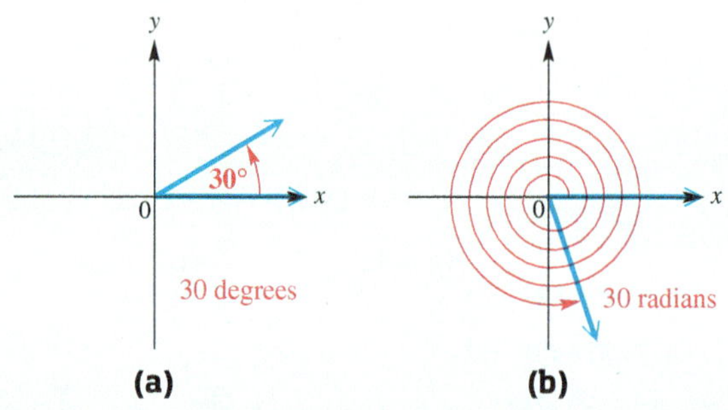

(a) **(b)**

Note the difference between an angle of 30 *degrees* and an angle of 30 *radians*.

Figure 3

The following table and **Figure 4** on the next page give some equivalent angle measures in degrees and radians. Keep in mind that

$$\mathbf{180° = \pi \text{ radians.}}$$

Equivalent Angle Measures

Degrees	Radians		Degrees	Radians	
	Exact	Approximate		Exact	Approximate
0°	0	0	90°	$\frac{\pi}{2}$	1.57
30°	$\frac{\pi}{6}$	0.52	180°	π	3.14
45°	$\frac{\pi}{4}$	0.79	270°	$\frac{3\pi}{2}$	4.71
60°	$\frac{\pi}{3}$	1.05	360°	2π	6.28

These exact values are *rational multiples of* π.

LOOKING AHEAD TO CALCULUS

In calculus, radian measure is much easier to work with than degree measure. If x is measured in radians, then the derivative of $f(x) = \sin x$ is

$$f'(x) = \cos x.$$

However, if x is measured in degrees, then the derivative of $f(x) = \sin x$ is

$$f'(x) = \frac{\pi}{180} \cos x.$$

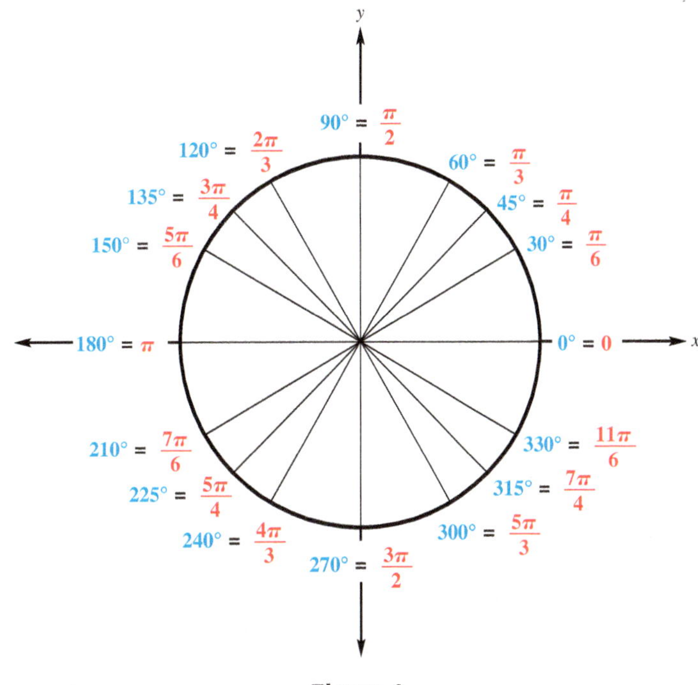

Figure 4

Learn the equivalences in* Figure 4. *They appear often in trigonometry.

Arc Length on a Circle The formula for finding the length of an arc of a circle follows directly from the definition of an angle θ in radians, where $\theta = \frac{s}{r}$.

In **Figure 5,** we see that angle QOP has measure 1 radian and intercepts an arc of length r on the circle. We also see that angle ROT has measure θ radians and intercepts an arc of length s on the circle. From plane geometry, we know that the lengths of the arcs are proportional to the measures of their central angles.

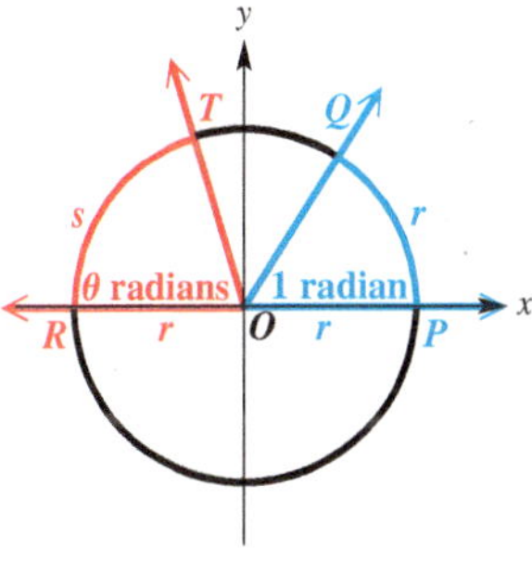

Figure 5

$$\frac{s}{r} = \frac{\theta}{1} \quad \text{Set up a proportion.}$$

Multiplying each side by r gives

$$s = r\theta. \quad \text{Solve for } s.$$

Arc Length

The length s of the arc intercepted on a circle of radius r by a central angle of measure θ radians is given by the product of the radius and the radian measure of the angle.

$$s = r\theta, \quad \text{where } \theta \text{ is in radians}$$

CAUTION *When the formula*

$$s = r\theta$$

is applied, the value of θ MUST *be expressed in radians, not degrees.*

EXAMPLE 3 Finding Arc Length Using $s = r\theta$

A circle has radius 18.20 cm. Find the length of the arc intercepted by a central angle having each of the following measures.

(a) $\frac{3\pi}{8}$ radians **(b)** 144°

SOLUTION

(a) As shown in **Figure 6,** $r = 18.20$ cm and $\theta = \frac{3\pi}{8}$.

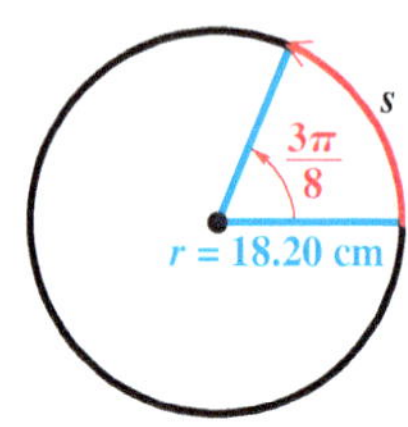

Figure 6

$s = r\theta$ Arc length formula

$s = 18.20\left(\frac{3\pi}{8}\right)$ Let $r = 18.20$ and $\theta = \frac{3\pi}{8}$.

$s \approx 21.44$ cm Use a calculator.

(b) The formula $s = r\theta$ requires that θ be measured in radians. First, convert θ to radians by multiplying 144° by $\frac{\pi}{180}$ radian.

$$144° = 144\left(\frac{\pi}{180}\right) = \frac{4\pi}{5} \text{ radians}$$ Convert from degrees to radians.

The length s is found using $s = r\theta$.

$$s = r\theta = 18.20\left(\frac{4\pi}{5}\right) \approx 45.74 \text{ cm}$$ Let $r = 18.20$ and $\theta = \frac{4\pi}{5}$.

Be sure to use radians for θ in $s = r\theta$.

✔ Now Try Exercises 67 and 71.

Latitude gives the measure of a central angle with vertex at Earth's center whose initial side goes through the equator and whose terminal side goes through the given location. As an example, see **Figure 7.**

EXAMPLE 4 Finding the Distance between Two Cities

Reno, Nevada, is approximately due north of Los Angeles. The latitude of Reno is 40° N, and that of Los Angeles is 34° N. (The N in 34° N means *north* of the equator.) The radius of Earth is 6400 km. Find the north-south distance between the two cities.

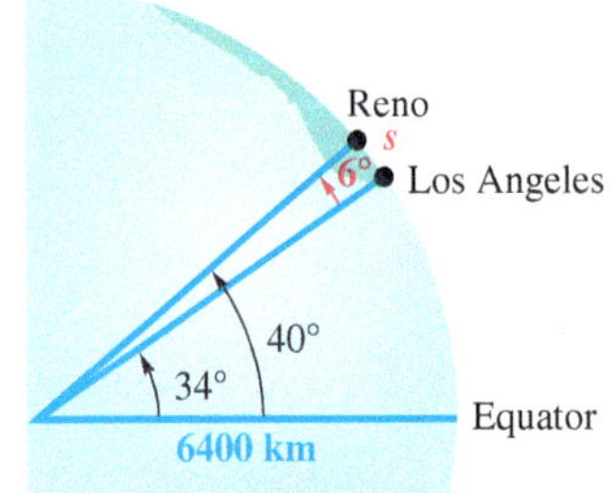

Figure 7

SOLUTION As shown in **Figure 7,** the central angle between Reno and Los Angeles is

$$40° - 34° = 6°.$$

The distance between the two cities can be found using the formula $s = r\theta$, after 6° is converted to radians.

$$6° = 6\left(\frac{\pi}{180}\right) = \frac{\pi}{30} \text{ radian}$$

The distance between the two cities is given by s.

$$s = r\theta = 6400\left(\frac{\pi}{30}\right) \approx 670 \text{ km}$$ Let $r = 6400$ and $\theta = \frac{\pi}{30}$.

✔ Now Try Exercise 75.

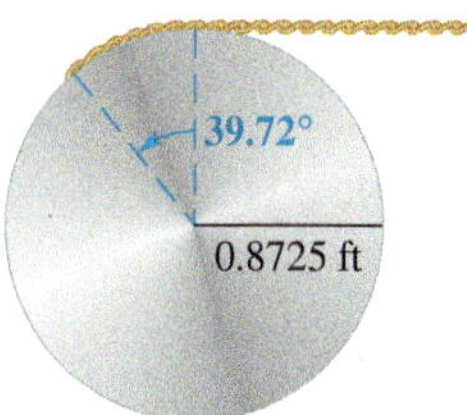

Figure 8

EXAMPLE 5 Finding a Length Using $s = r\theta$

A rope is being wound around a drum with radius 0.8725 ft. (See **Figure 8.**) How much rope will be wound around the drum if the drum is rotated through an angle of 39.72°?

SOLUTION The length of rope wound around the drum is the arc length for a circle of radius 0.8725 ft and a central angle of 39.72°. Use the formula $s = r\theta$, with the angle converted to radian measure. The length of the rope wound around the drum is approximated by s.

$$s = r\theta = 0.8725\left[39.72\left(\frac{\pi}{180}\right)\right] \approx 0.6049 \text{ ft}$$

Convert to radian measure.

✔ **Now Try Exercise 87(a).**

EXAMPLE 6 Finding an Angle Measure Using $s = r\theta$

Figure 9

Two gears are adjusted so that the smaller gear drives the larger one, as shown in **Figure 9.** If the smaller gear rotates through an angle of 225°, through how many degrees will the larger gear rotate?

SOLUTION First find the radian measure of the angle of rotation for the smaller gear, and then find the arc length on the smaller gear. This arc length will correspond to the arc length of the motion of the larger gear. Because $225° = \frac{5\pi}{4}$ radians, for the smaller gear we have arc length

$$s = r\theta = 2.5\left(\frac{5\pi}{4}\right) = \frac{12.5\pi}{4} = \frac{25\pi}{8} \text{ cm.}$$

The tips of the two mating gear teeth must move at the same linear speed, or the teeth will break. So we must have "equal arc lengths in equal times." An arc with this length s on the larger gear corresponds to an angle measure θ, in radians, where $s = r\theta$.

$$s = r\theta \quad \text{Arc length formula}$$

$$\frac{25\pi}{8} = 4.8\theta \quad \text{Let } s = \tfrac{25\pi}{8} \text{ and } r = 4.8 \text{ (for the larger gear).}$$

$$\frac{125\pi}{192} = \theta \quad 4.8 = \tfrac{48}{10} = \tfrac{24}{5}\text{; Multiply by } \tfrac{5}{24} \text{ to solve for } \theta.$$

Converting θ back to degrees shows that the larger gear rotates through

$$\frac{125\pi}{192}\left(\frac{180°}{\pi}\right) \approx 117°. \quad \text{Convert } \theta = \tfrac{125\pi}{192} \text{ to degrees.}$$

✔ **Now Try Exercise 81.**

Area of a Sector of a Circle

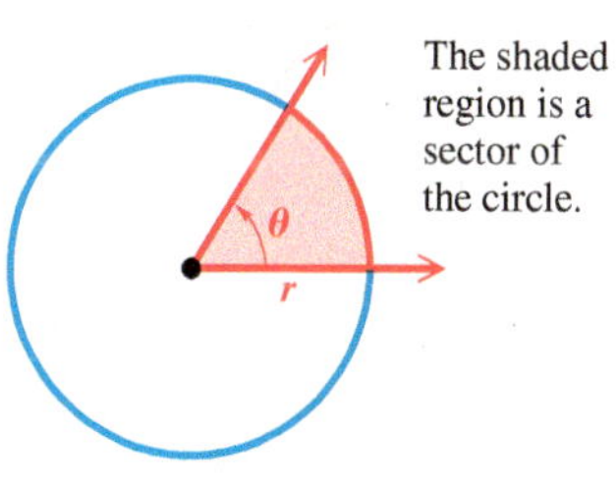

The shaded region is a sector of the circle.

Figure 10

A **sector of a circle** is the portion of the interior of a circle intercepted by a central angle. Think of it as a "piece of pie." See **Figure 10.** A complete circle can be thought of as an angle with measure 2π radians. If a central angle for a sector has measure θ radians, then the sector makes up the fraction $\frac{\theta}{2\pi}$ of a complete circle. The area $\mathcal{A}$ of a complete circle with radius r is $\mathcal{A} = \pi r^2$. Therefore, we have the following.

$$\text{Area } \mathcal{A} \text{ of a sector} = \frac{\theta}{2\pi}(\pi r^2) = \frac{1}{2}r^2\theta, \quad \text{where } \theta \text{ is in radians.}$$

Area of a Sector

The area $\mathscr{A}$ of a sector of a circle of radius r and central angle θ is given by the following formula.

$$\mathscr{A} = \frac{1}{2}r^2\theta, \quad \text{where } \theta \text{ is in radians}$$

CAUTION *As in the formula for arc length, the value of θ must be in radians when this formula is used to find the area of a sector.*

EXAMPLE 7 Finding the Area of a Sector-Shaped Field

Center-pivot irrigation system

A center-pivot irrigation system provides water to a sector-shaped field with the measures shown in **Figure 11.** Find the area of the field.

Figure 11

SOLUTION First, convert 15° to radians.

$$15° = 15\left(\frac{\pi}{180}\right) = \frac{\pi}{12} \text{ radian} \qquad \text{Convert to radians.}$$

Now find the area of a sector of a circle.

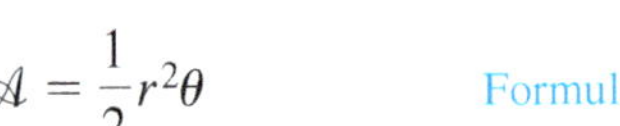

$$\mathscr{A} = \frac{1}{2}r^2\theta \qquad \text{Formula for area of a sector}$$

$$\mathscr{A} = \frac{1}{2}(321)^2\left(\frac{\pi}{12}\right) \qquad \text{Let } r = 321 \text{ and } \theta = \frac{\pi}{12}.$$

$$\mathscr{A} \approx 13{,}500 \text{ m}^2 \qquad \text{Multiply.}$$

✔ **Now Try Exercise 109.**

6.1 Exercises

CONCEPT PREVIEW *Fill in the blank(s) to correctly complete each sentence.*

1. An angle with its vertex at the center of a circle that intercepts an arc on the circle equal in length to the ________ of the circle has measure 1 radian.

2. 360° = ______ radians, and 180° = ______ radians.

3. To convert to radians, multiply a degree measure by ______ radian and simplify.

4. To convert to degrees, multiply a radian measure by ______ and simplify.

CONCEPT PREVIEW *Work each problem.*

5. Find the exact length of the arc intercepted by the given central angle.

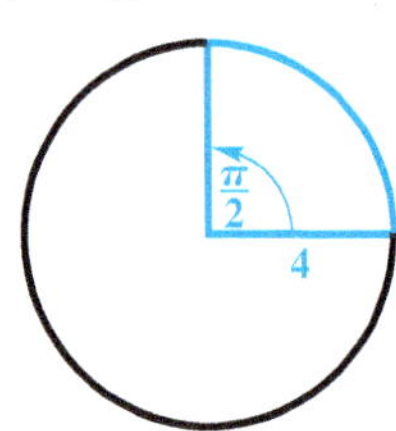

6. Find the radius of the circle.

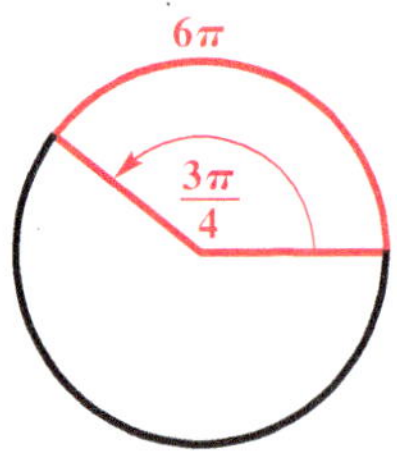

7. Find the measure of the central angle (in radians).

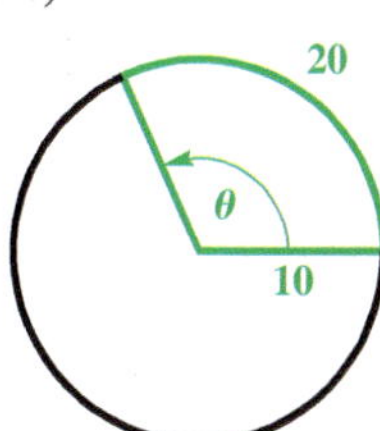

8. Find the area of the sector.

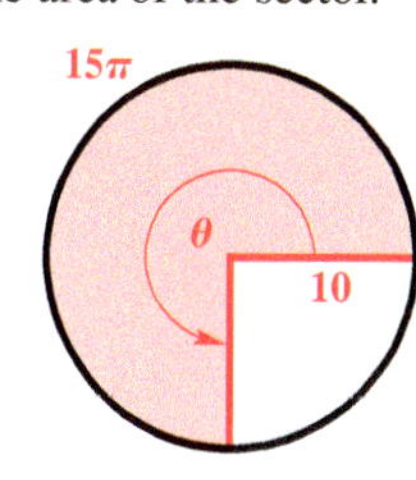

9. Find the measure (in radians) of the central angle. The number inside the sector is the area.

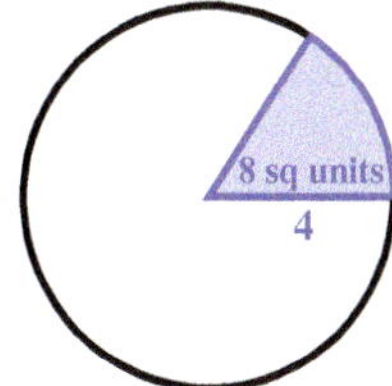

10. Find the measure (in degrees) of the central angle. The number inside the sector is the area.

96π sq units

12

Convert each degree measure to radians. Leave answers as multiples of π. ***See Examples 1(a) and 1(b).***

11. 60° **12.** 30° **13.** 90° **14.** 120°

15. 150° **16.** 270° **17.** −300° **18.** −315°

19. 450° **20.** 480° **21.** 1800° **22.** 3600°

23. 0° **24.** 180° **25.** −900° **26.** −1800°

Convert each radian measure to degrees. ***See Examples 2(a) and 2(b).***

27. $\frac{\pi}{3}$ **28.** $\frac{8\pi}{3}$ **29.** $\frac{7\pi}{4}$ **30.** $\frac{2\pi}{3}$

31. $\frac{11\pi}{6}$ **32.** $\frac{15\pi}{4}$ **33.** $-\frac{\pi}{6}$ **34.** $-\frac{8\pi}{5}$

35. $\frac{7\pi}{10}$ **36.** $\frac{11\pi}{15}$ **37.** $-\frac{4\pi}{15}$ **38.** $-\frac{7\pi}{20}$

39. $\frac{17\pi}{20}$ **40.** $\frac{11\pi}{30}$ **41.** -5π **42.** 15π

Convert each degree measure to radians. If applicable, round to the nearest thousandth. ***See Example 1(c).***

43. 39° **44.** 74° **45.** 42.5° **46.** 264.9°

47. 139° 10′ **48.** 174° 50′ **49.** 64.29° **50.** 85.04°

51. 56° 25′ **52.** 122° 37′ **53.** −47.69° **54.** −23.01°

Convert each radian measure to degrees. Write answers to the nearest minute. ***See Example 2(c).***

55. 2 **56.** 5 **57.** 1.74 **58.** 3.06

59. 0.3417 **60.** 9.84763 **61.** −5.01095 **62.** −3.47189

63. *Concept Check* The value of sin 30 is not $\frac{1}{2}$. Why is this true?

64. *Concept Check* What is meant by an angle of one radian?

65. *Concept Check* The figure shows the same angles measured in both degrees and radians. Complete the missing measures.

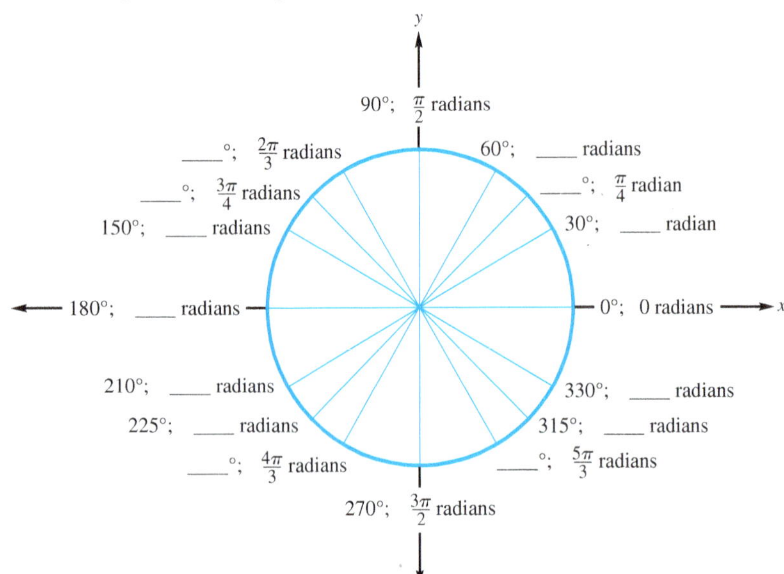

66. *Concept Check* What is the exact radian measure of an angle measuring π degrees?

Unless otherwise directed, give calculator approximations in answers in the rest of this exercise set.

Find the length to three significant digits of each arc intercepted by a central angle θ in a circle of radius r. ***See Example 3.***

67. $r = 12.3$ cm, $\theta = \frac{2\pi}{3}$ radians

68. $r = 0.892$ cm, $\theta = \frac{11\pi}{10}$ radians

69. $r = 1.38$ ft, $\theta = \frac{5\pi}{6}$ radians

70. $r = 3.24$ mi, $\theta = \frac{7\pi}{6}$ radians

71. $r = 4.82$ m, $\theta = 60°$

72. $r = 71.9$ cm, $\theta = 135°$

73. $r = 15.1$ in., $\theta = 210°$

74. $r = 12.4$ ft, $\theta = 330°$

Distance between Cities *Find the distance in kilometers between each pair of cities, assuming they lie on the same north-south line. Assume that the radius of Earth is* 6400 km. ***See Example 4.***

75. Panama City, Panama, 9° N, and Pittsburgh, Pennsylvania, 40° N

76. Farmersville, California, 36° N, and Penticton, British Columbia, 49° N

77. New York City, New York, 41° N, and Lima, Peru, 12° S

78. Halifax, Nova Scotia, 45° N, and Buenos Aires, Argentina, 34° S

79. *Latitude of Madison* Madison, South Dakota, and Dallas, Texas, are 1200 km apart and lie on the same north-south line. The latitude of Dallas is 33° N. What is the latitude of Madison?

80. *Latitude of Toronto* Charleston, South Carolina, and Toronto, Canada, are 1100 km apart and lie on the same north-south line. The latitude of Charleston is 33° N. What is the latitude of Toronto?

Work each problem. ***See Examples 5 and 6.***

81. *Gear Movement* Two gears are adjusted so that the smaller gear drives the larger one, as shown in the figure. If the smaller gear rotates through an angle of 300°, through how many degrees does the larger gear rotate?

82. *Gear Movement* Repeat **Exercise 81** for gear radii of 4.8 in. and 7.1 in. and for an angle of 315° for the smaller gear.

83. *Rotating Wheels* The rotation of the smaller wheel in the figure causes the larger wheel to rotate. Through how many degrees does the larger wheel rotate if the smaller one rotates through 60.0°?

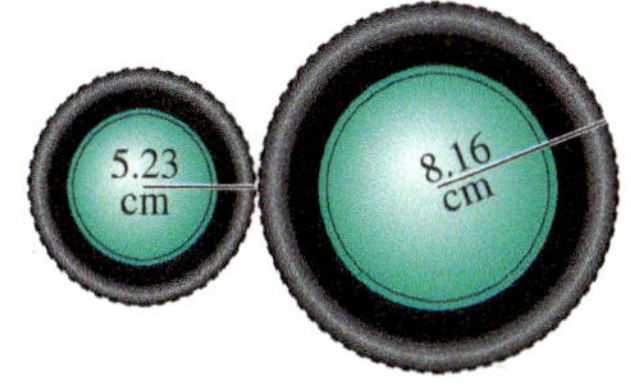

84. *Rotating Wheels* Repeat **Exercise 83** for wheel radii of 6.84 in. and 12.46 in. and an angle of 150.0° for the smaller wheel.

85. *Rotating Wheels* Find the radius of the larger wheel in the figure if the smaller wheel rotates 80.0° when the larger wheel rotates 50.0°.

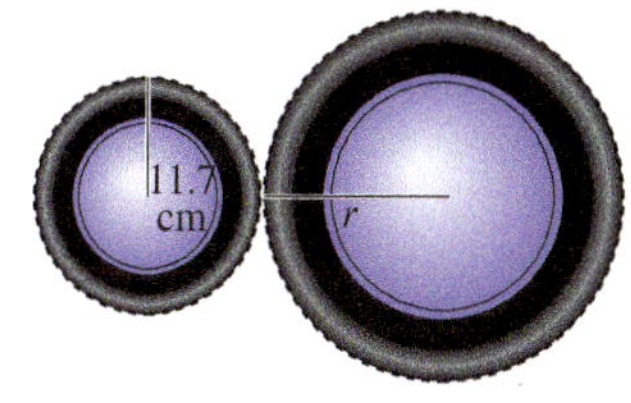

86. *Rotating Wheels* Repeat **Exercise 85** if the smaller wheel of radius 14.6 in. rotates 120.0° when the larger wheel rotates 60.0°.

87. *Pulley Raising a Weight* Refer to the figure.

(a) How many inches will the weight in the figure rise if the pulley is rotated through an angle of 71° 50′?

(b) Through what angle, to the nearest minute, must the pulley be rotated to raise the weight 6 in.?

88. *Pulley Raising a Weight* Find the radius of the pulley in the figure if a rotation of 51.6° raises the weight 11.4 cm.

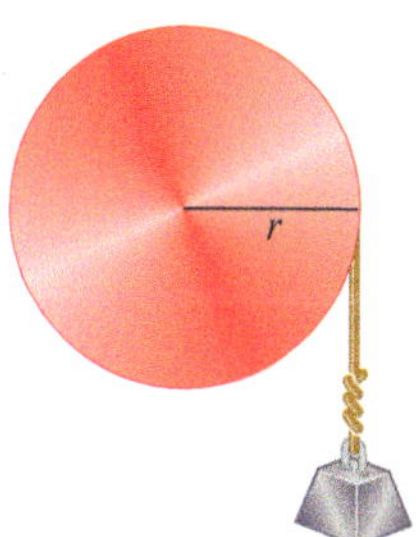

89. *Bicycle Chain Drive* The figure shows the chain drive of a bicycle. How far will the bicycle move if the pedals are rotated through 180.0°? Assume the radius of the bicycle wheel is 13.6 in.

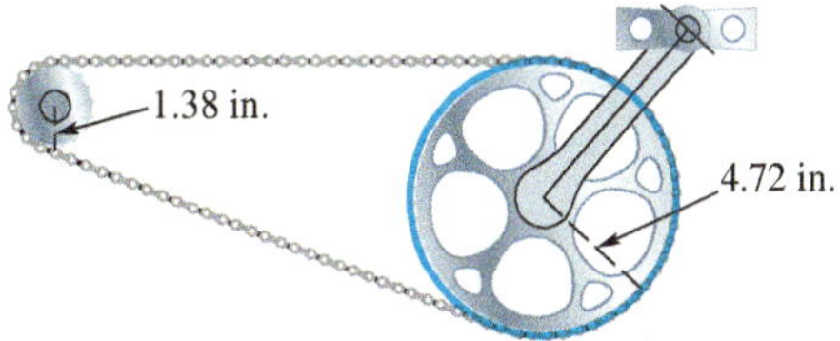

90. *Car Speedometer* The speedometer of Terry's Honda CR-V is designed to be accurate with tires of radius 14 in.

(a) Find the number of rotations of a tire in 1 hr if the car is driven at 55 mph.

(b) Suppose that oversize tires of radius 16 in. are placed on the car. If the car is now driven for 1 hr with the speedometer reading 55 mph, how far has the car gone? If the speed limit is 55 mph, does Terry deserve a speeding ticket?

Suppose the tip of the minute hand of a clock is 3 in. *from the center of the clock. For each duration, determine the distance traveled by the tip of the minute hand. Leave answers as multiples of* π.

91. 30 min

92. 40 min

93. 4.5 hr

94. $6\frac{1}{2}$ hr

If a central angle is very small, there is little difference in length between an arc and the inscribed chord. See the figure. Approximate each of the following lengths by finding the necessary arc length. (Note: When a central angle intercepts an arc, the arc is said to ***subtend*** *the angle.)*

Arc length ≈ length of inscribed chord

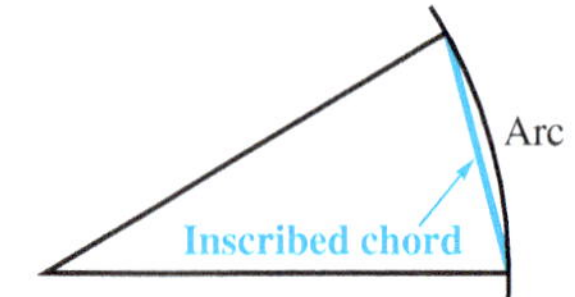

95. *Length of a Train* A railroad track in the desert is 3.5 km away. A train on the track subtends (horizontally) an angle of $3° 20'$. Find the length of the train.

96. Repeat **Exercise 95** for a railroad track 2.7 mi away and a train that subtends an angle of $2° 30'$.

97. *Distance to a Boat* The mast of a boat is 32.0 ft high. If it subtends an angle of $2° 11'$, how far away is it?

98. Repeat **Exercise 97** for a boat mast 11.0 m high that subtends an angle of $1° 45'$.

Find the area of a sector of a circle having radius r and central angle θ. *Express answers to the nearest tenth.* ***See Example 7.***

99. $r = 29.2$ m, $\theta = \frac{5\pi}{6}$ radians

100. $r = 59.8$ km, $\theta = \frac{2\pi}{3}$ radians

101. $r = 30.0$ ft, $\theta = \frac{\pi}{2}$ radians

102. $r = 90.0$ yd, $\theta = \frac{5\pi}{6}$ radians

103. $r = 12.7$ cm, $\theta = 81°$

104. $r = 18.3$ m, $\theta = 125°$

105. $r = 40.0$ mi, $\theta = 135°$

106. $r = 90.0$ km, $\theta = 270°$

Work each problem. ***See Example 7.***

107. *Angle Measure* Find the measure (in radians) of a central angle of a sector of area 16 in.^2 in a circle of radius 3.0 in.

108. *Radius Length* Find the radius of a circle in which a central angle of $\frac{\pi}{6}$ radian determines a sector of area 64 m^2.

109. *Irrigation Area* A center-pivot irrigation system provides water to a sector-shaped field as shown in the figure. Find the area of the field if $\theta = 40.0°$ and $r = 152$ yd.

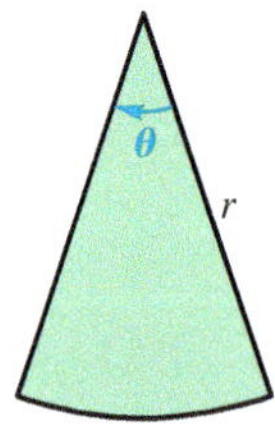

110. *Irrigation Area* Suppose that in **Exercise 109** the angle is halved and the radius length is doubled. How does the new area compare to the original area? Does this result hold in general for any values of θ and r?

111. *Arc Length* A circular sector has an area of 50 in.^2. The radius of the circle is 5 in. What is the arc length of the sector?

112. *Angle Measure* In a circle, a sector has an area of 16 cm^2 and an arc length of 6.0 cm. What is the measure of the central angle in degrees?

113. *Measures of a Structure* The figure illustrates Medicine Wheel, a Native American structure in northern Wyoming. There are 27 aboriginal spokes in the wheel, all equally spaced.

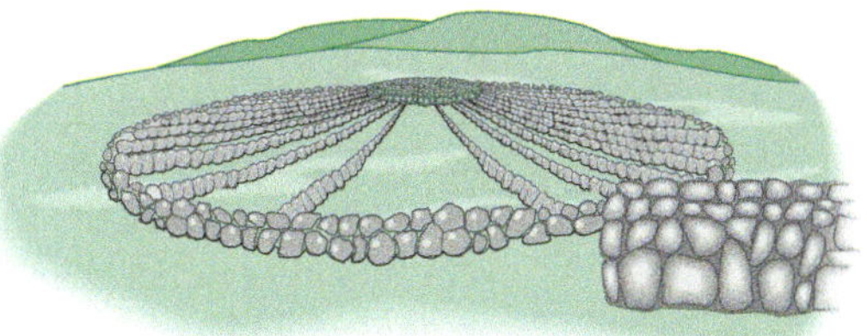

(a) Find the measure of each central angle in degrees and in radians in terms of π.

(b) If the radius of the wheel is 76.0 ft, find the circumference.

(c) Find the length of each arc intercepted by consecutive pairs of spokes.

(d) Find the area of each sector formed by consecutive spokes.

114. *Area Cleaned by a Windshield Wiper* The Ford Model A, built from 1928 to 1931, had a single windshield wiper on the driver's side. The total arm and blade was 10 in. long and rotated back and forth through an angle of 95°. The shaded region in the figure is the portion of the windshield cleaned by the 7-in. wiper blade. Find the area of the region cleaned to the nearest tenth.

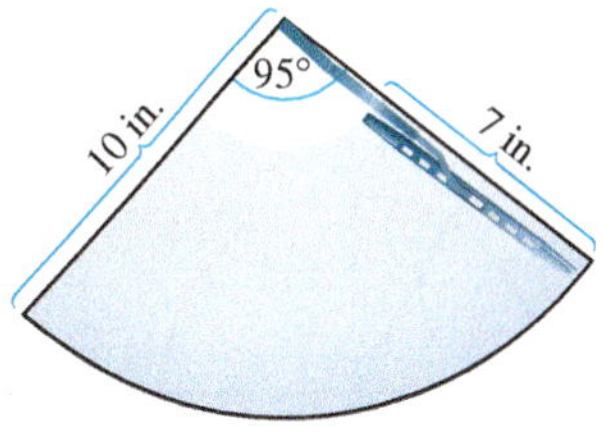

115. *Circular Railroad Curves* In the United States, circular railroad curves are designated by the **degree of curvature,** the central angle subtended by a chord of 100 ft. Suppose a portion of track has curvature 42.0°. (*Source:* Hay, W., *Railroad Engineering,* John Wiley and Sons.)

(a) What is the radius of the curve?

(b) What is the length of the arc determined by the 100-ft chord?

(c) What is the area of the portion of the circle bounded by the arc and the 100-ft chord?

116. *Land Required for a Solar-Power Plant* A 300-megawatt solar-power plant requires approximately 950,000 m^2 of land area to collect the required amount of energy from sunlight. If this land area is circular, what is its radius? If this land area is a 35° sector of a circle, what is its radius?

117. *Area of a Lot* A frequent problem in surveying city lots and rural lands adjacent to curves of highways and railways is that of finding the area when one or more of the boundary lines is the arc of a circle. Find the area (to two significant digits) of the lot shown in the figure. (*Source:* Anderson, J. and E. Michael, *Introduction to Surveying,* McGraw-Hill.)

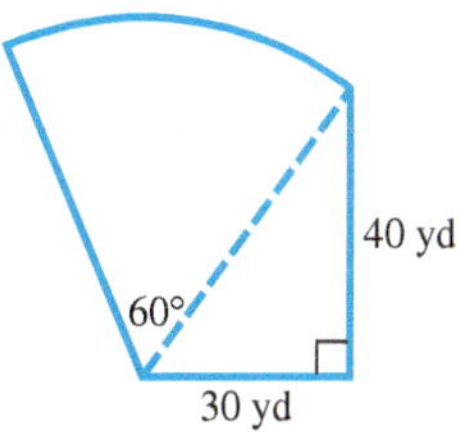

118. *Nautical Miles* **Nautical miles** are used by ships and airplanes. They are different from **statute miles,** where 1 mi = 5280 ft. A nautical mile is defined to be the arc length along the equator intercepted by a central angle *AOB* of 1′, as illustrated in the figure. If the equatorial radius of Earth is 3963 mi, use the arc length formula to approximate the number of statute miles in 1 nautical mile. Round the answer to two decimal places.

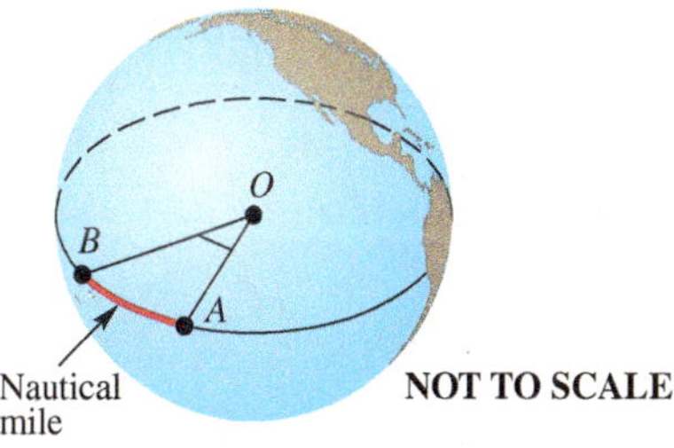

6.2 The Unit Circle and Circular Functions

- Circular Functions
- Values of the Circular Functions
- Determining a Number with a Given Circular Function Value
- Function Values as Lengths of Line Segments
- Linear and Angular Speed

We have defined the six trigonometric functions in such a way that the domain of each function was a set of *angles* in standard position. These angles can be measured in degrees or in radians. In advanced courses, such as calculus, it is necessary to modify the trigonometric functions so that their domains consist of *real numbers* rather than angles. We do this by using the relationship between an angle θ and an arc of length s on a circle.

Circular Functions

In **Figure 12,** we start at the point $(1, 0)$ and measure an arc of length s along the circle. If $s > 0$, then the arc is measured in a counterclockwise direction, and if $s < 0$, then the direction is clockwise. (If $s = 0$, then no arc is measured.) Let the endpoint of this arc be at the point (x, y). The circle in **Figure 12** is the **unit circle**—it has center at the origin and radius 1 unit (hence the name *unit circle*). Recall from algebra that the equation of this circle is

$$x^2 + y^2 = 1. \quad \text{The unit circle}$$

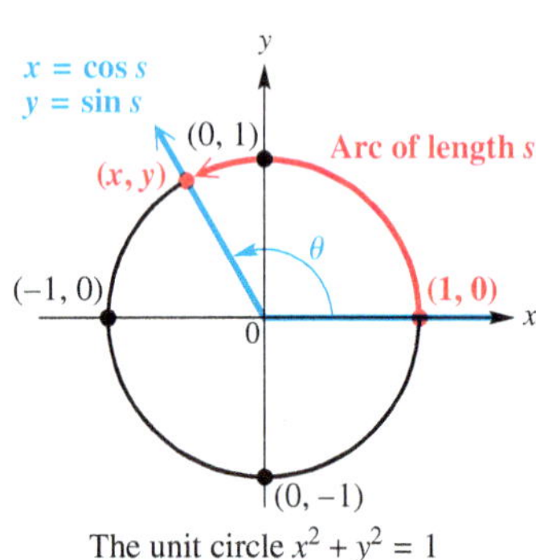

The unit circle $x^2 + y^2 = 1$

Figure 12

The radian measure of θ is related to the arc length s. For θ measured in radians and for r and s measured in the same linear units, we know that

$$s = r\theta.$$

When the radius has measure 1 unit, the formula $s = r\theta$ becomes $s = \theta$. Thus, the trigonometric functions of angle θ in radians found by choosing a point (x, y) on the unit circle can be rewritten as functions of the arc length s, a real number. When interpreted this way, they are called **circular functions.**

LOOKING AHEAD TO CALCULUS

If you plan to study calculus, you must become very familiar with radian measure. In calculus, the trigonometric or circular functions are always understood to have real number domains.

Circular Functions

The following functions are defined for any real number s represented by a directed arc on the unit circle.

$$\sin s = y \qquad \cos s = x \qquad \tan s = \frac{y}{x} \quad (x \neq 0)$$

$$\csc s = \frac{1}{y} \quad (y \neq 0) \qquad \sec s = \frac{1}{x} \quad (x \neq 0) \qquad \cot s = \frac{x}{y} \quad (y \neq 0)$$

The unit circle is symmetric with respect to the x-axis, the y-axis, and the origin. If a point (a, b) lies on the unit circle, so do $(a, -b)$, $(-a, b)$, and $(-a, -b)$. Furthermore, each of these points has a *reference arc* of equal magnitude. For a point on the unit circle, its **reference arc** is the shortest arc from the point itself to the nearest point on the x-axis. (This concept is analogous to the reference angle concept.) Using the concept of symmetry makes determining sines and cosines of the real numbers identified in **Figure 13*** on the next page a relatively simple procedure if we know the coordinates of the points labeled in quadrant I.

*The authors thank Professor Marvel Townsend of the University of Florida for her suggestion to include **Figure 13.**

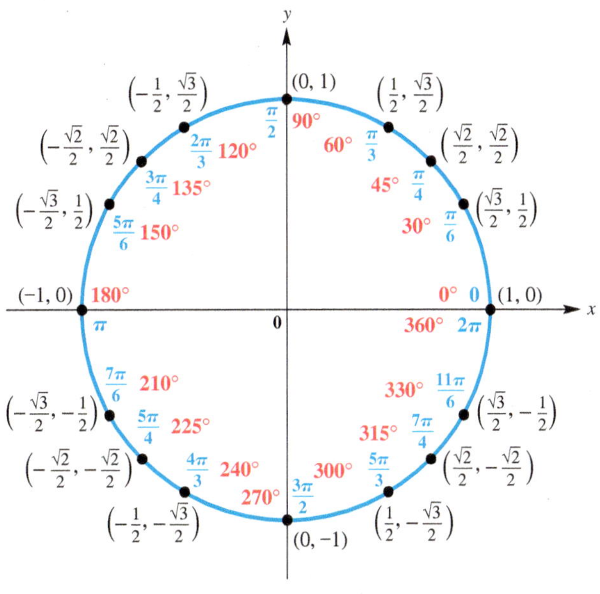

Figure 13

For example, the quadrant I real number $\frac{\pi}{3}$ is associated with the point $\left(\frac{1}{2}, \frac{\sqrt{3}}{2}\right)$ on the unit circle. Therefore, we can use symmetry to identify the coordinates of points having $\frac{\pi}{3}$ as reference arc.

Symmetry and Function Values for Real Numbers with Reference Arc $\frac{\pi}{3}$

s	Quadrant of s	Symmetry Type and Corresponding Point	$\cos s$	$\sin s$
$\frac{\pi}{3}$	I	not applicable; $\left(\frac{1}{2}, \frac{\sqrt{3}}{2}\right)$	$\frac{1}{2}$	$\frac{\sqrt{3}}{2}$
$\pi - \frac{\pi}{3} = \frac{2\pi}{3}$	II	y-axis; $\left(-\frac{1}{2}, \frac{\sqrt{3}}{2}\right)$	$-\frac{1}{2}$	$\frac{\sqrt{3}}{2}$
$\pi + \frac{\pi}{3} = \frac{4\pi}{3}$	III	origin; $\left(-\frac{1}{2}, -\frac{\sqrt{3}}{2}\right)$	$-\frac{1}{2}$	$-\frac{\sqrt{3}}{2}$
$2\pi - \frac{\pi}{3} = \frac{5\pi}{3}$	IV	x-axis; $\left(\frac{1}{2}, -\frac{\sqrt{3}}{2}\right)$	$\frac{1}{2}$	$-\frac{\sqrt{3}}{2}$

NOTE Because $\cos s = x$ and $\sin s = y$, we can replace x and y in the equation of the unit circle $x^2 + y^2 = 1$ and obtain the following.

$$\cos^2 s + \sin^2 s = 1 \quad \text{Pythagorean identity}$$

The ordered pair (x, y) represents a point on the unit circle, and therefore

$$-1 \le x \le 1 \quad \text{and} \quad -1 \le y \le 1,$$

$$\mathbf{-1 \le \cos s \le 1} \quad \text{and} \quad \mathbf{-1 \le \sin s \le 1.}$$

For any value of s, both $\sin s$ and $\cos s$ exist, so the domain of these functions is the set of all real numbers.

For tan s, defined as $\frac{y}{x}$, x must not equal 0. The only way x can equal 0 is when the arc length s is $\frac{\pi}{2}$, $-\frac{\pi}{2}$, $\frac{3\pi}{2}$, $-\frac{3\pi}{2}$, and so on. To avoid a 0 denominator, the domain of the tangent function must be restricted to those values of s that satisfy

$$s \neq (2n + 1)\frac{\pi}{2}, \quad \textbf{where } n \textbf{ is any integer.}$$

The definition of secant also has x in the denominator, so the domain of secant is the same as the domain of tangent. Both cotangent and cosecant are defined with a denominator of y. To guarantee that $y \neq 0$, the domain of these functions must be the set of all values of s that satisfy

$$s \neq n\pi, \quad \textbf{where } n \textbf{ is any integer.}$$

Domains of the Circular Functions

The domains of the circular functions are as follows.

Sine and Cosine Functions: $(-\infty, \infty)$

Tangent and Secant Functions:

$$\{s \mid s \neq (2n + 1)\frac{\pi}{2}, \quad \textbf{where } n \textbf{ is any integer}\}$$

Cotangent and Cosecant Functions:

$$\{s \mid s \neq n\pi, \quad \textbf{where } n \textbf{ is any integer}\}$$

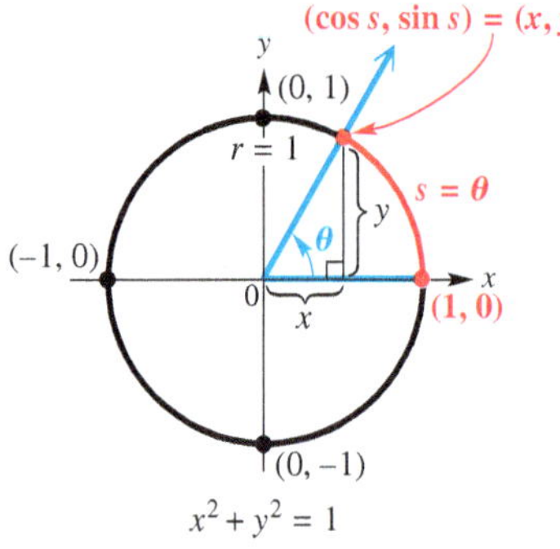

Figure 14

Values of the Circular Functions The circular functions of real numbers correspond to the trigonometric functions of angles measured in radians. Let us assume that angle θ is in standard position, superimposed on the unit circle. See **Figure 14.** Suppose that θ is the *radian* measure of this angle. Using the arc length formula

$$s = r\theta \quad \text{with } r = 1, \quad \text{we have} \quad s = \theta.$$

Thus, the length of the intercepted arc is the real number that corresponds to the radian measure of θ. We use the trigonometric function definitions to obtain the following.

$$\sin \theta = \frac{y}{r} = \frac{y}{1} = y = \sin s, \quad \cos \theta = \frac{x}{r} = \frac{x}{1} = x = \cos s, \quad \text{and so on.}$$

As shown here, the trigonometric functions and the circular functions lead to the same function values, provided that we think of the angles as being in radian measure. This leads to the following important result.

Evaluating a Circular Function

Circular function values of real numbers are obtained in the same manner as trigonometric function values of angles measured in radians. This applies both to methods of finding exact values (such as reference angle analysis) and to calculator approximations. ***Calculators must be in radian mode when they are used to find circular function values.***

EXAMPLE 1 Finding Exact Circular Function Values

Find the exact values of $\sin \frac{3\pi}{2}$, $\cos \frac{3\pi}{2}$, and $\tan \frac{3\pi}{2}$.

SOLUTION Evaluating a circular function at the real number $\frac{3\pi}{2}$ is equivalent to evaluating it at $\frac{3\pi}{2}$ radians. An angle of $\frac{3\pi}{2}$ radians intersects the unit circle at the point $(0, -1)$, as shown in **Figure 15.** Because

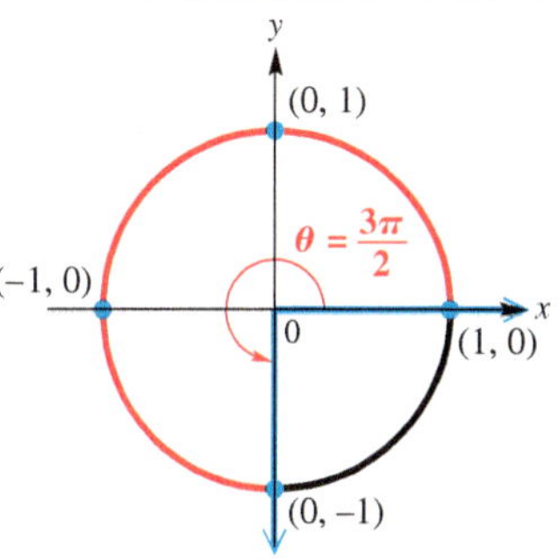

Figure 15

$$\sin s = y, \quad \cos s = x, \quad \text{and} \quad \tan s = \frac{y}{x},$$

it follows that

$$\sin \frac{3\pi}{2} = -1, \quad \cos \frac{3\pi}{2} = 0, \quad \text{and} \quad \tan \frac{3\pi}{2} \text{ is undefined.}$$

✔ **Now Try Exercises 13 and 15.**

EXAMPLE 2 Finding Exact Circular Function Values

Find each exact function value using the specified method.

(a) Use **Figure 13** to find the exact values of $\cos \frac{7\pi}{4}$ and $\sin \frac{7\pi}{4}$.

(b) Use **Figure 13** and the definition of the tangent to find the exact value of $\tan\left(-\frac{5\pi}{3}\right)$.

(c) Use reference angles and radian-to-degree conversion to find the exact value of $\cos \frac{2\pi}{3}$.

SOLUTION

(a) In **Figure 13,** we see that the real number $\frac{7\pi}{4}$ corresponds to the unit circle point $\left(\frac{\sqrt{2}}{2}, -\frac{\sqrt{2}}{2}\right)$.

$$\cos \frac{7\pi}{4} = \frac{\sqrt{2}}{2} \quad \text{and} \quad \sin \frac{7\pi}{4} = -\frac{\sqrt{2}}{2}$$

(b) Moving around the unit circle $\frac{5\pi}{3}$ units in the *negative* direction yields the same ending point as moving around $\frac{\pi}{3}$ units in the positive direction. Thus, $-\frac{5\pi}{3}$ corresponds to $\left(\frac{1}{2}, \frac{\sqrt{3}}{2}\right)$.

$$\tan\left(-\frac{5\pi}{3}\right) = \tan \frac{\pi}{3} = \frac{\frac{\sqrt{3}}{2}}{\frac{1}{2}} = \frac{\sqrt{3}}{2} \div \frac{1}{2} = \frac{\sqrt{3}}{2} \cdot \frac{2}{1} = \sqrt{3}$$

$\tan s = \frac{y}{x}$

Simplify this complex fraction.

(c) An angle of $\frac{2\pi}{3}$ radians corresponds to an angle of 120°. In standard position, 120° lies in quadrant II with a reference angle of 60°.

Cosine is negative in quadrant II.

$$\cos \frac{2\pi}{3} = \cos 120^\circ = -\cos 60^\circ = -\frac{1}{2}$$

Reference angle

✔ **Now Try Exercises 19, 25, 29, and 33.**

EXAMPLE 3 Approximating Circular Function Values

Find a calculator approximation for each circular function value.

(a) $\cos 1.85$ **(b)** $\cos 0.5149$ **(c)** $\cot 1.3209$ **(d)** $\sec(-2.9234)$

SOLUTION

```
NORMAL FIX4 AUTO REAL RADIAN MP
cos(1.85)
                    -.2756
cos(.5149)
                     .8703
1/tan(1.3209)
                     .2552
1/cos(-2.9234)
                   -1.0243
```

Radian mode

This is how the TI-84 Plus calculator displays the results of **Example 3**, fixed to four decimal places.

(a) $\cos 1.85 \approx -0.2756$ Use a calculator in radian mode.

(b) $\cos 0.5149 \approx 0.8703$ Use a calculator in radian mode.

(c) As before, to find cotangent, secant, and cosecant function values, we must use the appropriate reciprocal functions. To find $\cot 1.3209$, first find $\tan 1.3209$ and then find the reciprocal.

$$\cot 1.3209 = \frac{1}{\tan 1.3209} \approx 0.2552$$ Tangent and cotangent are reciprocals.

(d) $$\sec(-2.9234) = \frac{1}{\cos(-2.9234)} \approx -1.0243$$ Cosine and secant are reciprocals.

✔ **Now Try Exercises 35, 41, and 45.**

CAUTION ***Remember, when used to find a circular function value of a real number, a calculator must be in radian mode.***

Determining a Number with a Given Circular Function Value

We can reverse the process of **Example 3** and use a calculator to determine an angle measure, given a trigonometric function value of the angle. ***Remember that the keys marked $\sin^{-1}$, $\cos^{-1}$, and $\tan^{-1}$ do not represent reciprocal functions. They enable us to find inverse function values.***

For reasons explained in a later chapter, the following statements are true.

- For all x in $[-1, 1]$, a calculator in radian mode returns a single value in $\left[-\frac{\pi}{2}, \frac{\pi}{2}\right]$ for $\sin^{-1} x$.
- For all x in $[-1, 1]$, a calculator in radian mode returns a single value in $[0, \pi]$ for $\cos^{-1} x$.
- For all real numbers x, a calculator in radian mode returns a single value in $\left(-\frac{\pi}{2}, \frac{\pi}{2}\right)$ for $\tan^{-1} x$.

EXAMPLE 4 Finding Numbers Given Circular Function Values

Find each value as specified.

(a) Approximate the value of s in the interval $\left[0, \frac{\pi}{2}\right]$ if $\cos s = 0.9685$.

(b) Find the exact value of s in the interval $\left[\pi, \frac{3\pi}{2}\right]$ if $\tan s = 1$.

SOLUTION

(a) Because we are given a cosine value and want to determine the real number in $\left[0, \frac{\pi}{2}\right]$ that has this cosine value, we use the *inverse cosine* function of a calculator. With the calculator in radian mode, we find s as follows.

$$s = \cos^{-1}(0.9685) \approx 0.2517$$

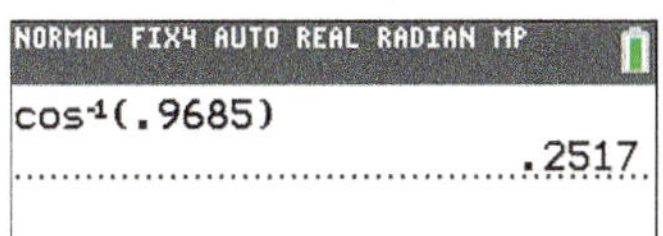

Radian mode

Figure 16

See **Figure 16.** The screen indicates that the real number in $\left[0, \frac{\pi}{2}\right]$ having cosine equal to 0.9685 is 0.2517.

(b) Recall that $\tan \frac{\pi}{4} = 1$, and in quadrant III $\tan s$ is positive.

$$\tan\left(\pi + \frac{\pi}{4}\right) = \tan \frac{5\pi}{4} = 1$$

Thus, $s = \frac{5\pi}{4}$. See **Figure 17.**

```
NORMAL FLOAT AUTO REAL RADIAN MP
tan⁻¹(1)
                .7853981634
Ans+π
                3.926990817
tan(Ans)
                          1
```

This screen supports the result in **Example 4(b)** with calculator approximations.

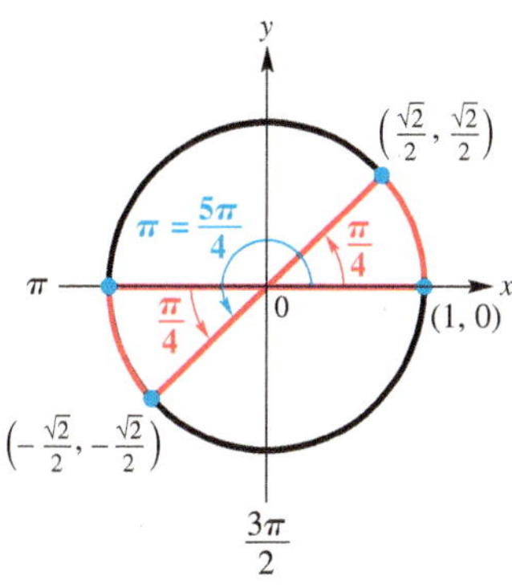

Figure 17

Now Try Exercises 65 and 73.

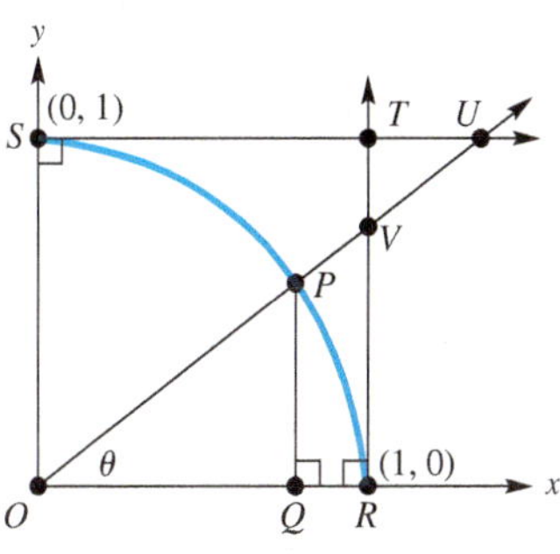

Figure 18

Function Values as Lengths of Line Segments The diagram shown in **Figure 18** illustrates a correspondence that ties together the right triangle ratio definitions of the trigonometric functions and the unit circle interpretation. The arc SR is the first-quadrant portion of the unit circle, and the standard-position angle POQ is designated θ. By definition, the coordinates of P are $(\cos \theta, \sin \theta)$. The six trigonometric functions of θ can be interpreted as lengths of line segments found in **Figure 18.**

For $\cos \theta$ and $\sin \theta$, use right triangle POQ and right triangle ratios.

$$\boldsymbol{\cos \theta} = \frac{\text{side adjacent to } \theta}{\text{hypotenuse}} = \frac{OQ}{OP} = \frac{OQ}{1} = \boldsymbol{OQ}$$

$$\boldsymbol{\sin \theta} = \frac{\text{side opposite } \theta}{\text{hypotenuse}} = \frac{PQ}{OP} = \frac{PQ}{1} = \boldsymbol{PQ}$$

For $\tan \theta$ and $\sec \theta$, use right triangle VOR in **Figure 18** and right triangle ratios.

$$\boldsymbol{\tan \theta} = \frac{\text{side opposite } \theta}{\text{side adjacent to } \theta} = \frac{VR}{OR} = \frac{VR}{1} = \boldsymbol{VR}$$

$$\boldsymbol{\sec \theta} = \frac{\text{hypotenuse}}{\text{side adjacent to } \theta} = \frac{OV}{OR} = \frac{OV}{1} = \boldsymbol{OV}$$

For $\csc \theta$ and $\cot \theta$, first note that US and OR are parallel. Thus angle SUO is equal to θ because it is an alternate interior angle to angle POQ, which is equal to θ. Use right triangle USO and right triangle ratios.

$$\csc SUO = \boldsymbol{\csc \theta} = \frac{\text{hypotenuse}}{\text{side opposite } \theta} = \frac{OU}{OS} = \frac{OU}{1} = \boldsymbol{OU}$$

$$\cot SUO = \boldsymbol{\cot \theta} = \frac{\text{side adjacent to } \theta}{\text{side opposite } \theta} = \frac{US}{OS} = \frac{US}{1} = \boldsymbol{US}$$

Figure 19 uses color to illustrate the results just found.

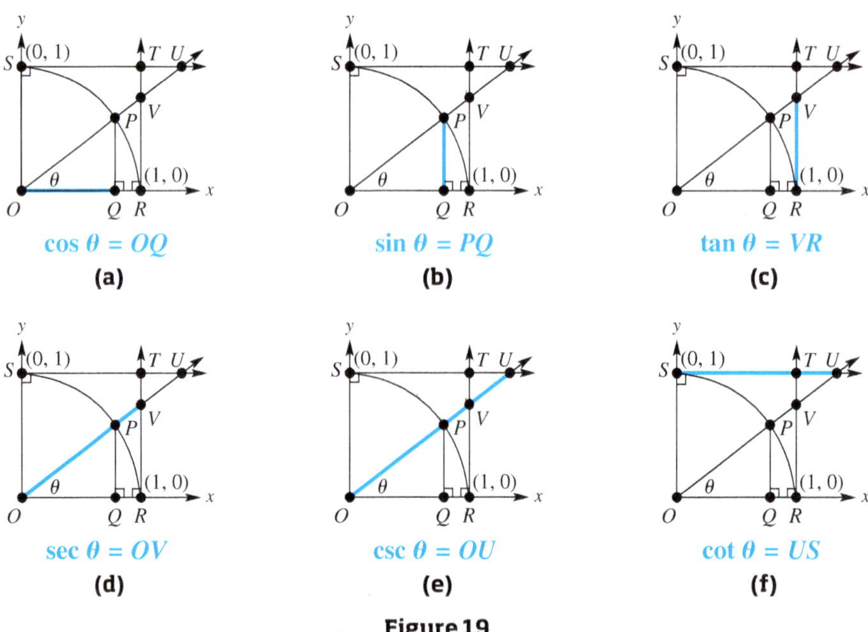

Figure 19

EXAMPLE 5 Finding Lengths of Line Segments

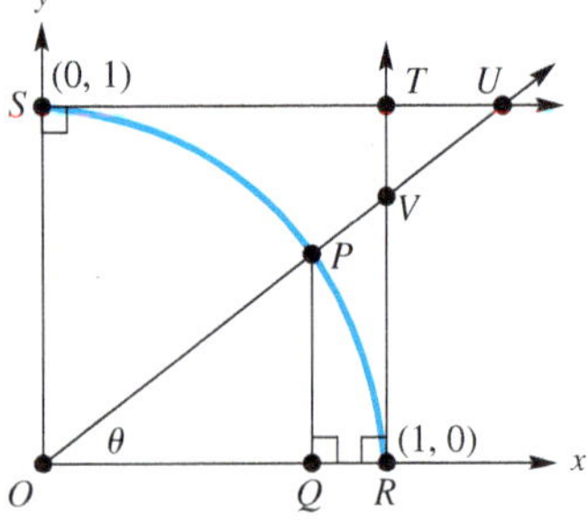

Figure 18 (repeated)

Figure 18 is repeated in the margin. Suppose that angle TVU measures 60°. Find the exact lengths of segments OQ, PQ, VR, OV, OU, and US.

SOLUTION Angle TVU has the same measure as angle OVR because they are vertical angles. Therefore, angle OVR measures 60°. Because it is one of the acute angles in right triangle VOR, θ must be its complement, measuring 30°.

$$OQ = \cos 30° = \frac{\sqrt{3}}{2} \qquad OV = \sec 30° = \frac{2\sqrt{3}}{3}$$

$$PQ = \sin 30° = \frac{1}{2} \qquad OU = \csc 30° = 2$$

$$VR = \tan 30° = \frac{\sqrt{3}}{3} \qquad US = \cot 30° = \sqrt{3}$$

Use the equations found in **Figure 19**, with $\theta = 30°$.

✔ **Now Try Exercise 81.**

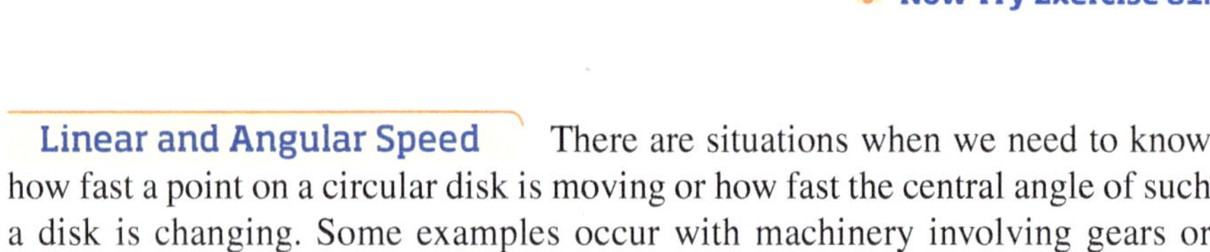

Linear and Angular Speed

There are situations when we need to know how fast a point on a circular disk is moving or how fast the central angle of such a disk is changing. Some examples occur with machinery involving gears or pulleys or the speed of a car around a curved portion of highway.

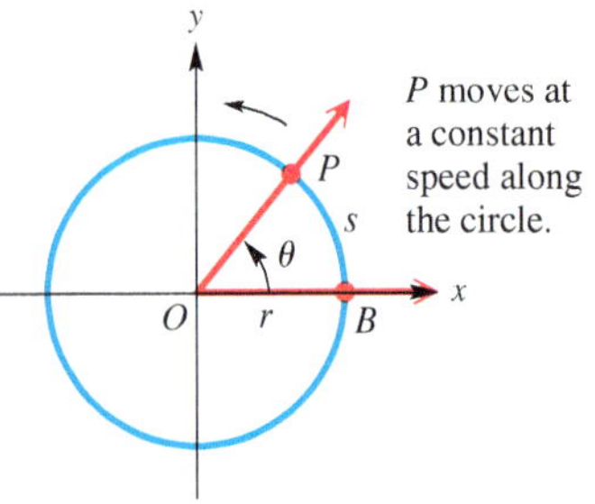

Figure 20

Suppose that point P moves at a constant speed along a circle of radius r and center O. See **Figure 20.** The measure of how fast the position of P is changing is the **linear speed.** If v represents linear speed, then

$$\text{speed} = \frac{\text{distance}}{\text{time}}, \quad \text{or} \quad v = \frac{s}{t},$$

where s is the length of the arc traced by point P at time t. (This formula is just a restatement of $r = \frac{d}{t}$ with s as distance, v as rate (speed), and t as time.)

Refer to **Figure 20** on the previous page. As point P in the figure moves along the circle, ray OP rotates around the origin. Because ray OP is the terminal side of angle POB, the measure of the angle changes as P moves along the circle. The measure of how fast angle POB is changing is its **angular speed.** Angular speed, symbolized ω, is given as

Formulas for Angular and Linear Speed

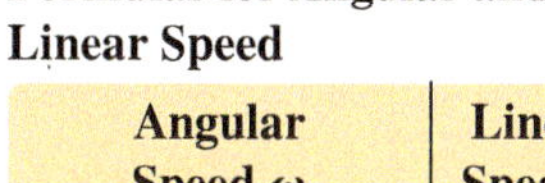

Angular Speed ω	Linear Speed v
$\omega = \frac{\theta}{t}$	$v = \frac{s}{t}$
(ω in radians per unit time t, θ in radians)	$v = \frac{r\theta}{t}$
	$v = r\omega$

$$\omega = \frac{\theta}{t}, \quad \textbf{where } \theta \textbf{ is in radians.}$$

Here θ is the measure of angle POB at time t. ***As with earlier formulas in this chapter, θ must be measured in radians, with ω expressed in radians per unit of time.***

The length s of the arc intercepted on a circle of radius r by a central angle of measure θ radians is $s = r\theta$. Using this formula, the formula for linear speed, $v = \frac{s}{t}$, can be written in several useful forms.

$$v = \frac{s}{t} \qquad \text{Formula for linear speed}$$

$$v = \frac{r\theta}{t} \qquad s = r\theta$$

$$v = r \cdot \frac{\theta}{t} \qquad \frac{ab}{c} = a \cdot \frac{b}{c}$$

$$v = r\omega \qquad \omega = \frac{\theta}{t}$$

As an example of linear and angular speeds, consider the following. The human joint that can be flexed the fastest is the wrist, which can rotate through 90°, or $\frac{\pi}{2}$ radians, in 0.045 sec while holding a tennis racket. The angular speed of a human wrist swinging a tennis racket is

$$\omega = \frac{\theta}{t} \qquad \text{Formula for angular speed}$$

$$\omega = \frac{\frac{\pi}{2}}{0.045} \qquad \text{Let } \theta = \frac{\pi}{2} \text{ and } t = 0.045.$$

$$\omega \approx 35 \text{ radians per sec.} \qquad \text{Use a calculator.}$$

If the radius (distance) from the tip of the racket to the wrist joint is 2 ft, then the speed at the tip of the racket is

$$v = r\omega \qquad \text{Formula for linear speed}$$

$$v \approx 2(35) \qquad \text{Let } r = 2 \text{ and } \omega = 35.$$

$$v = 70 \text{ ft per sec, or about 48 mph.} \qquad \text{Use a calculator.}$$

In a tennis serve the arm rotates at the shoulder, so the final speed of the racket is considerably greater. (*Source:* Cooper, J. and R. Glassow, *Kinesiology,* Second Edition, C.V. Mosby.)

EXAMPLE 6 Using Linear and Angular Speed Formulas

Suppose that point P is on a circle with radius 10 cm, and ray OP is rotating with angular speed $\frac{\pi}{18}$ radian per sec.

(a) Find the angle generated by P in 6 sec.

(b) Find the distance traveled by P along the circle in 6 sec.

(c) Find the linear speed of P in centimeters per second.

SOLUTION

(a) To find the angle generated by P, solve for θ in the angular speed formula $\omega = \frac{\theta}{t}$. Substitute the known quantities $\omega = \frac{\pi}{18}$ radian per sec and $t = 6$ sec in the formula.

$$\theta = \omega t \quad \text{Angular speed formula solved for } \theta$$

$$\theta = \frac{\pi}{18}(6) \quad \text{Let } \omega = \tfrac{\pi}{18} \text{ and } t = 6.$$

$$\theta = \frac{\pi}{3} \text{ radians} \quad \text{Multiply.}$$

(b) To find the distance traveled by P, use the arc length formula $s = r\theta$ with $r = 10$ cm and, from part (a), $\theta = \frac{\pi}{3}$ radians.

$$s = r\theta = 10\left(\frac{\pi}{3}\right) = \frac{10\pi}{3} \text{ cm} \quad \text{Let } r = 10 \text{ and } \theta = \tfrac{\pi}{3}.$$

(c) Use the formula for linear speed with $r = 10$ cm and $\omega = \frac{\pi}{18}$ radians per sec.

$$v = r\omega = 10\left(\frac{\pi}{18}\right) = \frac{5\pi}{9} \text{ cm per sec} \quad \text{Linear speed formula}$$

✔ **Now Try Exercise 83.**

EXAMPLE 7 Finding Angular Speed of a Pulley and Linear Speed of a Belt

A belt runs a pulley of radius 6 cm at 80 revolutions per min. See **Figure 21.**

(a) Find the angular speed of the pulley in radians per second.

(b) Find the linear speed of the belt in centimeters per second.

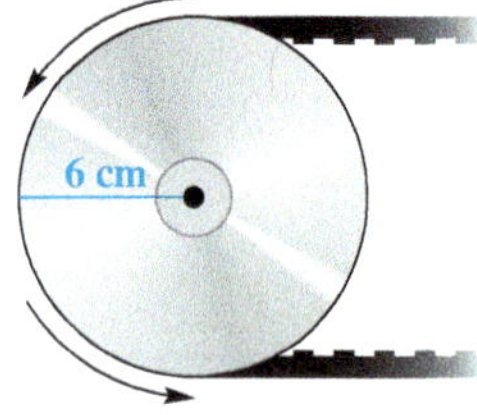

Figure 21

SOLUTION

(a) The angular speed 80 revolutions per min can be converted to radians per second using the following facts.

$$1 \text{ revolution} = 2\pi \text{ radians} \quad \text{and} \quad 1 \text{ min} = 60 \text{ sec}$$

We multiply by the corresponding **unit fractions.** Here, just as with the unit circle, the word *unit* means 1, so multiplying by a unit fraction is equivalent to multiplying by 1. We divide out common units in the same way that we divide out common factors.

$$\omega = \frac{80 \text{ revolutions}}{1 \text{ min}} \cdot \frac{2\pi \text{ radians}}{1 \text{ revolution}} \cdot \frac{1 \text{ min}}{60 \text{ sec}}$$

$$\omega = \frac{160\pi \text{ radians}}{60 \text{ sec}} \quad \text{Multiply. Divide out common units.}$$

$$\omega = \frac{8\pi}{3} \text{ radians per sec} \quad \text{Angular speed}$$

(b) The linear speed v of the belt will be the same as that of a point on the circumference of the pulley.

$$v = r\omega = 6\left(\frac{8\pi}{3}\right) = 16\pi \approx 50 \text{ cm per sec} \quad \text{Linear speed}$$

✔ **Now Try Exercise 123.**

6.2 Exercises

CONCEPT PREVIEW *Fill in the blanks to complete the coordinates for each point indicated in the first quadrant of the unit circle in Exercise 1. Then use it to find each exact circular function value in Exercises 2–5, and work Exercise 6.*

1.

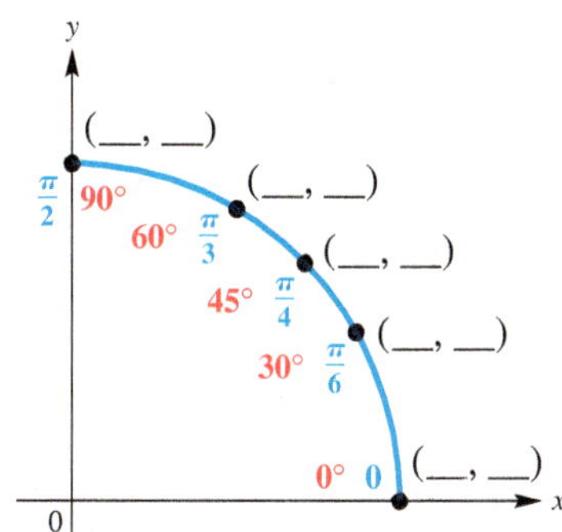

2. $\cos 0$

3. $\sin \frac{\pi}{4}$

4. $\sin \frac{\pi}{3}$

5. $\tan \frac{\pi}{4}$

6. Find s in the interval $\left[0, \frac{\pi}{2}\right]$ if $\cos s = \frac{1}{2}$.

CONCEPT PREVIEW *Fill in the blank to correctly complete each sentence. As necessary, refer to the figure that shows point P moving at a constant speed along the unit circle.*

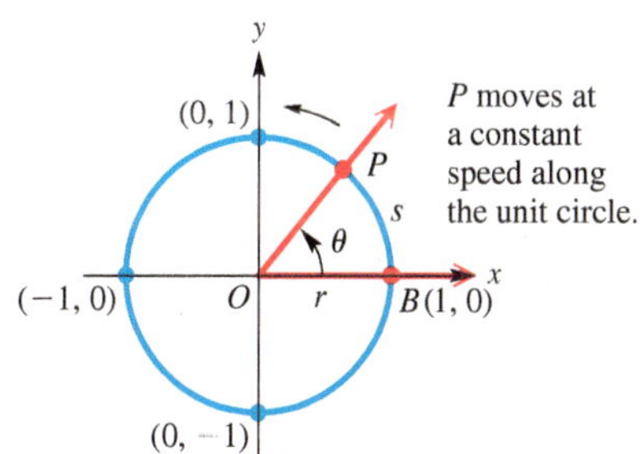

7. The measure of how fast the position of point P is changing is the ________.

8. The measure of how fast angle POB is changing is the ________.

9. If the angular speed of point P is 1 radian per sec, then P will move around the entire unit circle in ______ sec.

10. If the angular speed of point P is π radians per sec, then the linear speed is ______ unit(s) per sec.

11. An angular speed of 1 revolution per min on the unit circle is equivalent to an angular speed, ω, of ______ radians per min.

12. If P is rotating with angular speed $\frac{\pi}{2}$ radians per sec, then the distance traveled by P in 10 sec is ______ units.

Find the exact values of **(a)** sin *s*, **(b)** cos *s*, *and* **(c)** tan *s for each real number s.* ***See Example 1.***

13. $s = \frac{\pi}{2}$ **14.** $s = \pi$ **15.** $s = 2\pi$

16. $s = 3\pi$ **17.** $s = -\pi$ **18.** $s = -\frac{3\pi}{2}$

Find each exact function value. ***See Example 2.***

19. $\sin \frac{7\pi}{6}$ **20.** $\cos \frac{5\pi}{3}$ **21.** $\tan \frac{3\pi}{4}$ **22.** $\sec \frac{2\pi}{3}$

23. $\csc \frac{11\pi}{6}$ **24.** $\cot \frac{5\pi}{6}$ **25.** $\cos\left(-\frac{4\pi}{3}\right)$ **26.** $\tan\left(-\frac{17\pi}{3}\right)$

27. $\cos \frac{7\pi}{4}$ **28.** $\sec \frac{5\pi}{4}$ **29.** $\sin\left(-\frac{4\pi}{3}\right)$ **30.** $\sin\left(-\frac{5\pi}{6}\right)$

31. $\sec \frac{23\pi}{6}$ **32.** $\csc \frac{13\pi}{3}$ **33.** $\tan \frac{5\pi}{6}$ **34.** $\cos \frac{3\pi}{4}$

Find a calculator approximation to four decimal places for each circular function value. ***See Example 3.***

35. sin 0.6109 **36.** sin 0.8203 **37.** cos(−1.1519)

38. cos(−5.2825) **39.** tan 4.0203 **40.** tan 6.4752

41. csc(−9.4946) **42.** csc 1.3875 **43.** sec 2.8440

44. sec(−8.3429) **45.** cot 6.0301 **46.** cot 3.8426

Concept Check *The figure displays a unit circle and an angle of* 1 *radian. The tick marks on the circle are spaced at every two-tenths radian. Use the figure to estimate each value.*

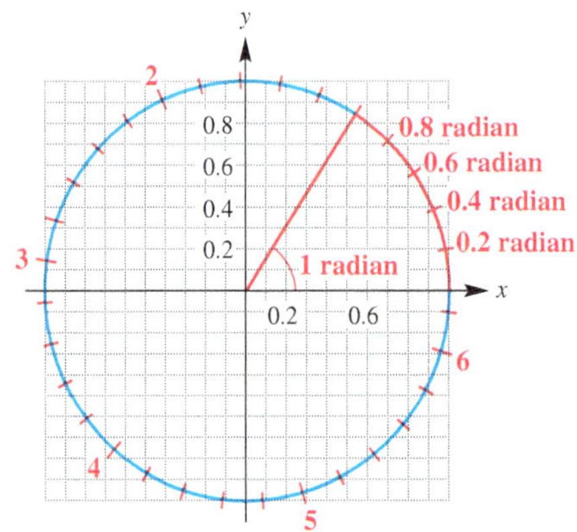

47. cos 0.8 **48.** cos 0.6 **49.** sin 2

50. sin 5.4 **51.** sin 3.8 **52.** cos 3.2

53. a positive angle whose cosine is −0.65

54. a positive angle whose sine is −0.95

55. a positive angle whose sine is 0.7

56. a positive angle whose cosine is 0.3

Concept Check *Without using a calculator, decide whether each function value is positive or negative. (Hint: Consider the radian measures of the quadrantal angles, and remember that $\pi \approx 3.14$.)*

57. $\cos 2$ **58.** $\sin(-1)$ **59.** $\sin 5$

60. $\cos 6$ **61.** $\tan 6.29$ **62.** $\tan(-6.29)$

Find the approximate value of s, to four decimal places, in the interval $\left[0, \frac{\pi}{2}\right]$ that makes each statement true. ***See Example 4(a).***

63. $\tan s = 0.2126$ **64.** $\cos s = 0.7826$ **65.** $\sin s = 0.9918$

66. $\cot s = 0.2994$ **67.** $\sec s = 1.0806$ **68.** $\csc s = 1.0219$

Find the exact value of s in the given interval that has the given circular function value. ***See Example 4(b).***

69. $\left[\frac{\pi}{2}, \pi\right]$; $\sin s = \frac{1}{2}$ **70.** $\left[\frac{\pi}{2}, \pi\right]$; $\cos s = -\frac{1}{2}$

71. $\left[\pi, \frac{3\pi}{2}\right]$; $\tan s = \sqrt{3}$ **72.** $\left[\pi, \frac{3\pi}{2}\right]$; $\sin s = -\frac{1}{2}$

73. $\left[\frac{3\pi}{2}, 2\pi\right]$; $\tan s = -1$ **74.** $\left[\frac{3\pi}{2}, 2\pi\right]$; $\cos s = \frac{\sqrt{3}}{2}$

Find the exact values of s in the given interval that satisfy the given condition.

75. $[0, 2\pi)$; $\sin s = -\frac{\sqrt{3}}{2}$ **76.** $[0, 2\pi)$; $\cos s = -\frac{1}{2}$

77. $[0, 2\pi)$; $\cos^2 s = \frac{1}{2}$ **78.** $[0, 2\pi)$; $\tan^2 s = 3$

79. $[-2\pi, \pi)$; $3 \tan^2 s = 1$ **80.** $[-\pi, \pi)$; $\sin^2 s = \frac{1}{2}$

Refer to **Figures 18 and 19,** *and work each problem.* ***See Example 5.***

81. Suppose that angle θ measures 60°. Find the exact length of each segment.

(a) OQ **(b)** PQ **(c)** VR

(d) OV **(e)** OU **(f)** US

82. Repeat **Exercise 81** for $\theta = 38°$. Give lengths as approximations to four significant digits.

Suppose that point P is on a circle with radius r, and ray OP is rotating with angular speed ω. Use the given values of r, ω, and t to do the following. ***See Example 6.***

(a) *Find the angle generated by P in time t.*

(b) *Find the distance traveled by P along the circle in time t.*

(c) *Find the linear speed of P.*

83. $r = 20$ cm, $\omega = \frac{\pi}{12}$ radian per sec, $t = 6$ sec

84. $r = 30$ cm, $\omega = \frac{\pi}{10}$ radian per sec, $t = 4$ sec

85. $r = 8$ in., $\omega = \frac{\pi}{3}$ radians per min, $t = 9$ min

86. $r = 12$ ft, $\omega = 8\pi$ radians per min, $t = 5$ min

Use the formula $\omega = \frac{\theta}{t}$ to find the value of the missing variable.

87. $\omega = \frac{2\pi}{3}$ radians per sec, $t = 3$ sec

88. $\omega = \frac{\pi}{4}$ radian per min, $t = 5$ min

89. $\omega = 0.91$ radian per min, $t = 8.1$ min

90. $\omega = 4.3$ radians per min, $t = 1.6$ min

91. $\theta = \frac{3\pi}{4}$ radians, $t = 8$ sec

92. $\theta = \frac{2\pi}{5}$ radians, $t = 10$ sec

93. $\theta = 3.871$ radians, $t = 21.47$ sec

94. $\theta = 5.225$ radians, $t = 2.515$ sec

95. $\theta = \frac{2\pi}{9}$ radian, $\omega = \frac{5\pi}{27}$ radian per min

96. $\theta = \frac{3\pi}{8}$ radians, $\omega = \frac{\pi}{24}$ radian per min

Use the formula $v = r\omega$ to find the value of the missing variable.

97. $r = 12$ m, $\omega = \frac{2\pi}{3}$ radians per sec

98. $r = 8$ cm, $\omega = \frac{9\pi}{5}$ radians per sec

99. $v = 9$ m per sec, $r = 5$ m

100. $v = 18$ ft per sec, $r = 3$ ft

101. $v = 12$ m per sec, $\omega = \frac{3\pi}{2}$ radians per sec

102. $v = 24.93$ cm per sec, $\omega = 0.3729$ radian per sec

The formula $\omega = \frac{\theta}{t}$ can be rewritten as $\theta = \omega t$. Substituting ωt for θ converts $s = r\theta$ to $s = r\omega t$. Use the formula $s = r\omega t$ to find the value of the missing variable.

103. $r = 6$ cm, $\omega = \frac{\pi}{3}$ radians per sec, $t = 9$ sec

104. $r = 9$ yd, $\omega = \frac{2\pi}{5}$ radians per sec, $t = 12$ sec

105. $s = 6\pi$ cm, $r = 2$ cm, $\omega = \frac{\pi}{4}$ radian per sec

106. $s = \frac{12\pi}{5}$ m, $r = \frac{3}{2}$ m, $\omega = \frac{2\pi}{5}$ radians per sec

107. $s = \frac{3\pi}{4}$ km, $r = 2$ km, $t = 4$ sec

108. $s = \frac{8\pi}{9}$ m, $r = \frac{4}{3}$ m, $t = 12$ sec

Find the angular speed ω for each of the following.

109. the hour hand of a clock

110. the second hand of a clock

111. the minute hand of a clock

112. a gear revolving 300 times per min

Find the linear speed v for each of the following.

113. the tip of the minute hand of a clock, if the hand is 7 cm long

114. the tip of the second hand of a clock, if the hand is 28 mm long

115. a point on the edge of a flywheel of radius 2 m, rotating 42 times per min

116. a point on the tread of a tire of radius 18 cm, rotating 35 times per min

117. the tip of a propeller 3 m long, rotating 500 times per min (*Hint*: $r = 1.5$ m)

118. a point on the edge of a gyroscope of radius 83 cm, rotating 680 times per min

Solve each problem. ***See Examples 6 and 7.***

119. ***Speed of a Bicycle*** The tires of a bicycle have radius 13.0 in. and are turning at the rate of 215 revolutions per min. See the figure. How fast is the bicycle traveling in miles per hour? (*Hint:* 5280 ft = 1 mi)

120. *Hours in a Martian Day* Mars rotates on its axis at the rate of about 0.2552 radian per hr. Approximately how many hours are in a Martian day (or *sol*)? (*Source: World Almanac and Book of Facts.*)

Opposite sides of Mars

121. *Angular and Linear Speeds of Earth* The orbit of Earth about the sun is almost circular. Assume that the orbit is a circle with radius 93,000,000 mi. Its angular and linear speeds are used in designing solar-power facilities.

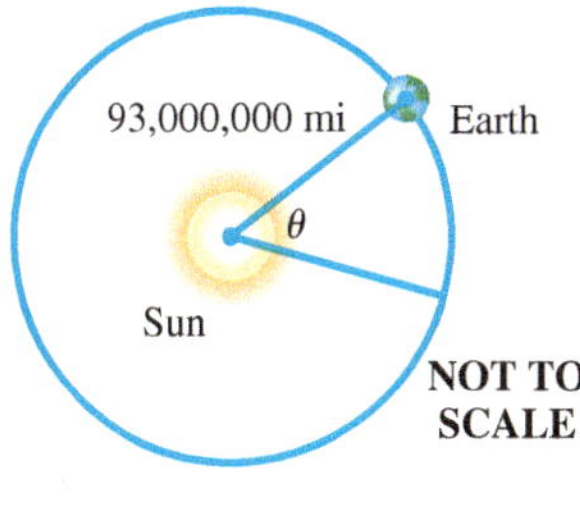

(a) Assume that a year is 365 days, and find the angle formed by Earth's movement in one day.

(b) Give the angular speed in radians per hour.

(c) Find the approximate linear speed of Earth in miles per hour.

122. *Angular and Linear Speeds of Earth* Earth revolves on its axis once every 24 hr. Assuming that Earth's radius is 6400 km, find the following.

(a) angular speed of Earth in radians per hour

(b) linear speed at the North Pole or South Pole

(c) approximate linear speed at Quito, Ecuador, a city on the equator

(d) approximate linear speed at Salem, Oregon (halfway from the equator to the North Pole)

123. *Speeds of a Pulley and a Belt* The pulley shown has a radius of 12.96 cm. Suppose it takes 18 sec for 56 cm of belt to go around the pulley.

(a) Find the linear speed of the belt in centimeters per second.

(b) Find the angular speed of the pulley in radians per second.

124. *Angular Speeds of Pulleys* The two pulleys in the figure have radii of 15 cm and 8 cm, respectively. The larger pulley rotates 25 times in 36 sec. Find the angular speed of each pulley in radians per second.

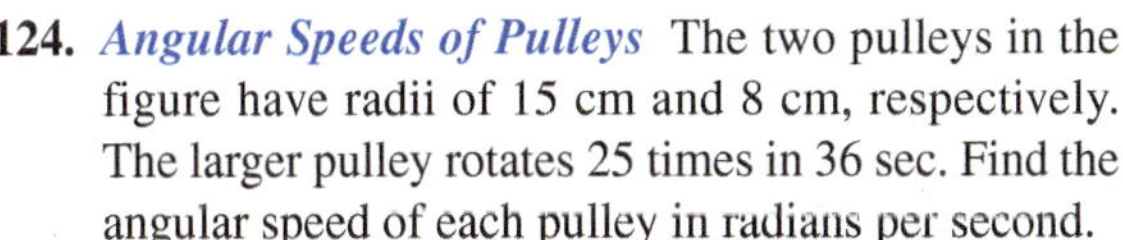

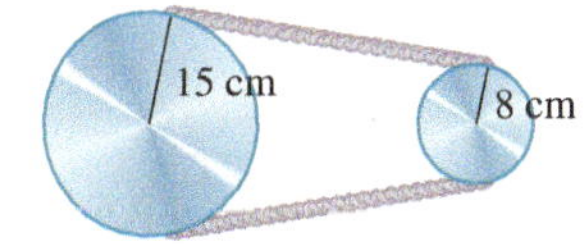

125. *Radius of a Spool of Thread* A thread is being pulled off a spool at the rate of 59.4 cm per sec. Find the radius of the spool if it makes 152 revolutions per min.

126. *Time to Move along a Railroad Track* A railroad track is laid along the arc of a circle of radius 1800 ft. The circular part of the track subtends a central angle of 40°. How long (in seconds) will it take a point on the front of a train traveling 30.0 mph to go around this portion of the track?

127. *Angular Speed of a Motor Propeller* The propeller of a 90-horsepower outboard motor at full throttle rotates at exactly 5000 revolutions per min. Find the angular speed of the propeller in radians per second.

128. *Linear Speed of a Golf Club* The shoulder joint can rotate at 25.0 radians per sec. If a golfer's arm is straight and the distance from the shoulder to the club head is 5.00 ft, find the linear speed of the club head from shoulder rotation. (*Source*: Cooper, J. and R. Glassow, *Kinesiology*, Second Edition, C.V. Mosby.)

6.3 Graphs of the Sine and Cosine Functions

- Periodic Functions
- Graph of the Sine Function
- Graph of the Cosine Function
- Techniques for Graphing, Amplitude, and Period
- Connecting Graphs with Equations
- A Trigonometric Model

Periodic Functions

Phenomena that repeat with a predictable pattern, such as tides, phases of the moon, and hours of daylight, can be modeled by sine and cosine functions. These functions are *periodic*. The periodic graph in **Figure 22** represents a normal heartbeat.

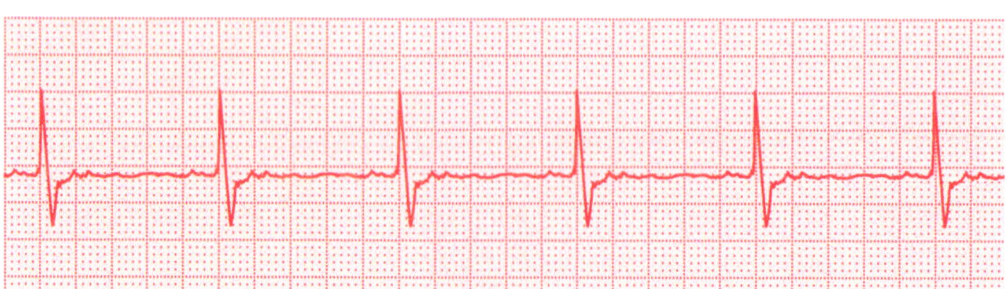

Figure 22

Periodic functions are defined as follows.

LOOKING AHEAD TO CALCULUS

Periodic functions are used throughout calculus, so it is important to know their characteristics. One use of these functions is to describe the location of a point in the plane using **polar coordinates,** an alternative to rectangular coordinates.

Periodic Function

A **periodic function** is a function f such that

$$f(x) = f(x + np),$$

for every real number x in the domain of f, every integer n, and some positive real number p. The least possible positive value of p is the **period** of the function.

The circumference of the unit circle is 2π, so the least value of p for which the sine and cosine functions repeat is 2π. ***Therefore, the sine and cosine functions are periodic functions with period 2π.*** For every positive integer n,

$$\sin x = \sin(x + n \cdot 2\pi) \quad \text{and} \quad \cos x = \cos(x + n \cdot 2\pi).$$

Graph of the Sine Function

We have seen that for a real number s, the point on the unit circle corresponding to s has coordinates $(\cos s, \sin s)$. See **Figure 23.** Trace along the circle to verify the results shown in the table.

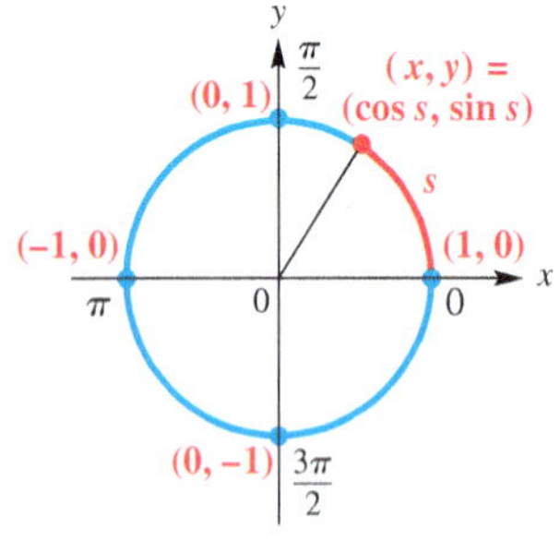

Figure 23

As s Increases from	$\sin s$	$\cos s$
0 to $\frac{\pi}{2}$	Increases from 0 to 1	Decreases from 1 to 0
$\frac{\pi}{2}$ to π	Decreases from 1 to 0	Decreases from 0 to -1
π to $\frac{3\pi}{2}$	Decreases from 0 to -1	Increases from -1 to 0
$\frac{3\pi}{2}$ to 2π	Increases from -1 to 0	Increases from 0 to 1

To avoid confusion when graphing the sine function, we use x rather than s. This corresponds to the letters in the xy-coordinate system. Selecting key values of x and finding the corresponding values of $\sin x$ leads to the table in **Figure 24** on the next page.

To obtain the traditional graph in **Figure 24,** we plot the points from the table, use symmetry, and join them with a smooth curve. Because $y = \sin x$ is periodic with period 2π and has domain $(-\infty, \infty)$, the graph continues in the same pattern in both directions. This graph is a **sine wave,** or **sinusoid.**

Sine Function $f(x) = \sin x$

Domain: $(-\infty, \infty)$ Range: $[-1, 1]$

x	y
0	0
$\frac{\pi}{6}$	$\frac{1}{2}$
$\frac{\pi}{4}$	$\frac{\sqrt{2}}{2}$
$\frac{\pi}{3}$	$\frac{\sqrt{3}}{2}$
$\frac{\pi}{2}$	1
π	0
$\frac{3\pi}{2}$	-1
2π	0

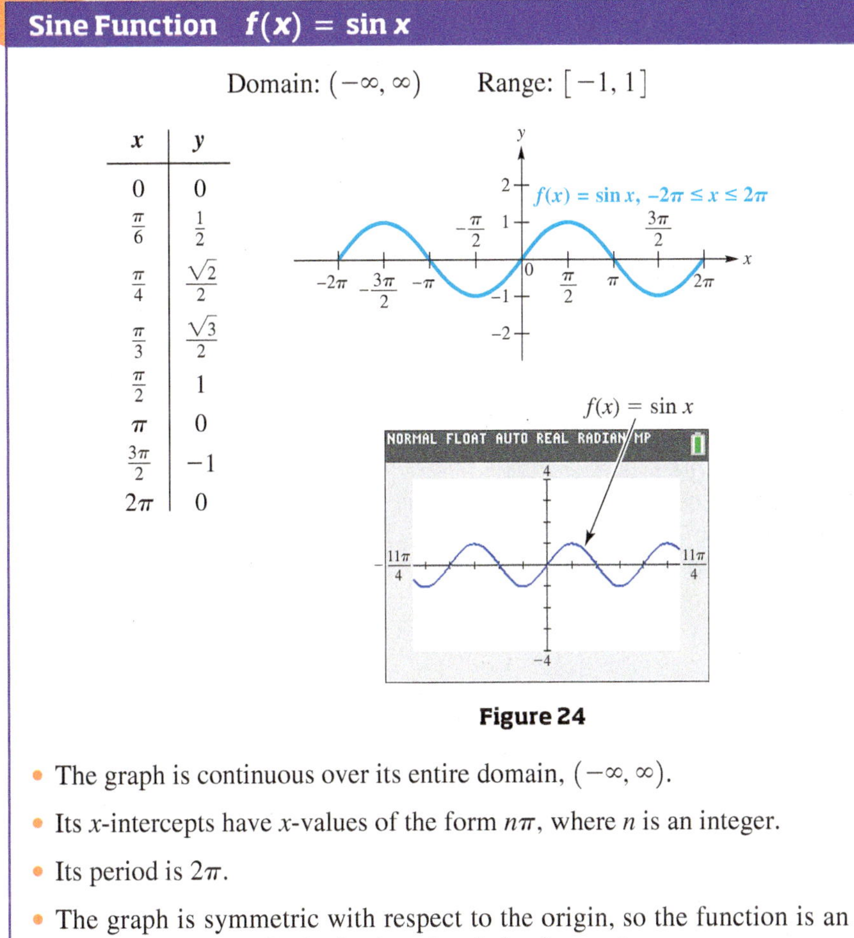

Figure 24

- The graph is continuous over its entire domain, $(-\infty, \infty)$.
- Its x-intercepts have x-values of the form $n\pi$, where n is an integer.
- Its period is 2π.
- The graph is symmetric with respect to the origin, so the function is an odd function. For all x in the domain, $\sin(-x) = -\sin x$.

NOTE A function f is an **odd function** if for all x in the domain of f,

$$f(-x) = -f(x).$$

The graph of an odd function is symmetric with respect to the origin. This means that if (x, y) belongs to the function, then $(-x, -y)$ also belongs to the function. For example, $\left(\frac{\pi}{2}, 1\right)$ and $\left(-\frac{\pi}{2}, -1\right)$ are points on the graph of $y = \sin x$, illustrating the property $\sin(-x) = -\sin x$.

The sine function is related to the unit circle. **Its domain consists of real numbers corresponding to angle measures (or arc lengths) on the unit circle. Its range corresponds to y-coordinates (or sine values) on the unit circle.**

Consider the unit circle in **Figure 23** and assume that the line from the origin to some point on the circle is part of the pedal of a bicycle, with a foot placed on the circle itself. As the pedal is rotated from 0 radians on the horizontal axis through various angles, the angle (or arc length) giving the pedal's location and its corresponding height from the horizontal axis given by $\sin x$ are used to create points on the sine graph. See **Figure 25** on the next page.

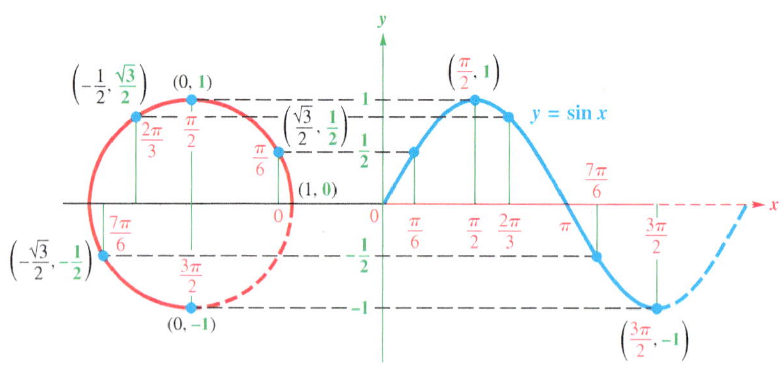

Figure 25

LOOKING AHEAD TO CALCULUS

The discussion of the derivative of a function in calculus shows that for the sine function, the slope of the tangent line at any point x is given by $\cos x$. For example, look at the graph of $y = \sin x$ and notice that a tangent line at $x = \pm\frac{\pi}{2}, \pm\frac{3\pi}{2}, \pm\frac{5\pi}{2}, \ldots$ will be horizontal and thus have slope 0. Now look at the graph of $y = \cos x$ and see that for these values, $\cos x = 0$.

Graph of the Cosine Function *The graph of* $y = \cos x$ *in* **Figure 26** *is the graph of the sine function shifted, or translated,* $\frac{\pi}{2}$ *units to the left.*

Cosine Function $f(x) = \cos x$

Domain: $(-\infty, \infty)$ Range: $[-1, 1]$

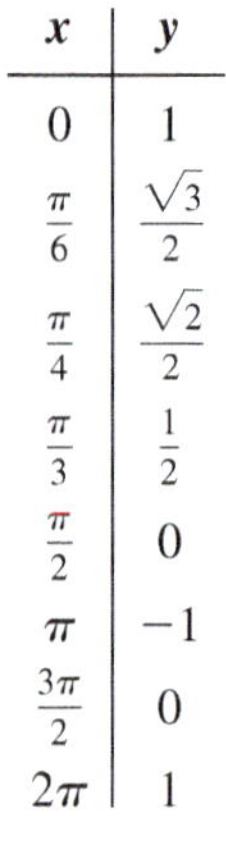

x	y
0	1
$\frac{\pi}{6}$	$\frac{\sqrt{3}}{2}$
$\frac{\pi}{4}$	$\frac{\sqrt{2}}{2}$
$\frac{\pi}{3}$	$\frac{1}{2}$
$\frac{\pi}{2}$	0
π	-1
$\frac{3\pi}{2}$	0
2π	1

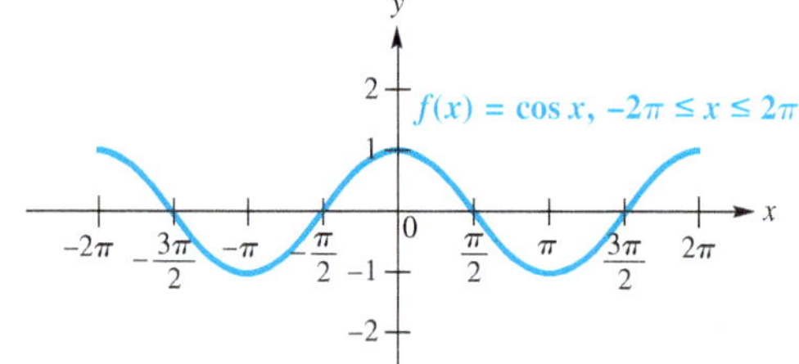

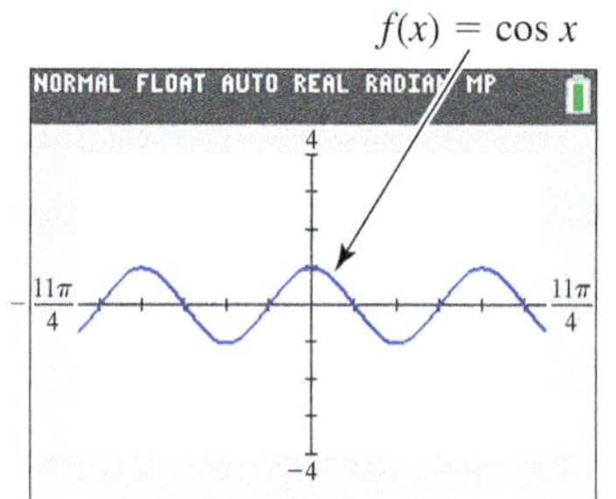

Figure 26

- The graph is continuous over its entire domain, $(-\infty, \infty)$.
- Its x-intercepts have x-values of the form $(2n + 1)\frac{\pi}{2}$, where n is an integer.
- Its period is 2π.
- The graph is symmetric with respect to the y-axis, so the function is an even function. For all x in the domain, $\cos(-x) = \cos x$.

NOTE A function f is an **even function** if for all x in the domain of f,

$$f(-x) = f(x).$$

The graph of an even function is symmetric with respect to the y-axis. This means that if (x, y) belongs to the function, then $(-x, y)$ also belongs to the function. For example, $\left(\frac{\pi}{2}, 0\right)$ and $\left(-\frac{\pi}{2}, 0\right)$ are points on the graph of $y = \cos x$, illustrating the property $\cos(-x) = \cos x$.

The calculator graphs of $f(x) = \sin x$ in **Figure 24** and $f(x) = \cos x$ in **Figure 26** are shown in the ZTrig viewing window

$$\left[-\frac{11\pi}{4}, \frac{11\pi}{4}\right] \text{ by } [-4, 4] \quad \left(\frac{11\pi}{4} \approx 8.639379797\right)$$

of the TI-84 Plus calculator, with Xscl $= \frac{\pi}{2}$ and Yscl $= 1$. (Other models have different trigonometry viewing windows.) ■

Techniques for Graphing, Amplitude, and Period The examples that follow show graphs that are "stretched" or "compressed" (shrunk) either vertically, horizontally, or both when compared with the graphs of $y = \sin x$ or $y = \cos x$.

EXAMPLE 1 Graphing $y = a \sin x$

Graph $y = 2 \sin x$, and compare to the graph of $y = \sin x$.

SOLUTION For a given value of x, the value of y is twice what it would be for $y = \sin x$. See the table of values. The change in the graph is the range, which becomes $[-2, 2]$. See **Figure 27,** which also includes a graph of $y = \sin x$.

x	0	$\frac{\pi}{2}$	π	$\frac{3\pi}{2}$	2π
$\sin x$	0	1	0	−1	0
$2 \sin x$	0	2	0	−2	0

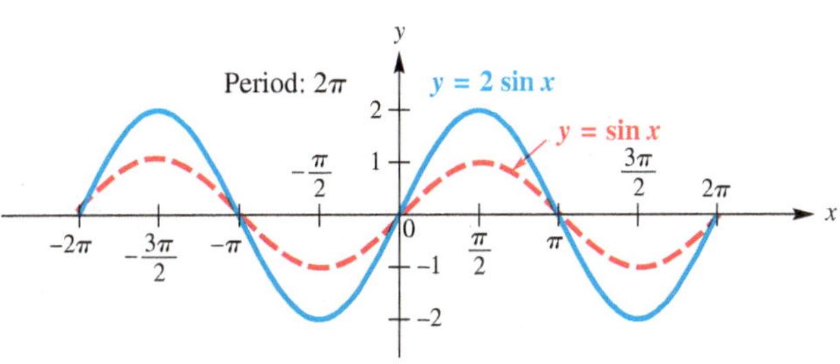

Figure 27

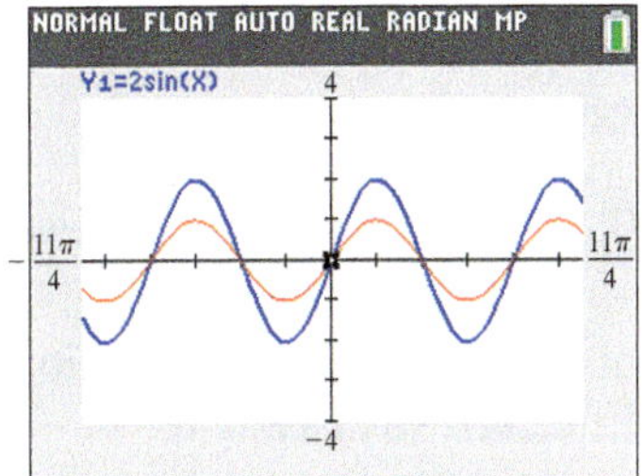

The graph of $y = 2 \sin x$ is shown in blue, and that of $y = \sin x$ is shown in red. Compare to **Figure 27.**

The **amplitude** of a periodic function is half the difference between the maximum and minimum values. It describes the height of the graph both above and below a horizontal line passing through the "middle" of the graph. Thus, for the basic sine function $y = \sin x$ (and also for the basic cosine function $y = \cos x$), the amplitude is computed as follows.

$$\frac{1}{2}[1 - (-1)] = \frac{1}{2}(2) = 1 \quad \text{Amplitude of } y = \sin x$$

For $y = 2 \sin x$, the amplitude is

$$\frac{1}{2}[2 - (-2)] = \frac{1}{2}(4) = 2. \quad \text{Amplitude of } y = 2 \sin x$$

We can think of the graph of $y = a \sin x$ as a vertical stretching of the graph of $y = \sin x$ when $a > 1$ and a vertical shrinking when $0 < a < 1$.

✔ **Now Try Exercise 15.**

Amplitude

The graph of $y = a \sin x$ or $y = a \cos x$, with $a \neq 0$, will have the same shape as the graph of $y = \sin x$ or $y = \cos x$, respectively, except with range $[-|a|, |a|]$. The amplitude is $|a|$.

While the coefficient a in $y = a \sin x$ or $y = a \cos x$ affects the amplitude of the graph, the coefficient of x in the argument affects the period. Consider $y = \sin 2x$. We can complete a table of values for the interval $[0, 2\pi]$.

x	0	$\frac{\pi}{4}$	$\frac{\pi}{2}$	$\frac{3\pi}{4}$	π	$\frac{5\pi}{4}$	$\frac{3\pi}{2}$	$\frac{7\pi}{4}$	2π
$\sin 2x$	0	1	0	-1	0	1	0	-1	0

Note that one complete cycle occurs in π units, not 2π units. Therefore, the period here is π, which equals $\frac{2\pi}{2}$. Now consider $y = \sin 4x$. Look at the next table.

x	0	$\frac{\pi}{8}$	$\frac{\pi}{4}$	$\frac{3\pi}{8}$	$\frac{\pi}{2}$	$\frac{5\pi}{8}$	$\frac{3\pi}{4}$	$\frac{7\pi}{8}$	π
$\sin 4x$	0	1	0	-1	0	1	0	-1	0

These values suggest that one complete cycle is achieved in $\frac{\pi}{2}$ or $\frac{2\pi}{4}$ units, which is reasonable because

$$\sin\left(4 \cdot \frac{\pi}{2}\right) = \sin 2\pi = 0.$$

In general, the graph of a function of the form $y = \sin bx$ ***or*** $y = \cos bx$***, for*** $b > 0$***, will have a period different from*** 2π ***when*** $b \neq 1$***.***

To see why this is so, remember that the values of $\sin bx$ or $\cos bx$ will take on all possible values as bx ranges from 0 to 2π. Therefore, to find the period of either of these functions, we must solve the following three-part inequality.

$$0 \le bx \le 2\pi \quad \text{bx ranges from 0 to } 2\pi.$$

$$0 \le x \le \frac{2\pi}{b} \quad \text{Divide each part by the positive number } b.$$

Thus, the period is $\frac{2\pi}{b}$***.*** By dividing the interval $\left[0, \frac{2\pi}{b}\right]$ into four equal parts, we obtain the values for which $\sin bx$ or $\cos bx$ is -1, 0, or 1. These values will give minimum points, x-intercepts, and maximum points on the graph. (If a function has $b < 0$, then identities can be used to rewrite the function so that $b > 0$.)

NOTE One method to divide an interval into four equal parts is as follows.

Step 1 Find the midpoint of the interval by adding the x-values of the endpoints and dividing by 2.

Step 2 Find the quarter points (the midpoints of the two intervals found in Step 1) using the same procedure.

EXAMPLE 2 Graphing $y = \sin bx$

Graph $y = \sin 2x$, and compare to the graph of $y = \sin x$.

SOLUTION In this function the coefficient of x is 2, so $b = 2$ and the period is $\frac{2\pi}{2} = \pi$. Therefore, the graph will complete one period over the interval $[0, \pi]$.

We can divide the interval $[0, \pi]$ into four equal parts by first finding its midpoint: $\frac{1}{2}(0 + \pi) = \frac{\pi}{2}$. The quarter points are found next by determining the midpoints of the two intervals $\left[0, \frac{\pi}{2}\right]$ and $\left[\frac{\pi}{2}, \pi\right]$.

$$\frac{1}{2}\left(\frac{\pi}{2} + \pi\right) = \frac{1}{2}\left(\frac{3\pi}{2}\right) = \frac{3\pi}{4}$$

$$\frac{1}{2}\left(0 + \frac{\pi}{2}\right) = \frac{\pi}{4} \quad \text{and} \quad \frac{1}{2}\left(\frac{\pi}{2} + \pi\right) = \frac{3\pi}{4} \quad \text{Quarter points}$$

The interval $[0, \pi]$ is divided into four equal parts using these x-values.

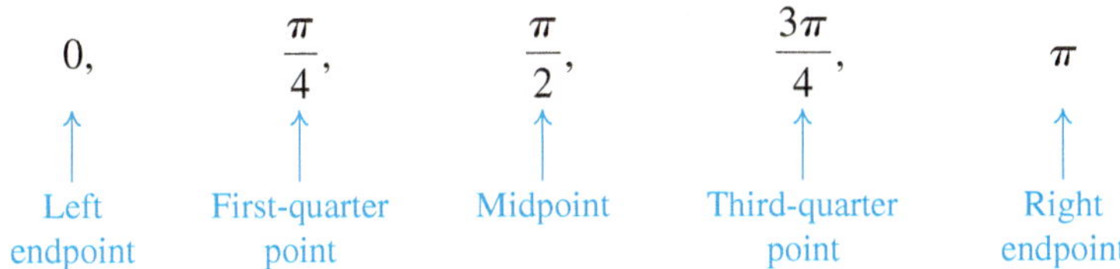

We plot the points from the table of values given at the top of the previous page, and join them with a smooth sinusoidal curve. More of the graph can be sketched by repeating this cycle, as shown in **Figure 28.** The amplitude is not changed.

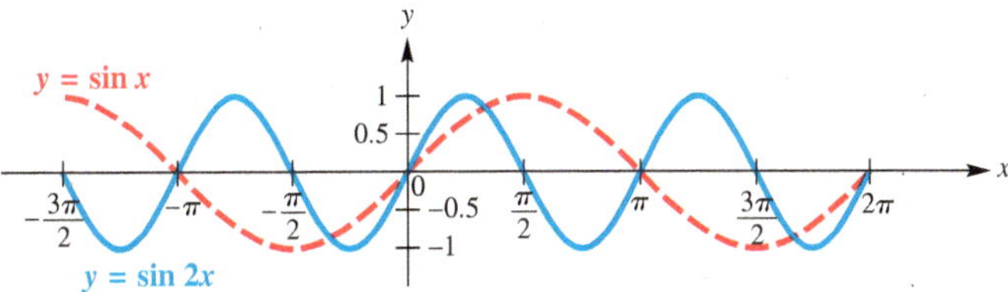

Figure 28

We can think of the graph of $y = \sin bx$ as a horizontal stretching of the graph of $y = \sin x$ when $0 < b < 1$ and a horizontal shrinking when $b > 1$.

✔ **Now Try Exercise 27.**

Period

For $b > 0$, the graph of $y = \sin bx$ will resemble that of $y = \sin x$, but with period $\frac{2\pi}{b}$. Also, the graph of $y = \cos bx$ will resemble that of $y = \cos x$, but with period $\frac{2\pi}{b}$.

EXAMPLE 3 Graphing $y = \cos bx$

Graph $y = \cos \frac{2}{3}x$ over one period.

SOLUTION The period is

$$\frac{2\pi}{\frac{2}{3}} = 2\pi \div \frac{2}{3} = 2\pi \cdot \frac{3}{2} = 3\pi.$$

To divide by a fraction, multiply by its reciprocal.

We divide the interval $[0, 3\pi]$ into four equal parts to obtain the x-values $0, \frac{3\pi}{4}, \frac{3\pi}{2}, \frac{9\pi}{4}$, and 3π that yield minimum points, maximum points, and x-intercepts. We use these values to obtain a table of key points for one period.

x	0	$\frac{3\pi}{4}$	$\frac{3\pi}{2}$	$\frac{9\pi}{4}$	3π
$\frac{2}{3}x$	0	$\frac{\pi}{2}$	π	$\frac{3\pi}{2}$	2π
$\cos \frac{2}{3}x$	1	0	-1	0	1

Figure 29

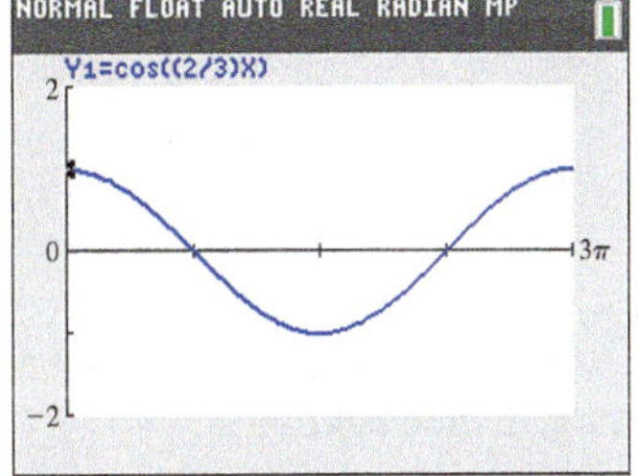

This screen shows a graph of the function in **Example 3.** By choosing Xscl $= \frac{3\pi}{4}$, the x-intercepts, maxima, and minima coincide with tick marks on the x-axis.

The amplitude is 1 because the maximum value is 1, the minimum value is -1, and $\frac{1}{2}[1 - (-1)] = \frac{1}{2}(2) = 1$. We plot these points and join them with a smooth curve. The graph is shown in **Figure 29.**

✔ **Now Try Exercise 25.**

NOTE Look at the middle row of the table in **Example 3.** Dividing $\left[0, \frac{2\pi}{b}\right]$ into four equal parts gives the values $0, \frac{\pi}{2}, \pi, \frac{3\pi}{2}$, and 2π for this row, resulting here in values of -1, 0, or 1. These values lead to key points on the graph, which can be plotted and joined with a smooth sinusoidal curve.

Guidelines for Sketching Graphs of Sine and Cosine Functions

To graph $y = a \sin bx$ or $y = a \cos bx$, with $b > 0$, follow these steps.

Step 1 Find the period, $\frac{2\pi}{b}$. Start at 0 on the x-axis, and lay off a distance of $\frac{2\pi}{b}$.

Step 2 Divide the interval into four equal parts. (See the Note preceding Example 2.)

Step 3 Evaluate the function for each of the five x-values resulting from Step 2. The points will be maximum points, minimum points, and x-intercepts.

Step 4 Plot the points found in Step 3, and join them with a sinusoidal curve having amplitude $|a|$.

Step 5 Draw the graph over additional periods as needed.

EXAMPLE 4 Graphing $y = a \sin bx$

Graph $y = -2 \sin 3x$ over one period using the preceding guidelines.

SOLUTION

Step 1 For this function, $b = 3$, so the period is $\frac{2\pi}{3}$. The function will be graphed over the interval $\left[0, \frac{2\pi}{3}\right]$.

Step 2 Divide the interval $\left[0, \frac{2\pi}{3}\right]$ into four equal parts to obtain the x-values $0, \frac{\pi}{6}, \frac{\pi}{3}, \frac{\pi}{2}$, and $\frac{2\pi}{3}$.

Step 3 Make a table of values determined by the x-values from Step 2.

x	0	$\frac{\pi}{6}$	$\frac{\pi}{3}$	$\frac{\pi}{2}$	$\frac{2\pi}{3}$
$3x$	0	$\frac{\pi}{2}$	π	$\frac{3\pi}{2}$	2π
$\sin 3x$	0	1	0	-1	0
$-2 \sin 3x$	0	-2	0	2	0

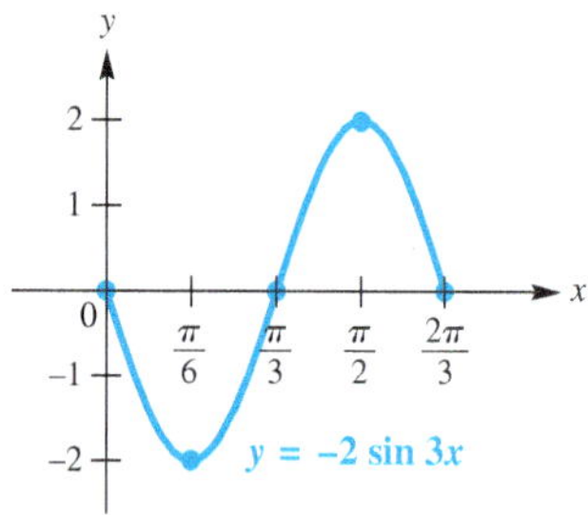

Figure 30

Step 4 Plot the points $(0, 0)$, $\left(\frac{\pi}{6}, -2\right)$, $\left(\frac{\pi}{3}, 0\right)$, $\left(\frac{\pi}{2}, 2\right)$, and $\left(\frac{2\pi}{3}, 0\right)$, and join them with a sinusoidal curve having amplitude 2. See **Figure 30.**

Step 5 The graph can be extended by repeating the cycle.

Notice that when a is negative, the graph of $y = a \sin bx$ *is a reflection across the x-axis of the graph of* $y = |a| \sin bx$.

✔ **Now Try Exercise 29.**

EXAMPLE 5 Graphing $y = a \cos bx$ (Where b Is a Multiple of π)

Graph $y = -3 \cos \pi x$ over one period.

SOLUTION

Step 1 Here $b = \pi$ and the period is $\frac{2\pi}{\pi} = 2$, so we will graph the function over the interval $[0, 2]$.

Step 2 Dividing $[0, 2]$ into four equal parts yields the x-values $0, \frac{1}{2}, 1, \frac{3}{2}$, and 2.

Step 3 Make a table using these x-values.

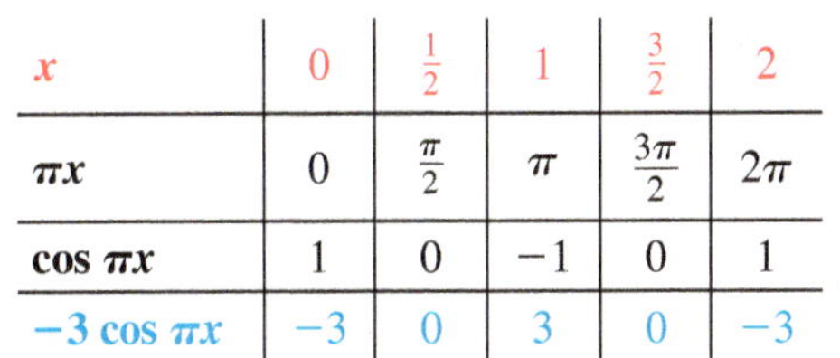

x	0	$\frac{1}{2}$	1	$\frac{3}{2}$	2
πx	0	$\frac{\pi}{2}$	π	$\frac{3\pi}{2}$	2π
$\cos \pi x$	1	0	-1	0	1
$-3 \cos \pi x$	-3	0	3	0	-3

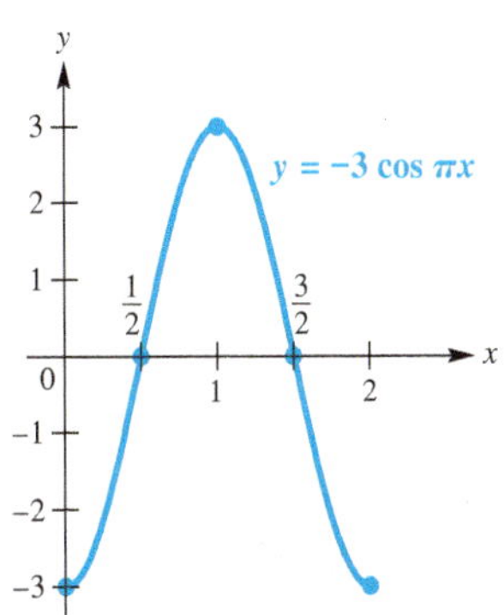

Figure 31

Step 4 Plot the points $(0, -3)$, $\left(\frac{1}{2}, 0\right)$, $(1, 3)$, $\left(\frac{3}{2}, 0\right)$, and $(2, -3)$, and join them with a sinusoidal curve having amplitude $|-3| = 3$. See **Figure 31.**

Step 5 The graph can be extended by repeating the cycle.

Notice that when b is an integer multiple of π, the first coordinates of the x-intercepts of the graph are rational numbers.

✔ **Now Try Exercise 37.**

Connecting Graphs with Equations

EXAMPLE 6 Determining an Equation for a Graph

Determine an equation of the form $y = a \cos bx$ or $y = a \sin bx$, where $b > 0$, for the given graph.

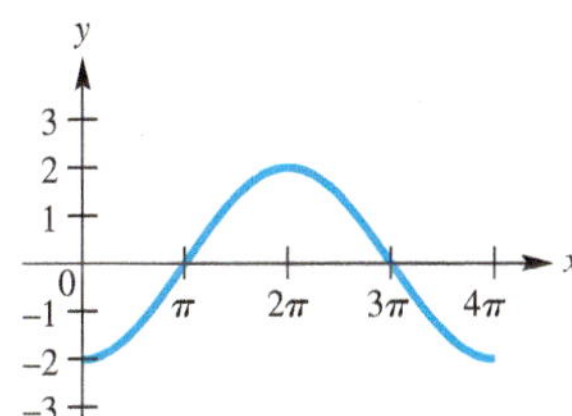

SOLUTION This graph is that of a cosine function that is reflected across its horizontal axis, the x-axis. The amplitude is half the distance between the maximum and minimum values.

$$\frac{1}{2}[2-(-2)] = \frac{1}{2}(4) = 2 \qquad \text{The amplitude } |a| \text{ is 2.}$$

Because the graph completes a cycle on the interval $[0, 4\pi]$, the period is 4π. We use this fact to solve for b.

$$4\pi = \frac{2\pi}{b} \qquad \text{Period} = \frac{2\pi}{b}$$

$$4\pi b = 2\pi \qquad \text{Multiply each side by } b.$$

$$b = \frac{1}{2} \qquad \text{Divide each side by } 4\pi.$$

An equation for the graph is

$$y = -2 \cos \frac{1}{2}x.$$

(-2: x-axis reflection; $\frac{1}{2}$: Horizontal stretch)

✔ **Now Try Exercise 41.**

A Trigonometric Model Sine and cosine functions may be used to model many real-life phenomena that repeat their values in a cyclical, or periodic, manner. Average temperature in a certain geographic location is one such example.

EXAMPLE 7 Interpreting a Sine Function Model

The average temperature (in °F) at Mould Bay, Canada, can be approximated by the function

$$f(x) = 34 \sin\left[\frac{\pi}{6}(x - 4.3)\right],$$

where x is the month and $x = 1$ corresponds to January, $x = 2$ to February, and so on.

(a) To observe the graph over a two-year interval, graph f in the window $[0, 25]$ by $[-45, 45]$.

(b) According to this model, what is the average temperature during the month of May?

(c) What would be an approximation for the average annual temperature at Mould Bay?

SOLUTION

(a) The graph of $f(x) = 34 \sin\left[\frac{\pi}{6}(x - 4.3)\right]$ is shown in **Figure 32.** Its amplitude is 34, and the period is

$$\frac{2\pi}{\frac{\pi}{6}} = 2\pi \div \frac{\pi}{6} = 2\pi \cdot \frac{6}{\pi} = 12. \quad \text{Simplify the complex fraction.}$$

Function f has a period of 12 months, or 1 year, which agrees with the changing of the seasons.

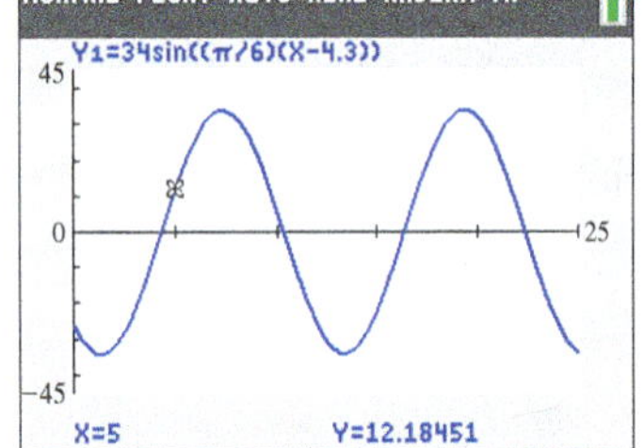

Figure 32

(b) May is the fifth month, so the average temperature during May is

$$f(5) = 34 \sin\left[\frac{\pi}{6}(5 - 4.3)\right] \approx 12°\text{F}. \quad \text{Let } x = 5 \text{ in the given function.}$$

See the display at the bottom of the screen in **Figure 32.**

(c) From the graph, it appears that the average annual temperature is about 0°F because the graph is centered vertically about the line $y = 0$.

✔ **Now Try Exercise 57.**

6.3 Exercises

CONCEPT PREVIEW *Fill in the blank(s) to correctly complete each sentence.*

1. The amplitude of the graphs of the sine and cosine functions is ______, and the period of each is ______.

2. For the x-values 0 to $\frac{\pi}{2}$, the graph of the sine function ____________ (rises/falls) and that of the cosine function ____________ (rises/falls).

3. The graph of the sine function crosses the x-axis for all numbers of the form ______, where n is an integer.

4. The domain of both the sine and cosine functions (in interval form) is ______, and the range is ______.

5. The least positive number x for which $\cos x = 0$ is ______.

6. On the interval $[\pi, 2\pi]$, the function values of the cosine function increase from ______ to ______.

Concept Check *Match each function with its graph in choices A–F.*

7. $y = -\sin x$ **8.** $y = -\cos x$ **9.** $y = \sin 2x$

10. $y = \cos 2x$ **11.** $y = 2\sin x$ **12.** $y = 2\cos x$

A. 

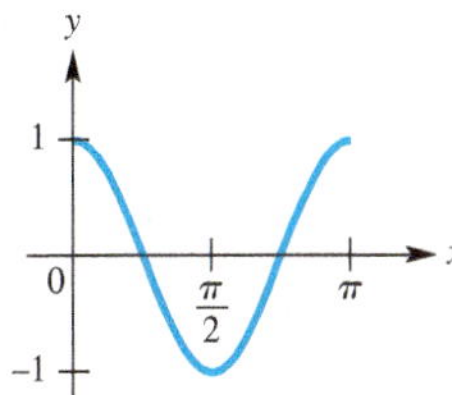

B. 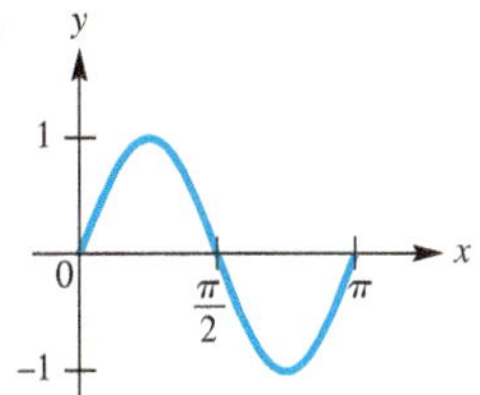

C. 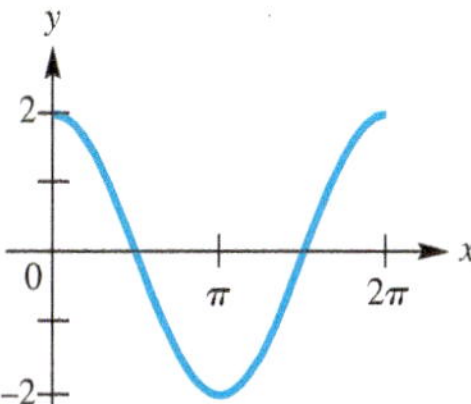

D. 

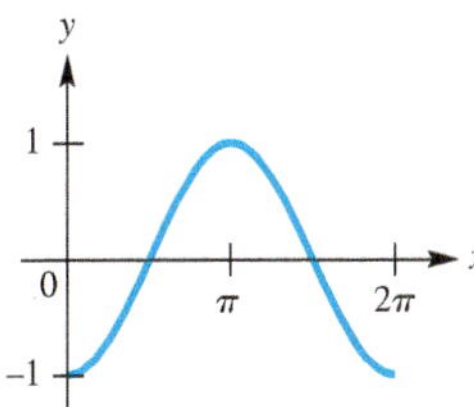

E. 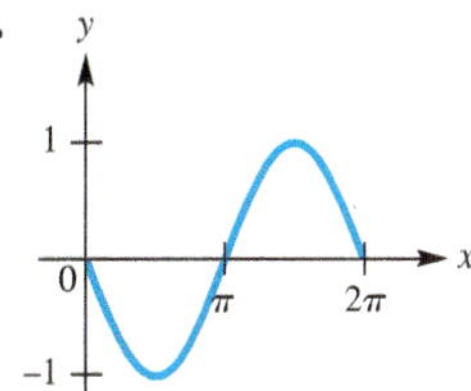

F. 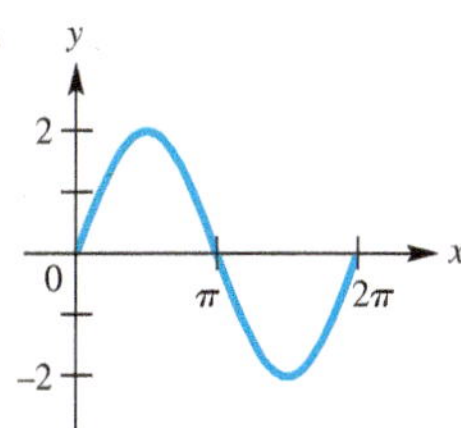

Graph each function over the interval $[-2\pi, 2\pi]$. Give the amplitude. ***See Example 1.***

13. $y = 2\cos x$ **14.** $y = 3\sin x$ **15.** $y = \frac{2}{3}\sin x$

16. $y = \frac{3}{4}\cos x$ **17.** $y = -\cos x$ **18.** $y = -\sin x$

19. $y = -2\sin x$ **20.** $y = -3\cos x$ **21.** $y = \sin(-x)$

22. ***Concept Check*** In **Exercise 21,** why is the graph the same as that of $y = -\sin x$?

Graph each function over a two-period interval. Give the period and amplitude. ***See Examples 2–5.***

23. $y = \sin \frac{1}{2}x$ **24.** $y = \sin \frac{2}{3}x$ **25.** $y = \cos \frac{3}{4}x$

26. $y = \cos \frac{1}{3}x$ **27.** $y = \sin 3x$ **28.** $y = \cos 2x$

29. $y = 2\sin \frac{1}{4}x$ **30.** $y = 3\sin 2x$ **31.** $y = -2\cos 3x$

32. $y = -5\cos 2x$ **33.** $y = \cos \pi x$ **34.** $y = -\sin \pi x$

35. $y = -2\sin 2\pi x$ **36.** $y = 3\cos 2\pi x$ **37.** $y = \frac{1}{2}\cos \frac{\pi}{2}x$

38. $y = -\frac{2}{3}\sin \frac{\pi}{4}x$ **39.** $y = \pi \sin \pi x$ **40.** $y = -\pi \cos \pi x$

Connecting Graphs with Equations *Determine an equation of the form* $y = a \cos bx$ *or* $y = a \sin bx$*, where* $b > 0$*, for the given graph.* ***See Example 6.***

41. **42.** **43.**

44. **45.** **46.**

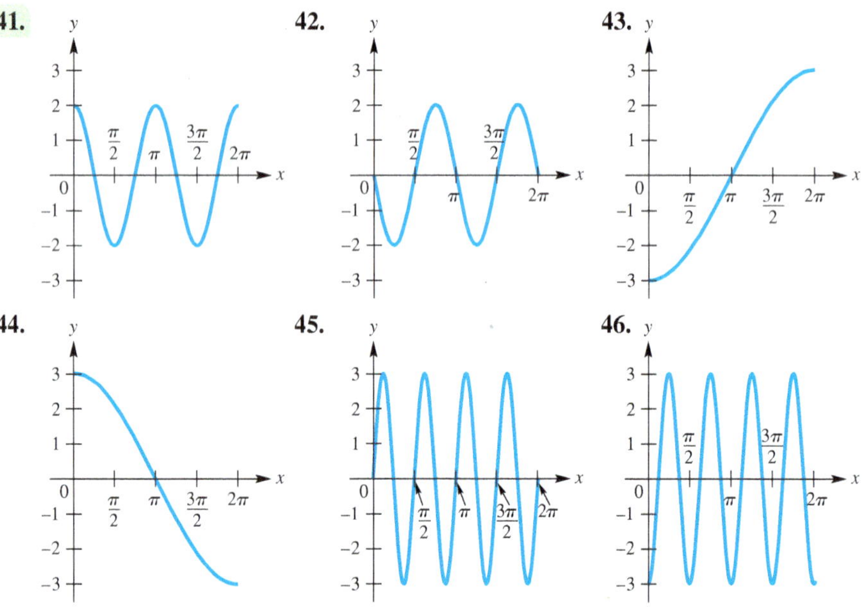

(Modeling) Solve each problem.

47. ***Average Annual Temperature*** Scientists believe that the average annual temperature in a given location is periodic. The average temperature at a given place during a given season fluctuates as time goes on, from colder to warmer, and back to colder. The graph shows an idealized description of the temperature (in °F) for approximately the last 150 thousand years of a particular location.

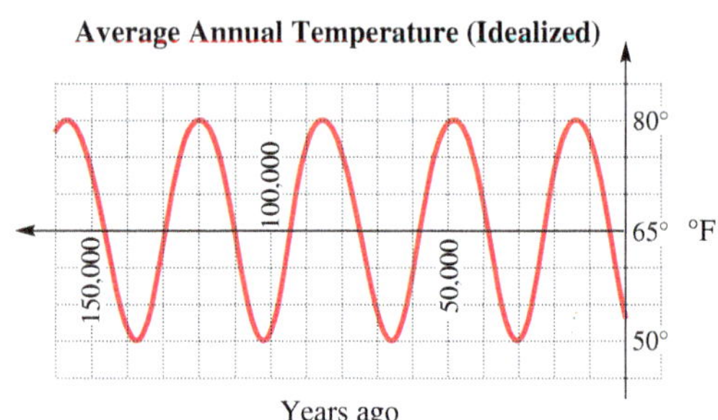

(a) Find the highest and lowest temperatures recorded.

(b) Use these two numbers to find the amplitude.

(c) Find the period of the function.

(d) What is the trend of the temperature now?

48. ***Blood Pressure Variation*** The graph gives the variation in blood pressure for a typical person. **Systolic** and **diastolic pressures** are the upper and lower limits of the periodic changes in pressure that produce the pulse. The length of time between peaks is the period of the pulse.

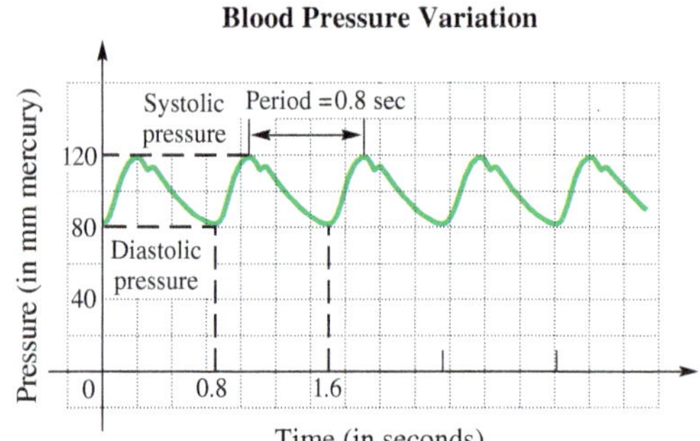

(a) Find the systolic and diastolic pressures.

(b) Find the amplitude of the graph.

(c) Find the pulse rate (the number of pulse beats in 1 min) for this person.

(Modeling) Tides for Kahului Harbor *The chart shows the tides for Kahului Harbor (on the island of Maui, Hawaii). To identify high and low tides and times for other Maui areas, the following adjustments must be made.*

Hana:	High, $+40$ min, $+0.1$ ft; Low, $+18$ min, -0.2 ft	Makena:	High, $+1{:}21$, -0.5 ft; Low, $+1{:}09$, -0.2 ft
Maalaea:	High, $+1{:}52$, -0.1 ft; Low, $+1{:}19$, -0.2 ft	Lahaina:	High, $+1{:}18$, -0.2 ft; Low, $+1{:}01$, -0.1 ft

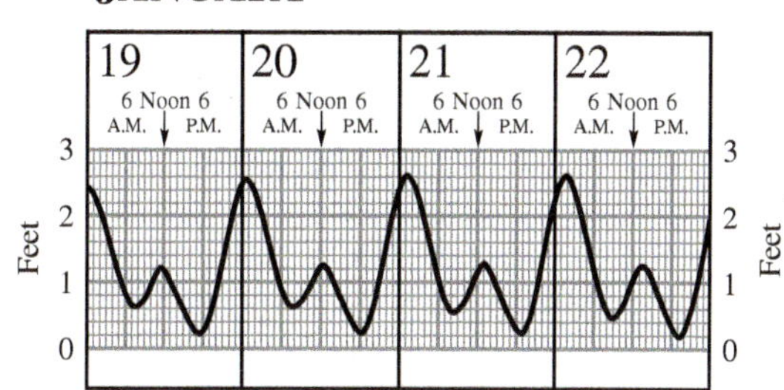

Source: Maui News. Original chart prepared by Edward K. Noda and Associates.

Use the graph to approximate each answer.

49. The graph is an example of a periodic function. What is the period (in hours)?

50. What is the amplitude?

51. At what time on January 20 was low tide at Kahului? What was the height then?

52. Repeat **Exercise 51** for Maalaea.

53. At what time on January 22 was high tide at Lahaina? What was the height then?

(Modeling) Solve each problem.

54. ***Activity of a Nocturnal Animal*** Many activities of living organisms are periodic. For example, the graph at the right below shows the time that a certain nocturnal animal begins its evening activity.

(a) Find the amplitude of this graph.

(b) Find the period.

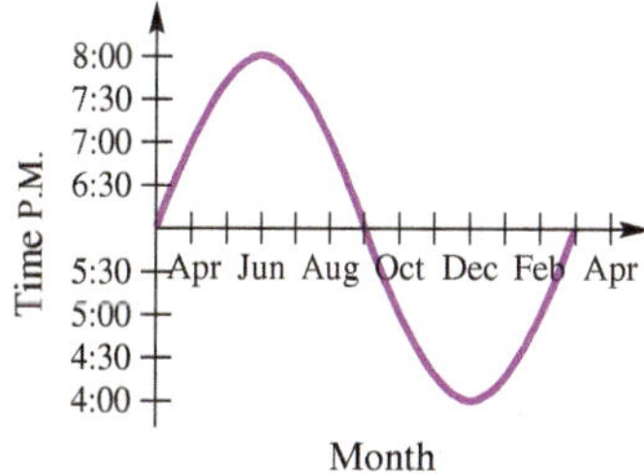

55. ***Atmospheric Carbon Dioxide*** At Mauna Loa, Hawaii, atmospheric carbon dioxide levels in parts per million (ppm) were measured regularly, beginning in 1958. The function

$$L(x) = 0.022x^2 + 0.55x + 316 + 3.5 \sin 2\pi x$$

can be used to model these levels, where x is in years and $x = 0$ corresponds to 1960. (*Source*: Nilsson, A., *Greenhouse Earth*, John Wiley and Sons.)

(a) Graph L in the window $[15, 45]$ by $[325, 385]$.

(b) When do the seasonal maximum and minimum carbon dioxide levels occur?

(c) L is the sum of a quadratic function and a sine function. What is the significance of each of these functions?

56. *Atmospheric Carbon Dioxide* Refer to **Exercise 55.** The carbon dioxide content in the atmosphere at Barrow, Alaska, in parts per million (ppm) can be modeled by the function

$$C(x) = 0.04x^2 + 0.6x + 330 + 7.5 \sin 2\pi x,$$

where $x = 0$ corresponds to 1970. (*Source*: Zeilik, M. and S. Gregory, *Introductory Astronomy and Astrophysics,* Brooks/Cole.)

(a) Graph C in the window $[5, 50]$ by $[320, 450]$.

(b) What part of the function causes the amplitude of the oscillations in the graph of C to be larger than the amplitude of the oscillations in the graph of L in **Exercise 55,** which models Hawaii?

57. *Average Daily Temperature* The temperature in Anchorage, Alaska, can be approximated by the function

$$T(x) = 37 + 21 \sin\left[\frac{2\pi}{365}(x - 91)\right],$$

where $T(x)$ is the temperature in degrees Fahrenheit on day x, with $x = 1$ corresponding to January 1 and $x = 365$ corresponding to December 31. Use a calculator to estimate the temperature on the following days. (*Source*: *World Almanac and Book of Facts.*)

(a) March 15 (day 74) **(b)** April 5 (day 95) **(c)** Day 200

(d) June 25 **(e)** October 1 **(f)** December 31

58. *Fluctuation in the Solar Constant* The **solar constant** S is the amount of energy per unit area that reaches Earth's atmosphere from the sun. It is equal to 1367 watts per m^2 but varies slightly throughout the seasons. This fluctuation ΔS in S can be calculated using the formula

$$\Delta S = 0.034S \sin\left[\frac{2\pi(82.5 - N)}{365.25}\right].$$

In this formula, N is the day number covering a four-year period, where $N = 1$ corresponds to January 1 of a leap year and $N = 1461$ corresponds to December 31 of the fourth year. (*Source*: Winter, C., R. Sizmann, and L. L. Vant-Hull, Editors, *Solar Power Plants,* Springer-Verlag.)

(a) Calculate ΔS for $N = 80$, which is the spring equinox in the first year.

(b) Calculate ΔS for $N = 1268$, which is the summer solstice in the fourth year.

(c) What is the maximum value of ΔS?

(d) Find a value for N where ΔS is equal to 0.

Musical Sound Waves *Pure sounds produce single sine waves on an oscilloscope. Find the amplitude and period of each sine wave graph. On the vertical scale, each square represents* 0.5. *On the horizontal scale, each square represents* 30° *or* $\frac{\pi}{6}$.

59.

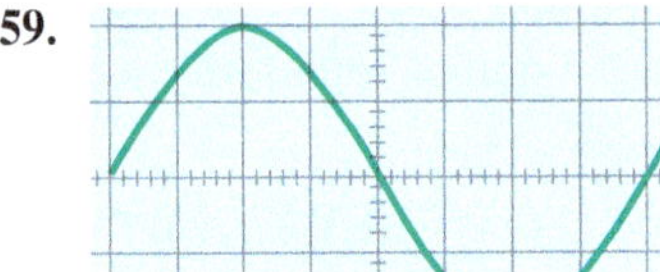

60.

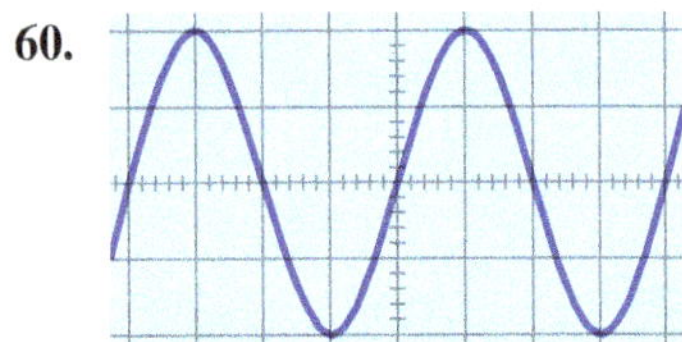

61. *Concept Check* Compare the graphs of $y = \sin 2x$ and $y = 2 \sin x$ over the interval $[0, 2\pi]$. Can we say that, in general, $\sin bx = b \sin x$ for $b > 0$? Explain.

62. *Concept Check* Compare the graphs of $y = \cos 3x$ and $y = 3 \cos x$ over the interval $[0, 2\pi]$. Can we say that, in general, $\cos bx = b \cos x$ for $b > 0$? Explain.

Relating Concepts

For individual or collaborative investigation *(Exercises 63-66)*

Connecting the Unit Circle and Sine Graph *Using a TI-84 Plus calculator, adjust the settings to correspond to the following screens.*

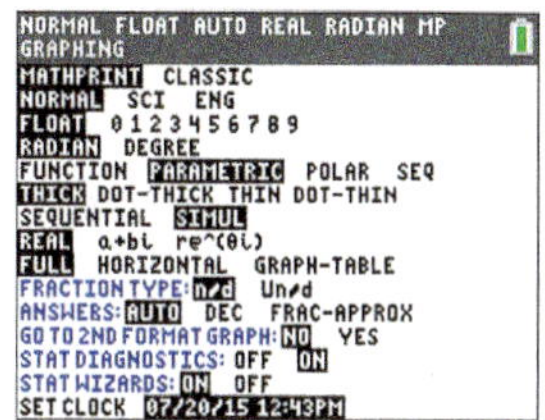

MODE

FORMAT

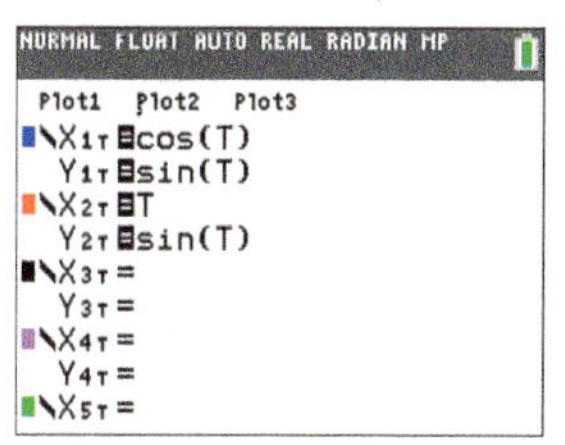

Y= EDITOR

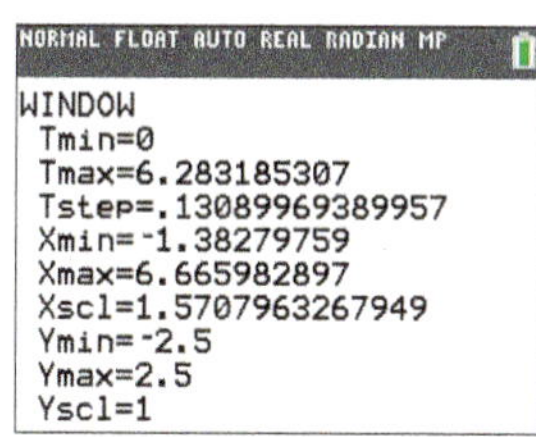

Graph the two equations (which are in ***parametric form****), and watch as the unit circle and the sine function are graphed simultaneously. Press the* TRACE *key once to obtain the screen shown on the left below. Then press the up-arrow key to obtain the screen shown on the right below. The screen on the left gives a unit circle interpretation of* $\cos 0 = 1$ *and* $\sin 0 = 0$*. The screen on the right gives a rectangular coordinate graph interpretation of* $\sin 0 = 0$.

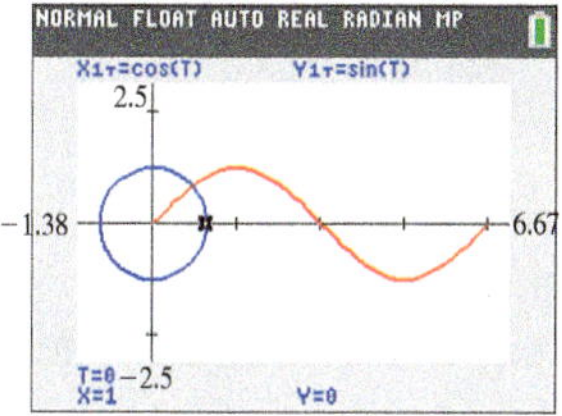

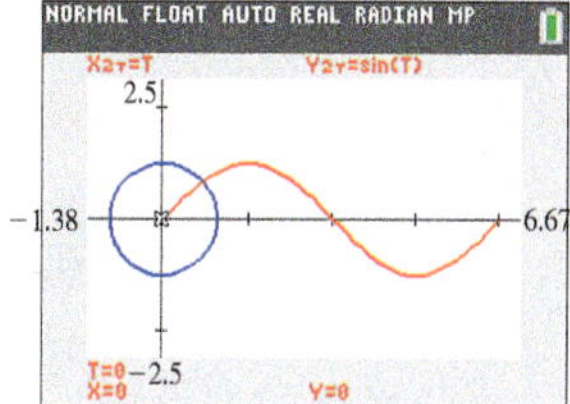

63. On the unit circle graph, let $T = 2$. Find X and Y, and interpret their values.

64. On the sine graph, let $T = 2$. What values of X and Y are displayed? Interpret these values with an equation in X and Y.

65. Now go back and redefine Y_{2T} as $\cos(T)$. Graph both equations. On the cosine graph, let $T = 2$. What values of X and Y are displayed? Interpret these values with an equation in X and Y.

66. Explain the relationship between the coordinates of the unit circle and the coordinates of the sine and cosine graphs.

6.4 Translations of the Graphs of the Sine and Cosine Functions

- Horizontal Translations
- Vertical Translations
- Combinations of Translations
- A Trigonometric Model

Horizontal Translations

The graph of the function

$$y = f(x - d)$$

is translated *horizontally* compared to the graph of $y = f(x)$. The translation is d units to the right if $d > 0$ and $|d|$ units to the left if $d < 0$. See **Figure 33**.

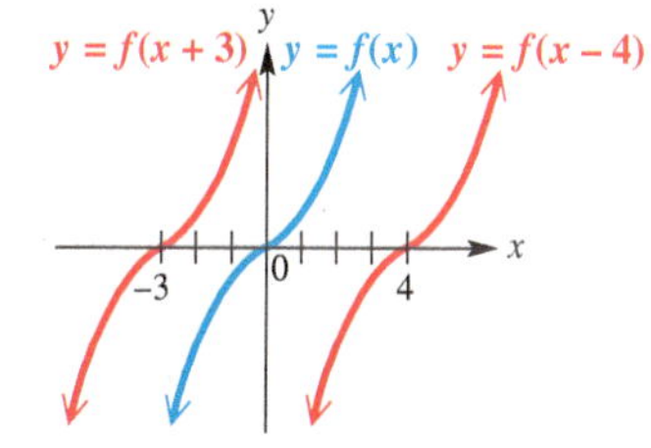

Horizontal translations of $y = f(x)$

Figure 33

With circular functions, a horizontal translation is a **phase shift.** In the function $y = f(x - d)$, the expression $x - d$ is the **argument.**

EXAMPLE 1 Graphing $y = \sin(x - d)$

Graph $y = \sin\left(x - \frac{\pi}{3}\right)$ over one period.

SOLUTION ***Method 1*** For the argument $x - \frac{\pi}{3}$ to result in all possible values throughout one period, it must take on all values between 0 and 2π, inclusive. To find an interval of one period, we solve the following three-part inequality.

$$0 \leq x - \frac{\pi}{3} \leq 2\pi \qquad \text{Three-part inequality}$$

$$\frac{\pi}{3} \leq x \leq \frac{7\pi}{3} \qquad \text{Add } \tfrac{\pi}{3} \text{ to each part.}$$

Use the method described in the previous section to divide the interval $\left[\frac{\pi}{3}, \frac{7\pi}{3}\right]$ into four equal parts, obtaining the following x-values.

$$\frac{\pi}{3}, \frac{5\pi}{6}, \frac{4\pi}{3}, \frac{11\pi}{6}, \frac{7\pi}{3}$$

These are *key* x-values.

A table of values using these x-values follows.

x	$\frac{\pi}{3}$	$\frac{5\pi}{6}$	$\frac{4\pi}{3}$	$\frac{11\pi}{6}$	$\frac{7\pi}{3}$
$x - \frac{\pi}{3}$	0	$\frac{\pi}{2}$	π	$\frac{3\pi}{2}$	2π
$\sin\left(x - \frac{\pi}{3}\right)$	0	1	0	-1	0

We join the corresponding points with a smooth curve to obtain the solid blue graph shown in **Figure 34.** The period is 2π, and the amplitude is 1.

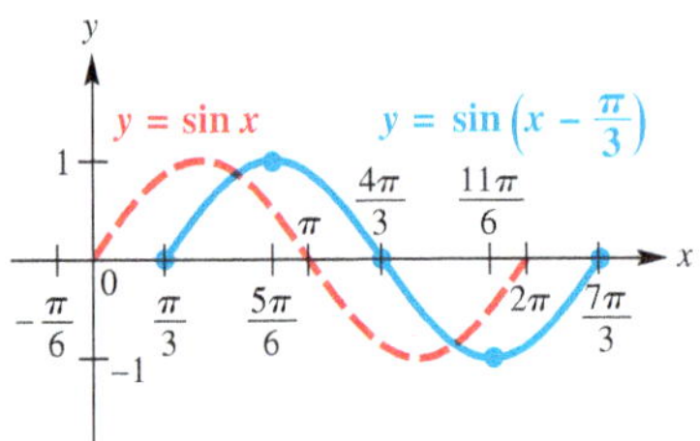

The graph can be extended through additional periods by repeating the given portion of the graph, as necessary.

Figure 34

Method 2 We can also graph $y = \sin\left(x - \frac{\pi}{3}\right)$ by using a horizontal translation of the graph of $y = \sin x$. The argument $x - \frac{\pi}{3}$ indicates that the graph will be translated $\frac{\pi}{3}$ units to the *right* (the phase shift) compared to the graph of $y = \sin x$. See **Figure 34.**

To graph a function using this method, first graph the basic circular function, and then graph the desired function using the appropriate translation.

✔ **Now Try Exercise 39.**

CAUTION In **Example 1,** the argument of the function is $\left(x - \frac{\pi}{3}\right)$. The parentheses are important here. If the function had been

$$y = \sin x - \frac{\pi}{3},$$

the graph would be that of $y = \sin x$ translated *vertically* $\frac{\pi}{3}$ units down.

EXAMPLE 2 Graphing $y = a\cos(x - d)$

Graph $y = 3\cos\left(x + \frac{\pi}{4}\right)$ over one period.

SOLUTION ***Method 1*** We first solve the following three-part inequality.

$$0 \le x + \frac{\pi}{4} \le 2\pi \quad \text{Three-part inequality}$$

$$-\frac{\pi}{4} \le x \le \frac{7\pi}{4} \quad \text{Subtract } \tfrac{\pi}{4} \text{ from each part.}$$

Dividing this interval into four equal parts gives the following x-values. We use them to make a table of key points.

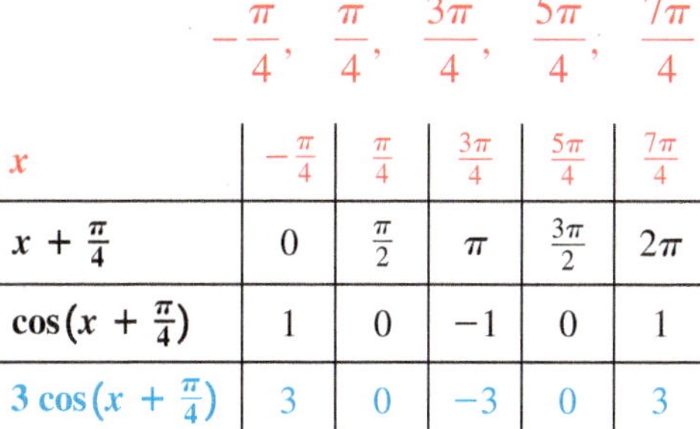

$$-\frac{\pi}{4}, \quad \frac{\pi}{4}, \quad \frac{3\pi}{4}, \quad \frac{5\pi}{4}, \quad \frac{7\pi}{4} \quad \text{Key } x\text{-values}$$

x	$-\frac{\pi}{4}$	$\frac{\pi}{4}$	$\frac{3\pi}{4}$	$\frac{5\pi}{4}$	$\frac{7\pi}{4}$
$x + \frac{\pi}{4}$	0	$\frac{\pi}{2}$	π	$\frac{3\pi}{2}$	2π
$\cos\left(x + \frac{\pi}{4}\right)$	1	0	-1	0	1
$3\cos\left(x + \frac{\pi}{4}\right)$	3	0	-3	0	3

These x-values lead to maximum points, minimum points, and x-intercepts.

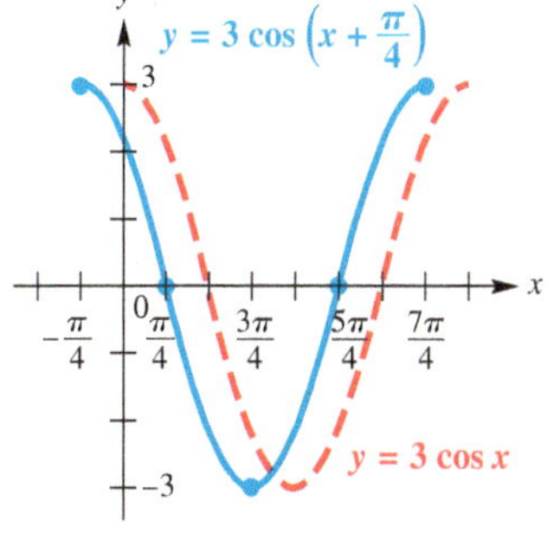

Figure 35

We join the corresponding points with a smooth curve to obtain the solid blue graph shown in **Figure 35.** The period is 2π, and the amplitude is 3.

Method 2 Write $y = 3\cos\left(x + \frac{\pi}{4}\right)$ in the form $y = a\cos(x - d)$.

$$y = 3\cos\left(x + \frac{\pi}{4}\right), \quad \text{or} \quad y = 3\cos\left[x - \left(-\frac{\pi}{4}\right)\right] \quad \text{Rewrite to subtract } -\tfrac{\pi}{4}.$$

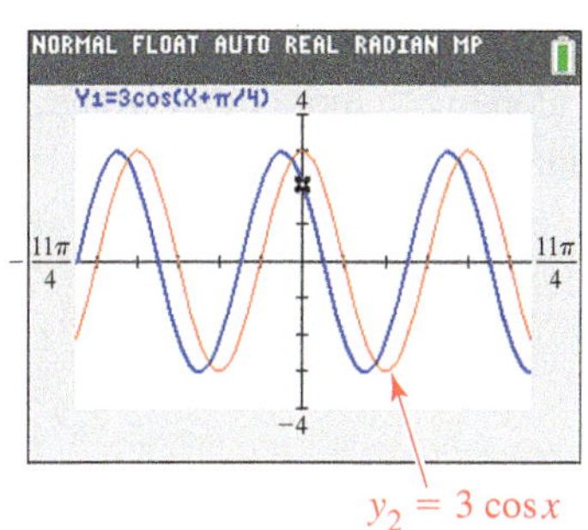

This result shows that $d = -\frac{\pi}{4}$. Because $-\frac{\pi}{4}$ is negative, the phase shift is $\left|-\frac{\pi}{4}\right| = \frac{\pi}{4}$ unit to the left. The graph is the same as that of $y = 3\cos x$ (the red graph in the calculator screen shown in the margin), except that it is translated $\frac{\pi}{4}$ unit to the left (the blue graph).

✔ **Now Try Exercise 41.**

EXAMPLE 3 Graphing $y = a\cos[b(x - d)]$

Graph $y = -2\cos(3x + \pi)$ over two periods.

SOLUTION ***Method 1*** We first solve the three-part inequality

$$0 \le 3x + \pi \le 2\pi$$

to find the interval $\left[-\frac{\pi}{3}, \frac{\pi}{3}\right]$. Dividing this interval into four equal parts gives the points $\left(-\frac{\pi}{3}, -2\right)$, $\left(-\frac{\pi}{6}, 0\right)$, $(0, 2)$, $\left(\frac{\pi}{6}, 0\right)$, and $\left(\frac{\pi}{3}, -2\right)$. We plot these points and join them with a smooth curve. By graphing an additional half period to the left and to the right, we obtain the graph shown in **Figure 36.**

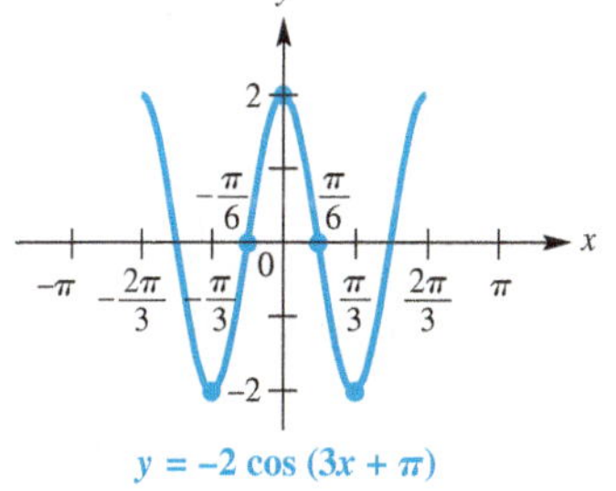

Figure 36

Method 2 First write the equation in the form $y = a\cos[b(x - d)]$.

$$y = -2\cos(3x + \pi), \quad \text{or} \quad y = -2\cos\left[3\left(x + \frac{\pi}{3}\right)\right] \quad \text{Rewrite by factoring out 3.}$$

Then $a = -2$, $b = 3$, and $d = -\frac{\pi}{3}$. The amplitude is $|-2| = 2$, and the period is $\frac{2\pi}{3}$ (because the value of b is 3). The phase shift is $\left|-\frac{\pi}{3}\right| = \frac{\pi}{3}$ units to the left compared to the graph of $y = -2\cos 3x$. Again, see **Figure 36.**

✔ **Now Try Exercise 47.**

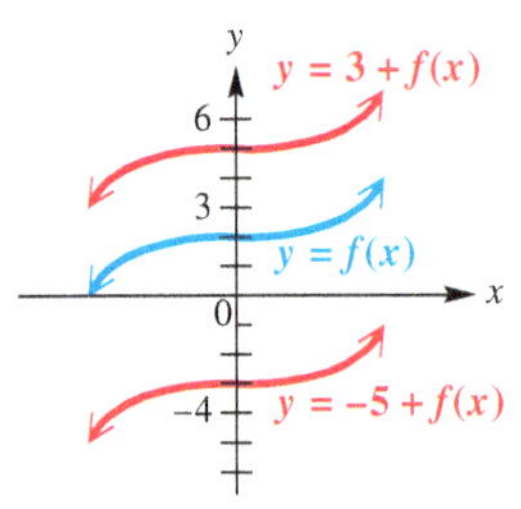

Vertical translations of $y = f(x)$

Figure 37

Vertical Translations The graph of a function of the form

$$y = c + f(x)$$

is translated *vertically* compared to the graph of $y = f(x)$. See **Figure 37.** The translation is c units up if $c > 0$ and is $|c|$ units down if $c < 0$.

EXAMPLE 4 Graphing $y = c + a \cos bx$

Graph $y = 3 - 2 \cos 3x$ over two periods.

SOLUTION We use Method 1. The values of y will be 3 greater than the corresponding values of y in $y = -2 \cos 3x$. This means that the graph of $y = 3 - 2 \cos 3x$ is the same as the graph of $y = -2 \cos 3x$, vertically translated 3 units up. The period of $y = -2 \cos 3x$ is $\frac{2\pi}{3}$, so the key points have these x-values.

$$0, \quad \frac{\pi}{6}, \quad \frac{\pi}{3}, \quad \frac{\pi}{2}, \quad \frac{2\pi}{3} \qquad \text{Key } x\text{-values}$$

Use these x-values to make a table of points.

x	0	$\frac{\pi}{6}$	$\frac{\pi}{3}$	$\frac{\pi}{2}$	$\frac{2\pi}{3}$
$\cos 3x$	1	0	−1	0	1
$2 \cos 3x$	2	0	−2	0	2
$3 - 2 \cos 3x$	1	3	5	3	1

The key points are shown on the graph in **Figure 38,** along with more of the graph, which is sketched using the fact that the function is periodic.

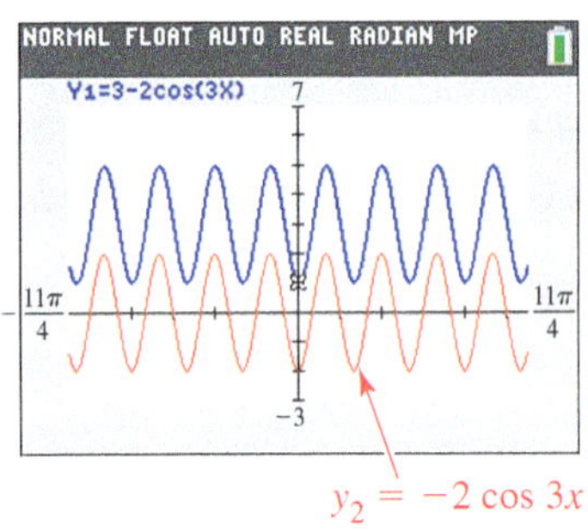

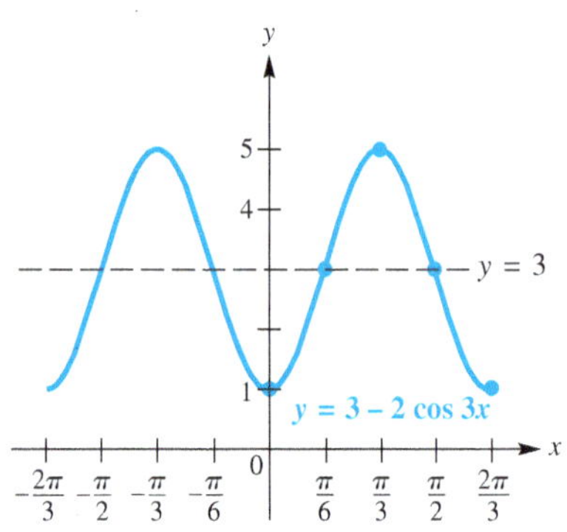

Figure 38

✔ **Now Try Exercise 51.**

CAUTION If we use ***Method 2*** to graph the function $y = 3 - 2 \cos 3x$ in Example 4, we must ***first*** graph

$$y = -2 \cos 3x$$

and ***then*** apply the vertical translation 3 units up. ***To begin,*** use the fact that $a = -2$ and $b = 3$ to determine that the amplitude is 2, the period is $\frac{2\pi}{3}$, and the graph is the reflection of the graph of $y = 2 \cos 3x$ across the x-axis. ***Then,*** because $c = 3$, translate the graph of $y = -2 \cos 3x$ up 3 units. See **Figure 38.**

If the vertical translation is applied first, then the reflection must be across the line $y = 3$, not across the x-axis.

Combinations of Translations

Further Guidelines for Sketching Graphs of Sine and Cosine Functions

To graph $\boldsymbol{y = c + a \sin[b(x - d)]}$ or $\boldsymbol{y = c + a \cos[b(x - d)]}$, with $b > 0$, follow these steps.

Method 1

Step 1 Find an interval whose length is one period $\frac{2\pi}{b}$ by solving the three-part inequality $0 \leq b(x - d) \leq 2\pi$.

Step 2 Divide the interval into four equal parts to obtain five key x-values.

Step 3 Evaluate the function for each of the five x-values resulting from Step 2. The points will be maximum points, minimum points, and points that intersect the line $y = c$ ("middle" points of the wave).

Step 4 Plot the points found in Step 3, and join them with a sinusoidal curve having amplitude $|a|$.

Step 5 Draw the graph over additional periods, as needed.

Method 2

Step 1 Graph $y = a \sin bx$ or $y = a \cos bx$. The amplitude of the function is $|a|$, and the period is $\frac{2\pi}{b}$.

Step 2 Use translations to graph the desired function. The vertical translation is c units up if $c > 0$ and $|c|$ units down if $c < 0$. The horizontal translation (phase shift) is d units to the right if $d > 0$ and $|d|$ units to the left if $d < 0$.

EXAMPLE 5 **Graphing $y = c + a\sin[b(x - d)]$**

Graph $y = -1 + 2\sin(4x + \pi)$ over two periods.

SOLUTION We use Method 1. We must first write the expression on the right side of the equation in the form $c + a \sin[b(x - d)]$.

$$y = -1 + 2\sin(4x + \pi), \quad \text{or} \quad y = -1 + 2\sin\left[4\left(x + \frac{\pi}{4}\right)\right] \qquad \text{Rewrite by factoring out 4.}$$

Step 1 Find an interval whose length is one period.

$$0 \leq 4\left(x + \frac{\pi}{4}\right) \leq 2\pi \qquad \text{Three-part inequality}$$

$$0 \leq x + \frac{\pi}{4} \leq \frac{\pi}{2} \qquad \text{Divide each part by 4.}$$

$$-\frac{\pi}{4} \leq x \leq \frac{\pi}{4} \qquad \text{Subtract } \tfrac{\pi}{4} \text{ from each part.}$$

Step 2 Divide the interval $\left[-\frac{\pi}{4}, \frac{\pi}{4}\right]$ into four equal parts to obtain these x-values.

$$-\frac{\pi}{4}, \quad -\frac{\pi}{8}, \quad 0, \quad \frac{\pi}{8}, \quad \frac{\pi}{4} \qquad \text{Key } x\text{-values}$$

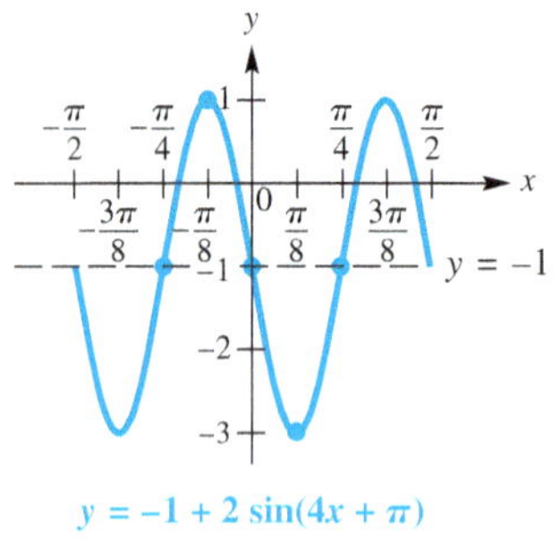

Figure 39

Step 3 Make a table of values.

x	$-\frac{\pi}{4}$	$-\frac{\pi}{8}$	0	$\frac{\pi}{8}$	$\frac{\pi}{4}$
$x + \frac{\pi}{4}$	0	$\frac{\pi}{8}$	$\frac{\pi}{4}$	$\frac{3\pi}{8}$	$\frac{\pi}{2}$
$4(x + \frac{\pi}{4})$	0	$\frac{\pi}{2}$	π	$\frac{3\pi}{2}$	2π
$\sin[4(x + \frac{\pi}{4})]$	0	1	0	-1	0
$2\sin[4(x + \frac{\pi}{4})]$	0	2	0	-2	0
$-1 + 2\sin(4x + \pi)$	-1	1	-1	-3	-1

Steps 4 and 5 Plot the points found in the table and join them with a sinusoidal curve. **Figure 39** shows the graph, extended to the right and left to include two full periods.

✔ **Now Try Exercise 57.**

A Trigonometric Model For natural phenomena that occur in periodic patterns (such as seasonal temperatures, phases of the moon, heights of tides) a sinusoidal function will provide a good approximation of a set of data points.

EXAMPLE 6 Modeling Temperature with a Sine Function

The maximum average monthly temperature in New Orleans, Louisiana, is 83°F, and the minimum is 53°F. The table shows the average monthly temperatures. The scatter diagram for a two-year interval in **Figure 40** strongly suggests that the temperatures can be modeled with a sine curve.

Month	°F	Month	°F
Jan	53	July	83
Feb	56	Aug	83
Mar	62	Sept	79
Apr	68	Oct	70
May	76	Nov	61
June	81	Dec	55

Source: World Almanac and Book of Facts.

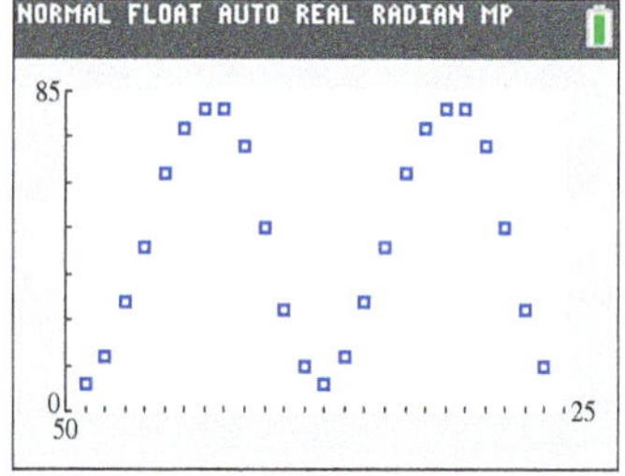

Figure 40

(a) Using only the maximum and minimum temperatures, determine a function of the form

$$f(x) = a\sin[b(x-d)] + c, \quad \text{where } a, b, c, \text{ and } d \text{ are constants,}$$

that models the average monthly temperature in New Orleans. Let x represent the month, with January corresponding to $x = 1$.

(b) On the same coordinate axes, graph f for a two-year period together with the actual data values found in the table.

(c) Use the **sine regression** feature of a graphing calculator to determine a second model for these data.

SOLUTION

(a) We use the maximum and minimum average monthly temperatures to find the amplitude a.

$$a = \frac{83 - 53}{2} = 15 \quad \text{Amplitude } a$$

The average of the maximum and minimum temperatures is a good choice for c. The average is

$$\frac{83 + 53}{2} = 68. \quad \text{Vertical translation } c$$

Because temperatures repeat every 12 months, b can be found as follows.

$$12 = \frac{2\pi}{b} \quad \text{Period} = \frac{2\pi}{b}$$

$$b = \frac{\pi}{6} \quad \text{Solve for } b.$$

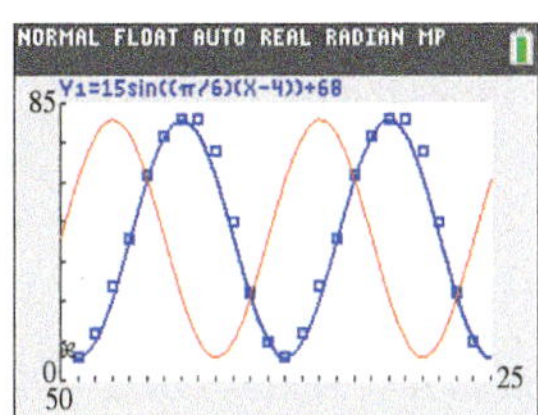

Figure 41

The coldest month is January, when $x = 1$, and the hottest month is July, when $x = 7$. A good choice for d is 4 because April, when $x = 4$, is located at the midpoint between January and July. Also, notice that the average monthly temperature in April is 68°F, which is the value of the vertical translation, c. The average monthly temperature in New Orleans is modeled closely by the following equation.

$$f(x) = a \sin[b(x - d)] + c$$

$$f(x) = 15 \sin\left[\frac{\pi}{6}(x - 4)\right] + 68 \quad \text{Substitute for } a, b, c, \text{ and } d.$$

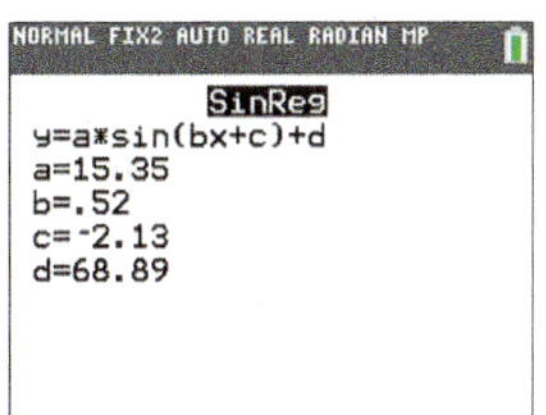

(a)

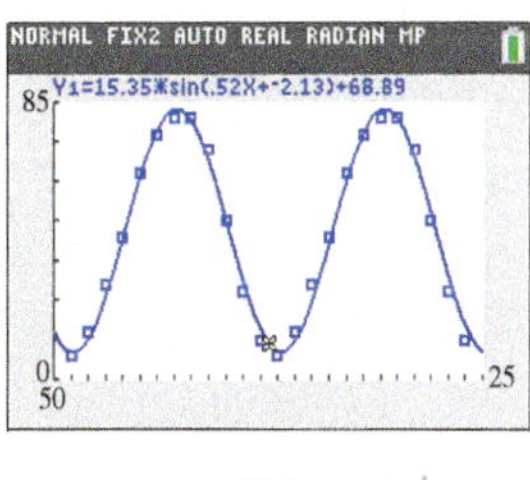

(b)

Figure 42

(b) **Figure 41** shows two iterations of the data points from the table, along with the graph of $y = 15 \sin\left[\frac{\pi}{6}(x - 4)\right] + 68$. The graph of $y = 15 \sin \frac{\pi}{6}x + 68$ is shown for comparison.

(c) We used the given data for a two-year period and the sine regression capability of a graphing calculator to produce the model

$$f(x) = 15.35 \sin(0.52x - 2.13) + 68.89$$

described in **Figure 42(a).** Its graph along with the data points is shown in **Figure 42(b).**

✔ **Now Try Exercise 61.**

6.4 Exercises

CONCEPT PREVIEW *Fill in the blank(s) to correctly complete each sentence.*

1. The graph of $y = \sin\left(x + \frac{\pi}{4}\right)$ is obtained by shifting the graph of $y = \sin x$ _____ unit(s) to the __________.
(right/left)

2. The graph of $y = \cos\left(x - \frac{\pi}{6}\right)$ is obtained by shifting the graph of $y = \cos x$ _____ unit(s) to the __________.
(right/left)

3. The graph of $y = 4 \sin x$ is obtained by stretching the graph of $y = \sin x$ vertically by a factor of ____.

4. The graph of $y = -3 \sin x$ is obtained by stretching the graph of $y = \sin x$ by a factor of ____ and reflecting across the ____-axis.

5. The graph of $y = 6 + 3 \sin x$ is obtained by shifting the graph of $y = 3 \sin x$ ____ unit(s) ________.
(up/down)

6. The graph of $y = -5 + 2 \cos x$ is obtained by shifting the graph of $y = 2 \cos x$ ____ unit(s) ________.
(up/down)

7. The graph of $y = 3 + 5 \cos\left(x + \frac{\pi}{5}\right)$ is obtained by shifting the graph of $y = \cos x$ ____ unit(s) horizontally to the ________, stretching it vertically by a factor
(right/left)
of ____, and then shifting it ____ unit(s) vertically ________.
(up/down)

8. Repeat **Exercise 7** for $y = -2 + 3 \cos\left(x - \frac{\pi}{6}\right)$.

Concept Check *Match each function with its graph in choices A–I. (One choice will not be used.)*

9. $y = \sin\left(x - \frac{\pi}{4}\right)$

10. $y = \sin\left(x + \frac{\pi}{4}\right)$

11. $y = \cos\left(x - \frac{\pi}{4}\right)$

12. $y = \cos\left(x + \frac{\pi}{4}\right)$

13. $y = 1 + \sin x$

14. $y = -1 + \sin x$

15. $y = 1 + \cos x$

16. $y = -1 + \cos x$

A.
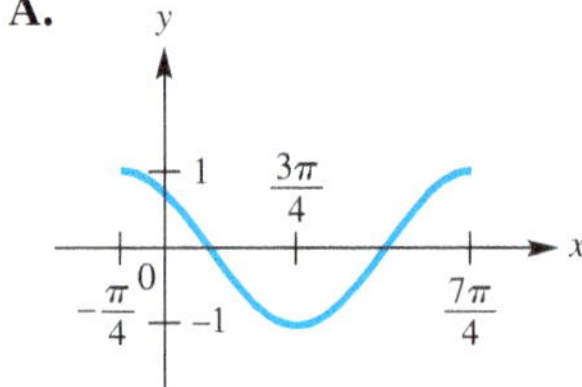

B.
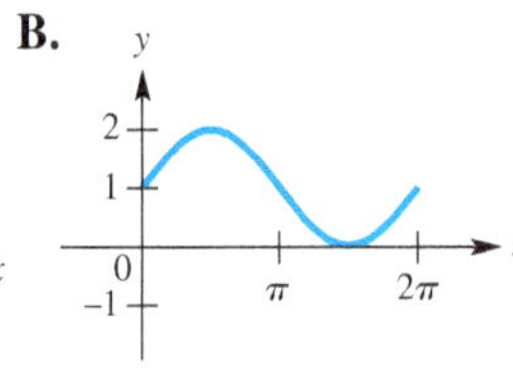

C.
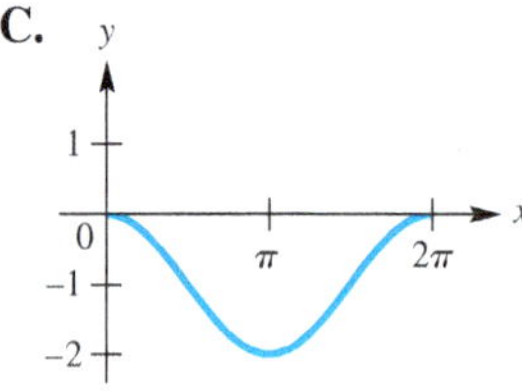

D.
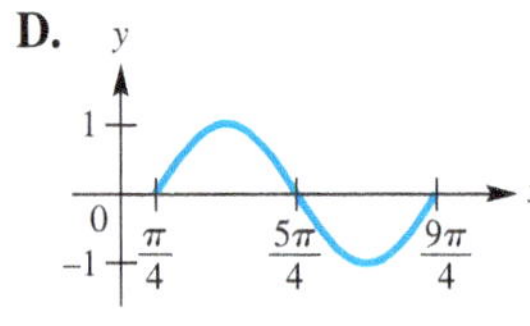

E.
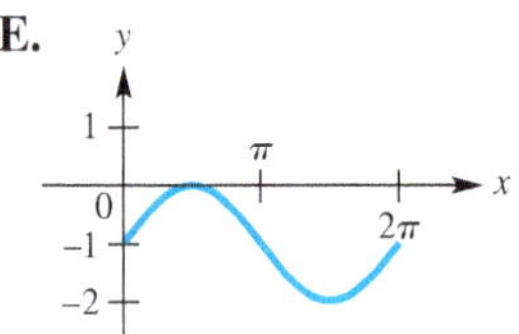

F.
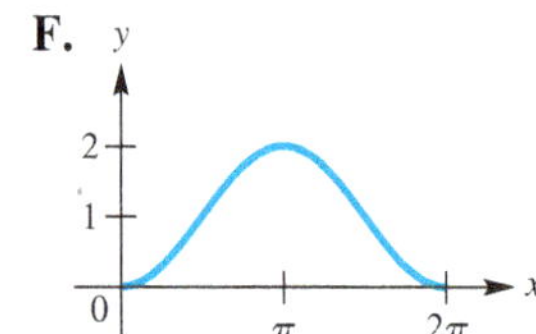

G.
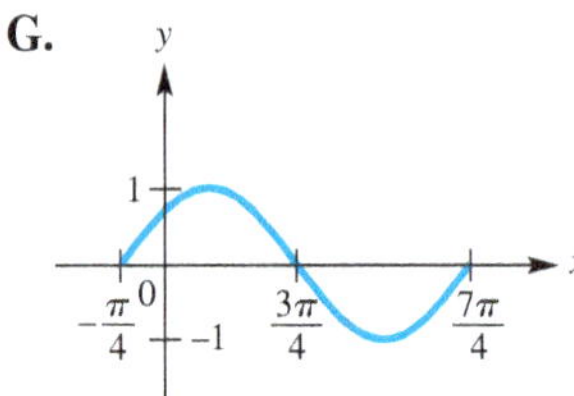

H.
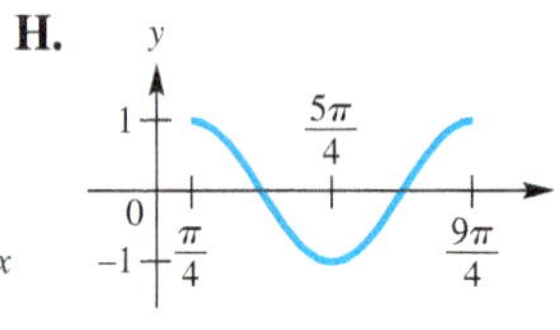

I.
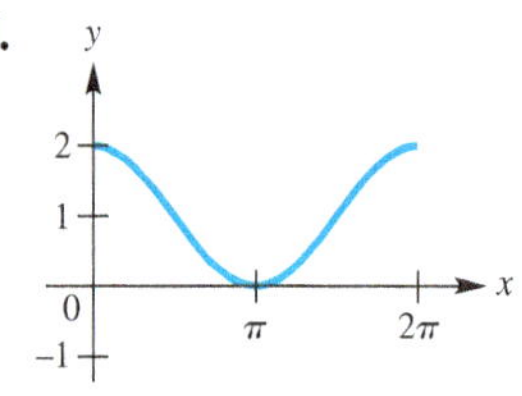

17. The graphs of $y = \sin x + 1$ and $y = \sin(x + 1)$ are **NOT** the same. Explain why this is so.

18. *Concept Check* Refer to **Exercise 17.** Which one of the two graphs is the same as that of $y = 1 + \sin x$?

Concept Check *Match each function in Column I with the appropriate description in Column II.*

I	II
19. $y = 3\sin(2x - 4)$	**A.** amplitude $= 2$, period $= \frac{\pi}{2}$, phase shift $= \frac{3}{4}$
20. $y = 2\sin(3x - 4)$	**B.** amplitude $= 3$, period $= \pi$, phase shift $= 2$
21. $y = -4\sin(3x - 2)$	**C.** amplitude $= 4$, period $= \frac{2\pi}{3}$, phase shift $= \frac{2}{3}$
22. $y = -2\sin(4x - 3)$	**D.** amplitude $= 2$, period $= \frac{2\pi}{3}$, phase shift $= \frac{4}{3}$

Concept Check *Fill in each blank with the word* right *or the word* left.

23. If the graph of $y = \cos x$ is translated $\frac{\pi}{2}$ units horizontally to the ________, it will coincide with the graph of $y = \sin x$.

24. If the graph of $y = \sin x$ is translated $\frac{\pi}{2}$ units horizontally to the ________, it will coincide with the graph of $y = \cos x$.

Connecting Graphs with Equations *Each function graphed is of the form $y = c + \cos x$, $y = c + \sin x$, $y = \cos(x - d)$, or $y = \sin(x - d)$, where d is the least possible positive value. Determine an equation of the graph.*

25.

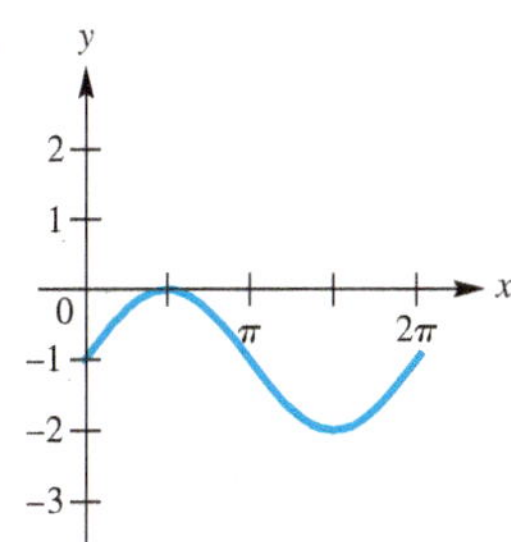

26.

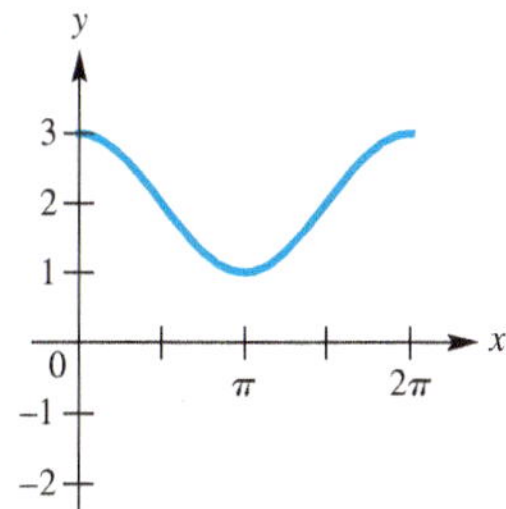

27.

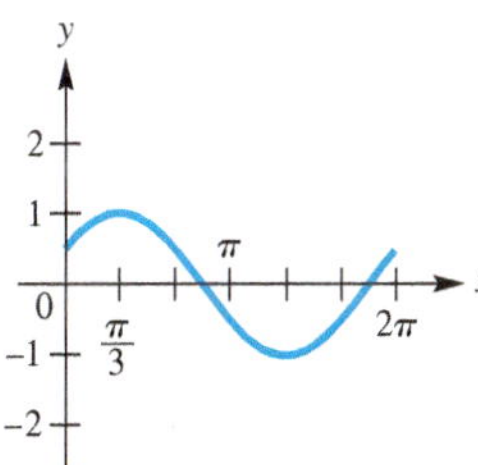

28.

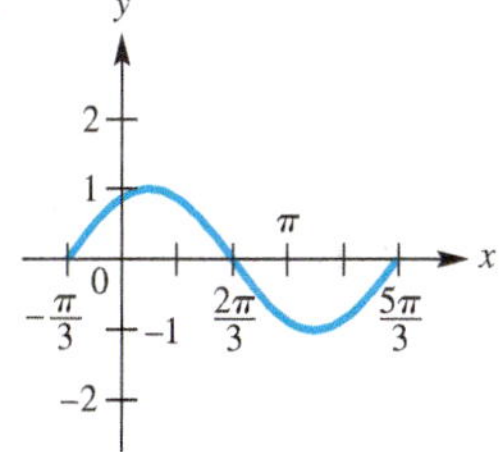

Find the amplitude, the period, any vertical translation, and any phase shift of the graph of each function. ***See Examples 1–5.***

29. $y = 2\sin(x + \pi)$

30. $y = 3\sin\left(x + \dfrac{\pi}{2}\right)$

31. $y = -\dfrac{1}{4}\cos\left(\dfrac{1}{2}x + \dfrac{\pi}{2}\right)$

32. $y = -\dfrac{1}{2}\sin\left(\dfrac{1}{2}x + \pi\right)$

33. $y = 3\cos\left[\dfrac{\pi}{2}\left(x - \dfrac{1}{2}\right)\right]$

34. $y = -\cos\left[\pi\left(x - \dfrac{1}{3}\right)\right]$

35. $y = 2 - \sin\left(3x - \dfrac{\pi}{5}\right)$

36. $y = -1 + \dfrac{1}{2}\cos(2x - 3\pi)$

Graph each function over a two-period interval. **See Examples 1 and 2.**

37. $y = \cos\left(x - \frac{\pi}{2}\right)$ **38.** $y = \sin\left(x - \frac{\pi}{4}\right)$ **39.** $y = \sin\left(x + \frac{\pi}{4}\right)$

40. $y = \cos\left(x + \frac{\pi}{3}\right)$ **41.** $y = 2\cos\left(x - \frac{\pi}{3}\right)$ **42.** $y = 3\sin\left(x - \frac{3\pi}{2}\right)$

Graph each function over a one-period interval. **See Example 3.**

43. $y = \frac{3}{2}\sin\left[2\left(x + \frac{\pi}{4}\right)\right]$ **44.** $y = -\frac{1}{2}\cos\left[4\left(x + \frac{\pi}{2}\right)\right]$

45. $y = -4\sin(2x - \pi)$ **46.** $y = 3\cos(4x + \pi)$

47. $y = \frac{1}{2}\cos\left(\frac{1}{2}x - \frac{\pi}{4}\right)$ **48.** $y = -\frac{1}{4}\sin\left(\frac{3}{4}x + \frac{\pi}{8}\right)$

Graph each function over a two-period interval. **See Example 4.**

49. $y = -3 + 2\sin x$ **50.** $y = 2 - 3\cos x$ **51.** $y = -1 - 2\cos 5x$

52. $y = 1 - \frac{2}{3}\sin\frac{3}{4}x$ **53.** $y = 1 - 2\cos\frac{1}{2}x$ **54.** $y = -3 + 3\sin\frac{1}{2}x$

55. $y = -2 + \frac{1}{2}\sin 3x$ **56.** $y = 1 + \frac{2}{3}\cos\frac{1}{2}x$

Graph each function over a one-period interval. **See Example 5.**

57. $y = -3 + 2\sin\left(x + \frac{\pi}{2}\right)$ **58.** $y = 4 - 3\cos(x - \pi)$

59. $y = \frac{1}{2} + \sin\left[2\left(x + \frac{\pi}{4}\right)\right]$ **60.** $y = -\frac{5}{2} + \cos\left[3\left(x - \frac{\pi}{6}\right)\right]$

(Modeling) Solve each problem. **See Example 6.**

61. *Average Monthly Temperature* The average monthly temperature (in °F) in Seattle, Washington, is shown in the table.

Month	°F	Month	°F
Jan	41	July	65
Feb	43	Aug	66
Mar	46	Sept	61
Apr	50	Oct	53
May	56	Nov	45
June	61	Dec	41

Source: World Almanac and Book of Facts.

(a) Plot the average monthly temperature over a two-year period, letting $x = 1$ correspond to January of the first year. Do the data seem to indicate a translated sine graph?

(b) The highest average monthly temperature is 66°F in August, and the lowest average monthly temperature is 41°F in January. Their average is 53.5°F. Graph the data together with the line $y = 53.5$. What does this line represent with regard to temperature in Seattle?

(c) Approximate the amplitude, period, and phase shift of the translated sine wave.

(d) Determine a function of the form $f(x) = a\sin[b(x - d)] + c$, where a, b, c, and d are constants, that models the data.

(e) Graph f together with the data on the same coordinate axes. How well does f model the given data?

(f) Use the sine regression capability of a graphing calculator to find the equation of a sine curve that fits these data.

62. *Average Monthly Temperature* The average monthly temperature (in °F) in Phoenix, Arizona, is shown in the table.

Month	°F	Month	°F
Jan	54	July	93
Feb	58	Aug	91
Mar	63	Sept	86
Apr	70	Oct	75
May	79	Nov	62
June	89	Dec	54

Source: World Almanac and Book of Facts.

(a) Predict the average annual temperature.

(b) Plot the average monthly temperature over a two-year period, letting $x = 1$ correspond to January of the first year.

(c) Determine a function of the form $f(x) = a\cos[b(x - d)] + c$, where a, b, c, and d are constants, that models the data.

(d) Graph f together with the data on the same coordinate axes. How well does f model the data?

(e) Use the sine regression capability of a graphing calculator to find the equation of a sine curve that fits these data (two years).

(Modeling) Monthly Temperatures A set of temperature data (in °F) is given for a particular location. (Source: www.weatherbase.com*)*

(a) Plot the data over a two-year interval.

(b) Use sine regression to determine a model for the two-year interval. Let $x = 1$ represent January of the first year.

(c) Graph the equation from part (b) together with the data on the same coordinate axes.

63. Average Monthly Temperature, Buenos Aires, Argentina

Jan	Feb	Mar	Apr	May	Jun	Jul	Aug	Sept	Oct	Nov	Dec
77.2	74.7	70.5	63.9	57.7	52.2	51.6	54.9	57.6	63.9	69.1	73.8

64. Average High Temperature, Buenos Aires, Argentina

Jan	Feb	Mar	Apr	May	Jun	Jul	Aug	Sept	Oct	Nov	Dec
86.7	83.7	79.5	72.9	66.2	60.1	58.8	63.1	66.0	72.5	77.5	82.6

(Modeling) Fractional Part of the Moon Illuminated The tables give the fractional part of the moon that is illuminated during the month indicated. (Source: http://aa.usno.navy.mil*)*

(a) Plot the data for the month.

(b) Use sine regression to determine a model for the data.

(c) Graph the equation from part (b) together with the data on the same coordinate axes.

65. January 2015

Day	1	2	3	4	5	6	7	8	9	10	11	12	13	14	15	16
Fraction	0.84	0.91	0.96	0.99	1.00	0.99	0.96	0.92	0.86	0.79	0.70	0.62	0.52	0.42	0.33	0.23

Day	17	18	19	20	21	22	23	24	25	26	27	28	29	30	31
Fraction	0.15	0.08	0.03	0.00	0.01	0.04	0.10	0.19	0.28	0.39	0.50	0.61	0.71	0.80	0.87

66. November 2015

Day	1	2	3	4	5	6	7	8	9	10	11	12	13	14
Fraction	0.73	0.63	0.53	0.43	0.34	0.25	0.18	0.11	0.06	0.02	0.00	0.00	0.02	0.06

Day	15	16	17	18	19	20	21	22	23	24	25	26	27	28	29	30
Fraction	0.12	0.19	0.28	0.39	0.49	0.61	0.71	0.81	0.90	0.96	0.99	1.00	0.98	0.93	0.87	0.79

Chapter 6 Quiz (Sections 6.1-6.4)

1. Convert 225° to radians.

2. Convert $-\frac{7\pi}{6}$ to degrees.

A central angle of a circle with radius 300 in. *intercepts an arc of* 450 in. *(These measures are accurate to the nearest inch.) Find each measure.*

3. the radian measure of the angle

4. the area of the sector

Find each exact circular function value.

5. $\cos \frac{7\pi}{4}$

6. $\sin\left(-\frac{5\pi}{6}\right)$

7. $\tan 3\pi$

8. Find the exact value of s in the interval $\left[\frac{\pi}{2}, \pi\right]$ if $\sin s = \frac{\sqrt{3}}{2}$.

9. Give the amplitude, period, vertical translation, and phase shift of the function $y = 3 - 4\sin\left(2x + \frac{\pi}{2}\right)$.

Graph each function over a two-period interval. Give the period and amplitude.

10. $y = -4 \sin x$

11. $y = -\frac{1}{2}\cos 2x$

12. $y = -2\cos\left(x + \frac{\pi}{4}\right)$

13. $y = 2 + \sin(2x - \pi)$

Connecting Graphs with Equations *Each function graphed is of the form* $y = a \cos bx$ *or* $y = a \sin bx$, *where* $b > 0$. *Determine an equation of the graph.*

14.

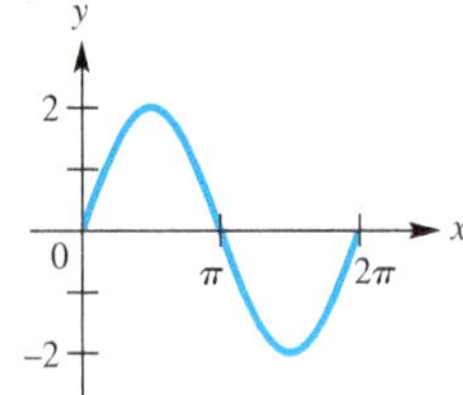

15.

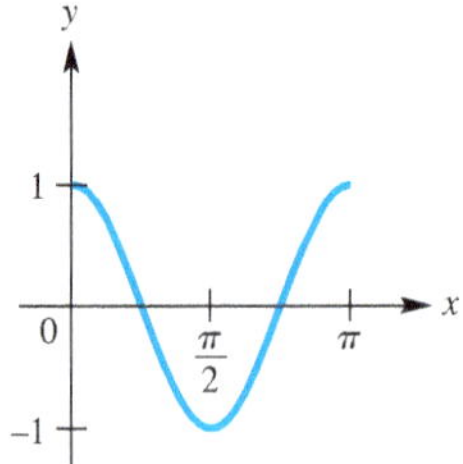

6.5 Graphs of the Tangent and Cotangent Functions

- Graph of the Tangent Function
- Graph of the Cotangent Function
- Techniques for Graphing
- Connecting Graphs with Equations

Graph of the Tangent Function Consider the table of selected points accompanying the graph of the tangent function in **Figure 43** on the next page. These points include special values between $-\frac{\pi}{2}$ and $\frac{\pi}{2}$. The tangent function is undefined for odd multiples of $\frac{\pi}{2}$ and, thus, has *vertical asymptotes* for such values. A **vertical asymptote** is a vertical line that the graph approaches but does not intersect. As the x-values get closer and closer to the line, the function values increase or decrease without bound. Furthermore, because

$$\tan(-x) = -\tan x, \quad \text{Odd function}$$

the graph of the tangent function is symmetric with respect to the origin.

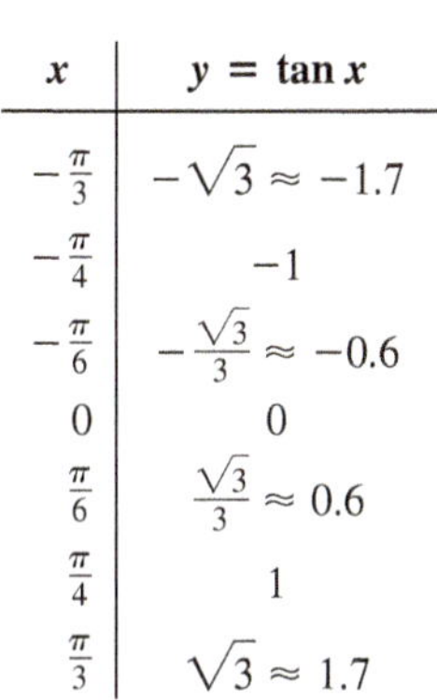

x	$y = \tan x$
$-\frac{\pi}{3}$	$-\sqrt{3} \approx -1.7$
$-\frac{\pi}{4}$	-1
$-\frac{\pi}{6}$	$-\frac{\sqrt{3}}{3} \approx -0.6$
0	0
$\frac{\pi}{6}$	$\frac{\sqrt{3}}{3} \approx 0.6$
$\frac{\pi}{4}$	1
$\frac{\pi}{3}$	$\sqrt{3} \approx 1.7$

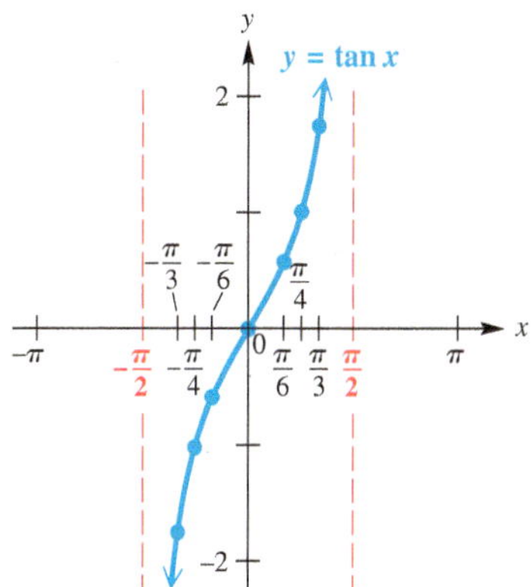

Figure 43

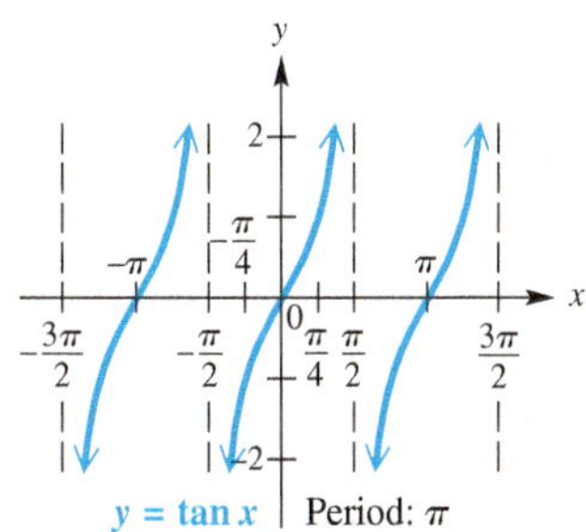

The graph continues in this pattern.

Figure 44

The tangent function has period π. Because $\tan x = \frac{\sin x}{\cos x}$, tangent values are 0 when sine values are 0, and are undefined when cosine values are 0. As x-values increase from $-\frac{\pi}{2}$ to $\frac{\pi}{2}$, tangent values range from $-\infty$ to ∞ and increase throughout the interval. Those same values are repeated as x increases from $\frac{\pi}{2}$ to $\frac{3\pi}{2}$, from $\frac{3\pi}{2}$ to $\frac{5\pi}{2}$, and so on. The graph of $y = \tan x$ from $-\frac{3\pi}{2}$ to $\frac{3\pi}{2}$ is shown in **Figure 44.**

Tangent Function $f(x) = \tan x$

Domain: $\{x \mid x \neq (2n+1)\frac{\pi}{2}$, where n is any integer$\}$ Range: $(-\infty, \infty)$

x	y
$-\frac{\pi}{2}$	undefined
$-\frac{\pi}{4}$	-1
0	0
$\frac{\pi}{4}$	1
$\frac{\pi}{2}$	undefined

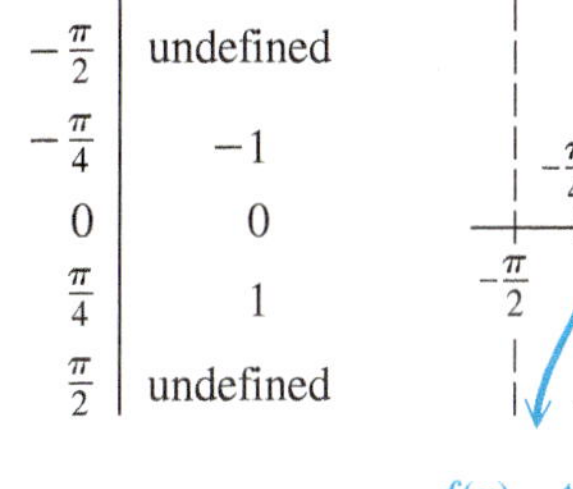

$f(x) = \tan x,\ -\frac{\pi}{2} < x < \frac{\pi}{2}$

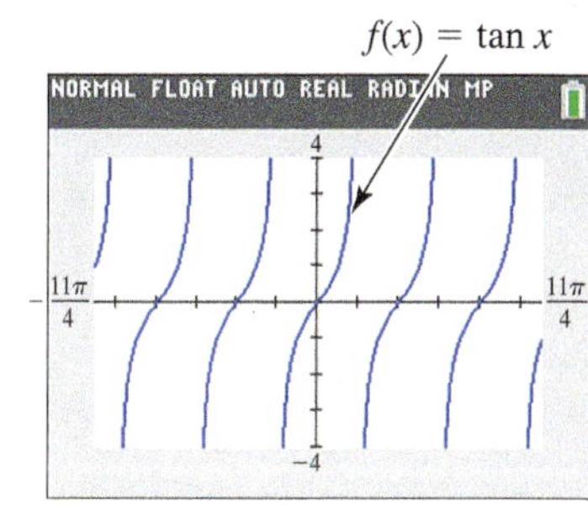

Figure 45

- The graph is discontinuous at values of x of the form $x = (2n+1)\frac{\pi}{2}$ and has vertical asymptotes at these values.
- Its x-intercepts have x-values of the form $n\pi$.
- Its period is π.
- There are no minimum or maximum values, so its graph has no amplitude.
- The graph is symmetric with respect to the origin, so the function is an odd function. For all x in the domain, $\tan(-x) = -\tan x$.

Graph of the Cotangent Function A similar analysis for selected points between 0 and π for the graph of the cotangent function yields the graph in **Figure 46** on the next page. Here the vertical asymptotes are at x-values that are integer multiples of π. Because

$$\cot(-x) = -\cot x, \quad \text{Odd function}$$

this graph is also symmetric with respect to the origin.

x	$y = \cot x$
$\frac{\pi}{6}$	$\sqrt{3} \approx 1.7$
$\frac{\pi}{4}$	1
$\frac{\pi}{3}$	$\frac{\sqrt{3}}{3} \approx 0.6$
$\frac{\pi}{2}$	0
$\frac{2\pi}{3}$	$-\frac{\sqrt{3}}{3} \approx -0.6$
$\frac{3\pi}{4}$	-1
$\frac{5\pi}{6}$	$-\sqrt{3} \approx -1.7$

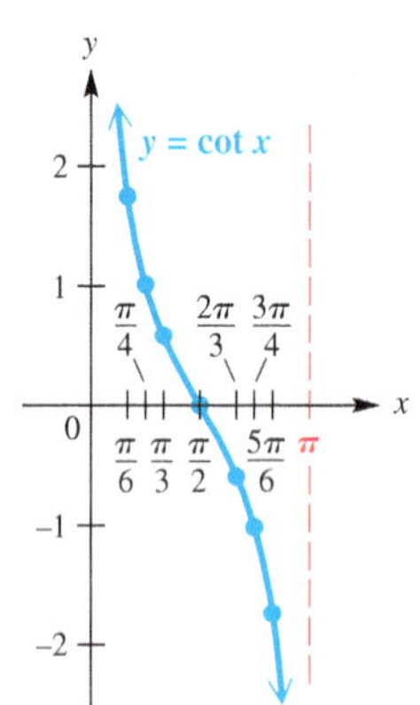

Figure 46

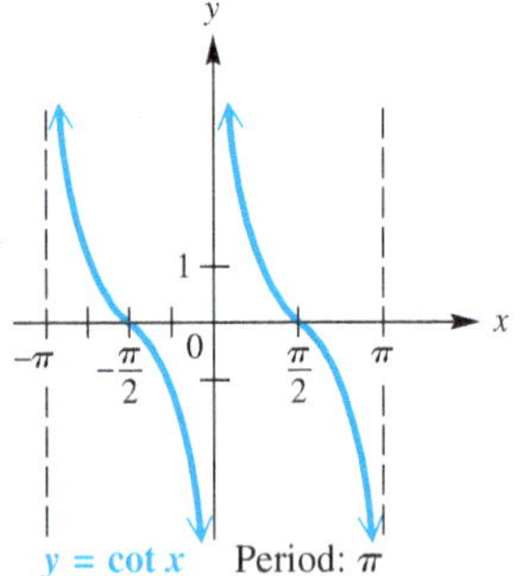

$y = \cot x$ Period: π

The graph continues in this pattern.

Figure 47

The cotangent function also has period π. Cotangent values are 0 when cosine values are 0, and are undefined when sine values are 0. As x-values increase from 0 to π, cotangent values range from ∞ to $-\infty$ and decrease throughout the interval. Those same values are repeated as x increases from π to 2π, from 2π to 3π, and so on. The graph of $y = \cot x$ from $-\pi$ to π is shown in **Figure 47.**

Cotangent Function $f(x) = \cot x$

Domain: $\{x \mid x \neq n\pi$, where n is any integer$\}$ Range: $(-\infty, \infty)$

x	y
0	undefined
$\frac{\pi}{4}$	1
$\frac{\pi}{2}$	0
$\frac{3\pi}{4}$	-1
π	undefined

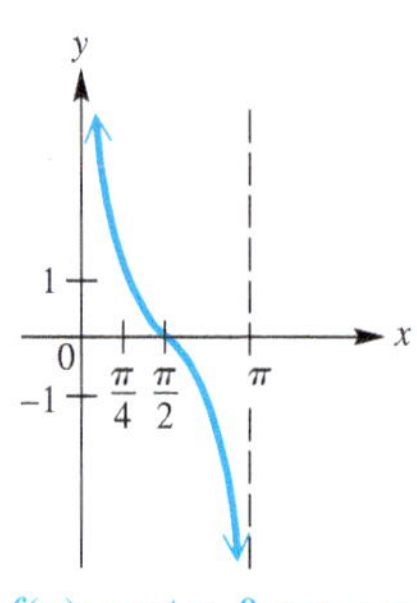

$f(x) = \cot x$, $0 < x < \pi$

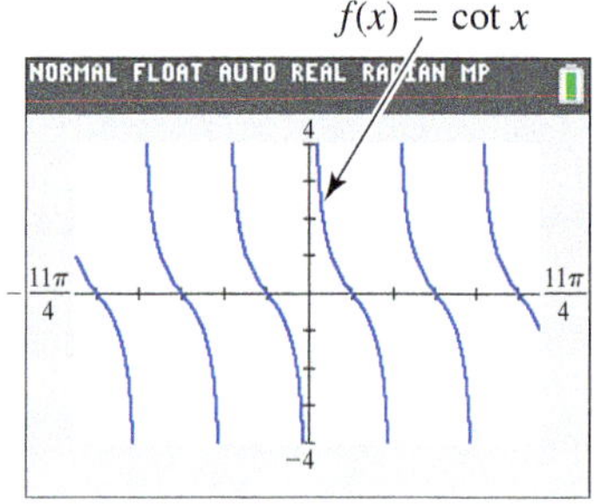

Figure 48

- The graph is discontinuous at values of x of the form $x = n\pi$ and has vertical asymptotes at these values.
- Its x-intercepts have x-values of the form $(2n + 1)\frac{\pi}{2}$.
- Its period is π.
- There are no minimum or maximum values, so its graph has no amplitude.
- The graph is symmetric with respect to the origin, so the function is an odd function. For all x in the domain, $\cot(-x) = -\cot x$.

The tangent function can be graphed directly with a graphing calculator, using the tangent key. To graph the cotangent function, however, we must use one of the identities

$$\cot x = \frac{1}{\tan x} \quad \text{or} \quad \cot x = \frac{\cos x}{\sin x}$$

because graphing calculators generally do not have cotangent keys. ■

Techniques for Graphing

Guidelines for Sketching Graphs of Tangent and Cotangent Functions

To graph $y = a \tan bx$ or $y = a \cot bx$, with $b > 0$, follow these steps.

Step 1 Determine the period, $\frac{\pi}{b}$. To locate two adjacent vertical asymptotes, solve the following equations for x:

For $y = a \tan bx$: $\quad bx = -\frac{\pi}{2}$ and $bx = \frac{\pi}{2}$.

For $y = a \cot bx$: $\quad bx = 0$ and $bx = \pi$.

Step 2 Sketch the two vertical asymptotes found in Step 1.

Step 3 Divide the interval formed by the vertical asymptotes into four equal parts.

Step 4 Evaluate the function for the first-quarter point, midpoint, and third-quarter point, using the x-values found in Step 3.

Step 5 Join the points with a smooth curve, approaching the vertical asymptotes. Indicate additional asymptotes and periods of the graph as necessary.

EXAMPLE 1 **Graphing $y = \tan bx$**

Graph $y = \tan 2x$.

SOLUTION

Step 1 The period of this function is $\frac{\pi}{2}$. To locate two adjacent vertical asymptotes, solve $2x = -\frac{\pi}{2}$ and $2x = \frac{\pi}{2}$ (because this is a tangent function). The two asymptotes have equations $x = -\frac{\pi}{4}$ and $x = \frac{\pi}{4}$.

Step 2 Sketch the two vertical asymptotes $x = \pm\frac{\pi}{4}$, as shown in **Figure 49.**

Step 3 Divide the interval $\left(-\frac{\pi}{4}, \frac{\pi}{4}\right)$ into four equal parts to find key x-values.

first-quarter value: $-\frac{\pi}{8}$, middle value: 0, third-quarter value: $\frac{\pi}{8}$ Key x-values

Step 4 Evaluate the function for the x-values found in Step 3.

x	$-\frac{\pi}{8}$	0	$\frac{\pi}{8}$
$2x$	$-\frac{\pi}{4}$	0	$\frac{\pi}{4}$
$\tan 2x$	-1	0	1

Another period has been graphed, one half period to the left and one half period to the right.

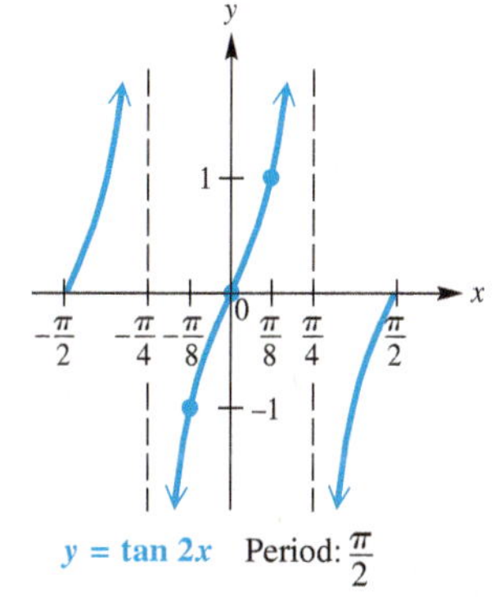

Figure 49

Step 5 Join these points with a smooth curve, approaching the vertical asymptotes. See **Figure 49.**

✔ **Now Try Exercise 13.**

EXAMPLE 2 Graphing $y = a \tan bx$

Graph $y = -3 \tan \frac{1}{2}x$.

SOLUTION The period is $\frac{\pi}{\frac{1}{2}} = \pi \div \frac{1}{2} = \pi \cdot \frac{2}{1} = 2\pi$. Adjacent asymptotes are at $x = -\pi$ and $x = \pi$. Dividing the interval $(-\pi, \pi)$ into four equal parts gives key x-values of $-\frac{\pi}{2}$, 0, and $\frac{\pi}{2}$. Evaluating the function at these x-values gives the following key points.

$$\left(-\frac{\pi}{2}, 3\right), \quad (0, 0), \quad \left(\frac{\pi}{2}, -3\right) \qquad \text{Key points}$$

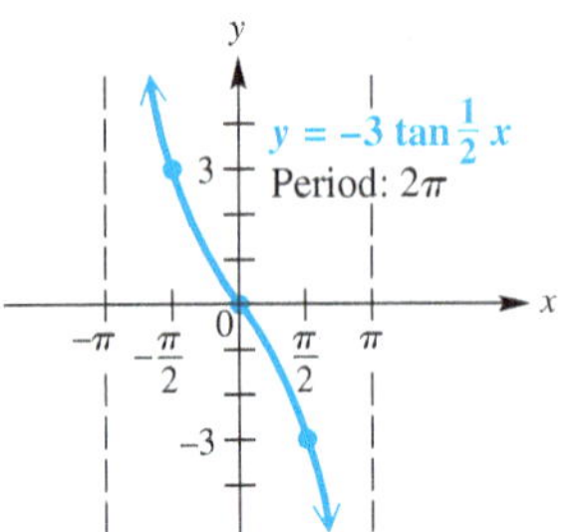

Figure 50

By plotting these points and joining them with a smooth curve, we obtain the graph shown in **Figure 50.** Because the coefficient -3 is negative, the graph is reflected across the x-axis compared to the graph of $y = 3 \tan \frac{1}{2}x$.

✔ **Now Try Exercise 21.**

NOTE The function $y = -3 \tan \frac{1}{2}x$ in Example 2, graphed in **Figure 50,** has a graph that compares to the graph of $y = \tan x$ as follows.

1. The period is larger because $b = \frac{1}{2}$, and $\frac{1}{2} < 1$.
2. The graph is stretched vertically because $a = -3$, and $|-3| > 1$.
3. Each branch of the graph falls from left to right (that is, the function decreases) between each pair of adjacent asymptotes because $a = -3$, and $-3 < 0$. When $a < 0$, the graph is reflected across the x-axis compared to the graph of $y = |a| \tan bx$.

EXAMPLE 3 Graphing $y = a \cot bx$

Graph $y = \frac{1}{2} \cot 2x$.

SOLUTION Because this function involves the cotangent, we can locate two adjacent asymptotes by solving the equations $2x = 0$ and $2x = \pi$. The lines $x = 0$ (the y-axis) and $x = \frac{\pi}{2}$ are two such asymptotes. We divide the interval $\left(0, \frac{\pi}{2}\right)$ into four equal parts, obtaining key x-values of $\frac{\pi}{8}$, $\frac{\pi}{4}$, and $\frac{3\pi}{8}$. Evaluating the function at these x-values gives the key points $\left(\frac{\pi}{8}, \frac{1}{2}\right)$, $\left(\frac{\pi}{4}, 0\right)$, $\left(\frac{3\pi}{8}, -\frac{1}{2}\right)$. We plot these points and join them with a smooth curve approaching the asymptotes to obtain the graph shown in **Figure 51.**

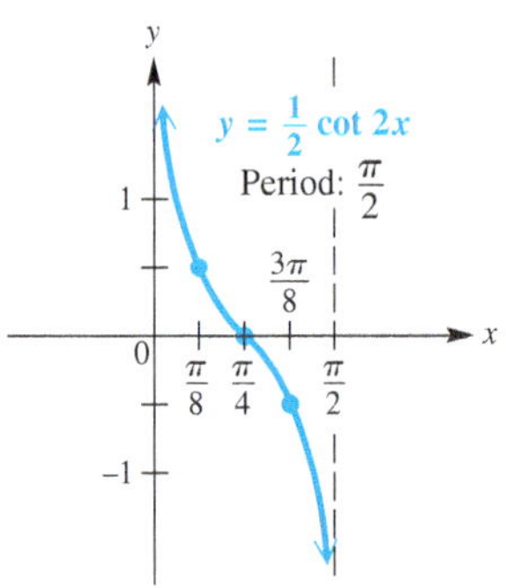

Figure 51

✔ **Now Try Exercise 23.**

Like the other circular functions, the graphs of the tangent and cotangent functions may be translated horizontally and vertically.

EXAMPLE 4 Graphing $y = c + \tan x$

Graph $y = 2 + \tan x$.

ANALYTIC SOLUTION

Every value of y for this function will be 2 units more than the corresponding value of y in $y = \tan x$, causing the graph of $y = 2 + \tan x$ to be translated 2 units up compared to the graph of $y = \tan x$. See **Figure 52.**

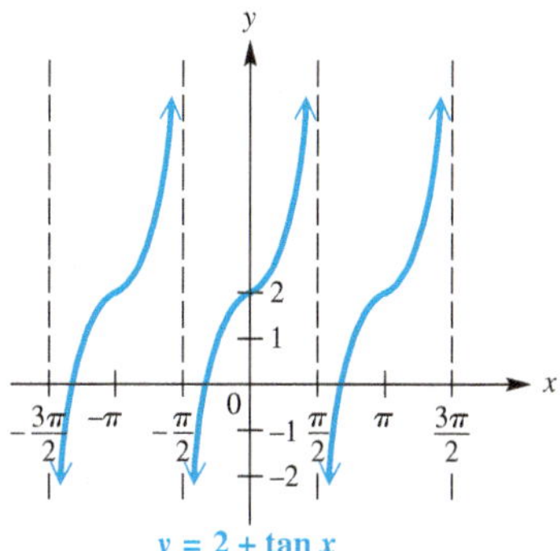

Figure 52

Three periods of the function are shown in **Figure 52.** Because the period of $y = 2 + \tan x$ is π, additional asymptotes and periods of the function can be drawn by repeating the basic graph every π units on the x-axis to the left or to the right of the graph shown.

GRAPHING CALCULATOR SOLUTION

Observe **Figures 53 and 54.** In these figures

$$y_2 = \tan x$$

is the red graph and

$$y_1 = 2 + \tan x$$

is the blue graph. Notice that for the arbitrarily-chosen value of $\frac{\pi}{4}$ (approximately 0.78539816), the difference in the y-values is

$$y_1 - y_2 = 3 - 1 = 2.$$

This illustrates the vertical translation 2 units up.

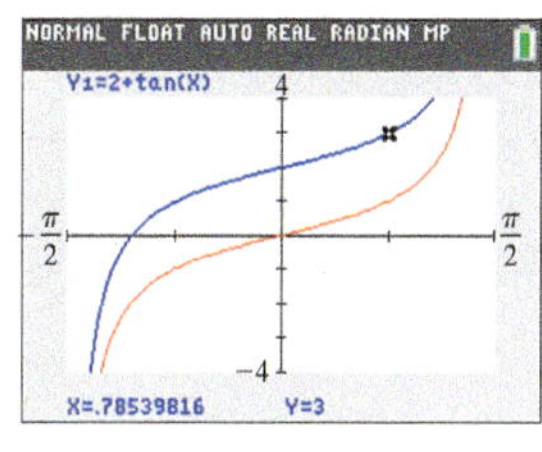

Figure 53

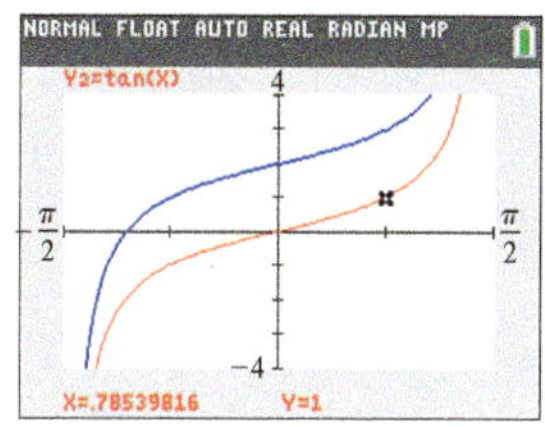

Figure 54

✔ **Now Try Exercise 29.**

EXAMPLE 5 Graphing $y = c + a\cot(x - d)$

Graph $y = -2 - \cot\left(x - \frac{\pi}{4}\right)$.

SOLUTION Here $b = 1$, so the period is π. The negative sign in front of the cotangent will cause the graph to be reflected across the x-axis, and the argument $\left(x - \frac{\pi}{4}\right)$ indicates a phase shift (horizontal shift) $\frac{\pi}{4}$ unit to the right. Because $c = -2$, the graph will then be translated 2 units down. To locate adjacent asymptotes, because this function involves the cotangent, we solve the following.

$$x - \frac{\pi}{4} = 0 \quad \text{and} \quad x - \frac{\pi}{4} = \pi$$

$$x = \frac{\pi}{4} \quad \text{and} \quad x = \frac{5\pi}{4} \quad \text{Add } \tfrac{\pi}{4}.$$

Dividing the interval $\left(\frac{\pi}{4}, \frac{5\pi}{4}\right)$ into four equal parts and evaluating the function at the three key x-values within the interval give these points.

$$\left(\frac{\pi}{2}, -3\right), \quad \left(\frac{3\pi}{4}, -2\right), \quad (\pi, -1) \quad \text{Key points}$$

We join these points with a smooth curve. This period of the graph, along with the one in the domain interval $\left(-\frac{3\pi}{4}, \frac{\pi}{4}\right)$, is shown in **Figure 55** on the next page.

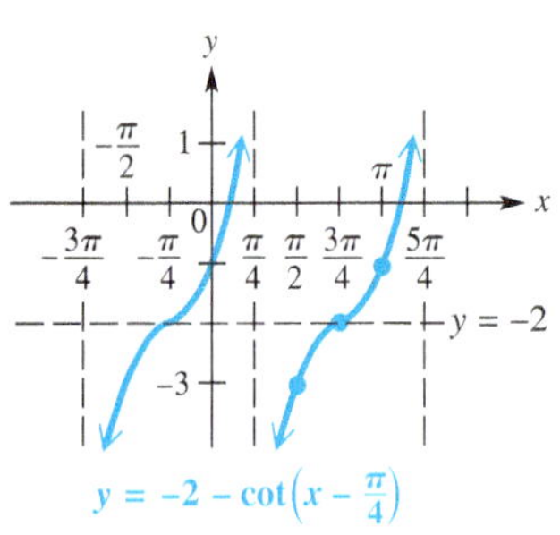

Figure 55

✔ **Now Try Exercise 37.**

Connecting Graphs with Equations

EXAMPLE 6 Determining an Equation for a Graph

Determine an equation for each graph.

(a)

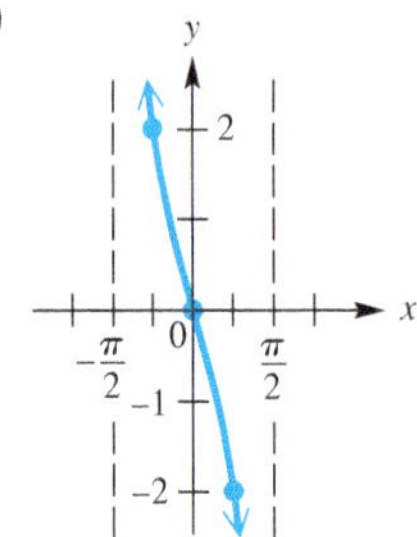

(b)

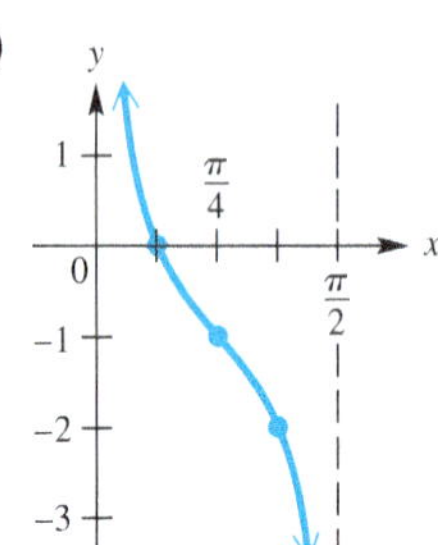

SOLUTION

(a) This graph is that of $y = \tan x$ but reflected across the x-axis and stretched vertically by a factor of 2. Therefore, an equation for this graph is

$$y = -2 \tan x.$$

x-axis reflection (points to −); Vertical stretch (points to 2)

(b) This is the graph of a cotangent function, but the period is $\frac{\pi}{2}$ rather than π. Therefore, the coefficient of x is 2. This graph is vertically translated 1 unit down compared to the graph of $y = \cot 2x$. An equation for this graph is

$$y = -1 + \cot 2x.$$

Vertical translation 1 unit down (points to −1); Period is $\frac{\pi}{2}$. (points to 2)

✔ **Now Try Exercises 39 and 43.**

NOTE Because the circular functions are periodic, there are infinitely many equations that correspond to each graph in **Example 6.** Confirm that both

$$y = -1 - \cot(-2x) \quad \text{and} \quad y = -1 - \tan\left(2x - \frac{\pi}{2}\right)$$

are equations for the graph in **Example 6(b).** When writing the equation from a graph, it is practical to write the simplest form. Therefore, we choose values of b where $b > 0$ and write the function without a phase shift when possible.

6.5 Exercises

CONCEPT PREVIEW *Fill in the blank to correctly complete each sentence.*

1. The least positive value x for which $\tan x = 0$ is ______.

2. The least positive value x for which $\cot x = 0$ is ______.

3. Between any two successive vertical asymptotes, the graph of $y = \tan x$ ____________.
   (increases/decreases)

4. Between any two successive vertical asymptotes, the graph of $y = \cot x$ ____________.
   (increases/decreases)

5. The negative value k with the greatest value for which $x = k$ is a vertical asymptote of the graph of $y = \tan x$ is ______.

6. The negative value k with the greatest value for which $x = k$ is a vertical asymptote of the graph of $y = \cot x$ is ______.

Concept Check Match each function with its graph from choices A–F.

7. $y = -\tan x$ **8.** $y = -\cot x$ **9.** $y = \tan\left(x - \frac{\pi}{4}\right)$

10. $y = \cot\left(x - \frac{\pi}{4}\right)$ **11.** $y = \cot\left(x + \frac{\pi}{4}\right)$ **12.** $y = \tan\left(x + \frac{\pi}{4}\right)$

A.

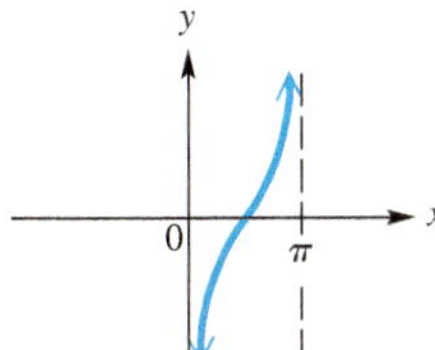

B.

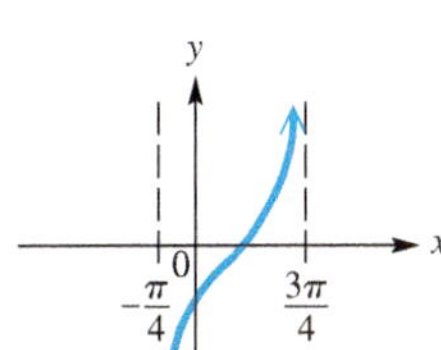

C.

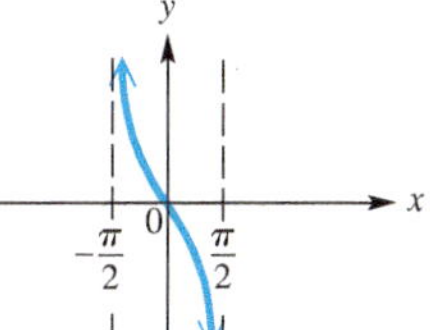

D.

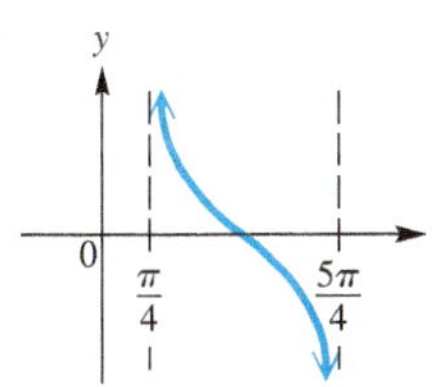

E.

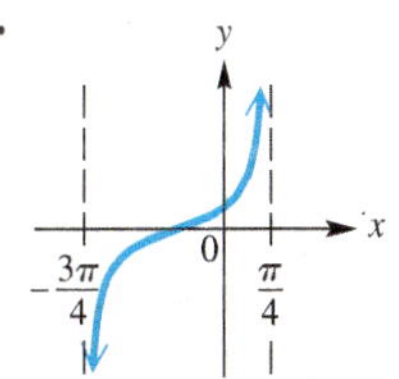

F.

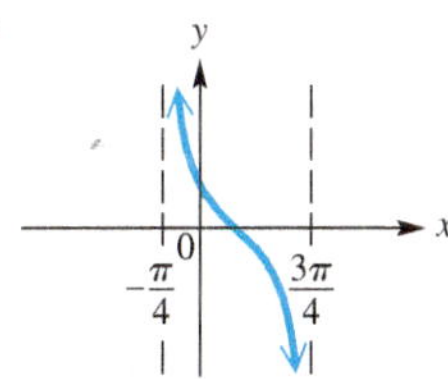

Graph each function over a one-period interval. ***See Examples 1–3.***

13. $y = \tan 4x$ **14.** $y = \tan \frac{1}{2}x$ **15.** $y = 2 \tan x$

16. $y = 2 \cot x$ **17.** $y = 2 \tan \frac{1}{4}x$ **18.** $y = \frac{1}{2} \cot x$

19. $y = \cot 3x$ **20.** $y = -\cot \frac{1}{2}x$ **21.** $y = -2 \tan \frac{1}{4}x$

22. $y = 3 \tan \frac{1}{2}x$ **23.** $y = \frac{1}{2} \cot 4x$ **24.** $y = -\frac{1}{2} \cot 2x$

Graph each function over a two-period interval. ***See Examples 4 and 5.***

25. $y = \tan(2x - \pi)$ **26.** $y = \tan\left(\frac{x}{2} + \pi\right)$ **27.** $y = \cot\left(3x + \frac{\pi}{4}\right)$

28. $y = \cot\left(2x - \frac{3\pi}{2}\right)$ **29.** $y = 1 + \tan x$ **30.** $y = 1 - \tan x$

31. $y = 1 - \cot x$

32. $y = -2 - \cot x$

33. $y = -1 + 2 \tan x$

34. $y = 3 + \frac{1}{2} \tan x$

35. $y = -1 + \frac{1}{2} \cot(2x - 3\pi)$

36. $y = -2 + 3 \tan(4x + \pi)$

37. $y = 1 - 2 \cot\left[2\left(x + \frac{\pi}{2}\right)\right]$

38. $y = -2 + \frac{2}{3} \tan\left(\frac{3}{4}x - \pi\right)$

Connecting Graphs with Equations *Determine the simplest form of an equation for each graph. Choose $b > 0$, and include no phase shifts. (Midpoints and quarter-points are identified by dots.)* ***See Example 6.***

39.

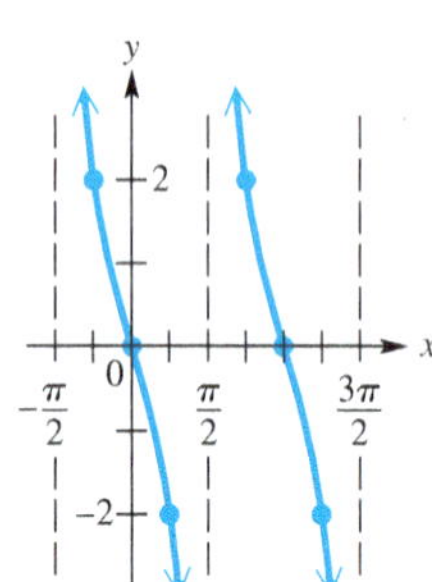

40.

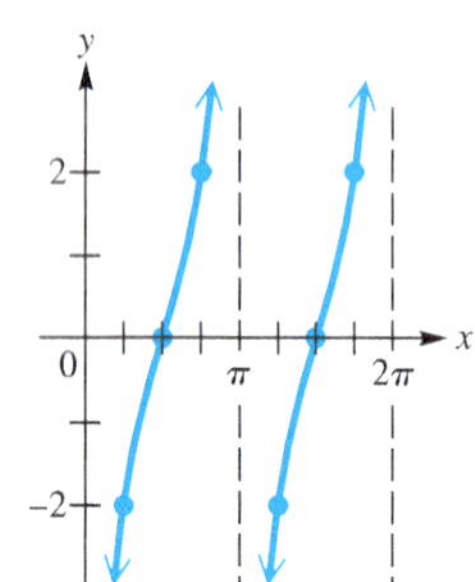

41.

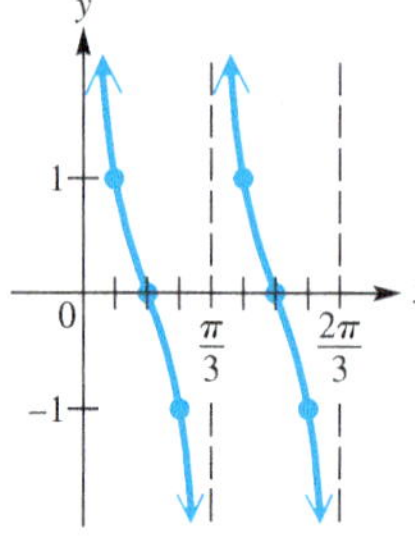

42.

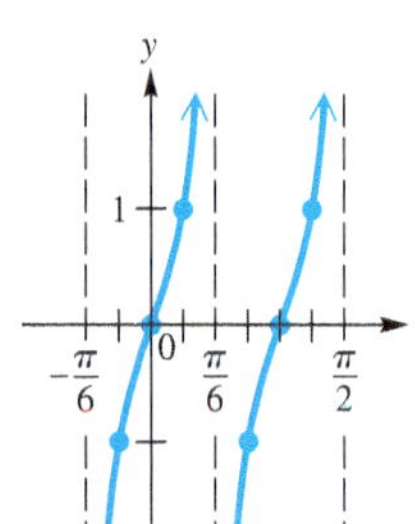

43.

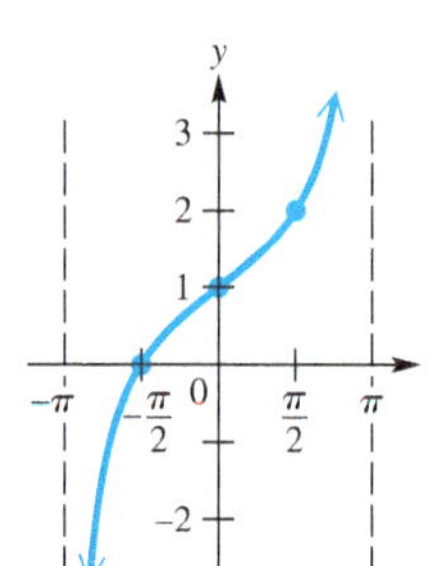

44.

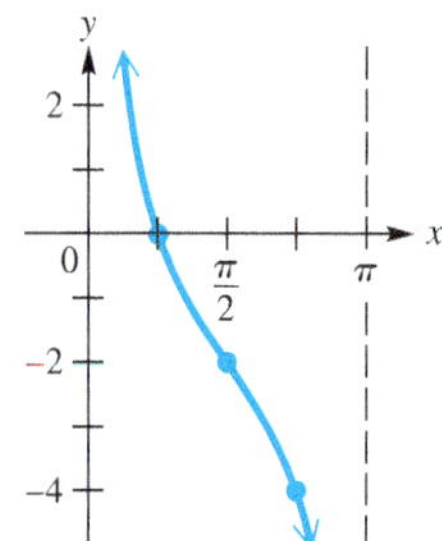

Concept Check *Decide whether each statement is* true *or* false. *If* false, *explain why.*

45. The least positive number k for which $x = k$ is an asymptote for the tangent function is $\frac{\pi}{2}$.

46. The least positive number k for which $x = k$ is an asymptote for the cotangent function is $\frac{\pi}{2}$.

47. The graph of $y = \tan x$ in **Figure 44** suggests that $\tan(-x) = \tan x$ for all x in the domain of $\tan x$.

48. The graph of $y = \cot x$ in **Figure 47** suggests that $\cot(-x) = -\cot x$ for all x in the domain of $\cot x$.

Work each problem.

49. ***Concept Check*** If c is any number, then how many solutions does the equation $c = \tan x$ have in the interval $(-2\pi, 2\pi]$?

50. ***Concept Check*** Consider the function defined by $f(x) = -4 \tan(2x + \pi)$. What is the domain of f? What is its range?

51. Show that $\tan(-x) = -\tan x$ by writing $\tan(-x)$ as $\frac{\sin(-x)}{\cos(-x)}$ and then using the relationships for $\sin(-x)$ and $\cos(-x)$.

52. Show that $\cot(-x) = -\cot x$ by writing $\cot(-x)$ as $\frac{\cos(-x)}{\sin(-x)}$ and then using the relationships for $\cos(-x)$ and $\sin(-x)$.

(Modeling) Distance of a Rotating Beacon *A rotating beacon is located at point A next to a long wall. The beacon is* 4 m *from the wall. The distance d is given by*

$$d = 4 \tan 2\pi t,$$

where t is time measured in seconds since the beacon started rotating. (When $t = 0$, the beacon is aimed at point R. When the beacon is aimed to the right of R, the value of d is positive; d is negative when the beacon is aimed to the left of R.) Find d for each time. Round to the nearest tenth if applicable.

53. $t = 0$

54. $t = 0.4$

55. $t = 1.2$

56. Why is 0.25 a meaningless value for t?

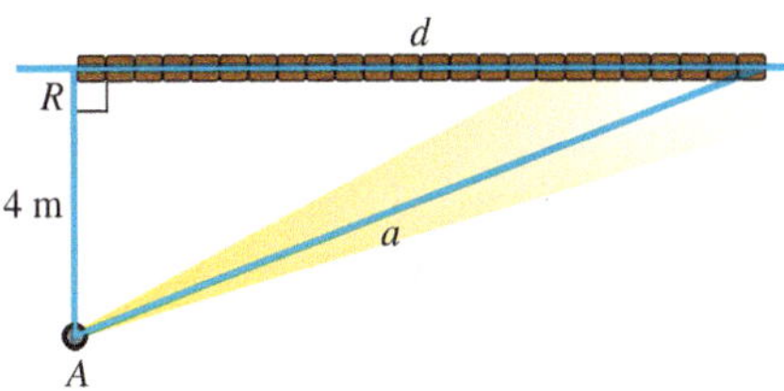

Relating Concepts

For individual or collaborative investigation *(Exercises 57–62)*

Consider the following function from **Example 5. Work these exercises in order.**

$$y = -2 - \cot\left(x - \frac{\pi}{4}\right)$$

57. What is the least positive number for which $y = \cot x$ is undefined?

58. Let k represent the number found in **Exercise 57.** Set $x - \frac{\pi}{4}$ equal to k, and solve to find a positive number for which $\cot\left(x - \frac{\pi}{4}\right)$ is undefined.

59. Based on the answer in **Exercise 58** and the fact that the cotangent function has period π, give the general form of the equations of the asymptotes of the graph of $y = -2 - \cot\left(x - \frac{\pi}{4}\right)$. Let n represent any integer.

60. Use the capabilities of a calculator to find the x-intercept with least positive x-value of the graph of this function. Round to the nearest hundredth.

61. Use the fact that the period of this function is π to find the next positive x-intercept. Round to the nearest hundredth.

62. Give the solution set of the equation $-2 - \cot\left(x - \frac{\pi}{4}\right) = 0$ over all real numbers. Let n represent any integer.

6.6 Graphs of the Secant and Cosecant Functions

- Graph of the Secant Function
- Graph of the Cosecant Function
- Techniques for Graphing
- Connecting Graphs with Equations
- Addition of Ordinates

Graph of the Secant Function Consider the table of selected points accompanying the graph of the secant function in **Figure 56** on the next page. These points include special values from $-\pi$ to π. The secant function is undefined for odd multiples of $\frac{\pi}{2}$ and thus, like the tangent function, has vertical asymptotes for such values. Furthermore, because

$$\sec(-x) = \sec x, \quad \text{Even function}$$

the graph of the secant function is symmetric with respect to the y-axis.

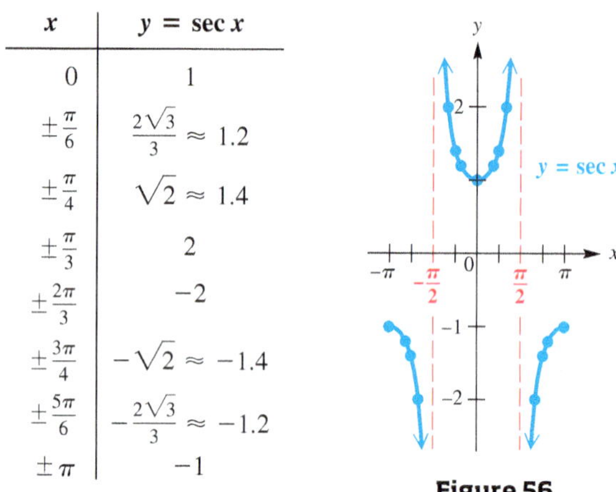

x	$y = \sec x$
0	1
$\pm\frac{\pi}{6}$	$\frac{2\sqrt{3}}{3} \approx 1.2$
$\pm\frac{\pi}{4}$	$\sqrt{2} \approx 1.4$
$\pm\frac{\pi}{3}$	2
$\pm\frac{2\pi}{3}$	-2
$\pm\frac{3\pi}{4}$	$-\sqrt{2} \approx -1.4$
$\pm\frac{5\pi}{6}$	$-\frac{2\sqrt{3}}{3} \approx -1.2$
$\pm\pi$	-1

Figure 56

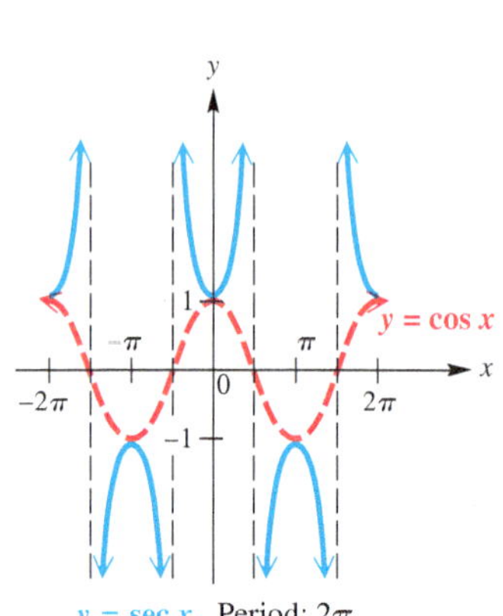

Figure 57

Because secant values are reciprocals of corresponding cosine values, the period of the secant function is 2π, the same as for $y = \cos x$. When $\cos x = 1$, the value of $\sec x$ is also 1. Likewise, when $\cos x = -1$, $\sec x = -1$. For all x, $-1 \leq \cos x \leq 1$, and thus, $|\sec x| \geq 1$ for all x in its domain. **Figure 57** shows how the graphs of $y = \cos x$ and $y = \sec x$ are related.

Secant Function $f(x) = \sec x$

Domain: $\left\{x \mid x \neq (2n+1)\frac{\pi}{2},\right.$ where n is any integer$\left.\right\}$ Range: $(-\infty, -1] \cup [1, \infty)$

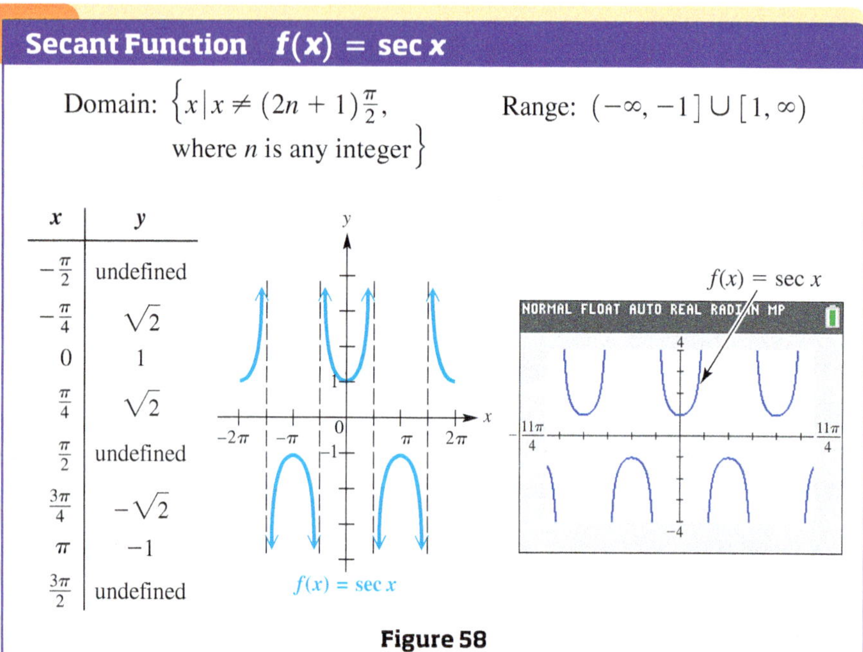

x	y
$-\frac{\pi}{2}$	undefined
$-\frac{\pi}{4}$	$\sqrt{2}$
0	1
$\frac{\pi}{4}$	$\sqrt{2}$
$\frac{\pi}{2}$	undefined
$\frac{3\pi}{4}$	$-\sqrt{2}$
π	-1
$\frac{3\pi}{2}$	undefined

Figure 58

- The graph is discontinuous at values of x of the form $x = (2n+1)\frac{\pi}{2}$ and has vertical asymptotes at these values.
- There are no x-intercepts.
- Its period is 2π.
- There are no minimum or maximum values, so its graph has no amplitude.
- The graph is symmetric with respect to the y-axis, so the function is an even function. For all x in the domain, $\sec(-x) = \sec x$.

As we shall see, locating the vertical asymptotes for the graph of a function involving the secant (as well as the cosecant) is helpful in sketching its graph.

Graph of the Cosecant function A similar analysis for selected points between $-\pi$ and π for the graph of the cosecant function yields the graph in **Figure 59.** The vertical asymptotes are at x-values that are integer multiples of π. This graph is symmetric with respect to the origin because

$$\csc(-x) = -\csc x. \quad \text{Odd function}$$

x	$y = \csc x$	x	$y = \csc x$
$\frac{\pi}{6}$	2	$-\frac{\pi}{6}$	-2
$\frac{\pi}{4}$	$\sqrt{2} \approx 1.4$	$-\frac{\pi}{4}$	$-\sqrt{2} \approx -1.4$
$\frac{\pi}{3}$	$\frac{2\sqrt{3}}{3} \approx 1.2$	$-\frac{\pi}{3}$	$-\frac{2\sqrt{3}}{3} \approx -1.2$
$\frac{\pi}{2}$	1	$-\frac{\pi}{2}$	-1
$\frac{2\pi}{3}$	$\frac{2\sqrt{3}}{3} \approx 1.2$	$-\frac{2\pi}{3}$	$-\frac{2\sqrt{3}}{3} \approx -1.2$
$\frac{3\pi}{4}$	$\sqrt{2} \approx 1.4$	$-\frac{3\pi}{4}$	$-\sqrt{2} \approx -1.4$
$\frac{5\pi}{6}$	2	$-\frac{5\pi}{6}$	-2

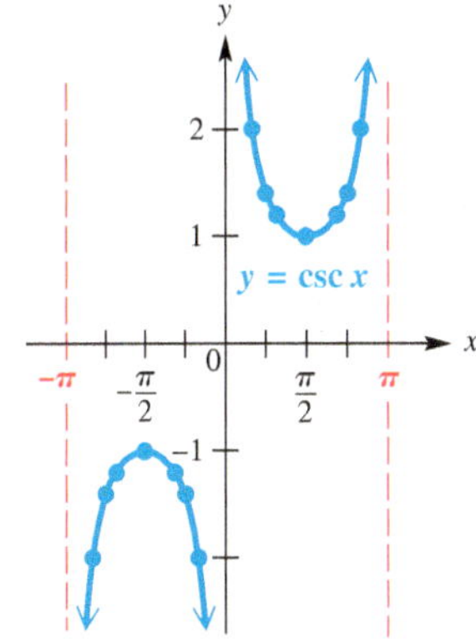

Figure 59

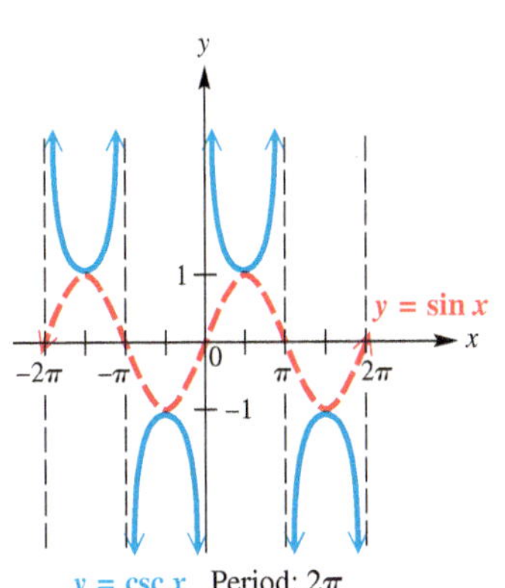

$y = \csc x$ Period: 2π

Figure 60

Because cosecant values are reciprocals of corresponding sine values, the period of the cosecant function is 2π, the same as for $y = \sin x$. When $\sin x = 1$, the value of $\csc x$ is also 1. Likewise, when $\sin x = -1$, $\csc x = -1$. For all x, $-1 \le \sin x \le 1$, and thus $|\csc x| \ge 1$ for all x in its domain. **Figure 60** shows how the graphs of $y = \sin x$ and $y = \csc x$ are related.

Cosecant Function $f(x) = \csc x$

Domain: $\{x \mid x \neq n\pi$, where n is any integer$\}$ Range: $(-\infty, -1] \cup [1, \infty)$

x	y
0	undefined
$\frac{\pi}{6}$	2
$\frac{\pi}{3}$	$\frac{2\sqrt{3}}{3}$
$\frac{\pi}{2}$	1
$\frac{2\pi}{3}$	$\frac{2\sqrt{3}}{3}$
π	undefined
$\frac{3\pi}{2}$	-1
2π	undefined

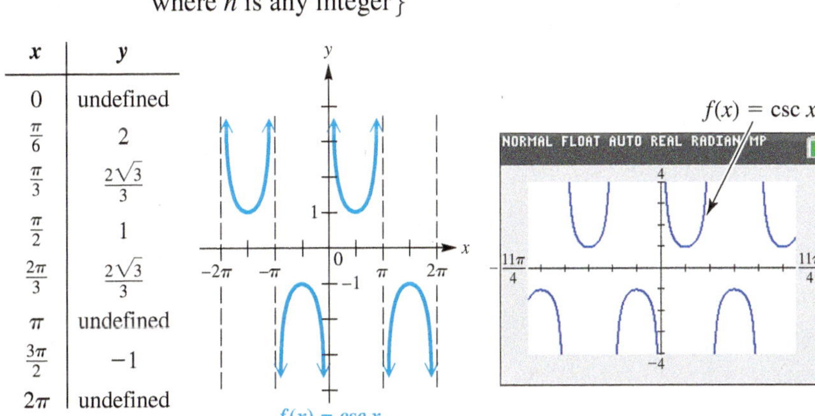

Figure 61

- The graph is discontinuous at values of x of the form $x = n\pi$ and has vertical asymptotes at these values.
- There are no x-intercepts.
- Its period is 2π.
- There are no minimum or maximum values, so its graph has no amplitude.
- The graph is symmetric with respect to the origin, so the function is an odd function. For all x in the domain, $\csc(-x) = -\csc x$.

Calculators do not have keys for the cosecant and secant functions. To graph them with a graphing calculator, use

$$\csc x = \frac{1}{\sin x} \quad \text{and} \quad \sec x = \frac{1}{\cos x}. \qquad \text{Reciprocal identities}$$

Techniques for Graphing

Guidelines for Sketching Graphs of Cosecant and Secant Functions

To graph $\boldsymbol{y = a \csc bx}$ or $\boldsymbol{y = a \sec bx}$, with $b > 0$, follow these steps.

Step 1 Graph the corresponding reciprocal function as a guide, using a dashed curve.

To Graph	Use as a Guide
$y = a \csc bx$	$y = a \sin bx$
$y = a \sec bx$	$y = a \cos bx$

Step 2 Sketch the vertical asymptotes. They will have equations of the form $x = k$, where k corresponds to an x-intercept of the graph of the guide function.

Step 3 Sketch the graph of the desired function by drawing the typical U-shaped branches between the adjacent asymptotes. The branches will be above the graph of the guide function when the guide function values are positive and below the graph of the guide function when the guide function values are negative. The graph will resemble those in **Figures 58 and 61** in the function boxes given earlier in this section.

Like graphs of the sine and cosine functions, graphs of the secant and cosecant functions may be translated vertically and horizontally. The period of both basic functions is 2π.

EXAMPLE 1 Graphing $y = a \sec bx$

Graph $y = 2 \sec \frac{1}{2}x$.

SOLUTION

Step 1 This function involves the secant, so the corresponding reciprocal function will involve the cosine. The guide function to graph is

$$y = 2 \cos \frac{1}{2}x.$$

Using the guidelines given earlier, we find that this guide function has amplitude 2 and that one period of the graph lies along the interval that satisfies the following inequality.

$$0 \le \frac{1}{2}x \le 2\pi$$

$$0 \le x \le 4\pi \qquad \text{Multiply each part by 2.}$$

Dividing the interval $[0, 4\pi]$ into four equal parts gives these key points.

$$(0, 2), \quad (\pi, 0), \quad (2\pi, -2), \quad (3\pi, 0), \quad (4\pi, 2) \qquad \text{Key points}$$

These key points are plotted and joined with a dashed red curve to indicate that this graph is only a guide. An additional period is graphed as shown in **Figure 62(a).**

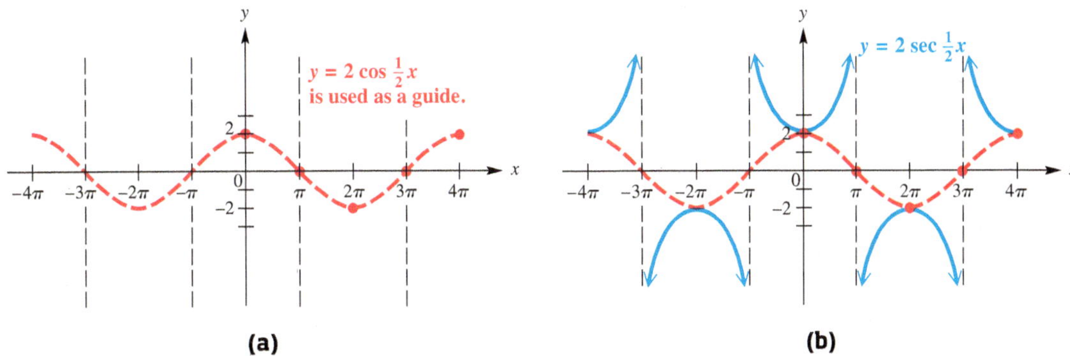

Figure 62

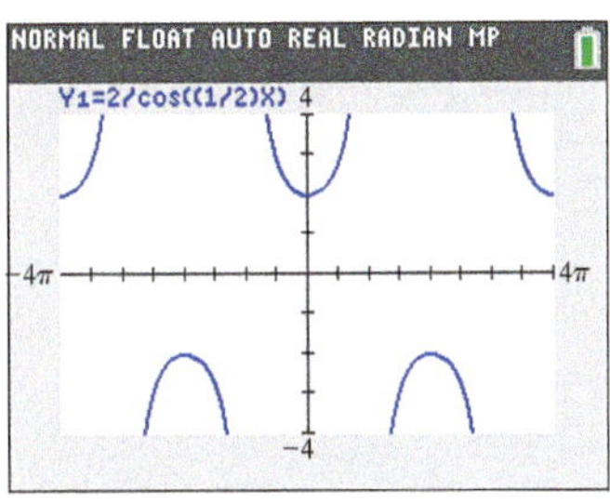

This is a calculator graph of the function in **Example 1.**

Step 2 Sketch the vertical asymptotes as shown in **Figure 62(a).** These occur at x-values for which the guide function equals 0, such as

$$x = -3\pi, \quad x = -\pi, \quad x = \pi, \quad x = 3\pi.$$

Step 3 Sketch the graph of $y = 2\sec\frac{1}{2}x$ by drawing typical U-shaped branches, approaching the asymptotes. See the solid blue graph in **Figure 62(b).**

Now Try Exercise 11.

EXAMPLE 2 Graphing $y = a\csc(x - d)$

Graph $y = \frac{3}{2}\csc\left(x - \frac{\pi}{2}\right)$.

SOLUTION

Step 1 Graph the corresponding reciprocal function

$$y = \frac{3}{2}\sin\left(x - \frac{\pi}{2}\right),$$

shown as a red dashed curve in **Figure 63.**

Step 2 Sketch the vertical asymptotes through the x-intercepts of the graph of $y = \frac{3}{2}\sin\left(x - \frac{\pi}{2}\right)$. These x-values have the form $(2n + 1)\frac{\pi}{2}$, where n is any integer. See the black dashed lines in **Figure 63.**

Step 3 Sketch the graph of $y = \frac{3}{2}\csc\left(x - \frac{\pi}{2}\right)$ by drawing the typical U-shaped branches between adjacent asymptotes. See the solid blue graph in **Figure 63.**

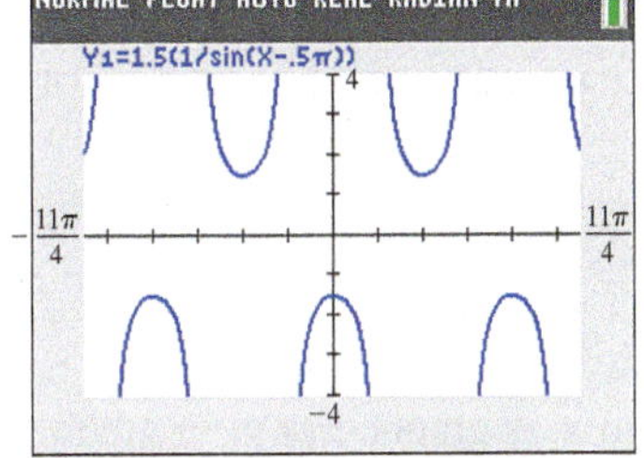

This is a calculator graph of the function in **Example 2.** (The use of decimal equivalents when defining y_1 eliminates the need for some parentheses.)

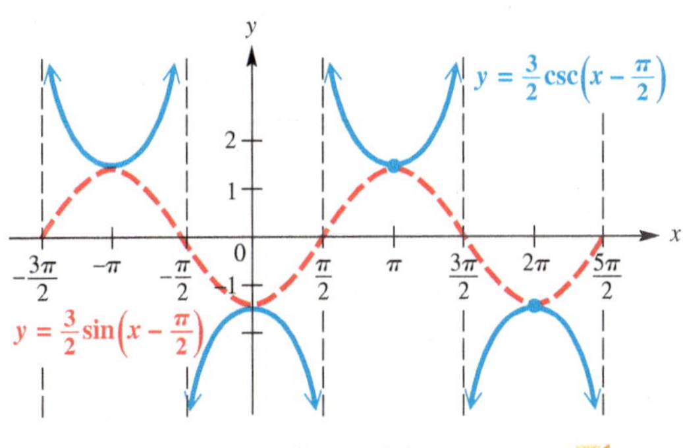

Figure 63

Now Try Exercise 13.

Connecting Graphs with Equations

EXAMPLE 3 Determining an Equation for a Graph

Determine an equation for each graph.

(a)

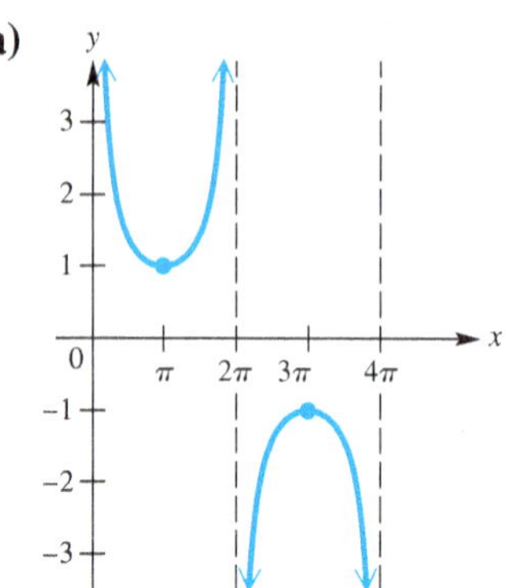

(b)

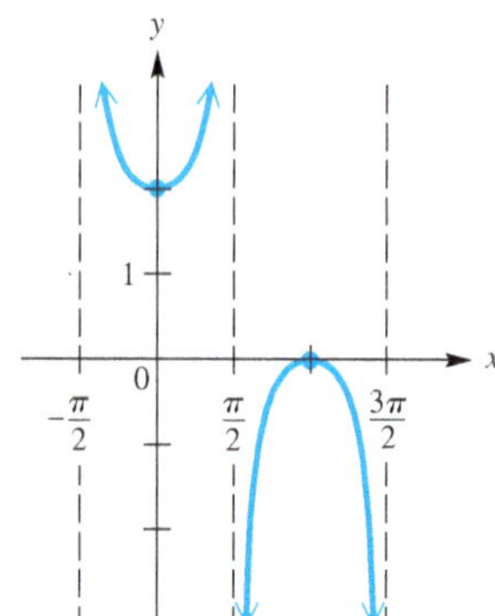

SOLUTION

(a) This graph is that of a cosecant function that is stretched horizontally having period 4π. If $y = \csc bx$, where $b > 0$, then we must have $b = \frac{1}{2}$ and

$$y = \csc \frac{1}{2}x. \qquad \frac{2\pi}{\frac{1}{2}} = 4\pi$$

↑ Horizontal stretch

(b) This is the graph of $y = \sec x$, translated 1 unit up. An equation is

$$y = 1 + \sec x.$$

↑ Vertical translation

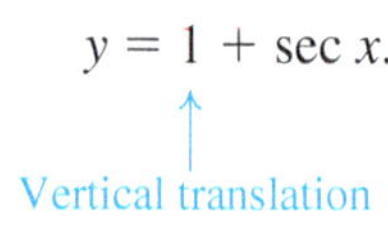

✔ **Now Try Exercises 25 and 27.**

Figure 64

Addition of Ordinates

A function formed by combining two other functions, such as

$$y_3 = y_1 + y_2,$$

has historically been graphed using a method known as **addition of ordinates.** (The x-value of a point is sometimes called its **abscissa,** while its y-value is called its **ordinate.**)

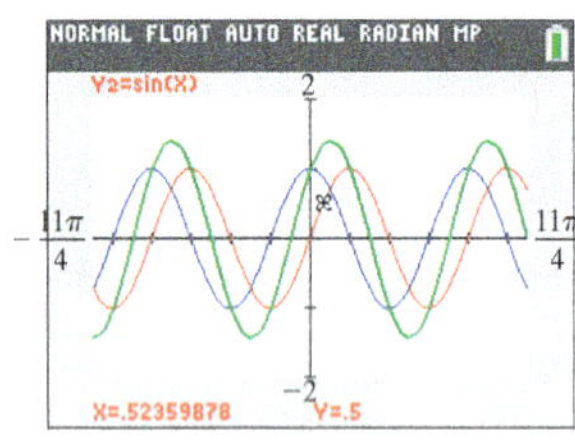

Figure 65

EXAMPLE 4 Illustrating Addition of Ordinates

Use the functions $y_1 = \cos x$ and $y_2 = \sin x$ to illustrate addition of ordinates for

$$y_3 = \cos x + \sin x$$

with the value $\frac{\pi}{6}$ for x.

SOLUTION In **Figures 64–66,** $y_1 = \cos x$ is graphed in blue, $y_2 = \sin x$ is graphed in red, and their sum, $y_1 + y_2 = \cos x + \sin x$, is graphed as y_3 in green. If the ordinates (y-values) for $x = \frac{\pi}{6}$ (approximately 0.52359878) in **Figures 64 and 65** are added, their sum is found in **Figure 66.** Verify that

$$0.8660254 + 0.5 = 1.3660254.$$

(This would occur for *any* value of x.)

Figure 66

✔ **Now Try Exercise 43.**

6.6 Exercises

CONCEPT PREVIEW *Match each description in Column I with the correct value in Column II. Refer to the basic graphs as needed.*

I	II
1. The least positive value k for which $x = k$ is a vertical asymptote for $y = \sec x$	**A.** $\frac{\pi}{2}$
2. The least positive value k for which $x = k$ is a vertical asymptote for $y = \csc x$	**B.** π
3. The least positive value that is in the range of $y = \sec x$	**C.** $-\pi$
4. The greatest negative value that is in the range of $y = \csc x$	**D.** 1
5. The greatest negative value of x for which $\sec x = -1$	**E.** $\frac{3\pi}{2}$
6. The least positive value of x for which $\csc x = -1$	**F.** -1

Concept Check Match each function with its graph from choices A–D.

7. $y = -\csc x$ **8.** $y = -\sec x$ **9.** $y = \sec\left(x - \frac{\pi}{2}\right)$ **10.** $y = \csc\left(x + \frac{\pi}{2}\right)$

A.

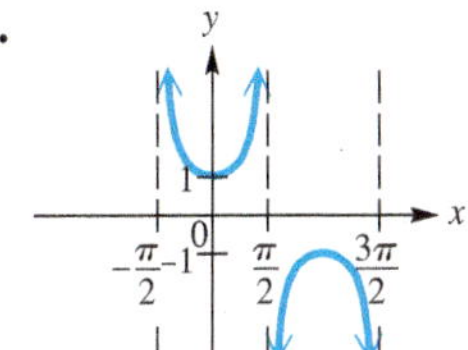

B.

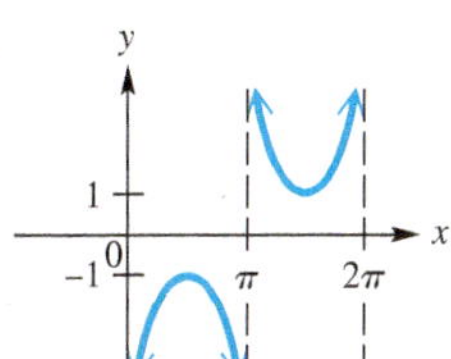

C.

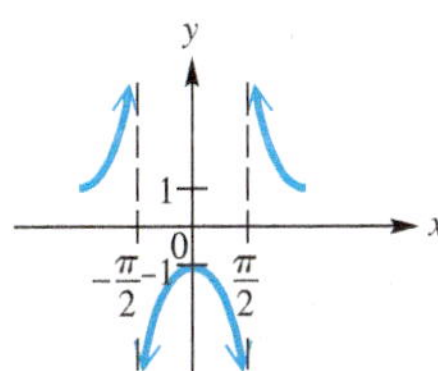

D.

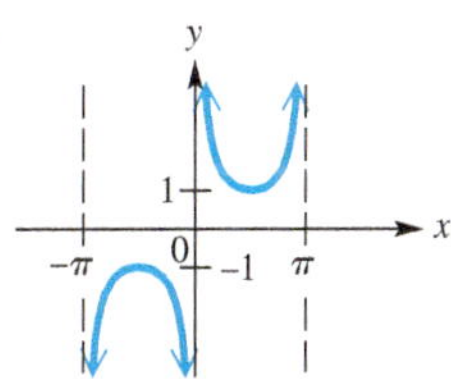

Graph each function over a one-period interval. ***See Examples 1 and 2.***

11. $y = 3 \sec \frac{1}{4}x$

12. $y = -2 \sec \frac{1}{2}x$

13. $y = -\frac{1}{2} \csc\left(x + \frac{\pi}{2}\right)$

14. $y = \frac{1}{2} \csc\left(x - \frac{\pi}{2}\right)$

15. $y = \csc\left(x - \frac{\pi}{4}\right)$

16. $y = \sec\left(x + \frac{3\pi}{4}\right)$

17. $y = \sec\left(x + \frac{\pi}{4}\right)$

18. $y = \csc\left(x + \frac{\pi}{3}\right)$

19. $y = \csc\left(\frac{1}{2}x - \frac{\pi}{4}\right)$

20. $y = \sec\left(\frac{1}{2}x + \frac{\pi}{3}\right)$

21. $y = 2 + 3 \sec(2x - \pi)$

22. $y = 1 - 2 \csc\left(x + \frac{\pi}{2}\right)$

23. $y = 1 - \frac{1}{2} \csc\left(x - \frac{3\pi}{4}\right)$

24. $y = 2 + \frac{1}{4} \sec\left(\frac{1}{2}x - \pi\right)$

Connecting Graphs with Equations *Determine an equation for each graph.* ***See Example 3.***

25.

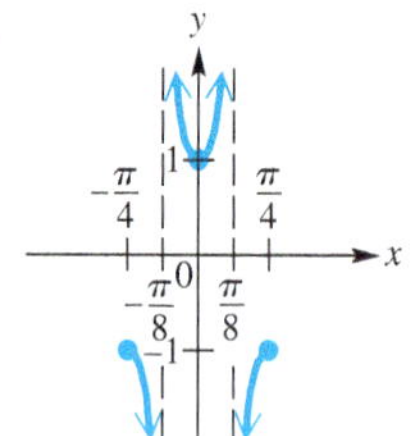

26.

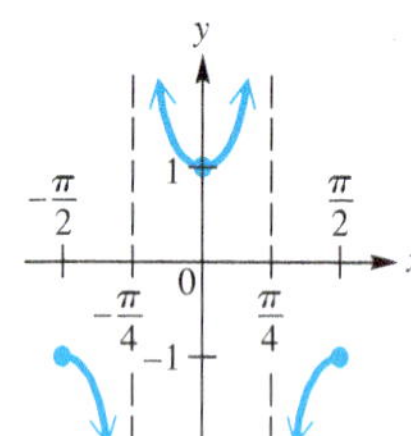

27.

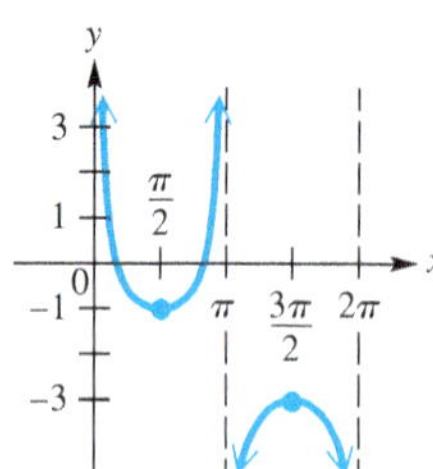

28.

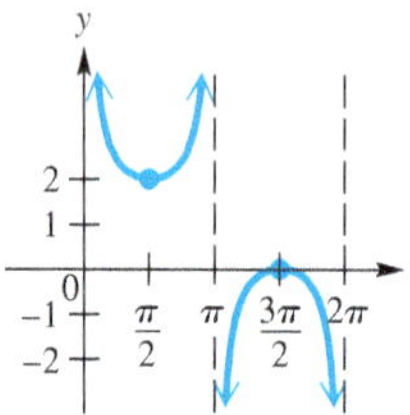

29.

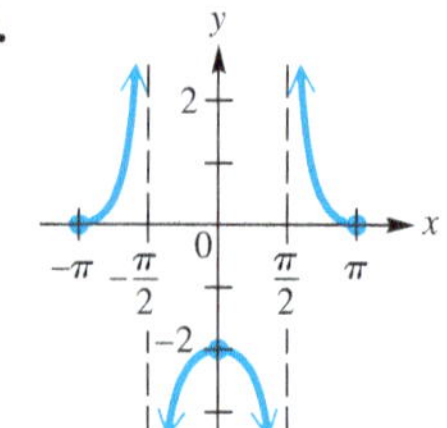

30.

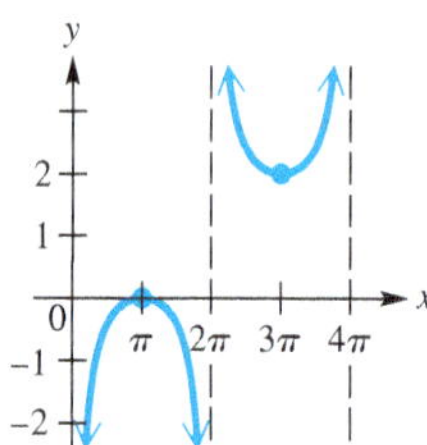

Concept Check *Decide whether each statement is* true *or* false. *If* false, *explain why.*

31. The tangent and secant functions are undefined for the same values.

32. The secant and cosecant functions are undefined for the same values.

33. The graph of $y = \sec x$ in **Figure 58** suggests that $\sec(-x) = \sec x$ for all x in the domain of $\sec x$.

34. The graph of $y = \csc x$ in **Figure 61** suggests that $\csc(-x) = -\csc x$ for all x in the domain of $\csc x$.

Work each problem.

35. ***Concept Check*** If c is any number such that $-1 < c < 1$, then how many solutions does the equation $c = \sec x$ have over the entire domain of the secant function?

36. ***Concept Check*** Consider the function $g(x) = -2 \csc(4x + \pi)$. What is the domain of g? What is its range?

37. Show that $\sec(-x) = \sec x$ by writing $\sec(-x)$ as $\frac{1}{\cos(-x)}$ and then using the relationship between $\cos(-x)$ and $\cos x$.

38. Show that $\csc(-x) = -\csc x$ by writing $\csc(-x)$ as $\frac{1}{\sin(-x)}$ and then using the relationship between $\sin(-x)$ and $\sin x$.

(Modeling) Distance of a Rotating Beacon *The distance a in the figure (repeated from the exercise set in the previous section) is given by*

$$a = 4|\sec 2\pi t|.$$

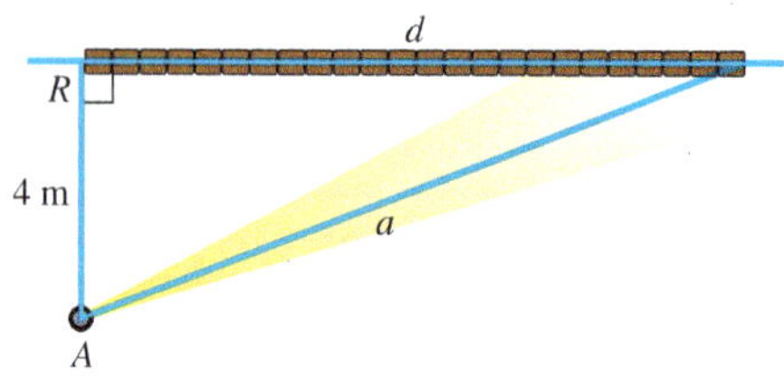

Find the value of a for each time t. Round to the nearest tenth if applicable.

39. $t = 0$ **40.** $t = 0.86$ **41.** $t = 1.24$ **42.** $t = 0.25$

Given y_1 and y_2, define their sum to be

$$y_3 = y_1 + y_2.$$

Evaluate y_1 and y_2 at the given value of x and show that their sum is equal to y_3 evaluated at x. Use the method of addition of ordinates. **See Example 4.**

43. $y_1 = \sin x,\ y_2 = \sin 2x;\quad x = \frac{\pi}{6}$

44. $y_1 = \cos x,\ y_2 = \cos 2x;\quad x = \frac{2\pi}{3}$

45. $y_1 = \tan x,\ y_2 = \sec x;\quad x = \frac{\pi}{4}$

46. $y_1 = \cot x,\ y_2 = \csc x;\quad x = \frac{\pi}{3}$

Summary Exercises on Graphing Circular Functions

These summary exercises provide practice with the various graphing techniques presented in this chapter. Graph each function over a one-period interval.

1. $y = 2 \sin \pi x$

2. $y = 4 \cos \frac{3}{2}x$

3. $y = -2 + \frac{1}{2} \cos \frac{\pi}{4}x$

4. $y = 3 \sec \frac{\pi}{2}x$

5. $y = -4 \csc \frac{1}{2}x$

6. $y = 3 \tan \left(\frac{\pi}{2}x + \pi\right)$

Graph each function over a two-period interval.

7. $y = -5 \sin \frac{x}{3}$

8. $y = 10 \cos \left(\frac{x}{4} + \frac{\pi}{2}\right)$

9. $y = 3 - 4 \sin \left(\frac{5}{2}x + \pi\right)$

10. $y = 2 - \sec[\pi(x - 3)]$

6.7 Harmonic Motion

- Simple Harmonic Motion
- Damped Oscillatory Motion

Simple Harmonic Motion In part A of **Figure 67,** a spring with a weight attached to its free end is in equilibrium (or rest) position. If the weight is pulled down a units and released (part B of the figure), the spring's elasticity causes the weight to rise a units $(a > 0)$ above the equilibrium position, as seen in part C, and then to oscillate about the equilibrium position.

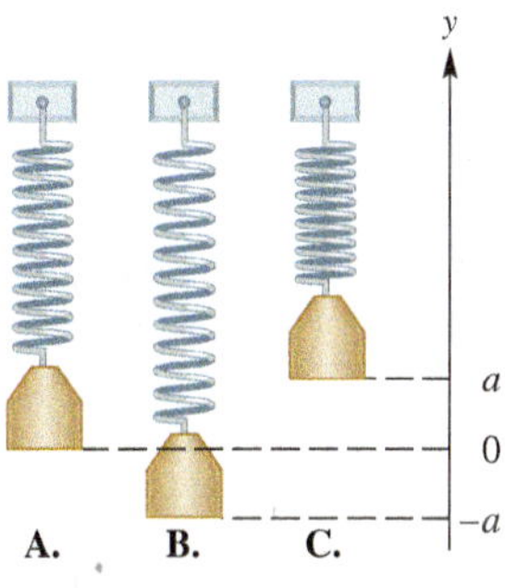

Figure 67

If friction is neglected, this oscillatory motion is described mathematically by a sinusoid. Other applications of this type of motion include sound, electric current, and electromagnetic waves.

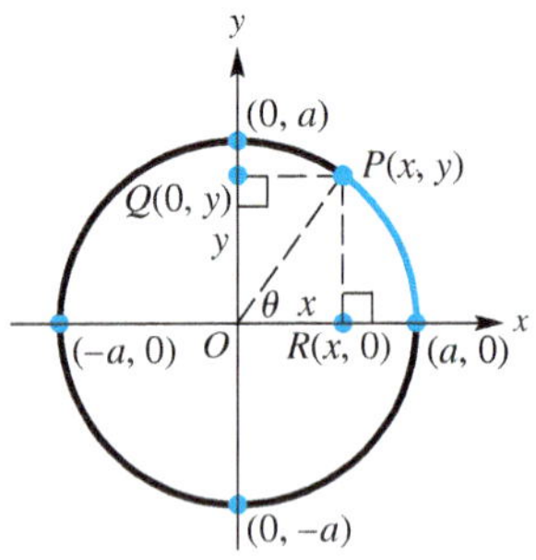

Figure 68

To develop a general equation for such motion, consider **Figure 68.** Suppose the point $P(x, y)$ moves around the circle counterclockwise at a uniform angular speed ω. Assume that at time $t = 0$, P is at $(a, 0)$. The angle swept out by ray OP at time t is given by $\theta = \omega t$. The coordinates of point P at time t are

$$x = a \cos \theta = a \cos \omega t \quad \text{and} \quad y = a \sin \theta = a \sin \omega t.$$

As P moves around the circle from the point $(a, 0)$, the point $Q(0, y)$ oscillates back and forth along the y-axis between the points $(0, a)$ and $(0, -a)$. Similarly, the point $R(x, 0)$ oscillates back and forth between $(a, 0)$ and $(-a, 0)$. This oscillatory motion is **simple harmonic motion.**

The amplitude of the motion is $|a|$, and the period is $\frac{2\pi}{\omega}$. The moving points P and Q or P and R complete one oscillation or cycle per period. The number of cycles per unit of time, called the **frequency,** is the reciprocal of the period, $\frac{\omega}{2\pi}$, where $\omega > 0$.

Simple Harmonic Motion

The position of a point oscillating about an equilibrium position at time t is modeled by either

$$s(t) = a \cos \omega t \quad \text{or} \quad s(t) = a \sin \omega t,$$

where a and ω are constants, with $\omega > 0$. The amplitude of the motion is $|a|$, the period is $\frac{2\pi}{\omega}$, and the frequency is $\frac{\omega}{2\pi}$ oscillations per time unit.

EXAMPLE 1 Modeling the Motion of a Spring

Suppose that an object is attached to a coiled spring such as the one in **Figure 67** on the preceding page. It is pulled down a distance of 5 in. from its equilibrium position and then released. The time for one complete oscillation is 4 sec.

(a) Give an equation that models the position of the object at time t.

(b) Determine the position at $t = 1.5$ sec.

(c) Find the frequency.

SOLUTION

(a) When the object is released at $t = 0$, the distance of the object from the equilibrium position is 5 in. below equilibrium. If $s(t)$ is to model the motion, then $s(0)$ must equal -5. We use

$$s(t) = a \cos \omega t, \quad \text{with } a = -5.$$

We choose the cosine function here because $\cos \omega(0) = \cos 0 = 1$, and $-5 \cdot 1 = -5$. (Had we chosen the sine function, a phase shift would have been required.) Use the fact that the period is 4 to solve for ω.

$$\frac{2\pi}{\omega} = 4 \quad \text{The period is } \frac{2\pi}{\omega}.$$

$$\omega = \frac{\pi}{2} \quad \text{Solve for } \omega.$$

Thus, the motion is modeled by $s(t) = -5 \cos \frac{\pi}{2} t$.

(b) Substitute the given value of t in the equation found in part (a).

$$s(t) = -5 \cos \frac{\pi}{2} t \qquad \text{Equation from part (a)}$$

$$s(1.5) = -5 \cos \left[\frac{\pi}{2}(1.5) \right] \qquad \text{Let } t = 1.5.$$

$$s(1.5) \approx 3.54 \text{ in.} \qquad \text{Use a calculator.}$$

Because $3.54 > 0$, the object is *above* the equilibrium position.

(c) The frequency is the reciprocal of the period, or $\frac{1}{4}$ oscillation per sec.

✔ **Now Try Exercise 7.**

EXAMPLE 2 Analyzing Harmonic Motion

Suppose that an object oscillates according to the model

$$s(t) = 8 \sin 3t,$$

where t is in seconds and $s(t)$ is in feet. Analyze the motion.

SOLUTION The motion is harmonic because the model is

$$s(t) = a \sin \omega t.$$

Because $a = 8$, the object oscillates 8 ft in either direction from its starting point. The period $\frac{2\pi}{3} \approx 2.1$ is the time, in seconds, it takes for one complete oscillation. The frequency is the reciprocal of the period, so the object completes $\frac{3}{2\pi} \approx 0.48$ oscillation per sec.

✔ **Now Try Exercise 17.**

Damped Oscillatory Motion In the example of the stretched spring, we disregard the effect of friction. Friction causes the amplitude of the motion to diminish gradually until the weight comes to rest. In this situation, we say that the motion has been *damped* by the force of friction. Most oscillatory motions are damped. For instance, shock absorbers are put on an automobile in order to damp oscillatory motion. Instead of oscillating up and down for a long while after hitting a bump or pothole, the oscillations of the car are quickly damped out for a smoother ride.

The decrease in amplitude of a **damped oscillatory motion** usually follows the pattern of exponential decay.

EXAMPLE 3 Analyzing Damped Oscillatory Motion

A typical example of damped oscillatory motion is provided by the function

$$s(x) = e^{-x} \cos 2\pi x.$$

(The number $e \approx 2.718$ is the base of the natural logarithm function.) We use x rather than t to match the variable for graphing calculators.

(a) Provide a calculator graph of $y_3 = e^{-x} \cos 2\pi x$, along with the graphs of $y_1 = e^{-x}$ and $y_2 = -e^{-x}$ for $0 \le x \le 3$.

(b) Describe the relationships among the three graphs drawn in part (a).

(c) For what values of x does the graph of y_3 touch the graph of y_1?

(d) For what values of x does the graph of y_3 intersect the x-axis?

SOLUTION

(a) **Figure 69** shows a TI-84 Plus graph of y_1, y_2, and y_3 in the window $[0, 3]$ by $[-1, 1]$.

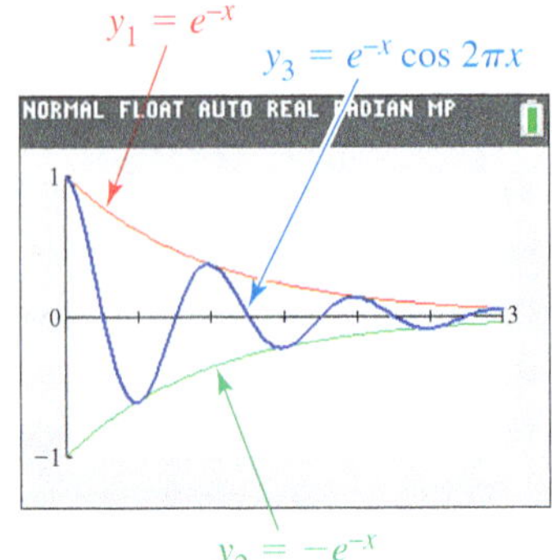

Figure 69

(b) The graph of y_3 is bounded above by the graph of y_1 and below by the graph of y_2. (The graphs of y_1 and y_2 are referred to as **envelopes** for the graph of y_3.)

(c) When $2\pi x = 0, 2\pi, 4\pi$, and 6π, $\cos 2\pi x = 1$. Thus, the value of $e^{-x} \cos 2\pi x$ is the same as the value of e^{-x} when $2\pi x = 0, 2\pi, 4\pi$, and 6π—that is, when $x = 0, 1, 2$, and 3.

(d) When $2\pi x = \frac{\pi}{2}, \frac{3\pi}{2}, \frac{5\pi}{2}, \frac{7\pi}{2}, \frac{9\pi}{2}$, and $\frac{11\pi}{2}$, $\cos 2\pi x = 0$. Thus, the graph of y_3 intersects the x-axis when $x = \frac{1}{4}, \frac{3}{4}, \frac{5}{4}, \frac{7}{4}, \frac{9}{4}$, and $\frac{11}{4}$.

✔ **Now Try Exercise 33.**

6.7 Exercises

CONCEPT PREVIEW *Refer to the equations in the definition of simple harmonic motion in this section, and consider the following equation.*

$$s(t) = 5 \cos 2t, \quad \text{where } t \text{ is time in seconds}$$

Answer each question.

1. What is the amplitude of this motion?

2. What is the period of this motion?

3. What is the frequency?

4. What is $s(0)$?

5. What is $s\left(\frac{\pi}{2}\right)$?

6. What is the range of the graph of this function? (*Hint*: See the answers to **Exercises 4 and 5.**)

(Modeling) Solve each problem. ***See Examples 1 and 2.***

7. *Spring Motion* A weight is attached to a coiled spring, as in **Figure 67** (repeated here). It is pulled down a distance of 4 units from its equilibrium position and then released. The time for one complete oscillation is 3 sec.

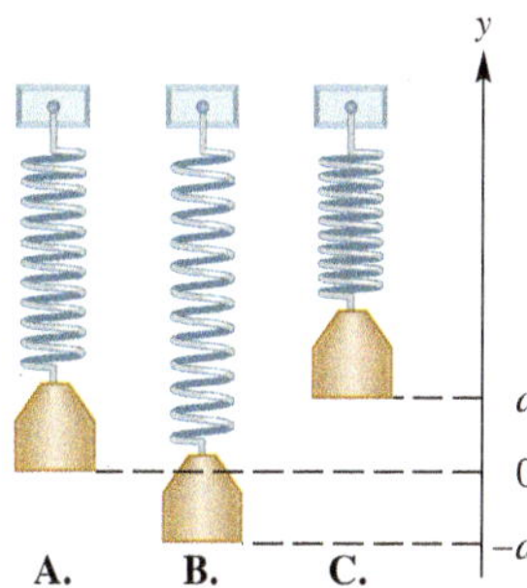

(a) Give an equation that models the position of the weight at time t.

(b) Determine the position at $t = 1.25$ sec to the nearest hundredth.

(c) Find the frequency.

8. *Spring Motion* Repeat **Exercise 7,** but assume that the object is pulled down a distance of 6 units and that the time for one complete oscillation is 4 sec.

9. *Voltage of an Electrical Circuit* The voltage E in an electrical circuit is modeled by

$$E = 5 \cos 120\pi t,$$

where t is time measured in seconds.

(a) Find the amplitude and the period.

(b) Find the frequency.

(c) Find E, to the nearest thousandth, when $t = 0, 0.03, 0.06, 0.09, 0.12$.

(d) Graph E for $0 \leq t \leq \frac{1}{30}$.

10. *Voltage of an Electrical Circuit* For another electrical circuit, the voltage E is modeled by

$$E = 3.8 \cos 40\pi t,$$

where t is time measured in seconds.

(a) Find the amplitude and the period.

(b) Find the frequency.

(c) Find E, to the nearest thousandth, when $t = 0.02, 0.04, 0.08, 0.12, 0.14$.

(d) Graph E for $0 \leq t \leq \frac{1}{20}$.

11. *Particle Movement* Write the equation and then determine the amplitude, period, and frequency of the simple harmonic motion of a particle moving uniformly around a circle of radius 2 units, with the given angular speed.

(a) 2 radians per sec **(b)** 4 radians per sec

12. *Spring Motion* The height attained by a weight attached to a spring set in motion is

$$s(t) = -4 \cos 8\pi t \text{ inches after } t \text{ seconds.}$$

(a) Find the maximum height that the weight rises above the equilibrium position of $s(t) = 0$.

(b) When does the weight first reach its maximum height if $t \geq 0$?

(c) What are the frequency and the period?

13. *Pendulum Motion* What are the period P and frequency T of oscillation of a pendulum of length $\frac{1}{2}$ ft? $\left(Hint: P = 2\pi\sqrt{\frac{L}{32}}\right.$, where L is the length of the pendulum in feet and the period P is in seconds.$\left.\right)$

14. *Pendulum Motion* In **Exercise 13,** how long should the pendulum be to have a period of 1 sec?

15. *Spring Motion* The formula for the up and down motion of a weight on a spring is given by

$$s(t) = a \sin \sqrt{\frac{k}{m}}t.$$

If the spring constant k is 4, what mass m must be used to produce a period of 1 sec?

16. *Spring Motion* (See **Exercise 15.**) A spring with spring constant $k = 2$ and a 1-unit mass m attached to it is stretched and then allowed to come to rest.

 (a) If the spring is stretched $\frac{1}{2}$ ft and released, what are the amplitude, period, and frequency of the resulting oscillatory motion?

 (b) What is the equation of the motion?

17. *Spring Motion* The position of a weight attached to a spring is

$$s(t) = -5 \cos 4\pi t \text{ inches after } t \text{ seconds.}$$

 (a) Find the maximum height that the weight rises above the equilibrium position of $s(t) = 0$.

 (b) What are the frequency and period?

 (c) When does the weight first reach its maximum height?

 (d) Calculate and interpret $s(1.3)$ to the nearest tenth.

18. *Spring Motion* The position of a weight attached to a spring is

$$s(t) = -4 \cos 10t \text{ inches after } t \text{ seconds.}$$

 (a) Find the maximum height that the weight rises above the equilibrium position of $s(t) = 0$.

 (b) What are the frequency and period?

 (c) When does the weight first reach its maximum height?

 (d) Calculate and interpret $s(1.466)$.

19. *Spring Motion* A weight attached to a spring is pulled down 3 in. below the equilibrium position.

 (a) Assuming that the frequency is $\frac{6}{\pi}$ cycles per sec, determine a model that gives the position of the weight at time t seconds.

 (b) What is the period?

20. *Spring Motion* A weight attached to a spring is pulled down 2 in. below the equilibrium position.

 (a) Assuming that the period is $\frac{1}{3}$ sec, determine a model that gives the position of the weight at time t seconds.

 (b) What is the frequency?

(Modeling) Springs *A weight on a spring has initial position* $s(0)$ *and period* P.

(a) *To model displacement of the weight, find a function s given by*

$$s(t) = a \cos \omega t.$$

(b) *Evaluate* $s(1)$. *Is the weight moving* upward, downward, *or* neither *when* $t = 1$? *Support the results graphically or numerically.*

21. $s(0) = 2$ in.; $P = 0.5$ sec

22. $s(0) = 5$ in.; $P = 1.5$ sec

23. $s(0) = -3$ in.; $P = 0.8$ sec

24. $s(0) = -4$ in.; $P = 1.2$ sec

(Modeling) Music *A note on a piano has given frequency F. Suppose the maximum displacement at the center of the piano wire is given by* $s(0)$. *Find constants a and* ω *so that the equation*

$$s(t) = a \cos \omega t$$

models this displacement. Graph s in the viewing window $[0, 0.05]$ *by* $[-0.3, 0.3]$.

25. $F = 27.5$; $s(0) = 0.21$

26. $F = 110$; $s(0) = 0.11$

27. $F = 55$; $s(0) = 0.14$

28. $F = 220$; $s(0) = 0.06$

(Modeling) Spring Motion *Consider the spring in the figure repeated with* ***Exercise 7,*** *but assume that because of friction and other resistive forces, the amplitude is decreasing over time, and that t seconds after the spring is released, its position in inches is given by the function*

$$s(t) = -11e^{-0.2t} \cos 0.5\pi t.$$

29. How far was the weight pulled down from the equilibrium position before it was released?

30. How far, to the nearest hundredth of an inch, is the weight from the equilibrium position after 6 sec?

31. Graph the function on the interval $[0, 12]$ by $[-12, 12]$, and determine the values for which the graph intersects the horizontal axis.

32. How many complete oscillations will the graph make during 12 sec?

(Modeling) Damped Oscillatory Motion *Work each problem.* ***See Example 3.***

33. Consider the damped oscillatory function

$$s(x) = 5e^{-0.3x} \cos \pi x.$$

(a) Graph the function $y_3 = 5e^{-0.3x} \cos \pi x$ in the window $[0, 3]$ by $[-5, 5]$.

(b) The graph of which function is the upper envelope of the graph of y_3?

(c) For what values of x does the graph of y_3 touch the graph of the function found in part (b)?

34. Consider the damped oscillatory function

$$s(x) = 10e^{-x} \sin 2\pi x.$$

(a) Graph the function $y_3 = 10e^{-x} \sin 2\pi x$ in the window $[0, 3]$ by $[-10, 10]$.

(b) The graph of which function is the lower envelope of the graph of y_3?

(c) For what values of x does the graph of y_3 touch the graph of the function found in part (b)?

Chapter 6 Test Prep

Key Terms

6.1 radian
circumference
latitude
sector of a circle
subtend
degree of curvature
nautical mile
statute mile
longitude

6.2 unit circle
circular functions
reference arc
linear speed v
angular speed ω
unit fraction

6.3 periodic function
period
sine wave (sinusoid)
amplitude

6.4 phase shift
argument

6.5 vertical asymptote

6.6 addition of ordinates

6.7 simple harmonic motion
frequency
damped oscillatory motion
envelope

Quick Review

Concepts	Examples

6.1 Radian Measure

An angle with its vertex at the center of a circle that intercepts an arc on the circle equal in length to the radius of the circle has a measure of **1 radian.**

$$180° = \pi \text{ radians} \quad \text{Degree/Radian Relationship}$$

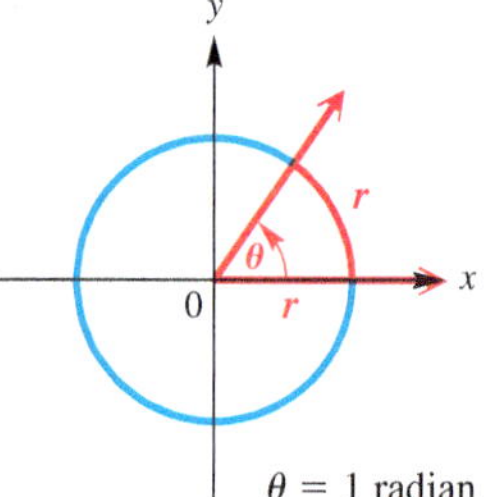

Converting between Degrees and Radians

- Multiply a degree measure by $\frac{\pi}{180}$ radian and simplify to convert to radians.
- Multiply a radian measure by $\frac{180°}{\pi}$ and simplify to convert to degrees.

Convert 135° to radians.

$$135° = 135\left(\frac{\pi}{180} \text{ radian}\right) = \frac{3\pi}{4} \text{ radians}$$

Convert $-\frac{5\pi}{3}$ radians to degrees.

$$-\frac{5\pi}{3} \text{ radians} = -\frac{5\pi}{3}\left(\frac{180°}{\pi}\right) = -300°$$

Arc Length

The length s of the arc intercepted on a circle of radius r by a central angle of measure θ radians is given by the product of the radius and the radian measure of the angle.

$$s = r\theta, \quad \text{where } \theta \text{ is in radians}$$

Find the central angle θ in the figure.

$$\theta = \frac{s}{r} = \frac{3}{4} \text{ radian}$$

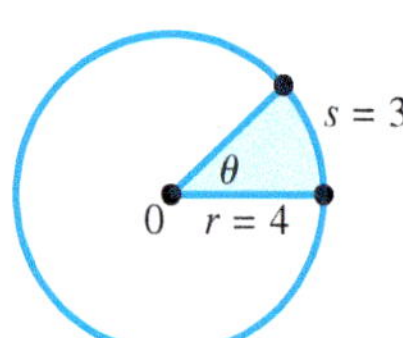

Area of a Sector

The area $\mathscr{A}$ of a sector of a circle of radius r and central angle θ is given by the following formula.

$$\mathscr{A} = \frac{1}{2}r^2\theta, \quad \text{where } \theta \text{ is in radians}$$

Find the area $\mathscr{A}$ of the sector in the figure above.

$$\mathscr{A} = \frac{1}{2}(4)^2\left(\frac{3}{4}\right) = 6 \text{ sq units}$$

Concepts	Examples

6.2 The Unit Circle and Circular Functions

Circular Functions

Start at the point $(1, 0)$ on the unit circle $x^2 + y^2 = 1$ and measure off an arc of length $|s|$ along the circle, moving counterclockwise if s is positive and clockwise if s is negative. Let the endpoint of the arc be at the point (x, y). The six circular functions of s are defined as follows. (Assume that no denominators are 0.)

$$\sin s = y \qquad \cos s = x \qquad \tan s = \frac{y}{x}$$

$$\csc s = \frac{1}{y} \qquad \sec s = \frac{1}{x} \qquad \cot s = \frac{x}{y}$$

Use the unit circle to find each value.

$$\sin \frac{5\pi}{6} = \frac{1}{2}$$

$$\cos \frac{3\pi}{2} = 0$$

$$\tan \frac{\pi}{4} = \frac{\frac{\sqrt{2}}{2}}{\frac{\sqrt{2}}{2}} = 1$$

$$\csc \frac{7\pi}{4} = \frac{1}{-\frac{\sqrt{2}}{2}} = -\sqrt{2}$$

$$\sec \frac{7\pi}{6} = \frac{1}{-\frac{\sqrt{3}}{2}} = -\frac{2\sqrt{3}}{3}$$

$$\cot \frac{\pi}{3} = \frac{\frac{1}{2}}{\frac{\sqrt{3}}{2}} = \frac{\sqrt{3}}{3}$$

$$\sin 0 = 0$$

$$\cos \frac{\pi}{2} = 0$$

The Unit Circle

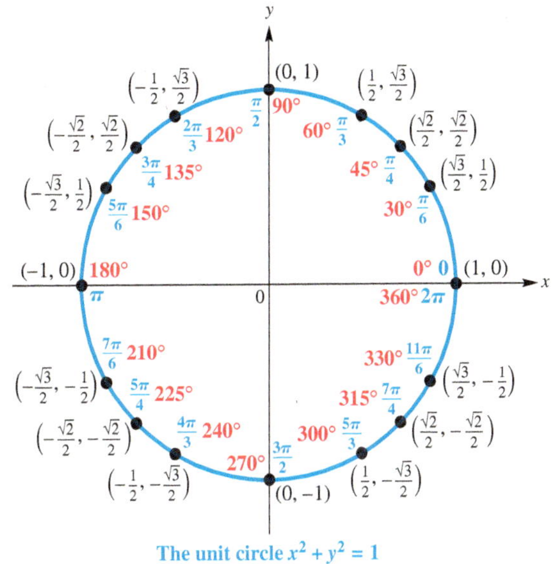

The unit circle $x^2 + y^2 = 1$

Find the exact value of s in $\left[0, \frac{\pi}{2}\right]$ if $\cos s = \frac{\sqrt{3}}{2}$.

In $\left[0, \frac{\pi}{2}\right]$, the arc length $s = \frac{\pi}{6}$ is associated with the point $\left(\frac{\sqrt{3}}{2}, \frac{1}{2}\right)$. The first coordinate is

$$\cos s = \cos \frac{\pi}{6} = \frac{\sqrt{3}}{2}.$$

Thus we have $s = \frac{\pi}{6}$.

Formulas for Angular and Linear Speed

Angular Speed ω	Linear Speed v
$\omega = \frac{\theta}{t}$ (ω in radians per unit time t, θ in radians)	$v = \frac{s}{t}$ $v = \frac{r\theta}{t}$ $v = r\omega$

A belt runs a machine pulley of radius 8 in. at 60 revolutions per min.

(a) Find the angular speed ω in radians per minute.

$$\omega = \frac{60 \text{ revolutions}}{1 \text{ min}} \cdot \frac{2\pi \text{ radians}}{1 \text{ revolution}}$$

$$\omega = 120\pi \text{ radians per min}$$

(b) Find the linear speed v in inches per minute.

$$v = r\omega$$

$$v = 8(120\pi)$$

$$v = 960\pi \text{ in. per min}$$

Concepts	Examples

6.3 Graphs of the Sine and Cosine Functions

6.4 Translations of the Graphs of the Sine and Cosine Functions

Sine and Cosine Functions

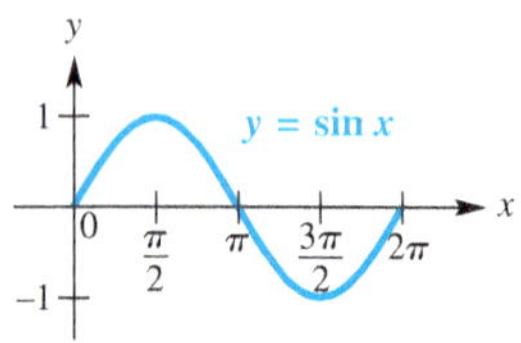

Domain: $(-\infty, \infty)$
Range: $[-1, 1]$
Amplitude: 1
Period: 2π

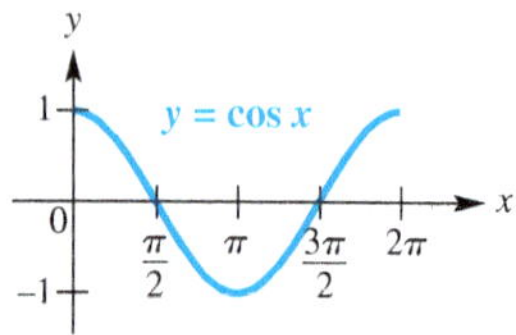

Domain: $(-\infty, \infty)$
Range: $[-1, 1]$
Amplitude: 1
Period: 2π

The graph of

$y = c + a\sin[b(x - d)]$ or $y = c + a\cos[b(x - d)]$,

with $b > 0$, has the following characteristics.

1. amplitude $|a|$
2. period $\frac{2\pi}{b}$
3. vertical translation c units up if $c > 0$ or $|c|$ units down if $c < 0$
4. phase shift d units to the right if $d > 0$ or $|d|$ units to the left if $d < 0$

Graph $y = 1 + \sin 3x$.

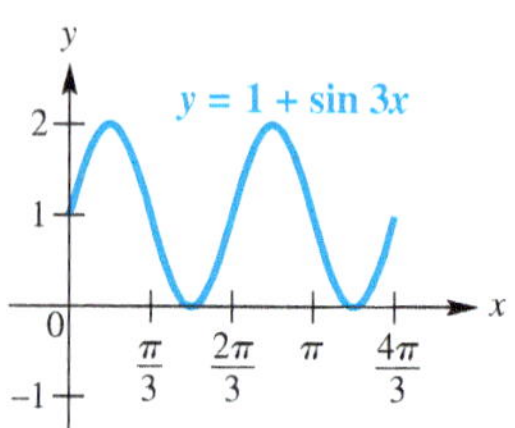

amplitude: 1
period: $\frac{2\pi}{3}$
vertical translation: 1 unit up
domain: $(-\infty, \infty)$
range: $[0, 2]$

Graph $y = -2\cos\left(x + \frac{\pi}{2}\right)$.

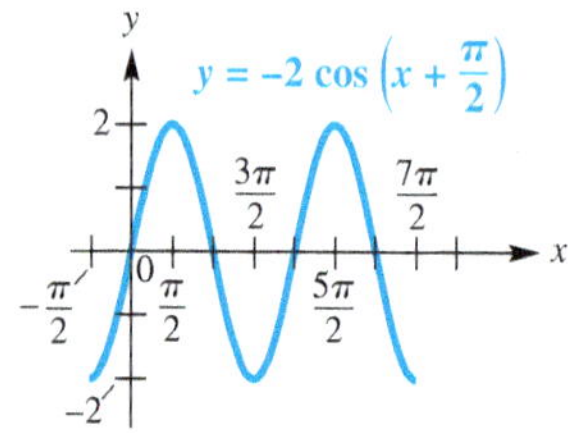

amplitude: 2
period: 2π
phase shift: $\frac{\pi}{2}$ units left
domain: $(-\infty, \infty)$
range: $[-2, 2]$

6.5 Graphs of the Tangent and Cotangent Functions

Tangent and Cotangent Functions

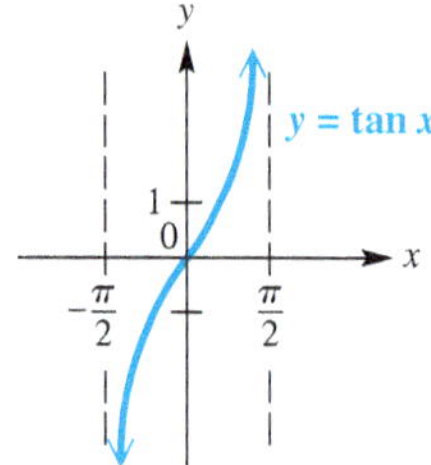

Domain: $\left\{x \,\middle|\, x \neq (2n + 1)\frac{\pi}{2},\right.$ where n is any integer$\left.\right\}$
Range: $(-\infty, \infty)$
Period: π

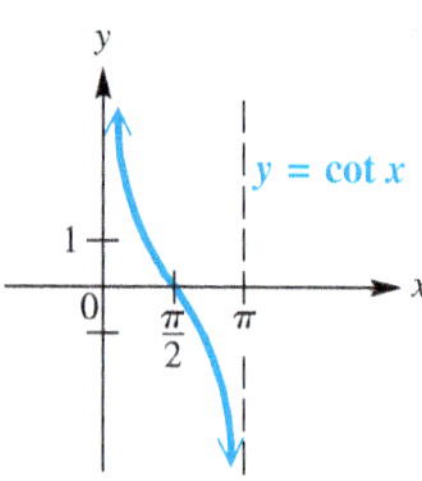

Domain: $\{x \,|\, x \neq n\pi,$ where n is any integer$\}$
Range: $(-\infty, \infty)$
Period: π

Graph $y = 2\tan x$ over a one-period interval.

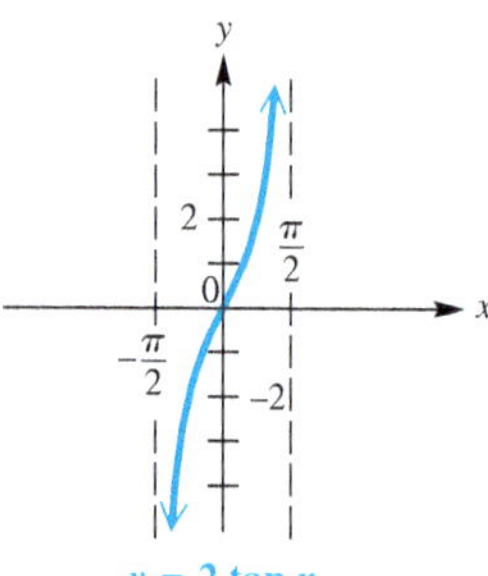

period: π
domain: $\left\{x \,\middle|\, x \neq (2n + 1)\frac{\pi}{2},\right.$ where n is any integer$\left.\right\}$
range: $(-\infty, \infty)$

Concepts	Examples

6.6 Graphs of the Secant and Cosecant Functions

Secant and Cosecant Functions

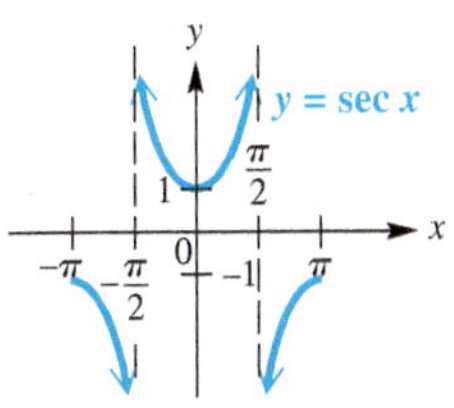

y = csc x

Domain: $\left\{x \mid x \neq (2n+1)\frac{\pi}{2},\right.$ where n is any integer$\left.\right\}$

Range: $(-\infty, -1] \cup [1, \infty)$

Period: 2π

Domain: $\{x \mid x \neq n\pi,$ where n is any integer$\}$

Range: $(-\infty, -1] \cup [1, \infty)$

Period: 2π

Graph $y = \sec\left(x + \frac{\pi}{4}\right)$ over a one-period interval.

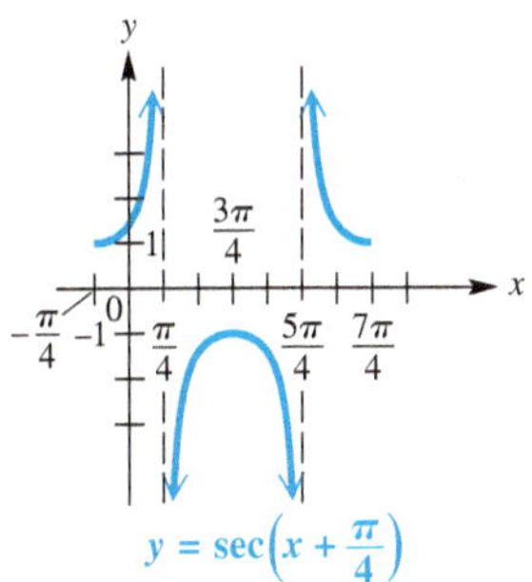

period: 2π

phase shift: $\frac{\pi}{4}$ unit left

domain: $\left\{x \mid x \neq \frac{\pi}{4} + n\pi,\right.$ where n is any integer$\left.\right\}$

range: $(-\infty, -1] \cup [1, \infty)$

6.7 Harmonic Motion

Simple Harmonic Motion

The position of a point oscillating about an equilibrium position at time t is modeled by either

$$s(t) = a \cos \omega t \quad \text{or} \quad s(t) = a \sin \omega t,$$

where a and ω are constants, with $\omega > 0$. The amplitude of the motion is $|a|$, the period is $\frac{2\pi}{\omega}$, and the frequency is $\frac{\omega}{2\pi}$ oscillations per time unit.

A spring oscillates according to

$$s(t) = -5 \cos 6t,$$

where t is in seconds and $s(t)$ is in inches. Find the amplitude, period, and frequency.

$$\text{amplitude} = |-5| = 5 \text{ in.} \qquad \text{period} = \frac{2\pi}{6} = \frac{\pi}{3} \text{ sec}$$

$$\text{frequency} = \frac{3}{\pi} \text{ oscillation per sec}$$

Chapter 6 Review Exercises

Concept Check *Work each problem.*

1. What is the meaning of "an angle with measure 2 radians"?

2. Consider each angle in standard position having the given radian measure. In what quadrant does the terminal side lie?

(a) 3 **(b)** 4 **(c)** -2 **(d)** 7

3. Find three angles coterminal with an angle of 1 radian.

4. Give an expression that generates all angles coterminal with an angle of $\frac{\pi}{6}$ radian. Let n represent any integer.

Convert each degree measure to radians. Leave answers as multiples of π.

5. $45°$ **6.** $120°$ **7.** $175°$ **8.** $330°$ **9.** $800°$ **10.** $1020°$

Convert each radian measure to degrees.

11. $\frac{5\pi}{4}$ **12.** $\frac{9\pi}{10}$ **13.** $\frac{8\pi}{3}$ **14.** $\frac{6\pi}{5}$ **15.** $-\frac{11\pi}{18}$ **16.** $-\frac{21\pi}{5}$

Suppose the tip of the minute hand of a clock is 2 in. *from the center of the clock. For each duration, determine the distance traveled by the tip of the minute hand. Leave answers as multiples of π.*

17. 15 min **18.** 20 min

19. 3 hr **20.** 8 hr

Solve each problem. Use a calculator as necessary.

21. ***Arc Length*** The radius of a circle is 15.2 cm. Find the length of an arc of the circle intercepted by a central angle of $\frac{3\pi}{4}$ radians.

22. ***Arc Length*** Find the length of an arc intercepted by a central angle of 0.769 radian on a circle with radius 11.4 cm.

23. ***Angle Measure*** Find the measure (in degrees) of a central angle that intercepts an arc of length 7.683 cm in a circle of radius 8.973 cm.

24. ***Angle Measure*** Find the measure (in radians) of a central angle whose sector has area $\frac{50\pi}{3}$ cm^2 in a circle of radius 10 cm.

25. ***Area of a Sector*** Find the area of a sector of a circle having a central angle of $21° 40'$ in a circle of radius 38.0 m.

26. ***Area of a Sector*** A central angle of $\frac{7\pi}{4}$ radians forms a sector of a circle. Find the area of the sector if the radius of the circle is 28.69 in.

Distance between Cities *Assume that the radius of Earth is* 6400 km.

27. Find the distance in kilometers between cities on a north-south line that are on latitudes 28° N and 12° S, respectively.

28. Two cities on the equator have longitudes of 72° E and 35° W, respectively. Find the distance between the cities.

Concept Check *Find the measure of each central angle θ (in radians) and the area of each sector.*

29.

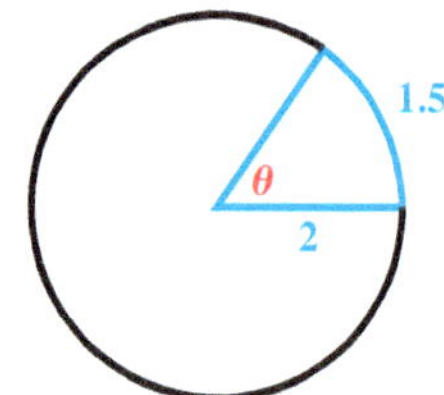

30.

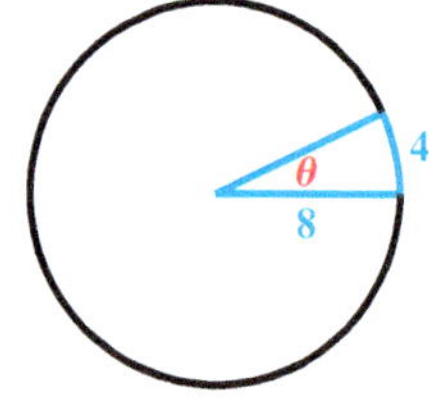

Find each exact function value.

31. $\tan \frac{\pi}{3}$ **32.** $\cos \frac{2\pi}{3}$ **33.** $\sin\left(-\frac{5\pi}{6}\right)$

34. $\tan\left(-\frac{7\pi}{3}\right)$ **35.** $\csc\left(-\frac{11\pi}{6}\right)$ **36.** $\cot(-13\pi)$

Find a calculator approximation to four decimal places for each circular function value.

37. sin 1.0472 **38.** tan 1.2275 **39.** $\cos(-0.2443)$

40. cot 3.0543 **41.** sec 7.3159 **42.** csc 4.8386

Find the approximate value of s, to four decimal places, in the interval $\left[0, \frac{\pi}{2}\right]$ that makes each statement true.

43. $\cos s = 0.9250$ **44.** $\tan s = 4.0112$ **45.** $\sin s = 0.4924$

46. $\csc s = 1.2361$ **47.** $\cot s = 0.5022$ **48.** $\sec s = 4.5600$

Find the exact value of s in the given interval that has the given circular function value.

49. $\left[0, \frac{\pi}{2}\right]$; $\cos s = \frac{\sqrt{2}}{2}$ **50.** $\left[\frac{\pi}{2}, \pi\right]$; $\tan s = -\sqrt{3}$

51. $\left[\pi, \frac{3\pi}{2}\right]$; $\sec s = -\frac{2\sqrt{3}}{3}$ **52.** $\left[\frac{3\pi}{2}, 2\pi\right]$; $\sin s = -\frac{1}{2}$

Suppose that point P is on a circle with radius r, and ray OP is rotating with angular speed ω. Use the given values of r, ω, and t to do the following.

(a) Find the angle generated by P in time t.

(b) Find the distance traveled by P along the circle in time t.

(c) Find the linear speed of P.

53. $r = 15$ cm, $\omega = \frac{2\pi}{3}$ radians per sec, $t = 30$ sec

54. $r = 45$ ft, $\omega = \frac{\pi}{36}$ radian per min, $t = 12$ min

Solve each problem.

55. ***Linear Speed of a Flywheel*** Find the linear speed of a point on the edge of a flywheel of radius 7 cm if the flywheel is rotating 90 times per sec.

56. ***Angular Speed of a Ferris Wheel*** A Ferris wheel has radius 25 ft. A person takes a seat, and then the wheel turns $\frac{5\pi}{6}$ radians.

(a) How far is the person above the ground to the nearest foot?

(b) If it takes 30 sec for the wheel to turn $\frac{5\pi}{6}$ radians, what is the angular speed of the wheel?

Figure A

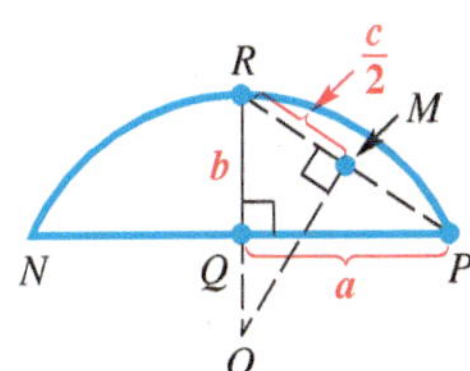

Figure B

57. ***(Modeling) Archaeology*** An archaeology professor believes that an unearthed fragment is a piece of the edge of a circular ceremonial plate and uses a formula that will give the radius of the original plate using measurements from the fragment, shown in **Figure A.** Measurements are in inches.

In **Figure B,** a is $\frac{1}{2}$ the length of chord NP, and b is the distance from the midpoint of chord NP to the circle. According to the formula, the radius r of the circle, OR, is given by

$$r = \frac{a^2 + b^2}{2b}.$$

What is the radius of the original plate from which the fragment came?

58. *(Modeling) Phase Angle of the Moon* Because the moon orbits Earth, we observe different phases of the moon during the period of a month. In the figure, t is the **phase angle.**

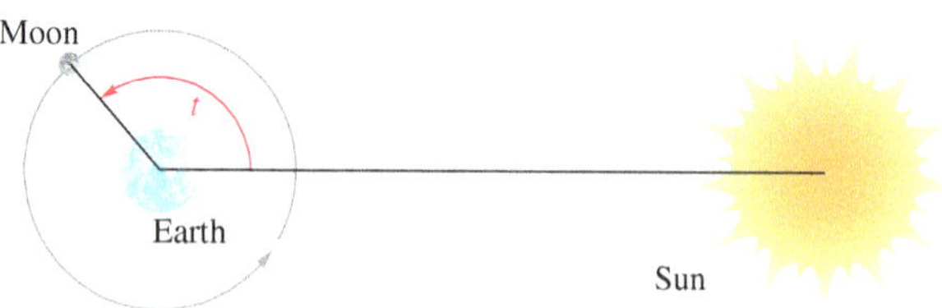

The **phase** F of the moon is modeled by

$$F(t) = \frac{1}{2}(1 - \cos t)$$

and gives the fraction of the moon's face that is illuminated by the sun. (*Source:* Duffet-Smith, P., *Practical Astronomy with Your Calculator,* Cambridge University Press.) Evaluate each expression and interpret the result.

(a) $F(0)$ **(b)** $F\left(\frac{\pi}{2}\right)$ **(c)** $F(\pi)$ **(d)** $F\left(\frac{3\pi}{2}\right)$

59. *Concept Check* Which one of the following statements is true about the graph of $y = 4 \sin 2x$?

A. It has amplitude 2 and period $\frac{\pi}{2}$.
B. It has amplitude 4 and period π.
C. Its range is $[0, 4]$.
D. Its range is $[-4, 0]$.

60. *Concept Check* Which one of the following statements is false about the graph of $y = -3 \cos \frac{1}{2}x$?

A. Its range is $[-3, 3]$.
B. Its domain is $(-\infty, \infty)$.
C. Its amplitude is 3, and its period is 4π.
D. Its amplitude is -3, and its period is π.

For each function, give the amplitude, period, vertical translation, and phase shift, as applicable.

61. $y = 2 \sin x$ **62.** $y = \tan 3x$ **63.** $y = -\frac{1}{2} \cos 3x$

64. $y = 2 \sin 5x$ **65.** $y = 1 + 2 \sin \frac{1}{4}x$ **66.** $y = 3 - \frac{1}{4} \cos \frac{2}{3}x$

67. $y = 3 \cos\left(x + \frac{\pi}{2}\right)$ **68.** $y = -\sin\left(x - \frac{3\pi}{4}\right)$ **69.** $y = \frac{1}{2} \csc\left(2x - \frac{\pi}{4}\right)$

70. $y = 2 \sec(\pi x - 2\pi)$ **71.** $y = \frac{1}{3} \tan\left(3x - \frac{\pi}{3}\right)$ **72.** $y = \cot\left(\frac{x}{2} + \frac{3\pi}{4}\right)$

Concept Check Identify the circular function that satisfies each description.

73. period is π; x-intercepts have x-values of the form $n\pi$, where n is any integer

74. period is 2π; graph passes through the origin

75. period is 2π; graph passes through the point $\left(\frac{\pi}{2}, 0\right)$

76. period is 2π; domain is $\{x \mid x \neq n\pi$, where n is any integer$\}$

77. period is π; function is decreasing on the interval $(0, \pi)$

78. period is 2π; has vertical asymptotes of the form $x = (2n + 1)\frac{\pi}{2}$, where n is any integer

Graph each function over a one-period interval.

79. $y = 3 \sin x$

80. $y = \frac{1}{2} \sec x$

81. $y = -\tan x$

82. $y = -2 \cos x$

83. $y = 2 + \cot x$

84. $y = -1 + \csc x$

85. $y = \sin 2x$

86. $y = \tan 3x$

87. $y = 3 \cos 2x$

88. $y = \frac{1}{2} \cot 3x$

89. $y = \cos\left(x - \frac{\pi}{4}\right)$

90. $y = \tan\left(x - \frac{\pi}{2}\right)$

91. $y = \sec\left(2x + \frac{\pi}{3}\right)$

92. $y = \sin\left(3x + \frac{\pi}{2}\right)$

93. $y = 1 + 2 \cos 3x$

94. $y = -1 - 3 \sin 2x$

95. $y = 2 \sin \pi x$

96. $y = -\frac{1}{2} \cos(\pi x - \pi)$

(Modeling) Monthly Temperatures *A set of temperature data (in °F) is given for a particular location.* (**Source:** *www.weatherbase.com*)

(a) Plot the data over a two-year interval.

(b) Use sine regression to determine a model for the two-year interval. Let $x = 1$ represent January of the first year.

(c) Graph the equation from part (b) together with the data on the same coordinate axes.

97. Average Monthly Temperature, Auckland, New Zealand

Jan	Feb	Mar	Apr	May	Jun	Jul	Aug	Sept	Oct	Nov	Dec
67.6	68.5	65.8	61.3	57.2	53.2	51.6	52.9	55.4	58.1	61.2	64.9

98. Average Low Temperature, Auckland, New Zealand

Jan	Feb	Mar	Apr	May	Jun	Jul	Aug	Sept	Oct	Nov	Dec
60.8	61.7	58.8	54.9	51.1	47.1	45.5	46.8	49.5	52.2	55.0	58.8

Connecting Graphs with Equations *Determine the simplest form of an equation for each graph. Choose $b > 0$, and include no phase shifts.*

99.

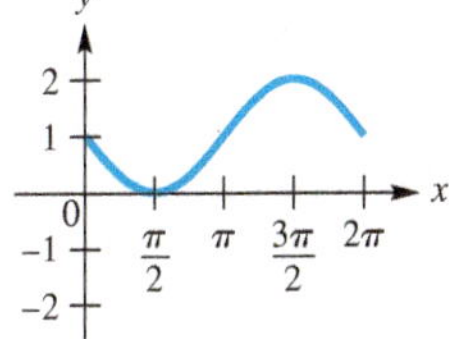

100.

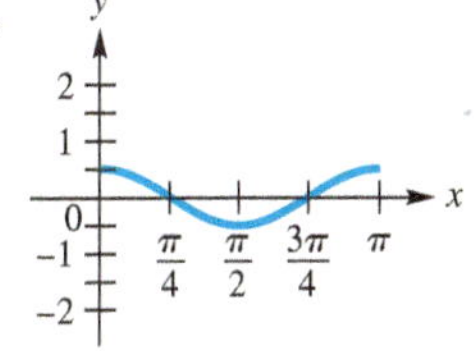

101.

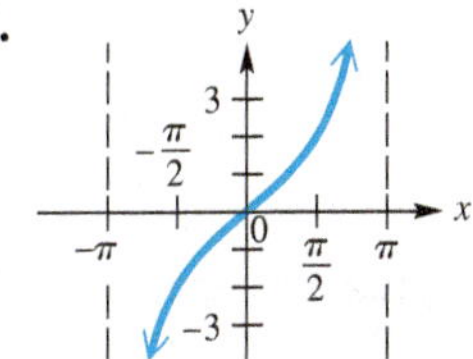

102.

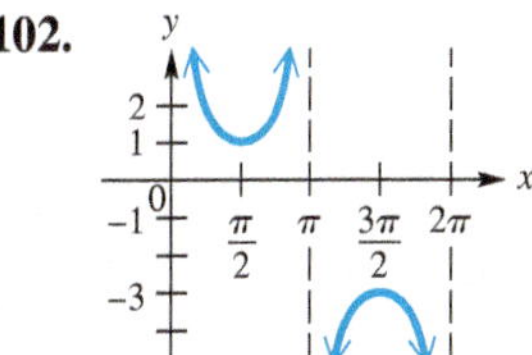

Solve each problem.

103. *Viewing Angle to an Object* Suppose that a person whose eyes are h_1 feet from the ground is standing d feet from an object h_2 feet tall, where $h_2 > h_1$. Let θ be the angle of elevation to the top of the object. See the figure.

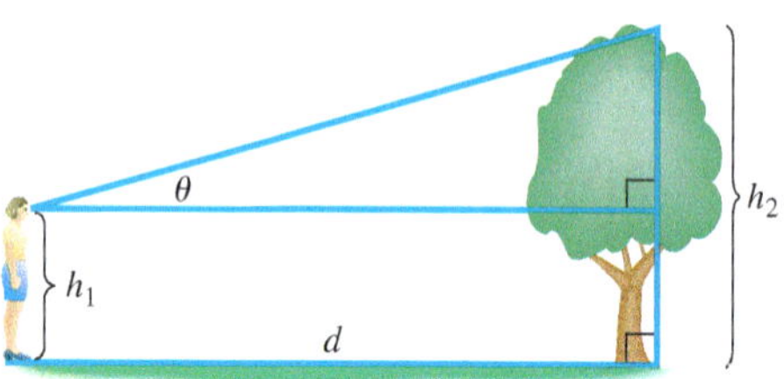

(a) Show that $d = (h_2 - h_1) \cot \theta$.

(b) Let $h_2 = 55$ and $h_1 = 5$. Graph d for the interval $0 < \theta \leq \frac{\pi}{2}$.

104. *(Modeling) Tides* The figure shows a function f that models the tides in feet at Clearwater Beach, Florida, x hours after midnight. (*Source*: Pentcheff, D., *WWW Tide and Current Predictor.*)

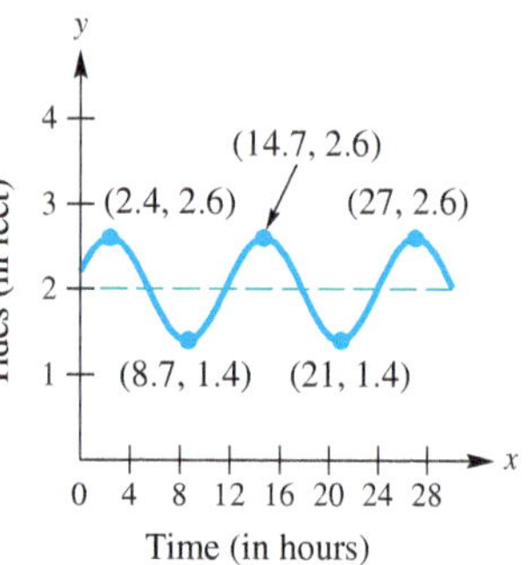

(a) Find the time between high tides.

(b) What is the difference in water levels between high tide and low tide?

(c) The tides can be modeled by

$$f(x) = 0.6 \cos[0.511(x - 2.4)] + 2.$$

Estimate the tides, to the nearest hundredth, when $x = 10$.

105. *(Modeling) Maximum Temperatures* The maximum afternoon temperature (in °F) in a given city can be modeled by

$$t = 60 - 30 \cos \frac{x\pi}{6},$$

where t represents the maximum afternoon temperature in month x, with $x = 0$ representing January, $x = 1$ representing February, and so on. Find the maximum afternoon temperature, to the nearest degree, for each month.

(a) January **(b)** April **(c)** May

(d) June **(e)** August **(f)** October

106. *(Modeling) Average Monthly Temperature* The average monthly temperature (in °F) in Chicago, Illinois, is shown in the table.

Month	°F	Month	°F
Jan	22	July	73
Feb	27	Aug	72
Mar	37	Sept	64
Apr	48	Oct	52
May	59	Nov	39
June	68	Dec	27

Source: *World Almanac and Book of Facts.*

(a) Plot the average monthly temperature over a two-year period. Let $x = 1$ correspond to January of the first year.

(b) To model the data, determine a function of the form

$$f(x) = a \sin[b(x - d)] + c,$$

where a, b, c, and d are constants.

(c) Graph f together with the data on the same coordinate axes. How well does f model the data?

(d) Use the sine regression capability of a graphing calculator to find the equation of a sine curve of the form $y = a \sin(bx + c) + d$ that fits these data.

107. *(Modeling) Pollution Trends* The amount of pollution in the air is lower after heavy spring rains and higher after periods of little rain. In addition to this seasonal fluctuation, the long-term trend is upward. An idealized graph of this situation is shown in the figure.

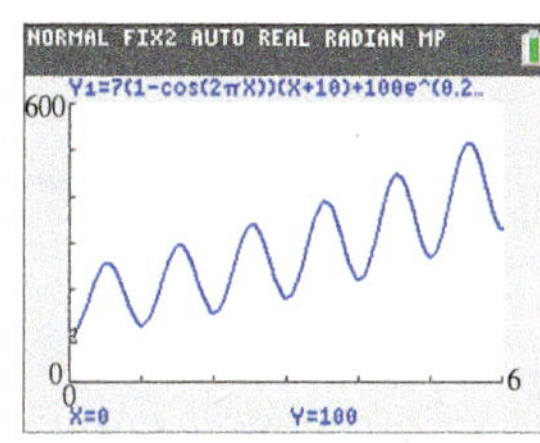

Circular functions can be used to model the fluctuating part of the pollution levels. Powers of the number e (e is the base of the natural logarithm; $e \approx 2.718282$) can be used to model long-term growth. The pollution level in a certain area might be given by

$$y = 7(1 - \cos 2\pi x)(x + 10) + 100e^{0.2x},$$

where x is the time in years, with $x = 0$ representing January 1 of the base year. July 1 of the same year would be represented by $x = 0.5$, October 1 of the following year would be represented by $x = 1.75$, and so on. Find the pollution levels on each date.

(a) January 1, base year (See the screen.) **(b)** July 1, base year

(c) January 1, following year **(d)** July 1, following year

108. *(Modeling) Lynx and Hare Populations* The figure shows the populations of lynx and hares in Canada for the years 1847–1903. The hares are food for the lynx. An increase in hare population causes an increase in lynx population some time later. The increasing lynx population then causes a decline in hare population. The two graphs have the same period.

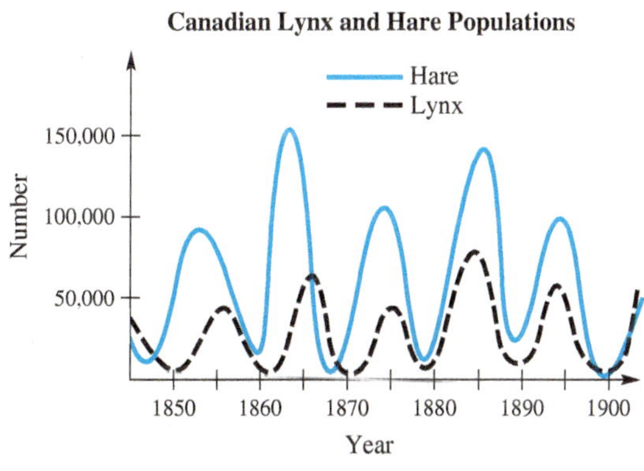

(a) Estimate the length of one period.

(b) Estimate the maximum and minimum hare populations.

An object in simple harmonic motion has position function $s(t)$ inches from an equilibrium point, where t is the time in seconds. Find the amplitude, period, and frequency.

109. $s(t) = 4 \sin \pi t$

110. $s(t) = 3 \cos 2t$

111. In **Exercise 109,** what does the frequency represent? Find the position of the object relative to the equilibrium point at 1.5 sec, 2 sec, and 3.25 sec.

112. In **Exercise 110,** what does the period represent? What does the amplitude represent?

Chapter 6 Test

Convert each degree measure to radians.

1. $120°$
2. $-45°$
3. $5°$ (to the nearest thousandth)

Convert each radian measure to degrees.

4. $\frac{3\pi}{4}$
5. $-\frac{7\pi}{6}$
6. 4 (to the nearest minute)

7. A central angle of a circle with radius 150 cm intercepts an arc of 200 cm. Find each measure.

   (a) the radian measure of the angle **(b)** the area of a sector with that central angle

8. ***Rotation of Gas Gauge Arrow*** The arrow on a car's gasoline gauge is $\frac{1}{2}$ in. long. See the figure. Through what angle does the arrow rotate when it moves 1 in. on the gauge?

Find each exact function value.

9. $\sin \frac{3\pi}{4}$
10. $\cos\left(-\frac{7\pi}{6}\right)$
11. $\tan \frac{3\pi}{2}$
12. $\sec \frac{8\pi}{3}$
13. $\tan \pi$
14. $\cos \frac{3\pi}{2}$

15. Determine the six exact circular function values of s in the figure.

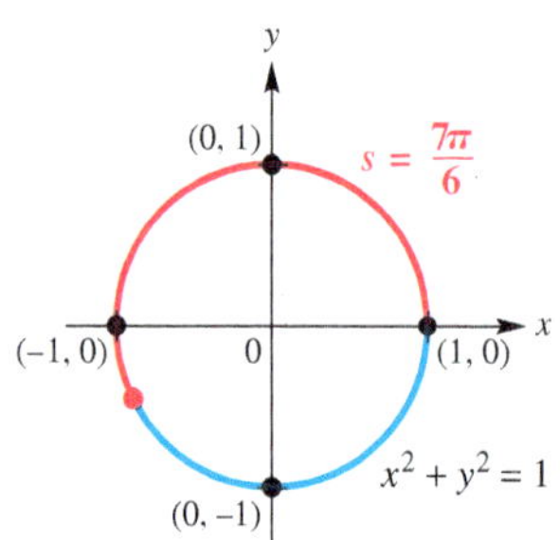

16. Do each of the following.

    (a) Use a calculator to approximate s in the interval $\left[0, \frac{\pi}{2}\right]$ if $\sin s = 0.8258$.

    (b) Find the exact value of s in the interval $\left[0, \frac{\pi}{2}\right]$ if $\cos s = \frac{1}{2}$.

17. ***Angular and Linear Speed of a Point*** Suppose that point P is on a circle with radius 60 cm, and ray OP is rotating with angular speed $\frac{\pi}{12}$ radian per sec.

    (a) Find the angle generated by P in 8 sec.

    (b) Find the distance traveled by P along the circle in 8 sec.

    (c) Find the linear speed of P.

18. ***Ferris Wheel*** A Ferris wheel has radius 50.0 ft. A person takes a seat, and then the wheel turns $\frac{2\pi}{3}$ radians.

    (a) How far is the person above the ground?

    (b) If it takes 30 sec for the wheel to turn $\frac{2\pi}{3}$ radians, what is the angular speed of the wheel?

19. Identify each of the following basic circular function graphs.

(a)

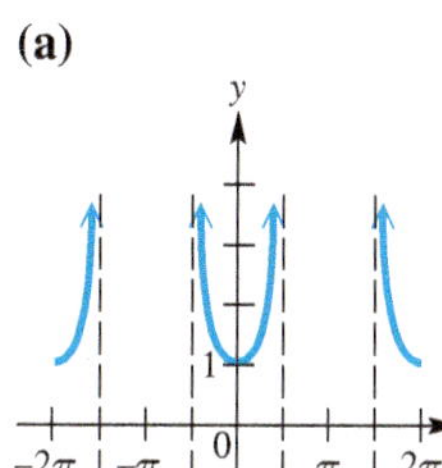

(b)

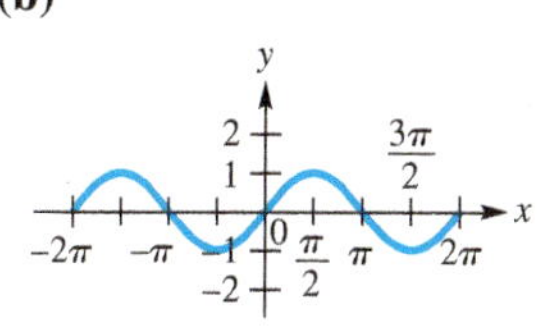

(c)

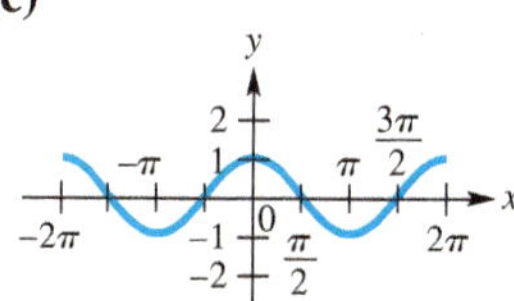

(d)

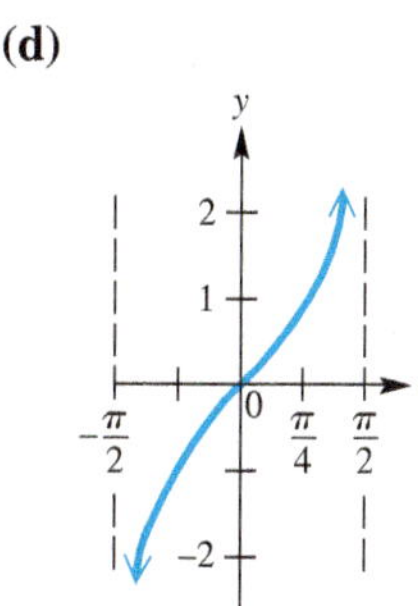

(e)

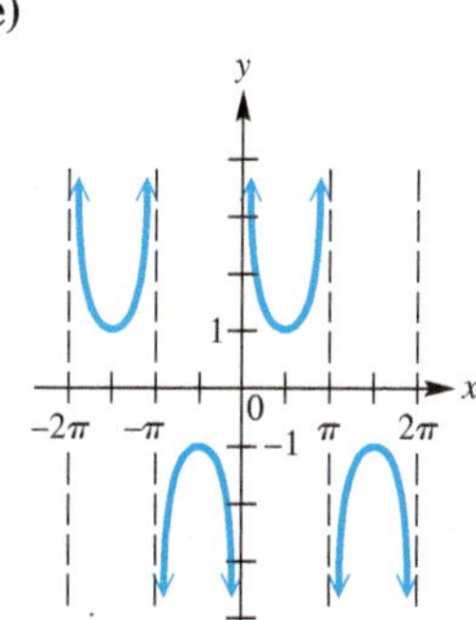

(f)

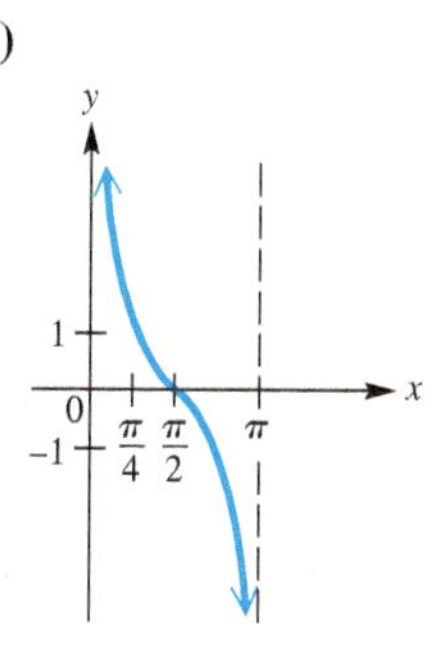

20. ***Connecting Graphs with Equations*** *Determine the simplest form of an equation for each graph. Choose $b > 0$, and include no phase shifts.*

(a)

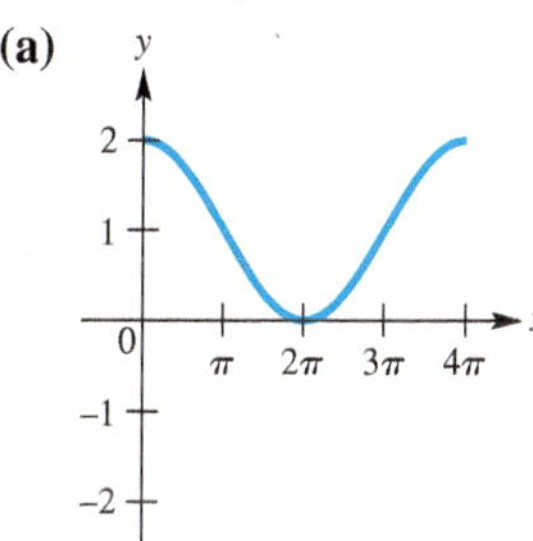

(b)

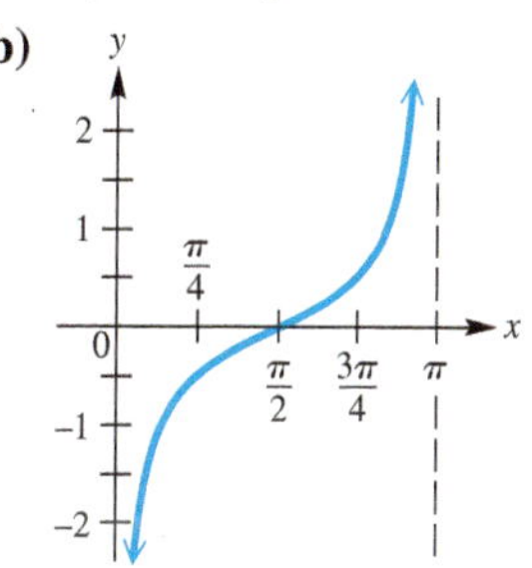

21. Answer each question.

(a) What is the domain of the cosine function?

(b) What is the range of the sine function?

(c) What is the least positive value for which the tangent function is undefined?

(d) What is the range of the secant function?

22. Consider the function $y = 3 - 6 \sin\left(2x + \frac{\pi}{2}\right)$.

(a) What is its period?

(b) What is the amplitude of its graph?

(c) What is its range?

(d) What is the y-intercept of its graph?

(e) What is its phase shift?

Graph each function over a two-period interval. Identify asymptotes when applicable.

23. $y = \sin(2x + \pi)$

24. $y = 2 + \cos x$

25. $y = -1 + 2 \sin(x + \pi)$

26. $y = \tan\left(x - \frac{\pi}{2}\right)$

27. $y = -2 - \cot\left(x - \frac{\pi}{2}\right)$

28. $y = -\csc 2x$

(Modeling) Solve each problem.

29. ***Average Monthly Temperature*** The average monthly temperature (in °F) in San Antonio, Texas, can be modeled by

$$f(x) = 16.5 \sin\left[\frac{\pi}{6}(x - 4)\right] + 67.5,$$

where x is the month and $x = 1$ corresponds to January. (*Source: World Almanac and Book of Facts.*)

(a) Graph f in the window $[0, 25]$ by $[40, 90]$.

(b) Determine the amplitude, period, phase shift, and vertical translation of f.

(c) What is the average monthly temperature for the month of December?

(d) Determine the minimum and maximum average monthly temperatures and the months when they occur.

(e) What would be an approximation for the average annual temperature in San Antonio? How is this related to the vertical translation of the sine function in the formula for f?

30. ***Spring Motion*** The position of a weight attached to a spring is

$$s(t) = -4 \cos 8\pi t \text{ inches after } t \text{ seconds.}$$

(a) Find the maximum height that the weight rises above the equilibrium position of $s(t) = 0$.

(b) When does the weight first reach its maximum height if $t \geq 0$?

(c) What are the frequency and period?

7 Trigonometric Identities and Equations

Electricity that passes through wires to homes and businesses alternates its direction on those wires and is modeled by *sine* and *cosine functions.*

7.1 Fundamental Identities

- Fundamental Identities
- Uses of the Fundamental Identities

Fundamental Identities

Recall that a function is **even** if $f(-x) = f(x)$ for all x in the domain of f, and a function is **odd** if $f(-x) = -f(x)$ for all x in the domain of f. We have used graphs to classify the trigonometric functions as even or odd. We can also use **Figure 1** to do this.

As suggested by the circle in **Figure 1,** an angle θ having the point (x, y) on its terminal side has a corresponding angle $-\theta$ with the point $(x, -y)$ on its terminal side.

$\sin(-\theta) = -\frac{y}{r} = -\sin\theta$

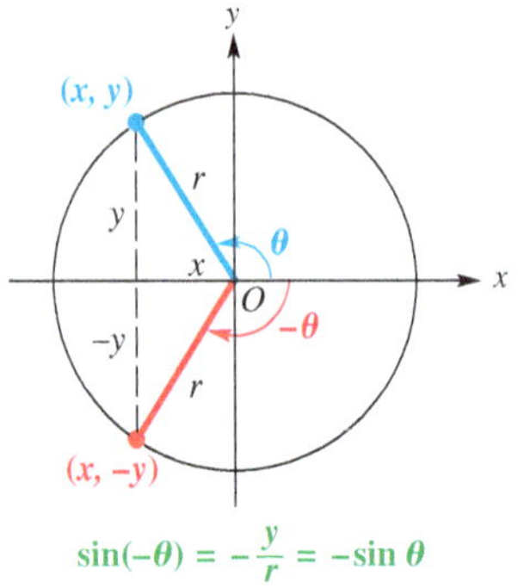

Figure 1

From the definition of sine, we see that $\sin(-\theta)$ and $\sin\theta$ are negatives of each other. That is,

$$\sin(-\theta) = \frac{-y}{r} \quad \text{and} \quad \sin\theta = \frac{y}{r},$$

so $$\mathbf{\sin(-\theta) = -\sin\theta}$$ Sine is an odd function.

This is an example of an **identity,** an equation that is satisfied by *every* value in the domain of its variable. Some examples from algebra follow.

$$x^2 - y^2 = (x + y)(x - y)$$
$$x(x + y) = x^2 + xy$$
$$x^2 + 2xy + y^2 = (x + y)^2$$

Identities

Figure 1 shows an angle θ in quadrant II, but the same result holds for θ in any quadrant. The figure also suggests the following identity for cosine.

$$\cos(-\theta) = \frac{x}{r} \quad \text{and} \quad \cos\theta = \frac{x}{r}$$

$$\mathbf{\cos(-\theta) = \cos\theta}$$ Cosine is an even function.

We use the identities for $\sin(-\theta)$ and $\cos(-\theta)$ to find $\tan(-\theta)$ in terms of $\tan\theta$.

$$\tan(-\theta) = \frac{\sin(-\theta)}{\cos(-\theta)} = \frac{-\sin\theta}{\cos\theta} = -\frac{\sin\theta}{\cos\theta}$$

$$\mathbf{\tan(-\theta) = -\tan\theta}$$ Tangent is an odd function.

The reciprocal identities are used to determine that cosecant and cotangent are odd functions and secant is an even function. These **even-odd identities** together with the reciprocal, quotient, and Pythagorean identities make up the **fundamental identities.**

NOTE In trigonometric identities, θ can represent an angle in degrees or radians, or a real number.

Fundamental Identities

Reciprocal Identities

$$\cot \theta = \frac{1}{\tan \theta} \qquad \sec \theta = \frac{1}{\cos \theta} \qquad \csc \theta = \frac{1}{\sin \theta}$$

Quotient Identities

$$\tan \theta = \frac{\sin \theta}{\cos \theta} \qquad \cot \theta = \frac{\cos \theta}{\sin \theta}$$

Pythagorean Identities

$$\sin^2 \theta + \cos^2 \theta = 1 \qquad \tan^2 \theta + 1 = \sec^2 \theta \qquad 1 + \cot^2 \theta = \csc^2 \theta$$

Even-Odd Identities

$$\sin(-\theta) = -\sin \theta \qquad \cos(-\theta) = \cos \theta \qquad \tan(-\theta) = -\tan \theta$$

$$\csc(-\theta) = -\csc \theta \qquad \sec(-\theta) = \sec \theta \qquad \cot(-\theta) = -\cot \theta$$

NOTE We will also use alternative forms of the fundamental identities. *For example, two other forms of* $\sin^2 \theta + \cos^2 \theta = 1$ *are*

$$\sin^2 \theta = 1 - \cos^2 \theta \quad \textit{and} \quad \cos^2 \theta = 1 - \sin^2 \theta.$$

Uses of the Fundamental Identities We can use these identities to find the values of other trigonometric functions from the value of a given trigonometric function.

EXAMPLE 1 Finding Trigonometric Function Values Given One Value and the Quadrant

If $\tan \theta = -\frac{5}{3}$ and θ is in quadrant II, find each function value.

(a) $\sec \theta$ **(b)** $\sin \theta$ **(c)** $\cot(-\theta)$

SOLUTION

(a) We use an identity that relates the tangent and secant functions.

$\tan^2 \theta + 1 = \sec^2 \theta$ Pythagorean identity

$\left(-\frac{5}{3}\right)^2 + 1 = \sec^2 \theta$ $\tan \theta = -\frac{5}{3}$

$\frac{25}{9} + 1 = \sec^2 \theta$ Square $-\frac{5}{3}$.

$\frac{34}{9} = \sec^2 \theta$ Add; $1 = \frac{9}{9}$

$-\sqrt{\frac{34}{9}} = \sec \theta$ Take the negative square root because θ is in quadrant II.

Choose the correct sign.

$\sec \theta = -\frac{\sqrt{34}}{3}$ Simplify the radical: $-\sqrt{\frac{34}{9}} = -\frac{\sqrt{34}}{\sqrt{9}} = -\frac{\sqrt{34}}{3}$, and rewrite.

(b)

$$\tan\theta = \frac{\sin\theta}{\cos\theta} \quad \text{Quotient identity}$$

$$\cos\theta\tan\theta = \sin\theta \quad \text{Multiply each side by } \cos\theta.$$

$$\left(\frac{1}{\sec\theta}\right)\tan\theta = \sin\theta \quad \text{Reciprocal identity}$$

$$\left(-\frac{3\sqrt{34}}{34}\right)\left(-\frac{5}{3}\right) = \sin\theta$$

$\tan\theta = -\frac{5}{3}$, and from part (a), $\frac{1}{\sec\theta} = \frac{1}{-\frac{\sqrt{34}}{3}} = -\frac{3}{\sqrt{34}} = -\frac{3}{\sqrt{34}} \cdot \frac{\sqrt{34}}{\sqrt{34}} = -\frac{3\sqrt{34}}{34}$.

$$\sin\theta = \frac{5\sqrt{34}}{34} \quad \text{Multiply and rewrite.}$$

(c)

$$\cot(-\theta) = \frac{1}{\tan(-\theta)} \quad \text{Reciprocal identity}$$

$$\cot(-\theta) = \frac{1}{-\tan\theta} \quad \text{Even-odd identity}$$

$$\cot(-\theta) = \frac{1}{-\left(-\frac{5}{3}\right)} \quad \tan\theta = -\frac{5}{3}$$

$$\cot(-\theta) = \frac{3}{5} \quad \frac{1}{-\left(-\frac{5}{3}\right)} = 1 \div \frac{5}{3} = 1 \cdot \frac{3}{5} = \frac{3}{5}$$

✔ **Now Try Exercises 11, 19, and 31.**

CAUTION ***When taking the square root, be sure to choose the sign based on the quadrant of θ and the function being evaluated.***

EXAMPLE 2 Writing One Trignometric Function in Terms of Another

Write $\cos x$ in terms of $\tan x$.

SOLUTION By identities, $\sec x$ is related to both $\cos x$ and $\tan x$.

$$1 + \tan^2 x = \sec^2 x \quad \text{Pythagorean identity}$$

$$\frac{1}{1+\tan^2 x} = \frac{1}{\sec^2 x} \quad \text{Take reciprocals.}$$

$$\frac{1}{1+\tan^2 x} = \cos^2 x \quad \text{The reciprocal of } \sec^2 x \text{ is } \cos^2 x.$$

Remember both the positive and negative roots.

$$\pm\sqrt{\frac{1}{1+\tan^2 x}} = \cos x \quad \text{Take the square root of each side.}$$

$$\cos x = \frac{\pm 1}{\sqrt{1+\tan^2 x}} \quad \text{Quotient rule for radicals: } \sqrt[n]{\frac{a}{b}} = \frac{\sqrt[n]{a}}{\sqrt[n]{b}}\text{; rewrite.}$$

$$\cos x = \frac{\pm\sqrt{1+\tan^2 x}}{1+\tan^2 x} \quad \text{Rationalize the denominator.}$$

The choice of the + sign or the − sign is made depending on the quadrant of x.

✔ **Now Try Exercise 47.**

$y_1 = \sin^2 x + \cos^2 x$
$y_2 = 1$

NORMAL FLOAT AUTO REAL RADIAN MP

4, −4, $-\frac{11\pi}{4}$, $\frac{11\pi}{4}$

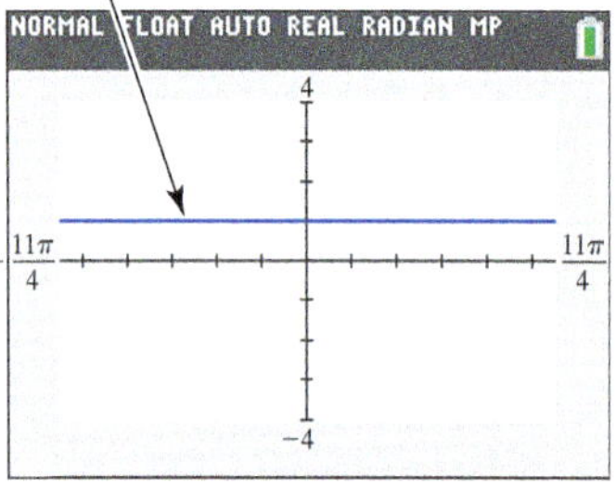

With an identity, there should be no difference between the two graphs.

Figure 2

Figure 2 supports the identity $\sin^2 x + \cos^2 x = 1$. ■

The functions tan θ, cot θ, sec θ, and csc θ can easily be expressed in terms of sin θ, cos θ, or both. We make such substitutions in an expression to simplify it.

EXAMPLE 3 Rewriting an Expression in Terms of Sine and Cosine

Write $\frac{1 + \cot^2 \theta}{1 - \csc^2 \theta}$ in terms of sin θ and cos θ, and then simplify the expression so that no quotients appear.

SOLUTION

$$\frac{1 + \cot^2 \theta}{1 - \csc^2 \theta} \quad \text{Given expression}$$

$$= \frac{1 + \frac{\cos^2 \theta}{\sin^2 \theta}}{1 - \frac{1}{\sin^2 \theta}} \quad \text{Quotient identities}$$

$$= \frac{\left(1 + \frac{\cos^2 \theta}{\sin^2 \theta}\right)\sin^2 \theta}{\left(1 - \frac{1}{\sin^2 \theta}\right)\sin^2 \theta} \quad \text{Simplify the complex fraction by multiplying both numerator and denominator by the LCD.}$$

$$= \frac{\sin^2 \theta + \cos^2 \theta}{\sin^2 \theta - 1} \quad \text{Distributive property: } (b + c)a = ba + ca$$

$$= \frac{1}{-\cos^2 \theta} \quad \text{Pythagorean identities}$$

$$= -\sec^2 \theta \quad \text{Reciprocal identity}$$

$y_1 = \frac{1 + \cot^2 x}{1 - \csc^2 x}$

$y_2 = -\sec^2 x$

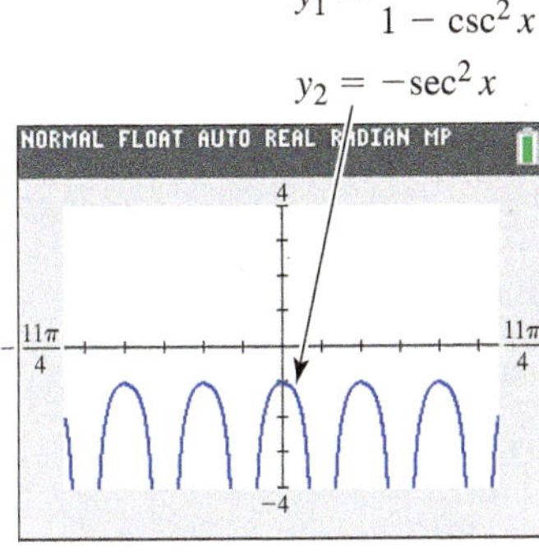

The graph supports the result in **Example 3.** The graphs of y_1 and y_2 coincide.

 **Now Try Exercise 59.**

CAUTION ***When working with trigonometric expressions and identities, be sure to write the argument of the function.*** For example, we would *not* write $\sin^2 + \cos^2 = 1$. An argument such as θ is necessary to write this correctly as $\sin^2 \theta + \cos^2 \theta = 1$.

7.1 Exercises

CONCEPT PREVIEW *For each expression in Column I, choose the expression from Column II that completes an identity.*

I	II
1. $\frac{\cos x}{\sin x} =$ _____	**A.** $\sin^2 x + \cos^2 x$
2. $\tan x =$ _____	**B.** $\cot x$
3. $\cos(-x) =$ _____	**C.** $\sec^2 x$
4. $\tan^2 x + 1 =$ _____	**D.** $\frac{\sin x}{\cos x}$
5. $1 =$ _____	**E.** $\cos x$

CONCEPT PREVIEW *Use identities to correctly complete each sentence.*

6. If $\tan \theta = 2.6$, then $\tan(-\theta) =$ ________.

7. If $\cos \theta = -0.65$, then $\cos(-\theta) =$ ________.

8. If $\tan \theta = 1.6$, then $\cot \theta =$ ________.

9. If $\cos \theta = 0.8$ and $\sin \theta = 0.6$, then $\tan(-\theta) =$ ________.

10. If $\sin \theta = \frac{2}{3}$, then $-\sin(-\theta) =$ ________.

Find $\sin \theta$. ***See Example 1.***

11. $\cos \theta = \frac{3}{4}$, θ in quadrant I

12. $\cos \theta = \frac{5}{6}$, θ in quadrant I

13. $\cot \theta = -\frac{1}{5}$, θ in quadrant IV

14. $\cot \theta = -\frac{1}{3}$, θ in quadrant IV

15. $\cos(-\theta) = \frac{\sqrt{5}}{5}$, $\tan \theta < 0$

16. $\cos(-\theta) = \frac{\sqrt{3}}{6}$, $\cot \theta < 0$

17. $\tan \theta = -\frac{\sqrt{6}}{2}$, $\cos \theta > 0$

18. $\tan \theta = -\frac{\sqrt{7}}{2}$, $\sec \theta > 0$

19. $\sec \theta = \frac{11}{4}$, $\cot \theta < 0$

20. $\sec \theta = \frac{7}{2}$, $\tan \theta < 0$

21. $\csc \theta = -\frac{9}{4}$

22. $\csc \theta = -\frac{8}{5}$

23. Why is it unnecessary to give the quadrant of θ in **Exercises 21 and 22?**

24. *Concept Check* What is **WRONG** with the statement of this problem?

Find $\cos(-\theta)$ *if* $\cos \theta = 3$.

Concept Check Find $f(-x)$ *to determine whether each function is* even *or* odd.

25. $f(x) = \dfrac{\sin x}{x}$

26. $f(x) = x \cos x$

Concept Check Identify the basic trigonometric function graphed and determine whether it is even *or* odd.

27.

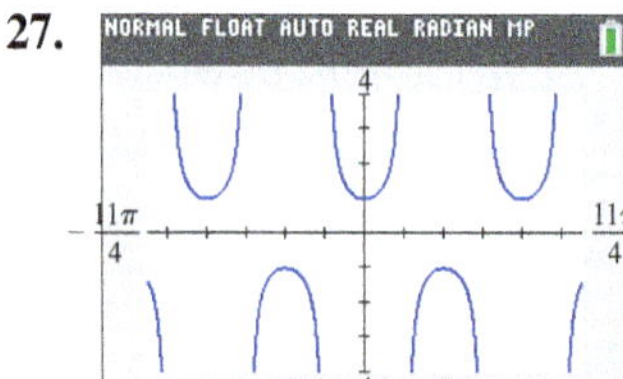

28.

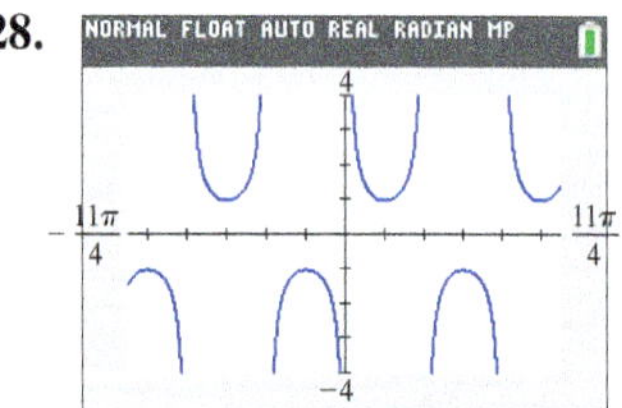

29.

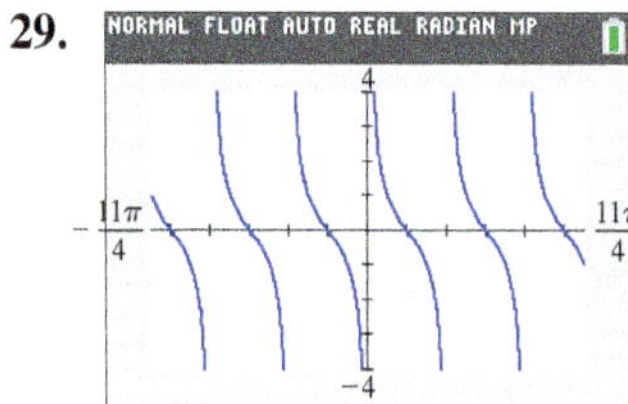

30.

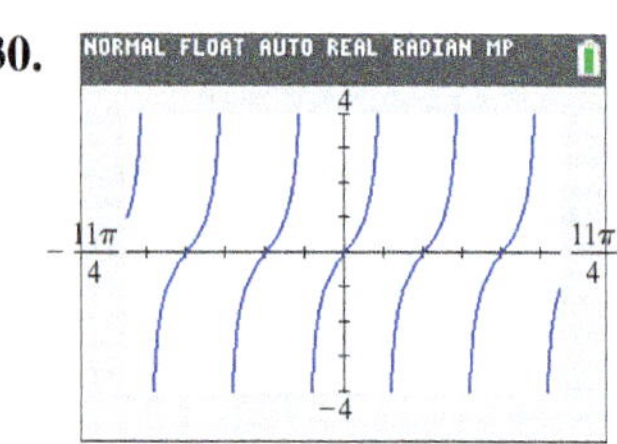

Find the remaining five trigonometric functions of θ. ***See Example 1.***

31. $\sin \theta = \frac{2}{3}$, θ in quadrant II

32. $\cos \theta = \frac{1}{5}$, θ in quadrant I

33. $\tan \theta = -\frac{1}{4}$, θ in quadrant IV

34. $\csc \theta = -\frac{5}{2}$, θ in quadrant III

35. $\cot \theta = \frac{4}{3}$, $\sin \theta > 0$

36. $\sin \theta = -\frac{4}{5}$, $\cos \theta < 0$

37. $\sec \theta = \frac{4}{3}$, $\sin \theta < 0$

38. $\cos \theta = -\frac{1}{4}$, $\sin \theta > 0$

Concept Check For each expression in Column I, choose the expression from Column II that completes an identity. One or both expressions may need to be rewritten.

I	II
39. $-\tan x \cos x =$ ____	**A.** $\dfrac{\sin^2 x}{\cos^2 x}$
40. $\sec^2 x - 1 =$ ____	**B.** $\dfrac{1}{\sec^2 x}$
41. $\dfrac{\sec x}{\csc x} =$ ____	**C.** $\sin(-x)$
42. $1 + \sin^2 x =$ ____	**D.** $\csc^2 x - \cot^2 x + \sin^2 x$
43. $\cos^2 x =$ ____	**E.** $\tan x$

44. A student writes "$1 + \cot^2 = \csc^2$." Comment on this student's work.

45. *Concept Check* Suppose that $\cos \theta = \frac{x}{x+1}$. Find an expression in x for $\sin \theta$.

46. *Concept Check* Suppose that $\sec \theta = \frac{x+4}{x}$. Find an expression in x for $\tan \theta$.

Perform each transformation. ***See Example 2.***

47. Write $\sin x$ in terms of $\cos x$.

48. Write $\cot x$ in terms of $\sin x$.

49. Write $\tan x$ in terms of $\sec x$.

50. Write $\cot x$ in terms of $\csc x$.

51. Write $\csc x$ in terms of $\cos x$.

52. Write $\sec x$ in terms of $\sin x$.

Write each expression in terms of sine and cosine, and then simplify the expression so that no quotients appear and all functions are of θ only. ***See Example 3.***

53. $\cot \theta \sin \theta$

54. $\tan \theta \cos \theta$

55. $\sec \theta \cot \theta \sin \theta$

56. $\csc \theta \cos \theta \tan \theta$

57. $\cos \theta \csc \theta$

58. $\sin \theta \sec \theta$

59. $\sin^2 \theta(\csc^2 \theta - 1)$

60. $\cot^2 \theta(1 + \tan^2 \theta)$

61. $(1 - \cos \theta)(1 + \sec \theta)$

62. $(\sec \theta - 1)(\sec \theta + 1)$

63. $\dfrac{1 + \tan(-\theta)}{\tan(-\theta)}$

64. $\dfrac{1 + \cot \theta}{\cot \theta}$

65. $\dfrac{1 - \cos^2(-\theta)}{1 + \tan^2(-\theta)}$

66. $\dfrac{1 - \sin^2(-\theta)}{1 + \cot^2(-\theta)}$

67. $\sec \theta - \cos \theta$

68. $\csc \theta - \sin \theta$

69. $(\sec \theta + \csc \theta)(\cos \theta - \sin \theta)$

70. $(\sin \theta - \cos \theta)(\csc \theta + \sec \theta)$

71. $\sin \theta(\csc \theta - \sin \theta)$

72. $\cos \theta(\cos \theta - \sec \theta)$

73. $\dfrac{1 + \tan^2 \theta}{1 + \cot^2 \theta}$

74. $\dfrac{\sec^2 \theta - 1}{\csc^2 \theta - 1}$

75. $\dfrac{\csc \theta}{\cot(-\theta)}$

76. $\dfrac{\tan(-\theta)}{\sec \theta}$

77. $\sin^2(-\theta) + \tan^2(-\theta) + \cos^2(-\theta)$

78. $-\sec^2(-\theta) + \sin^2(-\theta) + \cos^2(-\theta)$

Work each problem.

79. Let $\cos x = \frac{1}{5}$. Find all possible values of $\frac{\sec x - \tan x}{\sin x}$.

80. Let $\csc x = -3$. Find all possible values of $\frac{\sin x + \cos x}{\sec x}$.

Use a graphing calculator to make a conjecture about whether each equation is an identity.

81. $\cos 2x = 1 - 2 \sin^2 x$

82. $2 \sin x = \sin 2x$

83. $\sin x = \sqrt{1 - \cos^2 x}$

84. $\cos 2x = \cos^2 x - \sin^2 x$

Relating Concepts

For individual or collaborative investigation *(Exercises 85–90)*

Previously we graphed functions of the form

$$y = c + a \cdot f[b(x - d)]$$

with the assumption that $b > 0$. To see what happens when $b < 0$, ***work Exercises 85–90 in order.***

85. Use an even-odd identity to write $y = \sin(-2x)$ as a function of $2x$.

86. How is the answer to **Exercise 85** related to $y = \sin 2x$?

87. Use an even-odd identity to write $y = \cos(-4x)$ as a function of $4x$.

88. How is the answer to **Exercise 87** related to $y = \cos 4x$?

89. Use the results from **Exercises 85–88** to rewrite the following with a positive value of b.

(a) $y = \sin(-4x)$ **(b)** $y = \cos(-2x)$ **(c)** $y = -5 \sin(-3x)$

90. Write a short response to this statement, which is often used by one of the authors of this text in trigonometry classes:

Students who tend to ignore negative signs should enjoy graphing functions involving the cosine and the secant.

7.2 Verifying Trigonometric Identities

- Strategies
- Verifying Identities by Working with One Side
- Verifying Identities by Working with Both Sides

Strategies One of the skills required for more advanced work in mathematics, especially in calculus, is the ability to use identities to write expressions in alternative forms. We develop this skill by using the fundamental identities to verify that a trigonometric equation is an identity (for those values of the variable for which it is defined).

CAUTION ***The procedure for verifying identities is not the same as that for solving equations.*** Techniques used in solving equations, such as adding the same term to each side, and multiplying each side by the same term, should *not* be used when working with identities.

LOOKING AHEAD TO CALCULUS

Trigonometric identities are used in calculus to simplify trigonometric expressions, determine derivatives of trigonometric functions, and change the form of some integrals.

Hints for Verifying Identities

1. ***Learn the fundamental identities.*** Whenever you see either side of a fundamental identity, the other side should come to mind. ***Also, be aware of equivalent forms of the fundamental identities.*** For example,

$$\sin^2\theta = 1 - \cos^2\theta \quad \text{is an alternative form of} \quad \sin^2\theta + \cos^2\theta = 1.$$

2. ***Try to rewrite the more complicated side*** of the equation so that it is identical to the simpler side.

3. ***It is sometimes helpful to express all trigonometric functions in the equation in terms of sine and cosine*** and then simplify the result.

4. ***Usually, any factoring or indicated algebraic operations should be performed.*** These *algebraic* identities are often used in verifying trigonometric identities.

$$x^2 + 2xy + y^2 = (x + y)^2$$
$$x^2 - 2xy + y^2 = (x - y)^2$$
$$x^3 - y^3 = (x - y)(x^2 + xy + y^2)$$
$$x^3 + y^3 = (x + y)(x^2 - xy + y^2)$$
$$x^2 - y^2 = (x + y)(x - y)$$

For example, the expression

$$\sin^2 x + 2\sin x + 1 \quad \text{can be factored as} \quad (\sin x + 1)^2.$$

The sum or difference of two trigonometric expressions can be found in the same way as any other rational expression. For example,

$$\frac{1}{\sin\theta} + \frac{1}{\cos\theta}$$

$$= \frac{1 \cdot \cos\theta}{\sin\theta\cos\theta} + \frac{1 \cdot \sin\theta}{\cos\theta\sin\theta} \quad \text{Write with the LCD.}$$

$$= \frac{\cos\theta + \sin\theta}{\sin\theta\cos\theta}. \quad \frac{a}{c} + \frac{b}{c} = \frac{a+b}{c}$$

5. ***When selecting substitutions, keep in mind the side that is not changing, because it represents the goal.*** For example, to verify that the equation

$$\tan^2 x + 1 = \frac{1}{\cos^2 x}$$

is an identity, think of an identity that relates $\tan x$ to $\cos x$. In this case, because $\sec x = \frac{1}{\cos x}$ and $\sec^2 x = \tan^2 x + 1$, the secant function is the best link between the two sides.

6. If an expression contains $1 + \sin x$, ***multiplying both numerator and denominator*** by $1 - \sin x$ would give $1 - \sin^2 x$, which could be replaced with $\cos^2 x$. Similar procedures apply for $1 - \sin x$, $1 + \cos x$, and $1 - \cos x$.

Verifying Identities by Working with One Side Avoid the temptation to use algebraic properties of equations to verify identities.

One strategy is to work with one side and rewrite it to match the other side.

EXAMPLE 1 Verifying an Identity (Working with One Side)

Verify that the following equation is an identity.

$$\cot\theta + 1 = \csc\theta(\cos\theta + \sin\theta)$$

SOLUTION We use the fundamental identities to rewrite one side of the equation so that it is identical to the other side. The right side is more complicated, so we work with it, as suggested in Hint 2, and use Hint 3 to change all functions to expressions involving sine or cosine.

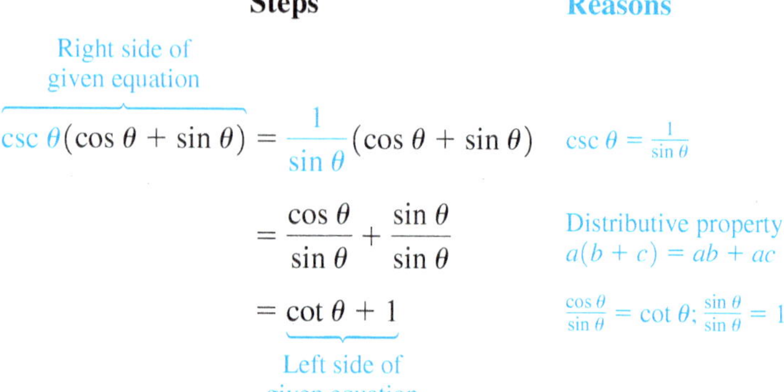

Steps	Reasons
Right side of given equation: $\csc\theta(\cos\theta + \sin\theta) = \frac{1}{\sin\theta}(\cos\theta + \sin\theta)$	$\csc\theta = \frac{1}{\sin\theta}$
$= \frac{\cos\theta}{\sin\theta} + \frac{\sin\theta}{\sin\theta}$	Distributive property: $a(b+c) = ab + ac$
$= \cot\theta + 1$ (Left side of given equation)	$\frac{\cos\theta}{\sin\theta} = \cot\theta$; $\frac{\sin\theta}{\sin\theta} = 1$

For $\theta = x$,
$y_1 = \cot x + 1$
$y_2 = \csc x(\cos x + \sin x)$

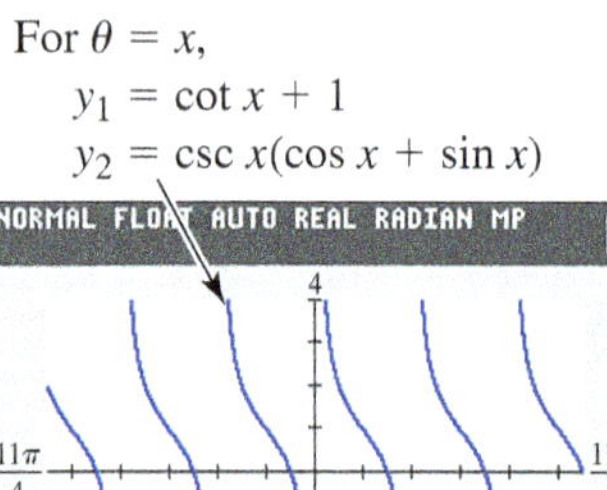

The graphs coincide, which supports the conclusion in **Example 1.**

The given equation is an identity. The right side of the equation is identical to the left side.

✔ **Now Try Exercise 45.**

EXAMPLE 2 Verifying an Identity (Working with One Side)

Verify that the following equation is an identity.

$$\tan^2 x(1 + \cot^2 x) = \frac{1}{1 - \sin^2 x}$$

SOLUTION We work with the more complicated left side, as suggested in Hint 2. Again, we use the fundamental identities.

Steps	Reasons
Left side of given equation: $\tan^2 x(1 + \cot^2 x) = \tan^2 x + \tan^2 x \cot^2 x$	Distributive property
$= \tan^2 x + \tan^2 x \cdot \frac{1}{\tan^2 x}$	$\cot^2 x = \frac{1}{\tan^2 x}$
$= \tan^2 x + 1$	$\tan^2 x \cdot \frac{1}{\tan^2 x} = 1$
$= \sec^2 x$	Pythagorean identity
$= \frac{1}{\cos^2 x}$	$\sec^2 x = \frac{1}{\cos^2 x}$
$= \frac{1}{1 - \sin^2 x}$ (Right side of given equation)	Pythagorean identity

$y_1 = \tan^2 x(1 + \cot^2 x)$
$y_2 = \frac{1}{1 - \sin^2 x}$

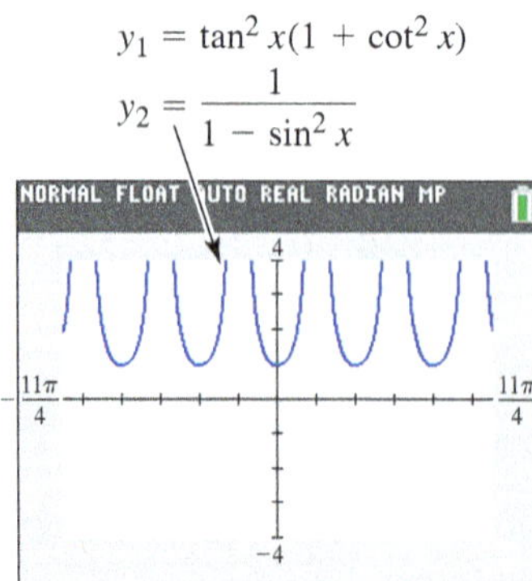

The screen supports the conclusion in **Example 2.**

Because the left side of the equation is identical to the right side, the given equation is an identity.

✔ **Now Try Exercise 49.**

EXAMPLE 3 Verifying an Identity (Working with One Side)

Verify that the following equation is an identity.

$$\frac{\tan t - \cot t}{\sin t \cos t} = \sec^2 t - \csc^2 t$$

SOLUTION We transform the more complicated left side to match the right side.

$$\begin{aligned}
\frac{\tan t - \cot t}{\sin t \cos t} &= \frac{\tan t}{\sin t \cos t} - \frac{\cot t}{\sin t \cos t} && \frac{a-b}{c} = \frac{a}{c} - \frac{b}{c} \\
&= \tan t \cdot \frac{1}{\sin t \cos t} - \cot t \cdot \frac{1}{\sin t \cos t} && \frac{a}{b} = a \cdot \frac{1}{b} \\
&= \frac{\sin t}{\cos t} \cdot \frac{1}{\sin t \cos t} - \frac{\cos t}{\sin t} \cdot \frac{1}{\sin t \cos t} && \tan t = \frac{\sin t}{\cos t};\ \cot t = \frac{\cos t}{\sin t} \\
&= \frac{1}{\cos^2 t} - \frac{1}{\sin^2 t} && \text{Multiply.} \\
&= \sec^2 t - \csc^2 t && \frac{1}{\cos^2 t} = \sec^2 t;\ \frac{1}{\sin^2 t} = \csc^2 t
\end{aligned}$$

Hint 3 about writing all trigonometric functions in terms of sine and cosine was used in the third line of the solution.

✔ **Now Try Exercise 53.**

EXAMPLE 4 Verifying an Identity (Working with One Side)

Verify that the following equation is an identity.

$$\frac{\cos x}{1 - \sin x} = \frac{1 + \sin x}{\cos x}$$

SOLUTION We work on the right side, using Hint 6 in the list given earlier to multiply the numerator and denominator on the right by $1 - \sin x$.

$$\begin{aligned}
\frac{1 + \sin x}{\cos x} &= \frac{(1 + \sin x)(1 - \sin x)}{\cos x(1 - \sin x)} && \text{Multiply by 1 in the form } \frac{1 - \sin x}{1 - \sin x}. \\
&= \frac{1 - \sin^2 x}{\cos x(1 - \sin x)} && (x + y)(x - y) = x^2 - y^2 \\
&= \frac{\cos^2 x}{\cos x(1 - \sin x)} && 1 - \sin^2 x = \cos^2 x \\
&= \frac{\cos x \cdot \cos x}{\cos x(1 - \sin x)} && a^2 = a \cdot a \\
&= \frac{\cos x}{1 - \sin x} && \text{Write in lowest terms.}
\end{aligned}$$

✔ **Now Try Exercise 59.**

Verifying Identities by Working with Both Sides If both sides of an identity appear to be equally complex, the identity can be verified by working independently on the left side and on the right side, until each side is changed into some common third result. ***Each step, on each side, must be reversible.*** With all steps reversible, the procedure is as shown in the margin. The left side leads to a common third expression, which leads back to the right side.

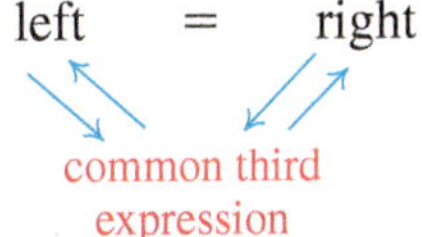

NOTE Working with both sides is often a good alternative for identities that are difficult. In practice, if working with one side does not seem to be effective, switch to the other side. Somewhere along the way it may happen that the same expression occurs on both sides.

EXAMPLE 5 Verifying an Identity (Working with Both Sides)

Verify that the following equation is an identity.

$$\frac{\sec \alpha + \tan \alpha}{\sec \alpha - \tan \alpha} = \frac{1 + 2 \sin \alpha + \sin^2 \alpha}{\cos^2 \alpha}$$

SOLUTION Both sides appear equally complex, so we verify the identity by changing each side into a common third expression. We work first on the left, multiplying the numerator and denominator by $\cos \alpha$.

$$\underbrace{\frac{\sec \alpha + \tan \alpha}{\sec \alpha - \tan \alpha}}_{\text{Left side of given equation}} = \frac{(\sec \alpha + \tan \alpha) \cos \alpha}{(\sec \alpha - \tan \alpha) \cos \alpha} \qquad \text{Multiply by 1 in the form } \frac{\cos \alpha}{\cos \alpha}.$$

$$= \frac{\sec \alpha \cos \alpha + \tan \alpha \cos \alpha}{\sec \alpha \cos \alpha - \tan \alpha \cos \alpha} \qquad \text{Distributive property}$$

$$= \frac{1 + \tan \alpha \cos \alpha}{1 - \tan \alpha \cos \alpha} \qquad \sec \alpha \cos \alpha = 1$$

$$= \frac{1 + \frac{\sin \alpha}{\cos \alpha} \cdot \cos \alpha}{1 - \frac{\sin \alpha}{\cos \alpha} \cdot \cos \alpha} \qquad \tan \alpha = \frac{\sin \alpha}{\cos \alpha}$$

$$= \frac{1 + \sin \alpha}{1 - \sin \alpha} \qquad \text{Simplify.}$$

On the right side of the original equation, we begin by factoring.

$$\underbrace{\frac{1 + 2 \sin \alpha + \sin^2 \alpha}{\cos^2 \alpha}}_{\text{Right side of given equation}} = \frac{(1 + \sin \alpha)^2}{\cos^2 \alpha} \qquad \text{Factor the numerator; } x^2 + 2xy + y^2 = (x + y)^2.$$

$$= \frac{(1 + \sin \alpha)^2}{1 - \sin^2 \alpha} \qquad \cos^2 \alpha = 1 - \sin^2 \alpha$$

$$= \frac{(1 + \sin \alpha)^2}{(1 + \sin \alpha)(1 - \sin \alpha)} \qquad \text{Factor the denominator; } x^2 - y^2 = (x + y)(x - y).$$

$$= \frac{1 + \sin \alpha}{1 - \sin \alpha} \qquad \text{Write in lowest terms.}$$

We have shown that

$$\overbrace{\frac{\sec \alpha + \tan \alpha}{\sec \alpha - \tan \alpha}}^{\text{Left side of given equation}} = \overbrace{\frac{1 + \sin \alpha}{1 - \sin \alpha}}^{\text{Common third expression}} = \overbrace{\frac{1 + 2 \sin \alpha + \sin^2 \alpha}{\cos^2 \alpha}}^{\text{Right side of given equation}},$$

and thus have verified that the given equation is an identity.

✔ **Now Try Exercise 75.**

CAUTION Use the method of **Example 5** *only* if the steps are reversible.

There are usually several ways to verify a given identity. Another way to begin verifying the identity in **Example 5** is to work on the left as follows.

$$\underbrace{\frac{\sec\alpha+\tan\alpha}{\sec\alpha-\tan\alpha}}_{\text{Left side of given equation in Example 5}}=\frac{\frac{1}{\cos\alpha}+\frac{\sin\alpha}{\cos\alpha}}{\frac{1}{\cos\alpha}-\frac{\sin\alpha}{\cos\alpha}} \quad \text{Fundamental identities}$$

$$=\frac{\frac{1+\sin\alpha}{\cos\alpha}}{\frac{1-\sin\alpha}{\cos\alpha}} \quad \text{Add and subtract fractions.}$$

$$=\frac{1+\sin\alpha}{\cos\alpha}\div\frac{1-\sin\alpha}{\cos\alpha} \quad \text{Simplify the complex fraction. Use the definition of division.}$$

$$=\frac{1+\sin\alpha}{\cos\alpha}\cdot\frac{\cos\alpha}{1-\sin\alpha} \quad \text{Multiply by the reciprocal.}$$

$$=\frac{1+\sin\alpha}{1-\sin\alpha} \quad \text{Multiply and write in lowest terms.}$$

Compare this with the result shown in **Example 5** for the right side to see that the two sides indeed agree.

EXAMPLE 6 Applying a Pythagorean Identity to Electronics

Tuners in radios select a radio station by adjusting the frequency. A tuner may contain an inductor L and a capacitor C, as illustrated in **Figure 3.** The energy stored in the inductor at time t is given by

$$L(t)=k\sin^2 2\pi Ft$$

and the energy stored in the capacitor is given by

$$C(t)=k\cos^2 2\pi Ft,$$

where F is the frequency of the radio station and k is a constant. The total energy E in the circuit is given by

$$E(t)=L(t)+C(t).$$

Show that E is a constant function. (*Source*: Weidner, R. and R. Sells, *Elementary Classical Physics,* Vol. 2, Allyn & Bacon.)

L C

An Inductor and a Capacitor

Figure 3

SOLUTION

$$\begin{aligned} E(t) &= L(t)+C(t) && \text{Given equation}\\ &= k\sin^2 2\pi Ft + k\cos^2 2\pi Ft && \text{Substitute.}\\ &= k[\sin^2 2\pi Ft+\cos^2 2\pi Ft] && \text{Factor out } k.\\ &= k(1) && \sin^2\theta+\cos^2\theta=1 \text{ (Here } \theta=2\pi Ft.)\\ &= k && \text{Identity property} \end{aligned}$$

Because k is a constant, $E(t)$ is a constant function. ✔ **Now Try Exercise 105.**

7.2 Exercises

To the student: **Exercises 1–44** are designed for practice in using the fundamental identities and applying algebraic techniques to trigonometric expressions. These skills are essential in verifying the identities that follow.

CONCEPT PREVIEW *Match each expression in Column I with its correct factorization in Column II.*

I	II
1. $x^2 - y^2$	**A.** $(x + y)(x^2 - xy + y^2)$
2. $x^3 - y^3$	**B.** $(x + y)(x - y)$
3. $x^3 + y^3$	**C.** $(x + y)^2$
4. $x^2 + 2xy + y^2$	**D.** $(x - y)(x^2 + xy + y^2)$

CONCEPT PREVIEW *Fill in the blank(s) to correctly complete each fundamental identity.*

5. $\sin^2 \theta + \cos^2 \theta =$ ______

6. $\tan^2 \theta + 1 =$ ______

7. $\sin(-\theta) =$ ______

8. $\sec(-\theta) =$ ______

9. $\tan \theta = \dfrac{1}{\_\_\_\_} = \dfrac{\sin \theta}{\_\_\_\_}$

10. $\cot \theta = \dfrac{1}{\_\_\_\_} = \dfrac{\cos \theta}{\_\_\_\_}$

Perform each indicated operation and simplify the result so that there are no quotients.

11. $\cot \theta + \dfrac{1}{\cot \theta}$

12. $\dfrac{\sec x}{\csc x} + \dfrac{\csc x}{\sec x}$

13. $\tan x(\cot x + \csc x)$

14. $\cos \beta(\sec \beta + \csc \beta)$

15. $\dfrac{1}{\csc^2 \theta} + \dfrac{1}{\sec^2 \theta}$

16. $\dfrac{\cos x}{\sec x} + \dfrac{\sin x}{\csc x}$

17. $(\sin \alpha - \cos \alpha)^2$

18. $(\tan x + \cot x)^2$

19. $(1 + \sin t)^2 + \cos^2 t$

20. $(1 + \tan \theta)^2 - 2 \tan \theta$

21. $\dfrac{1}{1 + \cos x} - \dfrac{1}{1 - \cos x}$

22. $\dfrac{1}{\sin \alpha - 1} - \dfrac{1}{\sin \alpha + 1}$

Factor each trigonometric expression.

23. $\sin^2 \theta - 1$

24. $\sec^2 \theta - 1$

25. $(\sin x + 1)^2 - (\sin x - 1)^2$

26. $(\tan x + \cot x)^2 - (\tan x - \cot x)^2$

27. $2 \sin^2 x + 3 \sin x + 1$

28. $4 \tan^2 \beta + \tan \beta - 3$

29. $\cos^4 x + 2 \cos^2 x + 1$

30. $\cot^4 x + 3 \cot^2 x + 2$

31. $\sin^3 x - \cos^3 x$

32. $\sin^3 \alpha + \cos^3 \alpha$

Each expression simplifies to a constant, a single function, or a power of a function. Use fundamental identities to simplify each expression.

33. $\tan \theta \cos \theta$

34. $\cot \alpha \sin \alpha$

35. $\sec r \cos r$

36. $\cot t \tan t$

37. $\dfrac{\sin \beta \tan \beta}{\cos \beta}$

38. $\dfrac{\csc \theta \sec \theta}{\cot \theta}$

39. $\sec^2 x - 1$

40. $\csc^2 t - 1$

41. $\dfrac{\sin^2 x}{\cos^2 x} + \sin x \csc x$

42. $\dfrac{1}{\tan^2 \alpha} + \cot \alpha \tan \alpha$

43. $1 - \dfrac{1}{\csc^2 x}$

44. $1 - \dfrac{1}{\sec^2 x}$

Verify that each equation is an identity. ***See Examples 1–5.***

45. $\dfrac{\cot \theta}{\csc \theta} = \cos \theta$

46. $\dfrac{\tan \alpha}{\sec \alpha} = \sin \alpha$

47. $\dfrac{1 - \sin^2 \beta}{\cos \beta} = \cos \beta$

48. $\dfrac{\tan^2 \alpha + 1}{\sec \alpha} = \sec \alpha$

49. $\cos^2 \theta(\tan^2 \theta + 1) = 1$

50. $\sin^2 \beta(1 + \cot^2 \beta) = 1$

51. $\cot \theta + \tan \theta = \sec \theta \csc \theta$

52. $\sin^2 \alpha + \tan^2 \alpha + \cos^2 \alpha = \sec^2 \alpha$

53. $\dfrac{\cos \alpha}{\sec \alpha} + \dfrac{\sin \alpha}{\csc \alpha} = \sec^2 \alpha - \tan^2 \alpha$

54. $\dfrac{\sin^2 \theta}{\cos \theta} = \sec \theta - \cos \theta$

55. $\sin^4 \theta - \cos^4 \theta = 2 \sin^2 \theta - 1$

56. $\sec^4 x - \sec^2 x = \tan^4 x + \tan^2 x$

57. $\dfrac{1 - \cos x}{1 + \cos x} = (\cot x - \csc x)^2$

58. $(\sec \alpha - \tan \alpha)^2 = \dfrac{1 - \sin \alpha}{1 + \sin \alpha}$

59. $\dfrac{\cos \theta + 1}{\tan^2 \theta} = \dfrac{\cos \theta}{\sec \theta - 1}$

60. $\dfrac{(\sec \theta - \tan \theta)^2 + 1}{\sec \theta \csc \theta - \tan \theta \csc \theta} = 2 \tan \theta$

61. $\dfrac{1}{1 - \sin \theta} + \dfrac{1}{1 + \sin \theta} = 2 \sec^2 \theta$

62. $\dfrac{1}{\sec \alpha - \tan \alpha} = \sec \alpha + \tan \alpha$

63. $\dfrac{\cot \alpha + 1}{\cot \alpha - 1} = \dfrac{1 + \tan \alpha}{1 - \tan \alpha}$

64. $\dfrac{\csc \theta + \cot \theta}{\tan \theta + \sin \theta} = \cot \theta \csc \theta$

65. $\dfrac{\cos \theta}{\sin \theta \cot \theta} = 1$

66. $\sin^2 \theta(1 + \cot^2 \theta) - 1 = 0$

67. $\dfrac{\sec^4 \theta - \tan^4 \theta}{\sec^2 \theta + \tan^2 \theta} = \sec^2 \theta - \tan^2 \theta$

68. $\dfrac{\sin^4 \alpha - \cos^4 \alpha}{\sin^2 \alpha - \cos^2 \alpha} = 1$

69. $\dfrac{\tan^2 t - 1}{\sec^2 t} = \dfrac{\tan t - \cot t}{\tan t + \cot t}$

70. $\dfrac{\cot^2 t - 1}{1 + \cot^2 t} = 1 - 2 \sin^2 t$

71. $\sin^2 \alpha \sec^2 \alpha + \sin^2 \alpha \csc^2 \alpha = \sec^2 \alpha$

72. $\tan^2 \alpha \sin^2 \alpha = \tan^2 \alpha + \cos^2 \alpha - 1$

73. $\dfrac{\tan x}{1 + \cos x} + \dfrac{\sin x}{1 - \cos x} = \cot x + \sec x \csc x$

74. $\dfrac{\sin \theta}{1 - \cos \theta} - \dfrac{\sin \theta \cos \theta}{1 + \cos \theta} = \csc \theta(1 + \cos^2 \theta)$

75. $\dfrac{1 + \cos x}{1 - \cos x} - \dfrac{1 - \cos x}{1 + \cos x} = 4 \cot x \csc x$

76. $\dfrac{1 + \sin \theta}{1 - \sin \theta} - \dfrac{1 - \sin \theta}{1 + \sin \theta} = 4 \tan \theta \sec \theta$

77. $\dfrac{1 - \sin \theta}{1 + \sin \theta} = \sec^2 \theta - 2 \sec \theta \tan \theta + \tan^2 \theta$

78. $\sin \theta + \cos \theta = \dfrac{\sin \theta}{1 - \cot \theta} + \dfrac{\cos \theta}{1 - \tan \theta}$

79. $\dfrac{-1}{\tan \alpha - \sec \alpha} + \dfrac{-1}{\tan \alpha + \sec \alpha} = 2 \tan \alpha$

80. $(1 + \sin x + \cos x)^2 = 2(1 + \sin x)(1 + \cos x)$

81. $(1 - \cos^2 \alpha)(1 + \cos^2 \alpha) = 2 \sin^2 \alpha - \sin^4 \alpha$

82. $(\sec \alpha + \csc \alpha)(\cos \alpha - \sin \alpha) = \cot \alpha - \tan \alpha$

83. $\dfrac{1 - \cos x}{1 + \cos x} = \csc^2 x - 2 \csc x \cot x + \cot^2 x$

84. $\dfrac{1 - \cos \theta}{1 + \cos \theta} = 2 \csc^2 \theta - 2 \csc \theta \cot \theta - 1$

85. $(2 \sin x + \cos x)^2 + (2 \cos x - \sin x)^2 = 5$

86. $\sin^2 x(1 + \cot x) + \cos^2 x(1 - \tan x) + \cot^2 x = \csc^2 x$

87. $\sec x - \cos x + \csc x - \sin x - \sin x \tan x = \cos x \cot x$

88. $\sin^3 \theta + \cos^3 \theta = (\cos \theta + \sin \theta)(1 - \cos \theta \sin \theta)$

Graph each expression and use the graph to make a conjecture, predicting what might be an identity. Then verify your conjecture algebraically.

89. $(\sec \theta + \tan \theta)(1 - \sin \theta)$

90. $(\csc \theta + \cot \theta)(\sec \theta - 1)$

91. $\dfrac{\cos \theta + 1}{\sin \theta + \tan \theta}$

92. $\tan \theta \sin \theta + \cos \theta$

Graph the expressions on each side of the equals symbol to determine whether the equation might be an identity. (Note: Use a domain whose length is at least 2π.) If the equation looks like an identity, then verify it algebraically. **See Example 1.**

93. $\dfrac{2 + 5 \cos x}{\sin x} = 2 \csc x + 5 \cot x$

94. $1 + \cot^2 x = \dfrac{\sec^2 x}{\sec^2 x - 1}$

95. $\dfrac{\tan x - \cot x}{\tan x + \cot x} = 2 \sin^2 x$

96. $\dfrac{1}{1 + \sin x} + \dfrac{1}{1 - \sin x} = \sec^2 x$

By substituting a number for t, show that the equation is not an identity.

97. $\sin(\csc t) = 1$

98. $\sqrt{\cos^2 t} = \cos t$

99. $\csc t = \sqrt{1 + \cot^2 t}$

100. $\cos t = \sqrt{1 - \sin^2 t}$

(Modeling) Work each problem.

101. ***Intensity of a Lamp*** According to **Lambert's law,** the intensity of light from a single source on a flat surface at point P is given by

$$I = k \cos^2 \theta,$$

where k is a constant. (*Source:* Winter, C., *Solar Power Plants,* Springer-Verlag.)

(a) Write I in terms of the sine function.

(b) Why does the maximum value of I occur when $\theta = 0$?

102. ***Oscillating Spring*** The distance or displacement y of a weight attached to an oscillating spring from its natural position is modeled by

$$y = 4 \cos 2\pi t,$$

where t is time in seconds. Potential energy is the energy of position and is given by

$$P = ky^2,$$

where k is a constant. The weight has the greatest potential energy when the spring is stretched the most. (*Source*: Weidner, R. and R. Sells, *Elementary Classical Physics,* Vol. 2, Allyn & Bacon.)

(a) Write an expression for P that involves the cosine function.

(b) Use a fundamental identity to write P in terms of $\sin 2\pi t$.

(Modeling) Radio Tuners See **Example 6.** Let the energy stored in the inductor be given by

$$L(t) = 3\cos^2 6{,}000{,}000t$$

and let the energy stored in the capacitor be given by

$$C(t) = 3\sin^2 6{,}000{,}000t,$$

where t is time in seconds. The total energy E in the circuit is given by

$$E(t) = L(t) + C(t).$$

103. Graph L, C, and E in the window $[0, 10^{-6}]$ by $[-1, 4]$, with Xscl $= 10^{-7}$ and Yscl $= 1$. Interpret the graph.

104. Make a table of values for L, C, and E starting at $t = 0$, incrementing by 10^{-7}. Interpret the results.

105. Use a fundamental identity to derive a simplified expression for $E(t)$.

7.3 Sum and Difference Identities

- Cosine Sum and Difference Identities
- Cofunction Identities
- Sine and Tangent Sum and Difference Identities
- Applications of the Sum and Difference Identities
- Verifying an Identity

Cosine Sum and Difference Identities Several examples presented earlier should have convinced you by now that

$$\cos(A - B) \textit{ does not equal } \cos A - \cos B.$$

For example, if $A = \frac{\pi}{2}$ and $B = 0$, then

$$\cos(A - B) = \cos\left(\frac{\pi}{2} - 0\right) = \cos\frac{\pi}{2} = 0,$$

while $$\cos A - \cos B = \cos\frac{\pi}{2} - \cos 0 = 0 - 1 = -1.$$

To derive a formula for $\cos(A - B)$, we start by locating angles A and B in standard position on a unit circle, with $B < A$. Let S and Q be the points where the terminal sides of angles A and B, respectively, intersect the circle. Let P be the point $(1, 0)$, and locate point R on the unit circle so that angle POR equals the difference $A - B$. See **Figure 4.**

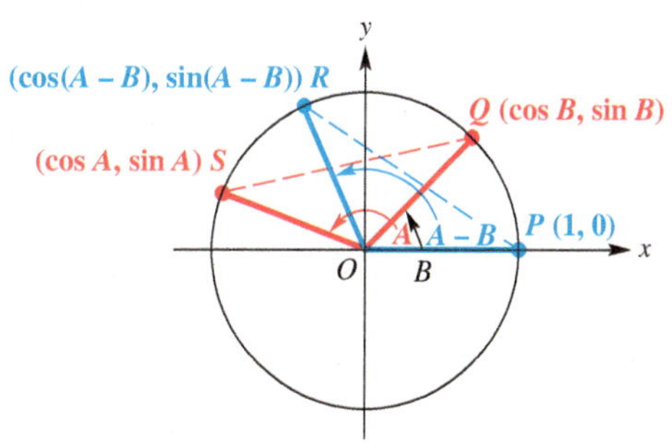

Figure 4

Because point Q is on the unit circle, the x-coordinate of Q is the cosine of angle B, while the y-coordinate of Q is the sine of angle B.

$$Q \text{ has coordinates } (\cos B, \sin B).$$

In the same way,

$$S \text{ has coordinates } (\cos A, \sin A),$$

and $$R \text{ has coordinates } (\cos(A-B), \sin(A-B)).$$

Angle SOQ also equals $A - B$. The central angles SOQ and POR are equal, so chords PR and SQ are equal. Because $PR = SQ$, by the distance formula,

$$\sqrt{[\cos(A-B)-1]^2 + [\sin(A-B)-0]^2} = \sqrt{(\cos A - \cos B)^2 + (\sin A - \sin B)^2}. \quad d = \sqrt{(x_2-x_1)^2 + (y_2-y_1)^2}$$

Square each side of this equation. Then square each expression, remembering that for any values of x and y, $(x-y)^2 = x^2 - 2xy + y^2$.

$$[\cos(A-B)-1]^2 + [\sin(A-B)-0]^2 = (\cos A - \cos B)^2 + (\sin A - \sin B)^2$$

$$\cos^2(A-B) - 2\cos(A-B) + 1 + \sin^2(A-B) = \cos^2 A - 2\cos A \cos B + \cos^2 B + \sin^2 A - 2\sin A \sin B + \sin^2 B$$

For any value of x, $\sin^2 x + \cos^2 x = 1$, so we can rewrite the equation.

$$2 - 2\cos(A-B) = 2 - 2\cos A \cos B - 2\sin A \sin B \quad \text{Use } \sin^2 x + \cos^2 x = 1 \text{ three times and add like terms.}$$

$$-2\cos(A-B) = -2\cos A \cos B - 2\sin A \sin B \quad \text{Subtract 2.}$$

$$\mathbf{\cos(A-B) = \cos A \cos B + \sin A \sin B} \quad \text{Divide by } -2.$$

This is the identity for $\cos(A-B)$. Although **Figure 4** on the previous page shows angles A and B in the second and first quadrants, respectively, this result is the same for any values of these angles.

To find a similar expression for $\cos(A+B)$, rewrite $A+B$ as $A-(-B)$ and use the identity for $\cos(A-B)$.

$$\cos(A+B) = \cos[A-(-B)] \quad \text{Definition of subtraction}$$

$$= \cos A \cos(-B) + \sin A \sin(-B) \quad \text{Cosine difference identity}$$

$$= \cos A \cos B + \sin A(-\sin B) \quad \text{Even-odd identities}$$

$$\mathbf{\cos(A+B) = \cos A \cos B - \sin A \sin B} \quad \text{Multiply.}$$

Cosine of a Sum or Difference

$$\cos(A+B) = \cos A \cos B - \sin A \sin B$$

$$\cos(A-B) = \cos A \cos B + \sin A \sin B$$

These identities are important in calculus and useful in certain applications. For example, the method shown in **Example 1** can be applied to find an exact value for $\cos 15°$.

EXAMPLE 1 Finding Exact Cosine Function Values

Find the *exact* value of each expression.

(a) $\cos 15^\circ$ **(b)** $\cos \frac{5\pi}{12}$ **(c)** $\cos 87^\circ \cos 93^\circ - \sin 87^\circ \sin 93^\circ$

SOLUTION

(a) To find cos 15°, we write 15° as the sum or difference of two angles with known function values, such as 45° and 30°, because

$$15^\circ = 45^\circ - 30^\circ. \quad \text{(We could also use } 60^\circ - 45^\circ.)$$

Then we use the cosine difference identity.

$\cos 15^\circ$

$= \cos(45^\circ - 30^\circ)$ — $15^\circ = 45^\circ - 30^\circ$

$= \cos 45^\circ \cos 30^\circ + \sin 45^\circ \sin 30^\circ$ — Cosine difference identity

$= \frac{\sqrt{2}}{2} \cdot \frac{\sqrt{3}}{2} + \frac{\sqrt{2}}{2} \cdot \frac{1}{2}$ — Substitute known values.

$= \frac{\sqrt{6} + \sqrt{2}}{4}$ — Multiply, and then add fractions.

(b) $\cos \frac{5\pi}{12}$

$= \cos\left(\frac{\pi}{6} + \frac{\pi}{4}\right)$ — $\frac{\pi}{6} = \frac{2\pi}{12}$ and $\frac{\pi}{4} = \frac{3\pi}{12}$

$= \cos \frac{\pi}{6} \cos \frac{\pi}{4} - \sin \frac{\pi}{6} \sin \frac{\pi}{4}$ — Cosine sum identity

$= \frac{\sqrt{3}}{2} \cdot \frac{\sqrt{2}}{2} - \frac{1}{2} \cdot \frac{\sqrt{2}}{2}$ — Substitute known values.

$= \frac{\sqrt{6} - \sqrt{2}}{4}$ — Multiply, and then subtract fractions.

```
NORMAL FLOAT AUTO REAL RADIAN MP
cos(5π/12)
                    .2588190451
(√6-√2)/4
                    .2588190451
```

This screen supports the solution in **Example 1(b)** by showing that the decimal approximations for $\cos \frac{5\pi}{12}$ and $\frac{\sqrt{6} - \sqrt{2}}{4}$ agree.

(c) $\cos 87^\circ \cos 93^\circ - \sin 87^\circ \sin 93^\circ$

$= \cos(87^\circ + 93^\circ)$ — Cosine sum identity

$= \cos 180^\circ$ — Add.

$= -1$ — $\cos 180^\circ = -1$

✔ **Now Try Exercises 11, 15, and 19.**

Cofunction Identities We can use the identity for the cosine of the difference of two angles and the fundamental identities to derive *cofunction identities,* presented previously for values of θ in the interval $[0^\circ, 90^\circ]$. For example, substituting 90° for A and θ for B in the identity for $\cos(A - B)$ gives the following.

$\mathbf{\cos(90^\circ - \theta)} = \cos 90^\circ \cos \theta + \sin 90^\circ \sin \theta$ — Cosine difference identity

$= 0 \cdot \cos \theta + 1 \cdot \sin \theta$ — $\cos 90^\circ = 0$ and $\sin 90^\circ = 1$

$= \mathbf{\sin \theta}$ — Simplify.

This result is true for *any* value of θ because the identity for $\cos(A - B)$ is true for any values of A and B.

Cofunction Identities

The following identities hold for any angle θ for which the functions are defined.

$$\cos(90° - \theta) = \sin\theta \qquad \cot(90° - \theta) = \tan\theta$$
$$\sin(90° - \theta) = \cos\theta \qquad \sec(90° - \theta) = \csc\theta$$
$$\tan(90° - \theta) = \cot\theta \qquad \csc(90° - \theta) = \sec\theta$$

The same identities can be obtained for a real number domain by replacing 90° with $\frac{\pi}{2}$.

EXAMPLE 2 Using Cofunction Identities to Find θ

Find one value of θ or x that satisfies each of the following.

(a) $\cot\theta = \tan 25°$ **(b)** $\sin\theta = \cos(-30°)$ **(c)** $\csc\frac{3\pi}{4} = \sec x$

SOLUTION

(a) Because tangent and cotangent are cofunctions, $\tan(90° - \theta) = \cot\theta$.

$$\cot\theta = \tan 25°$$
$$\tan(90° - \theta) = \tan 25° \quad \text{Cofunction identity}$$
$$90° - \theta = 25° \quad \text{Set angle measures equal.}$$
$$\theta = 65° \quad \text{Solve for } \theta.$$

(b)

$$\sin\theta = \cos(-30°)$$
$$\cos(90° - \theta) = \cos(-30°) \quad \text{Cofunction identity}$$
$$90° - \theta = -30° \quad \text{Set angle measures equal.}$$
$$\theta = 120° \quad \text{Solve for } \theta.$$

(c) $\csc\frac{3\pi}{4} = \sec x$

$$\csc\frac{3\pi}{4} = \csc\left(\frac{\pi}{2} - x\right) \quad \text{Cofunction identity}$$
$$\frac{3\pi}{4} = \frac{\pi}{2} - x \quad \text{Set angle measures equal.}$$
$$x = -\frac{\pi}{4} \quad \text{Solve for } x\text{: } \frac{\pi}{2} - \frac{3\pi}{4} = \frac{2\pi}{4} - \frac{3\pi}{4} = -\frac{\pi}{4}.$$

✔ **Now Try Exercises 35 and 39.**

Sine and Tangent Sum and Difference Identities We can derive identities for sine by replacing θ in $\sin\theta = \cos(90° - \theta)$ with $A + B$.

$$\sin(A + B) = \cos[90° - (A + B)] \quad \text{Cofunction identity}$$
$$= \cos[(90° - A) - B] \quad \text{Distribute negative sign and regroup.}$$
$$= \cos(90° - A)\cos B + \sin(90° - A)\sin B \quad \text{Cosine difference identity}$$
$$\mathbf{\sin(A + B) = \sin A\cos B + \cos A\sin B} \quad \text{Cofunction identities}$$

Now we write $\sin(A - B)$ as $\sin[A + (-B)]$ and use the identity just found for $\sin(A + B)$.

$$\sin(A - B) = \sin[A + (-B)] \qquad \text{Definition of subtraction}$$

$$= \sin A \cos(-B) + \cos A \sin(-B) \qquad \text{Sine sum identity}$$

$$\mathbf{\sin(A - B) = \sin A \cos B - \cos A \sin B} \qquad \text{Even-odd identities}$$

Sine of a Sum or Difference

$$\mathbf{\sin(A + B) = \sin A \cos B + \cos A \sin B}$$

$$\mathbf{\sin(A - B) = \sin A \cos B - \cos A \sin B}$$

We can derive an identity for $\tan(A + B)$ as follows.

$$\tan(A + B) = \frac{\sin(A + B)}{\cos(A + B)} \qquad \text{Fundamental identity}$$

We express this result in terms of the tangent function.

$$= \frac{\sin A \cos B + \cos A \sin B}{\cos A \cos B - \sin A \sin B} \qquad \text{Sum identities}$$

$$= \frac{\dfrac{\sin A \cos B + \cos A \sin B}{1}}{\dfrac{\cos A \cos B - \sin A \sin B}{1}} \cdot \frac{\dfrac{1}{\cos A \cos B}}{\dfrac{1}{\cos A \cos B}} \qquad \text{Multiply by 1, where } 1 = \frac{\frac{1}{\cos A \cos B}}{\frac{1}{\cos A \cos B}}.$$

$$= \frac{\dfrac{\sin A \cos B}{\cos A \cos B} + \dfrac{\cos A \sin B}{\cos A \cos B}}{\dfrac{\cos A \cos B}{\cos A \cos B} - \dfrac{\sin A \sin B}{\cos A \cos B}} \qquad \text{Multiply numerators. Multiply denominators.}$$

$$= \frac{\dfrac{\sin A}{\cos A} + \dfrac{\sin B}{\cos B}}{1 - \dfrac{\sin A}{\cos A} \cdot \dfrac{\sin B}{\cos B}} \qquad \text{Simplify.}$$

$$\mathbf{\tan(A + B) = \frac{\tan A + \tan B}{1 - \tan A \tan B}} \qquad \frac{\sin\theta}{\cos\theta} = \tan\theta$$

We can replace B with $-B$ and use the fact that

$$\tan(-B) = -\tan B$$

to obtain an identity for the tangent of the difference of two angles, as seen below.

Tangent of a Sum or Difference

$$\mathbf{\tan(A + B) = \frac{\tan A + \tan B}{1 - \tan A \tan B}} \qquad \mathbf{\tan(A - B) = \frac{\tan A - \tan B}{1 + \tan A \tan B}}$$

Applications of the Sum and Difference Identities

EXAMPLE 3 Finding Exact Sine and Tangent Function Values

Find the *exact* value of each expression.

(a) $\sin 75^\circ$ **(b)** $\tan \frac{7\pi}{12}$ **(c)** $\sin 40^\circ \cos 160^\circ - \cos 40^\circ \sin 160^\circ$

SOLUTION

(a) $\sin 75^\circ$

$= \sin(45^\circ + 30^\circ)$ — $75^\circ = 45^\circ + 30^\circ$

$= \sin 45^\circ \cos 30^\circ + \cos 45^\circ \sin 30^\circ$ — Sine sum identity

$= \frac{\sqrt{2}}{2} \cdot \frac{\sqrt{3}}{2} + \frac{\sqrt{2}}{2} \cdot \frac{1}{2}$ — Substitute known values.

$= \frac{\sqrt{6} + \sqrt{2}}{4}$ — Multiply, and then add fractions.

(b) $\tan \frac{7\pi}{12}$

$= \tan\left(\frac{\pi}{3} + \frac{\pi}{4}\right)$ — $\frac{\pi}{3} = \frac{4\pi}{12}$ and $\frac{\pi}{4} = \frac{3\pi}{12}$

$= \frac{\tan \frac{\pi}{3} + \tan \frac{\pi}{4}}{1 - \tan \frac{\pi}{3} \tan \frac{\pi}{4}}$ — Tangent sum identity

$= \frac{\sqrt{3} + 1}{1 - \sqrt{3} \cdot 1}$ — Substitute known values.

$= \frac{\sqrt{3} + 1}{1 - \sqrt{3}} \cdot \frac{1 + \sqrt{3}}{1 + \sqrt{3}}$ — Rationalize the denominator.

$= \frac{\sqrt{3} + 3 + 1 + \sqrt{3}}{1 - 3}$ — $(a + b)(c + d) = ac + ad + bc + bd$; $(x - y)(x + y) = x^2 - y^2$

$= \frac{4 + 2\sqrt{3}}{-2}$ — Combine like terms.

Factor first. Then divide out the common factor.

$= \frac{2(2 + \sqrt{3})}{2(-1)}$ — Factor out 2.

$= -2 - \sqrt{3}$ — Write in lowest terms.

(c) $\sin 40^\circ \cos 160^\circ - \cos 40^\circ \sin 160^\circ$

$= \sin(40^\circ - 160^\circ)$ — Sine difference identity

$= \sin(-120^\circ)$ — Subtract.

$= -\sin 120^\circ$ — Even-odd identity

$= -\frac{\sqrt{3}}{2}$ — Substitute the known value.

✔ **Now Try Exercises 41, 47, and 57.**

EXAMPLE 4 Writing Functions as Expressions Involving Functions of θ

Write each function as an expression involving functions of θ alone.

(a) $\cos(30° + \theta)$ **(b)** $\tan(45° - \theta)$ **(c)** $\sin(180° - \theta)$

SOLUTION

(a) $\cos(30° + \theta)$

$= \cos 30° \cos\theta - \sin 30° \sin\theta$ Cosine sum identity

$= \frac{\sqrt{3}}{2}\cos\theta - \frac{1}{2}\sin\theta$ $\cos 30° = \frac{\sqrt{3}}{2}$ and $\sin 30° = \frac{1}{2}$

$= \frac{\sqrt{3}\cos\theta - \sin\theta}{2}$ $\frac{a}{b} \cdot c = \frac{ac}{b}$; Subtract fractions.

(b) $\tan(45° - \theta)$

$= \frac{\tan 45° - \tan\theta}{1 + \tan 45° \tan\theta}$ Tangent difference identity

$= \frac{1 - \tan\theta}{1 + 1 \cdot \tan\theta}$ $\tan 45° = 1$

$= \frac{1 - \tan\theta}{1 + \tan\theta}$ Multiply.

(c) $\sin(180° - \theta)$

$= \sin 180° \cos\theta - \cos 180° \sin\theta$ Sine difference identity

$= 0 \cdot \cos\theta - (-1)\sin\theta$ $\sin 180° = 0$ and $\cos 180° = -1$

$= \sin\theta$ Simplify.

✔ **Now Try Exercises 65, 71, and 75.**

EXAMPLE 5 Finding Function Values and the Quadrant of $A + B$

Suppose that A and B are angles in standard position such that $\sin A = \frac{4}{5}$, $\frac{\pi}{2} < A < \pi$, and $\cos B = -\frac{5}{13}$, $\pi < B < \frac{3\pi}{2}$. Find each of the following.

(a) $\sin(A + B)$ **(b)** $\tan(A + B)$ **(c)** the quadrant of $A + B$

SOLUTION

(a) The identity for $\sin(A + B)$ involves $\sin A$, $\cos A$, $\sin B$, and $\cos B$. We are given values of $\sin A$ and $\cos B$. We must find values of $\cos A$ and $\sin B$.

$\sin^2 A + \cos^2 A = 1$ Fundamental identity

$\left(\frac{4}{5}\right)^2 + \cos^2 A = 1$ $\sin A = \frac{4}{5}$

$\frac{16}{25} + \cos^2 A = 1$ Square $\frac{4}{5}$.

$\cos^2 A = \frac{9}{25}$ Subtract $\frac{16}{25}$.

Pay attention to signs. $\cos A = -\frac{3}{5}$ Take square roots. Because A is in quadrant II, $\cos A < 0$.

In the same way, $\sin B = -\frac{12}{13}$. Now find $\sin(A + B)$.

$$\sin(A+B) = \sin A \cos B + \cos A \sin B \quad \text{Sine sum identity}$$

$$= \frac{4}{5}\left(-\frac{5}{13}\right) + \left(-\frac{3}{5}\right)\left(-\frac{12}{13}\right) \quad \text{Substitute the given values for } \sin A \text{ and } \cos B \text{ and the values found for } \cos A \text{ and } \sin B.$$

$$= -\frac{20}{65} + \frac{36}{65} \quad \text{Multiply.}$$

$$\sin(A+B) = \frac{16}{65} \quad \text{Add.}$$

(b) To find $\tan(A + B)$, use the values of sine and cosine from part (a), $\sin A = \frac{4}{5}$, $\cos A = -\frac{3}{5}$, $\sin B = -\frac{12}{13}$, and $\cos B = -\frac{5}{13}$, to obtain $\tan A$ and $\tan B$.

$$\tan A = \frac{\sin A}{\cos A} = \frac{\frac{4}{5}}{-\frac{3}{5}} = \frac{4}{5} \div \left(-\frac{3}{5}\right) = \frac{4}{5} \cdot \left(-\frac{5}{3}\right)$$

$$\tan A = -\frac{4}{3}$$

$$\tan B = \frac{\sin B}{\cos B} = \frac{-\frac{12}{13}}{-\frac{5}{13}} = -\frac{12}{13} \div \left(-\frac{5}{13}\right) = -\frac{12}{13} \cdot \left(-\frac{13}{5}\right)$$

$$\tan B = \frac{12}{5}$$

$$\tan(A+B) = \frac{\tan A + \tan B}{1 - \tan A \tan B} \quad \text{Tangent sum identity}$$

$$= \frac{\left(-\frac{4}{3}\right) + \frac{12}{5}}{1 - \left(-\frac{4}{3}\right)\left(\frac{12}{5}\right)} \quad \text{Substitute.}$$

$$= \frac{\frac{16}{15}}{1 + \frac{48}{15}} \quad \text{Perform the indicated operations.}$$

$$= \frac{\frac{16}{15}}{\frac{63}{15}} \quad \text{Add terms in the denominator.}$$

$$= \frac{16}{15} \div \frac{63}{15} \quad \text{Simplify the complex fraction.}$$

$$= \frac{16}{15} \cdot \frac{15}{63} \quad \text{Definition of division}$$

$$\tan(A+B) = \frac{16}{63} \quad \text{Multiply.}$$

(c) $\sin(A+B) = \frac{16}{65}$ and $\tan(A+B) = \frac{16}{63}$ See parts (a) and (b).

Both are positive. Therefore, $A + B$ must be in quadrant I, because it is the only quadrant in which both sine and tangent are positive.

✔ **Now Try Exercise 83.**

EXAMPLE 6 Applying the Cosine Difference Identity to Voltage

Common household electric current is called **alternating current** because the current alternates direction within the wires. The voltage V in a typical 115-volt outlet can be expressed by the function

$$V(t) = 163 \sin \omega t,$$

where ω is the angular speed (in radians per second) of the rotating generator at the electrical plant and t is time in seconds. (*Source*: Bell, D., *Fundamentals of Electric Circuits,* Fourth Edition, Prentice-Hall.)

(a) It is essential for electric generators to rotate at precisely 60 cycles per sec so household appliances and computers will function properly. Determine ω for these electric generators.

(b) Graph V in the window $[0, 0.05]$ by $[-200, 200]$.

(c) Determine a value of ϕ so that the graph of

$$V(t) = 163 \cos(\omega t - \phi) \quad \text{is the same as the graph of} \quad V(t) = 163 \sin \omega t.$$

SOLUTION

(a) We convert 60 cycles per sec to radians per second as follows.

$$\omega = \frac{60 \text{ cycles}}{1 \text{ sec}} \cdot \frac{2\pi \text{ radians}}{1 \text{ cycle}} = 120\pi \text{ radians per sec.}$$

(b)

$$V(t) = 163 \sin \omega t$$

$$V(t) = 163 \sin 120\pi t \quad \text{From part (a), } \omega = 120\pi \text{ radians per sec.}$$

Because the amplitude of the function $V(t)$ is 163, an appropriate interval for the range is $[-200, 200]$, as shown in the graph in **Figure 5.**

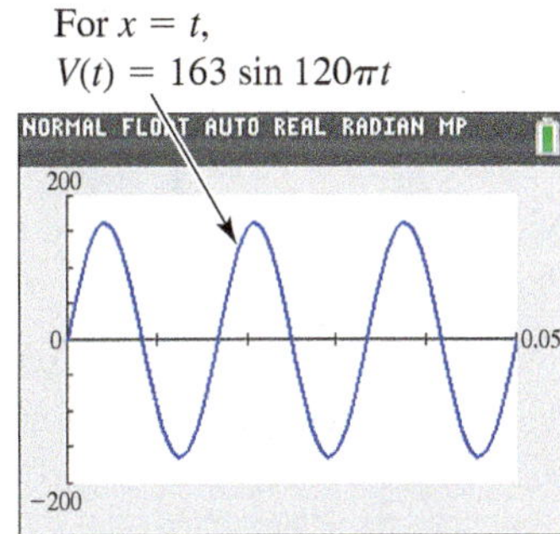

Figure 5

(c) Use the even-odd identity for cosine and a cofunction identity.

$$\cos\left(x - \frac{\pi}{2}\right) = \cos\left[-\left(\frac{\pi}{2} - x\right)\right] = \cos\left(\frac{\pi}{2} - x\right) = \sin x$$

Therefore, if $\phi = \frac{\pi}{2}$, then

$$V(t) = 163 \cos(\omega t - \phi) = 163 \cos\left(\omega t - \frac{\pi}{2}\right) = 163 \sin \omega t.$$

✔ **Now Try Exercise 103.**

Verifying an Identity

EXAMPLE 7 **Verifying an Identity**

Verify that the equation is an identity.

$$\sin\left(\frac{\pi}{6}+\theta\right)+\cos\left(\frac{\pi}{3}+\theta\right)=\cos\theta$$

SOLUTION Work on the left side, using the sine and cosine sum identities.

$$\sin\left(\frac{\pi}{6}+\theta\right)+\cos\left(\frac{\pi}{3}+\theta\right)$$

$$=\left(\sin\frac{\pi}{6}\cos\theta+\cos\frac{\pi}{6}\sin\theta\right)+\left(\cos\frac{\pi}{3}\cos\theta-\sin\frac{\pi}{3}\sin\theta\right)$$ Sine sum identity; cosine sum identity

$$=\left(\frac{1}{2}\cos\theta+\frac{\sqrt{3}}{2}\sin\theta\right)+\left(\frac{1}{2}\cos\theta-\frac{\sqrt{3}}{2}\sin\theta\right)$$ $\sin\frac{\pi}{6}=\frac{1}{2}$; $\cos\frac{\pi}{6}=\frac{\sqrt{3}}{2}$; $\cos\frac{\pi}{3}=\frac{1}{2}$; $\sin\frac{\pi}{3}=\frac{\sqrt{3}}{2}$

$$=\frac{1}{2}\cos\theta+\frac{1}{2}\cos\theta$$ Simplify.

$$=\cos\theta$$ Add.

The left side is identical to the right side, so the given equation is an identity.

✔ **Now Try Exercise 95.**

7.3 Exercises

CONCEPT PREVIEW *Match each expression in Column I with the correct expression in Column II to form an identity. Choices may be used once, more than once, or not at all.*

I	II
1. $\cos(x+y)=$ _____	**A.** $\cos x\cos y+\sin x\sin y$
2. $\cos(x-y)=$ _____	**B.** $-\cos x$
3. $\cos\left(\frac{\pi}{2}-x\right)=$ _____	**C.** $-\sin x$
4. $\sin\left(\frac{\pi}{2}-x\right)=$ _____	**D.** $\sin x$
5. $\cos\left(x-\frac{\pi}{2}\right)=$ _____	**E.** $\cos x\cos y-\sin x\sin y$
6. $\sin\left(x-\frac{\pi}{2}\right)=$ _____	**F.** $\cos x$

CONCEPT PREVIEW *Match each expression in Column I with the correct expression in Column II to form an identity.*

I	II
7. $\sin(A+B)$	**A.** $\sin A\cos B-\cos A\sin B$
8. $\sin(A-B)$	**B.** $\frac{\tan A+\tan B}{1-\tan A\tan B}$
9. $\tan(A+B)$	**C.** $\frac{\tan A-\tan B}{1+\tan A\tan B}$
10. $\tan(A-B)$	**D.** $\sin A\cos B+\cos A\sin B$

Find the exact value of each expression. (Do not use a calculator.) **See Example 1.**

11. $\cos 75°$

12. $\cos(-15°)$

13. $\cos(-105°)$
$(Hint: -105° = -60° + (-45°))$

14. $\cos 105°$
$(Hint: 105° = 60° + 45°)$

15. $\cos \frac{7\pi}{12}$

16. $\cos \frac{\pi}{12}$

17. $\cos\left(-\frac{\pi}{12}\right)$

18. $\cos\left(-\frac{7\pi}{12}\right)$

19. $\cos 40° \cos 50° - \sin 40° \sin 50°$

20. $\cos \frac{7\pi}{9} \cos \frac{2\pi}{9} - \sin \frac{7\pi}{9} \sin \frac{2\pi}{9}$

Write each function value in terms of the cofunction of a complementary angle. **See Example 2.**

21. $\tan 87°$

22. $\sin 15°$

23. $\cos \frac{\pi}{12}$

24. $\sin \frac{2\pi}{5}$

25. $\csc 14° 24'$

26. $\sin 142° 14'$

27. $\sin \frac{5\pi}{8}$

28. $\cot \frac{9\pi}{10}$

Use identities to fill in each blank with the appropriate trigonometric function name. **See Example 2.**

29. $\cot \frac{\pi}{3} = \underline{\qquad} \frac{\pi}{6}$

30. $\sin \frac{2\pi}{3} = \underline{\qquad}\left(-\frac{\pi}{6}\right)$

31. $\underline{\qquad} 33° = \sin 57°$

32. $\underline{\qquad} 72° = \cot 18°$

33. $\cos 70° = \frac{1}{\underline{\qquad} 20°}$

34. $\tan 24° = \frac{1}{\underline{\qquad} 66°}$

Find one value of θ or x that satisfies each of the following. **See Example 2.**

35. $\tan \theta = \cot(45° + 2\theta)$

36. $\sin \theta = \cos(2\theta + 30°)$

37. $\sec x = \csc \frac{2\pi}{3}$

38. $\cos x = \sin \frac{\pi}{12}$

39. $\sin(3\theta - 15°) = \cos(\theta + 25°)$

40. $\cot(\theta - 10°) = \tan(2\theta - 20°)$

Find the exact value of each expression. **See Example 3.**

41. $\sin 165°$

42. $\sin 255°$

43. $\tan 165°$

44. $\tan 285°$

45. $\sin \frac{5\pi}{12}$

46. $\sin \frac{13\pi}{12}$

47. $\tan \frac{\pi}{12}$

48. $\tan \frac{5\pi}{12}$

49. $\sin \frac{7\pi}{12}$

50. $\sin \frac{\pi}{12}$

51. $\sin\left(-\frac{7\pi}{12}\right)$

52. $\sin\left(-\frac{5\pi}{12}\right)$

53. $\tan\left(-\frac{5\pi}{12}\right)$

54. $\tan\left(-\frac{7\pi}{12}\right)$

55. $\tan \frac{11\pi}{12}$

56. $\sin\left(-\frac{13\pi}{12}\right)$

57. $\sin 76° \cos 31° - \cos 76° \sin 31°$

58. $\sin 40° \cos 50° + \cos 40° \sin 50°$

59. $\sin \frac{\pi}{5} \cos \frac{3\pi}{10} + \cos \frac{\pi}{5} \sin \frac{3\pi}{10}$

60. $\sin \frac{5\pi}{9} \cos \frac{\pi}{18} - \cos \frac{5\pi}{9} \sin \frac{\pi}{18}$

61. $\frac{\tan 80° + \tan 55°}{1 - \tan 80° \tan 55°}$

62. $\frac{\tan 80° - \tan(-55°)}{1 + \tan 80° \tan(-55°)}$

Write each function as an expression involving functions of θ or x alone. **See Example 4.**

63. $\cos(\theta - 180°)$ **64.** $\cos(\theta - 270°)$ **65.** $\cos(180° + \theta)$ **66.** $\cos(270° + \theta)$

67. $\cos(60° + \theta)$ **68.** $\cos(45° - \theta)$ **69.** $\cos\left(\frac{3\pi}{4} - x\right)$ **70.** $\sin(45° + \theta)$

71. $\tan(\theta + 30°)$ **72.** $\tan\left(\frac{\pi}{4} + x\right)$ **73.** $\sin\left(\frac{\pi}{4} + x\right)$ **74.** $\sin\left(\frac{3\pi}{4} - x\right)$

75. $\sin(270° - \theta)$ **76.** $\tan(180° + \theta)$ **77.** $\tan(2\pi - x)$ **78.** $\sin(\pi + x)$

Find $\cos(s + t)$ *and* $\cos(s - t)$. **See Example 5.**

79. $\sin s = \frac{3}{5}$ and $\sin t = -\frac{12}{13}$, s in quadrant I and t in quadrant III

80. $\cos s = -\frac{8}{17}$ and $\cos t = -\frac{3}{5}$, s and t in quadrant III

81. $\cos s = -\frac{1}{5}$ and $\sin t = \frac{3}{5}$, s and t in quadrant II

82. $\sin s = \frac{2}{3}$ and $\sin t = -\frac{1}{3}$, s in quadrant II and t in quadrant IV

Use the given information to find **(a)** $\sin(s + t)$, **(b)** $\tan(s + t)$, *and* **(c)** *the quadrant of* $s + t$. **See Example 5.**

83. $\cos s = \frac{3}{5}$ and $\sin t = \frac{5}{13}$, s and t in quadrant I

84. $\sin s = \frac{3}{5}$ and $\sin t = -\frac{12}{13}$, s in quadrant I and t in quadrant III

85. $\cos s = -\frac{8}{17}$ and $\cos t = -\frac{3}{5}$, s and t in quadrant III

86. $\cos s = -\frac{15}{17}$ and $\sin t = \frac{4}{5}$, s in quadrant II and t in quadrant I

87. $\sin s = \frac{2}{3}$ and $\sin t = -\frac{1}{3}$, s in quadrant II and t in quadrant IV

88. $\cos s = -\frac{1}{5}$ and $\sin t = \frac{3}{5}$, s and t in quadrant II

Graph each expression and use the graph to make a conjecture, predicting what might be an identity. Then verify your conjecture algebraically.

89. $\sin\left(\frac{\pi}{2} + \theta\right)$ **90.** $\sin\left(\frac{3\pi}{2} + \theta\right)$ **91.** $\tan\left(\frac{\pi}{2} + \theta\right)$ **92.** $\tan\left(\frac{\pi}{2} - \theta\right)$

Verify that each equation is an identity. **See Example 7.**

93. $\sin 2x = 2 \sin x \cos x$ (*Hint*: $\sin 2x = \sin(x + x)$)

94. $\sin(x + y) + \sin(x - y) = 2 \sin x \cos y$

95. $\sin\left(\frac{7\pi}{6} + x\right) - \cos\left(\frac{2\pi}{3} + x\right) = 0$

96. $\tan(x - y) - \tan(y - x) = \frac{2(\tan x - \tan y)}{1 + \tan x \tan y}$

97. $\frac{\cos(\alpha - \beta)}{\cos \alpha \sin \beta} = \tan \alpha + \cot \beta$ **98.** $\frac{\sin(s + t)}{\cos s \cos t} = \tan s + \tan t$

99. $\frac{\sin(x - y)}{\sin(x + y)} = \frac{\tan x - \tan y}{\tan x + \tan y}$ **100.** $\frac{\sin(x + y)}{\cos(x - y)} = \frac{\cot x + \cot y}{1 + \cot x \cot y}$

101. $\frac{\sin(s - t)}{\sin t} + \frac{\cos(s - t)}{\cos t} = \frac{\sin s}{\sin t \cos t}$ **102.** $\frac{\tan(\alpha + \beta) - \tan \beta}{1 + \tan(\alpha + \beta) \tan \beta} = \tan \alpha$

(Modeling) Solve each problem. **See Example 6.**

103. ***Electric Current*** The voltage V in a typical 115-volt outlet can be expressed by the function

$$V(t) = 163 \sin 120\pi t,$$

where 120π is the angular speed (in radians per second) of the rotating generator at an electrical power plant, and t is time in seconds. (*Source:* Bell, D., *Fundamentals of Electric Circuits,* Fourth Edition, Prentice-Hall.)

(a) How many times does the current oscillate in 0.05 sec?

(b) What are the maximum and minimum voltages in this outlet?

(c) Is the voltage always equal to 115 volts?

104. ***Sound Waves*** Sound is a result of waves applying pressure to a person's eardrum. For a pure sound wave radiating outward in a spherical shape, the trigonometric function

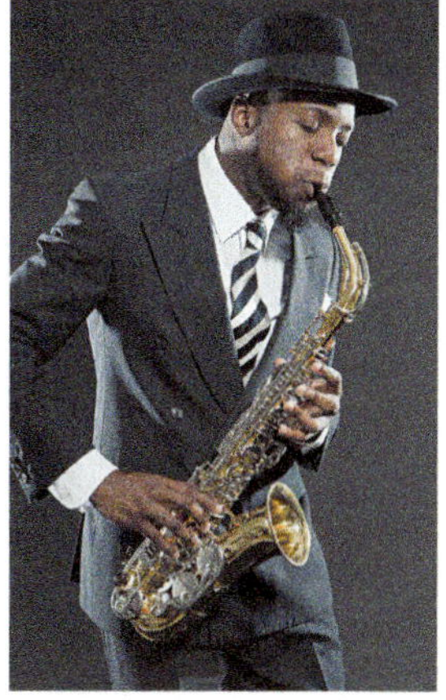

$$P = \frac{a}{r} \cos\left(\frac{2\pi r}{\lambda} - ct\right)$$

can be used to model the sound pressure at a radius of r feet from the source, where t is time in seconds, λ is length of the sound wave in feet, c is speed of sound in feet per second, and a is maximum sound pressure at the source measured in pounds per square foot. (*Source:* Beranek, L., *Noise and Vibration Control,* Institute of Noise Control Engineering, Washington, D.C.) Let $\lambda = 4.9$ ft and $c = 1026$ ft per sec.

(a) Let $a = 0.4$ lb per ft^2. Graph the sound pressure at distance $r = 10$ ft from its source in the window $[0, 0.05]$ by $[-0.05, 0.05]$. Describe P at this distance.

(b) Now let $a = 3$ and $t = 10$. Graph the sound pressure in the window $[0, 20]$ by $[-2, 2]$. What happens to pressure P as radius r increases?

(c) Suppose a person stands at a radius r so that $r = n\lambda$, where n is a positive integer. Use the difference identity for cosine to simplify P in this situation.

105. ***Back Stress*** If a person bends at the waist with a straight back making an angle of θ degrees with the horizontal, then the force F exerted on the back muscles can be modeled by the equation

$$F = \frac{0.6W \sin(\theta + 90^\circ)}{\sin 12^\circ},$$

where W is the weight of the person. (*Source:* Metcalf, H., *Topics in Classical Biophysics,* Prentice-Hall.)

(a) Calculate force F, to the nearest pound, for $W = 170$ lb and $\theta = 30^\circ$.

(b) Use an identity to show that F is approximately equal to $2.9W \cos \theta$.

(c) For what value of θ is F maximum?

106. ***Back Stress*** Refer to **Exercise 105.**

(a) Suppose a 200-lb person bends at the waist so that $\theta = 45^\circ$. Calculate the force, to the nearest pound, exerted on the person's back muscles.

(b) Approximate graphically the value of θ, to the nearest tenth, that results in the back muscles of a 200-lb person exerting a force of 400 lb.

Relating Concepts

For individual or collaborative investigation *(Exercises 107–112)*

(This discussion applies to functions of both angles and real numbers.) Consider the following.

$\cos(180° - \theta)$

$= \cos 180° \cos\theta + \sin 180° \sin\theta$ Cosine difference identity

$= (-1)\cos\theta + (0)\sin\theta$ $\cos 180° = -1$ and $\sin 180° = 0$

$= -\cos\theta$ Simplify.

$\mathbf{\cos(180° - \theta) = -\cos\theta}$ is an example of a **reduction formula,** which is an identity that *reduces* a function of a quadrantal angle plus or minus θ to a function of θ alone. Another example of a reduction formula is $\mathbf{\cos(270° + \theta) = \sin\theta}$.

Here is an interesting method for quickly determining a reduction formula for a trigonometric function f of the form $\boldsymbol{f(Q \pm \theta)}$, where Q is a quadrantal angle. ***There are two cases to consider, and in each case, think of θ as a small positive angle*** in order to determine the quadrant in which $Q \pm \theta$ will lie.

Case 1 **Q is a quadrantal angle whose terminal side lies along the x-axis.**

Determine the quadrant in which $Q \pm \theta$ will lie for a small positive angle θ. If the given function f is positive in that quadrant, use a + sign on the reduced form. If f is negative in that quadrant, use a − sign. The reduced form will have that sign, f as the function, and θ as the argument.

Example:

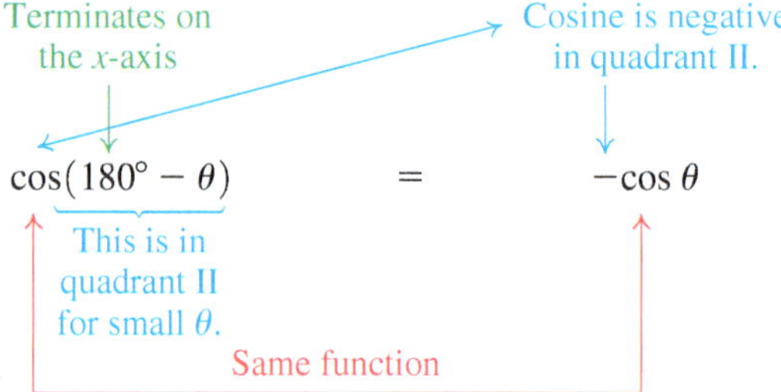

Case 2 **Q is a quadrantal angle whose terminal side lies along the y-axis.**

Determine the quadrant in which $Q \pm \theta$ will lie for a small positive angle θ. If the given function f is positive in that quadrant, use a + sign on the reduced form. If f is negative in that quadrant, use a − sign. The reduced form will have that sign, the ***cofunction of f*** as the function, and θ as the argument.

Example:

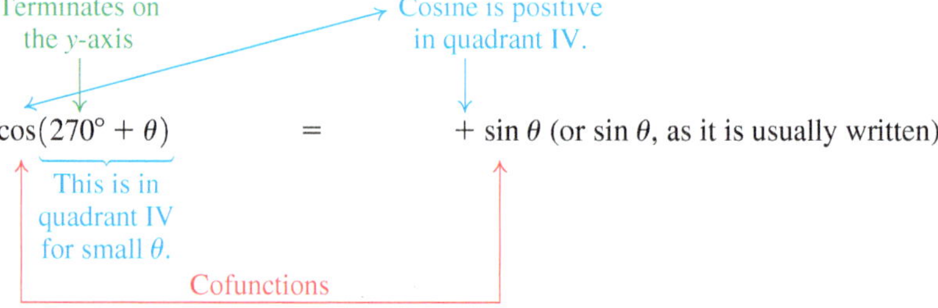

Use these ideas to write a reduction formula for each of the following.

107. $\cos(90° + \theta)$ **108.** $\cos(270° - \theta)$ **109.** $\cos(180° + \theta)$

110. $\cos(270° + \theta)$ **111.** $\sin(180° + \theta)$ **112.** $\tan(270° - \theta)$

Chapter 7 Quiz (Sections 7.1–7.3)

1. If $\sin\theta = -\frac{7}{25}$ and θ is in quadrant IV, find the remaining five trigonometric function values of θ.
2. Express $\cot^2 x + \csc^2 x$ in terms of $\sin x$ and $\cos x$, and simplify.
3. Find the exact value of $\sin\left(-\frac{7\pi}{12}\right)$.
4. Express $\cos(180° - \theta)$ as a function of θ alone.
5. If $\cos A = \frac{3}{5}$, $\sin B = -\frac{5}{13}$, $0 < A < \frac{\pi}{2}$, and $\pi < B < \frac{3\pi}{2}$, find each of the following.

   (a) $\cos(A + B)$ **(b)** $\sin(A + B)$ **(c)** the quadrant of $A + B$
6. Express $\tan\left(\frac{3\pi}{4} + x\right)$ as a function of x alone.

Verify that each equation is an identity.

7. $\dfrac{1 + \sin\theta}{\cot^2\theta} = \dfrac{\sin\theta}{\csc\theta - 1}$
8. $\sin\left(\dfrac{\pi}{3} + \theta\right) - \sin\left(\dfrac{\pi}{3} - \theta\right) = \sin\theta$
9. $\dfrac{\sin^2\theta - \cos^2\theta}{\sin^4\theta - \cos^4\theta} = 1$
10. $\dfrac{\cos(x + y) + \cos(x - y)}{\sin(x - y) + \sin(x + y)} = \cot x$

7.4 Double-Angle and Half-Angle Identities

- Double-Angle Identities
- An Application
- Product-to-Sum and Sum-to-Product Identities
- Half-Angle Identities
- Verifying an Identity

Double-Angle Identities When $A = B$ in the identities for the sum of two angles, the **double-angle identities** result. To derive an expression for $\cos 2A$, we let $B = A$ in the identity $\cos(A + B) = \cos A \cos B - \sin A \sin B$.

$$\begin{aligned} \cos 2A &= \cos(A + A) && 2A = A + A \\ &= \cos A \cos A - \sin A \sin A && \text{Cosine sum identity} \\ \mathbf{\cos 2A} &\mathbf{= \cos^2 A - \sin^2 A} && a \cdot a = a^2 \end{aligned}$$

Two other useful forms of this identity can be obtained by substituting

$$\cos^2 A = 1 - \sin^2 A \quad \text{or} \quad \sin^2 A = 1 - \cos^2 A.$$

Replacing $\cos^2 A$ with the expression $1 - \sin^2 A$ gives the following.

$$\begin{aligned} \cos 2A &= \cos^2 A - \sin^2 A && \text{Double-angle identity from above} \\ &= (1 - \sin^2 A) - \sin^2 A && \text{Fundamental identity} \\ \mathbf{\cos 2A} &\mathbf{= 1 - 2\sin^2 A} && \text{Subtract.} \end{aligned}$$

Replacing $\sin^2 A$ with $1 - \cos^2 A$ gives a third form.

$$\begin{aligned} \cos 2A &= \cos^2 A - \sin^2 A && \text{Double-angle identity from above} \\ &= \cos^2 A - (1 - \cos^2 A) && \text{Fundamental identity} \\ &= \cos^2 A - 1 + \cos^2 A && \text{Distributive property} \\ \mathbf{\cos 2A} &\mathbf{= 2\cos^2 A - 1} && \text{Add.} \end{aligned}$$

LOOKING AHEAD TO CALCULUS

The identities

$$\cos 2A = 1 - 2\sin^2 A$$

and $$\cos 2A = 2\cos^2 A - 1$$

can be rewritten as

$$\sin^2 A = \frac{1}{2}(1 - \cos 2A)$$

and $$\cos^2 A = \frac{1}{2}(1 + \cos 2A).$$

These identities are used to integrate the functions

$$f(A) = \sin^2 A$$

and $$g(A) = \cos^2 A.$$

We find $\sin 2A$ using $\sin(A + B) = \sin A \cos B + \cos A \sin B$, with $B = A$.

$$\sin 2A = \sin(A + A) \qquad 2A = A + A$$

$$= \sin A \cos A + \cos A \sin A \qquad \text{Sine sum identity}$$

$$\mathbf{\sin 2A = 2 \sin A \cos A} \qquad \text{Add.}$$

Using the identity for $\tan(A + B)$, we find $\tan 2A$.

$$\tan 2A = \tan(A + A) \qquad 2A = A + A$$

$$= \frac{\tan A + \tan A}{1 - \tan A \tan A} \qquad \text{Tangent sum identity}$$

$$\mathbf{\tan 2A = \frac{2 \tan A}{1 - \tan^2 A}} \qquad \text{Simplify.}$$

NOTE In general, for a trigonometric function f,

$$f(2A) \neq 2f(A).$$

Double-Angle Identities

$$\cos 2A = \cos^2 A - \sin^2 A \qquad \cos 2A = 1 - 2\sin^2 A$$

$$\cos 2A = 2\cos^2 A - 1 \qquad \sin 2A = 2 \sin A \cos A$$

$$\tan 2A = \frac{2 \tan A}{1 - \tan^2 A}$$

EXAMPLE 1 Finding Function Values of 2θ Given Information about θ

Given $\cos\theta = \frac{3}{5}$ and $\sin\theta < 0$, find $\sin 2\theta$, $\cos 2\theta$, and $\tan 2\theta$.

SOLUTION To find $\sin 2\theta$, we must first find the value of $\sin\theta$.

$$\sin^2\theta + \cos^2\theta = 1 \qquad \text{Pythagorean identity}$$

$$\sin^2\theta + \left(\frac{3}{5}\right)^2 = 1 \qquad \cos\theta = \tfrac{3}{5}$$

$$\sin^2\theta = \frac{16}{25} \qquad \left(\tfrac{3}{5}\right)^2 = \tfrac{9}{25};\ \text{Subtract } \tfrac{9}{25}.$$

Pay attention to signs here.

$$\sin\theta = -\frac{4}{5} \qquad \text{Take square roots. Choose the negative square root because } \sin\theta < 0.$$

Now use the double-angle identity for sine.

$$\sin 2\theta = 2 \sin\theta \cos\theta = 2\left(-\frac{4}{5}\right)\left(\frac{3}{5}\right) = -\frac{24}{25} \qquad \sin\theta = -\tfrac{4}{5} \text{ and } \cos\theta = \tfrac{3}{5}$$

Now we find $\cos 2\theta$, using the first of the double-angle identities for cosine.

Any of the three forms may be used.

$$\cos 2\theta = \cos^2\theta - \sin^2\theta = \frac{9}{25} - \frac{16}{25} = -\frac{7}{25} \qquad \cos\theta = \tfrac{3}{5} \text{ and } \left(\tfrac{3}{5}\right)^2 = \tfrac{9}{25};\ \sin\theta = -\tfrac{4}{5} \text{ and } \left(-\tfrac{4}{5}\right)^2 = \tfrac{16}{25}$$

The value of $\tan 2\theta$ can be found in either of two ways. We can use the double-angle identity and the fact that $\tan\theta = \frac{\sin\theta}{\cos\theta} = \frac{-\frac{4}{5}}{\frac{3}{5}} = -\frac{4}{5} \div \frac{3}{5} = -\frac{4}{5} \cdot \frac{5}{3} = -\frac{4}{3}$.

$$\tan 2\theta = \frac{2\tan\theta}{1-\tan^2\theta} = \frac{2\left(-\frac{4}{3}\right)}{1-\left(-\frac{4}{3}\right)^2} = \frac{-\frac{8}{3}}{-\frac{7}{9}} = \frac{24}{7}$$

Alternatively, we can find $\tan 2\theta$ by finding the quotient of $\sin 2\theta$ and $\cos 2\theta$.

$$\tan 2\theta = \frac{\sin 2\theta}{\cos 2\theta} = \frac{-\frac{24}{25}}{-\frac{7}{25}} = \frac{24}{7} \quad \text{Same result as above}$$

✔ **Now Try Exercise 15.**

EXAMPLE 2 Finding Function Values of θ Given Information about 2θ

Find the values of the six trigonometric functions of θ given $\cos 2\theta = \frac{4}{5}$ and $90° < \theta < 180°$.

SOLUTION We must obtain a trigonometric function value of θ alone.

$$\cos 2\theta = 1 - 2\sin^2\theta \quad \text{Double-angle identity}$$

$$\frac{4}{5} = 1 - 2\sin^2\theta \quad \cos 2\theta = \frac{4}{5}$$

$$-\frac{1}{5} = -2\sin^2\theta \quad \text{Subtract 1 from each side.}$$

$$\frac{1}{10} = \sin^2\theta \quad \text{Multiply by } -\frac{1}{2}.$$

$$\sin\theta = \sqrt{\frac{1}{10}} \quad \text{Take square roots. Choose the positive square root because } \theta \text{ terminates in quadrant II.}$$

$$\sin\theta = \frac{1}{\sqrt{10}} \cdot \frac{\sqrt{10}}{\sqrt{10}} \quad \text{Quotient rule for radicals; rationalize the denominator.}$$

$$\sin\theta = \frac{\sqrt{10}}{10} \quad \sqrt{a}\cdot\sqrt{a} = a$$

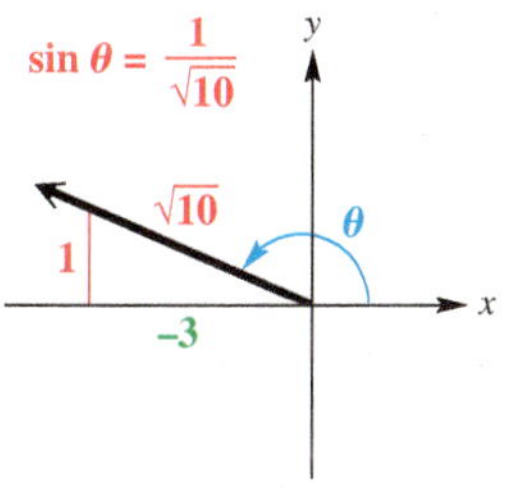

Figure 6

Now find values of $\cos\theta$ and $\tan\theta$ by sketching and labeling a right triangle in quadrant II. Because $\sin\theta = \frac{1}{\sqrt{10}}$, the triangle in **Figure 6** is labeled accordingly. The Pythagorean theorem is used to find the remaining leg.

$$\cos\theta = \frac{-3}{\sqrt{10}} = -\frac{3\sqrt{10}}{10} \quad \text{and} \quad \tan\theta = \frac{1}{-3} = -\frac{1}{3} \quad \cos\theta = \frac{x}{r} \text{ and } \tan\theta = \frac{y}{x}$$

We find the other three functions using reciprocals.

$$\csc\theta = \frac{1}{\sin\theta} = \sqrt{10}, \quad \sec\theta = \frac{1}{\cos\theta} = -\frac{\sqrt{10}}{3}, \quad \cot\theta = \frac{1}{\tan\theta} = -3$$

✔ **Now Try Exercise 19.**

EXAMPLE 3 Simplifying Expressions Using Double-Angle Identities

Simplify each expression.

(a) $\cos^2 7x - \sin^2 7x$ **(b)** $\sin 15° \cos 15°$

SOLUTION

(a) This expression suggests one of the double-angle identities for cosine: $\cos 2A = \cos^2 A - \sin^2 A$. Substitute $7x$ for A.

$$\cos^2 7x - \sin^2 7x = \cos 2(7x) = \cos 14x$$

(b) If the expression $\sin 15° \cos 15°$ were

$$2 \sin 15° \cos 15°,$$

we could apply the identity for $\sin 2A$ directly because $\sin 2A = 2 \sin A \cos A$.

$\sin 15° \cos 15°$

$= \frac{1}{2}(2) \sin 15° \cos 15°$ Multiply by 1 in the form $\frac{1}{2}(2)$.

This is not an obvious way to begin, but it is indeed valid.

$= \frac{1}{2}(2 \sin 15° \cos 15°)$ Associative property

$= \frac{1}{2}\sin(2 \cdot 15°)$ $2 \sin A \cos A = \sin 2A$, with $A = 15°$

$= \frac{1}{2}\sin 30°$ Multiply.

$= \frac{1}{2} \cdot \frac{1}{2}$ $\sin 30° = \frac{1}{2}$

$= \frac{1}{4}$ Multiply.

✔ **Now Try Exercises 21 and 23.**

Identities involving larger multiples of the variable can be derived by repeated use of the double-angle identities and other identities.

EXAMPLE 4 Deriving a Multiple-Angle Identity

Write $\sin 3x$ in terms of $\sin x$.

SOLUTION

$\sin 3x$

$= \sin(2x + x)$ $3x = 2x + x$

Use the simple fact that 3 = 2 + 1 here.

$= \sin 2x \cos x + \cos 2x \sin x$ Sine sum identity

$= (2 \sin x \cos x)\cos x + (\cos^2 x - \sin^2 x)\sin x$ Double-angle identities

$= 2 \sin x \cos^2 x + \cos^2 x \sin x - \sin^3 x$ Multiply.

$= 2 \sin x(1 - \sin^2 x) + (1 - \sin^2 x)\sin x - \sin^3 x$ $\cos^2 x = 1 - \sin^2 x$

$= 2 \sin x - 2 \sin^3 x + \sin x - \sin^3 x - \sin^3 x$ Distributive property

$= 3 \sin x - 4 \sin^3 x$ Combine like terms.

✔ **Now Try Exercise 33.**

An Application

EXAMPLE 5 Determining Wattage Consumption

If a toaster is plugged into a common household outlet, the wattage consumed is not constant. Instead, it varies at a high frequency according to the model

$$W = \frac{V^2}{R},$$

where V is the voltage and R is a constant that measures the resistance of the toaster in ohms. (*Source*: Bell, D., *Fundamentals of Electric Circuits,* Fourth Edition, Prentice-Hall.)

Graph the wattage W consumed by a toaster with $R = 15$ and $V = 163 \sin 120\pi t$ in the window $[0, 0.05]$ by $[-500, 2000]$. How many oscillations are there?

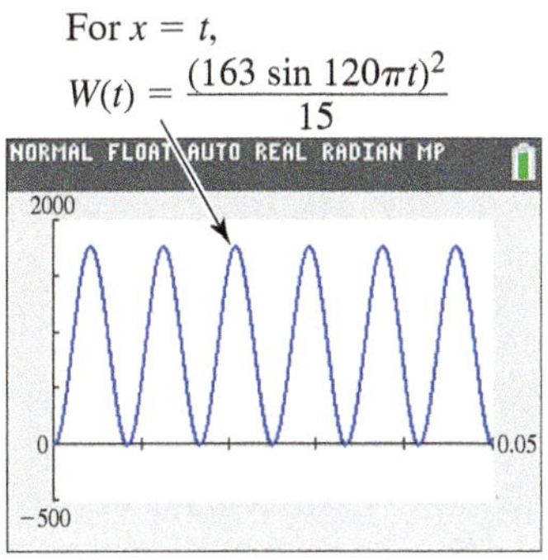

For $x = t$,
$W(t) = \frac{(163 \sin 120\pi t)^2}{15}$

NORMAL FLOAT AUTO REAL RADIAN MP
2000
0
−500
0.05

Figure 7

SOLUTION $W = \frac{V^2}{R} = \frac{(163 \sin 120\pi t)^2}{15}$ Substitute the given values into the wattage equation.

Note that $\sin 120\pi t$ has maximum value 1, so the expression for W has maximum value $\frac{163^2}{15} \approx 1771$. The minimum value is 0. The graph in **Figure 7** shows that there are six oscillations.

✔ **Now Try Exercise 107.**

Product-to-Sum and Sum-to-Product Identities We can add the corresponding sides of the identities for $\cos(A + B)$ and $\cos(A - B)$ to derive a product-to-sum identity that is useful in calculus.

$$\cos(A + B) = \cos A \cos B - \sin A \sin B$$
$$\cos(A - B) = \cos A \cos B + \sin A \sin B$$
$$\cos(A + B) + \cos(A - B) = 2 \cos A \cos B \quad \text{Add.}$$

$$\cos A \cos B = \frac{1}{2}[\cos(A + B) + \cos(A - B)]$$

Similarly, subtracting $\cos(A + B)$ from $\cos(A - B)$ gives

$$\sin A \sin B = \frac{1}{2}[\cos(A - B) - \cos(A + B)].$$

Using the identities for $\sin(A + B)$ and $\sin(A - B)$ in the same way, we obtain two more identities. Those and the previous ones are now summarized.

LOOKING AHEAD TO CALCULUS

The product-to-sum identities are used in calculus to find **integrals** of functions that are products of trigonometric functions. The classic calculus text by Earl Swokowski includes the following example:

Evaluate $\int \cos 5x \cos 3x \, dx$.

The first solution line reads:
"We may write

$\cos 5x \cos 3x = \frac{1}{2}[\cos 8x + \cos 2x]$."

Product-to-Sum Identities

$$\cos A \cos B = \frac{1}{2}[\cos(A + B) + \cos(A - B)]$$

$$\sin A \sin B = \frac{1}{2}[\cos(A - B) - \cos(A + B)]$$

$$\sin A \cos B = \frac{1}{2}[\sin(A + B) + \sin(A - B)]$$

$$\cos A \sin B = \frac{1}{2}[\sin(A + B) - \sin(A - B)]$$

EXAMPLE 6 Using a Product-to-Sum Identity

Write $4 \cos 75° \sin 25°$ as the sum or difference of two functions.

SOLUTION

$4 \cos 75° \sin 25°$

$$= 4\left[\frac{1}{2}(\sin(75° + 25°) - \sin(75° - 25°))\right]$$ Use the identity for $\cos A \sin B$, with $A = 75°$ and $B = 25°$.

$$= 2 \sin 100° - 2 \sin 50°$$ Simplify.

✔ **Now Try Exercise 37.**

We can transform the product-to-sum identities into equivalent useful forms—the sum-to-product identities—using substitution. Consider the product-to-sum identity for $\sin A \cos B$.

$$\sin A \cos B = \frac{1}{2}[\sin(A + B) + \sin(A - B)]$$ Product-to-sum identity

Let

$$u = A + B \quad \text{and} \quad v = A - B.$$

Then

$$u + v = 2A \quad \text{and} \quad u - v = 2B,$$

so

$$A = \frac{u + v}{2} \quad \text{and} \quad B = \frac{u - v}{2}.$$

Use substitution variables to write the product-to-sum identity in terms of u and v.

Substituting for A and B in the above product-to-sum identity gives the following.

$$\sin\left(\frac{u + v}{2}\right)\cos\left(\frac{u - v}{2}\right) = \frac{1}{2}(\sin u + \sin v)$$ Substitute.

$$\sin u + \sin v = 2 \sin\left(\frac{u + v}{2}\right)\cos\left(\frac{u - v}{2}\right)$$ Multiply by 2. Interchange sides.

The other three sum-to-product identities are derived using the same substitutions into the other three product-to-sum formulas.

Sum-to-Product Identities

$$\sin A + \sin B = 2 \sin\left(\frac{A + B}{2}\right) \cos\left(\frac{A - B}{2}\right)$$

$$\sin A - \sin B = 2 \cos\left(\frac{A + B}{2}\right) \sin\left(\frac{A - B}{2}\right)$$

$$\cos A + \cos B = 2 \cos\left(\frac{A + B}{2}\right) \cos\left(\frac{A - B}{2}\right)$$

$$\cos A - \cos B = -2 \sin\left(\frac{A + B}{2}\right) \sin\left(\frac{A - B}{2}\right)$$

EXAMPLE 7 Using a Sum-to-Product Identity

Write $\sin 2\theta - \sin 4\theta$ as a product of two functions.

SOLUTION $\sin 2\theta - \sin 4\theta$

$$= 2\cos\left(\frac{2\theta + 4\theta}{2}\right)\sin\left(\frac{2\theta - 4\theta}{2}\right)$$ Use the identity for $\sin A - \sin B$, with $A = 2\theta$ and $B = 4\theta$.

$$= 2\cos\frac{6\theta}{2}\sin\left(\frac{-2\theta}{2}\right)$$ Simplify the numerators.

$$= 2\cos 3\theta \sin(-\theta)$$ Divide.

$$= -2\cos 3\theta \sin\theta$$ $\sin(-\theta) = -\sin\theta$

✔ **Now Try Exercise 43.**

Half-Angle Identities

From alternative forms of the identity for $\cos 2A$, we derive identities for $\sin\frac{A}{2}$, $\cos\frac{A}{2}$, and $\tan\frac{A}{2}$, known as **half-angle identities.**

We derive the identity for $\sin\frac{A}{2}$ as follows.

$$\cos 2x = 1 - 2\sin^2 x$$ Cosine double-angle identity

$$2\sin^2 x = 1 - \cos 2x$$ Add $2\sin^2 x$ and subtract $\cos 2x$.

Remember both the positive and negative square roots.

$$\sin x = \pm\sqrt{\frac{1 - \cos 2x}{2}}$$ Divide by 2 and take square roots.

$$\boldsymbol{\sin\frac{A}{2} = \pm\sqrt{\frac{1 - \cos A}{2}}}$$ Let $2x = A$, so $x = \frac{A}{2}$. Substitute.

The $\pm$ symbol indicates that the appropriate sign is chosen depending on the quadrant of $\frac{A}{2}$. For example, if $\frac{A}{2}$ is a quadrant III angle, we choose the negative sign because the sine function is negative in quadrant III.

We derive the identity for $\cos\frac{A}{2}$ using another double-angle identity.

$$\cos 2x = 2\cos^2 x - 1$$ Cosine double-angle identity

$$1 + \cos 2x = 2\cos^2 x$$ Add 1.

$$\cos^2 x = \frac{1 + \cos 2x}{2}$$ Rewrite and divide by 2.

$$\cos x = \pm\sqrt{\frac{1 + \cos 2x}{2}}$$ Take square roots.

$$\boldsymbol{\cos\frac{A}{2} = \pm\sqrt{\frac{1 + \cos A}{2}}}$$ Replace x with $\frac{A}{2}$.

An identity for $\tan\frac{A}{2}$ comes from the identities for $\sin\frac{A}{2}$ and $\cos\frac{A}{2}$.

$$\tan\frac{A}{2} = \frac{\sin\frac{A}{2}}{\cos\frac{A}{2}} = \frac{\pm\sqrt{\frac{1 - \cos A}{2}}}{\pm\sqrt{\frac{1 + \cos A}{2}}} = \pm\sqrt{\frac{1 - \cos A}{1 + \cos A}}$$

We derive an alternative identity for $\tan \frac{A}{2}$ using double-angle identities.

$$\tan \frac{A}{2} = \frac{\sin \frac{A}{2}}{\cos \frac{A}{2}} \quad \text{Definition of tangent}$$

$$= \frac{2 \sin \frac{A}{2} \cos \frac{A}{2}}{2 \cos^2 \frac{A}{2}} \quad \text{Multiply by } 2\cos\tfrac{A}{2} \text{ in numerator and denominator.}$$

$$= \frac{\sin 2\left(\frac{A}{2}\right)}{1 + \cos 2\left(\frac{A}{2}\right)} \quad \text{Double-angle identities}$$

$$\tan \frac{A}{2} = \frac{\sin A}{1 + \cos A} \quad \text{Simplify.}$$

From the identity $\tan \frac{A}{2} = \frac{\sin A}{1 + \cos A}$, we can also derive an equivalent identity.

$$\tan \frac{A}{2} = \frac{1 - \cos A}{\sin A}$$

Half-Angle Identities

In the following identities, the $\pm$ symbol indicates that the sign is chosen based on the function under consideration and the quadrant of $\frac{A}{2}$.

$$\cos \frac{A}{2} = \pm\sqrt{\frac{1 + \cos A}{2}} \qquad \sin \frac{A}{2} = \pm\sqrt{\frac{1 - \cos A}{2}}$$

$$\tan \frac{A}{2} = \pm\sqrt{\frac{1 - \cos A}{1 + \cos A}} \qquad \tan \frac{A}{2} = \frac{\sin A}{1 + \cos A} \qquad \tan \frac{A}{2} = \frac{1 - \cos A}{\sin A}$$

Three of these identities require a sign choice. When using these identities, select the plus or minus sign according to the quadrant in which $\frac{A}{2}$ terminates. For example, if an angle $A = 324°$, then $\frac{A}{2} = 162°$, which lies in quadrant II. So when $A = 324°$, $\cos \frac{A}{2}$ and $\tan \frac{A}{2}$ are negative, and $\sin \frac{A}{2}$ is positive.

EXAMPLE 8 Using a Half-Angle Identity to Find an Exact Value

Find the exact value of $\cos 15°$ using the half-angle identity for cosine.

SOLUTION

$$\cos 15° = \cos \frac{30°}{2} = \sqrt{\frac{1 + \cos 30°}{2}}$$

Choose the positive square root.

$$= \sqrt{\frac{1 + \frac{\sqrt{3}}{2}}{2}} = \sqrt{\frac{\left(1 + \frac{\sqrt{3}}{2}\right) \cdot 2}{2 \cdot 2}} = \frac{\sqrt{2 + \sqrt{3}}}{2}$$

Simplify the radicals.

✔ Now Try Exercise 49.

EXAMPLE 9 Using a Half-Angle Identity to Find an Exact Value

Find the exact value of tan 22.5° using the identity $\tan \frac{A}{2} = \frac{\sin A}{1 + \cos A}$.

SOLUTION Because $22.5° = \frac{45°}{2}$, replace A with 45°.

$$\tan 22.5° = \tan \frac{45°}{2} = \frac{\sin 45°}{1 + \cos 45°} = \frac{\frac{\sqrt{2}}{2}}{1 + \frac{\sqrt{2}}{2}} = \frac{\frac{\sqrt{2}}{2}}{1 + \frac{\sqrt{2}}{2}} \cdot \frac{2}{2}$$

$$= \frac{\sqrt{2}}{2 + \sqrt{2}} = \frac{\sqrt{2}}{2 + \sqrt{2}} \cdot \frac{2 - \sqrt{2}}{2 - \sqrt{2}} = \frac{2\sqrt{2} - 2}{2} \quad \text{Rationalize the denominator.}$$

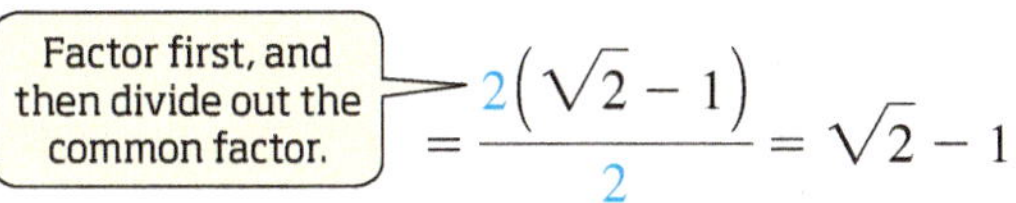

$$= \frac{2\left(\sqrt{2} - 1\right)}{2} = \sqrt{2} - 1$$

✔ **Now Try Exercise 51.**

EXAMPLE 10 Finding Function Values of $\frac{s}{2}$ Given Information about s

Given $\cos s = \frac{2}{3}$, with $\frac{3\pi}{2} < s < 2\pi$, find $\sin \frac{s}{2}$, $\cos \frac{s}{2}$, and $\tan \frac{s}{2}$.

SOLUTION The angle associated with $\frac{s}{2}$ terminates in quadrant II because

$$\frac{3\pi}{2} < s < 2\pi \quad \text{and} \quad \frac{3\pi}{4} < \frac{s}{2} < \pi. \quad \text{Divide by 2.}$$

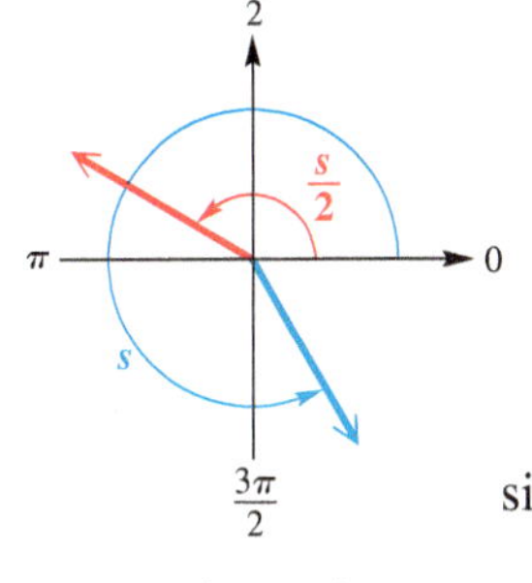

Figure 8

See **Figure 8.** In quadrant II, the values of $\cos \frac{s}{2}$ and $\tan \frac{s}{2}$ are negative and the value of $\sin \frac{s}{2}$ is positive. Use the appropriate half-angle identities and simplify.

$$\sin \frac{s}{2} = \sqrt{\frac{1 - \frac{2}{3}}{2}} = \sqrt{\frac{1}{6}} = \frac{\sqrt{1}}{\sqrt{6}} \cdot \frac{\sqrt{6}}{\sqrt{6}} = \frac{\sqrt{6}}{6}$$

Rationalize all denominators.

$$\cos \frac{s}{2} = -\sqrt{\frac{1 + \frac{2}{3}}{2}} = -\sqrt{\frac{5}{6}} = -\frac{\sqrt{5}}{\sqrt{6}} \cdot \frac{\sqrt{6}}{\sqrt{6}} = -\frac{\sqrt{30}}{6}$$

$$\tan \frac{s}{2} = \frac{\sin \frac{s}{2}}{\cos \frac{s}{2}} = \frac{\frac{\sqrt{6}}{6}}{-\frac{\sqrt{30}}{6}} = \frac{\sqrt{6}}{-\sqrt{30}} = -\frac{\sqrt{6}}{\sqrt{30}} \cdot \frac{\sqrt{30}}{\sqrt{30}} = -\frac{\sqrt{180}}{30} = -\frac{6\sqrt{5}}{6 \cdot 5} = -\frac{\sqrt{5}}{5}$$

Notice that it is not necessary to use a half-angle identity for $\tan \frac{s}{2}$ once we find $\sin \frac{s}{2}$ and $\cos \frac{s}{2}$. However, using this identity provides an excellent check.

✔ **Now Try Exercise 55.**

EXAMPLE 11 Simplifying Expressions Using Half-Angle Identities

Simplify each expression.

(a) $\pm\sqrt{\frac{1 + \cos 12x}{2}}$ **(b)** $\frac{1 - \cos 5\alpha}{\sin 5\alpha}$

SOLUTION

(a) This matches part of the identity for $\cos \frac{A}{2}$. Replace A with $12x$.

$$\cos \frac{A}{2} = \pm\sqrt{\frac{1 + \cos A}{2}} = \pm\sqrt{\frac{1 + \cos 12x}{2}} = \cos \frac{12x}{2} = \cos 6x$$

(b) Use the identity $\tan \frac{A}{2} = \frac{1 - \cos A}{\sin A}$ with $A = 5\alpha$.

$$\frac{1 - \cos 5\alpha}{\sin 5\alpha} = \tan \frac{5\alpha}{2}$$

✔ **Now Try Exercises 71 and 73.**

Verifying an Identity

EXAMPLE 12 Verifying an Identity

Verify that the following equation is an identity.

$$\left(\sin \frac{x}{2} + \cos \frac{x}{2}\right)^2 = 1 + \sin x$$

SOLUTION We work on the more complicated left side.

$$\left(\sin \frac{x}{2} + \cos \frac{x}{2}\right)^2$$

Remember the middle term when squaring a binomial.

$$= \sin^2 \frac{x}{2} + 2 \sin \frac{x}{2} \cos \frac{x}{2} + \cos^2 \frac{x}{2} \qquad (x + y)^2 = x^2 + 2xy + y^2$$

$$= 1 + 2 \sin \frac{x}{2} \cos \frac{x}{2} \qquad \sin^2 \tfrac{x}{2} + \cos^2 \tfrac{x}{2} = 1$$

$$= 1 + \sin 2\left(\frac{x}{2}\right) \qquad 2 \sin \tfrac{x}{2} \cos \tfrac{x}{2} = \sin 2\left(\tfrac{x}{2}\right)$$

$$= 1 + \sin x \qquad \text{Multiply.}$$

✔ **Now Try Exercise 93.**

7.4 Exercises

CONCEPT PREVIEW *Match each expression in Column I with its value in Column II.*

I

1. $2 \cos^2 15^\circ - 1$

2. $\dfrac{2 \tan 15^\circ}{1 - \tan^2 15^\circ}$

3. $2 \sin 22.5^\circ \cos 22.5^\circ$

4. $\cos^2 \dfrac{\pi}{6} - \sin^2 \dfrac{\pi}{6}$

5. $4 \sin \dfrac{\pi}{3} \cos \dfrac{\pi}{3}$

6. $\dfrac{2 \tan \frac{\pi}{3}}{1 - \tan^2 \frac{\pi}{3}}$

II

A. $\dfrac{1}{2}$

B. $\dfrac{\sqrt{2}}{2}$

C. $\dfrac{\sqrt{3}}{2}$

D. $-\sqrt{3}$

E. $\dfrac{\sqrt{3}}{3}$

F. $\sqrt{3}$

CONCEPT PREVIEW *Determine whether the positive or negative square root should be selected.*

7. $\sin 195^\circ = \pm\sqrt{\dfrac{1 - \cos 390^\circ}{2}}$

8. $\cos 58^\circ = \pm\sqrt{\dfrac{1 + \cos 116^\circ}{2}}$

9. $\tan 225^\circ = \pm\sqrt{\dfrac{1 - \cos 450^\circ}{1 + \cos 450^\circ}}$

10. $\sin(-10^\circ) = \pm\sqrt{\dfrac{1 - \cos(-20^\circ)}{2}}$

Find values of the sine and cosine functions for each angle measure. ***See Examples 1 and 2.***

11. 2θ, given $\sin\theta = \frac{2}{5}$ and $\cos\theta < 0$

12. 2θ, given $\cos\theta = -\frac{12}{13}$ and $\sin\theta > 0$

13. $2x$, given $\tan x = 2$ and $\cos x > 0$

14. $2x$, given $\tan x = \frac{5}{3}$ and $\sin x < 0$

15. 2θ, given $\sin\theta = -\frac{\sqrt{5}}{7}$ and $\cos\theta > 0$

16. 2θ, given $\cos\theta = \frac{\sqrt{3}}{5}$ and $\sin\theta > 0$

17. θ, given $\cos 2\theta = \frac{3}{5}$ and θ terminates in quadrant I

18. θ, given $\cos 2\theta = \frac{3}{4}$ and θ terminates in quadrant III

19. θ, given $\cos 2\theta = -\frac{5}{12}$ and $90° < \theta < 180°$

20. θ, given $\cos 2\theta = \frac{2}{3}$ and $90° < \theta < 180°$

Simplify each expression. ***See Example 3.***

21. $\cos^2 15° - \sin^2 15°$

22. $\dfrac{2\tan 15°}{1 - \tan^2 15°}$

23. $1 - 2\sin^2 15°$

24. $1 - 2\sin^2 22\dfrac{1°}{2}$

25. $2\cos^2 67\dfrac{1°}{2} - 1$

26. $\cos^2\dfrac{\pi}{8} - \dfrac{1}{2}$

27. $\dfrac{\tan 51°}{1 - \tan^2 51°}$

28. $\dfrac{\tan 34°}{2(1 - \tan^2 34°)}$

29. $\dfrac{1}{4} - \dfrac{1}{2}\sin^2 47.1°$

30. $\dfrac{1}{8}\sin 29.5° \cos 29.5°$

31. $\sin^2\dfrac{2\pi}{5} - \cos^2\dfrac{2\pi}{5}$

32. $\cos^2 2x - \sin^2 2x$

Express each function as a trigonometric function of x. ***See Example 4.***

33. $\sin 4x$

34. $\cos 3x$

35. $\tan 3x$

36. $\cos 4x$

Write each expression as a sum or difference of trigonometric functions. ***See Example 6.***

37. $2\sin 58° \cos 102°$

38. $2\cos 85° \sin 140°$

39. $2\sin\dfrac{\pi}{6}\cos\dfrac{\pi}{3}$

40. $5\cos 3x \cos 2x$

41. $6\sin 4x \sin 5x$

42. $8\sin 7x \sin 9x$

Write each expression as a product of trigonometric functions. ***See Example 7.***

43. $\cos 4x - \cos 2x$

44. $\cos 5x + \cos 8x$

45. $\sin 25° + \sin(-48°)$

46. $\sin 102° - \sin 95°$

47. $\cos 4x + \cos 8x$

48. $\sin 9x - \sin 3x$

Use a half-angle identity to find each exact value. ***See Examples 8 and 9.***

49. $\sin 67.5°$

50. $\sin 195°$

51. $\tan 195°$

52. $\cos 195°$

53. $\cos 165°$

54. $\sin 165°$

Use the given information to find each of the following. ***See Example 10.***

55. $\cos\frac{x}{2}$, given $\cos x = \frac{1}{4}$, with $0 < x < \frac{\pi}{2}$

56. $\sin\frac{x}{2}$, given $\cos x = -\frac{5}{8}$, with $\frac{\pi}{2} < x < \pi$

57. $\tan\frac{\theta}{2}$, given $\sin\theta = \frac{3}{5}$, with $90° < \theta < 180°$

58. $\cos\frac{\theta}{2}$, given $\sin\theta = -\frac{4}{5}$, with $180° < \theta < 270°$

59. $\sin\frac{x}{2}$, given $\tan x = 2$, with $0 < x < \frac{\pi}{2}$

60. $\cos\frac{x}{2}$, given $\cot x = -3$, with $\frac{\pi}{2} < x < \pi$

61. $\tan\frac{\theta}{2}$, given $\tan\theta = \frac{\sqrt{7}}{3}$, with $180° < \theta < 270°$

62. $\cot \frac{\theta}{2}$, given $\tan \theta = -\frac{\sqrt{5}}{2}$, with $90° < \theta < 180°$

63. $\sin \theta$, given $\cos 2\theta = \frac{3}{5}$ and θ terminates in quadrant I

64. $\cos \theta$, given $\cos 2\theta = \frac{1}{2}$ and θ terminates in quadrant II

65. $\cos x$, given $\cos 2x = -\frac{5}{12}$, with $\frac{\pi}{2} < x < \pi$

66. $\sin x$, given $\cos 2x = \frac{2}{3}$, with $\pi < x < \frac{3\pi}{2}$

Simplify each expression. ***See Example 11.***

67. $\sqrt{\dfrac{1 - \cos 40°}{2}}$

68. $\sqrt{\dfrac{1 + \cos 76°}{2}}$

69. $\sqrt{\dfrac{1 - \cos 147°}{1 + \cos 147°}}$

70. $\sqrt{\dfrac{1 + \cos 165°}{1 - \cos 165°}}$

71. $\dfrac{1 - \cos 59.74°}{\sin 59.74°}$

72. $\dfrac{\sin 158.2°}{1 + \cos 158.2°}$

73. $\pm\sqrt{\dfrac{1 + \cos 18x}{2}}$

74. $\pm\sqrt{\dfrac{1 + \cos 20\alpha}{2}}$

75. $\pm\sqrt{\dfrac{1 - \cos 8\theta}{1 + \cos 8\theta}}$

76. $\pm\sqrt{\dfrac{1 - \cos 5A}{1 + \cos 5A}}$

77. $\pm\sqrt{\dfrac{1 + \cos \frac{x}{4}}{2}}$

78. $\pm\sqrt{\dfrac{1 - \cos \frac{3\theta}{5}}{2}}$

Verify that each equation is an identity. ***See Example 12.***

79. $(\sin x + \cos x)^2 = \sin 2x + 1$

80. $\sec 2x = \dfrac{\sec^2 x + \sec^4 x}{2 + \sec^2 x - \sec^4 x}$

81. $(\cos 2x + \sin 2x)^2 = 1 + \sin 4x$

82. $(\cos 2x - \sin 2x)^2 = 1 - \sin 4x$

83. $\tan 8\theta - \tan 8\theta \tan^2 4\theta = 2 \tan 4\theta$

84. $\sin 2x = \dfrac{2 \tan x}{1 + \tan^2 x}$

85. $\cos 2\theta = \dfrac{2 - \sec^2 \theta}{\sec^2 \theta}$

86. $\tan 2\theta = \dfrac{-2 \tan \theta}{\sec^2 \theta - 2}$

87. $\sin 4x = 4 \sin x \cos x \cos 2x$

88. $\dfrac{1 + \cos 2x}{\sin 2x} = \cot x$

89. $\dfrac{2 \cos 2\theta}{\sin 2\theta} = \cot \theta - \tan \theta$

90. $\cot 4\theta = \dfrac{1 - \tan^2 2\theta}{2 \tan 2\theta}$

91. $\sec^2 \dfrac{x}{2} = \dfrac{2}{1 + \cos x}$

92. $\cot^2 \dfrac{x}{2} = \dfrac{(1 + \cos x)^2}{\sin^2 x}$

93. $\sin^2 \dfrac{x}{2} = \dfrac{\tan x - \sin x}{2 \tan x}$

94. $\dfrac{\sin 2x}{2 \sin x} = \cos^2 \dfrac{x}{2} - \sin^2 \dfrac{x}{2}$

95. $\dfrac{2}{1 + \cos x} - \tan^2 \dfrac{x}{2} = 1$

96. $\tan \dfrac{\theta}{2} = \csc \theta - \cot \theta$

97. $1 - \tan^2 \dfrac{\theta}{2} = \dfrac{2 \cos \theta}{1 + \cos \theta}$

98. $\cos x = \dfrac{1 - \tan^2 \frac{x}{2}}{1 + \tan^2 \frac{x}{2}}$

Graph each expression and use the graph to make a conjecture, predicting what might be an identity. Then verify your conjecture algebraically.

99. $\cos^4 x - \sin^4 x$

100. $\dfrac{4 \tan x \cos^2 x - 2 \tan x}{1 - \tan^2 x}$

101. $\dfrac{2 \tan x}{2 - \sec^2 x}$

102. $\dfrac{\cot^2 x - 1}{2 \cot x}$

103. $\dfrac{\sin x}{1 + \cos x}$

104. $\dfrac{1 - \cos x}{\sin x}$

105. $\dfrac{\tan \frac{x}{2} + \cot \frac{x}{2}}{\cot \frac{x}{2} - \tan \frac{x}{2}}$

106. $1 - 8 \sin^2 \dfrac{x}{2} \cos^2 \dfrac{x}{2}$

 *(Modeling) Solve each problem.* ***See Example 5.***

107. ***Wattage Consumption*** Use the identity $\cos 2\theta = 1 - 2 \sin^2 \theta$ to determine values of a, c, and ω so that the equation

$$W = \frac{(163 \sin 120\pi t)^2}{15} \quad \text{becomes} \quad W = a \cos(\omega t) + c.$$

Round to the nearest tenth as necessary. Check by graphing both expressions for W on the same coordinate axes.

108. ***Amperage, Wattage, and Voltage*** Amperage is a measure of the amount of electricity that is moving through a circuit, whereas voltage is a measure of the force pushing the electricity. The wattage W consumed by an electrical device can be determined by calculating the product of the amperage I and voltage V. (*Source*: Wilcox, G. and C. Hesselberth, *Electricity for Engineering Technology*, Allyn & Bacon.)

(a) A household circuit has voltage

$$V = 163 \sin 120\pi t$$

when an incandescent light bulb is turned on with amperage

$$I = 1.23 \sin 120\pi t.$$

Graph the wattage $W = VI$ consumed by the light bulb in the window $[0, 0.05]$ by $[-50, 300]$.

(b) Determine the maximum and minimum wattages used by the light bulb.

(c) Use identities to determine values for a, c, and ω so that $W = a \cos(\omega t) + c$.

(d) Check by graphing both expressions for W on the same coordinate axes.

(e) Use the graph to estimate the average wattage used by the light. For how many watts (to the nearest integer) would this incandescent light bulb be rated?

(Modeling) Mach Number *An airplane flying faster than the speed of sound sends out sound waves that form a cone, as shown in the figure. The cone intersects the ground to form a* ***hyperbola.*** *As this hyperbola passes over a particular point on the ground, a sonic boom is heard at that point. If θ is the angle at the vertex of the cone, then*

$$\sin \frac{\theta}{2} = \frac{1}{m},$$

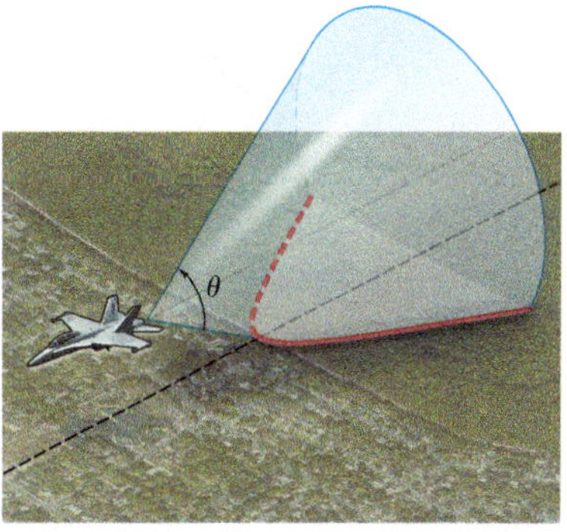

where m is the Mach number for the speed of the plane. (We assume $m > 1$.) The Mach number is the ratio of the speed of the plane to the speed of sound. Thus, a speed of Mach 1.4 means that the plane is flying at 1.4 times the speed of sound.

In each of the following exercises, θ or m is given. Find the other value (θ to the nearest degree and m to the nearest tenth as applicable).

109. $m = \dfrac{5}{4}$

110. $m = \dfrac{3}{2}$

111. $\theta = 60°$

112. $\theta = 30°$

Summary Exercises on Verifying Trigonometric Identities

These summary exercises provide practice with the various types of trigonometric identities presented in this chapter. Verify that each equation is an identity.

1. $\tan \theta + \cot \theta = \sec \theta \csc \theta$

2. $\csc \theta \cos^2 \theta + \sin \theta = \csc \theta$

3. $\tan \frac{x}{2} = \csc x - \cot x$

4. $\sec(\pi - x) = -\sec x$

5. $\frac{\sin t}{1 + \cos t} = \frac{1 - \cos t}{\sin t}$

6. $\frac{1 - \sin t}{\cos t} = \frac{1}{\sec t + \tan t}$

7. $\sin 2\theta = \frac{2 \tan \theta}{1 + \tan^2 \theta}$

8. $\frac{2}{1 + \cos x} - \tan^2 \frac{x}{2} = 1$

9. $\cot \theta - \tan \theta = \frac{2 \cos^2 \theta - 1}{\sin \theta \cos \theta}$

10. $\frac{1}{\sec t - 1} + \frac{1}{\sec t + 1} = 2 \cot t \csc t$

11. $\frac{\sin(x + y)}{\cos(x - y)} = \frac{\cot x + \cot y}{1 + \cot x \cot y}$

12. $1 - \tan^2 \frac{\theta}{2} = \frac{2 \cos \theta}{1 + \cos \theta}$

13. $\frac{\sin \theta + \tan \theta}{1 + \cos \theta} = \tan \theta$

14. $\csc^4 x - \cot^4 x = \frac{1 + \cos^2 x}{1 - \cos^2 x}$

15. $\cos x = \frac{1 - \tan^2 \frac{x}{2}}{1 + \tan^2 \frac{x}{2}}$

16. $\cos 2x = \frac{2 - \sec^2 x}{\sec^2 x}$

17. $\frac{\tan^2 t + 1}{\tan t \csc^2 t} = \tan t$

18. $\frac{\sin s}{1 + \cos s} + \frac{1 + \cos s}{\sin s} = 2 \csc s$

19. $\tan 4\theta = \frac{2 \tan 2\theta}{2 - \sec^2 2\theta}$

20. $\tan\left(\frac{x}{2} + \frac{\pi}{4}\right) = \sec x + \tan x$

21. $\frac{\cot s - \tan s}{\cos s + \sin s} = \frac{\cos s - \sin s}{\sin s \cos s}$

22. $\frac{\tan \theta - \cot \theta}{\tan \theta + \cot \theta} = 1 - 2 \cos^2 \theta$

23. $\frac{\tan(x + y) - \tan y}{1 + \tan(x + y) \tan y} = \tan x$

24. $2 \cos^2 \frac{x}{2} \tan x = \tan x + \sin x$

25. $\frac{\cos^4 x - \sin^4 x}{\cos^2 x} = 1 - \tan^2 x$

26. $\frac{\csc t + 1}{\csc t - 1} = (\sec t + \tan t)^2$

7.5 Inverse Circular Functions

- Review of Inverse Functions
- Inverse Sine Function
- Inverse Cosine Function
- Inverse Tangent Function
- Other Inverse Circular Functions
- Inverse Function Values

Review of Inverse Functions Recall that if a function is defined so that ***each range element is used only once,*** then it is a **one-to-one function.** For example, the function

$$f(x) = x^3 \text{ is a one-to-one function}$$

because every real number has exactly one real cube root. However,

$$g(x) = x^2 \text{ is not a one-to-one function}$$

because $g(2) = 4$ and $g(-2) = 4$. There are two domain elements, 2 and -2, that correspond to the range element 4.

By interchanging the components of the ordered pairs of a one-to-one function f, we obtain a new set of ordered pairs that satisfies the definition of a function. Recall that the **inverse function** of a one-to-one function f is defined as follows.

$$f^{-1} = \{(y, x) \mid (x, y) \text{ belongs to } f\}$$

The special notation used for inverse functions is f^{-1} (read "**f-inverse**"). It represents the function created by interchanging the input (domain) and the output (range) of a one-to-one function.

CAUTION ***Do not confuse the -1 in f^{-1} with a negative exponent.*** The symbol $f^{-1}(x)$ represents the inverse function of f, *not* $\frac{1}{f(x)}$.

The following statements summarize the concepts of inverse functions.

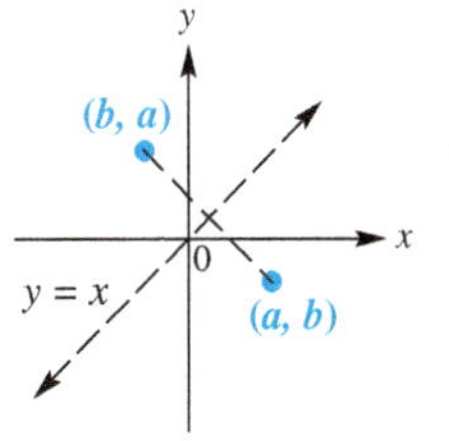

Review of Inverse Functions

1. In a one-to-one function, each x-value corresponds to only one y-value and each y-value corresponds to only one x-value.
2. If a function f is one-to-one, then f has an inverse function f^{-1}.
3. The domain of f is the range of f^{-1}, and the range of f is the domain of f^{-1}. That is, if the point (a, b) lies on the graph of f, then the point (b, a) lies on the graph of f^{-1}.
4. The graphs of f and f^{-1} are reflections of each other across the line $y = x$.
5. To find $f^{-1}(x)$ for $f(x)$, follow these steps.

   Step 1 Replace $f(x)$ with y and interchange x and y.

   Step 2 Solve for y.

   Step 3 Replace y with $f^{-1}(x)$.

(b, a) is the reflection of (a, b) across the line $y = x$.

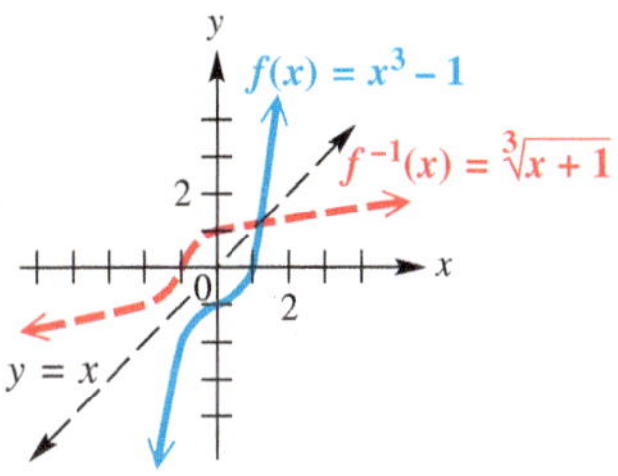

The graph of f^{-1} is the reflection of the graph of f across the line $y = x$.

Figure 9

Figure 9 illustrates some of these concepts.

NOTE Recall that we often restrict the domain of a function that is not one-to-one to make it one-to-one without changing the range. For example, the function $g(x) = x^2$, with its natural domain $(-\infty, \infty)$, is not one-to-one. However, if we restrict its domain to the set of nonnegative numbers $[0, \infty)$, we obtain a new function f that is one-to-one and has the same range as g, $[0, \infty)$. See **Figure 10.**

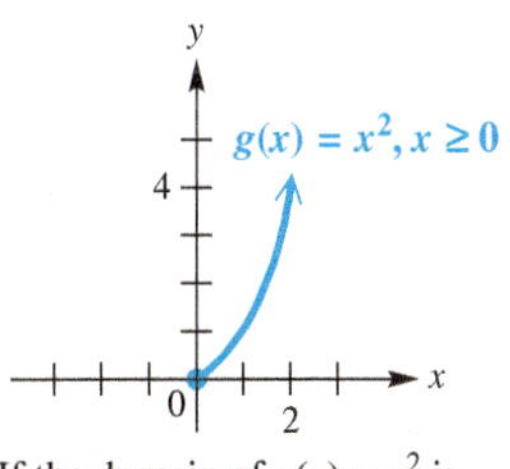

If the domain of $g(x) = x^2$ is restricted so that $x \geq 0$, then it is a one-to-one function.

Figure 10

Inverse Sine Function Refer to the graph of the sine function in **Figure 11** on the next page. Applying the horizontal line test, we see that $y = \sin x$ does not define a one-to-one function. If we restrict the domain to the interval $\left[-\frac{\pi}{2}, \frac{\pi}{2}\right]$, which is the part of the graph in **Figure 11** shown in color, this restricted function is one-to-one and has an inverse function. The range of $y = \sin x$ is $[-1, 1]$, so the domain of the inverse function will be $[-1, 1]$, and its range will be $\left[-\frac{\pi}{2}, \frac{\pi}{2}\right]$.

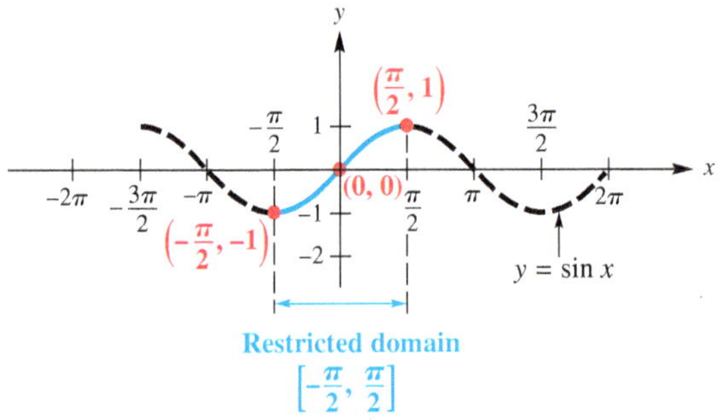

Figure 11

Reflecting the graph of $y = \sin x$ on the restricted domain, shown in **Figure 12(a),** across the line $y = x$ gives the graph of the inverse function, shown in **Figure 12(b).** Some key points are labeled on the graph. The equation of the inverse of $y = \sin x$ is found by interchanging x and y to obtain

$$x = \sin y.$$

This equation is solved for y by writing

$$\mathbf{y = \sin^{-1} x} \quad \text{(read “}\textbf{inverse sine of } \boldsymbol{x}\text{”).}$$

As **Figure 12(b)** shows, the domain of $y = \sin^{-1} x$ is $[-1, 1]$, while the restricted domain of $y = \sin x$, $\left[-\frac{\pi}{2}, \frac{\pi}{2}\right]$, is the range of $y = \sin^{-1} x$. ***An alternative notation for*** $\mathbf{\sin^{-1} x}$ ***is*** **arcsin *x*.**

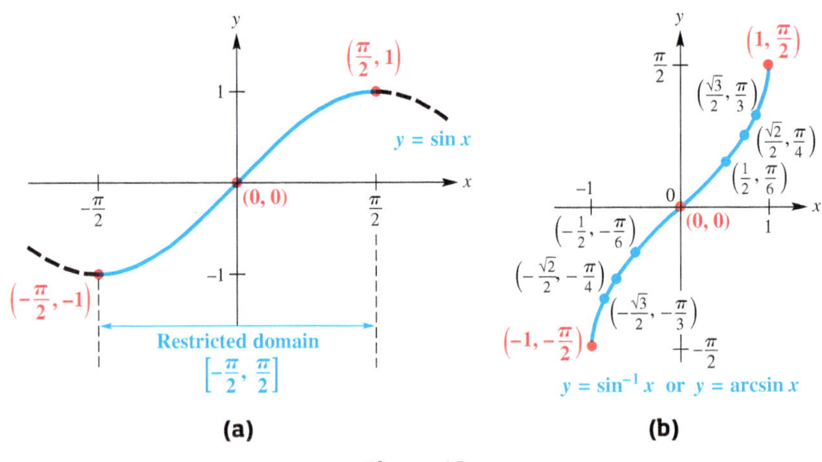

Figure 12

Inverse Sine Function

$\mathbf{y = \sin^{-1} x}$ or $\mathbf{y = \arcsin x}$ means that $x = \sin y$, for $-\frac{\pi}{2} \leq y \leq \frac{\pi}{2}$.

We can think of $\mathbf{y = \sin^{-1} x}$ ***or*** $\mathbf{y = \arcsin x}$ ***as***
“y is the number (angle) in the interval $\left[-\frac{\pi}{2}, \frac{\pi}{2}\right]$ ***whose sine is x.”***

Thus, we can write $y = \sin^{-1} x$ as $\sin y = x$ to evaluate it. We must pay close attention to the domain and range intervals.

EXAMPLE 1 Finding Inverse Sine Values

Find the value of each real number y if it exists.

(a) $y = \arcsin \frac{1}{2}$ **(b)** $y = \sin^{-1}(-1)$ **(c)** $y = \sin^{-1}(-2)$

ALGEBRAIC SOLUTION

(a) The graph of the function defined by $y = \arcsin x$ (**Figure 12(b)**) includes the point $\left(\frac{1}{2}, \frac{\pi}{6}\right)$. Therefore, $\arcsin \frac{1}{2} = \frac{\pi}{6}$.

Alternatively, we can think of $y = \arcsin \frac{1}{2}$ as "y is the number in $\left[-\frac{\pi}{2}, \frac{\pi}{2}\right]$ whose sine is $\frac{1}{2}$." Then we can write the given equation as $\sin y = \frac{1}{2}$. Because $\sin \frac{\pi}{6} = \frac{1}{2}$ and $\frac{\pi}{6}$ is in the range of the arcsine function, $y = \frac{\pi}{6}$.

(b) Writing the equation $y = \sin^{-1}(-1)$ in the form $\sin y = -1$ shows that $y = -\frac{\pi}{2}$. Notice that the point $\left(-1, -\frac{\pi}{2}\right)$ is on the graph of $y = \sin^{-1} x$.

(c) Because -2 is not in the domain of the inverse sine function, $\sin^{-1}(-2)$ does not exist.

GRAPHING CALCULATOR SOLUTION

(a)–(c) We graph the equation $y_1 = \sin^{-1} x$ and find the points with x-values $\frac{1}{2} = 0.5$ and -1. For these two x-values, **Figure 13** indicates that $y = \frac{\pi}{6} \approx 0.52359878$ and $y = -\frac{\pi}{2} \approx -1.570796$.

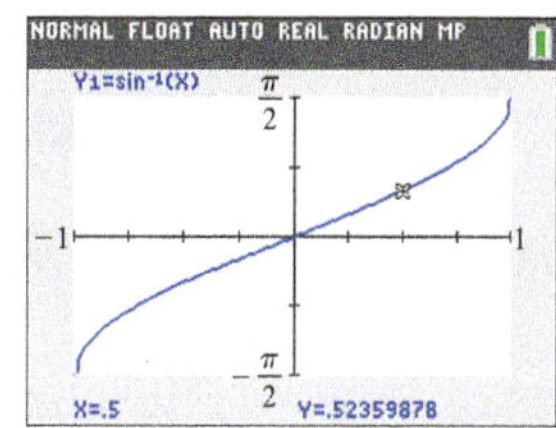

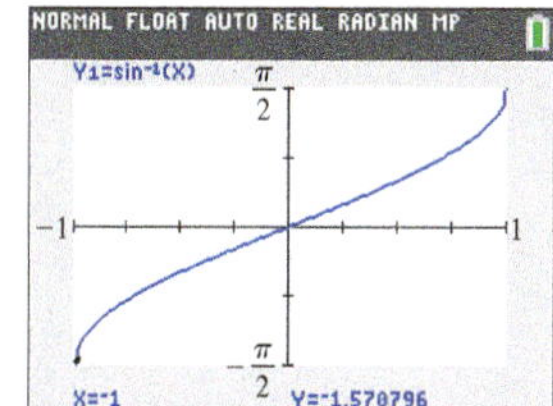

Figure 13

Because $\sin^{-1}(-2)$ does not exist, a calculator will give an error message for this input.

✔ **Now Try Exercises 13, 21, and 25.**

CAUTION In **Example 1(b),** it is tempting to give the value of $\sin^{-1}(-1)$ as $\frac{3\pi}{2}$ because $\sin \frac{3\pi}{2} = -1$. However, $\frac{3\pi}{2}$ is not in the range of the inverse sine function. ***Be certain that the number given for an inverse function value is in the range of the particular inverse function being considered.***

Inverse Sine Function $y = \sin^{-1} x$ or $y = \arcsin x$

Domain: $[-1, 1]$ Range: $\left[-\frac{\pi}{2}, \frac{\pi}{2}\right]$

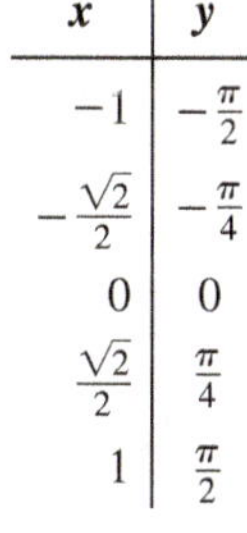

x	y
-1	$-\frac{\pi}{2}$
$-\frac{\sqrt{2}}{2}$	$-\frac{\pi}{4}$
0	0
$\frac{\sqrt{2}}{2}$	$\frac{\pi}{4}$
1	$\frac{\pi}{2}$

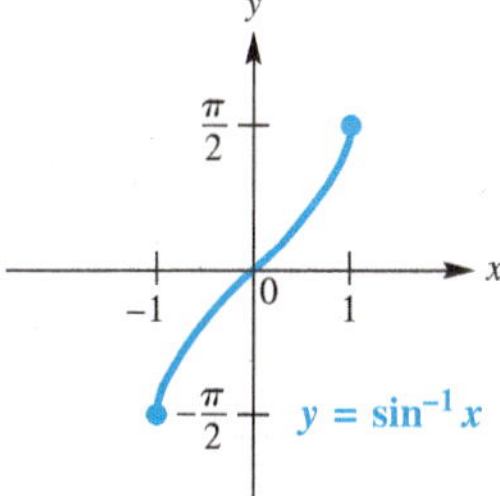

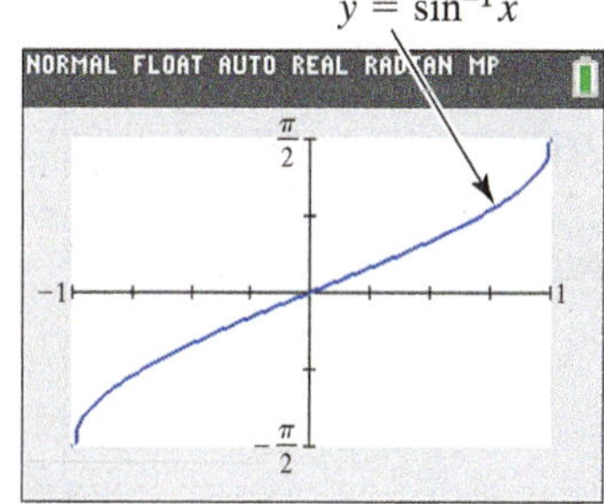

Figure 14

- The inverse sine function is increasing on the open interval $(-1, 1)$ and continuous on its domain $[-1, 1]$.
- Its x- and y-intercepts are both $(0, 0)$.
- Its graph is symmetric with respect to the origin, so the function is an odd function. For all x in the domain, $\sin^{-1}(-x) = -\sin^{-1} x$.

LOOKING AHEAD TO CALCULUS

The **inverse circular functions** are used in calculus to solve certain types of related-rates problems and to integrate certain rational functions.

Inverse Cosine Function The function

$$y = \cos^{-1} x \quad \text{or} \quad y = \arccos x$$

is defined by restricting the domain of the function $y = \cos x$ to the interval $[0, \pi]$ as in **Figure 15.** This restricted function, which is the part of the graph in **Figure 15** shown in color, is one-to-one and has an inverse function. The inverse function, $y = \cos^{-1} x$, is found by interchanging the roles of x and y. Reflecting the graph of $y = \cos x$ across the line $y = x$ gives the graph of the inverse function shown in **Figure 16.** Some key points are shown on the graph.

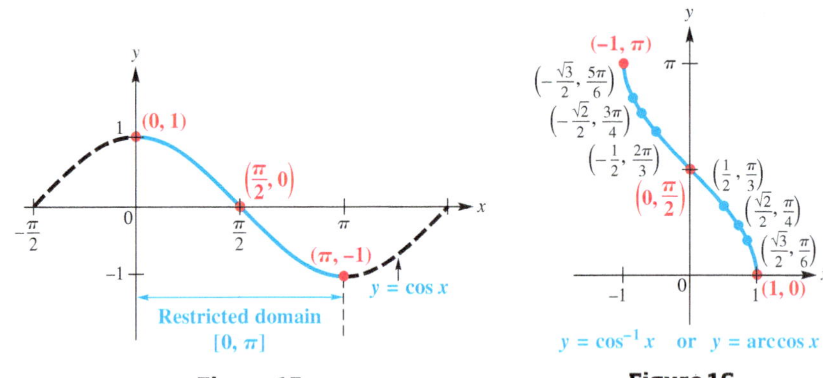

Figure 15 **Figure 16**

Inverse Cosine Function

$y = \cos^{-1} x$ or $y = \arccos x$ means that $x = \cos y$, for $0 \le y \le \pi$.

We can think of $y = \cos^{-1} x$ *or* $y = \arccos x$ *as*

"y is the number (angle) in the interval $[0, \pi]$ *whose cosine is x."*

EXAMPLE 2 Finding Inverse Cosine Values

Find the value of each real number y if it exists.

(a) $y = \arccos 1$ **(b)** $y = \cos^{-1}\left(-\frac{\sqrt{2}}{2}\right)$

SOLUTION

(a) Because the point $(1, 0)$ lies on the graph of $y = \arccos x$ in **Figure 16,** the value of y, or arccos 1, is 0. Alternatively, we can think of $y = \arccos 1$ as

"y is the number in $[0, \pi]$ whose cosine is 1," or $\cos y = 1$.

Thus, $y = 0$, since $\cos 0 = 1$ and 0 is in the range of the arccosine function.

(b) We must find the value of y that satisfies

$$\cos y = -\frac{\sqrt{2}}{2}, \quad \text{where } y \text{ is in the interval } [0, \pi],$$

which is the range of the function $y = \cos^{-1} x$. The only value for y that satisfies these conditions is $\frac{3\pi}{4}$. Again, this can be verified from the graph in **Figure 16.**

These screens support the results of **Example 2** because

$-\frac{\sqrt{2}}{2} \approx -0.7071068$

and $\frac{3\pi}{4} \approx 2.3561945.$

✔ **Now Try Exercises 15 and 23.**

Inverse Cosine Function $y = \cos^{-1} x$ or $y = \arccos x$

Domain: $[-1, 1]$ Range: $[0, \pi]$

x	y
-1	π
$-\frac{\sqrt{2}}{2}$	$\frac{3\pi}{4}$
0	$\frac{\pi}{2}$
$\frac{\sqrt{2}}{2}$	$\frac{\pi}{4}$
1	0

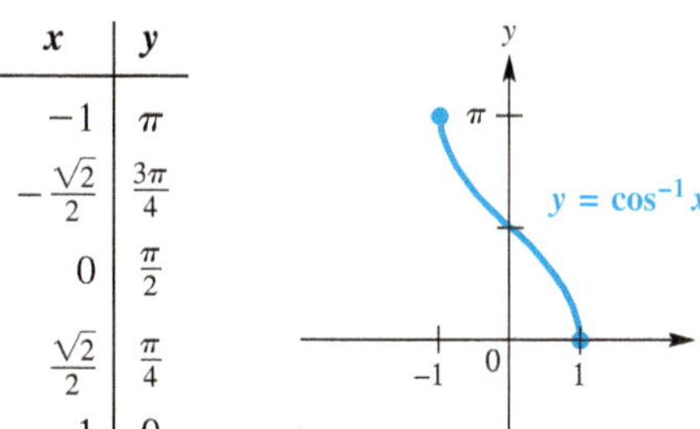

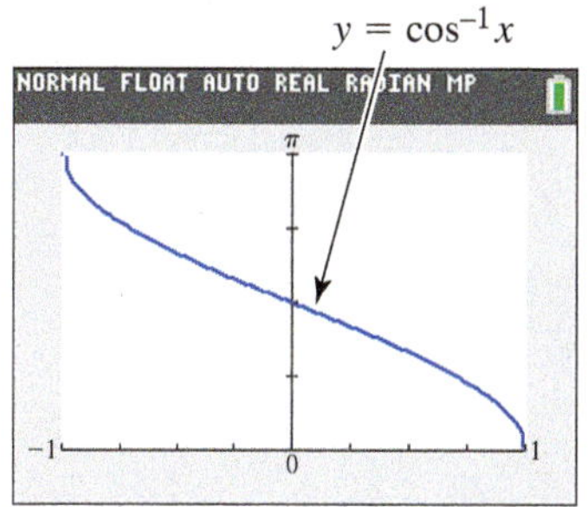

Figure 17

- The inverse cosine function is decreasing on the open interval $(-1, 1)$ and continuous on its domain $[-1, 1]$.
- Its x-intercept is $(1, 0)$ and its y-intercept is $\left(0, \frac{\pi}{2}\right)$.
- Its graph is not symmetric with respect to either the y-axis or the origin.

Inverse Tangent Function Restricting the domain of the function $y = \tan x$ to the open interval $\left(-\frac{\pi}{2}, \frac{\pi}{2}\right)$ yields a one-to-one function. By interchanging the roles of x and y, we obtain the inverse tangent function given by

$$y = \tan^{-1} x \quad \text{or} \quad y = \arctan x.$$

Figure 18 shows the graph of the restricted tangent function. **Figure 19** gives the graph of $y = \tan^{-1} x$.

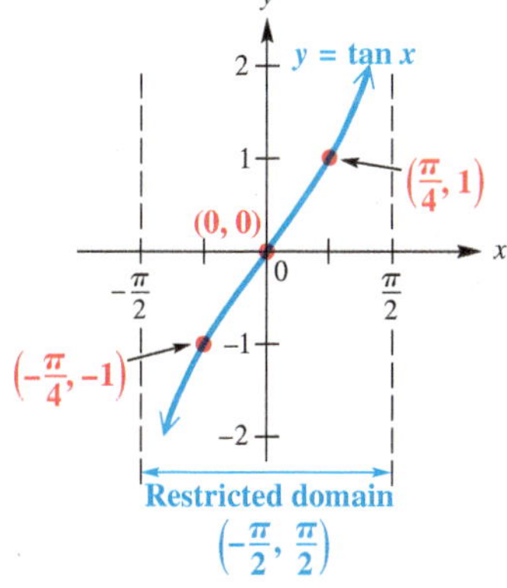

Figure 18

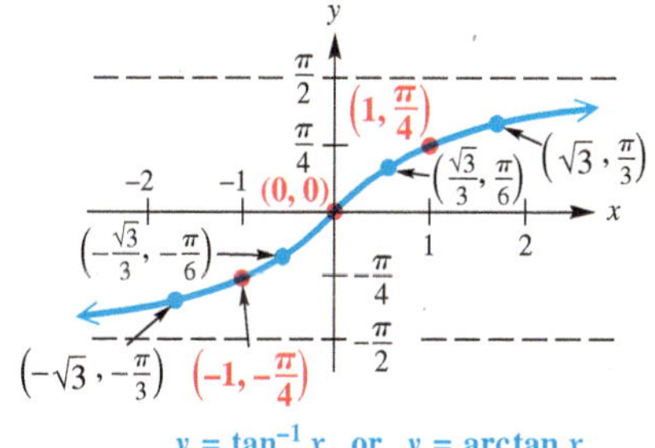

Figure 19

Inverse Tangent Function

$y = \tan^{-1} x$ or $y = \arctan x$ means that $x = \tan y$, for $-\frac{\pi}{2} < y < \frac{\pi}{2}$.

We can think of $y = \tan^{-1} x$ *or* $y = \arctan x$ *as*

"y is the number (angle) in the interval $\left(-\frac{\pi}{2}, \frac{\pi}{2}\right)$ *whose tangent is x."*

We summarize this discussion about the inverse tangent function as follows.

Inverse Tangent Function $y = \tan^{-1} x$ or $y = \arctan x$

Domain: $(-\infty, \infty)$ Range: $\left(-\frac{\pi}{2}, \frac{\pi}{2}\right)$

x	y
-1	$-\frac{\pi}{4}$
$-\frac{\sqrt{3}}{3}$	$-\frac{\pi}{6}$
0	0
$\frac{\sqrt{3}}{3}$	$\frac{\pi}{6}$
1	$\frac{\pi}{4}$

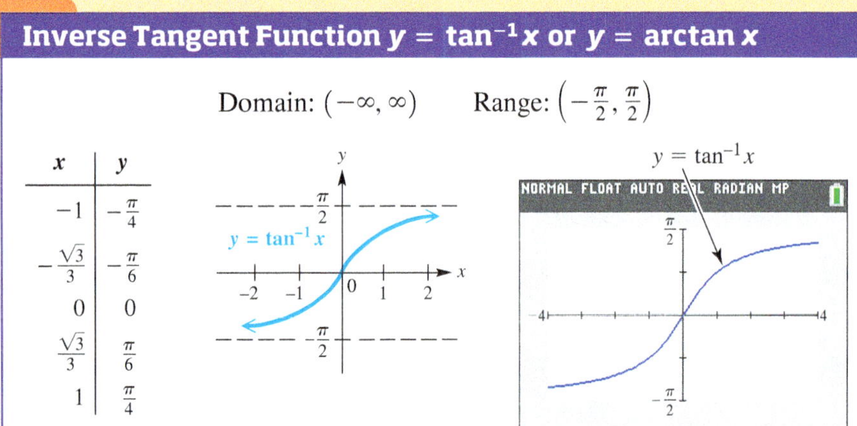

Figure 20

- The inverse tangent function is increasing on $(-\infty, \infty)$ and continuous on its domain $(-\infty, \infty)$.
- Its x- and y-intercepts are both $(0, 0)$.
- Its graph is symmetric with respect to the origin, so the function is an odd function. For all x in the domain, $\tan^{-1}(-x) = -\tan^{-1} x$.
- The lines $y = \frac{\pi}{2}$ and $y = -\frac{\pi}{2}$ are horizontal asymptotes.

Other Inverse Circular Functions The other three inverse trigonometric functions are defined similarly. Their graphs are shown in **Figure 21.**

Inverse Cotangent, Secant, and Cosecant Functions*

$\mathbf{y = \cot^{-1} x}$ or $\mathbf{y = \text{arccot}\, x}$ means that $x = \cot y$, for $0 < y < \pi$.

$\mathbf{y = \sec^{-1} x}$ or $\mathbf{y = \text{arcsec}\, x}$ means that $x = \sec y$, for $0 \le y \le \pi$, $y \neq \frac{\pi}{2}$.

$\mathbf{y = \csc^{-1} x}$ or $\mathbf{y = \text{arccsc}\, x}$ means that $x = \csc y$, for $-\frac{\pi}{2} \le y \le \frac{\pi}{2}$, $y \neq 0$.

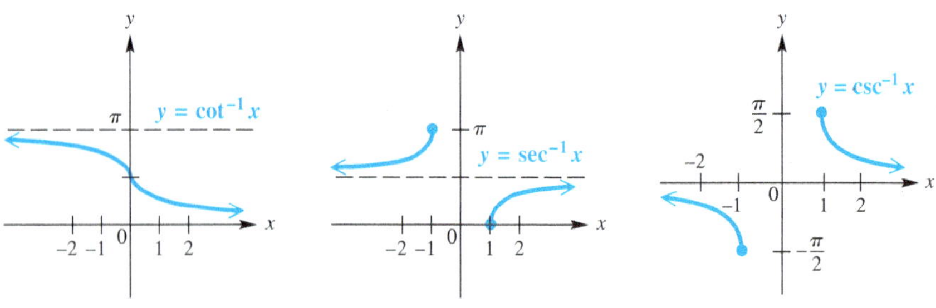

Figure 21

*The inverse secant and inverse cosecant functions are sometimes defined with different ranges. We use intervals that match those of the inverse cosine and inverse sine functions, respectively (except for one missing point).

The table gives all six inverse circular functions with their domains and ranges.

Summary of Inverse Circular Functions

Inverse Function	Domain	Range	
		Interval	Quadrants of the Unit Circle
$y = \sin^{-1} x$	$[-1, 1]$	$\left[-\frac{\pi}{2}, \frac{\pi}{2}\right]$	I and IV
$y = \cos^{-1} x$	$[-1, 1]$	$[0, \pi]$	I and II
$y = \tan^{-1} x$	$(-\infty, \infty)$	$\left(-\frac{\pi}{2}, \frac{\pi}{2}\right)$	I and IV
$y = \cot^{-1} x$	$(-\infty, \infty)$	$(0, \pi)$	I and II
$y = \sec^{-1} x$	$(-\infty, -1] \cup [1, \infty)$	$\left[0, \frac{\pi}{2}\right) \cup \left(\frac{\pi}{2}, \pi\right]$	I and II
$y = \csc^{-1} x$	$(-\infty, -1] \cup [1, \infty)$	$\left[-\frac{\pi}{2}, 0\right) \cup \left(0, \frac{\pi}{2}\right]$	I and IV

Inverse Function Values The inverse circular functions are formally defined with real number ranges. However, there are times when it may be convenient to find degree-measured angles equivalent to these real number values. It is also often convenient to think in terms of the unit circle and choose the inverse function values on the basis of the quadrants given in the preceding table.

EXAMPLE 3 Finding Inverse Function Values (Degree-Measured Angles)

Find the *degree measure* of θ if it exists.

(a) $\theta = \arctan 1$ **(b)** $\theta = \sec^{-1} 2$

SOLUTION

(a) Here θ must be in $(-90°, 90°)$, but because 1 is positive, θ must be in quadrant I. The alternative statement, $\tan \theta = 1$, leads to $\theta = 45°$.

(b) Write the equation as $\sec \theta = 2$. For $\sec^{-1} x$, θ is in quadrant I or II. Because 2 is positive, θ is in quadrant I and $\theta = 60°$, since $\sec 60° = 2$. Note that $60°$ (the degree equivalent of $\frac{\pi}{3}$) is in the range of the inverse secant function.

✔ **Now Try Exercises 37 and 45.**

The inverse trigonometric function keys on a calculator give correct results for the inverse sine, inverse cosine, and inverse tangent functions.

$$\sin^{-1} 0.5 = 30°, \quad \sin^{-1}(-0.5) = -30°,$$
$$\tan^{-1}(-1) = -45°, \quad \text{and} \quad \cos^{-1}(-0.5) = 120° \qquad \text{Degree mode}$$

However, finding $\cot^{-1} x$, $\sec^{-1} x$, and $\csc^{-1} x$ with a calculator is not as straightforward because these functions must first be expressed in terms of $\tan^{-1} x$, $\cos^{-1} x$, and $\sin^{-1} x$, respectively. If $y = \sec^{-1} x$, for example, then we have $\sec y = x$, which must be written in terms of cosine as follows.

$$\text{If} \quad \sec y = x, \quad \text{then} \quad \frac{1}{\cos y} = x, \quad \text{or} \quad \cos y = \frac{1}{x}, \quad \text{and} \quad y = \cos^{-1} \frac{1}{x}.$$

Use the following to evaluate these inverse functions on a calculator.

$$\sec^{-1} x \textit{ is evaluated as } \cos^{-1}\frac{1}{x}; \quad \csc^{-1} x \textit{ is evaluated as } \sin^{-1}\frac{1}{x};$$

$$\cot^{-1} x \textit{ is evaluated as } \begin{cases} \tan^{-1}\frac{1}{x} & \text{if } x > 0 \\ 180^\circ + \tan^{-1}\frac{1}{x} & \text{if } x < 0. \end{cases} \quad \text{Degree mode}$$

EXAMPLE 4 Finding Inverse Function Values with a Calculator

Use a calculator to approximate each value.

(a) Find y in radians if $y = \csc^{-1}(-3)$.

(b) Find θ in degrees if $\theta = \text{arccot}(-0.3541)$.

NORMAL FLOAT AUTO REAL RADIAN MP

$\sin^{-1}\left(\frac{1}{-3}\right)$ -.3398369095

(a)

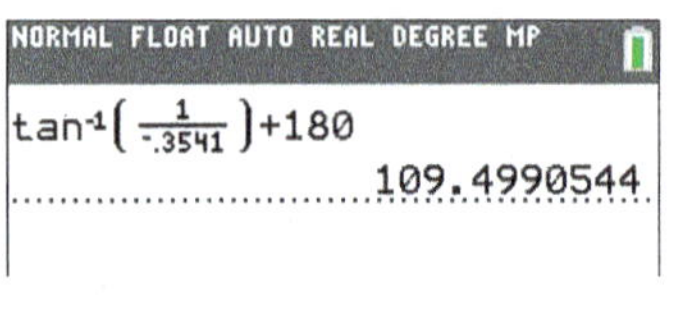

(b)

Figure 22

SOLUTION

(a) With the calculator in radian mode, enter $\csc^{-1}(-3)$ as $\sin^{-1}\left(\frac{1}{-3}\right)$ to obtain $y \approx -0.3398369095$. See **Figure 22(a).**

(b) A calculator in degree mode gives the inverse tangent value of a negative number as a quadrant IV angle. The restriction on the range of arccotangent implies that θ must be in quadrant II.

$$\text{arccot}(-0.3541) \quad \text{is entered as} \quad \tan^{-1}\left(\frac{1}{-0.3541}\right) + 180^\circ.$$

As shown in **Figure 22(b),**

$$\theta \approx 109.4990544^\circ.$$

✔ **Now Try Exercises 53 and 65.**

CAUTION ***Be careful when using a calculator to evaluate the inverse cotangent of a negative quantity.*** Enter the inverse tangent of the *reciprocal* of the negative quantity, which returns an angle in quadrant IV. Because inverse cotangent is negative in quadrant II, adjust the calculator result by adding π or 180° accordingly. (Note that $\cot^{-1} 0 = \frac{\pi}{2}$ or 90°.)

EXAMPLE 5 Finding Function Values Using Definitions of the Trigonometric Functions

Evaluate each expression without using a calculator.

(a) $\sin\left(\tan^{-1}\frac{3}{2}\right)$ **(b)** $\tan\left(\cos^{-1}\left(-\frac{5}{13}\right)\right)$

SOLUTION

(a) Let $\theta = \tan^{-1}\frac{3}{2}$, and thus $\tan\theta = \frac{3}{2}$. The inverse tangent function yields values only in quadrants I and IV, and because $\frac{3}{2}$ is positive, θ is in quadrant I. Sketch θ in quadrant I, and label a triangle, as shown in **Figure 23** on the next page. By the Pythagorean theorem, the hypotenuse is $\sqrt{13}$. The value of sine is the quotient of the side opposite and the hypotenuse.

$$\sin\left(\tan^{-1}\frac{3}{2}\right) = \sin\theta = \frac{3}{\sqrt{13}} = \frac{3}{\sqrt{13}} \cdot \frac{\sqrt{13}}{\sqrt{13}} = \frac{3\sqrt{13}}{13}$$

Rationalize the denominator.

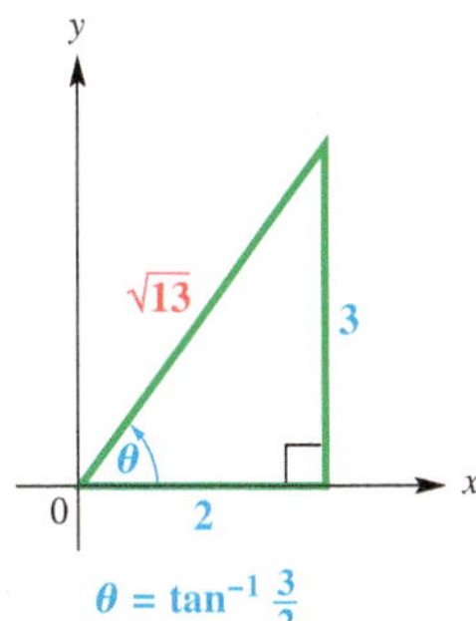

Figure 23

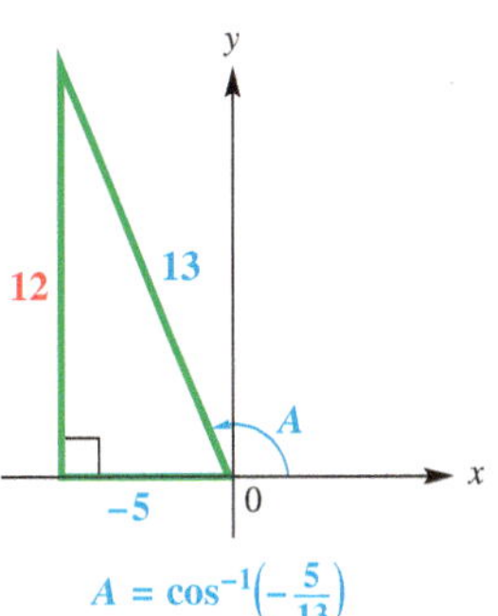

Figure 24

(b) Let $A = \cos^{-1}\left(-\frac{5}{13}\right)$. Then, $\cos A = -\frac{5}{13}$. Because $\cos^{-1} x$ for a negative value of x is in quadrant II, sketch A in quadrant II. See **Figure 24.**

$$\tan\left(\cos^{-1}\left(-\frac{5}{13}\right)\right) = \tan A = -\frac{12}{5}$$

✔ **Now Try Exercises 75 and 77.**

EXAMPLE 6 Finding Function Values Using Identities

Evaluate each expression without using a calculator.

(a) $\cos\left(\arctan\sqrt{3} + \arcsin\frac{1}{3}\right)$ **(b)** $\tan\left(2\arcsin\frac{2}{5}\right)$

SOLUTION

(a) Let $A = \arctan\sqrt{3}$ and $B = \arcsin\frac{1}{3}$. Therefore, $\tan A = \sqrt{3}$ and $\sin B = \frac{1}{3}$. Sketch both A and B in quadrant I, as shown in **Figure 25,** and use the Pythagorean theorem to find the unknown side in each triangle.

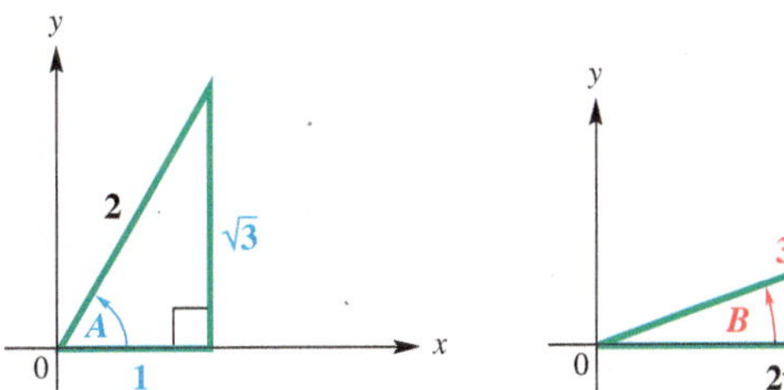

Figure 25

$$\cos\left(\arctan\sqrt{3} + \arcsin\frac{1}{3}\right) \qquad \text{Given expression}$$

$$= \cos(A + B) \qquad \text{Let } A = \arctan\sqrt{3} \text{ and } B = \arcsin\tfrac{1}{3}.$$

$$= \cos A \cos B - \sin A \sin B \qquad \text{Cosine sum identity}$$

$$= \frac{1}{2}\cdot\frac{2\sqrt{2}}{3} - \frac{\sqrt{3}}{2}\cdot\frac{1}{3} \qquad \text{Substitute values using \textbf{Figure 25.}}$$

$$= \frac{2\sqrt{2} - \sqrt{3}}{6} \qquad \text{Multiply and write as a single fraction.}$$

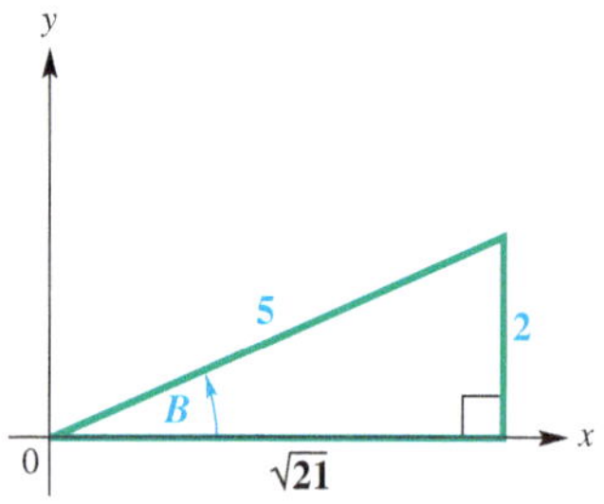

Figure 26

(b) Let $B = \arcsin \frac{2}{5}$, so that $\sin B = \frac{2}{5}$. Sketch angle B in quadrant I, and use the Pythagorean theorem to find the length of the third side of the triangle. See **Figure 26.**

$$\tan\left(2 \arcsin \frac{2}{5}\right) \quad \text{Given expression}$$

$$= \frac{2\left(\frac{2}{\sqrt{21}}\right)}{1 - \left(\frac{2}{\sqrt{21}}\right)^2} \quad \text{Use } \tan 2B = \frac{2 \tan B}{1 - \tan^2 B} \text{ with } \tan B = \frac{2}{\sqrt{21}} \text{ from Figure 26.}$$

$$= \frac{\frac{4}{\sqrt{21}}}{1 - \frac{4}{21}} \quad \text{Multiply and apply the exponent.}$$

$$= \frac{\frac{4}{\sqrt{21}} \cdot \frac{\sqrt{21}}{\sqrt{21}}}{\frac{17}{21}} \quad \text{Rationalize in the numerator. Subtract in the denominator.}$$

$$= \frac{\frac{4\sqrt{21}}{21}}{\frac{17}{21}} \quad \text{Multiply in the numerator.}$$

$$= \frac{4\sqrt{21}}{17} \quad \text{Divide; } \frac{\frac{a}{b}}{\frac{c}{d}} = \frac{a}{b} \div \frac{c}{d} = \frac{a}{b} \cdot \frac{d}{c}.$$

✔ **Now Try Exercises 79 and 87.**

While the work shown in **Examples 5 and 6** does not rely on a calculator, we can use one to support our algebraic work. By entering $\cos\left(\arctan \sqrt{3} + \arcsin \frac{1}{3}\right)$ from **Example 6(a)** into a calculator, we find the approximation 0.1827293862, the same approximation as when we enter $\frac{2\sqrt{2} - \sqrt{3}}{6}$ (the exact value we obtained algebraically). Similarly, we obtain the same approximation when we evaluate $\tan\left(2 \arcsin \frac{2}{5}\right)$ and $\frac{4\sqrt{21}}{17}$, supporting our answer in **Example 6(b).**

EXAMPLE 7 Writing Function Values in Terms of u

Write each trigonometric expression as an algebraic expression in u.

(a) $\sin(\tan^{-1} u)$ **(b)** $\cos(2 \sin^{-1} u)$

SOLUTION

(a) Let $\theta = \tan^{-1} u$, so $\tan \theta = u$. Because

$$-\frac{\pi}{2} < \tan^{-1} u < \frac{\pi}{2},$$

sketch θ in quadrants I and IV and label two triangles, as shown in **Figure 27.** Sine is given by the quotient of the side opposite and the hypotenuse, so we have the following.

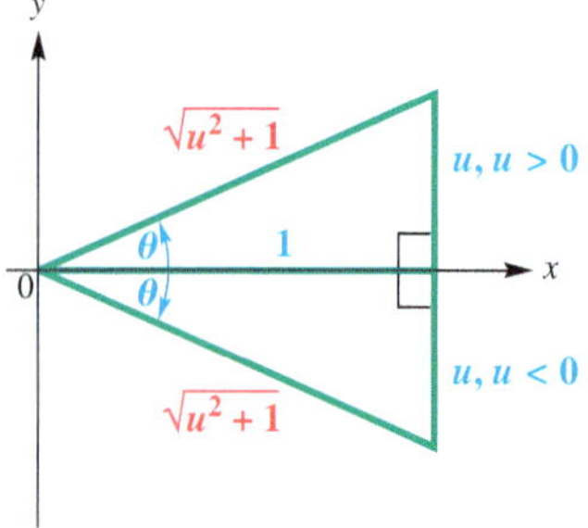

Figure 27

$$\sin(\tan^{-1} u) = \sin \theta = \frac{u}{\sqrt{u^2 + 1}} = \frac{u}{\sqrt{u^2 + 1}} \cdot \frac{\sqrt{u^2 + 1}}{\sqrt{u^2 + 1}} = \frac{u\sqrt{u^2 + 1}}{u^2 + 1}$$

Rationalize the denominator.

The result is positive when u is positive and negative when u is negative.

(b) Let $\theta = \sin^{-1} u$, so that $\sin \theta = u$. To find $\cos 2\theta$, use the double-angle identity $\cos 2\theta = 1 - 2 \sin^2 \theta$.

$$\cos(2 \sin^{-1} u) = \cos 2\theta = 1 - 2 \sin^2 \theta = 1 - 2u^2$$

✔ **Now Try Exercises 95 and 99.**

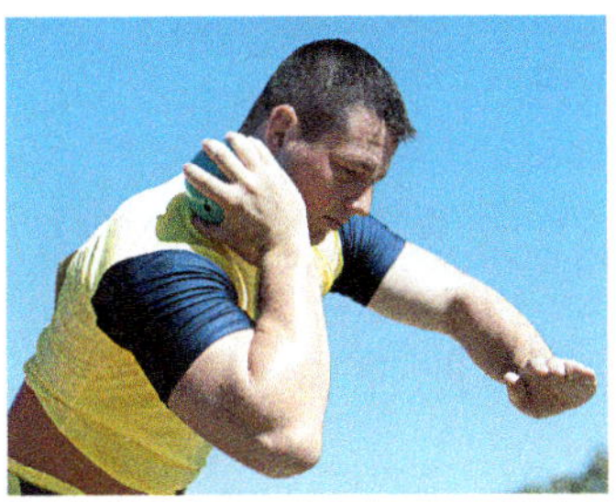

EXAMPLE 8 Finding Optimal Angle of Elevation of a Shot Put

The optimal angle of elevation θ for a shot-putter to achieve the greatest distance depends on the velocity v of the throw and the initial height h of the shot. See **Figure 28.** One model for θ that attains this greatest distance is

$$\theta = \arcsin\left(\sqrt{\frac{v^2}{2v^2 + 64h}}\right).$$

(*Source*: Townend, M. S., *Mathematics in Sport,* Chichester, Ellis Horwood Ltd.)

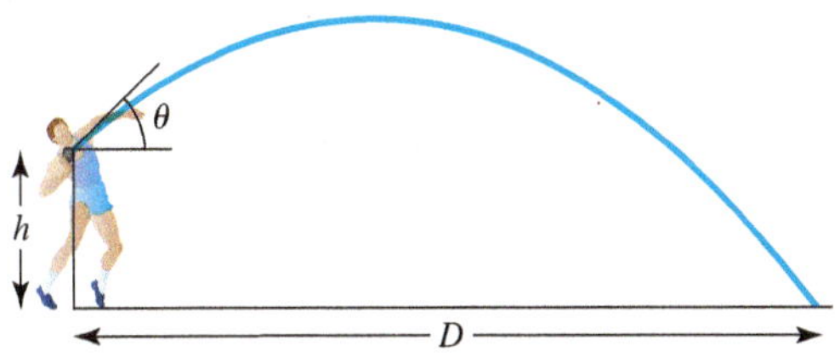

Figure 28

An athlete can consistently put the shot with $h = 6.6$ ft and $v = 42$ ft per sec. At what angle should he release the ball to maximize distance?

SOLUTION To find this angle, substitute and use a calculator in degree mode.

$$\theta = \arcsin\left(\sqrt{\frac{42^2}{2(42^2) + 64(6.6)}}\right) \approx 42° \quad \text{Use } h = 6.6,\ v = 42, \text{ and a calculator.}$$

✔ **Now Try Exercise 105.**

7.5 Exercises

CONCEPT PREVIEW *Fill in the blank(s) to correctly complete each sentence.*

1. For a function to have an inverse, it must be ________-to-________.
2. The domain of $y = \arcsin x$ equals the ________ of $y = \sin x$.
3. $y = \cos^{-1} x$ means that $x =$ ________ for $0 \le y \le \pi$.
4. The point $\left(\frac{\pi}{4}, 1\right)$ lies on the graph of $y = \tan x$. Therefore, the point ________ lies on the graph of $y = \tan^{-1} x$.
5. If a function f has an inverse and $f(\pi) = -1$, then $f^{-1}(-1) =$ ________.
6. To evaluate $\sec^{-1} x$, use the value of $\cos^{-1} \dfrac{1}{\_\_\_\_}$.

CONCEPT PREVIEW *Write a short answer for each of the following.*

7. Consider the inverse sine function $y = \sin^{-1} x$, or $y = \arcsin x$.
   (a) What is its domain?
   (b) What is its range?
   (c) Is this function increasing or decreasing?
   (d) Why is $\arcsin(-2)$ not defined?

8. Consider the inverse cosine function $y = \cos^{-1} x$, or $y = \arccos x$.

(a) What is its domain? **(b)** What is its range?

(c) Is this function increasing or decreasing?

(d) $\arccos\left(-\frac{1}{2}\right) = \frac{2\pi}{3}$. Why is $\arccos\left(-\frac{1}{2}\right)$ not equal to $-\frac{4\pi}{3}$?

9. Consider the inverse tangent function $y = \tan^{-1} x$, or $y = \arctan x$.

(a) What is its domain? **(b)** What is its range?

(c) Is this function increasing or decreasing?

(d) Is there any real number x for which $\arctan x$ is not defined? If so, what is it (or what are they)?

10. Give the domain and range of each inverse trigonometric function, as defined in this section.

(a) inverse cosecant function **(b)** inverse secant function

(c) inverse cotangent function

11. ***Concept Check*** Why are different intervals used when restricting the domains of the sine and cosine functions in the process of defining their inverse functions?

12. ***Concept Check*** For positive values of a, $\cot^{-1} a$ is calculated as $\tan^{-1} \frac{1}{a}$. How is $\cot^{-1} a$ calculated for negative values of a?

Find the exact value of each real number y if it exists. Do not use a calculator. ***See Examples 1 and 2.***

13. $y = \sin^{-1} 0$

14. $y = \sin^{-1}(-1)$

15. $y = \cos^{-1}(-1)$

16. $y = \arccos 0$

17. $y = \tan^{-1} 1$

18. $y = \arctan(-1)$

19. $y = \arctan 0$

20. $y = \tan^{-1}(-1)$

21. $y = \arcsin\left(-\frac{\sqrt{3}}{2}\right)$

22. $y = \sin^{-1} \frac{\sqrt{2}}{2}$

23. $y = \arccos\left(-\frac{\sqrt{3}}{2}\right)$

24. $y = \cos^{-1}\left(-\frac{1}{2}\right)$

25. $y = \sin^{-1} \sqrt{3}$

26. $y = \arcsin\left(-\sqrt{2}\right)$

27. $y = \cot^{-1}(-1)$

28. $y = \text{arccot}\left(-\sqrt{3}\right)$

29. $y = \csc^{-1}(-2)$

30. $y = \csc^{-1} \sqrt{2}$

31. $y = \text{arcsec} \frac{2\sqrt{3}}{3}$

32. $y = \sec^{-1}\left(-\sqrt{2}\right)$

33. $y = \sec^{-1} 1$

34. $y = \sec^{-1} 0$

35. $y = \csc^{-1} \frac{\sqrt{2}}{2}$

36. $y = \text{arccsc}\left(-\frac{1}{2}\right)$

Give the degree measure of θ if it exists. Do not use a calculator. ***See Example 3.***

37. $\theta = \arctan(-1)$

38. $\theta = \tan^{-1} \sqrt{3}$

39. $\theta = \arcsin\left(-\frac{\sqrt{3}}{2}\right)$

40. $\theta = \arcsin\left(-\frac{\sqrt{2}}{2}\right)$

41. $\theta = \arccos\left(-\frac{1}{2}\right)$

42. $\theta = \sec^{-1}(-2)$

43. $\theta = \cot^{-1}\left(-\frac{\sqrt{3}}{3}\right)$

44. $\theta = \cot^{-1} \frac{\sqrt{3}}{3}$

45. $\theta = \csc^{-1}(-2)$

46. $\theta = \csc^{-1}(-1)$

47. $\theta = \sin^{-1} 2$

48. $\theta = \cos^{-1}(-2)$

Use a calculator to approximate each value in decimal degrees. ***See Example 4.***

49. $\theta = \sin^{-1}(-0.13349122)$

50. $\theta = \arcsin 0.77900016$

51. $\theta = \arccos(-0.39876459)$

52. $\theta = \cos^{-1}(-0.13348816)$

53. $\theta = \csc^{-1} 1.9422833$

54. $\theta = \cot^{-1} 1.7670492$

55. $\theta = \cot^{-1}(-0.60724226)$

56. $\theta = \cot^{-1}(-2.7733744)$

57. $\theta = \tan^{-1}(-7.7828641)$

58. $\theta = \sec^{-1}(-5.1180378)$

Use a calculator to approximate each real number value. (Be sure the calculator is in radian mode.) ***See Example 4.***

59. $y = \arcsin 0.92837781$

60. $y = \arcsin 0.81926439$

61. $y = \cos^{-1}(-0.32647891)$

62. $y = \arccos 0.44624593$

63. $y = \arctan 1.1111111$

64. $y = \cot^{-1} 1.0036571$

65. $y = \cot^{-1}(-0.92170128)$

66. $y = \cot^{-1}(-36.874610)$

67. $y = \sec^{-1}(-1.2871684)$

68. $y = \sec^{-1} 4.7963825$

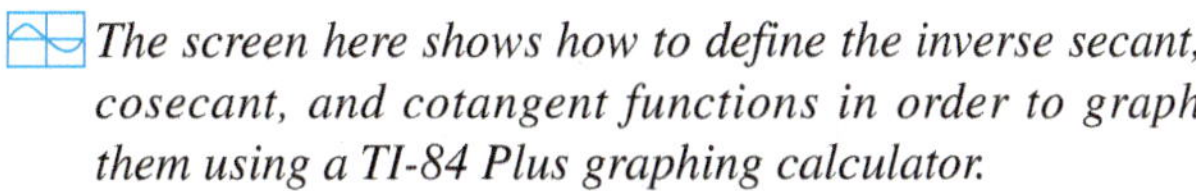

The screen here shows how to define the inverse secant, cosecant, and cotangent functions in order to graph them using a TI-84 Plus graphing calculator.

Use this information to graph each inverse circular function and compare the graphs to those in **Figure 21.**

NORMAL FLOAT AUTO REAL RADIAN MP
Plot1 Plot2 Plot3
$Y_1 = \cos^{-1}\left(\frac{1}{X}\right)$
$Y_2 = \sin^{-1}\left(\frac{1}{X}\right)$
$Y_3 = \frac{\pi}{2} - \tan^{-1}(X)$
$Y_4 =$
$Y_5 =$
$Y_6 =$
$Y_7 =$

69. $y = \sec^{-1} x$

70. $y = \csc^{-1} x$

71. $y = \cot^{-1} x$

Graph each inverse circular function by hand.

72. $y = \operatorname{arccsc} 2x$

73. $y = \operatorname{arcsec} \frac{1}{2}x$

74. $y = 2\cot^{-1} x$

Evaluate each expression without using a calculator. ***See Examples 5 and 6.***

75. $\tan\left(\arccos \frac{3}{4}\right)$

76. $\sin\left(\arccos \frac{1}{4}\right)$

77. $\cos(\tan^{-1}(-2))$

78. $\sec\left(\sin^{-1}\left(-\frac{1}{5}\right)\right)$

79. $\sin\left(2\tan^{-1}\frac{12}{5}\right)$

80. $\cos\left(2\sin^{-1}\frac{1}{4}\right)$

81. $\cos\left(2\arctan\frac{4}{3}\right)$

82. $\tan\left(2\cos^{-1}\frac{1}{4}\right)$

83. $\sin\left(2\cos^{-1}\frac{1}{5}\right)$

84. $\cos(2\tan^{-1}(-2))$

85. $\sec(\sec^{-1} 2)$

86. $\csc(\csc^{-1}\sqrt{2})$

87. $\cos\left(\tan^{-1}\frac{5}{12} - \tan^{-1}\frac{3}{4}\right)$

88. $\cos\left(\sin^{-1}\frac{3}{5} + \cos^{-1}\frac{5}{13}\right)$

89. $\sin\left(\sin^{-1}\frac{1}{2} + \tan^{-1}(-3)\right)$

90. $\tan\left(\cos^{-1}\frac{\sqrt{3}}{2} - \sin^{-1}\left(-\frac{3}{5}\right)\right)$

Use a calculator to find each value. Give answers as real numbers.

91. $\cos(\tan^{-1} 0.5)$

92. $\sin(\cos^{-1} 0.25)$

93. $\tan(\arcsin 0.12251014)$

94. $\cot(\arccos 0.58236841)$

Write each trigonometric expression as an algebraic expression in u, for $u > 0$. **See Example 7.**

95. $\sin(\arccos u)$

96. $\tan(\arccos u)$

97. $\cos(\arcsin u)$

98. $\cot(\arcsin u)$

99. $\sin\left(2 \sec^{-1} \frac{u}{2}\right)$

100. $\cos\left(2 \tan^{-1} \frac{3}{u}\right)$

101. $\tan\left(\sin^{-1} \frac{u}{\sqrt{u^2 + 2}}\right)$

102. $\sec\left(\cos^{-1} \frac{u}{\sqrt{u^2 + 5}}\right)$

103. $\sec\left(\text{arccot} \frac{\sqrt{4 - u^2}}{u}\right)$

104. $\csc\left(\arctan \frac{\sqrt{9 - u^2}}{u}\right)$

(Modeling) Solve each problem.

105. *Angle of Elevation of a Shot Put* Refer to **Example 8.** Suppose a shot-putter can consistently release the steel ball with velocity v of 32 ft per sec from an initial height h of 5.0 ft. What angle, to the nearest degree, will maximize the distance?

106. *Angle of Elevation of a Shot Put* Refer to **Example 8.**

(a) What is the optimal angle, to the nearest degree, when $h = 0$?

(b) Fix h at 6 ft and regard θ as a function of v. As v increases without bound, the graph approaches an asymptote. Find the equation of that asymptote.

107. *Observation of a Painting* A painting 1 m high and 3 m from the floor will cut off an angle θ to an observer, where

$$\theta = \tan^{-1}\left(\frac{x}{x^2 + 2}\right),$$

assuming that the observer is x meters from the wall where the painting is displayed and that the eyes of the observer are 2 m above the ground. (See the figure.) Find the value of θ for the following values of x. Round to the nearest degree.

(a) 1 **(b)** 2 **(c)** 3

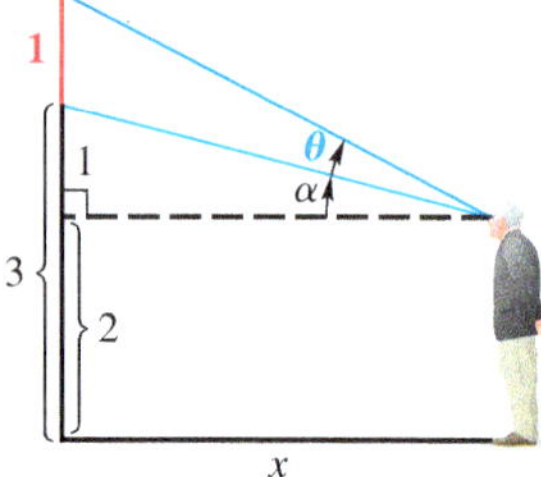

(d) Derive the formula given above. (*Hint*: Use the identity for $\tan(\theta + \alpha)$. Use right triangles.)

(e) Graph the function for θ with a graphing calculator, and determine the distance that maximizes the angle.

(f) The concept in part (e) was first investigated in 1471 by the astronomer Regiomontanus. (*Source*: Maor, E., *Trigonometric Delights*, Princeton University Press.) If the bottom of the picture is a meters above eye level and the top of the picture is b meters above eye level, then the optimum value of x is $\sqrt{ab}$ meters. Use this result to find the exact answer to part (e).

108. *Landscaping Formula* A shrub is planted in a 100-ft-wide space between buildings measuring 75 ft and 150 ft tall. The location of the shrub determines how much sun it receives each day. Show that if θ is the angle in the figure and x is the distance of the shrub from the taller building, then the value of θ (in radians) is given by

$$\theta = \pi - \arctan\left(\frac{75}{100 - x}\right) - \arctan\left(\frac{150}{x}\right).$$

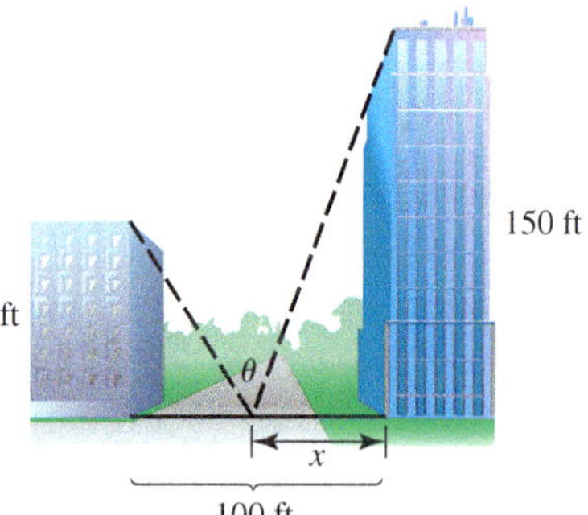

109. ***Communications Satellite Coverage*** The figure shows a stationary communications satellite positioned 20,000 mi above the equator. What percent, to the nearest tenth, of the equator can be seen from the satellite? The diameter of Earth is 7927 mi at the equator.

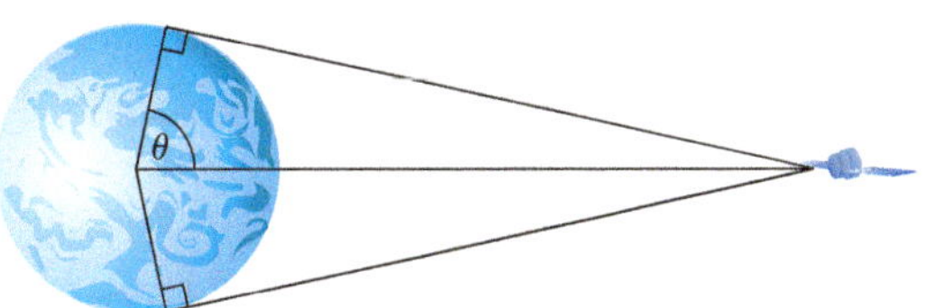

110. ***Oil in a Storage Tank*** The level of oil in a storage tank buried in the ground can be found in much the same way as a dipstick is used to determine the oil level in an automobile crankcase. Suppose the ends of the cylindrical storage tank in the figure are circles of radius 3 ft and the cylinder is 20 ft long. Determine the volume of oil in the tank to the nearest cubic foot if the rod shows a depth of 2 ft. (*Hint:* The volume will be 20 times the area of the shaded segment of the circle shown in the figure on the right.)

Relating Concepts

For individual or collaborative investigation *(Exercises 111–114)**

111. Consider the function

$$f(x) = 3x - 2 \quad \text{and its inverse} \quad f^{-1}(x) = \frac{1}{3}x + \frac{2}{3}.$$

Simplify $f(f^{-1}(x))$ and $f^{-1}(f(x))$. What do you notice in each case?

112. Now consider the general linear functions

$$f(x) = ax + b \quad \text{and} \quad f^{-1}(x) = \frac{1}{a}x - \frac{b}{a}, \quad \text{for } a \neq 0.$$

Simplify $f(f^{-1}(x))$ and $f^{-1}(f(x))$. What do you notice in each case? What is the graph in each case?

113. Use a graphing calculator to graph $y = \tan(\tan^{-1} x)$ in the standard viewing window, using radian mode. How does this compare to the graph you described in **Exercise 112?**

114. Use a graphing calculator to graph $y = \tan^{-1}(\tan x)$ in the standard viewing window, using radian and dot modes. Why does this graph not agree with the graph you found in **Exercise 113?**

*The authors wish to thank Carol Walker of Hinds Community College for making a suggestion on which these exercises are based.

7.6 Trigonometric Equations

Earlier we studied trigonometric equations that were identities. We now consider trigonometric equations that are *conditional*. These equations are satisfied by some values but not others.

Linear Methods

The most basic trigonometric equations are solved by first using properties of equality to isolate a trigonometric expression on one side of the equation.

EXAMPLE 1 Solving a Trigonometric Equation (Linear Methods)

Solve the equation $2 \sin \theta + 1 = 0$

(a) over the interval $[0°, 360°)$ **(b)** for all solutions.

ALGEBRAIC SOLUTION

(a) Because $\sin \theta$ is to the first power, we use the same method as we would to solve the linear equation $2x + 1 = 0$.

$$2 \sin \theta + 1 = 0 \quad \text{Original equation}$$

$$2 \sin \theta = -1 \quad \text{Subtract 1.}$$

$$\sin \theta = -\frac{1}{2} \quad \text{Divide by 2.}$$

To find values of θ that satisfy $\sin \theta = -\frac{1}{2}$, we observe that θ must be in either quadrant III or quadrant IV because the sine function is negative only in these two quadrants. Furthermore, the reference angle must be 30°. The graph of the unit circle in **Figure 29** shows the two possible values of θ. The solution set is $\{210°, 330°\}$.

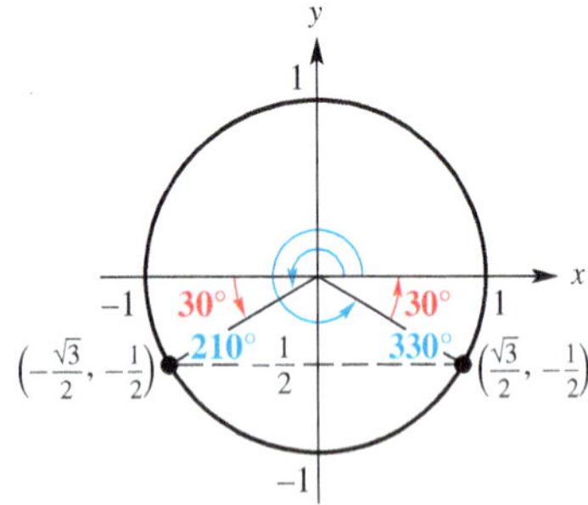

Figure 29

(b) To find all solutions, we add integer multiples of the period of the sine function, 360°, to each solution found in part (a). The solution set is written as follows.

$$\{210° + 360°n, 330° + 360°n, \text{ where } n \text{ is any integer}\}$$

GRAPHING CALCULATOR SOLUTION

(a) Consider the original equation.

$$2 \sin \theta + 1 = 0$$

We can find the solution set of this equation by graphing the function

$$y_1 = 2 \sin x + 1$$

and then determining its zeros. Because we are finding solutions over the interval $[0°, 360°)$, we use degree mode and choose this interval of values for the input x on the graph.

The screen in **Figure 30(a)** indicates that one solution is 210°, and the screen in **Figure 30(b)** indicates that the other solution is 330°. The solution set is $\{210°, 330°\}$, which agrees with the algebraic solution.

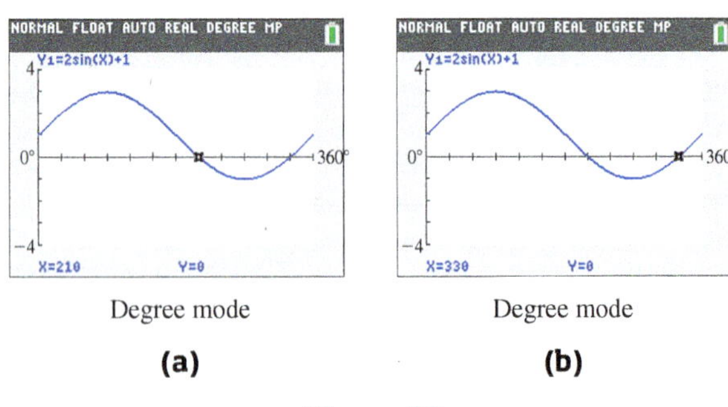

Figure 30

(b) Because the graph of

$$y_1 = 2 \sin x + 1$$

repeats the same y-values every 360°, all solutions are found by adding integer multiples of 360° to the solutions found in part (a). See the algebraic solution.

✓ **Now Try Exercises 15 and 47.**

Zero-Factor Property Method

EXAMPLE 2 Solving a Trigonometric Equation (Zero-Factor Property)

Solve $\sin\theta \tan\theta = \sin\theta$ over the interval $[0°, 360°)$.

SOLUTION

$\sin\theta \tan\theta = \sin\theta$ Original equation

$\sin\theta \tan\theta - \sin\theta = 0$ Subtract $\sin\theta$.

$\sin\theta(\tan\theta - 1) = 0$ Factor out $\sin\theta$.

$\sin\theta = 0$ or $\tan\theta - 1 = 0$ Zero-factor property

$\tan\theta = 1$

$\theta = 0°$ or $\theta = 180°$ $\theta = 45°$ or $\theta = 225°$ Apply the inverse function.

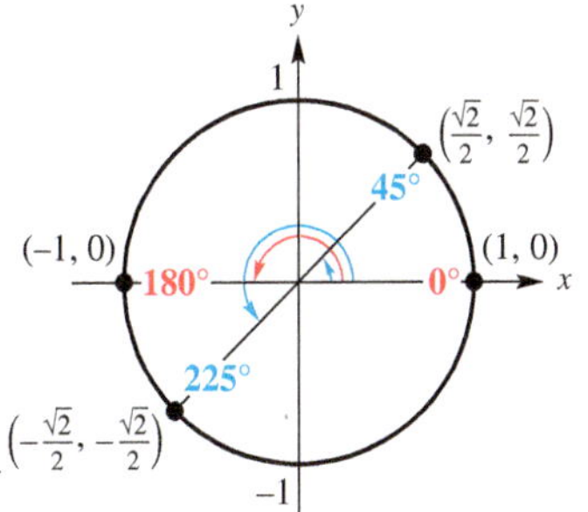

Figure 31

See **Figure 31.** The solution set is $\{0°, 45°, 180°, 225°\}$.

✔ **Now Try Exercise 35.**

CAUTION Trying to solve the equation in **Example 2** by dividing each side by $\sin\theta$ would lead to $\tan\theta = 1$, which would give $\theta = 45°$ or $\theta = 225°$. The missing two solutions are the ones that make the divisor, $\sin\theta$, equal 0. ***For this reason, we avoid dividing by a variable expression.***

Quadratic Methods

The equation $au^2 + bu + c = 0$, where u is an algebraic expression, is solved by quadratic methods. The expression u may be a trigonometric function.

EXAMPLE 3 Solving a Trigonometric Equation (Zero-Factor Property)

Solve $\tan^2 x + \tan x - 2 = 0$ over the interval $[0, 2\pi)$.

SOLUTION

$\tan^2 x + \tan x - 2 = 0$ This equation is quadratic in form.

$(\tan x - 1)(\tan x + 2) = 0$ Factor.

$\tan x - 1 = 0$ or $\tan x + 2 = 0$ Zero-factor property

$\tan x = 1$ or $\tan x = -2$ Solve each equation.

The solutions for $\tan x = 1$ over the interval $[0, 2\pi)$ are $x = \frac{\pi}{4}$ and $x = \frac{5\pi}{4}$.

To solve $\tan x = -2$ over that interval, we use a calculator set in *radian* mode. We find that $\tan^{-1}(-2) \approx -1.1071487$. This is a quadrant IV number, based on the range of the inverse tangent function. However, because we want solutions over the interval $[0, 2\pi)$, we must first add π to -1.1071487, and then add 2π. See **Figure 32.**

$$x \approx -1.1071487 + \pi \approx 2.0344439$$

$$x \approx -1.1071487 + 2\pi \approx 5.1760366$$

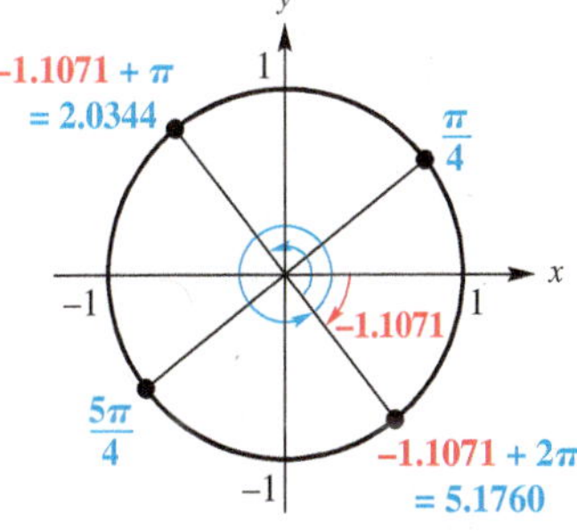

The solutions shown in blue represent angle measures, in radians, *and* their intercepted arc lengths on the unit circle.

Figure 32

The solutions over the required interval form the following solution set.

$$\left\{\underbrace{\frac{\pi}{4}, \quad \frac{5\pi}{4}}_{\text{Exact values}}, \quad \underbrace{2.0344, \quad 5.1760}_{\text{Approximate values to four decimal places}}\right\}$$

✔ **Now Try Exercise 25.**

EXAMPLE 4 Solving a Trigonometric Equation (Quadratic Formula)

Find all solutions of $\cot x(\cot x + 3) = 1$.

SOLUTION We multiply the factors on the left and subtract 1 to write the equation in standard quadratic form.

$$\cot x(\cot x + 3) = 1 \quad \text{Original equation}$$

$$\cot^2 x + 3 \cot x - 1 = 0 \quad \text{Distributive property; Subtract 1.}$$

This equation is quadratic in form, but cannot be solved using the zero-factor property. Therefore, we use the quadratic formula, with $a = 1$, $b = 3$, $c = -1$, and $\cot x$ as the variable.

$$\cot x = \frac{-b \pm \sqrt{b^2 - 4ac}}{2a} \quad \text{Quadratic formula}$$

$$= \frac{-3 \pm \sqrt{3^2 - 4(1)(-1)}}{2(1)} \quad a = 1, b = 3, c = -1$$

Be careful with signs.

$$= \frac{-3 \pm \sqrt{9 + 4}}{2} \quad \text{Simplify.}$$

$$= \frac{-3 \pm \sqrt{13}}{2} \quad \text{Add under the radical.}$$

$$\cot x \approx -3.302775638 \quad \text{or} \quad \cot x \approx 0.3027756377 \quad \text{Use a calculator.}$$

$$x \approx \cot^{-1}(-3.302775638) \quad \text{or} \quad x \approx \cot^{-1}(0.3027756377) \quad \text{Definition of inverse cotangent}$$

$$x \approx \tan^{-1}\left(\frac{1}{-3.302775638}\right) + \pi \quad \text{or} \quad x \approx \tan^{-1}\left(\frac{1}{0.3027756377}\right) \quad \text{Write inverse cotangent in terms of inverse tangent.}$$

$$x \approx -0.2940013018 + \pi \quad \text{or} \quad x \approx 1.276795025 \quad \text{Use a calculator in radian mode.}$$

$$x \approx 2.847591352$$

To find *all* solutions, we add integer multiples of the period of the tangent function, which is π, to each solution found previously. Although not unique, a common form of the solution set of the equation, written using the least possible nonnegative angle measures, is given as follows.

$$\{2.8476 + n\pi, 1.2768 + n\pi, \text{ where } n \text{ is any integer}\}$$

Round to four decimal places.

✔ **Now Try Exercise 57.**

LOOKING AHEAD TO CALCULUS

There are many instances in calculus where it is necessary to solve trigonometric equations. Examples include solving related-rates problems and optimization problems.

Trigonometric Identity Substitutions Recall that squaring each side of an equation, such as

$$\sqrt{x + 4} = x + 2,$$

will yield all solutions but may also give extraneous solutions—solutions that satisfy the final equation but *not* the original equation. As a result, all proposed solutions *must* be checked in the original equation as shown in **Example 5.**

EXAMPLE 5 Solving a Trigonometric Equation (Squaring)

Solve $\tan x + \sqrt{3} = \sec x$ over the interval $[0, 2\pi)$.

SOLUTION We must rewrite the equation in terms of a single trigonometric function. Because the tangent and secant functions are related by the identity $1 + \tan^2 x = \sec^2 x$, square each side and express $\sec^2 x$ in terms of $\tan^2 x$.

Don't forget the middle term.

$$\left(\tan x + \sqrt{3}\right)^2 = (\sec x)^2 \quad \text{Square each side.}$$

$$\tan^2 x + 2\sqrt{3}\tan x + 3 = \sec^2 x \quad (x+y)^2 = x^2 + 2xy + y^2$$

$$\tan^2 x + 2\sqrt{3}\tan x + 3 = 1 + \tan^2 x \quad \text{Pythagorean identity}$$

$$2\sqrt{3}\tan x = -2 \quad \text{Subtract } 3 + \tan^2 x.$$

$$\tan x = -\frac{1}{\sqrt{3}} \quad \text{Divide by } 2\sqrt{3}.$$

$$\tan x = -\frac{\sqrt{3}}{3} \quad \text{Rationalize the denominator.}$$

Solutions of $\tan x = -\frac{\sqrt{3}}{3}$ over $[0, 2\pi)$ are $\frac{5\pi}{6}$ and $\frac{11\pi}{6}$. These possible, or proposed, solutions must be checked to determine whether they are also solutions of the original equation.

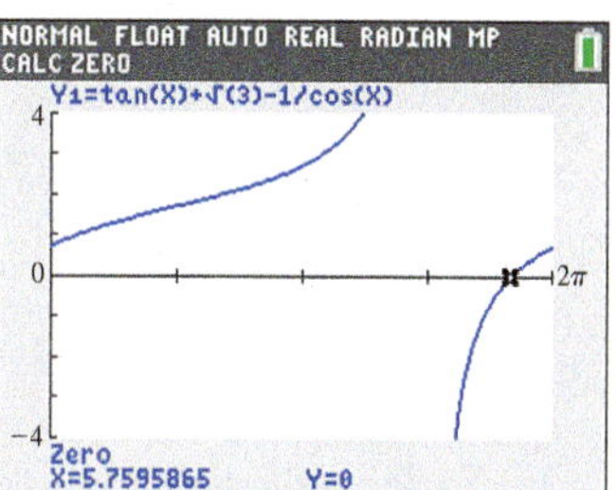

Radian mode

The graph shows that on the interval $[0, 2\pi)$, the only zero of the function $y = \tan x + \sqrt{3} - \sec x$ is 5.7595865, which is an approximation for $\frac{11\pi}{6}$, the solution found in **Example 5.**

CHECK $\tan x + \sqrt{3} = \sec x$ Original equation

$$\tan\left(\frac{5\pi}{6}\right) + \sqrt{3} \stackrel{?}{=} \sec\left(\frac{5\pi}{6}\right)$$

$$-\frac{\sqrt{3}}{3} + \frac{3\sqrt{3}}{3} \stackrel{?}{=} -\frac{2\sqrt{3}}{3}$$

$$\frac{2\sqrt{3}}{3} = -\frac{2\sqrt{3}}{3} \quad \text{False}$$

$$\tan\left(\frac{11\pi}{6}\right) + \sqrt{3} \stackrel{?}{=} \sec\left(\frac{11\pi}{6}\right)$$

$$-\frac{\sqrt{3}}{3} + \frac{3\sqrt{3}}{3} \stackrel{?}{=} \frac{2\sqrt{3}}{3}$$

$$\frac{2\sqrt{3}}{3} = \frac{2\sqrt{3}}{3} \checkmark \quad \text{True}$$

As the check shows, only $\frac{11\pi}{6}$ is a solution, so the solution set is $\left\{\frac{11\pi}{6}\right\}$.

✓ **Now Try Exercise 45.**

Solving a Trigonometric Equation

1. Decide whether the equation is linear or quadratic in form in order to determine the solution method.
2. If only one trigonometric function is present, solve the equation for that function.
3. If more than one trigonometric function is present, rewrite the equation so that one side equals 0. Then try to factor and apply the zero-factor property.
4. If the equation is quadratic in form, but not factorable, use the quadratic formula. Check that solutions are in the desired interval.
5. Try using identities to change the form of the equation. It may be helpful to square each side of the equation first. In this case, check for extraneous solutions.

Equations with Half-Angles

EXAMPLE 6 Solving an Equation with a Half-Angle

Solve $2 \sin \frac{x}{2} = 1$

(a) over the interval $[0, 2\pi)$ **(b)** for all solutions.

SOLUTION

(a) To solve over the interval $[0, 2\pi)$, we must have

$$0 \leq x < 2\pi.$$

The corresponding inequality for $\frac{x}{2}$ is

$$0 \leq \frac{x}{2} < \pi. \quad \text{Divide by 2.}$$

To find all values of $\frac{x}{2}$ over the interval $[0, \pi)$ that satisfy the given equation, first solve for $\sin \frac{x}{2}$.

$$2 \sin \frac{x}{2} = 1 \quad \text{Original equation}$$

$$\sin \frac{x}{2} = \frac{1}{2} \quad \text{Divide by 2.}$$

NORMAL FLOAT AUTO REAL RADIAN MP
CALC ZERO
Y1=2sin(X/2)-1
2
0
2π
−2
Zero
X=1.0471976 Y=0

The x-intercepts correspond to the solutions found in **Example 6(a).** Using Xscl $= \frac{\pi}{3}$ makes it possible to support the exact solutions by counting the tick marks from 0 on the graph.

The two numbers over the interval $[0, \pi)$ with sine value $\frac{1}{2}$ are $\frac{\pi}{6}$ and $\frac{5\pi}{6}$.

$$\frac{x}{2} = \frac{\pi}{6} \quad \text{or} \quad \frac{x}{2} = \frac{5\pi}{6} \quad \text{Definition of inverse sine}$$

$$x = \frac{\pi}{3} \quad \text{or} \quad x = \frac{5\pi}{3} \quad \text{Multiply by 2.}$$

The solution set over the given interval is $\left\{\frac{\pi}{3}, \frac{5\pi}{3}\right\}$.

(b) The argument $\frac{x}{2}$ in the expression $\sin \frac{x}{2}$ can also be written $\frac{1}{2}x$ to see that the value of b in $\sin bx$ is $\frac{1}{2}$. From earlier work we know that the period is $\frac{2\pi}{b}$, so we replace b with $\frac{1}{2}$ in this expression and perform the calculation. Here the period is

$$\frac{2\pi}{\frac{1}{2}} = 2\pi \div \frac{1}{2} = 2\pi \cdot 2 = 4\pi.$$

All solutions are found by adding integer multiples of 4π.

$$\left\{\frac{\pi}{3} + 4n\pi, \frac{5\pi}{3} + 4n\pi, \text{ where } n \text{ is any integer}\right\}$$

✔ **Now Try Exercises 77 and 91.**

CAUTION Because 2 is not a factor of $\cos 2x$, $\frac{\cos 2x}{2} \neq \cos x$. In **Example 7** on the next page, we change $\cos 2x$ to a function of x alone using an identity.

Equations with Multiple Angles

EXAMPLE 7 Solving an Equation Using a Double-Angle Identity

Solve $\cos 2x = \cos x$ over the interval $[0, 2\pi)$.

SOLUTION First convert $\cos 2x$ to a function of x alone. Use the identity $\cos 2x = 2\cos^2 x - 1$ so that the equation involves only $\cos x$. Then factor.

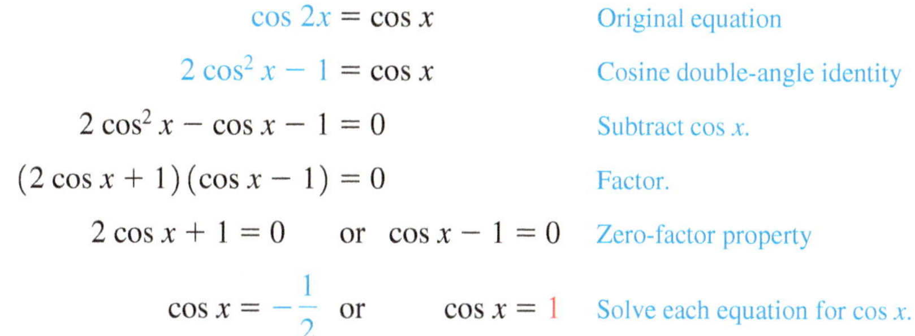

$$\cos 2x = \cos x \qquad \text{Original equation}$$

$$2\cos^2 x - 1 = \cos x \qquad \text{Cosine double-angle identity}$$

$$2\cos^2 x - \cos x - 1 = 0 \qquad \text{Subtract } \cos x.$$

$$(2\cos x + 1)(\cos x - 1) = 0 \qquad \text{Factor.}$$

$$2\cos x + 1 = 0 \quad \text{or} \quad \cos x - 1 = 0 \qquad \text{Zero-factor property}$$

$$\cos x = -\frac{1}{2} \quad \text{or} \quad \cos x = 1 \qquad \text{Solve each equation for } \cos x.$$

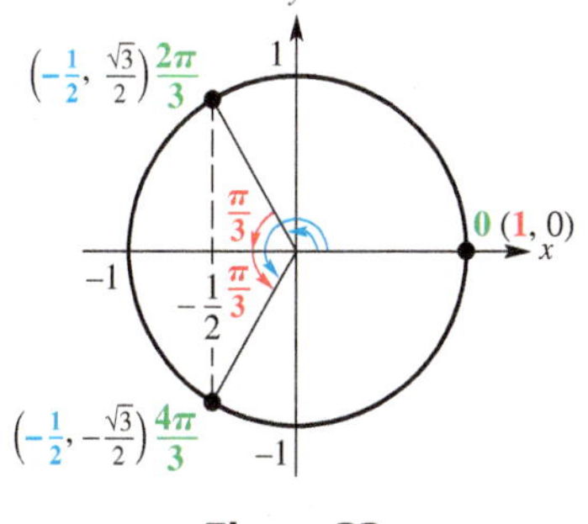

Figure 33

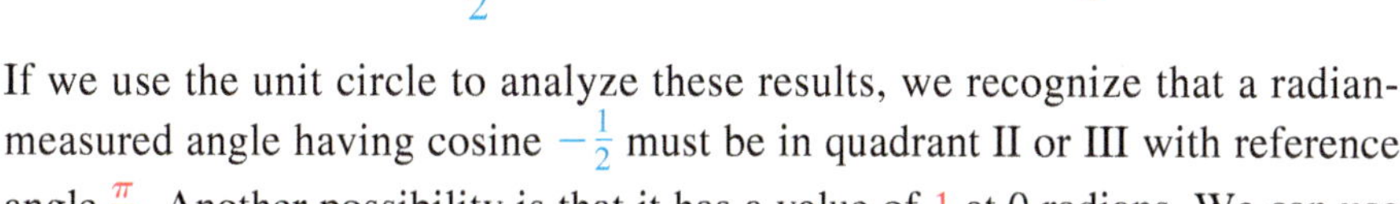

If we use the unit circle to analyze these results, we recognize that a radian-measured angle having cosine $-\frac{1}{2}$ must be in quadrant II or III with reference angle $\frac{\pi}{3}$. Another possibility is that it has a value of 1 at 0 radians. We can use **Figure 33** to determine that solutions over the required interval are as follows.

$$x = \frac{2\pi}{3} \quad \text{or} \quad x = \frac{4\pi}{3} \quad \text{or} \quad x = 0$$

The solution set is $\left\{0, \frac{2\pi}{3}, \frac{4\pi}{3}\right\}$.

✔ **Now Try Exercise 79.**

EXAMPLE 8 Solving an Equation Using a Double-Angle Identity

Solve $4 \sin\theta \cos\theta = \sqrt{3}$

(a) over the interval $[0°, 360°)$ **(b)** for all solutions.

SOLUTION

(a)

$$4\sin\theta\cos\theta = \sqrt{3} \qquad \text{Original equation}$$

$$2(2\sin\theta\cos\theta) = \sqrt{3} \qquad 4 = 2 \cdot 2$$

$$2\sin 2\theta = \sqrt{3} \qquad \text{Sine double-angle identity}$$

$$\sin 2\theta = \frac{\sqrt{3}}{2} \qquad \text{Divide by 2.}$$

From the given interval $0° \leq \theta < 360°$, the corresponding interval for 2θ is $0° \leq 2\theta < 720°$. Because the sine is positive in quadrants I and II, solutions over this interval are as follows.

$$2\theta = 60°, 120°, 420°, 480°, \qquad \text{Reference angle is } 60°.$$

$$\text{or} \quad \theta = 30°, 60°, 210°, 240° \qquad \text{Divide by 2.}$$

The final two solutions for 2θ were found by adding $360°$ to $60°$ and $120°$, respectively, which gives the solution set

$$\{30°, 60°, 210°, 240°\}.$$

(b) All angles 2θ that are solutions of the equation $\sin 2\theta = \frac{\sqrt{3}}{2}$ are found by adding integer multiples of 360° to the basic solution angles, 60° and 120°.

$$2\theta = 60° + 360°n \quad \text{and} \quad 2\theta = 120° + 360°n \qquad \text{Add integer multiples of } 360°.$$

$$\theta = 30° + 180°n \quad \text{and} \quad \theta = 60° + 180°n \qquad \text{Divide by 2.}$$

All solutions are given by the following set, where 180° represents the period of $\sin 2\theta$.

$$\{30° + 180°n, 60° + 180°n, \text{ where } n \text{ is any integer}\}$$

✔ **Now Try Exercises 75 and 99.**

Applications

EXAMPLE 9 Describing a Musical Tone from a Graph

A basic component of music is a pure tone. The graph in **Figure 34** models the sinusoidal pressure $y = P$ in pounds per square foot from a pure tone at time $x = t$ in seconds.

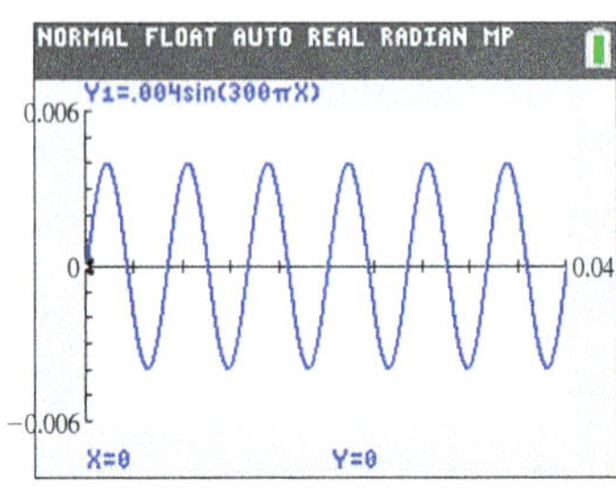

Figure 34

(a) The frequency of a pure tone is often measured in hertz. One hertz is equal to one cycle per second and is abbreviated Hz. What is the frequency f, in hertz, of the pure tone shown in the graph?

(b) The time for the tone to produce one complete cycle is the **period.** Approximate the period T, in seconds, of the pure tone.

(c) An equation for the graph is $y = 0.004 \sin 300\pi x$. Use a calculator to estimate all solutions that make $y = 0.004$ over the interval $[0, 0.02]$.

SOLUTION

(a) From **Figure 34,** we see that there are 6 cycles in 0.04 sec. This is equivalent to $\frac{6}{0.04} = 150$ cycles per sec. The pure tone has a frequency of $f = 150$ Hz.

(b) Six periods cover a time interval of 0.04 sec. One period would be equal to $T = \frac{0.04}{6} = \frac{1}{150}$, or $0.00\overline{6}$ sec.

(c) If we reproduce the graph in **Figure 34** on a calculator as y_1 and also graph a second function as $y_2 = 0.004$, we can determine that the approximate values of x at the points of intersection of the graphs over the interval $[0, 0.02]$ are

$$0.0017, \quad 0.0083, \quad \text{and} \quad 0.015.$$

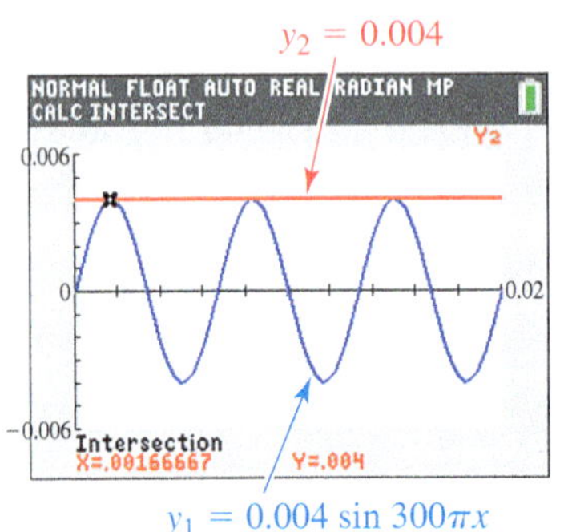

Figure 35

The first value is shown in **Figure 35.** These values represent time in seconds.

✔ **Now Try Exercise 105.**

A piano string can vibrate at more than one frequency when it is struck. It produces a complex wave that can mathematically be modeled by a sum of several pure tones. When a piano key with a frequency of f_1 is played, the corresponding string vibrates not only at f_1 but also at the higher frequencies of $2f_1, 3f_1, 4f_1, \ldots, nf_1$. f_1 is the **fundamental frequency** of the string, and higher frequencies are the **upper harmonics.** The human ear will hear the sum of these frequencies as one complex tone. (*Source*: Roederer, J., *Introduction to the Physics and Psychophysics of Music,* Second Edition, Springer-Verlag.)

EXAMPLE 10 Analyzing Pressures of Upper Harmonics

Suppose that the A key above middle C is played on a piano. Its fundamental frequency is $f_1 = 440$ Hz, and its associated pressure is expressed as

$$P_1 = 0.002 \sin 880\pi t.$$

The string will also vibrate at

$$f_2 = 880, \ f_3 = 1320, \ f_4 = 1760, \ f_5 = 2200, \ldots \text{ Hz}.$$

The corresponding pressures of these upper harmonics are as follows.

$$P_2 = \frac{0.002}{2} \sin 1760\pi t, \qquad P_3 = \frac{0.002}{3} \sin 2640\pi t,$$

$$P_4 = \frac{0.002}{4} \sin 3520\pi t, \qquad \text{and} \qquad P_5 = \frac{0.002}{5} \sin 4400\pi t$$

The graph of $P = P_1 + P_2 + P_3 + P_4 + P_5$ can be found by entering each P_i as a separate function y_i and graphing their sum. The graph, shown in **Figure 36,** is "saw-toothed."

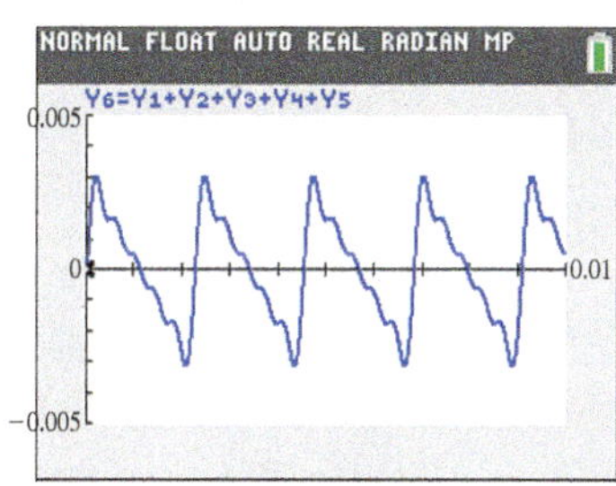

Figure 36

(a) Approximate the maximum value of P.

(b) At what values of $t = x$ does this maximum occur over $[0, 0.01]$?

SOLUTION

(a) A graphing calculator shows that the maximum value of P is approximately 0.00317. See **Figure 37.**

(b) The maximum occurs at

$$t = x \approx 0.000191, 0.00246, 0.00474, 0.00701, \text{ and } 0.00928.$$

Figure 37 shows how the second value is found. The other values are found similarly.

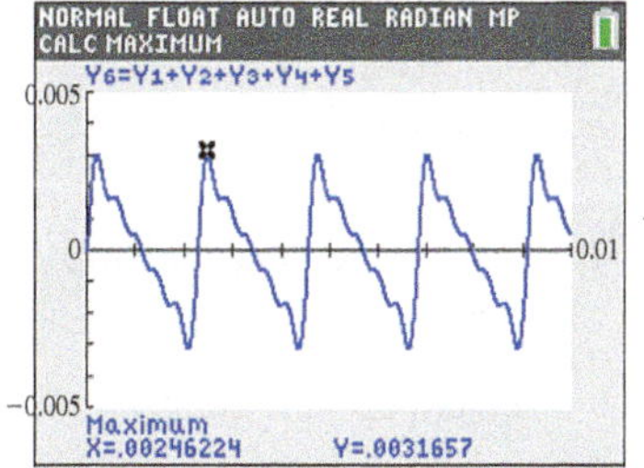

Figure 37

✔ **Now Try Exercise 109.**

7.6 Exercises

CONCEPT PREVIEW *Use the unit circle shown here to solve each simple trigonometric equation. If the variable is x, then solve over $[0, 2\pi)$. If the variable is θ, then solve over $[0°, 360°)$.*

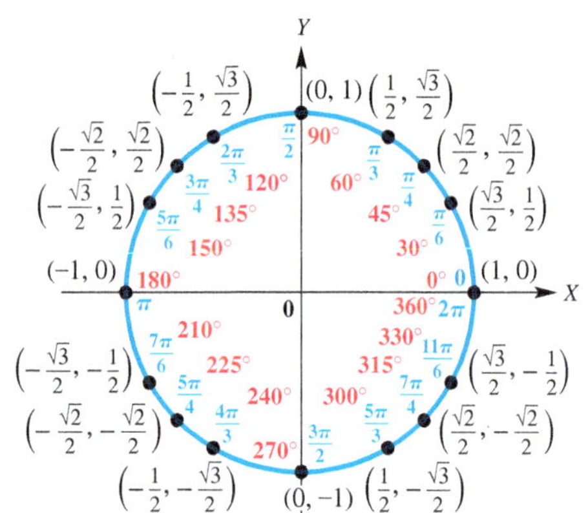

1. $\cos x = \frac{1}{2}$

2. $\cos x = \frac{\sqrt{3}}{2}$

3. $\sin x = -\frac{1}{2}$

4. $\sin \theta = 0$

5. $\sin \theta = -1$

6. $\cos \theta = -\frac{\sqrt{2}}{2}$

CONCEPT PREVIEW *Refer to* ***Exercises 1–3,*** *and use those results to solve each equation over the interval $[0, 2\pi)$.*

7. $\cos 2x = \frac{1}{2}$

8. $\cos 2x = \frac{\sqrt{3}}{2}$

9. $\sin 2x = -\frac{1}{2}$

CONCEPT PREVIEW *Refer to* ***Exercises 4–6,*** *and use those results to solve each equation over the interval $[0°, 360°)$.*

10. $\sin \frac{\theta}{2} = 0$

11. $\sin \frac{\theta}{2} = -1$

12. $\cos \frac{\theta}{2} = -\frac{\sqrt{2}}{2}$

13. *Concept Check* Suppose that in solving an equation over the interval $[0°, 360°)$, we reach the step $\sin \theta = -\frac{1}{2}$. Why is $-30°$ not a correct answer?

14. *Concept Check* Lindsay solved the equation $\sin x = 1 - \cos x$ by squaring each side to obtain

$$\sin^2 x = 1 - 2\cos x + \cos^2 x.$$

Several steps later, using correct algebra, she concluded that the solution set for solutions over the interval $[0, 2\pi)$ is $\left\{0, \frac{\pi}{2}, \frac{3\pi}{2}\right\}$. Explain why this is not correct.

Solve each equation for exact solutions over the interval $[0, 2\pi)$. ***See Examples 1–3.***

15. $2\cot x + 1 = -1$

16. $\sin x + 2 = 3$

17. $2\sin x + 3 = 4$

18. $2\sec x + 1 = \sec x + 3$

19. $\tan^2 x + 3 = 0$

20. $\sec^2 x + 2 = -1$

21. $(\cot x - 1)(\sqrt{3}\cot x + 1) = 0$

22. $(\csc x + 2)(\csc x - \sqrt{2}) = 0$

23. $\cos^2 x + 2\cos x + 1 = 0$

24. $2\cos^2 x - \sqrt{3}\cos x = 0$

25. $-2\sin^2 x = 3\sin x + 1$

26. $2\cos^2 x - \cos x = 1$

Solve each equation for solutions over the interval $[0°, 360°)$. Give solutions to the nearest tenth as appropriate. ***See Examples 2–5.***

27. $(\cot \theta - \sqrt{3})(2\sin \theta + \sqrt{3}) = 0$

28. $(\tan \theta - 1)(\cos \theta - 1) = 0$

29. $2\sin \theta - 1 = \csc \theta$

30. $\tan \theta + 1 = \sqrt{3} + \sqrt{3}\cot \theta$

31. $\tan\theta - \cot\theta = 0$

32. $\cos^2\theta = \sin^2\theta + 1$

33. $\csc^2\theta - 2\cot\theta = 0$

34. $\sin^2\theta\cos\theta = \cos\theta$

35. $2\tan^2\theta\sin\theta - \tan^2\theta = 0$

36. $\sin^2\theta\cos^2\theta = 0$

37. $\sec^2\theta\tan\theta = 2\tan\theta$

38. $\cos^2\theta - \sin^2\theta = 0$

39. $9\sin^2\theta - 6\sin\theta = 1$

40. $4\cos^2\theta + 4\cos\theta = 1$

41. $\tan^2\theta + 4\tan\theta + 2 = 0$

42. $3\cot^2\theta - 3\cot\theta - 1 = 0$

43. $\sin^2\theta - 2\sin\theta + 3 = 0$

44. $2\cos^2\theta + 2\cos\theta + 1 = 0$

45. $\cot\theta + 2\csc\theta = 3$

46. $2\sin\theta = 1 - 2\cos\theta$

Solve each equation (x in radians and θ in degrees) for all exact solutions where appropriate. Round approximate answers in radians to four decimal places and approximate answers in degrees to the nearest tenth. Write answers using the least possible nonnegative angle measures. ***See Examples 1–5.***

47. $\cos\theta + 1 = 0$

48. $\tan\theta + 1 = 0$

49. $3\csc x - 2\sqrt{3} = 0$

50. $\cot x + \sqrt{3} = 0$

51. $6\sin^2\theta + \sin\theta = 1$

52. $3\sin^2\theta - \sin\theta = 2$

53. $2\cos^2 x + \cos x - 1 = 0$

54. $4\cos^2 x - 1 = 0$

55. $\sin\theta\cos\theta - \sin\theta = 0$

56. $\tan\theta\csc\theta - \sqrt{3}\csc\theta = 0$

57. $\sin x(3\sin x - 1) = 1$

58. $\tan x(\tan x - 2) = 5$

59. $5 + 5\tan^2\theta = 6\sec\theta$

60. $\sec^2\theta = 2\tan\theta + 4$

61. $\dfrac{2\tan\theta}{3 - \tan^2\theta} = 1$

62. $\dfrac{2\cot^2\theta}{\cot\theta + 3} = 1$

The following equations cannot be solved by algebraic methods. Use a graphing calculator to find all solutions over the interval $[0, 2\pi)$. Express solutions to four decimal places.

63. $x^2 + \sin x - x^3 - \cos x = 0$

64. $x^3 - \cos^2 x = \frac{1}{2}x - 1$

Concept Check *Answer each question.*

65. Suppose solving a trigonometric equation for solutions over the interval $[0, 2\pi)$ leads to $2x = \frac{2\pi}{3}, 2\pi, \frac{8\pi}{3}$. What are the corresponding values of x?

66. Suppose solving a trigonometric equation for solutions over the interval $[0, 2\pi)$ leads to $\frac{1}{2}x = \frac{\pi}{16}, \frac{5\pi}{12}, \frac{5\pi}{8}$. What are the corresponding values of x?

67. Suppose solving a trigonometric equation for solutions over the interval $[0°, 360°)$ leads to $3\theta = 180°, 630°, 720°, 930°$. What are the corresponding values of θ?

68. Suppose solving a trigonometric equation for solutions over the interval $[0°, 360°)$ leads to $\frac{1}{3}\theta = 45°, 60°, 75°, 90°$. What are the corresponding values of θ?

Solve each equation in x for exact solutions over the interval $[0, 2\pi)$ and each equation in θ for exact solutions over the interval $[0°, 360°)$. ***See Examples 6–8.***

69. $2\cos 2x = \sqrt{3}$

70. $2\cos 2x = -1$

71. $\sin 3\theta = -1$

72. $\sin 3\theta = 0$

73. $3\tan 3x = \sqrt{3}$

74. $\cot 3x = \sqrt{3}$

75. $\sqrt{2}\cos 2\theta = -1$

76. $2\sqrt{3}\sin 2\theta = \sqrt{3}$

77. $\sin\frac{x}{2} = \sqrt{2} - \sin\frac{x}{2}$

78. $\tan 4x = 0$

79. $\sin x = \sin 2x$

80. $\cos 2x - \cos x = 0$

81. $8 \sec^2 \frac{x}{2} = 4$

82. $\sin^2 \frac{x}{2} - 2 = 0$

83. $\sin \frac{\theta}{2} = \csc \frac{\theta}{2}$

84. $\sec \frac{\theta}{2} = \cos \frac{\theta}{2}$

85. $\cos 2x + \cos x = 0$

86. $\sin x \cos x = \frac{1}{4}$

Solve each equation (x in radians and θ in degrees) for all exact solutions where appropriate. Round approximate answers in radians to four decimal places and approximate answers in degrees to the nearest tenth. Write answers using the least possible nonnegative angle measures. ***See Examples 6–8.***

87. $\sqrt{2} \sin 3x - 1 = 0$

88. $-2 \cos 2x = \sqrt{3}$

89. $\cos \frac{\theta}{2} = 1$

90. $\sin \frac{\theta}{2} = 1$

91. $2\sqrt{3} \sin \frac{x}{2} = 3$

92. $2\sqrt{3} \cos \frac{x}{2} = -3$

93. $2 \sin \theta = 2 \cos 2\theta$

94. $\cos \theta - 1 = \cos 2\theta$

95. $1 - \sin x = \cos 2x$

96. $\sin 2x = 2 \cos^2 x$

97. $3 \csc^2 \frac{x}{2} = 2 \sec x$

98. $\cos x = \sin^2 \frac{x}{2}$

99. $2 - \sin 2\theta = 4 \sin 2\theta$

100. $4 \cos 2\theta = 8 \sin \theta \cos \theta$

101. $2 \cos^2 2\theta = 1 - \cos 2\theta$

102. $\sin \theta - \sin 2\theta = 0$

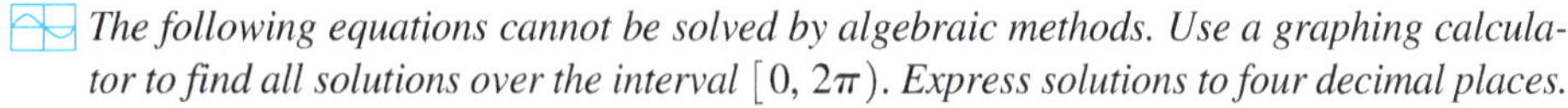

The following equations cannot be solved by algebraic methods. Use a graphing calculator to find all solutions over the interval $[0, 2\pi)$. Express solutions to four decimal places.

103. $2 \sin 2x - x^3 + 1 = 0$

104. $3 \cos \frac{x}{2} + \sqrt{x} - 2 = -\frac{1}{2}x + 2$

(Modeling) Solve each problem. ***See Examples 9 and 10.***

105. ***Pressure on the Eardrum*** A pure tone has a unique, constant frequency and amplitude that sounds rather dull. The pressures caused by pure tones on the eardrum are sinusoidal. The change in pressure P in pounds per square foot on a person's eardrum from a pure tone at time t in seconds can be modeled using the equation

$$P = A \sin(2\pi f t + \phi),$$

where f is the frequency in cycles per second, and ϕ is the phase angle. When P is positive, there is an increase in pressure and the eardrum is pushed inward. When P is negative, there is a decrease in pressure and the eardrum is pushed outward. (*Source*: Roederer, J., *Introduction to the Physics and Psychophysics of Music,* Second Edition, Springer-Verlag.)

(a) Determine algebraically the values of t for which $P = 0$ over $[0, 0.005]$.

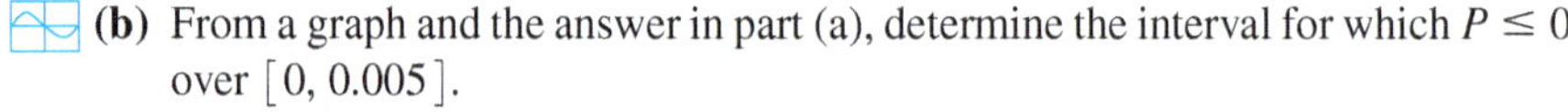

(b) From a graph and the answer in part (a), determine the interval for which $P \leq 0$ over $[0, 0.005]$.

(c) Would an eardrum hearing this tone be vibrating outward or inward when $P < 0$?

106. ***Accident Reconstruction*** To reconstruct accidents in which a vehicle vaults into the air after hitting an obstruction, the model

$$0.342D \cos \theta + h \cos^2 \theta = \frac{16D^2}{V_0^{\,2}}$$

can be used. V_0 is velocity in feet per second of the vehicle when it hits the obstruction, D is distance (in feet) from the obstruction to the landing point, and h is the difference in height (in feet) between landing point and takeoff point. Angle θ is the takeoff angle, the angle between the horizontal and the path of the vehicle. Find θ to the nearest degree if $V_0 = 60$, $D = 80$, and $h = 2$.

107. *Electromotive Force* In an electric circuit, suppose that the electromotive force in volts at t seconds can be modeled by

$$V = \cos 2\pi t.$$

Find the least value of t where $0 \le t \le \frac{1}{2}$ for each value of V.

(a) $V = 0$ **(b)** $V = 0.5$ **(c)** $V = 0.25$

108. *Voltage Induced by a Coil of Wire* A coil of wire rotating in a magnetic field induces a voltage modeled by

$$E = 20 \sin\left(\frac{\pi t}{4} - \frac{\pi}{2}\right),$$

where t is time in seconds. Find the least positive time to produce each voltage.

(a) 0 **(b)** $10\sqrt{3}$

109. *Pressure of a Plucked String* If a string with a fundamental frequency of 110 Hz is plucked in the middle, it will vibrate at the odd harmonics of 110, 330, 550, . . . Hz but not at the even harmonics of 220, 440, 660, . . . Hz. The resulting pressure P caused by the string is graphed below and can be modeled by the following equation.

$$P = 0.003 \sin 220\pi t + \frac{0.003}{3} \sin 660\pi t + \frac{0.003}{5} \sin 1100\pi t + \frac{0.003}{7} \sin 1540\pi t$$

(*Source:* Benade, A., *Fundamentals of Musical Acoustics,* Dover Publications. Roederer, J., *Introduction to the Physics and Psychophysics of Music,* Second Edition, Springer-Verlag.)

(a) Duplicate the graph shown here.

(b) Describe the shape of the sound wave that is produced.

(c) At lower frequencies, the inner ear will hear a tone only when the eardrum is moving outward. This occurs when P is negative. Determine the times over the interval $[0, 0.03]$ when this will occur.

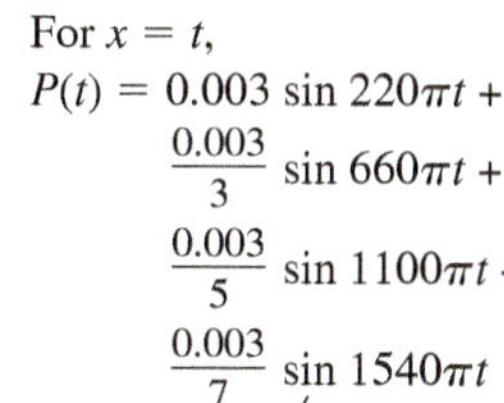

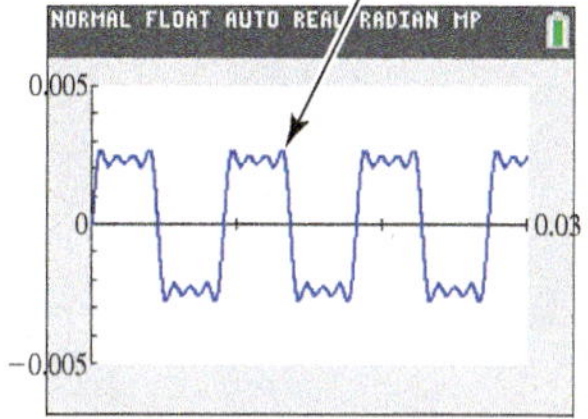

110. *Hearing Beats in Music* Musicians sometimes tune instruments by playing the same tone on two different instruments and listening for a phenomenon known as **beats.** Beats occur when two tones vary in frequency by only a few hertz. When the two instruments are in tune, the beats disappear. The ear hears beats because the pressure slowly rises and falls as a result of this slight variation in the frequency. (*Source:* Pierce, J., *The Science of Musical Sound,* Scientific American Books.)

(a) Consider the two tones with frequencies of 220 Hz and 223 Hz and pressures

$$P_1 = 0.005 \sin 440\pi t$$

and

$$P_2 = 0.005 \sin 446\pi t,$$

respectively. A graph of the pressure $P = P_1 + P_2$ felt by an eardrum over the 1-sec interval $[0.15, 1.15]$ is shown here. How many beats are there in 1 sec?

(b) Repeat part (a) with frequencies of 220 and 216 Hz.

(c) Determine a simple way to find the number of beats per second if the frequency of each tone is given.

For $x = t$,
$P(t) = 0.005 \sin 440\pi t + 0.005 \sin 446\pi t$

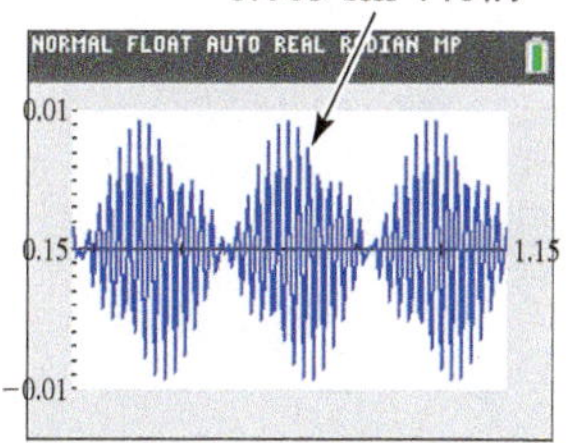

111. *Hearing Difference Tones* When a musical instrument creates a tone of 110 Hz, it also creates tones at 220, 330, 440, 550, 660, . . . Hz. A small speaker cannot reproduce the 110-Hz vibration but it can reproduce the higher frequencies, which are the **upper harmonics.** The low tones can still be heard because the speaker produces **difference tones** of the upper harmonics. The difference between consecutive frequencies is 110 Hz, and this difference tone will be heard by a listener. (*Source*: Benade, A., *Fundamentals of Musical Acoustics,* Dover Publications.)

(a) In the window $[0, 0.03]$ by $[-1, 1]$, graph the upper harmonics represented by the pressure

$$P = \frac{1}{2}\sin[2\pi(220)t] + \frac{1}{3}\sin[2\pi(330)t] + \frac{1}{4}\sin[2\pi(440)t].$$

(b) Estimate all t-coordinates where P is maximum.

(c) What does a person hear in addition to the frequencies of 220, 330, and 440 Hz?

(d) Graph the pressure produced by a speaker that can vibrate at 110 Hz and above.

112. *Daylight Hours in New Orleans* The seasonal variation in length of daylight can be modeled by a sine function. For example, the daily number of hours of daylight in New Orleans is given by

$$h = \frac{35}{3} + \frac{7}{3}\sin\frac{2\pi x}{365},$$

where x is the number of days after March 21 (disregarding leap year). (*Source:* Bushaw, D., et al., *A Sourcebook of Applications of School Mathematics*, Mathematical Association of America.)

(a) On what date will there be about 14 hr of daylight?

(b) What date has the least number of hours of daylight?

(c) When will there be about 10 hr of daylight?

113. *Average Monthly Temperature in Vancouver* The following function approximates average monthly temperature y (in °F) in Vancouver, Canada. Here x represents the month, where $x = 1$ corresponds to January, $x = 2$ corresponds to February, and so on. (*Source:* www.weather.com)

$$f(x) = 14\sin\left[\frac{\pi}{6}(x-4)\right] + 50$$

When is the average monthly temperature **(a)** 64°F **(b)** 39°F?

114. *Average Monthly Temperature in Phoenix* The following function approximates average monthly temperature y (in °F) in Phoenix, Arizona. Here x represents the month, where $x = 1$ corresponds to January, $x = 2$ corresponds to February, and so on. (*Source:* www.weather.com)

$$f(x) = 19.5\cos\left[\frac{\pi}{6}(x-7)\right] + 70.5$$

When is the average monthly temperature **(a)** 70.5°F **(b)** 55°F?

(Modeling) Alternating Electric Current The study of alternating electric current requires solving equations of the form

$$i = I_{\max}\sin 2\pi ft,$$

for time t in seconds, where i is instantaneous current in amperes, $I_{\max}$ is maximum current in amperes, and f is the number of cycles per second. (*Source:* Hannon, R. H., *Basic Technical Mathematics with Calculus,* W. B. Saunders Company.) *Find the least positive value of t, given the following data.*

115. $i = 40, I_{\max} = 100, f = 60$

116. $i = 50, I_{\max} = 100, f = 120$

117. $i = I_{\max}, f = 60$

118. $i = \frac{1}{2}I_{\max}, f = 60$

Chapter 7 Quiz (Sections 7.5–7.6)

1. Graph $y = \cos^{-1} x$, and indicate the coordinates of three points on the graph. Give the domain and range.

2. Find the exact value of each real number y. Do not use a calculator.

(a) $y = \sin^{-1}\left(-\frac{\sqrt{2}}{2}\right)$ **(b)** $y = \tan^{-1}\sqrt{3}$ **(c)** $y = \sec^{-1}\left(-\frac{2\sqrt{3}}{3}\right)$

3. Use a calculator to approximate each value in decimal degrees.

(a) $\theta = \arccos 0.92341853$ **(b)** $\theta = \cot^{-1}(-1.0886767)$

4. Evaluate each expression without using a calculator.

(a) $\cos\left(\tan^{-1}\frac{4}{5}\right)$ **(b)** $\sin\left(\cos^{-1}\left(-\frac{1}{2}\right) + \tan^{-1}\left(-\sqrt{3}\right)\right)$

Solve each equation for exact solutions over the interval $[0°, 360°)$.

5. $2\sin\theta - \sqrt{3} = 0$ **6.** $\cos\theta + 1 = 2\sin^2\theta$

7. *(Modeling) Electromotive Force* In an electric circuit, suppose that

$$V = \cos 2\pi t$$

models the electromotive force in volts at t seconds. Find the least value of t where $0 \le t \le \frac{1}{2}$ for each value of V.

(a) $V = 1$ **(b)** $V = 0.30$

Solve each equation for solutions over the interval $[0, 2\pi)$. *Round approximate answers to four decimal places.*

8. $\tan^2 x - 5\tan x + 3 = 0$ **9.** $3\cot 2x - \sqrt{3} = 0$

10. Solve $\cos\frac{x}{2} + \sqrt{3} = -\cos\frac{x}{2}$, giving all solutions in radians.

7.7 Equations Involving Inverse Trigonometric Functions

- Solution for x in Terms of y Using Inverse Functions
- Solution of Inverse Trigonometric Equations

Solution for x in Terms of y Using Inverse Functions

EXAMPLE 1 Solving an Equation for a Specified Variable

Solve $y = 3\cos 2x$ for x, where x is restricted to the interval $\left[0, \frac{\pi}{2}\right]$.

SOLUTION We want to isolate $\cos 2x$ on one side of the equation so that we can solve for $2x$, and then for x.

$$y = 3\cos 2x \quad \text{Our goal is to isolate } x.$$

$$\frac{y}{3} = \cos 2x \quad \text{Divide by 3.}$$

$$2x = \arccos\frac{y}{3} \quad \text{Definition of arccosine}$$

$$x = \frac{1}{2}\arccos\frac{y}{3} \quad \text{Multiply by } \tfrac{1}{2}.$$

An equivalent form of this answer is $x = \frac{1}{2}\cos^{-1}\frac{y}{3}$.

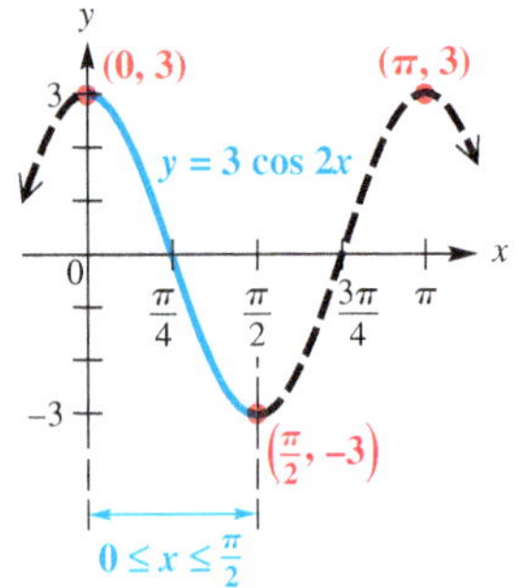

Figure 38

Because the function $y = 3 \cos 2x$ is periodic, with period π, there are infinitely many domain values (x-values) that will result in a given range value (y-value). For example, the x-values 0 and π both correspond to the y-value 3. See **Figure 38.** The restriction $0 \le x \le \frac{\pi}{2}$ given in the original problem ensures that this function is one-to-one, and, correspondingly, that

$$x = \frac{1}{2} \arccos \frac{y}{3}$$

has a one-to-one relationship. Thus, each y-value in $[-3, 3]$ substituted into this equation will lead to a single x-value.

✔ **Now Try Exercise 9.**

Solution of Inverse Trigonometric Equations

EXAMPLE 2 Solving an Equation Involving an Inverse Trigonometric Function

Solve $2 \arcsin x = \pi$.

SOLUTION First solve for arcsin x, and then for x.

$$2 \arcsin x = \pi \quad \text{Original equation}$$

$$\arcsin x = \frac{\pi}{2} \quad \text{Divide by 2.}$$

$$x = \sin \frac{\pi}{2} \quad \text{Definition of arcsine}$$

$$x = 1 \quad \arcsin 1 = \frac{\pi}{2}$$

CHECK

$$2 \arcsin x = \pi \quad \text{Original equation}$$

$$2 \arcsin 1 \stackrel{?}{=} \pi \quad \text{Let } x = 1.$$

$$2\left(\frac{\pi}{2}\right) \stackrel{?}{=} \pi \quad \text{Substitute the inverse value.}$$

$$\pi = \pi \ ✓ \quad \text{True}$$

The solution set is $\{1\}$.

✔ **Now Try Exercise 27.**

EXAMPLE 3 Solving an Equation Involving Inverse Trigonometric Functions

Solve $\cos^{-1} x = \sin^{-1} \frac{1}{2}$.

SOLUTION Let $\sin^{-1} \frac{1}{2} = u$. Then $\sin u = \frac{1}{2}$, and for u in quadrant I we have the following.

$$\cos^{-1} x = \sin^{-1} \frac{1}{2} \quad \text{Original equation}$$

$$\cos^{-1} x = u \quad \text{Substitute.}$$

$$\cos u = x \quad \text{Alternative form}$$

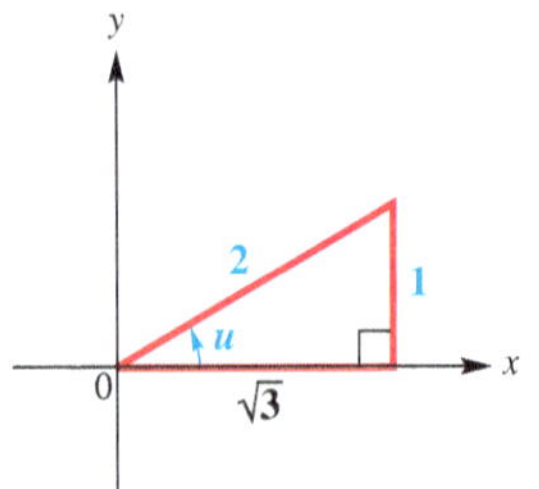

Figure 39

Sketch a triangle and label it using the facts that u is in quadrant I and $\sin u = \frac{1}{2}$. See **Figure 39.** Because $x = \cos u$, we have $x = \frac{\sqrt{3}}{2}$. The solution set is $\left\{\frac{\sqrt{3}}{2}\right\}$.

✔ **Now Try Exercise 35.**

EXAMPLE 4 Solving an Inverse Trigonometric Equation Using an Identity

Solve $\arcsin x - \arccos x = \frac{\pi}{6}$.

SOLUTION Isolate one inverse function on one side of the equation.

$$\arcsin x - \arccos x = \frac{\pi}{6} \quad \text{Original equation}$$

$$\arcsin x = \arccos x + \frac{\pi}{6} \quad \text{Add arccos } x. \quad (1)$$

$$x = \sin\left(\arccos x + \frac{\pi}{6}\right) \quad \text{Definition of arcsine}$$

Let $u = \arccos x$. The arccosine function yields angles in quadrants I and II, so $0 \le u \le \pi$ by definition.

$$x = \sin\left(u + \frac{\pi}{6}\right) \quad \text{Substitute.}$$

$$x = \sin u \cos \frac{\pi}{6} + \cos u \sin \frac{\pi}{6} \quad \text{Sine sum identity} \quad (2)$$

Use equation (1) and the definition of the arcsine function.

$$-\frac{\pi}{2} \le \arccos x + \frac{\pi}{6} \le \frac{\pi}{2} \quad \text{Range of arcsine is } \left[-\frac{\pi}{2}, \frac{\pi}{2}\right].$$

$$-\frac{2\pi}{3} \le \arccos x \le \frac{\pi}{3} \quad \text{Subtract } \frac{\pi}{6} \text{ from each part.}$$

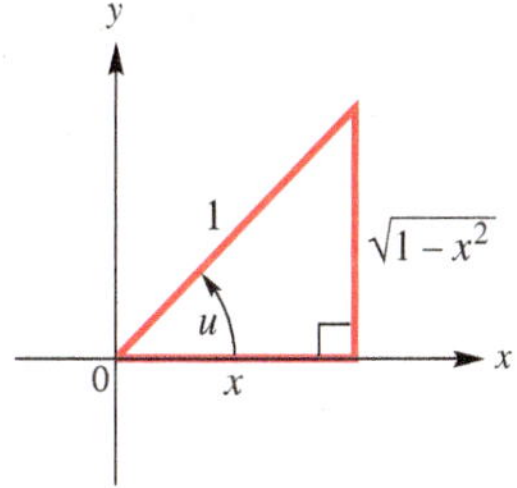

Figure 40

Because both $0 \le \arccos x \le \pi$ **and** $-\frac{2\pi}{3} \le \arccos x \le \frac{\pi}{3}$, the intersection yields $0 \le \arccos x \le \frac{\pi}{3}$. This places u in quadrant I, and we can sketch the triangle in **Figure 40.** From this triangle we find that $\sin u = \sqrt{1-x^2}$. Now substitute into equation (2) using $\sin u = \sqrt{1-x^2}$, $\sin \frac{\pi}{6} = \frac{1}{2}$, $\cos \frac{\pi}{6} = \frac{\sqrt{3}}{2}$, and $\cos u = x$.

$$x = \sin u \cos \frac{\pi}{6} + \cos u \sin \frac{\pi}{6} \quad (2)$$

$$x = \left(\sqrt{1-x^2}\right)\frac{\sqrt{3}}{2} + x \cdot \frac{1}{2} \quad \text{Substitute.}$$

$$2x = \left(\sqrt{1-x^2}\right)\sqrt{3} + x \quad \text{Multiply by 2.}$$

$$x = \left(\sqrt{3}\right)\sqrt{1-x^2} \quad \text{Subtract } x\text{; commutative property}$$

Square each factor.

$$x^2 = 3(1-x^2) \quad \text{Square each side; } (ab)^2 = a^2b^2$$

$$x^2 = 3 - 3x^2 \quad \text{Distributive property}$$

$$x^2 = \frac{3}{4} \quad \text{Add } 3x^2\text{. Divide by 4.}$$

Choose the positive square root, $x > 0$.

$$x = \sqrt{\frac{3}{4}} \quad \text{Take the square root on each side.}$$

$$x = \frac{\sqrt{3}}{2} \quad \text{Quotient rule: } \sqrt[n]{\frac{a}{b}} = \frac{\sqrt[n]{a}}{\sqrt[n]{b}}$$

CHECK A check is necessary because we squared each side when solving the equation.

$$\arcsin x - \arccos x = \frac{\pi}{6} \quad \text{Original equation}$$

$$\arcsin \frac{\sqrt{3}}{2} - \arccos \frac{\sqrt{3}}{2} \stackrel{?}{=} \frac{\pi}{6} \quad \text{Let } x = \tfrac{\sqrt{3}}{2}.$$

$$\frac{\pi}{3} - \frac{\pi}{6} \stackrel{?}{=} \frac{\pi}{6} \quad \text{Substitute inverse values.}$$

$$\frac{\pi}{6} = \frac{\pi}{6} \checkmark \quad \text{True}$$

The solution set is $\left\{\frac{\sqrt{3}}{2}\right\}$.

✔ **Now Try Exercise 37.**

7.7 Exercises

CONCEPT PREVIEW *Answer each question.*

1. Which one of the following equations has solution 0?

A. $\arctan 1 = x$ **B.** $\arccos 0 = x$ **C.** $\arcsin 0 = x$

2. Which one of the following equations has solution $\frac{\pi}{4}$?

A. $\arcsin \frac{\sqrt{2}}{2} = x$ **B.** $\arccos\left(-\frac{\sqrt{2}}{2}\right) = x$ **C.** $\arctan \frac{\sqrt{3}}{3} = x$

3. Which one of the following equations has solution $\frac{3\pi}{4}$?

A. $\arctan 1 = x$ **B.** $\arcsin \frac{\sqrt{2}}{2} = x$ **C.** $\arccos\left(-\frac{\sqrt{2}}{2}\right) = x$

4. Which one of the following equations has solution $-\frac{\pi}{6}$?

A. $\arctan \frac{\sqrt{3}}{3} = x$ **B.** $\arccos\left(-\frac{1}{2}\right) = x$ **C.** $\arcsin\left(-\frac{1}{2}\right) = x$

5. Which one of the following equations has solution π?

A. $\arccos(-1) = x$ **B.** $\arccos 1 = x$ **C.** $\arcsin(-1) = x$

6. Which one of the following equations has solution $-\frac{\pi}{2}$?

A. $\arctan(-1) = x$ **B.** $\arcsin(-1) = x$ **C.** $\arccos(-1) = x$

Solve each equation for x, where x is restricted to the given interval. ***See Example 1.***

7. $y = 5 \cos x$, for x in $[0, \pi]$

8. $y = \frac{1}{4} \sin x$, for x in $\left[-\frac{\pi}{2}, \frac{\pi}{2}\right]$

9. $y = 3 \tan 2x$, for x in $\left(-\frac{\pi}{4}, \frac{\pi}{4}\right)$

10. $y = 3 \sin \frac{x}{2}$, for x in $[-\pi, \pi]$

11. $y = 6 \cos \frac{x}{4}$, for x in $[0, 4\pi]$

12. $y = -\sin \frac{x}{3}$, for x in $\left[-\frac{3\pi}{2}, \frac{3\pi}{2}\right]$

13. $y = -2 \cos 5x$, for x in $\left[0, \frac{\pi}{5}\right]$

14. $y = 3 \cot 5x$, for x in $\left(0, \frac{\pi}{5}\right)$

15. $y = \sin x - 2$, for x in $\left[-\frac{\pi}{2}, \frac{\pi}{2}\right]$

16. $y = \cot x + 1$, for x in $(0, \pi)$

17. $y = -4 + 2 \sin x$, for x in $\left[-\frac{\pi}{2}, \frac{\pi}{2}\right]$

18. $y = 4 + 3 \cos x$, for x in $[0, \pi]$

19. $y = \frac{1}{2} \cot 3x$, for x in $\left(0, \frac{\pi}{3}\right)$

20. $y = \frac{1}{12} \sec x$, for x in $\left[0, \frac{\pi}{2}\right) \cup \left(\frac{\pi}{2}, \pi\right]$

21. $y = \cos(x + 3)$, for x in $[-3, \pi - 3]$

22. $y = \tan(2x - 1)$, for x in $\left(\frac{1}{2} - \frac{\pi}{4}, \frac{1}{2} + \frac{\pi}{4}\right)$

23. $y = \sqrt{2} + 3 \sec 2x$, for x in $\left[0, \frac{\pi}{4}\right) \cup \left(\frac{\pi}{4}, \frac{\pi}{2}\right]$

24. $y = -\sqrt{3} + 2 \csc \frac{x}{2}$, for x in $[-\pi, 0) \cup (0, \pi]$

25. Refer to **Exercise 15.** A student solving this equation wrote $y = \sin(x - 2)$ as the first step, inserting parentheses as shown. Explain why this is incorrect.

26. Explain why the equation $\sin^{-1} x = \cos^{-1} 2$ cannot have a solution. (No work is required.)

Solve each equation for exact solutions. ***See Examples 2 and 3.***

27. $-4 \arcsin x = \pi$

28. $6 \arccos x = 5\pi$

29. $\frac{4}{3} \cos^{-1} \frac{x}{4} = \pi$

30. $4 \tan^{-1} x = -3\pi$

31. $2 \arccos\left(\frac{x}{3} - \frac{\pi}{3}\right) = 2\pi$

32. $6 \arccos\left(x - \frac{\pi}{3}\right) = \pi$

33. $\arcsin x = \arctan \frac{3}{4}$

34. $\arctan x = \arccos \frac{5}{13}$

35. $\cos^{-1} x = \sin^{-1} \frac{3}{5}$

36. $\cot^{-1} x = \tan^{-1} \frac{4}{3}$

Solve each equation for exact solutions. ***See Example 4.***

37. $\sin^{-1} x - \tan^{-1} 1 = -\frac{\pi}{4}$

38. $\sin^{-1} x + \tan^{-1} \sqrt{3} = \frac{2\pi}{3}$

39. $\arccos x + 2 \arcsin \frac{\sqrt{3}}{2} = \pi$

40. $\arccos x + 2 \arcsin \frac{\sqrt{3}}{2} = \frac{\pi}{3}$

41. $\arcsin 2x + \arccos x = \frac{\pi}{6}$

42. $\arcsin 2x + \arcsin x = \frac{\pi}{2}$

43. $\cos^{-1} x + \tan^{-1} x = \frac{\pi}{2}$

44. $\sin^{-1} x + \tan^{-1} x = 0$

Use a graphing calculator in each of the following.

45. Provide graphical support for the solution in **Example 4** by showing that the graph of

$$y = \sin^{-1} x - \cos^{-1} x - \frac{\pi}{6} \quad \text{has a zero of} \quad \frac{\sqrt{3}}{2} \approx 0.8660254.$$

46. Provide graphical support for the solution in **Example 4** by showing that the x-coordinate of the point of intersection of the graphs of

$$y_1 = \sin^{-1} x - \cos^{-1} x \quad \text{and} \quad y_2 = \frac{\pi}{6} \quad \text{is} \quad \frac{\sqrt{3}}{2} \approx 0.8660254.$$

The following equations cannot be solved by algebraic methods. Use a graphing calculator to find all solutions over the interval $[0, 6]$. Express solutions to four decimal places.

47. $(\arctan x)^3 - x + 2 = 0$

48. $\pi \sin^{-1}(0.2x) - 3 = -\sqrt{x}$

(Modeling) Solve each problem.

49. ***Tone Heard by a Listener*** When two sources located at different positions produce the same pure tone, the human ear will often hear one sound that is equal to the sum of the individual tones. Because the sources are at different locations, they will have different phase angles ϕ. If two speakers located at different positions produce pure tones $P_1 = A_1 \sin(2\pi f t + \phi_1)$ and $P_2 = A_2 \sin(2\pi f t + \phi_2)$, where $-\frac{\pi}{4} \le \phi_1, \phi_2 \le \frac{\pi}{4}$, then the resulting tone heard by a listener can be written as $P = A \sin(2\pi f t + \phi)$, where

$$A = \sqrt{(A_1 \cos \phi_1 + A_2 \cos \phi_2)^2 + (A_1 \sin \phi_1 + A_2 \sin \phi_2)^2}$$

$$\text{and} \quad \phi = \arctan\left(\frac{A_1 \sin \phi_1 + A_2 \sin \phi_2}{A_1 \cos \phi_1 + A_2 \cos \phi_2}\right).$$

(*Source*: Fletcher, N. and T. Rossing, *The Physics of Musical Instruments,* Second Edition, Springer-Verlag.)

(a) Calculate A and ϕ if $A_1 = 0.0012$, $\phi_1 = 0.052$, $A_2 = 0.004$, and $\phi_2 = 0.61$. Also, if $f = 220$, find an expression for

$$P = A \sin(2\pi f t + \phi).$$

(b) Graph $Y_1 = P$ and $Y_2 = P_1 + P_2$ on the same coordinate axes over the interval $[0, 0.01]$. Are the two graphs the same?

50. ***Tone Heard by a Listener*** Repeat **Exercise 49.** Use $A_1 = 0.0025$, $\phi_1 = \frac{\pi}{7}$, $A_2 = 0.001$, $\phi_2 = \frac{\pi}{6}$, and $f = 300$.

51. ***Depth of Field*** When a large-view camera is used to take a picture of an object that is not parallel to the film, the lens board should be tilted so that the planes containing the subject, the lens board, and the film intersect in a line. This gives the best "depth of field." See the figure. (*Source*: Bushaw, D., et al., *A Sourcebook of Applications of School Mathematics,* Mathematical Association of America.)

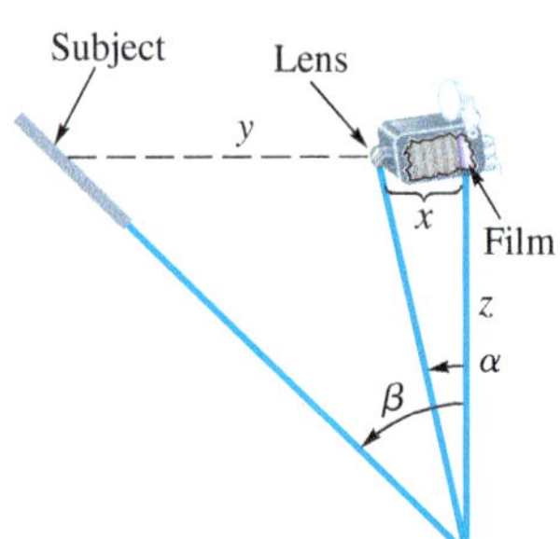

(a) Write two equations, one relating α, x, and z, and the other relating β, x, y, and z.

(b) Eliminate z from the equations in part (a) to get one equation relating α, β, x, and y.

(c) Solve the equation from part (b) for α.

(d) Solve the equation from part (b) for β.

52. ***Programming Language for Inverse Functions*** In some programming languages, the only inverse trigonometric function available is arctangent. The other inverse trigonometric functions can be expressed in terms of arctangent.

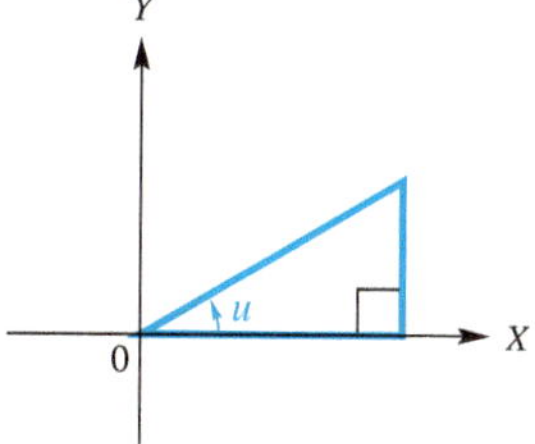

(a) Let $u = \arcsin x$. Solve the equation for x in terms of u.

(b) Use the result of part (a) to label the three sides of the triangle in the figure in terms of x.

(c) Use the triangle from part (b) to write an equation for $\tan u$ in terms of x.

(d) Solve the equation from part (c) for u.

53. ***Alternating Electric Current*** In the study of alternating electric current, instantaneous voltage is modeled by

$$E = E_{\max} \sin 2\pi f t,$$

where f is the number of cycles per second, $E_{\max}$ is the maximum voltage, and t is time in seconds.

(a) Solve the equation for t.

(b) Find the least positive value of t if $E_{\max} = 12$, $E = 5$, and $f = 100$. Use a calculator and round to two significant digits.

54. ***Viewing Angle of an Observer*** While visiting a museum, an observer views a painting that is 3 ft high and hangs 6 ft above the ground. See the figure. Assume her eyes are 5 ft above the ground, and let x be the distance from the spot where she is standing to the wall displaying the painting.

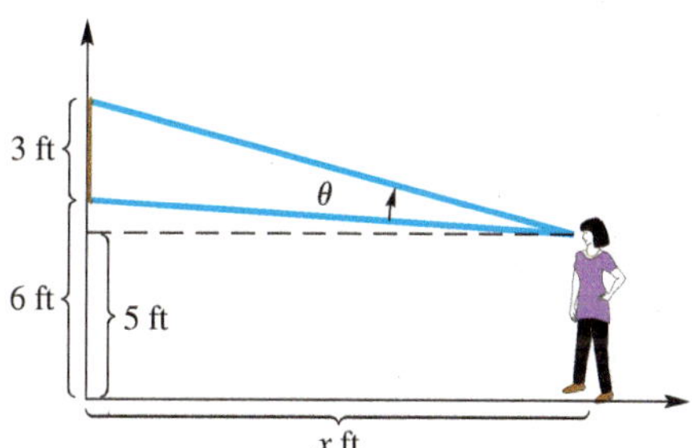

(a) Show that θ, the viewing angle subtended by the painting, is given by

$$\theta = \tan^{-1}\left(\frac{4}{x}\right) - \tan^{-1}\left(\frac{1}{x}\right).$$

(b) Find the value of x to the nearest hundredth for each value of θ.

(i) $\theta = \frac{\pi}{6}$ **(ii)** $\theta = \frac{\pi}{8}$

(c) Find the value of θ to the nearest hundredth for each value of x.

(i) $x = 4$ **(ii)** $x = 3$

55. ***Movement of an Arm*** In the equation below, t is time (in seconds) and y is the angle formed by a rhythmically moving arm.

$$y = \frac{1}{3}\sin\frac{4\pi t}{3}$$

(a) Solve the equation for t.

(b) At what time, to the nearest hundredth of a second, does the arm first form an angle of 0.3 radian?

Chapter 7 Test Prep

New Symbols

$\sin^{-1} x$ ($\arcsin x$)	inverse sine of x	$\cot^{-1} x$ ($\text{arccot}\, x$)	inverse cotangent of x
$\cos^{-1} x$ ($\arccos x$)	inverse cosine of x	$\sec^{-1} x$ ($\text{arcsec}\, x$)	inverse secant of x
$\tan^{-1} x$ ($\arctan x$)	inverse tangent of x	$\csc^{-1} x$ ($\text{arccsc}\, x$)	inverse cosecant of x

Quick Review

Concepts | **Examples**

7.1 Fundamental Identities

Reciprocal Identities

$$\cot\theta = \frac{1}{\tan\theta} \qquad \sec\theta = \frac{1}{\cos\theta} \qquad \csc\theta = \frac{1}{\sin\theta}$$

Quotient Identities

$$\tan\theta = \frac{\sin\theta}{\cos\theta} \qquad \cot\theta = \frac{\cos\theta}{\sin\theta}$$

Pythagorean Identities

$$\sin^2\theta + \cos^2\theta = 1 \qquad \tan^2\theta + 1 = \sec^2\theta$$

$$1 + \cot^2\theta = \csc^2\theta$$

Even-Odd Identities

$$\sin(-\theta) = -\sin\theta \quad \cos(-\theta) = \cos\theta \quad \tan(-\theta) = -\tan\theta$$

$$\csc(-\theta) = -\csc\theta \quad \sec(-\theta) = \sec\theta \quad \cot(-\theta) = -\cot\theta$$

If θ is in quadrant IV and $\sin\theta = -\frac{3}{5}$, find $\csc\theta$, $\cos\theta$, and $\sin(-\theta)$.

$\csc\theta = \frac{1}{\sin\theta} = \frac{1}{-\frac{3}{5}} = -\frac{5}{3}$ Reciprocal identity

$\sin^2\theta + \cos^2\theta = 1$ Pythagorean identity

$\left(-\frac{3}{5}\right)^2 + \cos^2\theta = 1$ Substitute.

$\cos^2\theta = \frac{16}{25}$ $\left(-\frac{3}{5}\right)^2 = \frac{9}{25}$; Subtract $\frac{9}{25}$.

$\cos\theta = +\sqrt{\frac{16}{25}}$ $\cos\theta$ is positive in quadrant IV.

$\cos\theta = \frac{4}{5}$

$\sin(-\theta) = -\sin\theta = -\left(-\frac{3}{5}\right) = \frac{3}{5}$

Even-odd identity

7.2 Verifying Trigonometric Identities

See the box titled Hints for Verifying Identities in **Section 7.2.**

7.3 Sum and Difference Identities

Cofunction Identities

$$\cos(90^\circ - \theta) = \sin\theta \qquad \cot(90^\circ - \theta) = \tan\theta$$

$$\sin(90^\circ - \theta) = \cos\theta \qquad \sec(90^\circ - \theta) = \csc\theta$$

$$\tan(90^\circ - \theta) = \cot\theta \qquad \csc(90^\circ - \theta) = \sec\theta$$

Find one value of θ such that $\tan\theta = \cot 78^\circ$.

$\tan\theta = \cot 78^\circ$

$\cot(90^\circ - \theta) = \cot 78^\circ$ Cofunction identity

$90^\circ - \theta = 78^\circ$ Set angles equal.

$\theta = 12^\circ$ Solve for θ.

Concepts	Examples
Sum and Difference Identities $\cos(A-B)=\cos A\cos B+\sin A\sin B$ $\cos(A+B)=\cos A\cos B-\sin A\sin B$ $\sin(A+B)=\sin A\cos B+\cos A\sin B$ $\sin(A-B)=\sin A\cos B-\cos A\sin B$ $\tan(A+B)=\dfrac{\tan A+\tan B}{1-\tan A\tan B}$ $\tan(A-B)=\dfrac{\tan A-\tan B}{1+\tan A\tan B}$	Find the exact value of $\cos(-15^\circ)$. $\cos(-15^\circ)$ $=\cos(30^\circ-45^\circ)$ $-15^\circ=30^\circ-45^\circ$ $=\cos 30^\circ\cos 45^\circ+\sin 30^\circ\sin 45^\circ$ Cosine difference identity $=\dfrac{\sqrt{3}}{2}\cdot\dfrac{\sqrt{2}}{2}+\dfrac{1}{2}\cdot\dfrac{\sqrt{2}}{2}$ Substitute known values. $=\dfrac{\sqrt{6}+\sqrt{2}}{4}$ Simplify.

7.4 Double-Angle and Half-Angle Identities

Concepts	Examples
Double-Angle Identities $\cos 2A=\cos^2 A-\sin^2 A$ $\cos 2A=1-2\sin^2 A$ $\cos 2A=2\cos^2 A-1$ $\sin 2A=2\sin A\cos A$ $\tan 2A=\dfrac{2\tan A}{1-\tan^2 A}$	Given $\cos\theta=-\frac{5}{13}$ and $\sin\theta>0$, find $\sin 2\theta$. Sketch a triangle in quadrant II because $\cos\theta<0$ and $\sin\theta>0$. Use it to find that $\sin\theta=\frac{12}{13}$. $\sin 2\theta=2\sin\theta\cos\theta$ $=2\left(\dfrac{12}{13}\right)\left(-\dfrac{5}{13}\right)$ $=-\dfrac{120}{169}$
Product-to-Sum Identities $\cos A\cos B=\frac{1}{2}[\cos(A+B)+\cos(A-B)]$ $\sin A\sin B=\frac{1}{2}[\cos(A-B)-\cos(A+B)]$ $\sin A\cos B=\frac{1}{2}[\sin(A+B)+\sin(A-B)]$ $\cos A\sin B=\frac{1}{2}[\sin(A+B)-\sin(A-B)]$	Write $\sin(-\theta)\sin 2\theta$ as the difference of two functions. $\sin(-\theta)\sin 2\theta$ $=\frac{1}{2}[\cos(-\theta-2\theta)-\cos(-\theta+2\theta)]$ $=\frac{1}{2}[\cos(-3\theta)-\cos\theta]$ $=\frac{1}{2}\cos(-3\theta)-\frac{1}{2}\cos\theta$ $=\frac{1}{2}\cos 3\theta-\frac{1}{2}\cos\theta$
Sum-to-Product Identities $\sin A+\sin B=2\sin\left(\dfrac{A+B}{2}\right)\cos\left(\dfrac{A-B}{2}\right)$ $\sin A-\sin B=2\cos\left(\dfrac{A+B}{2}\right)\sin\left(\dfrac{A-B}{2}\right)$ $\cos A+\cos B=2\cos\left(\dfrac{A+B}{2}\right)\cos\left(\dfrac{A-B}{2}\right)$ $\cos A-\cos B=-2\sin\left(\dfrac{A+B}{2}\right)\sin\left(\dfrac{A-B}{2}\right)$	Write $\cos\theta+\cos 3\theta$ as a product of two functions. $\cos\theta+\cos 3\theta$ $=2\cos\left(\dfrac{\theta+3\theta}{2}\right)\cos\left(\dfrac{\theta-3\theta}{2}\right)$ $=2\cos\left(\dfrac{4\theta}{2}\right)\cos\left(\dfrac{-2\theta}{2}\right)$ $=2\cos 2\theta\cos(-\theta)$ $=2\cos 2\theta\cos\theta$

Concepts	Examples

Half-Angle Identities

$$\cos\frac{A}{2} = \pm\sqrt{\frac{1+\cos A}{2}} \qquad \sin\frac{A}{2} = \pm\sqrt{\frac{1-\cos A}{2}}$$

$$\tan\frac{A}{2} = \pm\sqrt{\frac{1-\cos A}{1+\cos A}} \qquad \tan\frac{A}{2} = \frac{\sin A}{1+\cos A}$$

$$\tan\frac{A}{2} = \frac{1-\cos A}{\sin A}$$

(In the identities involving radicals, the sign is chosen based on the function under consideration and the quadrant of $\frac{A}{2}$.)

Find the exact value of $\tan 67.5^\circ$.

We choose the last form with $A = 135^\circ$.

$$\tan 67.5^\circ = \tan\frac{135^\circ}{2} = \frac{1-\cos 135^\circ}{\sin 135^\circ} = \frac{1-\left(-\frac{\sqrt{2}}{2}\right)}{\frac{\sqrt{2}}{2}}$$

$$= \frac{1+\frac{\sqrt{2}}{2}}{\frac{\sqrt{2}}{2}} \cdot \frac{2}{2} = \frac{2+\sqrt{2}}{\sqrt{2}} = \sqrt{2}+1$$

Rationalize the denominator and simplify.

7.5 Inverse Circular Functions

Inverse Function	Domain	Range: Interval	Range: Quadrants of the Unit Circle
$y = \sin^{-1} x$	$[-1, 1]$	$\left[-\frac{\pi}{2}, \frac{\pi}{2}\right]$	I and IV
$y = \cos^{-1} x$	$[-1, 1]$	$[0, \pi]$	I and II
$y = \tan^{-1} x$	$(-\infty, \infty)$	$\left(-\frac{\pi}{2}, \frac{\pi}{2}\right)$	I and IV
$y = \cot^{-1} x$	$(-\infty, \infty)$	$(0, \pi)$	I and II
$y = \sec^{-1} x$	$(-\infty, -1] \cup [1, \infty)$	$\left[0, \frac{\pi}{2}\right) \cup \left(\frac{\pi}{2}, \pi\right]$	I and II
$y = \csc^{-1} x$	$(-\infty, -1] \cup [1, \infty)$	$\left[-\frac{\pi}{2}, 0\right) \cup \left(0, \frac{\pi}{2}\right]$	I and IV

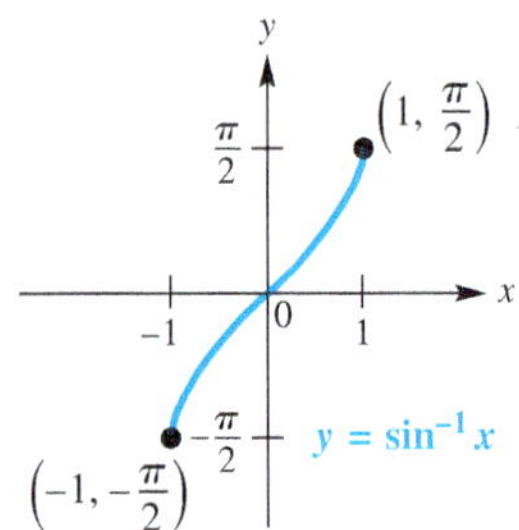

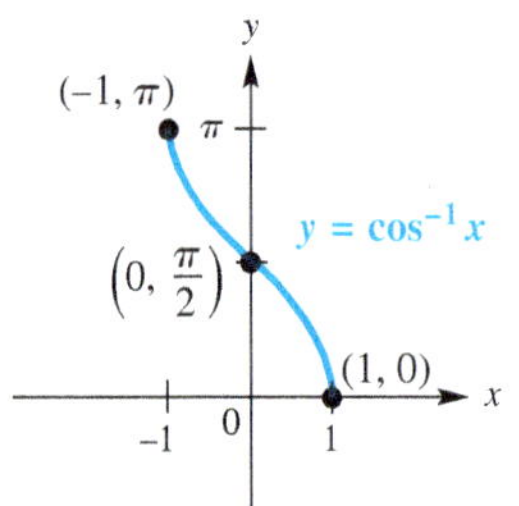

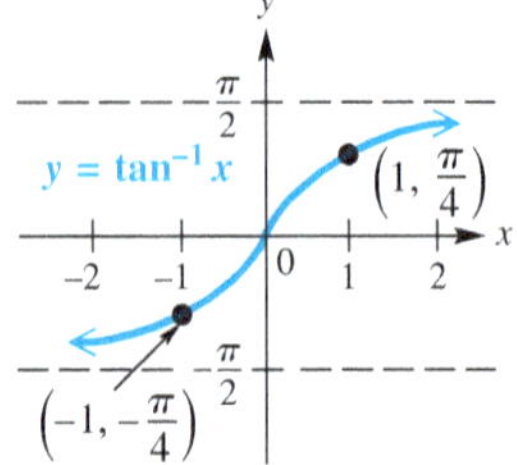

See the section for graphs of the other inverse circular (trigonometric) functions.

Evaluate $y = \cos^{-1} 0$.

Write $y = \cos^{-1} 0$ as $\cos y = 0$. Then

$$y = \frac{\pi}{2}$$

because $\cos\frac{\pi}{2} = 0$ and $\frac{\pi}{2}$ is in the range of $\cos^{-1} x$.

Use a calculator to find y in radians if $y = \sec^{-1}(-3)$.

With the calculator in radian mode, enter $\sec^{-1}(-3)$ as $\cos^{-1}\left(\frac{1}{-3}\right)$ to obtain

$$y \approx 1.9106332.$$

Evaluate $\sin\left(\tan^{-1}\left(-\frac{3}{4}\right)\right)$.

Let $u = \tan^{-1}\left(-\frac{3}{4}\right)$. Then $\tan u = -\frac{3}{4}$. Because $\tan u$ is negative when u is in quadrant IV, sketch a triangle as shown.

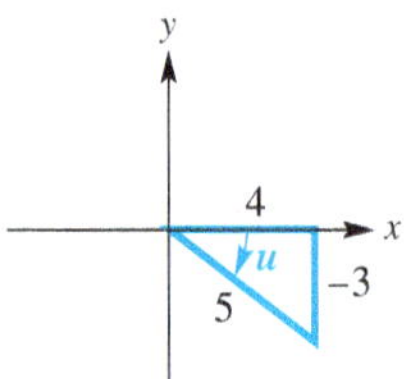

We want $\sin\left(\tan^{-1}\left(-\frac{3}{4}\right)\right) = \sin u$. From the triangle, we have the following.

$$\sin u = -\frac{3}{5}$$

Concepts	Examples

7.6 Trigonometric Equations

Solving a Trigonometric Equation

1. Decide whether the equation is linear or quadratic in form in order to determine the solution method.
2. If only one trigonometric function is present, solve the equation for that function.
3. If more than one trigonometric function is present, rewrite the equation so that one side equals 0. Then try to factor and apply the zero-factor property.
4. If the equation is quadratic in form, but not factorable, use the quadratic formula. Check that solutions are in the desired interval.
5. Try using identities to change the form of the equation. It may be helpful to square each side of the equation first. In this case, check for extraneous solutions.

Solve $\tan\theta + \sqrt{3} = 2\sqrt{3}$ over the interval $[0°, 360°)$.

$\tan\theta + \sqrt{3} = 2\sqrt{3}$ Original equation

$\tan\theta = \sqrt{3}$ Subtract $\sqrt{3}$.

$\theta = 60°$ Definition of inverse tangent

Another solution over $[0°, 360°)$ is

$$\theta = 60° + 180° = 240°.$$

The solution set is $\{60°, 240°\}$.

Solve $2\cos^2 x = 1$ for all solutions.

$2\cos^2 x = 1$ Original equation

$2\cos^2 x - 1 = 0$ Subtract 1.

$\cos 2x = 0$ Cosine double-angle identity

$2x = \frac{\pi}{2} + 2n\pi$ and $2x = \frac{3\pi}{2} + 2n\pi$ Add integer multiples of 2π.

$x = \frac{\pi}{4} + n\pi$ and $x = \frac{3\pi}{4} + n\pi$ Divide by 2.

The solution set, where π is the period of $\cos 2x$, is

$$\left\{\frac{\pi}{4} + n\pi, \frac{3\pi}{4} + n\pi, \text{ where } n \text{ is any integer}\right\}.$$

7.7 Equations Involving Inverse Trigonometric Functions

We solve equations of the form $y = f(x)$, where $f(x)$ involves a trigonometric function, using inverse trigonometric functions.

Solve $y = 2\sin 3x$ for x, where x is restricted to the interval $\left[-\frac{\pi}{6}, \frac{\pi}{6}\right]$.

$y = 2\sin 3x$ Original equation

$\frac{y}{2} = \sin 3x$ Divide by 2.

$3x = \arcsin\frac{y}{2}$ Definition of arcsine

$x = \frac{1}{3}\arcsin\frac{y}{2}$ Multiply by $\frac{1}{3}$.

Techniques introduced in this section also show how to solve equations that involve inverse functions.

Solve.

$4\tan^{-1} x = \pi$ Original equation

$\tan^{-1} x = \frac{\pi}{4}$ Divide by 4.

$x = \tan\frac{\pi}{4}$ Definition of arctangent

$x = 1$ Evaluate.

The solution set is $\{1\}$.

Chapter 7 Review Exercises

Concept Check For each expression in Column I, choose the expression from Column II that completes an identity.

I		II	
1. $\sec x =$ ____	**2.** $\csc x =$ ____	**A.** $\frac{1}{\sin x}$	**B.** $\frac{1}{\cos x}$
3. $\tan x =$ ____	**4.** $\cot x =$ ____	**C.** $\frac{\sin x}{\cos x}$	**D.** $\frac{1}{\cot^2 x}$
5. $\tan^2 x =$ ____	**6.** $\sec^2 x =$ ____	**E.** $\frac{1}{\cos^2 x}$	**F.** $\frac{\cos x}{\sin x}$

Use identities to write each expression in terms of $\sin \theta$ *and* $\cos \theta$, *and then simplify so that no quotients appear and all functions are of* θ *only.*

7. $\sec^2 \theta - \tan^2 \theta$

8. $\frac{\cot(-\theta)}{\sec(-\theta)}$

9. $\tan^2 \theta(1 + \cot^2 \theta)$

10. $\csc \theta - \sin \theta$

11. $\tan \theta - \sec \theta \csc \theta$

12. $\csc^2 \theta + \sec^2 \theta$

Work each problem.

13. Use the trigonometric identities to find $\sin x$, $\tan x$, and $\cot(-x)$, given $\cos x = \frac{3}{5}$ and x in quadrant IV.

14. Given $\tan x = -\frac{5}{4}$, where $\frac{\pi}{2} < x < \pi$, use the trigonometric identities to find $\cot x$, $\csc x$, and $\sec x$.

15. Find the exact values of the six trigonometric functions of 165°.

16. Find the exact values of $\sin x$, $\cos x$, and $\tan x$, for $x = \frac{\pi}{12}$, using

(a) difference identities **(b)** half-angle identities.

Concept Check For each expression in Column I, use an identity to choose an expression from Column II with the same value. Choices may be used once, more than once, or not at all.

I		II	
17. $\cos 210°$	**18.** $\sin 35°$	**A.** $\sin(-35°)$	**B.** $\cos 55°$
19. $\tan(-35°)$	**20.** $-\sin 35°$	**C.** $\sqrt{\frac{1 + \cos 150°}{2}}$	**D.** $2 \sin 150° \cos 150°$
21. $\cos 35°$	**22.** $\cos 75°$	**E.** $\cot(-35°)$	**F.** $\cos^2 150° - \sin^2 150°$
23. $\sin 75°$	**24.** $\sin 300°$	**G.** $\cos(-35°)$	**H.** $\cot 125°$
25. $\cos 300°$	**26.** $\cos(-55°)$	**I.** $\cos 150° \cos 60° - \sin 150° \sin 60°$	
		J. $\sin 15° \cos 60° + \cos 15° \sin 60°$	

Use the given information to find $\sin(x + y)$, $\cos(x - y)$, $\tan(x + y)$, *and the quadrant of* $x + y$.

27. $\sin x = -\frac{3}{5}$, $\cos y = -\frac{7}{25}$, x and y in quadrant III

28. $\sin x = \frac{3}{5}$, $\cos y = \frac{24}{25}$, x in quadrant I, y in quadrant IV

29. $\sin x = -\frac{1}{2}$, $\cos y = -\frac{2}{5}$, x and y in quadrant III

30. $\sin y = -\frac{2}{3}$, $\cos x = -\frac{1}{5}$, x in quadrant II, y in quadrant III

31. $\sin x = \frac{1}{10}$, $\cos y = \frac{4}{5}$, x in quadrant I, y in quadrant IV

32. $\cos x = \frac{2}{9}$, $\sin y = -\frac{1}{2}$, x in quadrant IV, y in quadrant III

Find values of the sine and cosine functions for each angle measure.

33. θ, given $\cos 2\theta = -\frac{3}{4}$, $90° < 2\theta < 180°$

34. B, given $\cos 2B = \frac{1}{8}$, $540° < 2B < 720°$

35. $2x$, given $\tan x = 3$, $\sin x < 0$

36. $2y$, given $\sec y = -\frac{5}{3}$, $\sin y > 0$

Use the given information to find each of the following.

37. $\cos \frac{\theta}{2}$, given $\cos \theta = -\frac{1}{2}$, $90° < \theta < 180°$

38. $\sin \frac{A}{2}$, given $\cos A = -\frac{3}{4}$, $90° < A < 180°$

39. $\tan x$, given $\tan 2x = 2$, $\pi < x < \frac{3\pi}{2}$

40. $\sin y$, given $\cos 2y = -\frac{1}{3}$, $\frac{\pi}{2} < y < \pi$

41. $\tan \frac{x}{2}$, given $\sin x = 0.8$, $0 < x < \frac{\pi}{2}$

42. $\sin 2x$, given $\sin x = 0.6$, $\frac{\pi}{2} < x < \pi$

Graph each expression and use the graph to make a conjecture, predicting what might be an identity. Then verify your conjecture algebraically.

43. $\dfrac{\sin 2x + \sin x}{\cos x - \cos 2x}$

44. $\dfrac{1 - \cos 2x}{\sin 2x}$

45. $\dfrac{\sin x}{1 - \cos x}$

46. $\dfrac{\cos x \sin 2x}{1 + \cos 2x}$

47. $\dfrac{2(\sin x - \sin^3 x)}{\cos x}$

48. $\csc x - \cot x$

Verify that each equation is an identity.

49. $\sin^2 x - \sin^2 y = \cos^2 y - \cos^2 x$

50. $2 \cos^3 x - \cos x = \dfrac{\cos^2 x - \sin^2 x}{\sec x}$

51. $\dfrac{\sin^2 x}{2 - 2 \cos x} = \cos^2 \dfrac{x}{2}$

52. $\dfrac{\sin 2x}{\sin x} = \dfrac{2}{\sec x}$

53. $2 \cos A - \sec A = \cos A - \dfrac{\tan A}{\csc A}$

54. $\dfrac{2 \tan B}{\sin 2B} = \sec^2 B$

55. $1 + \tan^2 \alpha = 2 \tan \alpha \csc 2\alpha$

56. $\dfrac{2 \cot x}{\tan 2x} = \csc^2 x - 2$

57. $\tan \theta \sin 2\theta = 2 - 2 \cos^2 \theta$

58. $\csc A \sin 2A - \sec A = \cos 2A \sec A$

59. $2 \tan x \csc 2x - \tan^2 x = 1$

60. $2 \cos^2 \theta - 1 = \dfrac{1 - \tan^2 \theta}{1 + \tan^2 \theta}$

61. $\tan \theta \cos^2 \theta = \dfrac{2 \tan \theta \cos^2 \theta - \tan \theta}{1 - \tan^2 \theta}$

62. $\sec^2 \alpha - 1 = \dfrac{\sec 2\alpha - 1}{\sec 2\alpha + 1}$

63. $\dfrac{\sin^2 x - \cos^2 x}{\csc x} = 2 \sin^3 x - \sin x$

64. $\sin^3 \theta = \sin \theta - \cos^2 \theta \sin \theta$

65. $\tan 4\theta = \dfrac{2 \tan 2\theta}{2 - \sec^2 2\theta}$

66. $2 \cos^2 \dfrac{x}{2} \tan x = \tan x + \sin x$

67. $\tan\left(\dfrac{x}{2} + \dfrac{\pi}{4}\right) = \sec x + \tan x$

68. $\dfrac{1}{2} \cot \dfrac{x}{2} - \dfrac{1}{2} \tan \dfrac{x}{2} = \cot x$

69. $-\cot \dfrac{x}{2} = \dfrac{\sin 2x + \sin x}{\cos 2x - \cos x}$

70. $\dfrac{\sin 3t + \sin 2t}{\sin 3t - \sin 2t} = \dfrac{\tan \frac{5t}{2}}{\tan \frac{t}{2}}$

71. Graph the inverse sine, cosine, and tangent functions, indicating the coordinates of three points on each graph. Give the domain and range for each.

Concept Check *Determine whether each statement is* true *or* false. *If false, tell why.*

72. The ranges of the inverse tangent and inverse cotangent functions are the same.

73. It is true that $\sin \frac{11\pi}{6} = -\frac{1}{2}$, and therefore $\arcsin\left(-\frac{1}{2}\right) = \frac{11\pi}{6}$.

74. For all x, $\tan(\tan^{-1} x) = x$.

Find the exact value of each real number y. Do not use a calculator.

75. $y = \sin^{-1} \frac{\sqrt{2}}{2}$

76. $y = \arccos\left(-\frac{1}{2}\right)$

77. $y = \tan^{-1}(-\sqrt{3})$

78. $y = \arcsin(-1)$

79. $y = \cos^{-1}\left(-\frac{\sqrt{2}}{2}\right)$

80. $y = \arctan \frac{\sqrt{3}}{3}$

81. $y = \sec^{-1}(-2)$

82. $y = \operatorname{arccsc} \frac{2\sqrt{3}}{3}$

83. $y = \operatorname{arccot}(-1)$

Give the degree measure of θ. Do not use a calculator.

84. $\theta = \arccos \frac{1}{2}$

85. $\theta = \arcsin\left(-\frac{\sqrt{3}}{2}\right)$

86. $\theta = \tan^{-1} 0$

Use a calculator to approximate each value in decimal degrees.

87. $\theta = \arctan 1.7804675$

88. $\theta = \sin^{-1}(-0.66045320)$

89. $\theta = \cos^{-1} 0.80396577$

90. $\theta = \cot^{-1} 4.5046388$

91. $\theta = \operatorname{arcsec} 3.4723155$

92. $\theta = \csc^{-1} 7.4890096$

Evaluate each expression without using a calculator.

93. $\cos(\arccos(-1))$

94. $\sin\left(\arcsin\left(-\frac{\sqrt{3}}{2}\right)\right)$

95. $\arccos\left(\cos \frac{3\pi}{4}\right)$

96. $\operatorname{arcsec}(\sec \pi)$

97. $\tan^{-1}\left(\tan \frac{\pi}{4}\right)$

98. $\cos^{-1}(\cos 0)$

99. $\sin\left(\arccos \frac{3}{4}\right)$

100. $\cos(\arctan 3)$

101. $\cos(\csc^{-1}(-2))$

102. $\sec\left(2 \sin^{-1}\left(-\frac{1}{3}\right)\right)$

103. $\tan\left(\arcsin \frac{3}{5} + \arccos \frac{5}{7}\right)$

Write each trigonometric expression as an algebraic expression in u, for $u > 0$.

104. $\cos\left(\arctan \frac{u}{\sqrt{1-u^2}}\right)$

105. $\tan\left(\operatorname{arcsec} \frac{\sqrt{u^2+1}}{u}\right)$

Solve each equation for exact solutions over the interval $[0, 2\pi)$ *where appropriate. Give approximate solutions to four decimal places.*

106. $\sin^2 x = 1$

107. $2 \tan x - 1 = 0$

108. $3 \sin^2 x - 5 \sin x + 2 = 0$

109. $\tan x = \cot x$

110. $\sec^2 2x = 2$

111. $\tan^2 2x - 1 = 0$

Solve each equation for all exact solutions, in radians.

112. $\sec \frac{x}{2} = \cos \frac{x}{2}$ **113.** $\cos 2x + \cos x = 0$ **114.** $4 \sin x \cos x = \sqrt{3}$

Solve each equation for exact solutions over the interval $[0° \; 360°)$ where appropriate. Give approximate solutions to the nearest tenth of a degree.

115. $\sin^2 \theta + 3 \sin \theta + 2 = 0$ **116.** $2 \tan^2 \theta = \tan \theta + 1$

117. $\sin 2\theta = \cos 2\theta + 1$ **118.** $2 \sin 2\theta = 1$

119. $3 \cos^2 \theta + 2 \cos \theta - 1 = 0$ **120.** $5 \cot^2 \theta - \cot \theta - 2 = 0$

Solve each equation for all exact solutions, in degrees.

121. $2\sqrt{3} \cos \frac{\theta}{2} = -3$ **122.** $\sin \theta - \cos 2\theta = 0$ **123.** $\tan \theta - \sec \theta = 1$

Solve each equation for x.

124. $4\pi - 4 \cot^{-1} x = \pi$ **125.** $\frac{4}{3} \arctan \frac{x}{2} = \pi$

126. $\arccos x = \arcsin \frac{2}{7}$ **127.** $\arccos x + \arctan 1 = \frac{11\pi}{12}$

128. $y = 3 \cos \frac{x}{2}$, for x in $[0, 2\pi]$ **129.** $y = \frac{1}{2} \sin x$, for x in $\left[-\frac{\pi}{2}, \frac{\pi}{2}\right]$

130. $y = \frac{4}{5} \sin x - \frac{3}{5}$, for x in $\left[-\frac{\pi}{2}, \frac{\pi}{2}\right]$

131. $y = \frac{1}{2} \tan(3x + 2)$, for x in $\left(-\frac{2}{3} - \frac{\pi}{6}, -\frac{2}{3} + \frac{\pi}{6}\right)$

132. Solve $d = 550 + 450 \cos\left(\frac{\pi}{50}t\right)$ for t, where t is in the interval $[0, 50]$.

(Modeling) Solve each problem.

133. ***Viewing Angle of an Observer*** A 10-ft-wide chalkboard is situated 5 ft from the left wall of a classroom. See the figure. A student sitting next to the wall x feet from the front of the classroom has a viewing angle of θ radians.

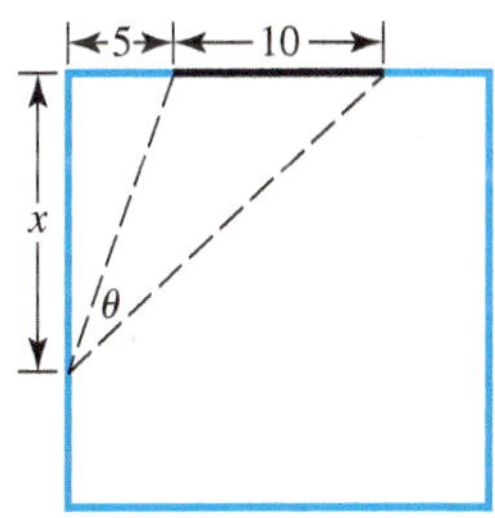

(a) Show that the value of θ is given by

$$y_1 = \tan^{-1}\left(\frac{15}{x}\right) - \tan^{-1}\left(\frac{5}{x}\right).$$

(b) Graph y_1 with a graphing calculator to estimate the value of x that maximizes the viewing angle.

134. ***Snell's Law*** Snell's law states that

$$\frac{c_1}{c_2} = \frac{\sin \theta_1}{\sin \theta_2},$$

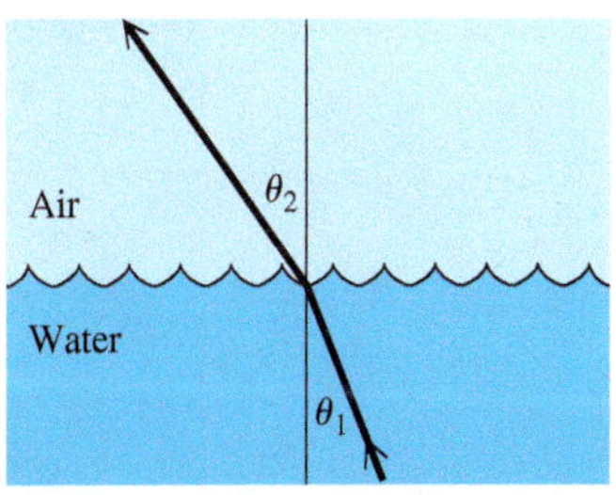

where c_1 is the speed of light in one medium, c_2 is the speed of light in a second medium, and θ_1 and θ_2 are the angles shown in the figure. Suppose a light is shining up through water into the air as in the figure. As θ_1 increases, θ_2 approaches 90°, at which point no light will emerge from the water. Assume the ratio $\frac{c_1}{c_2}$ in this case is 0.752. For what value of θ_1, to the nearest tenth, does $\theta_2 = 90°$? This value of θ_1 is the **critical angle** for water.

135. *Snell's Law* Refer to **Exercise 134.** What happens when θ_1 is greater than the critical angle?

136. *British Nautical Mile* The British nautical mile is defined as the length of a minute of arc of a meridian. Because Earth is flat at its poles, the nautical mile, in feet, is given by

$$L = 6077 - 31 \cos 2\theta,$$

where θ is the latitude in degrees. See the figure. (*Source:* Bushaw, D., et al., *A Sourcebook of Applications of School Mathematics,* National Council of Teachers of Mathematics.) Give answers to the nearest tenth if applicable.

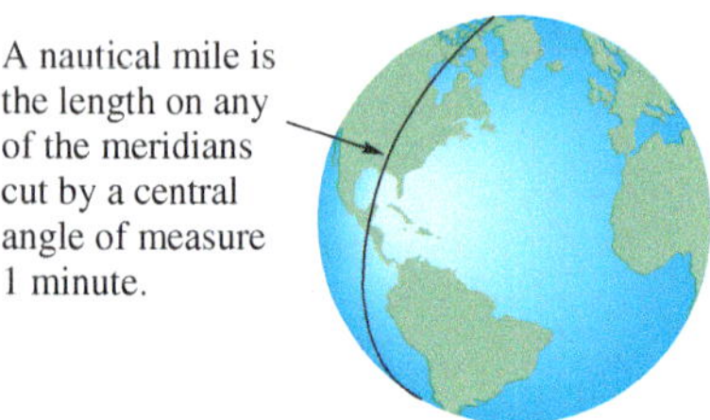

(a) Find the latitude between 0° and 90° at which the nautical mile is 6074 ft.

(b) At what latitude between 0° and 180° is the nautical mile 6108 ft?

(c) In the United States, the nautical mile is defined everywhere as 6080.2 ft. At what latitude between 0° and 90° does this agree with the British nautical mile?

Chapter 7 Test

Work each problem.

1. If $\cos \theta = \frac{24}{25}$ and θ is in quadrant IV, find the other five trigonometric functions of θ.

2. Express $\sec \theta - \sin \theta \tan \theta$ as a single function of θ.

3. Express $\tan^2 x - \sec^2 x$ in terms of $\sin x$ and $\cos x$, and simplify.

4. Find the exact value of $\cos \frac{5\pi}{12}$.

5. Express **(a)** $\cos(270° - \theta)$ and **(b)** $\tan(\pi + x)$ as functions of θ or x alone.

6. Use a half-angle identity to find the exact value of $\sin(-22.5°)$.

7. Given that $\sin A = \frac{5}{13}$, $\cos B = -\frac{3}{5}$, A is a quadrant I angle, and B is a quadrant II angle, find each of the following.

(a) $\sin(A + B)$ **(b)** $\cos(A + B)$ **(c)** $\tan(A - B)$ **(d)** the quadrant of $A + B$

8. Given that $\cos \theta = -\frac{3}{5}$ and $90° < \theta < 180°$, find each of the following.

(a) $\cos 2\theta$ **(b)** $\sin 2\theta$ **(c)** $\tan 2\theta$ **(d)** $\cos \frac{\theta}{2}$ **(e)** $\tan \frac{\theta}{2}$

Verify that each equation is an identity.

9. $\sec^2 B = \dfrac{1}{1 - \sin^2 B}$

10. $\cos 2A = \dfrac{\cot A - \tan A}{\csc A \sec A}$

11. $\tan^2 x - \sin^2 x = (\tan x \sin x)^2$

12. $\dfrac{\tan x - \cot x}{\tan x + \cot x} = 2 \sin^2 x - 1$

Work each problem.

13. *(Modeling) Voltage* The voltage in common household current is expressed as $V = 163 \sin \omega t$, where ω is the angular speed (in radians per second) of the generator at an electrical plant and t is time (in seconds).

(a) Use an identity to express V in terms of cosine.

(b) If $\omega = 120\pi$, what is the maximum voltage? Give the least positive value of t when the maximum voltage occurs.

14. Graph $y = \sin^{-1} x$, and indicate the coordinates of three points on the graph. Give the domain and range.

15. Find the exact value of each real number y. Do not use a calculator.

(a) $y = \arccos\left(-\frac{1}{2}\right)$ **(b)** $y = \sin^{-1}\left(-\frac{\sqrt{3}}{2}\right)$

(c) $y = \tan^{-1} 0$ **(d)** $y = \text{arcsec}(-2)$

16. Give the degree measure of θ. Do not use a calculator.

(a) $\theta = \arccos \frac{\sqrt{3}}{2}$ **(b)** $\theta = \tan^{-1}(-1)$

(c) $\theta = \cot^{-1}(-1)$ **(d)** $\theta = \csc^{-1}\left(-\frac{2\sqrt{3}}{3}\right)$

17. Use a calculator to approximate each value in decimal degrees to the nearest hundredth.

(a) $\sin^{-1} 0.69431882$ **(b)** $\sec^{-1} 1.0840880$

(c) $\cot^{-1}(-0.7125586)$

18. Evaluate each expression without using a calculator.

(a) $\cos\left(\arcsin \frac{2}{3}\right)$ **(b)** $\sin\left(2 \cos^{-1} \frac{1}{3}\right)$

19. Write $\tan(\arcsin u)$ as an algebraic expression in u, for $u > 0$.

Solve each equation for exact solutions over the interval $[0°, 360°)$ *where appropriate. Give approximate solutions to the nearest tenth of a degree.*

20. $-3 \sec \theta + 2\sqrt{3} = 0$ **21.** $\sin^2 \theta = \cos^2 \theta + 1$ **22.** $\csc^2 \theta - 2 \cot \theta = 4$

Solve each equation for exact solutions over the interval $[0, 2\pi)$ *where appropriate. Give approximate solutions to four decimal places.*

23. $\cos x = \cos 2x$ **24.** $\sqrt{2} \cos 3x - 1 = 0$ **25.** $\sin x \cos x = \frac{1}{3}$

Solve each equation for all exact solutions in radians (for x) or in degrees (for θ). Write answers using the least possible nonnegative angle measures.

26. $\sin^2 \theta = -\cos 2\theta$ **27.** $2\sqrt{3} \sin \frac{x}{2} = 3$

Work each problem.

28. Solve each equation for x, where x is restricted to the given interval.

(a) $y = \cos 3x$, for x in $\left[0, \frac{\pi}{3}\right]$ **(b)** $y = 4 + 3 \cot x$, for x in $(0, \pi)$

29. Solve each equation for exact solutions.

(a) $\arcsin x = \arctan \frac{4}{3}$ **(b)** $\text{arccot}\, x + 2 \arcsin \frac{\sqrt{3}}{2} = \pi$

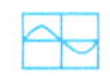 **30.** *(Modeling) Upper Harmonics Pressures* Suppose that the E key above middle C is played on a piano, and its fundamental frequency is $f_1 = 330$ Hz. Its associated pressure is expressed as

$$P_1 = 0.002 \sin 660\, \pi t.$$

The pressures associated with the next four frequencies are $P_2 = \frac{0.002}{2} \sin 1320\pi t$, $P_3 = \frac{0.002}{3} \sin 1980\pi t$, $P_4 = \frac{0.002}{4} \sin 2640\pi t$, and $P_5 = \frac{0.002}{5} \sin 3300\pi t$. Duplicate the graph shown below of

$$P = P_1 + P_2 + P_3 + P_4 + P_5.$$

Approximate the maximum value of P to four significant digits and the least positive value of t for which P reaches this maximum.

For $x = t$,
$y_6 = P_1 + P_2 + P_3 + P_4 + P_5 = P$

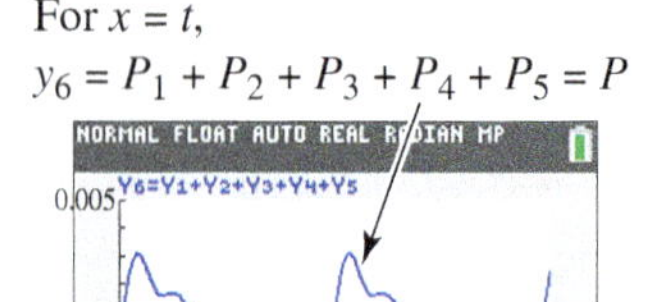

8 Applications of Trigonometry

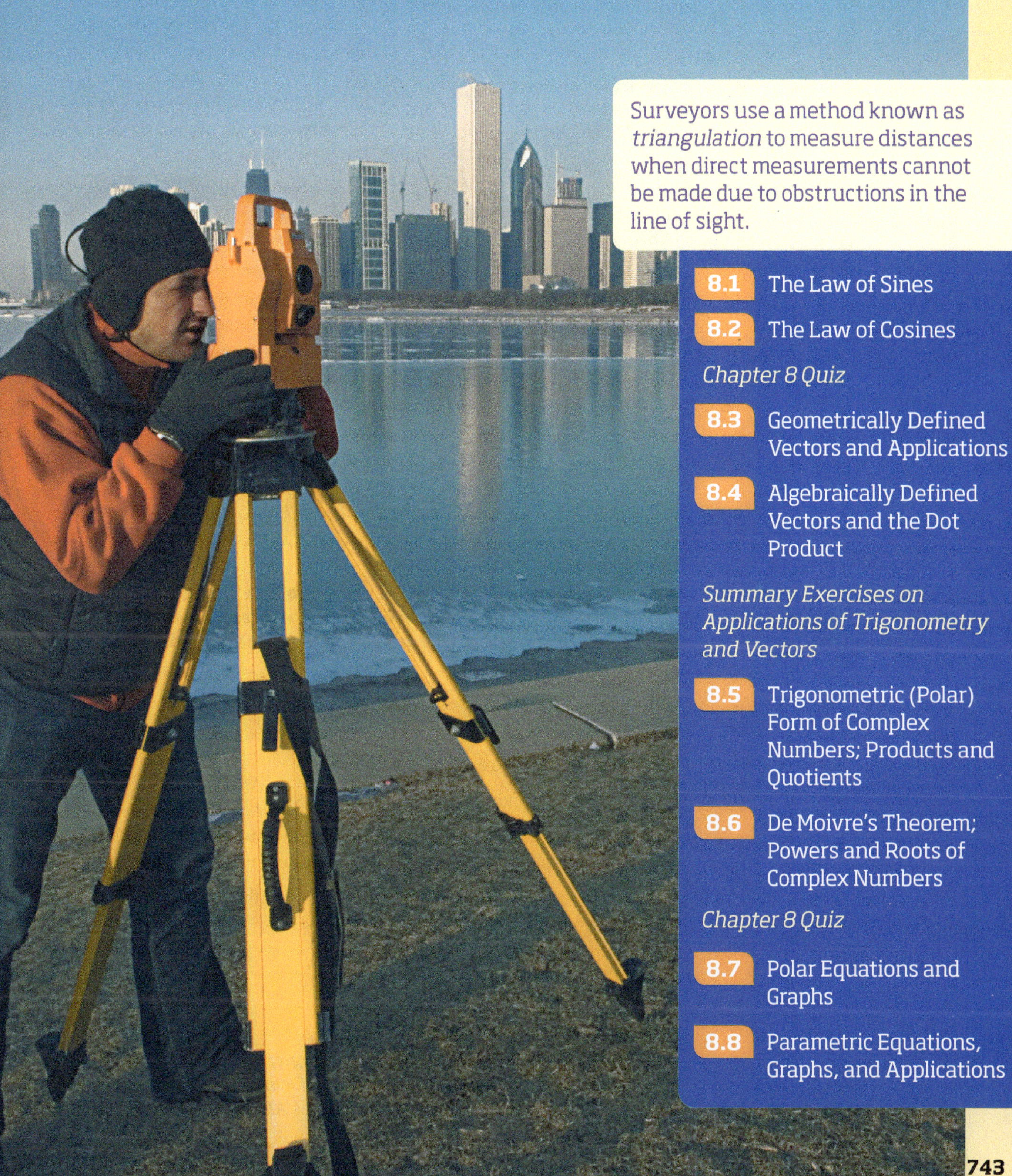

Surveyors use a method known as *triangulation* to measure distances when direct measurements cannot be made due to obstructions in the line of sight.

8.1 The Law of Sines

- Congruency and Oblique Triangles
- Derivation of the Law of Sines
- Using the Law of Sines
- Description of the Ambiguous Case
- Area of a Triangle

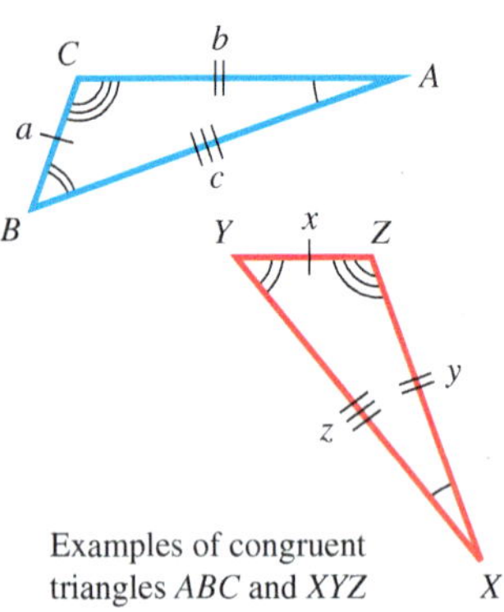

Examples of congruent triangles ABC and XYZ

Congruency and Oblique Triangles

We now turn our attention to solving triangles that are *not* right triangles. To do this we develop new relationships, or laws, that exist between the sides and angles of any triangle. The congruence axioms assist in this process. ***Recall that two triangles are congruent if their corresponding sides and angles are equal.***

Congruence Axioms

Side-Angle-Side (SAS) If two sides and the included angle of one triangle are equal, respectively, to two sides and the included angle of a second triangle, then the triangles are congruent.

Angle-Side-Angle (ASA) If two angles and the included side of one triangle are equal, respectively, to two angles and the included side of a second triangle, then the triangles are congruent.

Side-Side-Side (SSS) If three sides of one triangle are equal, respectively, to three sides of a second triangle, then the triangles are congruent.

If a side and *any* two angles are given **(SAA),** the third angle can be determined by the angle sum formula

$$A + B + C = 180°.$$

Then the ASA axiom can be applied. Whenever SAS, ASA, or SSS is given, the triangle is unique.

A triangle that is not a right triangle is an **oblique triangle.** ***Recall that a triangle can be solved—that is, the measures of the three sides and three angles can be found—if at least one side and any other two measures are known.***

Data Required for Solving Oblique Triangles

There are four possible cases.

Case 1 One side and two angles are known (SAA or ASA).

Case 2 Two sides and one angle not included between the two sides are known (SSA). This case may lead to more than one triangle.

Case 3 Two sides and the angle included between the two sides are known (SAS).

Case 4 Three sides are known (SSS).

NOTE ***If we know three angles of a triangle, we cannot find unique side lengths because AAA assures us only of similarity, not congruence.*** For example, there are infinitely many triangles ABC of different sizes with $A = 35°$, $B = 65°$, and $C = 80°$.

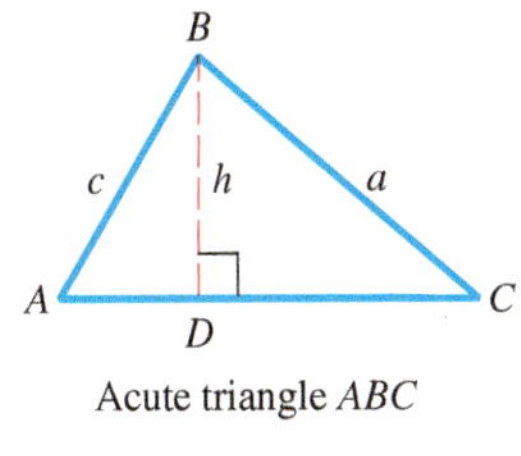

Acute triangle ABC

(a)

Obtuse triangle ABC

(b)

We label oblique triangles as we did right triangles: side a opposite angle A, side b opposite angle B, and side c opposite angle C.

Figure 1

Solving a triangle with given information matching Case 1 or Case 2 requires using the *law of sines,* while solving a triangle with given information matching Case 3 or Case 4 requires using the *law of cosines.*

Derivation of the Law of Sines To derive the law of sines, we start with an oblique triangle, such as the **acute triangle** in **Figure 1(a)** or the **obtuse triangle** in **Figure 1(b).** This discussion applies to both triangles. First, construct the perpendicular from B to side AC (or its extension). Let h be the length of this perpendicular. Then c is the hypotenuse of right triangle ADB, and a is the hypotenuse of right triangle BDC.

$$\text{In triangle } ADB, \quad \sin A = \frac{h}{c}, \quad \text{or} \quad h = c \sin A.$$

$$\text{In triangle } BDC, \quad \sin C = \frac{h}{a}, \quad \text{or} \quad h = a \sin C.$$

Because $h = c \sin A$ and $h = a \sin C$, we set these two expressions equal.

$$a \sin C = c \sin A$$

$$\frac{a}{\sin A} = \frac{c}{\sin C} \quad \text{Divide each side by } \sin A \sin C.$$

In a similar way, by constructing perpendicular lines from the other vertices, we can show that these two equations are also true.

$$\frac{a}{\sin A} = \frac{b}{\sin B} \quad \text{and} \quad \frac{b}{\sin B} = \frac{c}{\sin C}$$

This discussion proves the following theorem.

Law of Sines

In any triangle ABC, with sides a, b, and c, the following hold.

$$\frac{a}{\sin A} = \frac{b}{\sin B}, \quad \frac{a}{\sin A} = \frac{c}{\sin C}, \quad \text{and} \quad \frac{b}{\sin B} = \frac{c}{\sin C}$$

This can be written in compact form as follows.

$$\frac{a}{\sin A} = \frac{b}{\sin B} = \frac{c}{\sin C}$$

That is, according to the law of sines, the lengths of the sides in a triangle are proportional to the sines of the measures of the angles opposite them.

In practice we can also use an alternative form of the law of sines.

$$\frac{\sin A}{a} = \frac{\sin B}{b} = \frac{\sin C}{c} \quad \text{Alternative form of the law of sines}$$

NOTE When using the law of sines, a good strategy is to select a form that has the unknown variable in the numerator and where all other variables are known. This makes computation easier.

Using the Law of Sines

EXAMPLE 1 Applying the Law of Sines (SAA)

Solve triangle ABC if $A = 32.0°$, $B = 81.8°$, and $a = 42.9$ cm.

SOLUTION Start by drawing a triangle, roughly to scale, and labeling the given parts as in **Figure 2.** The values of A, B, and a are known, so use the form of the law of sines that involves these variables, and then solve for b.

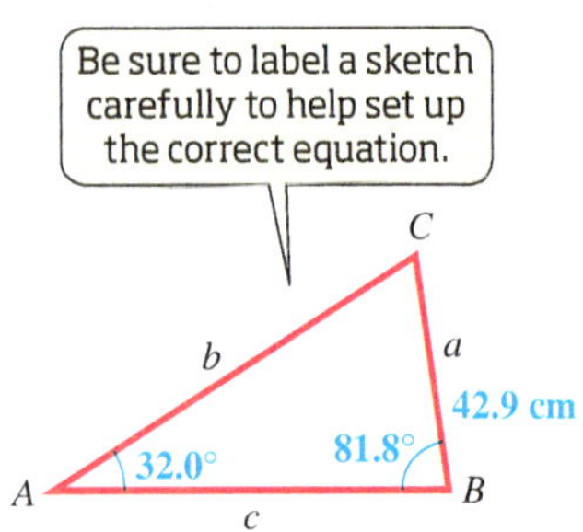

Figure 2

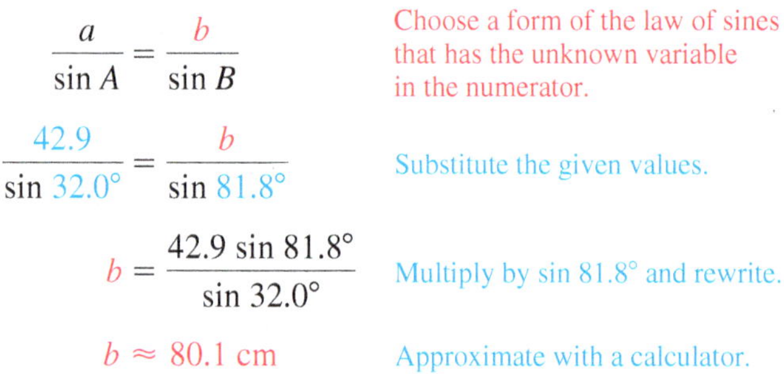

$$\frac{a}{\sin A} = \frac{b}{\sin B}$$ Choose a form of the law of sines that has the unknown variable in the numerator.

$$\frac{42.9}{\sin 32.0°} = \frac{b}{\sin 81.8°}$$ Substitute the given values.

$$b = \frac{42.9 \sin 81.8°}{\sin 32.0°}$$ Multiply by sin 81.8° and rewrite.

$$b \approx 80.1 \text{ cm}$$ Approximate with a calculator.

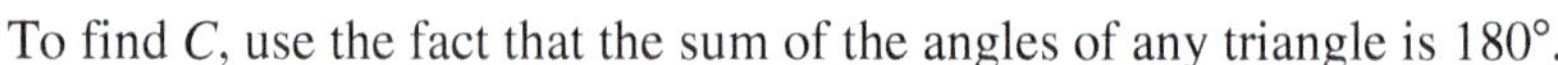

To find C, use the fact that the sum of the angles of any triangle is 180°.

$$A + B + C = 180°$$ Angle sum formula

$$C = 180° - A - B$$ Solve for C.

$$C = 180° - 32.0° - 81.8°$$ Substitute.

$$C = 66.2°$$ Subtract.

Now use the law of sines to find c. (The Pythagorean theorem does not apply because this is not a right triangle.)

$$\frac{a}{\sin A} = \frac{c}{\sin C}$$ Law of sines

$$\frac{42.9}{\sin 32.0°} = \frac{c}{\sin 66.2°}$$ Substitute known values.

$$c = \frac{42.9 \sin 66.2°}{\sin 32.0°}$$ Multiply by sin 66.2° and rewrite.

$$c \approx 74.1 \text{ cm}$$ Approximate with a calculator.

✔ **Now Try Exercise 19.**

CAUTION Whenever possible, use given values in solving triangles, rather than values obtained in intermediate steps, to avoid rounding errors.

EXAMPLE 2 Applying the Law of Sines (ASA)

An engineer wishes to measure the distance across a river. See **Figure 3.** He determines that $C = 112.90°$, $A = 31.10°$, and $b = 347.6$ ft. Find the distance a.

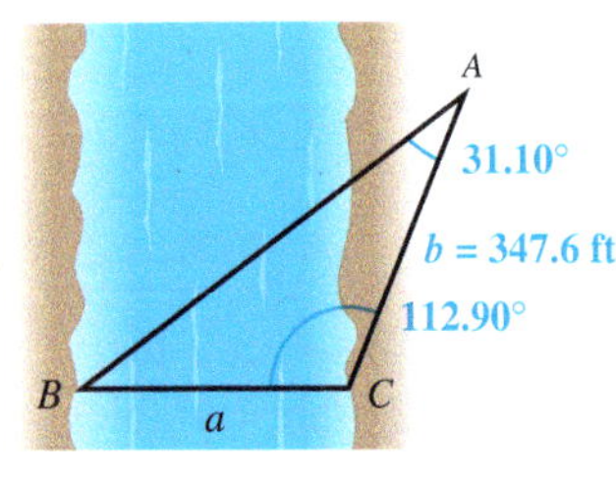

Figure 3

SOLUTION To use the law of sines, one side and the angle opposite it must be known. Here b is the only side whose length is given, so angle B must be found before the law of sines can be used.

$$B = 180° - A - C$$ Angle sum formula, solved for B

$$B = 180° - 31.10° - 112.90°$$ Substitute the given values.

$$B = 36.00°$$ Subtract.

Now use the form of the law of sines involving A, B, and b to find side a.

Solve for a.

$$\frac{a}{\sin A} = \frac{b}{\sin B} \quad \text{Law of sines}$$

$$\frac{a}{\sin 31.10°} = \frac{347.6}{\sin 36.00°} \quad \text{Substitute known values.}$$

$$a = \frac{347.6 \sin 31.10°}{\sin 36.00°} \quad \text{Multiply by } \sin 31.10°.$$

$$a \approx 305.5 \text{ ft} \quad \text{Use a calculator.}$$

Now Try Exercise 55.

Recall that **bearing** is used in navigation to refer to direction of motion or direction of a distant object relative to current course. We consider two methods for expressing bearing.

Method 1

When a single angle is given, such as 220°, this bearing is measured in a clockwise direction from north.

Example: 220°

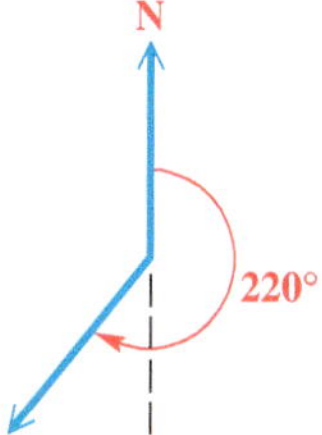

Method 2

Start with a north-south line and use an acute angle to show direction, either east or west, from this line.

Example: S 40° W

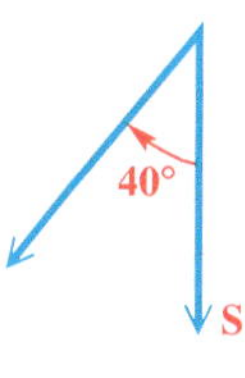

EXAMPLE 3 Applying the Law of Sines (ASA)

Two ranger stations are on an east-west line 110 mi apart. A forest fire is located on a bearing of N 42° E from the western station at A and a bearing of N 15° E from the eastern station at B. To the nearest ten miles, how far is the fire from the western station?

SOLUTION **Figure 4** shows the two ranger stations at points A and B and the fire at point C. Angle BAC measures $90° - 42° = 48°$, obtuse angle B measures $90° + 15° = 105°$, and the third angle, C, measures $180° - 105° - 48° = 27°$. We use the law of sines to find side b.

Solve for b.

$$\frac{b}{\sin B} = \frac{c}{\sin C} \quad \text{Law of sines}$$

$$\frac{b}{\sin 105°} = \frac{110}{\sin 27°} \quad \text{Substitute known values.}$$

$$b = \frac{110 \sin 105°}{\sin 27°} \quad \text{Multiply by } \sin 105°.$$

$$b \approx 230 \text{ mi} \quad \text{Use a calculator and give two significant digits.}$$

Now Try Exercise 57.

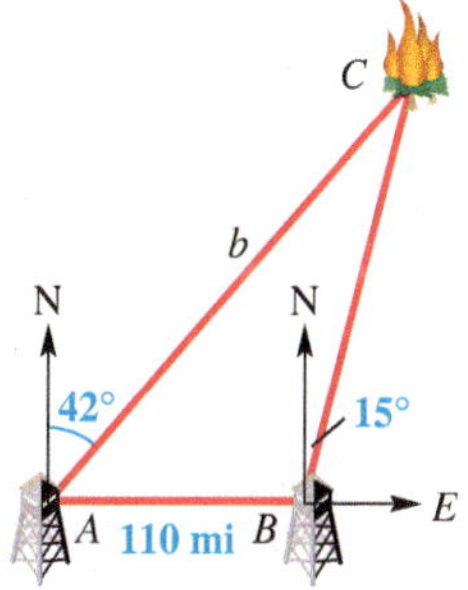

Figure 4

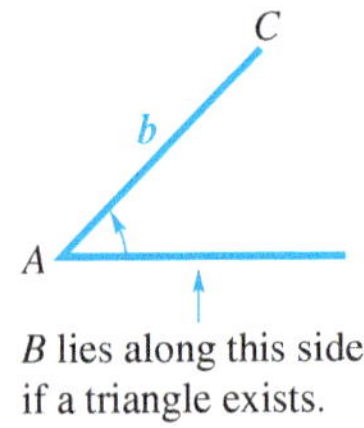

B lies along this side if a triangle exists.

Figure 5

Description of the Ambiguous Case We have used the law of sines to solve triangles involving Case 1, given SAA or ASA. If we are given the lengths of two sides and the angle opposite one of them (Case 2, SSA), then zero, one, or two such triangles may exist. (There is no SSA congruence axiom.)

Suppose we know the measure of acute angle A of triangle ABC, the length of side a, and the length of side b, as shown in **Figure 5.** We must draw the side of length a opposite angle A. The table shows possible outcomes. This situation (SSA) is called the **ambiguous case** of the law of sines.

As shown in the table, if angle A is acute, there are four possible outcomes. If A is obtuse, there are two possible outcomes.

Possible Outcomes for Applying the Law of Sines

Angle A is	Possible Number of Triangles	Sketch	Applying Law of Sines Leads to
Acute	0		$\sin B > 1$, $a < h < b$
Acute	1		$\sin B = 1$, $a = h$ and $h < b$
Acute	1		$0 < \sin B < 1$, $a \geq b$
Acute	2		$0 < \sin B_1 < 1$, $h < a < b$, $A + B_2 < 180°$
Obtuse	0		$\sin B \geq 1$, $a \leq b$
Obtuse	1		$0 < \sin B < 1$, $a > b$

The following basic facts help determine which situation applies.

Applying the Law of Sines

1. For any angle θ of a triangle, $0 < \sin \theta \leq 1$. If $\sin \theta = 1$, then $\theta = 90°$ and the triangle is a right triangle.
2. $\sin \theta = \sin(180° - \theta)$ (Supplementary angles have the same sine value.)
3. The smallest angle is opposite the shortest side, the largest angle is opposite the longest side, and the middle-valued angle is opposite the intermediate side (assuming the triangle has sides that are all of different lengths).

EXAMPLE 4 Solving the Ambiguous Case (No Such Triangle)

Solve triangle ABC if $B = 55° 40'$, $b = 8.94$ m, and $a = 25.1$ m.

SOLUTION We are given B, b, and a. We use the law of sines to find angle A.

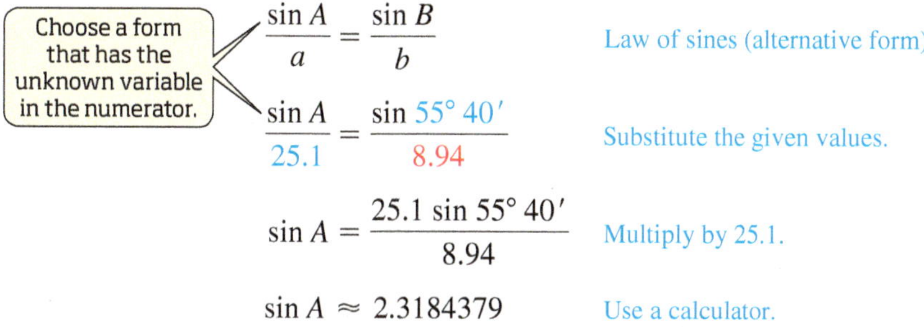

$$\frac{\sin A}{a} = \frac{\sin B}{b} \quad \text{Law of sines (alternative form)}$$

$$\frac{\sin A}{25.1} = \frac{\sin 55° 40'}{8.94} \quad \text{Substitute the given values.}$$

$$\sin A = \frac{25.1 \sin 55° 40'}{8.94} \quad \text{Multiply by 25.1.}$$

$$\sin A \approx 2.3184379 \quad \text{Use a calculator.}$$

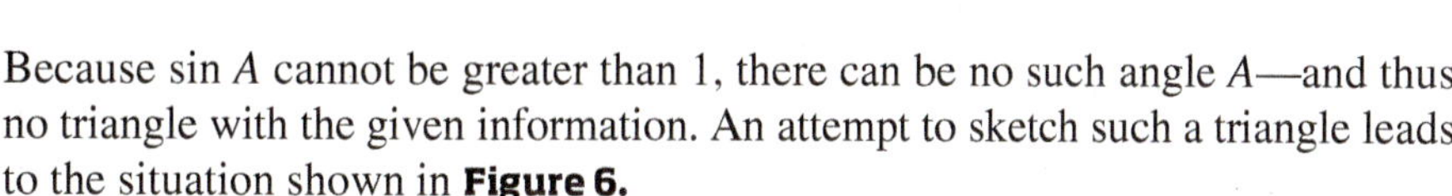

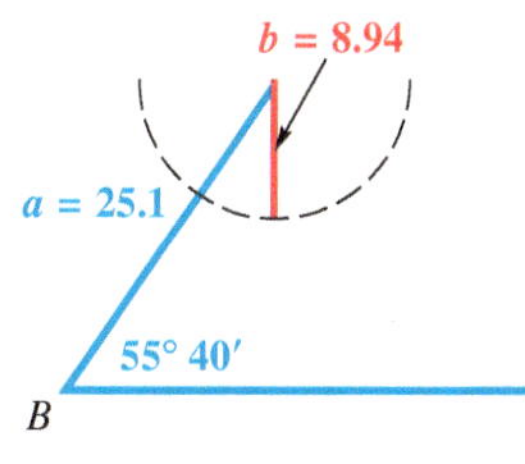

Figure 6

Because $\sin A$ cannot be greater than 1, there can be no such angle A—and thus no triangle with the given information. An attempt to sketch such a triangle leads to the situation shown in **Figure 6.**

✓ **Now Try Exercise 37.**

NOTE In the ambiguous case, we are given two sides and an angle opposite one of the sides (SSA). For example, suppose b, c, and angle C are given. This situation represents the ambiguous case because angle C is opposite side c.

EXAMPLE 5 Solving the Ambiguous Case (Two Triangles)

Solve triangle ABC if $A = 55.3°$, $a = 22.8$ ft, and $b = 24.9$ ft.

SOLUTION To begin, use the law of sines to find angle B.

$$\frac{\sin A}{a} = \frac{\sin B}{b} \quad \text{Solve for } \sin B.$$

$$\frac{\sin 55.3°}{22.8} = \frac{\sin B}{24.9} \quad \text{Substitute the given values.}$$

$$\sin B = \frac{24.9 \sin 55.3°}{22.8} \quad \text{Multiply by 24.9 and rewrite.}$$

$$\sin B \approx 0.8978678 \quad \text{Use a calculator.}$$

There are two angles B between 0° and 180° that satisfy this condition. Because $\sin B \approx 0.8978678$, one value of angle B, to the nearest tenth, is

$$B_1 = 63.9°. \quad \text{Use the inverse sine function.}$$

Supplementary angles have the same sine value, so another *possible* value of angle B is

$$B_2 = 180° - 63.9° = 116.1°.$$

To see whether $B_2 = 116.1°$ is a valid possibility, add 116.1° to the measure of A, 55.3°. Because

$$116.1° + 55.3° = 171.4°,$$

and this sum is less than 180°, it is a valid angle measure for this triangle.

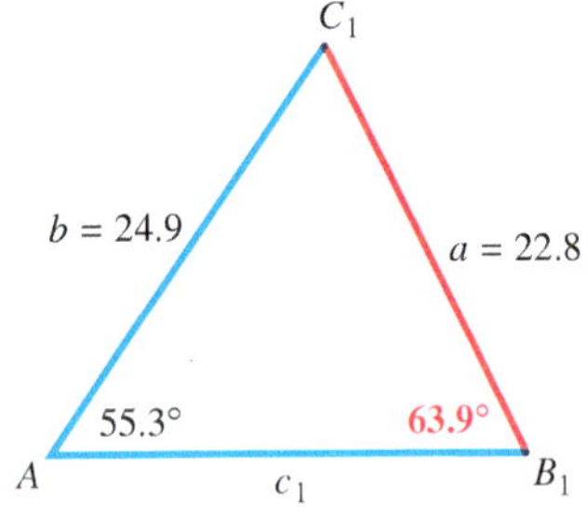

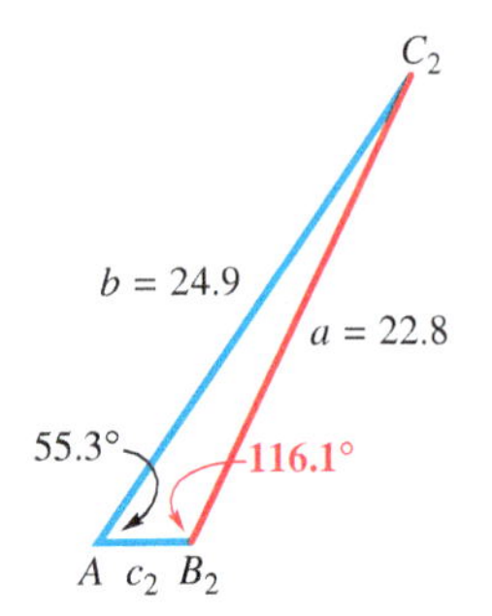

Figure 7

Now separately solve triangles AB_1C_1 and AB_2C_2 shown in **Figure 7.** Begin with AB_1C_1. Find angle C_1 first.

$$C_1 = 180° - A - B_1 \quad \text{Angle sum formula, solved for } C_1$$
$$C_1 = 180° - 55.3° - 63.9° \quad \text{Substitute.}$$
$$C_1 = 60.8° \quad \text{Subtract.}$$

Now, use the law of sines to find side c_1.

$$\frac{a}{\sin A} = \frac{c_1}{\sin C_1} \quad \text{Solve for } c_1.$$
$$\frac{22.8}{\sin 55.3°} = \frac{c_1}{\sin 60.8°} \quad \text{Substitute.}$$
$$c_1 = \frac{22.8 \sin 60.8°}{\sin 55.3°} \quad \text{Multiply by } \sin 60.8° \text{ and rewrite.}$$
$$c_1 \approx 24.2 \text{ ft} \quad \text{Use a calculator.}$$

To solve triangle AB_2C_2, first find angle C_2.

$$C_2 = 180° - A - B_2 \quad \text{Angle sum formula, solved for } C_2$$
$$C_2 = 180° - 55.3° - 116.1° \quad \text{Substitute.}$$
$$C_2 = 8.6° \quad \text{Subtract.}$$

Use the law of sines to find side c_2.

$$\frac{a}{\sin A} = \frac{c_2}{\sin C_2} \quad \text{Solve for } c_2.$$
$$\frac{22.8}{\sin 55.3°} = \frac{c_2}{\sin 8.6°} \quad \text{Substitute.}$$
$$c_2 = \frac{22.8 \sin 8.6°}{\sin 55.3°} \quad \text{Multiply by } \sin 8.6° \text{ and rewrite.}$$
$$c_2 \approx 4.15 \text{ ft} \quad \text{Use a calculator.}$$

✔ **Now Try Exercise 45.**

The ambiguous case results in zero, one, or two triangles. The following guidelines can be used to determine how many triangles there are.

Number of Triangles Satisfying the Ambiguous Case (SSA)

Let sides a and b and angle A be given in triangle ABC. (The law of sines can be used to calculate the value of $\sin B$.)

1. If applying the law of sines results in an equation having $\sin B > 1$, then *no triangle* satisfies the given conditions.

2. If $\sin B = 1$, then *one triangle* satisfies the given conditions and $B = 90°$.

3. If $0 < \sin B < 1$, then either *one or two triangles* satisfy the given conditions.

  (a) If $\sin B = k$, then let $B_1 = \sin^{-1} k$ and use B_1 for B in the first triangle.

  (b) Let $B_2 = 180° - B_1$. If $A + B_2 < 180°$, then a second triangle exists. In this case, use B_2 for B in the second triangle.

EXAMPLE 6 Solving the Ambiguous Case (One Triangle)

Solve triangle ABC, given $A = 43.5°$, $a = 10.7$ in., and $c = 7.2$ in.

SOLUTION Find angle C.

$$\frac{\sin C}{c} = \frac{\sin A}{a} \quad \text{Law of sines (alternative form)}$$

$$\frac{\sin C}{7.2} = \frac{\sin 43.5°}{10.7} \quad \text{Substitute the given values.}$$

$$\sin C = \frac{7.2 \sin 43.5°}{10.7} \quad \text{Multiply by 7.2.}$$

$$\sin C \approx 0.46319186 \quad \text{Use a calculator.}$$

$$C \approx 27.6° \quad \text{Use the inverse sine function.}$$

There is another angle C that has sine value 0.46319186. It is

$$C = 180° - 27.6° = 152.4°.$$

However, notice in the given information that $c < a$, meaning that in the triangle, angle C must have measure *less than* angle A. Notice also that when we add this obtuse value to the given angle $A = 43.5°$, we obtain

$$152.4° + 43.5° = 195.9°,$$

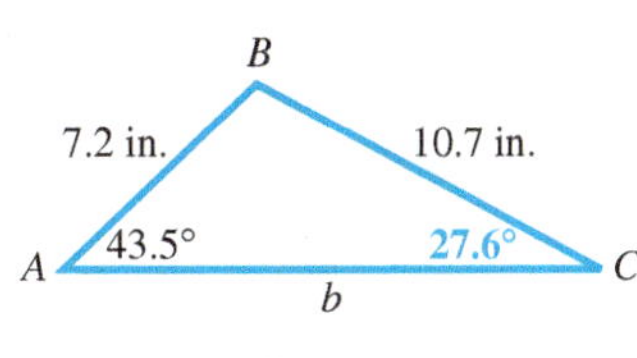

Figure 8

which is *greater than* 180°. Thus either of these approaches shows that there can be only one triangle. See **Figure 8.** The measure of angle B can be found next.

$$B = 180° - 27.6° - 43.5° \quad \text{Substitute.}$$

$$B = 108.9° \quad \text{Subtract.}$$

We can find side b with the law of sines.

$$\frac{b}{\sin B} = \frac{a}{\sin A} \quad \text{Law of sines}$$

$$\frac{b}{\sin 108.9°} = \frac{10.7}{\sin 43.5°} \quad \text{Substitute known values.}$$

$$b = \frac{10.7 \sin 108.9°}{\sin 43.5°} \quad \text{Multiply by } \sin 108.9°.$$

$$b \approx 14.7 \text{ in.} \quad \text{Use a calculator.}$$

✔ **Now Try Exercise 41.**

When solving triangles, it is important to analyze the given information to determine whether it forms a valid triangle.

EXAMPLE 7 Analyzing Data Involving an Obtuse Angle

Without using the law of sines, explain why $A = 104°$, $a = 26.8$ m, and $b = 31.3$ m cannot be valid for a triangle ABC.

SOLUTION Because A is an obtuse angle, it is the largest angle, and so the longest side of the triangle must be a. However, we are given $b > a$.

Thus, $B > A$, which is impossible if A is obtuse.

Therefore, no such triangle ABC exists. ✔ **Now Try Exercise 53.**

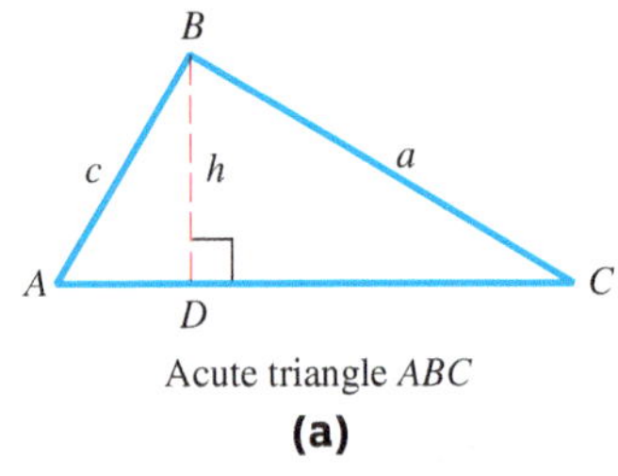

Acute triangle ABC

(a)

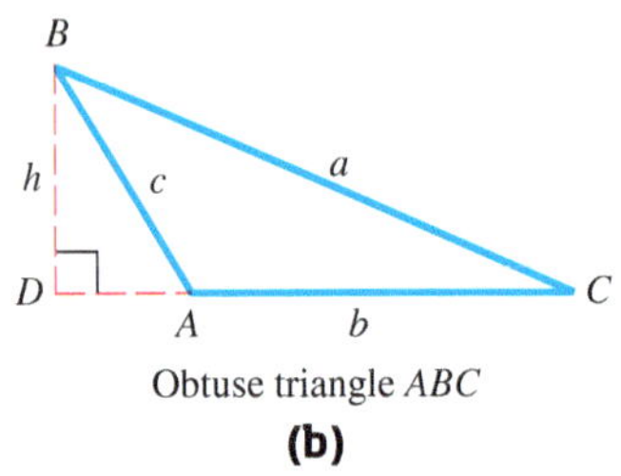

Obtuse triangle ABC

(b)

Figure 9

Area of a Triangle

A familiar formula for the area of a triangle is

$$\mathcal{A} = \frac{1}{2}bh, \quad \text{where } \mathcal{A} \text{ represents area, } b \text{ base, and } h \text{ height.}$$

This formula cannot always be used easily because in practice, h is often unknown. To find another formula, refer to acute triangle ABC in **Figure 9(a)** or obtuse triangle ABC in **Figure 9(b).**

A perpendicular has been drawn from B to the base of the triangle (or the extension of the base). Consider right triangle ADB in either figure.

$$\sin A = \frac{h}{c}, \quad \text{or} \quad h = c \sin A$$

Substitute into the formula for the area of a triangle.

$$\mathcal{A} = \frac{1}{2}bh = \frac{1}{2}bc \sin A$$

Any other pair of sides and the angle between them could have been used.

Area of a Triangle (SAS)

In any triangle ABC, the area $\mathcal{A}$ is given by the following formulas.

$$\mathcal{A} = \frac{1}{2}bc \sin A, \quad \mathcal{A} = \frac{1}{2}ab \sin C, \quad \text{and} \quad \mathcal{A} = \frac{1}{2}ac \sin B$$

That is, the area is half the product of the lengths of two sides and the sine of the angle included between them.

NOTE If the included angle measures 90°, its sine is 1 and the formula becomes the familiar $\mathcal{A} = \frac{1}{2}bh$.

EXAMPLE 8 Finding the Area of a Triangle (SAS)

Find the area of triangle ABC in **Figure 10.**

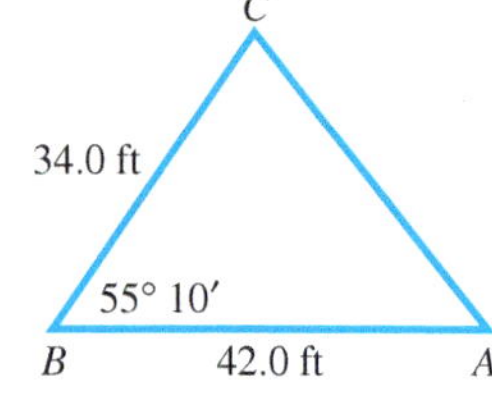

Figure 10

SOLUTION Substitute $B = 55° \, 10'$, $a = 34.0$ ft, and $c = 42.0$ ft into the area formula.

$$\mathcal{A} = \frac{1}{2}ac \sin B = \frac{1}{2}(34.0)(42.0) \sin 55° \, 10' \approx 586 \text{ ft}^2$$

✔ **Now Try Exercise 75.**

EXAMPLE 9 Finding the Area of a Triangle (ASA)

Find the area of triangle ABC in **Figure 11.**

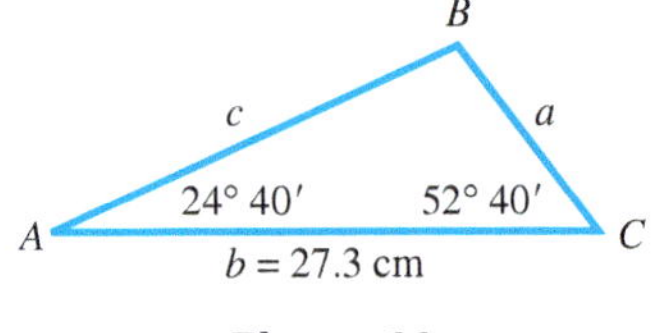

Figure 11

SOLUTION Before the area formula can be used, we must find side a or c.

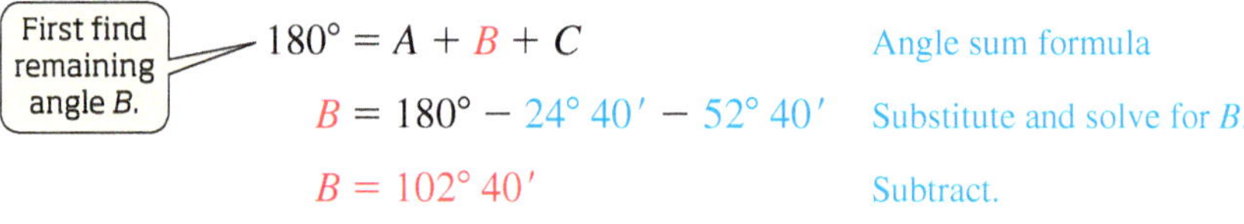

$$180° = A + B + C \qquad \text{Angle sum formula}$$

$$B = 180° - 24° \, 40' - 52° \, 40' \qquad \text{Substitute and solve for } B.$$

$$B = 102° \, 40' \qquad \text{Subtract.}$$

Next use the law of sines to find side a.

Solve for a.

$$\frac{a}{\sin A} = \frac{b}{\sin B} \qquad \text{Law of sines}$$

$$\frac{a}{\sin 24^\circ 40'} = \frac{27.3}{\sin 102^\circ 40'} \qquad \text{Substitute known values.}$$

$$a = \frac{27.3 \sin 24^\circ 40'}{\sin 102^\circ 40'} \qquad \text{Multiply by } \sin 24^\circ 40'.$$

$$a \approx 11.7 \text{ cm} \qquad \text{Use a calculator.}$$

Now that we know two sides, a and b, and their included angle C, we find the area.

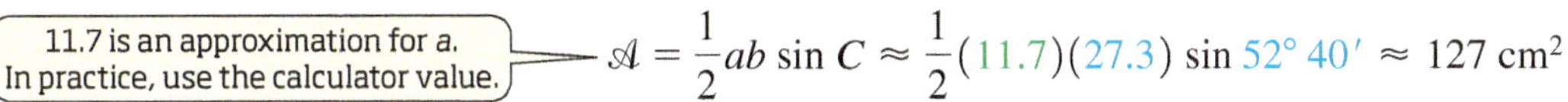

$$\mathcal{A} = \frac{1}{2}ab \sin C \approx \frac{1}{2}(11.7)(27.3) \sin 52^\circ 40' \approx 127 \text{ cm}^2$$

✔ **Now Try Exercise 81.**

8.1 Exercises

CONCEPT PREVIEW *Consider each case and determine whether there is sufficient information to solve the triangle using the law of sines.*

1. Two angles and the side included between them are known.
2. Two angles and a side opposite one of them are known.
3. Two sides and the angle included between them are known.
4. Three sides are known.
5. **CONCEPT PREVIEW** Which one of the following sets of data does *not* determine a unique triangle?

   A. $A = 50^\circ,\ b = 21,\ a = 19$ **B.** $A = 45^\circ,\ b = 10,\ a = 12$

   C. $A = 130^\circ,\ b = 4,\ a = 7$ **D.** $A = 30^\circ,\ b = 8,\ a = 4$

6. **CONCEPT PREVIEW** Which one of the following sets of data determines a unique triangle?

   A. $A = 50^\circ,\ B = 50^\circ,\ C = 80^\circ$ **B.** $a = 3,\ b = 5,\ c = 20$

   C. $A = 40^\circ,\ B = 20^\circ,\ C = 30^\circ$ **D.** $a = 7,\ b = 24,\ c = 25$

CONCEPT PREVIEW *In each figure, a line segment of length L is to be drawn from the given point to the positive x-axis in order to form a triangle. For what value(s) of L can we draw the following?*

(a) *two triangles* **(b)** *exactly one triangle* **(c)** *no triangle*

7.

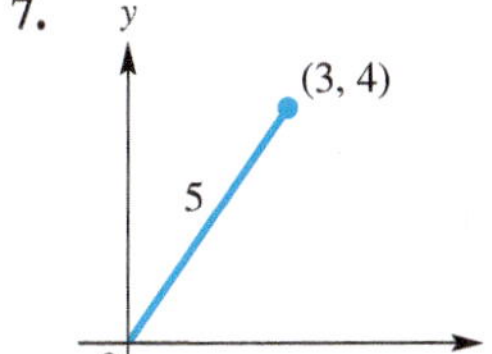

8.

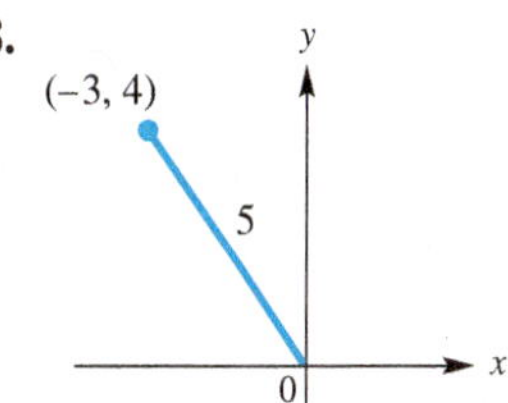

CONCEPT PREVIEW *Determine the number of triangles ABC possible with the given parts.*

9. $a = 50,\ b = 26,\ A = 95^\circ$ **10.** $a = 31,\ b = 26,\ B = 48^\circ$

11. $c = 50,\ b = 61,\ C = 58^\circ$ **12.** $a = 35,\ b = 30,\ A = 40^\circ$

Find the length of each side a. Do not use a calculator.

13.

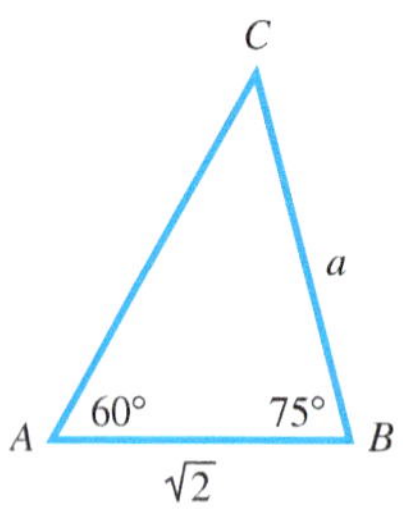

14.

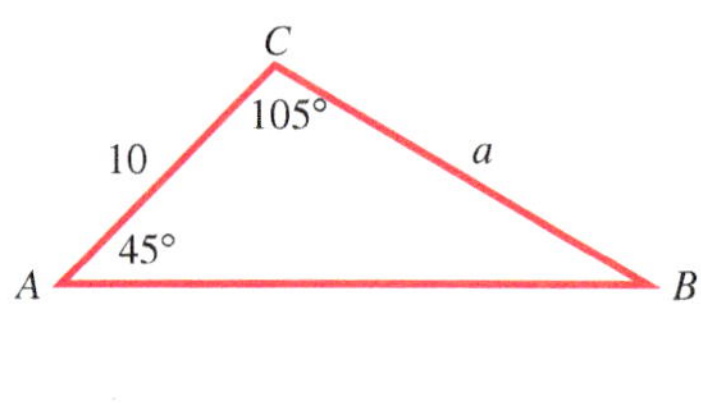

Determine the remaining sides and angles of each triangle ABC. **See Example 1.**

15.

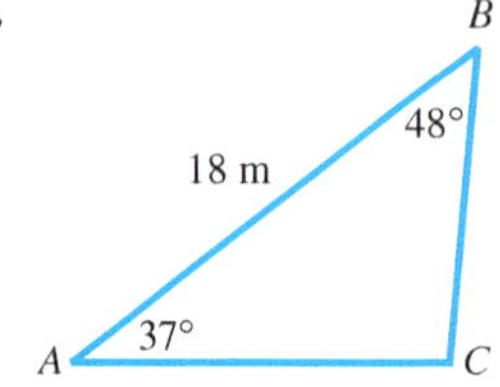

16.

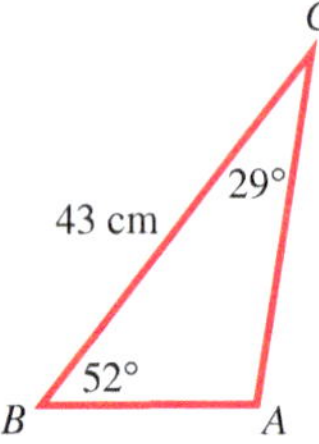

17.

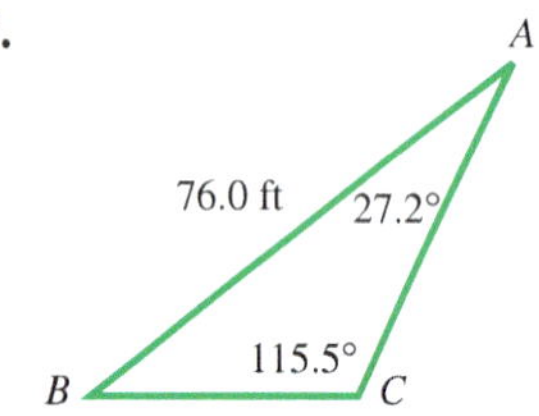

18.

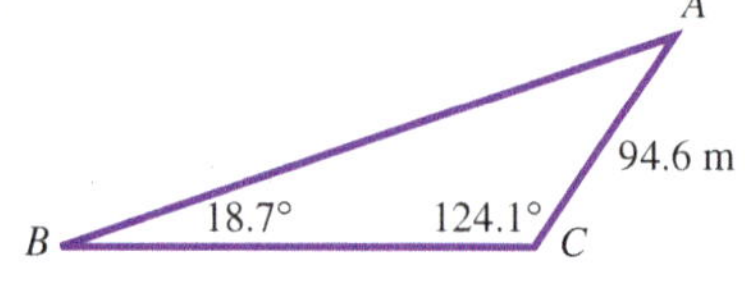

19. $A = 68.41°$, $B = 54.23°$, $a = 12.75$ ft

20. $C = 74.08°$, $B = 69.38°$, $c = 45.38$ m

21. $A = 87.2°$, $b = 75.9$ yd, $C = 74.3°$

22. $B = 38° \, 40'$, $a = 19.7$ cm, $C = 91° \, 40'$

23. $B = 20° \, 50'$, $C = 103° \, 10'$, $AC = 132$ ft

24. $A = 35.3°$, $B = 52.8°$, $AC = 675$ ft

25. $A = 39.70°$, $C = 30.35°$, $b = 39.74$ m

26. $C = 71.83°$, $B = 42.57°$, $a = 2.614$ cm

27. $B = 42.88°$, $C = 102.40°$, $b = 3974$ ft

28. $C = 50.15°$, $A = 106.1°$, $c = 3726$ yd

29. $A = 39° \, 54'$, $a = 268.7$ m, $B = 42° \, 32'$

30. $C = 79° \, 18'$, $c = 39.81$ mm, $A = 32° \, 57'$

Find the measure of each angle B. Do not use a calculator.

31.

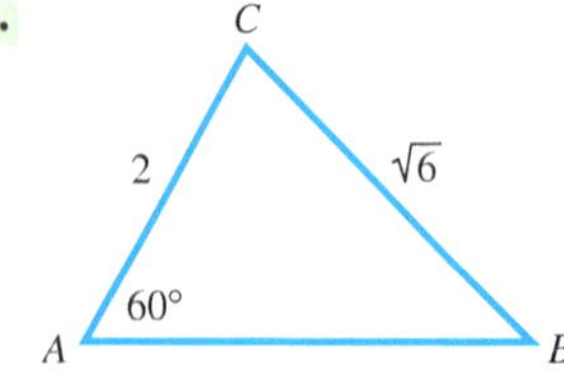

32.

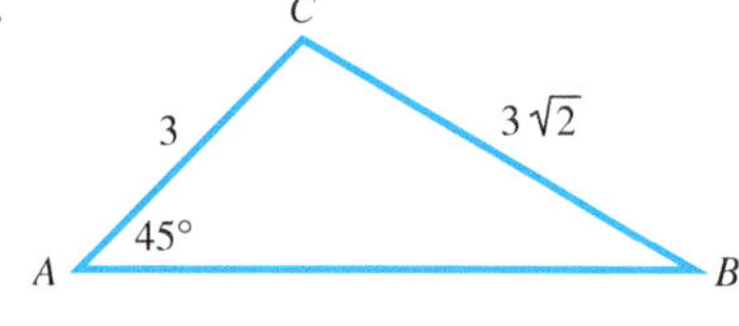

Find the unknown angles in triangle ABC for each triangle that exists. **See Examples 4–6.**

33. $A = 29.7°$, $b = 41.5$ ft, $a = 27.2$ ft

34. $B = 48.2°$, $a = 890$ cm, $b = 697$ cm

35. $C = 41° \, 20'$, $b = 25.9$ m, $c = 38.4$ m

36. $B = 48° \, 50'$, $a = 3850$ in., $b = 4730$ in.

37. $B = 74.3°$, $a = 859$ m, $b = 783$ m

38. $C = 82.2°$, $a = 10.9$ km, $c = 7.62$ km

39. $A = 142.13°$, $b = 5.432$ ft, $a = 7.297$ ft

40. $B = 113.72°$, $a = 189.6$ yd, $b = 243.8$ yd

Solve each triangle ABC that exists. ***See Examples 4–6.***

41. $A = 42.5°$, $a = 15.6$ ft, $b = 8.14$ ft

42. $C = 52.3°$, $a = 32.5$ yd, $c = 59.8$ yd

43. $B = 72.2°$, $b = 78.3$ m, $c = 145$ m

44. $C = 68.5°$, $c = 258$ cm, $b = 386$ cm

45. $A = 38° \, 40'$, $a = 9.72$ m, $b = 11.8$ m

46. $C = 29° \, 50'$, $a = 8.61$ m, $c = 5.21$ m

47. $A = 96.80°$, $b = 3.589$ ft, $a = 5.818$ ft

48. $C = 88.70°$, $b = 56.87$ m, $c = 112.4$ m

49. $B = 39.68°$, $a = 29.81$ m, $b = 23.76$ m

50. $A = 51.20°$, $c = 7986$ cm, $a = 7208$ cm

Concept Check Answer each question.

51. Apply the law of sines to the following: $a = \sqrt{5}$, $c = 2\sqrt{5}$, $A = 30°$. What is the value of $\sin C$? What is the measure of C? Based on its angle measures, what kind of triangle is triangle ABC?

52. What condition must exist to determine that there is no triangle satisfying the given values of a, b, and B, once the value of $\sin A$ is found by applying the law of sines?

53. Without using the law of sines, why can no triangle ABC exist that satisfies $A = 103° \, 20'$, $a = 14.6$ ft, $b = 20.4$ ft?

54. If the law of sines is applied to the data given in **Example 7,** what happens when we try to find the measure of angle B using a calculator?

Solve each problem. ***See Examples 2 and 3.***

55. ***Distance across a River*** To find the distance AB across a river, a surveyor laid off a distance $BC = 354$ m on one side of the river. It is found that $B = 112° \, 10'$ and $C = 15° \, 20'$. Find AB. See the figure.

56. ***Distance across a Canyon*** To determine the distance RS across a deep canyon, Rhonda lays off a distance $TR = 582$ yd. She then finds that $T = 32° \, 50'$ and $R = 102° \, 20'$. Find RS. See the figure.

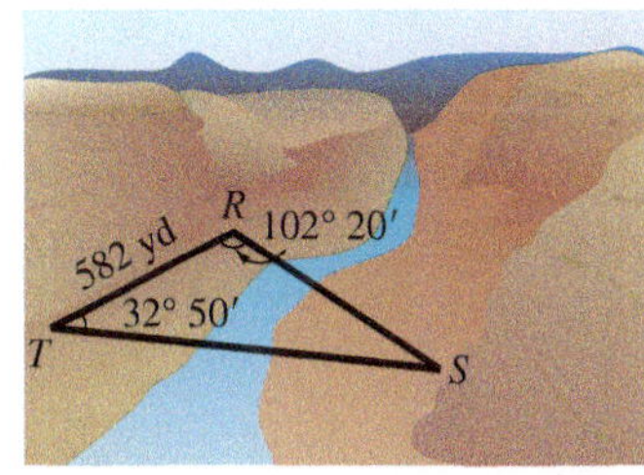

57. ***Distance a Ship Travels*** A ship is sailing due north. At a certain point the bearing of a lighthouse 12.5 km away is N 38.8° E. Later on, the captain notices that the bearing of the lighthouse has become S 44.2° E. How far did the ship travel between the two observations of the lighthouse?

58. ***Distance between Radio Direction Finders*** Radio direction finders are placed at points A and B, which are 3.46 mi apart on an east-west line, with A west of B. From A the bearing of a certain radio transmitter is 47.7°, and from B the bearing is 302.5°. Find the distance of the transmitter from A.

59. ***Distance between a Ship and a Lighthouse*** The bearing of a lighthouse from a ship was found to be N 37° E. After the ship sailed 2.5 mi due south, the new bearing was N 25° E. Find the distance between the ship and the lighthouse at each location.

60. ***Distance across a River*** Standing on one bank of a river flowing north, Mark notices a tree on the opposite bank at a bearing of 115.45°. Lisa is on the same bank as Mark, but 428.3 m away. She notices that the bearing of the tree is 45.47°. The two banks are parallel. What is the distance across the river?

61. ***Height of a Balloon*** A balloonist is directly above a straight road 1.5 mi long that joins two villages. She finds that the town closer to her is at an angle of depression of 35°, and the farther town is at an angle of depression of 31°. How high above the ground is the balloon?

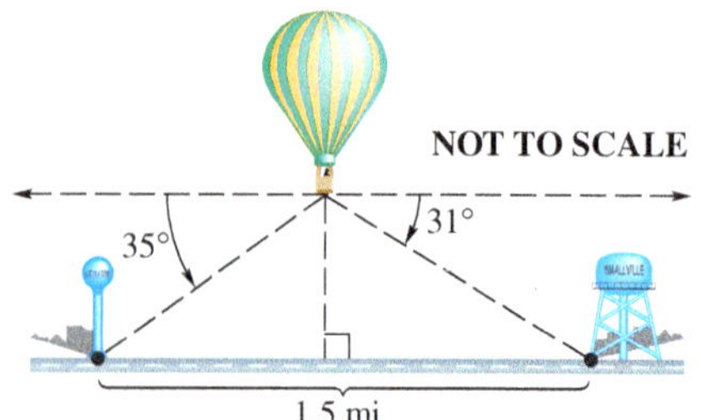

62. ***Measurement of a Folding Chair*** A folding chair is to have a seat 12.0 in. deep with angles as shown in the figure. How far down from the seat should the crossing legs be joined? (Find length x in the figure.)

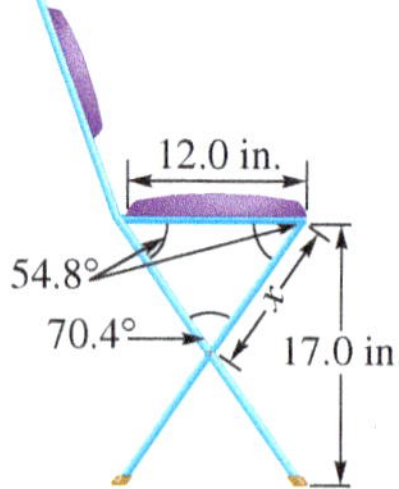

63. ***Angle Formed by Radii of Gears*** Three gears are arranged as shown in the figure. Find angle θ.

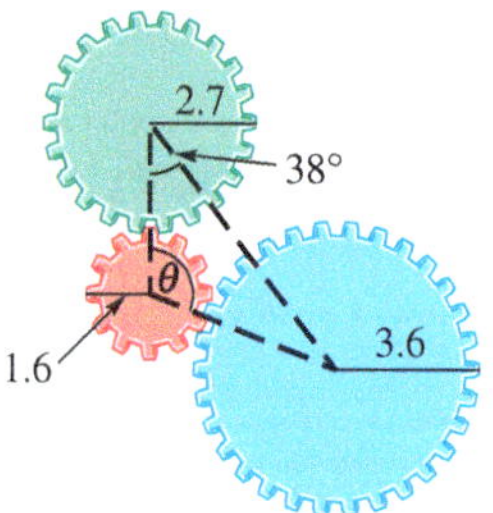

64. ***Distance between Atoms*** Three atoms with atomic radii of 2.0, 3.0, and 4.5 are arranged as in the figure. Find the distance between the centers of atoms A and C.

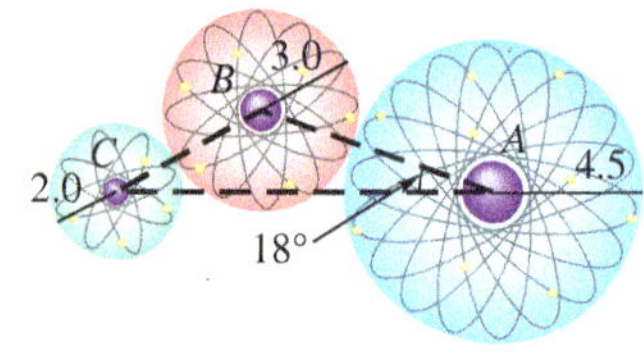

65. ***Distance between Inaccessible Points*** To find the distance between a point X and an inaccessible point Z, a line segment XY is constructed. It is found that $XY = 960$ m, angle $XYZ = 43° 30'$, and angle $YZX = 95° 30'$. Find the distance between X and Z to the nearest meter.

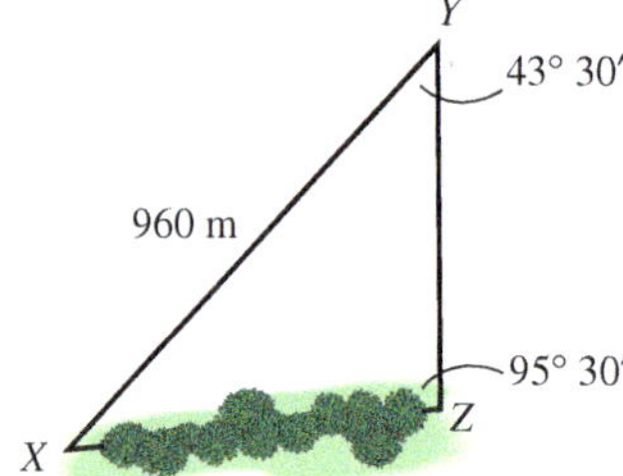

66. ***Height of an Antenna Tower*** The angle of elevation from the top of a building 45.0 ft high to the top of a nearby antenna tower is 15° 20′. From the base of the building, the angle of elevation of the tower is 29° 30′. Find the height of the tower.

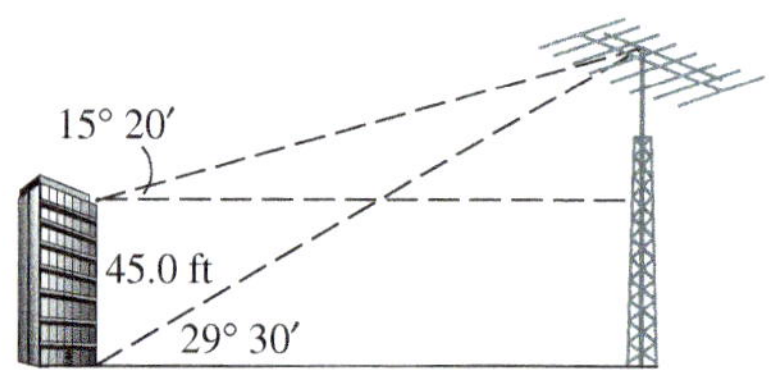

67. ***Height of a Building*** A flagpole 95.0 ft tall is on the top of a building. From a point on level ground, the angle of elevation of the top of the flagpole is 35.0°, and the angle of elevation of the bottom of the flagpole is 26.0°. Find the height of the building.

68. *Flight Path of a Plane* A pilot flies her plane on a bearing of 35° 00′ from point *X* to point *Y*, which is 400 mi from *X*. Then she turns and flies on a bearing of 145° 00′ to point *Z*, which is 400 mi from her starting point *X*. What is the bearing of *Z* from *X*, and what is the distance *YZ*?

69. *Distance to the Moon* The moon is a relatively close celestial object, so its distance can be measured directly by taking two different photographs at precisely the same time from two different locations. The moon will have a different angle of elevation at each location. On April 29, 1976, at 11:35 A.M., the lunar angles of elevation during a partial solar eclipse at Bochum in upper Germany and at Donaueschingen in lower Germany were measured as 52.6997° and 52.7430°, respectively. The two cities are 398 km apart.

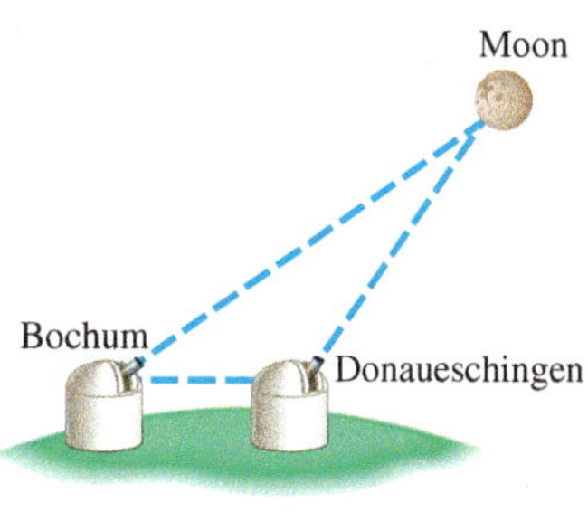

Calculate the distance to the moon, to the nearest thousand kilometers, from Bochum on this day, and compare it with the actual value of 406,000 km. Disregard the curvature of Earth in this calculation. (*Source:* Scholosser, W., T. Schmidt-Kaler, and E. Milone, *Challenges of Astronomy*, Springer-Verlag.)

70. *Ground Distances Measured by Aerial Photography* The distance covered by an aerial photograph is determined by both the focal length of the camera and the tilt of the camera from the perpendicular to the ground. A camera lens with a 12-in. focal length will have an angular coverage of 60°. If an aerial photograph is taken with this camera tilted $\theta = 35°$ at an altitude of 5000 ft, calculate to the nearest foot the ground distance *d* that will be shown in this photograph. (*Source:* Brooks, R. and D. Johannes, *Phytoarchaeology*, Dioscorides Press.)

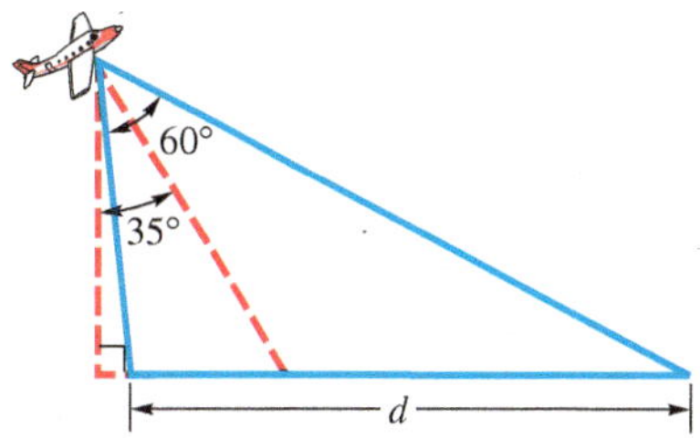

Find the area of each triangle using the formula $\mathcal{A} = \frac{1}{2}bh$, *and then verify that the formula* $\mathcal{A} = \frac{1}{2}ab \sin C$ *gives the same result.*

71.

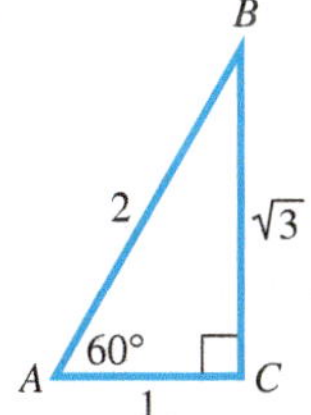

72.

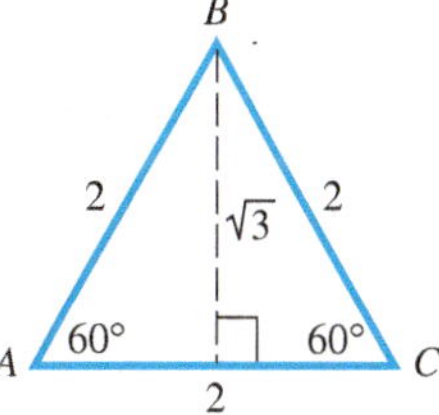

73.

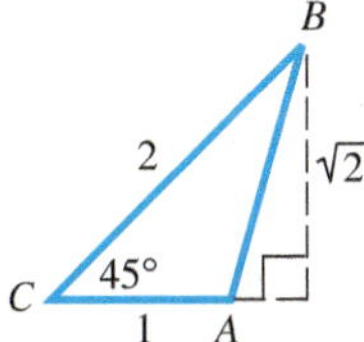

74.

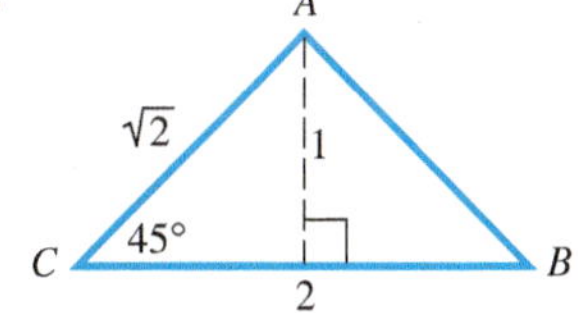

Find the area of each triangle ABC. ***See Examples 8 and 9.***

75. $A = 42.5°$, $b = 13.6$ m, $c = 10.1$ m

76. $C = 72.2°$, $b = 43.8$ ft, $a = 35.1$ ft

77. $B = 124.5°$, $a = 30.4$ cm, $c = 28.4$ cm

78. $C = 142.7°$, $a = 21.9$ km, $b = 24.6$ km

79. $A = 56.80°$, $b = 32.67$ in., $c = 52.89$ in.

80. $A = 34.97°$, $b = 35.29$ m, $c = 28.67$ m

81. $A = 30.50°$, $b = 13.00$ cm, $C = 112.60°$

82. $A = 59.80°$, $b = 15.00$ m, $C = 53.10°$

Solve each problem.

83. ***Area of a Metal Plate*** A painter is going to apply a special coating to a triangular metal plate on a new building. Two sides measure 16.1 m and 15.2 m. She knows that the angle between these sides is 125°. What is the area of the surface she plans to cover with the coating?

84. ***Area of a Triangular Lot*** A real estate agent wants to find the area of a triangular lot. A surveyor takes measurements and finds that two sides are 52.1 m and 21.3 m, and the angle between them is 42.2°. What is the area of the triangular lot?

85. ***Triangle Inscribed in a Circle*** For a triangle inscribed in a circle of radius r, the law of sines ratios

$$\frac{a}{\sin A}, \quad \frac{b}{\sin B}, \quad \text{and} \quad \frac{c}{\sin C} \quad \text{have value } 2r.$$

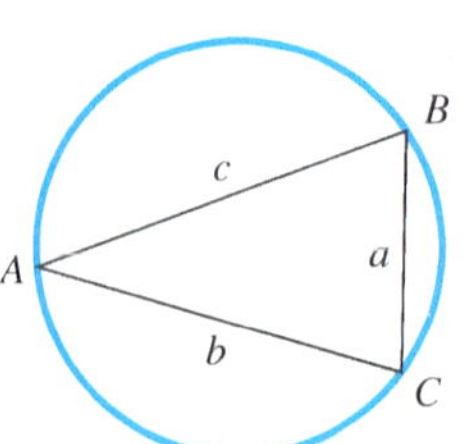

The circle in the figure has diameter 1. What are the values of a, b, and c? (*Note:* This result provides an alternative way to define the sine function for angles between 0° and 180°. It was used nearly 2000 yr ago by the mathematician Ptolemy to construct one of the earliest trigonometric tables.)

86. ***Theorem of Ptolemy*** The following theorem is also attributed to Ptolemy:

In a quadrilateral inscribed in a circle, the product of the diagonals is equal to the sum of the products of the opposite sides.

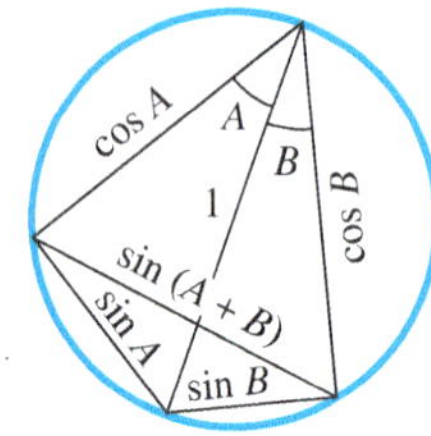

(*Source:* Eves, H., *An Introduction to the History of Mathematics,* Sixth Edition, Saunders College Publishing.) The circle in the figure has diameter 1. Use Ptolemy's theorem to derive the formula for the sine of the sum of two angles.

87. ***Law of Sines*** Several of the exercises on right triangle applications involved a figure similar to the one shown here, in which angles α and β and the length of line segment AB are known, and the length of side CD is to be determined. Use the law of sines to obtain x in terms of α, β, and d.

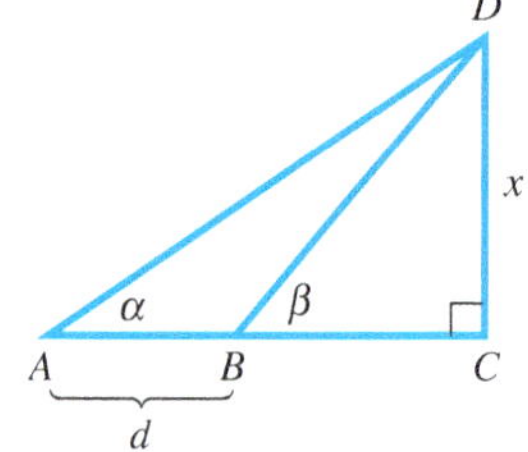

88. ***Aerial Photography*** Aerial photographs can be used to provide coordinates of ordered pairs to determine distances on the ground. Suppose we assign coordinates as shown in the figure. If an object's photographic coordinates are (x, y), then its ground coordinates (X, Y) in feet can be computed using the following formulas.

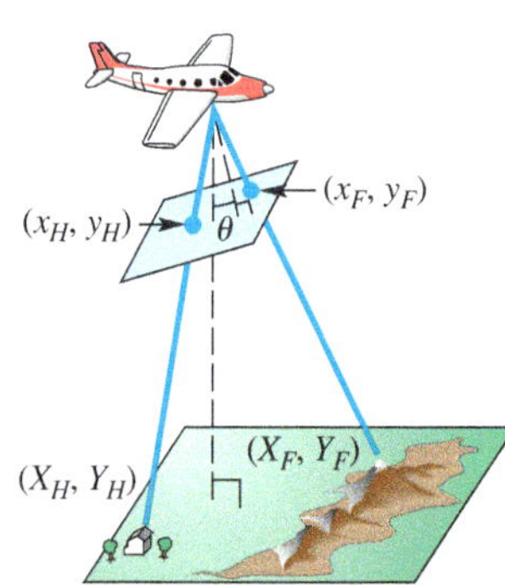

$$X = \frac{(a - h)x}{f \sec \theta - y \sin \theta}, \quad Y = \frac{(a - h)y \cos \theta}{f \sec \theta - y \sin \theta}$$

Here, f is focal length of the camera in inches, a is altitude in feet of the airplane, and h is elevation in feet of the object. Suppose that a house has photographic coordinates $(x_H, y_H) = (0.9, 3.5)$ with elevation 150 ft, and a nearby forest fire has photographic coordinates $(x_F, y_F) = (2.1, -2.4)$ and is at elevation 690 ft. Also suppose the photograph was taken at 7400 ft by a camera with focal length 6 in. and tilt angle $\theta = 4.1°$. (*Source:* Moffitt, F. and E. Mikhail, *Photogrammetry,* Third Edition, Harper & Row.)

(a) Use the formulas to find the ground coordinates of the house and the fire to the nearest tenth of a foot.

(b) Use the distance formula $d = \sqrt{(x_2 - x_1)^2 + (y_2 - y_1)^2}$ to find the distance on the ground between the house and the fire to the nearest tenth of a foot.

Relating Concepts

For individual or collaborative investigation. *(Exercises 89–92)*

Colors of the U.S. Flag *The flag of the United States includes the colors red, white, and blue.*

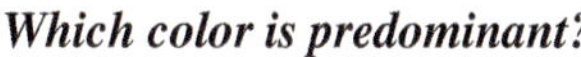

Which color is predominant?

Clearly the answer is either red or white. (It can be shown that only 18.73% *of the total area is blue.)* (Source: *Banks, R.,* Slicing Pizzas, Racing Turtles, and Further Adventures in Applied Mathematics, *Princeton University Press.)*

To answer this question, work ***Exercises 89–92 in order.***

89. Let R denote the radius of the circumscribing circle of a five-pointed star appearing on the American flag. The star can be decomposed into ten congruent triangles. In the figure, r is the radius of the circumscribing circle of the pentagon in the interior of the star. Show that the area of a star is

$$\mathcal{A} = \left[5\frac{\sin A \sin B}{\sin(A+B)}\right]R^2. \quad (\textit{Hint: } \sin C = \sin[180^\circ - (A+B)] = \sin(A+B).)$$

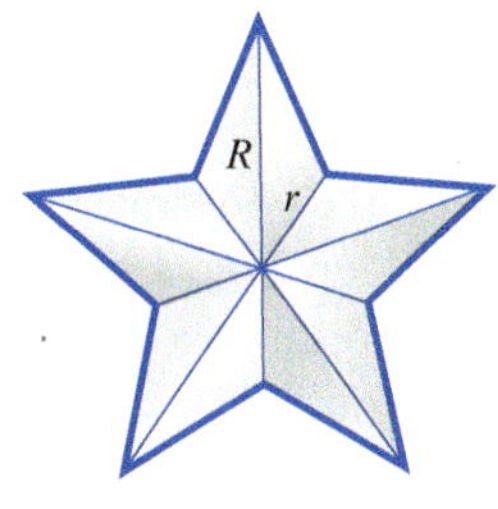

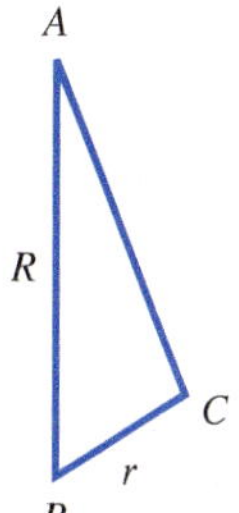

90. Angles A and B have values 18° and 36°, respectively. Express the area $\mathcal{A}$ of a star in terms of its radius, R.

91. To determine whether red or white is predominant, we must know the measurements of the flag. Consider a flag of width 10 in., length 19 in., length of each upper stripe 11.4 in., and radius R of the circumscribing circle of each star 0.308 in. The thirteen stripes consist of six matching pairs of red and white stripes and one additional red, upper stripe. Therefore, we must compare the area of a red, upper stripe with the total area of the 50 white stars.

(a) Compute the area of the red, upper stripe.

(b) Compute the total area of the 50 white stars.

92. Which color occupies the greatest area on the flag?

8.2 The Law of Cosines

- Derivation of the Law of Cosines
- Using the Law of Cosines
- Heron's Formula for the Area of a Triangle
- Derivation of Heron's Formula

If we are given two sides and the included angle (Case 3) or three sides (Case 4) of a triangle, then a unique triangle is determined. These are the SAS and SSS cases, respectively. Both require using the *law of cosines* to solve the triangle.

The following property is important when applying the law of cosines.

Triangle Side Length Restriction

In any triangle, the sum of the lengths of any two sides must be greater than the length of the remaining side.

As an example of this property, it would be impossible to construct a triangle with sides of lengths 3, 4, and 10. See **Figure 12.**

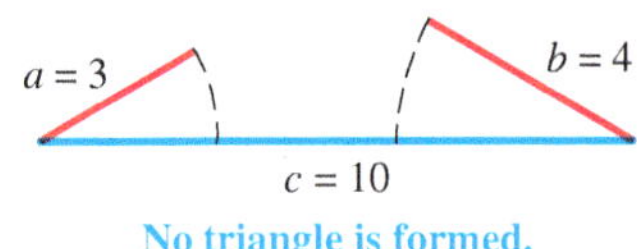

Figure 12

Derivation of the Law of Cosines To derive the law of cosines, let ABC be any oblique triangle. Choose a coordinate system so that vertex B is at the origin and side BC is along the positive x-axis. See **Figure 13.**

Let (x, y) be the coordinates of vertex A of the triangle. Then the following are true for angle B, whether obtuse or acute.

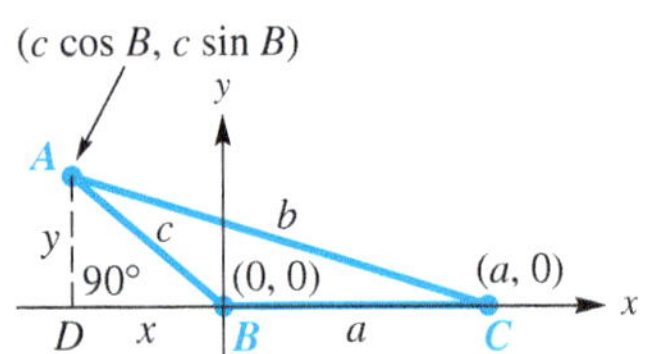

Figure 13

$$\sin B = \frac{y}{c} \quad \text{and} \quad \cos B = \frac{x}{c}$$ Definition of sine and cosine

$$y = c \sin B \quad \text{and} \quad x = c \cos B$$ Here x is negative when B is obtuse.

Thus, the coordinates of point A become $(c \cos B, c \sin B)$.

Point C in **Figure 13** has coordinates $(a, 0)$, AC has length b, and point A has coordinates $(c \cos B, c \sin B)$. We can use the distance formula to write an equation.

$$b = \sqrt{(c \cos B - a)^2 + (c \sin B - 0)^2}$$ $d = \sqrt{(x_2 - x_1)^2 + (y_2 - y_1)^2}$

$$b^2 = (c \cos B - a)^2 + (c \sin B)^2$$ Square each side.

$$b^2 = (c^2 \cos^2 B - 2ac \cos B + a^2) + c^2 \sin^2 B$$ Multiply; $(x - y)^2 = x^2 - 2xy + y^2$

$$b^2 = a^2 + c^2(\cos^2 B + \sin^2 B) - 2ac \cos B$$ Properties of real numbers

$$b^2 = a^2 + c^2(1) - 2ac \cos B$$ Fundamental identity

$$b^2 = a^2 + c^2 - 2ac \cos B$$ Law of cosines

This result is one of three possible forms of the law of cosines. In our work, we could just as easily have placed vertex A or C at the origin. This would have given the same result, but with the variables rearranged.

Law of Cosines

In any triangle ABC, with sides a, b, and c, the following hold.

$$a^2 = b^2 + c^2 - 2bc \cos A$$
$$b^2 = a^2 + c^2 - 2ac \cos B$$
$$c^2 = a^2 + b^2 - 2ab \cos C$$

That is, according to the law of cosines, the square of a side of a triangle is equal to the sum of the squares of the other two sides, minus twice the product of those two sides and the cosine of the angle included between them.

NOTE If we let $C = 90°$ in the third form of the law of cosines, then $\cos C = \cos 90° = 0$, and the formula becomes

$$c^2 = a^2 + b^2. \quad \text{Pythagorean theorem}$$

The Pythagorean theorem is a special case of the law of cosines.

Using the Law of Cosines

EXAMPLE 1 Applying the Law of Cosines (SAS)

A surveyor wishes to find the distance between two inaccessible points A and B on opposite sides of a lake. While standing at point C, she finds that $b = 259$ m, $a = 423$ m, and angle ACB measures $132°\ 40'$. Find the distance c. See **Figure 14.**

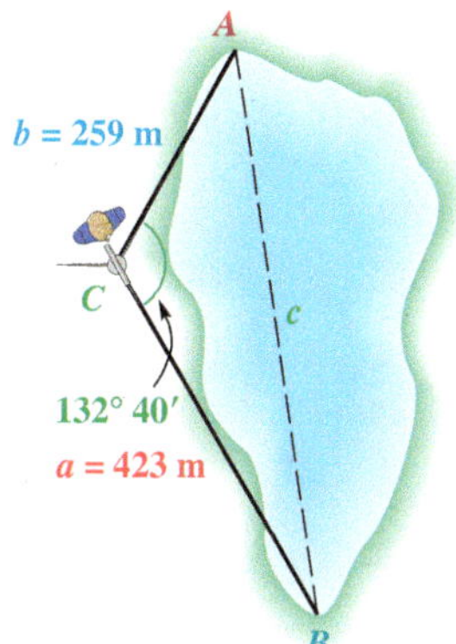

Figure 14

SOLUTION We can use the law of cosines here because we know the lengths of two sides of the triangle and the measure of the included angle.

$$c^2 = a^2 + b^2 - 2ab \cos C \qquad \text{Law of cosines}$$

$$c^2 = 423^2 + 259^2 - 2(423)(259) \cos 132°\ 40' \qquad \text{Substitute.}$$

$$c^2 \approx 394{,}510.6 \qquad \text{Use a calculator.}$$

$$c \approx 628 \qquad \text{Take the square root of each side. Choose the positive root.}$$

The distance between the points is approximately 628 m. ✔ **Now Try Exercise 39.**

EXAMPLE 2 Applying the Law of Cosines (SAS)

Solve triangle ABC if $A = 42.3°$, $b = 12.9$ m, and $c = 15.4$ m.

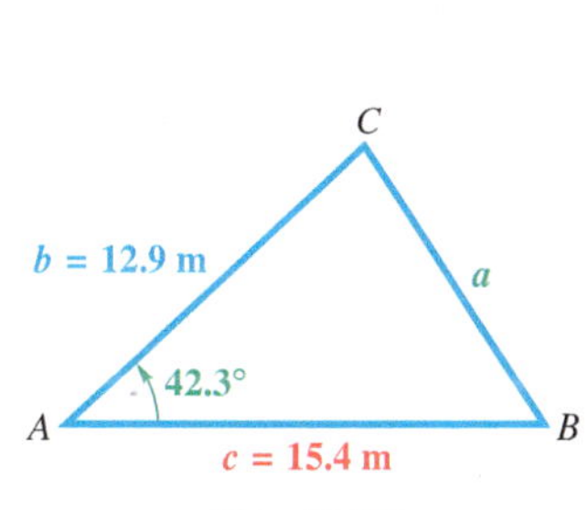

Figure 15

SOLUTION See **Figure 15.** We start by finding side a with the law of cosines.

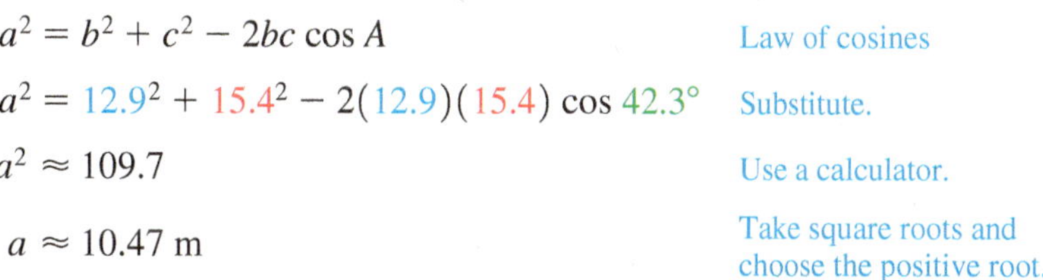

$$a^2 = b^2 + c^2 - 2bc \cos A \qquad \text{Law of cosines}$$

$$a^2 = 12.9^2 + 15.4^2 - 2(12.9)(15.4) \cos 42.3° \qquad \text{Substitute.}$$

$$a^2 \approx 109.7 \qquad \text{Use a calculator.}$$

$$a \approx 10.47 \text{ m} \qquad \text{Take square roots and choose the positive root.}$$

Of the two remaining angles B and C, B must be the smaller because it is opposite the shorter of the two sides b and c. Therefore, B cannot be obtuse.

$$\frac{\sin A}{a} = \frac{\sin B}{b} \qquad \text{Law of sines (alternative form)}$$

$$\frac{\sin 42.3°}{10.47} = \frac{\sin B}{12.9} \qquad \text{Substitute.}$$

$$\sin B = \frac{12.9 \sin 42.3°}{10.47} \qquad \text{Multiply by 12.9 and rewrite.}$$

$$B \approx 56.0° \qquad \text{Use the inverse sine function.}$$

The easiest way to find C is to subtract the measures of A and B from 180°.

$$C = 180° - A - B \qquad \text{Angle sum formula, solved for } C$$

$$C \approx 180° - 42.3° - 56.0° \qquad \text{Substitute.}$$

$$C \approx 81.7° \qquad \text{Subtract.}$$

✔ **Now Try Exercise 19.**

CAUTION Had we used the law of sines to find C rather than B in **Example 2,** we would not have known whether C was equal to $81.7°$ or to its supplement, $98.3°$.

EXAMPLE 3 Applying the Law of Cosines (SSS)

Solve triangle ABC if $a = 9.47$ ft, $b = 15.9$ ft, and $c = 21.1$ ft.

SOLUTION We can use the law of cosines to solve for any angle of the triangle. We solve for C, the largest angle. We will know that C is obtuse if $\cos C < 0$.

$$c^2 = a^2 + b^2 - 2ab\cos C \quad \text{Law of cosines}$$

$$\cos C = \frac{a^2 + b^2 - c^2}{2ab} \quad \text{Solve for } \cos C.$$

$$\cos C = \frac{9.47^2 + 15.9^2 - 21.1^2}{2(9.47)(15.9)} \quad \text{Substitute.}$$

$$\cos C \approx -0.34109402 \quad \text{Use a calculator.}$$

$$C \approx 109.9° \quad \text{Use the inverse cosine function.}$$

Now use the law of sines to find angle B.

$$\frac{\sin B}{b} = \frac{\sin C}{c} \quad \text{Law of sines (alternative form)}$$

$$\frac{\sin B}{15.9} = \frac{\sin 109.9°}{21.1} \quad \text{Substitute.}$$

$$\sin B = \frac{15.9 \sin 109.9°}{21.1} \quad \text{Multiply by 15.9.}$$

$$B \approx 45.1° \quad \text{Use the inverse sine function.}$$

Since $A = 180° - B - C$, we have $A \approx 180° - 45.1° - 109.9° \approx 25.0°$.

✔ **Now Try Exercise 23.**

Figure 16

Trusses are frequently used to support roofs on buildings, as illustrated in **Figure 16.** The simplest type of roof truss is a triangle, as shown in **Figure 17.** (*Source:* Riley, W., L. Sturges, and D. Morris, *Statics and Mechanics of Materials,* John Wiley and Sons.)

EXAMPLE 4 Designing a Roof Truss (SSS)

Find angle B to the nearest degree for the truss shown in **Figure 17.**

SOLUTION

$$b^2 = a^2 + c^2 - 2ac\cos B \quad \text{Law of cosines}$$

$$\cos B = \frac{a^2 + c^2 - b^2}{2ac} \quad \text{Solve for } \cos B.$$

$$\cos B = \frac{11^2 + 9^2 - 6^2}{2(11)(9)} \quad \text{Let } a = 11,\ b = 6, \text{ and } c = 9.$$

$$\cos B \approx 0.83838384 \quad \text{Use a calculator.}$$

$$B \approx 33° \quad \text{Use the inverse cosine function.}$$

A
9 ft
6 ft
B
C
11 ft

Figure 17

✔ **Now Try Exercise 49.**

Four possible cases can occur when we solve an oblique triangle. They are summarized in the following table. In all four cases, it is assumed that the given information actually produces a triangle.

Four Cases for Solving Oblique Triangles

Oblique Triangle	Suggested Procedure for Solving
Case 1: One side and two angles are known. **(SAA or ASA)**	*Step 1* Find the remaining angle using the angle sum formula $(A + B + C = 180°)$. *Step 2* Find the remaining sides using the law of sines.
Case 2: Two sides and one angle (not included between the two sides) are known. **(SSA)**	*This is the ambiguous case. There may be no triangle, one triangle, or two triangles.* *Step 1* Find an angle using the law of sines. *Step 2* Find the remaining angle using the angle sum formula. *Step 3* Find the remaining side using the law of sines. *If two triangles exist, repeat Steps 2 and 3.*
Case 3: Two sides and the included angle are known. **(SAS)**	*Step 1* Find the third side using the law of cosines. *Step 2* Find the smaller of the two remaining angles using the law of sines. *Step 3* Find the remaining angle using the angle sum formula.
Case 4: Three sides are known. **(SSS)**	*Step 1* Find the largest angle using the law of cosines. *Step 2* Find either remaining angle using the law of sines. *Step 3* Find the remaining angle using the angle sum formula.

Heron's Formula for the Area of a Triangle A formula for finding the area of a triangle given the lengths of the three sides, known as **Heron's formula,** is named after the Greek mathematician Heron of Alexandria. It is found in his work *Metrica*. Heron's formula can be used for the case SSS.

Heron of Alexandria (c. 62 CE)

Heron (also called Hero), a Greek geometer and inventor, produced writings that contain knowledge of the mathematics and engineering of Babylonia, ancient Egypt, and the Greco-Roman world.

Heron's Area Formula (SSS)

If a triangle has sides of lengths a, b, and c, with **semiperimeter**

$$s = \frac{1}{2}(a + b + c),$$

then the area $\mathcal{A}$ of the triangle is given by the following formula.

$$\mathcal{A} = \sqrt{s(s - a)(s - b)(s - c)}$$

That is, according to Heron's formula, the area of a triangle is the square root of the product of four factors: (1) the semiperimeter, (2) the semiperimeter minus the first side, (3) the semiperimeter minus the second side, and (4) the semiperimeter minus the third side.

EXAMPLE 5 Using Heron's Formula to Find an Area (SSS)

The distance "as the crow flies" from Los Angeles to New York is 2451 mi, from New York to Montreal is 331 mi, and from Montreal to Los Angeles is 2427 mi. What is the area of the triangular region having these three cities as vertices? (Ignore the curvature of Earth.)

SOLUTION In **Figure 18,** we let $a = 2451$, $b = 331$, and $c = 2427$.

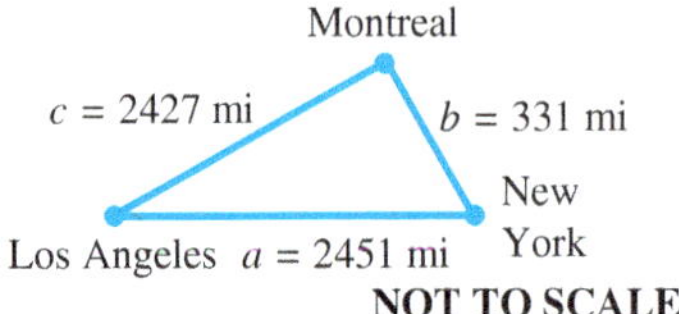

Figure 18

First, find the semiperimeter s.

$$s = \frac{1}{2}(a + b + c) \quad \text{Semiperimeter}$$

$$s = \frac{1}{2}(2451 + 331 + 2427) \quad \text{Substitute the given values.}$$

$$s = 2604.5 \quad \text{Add, and then multiply.}$$

Now use Heron's formula to find the area $\mathcal{A}$.

$$\mathcal{A} = \sqrt{s(s-a)(s-b)(s-c)}$$

Don't forget the factor s.

$$\mathcal{A} = \sqrt{2604.5(2604.5 - 2451)(2604.5 - 331)(2604.5 - 2427)}$$

$$\mathcal{A} \approx 401{,}700 \text{ mi}^2 \quad \text{Use a calculator.}$$

✔ **Now Try Exercise 73.**

Derivation of Heron's Formula A trigonometric derivation of Heron's formula illustrates some ingenious manipulation.

Let triangle ABC have sides of lengths a, b, and c. Apply the law of cosines.

$$a^2 = b^2 + c^2 - 2bc \cos A \quad \text{Law of cosines}$$

$$\cos A = \frac{b^2 + c^2 - a^2}{2bc} \quad \text{Solve for } \cos A. \quad (1)$$

The perimeter of the triangle is $a + b + c$, so half of the perimeter (the semiperimeter) is given by the formula in equation (2) below.

$$s = \frac{1}{2}(a + b + c) \quad (2)$$

$$2s = a + b + c \quad \text{Multiply by 2.} \quad (3)$$

$$b + c - a = 2s - 2a \quad \text{Subtract } 2a \text{ from each side and rewrite.}$$

$$b + c - a = 2(s - a) \quad \text{Factor.} \quad (4)$$

Subtract $2b$ and $2c$ in a similar way in equation (3) to obtain the following.

$$a - b + c = 2(s - b) \quad (5)$$

$$a + b - c = 2(s - c) \quad (6)$$

Now we obtain an expression for $1 - \cos A$.

$$1 - \cos A = 1 - \underbrace{\frac{b^2 + c^2 - a^2}{2bc}}_{\cos A, \text{ from } (1)}$$

$$= \frac{2bc + a^2 - b^2 - c^2}{2bc}$$ Find a common denominator, and distribute the − sign.

$$= \frac{a^2 - (b^2 - 2bc + c^2)}{2bc}$$ Regroup.

Pay attention to signs.

$$= \frac{a^2 - (b - c)^2}{2bc}$$ Factor the perfect square trinomial.

$$= \frac{[a - (b - c)][a + (b - c)]}{2bc}$$ Factor the difference of squares.

$$= \frac{(a - b + c)(a + b - c)}{2bc}$$ Distributive property

$$= \frac{2(s - b) \cdot 2(s - c)}{2bc}$$ Use equations (5) and (6).

$$1 - \cos A = \frac{2(s - b)(s - c)}{bc}$$ Lowest terms (7)

Similarly, it can be shown that

$$1 + \cos A = \frac{2s(s - a)}{bc}. \quad (8)$$

Recall the double-angle identities for $\cos 2\theta$.

$$\cos 2\theta = 2\cos^2\theta - 1$$

$$\cos A = 2\cos^2\left(\frac{A}{2}\right) - 1$$ Let $\theta = \frac{A}{2}$.

$$1 + \cos A = 2\cos^2\left(\frac{A}{2}\right)$$ Add 1.

$$\underbrace{\frac{2s(s - a)}{bc}}_{\text{From } (8)} = 2\cos^2\left(\frac{A}{2}\right)$$ Substitute.

$$\frac{s(s - a)}{bc} = \cos^2\left(\frac{A}{2}\right)$$ Divide by 2.

$$\cos\left(\frac{A}{2}\right) = \sqrt{\frac{s(s - a)}{bc}} \quad (9)$$

$$\cos 2\theta = 1 - 2\sin^2\theta$$

$$\cos A = 1 - 2\sin^2\left(\frac{A}{2}\right)$$ Let $\theta = \frac{A}{2}$.

$$1 - \cos A = 2\sin^2\left(\frac{A}{2}\right)$$ Subtract 1. Multiply by −1.

$$\underbrace{\frac{2(s - b)(s - c)}{bc}}_{\text{From } (7)} = 2\sin^2\left(\frac{A}{2}\right)$$ Substitute.

$$\frac{(s - b)(s - c)}{bc} = \sin^2\left(\frac{A}{2}\right)$$ Divide by 2.

$$\sin\left(\frac{A}{2}\right) = \sqrt{\frac{(s - b)(s - c)}{bc}} \quad (10)$$

The area of triangle ABC can be expressed as follows.

$$\mathscr{A} = \frac{1}{2}bc \sin A$$ Area formula

$$2\mathscr{A} = bc \sin A$$ Multiply by 2.

$$\frac{2\mathscr{A}}{bc} = \sin A$$ Divide by bc. (11)

Recall the double-angle identity for sin 2θ.

$$\sin 2\theta = 2 \sin \theta \cos \theta$$

$$\sin A = 2 \sin\left(\frac{A}{2}\right)\cos\left(\frac{A}{2}\right) \qquad \text{Let } \theta = \tfrac{A}{2}.$$

$$\frac{2\mathcal{A}}{bc} = 2 \sin\left(\frac{A}{2}\right)\cos\left(\frac{A}{2}\right) \qquad \text{Use equation (11).}$$

$$\frac{2\mathcal{A}}{bc} = 2\sqrt{\frac{(s-b)(s-c)}{bc}} \cdot \sqrt{\frac{s(s-a)}{bc}} \qquad \text{Use equations (9) and (10).}$$

$$\frac{2\mathcal{A}}{bc} = 2\sqrt{\frac{s(s-a)(s-b)(s-c)}{b^2c^2}} \qquad \text{Multiply.}$$

$$\frac{2\mathcal{A}}{bc} = \frac{2\sqrt{s(s-a)(s-b)(s-c)}}{bc} \qquad \text{Simplify the denominator.}$$

Heron's formula results. $\longrightarrow$

$$\mathcal{A} = \sqrt{s(s-a)(s-b)(s-c)} \qquad \text{Multiply by } bc. \text{ Divide by 2.}$$

8.2 Exercises

CONCEPT PREVIEW *Assume a triangle ABC has standard labeling.*

(a) *Determine whether* SAA, ASA, SSA, SAS, *or* SSS *is given.*

(b) *Decide whether the law of sines or the law of cosines should be used to begin solving the triangle.*

1. a, b, and C **2.** A, C, and c **3.** a, b, and A **4.** a, B, and C

5. A, B, and c **6.** a, c, and A **7.** a, b, and c **8.** b, c, and A

Find the length of the remaining side of each triangle. Do not use a calculator.

9.

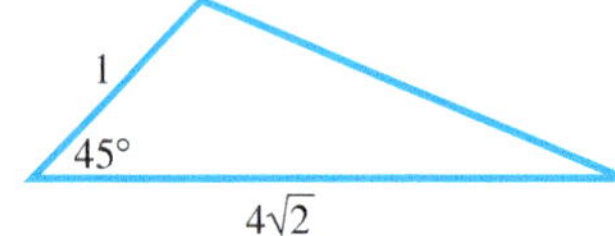

10.

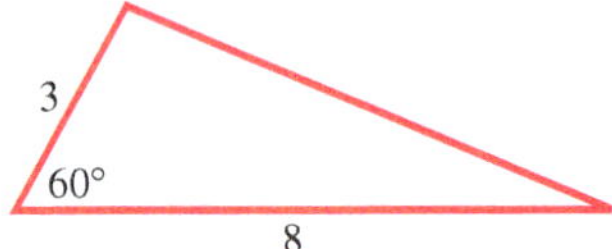

Find the measure of θ in each triangle. Do not use a calculator.

11.

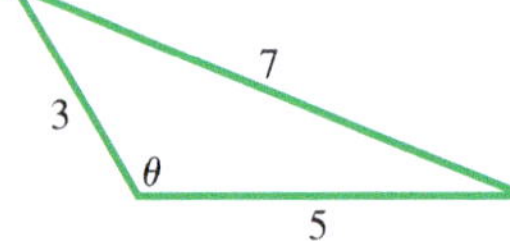

12.

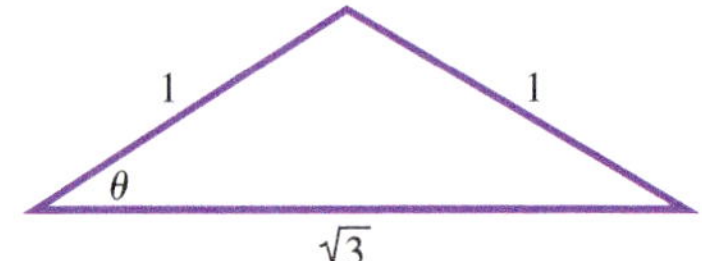

Solve each triangle. Approximate values to the nearest tenth.

13.

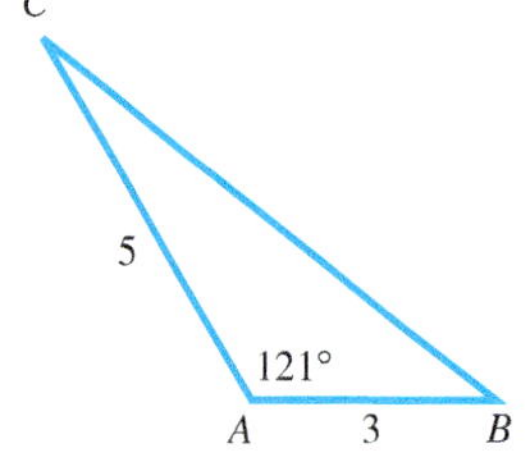

14.

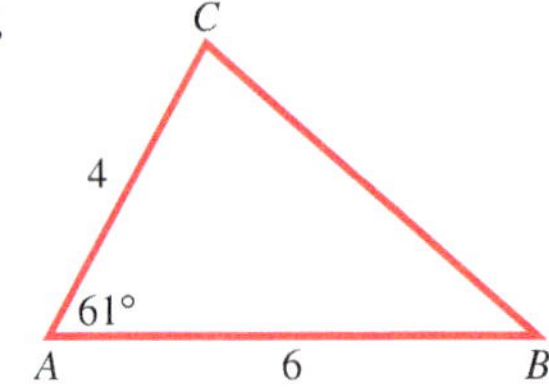

15.

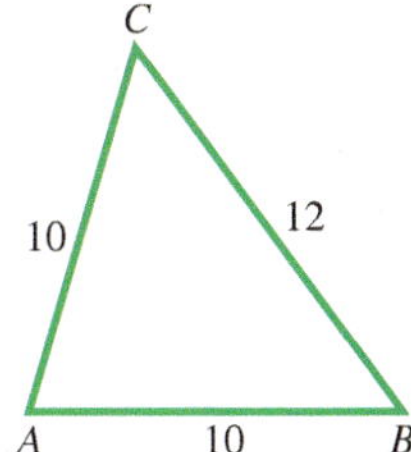

16.

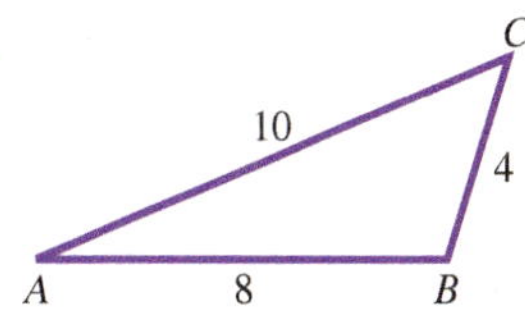

17.

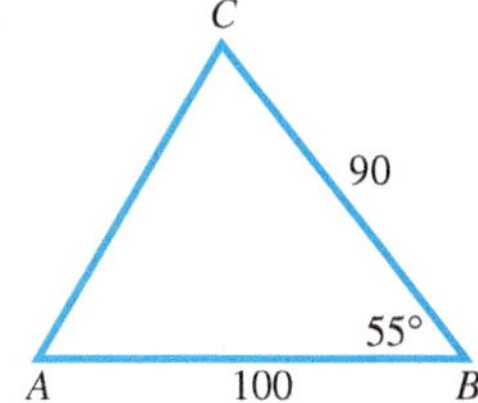

18.

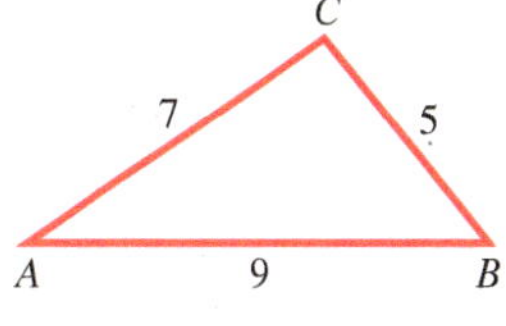

Solve each triangle. ***See Examples 2 and 3.***

19. $A = 41.4°$, $b = 2.78$ yd, $c = 3.92$ yd

20. $C = 28.3°$, $b = 5.71$ in., $a = 4.21$ in.

21. $C = 45.6°$, $b = 8.94$ m, $a = 7.23$ m

22. $A = 67.3°$, $b = 37.9$ km, $c = 40.8$ km

23. $a = 9.3$ cm, $b = 5.7$ cm, $c = 8.2$ cm

24. $a = 28$ ft, $b = 47$ ft, $c = 58$ ft

25. $a = 42.9$ m, $b = 37.6$ m, $c = 62.7$ m

26. $a = 189$ yd, $b = 214$ yd, $c = 325$ yd

27. $a = 965$ ft, $b = 876$ ft, $c = 1240$ ft

28. $a = 324$ m, $b = 421$ m, $c = 298$ m

29. $A = 80° 40'$, $b = 143$ cm, $c = 89.6$ cm

30. $C = 72° 40'$, $a = 327$ ft, $b = 251$ ft

31. $B = 74.8°$, $a = 8.92$ in., $c = 6.43$ in.

32. $C = 59.7°$, $a = 3.73$ mi, $b = 4.70$ mi

33. $A = 112.8°$, $b = 6.28$ m, $c = 12.2$ m

34. $B = 168.2°$, $a = 15.1$ cm, $c = 19.2$ cm

35. $a = 3.0$ ft, $b = 5.0$ ft, $c = 6.0$ ft

36. $a = 4.0$ ft, $b = 5.0$ ft, $c = 8.0$ ft

Concept Check *Answer each question.*

37. Refer to **Figure 12.** If we attempt to find any angle of a triangle with the values $a = 3$, $b = 4$, and $c = 10$ using the law of cosines, what happens?

38. "The shortest distance between two points is a straight line." How is this statement related to the geometric property that states that the sum of the lengths of any two sides of a triangle must be greater than the length of the remaining side?

Solve each problem. ***See Examples 1–4.***

39. ***Distance across a River*** Points A and B are on opposite sides of False River. From a third point, C, the angle between the lines of sight to A and B is 46.3°. If AC is 350 m long and BC is 286 m long, find AB.

40. *Distance across a Ravine* Points X and Y are on opposite sides of a ravine. From a third point Z, the angle between the lines of sight to X and Y is 37.7°. If XZ is 153 m long and YZ is 103 m long, find XY.

41. *Angle in a Parallelogram* A parallelogram has sides of length 25.9 cm and 32.5 cm. The longer diagonal has length 57.8 cm. Find the measure of the angle opposite the longer diagonal.

42. *Diagonals of a Parallelogram* The sides of a parallelogram are 4.0 cm and 6.0 cm. One angle is 58°, while another is 122°. Find the lengths of the diagonals of the parallelogram.

43. *Flight Distance* Airports A and B are 450 km apart, on an east-west line. Tom flies in a northeast direction from airport A to airport C. From C he flies 359 km on a bearing of 128° 40′ to B. How far is C from A?

44. *Distance Traveled by a Plane* An airplane flies 180 mi from point X at a bearing of 125°, and then turns and flies at a bearing of 230° for 100 mi. How far is the plane from point X?

45. *Distance between Ends of the Vietnam Memorial* The Vietnam Veterans Memorial in Washington, D.C., is V-shaped with equal sides of length 246.75 ft. The angle between these sides measures 125° 12′. Find the distance between the ends of the two sides. (*Source:* Pamphlet obtained at Vietnam Veterans Memorial.)

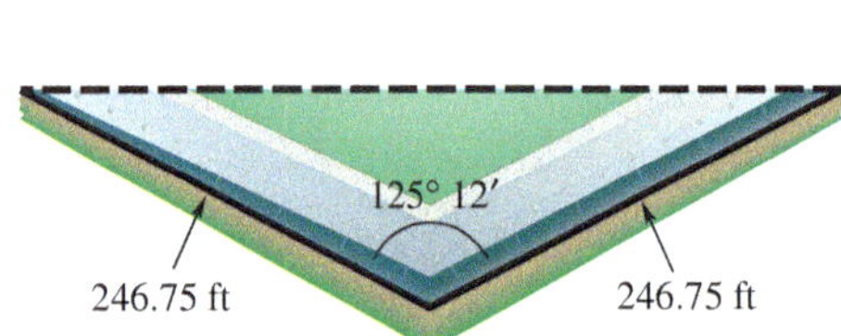

46. *Distance between Two Ships* Two ships leave a harbor together, traveling on courses that have an angle of 135° 40′ between them. If each travels 402 mi, how far apart are they?

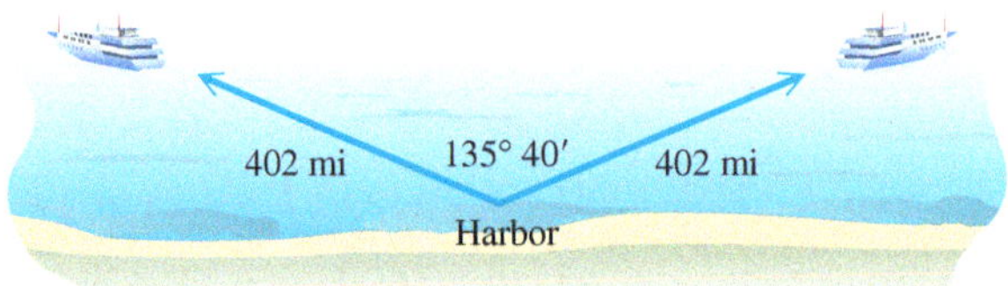

47. *Distance between a Ship and a Rock* A ship is sailing east. At one point, the bearing of a submerged rock is 45° 20′. After the ship has sailed 15.2 mi, the bearing of the rock has become 308° 40′. Find the distance of the ship from the rock at the latter point.

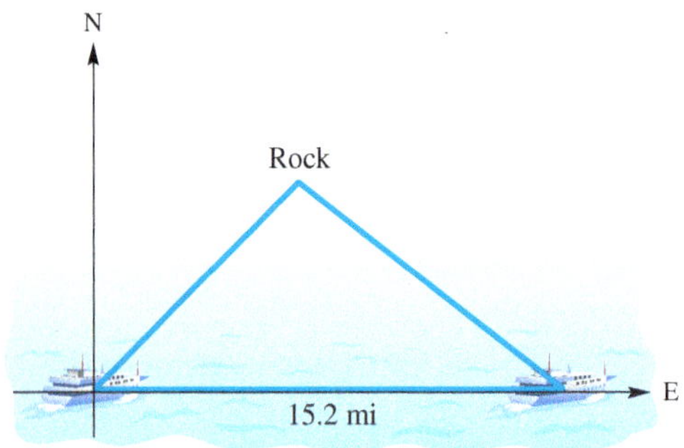

48. ***Distance between a Ship and a Submarine*** From an airplane flying over the ocean, the angle of depression to a submarine lying under the surface is 24° 10′. At the same moment, the angle of depression from the airplane to a battleship is 17° 30′. See the figure. The distance from the airplane to the battleship is 5120 ft. Find the distance between the battleship and the submarine. (Assume the airplane, submarine, and battleship are in a vertical plane.)

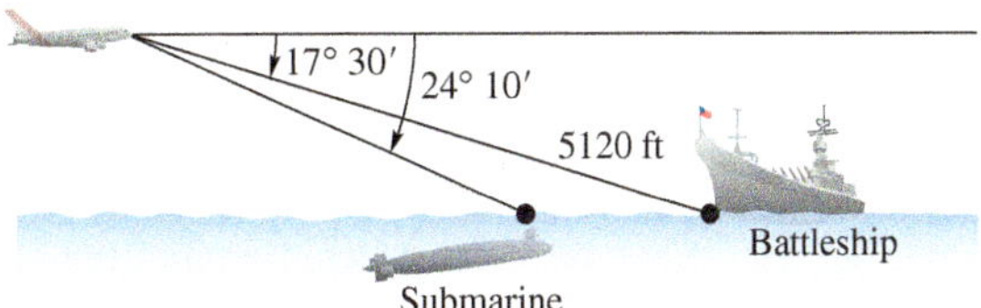

49. ***Truss Construction*** A triangular truss is shown in the figure. Find angle θ.

50. ***Truss Construction*** Find angle β in the truss shown in the figure.

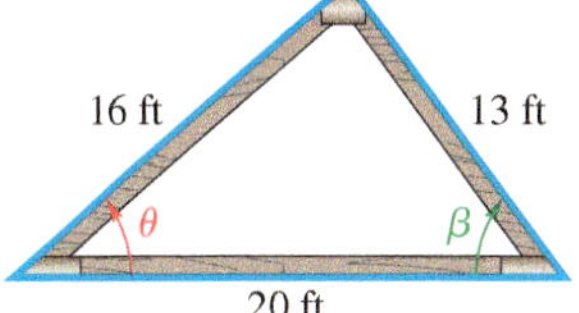

51. ***Distance between a Beam and Cables*** A weight is supported by cables attached to both ends of a balance beam, as shown in the figure. What angles are formed between the beam and the cables?

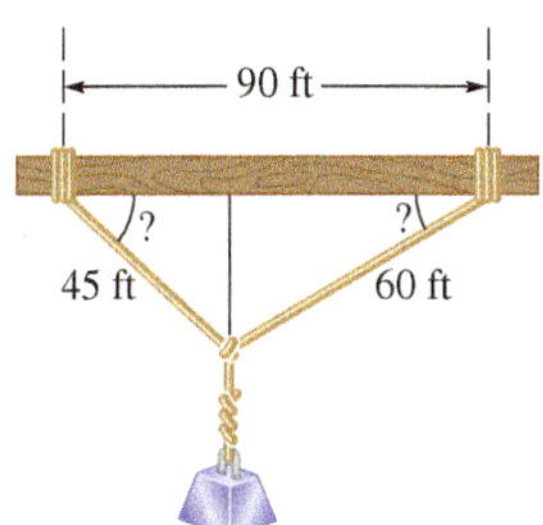

52. ***Distance between Points on a Crane*** A crane with a counterweight is shown in the figure. Find the horizontal distance between points A and B to the nearest foot.

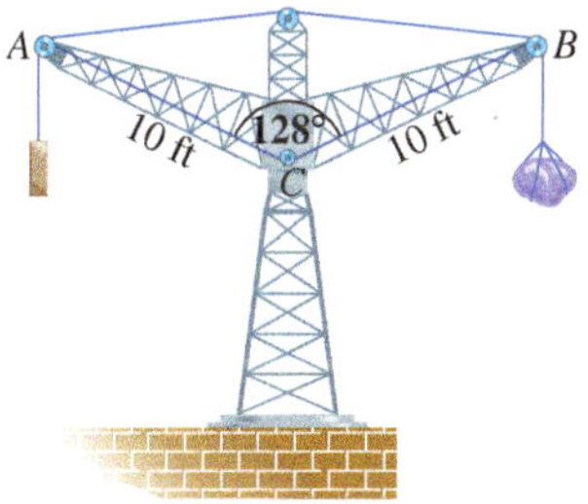

53. ***Distance on a Baseball Diamond*** A baseball diamond is a square, 90.0 ft on a side, with home plate and the three bases as vertices. The pitcher's position is 60.5 ft from home plate. Find the distance from the pitcher's position to each of the bases.

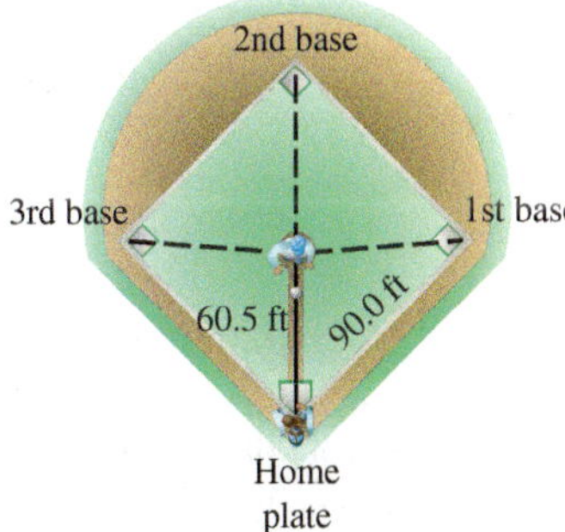

54. ***Distance on a Softball Diamond*** A softball diamond is a square, 60.0 ft on a side, with home plate and the three bases as vertices. The pitcher's position is 46.0 ft from home plate. Find the distance from the pitcher's position to each of the bases.

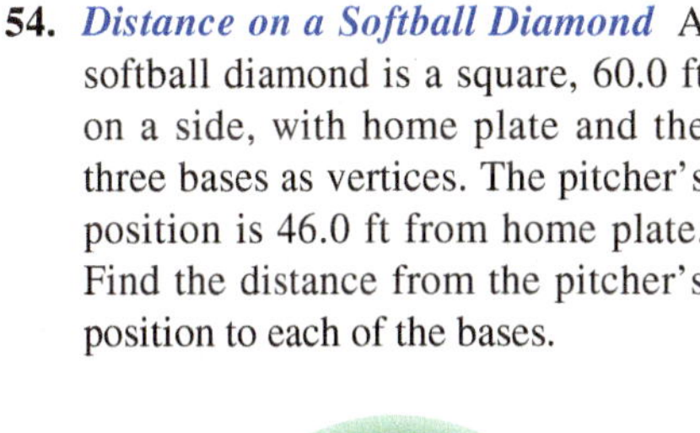

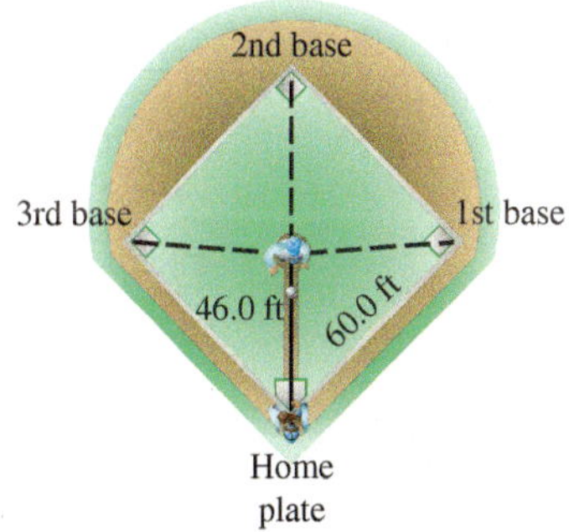

55. *Distance between a Ship and a Point* Starting at point A, a ship sails 18.5 km on a bearing of 189°, then turns and sails 47.8 km on a bearing of 317°. Find the distance of the ship from point A.

56. *Distance between Two Factories* Two factories blow their whistles at exactly 5:00. A man hears the two blasts at 3 sec and 6 sec after 5:00, respectively. The angle between his lines of sight to the two factories is 42.2°. If sound travels 344 m per sec, how far apart are the factories?

57. *Measurement Using Triangulation* Surveyors are often confronted with obstacles, such as trees, when measuring the boundary of a lot. One technique used to obtain an accurate measurement is the **triangulation method.** In this technique, a triangle is constructed around the obstacle and one angle and two sides of the triangle are measured. Use this technique to find the length of the property line (the straight line between the two markers) in the figure. (*Source:* Kavanagh, B., *Surveying Principles and Applications,* Sixth Edition, Prentice-Hall.)

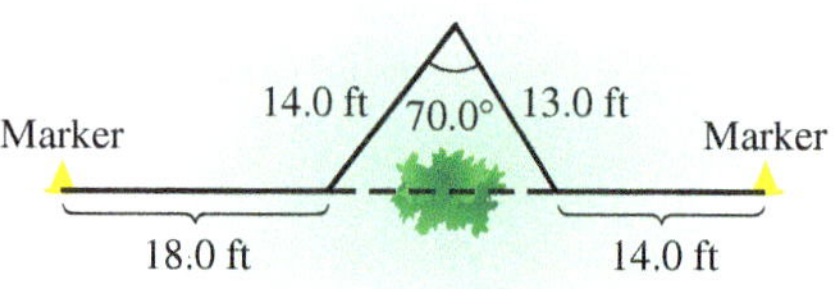

NOT TO SCALE

58. *Path of a Ship* A ship sailing due east in the North Atlantic has been warned to change course to avoid icebergs. The captain turns and sails on a bearing of 62°, then changes course again to a bearing of 115° until the ship reaches its original course. See the figure. How much farther did the ship have to travel to avoid the icebergs?

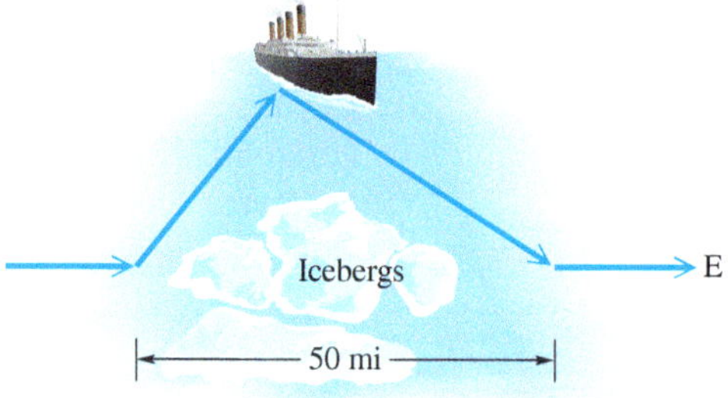

59. *Length of a Tunnel* To measure the distance through a mountain for a proposed tunnel, a point C is chosen that can be reached from each end of the tunnel. See the figure. If $AC = 3800$ m, $BC = 2900$ m, and angle $C = 110°$, find the length of the tunnel.

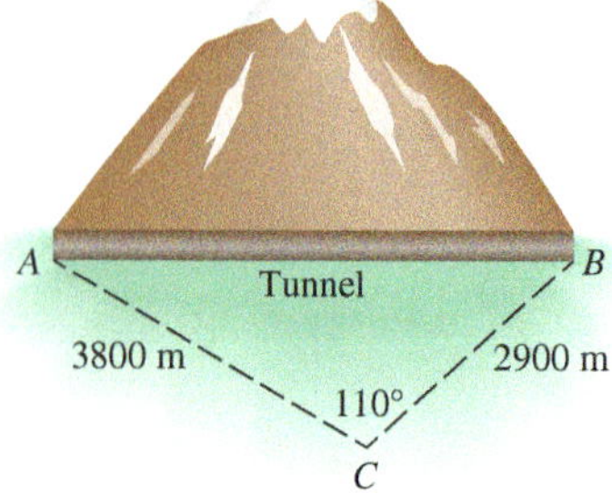

60. *Distance between an Airplane and a Mountain* A person in a plane flying straight north observes a mountain at a bearing of 24.1°. At that time, the plane is 7.92 km from the mountain. A short time later, the bearing to the mountain becomes 32.7°. How far is the airplane from the mountain when the second bearing is taken?

Find the measure of each angle θ to two decimal places.

61.

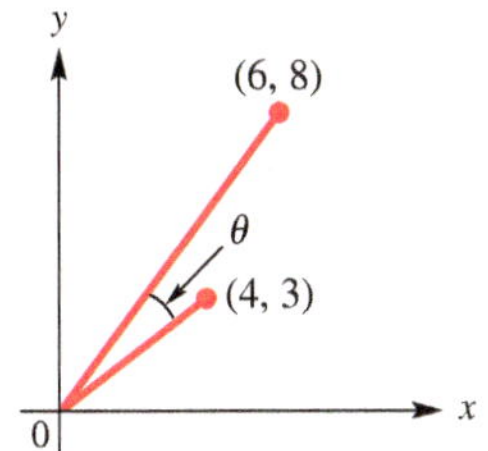

62.

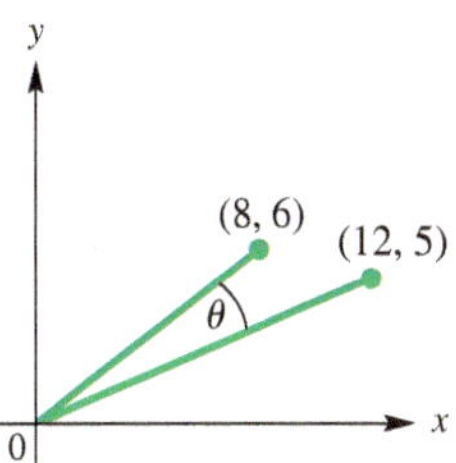

Find the exact area of each triangle using the formula $\mathcal{A} = \frac{1}{2}bh$, *and then verify that Heron's formula gives the same result.*

63.

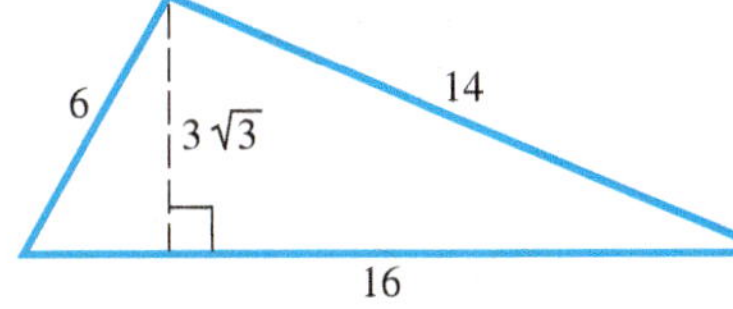

64.

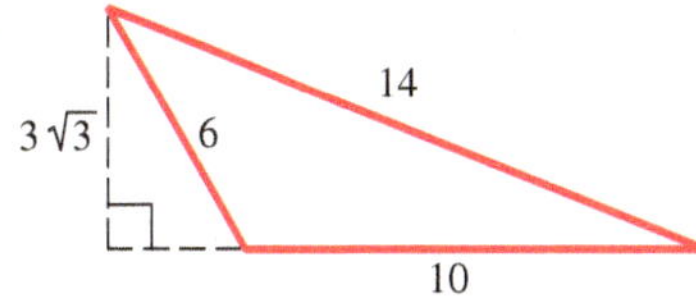

Find the area of each triangle ABC. ***See Example 5.***

65. $a = 12$ m, $b = 16$ m, $c = 25$ m

66. $a = 22$ in., $b = 45$ in., $c = 31$ in.

67. $a = 154$ cm, $b = 179$ cm, $c = 183$ cm

68. $a = 25.4$ yd, $b = 38.2$ yd, $c = 19.8$ yd

69. $a = 76.3$ ft, $b = 109$ ft, $c = 98.8$ ft

70. $a = 15.8$ m, $b = 21.7$ m, $c = 10.9$ m

Solve each problem. ***See Example 5.***

71. ***Perfect Triangles*** A **perfect triangle** is a triangle whose sides have whole number lengths and whose area is numerically equal to its perimeter. Show that the triangle with sides of length 9, 10, and 17 is perfect.

72. ***Heron Triangles*** A **Heron triangle** is a triangle having integer sides and area. Show that each of the following is a Heron triangle.

(a) $a = 11, b = 13, c = 20$

(b) $a = 13, b = 14, c = 15$

(c) $a = 7, b = 15, c = 20$

(d) $a = 9, b = 10, c = 17$

73. ***Area of the Bermuda Triangle*** Find the area of the Bermuda Triangle if the sides of the triangle have approximate lengths 850 mi, 925 mi, and 1300 mi.

74. ***Required Amount of Paint*** A painter needs to cover a triangular region 75 m by 68 m by 85 m. A can of paint covers 75 m^2 of area. How many cans (to the next higher number of cans) will be needed?

75. Consider triangle *ABC* shown here.

(a) Use the law of sines to find candidates for the value of angle *C*. Round angle measures to the nearest tenth of a degree.

(b) Rework part (a) using the law of cosines.

(c) Why is the law of cosines a better method in this case?

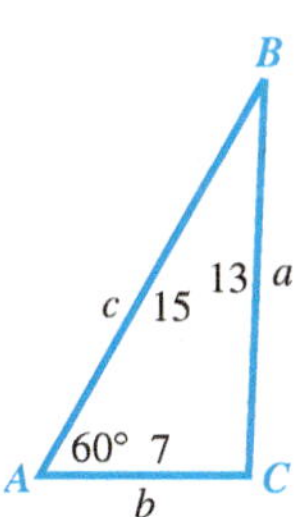

76. Show that the measure of angle *A* is twice the measure of angle *B*. (*Hint:* Use the law of cosines to find cos *A* and cos *B*, and then show that $\cos A = 2\cos^2 B - 1$.)

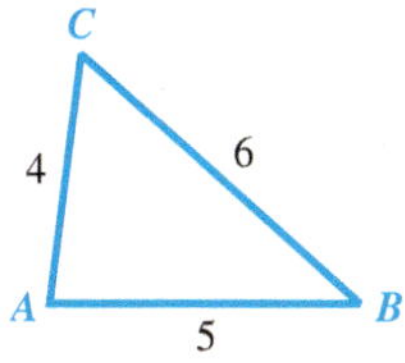

Relating Concepts

For individual or collaborative investigation *(Exercises 77–80)*

We have introduced two new formulas for the area of a triangle in this chapter. We can now find the area $\mathcal{A}$ of a triangle using one of three formulas.

(a) $\mathcal{A} = \frac{1}{2}bh$

(b) $\mathcal{A} = \frac{1}{2}ab \sin C \quad \left(\text{or } \mathcal{A} = \frac{1}{2}ac \sin B \text{ or } \mathcal{A} = \frac{1}{2}bc \sin A\right)$

(c) $\mathcal{A} = \sqrt{s(s-a)(s-b)(s-c)}$ (Heron's formula)

If the coordinates of the vertices of a triangle are given, then the following area formula is also valid.

(d) $\mathcal{A} = \frac{1}{2}\left|(x_1y_2 - y_1x_2 + x_2y_3 - y_2x_3 + x_3y_1 - y_3x_1)\right|$

The vertices are the ordered pairs (x_1, y_1), (x_2, y_2), and (x_3, y_3).

Work Exercises 77–80 in order, *showing that the various formulas all lead to the same area.*

77. Draw a triangle with vertices $A(2, 5)$, $B(-1, 3)$, and $C(4, 0)$, and use the distance formula to find the lengths of the sides a, b, and c.

78. Find the area of triangle ABC using formula (b). (First use the law of cosines to find the measure of an angle.)

79. Find the area of triangle ABC using formula (c)—that is, Heron's formula.

80. Find the area of triangle ABC using new formula (d).

Chapter 8 Quiz (Sections 8.1–8.2)

Find the indicated part of each triangle ABC.

1. Find A if $B = 30.6°$, $b = 7.42$ in., and $c = 4.54$ in.

2. Find a if $A = 144°$, $c = 135$ m, and $b = 75.0$ m.

3. Find C if $a = 28.4$ ft, $b = 16.9$ ft, and $c = 21.2$ ft.

Solve each problem.

4. Find the area of the triangle shown here.

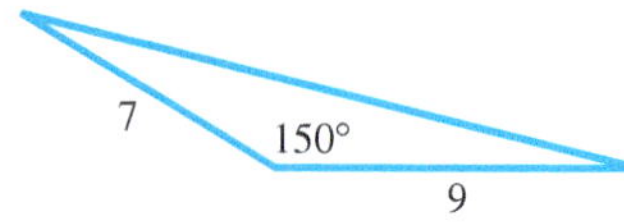

5. Find the area of triangle ABC if $a = 19.5$ km, $b = 21.0$ km, and $c = 22.5$ km.

6. For triangle ABC with $c = 345$, $a = 534$, and $C = 25.4°$, there are two possible values for angle A. What are they?

7. Solve triangle ABC if $c = 326$, $A = 111°$, and $B = 41.0°$.

8. ***Height of a Balloon*** The angles of elevation of a hot air balloon from two observation points X and Y on level ground are $42° \, 10'$ and $23° \, 30'$, respectively. As shown in the figure, points X, Y, and Z are in the same vertical plane and points X and Y are 12.2 mi apart. Approximate the height of the balloon to the nearest tenth of a mile.

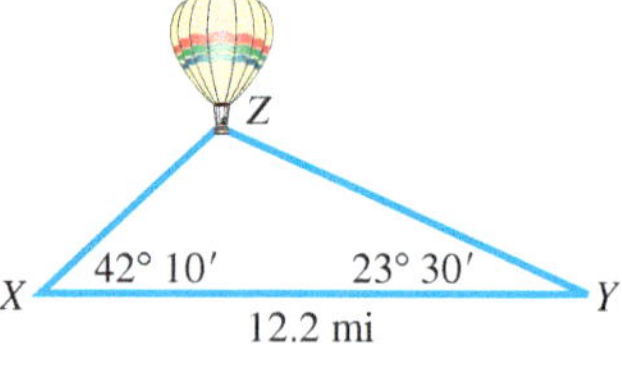

9. *Volcano Movement* To help predict eruptions from the volcano Mauna Loa on the island of Hawaii, scientists keep track of the volcano's movement by using a "super triangle" with vertices on the three volcanoes shown on the map at the right. Find BC given that $AB = 22.47928$ mi, $AC = 28.14276$ mi, and $A = 58.56989°$.

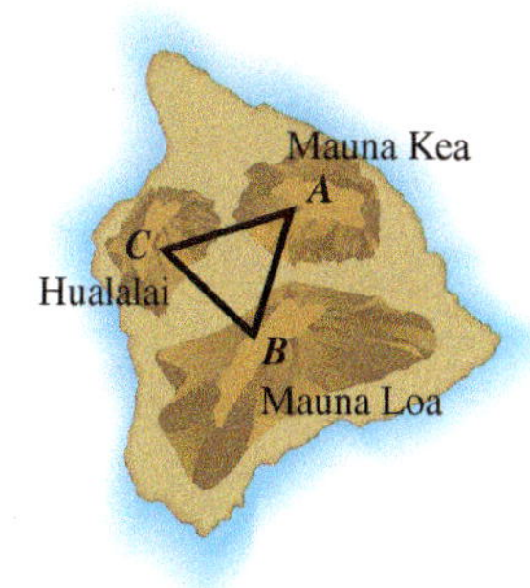

10. *Distance between Two Towns* To find the distance between two small towns, an electronic distance measuring (EDM) instrument is placed on a hill from which both towns are visible. The distance to each town from the EDM and the angle between the two lines of sight are measured. See the figure. Find the distance between the towns.

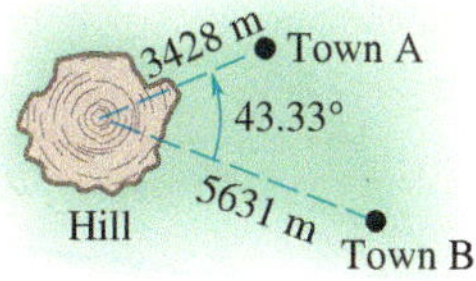

8.3 Geometrically Defined Vectors and Applications

- Basic Terminology
- The Equilibrant
- Incline Applications
- Navigation Applications

Basic Terminology Quantities that involve magnitudes, such as 45 lb or 60 mph, can be represented by real numbers called **scalars.** Other quantities, called **vector quantities,** involve both magnitude *and* direction. Typical vector quantities are velocity, acceleration, and force. For example, traveling 50 mph *east* represents a vector quantity.

A vector quantity can be represented with a directed line segment (a segment that uses an arrowhead to indicate direction) called a **vector.** The *length* of the vector represents the **magnitude** of the vector quantity. The *direction* of the vector, indicated by the arrowhead, represents the direction of the quantity. See **Figure 19.**

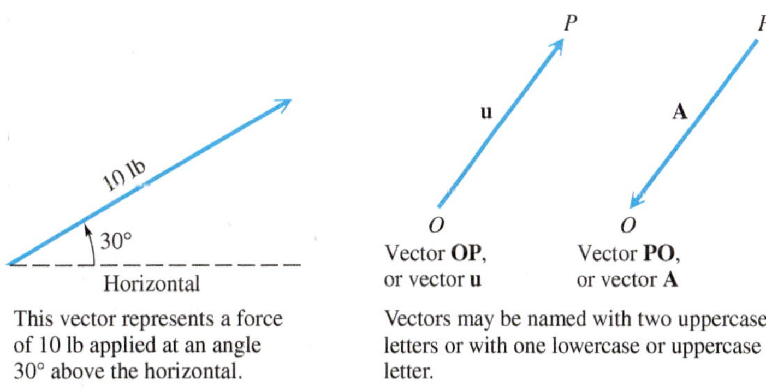

This vector represents a force of 10 lb applied at an angle 30° above the horizontal.

Figure 19

Vectors may be named with two uppercase letters or with one lowercase or uppercase letter.

Figure 20

When we indicate vectors in print, it is customary to use boldface type or an arrow over the letter or letters. Thus, **OP** and $\overrightarrow{OP}$ both represent the vector **OP**. When two letters name a vector, the first indicates the **initial point** and the second indicates the **terminal point** of the vector. Knowing these points gives the direction of the vector. For example, vectors **OP** and **PO** in **Figure 20** are not the same vector. They have the same magnitude but *opposite* directions. The magnitude of vector **OP** is written $|\mathbf{OP}|$.

Two vectors are equal if and only if they have the same direction and the same magnitude. In **Figure 21,** vectors **A** and **B** are equal, as are vectors **C** and **D**. As **Figure 21** shows, equal vectors need not coincide, but they must be parallel and in the same direction. Vectors **A** and **E** are unequal because they do not have the same direction, while $\mathbf{A} \neq \mathbf{F}$ because they have different magnitudes.

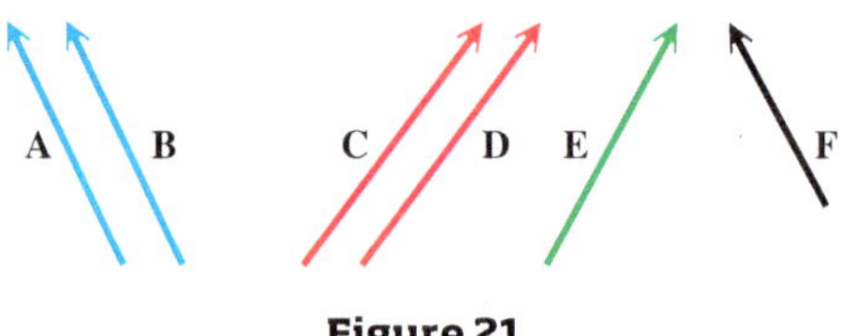

Figure 21

The sum of two vectors is also a vector. There are two ways to find the sum of two vectors **A** and **B** geometrically.

A + B
B
A
(a)

A + B
B
A
(b)

Figure 22

1. Place the initial point of vector **B** at the terminal point of vector **A**, as shown in **Figure 22(a).** The vector with the same initial point as **A** and the same terminal point as **B** is the sum $\mathbf{A} + \mathbf{B}$.
2. Use the **parallelogram rule.** Place vectors **A** and **B** so that their initial points coincide, as in **Figure 22(b).** Then, complete a parallelogram that has **A** and **B** as two sides. The diagonal of the parallelogram with the same initial point as **A** and **B** is the sum $\mathbf{A} + \mathbf{B}$.

Parallelograms can be used to show that vector $\mathbf{B} + \mathbf{A}$ is the same as vector $\mathbf{A} + \mathbf{B}$, or that $\mathbf{A} + \mathbf{B} = \mathbf{B} + \mathbf{A}$, so ***vector addition is commutative.*** The vector sum $\mathbf{A} + \mathbf{B}$ is the **resultant** of vectors **A** and **B**.

For every vector **v** there is a vector $-\mathbf{v}$ that has the same magnitude as **v** but opposite direction. Vector $-\mathbf{v}$ is the **opposite** of **v**. See **Figure 23.** The sum of **v** and $-\mathbf{v}$ has magnitude 0 and is the **zero vector.** As with real numbers, to subtract vector **B** from vector **A**, find the vector sum $\mathbf{A} + (-\mathbf{B})$. See **Figure 24.**

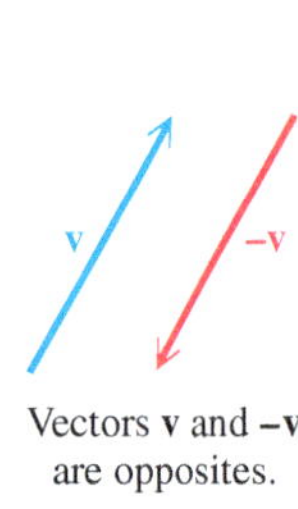

Vectors **v** and −**v** are opposites.

Figure 23

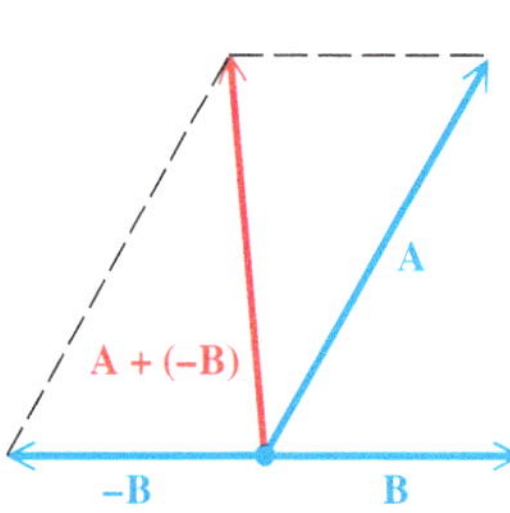

Figure 24

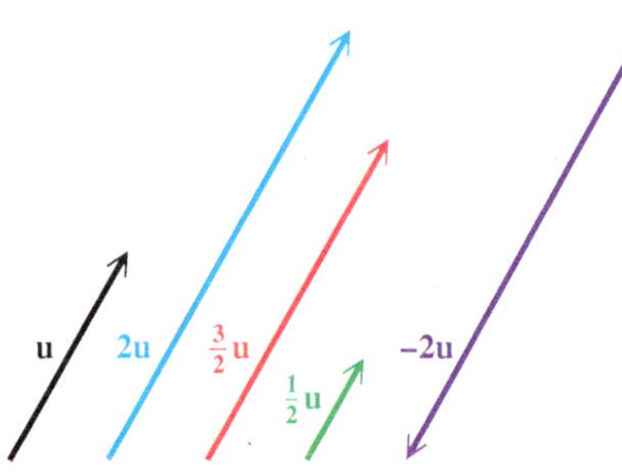

Figure 25

The product of a real number (or scalar) k and a vector **u** is the vector $k \cdot \mathbf{u}$, which has magnitude $|k|$ times the magnitude of **u**. The vector $k \cdot \mathbf{u}$ has the same direction as **u** if $k > 0$ and the opposite direction if $k < 0$. See **Figure 25.**

The following properties are helpful when solving vector applications.

Geometric Properties of Parallelograms

1. A parallelogram is a quadrilateral whose opposite sides are parallel.
2. The opposite sides and opposite angles of a parallelogram are equal, and adjacent angles of a parallelogram are supplementary.
3. The diagonals of a parallelogram bisect each other, but they do not necessarily bisect the angles of the parallelogram.

EXAMPLE 1 Finding the Magnitude of a Resultant

Two forces of 15 and 22 newtons act on a point in the plane. (A **newton** is a unit of force that equals 0.225 lb.) If the angle between the forces is 100°, find the magnitude of the resultant force.

SOLUTION As shown in **Figure 26,** a parallelogram that has the forces as adjacent sides can be formed. The angles of the parallelogram adjacent to angle P measure 80° because adjacent angles of a parallelogram are supplementary. Opposite sides of the parallelogram are equal in length. The resultant force divides the parallelogram into two triangles. Use the law of cosines with either triangle.

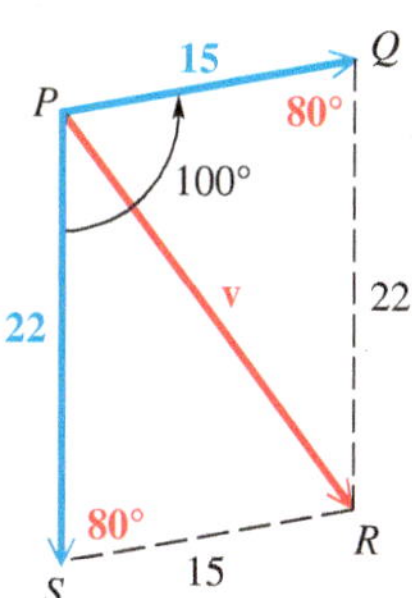

Figure 26

$|\mathbf{v}|^2 = 15^2 + 22^2 - 2(15)(22)\cos 80°$ Law of cosines

$|\mathbf{v}|^2 \approx 225 + 484 - 115$ Evaluate powers and cos 80°. Multiply.

$|\mathbf{v}|^2 \approx 594$ Add and subtract.

$|\mathbf{v}| \approx 24$ Take square roots and choose the positive square root.

To the nearest unit, the magnitude of the resultant force is 24 newtons.

✔ **Now Try Exercise 27.**

The Equilibrant The previous example showed a method for finding the resultant of two vectors. Sometimes it is necessary to find a vector that will counterbalance the resultant. This opposite vector is the **equilibrant.** That is, the equilibrant of vector **u** is the vector $-\mathbf{u}$.

EXAMPLE 2 Finding the Magnitude and Direction of an Equilibrant

Find the magnitude of the equilibrant of forces of 48 newtons and 60 newtons acting on a point A, if the angle between the forces is 50°. Then find the angle between the equilibrant and the 48-newton force.

SOLUTION As shown in **Figure 27,** the equilibrant is $-\mathbf{v}$.

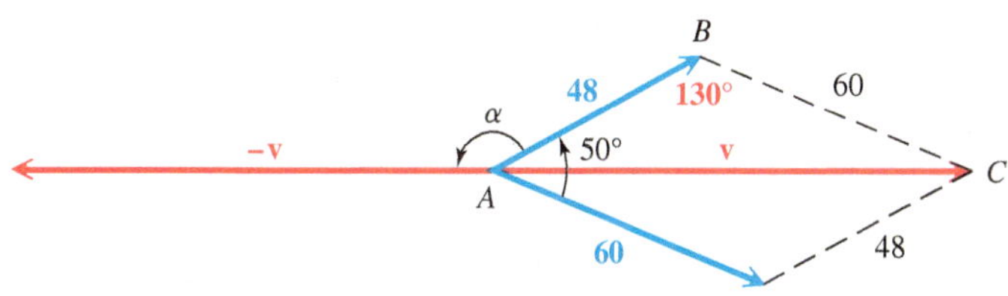

Figure 27

The magnitude of **v**, and hence of $-\mathbf{v}$, is found using triangle ABC and the law of cosines.

$|\mathbf{v}|^2 = 48^2 + 60^2 - 2(48)(60)\cos 130°$ Law of cosines

$|\mathbf{v}|^2 \approx 9606.5$ Use a calculator.

$|\mathbf{v}| \approx 98$ Square root property; Give two significant digits.

To the nearest unit, the magnitude is 98 newtons.

The required angle, labeled α in **Figure 27,** can be found by subtracting angle CAB from 180°. Use the law of sines to find angle CAB.

$$\frac{\sin CAB}{60} = \frac{\sin 130^\circ}{98}$$ Law of sines (alternative form)

$$\sin CAB \approx 0.46900680$$ Multiply by 60 and use a calculator.

$$CAB \approx 28^\circ$$ Use the inverse sine function.

Finally, $\alpha \approx 180^\circ - 28^\circ = 152^\circ$.

✔ Now Try Exercise 31.

Incline Applications

We can use vectors to solve incline problems.

EXAMPLE 3 Finding a Required Force

Find the force required to keep a 50-lb wagon from sliding down a ramp inclined at 20° to the horizontal. (Assume there is no friction.)

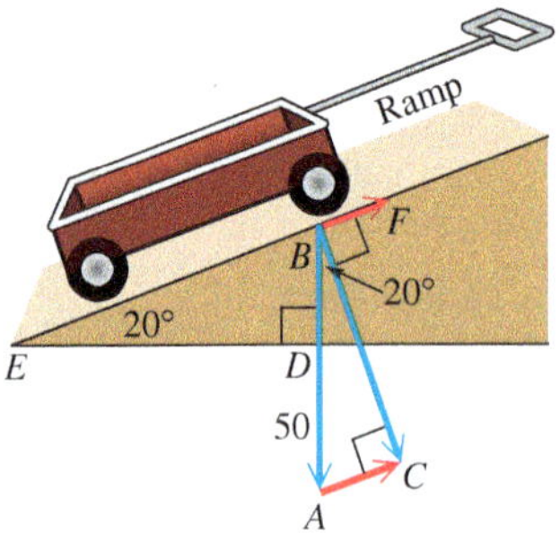

SOLUTION In **Figure 28,** the vertical 50-lb force **BA** represents the force of gravity. It is the sum of vectors **BC** and −**AC**. The vector **BC** represents the force with which the weight pushes against the ramp. The vector **BF** represents the force that would pull the weight up the ramp. Because vectors **BF** and **AC** are equal, $|\mathbf{AC}|$ gives the magnitude of the required force.

Vectors **BF** and **AC** are parallel, so angle EBD equals angle A by alternate interior angles. Because angle BDE and angle C are right angles, triangles CBA and DEB have two corresponding angles equal and, thus, are similar triangles. Therefore, angle ABC equals angle E, which is 20°. From right triangle ABC, we have the following.

Figure 28

$$\sin 20^\circ = \frac{|\mathbf{AC}|}{50}$$ $\sin B = \frac{\text{side opposite } B}{\text{hypotenuse}}$

$$|\mathbf{AC}| = 50 \sin 20^\circ$$ Multiply by 50 and rewrite.

$$|\mathbf{AC}| \approx 17$$ Use a calculator.

A force of approximately 17 lb will keep the wagon from sliding down the ramp.

✔ Now Try Exercise 39.

EXAMPLE 4 Finding an Incline Angle

A force of 16.0 lb is required to hold a 40.0-lb lawn mower on an incline. What angle does the incline make with the horizontal?

SOLUTION This situation is illustrated in **Figure 29.** Consider right triangle ABC. Angle B equals angle θ, the magnitude of vector **BA** represents the weight of the mower, and vector **AC** equals vector **BE**, which represents the force required to hold the mower on the incline.

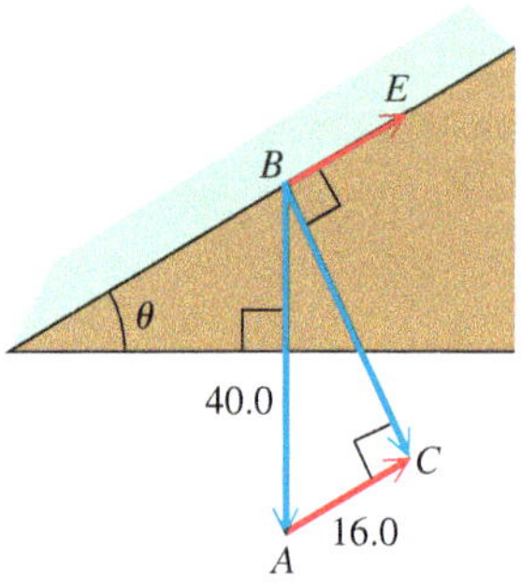

$$\sin B = \frac{16.0}{40.0}$$ $\sin B = \frac{\text{side opposite } B}{\text{hypotenuse}}$

$$\sin B = 0.4$$ Simplify.

$$B \approx 23.6^\circ$$ Use the inverse sine function.

Figure 29

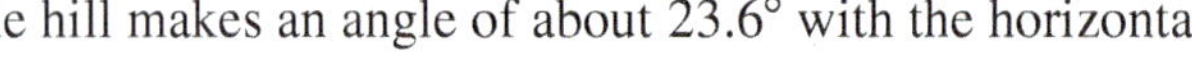

The hill makes an angle of about 23.6° with the horizontal.

✔ Now Try Exercise 41.

Navigation Applications Problems that involve bearing can also be solved using vectors.

EXAMPLE 5 Applying Vectors to a Navigation Problem

A ship leaves port on a bearing of 28.0° and travels 8.20 mi. The ship then turns due east and travels 4.30 mi. How far is the ship from port? What is its bearing from port?

SOLUTION In **Figure 30,** vectors **PA** and **AE** represent the ship's path. The magnitude and bearing of the resultant **PE** can be found as follows. Triangle *PNA* is a right triangle, so

$$\text{angle } NAP = 90° - 28.0° = 62.0°,$$

and $$\text{angle } PAE = 180° - 62.0° = 118.0°.$$

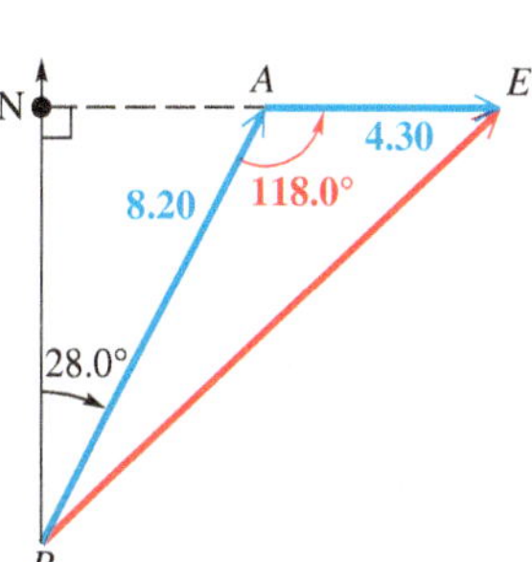

Figure 30

Use the law of cosines to find $|\mathbf{PE}|$, the magnitude of vector **PE**.

$$|\mathbf{PE}|^2 = 8.20^2 + 4.30^2 - 2(8.20)(4.30)\cos 118.0° \quad \text{Law of cosines}$$

$$|\mathbf{PE}|^2 \approx 118.84 \quad \text{Use a calculator.}$$

$$|\mathbf{PE}| \approx 10.9 \quad \text{Square root property}$$

The ship is about 10.9 mi from port.

To find the bearing of the ship from port, find angle *APE*.

$$\frac{\sin APE}{4.30} = \frac{\sin 118.0°}{10.9} \quad \text{Law of sines}$$

$$\sin APE = \frac{4.30 \sin 118.0°}{10.9} \quad \text{Multiply by 4.30.}$$

$$APE \approx 20.4° \quad \text{Use the inverse sine function.}$$

Finally, 28.0° + 20.4° = 48.4°, so the bearing is 48.4°.

✓ **Now Try Exercise 45.**

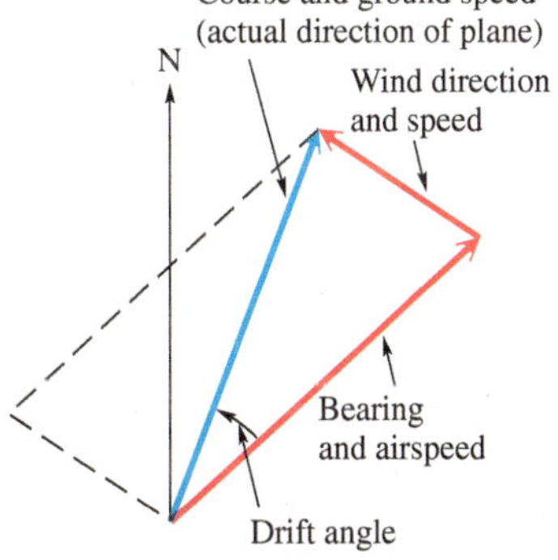

Figure 31

In air navigation, the **airspeed** of a plane is its speed relative to the air, and the **ground speed** is its speed relative to the ground. Because of wind, these two speeds are usually different. The ground speed of the plane is represented by the vector sum of the airspeed and windspeed vectors. See **Figure 31.**

EXAMPLE 6 Applying Vectors to a Navigation Problem

An airplane that is following a bearing of 239° at an airspeed of 425 mph encounters a wind blowing at 36.0 mph from a direction of 115°. Find the resulting bearing and ground speed of the plane.

SOLUTION An accurate sketch is essential to the solution of this problem. We have included two sets of geographical axes, which enable us to determine measures of necessary angles. Analyze **Figure 32** on the next page carefully.

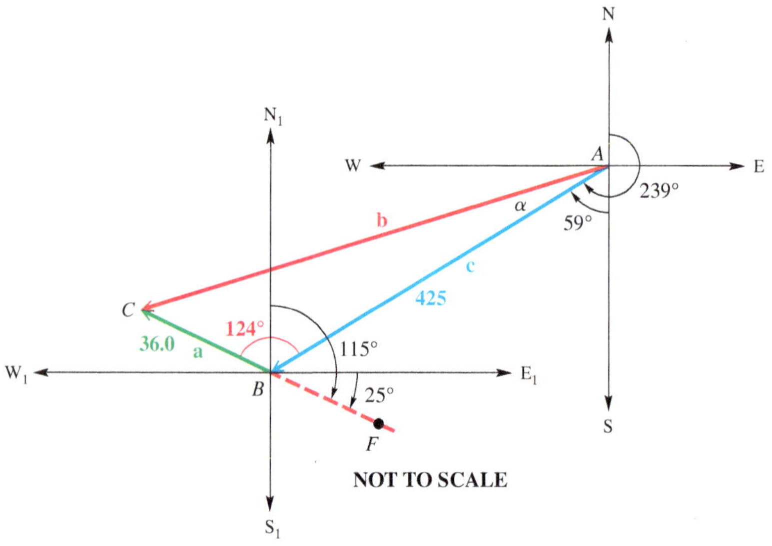

Figure 32

Vector **c** represents the airspeed and bearing of the plane, vector **a** represents the speed and direction of the wind, and vector **b** represents the resulting bearing and ground speed of the plane. Angle ABC has as its measure the sum of angle ABN_1 and angle N_1BC.

- Angle SAB measures $239° - 180° = 59°$. Because angle ABN_1 is an alternate interior angle to it, $ABN_1 = 59°$.
- Angle E_1BF measures $115° - 90° = 25°$. Thus, angle CBW_1 also measures 25° because it is a vertical angle. Angle N_1BC is the complement of 25°, which is $90° - 25° = 65°$.

By these results,

$$\text{angle } ABC = 59° + 65° = 124°.$$

To find $|\mathbf{b}|$, we use the law of cosines.

$$|\mathbf{b}|^2 = |\mathbf{a}|^2 + |\mathbf{c}|^2 - 2|\mathbf{a}||\mathbf{c}| \cos ABC \qquad \text{Law of cosines}$$

$$|\mathbf{b}|^2 = 36.0^2 + 425^2 - 2(36.0)(425) \cos 124° \qquad \text{Substitute.}$$

$$|\mathbf{b}|^2 \approx 199{,}032 \qquad \text{Use a calculator.}$$

$$|\mathbf{b}| \approx 446 \qquad \text{Square root property}$$

The ground speed is approximately 446 mph.

To find the resulting bearing of **b**, we must find the measure of angle α in **Figure 32** and then add it to 239°. To find α, we use the law of sines.

$$\frac{\sin \alpha}{36.0} = \frac{\sin 124°}{446}$$

To maintain accuracy, use all the significant digits that a calculator allows.

$$\sin \alpha = \frac{36.0 \sin 124°}{446} \qquad \text{Multiply by 36.0.}$$

$$\alpha = \sin^{-1}\left(\frac{36.0 \sin 124°}{446}\right) \qquad \text{Use the inverse sine function.}$$

$$\alpha \approx 4° \qquad \text{Use a calculator.}$$

Add 4° to 239° to find the resulting bearing of 243°. ✔ **Now Try Exercise 51.**

8.3 Exercises

CONCEPT PREVIEW *Refer to the vectors* **m** *through* **t** *below.*

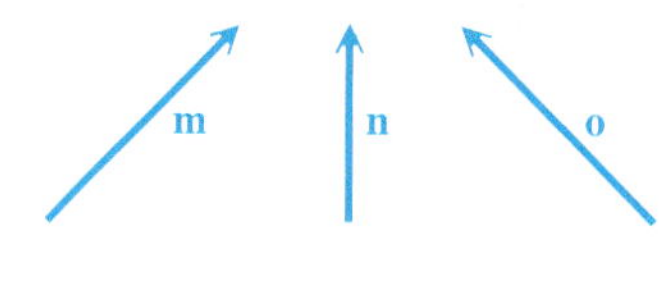

1. Name all pairs of vectors that appear to be equal.

2. Name all pairs of vectors that are opposites.

3. Name all pairs of vectors where the first is a scalar multiple of the other, with the scalar positive.

4. Name all pairs of vectors where the first is a scalar multiple of the other, with the scalar negative.

CONCEPT PREVIEW *Refer to vectors* **a** *through* **h** *below. Make a copy or a sketch of each vector, and then draw a sketch to represent each of the following. For example, find* **a** + **e** *by placing* **a** *and* **e** *so that their initial points coincide. Then use the parallelogram rule to find the resultant, as shown in the figure on the right.*

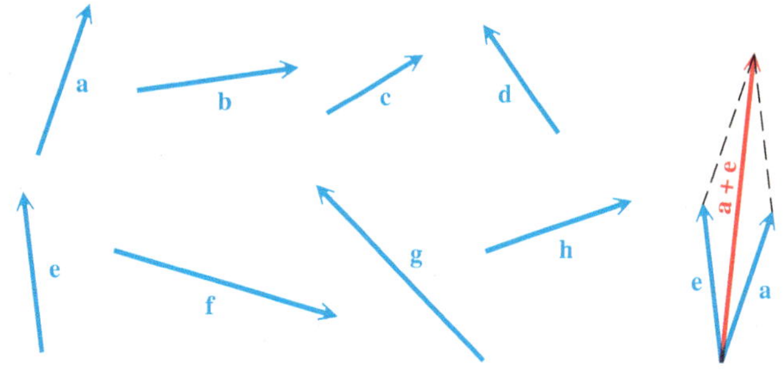

5. −**b**
6. −**g**
7. 2**c**
8. 2**h**
9. **a** + **b**
10. **h** + **g**
11. **a** − **c**
12. **d** − **e**
13. **a** + (**b** + **c**)
14. (**a** + **b**) + **c**
15. **c** + **d**
16. **d** + **c**
17. From the results of **Exercises 13 and 14,** does it appear that vector addition is associative?
18. From the results of **Exercises 15 and 16,** does it appear that vector addition is commutative?

For each pair of vectors **u** *and* **v** *with angle* θ *between them, sketch the resultant.*

19. $|\mathbf{u}| = 12$, $|\mathbf{v}| = 20$, $\theta = 27°$
20. $|\mathbf{u}| = 8$, $|\mathbf{v}| = 12$, $\theta = 20°$
21. $|\mathbf{u}| = 20$, $|\mathbf{v}| = 30$, $\theta = 30°$
22. $|\mathbf{u}| = 50$, $|\mathbf{v}| = 70$, $\theta = 40°$

Use the parallelogram rule to find the magnitude of the resultant force for the two forces shown in each figure. Round answers to the nearest tenth.

23.
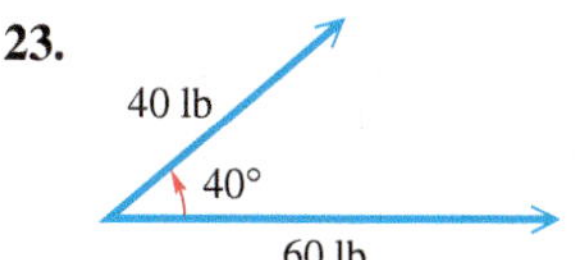

24.
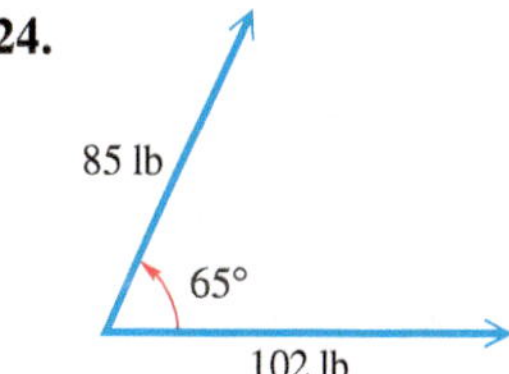

25.
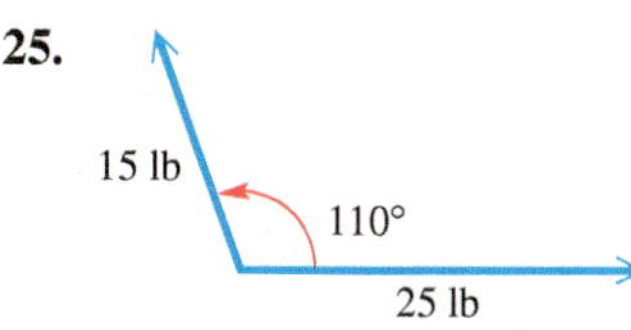

26.
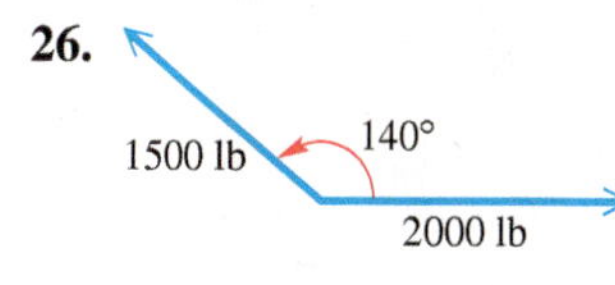

Two forces act at a point in the plane. The angle between the two forces is given. Find the magnitude of the resultant force. ***See Example 1.***

27. forces of 250 and 450 newtons, forming an angle of 85°

28. forces of 19 and 32 newtons, forming an angle of 118°

29. forces of 116 and 139 lb, forming an angle of 140° 50′

30. forces of 37.8 and 53.7 lb, forming an angle of 68.5°

Solve each problem. ***See Examples 1–4.***

31. ***Direction and Magnitude of an Equilibrant*** Two tugboats are pulling a disabled speedboat into port with forces of 1240 lb and 1480 lb. The angle between these forces is 28.2°. Find the direction and magnitude of the equilibrant.

32. ***Direction and Magnitude of an Equilibrant*** Two rescue vessels are pulling a broken-down motorboat toward a boathouse with forces of 840 lb and 960 lb. The angle between these forces is 24.5°. Find the direction and magnitude of the equilibrant.

33. ***Angle between Forces*** Two forces of 692 newtons and 423 newtons act at a point. The resultant force is 786 newtons. Find the angle between the forces.

34. ***Angle between Forces*** Two forces of 128 lb and 253 lb act at a point. The resultant force is 320 lb. Find the angle between the forces.

35. ***Magnitudes of Forces*** A force of 176 lb makes an angle of 78° 50′ with a second force. The resultant of the two forces makes an angle of 41° 10′ with the first force. Find the magnitudes of the second force and of the resultant.

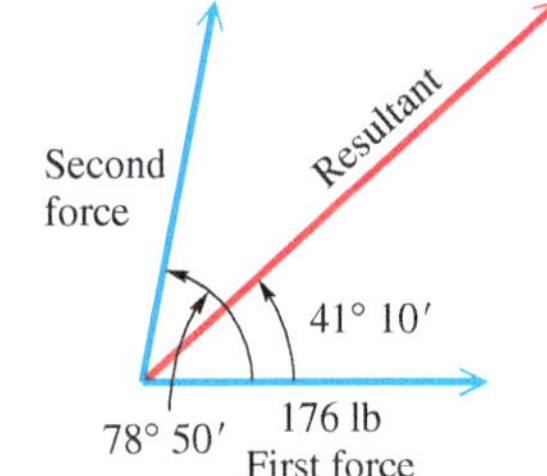

36. ***Magnitudes of Forces*** A force of 28.7 lb makes an angle of 42° 10′ with a second force. The resultant of the two forces makes an angle of 32° 40′ with the first force. Find the magnitudes of the second force and of the resultant.

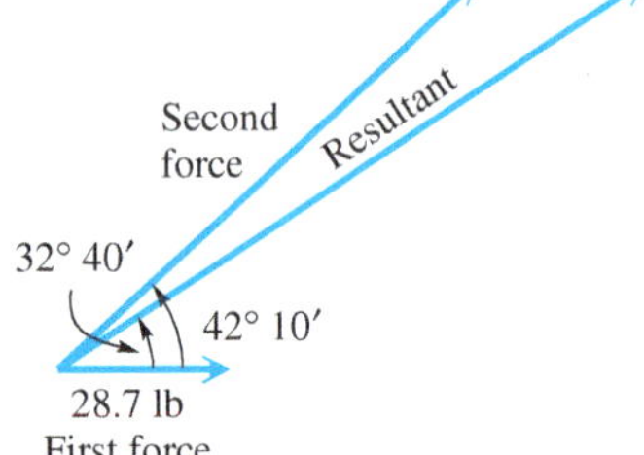

37. ***Angle of a Hill Slope*** A force of 25 lb is required to hold an 80-lb crate on a hill. What angle does the hill make with the horizontal?

38. ***Force Needed to Keep a Car Parked*** Find the force required to keep a 3000-lb car parked on a hill that makes an angle of 15° with the horizontal.

39. ***Force Needed for a Monolith*** To build the pyramids in Egypt, it is believed that giant causeways were constructed to transport the building materials to the site. One such causeway is said to have been 3000 ft long, with a slope of about 2.3°. How much force would be required to hold a 60-ton monolith on this causeway?

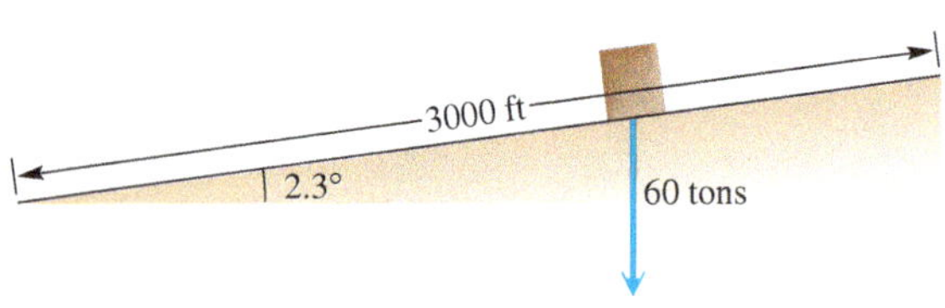

40. *Force Needed for a Monolith* If the causeway in **Exercise 39** were 500 ft longer and the monolith weighed 10 tons more, how much force would be required?

41. *Incline Angle* A force of 18.0 lb is required to hold a 60.0-lb stump grinder on an incline. What angle does the incline make with the horizontal?

42. *Incline Angle* A force of 30.0 lb is required to hold an 80.0-lb pressure washer on an incline. What angle does the incline make with the horizontal?

43. *Weight of a Box* Two people are carrying a box. One person exerts a force of 150 lb at an angle of 62.4° with the horizontal. The other person exerts a force of 114 lb at an angle of 54.9°. Find the weight of the box.

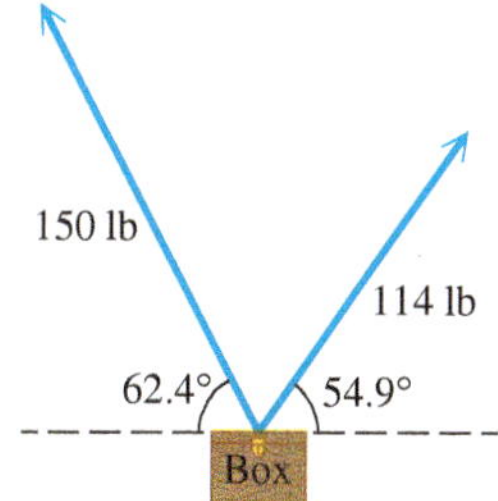

44. *Weight of a Crate and Tension of a Rope* A crate is supported by two ropes. One rope makes an angle of 46° 20′ with the horizontal and has a tension of 89.6 lb on it. The other rope is horizontal. Find the weight of the crate and the tension in the horizontal rope.

Solve each problem. ***See Examples 5 and 6.***

45. *Distance and Bearing of a Ship* A ship leaves port on a bearing of 34.0° and travels 10.4 mi. The ship then turns due east and travels 4.6 mi. How far is the ship from port, and what is its bearing from port?

46. *Distance and Bearing of a Luxury Liner* A luxury liner leaves port on a bearing of 110.0° and travels 8.8 mi. It then turns due west and travels 2.4 mi. How far is the liner from port, and what is its bearing from port?

47. *Distance of a Ship from Its Starting Point* Starting at point A, a ship sails 18.5 km on a bearing of 189°, then turns and sails 47.8 km on a bearing of 317°. Find the distance of the ship from point A.

48. *Distance of a Ship from Its Starting Point* Starting at point X, a ship sails 15.5 km on a bearing of 200°, then turns and sails 2.4 km on a bearing of 320°. Find the distance of the ship from point X.

49. *Distance and Direction of a Motorboat* A motorboat sets out in the direction N 80° 00′ E. The speed of the boat in still water is 20.0 mph. If the current is flowing directly south, and the actual direction of the motorboat is due east, find the speed of the current and the actual speed of the motorboat.

50. *Movement of a Motorboat* Suppose we would like to cross a 132-ft-wide river in a motorboat. Assume that the motorboat can travel at 7.0 mph relative to the water and that the current is flowing west at the rate of 3.0 mph. The bearing θ is chosen so that the motorboat will land at a point exactly across from the starting point.

(a) At what speed will the motorboat be traveling relative to the banks?

(b) How long will it take for the motorboat to make the crossing?

(c) What is the measure of angle θ?

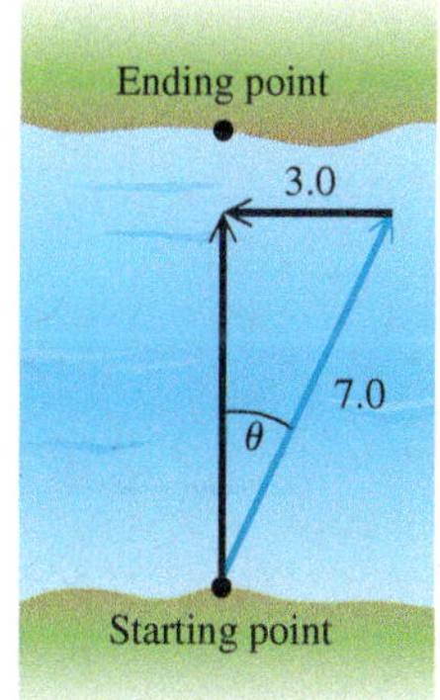

51. *Bearing and Ground Speed of a Plane* An airline route from San Francisco to Honolulu is on a bearing of 233.0°. A jet flying at 450 mph on that bearing encounters a wind blowing at 39.0 mph from a direction of 114.0°. Find the resulting bearing and ground speed of the plane.

52. *Path Traveled by a Plane* The aircraft carrier *Tallahassee* is traveling at sea on a steady course with a bearing of 30° at 32 mph. Patrol planes on the carrier have enough fuel for 2.6 hr of flight when traveling at a speed of 520 mph. One of the pilots takes off on a bearing of 338° and then turns and heads in a straight line, so as to be able to catch the carrier and land on the deck at the exact instant that his fuel runs out. If the pilot left at 2 P.M., at what time did he turn to head for the carrier?

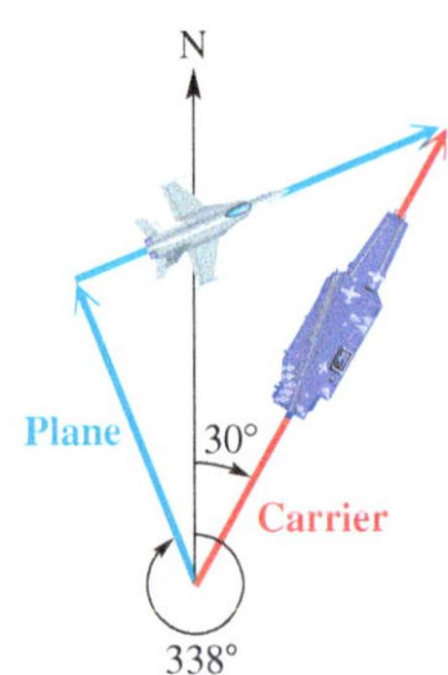

53. *Airspeed and Ground Speed* A pilot wants to fly on a bearing of 74.9°. By flying due east, he finds that a 42.0-mph wind, blowing from the south, puts him on course. Find the airspeed and the ground speed.

54. *Bearing of a Plane* A plane flies 650 mph on a bearing of 175.3°. A 25-mph wind, from a direction of 266.6°, blows against the plane. Find the resulting bearing of the plane.

55. *Bearing and Ground Speed of a Plane* A pilot is flying at 190.0 mph. He wants his flight path to be on a bearing of 64° 30′. A wind is blowing from the south at 35.0 mph. Find the bearing he should fly, and find the plane's ground speed.

56. *Bearing and Ground Speed of a Plane* A pilot is flying at 168 mph. She wants her flight path to be on a bearing of 57° 40′. A wind is blowing from the south at 27.1 mph. Find the bearing she should fly, and find the plane's ground speed.

57. *Bearing and Airspeed of a Plane* What bearing and airspeed are required for a plane to fly 400 mi due north in 2.5 hr if the wind is blowing from a direction of 328° at 11 mph?

58. *Ground Speed and Bearing of a Plane* A plane is headed due south with an airspeed of 192 mph. A wind from a direction of 78.0° is blowing at 23.0 mph. Find the ground speed and resulting bearing of the plane.

59. *Ground Speed and Bearing of a Plane* An airplane is headed on a bearing of 174° at an airspeed of 240 km per hr. A 30-km-per-hr wind is blowing from a direction of 245°. Find the ground speed and resulting bearing of the plane.

60. *Velocity of a Star* The space velocity $\mathbf{v}$ of a star relative to the sun can be expressed as the resultant vector of two perpendicular vectors—the radial velocity $\mathbf{v}_r$ and the tangential velocity $\mathbf{v}_t$, where $\mathbf{v} = \mathbf{v}_r + \mathbf{v}_t$. If a star is located near the sun and its space velocity is large, then its motion across the sky will also be large. Barnard's Star is a relatively close star with a distance of 35 trillion mi from the sun. It moves across the sky through an angle of 10.34″ per year, which is the largest motion of any known star. Its radial velocity $\mathbf{v}_r$ is 67 mi per sec toward the sun. (*Sources:* Zeilik, M., S. Gregory, and E. Smith, *Introductory Astronomy and Astrophysics,* Second Edition, Saunders College Publishing; Acker, A. and C. Jaschek, *Astronomical Methods and Calculations,* John Wiley and Sons.)

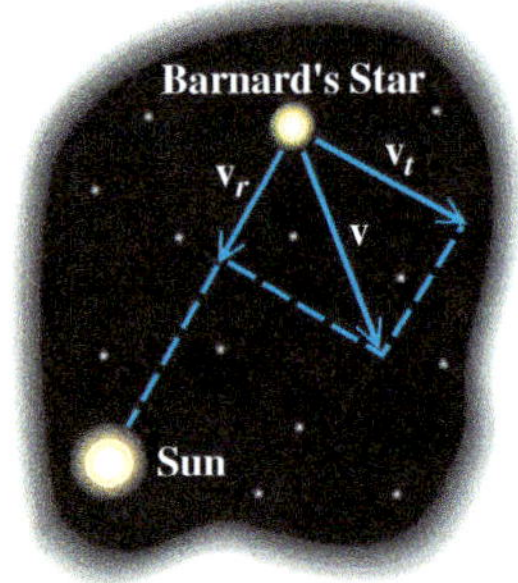

NOT TO SCALE

(a) Approximate the tangential velocity $\mathbf{v}_t$ of Barnard's Star. (*Hint:* Use the arc length formula $s = r\theta$.)

(b) Compute the magnitude of $\mathbf{v}$.

8.4 Algebraically Defined Vectors and the Dot Product

- Algebraic Interpretation of Vectors
- Operations with Vectors
- The Dot Product and the Angle between Vectors

LOOKING AHEAD TO CALCULUS

In addition to two-dimensional vectors in a plane, calculus courses introduce three-dimensional vectors in space. The magnitude of the two-dimensional vector $\langle a, b \rangle$ is given by

$$\sqrt{a^2 + b^2}.$$

If we extend this to the three-dimensional vector $\langle a, b, c \rangle$, the expression becomes

$$\sqrt{a^2 + b^2 + c^2}.$$

Similar extensions are made for other concepts.

Algebraic Interpretation of Vectors

A vector with initial point at the origin in a rectangular coordinate system is a **position vector.** A position vector **u** with endpoint at the point (a, b) is written $\langle a, b \rangle$, so

$$\mathbf{u} = \langle a, b \rangle.$$

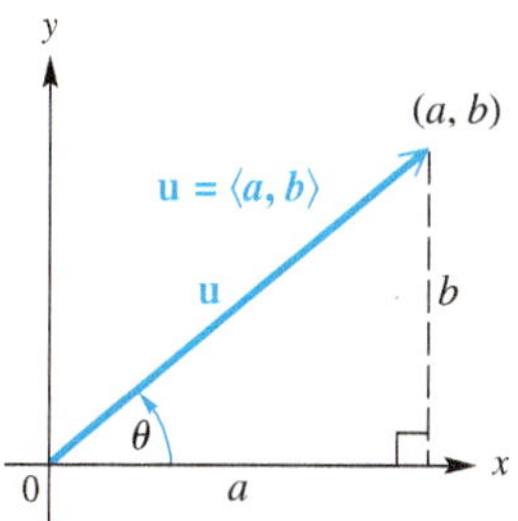

Figure 33

This means that every vector in the real plane corresponds to an ordered pair of real numbers. ***Thus, geometrically a vector is a directed line segment while algebraically it is an ordered pair.*** The numbers a and b are the **horizontal component** and the **vertical component,** respectively, of vector **u**.

Figure 33 shows the vector $\mathbf{u} = \langle a, b \rangle$. The positive angle between the x-axis and a position vector is the **direction angle** for the vector. In **Figure 33,** θ is the direction angle for vector **u.** The magnitude and direction angle of a vector are related to its horizontal and vertical components.

Magnitude and Direction Angle of a Vector $\langle a, b \rangle$

The magnitude (length) of vector $\mathbf{u} = \langle a, b \rangle$ is given by the following.

$$|\mathbf{u}| = \sqrt{a^2 + b^2}$$

The direction angle θ satisfies $\tan\theta = \frac{b}{a}$, where $a \neq 0$.

EXAMPLE 1 Finding Magnitude and Direction Angle

Find the magnitude and direction angle for $\mathbf{u} = \langle 3, -2 \rangle$.

ALGEBRAIC SOLUTION

The magnitude is $|\mathbf{u}| = \sqrt{3^2 + (-2)^2} = \sqrt{13}$. To find the direction angle θ, start with $\tan\theta = \frac{b}{a} = \frac{-2}{3} = -\frac{2}{3}$. Vector **u** has a positive horizontal component and a negative vertical component, which places the position vector in quadrant IV. A calculator then gives $\tan^{-1}\left(-\frac{2}{3}\right) \approx -33.7°$. Adding 360° yields the direction angle $\theta \approx 326.3°$. See **Figure 34.**

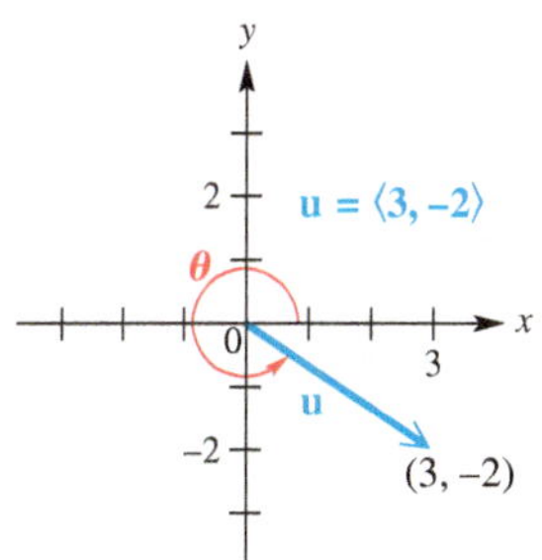

Figure 34

GRAPHING CALCULATOR SOLUTION

The TI-84 Plus calculator can find the magnitude and direction angle using rectangular to polar conversion (which is covered in detail in the next chapter). An approximation for $\sqrt{13}$ is given, and the TI-84 Plus gives the direction angle with the least possible absolute value. We must add 360° to the given value $-33.7°$ to obtain the positive direction angle $\theta \approx 326.3°$.

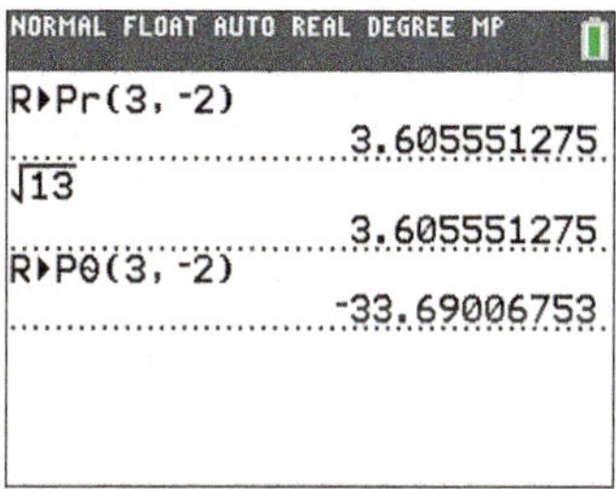

Figure 35

✔ **Now Try Exercise 9.**

Horizontal and Vertical Components

The horizontal and vertical components, respectively, of a vector **u** having magnitude $|\mathbf{u}|$ and direction angle θ are the following.

$$a = |\mathbf{u}| \cos \theta \quad \text{and} \quad b = |\mathbf{u}| \sin \theta$$

That is, $\mathbf{u} = \langle a, b \rangle = \langle |\mathbf{u}| \cos \theta, |\mathbf{u}| \sin \theta \rangle$.

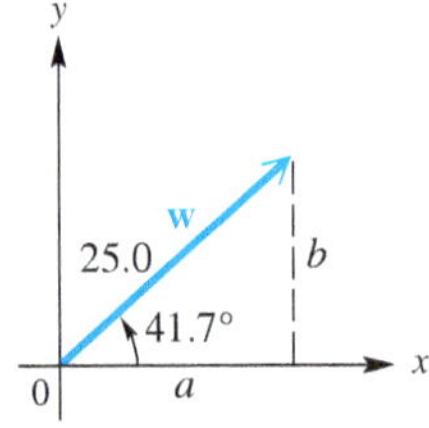

Figure 36

EXAMPLE 2 Finding Horizontal and Vertical Components

Vector **w** in **Figure 36** has magnitude 25.0 and direction angle 41.7°. Find the horizontal and vertical components.

ALGEBRAIC SOLUTION

Use the formulas below, with $|\mathbf{w}| = 25.0$ and $\theta = 41.7°$.

$a = \lvert\mathbf{w}\rvert \cos \theta$	$b = \lvert\mathbf{w}\rvert \sin \theta$
$a = 25.0 \cos 41.7°$	$b = 25.0 \sin 41.7°$
$a \approx 18.7$	$b \approx 16.6$

Therefore, $\mathbf{w} = \langle 18.7, 16.6 \rangle$. The horizontal component is 18.7, and the vertical component is 16.6 (rounded to the nearest tenth).

GRAPHING CALCULATOR SOLUTION

See **Figure 37.** The results support the algebraic solution.

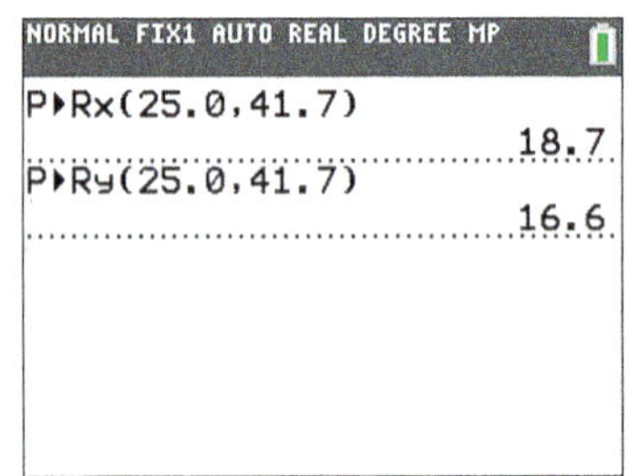

Figure 37

✔ **Now Try Exercise 13.**

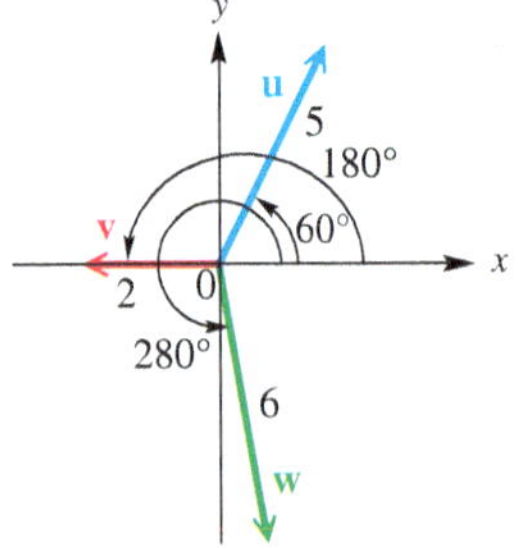

Figure 38

EXAMPLE 3 Writing Vectors in the Form $\langle a, b \rangle$

Write each vector in **Figure 38** in the form $\langle a, b \rangle$.

SOLUTION

$$\mathbf{u} = \langle 5 \cos 60°, 5 \sin 60° \rangle = \left\langle 5 \cdot \frac{1}{2}, 5 \cdot \frac{\sqrt{3}}{2} \right\rangle = \left\langle \frac{5}{2}, \frac{5\sqrt{3}}{2} \right\rangle$$

$$\mathbf{v} = \langle 2 \cos 180°, 2 \sin 180° \rangle = \langle 2(-1), 2(0) \rangle = \langle -2, 0 \rangle$$

$$\mathbf{w} = \langle 6 \cos 280°, 6 \sin 280° \rangle \approx \langle 1.0419, -5.9088 \rangle \quad \text{Use a calculator.}$$

✔ **Now Try Exercises 19 and 21.**

Operations with Vectors As shown in **Figure 39,**

$$\mathbf{m} = \langle a, b \rangle, \quad \mathbf{n} = \langle c, d \rangle, \quad \text{and} \quad \mathbf{p} = \langle a + c, b + d \rangle.$$

Using geometry, we can show that the endpoints of the three vectors and the origin form a parallelogram. A diagonal of this parallelogram gives the resultant of **m** and **n**, so we have $\mathbf{p} = \mathbf{m} + \mathbf{n}$ or

$$\langle a + c, b + d \rangle = \langle a, b \rangle + \langle c, d \rangle.$$

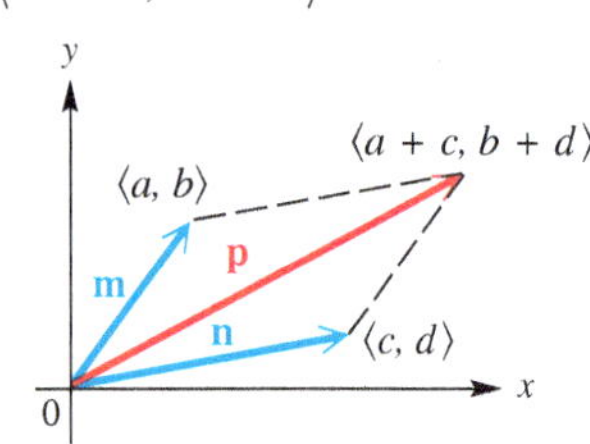

Figure 39

Similarly, we can verify the following operations.

Vector Operations

Let a, b, c, d, and k represent real numbers.

$$\langle a, b\rangle + \langle c, d\rangle = \langle a + c, b + d\rangle$$

$$k \cdot \langle a, b\rangle = \langle ka, kb\rangle$$

$$\text{If } \mathbf{u} = \langle a_1, a_2\rangle, \text{ then } -\mathbf{u} = \langle -a_1, -a_2\rangle.$$

$$\langle a, b\rangle - \langle c, d\rangle = \langle a, b\rangle + (-\langle c, d\rangle) = \langle a - c, b - d\rangle$$

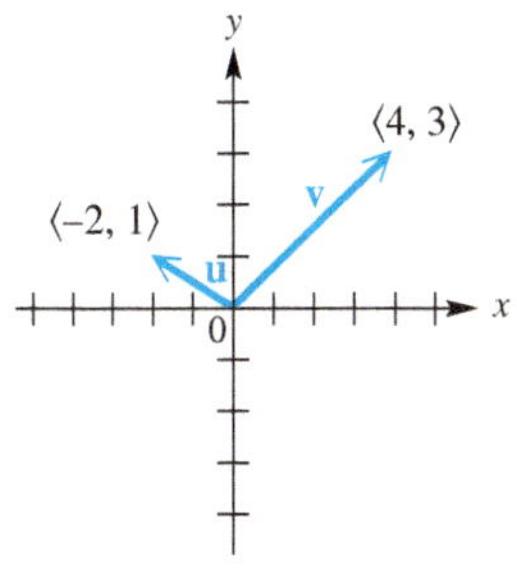

Figure 40

EXAMPLE 4 Performing Vector Operations

Let $\mathbf{u} = \langle -2, 1\rangle$ and $\mathbf{v} = \langle 4, 3\rangle$. See **Figure 40.** Find and illustrate each of the following.

(a) $\mathbf{u} + \mathbf{v}$ **(b)** $-2\mathbf{u}$ **(c)** $3\mathbf{u} - 2\mathbf{v}$

SOLUTION See **Figure 41.**

(a) $\mathbf{u} + \mathbf{v}$

$$= \langle -2, 1\rangle + \langle 4, 3\rangle$$
$$= \langle -2 + 4, 1 + 3\rangle$$
$$= \langle 2, 4\rangle$$

(b) $-2\mathbf{u}$

$$= -2 \cdot \langle -2, 1\rangle$$
$$= \langle -2(-2), -2(1)\rangle$$
$$= \langle 4, -2\rangle$$

(c) $3\mathbf{u} - 2\mathbf{v}$

$$= 3 \cdot \langle -2, 1\rangle - 2 \cdot \langle 4, 3\rangle$$
$$= \langle -6, 3\rangle - \langle 8, 6\rangle$$
$$= \langle -6 - 8, 3 - 6\rangle$$
$$= \langle -14, -3\rangle$$

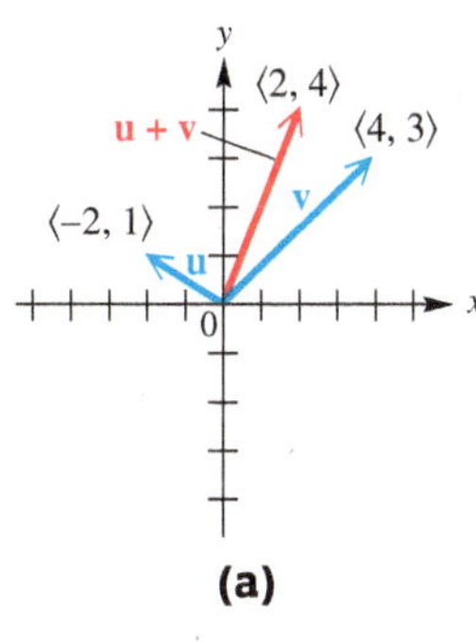

(a)

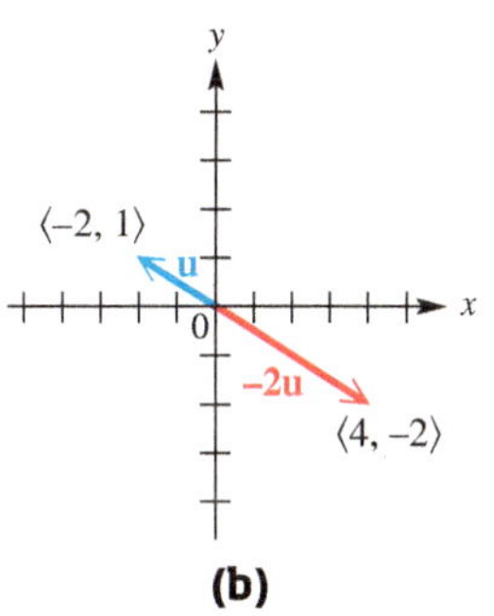

(b)

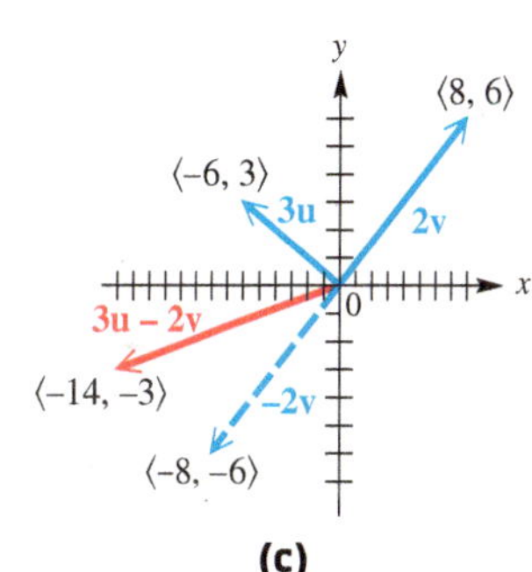

(c)

Figure 41

✔ Now Try Exercises 35, 37, and 39.

A **unit vector** is a vector that has magnitude 1. Two very important unit vectors are defined as follows and shown in **Figure 42(a).**

$$\mathbf{i} = \langle 1, 0\rangle \qquad \mathbf{j} = \langle 0, 1\rangle$$

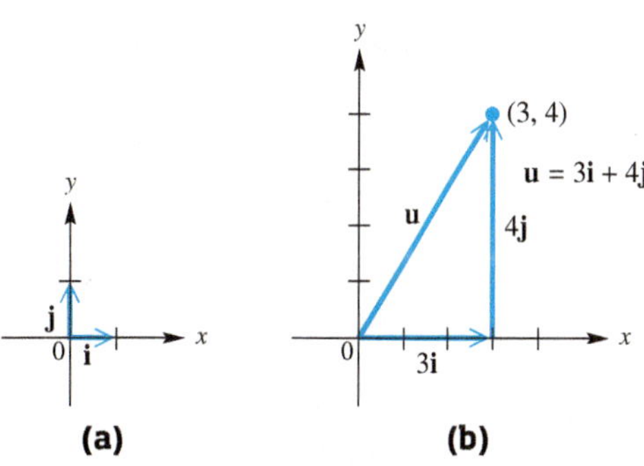

Figure 42

With the unit vectors **i** and **j**, we can express any other vector $\langle a, b \rangle$ in the form $a\mathbf{i} + b\mathbf{j}$, as shown in **Figure 42(b)** on the previous page, where $\langle 3, 4 \rangle = 3\mathbf{i} + 4\mathbf{j}$. The vector operations previously given can be restated, using $a\mathbf{i} + b\mathbf{j}$ notation.

i, j Form for Vectors

If $\mathbf{v} = \langle a, b \rangle$, then

$$\mathbf{v} = a\mathbf{i} + b\mathbf{j}, \quad \text{where } \mathbf{i} = \langle 1, 0 \rangle \text{ and } \mathbf{j} = \langle 0, 1 \rangle.$$

The Dot Product and the Angle between Vectors ***The* dot product *of two vectors is a real number, not a vector.*** It is also known as the *inner product*. Dot products are used to determine the angle between two vectors, to derive geometric theorems, and to solve physics problems.

Dot Product

The **dot product** of the two vectors $\mathbf{u} = \langle a, b \rangle$ and $\mathbf{v} = \langle c, d \rangle$ is denoted $\mathbf{u} \cdot \mathbf{v}$, read "**u** dot **v**," and given by the following.

$$\mathbf{u} \cdot \mathbf{v} = ac + bd$$

That is, the dot product of two vectors is the sum of the product of their first components and the product of their second components.

EXAMPLE 5 Finding Dot Products

Find each dot product.

(a) $\langle 2, 3 \rangle \cdot \langle 4, -1 \rangle$ **(b)** $\langle 6, 4 \rangle \cdot \langle -2, 3 \rangle$

SOLUTION

(a) $\langle 2, 3 \rangle \cdot \langle 4, -1 \rangle$

$= 2(4) + 3(-1)$

$= 5$

(b) $\langle 6, 4 \rangle \cdot \langle -2, 3 \rangle$

$= 6(-2) + 4(3)$

$= 0$

✔ **Now Try Exercises 47 and 49.**

The following properties of dot products can be verified using the definitions presented so far.

Properties of the Dot Product

For all vectors **u**, **v**, and **w** and real numbers k, the following hold.

(a) $\mathbf{u} \cdot \mathbf{v} = \mathbf{v} \cdot \mathbf{u}$ **(b)** $\mathbf{u} \cdot (\mathbf{v} + \mathbf{w}) = \mathbf{u} \cdot \mathbf{v} + \mathbf{u} \cdot \mathbf{w}$

(c) $(\mathbf{u} + \mathbf{v}) \cdot \mathbf{w} = \mathbf{u} \cdot \mathbf{w} + \mathbf{v} \cdot \mathbf{w}$ **(d)** $(k\mathbf{u}) \cdot \mathbf{v} = k(\mathbf{u} \cdot \mathbf{v}) = \mathbf{u} \cdot (k\mathbf{v})$

(e) $\mathbf{0} \cdot \mathbf{u} = 0$ **(f)** $\mathbf{u} \cdot \mathbf{u} = |\mathbf{u}|^2$

For example, to prove the first part of property (d),

$$(k\mathbf{u}) \cdot \mathbf{v} = k(\mathbf{u} \cdot \mathbf{v}),$$

we let $\mathbf{u} = \langle a, b \rangle$ and $\mathbf{v} = \langle c, d \rangle$.

$$\begin{aligned}
(k\mathbf{u}) \cdot \mathbf{v} &= \big(k\langle a, b \rangle\big) \cdot \langle c, d \rangle && \text{Substitute.} \\
&= \langle ka, kb \rangle \cdot \langle c, d \rangle && \text{Multiply by scalar } k. \\
&= kac + kbd && \text{Dot product} \\
&= k(ac + bd) && \text{Distributive property} \\
&= k\big(\langle a, b \rangle \cdot \langle c, d \rangle\big) && \text{Dot product} \\
&= k(\mathbf{u} \cdot \mathbf{v}) && \text{Substitute.}
\end{aligned}$$

The proofs of the remaining properties are similar.

The dot product of two vectors can be positive, 0, or negative. A geometric interpretation of the dot product explains when each of these cases occurs. This interpretation involves the angle between the two vectors.

Consider the two vectors $\mathbf{u} = \langle a_1, a_2 \rangle$ and $\mathbf{v} = \langle b_1, b_2 \rangle$, as shown in **Figure 43.** The **angle θ between u and v** is defined to be the angle having the two vectors as its sides for which $0° \le \theta \le 180°$.

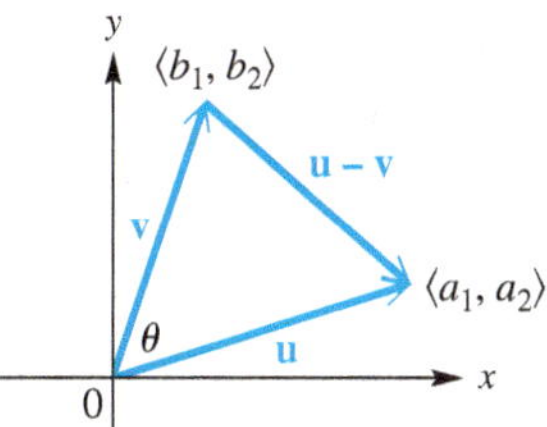

Figure 43

We can use the law of cosines to develop a formula to find angle θ in **Figure 43.**

$$|\mathbf{u} - \mathbf{v}|^2 = |\mathbf{u}|^2 + |\mathbf{v}|^2 - 2|\mathbf{u}||\mathbf{v}| \cos \theta \qquad \text{Law of cosines applied to Figure 43}$$

$$\left(\sqrt{(a_1 - b_1)^2 + (a_2 - b_2)^2}\right)^2 = \left(\sqrt{a_1^2 + a_2^2}\right)^2 + \left(\sqrt{b_1^2 + b_2^2}\right)^2 - 2|\mathbf{u}||\mathbf{v}| \cos \theta \qquad \text{Magnitude of a vector}$$

$$a_1^2 - 2a_1b_1 + b_1^2 + a_2^2 - 2a_2b_2 + b_2^2 = a_1^2 + a_2^2 + b_1^2 + b_2^2 - 2|\mathbf{u}||\mathbf{v}| \cos \theta \qquad \text{Square.}$$

$$-2a_1b_1 - 2a_2b_2 = -2|\mathbf{u}||\mathbf{v}| \cos \theta \qquad \text{Subtract like terms from each side.}$$

$$a_1b_1 + a_2b_2 = |\mathbf{u}||\mathbf{v}| \cos \theta \qquad \text{Divide by } -2.$$

$$\mathbf{u} \cdot \mathbf{v} = |\mathbf{u}||\mathbf{v}| \cos \theta \qquad \text{Definition of dot product}$$

$$\cos \theta = \frac{\mathbf{u} \cdot \mathbf{v}}{|\mathbf{u}||\mathbf{v}|} \qquad \text{Divide by } |\mathbf{u}||\mathbf{v}| \text{ and rewrite.}$$

Geometric Interpretation of Dot Product

If θ is the angle between the two nonzero vectors **u** and **v**, where $0° \le \theta \le 180°$, then the following holds.

$$\cos \theta = \frac{\mathbf{u} \cdot \mathbf{v}}{|\mathbf{u}||\mathbf{v}|}$$

EXAMPLE 6 Finding the Angle between Two Vectors

Find the angle θ between the two vectors.

(a) $\mathbf{u} = \langle 3, 4 \rangle$ and $\mathbf{v} = \langle 2, 1 \rangle$ **(b)** $\mathbf{u} = \langle 2, -6 \rangle$ and $\mathbf{v} = \langle 6, 2 \rangle$

SOLUTION

(a)

$$\cos\theta = \frac{\mathbf{u} \cdot \mathbf{v}}{|\mathbf{u}||\mathbf{v}|} \quad \text{Geometric interpretation of the dot product}$$

$$\cos\theta = \frac{\langle 3, 4 \rangle \cdot \langle 2, 1 \rangle}{|\langle 3, 4 \rangle||\langle 2, 1 \rangle|} \quad \text{Substitute values.}$$

$$\cos\theta = \frac{3(2) + 4(1)}{\sqrt{9+16} \cdot \sqrt{4+1}} \quad \text{Use the definitions.}$$

$$\cos\theta = \frac{10}{5\sqrt{5}} \quad \text{Simplify.}$$

$$\cos\theta \approx 0.894427191 \quad \text{Use a calculator.}$$

$$\theta \approx 26.57° \quad \text{Use the inverse cosine function.}$$

(b)

$$\cos\theta = \frac{\mathbf{u} \cdot \mathbf{v}}{|\mathbf{u}||\mathbf{v}|} \quad \text{Geometric interpretation of the dot product}$$

$$\cos\theta = \frac{\langle 2, -6 \rangle \cdot \langle 6, 2 \rangle}{|\langle 2, -6 \rangle||\langle 6, 2 \rangle|} \quad \text{Substitute values.}$$

$$\cos\theta = \frac{2(6) + (-6)(2)}{\sqrt{4+36} \cdot \sqrt{36+4}} \quad \text{Use the definitions.}$$

$$\cos\theta = 0 \quad \text{Evaluate. The numerator is equal to 0.}$$

$$\theta = 90° \quad \cos^{-1} 0 = 90°$$

✔ **Now Try Exercises 53 and 55.**

For angles θ between 0° and 180°, $\cos\theta$ is positive, 0, or negative when θ is less than, equal to, or greater than 90°, respectively. Therefore, the dot product of nonzero vectors is positive, 0, or negative according to this table.

Dot Product	Angle between Vectors
Positive	Acute
0	Right
Negative	Obtuse

Thus, in **Example 6,** the vectors in part (a) form an acute angle, and those in part (b) form a right angle. If $\mathbf{u} \cdot \mathbf{v} = 0$ for two nonzero vectors $\mathbf{u}$ and $\mathbf{v}$, then $\cos\theta = 0$ and $\theta = 90°$. Thus, $\mathbf{u}$ and $\mathbf{v}$ are perpendicular vectors, also called **orthogonal vectors.** See **Figure 44.**

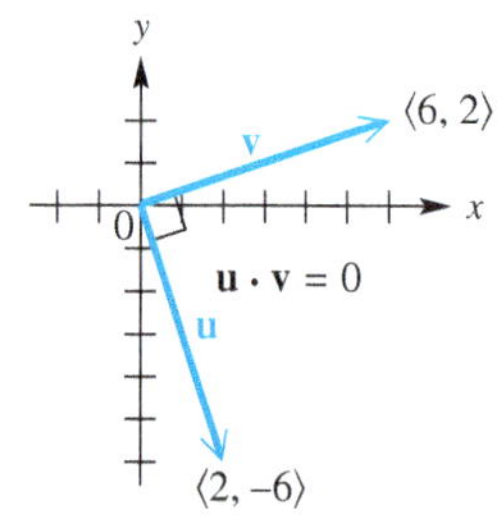

Orthogonal vectors

Figure 44

8.4 Exercises

CONCEPT PREVIEW *Fill in the blank to correctly complete each sentence.*

1. The magnitude of vector **u** is ________.

2. The direction angle of vector **u** is ________.

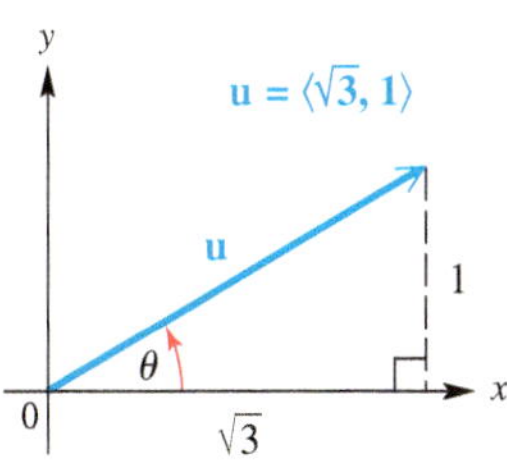

3. The horizontal component, a, of vector **v** is ________.

4. The vertical component, b, of vector **v** is ________.

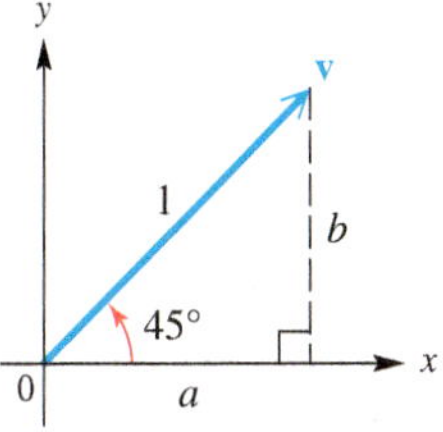

5. The sum of the vectors $\mathbf{u} = \langle -3, 5 \rangle$ and $\mathbf{v} = \langle 7, 4 \rangle$ is $\mathbf{u} + \mathbf{v} =$ ________.

6. The vector $\mathbf{u} = \langle 4, -2 \rangle$ is written in **i**, **j** form as ________.

7. The formula for the dot product of the two vectors $\mathbf{u} = \langle a, b \rangle$ and $\mathbf{v} = \langle c, d \rangle$ is

$$\mathbf{u} \cdot \mathbf{v} = \_\_\_\_\_\_\_\_.$$

8. If the dot product of two vectors is a positive number, then the angle between them is ________________.
(acute/obtuse)

Find the magnitude and direction angle for each vector. ***See Example 1.***

9. $\langle 15, -8 \rangle$

10. $\langle -7, 24 \rangle$

11. $\langle -4, 4\sqrt{3} \rangle$

12. $\langle 8\sqrt{2}, -8\sqrt{2} \rangle$

Vector **v** *has the given direction and magnitude. Find the horizontal and vertical components of* **v**, *if* θ *is the direction angle of* **v** *from the horizontal.* ***See Example 2.***

13. $\theta = 20°, \quad |\mathbf{v}| = 50$

14. $\theta = 50°, \quad |\mathbf{v}| = 26$

15. $\theta = 35° \, 50', \quad |\mathbf{v}| = 47.8$

16. $\theta = 27° \, 30', \quad |\mathbf{v}| = 15.4$

17. $\theta = 128.5°, \quad |\mathbf{v}| = 198$

18. $\theta = 146.3°, \quad |\mathbf{v}| = 238$

Write each vector in the form $\langle a, b \rangle$. *Round to four decimal places as applicable.* ***See Example 3.***

19. 

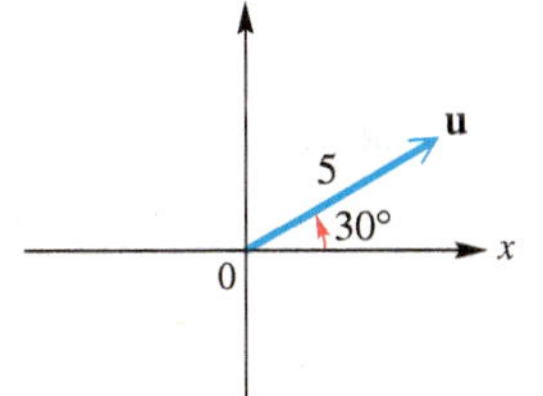

20.

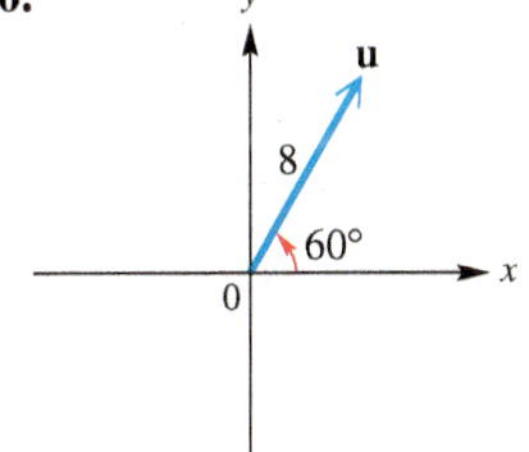

21.

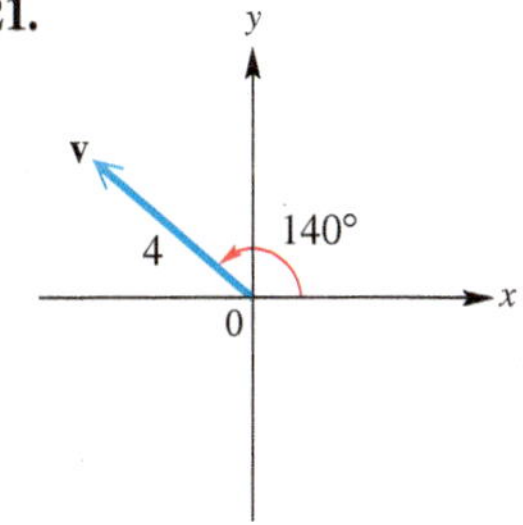

22. **23.** **24.**

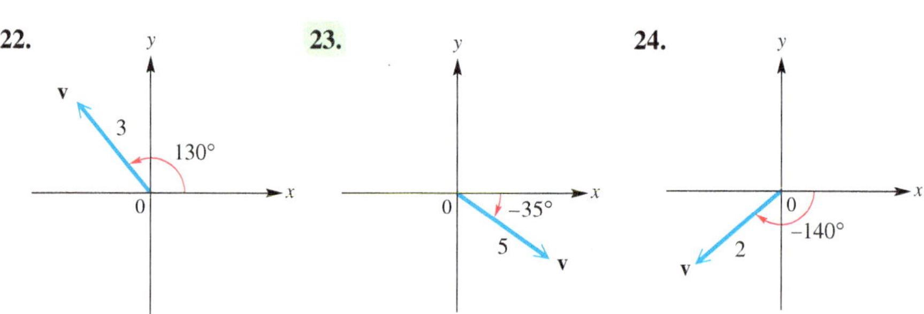

Use the figure to find each vector: ***(a)*** **u** + **v** ***(b)*** **u** − **v** ***(c)*** −**u**. *Use vector notation as in* ***Example 4.***

25. **26.** **27.** **28.** **29.** **30.**

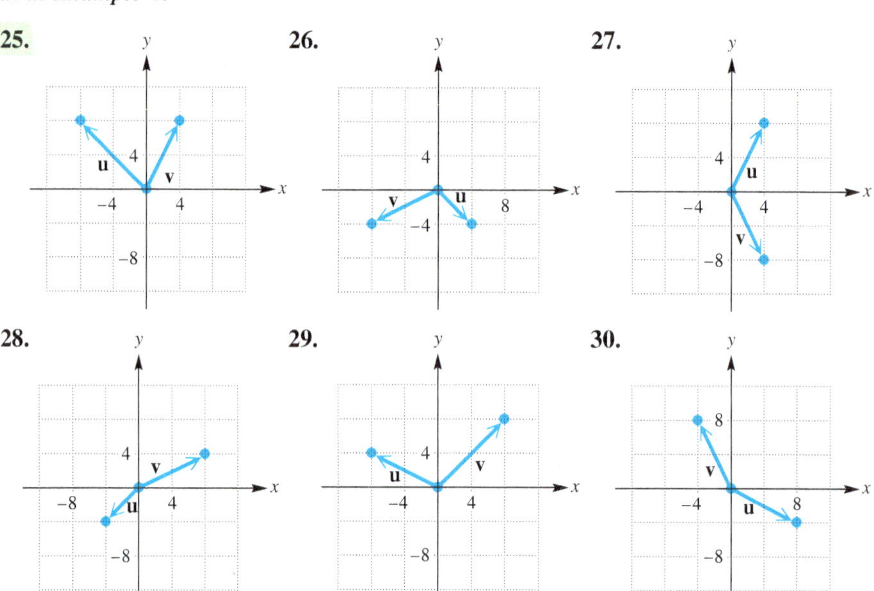

Given vectors **u** *and* **v**, *find:* ***(a)*** 2**u** ***(b)*** 2**u** + 3**v** ***(c)*** **v** − 3**u**. ***See Example 4.***

31. **u** = 2**i**, **v** = **i** + **j**

32. **u** = −**i** + 2**j**, **v** = **i** − **j**

33. **u** = $\langle -1, 2 \rangle$, **v** = $\langle 3, 0 \rangle$

34. **u** = $\langle -2, -1 \rangle$, **v** = $\langle -3, 2 \rangle$

Given **u** = $\langle -2, 5 \rangle$ *and* **v** = $\langle 4, 3 \rangle$, *find each of the following.* ***See Example 4.***

35. **u** − **v**

36. **v** − **u**

37. −4**u**

38. −5**v**

39. 3**u** − 6**v**

40. −2**u** + 4**v**

41. **u** + **v** − 3**u**

42. 2**u** + **v** − 6**v**

*Write each vector in the form a***i** + *b***j**.

43. $\langle -5, 8 \rangle$

44. $\langle 6, -3 \rangle$

45. $\langle 2, 0 \rangle$

46. $\langle 0, -4 \rangle$

Find the dot product for each pair of vectors. ***See Example 5.***

47. $\langle 6, -1 \rangle$, $\langle 2, 5 \rangle$

48. $\langle -3, 8 \rangle$, $\langle 7, -5 \rangle$

49. $\langle 5, 2 \rangle$, $\langle -4, 10 \rangle$

50. $\langle 7, -2 \rangle$, $\langle 4, 14 \rangle$

51. 4**i**, 5**i** − 9**j**

52. 2**i** + 4**j**, −**j**

Find the angle between each pair of vectors. ***See Example 6.***

53. $\langle 2, 1 \rangle$, $\langle -3, 1 \rangle$

54. $\langle 1, 7 \rangle$, $\langle 1, 1 \rangle$

55. $\langle 1, 2 \rangle$, $\langle -6, 3 \rangle$

56. $\langle 4, 0 \rangle$, $\langle 2, 2 \rangle$

57. 3**i** + 4**j**, **j**

58. −5**i** + 12**j**, 3**i** + 2**j**

Let **u** = $\langle -2, 1 \rangle$, **v** = $\langle 3, 4 \rangle$, *and* **w** = $\langle -5, 12 \rangle$. *Evaluate each expression.*

59. (3**u**) · **v**

60. **u** · (3**v**)

61. **u** · **v** − **u** · **w**

62. **u** · (**v** − **w**)

Determine whether each pair of vectors is orthogonal. ***See Example 6(b).***

63. $\langle 1, 2 \rangle, \langle -6, 3 \rangle$

64. $\langle 1, 1 \rangle, \langle 1, -1 \rangle$

65. $\langle 1, 0 \rangle, \langle \sqrt{2}, 0 \rangle$

66. $\langle 3, 4 \rangle, \langle 6, 8 \rangle$

67. $\sqrt{5}\,\mathbf{i} - 2\mathbf{j}, -5\mathbf{i} + 2\sqrt{5}\,\mathbf{j}$

68. $-4\mathbf{i} + 3\mathbf{j}, 8\mathbf{i} - 6\mathbf{j}$

69. *(Modeling) Measuring Rainfall* Suppose that vector **R** models the amount of rainfall in inches and the direction it falls, and vector **A** models the area in square inches and the orientation of the opening of a rain gauge, as illustrated in the figure. The total volume V of water collected in the rain gauge is given by

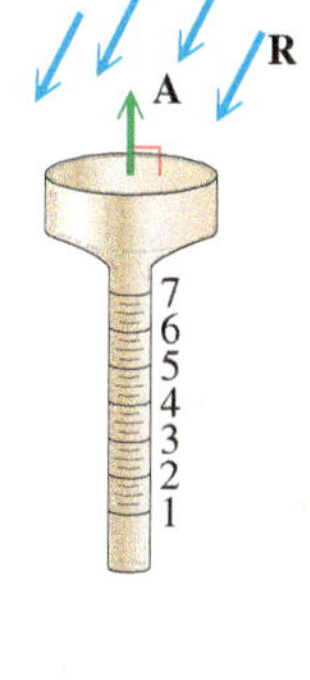

$$V = |\mathbf{R} \cdot \mathbf{A}|.$$

This formula calculates the volume of water collected even if the wind is blowing the rain in a slanted direction or the rain gauge is not exactly vertical. Let $\mathbf{R} = \mathbf{i} - 2\mathbf{j}$ and $\mathbf{A} = 0.5\mathbf{i} + \mathbf{j}$.

(a) Find $|\mathbf{R}|$ and $|\mathbf{A}|$ to the nearest tenth. Interpret the results.

(b) Calculate V to the nearest tenth, and interpret this result.

70. *Concept Check* In **Exercise 69,** for the rain gauge to collect the maximum amount of water, what should be true about vectors **R** and **A**?

Relating Concepts

For individual or collaborative investigation *(Exercises 71–76)*

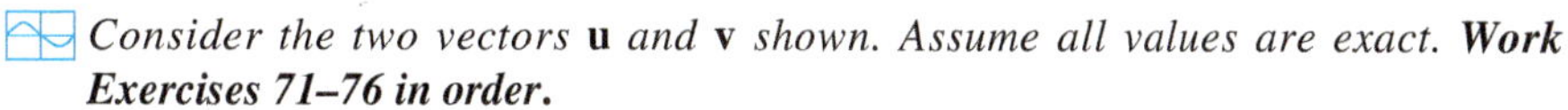

Consider the two vectors **u** *and* **v** *shown. Assume all values are exact.* ***Work Exercises 71–76 in order.***

71. Use trigonometry alone (without using vector notation) to find the magnitude and direction angle of $\mathbf{u} + \mathbf{v}$. Use the law of cosines and the law of sines in your work.

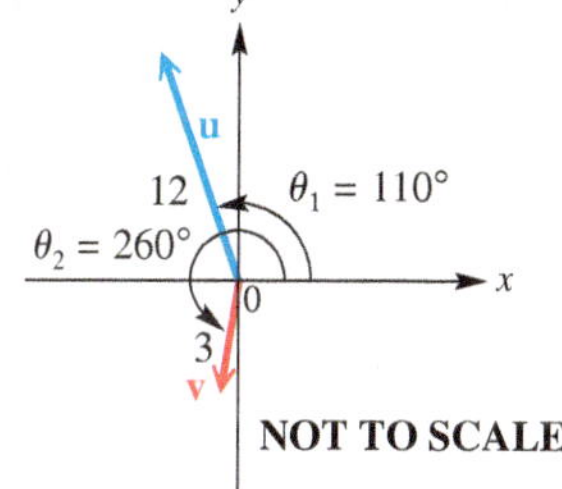

72. Find the horizontal and vertical components of **u**, using a calculator.

73. Find the horizontal and vertical components of **v**, using a calculator.

74. Find the horizontal and vertical components of $\mathbf{u} + \mathbf{v}$ by adding the results obtained in **Exercises 72 and 73.**

75. Use a calculator to find the magnitude and direction angle of the vector $\mathbf{u} + \mathbf{v}$.

76. Compare the answers in **Exercises 71 and 75.** What do you notice? Which method of solution do you prefer?

Summary Exercises on Applications of Trigonometry and Vectors

These summary exercises provide practice with applications that involve solving triangles and using vectors.

1. *Wires Supporting a Flagpole* A flagpole stands vertically on a hillside that makes an angle of 20° with the horizontal. Two supporting wires are attached as shown in the figure. What are the lengths of the supporting wires?

30 ft

15 ft

15 ft

20°

2. ***Distance between a Pin and a Rod*** A slider crank mechanism is shown in the figure. Find the distance between the wrist pin W and the connecting rod center C.

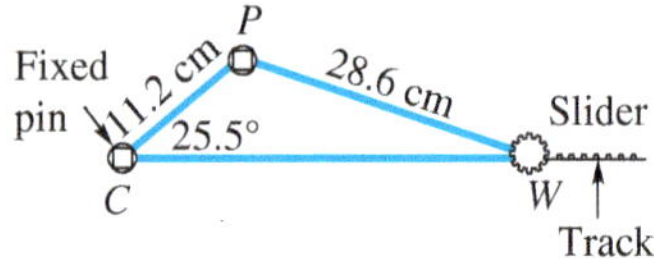

3. ***Distance between Two Lighthouses*** Two lighthouses are located on a north-south line. From lighthouse A, the bearing of a ship 3742 m away is 129° 43′. From lighthouse B, the bearing of the ship is 39° 43′. Find the distance between the lighthouses.

4. ***Hot-Air Balloon*** A hot-air balloon is rising straight up at the speed of 15.0 ft per sec. Then a wind starts blowing horizontally at 5.00 ft per sec. What will the new speed of the balloon be and what angle with the horizontal will the balloon's path make?

5. ***Playing on a Swing*** Mary is playing with her daughter Brittany on a swing. Starting from rest, Mary pulls the swing through an angle of 40° and holds it briefly before releasing the swing. If Brittany weighs 50 lb, what horizontal force, to the nearest pound, must Mary apply while holding the swing?

6. ***Height of an Airplane*** Two observation points A and B are 950 ft apart. From these points the angles of elevation of an airplane are 52° and 57°. See the figure. Find the height of the airplane.

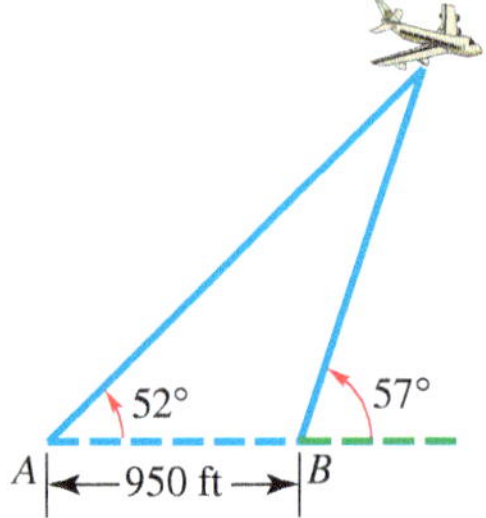

7. ***Wind and Vectors*** A wind can be described by $\mathbf{v} = 6\mathbf{i} + 8\mathbf{j}$, where vector $\mathbf{j}$ points north and represents a south wind of 1 mph.
   (a) What is the speed of the wind?
   (b) Find $3\mathbf{v}$ and interpret the result.
   (c) Interpret the direction and speed of the wind if it changes to $\mathbf{u} = -8\mathbf{i} + 8\mathbf{j}$.

8. ***Ground Speed and Bearing*** A plane with an airspeed of 355 mph is on a bearing of 62°. A wind is blowing from west to east at 28.5 mph. Find the ground speed and the actual bearing of the plane.

9. ***Property Survey*** A surveyor reported the following data about a piece of property: "The property is triangular in shape, with dimensions as shown in the figure." Use the law of sines to see whether such a piece of property could exist.

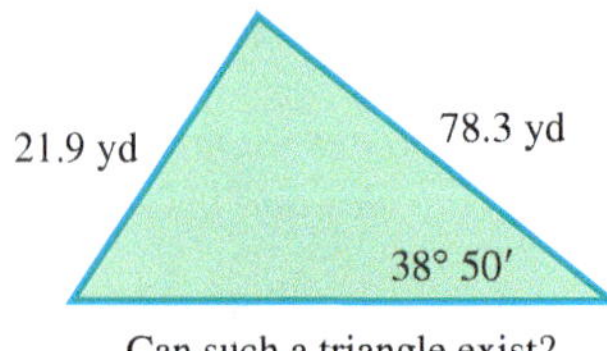

Can such a triangle exist?

10. ***Property Survey*** A triangular piece of property has the dimensions shown. It turns out that the surveyor did not consider every possible case. Use the law of sines to show why.

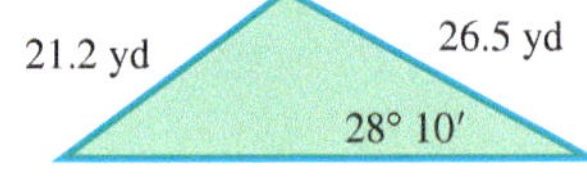

8.5 Trigonometric (Polar) Form of Complex Numbers; Products and Quotients

- The Complex Plane and Vector Representation
- Trigonometric (Polar) Form
- Converting between Rectangular and Trigonometric Forms
- An Application of Complex Numbers to Fractals
- Products of Complex Numbers in Trigonometric Form
- Quotients of Complex Numbers in Trigonometric Form

The Complex Plane and Vector Representation

Unlike real numbers, complex numbers cannot be ordered. One way to organize and illustrate them is by using a graph in a rectangular coordinate system.

To graph a complex number such as $2 - 3i$, we modify the coordinate system by calling the horizontal axis the **real axis** and the vertical axis the **imaginary axis.** Then complex numbers can be graphed in this **complex plane,** as shown in **Figure 45.** ***Each complex number a + bi determines a unique position vector with initial point (0, 0) and terminal point (a, b).***

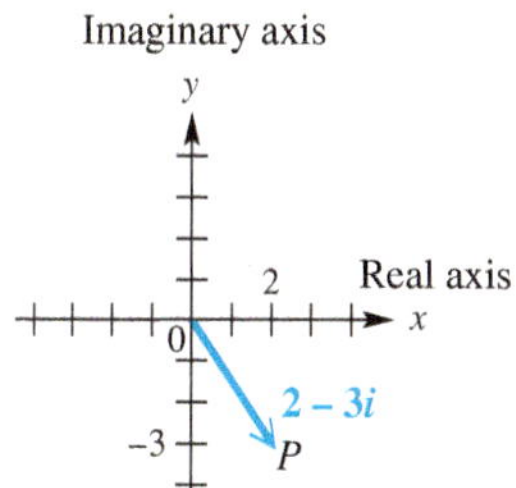

Figure 45

NOTE This geometric representation is the reason that $a + bi$ is called the **rectangular form** of a complex number. (*Rectangular form* is also known as *standard form.*)

Recall that $(4 + i) + (1 + 3i) = 5 + 4i$. Graphically, the sum of two complex numbers is represented by the vector that is the **resultant** of the vectors corresponding to the two numbers. See **Figure 46.**

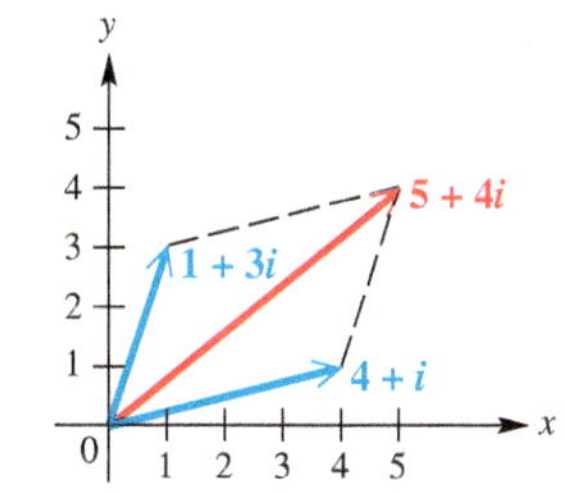

Figure 46

EXAMPLE 1 Expressing the Sum of Complex Numbers Graphically

Find the sum of $6 - 2i$ and $-4 - 3i$. Graph both complex numbers and their resultant.

SOLUTION The sum is found by adding the two numbers.

$$(6 - 2i) + (-4 - 3i) = 2 - 5i \quad \text{Add real parts, and add imaginary parts.}$$

The graphs are shown in **Figure 47.**

Now Try Exercise 21.

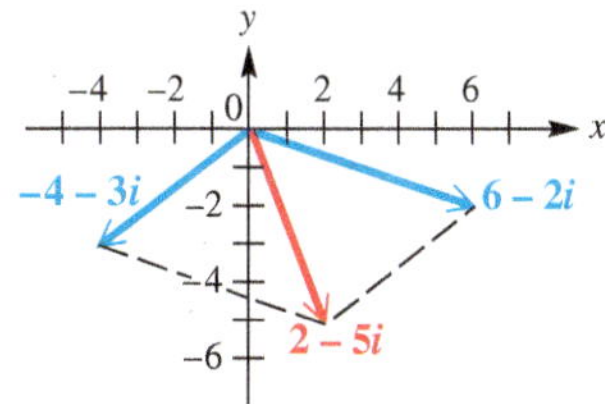

Figure 47

Trigonometric (Polar) Form

Figure 48 shows the complex number $x + yi$ that corresponds to a vector **OP** with direction angle θ and magnitude r. The following relationships among x, y, r, and θ can be verified from **Figure 48.**

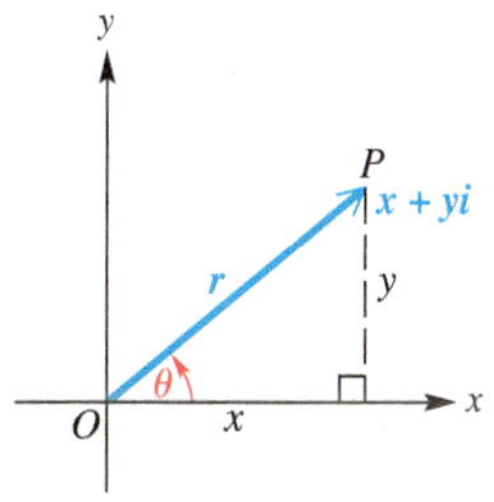

Figure 48

Relationships among x, y, r, and θ

$$x = r\cos\theta \qquad y = r\sin\theta$$

$$r = \sqrt{x^2 + y^2} \qquad \tan\theta = \frac{y}{x}, \quad \text{if } x \neq 0$$

Substituting $x = r\cos\theta$ and $y = r\sin\theta$ into $x + yi$ gives the following.

$$\begin{aligned} & x + yi \\ & = r\cos\theta + (r\sin\theta)i && \text{Substitute.} \\ & = r(\cos\theta + i\sin\theta) && \text{Factor out } r. \end{aligned}$$

Trigonometric (Polar) Form of a Complex Number

The **trigonometric form** (or **polar form**) of the complex number $x + yi$ is

$$r(\cos \theta + i \sin \theta).$$

The expression $\cos \theta + i \sin \theta$ is sometimes abbreviated **cis** $\boldsymbol{\theta}$. Using this notation, $\boldsymbol{r(\cos \theta + i \sin \theta)}$ is written $\boldsymbol{r}$ **cis** $\boldsymbol{\theta}$.

The number r is the **absolute value** (or **modulus**) of $x + yi$, and θ is the **argument** of $x + yi$. In this section, we choose the value of θ in the interval $[0°, 360°)$. Any angle coterminal with θ also could serve as the argument.

EXAMPLE 2 Converting from Trigonometric Form to Rectangular Form

Write $2(\cos 300° + i \sin 300°)$ in rectangular form.

ALGEBRAIC SOLUTION

$2(\cos 300° + i \sin 300°)$

$= 2\left(\frac{1}{2} - i\frac{\sqrt{3}}{2}\right)$ $\quad$ $\cos 300° = \frac{1}{2}$; $\sin 300° = -\frac{\sqrt{3}}{2}$

$= 1 - i\sqrt{3}$ $\quad$ Distributive property

Note that the real part is positive and the imaginary part is negative. This is consistent with 300° being a quadrant IV angle. For a 300° angle, the reference angle is 60°. Thus the function values cos 300° and sin 300° correspond *in absolute value* to those of cos 60° and sin 60°, with the first of these equal to $\frac{1}{2}$ and the second equal to $-\frac{\sqrt{3}}{2}$.

GRAPHING CALCULATOR SOLUTION

In **Figure 49,** the first result confirms the algebraic solution, where an approximation for $-\sqrt{3}$ is used for the imaginary part (from the second result). The TI-84 Plus also converts from polar to rectangular form, as seen in the third and fourth results.

```
NORMAL FLOAT AUTO a+bi DEGREE MP
2(cos(300)+isin(300))
                      1-1.732050808i
-√3
                        -1.732050808
P▸Rx(2,300)
                                   1
P▸Ry(2,300)
                        -1.732050808
```

Figure 49

✔ **Now Try Exercise 37.**

Converting between Rectangular and Trigonometric Forms To convert from rectangular form to trigonometric form, we use the following procedure.

Converting from Rectangular to Trigonometric Form

Step 1 Sketch a graph of the number $x + yi$ in the complex plane.

Step 2 Find r by using the equation $r = \sqrt{x^2 + y^2}$.

Step 3 Find θ by using the equation $\tan \theta = \frac{y}{x}$, where $x \neq 0$, choosing the quadrant indicated in Step 1.

CAUTION Errors often occur in Step 3. ***Be sure to choose the correct quadrant for θ by referring to the graph sketched in Step 1.***

EXAMPLE 3 Converting from Rectangular to Trigonometric Form

Write each complex number in trigonometric form.

(a) $-\sqrt{3}+i$ (Use radian measure.) **(b)** $-3i$ (Use degree measure.)

SOLUTION

(a) We start by sketching the graph of $-\sqrt{3}+i$ in the complex plane, as shown in **Figure 50.** Next, we use $x=-\sqrt{3}$ and $y=1$ to find r and θ.

$$r=\sqrt{x^2+y^2}=\sqrt{\left(-\sqrt{3}\right)^2+1^2}=\sqrt{3+1}=2$$

$$\tan\theta=\frac{y}{x}=\frac{1}{-\sqrt{3}}=-\frac{1}{\sqrt{3}}\cdot\frac{\sqrt{3}}{\sqrt{3}}=-\frac{\sqrt{3}}{3}$$

Rationalize the denominator.

```
NORMAL FLOAT AUTO a+bi RADIAN MP
R▸Pr(-√3,1)
                               2
R▸Pθ(-√3,1)
                     2.617993878
5π
6
                     2.617993878
```

See **Example 3(a).** The TI-84 Plus converts from rectangular form to polar form. The value of θ in the second result is an approximation for $\frac{5\pi}{6}$, as shown in the third result.

Because $\tan\theta=-\frac{\sqrt{3}}{3}$, the reference angle for θ in radians is $\frac{\pi}{6}$. From the graph, we see that θ is in quadrant II, so $\theta=\pi-\frac{\pi}{6}=\frac{5\pi}{6}$.

$$-\sqrt{3}+i=2\left(\cos\frac{5\pi}{6}+i\sin\frac{5\pi}{6}\right),\quad\text{or}\quad 2\text{ cis }\frac{5\pi}{6}$$

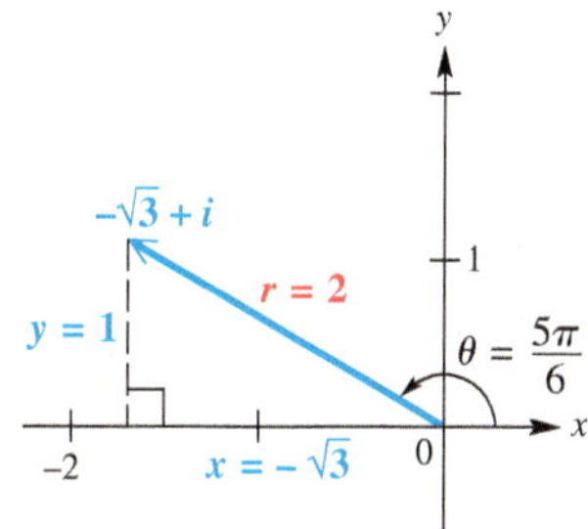

Figure 50

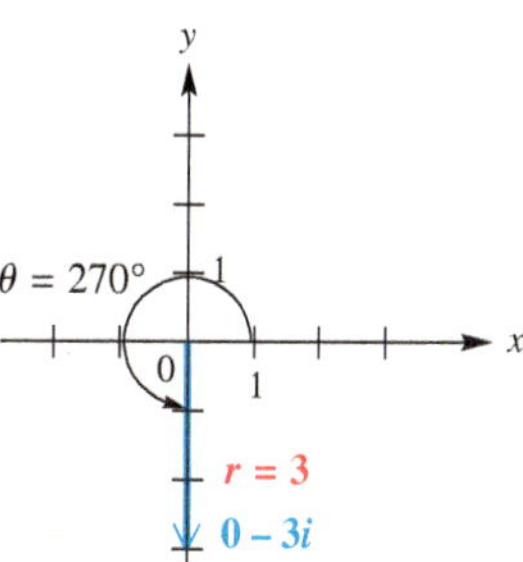

Figure 51

(b) See **Figure 51.** Because $-3i=0-3i$, we have $x=0$ and $y=-3$.

$$r=\sqrt{0^2+(-3)^2}=\sqrt{0+9}=\sqrt{9}=3 \quad \text{Substitute.}$$

```
NORMAL FLOAT AUTO a+bi DEGREE MP
R▸Pr(0,-3)
                               3
R▸Pθ(0,-3)
                             -90
```

Compare to the result in **Example 3(b).** The angle $-90°$ is coterminal with $270°$. The calculator returns θ values between $-180°$ and $180°$.

We cannot find θ by using $\tan\theta=\frac{y}{x}$ because $x=0$. However, the graph shows that the least positive value for θ is $270°$.

$$-3i=3(\cos 270°+i\sin 270°),\quad\text{or}\quad 3\text{ cis }270° \quad \text{Trigonometric form}$$

✔ **Now Try Exercises 49 and 55.**

EXAMPLE 4 Converting between Trigonometric and Rectangular Forms Using Calculator Approximations

Write each complex number in its alternative form, using calculator approximations as necessary.

(a) $6(\cos 125°+i\sin 125°)$ **(b)** $5-4i$

SOLUTION

(a) Because $125°$ does not have a special angle as a reference angle, we cannot find exact values for $\cos 125°$ and $\sin 125°$.

$$6(\cos 125°+i\sin 125°)$$

$$\approx 6(-0.5735764364+0.8191520443i) \quad \text{Use a calculator set to degree mode.}$$

$$\approx -3.4415+4.9149i \quad \text{Four decimal places}$$

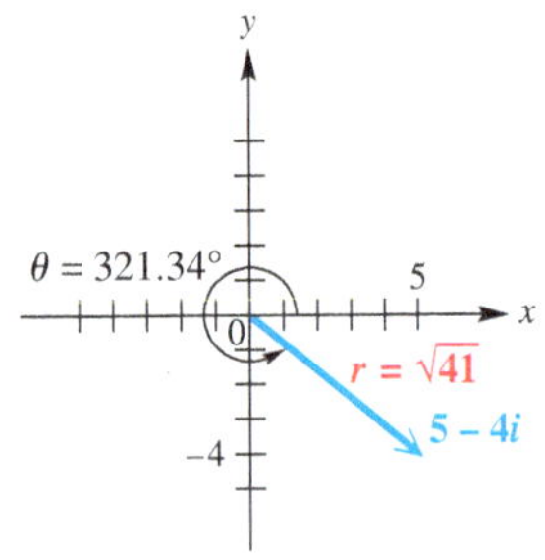

Figure 52

(b) A sketch of $5 - 4i$ shows that θ must be in quadrant IV. See **Figure 52.**

$$r = \sqrt{5^2 + (-4)^2} = \sqrt{41} \quad \text{and} \quad \tan\theta = -\frac{4}{5}$$

Use a calculator to find that one measure of θ is $-38.66°$. In order to express θ in the interval $[0, 360°)$, we find $\theta = 360° - 38.66° = 321.34°$.

$$5 - 4i = \sqrt{41}\ \text{cis}\ 321.34°$$

✔ **Now Try Exercises 61 and 65.**

An Application of Complex Numbers to Fractals At its basic level, a **fractal** is a unique, enchanting geometric figure with an endless self-similarity property. A fractal image repeats itself infinitely with ever-decreasing dimensions. If we look at smaller and smaller portions, we will continue to see the whole—it is much like looking into two parallel mirrors that are facing each other.

EXAMPLE 5 Deciding Whether a Complex Number Is in the Julia Set

The fractal called the **Julia set** is shown in **Figure 53.** To determine whether a complex number $z = a + bi$ is in this Julia set, perform the following sequence of calculations.

$$z^2 - 1, \quad (z^2 - 1)^2 - 1, \quad [(z^2 - 1)^2 - 1]^2 - 1, \quad \ldots$$

If the absolute values of any of the resulting complex numbers exceed 2, then the complex number z is not in the Julia set. Otherwise z is part of this set and the point (a, b) should be shaded in the graph.

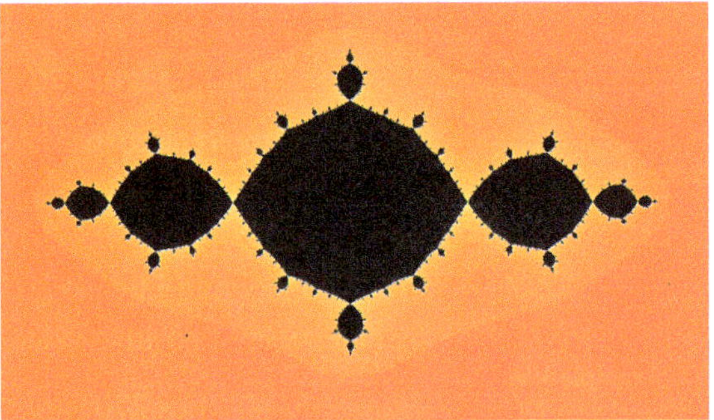

Figure 53

Determine whether each number belongs to the Julia set.

(a) $z = 0 + 0i$ **(b)** $z = 1 + 1i$

SOLUTION

(a) Here

$$\begin{aligned} z &= 0 + 0i = 0, \\ z^2 - 1 &= 0^2 - 1 = -1, \\ (z^2 - 1)^2 - 1 &= (-1)^2 - 1 = 0, \\ [(z^2 - 1)^2 - 1]^2 - 1 &= 0^2 - 1 = -1, \quad \text{and so on.} \end{aligned}$$

We see that the calculations repeat as $0, -1, 0, -1$, and so on. The absolute values are either 0 or 1, which do not exceed 2, so $0 + 0i$ is in the Julia set and the point $(0, 0)$ is part of the graph.

(b) For $z = 1 + 1i$, we have the following.

$$
\begin{aligned}
&z^2 - 1 \\
&\quad = (1 + i)^2 - 1 && \text{Substitute for } z;\ 1 + 1i = 1 + i. \\
&\quad = (1 + 2i + i^2) - 1 && \text{Square the binomial; } (x + y)^2 = x^2 + 2xy + y^2. \\
&\quad = -1 + 2i && i^2 = -1
\end{aligned}
$$

The absolute value is

$$\sqrt{(-1)^2 + 2^2} = \sqrt{5}.$$

Because $\sqrt{5}$ is greater than 2, the number $1 + 1i$ is not in the Julia set and $(1, 1)$ is not part of the graph.

✔ **Now Try Exercise 71.**

Products of Complex Numbers in Trigonometric Form Using the FOIL method to multiply complex numbers in rectangular form, we find the product of $1 + i\sqrt{3}$ and $-2\sqrt{3} + 2i$ as follows.

$$
\begin{aligned}
&\left(1 + i\sqrt{3}\right)\left(-2\sqrt{3} + 2i\right) \\
&\quad = -2\sqrt{3} + 2i - 2i(3) + 2i^2\sqrt{3} && \text{FOIL method} \\
&\quad = -2\sqrt{3} + 2i - 6i - 2\sqrt{3} && i^2 = -1 \\
&\quad = -4\sqrt{3} - 4i && \text{Combine like terms.}
\end{aligned}
$$

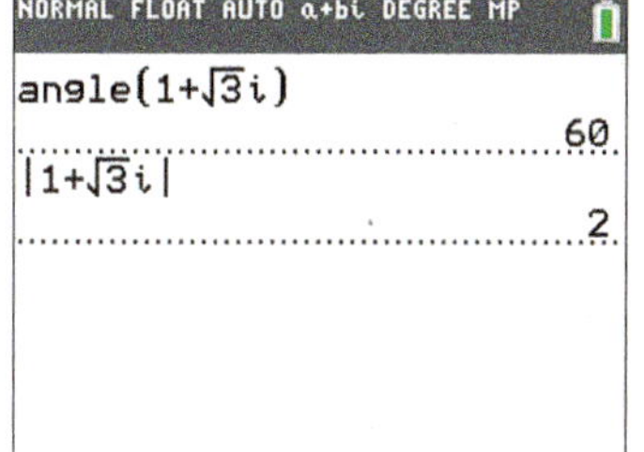

With the TI-84 Plus calculator in complex and degree modes, the MATH menu can be used to find the angle and the magnitude (absolute value) of a complex number.

We can also find this same product by first converting the complex numbers $1 + i\sqrt{3}$ and $-2\sqrt{3} + 2i$ to trigonometric form using the method explained earlier in this section.

$$1 + i\sqrt{3} = 2(\cos 60° + i \sin 60°)$$

$$-2\sqrt{3} + 2i = 4(\cos 150° + i \sin 150°)$$

If we multiply the trigonometric forms and use identities for the cosine and the sine of the sum of two angles, then the result is as follows.

$$
\begin{aligned}
&[2(\cos 60° + i \sin 60°)][4(\cos 150° + i \sin 150°)] \\
&\quad = 2 \cdot 4(\cos 60° \cdot \cos 150° + i \sin 60° \cdot \cos 150° \\
&\qquad + i \cos 60° \cdot \sin 150° + i^2 \sin 60° \cdot \sin 150°) && \text{Multiply the absolute values and the binomials.} \\
&\quad = 8[(\cos 60° \cdot \cos 150° - \sin 60° \cdot \sin 150°) \\
&\qquad + i(\sin 60° \cdot \cos 150° + \cos 60° \cdot \sin 150°)] && i^2 = -1;\ \text{Factor out } i. \\
&\quad = 8[\cos(60° + 150°) + i \sin(60° + 150°)] \\
&\qquad && \cos(A + B) = \cos A \cdot \cos B - \sin A \cdot \sin B; \\
&\qquad && \sin(A + B) = \sin A \cdot \cos B + \cos A \cdot \sin B \\
&\quad = 8(\cos 210° + i \sin 210°) && \text{Add.}
\end{aligned}
$$

Notice the following.

- The absolute value of the product, 8, is equal to the product of the absolute values of the factors, $2 \cdot 4$.
- The argument of the product, 210°, is equal to the sum of the arguments of the factors, $60° + 150°$.

The product obtained when multiplying by the first method is the rectangular form of the product obtained when multiplying by the second method.

$$8(\cos 210^\circ + i \sin 210^\circ)$$

$$= 8\left(-\frac{\sqrt{3}}{2} - \frac{1}{2}i\right) \quad \cos 210^\circ = -\frac{\sqrt{3}}{2};\ \sin 210^\circ = -\frac{1}{2}$$

$$= -4\sqrt{3} - 4i \quad \text{Rectangular form}$$

Product Theorem

If $r_1(\cos \theta_1 + i \sin \theta_1)$ and $r_2(\cos \theta_2 + i \sin \theta_2)$ are any two complex numbers, then the following holds.

$$[r_1(\cos \theta_1 + i \sin \theta_1)] \cdot [r_2(\cos \theta_2 + i \sin \theta_2)]$$
$$= r_1 r_2[\cos(\theta_1 + \theta_2) + i \sin(\theta_1 + \theta_2)]$$

In compact form, this is written

$$(r_1 \operatorname{cis} \theta_1)(r_2 \operatorname{cis} \theta_2) = r_1 r_2 \operatorname{cis}(\theta_1 + \theta_2).$$

That is, to multiply complex numbers in trigonometric form, multiply their absolute values and add their arguments.

EXAMPLE 6 Using the Product Theorem

Find the product of $3(\cos 45^\circ + i \sin 45^\circ)$ and $2(\cos 135^\circ + i \sin 135^\circ)$. Write the answer in rectangular form.

SOLUTION

$$[3(\cos 45^\circ + i \sin 45^\circ)][2(\cos 135^\circ + i \sin 135^\circ)] \quad \text{Write as a product.}$$

$$= 3 \cdot 2[\cos(45^\circ + 135^\circ) + i \sin(45^\circ + 135^\circ)] \quad \text{Product theorem}$$

$$= 6(\cos 180^\circ + i \sin 180^\circ) \quad \text{Multiply and add.}$$

$$= 6(-1 + i \cdot 0) \quad \cos 180^\circ = -1;\ \sin 180^\circ = 0$$

$$= -6 \quad \text{Rectangular form}$$

✔ **Now Try Exercise 77.**

Quotients of Complex Numbers in Trigonometric Form The rectangular form of the quotient of $1 + i\sqrt{3}$ and $-2\sqrt{3} + 2i$ is found as follows.

$$\frac{1 + i\sqrt{3}}{-2\sqrt{3} + 2i}$$

$$= \frac{(1 + i\sqrt{3})(-2\sqrt{3} - 2i)}{(-2\sqrt{3} + 2i)(-2\sqrt{3} - 2i)} \quad \text{Multiply both numerator and denominator by the conjugate of the denominator.}$$

$$= \frac{-2\sqrt{3} - 2i - 6i - 2i^2\sqrt{3}}{12 - 4i^2} \quad \text{FOIL method; } (x + y)(x - y) = x^2 - y^2$$

$$= \frac{-8i}{16} \quad \text{Simplify.}$$

$$= -\frac{1}{2}i \quad \text{Lowest terms}$$

Writing $1 + i\sqrt{3}$, $-2\sqrt{3} + 2i$, and $-\frac{1}{2}i$ in trigonometric form gives

$$1 + i\sqrt{3} = 2(\cos 60° + i \sin 60°),$$

$$-2\sqrt{3} + 2i = 4(\cos 150° + i \sin 150°), \quad \text{Use } r = \sqrt{x^2 + y^2} \text{ and } \tan\theta = \frac{y}{x}.$$

and $$-\frac{1}{2}i = \frac{1}{2}[\cos(-90°) + i \sin(-90°)]. \quad -90° \text{ can be replaced by } 270°.$$

Here, the absolute value of the quotient, $\frac{1}{2}$, is the quotient of the two absolute values, $\frac{2}{4} = \frac{1}{2}$. The argument of the quotient, $-90°$, is the difference of the two arguments,

$$60° - 150° = -90°.$$

Quotient Theorem

If $r_1(\cos\theta_1 + i \sin\theta_1)$ and $r_2(\cos\theta_2 + i\sin\theta_2)$ are any two complex numbers, where $r_2(\cos\theta_2 + i\sin\theta_2) \neq 0$, then the following holds.

$$\frac{r_1(\cos\theta_1 + i\sin\theta_1)}{r_2(\cos\theta_2 + i\sin\theta_2)} = \frac{r_1}{r_2}[\cos(\theta_1 - \theta_2) + i\sin(\theta_1 - \theta_2)]$$

In compact form, this is written

$$\frac{r_1 \text{ cis } \theta_1}{r_2 \text{ cis } \theta_2} = \frac{r_1}{r_2}\text{cis}(\theta_1 - \theta_2).$$

That is, to divide complex numbers in trigonometric form, divide their absolute values and subtract their arguments.

EXAMPLE 7 Using the Quotient Theorem

Find the quotient $\frac{10 \text{ cis}(-60°)}{5 \text{ cis } 150°}$. Write the answer in rectangular form.

SOLUTION

$$\frac{10 \text{ cis}(-60°)}{5 \text{ cis } 150°}$$

$= \frac{10}{5}\text{cis}(-60° - 150°)$ Quotient theorem

$= 2 \text{ cis}(-210°)$ Divide and subtract.

$= 2[\cos(-210°) + i \sin(-210°)]$ Rewrite.

$= 2\left[-\frac{\sqrt{3}}{2} + i\left(\frac{1}{2}\right)\right]$ $\cos(-210°) = -\frac{\sqrt{3}}{2}$; $\sin(-210°) = \frac{1}{2}$

$= -\sqrt{3} + i$ Distributive property

✔ Now Try Exercise 87.

8.5 Exercises

CONCEPT PREVIEW *For each complex number shown, give **(a)** its rectangular form and **(b)** its trigonometric (polar) form with $r > 0$, $0° \le \theta < 360°$.*

1. **2.** 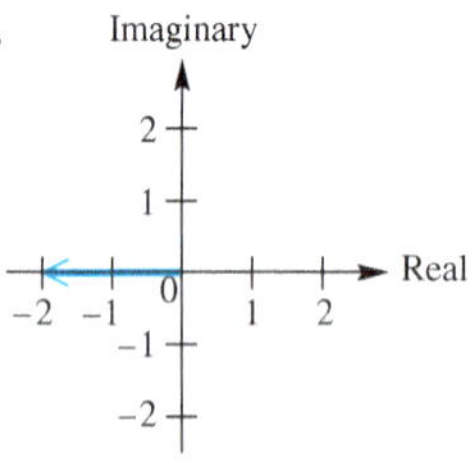 **3.** 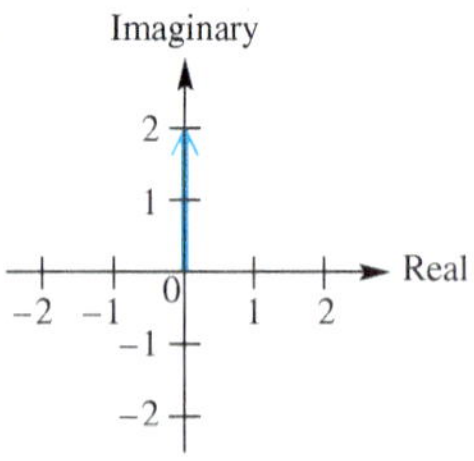 **4.** 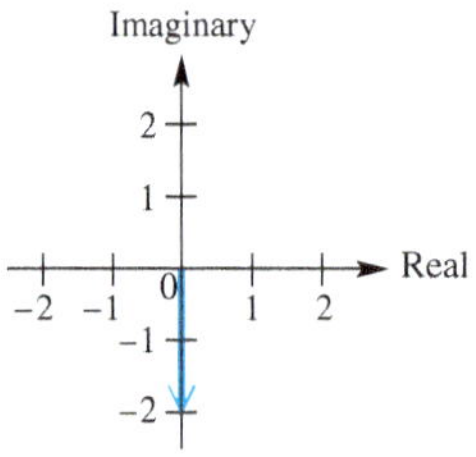 **5.** 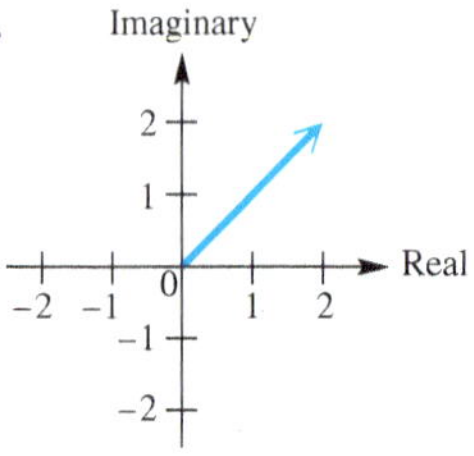 **6.**

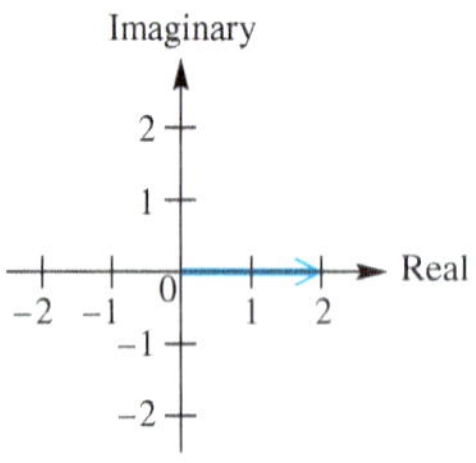

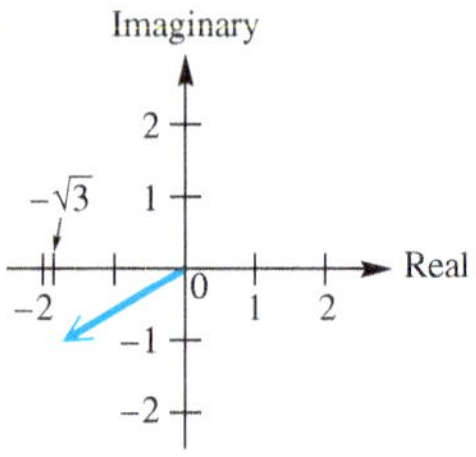

CONCEPT PREVIEW *Fill in the blanks to correctly complete each problem.*

7. When multiplying two complex numbers in trigonometric form, we ______ their absolute values and ______ their arguments.

8. When dividing two complex numbers in trigonometric form, we ______ their absolute values and ______ their arguments.

9. $[5(\cos 150° + i \sin 150°)][2(\cos 30° + i \sin 30°)]$

$= \_\_\_\_ (\cos \_\_\_\_ + i \sin \_\_\_\_)$

$= \_\_\_\_ + \_\_\_\_ i$

10. $\dfrac{6(\cos 120° + i \sin 120°)}{2(\cos 30° + i \sin 30°)}$

$= \_\_\_\_ (\cos \_\_\_\_ + i \sin \_\_\_\_)$

$= \_\_\_\_ + \_\_\_\_ i$

11. $\text{cis}(-1000°) \cdot \text{cis } 1000°$

$= \text{cis } \_\_\_\_$

$= \_\_\_\_ + \_\_\_\_ i$

12. $\dfrac{5 \text{ cis } 50{,}000°}{\text{cis } 50{,}000°}$

$= 5 \text{ cis } \_\_\_\_$

$= \_\_\_\_ + \_\_\_\_ i$

Graph each complex number. ***See Example 1.***

13. $-3 + 2i$ **14.** $6 - 5i$ **15.** $\sqrt{2} + \sqrt{2}i$ **16.** $2 - 2i\sqrt{3}$

17. $-4i$ **18.** $3i$ **19.** -8 **20.** 2

Find the sum of each pair of complex numbers. In Exercises 21–24, graph both complex numbers and their resultant. ***See Example 1.***

21. $4 - 3i,\ -1 + 2i$ **22.** $2 + 3i,\ -4 - i$ **23.** $5 - 6i,\ -5 + 3i$

24. $7 - 3i,\ -4 + 3i$ **25.** $-3,\ 3i$ **26.** $6,\ -2i$

27. $-5 - 8i,\ -1$ **28.** $4 - 2i,\ 5$ **29.** $7 + 6i,\ 3i$

30. $2 + 6i, -2i$ **31.** $\frac{1}{2} + \frac{2}{3}i, \frac{2}{3} + \frac{1}{2}i$ **32.** $-\frac{1}{5} + \frac{2}{7}i, \frac{3}{7} - \frac{3}{4}i$

Write each complex number in rectangular form. ***See Example 2.***

33. $2(\cos 45° + i \sin 45°)$ **34.** $4(\cos 60° + i \sin 60°)$

35. $10(\cos 90° + i \sin 90°)$ **36.** $8(\cos 270° + i \sin 270°)$

37. $4(\cos 240° + i \sin 240°)$ **38.** $2(\cos 330° + i \sin 330°)$

39. $3 \text{ cis } 150°$ **40.** $2 \text{ cis } 30°$ **41.** $5 \text{ cis } 300°$

42. $6 \text{ cis } 135°$ **43.** $\sqrt{2} \text{ cis } 225°$ **44.** $\sqrt{3} \text{ cis } 315°$

45. $4(\cos(-30°) + i \sin(-30°))$ **46.** $\sqrt{2}(\cos(-60°) + i \sin(-60°))$

Write each complex number in trigonometric form $r(\cos \theta + i \sin \theta)$, with θ in the interval $[0°, 360°)$. ***See Example 3.***

47. $-3 - 3i\sqrt{3}$ **48.** $1 + i\sqrt{3}$ **49.** $\sqrt{3} - i$ **50.** $4\sqrt{3} + 4i$

51. $-5 - 5i$ **52.** $-2 + 2i$ **53.** $2 + 2i$ **54.** $4 + 4i$

55. $5i$ **56.** $-2i$ **57.** -4 **58.** 7

Write each complex number in its alternative form, using a calculator to approximate answers to four decimal places as necessary. ***See Example 4.***

	Rectangular Form	Trigonometric Form
59.	$2 + 3i$	______
60.	______	$\cos 35° + i \sin 35°$
61.	______	$3(\cos 250° + i \sin 250°)$
62.	$-4 + i$	______
63.	$12i$	______
64.	______	$3 \text{ cis } 180°$
65.	$3 + 5i$	______
66.	______	$\text{cis } 110.5°$

Concept Check *The complex number z, where $z = x + yi$, can be graphed in the plane as (x, y). Describe the graphs of all complex numbers z satisfying the given conditions.*

67. The absolute value of z is 1. **68.** The real and imaginary parts of z are equal.

69. The real part of z is 1. **70.** The imaginary part of z is 1.

Julia Set *Refer to* ***Example 5.***

71. Is $z = -0.2i$ in the Julia set?

72. The graph of the Julia set in **Figure 53** appears to be symmetric with respect to both the x-axis and the y-axis. Complete the following to show that this is true.

(a) Show that complex conjugates have the same absolute value.

(b) Compute $z_1^2 - 1$ and $z_2^2 - 1$, where $z_1 = a + bi$ and $z_2 = a - bi$.

(c) Discuss why if (a, b) is in the Julia set, then so is $(a, -b)$.

(d) Conclude that the graph of the Julia set must be symmetric with respect to the x-axis.

(e) Using a similar argument, show that the Julia set must also be symmetric with respect to the y-axis.

Find each product. Write answers in rectangular form. ***See Example 6.***

73. $[3(\cos 60^\circ + i \sin 60^\circ)][2(\cos 90^\circ + i \sin 90^\circ)]$

74. $[4(\cos 30^\circ + i \sin 30^\circ)][5(\cos 120^\circ + i \sin 120^\circ)]$

75. $[4(\cos 60^\circ + i \sin 60^\circ)][6(\cos 330^\circ + i \sin 330^\circ)]$

76. $[8(\cos 300^\circ + i \sin 300^\circ)][5(\cos 120^\circ + i \sin 120^\circ)]$

77. $[2(\cos 135^\circ + i \sin 135^\circ)][2(\cos 225^\circ + i \sin 225^\circ)]$

78. $[8(\cos 210^\circ + i \sin 210^\circ)][2(\cos 330^\circ + i \sin 330^\circ)]$

79. $(\sqrt{3} \text{ cis } 45^\circ)(\sqrt{3} \text{ cis } 225^\circ)$

80. $(\sqrt{6} \text{ cis } 120^\circ)[\sqrt{6} \text{ cis}(-30^\circ)]$

81. $(5 \text{ cis } 90^\circ)(3 \text{ cis } 45^\circ)$

82. $(3 \text{ cis } 300^\circ)(7 \text{ cis } 270^\circ)$

Find each quotient. Write answers in rectangular form. In Exercises 89–94, first convert the numerator and the denominator to trigonometric form. ***See Example 7.***

83. $\dfrac{4(\cos 150^\circ + i \sin 150^\circ)}{2(\cos 120^\circ + i \sin 120^\circ)}$

84. $\dfrac{24(\cos 150^\circ + i \sin 150^\circ)}{2(\cos 30^\circ + i \sin 30^\circ)}$

85. $\dfrac{10(\cos 50^\circ + i \sin 50^\circ)}{5(\cos 230^\circ + i \sin 230^\circ)}$

86. $\dfrac{12(\cos 23^\circ + i \sin 23^\circ)}{6(\cos 293^\circ + i \sin 293^\circ)}$

87. $\dfrac{3 \text{ cis } 305^\circ}{9 \text{ cis } 65^\circ}$

88. $\dfrac{16 \text{ cis } 310^\circ}{8 \text{ cis } 70^\circ}$

89. $\dfrac{8}{\sqrt{3} + i}$

90. $\dfrac{2i}{-1 - i\sqrt{3}}$

91. $\dfrac{-i}{1 + i}$

92. $\dfrac{1}{2 - 2i}$

93. $\dfrac{2\sqrt{6} - 2i\sqrt{2}}{\sqrt{2} - i\sqrt{6}}$

94. $\dfrac{-3\sqrt{2} + 3i\sqrt{6}}{\sqrt{6} + i\sqrt{2}}$

Use a calculator to perform the indicated operations. Write answers in rectangular form, expressing real and imaginary parts to four decimal places.

95. $[2.5(\cos 35^\circ + i \sin 35^\circ)][3.0(\cos 50^\circ + i \sin 50^\circ)]$

96. $[4.6(\cos 12^\circ + i \sin 12^\circ)][2.0(\cos 13^\circ + i \sin 13^\circ)]$

97. $(12 \text{ cis } 18.5^\circ)(3 \text{ cis } 12.5^\circ)$

98. $(4 \text{ cis } 19.25^\circ)(7 \text{ cis } 41.75^\circ)$

99. $\dfrac{45\left(\cos \frac{2\pi}{3} + i \sin \frac{2\pi}{3}\right)}{22.5\left(\cos \frac{3\pi}{5} + i \sin \frac{3\pi}{5}\right)}$

100. $\dfrac{30\left(\cos \frac{2\pi}{5} + i \sin \frac{2\pi}{5}\right)}{10\left(\cos \frac{\pi}{7} + i \sin \frac{\pi}{7}\right)}$

101. $\left[2 \text{ cis } \dfrac{5\pi}{9}\right]^2$

102. $\left[24.3 \text{ cis } \dfrac{7\pi}{12}\right]^2$

Work each problem.

103. Note that $(r \text{ cis } \theta)^2 = (r \text{ cis } \theta)(r \text{ cis } \theta) = r^2 \text{ cis}(\theta + \theta) = r^2 \text{ cis } 2\theta$. Explain how we can square a complex number in trigonometric form. (In the next section, we will develop this idea more fully.)

104. Without actually performing the operations, state why the following products are the same.

$$[2(\cos 45^\circ + i \sin 45^\circ)] \cdot [5(\cos 90^\circ + i \sin 90^\circ)]$$

and $[2[\cos(-315^\circ) + i \sin(-315^\circ)]] \cdot [5[\cos(-270^\circ) + i \sin(-270^\circ)]]$

105. Show that $\frac{1}{z} = \frac{1}{r}(\cos \theta - i \sin \theta)$, where $z = r(\cos \theta + i \sin \theta)$.

106. The complex conjugate of $r(\cos\theta + i\sin\theta)$ is $r(\cos\theta - i\sin\theta)$. Use these trigonometric forms to show that the product of complex conjugates is always a real number.

(Modeling) Electrical Current *Solve each problem.*

107. The alternating current in an electric inductor is $I = \frac{E}{Z}$ amperes, where E is voltage and $Z = R + X_L i$ is impedance. If $E = 8(\cos 20° + i\sin 20°)$, $R = 6$, and $X_L = 3$, find the current. Give the answer in rectangular form, with real and imaginary parts to the nearest hundredth.

108. The current I in a circuit with voltage E, resistance R, capacitive reactance X_c, and inductive reactance X_L is

$$I = \frac{E}{R + (X_L - X_c)i}.$$

Find I if $E = 12(\cos 25° + i\sin 25°)$, $R = 3$, $X_L = 4$, and $X_c = 6$. Give the answer in rectangular form, with real and imaginary parts to the nearest tenth.

(Modeling) Impedance *In the parallel electrical circuit shown in the figure, the impedance Z can be calculated using the equation*

$$Z = \frac{1}{\frac{1}{Z_1} + \frac{1}{Z_2}},$$

where Z_1 and Z_2 are the impedances for the branches of the circuit.

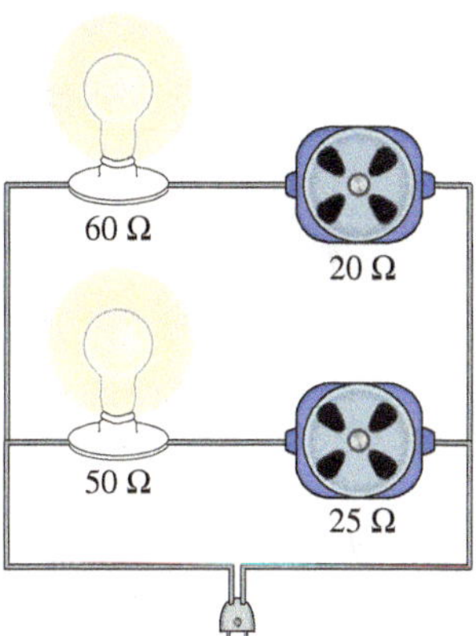

109. If $Z_1 = 50 + 25i$ and $Z_2 = 60 + 20i$, approximate Z to the nearest hundredth.

110. Determine the angle θ, to the nearest hundredth, for the value of Z found in **Exercise 109.**

Relating Concepts

For individual or collaborative investigation *(Exercises 111–118)*

Consider the following complex numbers, and ***work Exercises 111–118 in order.***

$$w = -1 + i \quad \text{and} \quad z = -1 - i$$

111. Multiply w and z using their rectangular forms and the FOIL method. Leave the product in rectangular form.

112. Find the trigonometric forms of w and z.

113. Multiply w and z using their trigonometric forms and the method described in this section.

114. Use the result of **Exercise 113** to find the rectangular form of wz. How does this compare to the result in **Exercise 111?**

115. Find the quotient $\frac{w}{z}$ using their rectangular forms and multiplying both the numerator and the denominator by the conjugate of the denominator. Leave the quotient in rectangular form.

116. Use the trigonometric forms of w and z, found in **Exercise 112,** to divide w by z using the method described in this section.

117. Use the result in **Exercise 116** to find the rectangular form of $\frac{w}{z}$.

118. How does the result in **Exercise 117** compare to the result in **Exercise 115?**

8.6 De Moivre's Theorem; Powers and Roots of Complex Numbers

- Powers of Complex Numbers (De Moivre's Theorem)
- Roots of Complex Numbers

Powers of Complex Numbers (De Moivre's Theorem)

Because raising a number to a positive integer power is a repeated application of the product rule, it would seem likely that a theorem for finding powers of complex numbers exists. Consider the following.

$$[r(\cos\theta + i\sin\theta)]^2$$
$$= [r(\cos\theta + i\sin\theta)][r(\cos\theta + i\sin\theta)] \quad a^2 = a \cdot a$$
$$= r \cdot r[\cos(\theta+\theta) + i\sin(\theta+\theta)] \quad \text{Product theorem}$$
$$= r^2(\cos 2\theta + i\sin 2\theta) \quad \text{Multiply and add.}$$

In the same way,

$$[r(\cos\theta + i\sin\theta)]^3 \text{ is equivalent to } r^3(\cos 3\theta + i\sin 3\theta).$$

These results suggest the following theorem for positive integer values of n. Although the theorem is stated and can be proved for all n, we use it only for positive integer values of n and their reciprocals.

Abraham De Moivre (1667–1754)

Named after this French expatriate friend of Isaac Newton, De Moivre's theorem relates complex numbers and trigonometry.

De Moivre's Theorem

If $r(\cos\theta + i\sin\theta)$ is a complex number, and if n is any real number, then the following holds.

$$[r(\cos\theta + i\sin\theta)]^n = r^n(\cos n\theta + i\sin n\theta)$$

In compact form, this is written

$$[r \operatorname{cis}\theta]^n = r^n(\operatorname{cis} n\theta).$$

EXAMPLE 1 Finding a Power of a Complex Number

Find $\left(1 + i\sqrt{3}\right)^8$ and write the answer in rectangular form.

SOLUTION Using earlier methods, write $1 + i\sqrt{3}$ in trigonometric form.

$$2(\cos 60° + i\sin 60°) \quad \text{Trigonometric form of } 1 + i\sqrt{3}$$

Now, apply De Moivre's theorem.

$$\left(1 + i\sqrt{3}\right)^8$$
$$= [2(\cos 60° + i\sin 60°)]^8 \quad \text{Trigonometric form}$$
$$= 2^8[\cos(8 \cdot 60°) + i\sin(8 \cdot 60°)] \quad \text{De Moivre's theorem}$$
$$= 256(\cos 480° + i\sin 480°) \quad \text{Apply the exponent and multiply.}$$
$$= 256(\cos 120° + i\sin 120°) \quad 480° \text{ and } 120° \text{ are coterminal.}$$
$$= 256\left(-\frac{1}{2} + i\frac{\sqrt{3}}{2}\right) \quad \cos 120° = -\tfrac{1}{2};\ \sin 120° = \tfrac{\sqrt{3}}{2}$$
$$= -128 + 128i\sqrt{3} \quad \text{Rectangular form}$$

✔ Now Try Exercise 13.

Roots of Complex Numbers Every nonzero complex number has exactly n distinct complex nth roots. De Moivre's theorem can be extended to find all nth roots of a complex number.

*n*th Root

For a positive integer n, the complex number $a + bi$ is an ***n*th root** of the complex number $x + yi$ if the following holds.

$$(a + bi)^n = x + yi$$

To find the three complex cube roots of $8(\cos 135° + i \sin 135°)$, for example, look for a complex number, say $r(\cos \alpha + i \sin \alpha)$, that will satisfy

$$[r(\cos \alpha + i \sin \alpha)]^3 = 8(\cos 135° + i \sin 135°).$$

By De Moivre's theorem, this equation becomes

$$r^3(\cos 3\alpha + i \sin 3\alpha) = 8(\cos 135° + i \sin 135°).$$

Set $r^3 = 8$ and $\cos 3\alpha + i \sin 3\alpha = \cos 135° + i \sin 135°$, to satisfy this equation. The first of these conditions implies that $r = 2$, and the second implies that

$$\cos 3\alpha = \cos 135° \quad \text{and} \quad \sin 3\alpha = \sin 135°.$$

For these equations to be satisfied, 3α must represent an angle that is coterminal with 135°. Therefore, we must have

$$3\alpha = 135° + 360° \cdot k, \quad k \text{ any integer}$$

or

$$\alpha = \frac{135° + 360° \cdot k}{3}, \quad k \text{ any integer.}$$

Now, let k take on the integer values 0, 1, and 2.

$$\text{If } k = 0, \text{ then} \quad \alpha = \frac{135° + 360° \cdot 0}{3} = 45°.$$

$$\text{If } k = 1, \text{ then} \quad \alpha = \frac{135° + 360° \cdot 1}{3} = \frac{495°}{3} = 165°.$$

$$\text{If } k = 2, \text{ then} \quad \alpha = \frac{135° + 360° \cdot 2}{3} = \frac{855°}{3} = 285°.$$

In the same way, $\alpha = 405°$ when $k = 3$. But note that $405° = 45° + 360°$, so $\sin 405° = \sin 45°$ and $\cos 405° = \cos 45°$. Similarly, if $k = 4$, then $\alpha = 525°$, which has the same sine and cosine values as 165°. Continuing with larger values of k would repeat solutions already found. Therefore, all of the cube roots (three of them) can be found by letting $k = 0$, 1, and 2, respectively.

When $k = 0$, the root is $2(\cos 45° + i \sin 45°)$.

When $k = 1$, the root is $2(\cos 165° + i \sin 165°)$.

When $k = 2$, the root is $2(\cos 285° + i \sin 285°)$.

In summary, we see that $2(\cos 45° + i \sin 45°)$, $2(\cos 165° + i \sin 165°)$, and $2(\cos 285° + i \sin 285°)$ are the three cube roots of $8(\cos 135° + i \sin 135°)$.

*n*th Root Theorem

If n is any positive integer, r is a positive real number, and θ is in degrees, then the nonzero complex number $r(\cos\theta + i\sin\theta)$ has exactly n distinct nth roots, given by the following.

$$\sqrt[n]{r}(\cos\alpha + i\sin\alpha) \quad \text{or} \quad \sqrt[n]{r}\,\text{cis}\,\alpha,$$

where

$$\alpha = \frac{\theta + 360^\circ \cdot k}{n}, \quad \text{or} \quad \alpha = \frac{\theta}{n} + \frac{360^\circ \cdot k}{n}, \quad k = 0, 1, 2, \ldots, n-1$$

If θ is in radians, then

$$\alpha = \frac{\theta + 2\pi k}{n}, \quad \text{or} \quad \alpha = \frac{\theta}{n} + \frac{2\pi k}{n}, \quad k = 0, 1, 2, \ldots, n-1.$$

EXAMPLE 2 Finding Complex Roots

Find the two square roots of $4i$. Write the roots in rectangular form.

SOLUTION First write $4i$ in trigonometric form.

$$4\left(\cos\frac{\pi}{2} + i\sin\frac{\pi}{2}\right) \quad \text{Trigonometric form (using radian measure)}$$

Here $r = 4$ and $\theta = \frac{\pi}{2}$. The square roots have absolute value $\sqrt{4} = 2$ and arguments as follows.

$$\alpha = \frac{\frac{\pi}{2}}{2} + \frac{2\pi k}{2} = \frac{\pi}{4} + \pi k$$

Be careful simplifying here.

Because there are two square roots, let $k = 0$ and 1.

$$\text{If } k = 0, \text{ then} \quad \alpha = \frac{\pi}{4} + \pi \cdot 0 = \frac{\pi}{4}.$$

$$\text{If } k = 1, \text{ then} \quad \alpha = \frac{\pi}{4} + \pi \cdot 1 = \frac{5\pi}{4}.$$

Using these values for α, the square roots are $2\,\text{cis}\,\frac{\pi}{4}$ and $2\,\text{cis}\,\frac{5\pi}{4}$, which can be written in rectangular form as

$$\sqrt{2} + i\sqrt{2} \quad \text{and} \quad -\sqrt{2} - i\sqrt{2}.$$

✔ **Now Try Exercise 23(a).**

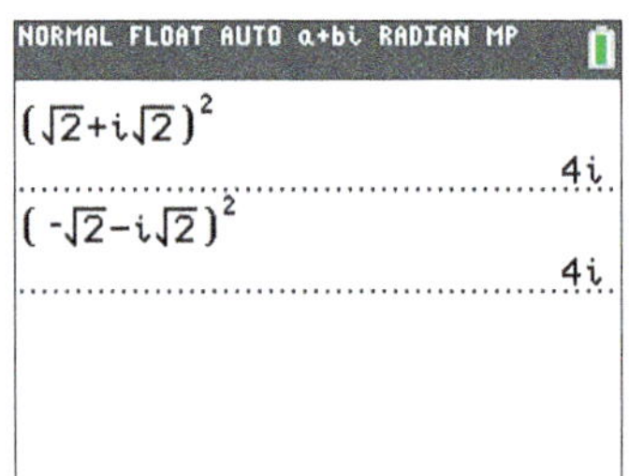

This screen confirms the results of **Example 2.**

NORMAL FLOAT AUTO a+bi DEGREE MP
R▸Pr(-8,8√3)
16
R▸Pθ(-8,8√3)
120

This screen shows how a calculator finds r and θ for the number in **Example 3.**

EXAMPLE 3 Finding Complex Roots

Find all fourth roots of $-8 + 8i\sqrt{3}$. Write the roots in rectangular form.

SOLUTION $$-8 + 8i\sqrt{3} = 16\,\text{cis}\,120^\circ \quad \text{Write in trigonometric form.}$$

Here $r = 16$ and $\theta = 120^\circ$. The fourth roots of this number have absolute value $\sqrt[4]{16} = 2$ and arguments as follows.

$$\alpha = \frac{120^\circ}{4} + \frac{360^\circ \cdot k}{4} = 30^\circ + 90^\circ \cdot k$$

Because there are four fourth roots, let $k = 0, 1, 2,$ and 3.

$$\text{If } k = 0, \text{ then} \quad \alpha = 30° + 90° \cdot 0 = 30°.$$
$$\text{If } k = 1, \text{ then} \quad \alpha = 30° + 90° \cdot 1 = 120°.$$
$$\text{If } k = 2, \text{ then} \quad \alpha = 30° + 90° \cdot 2 = 210°.$$
$$\text{If } k = 3, \text{ then} \quad \alpha = 30° + 90° \cdot 3 = 300°.$$

Using these angles, the fourth roots are

$$2 \text{ cis } 30°, \quad 2 \text{ cis } 120°, \quad 2 \text{ cis } 210°, \quad \text{and} \quad 2 \text{ cis } 300°.$$

These four roots can be written in rectangular form as

$$\sqrt{3} + i, \quad -1 + i\sqrt{3}, \quad -\sqrt{3} - i, \quad \text{and} \quad 1 - i\sqrt{3}.$$

The graphs of these roots lie on a circle with center at the origin and radius 2. See **Figure 54.** The roots are equally spaced about the circle, 90° apart. (For convenience, we label the real axis x and the imaginary axis y.)

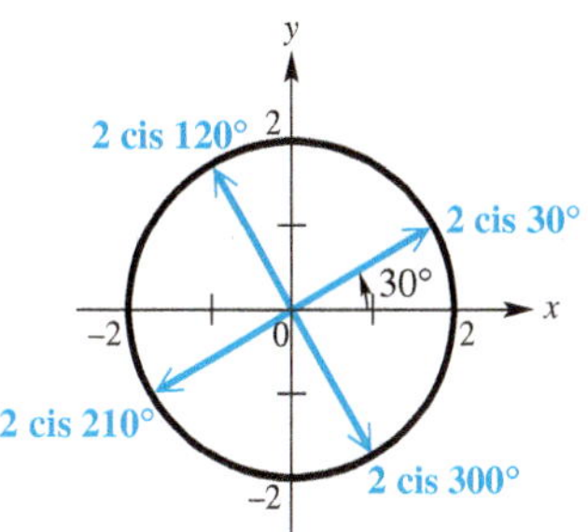

Figure 54

✔ **Now Try Exercise 29.**

EXAMPLE 4 Solving an Equation (Complex Roots)

Find all complex number solutions of $x^5 - 1 = 0$. Graph them as vectors in the complex plane.

SOLUTION Write the equation as

$$x^5 - 1 = 0, \quad \text{or} \quad x^5 = 1.$$

Because $1^5 = 1$, there is a real number solution, 1, and it is the only one. There are a total of five complex number solutions. To find these solutions, first write 1 in trigonometric form.

$$1 = 1 + 0i = 1(\cos 0° + i \sin 0°) \quad \text{Trigonometric form}$$

The absolute value of the fifth roots is $\sqrt[5]{1} = 1$. The arguments are given by

$$0° + 72° \cdot k, \quad k = 0, 1, 2, 3, \text{ and } 4.$$

By using these arguments, we find that the fifth roots are as follows.

$$\text{Real solution} \longrightarrow 1(\cos 0° + i \sin 0°), \quad k = 0$$
$$1(\cos 72° + i \sin 72°), \quad k = 1$$
$$1(\cos 144° + i \sin 144°), \quad k = 2$$
$$1(\cos 216° + i \sin 216°), \quad k = 3$$
$$1(\cos 288° + i \sin 288°) \quad k = 4$$

The solution set of the equation can be written as

$$\{\text{cis } 0°, \text{cis } 72°, \text{cis } 144°, \text{cis } 216°, \text{cis } 288°\}.$$

The first of these roots is the real number 1. The others cannot easily be expressed in rectangular form but can be approximated using a calculator.

The tips of the arrows representing the five fifth roots all lie on a unit circle and are equally spaced around it every 72°, as shown in **Figure 55.**

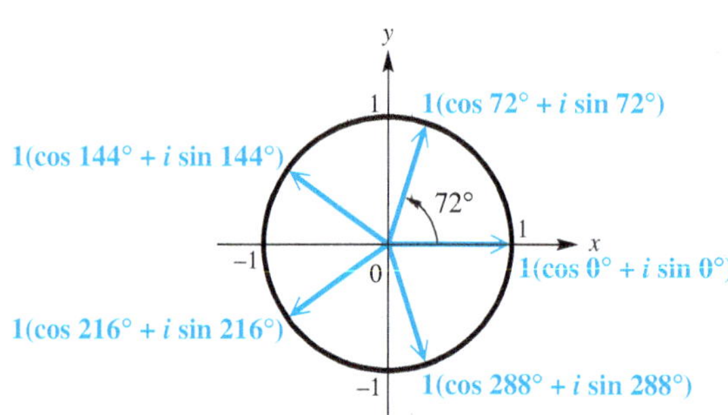

Figure 55

✔ **Now Try Exercise 41.**

8.6 Exercises

CONCEPT PREVIEW *Fill in the blanks to correctly complete each problem.*

1. If $z = 3(\cos 30° + i \sin 30°)$, it follows that

$$z^3 = \_\_\_(\cos \_\_\_ + i \sin \_\_\_)$$
$$= \_\_\_(\_\_\_ + i\_\_\_)$$
$$= \_\_\_ + \_\_\_ i, \text{ or simply } \_\_\_.$$

2. If we are given

$$z = 16(\cos 80° + i \sin 80°),$$

then any fourth root of z has $r =$ ___, and the fourth root with least positive argument has $\theta =$ ___.

3. $[\cos 6° + i \sin 6°]^{30}$

$$= \cos \_\_\_ + i \sin \_\_\_$$
$$= \_\_\_ + \_\_\_ i$$

4. Based on the result of **Exercise 3,**

$$\cos 6° + i \sin 6°$$

is a(n) ___ root of ___.

CONCEPT PREVIEW *Answer each question.*

5. How many real tenth roots of 1 exist?

6. How many nonreal complex tenth roots of 1 exist?

Find each power. Write answers in rectangular form. ***See Example 1.***

7. $[3(\cos 30° + i \sin 30°)]^3$

8. $[2(\cos 135° + i \sin 135°)]^4$

9. $(\cos 45° + i \sin 45°)^8$

10. $[2(\cos 120° + i \sin 120°)]^3$

11. $[3 \text{ cis } 100°]^3$

12. $[3 \text{ cis } 40°]^3$

13. $(\sqrt{3} + i)^5$

14. $(2 - 2i\sqrt{3})^4$

15. $(2\sqrt{2} - 2i\sqrt{2})^6$

16. $\left(\frac{\sqrt{2}}{2} - \frac{\sqrt{2}}{2}i\right)^8$

17. $(-2 - 2i)^5$

18. $(-1 + i)^7$

For each of the following, ***(a)*** *find all cube roots of each complex number. Write answers in trigonometric form.* ***(b)*** *Graph each cube root as a vector in the complex plane.* ***See Examples 2 and 3.***

19. $\cos 0° + i \sin 0°$ **20.** $\cos 90° + i \sin 90°$ **21.** $8 \text{ cis } 60°$

22. $27 \text{ cis } 300°$ **23.** $-8i$ **24.** $27i$

25. -64 **26.** 27 **27.** $1 + i\sqrt{3}$

28. $2 - 2i\sqrt{3}$ **29.** $-2\sqrt{3} + 2i$ **30.** $\sqrt{3} - i$

Find and graph all specified roots of 1.

31. second (square) **32.** fourth **33.** sixth

Find and graph all specified roots of i.

34. second (square) **35.** third (cube) **36.** fourth

Find all complex number solutions of each equation. Write answers in trigonometric form. ***See Example 4.***

37. $x^3 - 1 = 0$ **38.** $x^3 + 1 = 0$ **39.** $x^3 + i = 0$

40. $x^4 + i = 0$ **41.** $x^3 - 8 = 0$ **42.** $x^3 + 27 = 0$

43. $x^4 + 1 = 0$ **44.** $x^4 + 16 = 0$ **45.** $x^4 - i = 0$

46. $x^5 - i = 0$ **47.** $x^3 - (4 + 4i\sqrt{3}) = 0$ **48.** $x^4 - (8 + 8i\sqrt{3}) = 0$

Solve each problem.

49. Solve the cubic equation

$$x^3 = 1$$

by writing it as $x^3 - 1 = 0$, factoring the left side as the difference of two cubes, and using the zero-factor property. Apply the quadratic formula as needed. Then compare the solutions to those of **Exercise 37.**

50. Solve the cubic equation

$$x^3 = -27$$

by writing it as $x^3 + 27 = 0$, factoring the left side as the sum of two cubes, and using the zero-factor property. Apply the quadratic formula as needed. Then compare the solutions to those of **Exercise 42.**

51. *Mandelbrot Set* The fractal known as the **Mandelbrot set** is shown in the figure. To determine whether a complex number $z = a + bi$ is in this set, perform the following sequence of calculations. Repeatedly compute

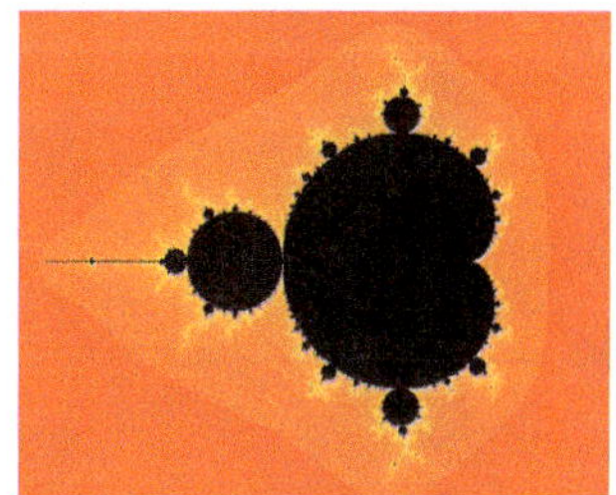

$$z, \quad z^2 + z, \quad (z^2 + z)^2 + z,$$
$$[(z^2 + z)^2 + z]^2 + z, \ldots.$$

In a manner analogous to the Julia set, the complex number z does not belong to the Mandelbrot set if any of the resulting absolute values exceeds 2. Otherwise z is in the set and the point (a, b) should be shaded in the graph. Determine whether the following numbers belong to the Mandelbrot set. (*Source:* Lauwerier, H., *Fractals,* Princeton University Press.)

(a) $z = 0 + 0i$ **(b)** $z = 1 - 1i$ **(c)** $z = -0.5i$

52. *Basins of Attraction* The fractal shown in the figure is the solution to Cayley's problem of determining the basins of attraction for the cube roots of unity. The three cube roots of unity are

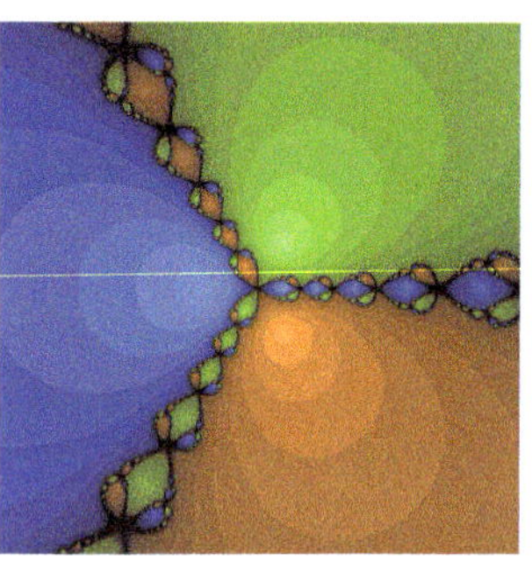

$$w_1 = 1, \quad w_2 = -\frac{1}{2} + \frac{\sqrt{3}}{2}i,$$

and $$w_3 = -\frac{1}{2} - \frac{\sqrt{3}}{2}i.$$

A fractal of this type can be generated by repeatedly evaluating the function

$$f(z) = \frac{2z^3 + 1}{3z^2},$$

where z is a complex number. We begin by picking $z_1 = a + bi$ and successively computing $z_2 = f(z_1)$, $z_3 = f(z_2)$, $z_4 = f(z_3)$, ... Suppose that if the resulting values of $f(z)$ approach w_1, we color the pixel at (a, b) red. If they approach w_2, we color it blue, and if they approach w_3, we color it yellow. If this process continues for a large number of different z_1, a fractal similar to the figure will appear. Determine the appropriate color of the pixel for each value of z_1. (*Source:* Crownover, R., *Introduction to Fractals and Chaos,* Jones and Bartlett Publishers.)

(a) $z_1 = i$ **(b)** $z_1 = 2 + i$ **(c)** $z_1 = -1 - i$

53. The screens here illustrate how a pentagon can be graphed using a graphing calculator. Note that a pentagon has five sides, and the Tstep is $\frac{360}{5} = 72$. The display at the bottom of the graph screen indicates that one fifth root of 1 is $1 + 0i = 1$. Use this technique to find all fifth roots of 1, and express the real and imaginary parts in decimal form.

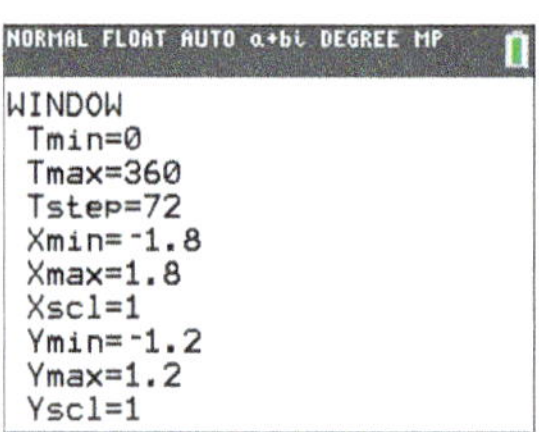

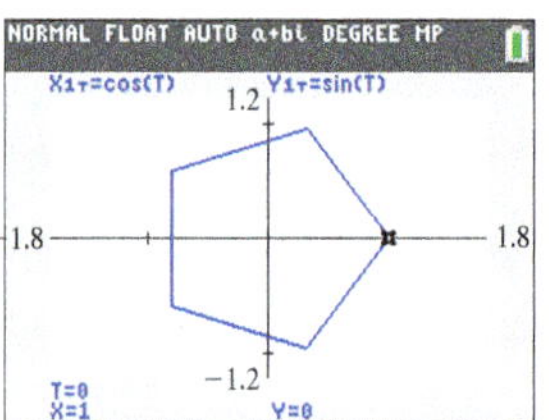

The calculator is in parametric, degree, and connected graph modes.

54. Use the method of **Exercise 53** to find the first three of the ten 10th roots of 1.

Use a calculator to find all solutions of each equation in rectangular form.

55. $x^2 - 3 + 2i = 0$

56. $x^2 + 2 - i = 0$

57. $x^5 + 2 + 3i = 0$

58. $x^3 + 4 - 5i = 0$

Relating Concepts

For individual or collaborative investigation *(Exercises 59–62)*

In earlier work we derived identities, or formulas, for cos 2θ *and* sin 2θ. *These identities can also be derived using De Moivre's theorem.* ***Work Exercises 59–62 in order, to see how this is done.***

59. De Moivre's theorem states that $(\cos\theta + i\sin\theta)^2 =$ ________.

60. Expand the left side of the equation in **Exercise 59** as a binomial and combine like terms to write the left side in the form $a + bi$.

61. Use the result of **Exercise 60** to obtain the double-angle formula for cosine.

62. Repeat **Exercise 61,** but find the double-angle formula for sine.

Chapter 8 Quiz (Sections 8.3–8.6)

1. Given vectors $\mathbf{a} = \langle -1, 4 \rangle$ and $\mathbf{b} = \langle 5, 2 \rangle$, find each of the following.

   (a) $3\mathbf{a}$ **(b)** $4\mathbf{a} - 2\mathbf{b}$ **(c)** $|\mathbf{a}|$

   (d) $\mathbf{a} \cdot \mathbf{b}$ **(e)** the angle between **a** and **b**

2. ***Walking Dogs on Leashes*** While Michael is walking his two dogs, Gus and Dotty, they reach a corner and must wait for a WALK sign. Michael is holding the two leashes in the same hand, and the dogs are pulling on their leashes at the angles and forces shown in the figure. Find the magnitude of the equilibrant force (to the nearest tenth of a pound) that Michael must apply to restrain the dogs.

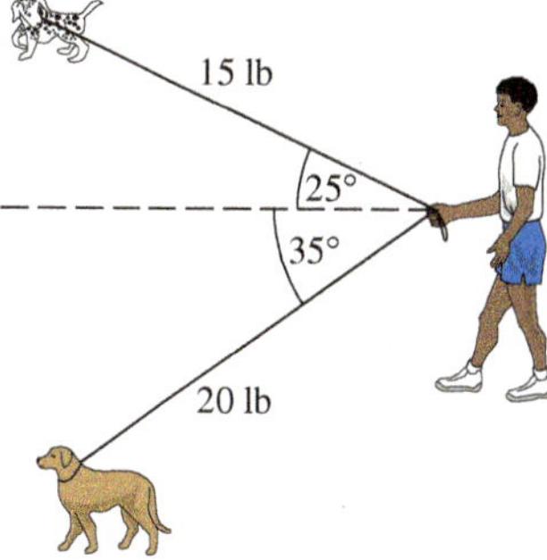

3. For the following complex numbers,

$$w = 3 + 5i \text{ and } z = -4 + i,$$

   find $w + z$ and give a geometric representation of the sum.

4. Express $(1 - i)^3$ in rectangular form.

5. Write each complex number in trigonometric (polar) form, where $0° \le \theta < 360°$.

   (a) $-4i$ **(b)** $1 - i\sqrt{3}$ **(c)** $-3 - i$

6. Write each complex number in rectangular form.

   (a) $4(\cos 60° + i \sin 60°)$ **(b)** $5 \text{ cis } 130°$ **(c)** $7(\cos 270° + i \sin 270°)$

7. Write each of the following in the form specified for the complex numbers

$$w = 12(\cos 80° + i \sin 80°) \quad \text{and} \quad z = 3(\cos 50° + i \sin 50°).$$

   (a) wz (trigonometric form) **(b)** $\frac{w}{z}$ (rectangular form) **(c)** z^3 (rectangular form)

8. Find the four complex fourth roots of -16. Write them in both trigonometric and rectangular forms.

8.7 Polar Equations and Graphs

- Polar Coordinate System
- Graphs of Polar Equations
- Conversion from Polar to Rectangular Equations
- Classification of Polar Equations

Polar Coordinate System Previously we have used the rectangular coordinate system to graph points and equations. In the rectangular coordinate system, each point in the plane is specified by giving two numbers (x, y). These represent the directed distances from a pair of perpendicular axes, the x-axis and the y-axis.

Now we consider the **polar coordinate system** which is based on a point, called the **pole,** and a ray, called the **polar axis.** The polar axis is usually drawn in the direction of the positive x-axis, as shown in **Figure 56.**

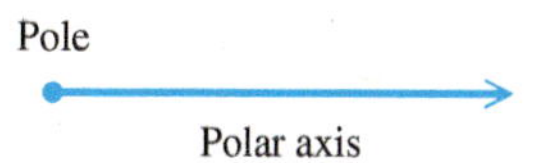

Figure 56

In **Figure 57** the pole has been placed at the origin of a rectangular coordinate system so that the polar axis coincides with the positive x-axis. Point P has rectangular coordinates (x, y). Point P can also be located by giving the directed angle θ from the positive x-axis to ray OP and the *directed* distance r from the pole to point P. The ordered pair (r, θ) gives the **polar coordinates** of point P. If $r > 0$ then point P lies on the terminal side of θ, and if $r < 0$ then point P lies on the ray pointing in the opposite direction of the terminal side of θ, a distance $|r|$ from the pole.

Figure 58 shows rectangular axes superimposed on a polar coordinate grid.

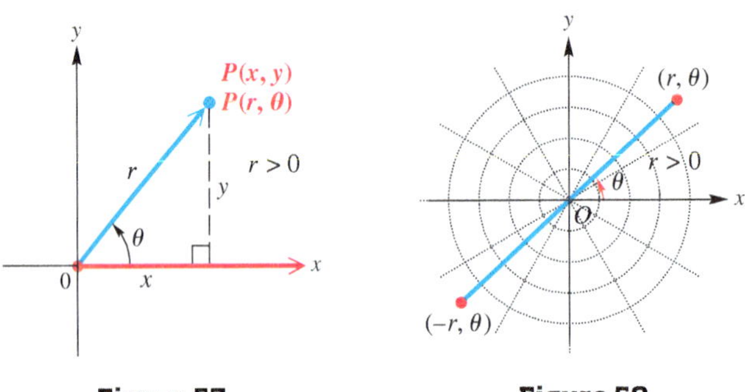

Figure 57 **Figure 58**

Rectangular and Polar Coordinates

If a point has rectangular coordinates (x, y) and polar coordinates (r, θ), then these coordinates are related as follows.

$$\boldsymbol{x = r\cos\theta} \qquad \boldsymbol{y = r\sin\theta}$$

$$\boldsymbol{r^2 = x^2 + y^2} \qquad \boldsymbol{\tan\theta = \frac{y}{x}}, \quad \text{if } x \neq 0$$

EXAMPLE 1 Plotting Points with Polar Coordinates

Plot each point in the polar coordinate system. Then determine the rectangular coordinates of each point.

(a) $P(2, 30°)$ **(b)** $Q\left(-4, \frac{2\pi}{3}\right)$ **(c)** $R\left(5, -\frac{\pi}{4}\right)$

SOLUTION

(a) In the point $P(2, 30°)$, $r = 2$ and $\theta = 30°$, so P is located 2 units from the origin in the positive direction on a ray making a 30° angle with the polar axis, as shown in **Figure 59.**

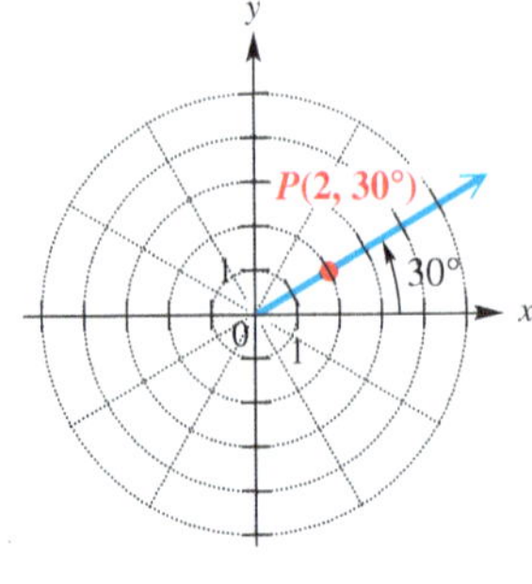

Figure 59

We find the rectangular coordinates as follows.

$x = r\cos\theta$	$y = r\sin\theta$	Conversion equations
$x = 2\cos 30°$	$y = 2\sin 30°$	Substitute.
$x = 2\left(\frac{\sqrt{3}}{2}\right)$	$y = 2\left(\frac{1}{2}\right)$	Evaluate the functions.
$x = \sqrt{3}$	$y = 1$	Multiply.

The rectangular coordinates are $\left(\sqrt{3}, 1\right)$.

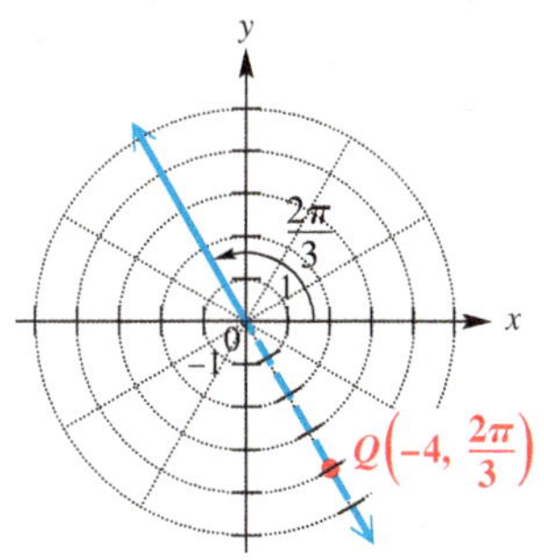

Figure 60

(b) In the point $Q\left(-4, \frac{2\pi}{3}\right)$, r is *negative*, so Q is 4 units in the *opposite* direction from the pole on an extension of the $\frac{2\pi}{3}$ ray. See **Figure 60.** The rectangular coordinates are

$$x = -4 \cos \frac{2\pi}{3} = -4\left(-\frac{1}{2}\right) = 2$$

and

$$y = -4 \sin \frac{2\pi}{3} = -4\left(\frac{\sqrt{3}}{2}\right) = -2\sqrt{3}.$$

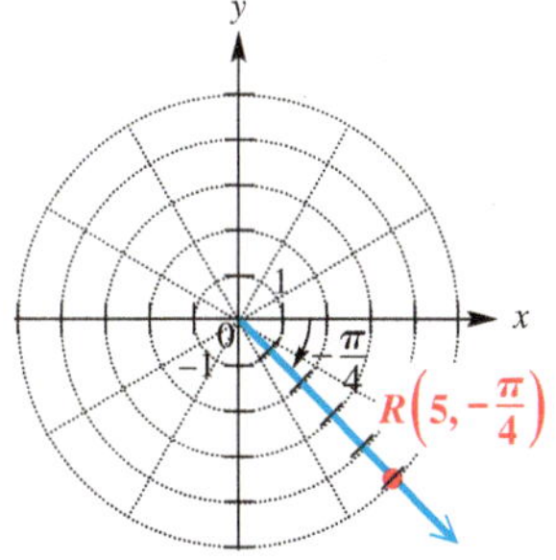

Figure 61

(c) Point $R\left(5, -\frac{\pi}{4}\right)$ is shown in **Figure 61.** Because θ is negative, the angle is measured in the clockwise direction.

$$x = 5 \cos\left(-\frac{\pi}{4}\right) = \frac{5\sqrt{2}}{2} \quad \text{and} \quad y = 5 \sin\left(-\frac{\pi}{4}\right) = -\frac{5\sqrt{2}}{2}$$

✔ **Now Try Exercises 13(a), (c), 15(a), (c), and 21(a), (c).**

While a given point in the plane can have only one pair of rectangular coordinates, this same point can have an infinite number of pairs of polar coordinates. For example, (2, 30°) locates the same point as

$$(2, 390°), \quad (2, -330°), \quad \text{and} \quad (-2, 210°).$$

EXAMPLE 2 Giving Alternative Forms for Coordinates of Points

Determine the following.

(a) Three other pairs of polar coordinates for the point $P(3, 140°)$

(b) Two pairs of polar coordinates for the point with rectangular coordinates $(-1, 1)$

SOLUTION

(a) Three pairs that could be used for the point are $(3, -220°)$, $(-3, 320°)$, and $(-3, -40°)$. See **Figure 62.**

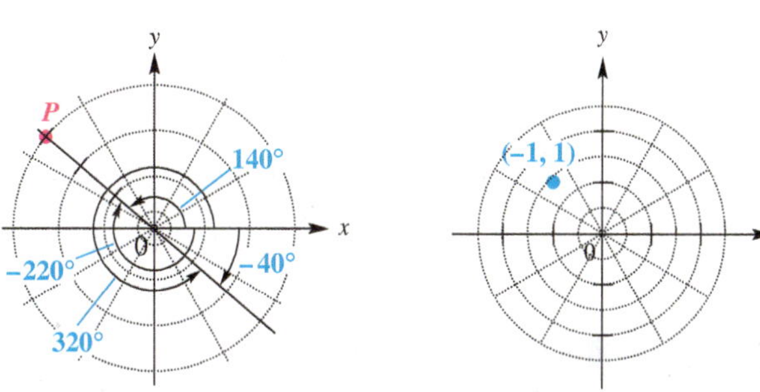

Figure 62 Figure 63

LOOKING AHEAD TO CALCULUS
Techniques studied in calculus associated with derivatives and integrals provide methods of finding slopes of tangent lines to polar curves, areas bounded by such curves, and lengths of their arcs.

(b) As shown in **Figure 63,** the point $(-1, 1)$ lies in the second quadrant. Because $\tan \theta = \frac{1}{-1} = -1$, one possible value for θ is 135°. Also,

$$r = \sqrt{x^2 + y^2} = \sqrt{(-1)^2 + 1^2} = \sqrt{2}.$$

Two pairs of polar coordinates are $\left(\sqrt{2}, 135°\right)$ and $\left(-\sqrt{2}, 315°\right)$.

✔ **Now Try Exercises 13(b), 15(b), 21(b), and 25.**

Graphs of Polar Equations An equation in the variables x and y is a **rectangular** (or **Cartesian**) **equation.** An equation in which r and θ are the variables instead of x and y is a **polar equation.**

$$r = 3 \sin \theta, \quad r = 2 + \cos \theta, \quad r = \theta \qquad \text{Polar equations}$$

Although the rectangular forms of lines and circles are the ones most often encountered, they can also be defined in terms of polar coordinates. The polar equation of the line $ax + by = c$ can be derived as follows.

Line:

$$ax + by = c \qquad \text{Rectangular equation of a line}$$

$$a(r \cos \theta) + b(r \sin \theta) = c \qquad \text{Convert to polar coordinates.}$$

$$r(a \cos \theta + b \sin \theta) = c \qquad \text{Factor out } r.$$

$$r = \frac{c}{a \cos \theta + b \sin \theta} \qquad \text{Polar equation of a line}$$

This is the polar equation of $ax + by = c$.

For the circle $x^2 + y^2 = a^2$, the polar equation can be found in a similar manner.

Circle:

$$x^2 + y^2 = a^2 \qquad \text{Rectangular equation of a circle}$$

$$r^2 = a^2 \qquad x^2 + y^2 = r^2$$

$$r = \pm a \qquad \text{Polar equation of a circle; } r \text{ can be negative in polar coordinates.}$$

These are polar equations of $x^2 + y^2 = a^2$.

We use these forms in the next example.

EXAMPLE 3 Finding Polar Equations of Lines and Circles

For each rectangular equation, give the equivalent polar equation and sketch its graph.

(a) $y = x - 3$ **(b)** $x^2 + y^2 = 4$

SOLUTION

(a) This is the equation of a line.

$$y = x - 3$$

$$x - y = 3 \qquad \text{Write in standard form } ax + by = c.$$

$$r \cos \theta - r \sin \theta = 3 \qquad \text{Substitute for } x \text{ and } y.$$

$$r(\cos \theta - \sin \theta) = 3 \qquad \text{Factor out } r.$$

$$r = \frac{3}{\cos \theta - \sin \theta} \qquad \text{Divide by } \cos \theta - \sin \theta.$$

A traditional graph is shown in **Figure 64(a),** and a calculator graph is shown in **Figure 64(b).**

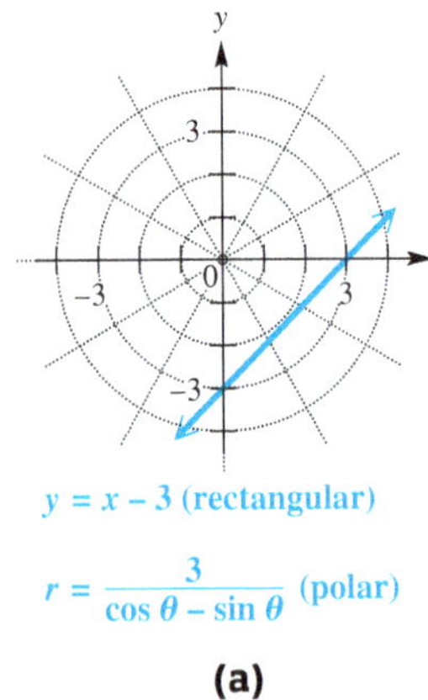

$y = x - 3$ (rectangular)

$r = \frac{3}{\cos \theta - \sin \theta}$ (polar)

(a)

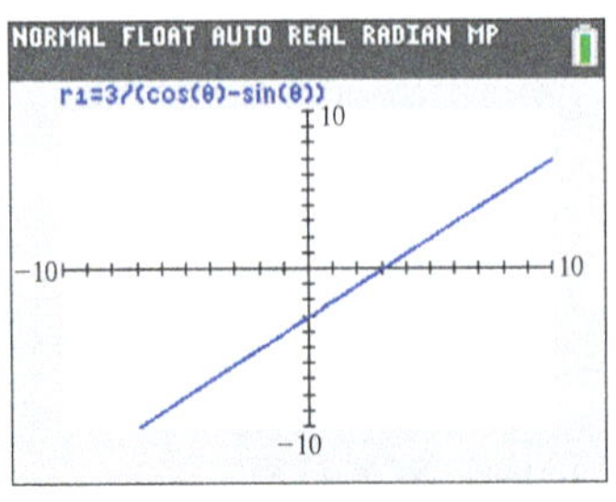

Polar graphing mode

(b)

Figure 64

(b) The graph of $x^2 + y^2 = 4$ is a circle with center at the origin and radius 2.

$$x^2 + y^2 = 4$$

$$r^2 = 4 \qquad x^2 + y^2 = r^2$$

$$r = 2 \quad \text{or} \quad r = -2$$

In polar coordinates, we may have $r < 0$.

The graphs of $r = 2$ and $r = -2$ coincide. See **Figure 65** on the next page.

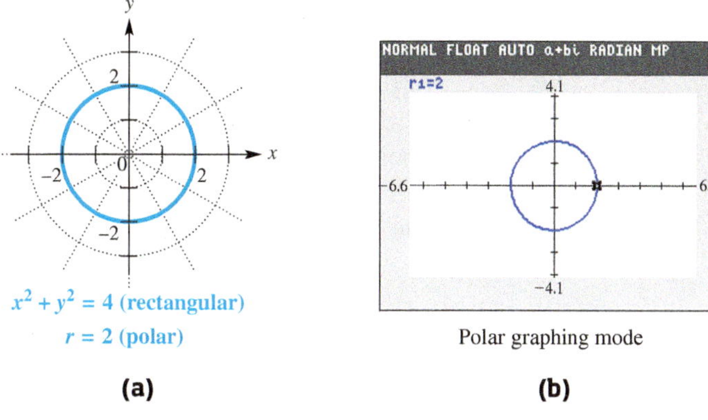

$x^2 + y^2 = 4$ (rectangular)
$r = 2$ (polar)

Polar graphing mode

(a) **(b)**

Figure 65

✔ **Now Try Exercises 37 and 39.**

To graph polar equations, evaluate r for various values of θ until a pattern appears, and then join the points with a smooth curve. The next four examples illustrate curves that are not usually discussed when rectangular coordinates are covered. (Using graphing calculators makes the task of graphing them quite a bit easier than using traditional point-plotting methods.)

EXAMPLE 4 Graphing a Polar Equation (Cardioid)

Graph $r = 1 + \cos \theta$.

ALGEBRAIC SOLUTION

To graph this equation, find some ordered pairs as in the table. Once the pattern of values of r becomes clear, it is not necessary to find more ordered pairs. The table includes approximate values for $\cos \theta$ and r.

θ	$\cos \theta$	$r = 1 + \cos \theta$	θ	$\cos \theta$	$r = 1 + \cos \theta$
0°	1	2	135°	−0.7	0.3
30°	0.9	1.9	150°	−0.9	0.1
45°	0.7	1.7	180°	−1	0
60°	0.5	1.5	270°	0	1
90°	0	1	315°	0.7	1.7
120°	−0.5	0.5	330°	0.9	1.9

Connect the points in order—from $(2, 0°)$ to $(1.9, 30°)$ to $(1.7, 45°)$ and so on. See **Figure 66.** This curve is called a **cardioid** because of its heart shape. The curve has been graphed on a **polar grid.**

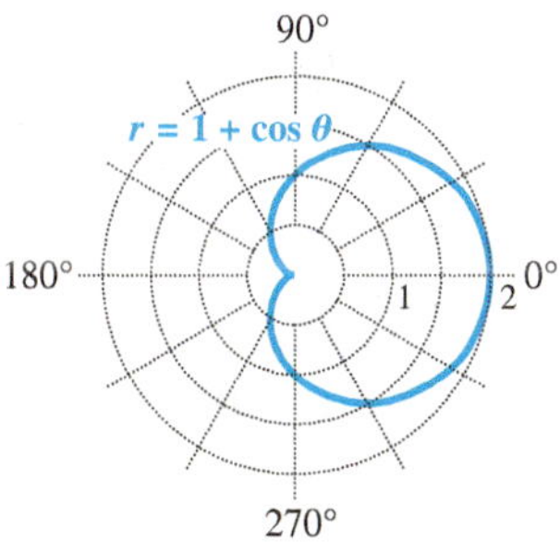

Figure 66

GRAPHING CALCULATOR SOLUTION

We choose degree mode and graph values of θ in the interval $[0°, 360°]$. The screen in **Figure 67(a)** shows the choices needed to generate the graph in **Figure 67(b).**

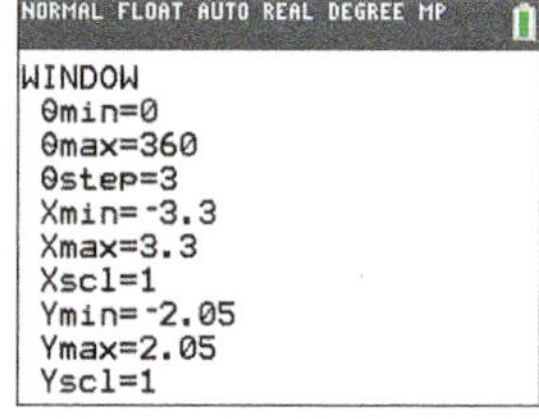

(a)

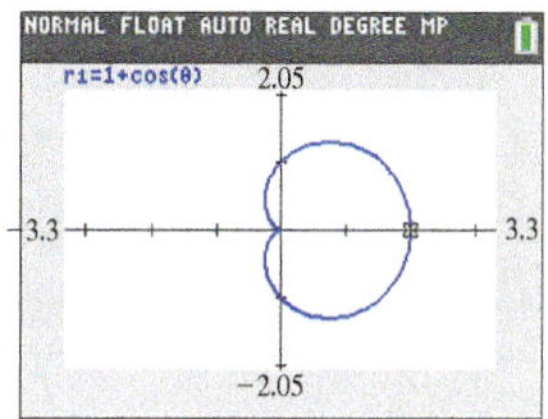

Polar graphing mode

(b)

Figure 67

✔ **Now Try Exercise 47.**

EXAMPLE 5 Graphing a Polar Equation (Rose)

Graph $r = 3 \cos 2\theta$.

SOLUTION Because the argument is 2θ, the graph requires a greater number of points than when the argument is just θ. We complete the table using selected angle measures through 360° in order to see the pattern of the graph. Approximate values in the table have been rounded to the nearest tenth.

θ	2θ	$\cos 2\theta$	$r = 3\cos 2\theta$	θ	2θ	$\cos 2\theta$	$r = 3\cos 2\theta$
0°	0°	1	3	120°	240°	−0.5	−1.5
15°	30°	0.9	2.6	135°	270°	0	0
30°	60°	0.5	1.5	180°	360°	1	3
45°	90°	0	0	225°	450°	0	0
60°	120°	−0.5	−1.5	270°	540°	−1	−3
75°	150°	−0.9	−2.6	315°	630°	0	0
90°	180°	−1	−3	360°	720°	1	3

Plotting these points in order gives the graph of a **four-leaved rose.** Note in **Figure 68(a)** how the graph is developed with a continuous curve, beginning with the upper half of the right horizontal leaf and ending with the lower half of that leaf. As the graph is traced, the curve goes through the pole four times. This can be seen as a calculator graphs the curve. See **Figure 68(b).**

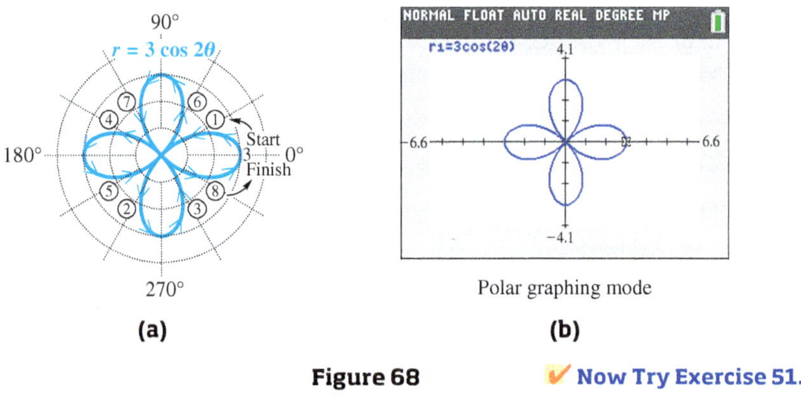

(a) (b)

Figure 68

✔ **Now Try Exercise 51.**

NOTE To sketch the graph of $r = 3 \cos 2\theta$ in polar coordinates, it may be helpful to first sketch the graph of $y = 3 \cos 2x$ in rectangular coordinates. The minimum and maximum values of this function may be used to determine the location of the tips of the leaves, and the x-intercepts of this function may be used to determine where the polar graph passes through the pole.

The equation $r = 3 \cos 2\theta$ in **Example 5** has a graph that belongs to a family of curves called **roses.**

$$r = a \sin n\theta \quad \text{and} \quad r = a \cos n\theta \qquad \text{Equations of roses}$$

- The graph has n leaves if n is odd, and $2n$ leaves if n is even.
- The absolute value of a determines the length of the leaves.

EXAMPLE 6 Graphing a Polar Equation (Lemniscate)

Graph $r^2 = \cos 2\theta$.

ALGEBRAIC SOLUTION

Complete a table of ordered pairs, and sketch the graph, as in **Figure 69.** The point $(-1, 0°)$, with r negative, may be plotted as $(1, 180°)$. Also, $(-0.7, 30°)$ may be plotted as $(0.7, 210°)$, and so on.

Values of θ for $45° < \theta < 135°$ are not included in the table because the corresponding values of $\cos 2\theta$ are negative (quadrants II and III) and so do not have real square roots. Values of θ greater than 180° give 2θ greater than 360° and would repeat the points already found. This curve is called a **lemniscate.**

θ	0°	30°	45°	135°	150°	180°
2θ	0°	60°	90°	270°	300°	360°
$\cos 2\theta$	1	0.5	0	0	0.5	1
$r = \pm\sqrt{\cos 2\theta}$	± 1	± 0.7	0	0	± 0.7	± 1

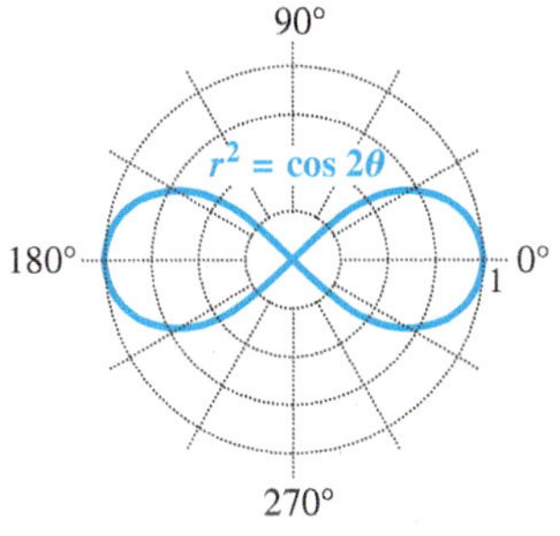

Figure 69

GRAPHING CALCULATOR SOLUTION

To graph $r^2 = \cos 2\theta$ with a graphing calculator, first solve for r by considering both square roots. Enter the two polar equations as

$$r_1 = \sqrt{\cos 2\theta} \quad \text{and} \quad r_2 = -\sqrt{\cos 2\theta}.$$

See **Figures 70(a) and (b).**

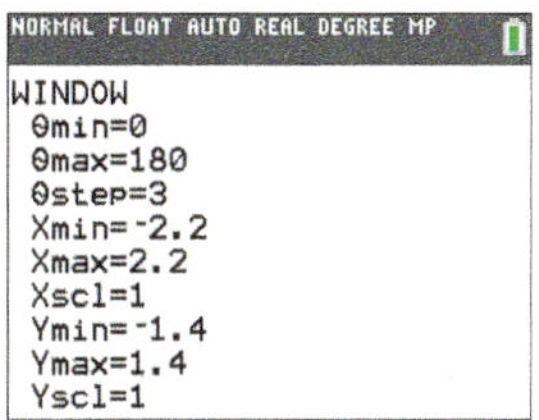

Settings for the graph below

(a)

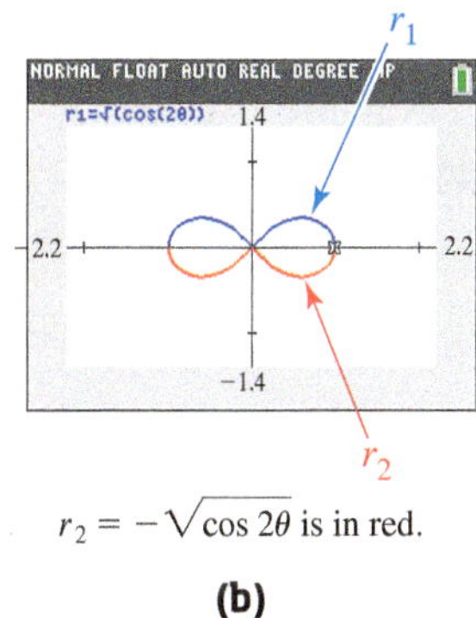

$r_2 = -\sqrt{\cos 2\theta}$ is in red.

(b)

Figure 70

✔ **Now Try Exercise 53.**

EXAMPLE 7 Graphing a Polar Equation (Spiral of Archimedes)

Graph $r = 2\theta$ (with θ measured in radians).

SOLUTION Some ordered pairs are shown in the table. Because $r = 2\theta$ does *not* involve a trigonometric function of θ, we must also consider negative values of θ. The graph in **Figure 71(a)** on the next page is a **spiral of Archimedes. Figure 71(b)** shows a calculator graph of this spiral.

θ (radians)	$r = 2\theta$	θ (radians)	$r = 2\theta$
$-\pi$	−6.3	$\frac{\pi}{3}$	2.1
$-\frac{\pi}{2}$	−3.1	$\frac{\pi}{2}$	3.1
$-\frac{\pi}{4}$	−1.6	π	6.3
0	0	$\frac{3\pi}{2}$	9.4
$\frac{\pi}{6}$	1	2π	12.6

Radian measures have been rounded.

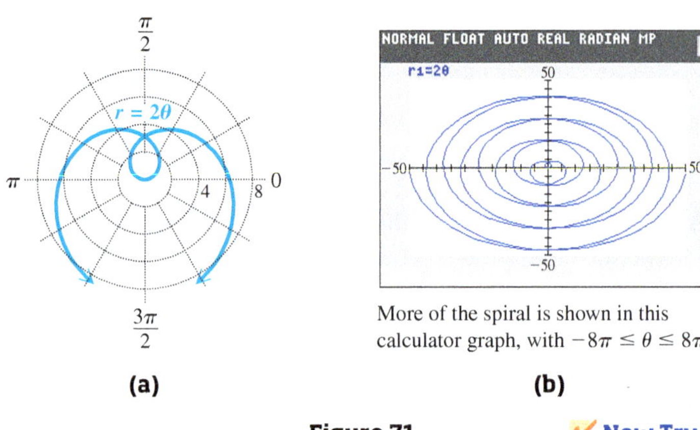

More of the spiral is shown in this calculator graph, with $-8\pi \le \theta \le 8\pi$.

(a) **(b)**

Figure 71

✔ **Now Try Exercise 59.**

Conversion from Polar to Rectangular Equations

EXAMPLE 8 Converting a Polar Equation to a Rectangular Equation

For the equation $r = \frac{4}{1 + \sin\theta}$, write an equivalent equation in rectangular coordinates, and graph.

SOLUTION

$$r = \frac{4}{1 + \sin\theta} \qquad \text{Polar equation}$$

$$r(1 + \sin\theta) = 4 \qquad \text{Multiply by } 1 + \sin\theta.$$

$$r + r\sin\theta = 4 \qquad \text{Distributive property}$$

$$\sqrt{x^2 + y^2} + y = 4 \qquad \text{Let } r = \sqrt{x^2 + y^2} \text{ and } r\sin\theta = y.$$

$$\sqrt{x^2 + y^2} = 4 - y \qquad \text{Subtract } y.$$

$$x^2 + y^2 = (4 - y)^2 \qquad \text{Square each side.}$$

$$x^2 + y^2 = 16 - 8y + y^2 \qquad \text{Expand the right side.}$$

$$x^2 = -8y + 16 \qquad \text{Subtract } y^2.$$

$$x^2 = -8(y - 2) \qquad \text{Rectangular equation}$$

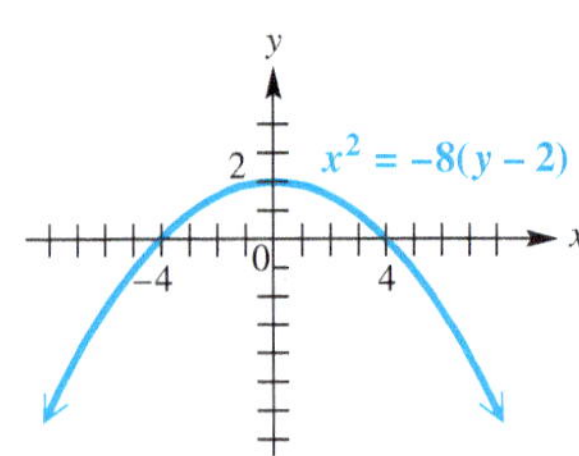

Figure 72

The final equation represents a parabola and is graphed in **Figure 72.**

✔ **Now Try Exercise 63.**

The conversion in **Example 8** is not necessary when using a graphing calculator. **Figure 73** shows the graph of $r = \frac{4}{1 + \sin\theta}$, graphed directly with the calculator in polar mode. ■

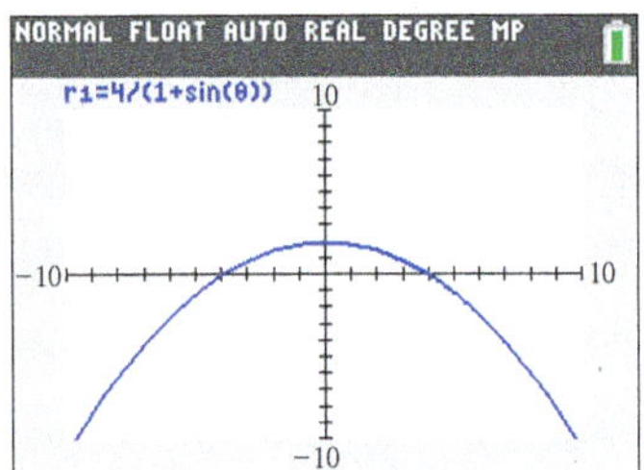

$0° \le \theta \le 360°$

Figure 73

Classification of Polar Equations The table on the next page summarizes common polar graphs and forms of their equations. In addition to circles, lemniscates, and roses, we include **limaçons.** Cardioids are a special case of limaçons, where $\left|\frac{a}{b}\right| = 1$.

NOTE Some other polar curves are the **cissoid, kappa curve, conchoid, trisectrix, cruciform, strophoid,** and **lituus.** Refer to older textbooks on analytic geometry or the Internet to investigate them.

Polar Graphs and Forms of Equations

Circles and Lemniscates

Circles

$r = a \cos \theta$

$r = a \sin \theta$

Lemniscates

$r^2 = a^2 \sin 2\theta$

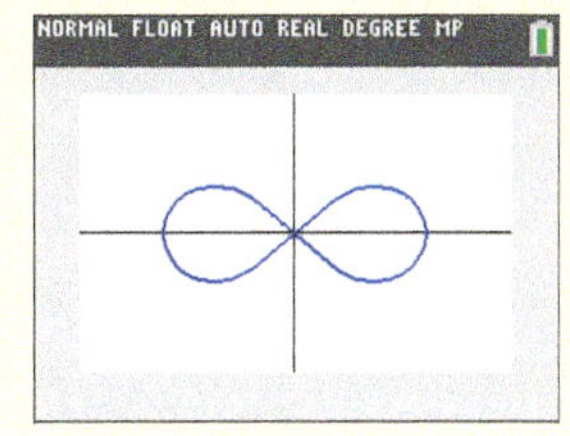

$r^2 = a^2 \cos 2\theta$

Limaçons

$r = a \pm b \sin \theta$ or $r = a \pm b \cos \theta$

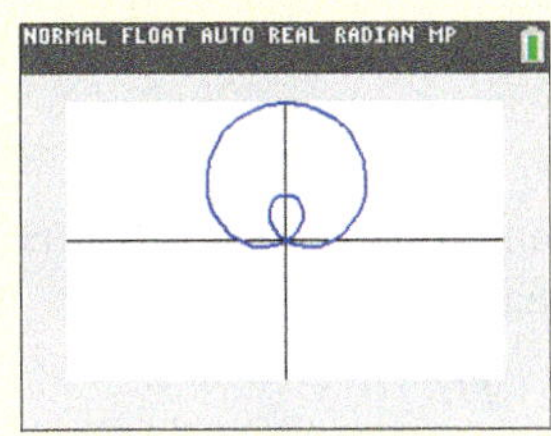

$\frac{a}{b} < 1$

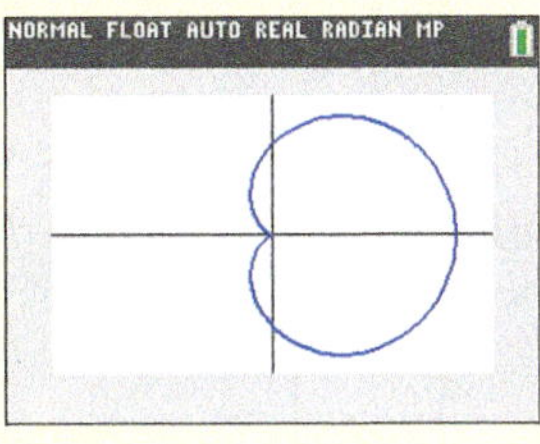

$\frac{a}{b} = 1$

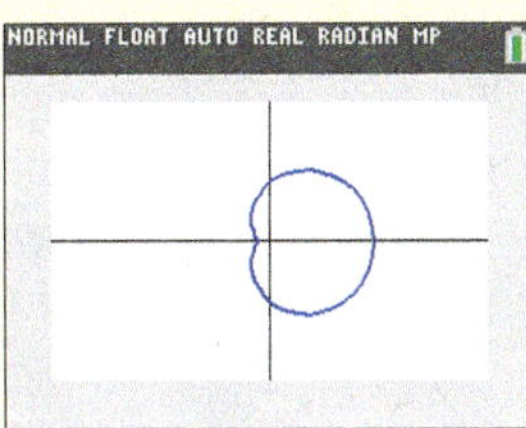

$1 < \frac{a}{b} < 2$

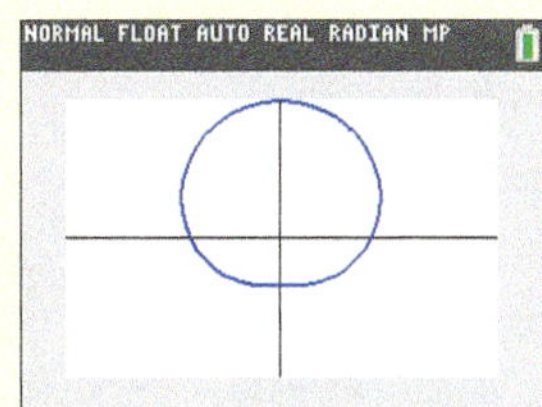

$\frac{a}{b} \geq 2$

Rose Curves

$2n$ leaves if n is even, $n \geq 2$

$n = 2$

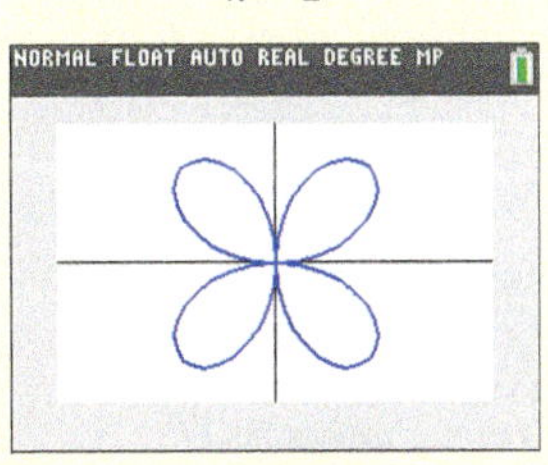

$r = a \sin n\theta$

$n = 4$

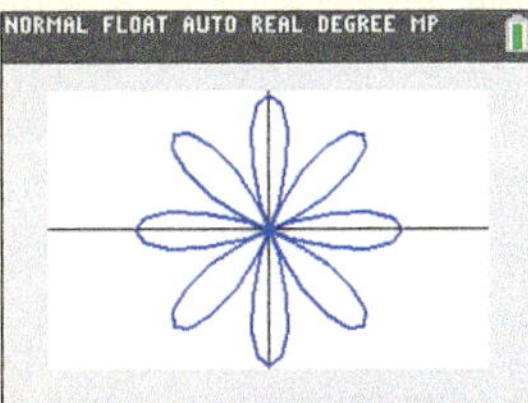

$r = a \cos n\theta$

n leaves if n is odd

$n = 3$

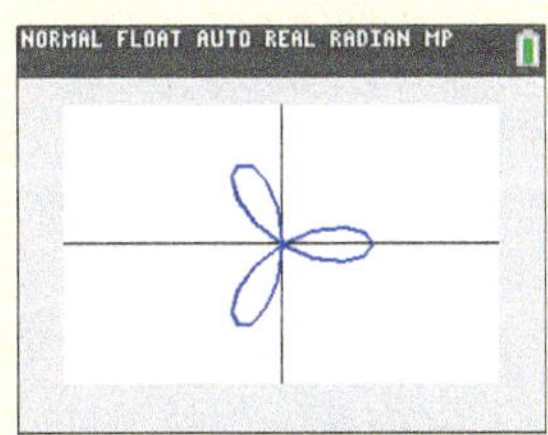

$r = a \cos n\theta$

$n = 5$

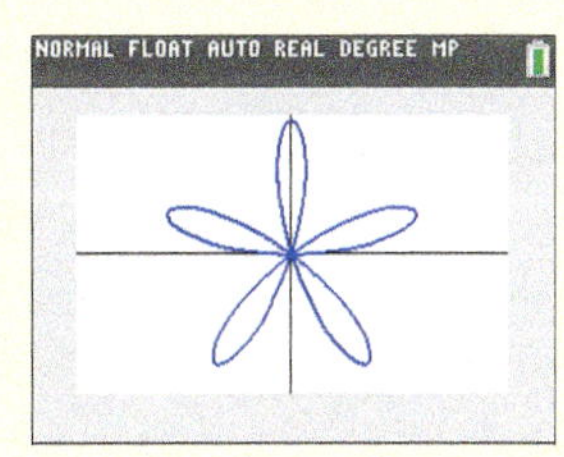

$r = a \sin n\theta$

8.7 Exercises

CONCEPT PREVIEW *Fill in the blank to correctly complete each sentence.*

1. For the polar equation $r = 3 \cos \theta$, if $\theta = 60°$, then $r =$ _____.

2. For the polar equation $r = 2 \sin 2\theta$, if $\theta = 15°$, then $r =$ _____.

3. For the polar equation $r^2 = 4 \sin 2\theta$, if $\theta = 15°$, then $r =$ _____.

4. For the polar equation $r^2 = -2 \cos 2\theta$, if $\theta = 60°$, then $r =$ _____.

CONCEPT PREVIEW *For each point given in polar coordinates, state the quadrant in which the point lies if it is graphed in a rectangular coordinate system.*

5. $(5, 135°)$ **6.** $(2, 60°)$ **7.** $(6, -30°)$ **8.** $(4.6, 213°)$

CONCEPT PREVIEW *For each point given in polar coordinates, state the axis on which the point lies if it is graphed in a rectangular coordinate system. Also state whether it is on the positive portion or the negative portion of the axis. (For example, $(5, 0°)$ lies on the positive x-axis.)*

9. $(7, 360°)$ **10.** $(4, 180°)$ **11.** $(2, -90°)$ **12.** $(8, 450°)$

For each pair of polar coordinates, ***(a)*** *plot the point,* ***(b)*** *give two other pairs of polar coordinates for the point, and* ***(c)*** *give the rectangular coordinates for the point.* ***See Examples 1 and 2.***

13. $(1, 45°)$ **14.** $(3, 120°)$ **15.** $(-2, 135°)$ **16.** $(-4, 30°)$

17. $(5, -60°)$ **18.** $(2, -45°)$ **19.** $(-3, -210°)$ **20.** $(-1, -120°)$

21. $\left(3, \frac{5\pi}{3}\right)$ **22.** $\left(4, \frac{3\pi}{2}\right)$ **23.** $\left(-2, \frac{\pi}{3}\right)$ **24.** $\left(-5, \frac{5\pi}{6}\right)$

For each pair of rectangular coordinates, ***(a)*** *plot the point and* ***(b)*** *give two pairs of polar coordinates for the point, where $0° \le \theta < 360°$.* ***See Example 2(b).***

25. $(1, -1)$ **26.** $(1, 1)$ **27.** $(0, 3)$

28. $(0, -3)$ **29.** $(\sqrt{2}, \sqrt{2})$ **30.** $(-\sqrt{2}, \sqrt{2})$

31. $\left(\frac{\sqrt{3}}{2}, \frac{3}{2}\right)$ **32.** $\left(-\frac{\sqrt{3}}{2}, -\frac{1}{2}\right)$ **33.** $(3, 0)$

34. $(-2, 0)$ **35.** $\left(-\frac{3}{2}, -\frac{3\sqrt{3}}{2}\right)$ **36.** $\left(\frac{1}{2}, -\frac{\sqrt{3}}{2}\right)$

For each rectangular equation, give the equivalent polar equation and sketch its graph. ***See Example 3.***

37. $x - y = 4$ **38.** $x + y = -7$ **39.** $x^2 + y^2 = 16$

40. $x^2 + y^2 = 9$ **41.** $2x + y = 5$ **42.** $3x - 2y = 6$

Concept Check *Match each equation with its polar graph from choices A–D.*

43. $r = 3$ **44.** $r = \cos 3\theta$ **45.** $r = \cos 2\theta$ **46.** $r = \frac{2}{\cos\theta + \sin\theta}$

A.

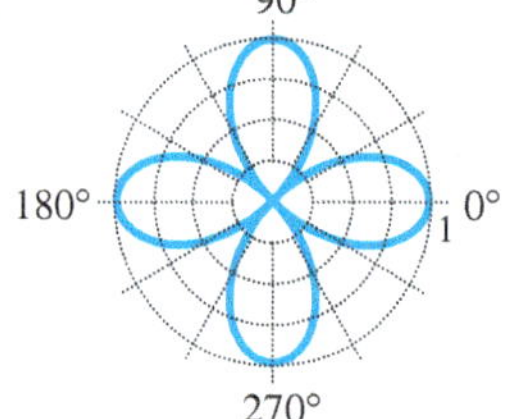

B.

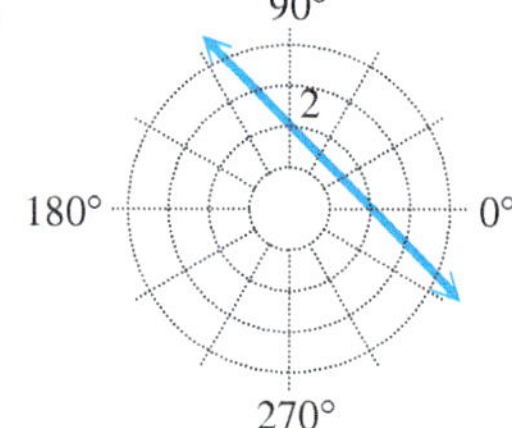

C.

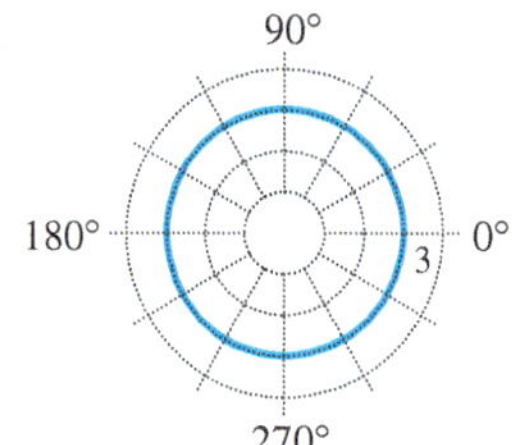

D.

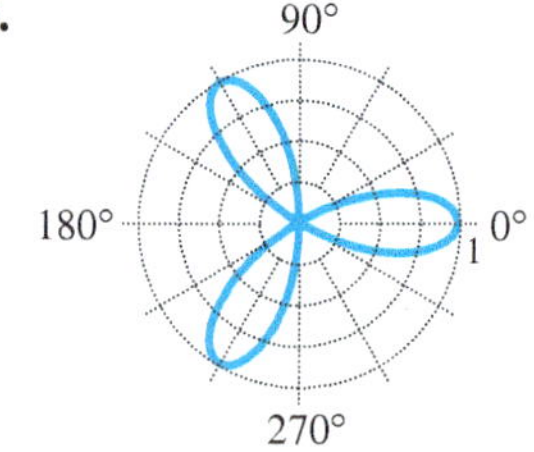

Graph each polar equation. In Exercises 47–56, also identify the type of polar graph. ***See Examples 4–6.***

47. $r = 2 + 2\cos\theta$

48. $r = 8 + 6\cos\theta$

49. $r = 3 + \cos\theta$

50. $r = 2 - \cos\theta$

51. $r = 4\cos 2\theta$

52. $r = 3\cos 5\theta$

53. $r^2 = 4\cos 2\theta$

54. $r^2 = 4\sin 2\theta$

55. $r = 4 - 4\cos\theta$

56. $r = 6 - 3\cos\theta$

57. $r = 2\sin\theta\tan\theta$
(This is a **cissoid.**)

58. $r = \dfrac{\cos 2\theta}{\cos\theta}$
(This is a **cissoid with a loop.**)

Graph each spiral of Archimedes. ***See Example 7.***

59. $r = \theta$ (Use both positive and nonpositive values.)

60. $r = -4\theta$ (Use a graphing calculator in a window of $[-30, 30]$ by $[-30, 30]$, in radian mode, and θ in $[-12\pi, 12\pi]$.)

For each equation, find an equivalent equation in rectangular coordinates, and graph. ***See Example 8.***

61. $r = 2\sin\theta$

62. $r = 2\cos\theta$

63. $r = \dfrac{2}{1 - \cos\theta}$

64. $r = \dfrac{3}{1 - \sin\theta}$

65. $r = -2\cos\theta - 2\sin\theta$

66. $r = \dfrac{3}{4\cos\theta - \sin\theta}$

67. $r = 2\sec\theta$

68. $r = -5\csc\theta$

69. $r = \dfrac{2}{\cos\theta + \sin\theta}$

70. $r = \dfrac{2}{2\cos\theta + \sin\theta}$

Solve each problem.

71. Find the polar equation of the line that passes through the points $(1, 0°)$ and $(2, 90°)$.

72. Explain how to plot a point (r, θ) in polar coordinates, if $r < 0$ and θ is in degrees.

Concept Check *The polar graphs in this section exhibit symmetry. Visualize an xy-plane superimposed on the polar coordinate system, with the pole at the origin and the polar axis on the positive x-axis. Then a polar graph may be symmetric with respect to the x-axis (the polar axis), the y-axis (the line $\theta = \frac{\pi}{2}$), or the origin (the pole).*

73. Complete the missing ordered pairs in the graphs below.

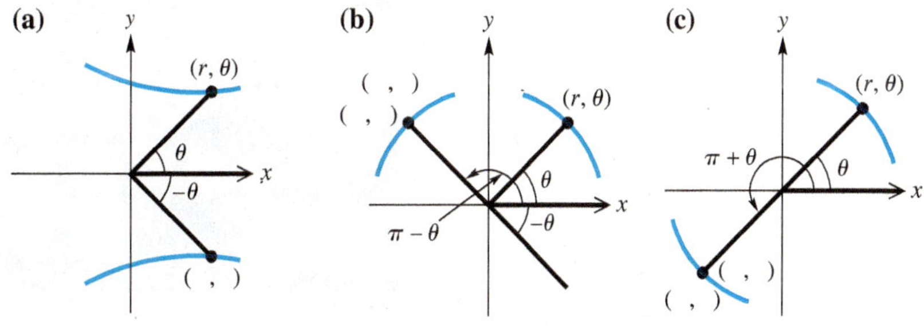

74. Based on the results in **Exercise 73,** fill in the blank(s) to correctly complete each sentence.

(a) The graph of $r = f(\theta)$ is symmetric with respect to the polar axis if substitution of ______ for θ leads to an equivalent equation.

(b) The graph of $r = f(\theta)$ is symmetric with respect to the vertical line $\theta = \frac{\pi}{2}$ if substitution of ______ for θ leads to an equivalent equation.

(c) Alternatively, the graph of $r = f(\theta)$ is symmetric with respect to the vertical line $\theta = \frac{\pi}{2}$ if substitution of ______ for r and ______ for θ leads to an equivalent equation.

(d) The graph of $r = f(\theta)$ is symmetric with respect to the pole if substitution of ______ for r leads to an equivalent equation.

(e) Alternatively, the graph of $r = f(\theta)$ is symmetric with respect to the pole if substitution of ______ for θ leads to an equivalent equation.

(f) In general, the completed statements in parts (a)–(e) mean that the graphs of polar equations of the form $r = a \pm b \cos \theta$ (where a may be 0) are symmetric with respect to ________.

(g) In general, the completed statements in parts (a)–(e) mean that the graphs of polar equations of the form $r = a \pm b \sin \theta$ (where a may be 0) are symmetric with respect to ________.

Spirals of Archimedes *The graph of $r = a\theta$ in polar coordinates is an example of a spiral of Archimedes. With a calculator set to radian mode, use the given value of a and interval of θ to graph the spiral in the window specified.*

75. $a = 1, 0 \le \theta \le 4\pi,$
$[-15, 15]$ by $[-15, 15]$

76. $a = 2, -4\pi \le \theta \le 4\pi,$
$[-30, 30]$ by $[-30, 30]$

77. $a = 1.5, -4\pi \le \theta \le 4\pi,$
$[-20, 20]$ by $[-20, 20]$

78. $a = -1, 0 \le \theta \le 12\pi,$
$[-40, 40]$ by $[-40, 40]$

Intersection of Polar Curves *Find the polar coordinates of the points of intersection of the given curves for the specified interval of θ.*

79. $r = 4 \sin \theta, \; r = 1 + 2 \sin \theta;$
$0 \le \theta < 2\pi$

80. $r = 3, \; r = 2 + 2 \cos \theta;$
$0° \le \theta < 360°$

81. $r = 2 + \sin \theta, \; r = 2 + \cos \theta;$
$0 \le \theta < 2\pi$

82. $r = \sin 2\theta, \; r = \sqrt{2} \cos \theta;$
$0 \le \theta < \pi$

(Modeling) *Solve each problem.*

83. ***Orbits of Satellites*** The polar equation

$$r = \frac{a(1 - e^2)}{1 + e \cos \theta}$$

can be used to graph the orbits of the satellites of our sun, where a is the average distance in astronomical units from the sun and e is a constant called the **eccentricity.** The sun will be located at the pole. The table lists the values of a and e.

Satellite	a	e
Mercury	0.39	0.206
Venus	0.78	0.007
Earth	1.00	0.017
Mars	1.52	0.093
Jupiter	5.20	0.048
Saturn	9.54	0.056
Uranus	19.20	0.047
Neptune	30.10	0.009
Pluto	39.40	0.249

Source: Karttunen, H., P. Kröger, H. Oja, M. Putannen, and K. Donners (Editors), *Fundamental Astronomy, 4th edition,* Springer-Verlag. Zeilik, M., S. Gregory, and E. Smith, *Introductory Astronomy and Astrophysics,* Saunders College Publishers.

(a) Graph the orbits of the four closest satellites on the same polar grid. Choose a viewing window that results in a graph with nearly circular orbits.

(b) Plot the orbits of Earth, Jupiter, Uranus, and Pluto on the same polar grid. How does Earth's distance from the sun compare to the others' distances from the sun?

(c) Use graphing to determine whether or not Pluto is always farthest from the sun.

84. *Radio Towers and Broadcasting Patterns* Radio stations do not always broadcast in all directions with the same intensity. To avoid interference with an existing station to the north, a new station may be licensed to broadcast only east and west. To create an east-west signal, two radio towers are sometimes used. See the figure. Locations where the radio signal is received correspond to the interior of the curve

$$r^2 = 40{,}000 \cos 2\theta,$$

where the polar axis (or positive x-axis) points east.

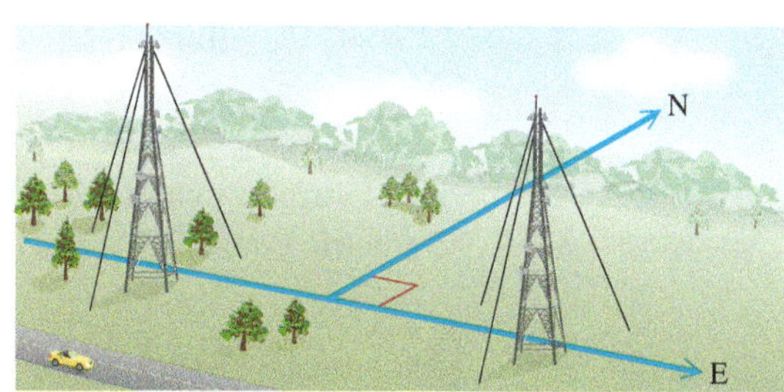

(a) Graph $r^2 = 40{,}000 \cos 2\theta$ for $0° \le \theta \le 360°$, where distances are in miles. Assuming the radio towers are located near the pole, use the graph to describe the regions where the signal can be received and where the signal cannot be received.

(b) Suppose a radio signal pattern is given by the following equation. Graph this pattern and interpret the results.

$$r^2 = 22{,}500 \sin 2\theta$$

Relating Concepts

For individual or collaborative investigation *(Exercises 85–92)*

In rectangular coordinates, the graph of

$$ax + by = c$$

is a horizontal line if $a = 0$ or a vertical line if $b = 0$. ***Work Exercises 85–92 in order,*** *to determine the general forms of polar equations for horizontal and vertical lines.*

85. Begin with the equation $y = k$, whose graph is a horizontal line. Make a trigonometric substitution for y using r and θ.

86. Solve the equation in **Exercise 85** for r.

87. Rewrite the equation in **Exercise 86** using the appropriate reciprocal function.

88. Sketch the graph of the equation $r = 3 \csc \theta$. What is the corresponding rectangular equation?

89. Begin with the equation $x = k$, whose graph is a vertical line. Make a trigonometric substitution for x using r and θ.

90. Solve the equation in **Exercise 89** for r.

91. Rewrite the equation in **Exercise 90** using the appropriate reciprocal function.

92. Sketch the graph of $r = 3 \sec \theta$. What is the corresponding rectangular equation?

8.8 Parametric Equations, Graphs, and Applications

Basic Concepts

We have graphed sets of ordered pairs that correspond to a function of the form $y = f(x)$ or $r = g(\theta)$. Another way to determine a set of ordered pairs involves the equations $x = f(t)$ and $y = g(t)$, where t is a real number in an interval I. Each value of t leads to a corresponding x-value and a corresponding y-value, and thus to an ordered pair (x, y).

Parametric Equations of a Plane Curve

A **plane curve** is a set of points (x, y) such that $x = f(t)$, $y = g(t)$, and f and g are both defined on an interval I. The equations $x = f(t)$ and $y = g(t)$ are **parametric equations** with **parameter** t.

Graphing calculators are capable of graphing plane curves defined by parametric equations. The calculator must be set to parametric mode. ■

Parametric Graphs and Their Rectangular Equivalents

EXAMPLE 1 Graphing a Plane Curve Defined Parametrically

Let $x = t^2$ and $y = 2t + 3$, for t in $[-3, 3]$. Graph the set of ordered pairs (x, y).

ALGEBRAIC SOLUTION

Make a table of corresponding values of t, x, and y over the domain of t. Plot the points as shown in **Figure 74.** The graph is a portion of a parabola with horizontal axis $y = 3$. The arrowheads indicate the direction the curve traces as t increases.

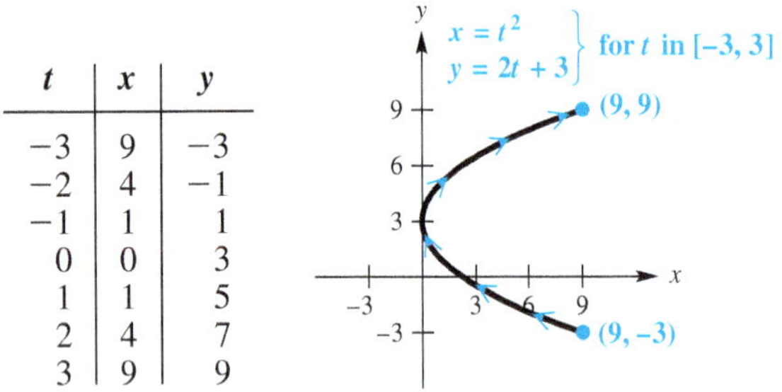

t	x	y
−3	9	−3
−2	4	−1
−1	1	1
0	0	3
1	1	5
2	4	7
3	9	9

Figure 74

GRAPHING CALCULATOR SOLUTION

We set the parameters of the TI-84 Plus as shown to obtain the graph. See **Figure 75.**

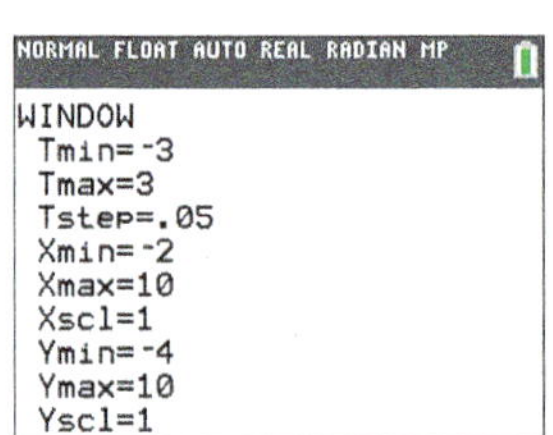

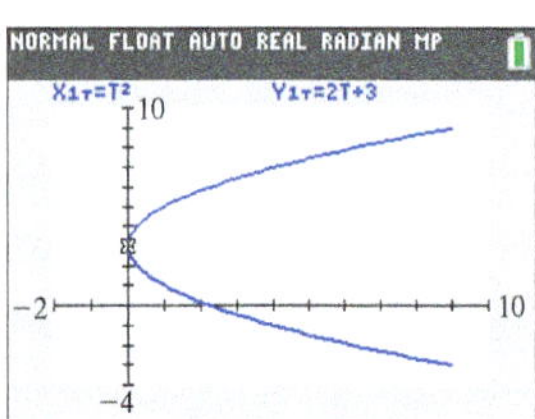

Figure 75

Duplicate this graph and observe how the curve is traced. It should match **Figure 74.**

✔ Now Try Exercise 9(a).

EXAMPLE 2 Finding an Equivalent Rectangular Equation

Find a rectangular equation for the plane curve of **Example 1,**

$$x = t^2, \quad y = 2t + 3, \quad \text{for } t \text{ in } [-3, 3].$$

SOLUTION To eliminate the parameter t, first solve either equation for t.

This equation leads to a unique solution for t.

$y = 2t + 3$ Choose the simpler equation.

$2t = y - 3$ Subtract 3 and rewrite.

$t = \dfrac{y - 3}{2}$ Divide by 2.

Now substitute this result into the first equation to eliminate the parameter t.

$$x = t^2$$

$$x = \left(\frac{y-3}{2}\right)^2 \quad \text{Substitute for } t.$$

$$x = \frac{(y-3)^2}{4} \quad \left(\frac{a}{b}\right)^2 = \frac{a^2}{b^2}$$

$$4x = (y-3)^2 \quad \text{Multiply by 4.}$$

This is the equation of a horizontal parabola opening to the right, which agrees with the graph given in **Figure 74.** Because t is in $[-3, 3]$, x is in $[0, 9]$ and y is in $[-3, 9]$. The rectangular equation must be given with restricted domain as

$$4x = (y-3)^2, \quad \text{for } x \text{ in } [0, 9].$$

✔ **Now Try Exercise 9(b).**

EXAMPLE 3 Graphing a Plane Curve Defined Parametrically

Graph the plane curve defined by $x = 2 \sin t$, $y = 3 \cos t$, for t in $[0, 2\pi]$.

SOLUTION To convert to a rectangular equation, it is not productive here to solve either equation for t. Instead, we use the fact that $\sin^2 t + \cos^2 t = 1$ to apply another approach.

$$x = 2 \sin t \qquad y = 3 \cos t \qquad \text{Given equations}$$

$$x^2 = 4 \sin^2 t \qquad y^2 = 9 \cos^2 t \qquad \text{Square each side.}$$

$$\frac{x^2}{4} = \sin^2 t \qquad \frac{y^2}{9} = \cos^2 t \qquad \text{Solve for } \sin^2 t \text{ and } \cos^2 t.$$

Now add corresponding sides of the two equations.

$$\frac{x^2}{4} + \frac{y^2}{9} = \sin^2 t + \cos^2 t$$

$$\frac{x^2}{4} + \frac{y^2}{9} = 1 \qquad \sin^2 t + \cos^2 t = 1$$

This is an equation of an **ellipse.** See **Figure 76.**

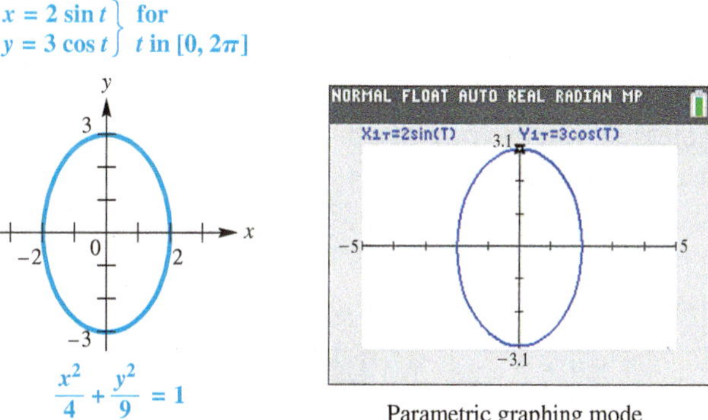

Parametric graphing mode

Figure 76

✔ **Now Try Exercise 31.**

Parametric representations of a curve are not unique. In fact, there are infinitely many parametric representations of a given curve. If the curve can be described by a rectangular equation $y = f(x)$, with domain X, then one simple parametric representation is

$$x = t, \quad y = f(t), \quad \text{for } t \text{ in } X.$$

EXAMPLE 4 Finding Alternative Parametric Equation Forms

Give two parametric representations for the equation of the parabola.

$$y = (x - 2)^2 + 1$$

SOLUTION The simplest choice is to let

$$x = t, \quad y = (t - 2)^2 + 1, \qquad \text{for } t \text{ in } (-\infty, \infty).$$

Another choice, which leads to a simpler equation for y, is

$$x = t + 2, \quad y = t^2 + 1, \qquad \text{for } t \text{ in } (-\infty, \infty).$$

✔ **Now Try Exercise 33.**

NOTE Verify that another choice in **Example 4** is

$$x = 2 + \tan t, \quad y = \sec^2 t, \qquad \text{for } t \text{ in } \left(-\frac{\pi}{2}, \frac{\pi}{2}\right).$$

Other choices are possible.

The Cycloid The *cycloid* is a special case of the **trochoid**—a curve traced out by a point at a given distance from the center of a circle as the circle rolls along a straight line. If the given point is on the *circumference* of the circle, then the path traced as the circle rolls along a straight line is a **cycloid,** which is defined parametrically as follows.

$$\boldsymbol{x = at - a \sin t, \quad y = a - a \cos t, \qquad \text{for } t \text{ in } (-\infty, \infty)}$$

Other curves related to trochoids are **hypotrochoids** and **epitrochoids,** which are traced out by a point that is a given distance from the center of a circle that rolls not on a straight line, but on the inside or outside, respectively, of another circle. The classic Spirograph toy can be used to draw these curves.

EXAMPLE 5 Graphing a Cycloid

Graph the cycloid.

$$x = t - \sin t, \quad y = 1 - \cos t, \qquad \text{for } t \text{ in } [0, 2\pi]$$

ALGEBRAIC SOLUTION

There is no simple way to find a rectangular equation for the cycloid from its parametric equations. Instead, begin with a table using selected values for t in $[0, 2\pi]$. Approximate values have been rounded as necessary.

t	0	$\frac{\pi}{4}$	$\frac{\pi}{2}$	π	$\frac{3\pi}{2}$	2π
x	0	0.08	0.6	π	5.7	2π
y	0	0.3	1	2	1	0

Figure 77

Plotting the ordered pairs (x, y) from the table of values leads to the portion of the graph in **Figure 77** from 0 to 2π.

GRAPHING CALCULATOR SOLUTION

It is easier to graph a cycloid with a graphing calculator in parametric mode than with traditional methods. See **Figure 78.**

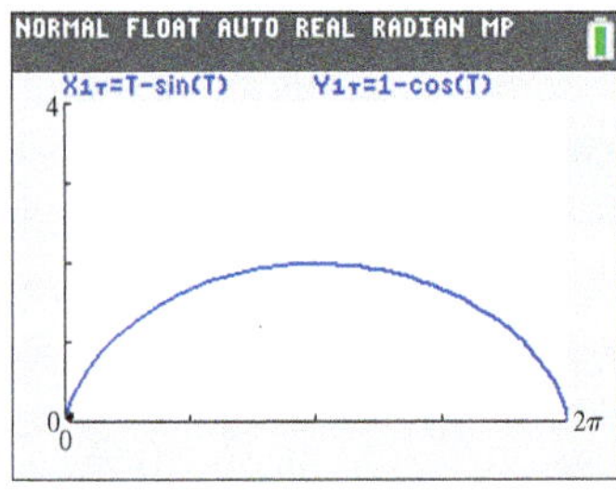

Figure 78

Using a larger interval for t would show that the cycloid repeats the pattern shown here every 2π units.

✔ **Now Try Exercise 37.**

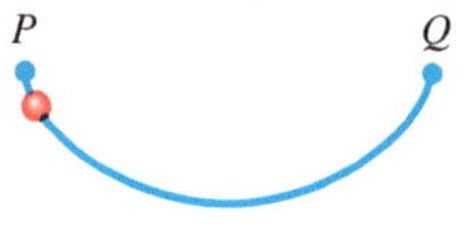

Figure 79

The cycloid has an interesting physical property. If a flexible cord or wire goes through points P and Q as in **Figure 79,** and a bead is allowed to slide due to the force of gravity without friction along this path from P to Q, the path that requires the shortest time takes the shape of the graph of an inverted cycloid.

Applications of Parametric Equations Parametric equations are used to simulate motion. If an object is thrown with a velocity of v feet per second at an angle θ with the horizontal, then its flight can be modeled by

$$x = (v \cos \theta)t \quad \text{and} \quad y = (v \sin \theta)t - 16t^2 + h,$$

where t is in seconds and h is the object's initial height in feet above the ground. Here, x gives the horizontal position information and y gives the vertical position information. The term $-16t^2$ occurs because gravity is pulling downward. See **Figure 80.** These equations ignore air resistance.

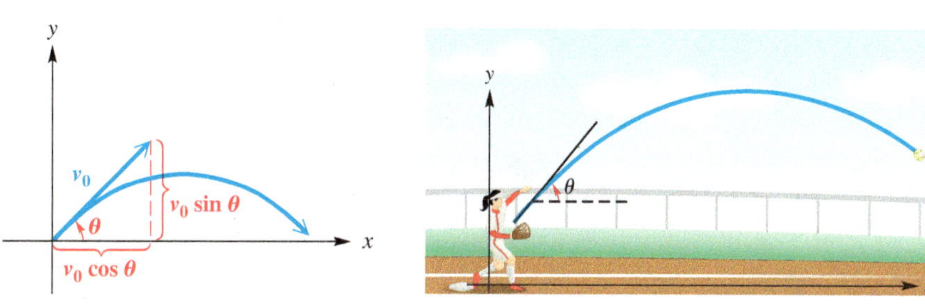

Figure 80

EXAMPLE 6 Simulating Motion with Parametric Equations

Three golf balls are hit simultaneously into the air at 132 ft per sec (90 mph) at angles of 30°, 50°, and 70° with the horizontal.

(a) Assuming the ground is level, determine graphically which ball travels the greatest distance. Estimate this distance.

(b) Which ball reaches the greatest height? Estimate this height.

SOLUTION

(a) Use the following parametric equations to model the flight of the golf balls.

$$x = (v \cos \theta)t \quad \text{and} \quad y = (v \sin \theta)t - 16t^2 + h$$

Write three sets of parametric equations.

$$x_1 = (132 \cos 30°)t, \quad y_1 = (132 \sin 30°)t - 16t^2$$
$$x_2 = (132 \cos 50°)t, \quad y_2 = (132 \sin 50°)t - 16t^2$$
$$x_3 = (132 \cos 70°)t, \quad y_3 = (132 \sin 70°)t - 16t^2$$

Substitute $h = 0$, $v = 132$ ft per sec, and $\theta = 30°, 50°$, and $70°$.

The graphs of the three sets of parametric equations are shown in **Figure 81(a),** where $0 \le t \le 3$. From the graph in **Figure 81(b),** where $0 \le t \le 9$, we see that the ball hit at 50° travels the greatest distance. Using the tracing feature of the TI-84 Plus calculator, we find that this distance is about 540 ft.

(b) Again, use the tracing feature to find that the ball hit at 70° reaches the greatest height, about 240 ft.

Now Try Exercise 43.

(a)

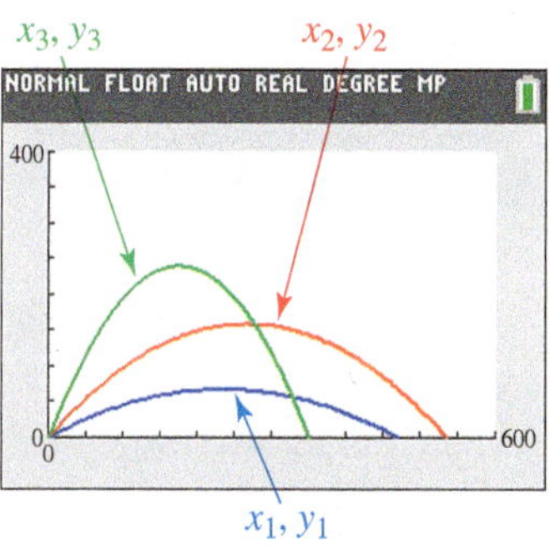

(b)

Figure 81

A TI-84 Plus calculator allows us to view the graphing of more than one equation either *sequentially* or *simultaneously*. By choosing the latter, the three golf balls in **Figure 81** can be viewed in flight at the same time. ■

EXAMPLE 7 Examining Parametric Equations of Flight

Jack launches a small rocket from a table that is 3.36 ft above the ground. Its initial velocity is 64 ft per sec, and it is launched at an angle of 30° with respect to the ground. Find the rectangular equation that models its path. What type of path does the rocket follow?

SOLUTION The path of the rocket is defined by the parametric equations

$$x = (64 \cos 30°)t \quad \text{and} \quad y = (64 \sin 30°)t - 16t^2 + 3.36$$

or, equivalently, $\quad x = 32\sqrt{3}t \quad$ and $\quad y = -16t^2 + 32t + 3.36.$

From $x = 32\sqrt{3}t$, we solve for t to obtain

$$t = \frac{x}{32\sqrt{3}}. \qquad \text{Divide by } 32\sqrt{3}.$$

Substituting for t in the other parametric equation yields the following.

$$y = -16t^2 + 32t + 3.36$$

$$y = -16\left(\frac{x}{32\sqrt{3}}\right)^2 + 32\left(\frac{x}{32\sqrt{3}}\right) + 3.36 \qquad \text{Let } t = \frac{x}{32\sqrt{3}}.$$

$$y = -\frac{1}{192}x^2 + \frac{\sqrt{3}}{3}x + 3.36 \qquad \text{Simplify.}$$

This equation defines a parabola. The rocket follows a parabolic path.

✔ **Now Try Exercise 47(a).**

EXAMPLE 8 Analyzing the Path of a Projectile

Determine the total flight time and the horizontal distance traveled by the rocket in **Example 7.**

ALGEBRAIC SOLUTION

The equation $y = -16t^2 + 32t + 3.36$ tells the vertical position of the rocket at time t. We need to determine the positive value of t for which $y = 0$ because this value corresponds to the rocket at ground level. This yields

$$0 = -16t^2 + 32t + 3.36.$$

Using the quadratic formula, the solutions are $t = -0.1$ or $t = 2.1$. Because t represents time, $t = -0.1$ is an unacceptable answer. Therefore, the flight time is 2.1 sec.

The rocket was in the air for 2.1 sec, so we can use $t = 2.1$ and the parametric equation that models the horizontal position, $x = 32\sqrt{3}t$, to obtain

$$x = 32\sqrt{3}(2.1) \approx 116.4 \text{ ft}.$$

GRAPHING CALCULATOR SOLUTION

Figure 82 shows that when $t = 2.1$, the horizontal distance x covered is approximately 116.4 ft, which agrees with the algebraic solution.

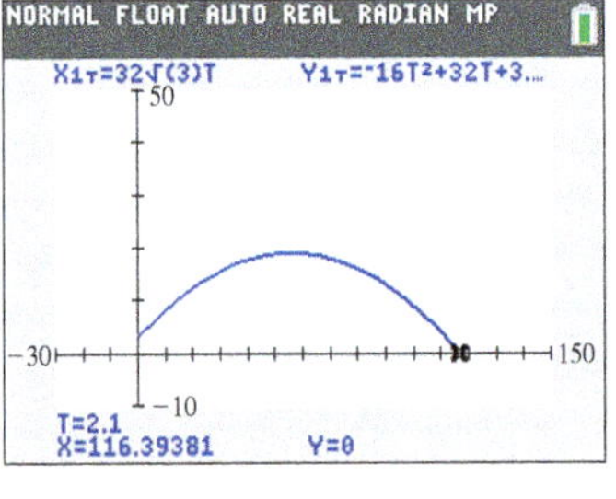

Figure 82

✔ **Now Try Exercise 47(b).**

8.8 Exercises

CONCEPT PREVIEW *Fill in the blank to correctly complete each sentence.*

1. For the plane curve defined by

$$x = t^2 + 1, \quad y = 2t + 3, \quad \text{for } t \text{ in } [-4, 4],$$

the ordered pair that corresponds to $t = -3$ is ______.

2. For the plane curve defined by

$$x = -3t + 6, \quad y = t^2 - 3, \quad \text{for } t \text{ in } [-5, 5],$$

the ordered pair that corresponds to $t = 4$ is ______.

3. For the plane curve defined by

$$x = \cos t, \quad y = 2 \sin t, \quad \text{for } t \text{ in } [0, 2\pi],$$

the ordered pair that corresponds to $t = \frac{\pi}{3}$ is ______.

4. For the plane curve defined by

$$x = \sqrt{t}, \quad y = t^2 + 3, \quad \text{for } t \text{ in } (0, \infty),$$

the ordered pair that corresponds to $t = 16$ is ______.

CONCEPT PREVIEW *Match the ordered pair from Column II with the pair of parametric equations in Column I on whose graph the point lies. In each case, consider the given value of t.*

I	II
5. $x = 3t + 6, \; y = -2t + 4; \quad t = 2$	**A.** $(5, 25)$
6. $x = \cos t, \; y = \sin t; \quad t = \frac{\pi}{4}$	**B.** $(7, 2)$
7. $x = t, \; y = t^2; \quad t = 5$	**C.** $(12, 0)$
8. $x = t^2 + 3, \; y = t^2 - 2; \quad t = 2$	**D.** $\left(\frac{\sqrt{2}}{2}, \frac{\sqrt{2}}{2}\right)$

For each plane curve, (a) graph the curve, and (b) find a rectangular equation for the curve. ***See Examples 1 and 2.***

9. $x = t + 2, \; y = t^2,$ for t in $[-1, 1]$

10. $x = 2t, \; y = t + 1,$ for t in $[-2, 3]$

11. $x = \sqrt{t}, \; y = 3t - 4,$ for t in $[0, 4]$

12. $x = t^2, \; y = \sqrt{t},$ for t in $[0, 4]$

13. $x = t^3 + 1, \; y = t^3 - 1,$ for t in $(-\infty, \infty)$

14. $x = 2t - 1, \; y = t^2 + 2,$ for t in $(-\infty, \infty)$

15. $x = 2 \sin t, \; y = 2 \cos t,$ for t in $[0, 2\pi]$

16. $x = \sqrt{5} \sin t, \; y = \sqrt{3} \cos t,$ for t in $[0, 2\pi]$

17. $x = 3 \tan t, \; y = 2 \sec t,$ for t in $\left(-\frac{\pi}{2}, \frac{\pi}{2}\right)$

18. $x = \cot t, \; y = \csc t,$ for t in $(0, \pi)$

19. $x = \sin t, \; y = \csc t,$ for t in $(0, \pi)$

20. $x = \tan t, \; y = \cot t,$ for t in $\left(0, \frac{\pi}{2}\right)$

21. $x = t, \; y = \sqrt{t^2 + 2},$ for t in $(-\infty, \infty)$

22. $x = \sqrt{t}, \; y = t^2 - 1,$ for t in $[0, \infty)$

23. $x = 2 + \sin t,\ y = 1 + \cos t,$ for t in $[0, 2\pi]$

24. $x = 1 + 2\sin t,\ y = 2 + 3\cos t,$ for t in $[0, 2\pi]$

25. $x = t + 2,\ y = \dfrac{1}{t+2},$ for $t \neq -2$

26. $x = t - 3,\ y = \dfrac{2}{t-3},$ for $t \neq 3$

27. $x = t + 2,\ y = t - 4,$ for t in $(-\infty, \infty)$

28. $x = t^2 + 2,\ y = t^2 - 4,$ for t in $(-\infty, \infty)$

Graph each plane curve defined by the parametric equations for t in $[0, 2\pi]$. Then find a rectangular equation for the plane curve. ***See Example 3.***

29. $x = 3\cos t,\ y = 3\sin t$

30. $x = 2\cos t,\ y = 2\sin t$

31. $x = 3\sin t,\ y = 2\cos t$

32. $x = 4\sin t,\ y = 3\cos t$

Give two parametric representations for the equation of each parabola. ***See Example 4.***

33. $y = (x + 3)^2 - 1$

34. $y = (x + 4)^2 + 2$

35. $y = x^2 - 2x + 3$

36. $y = x^2 - 4x + 6$

Graph each cycloid defined by the given equations for t in the specified interval. ***See Example 5.***

37. $x = 2t - 2\sin t,\ y = 2 - 2\cos t,$ for t in $[0, 4\pi]$

38. $x = t - \sin t,\ y = 1 - \cos t,$ for t in $[0, 4\pi]$

Lissajous Figures The screen shown here is an example of a ***Lissajous figure.*** *Such figures occur in electronics and may be used to find the frequency of an unknown voltage. Graph each Lissajous figure for t in $[0, 6.5]$ using the window $[-6, 6]$ by $[-4, 4]$.*

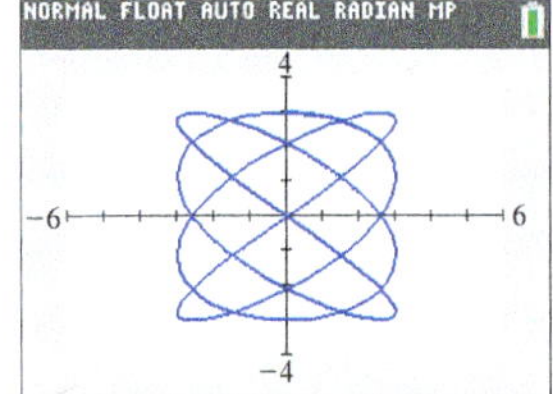

39. $x = 2\cos t,\ y = 3\sin 2t$

40. $x = 3\cos t,\ y = 2\sin 2t$

41. $x = 3\sin 4t,\ y = 3\cos 3t$

42. $x = 4\sin 4t,\ y = 3\sin 5t$

(Modeling) Do the following. ***See Examples 6–8.***

(a) Determine parametric equations that model the path of the projectile.

(b) Determine a rectangular equation that models the path of the projectile.

(c) Determine approximately how long the projectile is in flight and the horizontal distance it covers.

43. *Flight of a Model Rocket* A model rocket is launched from the ground with velocity 48 ft per sec at an angle of 60° with respect to the ground.

44. *Flight of a Golf Ball* Tyler is playing golf. He hits a golf ball from the ground at an angle of 60° with respect to the ground at velocity 150 ft per sec.

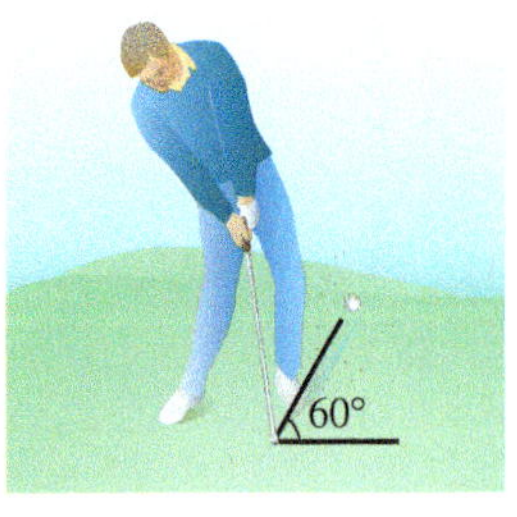

45. *Flight of a Softball* Sally hits a softball when it is 2 ft above the ground. The ball leaves her bat at an angle of 20° with respect to the ground at velocity 88 ft per sec.

46. *Flight of a Baseball* Francisco hits a baseball when it is 2.5 ft above the ground. The ball leaves his bat at an angle of 29° from the horizontal with velocity 136 ft per sec.

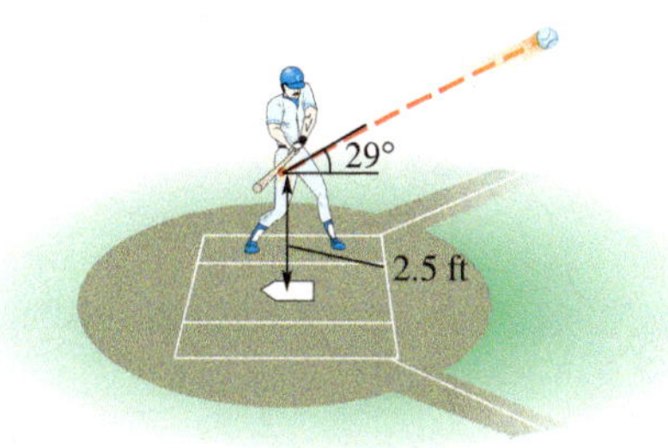

(Modeling) Solve each problem. ***See Examples 7 and 8.***

47. *Path of a Rocket* A rocket is launched from the top of an 8-ft platform. Its initial velocity is 128 ft per sec. It is launched at an angle of 60° with respect to the ground.

(a) Find the rectangular equation that models its path. What type of path does the rocket follow?

(b) Determine the total flight time, to the nearest second, and the horizontal distance the rocket travels, to the nearest foot.

48. *Simulating Gravity on the Moon* If an object is thrown on the moon, then the parametric equations of flight are

$$x = (v \cos \theta)t \quad \text{and} \quad y = (v \sin \theta)t - 2.66t^2 + h.$$

Estimate, to the nearest foot, the distance a golf ball hit at 88 ft per sec (60 mph) at an angle of 45° with the horizontal travels on the moon if the moon's surface is level.

49. *Flight of a Baseball* A baseball is hit from a height of 3 ft at a 60° angle above the horizontal. Its initial velocity is 64 ft per sec.

(a) Write parametric equations that model the flight of the baseball.

(b) Determine the horizontal distance, to the nearest tenth of a foot, traveled by the ball in the air. Assume that the ground is level.

(c) What is the maximum height of the baseball, to the nearest tenth of a foot? At that time, how far has the ball traveled horizontally?

(d) Would the ball clear a 5-ft-high fence that is 100 ft from the batter?

50. *Path of a Projectile* A projectile has been launched from the ground with initial velocity 88 ft per sec. The parametric equations

$$x = 82.7t \quad \text{and} \quad y = -16t^2 + 30.1t$$

model the path of the projectile, where t is in seconds.

(a) Approximate the angle θ that the projectile makes with the horizontal at the launch, to the nearest tenth of a degree.

(b) Write parametric equations for the path using the cosine and sine functions.

Work each problem.

51. Give two parametric representations of the parabola $y = a(x - h)^2 + k$.

52. Give a parametric representation of the rectangular equation $\frac{x^2}{a^2} - \frac{y^2}{b^2} = 1$.

53. Give a parametric representation of the rectangular equation $\frac{x^2}{a^2} + \frac{y^2}{b^2} = 1$.

54. The spiral of Archimedes has polar equation $r = a\theta$, where $r^2 = x^2 + y^2$. Show that a parametric representation of the spiral of Archimedes is

$$x = a\theta \cos \theta, \quad y = a\theta \sin \theta, \qquad \text{for } \theta \text{ in } (-\infty, \infty).$$

55. Show that the **hyperbolic spiral** $r\theta = a$, where $r^2 = x^2 + y^2$, is given parametrically by

$$x = \frac{a \cos \theta}{\theta}, \quad y = \frac{a \sin \theta}{\theta}, \qquad \text{for } \theta \text{ in } (-\infty, 0) \cup (0, \infty).$$

Chapter 8 Test Prep

Key Terms

8.1 Side-Angle-Side (SAS)
Angle-Side-Angle (ASA)
Side-Side-Side (SSS)
oblique triangle
Side-Angle-Angle (SAA)
ambiguous case

8.2 semiperimeter

8.3 scalar
vector quantity
vector
magnitude
initial point
terminal point
parallelogram rule
resultant
opposite (of a vector)
zero vector
equilibrant
airspeed
ground speed

8.4 position vector
horizontal component
vertical component
direction angle
unit vector
dot product (inner product)
angle between two vectors
orthogonal vectors

8.5 resultant
real axis
imaginary axis
complex plane
rectangular form of a complex number
trigonometric (polar) form of a complex number
absolute value (modulus)
argument

8.6 *n*th root of a complex number

8.7 polar coordinate system
pole
polar axis
polar coordinates
rectangular (Cartesian) equation
polar equation
cardioid
polar grid
rose curve
lemniscate
spiral of Archimedes
limaçon

8.8 plane curve
parametric equations of a plane curve
parameter
cycloid

New Symbols

OP or $\overrightarrow{\mathbf{OP}}$	vector **OP**
$\lvert\mathbf{OP}\rvert$	magnitude of vector **OP**
$\langle a, b\rangle$	position vector
i, **j**	unit vectors
cis θ	$\cos\theta + i\sin\theta$

Quick Review

Concepts	Examples

8.1 The Law of Sines

Law of Sines

In any triangle ABC, with sides a, b, and c, the following holds.

$$\frac{a}{\sin A} = \frac{b}{\sin B} = \frac{c}{\sin C}$$

$$\frac{\sin A}{a} = \frac{\sin B}{b} = \frac{\sin C}{c} \quad \text{Alternative form}$$

In triangle ABC, find c, to the nearest hundredth, if $A = 44°$, $C = 62°$, and $a = 12.00$ units.

$$\frac{a}{\sin A} = \frac{c}{\sin C} \quad \text{Law of sines}$$

$$\frac{12.00}{\sin 44°} = \frac{c}{\sin 62°} \quad \text{Substitute.}$$

$$c = \frac{12.00 \sin 62°}{\sin 44°} \quad \text{Multiply by } \sin 62° \text{ and rewrite.}$$

$$c \approx 15.25 \text{ units} \quad \text{Use a calculator.}$$

Area of a Triangle

In any triangle ABC, the area $\mathcal{A}$ is half the product of the lengths of two sides and the sine of the angle between them.

$$\mathcal{A} = \frac{1}{2}bc \sin A, \quad \mathcal{A} = \frac{1}{2}ab \sin C, \quad \mathcal{A} = \frac{1}{2}ac \sin B$$

For triangle ABC above, apply the appropriate area formula. Here, $B = 180° - 44° - 62° = 74°$.

$$\mathcal{A} = \frac{1}{2}ac \sin B = \frac{1}{2}(12.00)(15.25) \sin 74° \approx 87.96 \text{ sq units}$$

Concepts	Examples
Ambiguous Case If we are given the lengths of two sides and the angle opposite one of them (for example, A, a, and b in triangle ABC), then it is possible that zero, one, or two such triangles exist. If A is acute, h is the altitude from C, and • $a < h < b$, then there is no triangle. • $a = h$ and $h < b$, then there is one triangle (a right triangle). • $a \geq b$, then there is one triangle. • $h < a < b$, then there are two triangles. If A is obtuse and • $a \leq b$, then there is no triangle. • $a > b$, then there is one triangle. See the guidelines in this section that illustrate the possible outcomes.	Solve triangle ABC, given $A = 44.5°$, $a = 11.0$ in., and $c = 7.0$ in. Find angle C. $\frac{\sin C}{7.0} = \frac{\sin 44.5°}{11.0}$ Law of sines $\sin C \approx 0.4460$ Solve for sin C. $C \approx 26.5°$ Use the inverse sine function. Another angle with this sine value is $180° - 26.5° \approx 153.5°$. However, $153.5° + 44.5° > 180°$, so there is only one triangle. $B = 180° - 44.5° - 26.5°$ Angle sum formula $B = 109°$ Subtract. Use the law of sines again to solve for b. $b \approx 14.8$ in.

8.2 The Law of Cosines

Concepts	Examples
Law of Cosines In any triangle ABC, with sides a, b, and c, the following hold. $a^2 = b^2 + c^2 - 2bc \cos A$ $b^2 = a^2 + c^2 - 2ac \cos B$ $c^2 = a^2 + b^2 - 2ab \cos C$	In triangle ABC, find C if $a = 11$ units, $b = 13$ units, and $c = 20$ units. Then find its area. $c^2 = a^2 + b^2 - 2ab \cos C$ Law of cosines $20^2 = 11^2 + 13^2 - 2(11)(13) \cos C$ Substitute. $400 = 121 + 169 - 286 \cos C$ Square and multiply. $\cos C = \frac{400 - 121 - 169}{-286}$ Solve for cos C. $\cos C \approx -0.38461538$ Use a calculator. $C \approx 113°$ Use the inverse cosine function.
Heron's Area Formula If a triangle has sides of lengths a, b, and c, with semiperimeter $s = \frac{1}{2}(a + b + c)$, then the area $\mathcal{A}$ of the triangle is given by the following. $\mathcal{A} = \sqrt{s(s-a)(s-b)(s-c)}$	The semiperimeter s of the above triangle is $s = \frac{1}{2}(11 + 13 + 20) = 22$, so the area is $\mathcal{A} = \sqrt{22(22-11)(22-13)(22-20)}$ $\mathcal{A} = 66$ sq units.

Concepts | **Examples**

8.3 Geometrically Defined Vectors and Applications

Vector Sum

The sum of two vectors is also a vector. There are two ways to find the sum of two vectors **A** and **B** geometrically.

1. The vector with the same initial point as **A** and the same terminal point as **B** is the sum **A** + **B**.

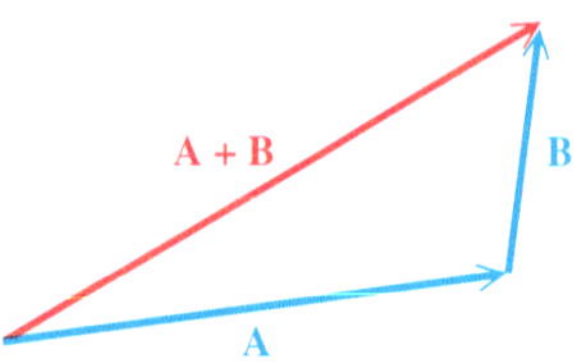

2. The diagonal of the parallelogram with the same initial point as **A** and **B** is the sum **A** + **B**. This is the **parallelogram rule.**

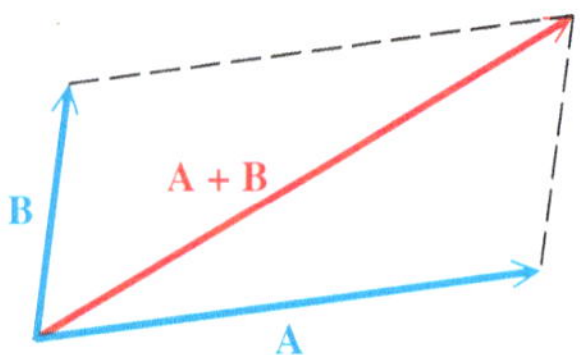

Two forces of 25 newtons and 32 newtons act on a point in a plane. If the angle between the forces is 62°, find the magnitude of the resultant force.

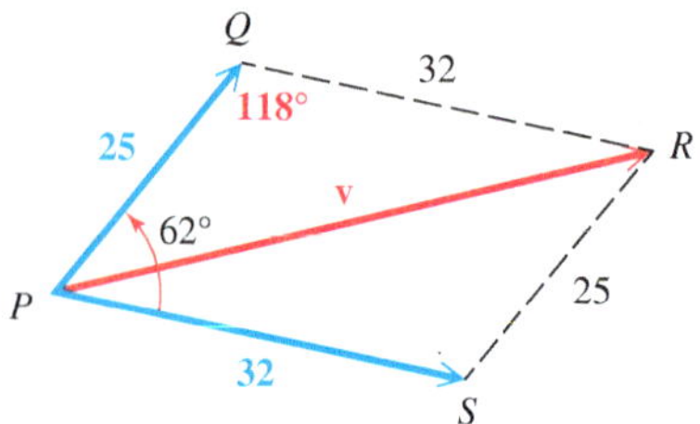

The resultant force divides a parallelogram into two triangles. The measure of angle Q in the figure is 118°. We use the law of cosines to find the desired magnitude.

$$|\mathbf{v}|^2 = 25^2 + 32^2 - 2(25)(32)\cos 118^\circ$$

$$|\mathbf{v}|^2 \approx 2400$$

$$|\mathbf{v}| \approx 49$$

The magnitude of the resultant force is 49 newtons.

8.4 Algebraically Defined Vectors and the Dot Product

Magnitude and Direction Angle of a Vector

The magnitude (length) of vector $\mathbf{u} = \langle a, b \rangle$ is given by the following.

$$|\mathbf{u}| = \sqrt{a^2 + b^2}$$

The direction angle θ satisfies $\tan\theta = \frac{b}{a}$, where $a \neq 0$.

Find the magnitude and direction angle of vector **u** in the figure.

$$|\mathbf{u}| = \sqrt{\left(2\sqrt{3}\right)^2 + 2^2} = \sqrt{16}$$

$$= 4 \leftarrow \text{Magnitude}$$

$$\tan\theta = \frac{2}{2\sqrt{3}} = \frac{1}{\sqrt{3}} \cdot \frac{\sqrt{3}}{\sqrt{3}} = \frac{\sqrt{3}}{3}, \text{ so } \theta = 30^\circ.$$

If $\mathbf{u} = \langle a, b \rangle$ has direction angle θ, then

$$\mathbf{u} = \langle |\mathbf{u}| \cos\theta, |\mathbf{u}| \sin\theta \rangle.$$

For **u** defined above,

$$\mathbf{u} = \langle 4\cos 30^\circ, 4\sin 30^\circ \rangle$$

$$= \langle 2\sqrt{3}, 2 \rangle. \quad \cos 30^\circ = \frac{\sqrt{3}}{2};\ \sin 30^\circ = \frac{1}{2}$$

Vector Operations

Let a, b, c, d, and k represent real numbers.

$$\langle a, b \rangle + \langle c, d \rangle = \langle a + c, b + d \rangle$$

$$k \cdot \langle a, b \rangle = \langle ka, kb \rangle$$

$$\text{If } \mathbf{u} = \langle a_1, a_2 \rangle, \text{ then } -\mathbf{u} = \langle -a_1, -a_2 \rangle.$$

$$\langle a, b \rangle - \langle c, d \rangle = \langle a, b \rangle + (-\langle c, d \rangle) = \langle a - c, b - d \rangle$$

Find each of the following.

$$\langle 4, 6 \rangle + \langle -8, 3 \rangle = \langle -4, 9 \rangle$$

$$5\langle -2, 1 \rangle = \langle -10, 5 \rangle$$

$$-\langle -9, 6 \rangle = \langle 9, -6 \rangle$$

$$\langle 4, 6 \rangle - \langle -8, 3 \rangle = \langle 12, 3 \rangle$$

i, j Form for Vectors

If $\mathbf{v} = \langle a, b \rangle$, then

$$\mathbf{v} = a\mathbf{i} + b\mathbf{j}, \quad \text{where } \mathbf{i} = \langle 1, 0 \rangle \text{ and } \mathbf{j} = \langle 0, 1 \rangle.$$

If $\mathbf{u} = \langle 2\sqrt{3}, 2 \rangle$ as above, then

$$\mathbf{u} = 2\sqrt{3}\,\mathbf{i} + 2\mathbf{j}.$$

Concepts	Examples
Dot Product The dot product of the two vectors $\mathbf{u} = \langle a, b\rangle$ and $\mathbf{v} = \langle c, d\rangle$, denoted $\mathbf{u} \cdot \mathbf{v}$, is given by the following. $$\mathbf{u} \cdot \mathbf{v} = ac + bd$$	Find the dot product. $$\langle 2, 1\rangle \cdot \langle 5, -2\rangle = 2 \cdot 5 + 1(-2) = 8$$
Geometric Interpretation of the Dot Product If θ is the angle between the two nonzero vectors **u** and **v**, where $0^\circ \le \theta \le 180^\circ$, then the following holds. $$\cos\theta = \frac{\mathbf{u} \cdot \mathbf{v}}{\lvert\mathbf{u}\rvert\,\lvert\mathbf{v}\rvert}$$	Find the angle θ between $\mathbf{u} = \langle 3, 1\rangle$ and $\mathbf{v} = \langle 2, -3\rangle$. $\cos\theta = \frac{\mathbf{u} \cdot \mathbf{v}}{\lvert\mathbf{u}\rvert\,\lvert\mathbf{v}\rvert}$ Geometric interpretation of the dot product $\cos\theta = \frac{3(2) + 1(-3)}{\sqrt{3^2 + 1^2} \cdot \sqrt{2^2 + (-3)^2}}$ Use the definitions. $\cos\theta = \frac{3}{\sqrt{130}}$ Simplify. $\cos\theta \approx 0.26311741$ Use a calculator. $\theta \approx 74.7^\circ$ Use the inverse cosine function.

8.5 Trigonometric (Polar) Form of Complex Numbers; Products and Quotients

Concepts	Examples
Trigonometric (Polar) Form of Complex Numbers Let the complex number $x + yi$ correspond to the vector with direction angle θ and magnitude r. $x = r\cos\theta \qquad y = r\sin\theta$ $r = \sqrt{x^2 + y^2} \qquad \tan\theta = \frac{y}{x}, \text{ if } x \neq 0$	Write $2(\cos 60^\circ + i\sin 60^\circ)$ in rectangular form. $$2(\cos 60^\circ + i\sin 60^\circ) = 2\left(\frac{1}{2} + i \cdot \frac{\sqrt{3}}{2}\right) = 1 + i\sqrt{3}$$
The trigonometric (polar) form of the expression $x + yi$ is $r(\cos\theta + i\sin\theta)$ or $r \text{ cis } \theta$.	Write $-\sqrt{2} + i\sqrt{2}$ in trigonometric form. 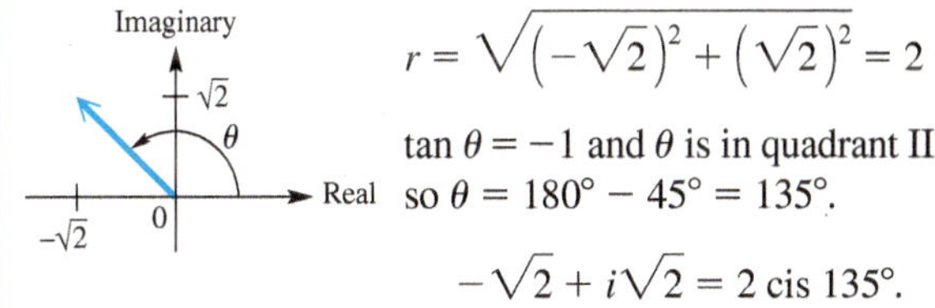 $r = \sqrt{(-\sqrt{2})^2 + (\sqrt{2})^2} = 2$ $\tan\theta = -1$ and θ is in quadrant II, so $\theta = 180^\circ - 45^\circ = 135^\circ$. $-\sqrt{2} + i\sqrt{2} = 2 \text{ cis } 135^\circ.$
Product and Quotient Theorems If $r_1(\cos\theta_1 + i\sin\theta_1)$ and $r_2(\cos\theta_2 + i\sin\theta_2)$ are any two complex numbers, then the following hold. $[r_1(\cos\theta_1 + i\sin\theta_1)] \cdot [r_2(\cos\theta_2 + i\sin\theta_2)] = r_1r_2[\cos(\theta_1 + \theta_2) + i\sin(\theta_1 + \theta_2)]$ and $\frac{r_1(\cos\theta_1 + i\sin\theta_1)}{r_2(\cos\theta_2 + i\sin\theta_2)} = \frac{r_1}{r_2}[\cos(\theta_1 - \theta_2) + i\sin(\theta_1 - \theta_2)]$, where $r_2(\cos\theta_2 + i\sin\theta_2) \neq 0$.	Let $z_1 = 4(\cos 135^\circ + i\sin 135^\circ)$ and $z_2 = 2(\cos 45^\circ + i\sin 45^\circ)$. $z_1z_2 = 8(\cos 180^\circ + i\sin 180^\circ)$ $4 \cdot 2 = 8$; $135^\circ + 45^\circ = 180^\circ$ $= 8(-1 + i \cdot 0)$ $= -8$ $\frac{z_1}{z_2} = 2(\cos 90^\circ + i\sin 90^\circ)$ $\frac{4}{2} = 2$; $135^\circ - 45^\circ = 90^\circ$ $= 2(0 + i \cdot 1)$ $= 2i$

Concepts | **Examples**

8.6 De Moivre's Theorem; Powers and Roots of Complex Numbers

De Moivre's Theorem

$$[r(\cos\theta + i\sin\theta)]^n = r^n(\cos n\theta + i\sin n\theta)$$

***n*th Root Theorem**

If n is any positive integer, r is a positive real number, and θ is in degrees, then the nonzero complex number $r(\cos\theta + i\sin\theta)$ has exactly n distinct nth roots, given by the following.

$$\sqrt[n]{r}\,(\cos\alpha + i\sin\alpha), \quad \text{or} \quad \sqrt[n]{r}\ \text{cis}\ \alpha,$$

where

$$\alpha = \frac{\theta + 360° \cdot k}{n}, \quad k = 0, 1, 2, \ldots, n-1.$$

If θ is in radians, then

$$\alpha = \frac{\theta + 2\pi k}{n}, \quad k = 0, 1, 2, \ldots, n-1.$$

Let $z = 4(\cos 180° + i\sin 180°)$. Find z^3 and the square roots of z.

$$[4(\cos 180° + i\sin 180°)]^3 \qquad \text{Find } z^3.$$
$$= 4^3(\cos 3 \cdot 180° + i\sin 3 \cdot 180°)$$
$$= 64(\cos 540° + i\sin 540°)$$
$$= 64(-1 + i \cdot 0)$$
$$= -64$$

For the given z, $r = 4$ and $\theta = 180°$. Its square roots are

$$\sqrt{4}\left(\cos\frac{180°}{2} + i\sin\frac{180°}{2}\right)$$
$$= 2(0 + i \cdot 1)$$
$$= 2i$$

and $\sqrt{4}\left(\cos\frac{180° + 360°}{2} + i\sin\frac{180° + 360°}{2}\right)$

$$= 2(0 + i(-1))$$
$$= -2i.$$

8.7 Polar Equations and Graphs

Rectangular and Polar Coordinates

If a point has rectangular coordinates (x, y) and polar coordinates (r, θ), then these coordinates are related as follows.

$$x = r\cos\theta \qquad y = r\sin\theta$$

$$r^2 = x^2 + y^2 \qquad \tan\theta = \frac{y}{x}, \ \text{if } x \neq 0$$

Find the rectangular coordinates for the point $(5, 60°)$ in polar coordinates.

$$x = 5\cos 60° = 5\left(\frac{1}{2}\right) = \frac{5}{2}$$

$$y = 5\sin 60° = 5\left(\frac{\sqrt{3}}{2}\right) = \frac{5\sqrt{3}}{2}$$

The rectangular coordinates are $\left(\frac{5}{2}, \frac{5\sqrt{3}}{2}\right)$.

Find polar coordinates for $(-1, -1)$ in rectangular coordinates.

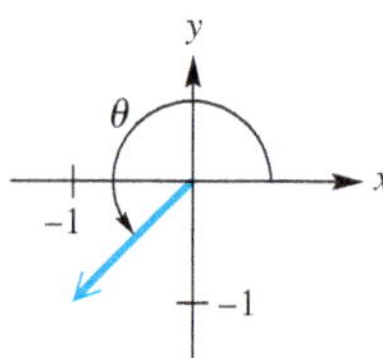

$r = \sqrt{(-1)^2 + (-1)^2} = \sqrt{2}$

$\tan\theta = 1$ and θ is in quadrant III, so $\theta = 225°$.

One pair of polar coordinates for $(-1, -1)$ is $\left(\sqrt{2}, 225°\right)$.

Polar Equations and Graphs

Equations	Graph
$r = a\cos\theta$, $r = a\sin\theta$	Circles
$r^2 = a^2\sin 2\theta$, $r^2 = a^2\cos 2\theta$	Lemniscates
$r = a \pm b\sin\theta$, $r = a \pm b\cos\theta$	Limaçons
$r = a\sin n\theta$, $r = a\cos n\theta$	Rose curves

Graph $r = 4\cos 2\theta$.

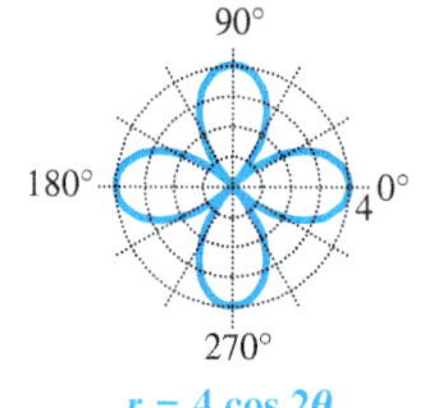

$r = 4\cos 2\theta$

Concepts	Examples

8.8 Parametric Equations, Graphs, and Applications

Parametric Equations of a Plane Curve
A **plane curve** is a set of points (x, y) such that $x = f(t)$, $y = g(t)$, and f and g are both defined on an interval I. The equations

$$x = f(t) \quad \text{and} \quad y = g(t)$$

are **parametric equations** with **parameter t.**

Graph $x = 2 - \sin t$, $y = \cos t - 1$, for $0 \le t \le 2\pi$.

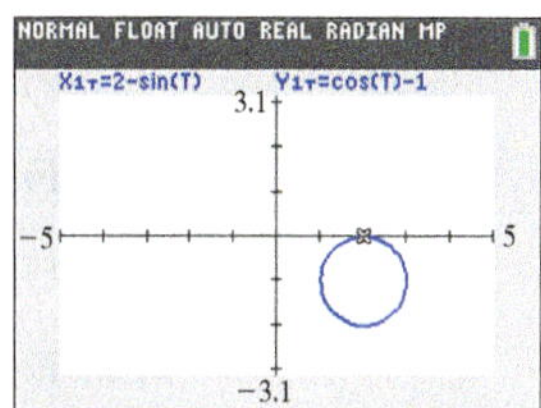

Flight of an Object
If an object has initial velocity v and initial height h, and travels such that its initial angle of elevation is θ, then its flight after t seconds can be modeled by the following parametric equations.

$$x = (v \cos \theta)t \quad \text{and} \quad y = (v \sin \theta)t - 16t^2 + h$$

Joe kicks a football from the ground at an angle of 45° with a velocity of 48 ft per sec. Give the parametric equations that model the path of the football and the distance it travels before hitting the ground.

$$x = (48 \cos 45°)t = 24\sqrt{2}\,t$$

$$y = (48 \sin 45°)t - 16t^2 = 24\sqrt{2}\,t - 16t^2$$

When the ball hits the ground, $y = 0$.

$$24\sqrt{2}\,t - 16t^2 = 0 \quad \text{Substitute } y = 0.$$

$$8t\left(3\sqrt{2} - 2t\right) = 0 \quad \text{Factor.}$$

$$t = 0 \text{ (Reject)} \quad \text{or} \quad t = \frac{3\sqrt{2}}{2} \quad \text{Zero-factor property}$$

The distance it travels is $x = 24\sqrt{2}\left(\frac{3\sqrt{2}}{2}\right) = 72$ ft.

Chapter 8 Review Exercises

Use the law of sines to find the indicated part of each triangle ABC.

1. Find b if $C = 74.2°$, $c = 96.3$ m, $B = 39.5°$.
2. Find B if $A = 129.7°$, $a = 127$ ft, $b = 69.8$ ft.
3. Find B if $C = 51.3°$, $c = 68.3$ m, $b = 58.2$ m.
4. Find b if $a = 165$ m, $A = 100.2°$, $B = 25.0°$.
5. Find A if $B = 39° \, 50'$, $b = 268$ m, $a = 340$ m.
6. Find A if $C = 79° \, 20'$, $c = 97.4$ mm, $a = 75.3$ mm.

Answer each question.

7. If we are given a, A, and C in a triangle ABC, does the possibility of the ambiguous case exist? If not, explain why.
8. Can triangle ABC exist if $a = 4.7$, $b = 2.3$, and $c = 7.0$? If not, explain why. Answer this question without using trigonometry.

9. Given $a = 10$ and $B = 30°$ in triangle ABC, for what values of b does A have

(a) exactly one value **(b)** two possible values **(c)** no value?

10. Why can there be no triangle ABC satisfying $A = 140°$, $a = 5$, and $b = 7$?

Use the law of cosines to find the indicated part of each triangle ABC.

11. Find A if $a = 86.14$ in., $b = 253.2$ in., $c = 241.9$ in.

12. Find b if $B = 120.7°$, $a = 127$ ft, $c = 69.8$ ft.

13. Find a if $A = 51° \, 20'$, $c = 68.3$ m, $b = 58.2$ m.

14. Find B if $a = 14.8$ m, $b = 19.7$ m, $c = 31.8$ m.

15. Find a if $A = 60°$, $b = 5.0$ cm, $c = 21$ cm.

16. Find A if $a = 13$ ft, $b = 17$ ft, $c = 8$ ft.

Solve each triangle ABC.

17. $A = 25.2°$, $a = 6.92$ yd, $b = 4.82$ yd

18. $A = 61.7°$, $a = 78.9$ m, $b = 86.4$ m

19. $a = 27.6$ cm, $b = 19.8$ cm, $C = 42° \, 30'$

20. $a = 94.6$ yd, $b = 123$ yd, $c = 109$ yd

Find the area of each triangle ABC.

21. $b = 840.6$ m, $c = 715.9$ m, $A = 149.3°$

22. $a = 6.90$ ft, $b = 10.2$ ft, $C = 35° \, 10'$

23. $a = 0.913$ km, $b = 0.816$ km, $c = 0.582$ km

24. $a = 43$ m, $b = 32$ m, $c = 51$ m

Solve each problem.

25. ***Distance across a Canyon*** To measure the distance AB across a canyon for a power line, a surveyor measures angles B and C and the distance BC, as shown in the figure. What is the distance from A to B?

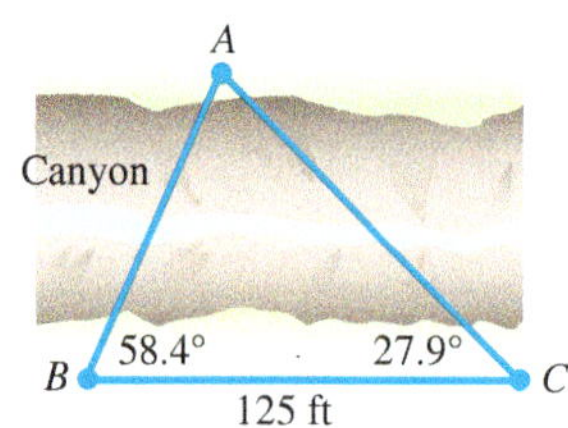

26. ***Length of a Brace*** A banner on an 8.0-ft pole is to be mounted on a building at an angle of 115°, as shown in the figure. Find the length of the brace.

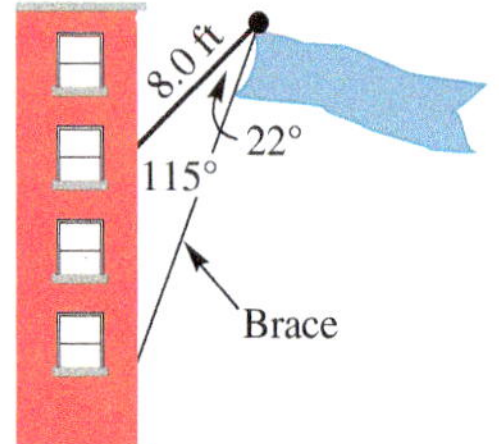

27. ***Height of a Tree*** A tree leans at an angle of 8.0° from the vertical. From a point 7.0 m from the bottom of the tree, the angle of elevation to the top of the tree is 68°. Find the slanted height x in the figure.

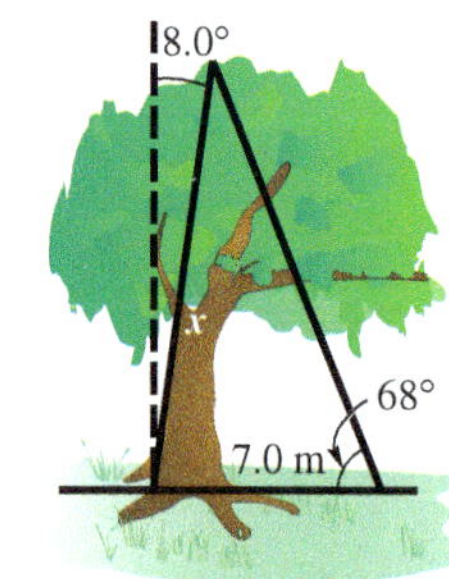

28. *Hanging Sculpture* A hanging sculpture is to be hung in an art gallery with two wires of lengths 15.0 ft and 12.2 ft so that the angle between them is 70.3°. How far apart should the ends of the wire be placed on the ceiling?

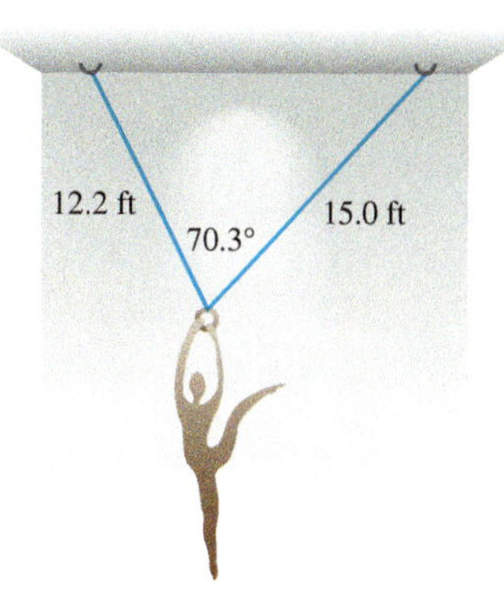

29. *Height of a Tree* A hill makes an angle of 14.3° with the horizontal. From the base of the hill, the angle of elevation to the top of a tree on top of the hill is 27.2°. The distance along the hill from the base to the tree is 212 ft. Find the height of the tree.

30. *Pipeline Position* A pipeline is to run between points A and B, which are separated by a protected wetlands area. To avoid the wetlands, the pipe will run from point A to C and then to B. The distances involved are $AB = 150$ km, $AC = 102$ km, and $BC = 135$ km. What angle should be used at point C?

31. *Distance between Two Boats* Two boats leave a dock together. Each travels in a straight line. The angle between their courses measures 54° 10′. One boat travels 36.2 km per hr, and the other travels 45.6 km per hr. How far apart will they be after 3 hr?

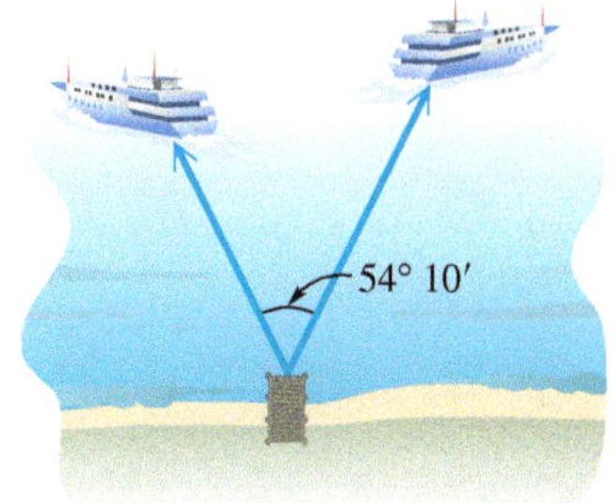

32. *Distance from a Ship to a Lighthouse* A ship sailing parallel to shore sights a lighthouse at an angle of 30° from its direction of travel. After the ship travels 2.0 mi farther, the angle has increased to 55°. At that time, how far is the ship from the lighthouse?

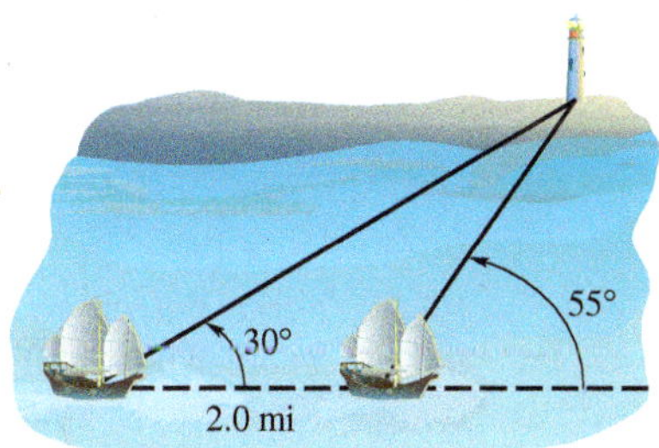

33. *Area of a Triangle* Find the area of the triangle shown in the figure using Heron's area formula.

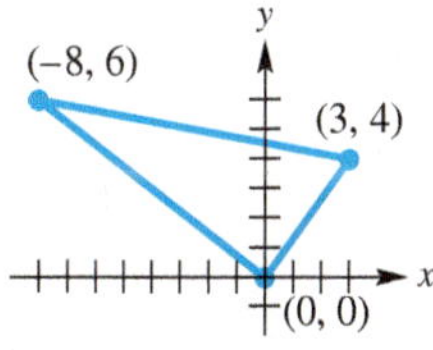

34. Show that the triangle in **Exercise 33** is a right triangle. Then use the formula $\mathcal{A} = \frac{1}{2}ac \sin B$, with $B = 90°$, to find the area.

Use the given vectors to sketch each of the following.

35. $\mathbf{a} - \mathbf{b}$

36. $\mathbf{a} + 3\mathbf{c}$

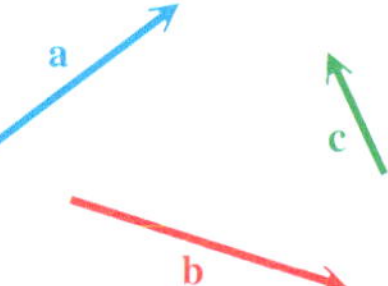

Given two forces and the angle between them, find the magnitude of the resultant force.

37.

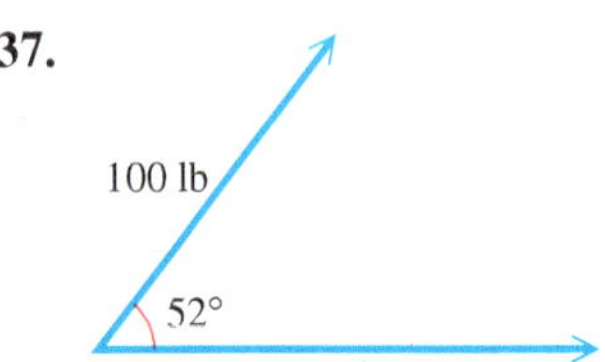

38. forces of 142 and 215 newtons, forming an angle of 112°

Vector **v** *has the given magnitude and direction angle. Find the horizontal and vertical components of* **v**.

39. $|\mathbf{v}| = 964$, $\theta = 154° \, 20'$

40. $|\mathbf{v}| = 50$, $\theta = 45°$ (Give exact values.)

Find the magnitude and direction angle for **u** *rounded to the nearest tenth.*

41. $\mathbf{u} = \langle -9, 12 \rangle$

42. $\mathbf{u} = \langle 21, -20 \rangle$

43. Let $\mathbf{v} = 2\mathbf{i} - \mathbf{j}$ and $\mathbf{u} = -3\mathbf{i} + 2\mathbf{j}$. Express each in terms of **i** and **j**.

(a) $2\mathbf{v} + \mathbf{u}$ **(b)** $2\mathbf{v}$ **(c)** $\mathbf{v} - 3\mathbf{u}$

Find the angle between the vectors. Round to the nearest tenth of a degree. If the vectors are orthogonal, say so.

44. $\langle 3, -2 \rangle$, $\langle -1, 3 \rangle$

45. $\langle 5, -3 \rangle$, $\langle 3, 5 \rangle$

46. $\langle 0, 4 \rangle$, $\langle -4, 4 \rangle$

Solve each problem.

47. ***Weight of a Sled and Passenger*** Paula and Steve are pulling their daughter Jessie on a sled. Steve pulls with a force of 18 lb at an angle of 10°. Paula pulls with a force of 12 lb at an angle of 15°. Find the magnitude of the resultant force on Jessie and the sled.

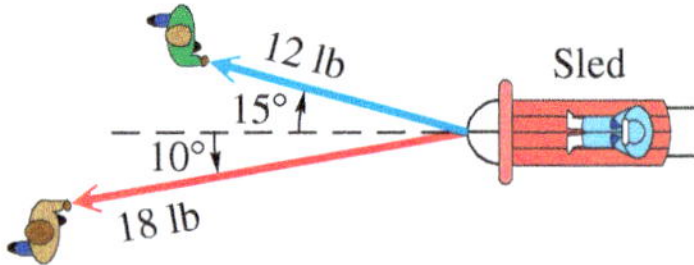

48. ***Force Placed on a Barge*** One boat pulls a barge with a force of 100 newtons. Another boat pulls the barge at an angle of 45° to the first force, with a force of 200 newtons. Find the resultant force acting on the barge, to the nearest unit, and the angle between the resultant and the first boat, to the nearest tenth.

49. ***Direction and Speed of a Plane*** A plane has an airspeed of 520 mph. The pilot wishes to fly on a bearing of 310°. A wind of 37 mph is blowing from a bearing of 212°. In what direction should the pilot fly, and what will be her ground speed?

50. ***Angle of a Hill*** A 186-lb force is required to hold a 2800-lb car on a hill. What angle does the hill make with the horizontal?

51. ***Incline Force*** Find the force required to keep a 75-lb sled from sliding down an incline that makes an angle of 27° with the horizontal. (Assume there is no friction.)

52. ***Speed and Direction of a Boat*** A boat travels 15 km per hr in still water. The boat is traveling across a large river, on a bearing of 130°. The current in the river, coming from the west, has a speed of 7 km per hr. Find the resulting speed of the boat and its resulting direction of travel.

Perform each operation. Write answers in rectangular form.

53. $[5(\cos 90° + i \sin 90°)][6(\cos 180° + i \sin 180°)]$

54. $[3 \text{ cis } 135°][2 \text{ cis } 105°]$

55. $\dfrac{2(\cos 60° + i \sin 60°)}{8(\cos 300° + i \sin 300°)}$

56. $\dfrac{4 \text{ cis } 270°}{2 \text{ cis } 90°}$

57. $\left(\sqrt{3} + i\right)^3$

58. $(2 - 2i)^5$

59. $(\cos 100° + i \sin 100°)^6$

60. *Concept Check* The vector representing a real number will lie on the ______-axis in the complex plane.

Graph each complex number.

61. $5i$

62. $-4 + 2i$

63. $3 - 3i\sqrt{3}$

64. Find the sum of $7 + 3i$ and $-2 + i$. Graph both complex numbers and their resultant.

Write each complex number in its alternative form, using a calculator to approximate answers to four decimal places as necessary.

	Rectangular Form	Trigonometric Form
65.	$-2 + 2i$	______
66.	______	$3(\cos 90° + i \sin 90°)$
67.	______	$2(\cos 225° + i \sin 225°)$
68.	$-4 + 4i\sqrt{3}$	______
69.	$1 - i$	______
70.	______	$4 \text{ cis } 240°$
71.	$-4i$	______
72.	______	$7 \text{ cis } 310°$

Concept Check *The complex number z, where $z = x + yi$, can be graphed in the plane as (x, y). Describe the graph of all complex numbers z satisfying the given conditions.*

73. The imaginary part of z is the negative of the real part of z.

74. The absolute value of z is 2.

Find all roots as indicated. Write answers in trigonometric form.

75. the cube roots of $1 - i$

76. the fifth roots of $-2 + 2i$

77. *Concept Check* How many real sixth roots does -64 have?

78. *Concept Check* How many real fifth roots does -32 have?

Find all complex number solutions. Write answers in trigonometric form.

79. $x^4 + 16 = 0$

80. $x^3 + 125 = 0$

81. $x^2 + i = 0$

82. Convert $(5, 315°)$ to rectangular coordinates.

83. Convert $\left(-1, \sqrt{3}\right)$ to polar coordinates, with $0° \le \theta < 360°$ and $r > 0$.

84. *Concept Check* Describe the graph of $r = k$ for $k > 0$.

Identify and graph each polar equation for θ in $[0°, 360°)$.

85. $r = 4 \cos \theta$

86. $r = -1 + \cos \theta$

87. $r = 2 \sin 4\theta$

88. $r = \dfrac{2}{2 \cos \theta - \sin \theta}$

Find an equivalent equation in rectangular coordinates.

89. $r = \frac{3}{1 + \cos \theta}$ **90.** $r = \sin \theta + \cos \theta$ **91.** $r = 2$

Find an equivalent equation in polar coordinates.

92. $y = x$ **93.** $y = x^2$ **94.** $x^2 + y^2 = 25$

Find a polar equation having the given graph.

95.

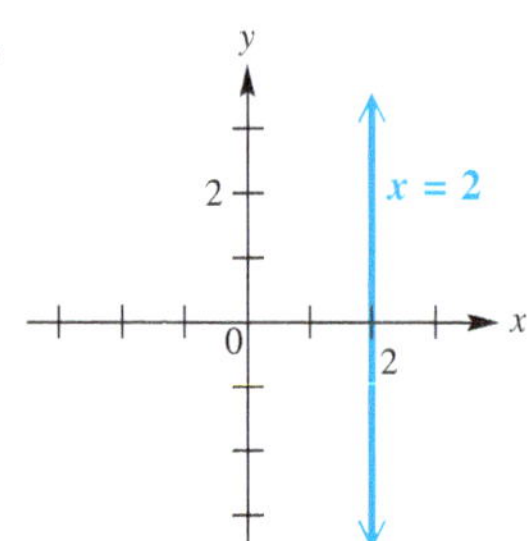

96.

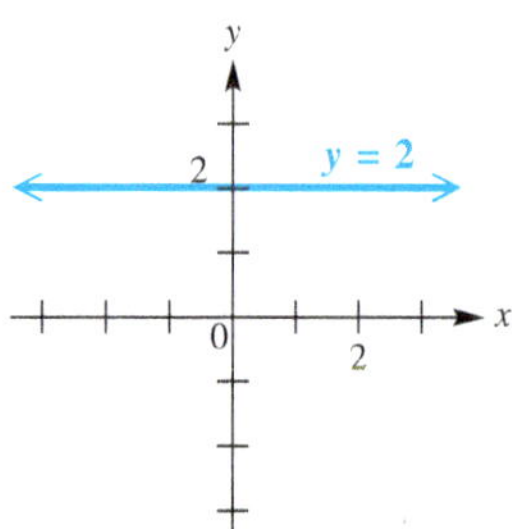

97.

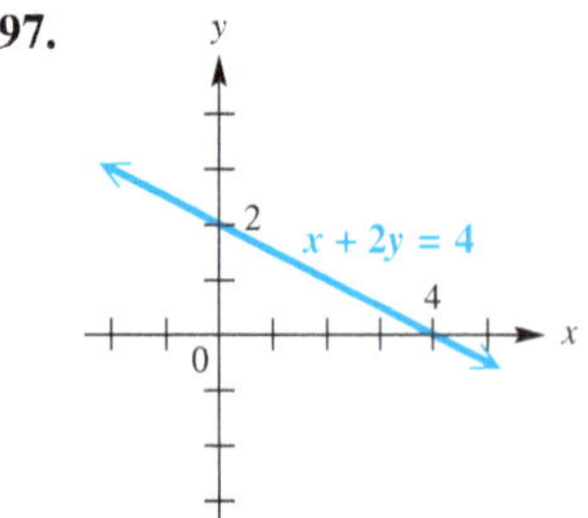

98.

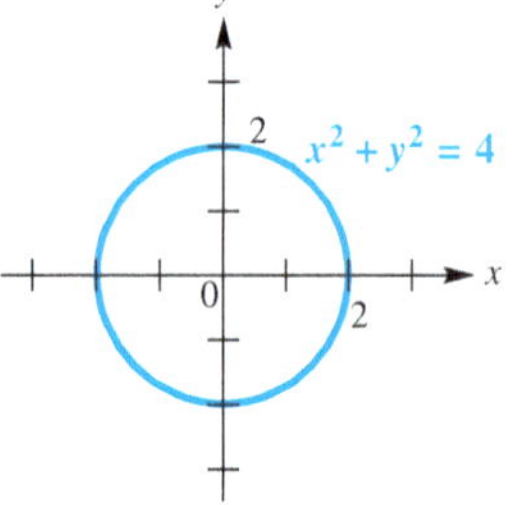

99. Graph the plane curve defined by the parametric equations $x = t + \cos t$, $y = \sin t$, for t in $[0, 2\pi]$.

100. Show that the distance between (r_1, θ_1) and (r_2, θ_2) in polar coordinates is given by

$$d = \sqrt{r_1{}^2 + r_2{}^2 - 2r_1r_2 \cos(\theta_1 - \theta_2)}.$$

Find a rectangular equation for each plane curve with the given parametric equations.

101. $x = \sqrt{t - 1}$, $y = \sqrt{t}$, for t in $[1, \infty)$

102. $x = 3t + 2$, $y = t - 1$, for t in $[-5, 5]$

103. $x = 5 \tan t$, $y = 3 \sec t$, for t in $\left(-\frac{\pi}{2}, \frac{\pi}{2}\right)$

104. $x = t^2 + 5$, $y = \frac{1}{t^2 + 1}$, for t in $(-\infty, \infty)$

105. $x = \cos 2t$, $y = \sin t$, for t in $(-\pi, \pi)$

106. Give a pair of parametric equations whose graph is the circle having center $(3, 4)$ and passing through the origin.

107. ***(Modeling) Flight of a Baseball*** A batter hits a baseball when it is 3.2 ft above the ground. It leaves the bat with velocity 118 ft per sec at an angle of 27° with respect to the ground.

(a) Determine parametric equations that model the path of the baseball.

(b) Determine a rectangular equation that models the path of the baseball.

(c) Determine approximately how long the baseball is in flight and the horizontal distance it covers.

108. ***Mandelbrot Set*** Consider the complex number $z = 1 + i$. Compute the value of $z^2 + z$, and show that its absolute value exceeds 2, indicating that $1 + i$ is not in the Mandelbrot set.

Chapter 8 Test

Find the indicated part of each triangle ABC.

1. Find C if $A = 25.2°$, $a = 6.92$ yd, and $b = 4.82$ yd.
2. Find c if $C = 118°$, $a = 75.0$ km, and $b = 131$ km.
3. Find B if $a = 17.3$ ft, $b = 22.6$ ft, $c = 29.8$ ft.

Solve each problem.

4. Find the area of triangle ABC if $a = 14$, $b = 30$, and $c = 40$.
5. Find the area of triangle XYZ shown here.

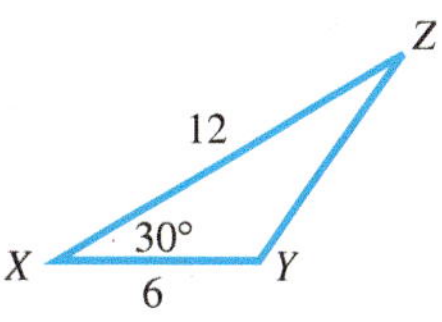

6. Given $a = 10$ and $B = 150°$ in triangle ABC, determine the values of b for which A has
   (a) exactly one value **(b)** two possible values **(c)** no value.

Solve each triangle ABC.

7. $A = 60°$, $b = 30$ m, $c = 45$ m
8. $b = 1075$ in., $c = 785$ in., $C = 38° 30'$

Work each problem.

9. Find the magnitude and the direction angle, to the nearest tenth, for the vector shown in the figure.

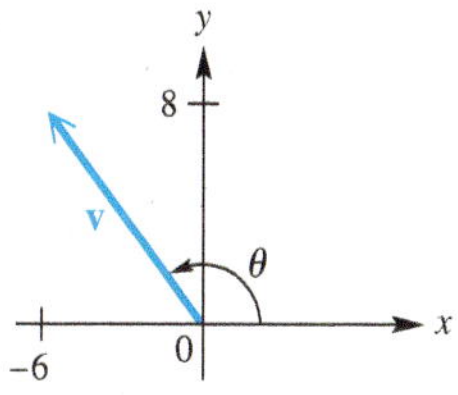

10. For the vectors $\mathbf{u} = \langle -1, 3 \rangle$ and $\mathbf{v} = \langle 2, -6 \rangle$, find each of the following.
    (a) $\mathbf{u} + \mathbf{v}$ **(b)** $-3\mathbf{v}$ **(c)** $\mathbf{u} \cdot \mathbf{v}$ **(d)** $|\mathbf{u}|$
11. Find the measure of the angle θ between $\mathbf{u} = \langle 4, 3 \rangle$ and $\mathbf{v} = \langle 1, 5 \rangle$ to the nearest tenth.

Solve each problem.

12. ***Height of a Balloon*** The angles of elevation of a balloon from two points A and B on level ground are $24° 50'$ and $47° 20'$, respectively. As shown in the figure, points A, B, and C are in the same vertical plane and points A and B are 8.4 mi apart. Approximate the height of the balloon above the ground to the nearest tenth of a mile.

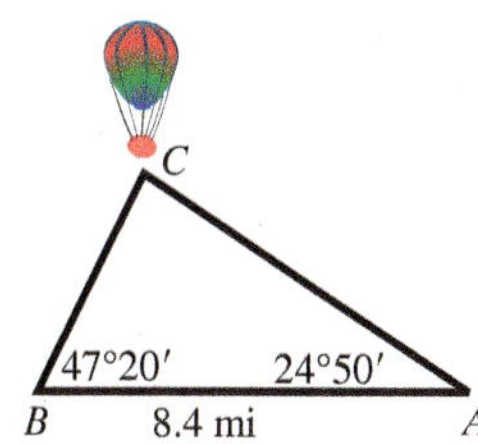

13. ***Horizontal and Vertical Components*** Find the horizontal and vertical components of the vector with magnitude 569 and direction angle 127.5° from the horizontal. Give your answer in the form $\langle a, b \rangle$ to the nearest unit.
14. ***Radio Direction Finders*** Radio direction finders are placed at points A and B, which are 3.46 mi apart on an east-west line, with A west of B. From A, the bearing of a certain illegal pirate radio transmitter is 48°, and from B the bearing is 302°. Find the distance between the transmitter and A to the nearest hundredth of a mile.

15. ***Height of a Tree*** A tree leans at an angle of 8.0° from the vertical, as shown in the figure. From a point 8.0 m from the bottom of the tree, the angle of elevation to the top of the tree is 66°. Find the slanted height x in the figure.

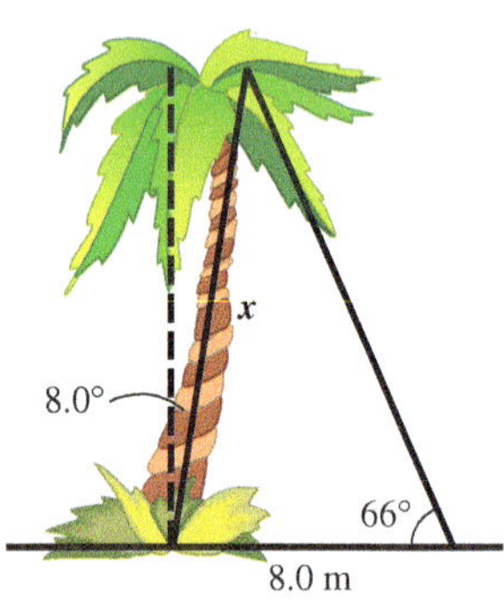

16. ***Bearing and Airspeed*** Find the bearing and airspeed required for a plane to fly 630 mi due north in 3.0 hr if the wind is blowing from a direction of 318° at 15 mph. Approximate the bearing to the nearest degree and the airspeed to the nearest 10 mph.

17. ***Incline Angle*** A force of 16.0 lb is required to hold a 50.0-lb wheelbarrow on an incline. What angle does the incline make with the horizontal?

18. ***Magnitude of a Force*** Two forces of 135 newtons and 260 newtons act on a point, forming an angle of 115°. Find the magnitude of the resultant force.

19. For the complex numbers $w = 2 - 4i$ and $z = 5 + i$, find $w + z$ in rectangular form and give a geometric representation of the sum.

20. Write each complex number in trigonometric (polar) form, where $0° \le \theta < 360°$.

(a) $3i$ **(b)** $1 + 2i$ **(c)** $-1 - i\sqrt{3}$

21. Write each complex number in rectangular form.

(a) $3(\cos 30° + i \sin 30°)$ **(b)** $4 \text{ cis } 40°$ **(c)** $3(\cos 90° + i \sin 90°)$

22. For the complex numbers

$$w = 8(\cos 40° + i \sin 40°) \quad \text{and} \quad z = 2(\cos 10° + i \sin 10°),$$

find each of the following in the form specified.

(a) wz (trigonometric form) **(b)** $\frac{w}{z}$ (rectangular form) **(c)** z^3 (rectangular form)

23. Find the four complex fourth roots of $-16i$. Write answers in trigonometric form.

24. Convert the given rectangular coordinates to polar coordinates. Give two pairs of polar coordinates for each point.

(a) $(0, 5)$ **(b)** $(-2, -2)$

25. Convert the given polar coordinates to rectangular coordinates.

(a) $(3, 315°)$ **(b)** $(-4, 90°)$

Identify and graph each polar equation for θ in $[0°, 360°)$.

26. $r = 1 - \cos \theta$

27. $r = 3 \cos 3\theta$

28. Convert each polar equation to a rectangular equation, and sketch its graph.

(a) $r = \frac{4}{2 \sin \theta - \cos \theta}$ **(b)** $r = 6$

Graph each pair of parametric equations.

29. $x = 4t - 3, \; y = t^2, \;$ for t in $[-3, 4]$

30. $x = 2 \cos 2t, \; y = 2 \sin 2t, \;$ for t in $[0, 2\pi]$

9 Systems and Matrices

These two linear trails of jet exhaust crossing each other illustrate the concept of two distinct, nonparallel lines intersecting in a single point—a geometric interpretation of a *system of linear equations* having a single solution.

9.1 Systems of Linear Equations

- Linear Systems
- Substitution Method
- Elimination Method
- Special Systems
- Application of Systems of Equations
- Linear Systems with Three Unknowns (Variables)
- Application of Systems to Model Data

Linear Systems

The definition of a linear equation can be extended to more than one variable. Any equation of the form

$$a_1x_1 + a_2x_2 + \cdots + a_nx_n = b,$$

for real numbers $a_1, a_2, \ldots, a_n$ (all nonzero) and b, is a **linear equation,** or a **first-degree equation, in n unknowns.**

A set of equations considered simultaneously is a **system of equations.** The **solutions** of a system of equations must satisfy every equation in the system. If all the equations in a system are linear, the system is a **system of linear equations,** or a **linear system.**

The solution set of a linear equation in two unknowns (or variables) is an infinite set of ordered pairs. The graph of such an equation is a straight line, so there are three possibilities for the number of elements in the solution set of a system of two linear equations in two unknowns. See **Figure 1.** The possible graphs of a linear system in two unknowns are as follows.

1. **The graphs intersect at exactly one point,** which gives the (single) ordered-pair solution of the system. The **system is consistent** and the **equations are independent.** See **Figure 1(a).**
2. **The graphs are parallel lines,** so there is no solution and the solution set is $\varnothing$. The **system is inconsistent** and the **equations are independent.** See **Figure 1(b).**
3. **The graphs are the same line,** and there are an infinite number of solutions. The **system is consistent** and the **equations are dependent.** See **Figure 1(c).**

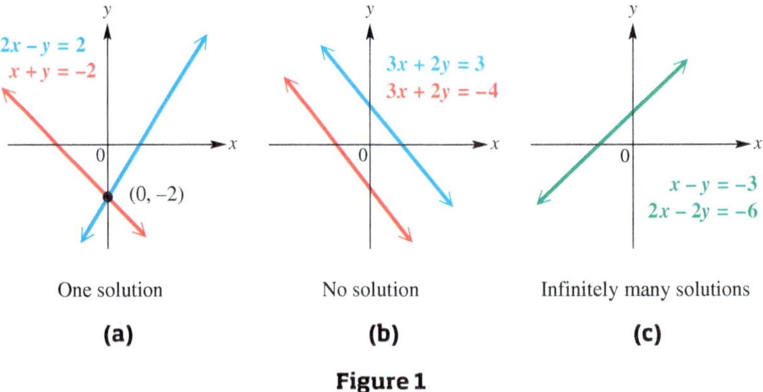

Figure 1

Using graphs to find the solution set of a linear system in two unknowns provides a good visual perspective, but may be inefficient when the solution set contains non-integer values. Thus, we introduce two algebraic methods for solving systems with two unknowns: *substitution* and *elimination.*

Substitution Method

In a system of two equations with two variables, the **substitution method** involves using one equation to find an expression for one variable in terms of the other, and then substituting this expression into the other equation of the system.

EXAMPLE 1 Solving a System (Substitution Method)

Solve the system.

$$3x + 2y = 11 \quad (1)$$
$$-x + y = 3 \quad (2)$$

SOLUTION Begin by solving one of the equations for one of the variables. We solve equation (2) for y.

$$-x + y = 3 \quad (2)$$
$$y = x + 3 \quad \text{Add } x. \quad (3)$$

Now replace y with $x + 3$ in equation (1), and solve for x.

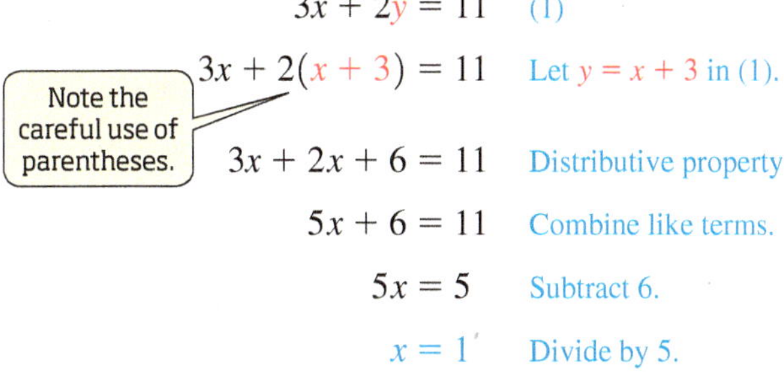

$$3x + 2y = 11 \quad (1)$$
$$3x + 2(x + 3) = 11 \quad \text{Let } y = x + 3 \text{ in (1).}$$
$$3x + 2x + 6 = 11 \quad \text{Distributive property}$$
$$5x + 6 = 11 \quad \text{Combine like terms.}$$
$$5x = 5 \quad \text{Subtract 6.}$$
$$x = 1 \quad \text{Divide by 5.}$$

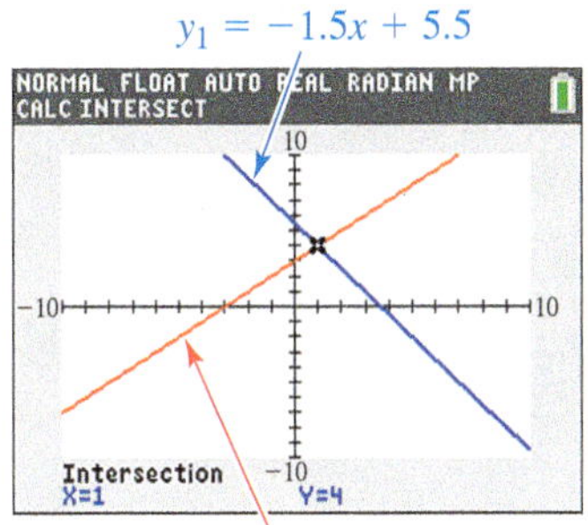

To solve the system in **Example 1** graphically, solve both equations for y.

$3x + 2y = 11$ leads to

$$y_1 = -1.5x + 5.5.$$

$-x + y = 3$ leads to

$$y_2 = x + 3.$$

Graph both y_1 and y_2 in the standard window to find that their point of intersection is $(1, 4)$.

Replace x with 1 in equation (3) to obtain

$$y = x + 3 = 1 + 3 = 4.$$

The solution of the system is the ordered pair $(1, 4)$. ***Check this solution in both equations (1) and (2).***

CHECK

$3x + 2y = 11$ (1)	$-x + y = 3$ (2)
$3(1) + 2(4) \stackrel{?}{=} 11$	$-1 + 4 \stackrel{?}{=} 3$
$11 = 11$ ✓ True	$3 = 3$ ✓ True

True statements result when the solution is substituted in both equations, confirming that the solution set is $\{(1, 4)\}$.

✔ **Now Try Exercise 7.**

Elimination Method The **elimination method** for solving a system of two equations uses multiplication and addition to eliminate a variable from one equation. To eliminate a variable, the coefficients of that variable in the two equations must be additive inverses. We use properties of algebra to change the system to an **equivalent system,** one with the same solution set.

The three transformations that produce an equivalent system are listed here.

Transformations of a Linear System

1. Interchange any two equations of the system.
2. Multiply or divide any equation of the system by a nonzero real number.
3. Replace any equation of the system by the sum of that equation and a multiple of another equation in the system.

EXAMPLE 2 Solving a System (Elimination Method)

Solve the system.

$$3x - 4y = 1 \quad (1)$$
$$2x + 3y = 12 \quad (2)$$

SOLUTION One way to eliminate a variable is to use the second transformation and multiply each side of equation (2) by -3, to obtain an equivalent system.

$$3x - 4y = 1 \quad (1)$$
$$-6x - 9y = -36 \quad \text{Multiply (2) by } -3. \quad (3)$$

Now multiply each side of equation (1) by 2, and use the third transformation to add the result to equation (3), eliminating x. Solve the result for y.

$$6x - 8y = 2 \quad \text{Multiply (1) by 2.}$$
$$-6x - 9y = -36 \quad (3)$$
$$-17y = -34 \quad \text{Add.}$$
$$y = 2 \quad \text{Solve for } y.$$

Substitute 2 for y in either of the original equations and solve for x.

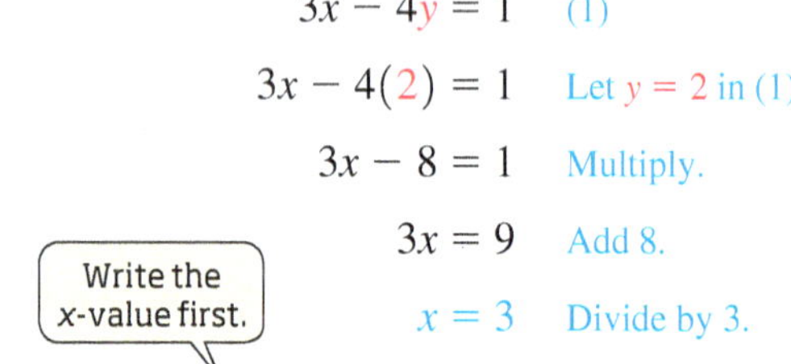

$$3x - 4y = 1 \quad (1)$$
$$3x - 4(2) = 1 \quad \text{Let } y = 2 \text{ in (1).}$$
$$3x - 8 = 1 \quad \text{Multiply.}$$
$$3x = 9 \quad \text{Add 8.}$$
$$x = 3 \quad \text{Divide by 3.}$$

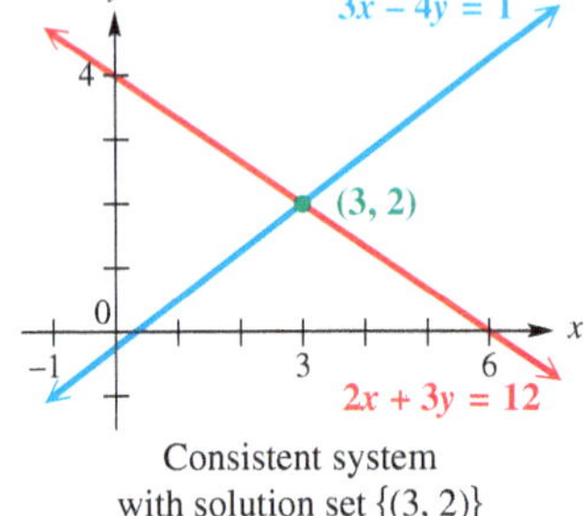

Consistent system with solution set {(3, 2)}

Figure 2

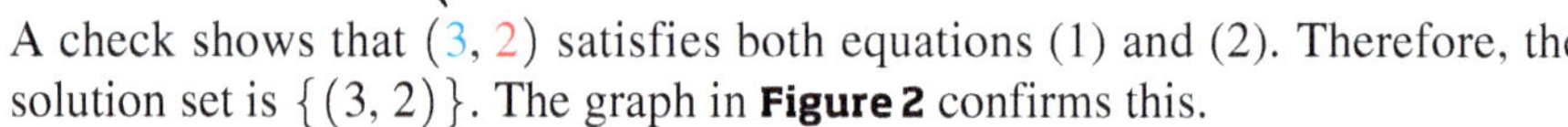

A check shows that (3, 2) satisfies both equations (1) and (2). Therefore, the solution set is $\{(3, 2)\}$. The graph in **Figure 2** confirms this.

✔ **Now Try Exercise 21.**

Special Systems The systems in **Examples 1 and 2** were both consistent, having a single solution. This is not always the case.

EXAMPLE 3 Solving an Inconsistent System

Solve the system.

$$3x - 2y = 4 \quad (1)$$
$$-6x + 4y = 7 \quad (2)$$

SOLUTION To eliminate the variable x, multiply each side of equation (1) by 2, and add the result to equation (2).

$$6x - 4y = 8 \quad \text{Multiply (1) by 2.}$$
$$-6x + 4y = 7 \quad (2)$$
$$0 = 15 \quad \text{False}$$

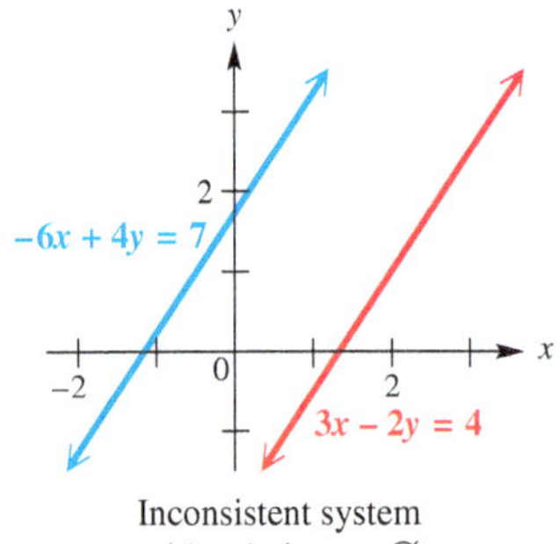

Inconsistent system with solution set ∅

Figure 3

Since $0 = 15$ is false, the system is inconsistent and has no solution. As suggested by **Figure 3,** this means that the graphs of the equations of the system never intersect. (The lines are parallel.) The solution set is ∅.

✔ **Now Try Exercise 31.**

EXAMPLE 4 Solving a System with Infinitely Many Solutions

Solve the system.

$$8x - 2y = -4 \quad (1)$$
$$-4x + y = 2 \quad (2)$$

ALGEBRAIC SOLUTION

Divide each side of equation (1) by 2, and add the result to equation (2).

$$\begin{aligned} 4x - y &= -2 && \text{Divide (1) by 2.} \\ -4x + y &= 2 && (2) \\ \hline 0 &= 0 && \text{True} \end{aligned}$$

The result, $0 = 0$, is a true statement, which indicates that the equations are equivalent. Any ordered pair (x, y) that satisfies either equation will satisfy the system. Solve equation (2) for y.

$$-4x + y = 2 \quad (2)$$
$$y = 4x + 2$$

The solutions of the system can be written in the form of a set of ordered pairs $(x, 4x + 2)$, for any real number x. Some ordered pairs in the solution set are $(0, 4 \cdot 0 + 2)$, or $(0, 2)$, and $(1, 4 \cdot 1 + 2)$, or $(1, 6)$, as well as $(3, 14)$, and $(-2, -6)$.

As shown in **Figure 4,** the equations of the original system are dependent and lead to the same straight-line graph. Using this method, the solution set can be written $\{(x, 4x + 2)\}$.

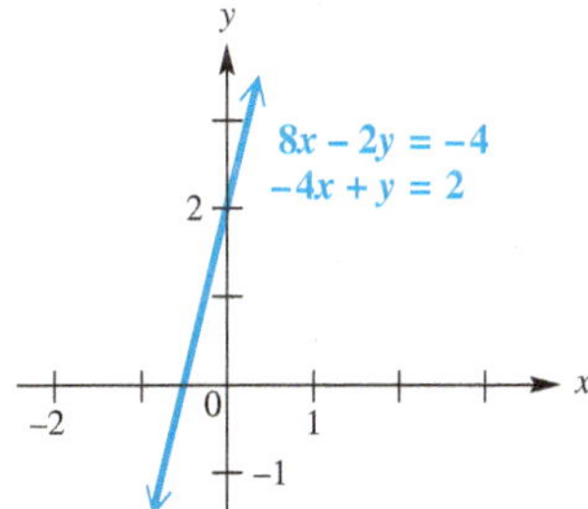

Consistent system with infinitely many solutions

Figure 4

GRAPHING CALCULATOR SOLUTION

Solving the equations for y gives

$$y_1 = 4x + 2 \quad (1)$$

and

$$y_2 = 4x + 2. \quad (2)$$

When written in this form, we can immediately determine that the equations are identical. Each has slope 4 and y-intercept $(0, 2)$.

As expected, the graphs coincide. See the top screen in **Figure 5.** The table indicates that

$$y_1 = y_2 \text{ for selected values of } x,$$

providing another way to show that the two equations lead to the same graph.

$8x - 2y = -4$
$-4x + y = 2$

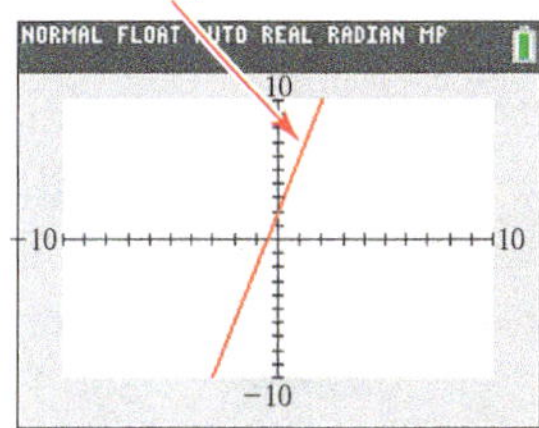

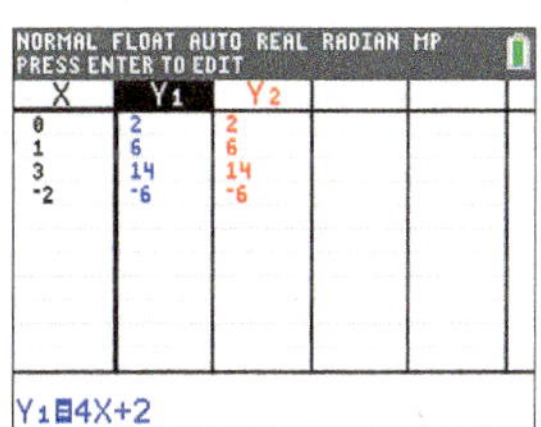

Figure 5

Refer to the algebraic solution to see how the solution set can be written using an arbitrary variable.

✔ **Now Try Exercise 33.**

NOTE In the algebraic solution for **Example 4,** we wrote the solution set with the variable x arbitrary. We could write the solution set with y arbitrary.

$$\left\{\left(\frac{y-2}{4}, y\right)\right\} \quad \text{Solve } -4x + y = 2 \text{ for } x.$$

By selecting values for y and solving for x in this ordered pair, we can find individual solutions. Verify again that $(0, 2)$ is a solution by letting $y = 2$ and solving for x to obtain $\frac{2-2}{4} = 0$.

Application of Systems of Equations Many applied problems involve more than one unknown quantity. Although some problems with two unknowns can be solved using just one variable, it is often easier to use two variables.

To solve a problem with two unknowns, we must write two equations that relate the unknown quantities. The system formed by the pair of equations can then be solved using the methods of this chapter. The following steps, based on the six-step problem-solving method introduced earlier, give a strategy for solving such applied problems.

Solving an Applied Problem by Writing a System of Equations

Step 1 **Read** the problem carefully until you understand what is given and what is to be found.

Step 2 **Assign variables** to represent the unknown values, using diagrams or tables as needed. *Write down* what each variable represents.

Step 3 **Write a system of equations** that relates the unknowns.

Step 4 **Solve** the system of equations.

Step 5 **State the answer** to the problem. Does it seem reasonable?

Step 6 **Check** the answer in the words of the original problem.

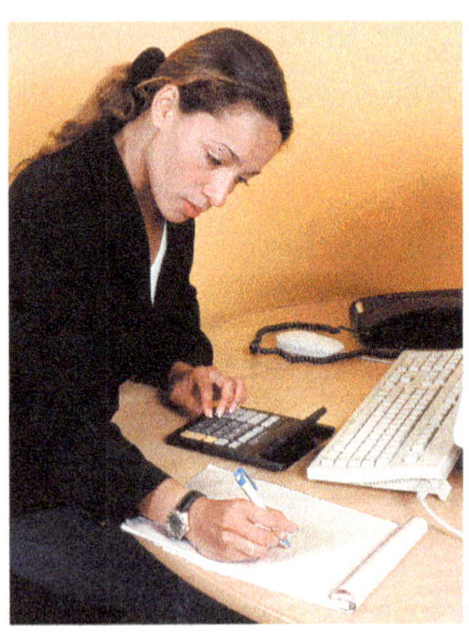

EXAMPLE 5 Using a Linear System to Solve an Application

Salaries for the same position can vary depending on the location. In 2015, the average of the median salaries for the position of Accountant I in San Diego, California, and Salt Lake City, Utah, was \$47,449.50. The median salary in San Diego, however, exceeded the median salary in Salt Lake City by \$5333. Determine the median salary for the Accountant I position in San Diego and in Salt Lake City. (*Source:* www.salary.com)

SOLUTION

Step 1 **Read** the problem. We must find the median salary of the Accountant I position in San Diego and in Salt Lake City.

Step 2 **Assign variables.** Let x represent the median salary of the Accountant I position in San Diego and y represent the median salary for the same position in Salt Lake City.

Step 3 **Write a system of equations.** Since the average of the two medians for the Accountant I position in San Diego and Salt Lake City was \$47,449.50, one equation is as follows.

$$\frac{x+y}{2} = 47{,}449.50$$

Multiply each side of this equation by 2 to clear the fraction and obtain an equivalent equation.

$$x + y = 94{,}899 \quad (1)$$

The median salary in San Diego exceeded the median salary in Salt Lake City by \$5333. Thus, $x - y = 5333$, which gives the following system of equations.

$$x + y = 94{,}899 \quad (1)$$
$$x - y = 5333 \quad (2)$$

Step 4 **Solve** the system. To eliminate y, add the two equations.

$$\begin{aligned} x + y &= 94{,}899 && (1) \\ x - y &= 5333 && (2) \\ \hline 2x \quad &= 100{,}232 && \text{Add.} \\ x &= 50{,}116 && \text{Solve for } x. \end{aligned}$$

To find y, substitute 50,116 for x in equation (2).

$$\begin{aligned} x - y &= 5333 && (2) \\ 50{,}116 - y &= 5333 && \text{Let } x = 50{,}116. \\ -y &= -44{,}783 && \text{Subtract } 50{,}116. \\ y &= 44{,}783 && \text{Multiply by } -1. \end{aligned}$$

Step 5 **State the answer.** The median salary for the position of Accountant I was $50,116 in San Diego and $44,783 in Salt Lake City.

Step 6 **Check.** The average of $50,116 and $44,783 is

$$\frac{\$50{,}116 + \$44{,}783}{2} = \$47{,}449.50.$$

Also, $\quad \$50{,}116 - \$44{,}783 = \$5333, \quad$ as required.

✔ **Now Try Exercise 101.**

Linear Systems with Three Unknowns (Variables) We have seen that the graph of a linear equation in two unknowns is a straight line. The graph of a linear equation in three unknowns requires a three-dimensional coordinate system. The three number lines are placed at right angles. The graph of a linear equation in three unknowns is a plane. Some possible intersections of planes representing three equations in three variables are shown in **Figure 6.**

In two dimensions we customarily label the axes x and y. When working in three dimensions they are usually labeled x, y, and z.

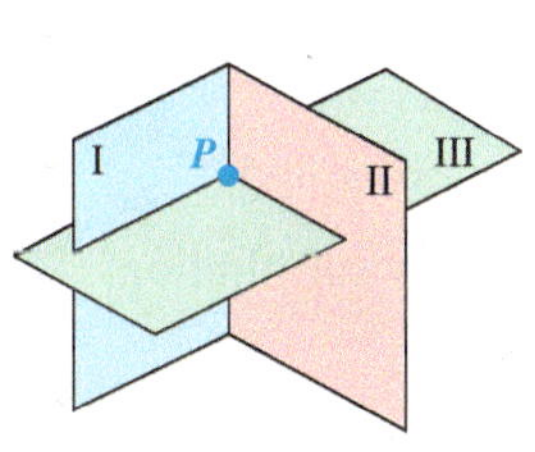

A single solution

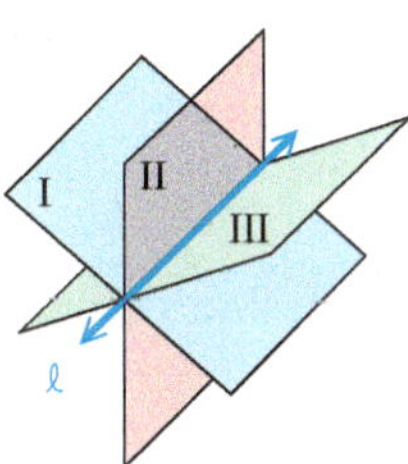

Points of a line in common

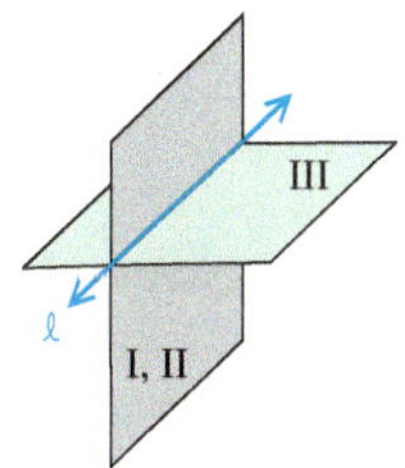

Points of a line in common

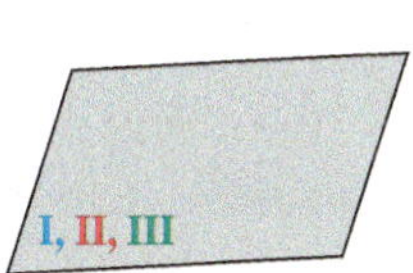

All points in common

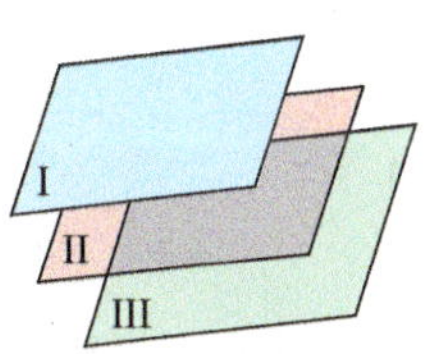

No points in common

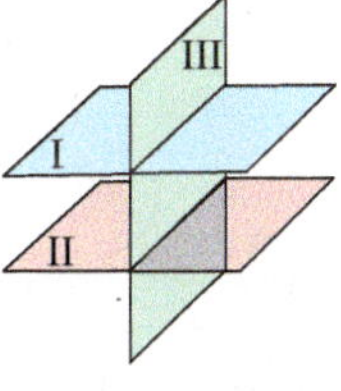

No points in common

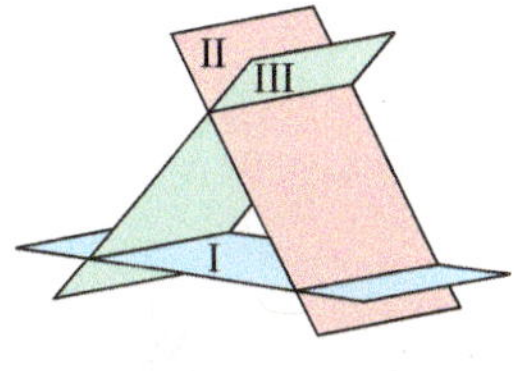

No points in common

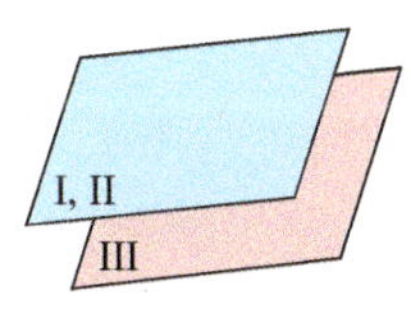

No points in common

Figure 6

Solving a Linear System with Three Unknowns

Step 1 Eliminate a variable from any two of the equations.

Step 2 Eliminate the ***same variable*** from a different pair of equations.

Step 3 Eliminate a second variable using the resulting two equations in two variables to obtain an equation with just one variable whose value we can now determine.

Step 4 Find the values of the remaining variables by substitution. Write the solution of the system as an **ordered triple.**

EXAMPLE 6 Solving a System of Three Equations with Three Variables

Solve the system.

$$3x + 9y + 6z = 3 \quad (1)$$
$$2x + y - z = 2 \quad (2)$$
$$x + y + z = 2 \quad (3)$$

SOLUTION

Step 1 Eliminate z by adding equations (2) and (3).

$$3x + 2y = 4 \quad (4)$$

Step 2 To eliminate z from another pair of equations, multiply each side of equation (2) by 6 and add the result to equation (1).

$$12x + 6y - 6z = 12 \quad \text{Multiply (2) by 6.}$$
$$3x + 9y + 6z = 3 \quad (1)$$
$$15x + 15y = 15 \quad (5)$$

Make sure equation (5) has the *same* two variables as equation (4).

Step 3 To eliminate x from equations (4) and (5), multiply each side of equation (4) by -5 and add the result to equation (5). Solve the resulting equation for y.

$$-15x - 10y = -20 \quad \text{Multiply (4) by } -5.$$
$$15x + 15y = 15 \quad (5)$$
$$5y = -5 \quad \text{Add.}$$
$$y = -1 \quad \text{Divide by 5.}$$

Step 4 Use $y = -1$ to find x from equation (4) by substitution.

$$3x + 2y = 4 \quad (4)$$
$$3x + 2(-1) = 4 \quad \text{Let } y = -1.$$
$$x = 2 \quad \text{Solve for } x.$$

Substitute 2 for x and -1 for y in equation (3) to find z.

$$x + y + z = 2 \quad (3)$$
$$2 + (-1) + z = 2 \quad \text{Let } x = 2, y = -1.$$
$$z = 1 \quad \text{Solve for } z.$$

Write the values of x, y, and z in the correct order.

Verify that the ordered triple $(2, -1, 1)$ satisfies all three equations in the *original* system. The solution set is $\{(2, -1, 1)\}$.

✓ Now Try Exercise 47.

EXAMPLE 7 Solving a System of Two Equations with Three Variables

Solve the system.

$$x + 2y + z = 4 \quad (1)$$
$$3x - y - 4z = -9 \quad (2)$$

SOLUTION Geometrically, the solution is the intersection of the two planes given by equations (1) and (2). The intersection of two different nonparallel planes is a line. Thus there will be an infinite number of ordered triples in the solution set, representing the points on the line of intersection.

To eliminate x, multiply both sides of equation (1) by -3 and add the result to equation (2). (Either y or z could have been eliminated instead.)

$$-3x - 6y - 3z = -12 \quad \text{Multiply (1) by } -3.$$
$$3x - y - 4z = -9 \quad (2)$$
$$-7y - 7z = -21 \quad (3)$$

Solve this equation for z.

$$-7z = 7y - 21 \quad \text{Add } 7y.$$
$$z = -y + 3 \quad \text{Divide } each \text{ term by } -7.$$

This gives z in terms of y. Express x also in terms of y by solving equation (1) for x and substituting $-y + 3$ for z in the result.

$$x + 2y + z = 4 \quad (1)$$
$$x = -2y - z + 4 \quad \text{Solve for } x.$$
$$x = -2y - (-y + 3) + 4 \quad \text{Substitute } (-y + 3) \text{ for } z.$$

Use parentheses around $-y + 3$.

$$x = -y + 1 \quad \text{Simplify.}$$

The system has an infinite number of solutions. For any value of y, the value of z is $-y + 3$ and the value of x is $-y + 1$. For example, if $y = 1$, then $x = -1 + 1 = 0$ and $z = -1 + 3 = 2$, giving the solution $(0, 1, 2)$. Verify that another solution is $(-1, 2, 1)$.

With y arbitrary, the solution set is of the form $\{(-y + 1, y, -y + 3)\}$.

✔ **Now Try Exercise 59.**

NOTE Had we solved equation (3) in Example 7 for y instead of z, the solution would have had a different form but would have led to the same set of solutions.

$$\{(z - 2, -z + 3, z)\} \quad \text{Solution set with } z \text{ arbitrary}$$

By choosing $z = 2$, one solution would be $(0, 1, 2)$, which was found above.

Application of Systems to Model Data Applications with three unknowns usually require solving a system of three equations. If we know three points on the graph, we can find the equation of a parabola in the form

$$y = ax^2 + bx + c$$

by solving a system of three equations with three variables.

EXAMPLE 8 **Using Modeling to Find an Equation through Three Points**

Find an equation of the parabola $y = ax^2 + bx + c$ that passes through the points $(2, 4)$, $(-1, 1)$, and $(-2, 5)$.

SOLUTION The three ordered pairs represent points that lie on the graph of the given equation $y = ax^2 + bx + c$, so they must all satisfy the equation. Substituting each ordered pair into the equation gives three equations with three unknowns.

$$4 = a(2)^2 + b(2) + c, \quad \text{or} \quad 4 = 4a + 2b + c \quad (1)$$
$$1 = a(-1)^2 + b(-1) + c, \quad \text{or} \quad 1 = a - b + c \quad (2)$$
$$5 = a(-2)^2 + b(-2) + c, \quad \text{or} \quad 5 = 4a - 2b + c \quad (3)$$

To solve this system, first eliminate c using equations (1) and (2).

$$4 = 4a + 2b + c \quad (1)$$
$$-1 = -a + b - c \quad \text{Multiply (2) by } -1.$$
$$3 = 3a + 3b \quad (4)$$

Now, use equations (2) and (3) to eliminate the same unknown, c.

$$1 = a - b + c \quad (2)$$
$$-5 = -4a + 2b - c \quad \text{Multiply (3) by } -1.$$
$$-4 = -3a + b \quad (5)$$

Equation (5) must have the *same* two unknowns as equation (4).

Solve the system of equations (4) and (5) in two unknowns by eliminating a.

$$3 = 3a + 3b \quad (4)$$
$$-4 = -3a + b \quad (5)$$
$$-1 = 4b \quad \text{Add.}$$
$$-\frac{1}{4} = b \quad \text{Divide by 4.}$$

Find a by substituting $-\frac{1}{4}$ for b in equation (4).

$$1 = a + b \quad \text{Equation (4) divided by 3}$$
$$1 = a + \left(-\frac{1}{4}\right) \quad \text{Let } b = -\tfrac{1}{4}.$$
$$\frac{5}{4} = a \quad \text{Add } \tfrac{1}{4}.$$

Finally, find c by substituting $a = \frac{5}{4}$ and $b = -\frac{1}{4}$ in equation (2).

$$1 = a - b + c \quad (2)$$
$$1 = \frac{5}{4} - \left(-\frac{1}{4}\right) + c \quad \text{Let } a = \tfrac{5}{4},\ b = -\tfrac{1}{4}.$$
$$1 = \frac{6}{4} + c \quad \text{Add.}$$
$$-\frac{1}{2} = c \quad \text{Subtract } \tfrac{6}{4}.$$

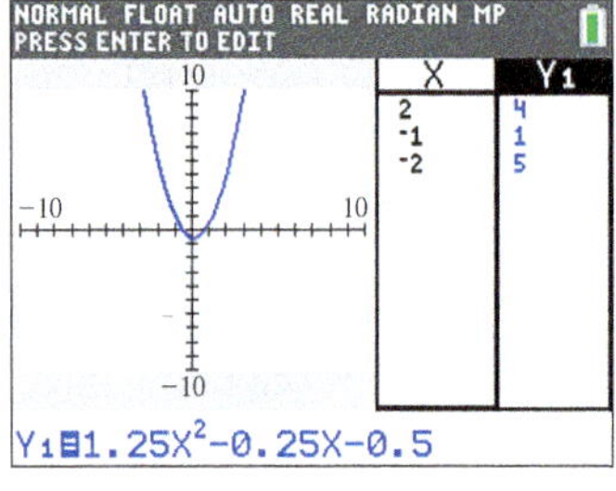

This graph/table screen shows that the points $(2, 4)$, $(-1, 1)$, and $(-2, 5)$ lie on the graph of $y_1 = 1.25x^2 - 0.25x - 0.5$. This supports the result of **Example 8.**

The required equation is $y = \frac{5}{4}x^2 - \frac{1}{4}x - \frac{1}{2}$, or $y = 1.25x^2 - 0.25x - 0.5$.

✔ **Now Try Exercise 79.**

EXAMPLE 9 Solving an Application Using a System of Three Equations

An animal feed is made from three ingredients: corn, soybeans, and cottonseed. One unit of each ingredient provides units of protein, fat, and fiber as shown in the table. How many units of each ingredient should be used to make a feed that contains 22 units of protein, 28 units of fat, and 18 units of fiber?

	Corn	Soybeans	Cottonseed	Total
Protein	0.25	0.4	0.2	22
Fat	0.4	0.2	0.3	28
Fiber	0.3	0.2	0.1	18

SOLUTION

Step 1 **Read** the problem. We must determine the number of units of corn, soybeans, and cottonseed.

Step 2 **Assign variables.** Let x represent the number of units of corn, y the number of units of soybeans, and z the number of units of cottonseed.

Step 3 **Write a system of equations.** The total amount of protein is to be 22 units, so we use the first row of the table to write equation (1).

$$0.25x + 0.4y + 0.2z = 22 \quad (1)$$

We use the second row of the table to obtain 28 units of fat.

$$0.4x + 0.2y + 0.3z = 28 \quad (2)$$

Finally, we use the third row of the table to obtain 18 units of fiber.

$$0.3x + 0.2y + 0.1z = 18 \quad (3)$$

Multiply equation (1) on each side by 100, and equations (2) and (3) by 10, to obtain an equivalent system.

$$\begin{aligned} 25x + 40y + 20z &= 2200 \quad (4) \\ 4x + 2y + 3z &= 280 \quad (5) \\ 3x + 2y + z &= 180 \quad (6) \end{aligned}$$

Eliminate the decimal points in equations (1), (2), and (3) by multiplying each equation by an appropriate power of 10.

Step 4 **Solve** the system. Using the methods described earlier in this section, we find the following.

$$x = 40, \quad y = 15, \quad \text{and} \quad z = 30$$

Step 5 **State the answer.** The feed should contain 40 units of corn, 15 units of soybeans, and 30 units of cottonseed.

Step 6 **Check.** Show that the ordered triple $(40, 15, 30)$ satisfies the system formed by equations (1), (2), and (3).

✔ **Now Try Exercise 107.**

NOTE Notice how the table in **Example 9** is used to set up the equations of the system. The coefficients in each equation are read from left to right. This idea is extended in the next section, where we introduce the solution of systems by matrices.

9.1 Exercises

CONCEPT PREVIEW *Fill in the blank(s) to correctly complete each sentence.*

1. The solution set of the following system is $\{(1, \_\_)\}$.

$$-2x + 5y = 18$$
$$x + y = 5$$

2. The solution set of the following system is $\{(\_\_, 0)\}$.

$$6x + y = -18$$
$$13x + y = -39$$

3. One way of solving the following system by elimination is to multiply equation (2) by the integer _____ to eliminate the y-terms by direct addition.

$$14x + 11y = 80 \quad (1)$$
$$2x + y = 19 \quad (2)$$

4. To solve the system

$$3x + y = 4 \quad (1)$$
$$7x + 8y = -2 \quad (2)$$

by substitution, it is easiest to begin by solving equation (1) for the variable _____ and then substituting into equation (2), because no fractions will appear in the algebraic work.

5. If a system of linear equations in two variables has two graphs that coincide, there is/are _________________ solutions to the system.
(one/no/infinitely many)

6. If a system of linear equations in two variables has two graphs that are parallel lines, there is/are _________________ solutions to the system.
(one/no/infinitely many)

Solve each system by substitution. ***See Example 1.***

7. $4x + 3y = -13$; $-x + y = 5$

8. $3x + 4y = 4$; $x - y = 13$

9. $x - 5y = 8$; $x = 6y$

10. $6x - y = 5$; $y = 11x$

11. $8x - 10y = -22$; $3x + y = 6$

12. $4x - 5y = -11$; $2x + y = 5$

13. $7x - y = -10$; $3y - x = 10$

14. $4x + 5y = 7$; $9y = 31 + 2x$

15. $-2x = 6y + 18$; $-29 = 5y - 3x$

16. $3x - 7y = 15$; $3x + 7y = 15$

17. $3y = 5x + 6$; $x + y = 2$

18. $4y = 2x - 4$; $x - y = 4$

Solve each system by elimination. In systems with fractions, first clear denominators. ***See Example 2.***

19. $3x - y = -4$; $x + 3y = 12$

20. $4x + y = -23$; $x - 2y = -17$

21. $2x - 3y = -7$; $5x + 4y = 17$

22. $4x + 3y = -1$; $2x + 5y = 3$

23. $5x + 7y = 6$; $10x - 3y = 46$

24. $12x - 5y = 9$; $3x - 8y = -18$

25. $6x + 7y + \ \ 2 = 0$
$7x - 6y - 26 = 0$

26. $5x + 4y + \ \ 2 = 0$
$4x - 5y - 23 = 0$

27. $\frac{x}{2} + \frac{y}{3} = 4$
$\frac{3x}{2} + \frac{3y}{2} = 15$

28. $\frac{3x}{2} + \frac{y}{2} = -2$
$\frac{x}{2} + \frac{y}{2} = 0$

29. $\frac{2x-1}{3} + \frac{y+2}{4} = 4$
$\frac{x+3}{2} - \frac{x-y}{3} = 3$

30. $\frac{x+6}{5} + \frac{2y-x}{10} = 1$
$\frac{x+2}{4} + \frac{3y+2}{5} = -3$

Solve each system of equations. State whether it is an inconsistent system *or has* infinitely many solutions. *If the system has infinitely many solutions, write the solution set with y arbitrary.* ***See Examples 3 and 4.***

31. $9x - \ \ 5y = 1$
$-18x + 10y = 1$

32. $3x + 2y = 5$
$6x + 4y = 8$

33. $4x - \ \ y = 9$
$-8x + 2y = -18$

34. $3x + \ \ 5y = -2$
$9x + 15y = -6$

35. $5x - 5y - \ \ 3 = 0$
$x - \ \ y - 12 = 0$

36. $2x - 3y - \ \ 7 = 0$
$-4x + 6y - 14 = 0$

37. $7x + 2y = 6$
$14x + 4y = 12$

38. $2x - 8y = 4$
$x - 4y = 2$

39. $2x - \ \ 6y = 0$
$-7x + 21y = 10$

40. *Concept Check* Which screen gives the correct graphical solution of the system? (*Hint:* Solve for y first in each equation and use the slope-intercept forms to answer the question.)

$$4x - 5y = -11$$
$$2x + \ \ y = 5$$

A.

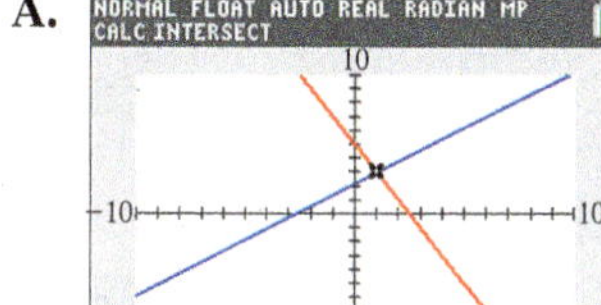

B.

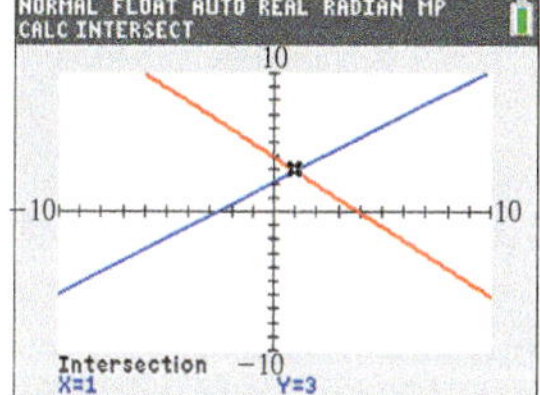

C.

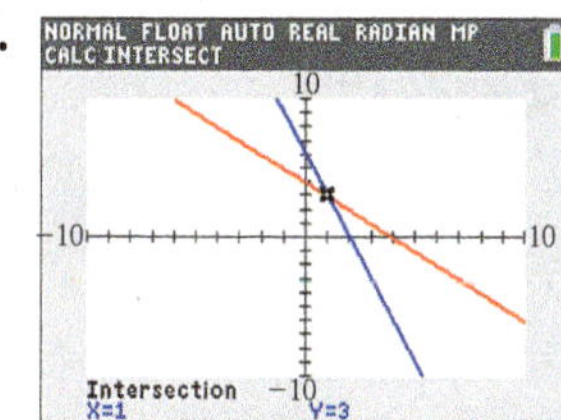

D.

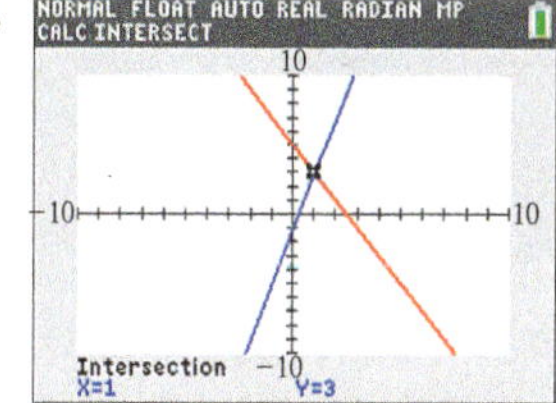

Connecting Graphs with Equations *Determine the system of equations illustrated in each graph. Write equations in standard form.*

41.

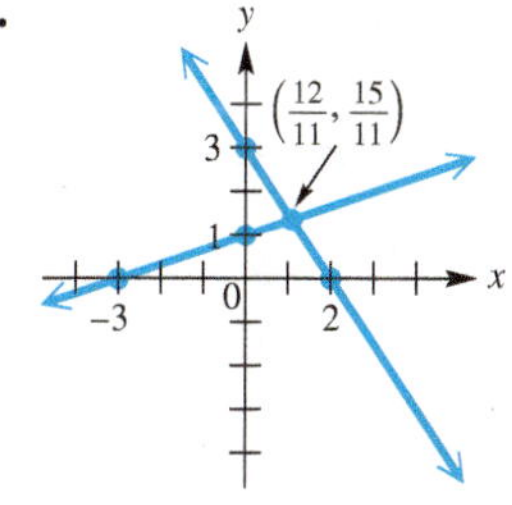

42.

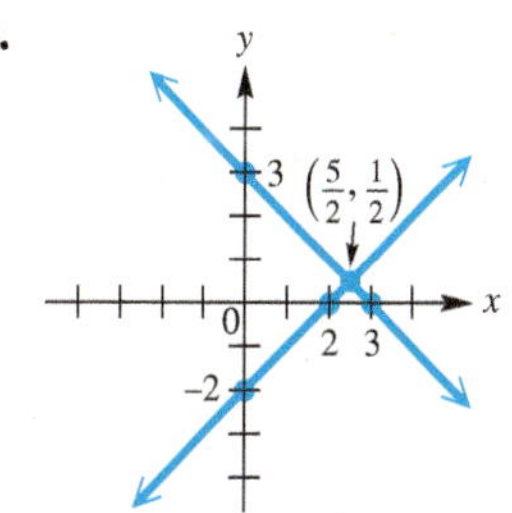

Use a graphing calculator to solve each system. Express solutions with approximations to the nearest thousandth.

43. $\frac{11}{3}x + y = 0.5$
$0.6x - y = 3$

44. $\sqrt{3}x - y = 5$
$100x + y = 9$

45. $\sqrt{7}x + \sqrt{2}y = 3$
$\sqrt{6}x - y = \sqrt{3}$

46. $0.2x + \sqrt{2}y = 1$
$\sqrt{5}x + 0.7y = 1$

Solve each system. ***See Example 6.***

47. $x + y + z = 2$
$2x + y - z = 5$
$x - y + z = -2$

48. $2x + y + z = 9$
$-x - y + z = 1$
$3x - y + z = 9$

49. $x + 3y + 4z = 14$
$2x - 3y + 2z = 10$
$3x - y + z = 9$

50. $4x - y + 3z = -2$
$3x + 5y - z = 15$
$-2x + y + 4z = 14$

51. $x + 4y - z = 6$
$2x - y + z = 3$
$3x + 2y + 3z = 16$

52. $4x - 3y + z = 9$
$3x + 2y - 2z = 4$
$x - y + 3z = 5$

53. $x - 3y - 2z = -3$
$3x + 2y - z = 12$
$-x - y + 4z = 3$

54. $x + y + z = 3$
$3x - 3y - 4z = -1$
$x + y + 3z = 11$

55. $2x + 6y - z = 6$
$4x - 3y + 5z = -5$
$6x + 9y - 2z = 11$

56. $8x - 3y + 6z = -2$
$4x + 9y + 4z = 18$
$12x - 3y + 8z = -2$

57. $2x - 3y + 2z - 3 = 0$
$4x + 8y + z - 2 = 0$
$-x - 7y + 3z - 14 = 0$

58. $-x + 2y - z - 1 = 0$
$-x - y - z + 2 = 0$
$x - y + 2z - 2 = 0$

Solve each system in terms of the arbitrary variable z. ***See Example 7.***

59. $x - 2y + 3z = 6$
$2x - y + 2z = 5$

60. $3x - 2y + z = 15$
$x + 4y - z = 11$

61. $5x - 4y + z = 9$
$y + z = 15$

62. $3x - 5y - 4z = -7$
$y - z = -13$

63. $3x + 4y - z = 13$
$x + y + 2z = 15$

64. $x - y + z = -6$
$4x + y + z = 7$

Solve each system. State whether it is an inconsistent system *or has* infinitely many solutions. *If the system has infinitely many solutions, write the solution set with z arbitrary.* ***See Examples 3, 4, 6, and 7.***

65. $3x + 5y - z = -2$
$4x - y + 2z = 1$
$-6x - 10y + 2z = 0$

66. $3x + y + 3z = 1$
$x + 2y - z = 2$
$2x - y + 4z = 4$

67. $5x - 4y + z = 0$
$x + y = 0$
$-10x + 8y - 2z = 0$

68. $2x + y - 3z = 0$
$4x + 2y - 6z = 0$
$x - y + z = 0$

Solve each system. (Hint: In Exercises 69–72, let $\frac{1}{x} = t$ and $\frac{1}{y} = u$.)

69. $\frac{2}{x} + \frac{1}{y} = \frac{3}{2}$
$\frac{3}{x} - \frac{1}{y} = 1$

70. $\frac{1}{x} + \frac{3}{y} = \frac{16}{5}$
$\frac{5}{x} + \frac{4}{y} = 5$

71. $\frac{2}{x} + \frac{1}{y} = 11$
$\frac{3}{x} - \frac{5}{y} = 10$

72. $\frac{2}{x} + \frac{3}{y} = 18$

$\frac{4}{x} - \frac{5}{y} = -8$

73. $\frac{2}{x} + \frac{3}{y} - \frac{2}{z} = -1$

$\frac{8}{x} - \frac{12}{y} + \frac{5}{z} = 5$

$\frac{6}{x} + \frac{3}{y} - \frac{1}{z} = 1$

74. $-\frac{5}{x} + \frac{4}{y} + \frac{3}{z} = 2$

$\frac{10}{x} + \frac{3}{y} - \frac{6}{z} = 7$

$\frac{5}{x} + \frac{2}{y} - \frac{9}{z} = 6$

75. *Concept Check* For what value(s) of k will the following system of linear equations have no solution? infinitely many solutions?

$$x - 2y = 3$$
$$-2x + 4y = k$$

76. *Concept Check* Consider the linear equation in three variables

$$x + y + z = 4.$$

Find a pair of linear equations in three variables that, when considered together with the given equation, form a system having **(a)** exactly one solution, **(b)** no solution, **(c)** infinitely many solutions.

(Modeling) Use a system of equations to solve each problem. ***See Example 8.***

77. Find an equation of the line $y = ax + b$ that passes through the points $(-2, 1)$ and $(-1, -2)$.

78. Find an equation of the line $y = ax + b$ that passes through the points $(3, -4)$ and $(-1, 4)$.

79. Find an equation of the parabola $y = ax^2 + bx + c$ that passes through the points $(2, 3)$, $(-1, 0)$, and $(-2, 2)$.

80. Find an equation of the parabola $y = ax^2 + bx + c$ that passes through the points $(-2, 4)$, $(2, 2)$, and $(4, 9)$.

81. *Connecting Graphs with Equations* Use a system to find an equation of the line through the given points.

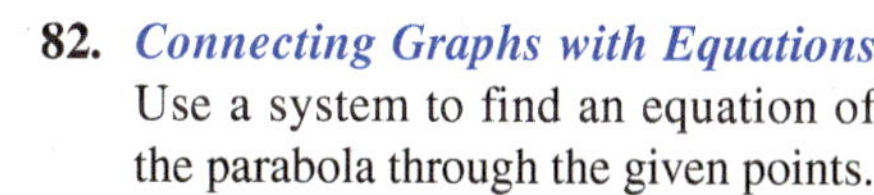

82. *Connecting Graphs with Equations* Use a system to find an equation of the parabola through the given points.

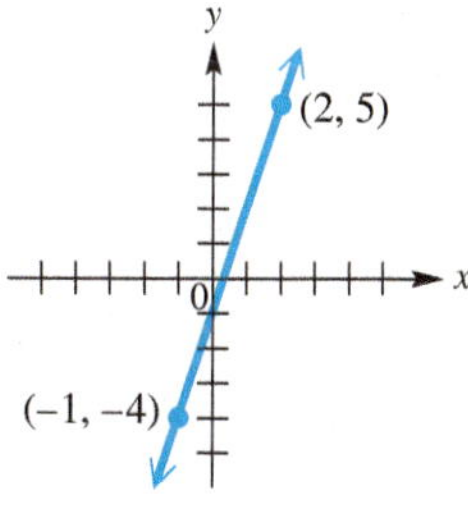

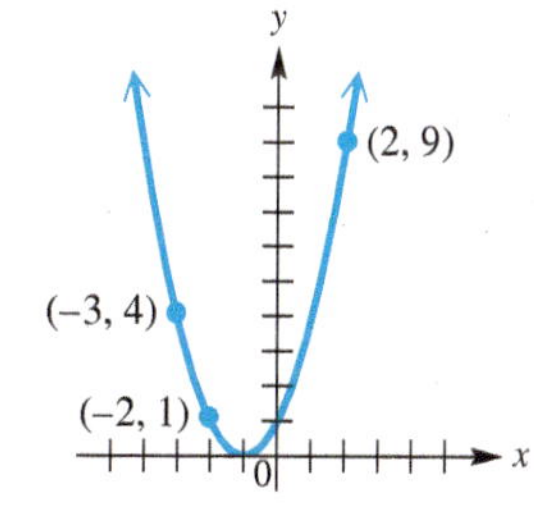

83. *Connecting Graphs with Equations* Find an equation of the parabola. Three views of the same curve are given.

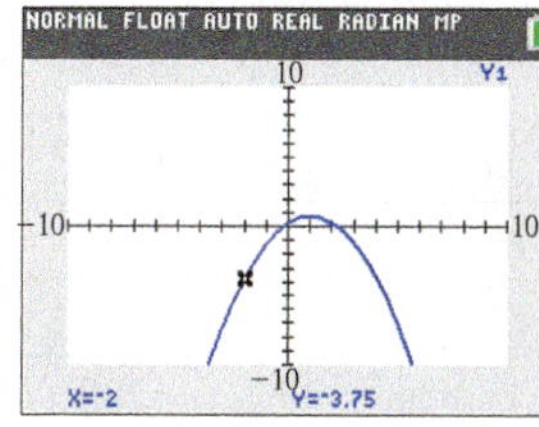

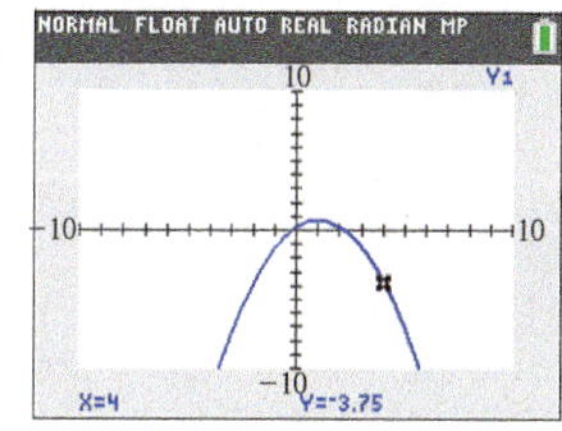

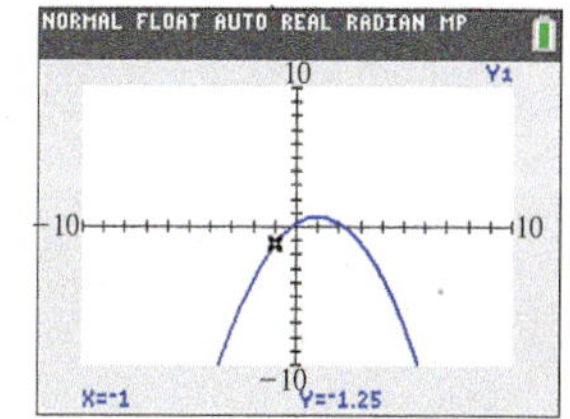

84. *(Modeling)* The table was generated using a function

$$y_1 = ax^2 + bx + c.$$

Use any three points from the table to find an equation for y_1.

NORMAL FLOAT AUTO REAL RADIAN MP
PRESS + FOR ΔTbl

X	Y1
-3	5.48
-2	2.9
-1	1.26
0	.56
1	.8
2	1.98
3	4.1
4	7.16
5	11.16
6	16.1
7	21.98

X= -3

(Modeling) Given three noncollinear points, there is one and only one circle that passes through them. Knowing that the equation of a circle may be written in the form

$$x^2 + y^2 + ax + by + c = 0,$$

find an equation of the circle passing through the given points.

85. $(-1, 3)$, $(6, 2)$, and $(-2, -4)$

86. $(-1, 5)$, $(6, 6)$, and $(7, -1)$

87. $(2, 1)$, $(-1, 0)$, and $(3, 3)$

88. $(-5, 0)$, $(2, -1)$, and $(4, 3)$

89. *Connecting Graphs with Equations*

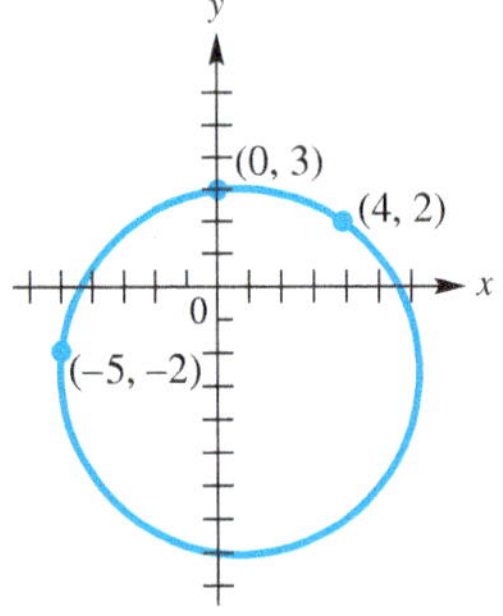

90. *Connecting Graphs with Equations*

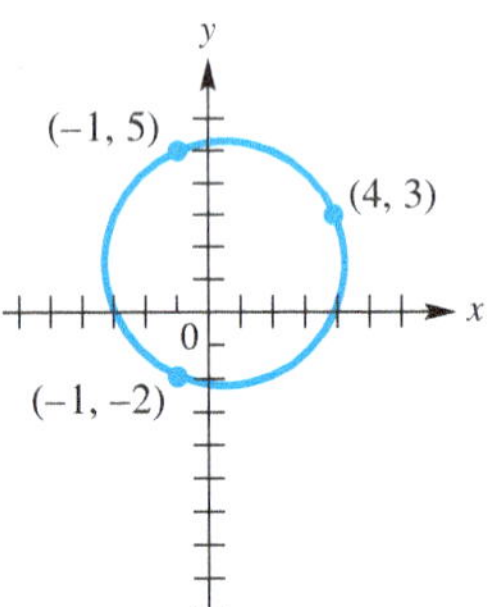

(Modeling) Use the method of ***Example 8*** *to work each problem.*

91. *Atmospheric Carbon Dioxide* Carbon dioxide concentrations (in parts per million) have been measured directly from the atmosphere since 1960. This concentration has increased quadratically. The table lists readings for three years.

Year	CO_2
1960	317
1980	339
2013	396

Source: U.S. Department of Energy; Carbon Dioxide Information Analysis Center.

(a) If the quadratic relationship between the carbon dioxide concentration C and the year t is expressed as

$$C = at^2 + bt + c,$$

where $t = 0$ corresponds to 1960, use a system of linear equations to determine the constants a, b, and c, and give the equation.

(b) Predict when the amount of carbon dioxide in the atmosphere will be double its 1960 level.

92. *Aircraft Speed and Altitude* For certain aircraft there exists a quadratic relationship between an airplane's maximum speed S (in knots) and its ceiling C, or highest altitude possible (in thousands of feet). The table lists three such airplanes.

Airplane	Max Speed (S)	Ceiling (C)
Hawkeye	320	33
Corsair	600	40
Tomcat	1283	50

Source: Sanders, D., *Statistics: A First Course*, Sixth Edition, McGraw Hill.

(a) If the quadratic relationship between C and S is written as

$$C = aS^2 + bS + c,$$

use a system of linear equations to determine the constants a, b and c, and give the equation.

(b) A new aircraft of this type has a ceiling of 45,000 ft. Predict its top speed to the nearest knot.

Changes in Population *The graph shows the populations of the New Orleans, LA, and the Jacksonville, FL, metropolitan areas over the years 2004–2013.*

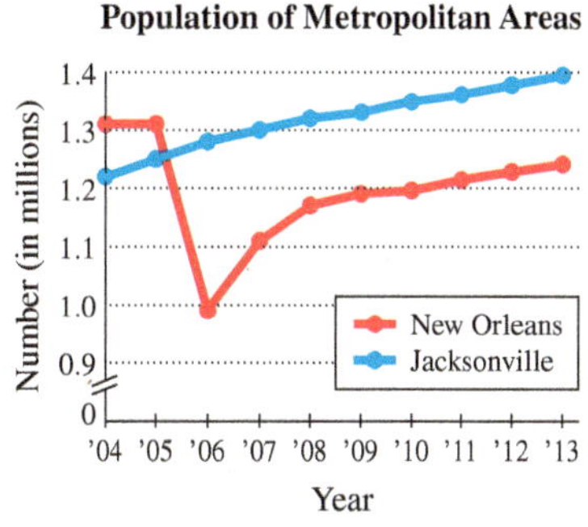

Source: U.S. Census Bureau.

93. In what years was the population of the Jacksonville metropolitan area greater than that of the New Orleans metropolitan area?

94. At the time when the populations of the two metropolitan areas were equal, what was the approximate population of each area? Round to the nearest hundredth million.

95. Express the solution of the system as an ordered pair to the nearest tenth of a year and the nearest hundredth million.

96. Use the terms *increasing, decreasing*, and *constant* to describe the trends for the population of the New Orleans metropolitan area.

97. If equations of the form $y = f(t)$ were determined that modeled either of the two graphs, then the variable t would represent ________ and the variable y would represent ________.

98. Why is each graph that of a function?

Solve each problem. ***See Examples 5 and 9.***

99. *Unknown Numbers* The sum of two numbers is 47, and the difference between the numbers is 1. Find the numbers.

100. *Costs of Goats and Sheep* At the Berger ranch, 6 goats and 5 sheep sell for \$305, while 2 goats and 9 sheep sell for \$285. Find the cost of a single goat and of a single sheep.

101. *Fan Cost Index* The Fan Cost Index (FCI) is a measure of how much it will cost a family of four to attend a professional sports event. In 2014, the FCI prices for Major League Baseball and the National Football League averaged \$345.53. The FCI for baseball was \$266.13 less than that for football. What were the FCIs for these sports? (*Source:* Team Marketing Report.)

102. *Money Denominations* A cashier has a total of 30 bills, made up of ones, fives, and twenties. The number of twenties is 9 more than the number of ones. The total value of the money is \$351. How many of each denomination of bill are there?

103. *Mixing Water* A sparkling-water distributor wants to make up 300 gal of sparkling water to sell for \$6.00 per gallon. She wishes to mix three grades of water selling for \$9.00, \$3.00, and \$4.50 per gallon, respectively. She must use twice as much of the \$4.50 water as of the \$3.00 water. How many gallons of each should she use?

104. *Mixing Glue* A glue company needs to make some glue that it can sell for \$120 per barrel. It wants to use 150 barrels of glue worth \$100 per barrel, along with some glue worth \$150 per barrel and some glue worth \$190 per barrel. It must use the same number of barrels of \$150 and \$190 glue. How much of the \$150 and \$190 glue will be needed? How many barrels of \$120 glue will be produced?

105. *Triangle Dimensions* The perimeter of a triangle is 59 in. The longest side is 11 in. longer than the medium side, and the medium side is 3 in. longer than the shortest side. Find the length of each side of the triangle.

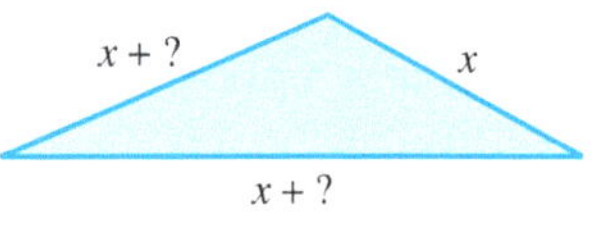

106. *Triangle Dimensions* The sum of the measures of the angles of any triangle is 180°. In a certain triangle, the largest angle measures 55° less than twice the medium angle, and the smallest angle measures 25° less than the medium angle. Find the measures of all three angles.

107. *Investment Decisions* Patrick wins \$200,000 in the Louisiana state lottery. He invests part of the money in real estate with an annual return of 3% and another part in a money market account at 2.5% interest. He invests the rest, which amounts to \$80,000 less than the sum of the other two parts, in certificates of deposit that pay 1.5%. If the total annual interest on the money is \$4900, how much was invested at each rate?

	Amount Invested	Rate (as a decimal)	Annual Interest
Real Estate		0.03	
Money Market		0.025	
CDs		0.015	

108. *Investment Decisions* Jane invests \$40,000 received as an inheritance in three parts. With one part she buys mutual funds that offer a return of 2% per year. The second part, which amounts to twice the first, is used to buy government bonds paying 2.5% per year. She puts the rest of the money into a savings account that pays 1.25% annual interest. During the first year, the total interest is \$825. How much did she invest at each rate?

	Amount Invested	Rate (as a decimal)	Annual Interest
Mutual Funds		0.02	
Government Bonds		0.025	
Savings Account		0.0125	

109. Solve the system of equations (4), (5), and (6) from **Example 9.**

$$\begin{aligned} 25x + 40y + 20z &= 2200 \quad (4) \\ 4x + 2y + 3z &= 280 \quad (5) \\ 3x + 2y + z &= 180 \quad (6) \end{aligned}$$

110. Check the solution in **Exercise 109,** showing that it satisfies all three equations of the system.

111. ***Blending Coffee Beans*** Three varieties of coffee—Arabian Mocha Sanani, Organic Shade Grown Mexico, and Guatemala Antigua—are combined and roasted, yielding a 50-lb batch of coffee beans. Twice as many pounds of Guatemala Antigua, which retails for \$10.19 per lb, are needed as of Arabian Mocha Sanani, which retails for \$15.99 per lb. Organic Shade Grown Mexico retails for \$12.99 per lb. How many pounds, to the nearest hundredth, of each coffee should be used in a blend that sells for \$12.37 per lb?

112. ***Blending Coffee Beans*** Rework **Exercise 111** if Guatemala Antigua retails for \$12.49 per lb instead of \$10.19 per lb. Does the answer seem reasonable?

Relating Concepts

For individual or collaborative investigation *(Exercises 113–118)*

Supply and Demand *In many applications of economics, as the price of an item goes up, demand for the item goes down and supply of the item goes up. The price where supply and demand are equal is the* ***equilibrium price,*** *and the resulting supply or demand is the* ***equilibrium supply*** *or* ***equilibrium demand.***

Suppose the supply of a product is related to its price by the equation

$$p = \frac{2}{3}q,$$

where p is in dollars and q is supply in appropriate units. (Here, q stands for quantity.) Furthermore, suppose demand and price for the same product are related by

$$p = -\frac{1}{3}q + 18,$$

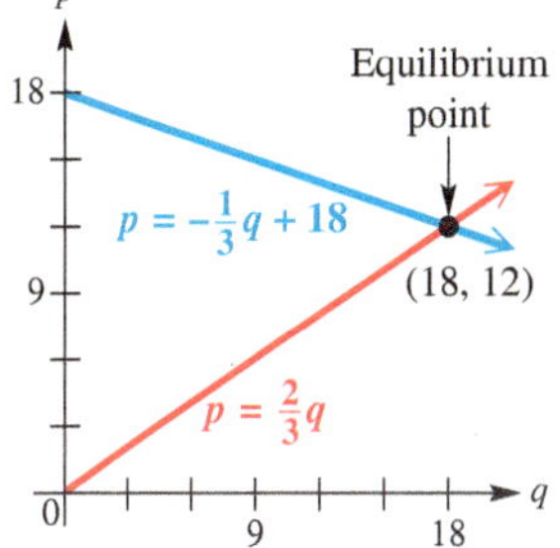

where p is price and q is demand. The system formed by these two equations has solution $(18, 12)$, *as seen in the graph. Use this information to* ***work Exercises 113–118 in order.***

113. Suppose the demand and price for a certain model of electric can opener are related by $p = 16 - \frac{5}{4}q$, where p is price, in dollars, and q is demand, in appropriate units. Find the price when the demand is at each level.

(a) 0 units **(b)** 4 units **(c)** 8 units

114. Find the demand for the electric can opener at each price.

(a) \$6 **(b)** \$11 **(c)** \$16

115. Graph $p = 16 - \frac{5}{4}q$.

116. Suppose the price and supply of the can opener are related by $p = \frac{3}{4}q$, where q represents the supply and p the price. Find the supply at each price.

(a) \$0 **(b)** \$10 **(c)** \$20

117. Graph $p = \frac{3}{4}q$ on the same axes used for **Exercise 115.**

118. Use the result of **Exercise 117** to find the equilibrium price and the equilibrium demand.

9.2 Matrix Solution of Linear Systems

- The Gauss-Jordan Method
- Special Systems

Systems of linear equations occur in many practical situations, and as a result, computer methods have been developed for efficiently solving linear systems. Computer solutions of linear systems depend on the idea of a **matrix** (plural **matrices**), a rectangular array of numbers enclosed in brackets. Each number is an **element** of the matrix.

$$\begin{bmatrix} 2 & 3 & 7 \\ 5 & -1 & 10 \end{bmatrix} \quad \text{Matrix}$$

The Gauss-Jordan Method

In this section, we develop a method for solving linear systems using matrices. We start with a system and write the coefficients of the variables and the constants as an **augmented matrix** of the system.

Linear system of equations

$$\begin{aligned} x + 3y + 2z &= 1 \\ 2x + y - z &= 2 \\ x + y + z &= 2 \end{aligned} \quad \text{can be written as} \quad \left[\begin{array}{ccc|c} 1 & 3 & 2 & 1 \\ 2 & 1 & -1 & 2 \\ 1 & 1 & 1 & 2 \end{array}\right]$$

Augmented matrix

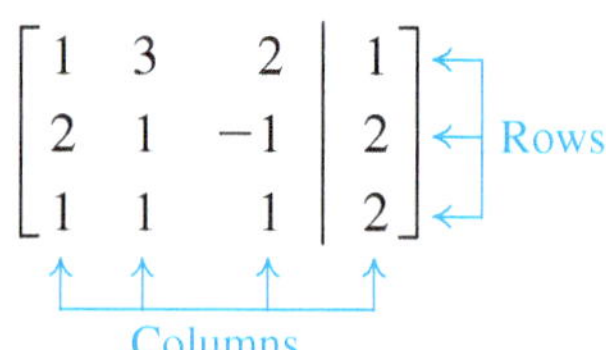

The vertical line, which is optional, separates the coefficients from the constants. Because this matrix has 3 rows (horizontal) and 4 columns (vertical), we say its **dimension*** is 3×4 (read "three by four"). ***The number of rows is always given first.*** To refer to a number in the matrix, use its row and column numbers. For example, the number 3 is in the first row, second column.

We can treat the rows of this matrix just like the equations of the corresponding system of linear equations. Because an augmented matrix is nothing more than a shorthand form of a system, any transformation of the matrix that results in an equivalent system of equations can be performed.

Matrix Row Transformations

For any augmented matrix of a system of linear equations, the following row transformations will result in the matrix of an equivalent system.

1. Interchange any two rows.
2. Multiply or divide the elements of any row by a nonzero real number.
3. Replace any row of the matrix by the sum of the elements of that row and a multiple of the elements of another row.

These transformations are restatements in matrix form of the transformations of systems discussed in the previous section. From now on, when referring to the third transformation, we will abbreviate "a multiple of the elements of a row" as "a multiple of a row."

Before matrices can be used to solve a linear system, the system must be arranged in the proper form, with variable terms on the left side of the equation and constant terms on the right. The variable terms must be in the same order in each of the equations.

*Other terms used to describe the dimension of a matrix are *order* and *size*.

The **Gauss-Jordan method** is a systematic technique for applying matrix row transformations in an attempt to reduce a matrix to **diagonal form,** with 1s along the diagonal, from which the solutions are easily obtained.

$$\left[\begin{array}{cc|c} 1 & 0 & a \\ 0 & 1 & b \end{array}\right] \quad \text{or} \quad \left[\begin{array}{ccc|c} 1 & 0 & 0 & a \\ 0 & 1 & 0 & b \\ 0 & 0 & 1 & c \end{array}\right]$$

Diagonal form, or reduced-row echelon form

This form is also called **reduced-row echelon form.**

Using the Gauss-Jordan Method to Transform a Matrix into Diagonal Form

Step 1 Obtain 1 as the first element of the first column.

Step 2 Use the first row to transform the remaining entries in the first column to 0.

Step 3 Obtain 1 as the second entry in the second column.

Step 4 Use the second row to transform the remaining entries in the second column to 0.

Step 5 Continue in this manner as far as possible.

NOTE ***The Gauss-Jordan method proceeds column by column, from left to right.*** In each column, we work to obtain 1 in the appropriate diagonal location, and then use it to transform the remaining elements in that column to 0s. When we are working with a particular column, no row operation should undo the form of a preceding column.

EXAMPLE 1 Using the Gauss-Jordan Method

Solve the system.

$$3x - 4y = 1$$

$$5x + 2y = 19$$

SOLUTION Both equations are in the same form, with variable terms in the same order on the left, and constant terms on the right.

$$\left[\begin{array}{cc|c} 3 & -4 & 1 \\ 5 & 2 & 19 \end{array}\right]$$

Write the augmented matrix.

The goal is to transform the augmented matrix into one in which the value of the variables will be easy to see. That is, because each of the first two columns in the matrix represents the coefficients of one variable, the augmented matrix should be transformed so that it is of the following form.

This form is our goal.

$$\left[\begin{array}{cc|c} \mathbf{1} & \mathbf{0} & k \\ \mathbf{0} & \mathbf{1} & j \end{array}\right]$$

Here k and j are real numbers.

In this form, the matrix can be rewritten as a linear system.

$$x = k$$

$$y = j$$

It is best to work in columns, beginning in each column with the element that is to become 1. In the augmented matrix

$$\left[\begin{array}{cc|c} 3 & -4 & 1 \\ 5 & 2 & 19 \end{array}\right],$$

3 is in the first row, first column position. Use transformation 2, multiplying each entry in the first row by $\frac{1}{3}$ (abbreviated $\frac{1}{3}$R1) to obtain 1 in this position.

$$\left[\begin{array}{cc|c} 1 & -\frac{4}{3} & \frac{1}{3} \\ 5 & 2 & 19 \end{array}\right] \quad \tfrac{1}{3}\text{R1}$$

Introduce 0 in the second row, first column by multiplying each element of the first row by -5 and adding the result to the corresponding element in the second row, using transformation 3.

$$\left[\begin{array}{cc|c} 1 & -\frac{4}{3} & \frac{1}{3} \\ 0 & \frac{26}{3} & \frac{52}{3} \end{array}\right] \quad -5\text{R1} + \text{R2}$$

Obtain 1 in the second row, second column by multiplying each element of the second row by $\frac{3}{26}$, using transformation 2.

$$\left[\begin{array}{cc|c} 1 & -\frac{4}{3} & \frac{1}{3} \\ 0 & 1 & 2 \end{array}\right] \quad \tfrac{3}{26}\text{R2}$$

Finally, obtain 0 in the first row, second column by multiplying each element of the second row by $\frac{4}{3}$ and adding the result to the corresponding element in the first row.

$$\left[\begin{array}{cc|c} 1 & 0 & 3 \\ 0 & 1 & 2 \end{array}\right] \quad \tfrac{4}{3}\text{R2} + \text{R1}$$

This last matrix corresponds to the system

$$x = 3$$
$$y = 2,$$

which indicates the solution $(3, 2)$. We can read this solution directly from the third column of the final matrix.

CHECK Substitute the solution in both equations of the *original* system.

$$3x - 4y = 1 \qquad\qquad 5x + 2y = 19$$
$$3(3) - 4(2) \stackrel{?}{=} 1 \qquad\qquad 5(3) + 2(2) \stackrel{?}{=} 19$$
$$9 - 8 \stackrel{?}{=} 1 \qquad\qquad 15 + 4 \stackrel{?}{=} 19$$
$$1 = 1 \checkmark \text{ True} \qquad\qquad 19 = 19 \checkmark \text{ True}$$

True statements result, so the solution set is $\{(3, 2)\}$.

✔ **Now Try Exercise 23.**

NOTE Using row operations to write a matrix in diagonal form requires effective use of the inverse properties of addition and multiplication.

A linear system with three equations is solved in a similar way. Row transformations are used to introduce 1s down the diagonal from left to right and 0s above and below each 1.

EXAMPLE 2 Using the Gauss-Jordan Method

Solve the system.

$$\begin{aligned} x - y + 5z &= -6 \\ 3x + 3y - z &= 10 \\ x + 3y + 2z &= 5 \end{aligned}$$

SOLUTION

$$\left[\begin{array}{ccc|c} 1 & -1 & 5 & -6 \\ 3 & 3 & -1 & 10 \\ 1 & 3 & 2 & 5 \end{array}\right] \quad \text{Write the augmented matrix.}$$

There is already a 1 in the first row, first column. Introduce 0 in the second row of the first column by multiplying each element in the first row by -3 and adding the result to the corresponding element in the second row.

$$\left[\begin{array}{ccc|c} 1 & -1 & 5 & -6 \\ 0 & 6 & -16 & 28 \\ 1 & 3 & 2 & 5 \end{array}\right] \quad -3\text{R1} + \text{R2}$$

To change the third element in the first column to 0, multiply each element of the first row by -1. Add the result to the corresponding element of the third row.

$$\left[\begin{array}{ccc|c} 1 & -1 & 5 & -6 \\ 0 & 6 & -16 & 28 \\ 0 & 4 & -3 & 11 \end{array}\right] \quad -1\text{R1} + \text{R3}$$

Use the same procedure to transform the second and third columns. Obtain 1 in the appropriate position of each column by multiplying the elements of the row by the reciprocal of the number in that position.

$$\left[\begin{array}{ccc|c} 1 & -1 & 5 & -6 \\ 0 & 1 & -\frac{8}{3} & \frac{14}{3} \\ 0 & 4 & -3 & 11 \end{array}\right] \quad \tfrac{1}{6}\text{R2}$$

$$\left[\begin{array}{ccc|c} 1 & 0 & \frac{7}{3} & -\frac{4}{3} \\ 0 & 1 & -\frac{8}{3} & \frac{14}{3} \\ 0 & 4 & -3 & 11 \end{array}\right] \quad \text{R2} + \text{R1}$$

$$\left[\begin{array}{ccc|c} 1 & 0 & \frac{7}{3} & -\frac{4}{3} \\ 0 & 1 & -\frac{8}{3} & \frac{14}{3} \\ 0 & 0 & \frac{23}{3} & -\frac{23}{3} \end{array}\right] \quad -4\text{R2} + \text{R3}$$

$$\left[\begin{array}{ccc|c} 1 & 0 & \frac{7}{3} & -\frac{4}{3} \\ 0 & 1 & -\frac{8}{3} & \frac{14}{3} \\ 0 & 0 & 1 & -1 \end{array}\right] \quad \tfrac{3}{23}\text{R3}$$

$$\left[\begin{array}{ccc|c} 1 & 0 & 0 & 1 \\ 0 & 1 & -\frac{8}{3} & \frac{14}{3} \\ 0 & 0 & 1 & -1 \end{array}\right] \quad -\tfrac{7}{3}\text{R3} + \text{R1}$$

$$\left[\begin{array}{ccc|c} 1 & 0 & 0 & 1 \\ 0 & 1 & 0 & 2 \\ 0 & 0 & 1 & -1 \end{array}\right] \quad \tfrac{8}{3}\text{R3} + \text{R2}$$

$$\left[\begin{array}{ccc|c} 1 & 0 & 0 & 1 \\ 0 & 1 & 0 & 2 \\ 0 & 0 & 1 & -1 \end{array}\right]$$

Final Matrix

The linear system associated with this final matrix is

$$x = 1$$
$$y = 2$$
$$z = -1.$$

The solution set is $\{(1, 2, -1)\}$. Check the solution in the original system.

✔ **Now Try Exercise 31.**

The TI-84 Plus graphing calculator is able to perform row operations. See **Figure 7(a).** The screen in **Figure 7(b)** shows typical entries for the matrix in the second step of the solution in **Example 2.** The entire Gauss-Jordan method can be carried out in one step with the rref (reduced-row echelon form) command, as shown in **Figure 7(c).**

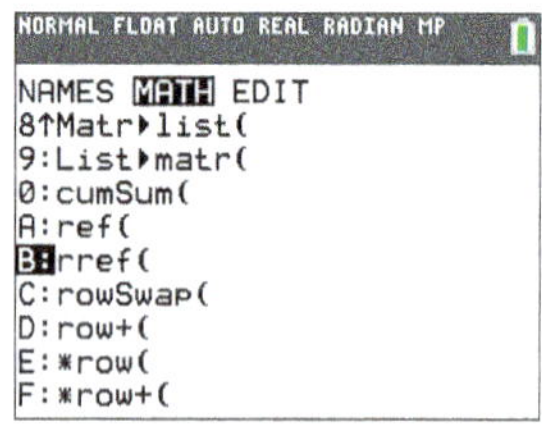

This menu shows various options for matrix row transformations in choices C–F.

(a)

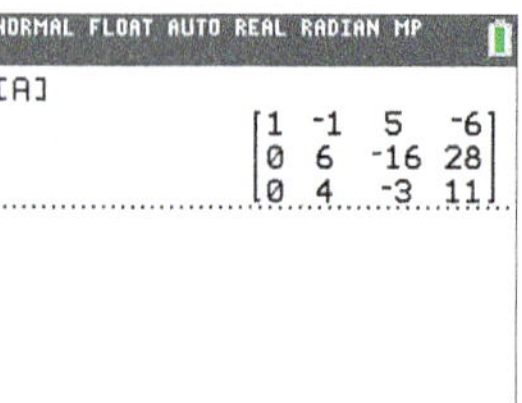

This is the matrix after column 1 has been transformed.

(b)

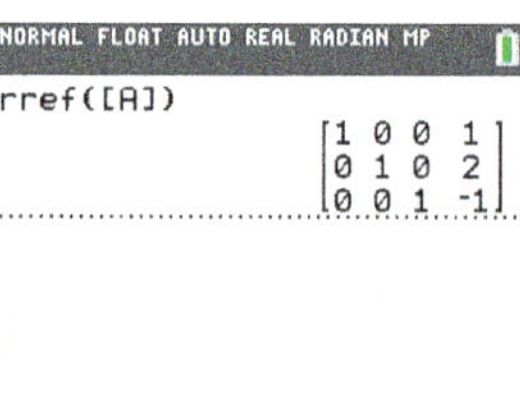

This screen shows the final matrix in the algebraic solution, found by using the rref command.

(c)

Figure 7

Special Systems

The next two examples show how to recognize inconsistent systems or systems with infinitely many solutions when solving such systems using row transformations.

EXAMPLE 3 Solving an Inconsistent System

Use the Gauss-Jordan method to solve the system.

$$x + y = 2$$
$$2x + 2y = 5$$

SOLUTION

$$\left[\begin{array}{cc|c} 1 & 1 & 2 \\ 2 & 2 & 5 \end{array}\right] \quad \text{Write the augmented matrix.}$$

$$\left[\begin{array}{cc|c} 1 & 1 & 2 \\ 0 & 0 & 1 \end{array}\right] \quad -2R1 + R2$$

The next step would be to introduce 1 in the second row, second column. Because of the 0 there, this is impossible. The second row corresponds to

$$0x + 0y = 1$$

which is false for all pairs of x and y, so the system has no solution. The system is inconsistent, and the solution set is $\emptyset$.

✔ **Now Try Exercise 27.**

EXAMPLE 4 Solving a System with Infinitely Many Solutions

Use the Gauss-Jordan method to solve the system. Write the solution set with z arbitrary.

$$2x - 5y + 3z = 1$$
$$x - 2y - 2z = 8$$

SOLUTION Recall from the previous section that a system with two equations in three variables usually has an infinite number of solutions. We can use the Gauss-Jordan method to give the solution with z arbitrary.

$$\left[\begin{array}{ccc|c} 2 & -5 & 3 & 1 \\ 1 & -2 & -2 & 8 \end{array}\right] \quad \text{Write the augmented matrix.}$$

$$\left[\begin{array}{ccc|c} 1 & -2 & -2 & 8 \\ 2 & -5 & 3 & 1 \end{array}\right] \quad \text{Interchange rows to obtain 1 in the first row, first column position.}$$

$$\left[\begin{array}{ccc|c} 1 & -2 & -2 & 8 \\ 0 & -1 & 7 & -15 \end{array}\right] \quad -2\text{R1} + \text{R2}$$

$$\left[\begin{array}{ccc|c} 1 & -2 & -2 & 8 \\ 0 & 1 & -7 & 15 \end{array}\right] \quad -1\text{R2}$$

$$\left[\begin{array}{ccc|c} 1 & 0 & -16 & 38 \\ 0 & 1 & -7 & 15 \end{array}\right] \quad 2\text{R2} + \text{R1}$$

It is not possible to go further with the Gauss-Jordan method. The equations that correspond to the final matrix are

$$x - 16z = 38 \quad \text{and} \quad y - 7z = 15.$$

Solve these equations for x and y, respectively.

$$x - 16z = 38 \qquad\qquad y - 7z = 15$$
$$x = 16z + 38 \quad \text{Add } 16z. \qquad\qquad y = 7z + 15 \quad \text{Add } 7z.$$

The solution set, written with z arbitrary, is $\{(16z + 38, 7z + 15, z)\}$.

✔ **Now Try Exercise 43.**

Summary of Possible Cases

When matrix methods are used to solve a system of linear equations and the resulting matrix is written in diagonal form (or as close as possible to diagonal form), there are three possible cases.

1. If the number of rows with nonzero elements to the left of the vertical line is equal to the number of variables in the system, then the system has a single solution. **See Examples 1 and 2.**
2. If one of the rows has the form $[0 \;\; 0 \cdots 0 \,|\, a]$ with $a \neq 0$, then the system has no solution. **See Example 3.**
3. If there are fewer rows in the matrix containing nonzero elements than the number of variables, then the system has either no solution or infinitely many solutions. If there are infinitely many solutions, give the solutions in terms of one or more arbitrary variables. **See Example 4.**

9.2 Exercises

CONCEPT PREVIEW *Answer each question.*

1. How many rows and how many columns does this matrix have? What is its dimension?

$$\begin{bmatrix} -2 & 5 & 8 & 0 \\ 1 & 13 & -6 & 9 \end{bmatrix}$$

2. What is the element in the second row, first column of the matrix in **Exercise 1?**

3. What is the augmented matrix of the following system?

$$\begin{aligned} -3x + 5y &= 2 \\ 6x + 2y &= 7 \end{aligned}$$

4. By what number must the first row of the augmented matrix of **Exercise 3** be multiplied so that when it is added to the second row, the element in the second row, first column becomes 0?

5. What is the augmented matrix of the following system?

$$\begin{aligned} 3x + 2y &= 5 \\ -9x + 6z &= 1 \\ -8y + z &= 4 \end{aligned}$$

6. By what number must the first row of the augmented matrix of **Exercise 5** be multiplied so that when it is added to the second row, the element in the second row, first column becomes 0?

Use the given row transformation to change each matrix as indicated. ***See Example 1.***

7. $\begin{bmatrix} 1 & 4 \\ 4 & 7 \end{bmatrix}$; -4 times row 1 added to row 2

8. $\begin{bmatrix} 1 & -4 \\ 7 & 0 \end{bmatrix}$; -7 times row 1 added to row 2

9. $\begin{bmatrix} 1 & 5 & 6 \\ -2 & 3 & -1 \\ 4 & 7 & 0 \end{bmatrix}$; 2 times row 1 added to row 2

10. $\begin{bmatrix} 1 & 5 & 6 \\ -4 & -1 & 2 \\ 3 & 7 & 1 \end{bmatrix}$; 4 times row 1 added to row 2

Concept Check *Write the augmented matrix for each system and give its dimension. Do not solve.*

11. $\begin{aligned} 2x + 3y &= 11 \\ x + 2y &= 8 \end{aligned}$

12. $\begin{aligned} 3x + 5y &= -13 \\ 2x + 3y &= -9 \end{aligned}$

13. $\begin{aligned} 2x + y + z - 3 &= 0 \\ 3x - 4y + 2z + 7 &= 0 \\ x + y + z - 2 &= 0 \end{aligned}$

14. $\begin{aligned} 4x - 2y + 3z - 4 &= 0 \\ 3x + 5y + z - 7 &= 0 \\ 5x - y + 4z - 7 &= 0 \end{aligned}$

Concept Check *Write the system of equations associated with each augmented matrix. Do not solve.*

15. $\left[\begin{array}{ccc|c} 3 & 2 & 1 & 1 \\ 0 & 2 & 4 & 22 \\ -1 & -2 & 3 & 15 \end{array}\right]$

16. $\left[\begin{array}{ccc|c} 2 & 1 & 3 & 12 \\ 4 & -3 & 0 & 10 \\ 5 & 0 & -4 & -11 \end{array}\right]$

17. $\left[\begin{array}{ccc|c} 1 & 0 & 0 & 2 \\ 0 & 1 & 0 & 3 \\ 0 & 0 & 1 & -2 \end{array}\right]$

18. $\left[\begin{array}{ccc|c} 1 & 0 & 0 & 4 \\ 0 & 1 & 0 & 2 \\ 0 & 0 & 1 & 3 \end{array}\right]$

19.

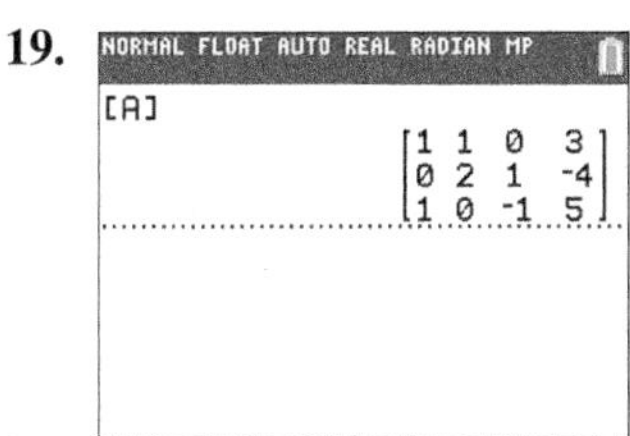

20.

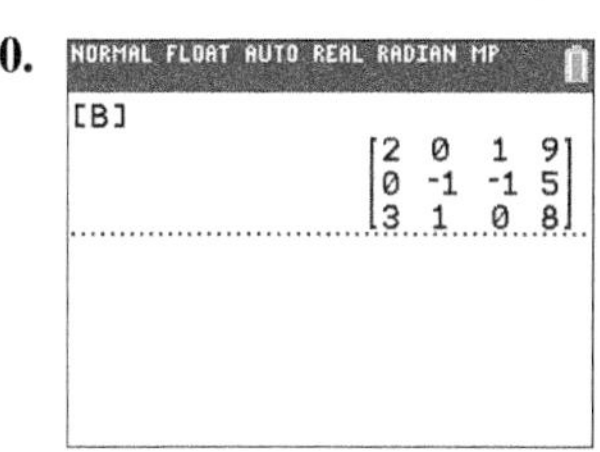

Use the Gauss-Jordan method to solve each system of equations. For systems in two variables with infinitely many solutions, give the solution with y arbitrary. For systems in three variables with infinitely many solutions, give the solution with z arbitrary. ***See Examples 1–4.***

21. $x + y = 5$
$x - y = -1$

22. $x + 2y = 5$
$2x + y = -2$

23. $3x + 2y = -9$
$2x - 5y = -6$

24. $2x - 5y = 10$
$3x + y = 15$

25. $6x - 3y - 4 = 0$
$3x + 6y - 7 = 0$

26. $2x - 3y - 10 = 0$
$2x + 2y - 5 = 0$

27. $2x - y = 6$
$4x - 2y = 0$

28. $3x - 2y = 1$
$6x - 4y = -1$

29. $\frac{3}{8}x - \frac{1}{2}y = \frac{7}{8}$
$-6x + 8y = -14$

30. $\frac{1}{2}x + \frac{3}{5}y = \frac{1}{4}$
$10x + 12y = 5$

31. $x + y - 5z = -18$
$3x - 3y + z = 6$
$x + 3y - 2z = -13$

32. $-x + 2y + 6z = 2$
$3x + 2y + 6z = 6$
$x + 4y - 3z = 1$

33. $x + y - z = 6$
$2x - y + z = -9$
$x - 2y + 3z = 1$

34. $x + 3y - 6z = 7$
$2x - y + z = 1$
$x + 2y + 2z = -1$

35. $x - z = -3$
$y + z = 9$
$x + z = 7$

36. $-x + y = -1$
$y - z = 6$
$x + z = -1$

37. $y = -2x - 2z + 1$
$x = -2y - z + 2$
$z = x - y$

38. $x = -y + 1$
$z = 2x$
$y = -2z - 2$

39. $2x - y + 3z = 0$
$x + 2y - z = 5$
$2y + z = 1$

40. $4x + 2y - 3z = 6$
$x - 4y + z = -4$
$-x + 2z = 2$

41. $3x + 5y - z + 2 = 0$
$4x - y + 2z - 1 = 0$
$-6x - 10y + 2z = 0$

42. $3x + y + 3z - 1 = 0$
$x + 2y - z - 2 = 0$
$2x - y + 4z - 4 = 0$

43. $x - 8y + z = 4$
$3x - y + 2z = -1$

44. $5x - 3y + z = 1$
$2x + y - z = 4$

45. $x - y + 2z + w = 4$
$y + z = 3$
$z - w = 2$
$x - y = 0$

46. $x + 2y + z - 3w = 7$
$y + z = 0$
$x - w = 4$
$-x + y = -3$

47. $x + 3y - 2z - w = 9$
$4x + y + z + 2w = 2$
$-3x - y + z - w = -5$
$x - y - 3z - 2w = 2$

48. $2x + y - z + 3w = 0$
$3x - 2y + z - 4w = -24$
$x + y - z + w = 2$
$x - y + 2z - 5w = -16$

Solve each system using a graphing calculator capable of performing row operations. Give solutions with values correct to the nearest thousandth.

49. $0.3x + 2.7y - \sqrt{2}z = 3$
$\sqrt{7}x - 20y + 12z = -2$
$4x + \sqrt{3}y - 1.2z = \frac{3}{4}$

50. $\sqrt{5}x - 1.2y + z = -3$
$\frac{1}{2}x - 3y + 4z = \frac{4}{3}$
$4x + 7y - 9z = \sqrt{2}$

Graph each system of three equations together on the same axes, and determine the number of solutions (exactly one, none, or infinitely many). If there is exactly one solution, estimate the solution. Then confirm the answer by solving the system using the Gauss-Jordan method.

51. $2x + 3y = 5$
$-3x + 5y = 22$
$2x + y = -1$

52. $3x - 2y = 3$
$-2x + 4y = 14$
$x + y = 11$

For each equation, determine the constants A and B that make the equation an identity. (Hint: Combine terms on the right, and set coefficients of corresponding terms in the numerators equal.)

53. $\frac{1}{(x-1)(x+1)} = \frac{A}{x-1} + \frac{B}{x+1}$

54. $\frac{x+4}{x^2} = \frac{A}{x} + \frac{B}{x^2}$

55. $\frac{x}{(x-a)(x+a)} = \frac{A}{x-a} + \frac{B}{x+a}$

56. $\frac{2x}{(x+2)(x-1)} = \frac{A}{x+2} + \frac{B}{x-1}$

Solve each problem using the Gauss-Jordan method.

57. *Daily Wages* Dan is a building contractor. If he hires 7 day laborers and 2 concrete finishers, his payroll for the day is $1384. If he hires 1 day laborer and 5 concrete finishers, his daily cost is $952. Find the daily wage for each type of worker.

58. *Mixing Nuts* At the Everglades Nut Company, 5 lb of peanuts and 6 lb of cashews cost $33.60, while 3 lb of peanuts and 7 lb of cashews cost $32.40. Find the cost of a single pound of peanuts and a single pound of cashews.

59. *Unknown Numbers* Find three numbers whose sum is 20, if the first number is three times the difference between the second and the third, and the second number is two more than twice the third.

60. *Car Sales Quota* To meet a sales quota, a car salesperson must sell 24 new cars, consisting of small, medium, and large cars. She must sell 3 more small cars than medium cars, and the same number of medium cars as large cars. How many of each size must she sell?

61. *Mixing Acid Solutions* A chemist has two prepared acid solutions, one of which is 2% acid by volume, the other 7% acid. How many cubic centimeters of each should the chemist mix together to obtain 40 cm^3 of a 3.2% acid solution?

62. *Borrowing Money* A small company took out three loans totaling $25,000. The company was able to borrow some of the money at 4% interest. It borrowed $2000 more than one-half the amount of the 4% loan at 6%, and the rest at 5%. The total annual interest was $1220. How much did the company borrow at each rate?

63. *Investing Money* An investor deposited some money at 1.5% annual interest, and two equal but larger amounts at 2.2% and 2.4%. The total amount invested was $25,000, and the total annual interest earned was $535. How much was invested at each rate?

64. *Investing Money* An investor deposited some money at 1.75% annual interest, some at 2.25%, and twice as much as the sum of the first two at 2.5%. The total amount invested was \$30,000, and the total annual interest earned was \$710. How much was invested at each rate?

65. *Planning a Diet* In a special diet for a hospital patient, the total amount per meal of food groups A, B, and C must equal 400 g. The diet should include one-third as much of group A as of group B. The sum of the amounts of group A and group C should equal twice the amount of group B. How many grams of each food group should be included? (Give answers to the nearest tenth.)

66. *Planning a Diet* In **Exercise 65,** suppose that, in addition to the conditions given there, foods A and B cost \$0.02 per gram, food C costs \$0.03 per gram, and a meal must cost \$8. Is a solution possible?

(Modeling) Age Distribution in the United States Use matrices to solve each problem. Let $x = 0$ represent 2015 and $x = 35$ represent 2050. Round values to four decimal places as necessary and percents to the nearest tenth.

67. In 2015, 14.8% of the population was 65 or older. By 2050, this percent is expected to be 20.9%. The percent of the population aged 25–39 in 2015 was 20.0%. That age group is expected to include 10.3% of the population in 2050. (*Source:* U.S. Census Bureau.)

(a) Assuming these population changes are linear, use the data for the 65-or-older age group to write a linear equation. Then do the same for the 25–39 age group.

(b) Solve the system of linear equations from part (a). In what year will the two age groups include the same percent of the population? What is that percent?

(c) Does the answer to part (b) suggest that the *number* of people in the U.S. population aged 25–39 is decreasing? Explain.

68. In 2015, 19.7% of the U.S. population was aged 40–54. This percent is expected to decrease to 18.7% in 2050. (*Source:* U.S. Census Bureau.)

(a) Write a linear equation representing this population change.

(b) Solve the system containing the equation from part (a) and the equation from **Exercise 67** for the 65-or-older age group. Give the year in which these two age groups will include the same percent of the population. What is that percent?

(Modeling) Solve each problem using matrices.

69. *Athlete's Weight and Height* The relationship between a professional basketball player's height H (in inches) and weight W (in pounds) was modeled using two different samples of players. The resulting equations that modeled the two samples were

$$W = 7.46H - 374$$

and $$W = 7.93H - 405.$$

(a) Use each equation to predict the weight of a 6 ft 11 in. professional basketball player to the nearest pound.

(b) According to each model, what change in weight, to the nearest hundredth pound, is associated with a 1-in. increase in height?

(c) Determine the weight and height, to the nearest unit, where the two models agree.

70. ***Traffic Congestion*** At rush hours, substantial traffic congestion is encountered at the traffic intersections shown in the figure. (All streets are one-way.) The city wishes to improve the signals at these corners to speed the flow of traffic. The traffic engineers first gather data. As the figure shows, 700 cars per hour come down M Street to intersection A, and 300 cars per hour come to intersection A on 10th Street. A total of x_1 of these cars leave A on M Street, while x_4 cars leave A on 10th Street. The number of cars entering A must equal the number leaving, which suggests the following equation.

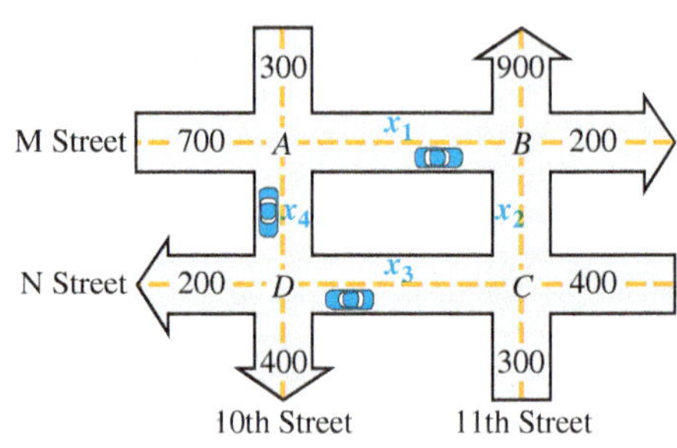

$$x_1 + x_4 = 700 + 300$$
$$x_1 + x_4 = 1000$$

For intersection B, x_1 cars enter B on M Street, and x_2 cars enter B on 11th Street. As the figure shows, 900 cars leave B on 11th Street, while 200 leave on M Street, which leads to the following equation.

$$x_1 + x_2 = 900 + 200$$
$$x_1 + x_2 = 1100$$

At intersection C, 400 cars enter on N Street and 300 on 11th Street, while x_2 cars leave on 11th Street and x_3 cars leave on N Street.

$$x_2 + x_3 = 400 + 300$$
$$x_2 + x_3 = 700$$

Finally, intersection D has x_3 cars entering on N Street and x_4 cars entering on 10th Street. There are 400 cars leaving D on 10th Street and 200 leaving on N Street.

(a) Set up an equation for intersection D.

(b) Use the four equations to write an augmented matrix, and then transform it so that 1s are on the diagonal and 0s are below. This is **triangular form.**

(c) Since there is a row of all 0s, the system of equations does not have a unique solution. Write three equations, corresponding to the three nonzero rows of the matrix. Solve each of the equations for x_4.

(d) One of the equations should have been $x_4 = 1000 - x_1$. What is the greatest possible value of x_1 so that x_4 is not negative?

(e) Another equation should have been $x_4 = x_2 - 100$. Find the least possible value of x_2 so that x_4 is not negative.

(f) Find the greatest possible values of x_3 and x_4 so that neither variable is negative.

(g) Use the results of parts (a)–(f) to give a solution for the problem in which all the equations are satisfied and all variables are nonnegative. Is the solution unique?

(Modeling) Number of Calculations *When computers are programmed to solve large linear systems involved in applications like designing aircraft or electrical circuits, they frequently use an algorithm that is similar to the Gauss-Jordan method presented in this section. Solving a linear system with n equations and n variables requires the computer to perform a total of*

$$T(n) = \frac{2}{3}n^3 + \frac{3}{2}n^2 - \frac{7}{6}n$$

arithmetic calculations, including additions, subtractions, multiplications, and divisions. Use this model to solve each problem. (*Source:* Burden, R. and J. Faires, *Numerical Analysis*, Sixth Edition, Brooks/Cole Publishing Company.)

71. Compute T for the following values of n. Write the results in a table.

$$n = 3, 6, 10, 29, 100, 200, 400, 1000, 5000, 10{,}000, 100{,}000$$

72. In 1940, John Atanasoff, a physicist from Iowa State University, wanted to solve a 29×29 linear system of equations. How many arithmetic operations would this have required? Is this too many to do by hand? (Atanasoff's work led to the invention of the first fully electronic digital computer.) (*Source: The Gazette.*)

Atanasoff-Berry Computer

73. If the number of equations and variables is doubled, does the number of arithmetic operations double?

74. Suppose that a supercomputer can execute up to 60 billion arithmetic operations per second. How many hours would be required to solve a linear system with 100,000 variables?

Relating Concepts

For individual or collaborative investigation *(Exercises 75-78)*

(Modeling) Number of Fawns *To model spring fawn count F from adult pronghorn population A, precipitation P, and severity of the winter W, environmentalists have used the equation*

$$F = a + bA + cP + dW,$$

where a, b, c, and d are constants that must be determined before using the equation. (Winter severity is scaled between 1 *and* 5, *with* 1 *being mild and* 5 *being severe.)* ***Work Exercises 75–78 in order.*** (*Source:* Brase, C. and C. Brase, *Understandable Statistics,* D.C. Heath and Company; Bureau of Land Management.)

75. Substitute the values for F, A, P, and W from the table for Years 1–4 into the equation

$$F = a + bA + cP + dW$$

and obtain four linear equations involving a, b, c, and d.

Year	Fawns	Adults	Precip. (in inches)	Winter Severity
1	239	871	11.5	3
2	234	847	12.2	2
3	192	685	10.6	5
4	343	969	14.2	1
5	?	960	12.6	3

76. Write an augmented matrix representing the system in **Exercise 75,** and solve for a, b, c, and d. Round coefficients to three decimal places.

77. Write the equation for F using the values found in **Exercise 76** for the coefficients.

78. Use the information in the table to predict the spring fawn count in Year 5. (Compare this with the actual count of 320.)

9.3 Determinant Solution of Linear Systems

- Determinants
- Cofactors
- $n \times n$ Determinants
- Determinant Theorems
- Cramer's Rule

Determinants

Every $n \times n$ matrix A is associated with a real number called the **determinant** of A, written $|A|$. In this section we show how to evaluate determinants of square matrices, providing mathematical justification as we proceed. Modern graphing calculators are programmed to evaluate determinants in their matrix menus.

The determinant of a 2×2 matrix is defined as follows.

Determinant of a 2 × 2 Matrix

$$\text{If } A = \begin{bmatrix} a_{11} & a_{12} \\ a_{21} & a_{22} \end{bmatrix}, \text{ then } |A| = \begin{vmatrix} a_{11} & a_{12} \\ a_{21} & a_{22} \end{vmatrix} = a_{11}a_{22} - a_{21}a_{12}.$$

NOTE ***Matrices are enclosed with square brackets, while determinants are denoted with vertical bars.*** A matrix is an *array* of numbers, but its determinant is a *single* number.

The arrows in the diagram below indicate which products to find when evaluating a 2×2 determinant.

$$\begin{vmatrix} a_{11} & a_{12} \\ a_{21} & a_{22} \end{vmatrix}$$

EXAMPLE 1 Evaluating a 2 × 2 Determinant

Let $A = \begin{bmatrix} -3 & 4 \\ 6 & 8 \end{bmatrix}$. Find $|A|$.

ALGEBRAIC SOLUTION

Use the definition with

$$a_{11} = -3, \quad a_{12} = 4, \quad a_{21} = 6, \quad a_{22} = 8.$$

$$|A| = \begin{vmatrix} -3 & 4 \\ 6 & 8 \end{vmatrix}$$

$$|A| = \underset{a_{11}}{-3} \cdot \underset{a_{22}}{8} - \underset{a_{21}}{6} \cdot \underset{a_{12}}{4}$$

$= -24 - 24$ Multiply.

$= -48$ Subtract.

GRAPHING CALCULATOR SOLUTION

We can define a matrix and then use the capability of a graphing calculator to find the determinant of the matrix. In the screen in **Figure 8,** the symbol det([A]) represents the determinant of [A].

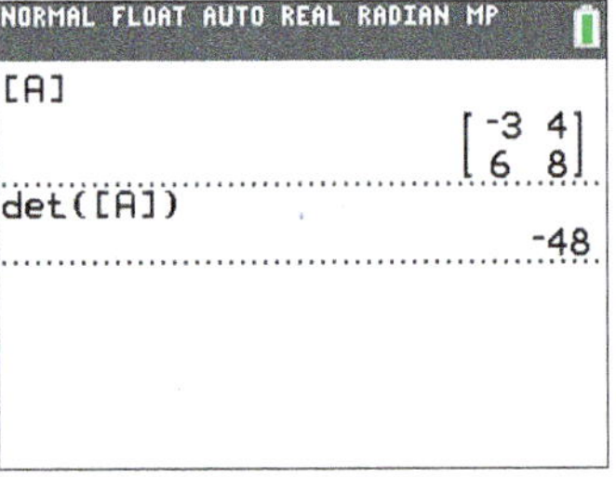

Figure 8

✔ Now Try Exercise 7.

LOOKING AHEAD TO CALCULUS

Determinants are used in calculus to find **vector cross products**, which are used to study the effect of forces in the plane or in space.

Determinant of a 3 × 3 Matrix

If $A = \begin{bmatrix} a_{11} & a_{12} & a_{13} \\ a_{21} & a_{22} & a_{23} \\ a_{31} & a_{32} & a_{33} \end{bmatrix}$, then the determinant of A, symbolized $|A|$, is defined as follows.

$$|A| = \begin{vmatrix} a_{11} & a_{12} & a_{13} \\ a_{21} & a_{22} & a_{23} \\ a_{31} & a_{32} & a_{33} \end{vmatrix} = (a_{11}a_{22}a_{33} + a_{12}a_{23}a_{31} + a_{13}a_{21}a_{32}) - (a_{31}a_{22}a_{13} + a_{32}a_{23}a_{11} + a_{33}a_{21}a_{12})$$

The terms on the right side of the equation in the definition of $|A|$ can be rearranged to obtain the following.

$$\begin{vmatrix} a_{11} & a_{12} & a_{13} \\ a_{21} & a_{22} & a_{23} \\ a_{31} & a_{32} & a_{33} \end{vmatrix} = a_{11}(a_{22}a_{33} - a_{32}a_{23}) - a_{21}(a_{12}a_{33} - a_{32}a_{13}) + a_{31}(a_{12}a_{23} - a_{22}a_{13})$$

Each quantity in parentheses represents the determinant of a 2 × 2 matrix that is the part of the 3 × 3 matrix remaining when the row and column of the multiplier are eliminated, as shown below.

$$a_{11}(a_{22}a_{33} - a_{32}a_{23}) \quad \begin{bmatrix} a_{11} & a_{12} & a_{13} \\ a_{21} & a_{22} & a_{23} \\ a_{31} & a_{32} & a_{33} \end{bmatrix}$$

$$a_{21}(a_{12}a_{33} - a_{32}a_{13}) \quad \begin{bmatrix} a_{11} & a_{12} & a_{13} \\ a_{21} & a_{22} & a_{23} \\ a_{31} & a_{32} & a_{33} \end{bmatrix}$$

$$a_{31}(a_{12}a_{23} - a_{22}a_{13}) \quad \begin{bmatrix} a_{11} & a_{12} & a_{13} \\ a_{21} & a_{22} & a_{23} \\ a_{31} & a_{32} & a_{33} \end{bmatrix}$$

Cofactors The determinant of each 2 × 2 matrix above is the **minor** of the associated element in the 3 × 3 matrix. The symbol M_{ij} represents the minor that results when row i and column j are eliminated. The following list gives some of the minors from the matrix above.

Element	Minor	Element	Minor
a_{11}	$M_{11} = \begin{vmatrix} a_{22} & a_{23} \\ a_{32} & a_{33} \end{vmatrix}$	a_{22}	$M_{22} = \begin{vmatrix} a_{11} & a_{13} \\ a_{31} & a_{33} \end{vmatrix}$
a_{21}	$M_{21} = \begin{vmatrix} a_{12} & a_{13} \\ a_{32} & a_{33} \end{vmatrix}$	a_{23}	$M_{23} = \begin{vmatrix} a_{11} & a_{12} \\ a_{31} & a_{32} \end{vmatrix}$
a_{31}	$M_{31} = \begin{vmatrix} a_{12} & a_{13} \\ a_{22} & a_{23} \end{vmatrix}$	a_{33}	$M_{33} = \begin{vmatrix} a_{11} & a_{12} \\ a_{21} & a_{22} \end{vmatrix}$

In a 4×4 matrix, the minors are determinants of 3×3 matrices. Similarly, an $n \times n$ matrix has minors that are determinants of $(n-1) \times (n-1)$ matrices.

To find the determinant of a 3×3 or larger matrix, first choose any row or column. Then the minor of each element in that row or column must be multiplied by $+1$ or -1, depending on whether the sum of the row number and column number is even or odd. The product of a minor and the number $+1$ or -1 is a *cofactor*.

Cofactor

Let M_{ij} be the minor for element a_{ij} in an $n \times n$ matrix. The **cofactor** of a_{ij}, written A_{ij}, is defined as follows.

$$A_{ij} = (-1)^{i+j} \cdot M_{ij}$$

EXAMPLE 2 Finding Cofactors of Elements

Find the cofactor of each of the following elements of the given matrix.

$$\begin{bmatrix} 6 & 2 & 4 \\ 8 & 9 & 3 \\ 1 & 2 & 0 \end{bmatrix}$$

(a) 6 **(b)** 3 **(c)** 8

SOLUTION

(a) The element 6 is in the first row, first column of the matrix, so $i = 1$ and $j = 1$. $M_{11} = \begin{vmatrix} 9 & 3 \\ 2 & 0 \end{vmatrix} = -6$. The cofactor is

$$(-1)^{1+1}(-6) = 1(-6) = -6.$$

(b) Here $i = 2$ and $j = 3$, so $M_{23} = \begin{vmatrix} 6 & 2 \\ 1 & 2 \end{vmatrix} = 10$. The cofactor is

$$(-1)^{2+3}(10) = -1(10) = -10.$$

(c) We have $i = 2$ and $j = 1$, and $M_{21} = \begin{vmatrix} 2 & 4 \\ 2 & 0 \end{vmatrix} = -8$. The cofactor is

$$(-1)^{2+1}(-8) = -1(-8) = 8.$$

✔ Now Try Exercise 17.

$n \times n$ Determinants The determinant of a 3×3 or larger matrix is found as follows.

Finding the Determinant of a Matrix

Multiply each element in any row or column of the matrix by its cofactor. The sum of these products gives the value of the determinant.

The process of forming this sum of products is called **expansion by a given row or column.**

EXAMPLE 3 Evaluating a 3 × 3 Determinant

Evaluate $\begin{vmatrix} 2 & -3 & -2 \\ -1 & -4 & -3 \\ -1 & 0 & 2 \end{vmatrix}$, expanding by the second column.

SOLUTION First find the minor of each element in the second column.

$$M_{12} = \begin{vmatrix} -1 & -3 \\ -1 & 2 \end{vmatrix} = -1(2) - (-1)(-3) = -5$$

Use parentheses, and keep track of all negative signs to avoid errors.

$$M_{22} = \begin{vmatrix} 2 & -2 \\ -1 & 2 \end{vmatrix} = 2(2) - (-1)(-2) = 2$$

$$M_{32} = \begin{vmatrix} 2 & -2 \\ -1 & -3 \end{vmatrix} = 2(-3) - (-1)(-2) = -8$$

Now find the cofactor of each element of these minors.

$$A_{12} = (-1)^{1+2} \cdot M_{12} = (-1)^3 \cdot (-5) = -1(-5) = 5$$

$$A_{22} = (-1)^{2+2} \cdot M_{22} = (-1)^4 \cdot 2 = 1 \cdot 2 = 2$$

$$A_{32} = (-1)^{3+2} \cdot M_{32} = (-1)^5 \cdot (-8) = -1(-8) = 8$$

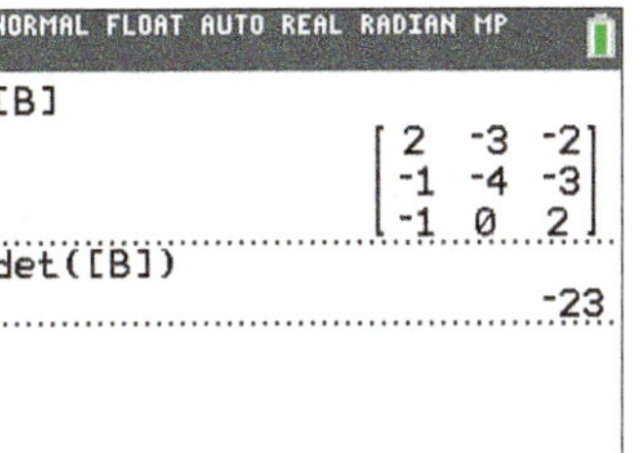

The matrix in **Example 3** can be entered as [B] and its determinant found using a graphing calculator, as shown in the screen above.

Find the determinant by multiplying each cofactor by its corresponding element in the matrix and finding the sum of these products.

$$\begin{vmatrix} 2 & -3 & -2 \\ -1 & -4 & -3 \\ -1 & 0 & 2 \end{vmatrix} = a_{12} \cdot A_{12} + a_{22} \cdot A_{22} + a_{32} \cdot A_{32}$$

$$= -3(5) + (-4)2 + 0(8)$$

$$= -15 + (-8) + 0$$

$$= -23$$

✔ **Now Try Exercise 21.**

In **Example 3,** we would have found the same answer using any row or column of the matrix. One reason we used column 2 is that it contains a 0 element, so it was not really necessary to calculate M_{32} and A_{32}.

Instead of calculating $(-1)^{i+j}$ for a given element, we can use the sign checkerboard shown below. The signs alternate for each row and column, beginning with + in the first row, first column position. If we expand a 3 × 3 matrix about row 3, for example, the first minor would have a + sign associated with it, the second minor a − sign, and the third minor a + sign.

Sign array for 3 × 3 matrices

$$\begin{matrix} + & - & + \\ - & + & - \\ + & - & + \end{matrix}$$

This sign array can be extended for determinants of larger matrices.

Determinant Theorems The following theorems are true for square matrices of any dimension and can be used to simplify finding determinants.

Determinant Theorems

1. If every element in a row (or column) of matrix A is 0, then $|A| = 0$.
2. If the rows of matrix A are the corresponding columns of matrix B, then $|B| = |A|$.
3. If any two rows (or columns) of matrix A are interchanged to form matrix B, then $|B| = -|A|$.
4. Suppose matrix B is formed by multiplying every element of a row (or column) of matrix A by the real number k. Then $|B| = k \cdot |A|$.
5. If two rows (or columns) of matrix A are identical, then $|A| = 0$.
6. Changing a row (or column) of a matrix by adding to it a constant times another row (or column) does not change the determinant of the matrix.
7. If matrix A is in triangular form, having only zeros either above or below the main diagonal, then $|A|$ is the product of the elements on the main diagonal of A.

EXAMPLE 4 Using the Determinant Theorems

Use the determinant theorems to evaluate each determinant.

(a) $\begin{vmatrix} -2 & 4 & 2 \\ 6 & 7 & 3 \\ 0 & 16 & 8 \end{vmatrix}$ **(b)** $\begin{vmatrix} 3 & -7 & 4 & 10 \\ 0 & 1 & 8 & 3 \\ 0 & 0 & -5 & 2 \\ 0 & 0 & 0 & 6 \end{vmatrix}$

SOLUTION

(a) Use determinant theorem 6 to obtain a 0 in the second row of the first column. Multiply each element in the first row by 3, and add the result to the corresponding element in the second row.

$$\begin{vmatrix} -2 & 4 & 2 \\ 0 & 19 & 9 \\ 0 & 16 & 8 \end{vmatrix} \quad 3R1 + R2$$

Now, find the determinant by expanding by column 1.

$$-2(-1)^{1+1}\begin{vmatrix} 19 & 9 \\ 16 & 8 \end{vmatrix} = -2(1)(8) = -16 \qquad 19(8) - 16(9) = 152 - 144 = 8$$

(b) Use determinant theorem 7 to find the determinant of this triangular matrix by multiplying the elements on the main diagonal.

$$\begin{vmatrix} 3 & -7 & 4 & 10 \\ 0 & 1 & 8 & 3 \\ 0 & 0 & -5 & 2 \\ 0 & 0 & 0 & 6 \end{vmatrix} = 3(1)(-5)(6) = -90$$

✔ Now Try Exercises 51 and 53.

Cramer's Rule The elimination method can be used to develop a process for solving a linear system in two unknowns using determinants. Consider the following system.

$$a_1x + b_1y = c_1 \quad (1)$$
$$a_2x + b_2y = c_2 \quad (2)$$

The variable y in this system of equations can be eliminated by using multiplication to create coefficients that are additive inverses and by adding the two equations.

$$a_1b_2x + b_1b_2y = c_1b_2 \qquad \text{Multiply (1) by } b_2.$$
$$-a_2b_1x - b_1b_2y = -c_2b_1 \qquad \text{Multiply (2) by } -b_1.$$
$$(a_1b_2 - a_2b_1)x = c_1b_2 - c_2b_1 \qquad \text{Add.}$$
$$x = \frac{c_1b_2 - c_2b_1}{a_1b_2 - a_2b_1}, \quad \text{if } a_1b_2 - a_2b_1 \neq 0$$

Similarly, the variable x can be eliminated.

$$-a_1a_2x - a_2b_1y = -a_2c_1 \qquad \text{Multiply (1) by } -a_2.$$
$$a_1a_2x + a_1b_2y = a_1c_2 \qquad \text{Multiply (2) by } a_1.$$
$$(a_1b_2 - a_2b_1)y = a_1c_2 - a_2c_1 \qquad \text{Add.}$$
$$y = \frac{a_1c_2 - a_2c_1}{a_1b_2 - a_2b_1}, \quad \text{if } a_1b_2 - a_2b_1 \neq 0$$

Both numerators and the common denominator of these values for x and y can be written as determinants.

$$c_1b_2 - c_2b_1 = \begin{vmatrix} c_1 & b_1 \\ c_2 & b_2 \end{vmatrix}, \quad a_1c_2 - a_2c_1 = \begin{vmatrix} a_1 & c_1 \\ a_2 & c_2 \end{vmatrix}, \quad \text{and} \quad a_1b_2 - a_2b_1 = \begin{vmatrix} a_1 & b_1 \\ a_2 & b_2 \end{vmatrix}$$

The solutions for x and y can be written using these determinants.

$$x = \frac{\begin{vmatrix} c_1 & b_1 \\ c_2 & b_2 \end{vmatrix}}{\begin{vmatrix} a_1 & b_1 \\ a_2 & b_2 \end{vmatrix}} \quad \text{and} \quad y = \frac{\begin{vmatrix} a_1 & c_1 \\ a_2 & c_2 \end{vmatrix}}{\begin{vmatrix} a_1 & b_1 \\ a_2 & b_2 \end{vmatrix}}, \quad \text{if} \quad \begin{vmatrix} a_1 & b_1 \\ a_2 & b_2 \end{vmatrix} \neq 0.$$

We denote the three determinants in the solution as follows.

$$\begin{vmatrix} a_1 & b_1 \\ a_2 & b_2 \end{vmatrix} = D, \quad \begin{vmatrix} c_1 & b_1 \\ c_2 & b_2 \end{vmatrix} = D_x, \quad \text{and} \quad \begin{vmatrix} a_1 & c_1 \\ a_2 & c_2 \end{vmatrix} = D_y$$

NOTE The elements of D are the four coefficients of the variables in the given system. The elements of D_x are obtained by replacing the coefficients of x in D by the respective constants, and the elements of D_y are obtained by replacing the coefficients of y in D by the respective constants.

These results are summarized as **Cramer's rule.**

Cramer's Rule for Two Equations in Two Variables

Given the system
$$a_1x + b_1y = c_1$$
$$a_2x + b_2y = c_2,$$
if $D \neq 0$, then the system has the unique solution
$$x = \frac{D_x}{D} \quad \text{and} \quad y = \frac{D_y}{D},$$
where $D = \begin{vmatrix} a_1 & b_1 \\ a_2 & b_2 \end{vmatrix}$, $D_x = \begin{vmatrix} c_1 & b_1 \\ c_2 & b_2 \end{vmatrix}$, and $D_y = \begin{vmatrix} a_1 & c_1 \\ a_2 & c_2 \end{vmatrix}$.

CAUTION Evaluate D first. ***If $D = 0$, then Cramer's rule does not apply.*** The system is inconsistent or has infinitely many solutions.

```
NORMAL FLOAT AUTO REAL RADIAN MP
[A]
                    [5 7]
                    [6 8]
[B]
                    [-1 7]
                    [ 1 8]
```

```
NORMAL FLOAT AUTO REAL RADIAN MP
[C]
                    [5 -1]
                    [6  1]
det([B])/det([A])▸Frac
                      15/2
det([C])/det([A])▸Frac
                     -11/2
```

Because graphing calculators can evaluate determinants, they can also be used to apply Cramer's rule to solve a system of linear equations. The screens above support the result in **Example 5.**

EXAMPLE 5 Applying Cramer's Rule to a 2 × 2 System

Use Cramer's rule to solve the system of equations.
$$5x + 7y = -1$$
$$6x + 8y = 1$$

SOLUTION First find D. If $D \neq 0$, then find D_x and D_y.

$$D = \begin{vmatrix} 5 & 7 \\ 6 & 8 \end{vmatrix} = 5(8) - 6(7) = -2$$

$$D_x = \begin{vmatrix} -1 & 7 \\ 1 & 8 \end{vmatrix} = -1(8) - 1(7) = -15$$

$$D_y = \begin{vmatrix} 5 & -1 \\ 6 & 1 \end{vmatrix} = 5(1) - 6(-1) = 11$$

$$x = \frac{D_x}{D} = \frac{-15}{-2} = \frac{15}{2} \quad \text{and} \quad y = \frac{D_y}{D} = \frac{11}{-2} = -\frac{11}{2} \qquad \text{Cramer's rule}$$

The solution set is $\left\{\left(\frac{15}{2}, -\frac{11}{2}\right)\right\}$. Verify by substituting in the given system.

✔ **Now Try Exercise 65.**

General Form of Cramer's Rule

Let an $n \times n$ system have linear equations of the following form.
$$a_1x_1 + a_2x_2 + a_3x_3 + \cdots + a_nx_n = b$$
Define D as the determinant of the $n \times n$ matrix of all coefficients of the variables. Define D_{x_1} as the determinant obtained from D by replacing the entries in column 1 of D with the constants of the system. Define D_{x_i} as the determinant obtained from D by replacing the entries in column i with the constants of the system. If $D \neq 0$, then the unique solution of the system is
$$x_1 = \frac{D_{x_1}}{D}, \quad x_2 = \frac{D_{x_2}}{D}, \quad x_3 = \frac{D_{x_3}}{D}, \quad \ldots, \quad x_n = \frac{D_{x_n}}{D}.$$

EXAMPLE 6 Applying Cramer's Rule to a 3 × 3 System

Use Cramer's rule to solve the system of equations.

$$\begin{aligned} x + y - z + 2 &= 0 \\ 2x - y + z + 5 &= 0 \\ x - 2y + 3z - 4 &= 0 \end{aligned}$$

SOLUTION

$$\begin{aligned} x + y - z &= -2 \\ 2x - y + z &= -5 \\ x - 2y + 3z &= 4 \end{aligned}$$

Rewrite each equation in the form $ax + by + cz + \cdots = k$.

Verify the required determinants.

$$D = \begin{vmatrix} 1 & 1 & -1 \\ 2 & -1 & 1 \\ 1 & -2 & 3 \end{vmatrix} = -3, \quad D_x = \begin{vmatrix} -2 & 1 & -1 \\ -5 & -1 & 1 \\ 4 & -2 & 3 \end{vmatrix} = 7,$$

$$D_y = \begin{vmatrix} 1 & -2 & -1 \\ 2 & -5 & 1 \\ 1 & 4 & 3 \end{vmatrix} = -22, \quad D_z = \begin{vmatrix} 1 & 1 & -2 \\ 2 & -1 & -5 \\ 1 & -2 & 4 \end{vmatrix} = -21$$

$$x = \frac{D_x}{D} = \frac{7}{-3} = -\frac{7}{3}, \quad y = \frac{D_y}{D} = \frac{-22}{-3} = \frac{22}{3}, \quad z = \frac{D_z}{D} = \frac{-21}{-3} = 7$$

The solution set is $\left\{\left(-\frac{7}{3}, \frac{22}{3}, 7\right)\right\}$.

✔ **Now Try Exercise 81.**

CAUTION As shown in **Example 6,** each equation in the system must be written in the form $ax + by + cz + \cdots = k$ before Cramer's rule is used.

EXAMPLE 7 Showing That Cramer's Rule Does Not Apply

Show that Cramer's rule does not apply to the following system.

$$\begin{aligned} 2x - 3y + 4z &= 10 \\ 6x - 9y + 12z &= 24 \\ x + 2y - 3z &= 5 \end{aligned}$$

SOLUTION We need to show that $D = 0$. Expand about column 1.

$$D = \begin{vmatrix} 2 & -3 & 4 \\ 6 & -9 & 12 \\ 1 & 2 & -3 \end{vmatrix} = 2\begin{vmatrix} -9 & 12 \\ 2 & -3 \end{vmatrix} - 6\begin{vmatrix} -3 & 4 \\ 2 & -3 \end{vmatrix} + 1\begin{vmatrix} -3 & 4 \\ -9 & 12 \end{vmatrix}$$

$$= 2(3) - 6(1) + 1(0)$$

$$= 0$$

Because $D = 0$, Cramer's rule does not apply.

✔ **Now Try Exercise 79.**

NOTE ***When $D = 0$, the system is either inconsistent or has infinitely many solutions.*** Use the elimination method to tell which is the case. Verify that the system in **Example 7** is inconsistent, and thus the solution set is ∅.

9.3 Exercises

CONCEPT PREVIEW *Answer each question.*

1. What is the value of $\begin{vmatrix} 4 & 0 \\ -2 & 0 \end{vmatrix}$?

2. What is the value of $\begin{vmatrix} 4 & 4 \\ -2 & -2 \end{vmatrix}$?

3. What expression in x represents $\begin{vmatrix} x & 4 \\ 3 & x \end{vmatrix}$?

4. What expression in x represents $\begin{vmatrix} 4 & 3 \\ x & x \end{vmatrix}$?

5. What is the value of x if $\begin{vmatrix} x & 0 \\ 0 & x \end{vmatrix} = 9$?

6. What is the value of x if $\begin{vmatrix} 0 & x \\ x & 0 \end{vmatrix} = -4$?

Evaluate each determinant. ***See Example 1.***

7. $\begin{vmatrix} -5 & 9 \\ 4 & -1 \end{vmatrix}$

8. $\begin{vmatrix} -1 & 3 \\ -2 & 9 \end{vmatrix}$

9. $\begin{vmatrix} -1 & -2 \\ 5 & 3 \end{vmatrix}$

10. $\begin{vmatrix} 6 & -4 \\ 0 & -1 \end{vmatrix}$

11. $\begin{vmatrix} 9 & 3 \\ -3 & -1 \end{vmatrix}$

12. $\begin{vmatrix} 0 & 2 \\ 1 & 5 \end{vmatrix}$

13. $\begin{vmatrix} 3 & 4 \\ 5 & -2 \end{vmatrix}$

14. $\begin{vmatrix} -9 & 7 \\ 2 & 6 \end{vmatrix}$

15. $\begin{vmatrix} -7 & 0 \\ 3 & 0 \end{vmatrix}$

16. ***Concept Check*** Refer to **Exercise 11.** Make a conjecture about the value of the determinant of a matrix in which one row is a multiple of another row.

Find the cofactor of each element in the second row of each matrix. ***See Example 2.***

17. $\begin{bmatrix} -2 & 0 & 1 \\ 1 & 2 & 0 \\ 4 & 2 & 1 \end{bmatrix}$

18. $\begin{bmatrix} 1 & -1 & 2 \\ 1 & 0 & 2 \\ 0 & -3 & 1 \end{bmatrix}$

19. $\begin{bmatrix} 1 & 2 & -1 \\ 2 & 3 & -2 \\ -1 & 4 & 1 \end{bmatrix}$

20. $\begin{bmatrix} 2 & -1 & 4 \\ 3 & 0 & 1 \\ -2 & 1 & 4 \end{bmatrix}$

Evaluate each determinant. ***See Example 3.***

21. $\begin{vmatrix} 4 & -7 & 8 \\ 2 & 1 & 3 \\ -6 & 3 & 0 \end{vmatrix}$

22. $\begin{vmatrix} 8 & -2 & -4 \\ 7 & 0 & 3 \\ 5 & -1 & 2 \end{vmatrix}$

23. $\begin{vmatrix} 1 & 2 & 0 \\ -1 & 2 & -1 \\ 0 & 1 & 4 \end{vmatrix}$

24. $\begin{vmatrix} 2 & 1 & -1 \\ 4 & 7 & -2 \\ 2 & 4 & 0 \end{vmatrix}$

25. $\begin{vmatrix} 10 & 2 & 1 \\ -1 & 4 & 3 \\ -3 & 8 & 10 \end{vmatrix}$

26. $\begin{vmatrix} 7 & -1 & 1 \\ 1 & -7 & 2 \\ -2 & 1 & 1 \end{vmatrix}$

27. $\begin{vmatrix} 1 & -2 & 3 \\ 0 & 0 & 0 \\ 1 & 10 & -12 \end{vmatrix}$

28. $\begin{vmatrix} 2 & 3 & 0 \\ 1 & 9 & 0 \\ -1 & -2 & 0 \end{vmatrix}$

29. $\begin{vmatrix} 3 & 3 & -1 \\ 2 & 6 & 0 \\ -6 & -6 & 2 \end{vmatrix}$

30. $\begin{vmatrix} 5 & -3 & 2 \\ -5 & 3 & -2 \\ 1 & 0 & 1 \end{vmatrix}$

31. $\begin{vmatrix} 1 & 0 & 0 \\ 0 & 1 & 0 \\ 0 & 0 & 1 \end{vmatrix}$

32. $\begin{vmatrix} 1 & 0 & 0 \\ 0 & -1 & 0 \\ 1 & 0 & 1 \end{vmatrix}$

33. $\begin{vmatrix} -2 & 0 & 1 \\ 0 & 1 & 0 \\ 0 & 0 & -1 \end{vmatrix}$

34. $\begin{vmatrix} 0 & 0 & -1 \\ -1 & 0 & 1 \\ 0 & -1 & 0 \end{vmatrix}$

35. $\begin{vmatrix} \sqrt{2} & 4 & 0 \\ 1 & -\sqrt{5} & 7 \\ -5 & \sqrt{5} & 1 \end{vmatrix}$

36. $\begin{vmatrix} \sqrt{3} & 1 & 0 \\ \sqrt{7} & 4 & -1 \\ 5 & 0 & -\sqrt{7} \end{vmatrix}$

37. $\begin{vmatrix} 0.4 & -0.8 & 0.6 \\ 0.3 & 0.9 & 0.7 \\ 3.1 & 4.1 & -2.8 \end{vmatrix}$

38. $\begin{vmatrix} -0.3 & -0.1 & 0.9 \\ 2.5 & 4.9 & -3.2 \\ -0.1 & 0.4 & 0.8 \end{vmatrix}$

Use the determinant theorems and the fact that $\begin{vmatrix} 1 & 2 & 3 \\ 4 & 5 & 6 \\ 7 & 9 & 10 \end{vmatrix} = 3$ *to evaluate each determinant.*

39. $\begin{vmatrix} 4 & 5 & 6 \\ 1 & 2 & 3 \\ 7 & 9 & 10 \end{vmatrix}$

40. $\begin{vmatrix} 3 & 2 & 1 \\ 6 & 5 & 4 \\ 10 & 9 & 7 \end{vmatrix}$

41. $\begin{vmatrix} 5 & 10 & 15 \\ 4 & 5 & 6 \\ 7 & 9 & 10 \end{vmatrix}$

42. $\begin{vmatrix} 1 & 20 & 3 \\ 4 & 50 & 6 \\ 7 & 90 & 10 \end{vmatrix}$

43. $\begin{vmatrix} 1 & 2 & 3 \\ 4 & 5 & 6 \\ 8 & 11 & 13 \end{vmatrix}$

44. $\begin{vmatrix} 1 & 2 & 0 \\ 4 & 5 & -6 \\ 7 & 9 & -11 \end{vmatrix}$

Use the determinant theorems to evaluate each determinant. ***See Example 4.***

45. $\begin{vmatrix} 1 & 0 & 0 \\ 1 & 0 & 1 \\ 3 & 0 & 0 \end{vmatrix}$

46. $\begin{vmatrix} -1 & 2 & 4 \\ 4 & -8 & -16 \\ 0 & 0 & 0 \end{vmatrix}$

47. $\begin{vmatrix} 6 & 8 & -12 \\ -1 & 0 & 2 \\ 4 & 0 & -8 \end{vmatrix}$

48. $\begin{vmatrix} 4 & 8 & 0 \\ -1 & -2 & 1 \\ 2 & 4 & 3 \end{vmatrix}$

49. $\begin{vmatrix} -4 & 1 & 4 \\ 2 & 0 & 1 \\ 0 & 2 & 4 \end{vmatrix}$

50. $\begin{vmatrix} 6 & 3 & 2 \\ 1 & 0 & 2 \\ 5 & 7 & 3 \end{vmatrix}$

51. $\begin{vmatrix} 0 & 1 & -3 \\ 7 & 5 & 2 \\ 1 & -2 & 6 \end{vmatrix}$

52. $\begin{vmatrix} 7 & 9 & -3 \\ 7 & -6 & 2 \\ 8 & 1 & 0 \end{vmatrix}$

53. $\begin{vmatrix} 1 & 6 & 7 \\ 0 & 6 & 7 \\ 0 & 0 & 9 \end{vmatrix}$

54. $\begin{vmatrix} 7 & 0 & 0 \\ 1 & 6 & 0 \\ 4 & 2 & 4 \end{vmatrix}$

55. $\begin{vmatrix} 2 & -1 & 3 \\ 6 & 4 & 10 \\ 4 & 5 & 7 \end{vmatrix}$

56. $\begin{vmatrix} 9 & 1 & 7 \\ 12 & 5 & 2 \\ 11 & 4 & 3 \end{vmatrix}$

57. $\begin{vmatrix} -1 & 0 & 2 & 3 \\ 5 & 4 & -3 & 7 \\ 8 & 2 & 9 & -5 \\ 4 & 4 & -1 & 10 \end{vmatrix}$

58. $\begin{vmatrix} 5 & 1 & 4 & 2 \\ 4 & -3 & 7 & -4 \\ 5 & 8 & -3 & 6 \\ 9 & 9 & 0 & 8 \end{vmatrix}$

59. $\begin{vmatrix} 4 & 0 & 0 & 2 \\ -1 & 0 & 3 & 0 \\ 2 & 4 & 0 & 1 \\ 0 & 0 & 1 & 2 \end{vmatrix}$

60. $\begin{vmatrix} -2 & 0 & 4 & 2 \\ 3 & 6 & 0 & 4 \\ 0 & 0 & 0 & 3 \\ 9 & 0 & 2 & -1 \end{vmatrix}$

61. $\begin{vmatrix} 3 & -6 & 5 & -1 \\ 0 & 2 & -1 & 3 \\ -6 & 4 & 2 & 0 \\ -7 & 3 & 1 & 1 \end{vmatrix}$

62. $\begin{vmatrix} 4 & 5 & -1 & -1 \\ 2 & -3 & 1 & 0 \\ -5 & 1 & 3 & 9 \\ 0 & -2 & 1 & 5 \end{vmatrix}$

Use Cramer's rule to solve each system of equations. If $D = 0$, then use another method to determine the solution set. ***See Examples 5–7.***

63. $x + y = 4$
$2x - y = 2$

64. $3x + 2y = -4$
$2x - y = -5$

65. $4x + 3y = -7$
$2x + 3y = -11$

66. $4x - y = 0$
$2x + 3y = 14$

67. $5x + 4y = 10$
$3x - 7y = 6$

68. $3x + 2y = -4$
$5x - y = 2$

69. $1.5x + 3y = 5$
$2x + 4y = 3$

70. $12x + 8y = 3$
$1.5x + y = 0.9$

71. $3x + 2y = 4$
$6x + 4y = 8$

72. $4x + 3y = 9$
$12x + 9y = 27$

73. $\frac{1}{2}x + \frac{1}{3}y = 2$
$\frac{3}{2}x - \frac{1}{2}y = -12$

74. $-\frac{3}{4}x + \frac{2}{3}y = 16$
$\frac{5}{2}x + \frac{1}{2}y = -37$

75. $2x - y + 4z = -2$
$3x + 2y - z = -3$
$x + 4y + 2z = 17$

76. $x + y + z = 4$
$2x - y + 3z = 4$
$4x + 2y - z = -15$

77. $x + 2y + 3z = 4$
$4x + 3y + 2z = 1$
$-x - 2y - 3z = 0$

78. $2x - y + 3z = 1$
$-2x + y - 3z = 2$
$5x - y + z = 2$

79. $-2x - 2y + 3z = 4$
$5x + 7y - z = 2$
$2x + 2y - 3z = -4$

80. $3x - 2y + 4z = 1$
$4x + y - 5z = 2$
$-6x + 4y - 8z = -2$

81. $4x - 3y + z + 1 = 0$
$5x + 7y + 2z + 2 = 0$
$3x - 5y - z - 1 = 0$

82. $2x - 3y + z - 8 = 0$
$-x - 5y + z + 4 = 0$
$3x - 5y + 2z - 12 = 0$

83. $5x - y = -4$
$3x + 2z = 4$
$4y + 3z = 22$

84. $3x + 5y = -7$
$2x + 7z = 2$
$4y + 3z = -8$

85. $x + 2y = 10$
$3x + 4z = 7$
$-y - z = 1$

86. $5x - 2y = 3$
$4y + z = 8$
$x + 2z = 4$

(Modeling) Solve each problem.

87. ***Roof Trusses*** The simplest type of roof truss is a triangle. The truss shown in the figure is used to frame roofs of small buildings. If a 100-pound force is applied at the peak of the truss, then the forces or weights W_1 and W_2 exerted parallel to each rafter of the truss are determined by the following linear system of equations.

$$\frac{\sqrt{3}}{2}(W_1 + W_2) = 100$$

$$W_1 - W_2 = 0$$

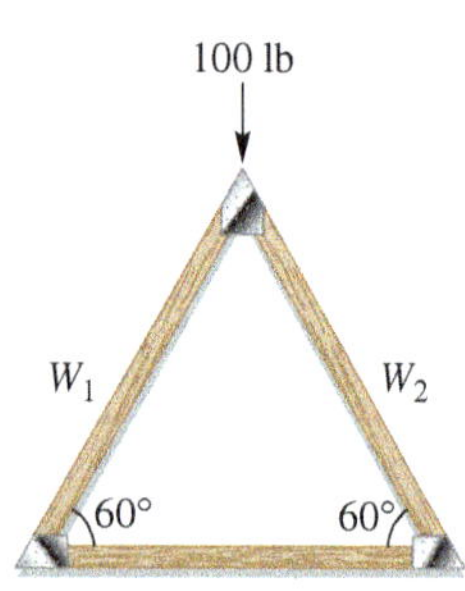

Solve the system to find W_1 and W_2. (*Source:* Hibbeler, R., *Structural Analysis,* Fourth Edition. Copyright © 1999. Reprinted by permission of Pearson Education, Inc., New York, NY.)

88. *Roof Trusses* **(Refer to Exercise 87.)** Use the following system of equations to determine the forces or weights W_1 and W_2 exerted on each rafter for the truss shown in the figure.

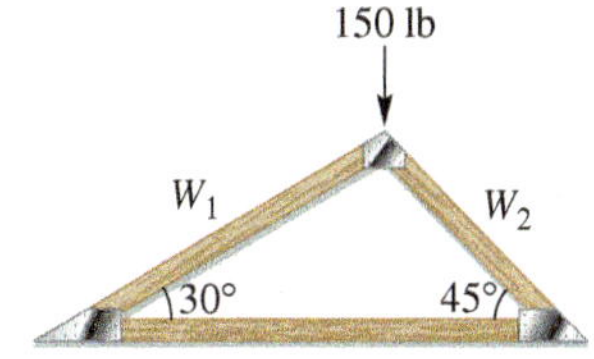

$$W_1 + \sqrt{2}W_2 = 300$$
$$\sqrt{3}W_1 - \sqrt{2}W_2 = 0$$

Area of a Triangle *A triangle with vertices at (x_1, y_1), (x_2, y_2), and (x_3, y_3), as shown in the figure, has area equal to the absolute value of D, where*

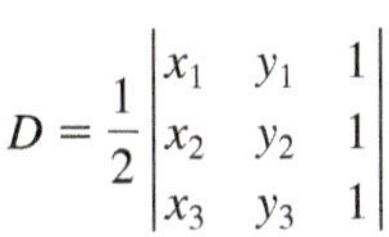

$$D = \frac{1}{2}\begin{vmatrix} x_1 & y_1 & 1 \\ x_2 & y_2 & 1 \\ x_3 & y_3 & 1 \end{vmatrix}.$$

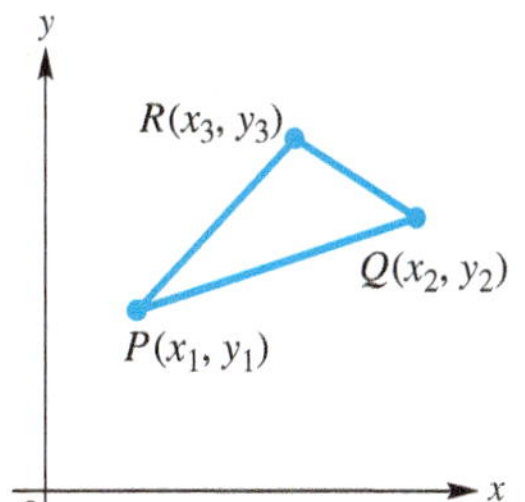

Find the area of each triangle having vertices at P, Q, and R.

89. $P(0, 0), Q(0, 2), R(1, 4)$

90. $P(0, 1), Q(2, 0), R(1, 5)$

91. $P(2, 5), Q(-1, 3), R(4, 0)$

92. $P(2, -2), Q(0, 0), R(-3, -4)$

93. *Area of a Triangle* Find the area of a triangular lot whose vertices have the following coordinates in feet. Round the answer to the nearest tenth of a foot.

$$(101.3, 52.7), \quad (117.2, 253.9), \quad \text{and} \quad (313.1, 301.6)$$

(*Source:* Al-Khafaji, A. and J. Tooley, *Numerical Methods in Engineering Practice,* Holt, Rinehart, and Winston.)

94. Let $A = \begin{bmatrix} a_{11} & a_{12} & a_{13} \\ a_{21} & a_{22} & a_{23} \\ a_{31} & a_{32} & a_{33} \end{bmatrix}$. Find $|A|$ by expansion about row 3 of the matrix. Show that the result is really equal to $|A|$ as given in the definition of the determinant of a 3×3 matrix at the beginning of this section.

To solve a ***determinant equation*** *such as*

$$\begin{vmatrix} 6 & 4 \\ -2 & x \end{vmatrix} = 2,$$

expand the determinant to obtain

$$6x - (-2)(4) = 2.$$

Then solve to obtain the solution set $\{-1\}$. Use this method to solve each equation.

95. $\begin{vmatrix} 5 & x \\ -3 & 2 \end{vmatrix} = 6$

96. $\begin{vmatrix} -0.5 & 2 \\ x & x \end{vmatrix} = 0$

97. $\begin{vmatrix} x & 3 \\ x & x \end{vmatrix} = 4$

98. $\begin{vmatrix} 2x & x \\ 11 & x \end{vmatrix} = 6$

99. $\begin{vmatrix} -2 & 0 & 1 \\ -1 & 3 & x \\ 5 & -2 & 0 \end{vmatrix} = 3$

100. $\begin{vmatrix} 4 & 3 & 0 \\ 2 & 0 & 1 \\ -3 & x & -1 \end{vmatrix} = 5$

101. $\begin{vmatrix} 5 & 3x & -3 \\ 0 & 2 & -1 \\ 4 & -1 & x \end{vmatrix} = -7$

102. $\begin{vmatrix} 2x & 1 & -1 \\ 0 & 4 & x \\ 3 & 0 & 2 \end{vmatrix} = x$

103. $\begin{vmatrix} x & 0 & -1 \\ 2 & -3 & x \\ x & 0 & 7 \end{vmatrix} = 12$

104. *Concept Check* Write the sign array representing $(-1)^{i+j}$ for each element of a 4×4 matrix.

Solve each system for x and y using Cramer's rule. Assume a and b are nonzero constants.

105. $bx + y = a^2$
$ax + y = b^2$

106. $ax + by = \frac{b}{a}$
$x + y = \frac{1}{b}$

107. $b^2x + a^2y = b^2$
$ax + by = a$

108. $x + by = b$
$ax + y = a$

109. Use Cramer's rule to find the solution set if $a, b, c, d, e,$ and f are consecutive integers.

$$ax + by = c$$
$$dx + ey = f$$

110. In the following system, $a, b, c, \ldots, l$ are consecutive integers. Express the solution set in terms of z.

$$ax + by + cz = d$$
$$ex + fy + gz = h$$
$$ix + jy + kz = l$$

Relating Concepts

For individual or collaborative investigation *(Exercises 111–114)*

The determinant of a 3×3 matrix A is defined as follows.

$$\text{If } A = \begin{bmatrix} a_{11} & a_{12} & a_{13} \\ a_{21} & a_{22} & a_{23} \\ a_{31} & a_{32} & a_{33} \end{bmatrix}, \text{ then } |A| = \begin{vmatrix} a_{11} & a_{12} & a_{13} \\ a_{21} & a_{22} & a_{23} \\ a_{31} & a_{32} & a_{33} \end{vmatrix}$$
$$= (a_{11}a_{22}a_{33} + a_{12}a_{23}a_{31} + a_{13}a_{21}a_{32}) - (a_{31}a_{22}a_{13} + a_{32}a_{23}a_{11} + a_{33}a_{21}a_{12}).$$

Work these exercises in order.

111. The determinant of a 3×3 matrix can also be found using the method of "diagonals."

Step 1 Rewrite columns 1 and 2 of matrix A to the right of matrix A.

Step 2 Identify the diagonals d_1 through d_6 and multiply their elements.

Step 3 Find the sum of the products from d_1, d_2, and d_3.

Step 4 Subtract the sum of the products from d_4, d_5, and d_6 from that sum:

$$(d_1 + d_2 + d_3) - (d_4 + d_5 + d_6).$$

Verify that this method produces the same results as the previous method given.

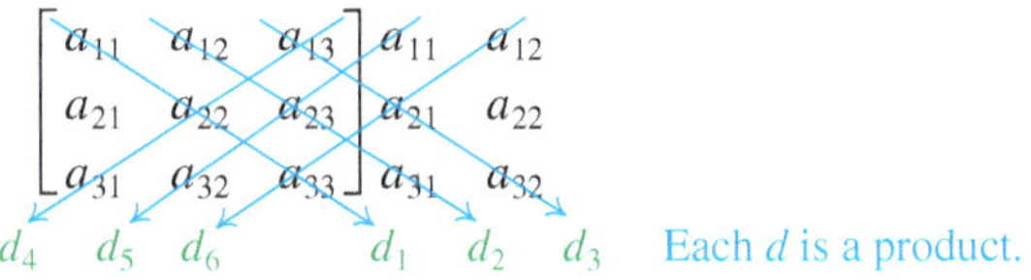

112. Evaluate the determinant $\begin{vmatrix} 1 & 3 & 2 \\ 0 & 2 & 6 \\ 7 & 1 & 5 \end{vmatrix}$ using the method of "diagonals."

113. See **Exercise 112.** Evaluate the determinant by expanding about column 1 and using the method of cofactors. Do these methods give the same determinant for 3×3 matrices?

114. *Concept Check* Does the method of evaluating a determinant using "diagonals" extend to 4×4 matrices?

9.4 Partial Fractions

- Decomposition of Rational Expressions
- Distinct Linear Factors
- Repeated Linear Factors
- Distinct Linear and Quadratic Factors
- Repeated Quadratic Factors

Decomposition of Rational Expressions

The sums of rational expressions are found by combining two or more rational expressions into one rational expression. Here, the reverse process is considered:

Given one rational expression, express it as the sum of two or more rational expressions.

A special type of sum involving rational expressions is a **partial fraction decomposition**—each term in the sum is a **partial fraction.**

Add rational expressions

$$\frac{2}{x+1} + \frac{3}{x} = \frac{5x+3}{x(x+1)}$$

Partial fraction decomposition

LOOKING AHEAD TO CALCULUS

In calculus, partial fraction decomposition provides a powerful technique for determining integrals of rational functions.

Partial Fraction Decomposition of $\frac{f(x)}{g(x)}$

To form a partial fraction decomposition of a rational expression, follow these steps.

Step 1 If $\frac{f(x)}{g(x)}$ is not a proper fraction (a fraction with the numerator of lesser degree than the denominator), divide $f(x)$ by $g(x)$. For example,

$$\frac{x^4 - 3x^3 + x^2 + 5x}{x^2 + 3} = x^2 - 3x - 2 + \frac{14x + 6}{x^2 + 3}.$$

Then apply the following steps to the remainder, which is a proper fraction.

Step 2 Factor the denominator $g(x)$ completely into factors of the form $(ax + b)^m$ or $(cx^2 + dx + e)^n$, where $cx^2 + dx + e$ is not factorable and m and n are positive integers.

Step 3 ***(a)*** For each distinct linear factor $(ax + b)$, the decomposition must include the term $\frac{A}{ax + b}$.

(b) For each repeated linear factor $(ax + b)^m$, the decomposition must include the terms

$$\frac{A_1}{ax + b} + \frac{A_2}{(ax + b)^2} + \cdots + \frac{A_m}{(ax + b)^m}.$$

Step 4 ***(a)*** For each distinct quadratic factor $(cx^2 + dx + e)$, the decomposition must include the term $\frac{Bx + C}{cx^2 + dx + e}$.

(b) For each repeated quadratic factor $(cx^2 + dx + e)^n$, the decomposition must include the terms

$$\frac{B_1x + C_1}{cx^2 + dx + e} + \frac{B_2x + C_2}{(cx^2 + dx + e)^2} + \cdots + \frac{B_nx + C_n}{(cx^2 + dx + e)^n}.$$

Step 5 Use algebraic techniques to solve for the constants in the numerators of the decomposition.

To find the constants in Step 5, the goal is to form a system of equations with as many equations as there are unknowns in the numerators.

Distinct Linear Factors

EXAMPLE 1 Finding a Partial Fraction Decomposition

Find the partial fraction decomposition of $\dfrac{2x^4 - 8x^2 + 5x - 2}{x^3 - 4x}$.

SOLUTION The given fraction is not a proper fraction—the numerator has greater degree than the denominator. Perform the division (Step 1).

$$\begin{array}{r} 2x \\ x^3 - 4x\overline{)2x^4 - 8x^2 + 5x - 2} \\ \underline{2x^4 - 8x^2 \qquad\qquad} \\ 5x - 2 \end{array}$$

The quotient is $\dfrac{2x^4 - 8x^2 + 5x - 2}{x^3 - 4x} = 2x + \dfrac{5x - 2}{x^3 - 4x}$.

Now, work with the remainder fraction. Factor the denominator (Step 2) as

$$x^3 - 4x = x(x + 2)(x - 2).$$

The factors are distinct linear factors (Step 3(a)). Write the decomposition as

$$\frac{5x - 2}{x^3 - 4x} = \frac{A}{x} + \frac{B}{x + 2} + \frac{C}{x - 2}, \quad (1)$$

where A, B, and C are constants that need to be found (Step 5). Multiply each side of equation (1) by $x(x + 2)(x - 2)$ to obtain

$$5x - 2 = A(x + 2)(x - 2) + Bx(x - 2) + Cx(x + 2). \quad (2)$$

Equation (1) is an identity because each side represents the same rational expression. Thus, equation (2) is also an identity. Equation (1) holds for all values of x except 0, -2, and 2. However, equation (2) holds for all values of x.

We can solve for A by letting $x = 0$ in equation (2).

$$5x - 2 = A(x + 2)(x - 2) + Bx(x - 2) + Cx(x + 2) \quad (2)$$

$$5(0) - 2 = A(0 + 2)(0 - 2) + B(0)(0 - 2) + C(0)(0 + 2) \quad \text{Let } x = 0.$$

$$-2 = -4A \quad \text{Simplify each term.}$$

$$A = \frac{1}{2} \quad \text{Divide by } -4.$$

Similarly, letting $x = -2$ in equation (2) enables us to solve for B.

$$5(-2) - 2 = A(-2 + 2)(-2 - 2) + B(-2)(-2 - 2) + C(-2)(-2 + 2) \quad \text{Let } x = -2 \text{ in (2).}$$

$$-12 = 8B \quad \text{Simplify each term.}$$

$$B = -\frac{3}{2} \quad \text{Divide by 8.}$$

Letting $x = 2$ gives the following for C.

$$5(2) - 2 = A(2 + 2)(2 - 2) + B(2)(2 - 2) + C(2)(2 + 2) \quad \text{Let } x = 2 \text{ in (2).}$$

$$8 = 8C \quad \text{Simplify each term.}$$

$$C = 1 \quad \text{Divide by 8.}$$

The remainder rational expression can be written as the following sum of partial fractions. Use $A = \frac{1}{2}$, $B = -\frac{3}{2}$, and $C = 1$ in equation (1).

$$\frac{5x - 2}{x^3 - 4x} = \frac{1}{2x} + \frac{-3}{2(x+2)} + \frac{1}{x-2} \qquad \frac{\frac{1}{2}}{x} = \frac{1}{2x} \text{ and } \frac{-\frac{3}{2}}{x+2} = \frac{-3}{2(x+2)}$$

The given rational expression can now be written as follows.

$$\frac{2x^4 - 8x^2 + 5x - 2}{x^3 - 4x} = 2x + \frac{1}{2x} + \frac{-3}{2(x+2)} + \frac{1}{x-2}$$

Check this result by combining the terms on the right.

✔ **Now Try Exercise 29.**

Repeated Linear Factors

EXAMPLE 2 Finding a Partial Fraction Decomposition

Find the partial fraction decomposition of $\frac{2x}{(x-1)^3}$.

SOLUTION This is a proper fraction. The denominator is already factored with repeated linear factors. Write the decomposition as shown using Step 3(b).

$$\frac{2x}{(x-1)^3} = \frac{A}{x-1} + \frac{B}{(x-1)^2} + \frac{C}{(x-1)^3}$$

Clear the denominators by multiplying each side of this equation by $(x-1)^3$.

$$2x = A(x-1)^2 + B(x-1) + C$$

Substituting 1 for x leads to $C = 2$.

$$2x = A(x-1)^2 + B(x-1) + 2 \qquad (1)$$

The only root has been substituted, and values for A and B still need to be found. However, *any* number can be substituted for x. For example, when we choose $x = -1$ (because it is easy to substitute), equation (1) becomes the following.

$$2(-1) = A(-1-1)^2 + B(-1-1) + 2 \qquad \text{Let } x = -1 \text{ in (1).}$$
$$-2 = 4A - 2B + 2 \qquad \text{Simplify each term.}$$
$$-4 = 4A - 2B \qquad \text{Subtract 2.}$$
$$-2 = 2A - B \qquad \text{Divide by 2.} \quad (2)$$

Substituting 0 for x in equation (1) gives another equation in A and B.

$$2(0) = A(0-1)^2 + B(0-1) + 2 \qquad \text{Let } x = 0 \text{ in (1).}$$
$$0 = A - B + 2 \qquad \text{Simplify each term.}$$
$$2 = -A + B \qquad \text{Subtract 2. Multiply by } -1. \quad (3)$$

Now, solve the system of equations (2) and (3) as shown in the margin to find $A = 0$ and $B = 2$. Substitute these values for A and B and 2 for C.

$$-2 = 2A - B \quad (2)$$
$$2 = -A + B \quad (3)$$
$$0 = A \quad \text{Add.}$$

If $A = 0$, then using equation (3),

$$2 = 0 + B$$
$$2 = B.$$

$$\frac{2x}{(x-1)^3} = \frac{2}{(x-1)^2} + \frac{2}{(x-1)^3} \qquad \text{Partial fraction decomposition}$$

We needed three substitutions because there were three constants to evaluate: A, B, and C. To check this result, we could combine the terms on the right.

✔ **Now Try Exercise 17.**

Distinct Linear and Quadratic Factors

EXAMPLE 3 Finding a Partial Fraction Decomposition

Find the partial fraction decomposition of $\frac{x^2 + 3x - 1}{(x+1)(x^2+2)}$.

SOLUTION The denominator $(x+1)(x^2+2)$ has distinct linear and quadratic factors, where neither is repeated. Because $x^2 + 2$ cannot be factored, it is irreducible. The partial fraction decomposition is of the following form.

$$\frac{x^2 + 3x - 1}{(x+1)(x^2+2)} = \frac{A}{x+1} + \frac{Bx + C}{x^2 + 2}$$

Multiply each side by $(x+1)(x^2+2)$.

$$x^2 + 3x - 1 = A(x^2 + 2) + (Bx + C)(x + 1) \quad (1)$$

First, substitute -1 for x.

Use parentheses around substituted values to avoid errors.

$$(-1)^2 + 3(-1) - 1 = A[(-1)^2 + 2] + [B(-1) + C](-1 + 1)$$
$$-3 = 3A$$
$$A = -1$$

Replace A with -1 in equation (1) and substitute any value for x. Let $x = 0$.

$$0^2 + 3(0) - 1 = -1(0^2 + 2) + (B \cdot 0 + C)(0 + 1)$$
$$-1 = -2 + C$$
$$C = 1$$

Now, letting $A = -1$ and $C = 1$, substitute again in equation (1), using another value for x. Let $x = 1$.

$$1^2 + 3(1) - 1 = -1(1^2 + 2) + [B(1) + 1](1 + 1)$$
$$3 = -3 + (B + 1)(2)$$
$$6 = 2B + 2$$
$$B = 2$$

Use $A = -1$, $B = 2$, and $C = 1$ to find the partial fraction decomposition.

$$\frac{x^2 + 3x - 1}{(x+1)(x^2+2)} = \frac{-1}{x+1} + \frac{2x + 1}{x^2 + 2}$$

Check by combining the terms on the right.

✔ **Now Try Exercise 25.**

For fractions with denominators that have quadratic factors, an alternative method is often more convenient. A system of equations is formed by equating coefficients of like terms on each side of the partial fraction decomposition. For instance, in **Example 3,** equation (1) was

$$x^2 + 3x - 1 = A(x^2 + 2) + (Bx + C)(x + 1). \quad (1)$$

Multiply on the right and collect like terms.

$$x^2 + 3x - 1 = Ax^2 + 2A + Bx^2 + Bx + Cx + C$$
$$1x^2 + 3x - 1 = (A + B)x^2 + (B + C)x + (C + 2A)$$

Now, equate the coefficients of like powers of x to obtain three equations.

$$1 = A + B$$
$$3 = B + C$$
$$-1 = C + 2A$$

Solving this system for A, B, and C gives the partial fraction decomposition.

Repeated Quadratic Factors

EXAMPLE 4 Finding a Partial Fraction Decomposition

Find the partial fraction decomposition of $\dfrac{2x}{(x^2+1)^2(x-1)}$.

SOLUTION This expression has both a linear factor and a repeated quadratic factor. Use Steps 3(a) and 4(b) from the box at the beginning of this section.

$$\frac{2x}{(x^2+1)^2(x-1)} = \frac{Ax+B}{x^2+1} + \frac{Cx+D}{(x^2+1)^2} + \frac{E}{x-1}$$

Multiply each side by $(x^2+1)^2(x-1)$.

$$2x = (Ax+B)(x^2+1)(x-1) + (Cx+D)(x-1) + E(x^2+1)^2 \qquad (1)$$

If $x = 1$, then equation (1) reduces to $2 = 4E$, or $E = \frac{1}{2}$. Substitute $\frac{1}{2}$ for E in equation (1), and expand and combine like terms on the right.

$$2x = Ax^4 - Ax^3 + Ax^2 - Ax + Bx^3 - Bx^2 + Bx - B$$
$$+ Cx^2 - Cx + Dx - D + \frac{1}{2}x^4 + x^2 + \frac{1}{2}$$

$$2x = \left(A + \frac{1}{2}\right)x^4 + (-A+B)x^3 + (A - B + C + 1)x^2$$
$$+ (-A + B - C + D)x + \left(-B - D + \frac{1}{2}\right) \qquad (2)$$

To obtain additional equations involving the unknowns, equate the coefficients of like powers of x on the two sides of equation (2). Setting corresponding coefficients of x^4 equal, $0 = A + \frac{1}{2}$, or $A = -\frac{1}{2}$. From the corresponding coefficients of x^3, $0 = -A + B$. Because $A = -\frac{1}{2}$, it follows that $B = -\frac{1}{2}$.

Using the coefficients of x^2, $0 = A - B + C + 1$. Since $A = -\frac{1}{2}$ and $B = -\frac{1}{2}$, it follows that $C = -1$. From the coefficients of x, $2 = -A + B - C + D$. Substituting for A, B, and C gives $D = 1$. With $A = -\frac{1}{2}$, $B = -\frac{1}{2}$, $C = -1$, $D = 1$, and $E = \frac{1}{2}$, the given fraction has partial fraction decomposition as follows.

$$\frac{2x}{(x^2+1)^2(x-1)} = \frac{-\frac{1}{2}x - \frac{1}{2}}{x^2+1} + \frac{-1x+1}{(x^2+1)^2} + \frac{\frac{1}{2}}{x-1} \qquad \text{Substitute for } A, B, C, D, \text{ and } E.$$

$$\frac{2x}{(x^2+1)^2(x-1)} = \frac{-(x+1)}{2(x^2+1)} + \frac{-x+1}{(x^2+1)^2} + \frac{1}{2(x-1)} \qquad \text{Simplify complex fractions.}$$

✔ **Now Try Exercise 31.**

In summary, to solve for the constants in the numerators of a partial fraction decomposition, use either of the following methods or a combination of the two.

Techniques for Decomposition into Partial Fractions

Method 1 **For Linear Factors**

Step 1 Multiply each side of the resulting rational equation by the common denominator.

Step 2 Substitute the zero of each factor in the resulting equation. For repeated linear factors, substitute as many other numbers as necessary to find all the constants in the numerators. The number of substitutions required will equal the number of constants $A, B, \ldots.$

Method 2 **For Quadratic Factors**

Step 1 Multiply each side of the resulting rational equation by the common denominator.

Step 2 Collect like terms on the right side of the equation.

Step 3 Equate the coefficients of like terms to form a system of equations.

Step 4 Solve the system to find the constants in the numerators.

9.4 Exercises

CONCEPT PREVIEW *Answer each question.*

1. By what expression should we multiply each side of

$$\frac{5}{3x(2x+1)} = \frac{A}{3x} + \frac{B}{2x+1}$$

so that there are no fractions in the equation?

2. In **Exercise 1,** after clearing fractions to decompose, the equation

$$A(2x+1) + B(3x) = 5$$

results. If we let $x = 0$, what is the value of A?

3. By what expression should we multiply each side of

$$\frac{3x-2}{(x+4)(3x^2+1)} = \frac{A}{x+4} + \frac{Bx+C}{3x^2+1}$$

so that there are no fractions in the equation?

4. In **Exercise 3,** after clearing fractions to decompose, the equation

$$3x - 2 = A(3x^2+1) + (Bx+C)(x+4)$$

results. If we let $x = -4$, what is the value of A?

5. By what expression should we multiply each side of

$$\frac{3x-1}{x(2x^2+1)^2} = \frac{A}{x} + \frac{Bx+C}{2x^2+1} + \frac{Dx+E}{(2x^2+1)^2}$$

so that there are no fractions in the equation?

6. In **Exercise 5,** after clearing fractions to decompose, the equation

$$3x - 1 = A(2x^2 + 1)^2 + (Bx + C)(x)(2x^2 + 1) + (Dx + E)(x)$$

results. If we let $x = 0$, what is the value of A?

Find the partial fraction decomposition for each rational expression. ***See Examples 1–4.***

7. $\dfrac{5}{3x(2x+1)}$

8. $\dfrac{3x-1}{x(x+1)}$

9. $\dfrac{4x+2}{(x+2)(2x-1)}$

10. $\dfrac{x+2}{(x+1)(x-1)}$

11. $\dfrac{x}{x^2+4x-5}$

12. $\dfrac{5x-3}{x^2-2x-3}$

13. $\dfrac{4}{x(1-x)}$

14. $\dfrac{9}{x(x-3)}$

15. $\dfrac{4x^2-x-15}{x(x+1)(x-1)}$

16. $\dfrac{3}{(x+1)(x+3)}$

17. $\dfrac{2x+1}{(x+2)^3}$

18. $\dfrac{2x}{(x+1)(x+2)^2}$

19. $\dfrac{x^2}{x^2+2x+1}$

20. $\dfrac{2}{x^2(x+3)}$

21. $\dfrac{x^3+4}{9x^3-4x}$

22. $\dfrac{x^3+2}{x^3-3x^2+2x}$

23. $\dfrac{-3}{x^2(x^2+5)}$

24. $\dfrac{1}{x^2(x^2-2)}$

25. $\dfrac{3x-2}{(x+4)(3x^2+1)}$

26. $\dfrac{2x+1}{(x+1)(x^2+2)}$

27. $\dfrac{1}{x(2x+1)(3x^2+4)}$

28. $\dfrac{3}{x(x+1)(x^2+1)}$

29. $\dfrac{2x^5+3x^4-3x^3-2x^2+x}{2x^2+5x+2}$

30. $\dfrac{6x^5+7x^4-x^2+2x}{3x^2+2x-1}$

31. $\dfrac{3x-1}{x(2x^2+1)^2}$

32. $\dfrac{x^4+1}{x(x^2+1)^2}$

33. $\dfrac{-x^4-8x^2+3x-10}{(x+2)(x^2+4)^2}$

34. $\dfrac{3x^4+x^3+5x^2-x+4}{(x-1)(x^2+1)^2}$

35. $\dfrac{5x^5+10x^4-15x^3+4x^2+13x-9}{x^3+2x^2-3x}$

36. $\dfrac{3x^6+3x^4+3x}{x^4+x^2}$

37. $\dfrac{x^2}{x^4-1}$

38. $\dfrac{-2x^2-24}{x^4-16}$

39. $\dfrac{4x^2-3x-4}{x^3+x^2-2x}$

40. $\dfrac{2x+4}{x^3-2x^2}$

Chapter 9 Quiz (Sections 9.1-9.4)

Solve each system, using the method indicated, if possible.

1. (Substitution)
$$2x + y = -4$$
$$-x + 2y = 2$$

2. (Substitution)
$$5x + 10y = 10$$
$$x + 2y = 2$$

3. (Elimination)
$$x - y = 6$$
$$x - y = 4$$

4. (Elimination)
$$2x - 3y = 18$$
$$5x + 2y = 7$$

5. (Gauss-Jordan)
$$3x + 5y = -5$$
$$-2x + 3y = 16$$

6. (Cramer's rule)
$$5x + 2y = -3$$
$$4x - 3y = -30$$

7. (Elimination)
$$x + y + z = 1$$
$$-x + y + z = 5$$
$$y + 2z = 5$$

8. (Gauss-Jordan)
$$2x + 4y + 4z = 4$$
$$x + 3y + z = 4$$
$$-x + 3y + 2z = -1$$

9. (Cramer's rule)
$$7x + y - z = 4$$
$$2x - 3y + z = 2$$
$$-6x + 9y - 3z = -6$$

Solve each problem.

10. ***Spending on Food*** In 2013, the amount spent by a typical American household on food was about \$6602. For every \$10 spent on food away from home, about \$15 was spent on food at home. Find the amount of household spending on food in each category. (*Source:* U.S. Bureau of Labor Statistics.)

11. ***Investments*** A sum of \$5000 is invested in three accounts that pay 2%, 3%, and 4% interest rates. The amount of money invested in the account paying 4% equals the total amount of money invested in the other two accounts, and the total annual interest from all three investments is \$165. Find the amount invested at each rate.

12. Let $A = \begin{bmatrix} -5 & 4 \\ 2 & -1 \end{bmatrix}$. Find $|A|$.

13. Evaluate $\begin{vmatrix} -1 & 2 & 4 \\ -3 & -2 & -3 \\ 2 & -1 & 5 \end{vmatrix}$. Use determinant theorems if desired.

Find the partial fraction decomposition for each rational expression.

14. $\dfrac{10x + 13}{x^2 - x - 20}$

15. $\dfrac{2x^2 - 15x - 32}{(x - 1)(x^2 + 6x + 8)}$

9.5 Nonlinear Systems of Equations

- Nonlinear Systems with Real Solutions
- Nonlinear Systems with Nonreal Complex Solutions
- An Application of Nonlinear Systems

Nonlinear Systems with Real Solutions

A system of equations in which at least one equation is *not* linear is a **nonlinear system.**

$$\begin{aligned} x^2 - y &= 4 \quad (1) \\ x + y &= -2 \quad (2) \end{aligned} \qquad \begin{aligned} x^2 + y^2 &= 16 \quad (1) \\ |x| + y &= 4 \quad (2) \end{aligned} \qquad \text{Nonlinear systems}$$

The substitution method works well for solving many such systems, particularly when one of the equations is linear, as in the next example.

EXAMPLE 1 Solving a Nonlinear System (Substitution Method)

Solve the system.

$$\begin{aligned} x^2 - y &= 4 \quad (1) \\ x + y &= -2 \quad (2) \end{aligned}$$

ALGEBRAIC SOLUTION

When one of the equations in a nonlinear system is linear, it is usually best to begin by solving the linear equation for one of the variables.

$$y = -2 - x \quad \text{Solve equation (2) for } y.$$

Substitute this result for y in equation (1).

$$\begin{aligned} x^2 - y &= 4 && (1) \\ x^2 - (-2 - x) &= 4 && \text{Let } y = -2 - x. \\ x^2 + 2 + x &= 4 && \text{Distributive property} \\ x^2 + x - 2 &= 0 && \text{Standard form} \\ (x + 2)(x - 1) &= 0 && \text{Factor.} \\ x + 2 = 0 \quad \text{or} \quad x - 1 &= 0 && \text{Zero-factor property} \\ x = -2 \quad \text{or} \quad x &= 1 && \text{Solve each equation.} \end{aligned}$$

Substituting -2 for x in equation (2) gives $y = 0$. If $x = 1$, then $y = -3$. The solution set of the given system is $\{(-2, 0), (1, -3)\}$. A graph of the system is shown in **Figure 9.**

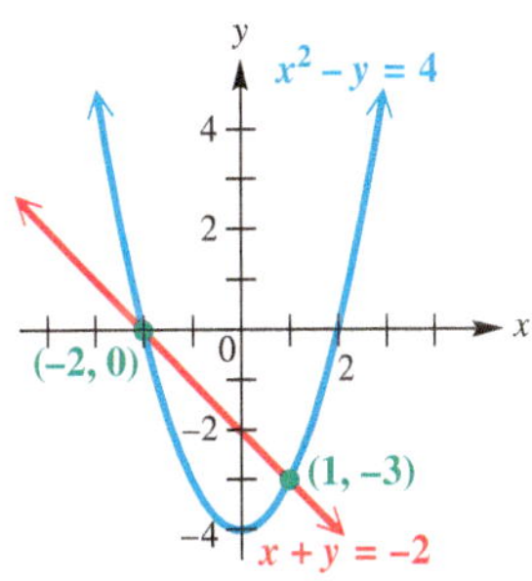

Figure 9

GRAPHING CALCULATOR SOLUTION

Solve each equation for y and graph them in the same viewing window. We obtain

$$y_1 = x^2 - 4 \quad \text{and} \quad y_2 = -x - 2.$$

The screens in **Figure 10,** which indicate that the points of intersection are

$$(-2, 0) \quad \text{and} \quad (1, -3),$$

support the solution found algebraically.

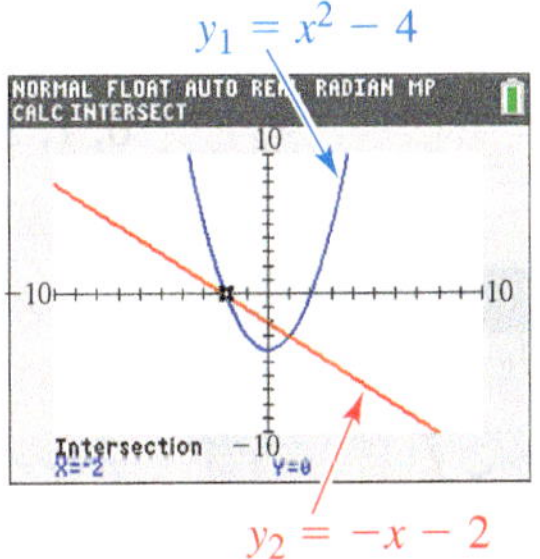

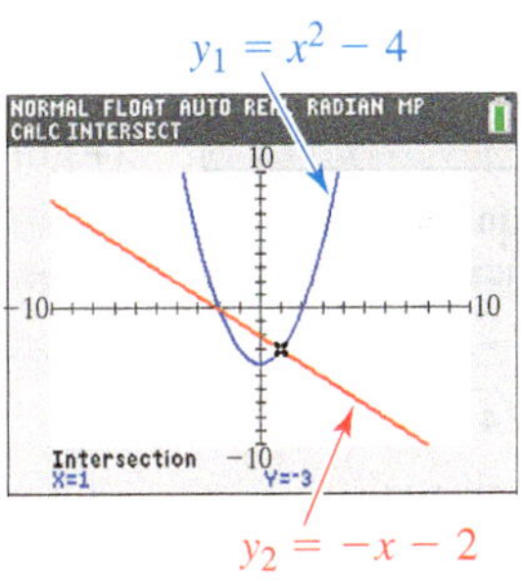

Figure 10

✔ **Now Try Exercise 15.**

The two equations form this system.

$$x^2y = 75 \quad (1)$$
$$x^2 + 4xy = 85 \quad (2)$$

Step 4 **Solve** the system. We solve equation (1) for y to obtain $y = \frac{75}{x^2}$.

$$x^2 + 4xy = 85 \quad (2)$$
$$x^2 + 4x\left(\frac{75}{x^2}\right) = 85 \quad \text{Let } y = \tfrac{75}{x^2}.$$
$$x^2 + \frac{300}{x} = 85 \quad \text{Multiply.}$$
$$x^3 + 300 = 85x \quad \text{Multiply by } x\text{, where } x \neq 0.$$
$$x^3 - 85x + 300 = 0 \quad \text{Subtract } 85x.$$

5)1	0	−85	300
	5	25	−300
1	5	−60	0

Coefficients of a quadratic polynomial factor

We are restricted to positive values for x, and considering the nature of the problem, any solution should be relatively small. By the rational zeros theorem, factors of 300 are the only possible rational solutions. Using synthetic division, as shown in the margin, we see that 5 is a solution. Therefore, one value of x is 5, and $y = \frac{75}{5^2} = 3$. We must now solve

$$x^2 + 5x - 60 = 0$$

for any other possible positive solutions. Use the quadratic formula to find the positive solution.

$$x = \frac{-5 + \sqrt{5^2 - 4(1)(-60)}}{2(1)} \approx 5.639 \quad \text{Quadratic formula with } a = 1,\ b = 5,\ c = -60$$

This value of x leads to $y \approx 2.359$.

Step 5 **State the answer.** There are two possible answers.

First answer: length = width = 5 in.; height = 3 in.

Second answer: length = width ≈ 5.639 in.; height ≈ 2.359 in.

Step 6 **Check.** See **Exercise 69.**

✔ **Now Try Exercises 67 and 69.**

9.5 Exercises

CONCEPT PREVIEW *Answer each of the following. When appropriate, fill in the blank to correctly complete the sentence.*

1. The following nonlinear system has two solutions, one of which is (3, ___).

$$x + y = 7$$
$$x^2 + y^2 = 25$$

2. The following nonlinear system has two solutions with real components, one of which is (2, ___).

$$y = x^2 + 6$$
$$x^2 - y^2 = -96$$

3. The following nonlinear system has two solutions, one of which is (___, 3).

$$2x + y = 1$$
$$x^2 + y^2 = 10$$

4. Refer to the system in **Exercise 2.** The other solution with real components has x-value -2. What is the y-value of this solution?

5. If we want to solve the following nonlinear system by substitution and we decide to solve equation (2) for y, what will be the resulting equation when the substitution is made into equation (1)?

$$x^2 + y = 2 \quad (1)$$
$$x - y = 0 \quad (2)$$

6. If we want to solve the following nonlinear system by eliminating the y^2 terms, by what number should we multiply equation (2)?

$$x^2 + 3y^2 = 4 \quad (1)$$
$$x^2 - y^2 = 0 \quad (2)$$

Concept Check *A nonlinear system is given, along with the graphs of both equations in the system. Verify that the points of intersection specified on the graph are solutions of the system by substituting directly into both equations.*

7. $x^2 = y - 1$
$y = 3x + 5$

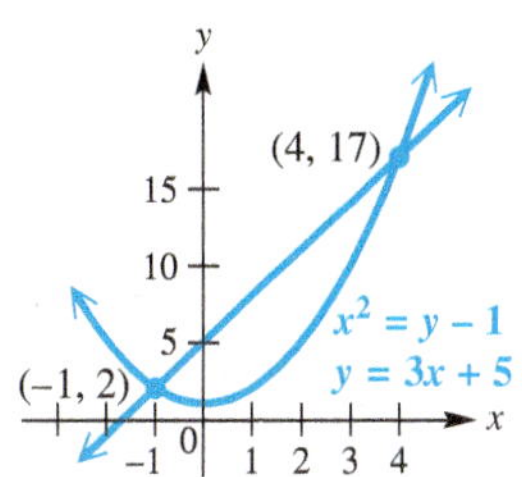

8. $2x^2 = 3y + 23$
$y = 2x - 5$

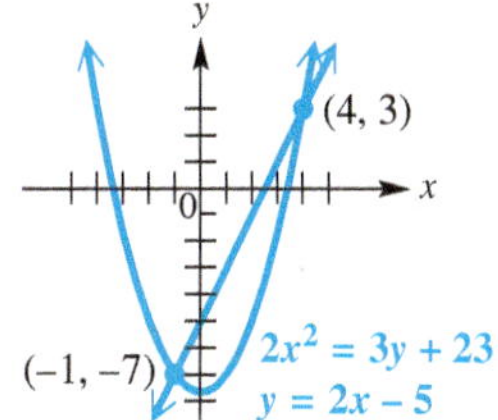

9. $x^2 + y^2 = 5$
$-3x + 4y = 2$

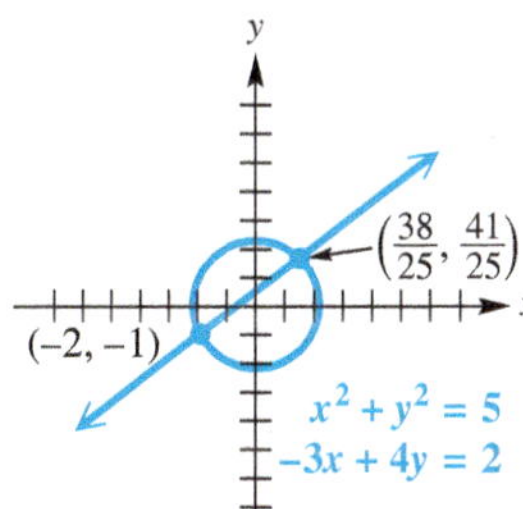

10. $x + y = -3$
$x^2 + y^2 = 45$

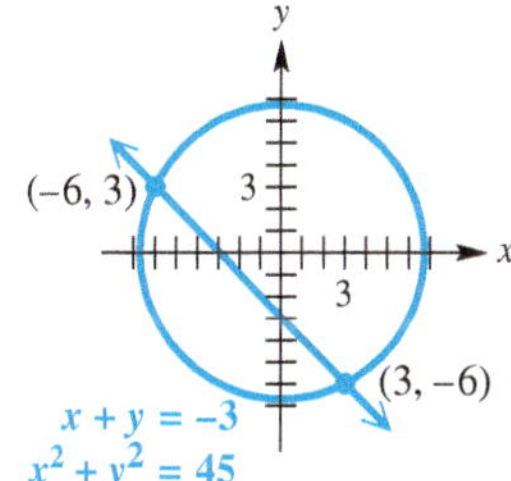

11. $y = 3x^2$
$x^2 + y^2 = 10$

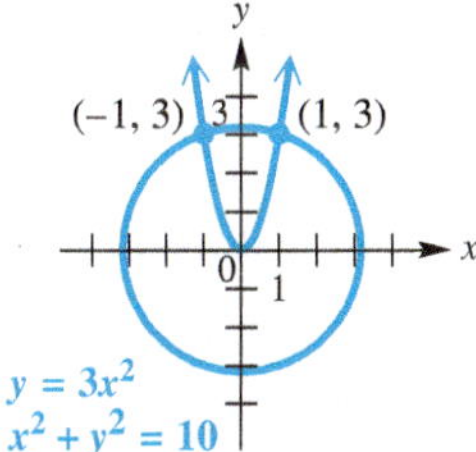

12. $y = -\frac{4}{9}x^2$
$x^2 + y^2 = 25$

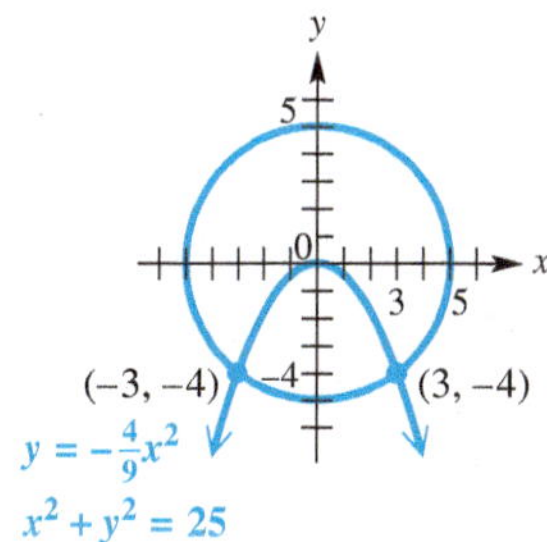

Concept Check *Answer each question.*

13. In **Example 1,** we solved the following system. How can we tell, before doing any work, that this system cannot have more than two solutions?

$$x^2 - y = 4$$
$$x + y = -2$$

14. In **Example 5,** there were four solutions to the system, but there were no points of intersection of the graphs. If a nonlinear system has nonreal complex numbers as components of its solutions, will they appear as intersection points of the graphs?

Solve each nonlinear system of equations. Give all solutions, including those with nonreal complex components. ***See Examples 1–5.***

15. $x^2 - y = 0$, $x + y = 2$

16. $x^2 + y = 2$, $x - y = 0$

17. $y = x^2 - 2x + 1$, $x - 3y = -1$

18. $y = x^2 + 6x + 9$, $x + 2y = -2$

19. $y = x^2 + 4x$, $2x - y = -8$

20. $y = 6x + x^2$, $4x - y = -3$

21. $3x^2 + 2y^2 = 5$, $x - y = -2$

22. $x^2 + y^2 = 5$, $-3x + 4y = 2$

23. $x^2 + y^2 = 8$, $x^2 - y^2 = 0$

24. $x^2 + y^2 = 10$, $2x^2 - y^2 = 17$

25. $5x^2 - y^2 = 0$, $3x^2 + 4y^2 = 0$

26. $x^2 + y^2 = 0$, $2x^2 - 3y^2 = 0$

27. $3x^2 + y^2 = 3$, $4x^2 + 5y^2 = 26$

28. $x^2 + 2y^2 = 9$, $x^2 + y^2 = 25$

29. $2x^2 + 3y^2 = 5$, $3x^2 - 4y^2 = -1$

30. $3x^2 + 5y^2 = 17$, $2x^2 - 3y^2 = 5$

31. $2x^2 + 2y^2 = 20$, $4x^2 + 4y^2 = 30$

32. $x^2 + y^2 = 4$, $5x^2 + 5y^2 = 28$

33. $2x^2 - 3y^2 = 12$, $6x^2 + 5y^2 = 36$

34. $5x^2 - 2y^2 = 25$, $10x^2 + y^2 = 50$

35. $xy = -15$, $4x + 3y = 3$

36. $xy = 8$, $3x + 2y = -16$

37. $2xy + 1 = 0$, $x + 16y = 2$

38. $-5xy + 2 = 0$, $x - 15y = 5$

39. $3x^2 - y^2 = 11$, $xy = 12$

40. $5x^2 - 2y^2 = 6$, $xy = 2$

41. $x^2 - xy + y^2 = 5$, $2x^2 + xy - y^2 = 10$

42. $3x^2 + 2xy - y^2 = 9$, $x^2 - xy + y^2 = 9$

43. $x^2 + 2xy - y^2 = 14$, $x^2 - y^2 = -16$

44. $x^2 + 3xy - y^2 = 12$, $x^2 - y^2 = -12$

45. $x^2 + y^2 = 25$, $|x| - y = 5$

46. $x^2 + y^2 = 9$, $|x| + y = 3$

47. $x = |y|$, $x^2 + y^2 = 18$

48. $2x + |y| = 4$, $x^2 + y^2 = 5$

49. $2x^2 - y^2 = 4$, $|x| = |y|$

50. $x^2 + y^2 = 9$, $|x| = |y|$

Many nonlinear systems cannot be solved algebraically, so graphical analysis is the only way to determine the solutions of such systems. Use a graphing calculator to solve each nonlinear system. Give x- and y-coordinates to the nearest hundredth.

51. $y = \log(x + 5)$, $y = x^2$

52. $y = 5^x$, $xy = 1$

53. $y = e^{x+1}$, $2x + y = 3$

54. $y = \sqrt[3]{x - 4}$, $x^2 + y^2 = 6$

Solve each problem using a system of equations in two variables. ***See Example 6.***

55. *Unknown Numbers* Find two numbers whose sum is −17 and whose product is 42.

56. *Unknown Numbers* Find two numbers whose sum is −10 and whose squares differ by 20.

57. *Unknown Numbers* Find two numbers whose squares have a sum of 100 and a difference of 28.

58. *Unknown Numbers* Find two numbers whose squares have a sum of 194 and a difference of 144.

59. *Unknown Numbers* Find two numbers whose ratio is 9 to 2 and whose product is 162.

60. *Unknown Numbers* Find two numbers whose ratio is 4 to 3 and are such that the sum of their squares is 100.

61. *Triangle Dimensions* The longest side of a right triangle is 13 m in length. One of the other sides is 7 m longer than the shortest side. Find the lengths of the two shorter sides of the triangle.

62. *Triangle Dimensions* The longest side of a right triangle is 29 ft in length. One of the other two sides is 1 ft longer than the shortest side. Find the lengths of the two shorter sides of the triangle.

Answer each question.

63. Does the straight line $3x - 2y = 9$ intersect the circle $x^2 + y^2 = 25$? (*Hint:* To find out, solve the system formed by these two equations.)

64. For what value(s) of b will the line $x + 2y = b$ touch the circle $x^2 + y^2 = 9$ in only one point?

65. A line passes through the points of intersection of the graphs of $y = x^2$ and $x^2 + y^2 = 90$. What is the equation of this line?

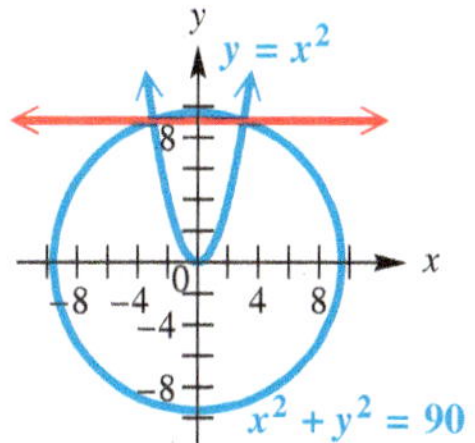

66. Suppose we are given the equations of two circles that are known to intersect in exactly two points. How would we find the equation of the only chord common to these circles?

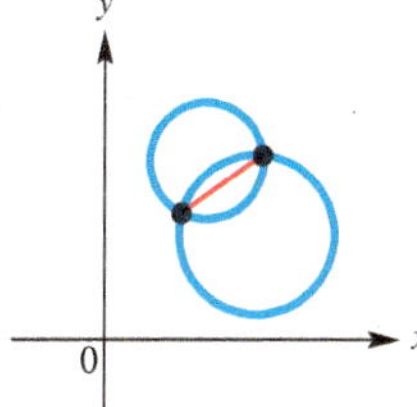

Solve each problem.

67. *Dimensions of a Box* A box with an open top has a square base and four sides of equal height. The volume of the box is 360 ft³. If the surface area is 276 ft², find the dimensions of the box. (Round answers to the nearest thousandth, if necessary.)

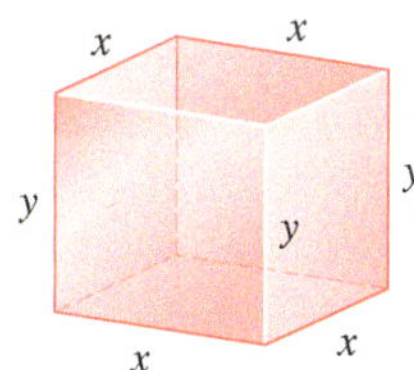

68. *Dimensions of a Cylinder* Find the radius and height (to the nearest thousandth) of an open-ended cylinder with volume 50 in.³ and lateral surface area 65 in.².

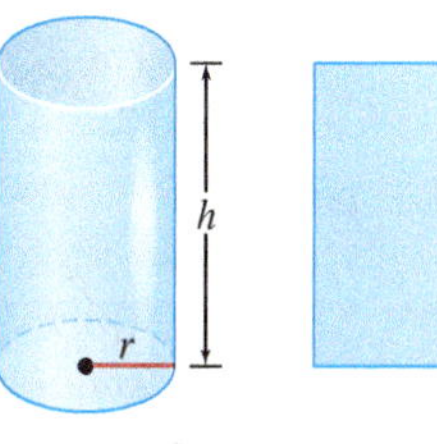

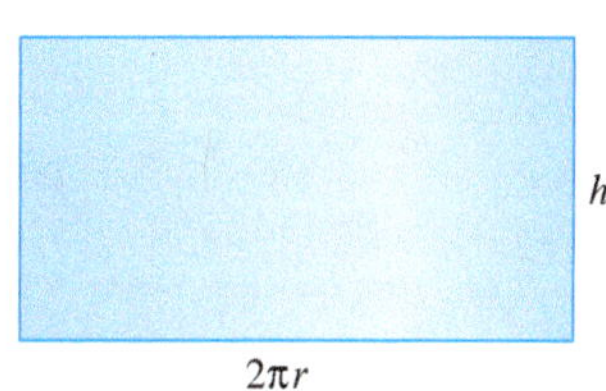

69. *Checking Answers* Check the two answers in **Example 6.**

70. *(Modeling) Equilibrium Demand and Price* The supply and demand equations for a certain commodity are given.

$$\text{supply: } p = \frac{2000}{2000 - q} \quad \text{and} \quad \text{demand: } p = \frac{7000 - 3q}{2q}$$

(a) Find the equilibrium demand.

(b) Find the equilibrium price (in dollars).

71. *(Modeling) Equilibrium Demand and Price* The supply and demand equations for a certain commodity are given.

$$\text{supply: } p = \sqrt{0.1q + 9} - 2 \quad \text{and} \quad \text{demand: } p = \sqrt{25 - 0.1q}$$

(a) Find the equilibrium demand. **(b)** Find the equilibrium price (in dollars).

72. *(Modeling) Circuit Gain* In electronics, circuit gain is modeled by

$$G = \frac{Bt}{R + R_t},$$

where R is the value of a resistor, t is temperature, R_t is the value of R at temperature t, and B is a constant. The sensitivity of the circuit to temperature is modeled by

$$S = \frac{BR}{(R + R_t)^2}.$$

If $B = 3.7$ and t is 90 K (kelvins), find the values of R and R_t that will result in the values $G = 0.4$ and $S = 0.001$. Round answers to the nearest whole number.

73. *(Modeling) Revenue for Public Colleges* The percents of revenue for public colleges from state sources and tuition are modeled in the accompanying graph.

(a) Interpret this graph. How are the sources of funding for public colleges changing with time?

(b) During what time period was the revenue from state sources increasing?

(c) Use the graph to estimate the year and the percent when the amounts from both sources were equal.

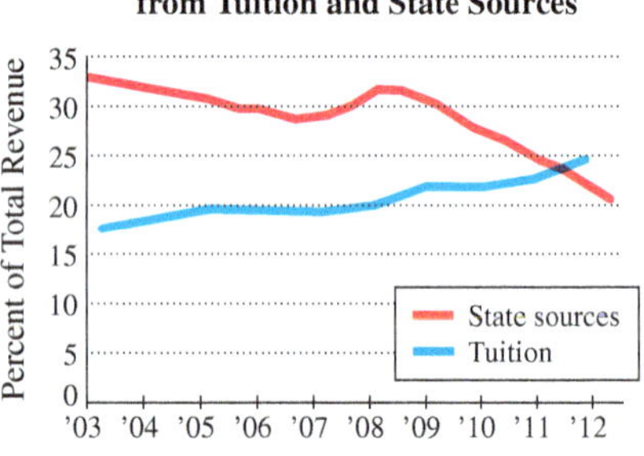

Source: Government Accountability Office.

74. *(Modeling) Revenue for Public Colleges* The following equations model the percents of revenue from both sources in **Exercise 73.** Use the equations to determine the year and percent when the amounts from both sources were equal.

$$S = 32(0.9637)^{x-2003} \quad \text{State sources}$$

$$T = 17(1.0438)^{x-2003} \quad \text{Tuition}$$

Relating Concepts

For individual or collaborative investigation *(Exercises 75–80)*

Consider the following nonlinear system.

$$y = |x - 1|$$

$$y = x^2 - 4$$

Work Exercises 75–80 in order, *to see how concepts of graphing are related to the solutions of this system.*

75. How is the graph of $y = |x - 1|$ obtained by transforming the graph of $y = |x|$?

76. How is the graph of $y = x^2 - 4$ obtained by transforming the graph of $y = x^2$?

77. Use the definition of absolute value to write $y = |x - 1|$ as a piecewise-defined function.

78. Write two quadratic equations that will be used to solve the system. (*Hint:* Set both parts of the piecewise-defined function in **Exercise 77** equal to $x^2 - 4$.)

79. Use the quadratic formula to solve both equations from **Exercise 78.** Pay close attention to the restriction on x.

80. Use the values of x found in **Exercise 79** to find the solution set of the system.

Summary Exercises on Systems of Equations

This chapter has introduced methods for solving systems of equations, including substitution and elimination, and matrix methods such as the Gauss-Jordan method and Cramer's rule. Use each method at least once when solving the systems below. Include solutions with nonreal complex number components. For systems with infinitely many solutions, write the solution set using an arbitrary variable.

1. $2x + 5y = 4$
$3x - 2y = -13$

2. $x - 3y = 7$
$-3x + 4y = -1$

3. $2x^2 + y^2 = 5$
$3x^2 + 2y^2 = 10$

4. $2x - 3y = -2$
$x + y = -16$
$3x - 2y + z = 7$

5. $6x - y = 5$
$xy = 4$

6. $4x + 2z = -12$
$x + y - 3z = 13$
$-3x + y - 2z = 13$

7. $x + 2y + z = 5$
$y + 3z = 9$

8. $x - 4 = 3y$
$x^2 + y^2 = 8$

9. $3x + 6y - 9z = 1$
$2x + 4y - 6z = 1$
$3x + 4y + 5z = 0$

10. $x + 2y + z = 0$
$x + 2y - z = 6$
$2x - y = -9$

11. $x^2 + y^2 = 4$
$y = x + 6$

12. $x + 5y = -23$
$4y - 3z = -29$
$2x + 5z = 19$

13. $y + 1 = x^2 + 2x$
$y + 2x = 4$

14. $3x + 6z = -3$
$y - z = 3$
$2x + 4z = -1$

15. $2x + 3y + 4z = 3$
$-4x + 2y - 6z = 2$
$4x + 3z = 0$

16. $\frac{3}{x} + \frac{4}{y} = 4$
$\frac{1}{x} + \frac{2}{y} = \frac{2}{3}$

17. $-5x + 2y + z = 5$
$-3x - 2y - z = 3$
$-x + 6y = 1$

18. $x + 5y + 3z = 9$
$2x + 9y + 7z = 5$

19. $2x^2 + y^2 = 9$
$3x - 2y = -6$

20. $2x - 4y - 6 = 0$
$-x + 2y + 3 = 0$

21. $x + y - z = 0$
$2y - z = 1$
$2x + 3y - 4z = -4$

22. $2y = 3x - x^2$
$x + 2y = 12$

23. $2x - 3y = -2$
$x + y - 4z = -16$
$3x - 2y + z = 7$

24. $x - y + 3z = 3$
$-2x + 3y - 11z = -4$
$x - 2y + 8z = 6$

25. $x^2 + 3y^2 = 28$
$y - x = -2$

26. $3x - y = -2$
$y + 5z = -4$
$-2x + 3y - z = -8$

27. $2x^2 + 3y^2 = 20$
$x^2 + 4y^2 = 5$

28. $x + y + z = -1$
$2x + 3y + 2z = 3$
$2x + y + 2z = -7$

29. $x + 2z = 9$
$y + z = 1$
$3x - 2y = 9$

30. $x^2 - y^2 = 15$
$x - 2y = 2$

31. $-x + y = -1$
$x + z = 4$
$6x - 3y + 2z = 10$

32. $2x - y - 2z = -1$
$-x + 2y + 13z = 12$
$3x + 9z = 6$

33. $xy = -3$
$x + y = -2$

34. $-3x + 2z = 1$
$4x + y - 2z = -6$
$x + y + 4z = 3$

35. $y = x^2 + 6x + 9$
$x + y = 3$

36. $5x - 2z = 8$
$4y + 3z = -9$
$\frac{1}{2}x + \frac{2}{3}y = -1$

9.6 Systems of Inequalities and Linear Programming

- Linear Inequalities in Two Variables
- Systems of Inequalities
- Linear Programming

Linear Inequalities in Two Variables A line divides a plane into three sets of points: the points of the line itself and the points belonging to the two regions determined by the line. Each of these two regions is a **half-plane.** In **Figure 14,** line r divides the plane into three different sets of points: line r, half-plane P, and half-plane Q. The points on r belong neither to P nor to Q. Line r is the **boundary** of each half-plane.

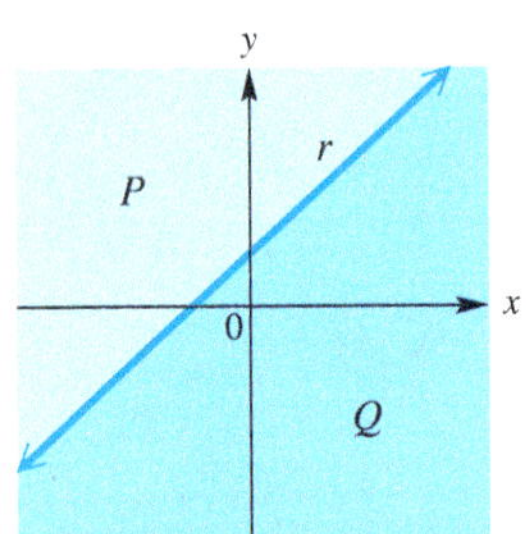

Figure 14

Linear Inequality in Two Variables

A **linear inequality in two variables** is an inequality of the form

$$Ax + By \le C,$$

where A, B, and C are real numbers, with A and B not both equal to 0. (The symbol $\le$ could be replaced with $\ge$, $<$, or $>$.)

The graph of a linear inequality is a half-plane, perhaps with its boundary.

EXAMPLE 1 Graphing a Linear Inequality

Graph $3x - 2y \le 6$.

SOLUTION First graph the boundary, $3x - 2y = 6$, as shown in **Figure 15.** Because the points of the line $3x - 2y = 6$ satisfy $3x - 2y \le 6$, this line is part of the solution set. To decide which half-plane (the one above the line $3x - 2y = 6$ or the one below the line) is part of the solution set, solve the original inequality for y.

$$3x - 2y \le 6$$

$$-2y \le -3x + 6 \quad \text{Subtract } 3x.$$

$$y \ge \frac{3}{2}x - 3 \quad \text{Divide by } -2, \text{ and change } \le \text{ to } \ge.$$

Reverse the inequality symbol when dividing by a negative number.

For a particular value of x, the inequality will be satisfied by all values of y that are *greater than or equal to* $\frac{3}{2}x - 3$. Thus, the solution set contains the half-plane *above* the line, as shown in **Figure 16.**

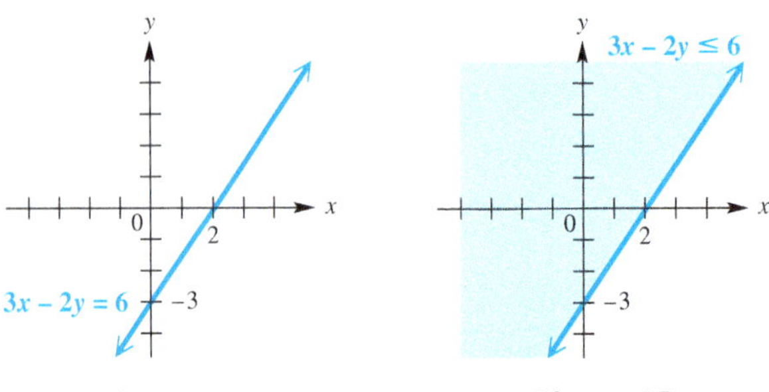

Figure 15 **Figure 16**

Coordinates for x and y from the solution set (the shaded region) satisfy the original inequality, while coordinates outside the solution set do not.

✔ Now Try Exercise 11.

CAUTION ***A linear inequality must be in slope-intercept form (solved for y) to determine, from the presence of a < symbol or a > symbol, whether to shade the lower or the upper half-plane.*** In **Figure 16,** the upper half-plane is shaded, even though the inequality is $3x - 2y \leq 6$ (with a $<$ symbol) in standard form. Only when we write the inequality as

$$y \geq \frac{3}{2}x - 3 \quad \text{Slope-intercept form}$$

does the $>$ symbol indicate to shade the upper half-plane.

EXAMPLE 2 Graphing a Linear Inequality

Graph $x + 4y < 4$.

SOLUTION The boundary of the graph is the straight line $x + 4y = 4$. Because points on this line do *not* satisfy $x + 4y < 4$, it is customary to make the line dashed, as in **Figure 17.**

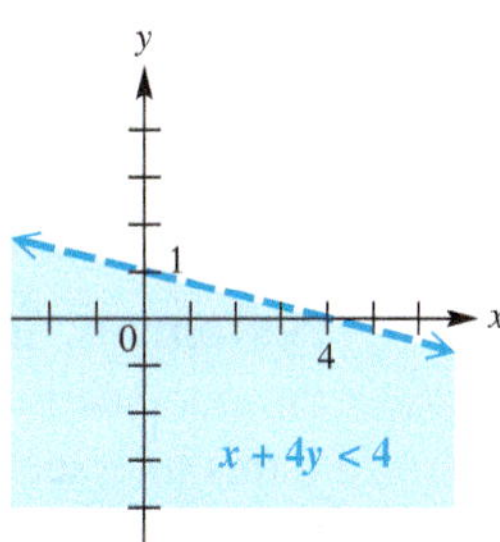

Figure 17

To decide which half-plane represents the solution set, solve for y.

$$x + 4y < 4$$

$$4y < -x + 4 \quad \text{Subtract } x.$$

$$y < -\frac{1}{4}x + 1 \quad \text{Divide by 4.}$$

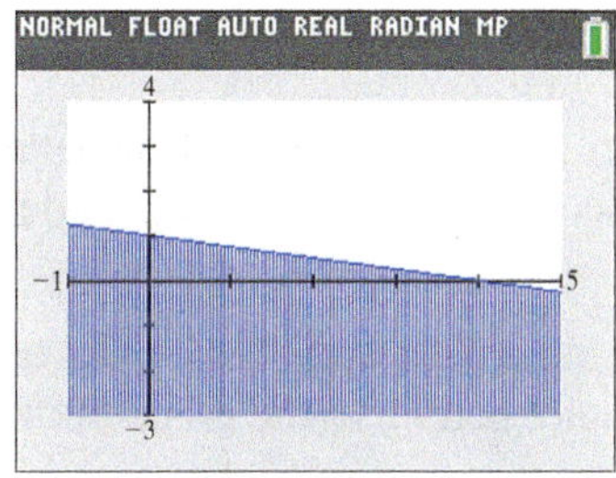

To graph the inequality in **Example 2** using a graphing calculator, solve for y, and then direct the calculator to shade below the boundary line, $y = -\frac{1}{4}x + 1$.

Because y is *less than* $-\frac{1}{4}x + 1$, the graph of the solution set is the half-plane *below* the boundary, as shown in **Figure 17.**

As an alternative method for deciding which half-plane to shade, or as a check, we can choose a test point not on the boundary line and substitute into the inequality. The point $(0, 0)$ is a good choice if it does not lie on the boundary because the substitution is easily done.

$$x + 4y < 4 \quad \text{Use the original inequality.}$$

$$0 + 4(0) \stackrel{?}{<} 4 \quad \text{Use } (0, 0) \text{ as a test point.}$$

$$0 < 4 \quad \text{True}$$

The test point $(0, 0)$ lies below the boundary, so all points that satisfy the inequality must also lie below the boundary. This agrees with the result above.

Now Try Exercise 15.

An inequality containing $<$ or $>$ is a **strict inequality** and does not include the boundary in its solution set. This is indicated with a dashed boundary, as shown in **Example 2.** A **nonstrict inequality** contains $\leq$ or $\geq$ and does include its boundary in the solution set. This is indicated with a solid boundary, as shown in **Example 1.**

Graphing an Inequality in Two Variables

Method 1 If the inequality is or can be solved for y, then the following hold.

- The graph of $y < f(x)$ consists of all the points that are *below* the graph of $y = f(x)$.
- The graph of $y > f(x)$ consists of all the points that are *above* the graph of $y = f(x)$.

Method 2 If the inequality is not or cannot be solved for y, then choose a test point not on the boundary.

- If the test point satisfies the inequality, then the graph includes all points on the *same* side of the boundary as the test point.
- If the test point does not satisfy the inequality, then the graph includes all points on the *other* side of the boundary.

For either method, use a solid boundary for a nonstrict inequality ($\leq$ or $\geq$) or a dashed boundary for a strict inequality ($<$ or $>$).

Systems of Inequalities The solution set of a **system of inequalities** is the intersection of the solution sets of all the inequalities in the system. We find this intersection by graphing the solution sets of all inequalities on the same coordinate axes and identifying, by shading, the region common to all graphs.

EXAMPLE 3 Graphing Systems of Inequalities

Graph the solution set of each system.

(a) $x > 6 - 2y$
$x^2 < 2y$

(b) $|x| \leq 3$
$y \leq 0$
$y \geq |x| + 1$

SOLUTION

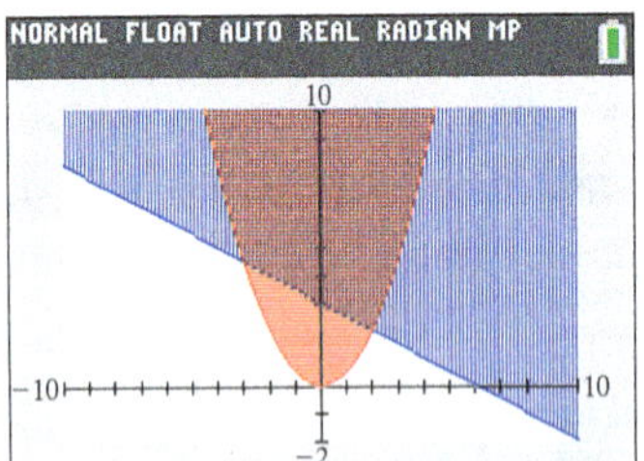

The region shaded twice is the solution set of the system in **Example 3(a).**

(a) **Figures 18(a) and (b)** show the graphs of $x > 6 - 2y$ and $x^2 < 2y$. The methods presented earlier in this chapter can be used to show that the boundaries intersect at the points $(2, 2)$ and $\left(-3, \frac{9}{2}\right)$.

The solution set of the system is shown in **Figure 18(c).** The points on the boundaries of $x > 6 - 2y$ and $x^2 < 2y$ do not belong to the graph of the solution set, so the boundaries are dashed.

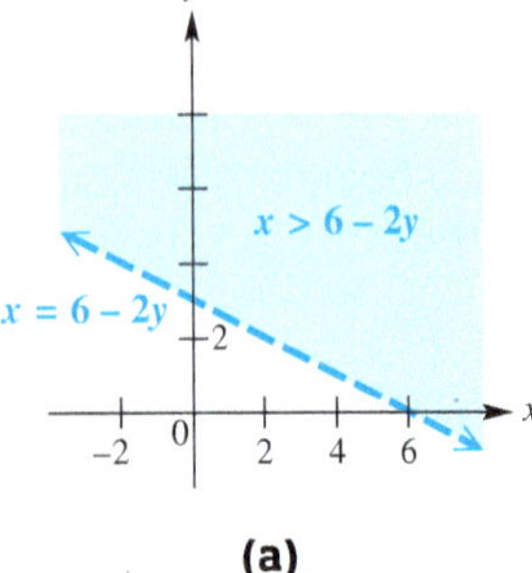

(a)

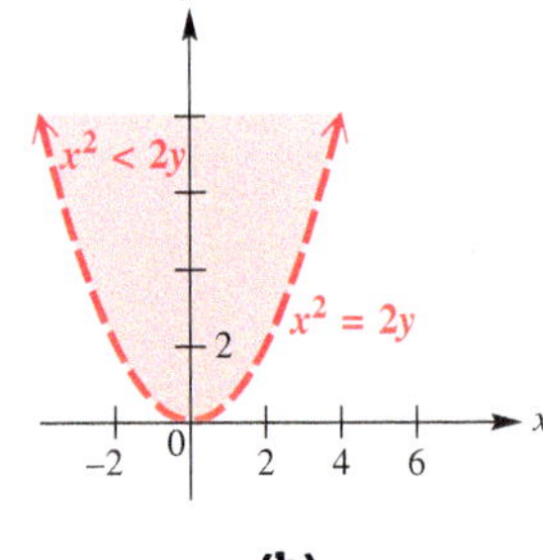

(b)

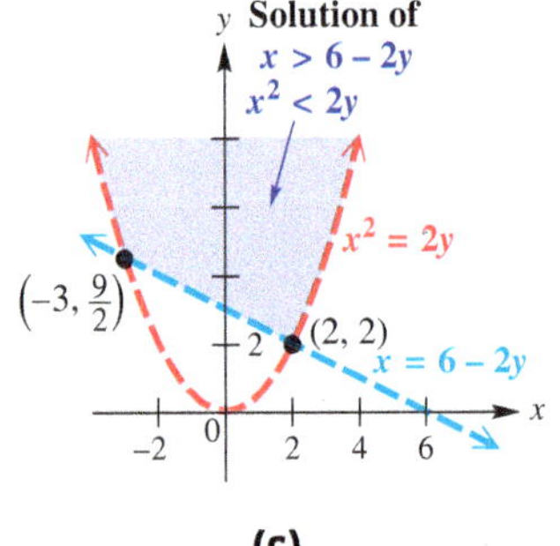

(c)

Figure 18

(b) Writing $|x| \leq 3$ as $-3 \leq x \leq 3$ shows that this inequality is satisfied by points in the region between and including

$$x = -3 \quad \text{and} \quad x = 3.$$

See **Figure 19(a).** The set of points that satisfies $y \leq 0$ includes the points below or on the x-axis. See **Figure 19(b).**

Graph $y = |x| + 1$ and use a test point to verify that the solutions of $y \geq |x| + 1$ are on or above the boundary. See **Figure 19(c).** Because the solution sets of $y \leq 0$ and $y \geq |x| + 1$ shown in **Figures 19(b) and (c)** have no points in common, ***the solution set of the system is*** $\varnothing$**.**

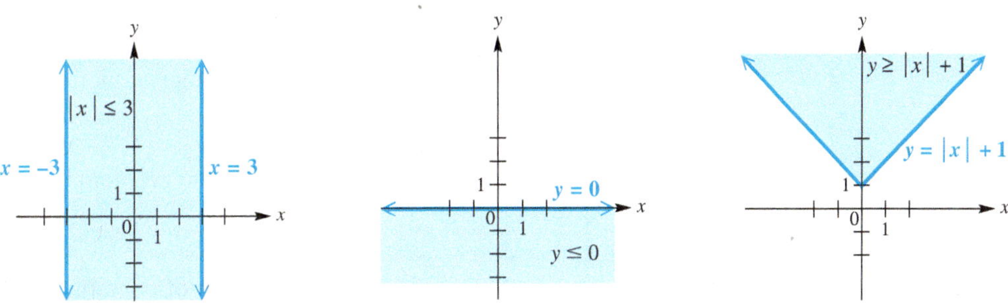

The solution set of the system is $\varnothing$ because there are no points common to all three regions simultaneously.

(a) **(b)** **(c)**

Figure 19

✔ **Now Try Exercises 43, 49, and 51.**

NOTE Although we gave three graphs in the solutions of **Example 3,** in practice we usually give only a final graph showing the solution set of the system. This would be an empty rectangular coordinate system in **Example 3(b).**

Linear Programming One important application of mathematics to business and social science is **linear programming.** Linear programming is used to find an optimum value—for example, minimum cost or maximum profit. Procedures for solving linear programming problems were developed in 1947 by George Dantzig while he was working on a problem of allocating supplies for the Air Force in a way that minimized total cost.

To solve a linear programming problem in general, use the following steps. (The italicized terms are defined in **Example 4.**)

Solving a Linear Programming Problem

Step 1 Write all necessary *constraints* and the *objective function.*

Step 2 Graph the *region of feasible solutions.*

Step 3 Identify all *vertices* (*corner points*).

Step 4 Find the value of the *objective function* at each *vertex.*

Step 5 The solution is given by the *vertex* producing the optimum value of the *objective function.*

In this procedure, Step 5 is an application of the following theorem.

Fundamental Theorem of Linear Programming

If an optimal value for a linear programming problem exists, then it occurs at a vertex of the region of feasible solutions.

EXAMPLE 4 Maximizing Rescue Efforts

Earthquake victims in China need medical supplies and bottled water. Each medical kit measures 1 ft^3 and weighs 10 lb. Each container of water is also 1 ft^3 and weighs 20 lb. The plane can carry only 80,000 lb with a total volume of 6000 ft^3. Each medical kit will aid 6 people, and each container of water will serve 10 people.

How many of each should be sent in order to maximize the number of victims aided? What is this maximum number of victims?

SOLUTION

Step 1 We translate the statements of the problem into symbols as follows.

Let x = the number of medical kits to be sent,

and y = the number of containers of water to be sent.

Because negative values of x and y are not valid for this problem, these two inequalities must be satisfied.

$$x \geq 0$$
$$y \geq 0$$

Each medical kit and each container of water will occupy 1 ft^3 of space, and there is a maximum of 6000 ft^3 available.

$$1x + 1y \leq 6000$$
$$x + y \leq 6000$$

Each medical kit weighs 10 lb, and each water container weighs 20 lb. The total weight cannot exceed 80,000 lb.

$$10x + 20y \leq 80{,}000$$
$$x + 2y \leq 8000 \quad \text{Divide by 10.}$$

The four inequalities in color form a system of linear inequalities.

$$x \geq 0$$
$$y \geq 0$$
$$x + y \leq 6000$$
$$x + 2y \leq 8000$$

These are the **constraints** on the variables in this application.

Because each medical kit will aid 6 victims and each container of water will serve 10 victims, the total number of victims served is represented by the following **objective function.**

$$\text{Number of victims served} = 6x + 10y \quad \text{Multiply the number of items by the number of victims served and add.}$$

Step 2 The maximum number of victims served, subject to these constraints, is found by sketching the graph of the solution set of the system. See **Figure 20.** The only feasible values of x and y are those that satisfy all constraints. These values correspond to points that lie on the boundary or in the shaded region, which is the **region of feasible solutions.**

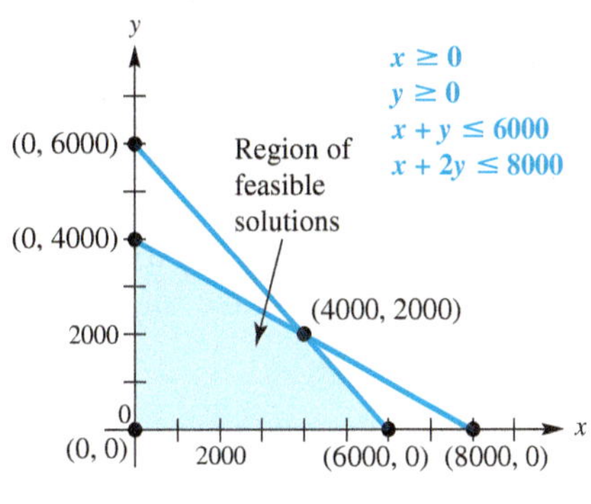

Figure 20

Step 3 The problem may now be stated as follows: Find values of x and y in the region of feasible solutions as shown in **Figure 20** that will produce the maximum possible value of $6x + 10y$. It can be shown that any optimum value (maximum or minimum) will always occur at a **vertex** (or **corner point)** of the region of feasible solutions. The vertices are

$$(0, 0), \quad (0, 4000), \quad (4000, 2000), \quad \text{and} \quad (6000, 0).$$

Step 4 To locate the point (x, y) that gives the maximum value, substitute the coordinates of the vertices into the objective function. See the table below. Find the number of victims served that corresponds to each coordinate pair.

Points are from **Figure 20.**

Point	Number of Victims Served = $6x + 10y$
$(0, 0)$	$6(0) + 10(0) = 0$
$(0, 4000)$	$6(0) + 10(4000) = 40{,}000$
$(4000, 2000)$	$6(4000) + 10(2000) = 44{,}000$
$(6000, 0)$	$6(6000) + 10(0) = 36{,}000$

44,000 is the maximum number.

Step 5 The vertex $(4000, 2000)$ gives the maximum number. The maximum number of victims served is 44,000, when 4000 medical kits and 2000 containers of water are sent.

✔ **Now Try Exercise 85.**

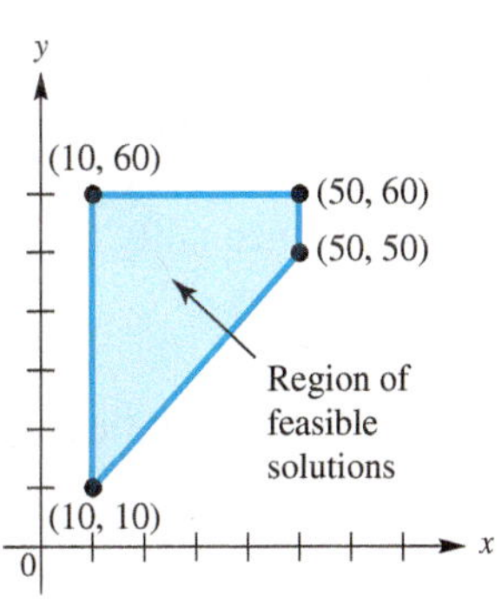

Figure 21

To justify the procedure used in a linear programming problem, suppose that we wish to determine the values of x and y that maximize the objective function

$$30x + 70y$$

with the constraints

$$10 \leq x \leq 50, \quad 10 \leq y \leq 60, \quad \text{and} \quad y \geq x.$$

Figure 21 shows this region of feasible solutions.

To locate the point (x, y) that gives the maximum objective function value, add to the graph of **Figure 21** lines corresponding to arbitrarily chosen values of 0, 1000, 3000, and 7000.

$$30x + 70y = 0$$
$$30x + 70y = 1000$$
$$30x + 70y = 3000$$
$$30x + 70y = 7000$$

Figure 22(a) shows the region of feasible solutions, together with these lines. The lines are parallel, and the higher the line, the greater the profit. The line $30x + 70y = 7000$ yields the greatest profit but does not contain any points of the region of feasible solutions. To find the feasible solution of greatest profit, lower the line $30x + 70y = 7000$ until it contains a feasible solution—that is, until it just touches the region of feasible solutions. This occurs at point A, a vertex of the region. The desired maximum value is 5700. See **Figure 22(b).**

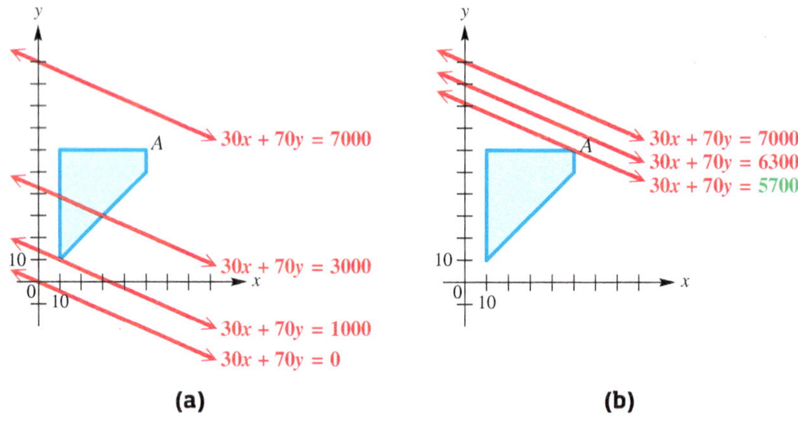

Figure 22

EXAMPLE 5 Minimizing Cost

Robin takes multivitamins each day. She wants at least 16 units of vitamin A, at least 5 units of vitamin B_1, and at least 20 units of vitamin C. Capsules, costing \$0.10 each, contain 8 units of A, 1 of B_1, and 2 of C. Chewable tablets, costing \$0.20 each, contain 2 units of A, 1 of B_1, and 7 of C.

How many of each should she take each day to minimize her cost and yet fulfill her daily requirements?

SOLUTION

Step 1 Let x represent the number of capsules to take each day, and let y represent the number of chewable tablets to take. Then the cost in *pennies* per day is

$$\text{cost} = 10x + 20y. \quad \text{Objective function}$$

Robin takes x of the \$0.10 capsules and y of the \$0.20 chewable tablets, and she gets 8 units of vitamin A from each capsule and 2 units of vitamin A from each tablet. Altogether she gets $8x + 2y$ units of A per day. She wants at least 16 units, which gives the following inequality for A.

$$8x + 2y \geq 16$$

Each capsule and each tablet supplies 1 unit of vitamin B_1. Robin wants at least 5 units per day, so the inequality for B is

$$x + y \geq 5.$$

For vitamin C, the inequality is

$$2x + 7y \geq 20.$$

Because Robin cannot take negative numbers of multivitamins,

$$x \geq 0 \quad \text{and} \quad y \geq 0.$$

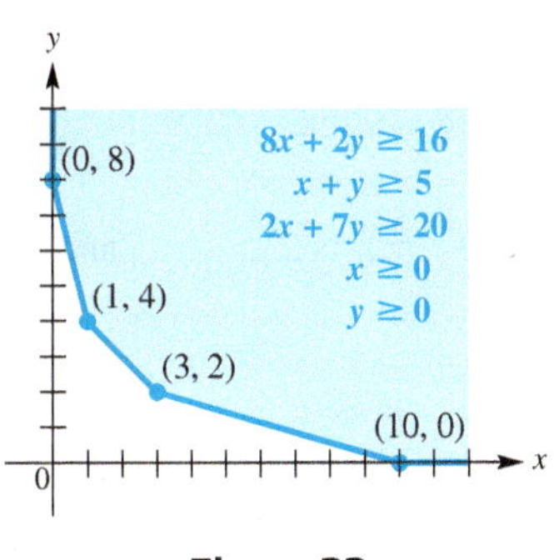

Figure 23

Step 2 **Figure 23** shows the region of feasible solutions formed by the inequalities

$$8x + 2y \geq 16, \quad x + y \geq 5, \quad 2x + 7y \geq 20, \quad x \geq 0, \quad \text{and} \quad y \geq 0.$$

Step 3 The vertices are $(0, 8)$, $(1, 4)$, $(3, 2)$, and $(10, 0)$.

Steps 4 and 5 See the table. The minimum cost of \$0.70 occurs at $(3, 2)$.

Points are from **Figure 23.**

Point	Cost $= 10x + 20y$
$(0, 8)$	$10(0) + 20(8) = 160$
$(1, 4)$	$10(1) + 20(4) = 90$
$(3, 2)$	$10(3) + 20(2) = 70$
$(10, 0)$	$10(10) + 20(0) = 100$

70 cents, or \$0.70, is the minimum cost.

Robin's best choice is to take 3 capsules and 2 chewable tablets each day, for a total cost of \$0.70 per day. She receives just the minimum amounts of vitamins B_1 and C, and an excess of vitamin A.

✔ Now Try Exercises 79 and 89.

9.6 Exercises

CONCEPT PREVIEW *Match each system of inequalities with the correct graph from choices A–D.*

1. $x \geq 5$
$y \leq -3$

2. $x \leq 5$
$y \geq -3$

3. $x > 5$
$y < -3$

4. $x < 5$
$y > -3$

A.

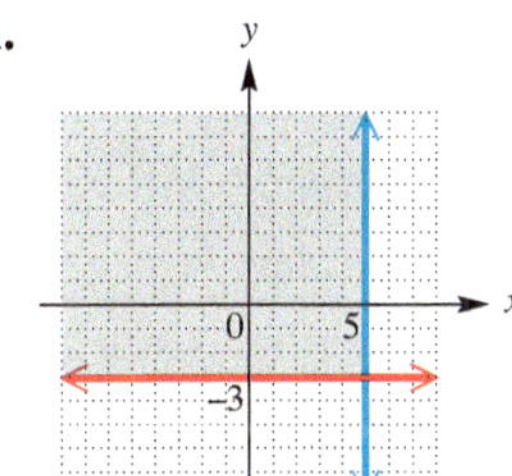

B.

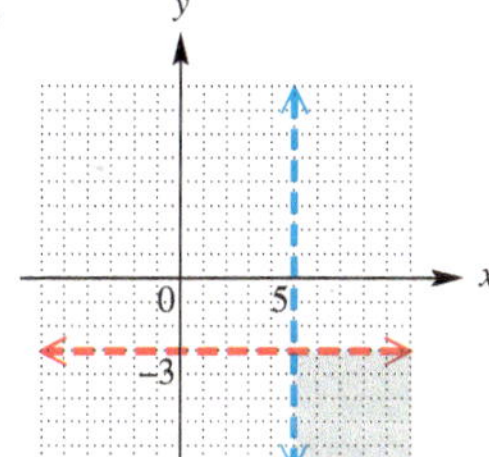

C.

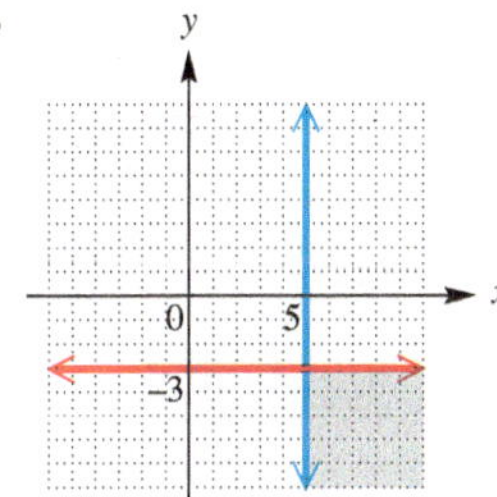

D.

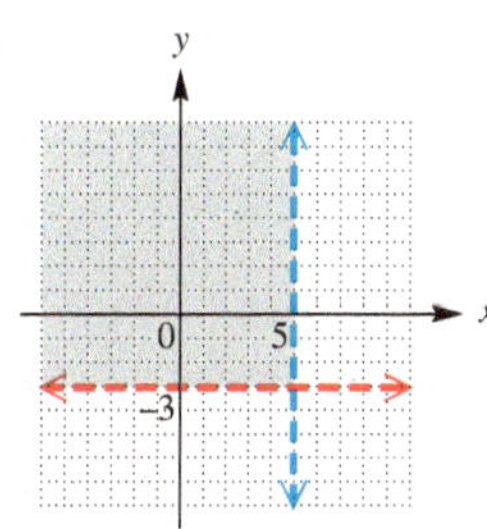

CONCEPT PREVIEW *Fill in the blank(s) to correctly complete the sentence, or answer the question.*

5. The test point $(0, 0)$ ____________ (does/does not) satisfy the inequality $-3x - 4y \geq 12$.

6. Any point that lies on the graph of $-3x - 4y = 12$ ____________ (does/does not) lie on the graph of $-3x - 4y > 12$.

7. What are the coordinates of the point of intersection of the boundary lines in the following system?

$$x \geq 2$$
$$y \leq -5$$

8. Does the point $(3, -8)$ satisfy the system in **Exercise 7?**

9. The graph of $4x - 7y < 28$ has a ______________ boundary line.
(solid/dashed)

10. When the inequality in **Exercise 9** is solved for y, the result is ________, and the points ______________ the boundary line are shaded.
(above/below)

Graph each inequality. ***See Examples 1–3.***

11. $x + 2y \leq 6$

12. $x - y \geq 2$

13. $2x + 3y \geq 4$

14. $4y - 3x \leq 5$

15. $3x - 5y > 6$

16. $x < 3 + 2y$

17. $5x < 4y - 2$

18. $2x > 3 - 4y$

19. $x \leq 3$

20. $y \leq -2$

21. $y < 3x^2 + 2$

22. $y \leq x^2 - 4$

23. $y > (x - 1)^2 + 2$

24. $y > 2(x + 3)^2 - 1$

25. $x^2 + (y + 3)^2 \leq 16$

26. $(x - 4)^2 + y^2 \geq 9$

27. $y > 2^x + 1$

28. $y \leq \log(x - 1) - 2$

Concept Check Work each problem.

29. For $Ax + By \geq C$, if $B > 0$, would the region above or below the line be shaded?

30. For $Ax + By \geq C$, if $B < 0$, would the region above or below the line be shaded?

31. Which one of the following is a description of the graph of the inequality

$$(x - 5)^2 + (y - 2)^2 < 4?$$

A. the region inside a circle with center $(-5, -2)$ and radius 2

B. the region inside a circle with center $(5, 2)$ and radius 2

C. the region inside a circle with center $(-5, -2)$ and radius 4

D. the region outside a circle with center $(5, 2)$ and radius 4

32. Write the inequality that represents the region inside a circle with center $(-5, -2)$ and radius 4.

Concept Check Match each inequality with the appropriate calculator graph in A–D. Do not use a calculator.

33. $y \leq 3x - 6$

34. $y \geq 3x - 6$

35. $y \leq -3x - 6$

36. $y \geq -3x - 6$

A.
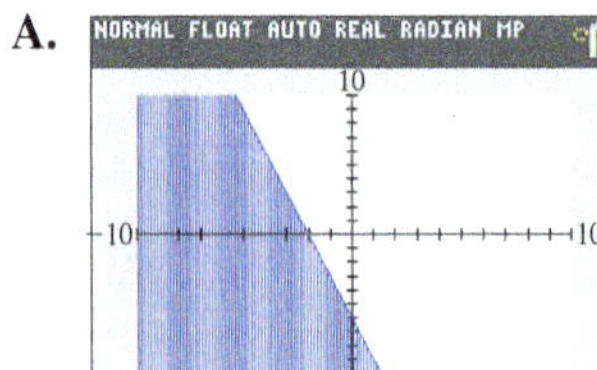

B.
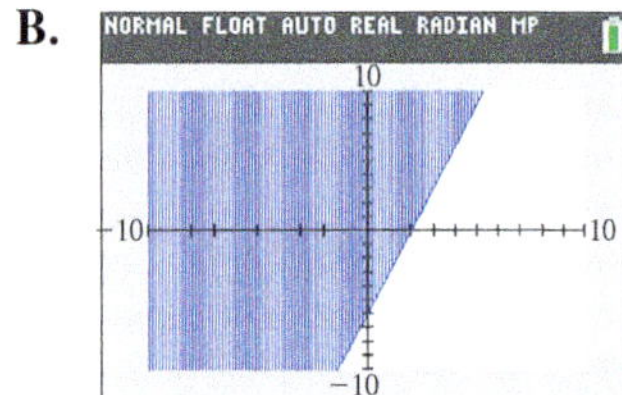

C.
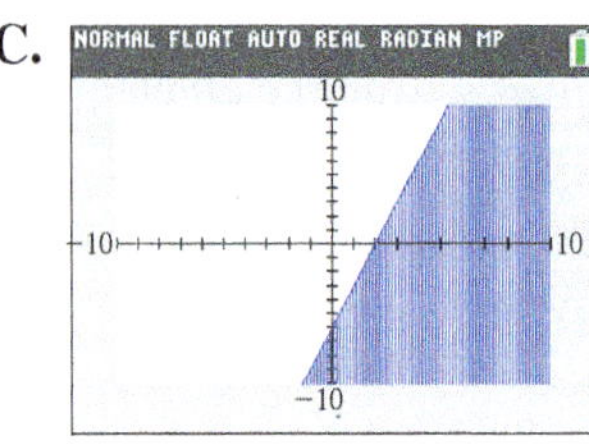

D.
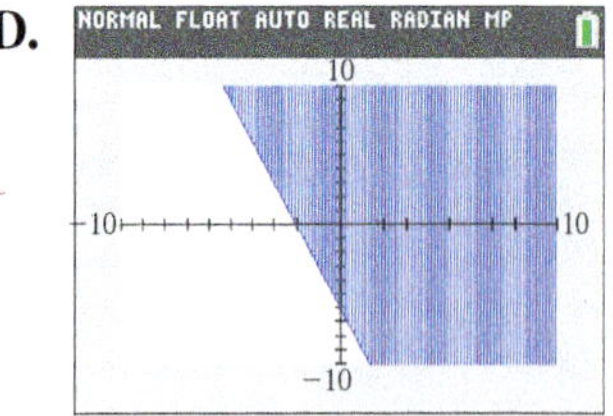

Graph the solution set of each system of inequalities. **See Example 3.**

37. $x + y \geq 0$
$2x - y \geq 3$

38. $x + y \leq 4$
$x - 2y \geq 6$

39. $2x + y > 2$
$x - 3y < 6$

40. $4x + 3y < 12$
$y + 4x > -4$

41. $3x + 5y \leq 15$
$x - 3y \geq 9$

42. $y \leq x$
$x^2 + y^2 < 1$

43. $4x - 3y \leq 12$
$y \leq x^2$

44. $y \leq -x^2$
$y \geq x^2 - 6$

45. $x + 2y \leq 4$
$y \geq x^2 - 1$

46. $x + y \leq 9$
$x \leq -y^2$

47. $y \leq (x + 2)^2$
$y \geq -2x^2$

48. $x - y < 1$
$-1 < y < 1$

49. $x + y \leq 36$
$-4 \leq x \leq 4$

50. $y \geq (x - 2)^2 + 3$
$y \leq -(x - 1)^2 + 6$

51. $y \geq x^2 + 4x + 4$
$y < -x^2$

52. $x \geq 0$
$x + y \leq 4$
$2x + y \leq 5$

53. $3x - 2y \geq 6$
$x + y \leq -5$
$y \leq 4$

54. $-2 < x < 3$
$-1 \leq y \leq 5$
$2x + y < 6$

55. $-2 < x < 2$
$y > 1$
$x - y > 0$

56. $x + y \leq 4$
$x - y \leq 5$
$4x + y \leq -4$

57. $x \leq 4$
$x \geq 0$
$y \geq 0$
$x + 2y \geq 2$

58. $2y + x \geq -5$
$y \leq 3 + x$
$x \leq 0$
$y \leq 0$

59. $2x + 3y \leq 12$
$2x + 3y > -6$
$3x + y < 4$
$x \geq 0$
$y \geq 0$

60. $y \geq 3^x$
$y \geq 2$

61. $y \leq \left(\frac{1}{2}\right)^x$
$y \geq 4$

62. $\ln x - y \geq 1$
$x^2 - 2x - y \leq 1$

63. $y \leq \log x$
$y \geq |x - 2|$

64. $e^{-x} - y \leq 1$
$x - 2y \geq 4$

65. $y > x^3 + 1$
$y \geq -1$

66. $y \leq x^3 - x$
$y > -3$

Use the shading capabilities of a graphing calculator to graph each inequality or system of inequalities.

67. $3x + 2y \geq 6$

68. $y \leq x^2 + 5$

69. $x + y \geq 2$
$x + y \leq 6$

70. $y \geq |x + 2|$
$y \leq 6$

Connecting Graphs with Equations *Determine the system of inequalities illustrated in each graph. Write each inequality in standard form.*

71.

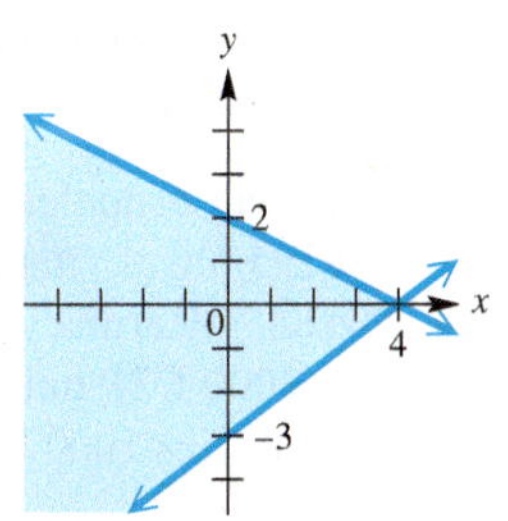

72.

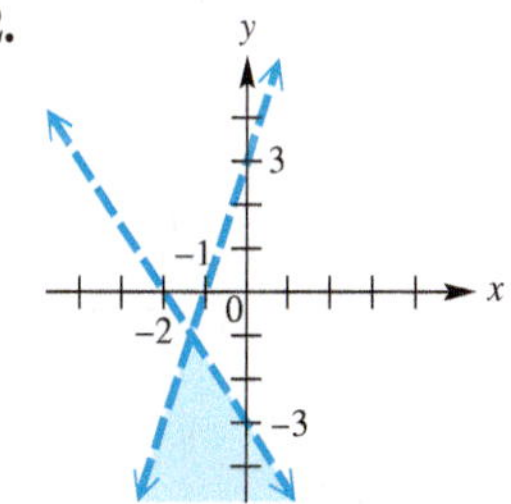

73.

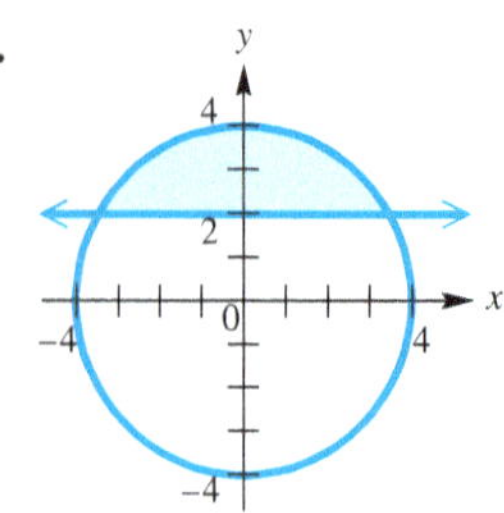

74.

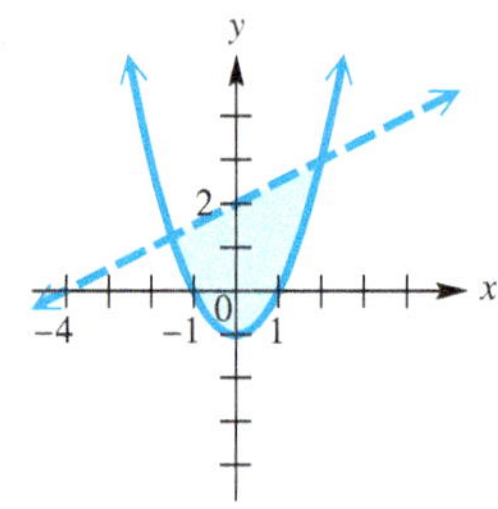

75. *Concept Check* Write a system of inequalities for which the graph is the region in the first quadrant inside and including the circle with radius 2 centered at the origin, and above (not including) the line that passes through the points $(0, -1)$ and $(2, 2)$.

76. *Cost of Vitamins* The figure shows the region of feasible solutions for the vitamin problem of **Example 5** and the straight-line graph of all combinations of capsules and chewable tablets for which the cost is \$0.40.

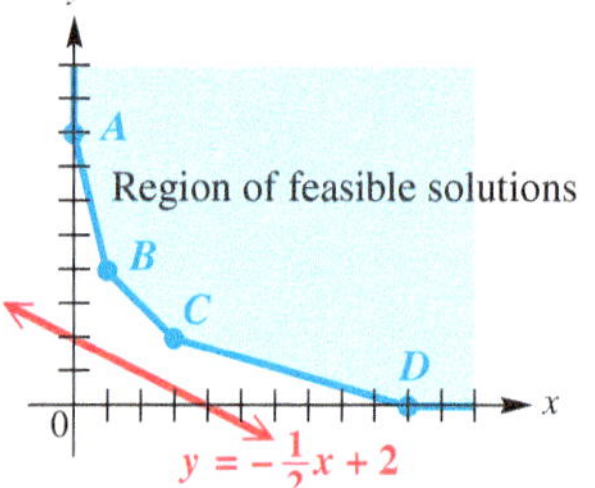

(a) The cost function is $10x + 20y$. Give the linear equation (in slope-intercept form) of the line of constant cost c.

(b) As c increases, does the line of constant cost move up or down?

(c) By inspection, find the vertex of the region of feasible solutions that gives the optimal value.

The graphs show regions of feasible solutions. Find the maximum and minimum values of each objective function. ***See Examples 4 and 5.***

77. objective function $= 3x + 5y$

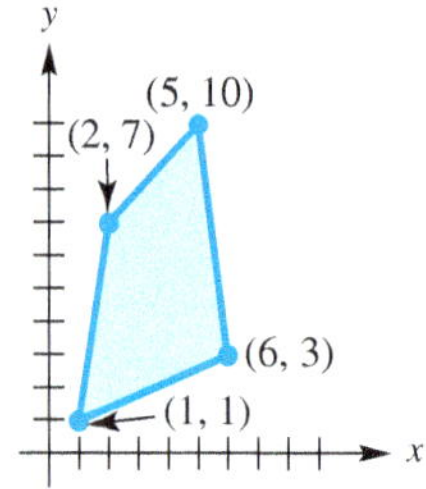

78. objective function $= 6x + y$

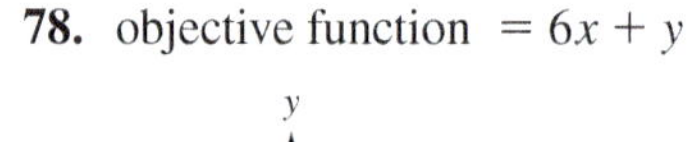

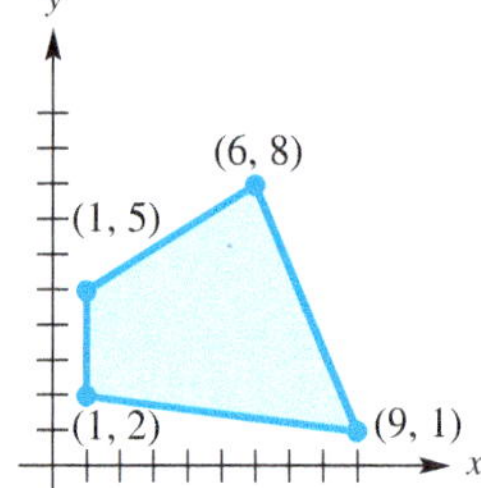

Find the maximum and minimum values of each objective function over the region of feasible solutions shown at the right. ***See Examples 4 and 5.***

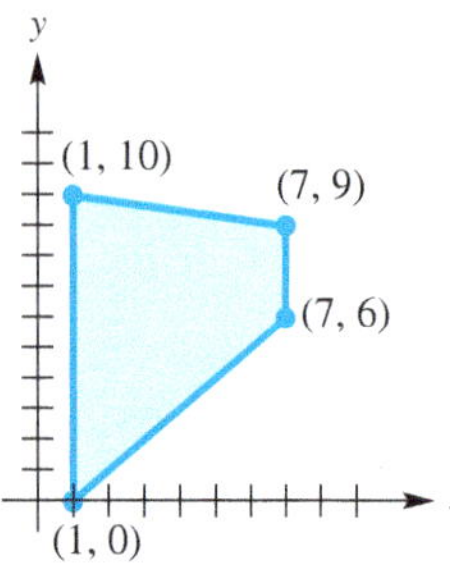

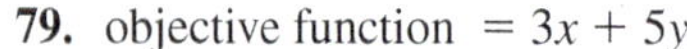

79. objective function $= 3x + 5y$

80. objective function $= 5x + 5y$

81. objective function $= 10y$

82. objective function $= 3x - y$

Write a system of inequalities for each problem, and then graph the region of feasible solutions of the system. ***See Examples 4 and 5.***

83. *Vitamin Requirements* Jane must supplement her daily diet with at least 6000 USP units of vitamin A, at least 195 mg of vitamin C, and at least 600 USP units of vitamin D. She finds that Mason's Pharmacy carries Brand X and Brand Y vitamins. Each Brand X pill contains 3000 USP units of A, 45 mg of C, and 75 USP units of D, while each Brand Y pill contains 1000 USP units of A, 50 mg of C, and 200 USP units of D.

84. *Shipping Requirements* The California Almond Growers have 2400 boxes of almonds to be shipped from their plant in Sacramento to Des Moines and San Antonio. The Des Moines market needs at least 1000 boxes, while the San Antonio market must have at least 800 boxes.

Solve each problem. ***See Examples 4 and 5.***

85. *Aid to Disaster Victims* An agency wants to ship food and clothing to tsunami victims in Japan. Commercial carriers have volunteered to transport the packages, provided they fit in the available cargo space. Each 20-ft^3 box of food weighs 40 lb and each 30-ft^3 box of clothing weighs 10 lb. The total weight cannot exceed 16,000 lb, and the total volume must be at most 18,000 ft^3. Each carton of food will feed 10 victims, and each carton of clothing will help 8 victims.

How many cartons of food and clothing should be sent to maximize the number of people assisted? What is the maximum number assisted?

86. *Aid to Disaster Victims* Refer to **Example 4.** Suppose that each medical kit aids 2 victims rather than 6, and each container of water serves 5 victims rather than 10.

How many of each should be sent in order to maximize the number of victims aided? What is this maximum number of victims?

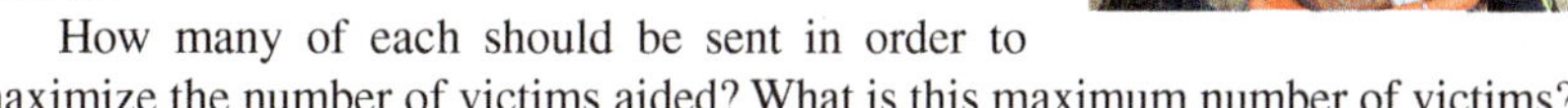

87. *Storage Capacity* An office manager wants to buy some filing cabinets. Cabinet A costs \$10 each, requires 6 ft^2 of floor space, and holds 8 ft^3 of files. Cabinet B costs \$20 each, requires 8 ft^2 of floor space, and holds 12 ft^3 of files. He can spend no more than \$140, and there is room for no more than 72 ft^2 of cabinets.

To maximize storage capacity within the limits imposed by funds and space, how many of each type of cabinet should he buy?

88. *Gasoline Revenues* The manufacturing process requires that oil refineries manufacture at least 2 gal of gasoline for each gallon of fuel oil. To meet the winter demand for fuel oil, at least 3 million gal per day must be produced. The demand for gasoline is no more than 6.4 million gal per day.

If the price of gasoline is \$2.90 per gal and the price of fuel oil is \$2.50 per gal, how much of each should be produced to maximize revenue?

89. *Diet Requirements* Theo requires two food supplements, I and II. He can get these supplements from two different products, A and B, as shown in the table. He must include at least 15 g of each supplement in his daily diet.

If product A costs \$0.25 per serving and product B costs \$0.40 per serving, how can he satisfy his requirements most economically?

Supplement (g/serving)	I	II
Product *A*	3	2
Product *B*	2	4

90. *Profit from Televisions* The GL company makes televisions. It produces a 32-inch screen that sells for \$100 profit and a 48-inch screen that sells for \$150 profit. On the assembly line, the 32-inch screen requires 3 hr, and the 48-inch screen takes 5 hr. The cabinet shop spends 1 hr on the cabinet for the 32-inch screen and 3 hr on the cabinet for the 48-inch screen. Both models require 2 hr of time for testing and packing. On one production run, the company has available 3900 work hours on the assembly line, 2100 work hours in the cabinet shop, and 2200 work hours in the testing and packing department.

How many of each model should it produce to maximize profit? What is the maximum profit?

9.7 Properties of Matrices

- Basic Definitions
- Matrix Addition
- Special Matrices
- Matrix Subtraction
- Scalar Multiplication
- Matrix Multiplication
- An Application of Matrix Algebra

Basic Definitions

In this section and the next, we discuss algebraic properties of matrices. It is customary to use capital letters to name matrices and subscript notation to name elements of a matrix, as in the following matrix A.

$$A = \begin{bmatrix} a_{11} & a_{12} & a_{13} & \cdots & a_{1n} \\ a_{21} & a_{22} & a_{23} & \cdots & a_{2n} \\ a_{31} & a_{32} & a_{33} & \cdots & a_{3n} \\ \vdots & \vdots & \vdots & & \vdots \\ a_{m1} & a_{m2} & a_{m3} & \cdots & a_{mn} \end{bmatrix}$$

The first row, first column element is a_{11} (read "a-sub-one-one"); the second row, third column element is a_{23}; and in general, the ith row, jth column element is a_{ij}.

An $n \times n$ matrix is a **square matrix of order n** because the number of rows is equal to the number of columns. A matrix with just one row is a **row matrix,** and a matrix with just one column is a **column matrix.**

Two matrices are equal if they have the same dimension and if corresponding elements, position by position, are equal. Using this definition, the matrices

$$\begin{bmatrix} 2 & 1 \\ 3 & -5 \end{bmatrix} \quad \text{and} \quad \begin{bmatrix} 1 & 2 \\ -5 & 3 \end{bmatrix} \quad \text{are } not \text{ equal}$$

(even though they contain the same elements and have the same dimension), because at least one pair of corresponding elements differ.

EXAMPLE 1 Finding Values to Make Two Matrices Equal

Find the values of the variables for which each statement is true, if possible.

(a) $\begin{bmatrix} 2 & 1 \\ p & q \end{bmatrix} = \begin{bmatrix} x & y \\ -1 & 0 \end{bmatrix}$ **(b)** $\begin{bmatrix} x \\ y \end{bmatrix} = \begin{bmatrix} 1 \\ 4 \\ 0 \end{bmatrix}$

SOLUTION

(a) From the definition of equality given above, the only way that the statement can be true is if $2 = x$, $1 = y$, $p = -1$, and $q = 0$.

(b) This statement can never be true because the two matrices have different dimensions. (One is 2×1 and the other is 3×1.)

✔ **Now Try Exercises 13 and 19.**

Matrix Addition

Addition of matrices is defined as follows.

Addition of Matrices

To add two matrices of the same dimension, add corresponding elements. ***Only matrices of the same dimension can be added.***

It can be shown that matrix addition satisfies the commutative, associative, closure, identity, and inverse properties. **(See Exercises 91 and 92.)**

EXAMPLE 2 Adding Matrices

Find each sum, if possible.

(a) $\begin{bmatrix} 5 & -6 \\ 8 & 9 \end{bmatrix} + \begin{bmatrix} -4 & 6 \\ 8 & -3 \end{bmatrix}$ **(b)** $\begin{bmatrix} 2 \\ 5 \\ 8 \end{bmatrix} + \begin{bmatrix} -6 \\ 3 \\ 12 \end{bmatrix}$

(c) $A + B$, if $A = \begin{bmatrix} 5 & 8 \\ 6 & 2 \end{bmatrix}$ and $B = \begin{bmatrix} 3 & 9 & 1 \\ 4 & 2 & 5 \end{bmatrix}$

ALGEBRAIC SOLUTION

(a) $\begin{bmatrix} 5 & -6 \\ 8 & 9 \end{bmatrix} + \begin{bmatrix} -4 & 6 \\ 8 & -3 \end{bmatrix}$

$= \begin{bmatrix} 5+(-4) & -6+6 \\ 8+8 & 9+(-3) \end{bmatrix}$

$= \begin{bmatrix} 1 & 0 \\ 16 & 6 \end{bmatrix}$

(b) $\begin{bmatrix} 2 \\ 5 \\ 8 \end{bmatrix} + \begin{bmatrix} -6 \\ 3 \\ 12 \end{bmatrix} = \begin{bmatrix} -4 \\ 8 \\ 20 \end{bmatrix}$

(c) The matrices

$$A = \begin{bmatrix} 5 & 8 \\ 6 & 2 \end{bmatrix}$$

and $B = \begin{bmatrix} 3 & 9 & 1 \\ 4 & 2 & 5 \end{bmatrix}$

have different dimensions, so A and B cannot be added. The sum $A + B$ does not exist.

GRAPHING CALCULATOR SOLUTION

(a) **Figure 24** shows the sum of matrices A and B.

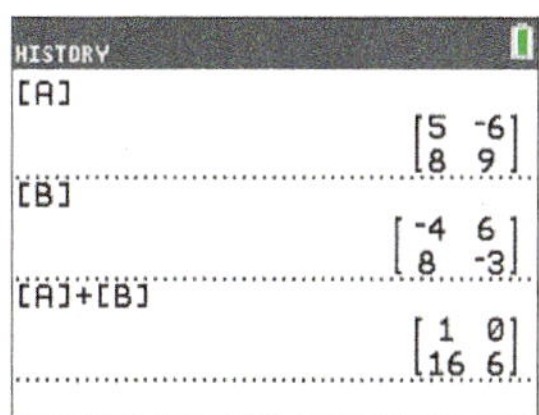

Figure 24

(b) The screen in **Figure 25** shows how the sum of two column matrices entered directly on the home screen is displayed.

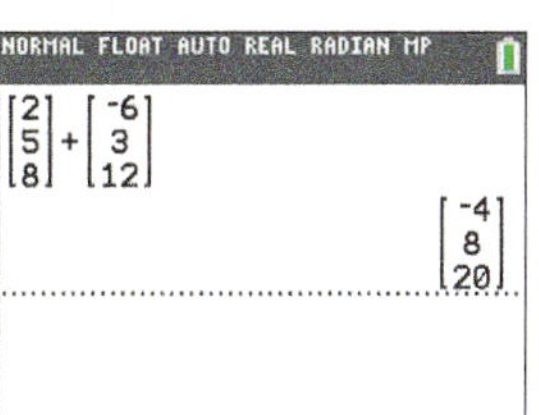

Figure 25

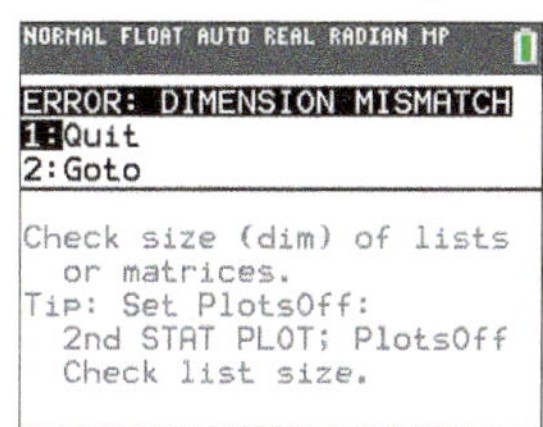

Figure 26

(c) A graphing calculator such as the TI-84 Plus will return an ERROR message if it is directed to perform an operation on matrices that is not possible due to dimension mismatch. See **Figure 26.**

✔ **Now Try Exercises 25, 27, and 29.**

Special Matrices

A matrix containing only zero elements is a **zero matrix.** A zero matrix can be written with any dimension.

$O = [0 \quad 0 \quad 0]$ 1×3 zero matrix $\qquad O = \begin{bmatrix} 0 & 0 & 0 \\ 0 & 0 & 0 \end{bmatrix}$ 2×3 zero matrix

By the additive inverse property, each real number has an additive inverse: If a is a real number, then there is a real number $-a$ such that

$$a + (-a) = 0 \quad \text{and} \quad -a + a = 0.$$

Given matrix A, there is a matrix $-A$ such that $A + (-A) = O$. The matrix $-A$ has as elements the additive inverses of the elements of A. (Remember, each element of A is a real number and therefore has an additive inverse.)

Example: If $A = \begin{bmatrix} -5 & 2 & -1 \\ 3 & 4 & -6 \end{bmatrix}$, then $-A = \begin{bmatrix} 5 & -2 & 1 \\ -3 & -4 & 6 \end{bmatrix}$.

CHECK Confirm that $A + (-A)$ equals the zero matrix, O.

$$A + (-A) = \begin{bmatrix} -5 & 2 & -1 \\ 3 & 4 & -6 \end{bmatrix} + \begin{bmatrix} 5 & -2 & 1 \\ -3 & -4 & 6 \end{bmatrix} = \begin{bmatrix} 0 & 0 & 0 \\ 0 & 0 & 0 \end{bmatrix} = O \checkmark$$

Matrix $-A$ is the **additive inverse,** or **negative,** of matrix A. Every matrix has an additive inverse.

Matrix Subtraction The real number b is subtracted from the real number a, written $a - b$, by adding a and the additive inverse of b.

$$a - b = a + (-b) \quad \text{Real number subtraction}$$

The same definition applies to subtraction of matrices.

Subtraction of Matrices

If A and B are two matrices of the same dimension, then the following holds.

$$A - B = A + (-B)$$

In practice, the difference of two matrices of the same dimension is found by subtracting corresponding elements.

EXAMPLE 3 Subtracting Matrices

Find each difference, if possible.

(a) $\begin{bmatrix} -5 & 6 \\ 2 & 4 \end{bmatrix} - \begin{bmatrix} -3 & 2 \\ 5 & -8 \end{bmatrix}$ **(b)** $\begin{bmatrix} 8 & 6 & -4 \end{bmatrix} - \begin{bmatrix} 3 & 5 & -8 \end{bmatrix}$

(c) $A - B$, if $A = \begin{bmatrix} -2 & 5 \\ 0 & 1 \end{bmatrix}$ and $B = \begin{bmatrix} 3 \\ 5 \end{bmatrix}$

SOLUTION

(a) $\begin{bmatrix} -5 & 6 \\ 2 & 4 \end{bmatrix} - \begin{bmatrix} -3 & 2 \\ 5 & -8 \end{bmatrix}$

$$= \begin{bmatrix} -5-(-3) & 6-2 \\ 2-5 & 4-(-8) \end{bmatrix} \quad \text{Subtract corresponding entries.}$$

$$= \begin{bmatrix} -2 & 4 \\ -3 & 12 \end{bmatrix} \quad \text{Simplify.}$$

```
HISTORY
[C]
                    [-5 6]
                    [ 2 4]
[D]
                    [-3  2]
                    [ 5 -8]
[C]-[D]
                    [-2  4]
                    [-3 12]
```

This screen supports the result in **Example 3(a).**

(b) $\begin{bmatrix} 8 & 6 & -4 \end{bmatrix} - \begin{bmatrix} 3 & 5 & -8 \end{bmatrix}$

$$= \begin{bmatrix} 5 & 1 & 4 \end{bmatrix} \quad \text{Subtract corresponding entries.}$$

(c) $A = \begin{bmatrix} -2 & 5 \\ 0 & 1 \end{bmatrix}$ and $B = \begin{bmatrix} 3 \\ 5 \end{bmatrix}$

These matrices have different dimensions and cannot be subtracted, so the difference $A - B$ does not exist.

✔ **Now Try Exercises 31, 33, and 35.**

Scalar Multiplication In work with matrices, a real number is called a **scalar** to distinguish it from a matrix.

The product of a scalar k and a matrix X is the matrix kX, each of whose elements is k times the corresponding element of X.

```
NORMAL FLOAT AUTO REAL RADIAN MP
[A]
                    [2 -3]
                    [0  4]
5[A]
                  [10 -15]
                  [0   20]
```

```
NORMAL FLOAT AUTO REAL RADIAN MP
[B]
                  [20  36]
                  [12 -16]
(3/4)[B]
                  [15  27]
                  [9  -12]
```

These screens support the results in **Example 4.**

EXAMPLE 4 Multiplying Matrices by Scalars

Find each product.

(a) $5\begin{bmatrix} 2 & -3 \\ 0 & 4 \end{bmatrix}$ **(b)** $\frac{3}{4}\begin{bmatrix} 20 & 36 \\ 12 & -16 \end{bmatrix}$

SOLUTION

(a) $5\begin{bmatrix} 2 & -3 \\ 0 & 4 \end{bmatrix}$

$= \begin{bmatrix} 5(2) & 5(-3) \\ 5(0) & 5(4) \end{bmatrix}$ Multiply each element of the matrix by the scalar 5.

$= \begin{bmatrix} 10 & -15 \\ 0 & 20 \end{bmatrix}$

(b) $\frac{3}{4}\begin{bmatrix} 20 & 36 \\ 12 & -16 \end{bmatrix}$

$= \begin{bmatrix} \frac{3}{4}(20) & \frac{3}{4}(36) \\ \frac{3}{4}(12) & \frac{3}{4}(-16) \end{bmatrix}$ Multiply each element of the matrix by the scalar $\frac{3}{4}$.

$= \begin{bmatrix} 15 & 27 \\ 9 & -12 \end{bmatrix}$

✔ **Now Try Exercises 41 and 43.**

The proofs of the following properties of scalar multiplication are left for **Exercises 95–98.**

Properties of Scalar Multiplication

Let A and B be matrices of the same dimension, and let c and d be scalars. Then these properties hold.

$$(c + d)A = cA + dA \qquad (cA)d = (cd)A$$
$$c(A + B) = cA + cB \qquad (cd)A = c(dA)$$

Matrix Multiplication We have seen how to multiply a real number (scalar) and a matrix. The product of two matrices can also be found. To illustrate, we multiply

$$A = \begin{bmatrix} -3 & 4 & 2 \\ 5 & 0 & 4 \end{bmatrix} \quad \text{and} \quad B = \begin{bmatrix} -6 & 4 \\ 2 & 3 \\ 3 & -2 \end{bmatrix}.$$

First locate *row* 1 of A and *column* 1 of B, which are shown shaded below.

$$A = \begin{bmatrix} -3 & 4 & 2 \\ 5 & 0 & 4 \end{bmatrix} \qquad B = \begin{bmatrix} -6 & 4 \\ 2 & 3 \\ 3 & -2 \end{bmatrix}$$

Multiply corresponding elements, and find the sum of the products.

$$-3(-6) + 4(2) + 2(3) = 32$$

This result is the element for row 1, column 1 of the product matrix.

Now use *row* 1 of A and *column* 2 of B to determine the element in row 1, column 2 of the product matrix.

$$\begin{bmatrix} -3 & 4 & 2 \\ 5 & 0 & 4 \end{bmatrix} \begin{bmatrix} -6 & 4 \\ 2 & 3 \\ 3 & -2 \end{bmatrix} \qquad -3(4) + 4(3) + 2(-2) = -4$$

Next, use *row* 2 of A and *column* 1 of B. This will give the row 2, column 1 element of the product matrix.

$$\begin{bmatrix} -3 & 4 & 2 \\ 5 & 0 & 4 \end{bmatrix} \begin{bmatrix} -6 & 4 \\ 2 & 3 \\ 3 & -2 \end{bmatrix} \qquad 5(-6) + 0(2) + 4(3) = -18$$

Finally, use *row* 2 of A and *column* 2 of B to find the element for row 2, column 2 of the product matrix.

$$\begin{bmatrix} -3 & 4 & 2 \\ 5 & 0 & 4 \end{bmatrix} \begin{bmatrix} -6 & 4 \\ 2 & 3 \\ 3 & -2 \end{bmatrix} \qquad 5(4) + 0(3) + 4(-2) = 12$$

The product matrix can be written using the four elements just found.

$$\begin{bmatrix} -3 & 4 & 2 \\ 5 & 0 & 4 \end{bmatrix} \begin{bmatrix} -6 & 4 \\ 2 & 3 \\ 3 & -2 \end{bmatrix} = \begin{bmatrix} 32 & -4 \\ -18 & 12 \end{bmatrix} \qquad \text{Product } AB$$

NOTE As seen here, the product of a 2×3 matrix and a 3×2 matrix is a 2×2 matrix. The dimension of a product matrix AB is given by the number of rows of A and the number of columns of B, respectively.

By definition, the product AB of an $m \times n$ matrix A and an $n \times p$ matrix B is found as follows.

> To find the ith row, jth column element of AB, multiply each element in the ith row of A by the corresponding element in the jth column of B. (Note the shaded areas in the matrices below.) The sum of these products will give the row i, column j element of AB.

$$A = \begin{bmatrix} a_{11} & a_{12} & a_{13} & \cdots & a_{1n} \\ a_{21} & a_{22} & a_{23} & \cdots & a_{2n} \\ \vdots & & & & \\ a_{i1} & a_{i2} & a_{i3} & \cdots & a_{in} \\ \vdots & & & & \\ a_{m1} & a_{m2} & a_{m3} & \cdots & a_{mn} \end{bmatrix} \qquad B = \begin{bmatrix} b_{11} & b_{12} & \cdots & b_{1j} & \cdots & b_{1p} \\ b_{21} & b_{22} & \cdots & b_{2j} & \cdots & b_{2p} \\ \vdots & & & & & \\ b_{n1} & b_{n2} & \cdots & b_{nj} & \cdots & b_{np} \end{bmatrix}$$

Matrix Multiplication

The number of columns of an $m \times n$ matrix A is the same as the number of rows of an $n \times p$ matrix B (i.e., both n). The element c_{ij} of the product matrix $C = AB$ is found as follows.

$$c_{ij} = a_{i1}b_{1j} + a_{i2}b_{2j} + \cdots + a_{in}b_{nj}$$

Matrix AB will be an $m \times p$ matrix.

EXAMPLE 5 Deciding Whether Two Matrices Can Be Multiplied

Suppose A is a 3×2 matrix, while B is a 2×4 matrix.

(a) Can the product AB be calculated?

(b) If AB can be calculated, what is its dimension?

(c) Can BA be calculated?

(d) If BA can be calculated, what is its dimension?

SOLUTION

(a) The following diagram shows that AB can be calculated because the number of columns of A is equal to the number of rows of B. (Both are 2.)

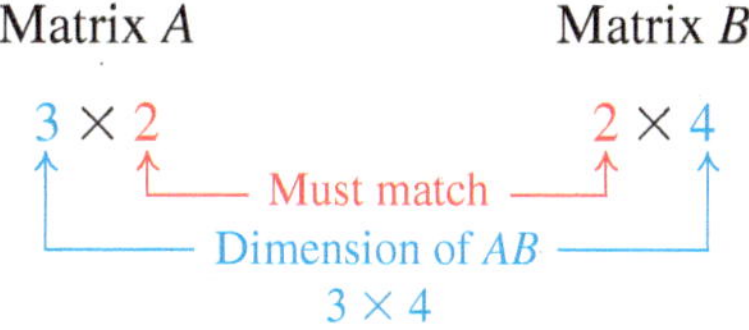

(b) As indicated in the diagram above, the product AB is a 3×4 matrix.

(c) The diagram below shows that BA cannot be calculated.

Matrix B Matrix A

2×4 3×2

Different

(d) The product BA cannot be calculated because B has 4 columns and A has only 3 rows.

✔ **Now Try Exercises 49 and 51.**

EXAMPLE 6 Multiplying Matrices

Let $A = \begin{bmatrix} 1 & -3 \\ 7 & 2 \end{bmatrix}$ and $B = \begin{bmatrix} 1 & 0 & -1 & 2 \\ 3 & 1 & 4 & -1 \end{bmatrix}$. Find each product, if possible.

(a) AB **(b)** BA

SOLUTION

(a) First decide whether AB can be found.

$$AB = \begin{bmatrix} 1 & -3 \\ 7 & 2 \end{bmatrix}\begin{bmatrix} 1 & 0 & -1 & 2 \\ 3 & 1 & 4 & -1 \end{bmatrix}$$ A is 2×2 and B is 2×4, so the product will be a 2×4 matrix.

$$= \begin{bmatrix} 1(1)+(-3)3 & 1(0)+(-3)1 & 1(-1)+(-3)4 & 1(2)+(-3)(-1) \\ 7(1)+2(3) & 7(0)+2(1) & 7(-1)+2(4) & 7(2)+2(-1) \end{bmatrix}$$

Use the definition of matrix multiplication.

$$= \begin{bmatrix} -8 & -3 & -13 & 5 \\ 13 & 2 & 1 & 12 \end{bmatrix}$$ Perform the operations.

(b) B is a 2×4 matrix, and A is a 2×2 matrix, so the number of columns of B (here 4) does not equal the number of rows of A (here 2). Therefore, the product BA cannot be calculated.

✔ **Now Try Exercises 69 and 73.**

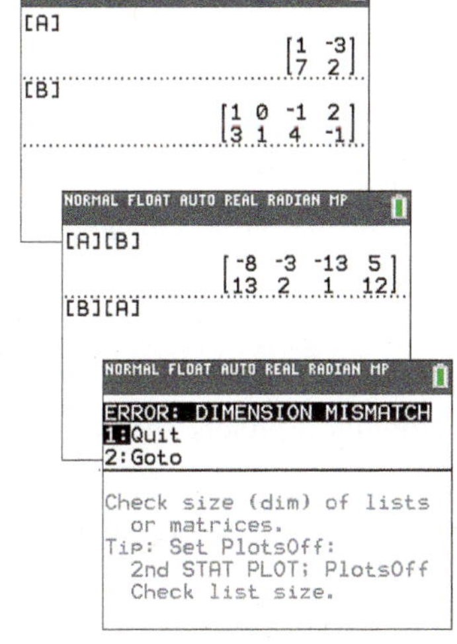

The three screens here support the results of the matrix multiplication in **Example 6.** The final screen indicates that the product BA cannot be found.

EXAMPLE 7 Multiplying Square Matrices in Different Orders

Let $A = \begin{bmatrix} 1 & 3 \\ -2 & 5 \end{bmatrix}$ and $B = \begin{bmatrix} -2 & 7 \\ 0 & 2 \end{bmatrix}$. Find each product.

(a) AB **(b)** BA

SOLUTION

(a) $AB = \begin{bmatrix} 1 & 3 \\ -2 & 5 \end{bmatrix}\begin{bmatrix} -2 & 7 \\ 0 & 2 \end{bmatrix}$

$$= \begin{bmatrix} 1(-2) + 3(0) & 1(7) + 3(2) \\ -2(-2) + 5(0) & -2(7) + 5(2) \end{bmatrix}$$ Multiply elements of each row of A by elements of each column of B.

$$= \begin{bmatrix} -2 & 13 \\ 4 & -4 \end{bmatrix}$$

(b) $BA = \begin{bmatrix} -2 & 7 \\ 0 & 2 \end{bmatrix}\begin{bmatrix} 1 & 3 \\ -2 & 5 \end{bmatrix}$

$$= \begin{bmatrix} -2(1) + 7(-2) & -2(3) + 7(5) \\ 0(1) + 2(-2) & 0(3) + 2(5) \end{bmatrix}$$ Multiply elements of each row of B by elements of each column of A.

$$= \begin{bmatrix} -16 & 29 \\ -4 & 10 \end{bmatrix}$$

Note that $AB \neq BA$.

✔ Now Try Exercise 79.

When multiplying matrices, it is important to pay special attention to the dimensions of the matrices as well as the order in which they are to be multiplied. **Examples 5 and 6** showed that the order in which two matrices are to be multiplied may determine whether their product can be found. **Example 7** showed that even when both products AB and BA can be found, they may not be equal.

Noncommutativity of Matrix Multiplication

In general, if A and B are matrices, then

$$AB \neq BA.$$

Matrix multiplication is not commutative.

Matrix multiplication does satisfy the associative and distributive properties.

Properties of Matrix Multiplication

If A, B, and C are matrices such that all the following products and sums exist, then these properties hold.

$$(AB)C = A(BC), \quad A(B + C) = AB + AC, \quad (B + C)A = BA + CA$$

For proofs of the first two results for the special cases when A, B, and C are square matrices, see **Exercises 93 and 94.** The identity and inverse properties for matrix multiplication are discussed later.

An Application of Matrix Algebra

EXAMPLE 8 Using Matrix Multiplication to Model Plans for a Subdivision

A contractor builds three kinds of houses, models A, B, and C, with a choice of two styles, colonial or ranch. Matrix P below shows the number of each kind of house the contractor is planning to build for a new 100-home subdivision. The amounts for each of the main materials used depend on the style of the house. These amounts are shown in matrix Q, while matrix R gives the cost in dollars for each kind of material. Concrete is measured here in cubic yards, lumber in 1000 board feet, brick in 1000s, and shingles in 100 square feet.

$$\begin{array}{r} \\ \text{Model A} \\ \text{Model B} \\ \text{Model C} \end{array} \begin{array}{c} \begin{array}{cc} \text{Colonial} & \text{Ranch} \end{array} \\ \begin{bmatrix} 0 & 30 \\ 10 & 20 \\ 20 & 20 \end{bmatrix} \end{array} = P$$

$$\begin{array}{r} \\ \text{Colonial} \\ \text{Ranch} \end{array} \begin{array}{c} \begin{array}{cccc} \text{Concrete} & \text{Lumber} & \text{Brick} & \text{Shingles} \end{array} \\ \begin{bmatrix} 10 & 2 & 0 & 2 \\ 50 & 1 & 20 & 2 \end{bmatrix} \end{array} = Q \qquad \begin{array}{r} \\ \text{Concrete} \\ \text{Lumber} \\ \text{Brick} \\ \text{Shingles} \end{array} \begin{array}{c} \text{Cost per Unit} \\ \begin{bmatrix} 20 \\ 180 \\ 60 \\ 25 \end{bmatrix} \end{array} = R$$

(a) What is the total cost of materials for all houses of each model?

(b) How much of each of the four kinds of material must be ordered?

(c) What is the total cost of the materials?

SOLUTION

(a) To find the materials cost for each model, first find matrix PQ, which will show the total amount of each material needed for all houses of each model.

$$PQ = \begin{bmatrix} 0 & 30 \\ 10 & 20 \\ 20 & 20 \end{bmatrix} \begin{bmatrix} 10 & 2 & 0 & 2 \\ 50 & 1 & 20 & 2 \end{bmatrix} = \begin{array}{c} \begin{array}{cccc} \text{Concrete} & \text{Lumber} & \text{Brick} & \text{Shingles} \end{array} \\ \begin{bmatrix} 1500 & 30 & 600 & 60 \\ 1100 & 40 & 400 & 60 \\ 1200 & 60 & 400 & 80 \end{bmatrix} \end{array} \begin{array}{l} \\ \text{Model A} \\ \text{Model B} \\ \text{Model C} \end{array}$$

Multiplying PQ and the cost matrix R gives the total cost of materials for each model.

$$(PQ)R = \begin{bmatrix} 1500 & 30 & 600 & 60 \\ 1100 & 40 & 400 & 60 \\ 1200 & 60 & 400 & 80 \end{bmatrix} \begin{bmatrix} 20 \\ 180 \\ 60 \\ 25 \end{bmatrix} = \begin{array}{c} \text{Cost} \\ \begin{bmatrix} 72{,}900 \\ 54{,}700 \\ 60{,}800 \end{bmatrix} \end{array} \begin{array}{l} \\ \text{Model A} \\ \text{Model B} \\ \text{Model C} \end{array}$$

(b) To find how much of each kind of material to order, refer to the columns of matrix PQ. The sums of the elements of the columns will give a matrix whose elements represent the total amounts of all materials needed for the subdivision. Call this matrix T, and write it as a row matrix.

$$T = \begin{bmatrix} 3800 & 130 & 1400 & 200 \end{bmatrix}$$

(c) The total cost of all the materials is given by the product of matrix T, the total amounts matrix, and matrix R, the cost matrix. To multiply these matrices and obtain a 1×1 matrix, representing the total cost, requires multiplying a 1×4 matrix and a 4×1 matrix. This is why in part (b) a row matrix was written rather than a column matrix.

The total materials cost is given by TR, so

$$TR = \begin{bmatrix} 3800 & 130 & 1400 & 200 \end{bmatrix} \begin{bmatrix} 20 \\ 180 \\ 60 \\ 25 \end{bmatrix} = \begin{bmatrix} 188{,}400 \end{bmatrix}.$$

The total cost of materials is \$188,400. This total may also be found by summing the elements of the column matrix $(PQ)R$.

✔ **Now Try Exercise 85.**

9.7 Exercises

CONCEPT PREVIEW *Fill in the blank(s) to correctly complete each sentence.*

1. For the following statement to be true, the value of x must be ______, and the value of y must be ______.

$$\begin{bmatrix} 3 & -6 \\ 5 & 1 \end{bmatrix} = \begin{bmatrix} x+1 & -6 \\ 5 & y+1 \end{bmatrix}$$

2. For the following sum to be true, we must have $w =$ ______, $x =$ ______, $y =$ ______, and $z =$ ______.

$$\begin{bmatrix} 0 & 5 \\ -3 & 10 \end{bmatrix} + \begin{bmatrix} 1 & 2 \\ 3 & 4 \end{bmatrix} = \begin{bmatrix} w & x \\ y & z \end{bmatrix}$$

3. For the following difference to be true, we must have $w =$ ______, $x =$ ______, $y =$ ______, and $z =$ ______.

$$\begin{bmatrix} 7 & 2 \\ 5 & 12 \end{bmatrix} - \begin{bmatrix} 1 & 3 \\ 6 & 0 \end{bmatrix} = \begin{bmatrix} w & x \\ y & z \end{bmatrix}$$

4. For the following scalar product to be true, we must have $w =$ ______, $x =$ ______, $y =$ ______, and $z =$ ______.

$$-2\begin{bmatrix} 5 & -1 \\ 6 & 0 \end{bmatrix} = \begin{bmatrix} w & x \\ y & z \end{bmatrix}$$

5. If the dimension of matrix A is 3×2 and the dimension of matrix B is 2×6, then the dimension of AB is ______.

6. For the following matrix product to be true, we must have $x =$ ______.

$$\begin{bmatrix} 2 & 1 \\ -4 & 3 \end{bmatrix}\begin{bmatrix} 3 & -1 \\ 2 & 7 \end{bmatrix} = \begin{bmatrix} x & 5 \\ -6 & 25 \end{bmatrix}$$

Concept Check Find the dimension of each matrix. Identify any square, column, or row matrices. ***See the discussion preceding Example 1.***

7. $\begin{bmatrix} -4 & 8 \\ 2 & 3 \end{bmatrix}$

8. $\begin{bmatrix} -9 & 6 & 2 \\ 4 & 1 & 8 \end{bmatrix}$

9. $\begin{bmatrix} -6 & 8 & 0 & 0 \\ 4 & 1 & 9 & 2 \\ 3 & -5 & 7 & 1 \end{bmatrix}$

10. $\begin{bmatrix} 8 & -2 & 4 & 6 & 3 \end{bmatrix}$

11. $\begin{bmatrix} 2 \\ 4 \end{bmatrix}$

12. $\begin{bmatrix} -9 \end{bmatrix}$

Find the values of the variables for which each statement is true, if possible. **See Examples 1 and 2.**

13. $\begin{bmatrix} -3 & a \\ b & 5 \end{bmatrix} = \begin{bmatrix} c & 0 \\ 4 & d \end{bmatrix}$

14. $\begin{bmatrix} w & x \\ 8 & -12 \end{bmatrix} = \begin{bmatrix} 9 & 17 \\ y & z \end{bmatrix}$

15. $\begin{bmatrix} x+2 & y-6 \\ z-3 & w+5 \end{bmatrix} = \begin{bmatrix} -2 & 8 \\ 0 & 3 \end{bmatrix}$

16. $\begin{bmatrix} 6 & a+3 \\ b+2 & 9 \end{bmatrix} = \begin{bmatrix} c-3 & 4 \\ -2 & d-4 \end{bmatrix}$

17. $\begin{bmatrix} x & y & z \end{bmatrix} = \begin{bmatrix} 21 & 6 \end{bmatrix}$

18. $\begin{bmatrix} p \\ q \\ r \end{bmatrix} = \begin{bmatrix} 3 \\ -9 \end{bmatrix}$

19. $\begin{bmatrix} 0 & 5 & x \\ -1 & 3 & y+2 \\ 4 & 1 & z \end{bmatrix} = \begin{bmatrix} 0 & w & 6 \\ -1 & 3 & 0 \\ 4 & 1 & 8 \end{bmatrix}$

20. $\begin{bmatrix} 5 & x-4 & 9 \\ 2 & -3 & 8 \\ 6 & 0 & 5 \end{bmatrix} = \begin{bmatrix} y+3 & 2 & 9 \\ z+4 & -3 & 8 \\ 6 & 0 & w \end{bmatrix}$

21. $\begin{bmatrix} -7+z & 4r & 8s \\ 6p & 2 & 5 \end{bmatrix} + \begin{bmatrix} -9 & 8r & 3 \\ 2 & 5 & 4 \end{bmatrix} = \begin{bmatrix} 2 & 36 & 27 \\ 20 & 7 & 12a \end{bmatrix}$

22. $\begin{bmatrix} a+2 & 3z+1 & 5m \\ 8k & 0 & 3 \end{bmatrix} + \begin{bmatrix} 3a & 2z & 5m \\ 2k & 5 & 6 \end{bmatrix} = \begin{bmatrix} 10 & -14 & 80 \\ 10 & 5 & 9 \end{bmatrix}$

23. Your friend missed the lecture on adding matrices. Explain to him how to add two matrices.

24. Explain how to subtract two matrices.

Find each sum or difference, if possible. **See Examples 2 and 3.**

25. $\begin{bmatrix} -4 & 3 \\ 12 & -6 \end{bmatrix} + \begin{bmatrix} 2 & -8 \\ 5 & 10 \end{bmatrix}$

26. $\begin{bmatrix} 9 & 4 \\ -8 & 2 \end{bmatrix} + \begin{bmatrix} -3 & 2 \\ -4 & 7 \end{bmatrix}$

27. $\begin{bmatrix} 6 & -9 & 2 \\ 4 & 1 & 3 \end{bmatrix} + \begin{bmatrix} -8 & 2 & 5 \\ 6 & -3 & 4 \end{bmatrix}$

28. $\begin{bmatrix} 4 & -3 \\ 7 & 2 \\ -6 & 8 \end{bmatrix} + \begin{bmatrix} 9 & -10 \\ 0 & 5 \\ -1 & 6 \end{bmatrix}$

29. $\begin{bmatrix} 2 & 4 & 6 \end{bmatrix} + \begin{bmatrix} -2 \\ -4 \\ -6 \end{bmatrix}$

30. $\begin{bmatrix} 3 \\ 1 \\ 0 \end{bmatrix} + \begin{bmatrix} 2 \\ -6 \end{bmatrix}$

31. $\begin{bmatrix} -6 & 8 \\ 0 & 0 \end{bmatrix} - \begin{bmatrix} 0 & 0 \\ -4 & -2 \end{bmatrix}$

32. $\begin{bmatrix} 11 & 0 \\ -4 & 0 \end{bmatrix} - \begin{bmatrix} 0 & 12 \\ 0 & -14 \end{bmatrix}$

33. $\begin{bmatrix} 12 \\ -1 \\ 3 \end{bmatrix} - \begin{bmatrix} 8 \\ 4 \\ -1 \end{bmatrix}$

34. $\begin{bmatrix} 10 & -4 & 6 \end{bmatrix} - \begin{bmatrix} -2 & 5 & 3 \end{bmatrix}$

35. $\begin{bmatrix} -4 & 3 \end{bmatrix} - \begin{bmatrix} 5 & 8 & 2 \end{bmatrix}$

36. $\begin{bmatrix} 4 & 6 \end{bmatrix} - \begin{bmatrix} 2 \\ 3 \end{bmatrix}$

37. $\begin{bmatrix} \sqrt{3} & -4 \\ 2 & -\sqrt{5} \\ -8 & \sqrt{8} \end{bmatrix} - \begin{bmatrix} 2\sqrt{3} & 9 \\ -2 & \sqrt{5} \\ -7 & 3\sqrt{2} \end{bmatrix}$

38. $\begin{bmatrix} 2 & \sqrt{7} \\ 3\sqrt{28} & -6 \end{bmatrix} - \begin{bmatrix} -1 & 5\sqrt{7} \\ 2\sqrt{7} & 2 \end{bmatrix}$

39. $\begin{bmatrix} 3x+y & -2y \\ x+2y & 3y \end{bmatrix} + \begin{bmatrix} 2x & 3y \\ 5x & x \end{bmatrix}$

40. $\begin{bmatrix} 4k-8y \\ 6z-3x \\ 2k+5a \\ -4m+2n \end{bmatrix} - \begin{bmatrix} 5k+6y \\ 2z+5x \\ 4k+6a \\ 4m-2n \end{bmatrix}$

Let $A = \begin{bmatrix} -2 & 4 \\ 0 & 3 \end{bmatrix}$ *and* $B = \begin{bmatrix} -6 & 2 \\ 4 & 0 \end{bmatrix}$. *Find each of the following.* ***See Examples 2–4.***

41. $2A$ **42.** $-3B$ **43.** $\frac{3}{2}B$ **44.** $-\frac{3}{2}A$

45. $2A - 3B$ **46.** $-2A + 4B$ **47.** $-A + \frac{1}{2}B$ **48.** $\frac{3}{4}A - B$

Suppose that matrix A has dimension 2×3, *B has dimension* 3×5, *and C has dimension* 5×2. *Decide whether the given product can be calculated. If it can, determine its dimension.* ***See Example 5.***

49. *AB* **50.** *CA* **51.** *BA* **52.** *AC* **53.** *BC* **54.** *CB*

Find each product, if possible. ***See Examples 5–7.***

55. $\begin{bmatrix} 1 & 2 \\ 3 & 4 \end{bmatrix}\begin{bmatrix} -1 \\ 7 \end{bmatrix}$

56. $\begin{bmatrix} -1 & 5 \\ 7 & 0 \end{bmatrix}\begin{bmatrix} 6 \\ 2 \end{bmatrix}$

57. $\begin{bmatrix} 3 & -4 & 1 \\ 5 & 0 & 2 \end{bmatrix}\begin{bmatrix} -1 \\ 4 \\ 2 \end{bmatrix}$

58. $\begin{bmatrix} -6 & 3 & 5 \\ 2 & 9 & 1 \end{bmatrix}\begin{bmatrix} -2 \\ 0 \\ 3 \end{bmatrix}$

59. $\begin{bmatrix} \sqrt{2} & \sqrt{2} & -\sqrt{18} \\ \sqrt{3} & \sqrt{27} & 0 \end{bmatrix}\begin{bmatrix} 8 & -10 \\ 9 & 12 \\ 0 & 2 \end{bmatrix}$

60. $\begin{bmatrix} -9 & 2 & 1 \\ 3 & 0 & 0 \end{bmatrix}\begin{bmatrix} \sqrt{5} \\ \sqrt{20} \\ -2\sqrt{5} \end{bmatrix}$

61. $\begin{bmatrix} \sqrt{3} & 1 \\ 2\sqrt{5} & 3\sqrt{2} \end{bmatrix}\begin{bmatrix} \sqrt{3} & -\sqrt{6} \\ 4\sqrt{3} & 0 \end{bmatrix}$

62. $\begin{bmatrix} \sqrt{7} & 0 \\ 2 & \sqrt{28} \end{bmatrix}\begin{bmatrix} 2\sqrt{3} & -\sqrt{7} \\ 0 & -6 \end{bmatrix}$

63. $\begin{bmatrix} -3 & 0 & 2 & 1 \\ 4 & 0 & 2 & 6 \end{bmatrix}\begin{bmatrix} -4 & 2 \\ 0 & 1 \end{bmatrix}$

64. $\begin{bmatrix} -1 & 2 & 4 & 1 \\ 0 & 2 & -3 & 5 \end{bmatrix}\begin{bmatrix} 1 & 2 & 4 \\ -2 & 5 & 1 \end{bmatrix}$

65. $\begin{bmatrix} -2 & 4 & 1 \end{bmatrix}\begin{bmatrix} 3 & -2 & 4 \\ 2 & 1 & 0 \\ 0 & -1 & 4 \end{bmatrix}$

66. $\begin{bmatrix} 0 & 3 & -4 \end{bmatrix}\begin{bmatrix} -2 & 6 & 3 \\ 0 & 4 & 2 \\ -1 & 1 & 4 \end{bmatrix}$

67. $\begin{bmatrix} -2 & -3 & -4 \\ 2 & -1 & 0 \\ 4 & -2 & 3 \end{bmatrix}\begin{bmatrix} 0 & 1 & 4 \\ 1 & 2 & -1 \\ 3 & 2 & -2 \end{bmatrix}$

68. $\begin{bmatrix} -1 & 2 & 0 \\ 0 & 3 & 2 \\ 0 & 1 & 4 \end{bmatrix}\begin{bmatrix} 2 & -1 & 2 \\ 0 & 2 & 1 \\ 3 & 0 & -1 \end{bmatrix}$

Given $A = \begin{bmatrix} 4 & -2 \\ 3 & 1 \end{bmatrix}$, $B = \begin{bmatrix} 5 & 1 \\ 0 & -2 \\ 3 & 7 \end{bmatrix}$, *and* $C = \begin{bmatrix} -5 & 4 & 1 \\ 0 & 3 & 6 \end{bmatrix}$, *find each product, if possible.* ***See Examples 5–7.***

69. *BA* **70.** *AC* **71.** *BC* **72.** *CB*

73. *AB* **74.** *CA* **75.** A^2 **76.** A^3 (*Hint:* $A^3 = A^2 \cdot A$)

77. *Concept Check* Compare the answers to **Exercises 69 and 73, 71 and 72,** and **70 and 74.** How do they show that matrix multiplication is not commutative?

78. *Concept Check* Why is it not possible to find C^2 for matrix *C* defined as follows.

$$C = \begin{bmatrix} -5 & 4 & 1 \\ 0 & 3 & 6 \end{bmatrix}$$

For each pair of matrices A and B, find (a) AB and (b) BA. ***See Example 7.***

79. $A = \begin{bmatrix} 3 & 4 \\ -2 & 1 \end{bmatrix}, B = \begin{bmatrix} 6 & 0 \\ 5 & -2 \end{bmatrix}$

80. $A = \begin{bmatrix} 0 & -5 \\ -4 & 2 \end{bmatrix}, B = \begin{bmatrix} 3 & -1 \\ -5 & 4 \end{bmatrix}$

81. $A = \begin{bmatrix} 0 & 1 & -1 \\ 0 & 1 & 0 \\ 0 & 0 & 1 \end{bmatrix}, B = \begin{bmatrix} 1 & 0 & 0 \\ 0 & 1 & 0 \\ 0 & 0 & 1 \end{bmatrix}$

82. $A = \begin{bmatrix} -1 & 0 & 1 \\ 0 & 1 & 1 \\ -1 & -1 & 0 \end{bmatrix}, B = \begin{bmatrix} 0 & 0 & 1 \\ 0 & 1 & 0 \\ 1 & 0 & 0 \end{bmatrix}$

83. ***Concept Check*** In **Exercise 81,** $AB = A$ and $BA = A$. For this pair of matrices, B acts the same way for matrix multiplication as the number ______ acts for multiplication of real numbers.

84. ***Concept Check*** Find AB and BA for the following matrices.

$$A = \begin{bmatrix} a & b \\ c & d \end{bmatrix} \quad \text{and} \quad B = \begin{bmatrix} 1 & 0 \\ 0 & 1 \end{bmatrix}$$

Matrix B acts as the multiplicative ______ element for 2×2 square matrices.

Solve each problem. ***See Example 8.***

85. ***Income from Yogurt*** Yagel's Yogurt sells three types of yogurt: nonfat, regular, and super creamy, at three locations. Location I sells 50 gal of nonfat, 100 gal of regular, and 30 gal of super creamy each day. Location II sells 10 gal of nonfat, and Location III sells 60 gal of nonfat each day. Daily sales of regular yogurt are 90 gal at Location II and 120 gal at Location III. At Location II, 50 gal of super creamy are sold each day, and 40 gal of super creamy are sold each day at Location III.

(a) Write a 3×3 matrix that shows the sales figures for the three locations, with the rows representing the three locations.

(b) The incomes per gallon for nonfat, regular, and super creamy are \$12, \$10, and \$15, respectively. Write a 1×3 or 3×1 matrix displaying the incomes.

(c) Find a matrix product that gives the daily income at each of the three locations.

(d) What is Yagel's Yogurt's total daily income from the three locations?

86. ***Purchasing Costs*** The Bread Box, a small neighborhood bakery, sells four main items: sweet rolls, bread, cakes, and pies. The amount of each ingredient (in cups, except for eggs) required for these items is given by matrix A.

	Eggs	Flour	Sugar	Shortening	Milk
Rolls (doz)	1	4	$\frac{1}{4}$	$\frac{1}{4}$	1
Bread (loaf)	0	3	0	$\frac{1}{4}$	0
Cake	4	3	2	1	1
Pie (crust)	0	1	0	$\frac{1}{3}$	0

$= A$

The cost (in cents) for each ingredient when purchased in large lots or small lots is given by matrix B.

	Cost	
	Large Lot	Small Lot
Eggs	5	5
Flour	8	10
Sugar	10	12
Shortening	12	15
Milk	5	6

$= B$

(a) Use matrix multiplication to find a matrix giving the comparative cost per bakery item for the two purchase options.

(b) Suppose a day's orders consist of 20 dozen sweet rolls, 200 loaves of bread, 50 cakes, and 60 pies. Write the orders as a 1×4 matrix, and, using matrix multiplication, write as a matrix the amount of each ingredient needed to fill the day's orders.

(c) Use matrix multiplication to find a matrix giving the costs under the two purchase options to fill the day's orders.

87. *(Modeling) Northern Spotted Owl Population* Mathematical ecologists created a model to analyze population dynamics of the endangered northern spotted owl in the Pacific Northwest. The ecologists divided the female owl population into three categories: juvenile (up to 1 yr old), subadult (1 to 2 yr old), and adult (over 2 yr old). They concluded that the change in the makeup of the northern spotted owl population in successive years could be described by the following matrix equation.

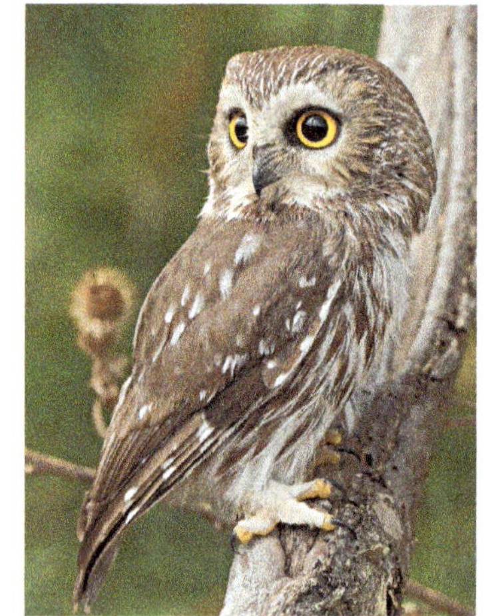

$$\begin{bmatrix} j_{n+1} \\ s_{n+1} \\ a_{n+1} \end{bmatrix} = \begin{bmatrix} 0 & 0 & 0.33 \\ 0.18 & 0 & 0 \\ 0 & 0.71 & 0.94 \end{bmatrix} \begin{bmatrix} j_n \\ s_n \\ a_n \end{bmatrix}$$

The numbers in the column matrices give the numbers of females in the three age groups after n years and $n + 1$ years. Multiplying the matrices yields the following.

$j_{n+1} = 0.33a_n$ — Each year 33 juvenile females are born for each 100 adult females.

$s_{n+1} = 0.18j_n$ — Each year 18% of the juvenile females survive to become subadults.

$a_{n+1} = 0.71s_n + 0.94a_n$ — Each year 71% of the subadults survive to become adults, and 94% of the adults survive.

(*Source:* Lamberson, R. H., R. McKelvey, B. R. Noon, and C. Voss, "A Dynamic Analysis of Northern Spotted Owl Viability in a Fragmented Forest Landscape," *Conservation Biology*, Vol. 6, No. 4.)

(a) Suppose there are currently 3000 female northern spotted owls made up of 690 juveniles, 210 subadults, and 2100 adults. Use the matrix equation to determine the total number of female owls for each of the next 5 yr.

(b) Using advanced techniques from linear algebra, we can show that in the long run,

$$\begin{bmatrix} j_{n+1} \\ s_{n+1} \\ a_{n+1} \end{bmatrix} \approx 0.98359 \begin{bmatrix} j_n \\ s_n \\ a_n \end{bmatrix}.$$

What can we conclude about the long-term fate of the northern spotted owl?

(c) In the model, the main impediment to the survival of the northern spotted owl is the number 0.18 in the second row of the 3×3 matrix. This number is low for two reasons.

- The first year of life is precarious for most animals living in the wild.
- Juvenile owls must eventually leave the nest and establish their own territory. If much of the forest near their original home has been cleared, then they are vulnerable to predators while searching for a new home.

Suppose that, thanks to better forest management, the number 0.18 can be increased to 0.3. Rework part (a) under this new assumption.

88. ***(Modeling) Predator-Prey Relationship*** In certain parts of the Rocky Mountains, deer provide the main food source for mountain lions. When the deer population is large, the mountain lions thrive. However, a large mountain lion population reduces the size of the deer population. Suppose the fluctuations of the two populations from year to year can be modeled with the matrix equation

$$\begin{bmatrix} m_{n+1} \\ d_{n+1} \end{bmatrix} = \begin{bmatrix} 0.51 & 0.4 \\ -0.05 & 1.05 \end{bmatrix} \begin{bmatrix} m_n \\ d_n \end{bmatrix}.$$

The numbers in the column matrices give the numbers of animals in the two populations after n years and $n + 1$ years, where the number of deer is measured in hundreds.

(a) Give the equation for d_{n+1} obtained from the second row of the square matrix. Use this equation to determine the rate at which the deer population will grow from year to year if there are no mountain lions.

(b) Suppose we start with a mountain lion population of 2000 and a deer population of 500,000 (that is, 5000 hundred deer). How large would each population be after 1 yr? 2 yr?

(c) Consider part (b) but change the initial mountain lion population to 4000. Show that the populations would both grow at a steady annual rate of 1.01.

89. ***Northern Spotted Owl Population*** Refer to **Exercise 87(b).** Show that the number 0.98359 is an approximate zero of the polynomial represented by

$$\begin{vmatrix} -x & 0 & 0.33 \\ 0.18 & -x & 0 \\ 0 & 0.71 & 0.94 - x \end{vmatrix}.$$

90. ***Predator-Prey Relationship*** Refer to **Exercise 88(c).** Show that the number 1.01 is a zero of the polynomial represented by

$$\begin{vmatrix} 0.51 - x & 0.4 \\ -0.05 & 1.05 - x \end{vmatrix}.$$

Use the following matrices, where all elements are real numbers, to show that each statement is true for 2×2 *matrices.*

$$A = \begin{bmatrix} a_{11} & a_{12} \\ a_{21} & a_{22} \end{bmatrix}, \quad B = \begin{bmatrix} b_{11} & b_{12} \\ b_{21} & b_{22} \end{bmatrix}, \quad \textit{and} \quad C = \begin{bmatrix} c_{11} & c_{12} \\ c_{21} & c_{22} \end{bmatrix},$$

91. $A + B = B + A$
(commutative property)

92. $A + (B + C) = (A + B) + C$
(associative property)

93. $(AB)C = A(BC)$
(associative property)

94. $A(B + C) = AB + AC$
(distributive property)

95. $c(A + B) = cA + cB$,
for any real number c.

96. $(c + d)A = cA + dA$,
for any real numbers c and d.

97. $(cA)d = (cd)A$, for any real numbers c and d.

98. $(cd)A = c(dA)$, for any real numbers c and d.

9.8 Matrix Inverses

- Identity Matrices
- Multiplicative Inverses
- Solution of Systems Using Inverse Matrices

We have seen several parallels between the set of real numbers and the set of matrices. Another similarity is that both sets have identity and inverse elements for multiplication.

Identity Matrices

By the identity property for real numbers,

$$a \cdot 1 = a \quad \text{and} \quad 1 \cdot a = a$$

for any real number a. If there is to be a multiplicative **identity matrix** I, such that

$$AI = A \quad \text{and} \quad IA = A,$$

for any matrix A, then A and I must be square matrices of the same dimension.

2 × 2 Identity Matrix

I_2 represents the 2×2 identity matrix.

$$I_2 = \begin{bmatrix} 1 & 0 \\ 0 & 1 \end{bmatrix}$$

To verify that I_2 is the 2×2 identity matrix, we must show that $AI = A$ and $IA = A$ for any 2×2 matrix A. Let

$$A = \begin{bmatrix} x & y \\ z & w \end{bmatrix}.$$

Then

$$AI = \begin{bmatrix} x & y \\ z & w \end{bmatrix}\begin{bmatrix} 1 & 0 \\ 0 & 1 \end{bmatrix} = \begin{bmatrix} x \cdot 1 + y \cdot 0 & x \cdot 0 + y \cdot 1 \\ z \cdot 1 + w \cdot 0 & z \cdot 0 + w \cdot 1 \end{bmatrix} = \begin{bmatrix} x & y \\ z & w \end{bmatrix} = A,$$

and

$$IA = \begin{bmatrix} 1 & 0 \\ 0 & 1 \end{bmatrix}\begin{bmatrix} x & y \\ z & w \end{bmatrix} = \begin{bmatrix} 1 \cdot x + 0 \cdot z & 1 \cdot y + 0 \cdot w \\ 0 \cdot x + 1 \cdot z & 0 \cdot y + 1 \cdot w \end{bmatrix} = \begin{bmatrix} x & y \\ z & w \end{bmatrix} = A.$$

Generalizing, there is an $n \times n$ identity matrix for every $n \times n$ square matrix. The $n \times n$ identity matrix has 1s on the main diagonal and 0s elsewhere.

n × n Identity Matrix

The $n \times n$ identity matrix is I_n.

$$I_n = \begin{bmatrix} 1 & 0 & \cdots & 0 \\ 0 & 1 & \cdots & 0 \\ \vdots & \vdots & a_{ij} & \vdots \\ 0 & 0 & \cdots & 1 \end{bmatrix}$$

The element $a_{ij} = 1$ when $i = j$ (the diagonal elements), and $a_{ij} = 0$ otherwise.

EXAMPLE 1 Verifying the Identity Property of I_3

Let $A = \begin{bmatrix} -2 & 4 & 0 \\ 3 & 5 & 9 \\ 0 & 8 & -6 \end{bmatrix}$. Give the 3×3 identity matrix I_3 and show that $AI_3 = A$.

ALGEBRAIC SOLUTION

The 3×3 identity matrix is

$$I_3 = \begin{bmatrix} 1 & 0 & 0 \\ 0 & 1 & 0 \\ 0 & 0 & 1 \end{bmatrix}.$$

Using matrix multiplication,

$$\begin{aligned} AI_3 &= \begin{bmatrix} -2 & 4 & 0 \\ 3 & 5 & 9 \\ 0 & 8 & -6 \end{bmatrix} \begin{bmatrix} 1 & 0 & 0 \\ 0 & 1 & 0 \\ 0 & 0 & 1 \end{bmatrix} \\ &= \begin{bmatrix} -2 & 4 & 0 \\ 3 & 5 & 9 \\ 0 & 8 & -6 \end{bmatrix} \\ &= A. \end{aligned}$$

GRAPHING CALCULATOR SOLUTION

The calculator screen in **Figure 27(a)** shows the identity matrix for $n = 3$. The screen in **Figure 27(b)** supports the algebraic result.

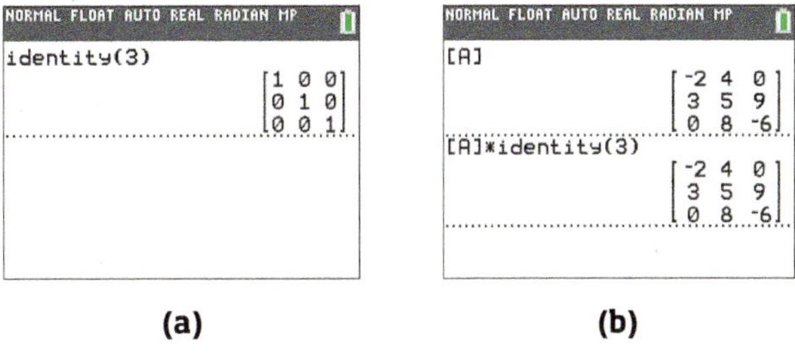

Figure 27

✔ **Now Try Exercise 7.**

Multiplicative Inverses For every *nonzero* real number a, there is a multiplicative inverse $\frac{1}{a}$ that satisfies both of the following.

$$a \cdot \frac{1}{a} = 1 \quad \text{and} \quad \frac{1}{a} \cdot a = 1$$

(Recall: $\frac{1}{a}$ is also written a^{-1}.) In a similar way, if A is an $n \times n$ matrix, then its **multiplicative inverse,** written A^{-1}, must satisfy both of the following.

$$AA^{-1} = I_n \quad \text{and} \quad A^{-1}A = I_n$$

This means that only a square matrix can have a multiplicative inverse.

NOTE Although $a^{-1} = \frac{1}{a}$ for any nonzero real number a, if A is a matrix, then $A^{-1} \neq \frac{1}{A}$. ***We do NOT use the symbol $\frac{1}{A}$ because 1 is a number and A is a matrix.***

To find the matrix A^{-1}, we use row transformations, introduced earlier in this chapter. As an example, we find the inverse of

$$A = \begin{bmatrix} 2 & 4 \\ 1 & -1 \end{bmatrix}.$$

Let the unknown inverse matrix be symbolized as follows.

$$A^{-1} = \begin{bmatrix} x & y \\ z & w \end{bmatrix}$$

By the definition of matrix inverse, $AA^{-1} = I_2$.

$$AA^{-1} = \begin{bmatrix} 2 & 4 \\ 1 & -1 \end{bmatrix} \begin{bmatrix} x & y \\ z & w \end{bmatrix} = \begin{bmatrix} 1 & 0 \\ 0 & 1 \end{bmatrix}$$

Use matrix multiplication for this product, which is repeated in the margin.

$$\begin{bmatrix} 2 & 4 \\ 1 & -1 \end{bmatrix}\begin{bmatrix} x & y \\ z & w \end{bmatrix} = \begin{bmatrix} 1 & 0 \\ 0 & 1 \end{bmatrix}$$

$$\begin{bmatrix} 2x + 4z & 2y + 4w \\ x - z & y - w \end{bmatrix} = \begin{bmatrix} 1 & 0 \\ 0 & 1 \end{bmatrix}$$

Set the corresponding elements equal to obtain a system of equations.

$$\begin{aligned} 2x + 4z &= 1 \quad (1) \\ 2y + 4w &= 0 \quad (2) \\ x - z &= 0 \quad (3) \\ y - w &= 1 \quad (4) \end{aligned}$$

Because equations (1) and (3) involve only x and z, while equations (2) and (4) involve only y and w, these four equations lead to two systems of equations.

$$\begin{aligned} 2x + 4z &= 1 \\ x - z &= 0 \end{aligned} \quad \text{and} \quad \begin{aligned} 2y + 4w &= 0 \\ y - w &= 1 \end{aligned}$$

Write the two systems as augmented matrices.

$$\left[\begin{array}{cc|c} 2 & 4 & 1 \\ 1 & -1 & 0 \end{array}\right] \quad \text{and} \quad \left[\begin{array}{cc|c} 2 & 4 & 0 \\ 1 & -1 & 1 \end{array}\right]$$

Each of these systems can be solved by the Gauss-Jordan method. However, since the elements to the left of the vertical bar are identical, the two systems can be combined into one matrix.

$$\left[\begin{array}{cc|c} 2 & 4 & 1 \\ 1 & -1 & 0 \end{array}\right] \quad \text{and} \quad \left[\begin{array}{cc|c} 2 & 4 & 0 \\ 1 & -1 & 1 \end{array}\right] \quad \text{yields} \quad \left[\begin{array}{cc|cc} 2 & 4 & 1 & 0 \\ 1 & -1 & 0 & 1 \end{array}\right]$$

We can solve simultaneously using matrix row transformations. We need to change the numbers on the left of the vertical bar to the 2×2 identity matrix.

$$\left[\begin{array}{cc|cc} 1 & -1 & 0 & 1 \\ 2 & 4 & 1 & 0 \end{array}\right] \quad \begin{array}{l}\text{Interchange R1 and R2 to introduce} \\ \text{1 in the upper left-hand corner.}\end{array}$$

$$\left[\begin{array}{cc|cc} 1 & -1 & 0 & 1 \\ 0 & 6 & 1 & -2 \end{array}\right] \quad -2\text{R1} + \text{R2}$$

$$\left[\begin{array}{cc|cc} 1 & -1 & 0 & 1 \\ 0 & 1 & \frac{1}{6} & -\frac{1}{3} \end{array}\right] \quad \tfrac{1}{6}\text{R2}$$

$$\left[\begin{array}{cc|cc} 1 & 0 & \frac{1}{6} & \frac{2}{3} \\ 0 & 1 & \frac{1}{6} & -\frac{1}{3} \end{array}\right] \quad \text{R2} + \text{R1}$$

The numbers in the first column to the right of the vertical bar in the final matrix give the values of x and z. The second column gives the values of y and w. That is,

$$\left[\begin{array}{cc|cc} 1 & 0 & x & y \\ 0 & 1 & z & w \end{array}\right] = \left[\begin{array}{cc|cc} 1 & 0 & \frac{1}{6} & \frac{2}{3} \\ 0 & 1 & \frac{1}{6} & -\frac{1}{3} \end{array}\right]$$

so that

$$A^{-1} = \begin{bmatrix} x & y \\ z & w \end{bmatrix} = \begin{bmatrix} \frac{1}{6} & \frac{2}{3} \\ \frac{1}{6} & -\frac{1}{3} \end{bmatrix}.$$

CHECK Multiply A by A^{-1}. The result should be I_2.

$$AA^{-1} = \begin{bmatrix} 2 & 4 \\ 1 & -1 \end{bmatrix} \begin{bmatrix} \frac{1}{6} & \frac{2}{3} \\ \frac{1}{6} & -\frac{1}{3} \end{bmatrix}$$

$$= \begin{bmatrix} \frac{1}{3} + \frac{2}{3} & \frac{4}{3} - \frac{4}{3} \\ \frac{1}{6} - \frac{1}{6} & \frac{2}{3} + \frac{1}{3} \end{bmatrix}$$

$$= \begin{bmatrix} 1 & 0 \\ 0 & 1 \end{bmatrix}$$

$$= I_2 \checkmark$$

Thus, $$A^{-1} = \begin{bmatrix} \frac{1}{6} & \frac{2}{3} \\ \frac{1}{6} & -\frac{1}{3} \end{bmatrix}.$$

This process is summarized below.

Finding an Inverse Matrix

To obtain A^{-1} for any $n \times n$ matrix A for which A^{-1} exists, follow these steps.

Step 1 Form the augmented matrix $[A \mid I_n]$, where I_n is the $n \times n$ identity matrix.

Step 2 Perform row transformations on $[A \mid I_n]$ to obtain a matrix of the form $[I_n \mid B]$.

Step 3 Matrix B is A^{-1}.

NOTE To confirm that two $n \times n$ matrices A and B are inverses of each other, it is sufficient to show that $AB = I_n$. It is not necessary to show also that $BA = I_n$.

As illustrated by the examples, the most efficient order for the transformations in Step 2 is to make the changes column by column from left to right, so that for each column the required 1 is the result of the first change. Next, perform the steps that obtain 0s in that column. Then proceed to the next column.

EXAMPLE 2 Finding the Inverse of a 3 × 3 Matrix

Find A^{-1} if $A = \begin{bmatrix} 1 & 0 & 1 \\ 2 & -2 & -1 \\ 3 & 0 & 0 \end{bmatrix}$.

SOLUTION Use row transformations as follows.

Step 1 Write the augmented matrix $[A \mid I_3]$.

$$\left[\begin{array}{ccc|ccc} 1 & 0 & 1 & 1 & 0 & 0 \\ 2 & -2 & -1 & 0 & 1 & 0 \\ 3 & 0 & 0 & 0 & 0 & 1 \end{array}\right]$$

Step 2 There is already a 1 in the upper left-hand corner as desired. Begin by using the row transformation that will result in 0 for the first element in the second row. Multiply the elements of the first row by -2, and add the result to the second row.

$$\left[\begin{array}{ccc|ccc} 1 & 0 & 1 & 1 & 0 & 0 \\ 0 & -2 & -3 & -2 & 1 & 0 \\ 3 & 0 & 0 & 0 & 0 & 1 \end{array}\right] \quad -2R1 + R2$$

We introduce 0 for the first element in the third row by multiplying the elements of the first row by -3 and adding to the third row.

$$\left[\begin{array}{ccc|ccc} 1 & 0 & 1 & 1 & 0 & 0 \\ 0 & -2 & -3 & -2 & 1 & 0 \\ 0 & 0 & -3 & -3 & 0 & 1 \end{array}\right] \quad -3R1 + R3$$

To obtain 1 for the second element in the second row, multiply the elements of the second row by $-\frac{1}{2}$.

$$\left[\begin{array}{ccc|ccc} 1 & 0 & 1 & 1 & 0 & 0 \\ 0 & 1 & \frac{3}{2} & 1 & -\frac{1}{2} & 0 \\ 0 & 0 & -3 & -3 & 0 & 1 \end{array}\right] \quad -\tfrac{1}{2}R2$$

We want 1 for the third element in the third row, so multiply the elements of the third row by $-\frac{1}{3}$.

$$\left[\begin{array}{ccc|ccc} 1 & 0 & 1 & 1 & 0 & 0 \\ 0 & 1 & \frac{3}{2} & 1 & -\frac{1}{2} & 0 \\ 0 & 0 & 1 & 1 & 0 & -\frac{1}{3} \end{array}\right] \quad -\tfrac{1}{3}R3$$

The third element in the first row should be 0, so multiply the elements of the third row by -1 and add to the first row.

$$\left[\begin{array}{ccc|ccc} 1 & 0 & 0 & 0 & 0 & \frac{1}{3} \\ 0 & 1 & \frac{3}{2} & 1 & -\frac{1}{2} & 0 \\ 0 & 0 & 1 & 1 & 0 & -\frac{1}{3} \end{array}\right] \quad -1R3 + R1$$

Finally, to introduce 0 as the third element in the second row, multiply the elements of the third row by $-\frac{3}{2}$ and add to the second row.

$$\left[\begin{array}{ccc|ccc} 1 & 0 & 0 & 0 & 0 & \frac{1}{3} \\ 0 & 1 & 0 & -\frac{1}{2} & -\frac{1}{2} & \frac{1}{2} \\ 0 & 0 & 1 & 1 & 0 & -\frac{1}{3} \end{array}\right] \quad -\tfrac{3}{2}R3 + R2$$

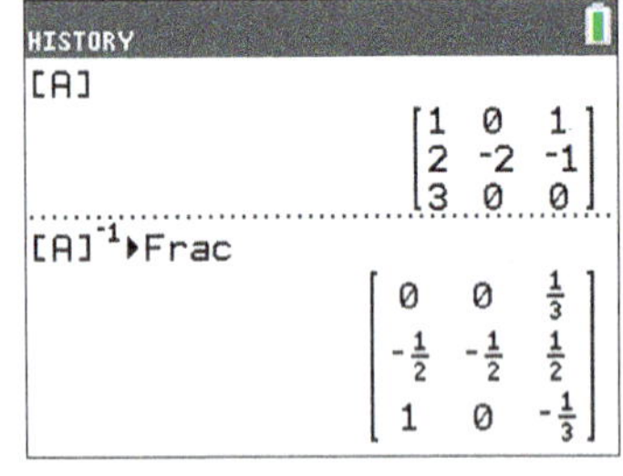

A graphing calculator can be used to find the inverse of a matrix. This screen supports the result in **Example 2.** The elements of the inverse are expressed as fractions, so it is easier to compare with the inverse matrix found in the example.

Step 3 The last transformation shows the inverse.

$$A^{-1} = \begin{bmatrix} 0 & 0 & \frac{1}{3} \\ -\frac{1}{2} & -\frac{1}{2} & \frac{1}{2} \\ 1 & 0 & -\frac{1}{3} \end{bmatrix}$$

Confirm this by forming the product $A^{-1}A$ or AA^{-1}, each of which should equal the matrix I_3.

✔ **Now Try Exercises 17 and 25.**

If the inverse of a matrix exists, it is unique. That is, any given square matrix has no more than one inverse. The proof of this is left as **Exercise 69.**

EXAMPLE 3 Identifying a Matrix with No Inverse

Find A^{-1}, if possible, given that $A = \begin{bmatrix} 2 & -4 \\ 1 & -2 \end{bmatrix}$.

ALGEBRAIC SOLUTION

Using row transformations to change the first column of the augmented matrix

$$\left[\begin{array}{cc|cc} 2 & -4 & 1 & 0 \\ 1 & -2 & 0 & 1 \end{array}\right]$$

results in the following matrices.

$$\left[\begin{array}{cc|cc} 1 & -2 & \frac{1}{2} & 0 \\ 1 & -2 & 0 & 1 \end{array}\right] \quad \frac{1}{2}\text{R1}$$

$$\left[\begin{array}{cc|cc} 1 & -2 & \frac{1}{2} & 0 \\ 0 & 0 & -\frac{1}{2} & 1 \end{array}\right] \quad -\text{R1} + \text{R2}$$

At this point, the matrix should be changed so that the second row, second element will be 1. Because that element is now 0, there is no way to complete the desired transformation, so A^{-1} does not exist for this matrix A.

Just as there is no multiplicative inverse for the real number 0, not every matrix has a multiplicative inverse. Matrix A is an example of such a matrix.

GRAPHING CALCULATOR SOLUTION

If the inverse of a matrix does not exist, the matrix is called **singular,** as shown in **Figure 28** for matrix [A]. This occurs when the determinant of the matrix is 0.

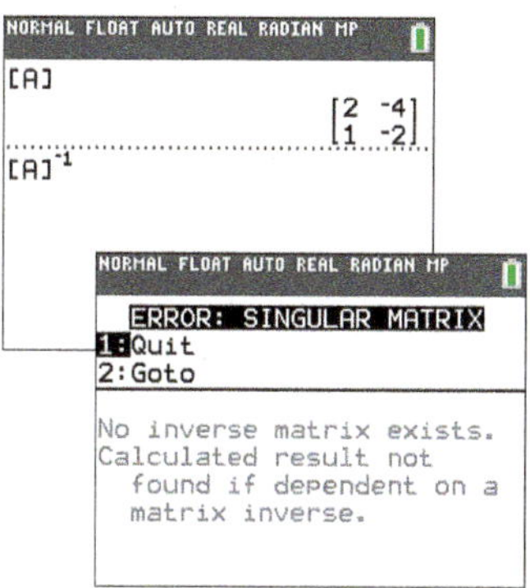

Figure 28

✔ **Now Try Exercise 21.**

Solution of Systems Using Inverse Matrices Matrix inverses can be used to solve *square linear systems of equations.* (A square system has the same number of equations as variables.) For example, consider the following linear system of three equations with three variables.

$$a_{11}x + a_{12}y + a_{13}z = b_1$$
$$a_{21}x + a_{22}y + a_{23}z = b_2$$
$$a_{31}x + a_{32}y + a_{33}z = b_3$$

The definition of matrix multiplication can be used to rewrite the system using matrices.

$$\begin{bmatrix} a_{11} & a_{12} & a_{13} \\ a_{21} & a_{22} & a_{23} \\ a_{31} & a_{32} & a_{33} \end{bmatrix} \cdot \begin{bmatrix} x \\ y \\ z \end{bmatrix} = \begin{bmatrix} b_1 \\ b_2 \\ b_3 \end{bmatrix} \quad (1)$$

(To see this, multiply the matrices on the left.)

$$\text{If} \quad A = \begin{bmatrix} a_{11} & a_{12} & a_{13} \\ a_{21} & a_{22} & a_{23} \\ a_{31} & a_{32} & a_{33} \end{bmatrix}, \quad X = \begin{bmatrix} x \\ y \\ z \end{bmatrix}, \quad \text{and} \quad B = \begin{bmatrix} b_1 \\ b_2 \\ b_3 \end{bmatrix},$$

then the system given in (1) becomes $AX = B$. If A^{-1} exists, then each side of $AX = B$ can be multiplied on the left as shown on the next page.

$$A^{-1}(AX) = A^{-1}B \quad \text{Multiply each side on the left by } A^{-1}.$$
$$(A^{-1}A)X = A^{-1}B \quad \text{Associative property}$$
$$I_3X = A^{-1}B \quad \text{Inverse property}$$
$$X = A^{-1}B \quad \text{Identity property}$$

Matrix $A^{-1}B$ gives the solution of the system.

Solution of the Matrix Equation $AX = B$

Suppose A is an $n \times n$ matrix with inverse A^{-1}, X is an $n \times 1$ matrix of variables, and B is an $n \times 1$ matrix. The matrix equation

$$\boldsymbol{AX = B}$$

has the solution $\quad \boldsymbol{X = A^{-1}B.}$

This method of using matrix inverses to solve systems of equations is useful when the inverse is already known or when many systems of the form $AX = B$ must be solved and only B changes.

EXAMPLE 4 Solving Systems of Equations Using Matrix Inverses

Solve each system using the inverse of the coefficient matrix.

(a) $2x - 3y = 4$
$\quad x + 5y = 2$

(b) $x + z = -1$
$\quad 2x - 2y - z = 5$
$\quad 3x = 6$

ALGEBRAIC SOLUTION

(a) The system can be written in matrix form as

$$\begin{bmatrix} 2 & -3 \\ 1 & 5 \end{bmatrix}\begin{bmatrix} x \\ y \end{bmatrix} = \begin{bmatrix} 4 \\ 2 \end{bmatrix},$$

where $\quad A = \begin{bmatrix} 2 & -3 \\ 1 & 5 \end{bmatrix}, \quad X = \begin{bmatrix} x \\ y \end{bmatrix}, \quad \text{and} \quad B = \begin{bmatrix} 4 \\ 2 \end{bmatrix}.$

An equivalent matrix equation is $AX = B$ with solution $X = A^{-1}B$. Use the methods described in this section to determine that

$$A^{-1} = \begin{bmatrix} \frac{5}{13} & \frac{3}{13} \\ -\frac{1}{13} & \frac{2}{13} \end{bmatrix},$$

and thus $A^{-1}B = \begin{bmatrix} \frac{5}{13} & \frac{3}{13} \\ -\frac{1}{13} & \frac{2}{13} \end{bmatrix}\begin{bmatrix} 4 \\ 2 \end{bmatrix} = \begin{bmatrix} 2 \\ 0 \end{bmatrix}.$

$X = A^{-1}B$, so

$$X = \begin{bmatrix} x \\ y \end{bmatrix} = \begin{bmatrix} 2 \\ 0 \end{bmatrix}.$$

The final matrix shows that the solution set of the system is $\{(2, 0)\}$.

GRAPHING CALCULATOR SOLUTION

(a) Enter [A] and [B] as defined in the algebraic solution, and then find the product $[A]^{-1}[B]$ as shown in **Figure 29.** The display indicates that the solution set is $\{(2, 0)\}$.

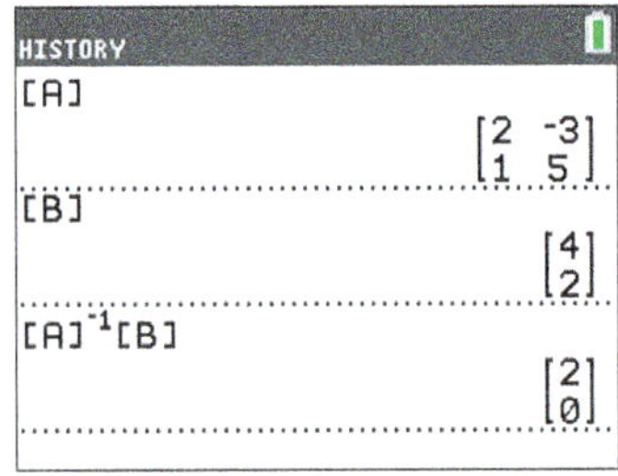

Figure 29

Notice that it is not necessary to actually compute $[A]^{-1}$ here. The calculator stores this inverse and then multiplies it by [B] to obtain the column matrix that represents the solution.

(b) The coefficient matrix A for this system is

$$A = \begin{bmatrix} 1 & 0 & 1 \\ 2 & -2 & -1 \\ 3 & 0 & 0 \end{bmatrix},$$

and its inverse A^{-1} was found in **Example 2.** Let

$$X = \begin{bmatrix} x \\ y \\ z \end{bmatrix} \quad \text{and} \quad B = \begin{bmatrix} -1 \\ 5 \\ 6 \end{bmatrix}.$$

Because $X = A^{-1}B$, we have

$$\begin{bmatrix} x \\ y \\ z \end{bmatrix} = \underbrace{\begin{bmatrix} 0 & 0 & \frac{1}{3} \\ -\frac{1}{2} & -\frac{1}{2} & \frac{1}{2} \\ 1 & 0 & -\frac{1}{3} \end{bmatrix}}_{A^{-1} \text{ from Example 2}} \begin{bmatrix} -1 \\ 5 \\ 6 \end{bmatrix}$$

$$= \begin{bmatrix} 2 \\ 1 \\ -3 \end{bmatrix}.$$

The solution set is $\{(2, 1, -3)\}$.

(b) Figure 30 shows the coefficient matrix $[A]$ and the column matrix of constants $[B]$. Be sure to enter the product of $[A]^{-1}$ and $[B]$ in the correct order. Remember that matrix multiplication is not commutative.

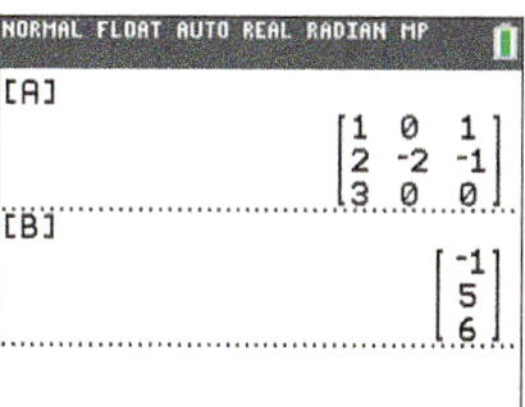

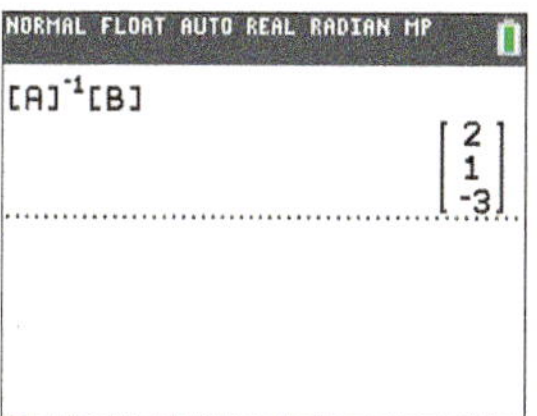

Figure 30

✔ **Now Try Exercises 35 and 49.**

9.8 Exercises

CONCEPT PREVIEW *Answer each question.*

1. What is the product of $\begin{bmatrix} 6 & 4 \\ -1 & 8 \end{bmatrix}$ and I_2 (in either order)?

2. What is the product $\begin{bmatrix} -5 & 7 & 4 \\ 2 & 3 & 0 \\ -1 & 6 & 6 \end{bmatrix}\begin{bmatrix} 1 & 0 & 0 \\ 0 & 1 & 0 \\ 0 & 0 & 1 \end{bmatrix}$?

3. It can be shown that the following matrices are inverses. What is their product (in either order)?

$$\begin{bmatrix} 5 & 1 \\ 4 & 1 \end{bmatrix} \quad \text{and} \quad \begin{bmatrix} 1 & -1 \\ -4 & 5 \end{bmatrix}$$

4. It can be shown that the following matrices are inverses. What is their product (in either order)?

$$\begin{bmatrix} 1 & 0 & 0 \\ 0 & -1 & 0 \\ -1 & 0 & 1 \end{bmatrix} \quad \text{and} \quad \begin{bmatrix} 1 & 0 & 0 \\ 0 & -1 & 0 \\ 1 & 0 & 1 \end{bmatrix}$$

5. What is the coefficient matrix of the following system?

$$3x - 6y = 8$$
$$-x + 3y = 4$$

6. What is the matrix equation form of the following system?

$$6x + 3y = 9$$
$$5x - \ \ y = 4$$

Provide a proof for each of the following.

7. Show that $I_3A = A$ for

$$A = \begin{bmatrix} -2 & 4 & 0 \\ 3 & 5 & 9 \\ 0 & 8 & -6 \end{bmatrix} \text{ and } I_3 = \begin{bmatrix} 1 & 0 & 0 \\ 0 & 1 & 0 \\ 0 & 0 & 1 \end{bmatrix}.$$

(This result, along with that of **Example 1,** illustrates that the commutative property holds when one of the matrices is an identity matrix.)

8. Let $A = \begin{bmatrix} a & b \\ c & d \end{bmatrix}$ and $I_2 = \begin{bmatrix} 1 & 0 \\ 0 & 1 \end{bmatrix}$. Show that $AI_2 = I_2A = A$, thus proving that I_2 is the identity element for matrix multiplication for 2×2 square matrices.

Decide whether or not the given matrices are inverses of each other. (Hint: Check to see whether their products are the identity matrix I_n.)

9. $\begin{bmatrix} 5 & 7 \\ 2 & 3 \end{bmatrix}$ and $\begin{bmatrix} 3 & -7 \\ -2 & 5 \end{bmatrix}$

10. $\begin{bmatrix} 2 & 3 \\ 1 & 1 \end{bmatrix}$ and $\begin{bmatrix} -1 & 3 \\ 1 & -2 \end{bmatrix}$

11. $\begin{bmatrix} -1 & 2 \\ 3 & -5 \end{bmatrix}$ and $\begin{bmatrix} -5 & -2 \\ -3 & -1 \end{bmatrix}$

12. $\begin{bmatrix} 2 & 1 \\ 3 & 2 \end{bmatrix}$ and $\begin{bmatrix} 2 & 1 \\ -3 & 2 \end{bmatrix}$

13. $\begin{bmatrix} 0 & 1 & 0 \\ 0 & 0 & -2 \\ 1 & -1 & 0 \end{bmatrix}$ and $\begin{bmatrix} 1 & 0 & 1 \\ 1 & 0 & 0 \\ 0 & -1 & 0 \end{bmatrix}$

14. $\begin{bmatrix} 1 & 2 & 0 \\ 0 & 1 & 0 \\ 0 & 1 & 0 \end{bmatrix}$ and $\begin{bmatrix} 1 & -2 & 0 \\ 0 & 1 & 0 \\ 0 & -1 & 1 \end{bmatrix}$

15. $\begin{bmatrix} 1 & 0 & 0 \\ 0 & -1 & 0 \\ 1 & 0 & 1 \end{bmatrix}$ and $\begin{bmatrix} 1 & 0 & 0 \\ 0 & -1 & 0 \\ -1 & 0 & 1 \end{bmatrix}$

16. $\begin{bmatrix} 1 & 3 & 3 \\ 1 & 4 & 3 \\ 1 & 3 & 4 \end{bmatrix}$ and $\begin{bmatrix} 7 & -3 & -3 \\ -1 & 1 & 0 \\ -1 & 0 & 1 \end{bmatrix}$

Find the inverse, if it exists, for each matrix. ***See Examples 2 and 3.***

17. $\begin{bmatrix} -1 & 2 \\ -2 & -1 \end{bmatrix}$

18. $\begin{bmatrix} 1 & -1 \\ 2 & 0 \end{bmatrix}$

19. $\begin{bmatrix} -1 & -2 \\ 3 & 4 \end{bmatrix}$

20. $\begin{bmatrix} 3 & -1 \\ -5 & 2 \end{bmatrix}$

21. $\begin{bmatrix} 5 & 10 \\ -3 & -6 \end{bmatrix}$

22. $\begin{bmatrix} -6 & 4 \\ -3 & 2 \end{bmatrix}$

23. $\begin{bmatrix} 1 & 0 & 1 \\ 0 & -1 & 0 \\ 2 & 1 & 1 \end{bmatrix}$

24. $\begin{bmatrix} 1 & 0 & 0 \\ 0 & -1 & 0 \\ 1 & 0 & 1 \end{bmatrix}$

25. $\begin{bmatrix} 2 & 3 & 3 \\ 1 & 4 & 3 \\ 1 & 3 & 4 \end{bmatrix}$

26. $\begin{bmatrix} -2 & 2 & 4 \\ -3 & 4 & 5 \\ 1 & 0 & 2 \end{bmatrix}$

27. $\begin{bmatrix} 2 & 2 & -4 \\ 2 & 6 & 0 \\ -3 & -3 & 5 \end{bmatrix}$

28. $\begin{bmatrix} 2 & 4 & 6 \\ -1 & -4 & -3 \\ 0 & 1 & -1 \end{bmatrix}$

29. $\begin{bmatrix} 1 & 1 & 0 & 2 \\ 2 & -1 & 1 & -1 \\ 3 & 3 & 2 & -2 \\ 1 & 2 & 1 & 0 \end{bmatrix}$

30. $\begin{bmatrix} 1 & -2 & 3 & 0 \\ 0 & 1 & -1 & 1 \\ -2 & 2 & -2 & 4 \\ 0 & 2 & -3 & 1 \end{bmatrix}$

31. $\begin{bmatrix} 3 & 2 & 0 & -1 \\ 2 & 0 & 1 & 2 \\ 1 & 2 & -1 & 0 \\ 2 & -1 & 1 & 1 \end{bmatrix}$

32. *Concept Check* Each graphing calculator screen shows A^{-1} for some matrix A. Find each matrix A. (*Hint:* $(A^{-1})^{-1} = A$)

(a) 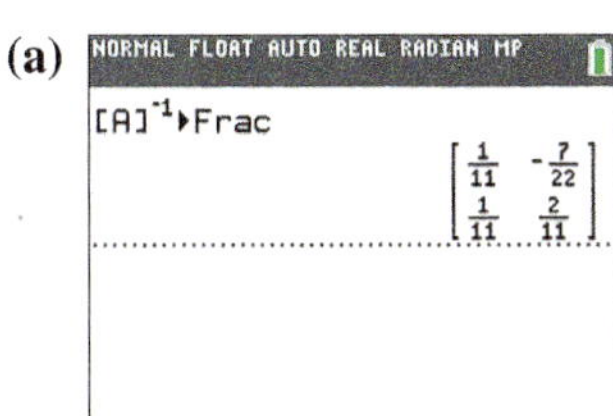

(b) 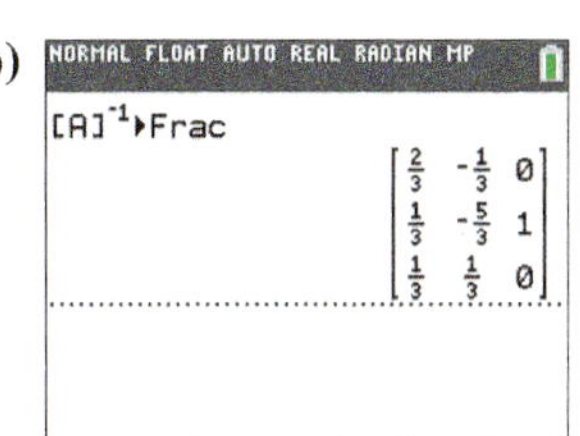

Solve each system using the inverse of the coefficient matrix. ***See Example 4.***

33. $-x + y = 1$
$2x - y = 1$

34. $x + y = 5$
$x - y = -1$

35. $2x - y = -8$
$3x + y = -2$

36. $x + 3y = -12$
$2x - y = 11$

37. $3x + 4y = -3$
$-5x + 8y = 16$

38. $2x - 3y = 10$
$2x + 2y = 5$

39. $6x + 9y = 3$
$-8x + 3y = 6$

40. $5x - 3y = 0$
$10x + 6y = -4$

41. $0.2x + 0.3y = -1.9$
$0.7x - 0.2y = 4.6$

42. $0.5x + 0.2y = 0.8$
$0.3x - 0.1y = 0.7$

43. $\frac{1}{2}x + \frac{1}{3}y = \frac{49}{18}$
$\frac{1}{2}x + 2y = \frac{4}{3}$

44. $\frac{1}{5}x + \frac{1}{7}y = \frac{12}{5}$
$\frac{1}{10}x + \frac{1}{3}y = \frac{5}{6}$

Concept Check Show that the matrix inverse method cannot be used to solve each system.

45. $7x - 2y = 3$
$14x - 4y = 1$

46. $x - 2y + 3z = 4$
$2x - 4y + 6z = 8$
$3x - 6y + 9z = 14$

Solve each system by using the inverse of the coefficient matrix. For Exercises 49–54, the inverses were found in ***Exercises 25–30. See Example 4.***

47. $x + y + z = 6$
$2x + 3y - z = 7$
$3x - y - z = 6$

48. $2x + 5y + 2z = 9$
$4x - 7y - 3z = 7$
$3x - 8y - 2z = 9$

49. $2x + 3y + 3z = 1$
$x + 4y + 3z = 0$
$x + 3y + 4z = -1$

50. $-2x + 2y + 4z = 3$
$-3x + 4y + 5z = 1$
$x + 2z = 2$

51. $2x + 2y - 4z = 12$
$2x + 6y = 16$
$-3x - 3y + 5z = -20$

52. $2x + 4y + 6z = 4$
$-x - 4y - 3z = 8$
$y - z = -4$

53. $x + y + 2w = 3$
$2x - y + z - w = 3$
$3x + 3y + 2z - 2w = 5$
$x + 2y + z = 3$

54. $x - 2y + 3z = 1$
$y - z + w = -1$
$-2x + 2y - 2z + 4w = 2$
$2y - 3z + w = -3$

(Modeling) Solve each problem.

55. *Plate-Glass Sales* The amount of plate-glass sales S (in millions of dollars) can be affected by the number of new building contracts B issued (in millions) and automobiles A produced (in millions). A plate-glass company in California wants to forecast future sales using the past three years of sales. The totals for the three years are given in the table.

S	A	B
602.7	5.543	37.14
656.7	6.933	41.30
778.5	7.638	45.62

To describe the relationship among these variables, we can use the equation

$$S = a + bA + cB,$$

where the coefficients a, b, and c are constants that must be determined before the equation can be used. (*Source:* Makridakis, S., and S. Wheelwright, *Forecasting Methods for Management*, John Wiley and Sons.)

(a) Substitute the values for S, A, and B for each year from the table into the equation $S = a + bA + cB$, and obtain three linear equations involving a, b, and c.

(b) Use a graphing calculator to solve this linear system for a, b, and c. Use matrix inverse methods.

(c) Write the equation for S using these values for the coefficients.

(d) For the next year it is estimated that $A = 7.752$ and $B = 47.38$. Predict S. (The actual value for S was 877.6.)

(e) It is predicted that in 6 yr, $A = 8.9$ and $B = 66.25$. Find the value of S in this situation and discuss its validity.

56. *Tire Sales* The number of automobile tire sales is dependent on several variables. In one study the relationship among annual tire sales S (in thousands of dollars), automobile registrations R (in millions), and personal disposable income I (in millions of dollars) was investigated. The results for three years are given in the table.

S	R	I
10,170	112.9	307.5
15,305	132.9	621.63
21,289	155.2	1937.13

To describe the relationship among these variables, we can use the equation

$$S = a + bR + cI,$$

where the coefficients a, b, and c are constants that must be determined before the equation can be used. (*Source:* Jarrett, J., *Business Forecasting Methods*, Basil Blackwell, Ltd.)

(a) Substitute the values for S, R, and I for each year from the table into the equation $S = a + bR + cI$, and obtain three linear equations involving a, b, and c.

(b) Use a graphing calculator to solve this linear system for a, b, and c. Use matrix inverse methods.

(c) Write the equation for S using these values for the coefficients.

(d) If $R = 117.6$ and $I = 310.73$, predict S. (The actual value for S was 11,314.)

(e) If $R = 143.8$ and $I = 829.06$, predict S. (The actual value for S was 18,481.)

57. ***Social Security Numbers*** It is possible to find a polynomial that goes through a given set of points in the plane by using a process called **polynomial interpolation.** Recall that three points define a second-degree polynomial, four points define a third-degree polynomial, and so on. The only restriction on the points, because polynomials define functions, is that no two distinct points can have the same x-coordinate.

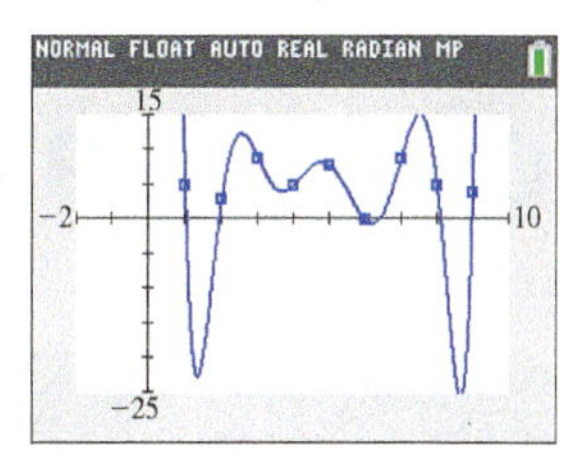

Using the SSN 539-58-0954, we can find an eighth-degree polynomial that lies on the nine points with x-coordinates 1 through 9 and y-coordinates that are digits of the SSN: $(1, 5)$, $(2, 3)$, $(3, 9)$, . . . , $(9, 4)$. This is done by writing a system of nine equations with nine variables, which is then solved by the inverse matrix method. The graph of this polynomial is shown. Find such a polynomial using your own SSN.

58. Repeat **Exercise 57** but use $-1, -2, \ldots, -9$ for the x-coordinates.

Use a graphing calculator to find the inverse of each matrix. Give as many decimal places as the calculator shows. ***See Example 2.***

59. $\begin{bmatrix} \frac{2}{3} & 0.7 \\ 22 & \sqrt{3} \end{bmatrix}$ **60.** $\begin{bmatrix} \sqrt{2} & 0.5 \\ -17 & \frac{1}{2} \end{bmatrix}$ **61.** $\begin{bmatrix} \frac{1}{2} & \frac{1}{4} & \frac{1}{3} \\ 0 & \frac{1}{4} & \frac{1}{3} \\ \frac{1}{2} & \frac{1}{2} & \frac{1}{3} \end{bmatrix}$ **62.** $\begin{bmatrix} 1.4 & 0.5 & 0.59 \\ 0.84 & 1.36 & 0.62 \\ 0.56 & 0.47 & 1.3 \end{bmatrix}$

Use a graphing calculator and the method of matrix inverses to solve each system. Give as many decimal places as the calculator shows. ***See Example 4.***

63. $\begin{aligned} 2.1x + y &= \sqrt{5} \\ \sqrt{2}x - 2y &= 5 \end{aligned}$

64. $\begin{aligned} x - \sqrt{2}y &= 2.6 \\ 0.75x + y &= -7 \end{aligned}$

65. $\begin{aligned} (\log 2)x + (\ln 3)y + (\ln 4)z &= 1 \\ (\ln 3)x + (\log 2)y + (\ln 8)z &= 5 \\ (\log 12)x + (\ln 4)y + (\ln 8)z &= 9 \end{aligned}$

66. $\begin{aligned} \pi x + ey + \sqrt{2}z &= 1 \\ ex + \pi y + \sqrt{2}z &= 2 \\ \sqrt{2}x + ey + \pi z &= 3 \end{aligned}$

Let $A = \begin{bmatrix} a & b \\ c & d \end{bmatrix}$, *and let O be the 2×2 zero matrix. Show that each statement is true.*

67. $A \cdot O = O \cdot A = O$

68. For square matrices A and B of the same dimension, if $AB = O$ and if A^{-1} exists, then $B = O$.

Work each problem.

69. Prove that any square matrix has no more than one inverse.

70. Give an example of two matrices A and B, where $(AB)^{-1} \neq A^{-1}B^{-1}$.

71. Suppose A and B are matrices, where A^{-1}, B^{-1}, and AB all exist. Show that $(AB)^{-1} = B^{-1}A^{-1}$.

72. Let $A = \begin{bmatrix} a & 0 & 0 \\ 0 & b & 0 \\ 0 & 0 & c \end{bmatrix}$, where a, b, and c are nonzero real numbers. Find A^{-1}.

73. Let $A = \begin{bmatrix} 1 & 0 & 0 \\ 0 & 0 & -1 \\ 0 & 1 & -1 \end{bmatrix}$. Show that $A^3 = I_3$, and use this result to find the inverse of A.

74. What are the inverses of I_n, $-A$ (in terms of A), and kA (k a scalar)?

Chapter 9 Test Prep

Key Terms

9.1 linear equation (first-degree equation) in n unknowns
system of equations
solutions of a system of equations
system of linear equations (linear system)
consistent system
independent equations
inconsistent system
dependent equations
equivalent system
ordered triple

9.2 matrix (matrices)
element (of a matrix)
augmented matrix
dimension (of a matrix)

9.3 determinant
minor
cofactor
expansion by a row or column
Cramer's rule

9.4 partial fraction decomposition
partial fraction

9.5 nonlinear system

9.6 half-plane
boundary
linear inequality in two variables
system of inequalities
linear programming
constraints
objective function
region of feasible solutions
vertex (corner point)

9.7 square matrix of order n
row matrix
column matrix
zero matrix
additive inverse (negative) of a matrix
scalar

9.8 identity matrix
multiplicative inverse (of a matrix)

New Symbols

(a, b, c)	ordered triple
$[A]$	matrix A (graphing calculator symbolism)
$\lvert A \rvert$	determinant of matrix A
a_{ij}	element in row i, column j of a matrix
D, D_x, D_y	determinants used in Cramer's rule
I_2, I_3	identity matrices
A^{-1}	multiplicative inverse of matrix A

Quick Review

Concepts	Examples

9.1 Systems of Linear Equations

Transformations of a Linear System

1. Interchange any two equations of the system.
2. Multiply or divide any equation of the system by a nonzero real number.
3. Replace any equation of the system by the sum of that equation and a multiple of another equation in the system.

A system may be solved by the substitution method, the elimination method, or a combination of the two methods.

Substitution Method

Use one equation to find an expression for one variable in terms of the other, and then substitute this expression into the other equation of the system.

Examples

Solve the system.

$$4x - y = 7 \quad (1)$$
$$3x + 2y = 30 \quad (2)$$

Solve for y in equation (1).

$$y = 4x - 7 \quad (3)$$

Substitute $4x - 7$ for y in equation (2), and solve for x.

$$3x + 2(4x - 7) = 30 \quad \text{(2) with } y = 4x - 7$$
$$3x + 8x - 14 = 30 \quad \text{Distributive property}$$
$$11x - 14 = 30 \quad \text{Combine like terms.}$$
$$11x = 44 \quad \text{Add 14.}$$
$$x = 4 \quad \text{Divide by 11.}$$

Substitute 4 for x in the equation $y = 4x - 7$ to find that $y = 9$. The solution set is $\{(4, 9)\}$.

Concepts	Examples
Elimination Method Use multiplication and addition to eliminate a variable from one equation. To eliminate a variable, the coefficients of that variable in the equations must be additive inverses.	Solve the system. $x + 2y - z = 6$ (1) $x + y + z = 6$ (2) $2x + y - z = 7$ (3)
Solving a Linear System with Three Unknowns ***Step 1*** Eliminate a variable from any two of the equations. ***Step 2*** Eliminate the ***same variable*** from a different pair of equations. ***Step 3*** Eliminate a second variable using the resulting two equations in two variables to obtain an equation with just one variable whose value we can now determine. ***Step 4*** Find the values of the remaining variables by substitution. Write the solution of the system as an ordered triple.	Add equations (1) and (2). The variable z is eliminated, and the result is $2x + 3y = 12$. Eliminate z again by adding equations (2) and (3) to obtain $3x + 2y = 13$. Solve the resulting system. $2x + 3y = 12$ (4) $3x + 2y = 13$ (5) $-6x - 9y = -36$ Multiply (4) by -3. $6x + 4y = 26$ Multiply (5) by 2. $-5y = -10$ Add. $y = 2$ Divide by -5. Substitute 2 for y in equation (4). $2x + 3(2) = 12$ (4) with $y = 2$ $2x + 6 = 12$ Multiply. $2x = 6$ Subtract 6. $x = 3$ Divide by 2. Let $y = 2$ and $x = 3$ in any of the original equations to find $z = 1$. The solution set is $\{(3, 2, 1)\}$.
9.2 Matrix Solution of Linear Systems **Matrix Row Transformations** For any augmented matrix of a system of linear equations, the following row transformations will result in the matrix of an equivalent system. **1.** Interchange any two rows. **2.** Multiply or divide the elements of any row by a nonzero real number. **3.** Replace any row of the matrix by the sum of the elements of that row and a multiple of the elements of another row. **Gauss-Jordan Method** The Gauss-Jordan method is a systematic technique for applying matrix row transformations in an attempt to reduce a matrix to diagonal form, with 1s along the diagonal.	Solve the system. $x + 3y = 7$ $2x + y = 4$ $\left[\begin{array}{cc\|c} 1 & 3 & 7 \\ 2 & 1 & 4 \end{array}\right]$ Augmented matrix $\left[\begin{array}{cc\|c} 1 & 3 & 7 \\ 0 & -5 & -10 \end{array}\right]$ $-2\text{R1} + \text{R2}$ $\left[\begin{array}{cc\|c} 1 & 3 & 7 \\ 0 & 1 & 2 \end{array}\right]$ $-\frac{1}{5}\text{R2}$ $\left[\begin{array}{cc\|c} 1 & 0 & 1 \\ 0 & 1 & 2 \end{array}\right]$ $-3\text{R2} + \text{R1}$ This leads to the system $x = 1$ $y = 2$. The solution set is $\{(1, 2)\}$.

Concepts | **Examples**

9.3 Determinant Solution of Linear Systems

Determinant of a 2 × 2 Matrix

If $A = \begin{bmatrix} a_{11} & a_{12} \\ a_{21} & a_{22} \end{bmatrix}$, then

$$|A| = \begin{vmatrix} a_{11} & a_{12} \\ a_{21} & a_{22} \end{vmatrix} = a_{11}a_{22} - a_{21}a_{12}.$$

Evaluate.

$$\begin{vmatrix} 3 & 5 \\ -2 & 6 \end{vmatrix} = 3(6) - (-2)5$$
$$= 28$$

Determinant of a 3 × 3 Matrix

If $A = \begin{bmatrix} a_{11} & a_{12} & a_{13} \\ a_{21} & a_{22} & a_{23} \\ a_{31} & a_{32} & a_{33} \end{bmatrix}$, then

$$|A| = \begin{vmatrix} a_{11} & a_{12} & a_{13} \\ a_{21} & a_{22} & a_{23} \\ a_{31} & a_{32} & a_{33} \end{vmatrix} = (a_{11}a_{22}a_{33} + a_{12}a_{23}a_{31} + a_{13}a_{21}a_{32}) - (a_{31}a_{22}a_{13} + a_{32}a_{23}a_{11} + a_{33}a_{21}a_{12})$$

In practice, we usually evaluate determinants by expansion by minors.

Evaluate by expanding about the second column.

$$\begin{vmatrix} 2 & -3 & -2 \\ -1 & -4 & -3 \\ -1 & 0 & 2 \end{vmatrix} = -(-3)\begin{vmatrix} -1 & -3 \\ -1 & 2 \end{vmatrix} + (-4)\begin{vmatrix} 2 & -2 \\ -1 & 2 \end{vmatrix}$$
$$- 0\begin{vmatrix} 2 & -2 \\ -1 & -3 \end{vmatrix}$$
$$= 3(-5) - 4(2) - 0(-8)$$
$$= -15 - 8 + 0$$
$$= -23$$

Cramer's Rule for Two Equations in Two Variables
Given the system

$$a_1x + b_1y = c_1$$
$$a_2x + b_2y = c_2,$$

if $D \neq 0$, then the system has the unique solution

$$x = \frac{D_x}{D} \quad \text{and} \quad y = \frac{D_y}{D},$$

where $D = \begin{vmatrix} a_1 & b_1 \\ a_2 & b_2 \end{vmatrix}$, $D_x = \begin{vmatrix} c_1 & b_1 \\ c_2 & b_2 \end{vmatrix}$, and $D_y = \begin{vmatrix} a_1 & c_1 \\ a_2 & c_2 \end{vmatrix}$.

Solve using Cramer's rule.

$$x - 2y = -1$$
$$2x + 5y = 16$$

$$x = \frac{D_x}{D} = \frac{\begin{vmatrix} -1 & -2 \\ 16 & 5 \end{vmatrix}}{\begin{vmatrix} 1 & -2 \\ 2 & 5 \end{vmatrix}} = \frac{-5 + 32}{5 + 4} = \frac{27}{9} = 3$$

$$y = \frac{D_y}{D} = \frac{\begin{vmatrix} 1 & -1 \\ 2 & 16 \end{vmatrix}}{\begin{vmatrix} 1 & -2 \\ 2 & 5 \end{vmatrix}} = \frac{16 + 2}{5 + 4} = \frac{18}{9} = 2$$

The solution set is $\{(3, 2)\}$.

General Form of Cramer's Rule
Let an $n \times n$ system have linear equations of the form

$$a_1x_1 + a_2x_2 + a_3x_3 + \cdots + a_nx_n = b.$$

Define D as the determinant of the $n \times n$ matrix of coefficients of the variables. Define D_{x_1} as the determinant obtained from D by replacing the entries in column 1 of D with the constants of the system. Define D_{x_i} as the determinant obtained from D by replacing the entries in column i with the constants of the system. If $D \neq 0$, then the unique solution of the system is

$$x_1 = \frac{D_{x_1}}{D}, \quad x_2 = \frac{D_{x_2}}{D}, \quad x_3 = \frac{D_{x_3}}{D}, \quad \ldots, \quad x_n = \frac{D_{x_n}}{D}.$$

Solve using Cramer's rule.

$$3x + 2y + z = -5$$
$$x - y + 3z = -5$$
$$2x + 3y + z = 0$$

Using the method of expansion by minors, it can be shown that $D_x = 45$, $D_y = -30$, $D_z = 0$, and $D = -15$.

$$x = \frac{D_x}{D} = \frac{45}{-15} = -3, \quad y = \frac{D_y}{D} = \frac{-30}{-15} = 2,$$
$$z = \frac{D_z}{D} = \frac{0}{-15} = 0$$

The solution set is $\{(-3, 2, 0)\}$.

Concepts	Examples
9.4 Partial Fractions To solve for the constants in the numerators of a partial fraction decomposition, use either of the following methods or a combination of the two. ***Method 1* For Linear Factors** *Step 1* Multiply each side by the common denominator. *Step 2* Substitute the zero of each factor in the resulting equation. For repeated linear factors, substitute as many other numbers as necessary to find all the constants in the numerators. The number of substitutions required will equal the number of constants $A, B, \ldots$. ***Method 2* For Quadratic Factors** *Step 1* Multiply each side by the common denominator. *Step 2* Collect like terms on the right side of the resulting equation. *Step 3* Equate the coefficients of like terms to form a system of equations. *Step 4* Solve the system to find the constants in the numerators.	Find the partial fraction decomposition of $\frac{9}{2x^2+9x+9}$. $\frac{9}{(2x+3)(x+3)} = \frac{A}{2x+3} + \frac{B}{x+3}$ (1) Multiply by $(2x+3)(x+3)$. $9 = A(x+3) + B(2x+3)$ $9 = Ax + 3A + 2Bx + 3B$ $9 = (A+2B)x + (3A+3B)$ Now solve the system $A + 2B = 0$ $3A + 3B = 9$ to obtain $A = 6$ and $B = -3$. $\frac{9}{2x^2+9x+9} = \frac{6}{2x+3} + \frac{-3}{x+3}$ Substitute into (1). Check this result by combining the terms on the right.
9.5 Nonlinear Systems of Equations **Solving a Nonlinear System of Equations** A nonlinear system can be solved by the substitution method, the elimination method, or a combination of the two methods.	Solve the system. $x^2 + 2xy - y^2 = 14$ (1) $x^2 - y^2 = -16$ (2) $x^2 + 2xy - y^2 = 14$ (1) $-x^2 + y^2 = 16$ Multiply (2) by -1. $2xy = 30$ Add to eliminate x^2 and y^2. $xy = 15$ Divide by 2. $y = \frac{15}{x}$ Solve for y. Substitute $\frac{15}{x}$ for y in equation (2). $x^2 - \left(\frac{15}{x}\right)^2 = -16$ (2) with $y = \frac{15}{x}$ $x^2 - \frac{225}{x^2} = -16$ Square. $x^4 + 16x^2 - 225 = 0$ Multiply by x^2. Add $16x^2$. $(x^2-9)(x^2+25) = 0$ Factor. $x^2 - 9 = 0$ or $x^2 + 25 = 0$ Zero-factor property $x = \pm 3$ or $x = \pm 5i$ Solve each equation. Find corresponding y-values. The solution set is $\{(3, 5), (-3, -5), (5i, -3i), (-5i, 3i)\}$.

Concepts	Examples

9.6 Systems of Inequalities and Linear Programming

Graphing an Inequality in Two Variables

Method 1

If the inequality is or can be solved for y, then the following hold.

- The graph of $y < f(x)$ consists of all the points that are *below* the graph of $y = f(x)$.
- The graph of $y > f(x)$ consists of all the points that are *above* the graph of $y = f(x)$.

Method 2

If the inequality is not or cannot be solved for y, then choose a test point not on the boundary.

- If the test point satisfies the inequality, then the graph includes all points on the *same* side of the boundary as the test point.
- If the test point does not satisfy the inequality, then the graph includes all points on the *other* side of the boundary.

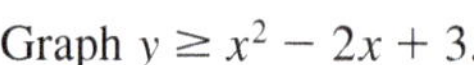

Graph $y \geq x^2 - 2x + 3$.

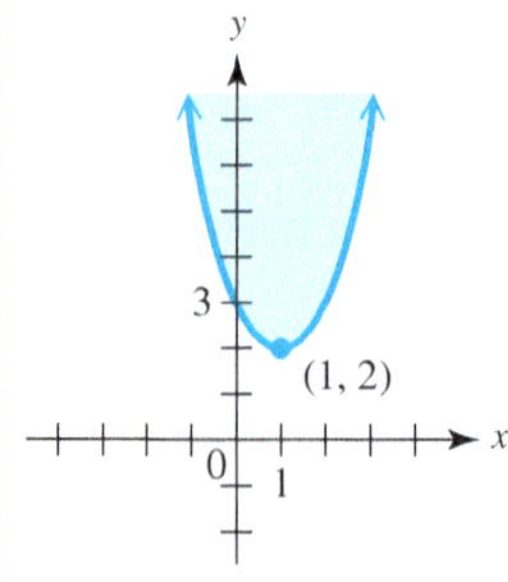

Graph the solution set of the system.

$$3x - 5y > -15$$
$$x^2 + y^2 \leq 25$$

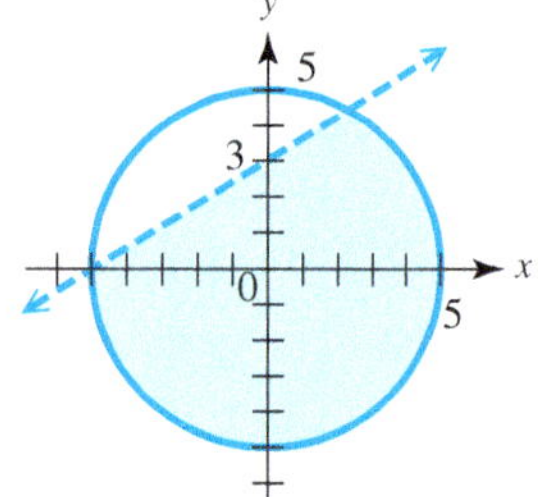

Solving Systems of Inequalities

To solve a system of inequalities, graph all inequalities on the same axes, and find the intersection of their solution sets.

Solving a Linear Programming Problem

Step 1 Write all necessary constraints and the objective function.

Step 2 Graph the region of feasible solutions.

Step 3 Identify all vertices (corner points).

Step 4 Find the value of the objective function at each vertex.

Step 5 The solution is given by the vertex producing the optimum value of the objective function.

Fundamental Theorem of Linear Programming

If an optimal value for a linear programming problem exists, then it occurs at a vertex of the region of feasible solutions.

The region of feasible solutions for the system below is given in the figure.

$$x + 2y \leq 14$$
$$3x + 4y \leq 36$$
$$x \geq 0$$
$$y \geq 0$$

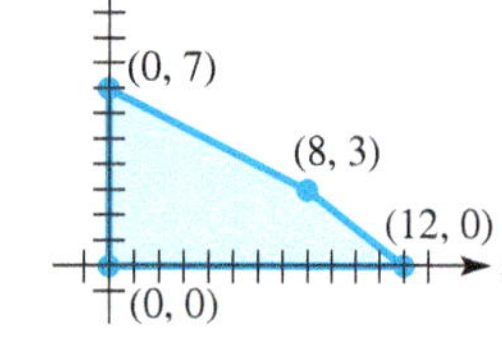

Maximize the objective function $8x + 12y$.

Vertex Point	Value of $8x + 12y$
(0, 0)	$8(0) + 12(0) = 0$
(0, 7)	$8(0) + 12(7) = 84$
(8, 3)	$8(8) + 12(3) = 100$ ← Maximum
(12, 0)	$8(12) + 12(0) = 96$

The maximum of 100 occurs at (8, 3).

9.7 Properties of Matrices

Addition and Subtraction of Matrices

To add (subtract) matrices of the same dimension, add (subtract) corresponding elements.

Find the sum or difference.

$$\begin{bmatrix} 2 & 3 & -1 \\ 0 & 4 & 9 \end{bmatrix} + \begin{bmatrix} -8 & 12 & 1 \\ 5 & 3 & -3 \end{bmatrix} = \begin{bmatrix} -6 & 15 & 0 \\ 5 & 7 & 6 \end{bmatrix}$$

$$\begin{bmatrix} 5 & -1 \\ -8 & 8 \end{bmatrix} - \begin{bmatrix} -2 & 4 \\ 3 & -6 \end{bmatrix} = \begin{bmatrix} 7 & -5 \\ -11 & 14 \end{bmatrix}$$

Concepts	Examples
Scalar Multiplication To multiply a matrix by a scalar, multiply each element of the matrix by the scalar.	Find the product. $3\begin{bmatrix} 6 & 2 \\ 1 & -2 \\ 0 & 8 \end{bmatrix} = \begin{bmatrix} 18 & 6 \\ 3 & -6 \\ 0 & 24 \end{bmatrix}$ Multiply each element by the scalar 3.
Matrix Multiplication The product AB of an $m \times n$ matrix A and an $n \times p$ matrix B is found as follows. To find the ith row, jth column element of matrix AB, multiply each element in the ith row of A by the corresponding element in the jth column of B. The sum of these products will give the row i, column j element of AB.	Find the matrix product. $\begin{bmatrix} 1 & -2 & 3 \\ 5 & 0 & 4 \\ -8 & 7 & -7 \end{bmatrix}\begin{bmatrix} 1 \\ -2 \\ 3 \end{bmatrix} = \begin{bmatrix} 1(1)+(-2)(-2)+3(3) \\ 5(1)+0(-2)+4(3) \\ -8(1)+7(-2)+(-7)(3) \end{bmatrix}$ 3×3 $\quad 3 \times 1$ $= \begin{bmatrix} 14 \\ 17 \\ -43 \end{bmatrix}$ 3×1
9.8 Matrix Inverses **Finding an Inverse Matrix** To obtain A^{-1} for any $n \times n$ matrix A for which A^{-1} exists, follow these steps. ***Step 1*** Form the augmented matrix $[A \mid I_n]$, where I_n is the $n \times n$ identity matrix. ***Step 2*** Perform row transformations on $[A \mid I_n]$ to obtain a matrix of the form $[I_n \mid B]$. ***Step 3*** Matrix B is A^{-1}.	Find A^{-1} if $A = \begin{bmatrix} 5 & 2 \\ 2 & 1 \end{bmatrix}$. $\left[\begin{array}{cc\|cc} 5 & 2 & 1 & 0 \\ 2 & 1 & 0 & 1 \end{array}\right]$ $\left[\begin{array}{cc\|cc} 1 & 0 & 1 & -2 \\ 2 & 1 & 0 & 1 \end{array}\right]$ $-2R2 + R1$ $\left[\begin{array}{cc\|cc} 1 & 0 & 1 & -2 \\ 0 & 1 & -2 & 5 \end{array}\right]$ $-2R1 + R2$ I_2 $\quad A^{-1}$ Therefore, $A^{-1} = \begin{bmatrix} 1 & -2 \\ -2 & 5 \end{bmatrix}$.

Chapter 9 Review Exercises

Use the substitution or elimination method to solve each system of equations. Identify any inconsistent systems *or systems with* infinitely many solutions. *If a system has infinitely many solutions, write the solution set with y arbitrary.*

1. $2x + 6y = 6$
$5x + 9y = 9$

2. $3x - 5y = 7$
$2x + 3y = 30$

3. $x + 5y = 9$
$2x + 10y = 18$

4. $\frac{1}{6}x + \frac{1}{3}y = 8$
$\frac{1}{4}x + \frac{1}{2}y = 12$

5. $y = -x + 3$
$2x + 2y = 1$

6. $0.2x + 0.5y = 6$
$0.4x + y = 9$

7. $3x - 2y = 0$
$9x + 8y = 7$

8. $6x + 10y = -11$
$9x + 6y = -3$

9. $2x - 5y + 3z = -1$
$x + 4y - 2z = 9$
$-x + 2y + 4z = 5$

10. $\begin{aligned} 4x + 3y + z &= -8 \\ 3x + y - z &= -6 \\ x + y + 2z &= -2 \end{aligned}$

11. $\begin{aligned} 5x - y &= 26 \\ 4y + 3z &= -4 \\ 3x + 3z &= 15 \end{aligned}$

12. $\begin{aligned} x + z &= 2 \\ 2y - z &= 2 \\ -x + 2y &= -4 \end{aligned}$

Solve each problem.

13. ***Concept Check*** Create an inconsistent system of two equations.

14. ***Connecting Graphs with Equations*** Determine the system of equations illustrated in the graph. Write equations in standard form.

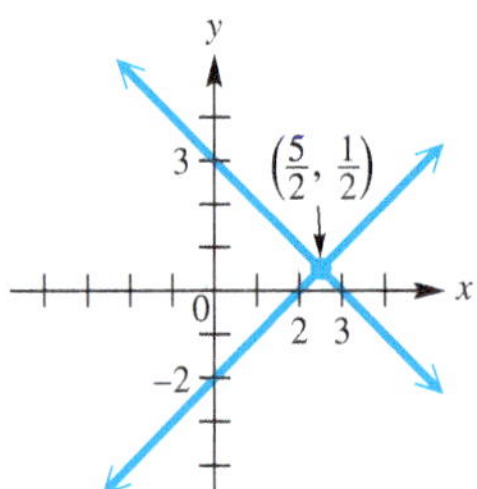

Solve each problem using a system of equations.

15. ***Meal Planning*** A cup of uncooked rice contains 15 g of protein and 810 calories. A cup of uncooked soybeans contains 22.5 g of protein and 270 calories. How many cups of each should be used for a meal containing 9.5 g of protein and 324 calories?

16. ***Order Quantities*** A company sells recordable CDs for \$0.80 each and play-only CDs for \$0.60 each. The company receives \$76.00 for an order of 100 CDs. However, the customer neglected to specify how many of each type to send. Determine the number of each type of CD that should be sent.

17. ***Indian Weavers*** The Waputi Indians make woven blankets, rugs, and skirts. Each blanket requires 24 hr for spinning the yarn, 4 hr for dyeing the yarn, and 15 hr for weaving. Rugs require 30, 5, and 18 hr and skirts 12, 3, and 9 hr, respectively. If there are 306, 59, and 201 hr available for spinning, dyeing, and weaving, respectively, how many of each item can be made?

18. ***(Modeling) Populations of Age Groups*** The estimated resident populations (in percent) of young people (age 14 and under) and seniors (age 65 and over) in the United States for the years 2015–2050 are modeled by the following linear functions.

$$y_1 = -0.04x + 19.3 \quad \text{Young people}$$
$$y_2 = 0.17x + 16.4 \quad \text{Seniors}$$

In each case, x represents the number of years since 2015. (*Source:* U.S. Census Bureau.)

(a) Solve the system to find the year when these population percents will be equal.

(b) What percent, to the nearest tenth, of the U.S. resident population will be young people or seniors in the year found in part (a)? Answer may vary due to rounding.

(c) Use a calculator graph of the system to support the algebraic solution.

(d) Which population is increasing? (*Hint:* Consider the slopes of the lines.)

19. ***(Modeling) Equilibrium Supply and Demand*** Let the supply and demand equations for units of backpacks be

$$\text{supply: } p = \frac{3}{2}q \quad \text{and} \quad \text{demand: } p = 81 - \frac{3}{4}q.$$

(a) Graph these equations on the same axes.

(b) Find the equilibrium demand.

(c) Find the equilibrium price.

20. ***(Modeling) Heart Rate*** In a study, a group of athletes was exercised to exhaustion. Let x represent an athlete's heart rate 5 sec after stopping exercise and y this rate 10 sec after stopping. It was found that the maximum heart rate H for these athletes satisfied the two equations

$$H = 0.491x + 0.468y + 11.2$$
$$H = -0.981x + 1.872y + 26.4.$$

If an athlete had maximum heart rate $H = 180$, determine x and y graphically. Round to the nearest tenth. Interpret the answer. (*Source:* Thomas, V., *Science and Sport*, Faber and Faber.)

21. *(Modeling)* The table was generated using a function $y_1 = ax^2 + bx + c$. Use any three points from the table to find the equation for y_1.

NORMAL FLOAT AUTO REAL RADIAN MP
PRESS + FOR △Tbl

X	Y1
0	1.5
1	-2.3
2	-1.3
3	4.5
4	15.1
5	30.5
6	50.7
7	75.7
8	105.5
9	140.1
10	179.5

X=0

22. *(Modeling)* The equation of a circle may be written in the form

$$x^2 + y^2 + ax + by + c = 0.$$

Find the equation of the circle passing through the points $(-3, -7)$, $(4, -8)$, and $(1, 1)$.

Solve each system in terms of the specified arbitrary variable.

23. $3x - 4y + z = 2$
$2x + y = 1$
(x arbitrary)

24. $2x - 6y + 4z = 5$
$5x + y - 3z = 1$
(z arbitrary)

Use the Gauss-Jordan method to solve each system.

25. $5x + 2y = -10$
$3x - 5y = -6$

26. $2x + 3y = 10$
$-3x + y = 18$

27. $3x + y = -7$
$x - y = -5$

28. $2x - y + 4z = -1$
$-3x + 5y - z = 5$
$2x + 3y + 2z = 3$

29. $x - z = -3$
$y + z = 6$
$2x - 3z = -9$

30. $2x - y + z = 4$
$x + 2y - z = 0$
$3x + y - 2z = 1$

Solve each problem using the Gauss-Jordan method to solve a system of equations.

31. ***Mixing Teas*** Three kinds of tea worth \$4.60, \$5.75, and \$6.50 per lb are to be mixed to get 20 lb of tea worth \$5.25 per lb. The amount of \$4.60 tea used is to be equal to the total amount of the other two kinds together. How many pounds of each tea should be used?

32. *Mixing Solutions* A 5% solution of a drug is to be mixed with some 15% solution and some 10% solution to make 20 ml of 8% solution. The amount of 5% solution used must be 2 ml more than the sum of the other two solutions. How many milliliters of each solution should be used?

33. *(Modeling) Master's Degrees* During the period 1975–2012, the numbers of master's degrees awarded to both males and females grew, but degrees earned by females grew at a greater rate. If $x = 0$ represents 1975 and $x = 37$ represents 2012, the number of master's degrees earned (in thousands) are closely modeled by the following system.

$$y = 3.79x + 128 \quad \text{Males}$$
$$y = 8.89x + 80.2 \quad \text{Females}$$

Solve the system to find the year in which males and females earned the same number of master's degrees. What was the total number, to the nearest thousand, of master's degrees earned in that year? (*Source:* U.S. Census Bureau.)

34. *(Modeling) Comparing Prices* One refrigerator sells for \$700 and uses \$85 worth of electricity per year. A second refrigerator is \$100 more expensive but costs only \$25 per year to run. Assuming that there are no repair costs, the costs to run the refrigerators over a 10-yr period are given by the following system of equations. Here, y represents the total cost in dollars, and x is time in years.

$$y = 700 + 85x$$
$$y = 800 + 25x$$

In how many years will the costs for the two refrigerators be equal? What are the equivalent costs at that time?

Evaluate each determinant.

35. $\begin{vmatrix} -1 & 8 \\ 2 & 9 \end{vmatrix}$

36. $\begin{vmatrix} -2 & 4 \\ 0 & 3 \end{vmatrix}$

37. $\begin{vmatrix} x & 4x \\ 2x & 8x \end{vmatrix}$

38. $\begin{vmatrix} -2 & 4 & 1 \\ 3 & 0 & 2 \\ -1 & 0 & 3 \end{vmatrix}$

39. $\begin{vmatrix} -1 & 2 & 3 \\ 4 & 0 & 3 \\ 5 & -1 & 2 \end{vmatrix}$

40. $\begin{vmatrix} -3 & 2 & 7 \\ 6 & -4 & -14 \\ 7 & 1 & 4 \end{vmatrix}$

Use Cramer's rule to solve each system of equations. If $D = 0$, use another method to determine the solution set.

41. $3x + 7y = 2$
$5x - y = -22$

42. $3x + y = -1$
$5x + 4y = 10$

43. $6x + y = -3$
$12x + 2y = 1$

44. $3x + 2y + z = 2$
$4x - y + 3z = -16$
$x + 3y - z = 12$

45. $x + y = -1$
$2y + z = 5$
$3x - 2z = -28$

46. $5x - 2y - z = 8$
$-5x + 2y + z = -8$
$x - 4y - 2z = 0$

Solve each equation.

47. $\begin{vmatrix} 3x & 7 \\ -x & 4 \end{vmatrix} = 8$

48. $\begin{vmatrix} 6x & 2 & 0 \\ 1 & 5 & 3 \\ x & 2 & -1 \end{vmatrix} = 2x$

Find the partial fraction decomposition for each rational expression.

49. $\dfrac{2}{3x^2 - 5x + 2}$

50. $\dfrac{11 - 2x}{x^2 - 8x + 16}$

51. $\dfrac{5 - 2x}{(x^2 + 2)(x - 1)}$

52. $\dfrac{x^3 + 2x^2 - 3}{x^4 - 4x^2 + 4}$

Solve each nonlinear system of equations.

53. $y = 2x + 10$
$x^2 + y = 13$

54. $x^2 = 2y - 3$
$x + y = 3$

55. $x^2 + y^2 = 17$
$2x^2 - y^2 = 31$

56. $2x^2 + 3y^2 = 30$
$x^2 + y^2 = 13$

57. $xy = -10$
$x + 2y = 1$

58. $xy + 2 = 0$
$y - x = 3$

59. $x^2 + 2xy + y^2 = 4$
$x - 3y = -2$

60. $x^2 + 2xy = 15 + 2x$
$xy - 3x + 3 = 0$

61. $2x^2 - 3y^2 = 18$
$2x^2 - 2y^2 = 14$

Solve each problem.

62. Find all values of b such that the straight line $3x - y = b$ touches the circle $x^2 + y^2 = 25$ at only one point.

63. Do the circle $x^2 + y^2 = 144$ and the line $x + 2y = 8$ have any points in common? If so, what are they?

64. Find the equation of the line passing through the points of intersection of the graphs of $x^2 + y^2 = 20$ and $x^2 - y = 0$.

Graph the solution set of each system of inequalities.

65. $x + y \le 6$
$2x - y \ge 3$

66. $y \le \dfrac{1}{3}x - 2$
$y^2 \le 16 - x^2$

67. Maximize the objective function $2x + 4y$ for the following constraints.

$$x \ge 0$$
$$y \ge 0$$
$$3x + 2y \le 12$$
$$5x + y \ge 5$$

68. ***Connecting Graphs with Equations*** Determine the system of inequalities illustrated in the graph. Write each inequality in standard form.

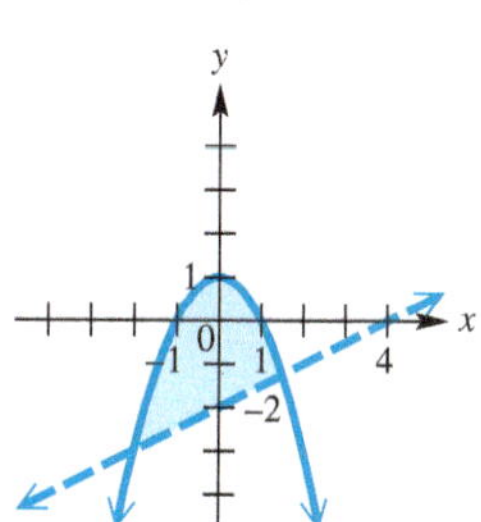

Solve each problem.

69. ***Cost of Nutrients*** Certain laboratory animals must have at least 30 g of protein and at least 20 g of fat per feeding period. These nutrients come from food A, which costs \$0.18 per unit and supplies 2 g of protein and 4 g of fat; and from food B, which costs \$0.12 per unit and has 6 g of protein and 2 g of fat. Food B is purchased under a long-term contract requiring that at least 2 units of B be used per serving. How much of each food must be purchased to produce the minimum cost per serving? What is the minimum cost?

70. *Profit from Farm Animals* A 4-H member raises only geese and pigs. She wants to raise no more than 16 animals, including no more than 10 geese. She spends \$5 to raise a goose and \$15 to raise a pig, and she has \$180 available for this project. Each goose produces \$6 in profit, and each pig produces \$20 in profit. How many of each animal should she raise to maximize her profit? What is her maximum profit?

Find the values of the variables for which each statement is true, if possible.

71. $\begin{bmatrix} 5 & x+2 \\ -6y & z \end{bmatrix} = \begin{bmatrix} a & 3x-1 \\ 5y & 9 \end{bmatrix}$

72. $\begin{bmatrix} -6+k & 2 & a+3 \\ -2+m & 3p & 2r \end{bmatrix} + \begin{bmatrix} 3-2k & 5 & 7 \\ 5 & 8p & 5r \end{bmatrix} = \begin{bmatrix} 5 & y & 6a \\ 2m & 11 & -35 \end{bmatrix}$

Perform each operation, if possible.

73. $\begin{bmatrix} 3 \\ 2 \\ 5 \end{bmatrix} - \begin{bmatrix} 8 \\ -4 \\ 6 \end{bmatrix} + \begin{bmatrix} 1 \\ 0 \\ 2 \end{bmatrix}$

74. $4\begin{bmatrix} 3 & -4 & 2 \\ 5 & -1 & 6 \end{bmatrix} + \begin{bmatrix} -3 & 2 & 5 \\ 1 & 0 & 4 \end{bmatrix}$

75. $\begin{bmatrix} 2 & 5 & 8 \\ 1 & 9 & 2 \end{bmatrix} - \begin{bmatrix} 3 & 4 \\ 7 & 1 \end{bmatrix}$

76. $\begin{bmatrix} -3 & 4 \\ 2 & 8 \end{bmatrix}\begin{bmatrix} -1 & 0 \\ 2 & 5 \end{bmatrix}$

77. $\begin{bmatrix} -1 & 0 \\ 2 & 5 \end{bmatrix}\begin{bmatrix} -3 & 4 \\ 2 & 8 \end{bmatrix}$

78. $\begin{bmatrix} 1 & 2 \\ 3 & 0 \\ -6 & 5 \end{bmatrix}\begin{bmatrix} 4 & 8 \\ -1 & 2 \end{bmatrix}$

79. $\begin{bmatrix} 3 & 2 & -1 \\ 4 & 0 & 6 \end{bmatrix}\begin{bmatrix} -2 & 0 \\ 0 & 2 \\ 3 & 1 \end{bmatrix}$

80. $\begin{bmatrix} 1 & -2 & 4 & 2 \\ 0 & 1 & -1 & 8 \end{bmatrix}\begin{bmatrix} -1 \\ 2 \\ 0 \\ 1 \end{bmatrix}$

81. $\begin{bmatrix} -2 & 5 & 5 \\ 0 & 1 & 4 \\ 3 & -4 & -1 \end{bmatrix}\begin{bmatrix} 1 & 0 & -1 \\ -1 & 0 & 0 \\ 1 & 1 & -1 \end{bmatrix}$

82. $\begin{bmatrix} 0 & 1 & 4 \\ 7 & -2 & 9 \\ 10 & 0 & 1 \end{bmatrix}\begin{bmatrix} 3 & 2 & 1 \\ -4 & 7 & 6 \end{bmatrix}$

Find the inverse, if it exists, for each matrix.

83. $\begin{bmatrix} 2 & 1 \\ 5 & 3 \end{bmatrix}$

84. $\begin{bmatrix} -4 & 2 \\ 0 & 3 \end{bmatrix}$

85. $\begin{bmatrix} 2 & -1 & 0 \\ 1 & 0 & 1 \\ 1 & -2 & 0 \end{bmatrix}$

86. $\begin{bmatrix} 2 & 3 & 5 \\ -2 & -3 & -5 \\ 1 & 4 & 2 \end{bmatrix}$

Solve each system using the inverse of the coefficient matrix.

87. $5x - 4y = 1$
$x + 4y = 3$

88. $2x + y = 5$
$3x - 2y = 4$

89. $3x + 2y + z = -5$
$x - y + 3z = -5$
$2x + 3y + z = 0$

90. $x + y + z = 1$
$2x - y = -2$
$3y + z = 2$

Chapter 9 Test

Use the substitution or elimination method to solve each system of equations. State whether it is an inconsistent system *or has* infinitely many solutions. *If a system has infinitely many solutions, write the solution set with y arbitrary.*

1. $3x - y = 9$
$x + 2y = 10$

2. $6x + 9y = -21$
$4x + 6y = -14$

3. $x - 2y = 4$
$-2x + 4y = 6$

4. $\frac{1}{4}x - \frac{1}{3}y = -\frac{5}{12}$
$\frac{1}{10}x + \frac{1}{5}y = \frac{1}{2}$

5. $2x + y + z = 3$
$x + 2y - z = 3$
$3x - y + z = 5$

Use the Gauss-Jordan method to solve each system.

6. $3x - 2y = 13$
$4x - y = 19$

7. $3x - 4y + 2z = 15$
$2x - y + z = 13$
$x + 2y - z = 5$

8. ***Connecting Graphs with Equations*** Find the equation $y = ax^2 + bx + c$ of the parabola through the given points.

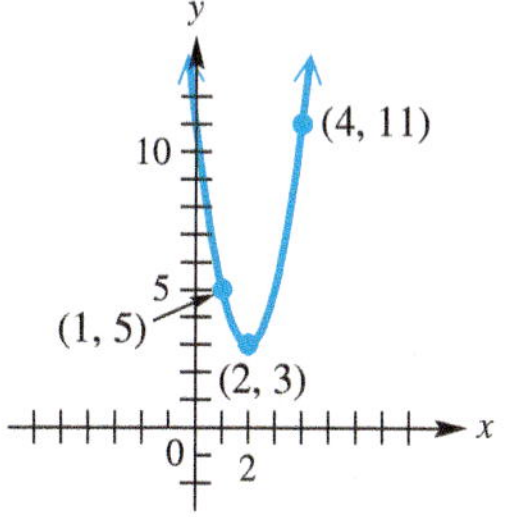

9. ***Ordering Supplies*** A knitting shop orders yarn from three suppliers in Toronto, Montreal, and Ottawa. One month the shop ordered a total of 100 units of yarn from these suppliers. The delivery costs were \$80, \$50, and \$65 per unit for the orders from Toronto, Montreal, and Ottawa, respectively, with total delivery costs of \$5990. The shop ordered the same amount from Toronto and Ottawa. How many units were ordered from each supplier?

Evaluate each determinant.

10. $\begin{vmatrix} 6 & 8 \\ 2 & -7 \end{vmatrix}$

11. $\begin{vmatrix} 2 & 0 & 8 \\ -1 & 7 & 9 \\ 12 & 5 & -3 \end{vmatrix}$

Use Cramer's rule to solve each system of equations.

12. $2x - 3y = -33$
$4x + 5y = 11$

13. $x + y - z = -4$
$2x - 3y - z = 5$
$x + 2y + 2z = 3$

Find the partial fraction decomposition for each rational expression.

14. $\frac{9x + 19}{x^2 + 2x - 3}$

15. $\frac{x + 2}{x^3 + 2x^2 + x}$

Solve each nonlinear system of equations.

16. $2x^2 + y^2 = 6$
$x^2 - 4y^2 = -15$

17. $x^2 + y^2 = 25$
$x + y = 7$

Work each problem.

18. ***Unknown Numbers*** Find two numbers such that their sum is -1 and the sum of their squares is 61.

19. Graph the solution set.

$$x - 3y \geq 6$$
$$y^2 \leq 16 - x^2$$

20. Maximize the objective function $2x + 3y$ for the following constraints.

$$x \geq 0$$
$$y \geq 0$$
$$x + 2y \leq 24$$
$$3x + 4y \leq 60$$

21. ***Jewelry Profits*** The Schwab Company designs and sells two types of rings: the VIP and the SST. The company can produce up to 24 rings each day using up to 60 total hours of labor. It takes 3 hr to make one VIP ring and 2 hr to make one SST ring. How many of each type of ring should be made daily in order to maximize the company's profit, if the profit on one VIP ring is \$30 and the profit on one SST ring is \$40? What is the maximum profit?

22. Find the value of each variable for which the statement is true.

$$\begin{bmatrix} 5 & x+6 \\ 0 & 4 \end{bmatrix} = \begin{bmatrix} y-2 & 4-x \\ 0 & w+7 \end{bmatrix}$$

Perform each operation, if possible.

23. $3\begin{bmatrix} 2 & 3 \\ 1 & -4 \\ 5 & 9 \end{bmatrix} - \begin{bmatrix} -2 & 6 \\ 3 & -1 \\ 0 & 8 \end{bmatrix}$

24. $\begin{bmatrix} 1 \\ 2 \end{bmatrix} + \begin{bmatrix} 4 \\ -6 \end{bmatrix} + \begin{bmatrix} 2 & 8 \\ -7 & 5 \end{bmatrix}$

25. $\begin{bmatrix} 2 & 1 & -3 \\ 4 & 0 & 5 \end{bmatrix}\begin{bmatrix} 1 & 3 \\ 2 & 4 \\ 3 & -2 \end{bmatrix}$

26. $\begin{bmatrix} 2 & -4 \\ 3 & 5 \end{bmatrix}\begin{bmatrix} 4 \\ 2 \\ 7 \end{bmatrix}$

27. ***Concept Check*** Which of the following properties does not apply to multiplication of matrices?

A. commutative **B.** associative **C.** distributive **D.** identity

Find the inverse, if it exists, for each matrix.

28. $\begin{bmatrix} -8 & 5 \\ 3 & -2 \end{bmatrix}$

29. $\begin{bmatrix} 4 & 12 \\ 2 & 6 \end{bmatrix}$

30. $\begin{bmatrix} 1 & 3 & 4 \\ 2 & 7 & 8 \\ -2 & -5 & -7 \end{bmatrix}$

Solve each system using the inverse of the coefficient matrix.

31. $2x + y = -6$
$3x - y = -29$

32. $x + y = 5$
$y - 2z = 23$
$x + 3z = -27$

10 Analytic Geometry

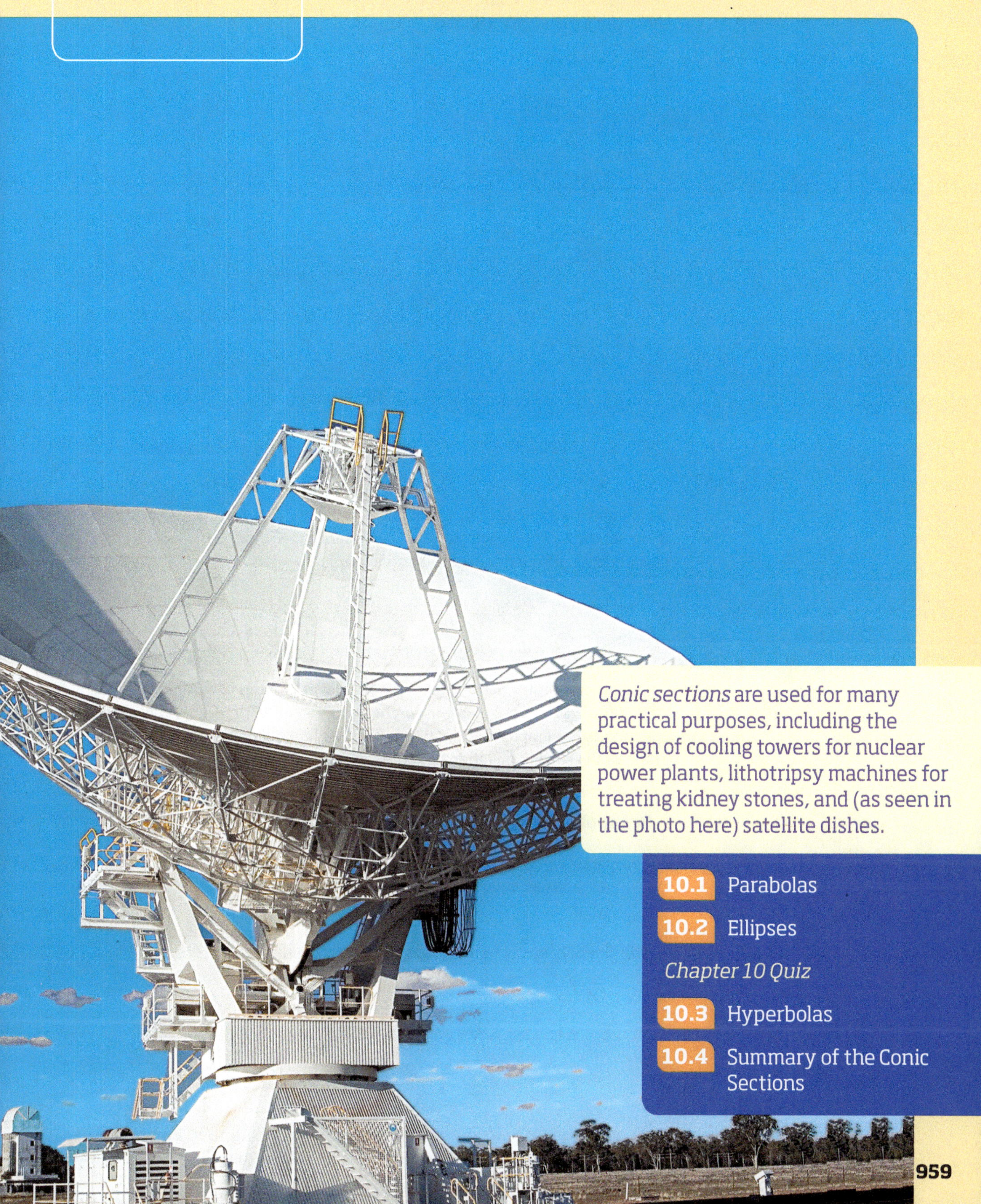

Conic sections are used for many practical purposes, including the design of cooling towers for nuclear power plants, lithotripsy machines for treating kidney stones, and (as seen in the photo here) satellite dishes.

10.1 Parabolas

- Conic Sections
- Horizontal Parabolas
- Geometric Definition and Equations of Parabolas
- An Application of Parabolas

Conic Sections

Parabolas, circles, ellipses, and *hyperbolas* form a group of curves known as **conic sections,** because they are the results of intersecting a cone with a plane. See **Figure 1.**

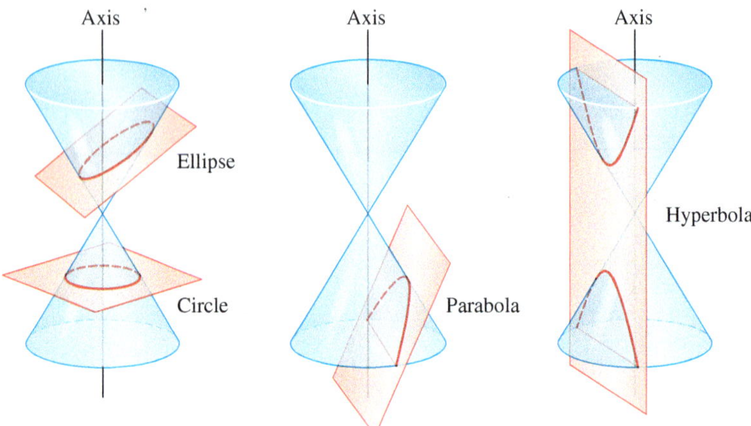

Figure 1

Horizontal Parabolas

We know that the graph of the equation

$$y = a(x - h)^2 + k$$

is a parabola with vertex (h, k) and the vertical line $x = h$ as its axis of symmetry. If we subtract k from each side and interchange the roles of $x - h$ and $y - k$, the new equation also has a parabola as its graph.

$$y - k = a(x - h)^2 \quad \text{Subtract } k. \quad (1)$$

$$x - h = a(y - k)^2 \quad \text{Interchange the roles of } x - h \text{ and } y - k. \quad (2)$$

While the graph of $y - k = a(x - h)^2$ has a *vertical* axis of symmetry, the graph of $x - h = a(y - k)^2$ has a *horizontal* axis of symmetry. The graph of the first equation is the graph of a function (specifically a quadratic function), while the graph of the second equation is not. Its graph fails the vertical line test.

Parabola with Horizontal Axis of Symmetry

The parabola with vertex (h, k) and the horizontal line $y = k$ as axis of symmetry has an equation of the following form.

$$\boldsymbol{x - h = a(y - k)^2}$$

The parabola opens to the right if $a > 0$ and to the left if $a < 0$.

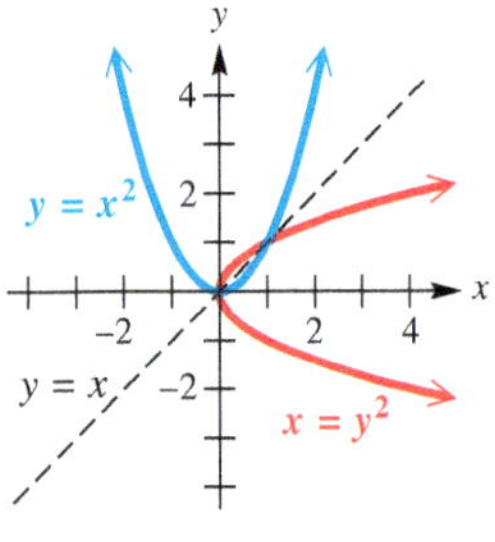

Figure 2

NOTE When the vertex (h, k) is $(0, 0)$ and $a = 1$ in

$$y - k = a(x - h)^2 \quad (1)$$

and

$$x - h = a(y - k)^2, \quad (2)$$

the equations $y = x^2$ and $x = y^2$, respectively, result. See **Figure 2.** The graphs are mirror images of each other with respect to the line $y = x$.

EXAMPLE 1 Graphing a Parabola (Horizontal Axis of Symmetry)

Graph $x + 3 = (y - 2)^2$. Give the domain and range.

SOLUTION The graph of $x + 3 = (y - 2)^2$, or $x - (-3) = (y - 2)^2$, has vertex $(-3, 2)$ and opens to the right because $a = 1$, and $1 > 0$. Plotting a few additional points gives the graph shown in **Figure 3.**

x	y
−3	2
−2	3
−2	1
1	4
1	0

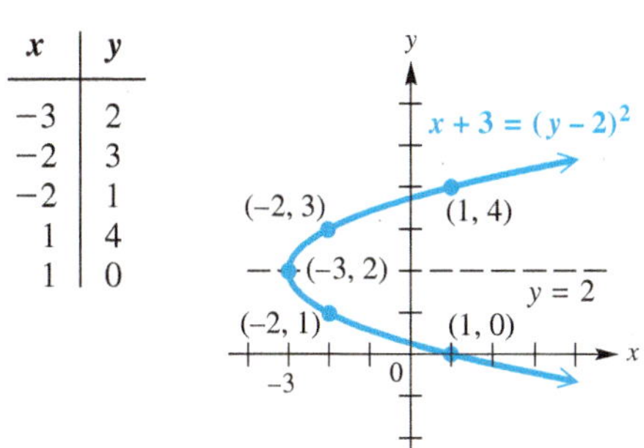

Figure 3

The graph is symmetric with respect to its axis, $y = 2$.
Domain: $[-3, \infty)$
Range: $(-\infty, \infty)$

✓ **Now Try Exercise 9.**

EXAMPLE 2 Graphing a Parabola (Horizontal Axis of Symmetry)

Graph $x = 2y^2 + 6y + 5$. Give the domain and range.

ALGEBRAIC SOLUTION

$$x = 2y^2 + 6y + 5$$

$$x = 2(y^2 + 3y \qquad) + 5 \quad \text{Factor out 2.}$$

$$x = 2\left(y^2 + 3y + \frac{9}{4} - \frac{9}{4}\right) + 5$$

Complete the square; $\left[\frac{1}{2}(3)\right]^2 = \frac{9}{4}$.

$$x = 2\left(y^2 + 3y + \frac{9}{4}\right) + 2\left(-\frac{9}{4}\right) + 5$$

Distributive property

$$x = 2\left(y + \frac{3}{2}\right)^2 + \frac{1}{2} \quad \text{Factor, and simplify.}$$

$$x - \frac{1}{2} = 2\left(y + \frac{3}{2}\right)^2 \quad \text{Subtract } \tfrac{1}{2}. \quad (*)$$

The vertex of the parabola is $\left(\frac{1}{2}, -\frac{3}{2}\right)$. The axis is the horizontal line $y = -\frac{3}{2}$. Using the vertex and the axis and plotting a few additional points gives the graph in **Figure 4.** If we let $y = 0$, we find that the x-intercept is $(5, 0)$, and because of symmetry, the point $(5, -3)$ also lies on the graph. The domain is $\left[\frac{1}{2}, \infty\right)$, and the range is $(-\infty, \infty)$.

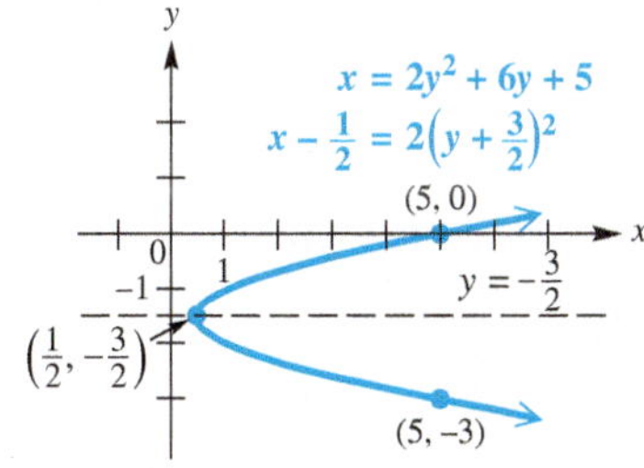

Figure 4

GRAPHING CALCULATOR SOLUTION

A horizontal parabola is *not* the graph of a function. To graph it using a graphing calculator in function mode, we must write two equations by solving for y.

$$x - \frac{1}{2} = 2\left(y + \frac{3}{2}\right)^2 \quad (*) \text{ from algebraic solution}$$

$$x - 0.5 = 2(y + 1.5)^2 \quad \text{Write with decimals.}$$

$$\frac{x - 0.5}{2} = (y + 1.5)^2 \quad \text{Divide by 2.}$$

$$\pm\sqrt{\frac{x - 0.5}{2}} = y + 1.5 \quad \text{Take the square root on each side.}$$

$$y = -1.5 \pm \sqrt{\frac{x - 0.5}{2}} \quad \text{Subtract 1.5, and rewrite.}$$

Figure 5 shows the graphs of the two functions defined in the final equation. Their union is the graph of $x = 2y^2 + 6y + 5$.

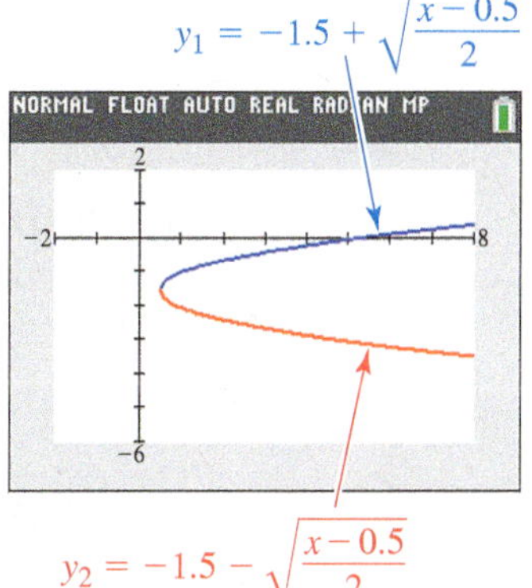

Figure 5

✓ **Now Try Exercise 17.**

Geometric Definition and Equations of Parabolas

The equation of a parabola comes from the geometric definition of a parabola as a set of points.

Parabola

A **parabola** is the set of points in a plane equidistant from a fixed point and a fixed line. The fixed point is the **focus,** and the fixed line is the **directrix** of the parabola.

As shown in **Figure 6,** the axis of symmetry of a parabola passes through the focus and is perpendicular to the directrix. The vertex is the midpoint of the line segment joining the focus and directrix on the axis.

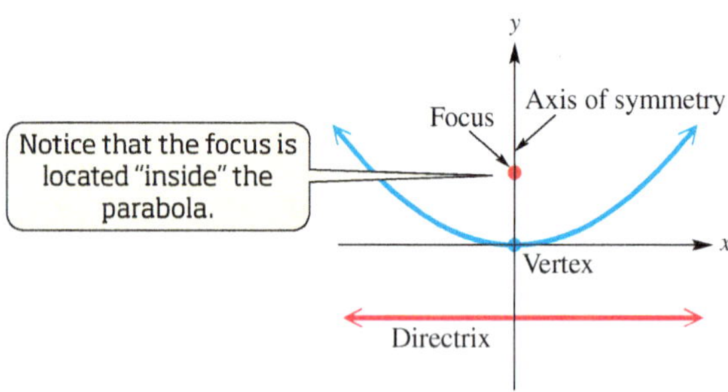

Figure 6

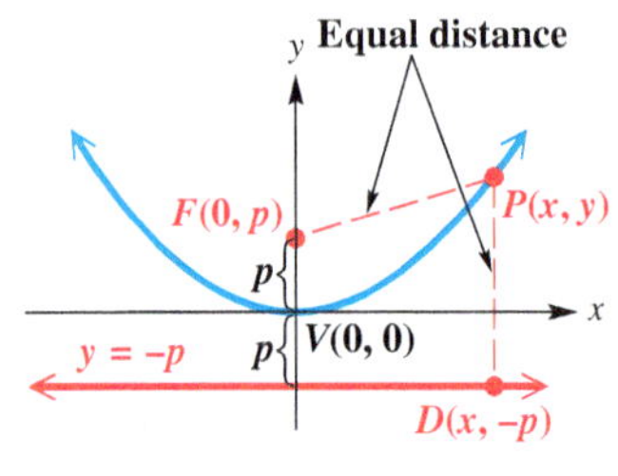

$d(P, F) = d(P, D)$ for all P on the parabola.

Figure 7

We can find an equation of a parabola from the preceding definition. **Let p represent the directed distance from the vertex to the focus.** Then the directrix is the line $y = -p$ and the focus is the point $F(0, p)$. See **Figure 7.** To find the equation of the set of points that are the same distance from the line $y = -p$ and the point $(0, p)$, choose one such point P with coordinates (x, y). Because $d(P, F)$ and $d(P, D)$ must be equal, using the distance formula gives the following.

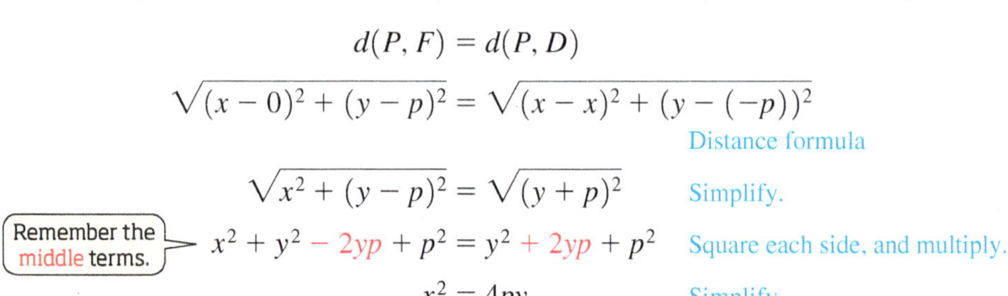

$$d(P, F) = d(P, D)$$

$$\sqrt{(x-0)^2 + (y-p)^2} = \sqrt{(x-x)^2 + (y-(-p))^2} \quad \text{Distance formula}$$

$$\sqrt{x^2 + (y-p)^2} = \sqrt{(y+p)^2} \quad \text{Simplify.}$$

$$x^2 + y^2 - 2yp + p^2 = y^2 + 2yp + p^2 \quad \text{Square each side, and multiply.}$$

$$x^2 = 4py \quad \text{Simplify.}$$

From this result, if the given form of the equation is $y = ax^2$, then $a = \frac{1}{4p}$.

Parabola with Vertical Axis of Symmetry and Vertex (0, 0)

The parabola with focus $(0, p)$ and directrix $y = -p$ has the following equation.

$$\mathbf{x^2 = 4py}$$

This parabola has vertical axis of symmetry $x = 0$ and opens up if $p > 0$ or down if $p < 0$.

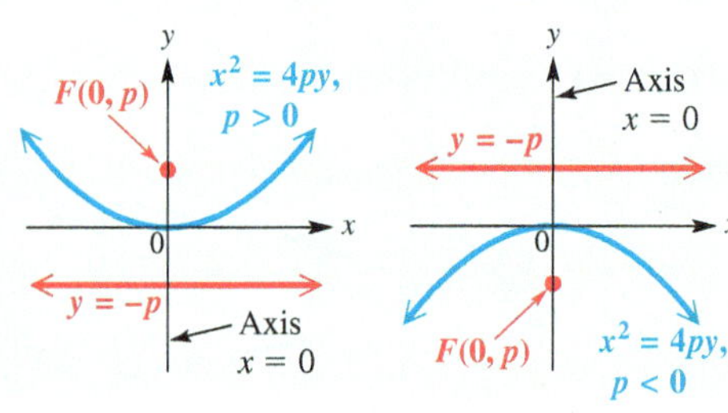

If the directrix is the line $x = -p$ and the focus is $(p, 0)$, a similar procedure leads to the equation of a parabola with a horizontal axis of symmetry.

Parabola with Horizontal Axis of Symmetry and Vertex (0, 0)

The parabola with focus $(p, 0)$ and directrix $x = -p$ has the following equation.

$$y^2 = 4px$$

This parabola has horizontal axis of symmetry $y = 0$ and opens to the right if $p > 0$ or to the left if $p < 0$.

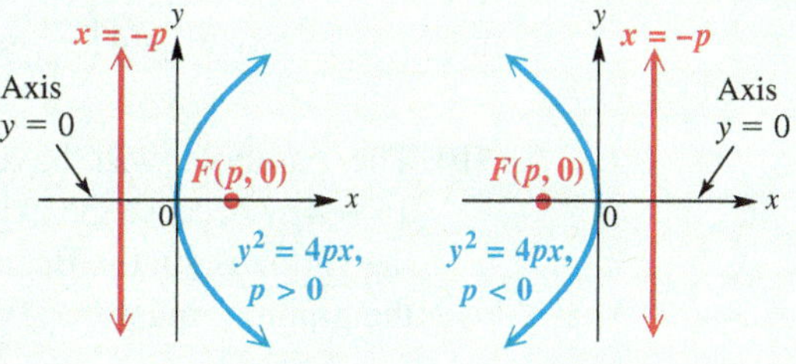

EXAMPLE 3 Graphing Parabolas

Give the focus, directrix, vertex, and axis of symmetry for each parabola. Then use this information to graph the parabola.

(a) $x^2 = 8y$ **(b)** $y^2 = -28x$

SOLUTION

(a) The equation $x^2 = 8y$ has the form

$$x^2 = 4py, \quad \text{with } 4p = 8, \text{ so } \quad p = 2.$$

The x-term is squared, so the parabola is vertical, with focus $(0, p) = (0, 2)$ and directrix $y = -p = -2$. The vertex is $(0, 0)$, and the axis of symmetry is the y-axis (that is, $x = 0$). See **Figure 8.**

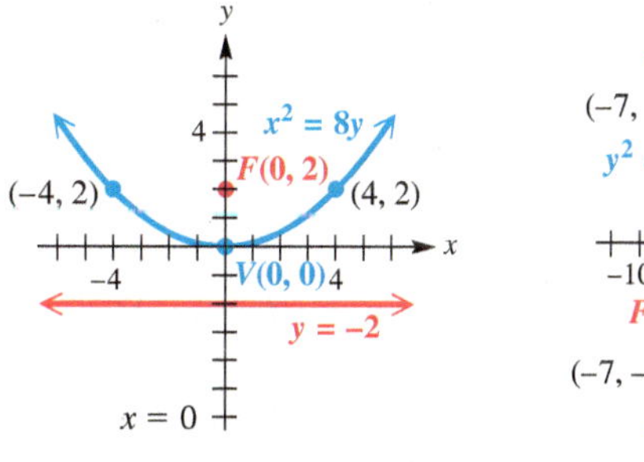

Figure 8

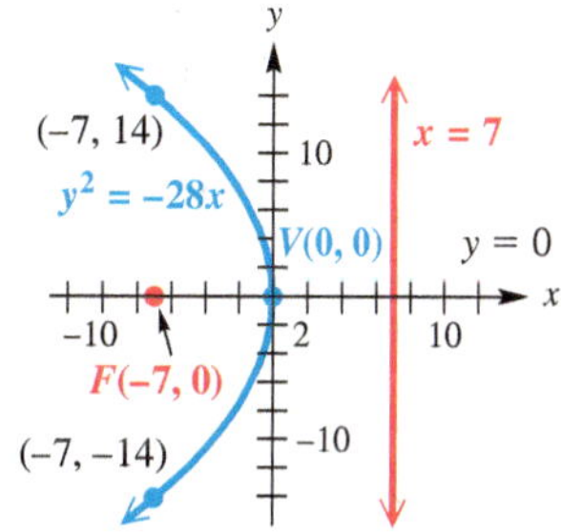

Figure 9

(b) The equation $y^2 = -28x$ has the form

$$y^2 = 4px, \quad \text{with } 4p = -28, \text{ so } \quad p = -7.$$

The parabola is horizontal, with focus $(-7, 0)$, directrix $x = 7$, vertex $(0, 0)$, and the x-axis (that is, $y = 0$) as axis of symmetry. Because p is negative, the graph opens to the left, as shown in **Figure 9.**

✔ **Now Try Exercises 23 and 27.**

EXAMPLE 4 Writing Equations of Parabolas (Vertex at the Origin)

Write an equation for each parabola with vertex at the origin.

(a) focus $\left(\frac{2}{3}, 0\right)$ **(b)** vertical axis of symmetry, through the point $(-2, 12)$

SOLUTION

(a) The focus $\left(\frac{2}{3}, 0\right)$ and the vertex $(0, 0)$ are both on the x-axis, so the parabola is horizontal. It opens to the right because $p = \frac{2}{3}$ is positive. See **Figure 10.** The equation will have the form $y^2 = 4px$.

$$y^2 = 4\left(\frac{2}{3}\right)x, \quad \text{or} \quad y^2 = \frac{8}{3}x$$

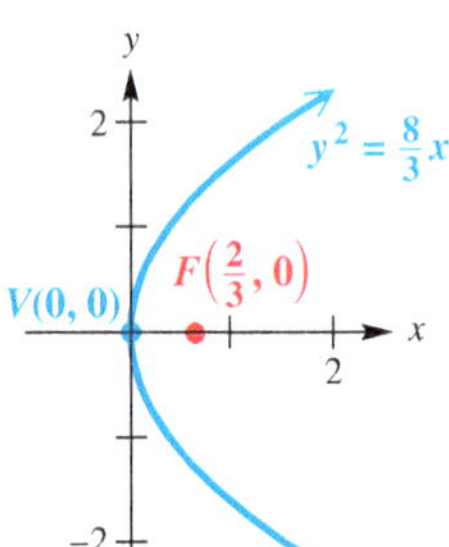

Figure 10

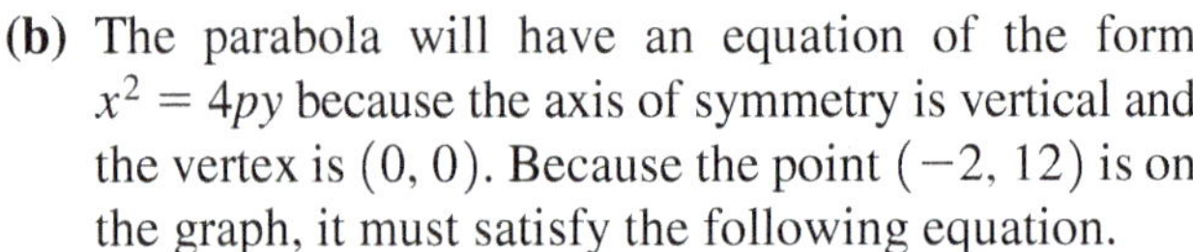

(b) The parabola will have an equation of the form $x^2 = 4py$ because the axis of symmetry is vertical and the vertex is $(0, 0)$. Because the point $(-2, 12)$ is on the graph, it must satisfy the following equation.

$$x^2 = 4py \qquad \text{Parabola with vertical axis of symmetry}$$

$$(-2)^2 = 4p(12) \qquad \text{Let } x = -2 \text{ and } y = 12.$$

$$4 = 48p \qquad \text{Apply the exponent, and multiply.}$$

$$p = \frac{1}{12} \qquad \text{Solve for } p.$$

Then, $$x^2 = 4\left(\frac{1}{12}\right)y \qquad \text{Let } p = \tfrac{1}{12} \text{ in the form } x^2 = 4py.$$

$$x^2 = \frac{1}{3}y, \quad \text{or} \quad y = 3x^2.$$

✔ **Now Try Exercises 35 and 39.**

The equations $x^2 = 4py$ and $y^2 = 4px$ can be extended to parabolas having vertex (h, k) by replacing x and y with $x - h$ and $y - k$, respectively.

Equation Forms for Translated Parabolas

A parabola with vertex (h, k) has an equation of the following form.

$$(x - h)^2 = 4p(y - k) \qquad \text{Vertical axis of symmetry}$$

or

$$(y - k)^2 = 4p(x - h) \qquad \text{Horizontal axis of symmetry}$$

The focus is distance $|p|$ from the vertex.

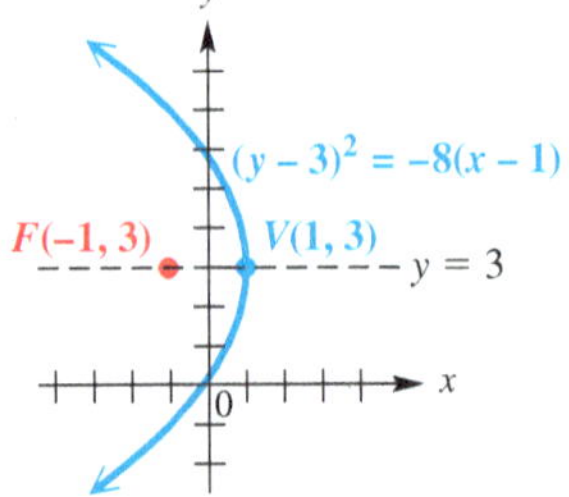

Figure 11

EXAMPLE 5 Writing an Equation of a Parabola

Write an equation for the parabola with vertex $(1, 3)$ and focus $(-1, 3)$, and graph it. Give the domain and range.

SOLUTION Because the focus is to the left of the vertex, the axis of symmetry is horizontal and the parabola opens to the left. See **Figure 11.** The directed distance between the vertex and the focus is $-1 - 1$, or -2, so $p = -2$ (because the parabola opens to the left).

The equation of the parabola is found as follows.

$$(y - k)^2 = 4p(x - h) \quad \text{Parabola with horizontal axis of symmetry}$$

$$(y - 3)^2 = 4(-2)(x - 1) \quad \text{Let } p = -2,\ h = 1, \text{ and } k = 3.$$

$$(y - 3)^2 = -8(x - 1) \quad \text{Multiply.}$$

The domain is $(-\infty, 1]$, and the range is $(-\infty, \infty)$. **Now Try Exercise 43.**

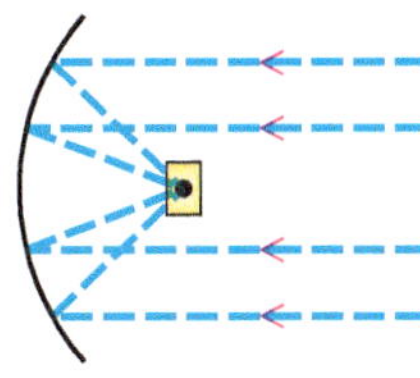

Solar furnace

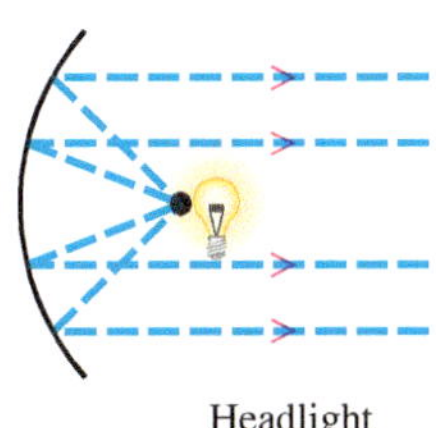

Headlight

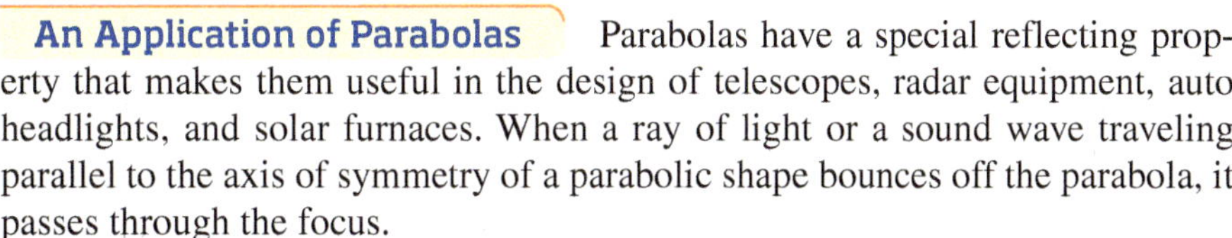

An Application of Parabolas Parabolas have a special reflecting property that makes them useful in the design of telescopes, radar equipment, auto headlights, and solar furnaces. When a ray of light or a sound wave traveling parallel to the axis of symmetry of a parabolic shape bounces off the parabola, it passes through the focus.

For example, in a solar furnace, a parabolic mirror collects light at the focus and thereby generates intense heat at that point. If a light source is placed at the focus, then the reflected light rays will be directed straight ahead.

EXAMPLE 6 Modeling the Reflective Property of Parabolas

The Parkes radio telescope has a parabolic dish shape with diameter 210 ft and depth 32 ft. Because of this parabolic shape, distant rays hitting the dish will be reflected directly toward the focus. A cross section of the dish is shown in **Figure 12.** (*Source:* Mar, J., and H. Liebowitz, *Structure Technology for Large Radio and Radar Telescope Systems,* The MIT Press, Massachusetts Institute of Technology.)

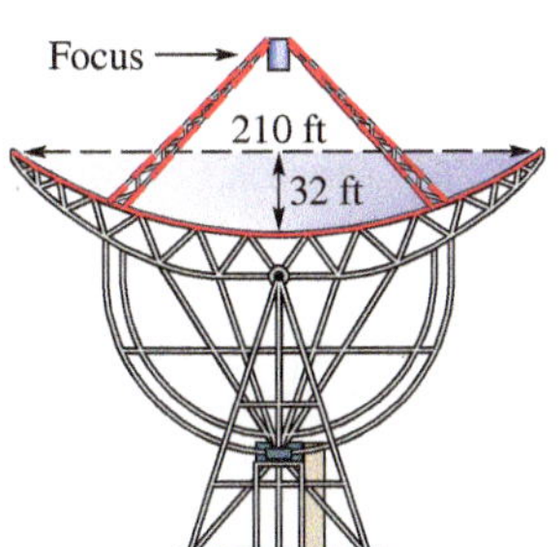

Figure 12

(a) Determine an equation that models this cross section by placing the vertex at the origin with the parabola opening up.

(b) The receiver must be placed at the focus of the parabola. How far from the vertex of the parabolic dish should the receiver be located?

SOLUTION

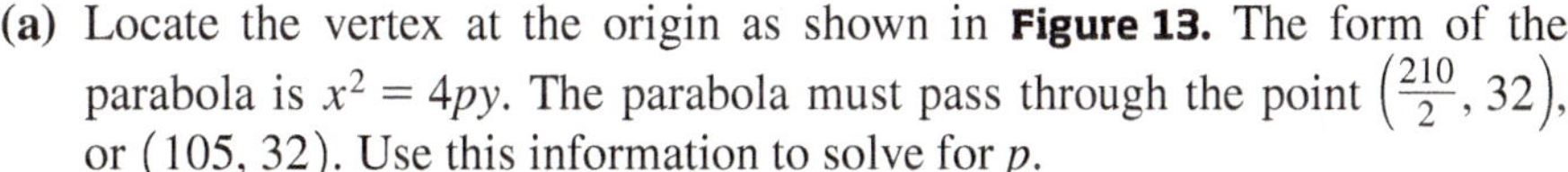

(a) Locate the vertex at the origin as shown in **Figure 13.** The form of the parabola is $x^2 = 4py$. The parabola must pass through the point $\left(\frac{210}{2}, 32\right)$, or $(105, 32)$. Use this information to solve for p.

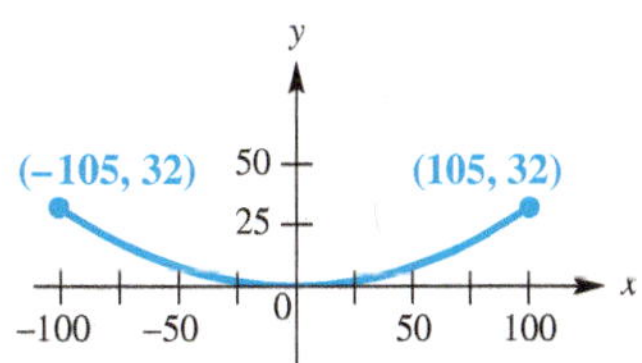

Figure 13

$$x^2 = 4py \quad \text{Parabola with vertical axis of symmetry}$$

$$(105)^2 = 4p(32) \quad \text{Let } x = 105 \text{ and } y = 32.$$

$$11{,}025 = 128p \quad \text{Multiply.}$$

$$p = \frac{11{,}025}{128} \quad \text{Solve for } p.$$

The cross section can be modeled by the following equation.

$$x^2 = 4py \quad \text{Parabola with vertical axis of symmetry}$$

$$x^2 = 4\left(\frac{11{,}025}{128}\right)y \quad \text{Substitute for } p.$$

$$x^2 = \frac{11{,}025}{32}y \quad \text{Simplify.}$$

(b) The distance between the vertex and the focus is p. In part (a), we found $p = \frac{11{,}025}{128} \approx 86.1$, so the receiver should be located at $(0, 86.1)$, or 86.1 ft above the vertex.

Now Try Exercise 51.

10.1 Exercises

1. **CONCEPT PREVIEW** Match each equation of a parabola in Column I with its description in Column II.

I	II
(a) $y - 2 = (x + 4)^2$	**A.** vertex $(-2, 4)$; opens down
(b) $y - 4 = (x + 2)^2$	**B.** vertex $(-2, 4)$; opens up
(c) $y - 2 = -(x + 4)^2$	**C.** vertex $(-4, 2)$; opens down
(d) $y - 4 = -(x + 2)^2$	**D.** vertex $(-4, 2)$; opens up
(e) $x - 2 = (y + 4)^2$	**E.** vertex $(2, -4)$; opens left
(f) $x - 4 = (y + 2)^2$	**F.** vertex $(2, -4)$; opens right
(g) $x - 2 = -(y + 4)^2$	**G.** vertex $(4, -2)$; opens left
(h) $x - 4 = -(y + 2)^2$	**H.** vertex $(4, -2)$; opens right

2. **CONCEPT PREVIEW** Match each equation of a parabola in Column I with its description in Column II.

I	II
(a) $y = 2x^2 + 3x + 9$	**A.** opens right
(b) $y = -3x^2 + 4x - 2$	**B.** opens up
(c) $x = 2y^2 - 3y + 9$	**C.** opens left
(d) $x = -3y^2 - 4y + 2$	**D.** opens down

CONCEPT PREVIEW *For each parabola, give the vertex, focus, directrix, axis of symmetry, domain, and range.*

3.

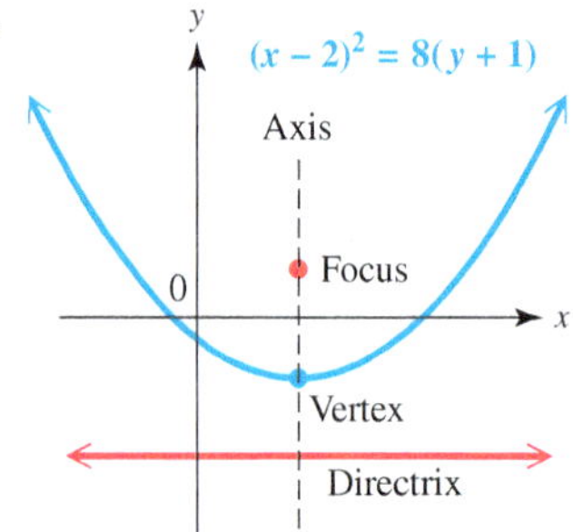

4.

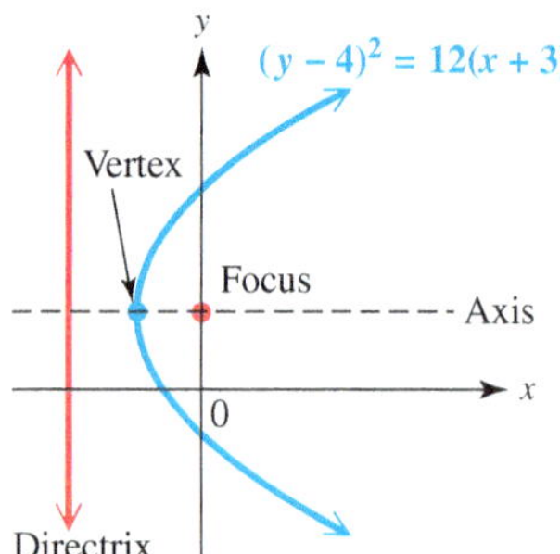

Graph each horizontal parabola, and give the domain and range. ***See Examples 1 and 2.***

5. $x + 4 = y^2$

6. $x - 2 = y^2$

7. $x = (y - 3)^2$

8. $x = (y + 1)^2$

9. $x - 2 = (y - 4)^2$

10. $x + 1 = (y + 2)^2$

11. $x - 2 = -3(y - 1)^2$

12. $x - 4 = \frac{1}{2}(y - 1)^2$

13. $-\frac{1}{2}x = (y + 3)^2$

14. $-\frac{1}{3}x = (y - 2)^2$

15. $x = y^2 + 4y + 2$

16. $x = 2y^2 - 4y + 6$

17. $x = -4y^2 - 4y + 3$

18. $x = -2y^2 + 2y - 3$

19. $2x - y^2 + 4y - 6 = 0$

20. $x + 3y^2 + 18y + 22 = 0$

21. $-x = 3y^2 + 6y + 2$

22. $-x = 2y^2 + 4y - 1$

Give the focus, directrix, and axis of symmetry for each parabola. **See Example 3.**

23. $x^2 = 24y$

24. $x^2 = \frac{1}{8}y$

25. $y = -4x^2$

26. $y = -\frac{1}{9}x^2$

27. $y^2 = -4x$

28. $y^2 = 16x$

29. $x = -32y^2$

30. $x = -16y^2$

31. $(y - 3)^2 = 12(x - 1)$

32. $(y - 2)^2 = 24(x - 3)$

33. $(x - 7)^2 = 16(y + 5)$

34. $(x + 2)^2 = 20(y - 2)$

Write an equation for each parabola with vertex at the origin. **See Example 4.**

35. focus $(5, 0)$

36. focus $\left(-\frac{1}{2}, 0\right)$

37. directrix $y = -\frac{1}{4}$

38. directrix $y = \frac{1}{3}$

39. through the point $\left(\sqrt{3}, 3\right)$, opens up

40. through the point $\left(-2, -2\sqrt{2}\right)$, opens left

41. through the point $(3, 2)$, symmetric with respect to the x-axis

42. through the point $(2, -4)$, symmetric with respect to the y-axis

Write an equation for each parabola. **See Example 5.**

43. vertex $(4, 3)$, focus $(4, 5)$

44. vertex $(-2, 1)$, focus $(-2, -3)$

45. vertex $(-5, 6)$, directrix $x = -12$

46. vertex $(1, 2)$, directrix $x = 4$

Determine the two equations necessary to graph each horizontal parabola using a graphing calculator, and graph it in the viewing window indicated. **See Example 2.**

47. $x = 3y^2 + 6y - 4$; $[-10, 2]$ by $[-4, 4]$

48. $x = -2y^2 + 4y + 3$; $[-10, 6.5]$ by $[-4, 4]$

49. $x + 2 = -(y + 1)^2$; $[-10, 2]$ by $[-4, 4]$

50. $x - 5 = 2(y - 2)^2$; $[-2, 12]$ by $[-2, 6]$

Solve each problem. **See Example 6.**

51. ***(Modeling) Radio Telescope Design*** The U.S. Naval Research Laboratory designed a giant radio telescope that had diameter 300 ft and maximum depth 44 ft. (*Source:* Mar, J., and H. Liebowitz, *Structure Technology for Large Radio and Radar Telescope Systems,* The MIT Press.)

(a) Write an equation of a parabola that models the cross section of the dish if the vertex is placed at the origin and the parabola opens up.

(b) The receiver must be placed at the focus of the parabola. How far from the vertex, to the nearest tenth of a foot, should the receiver be located?

52. *(Modeling) Radio Telescope Design* Suppose the telescope in **Exercise 51** had diameter 400 ft and maximum depth 50 ft.

(a) Write an equation of this parabola.

(b) The receiver must be placed at the focus of the parabola. How far from the vertex should the receiver be located?

53. *Parabolic Arch* An arch in the shape of a parabola has the dimensions shown in the figure. How wide is the arch 9 ft up?

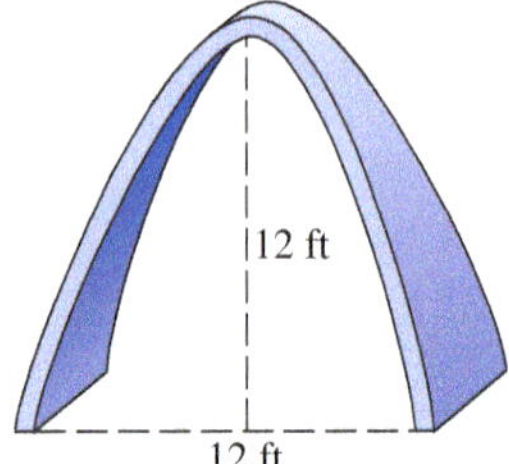

54. *Height of Bridge Cable Supports* The cable in the center portion of a bridge is supported as shown in the figure to form a parabola. The center vertical cable is 10 ft high, the supports are 210 ft high, and the distance between the two supports is 400 ft. Find the height of the remaining vertical cables, if the vertical cables are evenly spaced. (Ignore the width of the supports and cables.)

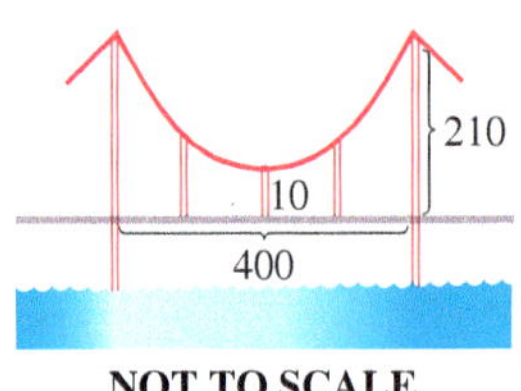

55. *(Modeling) Path of a Cannon Shell* The physicist Galileo observed that certain projectiles follow a parabolic path. For instance, if a cannon fires a shell at a 45° angle with a speed of v feet per second, then the path of the shell (see the figure on the left below) is modeled by the following equation.

$$y = x - \frac{32}{v^2}x^2$$

The figure on the right below shows the paths of shells all fired at the same speed but at different angles. The greatest distance is achieved with a 45° angle. The outline, or **envelope,** of this family of curves is another parabola with the cannon as focus. The horizontal line through the vertex of the envelope parabola is a directrix for all the other parabolas. Suppose all the shells are fired at a speed of 252.982 ft per sec.

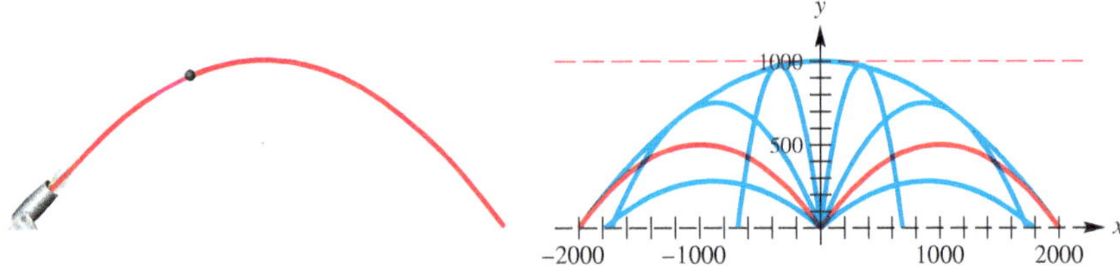

(a) What is the greatest distance, to the nearest foot, that a shell can be fired?

(b) What is the equation of the envelope parabola?

(c) Can a shell reach a helicopter 1500 ft due east of the cannon flying at a height of 450 ft?

56. *(Modeling) Path of a Projectile* When a projected object moves under the influence of a constant force (without air resistance), its path is parabolic. This occurs when a ball is thrown near the surface of a planet or other celestial body. Suppose two balls are simultaneously thrown upward at a 45° angle on two different planets. If their initial velocities are both 30 mph, then their paths can be modeled by the following equation.

$$y = x - \frac{g}{1936}x^2$$

Here g is the acceleration due to gravity, and x and y are the horizontal and vertical distances in feet, respectively. The value of g will vary depending on the mass and size of the planet.

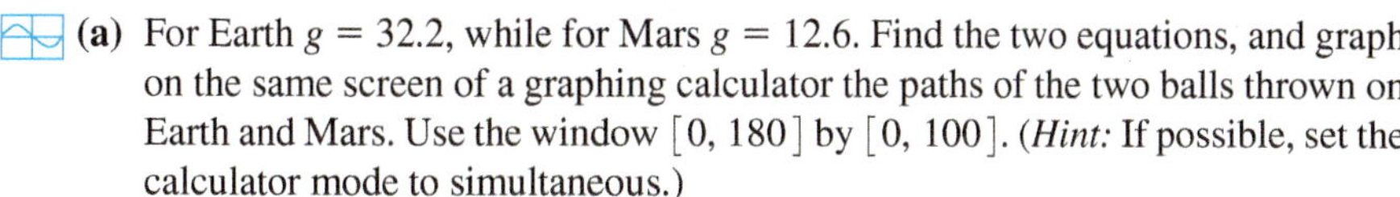

(a) For Earth $g = 32.2$, while for Mars $g = 12.6$. Find the two equations, and graph on the same screen of a graphing calculator the paths of the two balls thrown on Earth and Mars. Use the window $[0, 180]$ by $[0, 100]$. (*Hint:* If possible, set the calculator mode to simultaneous.)

(b) Determine the difference in the horizontal distances traveled by the two balls to the nearest foot.

57. ***(Modeling) Path of a Projectile*** (Refer to **Exercise 56.**) Suppose the two balls are now thrown upward at a 60° angle on Mars and the moon. If their initial velocity is 60 mph, then their paths in feet can be modeled by the following equation.

$$y = \sqrt{3}x - \frac{g}{3872}x^2$$

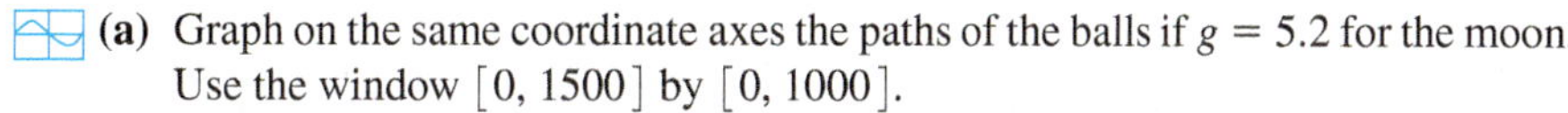

(a) Graph on the same coordinate axes the paths of the balls if $g = 5.2$ for the moon. Use the window $[0, 1500]$ by $[0, 1000]$.

(b) Determine the maximum height of each ball to the nearest foot.

58. Prove that the parabola with focus $(p, 0)$ and directrix $x = -p$ has the equation $y^2 = 4px$.

Relating Concepts

For individual or collaborative investigation *(Exercises 59–62)*

(Modeling) *Given three noncollinear points, we can find an equation of the form*

$$x = ay^2 + by + c$$

of the horizontal parabola joining them by solving a system of equations. ***Work Exercises 59–62 in order,*** *to find the equation of the horizontal parabola containing the points*

$$(-5, 1), \quad (-14, -2), \quad \textit{and} \quad (-10, 2).$$

59. Write three equations in a, b, and c, by substituting the given values of x and y into the equation $x = ay^2 + by + c$.

60. Solve the system of three equations determined in **Exercise 59.**

61. Does the horizontal parabola open to the left or to the right? Why?

62. Write the equation of the horizontal parabola.

10.2 Ellipses

- Equations and Graphs of Ellipses
- Translated Ellipses
- Eccentricity
- Applications of Ellipses

Equations and Graphs of Ellipses Like the circle and the parabola, the ellipse is defined as a set of points in a plane.

Ellipse

An **ellipse** is the set of all points in a plane the sum of whose distances from two fixed points is constant. Each fixed point is a **focus** (plural, **foci**) of the ellipse.

As shown in **Figure 14,** an ellipse has two axes of symmetry, the **major axis** (the longer one) and the **minor axis** (the shorter one). The foci are always located on the major axis. The midpoint of the major axis is the **center** of the ellipse, and the endpoints of the major axis are the **vertices** of the ellipse. ***The graph of an ellipse is not the graph of a function.*** It fails the vertical line test.

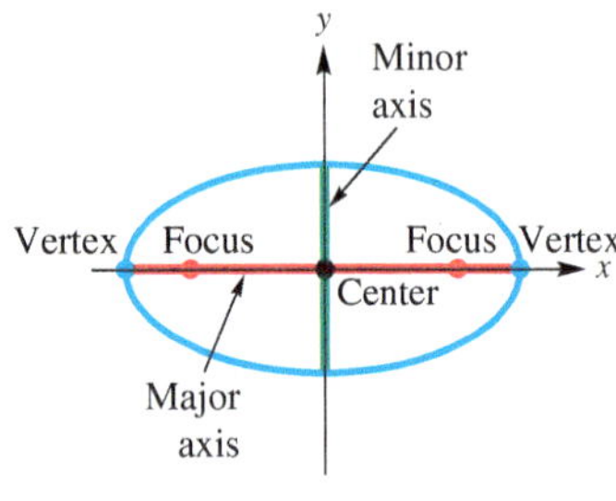

Figure 14

The ellipse in **Figure 15** has its center at the origin, foci $F(c, 0)$ and $F'(-c, 0)$, and vertices $V(a, 0)$ and $V'(-a, 0)$. From **Figure 15,** the distance from V to F is $a - c$ and the distance from V to F' is $a + c$. The sum of these distances is $2a$. Because V is on the ellipse, this sum is the constant referred to in the definition of an ellipse. Thus, for any point $P(x, y)$ on the ellipse,

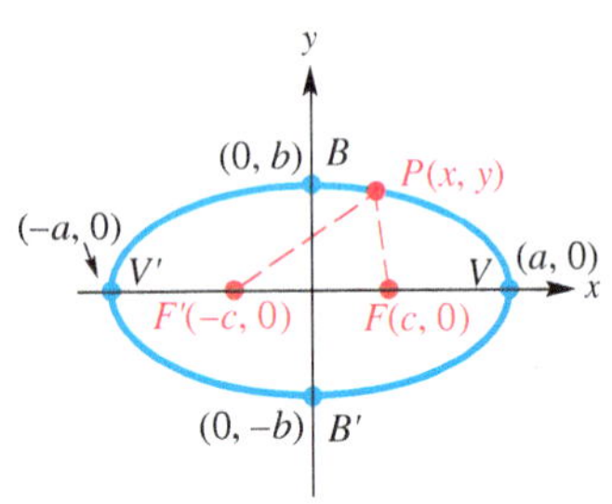

Ellipse centered at the origin

Figure 15

$$d(P, F) + d(P, F') = 2a.$$

By the distance formula,

$$d(P, F) = \sqrt{(x-c)^2 + y^2}$$

and $$d(P, F') = \sqrt{[x-(-c)]^2 + y^2} = \sqrt{(x+c)^2 + y^2}.$$

Thus, we have the following.

$$\sqrt{(x-c)^2 + y^2} + \sqrt{(x+c)^2 + y^2} = 2a \qquad d(P, F) + d(P, F') = 2a$$

$$\sqrt{(x-c)^2 + y^2} = 2a - \sqrt{(x+c)^2 + y^2} \qquad \text{Isolate } \sqrt{(x-c)^2 + y^2}.$$

$$(x-c)^2 + y^2 = 4a^2 - 4a\sqrt{(x+c)^2 + y^2} + (x+c)^2 + y^2 \qquad \text{Square each side.}$$

Be careful when squaring.

$$x^2 - 2cx + c^2 + y^2 = 4a^2 - 4a\sqrt{(x+c)^2 + y^2} + x^2 + 2cx + c^2 + y^2 \qquad \text{Square } x - c. \text{ Square } x + c.$$

$$4a\sqrt{(x+c)^2 + y^2} = 4a^2 + 4cx \qquad \text{Isolate } 4a\sqrt{(x+c)^2 + y^2}.$$

Divide each term by 4.

$$a\sqrt{(x+c)^2 + y^2} = a^2 + cx \qquad \text{Divide by 4.}$$

$$a^2(x^2 + 2cx + c^2 + y^2) = a^4 + 2ca^2x + c^2x^2 \qquad \text{Square each side. Square } x + c.$$

$$a^2x^2 + 2ca^2x + a^2c^2 + a^2y^2 = a^4 + 2ca^2x + c^2x^2 \qquad \text{Distributive property}$$

$$a^2x^2 + a^2c^2 + a^2y^2 = a^4 + c^2x^2 \qquad \text{Subtract } 2ca^2x.$$

$$a^2x^2 - c^2x^2 + a^2y^2 = a^4 - a^2c^2 \qquad \text{Rearrange terms.}$$

$$(a^2 - c^2)x^2 + a^2y^2 = a^2(a^2 - c^2) \qquad \text{Factor.}$$

$$\frac{x^2}{a^2} + \frac{y^2}{a^2 - c^2} = 1 \quad (1) \qquad \text{Divide by } a^2(a^2 - c^2).$$

The point $B(0, b)$ is on the ellipse in **Figure 15,** so we have the following.

$$d(B, F) + d(B, F') = 2a$$

$$\sqrt{(-c)^2 + b^2} + \sqrt{c^2 + b^2} = 2a \quad \text{Substitute.}$$

$$2\sqrt{c^2 + b^2} = 2a \quad \text{Combine like terms.}$$

$$\sqrt{c^2 + b^2} = a \quad \text{Divide by 2.}$$

$$c^2 + b^2 = a^2 \quad \text{Square each side.}$$

$$b^2 = a^2 - c^2 \quad \text{Subtract } c^2.$$

Replacing $a^2 - c^2$ with b^2 in equation (1) gives the standard form of the equation of an ellipse centered at the origin with foci on the x-axis.

$$\frac{x^2}{a^2} + \frac{y^2}{b^2} = 1$$

If the vertices and foci were on the y-axis, an almost identical derivation could be used to obtain the following standard form.

$$\frac{x^2}{b^2} + \frac{y^2}{a^2} = 1$$

Standard Forms of Equations for Ellipses

The ellipse with center at the origin and equation

$$\frac{x^2}{a^2} + \frac{y^2}{b^2} = 1 \quad (\text{where } a > b)$$

has vertices $(\pm a, 0)$, endpoints of the minor axis $(0, \pm b)$, and foci $(\pm c, 0)$, where $c^2 = a^2 - b^2$.

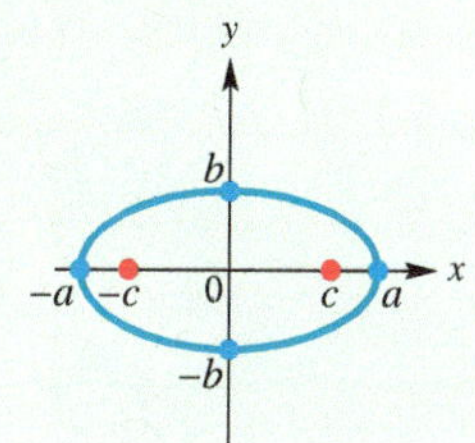

Major axis on x-axis

The ellipse with center at the origin and equation

$$\frac{x^2}{b^2} + \frac{y^2}{a^2} = 1 \quad (\text{where } a > b)$$

has vertices $(0, \pm a)$, endpoints of the minor axis $(\pm b, 0)$, and foci $(0, \pm c)$, where $c^2 = a^2 - b^2$.

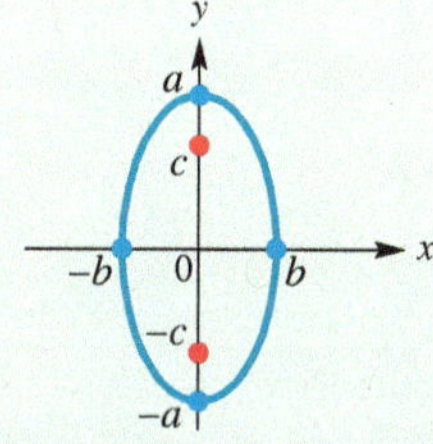

Major axis on y-axis

LOOKING AHEAD TO CALCULUS

Methods of calculus can be used to solve problems involving ellipses. For example, differentiation is used to find the slope of the tangent line at a point on the ellipse, and integration is used to find the length of any arc of the ellipse.

Do not be confused by the two standard forms.

- In the first form, a^2 is associated with x^2.
- In the second form, a^2 is associated with y^2.

In practice it is necessary only to find the intercepts of the graph—if the x-intercepts are farther from the center of the ellipse than the y-intercepts, then the major axis is horizontal; otherwise, it is vertical.

When using the relationship $c^2 = a^2 - b^2$, choose a^2 and b^2 so that $a^2 > b^2$.

EXAMPLE 1 Graphing Ellipses Centered at the Origin

Graph each ellipse, and find the coordinates of the foci. Give the domain and range.

(a) $4x^2 + 9y^2 = 36$ **(b)** $4x^2 = 64 - y^2$

SOLUTION

(a) Divide each side of $4x^2 + 9y^2 = 36$ by 36 to write the equation in standard form.

$$\frac{x^2}{9} + \frac{y^2}{4} = 1 \quad \text{Standard form of an ellipse}$$

(Each term was divided by 36.)

Thus, the x-intercepts are $(\pm 3, 0)$ and the y-intercepts are $(0, \pm 2)$. The graph of the ellipse is shown in **Figure 16.**

Because $9 > 4$, we find the foci of the ellipse by letting $a^2 = 9$ and $b^2 = 4$ in $c^2 = a^2 - b^2$.

$$c^2 = 9 - 4 = 5, \quad \text{so} \quad c = \sqrt{5} \quad \text{By definition, } c > 0.$$

The major axis is along the x-axis. Thus, the foci have coordinates $\left(-\sqrt{5}, 0\right)$ and $\left(\sqrt{5}, 0\right)$. The domain of this relation is $[-3, 3]$, and the range is $[-2, 2]$.

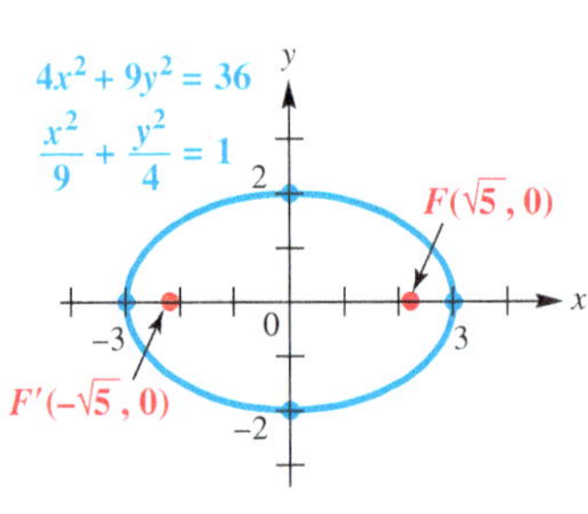

Figure 16

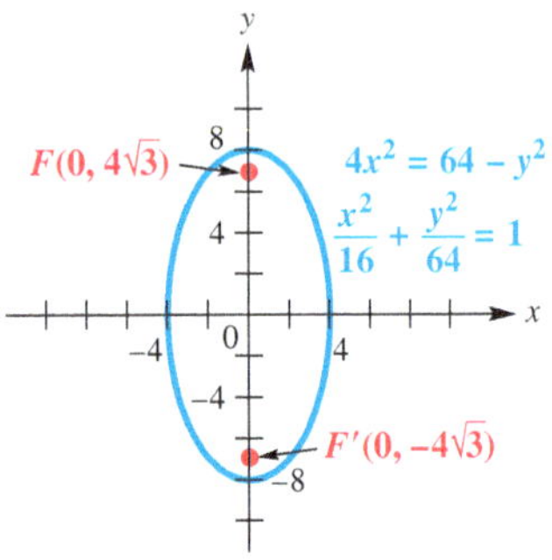

Figure 17

(b) Write the equation $4x^2 = 64 - y^2$ as $4x^2 + y^2 = 64$. Then divide each side by 64 to express it in standard form.

$$\frac{x^2}{16} + \frac{y^2}{64} = 1 \quad \text{Standard form of an ellipse}$$

(Each term was divided by 64.)

The x-intercepts are $(\pm 4, 0)$ and the y-intercepts are $(0, \pm 8)$. See **Figure 17.** Here $64 > 16$, so $a^2 = 64$ and $b^2 = 16$. Use $c^2 = a^2 - b^2$.

$$c^2 = 64 - 16 = 48, \quad \text{so} \quad c = \sqrt{48} = 4\sqrt{3} \quad \sqrt{48} = \sqrt{16 \cdot 3} = 4\sqrt{3}$$

The major axis is on the y-axis, which means the coordinates of the foci are $\left(0, -4\sqrt{3}\right)$ and $\left(0, 4\sqrt{3}\right)$. The domain of the relation is $[-4, 4]$, and the range is $[-8, 8]$.

✔ **Now Try Exercises 11 and 13.**

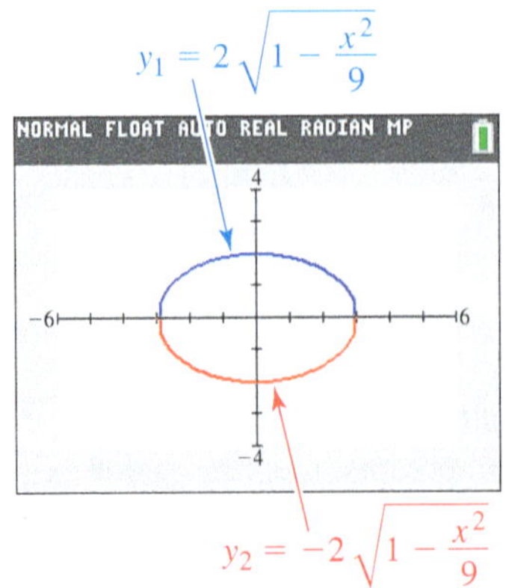

Figure 18

The graph of an ellipse is *not* the graph of a function. To graph the ellipse in **Example 1(a)** with a graphing calculator in function mode, solve for y in $4x^2 + 9y^2 = 36$ to obtain equations of the two functions shown in **Figure 18.**

$$y = 2\sqrt{1 - \frac{x^2}{9}} \quad \text{and} \quad y = -2\sqrt{1 - \frac{x^2}{9}}$$

(Their union is the graph of $4x^2 + 9y^2 = 36$.) ■

EXAMPLE 2 Writing an Equation of an Ellipse

Write an equation of the ellipse having center at the origin, foci at $(0, 3)$ and $(0, -3)$, and major axis of length 8 units.

SOLUTION Because the major axis is 8 units long, $2a = 8$ and thus $a = 4$. To find b^2, use the relationship $c^2 = a^2 - b^2$, with $a = 4$ and $c = 3$.

$$c^2 = a^2 - b^2 \quad \text{Relationship for ellipses}$$

$$3^2 = 4^2 - b^2 \quad \text{Let } c = 3 \text{ and } a = 4.$$

$$9 = 16 - b^2 \quad \text{Apply the exponents.}$$

$$b^2 = 7 \quad \text{Solve for } b^2.$$

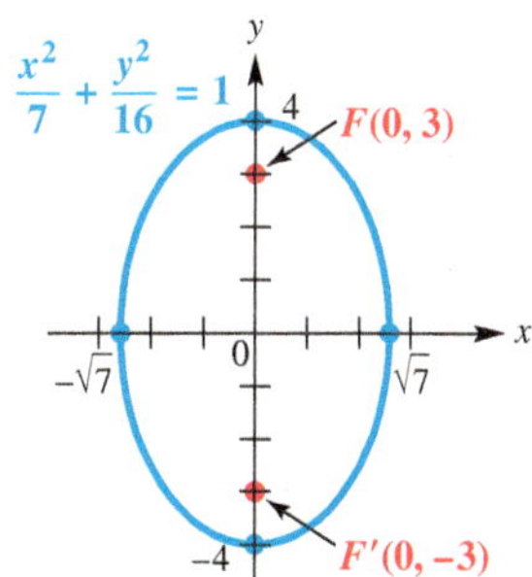

Figure 19

The foci are on the y-axis, so we use the larger value, $a = 4$, to find the denominator for y^2, giving the equation in standard form.

$$\frac{x^2}{7} + \frac{y^2}{16} = 1 \quad \text{Use } \frac{x^2}{b^2} + \frac{y^2}{a^2} = 1.$$

A graph of this ellipse is shown in **Figure 19.** This relation has

$$\text{domain } \left[-\sqrt{7}, \sqrt{7}\right] \quad \text{and} \quad \text{range } [-4, 4].$$

✔ **Now Try Exercise 21.**

EXAMPLE 3 Graphing a Half-Ellipse

Graph $\frac{y}{4} = \sqrt{1 - \frac{x^2}{25}}$. Give the domain and range.

SOLUTION We transform this equation to see that its graph is part of an ellipse.

$$\frac{y}{4} = \sqrt{1 - \frac{x^2}{25}} \quad \text{Given equation}$$

$$\frac{y^2}{16} = 1 - \frac{x^2}{25} \quad \text{Square each side.}$$

$$\frac{x^2}{25} + \frac{y^2}{16} = 1 \quad \text{Write in standard form.}$$

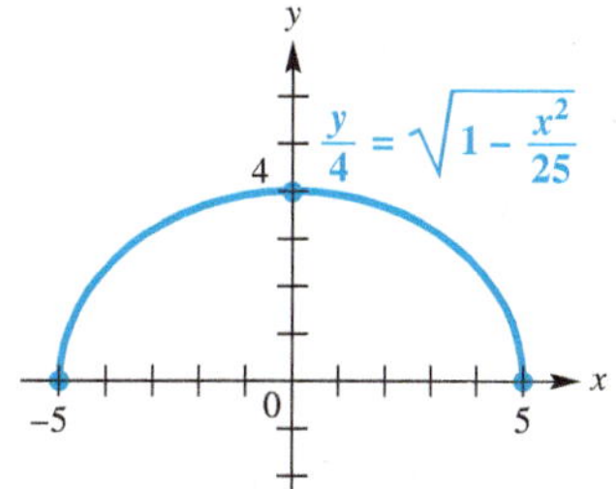

Figure 20

This is the equation of an ellipse with x-intercepts $(\pm 5, 0)$ and y-intercepts $(0, \pm 4)$. In the original equation, the radical expression

$$\sqrt{1 - \frac{x^2}{25}}$$

represents a nonnegative number, so the only possible values of y are those that give the half-ellipse shown in **Figure 20.** This is the graph of a function with

$$\text{domain } [-5, 5] \quad \text{and} \quad \text{range } [0, 4].$$

✔ **Now Try Exercise 33.**

Translated Ellipses An ellipse may have its center translated away from the origin by replacing x and y with $x - h$ and $y - k$, respectively.

Standard Forms for Ellipses Centered at (h, k)

An ellipse with center at (h, k) and either a horizontal or a vertical major axis of length $2a$ satisfies one of the following equations, where $a > b > 0$ and $c^2 = a^2 - b^2$ with $c > 0$.

$$\frac{(x-h)^2}{a^2} + \frac{(y-k)^2}{b^2} = 1$$

Major axis: horizontal; vertices: $(h \pm a, k)$; foci: $(h \pm c, k)$

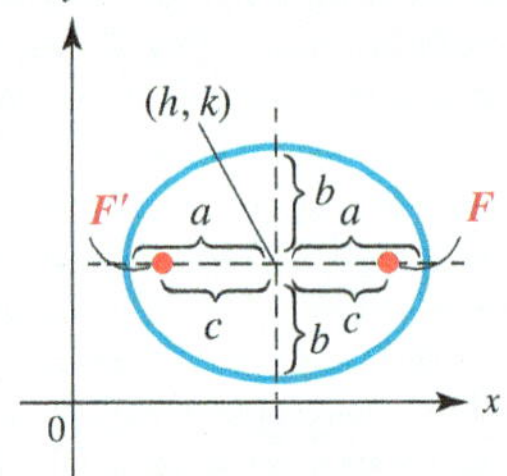

Horizontal major axis

$$\frac{(x-h)^2}{b^2} + \frac{(y-k)^2}{a^2} = 1$$

Major axis: vertical; vertices: $(h, k \pm a)$; foci: $(h, k \pm c)$

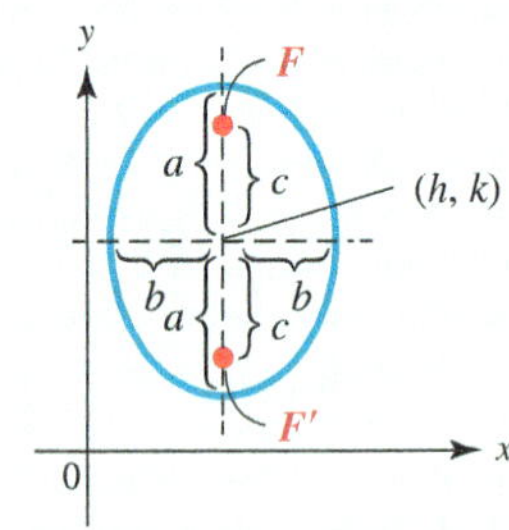

Vertical major axis

When graphing ellipses, remember that the location of a^2 (the greater denominator) determines whether the ellipse has a horizontal or a vertical major axis.

EXAMPLE 4 Graphing an Ellipse Translated Away from the Origin

Graph $\frac{(x-2)^2}{9} + \frac{(y+1)^2}{16} = 1$. Give the foci, domain, and range.

SOLUTION The graph of this equation is an ellipse centered at $(2, -1)$. Because $a > b$ for ellipses, $a = 4$ and $b = 3$. This ellipse has a vertical major axis because $a^2 = 16$ is associated with y^2.

The vertices are located a distance of $a = 4$ units directly above and below the center, at $(2, 3)$ and $(2, -5)$. Two other points on the ellipse, located a distance of $b = 3$ units to the left and right of the center, are $(-1, -1)$ and $(5, -1)$.

The foci are found using the following equation.

$$c^2 = a^2 - b^2 \quad \text{Relationship for ellipses}$$

$$c^2 = 16 - 9 \quad \text{Let } a^2 = 16 \text{ and } b^2 = 9.$$

$$c^2 = 7 \quad \text{Subtract.}$$

$$c = \sqrt{7} \quad \text{Take the positive square root because } c > 0.$$

The foci are located on the major axis a distance of $c = \sqrt{7}$ (approximately 2.6) units above and below the center $(2, -1)$, at $\left(2, -1 + \sqrt{7}\right)$ and $\left(2, -1 - \sqrt{7}\right)$. See the graph in **Figure 21.** The domain is $[-1, 5]$, and the range is $[-5, 3]$.

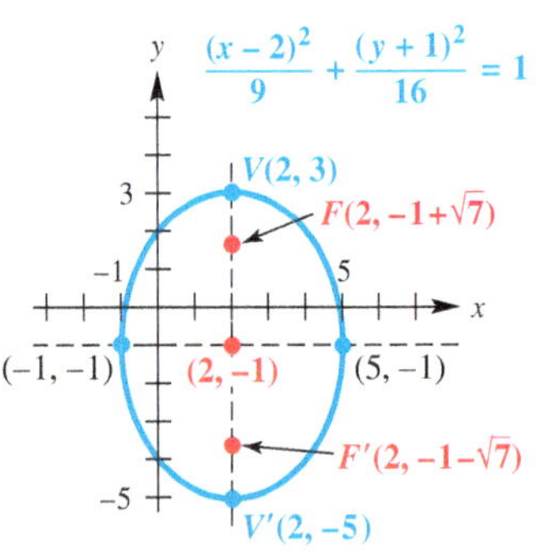

Figure 21

✔ **Now Try Exercise 17.**

NOTE As suggested by the graphs in this section, an ellipse is symmetric with respect to its major axis, its minor axis, and its center. ***If $a = b$ in the equation of an ellipse, then its graph is a circle.***

Eccentricity

All conics can be characterized by one general definition.

Conic

A **conic** is the set of all points $P(x, y)$ in a plane such that the ratio of the distance from P to a fixed point and the distance from P to a fixed line is constant.

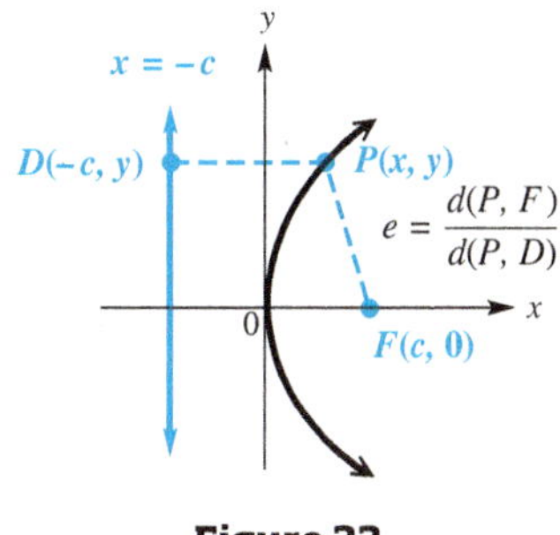

Figure 22

For a parabola, the fixed line is the directrix, and the fixed point is the focus. In **Figure 22,** the focus is $F(c, 0)$, and the directrix is the line $x = -c$. The constant ratio is the **eccentricity** of the conic, written $\boldsymbol{e}$. (*This is not the same e as the base of natural logarithms.*) If the conic is a parabola, then by definition, the distances $d(P, F)$ and $d(P, D)$ in **Figure 22** are equal. ***Thus, every parabola has eccentricity 1.***

For an ellipse, eccentricity is a measure of its "roundness." The constant ratio in the definition is $\boldsymbol{e} = \frac{c}{a}$, where (as before) c is the distance from the center of the figure to a focus, and a is the distance from the center to a vertex. By the definition of an ellipse, $a^2 > b^2$ and $c = \sqrt{a^2 - b^2}$. Thus, for the ellipse, we have the following.

$$0 < c < a \quad \text{Inequality fact for an ellipse}$$

$$0 < \frac{c}{a} < 1 \quad \text{Divide by } a.$$

$$0 < \boldsymbol{e} < 1 \quad \boldsymbol{e} = \frac{c}{a}$$

Thus, every ellipse has eccentricity between 0 and 1.

If a is constant, letting c approach 0 would force the ratio $\frac{c}{a}$ to approach 0, which also forces b to approach a (so that $\sqrt{a^2 - b^2} = c$ would approach 0). Because b determines the endpoints of the minor axis, this means that the lengths of the major and minor axes are almost the same, producing an ellipse very close in shape to a circle when $\boldsymbol{e}$ is very close to 0. In a similar manner, if $\boldsymbol{e}$ approaches 1, then b will approach 0.

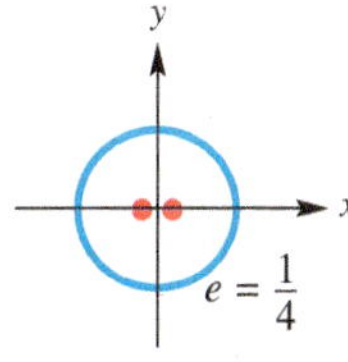

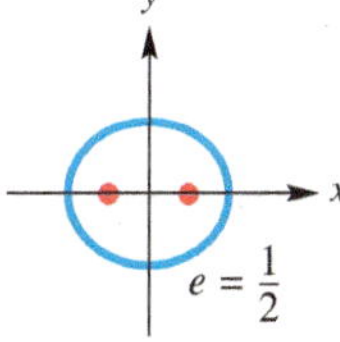

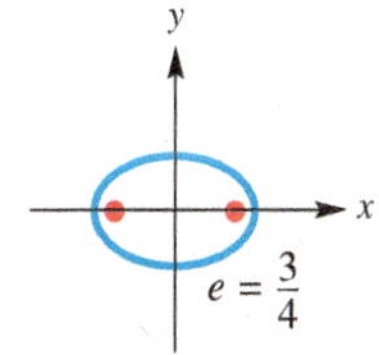

Eccentricities of ellipses

Figure 23

The path of Earth around the sun is an ellipse that is very nearly circular. In fact, for this ellipse, $\boldsymbol{e} \approx 0.017$. On the other hand, the path of Halley's comet is a very flat ellipse, with $\boldsymbol{e} \approx 0.97$. **Figure 23** compares ellipses with different eccentricities. The locations of the foci are shown in each case.

The equation of a circle with center (h, k) and radius r can be written as follows.

$$(x-h)^2 + (y-k)^2 = r^2 \quad \text{Center-radius form of a circle}$$

$$\frac{(x-h)^2}{r^2} + \frac{(y-k)^2}{r^2} = 1 \quad \text{Divide by } r^2.$$

In a circle, the foci coincide with the center, so $a = b$, $c = \sqrt{a^2 - b^2} = 0$, and $\boldsymbol{e} = 0$. ***Thus, every circle has eccentricity 0.***

EXAMPLE 5 Finding Eccentricity from Equations of Ellipses

Find the eccentricity e of each ellipse.

(a) $\frac{x^2}{9} + \frac{y^2}{16} = 1$ **(b)** $5x^2 + 10y^2 = 50$

SOLUTION

(a) Because $16 > 9$, $a^2 = 16$ and thus $a = 4$. To find c, use $c^2 = a^2 - b^2$.

$c^2 = a^2 - b^2$ Relationship for ellipses

$c^2 = 16 - 9$ Let $a^2 = 16$ and $b^2 = 9$.

$c = \sqrt{7}$ Subtract. Take the positive square root.

To find the eccentricity e, use $e = \frac{c}{a}$.

$e = \frac{\sqrt{7}}{4} \approx 0.66$ Substitute for c and a.

(b) Start by dividing each term of the given equation by 50.

$5x^2 + 10y^2 = 50$ Given equation

$\frac{5x^2}{50} + \frac{10y^2}{50} = \frac{50}{50}$ Divide by 50.

$\frac{x^2}{10} + \frac{y^2}{5} = 1$ Write in lowest terms.

For this ellipse, $a^2 = 10$ and thus $a = \sqrt{10}$. Find c as in part (a).

$c^2 = a^2 - b^2$ Relationship for ellipses

$c^2 = 10 - 5$ Let $a^2 = 10$ and $b^2 = 5$.

$c = \sqrt{5}$ Subtract. Take the positive square root.

Now find e.

$e = \frac{\sqrt{5}}{\sqrt{10}} \approx 0.71$ $e = \frac{c}{a}$

✔ **Now Try Exercises 41 and 43.**

Applications of Ellipses

EXAMPLE 6 Applying the Equation of an Ellipse to the Orbit of a Planet

The orbit of the planet Mars is an ellipse with the sun at one focus. The eccentricity of the ellipse is 0.0935, and the closest distance that Mars comes to the sun is 128.5 million mi. (*Source: World Almanac and Book of Facts.*) Find the maximum distance of Mars from the sun.

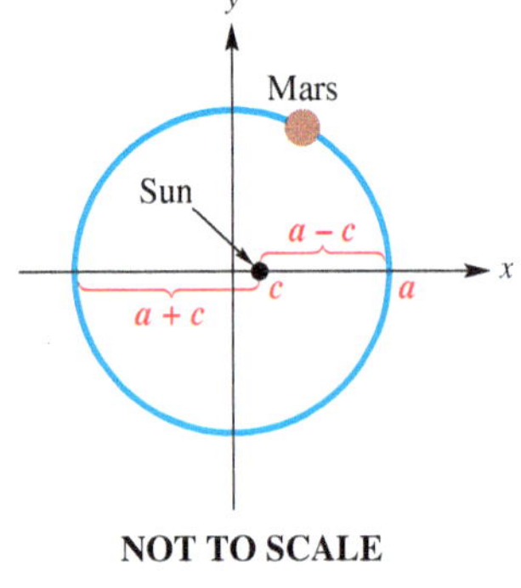

Figure 24

SOLUTION **Figure 24** shows the orbit of Mars with the origin at the center of the ellipse and the sun at one focus. Mars is closest to the sun when Mars is at the right endpoint of the major axis and farthest from the sun when Mars is at the left endpoint. Therefore, the least distance is $a - c$, and the greatest distance is $a + c$. Because $a - c = 128.5$, it follows that $c = a - 128.5$.

$$e = \frac{c}{a} \quad \text{Eccentricity formula for an ellipse}$$

$$0.0935 = \frac{a - 128.5}{a} \quad \text{Let } e = 0.0935 \text{ as given, and } c = a - 128.5.$$

$$0.0935a = a - 128.5 \quad \text{Multiply by } a.$$

$$a \approx 141.8 \quad \text{Solve for } a.$$

Then $c = 141.8 - 128.5 = 13.3$ and $a + c = 141.8 + 13.3 = 155.1$.

The maximum distance of Mars from the sun is about 155.1 million mi.

✔ **Now Try Exercise 49.**

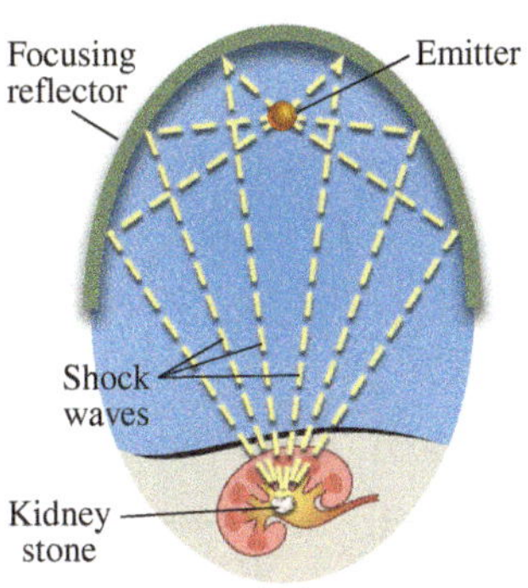

Aerial view of Old House Chamber

Figure 25

When a ray of light or sound emanating from one focus of an ellipse bounces off the ellipse, it passes through the other focus. See **Figure 25.** This reflecting property is responsible for "whispering galleries." In a whispering gallery, a person whispering at a certain point in the room can be heard clearly at another point across the room. The U.S. statesman John Quincy Adams was able to listen in on his opponents' conversations in the old House Chamber (Statuary Hall) because his desk was positioned at one of the foci beneath the ellipsoidal ceiling and his opponents were located across the room at the other focus.

Focusing reflector — Emitter — Shock waves — Kidney stone

The top of an ellipse is illustrated in this depiction of how a lithotripter crushes a kidney stone.

Figure 26

A lithotripter is a machine used to crush kidney stones using shock waves. The patient is placed in an elliptical tub with the kidney stone at one focus of the ellipse. A beam is projected from the other focus to the tub so that it reflects to hit the kidney stone. See **Figure 26.**

EXAMPLE 7 Modeling the Reflective Property of Ellipses

If a lithotripter is based on the ellipse

$$\frac{x^2}{36} + \frac{y^2}{27} = 1,$$

determine how many units both the kidney stone and the source of the beam must be placed from the center of the ellipse.

SOLUTION The kidney stone and the source of the beam must be placed at the foci, $(c, 0)$ and $(-c, 0)$. Here $a^2 = 36$ and $b^2 = 27$.

$$c^2 = a^2 - b^2 \quad \text{Relationship for ellipses}$$

$$c^2 = 36 - 27 \quad \text{Let } a^2 = 36 \text{ and } b^2 = 27.$$

$$c^2 = 9 \quad \text{Subtract.}$$

$$c = 3 \quad \text{Take the positive square root.}$$

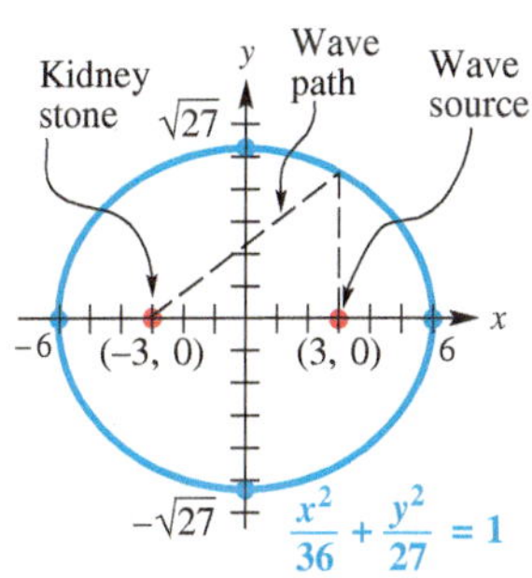

Figure 27

Thus, the foci are $(3, 0)$ and $(-3, 0)$. The kidney stone and the source both must be placed on the major axis 3 units from the center. See **Figure 27.**

✔ **Now Try Exercise 53.**

10.2 Exercises

1. **CONCEPT PREVIEW** Match each equation of an ellipse in Column I with the appropriate intercepts in Column II.

I	II
(a) $36x^2 + 9y^2 = 324$	**A.** $(-3, 0), (3, 0), (0, -6), (0, 6)$
(b) $9x^2 + 36y^2 = 324$	**B.** $(-4, 0), (4, 0), (0, -5), (0, 5)$
(c) $\frac{x^2}{25} + \frac{y^2}{16} = 1$	**C.** $(-6, 0), (6, 0), (0, -3), (0, 3)$
(d) $\frac{x^2}{16} + \frac{y^2}{25} = 1$	**D.** $(-5, 0), (5, 0), (0, -4), (0, 4)$

2. **CONCEPT PREVIEW** Determine whether or not each equation is that of an ellipse. If it is not, state the kind of graph the equation has.

(a) $x^2 + 4y^2 = 4$ **(b)** $x^2 + y^2 = 4$ **(c)** $x^2 + y = 4$ **(d)** $\frac{x}{4} + \frac{y}{25} = 1$

CONCEPT PREVIEW *For each ellipse, give the domain, range, center, vertices, and foci.*

3.

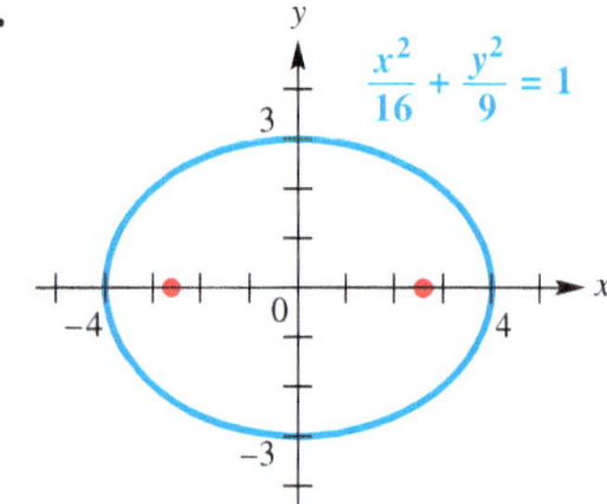

4.

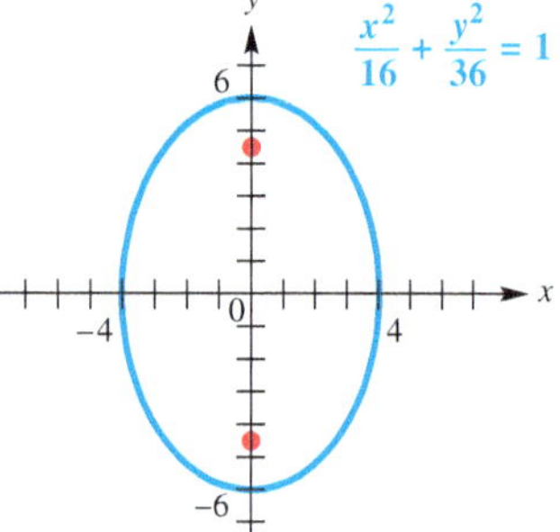

5.

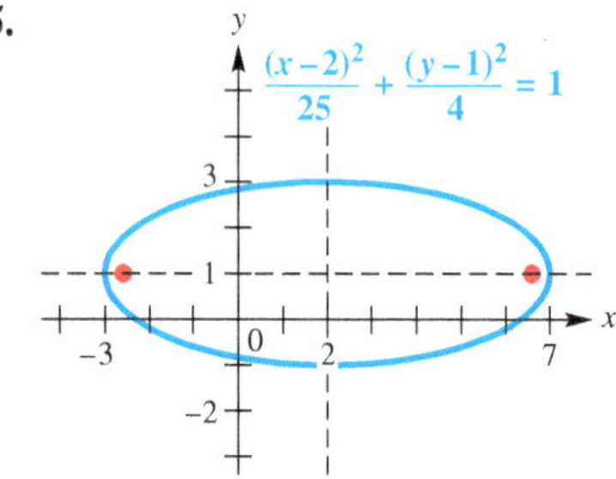

6.

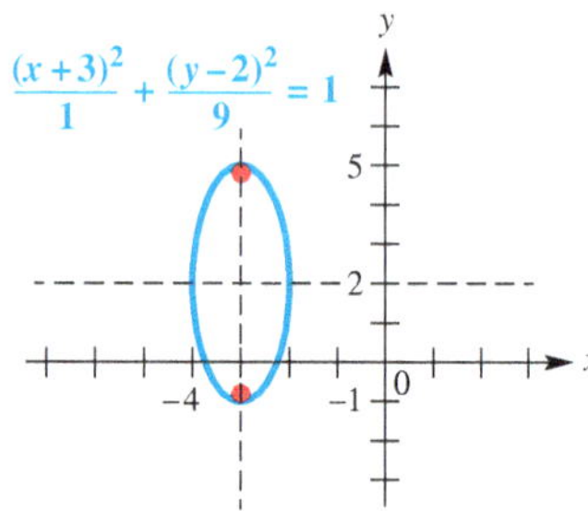

Graph each ellipse. Give the domain, range, center, vertices, endpoints of the minor axis, and foci. ***See Examples 1 and 4.***

7. $\frac{x^2}{25} + \frac{y^2}{9} = 1$

8. $\frac{x^2}{16} + \frac{y^2}{25} = 1$

9. $\frac{x^2}{9} + y^2 = 1$

10. $\frac{x^2}{36} + \frac{y^2}{16} = 1$

11. $9x^2 + y^2 = 81$

12. $4x^2 + 16y^2 = 64$

13. $4x^2 = 100 - 25y^2$

14. $4x^2 = 16 - y^2$

15. $\frac{(x-2)^2}{25} + \frac{(y-1)^2}{4} = 1$

16. $\frac{(x+2)^2}{16} + \frac{(y+1)^2}{9} = 1$

17. $\frac{(x+3)^2}{16} + \frac{(y-2)^2}{36} = 1$

18. $\frac{(x-1)^2}{9} + \frac{(y+3)^2}{25} = 1$

Write an equation for each ellipse. **See Example 2.**

19. x-intercepts $(\pm 5, 0)$; y-intercepts $(0, \pm 4)$

20. x-intercepts $(\pm \sqrt{15}, 0)$; y-intercepts $(0, \pm 4)$

21. major axis with length 6; foci at $(0, 2)$, $(0, -2)$

22. minor axis with length 4; foci at $(-5, 0)$, $(5, 0)$

23. center at $(3, 1)$; minor axis vertical, with length 8; $c = 3$

24. center at $(-2, 7)$; major axis vertical, with length 10; $c = 2$

25. foci at $(0, 4)$, $(0, -4)$; sum of distances from foci to point on ellipse is 10 (*Hint:* Consider one of the vertices.)

26. foci at $(-4, 0)$, $(4, 0)$; sum of distances from foci to point on ellipse is 9

27. foci at $(0, -3)$, $(0, 3)$; the point $(8, 3)$ on ellipse

28. foci at $(-3, -3)$, $(7, -3)$; the point $(2, -7)$ on ellipse

29. $e = \frac{4}{5}$; vertices at $(-5, 0)$, $(5, 0)$

30. $e = \frac{1}{2}$; vertices at $(-4, 0)$, $(4, 0)$

31. $e = \frac{3}{4}$; foci at $(0, -2)$, $(0, 2)$

32. $e = \frac{2}{3}$; foci at $(0, -9)$, $(0, 9)$

Graph each equation. Give the domain and range. Identify any that are functions. **See Example 3.**

33. $\frac{y}{2} = \sqrt{1 - \frac{x^2}{25}}$

34. $\frac{x}{4} = \sqrt{1 - \frac{y^2}{9}}$

35. $x = -\sqrt{1 - \frac{y^2}{64}}$

36. $y = -\sqrt{1 - \frac{x^2}{100}}$

Determine the two equations necessary to graph each ellipse using a graphing calculator, and graph it in the viewing window indicated. **See Figure 18.**

37. $\frac{x^2}{16} + \frac{y^2}{4} = 1$; $[-6.6, 6.6]$ by $[-4.1, 4.1]$

38. $\frac{x^2}{4} + \frac{y^2}{25} = 1$; $[-6.6, 6.6]$ by $[-5.2, 5.2]$

39. $\frac{(x-3)^2}{25} + \frac{y^2}{9} = 1$; $[-9.9, 9.9]$ by $[-8.2, 8.2]$

40. $\frac{x^2}{36} + \frac{(y+4)^2}{4} = 1$; $[-9.9, 9.9]$ by $[-8.2, 8.2]$

Find the eccentricity e *of each ellipse. Round to the nearest hundredth as needed.* **See Example 5.**

41. $\frac{x^2}{3} + \frac{y^2}{4} = 1$

42. $\frac{x^2}{8} + \frac{y^2}{4} = 1$

43. $4x^2 + 7y^2 = 28$

44. $x^2 + 25y^2 = 25$

45. *Concept Check* Draftspeople often use the method shown in the sketch to draw an ellipse. Why does this method work?

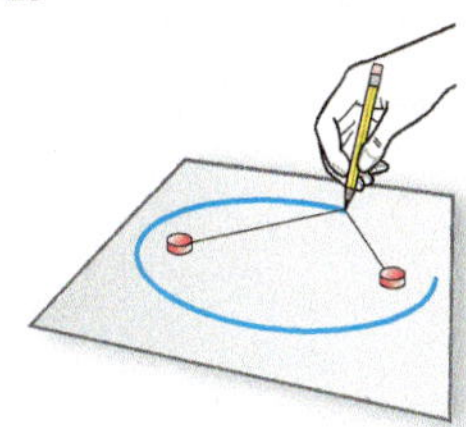

46. *Concept Check* How can the method of **Exercise 45** be modified to draw a circle?

Solve each problem. ***See Examples 6 and 7.***

47. ***Height of an Overpass*** A one-way road passes under an overpass in the shape of half an ellipse, 15 ft high at the center and 20 ft wide. Assuming a truck is 12 ft wide, what is the tallest truck that can pass under the overpass?

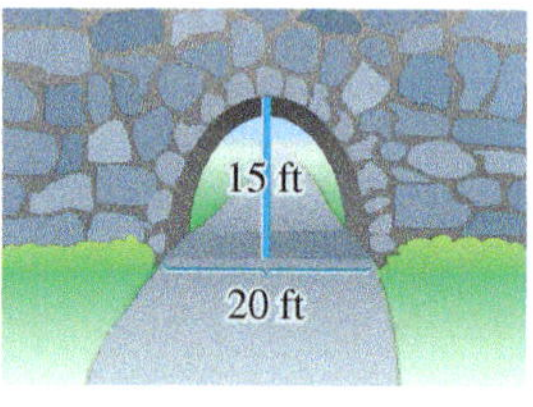

NOT TO SCALE

48. ***Height and Width of an Overpass*** An arch has the shape of half an ellipse. The equation of the ellipse is $100x^2 + 324y^2 = 32{,}400$, where x and y are in meters.

(a) How high is the center of the arch?

(b) How wide is the arch across the bottom?

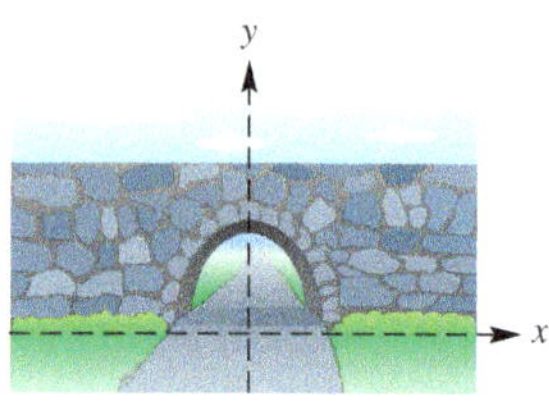

NOT TO SCALE

49. ***Orbit of Halley's Comet*** The famous Halley's comet last passed by Earth in February 1986 and will next return in 2062. It has an elliptical orbit of eccentricity 0.9673 with the sun at one focus. The greatest distance of the comet from the sun is 3281 million mi. Find the least distance between Halley's comet and the sun to the nearest million miles. (*Source: World Almanac and Book of Facts.*)

50. ***(Modeling) Orbit of a Satellite*** The coordinates in miles for the orbit of the artificial satellite Explorer VII can be modeled by the equation

$$\frac{x^2}{a^2} + \frac{y^2}{b^2} = 1,$$

where $a = 4465$ and $b = 4462$. Earth's center is located at one focus of the elliptical orbit. (*Source:* Loh, W., *Dynamics and Thermodynamics of Planetary Entry,* Prentice-Hall; Thomson, W., *Introduction to Space Dynamics,* John Wiley and Sons.)

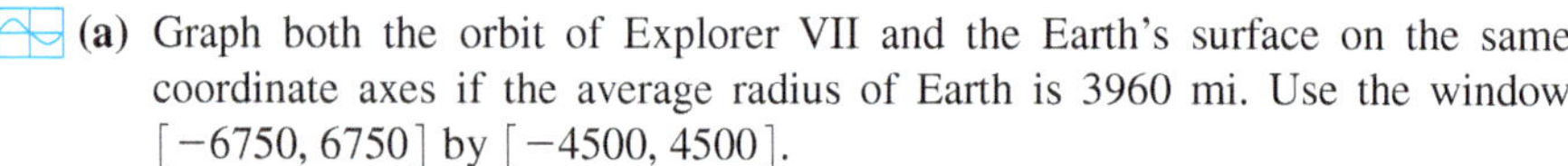

(a) Graph both the orbit of Explorer VII and the Earth's surface on the same coordinate axes if the average radius of Earth is 3960 mi. Use the window $[-6750, 6750]$ by $[-4500, 4500]$.

(b) Find the maximum and minimum heights of the satellite above Earth's surface to the nearest mile.

51. ***(Modeling) Orbits of Satellites*** Neptune and Pluto both have elliptical orbits with the sun at one focus. Neptune's orbit has $a = 30.1$ astronomical units (AU) with an eccentricity of $e = 0.009$, whereas Pluto's orbit has $a = 39.4$ and $e = 0.249$. (*Source:* Zeilik, M., S. Gregory, and E. Smith, *Introductory Astronomy and Astrophysics*, Fourth Edition, Saunders College Publishers.)

(a) Position the sun at the origin and determine equations that model each orbit.

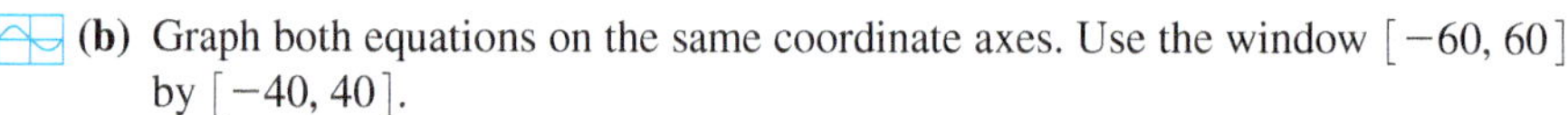

(b) Graph both equations on the same coordinate axes. Use the window $[-60, 60]$ by $[-40, 40]$.

52. ***(Modeling) The Roman Colosseum***

(a) The Roman Colosseum is an ellipse with major axis 620 ft and minor axis 513 ft. Find the distance between the foci of this ellipse to the nearest foot.

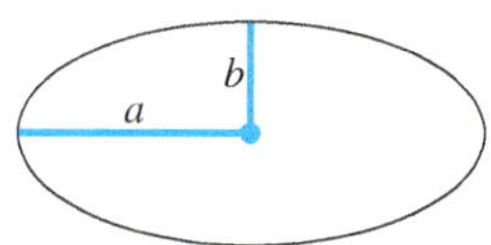

(b) A formula for the approximate perimeter of an ellipse is

$$P \approx 2\pi\sqrt{\frac{a^2 + b^2}{2}},$$

where a and b are the lengths shown in the figure. Use this formula to find the perimeter of the Roman Colosseum to the nearest foot.

53. *Design of a Lithotripter* Suppose a lithotripter is based on the ellipse with equation

$$\frac{x^2}{36} + \frac{y^2}{9} = 1.$$

How far from the center of the ellipse must the kidney stone and the source of the beam be placed? Give the exact answer.

54. *Design of a Lithotripter* Rework **Exercise 53** if the equation of the ellipse is $9x^2 + 4y^2 = 36$.

Chapter 10 Quiz (Sections 10.1–10.2)

1. *Concept Check* Match each equation of a conic section in Column I with the appropriate description in Column II.

I	II
(a) $x + 3 = 4(y - 1)^2$	**A.** circle; center $(-3, 1)$
(b) $(x + 3)^2 + (y - 1)^2 = 81$	**B.** parabola; opens right
(c) $25(x - 2)^2 + (y - 1)^2 = 100$	**C.** ellipse; major axis horizontal
(d) $\frac{(x-2)^2}{16} + \frac{(y-1)^2}{9} = 1$	**D.** parabola; opens down
(e) $-2(x + 3)^2 + 1 = y$	**E.** ellipse; major axis vertical

Write an equation for each conic section.

2. parabola with vertex $(-1, 2)$ and focus $(2, 2)$

3. parabola with vertex at the origin; through the point $\left(\sqrt{10}, -5\right)$; opens down

4. ellipse with center $(3, -2)$; $a = 5$; $c = 3$; major axis vertical

5. ellipse with foci at $(-3, 3)$ and $(-3, 11)$; major axis of length 10

Identify and then graph each conic section. If it is a parabola, give the vertex, focus, directrix, and axis of symmetry. If it is an ellipse, give the center, vertices, and foci.

6. $y + 4 = (x + 3)^2$ **7.** $4x^2 + 9y^2 = 36$ **8.** $8(x + 1) = (y + 3)^2$

9. $\frac{(x+3)^2}{25} + \frac{(y+2)^2}{36} = 1$ **10.** $x = -4y^2 - 4y - 3$

10.3 Hyperbolas

- Equations and Graphs of Hyperbolas
- Translated Hyperbolas
- Eccentricity

Equations and Graphs of Hyperbolas An ellipse was defined as the set of all points in a plane the sum of whose distances from two fixed points is a constant. A *hyperbola* is defined similarly.

Hyperbola

A **hyperbola** is the set of all points in a plane such that the absolute value of the difference of the distances from two fixed points is constant. The two fixed points are the **foci** of the hyperbola.

Suppose a hyperbola has center at the origin and foci at $F'(-c, 0)$ and $F(c, 0)$. See **Figure 28.** The midpoint of the segment $F'F$ is the center of the hyperbola and the points $V'(-a, 0)$ and $V(a, 0)$ are the **vertices** of the hyperbola. The line segment $V'V$ is the **transverse axis** of the hyperbola.

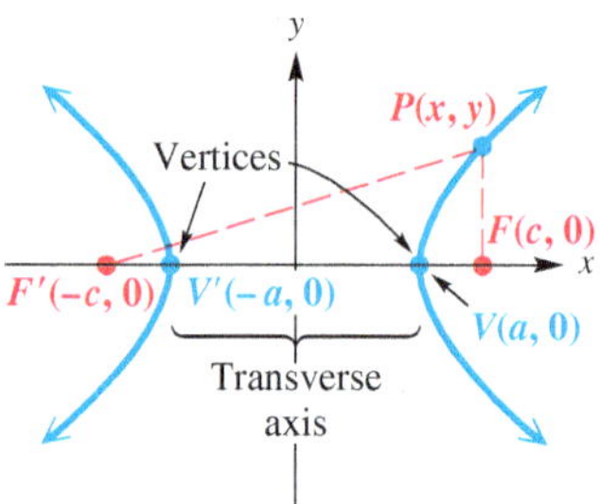

Figure 28

For a hyperbola,

$$d(V, F') - d(V, F) = (c + a) - (c - a) = 2a,$$

so the constant in the definition is $2a$, and

$$|d(P, F') - d(P, F)| = 2a$$

for any point $P(x, y)$ on the hyperbola. The distance formula and algebraic manipulation similar to that used for finding an equation for an ellipse produce the following result.

$$\frac{x^2}{a^2} - \frac{y^2}{c^2 - a^2} = 1$$

Replacing $c^2 - a^2$ with b^2 gives an equation of the hyperbola in **Figure 28.**

$$\mathbf{\frac{x^2}{a^2} - \frac{y^2}{b^2} = 1}$$

Letting $y = 0$ shows that the x-intercepts are $(\pm a, 0)$. If $x = 0$, the equation becomes $y^2 = -b^2$, which has no real number solutions, showing that this hyperbola has no y-intercepts.

To develop an aid for sketching the graph of a hyperbola, we start with the equation for a hyperbola and solve for y.

$$\frac{x^2}{a^2} - \frac{y^2}{b^2} = 1 \qquad \text{Hyperbola with transverse axis on the } x\text{-axis}$$

$$\frac{x^2}{a^2} - 1 = \frac{y^2}{b^2} \qquad \text{Subtract 1. Add } \frac{y^2}{b^2}.$$

$$\frac{x^2 - a^2}{a^2} = \frac{y^2}{b^2} \qquad \text{Write the left side as a single fraction.}$$

$$y = \pm\frac{b}{a}\sqrt{x^2 - a^2} \qquad \text{Take the square root on each side. Multiply by } b \text{, and rewrite.}$$

Remember both the positive and negative square roots.

If x^2 is very large in comparison to a^2, the difference $x^2 - a^2$ is very close to x^2. If this happens, then the points satisfying the final equation above are very close to one of the lines

$$\mathbf{y = \pm\frac{b}{a}x.}$$

Thus, as $|x|$ increases without bound, the points of the hyperbola $\frac{x^2}{a^2} - \frac{y^2}{b^2} = 1$ approach the lines $y = \pm\frac{b}{a}x$. These lines are **asymptotes** of the hyperbola and are useful when sketching the graph.

EXAMPLE 1 Using Asymptotes to Graph a Hyperbola

Graph $\frac{x^2}{25} - \frac{y^2}{49} = 1$. Sketch the asymptotes, and find the coordinates of the vertices and foci. Give the domain and range.

ALGEBRAIC SOLUTION

For this hyperbola, $a = 5$ and $b = 7$. With these values,

$$y = \pm\frac{b}{a}x \quad \text{becomes} \quad y = \pm\frac{7}{5}x. \quad \text{Asymptotes}$$

If we choose $x = 5$, then $y = \pm 7$. Choosing $x = -5$ also gives $y = \pm 7$. These four ordered pairs—$(5, 7)$, $(5, -7)$, $(-5, 7)$, and $(-5, -7)$—are the coordinates of the corners of the rectangle shown in **Figure 29.**

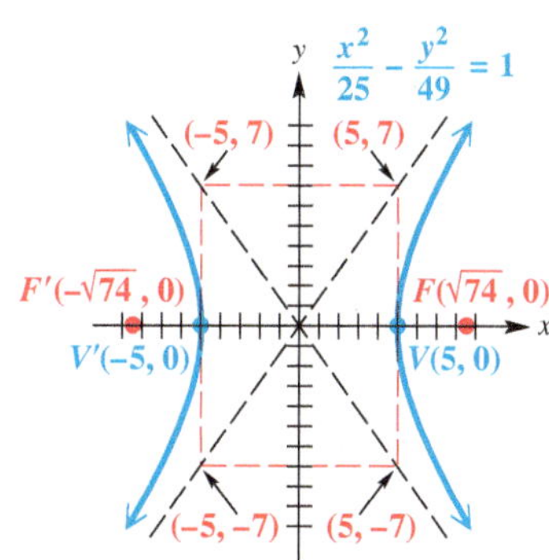

Figure 29

The extended diagonals of this rectangle, called the **fundamental rectangle,** are the asymptotes of the hyperbola. Because $a = 5$, the vertices of the hyperbola are $(5, 0)$ and $(-5, 0)$, as shown in **Figure 29.** We find the foci by letting

$$c^2 = a^2 + b^2 = 25 + 49 = 74, \quad \text{so} \quad c = \sqrt{74}.$$

Therefore, the foci are $(\sqrt{74}, 0)$ and $(-\sqrt{74}, 0)$. The domain is $(-\infty, -5] \cup [5, \infty)$, and the range is $(-\infty, \infty)$.

GRAPHING CALCULATOR SOLUTION

The graph of a hyperbola is not the graph of a function. We solve for y in $\frac{x^2}{25} - \frac{y^2}{49} = 1$ to obtain equations of the ***two*** functions

$$y = \pm\frac{7}{5}\sqrt{x^2 - 25}.$$

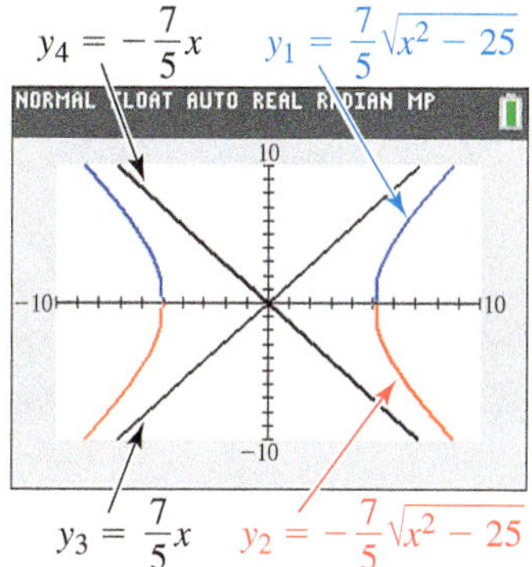

Figure 30

The graph of y_1 is the upper portion of each branch of the hyperbola shown in **Figure 30,** and the graph of y_2 is the lower portion of each branch. Alternatively, we could enter

$$y_2 = -y_1$$

to obtain the part of the graph below the x-axis.

The asymptotes are also shown. We can use tracing to observe how the branches of the hyperbola approach the asymptotes.

✔ **Now Try Exercise 9.**

NOTE When graphing hyperbolas, remember that the fundamental rectangle and the asymptotes are not actually parts of the graph. They are simply aids in sketching the graph.

While $a > b$ for an ellipse, examples show that for hyperbolas, it is possible that $a > b$, $a < b$, or $a = b$. If the foci of a hyperbola are on the y-axis, the equation of the hyperbola has the form

$$\frac{y^2}{a^2} - \frac{x^2}{b^2} = 1, \quad \text{with asymptotes} \quad y = \pm\frac{a}{b}x.$$

In either case, whether transverse axis on the x-axis or transverse axis on the y-axis, a^2 is chosen as the denominator of the leading term in the equation of a hyperbola written in *standard form.*

Standard Forms of Equations for Hyperbolas

The hyperbola with center at the origin and equation

$$\frac{x^2}{a^2} - \frac{y^2}{b^2} = 1$$

has vertices $(\pm a, 0)$, asymptotes $y = \pm \frac{b}{a}x$, and foci $(\pm c, 0)$, where $c^2 = a^2 + b^2$.

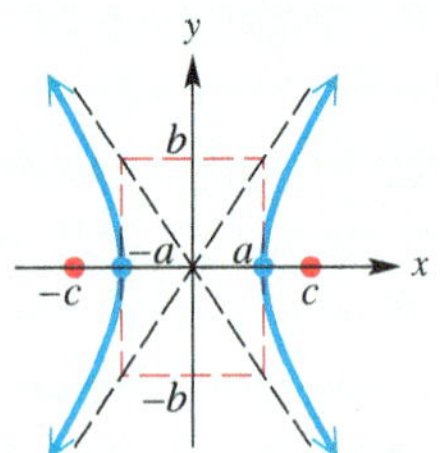

Transverse axis on *x*-axis

The hyperbola with center at the origin and equation

$$\frac{y^2}{a^2} - \frac{x^2}{b^2} = 1$$

has vertices $(0, \pm a)$, asymptotes $y = \pm \frac{a}{b}x$, and foci $(0, \pm c)$, where $c^2 = a^2 + b^2$.

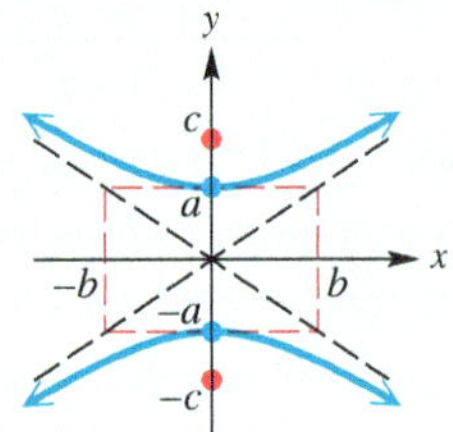

Transverse axis on *y*-axis

EXAMPLE 2 Graphing a Hyperbola

Graph $25y^2 - 4x^2 = 100$. Give the equations of the asymptotes, and the foci, domain, and range.

SOLUTION $\frac{y^2}{4} - \frac{x^2}{25} = 1$ Divide by 100, and write in standard form.

This hyperbola is centered at the origin, has foci on the *y*-axis, and has vertices $(0, 2)$ and $(0, -2)$. The equations of the asymptotes are found as follows.

$y = \pm \frac{a}{b}x$ Asymptotes for a hyperbola with vertical transverse axis

$y = \pm \frac{2}{5}x$ Let $a = 2$ and $b = 5$.

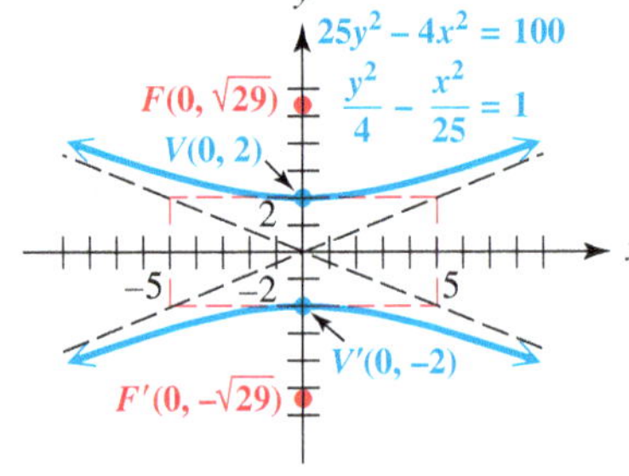

Figure 31

To graph the asymptotes, use the points $(5, 2)$, $(5, -2)$, $(-5, 2)$, and $(-5, -2)$ to determine the fundamental rectangle. The extended diagonals of this rectangle are the asymptotes for the graph, as shown in **Figure 31.**

The foci are located on the *y*-axis, *c* units above and below the origin.

$c^2 = a^2 + b^2$ Relationship for hyperbolas

$c^2 = 4 + 25$ Let $a^2 = 4$ and $b^2 = 25$.

$c^2 = 29$ Add.

$c = \sqrt{29}$ Take the positive square root because $c > 0$.

The coordinates of the foci are $(0, \sqrt{29})$ and $(0, -\sqrt{29})$. The domain of the relation is $(-\infty, \infty)$, and the range is $(-\infty, -2] \cup [2, \infty)$.

✔ **Now Try Exercise 17.**

Translated Hyperbolas Like an ellipse, a hyperbola can have its center translated away from the origin.

Standard Forms for Hyperbolas Centered at (h, k)

A hyperbola with center (h, k) and either a horizontal or vertical transverse axis satisfies one of the following equations, where $c^2 = a^2 + b^2$.

$$\frac{(x-h)^2}{a^2} - \frac{(y-k)^2}{b^2} = 1$$

Transverse axis: horizontal;
vertices: $(h \pm a, k)$;
foci: $(h \pm c, k)$;
asymptotes: $y = \pm\frac{b}{a}(x-h)+k$

Horizontal transverse axis

$$\frac{(y-k)^2}{a^2} - \frac{(x-h)^2}{b^2} = 1$$

Transverse axis: vertical;
vertices: $(h, k \pm a)$;
foci: $(h, k \pm c)$;
asymptotes: $y = \pm\frac{a}{b}(x-h)+k$

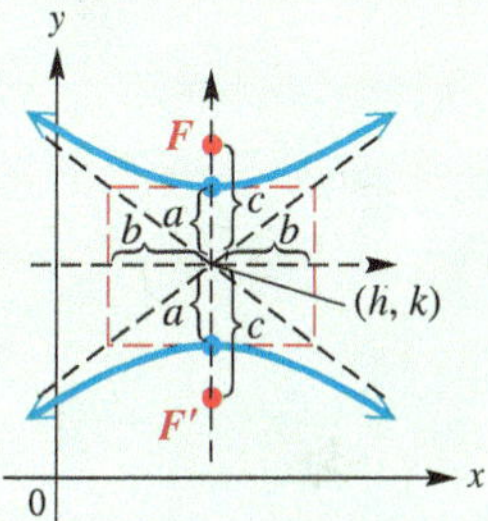

Vertical transverse axis

NOTE The asymptotes for a hyperbola *always* pass through the center (h, k). By the point-slope form of a line, the equation of any asymptote is $y = m(x-h)+k$. If the transverse axis is horizontal, then $m = \pm\frac{b}{a}$. If it is vertical, then $m = \pm\frac{a}{b}$.

EXAMPLE 3 Graphing a Hyperbola Translated Away from the Origin

Graph $\frac{(y+2)^2}{9} - \frac{(x+3)^2}{4} = 1$. Give the equations of the asymptotes, and the domain and range.

SOLUTION This equation represents a hyperbola centered at $(-3, -2)$. For this vertical hyperbola, $a = 3$ and $b = 2$. The x-values of the vertices are -3. Locate the y-values of the vertices by taking the y-value of the center, -2, and adding and subtracting 3. Thus, the vertices are $(-3, 1)$ and $(-3, -5)$.

The asymptotes have slopes $\pm\frac{3}{2}$ and pass through the center $(-3, -2)$. The equations of the asymptotes can be found using the point-slope form.

$$[y-(-2)] = \pm\frac{3}{2}[x-(-3)]$$ Point-slope form: $y - y_1 = m(x - x_1)$; Let $y_1 = -2$, $m = \pm\frac{3}{2}$, and $x_1 = -3$.

$$y = \pm\frac{3}{2}(x+3) - 2$$ Solve for y.

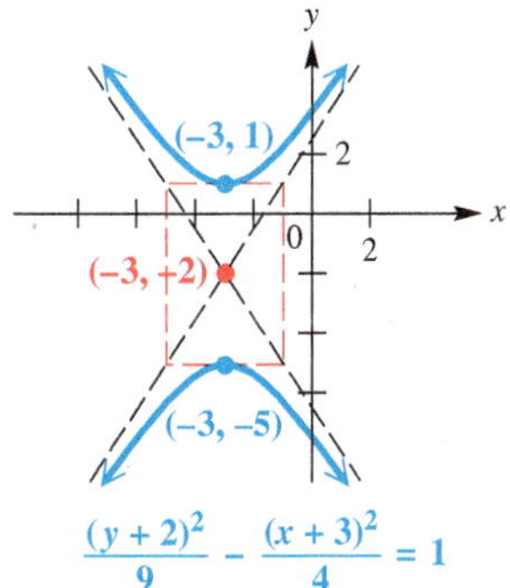

Figure 32

The graph is shown in **Figure 32.** The domain of the relation is $(-\infty, \infty)$, and the range is $(-\infty, -5] \cup [1, \infty)$.

Now Try Exercise 21.

Eccentricity If we apply the definition of eccentricity from the previous section to the hyperbola, we obtain the following.

$$e = \frac{\sqrt{a^2 + b^2}}{a} = \frac{c}{a} \quad \text{Eccentricity of a hyperbola}$$

Because $c > a$, we have $e > 1$. ***Thus, every hyperbola has eccentricity greater than 1.*** Narrow hyperbolas have e near 1, and wide hyperbolas have large e. See **Figure 33.**

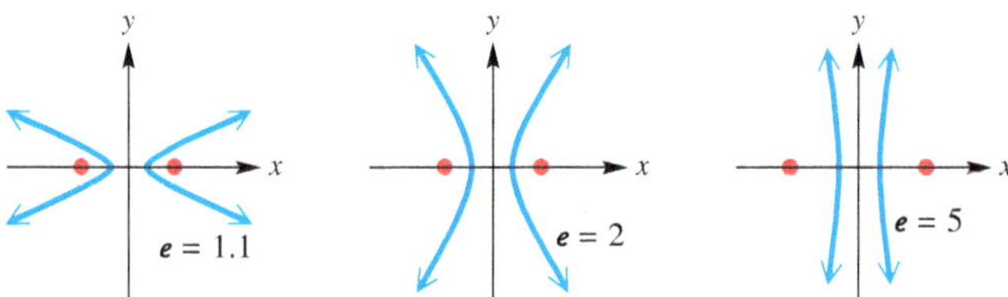

Eccentricities of hyperbolas

Figure 33

EXAMPLE 4 Finding Eccentricity from the Equation of a Hyperbola

Find the eccentricity of the hyperbola $\frac{x^2}{9} - \frac{y^2}{4} = 1$.

SOLUTION Here, $a^2 = 9$ and thus $a = 3$. Also, $b^2 = 4$.

$$c^2 = a^2 + b^2 \quad \text{Relationship for hyperbolas}$$

$$c^2 = 9 + 4 \quad \text{Let } a^2 = 9 \text{ and } b^2 = 4.$$

$$c^2 = 13 \quad \text{Add.}$$

$$c = \sqrt{13} \quad \text{Take the positive square root because } c > 0.$$

$$\text{eccentricity } e = \frac{c}{a} = \frac{\sqrt{13}}{3} \approx 1.2$$

✔ **Now Try Exercise 31.**

EXAMPLE 5 Writing an Equation of a Hyperbola

Write an equation for the hyperbola with $e = 2$ and foci at $(-9, 5)$ and $(-3, 5)$.

SOLUTION Because the foci have the same y-coordinate, the line through them, and therefore the hyperbola, is horizontal. The center of the hyperbola is halfway between the two foci at $(-6, 5)$. The distance from each focus to the center is $c = 3$, so $c^2 = 9$. Because $e = \frac{c}{a}$, we have $a = \frac{c}{e} = \frac{3}{2}$ and $a^2 = \frac{9}{4}$.

$$c^2 = a^2 + b^2 \quad \text{Relationship for hyperbolas}$$

$$9 = \frac{9}{4} + b^2 \quad \text{Let } c^2 = 9 \text{ and } a^2 = \frac{9}{4}.$$

$$b^2 = \frac{27}{4} \quad \text{Solve for } b^2;\ 9 - \frac{9}{4} = \frac{36}{4} - \frac{9}{4} = \frac{27}{4}.$$

The equation of the hyperbola is

$$\frac{(x+6)^2}{\frac{9}{4}} - \frac{(y-5)^2}{\frac{27}{4}} = 1, \quad \text{or} \quad \frac{4(x+6)^2}{9} - \frac{4(y-5)^2}{27} = 1.$$

Simplify complex fractions.

✔ **Now Try Exercise 49.**

The following chart summarizes our discussion of eccentricity in this chapter.

Summary of Eccentricity

Conic Section	Eccentricity e
Parabola	$e = 1$
Circle	$e = 0$
Ellipse	$e = \frac{c}{a}$ and $0 < e < 1$
Hyperbola	$e = \frac{c}{a}$ and $e > 1$

10.3 Exercises

CONCEPT PREVIEW *Match each equation of a hyperbola in Column I with its description in Column II.*

I	II
1. $\dfrac{(x-1)^2}{49} - \dfrac{(y-2)^2}{64} = 1$	**A.** center $(1, 2)$; horizontal transverse axis
2. $\dfrac{(x+1)^2}{64} - \dfrac{(y+2)^2}{49} = 1$	**B.** center $(-2, -1)$; vertical transverse axis
3. $\dfrac{(y-1)^2}{9} - \dfrac{(x-2)^2}{25} = 1$	**C.** center $(-1, -2)$; horizontal transverse axis
4. $\dfrac{(y+1)^2}{25} - \dfrac{(x+2)^2}{9} = 1$	**D.** center $(2, 1)$; vertical transverse axis

CONCEPT PREVIEW *Match each equation with the correct graph.*

5. $\dfrac{x^2}{25} + \dfrac{y^2}{9} = 1$

6. $\dfrac{x^2}{9} + \dfrac{y^2}{25} = 1$

7. $\dfrac{x^2}{9} - \dfrac{y^2}{25} = 1$

8. $\dfrac{x^2}{25} - \dfrac{y^2}{9} = 1$

A.

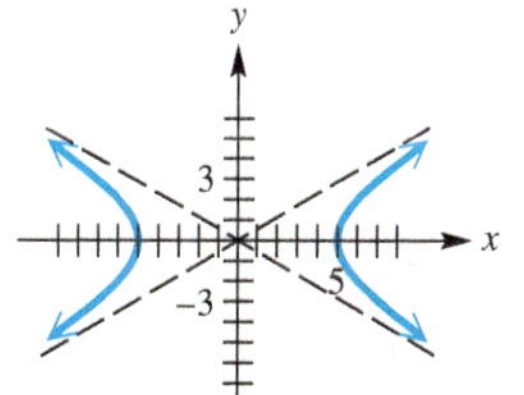

B.

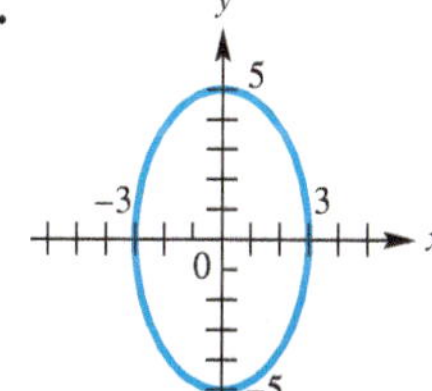

C.

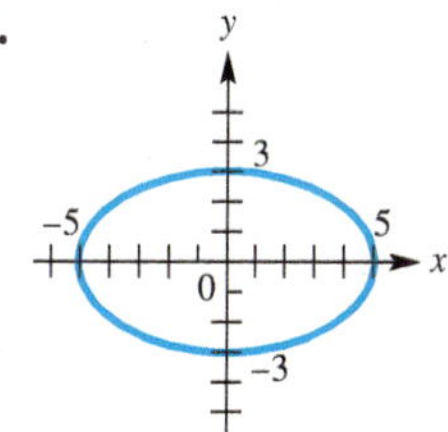

D.

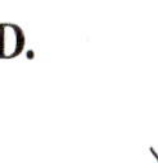

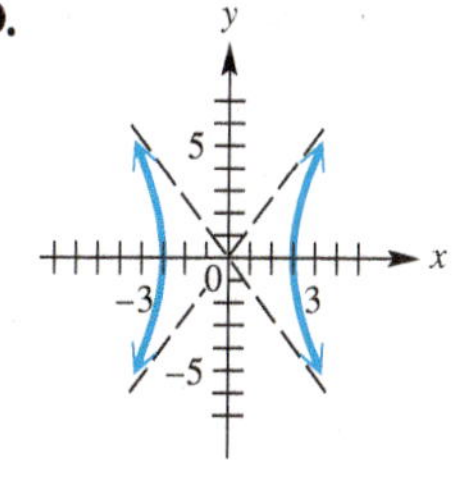

Graph each hyperbola. Give the domain, range, center, vertices, foci, and equations of the asymptotes. ***See Examples 1–3.***

9. $\dfrac{x^2}{16} - \dfrac{y^2}{9} = 1$

10. $\dfrac{x^2}{25} - \dfrac{y^2}{144} = 1$

11. $\dfrac{y^2}{25} - \dfrac{x^2}{49} = 1$

12. $\dfrac{y^2}{64} - \dfrac{x^2}{4} = 1$

13. $x^2 - y^2 = 9$

14. $x^2 - 4y^2 = 64$

15. $9x^2 - 25y^2 = 225$ **16.** $4y^2 - 16x^2 = 64$ **17.** $4y^2 - 25x^2 = 100$

18. $x^2 - 4y^2 = 16$ **19.** $9x^2 - 4y^2 = 1$ **20.** $25y^2 - 9x^2 = 1$

21. $\frac{(y-7)^2}{36} - \frac{(x-4)^2}{64} = 1$ **22.** $\frac{(x+6)^2}{144} - \frac{(y+4)^2}{81} = 1$

23. $\frac{(x+3)^2}{16} - \frac{(y-2)^2}{9} = 1$ **24.** $\frac{(y+5)^2}{4} - \frac{(x-1)^2}{16} = 1$

25. $16(x+5)^2 - (y-3)^2 = 1$ **26.** $4(x+9)^2 - 25(y+6)^2 = 100$

Graph each equation. Give the domain and range. Identify any that are functions.

27. $\frac{y}{3} = \sqrt{1 + \frac{x^2}{16}}$ **28.** $\frac{x}{3} = -\sqrt{1 + \frac{y^2}{25}}$

29. $5x = -\sqrt{1 + 4y^2}$ **30.** $3y = \sqrt{4x^2 - 16}$

*Find the eccentricity **e** of each hyperbola. Round to the nearest tenth.* ***See Example 4.***

31. $\frac{x^2}{8} - \frac{y^2}{8} = 1$ **32.** $\frac{x^2}{2} - \frac{y^2}{18} = 1$

33. $16y^2 - 8x^2 = 16$ **34.** $8y^2 - 2x^2 = 16$

Write an equation for each hyperbola. ***See Examples 4 and 5.***

35. x-intercepts $(\pm 3, 0)$; foci at $(-5, 0)$, $(5, 0)$

36. y-intercepts $(0, \pm 12)$; foci at $(0, -15)$, $(0, 15)$

37. vertices at $(0, 6)$, $(0, -6)$; asymptotes $y = \pm\frac{1}{2}x$

38. vertices at $(-10, 0)$, $(10, 0)$; asymptotes $y = \pm 5x$

39. vertices at $(-3, 0)$, $(3, 0)$; passing through the point $(-6, -1)$

40. vertices at $(0, 5)$, $(0, -5)$; passing through the point $(-3, 10)$

41. foci at $(0, \sqrt{13})$, $(0, -\sqrt{13})$; asymptotes $y = \pm 5x$

42. foci at $(-3\sqrt{5}, 0)$, $(3\sqrt{5}, 0)$; asymptotes $y = \pm 2x$

43. vertices at $(4, 5)$, $(4, 1)$; asymptotes $y = \pm 7(x - 4) + 3$

44. vertices at $(5, -2)$, $(1, -2)$; asymptotes $y = \pm\frac{3}{2}(x - 3) - 2$

45. center at $(1, -2)$; focus at $(-2, -2)$; vertex at $(-1, -2)$

46. center at $(9, -7)$; focus at $(9, -17)$; vertex at $(9, -13)$

47. $e = 3$; center at $(0, 0)$; vertex at $(0, 7)$ **48.** $e = \frac{5}{3}$; center at $(8, 7)$; focus at $(3, 7)$

49. $e = \frac{25}{9}$; foci at $(9, -1)$, $(-11, -1)$ **50.** $e = \frac{5}{4}$; vertices at $(2, 10)$, $(2, 2)$

Determine the two equations necessary to graph each hyperbola using a graphing calculator, and graph it in the viewing window indicated. ***See Example 1.***

51. $\frac{x^2}{4} - \frac{y^2}{16} = 1$; $[-6.6, 6.6]$ by $[-8, 8]$

52. $\frac{x^2}{25} - \frac{y^2}{49} = 1$; $[-10, 10]$ by $[-10, 10]$

53. $4y^2 - 36x^2 = 144$; $[-10, 10]$ by $[-15, 15]$

54. $y^2 - 9x^2 = 9$; $[-10, 10]$ by $[-10, 10]$

Solve each problem.

55. ***(Modeling) Atomic Structure*** In 1911, Ernest Rutherford discovered the basic structure of the atom by "shooting" positively charged alpha particles with a speed of 10^7 m per sec at a piece of gold foil 6×10^{-7} m thick. Only a small percentage of the alpha particles struck a gold nucleus head-on and were deflected directly back toward their source. The rest of the particles often followed a hyperbolic trajectory because they were repelled by positively charged gold nuclei. As a result of this famous experiment, Rutherford proposed that the atom was composed mostly of empty space with a small and dense nucleus.

Ernest Rutherford

The figure shows an alpha particle A initially approaching a gold nucleus N and being deflected at an angle $\theta = 90°$. N is located at a focus of the hyperbola, and the trajectory of A passes through a vertex of the hyperbola. (*Source:* Semat, H., and J. Albright, *Introduction to Atomic and Nuclear Physics,* Fifth Edition, International Thomson Publishing.)

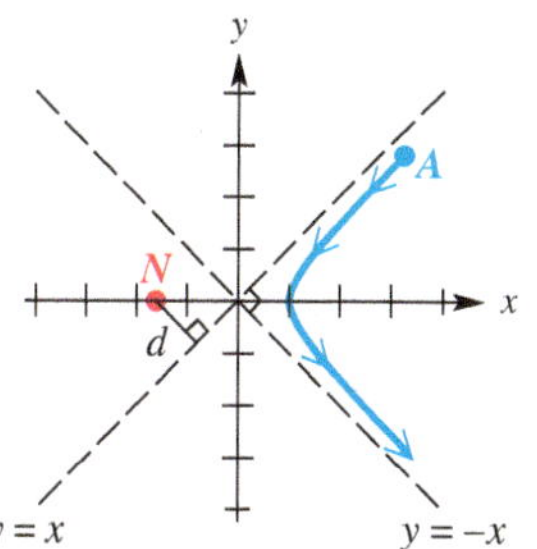

(a) Determine the equation of the trajectory of the alpha particle if $d = 5 \times 10^{-14}$ m.

(b) What was the minimum distance between the centers of the alpha particle and the gold nucleus? Write the answer using scientific notation. Round to the nearest tenth.

56. ***LORAN System*** Ships and planes often use a location-finding system called LORAN. With this system, a radio transmitter at M in the figure sends out a series of pulses. When each pulse is received at transmitter S, it then sends out a pulse. A ship at P receives pulses from both M and S. A receiver on the ship measures the difference in the arrival times of the pulses. The navigator then consults a special map showing hyperbolas that correspond to the differences in arrival times (which give the distances d_1 and d_2 in the figure). In this way the ship can be located as lying on a branch of a particular hyperbola.

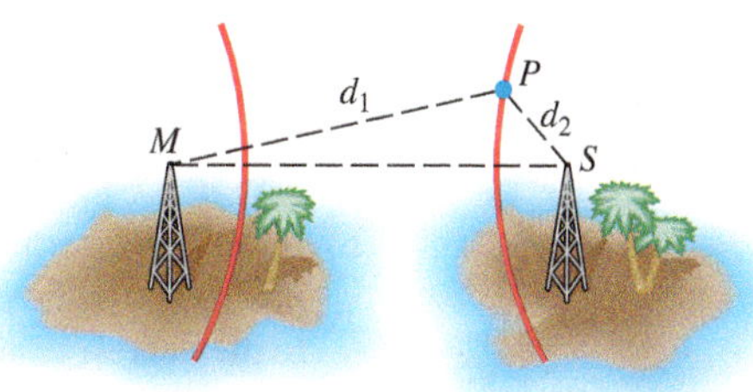

Suppose that in the figure, $d_1 = 80$ mi, $d_2 = 30$ mi, and the distance MS between the transmitters is 100 mi. Use the definition of a hyperbola to find an equation of the hyperbola on which the ship is located.

57. ***Sound Detection*** Microphones are placed at points $(-c, 0)$ and $(c, 0)$. An explosion occurs at point $P(x, y)$ having positive x-coordinate. See the figure. The sound is detected at the closer microphone t seconds before being detected at the farther microphone. Assume that sound travels at a speed of 330 m per sec, and show that P must be on the following hyperbola.

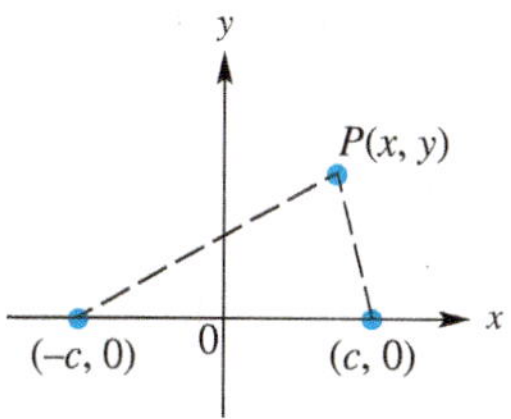

$$\frac{x^2}{330^2t^2} - \frac{y^2}{4c^2 - 330^2t^2} = \frac{1}{4}$$

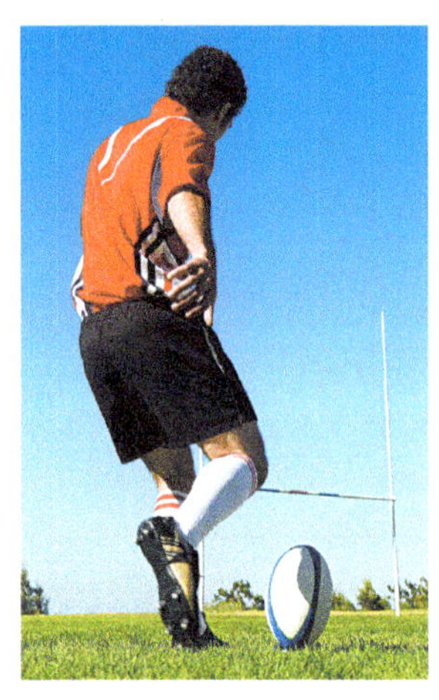

58. ***Rugby Algebra*** A rugby field is similar to a modern football field except that the goalpost, which is 18.5 ft wide, is located on the goal line instead of at the back of the endzone. The rugby equivalent of a touchdown, called a **try,** is scored by touching the ball down beyond the goal line. After a try is scored, the scoring team can earn extra points by kicking the ball through the goalposts. The ball must be placed somewhere on the line perpendicular to the goal line and passing through the point where the try was scored. See the figure below on the left.

If that line passes through the goalposts, then the kicker should place the ball at whatever distance is most comfortable. If the line passes outside the goalposts, then the player might choose the point on the line where angle θ in the figure on the left is as large as possible. The problem of determining this optimal point is similar to a problem posed in 1471 by the astronomer Regiomontanus. (*Source:* Maor, E., *Trigonometric Delights,* Princeton, NJ: Princeton University Press.)

The figure on the right below shows a vertical line segment AB, where A and B are a and b units above the horizontal axis, respectively. If point P is located on the axis at a distance of x units from point Q, then angle θ is greatest when $x = \sqrt{ab}$.

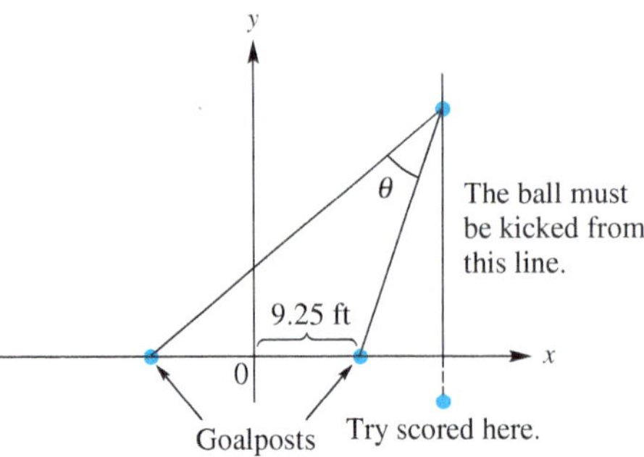

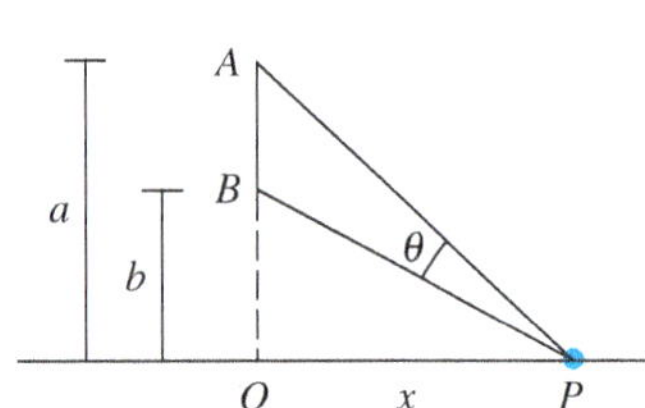

(a) Use the result from Regiomontanus' problem to show that when the line is outside the goalposts, the optimal location to kick the rugby ball lies on the following hyperbola.

$$x^2 - y^2 = 9.25^2$$

(b) If the line on which the ball must be kicked is 10 ft to the right of the goalpost, how far from the goal line should the ball be placed to maximize angle θ? Round to the nearest tenth.

(c) Rugby players find it easier to kick the ball from the hyperbola's asymptote. When the line on which the ball must be kicked is 10 ft to the right of the goalpost, how far will this point differ from the exact optimal location? Round to the nearest tenth.

59. ***(Modeling) Design of a Sports Complex*** Two buildings in a sports complex are shaped and positioned like a portion of the branches of the hyperbola

$$400x^2 - 625y^2 = 250{,}000,$$

where x and y are in meters.

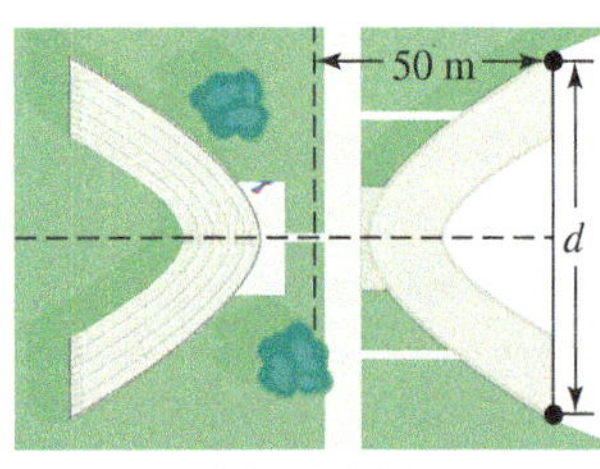

NOT TO SCALE

(a) How far apart are the buildings at their closest point?

(b) Find the distance d in the figure to the nearest tenth of a meter.

60. Suppose a hyperbola has center at the origin, foci at $F'(-c, 0)$ and $F(c, 0)$, and

$$|d(P, F') - d(P, F)| = 2a.$$

Let $b^2 = c^2 - a^2$, and show that an equation of the hyperbola is

$$\frac{x^2}{a^2} - \frac{y^2}{b^2} = 1.$$

Relating Concepts

For individual or collaborative investigation *(Exercises 61–66)*

The graph of $\frac{x^2}{4} - y^2 = 1$ is a hyperbola. We know that the graph of this hyperbola approaches its asymptotes as $|x|$ increases without bound. ***Work Exercises 61–66 in order,*** *to see the relationship between the hyperbola and one of its asymptotes.*

61. Solve $\frac{x^2}{4} - y^2 = 1$ for y, and choose the positive square root.

62. Find the equation of the asymptote with positive slope.

63. Use a calculator to evaluate the y-coordinate of the point where $x = 50$ on the graph of the portion of the hyperbola represented by the equation obtained in **Exercise 61.** Round the answer to the nearest hundredth.

64. Find the y-coordinate of the point where $x = 50$ on the graph of the asymptote found in **Exercise 62.**

65. Compare the results in **Exercises 63 and 64.** How do they support the following statement?

When $x = 50$, the graph of the function defined by the equation found in **Exercise 61** lies *below* the graph of the asymptote found in **Exercise 62.**

66. What happens if we choose x-values greater than 50?

10.4 Summary of the Conic Sections

- Characteristics
- Identifying Conic Sections
- Geometric Definition of Conic Sections

Characteristics The graphs of parabolas, circles, ellipses, and hyperbolas are called conic sections because each graph can be obtained by intersecting a cone with a plane, as suggested by **Figure 1** at the beginning of the chapter. All conic sections of the types presented in this chapter have equations of the general form

$$Ax^2 + Cy^2 + Dx + Ey + F = 0,$$

where either A or C must be nonzero.

Summary of Special Characteristics of Conic Sections

Conic Section	Characteristic	Example
Parabola	Either $A = 0$ or $C = 0$, but not both.	$x^2 - y - 4 = 0$ $y^2 - x - 4y = 0$
Circle	$A = C \neq 0$	$x^2 + y^2 - 16 = 0$
Ellipse	$A \neq C, AC > 0$	$25x^2 + 16y^2 - 400 = 0$
Hyperbola	$AC < 0$	$x^2 - y^2 - 1 = 0$

The following chart summarizes our work with conic sections.

Summary of Conic Sections

Equation	Graph	Description	Identification
$(x-h)^2 = 4p(y-k)$ or $y-k = a(x-h)^2$, where $a = \dfrac{1}{4p}$	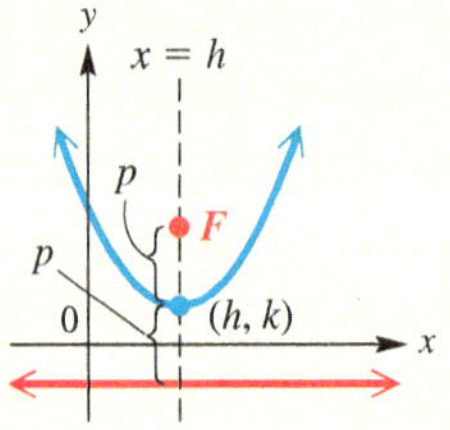 Parabola	Graph opens • up if $p > 0$ (or $a > 0$); • down if $p < 0$ (or $a < 0$). Vertex is (h, k). Axis of symmetry is $x = h$.	There is an x^2-term. y is not squared.
$(y-k)^2 = 4p(x-h)$ or $x-h = a(y-k)^2$, where $a = \dfrac{1}{4p}$	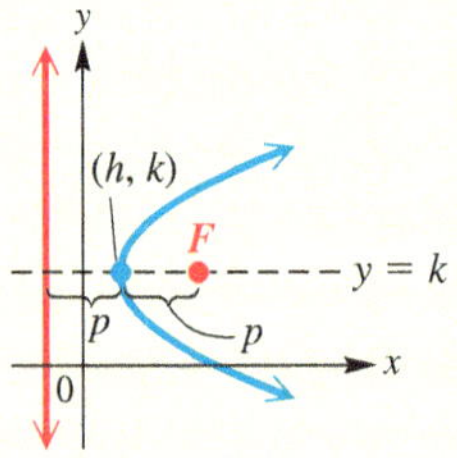 Parabola	Graph opens • to the right if $p > 0$ (or $a > 0$); • to the left if $p < 0$ (or $a < 0$). Vertex is (h, k). Axis of symmetry is $y = k$.	There is a y^2-term. x is not squared.
$(x-h)^2 + (y-k)^2 = r^2$	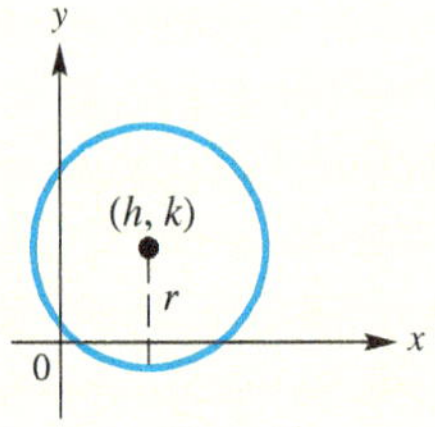 Circle	Center is (h, k). Radius is r.	x^2- and y^2-terms have the same positive coefficient.
$\dfrac{(x-h)^2}{a^2} + \dfrac{(y-k)^2}{b^2} = 1 \quad (a > b)$	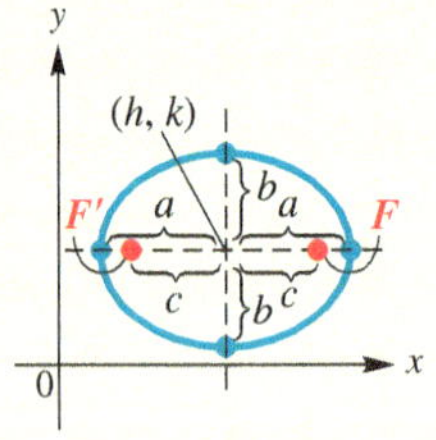 Ellipse	Horizontal major axis, length $= 2a$. $c^2 = a^2 - b^2$ Center is (h, k).	x^2- and y^2-terms have different positive coefficients.
$\dfrac{(x-h)^2}{b^2} + \dfrac{(y-k)^2}{a^2} = 1 \quad (a > b)$	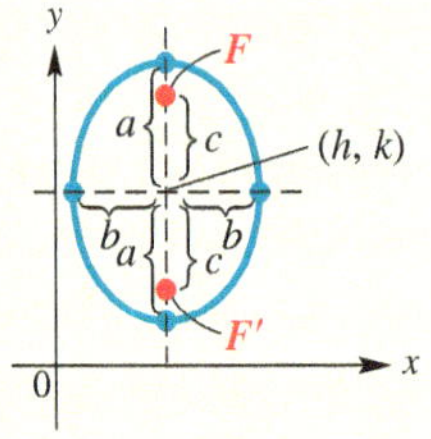 Ellipse	Vertical major axis, length $= 2a$. $c^2 = a^2 - b^2$ Center is (h, k).	x^2- and y^2-terms have different positive coefficients.
$\dfrac{(x-h)^2}{a^2} - \dfrac{(y-k)^2}{b^2} = 1$	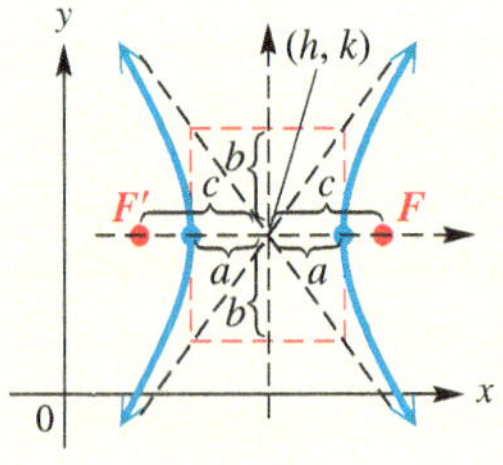 Hyperbola	Graph has horizontal transverse axis. $c^2 = a^2 + b^2$ Asymptotes are $y = \pm\frac{b}{a}(x-h) + k$. Center is (h, k).	x^2-term has a positive coefficient. y^2-term has a negative coefficient.

Equation	Graph	Description	Identification
$\dfrac{(y-k)^2}{a^2} - \dfrac{(x-h)^2}{b^2} = 1$	Hyperbola	Graph has vertical transverse axis. $c^2 = a^2 + b^2$ Asymptotes are $y = \pm \frac{a}{b}(x-h) + k$. Center is (h, k).	y^2-term has a positive coefficient. x^2-term has a negative coefficient.

Identifying Conic Sections To recognize the type of graph that a given conic section has, we may need to transform the equation into a more familiar form.

EXAMPLE 1 Determining Types of Conic Sections

Identify and sketch the graph of each relation.

(a) $x^2 = 25 + 5y^2$ **(b)** $x^2 - 8x + y^2 + 10y = -41$

(c) $4x^2 - 16x + 9y^2 + 54y = -61$ **(d)** $x^2 - 6x + 8y - 7 = 0$

SOLUTION

(a) This equation does *not* represent a parabola because both of its variables are squared.

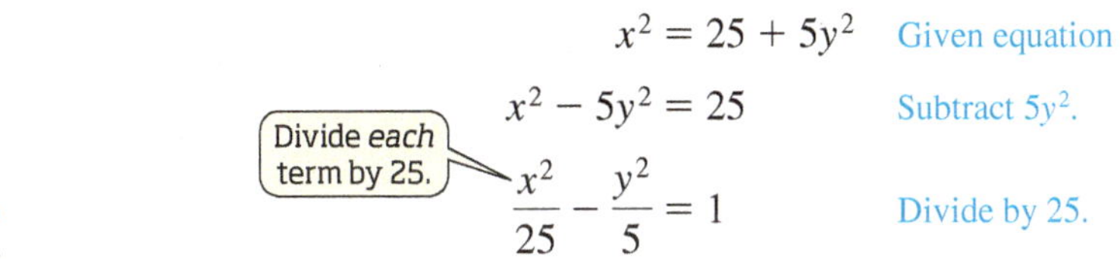

$$x^2 = 25 + 5y^2 \quad \text{Given equation}$$

$$x^2 - 5y^2 = 25 \quad \text{Subtract } 5y^2.$$

$$\frac{x^2}{25} - \frac{y^2}{5} = 1 \quad \text{Divide by 25.}$$

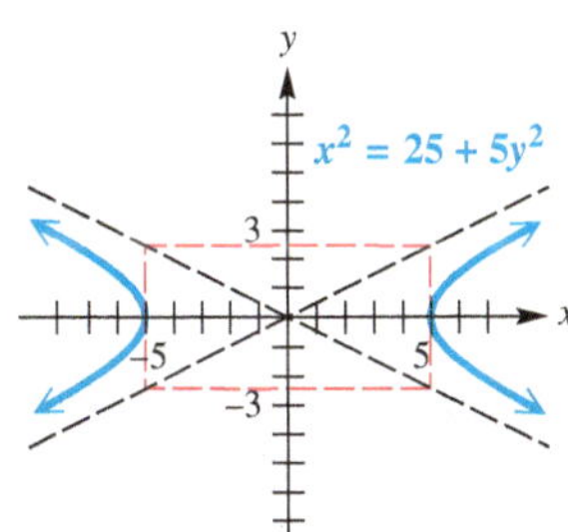

Figure 34

The equation represents a hyperbola centered at the origin, with asymptotes

$$y = \pm \frac{b}{a}x, \quad \text{or} \quad y = \pm \frac{\sqrt{5}}{5}x. \quad \text{Let } a = 5 \text{ and } b = \sqrt{5}.$$

The x-intercepts are $(\pm 5, 0)$. The graph is shown in **Figure 34.**

(b)

$$x^2 - 8x + y^2 + 10y = -41 \quad \text{Given equation}$$

$$(x^2 - 8x \qquad) + (y^2 + 10y \qquad) = -41 \quad \text{Rewrite in anticipation of completing the square.}$$

$$(x^2 - 8x + 16) + (y^2 + 10y + 25) = -41 + 16 + 25 \quad \text{Complete the square on both } x \text{ and } y.$$

$$(x-4)^2 + (y+5)^2 = 0 \quad \text{Factor, and add.}$$

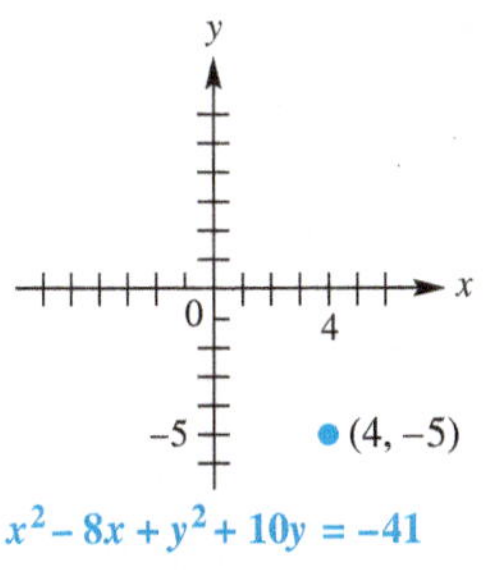

Figure 35

The resulting equation is that of a "circle" with radius 0. Its graph is the single point $(4, -5)$. See **Figure 35.** If we had obtained a negative number on the right (instead of 0), the equation would have no solution at all, and there would be no graph.

(c) In $4x^2 - 16x + 9y^2 + 54y = -61$, the coefficients of the x^2- and y^2-terms are unequal and both positive, so the equation might represent an ellipse but not a circle. (It might also represent a single point or no points at all.)

$$4x^2 - 16x + 9y^2 + 54y = -61 \quad \text{Given equation}$$

$$4(x^2 - 4x \qquad) + 9(y^2 + 6y \qquad) = -61 \quad \text{Factor out 4, and factor out 9.}$$

$$4(x^2 - 4x + 4 - 4) + 9(y^2 + 6y + 9 - 9) = -61 \quad \text{Complete the square.}$$

$$4(x^2 - 4x + 4) - 16 + 9(y^2 + 6y + 9) - 81 = -61 \quad \text{Distributive property}$$

Multiply $4(-4) = -16$ and $9(-9) = -81$.

$$4(x - 2)^2 + 9(y + 3)^2 = 36 \quad \text{Factor. Add 16 and add 81 on each side.}$$

$$\frac{(x-2)^2}{9} + \frac{(y+3)^2}{4} = 1 \quad \text{Divide by 36.}$$

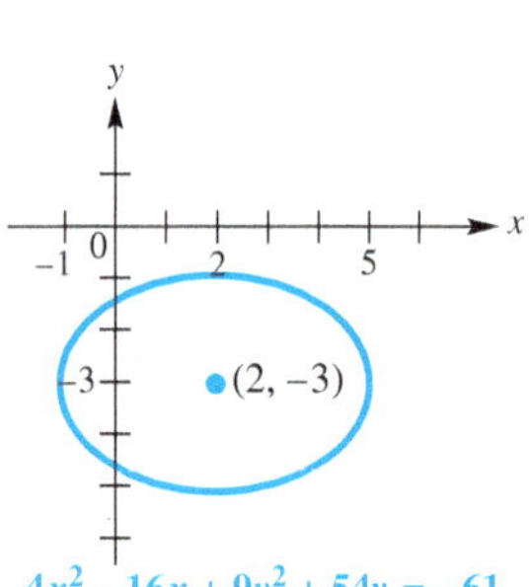

Figure 36

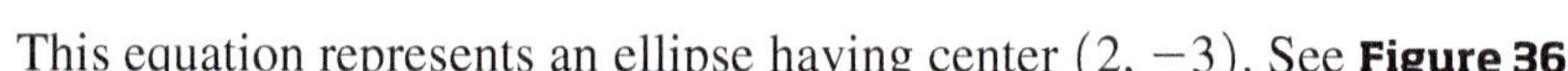

This equation represents an ellipse having center $(2, -3)$. See **Figure 36.**

(d) Since only one variable in $x^2 - 6x + 8y - 7 = 0$ is squared (x, and not y), the equation represents a parabola with a vertical axis of symmetry. Isolate the term with y (the variable that is not squared) on one side.

$$x^2 - 6x + 8y - 7 = 0 \quad \text{Given equation}$$

$$8y = -x^2 + 6x + 7 \quad \text{Isolate the } y\text{-term.}$$

$$8y = -(x^2 - 6x \qquad) + 7 \quad \text{Regroup terms, and factor out } -1.$$

$$8y = -(x^2 - 6x + 9 - 9) + 7 \quad \text{Complete the square.}$$

$$8y = -(x^2 - 6x + 9) + 9 + 7 \quad \text{Distributive property; } -(-9) = +9$$

$$8y = -(x - 3)^2 + 16 \quad \text{Factor, and add.}$$

Multiply *each* term by $\frac{1}{8}$. Here, $\frac{1}{8}(16) = 2$.

$$y = -\frac{1}{8}(x - 3)^2 + 2 \quad \text{Multiply by } \tfrac{1}{8}.$$

$$y - 2 = -\frac{1}{8}(x - 3)^2 \quad \text{Subtract 2.}$$

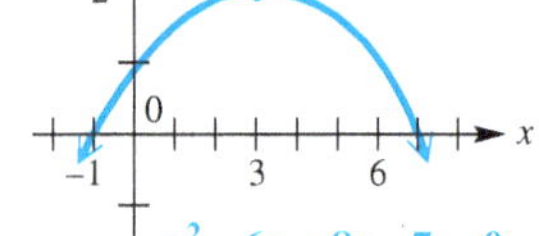

Figure 37

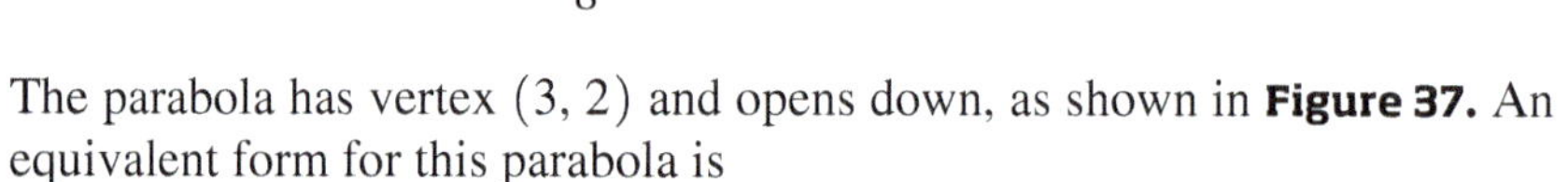

The parabola has vertex $(3, 2)$ and opens down, as shown in **Figure 37.** An equivalent form for this parabola is

$$(x - 3)^2 = -8(y - 2).$$

✔ **Now Try Exercises 33, 35, 37, and 41.**

EXAMPLE 2 Determining Type of Conic Section

Identify and sketch the graph of $4y^2 - 16y - 9x^2 + 18x = -43$.

SOLUTION

$$4y^2 - 16y - 9x^2 + 18x = -43 \quad \text{Given equation}$$

$$4(y^2 - 4y \qquad) - 9(x^2 - 2x \qquad) = -43 \quad \text{Factor out 4, and factor out } -9.$$

$$4(y^2 - 4y + 4 - 4) - 9(x^2 - 2x + 1 - 1) = -43 \quad \text{Complete the square.}$$

$$4(y^2 - 4y + 4) - 16 - 9(x^2 - 2x + 1) + 9 = -43 \quad \text{Distributive property}$$

Be careful with this step.

$$4(y - 2)^2 - 9(x - 1)^2 = -36 \quad \text{Factor. Add 16, and subtract 9.}$$

Because of the −36, we might think that this equation does not have a graph. However, dividing each side by −36 reveals that the graph is that of a hyperbola.

$$\frac{(x-1)^2}{4} - \frac{(y-2)^2}{9} = 1 \quad \text{Divide by } -36, \text{ and rearrange terms.}$$

Be careful here!

This hyperbola has center $(1, 2)$. The graph is shown in **Figure 38.**

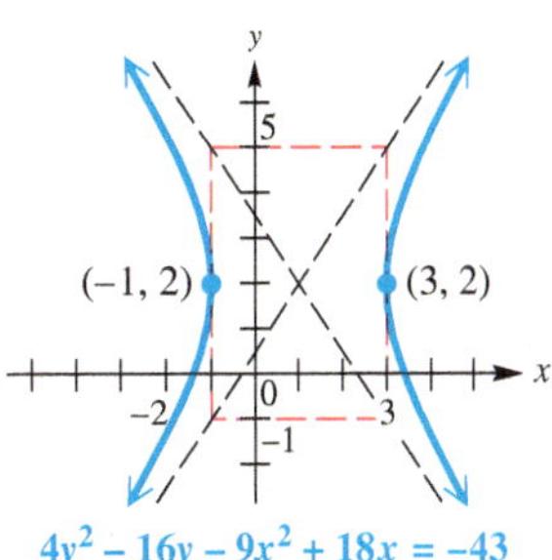

Figure 38

✔ **Now Try Exercise 43.**

Geometric Definition of Conic Sections A parabola was defined as the set of points in a plane equidistant from a fixed point (focus) and a fixed line (directrix). A parabola has eccentricity 1. This definition can be generalized to apply to ellipses and hyperbolas. **Figure 39** shows an ellipse with $a = 4$, $c = 2$, and $e = \frac{1}{2}$. The line $x = 8$ is shown also. For any point P on the ellipse,

$$\text{distance of } P \text{ from the focus} = \frac{1}{2}[\text{distance of } P \text{ from the line}].$$

Figure 40 shows a hyperbola with $a = 2$, $c = 4$, and $e = 2$, along with the line $x = 1$. For any point P on the hyperbola,

$$\text{distance of } P \text{ from the focus} = 2[\text{distance of } P \text{ from the line}].$$

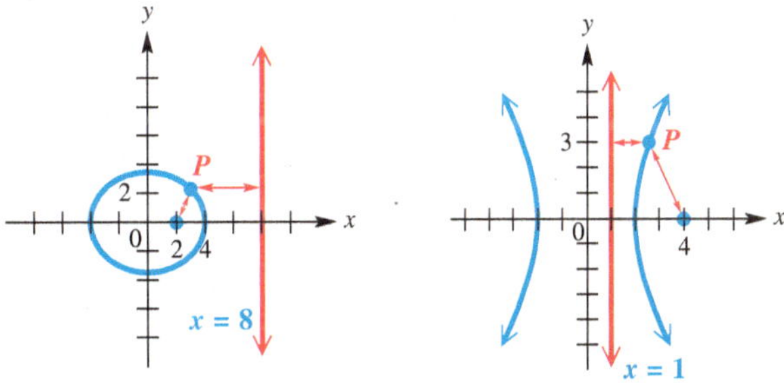

Figure 39 **Figure 40**

The following geometric definition applies to all conic sections except circles, which have $e = 0$.

Geometric Definition of a Conic Section

Given a fixed point F (focus), a fixed line L (directrix), and a positive number e, the set of all points P in the plane such that

$$\textbf{distance of } P \textbf{ from } F = e \cdot [\textbf{distance of } P \textbf{ from } L]$$

is a conic section of eccentricity e. ***The conic section is a parabola when*** $e = 1$, ***an ellipse when*** $0 < e < 1$, ***and a hyperbola when*** $e > 1$.

10.4 Exercises

CONCEPT PREVIEW *Identify the type of conic section described.*

1. The conic section consisting of the set of points in a plane that lie a given distance from a given point
2. The conic section consisting of the set of points in a plane that are equidistant from a fixed point and a fixed line
3. The conic section consisting of the set of points in a plane for which the distance from the point $(1, 3)$ is equal to the distance from the line $y = 1$
4. The conic section with eccentricity $e = 0$
5. The conic section consisting of the set of points in a plane for which the sum of the distances from the points $(5, 0)$ and $(-5, 0)$ is 14
6. The conic section consisting of the set of points in a plane for which the absolute value of the difference of the distances from the points $(3, 0)$ and $(-3, 0)$ is 2
7. The conic section consisting of the set of points in a plane for which the distance from the point $(3, 0)$ is one and one-half times the distance from the line $x = \frac{4}{3}$
8. The conic section consisting of the set of points in a plane for which the distance from the point $(2, 0)$ is one-third of the distance from the line $x = 10$

Identify the type of graph that each equation has, without actually graphing. ***See Examples 1 and 2.***

9. $x^2 + y^2 = 144$
10. $(x - 2)^2 + (y + 3)^2 = 25$
11. $y = 2x^2 + 3x - 4$
12. $x = 3y^2 + 5y - 6$
13. $x - 1 = -3(y - 4)^2$
14. $\frac{x^2}{25} + \frac{y^2}{36} = 1$
15. $\frac{x^2}{49} + \frac{y^2}{100} = 1$
16. $x^2 - y^2 = 1$
17. $\frac{x^2}{4} - \frac{y^2}{16} = 1$
18. $\frac{(x + 2)^2}{9} + \frac{(y - 4)^2}{16} = 1$
19. $\frac{x^2}{25} - \frac{y^2}{25} = 1$
20. $y + 7 = 4(x + 3)^2$
21. $\frac{x^2}{4} = 1 - \frac{y^2}{9}$
22. $\frac{x^2}{4} = 1 + \frac{y^2}{9}$
23. $\frac{(x + 3)^2}{16} + \frac{(y - 2)^2}{16} = 1$
24. $x^2 = 25 - y^2$
25. $x^2 - 6x + y = 0$
26. $11 - 3x = 2y^2 - 8y$
27. $4(x - 3)^2 + 3(y + 4)^2 = 0$
28. $2x^2 - 8x + 2y^2 + 20y = 12$
29. $x - 4y^2 - 8y = 0$
30. $x^2 + 2x = -4y$
31. $4x^2 - 24x + 5y^2 + 10y + 41 = 0$
32. $6x^2 - 12x + 6y^2 - 18y + 25 = 0$

Identify and sketch the graph of each relation. ***See Examples 1 and 2.***

33. $\frac{x^2}{4} + \frac{y^2}{4} = -1$
34. $\frac{x^2}{4} + \frac{y^2}{4} = 1$
35. $x^2 = 25 + y^2$
36. $9x^2 + 36y^2 = 36$

37. $x^2 = 4y - 8$

38. $\dfrac{(x-4)^2}{8} + \dfrac{(y+1)^2}{2} = 0$

39. $y^2 - 4y = x + 4$

40. $(x+7)^2 + (y-5)^2 + 4 = 0$

41. $3x^2 + 6x + 3y^2 - 12y = 12$

42. $-4x^2 + 8x + y^2 + 6y = -6$

43. $4x^2 - 8x + 9y^2 - 36y = -4$

44. $3x^2 + 12x + 3y^2 = 0$

Find the eccentricity ***e*** *of each conic section. The point shown on the x-axis is a focus, and the line shown is a directrix.*

45.

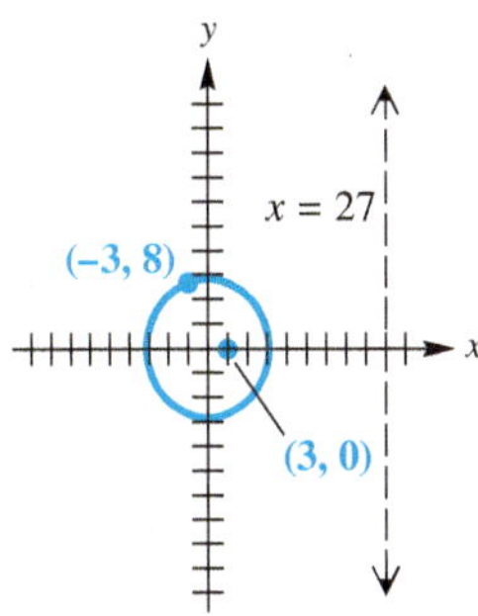

46.

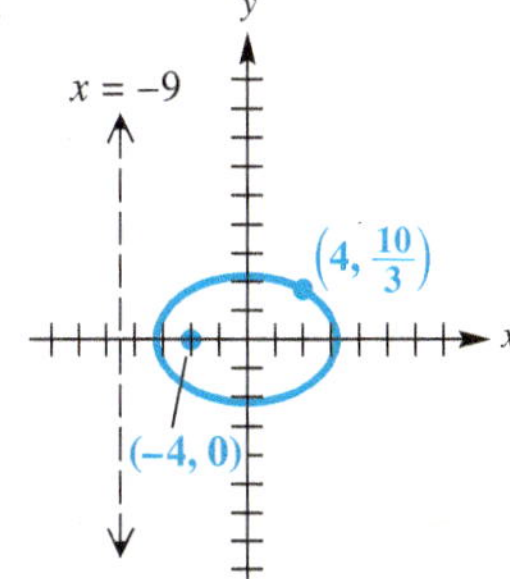

47.

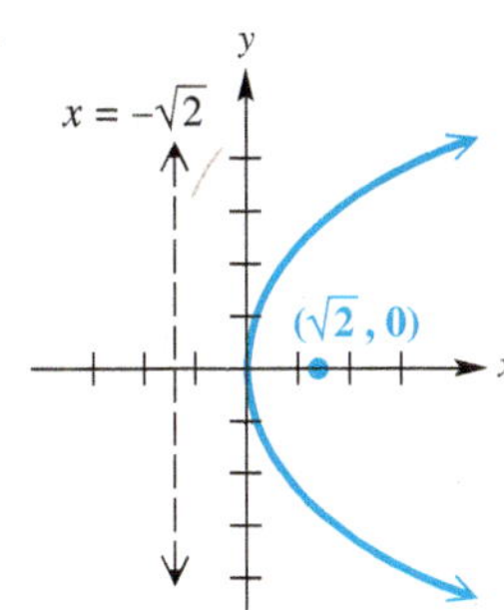

48.

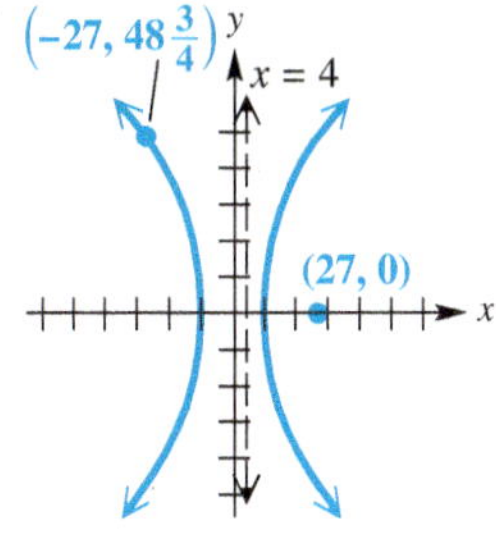

49.

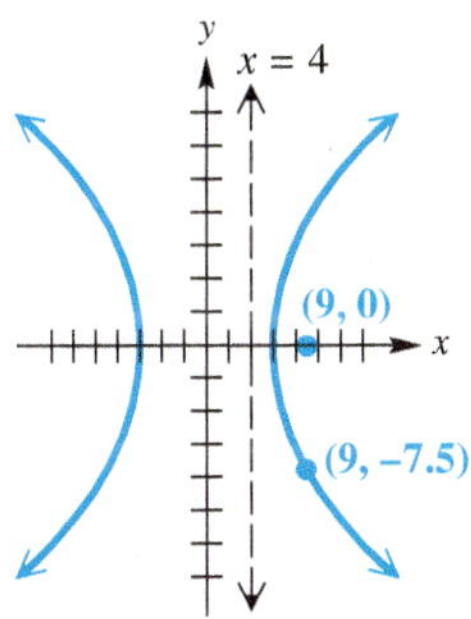

50.

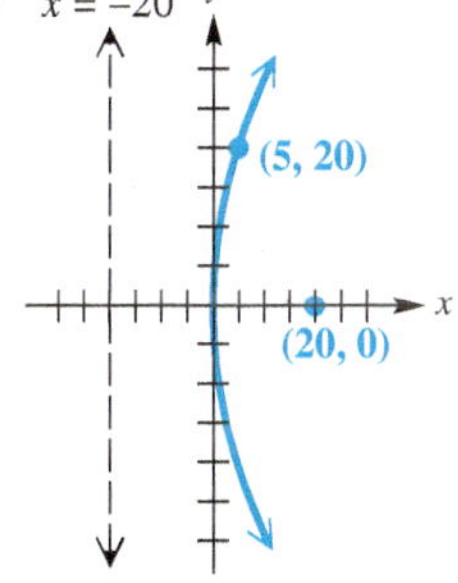

Satellite Trajectory *When a satellite is near Earth, its orbital trajectory may trace out a hyperbola, a parabola, or an ellipse. The type of trajectory depends on the satellite's velocity V in meters per second. It will be*

$$\text{hyperbolic if } V > \frac{k}{\sqrt{D}}, \quad \text{parabolic if } V = \frac{k}{\sqrt{D}}, \quad \text{or} \quad \text{elliptical if } V < \frac{k}{\sqrt{D}},$$

where $k = 2.82 \times 10^7$ *is a constant and D is the distance in meters from the satellite to the center of Earth.* (*Source:* Loh, W., *Dynamics and Thermodynamics of Planetary Entry,* Prentice-Hall, *and* Thomson, W., *Introduction to Space Dynamics,* John Wiley and Sons.)

51. When the artificial satellite Explorer IV was at a maximum distance D of 42.5×10^6 m from Earth's center, it had a velocity V of 2090 m per sec. Determine the shape of its trajectory.

52. If a satellite is scheduled to leave Earth's gravitational influence, its velocity must be increased so that its trajectory changes from elliptical to hyperbolic. Determine the minimum increase in velocity necessary for Explorer IV to escape Earth's gravitational influence when $D = 42.5 \times 10^6$ m. Round to the nearest whole number.

Solve each problem.

53. If $Ax^2 + Cy^2 + Dx + Ey + F = 0$ is the general equation of an ellipse, find the coordinates of its center point by completing the square.

54. Graph the hyperbola $\frac{x^2}{4} - \frac{y^2}{12} = 1$ using a graphing calculator. Trace to find the coordinates of several points on the hyperbola. For each of these points P, verify that

$$\text{distance of } P \text{ from } (4, 0) = 2[\text{distance of } P \text{ from the line } x = 1].$$

55. Graph the ellipse $\frac{x^2}{16} + \frac{y^2}{12} = 1$ using a graphing calculator. Trace to find the coordinates of several points on the ellipse. For each of these points P, verify that

$$\text{distance of } P \text{ from } (2, 0) = \frac{1}{2}[\text{distance of } P \text{ from the line } x = 8].$$

Chapter 10 Test Prep

Key Terms

10.1 conic sections, parabola, focus, directrix	**10.2** ellipse, foci, major axis, minor axis	center, vertices, conic, eccentricity	**10.3** hyperbola, transverse axis, asymptotes, fundamental rectangle

New Symbols

e eccentricity

Quick Review

Concepts	Examples

10.1 Parabolas

Parabola with Vertical Axis of Symmetry and Vertex (0, 0)

The parabola with focus $(0, p)$ and directrix $y = -p$ has equation $\boldsymbol{x^2 = 4py}$. The parabola has vertical axis of symmetry $x = 0$ and opens up if $p > 0$ or down if $p < 0$.

Parabola with Horizontal Axis of Symmetry and Vertex (0, 0)

The parabola with focus $(p, 0)$ and directrix $x = -p$ has equation $\boldsymbol{y^2 = 4px}$. The parabola has horizontal axis of symmetry $y = 0$ and opens to the right if $p > 0$ or to the left if $p < 0$.

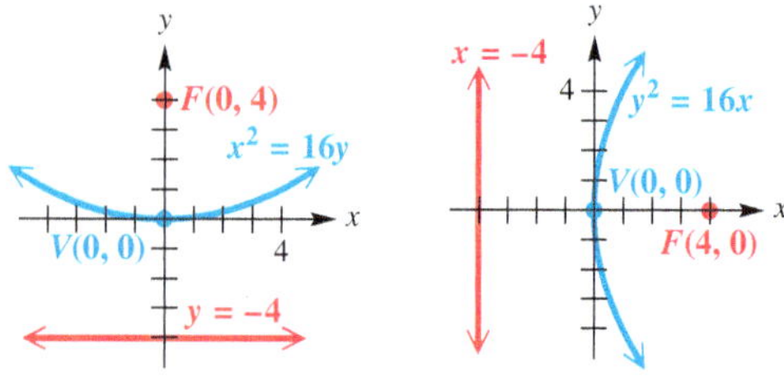

Concepts	Examples

Equation Forms for Translated Parabolas

A parabola with vertex (h, k) has an equation of the form

$$(x - h)^2 = 4p(y - k) \quad \text{Vertical axis of symmetry}$$

or $$(y - k)^2 = 4p(x - h). \quad \text{Horizontal axis of symmetry}$$

The focus is distance $|p|$ from the vertex.

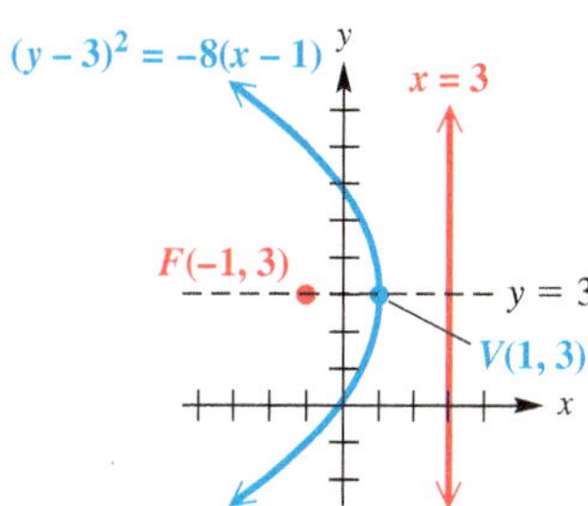

10.2 Ellipses

Standard Forms of Equations for Ellipses

The ellipse with center at the origin and equation

$$\frac{x^2}{a^2} + \frac{y^2}{b^2} = 1 \quad (\text{where } a > b)$$

has vertices $(\pm a, 0)$, endpoints of the minor axis $(0, \pm b)$, and foci $(\pm c, 0)$, where $c^2 = a^2 - b^2$.

The ellipse with center at the origin and equation

$$\frac{x^2}{b^2} + \frac{y^2}{a^2} = 1 \quad (\text{where } a > b)$$

has vertices $(0, \pm a)$, endpoints of the minor axis $(\pm b, 0)$, and foci $(0, \pm c)$, where $c^2 = a^2 - b^2$.

Translated Ellipses

The preceding equations can be extended to ellipses having center (h, k) by replacing x and y with $x - h$ and $y - k$, respectively.

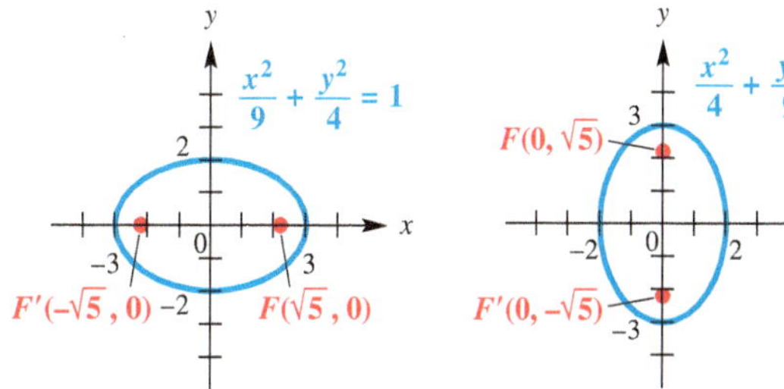

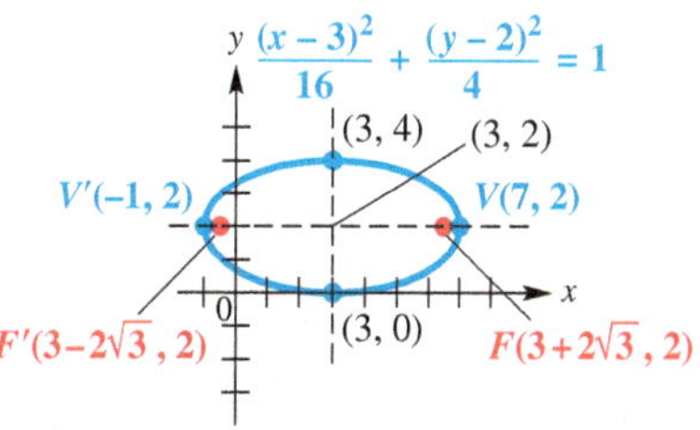

10.3 Hyperbolas

Standard Forms of Equations for Hyperbolas

The hyperbola with center at the origin and equation

$$\frac{x^2}{a^2} - \frac{y^2}{b^2} = 1$$

has vertices $(\pm a, 0)$, asymptotes $y = \pm \frac{b}{a}x$, and foci $(\pm c, 0)$, where $c^2 = a^2 + b^2$.

The hyperbola with center at the origin and equation

$$\frac{y^2}{a^2} - \frac{x^2}{b^2} = 1$$

has vertices $(0, \pm a)$, asymptotes $y = \pm \frac{a}{b}x$, and foci $(0, \pm c)$, where $c^2 = a^2 + b^2$.

Translated Hyperbolas

The preceding equations can be extended to hyperbolas having center (h, k) by replacing x and y with $x - h$ and $y - k$, respectively.

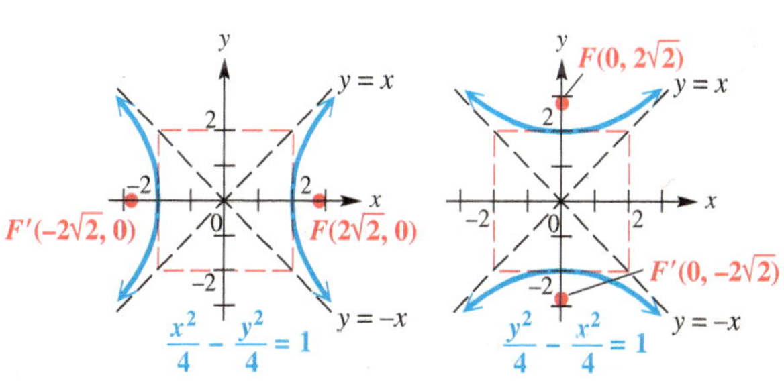

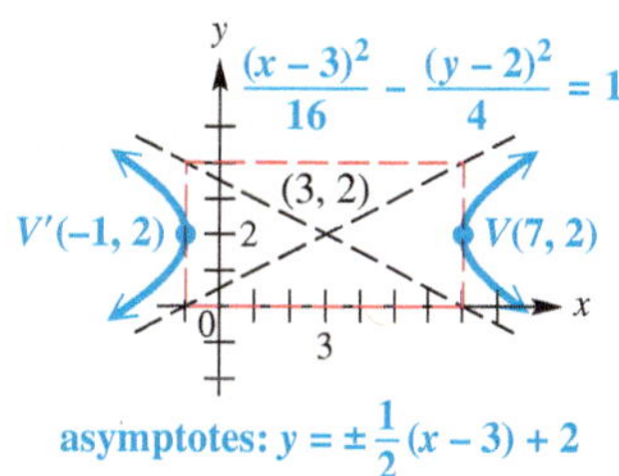

Concepts	Examples

10.4 Summary of the Conic Sections

Conic sections in this chapter have equations that can be written in the following form.

$$Ax^2 + Cy^2 + Dx + Ey + F = 0$$

Conic Section	Characteristic	Example
Parabola	Either $A = 0$ or $C = 0$, but not both.	$x^2 - y - 4 = 0$ $y^2 - x - 4y = 0$
Circle	$A = C \neq 0$	$x^2 + y^2 - 16 = 0$
Ellipse	$A \neq C, AC > 0$	$25x^2 + 16y^2 - 400 = 0$
Hyperbola	$AC < 0$	$x^2 - y^2 - 1 = 0$

See the summary chart in **Section 10.4.**

$y^2 - 4x - 10y + 21 = 0,$ or $(y - 5)^2 = 4(x + 1)$ Parabola; vertex: $(-1, 5)$; opens to the right

$x^2 - 4x + y^2 + 2y - 4 = 0,$ or $(x - 2)^2 + (y + 1)^2 = 9$ Circle; center: $(2, -1)$; radius: 3

$4x^2 + y^2 - 16 = 0,$ or $\frac{x^2}{4} + \frac{y^2}{16} = 1$ Ellipse; center: $(0, 0)$; major axis: vertical

$4x^2 - y^2 - 8x - 4y - 16 = 0,$ or $\frac{(x-1)^2}{4} - \frac{(y+2)^2}{16} = 1$ Hyperbola; center: $(1, -2)$; transverse axis: horizontal

Chapter 10 Review Exercises

Graph each parabola. In Exercises 1–4, give the domain, range, vertex, and axis of symmetry. In Exercises 5–8, give the domain, range, focus, directrix, and axis of symmetry.

1. $x = 4(y - 5)^2 + 2$

2. $x = -(y + 1)^2 - 7$

3. $x = 5y^2 - 5y + 3$

4. $x = 2y^2 - 4y + 1$

5. $y^2 = -\frac{2}{3}x$

6. $y^2 = 2x$

7. $3x^2 = y$

8. $x^2 + 2y = 0$

Write an equation for each parabola with vertex at the origin.

9. focus $(4, 0)$

10. focus $(0, -3)$

11. through the point $(-3, 4)$, opens up

12. through the point $(2, 5)$, opens right

Identify the type of graph that each equation has, without actually graphing.

13. $y^2 + 9x^2 = 9$

14. $9x^2 - 16y^2 = 144$

15. $3y^2 - 5x^2 = 30$

16. $y^2 + x = 4$

17. $4x^2 - y = 0$

18. $x^2 + y^2 = 25$

19. $4x^2 - 8x + 9y^2 + 36y = -4$

20. $9x^2 - 18x - 4y^2 - 16y - 43 = 0$

Concept Check Match each equation with its calculator graph in choices A–F. In all cases except choice B, Xscl = Yscl = 1.

21. $4x^2 + y^2 = 36$

22. $x = 2y^2 + 3$

23. $(x-2)^2 + (y+3)^2 = 36$

24. $\dfrac{x^2}{36} + \dfrac{y^2}{9} = 1$

25. $(y-1)^2 - (x-2)^2 = 36$

26. $y^2 = 36 + 4x^2$

A.

B.

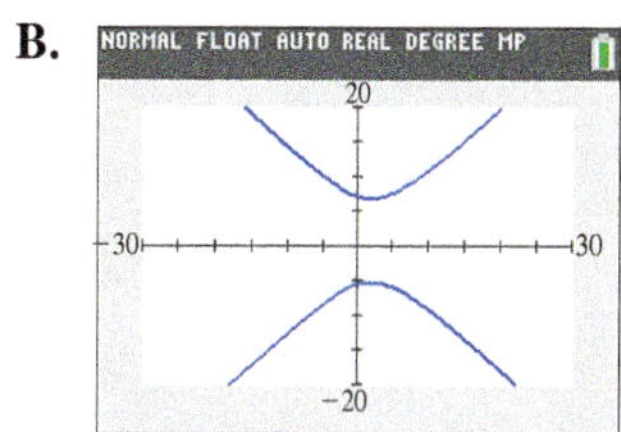

In this screen, Xscl = Yscl = 5.

C.

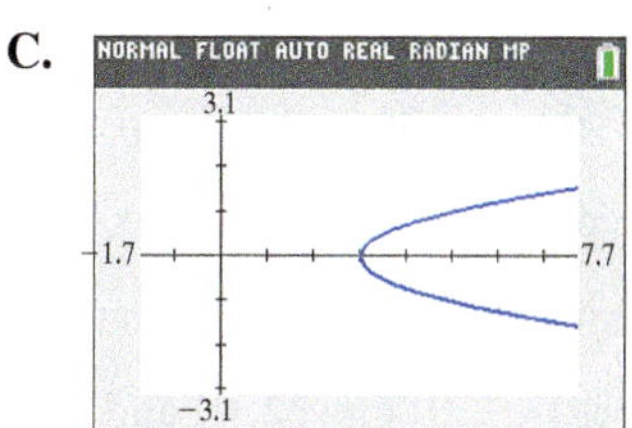

D.

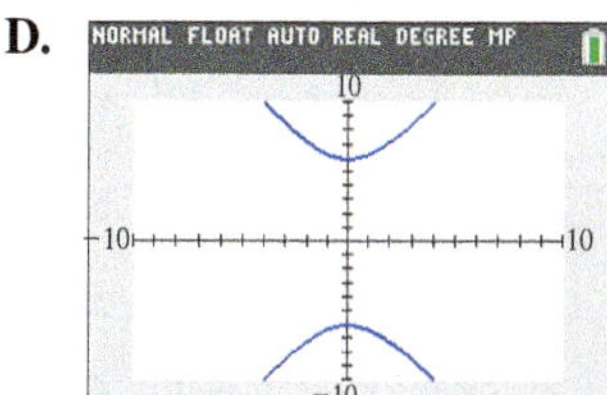

E.

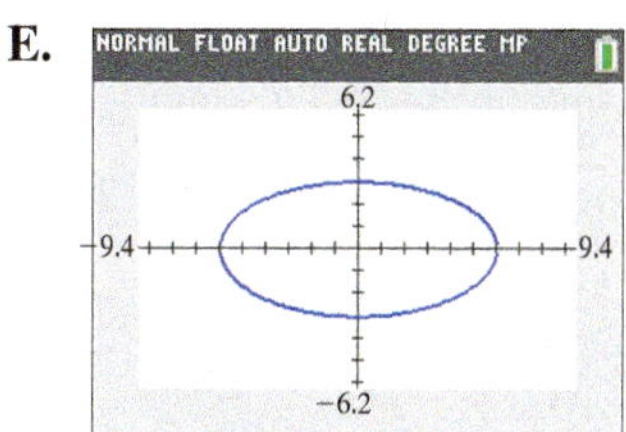

F.

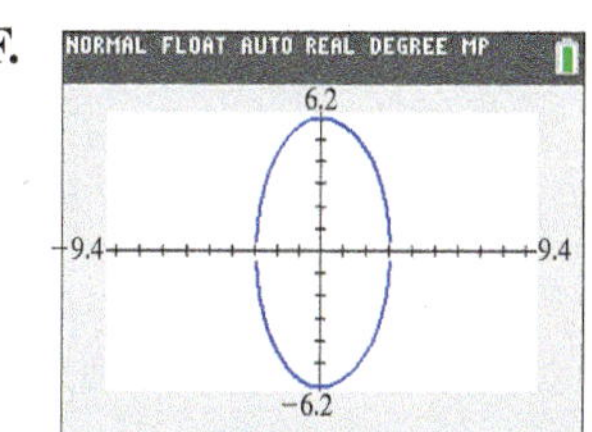

Identify and sketch the graph of each equation. Give the domain, range, coordinates of the vertices for each ellipse or hyperbola, and equations of the asymptotes for each hyperbola. Give the domain and range for each circle.

27. $\dfrac{x^2}{4} + \dfrac{y^2}{9} = 1$

28. $\dfrac{x^2}{16} + \dfrac{y^2}{4} = 1$

29. $\dfrac{x^2}{64} - \dfrac{y^2}{36} = 1$

30. $\dfrac{y^2}{25} - \dfrac{x^2}{9} = 1$

31. $\dfrac{(x+1)^2}{16} + \dfrac{(y-1)^2}{16} = 1$

32. $(x-3)^2 + (y+2)^2 = 9$

33. $4x^2 + 9y^2 = 36$

34. $x^2 = 16 + y^2$

35. $\dfrac{(x-3)^2}{4} + (y+1)^2 = 1$

36. $\dfrac{(x-2)^2}{9} + \dfrac{(y+3)^2}{4} = 1$

37. $\dfrac{(y+2)^2}{4} - \dfrac{(x+3)^2}{9} = 1$

38. $\dfrac{(x+1)^2}{16} - \dfrac{(y-2)^2}{4} = 1$

39. $x^2 - 4x + y^2 + 6y = -12$

40. $4x^2 + 8x + 25y^2 - 250y = -529$

41. $5x^2 + 20x + 2y^2 - 8y = -18$

42. $-4x^2 + 8x + 4y^2 + 8y = 16$

Graph each equation. Give the domain and range. Identify any that are functions.

43. $\frac{x}{3} = -\sqrt{1 - \frac{y^2}{16}}$

44. $x = -\sqrt{1 - \frac{y^2}{36}}$

45. $y = -\sqrt{1 + x^2}$

46. $y = -\sqrt{1 - \frac{x^2}{25}}$

Write an equation for each conic section with center at the origin.

47. ellipse; vertex at $(0, -4)$, focus at $(0, -2)$

48. ellipse; x-intercept $(6, 0)$, focus at $(2, 0)$

49. hyperbola; focus at $(0, 5)$, transverse axis with length 8

50. hyperbola; y-intercept $(0, -2)$, passing through the point $(2, 3)$

Write an equation for each conic section satisfying the given conditions.

51. parabola with focus at $(3, 2)$ and directrix $x = -3$

52. parabola with vertex at $(-3, 2)$ and y-intercepts $(0, 5)$ and $(0, -1)$

53. ellipse with foci at $(-2, 0)$ and $(2, 0)$ and major axis with length 10

54. ellipse with foci at $(0, 3)$ and $(0, -3)$ and vertex at $(0, -7)$

55. hyperbola with x-intercepts $(-3, 0)$ and $(3, 0)$ and foci at $(-5, 0)$ and $(5, 0)$

56. hyperbola with foci at $(0, 12)$ and $(0, -12)$ and asymptotes $y = \pm x$

Solve each problem.

57. Write the equation of an ellipse consisting of all points in the plane the sum of whose distances from $(0, 0)$ and $(4, 0)$ is 8.

58. Write the equation of a hyperbola consisting of all points in the plane for which the absolute value of the difference of the distances from $(0, 0)$ and $(0, 4)$ is 2.

59. Write the equation of a hyperbola consisting of all points in the plane for which the absolute value of the difference of the distances from $(-5, 0)$ and $(5, 0)$ is 8.

60. Calculator graphs are shown in choices A–D. Arrange the graphs so that their eccentricities are in increasing order.

A. 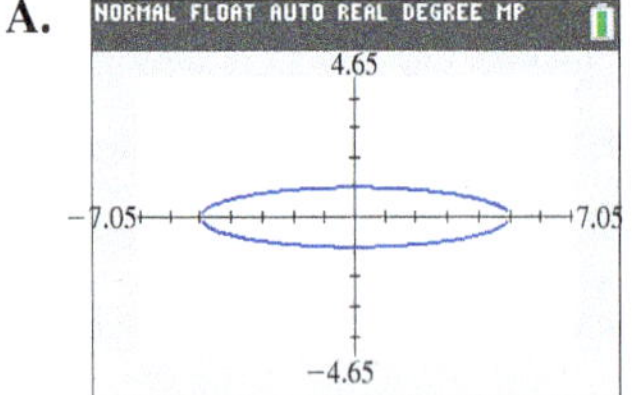

B. 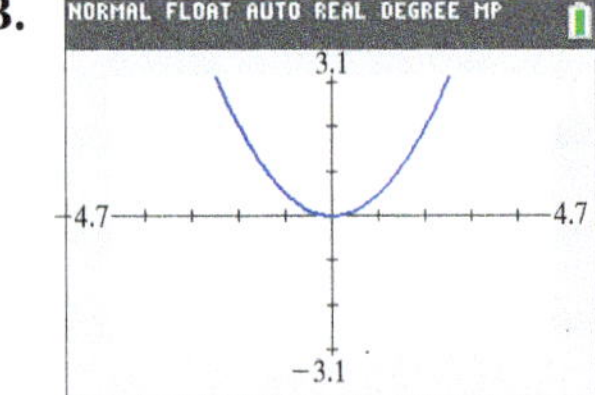

C. 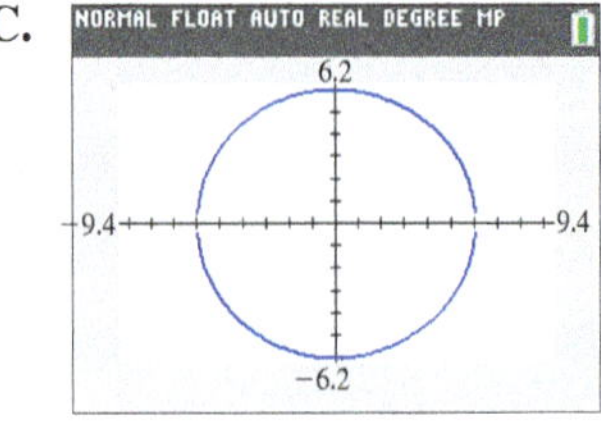

D. 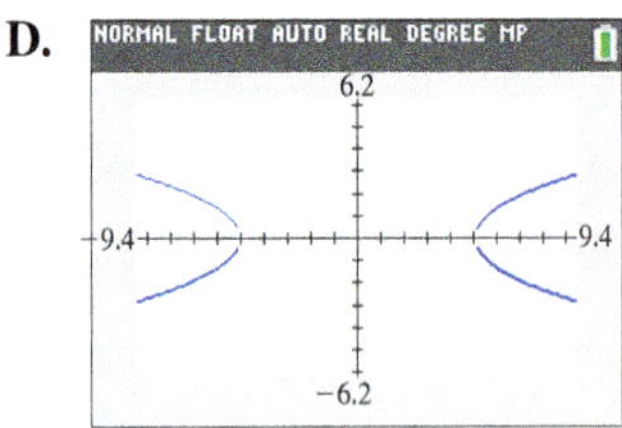

61. ***Orbit of a Comet*** The comet Swift-Tuttle has an elliptical orbit of eccentricity $e = 0.964$, with the sun at one focus. Find the equation of the comet given that the closest it comes to the sun is 89 million mi. Round values to the nearest million miles.

62. ***Orbit of Venus*** The orbit of Venus is an ellipse with the sun at one focus. The eccentricity of the orbit is $e = 0.006775$, and the major axis has length 134.5 million mi. (*Source: World Almanac and Book of Facts.*) Find the least and greatest distances of Venus from the sun to the nearest tenth of a million miles.

Chapter 10 Test

Graph each parabola. Give the domain, range, vertex, and axis of symmetry.

1. $y = -x^2 + 6x$

2. $x = 4y^2 + 8y$

3. Give the coordinates of the focus and the equation of the directrix for the parabola with equation $x = 8y^2$.

4. Write an equation for the parabola with vertex $(2, 3)$, passing through the point $(-18, 1)$, and opening to the left.

5. ***(Modeling) Radio Telescope Design*** A radio telescope has a diameter of 100 ft and a maximum depth of 15 ft.

(a) Write an equation of a parabola that models the cross section of the dish if the vertex is placed at the origin and the parabola opens up.

(b) The receiver must be placed at the focus of the parabola. How far from the vertex, to the nearest tenth of a foot, should the receiver be located?

Graph each ellipse. Give the domain and range.

6. $\dfrac{(x-8)^2}{100} + \dfrac{(y-5)^2}{49} = 1$

7. $16x^2 + 4y^2 = 64$

Solve each problem.

8. Graph $y = -\sqrt{1 - \dfrac{x^2}{36}}$. Tell whether the graph is that of a function.

9. Write the equation of an ellipse centered at the origin having horizontal major axis with length 6 and minor axis with length 4.

10. ***Height of the Arch of a Bridge*** An arch of a bridge has the shape of the top half of an ellipse. The arch is 40 ft wide and 12 ft high at the center. Write an equation of the complete ellipse. Find the height of the arch 10 ft from the center of the bottom. Round to the nearest tenth of a foot.

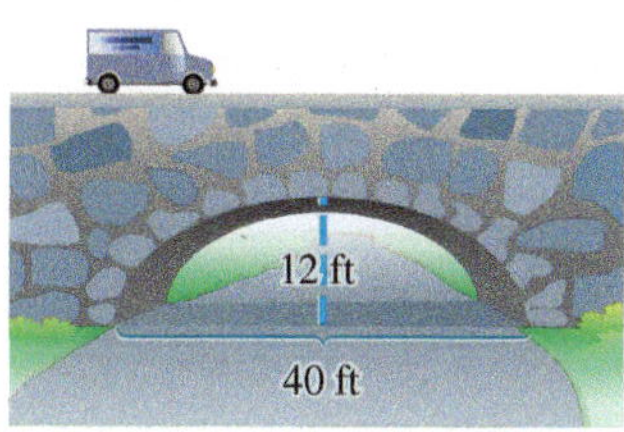

Graph each hyperbola. Give the domain, range, and equations of the asymptotes.

11. $\dfrac{x^2}{4} - \dfrac{y^2}{4} = 1$

12. $9x^2 - 4y^2 = 36$

13. Write the equation of a hyperbola with y-intercepts $(0, -5)$ and $(0, 5)$ and foci at $(0, -6)$ and $(0, 6)$.

Identify the type of graph that each equation has, without actually graphing.

14. $x^2 + 8x + y^2 - 4y + 2 = 0$

15. $5x^2 + 10x - 2y^2 - 12y - 23 = 0$

16. $3x^2 + 10y^2 - 30 = 0$

17. $x^2 - 4y = 0$

18. $(x + 9)^2 + (y - 3)^2 = 0$

19. $x^2 + 4x + y^2 - 6y + 30 = 0$

20. The screen shown here gives the graph of

$$\frac{x^2}{25} - \frac{y^2}{49} = 1$$

as generated by a graphing calculator. What two functions y_1 and y_2 were used to obtain the graph?

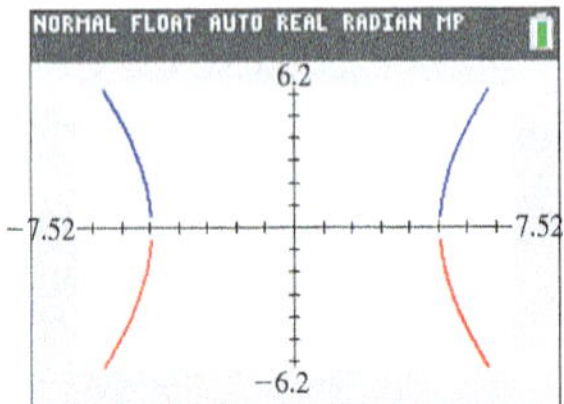

11 Further Topics in Algebra

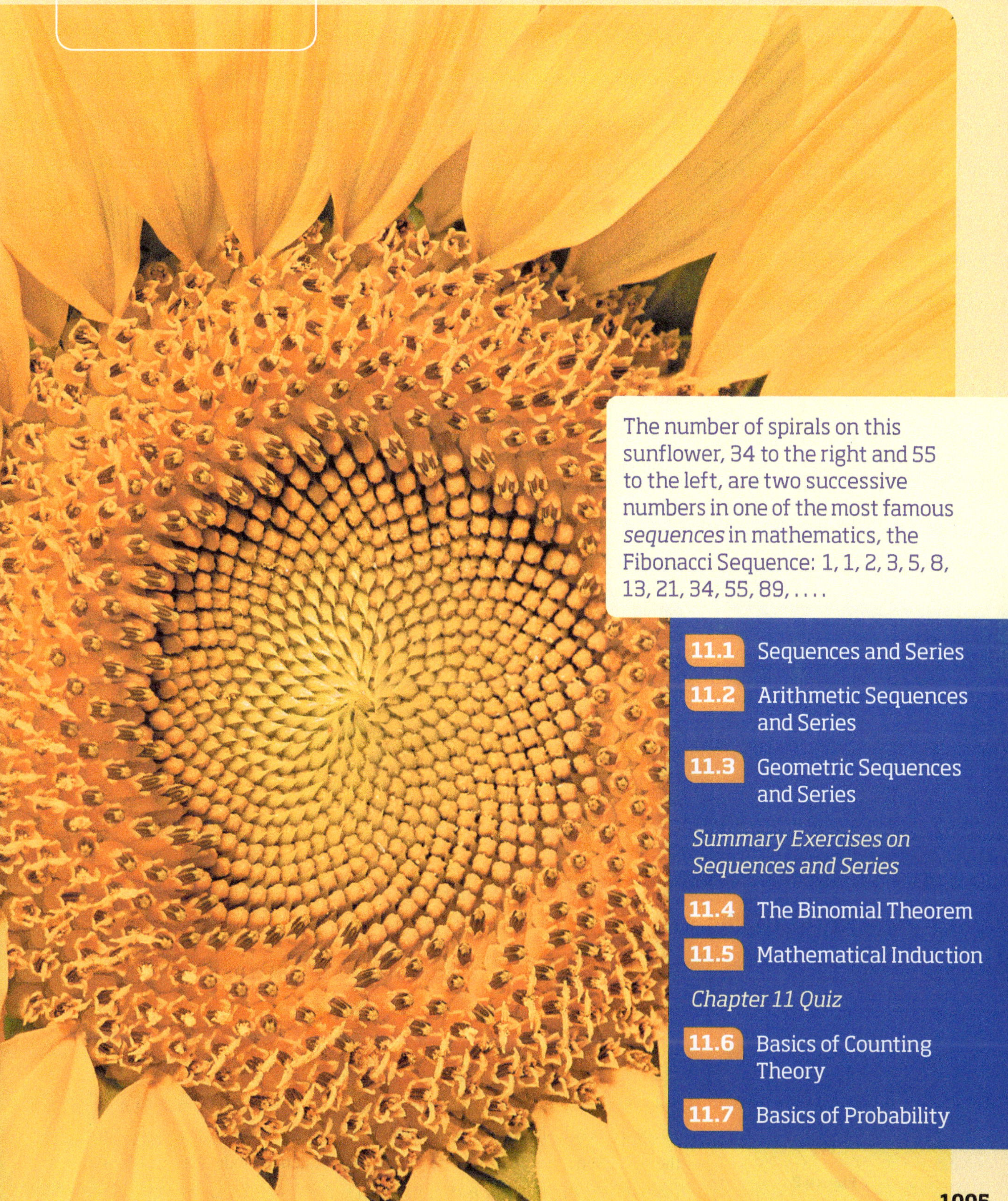

The number of spirals on this sunflower, 34 to the right and 55 to the left, are two successive numbers in one of the most famous *sequences* in mathematics, the Fibonacci Sequence: 1, 1, 2, 3, 5, 8, 13, 21, 34, 55, 89, ...

11.1 Sequences and Series

- Sequences
- Series and Summation Notation
- Summation Properties and Rules

Sequences

A *sequence* is a function that computes an ordered list. For example, the average person in the United States uses 100 gallons of water each day. The function $f(n) = 100n$ generates the terms of the sequence

$$100, 200, 300, 400, 500, 600, 700, \ldots,$$

when $n = 1, 2, 3, 4, 5, 6, 7, \ldots$. This function represents the number of gallons of water used by the average person after n days.

As another example, say \$100 is deposited into a savings account paying 3% interest compounded annually. The function $g(n) = 100(1.03)^n$ calculates the account balance after n years. The terms of the sequence are

$$g(1), g(2), g(3), g(4), g(5), g(6), g(7), \ldots,$$

and can be approximated as

$$103, 106.09, 109.27, 112.55, 115.93, 119.41, 122.99, \ldots.$$

Sequence

A **finite sequence** is a function that has a set of natural numbers of the form $\{1, 2, 3, \ldots, n\}$ as its domain. An **infinite sequence** has the set of natural numbers as its domain.

Instead of using function notation $f(x)$ to indicate a sequence, it is customary to use a_n, where $a_n = f(n)$. ***The letter n is used instead of x as a reminder that n represents a natural number.*** The elements in the range of a sequence, called the **terms** of the sequence, are $a_1, a_2, a_3, \ldots$. The elements of both the domain and the range of a sequence are *ordered.* The first term is found by letting $n = 1$, the second term is found by letting $n = 2$, and so on. The **general term,** or ***n*th term,** of the sequence is a_n.

Figure 1 shows graphs of $f(x) = 2x$ and $a_n = 2n$. Notice that $f(x)$ is a continuous function, and a_n consists of discrete points. To graph a_n, we plot points of the form $(n, 2n)$ for $n = 1, 2, 3, \ldots$. We show only the results for $n = 1, 2, 3, 4$, and 5.

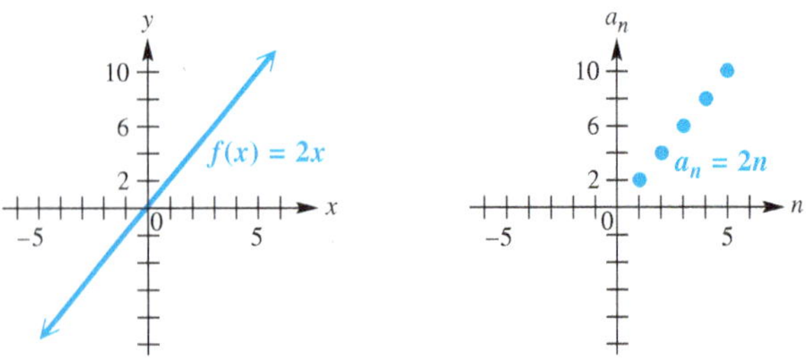

Figure 1

A graphing calculator can list the terms of a sequence. Using sequence mode to list the first 10 terms of the sequence with general term $a_n = n + \frac{1}{n}$ produces the result shown in **Figure 2(a).** The tenth term can be seen by scrolling to the right. Sequences can also be graphed in sequence mode. **Figure 2(b)** shows a graph of $a_n = n + \frac{1}{n}$. For $n = 5$, the term is $5 + \frac{1}{5} = 5.2$.

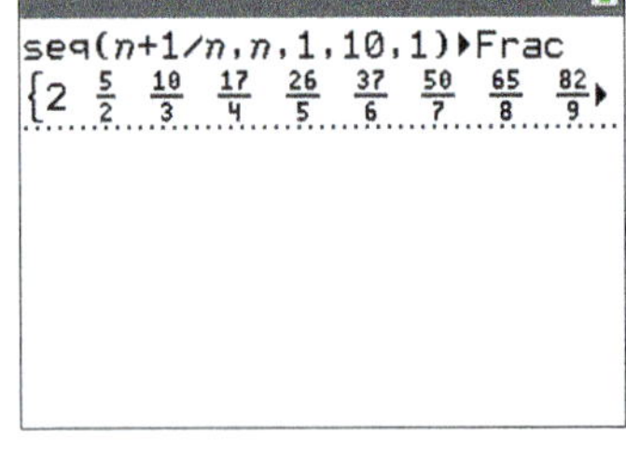

(a)

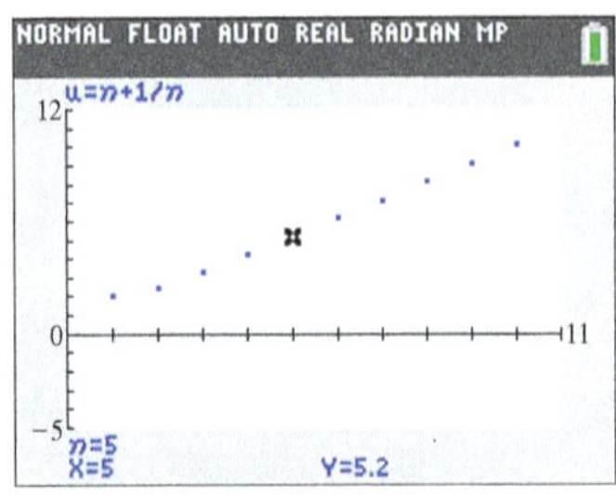

The fifth term is 5.2.

(b)

Figure 2

EXAMPLE 1 Finding Terms of Sequences

Write the first five terms of each sequence.

(a) $a_n = \frac{n+1}{n+2}$ **(b)** $a_n = (-1)^n \cdot n$ **(c)** $a_n = \frac{2n+1}{n^2+1}$

SOLUTION

(a) Replace n in $a_n = \frac{n+1}{n+2}$ with 1, 2, 3, 4, and 5.

$$n = 1: \quad a_1 = \frac{1+1}{1+2} = \frac{2}{3}$$

$$n = 2: \quad a_2 = \frac{2+1}{2+2} = \frac{3}{4}$$

$$n = 3: \quad a_3 = \frac{3+1}{3+2} = \frac{4}{5}$$

$$n = 4: \quad a_4 = \frac{4+1}{4+2} = \frac{5}{6}$$

$$n = 5: \quad a_5 = \frac{5+1}{5+2} = \frac{6}{7}$$

(b) Replace n in $a_n = (-1)^n \cdot n$ with 1, 2, 3, 4, and 5.

$$n = 1: \quad a_1 = (-1)^1 \cdot 1 = -1$$

$$n = 2: \quad a_2 = (-1)^2 \cdot 2 = 2$$

$$n = 3: \quad a_3 = (-1)^3 \cdot 3 = -3$$

$$n = 4: \quad a_4 = (-1)^4 \cdot 4 = 4$$

$$n = 5: \quad a_5 = (-1)^5 \cdot 5 = -5$$

(c) For $a_n = \frac{2n+1}{n^2+1}$, the first five terms are as follows.

$$a_1 = \frac{3}{2}, \quad a_2 = 1, \quad a_3 = \frac{7}{10}, \quad a_4 = \frac{9}{17}, \quad \text{and} \quad a_5 = \frac{11}{26} \qquad \text{Replace } n \text{ with } 1, 2, 3, 4, \text{ and } 5.$$

✔ **Now Try Exercises 13, 17, and 19.**

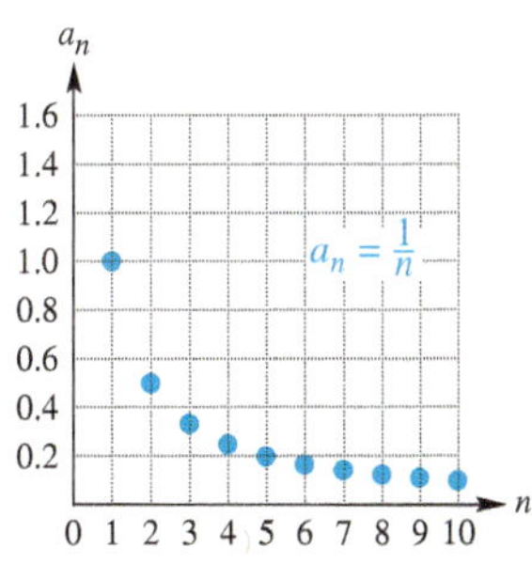

Figure 3

If the terms of an infinite sequence get closer and closer to some real number, the sequence is said to be **convergent** and to **converge** to that real number. For example, the sequence defined by $a_n = \frac{1}{n}$ approaches 0 as n becomes large. Thus, a_n is a convergent sequence that converges to 0. A graph of this sequence for $n = 1, 2, 3, \ldots, 10$ is shown in **Figure 3.** The terms of a_n approach the horizontal axis.

A sequence that does not converge to any number is **divergent.** The first nine terms of the sequence $a_n = n^2$ are

$$1, 4, 9, 16, 25, 36, 49, 64, 81, \ldots.$$

This sequence is divergent because as n becomes large, the values of a_n do not approach a fixed number—rather, they increase without bound.

Some sequences are defined by a **recursive definition,** one in which each term after the first term or first few terms is defined as an expression involving the previous term or terms. The sequences in **Example 1** were defined *explicitly,* with a formula for a_n that does not depend on a previous term.

Leonardo of Pisa (Fibonacci) (1170–1250)

One of the most famous sequences in mathematics is the **Fibonacci sequence,** 1, 1, 2, 3, 5, 8, 13, 21, 34, 55, . . . , named for the Italian mathematician Leonardo of Pisa, who was also known as Fibonacci. The Fibonacci sequence may be defined using a recursion formula. See **Exercise 33.**

EXAMPLE 2 Using a Recursion Formula

Find the first four terms of each sequence.

(a) $a_1 = 4$
$a_n = 2 \cdot a_{n-1} + 1, \text{ if } n > 1$

(b) $a_1 = 2$
$a_n = a_{n-1} + n - 1, \text{ if } n > 1$

SOLUTION

(a) This is a recursive definition. We know that $a_1 = 4$. Use $a_n = 2 \cdot a_{n-1} + 1$.

$$a_1 = 4$$
$$a_2 = 2 \cdot a_1 + 1 = 2 \cdot 4 + 1 = 9$$
$$a_3 = 2 \cdot a_2 + 1 = 2 \cdot 9 + 1 = 19$$
$$a_4 = 2 \cdot a_3 + 1 = 2 \cdot 19 + 1 = 39$$

(b) In this recursive definition, $a_1 = 2$ and $a_n = a_{n-1} + n - 1$.

$$a_1 = 2$$
$$a_2 = a_1 + 2 - 1 = 2 + 1 = 3$$
$$a_3 = a_2 + 3 - 1 = 3 + 2 = 5$$
$$a_4 = a_3 + 4 - 1 = 5 + 3 = 8$$

✔ **Now Try Exercises 31 and 35.**

EXAMPLE 3 Modeling Insect Population Growth

Frequently the population of a particular insect grows rapidly at first and then levels off because of competition for limited resources. In one study, the behavior of the winter moth was modeled with a sequence similar to the following, where a_n represents the population density, in thousands per acre, during year n. (*Source:* Varley, G. and G. Gradwell, "Population models for the winter moth," Symposium of the Royal Entomological Society of London.)

$$a_1 = 1$$
$$a_n = 2.85a_{n-1} - 0.19a_{n-1}^2, \quad \text{for } n \geq 2$$

(a) Give a table of values for $n = 1, 2, 3, \ldots, 10$.

(b) Graph the sequence. Describe what happens to the population density.

SOLUTION

(a) Evaluate $a_1, a_2, a_3, \ldots, a_{10}$ recursively. We are given $a_1 = 1$.

$$a_2 = 2.85a_1 - 0.19a_1^2 = 2.85(1) - 0.19(1)^2 = 2.66$$
$$a_3 = 2.85a_2 - 0.19a_2^2 = 2.85(2.66) - 0.19(2.66)^2 \approx 6.24$$

Approximate values for a_n are shown in the table. **Figure 4** shows computation of the sequence, denoted by $u(n)$ rather than a_n, using a calculator.

n	1	2	3	4	5	6	7	8	9	10
a_n	1	2.66	6.24	10.4	9.11	10.2	9.31	10.1	9.43	9.98

NORMAL FLOAT AUTO REAL RADIAN MP
PRESS ENTER TO EDIT

n	u(n)
1	1
2	2.66
3	6.2366
4	10.384
5	9.1069
6	10.197
7	9.3056
8	10.068
9	9.4345
10	9.9765

u(n)=2.85u(n−1)−.19u(n−1)²

Figure 4

(b) The graph of a sequence is a set of discrete points. Plot the points

$$(1, 1), \quad (2, 2.66), \quad (3, 6.24), \quad \ldots, \quad (10, 9.98),$$

as shown in **Figure 5(a).** At first, the insect population increases rapidly, and then it oscillates about the line $y = 9.7$. (See the Note following this example.) The oscillations become smaller as n increases, indicating that the population density converges to near 9.7 thousand per acre. In **Figure 5(b),** the first 20 terms have been plotted with a calculator.

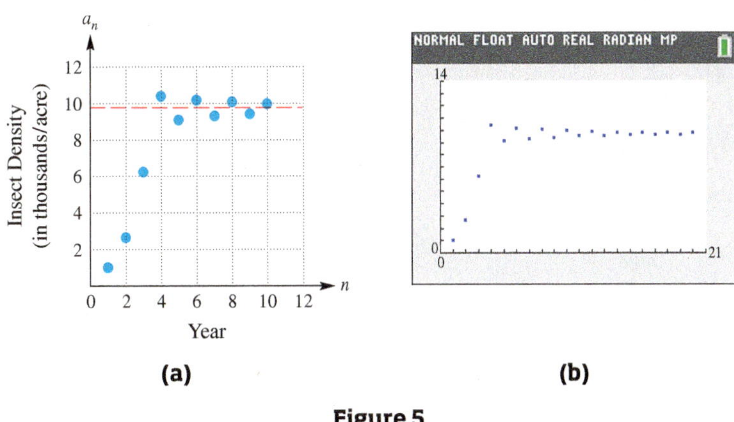

(a) **(b)**

Figure 5

✓ **Now Try Exercise 95.**

NOTE The insect population converges to the value $k = 9.7$ thousand per acre in **Example 3.** This value of k can be found by solving the quadratic equation $k = 2.85k - 0.19k^2$, which equates the values of a_n for consecutive years.

Series and Summation Notation Suppose a person has a starting salary of \$30,000 and receives a \$2000 raise each year. Then,

$$30{,}000, \quad 32{,}000, \quad 34{,}000, \quad 36{,}000, \quad 38{,}000$$

are terms of the sequence that describe this person's salaries over a 5-year period. The total earned is given by the *finite series*

$$30{,}000 + 32{,}000 + 34{,}000 + 36{,}000 + 38{,}000,$$

whose sum is \$170,000.

A sequence can be used to define a *series*. For example, the infinite sequence

$$1, \frac{1}{3}, \frac{1}{9}, \frac{1}{27}, \frac{1}{81}, \frac{1}{243}, \ldots$$

defines the terms of the *infinite series*

$$1 + \frac{1}{3} + \frac{1}{9} + \frac{1}{27} + \frac{1}{81} + \frac{1}{243} + \cdots.$$

If a sequence has terms $a_1, a_2, a_3, \ldots$, then S_n is defined as the sum of the first n terms. That is,

$$S_n = a_1 + a_2 + a_3 + \cdots + a_n.$$

LOOKING AHEAD TO CALCULUS

An infinite series converges if the sequence of partial sums $S_1, S_2, S_3, \ldots$ converges. For example, it can be shown that

$$1 + \frac{1}{2} + \frac{1}{3} + \frac{1}{4} + \cdots \text{ diverges,}$$

while

$$1 - \frac{1}{2} + \frac{1}{3} - \frac{1}{4} + \cdots \text{ converges.}$$

The sum of the terms of a sequence, called a **series,** is written using **summation notation.** The Greek capital letter **sigma** Σ is used to indicate a sum.

Series

A **finite series** is an expression of the form

$$S_n = a_1 + a_2 + a_3 + \cdots + a_n = \sum_{i=1}^{n} a_i,$$

and an **infinite series** is an expression of the form

$$S_\infty = a_1 + a_2 + a_3 + \cdots + a_n + \cdots = \sum_{i=1}^{\infty} a_i.$$

The letter i is the **index of summation.**

CAUTION ***Do not confuse this use of i with the use of i as the imaginary unit.*** Other letters, such as k and j, may be used for the index of summation.

EXAMPLE 4 Using Summation Notation

Evaluate $\sum_{k=1}^{6} (2^k + 1)$.

ALGEBRAIC SOLUTION

Write each of the six terms, and then evaluate the sum.

$$\sum_{k=1}^{6} (2^k + 1)$$

$$= (2^1 + 1) + (2^2 + 1) + (2^3 + 1) + (2^4 + 1) + (2^5 + 1) + (2^6 + 1)$$

$$= (2 + 1) + (4 + 1) + (8 + 1) + (16 + 1) + (32 + 1) + (64 + 1)$$

$$= 3 + 5 + 9 + 17 + 33 + 65$$

$$= 132$$

GRAPHING CALCULATOR SOLUTION

A graphing calculator can list the terms of a sequence and then compute the sum of the terms. See **Figure 6.**

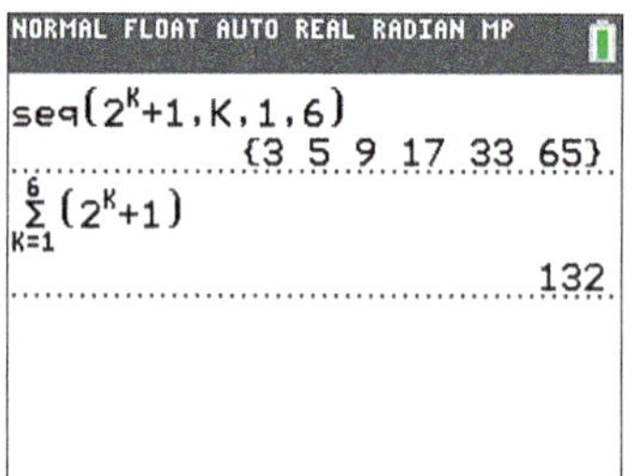

Figure 6

✔ **Now Try Exercise 51.**

EXAMPLE 5 Using Summation Notation with Subscripts

Write the terms for each series and, if possible, evaluate each sum.

(a) $\sum_{j=3}^{6} a_j$ **(b)** $\sum_{i=1}^{3} (6x_i - 2)$, if $x_1 = 2$, $x_2 = 4$, and $x_3 = 6$

(c) $\sum_{i=1}^{4} f(x_i)\Delta x$, if $f(x) = x^2$, $x_1 = 0$, $x_2 = 2$, $x_3 = 4$, $x_4 = 6$, and $\Delta x = 2$

SOLUTION

(a) $\sum_{j=3}^{6} a_j$

$= a_3 + a_4 + a_5 + a_6$ Replace j in a_j with 3, 4, 5, and 6.

(b) $\sum_{i=1}^{3}(6x_i - 2)$

$= (6x_1 - 2) + (6x_2 - 2) + (6x_3 - 2)$ Let $i = 1, 2,$ and 3, respectively.

$= (6 \cdot 2 - 2) + (6 \cdot 4 - 2) + (6 \cdot 6 - 2)$ Substitute the given values for x_1, x_2, and x_3.

$= 10 + 22 + 34$ Multiply and subtract inside the parentheses.

$= 66$ Add.

(c) $\sum_{i=1}^{4} f(x_i)\Delta x$

$= f(x_1)\Delta x + f(x_2)\Delta x + f(x_3)\Delta x + f(x_4)\Delta x$ Let $i = 1, 2, 3,$ and 4.

$= x_1{}^2\Delta x + x_2{}^2\Delta x + x_3{}^2\Delta x + x_4{}^2\Delta x$ $f(x) = x^2$

$= 0^2(2) + 2^2(2) + 4^2(2) + 6^2(2)$ Substitute the given values for x_1, x_2, x_3, and x_4, with $\Delta x = 2$.

$= 0 + 8 + 32 + 72$ Simplify.

$= 112$ Add.

✔ **Now Try Exercises 59, 61, and 71.**

LOOKING AHEAD TO CALCULUS

Summation notation is used in calculus to describe the area under a curve, to describe the volume of a figure rotated about an axis, and in many other applications, as well as in the definition of integral. In the definition of the definite integral, Σ is replaced with an elongated S:

$$\int_a^b f(x)\,dx = \lim_{n\to\infty} \sum_{i=1}^{n} f(x_i)\,\Delta x_i.$$

In some cases, the definite integral can be interpreted as the sum of the areas of rectangles.

Summation Properties and Rules

These provide useful shortcuts.

Summation Properties

If $a_1, a_2, a_3, \ldots, a_n$ and $b_1, b_2, b_3, \ldots, b_n$ are two sequences, and c is a constant, then for every positive integer n, the following hold.

(a) $\sum_{i=1}^{n} c = nc$ **(b)** $\sum_{i=1}^{n} ca_i = c\sum_{i=1}^{n} a_i$

(c) $\sum_{i=1}^{n} (a_i + b_i) = \sum_{i=1}^{n} a_i + \sum_{i=1}^{n} b_i$ **(d)** $\sum_{i=1}^{n} (a_i - b_i) = \sum_{i=1}^{n} a_i - \sum_{i=1}^{n} b_i$

To prove Property (a), expand the series.

$\sum_{i=1}^{n} c$

$= c + c + c + c + \cdots + c$ n terms of c

$= nc$

Property (c) also can be proved by first expanding the series.

$\sum_{i=1}^{n} (a_i + b_i)$

$= (a_1 + b_1) + (a_2 + b_2) + \cdots + (a_n + b_n)$

$= (a_1 + a_2 + \cdots + a_n) + (b_1 + b_2 + \cdots + b_n)$ Commutative and associative properties

$= \sum_{i=1}^{n} a_i + \sum_{i=1}^{n} b_i$

Proofs of the other two properties are similar.

The following rules, used in calculus, can be proved by *mathematical induction*.

Summation Rules

$$\sum_{i=1}^{n} i = 1 + 2 + \cdots + n = \frac{n(n+1)}{2}$$

$$\sum_{i=1}^{n} i^2 = 1^2 + 2^2 + \cdots + n^2 = \frac{n(n+1)(2n+1)}{6}$$

$$\sum_{i=1}^{n} i^3 = 1^3 + 2^3 + \cdots + n^3 = \frac{n^2(n+1)^2}{4}$$

EXAMPLE 6 Using the Summation Properties and Rules

Use the summation properties and rules to evaluate each series.

(a) $\sum_{i=1}^{40} 5$ **(b)** $\sum_{i=1}^{22} 2i$ **(c)** $\sum_{i=1}^{14} (2i^2 - 3)$

SOLUTION

(a) $\sum_{i=1}^{40} 5$

$= 40(5)$ Property (a) with $n = 40$ and $c = 5$

$= 200$ Multiply.

(b) $\sum_{i=1}^{22} 2i$

$= 2\sum_{i=1}^{22} i$ Property (b) with $c = 2$ and $a_i = i$

$= 2 \cdot \frac{22(22+1)}{2}$ Summation rule ($\sum_{i=1}^{n} i = \frac{n(n+1)}{2}$)

$= 506$ Evaluate.

(c) $\sum_{i=1}^{14} (2i^2 - 3)$

$= \sum_{i=1}^{14} 2i^2 - \sum_{i=1}^{14} 3$ Property (d) with $a_i = 2i^2$ and $b_i = 3$

$= 2\sum_{i=1}^{14} i^2 - \sum_{i=1}^{14} 3$ Property (b) with $c = 2$ and $a_i = i^2$

$= 2 \cdot \frac{14(14+1)(2 \cdot 14 + 1)}{6} - 14(3)$ Summation rule and Property (a) ($\sum_{i=1}^{n} i^2 = \frac{n(n+1)(2n+1)}{6}$)

$= 2030 - 42$ Simplify.

$= 1988$ Subtract.

✔ **Now Try Exercises 75, 77, and 79.**

EXAMPLE 7 Using the Summation Properties and Rules

Evaluate $\sum_{i=1}^{6} (i^2 + 3i + 5)$.

SOLUTION

$$\sum_{i=1}^{6} (i^2 + 3i + 5)$$

$$= \sum_{i=1}^{6} i^2 + \sum_{i=1}^{6} 3i + \sum_{i=1}^{6} 5 \quad \text{Property (c)}$$

$$= \sum_{i=1}^{6} i^2 + 3\sum_{i=1}^{6} i + \sum_{i=1}^{6} 5 \quad \text{Property (b)}$$

$$= \sum_{i=1}^{6} i^2 + 3\sum_{i=1}^{6} i + 6(5) \quad \text{Property (a)}$$

$$= \frac{6(6+1)(2 \cdot 6 + 1)}{6} + 3\left[\frac{6(6+1)}{2}\right] + 6(5) \quad \text{Summation rules}$$

$$= 91 + 63 + 30 \quad \text{Simplify.}$$

$$= 184 \quad \text{Add.}$$

✔ **Now Try Exercise 81.**

NOTE It is possible to evaluate the sums in **Examples 6 and 7** without using the summation properties and rules; however, this can be tedious.

11.1 Exercises

CONCEPT PREVIEW *Fill in the blank(s) to correctly complete each sentence.*

1. A(n) ________ is a function that computes an ordered list.

2. A(n) ________ sequence is a function that has the set of natural numbers of the form $\{1, 2, 3, \ldots, n\}$ as its domain.

3. Some sequences are defined by a(n) ________ definition, one in which each term after the first term or the first few terms is defined as an expression involving the previous term or terms.

4. The sum of the terms of a sequence is a(n) ________. It is written using the Greek capital letter symbol ________ to indicate a sum.

CONCEPT PREVIEW *Answer each of the following.*

5. Complete a table of values for the sequence $a_n = 5n + 2$ using $n = 1, 2, 3, 4, 5$.

6. Graph the sequence $a_n = 5n + 2$ using the values from **Exercise 5.**

7. Evaluate $\sum_{i=1}^{5} (5i + 2)$.

8. Find the first five terms of the sequence defined by the following recursive definition. How is the sequence related to the sequence in **Exercise 5?**

$$a_1 = 7$$
$$a_n = a_{n-1} + 5, \quad \text{if } n > 1$$

9. Find the first five terms of the sequence $a_n = 3(-3)^{n-1}$.

10. Evaluate $\sum_{i=1}^{5} 3(-3)^{i-1}$.

Write the first five terms of each sequence. ***See Example 1.***

11. $a_n = 4n + 10$

12. $a_n = 6n - 3$

13. $a_n = \frac{n+5}{n+4}$

14. $a_n = \frac{n-7}{n-6}$

15. $a_n = \left(\frac{1}{3}\right)^n (n-1)$

16. $a_n = \left(\frac{1}{2}\right)^n (n)$

17. $a_n = (-1)^n(2n)$

18. $a_n = (-1)^{n-1}(n+1)$

19. $a_n = \frac{4n-1}{n^2+2}$

20. $a_n = \frac{n^2-1}{n^2+1}$

21. $a_n = \frac{n^3+8}{n+2}$

22. $a_n = \frac{n^3+27}{n+3}$

Concept Check *Decide whether each sequence is* finite *or* infinite.

23. The sequence of days of the week

24. The sequence of pages in a book

25. 1, 2, 3, 4, 5

26. −1, −2, −3, −4, −5

27. 1, 2, 3, 4, 5, . . .

28. −1, −2, −3, −4, −5, . . .

29. $a_1 = 4$
$a_n = 4 \cdot a_{n-1}$, if $n \geq 2$

30. $a_1 = 2$
$a_2 = 5$
$a_n = a_{n-1} + a_{n-2}$, if $n \geq 3$

Find the first four terms of each sequence. ***See Example 2.***

31. $a_1 = -2$
$a_n = a_{n-1} + 3$, if $n > 1$

32. $a_1 = -1$
$a_n = a_{n-1} - 4$, if $n > 1$

33. $a_1 = 1$
$a_2 = 1$
$a_n = a_{n-1} + a_{n-2}$, if $n \geq 3$
(This is the Fibonacci sequence.)

34. $a_1 = 1$
$a_2 = 3$
$a_n = a_{n-1} + a_{n-2}$, if $n \geq 3$
(This is the Lucas sequence.)

35. $a_1 = 2$
$a_n = n \cdot a_{n-1}$, if $n > 1$

36. $a_1 = -3$
$a_n = 2n \cdot a_{n-1}$, if $n > 1$

Evaluate each series. ***See Example 4.***

37. $\sum_{i=1}^{5} (2i+1)$

38. $\sum_{i=1}^{6} (3i-2)$

39. $\sum_{j=1}^{4} j^{-1}$

40. $\sum_{i=1}^{5} (i+1)^{-1}$

41. $\sum_{i=1}^{4} i^i$

42. $\sum_{k=1}^{4} (k+1)^k$

43. $\sum_{k=1}^{6} (-1)^k \cdot k$

44. $\sum_{i=1}^{7} (-1)^{i+1} \cdot i^2$

45. $\sum_{i=2}^{5} (6-3i)$

46. $\sum_{i=3}^{7} (5i+2)$

47. $\sum_{i=-2}^{3} 2(3)^i$

48. $\sum_{i=-1}^{2} 5(2)^i$

49. $\sum_{i=-1}^{5} (i^2-2i)$

50. $\sum_{i=3}^{6} (2i^2+1)$

51. $\sum_{i=1}^{5} (3^i-4)$

52. $\sum_{i=1}^{4} [(-2)^i - 3]$

53. $\sum_{i=1}^{3} (i^3 - i)$

54. $\sum_{i=1}^{4} (i^4 - i^3)$

Use a graphing calculator to evaluate each series. **See Example 4.**

55. $\sum_{i=1}^{10} (4i^2 - 5)$

56. $\sum_{i=1}^{10} (i^3 - 6)$

57. $\sum_{j=3}^{9} (3j - j^2)$

58. $\sum_{k=5}^{10} (k^2 - 4k + 7)$

Write the terms for each series and evaluate the sum, given that $x_1 = -2$, $x_2 = -1$, $x_3 = 0$, $x_4 = 1$, and $x_5 = 2$. **See Examples 5(a) and 5(b).**

59. $\sum_{i=1}^{5} x_i$

60. $\sum_{i=1}^{5} -x_i$

61. $\sum_{i=1}^{5} (2x_i + 3)$

62. $\sum_{i=1}^{4} (-3x_i - 2)$

63. $\sum_{i=1}^{3} (3x_i - x_i^2)$

64. $\sum_{i=1}^{3} (x_i^2 + x_i)$

65. $\sum_{i=2}^{5} \frac{x_i + 1}{x_i + 2}$

66. $\sum_{i=1}^{5} \frac{x_i}{x_i + 3}$

67. $\sum_{i=1}^{4} \frac{x_i^3 + 1000}{x_i + 10}$

68. How can factoring make the work in **Exercises 21, 22, and 67** easier?

Write the terms of $\sum_{i=1}^{4} f(x_i)\Delta x$, with $x_1 = 0$, $x_2 = 2$, $x_3 = 4$, $x_4 = 6$, and $\Delta x = 0.5$, for each function. Evaluate the sum. **See Example 5(c).**

69. $f(x) = 4x - 7$

70. $f(x) = 6 + 2x$

71. $f(x) = 2x^2$

72. $f(x) = x^2 - 1$

73. $f(x) = \frac{-2}{x + 1}$

74. $f(x) = \frac{5}{2x - 1}$

Use the summation properties and rules to evaluate each series. **See Examples 6 and 7.**

75. $\sum_{i=1}^{100} 6$

76. $\sum_{i=1}^{20} 5$

77. $\sum_{i=1}^{15} i^2$

78. $\sum_{i=1}^{50} 2i^3$

79. $\sum_{i=1}^{5} (5i + 3)$

80. $\sum_{i=1}^{5} (8i - 1)$

81. $\sum_{i=1}^{5} (4i^2 - 2i + 6)$

82. $\sum_{i=1}^{6} (2 + i - i^2)$

83. $\sum_{i=1}^{4} (3i^3 + 2i - 4)$

84. $\sum_{i=1}^{6} (i^2 + 2i^3)$

Concept Check *Use summation notation to write each series.**

85. $\frac{1}{3(1)} + \frac{1}{3(2)} + \frac{1}{3(3)} + \cdots + \frac{1}{3(9)}$

86. $\frac{5}{1+1} + \frac{5}{1+2} + \frac{5}{1+3} + \cdots + \frac{5}{1+15}$

87. $1 - \frac{1}{2} + \frac{1}{4} - \frac{1}{8} + \cdots - \frac{1}{128}$

88. $1 - \frac{1}{4} + \frac{1}{9} - \frac{1}{16} + \cdots - \frac{1}{400}$

Use the sequence feature of a graphing calculator to graph the first ten terms of each sequence as defined. Use the graph to make a conjecture as to whether the sequence converges or diverges. If it converges, determine the number to which it converges.

89. $a_n = \frac{n + 4}{2n}$

90. $a_n = \frac{1 + 4n}{2n}$

91. $a_n = 2e^n$

92. $a_n = n(n + 2)$

93. $a_n = \left(1 + \frac{1}{n}\right)^n$

94. $a_n = (1 + n)^{1/n}$

---

*These exercises were suggested by Joe Lloyd Harris, Gulf Coast Community College.

Solve each problem. ***See Example 3.***

95. *(Modeling) Insect Population* Suppose an insect population density, in thousands per acre, during year n can be modeled by the recursively defined sequence

$$a_1 = 8$$
$$a_n = 2.9a_{n-1} - 0.2a_{n-1}^2, \quad \text{for } n > 1.$$

(a) Find the population for $n = 1, 2, 3$.

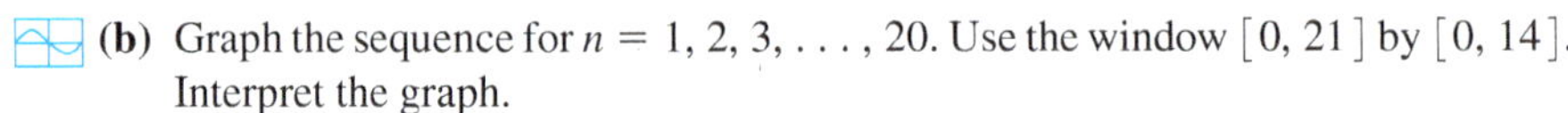

(b) Graph the sequence for $n = 1, 2, 3, \ldots, 20$. Use the window $[0, 21]$ by $[0, 14]$. Interpret the graph.

96. *Male Bee Ancestors* One of the most famous sequences in mathematics is the Fibonacci sequence,

$$1, 1, 2, 3, 5, 8, 13, 21, 34, 55, \ldots.$$

(Also see **Exercise 33.**) Male honeybees hatch from eggs that have not been fertilized, so a male bee has only one parent, a female. On the other hand, female honeybees hatch from fertilized eggs, so a female has two parents, one male and one female. The number of ancestors in consecutive generations of bees follows the Fibonacci sequence. Draw a tree showing the number of ancestors of a male bee in each generation following the description given above.

97. *(Modeling) Bacteria Growth* If certain bacteria are cultured in a medium with sufficient nutrients, they will double in size and then divide every 40 minutes. Let N_1 be the initial number of bacteria cells, N_2 the number after 40 minutes, N_3 the number after 80 minutes, and N_j the number after $40(j - 1)$ minutes. (*Source:* Hoppensteadt, F. and C. Peskin, *Mathematics in Medicine and the Life Sciences*, Springer-Verlag.)

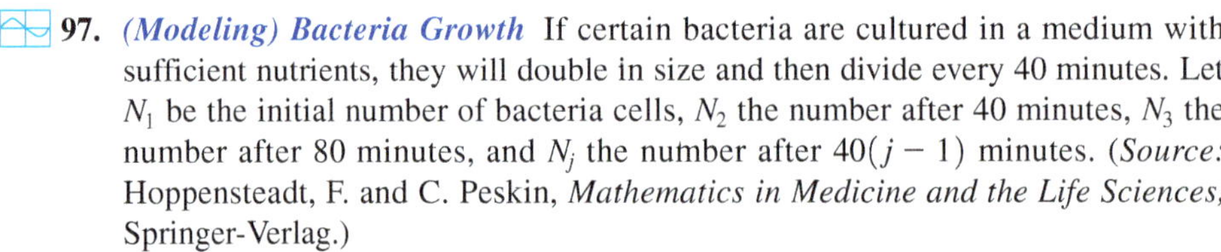

(a) Write N_{j+1} in terms of N_j for $j \geq 1$.

(b) Determine the number of bacteria after 2 hr if $N_1 = 230$.

(c) Graph the sequence N_j for $j = 1, 2, 3, \ldots, 7$, where $N_1 = 230$. Use the window $[0, 10]$ by $[0, 15{,}000]$.

(d) Describe the growth of these bacteria when there are unlimited nutrients.

98. *(Modeling) Verhulst's Model for Bacteria Growth* Refer to **Exercise 97.** If the bacteria are not cultured in a medium with sufficient nutrients, competition will ensue and growth will slow. According to Verhulst's model, the number of bacteria N_j at time $40(j - 1)$ in minutes can be determined by the sequence

$$N_{j+1} = \left[\frac{2}{1 + \frac{N_j}{K}}\right] N_j,$$

where K is a constant and $j \geq 1$. (*Source:* Hoppensteadt, F. and C. Peskin, *Mathematics in Medicine and the Life Sciences*, Springer-Verlag.)

(a) If $N_1 = 230$ and $K = 5000$, make a table of N_j for $j = 1, 2, 3, \ldots, 20$. Round values in the table to the nearest integer.

(b) Graph the sequence N_j for $j = 1, 2, 3, \ldots, 20$. Use the window $[0, 20]$ by $[0, 6000]$.

(c) Describe the growth of these bacteria when there are limited nutrients.

(d) Make a conjecture about why K is called the **saturation constant.** Test the conjecture by changing the value of K in the given formula.

99. *Approximating ln(1 + x)* The series

$$x - \frac{x^2}{2} + \frac{x^3}{3} - \frac{x^4}{4} + \cdots$$

can be used to approximate the value of $\ln(1 + x)$ for values of x in $(-1, 1]$. Use the first six terms of this series to approximate each expression. Compare this approximation with the value obtained on a calculator.

(a) $\ln 1.02 \; (x = 0.02)$ **(b)** $\ln 0.97 \; (x = -0.03)$

100. *Approximating π* Find the sum of the first six terms of the series

$$\frac{\pi^4}{90} = \frac{1}{1^4} + \frac{1}{2^4} + \frac{1}{3^4} + \frac{1}{4^4} + \frac{1}{5^4} + \cdots + \frac{1}{n^4} + \cdots.$$

Multiply this result by 90, and take the fourth root to obtain an approximation of π. Compare this answer to the actual decimal approximation of π.

101. *Approximating Powers of e* The series

$$e^a \approx 1 + a + \frac{a^2}{2!} + \frac{a^3}{3!} + \cdots + \frac{a^n}{n!},$$

where $n! = 1 \cdot 2 \cdot 3 \cdot 4 \cdot \cdots \cdot n$, can be used to approximate the value of e^a for any real number a. Use the first eight terms of this series to approximate each expression. Compare this approximation with the value obtained on a calculator.

(a) e **(b)** e^{-1}

102. *Approximating Square Roots* The recursively defined sequence

$$a_1 = k$$

$$a_n = \frac{1}{2}\left(a_{n-1} + \frac{k}{a_{n-1}}\right), \quad \text{if } n > 1$$

can be used to compute $\sqrt{k}$ for any positive number k. This sequence was known to Sumerian mathematicians 4000 years ago, and it is still used today. Use this sequence to approximate the given square root by finding a_6. Compare the result with the actual value. (*Source:* Heinz-Otto, P., *Chaos and Fractals,* Springer-Verlag.)

(a) $\sqrt{2}$ **(b)** $\sqrt{11}$

11.2 Arithmetic Sequences and Series

- Arithmetic Sequences
- Arithmetic Series

Arithmetic Sequences

An **arithmetic sequence** (or **arithmetic progression**) is a sequence in which each term after the first differs from the preceding term by a fixed constant, called the **common difference.** The sequence

$$5, 9, 13, 17, 21, \ldots$$

is an arithmetic sequence because each term after the first is obtained by adding 4 to the previous term. That is,

$$9 = 5 + 4, \quad 13 = 9 + 4, \quad 17 = 13 + 4, \quad 21 = 17 + 4, \quad \text{and so on.}$$

The common difference is 4.

If the common difference of an arithmetic sequence is d, then

$$\boldsymbol{d = a_{n+1} - a_n}, \qquad \text{Common difference } d$$

for every positive integer n in the domain of the sequence.

EXAMPLE 1 Finding the Common Difference

Determine the common difference, d, for the arithmetic sequence

$$-9, -7, -5, -3, -1, \ldots .$$

SOLUTION We find d by choosing any two adjacent terms and subtracting the first from the second. Choosing -7 and -5 gives the following.

$$d = a_{n+1} - a_n \quad \text{Common difference } d$$

$$d = -5 - (-7) \quad \text{Let } a_{n+1} = -5 \text{ and } a_n = -7.$$

$$d = 2 \quad \text{Subtract.}$$

Choosing -9 and -7 would give $d = -7 - (-9) = 2$, the same result.

✔ **Now Try Exercise 11.**

EXAMPLE 2 Finding Terms Given a_1 and d

Find the first five terms of each arithmetic sequence.

(a) The first term is 7, and the common difference is -3.

(b) $a_1 = -12, d = 5$

SOLUTION

(a)

$a_1 = 7$ Start with $a_1 = 7$.

$a_2 = 7 + (-3) = 4$ Add $d = -3$.

$a_3 = 4 + (-3) = 1$ Add -3.

$a_4 = 1 + (-3) = -2$ Add -3.

$a_5 = -2 + (-3) = -5$ Add -3.

(b)

$a_1 = -12$ Start with $a_1 = -12$.

$a_2 = -12 + 5 = -7$ Add $d = 5$.

$a_3 = -7 + 5 = -2$ Add 5.

$a_4 = -2 + 5 = 3$ Add 5.

$a_5 = 3 + 5 = 8$ Add 5.

✔ **Now Try Exercises 17 and 19.**

If a_1 is the first term of an arithmetic sequence and d is the common difference, then the terms of the sequence are given as follows.

$$a_1 = a_1$$

$$a_2 = a_1 + d$$

$$a_3 = a_2 + d = a_1 + d + d = a_1 + 2d$$

$$a_4 = a_3 + d = a_1 + 2d + d = a_1 + 3d$$

$$a_5 = a_1 + 4d$$

$$a_6 = a_1 + 5d$$

By this pattern, $a_n = a_1 + (n - 1)d$.

*n*th Term of an Arithmetic Sequence

In an arithmetic sequence with first term a_1 and common difference d, the nth term, a_n, is given by the following.

$$a_n = a_1 + (n - 1)d$$

EXAMPLE 3 Finding Terms of an Arithmetic Sequence

Determine a_{13} and a_n for the arithmetic sequence $-3, 1, 5, 9, \ldots$.

SOLUTION Here $a_1 = -3$ and $d = 1 - (-3) = 4$. To find a_{13}, substitute 13 for n in the formula for the nth term.

$a_n = a_1 + (n-1)d$	Formula for a_n
$a_{13} = -3 + (13-1)4$	Let $a_1 = -3$, $n = 13$, and $d = 4$.
$a_{13} = -3 + (12)4$	Subtract.
$a_{13} = -3 + 48$	Multiply.
$a_{13} = 45$	Add.

Find a_n by substituting values for a_1 and d in the formula for a_n.

$a_n = a_1 + (n-1)d$	Formula for a_n
$a_n = -3 + (n-1)4$	Let $a_1 = -3$ and $d = 4$.
$a_n = -3 + 4n - 4$	Distributive property
$a_n = 4n - 7$	Simplify.

✔ **Now Try Exercise 23.**

EXAMPLE 4 Finding Terms of an Arithmetic Sequence

Determine a_n and a_{18} for the arithmetic sequence having $a_2 = 9$ and $a_3 = 15$.

SOLUTION Subtract the given consecutive terms to obtain $d = 15 - 9 = 6$. The first term, a_1, is found as follows.

$a_2 = a_1 + d$	Definition of arithmetic sequence
$9 = a_1 + 6$	Let $a_2 = 9$ and $d = 6$.
$a_1 = 3$	Subtract 6 and interchange sides.

Find a_n by substituting values for a_1 and d in the formula for a_n.

$a_n = a_1 + (n-1)d$	Formula for a_n
$a_n = 3 + (n-1)6$	Let $a_1 = 3$ and $d = 6$.
$a_n = 3 + 6n - 6$	Distributive property
$a_n = 6n - 3$	Simplify.

Now, find a_{18}.

$a_{18} = 6(18) - 3$	Let $n = 18$.
$a_{18} = 105$	Multiply, and then subtract.

✔ **Now Try Exercise 27.**

EXAMPLE 5 Finding the First Term of an Arithmetic Sequence

An arithmetic sequence has $a_8 = -16$ and $a_{16} = -40$. Determine a_1.

SOLUTION We obtain a_{16} by adding the common difference to a_8 eight times.

$a_{16} = a_8 + 8d$	Definition of arithmetic sequence
$-40 = -16 + 8d$	Let $a_{16} = -40$ and $a_8 = -16$.
$-24 = 8d$	Add 16.
$d = -3$	Divide by 8 and interchange sides.

To find a_1, use the formula for a_n.

$$a_n = a_1 + (n-1)d \quad \text{Formula for } a_n$$
$$-16 = a_1 + (8-1)(-3) \quad \text{Let } a_n = -16,\ n = 8, \text{ and } d = -3.$$
$$-16 = a_1 - 21 \quad \text{Simplify.}$$
$$a_1 = 5 \quad \text{Add 21 and interchange sides.}$$

✔ **Now Try Exercise 33.**

To determine the characteristics of the graph of an arithmetic sequence, start by rewriting the formula for the nth term.

$$a_n = a_1 + (n-1)d \quad \text{Formula for the } n\text{th term}$$
$$a_n = a_1 + nd - d \quad \text{Distributive property}$$
$$a_n = dn + (a_1 - d) \quad \text{Commutative and associative properties}$$
$$a_n = dn + c \quad \text{Let } c = a_1 - d.$$

The points in the graph of an arithmetic sequence are determined by

$$a_n = dn + c,$$

where n is a natural number. Thus, the *discrete* points on the graph of the sequence must lie on the *continuous* linear graph

$$y = dx + c.$$

Slope (d); y-value of the y-intercept (c)

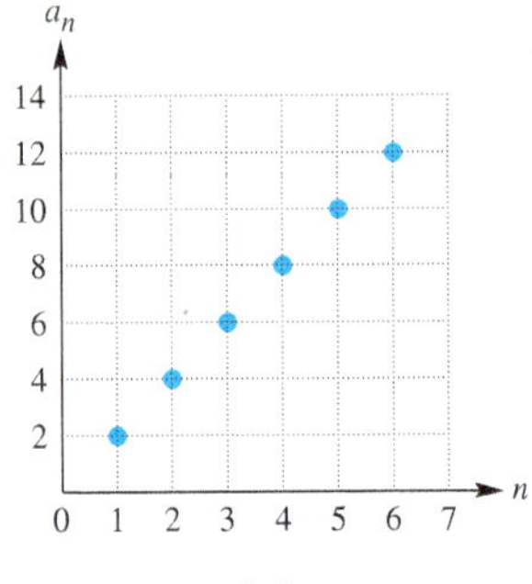

(a)

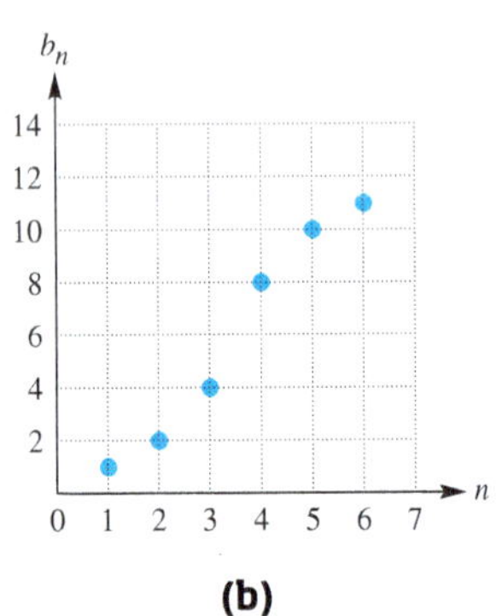

(b)

Figure 7

For example, the sequence a_n shown in **Figure 7(a)** is an arithmetic sequence because the points that make up its graph are collinear (lie on a line). The slope determined by these points is 2, so the common difference d equals 2. On the other hand, the sequence b_n shown in **Figure 7(b)** is not an arithmetic sequence because the points are not collinear.

EXAMPLE 6 Finding the *n*th Term from a Graph

Write a formula for the nth term of the sequence a_n shown in **Figure 8.** What are the domain and range of this sequence?

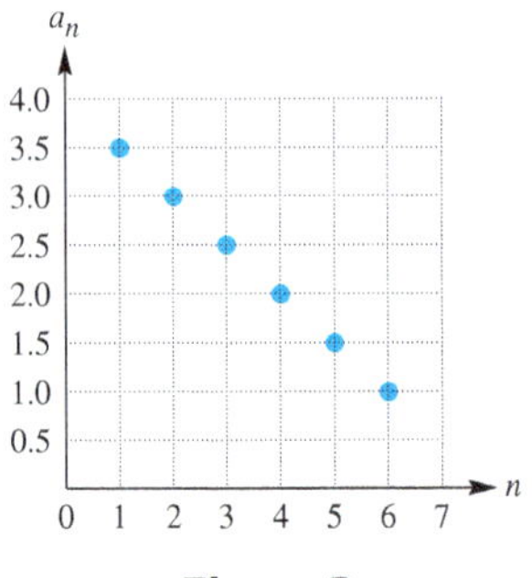

Figure 8

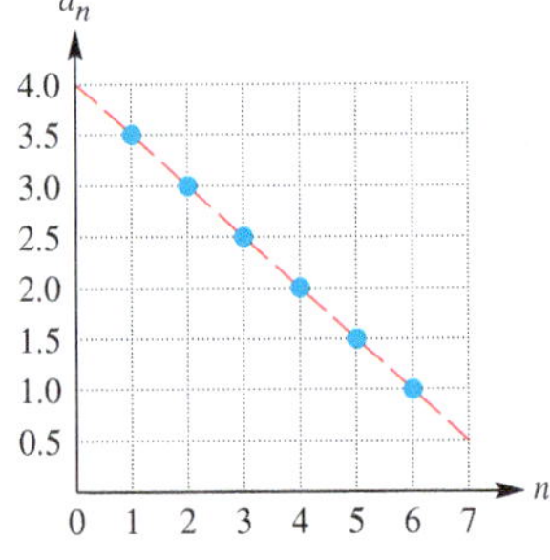

Figure 9

SOLUTION The points in **Figure 8** lie on a line, so the sequence is arithmetic. The dashed line in **Figure 9** has slope -0.5 and y-intercept $(0, 4)$, so its equation is $y = -0.5x + 4$. The nth term of this sequence is defined by

$$a_n = -0.5n + 4.$$

The sequence consists of the points

$$(1, 3.5),\ (2, 3),\ (3, 2.5),\ (4, 2),\ (5, 1.5),\ (6, 1).$$

Thus, the domain of the given sequence is $\{1, 2, 3, 4, 5, 6\}$, and the range is $\{3.5, 3, 2.5, 2, 1.5, 1\}$.

✔ **Now Try Exercise 39.**

Arithmetic Series An **arithmetic series** is the sum of the terms of an arithmetic sequence.

To illustrate, suppose that a person borrows \$3000 and agrees to pay \$100 per month plus interest of 1% per month on the unpaid balance until the loan is paid off. The first month, \$100 is paid to reduce the loan, plus interest of $(0.01)3000 = 30$ dollars. The second month, another \$100 is paid toward the loan, plus interest of $(0.01)2900 = 29$ dollars. The loan is reduced by \$100 each month. Interest payments decrease by $(0.01)100 = 1$ dollar each month, forming the arithmetic sequence

$$30, 29, 28, \ldots, 3, 2, 1.$$

The total amount of interest paid is given by the sum of the terms of this sequence. Now we develop a formula to find this sum without adding all 30 numbers directly. Because the sequence is arithmetic, we write the sum of the first n terms as follows.

$$S_n = a_1 + [a_1 + d] + [a_1 + 2d] + \cdots + [a_1 + (n-1)d]$$

Now we write the same sum in reverse order, beginning with a_n and *subtracting d.*

$$S_n = a_n + [a_n - d] + [a_n - 2d] + \cdots + [a_n - (n-1)d]$$

Adding the respective sides of these two equations term by term, we obtain the following.

$$S_n + S_n = (a_1 + a_n) + (a_1 + a_n) + \cdots + (a_1 + a_n)$$

$$2S_n = n(a_1 + a_n) \quad \text{There are } n \text{ terms of } a_1 + a_n \text{ on the right.}$$

$$S_n = \frac{n}{2}(a_1 + a_n) \quad \text{Solve for } S_n.$$

Using the formula $a_n = a_1 + (n-1)d$, we can also write this result as follows.

$$S_n = \frac{n}{2}[a_1 + a_1 + (n-1)d] \quad \text{Let } a_n = a_1 + (n-1)d.$$

or

$$S_n = \frac{n}{2}[2a_1 + (n-1)d] \quad \text{Alternative formula for the sum of the first } n \text{ terms}$$

Sum of the First *n* Terms of an Arithmetic Sequence

If an arithmetic sequence has first term a_1 and common difference d, then the sum S_n of the first n terms is given by the following.

$$S_n = \frac{n}{2}(a_1 + a_n), \quad \text{or} \quad S_n = \frac{n}{2}[2a_1 + (n-1)d]$$

The first formula is used when the first and last terms are known; otherwise, the second formula is used.

For example, in the sequence of interest payments discussed earlier, $n = 30$, $a_1 = 30$, and $a_n = 1$.

$$S_n = \frac{n}{2}(a_1 + a_n) \quad \text{First formula for } S_n$$

$$S_{30} = \frac{30}{2}(30 + 1) \quad \text{Let } n = 30,\ a_1 = 30, \text{ and } a_n = 1.$$

$$S_{30} = 15(31) \quad \text{Simplify.}$$

$$S_{30} = 465 \quad \text{Multiply.}$$

A total of \$465 interest will be paid over the 30 months.

EXAMPLE 7 Using the Sum Formulas

Consider the arithmetic sequence $-9, -5, -1, 3, 7, \ldots$.

(a) Evaluate S_{12}.

(b) Evaluate the sum of the first 60 positive integers.

SOLUTION

(a) We want the sum of the first 12 terms.

$$S_n = \frac{n}{2}[2a_1 + (n-1)d] \quad \text{Second formula for } S_n$$

$$S_{12} = \frac{12}{2}[2(-9) + (12-1)4] \quad \text{Let } n = 12,\ a_1 = -9, \text{ and } d = 4.$$

$$S_{12} = 156 \quad \text{Evaluate.}$$

(b) The first 60 positive integers form the arithmetic sequence $1, 2, 3, 4, \ldots, 60$.

$$S_n = \frac{n}{2}(a_1 + a_n) \quad \text{First formula for } S_n$$

$$S_{60} = \frac{60}{2}(1 + 60) \quad \text{Let } n = 60,\ a_1 = 1, \text{ and } a_{60} = 60.$$

$$S_{60} = 1830 \quad \text{Evaluate.}$$

✔ **Now Try Exercises 45 and 55.**

EXAMPLE 8 Using the Sum Formulas

The sum of the first 17 terms of an arithmetic sequence is 187. If $a_{17} = -13$, find a_1 and d.

SOLUTION

$$S_n = \frac{n}{2}(a_1 + a_n) \quad \text{Use the first formula for } S_n.$$

$$S_{17} = \frac{17}{2}(a_1 + a_{17}) \quad \text{Let } n = 17.$$

$$187 = \frac{17}{2}(a_1 - 13) \quad \text{Let } S_{17} = 187 \text{ and } a_{17} = -13.$$

$$22 = a_1 - 13 \quad \text{Multiply by } \tfrac{2}{17}.$$

$$a_1 = 35 \quad \text{Add 13 and interchange sides.}$$

Now find d.

$a_n = a_1 + (n-1)d$	Formula for the nth term
$a_{17} = a_1 + (17-1)d$	Let $n = 17$.
$-13 = 35 + 16d$	Let $a_{17} = -13$ and $a_1 = 35$; subtract.
$-48 = 16d$	Subtract 35.
$d = -3$	Divide by 16 and interchange sides.

✔ **Now Try Exercise 61.**

Any sum of the form

$$\sum_{i=1}^{n} (di + c), \quad \text{where } d \text{ and } c \text{ are real numbers,}$$

represents the sum of the terms of an arithmetic sequence having first term

$$a_1 = d(1) + c = d + c$$

and common difference d. These sums can be evaluated using the formulas in this section.

EXAMPLE 9 Using Summation Notation

Evaluate each sum.

(a) $\sum_{i=1}^{10} (4i + 8)$ **(b)** $\sum_{k=3}^{9} (4 - 3k)$

SOLUTION

(a) This sum contains the first 10 terms of the arithmetic sequence having

$$a_1 = 4 \cdot 1 + 8 = 12, \quad \text{First term}$$

and

$$a_{10} = 4 \cdot 10 + 8 = 48. \quad \text{Last term}$$

Using the formula $S_n = \frac{n}{2}(a_1 + a_n)$, we obtain the following sum.

$$\sum_{i=1}^{10} (4i + 8) = S_{10} = \frac{10}{2}(12 + 48) = 5(60) = 300$$

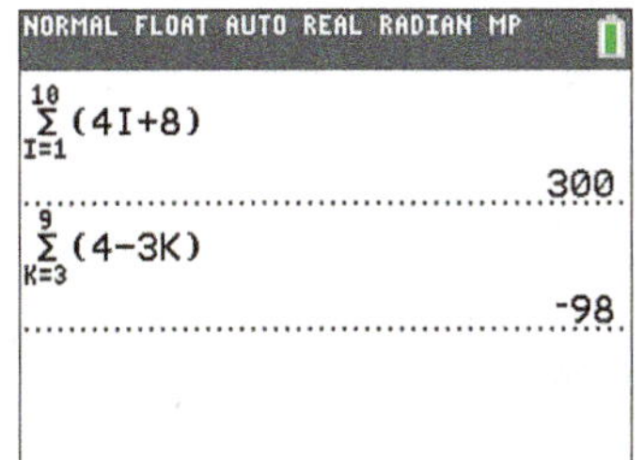

The TI-84 Plus will give the sum of a sequence without having to first store the sequence. The screen here illustrates this method for the sequences in **Example 9.**

(b) The first few terms are

$$[4 - 3(3)] + [4 - 3(4)] + [4 - 3(5)] + \cdots$$
$$= -5 + (-8) + (-11) + \cdots.$$

Thus, $a_1 = -5$ and $d = -3$. If the sequence started with $k = 1$, there would be nine terms. Because it starts at 3, two of those terms are missing, so there are seven terms and $n = 7$. Use the formula $S_n = \frac{n}{2}[2a_1 + (n-1)d]$.

$$\sum_{k=3}^{9} (4 - 3k) = \frac{7}{2}[2(-5) + (7-1)(-3)] = -98$$

✔ **Now Try Exercises 69 and 71.**

11.2 Exercises

CONCEPT PREVIEW *Fill in the blank to correctly complete each sentence.*

1. In an arithmetic sequence, each term after the first differs from the preceding term by a fixed constant called the common ________.

2. The common difference for the sequence $-25, -21, -17, -13, \ldots$ is ______.

3. For the arithmetic sequence having $a_1 = 10$ and $d = -2$, the term $a_3 =$ ______.

4. For the arithmetic sequence with nth term $a_n = 8n + 5$, the term $a_5 =$ ______.

CONCEPT PREVIEW *The figure shows the graph of a finite arithmetic sequence.*

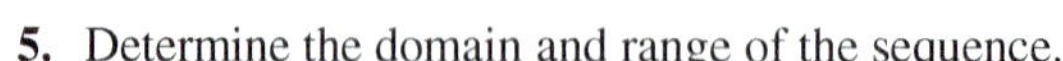

5. Determine the domain and range of the sequence.

6. What is the first term?

7. What is the common difference?

8. Write a formula for the nth term of the sequence.

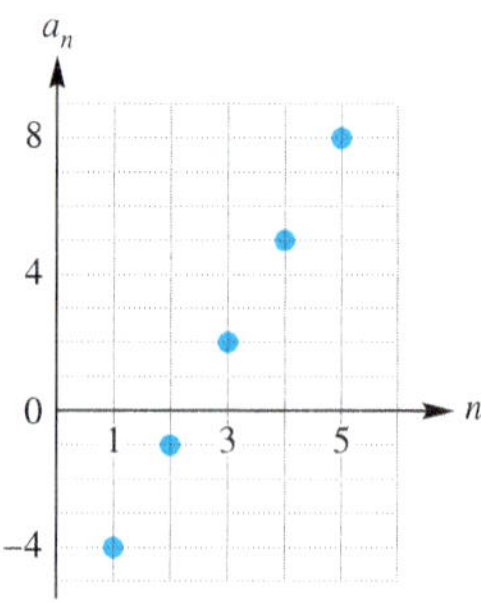

CONCEPT PREVIEW *Evaluate each sum.*

9. $\sum_{i=1}^{5} (4i - 1)$

10. S_{10} for the sequence $5, 10, 15, 20, \ldots$

Determine the common difference d for each arithmetic sequence **See Example 1.**

11. $2, 5, 8, 11, \ldots$

12. $4, 10, 16, 22, \ldots$

13. $3, -2, -7, -12, \ldots$

14. $-8, -12, -16, -20, \ldots$

15. $x + 3y, 2x + 5y, 3x + 7y, \ldots$

16. $t^2 + q, -4t^2 + 2q, -9t^2 + 3q, \ldots$

Find the first five terms of each arithmetic sequence. **See Example 2.**

17. The first term is 8, and the common difference is 6.

18. The first term is -2, and the common difference is 12.

19. $a_1 = 5, \ d = -2$

20. $a_1 = 4, \ d = 3$

21. $a_1 = 10 + \sqrt{7}, \ a_2 = 10$

22. $a_1 = 3 - \sqrt{2}, \ a_2 = 3$

Determine a_n and a_8 for each arithmetic sequence. **See Examples 3 and 4.**

23. $5, 7, 9, \ldots$

24. $-3, -7, -11, \ldots$

25. $a_1 = 5, \ a_4 = 15$

26. $a_1 = -4, \ a_5 = 16$

27. $a_{10} = 6, \ a_{11} = 10.5$

28. $a_{15} = 8, \ a_{16} = 5$

29. $a_1 = x, \ a_2 = x + 3$

30. $a_2 = y + 1, \ d = -3$

31. $a_4 = s + 6p, \ d = 2p$

32. *Concept Check* If a_1, a_2, a_3 represents an arithmetic sequence, express a_2 in terms of a_1 and a_3.

Determine a_1 for each arithmetic sequence. ***See Example 5.***

33. $a_5 = 27,\ a_{15} = 87$

34. $a_{12} = 60,\ a_{20} = 84$

35. $a_8 = -15,\ a_{18} = -85$

36. $a_6 = -72,\ a_{13} = 26$

37. $a_7 = 23.5,\ a_{20} = 69$

38. $a_{10} = -49.5,\ a_{14} = -73.5$

Write a formula for the nth term of the finite arithmetic sequence a_n shown in each graph. Then state the domain and range of the sequence. ***See Example 6.***

39.

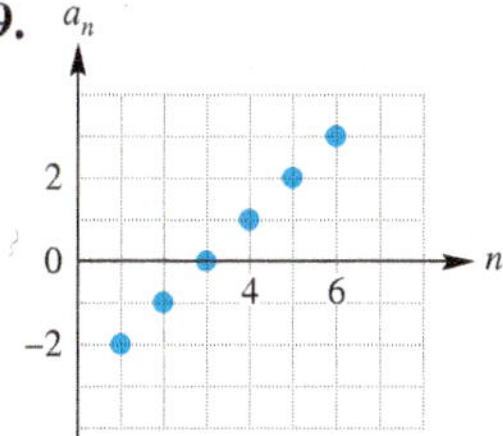

40.

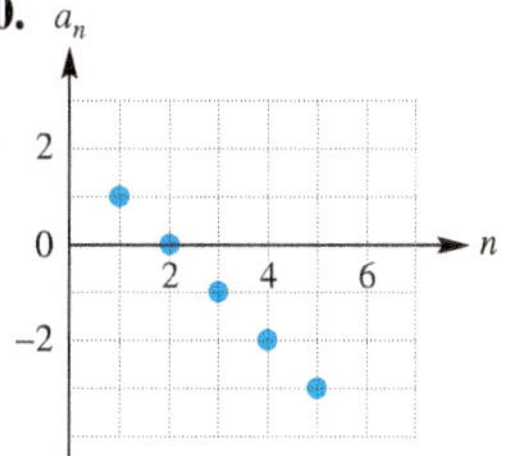

41.

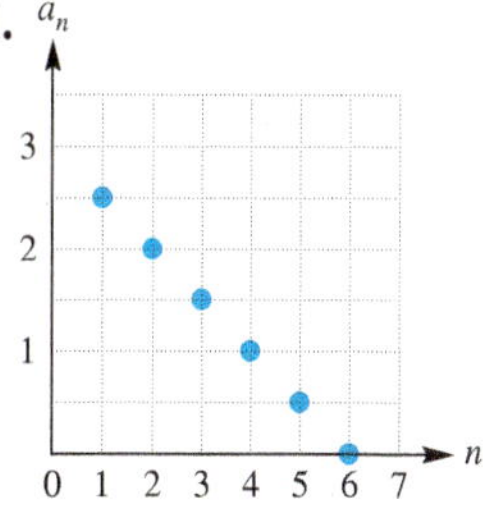

42.

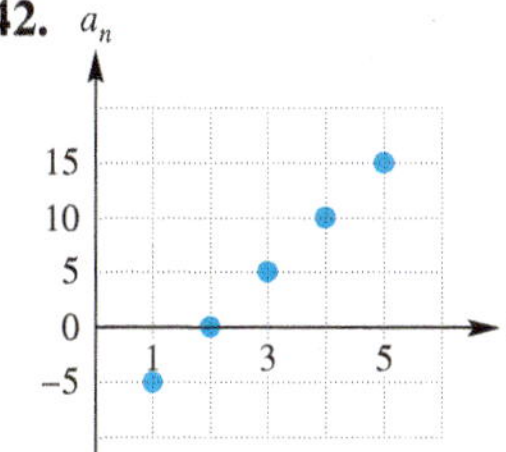

43.

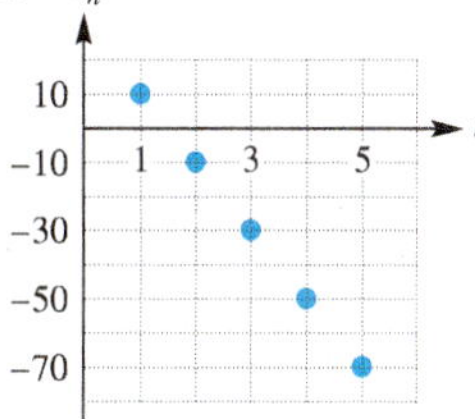

44.

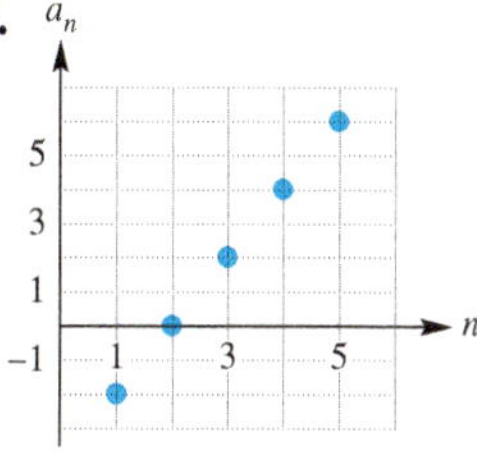

Evaluate S_{10}, the sum of the first ten terms, for each arithmetic sequence. ***See Example 7(a).***

45. 8, 11, 14, . . .

46. $-9, -5, -1, \ldots$

47. 5, 9, 13, . . .

48. 8, 6, 4, . . .

49. $a_2 = 9,\ a_4 = 13$

50. $a_3 = 5,\ a_4 = 8$

51. $a_1 = 10,\ a_{10} = 5.5$

52. $a_1 = -8,\ a_{10} = -1.25$

53. $a_1 = \pi,\ a_{10} = 10\pi$

54. ***Concept Check*** Is this statement accurate?

To find the sum of the first n positive integers, find half the product of n and $n + 1$.

Evaluate each sum as described. ***See Example 7(b).***

55. the sum of the first 80 positive integers

56. the sum of the first 120 positive integers

57. the sum of the first 50 positive odd integers

58. the sum of the first 90 positive odd integers

59. the sum of the first 60 positive even integers

60. the sum of the first 70 positive even integers

Find a_1 and d for each arithmetic series. ***See Example 8.***

61. $S_{20} = 1090,\ a_{20} = 102$

62. $S_{31} = 5580,\ a_{31} = 360$

63. $S_{16} = -160,\ a_{16} = -25$

64. $S_{25} = 650,\ a_{25} = 62$

65. $S_{12} = -108,\ a_{12} = -19$

66. $S_{31} = 620,\ a_{31} = 30$

Evaluate each sum. ***See Example 9.***

67. $\sum_{i=1}^{3}(i+4)$

68. $\sum_{i=1}^{5}(i-8)$

69. $\sum_{j=1}^{10}(2j+3)$

70. $\sum_{j=1}^{15}(5j-9)$

71. $\sum_{i=4}^{12}(-5-8i)$

72. $\sum_{k=5}^{19}(-3-4k)$

73. $\sum_{i=1}^{1000} i$

74. $\sum_{i=1}^{1000} -i$

75. $\sum_{k=1}^{100} 2k$

76. $\sum_{k=1}^{100}(2k-1)$

77. $\sum_{j=10}^{50} 5j$

78. $\sum_{j=20}^{80} 3j$

Use the summation feature of a graphing calculator to evaluate the sum of the first ten terms of each arithmetic series with a_n defined as shown. In Exercises 81 and 82, round to the nearest thousandth.

79. $a_n = 4.2n + 9.73$

80. $a_n = 8.42n + 36.18$

81. $a_n = \sqrt{8}n + \sqrt{3}$

82. $a_n = -\sqrt[3]{4}n + \sqrt{7}$

Solve each problem.

83. ***Integer Sum*** Find the sum of all the integers from 51 to 71.

84. ***Integer Sum*** Find the sum of all the integers from -8 to 30.

85. ***Clock Chimes*** If a clock strikes the proper number of chimes each hour on the hour, how many times will it chime in a month of 30 days?

86. ***Telephone Pole Stack*** A stack of telephone poles has 30 in the bottom row, 29 in the next, and so on, with one pole in the top row. How many poles are in the stack?

87. ***Population Growth*** Five years ago, the population of a city was 49,000. Each year the zoning commission permits an increase of 580 in the population. What will the maximum population be 5 yr from now?

88. ***Slide Supports*** A super slide of uniform slope is to be built on a level piece of land. There are to be 20 equally spaced vertical supports, with the longest support 15 m long and the shortest 2 m long. Find the total length of all the supports.

89. ***Rungs of a Ladder*** How much material will be needed for the rungs of a ladder of 31 rungs, if the rungs taper uniformly from 18 in. to 28 in.?

90. ***(Modeling) Children's Growth Pattern*** The normal growth pattern for children aged 3–11 follows that of an arithmetic sequence. An increase in height of about 6 cm per year is expected. Thus, 6 would be the common difference of the sequence.

For example, a child who measures 96 cm at age 3 would have his expected height in subsequent years represented by the sequence 102, 108, 114, 120, 126, 132, 138, 144. Each term differs from the adjacent terms by the common difference, 6.

(a) If a child measures 98.2 cm at age 3 and 109.8 cm at age 5, what would be the common difference of the arithmetic sequence describing his yearly height?

(b) What would we expect his height to be at age 8?

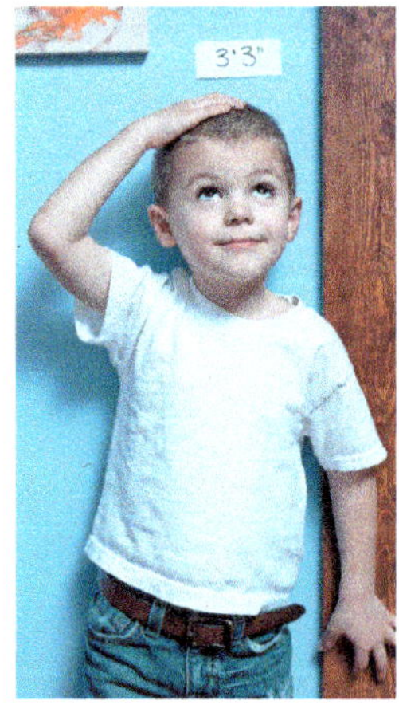

91. ***Concept Check*** Suppose that $a_1, a_2, a_3, a_4, a_5, \ldots$ is an arithmetic sequence. Is $a_1, a_3, a_5, \ldots$ an arithmetic sequence?

92. ***Concept Check*** Is the sequence log 2, log 4, log 8, log 16, . . . an arithmetic sequence?

11.3 Geometric Sequences and Series

Geometric Sequences

Suppose an employee agrees to work for \$0.01 the first day, \$0.02 the second day, \$0.04 the third day, \$0.08 the fourth day, and so on, with wages doubling each day. *How much is earned on day 20? How much is earned altogether in 20 days?* These questions will be answered in this section.

A **geometric sequence** (or **geometric progression**) is a sequence in which each term after the first is obtained by multiplying the preceding term by a fixed nonzero real number, called the **common ratio.** The sequence discussed above,

$$1, 2, 4, 8, 16, \ldots, \quad \text{In cents}$$

is a geometric sequence in which the first term is 1 and the common ratio is 2.

Notice that if we divide any term after the first term by the preceding term, we obtain the common ratio $r = 2$.

$$\frac{a_2}{a_1} = \frac{2}{1} = 2; \quad \frac{a_3}{a_2} = \frac{4}{2} = 2; \quad \frac{a_4}{a_3} = \frac{8}{4} = 2; \quad \frac{a_5}{a_4} = \frac{16}{8} = 2$$

If the common ratio of a geometric sequence is r, then

$$r = \frac{a_{n+1}}{a_n}, \quad \text{Common ratio } r$$

for every positive integer n. ***Therefore, we find the common ratio by choosing any term after the first and dividing it by the preceding term.***

In the geometric sequence 2, 8, 32, 128, . . . , $r = 4$. Notice that

$$8 = 2 \cdot 4$$
$$32 = 8 \cdot 4 = (2 \cdot 4) \cdot 4 = 2 \cdot 4^2$$
$$128 = 32 \cdot 4 = (2 \cdot 4^2) \cdot 4 = 2 \cdot 4^3.$$

To generalize this, assume that a geometric sequence has first term a_1 and common ratio r. The second term is $a_2 = a_1 r$, the third is $a_3 = a_2 r = (a_1 r)r = a_1 r^2$, and so on. Following this pattern, the nth term is $a_n = a_1 r^{n-1}$.

*n*th Term of a Geometric Sequence

In a geometric sequence with first term a_1 and common ratio r, the nth term, a_n, is given by the following.

$$a_n = a_1 r^{n-1}$$

EXAMPLE 1 Finding the *n*th Term of a Geometric Sequence

Use the formula for the nth term of a geometric sequence to answer the first question posed at the beginning of this section: *How much will be earned on day 20 if daily wages follow the sequence* 1, 2, 4, 8, 16, . . . *cents?*

SOLUTION

$$a_n = a_1 r^{n-1} \quad \text{Formula for } a_n$$

$$a_{20} = 1(2)^{20-1} \quad \text{Let } n = 20,\ a_1 = 1, \text{ and } r = 2.$$

$$a_{20} = 524{,}288 \text{ cents, or } \$5242.88 \quad \text{Evaluate.}$$

✔ **Now Try Exercise 11(a).**

EXAMPLE 2 Finding Terms of a Geometric Sequence

Determine a_5 and a_n for the geometric sequence 4, 12, 36, 108, ...

SOLUTION The first term, a_1, is 4. Find r by choosing any term after the first and dividing it by the preceding term. For example, $r = \frac{36}{12} = 3$.

$$a_5 = a_4 \cdot r \quad \text{Definition of geometric sequence}$$

$$a_5 = 108 \cdot 3 \quad \text{Let } a_4 = 108 \text{ and } r = 3.$$

$$a_5 = 324 \quad \text{Multiply.}$$

The nth term is found as follows.

$$a_n = a_1 r^{n-1} \quad \text{Formula for } a_n$$

$$a_n = 4(3)^{n-1} \quad \text{Let } a_1 = 4 \text{ and } r = 3.$$

We can also find the fifth term by replacing n with 5 in this formula.

$$a_5 = 4(3)^{5-1} = 4(3)^4 = 324$$

✔ **Now Try Exercise 21.**

EXAMPLE 3 Finding Terms of a Geometric Sequence

Determine r and a_1 for the geometric sequence with $a_3 = 20$ and $a_6 = 160$.

SOLUTION We obtain a_6 by multiplying a_3 by the common ratio three times.

$$a_6 = a_3 r^3 \quad \text{Definition of geometric sequence}$$

$$160 = 20r^3 \quad \text{Let } a_6 = 160 \text{ and } a_3 = 20.$$

$$8 = r^3 \quad \text{Divide by 20.}$$

$$r = 2 \quad \text{Take cube roots and interchange sides.}$$

Now use this value of r and the fact that $a_3 = 20$ to find the first term, a_1.

$$a_n = a_1 r^{n-1} \quad \text{Formula for } a_n$$

$$20 = a_1(2)^{3-1} \quad \text{Let } a_n = 20,\ r = 2, \text{ and } n = 3.$$

$$20 = a_1(4) \quad \text{Apply the exponent.}$$

$$a_1 = 5 \quad \text{Divide by 4 and interchange sides.}$$

✔ **Now Try Exercise 27.**

EXAMPLE 4 Modeling a Population of Fruit Flies

A population of fruit flies is growing in such a way that each generation is 1.5 times as large as the last generation. Suppose there are 100 insects in the first generation. How many would there be in the fourth generation? Round to the nearest whole number.

SOLUTION Consider the list of populations as a geometric sequence with a_1 as the first-generation population, a_2 the second-generation population, and so on. Then the fourth-generation population is a_4.

$$a_n = a_1 r^{n-1} \quad \text{Formula for } a_n$$

$$a_4 = 100(1.5)^{4-1} \quad \text{Let } a_1 = 100,\ r = 1.5, \text{ and } n = 4.$$

$$a_4 = 100(3.375) \quad \text{Apply the exponent.}$$

$$a_4 = 338 \quad \text{Multiply. Round to the nearest whole number.}$$

In the fourth generation, the population will number 338 insects.

✔ **Now Try Exercise 73.**

Geometric Series A **geometric series** is the sum of the terms of a geometric sequence. For example, a scientist might want to know the total number of insects in four generations of the population discussed in **Example 4.** This population would equal $a_1 + a_2 + a_3 + a_4$.

$$100 + 100(1.5) + 100(1.5)^2 + 100(1.5)^3 = 812.5 \approx 813 \text{ insects}$$

To find a formula for the sum of the first n terms of a geometric sequence, S_n, first write the sum as

$$S_n = a_1 + a_2 + a_3 + \cdots + a_n$$

or

$$S_n = a_1 + a_1r + a_1r^2 + \cdots + a_1r^{n-1}. \quad (1)$$

If $r = 1$, then $S_n = na_1$, which is a correct formula for this case. If $r \neq 1$, then multiply both sides of equation (1) by r to obtain

$$rS_n = a_1r + a_1r^2 + a_1r^3 + \cdots + a_1r^n. \quad (2)$$

Now subtract equation (2) from equation (1), and solve for S_n.

$$S_n = a_1 + a_1r + a_1r^2 + \cdots + a_1r^{n-1} \quad (1)$$

$$rS_n = \quad a_1r + a_1r^2 + \cdots + a_1r^{n-1} + a_1r^n \quad (2)$$

$$S_n - rS_n = a_1 \quad - a_1r^n \quad \text{Subtract.}$$

$$S_n(1 - r) = a_1(1 - r^n) \quad \text{Factor.}$$

$$S_n = \frac{a_1(1 - r^n)}{1 - r}, \quad \text{where } r \neq 1 \quad \text{Divide by } 1 - r.$$

Sum of the First *n* Terms of a Geometric Sequence

If a geometric sequence has first term a_1 and common ratio r, then the sum S_n of the first n terms is given by the following.

$$S_n = \frac{a_1(1 - r^n)}{1 - r}, \quad \text{where } r \neq 1$$

EXAMPLE 5 Finding the Sum of the First *n* Terms

At the beginning of this section, we found that an employee agreed to work for the following salary: \$0.01 the first day, \$0.02 the second day, \$0.04 the third day, \$0.08 the fourth day, and so on, with wages doubling each day. *How much is earned altogether in 20 days?*

SOLUTION We must find the total amount earned in 20 days with daily wages of 1, 2, 4, 8, . . . cents.

$$S_n = \frac{a_1(1 - r^n)}{1 - r} \quad \text{Formula for } S_n$$

$$S_{20} = \frac{1(1 - 2^{20})}{1 - 2} \quad \text{Let } n = 20,\ a_1 = 1, \text{ and } r = 2.$$

$$S_{20} = \frac{1 - 1{,}048{,}576}{-1} \quad \text{Evaluate } 2^{20}. \text{ Subtract in the denominator.}$$

$$S_{20} = 1{,}048{,}575 \text{ cents, or } \$10{,}485.75 \quad \text{Evaluate.}$$

✔ **Now Try Exercise 11(b).**

EXAMPLE 6 Finding the Sum of the First *n* Terms

Evaluate $\sum_{i=1}^{6} 2 \cdot 3^i$.

SOLUTION This series is the sum of the first six terms of a geometric sequence having $a_1 = 2 \cdot 3^1 = 6$ and $r = 3$.

$$S_n = \frac{a_1(1 - r^n)}{1 - r} \quad \text{Formula for } S_n$$

$$S_6 = \frac{6(1 - 3^6)}{1 - 3} \quad \text{Let } n = 6,\ a_1 = 6, \text{ and } r = 3.$$

$$S_6 = 2184 \quad \text{Evaluate.}$$

✔ **Now Try Exercise 41.**

Infinite Geometric Series

We extend our discussion of sums of sequences to include infinite geometric sequences such as

$$2,\ 1,\ \frac{1}{2},\ \frac{1}{4},\ \frac{1}{8},\ \frac{1}{16},\ \ldots, \quad \text{with first term 2 and common ratio } \frac{1}{2}.$$

Evaluating S_n gives the following sequence of sums.

$$S_1 = 2$$

$$S_2 = 2 + 1 = 3$$

$$S_3 = 2 + 1 + \frac{1}{2} = \frac{7}{2} = 3.5$$

$$S_4 = 2 + 1 + \frac{1}{2} + \frac{1}{4} = \frac{15}{4} = 3.75$$

$$S_5 = 2 + 1 + \frac{1}{2} + \frac{1}{4} + \frac{1}{8} = \frac{31}{8} = 3.875$$

$$S_6 = \frac{63}{16} = 3.9375, \quad \text{and so on}$$

The sums are getting closer and closer to the number 4.

LOOKING AHEAD TO CALCULUS

In the discussion of

$$\lim_{n \to \infty} S_n = 4,$$

we used the phrases "large enough" and "as close as desired." This description is made more precise in a standard calculus course.

For no value of n is $S_n = 4$. However, if n is large enough, then S_n is as close to 4 as desired. We say the sequence converges to 4. This is expressed as

$$\lim_{n \to \infty} S_n = 4.$$

Read this as:

"The limit of S_n as n increases without bound is 4."

Because $\lim_{n \to \infty} S_n = 4$, the number 4 is the *sum of the terms* of the infinite geometric sequence

$$2,\ 1,\ \frac{1}{2},\ \frac{1}{4},\ \ldots,$$

and

$$2 + 1 + \frac{1}{2} + \frac{1}{4} + \cdots = 4.$$

The calculator graph in **Figure 10** supports this.

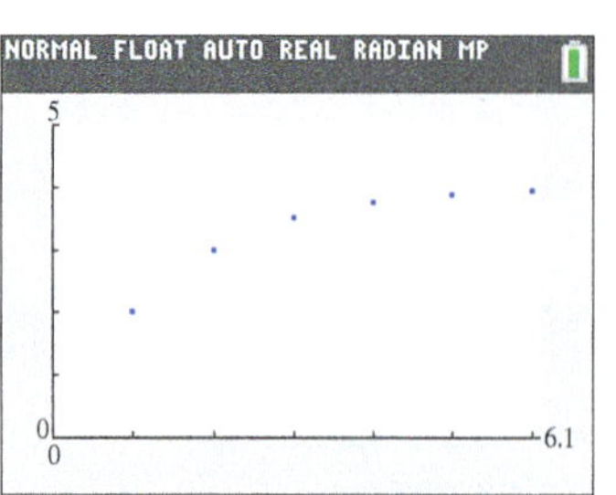

As n gets larger, S_n approaches 4.

Figure 10

EXAMPLE 7 Evaluating an Infinite Geometric Series

Evaluate $1 + \frac{1}{3} + \frac{1}{9} + \frac{1}{27} + \cdots$.

SOLUTION $S_1 = 1, \quad S_2 = \frac{4}{3}, \quad S_3 = \frac{13}{9}, \quad S_4 = \frac{40}{27}$

Using the formula for the sum of the first n terms of a geometric sequence, we obtain the following, in general.

$$S_n = \frac{a_1(1 - r^n)}{1 - r} \qquad \text{Formula for } S_n$$

$$S_n = \frac{1\left[1 - \left(\frac{1}{3}\right)^n\right]}{1 - \frac{1}{3}} \qquad \text{Let } a_1 = 1 \text{ and } r = \tfrac{1}{3}.$$

The table shows the value of $\left(\frac{1}{3}\right)^n$ for larger and larger values of n.

n	1	10	100	200
$\left(\frac{1}{3}\right)^n$	$\frac{1}{3}$	1.69×10^{-5}	1.94×10^{-48}	3.76×10^{-96}

NORMAL FLOAT AUTO REAL RADIAN MP

This graph of the first six values of S_n in **Example 7** suggests that its value is approaching $\frac{3}{2}$.

As n increases without bound, $\left(\frac{1}{3}\right)^n$ approaches 0. That is,

$$\lim_{n \to \infty} \left(\frac{1}{3}\right)^n = 0,$$

making it reasonable that

$$\lim_{n \to \infty} S_n = \lim_{n \to \infty} \frac{1\left[1 - \left(\frac{1}{3}\right)^n\right]}{1 - \frac{1}{3}} = \frac{1(1 - 0)}{1 - \frac{1}{3}} = \frac{1}{\frac{2}{3}} = \frac{3}{2}. \qquad \text{Simplify the complex fraction.}$$

Hence, $$1 + \frac{1}{3} + \frac{1}{9} + \frac{1}{27} + \cdots = \frac{3}{2}.$$

✔ **Now Try Exercise 51.**

LOOKING AHEAD TO CALCULUS

In calculus, functions are sometimes defined in terms of infinite series. Here are three functions we studied earlier in the text defined that way.

$$e^x = \frac{x^0}{0!} + \frac{x^1}{1!} + \frac{x^2}{2!} + \frac{x^3}{3!} + \cdots$$

$$\ln(1 + x) = x - \frac{x^2}{2} + \frac{x^3}{3} - \cdots$$

for x in $(-1, 1)$

$$\frac{1}{1 + x} = 1 - x + x^2 - x^3 + \cdots$$

for x in $(-1, 1)$

If a geometric series has first term a_1 and common ratio r, then

$$S_n = \frac{a_1(1 - r^n)}{1 - r} \quad (\text{where } r \neq 1),$$

for every positive integer n. If $|r| < 1$, then $\lim_{n \to \infty} r^n = 0$, and

$$\lim_{n \to \infty} S_n = \frac{a_1(1 - 0)}{1 - r} = \frac{a_1}{1 - r}.$$

The resulting quotient, $\frac{a_1}{1 - r}$, is the **sum of the terms of an infinite geometric sequence.**

The limit $\lim_{n \to \infty} S_n$ can be expressed as S_∞ or $\sum_{i=1}^{\infty} a_i$.

Sum of the Terms of an Infinite Geometric Sequence

The sum S_∞ of the terms of an infinite geometric sequence with first term a_1 and common ratio r, where $|r| < 1$, is given by the following.

$$S_\infty = \frac{a_1}{1 - r}$$

If $|r| > 1$, then the terms increase without bound in absolute value, so there is no limit as $n \to \infty$. ***Therefore, if*** $|r| > 1$***, then the terms of the sequence will not have a sum.***

EXAMPLE 8 Evaluating Infinite Geometric Series

Evaluate each sum.

(a) $\sum_{i=1}^{\infty}\left(-\frac{3}{4}\right)\left(-\frac{1}{2}\right)^{i-1}$ **(b)** $\sum_{i=1}^{\infty}\left(\frac{3}{5}\right)^{i}$

SOLUTION

(a) Here, $a_1 = -\frac{3}{4}$ and $r = -\frac{1}{2}$. Because $|r| < 1$, the sum converges.

$$S_\infty = \frac{a_1}{1 - r} = \frac{-\frac{3}{4}}{1 - \left(-\frac{1}{2}\right)} = \frac{-\frac{3}{4}}{\frac{3}{2}} = -\frac{3}{4} \div \frac{3}{2} = -\frac{3}{4} \cdot \frac{2}{3} = -\frac{1}{2}$$

(b) $$\sum_{i=1}^{\infty}\left(\frac{3}{5}\right)^{i} = \frac{\frac{3}{5}}{1 - \frac{3}{5}} = \frac{\frac{3}{5}}{\frac{2}{5}} = \frac{3}{5} \div \frac{2}{5} = \frac{3}{5} \cdot \frac{5}{2} = \frac{3}{2} \quad a_1 = \frac{3}{5},\ r = \frac{3}{5}$$

✔ **Now Try Exercises 61 and 63.**

Annuities A sequence of equal payments made after equal periods of time, such as car payments or house payments, is an **annuity.** If the payments are accumulated in an account (with no withdrawals), the sum of the payments and interest on the payments is the **future value** of the annuity.

EXAMPLE 9 Finding the Future Value of an Annuity

To save money for a trip, Jacqui deposited \$1000 at the *end* of each year for 4 yr in an account paying 2% interest, compounded annually. Find the future value of this annuity.

SOLUTION We use the formula for interest compounded annually.

$$A = P(1 + r)^t$$

The first payment earns interest for 3 yr, the second payment for 2 yr, and the third payment for 1 yr. The last payment earns no interest.

$$1000(1.02)^3 + 1000(1.02)^2 + 1000(1.02) + 1000 \quad \text{Total amount}$$

This is the sum of the terms of a geometric sequence with first term (starting at the end of the sum as written above) $a_1 = 1000$ and common ratio $r = 1.02$.

$$S_n = \frac{a_1(1 - r^n)}{1 - r} \quad \text{Formula for } S_n$$

$$S_4 = \frac{1000[1 - (1.02)^4]}{1 - 1.02} \quad \text{Substitute.}$$

$$S_4 \approx 4121.61 \quad \text{Evaluate.}$$

The future value of the annuity is $4121.61.

✔ Now Try Exercise 81.

Future Value of an Annuity

The formula for the future value of an annuity is given by the following.

$$S = R\left[\frac{(1 + i)^n - 1}{i}\right]$$

Here S is future value, R is payment at the end of each period, i is interest rate per period, and n is number of periods.

11.3 Exercises

CONCEPT PREVIEW *Fill in the blank to correctly complete each sentence.*

1. In a geometric sequence, each term after the first is obtained by multiplying the preceding term by a fixed nonzero real number called the common ________.
2. The common ratio for the sequence $-25, 5, -1, \frac{1}{5}, \ldots$ is ______.
3. For the geometric sequence having $a_1 = 6$ and $r = 2$, the term $a_3 =$ ______.
4. For the geometric sequence with nth term $a_n = 4\left(\frac{1}{2}\right)^{n-1}$, the term $a_5 =$ ______.
5. The sum of the first five terms of the geometric sequence 5, 10, 20, 40, . . . is ______.
6. When evaluated, $\sum_{i=1}^{5} 81\left(\frac{1}{3}\right)^{i-1}$ is ______.

CONCEPT PREVIEW *Determine whether each sequence is* arithmetic, geometric, *or* neither. *If it is arithmetic, give the common difference, d. If it is geometric, give the common ratio, r.*

7. $4, -8, 16, -32, \ldots$
8. $\frac{1}{3}, \frac{2}{3}, \frac{3}{3}, \frac{4}{3}, \ldots$
9. $5, 10, 20, 35, \ldots$
10. $8, 2, \frac{1}{2}, \frac{1}{8}, \ldots$

Recall from the beginning of this section that an employee agreed to work for the following salary: $0.01 *the first day,* $0.02 *the second day,* $0.04 *the third day,* $0.08 *the fourth day, and so on, with wages doubling each day. Determine* **(a)** *the amount earned on the day indicated and* **(b)** *the total amount earned altogether after wages are paid on the day indicated.* ***See Examples 1 and 5.***

11. day 10
12. day 12
13. day 15
14. day 18

Determine a_5 and a_n for each geometric sequence. ***See Example 2.***

15. $a_1 = 5,\ r = -2$

16. $a_1 = 8,\ r = -5$

17. $a_2 = -4,\ r = 3$

18. $a_3 = -2,\ r = 4$

19. $a_4 = 243,\ r = -3$

20. $a_4 = 18,\ r = 2$

21. $-4, -12, -36, -108, \ldots$

22. $-2, 6, -18, 54, \ldots$

23. $\frac{4}{5}, 2, 5, \frac{25}{2}, \ldots$

24. $\frac{1}{2}, \frac{2}{3}, \frac{8}{9}, \frac{32}{27}, \ldots$

25. $10, -5, \frac{5}{2}, -\frac{5}{4}, \ldots$

26. $3, -\frac{9}{4}, \frac{27}{16}, -\frac{81}{64}, \ldots$

Determine r and a_1 for each geometric sequence. ***See Example 3.***

27. $a_2 = -6,\ a_7 = -192$

28. $a_2 = -8,\ a_7 = 256$

29. $a_3 = 5,\ a_8 = \frac{1}{625}$

30. $a_4 = -\frac{1}{4},\ a_9 = -\frac{1}{128}$

31. $a_3 = 50,\ a_7 = 0.005$

32. $a_3 = 300,\ a_9 = \frac{100}{243}$

Use the formula for S_n to find the sum of the first five terms of each geometric sequence. In Exercises 37 and 38, round to the nearest hundredth. ***See Example 5.***

33. $2, 8, 32, 128, \ldots$

34. $4, 16, 64, 256, \ldots$

35. $18, -9, \frac{9}{2}, -\frac{9}{4}, \ldots$

36. $12, -4, \frac{4}{3}, -\frac{4}{9}, \ldots$

37. $a_1 = 8.423,\ r = 2.859$

38. $a_1 = -3.772,\ r = -1.553$

Evaluate each sum. ***See Example 6.***

39. $\sum_{i=1}^{5} (-3)^i$

40. $\sum_{i=1}^{4} (-2)^i$

41. $\sum_{j=1}^{6} 48\left(\frac{1}{2}\right)^j$

42. $\sum_{j=1}^{5} 243\left(\frac{2}{3}\right)^j$

43. $\sum_{k=4}^{10} 2^k$

44. $\sum_{k=3}^{9} 3^k$

45. *Concept Check* Under what conditions does the sum of an infinite geometric series exist?

46. The number 0.999 . . . can be written as the sum of the terms of an infinite geometric sequence: $0.9 + 0.09 + 0.009 + \cdots$. Here we have $a_1 = 0.9$ and $r = 0.1$. Use the formula for S_∞ to find this sum. Does intuition indicate that this answer is correct?

Find r for each infinite geometric sequence. Identify any whose sum diverges.

47. $12, 24, 48, 96, \ldots$

48. $2, -10, 50, -250, \ldots$

49. $-48, -24, -12, -6, \ldots$

50. $625, 125, 25, 5, \ldots$

Work each problem. ***See Examples 7 and 8.***

51. Use $\lim_{n\to\infty} S_n$ to show that $2 + 1 + \frac{1}{2} + \frac{1}{4} + \cdots$ converges to 4.

52. We determined that $1 + \frac{1}{3} + \frac{1}{9} + \frac{1}{27} + \cdots$ converges to $\frac{3}{2}$ using an argument involving limits. Use the formula for the sum of the terms of an infinite geometric sequence to obtain the same result.

Evaluate each sum. ***See Example 8.***

53. $18 + 6 + 2 + \frac{2}{3} + \cdots$

54. $100 + 10 + 1 + \cdots$

55. $\frac{1}{4} - \frac{1}{6} + \frac{1}{9} - \frac{2}{27} + \cdots$

56. $\frac{4}{3} + \frac{2}{3} + \frac{1}{3} + \cdots$

57. $\sum_{i=1}^{\infty} 3\left(\frac{1}{4}\right)^{i-1}$

58. $\sum_{i=1}^{\infty} 5\left(-\frac{1}{4}\right)^{i-1}$

59. $\sum_{k=1}^{\infty} 3^{-k}$

60. $\sum_{k=1}^{\infty} 10^{-k}$

61. $\sum_{i=1}^{\infty} \left(-\frac{2}{3}\right)\left(-\frac{1}{4}\right)^{i-1}$

62. $\sum_{i=1}^{\infty} \left(-\frac{1}{5}\right)\left(-\frac{2}{5}\right)^{i-1}$

63. $\sum_{i=1}^{\infty} \left(\frac{3}{7}\right)^{i}$

64. $\sum_{i=1}^{\infty} \left(\frac{5}{9}\right)^{i}$

Use the summation feature of a graphing calculator to evaluate each sum. Round to the nearest thousandth.

65. $\sum_{i=1}^{10} (1.4)^{i}$

66. $\sum_{j=1}^{6} -(3.6)^{j}$

67. $\sum_{j=3}^{8} 2(0.4)^{j}$

68. $\sum_{i=4}^{9} 3(0.25)^{i}$

Solve each problem. ***See Examples 1–8.***

69. ***(Modeling) Investment for Retirement*** According to T. Rowe Price Associates, a person with a moderate investment strategy and n years to retirement should have accumulated savings of a_n percent of his or her annual salary. The geometric sequence

$$a_n = 1276(0.916)^n$$

gives the appropriate percent for each year n.

(a) Find a_1 and r. Round a_1 to the nearest whole number.

(b) Find and interpret the terms a_{10} and a_{20}. Round to the nearest whole number.

70. ***(Modeling) Investment for Retirement*** Refer to **Exercise 69.** For someone who has a conservative investment strategy with n years to retirement, the geometric sequence is $a_n = 1278(0.935)^n$. (*Source:* T. Rowe Price Associates.)

(a) Repeat part (a) of **Exercise 69.**

(b) Repeat part (b) of **Exercise 69.**

(c) Why are the answers in parts (a) and (b) greater than those in **Exercise 69?**

71. ***(Modeling) Bacterial Growth*** Suppose that a strain of bacteria will double in size and then divide every 40 minutes. Let a_1 be the initial number of bacteria cells, a_2 the number after 40 minutes, and a_n the number after $40(n-1)$ minutes.

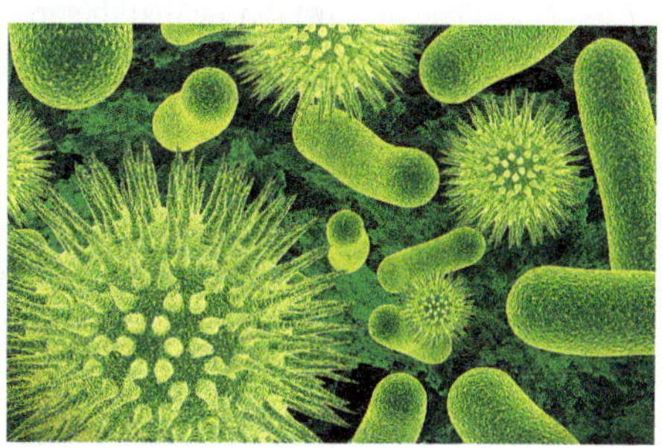

(a) Write a formula for the nth term a_n of the geometric sequence

$$a_1, a_2, a_3, \ldots, a_n, \ldots.$$

(b) Determine the first value for n where $a_n > 1{,}000{,}000$ if $a_1 = 100$.

(c) How long does it take for the number of bacteria to exceed one million?

72. *Depreciation* Each year a machine loses 20% of the value it had at the beginning of the year. Find the value of the machine at the end of 6 yr if it cost \$100,000 new.

73. *(Modeling) Fruit Flies Population* A population of fruit flies is growing in such a way that each generation is 1.25 times as large as the last generation. Suppose there were 200 insects in the first generation. How many, to the nearest whole number, would there be in the fifth generation?

74. *Height of a Dropped Ball* Alicia drops a ball from a height of 10 m and notices that on each bounce the ball returns to about $\frac{3}{4}$ of its previous height. About how far will the ball travel before it comes to rest? (*Hint:* Consider the sum of two sequences.)

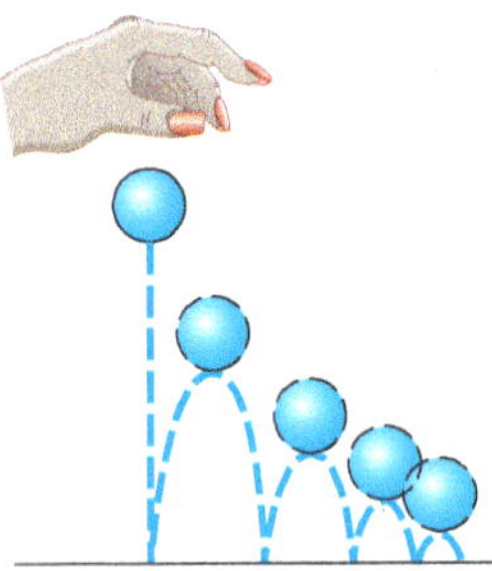

75. *Number of Ancestors* Each person has two parents, four grandparents, eight great-grandparents, and so on. What is the total number of ancestors a person has, going back five generations? ten generations?

76. *Drug Dosage* Certain medical conditions are treated with a fixed dose of a drug administered at regular intervals. Suppose a person is given 2 mg of a drug each day and that during each 24-hr period, the body utilizes 40% of the amount of drug that was present at the beginning of the period.

(a) Show that the amount of the drug present in the body at the end of n days is

$$\sum_{i=1}^{n} 2(0.6)^i.$$

(b) What will be the approximate quantity of the drug in the body at the end of each day after the treatment has been administered for a long period of time?

77. *Side Length of a Triangle* A sequence of equilateral triangles is constructed. The first triangle has sides 2 m in length. To construct the second triangle, midpoints of the sides of the original triangle are connected. What is the length of each side of the eighth such triangle? See the figure.

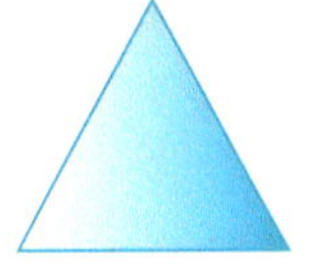 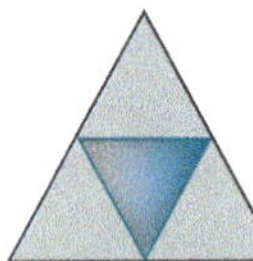 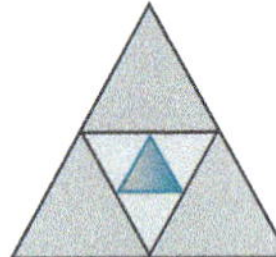 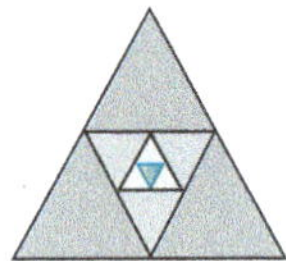

78. *Perimeter and Area of Triangles* In **Exercise 77,** if the process could be continued indefinitely, what would be the total perimeter of all the triangles? What would be the total area of all the triangles, disregarding the overlapping?

79. *Salaries* A student is offered a 6-week summer job and is asked to select one of the following salary options.

Option 1: \$5000 for the first day with a \$10,000 raise each day for the remaining 29 days (that is, \$15,000 for day 2, \$25,000 for day 3, and so on)

Option 2: \$0.01 for the first day with the pay doubled each day (that is, \$0.02 for day 2, \$0.04 for day 3, and so on)

Which option has a greater total salary?

80. *Number of Ancestors* A genealogical Web site allows a person to identify all of her or his ancestors who lived during the last 300 yr. Assuming that each generation spans about 25 yr, guess the number of ancestors that would be found during the 12 generations. Then use the formula for a geometric series to find the correct value.

Future Value of an Annuity *Find the future value of each annuity.* ***See Example 9.***

81. There are payments of \$1000 at the end of each year for 9 yr at 3% interest compounded annually.

82. There are payments of \$800 at the end of each year for 12 yr at 2% interest compounded annually.

83. There are payments of \$2430 at the end of each year for 10 yr at 1% interest compounded annually.

84. There are payments of \$1500 at the end of each year for 6 yr at 0.5% interest compounded annually.

85. Refer to **Exercise 83.** Use the answer and recursion to find the balance after 11 yr.

86. Refer to **Exercise 84.** Use the answer and recursion to find the balance after 7 yr.

87. *Individual Retirement Account* Starting on his 40th birthday, Michael deposits \$2000 per year in an Individual Retirement Account until age 65 (last payment at age 64). Find the total amount in the account if he had a guaranteed interest rate of 2% compounded annually.

88. *Retirement Savings* To save for retirement, Mort put \$3000 at the end of each year into an ordinary annuity for 20 yr at 1.5% annual interest. At the end of year 20, what was the amount of the annuity?

89. Show that the formula for future value of an annuity gives the correct answer when compared to the solution in **Example 9.**

90. The screen here shows how the TI-84 Plus calculator computes the future value of the annuity described in **Example 9.** Use a calculator with this capability to support the answers in **Exercises 81–88.**

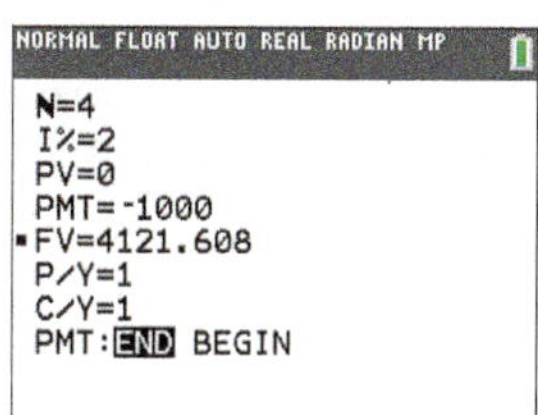

91. *Concept Check* Suppose that $a_1, a_2, a_3, a_4, a_5, \ldots$ is a geometric sequence. Is the sequence $a_1, a_3, a_5, \ldots$ geometric?

92. *Concept Check* Is the sequence log 6, log 36, log 1296, log 1,679,616, . . . a geometric sequence?

Summary Exercises on Sequences and Series

Use the following guidelines in the exercises that follow.

Given a sequence $a_1, a_2, a_3, a_4, a_5, \ldots,$

- If the differences

$$a_2 - a_1, \quad a_3 - a_2, \quad a_4 - a_3, \quad a_5 - a_4, \quad \ldots$$

are all equal to the same number d, then the sequence is *arithmetic,* and d is the *common difference.*

- If the ratios

$$\frac{a_2}{a_1}, \quad \frac{a_3}{a_2}, \quad \frac{a_4}{a_3}, \quad \frac{a_5}{a_4}, \quad \ldots$$

are all equal to the same number r, then the sequence is *geometric,* and r is the *common ratio.*

Determine whether each sequence is arithmetic, geometric, *or* neither. *If it is arithmetic, give the common difference, d. If it is geometric, give the common ratio, r.*

1. $2, 4, 8, 16, 32, \ldots$

2. $1, 4, 7, 10, 13, \ldots$

3. $3, \frac{1}{2}, -2, -\frac{9}{2}, -7, \ldots$

4. $1, 1, 2, 3, 5, 8, \ldots$

5. $\frac{3}{4}, 1, \frac{4}{3}, \frac{16}{9}, \frac{64}{27}, \ldots$

6. $4, -12, 36, -108, 324, \ldots$

7. $\frac{1}{2}, \frac{1}{3}, \frac{1}{4}, \frac{1}{5}, \frac{1}{6}, \ldots$

8. $5, 2, -1, -4, -7, \ldots$

9. $1, 9, 10, 19, 29, \ldots$

10. $-1, \sqrt{5}, -5, 5\sqrt{5}, -25, \ldots$

Determine whether each sequence is arithmetic *or* geometric. *Then find a_n and S_{10}.*

11. $3, 6, 12, 24, 48, \ldots$

12. $2, 6, 10, 14, 18, \ldots$

13. $4, \frac{5}{2}, 1, -\frac{1}{2}, -2, \ldots$

14. $\frac{3}{2}, 1, \frac{2}{3}, \frac{4}{9}, \frac{8}{27}, \ldots$

15. $3, -6, 12, -24, 48, \ldots$

16. $-5, -8, -11, -14, -17, \ldots$

Evaluate each sum that converges. Identify any that diverge.

17. $\sum_{i=1}^{\infty} \frac{1}{3}(-2)^{i-1}$

18. $\sum_{j=1}^{4} 2\left(\frac{1}{10}\right)^{j-1}$

19. $\sum_{i=1}^{25} (4 - 6i)$

20. $\sum_{i=1}^{6} 3^i$

21. $\sum_{i=1}^{\infty} 4\left(-\frac{1}{2}\right)^i$

22. $\sum_{i=1}^{\infty} (3i - 2)$

23. $\sum_{j=1}^{12} (2j - 1)$

24. $\sum_{k=1}^{\infty} 5^{-k}$

25. $\sum_{i=1}^{\infty} 1.0001^i$

26. Write $0.333\ldots$ as an infinite geometric series. Find the sum.

11.4 The Binomial Theorem

- A Binomial Expansion Pattern
- Pascal's Triangle
- *n*-Factorial
- Binomial Coefficients
- The Binomial Theorem
- *k*th Term of a Binomial Expansion

A Binomial Expansion Pattern

In this section, we introduce a method for writing the expansion of expressions of the form $(x+y)^n$, where n is a natural number. Some expansions for various nonnegative integer values of n follow.

$$(x+y)^0 = 1$$
$$(x+y)^1 = x+y$$
$$(x+y)^2 = x^2 + 2xy + y^2$$
$$(x+y)^3 = x^3 + 3x^2y + 3xy^2 + y^3$$
$$(x+y)^4 = x^4 + 4x^3y + 6x^2y^2 + 4xy^3 + y^4$$
$$(x+y)^5 = x^5 + 5x^4y + 10x^3y^2 + 10x^2y^3 + 5xy^4 + y^5$$

Notice that after the special case $(x+y)^0 = 1$, each expansion begins with x raised to the same power as the binomial itself. That is, the expansion of $(x+y)^1$ has a first term of x^1, $(x+y)^2$ has a first term of x^2, $(x+y)^3$ has a first term of x^3, and so on. Also, the last term in each expansion is y to the same power as the binomial. Thus, the expansion of $(x+y)^n$ should begin with the term x^n and end with the term y^n.

LOOKING AHEAD TO CALCULUS
Students taking calculus study the binomial series, which follows from Isaac Newton's extension to the case where the exponent is no longer a positive integer. His result led to a series for $(1+x)^k$, where k is a real number and $|x| < 1$.

Notice that the exponent on x decreases by 1 in each term after the first, while the exponent on y, beginning with y in the second term, increases by 1 in each succeeding term. That is, the *variables* in the terms of the expansion of $(x+y)^n$ have the following pattern.

$$x^n,\ x^{n-1}y,\ x^{n-2}y^2,\ x^{n-3}y^3,\ \dots,\ xy^{n-1},\ y^n$$

This pattern suggests that the sum of the exponents on x and y in each term is n. For example, the third term in the list above is $x^{n-2}y^2$, and the sum of the exponents is $n-2+2=n$.

Pascal's Triangle

Now, examine the *coefficients* in the terms of the expansion of $(x+y)^n$. Writing the coefficients alone gives the following pattern.

Pascal's Triangle

Triangle	Row
1	0
1 1	1
1 2 1	2
1 3 3 1	3
1 4 6 4 1	4
1 5 10 10 5 1	5

With the coefficients arranged in this way, each number in the triangle is the sum of the two numbers directly above it (one to the right and one to the left).

Blaise Pascal (1623–1662)

Pascal, a French mathematician, made mathematical contributions in the areas of calculus, geometry, and probability theory. At age 19, he invented the first adding machine, a precursor to our modern-day calculator.

For example, in row four of the triangle, 1 is the sum of 1 (the only number above it), 4 is the sum of 1 and 3, 6 is the sum of 3 and 3, and so on. This triangular array of numbers is called **Pascal's triangle,** in honor of the seventeenth-century mathematician Blaise Pascal. It was, however, known long before his time.

To find the coefficients for $(x + y)^6$, we need to include row six in Pascal's triangle. Adding adjacent numbers in row five, we find that row six is

$$1 \quad 6 \quad 15 \quad 20 \quad 15 \quad 6 \quad 1.$$

Using these coefficients, we obtain the expansion of $(x + y)^6$.

$$(x+y)^6 = x^6 + 6x^5y + 15x^4y^2 + 20x^3y^3 + 15x^2y^4 + 6xy^5 + y^6$$

n-Factorial

Although it is possible to use Pascal's triangle to find the coefficients of $(x + y)^n$ for any positive integer n, this calculation becomes impractical for large values of n because of the need to write all the preceding rows. A more efficient way of finding these coefficients uses **factorial notation.** The number $n!$ (read **"*n*-factorial"**) is defined as follows.

n-Factorial

For any positive integer n,

$$n! = n(n-1)(n-2)\cdots(3)(2)(1).$$

By definition,

$$0! = 1.$$

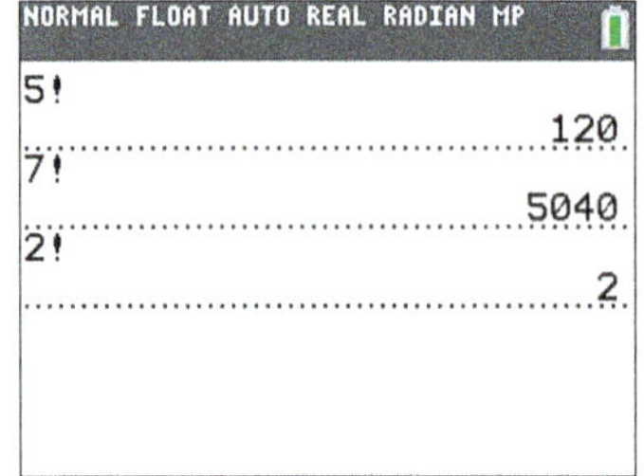

For example,

$$5! = 5 \cdot 4 \cdot 3 \cdot 2 \cdot 1 = 120,$$

$$7! = 7 \cdot 6 \cdot 5 \cdot 4 \cdot 3 \cdot 2 \cdot 1 = 5040,$$

and

$$2! = 2 \cdot 1 = 2.$$

Binomial Coefficients

Now look at the coefficients of the expansion

$$(x+y)^5 = x^5 + 5x^4y + 10x^3y^2 + 10x^2y^3 + 5xy^4 + y^5.$$

The coefficient of the second term, $5x^4y$, is 5, and the exponents on the variables are 4 and 1. Note that

$$5 = \frac{5!}{4!\,1!}.$$

The coefficient of the third term, $10x^3y^2$, is 10, with exponents of 3 and 2 on the variables, and

$$10 = \frac{5!}{3!\,2!}.$$

The last term (the sixth term) can be written as $y^5 = 1x^0y^5$, with coefficient 1 and exponents of 0 and 5. By definition $0! = 1$, so

$$1 = \frac{5!}{0!5!}.$$

Generalizing from these examples, the coefficient for the term of the expansion of $(x + y)^n$ in which the variable part is $x^r y^{n-r}$ (where $r \le n$) is

$$\frac{n!}{r!(n-r)!}.$$

This number, called a **binomial coefficient,** is often symbolized $\binom{n}{r}$ or ${}_nC_r$ (read **"n choose r"**).

Binomial Coefficient

For nonnegative integers n and r, with $r \le n$, the binomial coefficient is defined as follows.

$${}_nC_r = \binom{n}{r} = \frac{n!}{r!(n-r)!}$$

The binomial coefficients are numbers from Pascal's triangle. For example, $\binom{3}{0}$ is the first number in row three, and $\binom{7}{4}$ is the fifth number in row seven.

EXAMPLE 1 Evaluating Binomial Coefficients

Evaluate each binomial coefficient.

(a) $\binom{6}{2}$ **(b)** $\binom{8}{0}$ **(c)** $\binom{10}{10}$ **(d)** ${}_{12}C_{10}$

ALGEBRAIC SOLUTION

(a) $\binom{6}{2} = \frac{6!}{2!(6-2)!} = \frac{6!}{2!4!}$

$$= \frac{6 \cdot 5 \cdot 4 \cdot 3 \cdot 2 \cdot 1}{2 \cdot 1 \cdot 4 \cdot 3 \cdot 2 \cdot 1} = 15$$

(b) $\binom{8}{0} = \frac{8!}{0!(8-0)!} = \frac{8!}{0!8!} = \frac{8!}{1 \cdot 8!} = 1$ $\quad 0! = 1$

(c) $\binom{10}{10} = \frac{10!}{10!(10-10)!} = \frac{10!}{10!0!} = 1$ $\quad 0! = 1$

(d) ${}_{12}C_{10} = \frac{12!}{10!(12-10)!} = \frac{12!}{10!2!} = 66$

GRAPHING CALCULATOR SOLUTION

Graphing calculators calculate binomial coefficients using the notation ${}_nC_r$. For the TI-84 Plus, this function is found in the MATH menu. **Figure 11** shows the values of the binomial coefficients for parts (a)–(d). Compare the results to those in the algebraic solution.

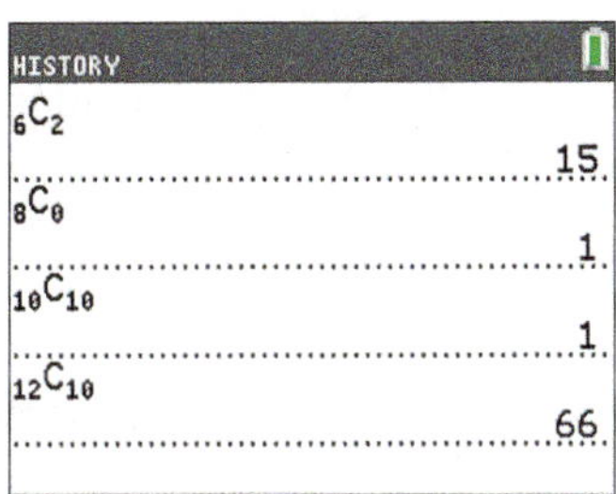

Figure 11

✔ **Now Try Exercises 15, 19, and 27.**

Refer again to Pascal's triangle. Notice the symmetry in each row. This suggests that binomial coefficients should have the same property. That is,

$$\binom{n}{r} = \binom{n}{n-r}.$$

This is true because the numerators are equal and the denominators are equal by the commutative property of multiplication.

$$\binom{n}{r} = \frac{n!}{r!(n-r)!} \quad \text{and} \quad \binom{n}{n-r} = \frac{n!}{(n-r)!r!}$$

The Binomial Theorem Our observations about the expansion of $(x+y)^n$ are summarized as follows.

1. There are $n+1$ terms in the expansion.
2. The first term is x^n, and the last term is y^n.
3. In each succeeding term, the exponent on x decreases by 1 and the exponent on y increases by 1.
4. The sum of the exponents on x and y in any term is n.
5. The coefficient of the term with $x^r y^{n-r}$ or $x^{n-r} y^r$ is $\binom{n}{r}$.

These observations suggest the **binomial theorem.**

LOOKING AHEAD TO CALCULUS

The binomial theorem is used to show that the derivative of $f(x) = x^n$ is given by the function $f'(x) = nx^{n-1}$. This fact is used extensively in calculus.

Binomial Theorem

For any positive integer n and any complex numbers x and y, $(x+y)^n$ is expanded as follows.

$$\begin{aligned}(x+y)^n &= x^n + \binom{n}{1}x^{n-1}y + \binom{n}{2}x^{n-2}y^2 + \binom{n}{3}x^{n-3}y^3 + \cdots \\ &\quad + \binom{n}{r}x^{n-r}y^r + \cdots + \binom{n}{n-1}xy^{n-1} + y^n\end{aligned}$$

NOTE The binomial theorem may also be written as a series using summation notation.

$$(x+y)^n = \sum_{r=0}^{n} \binom{n}{r} x^{n-r} y^r$$

In agreement with Pascal's triangle, the coefficients of the first and last terms are both 1. That is,

$$\binom{n}{0} = \binom{n}{n} = 1.$$

EXAMPLE 2 Applying the Binomial Theorem

Write the binomial expansion of $(x + y)^9$.

SOLUTION Apply the binomial theorem.

$$\begin{aligned}(x+y)^9 &= x^9 + \binom{9}{1}x^8y + \binom{9}{2}x^7y^2 + \binom{9}{3}x^6y^3 + \binom{9}{4}x^5y^4 + \binom{9}{5}x^4y^5 \\ &\quad + \binom{9}{6}x^3y^6 + \binom{9}{7}x^2y^7 + \binom{9}{8}xy^8 + y^9 \\ &= x^9 + \frac{9!}{1!8!}x^8y + \frac{9!}{2!7!}x^7y^2 + \frac{9!}{3!6!}x^6y^3 + \frac{9!}{4!5!}x^5y^4 + \frac{9!}{5!4!}x^4y^5 \\ &\quad + \frac{9!}{6!3!}x^3y^6 + \frac{9!}{7!2!}x^2y^7 + \frac{9!}{8!1!}xy^8 + y^9 \\ &= x^9 + 9x^8y + 36x^7y^2 + 84x^6y^3 + 126x^5y^4 + 126x^4y^5 \\ &\quad + 84x^3y^6 + 36x^2y^7 + 9xy^8 + y^9\end{aligned}$$

Evaluate each binomial coefficient.

✔ **Now Try Exercise 31.**

EXAMPLE 3 Applying the Binomial Theorem

Expand $\left(a - \frac{b}{2}\right)^5$.

SOLUTION Write the binomial as follows.

$$\left(a - \frac{b}{2}\right)^5 = \left(a + \left(-\frac{b}{2}\right)\right)^5$$

Now apply the binomial theorem with $x = a$, $y = -\frac{b}{2}$, and $n = 5$.

$$\begin{aligned}\left(a - \frac{b}{2}\right)^5 &= a^5 + \binom{5}{1}a^4\left(-\frac{b}{2}\right) + \binom{5}{2}a^3\left(-\frac{b}{2}\right)^2 + \binom{5}{3}a^2\left(-\frac{b}{2}\right)^3 + \binom{5}{4}a\left(-\frac{b}{2}\right)^4 + \left(-\frac{b}{2}\right)^5 \\ &= a^5 + 5a^4\left(-\frac{b}{2}\right) + 10a^3\left(-\frac{b}{2}\right)^2 + 10a^2\left(-\frac{b}{2}\right)^3 + 5a\left(-\frac{b}{2}\right)^4 + \left(-\frac{b}{2}\right)^5 \\ &= a^5 - \frac{5}{2}a^4b + \frac{5}{2}a^3b^2 - \frac{5}{4}a^2b^3 + \frac{5}{16}ab^4 - \frac{1}{32}b^5\end{aligned}$$

In this expansion, the signs of the terms alternate (as shown in color) because $y = -\frac{b}{2}$ has a negative sign.

✔ **Now Try Exercise 43.**

NOTE As **Example 3** illustrates, ***an expansion of the difference of two terms (for example, an expansion of $(x - y)^n$ for $n \geq 2$) has alternating signs.***

EXAMPLE 4 **Applying the Binomial Theorem**

Expand $\left(\frac{3}{m^2} - 2\sqrt{m}\right)^4$. (Assume $m > 0$.)

SOLUTION Apply the binomial theorem.

$$\left(\frac{3}{m^2} - 2\sqrt{m}\right)^4$$

$$= \left(\frac{3}{m^2}\right)^4 + \binom{4}{1}\left(\frac{3}{m^2}\right)^3\left(-2\sqrt{m}\right) + \binom{4}{2}\left(\frac{3}{m^2}\right)^2\left(-2\sqrt{m}\right)^2 + \binom{4}{3}\left(\frac{3}{m^2}\right)\left(-2\sqrt{m}\right)^3 + \left(-2\sqrt{m}\right)^4$$

$$= \frac{81}{m^8} + 4\left(\frac{27}{m^6}\right)(-2m^{1/2}) + 6\left(\frac{9}{m^4}\right)(4m) + 4\left(\frac{3}{m^2}\right)(-8m^{3/2}) + 16m^2$$

$\sqrt{m} = m^{1/2}$

$$= \frac{81}{m^8} - \frac{216}{m^{11/2}} + \frac{216}{m^3} - \frac{96}{m^{1/2}} + 16m^2$$

✔ **Now Try Exercise 45.**

*k*th Term of a Binomial Expansion

Earlier in this section, we wrote the binomial theorem in summation notation as $\sum_{r=0}^{n}\binom{n}{r}x^{n-r}y^r$, which gives the form of each term. We can use this form to write any particular term of a binomial expansion without writing out the entire expansion.

*k*th Term of the Binomial Expansion

The kth term of the binomial expansion of $(x + y)^n$, where $n \geq k - 1$, is given as follows.

$$\binom{n}{k-1}x^{n-(k-1)}y^{k-1}$$

To find the kth term of the binomial expansion, use the following steps.

Step 1 Find $k - 1$. This is the exponent on the second term of the binomial.

Step 2 Subtract the exponent found in Step 1 from n to obtain the exponent on the first term of the binomial.

Step 3 Determine the coefficient by using the exponents found in the first two steps and n.

EXAMPLE 5 **Finding a Particular Term of a Binomial Expansion**

Find the seventh term of the binominal expansion of $(a + 2b)^{10}$.

SOLUTION In the seventh term, $2b$ has an exponent of $7 - 1$, or 6, while a has an exponent of $10 - 6$, or 4.

$$\binom{10}{6}a^4(2b)^6 \quad \text{Seventh term of the binomial expansion}$$

$$= 210a^4(64b^6) \quad \text{Evaluate and apply the power rule for exponents.}$$

$$= 13{,}440a^4b^6 \quad \text{Multiply.}$$

✔ **Now Try Exercise 49.**

11.4 Exercises

CONCEPT PREVIEW *Fill in the blank(s) to correctly complete each sentence.*

1. Each number that is not a 1 in Pascal's triangle is the ______ of the two numbers directly above it (one to the right and one to the left).
2. The value of $8!$ is ______.
3. The value of $0!$ is ______.
4. The value of ${}_7C_3$ is ______.
5. ${}_{12}C_4 = {}_{12}C$___ (Do not use 4 in the blank.)
6. In the expansion of $(x + y)^5$, the number of terms is ______.
7. In the expansion of $(x + y)^8$, the first term is ______ and the last term is ______.
8. The sum of the exponents on x and y in any term of the expansion of $(x + y)^{10}$ is ______.
9. The second term in the expansion of $(p + q)^5$ is ______.
10. The fourth term in the expansion of $(2x - y)^7$ is ______.

Evaluate each binomial coefficient. In Exercises 21 and 22, leave answers in terms of n. ***See Example 1.***

11. $\frac{6!}{3!3!}$
12. $\frac{5!}{2!3!}$
13. $\frac{7!}{3!4!}$
14. $\frac{8!}{5!3!}$
15. $\binom{8}{5}$
16. $\binom{7}{3}$
17. $\binom{10}{2}$
18. $\binom{9}{3}$
19. $\binom{14}{14}$
20. $\binom{15}{15}$
21. $\binom{n}{n-1}$
22. $\binom{n}{n-2}$
23. ${}_8C_3$
24. ${}_9C_7$
25. ${}_{100}C_{98}$
26. ${}_{20}C_5$
27. ${}_9C_0$
28. ${}_4C_0$
29. ${}_{12}C_1$
30. ${}_5C_1$

Write the binomial expansion of each expression. ***See Examples 2–4.***

31. $(x + y)^6$
32. $(m + n)^4$
33. $(p - q)^5$
34. $(a - b)^7$
35. $(r^2 + s)^5$
36. $(m + n^2)^4$
37. $(p + 2q)^4$
38. $(3r + s)^6$
39. $(7p - 2q)^4$
40. $(4a - 5b)^5$
41. $(3x - 2y)^6$
42. $(7k - 9j)^4$
43. $\left(\frac{m}{2} - 1\right)^6$
44. $\left(3 - \frac{y}{3}\right)^5$
45. $\left(\sqrt{2}r + \frac{1}{m}\right)^4$
46. $\left(\frac{1}{k} + \sqrt{3}p\right)^3$
47. $\left(\frac{1}{x^4} + x^4\right)^4$
48. $\left(\frac{1}{y^5} + y^5\right)^5$

Find the indicated term of each binomial expansion. ***See Example 5.***

49. sixth term of $(4h - j)^8$

50. eighth term of $(2c - 3d)^{14}$

51. seventeenth term of $(a^2 + b)^{22}$

52. twelfth term of $(2x + y^2)^{16}$

53. fifteenth term of $(x - y^3)^{20}$

54. tenth term of $(a^3 + 3b)^{11}$

Concept Check *Work each problem.*

55. Find the middle term of $(3x^7 + 2y^3)^8$.

56. Find the two middle terms of $(-2m^{-1} + 3n^{-2})^{11}$.

57. Find the value of n for which the coefficients of the fifth and eighth terms in the expansion of $(x + y)^n$ are the same.

58. Find the term(s) in the expansion of $\left(3 + \sqrt{x}\right)^{11}$ that contain(s) x^4.

Relating Concepts

For individual or collaborative investigation *(Exercises 59–62)*

The factorial of a positive integer n can be computed as a product.

$$n! = 1 \cdot 2 \cdot 3 \cdot \cdots \cdot n$$

Calculators and computers can evaluate factorials very quickly. Before the days of modern technology, mathematicians developed ***Stirling's formula*** *for approximating large factorials. The formula involves the irrational numbers* π *and e.*

$$n! \approx \sqrt{2\pi n} \cdot n^n \cdot e^{-n}$$

As an example, the exact value of $5!$ *is* 120, *and Stirling's formula gives the approximation as* 118.019168 *with a graphing calculator. This is "off" by less than* 2, *an error of only* 1.65%. ***Work Exercises 59–62 in order.***

59. Use a calculator to find the exact value of $10!$ and its approximation, using Stirling's formula.

60. Subtract the smaller value from the larger value in **Exercise 59.** Divide it by $10!$ and convert to a percent. What is the percent error to three decimal places?

61. Repeat **Exercises 59 and 60** for $n = 12$.

62. Repeat **Exercises 59 and 60** for $n = 13$. What seems to happen as n gets larger?

11.5 Mathematical Induction

- Principle of Mathematical Induction
- Proofs of Statements
- Generalized Principle of Mathematical Induction
- Proof of the Binomial Theorem

Principle of Mathematical Induction

Many statements in mathematics are claimed true for every positive integer. Any of these statements could be checked for $n = 1$, $n = 2$, $n = 3$, and so on, but because the set of positive integers is infinite, it would be impossible to check every possible case.

For example, let S_n represent the statement that the sum of the first n positive integers is $\frac{n(n+1)}{2}$.

$$S_n\text{:}\quad 1 + 2 + 3 + \cdots + n = \frac{n(n+1)}{2}$$

The truth of this statement is easily verified for the first few values of n.

If $n = 1$, then S_1 is $\quad 1 = \frac{1(1 + 1)}{2}$. $\quad$ This is true because $1 = 1$.

If $n = 2$, then S_2 is $\quad 1 + 2 = \frac{2(2 + 1)}{2}$. $\quad$ This is true because $3 = 3$.

If $n = 3$, then S_3 is $\quad 1 + 2 + 3 = \frac{3(3 + 1)}{2}$. $\quad$ This is true because $6 = 6$.

If $n = 4$, then S_4 is $\quad 1 + 2 + 3 + 4 = \frac{4(4 + 1)}{2}$. $\quad$ This is true because $10 = 10$.

We cannot conclude that the statement is true for *every* positive integer n simply by observing a finite number of examples. To prove that a statement is true for every positive integer value of n, we use the following principle.

Principle of Mathematical Induction

Let S_n be a statement concerning the positive integer n. Suppose that both of the following are satisfied.

1. S_1 is true.

2. For any positive integer k, $k \leq n$, if S_k is true, then S_{k+1} is also true.

Then S_n is true for every positive integer value of n.

A proof by mathematical induction can be explained as follows. By assumption (1) above, the statement is true when $n = 1$. By assumption (2) above, the fact that the statement is true for $n = 1$ implies that it is true for

$$n = 1 + 1 = 2.$$

Using (2) again, the statement is thus true

$$\text{for } 2 + 1 = 3, \quad \text{for } 3 + 1 = 4, \quad \text{for } 4 + 1 = 5, \quad \text{and so on.}$$

Continuing in this way shows that the statement must be true for *every* positive integer.

The situation is similar to that of an infinite number of dominoes lined up as suggested in **Figure 12.** If the first domino is pushed over, it pushes the next, which pushes the next, and so on continuing indefinitely.

Figure 12

Another example of the principle of mathematical induction is the concept of an infinite ladder. Suppose the rungs are spaced so that whenever we are on a rung, we know we can move to the next rung. Then *if* we can get to the first rung, we can go as high up the ladder as we wish.

Proofs of Statements Two separate steps are required for a proof by mathematical induction.

Proof by Mathematical Induction

Step 1 Prove that the statement is true for $n = 1$.

Step 2 Show that, for any positive integer k, $k \leq n$, if S_k is true, then S_{k+1} is also true.

EXAMPLE 1 Proving an Equality Statement

Let S_n represent the following statement.

$$1 + 2 + 3 + \cdots + n = \frac{n(n+1)}{2}$$

Prove that S_n is true for every positive integer n.

SOLUTION

Step 1 Show that the statement is true when $n = 1$. If $n = 1$, S_1 becomes

$$1 = \frac{1(1+1)}{2}, \quad \text{which is true.}$$

Step 2 Show that S_k implies S_{k+1}, where S_k is the statement

$$1 + 2 + 3 + \cdots + k = \frac{k(k+1)}{2},$$

and S_{k+1} is the statement

$$1 + 2 + 3 + \cdots + k + (k+1) = \frac{(k+1)[(k+1)+1]}{2}.$$

Start with S_k and assume it is a true statement.

$$1 + 2 + 3 + \cdots + k = \frac{k(k+1)}{2}$$

Add $k + 1$ to each side of this equation to obtain S_{k+1}.

$$1 + 2 + 3 + \cdots + k + (k+1) = \frac{k(k+1)}{2} + (k+1) \quad \text{Add } k+1 \text{ to each side.}$$

$$= (k+1)\left(\frac{k}{2} + 1\right) \quad \text{Factor out } k+1 \text{ on the right.}$$

$$= (k+1)\left(\frac{k+2}{2}\right) \quad \text{Add inside the parentheses.}$$

This is the statement S_{k+1}.

$$1 + 2 + 3 + \cdots + k + (k+1) = \frac{(k+1)[(k+1)+1]}{2} \quad \text{Multiply; } k+2 = (k+1)+1$$

This final result is the statement for $n = k + 1$. It has been shown that if S_k is true, then S_{k+1} is also true. The two steps required for a proof by mathematical induction have been completed, so the statement S_n is true for every positive integer value of n.

✔ **Now Try Exercise 7.**

CAUTION Notice that the left side of the statement S_n in **Example 1** includes *all* the terms up to the nth term, as well as the nth term.

EXAMPLE 2 Proving an Inequality Statement

Prove that if x is a real number strictly between 0 and 1, then for every positive integer n, it follows that

$$0 < x^n < 1.$$

SOLUTION

Step 1 Let S_n represent the given statement. Here S_1 is the statement

$$\text{if} \quad 0 < x < 1, \quad \text{then} \quad 0 < x^1 < 1, \quad \text{which is true.}$$

Step 2 S_k is the statement

$$\text{if} \quad 0 < x < 1, \quad \text{then} \quad 0 < x^k < 1.$$

To show that this implies that S_{k+1} is true, multiply each of the three parts of $0 < x^k < 1$ by x.

$$0 < x^k < 1$$

$$x \cdot 0 < x \cdot x^k < x \cdot 1 \qquad \text{Use the fact that } 0 < x.$$

$$0 < x^{k+1} < x \qquad \text{Simplify.}$$

We now use a technique that allows us to reach our desired goal. From Step 1 we know that $x < 1$, so in the inequality $0 < x^{k+1} < x$, we can replace x with *any* greater value and the inequality is preserved. Because 1 is greater than x, replace x with 1.

$$0 < x^{k+1} < 1$$

This is the statement S_{k+1}. This work shows that if S_k is true, then S_{k+1} is true. Therefore, the given statement S_n is true for every positive integer n.

✔ **Now Try Exercise 21.**

Generalized Principle of Mathematical Induction Some statements S_n are not true for the first few values of n but *are* true for all values of n that are greater than or equal to some fixed integer j. The following generalized form of the principle of mathematical induction covers these cases.

Generalized Principle of Mathematical Induction

Let S_n be a statement concerning the positive integer n. Let j be a fixed positive integer. Suppose that both of the following are satisfied.

Step 1 S_j is true.

Step 2 For any positive integer k, $k \geq j$, S_k implies S_{k+1}.

Then S_n is true for all positive integers n, where $n \geq j$.

EXAMPLE 3 Using the Generalized Principle

Let S_n represent the statement $2^n > 2n + 1$. Show that S_n is true for all values of n such that $n \geq 3$.

SOLUTION (Check that S_n is false for $n = 1$ and $n = 2$.)

Step 1 Show that S_n is true for $n = 3$.

$$2^3 > 2 \cdot 3 + 1 \quad \text{Let } n = 3.$$
$$8 > 7 \quad \text{True}$$

Thus, S_3 is true.

Step 2 Now show that S_k implies S_{k+1}, where $k \geq 3$, and where

$$S_k \text{ is } 2^k > 2k + 1, \quad \text{and} \quad S_{k+1} \text{ is } 2^{k+1} > 2(k + 1) + 1.$$

Start with S_k and assume it is a true statement.

$$2^k > 2k + 1$$
$$2 \cdot 2^k > 2(2k + 1) \quad \text{Multiply each side by 2.}$$
$$2^{k+1} > 4k + 2 \quad \text{Product rule; distributive property}$$
$$2^{k+1} > 2k + 2 + 2k \quad \text{Rewrite } 4k \text{ as } 2k + 2k.$$
$$2^{k+1} > 2(k + 1) + 2k \quad \text{Factor } 2k + 2 \text{ on the right.}$$

$2 = 2^1$, so add the exponents.

Because $2k > 1$ for positive integers $k \geq 3$, replacing $2k$ with 1 will maintain the truth value of this inequality.

$$2^{k+1} > 2(k + 1) + 1 \quad S_{k+1}$$

Thus, S_k implies S_{k+1}. Together with the fact that S_3 is true, this shows that S_n is true for every positive integer value of n greater than or equal to 3.

✔ **Now Try Exercise 27.**

Proof of the Binomial Theorem The binomial theorem can be proved by mathematical induction.

$$(x + y)^n = x^n + \binom{n}{1}x^{n-1}y + \binom{n}{2}x^{n-2}y^2 + \binom{n}{3}x^{n-3}y^3 \quad \text{For any positive integer } n \text{ and any complex numbers } x \text{ and } y$$

$$+ \cdots + \binom{n}{r}x^{n-r}y^r + \cdots + \binom{n}{n-1}xy^{n-1} + y^n \quad (1)$$

Proof Let S_n be statement (1). Begin by verifying S_n for $n = 1$.

$$S_1: \quad (x + y)^1 = x^1 + y^1 \quad \text{True}$$

Now assume that S_n is true for the positive integer k. Statement S_k becomes

$$S_k: \quad (x + y)^k = x^k + \frac{k!}{1!(k-1)!}x^{k-1}y + \frac{k!}{2!(k-2)!}x^{k-2}y^2 \quad \text{Definition of the binomial coefficient}$$

$$+ \cdots + \frac{k!}{(k-1)!1!}xy^{k-1} + y^k. \quad (2)$$

Multiply each side of equation (2) by $x + y$.

$$(x+y)^k \cdot (x+y)$$

$$= x(x+y)^k + y(x+y)^k \quad \text{Distributive property}$$

$$= \left[x \cdot x^k + \frac{k!}{1!(k-1)!}x^k y + \frac{k!}{2!(k-2)!}x^{k-1}y^2 + \cdots + \frac{k!}{(k-1)!1!}x^2y^{k-1} + xy^k\right]$$

$$+ \left[x^k \cdot y + \frac{k!}{1!(k-1)!}x^{k-1}y^2 + \cdots + \frac{k!}{(k-1)!1!}xy^k + y \cdot y^k\right]$$

Rearrange terms.

$$(x+y)^{k+1}$$

$$= x^{k+1} + \left[\frac{k!}{1!(k-1)!} + 1\right]x^k y + \left[\frac{k!}{2!(k-2)!} + \frac{k!}{1!(k-1)!}\right]x^{k-1}y^2 + \cdots$$

$$+ \left[1 + \frac{k!}{(k-1)!1!}\right]xy^k + y^{k+1} \quad (3)$$

The first expression in brackets in equation (3) simplifies to $\binom{k+1}{1}$. To see this, note the following.

$$\binom{k+1}{1} = \frac{(k+1)(k)(k-1)(k-2)\cdots 1}{1 \cdot (k)(k-1)(k-2)\cdots 1} = k+1$$

Also,

$$\frac{k!}{1!(k-1)!} + 1 = \frac{k(k-1)!}{1(k-1)!} + 1 = k+1.$$

The second expression becomes $\binom{k+1}{2}$, the last $\binom{k+1}{k}$, and so on. The result of equation (3) is just equation (2) with every k replaced by $k+1$.

The truth of S_k implies the truth of S_{k+1}, which completes the proof of the theorem by mathematical induction.

11.5 Exercises

CONCEPT PREVIEW *Write out S_4 for each of the following, and decide whether it is* true *or* false.

1. S_n: $3 + 6 + 9 + \cdots + 3n = \dfrac{3n(n+1)}{2}$

2. S_n: $1^2 + 2^2 + 3^2 + \cdots + n^2 = \dfrac{n(n+1)(2n+1)}{6}$

3. S_n: $\dfrac{1}{2} + \dfrac{1}{2^2} + \dfrac{1}{2^3} + \cdots + \dfrac{1}{2^n} = \dfrac{2^n - 1}{2^n}$

4. S_n: $6 + 12 + 18 + \cdots + 6n = 3n^2 + 3n$

5. S_n: $2^n < 2n$

6. S_n: $n! > 6n$

Write out in full and verify the statements S_1, S_2, S_3, S_4, and S_5 for the following. Then use mathematical induction to prove that each statement is true for every positive integer n. ***See Example 1.***

7. $1 + 3 + 5 + \cdots + (2n - 1) = n^2$

8. $2 + 4 + 6 + \cdots + 2n = n(n + 1)$

Let S_n represent the given statement, and use mathematical induction to prove that S_n is true for every positive integer n. ***See Example 1.*** *Follow these steps.*

(a) *Verify S_1.* ***(b)*** *Write S_k.* ***(c)*** *Write S_{k+1}.*

(d) *Assume that S_k is true and use algebra to change S_k to S_{k+1}.*

(e) *Write a conclusion based on Steps (a)–(d).*

9. $3 + 6 + 9 + \cdots + 3n = \frac{3n(n + 1)}{2}$

10. $5 + 10 + 15 + \cdots + 5n = \frac{5n(n + 1)}{2}$

11. $2 + 4 + 8 + \cdots + 2^n = 2^{n+1} - 2$

12. $3 + 9 + 27 + \cdots + 3^n = \frac{1}{2}(3^{n+1} - 3)$

13. $1^2 + 2^2 + 3^2 + \cdots + n^2 = \frac{n(n + 1)(2n + 1)}{6}$

14. $1^3 + 2^3 + 3^3 + \cdots + n^3 = \frac{n^2(n + 1)^2}{4}$

15. $5 \cdot 6 + 5 \cdot 6^2 + 5 \cdot 6^3 + \cdots + 5 \cdot 6^n = 6(6^n - 1)$

16. $7 \cdot 8 + 7 \cdot 8^2 + 7 \cdot 8^3 + \cdots + 7 \cdot 8^n = 8(8^n - 1)$

See the text that illustrates the principle of mathematical induction using an infinite ladder.

17. $\frac{1}{1 \cdot 2} + \frac{1}{2 \cdot 3} + \frac{1}{3 \cdot 4} + \cdots + \frac{1}{n(n + 1)} = \frac{n}{n + 1}$

18. $\frac{1}{1 \cdot 4} + \frac{1}{4 \cdot 7} + \frac{1}{7 \cdot 10} + \cdots + \frac{1}{(3n - 2)(3n + 1)} = \frac{n}{3n + 1}$

19. $\frac{1}{2} + \frac{1}{2^2} + \frac{1}{2^3} + \cdots + \frac{1}{2^n} = 1 - \frac{1}{2^n}$

20. $\frac{4}{5} + \frac{4}{5^2} + \frac{4}{5^3} + \cdots + \frac{4}{5^n} = 1 - \frac{1}{5^n}$

Prove each of the following for every positive integer n. Use steps (a)–(e) as in Exercises 9–20. ***See Example 2.***

21. If $a > 1$, then $a^n > 1$.

22. If $a > 1$, then $a^n > a^{n-1}$.

23. If $0 < a < 1$, then $a^n < a^{n-1}$.

24. The bionomial $(x - y)$ is a factor of $x^{2n} - y^{2n}$.

25. $(a^m)^n = a^{mn}$
(Assume a and m are constant.)

26. $(ab)^n = a^n b^n$
(Assume a and b are constant.)

Let S_n represent the given statement. Show that S_n is true for the natural numbers n specified. ***See Example 3.***

27. $2^n > 2n$, for all n such that $n \geq 3$

28. $3^n > 2n + 1$, for all n such that $n \geq 2$

29. $2^n > n^2$, for all n such that $n \geq 5$

30. $4^n > n^4$, for all n such that $n \geq 5$

31. $n! > 2^n$, for all n such that $n \geq 4$

32. $n! > 3^n$, for all n such that $n \geq 7$

Solve each problem.

33. *Number of Handshakes* Suppose that each of the n (for $n \geq 2$) people in a room shakes hands with everyone else, but not with himself or herself. Show that the number of handshakes is $\frac{n^2 - n}{2}$.

34. *Sides of a Polygon* The series of sketches below starts with an equilateral triangle having sides of length 1. In the following steps, equilateral triangles are constructed on each side of the preceding figure. The length of the sides of each new triangle is $\frac{1}{3}$ the length of the sides of the preceding triangles. Develop a formula for the number of sides of the nth figure. Use mathematical induction to prove your answer.

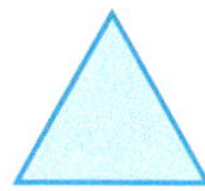   

35. *Perimeter* Find the perimeter of the nth figure in **Exercise 34.**

36. *Area* Show that the area of the nth figure in **Exercise 34** is

$$\sqrt{3}\left[\frac{2}{5}-\frac{3}{20}\left(\frac{4}{9}\right)^{n-1}\right].$$

37. *Tower of Hanoi* A pile of n rings, each ring smaller than the one below it, is on a peg. Two other pegs are attached to a board with this peg. In the game called the *Tower of Hanoi* puzzle, all the rings must be moved to a different peg, with only one ring moved at a time, and with no ring ever placed on top of a smaller ring. Find the least number of moves (in terms of n) that would be required.

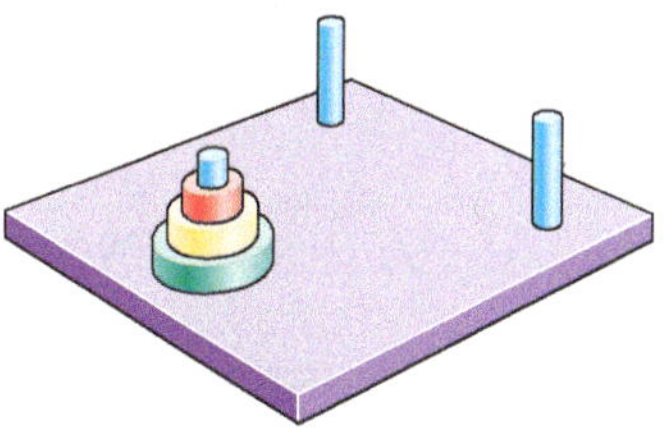

38. *Tower of Hanoi* Prove the result of **Exercise 37** using mathematical induction.

Chapter 11 Quiz (Sections 11.1–11.5)

Write the first five terms of each sequence. State whether the sequence is arithmetic, geometric, *or* neither.

1. $a_n = -4n + 2$

2. $a_n = -2\left(-\frac{1}{2}\right)^n$

3. $a_1 = 5, a_2 = 3, a_n = a_{n-1} + 3a_{n-2},\quad$ for $n \geq 3$

Solve each problem.

4. An arithmetic sequence has $a_1 = -6$ and $a_9 = 18$. Find a_7.

5. Find the sum of the first ten terms of each series described.

(a) arithmetic, $a_1 = -20, d = 14$ **(b)** geometric, $a_1 = -20, r = -\frac{1}{2}$

6. Evaluate each sum that converges. Identify any that diverge.

(a) $\sum_{i=1}^{30}(-3i + 6)$ **(b)** $\sum_{i=1}^{\infty}2^i$ **(c)** $\sum_{i=1}^{\infty}\left(\frac{3}{4}\right)^i$

7. Write the binomial expansion of $(x - 3y)^5$.

8. Find the fifth term of the binomial expansion of $\left(4x - \frac{1}{2}y\right)^5$.

9. Evaluate each expression.

(a) $9!$ **(b)** $\binom{10}{4}$

10. Let S_n represent the following statement, and use mathematical induction to prove that S_n is true for every positive integer n.

$$6 + 12 + 18 + \cdots + 6n = 3n(n + 1)$$

11.6 Basics of Counting Theory

- Fundamental Principle of Counting
- Permutations
- Combinations
- Characteristics That Distinguish Permutations from Combinations

Fundamental Principle of Counting

Consider the following problem.

If there are 3 roads from Albany to Baker and 2 roads from Baker to Creswich, in how many ways can one travel from Albany to Creswich by way of Baker?

For each of the 3 roads from Albany to Baker, there are 2 different roads from Baker to Creswich. Hence, there are

$$3 \cdot 2 = 6 \text{ different ways}$$

to make the trip, as shown in the **tree diagram** in **Figure 13.**

Here, each choice of road is an example of an *event*. Two events are **independent events** if neither influences the outcome of the other.

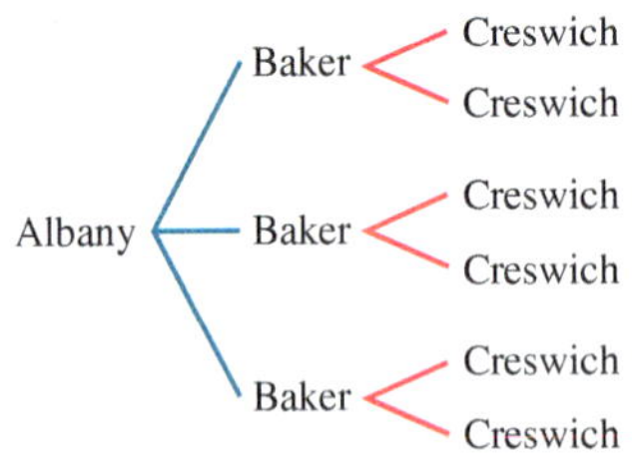

Figure 13

Fundamental Principle of Counting

If n independent events occur, with m_1 ways for event 1 to occur, m_2 ways for event 2 to occur, . . . and m_n ways for event n to occur, then there are

$$m_1 \cdot m_2 \cdot \cdots \cdot m_n \text{ different ways for all } n \text{ events to occur.}$$

The opening example illustrates the fundamental principle of counting with independent events.

EXAMPLE 1 Using the Fundamental Principle of Counting

A restaurant offers a choice of 3 salads, 5 main dishes, and 2 desserts. Use the fundamental principle of counting to find the number of different 3-course meals that can be selected.

SOLUTION Three independent events are involved: selecting a salad, selecting a main dish, and selecting a dessert. The first event can occur in 3 ways, the second event can occur in 5 ways, and the third event can occur in 2 ways.

$$3 \cdot 5 \cdot 2 = 30 \text{ possible meals}$$

✔ **Now Try Exercise 7.**

EXAMPLE 2 Using the Fundamental Principle of Counting

A teacher has 5 different books that he wishes to arrange in a row. How many different arrangements are possible?

SOLUTION Five events are involved: selecting a book for the first spot, selecting a book for the second spot, and so on. For the first spot the teacher has 5 choices. Here the outcome of the first event *does* influence the outcome of the second event, because one book has already been chosen. Thus the teacher has 4 choices for the second spot. Continuing in this manner, there are 3 choices for the third spot, 2 for the fourth spot, and 1 for the fifth spot. We use the fundamental principle of counting.

$$5 \cdot 4 \cdot 3 \cdot 2 \cdot 1 = 120 \text{ different arrangements}$$

✔ **Now Try Exercise 11.**

In using the fundamental principle of counting, products such as

$$5 \cdot 4 \cdot 3 \cdot 2 \cdot 1$$

occur often. We use the symbol $\boldsymbol{n!}$ (read **"n-factorial"**), for any counting number n, as follows.

$$\boldsymbol{n! = n(n-1)(n-2) \cdots (3)(2)(1)}$$

Examples: $5 \cdot 4 \cdot 3 \cdot 2 \cdot 1 = 5!$ and $3 \cdot 2 \cdot 1 = 3!$

By the definition of $n!$, $n[(n-1)!] = n!$ for all natural numbers $n \geq 2$. It is convenient to have this relation hold also for $n = 1$, and so, by definition,

$$\mathbf{0! = 1.}$$

EXAMPLE 3 Arranging r of n Items ($r < n$)

Suppose the teacher in **Example 2** wishes to place only 3 of the 5 books in a row. How many arrangements of 3 books are possible?

SOLUTION The teacher still has 5 ways to fill the first spot, 4 ways to fill the second spot, and 3 ways to fill the third. Only 3 books will be used, so there are only 3 spots to be filled (3 events) instead of 5. Again, we use the fundamental principle of counting.

$$5 \cdot 4 \cdot 3 = 60 \text{ arrangements}$$

✔ **Now Try Exercise 13.**

Permutations Because each ordering of three books is considered a different *arrangement*, the number 60 in the preceding example is called the number of *permutations* of 5 things taken 3 at a time, written

$$P(5, 3) = 60.$$

A **permutation** of n elements taken r at a time is one of the *arrangements* of r elements from a set of n elements. Generalizing, the number of permutations of n elements taken r at a time, denoted by $P(n, r)$, is given as follows.

$$P(n, r) = n(n-1)(n-2) \cdots (n-r+1)$$

$$P(n, r) = \frac{n(n-1)(n-2) \cdots (n-r+1)(n-r)(n-r-1) \cdots (2)(1)}{(n-r)(n-r-1) \cdots (2)(1)}$$

$$P(n, r) = \frac{n!}{(n-r)!}$$

Permutations of n Elements Taken r at a Time

If $P(n, r)$ denotes the number of permutations of n elements taken r at a time, with $r \leq n$, then the following holds.

$$\boldsymbol{P(n, r) = \frac{n!}{(n-r)!}}$$

An alternative notation for $\boldsymbol{P(n, r)}$ is $\boldsymbol{{}_nP_r}$.

EXAMPLE 4 Using the Permutations Formula

Evaluate.

(a) The number of permutations of the letters L, M, and N

(b) The number of permutations of 2 of the letters L, M, and N

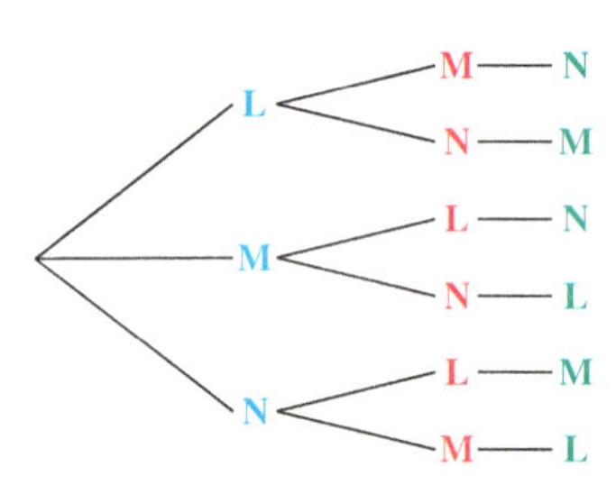

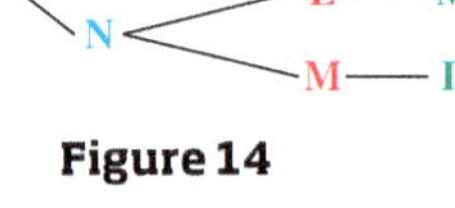

Figure 14

SOLUTION

(a) Use the formula for $P(n, r)$, with $n = 3$ and $r = 3$.

$$P(3, 3) = \frac{3!}{(3-3)!} = \frac{3!}{0!} = \frac{3 \cdot 2 \cdot 1}{1} = 6$$

As shown in the tree diagram in **Figure 14,** the 6 permutations of the letters are as follows.

LMN, LNM, MLN, MNL, NLM, NML

(b) Evaluate $P(3, 2)$.

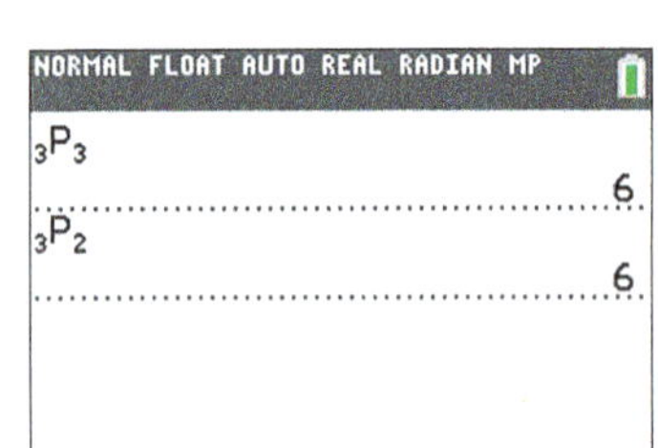

This screen shows how the TI-84 Plus calculates $P(3, 3)$ and $P(3, 2)$. **See Example 4.**

$$P(3, 2) = \frac{3!}{(3-2)!} = \frac{3!}{1!} = \frac{3 \cdot 2 \cdot 1}{1} = 6$$

This result is the same as the answer in part (a). After the first two letter choices are made, the third is already determined because only one letter is left.

✔ **Now Try Exercise 51.**

The result in **Example 4(a)** can be generalized for all n.

$$\boldsymbol{P(n, n) = n!}$$

EXAMPLE 5 Using the Permutations Formula

Suppose 8 people enter an event in a swim meet. In how many ways could the gold, silver, and bronze medals be awarded?

SOLUTION Using the fundamental principle of counting, there are 3 choices to be made, giving

$$8 \cdot 7 \cdot 6 = 336 \text{ ways.}$$

We can also use the formula for $P(n, r)$ to obtain the same result.

$P(8, 3)$

$= \frac{8!}{(8-3)!}$ Permutations formula with $n = 8$ and $r = 3$

$= \frac{8!}{5!}$ Subtract in the denominator.

$= \frac{8 \cdot 7 \cdot 6 \cdot 5 \cdot 4 \cdot 3 \cdot 2 \cdot 1}{5 \cdot 4 \cdot 3 \cdot 2 \cdot 1}$ Definition of $n!$

$= 8 \cdot 7 \cdot 6$ Divide out the common factors.

$= 336$ ways Multiply.

✔ **Now Try Exercise 49.**

EXAMPLE 6 **Using the Permutations Formula**

In how many ways can 6 students be seated in a row of 6 desks?

SOLUTION

$P(6, 6)$ ← A seating arrangement is a permutation.

$= \dfrac{6!}{(6-6)!}$ Permutations formula with $n = r = 6$

$= \dfrac{6!}{0!}$ Subtract; $0! = 1$

$= 6 \cdot 5 \cdot 4 \cdot 3 \cdot 2 \cdot 1$ Definition of 6!

$= 720$ ways Multiply.

✔ **Now Try Exercise 45.**

Combinations

In **Example 3** we saw that there are 60 ways in which a teacher can arrange 3 of 5 different books in a row. That is, there are 60 permutations of 5 things taken 3 at a time. Suppose now that the teacher does not wish to arrange the books in a row but rather wishes to choose, without regard to order, any 3 of the 5 books to donate to a book sale. In how many ways can the teacher do this?

The number 60 counts all possible *arrangements* of 3 books chosen from 5. The following 6 arrangements, however, would all lead to the same set of 3 books being given to the book sale.

mystery-biography-textbook	biography-textbook-mystery
mystery-textbook-biography	textbook-biography-mystery
biography-mystery-textbook	textbook-mystery-biography

The list shows 6 different *arrangements* of 3 books but only one *set* of 3 books. A subset of items selected *without regard to order* is a **combination.** The number of combinations of 5 things taken 3 at a time is written $C(5, 3)$ or ${}_5C_3$.

To evaluate $C(5, 3)$, start with the $5 \cdot 4 \cdot 3$ *permutations* of 5 things taken 3 at a time. Because order does not matter, and each subset of 3 items from the set of 5 items can have its elements rearranged in $3 \cdot 2 \cdot 1 = 3!$ ways, we find $C(5, 3)$ by dividing the number of permutations by 3!.

$$C(5, 3) = \frac{5 \cdot 4 \cdot 3}{3!} = \frac{5 \cdot 4 \cdot 3}{3 \cdot 2 \cdot 1} = 10$$

The teacher can choose 3 books for the book sale in 10 ways.

Generalizing this discussion gives the following formula for the number of combinations of n elements taken r at a time.

$$C(n, r) = \frac{P(n, r)}{r!}$$

An alternative version of this formula is found as follows.

$$C(n, r) = \frac{P(n, r)}{r!} = \frac{n!}{(n-r)!} \cdot \frac{1}{r!} = \frac{n!}{(n-r)!\,r!}$$

NOTE The formula for $C(n, r)$ given above is equivalent to the binomial coefficient formula, symbolized $\binom{n}{r}$, studied earlier in the chapter.

Combinations of *n* Elements Taken *r* at a Time

If $C(n, r)$ represents the number of combinations of n elements taken r at a time, with $r \leq n$, then the following holds.

$$C(n, r) = \frac{n!}{(n-r)!r!}, \quad \text{or} \quad C(n, r) = \frac{n!}{r!(n-r)!}$$

Alternative notations for $C(n, r)$ are $\binom{n}{r}$ and ${}_nC_r$.

EXAMPLE 7 Using the Combinations Formula

How many different committees of 3 people can be chosen from a group of 8 people?

SOLUTION A committee is an unordered set, so use the combinations formula with $n = 8$ and $r = 3$.

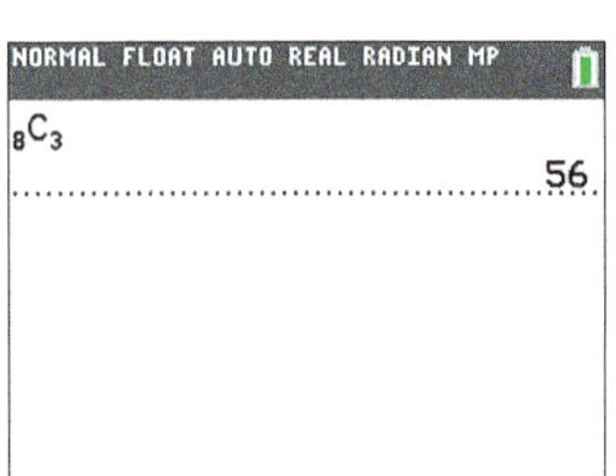

This screen shows how the TI-84 Plus calculates $C(8, 3)$. **See Example 7.**

$C(8, 3)$

$= \frac{8!}{3!(8-3)!}$ Combinations formula with $n = 8$ and $r = 3$

$= \frac{8!}{3!5!}$ Subtract in the denominator.

$= \frac{8 \cdot 7 \cdot 6 \cdot 5!}{3!5!}$ Definition of factorial

$= \frac{8 \cdot 7 \cdot 6}{6}$ $3! = 6$

$= 56$ committees Divide out the common factor; Multiply.

✔ **Now Try Exercise 53.**

EXAMPLE 8 Using the Combinations Formula

A group of stockbrokers consists of 11 women and 19 men. Four will be selected to work on a special project.

(a) In how many different ways can the stockbrokers be selected?

(b) In how many ways can the group of 4 be selected if 2 must be women and 2 must be men?

SOLUTION

(a) Here we wish to know the number of 4-element combinations that can be formed from a set of

$$11 + 19 = 30 \text{ elements.}$$

(We want combinations, not permutations, because order within the group does not matter.)

$$C(30, 4) = \frac{30!}{4!26!} = 27{,}405$$

There are 27,405 ways to select the project group.

(b) Order is not important, so we use combinations to select 2 of the 11 women and 2 of the 19 men.

$$\begin{aligned} &C(11, 2) \cdot C(19, 2) \\ &= \frac{11!}{2!9!} \cdot \frac{19!}{2!17!} && \text{Use combinations and the fundamental principle of counting.} \\ &= 55 \cdot 171 && \text{Evaluate.} \\ &= 9405 && \text{Multiply.} \end{aligned}$$

In this case, the project group can be selected in 9405 ways.

✔ **Now Try Exercise 61.**

Characteristics That Distinguish Permutations from Combinations

Consider the following table.

These are combinations. The order of the cards in the hands is *not* important.

Characteristics of Permutations and Combinations

Permutations	Combinations
These are selections of r items from n items. Repetitions are not allowed.	
Order is important.	Order is not important.
These are *arrangements* of r items from a set of n items.	These are *subsets* of r items from a set of n items.
$P(n, r) = \frac{n!}{(n-r)!}$	$C(n, r) = \binom{n}{r} = \frac{n!}{r!(n-r)!}$
Clue words: *arrangement, schedule, order*	Clue words: *group, committee, sample, selection*

EXAMPLE 9 Distinguishing Permutations and Combinations

Determine whether *permutations* or *combinations* should be used to solve each problem.

(a) How many 4-digit codes are possible if no digits are repeated?

(b) A sample of 4 light bulbs is randomly selected from a batch of 15 bulbs to be packaged and sold. How many different samples are possible?

(c) In a basketball tournament with 8 teams, how many games must be played so that each team plays every other team exactly once?

(d) In how many ways can 4 stockbrokers be assigned to 6 offices so that each broker has a private office?

SOLUTION

(a) Changing the order of the 4 digits results in a different code, so *permutations* should be used.

(b) The order in which the 4 light bulbs are selected is not important. The sample is unchanged if the items are rearranged, so *combinations* should be used.

(c) Selection of 2 teams for a game creates an *unordered* subset of 2 from the set of 8 teams. Use *combinations*.

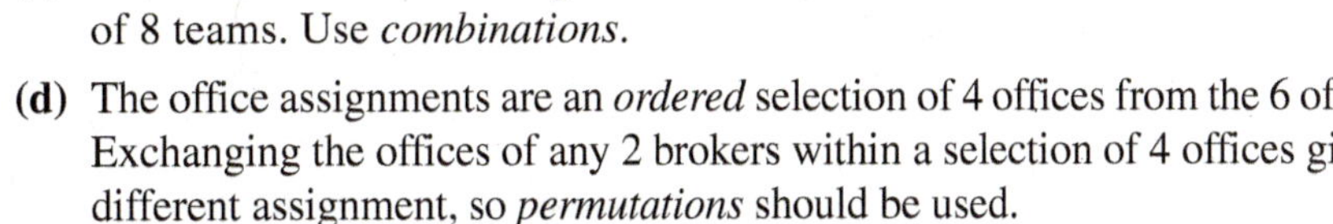

(d) The office assignments are an *ordered* selection of 4 offices from the 6 offices. Exchanging the offices of any 2 brokers within a selection of 4 offices gives a different assignment, so *permutations* should be used.

✔ **Now Try Exercise 35.**

To further illustrate the distinctions between permutations and combinations using tree diagrams, suppose we want to select 2 cans of soup from 4 cans.

noodle (N), bean (B), mushroom (M), and tomato (T)

As shown in **Figure 15(a),** there are 12 ways to select 2 cans from the 4 cans if order matters (if noodle first and bean second is considered different from bean, then noodle, for example). On the other hand, if order is unimportant, then there are 6 ways to choose 2 cans of soup from the 4 cans, as illustrated in **Figure 15(b).**

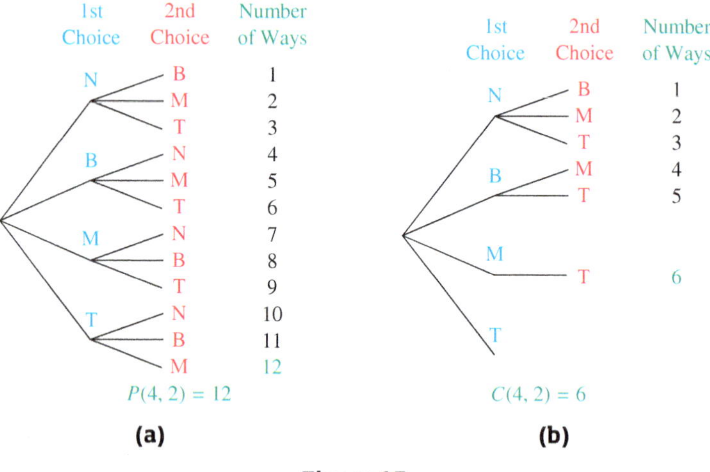

Figure 15

CAUTION Not all counting problems lend themselves to either permutations or combinations. Whenever the fundamental principle of counting or a tree diagram can be used directly, as in the soup example, use it.

11.6 Exercises

CONCEPT PREVIEW *Fill in the blank(s) to correctly complete each sentence.*

1. From the two choices *permutation* and *combination,* a computer password is an example of a ________ and a hand of cards is an example of a ________.

2. If there are 3 ways to choose a salad, 5 ways to choose an entrée, and 4 ways to choose a dessert, then there are ______ ways to form a meal consisting of these three choices.

3. There are ______ ways to form a three-digit number consisting of the digits 4, 5, and 9.

4. If there are 3 people to choose from, there are ______ ways to choose a pair of them.

5. When a fair die is rolled and a fair coin is tossed*, there are ______ possible outcomes.

6. A monogram consisting of three letters from the English alphabet can occur in ______ different ways.

*A fair die has 6 faces with a different number of dots 1–6 on each face. A fair coin has 2 sides with heads on one side, tails on the other. In both cases, all outcomes have the same chance of occurring.

Use the fundamental principle of counting to solve each problem. ***See Examples 1–3.***

7. On a business trip, Terry took 3 pairs of pants, 4 shirts, 1 jacket, and two pairs of shoes. Determine the number of outfits that Terry can choose.

8. When saddling her horse, Callie can choose from 2 saddles, 3 blankets, and 2 cinches. Find the number of possible choices for saddling Callie's horse.

9. A conference schedule offers 2 main sessions, 20 break-out sessions, and 4 mini-courses. In how many ways can an attendee choose 1 of each to attend?

10. A convenience store offers 16 types of soda with 4 options for flavoring and either crushed or cubed ice. Determine the total number of drink options available for selecting 1 soda with 1 flavor and 1 type of ice.

11. A college has 7 portraits of past college presidents to arrange in a row on a wall. How many different arrangements are possible?

12. A telephone messaging system requires a 4-digit security code. How many security codes are possible if numbers may be repeated?

13. In how many ways can judges select a 1st-place winner, a 2nd-place winner, and a 3rd-place winner from 16 desserts entered in a cooking contest?

14. In how many different ways can 4 different boys be selected from a group of 25 boys on a track team to receive 4 different awards?

Evaluate each expression. ***See Examples 4–8.***

15. $P(12, 2)$ **16.** $P(5, 2)$ **17.** $P(9, 2)$ **18.** $P(10, 4)$

19. $P(5, 1)$ **20.** $P(6, 1)$ **21.** $C(4, 2)$ **22.** $C(9, 3)$

23. $C(6, 0)$ **24.** $C(8, 0)$ **25.** $C(12, 4)$ **26.** $C(16, 3)$

Use a calculator to evaluate each expression. ***See Examples 4 and 7.***

27. ${}_{20}P_5$ **28.** ${}_{100}P_5$ **29.** ${}_{15}P_8$ **30.** ${}_{32}P_4$

31. ${}_{20}C_5$ **32.** ${}_{100}C_5$ **33.** ${}_{15}C_8$ **34.** ${}_{32}C_4$

35. Decide whether the situation described involves a permutation or a combination of objects. **See Example 9.**

(a) a telephone number **(b)** a Social Security number

(c) a hand of cards in poker **(d)** a committee of politicians

(e) the "combination" on a padlock **(f)** an automobile license plate

(g) a lottery choice of six numbers where order does not matter

36. ***Concept Check*** What is the difference between a permutation and a combination? Give an example of each.

Use the fundamental principle of counting or permutations to solve each problem. ***See Examples 1–6.***

37. ***Home Plan Choices*** How many different types of homes are available if a builder offers a choice of 5 basic plans, 4 roof styles, and 2 exterior finishes?

38. ***Auto Varieties*** An auto manufacturer produces 7 models, each available in 6 different colors, with 4 different upholstery fabrics, and 5 interior colors. How many varieties of the auto are available?

39. ***Radio-Station Call Letters*** How many different 4-letter radio-station call letters can be made under the following conditions? (Disregard the fact that some may be unacceptable for various reasons.)

(a) The first letter must be K or W, and no letter may be repeated.

(b) Repetitions are allowed (but the first letter is K or W).

(c) The first letter must be K or W, the last letter must be R, and repetitions are not allowed.

40. *Meal Choices* A menu offers a choice of 3 salads, 8 main dishes, and 5 desserts. How many different 3-course meals (salad, main dish, dessert) are possible?

41. *Arranging Blocks* Baby Finley is arranging 7 blocks in a row. How many different arrangements can he make?

42. *Names for a Baby* A couple has narrowed down the choice of a name for their new baby to 5 first names and 3 middle names. How many different first- and middle-name combinations are possible?

Finley

43. *License Plates* For many years, the state of California used 3 letters followed by 3 digits on its automobile license plates.

(a) How many different license plates are possible with this arrangement?

(b) When the state ran out of new plates, the order was reversed to 3 digits followed by 3 letters. How many additional plates were then possible?

(c) When the plates described in part (b) were also used up, the state then issued plates with 1 letter followed by 3 digits and then 3 letters. How many plates does this scheme provide?

44. *Telephone Numbers* How many 7-digit telephone numbers are possible if the first digit cannot be 0 and the following conditions apply?

(a) Only odd digits may be used.

(b) The telephone number must be a multiple of 10 (that is, it must end in 0).

(c) The telephone number must be a multiple of 100.

(d) The first 3 digits are 481.

(e) No repetitions are allowed.

45. *Seating People in a Row* In an experiment on social interaction, 9 people will sit in 9 seats in a row. In how many ways can this be done?

46. *Genetics Experiment* In how many ways can 7 of 10 rats be arranged in a row for a genetics experiment?

47. *Course Schedule Arrangement* A business school offers courses in keyboarding, spreadsheets, transcription, business English, technical writing, and accounting. In how many ways can a student arrange a schedule if 3 courses are taken?

48. *Course Schedule Arrangement* If your college offers 400 courses, 20 of which are in mathematics, and your counselor arranges your schedule of 4 courses by random selection, how many schedules are possible that do not include a math course?

49. *Club Officer Choices* In a club with 15 members, in how many ways can a slate of 3 officers consisting of president, vice-president, and secretary/treasurer be chosen?

50. *Batting Orders* A baseball team has 20 players. How many 9-player batting orders are possible?

51. *Letter Arrangement* Consider the word BRUCE.

(a) In how many ways can all the letters of the word BRUCE be arranged?

(b) In how many ways can all the first 3 letters of the word BRUCE be arranged?

52. *Basketball Positions* In how many ways can 5 players be assigned to the 5 positions on a basketball team, assuming that any player can play any position? In how many ways can 10 players be assigned to the 5 positions?

Solve each problem involving combinations. ***See Examples 7 and 8.***

53. *Seminar Presenters* A banker's association has 40 members. If 6 members are selected at random to present a seminar, how many different groups of 6 are possible?

54. *Financial Planners* Four financial planners are to be selected from a group of 12 to participate in a special program. In how many ways can this be done? In how many ways can the group that will not participate be selected?

55. *Apple Samples* How many different samples of 4 apples can be drawn from a crate of 25 apples?

56. *Apple Samples* Suppose that in **Exercise 55** there are 5 rotten apples in the crate.

(a) How many samples of 3 could be drawn in which all 3 are rotten?

(b) How many samples of 3 could be drawn in which there are 2 good apples and 1 rotten apple?

57. *Hamburger Choices* Howard's Hamburger Heaven sells hamburgers with cheese, relish, lettuce, tomato, mustard, or ketchup.

(a) How many different hamburgers can be made that use any 4 of the extras?

(b) How many different hamburgers can be made if one of the 4 extras must be cheese?

58. *Card Combinations* Five playing cards having the numbers 2, 3, 4, 5, and 6 are shuffled and 2 cards are then drawn. How many different 2-card hands are possible?

59. *Marble Samples* If a bag contains 15 marbles, how many samples of 2 marbles can be drawn from it? How many samples of 4 marbles can be drawn?

60. *Marble Samples* In **Exercise 59,** if the bag contains 3 yellow, 4 white, and 8 blue marbles, how many samples of 2 can be drawn in which both marbles are blue?

61. *Convention Delegation Choices* A city council is composed of 5 liberals and 4 conservatives. Three members are to be selected randomly as delegates to a convention.

(a) How many delegations are possible?

(b) How many delegations could have all liberals?

(c) How many delegations could have 2 liberals and 1 conservative?

(d) If 1 member of the council serves as mayor, how many delegations are possible that include the mayor?

62. *Delegation Choices* Seven workers decide to send a delegation of 2 to their supervisor to discuss their grievances.

(a) How many different delegations are possible?

(b) If it is decided that a certain employee must be in the delegation, how many different delegations are possible?

(c) If there are 2 women and 5 men in the group, how many delegations would include at least 1 woman?

Use any method described in this section to solve each problem. ***See Examples 1–9.***

63. *Course Schedule Arrangement* If Dwight has 8 courses to choose from, how many ways can he arrange his schedule if he must pick 4 of them?

64. *Pineapple Samples* How many samples of 9 pineapples can be drawn from a crate of 12?

65. *Soup Ingredients* Velma specializes in making different vegetable soups with carrots, celery, beans, peas, mushrooms, and potatoes. How many different soups can she make with any 4 ingredients?

66. *Secretary/Manager Assignments* From a pool of 7 secretaries, 3 are selected to be assigned to 3 managers, 1 secretary to each manager. In how many ways can this be done?

67. *Musical Chairs Seatings* In a game of musical chairs, 13 children will sit in 12 chairs. (1 will be left out.) How many seating arrangements are possible?

68. *Plant Samples* In an experiment on plant hardiness, a researcher gathers 6 wheat plants, 3 barley plants, and 2 rye plants. She wishes to select 4 plants at random.

(a) In how many ways can this be done?

(b) In how many ways can this be done if exactly 2 wheat plants must be included?

69. *Committee Choices* In a club with 8 women and 11 men members, how many 5-member committees can be chosen that satisfy the following conditions?

(a) All are women.
(b) All are men.
(c) There are 3 women and 2 men.
(d) There are no more than 3 men.

70. *Committee Choices* From 10 names on a ballot, 4 will be elected to a political party committee. In how many ways can the committee of 4 be formed if each person will have a different responsibility?

71. *Combination Lock* A briefcase has 2 locks. The combination to each lock consists of a 3-digit number, where digits may be repeated. How many combinations are possible? (*Hint:* The word *combination* is a misnomer. Lock combinations are permutations where the arrangement of the numbers is important.)

72. *Combination Lock* A typical "combination" for a padlock consists of 3 numbers from 0 to 39. Find the number of "combinations" that are possible with this type of lock, if a number may be repeated.

73. *Garage Door Openers* The code for some garage door openers consists of 12 electrical switches that can be set to either 0 or 1 by the owner. With this type of opener, how many codes are possible? (*Source:* Promax.)

74. *Lottery* To win the jackpot in a lottery game, a person must pick 4 numbers from 0 to 9 in the correct order. If a number can be repeated, how many ways are there to play the game?

75. *Keys* In how many distinguishable ways can 4 keys be put on a circular key ring?

76. *Sitting at a Round Table* In how many different ways can 8 people sit at a round table? Assume that "a different way" means that at least 1 person is sitting next to someone different.

Prove each statement for positive integers n and r, with $r \leq n$. (Hint: Use the definitions of permutations and combinations.)

77. $P(n, n-1) = P(n, n)$
78. $P(n, 1) = n$
79. $P(n, 0) = 1$
80. $P(n, n) = n!$
81. $C(n, n) = 1$
82. $C(n, 0) = 1$
83. $C(0, 0) = 1$
84. $C(n, n-1) = n$
85. $C(n, n-r) = C(n, r)$

86. Explain why the restriction $r \leq n$ is needed in the formulas for $C(n, r)$ and $P(n, r)$.

11.7 Basics of Probability

- Basic Concepts
- Complements and Venn Diagrams
- Odds
- Compound Events
- Binomial Probability

Basic Concepts

In probability, each repetition of an experiment is a **trial.** The possible results of each trial are **outcomes** of the experiment. In this section, we are concerned with outcomes that are equally likely to occur. (We assume that a die has 6 faces and a coin has 2 sides.)

For example, the experiment of tossing a fair coin has two equally likely outcomes:

$$\text{landing heads up } (H) \quad \text{or} \quad \text{landing tails up } (T).$$

Also, the experiment of rolling a fair die has 6 equally likely outcomes:

landing so the face that is up shows 1, 2, 3, 4, 5, or 6 dots.

The set S of all possible outcomes of a given experiment is the **sample space** for the experiment. (In this section, all sample spaces are finite.) A sample space S can be written in set notation.

Experiment	Sample Space
Toss a coin.	$S = \{H, T\}$
Roll a die.	$S = \{1, 2, 3, 4, 5, 6\}$
Toss two coins.	$S = \{(H, H), (H, T), (T, H), (T, T)\}$
Answer a true/false question.	$S = \{\text{true, false}\}$

Use set notation for a sample space.

Any subset of a sample space is an **event.** In the experiment with the die, for example, "the number showing is a 3" is an event, say E_1, such that $E_1 = \{3\}$. "The number showing is greater than 3" is also an event, say E_2, such that $E_2 = \{4, 5, 6\}$. To represent the number of outcomes that belong to event E, the notation $\boldsymbol{n(E)}$ is used. Then

$$n(E_1) = 1 \quad \text{and} \quad n(E_2) = 3.$$

The notation $\boldsymbol{P(E)}$ is used for the *probability* of an event E. If the outcomes in the sample space for an experiment are equally likely, then the probability of event E occurring is found as follows.

Probability of Event *E*

In a sample space with equally likely outcomes, the **probability** of event E, written $\boldsymbol{P(E)}$**,** is the ratio of the number of outcomes in sample space S that belong to event E, $n(E)$, to the total number of outcomes in sample space S, $n(S)$.

$$P(E) = \frac{n(E)}{n(S)}$$

To find the probability of event E_1 in the die experiment, start with the sample space, $S = \{1, 2, 3, 4, 5, 6\}$, and the desired event, $E_1 = \{3\}$.

$$P(E_1) = \frac{n(E_1)}{n(S)} = \frac{1}{6} \qquad \text{Use } n(E_1) = 1 \text{ and } n(S) = 6.$$

EXAMPLE 1 **Finding Probabilities of Events**

A single fair die is rolled. Write each event in set notation and give the probability of the event.

(a) E_3: the number showing is even

(b) E_4: the number showing is greater than 4

(c) E_5: the number showing is less than 7

(d) E_6: the number showing is 7

SOLUTION

(a) Because $E_3 = \{2, 4, 6\}$, we have $n(E_3) = 3$. As given earlier, $n(S) = 6$.

$$P(E_3) = \frac{3}{6} = \frac{1}{2}$$

(b) Again we have $n(S) = 6$. Event $E_4 = \{5, 6\}$, and thus $n(E_4) = 2$.

$$P(E_4) = \frac{2}{6} = \frac{1}{3}$$

(c) $E_5 = \{1, 2, 3, 4, 5, 6\}$ and $P(E_5) = \frac{6}{6} = 1$

(d) $E_6 = \varnothing$ and $P(E_6) = \frac{0}{6} = 0$

✔ **Now Try Exercises 7, 9, 13, and 15.**

In **Example 1(c),** $E_5 = S$. Therefore, event E_5 is *certain* to occur every time the experiment is performed. On the other hand, in **Example 1(d),** $E_6 = \varnothing$ and $P(E_6) = 0$, so E_6 is *impossible.*

Probability Values and Terminology

- A **certain event**—that is, an event that is certain to occur—always has probability 1.
- The probability of an **impossible event** is always 0 because none of the outcomes in the sample space satisfies the event.
- For any event E, $\boldsymbol{P(E)}$ **is between 0 and 1 inclusive of both.**

Complements and Venn Diagrams The set of all outcomes in the sample space that do *not* belong to event E is the **complement** of E, written $\boldsymbol{E'}$. For example, in the experiment of drawing a single card from a standard deck of 52 cards, let E be the event "the card is an ace." Then E' is the event "the card is not an ace." From the definition of E', for an event E,

$$E \cup E' = S \quad \text{and} \quad E \cap E' = \varnothing.*$$

*The **union** of two sets A and B is the set $\boldsymbol{A \cup B}$ of all elements from either A or B, or both. The **intersection** of sets A and B, written $\boldsymbol{A \cap B}$, includes all elements that belong to both sets.

NOTE A standard deck of 52 cards has four suits: hearts ♥, diamonds ♦, spades ♠, and clubs ♣. There are 13 cards in each suit, including a jack, a queen, and a king (sometimes called the "face cards"), an ace, and cards numbered from 2 to 10. The hearts and diamonds are red, and the spades and clubs are black. We refer to this standard deck of cards in this section.

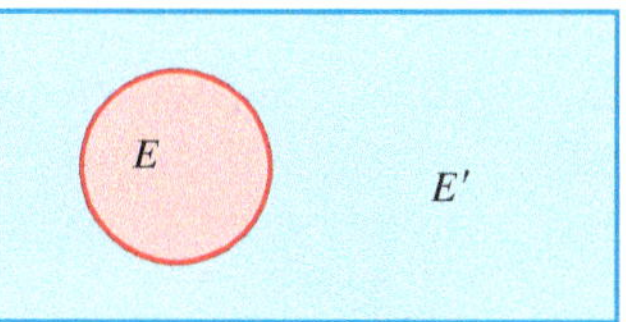

Figure 16

Probability concepts can be illustrated using **Venn diagrams,** as shown in **Figure 16.** The rectangle there represents the sample space in an experiment. The area inside the circle represents event E, and the area inside the rectangle, but outside the circle, represents event E'.

EXAMPLE 2 Using the Complement of an Event

In the experiment of drawing a card from a standard deck, find the probabilities of event E, "the card is an ace," and of event E'.

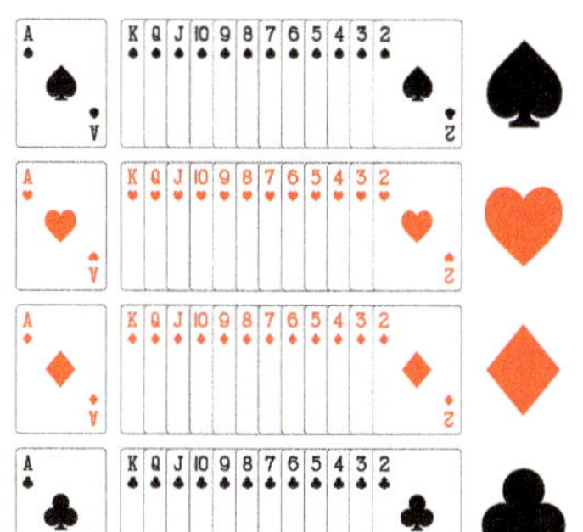

Standard deck of 52 cards

SOLUTION There are 4 aces in a standard deck of 52 cards, so $n(E) = 4$ and $n(S) = 52$.

$$P(E) = \frac{n(E)}{n(S)} = \frac{4}{52} = \frac{1}{13} \quad \text{Write in lowest terms.}$$

Of the 52 cards, 48 are not aces, so $n(E') = 48$.

$$P(E') = \frac{n(E')}{n(S)} = \frac{48}{52} = \frac{12}{13} \quad \text{Write in lowest terms.}$$

✔ **Now Try Exercises 23(a) and (b).**

In **Example 2,** $P(E) + P(E') = \frac{1}{13} + \frac{12}{13} = 1$. This is always true for any event E and its complement E'.

Rules for Complementary Events

If events E and E' are complements, then all of the following hold true.

$$\mathbf{P(E) + P(E') = 1 \qquad P(E) = 1 - P(E') \qquad P(E') = 1 - P(E)}$$

These equations suggest an alternative way to compute the probability of an event. For example, if it is known that $P(E) = \frac{1}{13}$, then

$$P(E') = 1 - \frac{1}{13} = \frac{12}{13}.$$

Odds Probability statements can be expressed in terms of *odds*, a comparison of $P(E)$ with $P(E')$. The **odds** in favor of an event E are expressed as

the ratio of $P(E)$ to $P(E')$, or as the quotient $\dfrac{P(E)}{P(E')}$.

For example, if the probability of rain can be established as $\frac{1}{3}$, the odds that it will rain are

$$P(\text{rain}) \text{ to } P(\text{no rain}) = \frac{1}{3} \text{ to } \frac{2}{3} = \frac{\frac{1}{3}}{\frac{2}{3}} = \frac{1}{3} \div \frac{2}{3} = \frac{1}{3} \cdot \frac{3}{2} = \frac{1}{2}, \quad \text{or} \quad 1 \text{ to } 2.$$

On the other hand, the odds *against* rain are 2 to 1 $\left(\text{or } \frac{2}{3} \text{ to } \frac{1}{3}\right)$. If the odds in favor of an event are, say, 3 to 5, then the probability of the event is $\frac{3}{8}$, and the probability of the complement of the event is $\frac{5}{8}$.

Rules for Odds

If m represents the number of outcomes in event E and n represents the number of outcomes in event E', then the following hold true.

$$P(E) = \frac{m}{m+n} \quad \text{and} \quad P(E') = \frac{n}{m+n}$$

The odds in favor of event E are $\frac{P(E)}{P(E')} = \frac{m}{n}$, or **$m$ to n.**

The odds against event E are $\frac{P(E')}{P(E)} = \frac{n}{m}$, or **$n$ to m.**

EXAMPLE 3 Finding Odds in Favor of an Event

A shirt is selected at random from a dark closet containing 6 blue shirts and 4 shirts that are not blue. Find the odds in favor of a blue shirt being selected.

SOLUTION Let E represent "a blue shirt is selected." Then

$$P(E) = \frac{6}{6+4} = \frac{6}{10} = \frac{3}{5} \quad \text{and} \quad P(E') = \frac{4}{6+4} = \frac{4}{10} = \frac{2}{5}.$$

Therefore, the odds in favor of a blue shirt being selected are found as follows.

$$\frac{P(E)}{P(E')} = \frac{\frac{3}{5}}{\frac{2}{5}} = \frac{3}{5} \div \frac{2}{5} = \frac{3}{5} \cdot \frac{5}{2} = \frac{3}{2}, \quad \text{or} \quad 3 \text{ to } 2$$

The odds *in favor of* a blue shirt being selected are 3 to 2, so we can quickly determine that the odds *against* selecting a blue shirt are 2 to 3.

✔ Now Try Exercise 23(e).

Compound Events A **compound event** involves an *alternative*, as in "H or K," where H and K are events. For example, suppose a fair die is rolled. Let H be the event "the result is a 3," and K the event "the result is an even number." From earlier in this section, we have the following.

$$H = \{3\} \qquad K = \{2, 4, 6\} \qquad H \cup K = \{2, 3, 4, 6\}$$

$$P(H) = \frac{1}{6} \qquad P(K) = \frac{3}{6} = \frac{1}{2} \qquad P(H \cup K) = \frac{4}{6} = \frac{2}{3}$$

Notice that in this case, $P(H) + P(K) = P(H \cup K)$.

Consider another event G for this experiment, "the result is a 2."

$$G = \{2\} \qquad K = \{2, 4, 6\} \qquad G \cup K = \{2, 4, 6\}$$

$$P(G) = \frac{1}{6} \qquad P(K) = \frac{3}{6} = \frac{1}{2} \qquad P(G \cup K) = \frac{3}{6} = \frac{1}{2}$$

In this case, $P(G) + P(K) \neq P(G \cup K)$.

As **Figure 17** suggests, the difference in the two preceding examples comes from the fact that events H and K cannot occur simultaneously. Such events are **mutually exclusive events.** In fact,

$$H \cap K = \emptyset, \quad \text{which is true for any two mutually exclusive events.}$$

Events G and K, however, can occur simultaneously. Both are satisfied if the result of the roll is a 2, the element in their intersection $(G \cap K = \{2\})$.

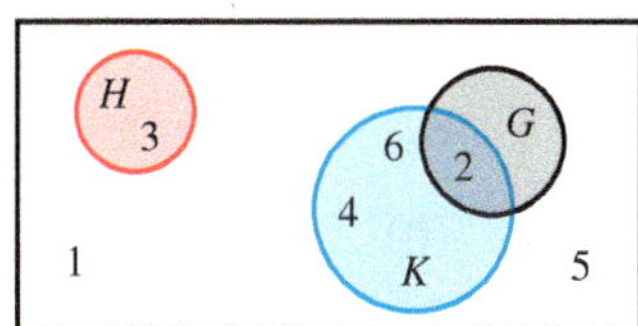

Figure 17

Probability of Compound Events

For any events E and F, the following holds.

$$\begin{aligned} P(E \text{ or } F) &= P(E \cup F) \\ &= P(E) + P(F) - P(E \cap F) \end{aligned}$$

CAUTION ***When finding the probability of a union, remember to subtract the probability of the intersection from the sum of the probabilities of the individual events.***

EXAMPLE 4 Finding Probabilities of Compound Events

One card is drawn from a standard deck of 52 cards. What is the probability of the following compound events?

(a) The card is an ace or a spade. **(b)** The card is a 3 or a king.

SOLUTION

(a) The events "drawing an ace" and "drawing a spade" are not mutually exclusive. It is possible to draw the ace of spades, an outcome satisfying both events.

$$P(\text{ace or spade}) = P(\text{ace}) + P(\text{spade}) - P(\text{ace and spade})$$ Probability of compound events

There are 4 aces, 13 spades, and 1 ace of spades.

$$= \frac{4}{52} + \frac{13}{52} - \frac{1}{52}$$ Find and substitute known probabilities.

$$= \frac{16}{52}$$ Add and subtract fractions.

$$= \frac{4}{13}$$ Write in lowest terms.

(b) "Drawing a 3" and "drawing a king" are mutually exclusive events because it is impossible to draw one card that is both a 3 and a king.

$$P(3 \text{ or } K) = P(3) + P(K) - P(3 \text{ and } K) \quad \text{Probability of compound events}$$

$$= \frac{4}{52} + \frac{4}{52} - 0 \quad \text{Find and substitute known probabilities.}$$

$$= \frac{8}{52} \quad \text{Add and subtract fractions.}$$

$$= \frac{2}{13} \quad \text{Write in lowest terms.}$$

✔ **Now Try Exercise 23(d).**

EXAMPLE 5 Finding Probabilities of Compound Events

Suppose two fair dice are rolled. Find each probability.

(a) The first die shows a 2, or the sum of the two dice is 6 or 7.

(b) The sum of the dots showing is at most 4.

SOLUTION

(a) Think of the two dice as being distinguishable—one red and one green, for example. (Actually, the sample space is the same even if they are not apparently distinguishable.) A sample space with equally likely outcomes is shown in **Figure 18,** where $(1, 1)$ represents the event "the first die (red) shows a 1 and the second die (green) shows a 1," $(1, 2)$ represents "the first die shows a 1 and the second die shows a 2," and so on.

(1, 1)	(1, 2)	(1, 3)	(1, 4)	(1, 5)	(1, 6)
(2, 1)	(2, 2)	(2, 3)	(2, 4)	(2, 5)	(2, 6)
(3, 1)	(3, 2)	(3, 3)	(3, 4)	(3, 5)	(3, 6)
(4, 1)	(4, 2)	(4, 3)	(4, 4)	(4, 5)	(4, 6)
(5, 1)	(5, 2)	(5, 3)	(5, 4)	(5, 5)	(5, 6)
(6, 1)	(6, 2)	(6, 3)	(6, 4)	(6, 5)	(6, 6)

Event *A* (row 2)

Event *B* (diagonal band)

Figure 18

Let *A* represent the event "the first die shows a 2," and *B* represent the event "the sum of the two dice is 6 or 7." See **Figure 18.** Event *A* has 6 elements, event *B* has 11 elements, and the sample space has 36 elements.

$$P(A) = \frac{6}{36}, \quad P(B) = \frac{11}{36}, \quad \text{and} \quad P(A \cap B) = \frac{2}{36}$$

$$P(A \cup B) = P(A) + P(B) - P(A \cap B) \quad \text{Probability of compound events}$$

$$= \frac{6}{36} + \frac{11}{36} - \frac{2}{36} \quad \text{Substitute known probabilities.}$$

$$= \frac{15}{36} \quad \text{Add and subtract fractions.}$$

$$= \frac{5}{12} \quad \text{Write in lowest terms.}$$

(b) "At most 4" can be written as "2 or 3 or 4." (A sum of 1 is meaningless here.) The events represented by "2," "3," and "4" are mutually exclusive.

$$P(\text{at most } 4) = P(2 \text{ or } 3 \text{ or } 4) = P(2) + P(3) + P(4) \quad (1)$$

The sample space for this experiment includes the 36 possible pairs of numbers shown in **Figure 18.** The pair $(1, 1)$ is the only one with a sum of 2, so $P(2) = \frac{1}{36}$. Also $P(3) = \frac{2}{36}$ because both $(1, 2)$ and $(2, 1)$ give a sum of 3. The pairs $(1, 3)$, $(2, 2)$, and $(3, 1)$ have a sum of 4, so $P(4) = \frac{3}{36}$.

$$P(\text{at most } 4) = \frac{1}{36} + \frac{2}{36} + \frac{3}{36} \quad \text{Substitute into equation (1).}$$

$$= \frac{6}{36} \quad \text{Add fractions.}$$

$$= \frac{1}{6} \quad \text{Write in lowest terms.}$$

✔ **Now Try Exercise 23(c).**

Summary of Properties of Probability

For any events E and F, the following hold true.

1. $0 \le P(E) \le 1$

2. $P(\text{a certain event}) = 1$

3. $P(\text{an impossible event}) = 0$

4. $P(E') = 1 - P(E)$

5. $P(E \text{ or } F) = P(E \cup F)$
$\quad = P(E) + P(F) - P(E \cap F)$

Binomial Probability A probability experiment may consist of a repeated number of independent trials (n) with only two possible outcomes.

Consider the example of tossing a coin 5 times and observing the number of tails. In this experiment there are $n = 5$ independent trials, or coin tosses, and there are two possible outcomes, head or tail, for each trial. It is common to consider "obtaining a tail" as a success because it is the outcome of interest, so "obtaining a head" would be considered a failure.

If a probability experiment consists of n independent trials with two possible outcomes for each trial, and the probabilities remain constant for each trial, then it is a **binomial experiment.** Recall that the expression $C(n, r)$ is equivalent to the binomial coefficient $\binom{n}{r}$.

Binomial Probability

Let p represent the probability of a success, and let $q = 1 - p$ represent the probability of a failure. In a binomial experiment, the probability of obtaining exactly r successes in n trials is found as follows.

$$P(r \text{ successes in } n \text{ trials}) = \binom{n}{r} p^r q^{n-r}$$

Suppose that we want to determine the probability of getting exactly 3 tails in 5 coin tosses. Here $n = 5$, $r = 3$, $p = P(\text{tail}) = \frac{1}{2}$, and $q = P(\text{head}) = 1 - \frac{1}{2} = \frac{1}{2}$.

$P(3 \text{ tails in } 5 \text{ coin tosses})$

$$= \binom{5}{3}\left(\frac{1}{2}\right)^3\left(\frac{1}{2}\right)^{5-3}$$ Use the binomial probability formula.

$$= \frac{5!}{3!\,2!}\left(\frac{1}{2}\right)^3\left(\frac{1}{2}\right)^2$$ Apply the formula for $C(n, r) = \binom{n}{r}$ and subtract.

$$= 10\left(\frac{1}{2}\right)^3\left(\frac{1}{2}\right)^2$$ Evaluate factorials and divide.

$$= 0.3125$$ Apply the exponents and multiply.

EXAMPLE 6 Finding Probabilities in a Binomial Experiment

An experiment consists of rolling a fair die 10 times and observing the number of 3s.

(a) Find the probability of getting exactly 4 threes.

(b) Find the probability that the result is not a 3 in exactly 9 of the rolls.

ALGEBRAIC SOLUTION

(a) There are $n = 10$ independent trials with $p = P(3) = \frac{1}{6}$ and $q = 1 - \frac{1}{6} = \frac{5}{6}$.

$P(4 \text{ threes in } 10 \text{ rolls})$

$$= \binom{10}{4}\left(\frac{1}{6}\right)^4\left(\frac{5}{6}\right)^{10-4}$$

$$= 210\left(\frac{1}{6}\right)^4\left(\frac{5}{6}\right)^6$$

$$\approx 0.054$$

(b) Here $n = 10$, $p = P(\text{not a } 3) = \frac{5}{6}$, and $q = \frac{1}{6}$.

$P(9 \text{ non-threes in } 10 \text{ rolls})$

$$= \binom{10}{9}\left(\frac{5}{6}\right)^9\left(\frac{1}{6}\right)^1$$

$$\approx 0.323$$

GRAPHING CALCULATOR SOLUTION

Graphing calculators, such as the TI-84 Plus, that have statistical distribution functions give binomial probabilities. **Figure 19** shows the results for parts (a) and (b). The numbers in parentheses separated by commas represent n, p, and r, respectively.

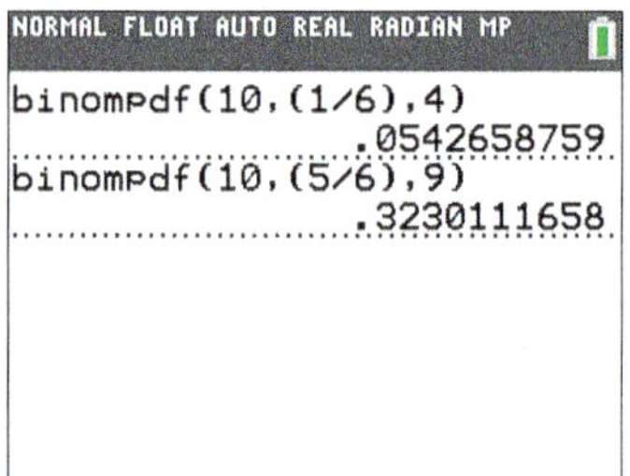

Figure 19

✔ **Now Try Exercise 41.**

11.7 Exercises

CONCEPT PREVIEW *Fill in the blank(s) to correctly complete each sentence.*

1. When a fair coin is tossed, there are ______ possible outcomes, and the probability of each outcome is ______.

2. When a fair die is rolled, there are ______ possible outcomes, and the probability of each outcome is ______.

3. When two different denominations of fair coins are tossed, there are ______ possible outcomes, and the probability of each outcome is ______.

4. When two distinct fair dice are rolled, there are ______ possible outcomes, and the probability of each outcome is ______.

5. When a fair coin is tossed 4 times, the probability of obtaining heads on all tosses is ______.

6. When a fair coin is tossed and a fair die is rolled, the probability of obtaining a "head" and a "3" is ______.

Concept Check *Write a sample space with equally likely outcomes for each experiment.*

7. Two fair coins are tossed.

8. A two-headed coin is tossed once.

9. Three fair coins are tossed.

10. Slips of paper marked with the numbers 1, 2, 3, and 4 are placed in a box. A slip is drawn and set aside, its number is recorded, and then a second slip is drawn.

11. The spinner shown here is spun twice.

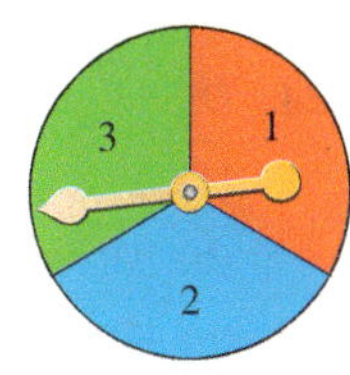

12. A fair die is rolled and then a fair coin is tossed.

Write each event in set notation and give the probability of the event. ***See Example 1.***

13. Refer to **Exercise 7.**

    (a) Both coins show the same face. **(b)** At least one coin is a head.

14. Refer to **Exercise 8.**

    (a) The result of the toss is heads. **(b)** The result of the toss is tails.

15. Refer to **Exercise 9.**

    (a) All three coins show the same face.

    (b) At least two coins are tails.

16. Refer to **Exercise 10.**

    (a) Both slips are marked with even numbers.

    (b) Both slips are marked with odd numbers.

    (c) Both slips are marked with the same number.

    (d) One slip is marked with an odd number, the other with an even number.

17. Refer to **Exercise 11.**

    (a) The result is a repeated number. **(b)** The second number is 1 or 3.

    (c) The first number is even and the second number is odd.

18. *Concept Check* A student gives the probability of an event in a problem as $\frac{6}{5}$. Why must this answer be incorrect?

19. *Concept Check* If the probability of an event is 0.857, what is the probability that the event will not occur?

20. *Concept Check* Associate each probability in parts (a)–(e) with one of the statements in choices A–E.

(a) $P(E) = 0.01$ **(b)** $P(E) = 1$ **(c)** $P(E) = 0.99$

(d) $P(E) = 0$ **(e)** $P(E) = 0.5$

A. The event is certain. **B.** The event is impossible.

C. The event is very likely to occur. **D.** The event is very unlikely to occur.

E. The event is just as likely to occur as not to occur.

Work each problem. ***See Examples 1–6.***

21. *Batting Average* A baseball player with a batting average of .300 comes to bat. What are the odds in favor of the ball player getting a hit?

22. *Small Business Loan* The probability that a bank with assets greater than or equal to \$30 billion will make a loan to a small business is 0.002. What are the odds against such a bank making a small business loan?

23. *Drawing a Card* A card is drawn at random from a standard deck of 52 cards. Find the probabilities in parts (a)–(d).

(a) The card is a spade. **(b)** The card is not a spade.

(c) The card is a spade or a heart. **(d)** The card is a spade or a face card.

(e) What are the odds in favor of drawing a spade?

24. *Dice Rolls* Two fair dice are rolled. Find the probabilities in parts (a)–(d).

(a) The sum of the dots is at least 10.

(b) The sum of the dots is less than 10.

(c) The sum of the dots is either 7 or at least 10.

(d) The sum of the dots is 2, or the dice both show the same number.

(e) What are the odds against rolling a 7?

25. *Origins of Foreign-Born Population* The numbers (in thousands) of foreign-born people who were living in the United States in 2012, according to region of birth, are given in the table. Find the probability, to the nearest thousandth, that a foreign-born U.S. resident in 2012 satisfied the following in parts (a)–(c).

Region	Number (in thousands)
Asia	11,587
Europe	11,596
Latin America	21,034
Other	2809

Source: U.S. Census Bureau.

(a) born in Asia **(b)** not born in Europe

(c) born in Asia or Europe

(d) What are the odds that a randomly selected foreign-born U.S. resident was born in Latin America?

26. *U.S. Population by Region* The U.S. resident population by region (in millions) for selected years is given in the table. Find the probability, to the nearest thousandth, that a U.S. resident selected at random satisfied the following in parts (a)–(d).

Region	2000	2006	2013
Northeast	53.6	54.5	55.9
Midwest	64.4	66.0	67.5
South	100.2	109.1	118.4
West	63.2	68.8	74.3

Source: U.S. Census Bureau.

(a) lived in the West in 2006

(b) lived in the Midwest in 2000

(c) lived in the Northeast or Midwest in 2006

(d) lived in the South or West in 2013

(e) What are the odds that a randomly selected U.S. resident in 2013 was not from the South?

27. *State Lottery* One game in a state lottery requires you to pick 1 heart, 1 club, 1 diamond, and 1 spade, in that order, from the 13 cards in each suit. What is the probability of getting all four picks correct and winning \$5000?

28. *State Lottery* If three of the four selections in **Exercise 27** are correct, the player wins \$200. Find the probability of this occurring.

29. *Male Life Table* The table is an abbreviated version of the 2010 **period life table** used by the Office of the Chief Actuary of the Social Security Administration. (The actual table includes every age, not just every tenth age.) Theoretically, this table follows a group of 100,000 males at birth and gives the number still alive at each age. Round answers to the questions that follow to the nearest thousandth.

Exact Age	Number of Lives	Exact Age	Number of Lives
0	100,000	60	86,010
10	99,155	70	73,355
20	98,731	80	49,939
30	97,450	90	17,152
40	95,889	100	873
50	92,820	110	2

Source: Office of the Actuary, Social Security Administration.

(a) What is the probability that a 40-year-old man will live 30 more years?

(b) What is the probability that a 40-year-old man will not live 30 more years?

(c) Consider a group of five 40-year-old men. What is the probability that exactly three of them survive to age 70? (*Hint:* The longevities of the individual men can be considered as independent trials.)

(d) Consider two 40-year-old men. What is the probability that at least one of them survives to age 70? (*Hint:* The complement of *at least one* is *none.*)

30. *Opinion Survey* The management of a firm wishes to survey the opinions of its workers, classified as follows for the purpose of an interview:

30% have worked for the company 5 or more years,

28% are female,

65% contribute to a voluntary retirement plan, and 50% of the female workers contribute to the retirement plan.

Find each probability if a worker is selected at random.

(a) A male worker is selected.

(b) A worker is selected who has worked for the company less than 5 yr.

(c) A worker is selected who contributes to the retirement plan or is female.

31. *Growth in Stock Value* A financial analyst has determined the possibilities (and their probabilities) for the growth in value of a certain stock during the next year. (Assume these are the only possibilities.) See the table. For instance, the probability of a 5% growth is 0.15. If you invest \$10,000 in the stock, what is the probability that the stock will be worth at least \$11,400 by the end of the year?

Percent Growth	Probability
5	0.15
8	0.20
10	0.35
14	0.20
18	0.10

32. *Growth in Stock Value* Refer to **Exercise 31.** Suppose the percents and probabilities in the table are estimates of annual growth during the next 3 yr. What is the probability that an investment of \$10,000 will grow in value to *at least* \$15,000 during the next 3 yr? (*Hint:* Use the formula for (annual) compound interest.)

U.S. House of Representatives In the U.S. House of Representatives, the number of representatives from each state is proportional to the state's population. California (the most populous state) has 53 representatives, whereas Wyoming (the least populous state) has just 1 representative. The table gives the percentage of states having various numbers of representatives in the current House of Representatives.

Number of Representatives	1	2–7	8–15	>15
Percent of States (as a decimal)	0.14	0.44	0.28	0.14

Source: www.house.gov

Using the percents as probabilities, find the probability that, out of 10 states selected at random, the following are true.

33. Exactly 4 states have 2–7 representatives.

34. Exactly 2 states have just one representative.

35. Fewer than 2 states have 8 or more representatives.

36. No more than 3 states have 2–15 representatives.

College Applications The table gives the results of a survey of 153,015 first-year students from the class of 2018 at 227 of the nation's four-year colleges and universities.

Number of Colleges Applied to	1	2 or 3	4–6	7 or more
Percent (as a decimal)	0.10	0.18	0.37	0.35

Source: Higher Education Research Institute, UCLA, 2014.

Using the percents as probabilities, find the probability of each event for a randomly selected student.

37. The student applied to fewer than 4 colleges.

38. The student applied to at least 2 colleges.

39. The student applied to more than 3 colleges.

40. The student applied to no colleges.

Solve each problem.

41. *Color-Blind Males* The probability that a male will be color-blind is 0.042. Find the probabilities that in a group of 53 men, the following are true.

(a) Exactly 5 are color-blind.

(b) No more than 5 are color-blind.

(c) None are color-blind.

(d) At least 1 is color-blind.

42. The screens illustrate how the table feature of a graphing calculator can be used to find the probabilities of having 0, 1, 2, 3, or 4 girls in a family of 4 children. (Note that 0 appears for values of x greater than 4 because these events are impossible.)

NORMAL FLOAT AUTO REAL RADIAN MP
Plot1 Plot2 Plot3
$Y_1 = ({}_4C_X)*(0.5^X)*(0.5^{4-X})$
$Y_2=$
$Y_3=$
$Y_4=$
$Y_5=$
$Y_6=$
$Y_7=$
$Y_8=$

NORMAL FLOAT AUTO REAL RADIAN MP
PRESS ENTER TO EDIT

X	Y1
0	.0625
1	.25
2	.375
3	.25
4	.0625
5	0
6	0
7	0
8	0
9	0
10	0

$Y_1 = ({}_4C_X)*(0.5^X)*(0.5^{4-X})$

Use the approach given on the previous page for the following.

(a) Find the probabilities of having 0, 1, 2, or 3 boys in a family of 3 children.

(b) Find the probabilities of having 0, 1, 2, 3, 4, 5, or 6 girls in a family of 6 children.

43. *(Modeling) Spread of Disease* What will happen when an infectious disease is introduced into a family? Suppose a family has I infected members and S members who are not infected but are susceptible to contracting the disease. The probability P of exactly k people not contracting the disease during a 1-week period can be calculated by the formula

$$P = \binom{S}{k} q^k (1 - q)^{S-k},$$

where $q = (1 - p)^I$, and p is the probability that a susceptible person contracts the disease from an infected person. For example, if $p = 0.5$, then there is a 50% chance that a susceptible person exposed to 1 infected person for 1 week will contract the disease. (*Source:* Hoppensteadt, F. and C. Peskin, *Mathematics in Medicine and the Life Sciences,* Springer-Verlag.) Give all answers to the nearest thousandth.

(a) Compute the probability P of 3 family members not becoming infected within 1 week if there are currently 2 infected and 4 susceptible members. Assume that $p = 0.1$. (*Hint:* To use the formula, first determine the values of k, I, S, and q.)

(b) A highly infectious disease can have $p = 0.5$. Repeat part (a) with this value of p.

(c) Determine the probability that everyone will become sick in a large family if, initially, $I = 1$, $S = 9$, and $p = 0.5$.

44. *(Modeling) Spread of Disease* (Refer to **Exercise 43.**) Suppose that in a family $I = 2$ and $S = 4$. If the probability P is 0.25 of there being $k = 2$ uninfected members after 1 week, find the possible values of p to the nearest thousandth. (*Hint:* Write P as a function of p.)

Chapter 11 Test Prep

Key Terms

11.1 finite sequence; infinite sequence; terms of a sequence; general term (*n*th term); convergent sequence; divergent sequence; recursive definition; Fibonacci sequence; series; summation notation; finite series; infinite series; index of summation

11.2 arithmetic sequence (arithmetic progression); common difference; arithmetic series

11.3 geometric sequence (geometric progression); common ratio; geometric series; annuity; future value of an annuity

11.4 Pascal's triangle; factorial notation; binomial coefficient; binomial theorem (general binomial expansion)

11.6 tree diagram; independent events; permutation; combination

11.7 trial; outcome; sample space; event; probability; certain event; impossible event; complement; Venn diagram; odds; compound event; mutually exclusive events; binomial experiment

New Symbols

a_n	nth term of a sequence	${}_nC_r$, $C(n, r)$, or $\binom{n}{r}$	binomial coefficient (combinations of n elements taken r at a time)
$\sum_{i=1}^{n} a_i$	summation notation; sum of n terms	${}_nP_r$ or $P(n, r)$	permutations of n elements taken r at a time
i	index of summation	$n(E)$	number of outcomes that belong to event E
S_n	sum of first n terms of a sequence	$P(E)$	probability of event E
Σ	Greek letter sigma	E'	complement of event E
$\sum_{i=1}^{\infty} a_i$	sum of an infinite number of terms		
$\lim_{n\to\infty} S_n$	limit of S_n as n increases without bound		
$n!$	n-factorial		

Quick Review

Concepts | **Examples**

11.1 Sequences and Series

A finite sequence is a function that has a set of natural numbers of the form $\{1, 2, 3, \ldots, n\}$ as its domain.

An infinite sequence has the set of natural numbers as its domain. The nth term of a sequence is symbolized a_n.

A series is an indicated sum of the terms of a sequence.

The sequence $1, \frac{1}{2}, \frac{1}{3}, \frac{1}{4}, \ldots, \frac{1}{n}$ has general term $a_n = \frac{1}{n}$. The corresponding series is the *sum*

$$1 + \frac{1}{2} + \frac{1}{3} + \frac{1}{4} + \cdots + \frac{1}{n}.$$

Summation Properties

If $a_1, a_2, a_3, \ldots, a_n$ and $b_1, b_2, b_3, \ldots, b_n$ are two sequences and c is a constant, then for every positive integer n, the following hold.

(a) $\sum_{i=1}^{n} c = nc$ **(b)** $\sum_{i=1}^{n} ca_i = c\sum_{i=1}^{n} a_i$

(c) $\sum_{i=1}^{n} (a_i \pm b_i) = \sum_{i=1}^{n} a_i \pm \sum_{i=1}^{n} b_i$

$$\sum_{i=1}^{6} 5 = 6 \cdot 5 = 30$$

$$\sum_{i=1}^{4} 3(2i + 1) = 3\sum_{i=1}^{4}(2i + 1) = 3(3 + 5 + 7 + 9) = 72$$

Summation Rules

(a) $\sum_{i=1}^{n} i = 1 + 2 + \cdots + n = \frac{n(n + 1)}{2}$

(b) $\sum_{i=1}^{n} i^2 = 1^2 + 2^2 + \cdots + n^2 = \frac{n(n + 1)(2n + 1)}{6}$

(c) $\sum_{i=1}^{n} i^3 = 1^3 + 2^3 + \cdots + n^3 = \frac{n^2(n + 1)^2}{4}$

$$\sum_{i=1}^{3} (5i + 6i^2)$$
$$= \sum_{i=1}^{3} 5i + \sum_{i=1}^{3} 6i^2$$
$$= (5 + 10 + 15) + (6 + 24 + 54)$$
$$= 30 + 84$$
$$= 114$$

11.2 Arithmetic Sequences and Series

Assume a_1 is the first term, a_n is the nth term, and d is the common difference in an arithmetic sequence.

Common Difference

$$d = a_{n+1} - a_n$$

The arithmetic sequence $2, 5, 8, 11, \ldots$ has $a_1 = 2$.

$d = 5 - 2 = 3$ Common difference

(Any two successive terms could have been used.)

Concepts	Examples
nth Term $a_n = a_1 + (n-1)d$	For the arithmetic sequence 2, 5, 8, 11, . . . , suppose that $n = 10$. The 10th term is found as follows. $a_{10} = 2 + (10-1)3 \quad a_1 = 2, d = 3$ $a_{10} = 2 + 9 \cdot 3$ $a_{10} = 29$
Sum of the First n Terms $S_n = \frac{n}{2}(a_1 + a_n)$ or $S_n = \frac{n}{2}[2a_1 + (n-1)d]$	The sum of the first 10 terms is found as follows. $S_{10} = \frac{10}{2}(a_1 + a_{10})$ $S_{10} = 5(2 + 29)$ $S_{10} = 155$ or $S_{10} = \frac{10}{2}[2(2) + (10-1)3]$ $S_{10} = 5(4 + 9 \cdot 3)$ $S_{10} = 5(4 + 27)$ $S_{10} = 155$
11.3 Geometric Sequences and Series Assume a_1 is the first term, a_n is the nth term, and r is the common ratio in a geometric sequence.	For the geometric sequence 1, 2, 4, 8, . . . , $a_1 = 1$.
Common Ratio $r = \frac{a_{n+1}}{a_n}$	$r = \frac{8}{4} = 2 \quad$ Common ratio (Any two successive terms could have been used.)
nth Term $a_n = a_1 r^{n-1}$	Suppose that $n = 6$. Then the sixth term is $a_6 = (1)(2)^{6-1} = 1(2)^5 = 32.$
Sum of the First n Terms $S_n = \frac{a_1(1-r^n)}{1-r} \quad (\text{where } r \neq 1)$	The sum of the first six terms is found as follows. $S_6 = \frac{1(1-2^6)}{1-2} = \frac{1-64}{-1} = 63$
Sum of the Terms of an Infinite Geometric Sequence $S_\infty = \frac{a_1}{1-r} \quad (\text{where } \lvert r \rvert < 1)$	The sum of the terms of the infinite geometric sequence $\sum_{k=0}^{\infty} \left(\frac{1}{2}\right)^k = 1 + \frac{1}{2} + \frac{1}{4} + \cdots$ is found as follows. $S_\infty = \frac{1}{1-\frac{1}{2}} = \frac{1}{\frac{1}{2}} = 2$
11.4 The Binomial Theorem **n-Factorial** For any positive integer n, $n! = n(n-1)(n-2)\cdots(3)(2)(1).$ By definition, $0! = 1.$	$4! = 4 \cdot 3 \cdot 2 \cdot 1 = 24$

Concepts	Examples
Binomial Coefficient For nonnegative integers n and r, with $r \le n$, $${}_nC_r = \binom{n}{r} = \frac{n!}{r!(n-r)!}.$$	$${}_5C_3 = \frac{5!}{3!(5-3)!} = \frac{5!}{3!2!} = \frac{5 \cdot 4 \cdot 3 \cdot 2 \cdot 1}{3 \cdot 2 \cdot 1 \cdot 2 \cdot 1} = 10$$
Binomial Theorem For any positive integer n and any complex numbers x and y, $(x+y)^n$ is expanded as follows. $$(x+y)^n = x^n + \binom{n}{1}x^{n-1}y + \binom{n}{2}x^{n-2}y^2 + \binom{n}{3}x^{n-3}y^3 + \cdots + \binom{n}{r}x^{n-r}y^r + \cdots + \binom{n}{n-1}xy^{n-1} + y^n$$	$$(2m+3)^4 = (2m)^4 + \frac{4!}{1!3!}(2m)^3(3) + \frac{4!}{2!2!}(2m)^2(3)^2 + \frac{4!}{3!1!}(2m)(3)^3 + 3^4$$ $$= 2^4m^4 + 4(2)^3m^3(3) + 6(2)^2m^2(9) + 4(2m)(27) + 81$$ $$= 16m^4 + 12(8)m^3 + 54(4)m^2 + 216m + 81$$ $$= 16m^4 + 96m^3 + 216m^2 + 216m + 81$$
***k*th Term of the Binomial Expansion of $(x+y)^n$** $$\binom{n}{k-1}x^{n-(k-1)}y^{k-1} \quad (\text{where } n \ge k-1)$$	The eighth term of $(a-2b)^{10}$ is found as follows. $$\binom{10}{7}a^3(-2b)^7 = \frac{10!}{7!3!}a^3(-2)^7b^7 = 120(-128)a^3b^7 = -15{,}360a^3b^7$$
11.5 Mathematical Induction	
Principle of Mathematical Induction Let S_n be a statement concerning the positive integer n. Suppose that both of the following are satisfied. **1.** S_1 is true. **2.** For any positive integer k, $k \le n$, if S_k is true, then S_{k+1} is also true. Then S_n is true for every positive integer value of n.	See **Examples 1 and 2** in **Section 11.5.** **Example 3** in **Section 11.5** illustrates the generalized principle of mathematical induction.
11.6 Basics of Counting Theory	
Fundamental Principle of Counting If n independent events occur, with m_1 ways for event 1 to occur, m_2 ways for event 2 to occur, . . . , and m_n ways for event n to occur, then there are $m_1 \cdot m_2 \cdot \cdots \cdot m_n$ different ways for all n events to occur.	If there are 2 ways to choose a pair of socks and 5 ways to choose a pair of shoes, then by the fundamental principle of counting there are $2 \cdot 5 = 10$ ways to choose socks and shoes.
Permutations Formula If $P(n, r)$ denotes the number of permutations of n elements taken r at a time, with $r \le n$, then the following holds. $$P(n,r) = \frac{n!}{(n-r)!}$$	How many ways are there to arrange the letters of the word TRIANGLE using 5 letters at a time? This is an arrangement. Use permutations. $$P(8,5) = \frac{8!}{(8-5)!} = \frac{8!}{3!} = 6720 \quad \text{Let } n = 8 \text{ and } r = 5.$$

Concepts	Examples
Combinations Formula The number of combinations of n elements taken r at a time, with $r \le n$, is determined as follows. $$C(n, r) = \frac{n!}{(n-r)!r!}, \quad \text{or} \quad C(n, r) = \frac{n!}{r!(n-r)!}$$	How many committees of 4 senators can be formed from a group of 9 senators? The arrangement of senators does not matter, so this is a combinations problem. $C(9, 4) = \frac{9!}{4!(9-4)!} = 126$ committees Let $n = 9$ and $r = 4$.
## 11.7 Basics of Probability **Probability of Event E** In a sample space S with equally likely outcomes, the probability of event E is determined as follows. $$P(E) = \frac{n(E)}{n(S)}$$	A number is chosen at random from $S = \{1, 2, 3, 4, 5, 6\}$. What is the probability that the number is less than 3? The event is $E = \{1, 2\}$, $n(S) = 6$, and $n(E) = 2$. $$P(E) = \frac{2}{6} = \frac{1}{3}$$
Properties of Probability For any events E and F, the following hold true. **1.** $0 \le P(E) \le 1$ **2.** $P(\text{a certain event}) = 1$ **3.** $P(\text{an impossible event}) = 0$ **4.** $P(E') = 1 - P(E)$ **5.** $P(E \text{ or } F) = P(E \cup F)$ $= P(E) + P(F) - P(E \cap F)$	What is the probability that the number is 3 or more? This event is E'. $$P(E') = 1 - \frac{1}{3} = \frac{2}{3}$$
Binomial Probability In a binomial experiment, let p represent the probability of a success, and let $q = 1 - p$ represent the probability of a failure. Then the probability of obtaining exactly r successes in n trials is found as follows. $$P(r \text{ successes in } n \text{ trials}) = \binom{n}{r} p^r q^{n-r}$$	An experiment consists of rolling a fair die 8 times. Find the probability that exactly 5 rolls result in a 2. $P(5 \text{ twos in } 8 \text{ rolls})$ $= \binom{8}{5}\left(\frac{1}{6}\right)^5\left(\frac{5}{6}\right)^{8-5}$ Let $n = 8$, $r = 5$, $p = \frac{1}{6}$, and $q = \frac{5}{6}$. $= 56\left(\frac{1}{6}\right)^5\left(\frac{5}{6}\right)^3$ ≈ 0.004

Chapter 11 Review Exercises

Write the first five terms of each sequence. State whether the sequence is arithmetic, geometric, *or* neither.

1. $a_n = \frac{n}{n+1}$

2. $a_n = (-2)^n$

3. $a_n = 2(n+3)$

4. $a_n = n(n+1)$

5. $a_1 = 5$
$a_n = a_{n-1} - 3$, if $n \ge 2$

6. $a_1 = 1$, $a_2 = 3$,
$a_n = a_{n-2} + a_{n-1}$, if $n \ge 3$

7. ***Concept Check*** Write an arithmetic sequence that consists of five terms, with first term 4, having the sum of the five terms equal to 25.

Write the first five terms of each sequence described.

8. arithmetic; $a_2 = 10$, $d = -2$

9. arithmetic; $a_3 = \pi$, $a_4 = 1$

10. geometric; $a_1 = 6$, $r = 2$

11. geometric; $a_1 = -5$, $a_2 = -1$

Determine the indicated terms for each sequence described.

12. An arithmetic sequence has $a_5 = -3$ and $a_{15} = 17$. Find a_1 and a_n.

13. A geometric sequence has $a_1 = -8$ and $a_7 = -\frac{1}{8}$. Find a_4 and a_n.

Determine a_8 for each arithmetic sequence.

14. $a_1 = 6, \; d = 2$

15. $a_1 = 6x - 9, \; a_2 = 5x + 1$

Determine S_{12} for each arithmetic sequence.

16. $a_1 = 2, \; d = 3$

17. $a_2 = 6, \; d = 10$

Determine a_5 for each geometric sequence.

18. $a_1 = -2, \; r = 3$

19. $a_3 = 4, \; r = \frac{1}{5}$

Determine S_4 for each geometric sequence.

20. $a_1 = 3, \; r = 2$

21. $a_1 = -1, \; r = 3$

22. $\frac{3}{4}, -\frac{1}{2}, \frac{1}{3}, \ldots$

Evaluate each sum that exists.

23. $\sum_{i=1}^{7} (-1)^{i-1}$

24. $\sum_{i=1}^{5} (i^2 + i)$

25. $\sum_{i=1}^{4} \frac{i+1}{i}$

26. $\sum_{j=1}^{10} (3j - 4)$

27. $\sum_{j=1}^{2500} j$

28. $\sum_{i=1}^{5} 4 \cdot 2^i$

29. $\sum_{i=1}^{\infty} \left(\frac{4}{7}\right)^i$

30. $\sum_{i=1}^{\infty} -2\left(\frac{6}{5}\right)^i$

31. $\sum_{i=1}^{\infty} 2\left(-\frac{2}{3}\right)^i$

32. ***Concept Check*** Find an infinite geometric series having common ratio $\frac{3}{4}$ and sum 6.

Evaluate each series that converges. Identify any that diverge.

33. $24 + 8 + \frac{8}{3} + \frac{8}{9} + \cdots$

34. $-\frac{3}{4} + \frac{1}{2} - \frac{1}{3} + \frac{2}{9} - \cdots$

35. $\frac{1}{12} + \frac{1}{6} + \frac{1}{3} + \frac{2}{3} + \cdots$

36. $0.9 + 0.09 + 0.009 + 0.0009 + \cdots$

Evaluate each sum where $x_1 = 0, \; x_2 = 1, \; x_3 = 2, \; x_4 = 3, \; x_5 = 4,$ and $x_6 = 5$.

37. $\sum_{i=1}^{4} (x_i^2 - 6)$

38. $\sum_{i=1}^{6} f(x_i)\Delta x; \; f(x) = (x-2)^3, \; \Delta x = 0.1$

Write each sum using summation notation.

39. $4 - 1 - 6 - \cdots - 66$

40. $10 + 14 + 18 + \cdots + 86$

41. $4 + 12 + 36 + \cdots + 972$

42. $\frac{5}{6} + \frac{6}{7} + \frac{7}{8} + \cdots + \frac{12}{13}$

Write the binomial expansion of each expression.

43. $(x + 2y)^4$

44. $(3z - 5w)^3$

45. $\left(3\sqrt{x} - \frac{1}{\sqrt{x}}\right)^5$

46. $(m^3 - m^{-2})^4$

Find the indicated term or terms of each expansion.

47. sixth term of $(4x - y)^8$

48. seventh term of $(m - 3n)^{14}$

49. first four terms of $(x + 2)^{12}$

50. last three terms of $(2a + 5b)^{16}$

Let S_n represent the statement, and use mathematical induction to prove that S_n is true for every positive integer n.

51. $1 + 3 + 5 + 7 + \cdots + (2n - 1) = n^2$

52. $2 + 6 + 10 + 14 + \cdots + (4n - 2) = 2n^2$

53. $2 + 2^2 + 2^3 + \cdots + 2^n = 2(2^n - 1)$

54. $1^3 + 3^3 + 5^3 + \cdots + (2n - 1)^3 = n^2(2n^2 - 1)$

Evaluate each expression.

55. $P(9, 2)$

56. $P(6, 0)$

57. $C(8, 3)$

58. $9!$

59. $C(10, 5)$

60. $10 \cdot 9!$

Solve each problem.

61. ***Median Annual Earnings*** In 2012 the median annual earnings of a high school graduate with no college attendance was \$30,000. This amount is expected to increase by about \$268 per year. How much will a person earning the median amount earn until retirement if he or she joins the work force at age 18 and works until age 66? (*Source:* U.S. Bureau of Labor Statistics.)

62. ***Median Annual Earnings*** In 2012 the median annual earnings of a person with 4 yr of college was \$46,900. This amount is expected to increase by about \$813 per year. How much will a person earning the median amount earn until retirement if he or she joins the work force at age 22 and works until age 66? (*Source:* U.S. Bureau of Labor Statistics.)

63. ***Median Annual Earnings*** Refer to **Exercises 61 and 62.** How much more will a person with 4 yr of college who earns the median amount make during his or her career than a person with no college attendance who earns the median amount during his or her career? If the expenses of a 4-yr college degree are estimated at \$95,780, is earning a 4-yr college degree worth it? (*Source:* U.S. Bureau of Labor Statistics.)

64. ***Wedding Plans*** Two people are planning their wedding. They can select from 2 different chapels, 4 soloists, 3 organists, and 2 ministers. How many different wedding arrangements are possible?

65. ***Couch Styles*** Bob is furnishing his apartment and wants to buy a new couch. He can select from 5 different styles, each available in 3 different fabrics, with 6 color choices. How many different couches are available?

66. ***Summer Job Assignments*** Four students are to be assigned to 4 different summer jobs. Each student is qualified for all 4 jobs. In how many ways can the jobs be assigned?

67. ***Conference Delegations*** A student council consists of 6 seniors and 3 juniors. Three members are to be selected to attend a conference.

(a) How many different such delegations are possible?

(b) How many are possible if 2 seniors and 1 junior must attend?

68. ***Tournament Outcomes*** Nine football teams are competing for first-, second-, and third-place titles in a statewide tournament. In how many ways can the winners be determined?

69. ***License Plates*** How many different license plates can be formed with a letter followed by 3 digits and then 3 letters? How many such license plates have no repeats?

70. *Racetrack Bets* Most racetracks have "compound" bets on 2 or more horses. An *exacta* is a bet in which the first and second finishers in a race are specified in order. A *quinella* is a bet on the first 2 finishers in a race, with order not specified.

(a) In a field of 9 horses, how many different exacta bets can be placed?

(b) How many different quinella bets can be placed in a field of 9 horses?

71. *Drawing a Marble* A marble is drawn at random from a box containing 4 green, 5 black, and 6 white marbles. Find the following probabilities.

(a) A green marble is drawn. **(b)** A marble that is not black is drawn.

(c) A blue marble is drawn.

(d) What are the odds in favor of drawing a marble that is not white?

72. *Drawing a Card* A card is drawn from a standard deck of 52 cards. Find the following probabilities.

(a) A black king is drawn. **(b)** A face card or an ace is drawn.

(c) An ace or a diamond is drawn. **(d)** A card that is not a diamond is drawn.

(e) What are the odds in favor of drawing an ace?

73. *Master's Degrees* There were 754,299 master's degrees awarded in the United States in 2012. The table shows the numbers of degrees awarded in several fields of study.

Field of Study	Number of Master's Degrees
Business	103,253
Education	178,062
Health professions and related clinical studies	83,893
Visual and performing arts	17,331
Other	371,760

Source: U.S. National Center for Education Statistics.

(a) What is the probability that a randomly selected student who earned a master's degree in 2012 earned a degree in business?

(b) What is the probability that a randomly selected student who earned a master's degree in 2012 earned a degree in either health professions and related clinical studies or the visual and performing arts?

(c) What is the probability that a randomly selected student who earned a master's degree in 2012 earned a degree that was not in education?

74. *Defective Toaster Ovens* A sample shipment of 5 toaster ovens is chosen. The probability of exactly 0, 1, 2, 3, 4, or 5 toaster ovens being defective is given in the table.

Number Defective	0	1	2	3	4	5
Probability	0.31	0.25	0.18	0.12	0.08	0.06

Find the probability that the given number of toaster ovens are defective.

(a) no more than 3 **(b)** at least 2 **(c)** more than 5

75. *Rolling a Die* A fair die is rolled 12 times. Find the probability (to three decimal places) that exactly 2 of the rolls result in a 5.

76. *Tossing a Coin* A fair coin is tossed 10 times. Find the probability (to three decimal places) that exactly 4 of the tosses result in a tail.

Chapter 11 Test

Write the first five terms of each sequence. State whether the sequence is arithmetic, geometric, *or* neither.

1. $a_n = (-1)^n(n^2 + 2)$

2. $a_n = -3\left(\frac{1}{2}\right)^n$

3. $a_1 = 2, \ a_2 = 3, \ a_n = a_{n-1} + 2a_{n-2}$, for $n \geq 3$

Determine the indicated term for each sequence described.

4. An arithmetic sequence has $a_1 = 1$ and $a_3 = 25$. Find a_5.

5. A geometric sequence has $a_1 = 81$ and $r = -\frac{2}{3}$. Find a_6.

Find the sum of the first ten terms of each series.

6. arithmetic; $a_1 = -43, \ d = 12$

7. geometric; $a_1 = 5, \ r = -2$

Evaluate each sum that exists.

8. $\sum_{i=1}^{30} (5i + 2)$

9. $\sum_{i=1}^{5} (-3 \cdot 2^i)$

10. $\sum_{i=1}^{\infty} (2^i) \cdot 4$

11. $\sum_{i=1}^{\infty} 54\left(\frac{2}{9}\right)^i$

Write the binomial expansion of each expression.

12. $(x + y)^6$

13. $(2x - 3y)^4$

14. Find the third term in the expansion of $(w - 2y)^6$.

Evaluate each expression.

15. $8!$

16. $C(10, 2)$

17. $C(7, 3)$

18. $P(11, 3)$

19. Let S_n represent the statement, and use mathematical induction to prove that S_n is true for every positive integer n.

$$1 + 7 + 13 + \cdots + (6n - 5) = n(3n - 2)$$

Solve each problem.

20. ***Athletic Shoe Styles*** A shoe manufacturer makes athletic shoes in 4 different styles. Each style comes in 3 different colors, and each color comes in 2 different shades. How many different types of shoes can be made?

21. ***Seminar Attendees*** A mortgage company has 10 loan officers: 4 women and 6 men. In how many ways can 4 of these officers be selected to attend a seminar? In how many ways can 2 women and 2 men be selected to attend the seminar?

22. ***Course Schedule Arrangement*** A student must select 4 courses from 15 that are offered in a semester. How many different arrangements of the 4 courses are possible?

23. ***Drawing Cards*** A card is drawn from a standard deck of 52 cards. Find the following probabilities in parts (a)–(c).

(a) A red three is drawn.

(b) A card that is not a face card is drawn.

(c) A king or a spade is drawn.

(d) What are the odds in favor of drawing a face card?

24. *Defective Light Bulbs* A sample of 4 light bulbs is chosen. The probability of exactly 0, 1, 2, 3, or 4 light bulbs being defective is given in the table. Find the probability that at most 2 are defective.

Number Defective	0	1	2	3	4
Probability	0.19	0.43	0.30	0.07	0.01

25. *Rolling a Die* Find the probability (to three decimal places) of obtaining 5 on exactly two of six rolls of a single fair die.

Appendices

A Polar Form of Conic Sections

- Equations and Graphs
- Conversion from Polar to Rectangular Form

Equations and Graphs

Until now we have worked with equations of conic sections in rectangular form. If the focus of a conic section is at the pole, the polar form of its equation is

$$r = \frac{ep}{1 \pm e \cdot f(\theta)},$$

where f is either the sine or cosine function.

Polar Forms of Conic Sections

A polar equation of the form

$$r = \frac{ep}{1 \pm e\cos\theta} \quad \text{or} \quad r = \frac{ep}{1 \pm e\sin\theta}$$

has a conic section as its graph. The eccentricity is e (where $e > 0$), and $|p|$ is the distance between the pole (focus) and the directrix.

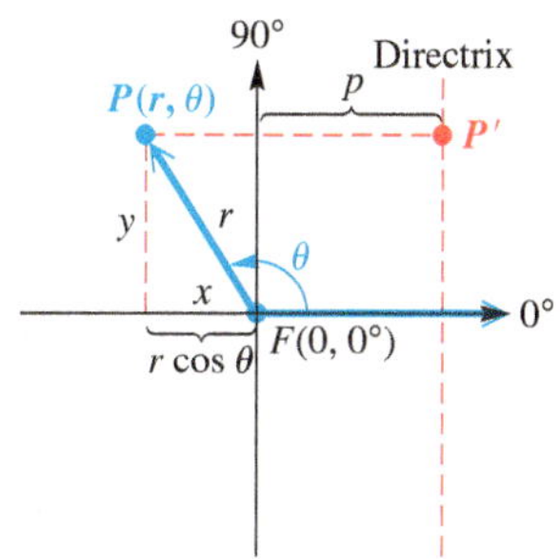

Figure 1

We can verify that $r = \frac{ep}{1 + e\cos\theta}$ does indeed satisfy the definition of a conic section. Consider **Figure 1**, where the directrix is vertical and is p units (where $p > 0$) to the right of the focus $F(0, 0°)$. If we let $P(r, \theta)$ be a point on the graph, then the distance between P and the directrix is found as follows.

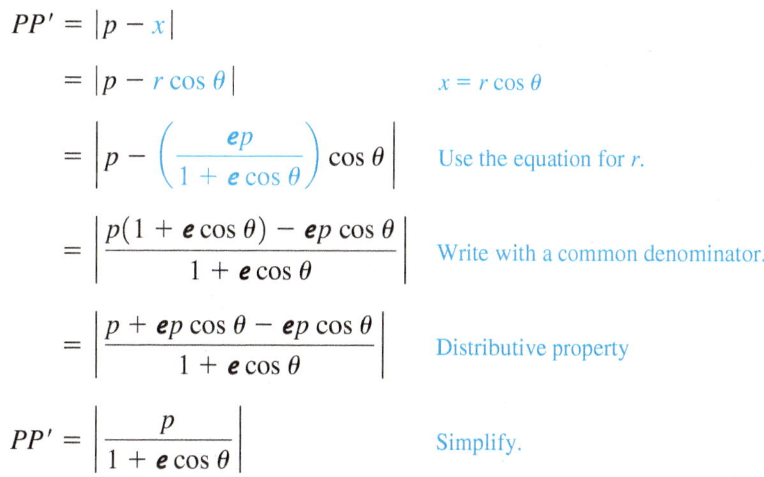

$$PP' = |p - x|$$

$$= |p - r\cos\theta| \qquad x = r\cos\theta$$

$$= \left|p - \left(\frac{ep}{1 + e\cos\theta}\right)\cos\theta\right| \qquad \text{Use the equation for } r.$$

$$= \left|\frac{p(1 + e\cos\theta) - ep\cos\theta}{1 + e\cos\theta}\right| \qquad \text{Write with a common denominator.}$$

$$= \left|\frac{p + ep\cos\theta - ep\cos\theta}{1 + e\cos\theta}\right| \qquad \text{Distributive property}$$

$$PP' = \left|\frac{p}{1 + e\cos\theta}\right| \qquad \text{Simplify.}$$

Because

$$r = \frac{ep}{1 + e\cos\theta},$$

we multiply each side by $\frac{1}{e}$.

$$\frac{r}{e} = \frac{p}{1 + e\cos\theta}$$

We can substitute $\frac{r}{e}$ for the expression in the absolute value bars for PP'.

$$PP' = \left|\frac{p}{1 + e\cos\theta}\right| = \left|\frac{r}{e}\right| = \frac{|r|}{|e|} = \frac{|r|}{e}$$

The distance between the pole and P is $PF = |r|$, so the ratio of PF to PP' is

$$\frac{PF}{PP'} = \frac{|r|}{\frac{|r|}{e}} = |r| \div \frac{|r|}{e} = |r| \cdot \frac{e}{|r|} = e.$$ Simplify the complex fraction.

Thus, by the definition, the graph has eccentricity e and must be a conic.

In the preceding discussion, we assumed a vertical directrix to the right of the pole. There are three other possible situations.

Location of the Directrix of a Conic Section

If the equation is:	then the directrix is:
$r = \frac{ep}{1 + e\cos\theta}$	*vertical*, p units to the *right* of the pole.
$r = \frac{ep}{1 - e\cos\theta}$	*vertical*, p units to the *left* of the pole.
$r = \frac{ep}{1 + e\sin\theta}$	*horizontal*, p units *above* the pole.
$r = \frac{ep}{1 - e\sin\theta}$	*horizontal*, p units *below* the pole.

EXAMPLE 1 Graphing a Conic in Polar Form

Graph $r = \frac{8}{4 + 4\sin\theta}$.

ALGEBRAIC SOLUTION

Divide both numerator and denominator by 4 to obtain

$$r = \frac{2}{1 + \sin\theta}.$$

Based on the preceding table, this is the equation of a conic with $ep = 2$ and $e = 1$. Thus, $p = 2$. Because $e = 1$, the graph is a parabola. The focus is at the pole, and the directrix is horizontal, 2 units *above* the pole.

The vertex must have polar coordinates $(1, 90°)$. Letting $\theta = 0°$ and $\theta = 180°$ gives the additional points $(2, 0°)$ and $(2, 180°)$. See **Figure 2.**

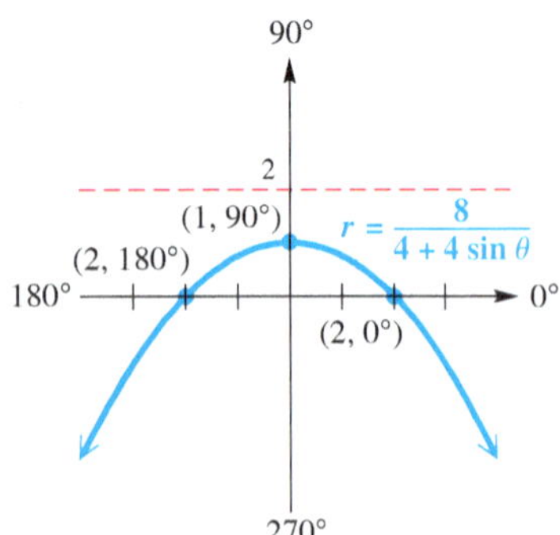

Figure 2

As expected, the graph is a parabola, and it opens downward because the directrix is above the pole.

GRAPHING CALCULATOR SOLUTION

Enter

$$r_1 = \frac{8}{4 + 4\sin\theta},$$

where the calculator is in polar and degree modes with polar coordinate displays. **Figure 3(a)** shows the window settings, and **Figure 3(b)** shows the graph.

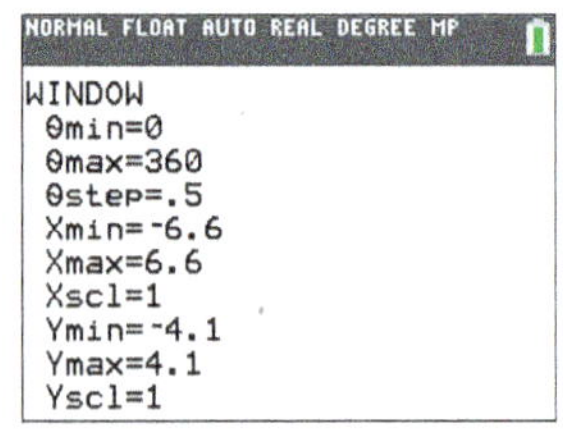

(a)

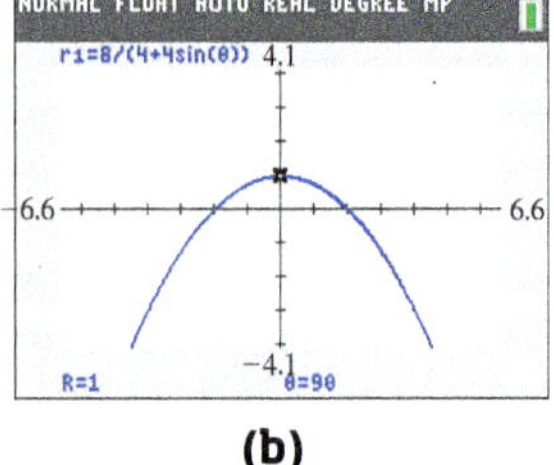

(b)

Figure 3

Notice that the point $(1, 90°)$ is indicated at the bottom.

✔ **Now Try Exercise 1.**

EXAMPLE 2 Finding a Polar Equation

Find the polar equation of a parabola with focus at the pole and vertical directrix 3 units to the left of the pole.

SOLUTION The eccentricity $\boldsymbol{e}$ must be 1, p must equal 3, and the equation must be of the following form.

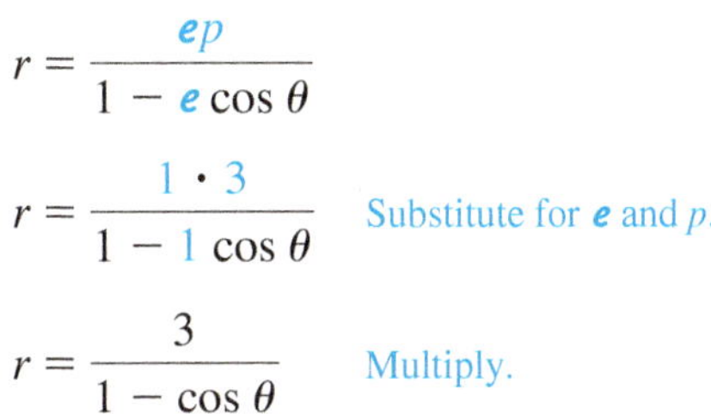

$$r = \frac{ep}{1 - e\cos\theta}$$

$$r = \frac{1 \cdot 3}{1 - 1\cos\theta} \quad \text{Substitute for } e \text{ and } p.$$

$$r = \frac{3}{1 - \cos\theta} \quad \text{Multiply.}$$

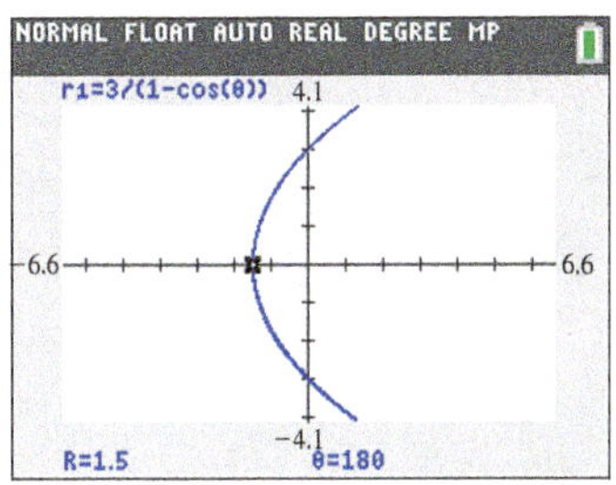

Degree mode

Figure 4

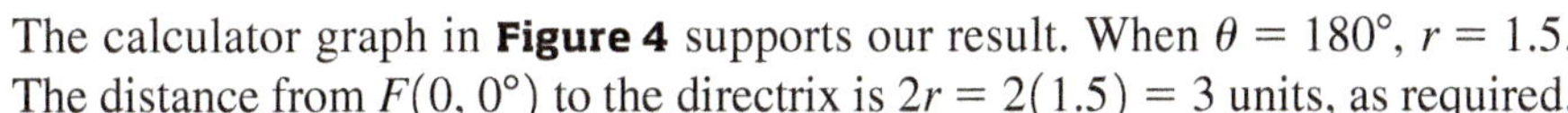

The calculator graph in **Figure 4** supports our result. When $\theta = 180°$, $r = 1.5$. The distance from $F(0, 0°)$ to the directrix is $2r = 2(1.5) = 3$ units, as required.

✔ Now Try Exercise 11.

Conversion from Polar to Rectangular Form

EXAMPLE 3 Identifying and Converting from Polar Form to Rectangular Form

Identify the type of conic represented by $r = \frac{8}{2 - \cos\theta}$. Then convert the equation to rectangular form.

SOLUTION To identify the type of conic, we divide both the numerator and the denominator on the right side of the equation by 2.

$$r = \frac{4}{1 - \frac{1}{2}\cos\theta}$$

From the table, we see that this is a conic that has a vertical directrix, with $\boldsymbol{e} = \frac{1}{2}$, making it an ellipse. To convert to rectangular form, we start with the given equation.

$$r = \frac{8}{2 - \cos\theta} \quad \text{Given equation}$$

$$r(2 - \cos\theta) = 8 \quad \text{Multiply by } 2 - \cos\theta.$$

$$2r - r\cos\theta = 8 \quad \text{Distributive property}$$

$$2r = r\cos\theta + 8 \quad \text{Add } r\cos\theta \text{ to each side.}$$

$$(2r)^2 = (r\cos\theta + 8)^2 \quad \text{Square each side.}$$

$$(2r)^2 = (x + 8)^2 \quad r\cos\theta = x$$

$$4r^2 = x^2 + 16x + 64 \quad \text{Multiply.}$$

$$4(x^2 + y^2) = x^2 + 16x + 64 \quad r^2 = x^2 + y^2$$

$$4x^2 + 4y^2 = x^2 + 16x + 64 \quad \text{Distributive property}$$

$$3x^2 + 4y^2 - 16x - 64 = 0 \quad \text{Standard form}$$

The coefficients of x^2 and y^2 are both positive and are not equal, further supporting our assertion that the graph is an ellipse.

✔ Now Try Exercise 19.

Appendix A Exercises

Graph each conic whose equation is given in polar form. **See Example 1.**

1. $r = \dfrac{6}{3 + 3 \sin \theta}$

2. $r = \dfrac{9}{3 - 3 \sin \theta}$

3. $r = \dfrac{-4}{6 + 2 \cos \theta}$

4. $r = \dfrac{-8}{4 + 2 \cos \theta}$

5. $r = \dfrac{2}{2 - 4 \sin \theta}$

6. $r = \dfrac{6}{2 - 4 \sin \theta}$

7. $r = \dfrac{-1}{1 + 2 \cos \theta}$

8. $r = \dfrac{-1}{1 - 2 \cos \theta}$

9. $r = \dfrac{1}{2 + \cos \theta}$

10. $r = \dfrac{1}{2 - \cos \theta}$

Find the polar equation of a parabola with focus at the pole, satisfying the given conditions. **See Example 2.**

11. The vertical directrix is 3 units to the right of the pole.

12. The vertical directrix is 4 units to the left of the pole.

13. The horizontal directrix is 5 units below the pole.

14. The horizontal directrix is 6 units above the pole.

Find a polar equation for the conic with focus at the pole, satisfying the given conditions. Also identify the type of conic represented. **See Example 2.**

15. $e = \frac{4}{5}$, and the vertical directrix is 5 units to the right of the pole.

16. $e = \frac{2}{3}$, and the vertical directrix is 6 units to the left of the pole.

17. $e = \frac{5}{4}$, and the horizontal directrix is 8 units below the pole.

18. $e = \frac{3}{2}$, and the horizontal directrix is 4 units above the pole.

Identify the type of conic represented by each equation. Then convert the equation to rectangular form. **See Example 3.**

19. $r = \dfrac{6}{3 - \cos \theta}$

20. $r = \dfrac{8}{4 - \cos \theta}$

21. $r = \dfrac{-2}{1 + 2 \cos \theta}$

22. $r = \dfrac{-3}{1 + 3 \cos \theta}$

23. $r = \dfrac{-6}{4 + 2 \sin \theta}$

24. $r = \dfrac{-12}{6 + 3 \sin \theta}$

25. $r = \dfrac{10}{2 - 2 \sin \theta}$

26. $r = \dfrac{12}{4 - 4 \sin \theta}$

B Rotation of Axes

- Derivation of Rotation Equations
- Application of a Rotation Equation

Derivation of Rotation Equations

If we begin with an xy-coordinate system having origin O and rotate the axes about O through an angle θ, the new coordinate system is a **rotation** of the xy-system. Trigonometric identities can be used to obtain equations for converting the coordinates of a point from the xy-system to the rotated $x'y'$-system.

Let P be any point other than the origin, with coordinates (x, y) in the xy-system and (x', y') in the $x'y'$-system. See **Figure 1.**

LOOKING AHEAD TO CALCULUS

Rotation of axes is a topic traditionally covered in calculus texts, in conjunction with parametric equations and polar coordinates. The coverage in calculus is typically the same as that seen in this section.

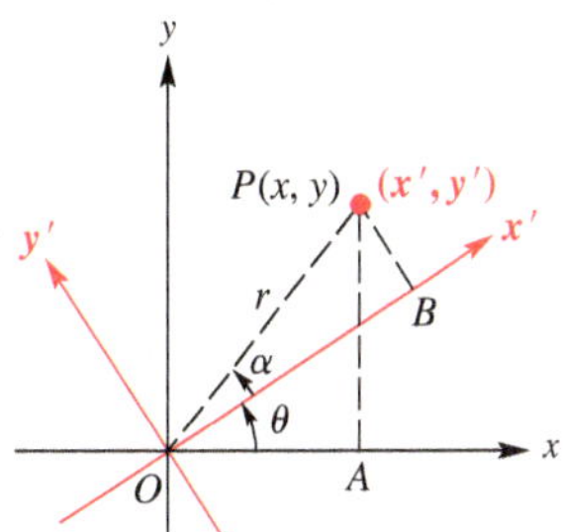

Figure 1

Let $OP = r$, and let α represent the angle made by OP and the x'-axis. **Figure 1** suggests that the following hold.

$$\cos(\theta + \alpha) = \frac{OA}{r} = \frac{x}{r}, \qquad \sin(\theta + \alpha) = \frac{AP}{r} = \frac{y}{r},$$

$$\cos \alpha = \frac{OB}{r} = \frac{x'}{r}, \qquad \sin \alpha = \frac{BP}{r} = \frac{y'}{r}$$

These four statements can be written as follows.

$$x = r\cos(\theta + \alpha), \qquad y = r\sin(\theta + \alpha),$$

$$x' = r\cos \alpha, \qquad y' = r\sin \alpha$$

The trigonometric identity for the cosine of the sum of two angles gives the following equation.

$$x = r\cos(\theta + \alpha)$$

$$x = r(\cos\theta\cos\alpha - \sin\theta\sin\alpha)$$

$$x = (r\cos\alpha)\cos\theta - (r\sin\alpha)\sin\theta \quad \text{Distributive property}$$

$$x = x'\cos\theta - y'\sin\theta \quad \text{Substitute.}$$

In the same way, the identity for the sine of the sum of two angles gives

$$y = x'\sin\theta + y'\cos\theta.$$

Rotation Equations

If the rectangular coordinate axes are rotated about the origin through an angle θ, and if the coordinates of a point P are (x, y) and (x', y') with respect to the xy-system and the $x'y'$-system, respectively, then the **rotation equations** are as follows.

$$\boldsymbol{x = x'\cos\theta - y'\sin\theta} \quad \text{and} \quad \boldsymbol{y = x'\sin\theta + y'\cos\theta}$$

Application of a Rotation Equation

EXAMPLE 1 **Finding an Equation after a Rotation**

The equation of a curve is

$$x^2 + y^2 + 2xy + 2\sqrt{2}x - 2\sqrt{2}y = 0.$$

Find the resulting equation if the axes are rotated 45°. Graph the equation.

SOLUTION If $\theta = 45°$, then $\sin\theta = \frac{\sqrt{2}}{2}$ and $\cos\theta = \frac{\sqrt{2}}{2}$, and the rotation equations become

$$x = \frac{\sqrt{2}}{2}x' - \frac{\sqrt{2}}{2}y' \quad \text{and} \quad y = \frac{\sqrt{2}}{2}x' + \frac{\sqrt{2}}{2}y'.$$

Substitute these values into the given equation.

$$x^2 + y^2 + 2xy + 2\sqrt{2}x - 2\sqrt{2}y = 0$$

$$\left[\frac{\sqrt{2}}{2}x' - \frac{\sqrt{2}}{2}y'\right]^2 + \left[\frac{\sqrt{2}}{2}x' + \frac{\sqrt{2}}{2}y'\right]^2 + 2\left[\frac{\sqrt{2}}{2}x' - \frac{\sqrt{2}}{2}y'\right]\left[\frac{\sqrt{2}}{2}x' + \frac{\sqrt{2}}{2}y'\right]$$

$$+ 2\sqrt{2}\left[\frac{\sqrt{2}}{2}x' - \frac{\sqrt{2}}{2}y'\right] - 2\sqrt{2}\left[\frac{\sqrt{2}}{2}x' + \frac{\sqrt{2}}{2}y'\right] = 0$$

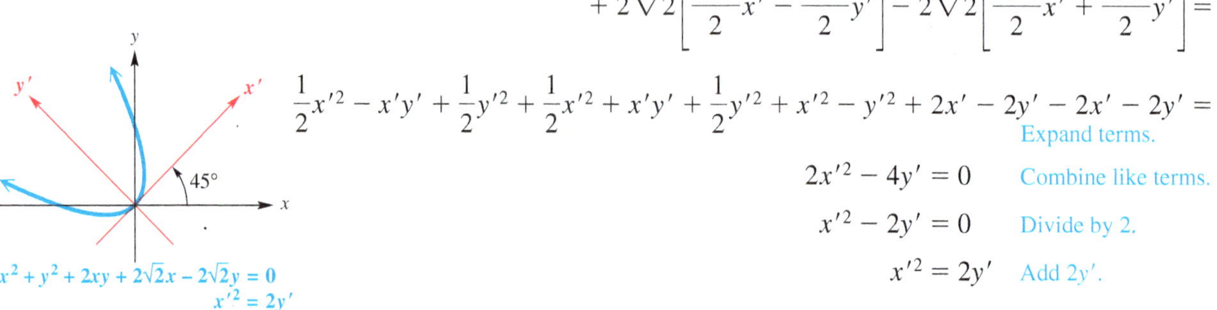

Figure 2

$$\frac{1}{2}x'^2 - x'y' + \frac{1}{2}y'^2 + \frac{1}{2}x'^2 + x'y' + \frac{1}{2}y'^2 + x'^2 - y'^2 + 2x' - 2y' - 2x' - 2y' = 0$$ Expand terms.

$$2x'^2 - 4y' = 0$$ Combine like terms.

$$x'^2 - 2y' = 0$$ Divide by 2.

$$x'^2 = 2y'$$ Add $2y'$.

This is the equation of a parabola. See **Figure 2.** ✔ **Now Try Exercise 13.**

We have graphed equations written in the general form

$$Ax^2 + Cy^2 + Dx + Ey + F = 0.$$

To graph an equation that has an xy-term by hand, it is necessary to find an appropriate **angle of rotation** to eliminate the xy-term. The necessary angle of rotation can be determined by using the following result. The proof is quite lengthy and is not presented here.

Angle of Rotation

The xy-term is removed from the general equation

$$Ax^2 + Bxy + Cy^2 + Dx + Ey + F = 0$$

by a rotation of the axes through an angle θ, $0° < \theta < 90°$, where

$$\cot 2\theta = \frac{A - C}{B}.$$

To find the rotation equations, first find $\sin\theta$ and $\cos\theta$. **Example 2** illustrates a way to obtain $\sin\theta$ and $\cos\theta$ from $\cot 2\theta$ without first identifying angle θ.

EXAMPLE 2 Rotating and Graphing

Remove the xy-term from $52x^2 - 72xy + 73y^2 = 200$ by performing a suitable rotation. Then graph the equation.

SOLUTION Here $A = 52$, $B = -72$, and $C = 73$.

$$\cot 2\theta = \frac{A - C}{B} = \frac{52 - 73}{-72} = \frac{-21}{-72} = \frac{7}{24}$$ Substitute into the angle of rotation equation, and simplify.

To find $\sin \theta$ and $\cos \theta$, use these trigonometric identities.

$$\sin \theta = \sqrt{\frac{1 - \cos 2\theta}{2}} \quad \text{and} \quad \cos \theta = \sqrt{\frac{1 + \cos 2\theta}{2}}$$

Sketch a right triangle as in **Figure 3**, to see that $\cos 2\theta = \frac{7}{25}$. (In the two quadrants for which we are concerned, cosine and cotangent have the same sign.)

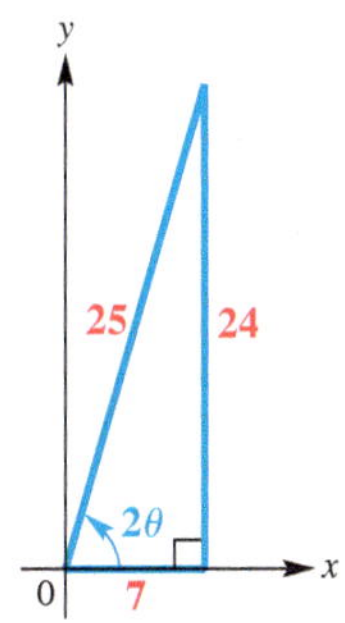

Figure 3

$$\sin \theta = \sqrt{\frac{1 - \frac{7}{25}}{2}} = \sqrt{\frac{9}{25}} = \frac{3}{5} \quad \text{and} \quad \cos \theta = \sqrt{\frac{1 + \frac{7}{25}}{2}} = \sqrt{\frac{16}{25}} = \frac{4}{5}$$

Use these values for $\sin \theta$ and $\cos \theta$ to obtain the following.

$$x = \frac{4}{5}x' - \frac{3}{5}y' \quad \text{and} \quad y = \frac{3}{5}x' + \frac{4}{5}y'$$

Substitute these expressions for x and y into the original equation.

$$52\left[\frac{4}{5}x' - \frac{3}{5}y'\right]^2 - 72\left[\frac{4}{5}x' - \frac{3}{5}y'\right]\left[\frac{3}{5}x' + \frac{4}{5}y'\right] + 73\left[\frac{3}{5}x' + \frac{4}{5}y'\right]^2 = 200$$

$$52\left[\frac{16}{25}x'^2 - \frac{24}{25}x'y' + \frac{9}{25}y'^2\right] - 72\left[\frac{12}{25}x'^2 + \frac{7}{25}x'y' - \frac{12}{25}y'^2\right] + 73\left[\frac{9}{25}x'^2 + \frac{24}{25}x'y' + \frac{16}{25}y'^2\right] = 200$$

$$25x'^2 + 100y'^2 = 200$$ Combine like terms.

$$\frac{x'^2}{8} + \frac{y'^2}{2} = 1$$ Divide by 200.

This is an equation of an ellipse having x'-intercepts $(\pm 2\sqrt{2}, 0)$ and y'-intercepts $(0, \pm\sqrt{2})$. The graph is shown in **Figure 4.** To find θ, use the following.

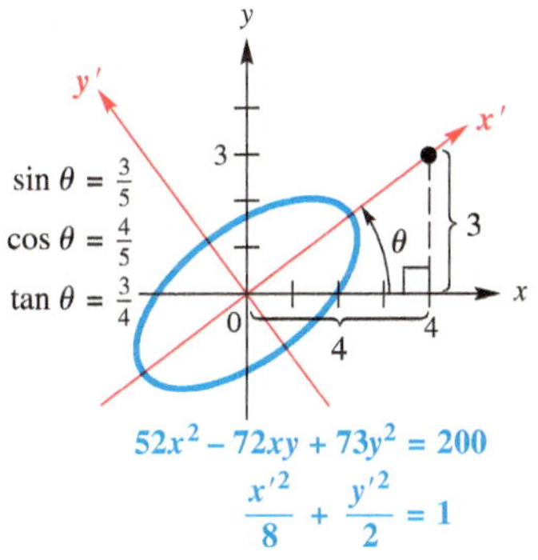

$$\frac{\sin \theta}{\cos \theta} = \frac{\frac{3}{5}}{\frac{4}{5}} = \frac{3}{5} \div \frac{4}{5} = \frac{3}{5} \cdot \frac{5}{4} = \frac{3}{4} = \tan \theta$$

$\sin \theta = \frac{3}{5}$
$\cos \theta = \frac{4}{5}$
$\tan \theta = \frac{3}{4}$

$52x^2 - 72xy + 73y^2 = 200$
$\frac{x'^2}{8} + \frac{y'^2}{2} = 1$

Figure 4

Use a calculator to find $\tan^{-1}\frac{3}{4} \approx 36.87°$. ✔ **Now Try Exercise 17.**

Equation of a Conic with an xy-Term

If the general second-degree equation

$$Ax^2 + Bxy + Cy^2 + Dx + Ey + F = 0$$

has a graph, it will be one of the following:

(a) a circle or an ellipse or a point if $B^2 - 4AC < 0$;

(b) a parabola or one line or two parallel lines if $B^2 - 4AC = 0$;

(c) a hyperbola or two intersecting lines if $B^2 - 4AC > 0$;

(d) a straight line if $A = B = C = 0$, and $D \neq 0$ or $E \neq 0$.

Appendix B Exercises

Concept Check *Use the summary at the end of the section to predict the type of graph of each second-degree equation.*

1. $4x^2 + 3y^2 + 2xy - 5x = 8$
2. $x^2 + 2xy - 3y^2 + 2y = 12$
3. $2x^2 + 3xy - 4y^2 = 0$
4. $x^2 - 2xy + y^2 + 4x - 8y = 0$
5. $4x^2 + 4xy + y^2 + 15 = 0$
6. $x^2 - 2xy + y^2 - 16 = 0$

Concept Check *Find the angle of rotation θ that will remove the xy-term in each equation.*

7. $2x^2 + \sqrt{3}xy + y^2 + x = 5$
8. $4\sqrt{3}x^2 + xy + 3\sqrt{3}y^2 = 10$
9. $3x^2 + \sqrt{3}xy + 4y^2 + 2x - 3y = 12$
10. $4x^2 + 2xy + 2y^2 + x = 7$
11. $x^2 - 4xy + 5y^2 = 18$
12. $3\sqrt{3}x^2 - 2xy + \sqrt{3}y^2 = 25$

Find the resulting equation if the axes are rotated through angle θ. Graph the equation. ***See Example 1.***

13. $x^2 - xy + y^2 = 6;\ \theta = 45°$
14. $5y^2 + 12xy = 10;\ \theta = \sin^{-1}\left(\frac{3\sqrt{13}}{13}\right)$

Remove the xy-term from each equation by performing a suitable rotation. Then graph each equation. ***See Example 2.***

15. $x^2 - 4xy + y^2 = -5$
16. $3x^2 - 2xy + 3y^2 = 8$
17. $7x^2 + 6\sqrt{3}xy + 13y^2 = 64$
18. $x^2 + 2xy + y^2 + 4\sqrt{2}x = 4\sqrt{2}y$
19. $3x^2 - 2\sqrt{3}xy + y^2 - 2x = 2\sqrt{3}y$
20. $7x^2 + 2\sqrt{3}xy + 5y^2 = 24$

Remove the xy-term by rotation. Then translate the axes and sketch the graph.

21. $x^2 + 3xy + y^2 - 5\sqrt{2}y - 15 = 0$
22. $x^2 - \sqrt{3}xy + 2\sqrt{3}x - 3y - 3 = 0$
23. $4x^2 + 4xy + y^2 - 24x + 38y - 19 = 0$
24. $12x^2 + 24xy + 19y^2 - 12x - 40y + 31 = 0$
25. $16x^2 + 24xy + 9y^2 - 130x + 90y = 0$
26. $9x^2 - 6xy + y^2 - 12\sqrt{10}x - 36\sqrt{10}y = 0$

C Geometry Formulas

Square
Perimeter: $P = 4s$
Area: $\mathcal{A} = s^2$

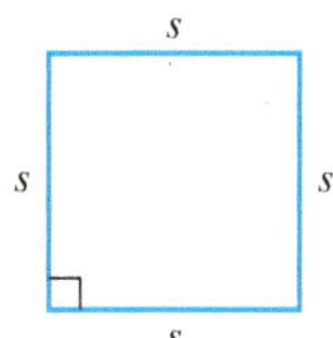

Rectangle
Perimeter: $P = 2L + 2W$
Area: $\mathcal{A} = LW$

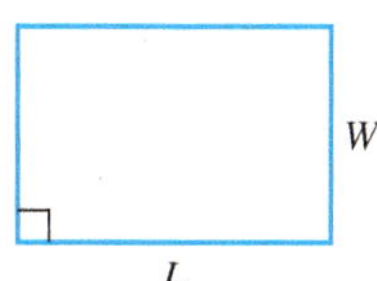

Triangle
Perimeter: $P = a + b + c$
Area: $\mathcal{A} = \frac{1}{2}bh$

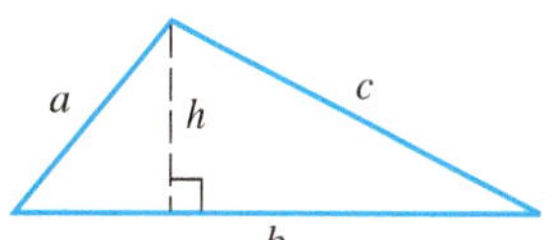

Parallelogram
Perimeter: $P = 2a + 2b$
Area: $\mathcal{A} = bh$

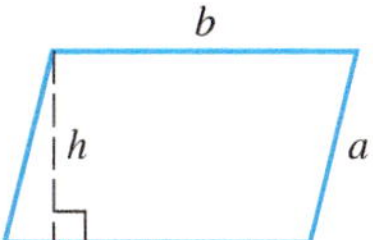

Trapezoid
Perimeter: $P = a + b + c + B$
Area: $\mathcal{A} = \frac{1}{2}h(B + b)$

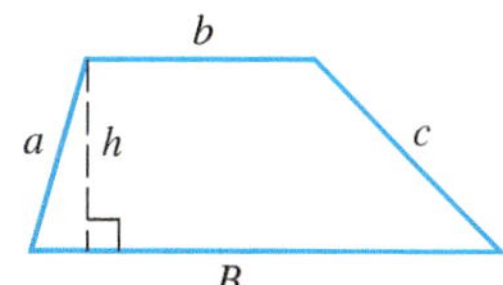

Circle
Diameter: $d = 2r$
Circumference: $C = 2\pi r = \pi d$
Area: $\mathcal{A} = \pi r^2$

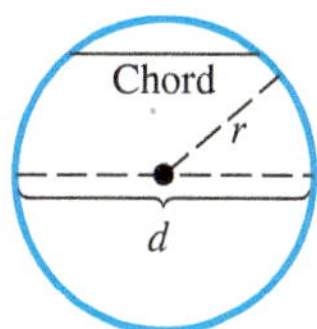

Cube
Volume: $V = e^3$
Surface area: $S = 6e^2$

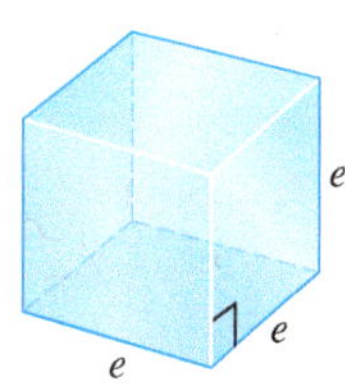

Rectangular Solid
Volume: $V = LWH$
Surface area: $S = 2HW + 2LW + 2LH$

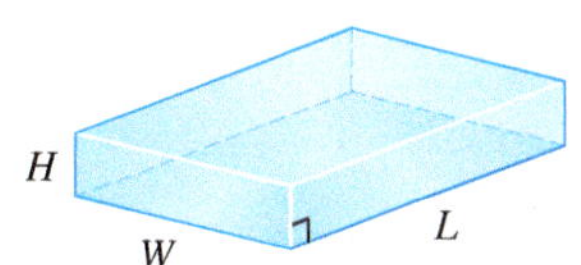

Sphere
Volume: $V = \frac{4}{3}\pi r^3$
Surface area: $S = 4\pi r^2$

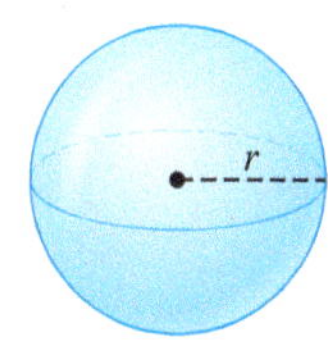

Cone
Volume: $V = \frac{1}{3}\pi r^2 h$
Surface area: $S = \pi r\sqrt{r^2 + h^2}$ (excludes the base)

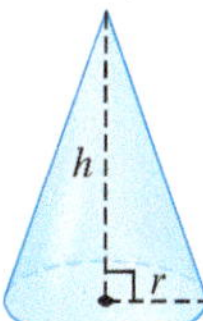

Right Circular Cylinder
Volume: $V = \pi r^2 h$
Surface area: $S = 2\pi rh + 2\pi r^2$ (includes top and bottom)

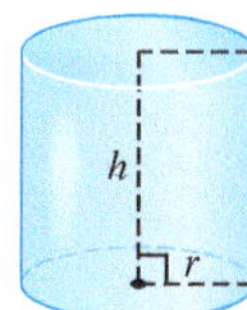

Right Pyramid
Volume: $V = \frac{1}{3}Bh$
B = area of the base

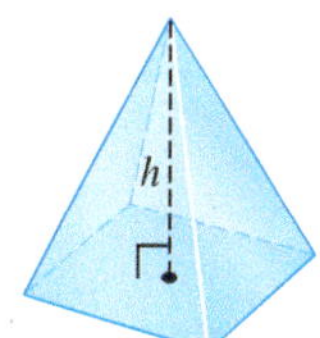

Answers to Selected Exercises

To The Student

In this section we provide the answers that we think most students will obtain when they work the exercises using the methods explained in the text. If your answer does not look exactly like the one given here, it is not necessarily wrong. In many cases there are equivalent forms of the answer. For example, if the answer section shows $\frac{3}{4}$ and your answer is 0.75, you have obtained the correct answer but written it in a different (yet equivalent) form. Unless the directions specify otherwise, 0.75 is just as valid an answer as $\frac{3}{4}$. In general, if your answer does not agree with the one given in the text, see whether it can be transformed into the other form. If it can, then it is the correct answer. If you still have doubts, talk with your instructor.

Chapter R Review of Basic Concepts

R.1 Exercises

1. $\{1, 2, 3, 4, \ldots\}$ **3.** complement **5.** union **7.** $\{1, 2, 3, 4, 5\}$ **9.** $\{16, 18\}$ **11.** finite; yes **13.** infinite; no **15.** infinite; no **17.** infinite; no **19.** $\{12, 13, 14, 15, 16, 17, 18, 19, 20\}$ **21.** $\left\{1, \frac{1}{2}, \frac{1}{4}, \frac{1}{8}, \frac{1}{16}, \frac{1}{32}\right\}$ **23.** $\{17, 22, 27, 32, 37, 42, 47\}$ **25.** $\{9, 10, 11, 12, 13, 14\}$ **27.** $\in$ **29.** $\notin$ **31.** $\in$ **33.** $\notin$ **35.** $\notin$ **37.** $\notin$ **39.** false **41.** true **43.** true **45.** true **47.** false **49.** true **51.** true **53.** true **55.** false **57.** true **59.** true **61.** false **63.** $\subseteq$ **65.** $\nsubseteq$ **67.** $\subseteq$ **69.** true **71.** false **73.** true **75.** false **77.** true **79.** true **81.** $\{0, 2, 4\}$ **83.** $\{0, 1, 2, 3, 4, 5, 6, 7, 8, 9, 11, 13\}$ **85.** $\varnothing$; M and N are disjoint sets. **87.** $\{0, 1, 2, 3, 4, 5, 7, 9, 11, 13\}$ **89.** Q, or $\{0, 2, 4, 6, 8, 10, 12\}$ **91.** $\{10, 12\}$ **93.** $\varnothing$; $\varnothing$ and R are disjoint. **95.** N, or $\{1, 3, 5, 7, 9, 11, 13\}$; N and $\varnothing$ are disjoint. **97.** R, or $\{0, 1, 2, 3, 4\}$; M and N are disjoint. $(M \cap N)$ and R are disjoint. **99.** $\{0, 1, 2, 3, 4, 6, 8\}$ **101.** R, or $\{0, 1, 2, 3, 4\}$ **103.** $\varnothing$; Q' and $(N' \cap U)$ are disjoint. **105.** $\{1, 3, 5, 7, 9, 10, 11, 12, 13\}$ **107.** M, or $\{0, 2, 4, 6, 8\}$ **109.** Q, or $\{0, 2, 4, 6, 8, 10, 12\}$

R.2 Exercises

1. whole numbers **3.** base; exponent **5.** absolute value **7.** 1000 **9.** 4 **11.** 1, 3 **13.** $-6, -\frac{12}{4}$ (or -3), 0, 1, 3 **15.** $-\sqrt{3}, 2\pi, \sqrt{12}$ **17.** -16 **19.** 16 **21.** -243 **23.** -162 **25.** -6 **27.** -60 **29.** -12 **31.** $-\frac{25}{36}$ **33.** $-\frac{6}{7}$ **35.** 28 **37.** $-\frac{1}{2}$ **39.** $-\frac{23}{20}$ **41.** $-\frac{25}{11}$ **43.** $-\frac{11}{3}$ **45.** $-\frac{13}{3}$ **47.** 6 **49.** distributive **51.** inverse **53.** identity **55.** commutative **57.** associative **59.** closure **61.** no; For example, $3 - 5 \neq 5 - 3$. **63.** $20z$ **65.** $m + 11$ **67.** $\frac{2}{3}y + \frac{4}{9}z - \frac{5}{3}$ **69.** $(8 - 14)p = -6p$ **71.** $-4z + 4y$ **73.** 1700 **75.** 150 **77.** false; $|6 - 8| = |8| - |6|$ **79.** true **81.** false; $|a - b| = |b| - |a|$ **83.** 10 **85.** $-\frac{4}{7}$ **87.** -8 **89.** 4 **91.** 16 **93.** 20 **95.** -1 **97.** -5 **99.** true **101.** false **103.** true **105.** 3 **107.** 9 **109.** x and y have the same sign. **111.** x and y have different signs. **113.** x and y have the same sign. **115.** 19; This represents the number of strokes between their scores. **117.** 0.031 **119.** 0.026; Increased weight results in lower BAC. **121.** 113.2 **123.** 103.3 **125.** 97.4 **127.** 96.5 **129.** 95.3 **131.** 93.9 **133.** 122.5 **135.** 111.0

R.3 Exercises

1. 5 **3.** binomial **5.** FOIL **7.** true **9.** false; $(x + y)^2 = x^2 + 2xy + y^2$ **11.** $-16x^7$ **13.** n^{11} **15.** 9^8 **17.** $72m^{11}$ **19.** $-15x^5y^5$ **21.** $4m^3n^3$ **23.** 2^{10} **25.** $-216x^6$ **27.** $-16m^6$ **29.** $\frac{r^{24}}{s^6}$ **31.** $\frac{256m^8}{t^4p^8}$ **33.** -1 **35.** **(a)** B **(b)** C **(c)** B **(d)** C **37.** polynomial; degree 11; monomial **39.** polynomial; degree 4; binomial **41.** polynomial; degree 5; trinomial **43.** polynomial; degree 11; none of these **45.** not a polynomial **47.** polynomial; degree 0; monomial **49.** $x^2 - x + 2$ **51.** $12y^2 + 4$ **53.** $6m^4 - 2m^3 - 7m^2 - 4m$ **55.** $28r^2 + r - 2$ **57.** $15x^4 - \frac{7}{3}x^3 - \frac{2}{9}x^2$ **59.** $12x^5 + 8x^4 - 20x^3 + 4x^2$ **61.** $-2z^3 + 7z^2 - 11z + 4$ **63.** $m^2 + mn - 2n^2 - 2km + 5kn - 3k^2$ **65.** $16x^4 - 72x^2 + 81$ **67.** $x^4 - 2x^2 + 1$ **69.** $4m^2 - 9$ **71.** $16x^4 - 25y^2$ **73.** $16m^2 + 16mn + 4n^2$ **75.** $25r^2 - 30rt^2 + 9t^4$ **77.** $4p^2 - 12p + 9 + 4pq - 6q + q^2$ **79.** $9q^2 + 30q + 25 - p^2$ **81.** $9a^2 + 6ab + b^2 - 6a - 2b + 1$ **83.** $y^3 + 6y^2 + 12y + 8$ **85.** $q^4 - 8q^3 + 24q^2 - 32q + 16$ **87.** $p^3 - 7p^2 - p - 7$ **89.** $49m^2 - 4n^2$ **91.** $-14q^2 + 11q - 14$ **93.** $4p^2 - 16$ **95.** $11y^3 - 18y^2 + 4y$ **97.** $2x^5 + 7x^4 - 5x^2 + 7$ **99.** $4x^2 + 5x + 10 + \frac{21}{x - 2}$ **101.** $2m^2 + m - 2 + \frac{6}{3m + 2}$ **103.** $x^2 + 2 + \frac{5x + 21}{x^2 + 3}$

105. **(a)** $(x+y)^2$ **(b)** $x^2+2xy+y^2$ **(c)** The expressions are equivalent because they represent the same area. **(d)** special product for squaring a binomial
107. **(a)** 60,501,000 ft^3 **(b)** The shape becomes a rectangular box with a square base, with volume $V=b^2h$.
(c) If we let $a=b$, then $V=\frac{1}{3}h(a^2+ab+b^2)$ becomes $V=\frac{1}{3}h(b^2+bb+b^2)$, which simplifies to $V=hb^2$. Yes, the Egyptian formula gives the same result.
109. 5.3; 0.3 low **111.** 2.2; 0.2 low **113.** 1,000,000
115. 32 **117.** 9999 **118.** 3591 **119.** 10,404
120. 5041

R.4 Exercises

1. factoring **3.** multiplying **5.** sum of squares
7. **(a)** B **(b)** C **(c)** A **9.** B **11.** $12(m+5)$
13. $8k(k^2+3)$ **15.** $xy(1-5y)$ **17.** $-2p^2q^4(2p+q)$
19. $4k^2m^3(1+2k^2-3m)$ **21.** $2(a+b)(1+2m)$
23. $(r+3)(3r-5)$ **25.** $(m-1)(2m^2-7m+7)$
27. The *completely* factored form is $4xy^3(xy^2-2)$.
29. $(2s+3)(3t-5)$ **31.** $(m^4+3)(2-a)$
33. $(p^2-2)(q^2+5)$ **35.** $(2a-1)(3a-4)$
37. $(3m+2)(m+4)$ **39.** prime
41. $2a(3a+7)(2a-3)$ **43.** $(3k-2p)(2k+3p)$
45. $(5a+3b)(a-2b)$ **47.** $(4x+y)(3x-y)$
49. $2a^2(4a-b)(3a+2b)$ **51.** $(3m-2)^2$
53. $2(4a+3b)^2$ **55.** $(2xy+7)^2$ **57.** $(a-3b-3)^2$
59. $(3a+4)(3a-4)$ **61.** $(x^2+4)(x+2)(x-2)$
63. $(5s^2+3t)(5s^2-3t)$ **65.** $(a+b+4)(a+b-4)$
67. $(p^2+25)(p+5)(p-5)$ **69.** $(x-4+y)(x-4-y)$
71. $(y+x-6)(y-x+6)$ **73.** $(2-a)(4+2a+a^2)$
75. $(5x-3)(25x^2+15x+9)$
77. $(3y^3+5z^2)(9y^6-15y^3z^2+25z^4)$
79. $r(r^2+18r+108)$
81. $(3-m-2n)(9+3m+6n+m^2+4mn+4n^2)$
83. $9(7k-3)(k+1)$ **85.** $(3a-7)^2$
87. $(a+4)(a^2-a+7)$ **89.** $9(x+1)(3x^2+9x+7)$
91. $(m^2-5)(m^2+2)$ **93.** $(3t^2+5)(4t^2-7)$
95. $(2b+c+4)(2b+c-4)$ **97.** $(x+y)(x-5)$
99. $(m-2n)(p^4+q)$ **101.** $(2z+7)^2$
103. $(10x+7y)(100x^2-70xy+49y^2)$
105. $(5m^2-6)(25m^4+30m^2+36)$
107. $9(x+2)(3x^2+4)$ **109.** $2y(3x^2+y^2)$ **111.** prime
113. $4xy$ **115.** In general, a sum of squares is not factorable over the real number system. If there is a greatest common factor, as in $4x^2+16$, it may be factored out, as, here, to obtain $4(x^2+4)$. **117.** $\left(7x+\frac{1}{5}\right)\left(7x-\frac{1}{5}\right)$
119. $\left(\frac{5}{3}x^2+3y\right)\left(\frac{5}{3}x^2-3y\right)$ **121.** ± 36 **123.** 9
125. $(x-1)(x^2+x+1)(x+1)(x^2-x+1)$
126. $(x-1)(x+1)(x^4+x^2+1)$
127. $(x^2-x+1)(x^2+x+1)$ **128.** additive inverse property (0 in the form x^2-x^2 was added on the right.); associative property of addition; factoring a perfect square trinomial; factoring a difference of squares; commutative property of addition **129.** They are the same.
130. $(x^4-x^2+1)(x^2+x+1)(x^2-x+1)$

R.5 Exercises

1. rational expression **3.** 5 **5.** $\frac{4}{x}$ **7.** $\frac{10}{x}$
9. $\frac{13x}{20}$ **11.** $\{x\,|\,x\neq 6\}$ **13.** $\left\{x\,|\,x\neq -\frac{1}{2}, 1\right\}$
15. $\{x\,|\,x\neq -2, -3\}$ **17.** $\{x\,|\,x\neq -1\}$ **19.** $\{x\,|\,x\neq 1\}$
21. $\frac{2x+4}{x}$ **23.** $\frac{-3}{t+5}$ **25.** $\frac{8}{9}$ **27.** $\frac{m-2}{m+3}$ **29.** $\frac{2m+3}{4m+3}$
31. $x^2-4x+16$ **33.** $\frac{25p^2}{9}$ **35.** $\frac{2}{9}$ **37.** $\frac{5x}{y}$
39. $\frac{2a+8}{a-3}$, or $\frac{2(a+4)}{a-3}$ **41.** 1 **43.** $\frac{m+6}{m+3}$ **45.** $\frac{x^2-xy+y^2}{x^2+xy+y^2}$
47. $\frac{x+2y}{4-x}$ **49.** B, C **51.** $\frac{19}{6k}$ **53.** $\frac{137}{30m}$ **55.** $\frac{a-b}{a^2}$
57. $\frac{5-22x}{12x^2y}$ **59.** 3 **61.** $\frac{2x}{(x+z)(x-z)}$ **63.** $\frac{4}{a-2}$, or $\frac{-4}{2-a}$
65. $\frac{3x+y}{2x-y}$, or $\frac{-3x-y}{y-2x}$ **67.** $\frac{4x-7}{x^2-x+1}$
69. $\frac{2x^2-9x}{(x-3)(x+4)(x-4)}$ **71.** $\frac{x+1}{x-1}$ **73.** $\frac{-1}{x+1}$
75. $\frac{(2-b)(1+b)}{b(1-b)}$ **77.** $\frac{a+b}{a^2-ab+b^2}$ **79.** $\frac{m^3-4m-1}{m-2}$
81. $\frac{p^3-16p+3}{p+4}$ **83.** $\frac{y^2-2y-3}{y^2+y-1}$ **85.** $\frac{-1}{x(x+h)}$
87. $\frac{-2x-h}{(x^2+9)[(x+h)^2+9]}$ **89.** 0 mi
91. 20.1 (thousand dollars)

R.6 Exercises

1. true **3.** false; $\frac{a^6}{a^4}=a^2$ **5.** false; $(x+y)^{-1}=\frac{1}{x+y}$
7. **(a)** B **(b)** D **(c)** B **(d)** D **9.** **(a)** E **(b)** G
(c) F **(d)** F **11.** $-\frac{1}{64}$ **13.** $-\frac{1}{625}$ **15.** 9 **17.** $\frac{1}{16x^2}$
19. $\frac{4}{x^2}$ **21.** $-\frac{1}{a^3}$ **23.** 16 **25.** x^4 **27.** $\frac{1}{r^3}$ **29.** 6^6
31. $\frac{2r^3}{3}$ **33.** $\frac{4n^7}{3m^7}$ **35.** $-4r^6$ **37.** $\frac{625}{a^{10}}$ **39.** $\frac{p^4}{5}$
41. $\frac{1}{2pq}$ **43.** $\frac{4}{a^2}$ **45.** $\frac{5}{x^2}$ **47.** 13 **49.** 2 **51.** $-\frac{4}{3}$
53. This expression is not a real number. **55.** 4
57. 1000 **59.** -27 **61.** $\frac{256}{81}$ **63.** 9 **65.** 4 **67.** y
69. $k^{2/3}$ **71.** x^3y^8 **73.** $\frac{1}{x^{10/3}}$ **75.** $\frac{6}{m^{1/4}n^{3/4}}$ **77.** p^2
79. **(a)** 250 sec **(b)** $2^{-1.5}\approx 0.3536$
81. $y-10y^2$ **83.** $-4k^{10/3}+24k^{4/3}$ **85.** x^2-x
87. $r-2+\frac{1}{r}$ **89.** $k^{-2}(4k+1)$ **91.** $4t^{-4}(t^2+2)$
93. $z^{-1/2}(9+2z)$ **95.** $p^{-7/4}(p-2)$ **97.** $4a^{-7/5}(-a+4)$
99. $(p+4)^{-3/2}(p^2+9p+21)$
101. $6(3x+1)^{-3/2}(9x^2+8x+2)$
103. $2x(2x+3)^{-5/9}(-16x^4-48x^3-30x^2+9x+2)$
105. $b+a$ **107.** -1 **109.** $\frac{y(xy-9)}{x^2y^2-9}$ **111.** $\frac{2x(1-3x^2)}{(x^2+1)^5}$
113. $\frac{1+2x^3-2x}{4}$ **115.** $\frac{3x-5}{(2x-3)^{4/3}}$ **117.** 27 **119.** 4
121. $\frac{1}{100}$

R.7 Exercises

1. $64^{1/3}$; 4 **3.** (a) F (b) H (c) G (d) C **5.** t **7.** $5\sqrt{2}$ **9.** $-5\sqrt{xy}$ **11.** 5 **13.** 3 **15.** -5 **17.** This expression is not a real number. **19.** 2 **21.** 2 **23.** $\sqrt[3]{m^2}$, or $\left(\sqrt[3]{m}\right)^2$ **25.** $\sqrt[3]{(2m+p)^2}$, or $\left(\sqrt[3]{2m+p}\right)^2$ **27.** $k^{2/5}$ **29.** $-3 \cdot 5^{1/2}p^{3/2}$ **31.** A **33.** $x \geq 0$ **35.** $|x|$ **37.** $5k^2|m|$ **39.** $|4x-y|$ **41.** $3\sqrt[3]{3}$ **43.** $-2\sqrt[4]{2}$ **45.** $\sqrt{42pqr}$ **47.** $\sqrt[3]{14xy}$ **49.** $-\frac{3}{5}$ **51.** $-\frac{\sqrt[3]{5}}{2}$ **53.** $\frac{\sqrt[4]{m}}{n}$ **55.** -15 **57.** $32\sqrt[3]{2}$ **59.** $2x^2z^4\sqrt{2x}$ **61.** This expression cannot be simplified further. **63.** $\frac{\sqrt{6x}}{3x}$ **65.** $\frac{x^2y\sqrt{xy}}{z}$ **67.** $\frac{2\sqrt[3]{x^2}}{x^2}$ **69.** $\frac{h\sqrt[4]{9g^3hr^2}}{3r^2}$ **71.** $\sqrt{3}$ **73.** $\sqrt[3]{2}$ **75.** $\sqrt[12]{2}$ **77.** $12\sqrt{2x}$ **79.** $7\sqrt[3]{3}$ **81.** $3x\sqrt[4]{x^2y^3} - 2x^2\sqrt[4]{x^2y^3}$ **83.** This expression cannot be simplified further. **85.** -7 **87.** 10 **89.** $11 + 4\sqrt{6}$ **91.** $5\sqrt{6}$ **93.** $\frac{m\sqrt[3]{n^2}}{n}$ **95.** $\frac{x\sqrt[3]{2} - \sqrt[3]{5}}{x^3}$ **97.** $\frac{11\sqrt{2}}{8}$ **99.** $-\frac{25\sqrt[3]{9}}{18}$ **101.** $\frac{\sqrt{15}-3}{2}$ **103.** $\frac{-7+2\sqrt{14}+\sqrt{7}-2\sqrt{2}}{2}$ **105.** $\sqrt{p}-2$ **107.** $\frac{3m\left(2-\sqrt{m+n}\right)}{4-m-n}$ **109.** $\frac{5\sqrt{x}\left(2\sqrt{x}-\sqrt{y}\right)}{4x-y}$ **111.** 17.7 ft per sec **113.** -12°F **115.** 2 **117.** 2 **119.** 3 **121.** It gives six decimal places of accuracy. **123.** It first differs in the fourth decimal place.

Chapter R Review Exercises

1. $\{6, 8, 10, 12, 14, 16, 18, 20\}$ **3.** true **5.** true **7.** false **9.** true **11.** true **13.** $\{2, 6, 9, 10\}$ **15.** $\varnothing$ **17.** $\varnothing$ **19.** $\{1, 2, 3, 4, 6, 8\}$ **21.** $\{1, 2, 3, 4, 5, 6, 7, 8, 9, 10\}$, or U **23.** $-12, -6, -\sqrt{4}$ (or -2), 0, 6 **25.** irrational number, real number **27.** whole number, integer, rational number, real number **29.** The reciprocal of a product is the product of the reciprocals. **31.** A product raised to a power is equal to the product of the factors to that power. **33.** A quotient raised to a power is equal to the quotient of the numerator and the denominator to that power. **35.** commutative **37.** associative **39.** identity **41.** 7.296 million **43.** 32 **45.** $-\frac{37}{18}$ **47.** $-\frac{12}{5}$ **49.** -32 **51.** -13 **53.** $7q^3 - 9q^2 - 8q + 9$ **55.** $16y^3 + 42y^2 - 73y + 21$ **57.** $9k^2 - 30km + 25m^2$ **59.** $6m^2 - 3m + 5$ **61.** $3b - 8 + \frac{2}{b^2+4}$ **63.** $3(z-4)^2(3z-11)$ **65.** $(z-8k)(z+2k)$ **67.** $6a^6(4a+5b)(2a-3b)$ **69.** $(7m^4+3n)(7m^4-3n)$ **71.** $3(9r-10)(2r+1)$ **73.** $(x-1)(y+2)$ **75.** $(3x-4)(9x-34)$ **77.** $\frac{1}{2k^2(k-1)}$ **79.** $\frac{x+1}{x+4}$ **81.** $\frac{p+6q}{p+q}$ **83.** $\frac{2m}{m-4}$, or $\frac{-2m}{4-m}$ **85.** $\frac{q+p}{pq-1}$ **87.** $\frac{16}{25}$ **89.** $-10z^8$ **91.** 1 **93.** $-8y^{11}p$ **95.** $\frac{1}{(p+q)^5}$ **97.** $-14r^{17/12}$ **99.** $y^{1/2}$ **101.** $\frac{1}{p^2q^4}$ **103.** $10\sqrt{2}$ **105.** $5\sqrt[4]{2}$ **107.** $-\frac{\sqrt[3]{50p}}{5p}$ **109.** $\sqrt[12]{m}$ **111.** 66 **113.** $-9m\sqrt{2m} + 5m\sqrt{m}$, or $m\left(-9\sqrt{2m}+5\sqrt{m}\right)$ **115.** $\frac{6\left(3+\sqrt{2}\right)}{7}$

In Exercises 117–125, we give only the corrected right sides of the equations.

117. $x^3 + 5x$ **119.** m^6 **121.** $\frac{a}{2b}$ **123.** $(-2)^{-3}$ **125.** $\frac{7b}{8b+7a}$

Chapter R Test

[R.1] **1.** false **2.** true **3.** false **4.** true [R.2] **5.** (a) $-13, -\frac{12}{4}$ (or -3), 0, $\sqrt{49}$ (or 7) (b) $-13, -\frac{12}{4}$ (or -3), 0, $\frac{3}{5}$, 5.9, $\sqrt{49}$ (or 7) (c) All are real numbers. **6.** 4 **7.** (a) associative (b) commutative (c) distributive (d) inverse **8.** 104.7 [R.3] **9.** $11x^2 - x + 2$ **10.** $36r^2 - 60r + 25$ **11.** $3t^3 + 5t^2 + 2t + 8$ **12.** $2x^2 - x - 5 + \frac{3}{x-5}$ **13.** \$9992 **14.** \$11,727 [R.4] **15.** $(3x-7)(2x-1)$ **16.** $(x^2+4)(x+2)(x-2)$ **17.** $2m(4m+3)(3m-4)$ **18.** $(x-2)(x^2+2x+4)(y+3)(y-3)$ **19.** $(a-b)(a-b+2)$ **20.** $(1-3x^2)(1+3x^2+9x^4)$ [R.5] **21.** $\frac{x^4(x+1)}{3(x^2+1)}$ **22.** $\frac{x(4x+1)}{(x+2)(x+1)(2x-3)}$ **23.** $\frac{2a}{2a-3}$, or $\frac{-2a}{3-2a}$ **24.** $\frac{y}{y+2}$ [R.6, R.7] **25.** $3x^2y^4\sqrt{2x}$ **26.** $2\sqrt{2x}$ **27.** $x-y$ **28.** $\frac{7\left(\sqrt{11}+\sqrt{7}\right)}{2}$ **29.** $\frac{y}{x}$ **30.** $\frac{9}{16}$ **31.** false **32.** 2.1 sec

Chapter 1 Equations and Inequalities

1.1 Exercises

1. equation **3.** first-degree equation **5.** contradiction **7.** true **9.** false **11.** $\{-4\}$ **13.** $\{1\}$ **15.** $\left\{-\frac{2}{7}\right\}$ **17.** $\left\{-\frac{7}{8}\right\}$ **19.** $\{-1\}$ **21.** $\{10\}$ **23.** $\{75\}$ **25.** $\{0\}$ **27.** $\{12\}$ **29.** $\{50\}$ **31.** identity; {all real numbers} **33.** conditional equation; $\{0\}$ **35.** contradiction; $\varnothing$ **37.** identity; {all real numbers} **39.** $l = \frac{V}{wh}$ **41.** $c = P - a - b$ **43.** $B = \frac{2\mathcal{A} - hb}{h}$, or $B = \frac{2\mathcal{A}}{h} - b$ **45.** $h = \frac{S - 2\pi r^2}{2\pi r}$, or $h = \frac{S}{2\pi r} - r$ **47.** $h = \frac{S - 2lw}{2w + 2l}$

Answers in Exercises 49–57 exist in equivalent forms.

49. $x = -3a + b$ **51.** $x = \frac{3a+b}{3-a}$ **53.** $x = \frac{3-3a}{a^2-a-1}$ **55.** $x = \frac{2a^2}{a^2+3}$ **57.** $x = \frac{m+4}{2m+5}$ **59.** (a) \$63 (b) \$3213 **61.** 68°F **63.** 10°C **65.** 37.8°C **67.** 463.9°C **69.** 45°C

1.2 Exercises

1. 8 hr **3.** \$40 **5.** 90 L **7.** A **9.** D **11.** 90 cm **13.** 6 cm **15.** 600 ft, 800 ft, 1000 ft **17.** 4 ft **19.** 50 mi **21.** 2.7 mi **23.** 45 min **25.** 1 hr, 7 min, 34 sec; It is about $\frac{1}{2}$ the world record time. **27.** 35 km per hr **29.** $7\frac{1}{2}$ gal **31.** 2 L **33.** 4 mL **35.** short-term note: \$100,000; long-term note: \$140,000 **37.** \$10,000 at 2.5%; \$20,000 at 3% **39.** \$50,000 at 1.5%; \$90,000 at 4% **41.** **(a)** \$52 **(b)** \$2500 **(c)** \$5000 **43.** **(a)** $F = 14{,}000x$ **(b)** 1.9 hr **45.** **(a)** 23.2 million **(b)** 2020 **(c)** They are quite close. **(d)** 17.5 million **(e)** When using the model for predictions, it is best to stay within the scope of the sample data.

1.3 Exercises

1. $\sqrt{-1}; -1$ **3.** complex conjugates **5.** denominator **7.** true **9.** false; $-12 + 13i$ **11.** real, complex **13.** pure imaginary, nonreal complex, complex **15.** nonreal complex, complex **17.** real, complex **19.** pure imaginary, nonreal complex, complex **21.** $5i$ **23.** $i\sqrt{10}$ **25.** $12i\sqrt{2}$ **27.** $-3i\sqrt{2}$ **29.** -13 **31.** $-2\sqrt{6}$ **33.** $\sqrt{3}$ **35.** $i\sqrt{3}$ **37.** $\frac{1}{2}$ **39.** -2 **41.** $-3 - i\sqrt{6}$ **43.** $2 + 2i\sqrt{2}$ **45.** $-\frac{1}{8} + \frac{\sqrt{2}}{8}i$ **47.** $12 - i$ **49.** 2 **51.** $1 - 10i$ **53.** $-13 + 4i\sqrt{2}$ **55.** $8 - i$ **57.** $-14 + 2i$ **59.** $5 - 12i$ **61.** 10 **63.** 13 **65.** 7 **67.** $25i$ **69.** $12 + 9i$ **71.** $20 + 15i$ **73.** $2 - 2i$ **75.** $\frac{3}{5} - \frac{4}{5}i$ **77.** $-1 - 2i$ **79.** $5i$ **81.** $8i$ **83.** $-\frac{2}{3}i$ **85.** $E = 2 + 62i$ **87.** $Z = 12 + 8i$ **89.** i **91.** -1 **93.** $-i$ **95.** 1 **97.** $-i$ **99.** $-i$

1.4 Exercises

1. G **3.** C **5.** H **7.** D **9.** D; $\left\{\frac{1}{3}, 7\right\}$ **11.** C; $\{-4, 3\}$ **13.** $\{2, 3\}$ **15.** $\left\{-\frac{2}{5}, 1\right\}$ **17.** $\left\{-\frac{3}{4}, 1\right\}$ **19.** $\{\pm 10\}$ **21.** $\left\{\frac{1}{2}\right\}$ **23.** $\left\{-\frac{3}{5}\right\}$ **25.** $\{\pm 4\}$ **27.** $\{\pm 3\sqrt{3}\}$ **29.** $\{\pm 9i\}$ **31.** $\left\{\frac{1 \pm 2\sqrt{3}}{3}\right\}$ **33.** $\{-5 \pm i\sqrt{3}\}$ **35.** $\left\{\frac{3}{5} \pm \frac{\sqrt{3}}{5}i\right\}$ **37.** $\{1, 3\}$ **39.** $\left\{-\frac{7}{2}, 4\right\}$ **41.** $\{1 \pm \sqrt{3}\}$ **43.** $\left\{-\frac{5}{2}, 2\right\}$ **45.** $\left\{\frac{2 \pm \sqrt{10}}{2}\right\}$ **47.** $\left\{1 \pm \frac{\sqrt{3}}{2}i\right\}$ **49.** He is incorrect because $c = 0$. **51.** $\left\{\frac{1 \pm \sqrt{5}}{2}\right\}$ **53.** $\{3 \pm \sqrt{2}\}$ **55.** $\{1 \pm 2i\}$ **57.** $\left\{\frac{3}{2} \pm \frac{\sqrt{2}}{2}i\right\}$ **59.** $\left\{\frac{-1 \pm \sqrt{97}}{4}\right\}$ **61.** $\left\{\frac{-2 \pm \sqrt{10}}{2}\right\}$ **63.** $\left\{\frac{-3 \pm \sqrt{41}}{8}\right\}$ **65.** $\{5\}$ **67.** $\{2, -1 \pm i\sqrt{3}\}$ **69.** $\left\{-3, \frac{3}{2} \pm \frac{3\sqrt{3}}{2}i\right\}$ **71.** $t = \frac{\pm\sqrt{2sg}}{g}$ **73.** $v = \frac{\pm\sqrt{FrkM}}{kM}$ **75.** $t = \frac{\pm\sqrt{2a(r - r_0)}}{a}$ **77.** $t = \frac{v_0 \pm \sqrt{v_0^2 - 64h + 64s_0}}{32}$ **79.** **(a)** $x = \frac{y \pm \sqrt{8 - 11y^2}}{4}$ **(b)** $y = \frac{x \pm \sqrt{6 - 11x^2}}{3}$ **81.** **(a)** $x = \frac{-2y \pm \sqrt{10y^2 + 4}}{2}$ **(b)** $y = \frac{2x \pm \sqrt{10x^2 - 6}}{3}$ **83.** 0; one rational solution (a double solution) **85.** 1; two distinct rational solutions **87.** 84; two distinct irrational solutions **89.** -23; two distinct nonreal complex solutions **91.** 2304; two distinct rational solutions **93.** no

In Exercises 95 and 97, there are other possible answers.

95. $a = 1, b = -9, c = 20$ **97.** $a = 1, b = -2, c = -1$

Chapter 1 Quiz

[1.1] **1.** $\{2\}$ **2.** **(a)** contradiction; $\emptyset$ **(b)** identity; {all real numbers} **(c)** conditional equation; $\left\{\frac{11}{4}\right\}$ **3.** $y = \frac{3x}{a - 1}$ [1.2] **4.** \$10,000 at 2.5%; \$20,000 at 3% **5.** \$6.59; The model predicts a wage that is \$0.04 greater than the actual wage. [1.3] **6.** $-\frac{1}{2} + \frac{\sqrt{6}}{4}i$ **7.** $\frac{3}{10} - \frac{8}{5}i$ [1.4] **8.** $\left\{\frac{1}{6} \pm \frac{\sqrt{11}}{6}i\right\}$ **9.** $\{\pm\sqrt{29}\}$ **10.** $r = \frac{\pm\sqrt{2s A\theta}}{\theta}$

1.5 Exercises

1. A **3.** D **5.** A **7.** B **9.** 7, 8 or $-8, -7$ **11.** 12, 14 or $-14, -12$ **13.** 7, 9 or $-9, -7$ **15.** 9, 11 or $-11, -9$ **17.** 20, 22 **19.** 6, 8, 10 **21.** 7 in., 10 in. **23.** 100 yd by 400 yd **25.** 9 ft by 12 ft **27.** 20 in. by 30 in. **29.** 1 ft **31.** 4 **33.** 3.75 cm **35.** 5 ft **37.** $10\sqrt{2}$ ft **39.** 16.4 ft **41.** 3000 yd **43.** **(a)** 1 sec, 5 sec **(b)** 6 sec **45.** **(a)** It will not reach 80 ft. **(b)** 2 sec **47.** **(a)** 0.19 sec, 10.92 sec **(b)** 11.32 sec **49.** **(a)** \$108.1 million **(b)** 2004 **51.** **(a)** 19.2 hr **(b)** 84.3 ppm (109.8 is not in the interval [50, 100].) **53.** **(a)** 549.2 million metric tons **(b)** 2019 **55.** \$42,795 million **57.** $80 - x$ **58.** $300 + 20x$ **59.** $R = (80 - x)(300 + 20x) = 24{,}000 + 1300x - 20x^2$ **60.** 10, 55; Because of the restriction, only $x = 10$ is valid. The number of apartments rented is 70. **61.** 80 **63.** 4

1.6 Exercises

1. rational equation **3.** $\frac{1}{4}$ **5.** rational exponent **7.** D **9.** E **11.** $-\frac{3}{2}, 6$ **13.** 2, -1 **15.** 0 **17.** $\{-10\}$ **19.** $\emptyset$ **21.** $\emptyset$ **23.** $\{-9\}$ **25.** $\{-2\}$ **27.** $\emptyset$ **29.** $\left\{-\frac{5}{2}, \frac{1}{9}\right\}$ **31.** $\left\{\frac{3}{4}, 1\right\}$ **33.** $\{3, 5\}$ **35.** $\left\{-2, \frac{5}{4}\right\}$ **37.** $1\frac{7}{8}$ hr **39.** 78 hr **41.** $13\frac{1}{3}$ hr **43.** 10 min **45.** $\{3\}$ **47.** $\{-1\}$ **49.** $\{5\}$ **51.** $\{9\}$ **53.** $\{9\}$ **55.** $\emptyset$ **57.** $\{\pm 2\}$ **59.** $\{0, 3\}$ **61.** $\{-2\}$ **63.** $\left\{-\frac{2}{9}, 2\right\}$ **65.** $\{4\}$ **67.** $\{-2\}$ **69.** $\left\{\frac{2}{5}, 1\right\}$ **71.** $\{31\}$ **73.** $\{-3, 1\}$ **75.** $\{25\}$ **77.** $\{-27, 3\}$ **79.** $\{-29, 35\}$ **81.** $\left\{\frac{3}{2}\right\}$ **83.** $\left\{\frac{1}{4}, 1\right\}$ **85.** $\{0, 8\}$ **87.** $\left\{\pm 1, \pm\frac{\sqrt{10}}{2}\right\}$ **89.** $\{\pm\sqrt{3}, \pm i\sqrt{5}\}$ **91.** $\{-63, 28\}$ **93.** $\{0, 31\}$ **95.** $\left\{-\frac{5}{2}, -2, 0, \frac{1}{2}\right\}$ **97.** $\left\{\frac{-6 \pm 2\sqrt{3}}{3}, \frac{-4 \pm \sqrt{2}}{2}\right\}$

99. $\left\{-\frac{2}{7}, 5\right\}$ **101.** $\left\{-\frac{1}{27}, \frac{1}{8}\right\}$ **103.** $\left\{\pm\frac{1}{2}, \pm 4\right\}$
105. $h = \frac{d^2}{k^2}$ **107.** $m = (1 - n^{3/4})^{4/3}$ **109.** $e = \frac{Er}{R + r}$
111. $\{16\}$ **112.** $\{16\}$ **113.** Answers may vary.
114. $\{4\}$

Summary Exercises on Solving Equations

1. $\{3\}$ **2.** $\{-1\}$ **3.** $\left\{-3 \pm 3\sqrt{2}\right\}$ **4.** $\{2, 6\}$
5. $\varnothing$ **6.** $\{-31\}$ **7.** $\{-6\}$ **8.** $\{6\}$ **9.** $\left\{\frac{1}{5} \pm \frac{2}{5}i\right\}$
10. $\{-2, 1\}$ **11.** $\left\{-\frac{1}{243}, \frac{1}{3125}\right\}$ **12.** $\{-1\}$
13. $\{\pm i, \pm 2\}$ **14.** $\{-2.4\}$ **15.** $\{4\}$
16. $\left\{\frac{1}{3} \pm \frac{\sqrt{2}}{3}i\right\}$ **17.** $\left\{\frac{15}{7}\right\}$ **18.** $\{4\}$ **19.** $\{3, 11\}$
20. $\{1\}$ **21.** $\{x \mid x \neq 3\}$ **22.** $a = \pm\sqrt{c^2 - b^2}$

1.7 Exercises

1. F **3.** A **5.** I **7.** B **9.** E **11.** A square bracket is used to show that a number is part of the solution set, and a parenthesis is used to indicate that a number is not part of the solution set. **13.** $[-4, \infty)$ **15.** $[-1, \infty)$ **17.** $(-\infty, \infty)$
19. $(-\infty, 4)$ **21.** $\left[-\frac{11}{5}, \infty\right)$ **23.** $\left(-\infty, \frac{48}{7}\right]$ **25.** $[500, \infty)$
27. The product will never break even. **29.** $(-5, 3)$
31. $[3, 6]$ **33.** $(4, 6)$ **35.** $[-9, 9]$ **37.** $\left(-\frac{16}{3}, \frac{19}{3}\right]$
39. $(-\infty, -2) \cup (3, \infty)$ **41.** $\left[-\frac{3}{2}, 6\right]$
43. $(-\infty, -3] \cup [-1, \infty)$ **45.** $[-2, 3]$ **47.** $[-3, 3]$
49. $\varnothing$ **51.** $\left[1 - \sqrt{2}, 1 + \sqrt{2}\right]$ **53.** A **55.** $(-5, 3]$
57. $(-\infty, -2)$ **59.** $(-\infty, 6) \cup \left[\frac{15}{2}, \infty\right)$
61. $(-\infty, 1) \cup \left(\frac{9}{5}, \infty\right)$ **63.** $\left(-\infty, -\frac{3}{2}\right) \cup \left[-\frac{1}{2}, \infty\right)$
65. $(-2, \infty)$ **67.** $\left(0, \frac{4}{11}\right) \cup \left(\frac{1}{2}, \infty\right)$
69. $(-\infty, -2] \cup (1, 2)$ **71.** $(-\infty, 5)$ **73.** $\left[\frac{3}{2}, \infty\right)$
75. $\left(\frac{5}{2}, \infty\right)$ **77.** $\left[-\frac{8}{3}, \frac{3}{2}\right] \cup (6, \infty)$ **79.** **(a)** 2000 **(b)** 2008 **81.** between (and inclusive of) 4 sec and 9.75 sec
83. between (and inclusive of) 1 sec and 1.75 sec
85. between -1.5 sec and 4 sec **87.** $\left\{\frac{4}{3}, -2, -6\right\}$

88. 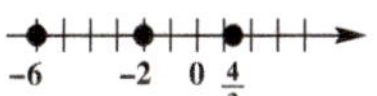

89. In the interval $(-\infty, -6)$, choose $x = -10$, for example. It satisfies the original inequality. In the interval $(-6, -2)$, choose $x = -4$, for example. It does not satisfy the inequality. In the interval $\left(-2, \frac{4}{3}\right)$, choose $x = 0$, for example. It satisfies the original inequality. In the interval $\left(\frac{4}{3}, \infty\right)$, choose $x = 4$, for example. It does not satisfy the original inequality.

90. **(a)** 

(b) $(-\infty, -6] \cup \left[-2, \frac{4}{3}\right]$

91. $\left[-2, \frac{3}{2}\right] \cup [3, \infty)$ **93.** $(-\infty, -2] \cup [0, 2]$

95. $(-\infty, -1) \cup (-1, 3)$ **97.** $[-4, -3] \cup [3, \infty)$
99. $(-\infty, \infty)$

1.8 Exercises

1. F **3.** D **5.** G **7.** C **9.** $\left\{-\frac{1}{3}, 1\right\}$ **11.** $\left\{\frac{2}{3}, \frac{8}{3}\right\}$
13. $\{-6, 14\}$ **15.** $\left\{\frac{5}{2}, \frac{7}{2}\right\}$ **17.** $\left\{-\frac{4}{3}, \frac{2}{9}\right\}$ **19.** $\left\{-\frac{7}{3}, -\frac{1}{7}\right\}$
21. $\{1\}$ **23.** $(-\infty, \infty)$ **25.** A positive solution will return a negative quantity for $-5x$, and the absolute value of an expression cannot be negative. **27.** $(-4, -1)$
29. $(-\infty, -4] \cup [-1, \infty)$ **31.** $\left(-\frac{3}{2}, \frac{5}{2}\right)$
33. $(-\infty, 0) \cup (6, \infty)$ **35.** $\left(-\infty, -\frac{2}{3}\right) \cup (4, \infty)$
37. $\left[-\frac{2}{3}, 4\right]$ **39.** $\left[-1, -\frac{1}{2}\right]$ **41.** $(-101, -99)$
43. $\left\{-1, -\frac{1}{2}\right\}$ **45.** $\{2, 4\}$ **47.** $\left(-\frac{4}{3}, \frac{2}{3}\right)$ **49.** $\left(-\frac{3}{2}, \frac{13}{10}\right)$
51. $\left(-\infty, \frac{3}{2}\right] \cup \left[\frac{7}{2}, \infty\right)$ **53.** $\varnothing$ **55.** $(-\infty, \infty)$ **57.** $\varnothing$
59. $\left\{-\frac{5}{8}\right\}$ **61.** $\varnothing$ **63.** $\left\{-\frac{1}{2}\right\}$ **65.** $\left(-\infty, -\frac{2}{3}\right) \cup \left(-\frac{2}{3}, \infty\right)$

In Exercises 67–73, the expression in absolute value bars may be replaced by its additive inverse. For example, in Exercise 67, $p - q$ may be written $q - p$.

67. $|p - q| = 2$ **69.** $|m - 7| \leq 2$ **71.** $|p - 9| < 0.0001$
73. $|r - 29| \geq 1$ **75.** $(0.9996, 1.0004)$ **77.** $[6.7, 9.7]$
79. $|F - 730| \leq 50$ **81.** $25.33 \leq R_L \leq 28.17$; $36.58 \leq R_E \leq 40.92$ **83.** -6 or 6 **84.** $x^2 - x = 6$; $\{-2, 3\}$ **85.** $x^2 - x = -6$; $\left\{\frac{1}{2} \pm \frac{\sqrt{23}}{2}i\right\}$
86. $\left\{-2, 3, \frac{1}{2} \pm \frac{\sqrt{23}}{2}i\right\}$ **87.** $\left\{-\frac{7}{3}, 2, -\frac{1}{6} \pm \frac{\sqrt{167}}{6}i\right\}$
89. $\left\{-\frac{1}{4}, 6\right\}$ **91.** $\{-1, 1\}$ **93.** $\varnothing$
95. $\left(-\infty, -\frac{1}{3}\right) \cup \left(-\frac{1}{3}, \infty\right)$

Chapter 1 Review Exercises

1. $\{6\}$ **3.** $\left\{-\frac{11}{3}\right\}$ **5.** $f = \frac{AB(p + 1)}{24}$ **7.** A, B
9. 13 in. on each side **11.** $3\frac{3}{7}$ L **13.** 560 km per hr
15. **(a)** $A = 36.525x$ **(b)** 2629.8 mg **17.** **(a)** \$4.31; The model gives a figure that is \$0.51 more than the actual figure of \$3.80. **(b)** 47.6 yr after 1956, which is mid-2003. This is close to the minimum wage changing to \$5.85 in 2007.
19. $13 - 3i$ **21.** $-14 + 13i$ **23.** $19 + 17i$ **25.** 146
27. $-30 - 40i$ **29.** $1 - 2i$ **31.** $-i$ **33.** i **35.** i
37. $\left\{-7 \pm \sqrt{5}\right\}$ **39.** $\left\{-3, \frac{5}{2}\right\}$ **41.** $\left\{-\frac{3}{2}, 7\right\}$
43. $\left\{2 \pm \sqrt{6}\right\}$ **45.** $\left\{\frac{\sqrt{5} \pm 3}{2}\right\}$ **47.** D
49. 76; two distinct irrational solutions
51. -124; two distinct nonreal complex solutions
53. 0; one rational solution (a double solution)
55. 6.25 sec and 7.5 sec **57.** $\frac{1}{2}$ ft **59.** \$801.9 billion
61. $\left\{\pm i, \pm\frac{1}{2}\right\}$ **63.** $\left\{-\frac{7}{24}\right\}$ **65.** $\varnothing$ **67.** $\{-239, 247\}$

69. $\{1, 4\}$ **71.** $\{-2, -1\}$ **73.** $\{3\}$ **75.** $\varnothing$
77. $\{-1\}$ **79.** $\left\{-\frac{7}{4}\right\}$ **81.** $\left\{-15, \frac{5}{2}\right\}$ **83.** $\left(-\frac{7}{13}, \infty\right)$
85. $(-\infty, 1]$ **87.** $[4, 5]$ **89.** $[-4, 1]$ **91.** $\left(-\frac{2}{3}, \frac{5}{2}\right)$
93. $(-\infty, -4] \cup [0, 4]$ **95.** $(-\infty, -2) \cup (5, \infty)$
97. $(-2, 0)$ **99.** $(-3, 1) \cup [7, \infty)$ **101. (a)** 79.8 ppb **(b)** 87.7 ppb **103. (a)** 20 sec **(b)** between 2 sec and 18 sec **105.** The value 3 makes the denominator 0. Therefore, it must be excluded from the solution set.
107. $\{-11, 3\}$ **109.** $\left\{\frac{11}{27}, \frac{25}{27}\right\}$ **111.** $\left\{-\frac{2}{7}, \frac{4}{3}\right\}$
113. $[-6, -3]$ **115.** $\left(-\infty, -\frac{1}{7}\right) \cup (1, \infty)$
117. $\left(-\frac{17}{3}, 1\right)$ **119.** $(-\infty, \infty)$ **121.** $\{0, -4\}$
123. $|k - 6| = 12$ (or $|6 - k| = 12$)
125. $|t - 5| \geq 0.01$ (or $|5 - t| \geq 0.01$)

Chapter 1 Test

[1.1] **1.** $\{0\}$ **2.** $\{-12\}$ [1.4] **3.** $\left\{-\frac{1}{2}, \frac{7}{3}\right\}$
4. $\left\{\frac{-1 \pm 2\sqrt{2}}{3}\right\}$ **5.** $\left\{-\frac{1}{3} \pm \frac{\sqrt{5}}{3}i\right\}$ [1.6] **6.** $\varnothing$
7. $\left\{-\frac{3}{4}\right\}$ **8.** $\{4\}$ **9.** $\{-3, 1\}$ **10.** $\{-2\}$
11. $\{\pm 1, \pm 4\}$ **12.** $\{-30, 5\}$ [1.8] **13.** $\left\{-\frac{5}{2}, 1\right\}$
14. $\left\{-6, \frac{4}{3}\right\}$ [1.1] **15.** $W = \frac{S - 2LH}{2H + 2L}$
[1.3] **16. (a)** $5 - 8i$ **(b)** $-29 - 3i$ **(c)** $55 + 48i$ **(d)** $6 + i$ **17. (a)** -1 **(b)** i **(c)** i
[1.2] **18. (a)** $A = 806{,}400x$ **(b)** 24,192,000 gal **(c)** $P = 40.32x$; 40 pools **(d)** 25 days
19. length: 200 m; width: 110 m **20.** cashews: $23\frac{1}{3}$ lb; walnuts: $11\frac{2}{3}$ lb **21.** 15 mph [1.2] **22. (a)** 1.7% **(b)** 2007
[1.5] **23. (a)** 1 sec and 5 sec **(b)** 6 sec
[1.7] **24.** $(-3, \infty)$ **25.** $[-10, 2]$
26. $(-\infty, -1] \cup \left[\frac{3}{2}, \infty\right)$ **27.** $(-\infty, 3) \cup (4, \infty)$
[1.8] **28.** $(-2, 7)$ **29.** $(-\infty, -6] \cup [5, \infty)$ **30.** $\left\{-\frac{7}{3}\right\}$

Chapter 2 Graphs and Functions

2.1 Exercises

1. II **3.** 0 **5.** $(5, 0)$ **7.** true **9.** false; The midpoint is a point with coordinates $(2, 2)$. **11.** any three of the following: $(2, -5)$, $(-1, 7)$, $(3, -9)$, $(5, -17)$, $(6, -21)$
13. any three of the following: $(1999, 35)$, $(2001, 29)$, $(2003, 22)$, $(2005, 23)$, $(2007, 20)$, $(2009, 20)$
15. (a) 13 **(b)** $\left(1, -\frac{7}{2}\right)$ **17. (a)** $\sqrt{34}$ **(b)** $\left(\frac{11}{2}, \frac{7}{2}\right)$
19. (a) $3\sqrt{41}$ **(b)** $\left(0, \frac{5}{2}\right)$ **21. (a)** $\sqrt{133}$
(b) $\left(2\sqrt{2}, \frac{3\sqrt{5}}{2}\right)$ **23.** yes **25.** no **27.** yes **29.** yes
31. no **33.** no **35.** $(-3, 6)$ **37.** $(5, -4)$
39. $(2a - p, 2b - q)$ **41.** 26.1%; This estimate is very close to the actual figure. **43.** \$23,428

Other ordered pairs are possible in Exercises 47–57.

47. (a)

x	y
0	-2
4	0
2	-1

(b)

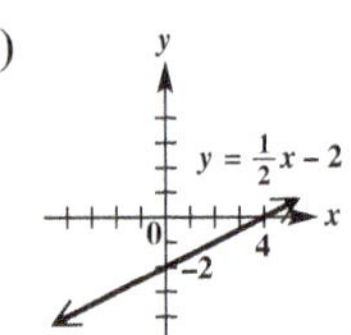

49. (a)

x	y
0	$\frac{5}{3}$
$\frac{5}{2}$	0
4	-1

(b)

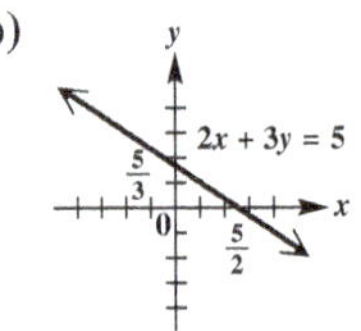

51. (a)

x	y
0	0
1	1
-2	4

(b)

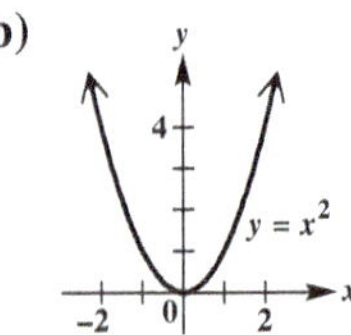

53. (a)

x	y
3	0
4	1
7	2

(b)

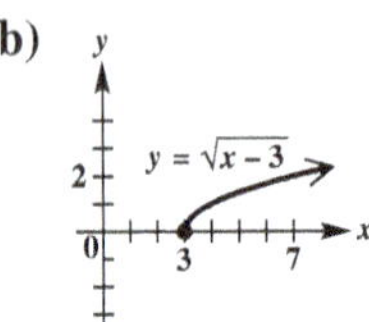

55. (a)

x	y
4	2
-2	4
0	2

(b)

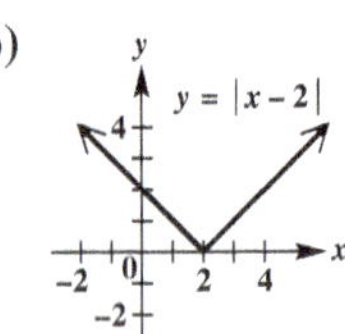

57. (a)

x	y
0	0
-1	-1
2	8

(b)

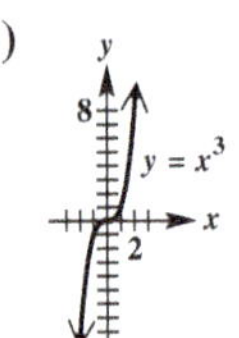

59. $(4, 0)$ **61.** III; I; IV; IV **63.** yes; no

2.2 Exercises

1. $(0, 0)$; 7 **3.** $(4, -7)$ **5.** B **7.** D
9. one (The point is $(0, 0)$.)
11. (a) $x^2 + y^2 = 36$ **(b)**

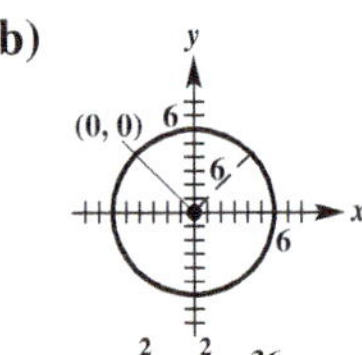

13. **(a)** $(x-2)^2+y^2=36$ **(b)**

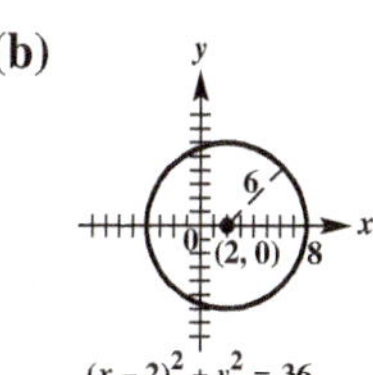

15. **(a)** $x^2+(y-4)^2=16$ **(b)**

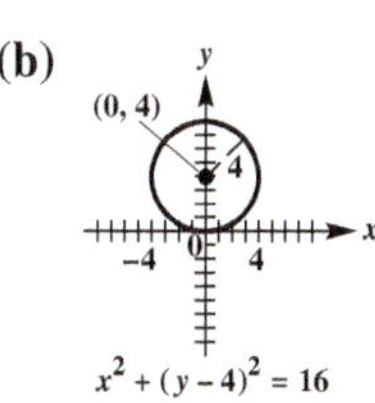

17. **(a)** $(x+2)^2+(y-5)^2=16$ **(b)**

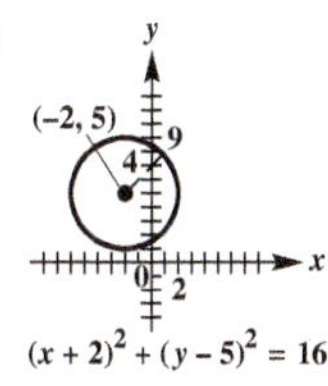

19. **(a)** $(x-5)^2+(y+4)^2=49$ **(b)**

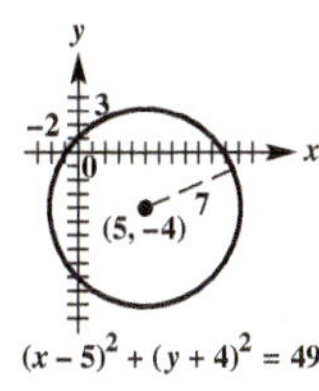

21. **(a)** $\left(x-\sqrt{2}\right)^2+\left(y-\sqrt{2}\right)^2=2$
(b)

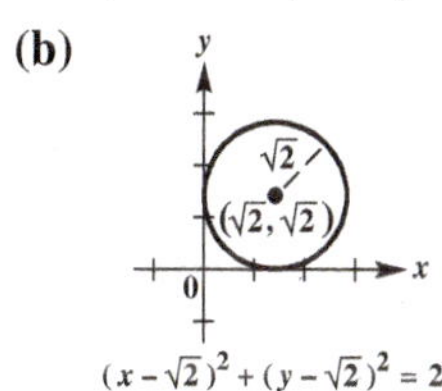

23. **(a)** $(x-3)^2+(y-1)^2=4$
(b) $x^2+y^2-6x-2y+6=0$
25. **(a)** $(x+2)^2+(y-2)^2=4$
(b) $x^2+y^2+4x-4y+4=0$
27. yes; center: $(-3,-4)$; radius: 4 **29.** yes; center: $(2,-6)$; radius: 6 **31.** yes; center: $\left(-\frac{1}{2},2\right)$; radius: 3 **33.** no; The graph is nonexistent.
35. no; The graph is the point $(3,3)$. **37.** yes; center: $\left(\frac{1}{3},-\frac{1}{3}\right)$; radius: $\frac{5}{3}$ **39.** $(3,1)$ **41.** $(-2,-2)$
43. $(x-3)^2+(y-2)^2=4$
45. $\left(2+\sqrt{7},2+\sqrt{7}\right),\left(2-\sqrt{7},2-\sqrt{7}\right)$
47. $(2,3)$ and $(4,1)$ **49.** $9+\sqrt{119}, 9-\sqrt{119}$
51. $\sqrt{113}-5$ **53.** $(2,-3)$ **54.** $3\sqrt{5}$ **55.** $3\sqrt{5}$
56. $3\sqrt{5}$ **57.** $(x-2)^2+(y+3)^2=45$
58. $(x+2)^2+(y+1)^2=41$
59. $(x-5)^2+\left(y-\frac{9}{2}\right)^2=\frac{169}{4}$
61. $(x-3)^2+\left(y-\frac{5}{2}\right)^2=\frac{25}{4}$

2.3 Exercises

1. $\{3,4,10\}$ **3.** x; y **5.** 10 **7.** $[0,\infty)$ **9.** $(-\infty,3)$
11. function **13.** not a function **15.** function
17. function **19.** not a function; domain: $\{0,1,2\}$; range: $\{-4,-1,0,1,4\}$ **21.** function; domain: $\{2,3,5,11,17\}$; range: $\{1,7,20\}$ **23.** function; domain: $\{0,-1,-2\}$; range: $\{0,1,2\}$ **25.** function; domain: $\{2010, 2011, 2012, 2013\}$; range: $\{64.9, 63.0, 65.1, 63.5\}$
27. function; domain: $(-\infty,\infty)$; range: $(-\infty,\infty)$
29. not a function; domain: $[3,\infty)$; range: $(-\infty,\infty)$
31. function; domain: $(-\infty,\infty)$; range: $(-\infty,\infty)$
33. function; domain: $(-\infty,\infty)$; range: $[0,\infty)$
35. not a function; domain: $[0,\infty)$; range: $(-\infty,\infty)$
37. function; domain: $(-\infty,\infty)$; range: $(-\infty,\infty)$
39. not a function; domain: $(-\infty,\infty)$; range: $(-\infty,\infty)$
41. function; domain: $[0,\infty)$; range: $[0,\infty)$
43. function; domain: $(-\infty,0)\cup(0,\infty)$; range: $(-\infty,0)\cup(0,\infty)$ **45.** function; domain: $\left[-\frac{1}{4},\infty\right)$; range: $[0,\infty)$ **47.** function; domain: $(-\infty,3)\cup(3,\infty)$; range: $(-\infty,0)\cup(0,\infty)$ **49.** B **51.** 4 **53.** -11
55. 3 **57.** $\frac{11}{4}$ **59.** $-3p+4$ **61.** $3x+4$ **63.** $-3x-2$
65. $-6m+13$ **67.** **(a)** 2 **(b)** 3 **69.** **(a)** 15 **(b)** 10
71. **(a)** 3 **(b)** -3 **73.** **(a)** 0 **(b)** 4 **(c)** 2 **(d)** 4
75. **(a)** -3 **(b)** -2 **(c)** 0 **(d)** 2
77. **(a)** $f(x)=-\frac{1}{3}x+4$ **(b)** 3
79. **(a)** $f(x)=-2x^2-x+3$ **(b)** -18
81. **(a)** $f(x)=\frac{4}{3}x-\frac{8}{3}$ **(b)** $\frac{4}{3}$ **83.** $f(3)=4$ **85.** -4
87. **(a)** $(-2,0)$ **(b)** $(-\infty,-2)$ **(c)** $(0,\infty)$
89. **(a)** $(-\infty,-2)$; $(2,\infty)$ **(b)** $(-2,2)$ **(c)** none
91. **(a)** $(-1,0)$; $(1,\infty)$ **(b)** $(-\infty,-1)$; $(0,1)$ **(c)** none
93. **(a)** yes **(b)** $[0,24]$ **(c)** 1200 megawatts
(d) at 17 hr or 5 p.m.; at 4 a.m. **(e)** $f(12)=1900$; At 12 noon, electricity use is 1900 megawatts.
(f) increasing from 4 a.m. to 5 p.m.; decreasing from midnight to 4 a.m. and from 5 p.m. to midnight
95. **(a)** 12 noon to 8 p.m. **(b)** from midnight until about 6 a.m. and after 10 p.m. **(c)** about 10 a.m. and 8:30 p.m.
(d) The temperature is 40° from midnight to 6 a.m., when it begins to rise until it reaches a maximum of just below 65° at 4 p.m. It then begins to fall until it reaches just under 40° at midnight.

2.4 Exercises

1. B **3.** C **5.** A **7.** C **9.** D

In Exercises 11–29, we give the domain first and then the range.

11.

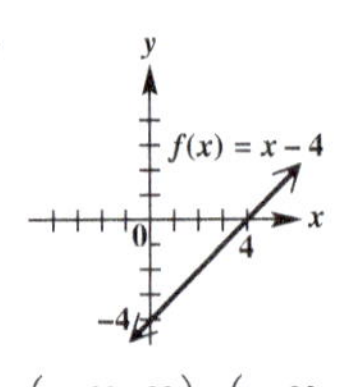

$(-\infty, \infty)$; $(-\infty, \infty)$

13.

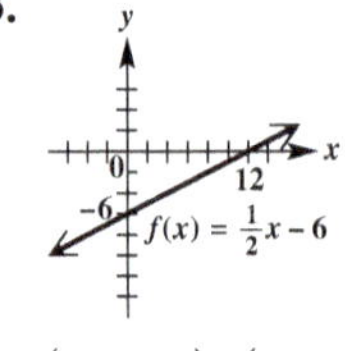

$(-\infty, \infty)$; $(-\infty, \infty)$

15.

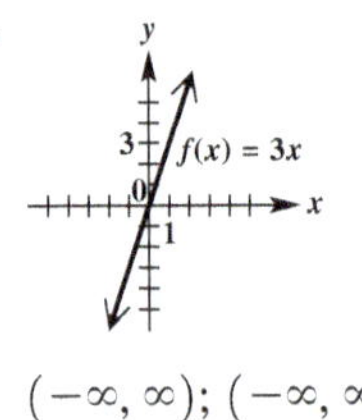

$(-\infty, \infty)$; $(-\infty, \infty)$

17.

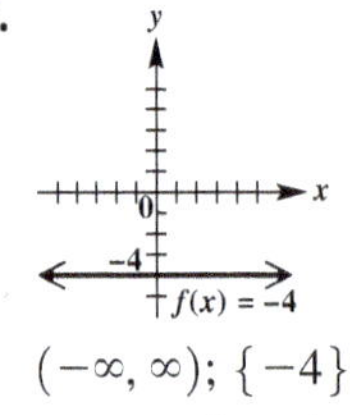

$(-\infty, \infty)$; $\{-4\}$; constant function

19.

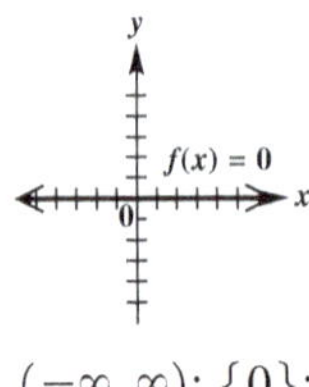

$(-\infty, \infty)$; $\{0\}$; constant function

21.

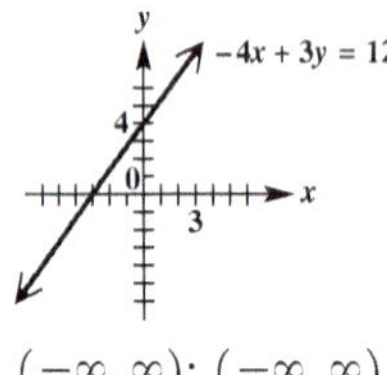

$(-\infty, \infty)$; $(-\infty, \infty)$

23.

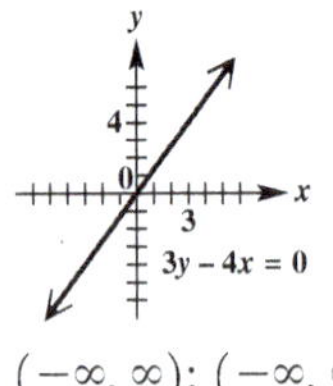

$(-\infty, \infty)$; $(-\infty, \infty)$

25.

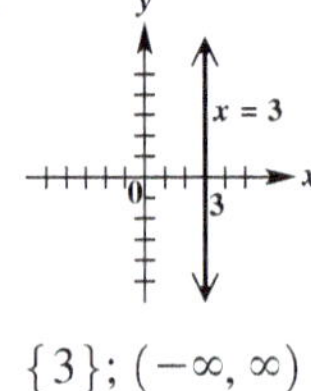

$\{3\}$; $(-\infty, \infty)$

27.

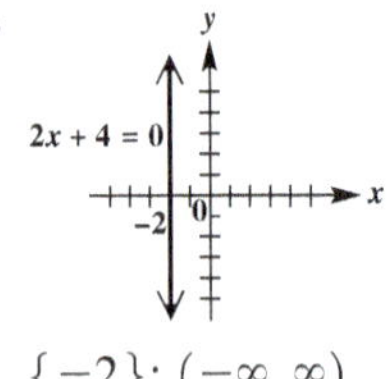

$\{-2\}$; $(-\infty, \infty)$

29.

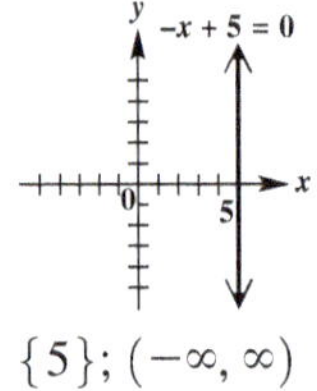

$\{5\}$; $(-\infty, \infty)$

31. A **33.** D

35.

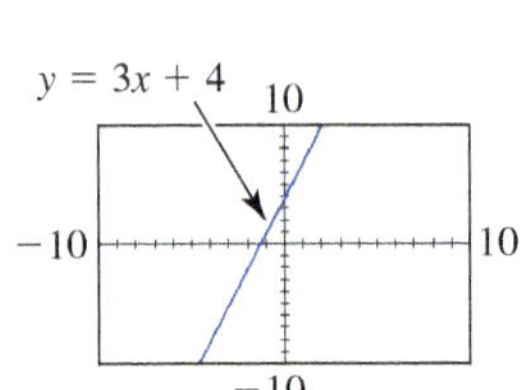

37.

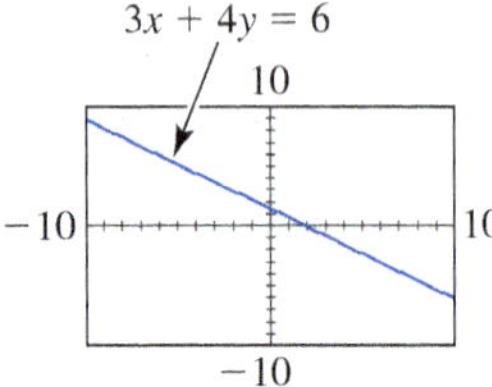

39. A, C, D, E **41.** $\frac{2}{5}$ **43.** -2 **45.** 0 **47.** 0 **49.** undefined

51. **(a)** $m = 3$ **(b)**

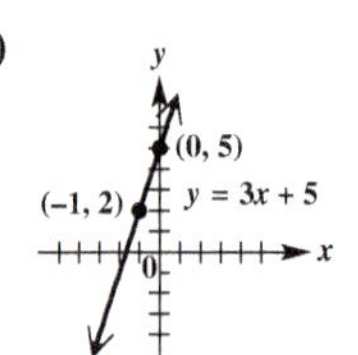

53. **(a)** $m = -\frac{3}{2}$ **(b)**

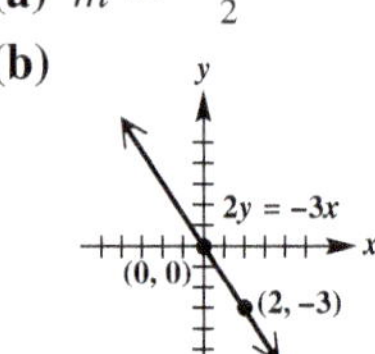

55. **(a)** $m = \frac{5}{2}$ **(b)**

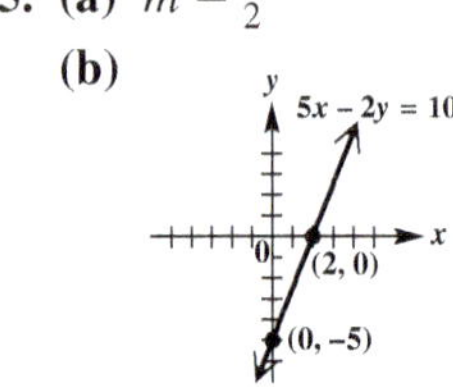

57.

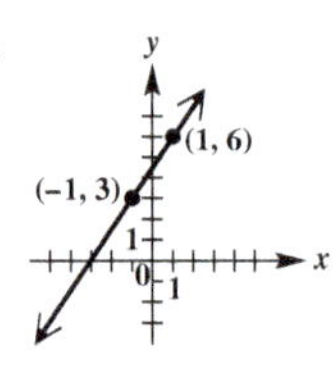

59.

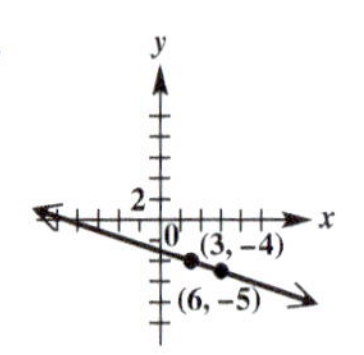

61.

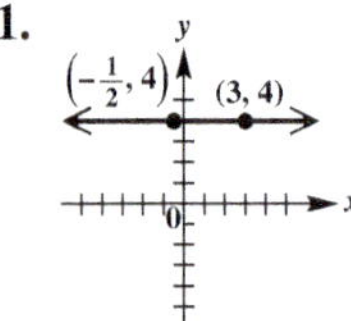

63.

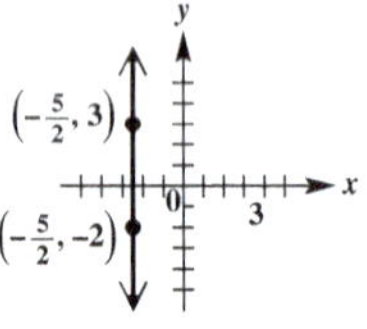

65. $-\$4000$ per year; The value of the machine is decreasing \$4000 each year during these years. **67.** 0% per year (or no change); The percent of pay raise is not changing—it is 3% each year during these years. **69.** -78.8 thousand per year; The number of high school dropouts decreased by an average of 78.8 thousand per year from 1980 to 2012. **71.** **(a)** The slope -0.0167 indicates that the average rate of change per year of the winning time for the 5000-m run is 0.0167 min less. It is negative because the times are generally decreasing as time progresses. **(b)** The Olympics were not held during World Wars I and II. **(c)** 13.05 min; The times differ by 0.30 min. **73.** 13,064 (thousands) **75.** **(a)** -8.17 thousand mobile homes per year **(b)** The negative slope means that the number of mobile homes *decreased* by an average of 8.17 thousand each year from 2003 to 2013. **77.** **(a)** $C(x) = 10x + 500$ **(b)** $R(x) = 35x$ **(c)** $P(x) = 25x - 500$ **(d)** 20 units; do not produce **79.** **(a)** $C(x) = 400x + 1650$ **(b)** $R(x) = 305x$ **(c)** $P(x) = -95x - 1650$ **(d)** $R(x) < C(x)$ for all positive x; don't produce, impossible to make a profit **81.** 25 units; \$6000 **83.** 3 **84.** 3 **85.** the same **86.** $\sqrt{10}$ **87.** $2\sqrt{10}$ **88.** $3\sqrt{10}$ **89.** The sum is $3\sqrt{10}$, which is equal to the answer in **Exercise 88.** **90.** *B*; *C*; *A*; *C* **91.** The midpoint is $(3, 3)$, which is the same as the middle entry in the table. **92.** 7.5

Chapter 2 Quiz

[2.1] **1.** $\sqrt{41}$ **2.** 2006: 6.76 million; 2010: 7.24 million

3.

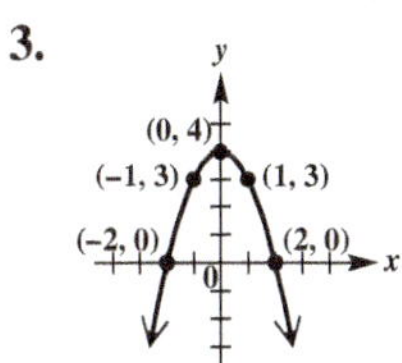

[2.2] **4.**

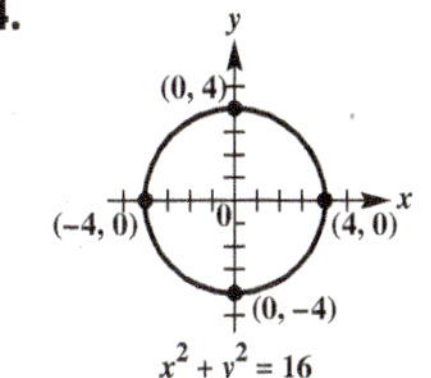

5. radius: $\sqrt{17}$; center: $(2, -4)$ [2.3] **6.** 2 **7.** domain: $(-\infty, \infty)$; range: $[0, \infty)$ **8.** **(a)** $(-\infty, -3)$ **(b)** $(-3, \infty)$ **(c)** none [2.4] **9.** **(a)** $\frac{3}{2}$ **(b)** 0 **(c)** undefined **10.** 1320.5 thousand per year; The number of new motor vehicles sold in the United States increased an average of 1320.5 thousand per year from 2009 to 2013.

2.5 Exercises

1. 4; 3 **3.** x **5.** $-\frac{6}{7}$ **7.** D **9.** C **11.** $2x + y = 5$ **13.** $3x + 2y = -7$ **15.** $x = -8$ **17.** $y = -8$ **19.** $x - 4y = -13$ **21.** $y = \frac{2}{3}x - 2$ **23.** $x = -6$ (cannot be written in slope-intercept form) **25.** $y = 4$ **27.** $y = 5x + 15$ **29.** $y = -4x - 3$ **31.** $y = \frac{3}{2}$ **33.** $(-2, 0)$; does not; undefined; $\left(0, \frac{1}{2}\right)$; does not; 0

35. slope: 3; y-intercept: $(0, -1)$

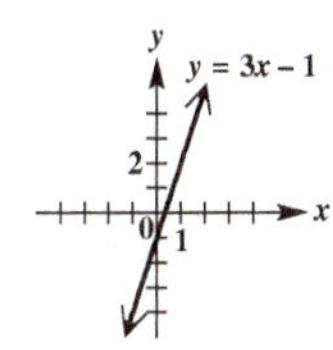

37. slope: 4; y-intercept: $(0, -7)$

39. slope: $-\frac{3}{4}$; y-intercept: $(0, 0)$

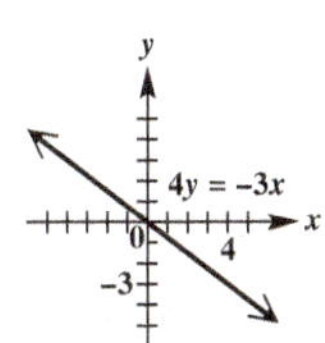

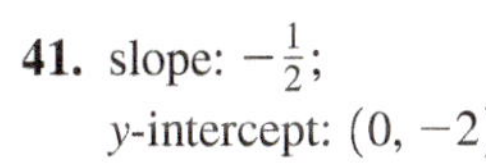

41. slope: $-\frac{1}{2}$; y-intercept: $(0, -2)$

43. slope: $\frac{3}{2}$; y-intercept: $(0, 1)$

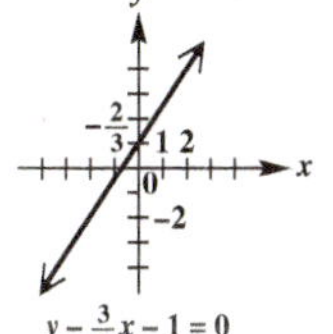

45. **(a)** -2; $(0, 1)$; $\left(\frac{1}{2}, 0\right)$ **(b)** $f(x) = -2x + 1$
47. **(a)** $-\frac{1}{3}$; $(0, 2)$; $(6, 0)$ **(b)** $f(x) = -\frac{1}{3}x + 2$
49. **(a)** -200; $(0, 300)$; $\left(\frac{3}{2}, 0\right)$ **(b)** $f(x) = -200x + 300$
51. **(a)** $x + 3y = 11$ **(b)** $y = -\frac{1}{3}x + \frac{11}{3}$
53. **(a)** $5x - 3y = -13$ **(b)** $y = \frac{5}{3}x + \frac{13}{3}$ **55.** **(a)** $y = 1$ **(b)** $y = 1$ **57.** **(a)** $y = 6$ **(b)** $y = 6$ **59.** **(a)** $-\frac{1}{2}$ **(b)** $-\frac{7}{2}$ **61.** **(a)** $y = 463.67x + 6312$ **(b)** \$8166.68; The result is \$96.68 more than the actual figure.

63. **(a)** $f(x) = 622.25x + 22{,}036$

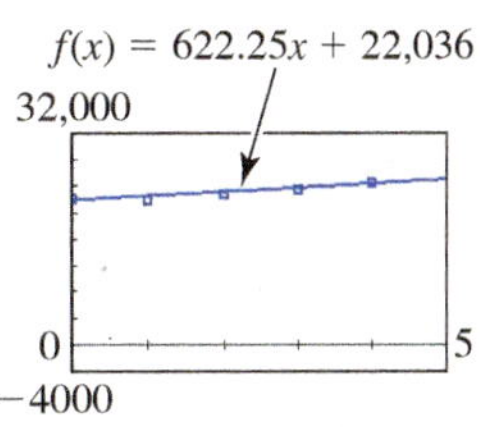

To the nearest dollar, the average tuition increase is \$622 per year for the period because this is the slope of the line.

(b) $f(3) = \$23{,}903$; This is a fairly good approximation. **(c)** $f(x) = 653x + 21{,}634$
65. **(a)** $F = \frac{9}{5}C + 32$ **(b)** $C = \frac{5}{9}(F - 32)$ **(c)** $-40°$
67. **(a)** $C = 0.6516I + 2253$ **(b)** 0.6516 **69.** $\{3\}$ **71.** $\{-0.5\}$ **73.** **(a)** $\{12\}$ **(b)** The solution does not appear in the x-values interval $[-10, 10]$. The minimum and maximum values must include 12. **75.** yes **77.** no **79.** $\sqrt{x_1{}^2 + m_1{}^2x_1{}^2}$ **80.** $\sqrt{x_2{}^2 + m_2{}^2x_2{}^2}$ **81.** $\sqrt{(x_2 - x_1)^2 + (m_2x_2 - m_1x_1)^2}$
83. $-2x_1x_2(m_1m_2 + 1) = 0$ **84.** Because $x_1 \neq 0$, $x_2 \neq 0$, we have $m_1m_2 + 1 = 0$, implying that $m_1m_2 = -1$.
85. If two nonvertical lines are perpendicular, then the product of the slopes of these lines is -1.

Summary Exercises on Graphs, Circles, Functions, and Equations

1. **(a)** $\sqrt{65}$ **(b)** $\left(\frac{5}{2}, 1\right)$ **(c)** $y = 8x - 19$ **2.** **(a)** $\sqrt{29}$ **(b)** $\left(\frac{3}{2}, -1\right)$ **(c)** $y = -\frac{2}{5}x - \frac{2}{5}$ **3.** **(a)** 5 **(b)** $\left(\frac{1}{2}, 2\right)$ **(c)** $y = 2$ **4.** **(a)** $\sqrt{10}$ **(b)** $\left(\frac{3\sqrt{2}}{2}, 2\sqrt{2}\right)$ **(c)** $y = -2x + 5\sqrt{2}$ **5.** **(a)** 2 **(b)** $(5, 0)$ **(c)** $x = 5$ **6.** **(a)** $4\sqrt{2}$ **(b)** $(-1, -1)$ **(c)** $y = x$ **7.** **(a)** $4\sqrt{3}$ **(b)** $\left(4\sqrt{3}, 3\sqrt{5}\right)$ **(c)** $y = 3\sqrt{5}$ **8.** **(a)** $\sqrt{34}$ **(b)** $\left(\frac{3}{2}, -\frac{3}{2}\right)$ **(c)** $y = \frac{5}{3}x - 4$

9. $y = -\frac{1}{3}x + \frac{1}{3}$

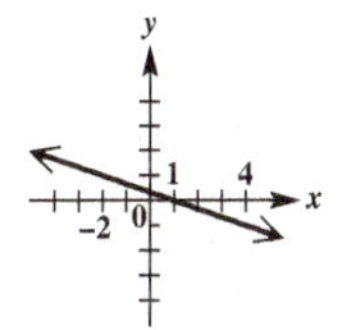

10. $y = 3$

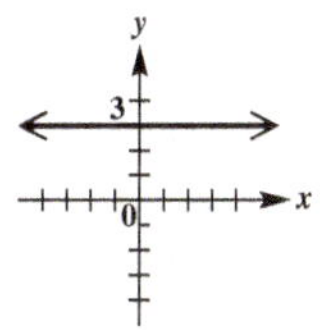

11. $(x - 2)^2 + (y + 1)^2 = 9$

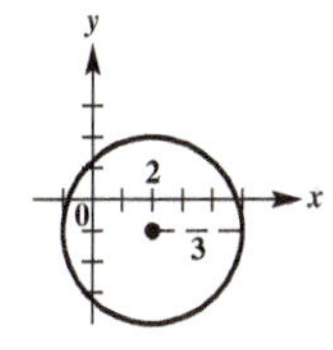

12. $x^2 + (y - 2)^2 = 4$

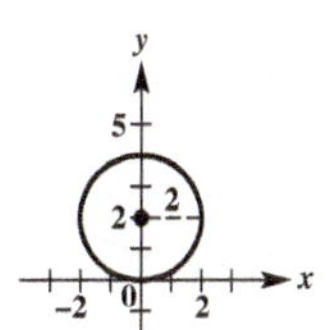

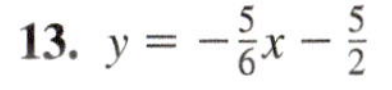

13. $y = -\frac{5}{6}x - \frac{5}{2}$

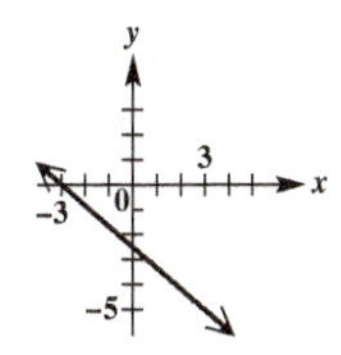

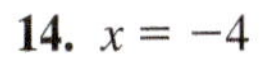

14. $x = -4$

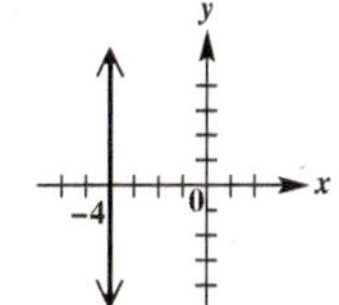

15. $y = -\frac{2}{3}x$

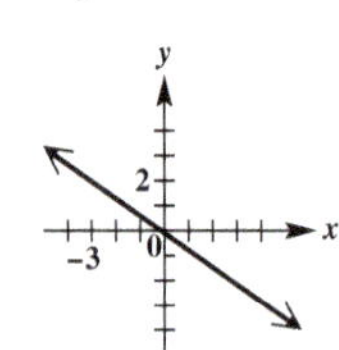

16. $y = -\frac{4}{3}x$

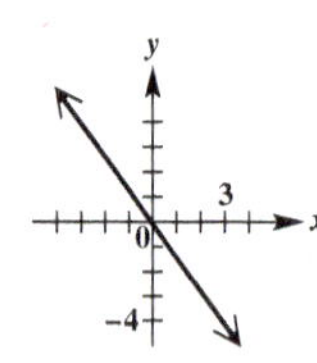

17. yes; center: $(2, -1)$; radius: 3 **18.** no **19.** yes; center: $(6, 0)$; radius: 4 **20.** yes; center: $(-1, -8)$; radius: 2 **21.** no **22.** yes; center: $(0, 4)$; radius: 5
23. $\left(4 - \sqrt{7}, 2\right), \left(4 + \sqrt{7}, 2\right)$ **24.** 8
25. **(a)** domain: $(-\infty, \infty)$; range: $(-\infty, \infty)$
(b) $f(x) = \frac{1}{4}x + \frac{3}{2}$; 1 **26.** **(a)** domain: $[-5, \infty)$; range: $(-\infty, \infty)$ **(b)** y is not a function of x.
27. **(a)** domain: $[-7, 3]$; range: $[-5, 5]$
(b) y is not a function of x. **28.** **(a)** domain: $(-\infty, \infty)$; range: $\left[-\frac{3}{2}, \infty\right)$ **(b)** $f(x) = \frac{1}{2}x^2 - \frac{3}{2}$; $\frac{1}{2}$

2.6 Exercises

1. E; $(-\infty, \infty)$ **3.** A; $(-\infty, \infty)$ **5.** F; $f(x) = x$
7. H; no **9.** B; $\{\dots, -3, -2, -1, 0, 1, 2, 3, \dots\}$
11. $(-\infty, \infty)$ **13.** $[0, \infty)$ **15.** $(-\infty, 3)$; $(3, \infty)$
17. **(a)** -10 **(b)** -2 **(c)** -1 **(d)** 2
19. **(a)** -3 **(b)** 1 **(c)** 0 **(d)** 9

21.

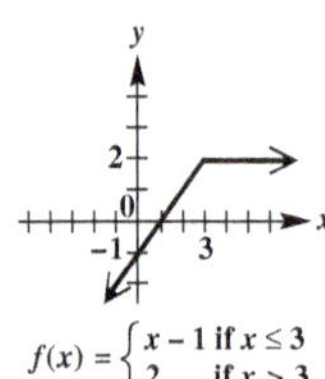

23.

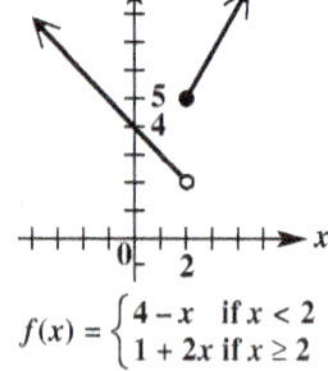

25.

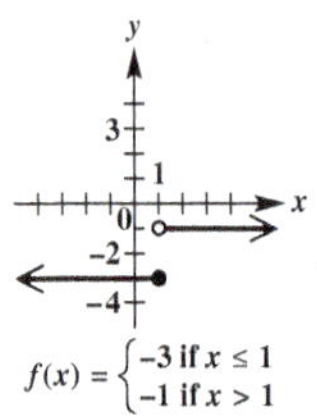

27.

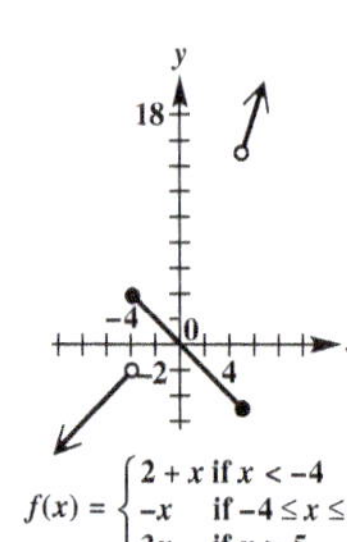

29.

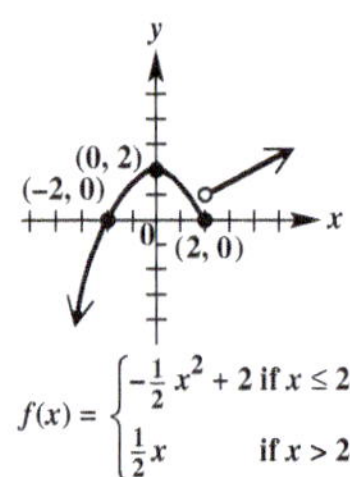

31.

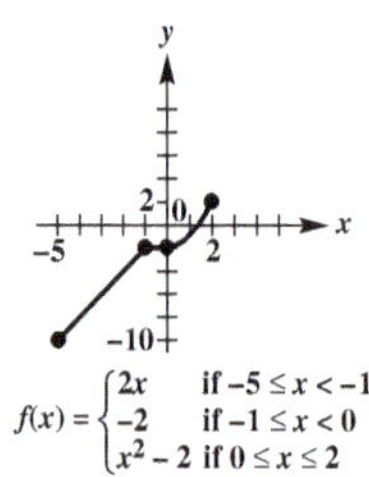

33.

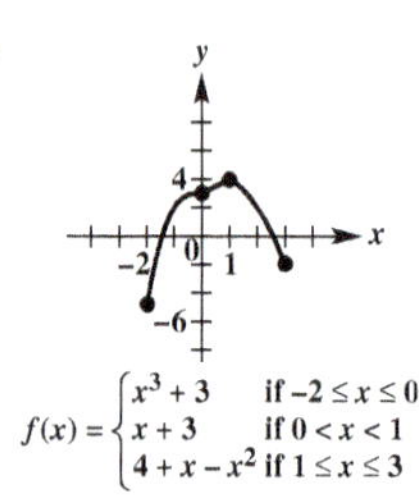

In Exercises 35–41, we give one of the possible rules, then the domain, and then the range.

35. $f(x) = \begin{cases} -1 & \text{if } x \le 0 \\ 1 & \text{if } x > 0 \end{cases}$; $(-\infty, \infty)$; $\{-1, 1\}$

37. $f(x) = \begin{cases} 2 & \text{if } x \le 0 \\ -1 & \text{if } x > 1 \end{cases}$; $(-\infty, 0] \cup (1, \infty)$; $\{-1, 2\}$

39. $f(x) = \begin{cases} x & \text{if } x \le 0 \\ 2 & \text{if } x > 0 \end{cases}$; $(-\infty, \infty)$; $(-\infty, 0] \cup \{2\}$

41. $f(x) = \begin{cases} \sqrt[3]{x} & \text{if } x < 1 \\ x + 1 & \text{if } x \ge 1 \end{cases}$; $(-\infty, \infty)$; $(-\infty, 1) \cup [2, \infty)$

43. $(-\infty, \infty)$; $\{\dots, -2, -1, 0, 1, 2, \dots\}$

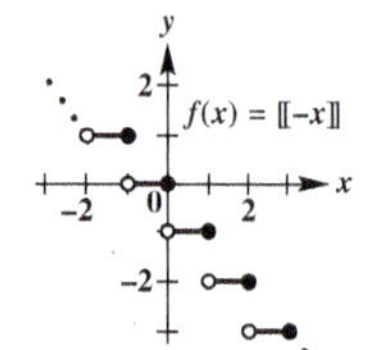

45. $(-\infty, \infty)$; $\{\dots, -2, -1, 0, 1, 2, \dots\}$

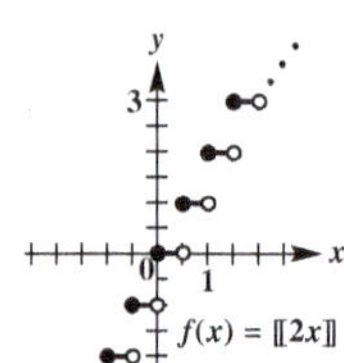

47.

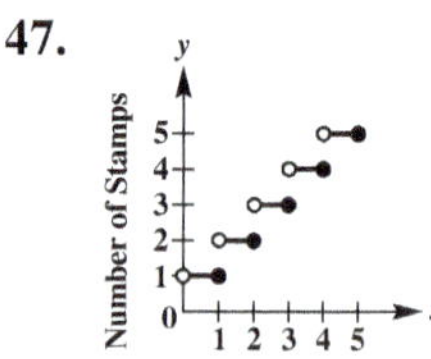

49.

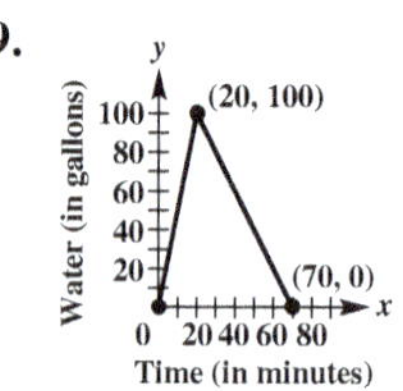

51. **(a)** for $[0, 8]$: $y = 1.95x + 34.2$; for $(8, 13]$: $y = 0.48x + 45.96$

(b) $f(x) = \begin{cases} 1.95x + 34.2 & \text{if } 0 \le x \le 8 \\ 0.48x + 45.96 & \text{if } 8 < x \le 13 \end{cases}$

53. **(a)** 50,000 gal; 30,000 gal **(b)** during the first and fourth days **(c)** 45,000; 40,000 **(d)** 5000 gal per day
55. **(a)** $f(x) = 0.80[\![\frac{x}{2}]\!]$ if $6 \le x \le 18$ **(b)** \$3.20; \$5.60

2.7 Exercises

1. 3 **3.** left **5.** x **7.** 2; 3 **9.** y **11.** **(a)** B **(b)** D **(c)** E **(d)** A **(e)** C **13.** **(a)** B **(b)** A **(c)** G **(d)** C **(e)** F **(f)** D **(g)** H **(h)** E **(i)** I
15. **(a)** F **(b)** C **(c)** H **(d)** D **(e)** G **(f)** A **(g)** E **(h)** I **(i)** B

17.

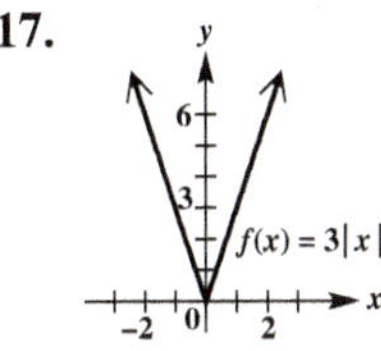

19.

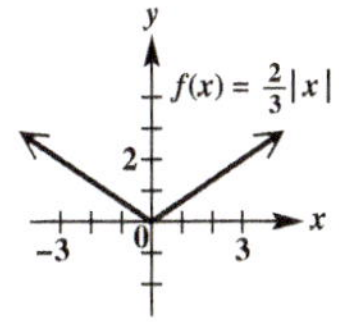

21.

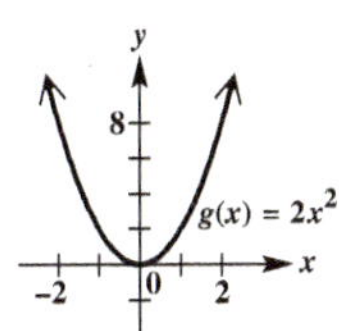

23.

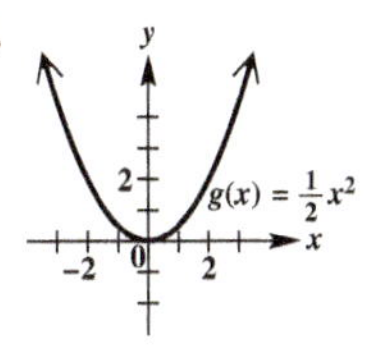

25.

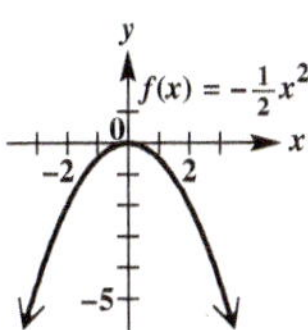

27.

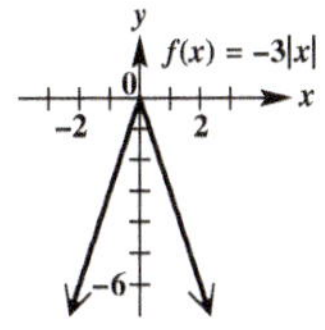

29.

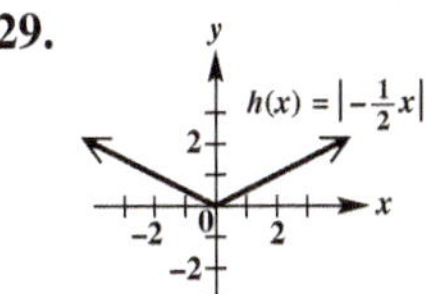

31.

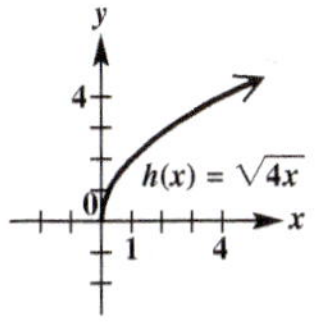

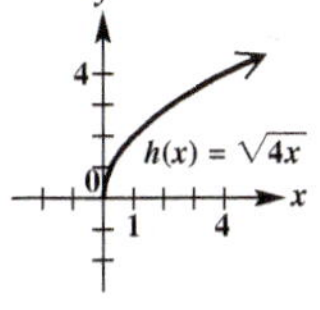

33.

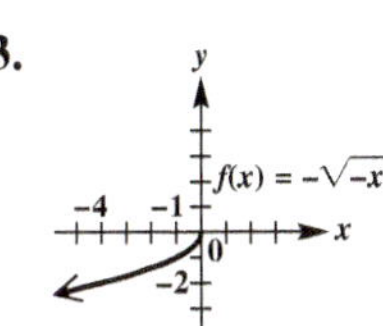

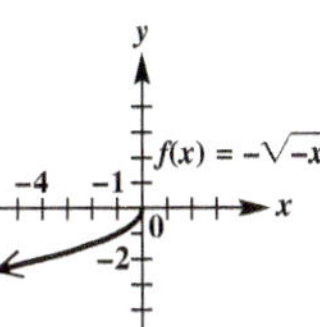

35. (a) $(4, 12)$ **(b)** $(8, 16)$

37. (a) $(2, 12)$ **(b)** $(32, 12)$

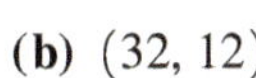

39.

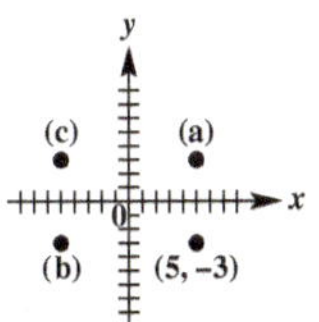

41.

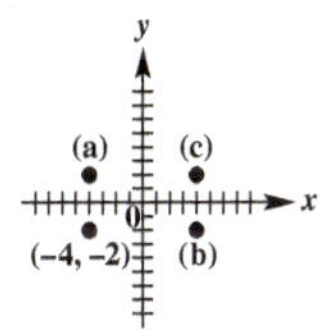

43. $x = 2$ **45.** y-axis **47.** x-axis, y-axis, origin **49.** origin **51.** none of these **53.** odd **55.** even **57.** neither

59.

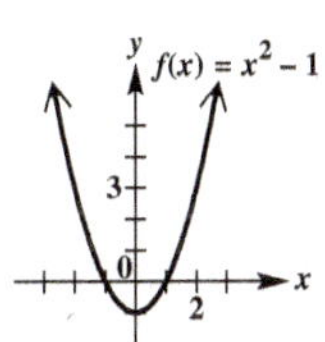

61.

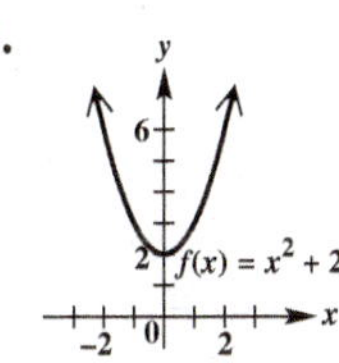

63.

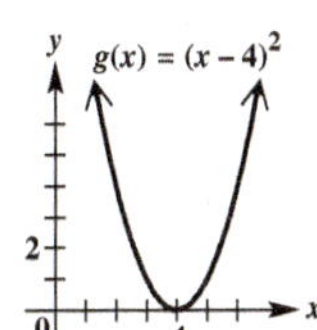

65.

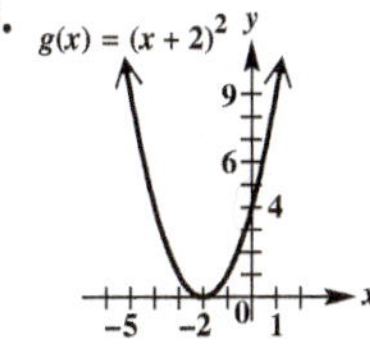

67.

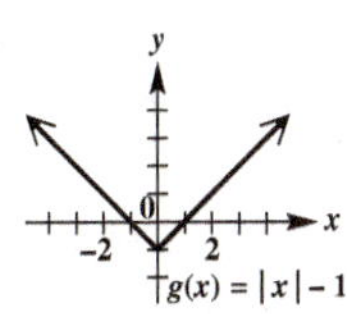

69.

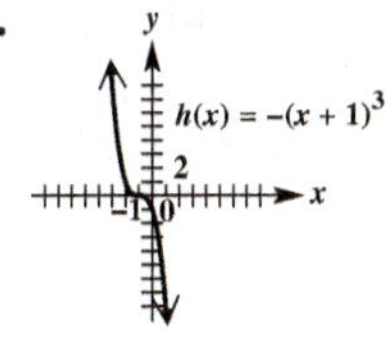

71.

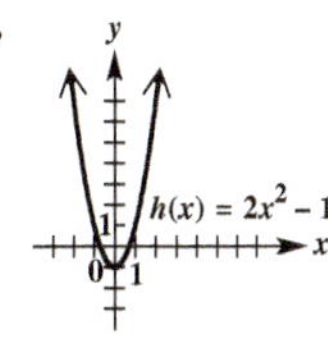

73.

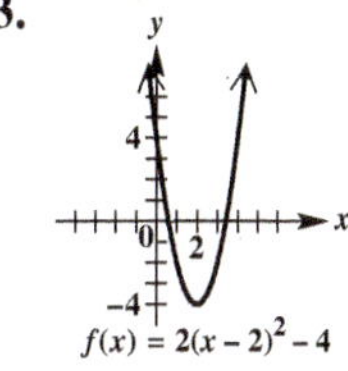

75.

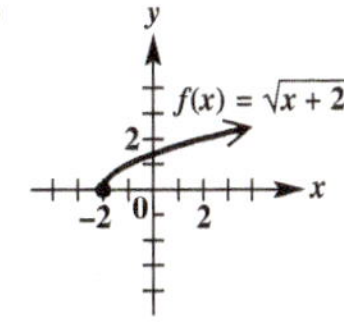

77.

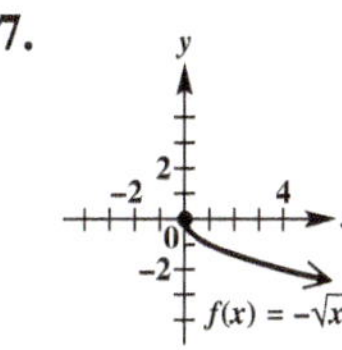

79.

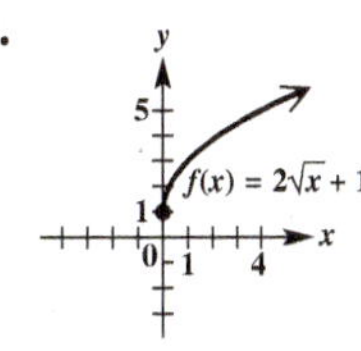

81.

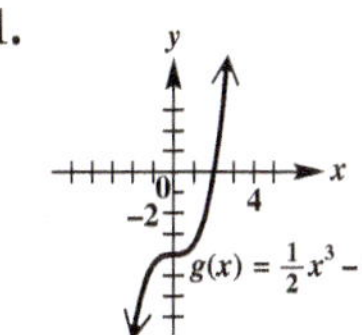

83.

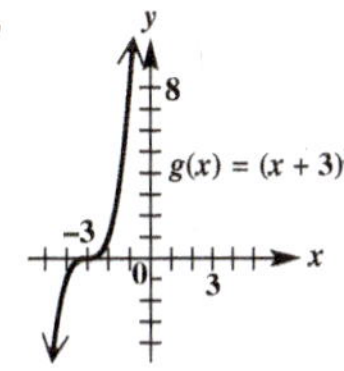

85.

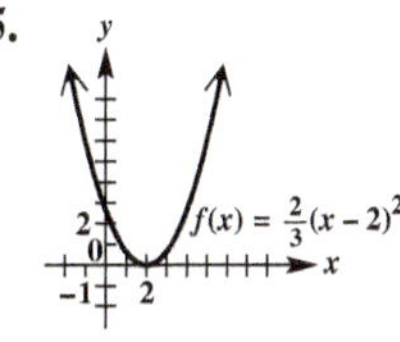

87. (a)

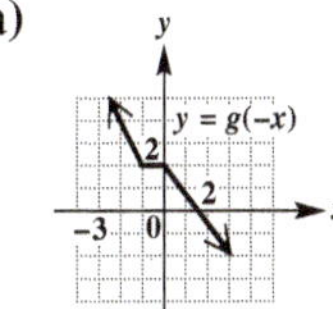

The graph of $g(x)$ is reflected across the y-axis.

(b)

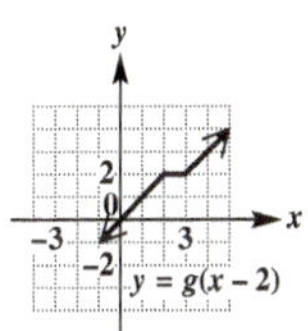

The graph of $g(x)$ is translated 2 units to the right.

(c)

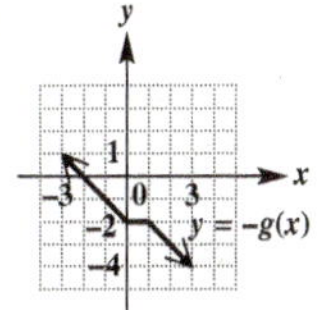

The graph of $g(x)$ is reflected across the x-axis.

(d)

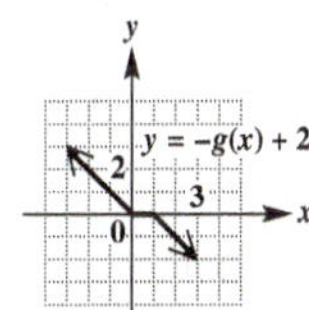

The graph of $g(x)$ is reflected across the x-axis and translated 2 units up.

89. It is the graph of $f(x) = |x|$ translated 1 unit to the left, reflected across the x-axis, and translated 3 units up. The equation is $y = -|x + 1| + 3$. **91.** It is the graph of $g(x) = \sqrt{x}$ translated 1 unit to the right and translated 3 units down. The equation is $y = \sqrt{x - 1} - 3$. **93.** It is the graph of $g(x) = \sqrt{x}$ translated 4 units to the left, stretched vertically by a factor of 2, and translated 4 units down. The equation is $y = 2\sqrt{x + 4} - 4$. **95.** $f(-3) = -6$

97. $f(9) = 6$ **99.** $f(-3) = -6$ **101.** $g(x) = 2x + 13$

103. (a)

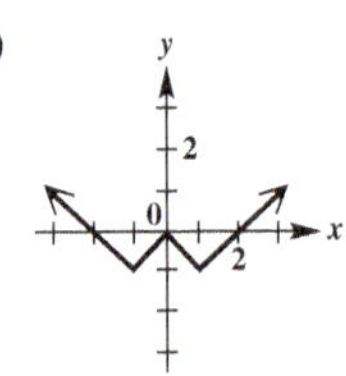

(b)

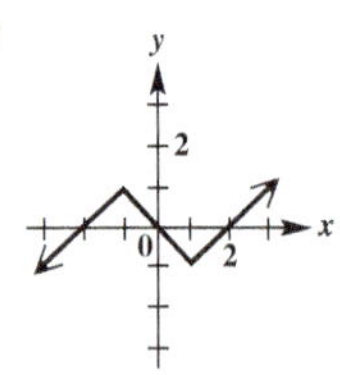

Chapter 2 Quiz

[2.5] **1. (a)** $y = 2x + 11$ **(b)** $\left(-\frac{11}{2}, 0\right)$ **2.** $y = -\frac{2}{3}x$ **3. (a)** $x = -8$ **(b)** $y = 5$ [2.6] **4. (a)** cubing function; domain: $(-\infty, \infty)$; range: $(-\infty, \infty)$; increasing over $(-\infty, \infty)$ **(b)** absolute value function; domain: $(-\infty, \infty)$; range: $[0, \infty)$; decreasing over $(-\infty, 0)$; increasing over $(0, \infty)$ **(c)** cube root function; domain: $(-\infty, \infty)$; range: $(-\infty, \infty)$; increasing over $(-\infty, \infty)$ **5.** \$2.75

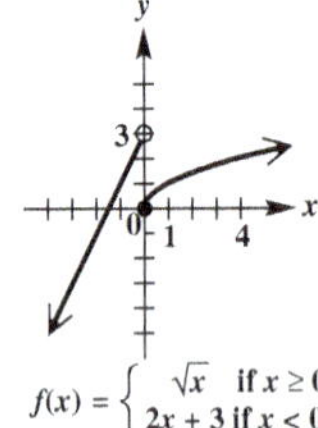

[2.7] **7.**

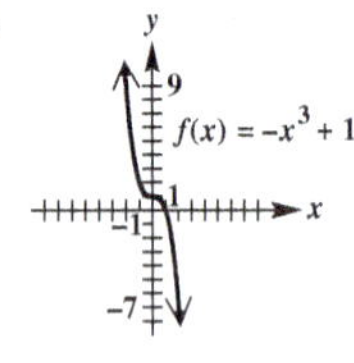

8.

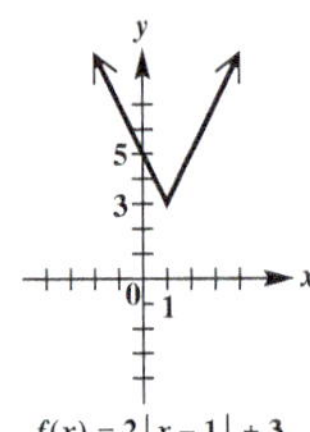

9. $g(x) = -\sqrt{x + 4} - 2$
10. (a) even **(b)** neither **(c)** odd

2.8 Exercises

1. 7 **3.** 12 **5.** 5 **7.** $(-\infty, \infty)$ **9.** $(-\infty, \infty)$ **11.** 12 **13.** -4 **15.** -38 **17.** $\frac{1}{2}$ **19.** $5x - 1$; $x + 9$; $6x^2 - 7x - 20$; $\frac{3x + 4}{2x - 5}$; All domains are $(-\infty, \infty)$ except for that of $\frac{f}{g}$, which is $\left(-\infty, \frac{5}{2}\right) \cup \left(\frac{5}{2}, \infty\right)$.
21. $3x^2 - 4x + 3$; $x^2 - 2x - 3$; $2x^4 - 5x^3 + 9x^2 - 9x$; $\frac{2x^2 - 3x}{x^2 - x + 3}$; All domains are $(-\infty, \infty)$. **23.** $\sqrt{4x - 1} + \frac{1}{x}$; $\sqrt{4x - 1} - \frac{1}{x}$; $\frac{\sqrt{4x - 1}}{x}$; $x\sqrt{4x - 1}$; All domains are $\left[\frac{1}{4}, \infty\right)$. **25.** 280; 470; 750 (all in thousands)
27. 2008–2012 **29.** 6; It represents the dollars (in billions) spent for general science in 2000. **31.** space and other technologies; 1995–2000 and 2010–2015
33. (a) 2 **(b)** 4 **(c)** 0 **(d)** $-\frac{1}{3}$ **35. (a)** 3 **(b)** -5 **(c)** 2 **(d)** undefined **37. (a)** 5 **(b)** 5 **(c)** 0 **(d)** undefined

39.

x	$(f + g)(x)$	$(f - g)(x)$	$(fg)(x)$	$\left(\frac{f}{g}\right)(x)$
-2	6	-6	0	0
0	5	5	0	undefined
2	5	9	-14	-3.5
4	15	5	50	2

41. Both the slope formula and the difference quotient represent the ratio of the vertical change to the horizontal change. The slope formula is stated for a line, while the difference quotient is stated for a function f.
43. (a) $2 - x - h$ **(b)** $-h$ **(c)** -1
45. (a) $6x + 6h + 2$ **(b)** $6h$ **(c)** 6
47. (a) $-2x - 2h + 5$ **(b)** $-2h$ **(c)** -2
49. (a) $\frac{1}{x + h}$ **(b)** $\frac{-h}{x(x + h)}$ **(c)** $\frac{-1}{x(x + h)}$
51. (a) $x^2 + 2xh + h^2$ **(b)** $2xh + h^2$ **(c)** $2x + h$
53. (a) $1 - x^2 - 2xh - h^2$ **(b)** $-2xh - h^2$ **(c)** $-2x - h$
55. (a) $x^2 + 2xh + h^2 + 3x + 3h + 1$ **(b)** $2xh + h^2 + 3h$ **(c)** $2x + h + 3$ **57.** -5 **59.** 7 **61.** 6 **63.** -1
65. 1 **67.** 9 **69.** 1 **71.** $g(1) = 9$, and $f(9)$ cannot be determined from the table given.
73. (a) $-30x - 33$; $(-\infty, \infty)$ **(b)** $-30x + 52$; $(-\infty, \infty)$
75. (a) $\sqrt{x + 3}$; $[-3, \infty)$ **(b)** $\sqrt{x} + 3$; $[0, \infty)$
77. (a) $(x^2 + 3x - 1)^3$; $(-\infty, \infty)$ **(b)** $x^6 + 3x^3 - 1$; $(-\infty, \infty)$ **79. (a)** $\sqrt{3x - 1}$; $\left[\frac{1}{3}, \infty\right)$ **(b)** $3\sqrt{x - 1}$; $[1, \infty)$
81. (a) $\frac{2}{x + 1}$; $(-\infty, -1) \cup (-1, \infty)$ **(b)** $\frac{2}{x} + 1$; $(-\infty, 0) \cup (0, \infty)$
83. (a) $\sqrt{-\frac{1}{x} + 2}$; $(-\infty, 0) \cup \left[\frac{1}{2}, \infty\right)$ **(b)** $-\frac{1}{\sqrt{x + 2}}$; $(-2, \infty)$
85. (a) $\sqrt{\frac{1}{x + 5}}$; $(-5, \infty)$ **(b)** $\frac{1}{\sqrt{x} + 5}$; $[0, \infty)$
87. (a) $\frac{x}{1 - 2x}$; $(-\infty, 0) \cup \left(0, \frac{1}{2}\right) \cup \left(\frac{1}{2}, \infty\right)$ **(b)** $x - 2$; $(-\infty, 2) \cup (2, \infty)$

89.

x	$f(x)$	$g(x)$	$g(f(x))$
1	3	2	7
2	1	5	2
3	2	7	5

In Exercises 97–101, we give only one of the many possible ways.

97. $g(x) = 6x - 2$, $f(x) = x^2$ **99.** $g(x) = x^2 - 1$, $f(x) = \sqrt{x}$ **101.** $g(x) = 6x$, $f(x) = \sqrt{x} + 12$
103. $(f \circ g)(x) = 63{,}360x$; It computes the number of inches in x miles. **105. (a)** $\mathscr{A}(2x) = \sqrt{3}x^2$ **(b)** $64\sqrt{3}$ square units **107. (a)** $(\mathscr{A} \circ r)(t) = 16\pi t^2$ **(b)** It defines the area of the leak in terms of the time t, in minutes. **(c)** 144π ft^2 **109. (a)** $N(x) = 100 - x$

(b) $G(x) = 20 + 5x$ **(c)** $C(x) = (100 - x)(20 + 5x)$ **(d)** \$9600 **111. (a)** $g(x) = \frac{1}{2}x$ **(b)** $f(x) = x + 1$ **(c)** $(f \circ g)(x) = f(g(x)) = g(x) + 1 = \frac{1}{2}x + 1$ **(d)** $(f \circ g)(60) = \frac{1}{2}(60) + 1 = 31$ (dollars)

Chapter 2 Review Exercises

1. $\sqrt{85}; \left(-\frac{1}{2}, 2\right)$ **3.** $5; \left(-6, \frac{11}{2}\right)$ **5.** $(22, -6)$ **7.** $(x + 2)^2 + (y - 3)^2 = 225$ **9.** $(x + 8)^2 + (y - 1)^2 = 289$ **11.** $x^2 + y^2 = 34$ **13.** $x^2 + (y - 3)^2 = 13$ **15.** $(2, -3); 1$ **17.** $\left(-\frac{7}{2}, -\frac{3}{2}\right); \frac{3\sqrt{6}}{2}$ **19.** no; $[-6, 6]$; $[-6, 6]$ **21.** no; $(-\infty, \infty)$; $(-\infty, -1] \cup [1, \infty)$ **23.** no; $[0, \infty)$; $(-\infty, \infty)$ **25.** function of x **27.** not a function of x **29.** $(-\infty, \infty)$ **31.** $(-\infty, 2]$ **33.** -15 **35.** -6

37.

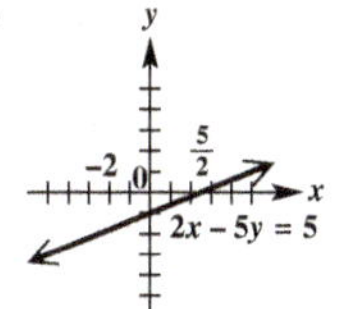

39.

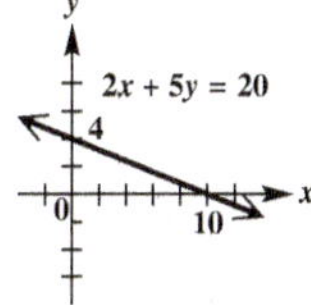

41.

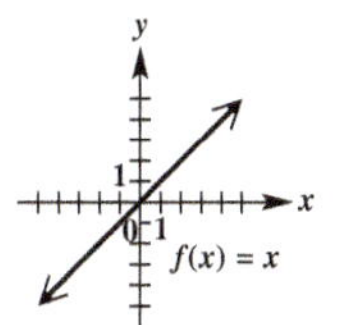

43.

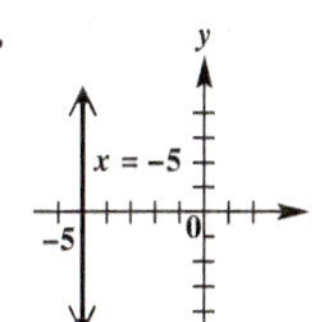

45.

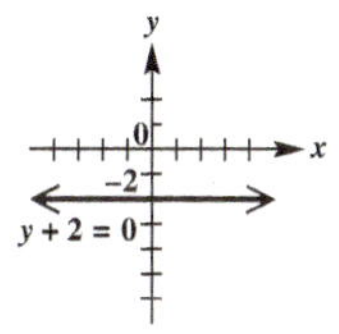

47.

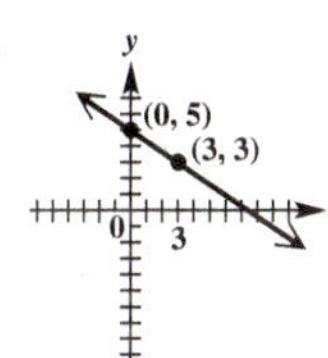

49. -2 **51.** 0 **53.** $-\frac{11}{2}$ **55.** undefined **57.** Initially, the car is at home. After traveling 30 mph for 1 hr, the car is 30 mi away from home. During the second hour, the car travels 20 mph until it is 50 mi away. During the third hour, the car travels toward home at 30 mph until it is 20 mi away. During the fourth hour, the car travels away from home at 40 mph until it is 60 mi away from home. During the last hour, the car travels 60 mi at 60 mph until it arrives home. **59. (a)** $y = 4.35x + 30.7$; The slope 4.35 indicates that the percent of returns filed electronically rose an average of 4.35% per year during this period. **(b)** 65.5% **61. (a)** $y = -2x + 1$ **(b)** $2x + y = 1$ **63. (a)** $y = 3x - 7$ **(b)** $3x - y = 7$ **65. (a)** $y = -10$ **(b)** $y = -10$ **67. (a)** not possible **(b)** $x = -7$

69.

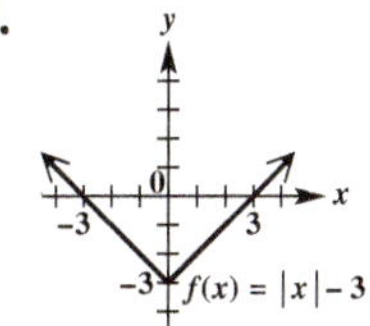

71.

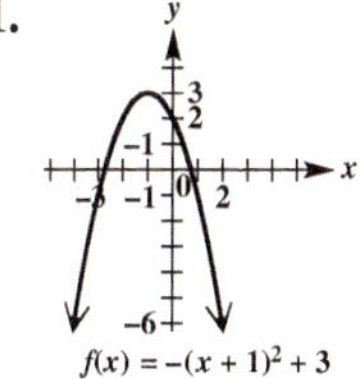

73.

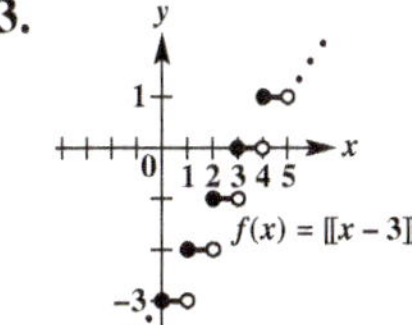

75.

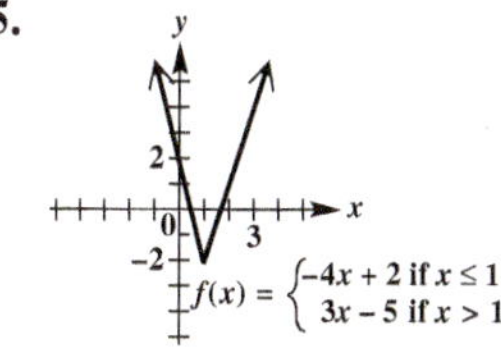

77.

79. true **81.** false; For example, $f(x) = x^2$ is even, and $(2, 4)$ is on the graph but $(2, -4)$ is not. **83.** true **85.** x-axis **87.** y-axis **89.** none of these **91.** y-axis **93.** x-axis, y-axis, origin **95.** Reflect the graph of $f(x) = |x|$ across the x-axis. **97.** Translate the graph of $f(x) = |x|$ to the right 4 units and stretch it vertically by a factor of 2. **99.** $y = -3x - 4$

101. (a)

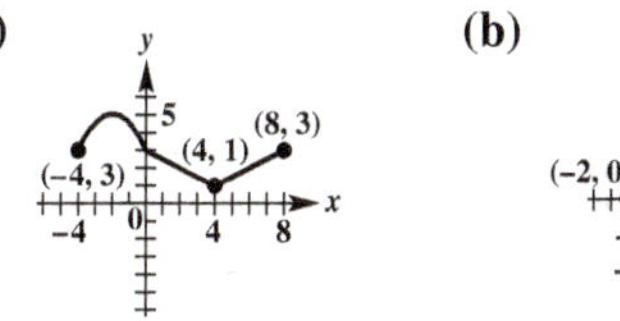

(b)

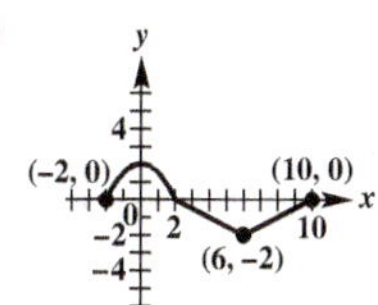

(c)

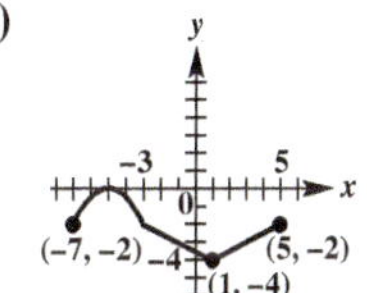

(d)

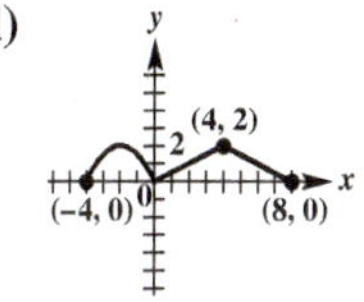

103. $3x^4 - 9x^3 - 16x^2 + 12x + 16$ **105.** 68 **107.** $-\frac{23}{4}$ **109.** $(-\infty, \infty)$ **111.** 2 **113.** $x - 2$ **115.** 1 **117.** undefined **119.** 8 **121.** -6 **123.** 2 **125.** 1 **127.** $f(x) = 36x$; $g(x) = 1760x$; $(f \circ g)(x) = f(g(x)) = f(1760x) = 36(1760x) = 63{,}360x$ **129.** $V(r) = \frac{4}{3}\pi(r + 3)^3 - \frac{4}{3}\pi r^3$

Chapter 2 Test

[2.3] **1. (a)** D **(b)** D **(c)** C **(d)** B **(e)** C **(f)** C **(g)** C **(h)** D **(i)** D **(j)** C [2.4] **2.** $\frac{3}{5}$ [2.1] **3.** $\sqrt{34}$ **4.** $\left(\frac{1}{2}, \frac{5}{2}\right)$ [2.5] **5.** $3x - 5y = -11$ **6.** $f(x) = \frac{3}{5}x + \frac{11}{5}$ [2.2] **7. (a)** $x^2 + y^2 = 4$ **(b)** $(x - 1)^2 + (y - 4)^2 = 1$

8.

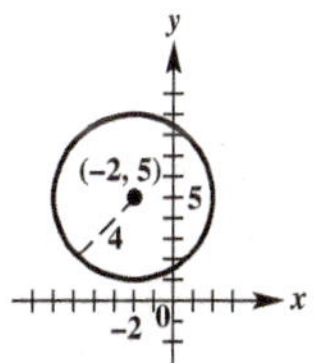

[2.3] **9. (a)** not a function; domain: $[0, 4]$; range: $[-4, 4]$ **(b)** function; domain: $(-\infty, -1) \cup (-1, \infty)$; range: $(-\infty, 0) \cup (0, \infty)$; decreasing on $(-\infty, -1)$ and $(-1, \infty)$

[2.5] **10. (a)** $x = 5$ **(b)** $y = -3$

11. **(a)** $y = -3x + 9$ **(b)** $y = \frac{1}{3}x + \frac{7}{3}$

[2.3] **12.** **(a)** $(2, \infty)$ **(b)** $(0, 2)$ **(c)** $(-\infty, 0)$ **(d)** $(-\infty, \infty)$ **(e)** $(-\infty, \infty)$ **(f)** $(-1, \infty)$

[2.6, 2.7]

13.

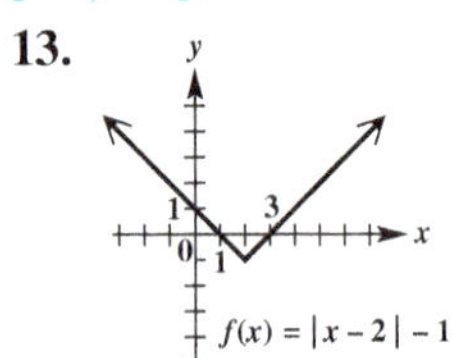

14.

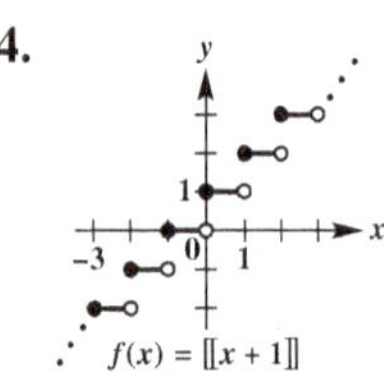

15.

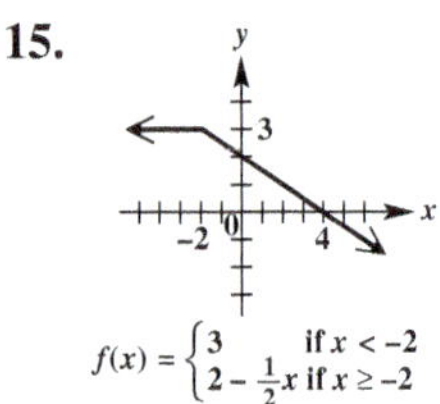

16. **(a)**

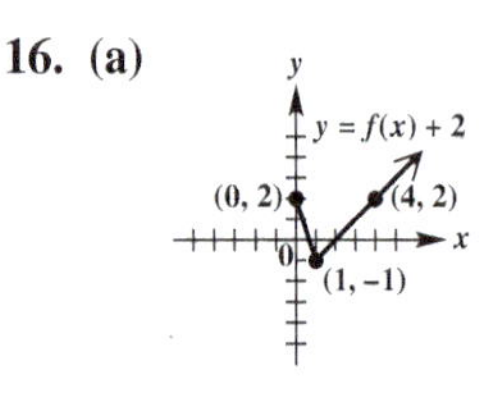

(b)

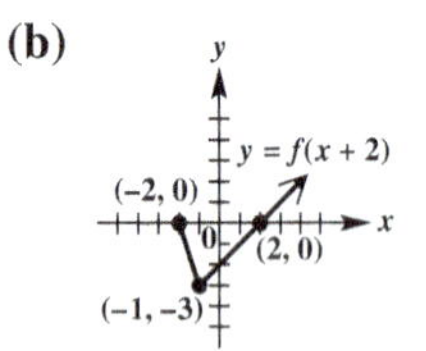

(c)

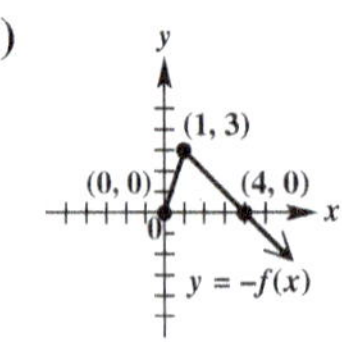

(d)

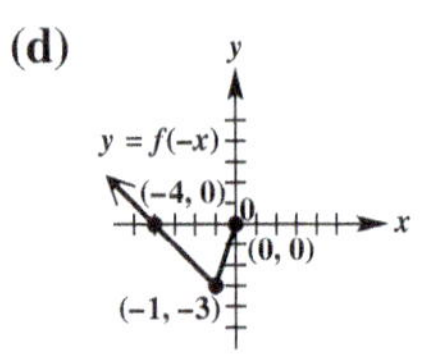

(e)

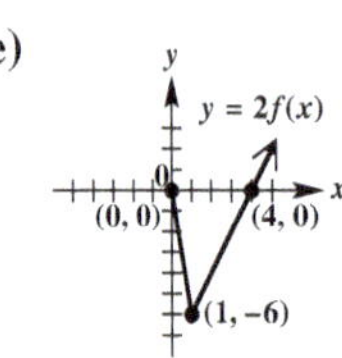

17. It is translated 2 units to the left, stretched vertically by a factor of 2, reflected across the x-axis, and translated 3 units down. **18.** **(a)** yes **(b)** yes **(c)** yes

[2.8] **19.** **(a)** $2x^2 - x + 1$ **(b)** $\frac{2x^2 - 3x + 2}{-2x + 1}$ **(c)** $\left(-\infty, \frac{1}{2}\right) \cup \left(\frac{1}{2}, \infty\right)$ **(d)** $4x + 2h - 3$ **(e)** 0 **(f)** -12 **(g)** 1 **20.** $\sqrt{2x - 6}$; $[3, \infty)$

21. $2\sqrt{x + 1} - 7$; $[-1, \infty)$

[2.4] **22.** **(a)** $C(x) = 3300 + 4.50x$ **(b)** $R(x) = 10.50x$ **(c)** $R(x) - C(x) = 6.00x - 3300$ **(d)** 551

Chapter 3 Polynomial and Rational Functions

3.1 Exercises

1. 5 **3.** vertex **5.** -1 **7.** C **9.** D

11. **(a)** domain: $(-\infty, \infty)$; range: $[-4, \infty)$ **(b)** $(-3, -4)$ **(c)** $x = -3$ **(d)** $(0, 5)$ **(e)** $(-5, 0), (-1, 0)$

13. **(a)** domain: $(-\infty, \infty)$; range: $(-\infty, 2]$ **(b)** $(-3, 2)$ **(c)** $x = -3$ **(d)** $(0, -16)$ **(e)** $(-4, 0), (-2, 0)$ **15.** B **17.** D

19. **(d)** The greater $|a|$ is, the narrower the parabola will be. The smaller $|a|$ is, the wider the parabola will be.

21. **(d)** The graph of $y = (x - h)^2$ is translated h units to the right if h is positive and $|h|$ units to the left if h is negative.

23. $f(x) = (x - 2)^2$

(a) $(2, 0)$ **(b)** $x = 2$ **(c)** $(-\infty, \infty)$ **(d)** $[0, \infty)$ **(e)** $(2, \infty)$ **(f)** $(-\infty, 2)$

25.

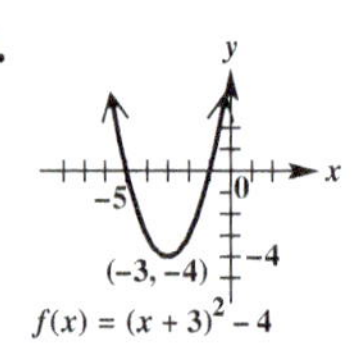

(a) $(-3, -4)$ **(b)** $x = -3$ **(c)** $(-\infty, \infty)$ **(d)** $[-4, \infty)$ **(e)** $(-3, \infty)$ **(f)** $(-\infty, -3)$

27.

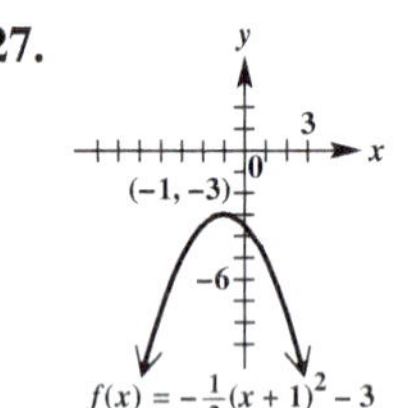

(a) $(-1, -3)$ **(b)** $x = -1$ **(c)** $(-\infty, \infty)$ **(d)** $(-\infty, -3]$ **(e)** $(-\infty, -1)$ **(f)** $(-1, \infty)$

29.

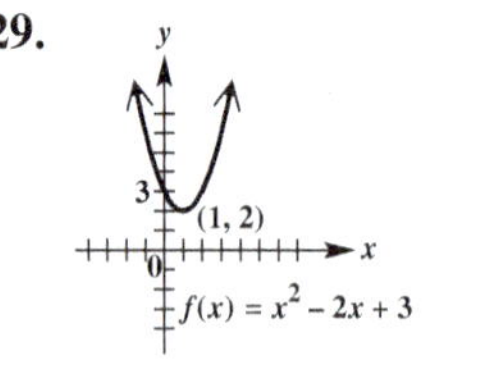

(a) $(1, 2)$ **(b)** $x = 1$ **(c)** $(-\infty, \infty)$ **(d)** $[2, \infty)$ **(e)** $(1, \infty)$ **(f)** $(-\infty, 1)$

31. $f(x) = x^2 - 10x + 21$

(a) $(5, -4)$ **(b)** $x = 5$ **(c)** $(-\infty, \infty)$ **(d)** $[-4, \infty)$ **(e)** $(5, \infty)$ **(f)** $(-\infty, 5)$

33.

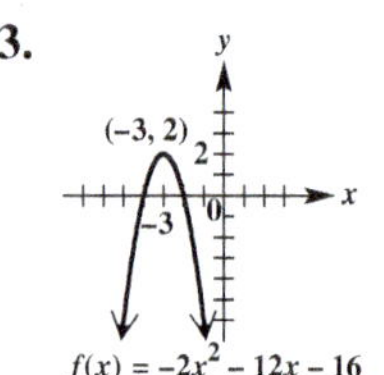

(a) $(-3, 2)$ **(b)** $x = -3$ **(c)** $(-\infty, \infty)$ **(d)** $(-\infty, 2]$ **(e)** $(-\infty, -3)$ **(f)** $(-3, \infty)$

35. $f(x) = -\frac{1}{2}x^2 - 3x - \frac{1}{2}$

(a) $(-3, 4)$ **(b)** $x = -3$ **(c)** $(-\infty, \infty)$ **(d)** $(-\infty, 4]$ **(e)** $(-\infty, -3)$ **(f)** $(-3, \infty)$

37. 3 **39.** none **41.** E **43.** D **45.** C

47. $f(x) = \frac{1}{4}(x - 2)^2 - 1$, or $f(x) = \frac{1}{4}x^2 - x$

49. $f(x) = -2(x - 1)^2 + 4$, or $f(x) = -2x^2 + 4x + 2$

51. linear; positive **53.** quadratic; positive

55. quadratic; negative **57.** **(a)** $f(t) = -16t^2 + 200t + 50$ **(b)** 6.25 sec; 675 ft **(c)** between 1.41 and 11.09 sec **(d)** 12.75 sec **59.** **(a)** $640 - 2x$ **(b)** $0 < x < 320$ **(c)** $\mathcal{A}(x) = -2x^2 + 640x$ **(d)** between 57.04 ft and 85.17 ft or 234.83 ft and 262.96 ft **(e)** 160 ft by 320 ft; The maximum area is 51,200 ft^2. **61.** **(a)** $2x$ **(b)** length: $2x - 4$; width: $x - 4$; $x > 4$ **(c)** $V(x) = 4x^2 - 24x + 32$ **(d)** 8 in. by 20 in. **(e)** 13.05 in. to 14.22 in. **63.** **(a)** 23.32 ft per sec

(b) 12.88 ft **65.** 10 and 10 **67.** 1 sec; 16 ft
69. \$4.97 **71.** 2002
73. (a)

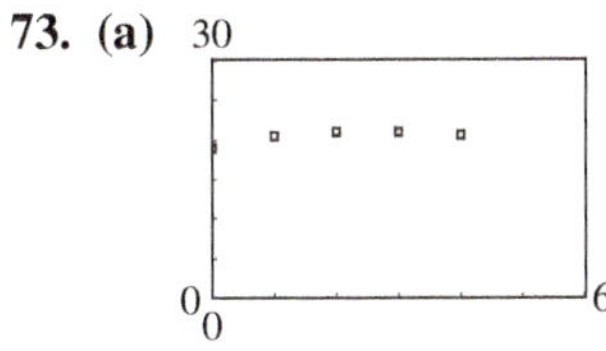

(b) $f(x) = -0.2857x^2 + 1.503x + 19.13$
(c)

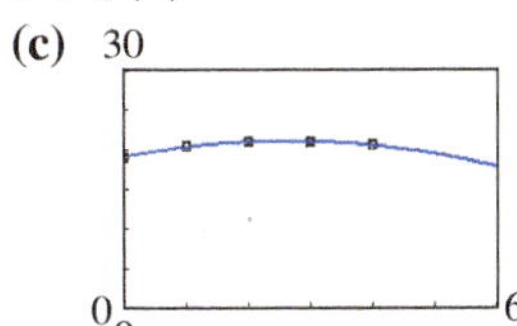

The graph of f models the data closely.

(d) 19.5 million
75. (a)

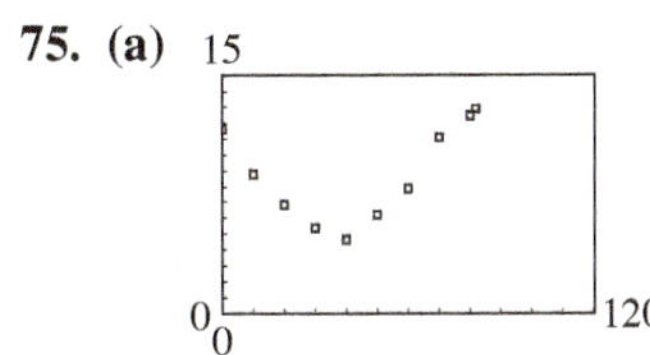

(b) $f(x) = 0.0055(x - 40)^2 + 4.7$
(c)

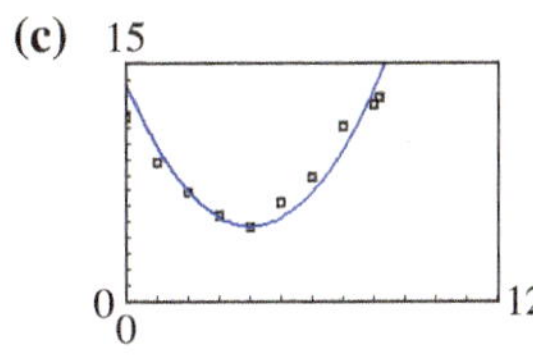

The graph of f models the data closely.

(d) $g(x) = 0.0041x^2 - 0.3130x + 11.43$
(e) $f(89) = 17.9\%$; $g(89) = 16.0\%$
77. $c = 25$ **79.** $f(x) = \frac{1}{2}x^2 - \frac{7}{2}x + 5$ **81.** $(3, 6)$
83.

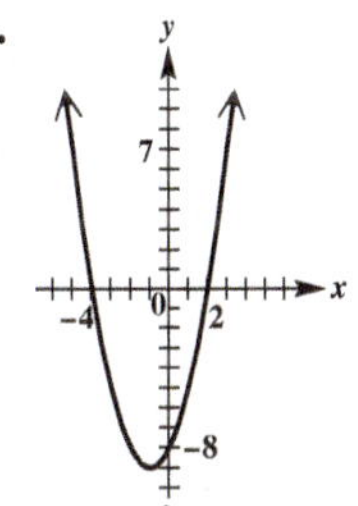

84. -4 and 2; $\{-4, 2\}$
85. $(-4, 2)$
86. $(-\infty, -4) \cup (2, \infty)$
87. $(-2, 3)$
89. $\left(-\infty, -\frac{3}{2}\right] \cup [6, \infty)$
91. $\left[2 - \sqrt{5}, 2 + \sqrt{5}\right]$

3.2 Exercises

1. 3; 4 **3.** 1; 1 **5.** 3; 6; 0 **7.** $x^2 + 2x + 9$
9. $5x^3 + 2x - 3$ **11.** $x^3 + 2x + 1$
13. $x^4 + x^3 + 2x - 1 + \frac{3}{x + 2}$
15. $-9x^2 - 10x - 27 + \frac{-52}{x - 2}$ **17.** $\frac{1}{3}x^2 - \frac{1}{9}x + \frac{1}{27}$
19. $x^3 - x^2 - 6x$ **21.** $x^2 + x + 1$
23. $x^4 - x^3 + x^2 - x + 1$
25. $f(x) = (x + 1)(2x^2 - x + 2) - 10$
27. $f(x) = (x + 2)(x^2 + 2x + 1) + 0$
29. $f(x) = (x - 3)(4x^3 + 9x^2 + 7x + 20) + 60$
31. $f(x) = (x + 1)(3x^3 + x^2 - 11x + 11) + 4$
33. 0 **35.** -1 **37.** -6 **39.** $-6 - i$ **41.** 0 **43.** -5
45. 7 **47.** yes **49.** yes **51.** no; -9 **53.** yes
55. no; $\frac{357}{125}$ **57.** yes **59.** no; $13 + 7i$ **61.** yes
63. no; $-2 + 7i$ **65.** -12; $(-2, -12)$ **66.** 0; $(-1, 0)$
67. $\frac{15}{8}$; $\left(-\frac{1}{2}, \frac{15}{8}\right)$ **68.** 2; $(0, 2)$ **69.** 0; $(1, 0)$
70. $-\frac{5}{8}$; $\left(\frac{3}{2}, -\frac{5}{8}\right)$ **71.** 0; $(2, 0)$ **72.** 8; $(3, 8)$
73. **74.**

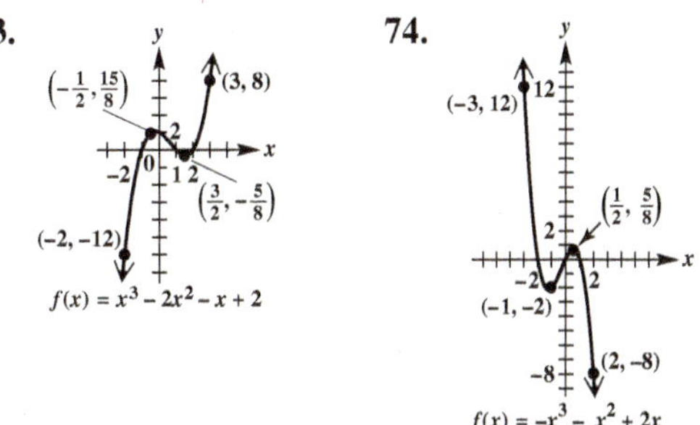

3.3 Exercises

1. true **3.** false; -2 is a zero of multiplicity 4. The number 2 is *not* a zero. **5.** true **7.** false; $\bar{z} = 7 + 6i$
9. yes **11.** no **13.** yes **15.** no **17.** no **19.** yes
21. $f(x) = (x - 2)(2x - 5)(x + 3)$
23. $f(x) = (x + 3)(3x - 1)(2x - 1)$
25. $f(x) = (x + 4)(3x - 1)(2x + 1)$
27. $f(x) = (x - 3i)(x + 4)(x + 3)$
29. $f(x) = [x - (1 + i)](2x - 1)(x + 3)$
31. $f(x) = (x + 2)^2(x + 1)(x - 3)$
33. $-1 \pm i$ **35.** $3, 2 + i$ **37.** $i, \pm 2i$
39. (a) $\pm 1, \pm 2, \pm 5, \pm 10$ **(b)** $-1, -2, 5$
(c) $f(x) = (x + 1)(x + 2)(x - 5)$ **41. (a)** $\pm 1, \pm 2, \pm 3, \pm 5, \pm 6, \pm 10, \pm 15, \pm 30$ **(b)** $-5, -3, 2$
(c) $f(x) = (x + 5)(x + 3)(x - 2)$ **43. (a)** $\pm 1, \pm 2, \pm 3, \pm 4, \pm 6, \pm 12, \pm\frac{1}{2}, \pm\frac{3}{2}, \pm\frac{1}{3}, \pm\frac{2}{3}, \pm\frac{4}{3}, \pm\frac{1}{6}$
(b) $-4, -\frac{1}{3}, \frac{3}{2}$ **(c)** $f(x) = (x + 4)(3x + 1)(2x - 3)$
45. (a) $\pm 1, \pm 2, \pm 3, \pm 4, \pm 6, \pm 12, \pm\frac{1}{2}, \pm\frac{3}{2}, \pm\frac{1}{3}, \pm\frac{2}{3}, \pm\frac{4}{3}, \pm\frac{1}{4}, \pm\frac{3}{4}, \pm\frac{1}{6}, \pm\frac{1}{8}, \pm\frac{3}{8}, \pm\frac{1}{12}, \pm\frac{1}{24}$
(b) $-\frac{3}{2}, -\frac{2}{3}, \frac{1}{2}$ **(c)** $f(x) = 2(2x + 3)(3x + 2)(2x - 1)$
47. 2 (multiplicity 3), $\pm\sqrt{7}$ **49.** $0, 2, -3, 1, -1$
51. -2 (multiplicity 5), 1 (multiplicity 5), $1 - \sqrt{3}$ (multiplicity 2) **53.** $f(x) = -3x^3 + 6x^2 + 33x - 36$
55. $f(x) = -\frac{1}{2}x^3 - \frac{1}{2}x^2 + x$
57. $f(x) = \frac{1}{6}x^3 + \frac{3}{2}x^2 + \frac{9}{2}x + \frac{9}{2}$
59. $f(x) = 5x^3 - 10x^2 + 5x$

In Exercises 61–77, we give only one possible answer.

61. $f(x) = x^2 - 10x + 26$
63. $f(x) = x^5 - 2x^4 + 3x^3 - 2x^2 + 2x$
65. $f(x) = x^3 - 3x^2 + x + 1$
67. $f(x) = x^4 - 6x^3 + 10x^2 + 2x - 15$
69. $f(x) = x^3 - 8x^2 + 22x - 20$

71. $f(x) = x^4 - 4x^3 + 5x^2 - 2x - 2$

73. $f(x) = x^4 - 16x^3 + 98x^2 - 240x + 225$

75. $f(x) = x^5 - 12x^4 + 74x^3 - 248x^2 + 445x - 500$

77. $f(x) = x^4 - 6x^3 + 17x^2 - 28x + 20$

79.

Positive	Negative	Nonreal Complex
2	1	0
0	1	2

81.

Positive	Negative	Nonreal Complex
3	0	0
1	0	2

83.

Positive	Negative	Nonreal Complex
1	1	2

85.

Positive	Negative	Nonreal Complex
4	0	0
2	0	2
0	0	4

87.

Positive	Negative	Nonreal Complex
2	3	0
2	1	2
0	3	2
0	1	4

89.

Positive	Negative	Nonreal Complex
2	3	0
2	1	2
0	3	2
0	1	4

91.

Positive	Negative	Nonreal Complex
4	2	0
4	0	2
2	2	2
2	0	4
0	2	4
0	0	6

93.

Positive	Negative	Nonreal Complex
0	5	0
0	3	2
0	1	4

95. $-5, 3, \pm 2i\sqrt{3}$ **97.** $1, 1, 1, -4$ **99.** $-3, -3, 0, \frac{1 \pm i\sqrt{31}}{4}$ **101.** $2, 2, 2, \pm i\sqrt{2}$ **103.** $-\frac{1}{2}, 1, \pm 2i$ **105.** $-\frac{1}{5}, 1 \pm i\sqrt{5}$ **107.** $\pm 2i, \pm 5i$ **109.** $\pm i, \pm i,$ **111.** $0, 0, 3 \pm \sqrt{2}$ **113.** $3, 3, 1 \pm i\sqrt{7}$ **115.** $\pm 2, \pm 3, \pm 2i$ **121.** 2

3.4 Exercises

1. A **3.** one **5.** B and D **7.** $f(x) = x(x+5)^2(x-3)$

9. $f(x) = 2x^4$

(a) $(0, \infty)$ (b) $(-\infty, 0)$

11. $f(x) = -\frac{2}{3}x^5$

(a) none (b) $(-\infty, \infty)$

13. $f(x) = \frac{1}{2}x^3 + 1$

(a) $(-\infty, \infty)$ (b) none

15.

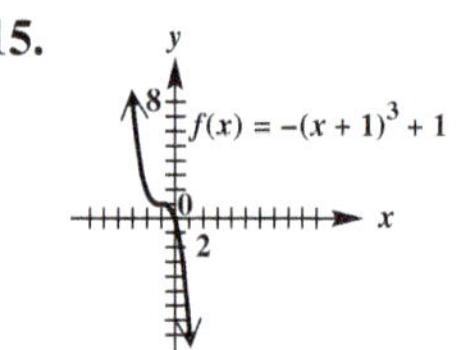

(a) none (b) $(-\infty, \infty)$

17. $f(x) = (x-1)^4 + 2$

(a) $(1, \infty)$ (b) $(-\infty, 1)$

19.

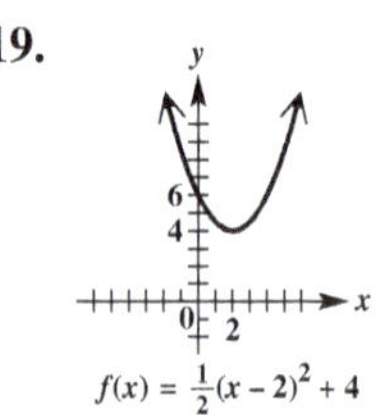

(a) $(2, \infty)$ (b) $(-\infty, 2)$

21. **23.** **25.** **27.**

29.

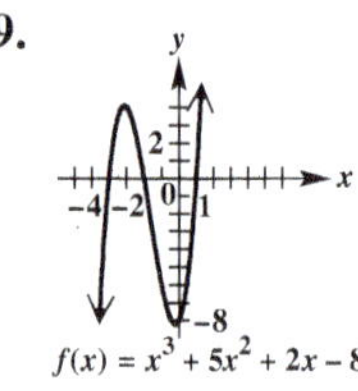

31. $f(x) = 2x(x-3)(x+2)$

33. $f(x) = x^2(x-2)(x+3)^2$

35. $f(x) = (3x-1)(x+2)^2$

37. $f(x) = x^3 + 5x^2 - x - 5$

39. $f(x) = x^3 - x^2 - 2x$

41. $f(x) = 2x^3(x^2-4)(x-1)$

43. $f(x) = 2x^3 - 5x^2 - x + 6$

45.

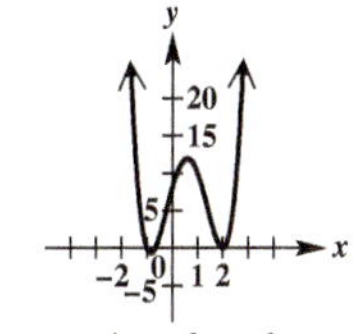

$f(x) = 3x^4 - 7x^3 - 6x^2 + 12x + 8$

47. $f(2) = -2 < 0$; $f(3) = 1 > 0$
49. $f(0) = 7 > 0$; $f(1) = -1 < 0$
51. $f(1) = -6 < 0$; $f(2) = 16 > 0$
53. $f(3.2) = -3.8144 < 0$; $f(3.3) = 7.1891 > 0$
55. $f(-1) = -35 < 0$; $f(0) = 12 > 0$
65. $f(x) = \frac{1}{2}(x+6)(x-2)(x-5)$, or $f(x) = \frac{1}{2}x^3 - \frac{1}{2}x^2 - 16x + 30$
67. $f(x) = (x-1)^3(x+1)^3$, or $f(x) = x^6 - 3x^4 + 3x^2 - 1$
69. $f(x) = (x-3)^2(x+3)^2$, or $f(x) = x^4 - 18x^2 + 81$
71. $f(x) = 2x(x-3)(x+2)$ **73.** $f(x) = (3x-1)(x+2)^2$

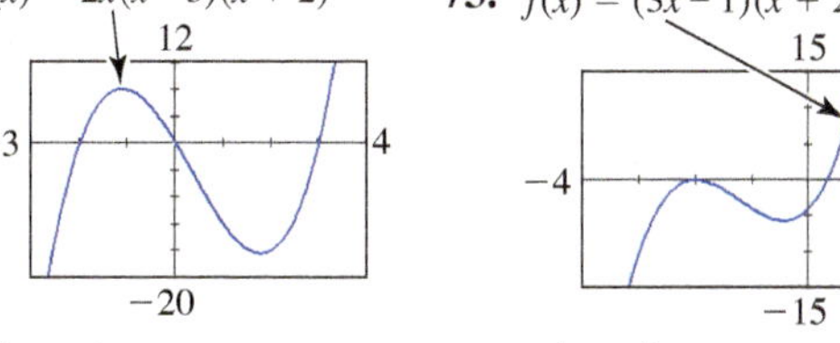

$f(1.25) = -14.21875$ $f(1.25) = 29.046875$

75. 2.7807764 **77.** 1.543689 **79.** $-3.0, -1.4, 1.4$
81. $-1.1, 1.2$ **83.** $(-0.09, 1.05)$ **85.** $(1.76, -5.34)$
87. $(-3.44, 26.15)$ **89.** Answers will vary.
91. **(a)** $0 < x < 6$ **(b)** $V(x) = x(18 - 2x)(12 - 2x)$, or $V(x) = 4x^3 - 60x^2 + 216x$ **(c)** $x \approx 2.35$; 228.16 in.^{3} **(d)** $0.42 < x < 5$ **93.** **(a)** $x - 1$; $(1, \infty)$ **(b)** $\sqrt{x^2 - (x-1)^2}$ **(c)** $2x^3 - 5x^2 + 4x - 28{,}225 = 0$ **(d)** hypotenuse: 25 in.; legs: 24 in. and 7 in. **95.** 3 ft
97. **(a)** 7.13 cm; The ball floats partly above the surface. **(b)** The sphere is more dense than water and sinks below the surface. **(c)** 10 cm; The balloon is submerged with its top even with the surface.
99. **(a)**

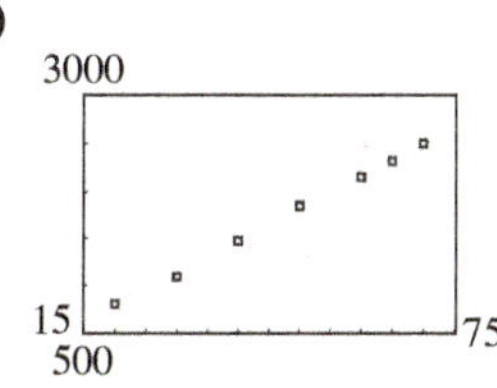

(b) $y = 33.93x + 113.4$

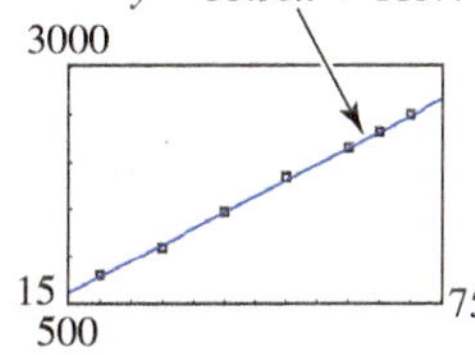

(c) $y = -0.0032x^3 + 0.4245x^2 + 16.64x + 323.1$

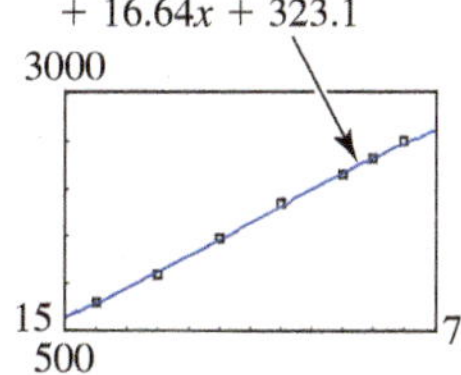

(d) linear: 1572 ft; cubic: 1569 ft **(e)** The cubic function is a slightly better fit because only one data point is not on the curve. **101.** B

103.

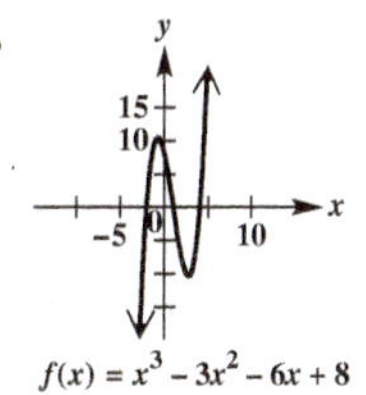

$f(x) = x^3 - 3x^2 - 6x + 8$
$f(x) = (x-4)(x-1)(x+2)$

(a) $\{-2, 1, 4\}$
(b) $(-\infty, -2) \cup (1, 4)$
(c) $(-2, 1) \cup (4, \infty)$

104.

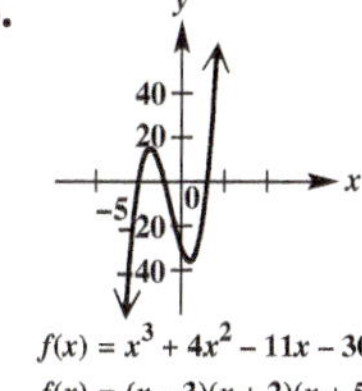

$f(x) = x^3 + 4x^2 - 11x - 30$
$f(x) = (x-3)(x+2)(x+5)$

(a) $\{-5, -2, 3\}$
(b) $(-\infty, -5) \cup (-2, 3)$
(c) $(-5, -2) \cup (3, \infty)$

105.

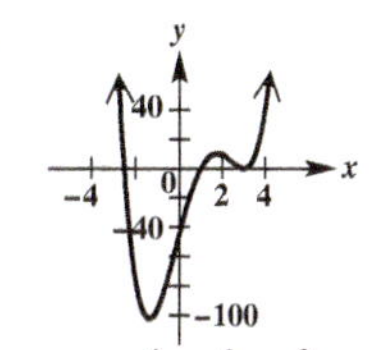

$f(x) = 2x^4 - 9x^3 - 5x^2 + 57x - 45$
$f(x) = (x-3)^2(2x+5)(x-1)$

(a) $\{-2.5, 1, 3 \text{ (multiplicity 2)}\}$ **(b)** $(-2.5, 1)$
(c) $(-\infty, -2.5) \cup (1, 3) \cup (3, \infty)$

106.

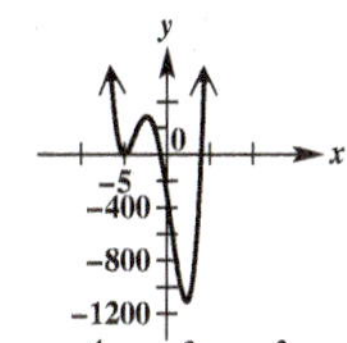

$f(x) = 4x^4 + 27x^3 - 42x^2 - 445x - 300$
$f(x) = (x+5)^2(4x+3)(x-4)$

(a) $\{-5 \text{ (multiplicity 2)}, -0.75, 4\}$ **(b)** $(-0.75, 4)$
(c) $(-\infty, -5) \cup (-5, -0.75) \cup (4, \infty)$

107.

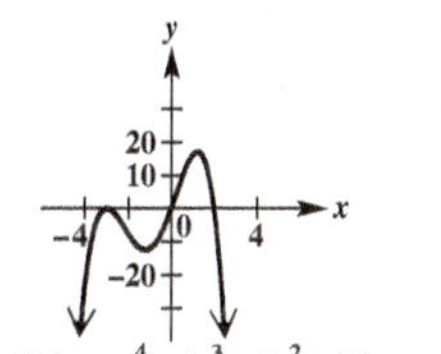

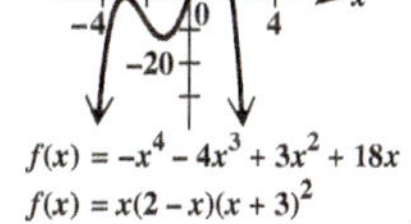

$f(x) = -x^4 - 4x^3 + 3x^2 + 18x$
$f(x) = x(2-x)(x+3)^2$

(a) $\{-3 \text{ (multiplicity 2)}, 0, 2\}$
(b) $\{-3\} \cup [0, 2]$
(c) $(-\infty, 0] \cup [2, \infty)$

108.

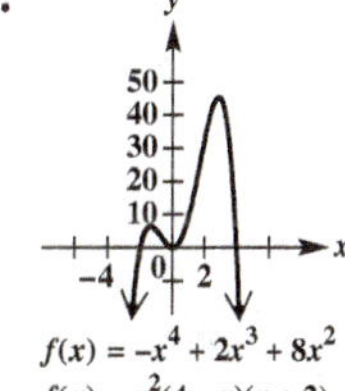

$f(x) = -x^4 + 2x^3 + 8x^2$
$f(x) = x^2(4-x)(x+2)$

(a) $\{-2, 0 \text{ (multiplicity 2)}, 4\}$
(b) $[-2, 4]$
(c) $(-\infty, -2] \cup \{0\} \cup [4, \infty)$

Summary Exercises on Polynomial Functions, Zeros, and Graphs

1. **(a)**

Positive	Negative	Nonreal Complex
2	1	0
0	1	2

(b) $\pm 1, \pm 2, \pm 3, \pm 4, \pm 6, \pm 8, \pm 12, \pm 24, \pm\frac{1}{2}, \pm\frac{3}{2}, \pm\frac{1}{3}, \pm\frac{2}{3}, \pm\frac{4}{3}, \pm\frac{8}{3}, \pm\frac{1}{6}$
(c) $-\frac{1}{2}, \frac{4}{3}, 6$ **(d)** no other complex zeros

2. (a)

Positive	Negative	Nonreal Complex
2	1	0
0	1	2

(b) $\pm 1, \pm 3, \pm \frac{1}{2}, \pm \frac{3}{2}$ **(c)** $-1, \frac{1}{2}, 3$
(d) no other complex zeros

3. (a)

Positive	Negative	Nonreal Complex
4	0	0
2	0	2
0	0	4

(b) $\pm 1, \pm 2, \pm 4, \pm 8, \pm \frac{1}{3}, \pm \frac{2}{3}, \pm \frac{4}{3}, \pm \frac{8}{3}$
(c) $\frac{2}{3}, 1$ **(d)** $-2i, 2i$

4. (a)

Positive	Negative	Nonreal Complex
3	1	0
1	1	2

(b) $\pm 1, \pm 2, \pm 3, \pm 6, \pm 9, \pm 18, \pm \frac{1}{2}, \pm \frac{3}{2}, \pm \frac{9}{2}$
(c) $-\frac{1}{2}, 2$ **(d)** $-3i, 3i$

5. (a)

Positive	Negative	Nonreal Complex
3	1	0
1	1	2

(b) $\pm 1, \pm 2, \pm \frac{1}{2}, \pm \frac{1}{3}, \pm \frac{2}{3}, \pm \frac{1}{6}$ **(c)** $\frac{1}{3}, \frac{1}{2}$
(d) $-\sqrt{2}, \sqrt{2}$

6. (a)

Positive	Negative	Nonreal Complex
2	2	0
2	0	2
0	2	2
0	0	4

(b) $\pm 1, \pm 2, \pm 3, \pm 4, \pm 6, \pm 12, \pm \frac{1}{5}, \pm \frac{2}{5}, \pm \frac{3}{5}, \pm \frac{4}{5}, \pm \frac{6}{5}, \pm \frac{12}{5}$ **(c)** $-2, \frac{2}{5}$ **(d)** $-\sqrt{3}, \sqrt{3}$

7. (a)

Positive	Negative	Nonreal Complex
4	0	0
2	0	2
0	0	4

(b) $0, \pm 1, \pm 2, \pm 4, \pm 8, \pm 16$
(c) 0, 2 (multiplicity 2) **(d)** $1 - i\sqrt{3}, 1 + i\sqrt{3}$

8. (a)

Positive	Negative	Nonreal Complex
1	3	0
1	1	2

(b) $\pm 1, \pm 3, \pm 9, \pm \frac{1}{2}, \pm \frac{3}{2}, \pm \frac{9}{2}$ **(c)** -3 (multiplicity 2)
(d) $\frac{2 - \sqrt{6}}{2}, \frac{2 + \sqrt{6}}{2}$

9. (a)

Positive	Negative	Nonreal Complex
1	3	0
1	1	2

(b) $\pm 1, \pm \frac{1}{2}, \pm \frac{1}{4}, \pm \frac{1}{8}$ **(c)** $-1, \frac{1}{2}$
(d) $-\frac{1}{4} - \frac{\sqrt{3}}{4}i, -\frac{1}{4} + \frac{\sqrt{3}}{4}i$

10. (a)

Positive	Negative	Nonreal Complex
3	2	0
3	0	2
1	2	2
1	0	4

(b) $\pm 1, \pm 2, \pm 3, \pm 6, \pm \frac{1}{2}, \pm \frac{3}{2}$ **(c)** $-3, -2, \frac{1}{2}$, 1 (multiplicity 2) **(d)** no other complex zeros

11. (a)

Positive	Negative	Nonreal Complex
1	3	0
1	1	2

(b) $\pm 1, \pm 2, \pm 3, \pm 6,$ **(c)** $-3, -1$ (multiplicity 2), 2
(d) no other real zeros **(e)** no other complex zeros
(f) $(-3, 0), (-1, 0), (2, 0)$ **(g)** $(0, -6)$
(h) $f(4) = 350$; $(4, 350)$ **(i)**  ↖ ↗
(j)

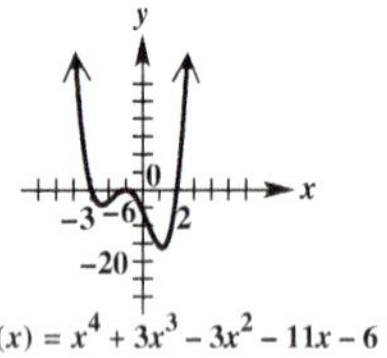

$f(x) = x^4 + 3x^3 - 3x^2 - 11x - 6$

12. (a)

Positive	Negative	Nonreal Complex
3	2	0
3	0	2
1	2	2
1	0	4

(b) $\pm 1, \pm 3, \pm 5, \pm 9, \pm 15, \pm 45, \pm \frac{1}{2}, \pm \frac{3}{2}, \pm \frac{5}{2}, \pm \frac{9}{2}, \pm \frac{15}{2}, \pm \frac{45}{2}$ **(c)** $-3, \frac{1}{2}, 5$ **(d)** $-\sqrt{3}, \sqrt{3}$
(e) no other complex zeros **(f)** $(-3, 0), \left(\frac{1}{2}, 0\right), (5, 0), \left(-\sqrt{3}, 0\right), \left(\sqrt{3}, 0\right)$ **(g)** $(0, 45)$ **(h)** $f(4) = 637$; $(4, 637)$ **(i)**  ↖ ↘ **(j)**

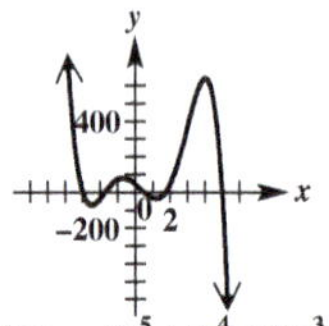

$f(x) = -2x^5 + 5x^4 + 34x^3 - 30x^2 - 84x + 45$

13. (a)

Positive	Negative	Nonreal Complex
4	1	0
2	1	2
0	1	4

(b) $\pm 1, \pm 5, \pm \frac{1}{2}, \pm \frac{5}{2}$ **(c)** 5 **(d)** $-\frac{\sqrt{2}}{2}, \frac{\sqrt{2}}{2}$
(e) $-i, i$ **(f)** $\left(-\frac{\sqrt{2}}{2}, 0\right), \left(\frac{\sqrt{2}}{2}, 0\right), (5, 0)$ **(g)** $(0, 5)$
(h) $f(4) = -527$; $(4, -527)$ **(i)**  ↙ ↗
(j)

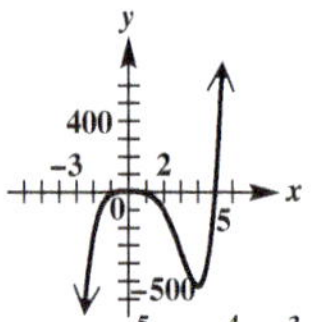

$f(x) = 2x^5 - 10x^4 + x^3 - 5x^2 - x + 5$

14. (a)

Positive	Negative	Nonreal Complex
2	2	0
2	0	2
0	2	2
0	0	4

(b) $\pm 1, \pm 2, \pm 3, \pm 6, \pm 9, \pm 18, \pm \frac{1}{3}, \pm \frac{2}{3}$ **(c)** $-\frac{2}{3}, 3$ **(d)** $\frac{-1+\sqrt{13}}{2}, \frac{-1-\sqrt{13}}{2}$ **(e)** no other complex zeros **(f)** $\left(-\frac{2}{3}, 0\right), (3, 0), \left(\frac{-1 \pm \sqrt{13}}{2}, 0\right)$ **(g)** $(0, 18)$ **(h)** $f(4) = 238; (4, 238)$ **(i)** **(j)**

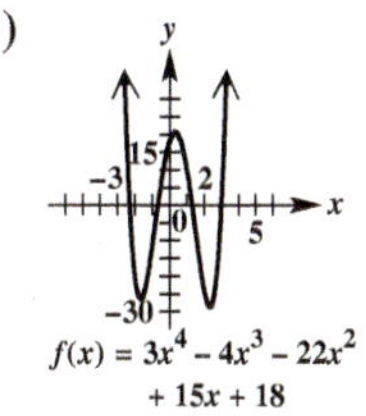

15. (a)

Positive	Negative	Nonreal Complex
1	3	0
1	1	2

(b) $\pm 1, \pm 2, \pm \frac{1}{2}$ **(c)** $-1, 1$ **(d)** no other real zeros **(e)** $-\frac{1}{4} + \frac{\sqrt{15}}{4}i, -\frac{1}{4} - \frac{\sqrt{15}}{4}i$ **(f)** $(-1, 0), (1, 0)$ **(g)** $(0, 2)$ **(h)** $f(4) = -570; (4, -570)$ **(i)** **(j)**

$f(x) = -2x^4 - x^3 + x + 2$

16. (a)

Positive	Negative	Nonreal Complex
0	4	0
0	2	2
0	0	4

(b) $0, \pm 1, \pm 3, \pm 9, \pm 27, \pm \frac{1}{2}, \pm \frac{3}{2}, \pm \frac{9}{2}, \pm \frac{27}{2}, \pm \frac{1}{4}, \pm \frac{3}{4}, \pm \frac{9}{4}, \pm \frac{27}{4}$ **(c)** $0, -\frac{3}{2}$ (multiplicity 2) **(d)** no other real zeros **(e)** $\frac{1}{2} + \frac{\sqrt{11}}{2}i, \frac{1}{2} - \frac{\sqrt{11}}{2}i$ **(f)** $\left(-\frac{3}{2}, 0\right), (0, 0)$ **(g)** $(0, 0)$ **(h)** $f(4) = 7260; (4, 7260)$ **(i)** **(j)**

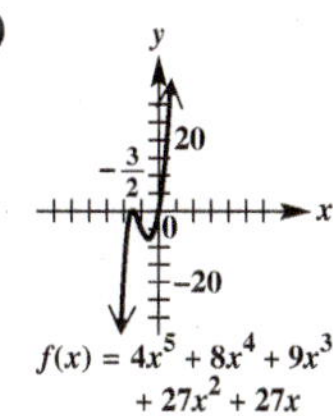

17. (a)

Positive	Negative	Nonreal Complex
1	1	2

(b) $\pm 1, \pm 5, \pm \frac{1}{3}, \pm \frac{5}{3}$ **(c)** no rational zeros **(d)** $-\sqrt{5}, \sqrt{5}$ **(e)** $-\frac{\sqrt{3}}{3}i, \frac{\sqrt{3}}{3}i$ **(f)** $\left(-\sqrt{5}, 0\right), \left(\sqrt{5}, 0\right)$ **(g)** $(0, -5)$ **(h)** $f(4) = 539; (4, 539)$ **(i)** **(j)**

$f(x) = 3x^4 - 14x^2 - 5$

18. (a)

Positive	Negative	Nonreal Complex
2	3	0
2	1	2
0	3	2
0	1	4

(b) $\pm 1, \pm 3, \pm 9$ **(c)** $-3, -1$ (multiplicity 2), $1, 3$ **(d)** no other real zeros **(e)** no other complex zeros **(f)** $(-3, 0), (-1, 0), (1, 0), (3, 0)$ **(g)** $(0, -9)$ **(h)** $f(4) = -525; (4, -525)$ **(i)**

(j)

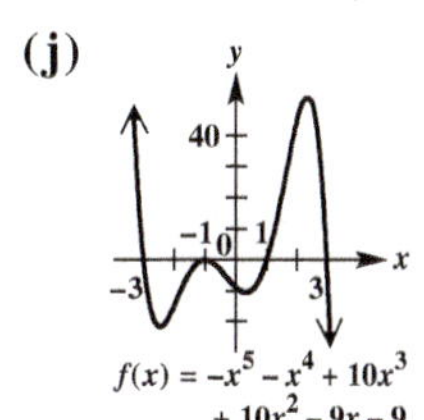

19. (a)

Positive	Negative	Nonreal Complex
4	0	0
2	0	2
0	0	4

(b) $\pm 1, \pm 2, \pm 3, \pm 4, \pm 6, \pm 12, \pm \frac{1}{3}, \pm \frac{2}{3}, \pm \frac{4}{3}$ **(c)** $\frac{1}{3}$, 2 (multiplicity 2), 3 **(d)** no other real zeros **(e)** no other complex zeros **(f)** $\left(\frac{1}{3}, 0\right), (2, 0), (3, 0)$ **(g)** $(0, -12)$ **(h)** $f(4) = -44; (4, -44)$ **(i)** **(j)**

$f(x) = -3x^4 + 22x^3 - 55x^2 + 52x - 12$

20. For the function in **Exercise 12:** ± 1.732; for the function in **Exercise 13:** ± 0.707; for the function in **Exercise 14:** $-2.303, 1.303$; for the function in **Exercise 17:** ± 2.236

3.5 Exercises

1. $(-\infty, 0) \cup (0, \infty)$; $(-\infty, 0) \cup (0, \infty)$ **3.** none; $(-\infty, 0)$ and $(0, \infty)$; none **5.** $x = 3$; $y = 2$ **7.** even; symmetry with respect to the y-axis **9.** A, B, C **11.** A **13.** A **15.** A, C, D

17. To obtain the graph of f, stretch the graph of $y = \frac{1}{x}$ vertically by a factor of 2.
(a) $(-\infty, 0) \cup (0, \infty)$
(b) $(-\infty, 0) \cup (0, \infty)$ **(c)** none
(d) $(-\infty, 0)$ and $(0, \infty)$

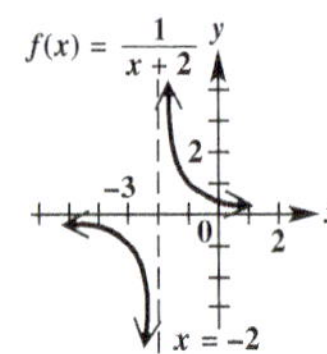

19. To obtain the graph of f, shift the graph of $y = \frac{1}{x}$ to the left 2 units.
(a) $(-\infty, -2) \cup (-2, \infty)$
(b) $(-\infty, 0) \cup (0, \infty)$ **(c)** none
(d) $(-\infty, -2)$ and $(-2, \infty)$

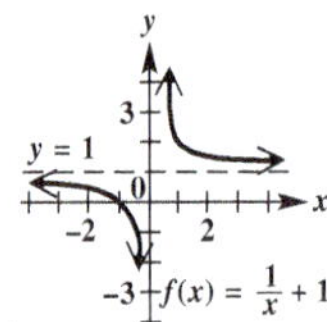

21. To obtain the graph of f, shift the graph of $y = \frac{1}{x}$ up 1 unit.
(a) $(-\infty, 0) \cup (0, \infty)$
(b) $(-\infty, 1) \cup (1, \infty)$ **(c)** none
(d) $(-\infty, 0)$ and $(0, \infty)$

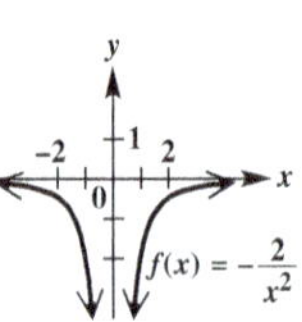

23. To obtain the graph of f, stretch the graph of $y = \frac{1}{x^2}$ vertically by a factor of 2 and reflect across the x-axis.
(a) $(-\infty, 0) \cup (0, \infty)$ **(b)** $(-\infty, 0)$
(c) $(0, \infty)$ **(d)** $(-\infty, 0)$

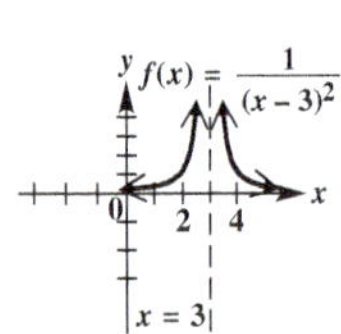

25. To obtain the graph of f, shift the graph of $y = \frac{1}{x^2}$ to the right 3 units.
(a) $(-\infty, 3) \cup (3, \infty)$ **(b)** $(0, \infty)$
(c) $(-\infty, 3)$ **(d)** $(3, \infty)$

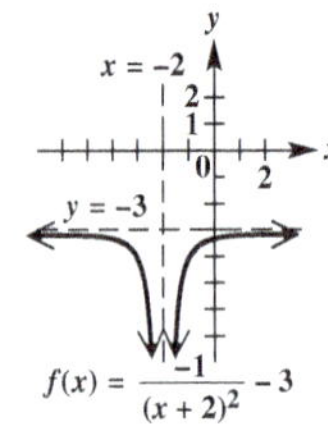

27. To obtain the graph of f, shift the graph of $y = \frac{1}{x^2}$ to the left 2 units, reflect across the x-axis, and shift 3 units down.
(a) $(-\infty, -2) \cup (-2, \infty)$
(b) $(-\infty, -3)$ **(c)** $(-2, \infty)$
(d) $(-\infty, -2)$

29. D **31.** G **33.** E **35.** F

In selected Exercises 37–59, V.A. represents vertical asymptote, H.A. represents horizontal asymptote, and O.A. represents oblique asymptote.

37. V.A.: $x = 5$; H.A.: $y = 0$
39. V.A.: $x = -\frac{1}{2}$; H.A.: $y = -\frac{3}{2}$
41. V.A.: $x = -3$; O.A.: $y = x - 3$
43. V.A.: $x = -2$, $x = \frac{5}{2}$; H.A.: $y = \frac{1}{2}$
45. V.A.: none; H.A.: $y = 1$
47. **(a)** $f(x) = \frac{2x-5}{x-3}$ **(b)** $\frac{5}{2}$ **(c)** H.A.: $y = 2$; V.A.: $x = 3$
49. **(a)** $y = x + 1$ **(b)** at $x = 0$ and $x = 1$ **(c)** above
51. A **53.** V.A.: $x = 2$; H.A.: $y = 4$; $(-\infty, 2) \cup (2, \infty)$
55. V.A.: $x = \pm 2$; H.A.: $y = -4$; $(-\infty, -2) \cup (-2, 2) \cup (2, \infty)$ **57.** V.A.: none; H.A.: $y = 0$; $(-\infty, \infty)$
59. V.A.: $x = -1$; O.A.: $y = x - 1$; $(-\infty, -1) \cup (-1, \infty)$

61.

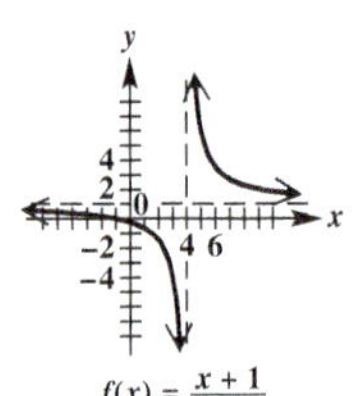

63.

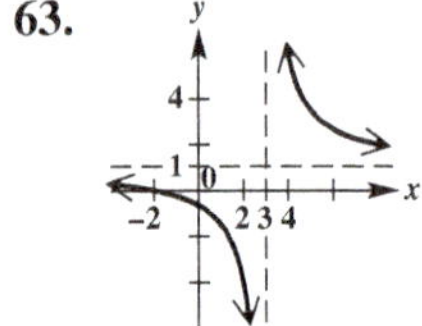

65.

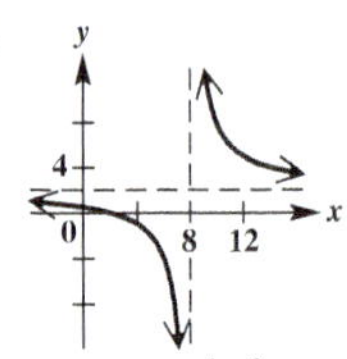

67.

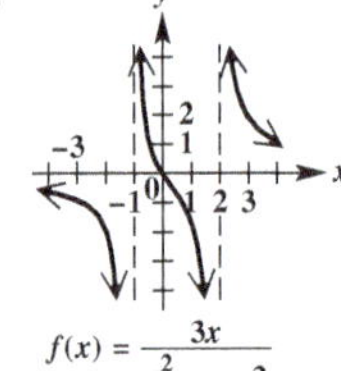

69.

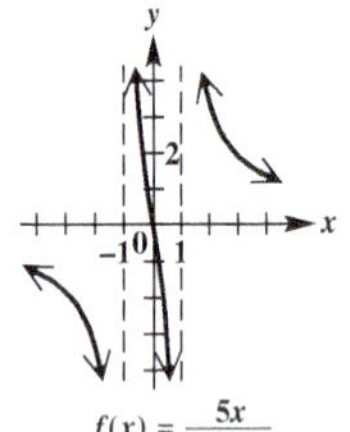

71.

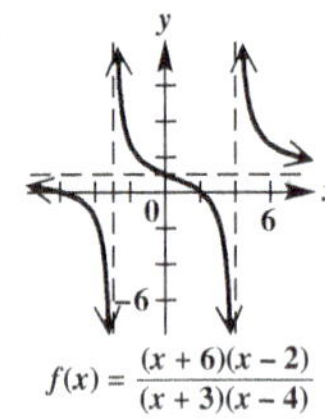

73.

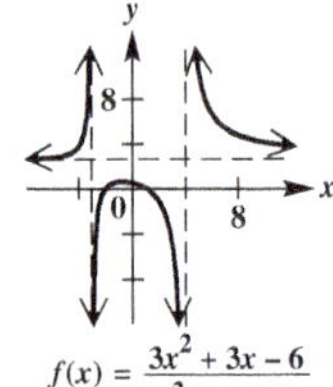

75.

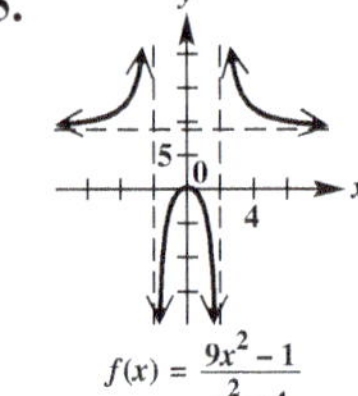

77.

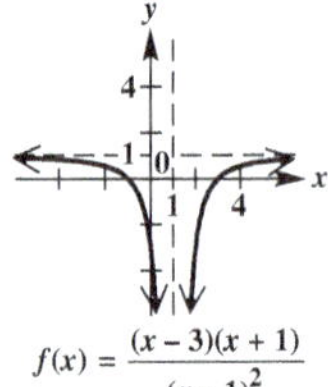

79.

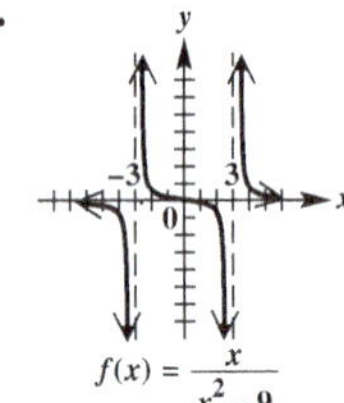

81.

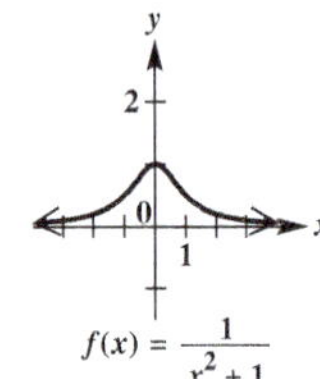

83.

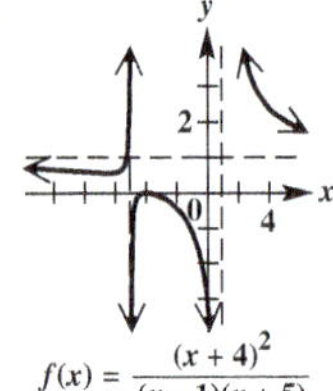

85.

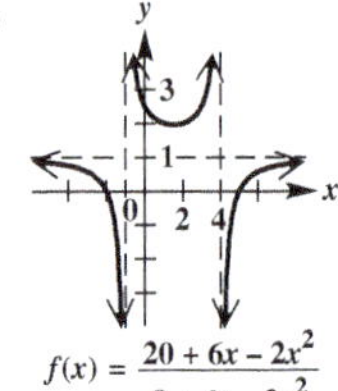

87.

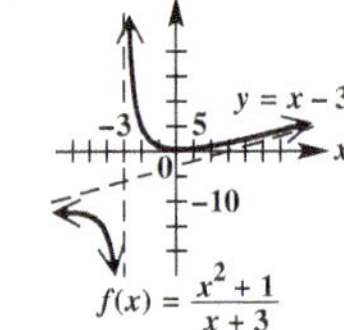

89.

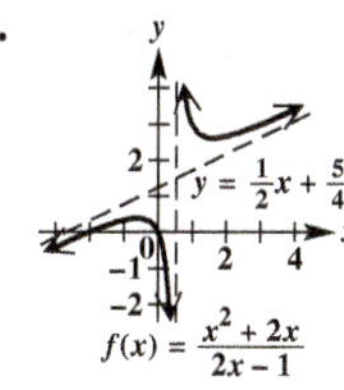

91.

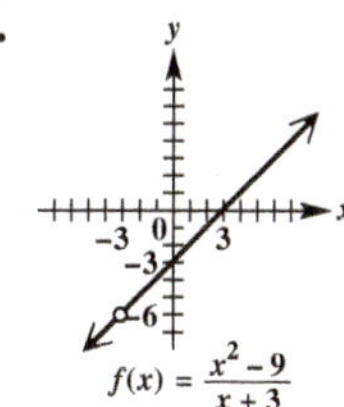

93.

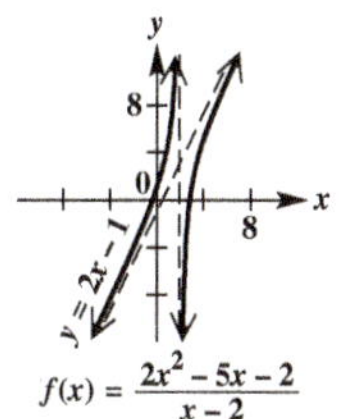

95.

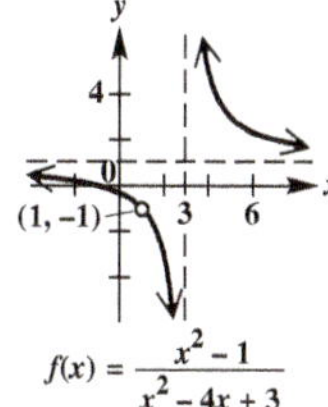

97.

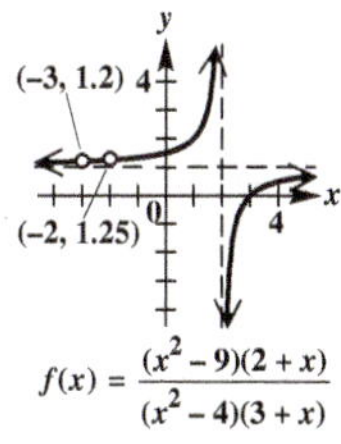

99.

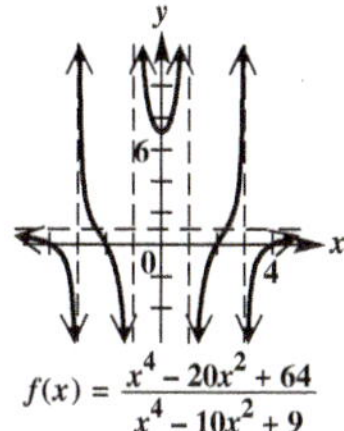

101. $f(x)=\frac{(x-3)(x+2)}{(x-2)(x+2)}$, or $f(x)=\frac{x^2-x-6}{x^2-4}$

103. $f(x)=\frac{x-2}{x(x-4)}$, or $f(x)=\frac{x-2}{x^2-4x}$

105. $f(x)=\frac{-x(x-2)}{(x-1)^2}$, or $f(x)=\frac{-x^2+2x}{x^2-2x+1}$

107. Several answers are possible. One answer is $f(x)=\frac{(x-3)(x+1)}{(x-1)^2}$.

109.

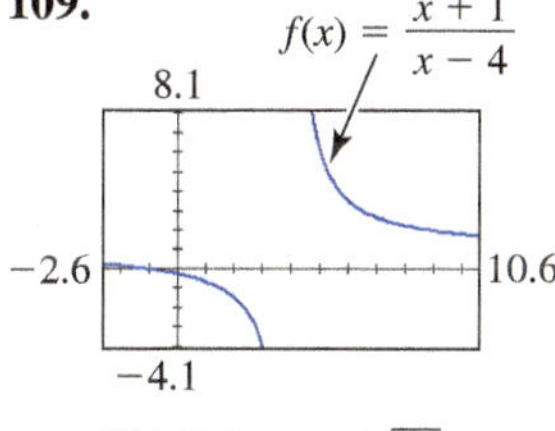

$f(1.25)=-0.\overline{81}$

111.

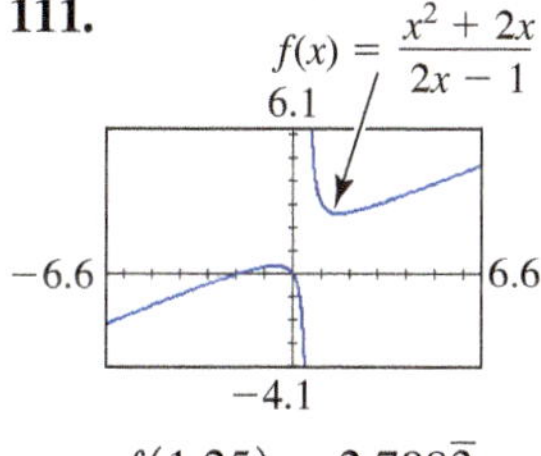

$f(1.25)=2.708\overline{3}$

113. **(a)** 26 per min **(b)** 5 park attendants

For $r=x$, $y=T(x)=\frac{2x-25}{2x^2-50x}$

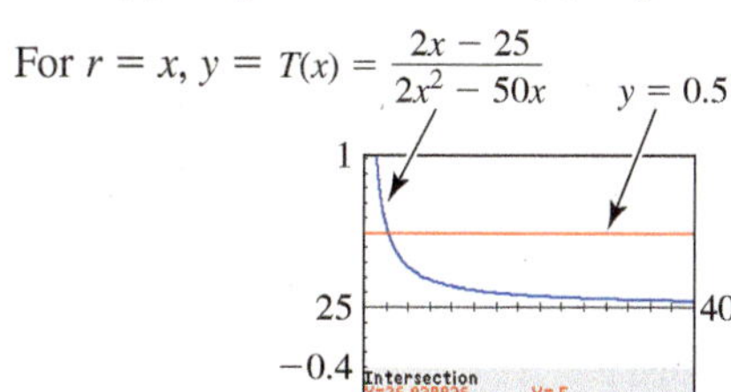

115. **(a)** 52 mph

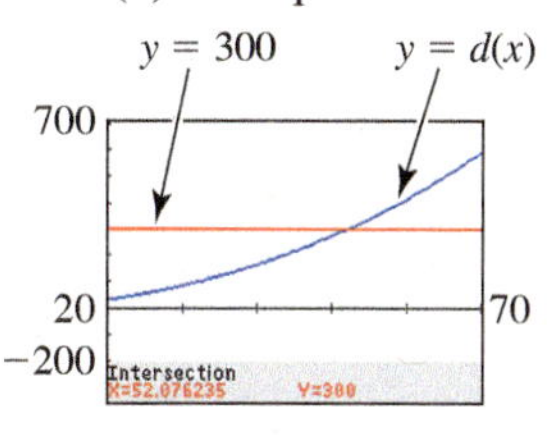

(b)

x	$d(x)$	x	$d(x)$
20	**34**	50	**273**
25	**56**	55	**340**
30	**85**	60	**415**
35	**121**	65	**499**
40	**164**	70	**591**
45	**215**		

(c) It more than doubles, compare the values of $d(20)$ and $d(40)$, for example. **(d)** It would be linear.

117. All answers are given in tens of millions.
(a) \$65.5 **(b)** \$64 **(c)** \$60 **(d)** \$40 **(e)** \$0

119. $y=1$ **120.** $(x+4)(x+1)(x-3)(x-5)$

121. **(a)** $(x-1)(x-2)(x+2)(x-5)$

(b) $f(x)=\frac{(x+4)(x+1)(x-3)(x-5)}{(x-1)(x-2)(x+2)(x-5)}$

122. **(a)** $x-5$ **(b)** 5 **123.** $(-4,0), (-1,0), (3,0)$

124. $(0,-3)$ **125.** $x=1, x=2, x=-2$

126. $\left(\frac{7\pm\sqrt{241}}{6}, 1\right)$

127.

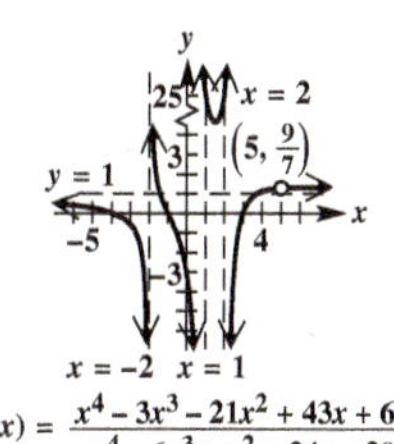

128. **(a)** $(-4,-2)\cup(-1,1)\cup(2,3)$
(b) $(-\infty,-4)\cup(-2,-1)\cup(1,2)\cup(3,5)\cup(5,\infty)$

Chapter 3 Quiz

[3.1] **1.** **(a)**

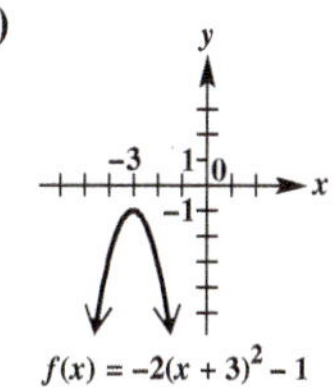

vertex: $(-3,-1)$; axis: $x=-3$; domain: $(-\infty,\infty)$; range: $(-\infty,-1]$; increasing on $(-\infty,-3)$; decreasing on $(-3,\infty)$

(b) $f(x)=2x^2-8x+3$

vertex: $(2,-5)$; axis: $x=2$; domain: $(-\infty,\infty)$; range: $[-5,\infty)$; increasing on $(2,\infty)$; decreasing on $(-\infty,2)$

2. **(a)** $s(t)=-16t^2+64t+200$
(b) between 0.78 sec and 3.22 sec

[3.2] **3.** no; 38 **4.** yes

[3.3] **5.** $f(x)=x^4-7x^3+10x^2+26x-60$

[3.4] **6.**

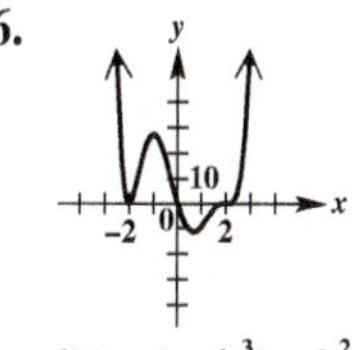

7.

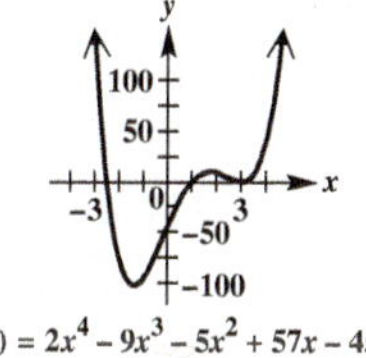

8.

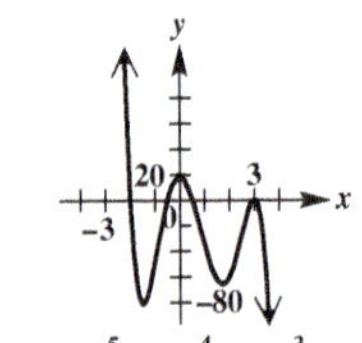

$f(x) = -4x^5 + 16x^4 + 13x^3 - 76x^2 - 3x + 18$
$f(x) = -(x + 2)(2x + 1)(2x - 1)(x - 3)^2$

[3.5] **9.**

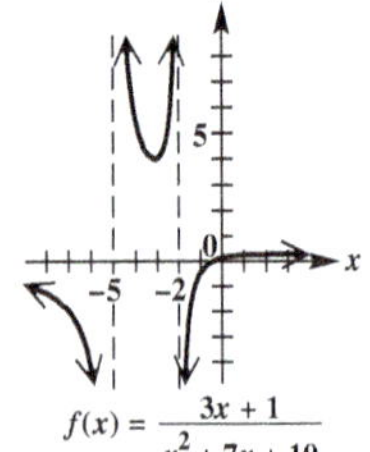

$f(x) = \dfrac{3x + 1}{x^2 + 7x + 10}$

10.

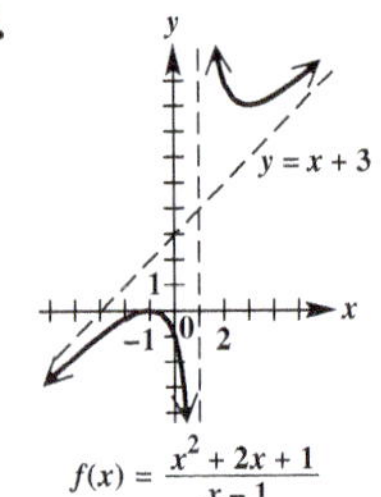

$f(x) = \dfrac{x^2 + 2x + 1}{x - 1}$

Summary Exercises on Solving Equations and Inequalities

1. **(a)** $(-1, 1) \cup (1, \infty)$ **(b)** $(-\infty, -1] \cup \{1\}$
2. **(a)** $(-\infty, -1) \cup (-1, 1) \cup (1, \infty)$ **(b)** $\varnothing$
3. **(a)** $\{0\}$ **(b)** $(-3, 0) \cup (3, \infty)$
4. **(a)** $\{1, 3\}$ **(b)** $[1, 2) \cup (2, 3]$
5. $\left\{\frac{4}{3}\right\}$ **6.** $\{20\}$ **7.** $\left\{\frac{9}{2}, \frac{14}{3}, \frac{16}{3}, \frac{11}{2}\right\}$ **8.** $\{25, 64\}$
9. $\{3, 7\}$ **10.** $\{2\}$ **11.** $\left\{\pm\frac{1}{8}\right\}$ **12.** $\{13\}$
13. inequality; $\left(-\infty, \frac{2}{5}\right) \cup \left(\frac{2}{5}, \infty\right)$ **14.** inequality; $(-\infty, -1] \cup [4, \infty)$ **15.** equation; $\{-1, 2\}$
16. inequality; $\{-3\} \cup [1, \infty)$ **17.** equation; $\{-4, 0, 3\}$
18. inequality; $\left(-\sqrt{3}, -\frac{1}{2}\right) \cup \left(\frac{1}{2}, \sqrt{3}\right)$
19. inequality; $(-\infty, 0) \cup \left[\frac{3}{2}, 3\right) \cup [5, \infty)$
20. inequality; $(-2, -1) \cup (2, \infty)$ **21.** equation; $\{1\}$
22. inequality; $(-2, 0) \cup (2, \infty)$

3.6 Exercises

1. increases; decreases **3.** 60 **5.** 6 **7.** −30 **9.** 60 **11.** $\frac{3}{2}$ **13.** $\frac{12}{5}$ **15.** $\frac{18}{125}$ **17.** C **19.** A **21.** The circumference of a circle varies directly as (or is proportional to) its radius. **23.** The speed varies directly as (or is proportional to) the distance traveled and inversely as the time.
25. The strength of a muscle varies directly as (or is proportional to) the cube of its length. **27.** 69.08 in. **29.** 850 ohms **31.** 12 km **33.** 16 in. **35.** 90 revolutions per minute **37.** 0.0444 ohm **39.** \$875 **41.** 800 lb **43.** $\frac{8}{9}$ metric ton **45.** $\frac{66\pi}{17}$ sec **47.** 21 **49.** 365.24 **51.** 92; undernourished **53.** y is half as large as before. **55.** y is one-third as large as before. **57.** p is $\frac{1}{32}$ as large as before.

Chapter 3 Review Exercises

1.

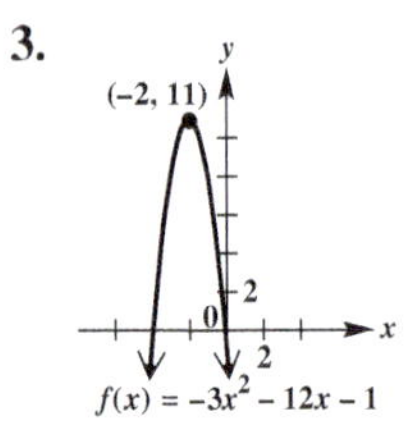

vertex: $(-4, -5)$; axis: $x = -4$; x-intercepts: $\left(\frac{-12 \pm \sqrt{15}}{3}, 0\right)$; y-intercept: $(0, 43)$; domain: $(-\infty, \infty)$; range: $[-5, \infty)$; increasing on $(-4, \infty)$; decreasing on $(-\infty, -4)$

3.

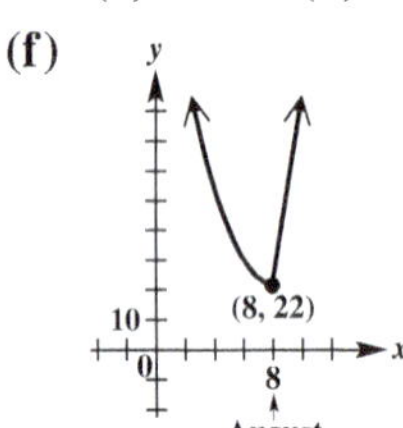

vertex: $(-2, 11)$; axis: $x = -2$; x-intercepts: $\left(\frac{-6 \pm \sqrt{33}}{3}, 0\right)$; y-intercept: $(0, -1)$; domain: $(-\infty, \infty)$; range: $(-\infty, 11]$; increasing on $(-\infty, -2)$; decreasing on $(-2, \infty)$

5. (h, k) **7.** $k \le 0$; $\left(h \pm \sqrt{\frac{-k}{a}}, 0\right)$ **9.** 90 m by 45 m
11. **(a)** 120 **(b)** 40 **(c)** 22 **(d)** 84 **(e)** 146
(f)

y
10
0
(8, 22)
8
August
x

The minimum occurs in August when $x = 8$.

13. Because the discriminant is 67.3033, a positive number, there are two x-intercepts. **15.** **(a)** the open interval $(-0.52, 2.59)$ **(b)** $(-\infty, -0.52) \cup (2.59, \infty)$
17. $x^2 + 4x + 1 + \frac{-7}{x - 3}$ **19.** $2x^2 - 8x + 31 + \frac{-118}{x + 4}$
21. $(x - 2)(5x^2 + 7x + 16) + 26$ **23.** −1 **25.** 28
27. yes **29.** $7 - 2i$

In Exercises 31–35, other answers are possible.

31. $f(x) = x^3 - 10x^2 + 17x + 28$
33. $f(x) = x^4 - 5x^3 + 3x^2 + 15x - 18$
35. $f(x) = x^4 - 6x^3 + 9x^2 - 6x + 8$ **37.** $\frac{1}{2}, -1, 5$
39. **(a)** $f(-1) = -10 < 0$; $f(0) = 2 > 0$ **(b)** $f(2) = -4 < 0$; $f(3) = 14 > 0$ **(c)** 2.414
41. **(a)** The numbers in the bottom row of synthetic division are all positive. **(b)** The numbers in the bottom row of synthetic division alternate positive and negative.
43. yes **45.** $f(x) = -2x^3 + 6x^2 + 12x - 16$
47. $1, -\frac{1}{2}, \pm 2i$ **49.** $\frac{13}{2}$
51. Any polynomial that can be factored into $a(x - b)^3$ satisfies the conditions. One example is $f(x) = 2(x - 1)^3$.

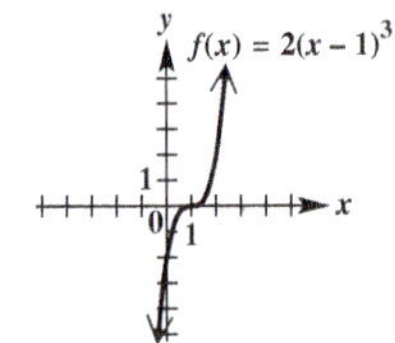

53. **(a)** $(-\infty, \infty)$ **(b)** $(-\infty, \infty)$ **(c)** $f(x) \to \infty$ as $x \to \infty$, $f(x) \to -\infty$ as $x \to -\infty$;
(d) at most seven **(e)** at most six

55.

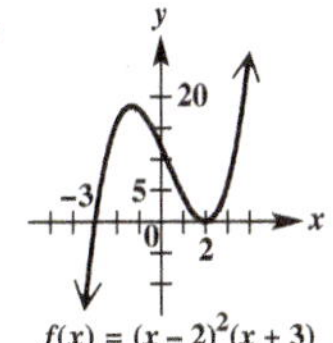

57.

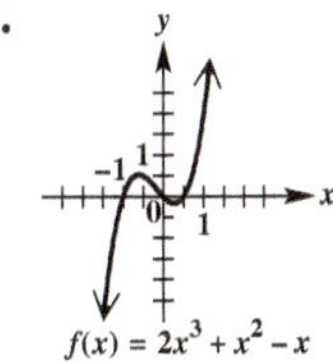

59.

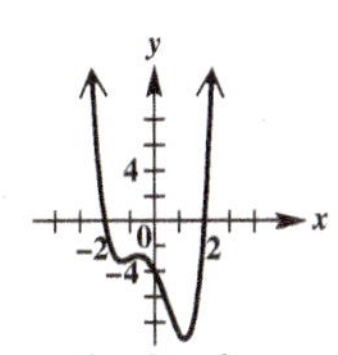

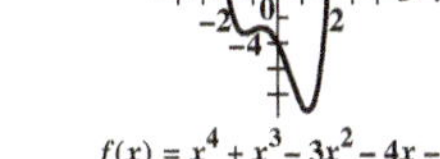

61. C **63.** E **65.** B

67. 7.6533119, 1, −0.6533119

69. (a)

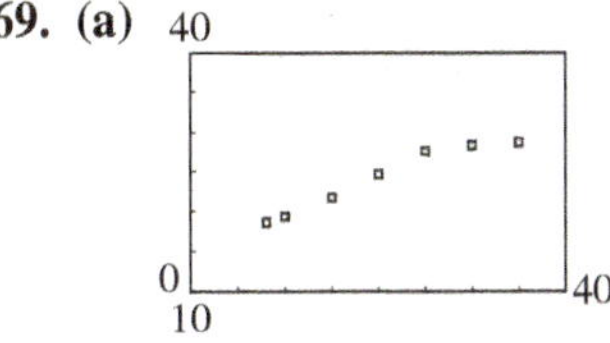

(b) $f(x) = -0.0109x^2 + 0.8693x + 11.85$

(c) $f(x) = -0.00087x^3 + 0.0456x^2 - 0.2191x + 17.83$

(d) $f(x) = -0.0109x^2 + 0.8693x + 11.85$

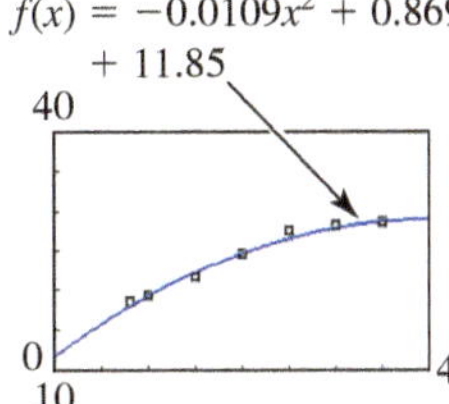

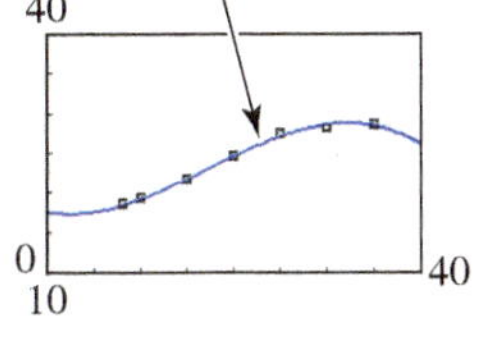

(e) Both functions approximate the data well. The quadratic function is probably better for prediction, because it is unlikely that the percent of out-of-pocket spending would decrease after 2025 (as the cubic function shows) unless changes were made in Medicare law.

71. 12 in. × 4 in. × 15 in.

73.

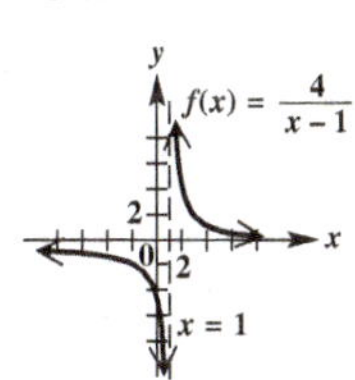

75.

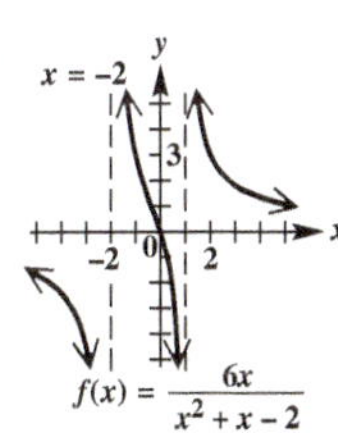

77.

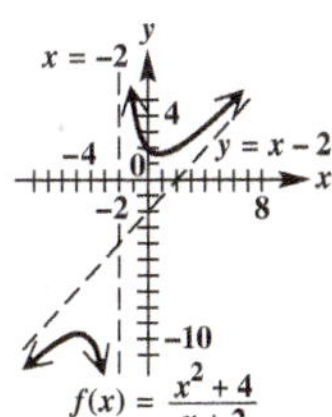

79.

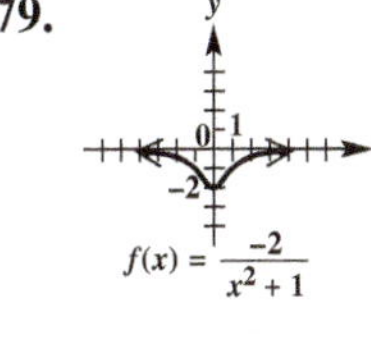

81. (a)

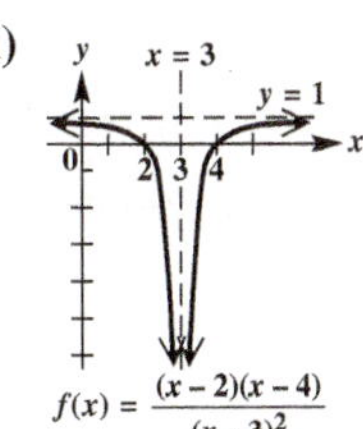

(b) One possibility is $f(x) = \frac{(x-2)(x-4)}{(x-3)^2}$.

83. $f(x) = \frac{-3x + 6}{x - 1}$

85. (a)

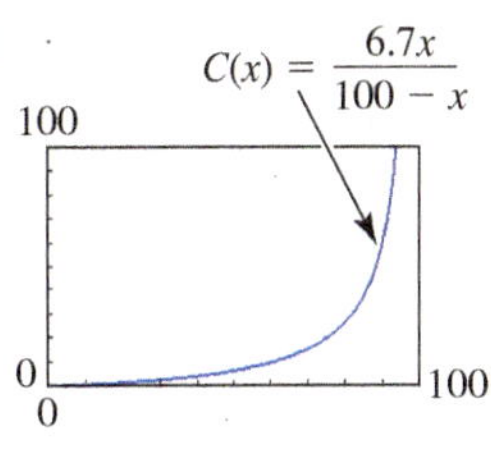

(b) \$127.3 thousand

87. 35 **89.** 0.75 **91.** 27 **93.** 33,750 units

Chapter 3 Test

[3.1] **1.**

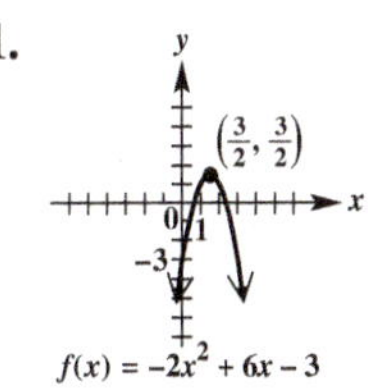

x-intercepts: $\left(\frac{-3 \pm \sqrt{3}}{-2}, 0\right)$ $\left(\text{or } \left(\frac{3 \pm \sqrt{3}}{2}, 0\right)\right)$; y-intercept: $(0, -3)$; vertex: $\left(\frac{3}{2}, \frac{3}{2}\right)$; axis: $x = \frac{3}{2}$; domain: $(-\infty, \infty)$; range: $\left(-\infty, \frac{3}{2}\right]$; increasing on $\left(-\infty, \frac{3}{2}\right)$; decreasing on $\left(\frac{3}{2}, \infty\right)$

2. (a) 2.75 sec **(b)** 169 ft **(c)** 0.7 sec and 4.8 sec **(d)** 6 sec [3.2] **3.** $3x^2 - 2x - 5 + \frac{16}{x+2}$

4. $2x^2 - x - 5$ **5.** 53 [3.3] **6.** It is a factor. The other factor is $6x^3 + 7x^2 - 14x - 8$. **7.** $-2, -3 - 2i, -3 + 2i$

8. $f(x) = 2x^4 - 2x^3 - 2x^2 - 2x - 4$ **9.** Because $f(x) > 0$ for all x, the graph never intersects or touches the x-axis. Therefore, $f(x)$ has no real zeros.

[3.3, 3.4] **10. (a)** $f(1) = 5 > 0$; $f(2) = -1 < 0$

(b)

Positive	Negative	Nonreal Complex
2	1	0
0	1	2

(c) 4.0937635, 1.8370381, −0.9308016

[3.4] **11.**

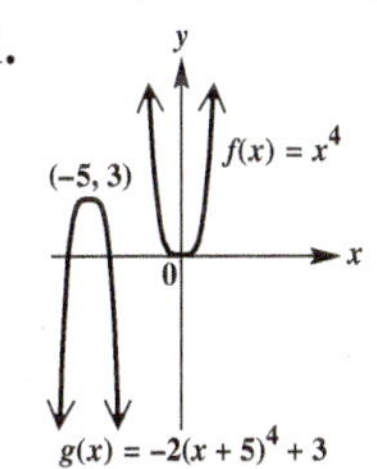

To obtain the graph of g, translate the graph of f 5 units to the left, stretch by a factor of 2, reflect across the x-axis, and translate 3 units up.

12. C **13.**

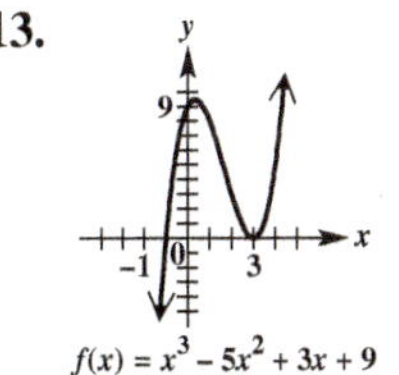

14.

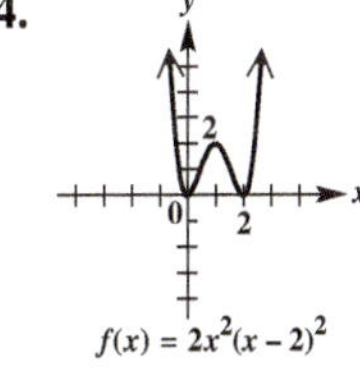

15.

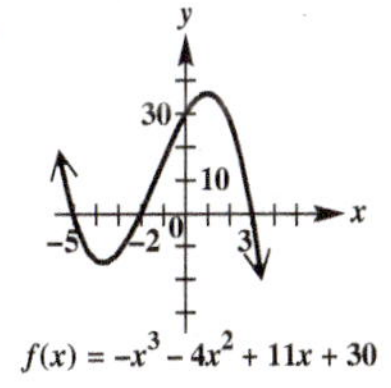

16. $f(x) = 2(x-2)^2(x+3)$, or $f(x) = 2x^3 - 2x^2 - 16x + 24$

17. (a) 270.08 **(b)** increasing from $t = 0$ to $t = 5.9$ and $t = 9.5$ to $t = 15$; decreasing from $t = 5.9$ to $t = 9.5$

[3.5] **18.**

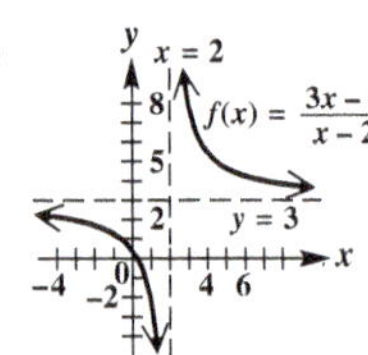

19.

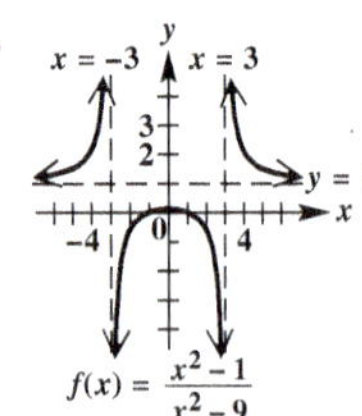

20. (a) $y = 2x + 3$ **(e)**

(b) $(-2, 0), \left(\frac{3}{2}, 0\right)$

(c) $(0, 6)$

(d) $x = 1$

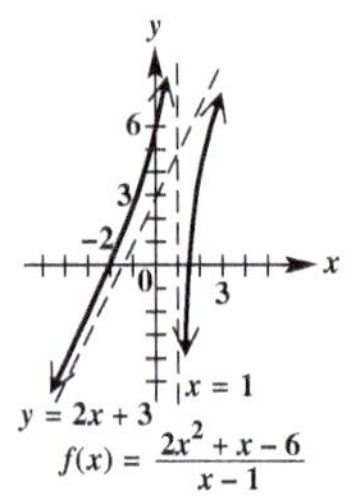

[3.6] **21.** 60 **22.** $\frac{640}{9}$ kg

Chapter 4 Inverse, Exponential, and Logarithmic Functions

4.1 Exercises

1. one-to-one **3.** one-to-one **5.** range; domain **7.** $\sqrt[3]{x}$ **9.** -3 **11.** one-to-one **13.** not one-to-one **15.** one-to-one **17.** one-to-one **19.** not one-to-one **21.** one-to-one **23.** one-to-one **25.** not one-to-one **27.** one-to-one **29.** no **31.** untying your shoelaces **33.** leaving a room **35.** unscrewing a light bulb **37.** inverses **39.** not inverses **41.** inverses **43.** not inverses **45.** inverses **47.** not inverses **49.** inverses **51.** $\{(6, -3), (1, 2), (8, 5)\}$ **53.** not one-to-one **55.** inverses **57.** not inverses

59. (a) $f^{-1}(x) = \frac{1}{3}x + \frac{4}{3}$ **(b)**

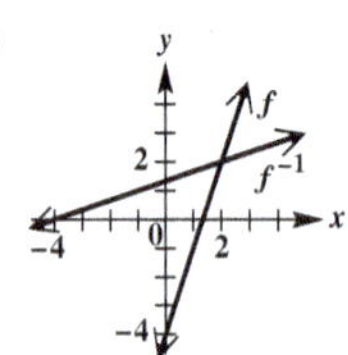

(c) Domains and ranges of both f and f^{-1} are $(-\infty, \infty)$.

61. (a) $f^{-1}(x) = -\frac{1}{4}x + \frac{3}{4}$ **(b)**

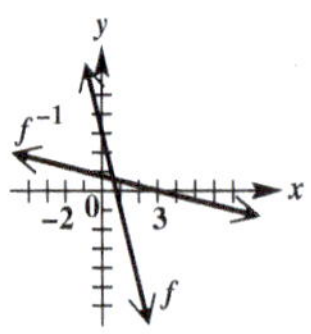

(c) Domains and ranges of both f and f^{-1} are $(-\infty, \infty)$.

63. (a) $f^{-1}(x) = \sqrt[3]{x - 1}$ **(b)**

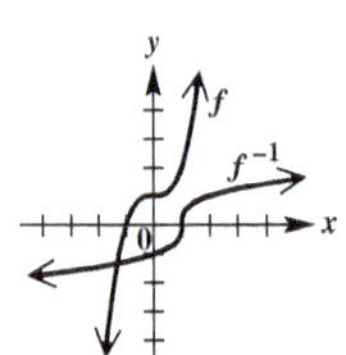

(c) Domains and ranges of both f and f^{-1} are $(-\infty, \infty)$.

65. not one-to-one

67. (a) $f^{-1}(x) = \frac{1}{x}, x \neq 0$ **(b)**

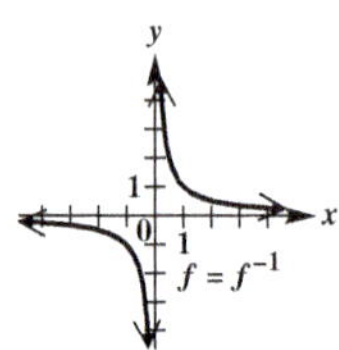

(c) Domains and ranges of both f and f^{-1} are $(-\infty, 0) \cup (0, \infty)$.

69. (a) $f^{-1}(x) = \frac{1 + 3x}{x}, x \neq 0$ **(b)**

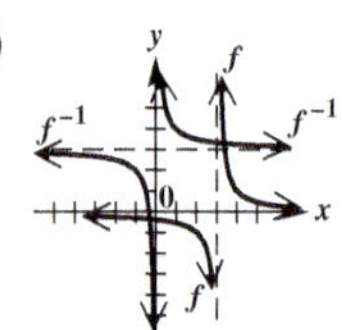

(c) Domain of f = range of $f^{-1} = (-\infty, 3) \cup (3, \infty)$.
Domain of f^{-1} = range of $f = (-\infty, 0) \cup (0, \infty)$.

71. (a) $f^{-1}(x) = \frac{3x + 1}{x - 1}, x \neq 1$ **(b)**

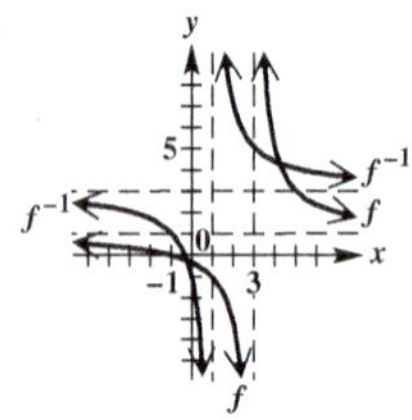

(c) Domain of f = range of $f^{-1} = (-\infty, 3) \cup (3, \infty)$.
Domain of f^{-1} = range of $f = (-\infty, 1) \cup (1, \infty)$.

73. (a) $f^{-1}(x) = \frac{3x + 6}{x - 2}, x \neq 2$ **(b)**

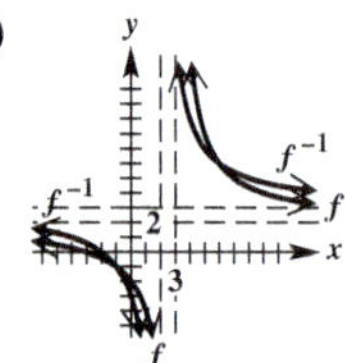

(c) Domain of f = range of $f^{-1} = (-\infty, 3) \cup (3, \infty)$.
Domain of f^{-1} = range of $f = (-\infty, 2) \cup (2, \infty)$.

75. (a) $f^{-1}(x) = x^2 - 6, x \geq 0$ **(b)**

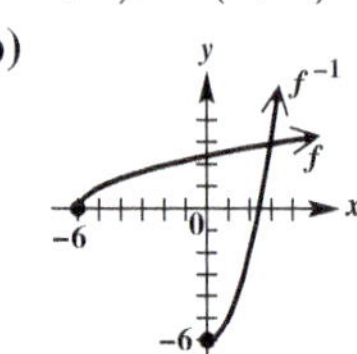

(c) Domain of f = range of $f^{-1} = [-6, \infty)$.
Domain of f^{-1} = range of $f = [0, \infty)$.

77.

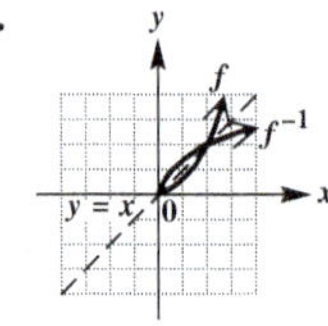

79.

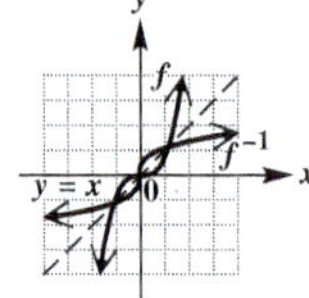

81.

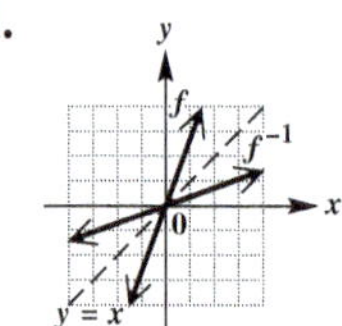

83. 4 **85.** 2 **87.** -2 **89.** It represents the cost, in dollars, of building 1000 cars. **91.** $\frac{1}{a}$ **93.** not one-to-one

95. one-to-one; $f^{-1}(x) = \frac{-5 - 3x}{x - 1}, x \neq 1$

97. $f^{-1}(x) = \frac{1}{3}x + \frac{2}{3}$; MIGUEL HAS ARRIVED

99. 6858 124 2743 63 511 124 1727 4095; $f^{-1}(x) = \sqrt[3]{x+1}$

4.2 Exercises

1. 16; $\frac{1}{16}$ **3.** falls **5.** $\frac{1}{8}$; 1; 8 **7.** $\{-3\}$ **9.** $\{2540.22\}$

11. 9 **13.** $\frac{1}{9}$ **15.** $\frac{1}{16}$ **17.** 16 **19.** $3\sqrt{3}$ **21.** $\frac{1}{8}$

23. 13.076 **25.** 10.267

27.

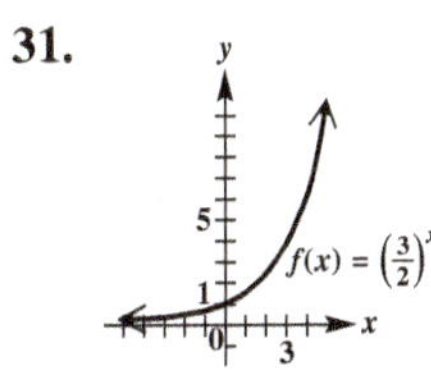

29.

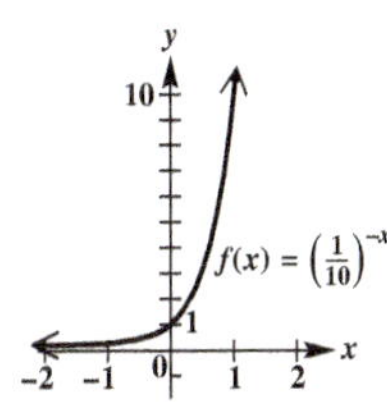

31.

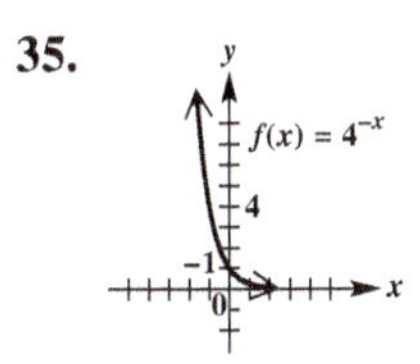

33.

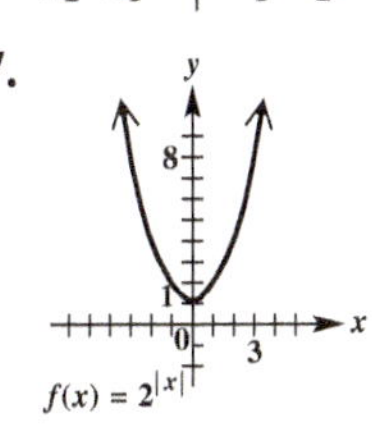

35.

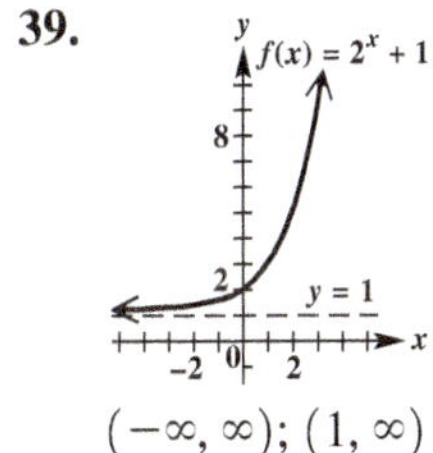

37.

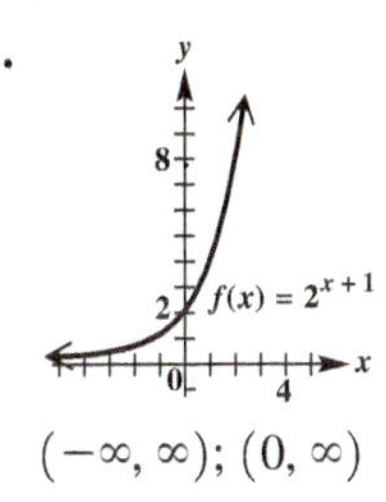

39.

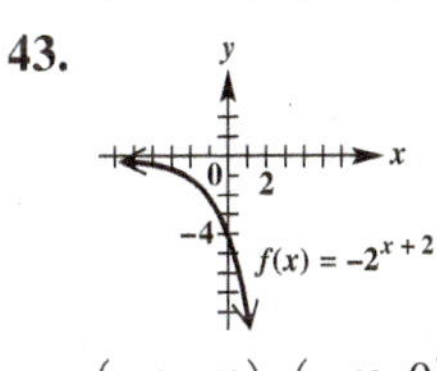

$(-\infty, \infty)$; $(1, \infty)$

41.

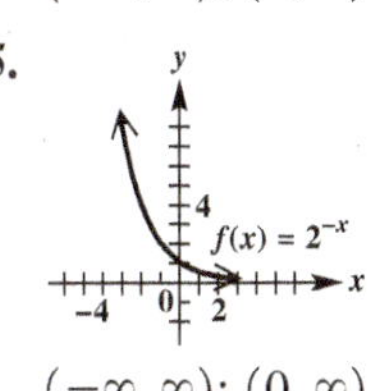

$(-\infty, \infty)$; $(0, \infty)$

43.

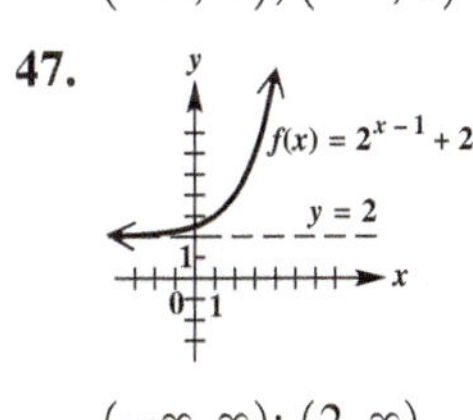

$(-\infty, \infty)$; $(-\infty, 0)$

45.

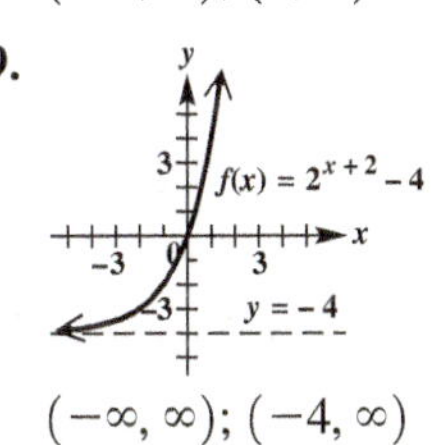

$(-\infty, \infty)$; $(0, \infty)$

47.

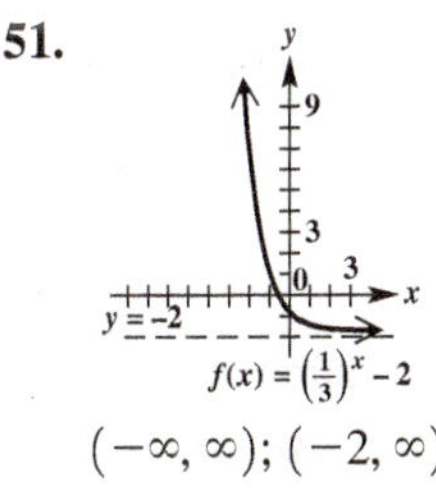

$(-\infty, \infty)$; $(2, \infty)$

49.

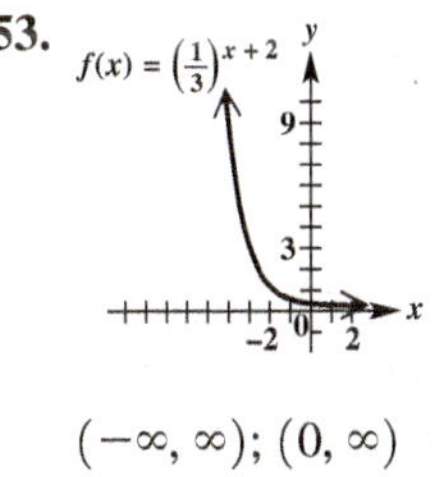

$(-\infty, \infty)$; $(-4, \infty)$

51.

$(-\infty, \infty)$; $(-2, \infty)$

53.

$(-\infty, \infty)$; $(0, \infty)$

55.

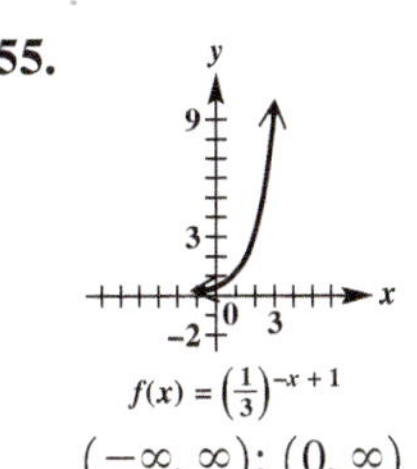

$(-\infty, \infty)$; $(0, \infty)$

57.

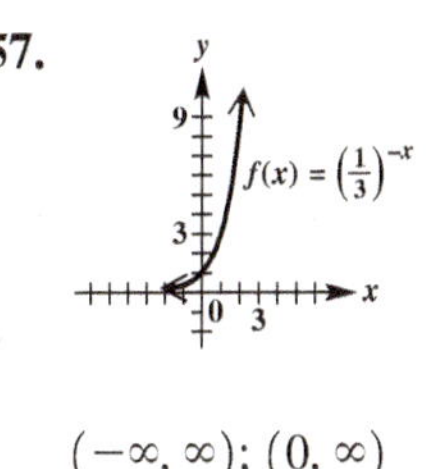

$(-\infty, \infty)$; $(0, \infty)$

59.

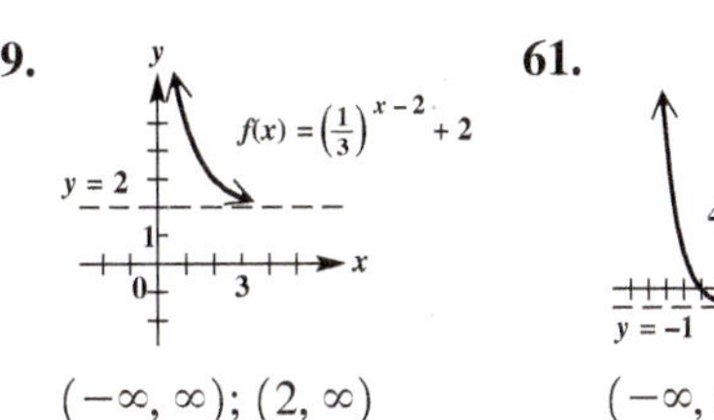

$(-\infty, \infty)$; $(2, \infty)$

61.

$(-\infty, \infty)$; $(-1, \infty)$

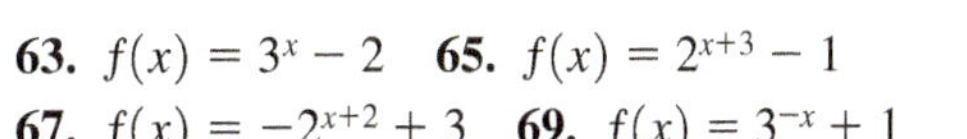

63. $f(x) = 3^x - 2$ **65.** $f(x) = 2^{x+3} - 1$

67. $f(x) = -2^{x+2} + 3$ **69.** $f(x) = 3^{-x} + 1$

71. $\left\{\frac{1}{2}\right\}$ **73.** $\{-2\}$ **75.** $\{0\}$ **77.** $\left\{\frac{1}{2}\right\}$

79. $\left\{\frac{1}{5}\right\}$ **81.** $\{-7\}$ **83.** $\{\pm 8\}$ **85.** $\{4\}$

87. $\{\pm 2\}$ **89.** $\{-27\}$ **91.** $\left\{-\frac{2}{3}\right\}$ **93.** $\left\{\frac{4}{3}\right\}$

95. $\{3\}$ **97.** **(a)** \$11,643.88; \$2737.34 **(b)** \$11,667.25; \$2760.71 **99.** \$22,902.04

101. \$3528.81 **103.** 2.5% **105.** Bank A (even though it has the greatest stated rate)

107. **(a)**

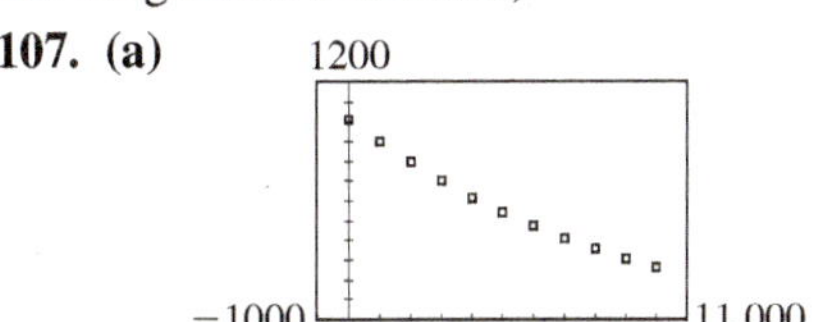

(b) exponential

(c)

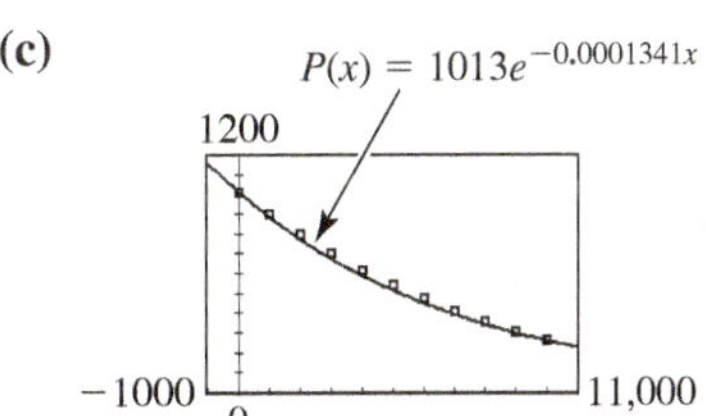

(d) $P(1500) \approx 828$ mb; $P(11{,}000) \approx 232$ mb

109. **(a)** 63,000 **(b)** 42,000 **(c)** 21,000

111. $\{0.9\}$ **113.** $\{-0.5, 1.3\}$ **115.** The variable is located in the base of a power function and in the exponent of an exponential function. **117.** $f(x) = 2^x$

119. $f(x) = \left(\frac{1}{4}\right)^x$ **121.** $f(t) = 27 \cdot 9^t$ **123.** $f(t) = \left(\frac{1}{3}\right)9^t$

125. 2.717 (A calculator gives 2.718.)

127. yes; an inverse function

128.

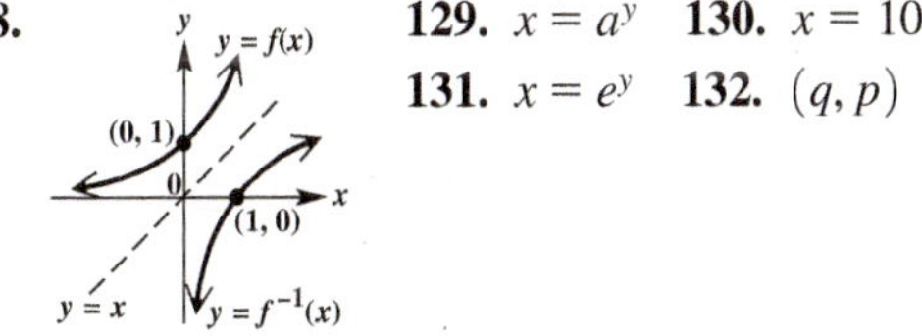

129. $x = a^y$ **130.** $x = 10^y$

131. $x = e^y$ **132.** (q, p)

4.3 Exercises

1. (a) C **(b)** A **(c)** E **(d)** B **(e)** F **(f)** D

3. $2^3 = 8$ **5.** $\left\{\frac{4}{9}\right\}$ **7.** $f(x) = \log_5 x$; $(0, \infty)$; $(-\infty, \infty)$

9. $\log_{10} 2 + \log_{10} x - \log_{10} 7$ **11.** $\log_3 81 = 4$

13. $\log_{2/3} \frac{27}{8} = -3$ **15.** $6^2 = 36$ **17.** $\left(\sqrt{3}\right)^8 = 81$

19. $\{-4\}$ **21.** $\left\{\frac{1}{2}\right\}$ **23.** $\left\{\frac{1}{4}\right\}$ **25.** $\{8\}$

27. $\{9\}$ **29.** $\left\{\frac{1}{5}\right\}$ **31.** $\{64\}$ **33.** $\left\{\frac{2}{3}\right\}$ **35.** $\{243\}$

37. $\{13\}$ **39.** $\{3\}$ **41.** $\{5\}$

43. $f(x) = (\log_2 x) + 3$; $(0, \infty)$; $(-\infty, \infty)$

45. $f(x) = |\log_2(x+3)|$, $x = -3$; $(-3, \infty)$; $[0, \infty)$

47. $f(x) = \log_{1/2}(x-2)$, $x = 2$; $(2, \infty)$; $(-\infty, \infty)$

49. E **51.** B **53.** F

55. $f(x) = \log_5 x$ **57.** $f(x) = \log_5 (x+1)$, $x = -1$

59. $f(x) = \log_{1/2}(1-x)$, $x = 1$ **61.** $f(x) = \log_3 (x-1) + 2$, $x = 1$

63. $f(x) = \log_{1/2} (x+3) - 2$, $x = -3$

65. $f(x) = \log_2(x + 1) - 3$

67. $f(x) = \log_2(-x + 3) - 2$

69. $f(x) = -\log_3(x - 1)$

71. $\log_2 6 + \log_2 x - \log_2 y$

73. $1 + \frac{1}{2} \log_5 7 - \log_5 3$

75. This cannot be simplified.

77. $\frac{1}{2}(\log_2 5 + 3 \log_2 r - 5 \log_2 z)$

79. $\log_2 a + \log_2 b - \log_2 c - \log_2 d$

81. $\frac{1}{2} \log_3 x + \frac{1}{3} \log_3 y - 2 \log_3 w - \frac{1}{2} \log_3 z$

83. $\log_a \frac{xy}{m}$ **85.** $\log_a \frac{m}{nt}$ **87.** $\log_b(x^{-1/6}y^{11/12})$

89. $\log_a[(z + 1)^2(3z + 2)]$ **91.** $\log_5 \frac{5^{1/3}}{m^{1/3}}$, or $\log_5 \sqrt[3]{\frac{5}{m}}$

93. 0.7781 **95.** 0.1761 **97.** 0.3522 **99.** 0.7386

101. (a)

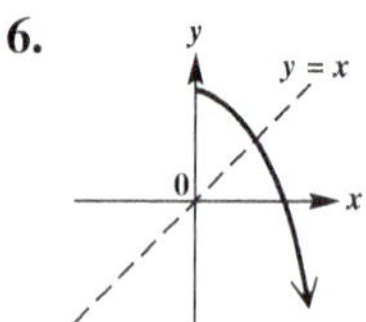

(b) logarithmic

103. (a) -4 **(b)** 6 **(c)** 4 **(d)** -1

107. $\{0.01, 2.38\}$

Summary Exercises on Inverse, Exponential, and Logarithmic Functions

1. They are inverses. **2.** They are not inverses.

3. They are inverses. **4.** They are inverses.

5. $y = x$ **6.** $y = x$

7. It is not one-to-one.

8. $y = x$ **9.** B **10.** D **11.** C **12.** A

13. The functions in **Exercises 9 and 12** are inverses of one another. The functions in **Exercises 10 and 11** are inverses of one another. **14.** $f^{-1}(x) = 5^x$

15. $f^{-1}(x) = \frac{1}{3}x + 2$; Domains and ranges of both f and f^{-1} are $(-\infty, \infty)$. **16.** $f^{-1}(x) = \sqrt[3]{\frac{x}{2}} - 1$; Domains and ranges of both f and f^{-1} are $(-\infty, \infty)$.

17. f is not one-to-one. **18.** $f^{-1}(x) = \frac{5x + 1}{2 + 3x}$; Domain of f = range of $f^{-1} = \left(-\infty, \frac{5}{3}\right) \cup \left(\frac{5}{3}, \infty\right)$. Domain of f^{-1} = range of $f = \left(-\infty, -\frac{2}{3}\right) \cup \left(-\frac{2}{3}, \infty\right)$.

19. f is not one-to-one. **20.** $f^{-1}(x) = \sqrt{x^2 + 9}$, $x \geq 0$; Domain of f = range of $f^{-1} = [3, \infty)$. Domain of f^{-1} = range of $f = [0, \infty)$. **21.** $\log_{1/10} 1000 = -3$

22. $\log_a c = b$ **23.** $\log_{\sqrt{3}} 9 = 4$ **24.** $\log_4 \frac{1}{8} = -\frac{3}{2}$

25. $\log_2 32 = x$ **26.** $\log_{27} 81 = \frac{4}{3}$ **27.** $\{2\}$ **28.** $\{-3\}$

29. $\{-3\}$ **30.** $\{25\}$ **31.** $\{-2\}$ **32.** $\left\{\frac{1}{3}\right\}$

33. $(0, 1) \cup (1, \infty)$ **34.** $\left\{\frac{3}{2}\right\}$ **35.** $\{5\}$ **36.** $\{243\}$

37. $\{1\}$ **38.** $\{-2\}$ **39.** $\{1\}$ **40.** $\{2\}$ **41.** $\{2\}$

42. $\left\{\frac{1}{9}\right\}$ **43.** $\left\{-\frac{1}{3}\right\}$ **44.** $(-\infty, \infty)$

4.4 Exercises

1. increasing **3.** $f^{-1}(x) = \log_5 x$ **5.** natural; common

7. There is no power of 2 that yields a result of 0.

9. $\log 8 = 0.90308999$ **11.** 12 **13.** -1 **15.** 1.7993 **17.** -2.6576 **19.** 3.9494 **21.** 0.1803 **23.** 3.9494 **25.** 0.1803 **27.** The logarithm of the product of two numbers is equal to the sum of the logarithms of the numbers. **29.** 3.2 **31.** 8.4 **33.** 2.0×10^{-3} **35.** 1.6×10^{-5} **37.** poor fen **39.** bog **41.** rich fen **43. (a)** 2.60031933 **(b)** 1.60031933 **(c)** 0.6003193298 **(d)** The whole number parts will vary, but the decimal parts will be the same. **45.** 1.6 **47.** -2 **49.** $\frac{1}{2}$ **51.** 3.3322 **53.** -8.9480 **55.** 10.1449 **57.** 2.0200 **59.** 10.1449 **61.** 2.0200 **63. (a)** 20 **(b)** 30 **(c)** 50 **(d)** 60 **(e)** 3 decibels **65. (a)** 3 **(b)** 6 **(c)** 8 **67.** $631{,}000{,}000 I_0$ **69.** 106.6 thousand; We must assume that the model continues to be logarithmic. **71. (a)** 2 **(b)** 2 **(c)** 2 **(d)** 1 **73.** 1 **75.** between 7°F and 11°F **77.** 1.13 billion yr **79.** 2.3219 **81.** -0.2537 **83.** -1.5850 **85.** 0.8736 **87.** 1.9376 **89.** -1.4125 **91.** $4v + \frac{1}{2}u$ **93.** $\frac{3}{2}u - \frac{5}{2}v$ **95. (a)** 4 **(b)** 25 **(c)** $\frac{1}{e}$ **97. (a)** 6 **(b)** $\ln 3$ **(c)** $\ln 9$ **99.** D **101.** domain: $(-\infty, 0) \cup (0, \infty)$; range: $(-\infty, \infty)$; symmetric with respect to the y-axis **103.** $f(x) = 2 + \ln x$, so it is the graph of $g(x) = \ln x$ translated 2 units up. **105.** $f(x) = \ln x - 2$, so it is the graph of $g(x) = \ln x$ translated 2 units down.

Chapter 4 Quiz

[4.1] **1.** $f^{-1}(x) = \frac{x^3 + 6}{3}$ [4.2] **2.** $\{4\}$

3.

$f(x) = -3^x$

domain: $(-\infty, \infty)$; range: $(-\infty, 0)$

[4.3] **4.**

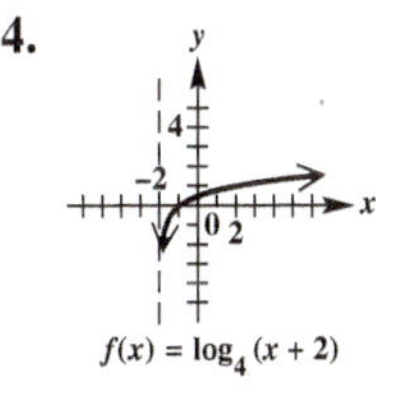

domain: $(-2, \infty)$; range: $(-\infty, \infty)$

[4.2] **5. (a)** \$18,563.28 **(b)** \$18,603.03 **(c)** \$18,612.02 **(d)** \$18,616.39 [4.4] **6. (a)** 1.5386 **(b)** 3.5427 [4.3] **7.** The expression $\log_6 25$ represents the exponent to which 6 must be raised to obtain 25. **8. (a)** $\{4\}$ **(b)** $\{5\}$ **(c)** $\left\{\frac{1}{16}\right\}$ [4.4] **9.** $\frac{1}{2}\log_3 x + \log_3 y - \log_3 p - 4\log_3 q$ **10.** 7.8137 **11.** 3.3578 **12.** 12

4.5 Exercises

1. B **3.** E **5.** D **7.** $\log_7 19$; $\frac{\log 19}{\log 7}$; $\frac{\ln 19}{\ln 7}$ **9.** $\log_{1/2} 12$; $\frac{\log 12}{\log \frac{1}{2}}$; $\frac{\ln 12}{\ln \frac{1}{2}}$ **11.** $\{1.771\}$ **13.** $\{-2.322\}$ **15.** $\{-6.213\}$ **17.** $\{-1.710\}$ **19.** $\{3.240\}$ **21.** $\{\pm 2.146\}$ **23.** $\{9.386\}$ **25.** $\varnothing$ **27.** $\{32.950\}$ **29.** $\{7.044\}$ **31.** $\{25.677\}$ **33.** $\{2011.568\}$ **35.** $\{\ln 2, \ln 4\}$ **37.** $\left\{\ln \frac{3}{2}\right\}$ **39.** $\{\log_5 4\}$ **41.** $\{e^2\}$ **43.** $\left\{\frac{e^{1.5}}{4}\right\}$ **45.** $\left\{2 - \sqrt{10}\right\}$ **47.** $\{16\}$ **49.** $\{3\}$ **51.** $\{e\}$ **53.** $\{-6, 4\}$ **55.** $\{-8, 0\}$ **57.** $\{5\}$ **59.** $\{-5\}$ **61.** $\varnothing$ **63.** $\{-2\}$ **65.** $\{0\}$ **67.** $\{12\}$ **69.** $\{25\}$ **71.** $\varnothing$ **73.** $\left\{\frac{5}{2}\right\}$ **75.** $\{3\}$ **77.** $\left\{\frac{1 + \sqrt{41}}{4}\right\}$ **79.** $\{6\}$ **81.** $\{4\}$ **83.** $\{1, 100\}$ **85.** Proposed solutions that cause *any argument of a logarithm* to be negative or zero must be rejected. The statement is not correct. For example, the solution set of $\log(-x + 99) = 2$ is $\{-1\}$. **87.** $x = e^{k/(p-a)}$ **89.** $t = -\frac{1}{k}\log\left(\frac{T - T_0}{T_1 - T_0}\right)$ **91.** $t = -\frac{2}{R}\ln\left(1 - \frac{RI}{E}\right)$ **93.** $x = \frac{\ln\left(\frac{A + B - y}{B}\right)}{-C}$ **95.** $A = \frac{B}{x^C}$ **97.** $t = \frac{\log \frac{A}{P}}{n \log\left(1 + \frac{r}{n}\right)}$ **99.** \$11,611.84 **101.** 2.6 yr **103.** 2.64% **105. (a)** 10.9% **(b)** 35.8% **(c)** 84.1% **107.** 2019 **109. (a)** 62% **(b)** 1989 **111. (a)** $P(x) = 1 - e^{-0.0034 - 0.0053x}$

(b)

$P(x) = 1 - e^{-0.0034 - 0.0053x}$

(c) $P(60) \approx 0.275$, or 27.5%; The reduction in carbon emissions from a tax of \$60 per ton of carbon is 27.5%. **(d)** $x = \$130.14$

113. $f^{-1}(x) = \ln x + 5$; domain: $(0, \infty)$; range: $(-\infty, \infty)$ **115.** $f^{-1}(x) = \ln(x + 4) - 1$; domain: $(-4, \infty)$; range: $(-\infty, \infty)$ **117.** $f^{-1}(x) = \frac{1}{3}e^{x/2}$; domain: $(-\infty, \infty)$; range: $(0, \infty)$ **119.** $\{1.52\}$ **121.** $\{0\}$ **123.** $\{2.45, 5.66\}$ **125.** When dividing each side by $\log \frac{1}{3}$, the direction of the inequality symbol should be reversed because $\log \frac{1}{3}$ is negative.

4.6 Exercises

1. B **3.** C **5.** B **7.** C **9.** $\frac{1}{3}\ln\frac{1}{3}$ **11.** $\frac{1}{100}\ln\frac{1}{2}$ **13.** $\frac{1}{2}\ln\frac{1}{4}$ **15. (a)** 440 g **(b)** 387 g **(c)** 264 g **(d)** 22 yr **17.** 1600 yr **19.** 3.6 g **21.** 16 days **23.** 9000 yr **25.** 15,600 yr **27.** 6.25°C **29. (a)** 3% compounded quarterly **(b)** \$826.95 **31.** 27.73 yr **33.** 36.62 yr **35. (a)** 315 **(b)** 229 **(c)** 142 **37. (a)** $P_0 = 1$; $a \approx 1.01355$ **(b)** 1.3 billion **(c)** 2030 **39. (a)** \$8412 **(b)** 2010 **41. (a)** \$14,542 **(b)** \$16,162 **(c)** \$17,494 **43. (a)** $y = 14{,}225e^{0.034x}$ **(b)** 2025–26 **45. (a)** 15,000 **(b)** 9098 **(c)** 5249 **47. (a)** 611 million **(b)** 746 million **(c)** 1007 million **49.** 13.2 hr **51.** 2020 **53.** 6.9 yr **55.** 11.6 yr **57. (a)** 0.065; 0.82; Among people age 25, 6.5% have some CHD, while among people age 65, 82% have some CHD. **(b)** 48 yr

Summary Exercises on Functions: Domains and Defining Equations

1. $(-\infty, \infty)$ **2.** $\left[\frac{7}{2}, \infty\right)$ **3.** $(-\infty, \infty)$
4. $(-\infty, 6) \cup (6, \infty)$ **5.** $(-\infty, \infty)$
6. $(-\infty, -3] \cup [3, \infty)$ **7.** $(-\infty, -3) \cup (-3, 3) \cup (3, \infty)$
8. $(-\infty, \infty)$ **9.** $(-4, 4)$ **10.** $(-\infty, -7) \cup (3, \infty)$
11. $(-\infty, -1] \cup [8, \infty)$ **12.** $(-\infty, 0) \cup (0, \infty)$
13. $(-\infty, \infty)$ **14.** $(-\infty, -5) \cup (-5, \infty)$ **15.** $[1, \infty)$
16. $\left(-\infty, -\sqrt{5}\right) \cup \left(-\sqrt{5}, \sqrt{5}\right) \cup \left(\sqrt{5}, \infty\right)$
17. $(-\infty, \infty)$ **18.** $(-\infty, -1) \cup (-1, 1) \cup (1, \infty)$
19. $(-\infty, 1)$ **20.** $(-\infty, 2) \cup (2, \infty)$
21. $(-\infty, \infty)$ **22.** $[-2, 3] \cup [4, \infty)$
23. $(-\infty, -2) \cup (-2, 3) \cup (3, \infty)$
24. $[-3, \infty)$ **25.** $(-\infty, 0) \cup (0, \infty)$
26. $\left(-\infty, -\sqrt{7}\right) \cup \left(-\sqrt{7}, \sqrt{7}\right) \cup \left(\sqrt{7}, \infty\right)$
27. $(-\infty, \infty)$ **28.** $\emptyset$ **29.** $[-2, 2]$ **30.** $(-\infty, \infty)$
31. $(-\infty, -7] \cup (-4, 3) \cup [9, \infty)$ **32.** $(-\infty, \infty)$
33. $(-\infty, 5]$ **34.** $(-\infty, 3)$ **35.** $(-\infty, 4) \cup (4, \infty)$
36. $(-\infty, \infty)$ **37.** $(-\infty, -5] \cup [5, \infty)$ **38.** $(-\infty, \infty)$
39. $(-2, 6)$ **40.** $(0, 1) \cup (1, \infty)$ **41.** A **42.** B
43. C **44.** D **45.** A **46.** B **47.** D **48.** C
49. C **50.** B

Chapter 4 Review Exercises

1. not one-to-one **3.** one-to-one **5.** not one-to-one
7. $f^{-1}(x) = \sqrt[3]{x+3}$ **9.** It represents the number of years after 2004 for the investment to reach \$50,000.
11. one-to-one **13.** B **15.** C **17.** $\log_2 32 = 5$
19. $\log_{3/4} \frac{4}{3} = -1$ **21.** $10^3 = 1000$ **23.** $e^{1/2} = \sqrt{e}$
25. 2 **27.** $\log_3 m + \log_3 n - \log_3 5 - \log_3 r$
29. -1.3862 **31.** 11.8776 **33.** 1.1592 **35.** $\left\{\frac{22}{5}\right\}$
37. $\{3.667\}$ **39.** $\{-13.257\}$ **41.** $\{-0.485\}$
43. $\{2.102\}$ **45.** $\{-2.487\}$ **47.** $\{3\}$ **49.** $\{\ln 3\}$
51. $\{6.959\}$ **53.** $\{e^{13/3}\}$ **55.** $\left\{\frac{\sqrt[4]{10}-7}{2}\right\}$ **57.** $\{3\}$
59. $\left\{-\frac{4}{3}, 5\right\}$ **61.** $\left\{1, \frac{10}{3}\right\}$ **63.** $\{2\}$
65. $\{-3\}$ **67.** $I_0 = \frac{I}{10^{d/10}}$ **69.** $\{1.315\}$
71. **(a)** $1{,}000{,}000 I_0$ **(b)** $158{,}500{,}000 I_0$
(c) 158.5 times greater **73.** 3.0% **75.** \$20,363.38
77. 17.3 yr **79.** 2016 **81.** **(a)** \$15,207 **(b)** \$10,716
(c) \$4491 **(d)** They are the same.

Chapter 4 Test

[4.1] **1.** **(a)** $(-\infty, \infty)$; $(-\infty, \infty)$ **(b)** The graph is a stretched translation of $y = \sqrt[3]{x}$, which passes the horizontal line test and is thus a one-to-one function.
(c) $f^{-1}(x) = \frac{x^3+7}{2}$ **(d)** $(-\infty, \infty)$; $(-\infty, \infty)$
(e)

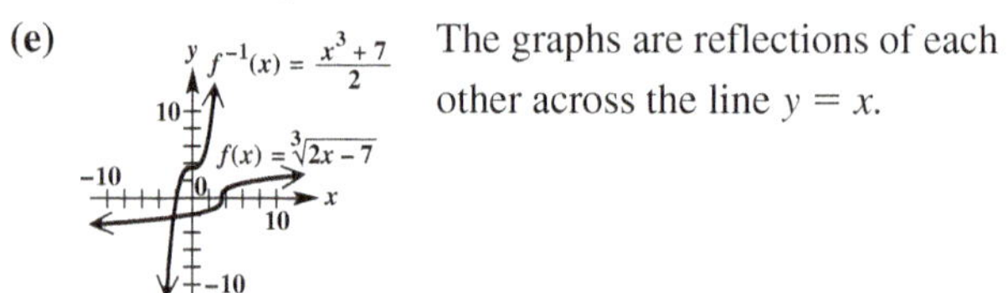

The graphs are reflections of each other across the line $y = x$.

[4.2, 4.3] **2.** **(a)** B **(b)** A **(c)** C **(d)** D
3. $\left\{\frac{1}{2}\right\}$ **4.** **(a)** $\log_4 8 = \frac{3}{2}$ **(b)** $8^{2/3} = 4$
[4.1–4.3] **5.**

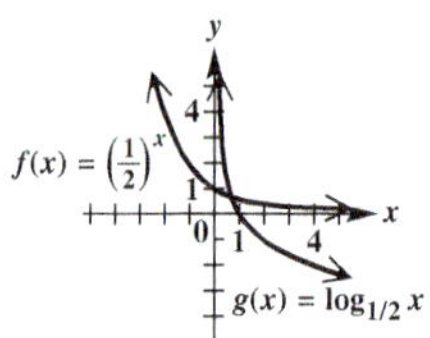

They are inverses.

[4.3] **6.** $2 \log_7 x + \frac{1}{4} \log_7 y - 3 \log_7 z$ [4.4] **7.** 3.3780
8. 7.7782 **9.** 1.1674 [4.2] **10.** $\{\pm 125\}$ **11.** $\{0\}$
[4.5] **12.** $\{0.631\}$ [4.2] **13.** $\{4\}$ [4.5] **14.** $\{12.548\}$
15. $\{2.811\}$ **16.** $\left\{0, \ln \frac{3}{2}\right\}$; $\{0, 0.405\}$ [4.3] **17.** $\left\{\frac{3}{4}\right\}$
[4.5] **18.** $\{0, 6\}$ **19.** $\{2\}$ **20.** $\emptyset$ **21.** $\left\{\frac{7}{2}\right\}$
[4.4] **22.** The expression $\log_5 27$ represents the exponent to which 5 must be raised in order to obtain 27. To approximate it with a calculator, use the change-of-base theorem. $\log_5 27 = \frac{\log 27}{\log 5} \approx 2.0478$ [4.6] **23.** 10 sec
24. **(a)** 42.8 yr **(b)** 42.7 yr **25.** 39.2 yr
26. **(a)** 329.3 g **(b)** 13.9 days

Chapter 5 Trigonometric Functions

5.1 Exercises

1. $\frac{1}{360}$ **3.** 180° **5.** 90° **7.** $\frac{1}{60}$ **9.** 55° 15′ **11.** **(a)** 60° **(b)** 150° **13.** **(a)** 45° **(b)** 135° **15.** **(a)** 36° **(b)** 126°
17. **(a)** 89° **(b)** 179° **19.** **(a)** 75° 40′ **(b)** 165° 40′
21. **(a)** 69° 49′ 30″ **(b)** 159° 49′ 30″ **23.** 70°; 110°
25. 30°; 60° **27.** 40°; 140° **29.** 107°; 73° **31.** 69°; 21°
33. 150° **35.** 7° 30′ **37.** 130° **39.** 83° 59′
41. 179° 19′ **43.** −23° 49′ **45.** 38° 32′ **47.** 60° 34′
49. 17° 01′ 49″ **51.** 30° 27′ **53.** 35.5° **55.** 112.25°
57. −60.2° **59.** 20.91° **61.** 91.598° **63.** 274.316°
65. 39° 15′ 00″ **67.** 126° 45′ 36″ **69.** −18° 30′ 54″
71. 31° 25′ 47″ **73.** 89° 54′ 01″ **75.** 178° 35′ 58″
77. 392° **79.** 386° 30′ **81.** 320° **83.** 234° 30′
85. 1° **87.** 359° **89.** 179° **91.** 130° **93.** 240°
95. 120°

In Exercises 97 and 99, answers may vary.
97. 450°, 810°; −270°, −630° **99.** 360°, 720°; −360°, −720° **101.** $30° + n \cdot 360°$ **103.** $135° + n \cdot 360°$
105. $-90° + n \cdot 360°$ **107.** $0° + n \cdot 360°$, or $n \cdot 360°$
109. 0° and 360° are coterminal angles.

Angles other than those given are possible in Exercises 111–121.

111.

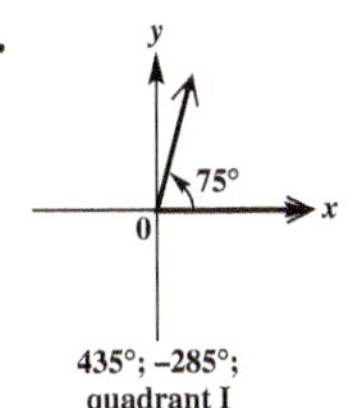

435°; −285°; quadrant I

113.

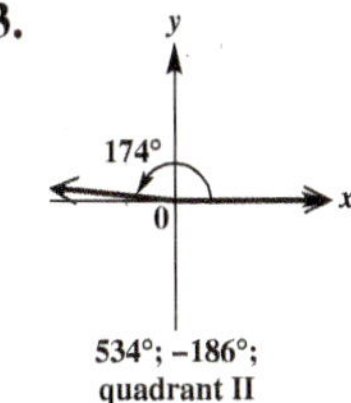

534°; −186°; quadrant II

115.

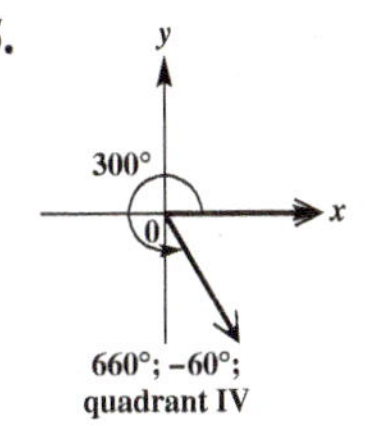

660°; −60°; quadrant IV

117.

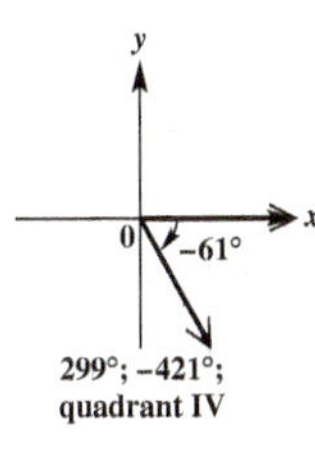

299°; −421°; quadrant IV

119.

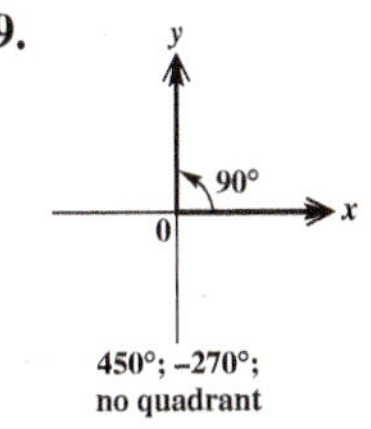

450°; −270°; no quadrant

121.

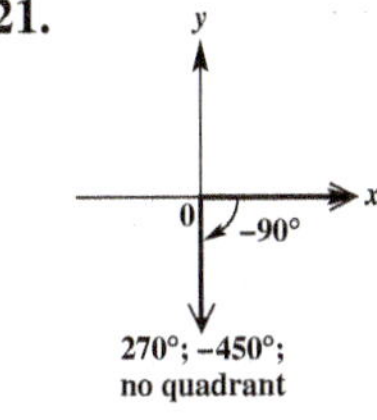

270°; −450°; no quadrant

123. $\frac{3}{4}$ **125.** 1800° **127.** 12.5 rotations per hr **129.** 4 sec

5.2 Exercises

1. $3\sqrt{2}$ **3.** $-\frac{\sqrt{2}}{2}$ **5.** possible **7.** impossible **9.** possible

In Exercises 11–25 and 43–49, we give, in order, sine, cosine, tangent, cotangent, secant, and cosecant.

11.

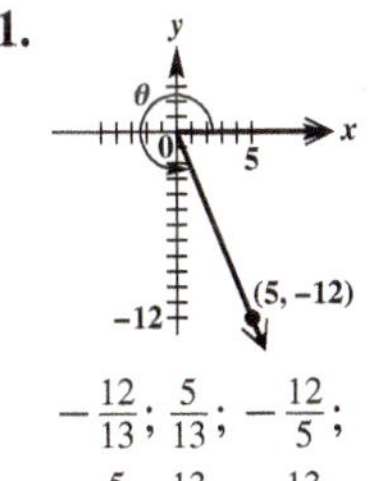

$-\frac{12}{13}$; $\frac{5}{13}$; $-\frac{12}{5}$; $-\frac{5}{12}$; $\frac{13}{5}$; $-\frac{13}{12}$

13.

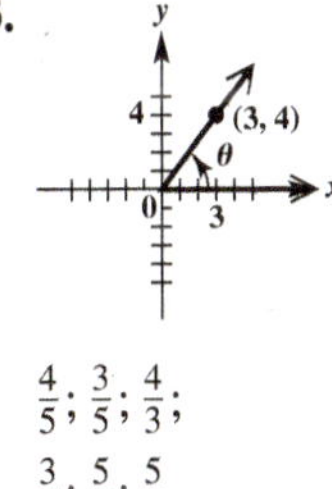

$\frac{4}{5}$; $\frac{3}{5}$; $\frac{4}{3}$; $\frac{3}{4}$; $\frac{5}{3}$; $\frac{5}{4}$

15.

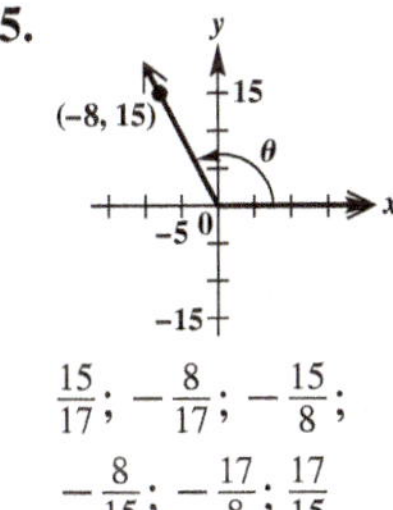

$\frac{15}{17}$; $-\frac{8}{17}$; $-\frac{15}{8}$; $-\frac{8}{15}$; $-\frac{17}{8}$; $\frac{17}{15}$

17.

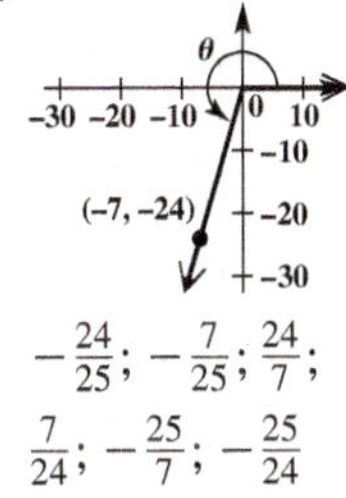

$-\frac{24}{25}$; $-\frac{7}{25}$; $\frac{24}{7}$; $\frac{7}{24}$; $-\frac{25}{7}$; $-\frac{25}{24}$

19.

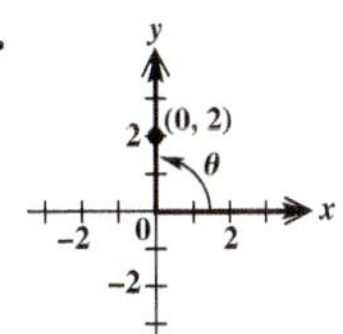

1; 0; undefined; 0; undefined; 1

21.

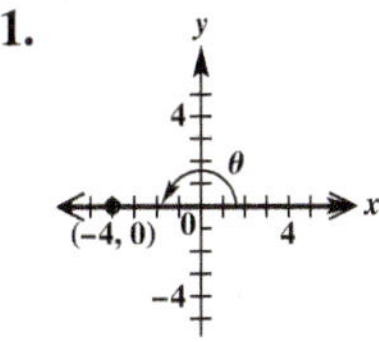

0; −1; 0; undefined; −1; undefined

23.

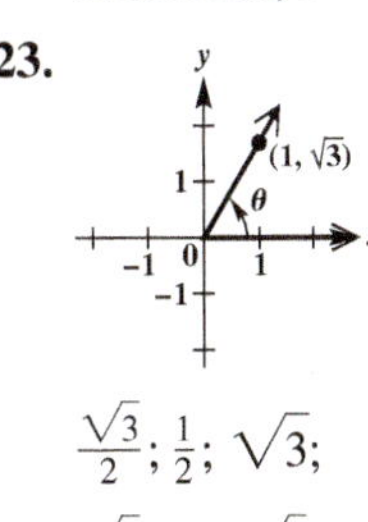

$\frac{\sqrt{3}}{2}$; $\frac{1}{2}$; $\sqrt{3}$; $\frac{\sqrt{3}}{3}$; 2; $\frac{2\sqrt{3}}{3}$

25.

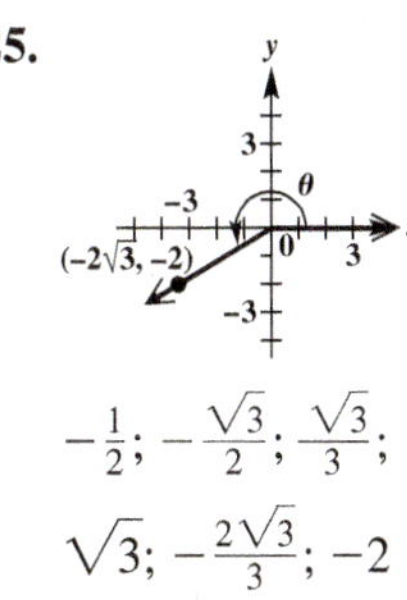

$-\frac{1}{2}$; $-\frac{\sqrt{3}}{2}$; $\frac{\sqrt{3}}{3}$; $\sqrt{3}$; $-\frac{2\sqrt{3}}{3}$; −2

27. negative **29.** negative **31.** positive **33.** positive **35.** negative **37.** positive **39.** negative **41.** positive

43.

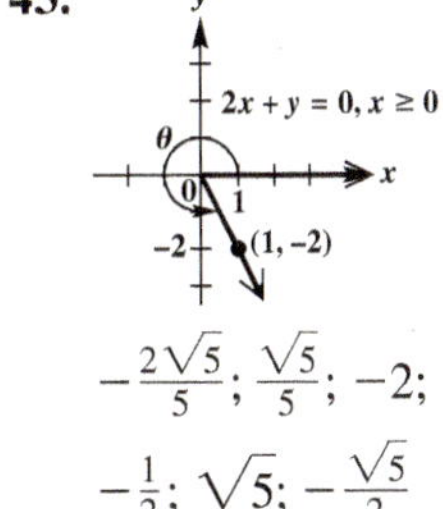

$-\frac{2\sqrt{5}}{5}$; $\frac{\sqrt{5}}{5}$; −2; $-\frac{1}{2}$; $\sqrt{5}$; $-\frac{\sqrt{5}}{2}$

45.

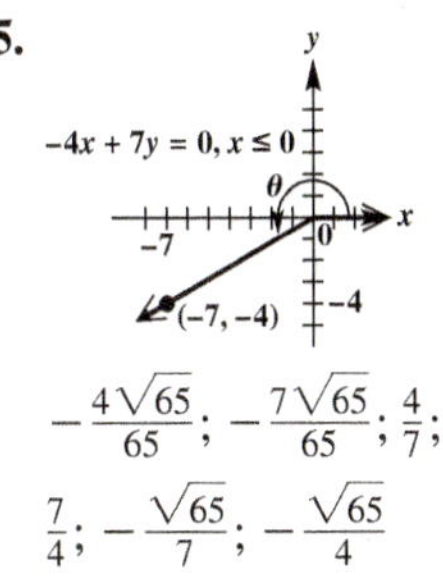

$-\frac{4\sqrt{65}}{65}$; $-\frac{7\sqrt{65}}{65}$; $\frac{4}{7}$; $\frac{7}{4}$; $-\frac{\sqrt{65}}{7}$; $-\frac{\sqrt{65}}{4}$

47.

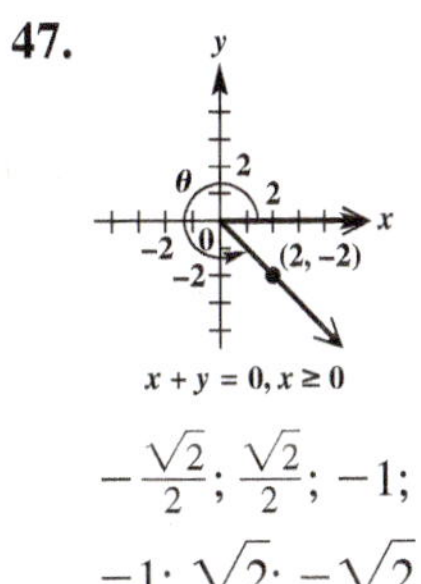

$-\frac{\sqrt{2}}{2}$; $\frac{\sqrt{2}}{2}$; −1; −1; $\sqrt{2}$; $-\sqrt{2}$

49.

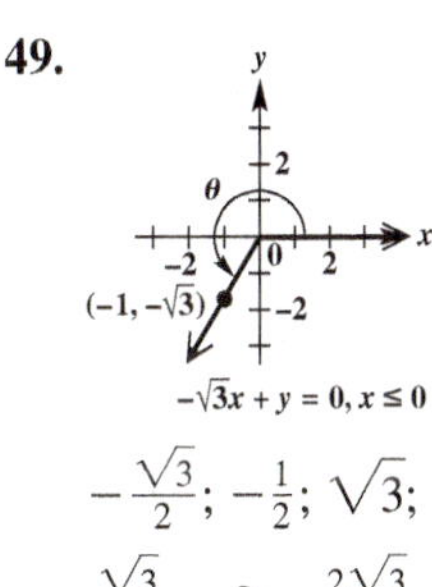

$-\frac{\sqrt{3}}{2}$; $-\frac{1}{2}$; $\sqrt{3}$; $\frac{\sqrt{3}}{3}$; −2; $-\frac{2\sqrt{3}}{3}$

51. 0 **53.** 0 **55.** −1 **57.** 1 **59.** undefined **61.** −1 **63.** 0 **65.** undefined **67.** 1 **69.** −1 **71.** −3 **73.** −3 **75.** 5 **77.** 0 **79.** 1 **81.** 0 **83.** 0 **85.** undefined **87.** $\frac{3}{2}$ **89.** $-\frac{7}{3}$ **91.** $\frac{1}{5}$ **93.** $\frac{\sqrt{2}}{2}$ **95.** 0.8 **97.** All are positive. **99.** Tangent and cotangent are positive. All others are negative. **101.** Sine and cosecant are positive. All others are negative. **103.** Cosine and secant are positive. All others are negative. **105.** Sine and cosecant are positive. All others are negative. **107.** All are positive. **109.** I, II **111.** I **113.** II **115.** I **117.** III **119.** III, IV **121.** impossible **123.** possible **125.** possible **127.** impossible **129.** possible **131.** $-\frac{4}{5}$ **133.** $-\frac{\sqrt{5}}{2}$ **135.** $-\frac{\sqrt{3}}{3}$ **137.** 1.05

In Exercises 139–149, we give, in order, sine, cosine, tangent, cotangent, secant, and cosecant.

139. $\frac{15}{17}$; $-\frac{8}{17}$; $-\frac{15}{8}$; $-\frac{8}{15}$; $-\frac{17}{8}$; $\frac{17}{15}$

141. $\frac{\sqrt{5}}{7}$; $\frac{2\sqrt{11}}{7}$; $\frac{\sqrt{55}}{22}$; $\frac{2\sqrt{55}}{5}$; $\frac{7\sqrt{11}}{22}$; $\frac{7\sqrt{5}}{5}$

143. $\frac{8\sqrt{67}}{67}$; $\frac{\sqrt{201}}{67}$; $\frac{8\sqrt{3}}{3}$; $\frac{\sqrt{3}}{8}$; $\frac{\sqrt{201}}{3}$; $\frac{\sqrt{67}}{8}$

145. $\frac{\sqrt{2}}{6}$; $-\frac{\sqrt{34}}{6}$; $-\frac{\sqrt{17}}{17}$; $-\sqrt{17}$; $-\frac{3\sqrt{34}}{17}$; $3\sqrt{2}$

147. $\frac{\sqrt{15}}{4}$; $-\frac{1}{4}$; $-\sqrt{15}$; $-\frac{\sqrt{15}}{15}$; -4; $\frac{4\sqrt{15}}{15}$

149. 1; 0; undefined; 0; undefined; 1 **153.** negative **155.** positive **157.** positive **159.** positive

5.3 Exercises

1. C **3.** B **5.** E **7.** negative; III; 60°; $-\frac{\sqrt{3}}{2}$
9. positive; III; 30°; $\frac{\sqrt{3}}{3}$

In Exercises 11 and 13, we give, in order, sine, cosine, and tangent.

11. $\frac{21}{29}$; $\frac{20}{29}$; $\frac{21}{20}$ **13.** $\frac{n}{p}$; $\frac{m}{p}$; $\frac{n}{m}$

In Exercises 15–21, we give, in order, the unknown side, sine, cosine, tangent, cotangent, secant, and cosecant.

15. $c = 13$; $\frac{12}{13}$; $\frac{5}{13}$; $\frac{12}{5}$; $\frac{5}{12}$; $\frac{13}{5}$; $\frac{13}{12}$

17. $b = \sqrt{13}$; $\frac{\sqrt{13}}{7}$; $\frac{6}{7}$; $\frac{\sqrt{13}}{6}$; $\frac{6\sqrt{13}}{13}$; $\frac{7}{6}$; $\frac{7\sqrt{13}}{13}$

19. $b = \sqrt{91}$; $\frac{\sqrt{91}}{10}$; $\frac{3}{10}$; $\frac{\sqrt{91}}{3}$; $\frac{3\sqrt{91}}{91}$; $\frac{10}{3}$; $\frac{10\sqrt{91}}{91}$

21. $b = \sqrt{3}$; $\frac{\sqrt{3}}{2}$; $\frac{1}{2}$; $\sqrt{3}$; $\frac{\sqrt{3}}{3}$; 2; $\frac{2\sqrt{3}}{3}$

23. sin 60° **25.** sec 30° **27.** csc 51° **29.** cos 51.3°

31. $\frac{\sqrt{3}}{3}$ **33.** $\frac{1}{2}$ **35.** $\frac{2\sqrt{3}}{3}$ **37.** $\sqrt{2}$ **39.** $\frac{\sqrt{2}}{2}$ **41.** 1

43. $\frac{\sqrt{3}}{2}$ **45.** $\sqrt{3}$ **47.** $y = \frac{\sqrt{3}}{3}x$ **49.** 60°

51. $x = \frac{9\sqrt{3}}{2}$; $y = \frac{9}{2}$; $z = \frac{3\sqrt{3}}{2}$; $w = 3\sqrt{3}$

53. $p = 15$; $r = 15\sqrt{2}$; $q = 5\sqrt{6}$; $t = 10\sqrt{6}$

55. $\mathcal{A} = \frac{s^2}{2}$ **57.** C **59.** A **61.** D **63.** $\frac{\sqrt{3}}{3}$; $\sqrt{3}$

65. $\frac{\sqrt{3}}{2}$; $\frac{\sqrt{3}}{3}$; $\frac{2\sqrt{3}}{3}$ **67.** -1; -1 **69.** $-\frac{\sqrt{3}}{2}$; $-\frac{2\sqrt{3}}{3}$

In Exercises 71–87, we give, in order, sine, cosine, tangent, cotangent, secant, and cosecant.

71. $-\frac{\sqrt{3}}{2}$; $\frac{1}{2}$; $-\sqrt{3}$; $-\frac{\sqrt{3}}{3}$; 2; $-\frac{2\sqrt{3}}{3}$

73. $\frac{\sqrt{2}}{2}$; $\frac{\sqrt{2}}{2}$; 1; 1; $\sqrt{2}$; $\sqrt{2}$

75. $\frac{\sqrt{3}}{2}$; $-\frac{1}{2}$; $-\sqrt{3}$; $-\frac{\sqrt{3}}{3}$; -2; $\frac{2\sqrt{3}}{3}$

77. $-\frac{1}{2}$; $-\frac{\sqrt{3}}{2}$; $\frac{\sqrt{3}}{3}$; $\sqrt{3}$; $-\frac{2\sqrt{3}}{3}$; -2

79. $-\frac{\sqrt{2}}{2}$; $-\frac{\sqrt{2}}{2}$; 1; 1; $-\sqrt{2}$; $-\sqrt{2}$

81. $\frac{\sqrt{3}}{2}$; $\frac{1}{2}$; $\sqrt{3}$; $\frac{\sqrt{3}}{3}$; 2; $\frac{2\sqrt{3}}{3}$

83. $-\frac{1}{2}$; $-\frac{\sqrt{3}}{2}$; $\frac{\sqrt{3}}{3}$; $\sqrt{3}$; $-\frac{2\sqrt{3}}{3}$; -2

85. $\frac{1}{2}$; $-\frac{\sqrt{3}}{2}$; $-\frac{\sqrt{3}}{3}$; $-\sqrt{3}$; $-\frac{2\sqrt{3}}{3}$; 2

87. $-\frac{\sqrt{3}}{2}$; $\frac{1}{2}$; $-\sqrt{3}$; $-\frac{\sqrt{3}}{3}$; 2; $-\frac{2\sqrt{3}}{3}$ **89.** $-\frac{\sqrt{2}}{2}$

91. $-\frac{\sqrt{3}}{2}$ **93.** $-\sqrt{2}$ **95.** -1 **97.** 30°; 150°

99. 120°; 300° **101.** 45°; 315° **103.** 210°; 330°
105. 30°; 210° **107.** 225°; 315°

In Exercises 109–121, the number of decimal places may vary depending on the calculator used. We show six places.

109. 0.625243 **111.** 1.027349 **113.** 15.055723
115. 0.740805 **117.** 1.483014
119. tan 23.4° ≈ 0.432739 **121.** cot 77° ≈ 0.230868
123. 55.845496° **125.** 16.166641° **127.** 38.491580°
129. 68.673241° **131.** 45.526434° **133.** 12.227282°
135. 70 lb **137.** −2.9° **139.** 2500 lb **141.** A 2200-lb car on a 2° uphill grade has greater grade resistance.
143. 703 ft **145.** *R* would decrease; 644 ft, 1559 ft
147. A: 69 mph; B: 66 mph

Chapter 5 Quiz

[5.1] **1.** **(a)** 71° **(b)** 161° **2.** 65°; 115° **3.** 26°; 64°
4. **(a)** 77.2025° **(b)** 22° 01′ 30″ **5.** **(a)** 50°
(b) 300° **(c)** 170° **(d)** 417° **6.** 1800°

[5.2] **7.** $\sin\theta = \frac{7}{25}$; $\cos\theta = -\frac{24}{25}$; $\tan\theta = -\frac{7}{24}$; $\cot\theta = -\frac{24}{7}$; $\sec\theta = -\frac{25}{24}$; $\csc\theta = \frac{25}{7}$

[5.3] **8.** $\sin A = \frac{3}{5}$; $\cos A = \frac{4}{5}$; $\tan A = \frac{3}{4}$; $\cot A = \frac{4}{3}$; $\sec A = \frac{5}{4}$; $\csc A = \frac{5}{3}$

9.

θ	$\sin\theta$	$\cos\theta$	$\tan\theta$	$\cot\theta$	$\sec\theta$	$\csc\theta$
30°	$\frac{1}{2}$	$\frac{\sqrt{3}}{2}$	$\frac{\sqrt{3}}{3}$	$\sqrt{3}$	$\frac{2\sqrt{3}}{3}$	2
45°	$\frac{\sqrt{2}}{2}$	$\frac{\sqrt{2}}{2}$	1	1	$\sqrt{2}$	$\sqrt{2}$
60°	$\frac{\sqrt{3}}{2}$	$\frac{1}{2}$	$\sqrt{3}$	$\frac{\sqrt{3}}{3}$	2	$\frac{2\sqrt{3}}{3}$

10. $\mathcal{A} = 3x^2 \sin\theta$ **11.** $w = 18$; $x = 18\sqrt{3}$; $y = 18$; $z = 18\sqrt{2}$

In Exercises 12–14, we give, in order, sine, cosine, tangent, cotangent, secant, and cosecant.

12. $\frac{\sqrt{2}}{2}$; $-\frac{\sqrt{2}}{2}$; -1; -1; $-\sqrt{2}$; $\sqrt{2}$

13. $-\frac{1}{2}$; $-\frac{\sqrt{3}}{2}$; $\frac{\sqrt{3}}{3}$; $\sqrt{3}$; $-\frac{2\sqrt{3}}{3}$; -2

14. $-\frac{\sqrt{3}}{2}$; $\frac{1}{2}$; $-\sqrt{3}$; $-\frac{\sqrt{3}}{3}$; 2; $-\frac{2\sqrt{3}}{3}$

15. 60°; 120° **16.** 135°; 225° **17.** 0.673013
18. −1.181763 **19.** 69.497888° **20.** 24.777233°

5.4 Exercises

1. B **3.** A **5.** C **7.** C **9.** A **11.** E
13. 23.825 to 23.835

Note to student: While most of the measures resulting from solving triangles in this chapter are approximations, for convenience we use = rather than ≈.

15. $B = 53° 40'$; $a = 571$ m; $b = 777$ m
17. $M = 38.8°$; $n = 154$ m; $p = 198$ m
19. $A = 47.9108°$; $c = 84.816$ cm; $a = 62.942$ cm
21. $A = 37° 40'$; $B = 52° 20'$; $c = 20.5$ ft
23. $B = 62.0°$; $a = 8.17$ ft; $b = 15.4$ ft
25. $A = 17.0°$; $a = 39.1$ in.; $c = 134$ in.
27. $B = 29.0°$; $a = 70.7$ cm; $c = 80.9$ cm
29. $A = 36°$; $B = 54°$; $b = 18$ m
31. $c = 85.9$ yd; $A = 62° 50'$; $B = 27° 10'$
33. $b = 42.3$ cm; $A = 24° 10'$; $B = 65° 50'$
35. $B = 36° 36'$; $a = 310.8$ ft; $b = 230.8$ ft
37. $A = 50° 51'$; $a = 0.4832$ m; $b = 0.3934$ m
39. 9.35 m **41.** 128 ft **43.** 26.92 in. **45.** 28.0 m **47.** 13.3 ft **49.** 37° 35′ **51.** 42.18° **53.** 22° **55.** **(a)** 29,000 ft **(b)** shorter **57.** 270°; N 90° W, or S 90° W **59.** 0°; N 0° E, or N 0° W **61.** 315°; N 45° W **63.** 135°; S 45° E **65.** 220 mi **67.** 47 nautical mi **69.** 2203 ft **71.** 148 mi **73.** 430 mi **75.** 140 mi **77.** 114 ft **79.** 5.18 m **81.** 433 ft **83.** 10.8 ft **85.** 1.95 mi **87.** **(a)** $d = \frac{b}{2}\left(\cot\frac{\alpha}{2} + \cot\frac{\beta}{2}\right)$ **(b)** 345.4 cm **89.** **(a)** 320 ft **(b)** $R\left(1 - \cos\frac{\theta}{2}\right)$

Chapter 5 Review Exercises

1. complement: 55°; supplement: 145° **3.** 186° **5.** 9360° **7.** 119.134° **9.** 275° 06′ 02″

In Exercises 11–35, we give, in order, sine, cosine, tangent, cotangent, secant, and cosecant.

11. $-\frac{\sqrt{3}}{2}$; $\frac{1}{2}$; $-\sqrt{3}$; $-\frac{\sqrt{3}}{3}$; 2; $-\frac{2\sqrt{3}}{3}$

13. 0; −1; 0; undefined; −1; undefined

15. $-\frac{4}{5}$; $\frac{3}{5}$; $-\frac{4}{3}$; $-\frac{3}{4}$; $\frac{5}{3}$; $-\frac{5}{4}$

17. $\frac{15}{17}$; $-\frac{8}{17}$; $-\frac{15}{8}$; $-\frac{8}{15}$; $-\frac{17}{8}$; $\frac{17}{15}$

19. $\frac{\sqrt{2}}{2}$; $-\frac{\sqrt{2}}{2}$; −1; −1; $-\sqrt{2}$; $\sqrt{2}$

21. $\frac{5\sqrt{34}}{34}$; $\frac{3\sqrt{34}}{34}$; $\frac{5}{3}$; $\frac{3}{5}$; $\frac{\sqrt{34}}{3}$; $\frac{\sqrt{34}}{5}$

23. 0; −1; 0; undefined; −1; undefined

25. $-\frac{\sqrt{39}}{8}$; $-\frac{5}{8}$; $\frac{\sqrt{39}}{5}$; $\frac{5\sqrt{39}}{39}$; $-\frac{8}{5}$; $-\frac{8\sqrt{39}}{39}$

27. $\frac{2\sqrt{5}}{5}$; $-\frac{\sqrt{5}}{5}$; −2; $-\frac{1}{2}$; $-\sqrt{5}$; $\frac{\sqrt{5}}{2}$

29. $-\frac{3}{5}$; $\frac{4}{5}$; $-\frac{3}{4}$; $-\frac{4}{3}$; $\frac{5}{4}$; $-\frac{5}{3}$

31. $\frac{60}{61}$; $\frac{11}{61}$; $\frac{60}{11}$; $\frac{11}{60}$; $\frac{61}{11}$; $\frac{61}{60}$ **33.** $-\frac{\sqrt{3}}{2}$; $\frac{1}{2}$; $-\sqrt{3}$; $-\frac{\sqrt{3}}{3}$; 2; $-\frac{2\sqrt{3}}{3}$ **35.** $-\frac{1}{2}$; $\frac{\sqrt{3}}{2}$; $-\frac{\sqrt{3}}{3}$; $-\sqrt{3}$; $\frac{2\sqrt{3}}{3}$; −2

37. 120°; 240° **39.** 150°; 210° **41.** $-\frac{\sqrt{2}}{2}$; $-\frac{\sqrt{2}}{2}$; 1 **43.** −1.356342 **45.** 1.021034 **47.** 0.208344 **49.** 55.673870° **51.** 12.733938° **53.** 63.008286° **55.** 47°; 133° **57.** No, this will result in an angle having tangent equal to 25. The function $\tan^{-1}$ is not the reciprocal of the tangent (cotangent) but is, rather, the *inverse tangent function*. To find cot 25°, the student must find the *reciprocal* of tan 25°. **59.** $B = 31° 30'$; $a = 638$; $b = 391$ **61.** $B = 50.28°$; $a = 32.38$ m; $c = 50.66$ m **63.** 73.7 ft **65.** 18.75 cm **67.** 1200 m **69.** 140 mi

Chapter 5 Test

[5.1] **1.** 23°; 113° **2.** 145°; 35° **3.** 20°; 70° **4.** 74.31° **5.** 45° 12′ 09″ **6.** **(a)** 30° **(b)** 280° **(c)** 90° **7.** 2700°

[5.2] **8.** $\sin\theta = -\frac{7\sqrt{53}}{53}$; $\cos\theta = \frac{2\sqrt{53}}{53}$; $\tan\theta = -\frac{7}{2}$; $\cot\theta = -\frac{2}{7}$; $\sec\theta = \frac{\sqrt{53}}{2}$; $\csc\theta = -\frac{\sqrt{53}}{7}$

9. $\sin\theta = -1$; $\cos\theta = 0$; $\tan\theta$ is undefined; $\cot\theta = 0$; $\sec\theta$ is undefined; $\csc\theta = -1$

10.

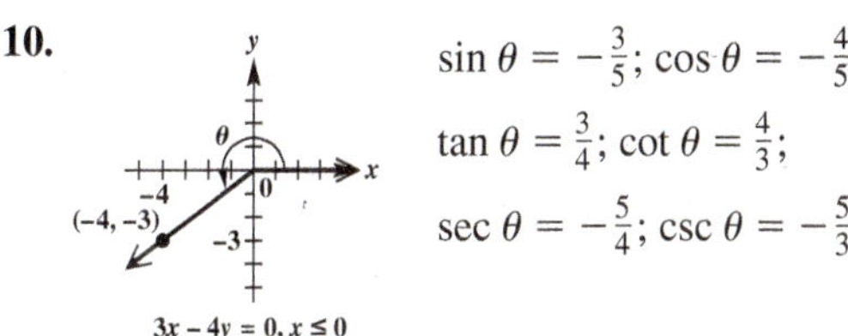

$\sin\theta = -\frac{3}{5}$; $\cos\theta = -\frac{4}{5}$; $\tan\theta = \frac{3}{4}$; $\cot\theta = \frac{4}{3}$; $\sec\theta = -\frac{5}{4}$; $\csc\theta = -\frac{5}{3}$

11. row 1: 1, 0, undefined, 0, undefined, 1; row 2: 0, 1, 0, undefined, 1, undefined; row 3: −1, 0, undefined, 0, undefined, −1 **12.** cosecant and cotangent **13.** **(a)** I **(b)** III, IV **(c)** III

14. $\cos\theta = -\frac{2\sqrt{10}}{7}$; $\tan\theta = -\frac{3\sqrt{10}}{20}$; $\cot\theta = -\frac{2\sqrt{10}}{3}$; $\sec\theta = -\frac{7\sqrt{10}}{20}$; $\csc\theta = \frac{7}{3}$

[5.3] **15.** $\sin A = \frac{12}{13}$; $\cos A = \frac{5}{13}$; $\tan A = \frac{12}{5}$; $\cot A = \frac{5}{12}$; $\sec A = \frac{13}{5}$; $\csc A = \frac{13}{12}$ **16.** $x = 4$; $y = 4\sqrt{3}$; $z = 4\sqrt{2}$; $w = 8$

In Exercises 17–19, we give, in order, sine, cosine, tangent, cotangent, secant, and cosecant.

17. $-\frac{\sqrt{3}}{2}$; $-\frac{1}{2}$; $\sqrt{3}$; $\frac{\sqrt{3}}{3}$; −2; $-\frac{2\sqrt{3}}{3}$

18. $-\frac{\sqrt{2}}{2}$; $-\frac{\sqrt{2}}{2}$; 1; 1; $-\sqrt{2}$; $-\sqrt{2}$

19. −1; 0; undefined; 0; undefined; −1 **20.** 135°; 225°

21. 240°; 300° **22.** 45°; 225° **23.** Take the reciprocal of $\tan\theta$ to obtain $\cot\theta = 0.59600119$. **24. (a)** 0.979399 **(b)** −1.905608 **(c)** 1.936213 **25.** 16.166641°

[5.4] **26.** $B = 31°\,30'$; $c = 877$; $b = 458$ **27.** 67.1°, or $67°\,10'$ **28.** 15.5 ft **29.** 8800 ft **30.** 72 nautical mi **31.** 92 km **32.** 448 m

Chapter 6 The Circular Functions and Their Graphs

6.1 Exercises

1. radius **3.** $\frac{\pi}{180}$ **5.** 2π **7.** 2 **9.** 1 **11.** $\frac{\pi}{3}$ **13.** $\frac{\pi}{2}$ **15.** $\frac{5\pi}{6}$ **17.** $-\frac{5\pi}{3}$ **19.** $\frac{5\pi}{2}$ **21.** 10π **23.** 0 **25.** -5π **27.** 60° **29.** 315° **31.** 330° **33.** −30° **35.** 126° **37.** −48° **39.** 153° **41.** −900° **43.** 0.681 **45.** 0.742 **47.** 2.429 **49.** 1.122 **51.** 0.985 **53.** −0.832 **55.** $114°\,35'$ **57.** $99°\,42'$ **59.** $19°\,35'$ **61.** $-287°\,06'$ **63.** In the expression "sin 30," 30 means 30 radians; $\sin 30° = \frac{1}{2}$, while $\sin 30 \approx -0.9880$. **65.** We begin the answers with the blank next to 30°, and then proceed counterclockwise from there: $\frac{\pi}{6}$; 45; $\frac{\pi}{3}$; 120; 135; $\frac{5\pi}{6}$; π; $\frac{7\pi}{6}$; $\frac{5\pi}{4}$; 240; 300; $\frac{7\pi}{4}$; $\frac{11\pi}{6}$. **67.** 25.8 cm **69.** 3.61 ft **71.** 5.05 m **73.** 55.3 in. **75.** 3500 km **77.** 5900 km **79.** 44° N **81.** 156° **83.** 38.5° **85.** 18.7 cm **87. (a)** 11.6 in. **(b)** $37°\,05'$ **89.** 146 in. **91.** 3π in. **93.** 27π in. **95.** 0.20 km **97.** 840 ft **99.** 1116.1 m^2 **101.** 706.9 ft^2 **103.** 114.0 cm^2 **105.** 1885.0 mi^2 **107.** 3.6 **109.** 8060 yd^2 **111.** 20 in. **113. (a)** $13\frac{1}{3}°$; $\frac{2\pi}{27}$ **(b)** 478 ft **(c)** 17.7 ft **(d)** 672 ft^2 **115. (a)** 140 ft **(b)** 102 ft **(c)** 622 ft^2 **117.** 1900 yd^2

6.2 Exercises

1. Counterclockwise from 0 radians, the coordinates are $(1, 0)$, $\left(\frac{\sqrt{3}}{2}, \frac{1}{2}\right)$, $\left(\frac{\sqrt{2}}{2}, \frac{\sqrt{2}}{2}\right)$, $\left(\frac{1}{2}, \frac{\sqrt{3}}{2}\right)$, and $(0, 1)$. **3.** $\frac{\sqrt{2}}{2}$ **5.** 1 **7.** linear speed (or linear velocity) **9.** 2π **11.** 2π **13. (a)** 1 **(b)** 0 **(c)** undefined **15. (a)** 0 **(b)** 1 **(c)** 0 **17. (a)** 0 **(b)** −1 **(c)** 0 **19.** $-\frac{1}{2}$ **21.** −1 **23.** −2 **25.** $-\frac{1}{2}$ **27.** $\frac{\sqrt{2}}{2}$ **29.** $\frac{\sqrt{3}}{2}$ **31.** $\frac{2\sqrt{3}}{3}$ **33.** $-\frac{\sqrt{3}}{3}$ **35.** 0.5736 **37.** 0.4068 **39.** 1.2065 **41.** 14.3338 **43.** −1.0460 **45.** −3.8665 **47.** 0.7 **49.** 0.9 **51.** −0.6 **53.** 2.3 or 4.0 **55.** 0.8 or 2.4 **57.** negative **59.** negative **61.** positive **63.** 0.2095 **65.** 1.4426 **67.** 0.3887 **69.** $\frac{5\pi}{6}$ **71.** $\frac{4\pi}{3}$ **73.** $\frac{7\pi}{4}$ **75.** $\frac{4\pi}{3}, \frac{5\pi}{3}$ **77.** $\frac{\pi}{4}, \frac{3\pi}{4}, \frac{5\pi}{4}, \frac{7\pi}{4}$ **79.** $-\frac{11\pi}{6}, -\frac{7\pi}{6}, -\frac{5\pi}{6}, -\frac{\pi}{6}, \frac{\pi}{6}, \frac{5\pi}{6}$ **81. (a)** $\frac{1}{2}$ **(b)** $\frac{\sqrt{3}}{2}$ **(c)** $\sqrt{3}$ **(d)** 2 **(e)** $\frac{2\sqrt{3}}{3}$ **(f)** $\frac{\sqrt{3}}{3}$ **83. (a)** $\frac{\pi}{2}$ radians **(b)** 10π cm **(c)** $\frac{5\pi}{3}$ cm per sec **85. (a)** 3π radians **(b)** 24π in. **(c)** $\frac{8\pi}{3}$ in. per min **87.** 2π radians **89.** 7.4 radians **91.** $\frac{3\pi}{32}$ radian per sec **93.** 0.1803 radian per sec **95.** $\frac{6}{5}$ min **97.** 8π m per sec **99.** $\frac{9}{5}$ radians per sec **101.** $\frac{8}{\pi}$ m **103.** 18π cm **105.** 12 sec **107.** $\frac{3\pi}{32}$ radian per sec **109.** $\frac{\pi}{6}$ radian per hr **111.** $\frac{\pi}{30}$ radian per min **113.** $\frac{7\pi}{30}$ cm per min **115.** 168π m per min **117.** 1500π m per min **119.** 16.6 mph **121. (a)** $\frac{2\pi}{365}$ radian **(b)** $\frac{\pi}{4380}$ radian per hr **(c)** 67,000 mph **123. (a)** 3.1 cm per sec **(b)** 0.24 radian per sec **125.** 3.73 cm **127.** 523.6 radians per sec

6.3 Exercises

1. 1; 2π **3.** $n\pi$ **5.** $\frac{\pi}{2}$ **7.** E **9.** B **11.** F

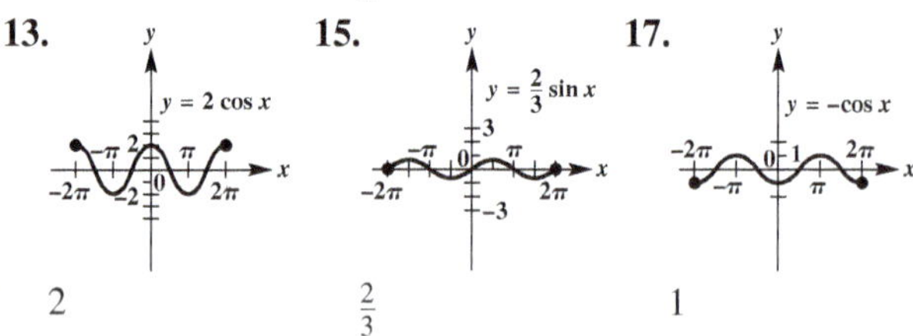

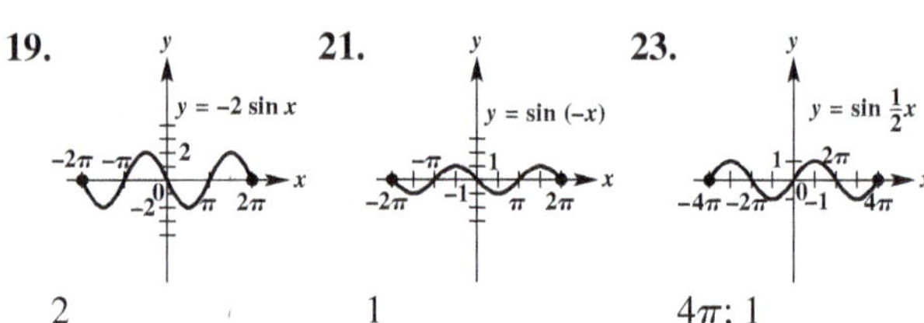

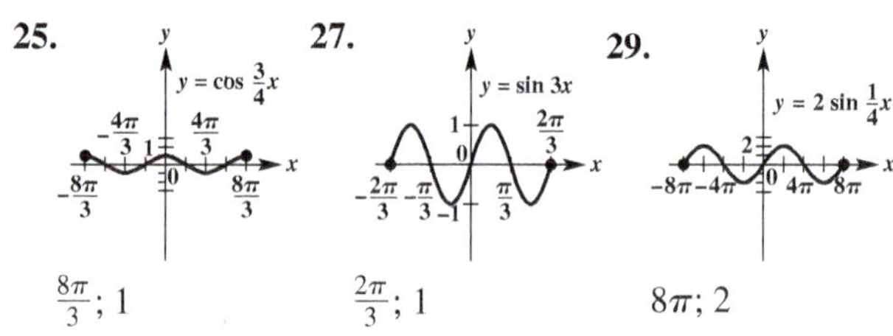

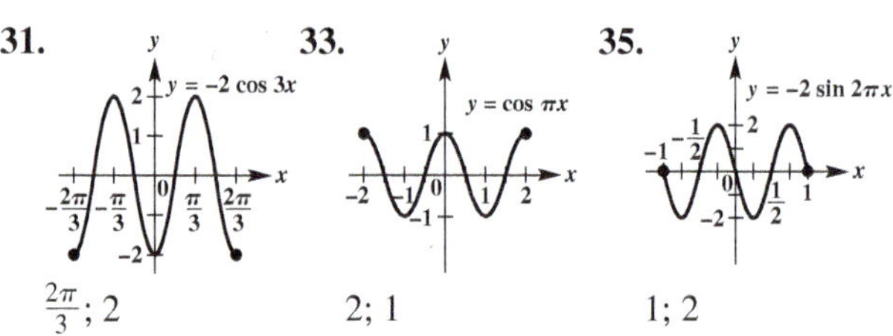

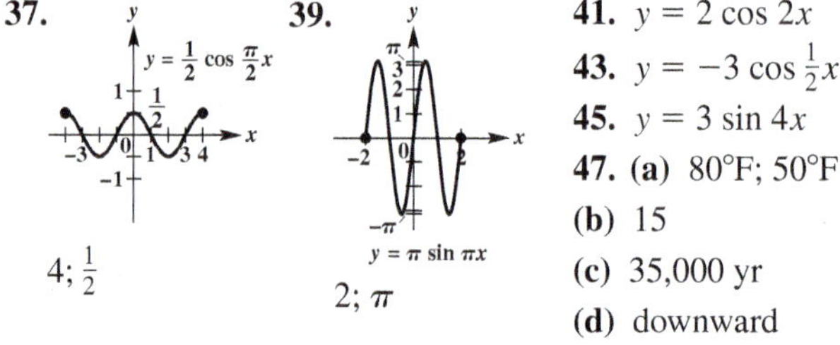

41. $y = 2\cos 2x$ **43.** $y = -3\cos\frac{1}{2}x$ **45.** $y = 3\sin 4x$ **47. (a)** 80°F; 50°F **(b)** 15 **(c)** 35,000 yr **(d)** downward **49.** 24 hr **51.** 6:00 P.M.; 0.2 ft **53.** 3:18 A.M.; 2.4 ft

55. (a) $L(x) = 0.022x^2 + 0.55x + 316 + 3.5 \sin 2\pi x$

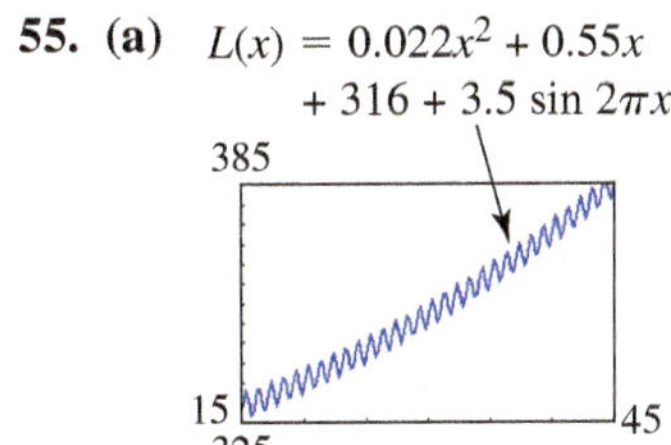

(b) maxima: $x = \frac{1}{4}, \frac{5}{4}, \frac{9}{4}, \ldots$; minima: $x = \frac{3}{4}, \frac{7}{4}, \frac{11}{4}, \ldots$ **(c)** The quadratic function provides the general increasing nature of the level, while the sine function provides the fluctuations as the years go by.

57. (a) 31°F **(b)** 38°F **(c)** 57°F **(d)** 58°F **(e)** 37°F **(f)** 16°F **59.** 1; 240°, or $\frac{4\pi}{3}$ **61.** No. For $b > 0$, $b \neq 1$, the graph of $y = \sin bx$ has amplitude 1 and period $\frac{2\pi}{b}$, while that of $y = b \sin x$ has amplitude b and period 2π.

63. X = −0.4161468, Y = 0.90929743; X is cos 2 and Y is sin 2. **64.** X = 2, Y = 0.90929743; sin 2 = 0.90929743 **65.** X = 2, Y = −0.4161468; cos 2 = −0.4161468 **66.** For an arc length T on the unit circle, X = cos T and Y = sin T.

6.4 Exercises

1. $\frac{\pi}{4}$; left **3.** 4 **5.** 6; up **7.** $\frac{\pi}{5}$; left; 5; 3; up **9.** D **11.** H **13.** B **15.** I **17.** The graph of $y = \sin x + 1$ is obtained by shifting the graph of $y = \sin x$ up 1 unit. The graph of $y = \sin(x + 1)$ is obtained by shifting the graph of $y = \sin x$ to the left 1 unit. **19.** B **21.** C **23.** right **25.** $y = -1 + \sin x$ **27.** $y = \cos\left(x - \frac{\pi}{3}\right)$ **29.** 2; 2π; none; π to the left **31.** $\frac{1}{4}$; 4π; none; π to the left **33.** 3; 4; none; $\frac{1}{2}$ to the right **35.** 1; $\frac{2\pi}{3}$; up 2; $\frac{\pi}{15}$ to the right

37.

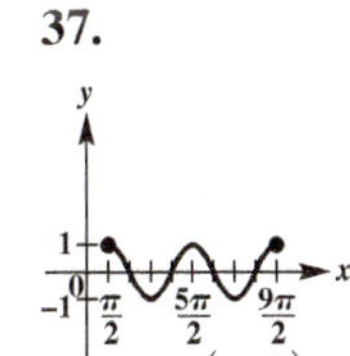

39.

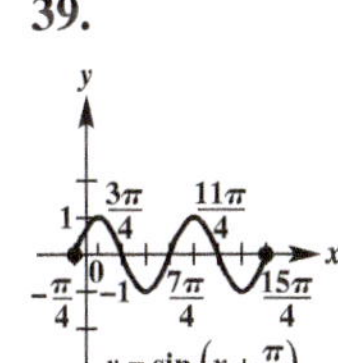

41.

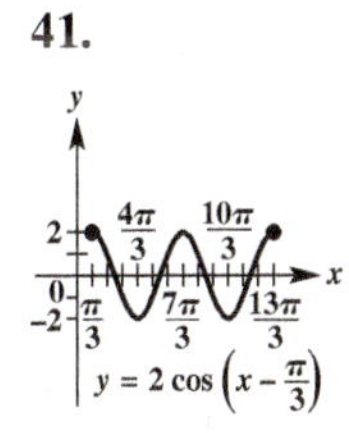

43.

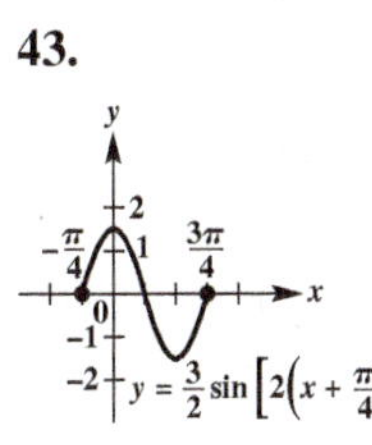

45.

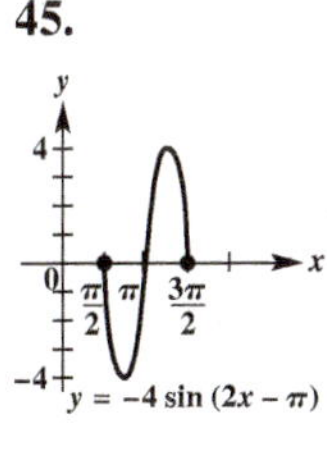

47.

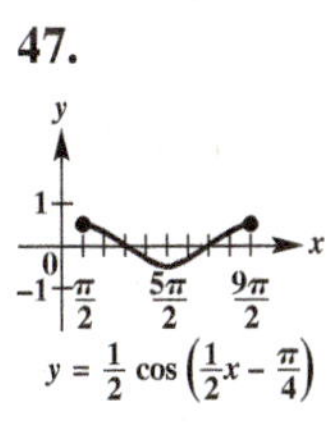

49.

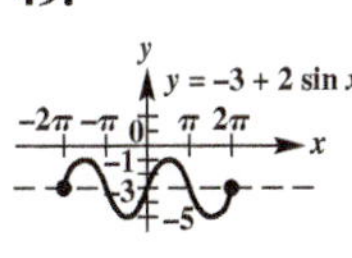

51.

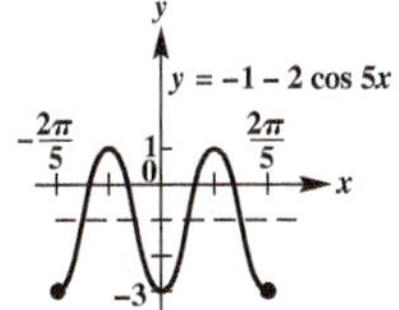

53.

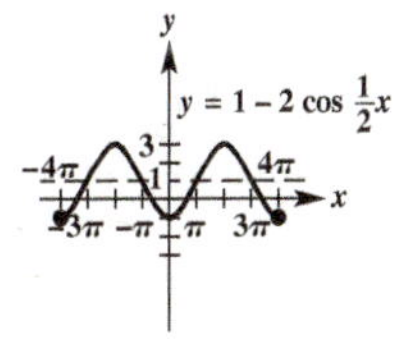

55.

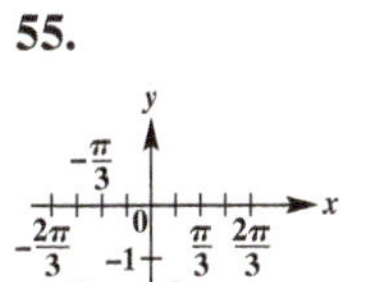

57.

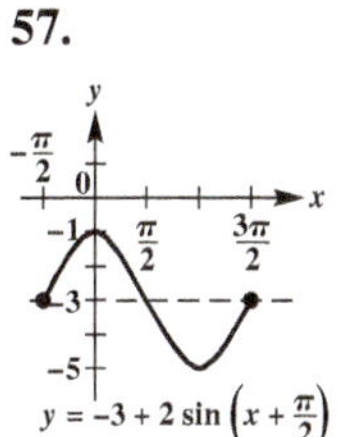

59.

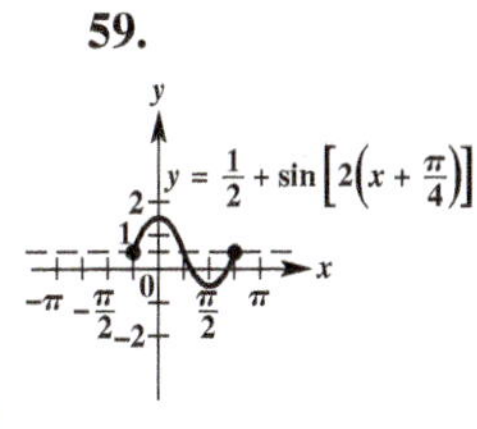

61. (a)

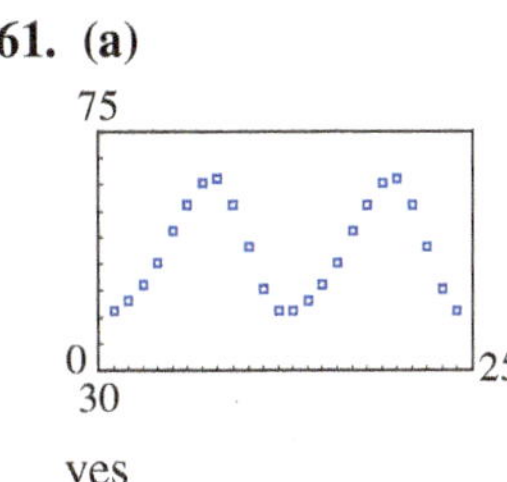

yes

(b)

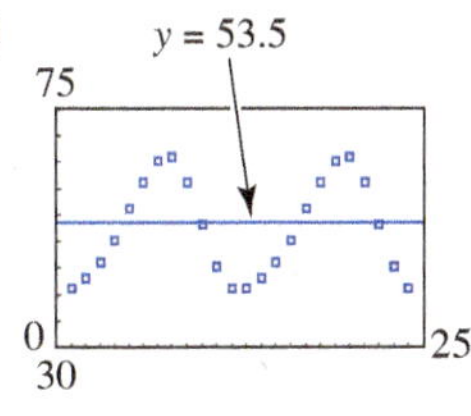

It represents the average annual temperature.

(c) 12.5; 12; 4.5 **(d)** $f(x) = 12.5 \sin\left[\frac{\pi}{6}(x - 4.5)\right] + 53.5$

(e)

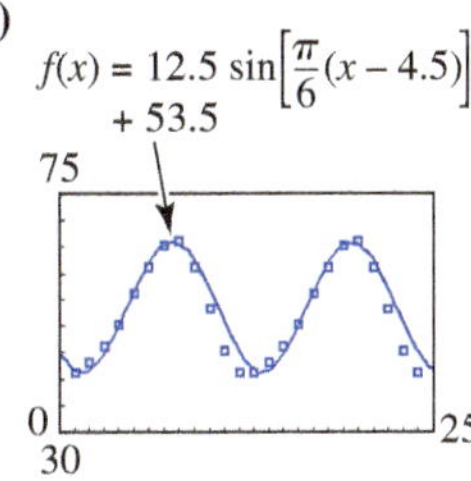

The function gives a good model for the given data.

(f)

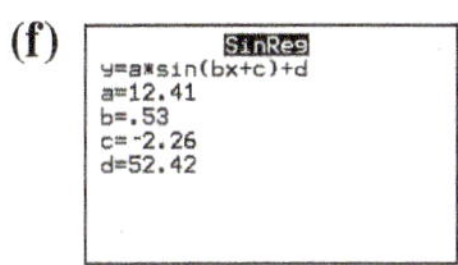

TI-84 Plus fixed to the nearest hundredth

In the answers to Exercises 63 and 65, we give the model and one graph of the data and equation.

63. (a) See the graph in part (c).
(b) $y = 12.28 \sin(0.52x + 1.06) + 63.96$
(c)

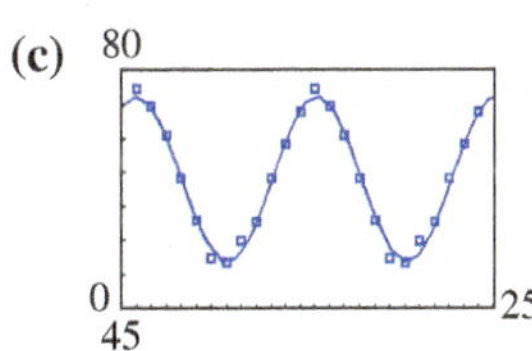

65. (a) See the graph in part (c).
(b) $y = 0.49 \sin(0.21x + 0.41) + 0.52$
(c)

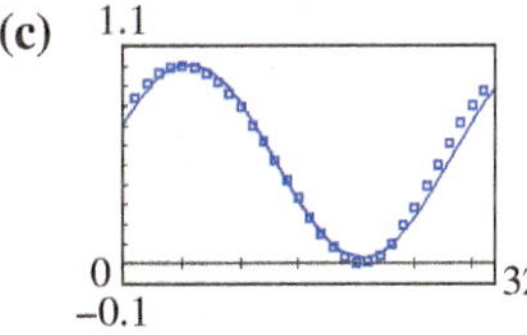

Chapter 6 Quiz

[6.1] **1.** $\frac{5\pi}{4}$ **2.** −210° **3.** 1.5 **4.** 67,500 in.^{2}

[6.2] **5.** $\frac{\sqrt{2}}{2}$ **6.** $-\frac{1}{2}$ **7.** 0 **8.** $\frac{2\pi}{3}$

[6.3, 6.4] **9.** 4; π; 3 up; $\frac{\pi}{4}$ to the left

10.

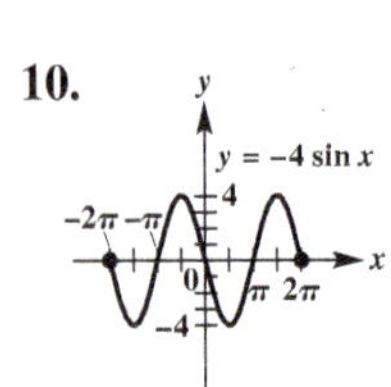

2π; 4

11.

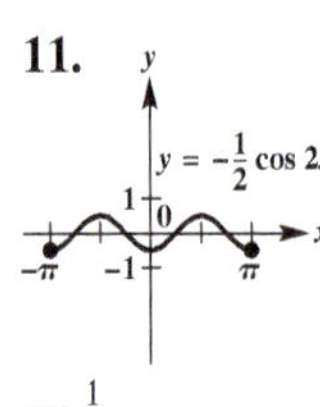

π; $\frac{1}{2}$

12.

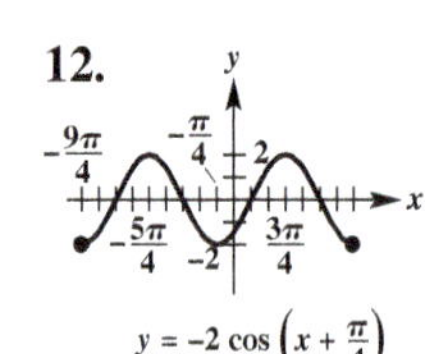

2π; 2

13.

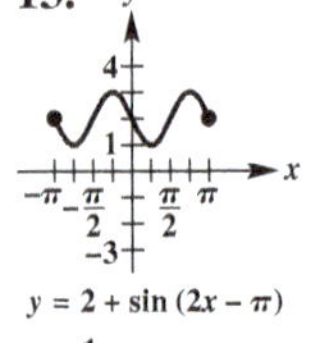

π; 1

14. $y = 2 \sin x$ **15.** $y = \cos 2x$

6.5 Exercises

1. π **3.** increases **5.** $-\frac{\pi}{2}$ **7.** C **9.** B **11.** F

13. $y = \tan 4x$

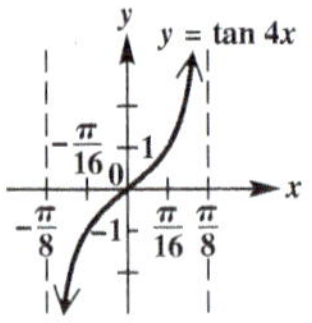

15. $y = 2 \tan x$

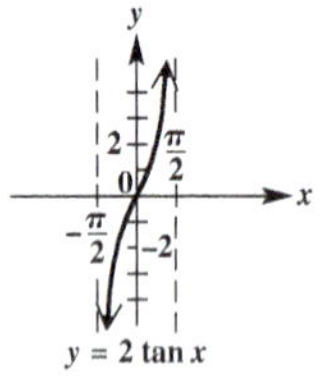

17. $y = 2 \tan \frac{1}{4}x$

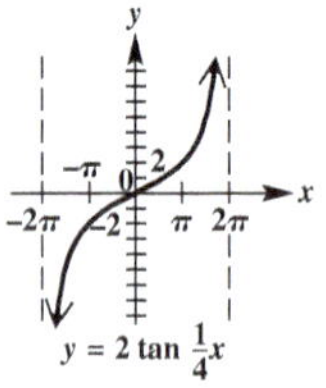

19.

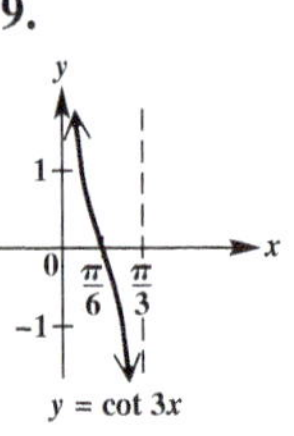

21. $y = -2 \tan \frac{1}{4}x$

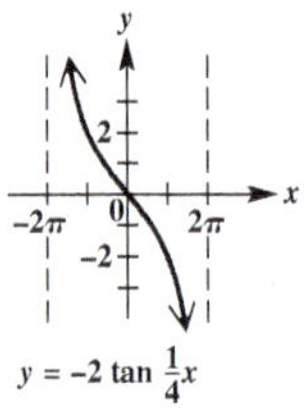

23. $y = \frac{1}{2}\cot 4x$

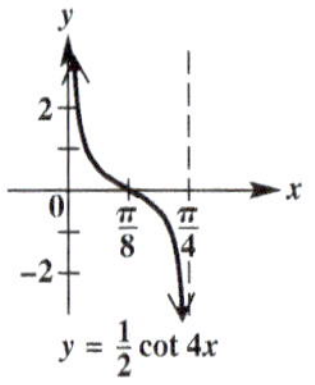

25. $y = \tan(2x - \pi)$

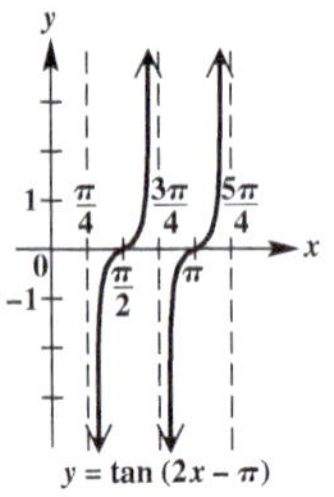

27. $y = \cot\left(3x + \frac{\pi}{4}\right)$

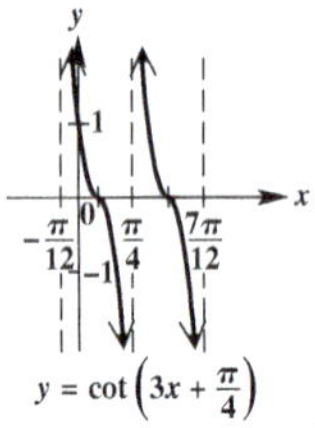

29. $y = 1 + \tan x$

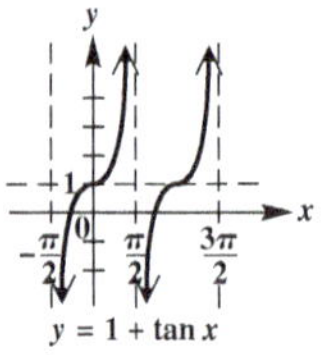

31. $y = 1 - \cot x$

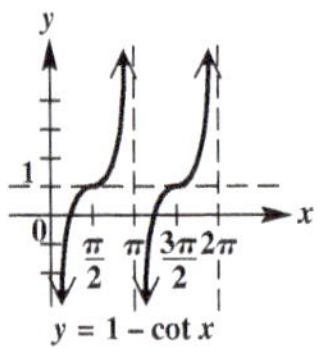

33. $y = -1 + 2\tan x$

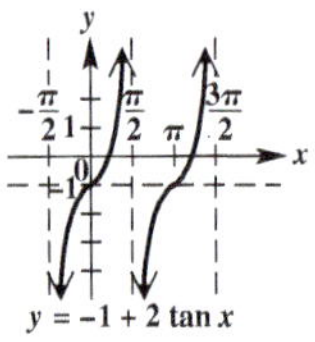

35. $y = -1 + \frac{1}{2}\cot(2x - 3\pi)$

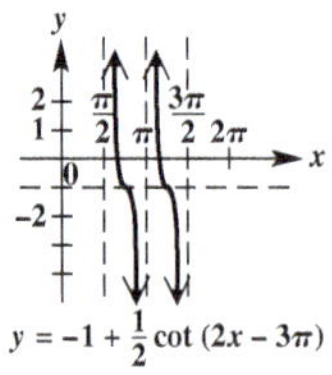

37. $y = 1 - 2\cot\left[2\left(x + \frac{\pi}{2}\right)\right]$

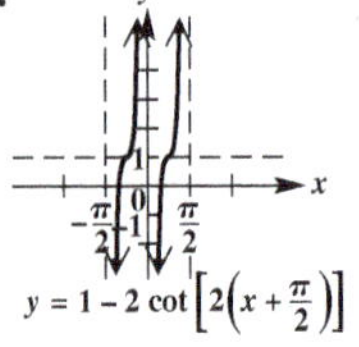

39. $y = -2 \tan x$ **41.** $y = \cot 3x$ **43.** $y = 1 + \tan \frac{1}{2}x$ **45.** true **47.** false; $\tan(-x) = -\tan x$ for all x in the domain. **49.** four **53.** 0 m **55.** 12.3 m **57.** π **58.** $\frac{5\pi}{4}$ **59.** $x = \frac{5\pi}{4} + n\pi$ **60.** (0.32, 0) **61.** (3.46, 0) **62.** $\{x \mid x = 0.32 + n\pi\}$

6.6 Exercises

1. A **3.** D **5.** C **7.** B **9.** D

11.

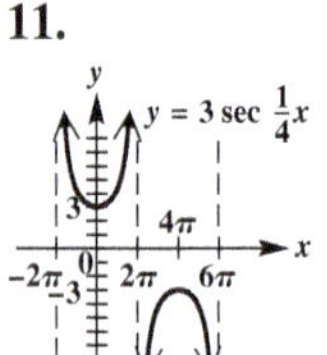

13. $y = -\frac{1}{2}\csc\left(x + \frac{\pi}{2}\right)$

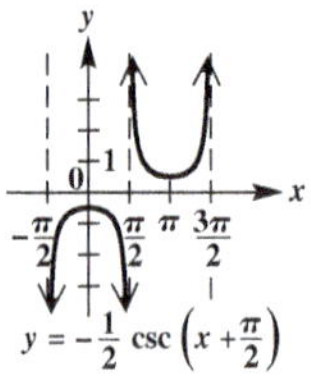

15.

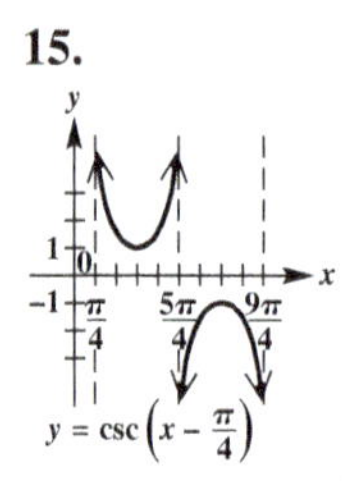

17. $y = \sec\left(x + \frac{\pi}{4}\right)$

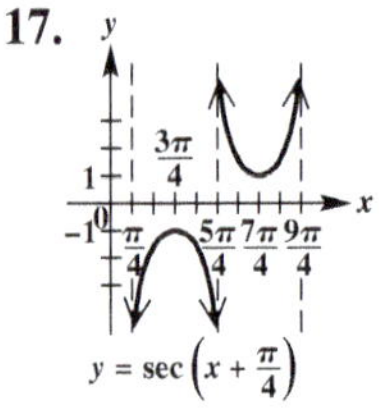

19. $y = \csc\left(\frac{1}{2}x - \frac{\pi}{4}\right)$

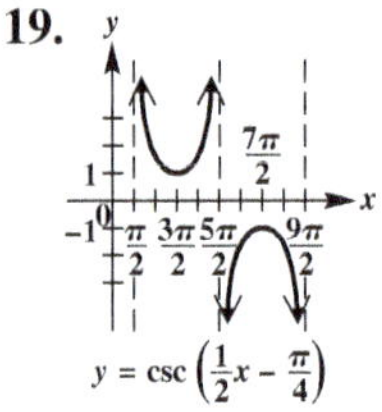

21. $y = 2 + 3\sec(2x - \pi)$

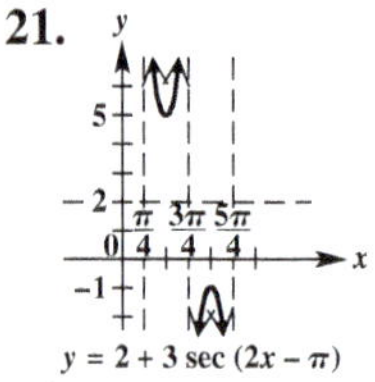

23. $y = 1 - \frac{1}{2}\csc\left(x - \frac{3\pi}{4}\right)$

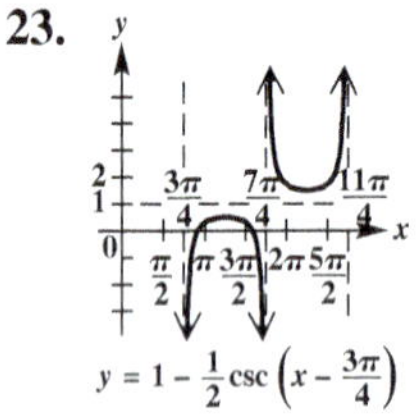

25. $y = \sec 4x$ **27.** $y = -2 + \csc x$ **29.** $y = -1 - \sec x$ **31.** true **33.** true **35.** none **39.** 4 m **41.** 63.7 m **43.** The value is 1.3660254 in both cases. **45.** The value is 2.4142136 in both cases.

Summary Exercises on Graphing Circular Functions

1.

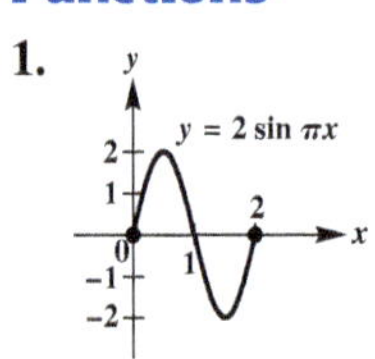

2.

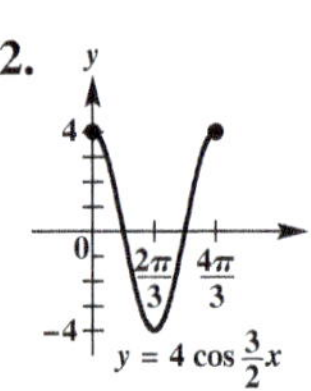

3.

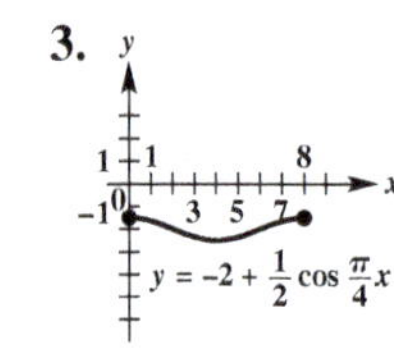

4.

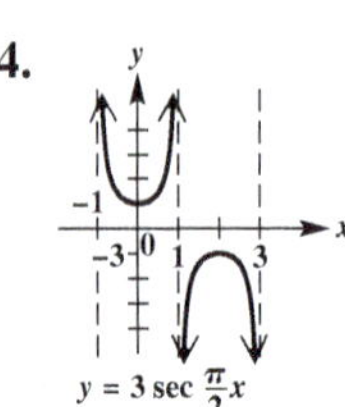

5.

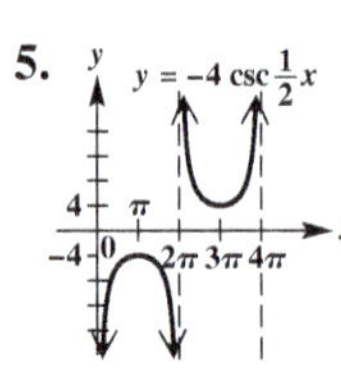

6.

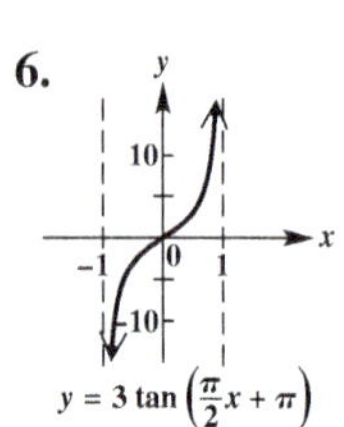

7.

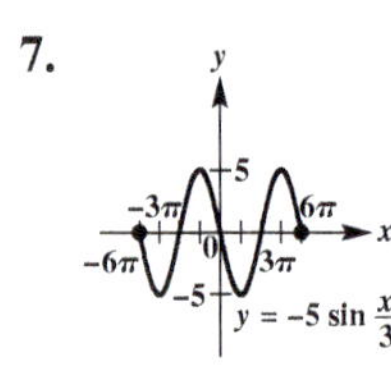

8.

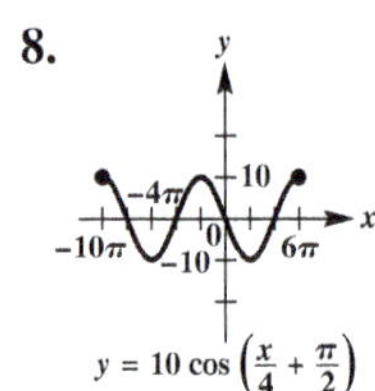

9.

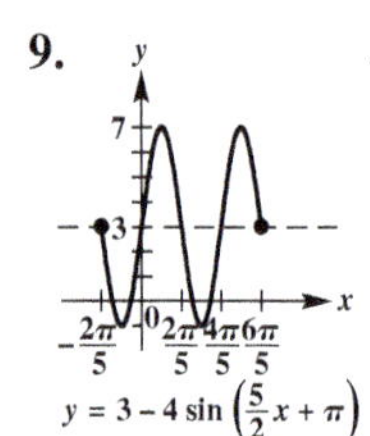

10.

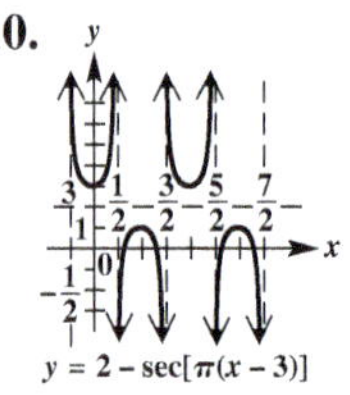

6.7 Exercises

1. 5 **3.** $\frac{1}{\pi}$ oscillation per sec **5.** -5

7. (a) $s(t) = -4 \cos \frac{2\pi}{3}t$ **(b)** 3.46 units **(c)** $\frac{1}{3}$ oscillation per sec **9. (a)** 5; $\frac{1}{60}$ **(b)** 60 oscillations per sec **(c)** 5; 1.545; -4.045; -4.045; 1.545

(d)

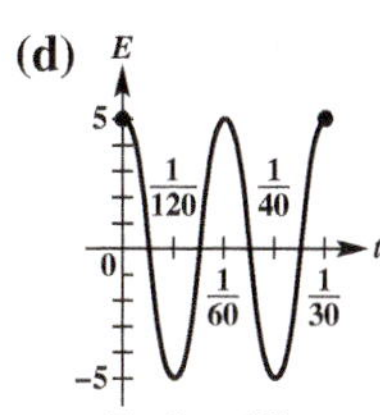

11. (a) $s(t) = 2 \sin 2t$; amplitude: 2; period: π; frequency: $\frac{1}{\pi}$ rotation per sec **(b)** $s(t) = 2 \sin 4t$; amplitude: 2; period: $\frac{\pi}{2}$; frequency: $\frac{2}{\pi}$ rotation per sec **13.** period: $\frac{\pi}{4}$; frequency: $\frac{4}{\pi}$ oscillations per sec **15.** $\frac{1}{\pi^2}$ **17. (a)** 5 in. **(b)** 2 cycles per sec; $\frac{1}{2}$ sec **(c)** after $\frac{1}{4}$ sec **(d)** 4.0; After 1.3 sec, the weight is about 4 in. above the equilibrium position.

19. (a) $s(t) = -3 \cos 12t$ **(b)** $\frac{\pi}{6}$ sec

21. (a) $s(t) = 2 \cos 4\pi t$ **(b)** $s(1) = 2$; The weight is moving neither upward nor downward. At $t = 1$, the motion of the weight is changing from up to down.

23. (a) $s(t) = -3 \cos 2.5\pi t$ **(b)** $s(1) = 0$; upward

25. $s(t) = 0.21 \cos 55\pi t$ **27.** $s(t) = 0.14 \cos 110\pi t$

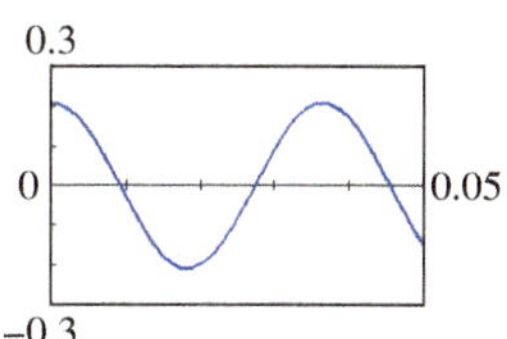

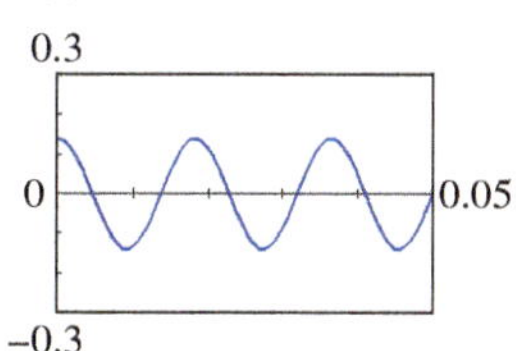

29. 11 in. **31.** 1, 3, 5, 7, 9, 11

33. (a)

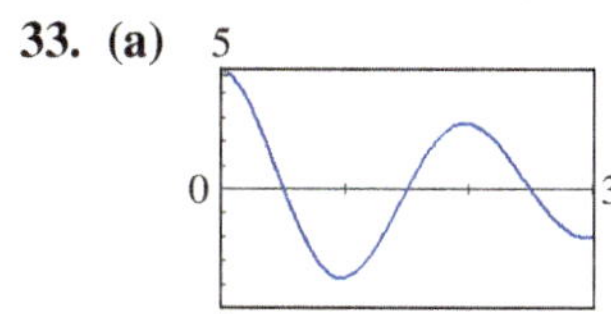

(b) $y_1 = 5e^{-0.3x}$ **(c)** 0, 2

Chapter 6 Review Exercises

1. A central angle of a circle that intercepts an arc of length 2 times the radius of the circle has a measure of 2 radians.

3. Three of many possible answers are $1 + 2\pi$, $1 + 4\pi$, and $1 + 6\pi$. **5.** $\frac{\pi}{4}$ **7.** $\frac{35\pi}{36}$ **9.** $\frac{40\pi}{9}$ **11.** 225° **13.** 480°

15. $-110°$ **17.** π in. **19.** 12π in. **21.** 35.8 cm

23. 49.06° **25.** 273 m^2 **27.** 4500 km **29.** $\frac{3}{4}$; 1.5 sq units

31. $\sqrt{3}$ **33.** $-\frac{1}{2}$ **35.** 2 **37.** 0.8660 **39.** 0.9703

41. 1.9513 **43.** 0.3898 **45.** 0.5148 **47.** 1.1054

49. $\frac{\pi}{4}$ **51.** $\frac{7\pi}{6}$ **53. (a)** 20π radians **(b)** 300π cm **(c)** 10π cm per sec **55.** 1260π cm per sec **57.** 5 in.

59. B **61.** 2; 2π; none; none **63.** $\frac{1}{2}$; $\frac{2\pi}{3}$; none; none

65. 2; 8π; 1 up; none **67.** 3; 2π; none; $\frac{\pi}{2}$ to the left

69. not applicable; π; none; $\frac{\pi}{8}$ to the right **71.** not applicable; $\frac{\pi}{3}$; none; $\frac{\pi}{9}$ to the right **73.** tangent **75.** cosine

77. cotangent

79.

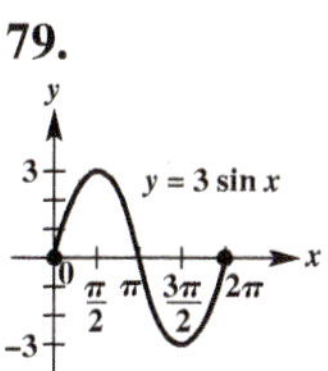

81.

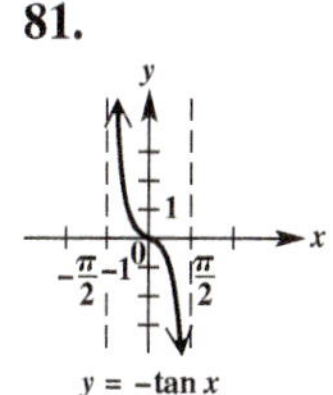

83.

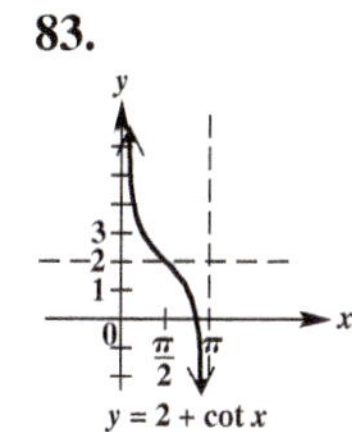

85.

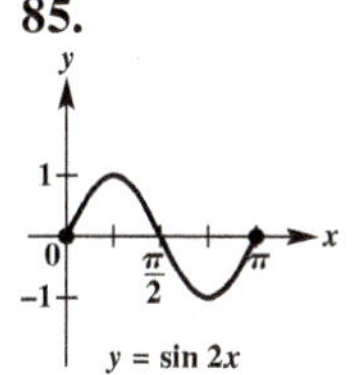

87.

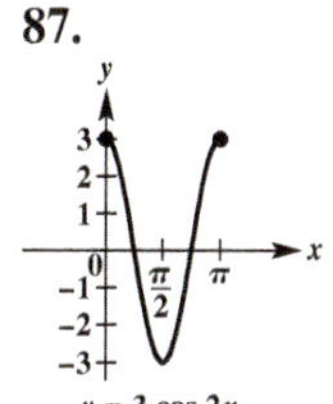

89.

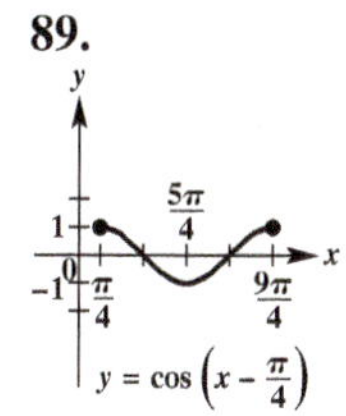

91.

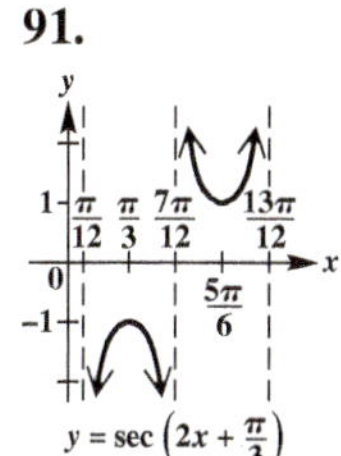

93.

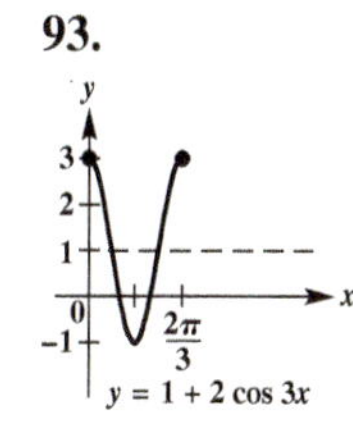

95.

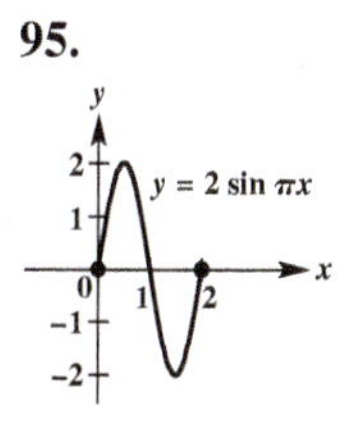

97. (a) See the graph in part (c).

(b) $y = 8.02 \sin(0.52x + 0.84) + 59.83$

(c)

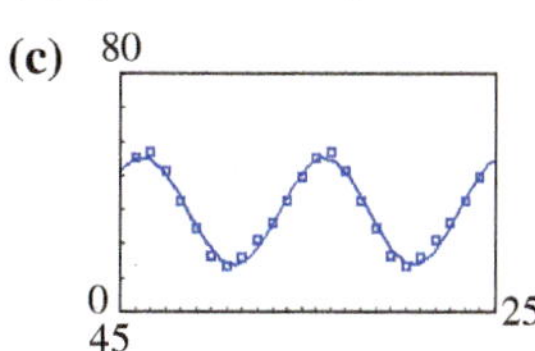

99. $y = 1 - \sin x$ **101.** $y = 2 \tan \frac{1}{2}x$

103. (b)

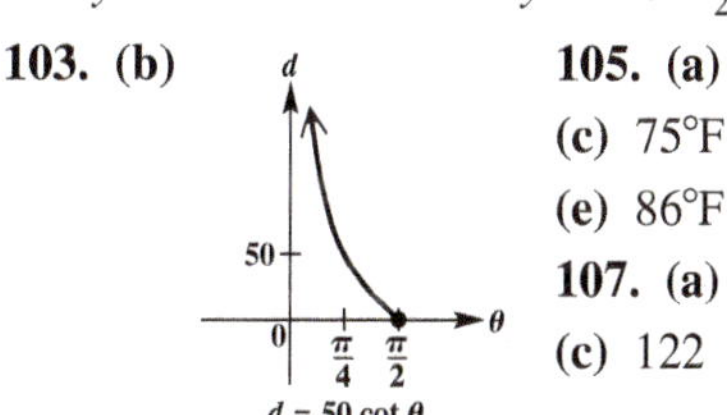

105. (a) 30°F **(b)** 60°F **(c)** 75°F **(d)** 86°F **(e)** 86°F **(f)** 60°F

107. (a) 100 **(b)** 258 **(c)** 122 **(d)** 296

109. amplitude: 4; period: 2; frequency: $\frac{1}{2}$ cycle per sec

111. The frequency is the number of cycles in one unit of time; -4; 0; $-2\sqrt{2}$

Chapter 6 Test

[6.1] **1.** $\frac{2\pi}{3}$ **2.** $-\frac{\pi}{4}$ **3.** 0.087 **4.** 135° **5.** $-210°$

6. 229° 11′ **7. (a)** $\frac{4}{3}$ **(b)** 15,000 cm^2 **8.** 2 radians

[6.2] **9.** $\frac{\sqrt{2}}{2}$ **10.** $-\frac{\sqrt{3}}{2}$ **11.** undefined **12.** -2 **13.** 0

14. 0 **15.** $\sin \frac{7\pi}{6} = -\frac{1}{2}$; $\cos \frac{7\pi}{6} = -\frac{\sqrt{3}}{2}$; $\tan \frac{7\pi}{6} = \frac{\sqrt{3}}{3}$; $\csc \frac{7\pi}{6} = -2$; $\sec \frac{7\pi}{6} = -\frac{2\sqrt{3}}{3}$; $\cot \frac{7\pi}{6} = \sqrt{3}$

16. (a) 0.9716 **(b)** $\frac{\pi}{3}$ **17. (a)** $\frac{2\pi}{3}$ radians **(b)** 40π cm **(c)** 5π cm per sec **18. (a)** 75 ft **(b)** $\frac{\pi}{45}$ radian per sec

[6.3–6.6] **19. (a)** $y = \sec x$ **(b)** $y = \sin x$ **(c)** $y = \cos x$ **(d)** $y = \tan x$ **(e)** $y = \csc x$ **(f)** $y = \cot x$

20. **(a)** $y = 1 + \cos \frac{1}{2}x$ **(b)** $y = -\frac{1}{2}\cot x$

[6.3, 6.5, 6.6] **21.** **(a)** $(-\infty, \infty)$ **(b)** $[-1, 1]$ **(c)** $\frac{\pi}{2}$ **(d)** $(-\infty, -1] \cup [1, \infty)$ [6.4] **22.** **(a)** π **(b)** 6 **(c)** $[-3, 9]$ **(d)** $(0, -3)$ **(e)** $\frac{\pi}{4}$ to the left $\left(\text{that is, } -\frac{\pi}{4}\right)$

23.

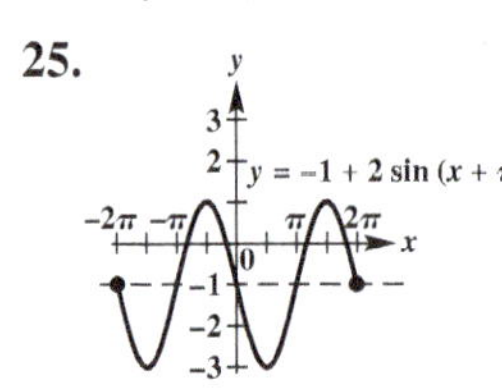

24.

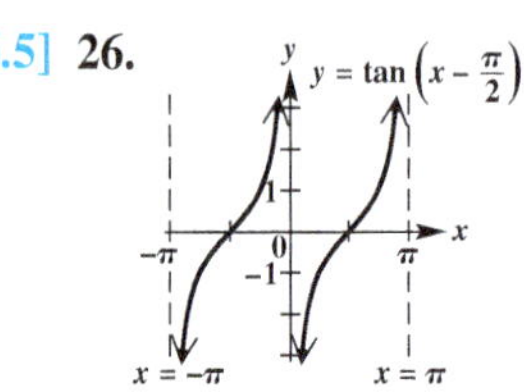

25.

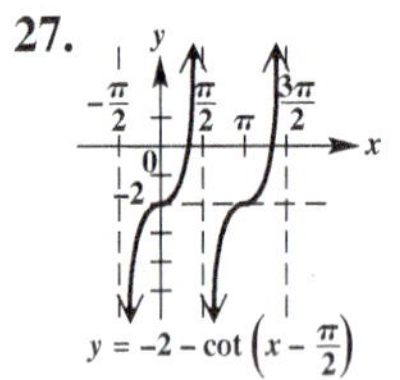

[6.5] **26.**

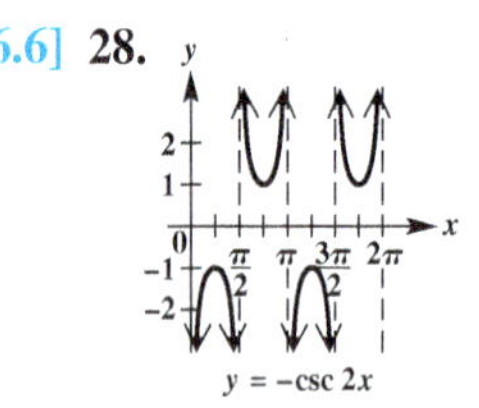

27. $y = -2 - \cot\left(x - \frac{\pi}{2}\right)$

[6.6] **28.** $y = -\csc 2x$

[6.3, 6.4]

29. **(a)**

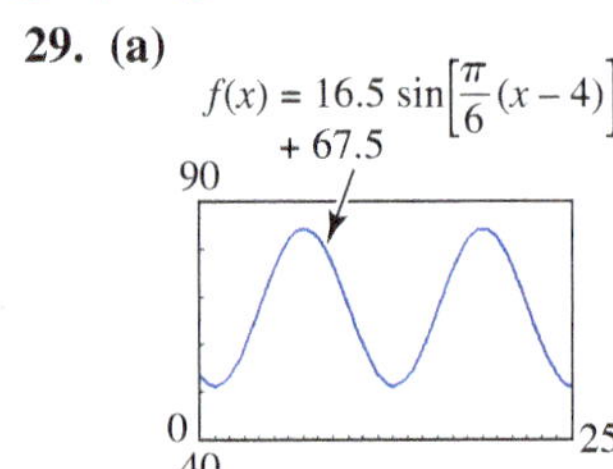

(b) 16.5; 12; 4 to the right; 67.5 up **(c)** 53°F **(d)** 51°F in January; 84°F in July **(e)** 67.5°F; This is the vertical translation.

[6.7] **30.** **(a)** 4 in. **(b)** after $\frac{1}{8}$ sec **(c)** 4 cycles per sec; $\frac{1}{4}$ sec

Chapter 7 Trigonometric Identities and Equations

7.1 Exercises

1. B **3.** E **5.** A **7.** −0.65 **9.** −0.75 **11.** $\frac{\sqrt{7}}{4}$ **13.** $-\frac{5\sqrt{26}}{26}$ **15.** $-\frac{2\sqrt{5}}{5}$ **17.** $-\frac{\sqrt{15}}{5}$ **19.** $-\frac{\sqrt{105}}{11}$ **21.** $-\frac{4}{9}$ **23.** $\sin\theta$ is the reciprocal of $\csc\theta$ and therefore has the same sign. **25.** $f(-x) = \frac{-\sin x}{-x} = \frac{\sin x}{x} = f(x)$; even **27.** $f(x) = \sec x$; even **29.** $f(x) = \cot x$; odd **31.** $\cos\theta = -\frac{\sqrt{5}}{3}$; $\tan\theta = -\frac{2\sqrt{5}}{5}$; $\cot\theta = -\frac{\sqrt{5}}{2}$; $\sec\theta = -\frac{3\sqrt{5}}{5}$; $\csc\theta = \frac{3}{2}$ **33.** $\sin\theta = -\frac{\sqrt{17}}{17}$; $\cos\theta = \frac{4\sqrt{17}}{17}$; $\cot\theta = -4$; $\sec\theta = \frac{\sqrt{17}}{4}$; $\csc\theta = -\sqrt{17}$ **35.** $\sin\theta = \frac{3}{5}$; $\cos\theta = \frac{4}{5}$; $\tan\theta = \frac{3}{4}$; $\sec\theta = \frac{5}{4}$; $\csc\theta = \frac{5}{3}$ **37.** $\sin\theta = -\frac{\sqrt{7}}{4}$; $\cos\theta = \frac{3}{4}$; $\tan\theta = -\frac{\sqrt{7}}{3}$; $\cot\theta = -\frac{3\sqrt{7}}{7}$; $\csc\theta = -\frac{4\sqrt{7}}{7}$ **39.** C **41.** E **43.** B **45.** $\sin\theta = \frac{\pm\sqrt{2x+1}}{x+1}$ **47.** $\sin x = \pm\sqrt{1 - \cos^2 x}$ **49.** $\tan x = \pm\sqrt{\sec^2 x - 1}$ **51.** $\csc x = \frac{\pm\sqrt{1-\cos^2 x}}{1 - \cos^2 x}$

In Exercises 53–77, there may be more than one possible answer.

53. $\cos\theta$ **55.** 1 **57.** $\cot\theta$ **59.** $\cos^2\theta$ **61.** $\sec\theta - \cos\theta$ **63.** $-\cot\theta + 1$ **65.** $\sin^2\theta\cos^2\theta$ **67.** $\tan\theta\sin\theta$ **69.** $\cot\theta - \tan\theta$ **71.** $\cos^2\theta$ **73.** $\tan^2\theta$ **75.** $-\sec\theta$ **77.** $\sec^2\theta$ **79.** $\frac{25\sqrt{6}-60}{12}$; $\frac{-25\sqrt{6}-60}{12}$ **81.** identity **83.** not an identity **85.** $y = -\sin 2x$ **86.** It is the negative of $y = \sin 2x$. **87.** $y = \cos 4x$ **88.** It is the same function. **89.** **(a)** $y = -\sin 4x$ **(b)** $y = \cos 2x$ **(c)** $y = 5\sin 3x$ **90.** Students who ignore negative signs will enjoy graphing cosine and secant functions containing a negative coefficient of x in the argument, because it can be ignored and the graph will still be correct.

7.2 Exercises

1. B **3.** A **5.** 1 **7.** $-\sin\theta$ **9.** $\cot\theta$; $\cos\theta$ **11.** $\csc\theta\sec\theta$ **13.** $1 + \sec x$ **15.** 1 **17.** $1 - 2\sin\alpha\cos\alpha$ **19.** $2 + 2\sin t$ **21.** $-2\cot x\csc x$ **23.** $(\sin\theta + 1)(\sin\theta - 1)$ **25.** $4\sin x$ **27.** $(2\sin x + 1)(\sin x + 1)$ **29.** $(\cos^2 x + 1)^2$ **31.** $(\sin x - \cos x)(1 + \sin x\cos x)$ **33.** $\sin\theta$ **35.** 1 **37.** $\tan^2\beta$ **39.** $\tan^2 x$ **41.** $\sec^2 x$ **43.** $\cos^2 x$ **89.** $(\sec\theta + \tan\theta)(1 - \sin\theta) = \cos\theta$ **91.** $\frac{\cos\theta + 1}{\sin\theta + \tan\theta} = \cot\theta$ **93.** identity **95.** not an identity **101.** **(a)** $I = k(1 - \sin^2\theta)$ **(b)** When $\theta = 0$, $\cos\theta = 1$, its maximum value. Thus, $\cos^2\theta$ will be a maximum and, as a result, I will be maximized if k is a positive constant.

103.

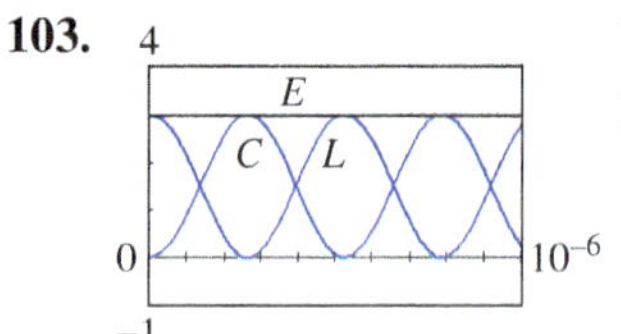

The sum of L and C equals 3. **105.** $E(t) = 3$

7.3 Exercises

1. E **3.** D **5.** D **7.** D **9.** B **11.** $\frac{\sqrt{6}-\sqrt{2}}{4}$ **13.** $\frac{\sqrt{2}-\sqrt{6}}{4}$ **15.** $\frac{\sqrt{2}-\sqrt{6}}{4}$ **17.** $\frac{\sqrt{6}+\sqrt{2}}{4}$ **19.** 0 **21.** $\cot 3°$ **23.** $\sin\frac{5\pi}{12}$ **25.** $\sec 75° 36'$ **27.** $\cos\left(-\frac{\pi}{8}\right)$ **29.** tan **31.** cos **33.** csc

For Exercises 35–39, other answers are possible. We give the most obvious one.

35. 15° **37.** $-\frac{\pi}{6}$ **39.** 20° **41.** $\frac{\sqrt{6}-\sqrt{2}}{4}$ **43.** $-2 + \sqrt{3}$ **45.** $\frac{\sqrt{6}+\sqrt{2}}{4}$ **47.** $2 - \sqrt{3}$ **49.** $\frac{\sqrt{6}+\sqrt{2}}{4}$ **51.** $\frac{-\sqrt{6}-\sqrt{2}}{4}$ **53.** $-2 - \sqrt{3}$ **55.** $-2 + \sqrt{3}$ **57.** $\frac{\sqrt{2}}{2}$ **59.** 1 **61.** −1 **63.** $-\cos\theta$ **65.** $-\cos\theta$ **67.** $\frac{\cos\theta - \sqrt{3}\sin\theta}{2}$ **69.** $\frac{\sqrt{2}(\sin x - \cos x)}{2}$ **71.** $\frac{\sqrt{3}\tan\theta + 1}{\sqrt{3} - \tan\theta}$ **73.** $\frac{\sqrt{2}(\cos x + \sin x)}{2}$ **75.** $-\cos\theta$ **77.** $-\tan x$ **79.** $\frac{16}{65}$; $-\frac{56}{65}$ **81.** $\frac{4-6\sqrt{6}}{25}$; $\frac{4+6\sqrt{6}}{25}$

83. **(a)** $\frac{63}{65}$ **(b)** $\frac{63}{16}$ **(c)** I **85.** **(a)** $\frac{77}{85}$ **(b)** $-\frac{77}{36}$ **(c)** II

87. **(a)** $\frac{4\sqrt{2}+\sqrt{5}}{9}$ **(b)** $\frac{-\sqrt{5}-\sqrt{2}}{2}$ **(c)** II

89. $\sin\left(\frac{\pi}{2}+\theta\right)=\cos\theta$ **91.** $\tan\left(\frac{\pi}{2}+\theta\right)=-\cot\theta$

103. **(a)** 3 **(b)** 163 and -163 **(c)** no

105. **(a)** 425 lb **(c)** 0° **107.** $\cos(90°+\theta)=-\sin\theta$

108. $\cos(270°-\theta)=-\sin\theta$

109. $\cos(180°+\theta)=-\cos\theta$

110. $\cos(270°+\theta)=\sin\theta$

111. $\sin(180°+\theta)=-\sin\theta$

112. $\tan(270°-\theta)=\cot\theta$

Chapter 7 Quiz

[7.1] **1.** $\cos\theta=\frac{24}{25}$; $\tan\theta=-\frac{7}{24}$; $\cot\theta=-\frac{24}{7}$; $\sec\theta=\frac{25}{24}$; $\csc\theta=-\frac{25}{7}$ **2.** $\frac{\cos^2 x+1}{\sin^2 x}$ [7.3] **3.** $\frac{-\sqrt{6}-\sqrt{2}}{4}$

4. $-\cos\theta$ **5.** **(a)** $-\frac{16}{65}$ **(b)** $-\frac{63}{65}$ **(c)** III

[7.1–7.3] **6.** $\frac{-1+\tan x}{1+\tan x}$

7.4 Exercises

1. C **3.** B **5.** F **7.** − **9.** + **11.** $\cos 2\theta=\frac{17}{25}$; $\sin 2\theta=-\frac{4\sqrt{21}}{25}$ **13.** $\cos 2x=-\frac{3}{5}$; $\sin 2x=\frac{4}{5}$

15. $\cos 2\theta=\frac{39}{49}$; $\sin 2\theta=-\frac{4\sqrt{55}}{49}$ **17.** $\cos\theta=\frac{2\sqrt{5}}{5}$; $\sin\theta=\frac{\sqrt{5}}{5}$ **19.** $\cos\theta=-\frac{\sqrt{42}}{12}$; $\sin\theta=\frac{\sqrt{102}}{12}$ **21.** $\frac{\sqrt{3}}{2}$

23. $\frac{\sqrt{3}}{2}$ **25.** $-\frac{\sqrt{2}}{2}$ **27.** $\frac{1}{2}\tan 102°$ **29.** $\frac{1}{4}\cos 94.2°$

31. $-\cos\frac{4\pi}{5}$ **33.** $\sin 4x=4\sin x\cos^3 x-4\sin^3 x\cos x$

35. $\tan 3x=\frac{3\tan x-\tan^3 x}{1-3\tan^2 x}$ **37.** $\sin 160°-\sin 44°$

39. $\sin\frac{\pi}{2}-\sin\frac{\pi}{6}$ **41.** $3\cos x-3\cos 9x$

43. $-2\sin 3x\sin x$ **45.** $-2\sin 11.5°\cos 36.5°$

47. $2\cos 6x\cos 2x$ **49.** $\frac{\sqrt{2+\sqrt{2}}}{2}$ **51.** $2-\sqrt{3}$

53. $-\frac{\sqrt{2+\sqrt{3}}}{2}$ **55.** $\frac{\sqrt{10}}{4}$ **57.** 3 **59.** $\frac{\sqrt{50-10\sqrt{5}}}{10}$

61. $-\sqrt{7}$ **63.** $\frac{\sqrt{5}}{5}$ **65.** $-\frac{\sqrt{42}}{12}$ **67.** $\sin 20°$

69. $\tan 73.5°$ **71.** $\tan 29.87°$ **73.** $\cos 9x$ **75.** $\tan 4\theta$

77. $\cos\frac{x}{8}$ **99.** $\cos^4 x-\sin^4 x=\cos 2x$

101. $\frac{2\tan x}{2-\sec^2 x}=\tan 2x$ **103.** $\frac{\sin x}{1+\cos x}=\tan\frac{x}{2}$

105. $\frac{\tan\frac{x}{2}+\cot\frac{x}{2}}{\cot\frac{x}{2}-\tan\frac{x}{2}}=\sec x$ **107.** $a=-885.6$; $c=885.6$; $\omega=240\pi$ **109.** 106° **111.** 2

7.5 Exercises

1. one; one **3.** $\cos y$ **5.** π **7.** **(a)** $[-1, 1]$ **(b)** $\left[-\frac{\pi}{2},\frac{\pi}{2}\right]$ **(c)** increasing **(d)** -2 is not in the domain.

9. **(a)** $(-\infty,\infty)$ **(b)** $\left(-\frac{\pi}{2},\frac{\pi}{2}\right)$ **(c)** increasing **(d)** no

11. The interval must be chosen so that the function is one-to-one, and the sine and cosine functions are not one-to-one on the same intervals. **13.** 0 **15.** π **17.** $\frac{\pi}{4}$ **19.** 0

21. $-\frac{\pi}{3}$ **23.** $\frac{5\pi}{6}$ **25.** $\sin^{-1}\sqrt{3}$ does not exist. **27.** $\frac{3\pi}{4}$

29. $-\frac{\pi}{6}$ **31.** $\frac{\pi}{6}$ **33.** 0 **35.** $\csc^{-1}\frac{\sqrt{2}}{2}$ does not exist.

37. −45° **39.** −60° **41.** 120° **43.** 120° **45.** −30°

47. $\sin^{-1} 2$ does not exist. **49.** −7.6713835°

51. 113.500970° **53.** 30.987961° **55.** 121.267893°

57. −82.678329° **59.** 1.1900238 **61.** 1.9033723

63. 0.83798122 **65.** 2.3154725 **67.** 2.4605221

69. **71.**

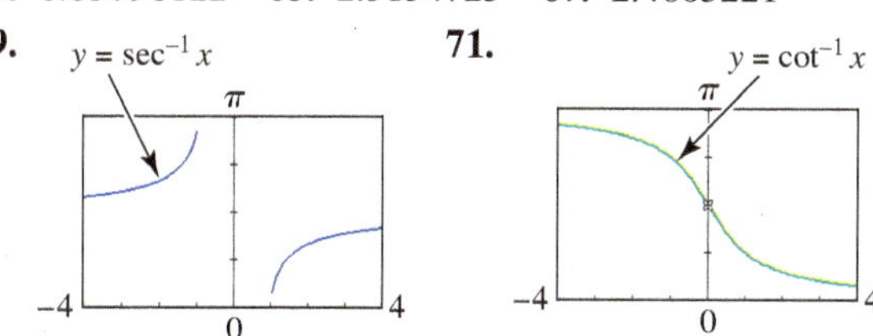

73.

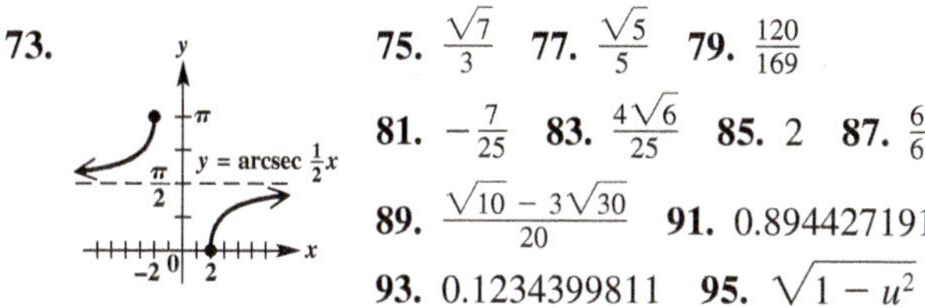

75. $\frac{\sqrt{7}}{3}$ **77.** $\frac{\sqrt{5}}{5}$ **79.** $\frac{120}{169}$

81. $-\frac{7}{25}$ **83.** $\frac{4\sqrt{6}}{25}$ **85.** 2 **87.** $\frac{63}{65}$

89. $\frac{\sqrt{10}-3\sqrt{30}}{20}$ **91.** 0.894427191

93. 0.1234399811 **95.** $\sqrt{1-u^2}$

97. $\sqrt{1-u^2}$ **99.** $\frac{4\sqrt{u^2-4}}{u^2}$ **101.** $\frac{u\sqrt{2}}{2}$ **103.** $\frac{2\sqrt{4-u^2}}{4-u^2}$

105. 41° **107.** **(a)** 18° **(b)** 18° **(c)** 15°

(e)

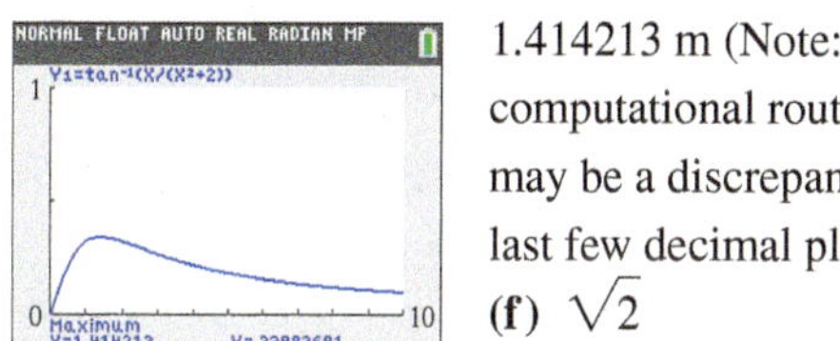

1.414213 m (Note: Due to the computational routine, there may be a discrepancy in the last few decimal places.) **(f)** $\sqrt{2}$

109. 44.7% **111.** In each case, the result is x. **112.** In each case, the result is x. The graph is that of the line $y=x$.

113.

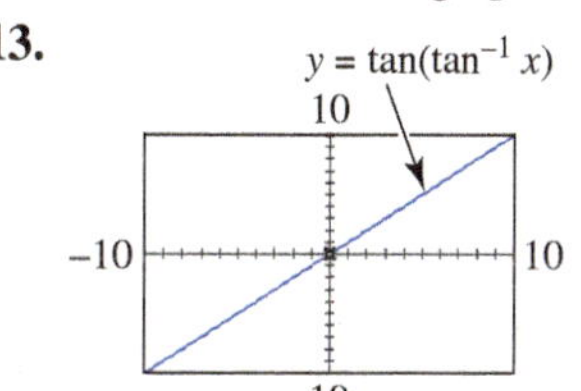

It is also the graph of $y=x$.

114.

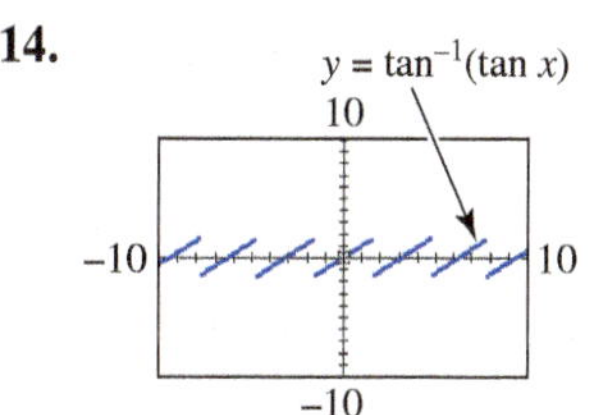

It does not agree because the range of the inverse tangent function is $\left(-\frac{\pi}{2},\frac{\pi}{2}\right)$, not $(-\infty,\infty)$, as was the case in **Exercise 113.**

7.6 Exercises

1. $\left\{\frac{\pi}{3}, \frac{5\pi}{3}\right\}$ **3.** $\left\{\frac{7\pi}{6}, \frac{11\pi}{6}\right\}$ **5.** $\{270°\}$ **7.** $\left\{\frac{\pi}{6}, \frac{5\pi}{6}, \frac{7\pi}{6}, \frac{11\pi}{6}\right\}$ **9.** $\left\{\frac{7\pi}{12}, \frac{11\pi}{12}, \frac{19\pi}{12}, \frac{23\pi}{12}\right\}$ **11.** $\varnothing$ **13.** $-30°$ is not in the interval $[0°, 360°)$. **15.** $\left\{\frac{3\pi}{4}, \frac{7\pi}{4}\right\}$ **17.** $\left\{\frac{\pi}{6}, \frac{5\pi}{6}\right\}$ **19.** $\varnothing$ **21.** $\left\{\frac{\pi}{4}, \frac{2\pi}{3}, \frac{5\pi}{4}, \frac{5\pi}{3}\right\}$ **23.** $\{\pi\}$ **25.** $\left\{\frac{7\pi}{6}, \frac{3\pi}{2}, \frac{11\pi}{6}\right\}$ **27.** $\{30°, 210°, 240°, 300°\}$ **29.** $\{90°, 210°, 330°\}$ **31.** $\{45°, 135°, 225°, 315°\}$ **33.** $\{45°, 225°\}$ **35.** $\{0°, 30°, 150°, 180°\}$ **37.** $\{0°, 45°, 135°, 180°, 225°, 315°\}$ **39.** $\{53.6°, 126.4°, 187.9°, 352.1°\}$ **41.** $\{149.6°, 329.6°, 106.3°, 286.3°\}$ **43.** $\varnothing$ **45.** $\{57.7°, 159.2°\}$ **47.** $\{180° + 360°n$, where n is any integer$\}$ **49.** $\left\{\frac{\pi}{3} + 2n\pi, \frac{2\pi}{3} + 2n\pi\right.$, where n is any integer$\left.\right\}$ **51.** $\{19.5° + 360°n, 160.5° + 360°n, 210° + 360°n, 330° + 360°n$, where n is any integer$\}$ **53.** $\left\{\frac{\pi}{3} + 2n\pi, \pi + 2n\pi, \frac{5\pi}{3} + 2n\pi\right.$, where n is any integer$\left.\right\}$ **55.** $\{180°n$, where n is any integer$\}$ **57.** $\{0.8751 + 2n\pi, 2.2665 + 2n\pi, 3.5908 + 2n\pi, 5.8340 + 2n\pi$, where n is any integer$\}$ **59.** $\{33.6° + 360°n, 326.4° + 360°n$, where n is any integer$\}$ **61.** $\{45° + 180°n, 108.4° + 180°n$, where n is any integer$\}$ **63.** $\{0.6806, 1.4159\}$ **65.** $\left\{\frac{\pi}{3}, \pi, \frac{4\pi}{3}\right\}$ **67.** $\{60°, 210°, 240°, 310°\}$ **69.** $\left\{\frac{\pi}{12}, \frac{11\pi}{12}, \frac{13\pi}{12}, \frac{23\pi}{12}\right\}$ **71.** $\{90°, 210°, 330°\}$ **73.** $\left\{\frac{\pi}{18}, \frac{7\pi}{18}, \frac{13\pi}{18}, \frac{19\pi}{18}, \frac{25\pi}{18}, \frac{31\pi}{18}\right\}$ **75.** $\{67.5°, 112.5°, 247.5°, 292.5°\}$ **77.** $\left\{\frac{\pi}{2}, \frac{3\pi}{2}\right\}$ **79.** $\left\{0, \frac{\pi}{3}, \pi, \frac{5\pi}{3}\right\}$ **81.** $\varnothing$ **83.** $\{180°\}$ **85.** $\left\{\frac{\pi}{3}, \pi, \frac{5\pi}{3}\right\}$ **87.** $\left\{\frac{\pi}{12} + \frac{2n\pi}{3}, \frac{\pi}{4} + \frac{2n\pi}{3}\right.$, where n is any integer$\left.\right\}$ **89.** $\{720°n$, where n is any integer$\}$ **91.** $\left\{\frac{2\pi}{3} + 4n\pi, \frac{4\pi}{3} + 4n\pi\right.$, where n is any integer$\left.\right\}$ **93.** $\{30° + 360°n, 150° + 360°n, 270° + 360°n$, where n is any integer$\}$ **95.** $\left\{n\pi, \frac{\pi}{6} + 2n\pi, \frac{5\pi}{6} + 2n\pi\right.$, where n is any integer$\left.\right\}$ **97.** $\{1.3181 + 2n\pi, 4.9651 + 2n\pi$, where n is any integer$\}$ **99.** $\{11.8° + 180°n, 78.2° + 180°n$, where n is any integer$\}$ **101.** $\{30° + 180°n, 90° + 180°n, 150° + 180°n$, where n is any integer$\}$ **103.** $\{1.2802\}$ **105.** **(a)** 0.00164 and 0.00355 **(b)** $[0.00164, 0.00355]$ **(c)** outward **107.** **(a)** $\frac{1}{4}$ sec **(b)** $\frac{1}{6}$ sec **(c)** 0.21 sec **109.** **(a)** See the graph in the text. **(b)** The graph is periodic, and the wave has "jagged square" tops and bottoms. **(c)** This will occur when t is in one of these intervals: $(0.0045, 0.0091), (0.0136, 0.0182), (0.0227, 0.0273)$.

111. **(a)** For $x = t$, $P(t) = \frac{1}{2}\sin[2\pi(220)t] + \frac{1}{3}\sin[2\pi(330)t] + \frac{1}{4}\sin[2\pi(440)t]$

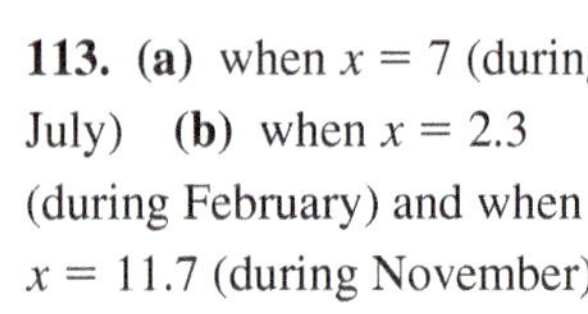

(b) 0.0007576, 0.009847, 0.01894, 0.02803 **(c)** 110 Hz

(d) For $x = t$, $P(t) = \sin[2\pi(110)t] + \frac{1}{2}\sin[2\pi(220)t] + \frac{1}{3}\sin[2\pi(330)t] + \frac{1}{4}\sin[2\pi(440)t]$

2
0
0.03
−2

113. **(a)** when $x = 7$ (during July) **(b)** when $x = 2.3$ (during February) and when $x = 11.7$ (during November) **115.** 0.001 sec **117.** 0.004 sec

Chapter 7 Quiz

[7.5] **1.**

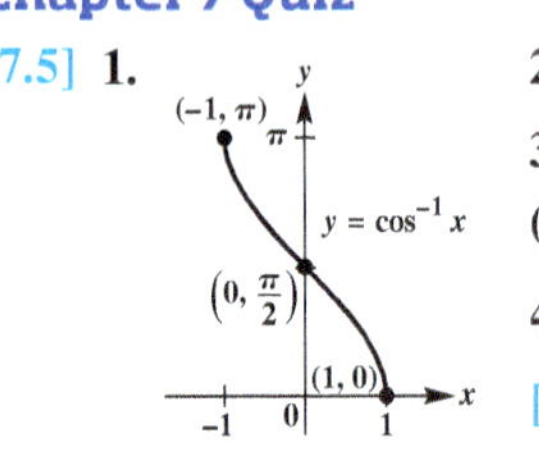

$[-1, 1]; [0, \pi]$

2. **(a)** $-\frac{\pi}{4}$ **(b)** $\frac{\pi}{3}$ **(c)** $\frac{5\pi}{6}$ **3.** **(a)** 22.568922° **(b)** 137.431085° **4.** **(a)** $\frac{5\sqrt{41}}{41}$ **(b)** $\frac{\sqrt{3}}{2}$ [7.6] **5.** $\{60°, 120°\}$ **6.** $\{60°, 180°, 300°\}$ **7.** **(a)** 0 sec **(b)** 0.20 sec **8.** $\{0.6089, 1.3424, 3.7505, 4.4840\}$ **9.** $\left\{\frac{\pi}{6}, \frac{2\pi}{3}, \frac{7\pi}{6}, \frac{5\pi}{3}\right\}$ **10.** $\left\{\frac{5\pi}{3} + 4n\pi, \frac{7\pi}{3} + 4n\pi\right.$, where n is any integer$\left.\right\}$

7.7 Exercises

1. C **3.** C **5.** A **7.** $x = \arccos \frac{y}{5}$ **9.** $x = \frac{1}{2}\arctan \frac{y}{3}$ **11.** $x = 4 \arccos \frac{y}{6}$ **13.** $x = \frac{1}{5}\arccos\left(-\frac{y}{2}\right)$ **15.** $x = \arcsin(y + 2)$ **17.** $x = \arcsin\left(\frac{y+4}{2}\right)$ **19.** $x = \frac{1}{3}\operatorname{arccot} 2y$ **21.** $x = -3 + \arccos y$ **23.** $x = \frac{1}{2}\sec^{-1}\left(\frac{y - \sqrt{2}}{3}\right)$ **25.** The argument of the sine function is x, not $x - 2$. To solve for x, first add 2 and then use the definition of arcsine. **27.** $\left\{-\frac{\sqrt{2}}{2}\right\}$ **29.** $\left\{-2\sqrt{2}\right\}$ **31.** $\{\pi - 3\}$ **33.** $\left\{\frac{3}{5}\right\}$ **35.** $\left\{\frac{4}{5}\right\}$ **37.** $\{0\}$ **39.** $\left\{\frac{1}{2}\right\}$ **41.** $\left\{-\frac{1}{2}\right\}$ **43.** $\{0\}$ **45.**

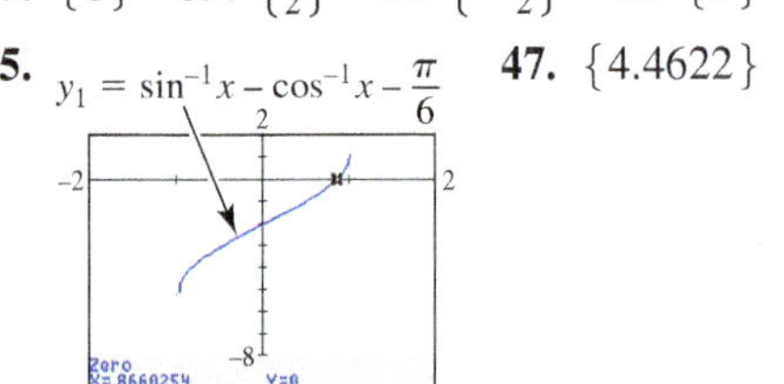

47. $\{4.4622\}$

49. **(a)** $A \approx 0.00506$, $\phi \approx 0.484$; $P = 0.00506 \sin(440\pi t + 0.484)$

(b) For $x = t$,
$P(t) = 0.00506 \sin(440\pi t + 0.484)$
$P_1(t) + P_2(t) = 0.0012 \sin(440\pi t + 0.052) + 0.004 \sin(440\pi t + 0.61)$

0.006
0 0.01
−0.006

The two graphs are the same.

51. **(a)** $\tan \alpha = \frac{x}{z}$; $\tan \beta = \frac{x+y}{z}$ **(b)** $\frac{x}{\tan \alpha} = \frac{x+y}{\tan \beta}$ **(c)** $\alpha = \arctan\left(\frac{x \tan \beta}{x+y}\right)$ **(d)** $\beta = \arctan\left(\frac{(x+y)\tan \alpha}{x}\right)$

53. **(a)** $t = \frac{1}{2\pi f} \arcsin \frac{E}{E_{max}}$ **(b)** 0.00068 sec

55. **(a)** $t = \frac{3}{4\pi} \arcsin 3y$ **(b)** 0.27 sec

Chapter 7 Review Exercises

1. B **3.** C **5.** D

In Exercises 7–11, there may be more than one possible answer.

7. 1 **9.** $\sec^2 \theta$ **11.** $-\cot \theta$ **13.** $\sin x = -\frac{4}{5}$; $\tan x = -\frac{4}{3}$; $\cot(-x) = \frac{3}{4}$ **15.** $\sin 165° = \frac{\sqrt{6} - \sqrt{2}}{4}$; $\cos 165° = \frac{-\sqrt{6} - \sqrt{2}}{4}$; $\tan 165° = -2 + \sqrt{3}$; $\csc 165° = \sqrt{6} + \sqrt{2}$; $\sec 165° = -\sqrt{6} + \sqrt{2}$; $\cot 165° = -2 - \sqrt{3}$ **17.** I **19.** H **21.** G **23.** J

25. F **27.** $\frac{117}{125}$; $\frac{4}{5}$; $-\frac{117}{44}$; II **29.** $\frac{2 + 3\sqrt{7}}{10}$; $\frac{2\sqrt{3} + \sqrt{21}}{10}$; $\frac{-25\sqrt{3} - 8\sqrt{21}}{9}$; II **31.** $\frac{4 - 9\sqrt{11}}{50}$; $\frac{12\sqrt{11} - 3}{50}$; $\frac{\sqrt{11} - 16}{21}$; IV

33. $\sin \theta = \frac{\sqrt{14}}{4}$; $\cos \theta = \frac{\sqrt{2}}{4}$ **35.** $\sin 2x = \frac{3}{5}$; $\cos 2x = -\frac{4}{5}$ **37.** $\frac{1}{2}$ **39.** $\frac{\sqrt{5} - 1}{2}$ **41.** 0.5

43. $\frac{\sin 2x + \sin x}{\cos x - \cos 2x} = \cot \frac{x}{2}$ **45.** $\frac{\sin x}{1 - \cos x} = \cot \frac{x}{2}$

47. $\frac{2(\sin x - \sin^3 x)}{\cos x} = \sin 2x$

71.

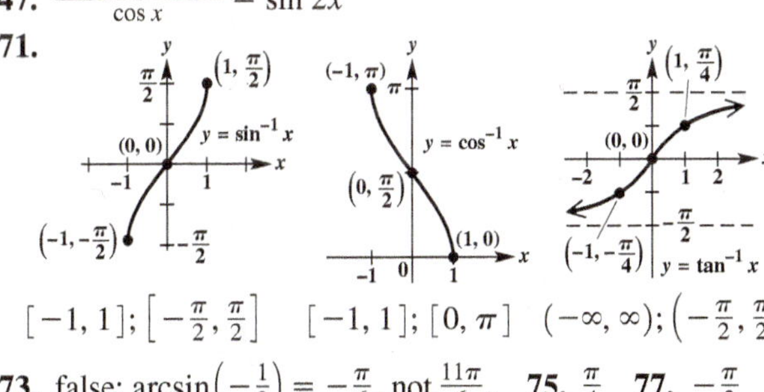

$[-1, 1]$; $\left[-\frac{\pi}{2}, \frac{\pi}{2}\right]$ $[-1, 1]$; $[0, \pi]$ $(-\infty, \infty)$; $\left(-\frac{\pi}{2}, \frac{\pi}{2}\right)$

73. false; $\arcsin\left(-\frac{1}{2}\right) = -\frac{\pi}{6}$, not $\frac{11\pi}{6}$. **75.** $\frac{\pi}{4}$ **77.** $-\frac{\pi}{3}$

79. $\frac{3\pi}{4}$ **81.** $\frac{2\pi}{3}$ **83.** $\frac{3\pi}{4}$ **85.** −60° **87.** 60.67924514°

89. 36.4895081° **91.** 73.26220613° **93.** −1 **95.** $\frac{3\pi}{4}$

97. $\frac{\pi}{4}$ **99.** $\frac{\sqrt{7}}{4}$ **101.** $\frac{\sqrt{3}}{2}$ **103.** $\frac{294 + 125\sqrt{6}}{92}$ **105.** $\frac{1}{u}$

107. $\{0.4636, 3.6052\}$ **109.** $\left\{\frac{\pi}{4}, \frac{3\pi}{4}, \frac{5\pi}{4}, \frac{7\pi}{4}\right\}$

111. $\left\{\frac{\pi}{8}, \frac{3\pi}{8}, \frac{5\pi}{8}, \frac{7\pi}{8}, \frac{9\pi}{8}, \frac{11\pi}{8}, \frac{13\pi}{8}, \frac{15\pi}{8}\right\}$ **113.** $\left\{\frac{\pi}{3} + 2n\pi, \pi + 2n\pi, \frac{5\pi}{3} + 2n\pi\text{, where } n \text{ is any integer}\right\}$ **115.** $\{270°\}$

117. $\{45°, 90°, 225°, 270°\}$ **119.** $\{70.5°, 180°, 289.5°\}$

121. $\{300° + 720°n, 420° + 720°n\text{, where } n \text{ is any integer}\}$

123. $\{180° + 360°n\text{, where } n \text{ is any integer}\}$

125. $\varnothing$ **127.** $\left\{-\frac{1}{2}\right\}$ **129.** $x = \arcsin 2y$

131. $x = \left(\frac{1}{3} \arctan 2y\right) - \frac{2}{3}$

133. **(b)**

$y_1 = \tan^{-1}\left(\frac{15}{x}\right) - \tan^{-1}\left(\frac{5}{x}\right)$
1
0 20
−0.5
Maximum X=8.6602534 Y=.52359878

8.6602534 ft; There may be a discrepancy in the final digits.

135. No light will emerge from the water.

Chapter 7 Test

[7.1] **1.** $\sin \theta = -\frac{7}{25}$; $\tan \theta = -\frac{7}{24}$; $\cot \theta = -\frac{24}{7}$; $\sec \theta = \frac{25}{24}$; $\csc \theta = -\frac{25}{7}$ **2.** $\cos \theta$ **3.** −1

[7.3] **4.** $\frac{\sqrt{6} - \sqrt{2}}{4}$ **5.** **(a)** $-\sin \theta$ **(b)** $\tan x$

[7.4] **6.** $-\frac{\sqrt{2 - \sqrt{2}}}{2}$ [7.3] **7.** **(a)** $\frac{33}{65}$ **(b)** $-\frac{56}{65}$ **(c)** $\frac{63}{16}$ **(d)** II [7.4] **8.** **(a)** $-\frac{7}{25}$ **(b)** $-\frac{24}{25}$ **(c)** $\frac{24}{7}$ **(d)** $\frac{\sqrt{5}}{5}$ **(e)** 2 [7.3] **13.** **(a)** $V = 163 \cos\left(\frac{\pi}{2} - \omega t\right)$

(b) 163 volts; $\frac{1}{240}$ sec

[7.5] **14.**

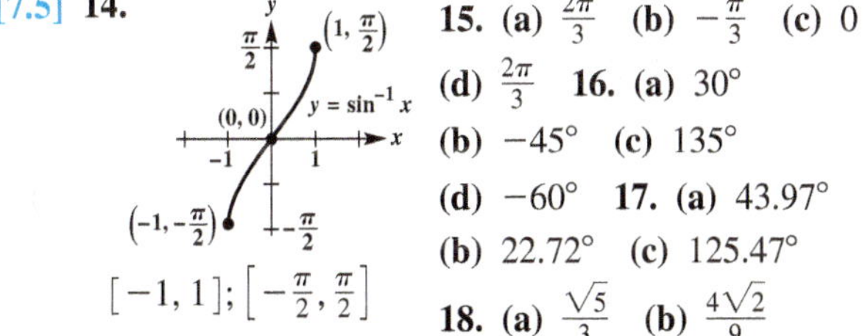

$[-1, 1]$; $\left[-\frac{\pi}{2}, \frac{\pi}{2}\right]$

15. **(a)** $\frac{2\pi}{3}$ **(b)** $-\frac{\pi}{3}$ **(c)** 0 **(d)** $\frac{2\pi}{3}$ **16.** **(a)** 30° **(b)** −45° **(c)** 135° **(d)** −60° **17.** **(a)** 43.97° **(b)** 22.72° **(c)** 125.47° **18.** **(a)** $\frac{\sqrt{5}}{3}$ **(b)** $\frac{4\sqrt{2}}{9}$

19. $\frac{u\sqrt{1 - u^2}}{1 - u^2}$ [7.6] **20.** $\{30°, 330°\}$ **21.** $\{90°, 270°\}$

22. $\{18.4°, 135°, 198.4°, 315°\}$ **23.** $\left\{0, \frac{2\pi}{3}, \frac{4\pi}{3}\right\}$

24. $\left\{\frac{\pi}{12}, \frac{7\pi}{12}, \frac{3\pi}{4}, \frac{5\pi}{4}, \frac{17\pi}{12}, \frac{23\pi}{12}\right\}$ **25.** $\{0.3649, 1.2059, 3.5065, 4.3475\}$ **26.** $\{90° + 180°n\text{, where } n \text{ is any integer}\}$ **27.** $\left\{\frac{2\pi}{3} + 4n\pi, \frac{4\pi}{3} + 4n\pi\text{, where } n \text{ is any integer}\right\}$ [7.7] **28.** **(a)** $x = \frac{1}{3} \arccos y$

(b) $x = \operatorname{arccot}\left(\frac{y - 4}{3}\right)$ **29.** **(a)** $\left\{\frac{4}{5}\right\}$ **(b)** $\left\{\frac{\sqrt{3}}{3}\right\}$

[7.6] **30.** P first reaches its maximum at approximately 2.5×10^{-4}. The maximum is approximately 0.003166.

Chapter 8 Applications of Trigonometry

Note to student: Although most of the measures resulting from solving triangles in this chapter are approximations, for convenience we use = rather than ≈ in the answers.

8.1 Exercises

1. The law of sines may be used. **3.** There is not sufficient information to use the law of sines. **5.** A
7. **(a)** $4 < L < 5$ **(b)** $L = 4$ or $L > 5$ **(c)** $L < 4$
9. 1 **11.** 0 **13.** $\sqrt{3}$ **15.** $C = 95°$, $b = 13$ m, $a = 11$ m
17. $B = 37.3°$, $a = 38.5$ ft, $b = 51.0$ ft
19. $C = 57.36°$, $b = 11.13$ ft, $c = 11.55$ ft
21. $B = 18.5°$, $a = 239$ yd, $c = 230$ yd
23. $A = 56° 00'$, $AB = 361$ ft, $BC = 308$ ft
25. $B = 110.0°$, $a = 27.01$ m, $c = 21.36$ m
27. $A = 34.72°$, $a = 3326$ ft, $c = 5704$ ft
29. $C = 97° 34'$, $b = 283.2$ m, $c = 415.2$ m **31.** 45°
33. $B_1 = 49.1°$, $C_1 = 101.2°$, $B_2 = 130.9°$, $C_2 = 19.4°$
35. $B = 26° 30'$, $A = 112° 10'$ **37.** No such triangle exists. **39.** $B = 27.19°$, $C = 10.68°$ **41.** $B = 20.6°$, $C = 116.9°$, $c = 20.6$ ft **43.** No such triangle exists.
45. $B_1 = 49° 20'$, $C_1 = 92° 00'$, $c_1 = 15.5$ m; $B_2 = 130° 40'$, $C_2 = 10° 40'$, $c_2 = 2.88$ m **47.** $B = 37.77°$, $C = 45.43°$, $c = 4.174$ ft **49.** $A_1 = 53.23°$, $C_1 = 87.09°$, $c_1 = 37.16$ m; $A_2 = 126.77°$, $C_2 = 13.55°$, $c_2 = 8.719$ m **51.** 1; 90°; a right triangle **53.** Because *A* is obtuse, it is the largest angle. Thus side *a* should be the longest side, but it is not. Therefore, no such triangle exists. **55.** 118 m
57. 17.8 km **59.** first location: 5.1 mi; second location: 7.2 mi **61.** 0.49 mi **63.** 111° **65.** 664 m **67.** 218 ft
69. The distance is 419,000 km, which compares favorably to the actual value. **71.** $\frac{\sqrt{3}}{2}$ sq unit **73.** $\frac{\sqrt{2}}{2}$ sq unit
75. 46.4 m² **77.** 356 cm² **79.** 722.9 in.²
81. 65.94 cm² **83.** 100 m² **85.** $a = \sin A$, $b = \sin B$, $c = \sin C$ **87.** $x = \frac{d \sin \alpha \sin \beta}{\sin(\beta - \alpha)}$ **90.** $\mathcal{A} = 1.12257R^2$
91. **(a)** 8.77 in.² **(b)** 5.32 in.² **92.** red

8.2 Exercises

1. **(a)** SAS **(b)** law of cosines **3.** **(a)** SSA **(b)** law of sines **5.** **(a)** ASA **(b)** law of sines
7. **(a)** SSS **(b)** law of cosines **9.** 5 **11.** 120°
13. $a = 7.0$, $B = 37.6°$, $C = 21.4°$ **15.** $A = 73.7°$, $B = 53.1°$, $C = 53.1°$ (The angles do not sum to 180° due to rounding.) **17.** $b = 88.2$, $A = 56.7°$, $C = 68.3°$
19. $a = 2.60$ yd, $B = 45.1°$, $C = 93.5°$ **21.** $c = 6.46$ m, $A = 53.1°$, $B = 81.3°$ **23.** $A = 82°$, $B = 37°$, $C = 61°$
25. $C = 102° 10'$, $B = 35° 50'$, $A = 42° 00'$
27. $C = 84° 30'$, $B = 44° 40'$, $A = 50° 50'$
29. $a = 156$ cm, $B = 64° 50'$, $C = 34° 30'$
31. $b = 9.53$ in., $A = 64.6°$, $C = 40.6°$ **33.** $a = 15.7$ m, $B = 21.6°$, $C = 45.6°$ **35.** $A = 30°$, $B = 56°$, $C = 94°$
37. The value of $\cos \theta$ will be greater than 1. A calculator will give an error message (or a nonreal complex number) when using the inverse cosine function. **39.** 257 m
41. 163.5° **43.** 281 km **45.** 438.14 ft **47.** 10.8 mi
49. 40° **51.** 26° and 36° **53.** second base: 66.8 ft; first and third bases: 63.7 ft **55.** 39.2 km **57.** 47.5 ft
59. 5500 m **61.** 16.26° **63.** $24\sqrt{3}$ sq units **65.** 78 m²
67. 12,600 cm² **69.** 3650 ft² **71.** Area and perimeter are both 36. **73.** 390,000 mi² **75.** **(a)** 87.8° and 92.2° are possible angle measures. **(b)** 92.2° **(c)** With the law of cosines we are required to find the inverse cosine of a negative number. Therefore, we know that angle *C* is greater than 90°.
77.

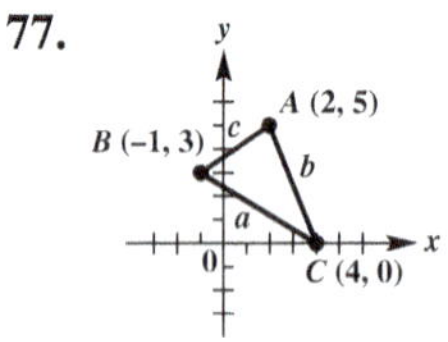

$a = \sqrt{34}$, $b = \sqrt{29}$, $c = \sqrt{13}$

78. 9.5 sq units
79. 9.5 sq units
80. 9.5 sq units

Chapter 8 Quiz

[8.1] **1.** 131° [8.2] **2.** 201 m **3.** 48.0°
[8.1] **4.** 15.75 sq units [8.2] **5.** 189 km² [8.1] **6.** 41.6°, 138.4° **7.** $a = 648$, $b = 456$, $C = 28°$ **8.** 3.6 mi
[8.2] **9.** 25.24983 mi **10.** 3921 m

8.3 Exercises

1. **m** and **p**; **n** and **r** **3.** **m** and **p** equal 2**t**, or **t** equals $\frac{1}{2}$**m** and $\frac{1}{2}$**p**. Also **m** = 1**p** and **n** = 1**r**.

5.

7.

9.

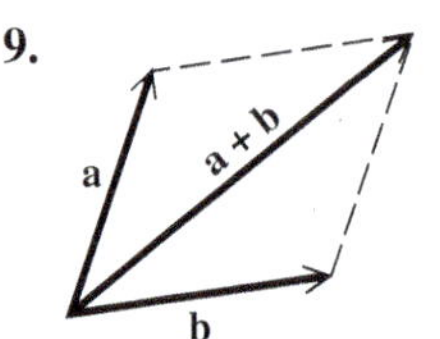

11.

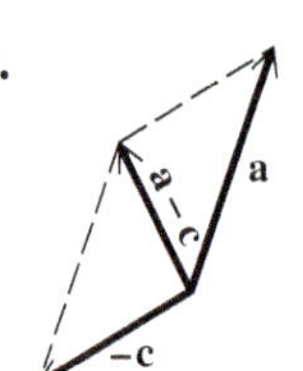

13.

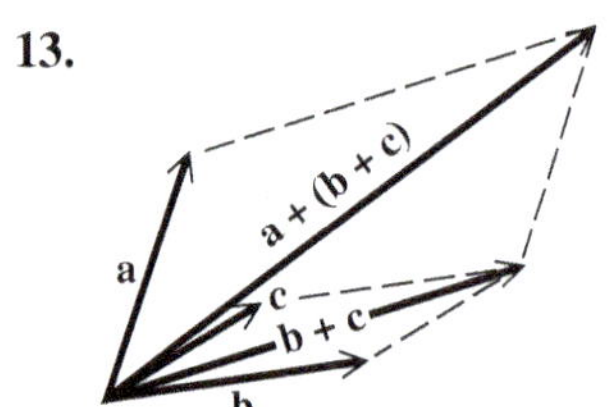

15.

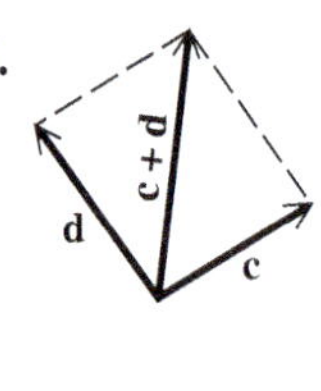

17. Yes, it appears that vector addition is associative (and this is true, in general).

19. 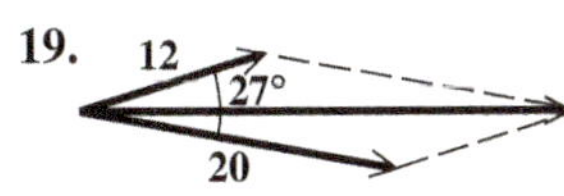

21. 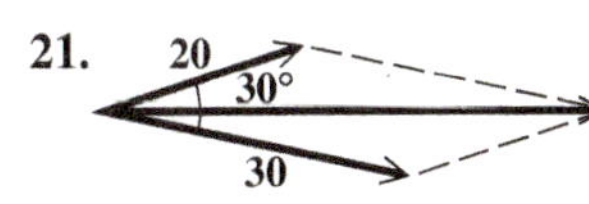

23. 94.2 lb **25.** 24.4 lb **27.** 530 newtons **29.** 88.2 lb **31.** 2640 lb at an angle of 167.2° with the 1480-lb force **33.** 93.9° **35.** 190 lb and 283 lb, respectively **37.** 18° **39.** 2.4 tons **41.** 17.5° **43.** 226 lb **45.** 13.5 mi; 50.4° **47.** 39.2 km **49.** current: 3.5 mph; motorboat: 19.7 mph **51.** bearing: 237°; ground speed: 470 mph **53.** ground speed: 161 mph; airspeed: 156 mph **55.** bearing: 74°; ground speed: 202 mph **57.** bearing: 358°; airspeed: 170 mph **59.** ground speed: 230 km per hr; bearing: 167°

8.4 Exercises

1. 2 **3.** $\frac{\sqrt{2}}{2}$ **5.** $\langle 4, 9 \rangle$ **7.** $ac + bd$ **9.** 17; 331.9° **11.** 8; 120° **13.** 47, 17 **15.** 38.8, 28.0 **17.** −123, 155 **19.** $\left\langle \frac{5\sqrt{3}}{2}, \frac{5}{2} \right\rangle$ **21.** $\langle -3.0642, 2.5712 \rangle$ **23.** $\langle 4.0958, -2.8679 \rangle$ **25. (a)** $\langle -4, 16 \rangle$ **(b)** $\langle -12, 0 \rangle$ **(c)** $\langle 8, -8 \rangle$ **27. (a)** $\langle 8, 0 \rangle$ **(b)** $\langle 0, 16 \rangle$ **(c)** $\langle -4, -8 \rangle$ **29. (a)** $\langle 0, 12 \rangle$ **(b)** $\langle -16, -4 \rangle$ **(c)** $\langle 8, -4 \rangle$ **31. (a)** $4\mathbf{i}$ **(b)** $7\mathbf{i} + 3\mathbf{j}$ **(c)** $-5\mathbf{i} + \mathbf{j}$ **33. (a)** $\langle -2, 4 \rangle$ **(b)** $\langle 7, 4 \rangle$ **(c)** $\langle 6, -6 \rangle$ **35.** $\langle -6, 2 \rangle$ **37.** $\langle 8, -20 \rangle$ **39.** $\langle -30, -3 \rangle$ **41.** $\langle 8, -7 \rangle$ **43.** $-5\mathbf{i} + 8\mathbf{j}$ **45.** $2\mathbf{i}$, or $2\mathbf{i} + 0\mathbf{j}$ **47.** 7 **49.** 0 **51.** 20 **53.** 135° **55.** 90° **57.** 36.87° **59.** −6 **61.** −24 **63.** orthogonal **65.** not orthogonal **67.** not orthogonal **69. (a)** $|\mathbf{R}| = \sqrt{5} \approx 2.2$, $|\mathbf{A}| = \sqrt{1.25} \approx 1.1$; 2.2 in. of rain fell. The area of the opening of the rain gauge is 1.1 in.^{2}. **(b)** $V = 1.5$; The volume of rain was 1.5 in.^{3}.

In Exercises 71–75, answers may vary due to rounding.
71. magnitude: 9.5208; direction angle: 119.0647° **72.** $\langle -4.1042, 11.2763 \rangle$ **73.** $\langle -0.5209, -2.9544 \rangle$ **74.** $\langle -4.6252, 8.3219 \rangle$ **75.** magnitude: 9.5208; direction angle: 119.0647° **76.** They are the same. Preference of method is an individual choice.

Summary Exercises on Applications of Trigonometry and Vectors

1. 29 ft; 38 ft **2.** 38.3 cm **3.** 5856 m **4.** 15.8 ft per sec; 71.6° **5.** 42 lb **6.** 7200 ft **7. (a)** 10 mph **(b)** $3\mathbf{v} = 18\mathbf{i} + 24\mathbf{j}$; This represents a 30-mph wind in the direction of **v**. **(c)** **u** represents a southeast wind of $\sqrt{128} \approx 11.3$ mph. **8.** 380 mph; 64° **9.** It cannot exist. **10.** Other angles can be 36° 10′, 115° 40′, third side 40.5, or other angles can be 143° 50′, 8° 00′, third side 6.25. (Lengths are in yards.)

8.5 Exercises

1. (a) 2 **(b)** $2(\cos 0° + i \sin 0°)$ **3. (a)** $2i$ **(b)** $2(\cos 90° + i \sin 90°)$ **5. (a)** $2 + 2i$ **(b)** $2\sqrt{2}(\cos 45° + i \sin 45°)$ **7.** multiply; add **9.** 10; 180°; 180°; −10; 0 **11.** 0°; 1; 0

13. 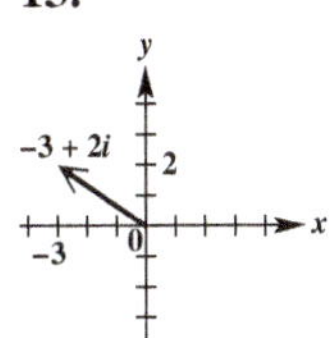

15. 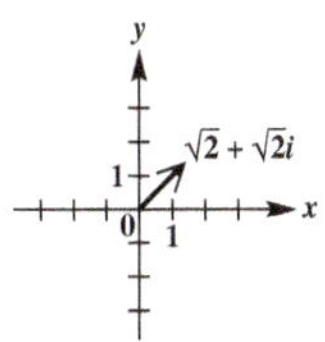

17. 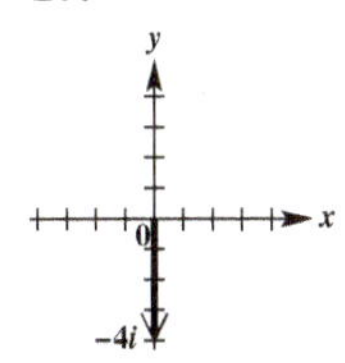

19. 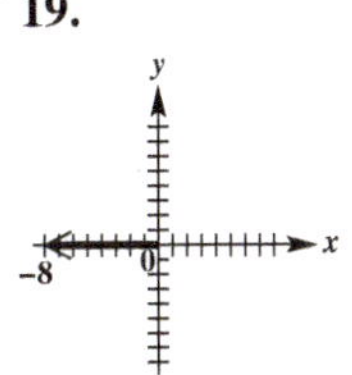

21. $3 - i$ 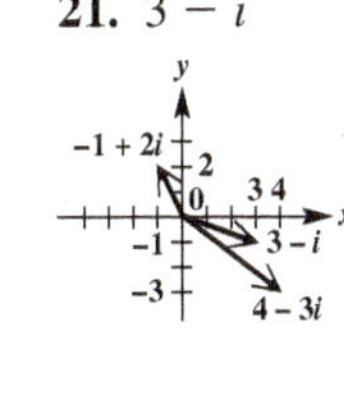

23. $-3i$ 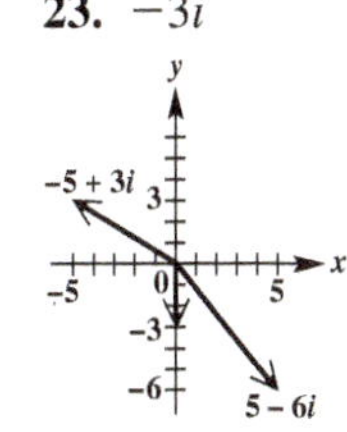

25. $-3 + 3i$ **27.** $-6 - 8i$ **29.** $7 + 9i$ **31.** $\frac{7}{6} + \frac{7}{6}i$ **33.** $\sqrt{2} + i\sqrt{2}$ **35.** $10i$ **37.** $-2 - 2i\sqrt{3}$ **39.** $-\frac{3\sqrt{3}}{2} + \frac{3}{2}i$ **41.** $\frac{5}{2} - \frac{5\sqrt{3}}{2}i$ **43.** $-1 - i$ **45.** $2\sqrt{3} - 2i$ **47.** $6(\cos 240° + i \sin 240°)$ **49.** $2(\cos 330° + i \sin 330°)$ **51.** $5\sqrt{2}(\cos 225° + i \sin 225°)$ **53.** $2\sqrt{2}(\cos 45° + i \sin 45°)$ **55.** $5(\cos 90° + i \sin 90°)$ **57.** $4(\cos 180° + i \sin 180°)$ **59.** $\sqrt{13}(\cos 56.31° + i \sin 56.31°)$ **61.** $-1.0261 - 2.8191i$ **63.** $12(\cos 90° + i \sin 90°)$ **65.** $\sqrt{34}(\cos 59.04° + i \sin 59.04°)$ **67.** It is the circle of radius 1 centered at the origin. **69.** It is the vertical line $x = 1$. **71.** yes **73.** $-3\sqrt{3} + 3i$ **75.** $12\sqrt{3} + 12i$ **77.** 4 **79.** $-3i$ **81.** $-\frac{15\sqrt{2}}{2} + \frac{15\sqrt{2}}{2}i$ **83.** $\sqrt{3} + i$ **85.** −2 **87.** $-\frac{1}{6} - \frac{\sqrt{3}}{6}i$ **89.** $2\sqrt{3} - 2i$ **91.** $-\frac{1}{2} - \frac{1}{2}i$ **93.** $\sqrt{3} + i$ **95.** $0.6537 + 7.4715i$ **97.** $30.8580 + 18.5414i$ **99.** $1.9563 + 0.4158i$ **101.** $-3.7588 - 1.3681i$ **103.** To square a complex number in trigonometric form, square its absolute value and double its argument. **107.** $1.18 - 0.14i$ **109.** $27.43 + 11.50i$ **111.** 2 **112.** $w = \sqrt{2}$ cis 135°; $z = \sqrt{2}$ cis 225° **113.** 2 cis 0° **114.** 2; It is the same. **115.** $-i$ **116.** cis(−90°) **117.** $-i$ **118.** It is the same.

8.6 Exercises

1. 27; 90°; 90°; 27; 0; 1; 0; 27; $27i$ **3.** 180°; 180°; −1; 0 **5.** two **7.** $27i$ **9.** 1 **11.** $\frac{27}{2} - \frac{27\sqrt{3}}{2}i$ **13.** $-16\sqrt{3} + 16i$ **15.** $4096i$ **17.** $128 + 128i$

19. **(a)** $\cos 0° + i \sin 0°$, $\cos 120° + i \sin 120°$, $\cos 240° + i \sin 240°$
(b)

21. **(a)** $2 \text{ cis } 20°$, $2 \text{ cis } 140°$, $2 \text{ cis } 260°$
(b)

23. **(a)** $2(\cos 90° + i \sin 90°)$, $2(\cos 210° + i \sin 210°)$, $2(\cos 330° + i \sin 330°)$
(b)

25. **(a)** $4(\cos 60° + i \sin 60°)$, $4(\cos 180° + i \sin 180°)$, $4(\cos 300° + i \sin 300°)$
(b)

27. **(a)** $\sqrt[3]{2}(\cos 20° + i \sin 20°)$, $\sqrt[3]{2}(\cos 140° + i \sin 140°)$, $\sqrt[3]{2}(\cos 260° + i \sin 260°)$
(b)

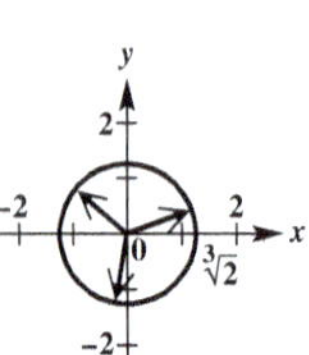

29. **(a)** $\sqrt[3]{4}(\cos 50° + i \sin 50°)$, $\sqrt[3]{4}(\cos 170° + i \sin 170°)$, $\sqrt[3]{4}(\cos 290° + i \sin 290°)$
(b)

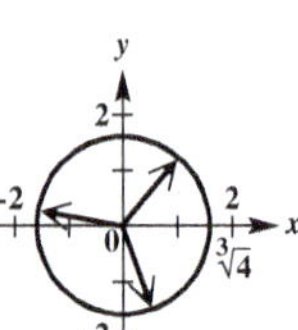

31. $\cos 0° + i \sin 0°$, $\cos 180° + i \sin 180°$

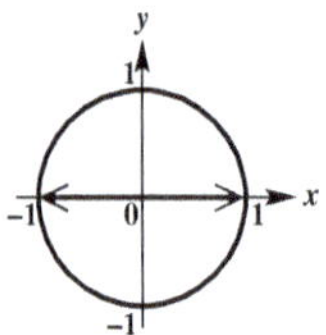

33. $\cos 0° + i \sin 0°$, $\cos 60° + i \sin 60°$, $\cos 120° + i \sin 120°$, $\cos 180° + i \sin 180°$, $\cos 240° + i \sin 240°$, $\cos 300° + i \sin 300°$

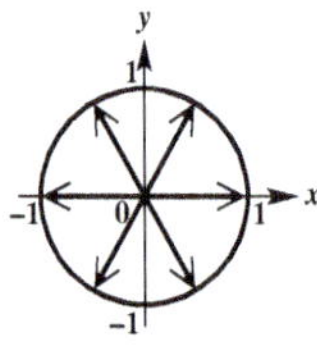

35. $\cos 30° + i \sin 30°$, $\cos 150° + i \sin 150°$, $\cos 270° + i \sin 270°$

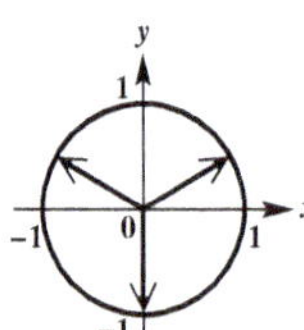

37. $\{\cos 0° + i \sin 0°, \cos 120° + i \sin 120°, \cos 240° + i \sin 240°\}$ **39.** $\{\cos 90° + i \sin 90°, \cos 210° + i \sin 210°, \cos 330° + i \sin 330°\}$
41. $\{2(\cos 0° + i \sin 0°), 2(\cos 120° + i \sin 120°), 2(\cos 240° + i \sin 240°)\}$

43. $\{\cos 45° + i \sin 45°, \cos 135° + i \sin 135°, \cos 225° + i \sin 225°, \cos 315° + i \sin 315°\}$
45. $\{\cos 22.5° + i \sin 22.5°, \cos 112.5° + i \sin 112.5°, \cos 202.5° + i \sin 202.5°, \cos 292.5° + i \sin 292.5°\}$
47. $\{2(\cos 20° + i \sin 20°), 2(\cos 140° + i \sin 140°), 2(\cos 260° + i \sin 260°)\}$
49. $1, -\frac{1}{2} + \frac{\sqrt{3}}{2}i, -\frac{1}{2} - \frac{\sqrt{3}}{2}i$ **51.** **(a)** yes **(b)** no **(c)** yes **53.** $1, 0.30901699 + 0.95105652i, -0.809017 + 0.58778525i, -0.809017 - 0.5877853i, 0.30901699 - 0.9510565i$ **55.** $\{-1.8174 + 0.5503i, 1.8174 - 0.5503i\}$ **57.** $\{0.8771 + 0.9492i, -0.6317 + 1.1275i, -1.2675 - 0.2524i, -0.1516 - 1.2835i, 1.1738 - 0.5408i\}$
59. $\cos 2\theta + i \sin 2\theta$
60. $(\cos^2 \theta - \sin^2 \theta) + i(2 \cos \theta \sin \theta) = \cos 2\theta + i \sin 2\theta$
61. $\cos 2\theta = \cos^2 \theta - \sin^2 \theta$ **62.** $\sin 2\theta = 2 \sin \theta \cos \theta$

Chapter 8 Quiz

[8.4] **1.** **(a)** $\langle -3, 12 \rangle$ **(b)** $\langle -14, 12 \rangle$ **(c)** $\sqrt{17}$ **(d)** 3 **(e)** 82.23° [8.3] **2.** 30 lb
[8.5]
3. $-1 + 6i$

[8.5, 8.6] **4.** $-2 - 2i$
5. **(a)** $4(\cos 270° + i \sin 270°)$
(b) $2(\cos 300° + i \sin 300°)$
(c) $\sqrt{10}(\cos 198.4° + i \sin 198.4°)$
6. **(a)** $2 + 2i\sqrt{3}$
(b) $-3.2139 + 3.8302i$
(c) $-7i$, or $0 - 7i$
7. **(a)** $36(\cos 130° + i \sin 130°)$ **(b)** $2\sqrt{3} + 2i$ **(c)** $-\frac{27\sqrt{3}}{2} + \frac{27}{2}i$ [8.6] **8.** $2(\cos 45° + i \sin 45°)$, $2(\cos 135° + i \sin 135°)$, $2(\cos 225° + i \sin 225°)$, $2(\cos 315° + i \sin 315°)$; $\sqrt{2} + i\sqrt{2}, -\sqrt{2} + i\sqrt{2}, -\sqrt{2} - i\sqrt{2}, \sqrt{2} - i\sqrt{2}$

8.7 Exercises

1. $\frac{3}{2}$ **3.** $\pm\sqrt{2}$ **5.** II **7.** IV **9.** positive x-axis
11. negative y-axis

Graphs for Exercises 13(a)–23(a)

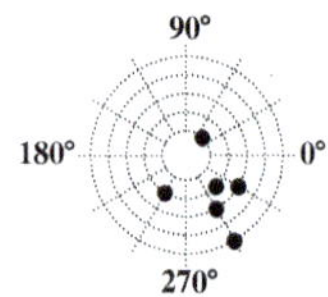

Answers may vary in Exercises 13(b)–23(b).

13. **(b)** $(1, 405°), (-1, 225°)$ **(c)** $\left(\frac{\sqrt{2}}{2}, \frac{\sqrt{2}}{2}\right)$
15. **(b)** $(-2, 495°), (2, 315°)$ **(c)** $\left(\sqrt{2}, -\sqrt{2}\right)$
17. **(b)** $(5, 300°), (-5, 120°)$ **(c)** $\left(\frac{5}{2}, -\frac{5\sqrt{3}}{2}\right)$

19. **(b)** $(-3, 150°), (3, -30°)$ **(c)** $\left(\frac{3\sqrt{3}}{2}, -\frac{3}{2}\right)$

21. **(b)** $\left(3, \frac{11\pi}{3}\right), \left(-3, \frac{2\pi}{3}\right)$ **(c)** $\left(\frac{3}{2}, -\frac{3\sqrt{3}}{2}\right)$

23. **(b)** $\left(-2, \frac{7\pi}{3}\right), \left(2, \frac{4\pi}{3}\right)$ **(c)** $\left(-1, -\sqrt{3}\right)$

Graphs for Exercises 25(a)–35(a)

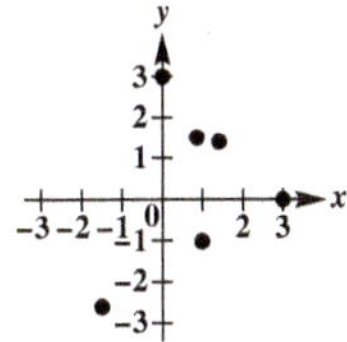

Answers may vary in Exercises 25(b)–35(b).

25. **(b)** $\left(\sqrt{2}, 315°\right), \left(-\sqrt{2}, 135°\right)$ **27.** **(b)** $(3, 90°)$, $(-3, 270°)$ **29.** **(b)** $(2, 45°), (-2, 225°)$

31. **(b)** $\left(\sqrt{3}, 60°\right), \left(-\sqrt{3}, 240°\right)$ **33.** **(b)** $(3, 0°)$, $(-3, 180°)$ **35.** **(b)** $(3, 240°), (-3, 60°)$

37. $r = \frac{4}{\cos\theta - \sin\theta}$

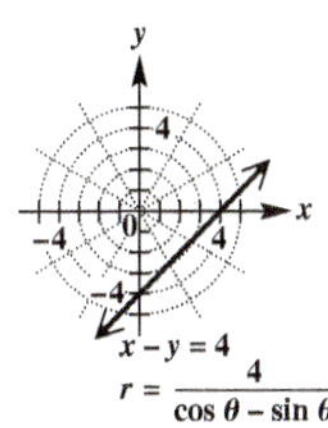

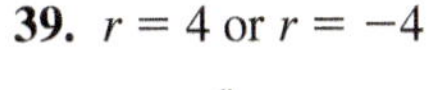

39. $r = 4$ or $r = -4$

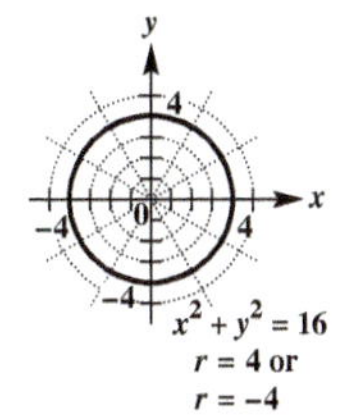

41. $r = \frac{5}{2\cos\theta + \sin\theta}$ **43.** C **45.** A

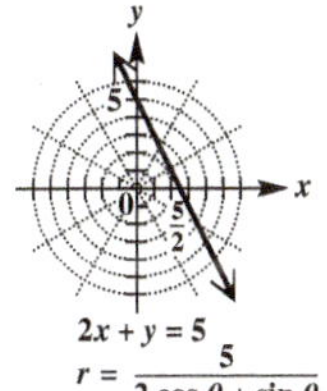

47.

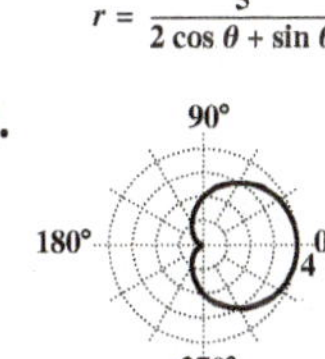

cardioid

49. $r = 3 + \cos\theta$

limaçon

51.

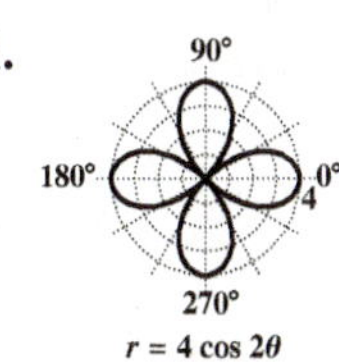

four-leaved rose

53. $r^2 = 4\cos 2\theta$

lemniscate

55.

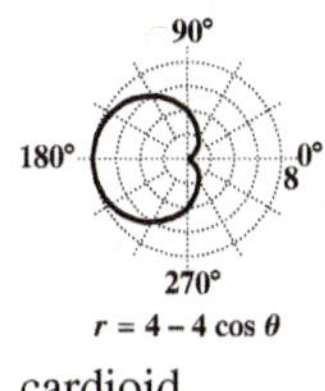

cardioid

57.

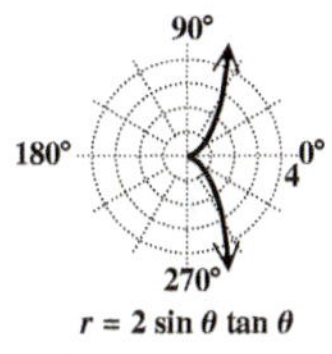

59.

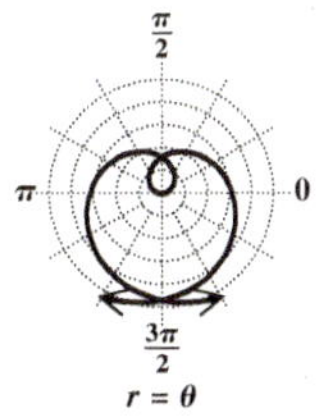

61. $x^2 + (y-1)^2 = 1$

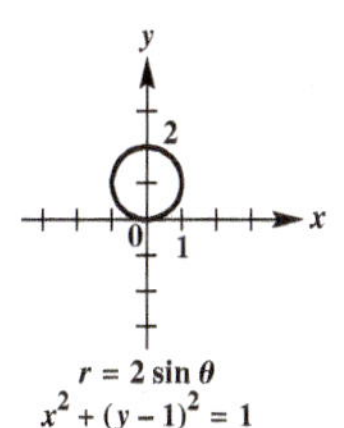

63. $y^2 = 4(x + 1)$

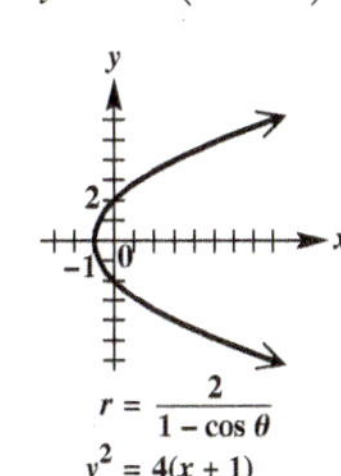

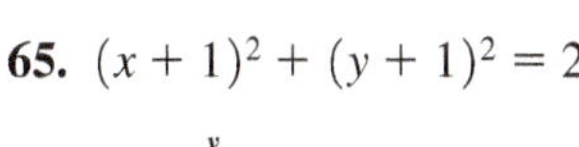

65. $(x + 1)^2 + (y + 1)^2 = 2$

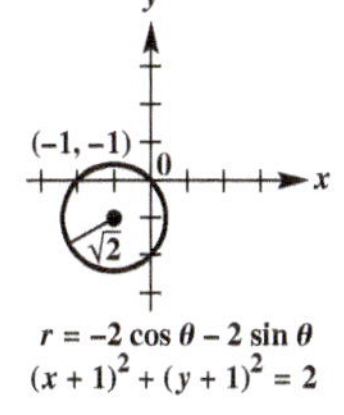

67. $x = 2$

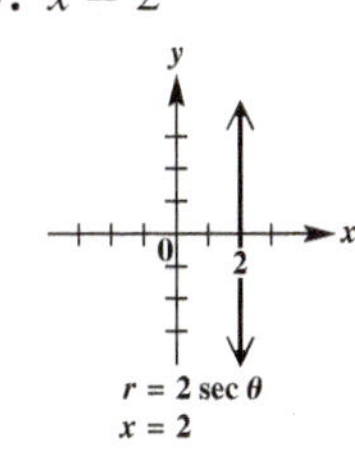

69. $x + y = 2$

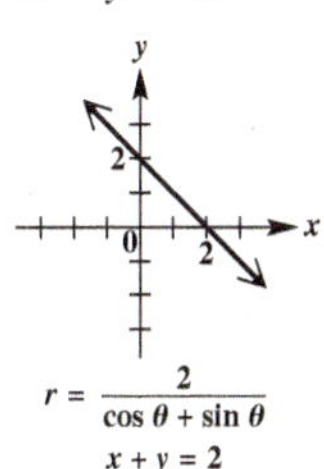

71. $r = \frac{2}{2\cos\theta + \sin\theta}$ **73.** **(a)** $(r, -\theta)$ **(b)** $(r, \pi - \theta)$ or $(-r, -\theta)$ **(c)** $(r, \pi + \theta)$ or $(-r, \theta)$

75.

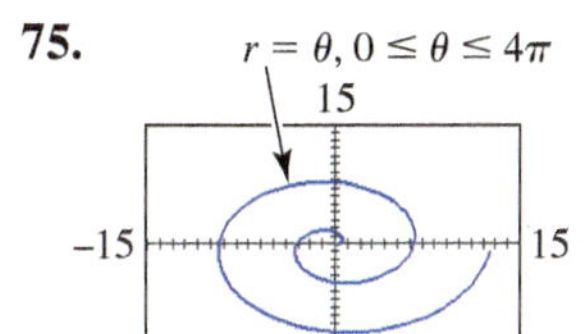

77.

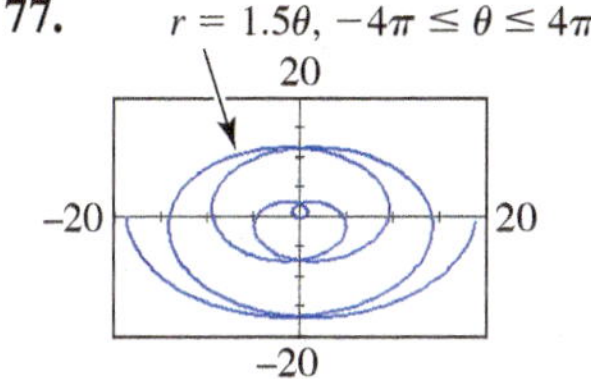

79. $\left(2, \frac{\pi}{6}\right), \left(2, \frac{5\pi}{6}\right), (0, 0)$ **81.** $\left(\frac{4+\sqrt{2}}{2}, \frac{\pi}{4}\right), \left(\frac{4-\sqrt{2}}{2}, \frac{5\pi}{4}\right)$

83. **(a)**

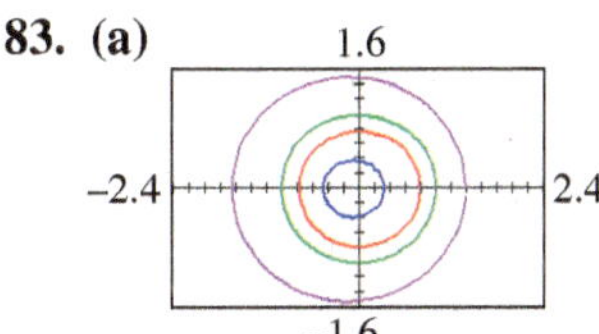

(b)

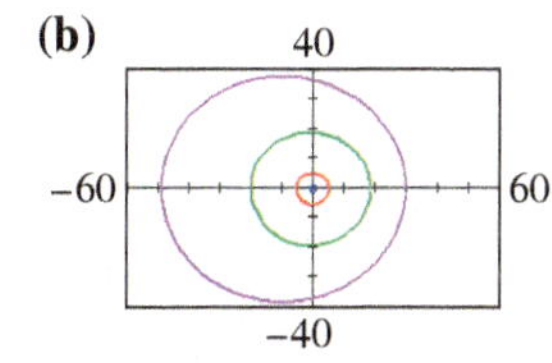

Earth is closest to the sun.

(c) no **85.** $r\sin\theta = k$ **86.** $r = \frac{k}{\sin\theta}$ **87.** $r = k\csc\theta$

88.

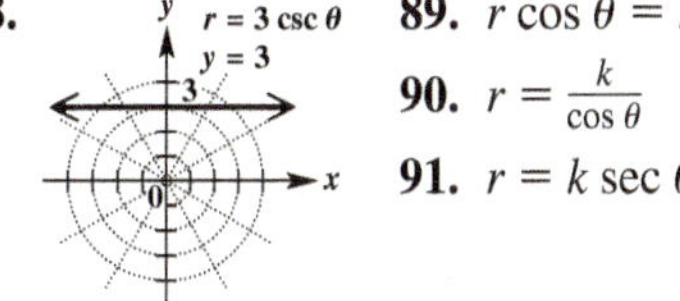

$y = 3$

89. $r\cos\theta = k$

90. $r = \frac{k}{\cos\theta}$

91. $r = k\sec\theta$

92.

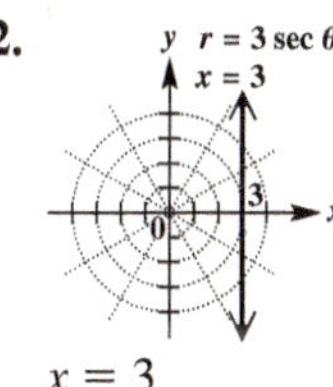

$x = 3$

8.8 Exercises

1. $(10, -3)$ **3.** $\left(\frac{1}{2}, \sqrt{3}\right)$ **5.** C **7.** A

9. (a)

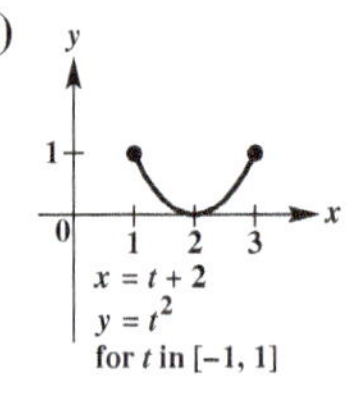

(b) $y = x^2 - 4x + 4$, for x in $[1, 3]$

11. (a)

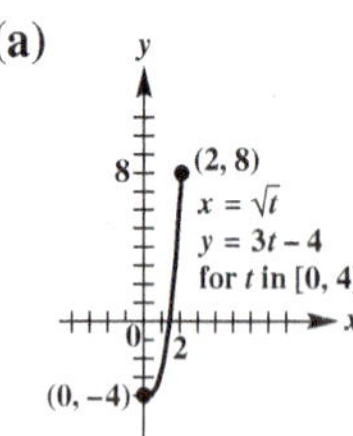

(b) $y = 3x^2 - 4$, for x in $[0, 2]$

13. (a) $x = t^3 + 1$, $y = t^3 - 1$, for t in $(-\infty, \infty)$

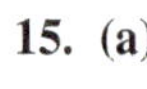

(b) $y = x - 2$, for x in $(-\infty, \infty)$

15. (a)

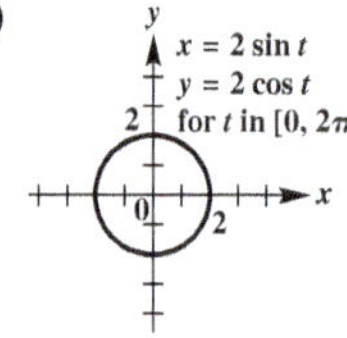

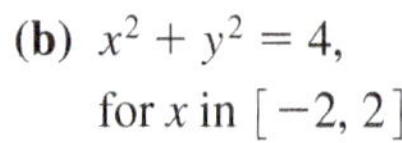

(b) $x^2 + y^2 = 4$, for x in $[-2, 2]$

17. (a)

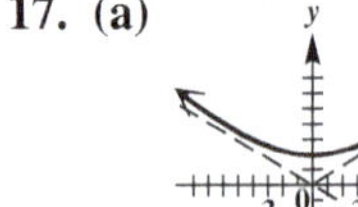

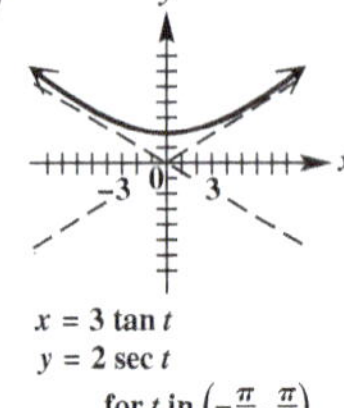

(b) $y = 2\sqrt{1 + \frac{x^2}{9}}$, for x in $(-\infty, \infty)$

19. (a)

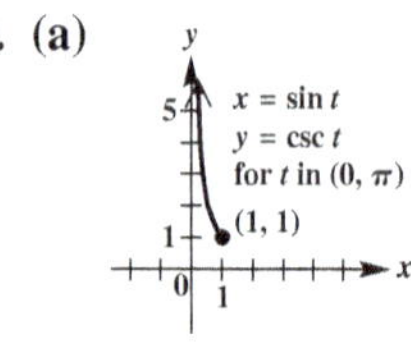

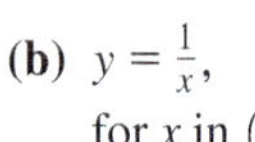

(b) $y = \frac{1}{x}$, for x in $(0, 1]$

21. (a)

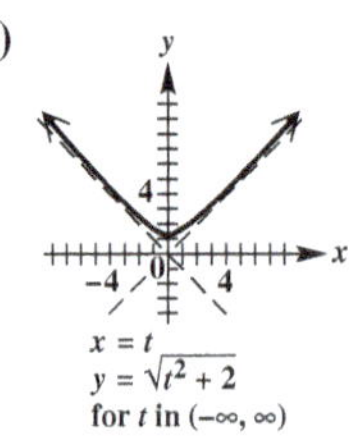

(b) $y = \sqrt{x^2 + 2}$, for x in $(-\infty, \infty)$

23. (a)

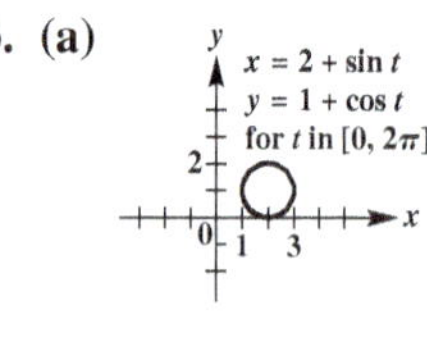

(b) $(x - 2)^2 + (y - 1)^2 = 1$, for x in $[1, 3]$

25. (a)

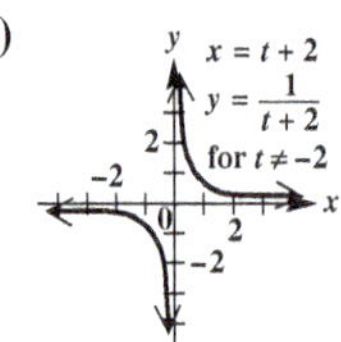

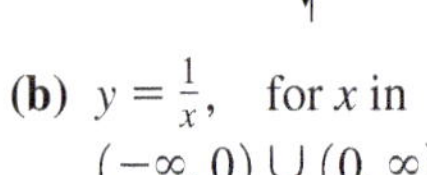

(b) $y = \frac{1}{x}$, for x in $(-\infty, 0) \cup (0, \infty)$

27. (a)

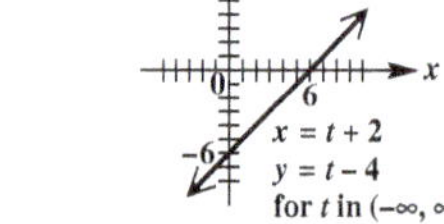

(b) $y = x - 6$, for x in $(-\infty, \infty)$

29.

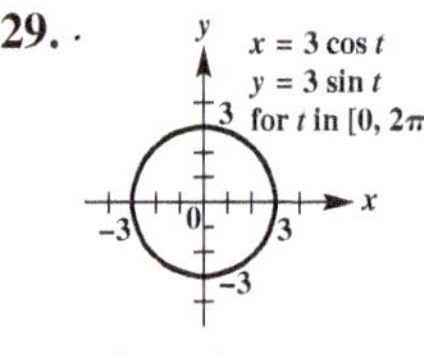

$x^2 + y^2 = 9$

31.

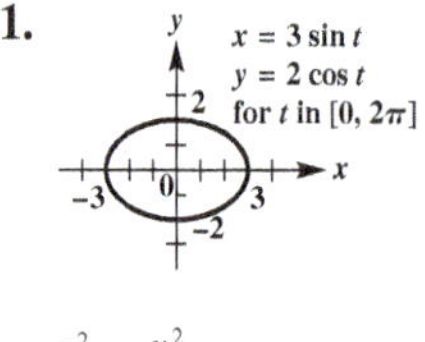

$\frac{x^2}{9} + \frac{y^2}{4} = 1$

Answers may vary for Exercises 33 and 35.

33. $x = t$, $y = (t + 3)^2 - 1$, for t in $(-\infty, \infty)$; $x = t - 3$, $y = t^2 - 1$, for t in $(-\infty, \infty)$ **35.** $x = t$, $y = t^2 - 2t + 3$, for t in $(-\infty, \infty)$; $x = t + 1$, $y = t^2 + 2$, for t in $(-\infty, \infty)$

37. $x = 2t - 2 \sin t$, $y = 2 - 2 \cos t$, for t in $[0, 4\pi]$

39.

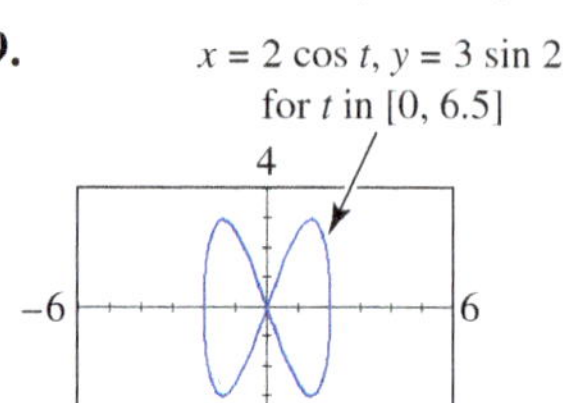

41.

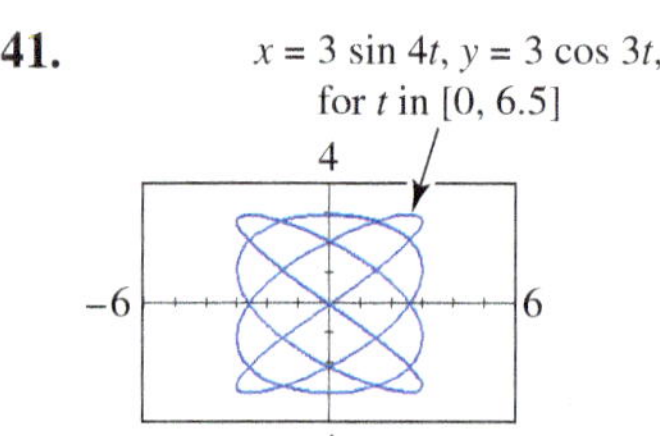

43. (a) $x = 24t$, $y = -16t^2 + 24\sqrt{3}t$

(b) $y = -\frac{1}{36}x^2 + \sqrt{3}x$ **(c)** 2.6 sec; 62 ft

45. (a) $x = (88 \cos 20°)t$, $y = 2 - 16t^2 + (88 \sin 20°)t$

(b) $y = 2 - \frac{x^2}{484 \cos^2 20°} + (\tan 20°)x$ **(c)** 1.9 sec; 161 ft

47. (a) $y = -\frac{1}{256}x^2 + \sqrt{3}x + 8$; parabolic path **(b)** 7 sec; 448 ft **49. (a)** $x = 32t$, $y = 32\sqrt{3}t - 16t^2 + 3$ **(b)** 112.6 ft **(c)** 51 ft maximum height; The ball had traveled horizontally 55.4 ft. **(d)** yes **51.** Many answers are possible; for example, $y = a(t - h)^2 + k$, $x = t$ and $y = at^2 + k$, $x = t + h$. **53.** Many answers are possible; for example, $x = a \sin t$, $y = b \cos t$ and $x = t$, $y^2 = b^2\left(1 - \frac{t^2}{a^2}\right)$.

Chapter 8 Review Exercises

1. 63.7 m **3.** 41.7° **5.** 54° 20′ or 125° 40′

7. If one side and two angles are given, the third angle can be determined using the angle sum formula, and then the ASA axiom can be applied. This is not the ambiguous case.

9. (a) $b = 5$, $b \geq 10$ **(b)** $5 < b < 10$ **(c)** $b < 5$

11. 19.87°, or 19° 52′ **13.** 55.5 m **15.** 19 cm

17. $B = 17.3°$, $C = 137.5°$, $c = 11.0$ yd **19.** $c = 18.7$ cm, $A = 91° 40'$, $B = 45° 50'$ **21.** 153,600 m^2

23. 0.234 km^2 **25.** 58.6 ft **27.** 13 m **29.** 53.2 ft

31. 115 km **33.** 25 sq units

35.

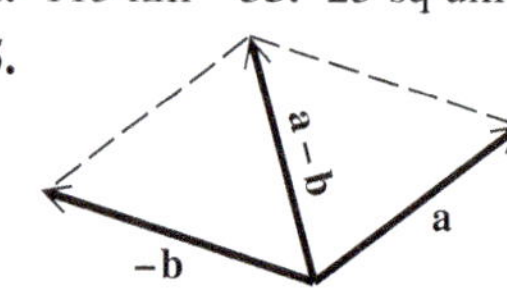

37. 207 lb **39.** −869; 418 **41.** 15; 126.9° **43. (a)** **i** **(b)** 4**i** − 2**j** **(c)** 11**i** − 7**j** **45.** 90°; orthogonal

47. 29 lb **49.** bearing: 306°; ground speed: 524 mph

51. 34 lb **53.** $-30i$ **55.** $-\frac{1}{8} + \frac{\sqrt{3}}{8}i$ **57.** $8i$

59. $-\frac{1}{2} - \frac{\sqrt{3}}{2}i$

61.

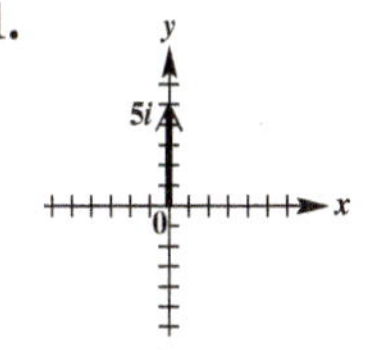

63.

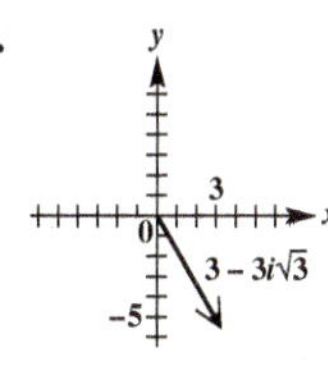

65. $2\sqrt{2}(\cos 135° + i \sin 135°)$ **67.** $-\sqrt{2} - i\sqrt{2}$

69. $\sqrt{2}(\cos 315° + i \sin 315°)$

71. $4(\cos 270° + i \sin 270°)$ **73.** It is the line $y = -x$.

75. $\sqrt[6]{2}(\cos 105° + i \sin 105°)$, $\sqrt[6]{2}(\cos 225° + i \sin 225°)$, $\sqrt[6]{2}(\cos 345° + i \sin 345°)$ **77.** none

79. $\{2(\cos 45° + i \sin 45°), 2(\cos 135° + i \sin 135°), 2(\cos 225° + i \sin 225°), 2(\cos 315° + i \sin 315°)\}$

81. $\{\cos 135° + i \sin 135°, \cos 315° + i \sin 315°\}$

83. $(2, 120°)$

85. circle

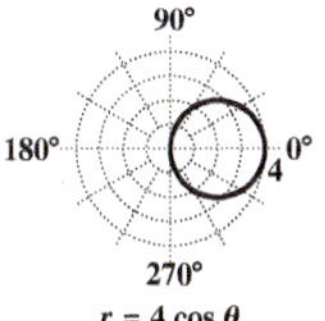

87. eight-leaved rose

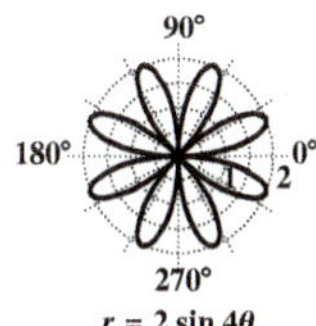

89. $y^2 = -6\left(x - \frac{3}{2}\right)$, or $y^2 + 6x - 9 = 0$ **91.** $x^2 + y^2 = 4$

93. $r = \tan\theta \sec\theta$, or $r = \frac{\tan\theta}{\cos\theta}$ **95.** $r = 2\sec\theta$, or $r = \frac{2}{\cos\theta}$ **97.** $r = \frac{4}{\cos\theta + 2\sin\theta}$

99.

$\left(\frac{\pi}{2}, 1\right)$, $(\pi - 1, 0)$, $(2\pi + 1, 0)$, $\left(\frac{3\pi}{2}, -1\right)$

$x = t + \cos t$
$y = \sin t$
for t in $[0, 2\pi]$

101. $y = \sqrt{x^2 + 1}$, for x in $[0, \infty)$

103. $y = 3\sqrt{1 + \frac{x^2}{25}}$, for x in $(-\infty, \infty)$

105. $y^2 = -\frac{1}{2}(x - 1)$, or $2y^2 + x - 1 = 0$, for x in $[-1, 1]$

107. **(a)** $x = (118 \cos 27°)t$, $y = 3.2 - 16t^2 + (118 \sin 27°)t$ **(b)** $y = 3.2 - \frac{4x^2}{3481 \cos^2 27°} + (\tan 27°)x$ **(c)** 3.4 sec; 358 ft

Chapter 8 Test

[8.1] **1.** 137.5° [8.2] **2.** 179 km **3.** 49.0° **4.** 168 sq units [8.1] **5.** 18 sq units **6.** **(a)** $b > 10$ **(b)** none **(c)** $b \leq 10$ [8.1–8.2] **7.** $a = 40$ m, $B = 41°$, $C = 79°$ **8.** $B_1 = 58° 30'$, $A_1 = 83° 00'$, $a_1 = 1250$ in.; $B_2 = 121° 30'$, $A_2 = 20° 00'$, $a_2 = 431$ in. [8.4] **9.** $|\mathbf{v}| = 10$; $\theta = 126.9°$ **10.** **(a)** $\langle 1, -3 \rangle$ **(b)** $\langle -6, 18 \rangle$ **(c)** -20 **(d)** $\sqrt{10}$ **11.** 41.8° [8.1] **12.** 2.7 mi [8.4] **13.** $\langle -346, 451 \rangle$ [8.3] **14.** 1.91 mi [8.1] **15.** 14 m [8.3] **16.** bearing: 357°; airspeed: 220 mph **17.** 18.7° **18.** 237 newtons

[8.5]

19. $7 - 3i$

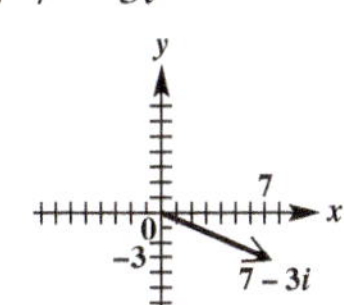

20. **(a)** $3(\cos 90° + i \sin 90°)$ **(b)** $\sqrt{5}$ cis 63.43° **(c)** $2(\cos 240° + i \sin 240°)$

21. **(a)** $\frac{3\sqrt{3}}{2} + \frac{3}{2}i$ **(b)** $3.06 + 2.57i$ **(c)** $3i$

[8.5, 8.6] **22.** **(a)** $16(\cos 50° + i \sin 50°)$ **(b)** $2\sqrt{3} + 2i$ **(c)** $4\sqrt{3} + 4i$ [8.6] **23.** 2 cis 67.5°, 2 cis 157.5°, 2 cis 247.5°, 2 cis 337.5° [8.7] **24.** Answers may vary. **(a)** $(5, 90°)$, $(5, -270°)$ **(b)** $\left(2\sqrt{2}, 225°\right)$, $\left(2\sqrt{2}, -135°\right)$

25. **(a)** $\left(\frac{3\sqrt{2}}{2}, -\frac{3\sqrt{2}}{2}\right)$ **(b)** $(0, -4)$

26. cardioid

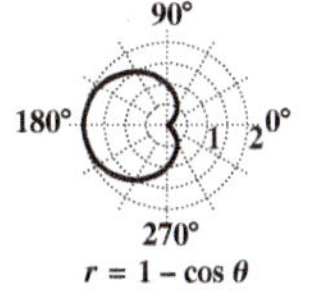

27. three-leaved rose

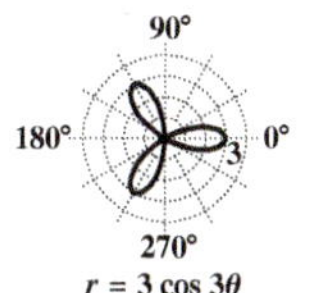

28. **(a)** $x - 2y = -4$

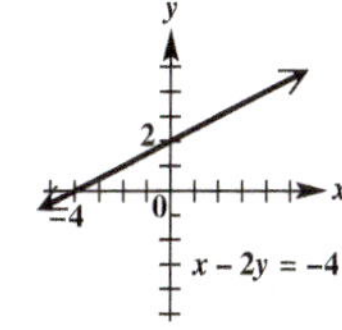

(b) $x^2 + y^2 = 36$

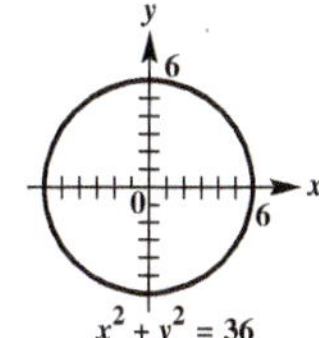

[8.8] **29.**

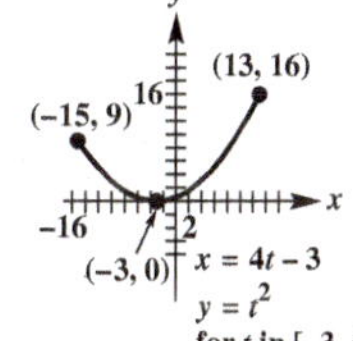

30.

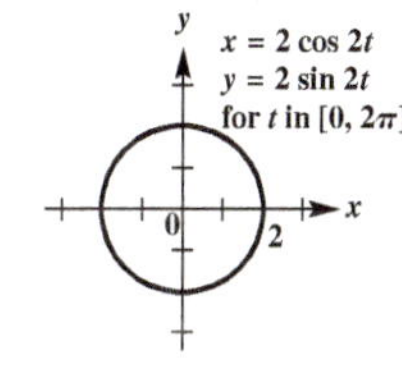

Chapter 9 Systems and Matrices

9.1 Exercises

1. 4 **3.** -11 **5.** infinitely many **7.** $\{(-4, 1)\}$ **9.** $\{(48, 8)\}$ **11.** $\{(1, 3)\}$ **13.** $\{(-1, 3)\}$ **15.** $\{(3, -4)\}$ **17.** $\{(0, 2)\}$ **19.** $\{(0, 4)\}$ **21.** $\{(1, 3)\}$ **23.** $\{(4, -2)\}$ **25.** $\{(2, -2)\}$ **27.** $\{(4, 6)\}$ **29.** $\{(5, 2)\}$ **31.** $\emptyset$; inconsistent system **33.** $\left\{\left(\frac{y + 9}{4}, y\right)\right\}$; infinitely many solutions **35.** $\emptyset$; inconsistent system **37.** $\left\{\left(\frac{-2y + 6}{7}, y\right)\right\}$; infinitely many solutions **39.** $\emptyset$; inconsistent system **41.** $x - 3y = -3$
$3x + 2y = 6$

43. $\{(0.820, -2.508)\}$ **45.** $\{(0.892, 0.453)\}$ **47.** $\{(1, 2, -1)\}$ **49.** $\{(2, 0, 3)\}$

51. $\{(1, 2, 3)\}$ **53.** $\{(4, 1, 2)\}$ **55.** $\left\{\left(\frac{1}{2}, \frac{2}{3}, -1\right)\right\}$

57. $\{(-3, 1, 6)\}$ **59.** $\left\{\left(\frac{-z+4}{3}, \frac{4z-7}{3}, z\right)\right\}$

61. $\left\{\left(\frac{-5z+69}{5}, -z+15, z\right)\right\}$

63. $\{(-9z+47, 7z-32, z)\}$ **65.** $\varnothing$; inconsistent system

67. $\left\{\left(-\frac{z}{9}, \frac{z}{9}, z\right)\right\}$; infinitely many solutions

69. $\{(2, 2)\}$ **71.** $\left\{\left(\frac{1}{5}, 1\right)\right\}$ **73.** $\{(4, 6, 1)\}$

75. $k \neq -6$; $k = -6$ **77.** $y = -3x - 5$

79. $y = \frac{3}{4}x^2 + \frac{1}{4}x - \frac{1}{2}$ **81.** $y = 3x - 1$

83. $y = -\frac{1}{2}x^2 + x + \frac{1}{4}$ **85.** $x^2 + y^2 - 4x + 2y - 20 = 0$

87. $x^2 + y^2 + x - 7y = 0$

89. $x^2 + y^2 - \frac{7}{5}x + \frac{27}{5}y - \frac{126}{5} = 0$

91. (a) $a = 0.0118, b = 0.8633, c = 317$; $C = 0.0118x^2 + 0.8633x + 317$ **(b)** 2091

93. 2006–2013 **95.** (2005.2, 1.26) **97.** year; population (in millions) **99.** 23, 24 **101.** baseball: \$212.47; football: \$478.60 **103.** 120 gal of \$9.00; 60 gal of \$3.00; 120 gal of \$4.50 **105.** 28 in.; 17 in.; 14 in. **107.** \$100,000 at 3%; \$40,000 at 2.5%; \$60,000 at 1.5% **109.** $\{(40, 15, 30)\}$

111. 11.92 lb of Arabian Mocha Sanani; 14.23 lb of Organic Shade Grown Mexico; 23.85 lb of Guatemala Antigua **113. (a)** \$16 **(b)** \$11 **(c)** \$6 **114. (a)** 8 **(b)** 4 **(c)** 0 **115.** See the answer to **Exercise 117.**

116. (a) 0 **(b)** $\frac{40}{3}$ **(c)** $\frac{80}{3}$

117.

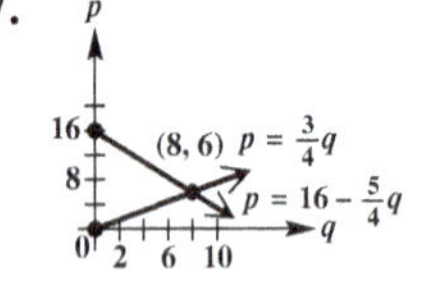

118. price: \$6; demand: 8

9.2 Exercises

1. 2 rows, 4 columns; 2×4 **3.** $\left[\begin{array}{cc|c} -3 & 5 & 2 \\ 6 & 2 & 7 \end{array}\right]$

5. $\left[\begin{array}{ccc|c} 3 & 2 & 0 & 5 \\ -9 & 0 & 6 & 1 \\ 0 & -8 & 1 & 4 \end{array}\right]$ **7.** $\begin{bmatrix} 1 & 4 \\ 0 & -9 \end{bmatrix}$ **9.** $\begin{bmatrix} 1 & 5 & 6 \\ 0 & 13 & 11 \\ 4 & 7 & 0 \end{bmatrix}$

11. $\left[\begin{array}{cc|c} 2 & 3 & 11 \\ 1 & 2 & 8 \end{array}\right]$; 2×3 **13.** $\left[\begin{array}{ccc|c} 2 & 1 & 1 & 3 \\ 3 & -4 & 2 & -7 \\ 1 & 1 & 1 & 2 \end{array}\right]$; 3×4

15. $\begin{aligned} 3x + 2y + z &= 1 \\ 2y + 4z &= 22 \\ -x - 2y + 3z &= 15 \end{aligned}$ **17.** $\begin{aligned} x &= 2 \\ y &= 3 \\ z &= -2 \end{aligned}$ **19.** $\begin{aligned} x + y &= 3 \\ 2y + z &= -4 \\ x - z &= 5 \end{aligned}$

21. $\{(2, 3)\}$ **23.** $\{(-3, 0)\}$ **25.** $\left\{\left(1, \frac{2}{3}\right)\right\}$

27. $\varnothing$ **29.** $\left\{\left(\frac{4y+7}{3}, y\right)\right\}$ **31.** $\{(-1, -2, 3)\}$

33. $\{(-1, 23, 16)\}$ **35.** $\{(2, 4, 5)\}$ **37.** $\left\{\left(\frac{1}{2}, 1, -\frac{1}{2}\right)\right\}$

39. $\{(2, 1, -1)\}$ **41.** $\varnothing$ **43.** $\left\{\left(\frac{-15z-12}{23}, \frac{z-13}{23}, z\right)\right\}$

45. $\{(1, 1, 2, 0)\}$ **47.** $\{(0, 2, -2, 1)\}$

49. $\{(0.571, 7.041, 11.442)\}$ **51.** none **53.** $A = \frac{1}{2}$, $B = -\frac{1}{2}$ **55.** $A = \frac{1}{2}, B = \frac{1}{2}$ **57.** day laborer: \$152; concrete finisher: \$160 **59.** 12, 6, 2 **61.** 9.6 cm^3 of 7%; 30.4 cm^3 of 2% **63.** \$5000 at 1.5%; \$10,000 at 2.2%; \$10,000 at 2.4% **65.** 44.4 g of A; 133.3 g of B; 222.2 g of C

67. (a) 65 or older: $y = 0.0017x + 0.148$; ages 25–39: $y = -0.0028x + 0.2$ **(b)** $\{(11.5556, 0.1676)\}$; 2026; 16.8% **(c)** The *percent* of people in the U.S. population aged 25–39 is decreasing, but not necessarily the *number* of people in this category. **69. (a)** using the first equation, 245 lb; using the second equation, 253 lb **(b)** for the first, 7.46 lb; for the second, 7.93 lb **(c)** 118 lb and 66 in.

71.

n	T
3	28
6	191
10	805
29	17,487
100	681,550
200	5,393,100
400	42,906,200
1000	668,165,500
5000	8.3×10^{10}
10,000	6.7×10^{11}
100,000	6.7×10^{14}

73. no; It increases by almost a factor of 8.

75. $\begin{aligned} a + 871b + 11.5c + 3d &= 239 \\ a + 847b + 12.2c + 2d &= 234 \\ a + 685b + 10.6c + 5d &= 192 \\ a + 969b + 14.2c + 1d &= 343 \end{aligned}$

76. $\left[\begin{array}{cccc|c} 1 & 871 & 11.5 & 3 & 239 \\ 1 & 847 & 12.2 & 2 & 234 \\ 1 & 685 & 10.6 & 5 & 192 \\ 1 & 969 & 14.2 & 1 & 343 \end{array}\right]$; $a \approx -715.457$, $b \approx 0.348$, $c \approx 48.659$, $d \approx 30.720$

77. $F = -715.457 + 0.348A + 48.659P + 30.720W$

78. 323; This estimate is very close to the actual value of 320.

9.3 Exercises

1. 0 **3.** $x^2 - 12$ **5.** -3 or 3 **7.** -31 **9.** 7 **11.** 0 **13.** -26 **15.** 0 **17.** 2, -6, 4 **19.** -6, 0, -6 **21.** 186 **23.** 17 **25.** 166 **27.** 0 **29.** 0 **31.** 1 **33.** 2 **35.** $-144 - 8\sqrt{10}$ **37.** -5.5 **39.** -3 **41.** 15 **43.** 3 **45.** 0 **47.** 0 **49.** 16 **51.** 17 **53.** 54 **55.** 0 **57.** 0 **59.** -88 **61.** 298 **63.** $\{(2, 2)\}$ **65.** $\{(2, -5)\}$ **67.** $\{(2, 0)\}$ **69.** Cramer's rule does not apply because $D = 0$; $\varnothing$ **71.** Cramer's rule does not apply because $D = 0$; $\left\{\left(\frac{-2y+4}{3}, y\right)\right\}$ **73.** $\{(-4, 12)\}$ **75.** $\{(-3, 4, 2)\}$

77. Cramer's rule does not apply because $D = 0$; $\emptyset$ **79.** Cramer's rule does not apply because $D = 0$; $\left\{\left(\frac{19z-32}{4}, \frac{-13z+24}{4}, z\right)\right\}$ **81.** $\{(0, 0, -1)\}$ **83.** $\{(0, 4, 2)\}$ **85.** $\left\{\left(\frac{31}{5}, \frac{19}{10}, -\frac{29}{10}\right)\right\}$ **87.** $W_1 = W_2 = \frac{100\sqrt{3}}{3} \approx 58$ lb **89.** 1 unit^2 **91.** 9.5 units^2 **93.** 19,328.3 ft^2 **95.** $\left\{-\frac{4}{3}\right\}$ **97.** $\{-1, 4\}$ **99.** $\{-4\}$ **101.** $\{13\}$ **103.** $\left\{-\frac{1}{2}\right\}$ **105.** $\{(-a-b, a^2+ab+b^2)\}$ **107.** $\{(1, 0)\}$ **109.** $\{(-1, 2)\}$ **112.** 102 **113.** 102; yes **114.** no

9.4 Exercises

1. $3x(2x+1)$ **3.** $(x+4)(3x^2+1)$ **5.** $x(2x^2+1)^2$ **7.** $\frac{5}{3x} + \frac{-10}{3(2x+1)}$ **9.** $\frac{6}{5(x+2)} + \frac{8}{5(2x-1)}$ **11.** $\frac{5}{6(x+5)} + \frac{1}{6(x-1)}$ **13.** $\frac{4}{x} + \frac{4}{1-x}$ **15.** $\frac{15}{x} + \frac{-5}{x+1} + \frac{-6}{x-1}$ **17.** $\frac{2}{(x+2)^2} + \frac{-3}{(x+2)^3}$ **19.** $1 + \frac{-2}{x+1} + \frac{1}{(x+1)^2}$ **21.** $\frac{1}{9} + \frac{-1}{x} + \frac{25}{18(3x+2)} + \frac{29}{18(3x-2)}$ **23.** $\frac{-3}{5x^2} + \frac{3}{5(x^2+5)}$ **25.** $\frac{-2}{7(x+4)} + \frac{6x-3}{7(3x^2+1)}$ **27.** $\frac{1}{4x} + \frac{-8}{19(2x+1)} + \frac{-9x-24}{76(3x^2+4)}$ **29.** $x^3 - x^2 + \frac{-1}{3(2x+1)} + \frac{2}{3(x+2)}$ **31.** $\frac{-1}{x} + \frac{2x}{2x^2+1} + \frac{2x+3}{(2x^2+1)^2}$ **33.** $\frac{-1}{x+2} + \frac{3}{(x^2+4)^2}$ **35.** $5x^2 + \frac{3}{x} + \frac{-1}{x+3} + \frac{2}{x-1}$ **37.** $\frac{-1}{4(x+1)} + \frac{1}{4(x-1)} + \frac{1}{2(x^2+1)}$ **39.** $\frac{2}{x} + \frac{-1}{x-1} + \frac{3}{x+2}$

Chapter 9 Quiz

[9.1] **1.** $\{(-2, 0)\}$ **2.** $\left\{\left(x, \frac{-x+2}{2}\right)\right\}$, or $\{(-2y+2, y)\}$ **3.** $\emptyset$ **4.** $\{(3, -4)\}$ [9.2] **5.** $\{(-5, 2)\}$ [9.3] **6.** $\{(-3, 6)\}$ [9.1] **7.** $\{(-2, 1, 2)\}$ [9.2] **8.** $\{(2, 1, -1)\}$ [9.3] **9.** Cramer's rule does not apply because $D = 0$; $\left\{\left(\frac{2y+6}{9}, y, \frac{23y+6}{9}\right)\right\}$ [9.1] **10.** at home: \$3961.20; away from home: \$2640.80 **11.** \$1000 at 2%; \$1500 at 3%; \$2500 at 4% [9.3] **12.** -3 **13.** 59 [9.4] **14.** $\frac{7}{x-5} + \frac{3}{x+4}$ **15.** $\frac{-1}{x+2} + \frac{6}{x+4} + \frac{-3}{x-1}$

9.5 Exercises

1. 4 **3.** -1 **5.** $x^2 + x = 2$ **13.** Consider the graphs. A line and a parabola cannot intersect in more than two points. **15.** $\{(1, 1), (-2, 4)\}$ **17.** $\left\{(2, 1), \left(\frac{1}{3}, \frac{4}{9}\right)\right\}$ **19.** $\{(2, 12), (-4, 0)\}$ **21.** $\left\{\left(-\frac{3}{5}, \frac{7}{5}\right), (-1, 1)\right\}$ **23.** $\{(2, 2), (2, -2), (-2, 2), (-2, -2)\}$ **25.** $\{(0, 0)\}$ **27.** $\left\{\left(i, \sqrt{6}\right), \left(-i, \sqrt{6}\right), \left(i, -\sqrt{6}\right), \left(-i, -\sqrt{6}\right)\right\}$ **29.** $\{(1, -1), (-1, 1), (1, 1), (-1, -1)\}$ **31.** $\emptyset$ **33.** $\left\{\left(\sqrt{6}, 0\right), \left(-\sqrt{6}, 0\right)\right\}$ **35.** $\left\{(-3, 5), \left(\frac{15}{4}, -4\right)\right\}$ **37.** $\left\{\left(4, -\frac{1}{8}\right), \left(-2, \frac{1}{4}\right)\right\}$ **39.** $\left\{(3, 4), (-3, -4), \left(\frac{4\sqrt{3}}{3}i, -3i\sqrt{3}\right), \left(-\frac{4\sqrt{3}}{3}i, 3i\sqrt{3}\right)\right\}$ **41.** $\left\{\left(\sqrt{5}, 0\right), \left(-\sqrt{5}, 0\right), \left(\sqrt{5}, \sqrt{5}\right), \left(-\sqrt{5}, -\sqrt{5}\right)\right\}$ **43.** $\{(3, 5), (-3, -5), (5i, -3i), (-5i, 3i)\}$ **45.** $\{(5, 0), (-5, 0), (0, -5)\}$ **47.** $\{(3, -3), (3, 3)\}$ **49.** $\{(2, 2), (-2, -2), (2, -2), (-2, 2)\}$ **51.** $\{(-0.79, 0.62), (0.88, 0.77)\}$ **53.** $\{(0.06, 2.88)\}$ **55.** -14 and -3 **57.** 8 and 6, 8 and -6, -8 and 6, -8 and -6 **59.** 27 and 6, -27 and -6 **61.** 5 m and 12 m **63.** yes **65.** $y = 9$ **67.** *First answer:* length = width: 6 ft; height: 10 ft *Second answer:* length = width: 12.780 ft; height: 2.204 ft **71. (a)** 160 units **(b)** \$3 **73. (a)** Revenue for public colleges from state sources is decreasing, and revenue from tuition is increasing. **(b)** 2007 to 2008 **(c)** The sources of revenue were equal in 2011, when they each contributed about 24%. **75.** Translate the graph of $y = |x|$ one unit to the right. **76.** Translate the graph of $y = x^2$ four units down. **77.** $y = \begin{cases} x-1 & \text{if } x \geq 1 \\ 1-x & \text{if } x < 1 \end{cases}$ **78.** $x^2 - 4 = x - 1\ (x \geq 1)$; $x^2 - 4 = 1 - x\ (x < 1)$ **79.** $\frac{1+\sqrt{13}}{2}$; $\frac{-1-\sqrt{21}}{2}$ **80.** $\left\{\left(\frac{1+\sqrt{13}}{2}, \frac{-1+\sqrt{13}}{2}\right), \left(\frac{-1-\sqrt{21}}{2}, \frac{3+\sqrt{21}}{2}\right)\right\}$

Summary Exercises on Systems of Equations

1. $\{(-3, 2)\}$ **2.** $\{(-5, -4)\}$ **3.** $\left\{\left(0, \sqrt{5}\right), \left(0, -\sqrt{5}\right)\right\}$ **4.** $\{(-10, -6, 25)\}$ **5.** $\left\{\left(\frac{4}{3}, 3\right), \left(-\frac{1}{2}, -8\right)\right\}$ **6.** $\{(-1, 2, -4)\}$ **7.** $\{(5z-13, -3z+9, z)\}$ **8.** $\left\{(-2, -2), \left(\frac{14}{5}, -\frac{2}{5}\right)\right\}$ **9.** $\emptyset$ **10.** $\{(-3, 3, -3)\}$ **11.** $\left\{\left(-3+i\sqrt{7}, 3+i\sqrt{7}\right), \left(-3-i\sqrt{7}, 3-i\sqrt{7}\right)\right\}$ **12.** $\{(2, -5, 3)\}$ **13.** $\{(1, 2), (-5, 14)\}$ **14.** $\emptyset$ **15.** $\{(0, 1, 0)\}$ **16.** $\left\{\left(\frac{3}{8}, -1\right)\right\}$ **17.** $\{(-1, 0, 0)\}$ **18.** $\{(-8z-56, z+13, z)\}$ **19.** $\left\{(0, 3), \left(-\frac{36}{17}, -\frac{3}{17}\right)\right\}$ **20.** $\{(2y+3, y)\}$ **21.** $\{(1, 2, 3)\}$ **22.** $\left\{\left(2+2i\sqrt{2}, 5-i\sqrt{2}\right), \left(2-2i\sqrt{2}, 5+i\sqrt{2}\right)\right\}$ **23.** $\{(2, 2, 5)\}$ **24.** $\emptyset$ **25.** $\{(4, 2), (-1, -3)\}$ **26.** $\{(-2, -4, 0)\}$ **27.** $\left\{\left(\sqrt{13}, i\sqrt{2}\right), \left(-\sqrt{13}, i\sqrt{2}\right), \left(\sqrt{13}, -i\sqrt{2}\right), \left(-\sqrt{13}, -i\sqrt{2}\right)\right\}$ **28.** $\{(-z-6, 5, z)\}$ **29.** $\{(1, -3, 4)\}$ **30.** $\left\{\left(-\frac{16}{3}, -\frac{11}{3}\right), (4, 1)\right\}$ **31.** $\{(-1, -2, 5)\}$ **32.** $\emptyset$ **33.** $\{(1, -3), (-3, 1)\}$ **34.** $\{(1, -6, 2)\}$ **35.** $\{(-6, 9), (-1, 4)\}$ **36.** $\{(2, -3, 1)\}$

9.6 Exercises

1. C **3.** B **5.** does not **7.** $(2, -5)$ **9.** dashed

11.

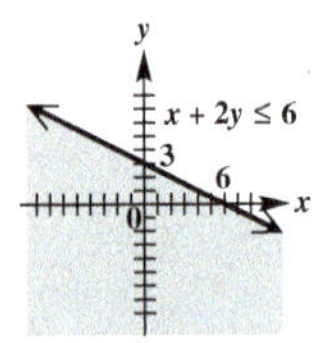

13.

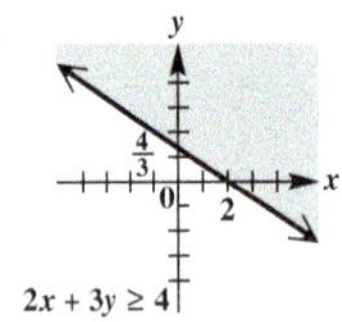

15.

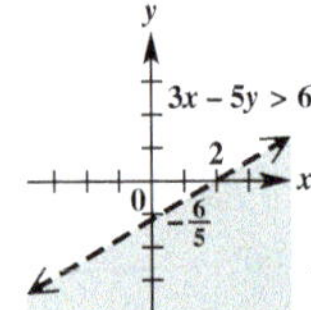

17.

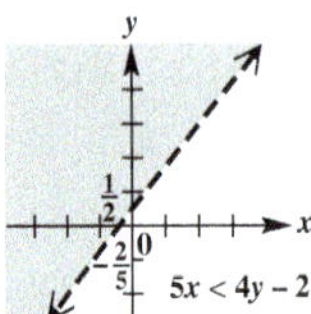

19.

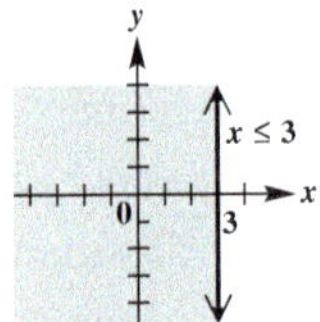

21.

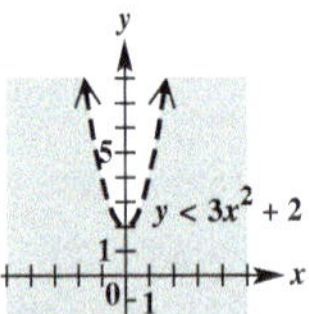

23.

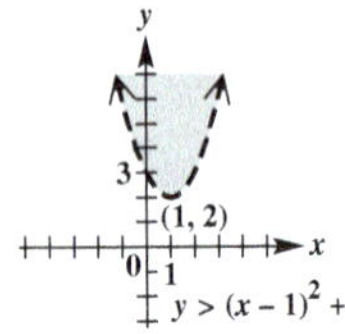

25.

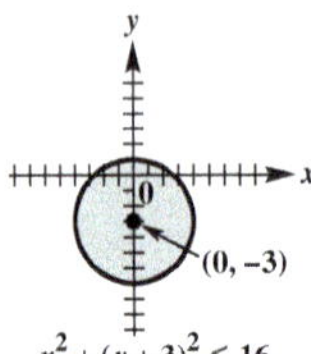

27.

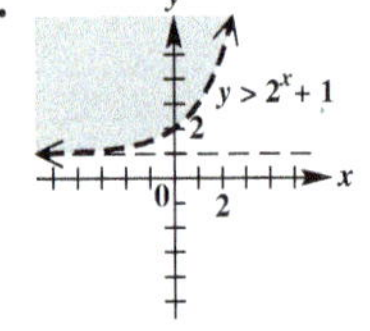

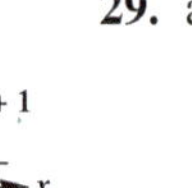

29. above **31.** B **33.** C **35.** A

37.

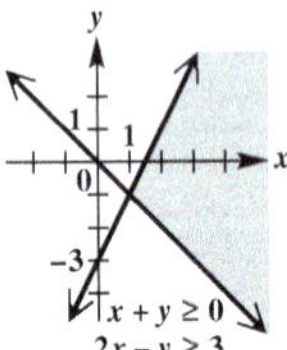

39.

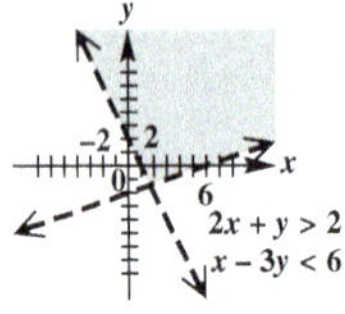

41.

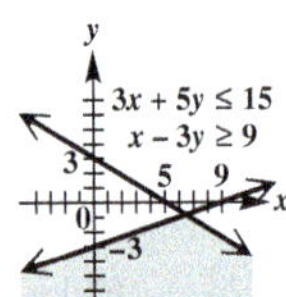

43.

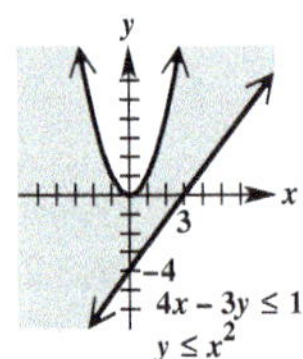

45.

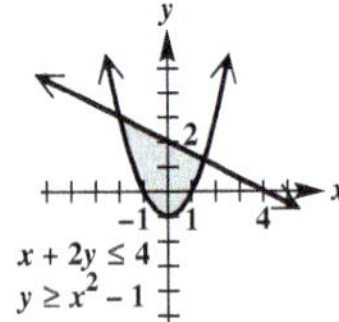

47.

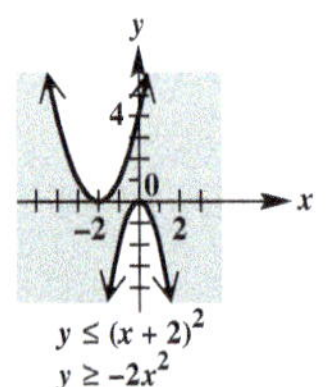

49.

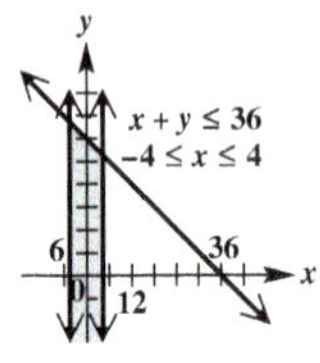

51. The solution set is ∅.

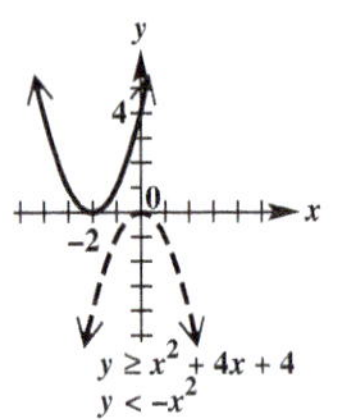

53.

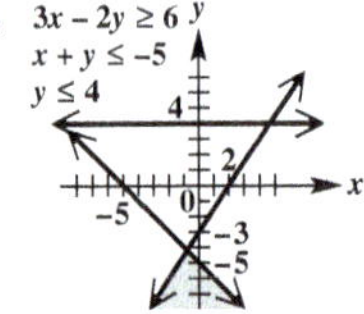

55.

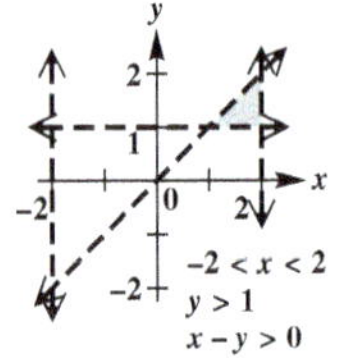

57.

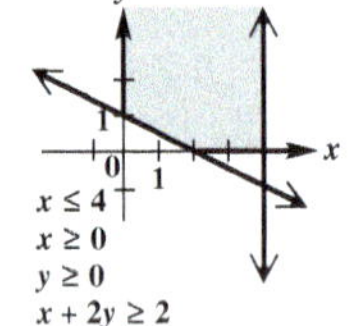

59.

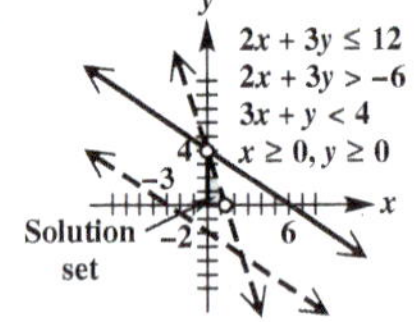

61.

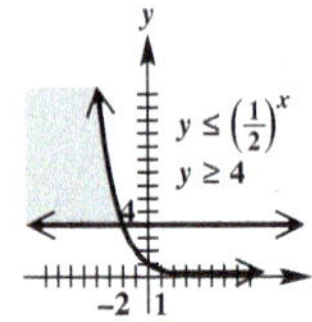

63.

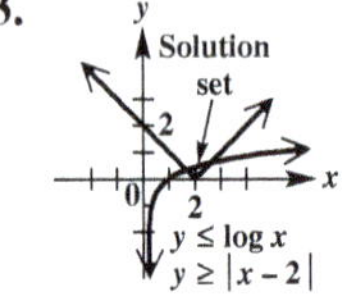

65.

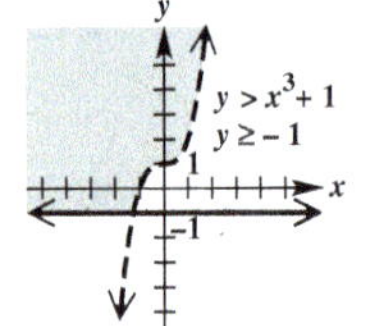

67.

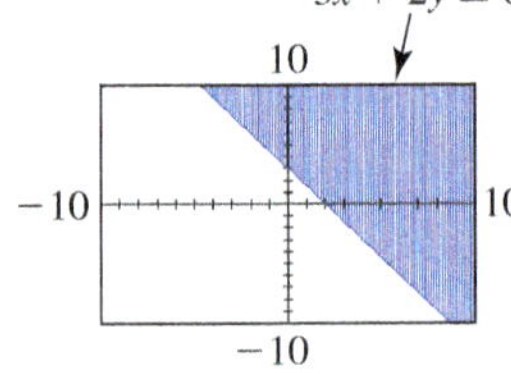

69.

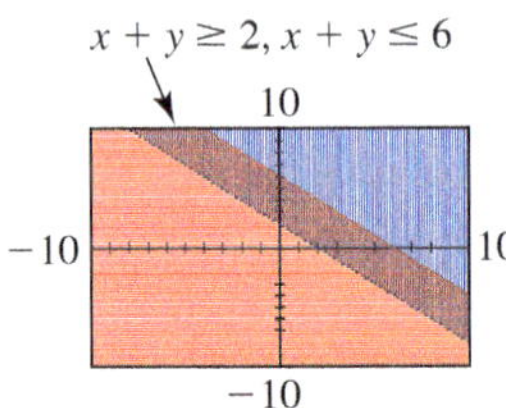

71. $x + 2y \leq 4$
$3x - 4y \leq 12$

73. $x^2 + y^2 \leq 16$
$y \geq 2$

75. $x > 0$
$y > 0$
$x^2 + y^2 \leq 4$
$y > \frac{3}{2}x - 1$

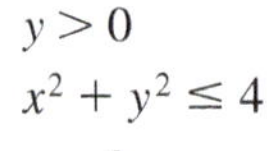

77. maximum of 65 at $(5, 10)$; minimum of 8 at $(1, 1)$

79. maximum of 66 at $(7, 9)$; minimum of 3 at $(1, 0)$

81. maximum of 100 at $(1, 10)$; minimum of 0 at $(1, 0)$

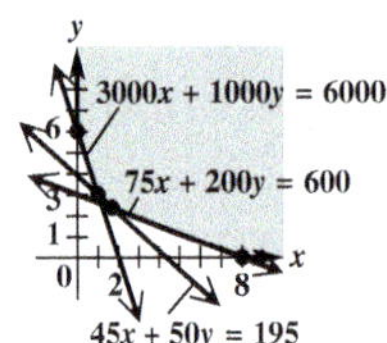

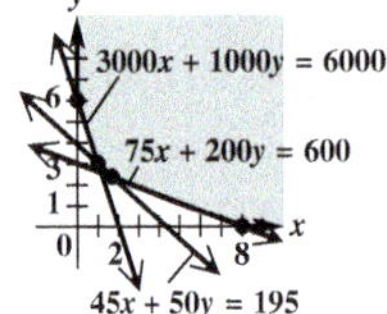

83. Let $x =$ the number of Brand X pills and $y =$ the number of Brand Y pills. Then $3000x + 1000y \geq 6000$, $45x + 50y \geq 195$, $75x + 200y \geq 600$, $x \geq 0$, $y \geq 0$.

85. 300 cartons of food and 400 cartons of clothes; 6200 people **87.** 8 of A and 3 of B, for a maximum storage capacity of 100 ft^3 **89.** $3\frac{3}{4}$ servings of A and $1\frac{7}{8}$ servings of B, for a minimum cost of $1.69

9.7 Exercises

1. 2; 0 **3.** 6; −1; −1; 12 **5.** 3×6 **7.** 2×2; square **9.** 3×4 **11.** 2×1; column **13.** $a = 0, b = 4, c = -3, d = 5$ **15.** $x = -4, y = 14, z = 3, w = -2$ **17.** This cannot be true. **19.** $w = 5, x = 6, y = -2, z = 8$ **21.** $z = 18, r = 3, s = 3, p = 3, a = \frac{3}{4}$ **23.** Be sure that the two matrices have the same dimension. The sum will have this dimension as well. To find the elements of the sum, add the corresponding elements of the two matrices.

25. $\begin{bmatrix} -2 & -5 \\ 17 & 4 \end{bmatrix}$ **27.** $\begin{bmatrix} -2 & -7 & 7 \\ 10 & -2 & 7 \end{bmatrix}$

29. They cannot be added. **31.** $\begin{bmatrix} -6 & 8 \\ 4 & 2 \end{bmatrix}$

33. $\begin{bmatrix} 4 \\ -5 \\ 4 \end{bmatrix}$ **35.** They cannot be subtracted.

37. $\begin{bmatrix} -\sqrt{3} & -13 \\ 4 & -2\sqrt{5} \\ -1 & -\sqrt{2} \end{bmatrix}$ **39.** $\begin{bmatrix} 5x + y & y \\ 6x + 2y & x + 3y \end{bmatrix}$

41. $\begin{bmatrix} -4 & 8 \\ 0 & 6 \end{bmatrix}$ **43.** $\begin{bmatrix} -9 & 3 \\ 6 & 0 \end{bmatrix}$ **45.** $\begin{bmatrix} 14 & 2 \\ -12 & 6 \end{bmatrix}$

47. $\begin{bmatrix} -1 & -3 \\ 2 & -3 \end{bmatrix}$ **49.** yes; 2×5 **51.** no **53.** yes; 3×2

55. $\begin{bmatrix} 13 \\ 25 \end{bmatrix}$ **57.** $\begin{bmatrix} -17 \\ -1 \end{bmatrix}$ **59.** $\begin{bmatrix} 17\sqrt{2} & -4\sqrt{2} \\ 35\sqrt{3} & 26\sqrt{3} \end{bmatrix}$

61. $\begin{bmatrix} 3 + 4\sqrt{4} & -3\sqrt{2} \\ 2\sqrt{15} + 12\sqrt{6} & -2\sqrt{30} \end{bmatrix}$

63. They cannot be multiplied. **65.** $\begin{bmatrix} 2 & 7 & -4 \end{bmatrix}$

67. $\begin{bmatrix} -15 & -16 & 3 \\ -1 & 0 & 9 \\ 7 & 6 & 12 \end{bmatrix}$ **69.** $\begin{bmatrix} 23 & -9 \\ -6 & -2 \\ 33 & 1 \end{bmatrix}$

71. $\begin{bmatrix} -25 & 23 & 11 \\ 0 & -6 & -12 \\ -15 & 33 & 45 \end{bmatrix}$ **73.** They cannot be multiplied.

75. $\begin{bmatrix} 10 & -10 \\ 15 & -5 \end{bmatrix}$ **77.** $BA \neq AB, BC \neq CB, AC \neq CA$

79. **(a)** $\begin{bmatrix} 38 & -8 \\ -7 & -2 \end{bmatrix}$ **(b)** $\begin{bmatrix} 18 & 24 \\ 19 & 18 \end{bmatrix}$

81. **(a)** $\begin{bmatrix} 0 & 1 & -1 \\ 0 & 1 & 0 \\ 0 & 0 & 1 \end{bmatrix}$ **(b)** $\begin{bmatrix} 0 & 1 & -1 \\ 0 & 1 & 0 \\ 0 & 0 & 1 \end{bmatrix}$ **83.** 1

85. **(a)** $\begin{bmatrix} 50 & 100 & 30 \\ 10 & 90 & 50 \\ 60 & 120 & 40 \end{bmatrix}$ **(b)** $\begin{bmatrix} 12 \\ 10 \\ 15 \end{bmatrix}$ **(c)** $\begin{bmatrix} 2050 \\ 1770 \\ 2520 \end{bmatrix}$

(d) $6340 **87.** Answers will vary a little if intermediate steps are rounded. **(a)** 2940, 2909, 2861, 2814, 2767 **(b)** The northern spotted owl will become extinct. **(c)** 3023, 3051, 3079, 3107, 3135

9.8 Exercises

1. $\begin{bmatrix} 6 & 4 \\ -1 & 8 \end{bmatrix}$ **3.** $\begin{bmatrix} 1 & 0 \\ 0 & 1 \end{bmatrix}$, or I_2 **5.** $\begin{bmatrix} 3 & -6 \\ -1 & 3 \end{bmatrix}$

9. yes **11.** no **13.** no **15.** yes

17. $\begin{bmatrix} -\frac{1}{5} & -\frac{2}{5} \\ \frac{2}{5} & -\frac{1}{5} \end{bmatrix}$ **19.** $\begin{bmatrix} 2 & 1 \\ -\frac{3}{2} & -\frac{1}{2} \end{bmatrix}$ **21.** The inverse does not exist.

23. $\begin{bmatrix} -1 & 1 & 1 \\ 0 & -1 & 0 \\ 2 & -1 & -1 \end{bmatrix}$ **25.** $\begin{bmatrix} \frac{7}{8} & -\frac{3}{8} & -\frac{3}{8} \\ -\frac{1}{8} & \frac{5}{8} & -\frac{3}{8} \\ -\frac{1}{8} & -\frac{3}{8} & \frac{5}{8} \end{bmatrix}$

27. $\begin{bmatrix} -\frac{15}{4} & -\frac{1}{4} & -3 \\ \frac{5}{4} & \frac{1}{4} & 1 \\ -\frac{3}{2} & 0 & -1 \end{bmatrix}$ **29.** $\begin{bmatrix} \frac{1}{2} & 0 & \frac{1}{2} & -1 \\ \frac{1}{10} & -\frac{2}{5} & \frac{3}{10} & -\frac{1}{5} \\ -\frac{7}{10} & \frac{4}{5} & -\frac{11}{10} & \frac{12}{5} \\ \frac{1}{5} & \frac{1}{5} & -\frac{2}{5} & \frac{3}{5} \end{bmatrix}$

31. $\begin{bmatrix} 0 & -\frac{1}{3} & \frac{1}{3} & \frac{2}{3} \\ \frac{1}{3} & \frac{2}{3} & -\frac{1}{3} & -1 \\ \frac{2}{3} & 1 & -\frac{4}{3} & -\frac{4}{3} \\ -\frac{1}{3} & \frac{1}{3} & \frac{1}{3} & 0 \end{bmatrix}$ **33.** $\{(2, 3)\}$

35. $\{(-2, 4)\}$ **37.** $\left\{\left(-2, \frac{3}{4}\right)\right\}$ **39.** $\left\{\left(-\frac{1}{2}, \frac{2}{3}\right)\right\}$ **41.** $\{(4, -9)\}$ **43.** $\left\{\left(6, -\frac{5}{6}\right)\right\}$ **45.** The inverse of $\begin{bmatrix} 7 & -2 \\ 14 & -4 \end{bmatrix}$ does not exist. **47.** $\{(3, 1, 2)\}$

49. $\left\{\left(\frac{5}{4}, \frac{1}{4}, -\frac{3}{4}\right)\right\}$ **51.** $\{(11, -1, 2)\}$ **53.** $\{(1, 0, 2, 1)\}$

55. **(a)** $602.7 = a + 5.543b + 37.14c$
$656.7 = a + 6.933b + 41.30c$
$778.5 = a + 7.638b + 45.62c$
(b) $a = -490.5, b = -89, c = 42.72$
(c) $S = -490.5 - 89A + 42.72B$ **(d)** $S = 843.6$
(e) $S = 1547.6$; Predictions made beyond the scope of the data are valid only if current trends continue.

57. Answers will vary.

59. $\begin{bmatrix} -0.1215875322 & 0.0491390161 \\ 1.544369078 & -0.046799063 \end{bmatrix}$

61. $\begin{bmatrix} 2 & -2 & 0 \\ -4 & 0 & 4 \\ 3 & 3 & -3 \end{bmatrix}$

63. $\{(1.68717058, -1.306990242)\}$

65. $\{(13.58736702, 3.929011993, -5.342780076)\}$

73. $A^{-1} = A^2 = \begin{bmatrix} 1 & 0 & 0 \\ 0 & -1 & 1 \\ 0 & -1 & 0 \end{bmatrix}$

Chapter 9 Review Exercises

1. $\{(0, 1)\}$ **3.** $\{(-5y + 9, y)\}$; infinitely many solutions **5.** $\varnothing$; inconsistent system **7.** $\left\{\left(\frac{1}{3}, \frac{1}{2}\right)\right\}$ **9.** $\{(3, 2, 1)\}$ **11.** $\{(5, -1, 0)\}$ **13.** One possible answer is $\begin{matrix} x + y = 2 \\ x + y = 3 \end{matrix}$. **15.** rice: $\frac{1}{3}$ cup; soybeans: $\frac{1}{5}$ cup **17.** 5 blankets; 3 rugs; 8 skirts **19. (a)**

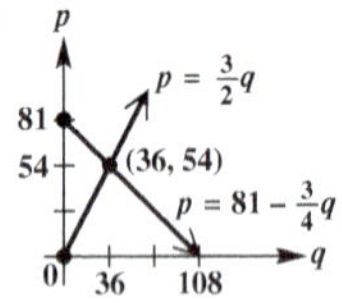

(b) 36 **(c)** $54

21. $y_1 = \frac{12}{5}x^2 - \frac{31}{5}x + \frac{3}{2}$, or $y_1 = 2.4x^2 - 6.2x + 1.5$ **23.** $\{(x, -2x + 1, -11x + 6)\}$ **25.** $\{(-2, 0)\}$ **27.** $\{(-3, 2)\}$ **29.** $\{(0, 3, 3)\}$ **31.** 10 lb of $4.60 tea; 8 lb of $5.75 tea; 2 lb of $6.50 tea **33.** 1984; 327 thousand **35.** -25 **37.** 0 **39.** -1 **41.** $\{(-4, 2)\}$ **43.** Cramer's rule does not apply because $D = 0$; $\varnothing$ **45.** $\{(14, -15, 35)\}$ **47.** $\left\{\frac{8}{19}\right\}$ **49.** $\frac{2}{x-1} - \frac{6}{3x-2}$ **51.** $\frac{1}{x-1} - \frac{x+3}{x^2+2}$ **53.** $\{(-3, 4), (1, 12)\}$ **55.** $\{(-4, 1), (-4, -1), (4, -1), (4, 1)\}$ **57.** $\left\{(5, -2), \left(-4, \frac{5}{2}\right)\right\}$ **59.** $\{(-2, 0), (1, 1)\}$ **61.** $\left\{\left(\sqrt{3}, 2i\right), \left(\sqrt{3}, -2i\right), \left(-\sqrt{3}, 2i\right), \left(-\sqrt{3}, -2i\right)\right\}$ **63.** yes; $\left\{\left(\frac{8 - 8\sqrt{41}}{5}, \frac{16 + 4\sqrt{41}}{5}\right), \left(\frac{8 + 8\sqrt{41}}{5}, \frac{16 - 4\sqrt{41}}{5}\right)\right\}$

65.

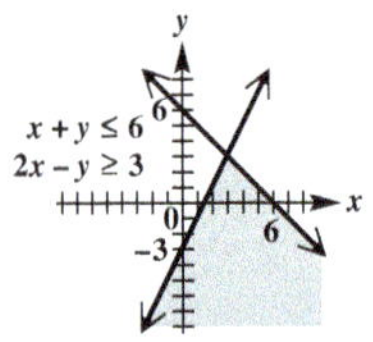

67. maximum of 24 at (0, 6) **69.** 3 units of food A and 4 units of food B; Minimum cost is $1.02 per serving. **71.** $a = 5$, $x = \frac{3}{2}$, $y = 0$, $z = 9$

73. $\begin{bmatrix} -4 \\ 6 \\ 1 \end{bmatrix}$ **75.** They cannot be subtracted.

77. $\begin{bmatrix} 3 & -4 \\ 4 & 48 \end{bmatrix}$ **79.** $\begin{bmatrix} -9 & 3 \\ 10 & 6 \end{bmatrix}$ **81.** $\begin{bmatrix} -2 & 5 & -3 \\ 3 & 4 & -4 \\ 6 & -1 & -2 \end{bmatrix}$

83. $\begin{bmatrix} 3 & -1 \\ -5 & 2 \end{bmatrix}$ **85.** $\begin{bmatrix} \frac{2}{3} & 0 & -\frac{1}{3} \\ \frac{1}{3} & 0 & -\frac{2}{3} \\ -\frac{2}{3} & 1 & \frac{1}{3} \end{bmatrix}$

87. $\left\{\left(\frac{2}{3}, \frac{7}{12}\right)\right\}$ **89.** $\{(-3, 2, 0)\}$

Chapter 9 Test

[9.1] **1.** $\{(4, 3)\}$ **2.** $\left\{\left(\frac{-3y - 7}{2}, y\right)\right\}$; infinitely many solutions **3.** $\varnothing$; inconsistent system **4.** $\{(1, 2)\}$ **5.** $\{(2, 0, -1)\}$ [9.2] **6.** $\{(5, 1)\}$ **7.** $\{(5, 3, 6)\}$ [9.1] **8.** $y = 2x^2 - 8x + 11$ **9.** 22 units from Toronto; 56 units from Montreal; 22 units from Ottawa [9.3] **10.** -58 **11.** -844 **12.** $\{(-6, 7)\}$ **13.** $\{(1, -2, 3)\}$ [9.4] **14.** $\frac{2}{x+3} + \frac{7}{x-1}$ **15.** $\frac{2}{x} + \frac{-2}{x+1} + \frac{-1}{(x+1)^2}$ [9.5] **16.** $\{(1, 2), (-1, 2), (1, -2), (-1, -2)\}$ **17.** $\{(3, 4), (4, 3)\}$ **18.** 5 and -6 [9.6] **19.**

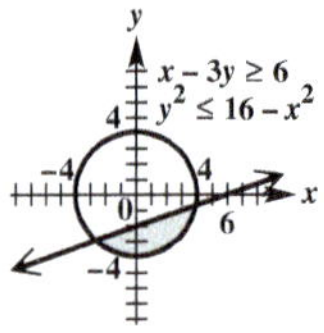

20. maximum of 42 at (12, 6) **21.** 0 VIP rings and 24 SST rings; Maximum profit is $960.

[9.7] **22.** $x = -1$, $y = 7$, $w = -3$ **23.** $\begin{bmatrix} 8 & 3 \\ 0 & -11 \\ 15 & 19 \end{bmatrix}$

24. The matrices cannot be added. **25.** $\begin{bmatrix} -5 & 16 \\ 19 & 2 \end{bmatrix}$

26. The matrices cannot be multiplied. **27.** A

[9.8] **28.** $\begin{bmatrix} -2 & -5 \\ -3 & -8 \end{bmatrix}$ **29.** The inverse does not exist.

30. $\begin{bmatrix} -9 & 1 & -4 \\ -2 & 1 & 0 \\ 4 & -1 & 1 \end{bmatrix}$ **31.** $\{(-7, 8)\}$ **32.** $\{(0, 5, -9)\}$

Chapter 10 Analytic Geometry

10.1 Exercises

1. (a) D **(b)** B **(c)** C **(d)** A **(e)** F **(f)** H **(g)** E **(h)** G **3.** $(2, -1)$; $(2, 1)$; $y = -3$; $x = 2$; $(-\infty, \infty)$; $[-1, \infty)$

In Exercises 5–21, we give the domain and then the range.

5.

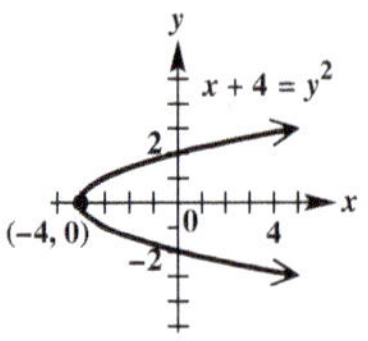

$[-4, \infty)$; $(-\infty, \infty)$

7.

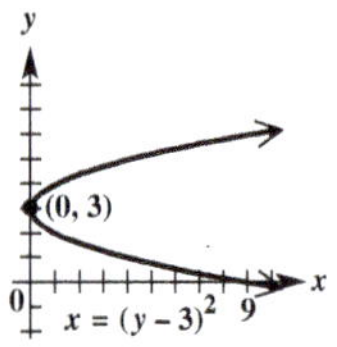

$[0, \infty)$; $(-\infty, \infty)$

9.

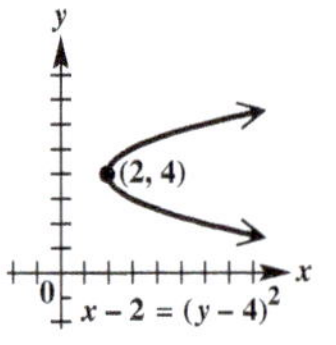

$[2, \infty)$; $(-\infty, \infty)$

11.

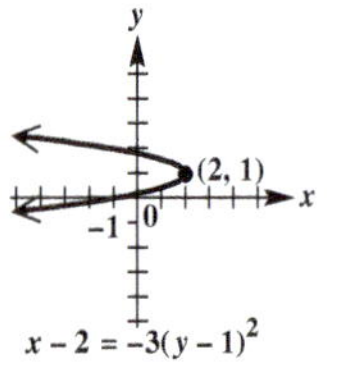

$(-\infty, 2]$; $(-\infty, \infty)$

13.

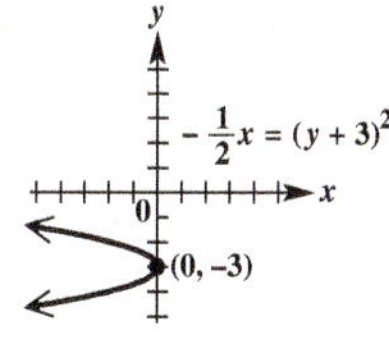

$(-\infty, 0]$; $(-\infty, \infty)$

15.

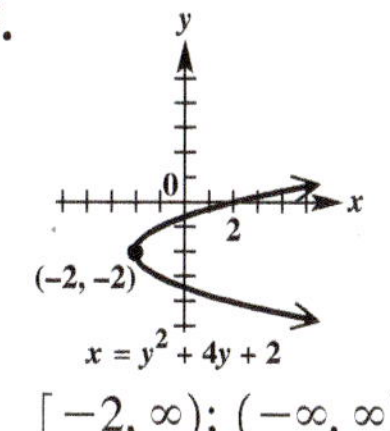

$[-2, \infty)$; $(-\infty, \infty)$

17.

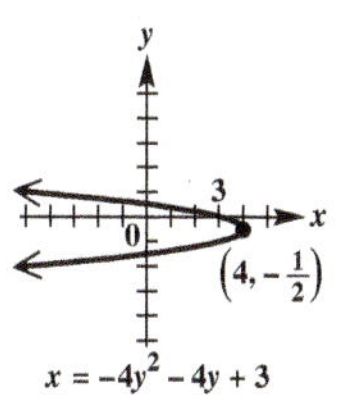

$(-\infty, 4]$; $(-\infty, \infty)$

19.

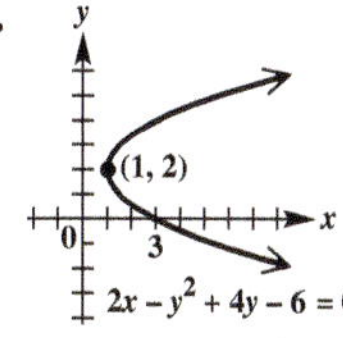

$[1, \infty)$; $(-\infty, \infty)$

21.

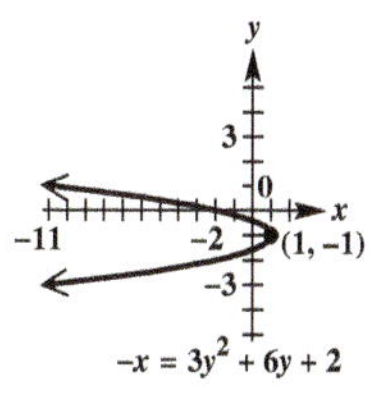

$(-\infty, 1]$; $(-\infty, \infty)$

23. $(0, 6)$; $y = -6$; $x = 0$ **25.** $\left(0, -\frac{1}{16}\right)$; $y = \frac{1}{16}$; $x = 0$ **27.** $(-1, 0)$; $x = 1$; $y = 0$ **29.** $\left(-\frac{1}{128}, 0\right)$; $x = \frac{1}{128}$; $y = 0$ **31.** $(4, 3)$; $x = -2$; $y = 3$ **33.** $(7, -1)$; $y = -9$; $x = 7$

35. $y^2 = 20x$ **37.** $x^2 = y$ **39.** $x^2 = y$ **41.** $y^2 = \frac{4}{3}x$ **43.** $(x-4)^2 = 8(y-3)$ **45.** $(y-6)^2 = 28(x+5)$

47.

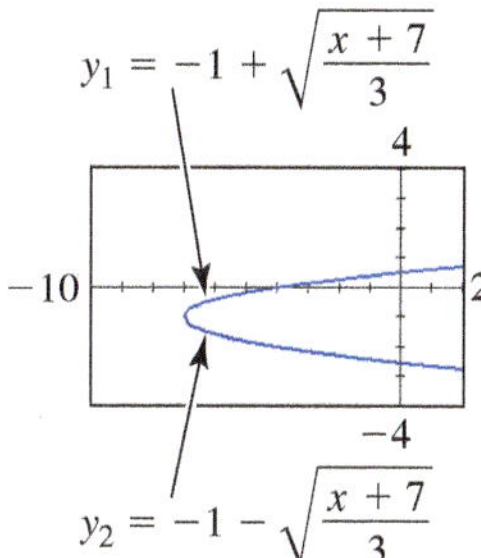

49.

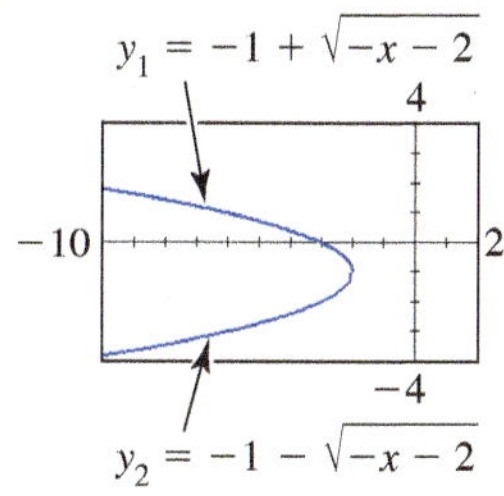

51. **(a)** $y = \frac{11}{5625}x^2$ **(b)** 127.8 ft **53.** 6 ft

55. **(a)** 2000 ft **(b)** $y = 1000 - 0.00025x^2$ **(c)** no

57. **(a)**

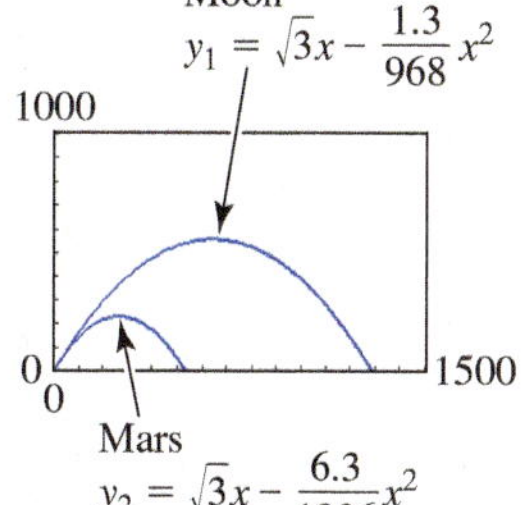

(b) Mars: 230 ft; moon: 558 ft

59. $a + b + c = -5$; $4a - 2b + c = -14$; $4a + 2b + c = -10$ **60.** $\{(-2, 1, -4)\}$

61. It opens to the left because $a = -2$ and $-2 < 0$.

62. $x = -2y^2 + y - 4$

10.2 Exercises

1. **(a)** A **(b)** C **(c)** D **(d)** B

3. $[-4, 4]$; $[-3, 3]$; $(0, 0)$; $(-4, 0)$, $(4, 0)$; $\left(-\sqrt{7}, 0\right)$, $\left(\sqrt{7}, 0\right)$ **5.** $[-3, 7]$; $[-1, 3]$; $(2, 1)$; $(-3, 1)$, $(7, 1)$; $\left(2 - \sqrt{21}, 1\right)$, $\left(2 + \sqrt{21}, 1\right)$

7.

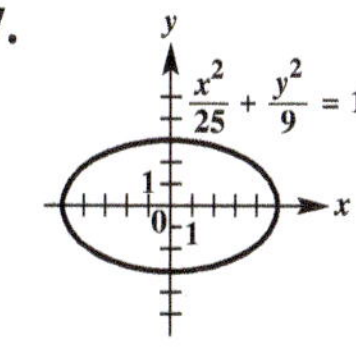

$[-5, 5]$; $[-3, 3]$; $(0, 0)$; $(-5, 0)$, $(5, 0)$; $(0, -3)$, $(0, 3)$; $(-4, 0)$, $(4, 0)$

9.

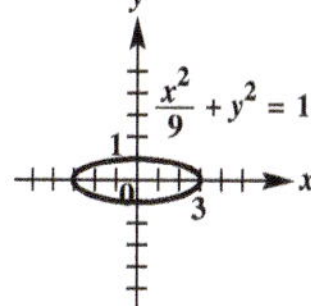

$[-3, 3]$; $[-1, 1]$; $(0, 0)$; $(-3, 0)$, $(3, 0)$; $(0, -1)$, $(0, 1)$; $\left(-2\sqrt{2}, 0\right)$, $\left(2\sqrt{2}, 0\right)$

11.

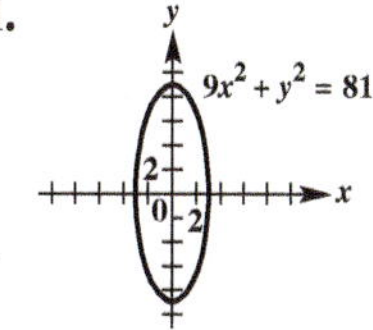

$[-3, 3]$; $[-9, 9]$; $(0, 0)$; $(0, -9)$, $(0, 9)$; $(-3, 0)$, $(3, 0)$; $\left(0, -6\sqrt{2}\right)$, $\left(0, 6\sqrt{2}\right)$

13.

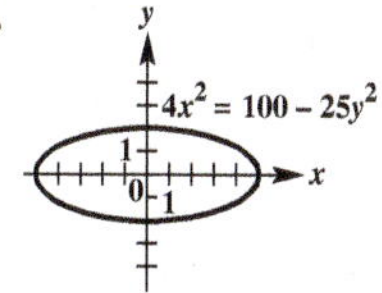

$[-5, 5]$; $[-2, 2]$; $(0, 0)$; $(-5, 0)$, $(5, 0)$; $(0, -2)$, $(0, 2)$; $\left(-\sqrt{21}, 0\right)$, $\left(\sqrt{21}, 0\right)$

15.

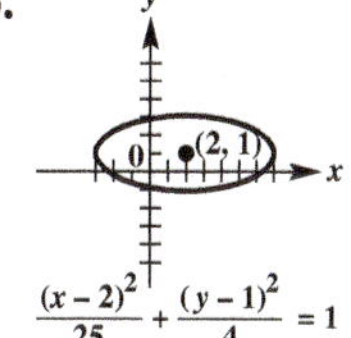

$[-3, 7]$; $[-1, 3]$; $(2, 1)$; $(-3, 1)$, $(7, 1)$; $(2, -1)$, $(2, 3)$; $\left(2 - \sqrt{21}, 1\right)$, $\left(2 + \sqrt{21}, 1\right)$

17.

y
8
(−3, 2)
x
0
2
$\frac{(x+3)^2}{16} + \frac{(y-2)^2}{36} = 1$

$[-7, 1]$; $[-4, 8]$; $(-3, 2)$; $(-3, -4)$, $(-3, 8)$; $(-7, 2)$, $(1, 2)$; $\left(-3, 2 - 2\sqrt{5}\right)$, $\left(-3, 2 + 2\sqrt{5}\right)$

19. $\frac{x^2}{25} + \frac{y^2}{16} = 1$ **21.** $\frac{x^2}{5} + \frac{y^2}{9} = 1$ **23.** $\frac{(x-3)^2}{25} + \frac{(y-1)^2}{16} = 1$ **25.** $\frac{x^2}{9} + \frac{y^2}{25} = 1$ **27.** $\frac{x^2}{72} + \frac{y^2}{81} = 1$ **29.** $\frac{x^2}{25} + \frac{y^2}{9} = 1$ **31.** $\frac{9x^2}{28} + \frac{9y^2}{64} = 1$

33.

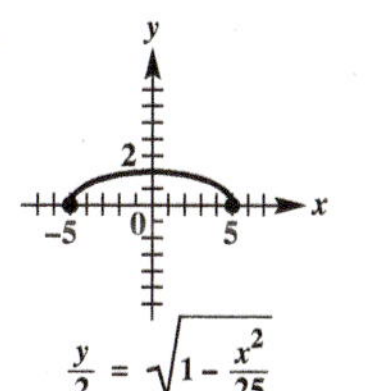

$[-5, 5]$; $[0, 2]$; function

35.

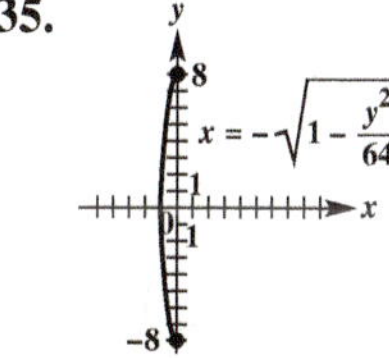

$[-1, 0]$; $[-8, 8]$

37.

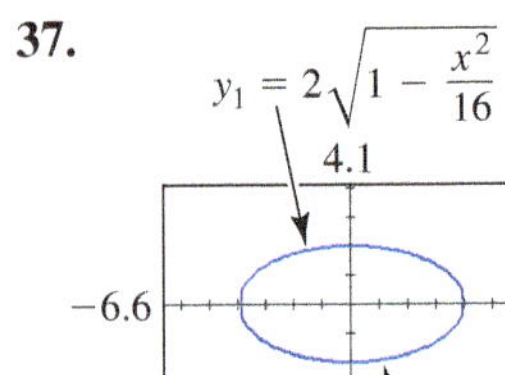

39.

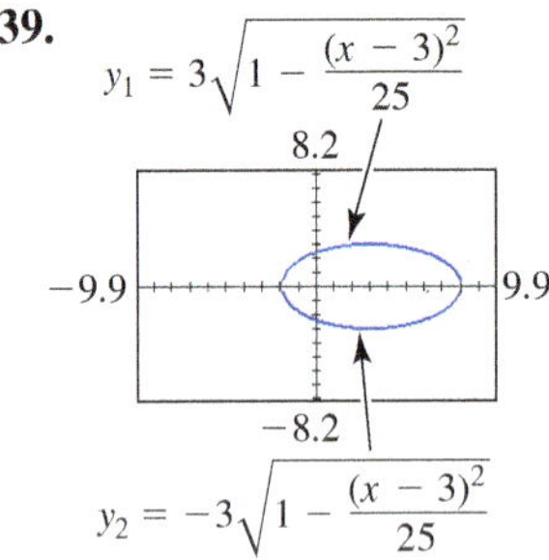

41. $\frac{1}{2}$ **43.** 0.65 **45.** Fixing a string to two points (foci) and sweeping a curve as shown will sketch an ellipse, because the sum of the distances from the two fixed points to the pencil's tip remains constant. **47.** 12 ft tall **49.** 55 million mi

51. (a) Neptune: $\frac{(x - 0.2709)^2}{30.1^2} + \frac{y^2}{30.1^2} = 1$; Pluto: $\frac{(x - 9.8106)^2}{39.4^2} + \frac{y^2}{38.2^2} = 1$

(b)

$y_3 = 38.2\sqrt{1 - \frac{(x - 9.8106)^2}{39.4^2}}$

$y_1 = \sqrt{30.1^2 - (x - 0.2709)^2}$

40; −60; 60; −40

$y_2 = -y_1$ $\quad y_4 = -y_3$

53. $3\sqrt{3}$ units

Chapter 10 Quiz

[10.1, 10.2] **1. (a)** B **(b)** A **(c)** E **(d)** C **(e)** D

[10.1] **2.** $(y - 2)^2 = 12(x + 1)$ **3.** $x^2 = -2y$

[10.2] **4.** $\frac{(x-3)^2}{16} + \frac{(y+2)^2}{25} = 1$ **5.** $\frac{(x+3)^2}{9} + \frac{(y-7)^2}{25} = 1$

[10.1, 10.2]

6. parabola

$y + 4 = (x + 3)^2$

vertex: $(-3, -4)$; focus: $\left(-3, -\frac{15}{4}\right)$; directrix: $y = -\frac{17}{4}$; axis: $x = -3$

7. ellipse

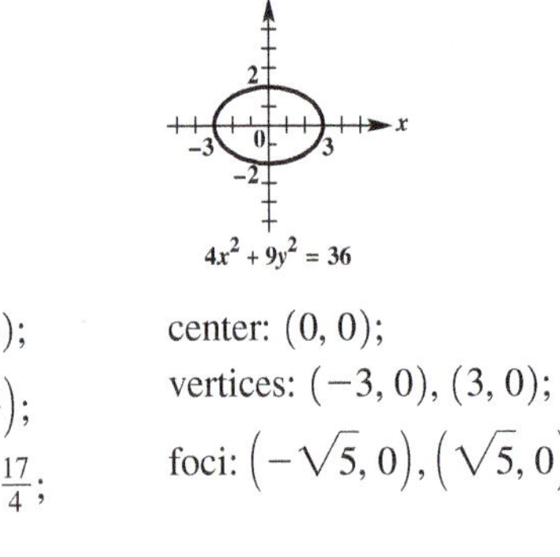

center: $(0, 0)$; vertices: $(-3, 0)$, $(3, 0)$; foci: $\left(-\sqrt{5}, 0\right)$, $\left(\sqrt{5}, 0\right)$

8. parabola

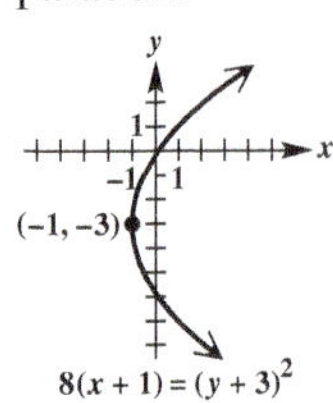

vertex: $(-1, -3)$; focus: $(1, -3)$; directrix: $x = -3$; axis: $y = -3$

9. ellipse

$\frac{(x+3)^2}{25} + \frac{(y+2)^2}{36} = 1$

center: $(-3, -2)$; vertices: $(-3, 4)$, $(-3, -8)$; foci: $\left(-3, -2 + \sqrt{11}\right)$, $\left(-3, -2 - \sqrt{11}\right)$

10. parabola

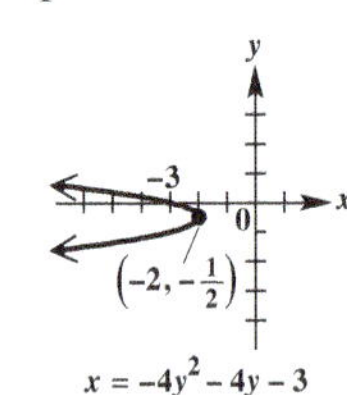

vertex: $\left(-2, -\frac{1}{2}\right)$; focus: $\left(-\frac{33}{16}, -\frac{1}{2}\right)$; directrix: $x = -\frac{31}{16}$; axis: $y = -\frac{1}{2}$

10.3 Exercises

1. A **3.** D **5.** C **7.** D

9. $\frac{x^2}{16} - \frac{y^2}{9} = 1$

$(-\infty, -4] \cup [4, \infty)$; $(-\infty, \infty)$; $(0, 0)$; $(-4, 0)$, $(4, 0)$; $(-5, 0)$, $(5, 0)$; $y = \pm\frac{3}{4}x$

11. $\frac{y^2}{25} - \frac{x^2}{49} = 1$

$(-\infty, \infty)$; $(-\infty, -5] \cup [5, \infty)$; $(0, 0)$; $(0, -5)$, $(0, 5)$; $\left(0, -\sqrt{74}\right)$, $\left(0, \sqrt{74}\right)$; $y = \pm\frac{5}{7}x$

13.

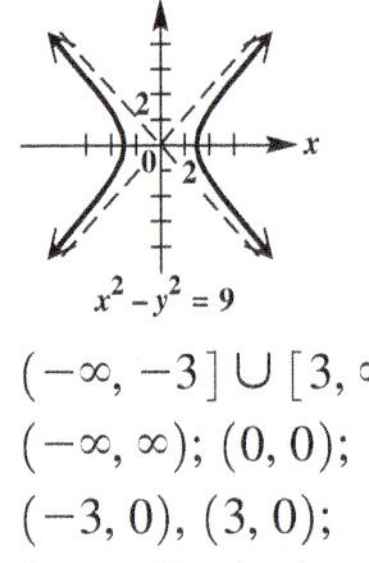

$(-\infty, -3] \cup [3, \infty)$; $(-\infty, \infty)$; $(0, 0)$; $(-3, 0)$, $(3, 0)$; $\left(-3\sqrt{2}, 0\right)$, $\left(3\sqrt{2}, 0\right)$; $y = \pm x$

15.

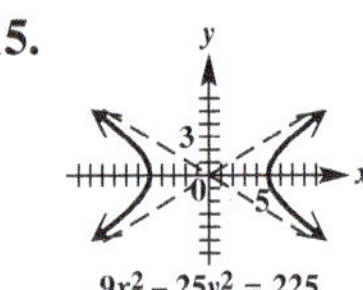

$(-\infty, -5] \cup [5, \infty)$; $(-\infty, \infty)$; $(0, 0)$; $(-5, 0)$, $(5, 0)$; $\left(-\sqrt{34}, 0\right)$, $\left(\sqrt{34}, 0\right)$; $y = \pm\frac{3}{5}x$

17.

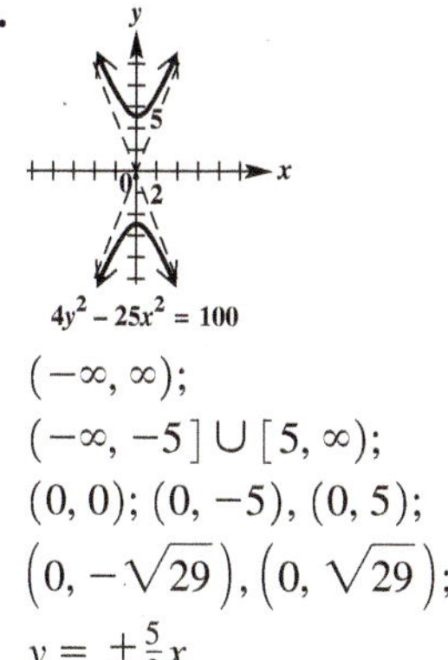

$(-\infty, \infty)$; $(-\infty, -5] \cup [5, \infty)$; $(0, 0)$; $(0, -5)$, $(0, 5)$; $\left(0, -\sqrt{29}\right)$, $\left(0, \sqrt{29}\right)$; $y = \pm\frac{5}{2}x$

19.

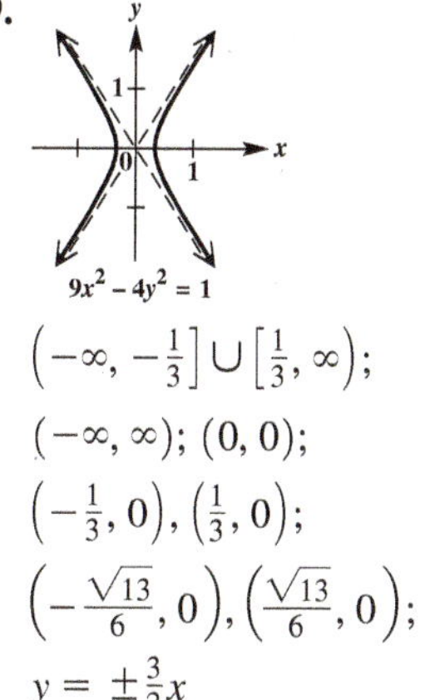

$\left(-\infty, -\frac{1}{3}\right] \cup \left[\frac{1}{3}, \infty\right)$; $(-\infty, \infty)$; $(0, 0)$; $\left(-\frac{1}{3}, 0\right)$, $\left(\frac{1}{3}, 0\right)$; $\left(-\frac{\sqrt{13}}{6}, 0\right)$, $\left(\frac{\sqrt{13}}{6}, 0\right)$; $y = \pm\frac{3}{2}x$

21.

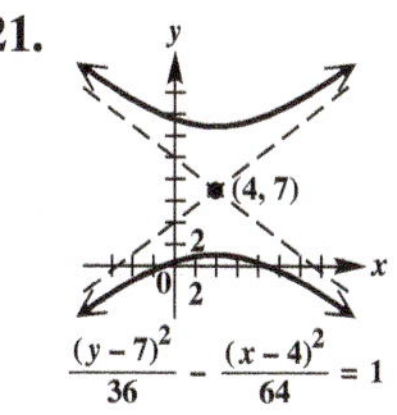

$(-\infty, \infty)$; $(-\infty, 1] \cup [13, \infty)$; $(4, 7)$; $(4, 1)$, $(4, 13)$; $(4, -3)$, $(4, 17)$; $y = \pm\frac{3}{4}(x - 4) + 7$

23.

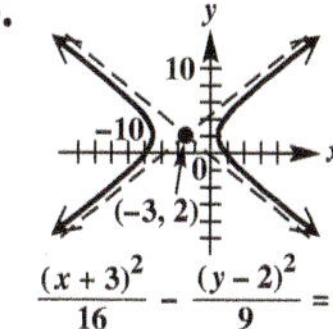

$(-\infty, -7] \cup [1, \infty)$; $(-\infty, \infty)$; $(-3, 2)$; $(-7, 2)$, $(1, 2)$; $(-8, 2)$, $(2, 2)$; $y = \pm\frac{3}{4}(x + 3) + 2$

25.

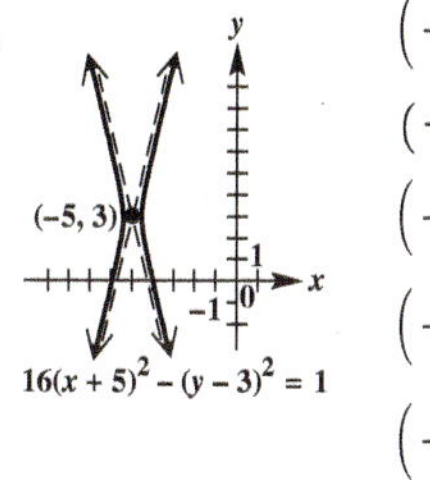

$\left(-\infty, -\frac{21}{4}\right] \cup \left[-\frac{19}{4}, \infty\right)$; $(-\infty, \infty)$; $(-5, 3)$; $\left(-\frac{21}{4}, 3\right)$, $\left(-\frac{19}{4}, 3\right)$; $\left(-5 - \frac{\sqrt{17}}{4}, 3\right)$, $\left(-5 + \frac{\sqrt{17}}{4}, 3\right)$; $y = \pm 4(x + 5) + 3$

27.

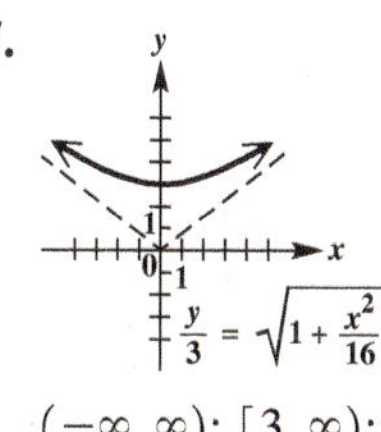

$(-\infty, \infty)$; $[3, \infty)$; function

29. $5x = -\sqrt{1 + 4y^2}$

$\left(-\infty, -\frac{1}{5}\right]$; $(-\infty, \infty)$

31. 1.4 **33.** 1.7 **35.** $\frac{x^2}{9} - \frac{y^2}{16} = 1$ **37.** $\frac{y^2}{36} - \frac{x^2}{144} = 1$ **39.** $\frac{x^2}{9} - 3y^2 = 1$ **41.** $\frac{2y^2}{25} - 2x^2 = 1$ **43.** $\frac{(y-3)^2}{4} - \frac{49(x-4)^2}{4} = 1$ **45.** $\frac{(x-1)^2}{4} - \frac{(y+2)^2}{5} = 1$ **47.** $\frac{y^2}{49} - \frac{x^2}{392} = 1$ **49.** $\frac{25(x+1)^2}{324} - \frac{25(y+1)^2}{2176} = 1$

51.

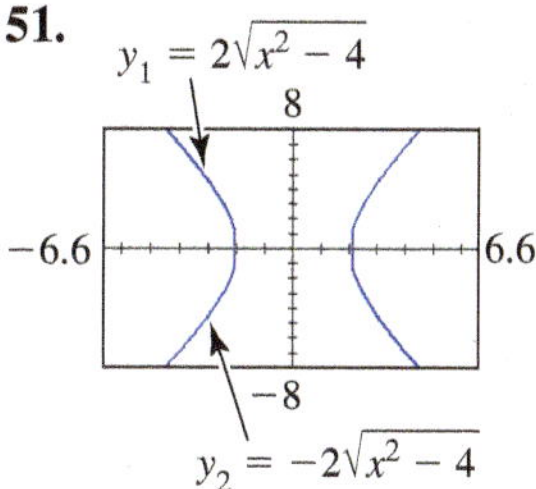

53.

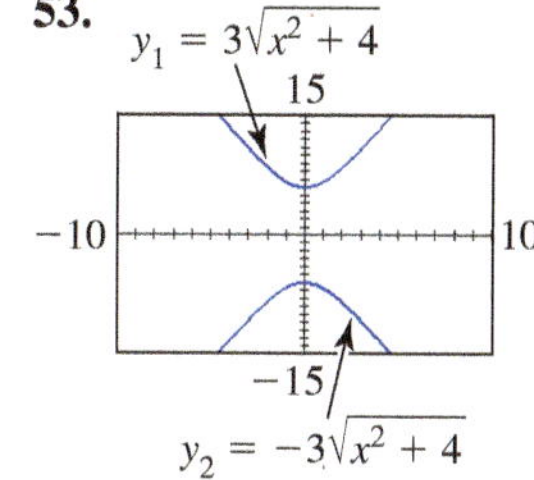

55. **(a)** $x = \sqrt{y^2 + 2.5 \times 10^{-27}}$ **(b)** 1.2×10^{-13} m **59.** **(a)** 50 m **(b)** 69.3 m **61.** $y = \frac{1}{2}\sqrt{x^2 - 4}$ **62.** $y = \frac{1}{2}x$ **63.** $y \approx 24.98$ **64.** $y = 25$ **65.** Because $24.98 < 25$, the graph of $y = \frac{1}{2}\sqrt{x^2 - 4}$ lies below the graph of $y = \frac{1}{2}x$ when $x = 50$. **66.** The y-values on the hyperbola will approach the y-values on the asymptote.

10.4 Exercises

1. circle **3.** parabola **5.** ellipse **7.** hyperbola **9.** circle **11.** parabola **13.** parabola **15.** ellipse **17.** hyperbola **19.** hyperbola **21.** ellipse **23.** circle **25.** parabola **27.** point **29.** parabola **31.** point **33.** no graph

35. hyperbola

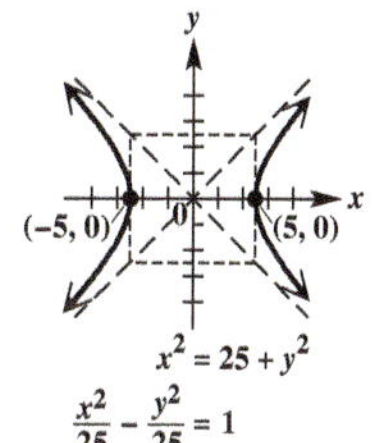

37. parabola

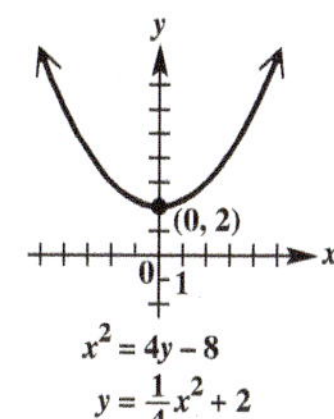

39. parabola

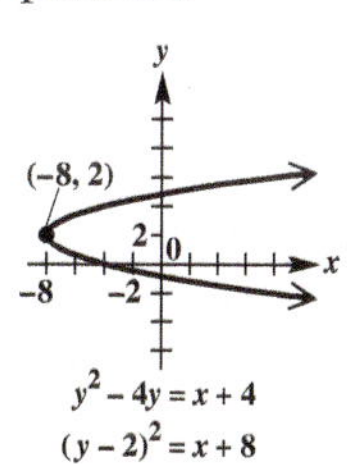

41. circle

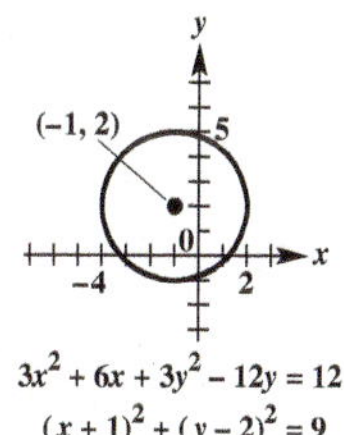

43. ellipse

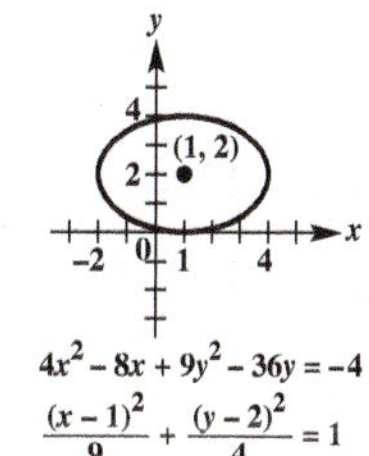

45. $\frac{1}{3}$ **47.** 1 **49.** 1.5 **51.** elliptical **53.** $\left(-\frac{D}{2A}, -\frac{E}{2C}\right)$

Chapter 10 Review Exercises

1.

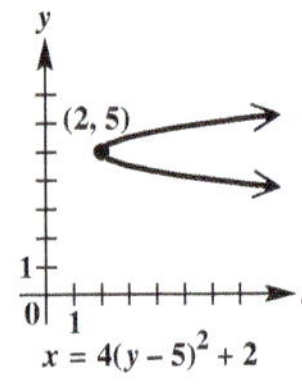

$[2, \infty)$; $(-\infty, \infty)$; $(2, 5)$; $y = 5$

3.

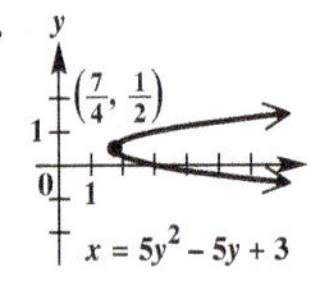

$\left[\frac{7}{4}, \infty\right)$; $(-\infty, \infty)$; $\left(\frac{7}{4}, \frac{1}{2}\right)$; $y = \frac{1}{2}$

5.

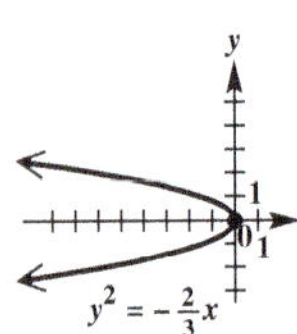

$(-\infty, 0]$; $(-\infty, \infty)$; $\left(-\frac{1}{6}, 0\right)$; $x = \frac{1}{6}$; x-axis

7.

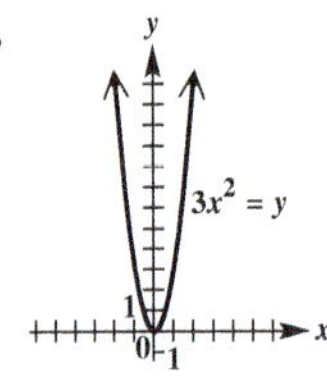

$(-\infty, \infty)$; $[0, \infty)$; $\left(0, \frac{1}{12}\right)$; $y = -\frac{1}{12}$; y-axis

9. $y^2 = 16x$ **11.** $x^2 = \frac{9}{4}y$ **13.** ellipse **15.** hyperbola **17.** parabola **19.** ellipse **21.** F **23.** A **25.** B

27. ellipse

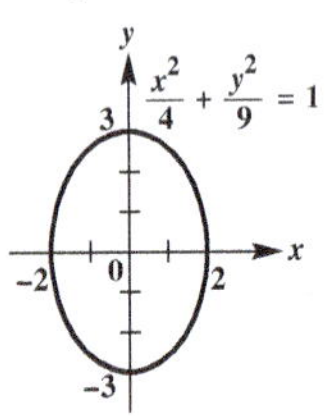

$[-2, 2]$; $[-3, 3]$; $(0, -3)$, $(0, 3)$

29. hyperbola

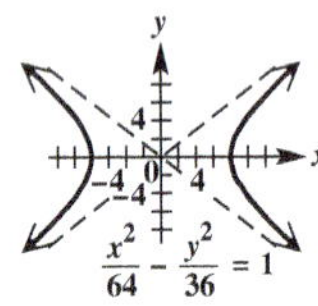

$(-\infty, -8] \cup [8, \infty)$; $(-\infty, \infty)$; $(-8, 0)$, $(8, 0)$; $y = \pm\frac{3}{4}x$

31. circle

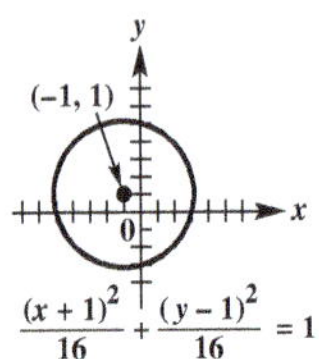

$[-5, 3]$; $[-3, 5]$

33. ellipse

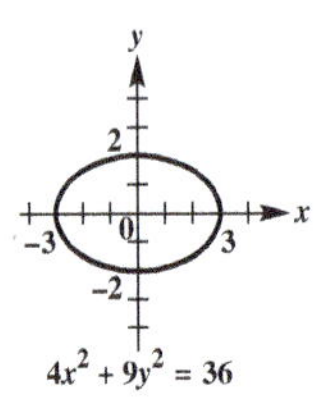

$[-3, 3]$; $[-2, 2]$; $(-3, 0)$, $(3, 0)$

35. ellipse

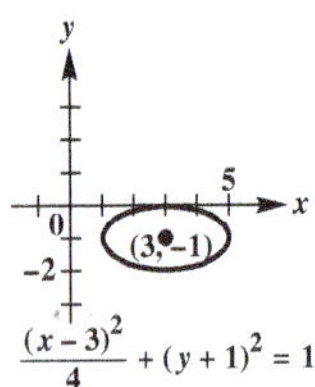

$[1, 5]$; $[-2, 0]$; $(1, -1)$, $(5, -1)$

37. hyperbola

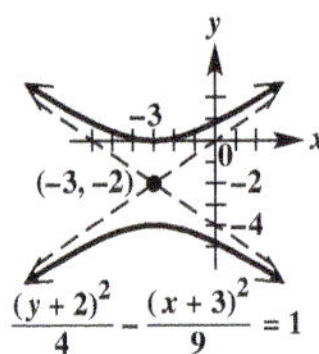

$(-\infty, \infty)$; $(-\infty, -4] \cup [0, \infty)$; $(-3, -4)$, $(-3, 0)$; $y = \pm\frac{2}{3}(x + 3) - 2$

39. circle

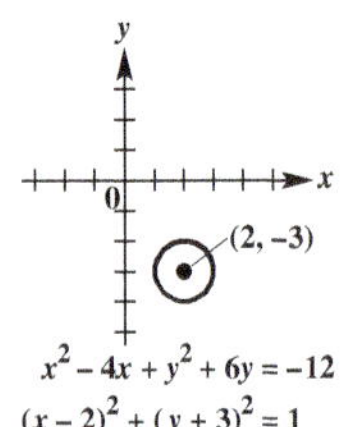

$[1, 3]$; $[-4, -2]$

41. ellipse

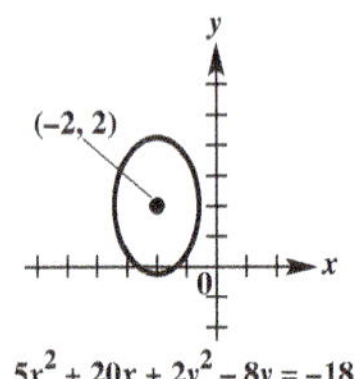

$\left[-2 - \sqrt{2}, -2 + \sqrt{2}\right]$; $\left[2 - \sqrt{5}, 2 + \sqrt{5}\right]$; $\left(-2, 2 - \sqrt{5}\right)$, $\left(-2, 2 + \sqrt{5}\right)$

43.

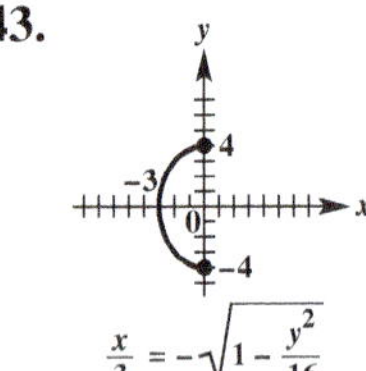

$[-3, 0]$; $[-4, 4]$

45.

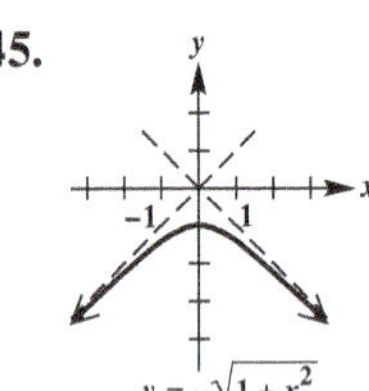

$(-\infty, \infty)$; $(-\infty, -1]$; function

47. $\frac{x^2}{12} + \frac{y^2}{16} = 1$ **49.** $\frac{y^2}{16} - \frac{x^2}{9} = 1$ **51.** $(y - 2)^2 = 12x$ **53.** $\frac{x^2}{25} + \frac{y^2}{21} = 1$ **55.** $\frac{x^2}{9} - \frac{y^2}{16} = 1$ **57.** $\frac{(x-2)^2}{16} + \frac{y^2}{12} = 1$ **59.** $\frac{x^2}{16} - \frac{y^2}{9} = 1$ **61.** $\frac{x^2}{6{,}111{,}883} + \frac{y^2}{432{,}135} = 1$

Chapter 10 Test

[10.1]

1.

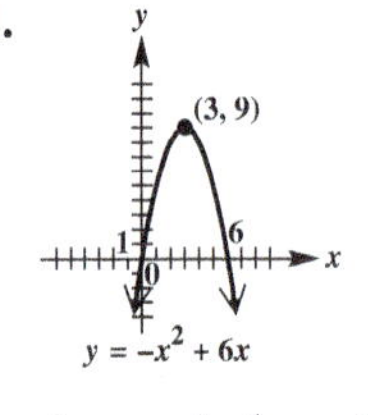

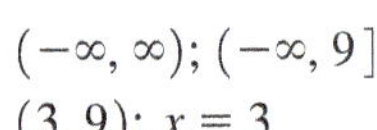

$(-\infty, \infty)$; $(-\infty, 9]$; $(3, 9)$; $x = 3$

2.

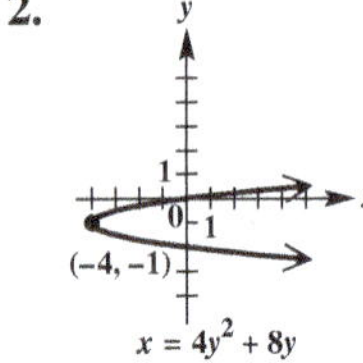

$[-4, \infty)$; $(-\infty, \infty)$; $(-4, -1)$; $y = -1$

3. $\left(\frac{1}{32}, 0\right)$; $x = -\frac{1}{32}$ **4.** $(y - 3)^2 = -\frac{1}{5}(x - 2)$

5. (a) $y = \frac{3}{500}x^2$ **(b)** 41.7 ft

[10.2]

6.

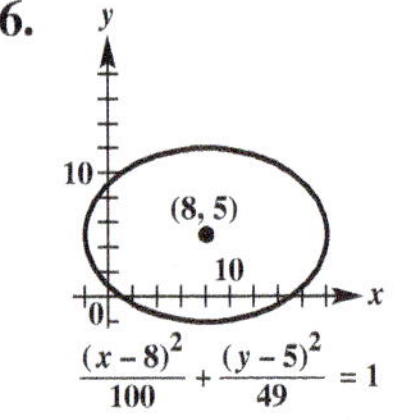

$[-2, 18]$; $[-2, 12]$

7.

$16x^2 + 4y^2 = 64$

$[-2, 2]$; $[-4, 4]$

8.

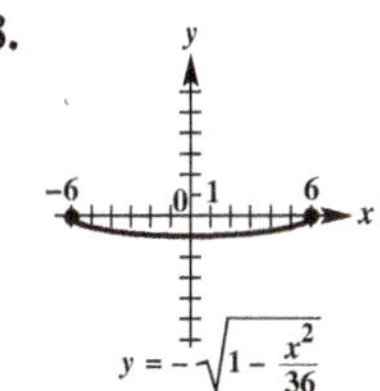

function

9. $\frac{x^2}{9} + \frac{y^2}{4} = 1$

10. $\frac{x^2}{400} + \frac{y^2}{144} = 1$; 10.4 ft

[10.3]

11.

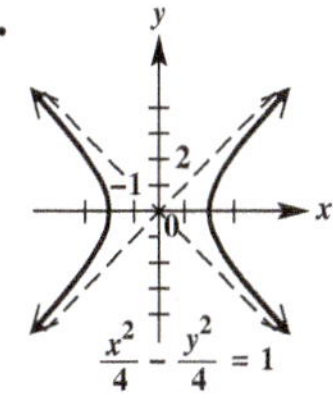

$(-\infty, -2] \cup [2, \infty)$; $(-\infty, \infty)$; $y = \pm x$

12.

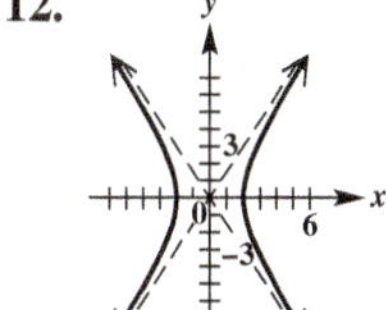

$(-\infty, -2] \cup [2, \infty)$; $(-\infty, \infty)$; $y = \pm\frac{3}{2}x$

13. $\frac{y^2}{25} - \frac{x^2}{11} = 1$ [10.4] **14.** circle **15.** hyperbola **16.** ellipse **17.** parabola **18.** point **19.** no graph

[10.3] **20.** $y_1 = 7\sqrt{\frac{x^2}{25} - 1}$, $y_2 = -7\sqrt{\frac{x^2}{25} - 1}$

Chapter 11 Further Topics in Algebra

11.1 Exercises

1. sequence **3.** recursive

5.

n	a_n
1	7
2	12
3	17
4	22
5	27

7. 85 **9.** 3, −9, 27, −81, 243 **11.** 14, 18, 22, 26, 30 **13.** $\frac{6}{5}, \frac{7}{6}, \frac{8}{7}, \frac{9}{8}, \frac{10}{9}$ **15.** $0, \frac{1}{9}, \frac{2}{27}, \frac{1}{27}, \frac{4}{243}$ **17.** −2, 4, −6, 8, −10 **19.** $1, \frac{7}{6}, 1, \frac{5}{6}, \frac{19}{27}$ **21.** 3, 4, 7, 12, 19 **23.** finite **25.** finite

27. infinite **29.** infinite **31.** −2, 1, 4, 7 **33.** 1, 1, 2, 3 **35.** 2, 4, 12, 48 **37.** 35 **39.** $\frac{25}{12}$ **41.** 288 **43.** 3 **45.** −18 **47.** $\frac{728}{9}$ **49.** 28 **51.** 343 **53.** 30 **55.** 1490 **57.** −154 **59.** $-2 + (-1) + 0 + 1 + 2$; 0 **61.** $-1 + 1 + 3 + 5 + 7$; 15 **63.** $-10 - 4 + 0$; −14 **65.** $0 + \frac{1}{2} + \frac{2}{3} + \frac{3}{4}$; $\frac{23}{12}$ **67.** $124 + 111 + 100 + 91$; 426 **69.** $-3.5 + 0.5 + 4.5 + 8.5$; 10 **71.** $0 + 4 + 16 + 36$; 56 **73.** $-1 - \frac{1}{3} - \frac{1}{5} - \frac{1}{7}$; $-\frac{176}{105}$ **75.** 600 **77.** 1240 **79.** 90 **81.** 220 **83.** 304

There are other acceptable forms of the answers in Exercises 85 and 87.

85. $\sum_{i=1}^{9} \frac{1}{3i}$ **87.** $\sum_{k=1}^{8} \left(-\frac{1}{2}\right)^{k-1}$ **89.** converges to $\frac{1}{2}$ **91.** diverges **93.** converges to $e \approx 2.71828$

95. **(a)** $a_1 = 8$ thousand per acre, $a_2 = 10.4$ thousand per acre, $a_3 = 8.528$ thousand per acre

(b)

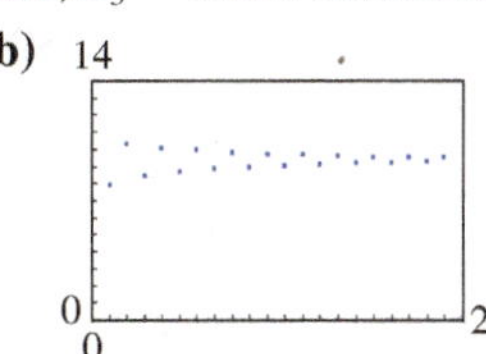

The population density converges to 9.5 thousand per acre.

97. **(a)** $N_{j+1} = 2N_j$ for $j \geq 1$ **(b)** 1840

(c) 15,000; 0; 0; 10

(d) As j becomes large, the values of N_j increase without bound.

99. **(a)** 0.0198026273; $\ln 1.02 \approx 0.0198026273$ **(b)** −0.0304592075; $\ln 0.97 \approx -0.0304592075$

101. **(a)** 2.718254; $e \approx 2.718282$ **(b)** 0.367857; $e^{-1} \approx 0.367879$

11.2 Exercises

1. difference **3.** 6 **5.** D: {1, 2, 3, 4, 5}; R: {−4, −1, 2, 5, 8} **7.** 3 **9.** 55 **11.** 3 **13.** −5 **15.** $x + 2y$ **17.** 8, 14, 20, 26, 32 **19.** 5, 3, 1, −1, −3 **21.** $10 + \sqrt{7}, 10, 10 - \sqrt{7}, 10 - 2\sqrt{7}, 10 - 3\sqrt{7}$ **23.** $a_n = 2n + 3$; $a_8 = 19$ **25.** $a_n = \frac{10}{3}n + \frac{5}{3}$; $a_8 = \frac{85}{3}$ **27.** $a_n = \frac{9}{2}n - 39$; $a_8 = -3$ **29.** $a_n = x + 3n - 3$; $a_8 = x + 21$ **31.** $a_n = s + 2pn - 2p$; $a_8 = s + 14p$ **33.** 3 **35.** 34 **37.** 2.5

In Exercises 39–43, D is the domain and R is the range.

39. $a_n = n - 3$; D: {1, 2, 3, 4, 5, 6}; R: {−2, −1, 0, 1, 2, 3} **41.** $a_n = -\frac{1}{2}n + 3$; D: {1, 2, 3, 4, 5, 6}; R: {0, 0.5, 1, 1.5, 2, 2.5} **43.** $a_n = -20n + 30$; D: {1, 2, 3, 4, 5}; R: {−70, −50, −30, −10, 10} **45.** 215 **47.** 230 **49.** 160 **51.** 77.5 **53.** 55π **55.** 3240 **57.** 2500 **59.** 3660 **61.** $a_1 = 7, d = 5$ **63.** $a_1 = 5, d = -2$ **65.** $a_1 = 1, d = -\frac{20}{11}$ **67.** 18 **69.** 140 **71.** −621 **73.** 500,500 **75.** 10,100 **77.** 6150 **79.** 328.3 **81.** 172.884 **83.** 1281 **85.** 4680 **87.** 54,800 **89.** 713 in. **91.** yes

11.3 Exercises

1. ratio **3.** 24 **5.** 155 **7.** geometric; $r = -2$ **9.** neither **11.** **(a)** \$5.12 **(b)** \$10.23 **13.** **(a)** \$163.84 **(b)** \$327.67

In Exercises 15–25, there may be other ways to express a_n.

15. $a_5 = 80$; $a_n = 5(-2)^{n-1}$ **17.** $a_5 = -108$; $a_n = -\frac{4}{3}(3)^{n-1}$ **19.** $a_5 = -729$; $a_n = -9(-3)^{n-1}$ **21.** $a_5 = -324$; $a_n = -4(3)^{n-1}$ **23.** $a_5 = \frac{125}{4}$; $a_n = \frac{4}{5}\left(\frac{5}{2}\right)^{n-1}$ **25.** $a_5 = \frac{5}{8}$; $a_n = 10\left(-\frac{1}{2}\right)^{n-1}$ **27.** 2; −3 **29.** $\frac{1}{5}$; 125 **31.** ±0.1; 5000 **33.** 682 **35.** $\frac{99}{8}$ **37.** 860.95 **39.** −183 **41.** $\frac{189}{4}$ **43.** 2032 **45.** The sum exists if $|r| < 1$. **47.** 2; diverges **49.** $\frac{1}{2}$ **53.** 27 **55.** $\frac{3}{20}$ **57.** 4 **59.** $\frac{1}{2}$ **61.** $-\frac{8}{15}$ **63.** $\frac{3}{4}$ **65.** 97.739 **67.** 0.212 **69.** **(a)** $a_1 = 1169$; $r = 0.916$ **(b)** $a_{10} = 531$; $a_{20} = 221$; A person who is 10 yr from retirement should have savings of 531% of his

or her annual salary. A person 20 yr from retirement should have savings of 221% of his or her annual salary.
71. **(a)** $a_n = a_1 \cdot 2^{n-1}$ **(b)** 15 (rounded from 14.29) **(c)** 560 min, or 9 hr, 20 min **73.** 488 **75.** 62; 2046 **77.** $\frac{1}{64}$ m **79.** Option 2 **81.** $10,159.11 **83.** $25,423.18 **85.** $28,107.41 **87.** $64,060.60 **91.** yes

Summary Exercises on Sequences and Series

1. geometric; $r = 2$ **2.** arithmetic; $d = 3$ **3.** arithmetic; $d = -\frac{5}{2}$ **4.** neither **5.** geometric; $r = \frac{4}{3}$ **6.** geometric; $r = -3$ **7.** neither **8.** arithmetic; $d = -3$ **9.** neither **10.** geometric; $r = -\sqrt{5}$ **11.** geometric; $3(2)^{n-1}$; 3069 **12.** arithmetic; $4n - 2$; 200 **13.** arithmetic; $-\frac{3}{2}n + \frac{11}{2}$; $-\frac{55}{2}$ **14.** geometric; $\frac{3}{2}\left(\frac{2}{3}\right)^{n-1}$, or $\left(\frac{2}{3}\right)^{n-2}$; $\frac{58{,}025}{13{,}122}$ **15.** geometric; $3(-2)^{n-1}$; -1023 **16.** arithmetic; $-3n - 2$; -185 **17.** diverges **18.** $\frac{1111}{500}$ **19.** -1850 **20.** 1092 **21.** $-\frac{4}{3}$ **22.** diverges **23.** 144 **24.** $\frac{1}{4}$ **25.** diverges **26.** $0.3 + 0.03 + 0.003 + \cdots$; $\frac{1}{3}$

11.4 Exercises

1. sum **3.** 1 **5.** 8 **7.** x^8; y^8 **9.** $5p^4q$ **11.** 20 **13.** 35 **15.** 56 **17.** 45 **19.** 1 **21.** n **23.** 56 **25.** 4950 **27.** 1 **29.** 12 **31.** $x^6 + 6x^5y + 15x^4y^2 + 20x^3y^3 + 15x^2y^4 + 6xy^5 + y^6$ **33.** $p^5 - 5p^4q + 10p^3q^2 - 10p^2q^3 + 5pq^4 - q^5$ **35.** $r^{10} + 5r^8s + 10r^6s^2 + 10r^4s^3 + 5r^2s^4 + s^5$ **37.** $p^4 + 8p^3q + 24p^2q^2 + 32pq^3 + 16q^4$ **39.** $2401p^4 - 2744p^3q + 1176p^2q^2 - 224pq^3 + 16q^4$ **41.** $729x^6 - 2916x^5y + 4860x^4y^2 - 4320x^3y^3 + 2160x^2y^4 - 576xy^5 + 64y^6$ **43.** $\frac{m^6}{64} - \frac{3m^5}{16} + \frac{15m^4}{16} - \frac{5m^3}{2} + \frac{15m^2}{4} - 3m + 1$ **45.** $4r^4 + \frac{8\sqrt{2}r^3}{m} + \frac{12r^2}{m^2} + \frac{4\sqrt{2}r}{m^3} + \frac{1}{m^4}$ **47.** $\frac{1}{x^{16}} + \frac{4}{x^8} + 6 + 4x^8 + x^{16}$ **49.** $-3584h^3j^5$ **51.** $74{,}613a^{12}b^{16}$ **53.** $38{,}760x^6y^{42}$ **55.** $90{,}720x^{28}y^{12}$ **57.** 11 **59.** exact: 3,628,800; approximate: 3,598,695.619 **60.** 0.830% **61.** exact: 479,001,600; approximate: 475,687,486.5; 0.692% **62.** exact: 6,227,020,800; approximate: 6,187,239,475; 0.639%; As n gets larger, the percent error decreases.

11.5 Exercises

1. $3 + 6 + 9 + 12 = \frac{3(4)(4+1)}{2}$; true **3.** $\frac{1}{2} + \frac{1}{2^2} + \frac{1}{2^3} + \frac{1}{2^4} = \frac{2^4 - 1}{2^4}$; true **5.** $2^4 < 2(4)$; false **7.** S_1: $1 = 1^2$; S_2: $1 + 3 = 2^2$; S_3: $1 + 3 + 5 = 3^2$; S_4: $1 + 3 + 5 + 7 = 4^2$; S_5: $1 + 3 + 5 + 7 + 9 = 5^2$

Although we do not usually give proofs, the answers for Exercises 9 and 17 are given here.

9. **(a)** $3(1) = 3$ and $\frac{3(1)(1+1)}{2} = \frac{6}{2} = 3$, so S_n is true for $n = 1$. **(b)** $3 + 6 + 9 + \cdots + 3k = \frac{3(k)(k+1)}{2}$ **(c)** $3 + 6 + 9 + \cdots + 3(k+1) = \frac{3(k+1)[(k+1)+1]}{2}$ **(d)** Add $3(k+1)$ to each side of the equation in part (b). Simplify the expression on the right side to match the right side of the equation in part (c). **(e)** Because S_n is true for $n = 1$ and S_n is true for $n = k + 1$ when it is true for $n = k$, S_n is true for every positive integer n.

17. **(a)** $\frac{1}{1 \cdot 2} = \frac{1}{2}$ and $\frac{1}{1+1} = \frac{1}{2}$, so S_n is true for $n = 1$. **(b)** $\frac{1}{1 \cdot 2} + \frac{1}{2 \cdot 3} + \frac{1}{3 \cdot 4} + \cdots + \frac{1}{k(k+1)} = \frac{k}{k+1}$ **(c)** $\frac{1}{1 \cdot 2} + \frac{1}{2 \cdot 3} + \cdots + \frac{1}{(k+1)[(k+1)+1]} = \frac{k+1}{(k+1)+1}$ **(d)** Add the last term on the left of the equation in part (c) to each side of the equation in part (b). Simplify the right side until it matches the right side in part (c). **(e)** Because S_n is true for $n = 1$ and S_n is true for $n = k + 1$ when it is true for $n = k$, S_n is true for every positive integer n.

For Exercises 21–31, we show only the proof for Exercise 25.

25. **(a)** $(a^m)^1 = a^m$ and $a^{m(1)} = a^m$, so S_n is true for $n = 1$. **(b)** $(a^m)^k = a^{mk}$ **(c)** $(a^m)^{(k+1)} = a^{m(k+1)}$ **(d)** $(a^m)^k \cdot (a^m)^1 = a^{mk} \cdot (a^m)^1$

$(a^m)^{(k+1)} = a^{(mk+m)}$ Product rule for exponents

$(a^m)^{(k+1)} = a^{m(k+1)}$ Factor.

(e) Because S_n is true for $n = 1$ and S_n is true for $n = k + 1$ when it is true for $n = k$, S_n is true for every positive integer n.

35. $\frac{4^{n-1}}{3^{n-2}}$, or $3\left(\frac{4}{3}\right)^{n-1}$ **37.** $2^n - 1$

Chapter 11 Quiz

[11.1–11.3] **1.** $-2, -6, -10, -14, -18$; arithmetic **2.** $1, -\frac{1}{2}, \frac{1}{4}, -\frac{1}{8}, \frac{1}{16}$; geometric **3.** 5, 3, 18, 27, 81; neither [11.2] **4.** 12 [11.2, 11.3] **5.** **(a)** 430 **(b)** $-\frac{1705}{128}$ **6.** **(a)** -1215 **(b)** diverges **(c)** 3 [11.4] **7.** $x^5 - 15x^4y + 90x^3y^2 - 270x^2y^3 + 405xy^4 - 243y^5$ **8.** $\frac{5}{4}xy^4$ **9.** **(a)** 362,880 **(b)** 210

11.6 Exercises

1. permutation; combination **3.** 6 **5.** 12 **7.** 24 **9.** 160 **11.** 5040 **13.** 3360 **15.** 132 **17.** 72 **19.** 5 **21.** 6 **23.** 1 **25.** 495 **27.** 1,860,480 **29.** 259,459,200 **31.** 15,504 **33.** 6435 **35.** **(a)** permutation **(b)** permutation **(c)** combination **(d)** combination **(e)** permutation **(f)** permutation **(g)** combination **37.** 40 **39.** **(a)** 27,600 **(b)** 35,152 **(c)** 1104 **41.** 5040 **43.** **(a)** 17,576,000 **(b)** 17,576,000 **(c)** 456,976,000 **45.** 362,880 **47.** 120 **49.** 2730 **51.** **(a)** 120 **(b)** 6 **53.** 3,838,380 **55.** 12,650 **57.** **(a)** 15 **(b)** 10 **59.** 105; 1365 **61.** **(a)** 84 **(b)** 10 **(c)** 40 **(d)** 28 **63.** 1680 **65.** 15 **67.** 6,227,020,800 **69.** **(a)** 56 **(b)** 462 **(c)** 3080 **(d)** 8526 **71.** 1,000,000 **73.** 4096 **75.** 6

11.7 Exercises

1. 2; $\frac{1}{2}$ **3.** 4; $\frac{1}{4}$ **5.** $\frac{1}{16}$ **7.** $\{(H, H), (H, T), (T, H), (T, T)\}$ **9.** $\{(H, H, H), (H, H, T), (H, T, H), (T, H, H), (H, T, T), (T, H, T), (T, T, H), (T, T, T)\}$ **11.** $\{(1, 1), (1, 2), (1, 3), (2, 1), (2, 2), (2, 3), (3, 1), (3, 2), (3, 3)\}$ **13.** **(a)** $\{(H, H), (T, T)\}; \frac{1}{2}$ **(b)** $\{(H, H), (H, T), (T, H)\}; \frac{3}{4}$ **15.** **(a)** $\{(H, H, H), (T, T, T)\}; \frac{1}{4}$ **(b)** $\{(H, T, T), (T, H, T), (T, T, H), (T, T, T)\}; \frac{1}{2}$ **17.** **(a)** $\{(1, 1), (2, 2), (3, 3)\}; \frac{1}{3}$ **(b)** $\{(1, 1), (1, 3), (2, 1), (2, 3), (3, 1), (3, 3)\}; \frac{2}{3}$ **(c)** $\{(2, 1), (2, 3)\}; \frac{2}{9}$ **19.** 0.143 **21.** 3 to 7 **23.** **(a)** $\frac{1}{4}$ **(b)** $\frac{3}{4}$ **(c)** $\frac{1}{2}$ **(d)** $\frac{11}{26}$ **(e)** 1 to 3 **25.** **(a)** 0.246 **(b)** 0.753 **(c)** 0.493 **(d)** 10,517 to 12,996 **27.** $\frac{1}{28{,}561} \approx 0.0000350$ **29.** **(a)** 0.765 **(b)** 0.235 **(c)** 0.247 **(d)** 0.945 **31.** 0.3 **33.** 0.243 **35.** 0.036 **37.** 0.28 **39.** 0.72 **41.** **(a)** 0.048 **(b)** 0.977 **(c)** 0.103 **(d)** 0.897 **43.** **(a)** 0.404 **(b)** 0.047 **(c)** 0.002

Chapter 11 Review Exercises

1. $\frac{1}{2}, \frac{2}{3}, \frac{3}{4}, \frac{4}{5}, \frac{5}{6}$; neither **3.** 8, 10, 12, 14, 16; arithmetic **5.** 5, 2, −1, −4, −7; arithmetic **7.** 4, 4.5, 5, 5.5, 6 **9.** $3\pi - 2, 2\pi - 1, \pi, 1, -\pi + 2$ **11.** $-5, -1, -\frac{1}{5}, -\frac{1}{25}, -\frac{1}{125}$ **13.** $\pm 1; -8\left(\frac{1}{2}\right)^{n-1} = -\left(\frac{1}{2}\right)^{n-4}$ or $-8\left(-\frac{1}{2}\right)^{n-1} = \left(-\frac{1}{2}\right)^{n-4}$ **15.** $-x + 61$ **17.** 612 **19.** $\frac{4}{25}$ **21.** −40 **23.** 1 **25.** $\frac{73}{12}$ **27.** 3,126,250 **29.** $\frac{4}{3}$ **31.** $-\frac{4}{5}$ **33.** 36 **35.** diverges **37.** −10

In Exercises 39 and 41, other answers are possible.

39. $\sum_{i=1}^{15}(-5i + 9)$ **41.** $\sum_{i=1}^{6} 4(3)^{i-1}$ **43.** $x^4 + 8x^3y + 24x^2y^2 + 32xy^3 + 16y^4$ **45.** $243x^{5/2} - 405x^{3/2} + 270x^{1/2} - 90x^{-1/2} + 15x^{-3/2} - x^{-5/2}$ **47.** $-3584x^3y^5$ **49.** $x^{12} + 24x^{11} + 264x^{10} + 1760x^9$ **55.** 72 **57.** 56 **59.** 252 **61.** $1,742,304 **63.** $1,090,394; Answers will vary. **65.** 90 **67.** **(a)** 84 **(b)** 45 **69.** 456,976,000; 258,336,000 **71.** **(a)** $\frac{4}{15}$ **(b)** $\frac{2}{3}$ **(c)** 0 **(d)** 3 to 2 **73.** **(a)** 0.137 **(b)** 0.134 **(c)** 0.764 **75.** 0.296

Chapter 11 Test

[11.1–11.3] **1.** −3, 6, −11, 18, −27; neither **2.** $-\frac{3}{2}, -\frac{3}{4}, -\frac{3}{8}, -\frac{3}{16}, -\frac{3}{32}$; geometric **3.** 2, 3, 7, 13, 27; neither [11.2] **4.** 49 [11.3] **5.** $-\frac{32}{3}$ [11.2] **6.** 110 [11.3] **7.** −1705 [11.2] **8.** 2385 [11.3] **9.** −186 **10.** The sum does not exist. **11.** $\frac{108}{7}$ [11.4] **12.** $x^6 + 6x^5y + 15x^4y^2 + 20x^3y^3 + 15x^2y^4 + 6xy^5 + y^6$ **13.** $16x^4 - 96x^3y + 216x^2y^2 - 216xy^3 + 81y^4$ **14.** $60w^4y^2$ [11.4, 11.6] **15.** 40,320 **16.** 45 **17.** 35 **18.** 990 [11.6] **20.** 24 **21.** 210; 90 **22.** 32,760 [11.7] **23.** **(a)** $\frac{1}{26}$ **(b)** $\frac{10}{13}$ **(c)** $\frac{4}{13}$ **(d)** 3 to 10 **24.** 0.92 **25.** 0.201

Appendices

Appendix A Exercises

1.

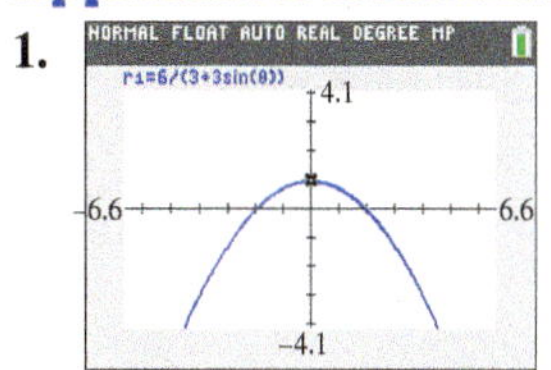

3.

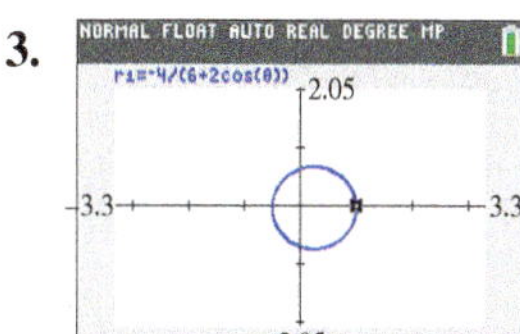

5.

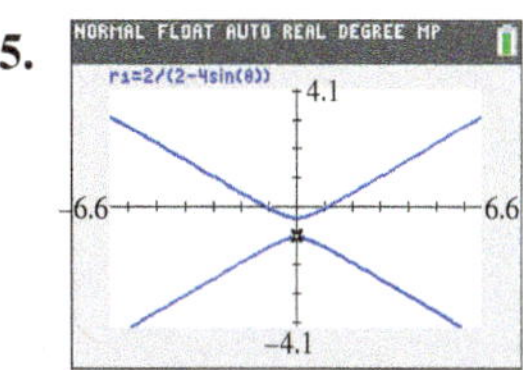

7.

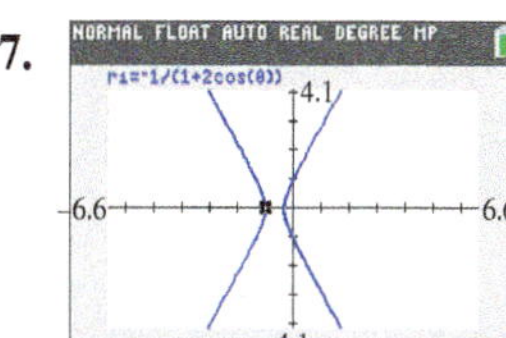

9.

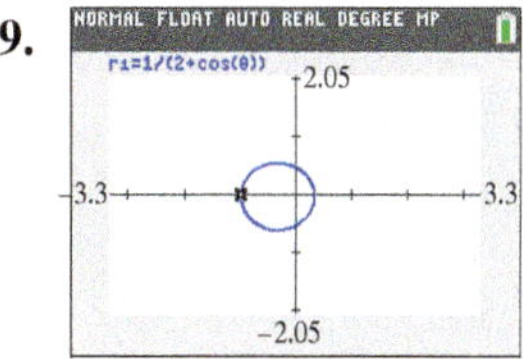

11. $r = \frac{3}{1 + \cos\theta}$ **13.** $r = \frac{5}{1 - \sin\theta}$ **15.** $r = \frac{20}{5 + 4\cos\theta}$; ellipse **17.** $r = \frac{40}{4 - 5\sin\theta}$; hyperbola **19.** ellipse; $8x^2 + 9y^2 - 12x - 36 = 0$ **21.** hyperbola; $3x^2 - y^2 + 8x + 4 = 0$ **23.** ellipse; $4x^2 + 3y^2 - 6y - 9 = 0$ **25.** parabola; $x^2 - 10y - 25 = 0$

Appendix B Exercises

1. circle or ellipse or a point **3.** hyperbola or two intersecting lines **5.** parabola or one line or two parallel lines **7.** 30° **9.** 60° **11.** 22.5°

13.

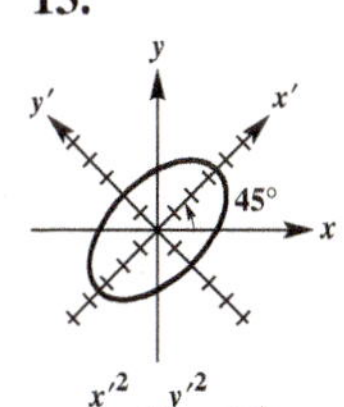

15.

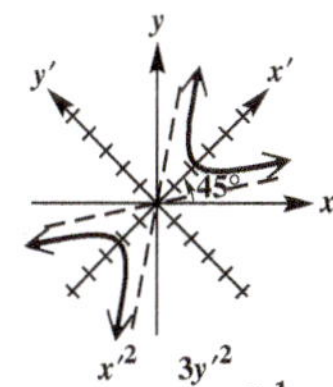

17.

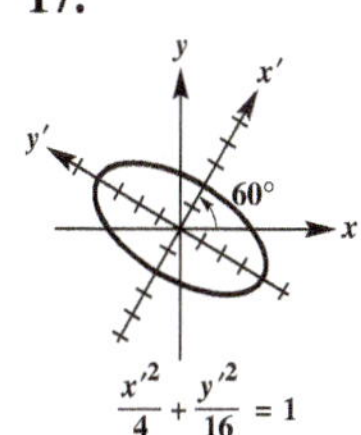

19.

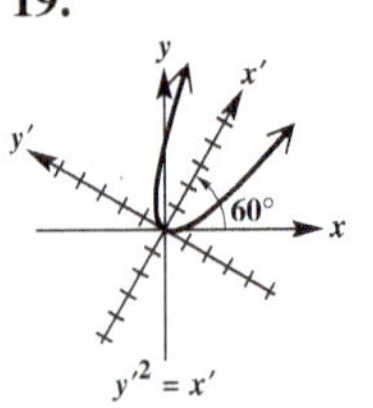
y
x′
y′
60°
x
$y'^2 = x'$

21.

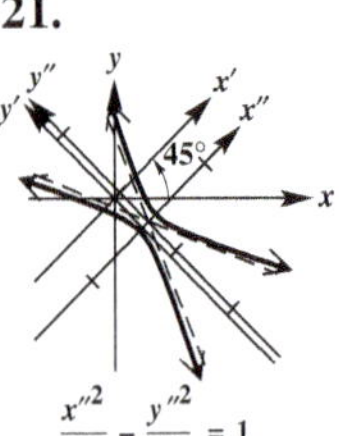
y′
y″
y
x′
x″
45°
x
$\frac{x''^2}{2} - \frac{y''^2}{10} = 1$

23.

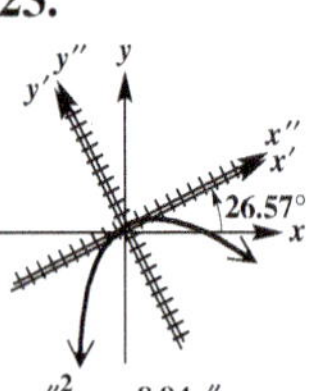
y′
y″
y
x″
x′
26.57°
x
$x''^2 \approx -8.94y''$

25.

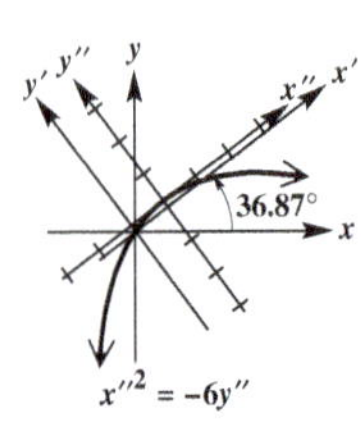
y′
y″
y
x″
x′
36.87°
x
$x''^2 = -6y''$

Photo Credits

CHAPTER R: **1** 123RF **17** Flashpics/Fotolia **22** McClatchy-Tribune/Tribune Content Agency LLC/Alamy **23** Courtesy of John Hornsby **35** Courtesy of Callie Daniels **55** Bestimagesever.com/Shutterstock **64** Satyrenko/Fotolia **76** Danshutter/Shutterstock **82** David Anderson/Shutterstock

CHAPTER 1: **87** Gunter Marx/Dorling Kindersley **94** Courtesy of Callie Daniels **96** AP Images **99** Michaeljung/Shutterstock **101** (left) DC5/WENN.com/Newscom; (right) Scanrail/Fotolia **103** 06photo/Shutterstock **104** Armadillo Stock/Shutterstock **105** (top) Bikeriderlondon/Shutterstock; (bottom) Kadmy/Fotolia **125** Jamie Roach/Shutterstock **126** Photos/Getty Images **127** Cphoto/Fotolia **128** (top) David Phillips/Shutterstock; (bottom) Courtesy of Callie Daniels **130** Marco Mayer/Shutterstock **131** Courtesy of Callie Daniels **133** Rich Kane/Icon Sportswire 781/Newscom **134** Vladislav Gajic/Fotolia **135** Courtesy of Callie Daniels **147** Columbia Pictures/Everett Collection **160** Snw/Fotolia **161** Picsfive/Shutterstock **166** Sarymsakov.com/Fotolia **168** My Good Images/Shutterstock **176** Monticellllo/Fotolia **177** Jules Selmes/Pearson Education, Inc. **182** Sergei Butorin/Shutterstock

CHAPTER 2: **183** Nancy Bauer/Shutterstock **185** Library of Congress Prints and Photographs Division [LC-USZ62-61365] **188** Foodpics/Shutterstock **199** Baloncici/Shutterstock **202** From the collection of John Hornsby **205** Courtesy of John Hornsby **240** Bikeriderlondon/Shutterstock **245** IvanRu/Shutterstock **246** Christian Martinez Kempin/123RF **254** Arsel/Fotolia **282** Jamdesign/Fotolia **291** (top) Catalin Petolea/Shutterstock; (bottom) Courtesy of Terry McGinnis

CHAPTER 3: **303** Jack Z Young/Shutterstock **311** Avava/Shutterstock **330** Bilwissedition Ltd. & Co./KG/Alamy **349** EML/Shutterstock **372** Robert Wilson/Fotolia **388** JonMilnes/Shutterstock **389** John Foxx Collection/Imagestate **391** Izf/Shutterstock **397** Monkey Business/Fotolia **402** Courtesy of Callie Daniels

CHAPTER 4: **405** Shi Yali/Shutterstock **434** Critterbiz/Shutterstock **443** National Geographic Image Collection/Alamy **451** Daizuoxin/Shutterstock **455** Kletr/Shutterstock **458** Courtesy of Callie Daniels **470** Sasha Buzko/Shutterstock **475** Rafael Ramirez Lee/Shutterstock **477** Mark Bassett/Pearson Education, Inc. **480** Mironov/Shutterstock **481** Courtesy of Callie Daniels **482** Crok Photography/Shutterstock **485** Courtesy of Callie Daniels **494** Dgareri/Dreamstime

CHAPTER 5: **497** Sergii Mostovyi/Fotolia **534** Brad Pict/Fotolia **535** VladKol/Shutterstock **536** Smileus/Shutterstock **537** (left) Unknown photographer; (right) Copyright 1899, 1900 by Elmer A. Lyman and Edwin C. Goddard **538** Africa Studio/Fotolia **543** 123RF **546** Courtesy of John Hornsby **549** F9photos/Shutterstock **554** Courtesy of Callie Daniels

CHAPTER 6: **565** Victoria Kalinina/Shutterstock **572** Cecilia Lim H M/Shutterstock **585** Andres Rodriguez/Fotolia **591** J. Bell, Cornell University/M. Wolff, Space Science Center/NASA **610** Gibleho/Fotolia **638** Alexander Sakhatovsky/Shutterstock **645** Selezeni/Fotolia **649** Caroline K Smith MD/Fotolia

CHAPTER 7: **653** Plampy/Shutterstock **665** Serggod/Shutterstock **681** (top) Ysbrand Cosijn/Shutterstock; (bottom) Mimagephotos/Fotolia **695** (top) Lithian/123RF; (bottom) US Navy Photo/Alamy **707** Robert Daly/OJO Images/Getty Images **719** MindStudio/Pearson Education Ltd **722** Svetoslav Radkov/Fotolia

CHAPTER 8: **743** Henryk Sadura/Shutterstock **759** Stephanie Kennedy/Shutterstock **763** Liszt collection/Alamy **780** Sculpies/Shutterstock **781** Gary Blakeley/Shutterstock **796** Ross Anderson **804** Pearson Education, Inc. **809** Ross Anderson **810** Ross Anderson **822** NatUlrich/Shutterstock

CHAPTER 9: **845** Jon Erickson/Flickr Flash/Getty Images **850** John Foxx Collection/Imagestate **863** Sozaijiten **872** 123RF **873** (top) Blend Images/Shutterstock; (bottom) Yuri Arcurs/Shutterstock **875** (top) U.S. Department of Energy; (bottom) Jeff Banke/Shutterstock **896** Hurst Photo/Shutterstock **919** Gautier Stephane/Sagaphoto.com/Alamy **927** Sue Ashe/Shutterstock **932** Tania Thomson/Shutterstock **933** Tom Reichner/Shutterstock **953** Andrey Yurlov/Shutterstock **954** Baloncici/Shutterstock

CHAPTER 10: **959** Cbpix/Shutterstock **967** Ilya Genkin/Shutterstock **977** Efrain Padro/Alamy **980** Vlad G/Shutterstock **989** Library of Congress Prints and Photographs Division [LC-DIG-ggbain-36570] **990** Jupiterimages/Pixland/Getty Images

CHAPTER 11: **1005** Courtesy of Callie Daniels **1008** (left) Pearson Education, Inc.; (right) Henrik Larsson/Fotolia **1016** Photo Fun/Shutterstock **1026** Michael Drager/Shutterstock **1028** Liew Weng Keong/Shutterstock **1035** Tischenko Irina/Shutterstock **1040** Georgios Kollidas/Fotolia **1057** Courtesy of John Hornsby **1059** Design56/Fotolia **1060** Courtesy of John Hornsby **1062** Courtesy of Margaret Westmoreland **1063** Courtesy of Callie Daniels **1065** Courtesy of Callie Daniels **1067** Xiver/Shutterstock **1083** Andresr/Shutterstock

Index

D

E

F

M

N

O

P

Q

R

S

T

U

V

W

X

Y

Z

Identities and Formulas

6.1 Conversion of Angular Measure

Degree/Radian Relationship: $180° = \pi$ radians

Conversion Formulas:

From	To	Multiply by
Degrees	Radians	$\frac{\pi}{180}$
Radians	Degrees	$\frac{180°}{\pi}$

6.1 Applications of Radian Measure

Arc Length: $s = r\theta$, θ in radians

Area of Sector: $\mathcal{A} = \frac{1}{2}r^2\theta$, θ in radians

6.2

Angular Speed ω	Linear Speed v
$\omega = \frac{\theta}{t}$ (ω in radians per unit time, θ in radians)	$v = \frac{s}{t}$ $v = \frac{r\theta}{t}$ $v = r\omega$

7.1 Fundamental Identities

$\cot\theta = \frac{1}{\tan\theta}$ $\quad \sec\theta = \frac{1}{\cos\theta}$ $\quad \csc\theta = \frac{1}{\sin\theta}$

$\tan\theta = \frac{\sin\theta}{\cos\theta}$ $\quad \cot\theta = \frac{\cos\theta}{\sin\theta}$

$\sin^2\theta + \cos^2\theta = 1$ $\quad \tan^2\theta + 1 = \sec^2\theta$ $\quad 1 + \cot^2\theta = \csc^2\theta$

$\sin(-\theta) = -\sin\theta$ $\quad \cos(-\theta) = \cos\theta$ $\quad \tan(-\theta) = -\tan\theta$

$\csc(-\theta) = -\csc\theta$ $\quad \sec(-\theta) = \sec\theta$ $\quad \cot(-\theta) = -\cot\theta$

7.3 Sum and Difference Identities

$\cos(A + B) = \cos A\cos B - \sin A\sin B$

$\cos(A - B) = \cos A\cos B + \sin A\sin B$

$\sin(A + B) = \sin A\cos B + \cos A\sin B$

$\sin(A - B) = \sin A\cos B - \cos A\sin B$

$\tan(A + B) = \frac{\tan A + \tan B}{1 - \tan A\tan B}$

$\tan(A - B) = \frac{\tan A - \tan B}{1 + \tan A\tan B}$

7.4 Product-to-Sum and Sum-to-Product Identities

$\cos A\cos B = \frac{1}{2}[\cos(A + B) + \cos(A - B)]$

$\sin A\sin B = \frac{1}{2}[\cos(A - B) - \cos(A + B)]$

$\sin A\cos B = \frac{1}{2}[\sin(A + B) + \sin(A - B)]$

$\cos A\sin B = \frac{1}{2}[\sin(A + B) - \sin(A - B)]$

$\sin A + \sin B = 2\sin\left(\frac{A + B}{2}\right)\cos\left(\frac{A - B}{2}\right)$

$\sin A - \sin B = 2\cos\left(\frac{A + B}{2}\right)\sin\left(\frac{A - B}{2}\right)$

$\cos A + \cos B = 2\cos\left(\frac{A + B}{2}\right)\cos\left(\frac{A - B}{2}\right)$

$\cos A - \cos B = -2\sin\left(\frac{A + B}{2}\right)\sin\left(\frac{A - B}{2}\right)$

7.3 Cofunction Identities

$\cos(90° - \theta) = \sin\theta$

$\sin(90° - \theta) = \cos\theta$

$\tan(90° - \theta) = \cot\theta$

$\cot(90° - \theta) = \tan\theta$

$\sec(90° - \theta) = \csc\theta$

$\csc(90° - \theta) = \sec\theta$

7.4 Double-Angle and Half-Angle Identities

$\cos 2A = \cos^2 A - \sin^2 A$ $\quad \cos 2A = 1 - 2\sin^2 A$

$\cos 2A = 2\cos^2 A - 1$ $\quad \sin 2A = 2\sin A\cos A$

$\tan 2A = \frac{2\tan A}{1 - \tan^2 A}$ $\quad \cos\frac{A}{2} = \pm\sqrt{\frac{1 + \cos A}{2}}$

$\sin\frac{A}{2} = \pm\sqrt{\frac{1 - \cos A}{2}}$ $\quad \tan\frac{A}{2} = \pm\sqrt{\frac{1 - \cos A}{1 + \cos A}}$

$\tan\frac{A}{2} = \frac{\sin A}{1 + \cos A}$ $\quad \tan\frac{A}{2} = \frac{1 - \cos A}{\sin A}$

8.1 Law of Sines

In any triangle ABC, with sides a, b, and c,

$$\frac{a}{\sin A} = \frac{b}{\sin B}, \quad \frac{a}{\sin A} = \frac{c}{\sin C}, \quad \text{and} \quad \frac{b}{\sin B} = \frac{c}{\sin C}.$$

Area of a Triangle

The area $\mathcal{A}$ of a triangle is given by half the product of the lengths of two sides and the sine of the angle between the two sides.

$$\mathcal{A} = \frac{1}{2}bc\sin A, \quad \mathcal{A} = \frac{1}{2}ab\sin C, \quad \mathcal{A} = \frac{1}{2}ac\sin B$$

8.2 Law of Cosines

In any triangle ABC, with sides a, b, and c,

$$a^2 = b^2 + c^2 - 2bc\cos A, \quad b^2 = a^2 + c^2 - 2ac\cos B,$$

and

$$c^2 = a^2 + b^2 - 2ab\cos C.$$

Heron's Area Formula

If a triangle has sides of lengths a, b, and c, with semiperimeter $s = \frac{1}{2}(a + b + c)$, then the area $\mathcal{A}$ of the triangle is

$$\mathcal{A} = \sqrt{s(s - a)(s - b)(s - c)}.$$

Graphs of Functions

2.6 Identity Function

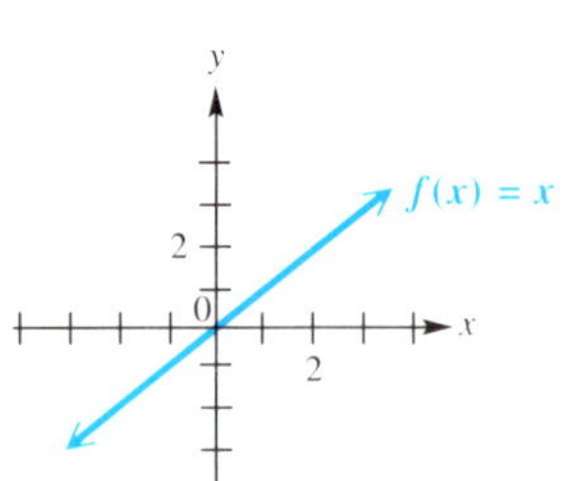

2.6 Squaring Function

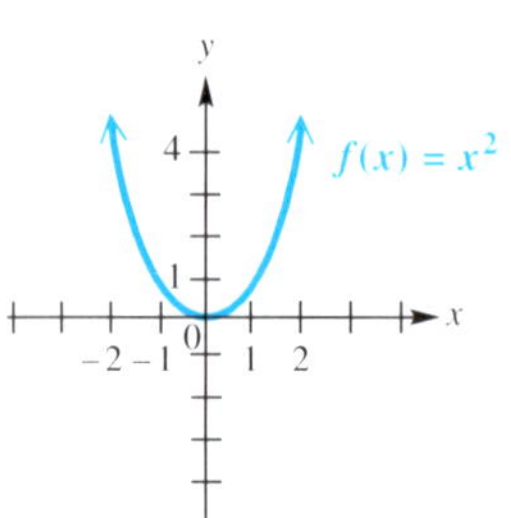

2.6 Cubing Function

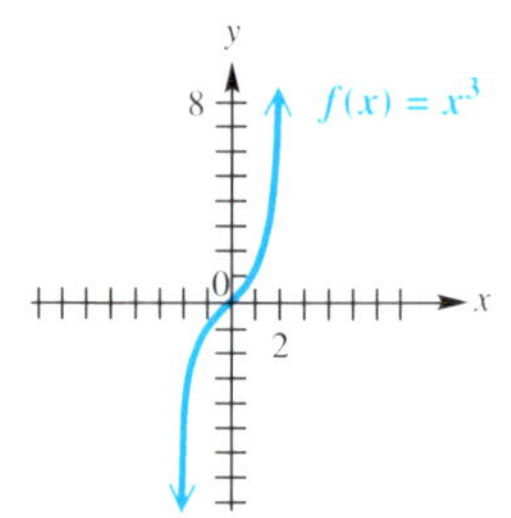

2.6 Square Root Function

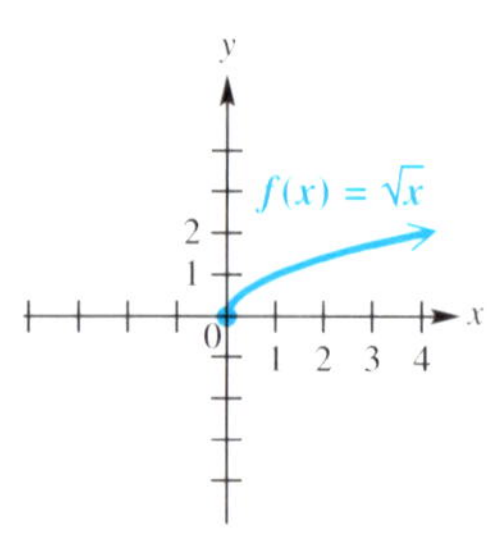

2.6 Cube Root Function

2.6 Absolute Value Function

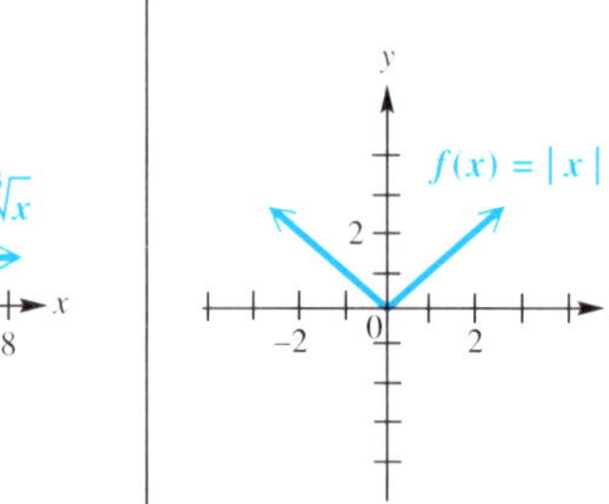

2.6 Greatest Integer Function

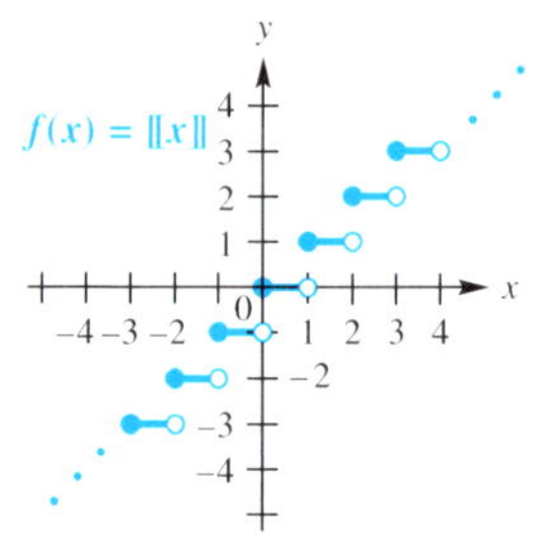

3.5 Reciprocal Function

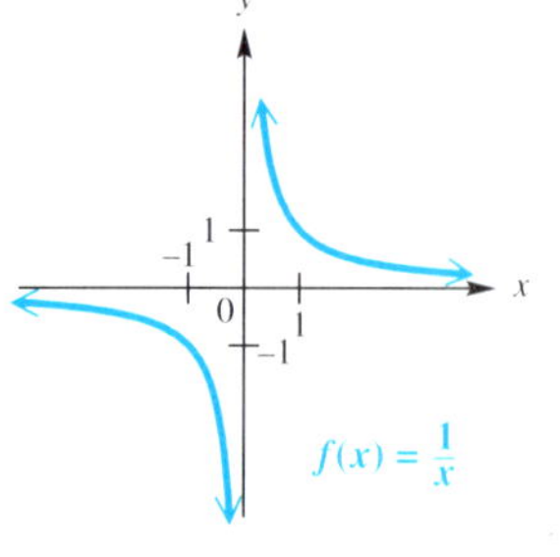

3.4 Polynomial Functions

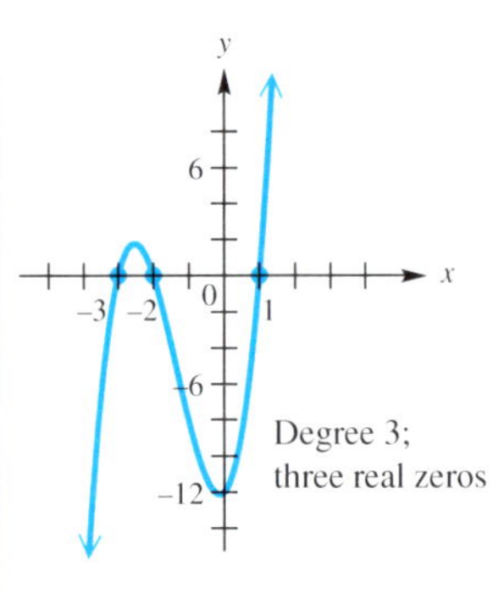

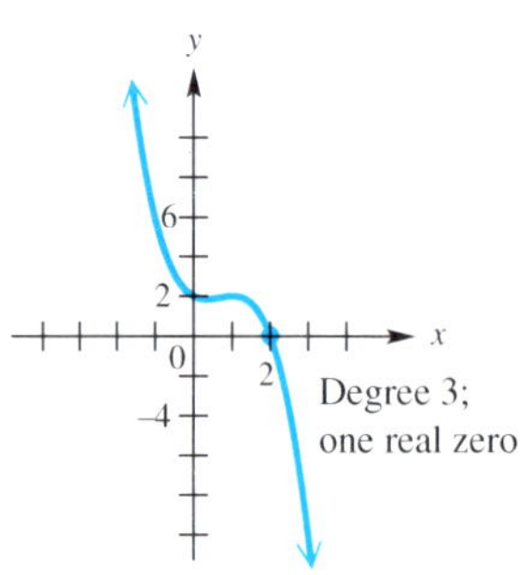

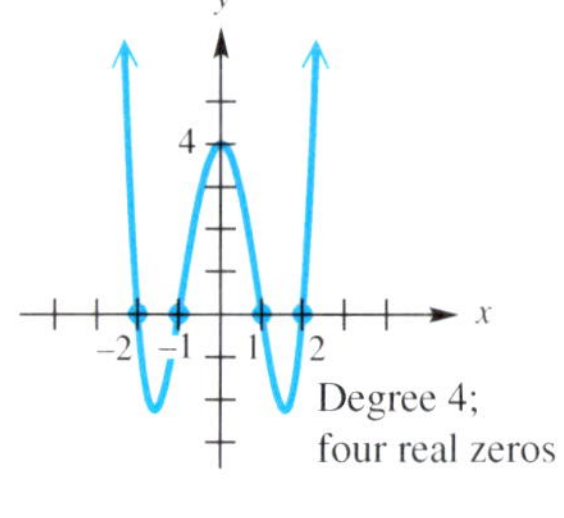

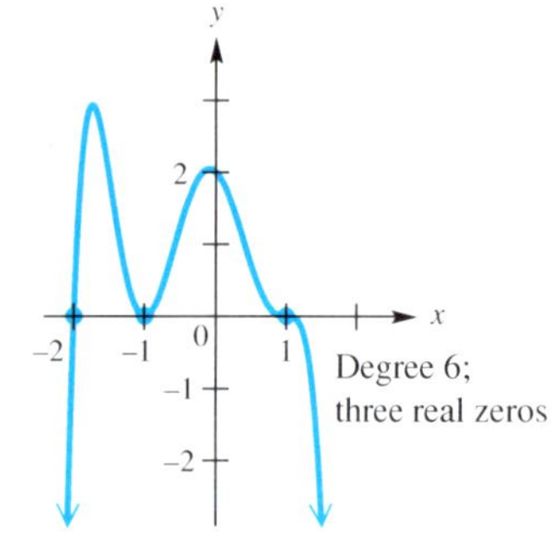

4.2 Exponential Functions

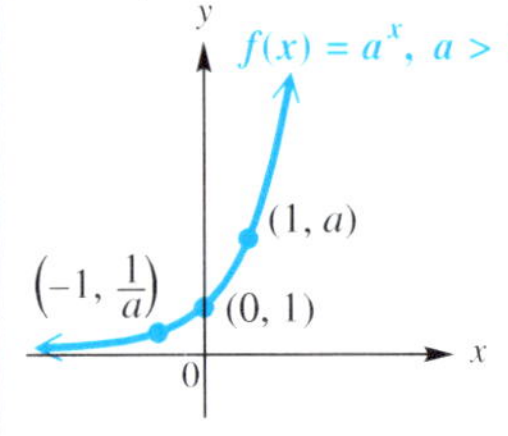

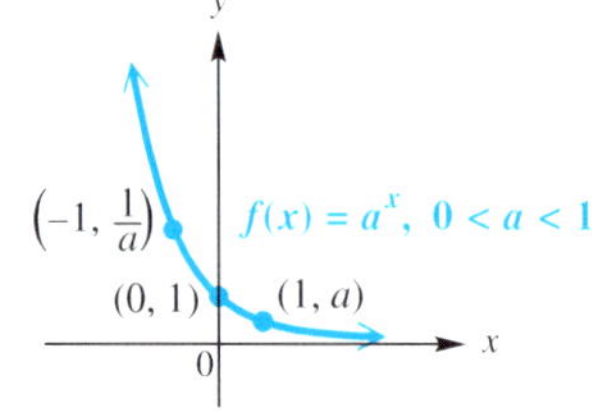

4.3 Logarithmic Functions

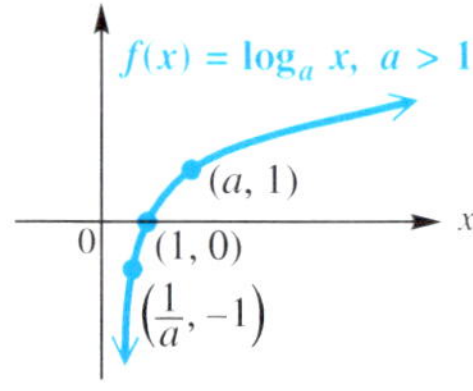

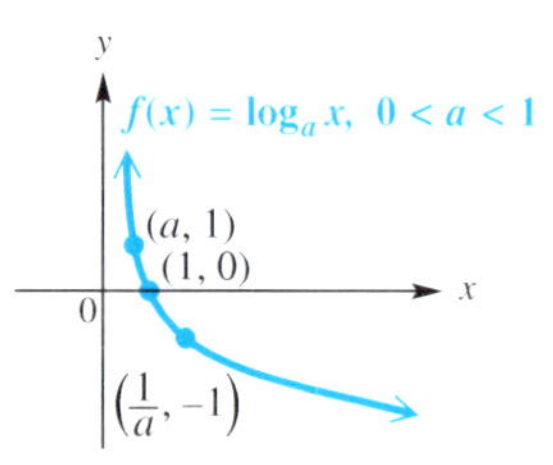

6.3–6.6 Trigonometric (Circular) Functions

The graph of $y = c + a \sin [b(x - d)]$ or $y = c + a \cos [b(x - d)]$, where $b > 0$, has amplitude $|a|$, period $\frac{2\pi}{b}$, a vertical translation c units up if $c > 0$ or $|c|$ units down if $c < 0$, and a phase shift d units to the right if $d > 0$ or $|d|$ units to the left if $d < 0$. The graph of $y = a \tan bx$ or $y = a \cot bx$ has period $\frac{\pi}{b}$, where $b > 0$.

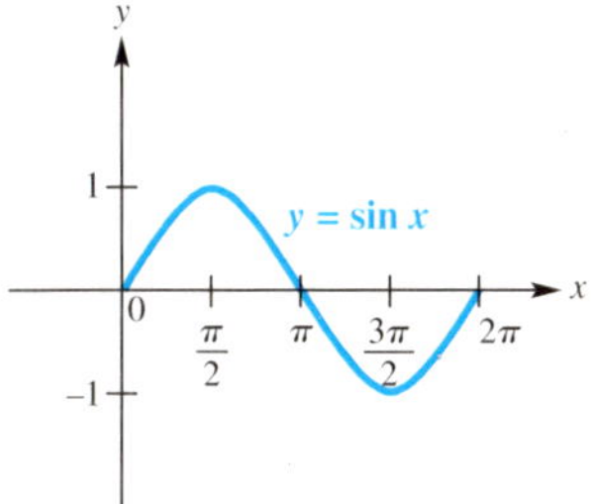

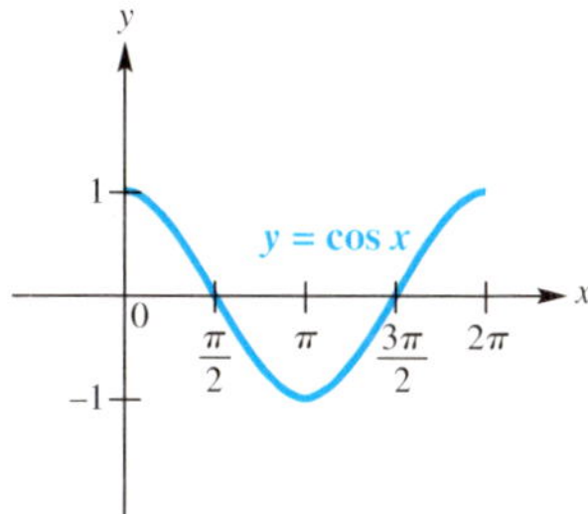

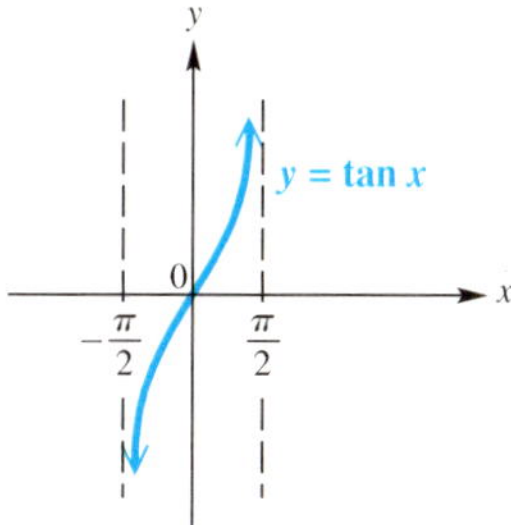

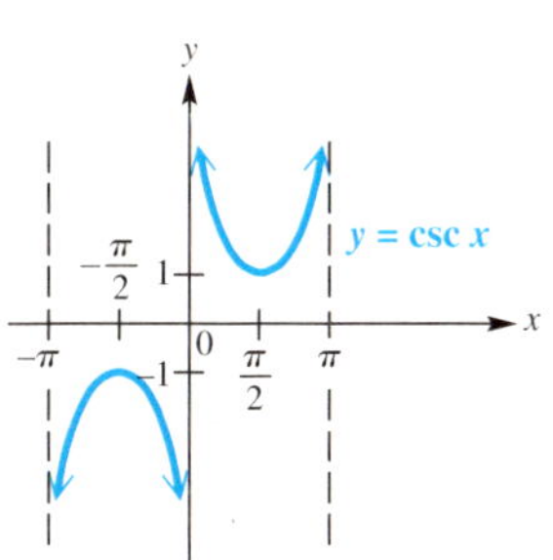

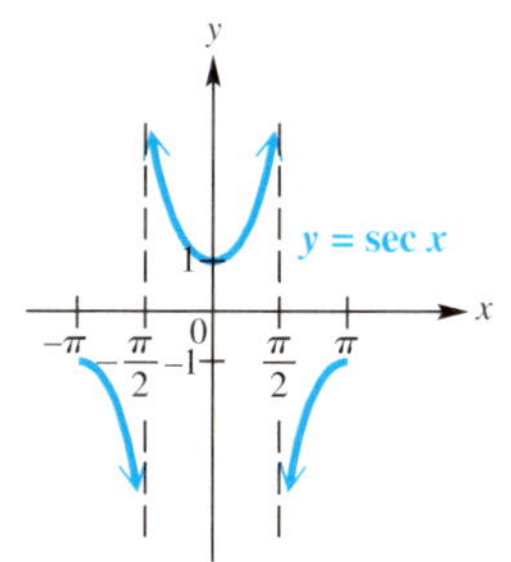

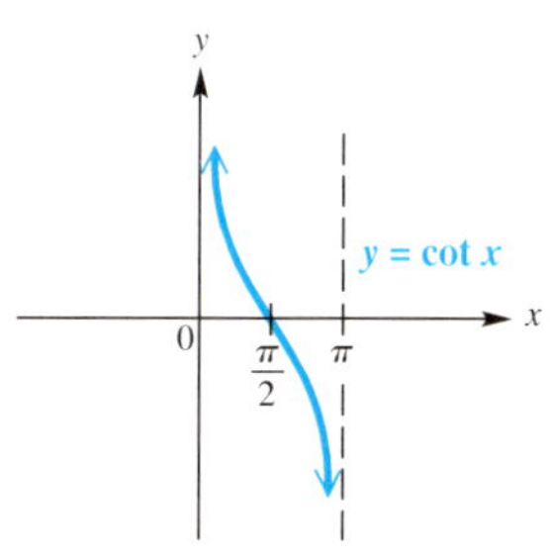

7.5 Inverse Trigonometric (Circular) Functions

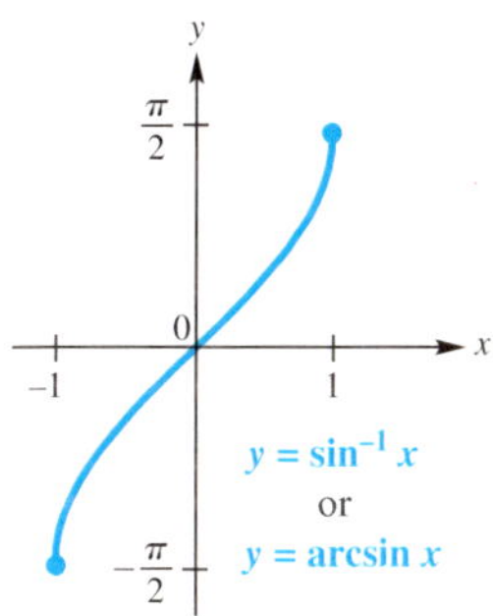

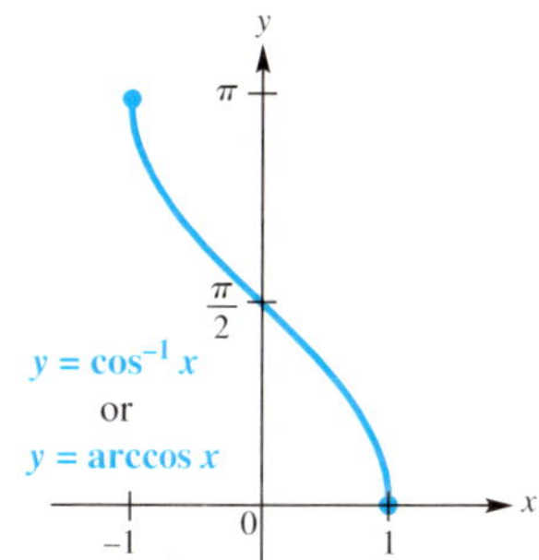

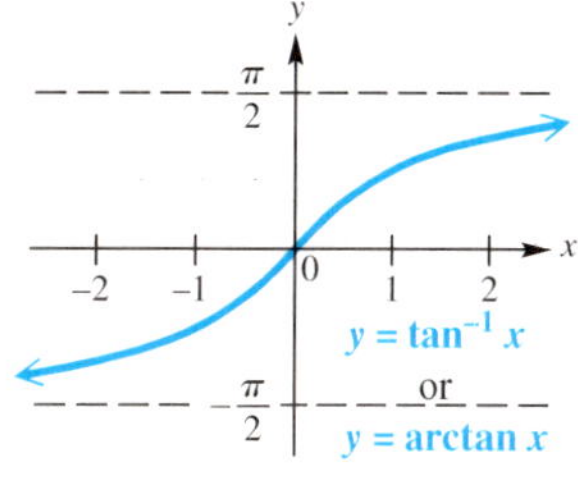

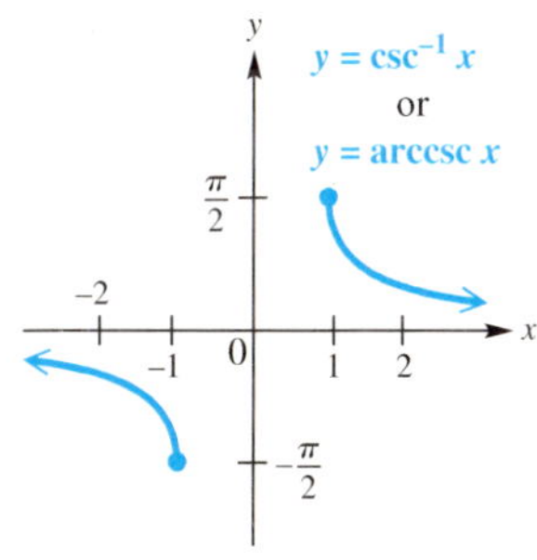

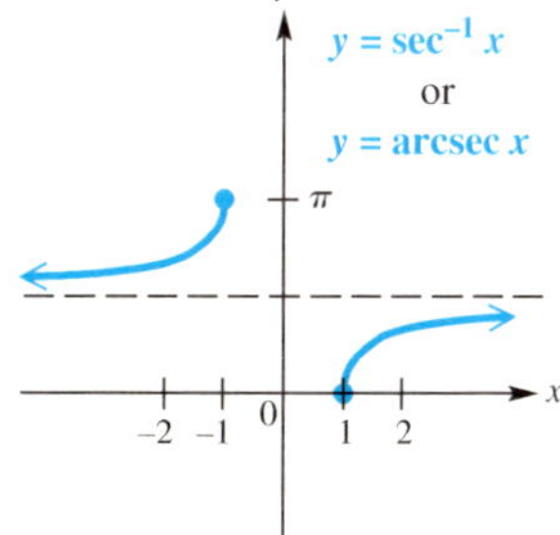

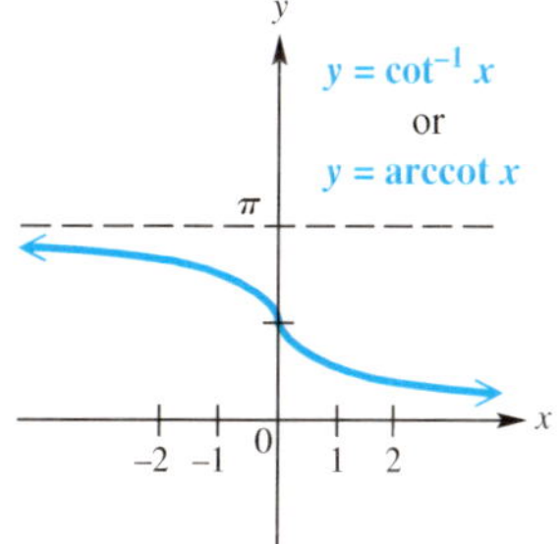